# *Handbook of* ECOTOXICOLOGY

**Second Edition**

**Edited by**
**David J. Hoffman**
**Barnett A. Rattner**
**G. Allen Burton, Jr.**
**John Cairns, Jr.**

LEWIS PUBLISHERS

A CRC Press Company
Boca Raton London New York Washington, D.C.

**Cover photograph of the California red-legged frog courtesy of Gary Fellers, U.S. Geological Survey.**

**Cover photograph of the American alligator courtesy of Heath Rauschenberger, U.S. Geological Survey.**


**Library of Congress Cataloging-in-Publication Data**

Handbook of ecotoxicology / David J. Hoffman ... [et al.] — 2nd ed.
p. cm.
Includes bibliographical references and index.
ISBN 1-56670-546-0 (alk. paper)
1. Environmental toxicology. I. Hoffman, David J. (David John), 1944-

RA1226 .H36 2002
615.9'02—dc21 2002075228





Direct all inquiries to CRC Press LLC, 2000 N.W. Corporate Blvd., Boca Raton, Florida 33431.




**Visit the CRC Press Web site at www.crcpress.com**




Lewis Publishers is an imprint of CRC Press LLC





International Standard Book Number 1-56670-546-0
Library of Congress Card Number 2002075228
Printed in the United States of America 2 3 4 5 6 7 8 9 0
Printed on acid-free paper

# Preface

The first edition of this book, a bestseller for Lewis Publishers/CRC Press, evolved from a series of articles on ecotoxicology authored by the editors and published in the journal *Environmental Science and Technology*. Ecotoxicology remains a rapidly growing field, with many components periodically being redefined or open to further interpretation. Therefore, this second edition of the *Handbook of Ecotoxicology* has expanded considerably in both concept and content over the first edition. The second edition contains 45 chapters with contributions from over 75 international experts. Eighteen new chapters have been introduced, and the original chapters have been substantially revised and updated. All of the content has been reviewed by a board of experts.

This edition is divided into five major sections: I. Quantifying and Measuring Ecotoxicological Effects, II. Contaminant Sources and Effects, III. Case Histories and Ecosystem Surveys, IV. Methods for Making Estimates, Predictability, and Risk Assessment in Ecotoxicology, and V. Special Issues in Ecotoxicology. In the first section, concepts and current methodologies for testing are provided for aquatic toxicology, wildlife toxicology, sediment toxicity, soil ecotoxicology, algal and plant toxicity, and landscape ecotoxicology. Biomonitoring programs and current use of bioindicators for aquatic and terrestrial monitoring are described. The second section contains chapters on major environmental contaminants and other anthropogenic processes capable of disrupting ecosystems including pesticides, petroleum and PAHs, heavy metals, selenium, polyhalogenated aromatic hydrocarbons, urban runoff, nuclear and thermal contamination, global effects of deforestation, pathogens and disease, and abiotic factors that interact with contaminants.

In order to illustrate the full impact of different environmental contaminants on diverse ecosystems, seven case histories and ecosystem surveys are described in the third section. The fourth section discusses methods and approaches used for estimating and predicting exposure and effects for purposes of risk assessment. These include global disposition of contaminants, bioaccumulation and bioconcentration, use of quantitative structure activity relationships (QSARs), population modeling, current guidelines and future directions for ecological risk assessment, and restoration ecology. The fifth section of this book identifies and describes a number of new and significant issues in ecotoxicology, most of which have come to the forefront of the field since the publication of the first edition. These include endocrine-disrupting chemicals in the environment, the possible role of contaminants in the worldwide decline of amphibian populations, potential genetic effects of contaminants on animal populations, the role of ecotoxicology in industrial ecology and natural capitalism, the consequences of indirect effects of agricultural pesticides on wildlife, the role of nutrition on trace element toxicity, and the role of environmental contaminants on endangered species.

This edition was designed to serve as a reference book for students entering the fields of ecotoxicology and other environmental sciences. Many portions of this handbook will serve as a convenient reference text for established investigators, resource managers, and those involved in risk assessment and management within regulatory agencies and the private sector.

**David J. Hoffman**
**Barnett A. Rattner**
**G. Allen Burton, Jr.**
**John Cairns, Jr.**

# The Editors

## *David J. Hoffman*

**David J. Hoffman** is a research physiologist in the field of environmental contaminants at the Patuxent Wildlife Research Center, U.S. Geological Survey of the Department of the Interior. He is also an Adjunct Professor in the Department of Biology, University of Maryland at Frostburg. Dr. Hoffman earned a Bachelor of Science degree in Zoology from McGill University in 1966 and his Doctor of Philosophy Degree in Zoology (developmental physiology) from the University of Maryland in 1971. He was an NIH Postdoctoral Fellow in the Biochemistry Section of Oak Ridge National Laboratory from 1971 to 1973. Other positions included teaching at Boston College during 1974 and research as a Senior Staff Physiologist with the Health Effects Research Laboratory of the U.S. Environmental Protection Agency in Cincinnati from 1974 to 1976 before joining the Environmental Contaminants Evaluation Program of the Patuxent Wildlife Research Center.

Dr. Hoffman's research over the past 20 years has focused on morphological and biochemical effects of environmental contaminants including bioindicators of developmental toxicity in birds in the laboratory and in natural ecosystems. Key areas of study have included sublethal indicators of exposure to planar PCBs, lead, selenium, and mercury; embryotoxicity and teratogenicity of pesticides and petroleum to birds and impact on nestlings; interactive toxicant and nutritional factors affecting agricultural drainwater and heavy-metal toxicity; and measurements of oxidative stress for monitoring contaminant exposure in wildlife.

Dr. Hoffman has published over 120 scientific papers including book chapters and review papers and has served on eight editorial boards.

## *Barnett A. Rattner*

**Barnett A. Rattner** is a research physiologist at the Patuxent Wildlife Research Center, U.S. Geological Survey of the Department of the Interior. He is also Adjunct Professor of the Department of Animal and Avian Science Sciences, University of Maryland. Dr. Rattner attended the University of Maryland, earning his Doctor of Philosophy degree in 1977. He was a National Research Council Postdoctoral Associate at the Naval Medical Research Institute in 1978 before joining the Environmental Contaminants Evaluation Program of the Patuxent Wildlife Research Center.

Dr. Rattner's research activities during the past 20 years have included evaluation of sublethal biochemical, endocrine, and physiological responses of wildlife to petroleum crude oil, various pesticides, industrial contaminants, and metals. He has investigated the interactive effects of natural stressors and toxic environmental pollutants, developed and applied cytochrome P450 assays as a biomarker of contaminant exposure, conducted risk assessments on potential substitutes for lead shot used in hunting, and compiled several large World Wide Web-accessible ecotoxicological databases for terrestrial vertebrates.

Dr. Rattner has published over 65 scientific articles and serves on four editorial boards and several federal committees including the Toxic Substances Control Act Interagency Testing Committee of the U.S. Environmental Protection Agency.

## G. Allen Burton, Jr.

**G. Allen Burton, Jr.** is the Brage Golding Distinguished Professor of Research and Director of the Institute for Environmental Quality at Wright State University. He earned a Ph.D. degree in Environmental Science from the University of Texas at Dallas in 1984. From 1980 until 1985 he was a Life Scientist with the U.S. Environmental Protection Agency. He was a Postdoctoral Fellow at the National Oceanic and Atmospheric Administration's Cooperative Institute for Research in Environmental Sciences at the University of Colorado. Since then he has had positions as a NATO Senior Research Fellow in Portugal and Visiting Senior Scientist in Italy and New Zealand. Dr. Burton's research during the past 20 years has focused on developing effective methods for identifying significant effects and stressors in aquatic systems where sediment and stormwater contamination is a concern. His ecosystem risk assessments have evaluated multiple levels of biological organization, ranging from microbial to amphibian effects. He has been active in the development and standardization of toxicity methods for the U.S. EPA, American Society for Testing and Materials (ASTM), Environment Canada, and the Organization of Economic Cooperation and Development (OECD). Dr. Burton has served on numerous national and international scientific committees and review panels and has had over $4 million in grants and contracts and more than 100 publications dealing with aquatic systems.

## John Cairns, Jr.

**John Cairns, Jr.** is University Distinguished Professor of Environmental Biology Emeritus in the Department of Biology at Virginia Polytechnic Institute and State University in Blacksburg, Virginia. His professional career includes 18 years as Curator of Limnology at the Academy of Natural Sciences of Philadelphia, 2 years at the University of Kansas, and over 34 years at his present institution. He has also taught periodically at various field stations.

Among his honors are Member, National Academy of Sciences; Member, American Philosophical Society; Fellow, American Academy of Arts and Sciences; Fellow, American Association for the Advancement of Science; the Founder's Award of the Society for Environmental Toxicology and Chemistry; the United Nations Environmental Programme Medal; Fellow, Association for Women in Science; U.S. Presidential Commendation for Environmental Activities; the Icko Iben Award for Interdisciplinary Activities from the American Water Resources Association; Phi Beta Kappa; the B. Y. Morrison Medal (awarded at the Pacific Rim Conference of the American Chemical Society); Distinguished Service Award, American Institute of Biological Sciences; Superior Achievement Award, U.S. Environmental Protection Agency; the Charles B. Dudley Award for excellence in publications from the American Society for Testing and Materials; the Life Achievement Award in Science from the Commonwealth of Virginia and the Science Museum of Virginia; the American Fisheries Society Award of Excellence; Doctor of Science, State University of New York at Binghamton; Fellow, Virginia Academy of Sciences; Fellow, Eco-Ethics International Union; Twentieth Century Distinguished Service Award, Ninth Lukacs Symposium; 2001 Ruth Patrick Award for

Environmental Problem Solving, American Society of Limnology and Oceanography; 2001 Sustained Achievement Award, Renewable Natural Resources Foundation, 2001.

Professor Cairns has served as both vice president and president of the American Microscopical Society, has served on 18 National Research Council committees (two as chair), is presently serving on 14 editorial boards, and has served on the Science Advisory Board of the International Joint Commission (United States and Canada) and on the U.S. EPA Science Advisory Board. The most recent of his 57 books are *Goals and Conditions for a Sustainable Planet, 2002* and the Japanese edition of *Restoration of Aquatic Ecosystems: Science, Technology, and Public Policy*, 1999.

# REVIEW BOARD
## *Handbook of Ecotoxicology* 2nd Edition

# Contributors

**William J. Adams**
Rio Tinto Corporation
Murray, Utah

**Peter H. Albers**
U.S. Geological Survey
Patuxent Wildlife Research Center
Laurel, Maryland

**Patrick J. Anderson**
U.S. Geological Survey
Mid-Continent Ecological Center
Fort Collins, Colorado

**Andrew S. Archuleta**
U.S. Fish and Wildlife Service
Colorado Field Office
Denver, Colorado

**Beverly S. Arnold**
U.S. Geological Survey
Florida Caribbean Science Center
Gainesville, Florida

**Pinar Balci**
University of North Texas
Institute of Applied Sciences
Denton, Texas

**Mace G. Barron**
P.E.A.K. Research
Longmont, Colorado

**Timothy M. Bartish**
U.S. Geological Survey
Mid-Continent Ecological Center
Fort Collins, Colorado

**Sally M. Benson**
Lawrence Berkeley National Laboratory
Berkeley, California

**W. Nelson Beyer**
U.S. Geological Survey
Patuxent Wildlife Research Center
Laurel, Maryland

**Amy M. Bickham**
Texas Tech University
Lubbock, Texas

**John W. Bickham**
Texas A & M University
College Station, Texas

**Lynn Blake-Hedges**
U.S. Environmental Protection Agency
Office of Pesticides, Prevention and Toxic Substances
Washington, D.C.

**Lawrence J. Blus**
U.S. Geological Survey
Forest and Rangeland Ecosystem Science Center
Corvallis, Oregon

**Dixie L. Bounds**
U.S. Geological Survey
Maryland Cooperative Fish and Wildlife Research Unit
Princess Anne, Maryland

**Robert P. Breckenridge**
Idaho National Engineering and Environmental Laboratory
Ecological and Cultural Resources
Idaho Falls, Idaho

**G. Allen Burton, Jr.**
Wright State University
Institute for Environmental Quality
Dayton, Ohio

**Earl R. Byron**
CH2M HILL
Sacramento, California

**John Cairns, Jr.**
Virginia Polytechnic Institute and State University
Blacksburg, Virginia

**Patricia A. Cirone**
U.S. Environmental Protection Agency
Seattle, Washington

**Laura C. Coppock**
U.S. Fish and Wildlife Service
Denver, Colorado

**Christine M. Custer**
U.S. Geological Survey
Upper Midwest Environmental Sciences Center
La Crosse, Wisconsin

**Thomas W. Custer**
U.S. Geological Survey
Upper Midwest Environmental Sciences Center
La Crosse, Wisconsin

**Michael Delamore**
U.S. Bureau of Reclamation
Fresno, California

**Debra L. Denton**
U.S. Environmental Protection Agency
Division of Water Quality
Sacramento, California

**Ronald Eisler**
U.S. Geological Survey
Patuxent Wildlife Research Center
Laurel, Maryland

**Valerie L. Fellows**
U.S. Fish and Wildlife Service
Annapolis, Maryland

**George F. Fries**
U.S. Department of Agriculture
Beltsville, Maryland

**Timothy S. Gross**
U.S. Geological Survey
Florida Caribbean Science Center
Gainesville, Florida

**Steven J. Hamilton**
U.S. Geological Survey
Columbia Environmental Research Center
Yankton, South Dakota

**Stuart Harrad**
University of Birmingham
School of Geography & Environmental Science
Edgbaston, Birmingham, United Kingdom

**Roy M. Harrison**
University of Birmingham
School of Geography & Environmental Science
Edgbaston, Birmingham, United Kingdom

**Alan G. Heath**
Virginia Polytechnic Institute and State University
Blacksburg, Virginia

**Gary H. Heinz**
U.S. Geological Survey
Patuxent Wildlife Research Center
Laurel, Maryland

**Gray Henderson**
University of Missouri
Columbia, Missouri

**Charles J. Henny**
U.S. Geological Survey
Forest and Rangeland Ecosystem Science Center
Corvallis, Oregon

**Elwood F. Hill**
U.S. Geological Survey
Patuxent Wildlife Research Center
Laurel, Maryland

**Kay Ho**
U.S. Environmental Protection Agency
Atlantic Ecology Division
Narragansett, Rhode Island

**David J. Hoffman**
U.S. Geological Survey
Patuxent Wildlife Research Center
Laurel, Maryland

**Karen D. Holl**
University of California
Department of Environmental Studies
Santa Cruz, California

**John Holland**
The Game Conservancy Trust
Fordingbridge
Hampshire, United Kingdom

**Michael J. Hooper**
Texas Tech University
Institute of Environmental and Human Health
Lubbock, Texas

**Richard A. Houghton**
The Woods Hole Research Center
Woods Hole, Massachusetts

**Elaine R. Ingham**
Soil FoodWeb, Inc.
Corvallis, Oregon

**D. Scott Ireland**
U.S. Environmental Protection Agency
Washington, D.C.

**Zane B. Johnson**
U.S. Geological Survey
Leetown Science Center
Kearneysville, West Virginia

**James H. Kennedy**
University of North Texas
Denton, Texas

**Stephen J. Klaine**
Clemson University
Pendleton, South Carolina

**Sandra L. Knuteson**
Clemson University
Pendleton, South Carolina

**David P. Krabbenhoft**
U.S. Geological Survey
Middleton, Wisconsin

**Timothy J. Kubiak**
U.S. Fish & Wildlife Service
Pleasantville, New Jersey

**Thomas W. La Point**
University of North Texas
Institute of Applied Sciences
Denton, Texas

**Jamie Lead**
University of Birmingham
School of Geography & Environmental Science
Edgbaston, Birmingham, United Kingdom

**Frederick A. Leighton**
University of Saskatchewan
Canadian Cooperative Wildlife Health Centre
Saskatoon, Saskatchewan, Canada

**Michael A. Lewis**
U.S. Environmental Protection Agency
Gulf Breeze, Florida

**Greg Linder**
U.S. Geological Survey
Columbia Environmental Research Center
Brooks, Oregon

**Marilynne Manguba**
Idaho National Engineering and Environmental Laboratory
Idaho Falls, Idaho

**Suzanne M. Marcy**
U.S. Environmental Protection Agency
Anchorage, Alaska

**John P. McCarty**
University of Nebraska-Omaha
Omaha, Nebraska

**Kelly McDonald**
U.S. Geological Survey
Florida Caribbean Science Center
Gainesville, Florida

**Mark J. Melancon**
U.S. Geological Survey
Patuxent Wildlife Research Center
Laurel, Maryland

**Linda Meyers-Schöne**
AMEC
Albuquerque, New Mexico

**Pierre Mineau**
Environment Canada
Canadian Wildlife Service
Hull, Quebec, Canada

**B. R. Niederlehner**
Virginia Polytechnic Institute and State University
Blacksburg, Virginia

**Susan B. Norton**
U.S. Environmental Protection Agency
Office of Research and Development
Washington, D.C.

**Harry M. Ohlendorf**
CH2M HILL
Sacramento, California

**Patrick W. O'Keefe**
NY State Health Department
Wadsworth Center for Laboratories and Research
Albany, New York

**Richard L. Orr**
U.S. Department of Agriculture
Animal and Plant Health Inspection Service
Riverdale, Maryland

**Deborah J. Pain**
Royal Society for the Protection of Birds
The Lodge Sandy
Bedfordshire, United Kingdom

**Gary Pascoe**
EA Engineering, Science & Technology, Inc.
Port Townsend, Washington

**Oliver H. Pattee**
U.S. Geological Survey
Patuxent Wildlife Research Center
Laurel, Maryland

**Grey W. Pendleton**
Alaska Department of Fish & Game
Douglas, Alaska

**Robert E. Pitt**
University of Alabama
Tuscaloosa, Alabama

**Barnett A. Rattner**
U.S. Geological Survey
Patuxent Wildlife Research Center
Laurel, Maryland

**Clifford P. Rice**
U.S. Department of Agriculture
Environmental Quality Laboratory
Beltsville, Maryland

**Donald J. Rodier**
U.S. Environmental Protection Agency
Office of Pesticides, Prevention and Toxic Substances
Washington, D.C.

**Carolyn D. Rowland**
Wright State University
Dayton, Ohio

**Gary M. Santolo**
CH2M HILL
Sacramento, California

**John R. Sauer**
U.S. Geological Survey
Patuxent Wildlife Research Center
Laurel, Maryland

**Anton M. Scheuhammer**
Environment Canada
Canadian Wildlife Service
Hull, Quebec, Canada

**T. Wayne Schultz**
University of Tennessee
College of Veterinary Medicine
Knoxville, Tennessee

**María S. Sepúlveda**
U.S. Geological Survey
Florida Caribbean Science Center
Gainesville, Florida

**Anne Sergeant**
U.S. Environmental Protection Agency
Office of Research and Development
Washington, D.C.

**Victor B. Serveiss**
U.S. Environmental Protection Agency
Office of Research and Development
Washington, D.C.

**Lee R. Shugart**
LR Shugart & Associates, Inc.
Oak Ridge, Tennessee

**Nick Sotherton**
The Game Conservancy Trust
Fordingbridge
Hampshire, United Kingdom

**Donald W. Sparling**
U.S. Geological Survey
Patuxent Wildlife Research Center
Laurel, Maryland

**Jacob K. Stanley**
University of North Texas
Denton, Texas

**Carol D. Swartz**
National Institute of Environmental Health Sciences
Research Triangle Park, North Carolina

**Sylvia S. Talmage**
Oak Ridge National Laboratory
Oak Ridge, Tennessee

**Christopher Theodorakis**
Texas Tech University
Lubbock, Texas

**Tetsu K. Tokunaga**
Lawrence Berkeley National Laboratory
Berkeley, California

**William H. van der Schalie**
U.S. Army
Center for Environmental Health Research
Fort Detrick, Maryland

**John D. Walker**
TSCA Interagency Testing Committee
U.S. Environmental Protection Agency
Washington, D.C.

**Randall Wentsel**
U.S. Environmental Protection Agency
Office of Research and Development
Washington, D.C.

**Steven Wharton**
U.S. Environmental Protection Agency
Denver, Colorado

**James G. Wiener**
University of Wisconsin-La Crosse
La Crosse, Wisconsin

**Daniel F. Woodward**
U.S. Geological Survey
Jackson, Wyoming

**Brian Woodbridge**
U.S. Forest Service
1312 Fairlane Road
Yreka, California

**María Elena Zaccagnini**
National Institute of Agricultural Technology
Agroecology and Wildlife Management
Parana, Argentina

**Peter T. Zawislanski**
Lawrence Berkeley National Laboratory
Berkeley, California

# Contents

# CHAPTER 1

# Introduction

**David J. Hoffman, Barnett A. Rattner, G. Allen Burton, Jr., and John Cairns, Jr.,**

## CONTENTS



## 1.1 HISTORY

The term *ecotoxicology* was first coined by Truhaut in 1969 as a natural extension from toxicology, the science of the effects of poisons on individual organisms, to the ecological effects of pollutants.[1] In the broadest sense ecotoxicology has been described as toxicity testing on one or more components of any ecosystem, as stated by Cairns.[2] This definition of ecotoxicology can be further expanded as the science of predicting effects of potentially toxic agents on natural ecosystems and on nontarget species. Ecotoxicology has not generally included the fields of industrial and human health toxicology or domestic animal and agricultural crop toxicology, which are not part of natural ecosystems, but are rather imposed upon them. Yet there is a growing belief by some that humanity and its artifacts should be regarded as components of natural systems, not apart from them. More recently, Newman has defined ecotoxicology as the science of contaminants in the biosphere and their effects on constituents of the biosphere, which includes humans.[3] Ecotoxicology employs ecological parameters to assess toxicity. In a more restrictive but useful sense, it can be defined as the science of assessing the effects of toxic substances on ecosystems with the goal of protecting entire ecosystems, and not merely isolated components.

Historically, some of the earliest observations of anthropogenic ecotoxic effects, such as industrial melanism of moths, date back to the industrial revolution of the 1850s (see Table 1.1). In the field of aquatic toxicology Forbes was one of the first researchers to recognize the significance of the presence or absence of species and communities within an aquatic ecosystem and to report approaches for classifying rivers into zones of pollution based on species tolerance.[4] At the same

**Table 1.1 Historical Overview: First Observations of Ecotoxic Effects of Different Classes of Environmental Contaminants**

| Date | Contaminant(s) | Effects |
|---|---|---|
| 1850s | Industrial revolution; soot from coal burning | Industrial melanism of moths |
| 1863 | Industrial wastewater | Toxicity to aquatic organisms; first acute toxicity tests |
| 1874 | Spent lead shot | Ingestion resulted in death of waterfowl and pheasants |
| 1887 | Industrial wastewater | Zones of pollution in rivers established by species tolerance |
| 1887 | Arsenic emissions from metal smelters | Death of fallow deer and foxes |
| 1907 | Crude oil spill | Death of thousands of puffins |
| 1924 | Lead and zinc mine runoff | Toxicity of metal ions to fish |
| 1927 | Hydrogen sulfide fumes in oil field | Large die-off of both wild birds and mammals |
| 1950s | DDT and organochlorines | Decline in American robins linked to DDT use for Dutch Elm disease; eggshell thinning in bald eagles, osprey, and brown pelicans linked to DDT; and fish-eating mammals at risk |
| 1960s | Anticholinesterase pesticides | Die-offs of wild birds, mammals, and other vertebrate species |
| 1970s | Mixtures of toxic wastes, including dioxins at hazardous waste sites | Human, aquatic, and wildlife health at risk |
| 1980s | Agricultural drainwater containing selenium and other contaminants | Multiple malformations and impaired reproduction in aquatic birds in central California |
| 1986 | Radioactive substances from Chernobyl nuclear power station | Worst nuclear incident in peacetime, affecting a wide variety of organisms and ecosystems |
| 1990s | Complex mixtures of potential endocrine disrupting chemicals, including PCBs and organochlorine pesticides | Abnormally developed reproductive organs, altered serum hormone concentrations, and decreased egg viability in alligators from contaminated lakes in Florida |

*Source*: Adapted from: Hoffman, D. J., Rattner, B. A., Burton, G. A. Jr., and Lavoie, D. R., Ecotoxicology, in *Handbook of Toxicology*, Derelanko, M. J., and Hollinger, M. A., Eds., CRC Press, Boca Raton, FL, 2002.

time some of the earliest acute aquatic toxicity tests were first performed by Penny and Adams (1863)[5] and Weigelt, Saare, and Schwab (1885),[6] who were concerned with toxic chemicals in industrial wastewater. The first "standard method" was published by Hart et al. in 1945 and subsequently adopted by the American Society for Testing and Materials.[7] In this manner it has become generally recognized that the presence or absence of species (especially populations or communities) in a given aquatic ecosystem provides a more sensitive and reliable indicator of the suitability of environmental conditions than do chemical and physical measurements alone.

In the field of terrestrial toxicology reports of anthropogenic contaminants affecting free-ranging wildlife first began to accumulate during the industrial revolution of the 1850s. These included cases of arsenic pollution and industrial smoke stack emission toxicity. One early report described the death of fallow deer (*Dama dama*) due to arsenic emissions from a silver foundry in Germany in 1887, and another described hydrogen sulfide fumes near a Texas oil field that resulted in a large die-off of many species of wild birds and mammals,[8] thus affecting multiple species within an ecosystem. With the advent of modern pesticides, most notably the introduction of dichlorodiphenyltrichloroethane (DDT) in 1943, a marked decline in the population of American robins (*Turdus migratorius*) was linked by the early 1950s to DDT spraying to control Dutch Elm disease. It soon became evident that ecosystems with bald eagles (*Haliaetus leucocephalus*), osprey (*Pandion haliaetus*), brown pelicans (*Pelecanus occidentalis*), and populations of fish-eating mammals were at risk.[9,10]

More recent observations of adverse effects of environmental contaminants and other anthropogenic processes capable of disrupting ecosystems will be covered in subsequent chapters of this book. Exposure and adverse effects, sometimes indirect, of anticholinesterase and other pesticides used in agriculture, petroleum and polycyclic aromatic hydrocarbons (PAHs), manufactured and waste polyhalogenated aromatic hydrocarbons, heavy metals, selenium and other trace elements are included. Other processes and contaminants include nuclear and thermal processes, urban runoff, pathogens and disease, deforestation and global warming, mining and smelting operations, waste and spent munitions, and released genotoxic and endocrine disruptive chemicals will be presented and discussed in detail.

This book is divided into five sections (I. Quantifying and Measuring Ecotoxicological Effects; II. Contaminant Sources and Effects; III. Case Histories and Ecosystem Surveys; IV. Methods for Making Estimates, Predictability, and Risk Assessment in Ecotoxicology; V. Special Issues in Ecotoxicology) in order to provide adequate coverage of the following general areas of ecotoxicology: (1) methods of quantifying and measuring ecotoxicological effects under controlled laboratory conditions and under natural or manipulated conditions in the field; (2) exposure to and effects of major classes of environmental contaminants and other ecological perturbations capable of altering ecosystems; (3) case histories involving disruption of natural ecosystems by environmental contaminants; (4) methods used for making estimates, predictions, models, and risk assessments; and (5) identification and description of a number of new and significant issues and methodologies, most of which have come to light since publication of the first edition of this book in 1995. The rationale and some of the key points and concepts presented in each of the five sections are presented below.

## 1.2 QUANTIFYING AND MEASURING ECOTOXICOLOGICAL EFFECTS

Current methodologies for testing and interpretation are provided for aquatic toxicology and design of model aquatic ecosystems, wildlife toxicology, sediment toxicity, soil ecotoxicology, algal and plant toxicity, and the concept of landscape. Identification of biomonitoring programs and current use of biomarkers and bioindicators in aquatic and terrestrial monitoring are also important chapters in this section.

Chapter 2, by Adams and Rowland, provides a comprehensive overview of aquatic toxicology with an emphasis on test methods to meet the requirements of various regulatory guidelines. The chapter describes recent efforts to develop protocols and identify species that permit full-life cycle studies to be performed over shorter durations (e.g., 7-day *Ceriodaphnia dubia* life cycle tests, two-dimensional rotifer tests) and to establish protocols that use sensitive species and life stages that generate accurate estimates of chronic no-effect levels. There has been an increasing need to assess the toxicity of various types of suspect samples in minutes to hours instead of days. The use of rapid assays during on-site effluent biomonitoring allows for the collection of extensive data sets. The expanded use of biomarkers in natural environments, where organisms are exposed to multiple stressors (natural and anthropogenic) over time, will allow better detection of stress and provide an early indication of the potential for population-level effects. Model aquatic ecosystems, known as microcosms and mesocosms, were designed to simulate ecosystems or portions of ecosystems in order to study and evaluate the fate and effects of contaminants. Microcosms are defined by Giesy and Odum[11] as artificially bounded subsets of naturally occurring environments that are replicable, contain several trophic levels, and exhibit system-level properties. Mesocosms are defined as larger, physically enclosed portions of natural ecosystems or man-made structures, such as ponds or stream channels, that may be self-sustaining for long durations. Chapter 3 by Kennedy et al. focuses on key factors in the experimental design of microcosm and mesocosm studies to increase their realism, reduce variability, and assess their ability to detect changes. The success in using such systems depends on the establishment of appropriate temporal and spatial scales of sampling. Emphasis is placed on the need to measure exposure as a function of life history using parameters of size, generation time, habitat, and food requirements. This chapter also addresses the utility of employing a suite of laboratory-to-field experiments and verification monitoring to more fully understand the consequences of single and multiple pollutants entering aquatic ecosystems.

With the advent of modern insecticides and the consequent wildlife losses, screening of pesticides for adverse effects has become an integral part of wildlife toxicology. Avian testing protocols developed by the U.S. Fish and Wildlife Service and other entities include protocols required for regulatory and other purposes. These are described with respect to acute, subacute, subchronic, chronic, developmental, field, and behavioral aspects of avian wildlife toxicity (Hoffman, Chapter 4). Several unique developmental toxicity tests assess the potential hazard of topical contaminant exposure to bird eggs and the sensitivity of "neonatal" nestlings to contaminants, including chemicals used for the control of aquatic weeds, mosquitoes, and wild fires. Coverage of toxicity testing for wild mammals, amphibians, and reptiles is provided as well, although in somewhat less detail since development of such tests has been more limited in scope and requirement.

Sediments serve as both a sink and a source of organic and inorganic materials in aquatic ecosystems, where cycling processes for organic matter and the critical elements occur. Since many potentially toxic chemicals of anthropogenic origin tend to sorb to sediments and organic materials, they become highly concentrated. Sediment toxicity testing (Burton et al., Chapter 5) is an expanding but still relatively new field in ecological assessments. The U.S. Environmental Protection Agency has initiated new efforts in managing contaminated sediments and method standardization that will result in an even greater degree of sediment toxicity testing, regulation, and research in the near future. A number of useful assays have been evaluated in freshwater and marine studies in which the importance of testing multiple species becomes apparent in order to protect the ecosystem. The assay methods described are sensitive to a wide variety of contaminants, discriminate differing levels of contamination, use relevant species, address critical levels of biological organization, and have been used successfully in sediment studies.

The importance of soil ingestion in estimating exposure to environmental contaminants has been best documented in assessments of pesticides or wastes applied to land supporting farm animals. Soil ingestion tends to be most important for those environmental contaminants that are found at relatively high concentrations compared to concentrations in a soil-free diet. Chapter 6, by Beyer and Fries, is designed to relate the toxicological significance of soil ingestion by wild and domestic animals. Concepts covered include methods for determining soil intake, intentional geophagy in animals, soil ingestion by both domestic animals and wildlife, toxicity of environmental contaminants in soil or sediment to animals, relation of particle size of ingested soil to exposure to contaminants, bioavailability of organic and inorganic contaminants in soil, and applications to risk assessments.

Chapter 7, by Linder, Henderson, and Ingham, focuses on applications of ecological risk assessment (ERA) of contaminated soils on wildlife and habitat restoration, since at present there is little or no federal, state, or other guidance to derive soil cleanup values or ecological-based remedial goal options. Three components of this chapter include ERA tools used to characterize a lower bound, the role of bioavailability in critically evaluating these lower bound preliminary remedial goals, and remediation measures intended to address field conditions and modify soil in order to decrease a chemical's immediate bioavailability, while increasing the likelihood of recovery to habitats suitable for future use by fish and wildlife.

Evaluation of the phytotoxicity of a chemical is an essential component of the ecological risk assessment, since primary producers form an essential trophic level of any ecosystem. Since most chemicals introduced into the environment ultimately find their way into aquatic ecosystems, aquatic algal and plant toxicity evaluations are particularly critical. Klaine, Lewis, and Knuteson (Chapter 8) discuss the current state of phytotoxicity testing with particular attention to algal and vascular (both aquatic and terrestrial) plant bioassays. The algal bioassay section not only focuses on test methods developed over the relatively long history of algal toxicity testing, but also includes many adaptations to traditional laboratory methods to provide more realistic phytotoxicity estimates. The vascular plant section focuses on different species used for bioassays and the various endpoints used. Bioassay systems described include soil, hydroponics, foliar, petri dish, and tissue culture.

In recent years ecologists have established a need for studying natural processes not only at the individual, community, or ecosystem level, but over the entire landscape,[12–14] since quite often ecological studies may be too small both spatially and temporally to detect certain important natural processes and the movement of pollutants across multiple ecosystems. Holl and Cairns (Chapter 9) discuss the concept of landscape ecology with a focus on (a) landscape structure, that is, spatial arrangement of ecosystems within landscapes; (b) landscape function, or the interaction among these ecosystems through flow of energy, materials, and organisms; and (c) alterations of this structure and function. Different types of landscape indicators in ecotoxicology are presented.

Biomonitoring data form the basis upon which most long-term stewardship decisions are made. These data often provide the critical linkage between field personnel and decision-makers. Data from biomonitoring programs have been very useful in identifying local, regional, and national ecotoxicological problems. Natural resource management decisions are being made that annually cost millions of dollars. These decisions should be supported by scientifically sound data. Chapter 10, by Breckenridge et al., discusses why monitoring programs are needed and how to design a program that is based on sound scientific principles and objectives. This chapter identifies many of the large-scale monitoring programs in the United States, how to access the information from the programs, and how this information can be used to improve long-term management of natural resources. Bioindicators are an important part of biomonitoring and reflect the bioavailability of contaminants, provide a rapid and inexpensive means for toxicity assessment, may serve as markers of specific classes of chemicals, and serve as an early warning of population and community stress. Melancon (Chapter 11) defines bioindicators as biomarkers (biochemical, physiological, or morphological responses) used to study the status of one or more species typical of a particular ecosystem. Systematically, the responses can range from minor biochemical or physiological homeostatic responses in individual organisms to major toxicity responses in an individual, a species, a community, or an ecosystem. Many currently used bioindicators of contaminant exposure/effect for environmental monitoring are discussed. Some of these bioindicators (e.g., inhibition of cholinesterase by pesticides, induction of hepatic microsomal cytochromes P450 by PAHs and polychlorinated biphenyls (PCBs), reproductive problems such as terata and eggshell thinning, aberrations of hemoglobin synthesis including the effects of lead on ALAD, and porphyria caused by chlorinated hydrocarbons) have been extensively field-validated. Other potentially valuable bioindicators undergoing further validation are discussed and include bile metabolite analysis, oxidative damage and immune competence, metallothioneins, stress proteins, gene arrays, and proteomics.

## 1.3 CONTAMINANT SOURCES AND EFFECTS

The purpose of this section is to identify and describe the effects of significant environmental contaminants and other anthropogenic processes capable of disrupting ecosystems. We have focused on major pesticides (including organophosphorus and carbamate anticholinesterases and persistent organochlorines), petroleum and PAHs, heavy metals (lead and mercury), selenium, polyhalogenated aromatic hydrocarbons, and urban runoff. Toxicity of other metals and trace elements is included in Chapter 40 on amphibian declines, Chapter 44 on trace element interactions, and in three of the case history chapters. Chapters in this section on other important anthropogenic processes include nuclear and thermal contamination, global effects of deforestation, pathogens and disease, and abiotic factors that interact with contaminants.

About 200 organophosphorus (OP) and 50 carbamate (CB) pesticides have been formulated into thousands of products that are available in the world's marketplace for control of fungi, insects, herbaceous plants, and terrestrial vertebrates following application to forests, rangelands, wetlands, cultivated crops, cities, and towns.[15,16] Though most applications are on field crops and other terrestrial habitats, the chemicals often drift or otherwise translocate into nontarget aquatic systems and affect a much larger number of species than originally intended. Hill (Chapter 12) provides an overview of the fate and toxicology of organophosphorus and carbamate pesticides. More attention is given to practical environmental considerations than interpretation of laboratory studies, which were detailed in the first edition of this book. Invertebrates, fish, amphibians, and reptiles are exemplified as ecosystem components and for comparison with birds and mammals. The focus is on concepts of ecological toxicology of birds and mammals related to natural systems as affected by pesticidal application in agriculture and public health. The environmental fate of representative OP and CB pesticides, their availability to wildlife, and toxicology as related to ambient factors, physiological cycles and status, product formulations and sources of exposure are discussed.

It is unlikely that any other group of contaminants has exerted such a heavy toll on the environment as have the organochlorine (OC) pesticides. Blus (Chapter 13) discusses the nature and extent of ecotoxicological problems resulting from the use of organochlorine pesticides for over a half century as well as the future relevance of these problems. Toxicity of OCs is described as influenced by species, sex, age, stress of various kinds, formulations used, and numerous other factors. The eggshell-thinning phenomenon, depressed productivity, and mortality of birds in the field led to experimental studies with OCs, clearly demonstrating their role in environmental problems. An assessment of the environmental impact of OCs leads to the conclusion that the ecotoxicologist must integrate data obtained from controlled experiments with those obtained from the field. In this manner through the use of the "sample egg technique" and other such innovative procedures, controversies over whether DDE or dieldrin were more important in causing a decline of peregrine falcons and other raptors in Great Britain could have been resolved. Although most of the problem OCs have been banned in a number of countries, exposure, bioaccumulation, and ecotoxicological effects will linger far into the future because of the environmental persistence of many compounds and their continued use in a fairly large area.

Petroleum and individual PAHs from anthropogenic sources are found throughout the world in all components of ecosystems. Chapter 14 (Albers) discusses sources and effects of petroleum in the environment. Less than half of the petroleum in the environment originates from spills and discharges associated with petroleum transportation; most comes from industrial, municipal, and household discharges, motorized vehicles, and natural oil seeps. Recovery from the effects of oil spills requires up to 5 years for many wetland plants. Sublethal effects of oil and PAHs on sensitive larval and early juvenile stages of fish, embryotoxic effects through direct exposure of bird eggs, and acute effects in vertebrates are discussed. Evidence linking environmental concentrations of PAHs to induction of cancer in wild animals is strongest for fish. Although concentrations of individual PAHs in aquatic environments are usually much lower than concentrations that are acutely toxic to aquatic organisms, sublethal effects can be produced. Effects of spills on populations of mobile species have been difficult to determine beyond an accounting of immediate losses and, sometimes, short-term changes in local populations.

Lead (Pb) is a nonessential, highly toxic heavy metal, and all known effects of lead on biological systems are deleterious. According to Pattee and Pain (Chapter 15), present anthropogenic lead emissions have resulted in soil and water lead concentrations of up to several orders of magnitude higher than estimated natural concentrations. Consequently, lead concentrations in many living organisms, including vertebrates, may be approaching adverse-effect thresholds. The influence of the chemical and physical form of lead on its distribution within the environment and recent technology to accurately quantify low lead concentrations are described. The chapter also discusses the most significant sources of lead related to direct wildlife mortality and physiological and behavioral effects detected at tissue lead concentrations below those previously considered safe for humans.

The widespread geographic extent and adverse consequences of mercury pollution continue to prompt considerable scientific investigation. Globally increasing concentrations of methylmercury are found in aquatic biota, even at remote sites, as a consequence of multiple anthropogenic sources and their releases of mercury into the environment. For example, in the marine food web of the North Atlantic Ocean, analysis of feathers of fish-eating seabirds sampled from 1885 through 1994 have shown a steady long-term increase in concentration of methylmercury.[17,18] Wiener et al. (Chapter 16) characterize the environmental mercury problem, critically review the ecotoxicology of mercury, and describe the consequences of methylmercury contamination of food webs. Topics include processes and factors that influence exposure to methylmercury, the highly neurotoxic form.

This form readily accumulates in exposed organisms and can biomagnify in food webs to concentrations that can adversely affect organisms in upper trophic levels. Emphasis is given to aquatic food webs, where methylmercury contamination is greatest.

Reproductive impairment due to bioaccumulation of selenium in fish and aquatic birds has been an ongoing focus of fish and wildlife research, not only in the western United States but also in other parts of the world. Selenium is a naturally occurring semimetallic trace element that is essential for animal nutrition in small quantities, but becomes toxic at dietary concentrations that are not much higher than those required for good health. Thus, dietary selenium concentrations that are either below or above the optimal range are of concern. Chapter 17, by Ohlendorf, summarizes the ecotoxicology of excessive selenium exposure for animals, especially as reported during the last 15 years. Focus is primarily on freshwater fish and aquatic birds, because fish and birds are the groups of animals for which most toxic effects have been reported in the wild. However, information related to bioaccumulation by plants and animals as well as to effects in invertebrates, amphibians, reptiles, and mammals is also presented.

PCBs, dioxins (PCDDs), and dibenzofurans (PCDFs) are all similar in their chemistry and manifestation of toxicity, including a high capacity for biomagnification within ecosystems. Mammals, birds, and fish all have representative species that are highly sensitive, as well as highly resistant, to dioxin-like adverse effects, especially chronic reproductive and developmental/endocrine effects. Aquatic food chain species (seals, dolphins, polar bears, fish-eating birds, and cold-water fish species) with high exposure potential through biomagnification are particularly vulnerable. Rice, O'Keefe, and Kubiak (Chapter 18) review the fate of these environmentally persistent compounds and their toxicity, which is complex and often chronic rather than acute. As for PCBs the complexity begins with the large number of compounds, with varying toxicities, that are regularly detected in the environment (100 to 150). With dioxin- and dibenzofuran-related compounds there are fewer commonly measured residues (< 20). However, environmental problems are confounded since they are not directly manufactured but occur as unwanted impurities in manufacturing and incineration.

Urban runoff investigations, which have examined mass balances of pollutants, have concluded that this process is a significant pollutant source. Some studies have even shown important aquatic life impacts for streams in watersheds that are less than 10% urbanized. In general, monitoring of urban stormwater runoff has indicated that the biological beneficial uses of urban receiving waters are most likely affected by habitat destruction and long-term exposures to contaminants (especially to macroinvertebrates via contaminated sediment), while documented effects associated with acute exposures of toxicants in the water column are less likely. Pitt (Chapter 19) recommends longer-term biological monitoring on a site-specific basis, using a variety of techniques, and sediment-quality analyses to best identify and understand these impacts, since water column testing alone has been shown to be very misleading. Most aquatic life impacts associated with urbanization are probably related to long-term problems caused by polluted sediments and food web disruption.

In addition to natural background radiation, irradiation occurs from the normal operation of nuclear power plants and plutonium production reactors, nuclear plant accidents, nuclear weapons testing, and contact with or leakage from radioactive waste storage sites. Assessing the impacts of nuclear power facilities on the environment from routine and accidental releases of radionuclides to aquatic and terrestrial ecosystems is important for the protection of these ecosystems and their species component. The impacts of power-plant cooling systems — impingement, entrainment, elevated water temperatures, heat shock, and cold shock — on aquatic populations and communities have been intensively studied as well. Discussion in Chapter 20 (Meyers-Schone and Talmage) focuses on basic radiological concepts and sources as well as the effects of radiation on terrestrial and aquatic populations and communities of plants and animals. Radiation effects in this chapter focus on field studies, with supporting information from relevant laboratory investigations. Selected examples attempt to relate estimated doses or tissue levels to potential effects; however, dose estimates in the field are often imprecise, and observations are further confounded by the presence of other contaminants or stressors. Thermal toxicity is related to power-plant cooling systems.

Nearly 17 million ha of tropical forests are being cleared each year for new agricultural lands, equivalent to clearing an area the size of the state of Georgia or Wisconsin annually.[19] Global effects of deforestation include irreplaceable loss of species, emissions into the atmosphere of chemically active and heat-trapping trace gases (carbon dioxide, methane, nitrous oxide, and carbon monoxide), and consequent global warming. Current emissions of greenhouse gases from deforestation account for about 25% of the global warming calculated to result from all anthropogenic emissions of greenhouse gases. Continued emissions of greenhouse gases from both deforestation and industrial sources will raise global mean temperature by an estimated 1 to 3.5°C by the end of this century. Houghton (Chapter 21) reviews the contribution of deforestation and subsequent land use with respect to the increasing concentrations of greenhouse gases in the atmosphere and projected global warming. Suggested remedial and preventative actions include (1) a large (≥60%) reduction in the use of fossil fuels through increased efficiency of energy use and a much expanded use of renewable energy sources, (2) the elimination of deforestation, and (3) reforestation of large areas of land, either to store carbon or to provide renewable fuels to replace fossil fuels.

Pathogenic organisms are life forms that cause disease in other life forms; they are components of all ecosystems. Although ecotoxicology is often considered to be the study of chemical pollutants in ecosystems, pathogenic organisms and their diseases are relevant in this context in at least several different ways, as described by Leighton (Chapter 22). Pathogens can be regarded as pollutants when they are released by humans into ecosystems for the first time or when they are concentrated in certain areas by human activity. Four situations in which human activities can alter the occurrence of diseases in the environment include: (1) translocation of pathogens, including manmade ones, host species, and vectors, to new environments; (2) concentration of pathogens or host species in particular areas; (3) changes in the environment that can alter host-pathogen relationships; and (4) creation of new pathogens by intentional genetic modification of organisms.

Environmental factors have long been shown to influence the toxicity of pollutants in living organisms. Drawing upon controlled experiments and field observations, Rattner and Heath (Chapter 23) provide an overview of abiotic environmental factors and perspective on their ecotoxicological significance. Factors discussed include temperature, salinity, water hardness, pH, oxygen tension, nonionizing radiation, photoperiod, and season. Free-ranging animals simultaneously encounter a combination of environmental variables that may influence, and even act synergistically, to alter contaminant toxicity. It is not possible to rank these factors, particularly since they are oftentimes interrelated (e.g., temperature and seasonal rhythms). However, it is clear that environmental factors (particularly temperature) may alter contaminant exposure and toxicity (accumulation, sublethal effects, and lethality) by more than an order of magnitude in some species. Accordingly, it is concluded that effects of abiotic environmental variables should be considered and factored into risk assessments of anthropogenic pollutants.

## 1.4 CASE HISTORIES AND ECOSYSTEM SURVEYS

To illustrate the full impact of different environmental contaminants on diverse ecosystems, seven case histories and ecosystem surveys are presented. These include effects of the nuclear meltdown of Chernobyl, agricultural pesticides on migratory birds in Argentina and Venezuela, impact of mining and smelting on several river basins in the western United States, white phosphorus from spent munitions on waterfowl, and effects of PCBs on the Hudson River.

The partial meltdown of the 1000 Mw reactor at Chernobyl in the Ukraine released large amounts of radiocesium and other radionuclides into the environment, causing widespread contamination of the northern hemisphere, particularly Europe and the former Soviet Union. Eisler (Chapter 24) provides a concise review of the ecological and toxicological aspects of the Chernobyl accident, with an emphasis on natural resources. The most sensitive local ecosystems and organisms are discussed, including soil fauna, pine forest communities, and certain populations of rodents. Elsewhere, reindeer in Scandinavia were among the most seriously afflicted by fallout since they are dependent on lichens, which absorb airborne particles containing radiocesium. Some reindeer calves contaminated with $^{137}Cs$ from Chernobyl showed $^{137}Cs$-dependent decreases in survival and increases in frequency of chromosomal aberrations. The full effect of the Chernobyl nuclear reactor accident on natural resources will probably not be known for at least several decades because of gaps in data on long-term genetic and reproductive effects and on radiocesium cycling and toxicokinetics.

Hooper and co-authors (Chapter 25) describe how recent events in Argentina and Venezuela have shown that pesticide effects transcend national borders with respect to migratory birds exposed to potentially damaging pesticides throughout their range. Despite its withdrawal from the United States, monocrotophos, one of the most acutely toxic pesticides to birds, remained the second highest use OP throughout the world through the mid-1990s, resulting in the death of an estimated 20,000 Swainson's hawks (*Buteo swainsoni*) in Argentina. What has been learned since, as demonstrated by the risks that face many trans-border avian migrants, has clarified the need for greater international cooperation and harmonization of pesticide use. Where a large portion of a species population occupies a small geographical area, either in the course of its migration or on wintering grounds, any localized contaminant or noncontaminant impact can have potentially serious consequences for that population.

Studies conducted in the vicinity of mining operations and smelters have provided some of the most revealing examples of environmental damage caused by metals and associated contaminants. Metal-contaminated soils eroded from exposed and disturbed landscapes and tailings generated during processing may be released to the environment and are associated with increased metal concentrations in surface water and groundwater. Similarly, dispersed sediments often become deposited as alluvial materials in riparian areas and can result in soil metal concentrations greatly exceeding predepositional conditions. Henny (Chapter 26) reviews the history and cause of waterfowl mortality in the Coeur d'Alene (CDA) River Basin of Idaho related to mining sediment containing high concentrations of lead and other metals. Diagnostic procedures and techniques to assess lead poisoning are discussed. Beyer and co-workers[20] concluded that exposure of waterfowl to lead in the CDA River Basin was principally related to the amount of ingested sediment, since the relative amount of lead in vegetation and prey was small. Following this logic and the fact that most raptors neither ingest sediment nor digest bones of prey species that are a major storage site for lead, it becomes clear why the ospreys, hawks, and owls in the CDA River Basin were less contaminated than waterfowl with lead from mining sources. Along the Arkansas River, lead concentrations in livers of nestling tree swallows conclusively demonstrated that lead from sediments is bioavailable to this species prior to fledging (Custer and co-authors, Chapter 28). Lead was detected in most tree swallow livers at two sites along an 11-mile stretch, the most sediment-contaminated section of the Arkansas River. The proportion of livers with detectable lead was less both downstream and upstream of the 11-mile stretch, but with a site-related upstream/downstream gradient in lead concentrations. Additionally, the mean half-peak coefficient of variation of DNA content (HPCV) indicative of possible chromosomal damage was positively correlated with both liver and carcass cadmium concentrations. Linder, Woodword, and Pascoe (Chapter 30) summarize ecological risk-assessment studies focused on metal-contaminated soil, sediment, and surface water for a series of Superfund sites located in the Clark Fork River (CFR) watershed of western Montana. Aquatic, terrestrial, and wetland resources at risk including benthic invertebrates, fish, earthworms, plants, and animals are evaluated.

Eagle River Flats (ERF), located on the Cook Inlet of Anchorage, Alaska, is used by waterfowl and shorebirds throughout the spring and summer and is particularly important as a spring and fall staging or stopover area for more than 75 species of migratory ducks, geese, swans, raptors, gulls, shorebirds, and passerines. Massive waterfowl mortalities due to ingestion of particles of white phosphorus (P4) originating from the firing of munitions into the area occurred and involved over

2000 ducks and swans each year. Sparling (Chapter 27) provides a discussion on toxicity of white phosphorus, a hazard assessment model, efforts to remediate the problem, and how ERF serves as a model for the effects and remediation of P4 on at least 71 other sites where the contaminant has been found.

The Hudson River dominates the history and landscape of eastern New York and surrounding states and has come to symbolize the difficulties associated with finding solutions to the problems of widespread contamination by persistent organic compounds. McCarty (Chapter 29) reviews the history of PCB contamination in the context of the Hudson River ecosystem, describing the patterns of PCB contamination in the biota of the Hudson, the risk contaminants pose to fish and wildlife, and attempts to mitigate those risks.

## 1.5 METHODS FOR MAKING ESTIMATES, PREDICTABILITY, AND RISK ASSESSMENT IN ECOTOXICOLOGY

Ecological risk assessment is a process that evaluates the likelihood that adverse ecological effects will occur as a result of exposure to one or more stressors; it is receiving increasing emphasis in ecotoxicology. Suter[21] defines risk assessment as the process of assigning magnitudes and probabilities to the adverse effects of human activities or natural catastrophes. Examples and uses of ecological risk assessment occur in chapters throughout all sections of this book. However, the fourth section of this book is focused on describing methods and approaches used for estimating and predicting the outcome of potentially ecotoxic events for purposes of risk assessment. These include global disposition of contaminants, bioaccumulation and bioconcentration of contaminants, use of quantitative structure activity relationships (QSARs) for predicting ecological effects of organic chemicals, and population modeling in contaminant studies. Another important part of this section is the current guidelines and future directions of the U.S. Environmental Protection Agency for ecological risk assessment, followed by an exemplary chapter of an ecological risk assessment. The final chapter of this section describes the relationship between restoration ecology and ecotoxicology and quantifies how damaged ecosystems can be restored.

The sources of many environmental contaminants are relatively easy to identify. While short-lived contaminants are most readily identified close to the source, the more persistent substances, such as heavy metals and PCBs, may achieve a truly global distribution due to atmospheric transport and deposition to soils and surface waters. The interim period between emission or discharge of an environmental contaminant and ultimate contact with a specific ecosystem or representative species often contains many varied and interesting processes. Harrison, Harrad, and Lead (Chapter 31) describe some of the more important processes involved in pollutant transport and removal from the environment and discuss how such processes influence the distribution of pollutants. Included are processes leading to the transfer of chemical substances between environmental compartments such as water to air and air to soil. Bioaccumulation and bioconcentration are terms describing the transfer of contaminants from the external environment to an organism. In aquatic organisms bioaccumulation can occur from exposure to sediment (including pore water) or via the food chain (termed trophic transfer). Bioconcentration is the accumulation of waterborne contaminants by aquatic animals through nondietary exposure routes. Biomagnification is defined as the increase in contaminant concentration in an organism in excess of bioconcentration. Biomagnification appears most significant for benthic-based food webs and for very hydrophobic contaminants resistant to biotransformation and biodegradation. As reviewed by Barron (Chapter 32) concern for the bioaccumulation of contaminants arose in the 1960s because of incidents such as toxicity from methyl mercury residues in shellfish and avian reproductive failure due to chlorinated pesticide residues in aquatic species. Bioaccumulation models were first developed in the 1970s to account for processes such as the partitioning of hydrophobic chemicals from water to aquatic organisms. To regulate new and existing chemicals laws such as the Federal Insecticide,

Fungicide and Rodenticide Act contain stringent requirements for bioaccumulation testing. This chapter presents an overview of the principles and determinants governing bioaccumulation from sediments and water and biomagnification in aquatic-based food webs. Organic and metal contaminants are discussed, with an emphasis on hydrophobic organics. The objective of this chapter is to elucidate concepts relating to bioaccumulation, rather than simply present an exhaustive review of the literature.

Structure activity relationships (SARs) are comparisons or relationships between a chemical structure, chemical substructure, or some physical or chemical property associated with that structure or substructure and a biological (e.g., acute toxicity) or chemical (e.g., hydrolysis) activity. When the result is expressed quantitatively, the relationship is a quantitative structure activity relationship (QSAR). Most SARs have been developed for predicting ecological effects of organic chemicals. For SARs that predict ecological effects Walker and Schulz (Chapter 33) provide examples, developmental approaches, universal principles, applications, and recommendations for new QSARs to predict ecotoxicological effects. Since most QSARS have been developed for freshwater aquatic organisms, it is recommended that additional QSARs be developed for predicting effects of chemicals on terrestrial and sediment-dwelling organisms. There is a critical need for these QSARs, especially given the high exposure potential of terrestrial wildlife to pesticides that are intentionally dispersed and to persistent industrial chemicals that are toxic and undergo long-range transport.

The inconvenience and hardship resulting from ecological failures in ecosystem services (e.g., waste processing, provision of potable water, and food production) motivated early attempts at proactive management of the environment, including prediction and mitigation of damage. Cairns and Niederlehner (Chapter 34) discuss the science of predictive ecotoxicology, emphasizing that prediction of environmental outcome is different from appraisal of existing damage and that prediction is solely dependent upon modeling. There is often no way to verify the accuracy of prediction through field survey, yet accuracy checks are essential in assuring that predictive techniques are adequate to management needs. Validation studies compare predictions to appraisals of damage in natural systems. Through these comparisons the magnitude and significance of predictive errors can be evaluated. Ways in which predictive models can be improved are discussed.

A population model is a set of rules or assumptions, expressed as mathematical equations, that describe how animals survive and reproduce. Ecology has a rich history of using models to gain insights into population dynamics. Population models provide a means for evaluating the effects of toxicants in the context of the life cycle of an organism. By developing a model and estimating demographic parameters effects of toxicants on demographic parameters of population growth rates and model stability can be assessed. Also, modeling allows one to identify what portions of the life cycle are most sensitive to toxicants and can guide future data collection and field experiments.

Chapter 35 (Sauer and Pendleton) reviews how population modeling has been used to provide insights into theoretical aspects of ecology and addresses practical questions for resource managers about how population dynamics are affected by changes in the environment. Specific concepts include (1) use of modeling procedures that group populations into discernible age classes, with survival and fecundity rates measured at various intervals for these groups; (2) methods for analyzing the stable population attributes of these models; (3) methods for assessing the effects of changes in the parameters of the models; and (4) applications of the models in evaluating the effects of changes in the demographic parameters.

ERA has been used to evaluate a wide variety of environmental issues of interest such as water-regime management, chemical and biological stressors used to control harmful insects, and toxic chemicals used in industrial processes or present at hazardous waste sites. Chapter 36 (Norton and co-authors) describes recent and ongoing guidelines established by the U.S. Environmental Protection Agency for ERA. Taken into consideration are effects on multiple populations, communities, and ecosystems and the need to consider nonchemical as well as chemical stressors. Problem formulation, characterization of exposure and ecological effects, risk characterization, and case studies and interaction between risk assessment and risk management are discussed. Applications are described in terms of assessing ecological risks from chemicals using probabilistic methods, conducting ecological risk assessment of biological stressors, expanding concepts of exposure, developing management objectives for ERA (including watershed management), and integrating ecological risk assessment with economic, human health, and cultural assessments. This chapter is followed by an exemplary chapter (Byron and co-authors, Chapter 37) describing an actual ERA involving selenium exposure in waterfowl and shorebirds feeding in ephemeral pools in the vicinity of Kesterson Reservoir in central California.

Cairns (Chapter 38) concludes that the adoption of a no-net-ecological-loss policy will require ecological restoration of systems damaged by accidental spills, cumulative impact of anthropogenic stresses over a long period of time, and even ecoterrorism. Ecotoxicology as a scientific discipline will remain essential to ensure that the management practices associated with potentially toxic materials are well understood. Illustrative examples of where the relationship between restoration ecology and ecotoxicology might be most effective include rivers chronically and cumulatively impacted by hazardous materials or by unexpected spills of hazardous materials. For terrestrial systems the Superfund sites in the United States, where accumulations of hazardous materials pose a threat to the surrounding environment and human health, provide an example. Although this chapter emphasizes the relationship between restoration ecology and ecotoxicology, other disciplines should be engaged in order to generate a long-term solution to a complex multivariate problem.

## 1.6 SPECIAL ISSUES IN ECOTOXICOLOGY

The purpose of the fifth section of this book is to identify and describe a number of new and significant issues and approaches in ecotoxicology, most of which have come into focus since the publication of the first edition of this book. These include endocrine-disrupting chemicals and endocrine active agents in the environment, the possible role of contaminants in the worldwide decline of amphibian populations, potential genetic effects of contaminants on animal populations, the role of ecotoxicology in industrial ecology and natural capitalism, the consequences of indirect effects of agricultural pesticides on wildlife, the role of nutrition on trace element toxicity in fish and wildlife, and the role of environmental contaminants on endangered species.

Over the last 5 years there has been a surge of reports in wildlife of suspect endocrine-disruptor-related effects based primarily on adverse reproductive and developmental effects.[22–24] Collectively, there is some evidence of altered reproductive and developmental processes in wildlife exposed to endocrine disruptors, and in the United States, Congress has passed legislation requiring the Environmental Protection Agency to develop, validate, and implement an Endocrine Disruptor Screening Program (EDSP) for identifying potential endocrine-disrupting chemicals. A wide variety of chemicals have been reported as potential endocrine disruptors and are described by Gross and co-workers in Chapter 39 of this book. This chapter reviews and selectively summarizes methods for screening and monitoring of potential endocrine disruptors, the current evidence for endocrine-disrupting effects, and chemical classes in vertebrate wildlife and their potential modes of action. Classes of chemicals include polycyclic aromatic hydrocarbons; polychlorinated and polybrominated biphenyls, dibenzo-*p*-dioxins, dibenzo-*p*-furans; organochlorine pesticides and fungicides; some nonorganochlorine pesticides; complex environmental mixtures; and a few metals.

Over the past decade widespread population declines of amphibians have been documented in North America, Europe, Australia, and Central and South America.[25–27] Population declines in eastern Europe, Asia, and Africa have been suggested as well but are not as well documented. Based on comparative toxicities of organic compounds and metals between amphibians and fish, the overall conclusions were that there was great variation among amphibian species in their sensitivity to metal and organic contaminants, that amphibians generally were more sensitive than fish, and that water-quality criteria established for fish may not be protective of amphibians.

Contaminants may be involved with amphibian population declines, including their possible interaction with other factors discussed by Sparling in Chapter 40.

Understanding changes to the genetic apparatus of an organism exposed to contaminants in the environment is essential to demonstrating an impact on parameters of ecological significance such as population effects. Genetic ecotoxicology attempts to identify changes in the genetic material of natural biota that may be induced by exposure to genotoxicants in their environment and the consequences at various levels of biological organization (molecular, cellular, individual, population, etc.) that may result from this exposure. Shugart, Theodorakis, and Bickham (Chapter 41) describe two major classes of effects studied in genetic ecotoxicology: (1) direct exposure to genotoxicants that have the potential to lead to somatic or heritable (genotoxicological) disease states and that could lower the reproductive output of an affected population, and (2) indirect effects from contaminant stress on populations that lead to alterations in the genetic makeup, a process termed evolutionary toxicology.[28] These latter types of effects alter the inclusive fitness of populations such as by the reduction of genetic variability.

Industrial ecology is the study of the flows of materials and energy in the industrial environment and the effects of those flows on natural systems. Natural capitalism refers to the increasingly critical relationship between natural capital (i.e., natural resources), living systems and the ecosystem services they provide, and manmade capital. Natural capitalism and two of its subdisciplines, industrial and municipal ecology, are essential components in developing a sustainable relationship with natural systems and protecting both natural capital and the delivery of ecosystem services. Cairns (Chapter 42) discusses the role of ecotoxicologists in this sustainability.

Agricultural pesticides have been identified as contributing to the decline of farmland wildlife, although the impact is often exacerbated by other farm practices associated with intensive agriculture. Many species of farmland birds are in decline in the United Kingdom, and there is considerable evidence for the indirect effects of pesticides as the cause. Sotherton and Holland (Chapter 43) discuss how changes in chick survival drive the population size of the once common United Kingdom farmland grey partridge, and conclude that the timing and magnitude of changes in population size and chick survival are consistent with having been caused by the increased use of pesticides, which reduce the insect foods available for partridge chicks. Indirect effects are also likely to impact upon a wide variety of farmland wildlife that are dependent on the same food chain as the grey partridge, and evidence of this is starting to appear for some passerines.

Nutrition of test organisms is one of the most important variables in the conduct of any biological experiment. Deficiencies of vitamins, minerals, and other nutrients in the diet of captive and free-ranging fish and wildlife can result in skeletal deformities, cataracts, histological lesions, abnormal behavior, and many other abnormalities. Excessive amounts of vitamins and minerals have also resulted in abnormalities. The quality of commercial or experimentally prepared diets of captive animals as well as diets consumed by wild animals can influence the acute and chronic toxicity of test compounds. Chapter 44 (Hamilton and Hoffman) examines interactions between nutrition and potentially toxic trace elements and interactions among trace elements. Limited information from dietary studies with trace elements, especially selenium, reveals that diet can have a profound effect on toxicity observed in contaminated ecosystems, yet water-quality standards are rarely derived taking this factor into account. Incorporation of dietary criteria into national criteria for trace elements will occur only after a sufficient database of information is generated from dietary toxicity studies. Recent findings with environmentally relevant forms of mercury (methylmercury) and selenium (selenomethionine) in birds have shown that mercury and selenium may be mutually protective to the toxicity of each other in adult birds but synergistic in combination to the reproductive process in embryos. Further studies are needed to examine the relationship between selenium, nutrition, and other trace elements that may be toxic by compromising cellular antioxidative defense mechanisms. There is also a need for comparative interaction studies in species of wild mammals.

The speed, severity, and taxonomic diversity of declining species is a major concern to ecologists because extinctions are taking place at a rate of approximately 100 species per day.[29] Previously,

Wilson[30] projected the loss of species at more than 20% of the planet's total biodiversity in 20 years. The last chapter (45) of this section by Pattee, Fellows, and Bounds examines the role of contaminants/pollution in the decline of species. Habitat destruction is the primary factor that threatens species and is listed as a significant factor affecting 73% of endangered species. The second major factor causing species decline is the introduction of nonnative species. This affects 68% of endangered species. Pollution and overharvesting were identified as impacting 38% and 15%, respectively, of endangered species. No other contaminant has impacted animal survival to the extent of DDT, which remains one of the few examples of pollution actually extirpating animal species over a significant portion of DDT's range. Once a species is reduced to a remnant of its former population size and distribution, its vulnerability to catastrophic pollution events increases, frequently exceeding or replacing the factors responsible for the initial decline. Therefore, large-scale environmental events, such as global warming, acid rain, and sea-level rise, attract considerable attention as speciation events, adversely affecting some species populations while causing other species to flourish.

The editors of this book conclude that with increasing loss of habitat the quality and fate of the remaining habitat becomes increasingly critical to the survival of species and ecosystems. Species that are endangered or at risk and that occupy a very limited geographical area could be easily decimated by a single event such as an oil or chemical spill or misapplication of pesticides. Furthermore, on a temporal basis where a large portion of a species population occupies a small geographical area, either in the course of its migration or on wintering grounds, any localized impact, whether pesticide-related (e.g., as reported by Hooper and co-authors, Chapter 25) or not, has the potential for serious consequences to populations. For these reasons the balance between shrinking habitat and anthropogenic stressors becomes increasingly crucial to sustain both ecosystems and species diversity.

## REFERENCES


1. Moriarity, F., *Ecotoxicology: The Study of Pollutants in Ecosystems*, 2nd ed., Academic Press, San Diego, 1988.
2. Cairns, J., Jr., Will the real ecotoxicologist please stand up?, *Environ. Toxicol. Chem.*, 8, 843, 1989.
3. Newman, M. C., *Fundamentals of Ecotoxicology*, Sleeping Bear/Ann Arbor Press, Chelsea, Michigan, 1998.
4. Forbes, S. A., The lake as a microcosm, *Bulletin of the Peoria Scientific Association*, 1887, reprinted in *Bull. Ill. State Nat. History Survey*, 15, 537–550, 1925.
5. Penny, C. and Adams, C., Fourth report of the Royal Commission on Pollution in Scotland, *London*, 2, 377, 1863.
6. Weigelt, C., Saare, O., and Schwab, L., Die Schädigung von Fischerei und Fischzueht durch Industrie und Hausabwasser, *Archiv für Hygiene*, 3, 39, 1885.
7. Hart, W. B., Doudoroff, P., and Greenbank, J., *The Evaluation of the Toxicity of Industrial Wastes, Chemicals, and Other Substances to Freshwater Fishes*. Waste Control Laboratory, Atlantic Refining Co., Philadelphia, PA, 1945.
8. Newman, J. R., Effects of industrial air pollution on wildlife, *Biol. Conserv.*, 15, 181, 1979.
9. Carson, R., *Silent Spring*, Houghton Mifflin, Boston, 1962, p. 103.
10. Blus, L. J., Gish, C. D., Belisle, A. A., and Prouty, R. M., Logarithmic relationship of DDE residues to eggshell thinning, *Nature*, 235, 376, 1972.
11. Giesy, J. P., Jr. and Odum, E. P., Microcosmology: Introductory comments, in *Microcosms in Ecological Research*, Giesy, J. P., Jr., Ed., Dept. of Energy Symposium Series 52, Conf. 781101, National Technical Information Service, Springfield, VA, 1990.
12. Forman, R. T. T. and Godron, M., *Landscape Ecology*, John Wiley and Sons, New York, 1986.
13. Cairns, J., Jr. and Niederlehner, B. R., Developing a field of landscape ecotoxicology, *Ecol. Appl.*, 6, 790, 1996.

14. O'Neill, R. V. et al., Monitoring environmental quality at the landscape scale, *Bioscience*, 47, 513, 1997.
15. Smith, G. J., Pesticide Use and Toxicology in Relation to Wildlife: Organophosphorus and Carbamate Compounds, U.S. Fish Wildl. Serv., Resour. Pub. 170, Washington, D.C., 1987.
16. Ecobichon, D. J., Toxic effects of pesticides, in *Casarett and Doull's Toxicology: The Basic Science of Poisons*, 5th ed., Klaassen, C. D., Amdur, M. O., and Doull, J., Eds., McGraw-Hill, New York, 1996, p. 643.
17. FAO, State of the World's Forests 1997, FAO, Rome, 1997.
18. Monteiro, L. R. and Furness, R. W., Accelerated increase in mercury contamination in North Atlantic mesopelagic food chains as indicated by time series of seabird feathers, *Environ. Toxicol. Chem.*, 16, 2489, 1997.
19. Thompson, D. R., Furness, R. W., and Walsh, P. M., Historical changes in mercury concentrations in the marine ecosystem of the north and north-east Atlantic ocean as indicated by seabird feathers, *J. Appl. Ecol.*, 29, 79, 1992.
20. Beyer, W. N., Audet, D. J., Heinz, G. H., Hoffman, D. J., and Day, D., Relation of waterfowl poisoning to sediment lead concentrations in the Coeur d'Alene River Basin, *Ecotoxicology*, 9, 207, 2000
21. Suter, G. W., II, *Ecological Risk Assessment*, Lewis Publishers, Boca Raton, FL, 1993.
22. Colborn, T., von Saal, F. S., and Soto, A. M., Developmental effects of endocrine-disrupting chemicals in wildlife and humans, *Environ. Health Perspect.*, 101, 378, 1993.
23. Kavlock, R. J. and Ankley, G. T., A perspective on the risk assessment process for endocrine-disruptive effects on wildlife and human health, *Risk Analysis*, 16, 731, 1996.
24. Tyler, C. R., Jobling, S., and Sumpter, J. P., Endocrine disruption in wildlife: A critical review of the evidence, *Crit. Rev. Toxicol.*, 28, 319, 1998.
25. Wake, D. B., Declining amphibian populations, *Science*, 253, 860, 1991.
26. Houlahan, J. E., Findlay, C. S., Schmidt, B. R., Meyers, A. H., and Kuzmin, S. L., Quantitative evidence for global amphibian population declines, *Nature*, 404, 752, 2000.
27. Alford, R. A. and Richards, S. J., Global amphibian declines: A problem in applied ecology, *Annu. Rev. Ecol. Sys.*, 30, 133, 1999.
28. Bickham, J. W. and Smolen, M. J., Somatic and heritable effects of environmental genotoxins and the emergence of evolutionary toxicology, *Environ. Health Perspec.*, 102, 25, 1994.
29. Clark, T. W., Reading, R. P., and Clarke, A. L., *Endangered Species Recovery: Finding the Lessons, Improving the Process*, Island Press, Washington, D.C., 1994.
30. Wilson, E.O., Threats to biodiversity, *Sci. Amer.*, 261,108, 1989.

SECTION I

# Quantifying and Measuring Ecotoxicological Effects

CHAPTER 2

# Aquatic Toxicology Test Methods

**William J. Adams and Carolyn D. Rowland**

## CONTENTS



## 2.1 INTRODUCTION

Aquatic toxicology is the study of the effects of toxic agents on aquatic organisms. This broad definition includes the study of toxic effects at the cellular, individual, population, and community levels. The vast majority of studies performed to date have been at the individual level. The intention of this chapter is to provide an overview of aquatic toxicology with an emphasis on reviewing test methods and data collection to meet the requirements of various regulatory guidance.

## 2.2 HISTORICAL REVIEW OF THE DEVELOPMENT OF AQUATIC TOXICOLOGY

Aquatic toxicology grew primarily out of two disciplines: water pollution biology and limnology. The development of these disciplines spanned about 130 years in Europe and the United States. Early studies included basic research to define and identify the biology and morphology of lakes, streams, and rivers. These studies included investigations on how plants, animals, and microorganisms interact to biologically treat sewage and thus reduce organic pollution. For example, the role of bacteria in the nitrification process was demonstrated in 1877 by Schoesing and Muntz. Stephen Forbes is generally credited as one of the earliest researchers of integrated biological communities (Forbes, 1887).[1] Kolwitz and Marsson[2,3] and Forbes and Richardson[4] published approaches for classifying rivers into zones of pollution based on species tolerance and their presence or absence. It has become an accepted belief that the presence or absence of species (especially populations or communities) living in a given aquatic ecosystem provides a more sensitive and reliable indicator of the suitability of environmental conditions than do chemical and physical measurements. Thus, a great deal of effort has been expended over many years in the search for organisms that are sensitive to environmental factors and changes in these parameters. This effort has been paralleled by similar attempts to culture and test sensitive organisms in laboratory settings. The underlying belief has been that organisms tested under controlled laboratory conditions provide a means to evaluate observed effects in natural ecosystems and to predict possible future effects from human-made and natural perturbations. The science of aquatic toxicology evolved out of these studies and has concentrated on studying the effects of toxic agents (chemicals, temperature, dissolved oxygen, pH, salinity, etc.) on aquatic life.

The historical development of aquatic toxicology up to 1970 has been summarized by Warren.[5] Most early toxicity tests consisted of short-term exposure of chemicals or effluents to a limited number of species. Tests ranged from a few minutes to several hours and occasionally 2 to 4 days. There were no standardized procedures. Some of the earliest acute toxicity tests were performed by Penny and Adams (1863)[6] and Weigelt, Saare, and Schwab (1885),[7] who were concerned with toxic chemicals in industrial wastewaters. In 1924 Kathleen Carpenter published the first of her notable papers on the toxicity to fish of heavy metal ions from lead and zinc mines.[8] This was expanded by the work of Jones (1939)[9] and has been followed by thousands of publications over the years on the toxicity of various metals to a wide variety of organisms.

Much of the work conducted in the 1930s and 1940s was done to provide insight into the interpretation of chemical tests as a first step into the incorporation of biological effects testing into the wastewater treatment process or to expand the basic information available on species tolerances, metabolism, and energetics. In 1947 F.E.J. Fry published a classical paper entitled *Effects of the Environment on Animal Activity.*[10] This study investigated the metabolic rate of fish as an integrated response of the whole organism and conceptualized how temperature and oxygen interact to control metabolic rate and hence the scope for activity and growth. Ellis (1937)[11] conducted some of the earliest studies with *Daphnia magna* as a species for evaluating stream pollution. Anderson (1944, 1946)[12,13] expanded this work and laid the groundwork for standardizing procedures for toxicity testing with *Daphnia magna*. Biologists became increasingly aware during this time that chemical analyses could not measure toxicity but only predict it. Hart, Doudoroff, and Greenbank (1945)[14] and Doudoroff (1951)[15] advocated using toxicity tests with fish to evaluate effluent toxicity and supported the development of standardized methods. Using aquatic organisms as reagents to assay effluents led to their description as aquatic bioassays. Doudoroff's 1951 publication[15] led to the first standard procedures, which were eventually included in *Standard Methods for the Examination of Water and Wastewater.*[16] Efforts to standardize aquatic tests were renewed, and the Environmental

Protection Agency (EPA) sponsored a workshop that resulted in a document entitled *Standard Methods for Acute Toxicity Test for Fish and Invertebrates*.[17] This important publication has been the primer for subsequent aquatic standards development and has been used worldwide.

The concept of water quality criteria (WQC) was formulated shortly after World War II. McKee (1952)[18] published a report entitled *Water Quality Criteria* that provided guidance on chemical concentrations not to be exceeded for the protection of aquatic life for the State of California. A second well-known edition by McKee and Wolf (1963)[19] expanded the list of chemicals and the toxicity database. WQC are defined as the scientific data used to judge what limits of variation or alteration of water will not have an adverse effect on the use of water by man or aquatic organisms.[1] An aquatic water quality criterion is usually referred to as a chemical concentration in water derived from a set of toxicity data (criteria) that should not be exceeded (often for a specified period of time) to protect aquatic life. Water quality standards are enforceable limits (concentration in water) not to be exceeded that are adopted by states and approved by the U. S. federal government. Water quality standards consist of WQC in conjunction with plans for their implementation.

In 1976 the EPA published formal guidelines for establishing WQC for aquatic life that were subsequently revised in 1985.[20] Prior to this time WQC were derived by assessing available acute and chronic aquatic toxicity data and selecting levels deemed to protect aquatic life based on the best available data and on good scientific judgment. These national WQC were published at various intervals in books termed the Green Book (1972),[21] the Blue Book (1976),[22] the Red Book (1977),[23] and the Gold Book (1986).[24] In some cases WQC were derived without chronic or partial life-cycle test data. Acute toxicity test results ($LC_{50}$ — lethal concentration to 50% of the test organisms) were used to predict chronic no-effect levels by means of an application factor (AF). The acute value was typically divided by 10 to provide a margin of safety, and the resulting chronic estimate was used as the water quality criterion. It was not until the mid-1960s that chronic test methods were developed and the first full life-cycle chronic toxicity test (with fathead minnows) was performed.[25]

The AF concept emerged in the 1950s as an approach for estimating chronic toxicity from acute data.[26] Stephan and Mount (1967)[27] formalized this AF approach, which was revised by Stephan (1987)[28] and termed the acute-to-chronic ratio (ACR). This approach provides a method for calculating a chronic-effects threshold for a given species when the $LC_{50}$ for that species is known and the average acute-to-chronic ratio for two or more similar species is also available. Dividing the $LC_{50}$ by the ACR provides an estimate of the chronic threshold for the additional species. The approach has generally been calculated as the $LC_{50} \div GMCV$, where GMCV = the geometric mean of the no-observed effect concentration (NOEC) and the lowest observed effect level (LOEC), termed the chronic value (CV). Before the ACR method was published, the AF concept was not used consistently. Arbitrary or "best judgment" values were often used as AFs to estimate chronic thresholds (CVs). Values in the range of 10 to 100 were most often used, but there was no consistent approach. The chronic value has also been alternatively referred to as the geometric mean maximum acceptable toxicant threshold (GM-MATC).

The passage of the Federal Insecticide, Fungicide and Rodenticide Act (FIFRA, 1972), the Toxic Substances Control Act (TSCA, 1976), and the Comprehensive Environmental Compensation Liabilities Act (CERCLA, 1980) as well as the incorporation of toxicity testing (termed biomonitoring) as part of the National Pollution Discharge Elimination System (NPDES, 1989)[29] have increased the need for aquatic toxicological information. Standard methods now exist for numerous freshwater and marine species, including fishes, invertebrates, and algae, that occupy water and sediment environments.

## 2.3 TEST METHODS

The fundamental principle upon which all toxicity tests are based is the recognition that the response of living organisms to the presence (exposure) of toxic agents is dependent upon the dose (exposure level) of the toxic agent. Using this principle, aquatic toxicity tests are designed to describe a concentration-response relationship, referred to as the concentration-response curve when the measured effect is plotted graphically with the concentration. Acute toxicity tests are usually designed to evaluate the concentration-response relationship for survival, whereas chronic studies evaluate sublethal effects such as growth, reproduction, behavior, tissue residues, or biochemical effects and are usually designed to provide an estimate of the concentration that produces no adverse effects.

### 2.3.1 Acute Toxicity Tests

Acute toxicity tests are short-term tests designed to measure the effects of toxic agents on aquatic species during a short period of their life span. Acute toxicity tests evaluate effects on survival over a 24- to 96-hour period. The American Society for Testing and Materials (ASTM), Environment Canada, and the U.S. EPA have published standard guides on how to perform acute toxicity tests for water column and sediment-dwelling species for both freshwater and marine invertebrates and fishes. A list of the standard methods and practices for water-column tests for several species is presented in Table 2.1. The species most often used in North America include the fathead minnow (*Pimephales promelas*), rainbow trout (*Oncorhynchus mykiss*), bluegill (*Lepomis macrochirus*), channel catfish (*Ictalurus punctatus*), sheepshead minnows (*Cyprinodon variegatus*), *Daphnia magna, Ceriodaphnia dubia*, amphipods (*Hyalella azteca*), midges (*Chironomus* sp.), duckweed (*Lemna* sp.), green algae (*Selenastrum capricornutum*), marine algae (*Skeletonema costatum*), mayflies (*Hexagenia* sp.), mysid shrimp (*Mysidopsis bahia*), penaid shrimp (*Penaeus* sp.), grass shrimp (*Palaemonetes pugio*), marine amphipods (*Rhepoynius aboronius and Ampleisca abdita*), marine worms (*Nereis virens*), oysters (*Crassotrea virginica),* marine mussel *(Mytilus edulis),* and marine clams (*Macoma* sp.). Use of particular species for different tests, environmental compartments, and regulations is discussed in the following sections.

Acute toxicity tests are usually performed by using five concentrations, a control, and a vehicle (i.e., solvent) control if a vehicle is needed, generally with 10 to 20 organisms per concentration. Most regulatory guidelines require duplicate exposure levels, although this is not required for pesticide registration. Overlying water quality parameters are generally required to fall within the following range: temperature, ±1°C; pH, 6.5 to 8.5; dissolved oxygen, greater than 60% of saturation; hardness (moderately hard), 140 to 160 mg/L as $CaCO_3$. For marine testing, salinity is controlled to appropriate specified levels. All of the above variables, as well as the test concentration, are typically measured at the beginning and end of the study and occasionally more often. This basic experimental design applies for most regulations and species.

### 2.3.2 Chronic Toxicity Tests

Chronic toxicity tests are designed to measure the effects of toxicants to aquatic species over a significant portion of the organism's life cycle, typically one tenth or more of the organism's lifetime. Chronic studies evaluate the sublethal effects of toxicants on reproduction, growth, and behavior due to physiological and biochemical disruptions. Effects on survival are most frequently evaluated, but they are not always the main objective of the study. Examples of chronic aquatic toxicity studies have included: brook trout (*Salvelinus fontinalis*), fathead and sheepshead minnow, daphnids, (*Daphnia magna), (Ceriodaphnia dubia*), oligochaete (*Lumbriculus variegatus), midge (Chironomus tentans),* freshwater amphipod *(Hyalella azteca),* zebrafish (*Brachydanio rerio*), and mysid shrimp (*Americamysis bahia*). Algal tests are typically 3 to 4 days in length and are often reported as acute tests. However, algal species reproduce fast enough that several generations are exposed during a typical study, and therefore these studies should be classified as chronic studies. Currently, many regulatory agencies regard an algal $EC_{50}$ as an acute test result and the NOEC or the $EC_{10}$ as a chronic test result.

**Table 2.1 Summary of Published U.S. Environmental Protection Agency (U.S. EPA), the American Society for Testing and Materials (ASTM), and Environment Canada (EC) Methods for Conducting Aquatic Toxicity Tests**

| Test Description | Reference |
|---|---|
| Methods for Acute Toxicity Tests with Fish, Macroinvertebrates, and Amphibians | EPA-660/3-75-009 |
| Methods for Measuring the Acute Toxicity of Effluents and Receiving Waters to Freshwater and Marine Organisms | EPA/600/4-90/027F |
| Short-Term Methods for Estimating the Chronic Toxicity of Effluents and Receiving Waters to Freshwater Organisms | EPA/600/4-91/002 |
| Short-Term Methods for Estimating the Chronic Toxicity of Effluents and Receiving Waters to West Coast and Marine and Estuarine Organisms | EPA/600/R-95/136 |
| Short-Term Methods for Estimating the Chronic Toxicity of Effluents and Receiving Waters to Marine and Estuarine Organisms | EPA/600/4-91/003 |
| Methods Guidance and Recommendations for Whole Effluent Toxicity (WET) Testing (40 CFR Part 136) | EPA/821/B-00/004 |
| Methods for Aquatic Toxicity Identification Evaluations: Phase I. Toxicity Characterization Procedures | EPA-600/6-91/003 |
| Methods for Aquatic Toxicity Identification Evaluations: Phase II. Toxicity Identification Procedures for Samples Exhibiting Acute and Chronic Toxicity. | EPA-600/R-92/080 |
| Methods for Aquatic Toxicity Identification Evaluations: Phase III. Toxicity Confirmation Procedures for Samples Exhibiting Acute and Chronic Toxicity. | EPA-600/R-92/081 |
| Toxicity Identification Evaluation: Characterization of Chronically Toxic Effluents, Phase I | EPA-600/6-91/005F |
| Conducting Static Acute Toxicity Tests Starting with Embryos of Four Species of Saltwater Bivalve Mollusks | ASTM E 724-98 |
| Conducting Acute Toxicity Tests on Materials with Fishes, Macroinvertebrates, and Amphibians | ASTM E 729-96 |
| Guide for Conducting Acute Toxicity Tests with Fishes, Macroinvertebrates, and Amphibians | ASTM E 729-88 |
| Conducting Bioconcentration Tests with Fishes and Saltwater Bivalve Mollusks | ASTM E 1022-94 |
| Assessing the Hazard of a Material to Aquatic Organisms and Their Uses | ASTM E 1023-84 |
| Life-Cycle Toxicity Tests with Saltwater Mysids | ASTM E 1191-97 |
| Conducting Acute Toxicity Tests on Aqueous Ambient Samples and Effluents with Fishes, Macroinvertebrates, and Amphibians | ASTM E 1192-97 |
| Conducting *Daphnia magna* Life Cycle Toxicity Tests | ASTM E 1193-97 |
| Using Brine Shrimp Nauplii as Food for Test Animals in Aquatic Toxicology | ASTM E 1203-98 |
| Conducting Static 96-h Toxicity Tests with Microalgae | ASTM E 1218-97a |
| Conducting Early Life-Stage Toxicity Tests with Fishes | ASTM E 1241-97 |
| Using Octanol-Water Partition Coefficient to Estimate Median Lethal Concentrations for Fish Due to Narcosis | ASTM E 1242-88 |
| Three-Brood, Renewal Toxicity Tests with *Ceriodaphnia dubia* | ASTM E 1295-89 |
| Standardized Aquatic Microcosm: Fresh Water | ASTM E 1366-96 |
| Conducting Static Toxicity Tests with *Lemna gibba* G3 | ASTM E 1415-91 |
| Conducting the Frog Embryo Teratogenesis Assay-Xenopus (FETAX) | ASTM E 1439-98 |
| Acute Toxicity Tests with the Rotifer Brachionus | ASTM E 1440-91 |
| Conducting Static and Flow-Through Acute Toxicity Tests with Mysids from the West Coast of the United States | ASTM E 1463-92 |
| Conducting Sexual Reproduction Tests with Seaweeds | ASTM E 1498-92 |
| Conducting Acute, Chronic and Life-Cycle Aquatic Toxicity Tests with Polychaetous Annelids | ASTM E 1562-94 |
| Conducting Static Acute Toxicity Tests with Echinoid Embryos | ASTM E 1563-98 |
| Conducting Renewal Phytotoxicity Tests with Freshwater Emergent Macrophytes | ASTM E 1841-96 |
| Conducting Static, Axenic, 14-day Phytotoxicity Tests in Test Tubes with the Submersed Aquatic Macrophyte *Myriophyllum sibiricum* Komarov | ASTM E 1913-97 |
| Conducting Toxicity Tests with Bioluminescent Dinoflagellates | ASTM E 1924-97 |
| Algal Growth Potential Testing with *Selenastrum capricornutum* | ASTM D 3978-80 |
| Acute Lethality Test Using Rainbow Trout | EPS 1/RM/9 |
| Acute Lethality Test Using Threespine Stickleback | EPS 1 RM/10 |
| Acute Lethality Test Using *Daphnia* ssp. | EPS 1/RM/11 |
| Test of Reproduction and Survival Using the Cladoceran *Ceriodaphnia dubia* | EPS 1/RM/21 |
| Test of Larval Growth and Survival Using Fathead Minnows | EPS 1/RM/22 |
| Toxicity Test Using Luminescent Bacteria (*Photobacterium phosphoreum*) | EPS 1/RM/24 |

**Table 2.1 Summary of Published U.S. Environmental Protection Agency (U.S. EPA), the American Society for Testing and Materials (ASTM), and Environment Canada (EC) Methods for Conducting Aquatic Toxicity Tests *(Continued)***

| Test Description | Reference |
|---|---|
| Growth Inhibition Test Using the Freshwater Alga (*Selenastrum capricornutum*) | EPS 1/RM/25 |
| Fertilization Assay with Echinoids (Sea Urchin and Sand Dollars) | EPS 1/RM/27 |
| Toxicity Testing Using Early Life Stages of Salmonid Fish (Rainbow Trout) – Second Edition | EPS 1/RM/28 |
| Test for Measuring the Inhibition of Growth Using the Freshwater Macrophyte – *Lemna minor* | EPS 1/RM/37 |
| Reference Method of Determining Acute Lethality of Effluents to Rainbow Trout | EPS 1/RM/13 |
| Reference Method for Determining Acute Lethality of Effluents to *Daphnia magna* | EPS 1/RM/1 |

*Note:* EPS = Environmental Protection Series (Environment Canada).

Partial life-cycle studies are often referred to as chronic studies; however, frequently only the most sensitive life stages are utilized for exposure in these studies and they should therefore not be considered true chronic studies. Hence, they are often referred to as partial chronic or subchronic studies. Common examples of partial life-cycle studies are the fish early-life-stage studies with fathead and sheepshead minnows, zebrafish, and rainbow trout. These studies generally expose the most vulnerable developmental stage, the embryo and larval stage (30 to 60 days post-hatch), to a toxicant and evaluate the effects on survival, growth, and sometimes behavior. Recently, procedures have been developed for an abbreviated fathead minnow life-cycle test to assess the potential of substances to affect reproduction.[30] This test was developed in response to a need to screen for endocrine-disrupting chemicals. Likewise, a partial life-cycle test with *Xenopus laevis* that evaluates tail resorption as a screen for thyroid active substances was recently developed.[31]

### 2.3.3 Static Toxicity Tests

Effluent, sediment, and dredged-materials tests are often performed in static or static renewal systems. Static toxicity tests are assays in which the water or toxicant in test beakers is not renewed during the exposure period. Static toxicity tests are most frequently associated with acute testing. The most common static acute tests are those performed with daphnids, mysids, amphipods, and various fishes. Renewal tests (sometimes called static renewal tests) refer to tests where the toxicant and dilution water is replaced periodically (usually daily or every other day). Renewal tests are often used for daphnid life-cycle studies with *Ceriodaphnia dubia* and *Daphnia magna* that are conducted for 7 and 21 days, respectively. Renewal tests have also been standardized for abbreviated early-life-stage studies or partial life-cycle studies with several species (e.g., 7- to 10- day fathead minnow early-life-stage studies).

Static and renewal tests are usually not an appropriate choice if the test material is unstable, sorbs to the test vessel, is highly volatile, or exerts a large oxygen demand. When any of these situations is apparent, a flow-through system is preferable. Static-test systems are usually limited to 1.0 g of biomass per liter of test solution so as not to deplete the oxygen in the test solution. More detail on fundamental procedures for conducting aquatic toxicity bioassays can be found in Sprague, 1969, 1973 and Rand, 1995.[32–34]

### 2.3.4 Flow-Through Toxicity Tests

Flow-through tests are designed to replace toxicant and the dilution water either continuously (continuous-flow tests) or at regular intermittent intervals (intermittent-flow tests). Longer-term studies are usually performed in this manner. Flow-through tests are generally thought of as being superior to static tests as they are much more efficient at maintaining a higher-level of water quality, ensuring the health of the test organisms. Static tests designed to provide the same organism mass to total water test volume as used in a flow-through study can maintain approximately the same water quality. Flow-through tests usually eliminate concerns related to ammonia buildup and dissolved oxygen usage as well as ensure that the toxicant concentration remains constant. This approach allows for more test organisms to be used in a similar size test system (number of organisms/standing volume/unit time) than do static tests.

There are many types of intermittent-flow diluter systems that have been designed to deliver dilution water and test for chemical presence in intermittent-flow toxicity tests. The most common system is that published by Mount and Brungs.[35] Continuous-flow systems provide a steady supply of dilution water and toxicant to the test vessels. This is achieved with a diluter system that utilizes flow meters to accurately control the delivery of water and metering pumps or syringes to deliver the toxicant.[36]

### 2.3.5 Sediment Tests

The science of sediment-toxicity testing has rapidly expanded during the past decade. Sediments in natural systems and in test systems often act as a sink for environmental contaminants, frequently reducing their bioavailability. Bioavailability refers to that fraction of a contaminant present that is available for uptake by aquatic organisms and capable of exerting a toxic effect. The extent to which the bioavailability is reduced by sediments is dependent upon the physical-chemical properties of the test chemical and the properties of the sediment. Past studies have demonstrated that chemical concentrations that produce biological effects in one sediment type often do not produce effects in other sediments even when the concentration is a factor of 10 or higher. The difference is due to the bioavailability of the sediment-sorbed chemical.

The ability to estimate bioavailability is a key factor in ultimately assessing the hazard of chemicals associated with sediments. Much progress has been made in this area recently. It is now widely recognized that the organic carbon content of the sediment is the component most responsible for controlling the bioavailability of nonionic (nonpolar) organic chemicals.[37, 38] This concept has been incorporated into an approach termed the "Equilibrium Partitioning Approach" and is being used by the EPA for establishing sediment quality criteria.[39] For some metals (cadmium, copper, nickel, and lead, silver, and zinc) the acid volatile sulfide (AVS) content of the sediments has recently been shown to control metal bioavailability in sediments with sufficient sulfide. AVS is a measure of the easily extractable fraction of the total sulfide content associated with sediment mineral surfaces. Metal-sulfide complexes are highly insoluble, which limits the bioavailability of certain metals. When the AVS content of the sediment is exceeded by the metal concentration (on a molar ratio of 1:1), free metal ion toxicity may be expressed.[40] Recent research shows that toxicity is frequently not expressed when SEM exceeds AVS due to the fact that metal ions are sorbed to sediment organic carbon or other reactive surfaces such as iron and manganese oxides.[41] Approaches for additional classes of compounds such as polar ionic chemicals have been proposed.[42] Recently, an approach was developed for assessing the combined effects of multiple PAHs sorbed to sediments based on equilibrium partitioning, narcosis toxicity theory, and the concept that chemicals within a given class of compounds with the same mode of action act in a predictive and additive manner.[43, 44]

The recognition that sediments are both a sink and a source for chemicals in natural environments has led to increased interest in sediments and to the development of standard testing methods for sediment-dwelling organisms. Until recently, most sediment tests were acute studies. Greater emphasis is now placed on chronic sediment-toxicity tests with sensitive organisms and sensitive life stages. For example, partial life-cycle test procedures are available for several species of amphipods and the sea urchin. Full life-cycle tests can be performed with the marine worm *Nereis virens,* freshwater midges (*C. tentans* and *Paratanytarsus disimilis*), and freshwater amphipods (*H. azteca*) (Table 2.2). Partial and full life-cycle tests can be performed with epibenthic species such

**Table 2.2 Summary of Published U.S. Environmental Protection Agency (U.S. EPA), the American Society for Testing and Materials (ASTM) and Environment Canada (EC) Methods for Conducting Sediment Toxicity Tests**

| Test Description | Reference |
|---|---|
| Methods for Measuring the Toxicity and Bioaccumulation of Sediment-Associated Contaminants with Freshwater Invertebrates. | EPA/600/R-99/064 |
| Standard Guide for Conduction of 10-day Static Sediment Toxicity Tests with Marine and Estuarine Amphipods | ASTM E 1367-92 |
| Standard Guide for Collection, Storage, Characterization, and Manipulation of Sediments for Toxicological Testing | ASTM E 1391-94 |
| Standard Guide for Designing Biological Test with Sediments | ASTM E 1525-94a |
| Standard Test Methods for Measuring the Toxicity of Sediment-Associated Contaminants with Freshwater invertebrates | ASTM E 1706-95b |
| Standard Guide for Conduction of Sediment Toxicity Tests with Marine and Estuarine Polychaetous Annelids | ASTM E 1611 |
| Standard Guide for Determination of Bioaccumulation of Sediment-Associated Contaminants by Benthic Invertebrates | ASTM E 1688-00 |
| Acute Test for Sediment Toxicity Using Marine and Estuarine Amphipods | EPS 1/RM/26 |
| Test for Survival and Growth in Sediment Using Freshwater Midge Larvae *Chironomus tentans* or *riparius* | EPS 1/RM/32 |
| Test for Survival and Growth in Sediment Using Freshwater Amphipod *Hyalella azteca* I | EPS 1/RM/33 |
| Test for Survival and Growth for Sediment Using a Marine Polychaete Worm | EPS 1/RM/* |
| Reference Method for Determining Acute Lethality of Sediments to Estuarine or Marine Amphipods | EPS 1/RM/35 |
| Reference Method of Determining Sediment Toxicity Using Luminescent Bacteria | EPS 1/RM/* |
| Sediment-Water Chironomid Toxicity Test Using Spiked Sediment | 218 |
| Sediment-Water Chironomid Toxicity Test Using Spiked Water | 219 |

*Note:* EPS = Environmental Protection Series (Environment Canada).

* Document in preparation.

as *D. magna* and *C. dubia*. These species can be tested with sediments present in the test vessels. Porewater (interstitial water) exposures offer a potentially sensitive approach to the toxicity of the freely dissolved fractions of contaminants. The interstitial water is extracted from the sediment, usually by centrifugation, and subsequently used in toxicity tests with a wide variety of test organisms and life stages. The use of porewater allows for the testing of fish early-life stages as well as invertebrates. An extensive review of porewater testing methods and utility of the data was recently summarized at a SETAC Pellston Workshop.[45] Available sediment-assessment methods have been reviewed by Adams et al.[46] Guidance for conducting sediment bioassays for evaluating the potential to dispose of dredge sediment via open ocean disposal has been summarized in the EPA-Corps of Engineers (COE) Green Book.[47]

Typical sediment bioassays are used to evaluate the potential toxicity or bioaccumulation of chemicals in whole sediments. Sediments may be collected from the field or spiked with compounds in the laboratory. Spiked and unspiked sediment tests are performed in either static or flow-through systems, depending on the organism and the test design. Flow-through procedures are most often preferred. Between 2 and 16 replicates are used, and the number of organisms varies from 10 to 30 per test vessel. Sediment depth in the test vessels often ranges from 2 to 6 cm and occasionally as deep as 10 cm. Test vessels often range from 100 to 4000 mL in volume. Sediment tests for field projects are not based on a set number of test concentrations but rely on a comparison of control and reference samples with sediments from sites of interest. Care must be exercised in selecting sites for testing, collecting, handling, and storing the sediments. [48,49] Likewise, special procedures have been devised for spiking sediments with test substances. A reference sediment from an area known to be contaminant-free and that provides for good survival and growth of the test organisms is often included as an additional control in the test design. Guidance for selecting reference samples and sites can be found in the EPA-COE Green Book.[47]

### 2.3.6 Bioconcentration Studies

Bioconcentration is defined as the net accumulation of a material from water into and onto an aquatic organism resulting from simultaneous uptake and depuration. Bioconcentration studies are performed to evaluate the potential for a chemical to accumulate in aquatic organisms, which may subsequently be consumed by higher trophic-level organisms including man (ASTM E 1022–94, Table 2.1). The extent to which a chemical is concentrated in tissue above the level in water is referred to as the bioconcentration factor (BCF). It is widely recognized that the octanol/water partition coefficient — referred to as $K_{ow,}$ Log $K_{ow}$ or Log *P* — can be used to estimate the potential for nonionizable organic chemicals to bioconcentrate in aquatic organisms. Octanol is used as a surrogate for tissue lipid in the estimation procedure. Equations used to predict BCFs have been summarized by Boethling and Mackay.[50] While the use of $K_{ow}$ is useful for estimating the bioconcentration potential of nonpolar organics, it is not useful for metals or ionizable or polar substances. Additionally, it should be recognized that the use of BCFs have limited utility for metals and other inorganic substances that may be regulated to some extent and that typically have BCFs that are inversely related to exposure concentration. For these substances the BCF value is not an intrinsic property of the substance.[51,52]

Methods for conducting bioconcentration studies have been described and summarized for fishes and saltwater bivalves by ASTM (Table 2.1) and TSCA (Table 2.3). To date, the scientific community has focused its efforts on developing methods for fishes and bivalves because these species are higher trophic-level organisms and are most often consumed by man. In general, the approach for determining the BCF for a given chemical and species is to expose several organisms to an environmentally relevant chemical of interest that is no more than one tenth of the $LC_{50}$ (lethal concentration) for the species being tested. At this exposure level mortality due to the test chemical can usually be avoided. The test population is sampled repeatedly, and tissue residues (usually in the fillet, viscera, and whole fish) are measured. This is most often done with $C^{14}$ chemicals to facilitate tissue residue measurements. The study continues until apparent steady state is reached (a plot of tissue chemical concentrations becomes asymptotic with time) or for 28 days. At this point the remaining fish are placed in clean water, and the elimination (depuration) of the chemical from the test species is measured by analyzing tissues at several time intervals.

Apparent steady state can be defined as that point in the experiment where tissue residue levels are no longer increasing. Three successive measurements over 2 to 4 days showing similar tissue concentrations are usually indicative of steady state. When steady state has been achieved, the uptake and depuration rates are approximately equal. It has been shown that 28 days is adequate for most chemicals to reach steady state. However, this is not true for chemicals with a large $K_{ow}$ (e.g., DDT, PCBs). An estimate of the time required to reach apparent steady state can be made for a given species based on previous experiments with a similar chemical or using $K_{ow}$ for nonionizable chemicals that follow a two-compartment, two-parameter model for uptake and depuration. The following equation is used: $S = \{\ln[1/(1.00 - 0.95)]\}/k_2 = 3.0/k_2$, where: S = number of days, ln = logarithm to the base e, $k_2$ = the first-order depuration constant ($day^{-1}$) and where $k_2$ for fishes is estimated as antilog ($1.47 - 0.414 \log K_{ow}$).[53] The use of $K_{ow}$ for estimating the BCF or time to equilibrium is not useful for polar substances or inorganic substances such as metals.

Two additional terms of interest are bioaccumulation and biomagnification. The first refers to chemical uptake and accumulation in tissues by an organism from any external phase (water, food, or sediment). Biomagnification is the process whereby a chemical is passed from a lower to successively higher trophic levels, resulting in successively higher residue at each trophic level. Biomagnification is said to occur when the trophic transfer factor exceeds 1.0 for two successive trophic levels (e.g., algae to invertebrates to fish). Biomagnification is generally thought to occur only with chemicals with a large $K_{ow}$ (>4.0) and does not occur for inorganic substances.[54] Specific tests and standard guidelines have been developed for measuring bioaccumulation of sediment associated contaminants in the freshwater oligochaete *L. variegatus* (EPA and ASTM).[55, 56]

**Table 2.3 Summary of the Toxicity Test Requirements by Regulatory Guideline**

| Regulatory Guideline | Type of Testing Required |
|---|---|
| **Clean Water Act (CWA)** | Aquatic Tests for the Protection of Surface Waters |
| U.S. EPA NPDES Regulations | Effluent Biomonitoring Studies |
| | Toxicity Identification and Reduction Evaluations |
| **Water Quality Standards** | Aquatic Tests for the Development of Water Quality Criteria (WQC) |
| **Toxic Substances Control Act (TSCA)** | Industrial and Specialty Chemicals: Aquatic Assessments |
| **Federal Insecticide, Fungicide and Rodenticide Act (FIFRA)** | |
| Premanufacture Notification, PMN | Algae, daphnid, and one fish species |
| Section Four Test Rule | Data set requirements may include multiple acutes with fish algae and invertebrates, freshwater and marine; followed by 1–3 chronic or partial life-cycle studies. A sediment study with midge and a bioconcentration study may be required if low $K_{ow} > 3.0$. |
| Adams et al. (1985) | Midge partial life cycle test with sediments |
| **TSCA and FIFRA Aquatic Test Guideline Numbers** | |
| Series 850 OPPTS Ecological Effects Test Guidelines | |
| (Aquatic Test Guideline Number): | |
| 850.1010 | Aquatic invertebrate acute toxicity test, freshwater daphnids |
| 850.1012 | Gammarid acute toxicity test |
| 850.1025 | Oyster acute toxicity test (shell deposition) |
| 850.1035 | Mysid acute toxicity test |
| 850.1045 | Penaeid acute toxicity test |
| 850.1055 | Bivalve acute toxicity test (embryo larval) |
| 850.1075 | Fish acute toxicity test freshwater and marine |
| 850.1085 | Fish acute toxicity test mitigated by humic acid |
| 850.1300 | Daphid chronic toxicity test |
| 850.1350 | Mysid chronic toxicity test |
| 850.1400 | Fish early-life state toxicity test |
| 850.1500 | Fish life cycle toxicity test |
| 850.1710 | Oyster BCF |
| 850.1730 | Fish BCF |
| 850.1735 | Whole sediment acute toxicity invertebrates, freshwater |
| 850.1740 | Whole sediment acute toxicity invertebrates, marine |
| 850.1790 | Chironomid sediment toxicity test |
| 850.1800 | Tadpole/sediment subchronic toxicity test |
| 850.1850 | Aquatic food chain transfer |
| 850.1900 | Generic freshwater microcosm test, laboratory |
| 850.1925 | Site-specific microcosm test, laboratory |
| 850.1950 | Field testing for aquatic organisms |
| 850.4400 | Aquatic plant toxicity test using *Lemna* spp. |
| 850.4450 | Aquatic plants field study, Tier III |
| 850.5100 | Soil microbial community toxicity test |
| 850.5400 | Algal toxicity, Tiers I and II |
| 850.6200 | Earthworm subchronic toxicity test |
| Adams et al. (1985) | Midge partial life cycle test with sediments |
| **Food and Drug Administration (FDA)** | New Drug Environmental Assessments |
| Environmental Effects Test Number: | |
| 4.01 | Algal test |
| 4.08 | *Daphnia magna* acute toxicity |
| 4.09 | *Daphnia magna* chronic toxicity |
| 4.10 | *Hyalella azteca* acute toxicity |
| 4.11 | Freshwater fish acute toxicity |
| 4.12 | Earthworm subacute toxicity |

**Table 2.3 Summary of the Toxicity Test Requirements by Regulatory Guideline *(Continued)***

| Regulatory Guideline | Type of Testing Required |
|---|---|
| **Organization of Economic Cooperation and Development (OEDC) and European Economic Community (EEC)** | European Community Aquatic Testing Requirements |
| Aquatic Effects Testing: | |
| 201 | Algal growth inhibition test |
| 202 | *Daphnia magna* Acute Immobilization Test and Reproduction Test |
| 203 | Fish, Acute Toxicity Test: 14-Day Study |
| 204 | Fish, Prolonged Toxicity Test: 14-Day Study |
| 210 | Fish, Early Life-Stage Toxicity Test |
| 211 | *Daphnia magna* Reproduction Test |
| 212 | Fish, Short-Term Toxicity Test on Embryo and Sac-Fry Stages |
| 215 | Fish, Juvenile Growth Test |
| 221 | *Lemna* sp. Growth Inhibition Test |
| 305 | Bioconcentration: Flow-Through Fish Test |
| **PARCOM European Community: Paris Commission** | Offshore Chemical Notification/Evaluation |
| — | Algal growth inhibition test (*Skeletonema costatum* or *Phaeodactylum tricornutum*) |
| — | Invertebrate acute toxicity test (*Acartia tonsa*, *Mysidopsid* sp., *Tisbe* sp.) |
| — | Fish Acute Toxicity Test (*Scophthalmus* sp.) |
| | Sediment Reworker Test (*Corophium volutator, Nereis virens*, and *Abra alba*) |

*Note:* — Indicates no guideline number available.

## 2.4 TOXICOLOGICAL ENDPOINTS

Toxicological endpoints are values derived from toxicity tests that are the results of specific measurements made during or at the conclusion of the test. Two broad categories of endpoints widely used are assessment and measures of effect. Assessment endpoints refer to the population, community, or ecosystem parameters that are to be protected (e.g., population growth rate, sustainable yield). Measures of effect refer to the variables measured, often at the individual level, that are used to evaluate the assessment endpoints. The measures of effect describe the variables of interest for a given test. The most common measures of effect include descriptions of the effects of toxic agents on survival, growth, and reproduction of a single species. Other measures of effect include descriptions of community effects (respiration, photosynthesis, diversity) or cellular effects such as physiological/histopathological effects (backbone collagen levels, ATP/ADP levels, RNA/DNA ratios, biomarkers, etc.). In each case the endpoint is a variable that can be quantitatively measured and used to evaluate the effects of a toxic agent on a given individual, population, or community. The underlying assumption in making toxicological endpoint measurements is that the endpoints can be used to evaluate or predict the effects of toxic agents in natural environments. EPA risk-assessment guidelines provide information on how endpoints can be used in the environmental risk-assessment process.[56]

### 2.4.1 Acute Toxicity Tests

Endpoints most often measured in acute toxicity tests include a determination of the LC or $EC_{50}$ (median effective concentration), an estimate of the acute no-observed effect concentration (NOEC), and behavioral observations. The primary endpoint is the LC or $EC_{50}$. The $LC_{50}$ is a lethal concentration that is estimated to kill 50% of a test population. An $EC_{50}$ measures immobilization or an endpoint other than death. The LC and $EC_{50}$ values are measures of central tendency and can be determined by a number of statistical approaches.[57] The Litchfield-Wilcoxen approach is most often used[58] and consists of plotting the survival and test chemical concentration data on log-probability paper, drawing a straight line through the data, checking the goodness of fit of the line with a chi-squared test, and reading the LC or $EC_{50}$ directly off the graph. Various computer packages are also available to perform this calculation. Other common methods include the moving-average and binomial methods. The latter is most often used with data sets where the dose-response curve is steep and no mortality was observed between the concentrations where zero and 100% mortality was observed.

The NOEC (no-observed effect concentration — acute and chronic tests) is the highest concentration in which there is no significant difference from the control treatment. The LOEC (lowest observed effect concentration — acute and chronic tests) is the lowest concentration in which there is a significant difference from the control treatment. The NOEC and LOEC are determined by examining the data and comparing treatments against the control in order to detect significant differences via hypothesis testing. The effects can be mortality, immobilization, reduced cell count (algae), or behavioral observations. These endpoints are typically determined using t-tests and analysis of variance (ANOVA) and are most often associated with chronic tests. NOECs/LOECs are concentration-dependent and do not have associated confidence intervals. Sebaugh et al. demonstrated that the $LC_{10}$ could be used as a substitute for the observed no-effect concentration for acute tests.[59] This provides a statistically valid approach for calculating the endpoint and makes it possible to estimate when the lowest concentration results in greater than 10% effects. It should be noted, however, that the confidence in the estimated LC decreases as one moves away from 50%. Regression analysis, as opposed to hypothesis testing, is gaining favor as a technique for evaluating both acute and chronic data. The advantage is that it allows for the calculation of a percentage of the population of test organisms affected, as opposed to ANOVA, which simply determines whether or not a given response varies significantly from the control organisms. EC and LC values are readily incorporated into risk-assessment models and are particularly useful in probabilistic risk assessments.[60,61]

### 2.4.2 Partial Life-Cycle and Chronic Toxicity Tests

In partial life-cycle studies the endpoints most often measured include egg hatchability (%), growth (both length and weight), and survival (%). Hatchability is observed visually; growth is determined by weighing and measuring the organisms physically at the termination of the study. Computer systems are available that allow the organisms to be weighed and measured electronically and the data to be automatically placed in a computer spreadsheet for analysis. In chronic studies, reproduction is also evaluated. Endpoints include all the parameters of interest, i.e., egg hatchability, length, weight, behavior, total number of young produced, number of young produced per adult, number of spawns or broods released per treatment group or spawning pair, physiological effects, and survival. In partial and full life-cycle studies, the endpoints of interest are expressed as NOEC/LOEC or $LC_x$ values. The geometric mean of these two values has traditionally been referred to as the maximum acceptable toxicant concentration (MATC). More recently, the term MATC has been referred to as the chronic value (CV), which is defined as the concentration (threshold) at which chronic effects are first observed. Other endpoints (LC or $EC_{50}$) may be estimated in chronic and subchronic studies, but they are of lesser interest. It is the CV that is compared to the LC or $EC_{50}$ to determine the acute-to-chronic ratio for a given species and toxicant.

The approach for assessing the aforementioned endpoints is based upon selecting the appropriate statistical model for comparing each concentration level to the control. Dichotomous data (hatchability or survival expressed as number dead and alive) require the Fisher's exact or chi-square test.[62] For continuous data (growth variables, e.g., length and weight; reproductive variables, e.g., number of spawns; hatchability or survival data expressed as percentages) the Dunnett's means-comparison procedure would be used based on an analysis of variance (ANOVA).[62] The type of ANOVA, such as one-way or nested, and the error term used, such as between chambers or between aquaria, should correspond to the experimental design and an evaluation of the appropriate experimental unit. Typically, a one-tailed test is used because primary interest is in the detrimental negative effects of the compound being tested and not on both a negative and a positive effect (two tails). Although parametric ANOVA procedures are robust, a nonparametric Dunnett's test should be performed if there are large departures from normality within treatment groups or large departures from homogeneous variance across treatment groups.

For studies that provide continuous data that are analyzed by calculating a percent change from the control, the most appropriate approach is to plot the percent change against the logarithm of the test concentration. The resulting regression line can be used to calculate a percent reduction of choice along with its corresponding confidence interval. It is common to calculate a 25% reduction and express the value either as an ICp (inhibition concentration for a percent effect) or as an EC. Probit analysis of these data is not appropriate. Expressing the data as an ICp, as opposed to an EC, is probably a better approach because it does not have as its basis the concept of a median effect concentration, which is dependent on dichotomous data as opposed to continuous data.

## 2.5 REGULATORY ASPECTS OF AQUATIC TOXICOLOGY IN THE UNITED STATES

### 2.5.1 Clean Water Act (CWA)

The CWA was passed in 1972 and has been amended several times since then. A primary goal of this regulation was to ensure that toxic chemicals were not allowed in U.S. surface waters in toxic amounts. The passage of this act had a major impact on environmental engineering and aquatic toxicology, which led to formalized guidelines for deriving water quality criteria.[20] These criteria were used to develop federal water quality standards that all states adopted and enforced. To date, 24 WQC have been developed in the United States.[63] The aquatic tests required to derive WQC are listed in Table 2.4. Additionally, 129 priority pollutants have been identified, and discharge enforceable limits that cannot be exceeded have been set.

Under the authority of the Clean Water Act, the EPA, Office of Water, Enforcement Branch, established a system of permits for industrial and municipal dischargers (effluents) into surface waters. This permit system is termed the National Pollutant Discharge Elimination System (NPDES). Chemical producers are classified according to the type of chemicals they produce (organic chemicals, plastics, textiles, pesticides, etc.). Each chemical industry category has a list of chemicals and corresponding concentrations that are not to be exceeded in the industry's wastewater effluent. These chemical lists apply to all producers for a given category and are part of each producer's NPDES permit. Each producer also has other water-quality-parameter requirements built into their permit that are specific to their operations. These usually include limitations on the amount (pounds) of chemical that is permitted to be discharged per month and may include items such as total organic carbon, biochemical oxygen demand, suspended solids, ammonia, and process-specific chemicals.

The NPDES permit system incorporates biomonitoring of effluents, usually on a monthly, quarterly, or yearly basis.[29] A toxicity limit is built into the discharger's permit for both industrial and municipal dischargers that must be achieved. If the toxicity limit is exceeded, the permittee is required to identify the chemical responsible for the excess toxicity and take steps to eliminate the chemical, reduce the toxicity, or both. Effluent biomonitoring most often consists of acute toxicity tests with *daphnia* (*Ceriodaphnia dubia*) and fathead minnow (*Pimephales promelas*). Seven-day life-cycle and partial life-cycle studies are required in some cases. An extensive set of procedures (toxicity identification evaluation, TIE) for identifying the substance or substances responsible for

**Table 2.4 Aquatic Toxicity Tests Required by U.S. EPA for the Development of Water Quality Criteria**

| Type of Testing | Recommended Aquatic Tests |
|---|---|
| Acute Toxicity Tests | Eight different families must be tested for both freshwater and marine species (16 acute tests):<br>*Freshwater*<br>1. A species in the family Salmonidae<br>2. A species in another family of the class Osteichthyes<br>3. A species in another family of the phylum Chordata<br>4. A plankton species in class Crustacea<br>5. A benthic species in class Crustacea<br>6. A species in class Insecta<br>7. A species in a phylum other than Chordata or Arthropoda<br>8. A species in another order of Insecta or in another phylum<br>*Marine*<br>1. Two families in the phylum Chordata<br>2. A family in a phylum other than Arthropoda or<br>3. Chordata<br>4. Either Family Mysidae or Penaeidae<br>5. Three other families not in the phylum Chordata (may include Mysidae or Penaeidae, whichever was not used above)<br>6. Any other family |
| Chronic Toxicity Tests | Three chronic or partial life-cycle studies are required:<br>One invertebrate and one fish<br>One freshwater and one marine species |
| Plant Testing | At least one algal or vascular plant test must be performed with a freshwater and marine species. |
| Bioconcentration Testing | At least one bioconcentration study with an appropriate freshwater and saltwater species is required. |

toxicity in effluents and sediments has been developed over the past decade.[64–75] Present tie research efforts are focused primarily on freshwater and marine sediments.

### 2.5.2 Toxic Substances Control Act (TSCA)

The Toxic Substances Control Act (TSCA) was established by Congress on October 8, 1976 as public law 94–469 to regulate toxic industrial chemicals and mixtures. The goal of Congress was to establish specific requirements and authorities to identify and control chemical hazards to both human health and the environment. The office of Pollution Prevention and Toxics (OPPT) is the lead office responsible for implementing the Toxic Substances Control Act, which was established to reduce the risk of new and existing chemicals in the marketplace.

There are approximately 80,000+ compounds listed on the TSCA Chemical Inventory that are approved for use in the United States.[76] The detection of polychlorinated biphenyls (PCBs), an industrial heat transformer and dielectric fluid found in aquatic and terrestrial ecosystems in many parts of the United States, emphasized the need for controlling industrial chemicals not regulated by pesticide or food and drug regulations. From the viewpoint of aquatic testing, this regulation has focused on two areas: test requirements for new chemicals and existing chemicals. Under TSCA Section 5, notice must be given to the Office of Pollution Prevention and Toxics (OPPT) prior to manufacture or importation of any new or existing chemical. No toxicity information is required for the Premanufacture Notification (PMN). OPPT has 90 days to conduct a hazard/risk assessment and may require generation of toxicity information. Toxicity testing is required only if a potential hazard or risk is demonstrated.[76]

Existing chemicals listed on the TSCA inventory register prior to 1976 are not required to undergo a PMN review. However, the EPA, through the Interagency Testing Committee (ITC), reviews individual chemicals and classes of chemicals to determine the need for environmental and human health data to assess the safety of the chemicals. If the ITC determines that a potential exists for significant chemical exposure to humans or the environment, they can require the manufacturers to provide additional data for the chemicals to help assess the risk associated with the manufacture and use of the product. TSCA empowers the EPA to restrict chemical production and usage when the risk is considered severe enough. Data collection is accomplished through Section 4 of TSCA by means of developing a legally binding consent order on a Test Rule. The Test Rule spells out the reasons for the testing and identifies which tests are required. Aquatic tests that are most often required for PMNs or by Test Rules are listed in Table 2.3.

The Chemical Right-to-Know Initiative was begun in 1998 in response to the finding that very little toxicity information is publicly available for most of the high production volume (HPV) commercial chemicals made and used (more than 1 million lbs/yr) in the United States. Without this basic hazard information, it is difficult to make sound judgments about what potential risks these chemicals could present to people and the environment. An ambitious testing program has been established, especially for those chemicals that are persistent, bioaccumulative, and toxic (PBT), or which are of particular concern to children's health.[77]

### 2.5.3 Federal Insecticide, Fungicide and Rodenticide Act (FIFRA)

Under FIFRA the EPA is responsible for protecting the environment from unreasonable adverse effects of pesticides.[78] This legislation is unique among environmental protection statutes in that it licenses chemicals known to be toxic for intentional release into the environment for the benefit of mankind. Most regulatory statutes are designed to limit or prevent the release of chemicals into the environment. The FIFRA licensing process regulates three distinct areas: (1) labeling, (2) classification, and (3) registration. To fulfill its responsibility the EPA requires disclosure of scientific data regarding the effects of pesticides on humans, wildlife, and aquatic species.

By statutory authority the EPA assumes that pesticides present a risk to humans or to the environment. The pesticide registrant is responsible for rebutting the EPA's presumption of risk. To accomplish this the EPA recommends a four-tiered testing series.[79] The tests become progressively more complex, lengthy, and costly, going from Tier I to Tier IV.[80] Studies in Tiers I and II evaluate a substance for acute toxicity and significant chronic effects, respectively. Higher-tier tests evaluate long-term chronic and subchronic effects. In Tier IV, field and mesocosm studies can be required. The need to perform successively higher-tiered tests is triggered by the degree of risk the pesticide presents to the environment. Risk is determined by the quotient method, i.e., by comparing the expected environmental concentration with the measured levels of biological effect. The difference between the two levels is referred to as the margin of safety. When the margin of safety is small in Tiers I and II, additional higher-tier studies are required to rebut the presumption of risk to the environment.

### 2.5.4 Federal Food, Drug, and Cosmetics Act (FFDCA)

The FFDCA of 1980 is administrated by the Food and Drug Administration (FDA). This act empowers the FDA to regulate food additives, pharmaceuticals, and cosmetics that are shipped between states. The intent of this act is to protect the human food supply and to ensure that all drugs are properly tested and safe for use. The FDA enforces pesticide tolerance and action levels set by the EPA. This can result in a ban or food consumption advisory for fish and seafood from certain areas. The FDA also regulates drugs that are used for animals, including fish, as well as human drugs. The use of drugs to treat fish diseases has drawn national and international attention since the FDA has begun to restrict the use of certain drugs that have not been properly tested for potential environmental effects. These drugs are used in significant quantities in commercial fishery operations.

The U.S. FDA is responsible for reviewing the potential environmental impact from the intended use of human and veterinary pharmaceuticals, food or color additives, Class III medical devices, and biological products. To evaluate the potential effects of a proposed compound the FDA requires the submission of an Environmental Assessment (EA). The National Environmental Protection Act, passed by Congress in 1969, provides the statutory authority for the FDA to conduct EA requirements.

EAs are required for all New Drug Applications as well as for some supplementary submissions and communications. Previously, an EA required little more than a statement that a compound had no potential environmental impact; however, changes within the FDA have increased and intensified the EA review and approval process. Under current policy the FDA requests quantitative documentation of a compound's potential environmental impact. The EA must contain statistically sound conclusions based on scientific data obtained through studies conducted under Good Laboratory Practices (GLPs). These changes have significantly impacted the content, manner of data acquisition, and preparation of an EA. Details relative to the FDA statutory authority and information required for an EA submission are contained in the Code of Federal Regulations.[81] The specific aquatic toxicity tests recommended by the FDA for inclusion with the EA submission are shown in Table 2.3.

### 2.5.5 Comprehensive Environmental Response, Compensation, Liability Act

Superfund is the name synonymous with the Comprehensive Environmental Response, Compensation, Liability Act (CERCLA, 1980). This act requires the EPA to clean up uncontrolled hazardous waste sites to protect both human health and the environment. CERCLA provides the statutory authority for the EPA to require environmental risk assessment as part of the Superfund site assessment process. Part of risk assessment includes evaluating the potential for risk to aquatic species, if appropriate, for a given site. Additional authority comes from the National Oil and Hazardous Materials Contingency Plan, which specifies that environmental evaluations shall be performed to assess threats to the environment, especially sensitive habitats and critical habitats of species protected under the Endangered Species Act.

The Superfund program provides (1) the EPA with the authority to force polluters to take responsibility for cleaning up their own wastes; (2) the EPA with the authority to take action to protect human health and the environment, including cleaning up waste sites, if responsible parties do not take timely and adequate action; and (3) a Hazardous Substance Response Trust Fund to cover the cost of EPA enforcement and cleanup activities. The Superfund process consists of: site discovery, preliminary assessment (PA)/site assessment (SA), hazard ranking/nomination to National Priorities List (NPL), remedial investigation (RI)/feasibility study (FS), selection of remedy, remedial design, remedial action, operation and maintenance, and NPL deletion.[82]

Environmental risk assessment is conducted as part of the PA/SA investigation and as part of the RI/FS studies. Sites that have the potential for contaminants to migrate to surface waters and sediments require aquatic assessment. Risk assessment procedures have been evolving, and guidance in the selection of tests and species is available.[83–85] Many of the tests for TSCA and FIFRA assessments are acceptable (Table 2.3). Most often, aquatic tests are performed on soils/sediments, which are shipped to an aquatic testing facility for studies with amphipods, midges, and earthworms. Most studies are static acute or static renewal partial life-cycle studies.

### 2.5.6 Marine Protection, Research and Sanctuaries Act (MPRSA)

The MPRSA of 1972 requires that dredged material be evaluated for its suitability for ocean disposal according to criteria published by the EPA (40 CFR 220–228) before disposal is approved. The maintenance of navigation channels requires dredging, and the disposal of that dredged material is a concern. For ocean disposal the dredged material must be evaluated to determine its potential for impact to the water column at the disposal site. In 1977 the EPA and COE developed the manual, "Ecological Evaluation of Proposed Discharge of Dredged Material into Ocean Waters," which contains technical guidance on chemical, physical, and biological procedures to evaluate the acceptability of dredged material for ocean disposal.[86] A similar manual was developed in 1998 for freshwater systems entitled "Evaluation of Dredged Material Proposed for Discharge in Waters of the U.S."[87]

The manual outlines a tiered testing procedure for evaluating compliance with the limiting permissible concentration (LPC) as defined by ocean dumping regulations. The liquid-phase or water-column LPC must not exceed applicable marine water quality criteria or a toxicity threshold (0.01 times the acutely toxic concentration). The suspended and solid-phase LPCs must not cause unreasonable toxicity or bioaccumulation. The document describes four levels (tiers) of evaluation. Tiers I and II utilize existing information, which is often available for recurring disposals of dredged materials from channel maintenance, to determine the appropriateness for ocean disposal. Tier III contains most laboratory bioassays, and Tier IV includes some tests of bioaccumulation and a range of possible field investigations. The evaluation also recommends using a reference site, specifically a site that is free of contamination, as a source of sediments for comparison testing with the dredged materials. Examples of aquatic marine species that are acceptable to evaluate the suitability of dredge materials for ocean disposal via water-column, solid-phase, and bioaccumulation tests are presented in Table 2.5 (taken from the Green Book).[47]

### 2.5.7 European Community (EC) Aquatic Test Requirements

The European Community (EC) also requires toxicity testing as part of their chemical environmental assessment process. The EC is managed by four institutions — the Commission, the Council of Ministers, the Parliament, and the Court of Justice. The Commission proposes regulations to the Council of Ministers, who make final rulings. Actions taken by the Council have the force of law and are referred to as regulations, directives, decisions, recommendations, and opinions. Most actions taken relative to chemical environmental assessment have taken the form of directives. Directives are binding on member countries; however, member countries may choose the method of implementation.

Critical directives that mandate aquatic toxicity tests are the Pesticide Registration Directive[88] and the Sixth and Draft Seventh Amendments of Directive 67/548/CEE, Classification, Packaging, and Labeling of Dangerous Substances. Additionally, the Paris Commission was established to prepare guidelines to ensure that offshore (North Sea) oil exploration would not endanger the marine environment. The directives of the Paris Commission as well as the previously mentioned directives require aquatic toxicity tests as part of environmental assessments.

### 2.5.8 Organization for Economic Cooperation and Development (OECD)

The published list of aquatic test methods and species required to be used when fulfilling the data requirements of EC directives is shown in Table 2.3. Test guidelines are listed as EEC (European Economic Community) or OECD (Organization of Economic Cooperation and Development). The OECD operates as a methods-generating and standardization body, whereas the EEC formally adopts test guidelines that become the legally binding method to be used. Relevant internationally agreed-upon OECD test methods used by government, industry, and independent laboratories have been published and are available as a compendium of guidelines[89, 90] (Table 2.6).

## 2.6 SUMMARY AND FUTURE DIRECTION OF AQUATIC TOXICOLOGY

The field of aquatic toxicology has grown out of the disciplines of water pollution biology and limnology. Aquatic toxicology studies have been performed for the past 120 years. Studies evolved from simple tests conducted over intervals as short as a few hours to standard acute lethality tests lasting 48 or 96 hours, depending on the species. Acute toxicity tests were followed by the

**Table 2.5 Examples of Appropriate Test Species for Use with Dredge Material when Performing Water Column, Solid-Phase Benthic, and Bioaccumulation Effects Testing**

| Type of Testing and Recommended Species | |
|---|---|
| **Water Column Toxicity Tests** | |
| ***Crustaceans*** | ***Zooplankton*** |
| Mysid shrimp, *Americamysis bahia* sp.* | Copepods, *Acartia* sp.* |
| *Neomysis* sp.* | Larvae of: |
| *Holmesimysis* sp.* | Mussels, *Mytilus edulis** |
| Grass shrimp, *Palaemonetes* sp. | Oysters, *Crassostrea virginica** |
| Oysters, *Crassostrea virginica** | *Ostrea* sp.* |
| Commercial shrimp, *Penaeus* sp. | Sea urchin, *Stronglyocentrotus purpuratus* |
| Oceanic shrimp, *Pandalus* sp. | *Lyetechinus pictus* |
| Blue crab, *Callinectes sapidus* | |
| Cancer crab, *Cancer* sp. | |
| ***Fish*** | ***Bivalves*** |
| Silversides, *Menidia* sp.* | Mussel, *Mytilus* sp. |
| Shiner perch, *Cymatogaster aggregata** | Oyster, *Crassostrea* sp. |
| Sheepshead minnow, *Cyprinodon variegatus* | |
| Pinfish, *Lagodon rhomboides* | |
| Spot, *Leiostomus xamthurus* | |
| Sanddab, *Citharichys stigmaeus* | |
| Grunion, *Leuresthes tenuis* | |
| Dolphinfish, *Coryphaena hippurus* | |
| **Benthic Solid-Phase Toxicity Tests** | |
| ***Infaunal Amphipods*** | ***Crustaceans*** |
| *Ampelisca* sp.* | Mysid shrimp, *Americamysis bahia* sp. |
| *Rhepoxynius* sp.* | *Neomysis* sp. |
| *Eohaustorius* sp.* | *Holmesimysis* sp. |
| *Grandiderella japonica* | Commercial shrimp, *Penaeus* sp. |
| *Corophium insidiosum* | Grass shrimp, *Palaemonetes* sp. |
| | Sand shrimp, *Crangon* sp. |
| | Blue crab, *Callinectes sapidus* |
| | Cancer crab, *Cancer* sp. |
| | Ridge-back prawn, *Sicyonia ingentis* |
| ***Burrowing Polychaetes*** | ***Fish*** |
| *Neanthes* sp.* | Arrow gobi, *Clevelandia ios* |
| *Nereis* sp.* | |
| *Nephthys* sp.* | |
| *Glycera* sp.* | |
| *Arenicola* sp.* | |
| *Abarenicola* sp.* | |
| ***Mollusks*** | |
| Yoldia clam, *Yoldia limatula* sp. | |
| Littleneck clam, *Protothaca staminea* | |
| Japanese clam, *Tapes japonica* | |
| **Bioaccumulation Tests** | |
| ***Polychaetes*** | ***Mollusks*** |
| *Neanthes* sp.* | Macoma clam, *Macoma* sp. |
| *Nereis* sp.* | Yoldia clam, *Yoldia limatula* sp. |

**Table 2.5 Examples of Appropriate Test Species for Use with Dredge Material when Performing Water Column, Solid-Phase Benthic, and Bioaccumulation Effects Testing *(Continued)***

| Type of Testing and Recommended Species | |
|---|---|
| ***Polychaetes*** | ***Mollusks*** |
| *Nephthys* sp.* | Nucula clam, *Nucula* sp. |
| *Arenicola* sp.* | Littleneck clam, *Protothaca staminea* |
| *Abarenicola* sp.* | Japanese clam, *Tapes japonica* |
| | Quahog clam, *Mercenaria mercenaria* |
| ***Fish*** | ***Crustaceans*** |
| Arrow gobi, *Clevelandia ios* | Ridge-back prawn, *Sicyonia ingentis* |
| Topsmelt, *Atherinops affinis* | Shrimp, *Peneaus* sp. |

*Note:* Information is taken from the EPA-COE Green Book.[47]

* Recommended species.

development of various short sublethal tests (e.g., behavior or biochemical studies) and tests with prolonged exposures such as partial life-cycle studies and full life-cycle studies. Early studies were performed in the absence of regulatory requirements by individuals with a high degree of scientific curiosity. Today, aquatic toxicology studies are done for research purposes or environmental risk assessments and are required by many regulatory agencies for product registration, labeling, shipping, or waste disposal.

The cost and length of time required to perform full life-cycle tests have encouraged scientists to search for sensitive test species and sensitive life stages. Full life-cycle fish studies have, for the most part, been replaced by embryo-larval studies (partial life-cycle studies).[91] A major effort has been expended to identify species that allow full life-cycle studies to be performed in much shorter periods (e.g., 7-day *Ceriodaphnia dubia* life cycle tests,[92] two-dimensional rotifer tests[93]) or tests that use sensitive species and sensitive life stages. For example, a 7-day fathead minnow embryo-larval growth and survival study is used to evaluate effluents.[94] The goal of these tests is to quickly provide accurate estimates of chronic no-effect levels. It is important to remember that these tests estimate chronic results, not duplicate them. The estimated value is often within a factor of 2 to 4 of the chronic value and, depending on the use of these data, may provide adequate accuracy.

During the last decade significant effort has been expended in developing rapid toxicity assays. There has been an increasing need to assess the toxicity of various sample types in minutes to hours instead of days. For example, effluent toxicity identification evaluation (TIE) procedures require multiple toxicity tests on successive days. The use of assays (such as the Microtox [95]assay) can speed up the TIE process considerably. The use of rapid assays during on-site effluent biomonitoring allows for collection of a more extensive data set during the limited testing time available.

**Table 2.6 Adopted and Draft OECD Test Guidelines Harmonized with OPPTS since 1990**

| Test Guideline No. | Guideline Title | Date of Adoption as an Original or as an Updated Version and Draft Date |
|---|---|---|
| 203 | Fish, Acute Toxicity Test. | July 17, 1992 |
| 210 | Fish, Early-Life Stage Toxicity Test | July 17, 1992 |
| 211 | *Daphnia magna* Reproduction Test | September 21, 1998 |
| 212 | Fish, Short-term Toxicity Test on Embryo and Sac-Fry Stages | September 21, 1998 |
| 215 | Fish, Juvenile Growth Test | January 21, 2000 Draft Guideline, July 1999 |
| 202 | *Daphnia* sp., Acute Immobilization Test | Draft |
| 305 | Bioconcentration; Flow-Through Fish Test | June 14, 1996 |

In recent years the increasing desire to link exposure to effect has drawn considerable attention to the "biomarker approach." Because chemical contaminants are known to evoke distinct measurable biological responses in exposed organisms, biomarker-based techniques are currently being investigated to assess toxicant-induced changes at the biological and ecological levels.[96] Collectively, the term biomarker refers to the use of physiological, biochemical, and histological changes as "indicators" of exposure and effects of xenobiotics at the suborganism or organism level.[97] However, indicators or biomarkers can be defined at any level of biological organization, including changes manifested as enzyme content or activity, DNA adducts, chromosomal aberrations, histopathological alterations, immune-system effects, reproductive effects, physiological effects, and fertility at the molecular and individual level, as well as size distributions, diversity indices, and functional parameters at the population and ecosystem level. In the field of ecotoxicology, the use of biomarkers has emerged as a new and very powerful tool for detecting both exposure and effects resulting from environmental contaminants.[97–104] Unlike most chemical monitoring, biomarker endpoints have the potential to reflect and assess the bioavailability of complex mixtures present in the environment as well as render biological significance. Biomarkers provide rapid toxicity assessment and early indication of population and community stress and offer the potential to be used as markers of specific chemicals.

Chemical effects are thought to be the result of the interaction between toxicant and biochemical receptor. Therefore, biochemical responses are expected to occur before effects are observed at higher levels of biological organization. Biomarker response frequently provides a high degree of sensitivity to environmental impacts, thereby providing an "early warning" to potential problems or irreversible effects. In natural environments, where organisms are exposed to multiple stresses (natural and anthropogenic) over time, biomarkers reflect this integrated exposure of cumulative, synergistic, or antagonistic effects of complex mixtures. A myriad of recent studies have demonstrated the utility of biomarker techniques in the assessment of contaminants ranging from single compounds to complex mixtures in both the laboratory and the field.[105–109]

To date, biomarker assays have not been standardized or incorporated into regulatory guidelines as part of chemical environmental risk assessments. It is expected that in the future a variety of specific biomarkers will be sufficiently validated as predictors of whole organism and population effects; however, it is unlikely that they will therefore tell us if an ecosystem is in danger of losing its integrity or if compensation to a particular insult is possible. A more reasonable application would be use as either part of a tiered assessment or as measurement by some standard of predefined ecological health. The trend toward more sensitive, biologically relevant test methods predictive of early ecosystem stress will continue, and biomarkers are expected to play a role as surrogate measures or predictors of ecosystem well-being.

## ACKNOWLEDGMENTS

We wish to thank Jerry Smrchek for critically reviewing this manuscript.

## REFERENCES

1. Forbes, S. A., The lake as a microcosm, *Bulletin of the Peoria Scientific Association*, 1887, reprinted in *Bulletin of the Illinois State Natural History Survey*, 15, 537–550, 1925.
2. Kolwitz, R. and Marsson, M., Ecology of plant saprobia, in *Biology of Water Pollution, Federal Water Pollution Administration*, Keup, L. E., Ingram, W. M., and Mackenthun, K. M., Eds., U.S. Department of the Interior, 1908, 47.
3. Kolwitz, R. and Marsson, M., Ecology of animal saprobia, in *Biology of Water Pollution, Federal Water Pollution Administration*, Keup, L. E., Ingram, W. M., and Mackenthun, K. M., Eds., U.S. Department of the Interior, 1909, 85.

4. Forbes, S. A. and Richardson, R. E., Studies on the biology of the upper Illinois River, *Bulletin of the Illinois State Laboratory of Natural History*, 9, 481, 1913.
5. Warren, C. E., *Biology and Water Pollution Control*, W. B. Saunders, Philadelphia, 1971, Chap. 1.
6. Penny C. and Adams, C., Fourth report of the royal commission on pollution in Scotland, *London*, 2, 377, 1863.
7. Weigelt, C., Saare, O., and Schwab, L., Die Schädigung von Fischerei und Fischzueht durch Industrie und Hausabwasser, *Archiv für Hygiene*, 3, 39, 1885.
8. Carpenter, K. E., A study of the fauna of rivers polluted by lead mining in the Aberystwth district of Cardiganshire, *Ann. Applied Biol.*, 11, 1, 1924.
9. Jones, J. R. E., The relationship between the electrolytic solution pressures of the metals and their toxicity to the stickleback (*Gasterosterus aculeatus* L.), *J. Exp. Biol.*, 16, 425, 1939.
10. Fry, F. E. J., Effects of the environment on animal activity, University of Toronto Studies Biological Series 55, Ontario Fisheries Research Laboratory Publication, 68, 1, 1947.
11. Ellis, M. M., Detection and measurement of stream pollution, *Bulletin of the U.S. Bureau of Fisheries*, 48, 365, 1937.
12. Anderson, B. G., The toxicity thresholds of various substances found in *Daphnia magna*, *Sewage Works J.*, 16, 1156, 1944.
13. Anderson, B. G., The toxicity thresholds of various salts determined by the use of *Daphnia magna*, *Sewage Works J.*, 18, 82, 1946.
14. Hart, W. B., Doudoroff, P., and Greenbank, J., *The Evaluation of the Toxicity of Industrial Wastes, Chemicals and Other Substances to Freshwater Fishes*, Waste Control Laboratory, The Atlantic Refining Company, 1945, 1.
15. Doudoroff, P., Anderson, B. G., Burdick, G. E., Galtsoff, P. S., Hart, W. B., Patrick R., Strong, E. R., Surber, E. W., and Van Horn, W. M., Bio-assay methods for the evaluation of acute toxicity of industrial wastes to fish, *Sewage and Industrial Wastes*, 23, 1381, 1951.
16. American Public Health Association, American Water Works Association, and Water Pollution Control Federation, *Standard Methods for the Examination of Water and Wastewater*, 17th ed., Washington, D.C., American Public Health Association, 1989.
17. Stephan, C. E., *Methods for Acute Toxicity Tests with Fish, Macroinvertebrates, and Amphibians*, U.S. Environmental Protection Agency, Ecological Research Series, EPA-660/3–75–009, 1972.
18. McKee, J. E., *Water Quality Criteria*, California State Water Pollution Control Board Publication 3, 1, 1952.
19. McKee, J. E. and Wolf, H. W., *Water Quality Criteria 2nd ed.*, California State Water Pollution Control Board Publication, 3-A, 1963, 1.
20. Stephan, C. E., Mount, D. I., Hansen, D. J., Gentile, J. H., Chapman, G. A., and Brungs, W. A., *Guidelines for Deriving Numerical National Water Quality Criteria for the Protection of Aquatic Organisms and Their Uses*, PB85–227049, National Technical Information Service, Springfield, VA, 1985, 1.
21. National Technical Advisory Committee, *Quality Criteria for Water*, reprinted by U.S. Environmental Protection Agency, Superintendent of Documents, U.S. Government Printing Office, Washington, D.C., 1972, 1.
22. U.S. Environmental Protection Agency, *Quality Criteria for Water*, U.S. Environmental Protection Agency, Washington, D.C., EPA-440/9–76–023, 1976, 1.
23. U.S. Environmental Protection Agency, *Quality Criteria for Water*, Office of Water and Hazardous Materials, U.S. Environmental Protection Agency, Washington, D.C., 1977, 1.
24. U.S. Environmental Protection Agency, *Quality Criteria for Water*, 1986, Office of Water Regulations and Standards, U.S. Environmental Protection Agency, Washington, D.C., EPA 440/5–86–001, 1986, 1.
25. Mount, D. I., Present approaches to toxicity testing: A perspective, in *Aquatic Toxicology and Hazardous Evaluation*, Mayer, F. L. and Hamelink, J. L., Eds., American Society for Testing and Materials, Philadelphia, Special Technical Publication, 5, 634, 1977.
26. Henderson, C. and Tarzwell, C. M., Bioassays for the control of industrial effluents, *Sewage and Industrial Wastes*, 29, 1002, 1957.
27. Mount, D. I. and Stephan, C. E., A method for establishing acceptable toxicant limits for fish: Malathion and the butoxyethanol ester of 2,4,-D, *Trans. Amer. Fish. Soc.*, 96, 185, 1967.

28. Stephan, C.E., Topics on expressing and predicting results of life-cycle tests, in *Aquatic Toxicology and Environmental Fate: Eleventh Volume*, Suter, G. W., II, and Lewis, M. A., Eds., American Society for Testing and Materials, Philadelphia, Special Technical Publication, 1007, 1988, 263.
29. U.S. Environmental Protection Agency, *Technical Support Document for Water Quality-Based Toxics Control*, Office of Water Regulation and Standards, Environmental Protection Agency, Washington, D.C., EPA/505/2–90–001, 1991.
30. Ankley, G. T., Jensen, K. M., Kahl, M. D., Korte, J. J., and Makynen, E. A., Description and evaluation of a short-term reproduction t-test with the fathead minnow (*Pimephales promelas), Environ. Toxicol. Chem.,* 20,1276–1290, 2001.
31. Tietge, J. E. G. T., Ankley, G. T., and Degitz, S. J., Report on the Xenopus tail resorption assay as a Tier I screen for thyroid active chemicals, U.S. EPA, Office of Research and Development, Duluth, MN, 2000.
32. Sprague, J. B., The ABC's of pollutant bioassay using fish, in *Biological Methods for the Assessment of Water Quality*, Cairns, J., Jr., and Dickson, K. L., Eds., American Society for Testing and Materials, Philadelphia, Special Technical Publication, 528, 1973, 6.
33. Sprague, J. B., Measurement of pollutant toxicity to fish I. Bioassay methods for acute toxicity, *Water Res.*, 3, 793, 1969.
34. Rand, G. M., *Fundamentals of Aquatic Toxicology*, Effects, Environmental Fate, and Risk Assessment. 2nd ed., Taylor and Francis, Washington, D.C., 1995.
35. Mount, D. I. and Brungs, W. A., A simplified dosing apparatus for fish toxicology studies, *Water Res.*, 1, 21, 1967.
36. American Society for Testing and Materials, Standard Guide for Conducting *Daphnia magna* Life Cycle Toxicity Tests, Annual Book of ASTM Standards, American Society for Testing and Materials, West Conshohocken, PA, E 1193–97, 2001, 1.
37. Adams, W. J., Kimerle, R. A., and Mosher, R. G., An aquatic safety assessment of chemicals sorbed to sediments, in *Aquatic Toxicology and Hazard Assessment: Seventh Symposium*, Cardwell, R. D., Purdy, R., and Bahner, R. C., Eds., American Society for Testing and Materials, Philadelphia, Special Technical Publication, 854, 1985, 429.
38. DiToro, D. M., Zarba, C. S., Hansen, D. J., Berry, W. J., Swartz, R. C., Cowan, C. C., Pavlou, S. P., Allen, H. E., Thomas, N. A., and Paquin, P. R., Technical basis for establishing sediment quality criteria for nonionic organic chemicals using equilibrium partitioning, *Environ. Toxicol. Chem.*, 10, 1541, 1991.
39. U.S. Environmental Protection Agency, Briefing Report to the EPA Science Advisory Board on the Equilibrium Partitioning Approach to Generating Sediment Quality Criteria, Office of Water Regulations and Standards, U.S. Environmental Protection Agency, Washington, D.C., EPA 440/5–89–002, 1989.
40. Ankley, G. T., DiToro, D. M., and Hansen, D. J., Technical basis and proposal for deriving sediment quality criteria for metals, *Environ. Toxicol. Chem.*, 15, 2056–2066, 1996.
41. Allen, H. E., Bell, H. E., Berry, W. J., DiToro, D. M., Hansen, D. J., Meyer, J. S., Mitchell, J. L., Paquin, P. R., Reiley, M. C., and Santore, R. C. *Integrated approach to assessing the bioavailability and toxicity of metals in surface waters and sediments*, Briefing document presented to EPA Science Advisory Board, U.S. EPA Office of Water and Research and Development, Washington, D.C., May 13, 1999.
42. DiToro, D. M., Dodge, L. J., and Hand, V. C., A model for anionic surfactant sorption, *Environ. Sci. Technol.*, 24, 1013, 1990.
43. Bell, H. M., DiToro, D. M., Hansen, D. J., McGrath, J. A., Mount, D. R., Reiley, M. C., and Swartz, R. C., *Assessing the toxicity and bioavailability of PAH mixtures in sediments*, Briefing document presented to EPA Science Advisory Board, U.S. EPA Office of Water, Washington, D.C., May 13, 1997.
44. Erickson, R. J., Ankley, G. T., DeFoe, D. L., Kosian, P. A., and Makynen, E. A., Additive toxicity of binary mixtures of phototoxic polycyclic aromatic hydrocarbons to the oligochaeta *Lumbriculus variegatus*, *Toxicol. Appl. Pharm.*, 154, 97–105, 1999.
45. Carr, R. S. and Nipper, M., Porewater Toxicity Testing: Biological, Chemical, and Ecological Considerations with a Review of Methods and Applications, and Recommendations for Future Areas of Research – Summary of a SETAC Technical Workshop, SETAC Press, Pensacola, FL, 2001.
46. Adams, W. J., Kimerle, R. A., and Barnett, J. W., Jr., Sediment quality and aquatic life assessment, *Environ. Sci. Technol.*, 25, 1965, 1992.

47. U.S. Environmental Protection Agency, *Evaluation of Dredged Material Proposed for Ocean Disposal, Testing Manual*, U.S. Environmental Protection Agency, Marine Protection Branch, Washington, EPA-503/8–91/001, 1991, 1.
48. American Society for Testing and Materials, *Standard Practice for Storage, Characterization, and Manipulation of Sediments for Toxicological Testing*, in Volume 11.05, Annual Book of ASTM Standards, American Society for Testing and Materials, West Conshohocken, PA, E1391–94, 2001, 1.
49. U.S. Environmental Protection Agency, Methods for Collection, Storage, and Manipulation of Sediments for Chemical and Toxicological Analyses, Technical manual, U.S. EPA, Office of Water, Washington, D.C., EPA-828-F-01–023, 2001.
50. Boethling, R. S. and Mackay, D., *Property Estimation Methods for Chemicals*, Environmental and Health Sciences, Lewis Publishers, Boca Raton, FL, 2000, 189.
51. Adams, W. J., Conard, B., Ethier, G., Brix, K. V., Paquin, P. R., and DiToro, D. M., The challenges of hazard identification and classification of insoluble metals and metal substances for the aquatic environment, *Hum. Ecol. Risk Assess.* 6(6), 1019–1038, 2000.
52. Brix, K. V. and DeForest, D. K., Critical review of the use of bioconcentration factors for hazard classification of metals and metal compounds, OECD (Organization for Economic Cooperation and Development) Aquatic Hazards Extended Workgroup Meeting, Paris, France, May 15, 2000.
53. American Society for Testing and Materials, *Standard Practice for Conducting Bioconcentration Tests With Fishes and Saltwater Bivalve Mollusks*, in Volume 11.05, Annual Book of ASTM Standards, American Society for Testing and Materials, West Conshohocken, PA, E1022–94, 2001, 1.
54. Thomann, R. V., Connolly, J. P., and Parkerton, T. F., An equilibrium model of organic chemical accumulation in aquatic food webs with sediment interaction, *Environ. Toxicol. Chem.*, 11, 615, 1992.
55. U.S. Environmental Protection Agency, Methods for Measuring the Bioaccumulation and Toxicity of Sediment-Associated Contaminants with Freshwater Invertebrates, U.S. EPA, Office of Research and Development, Washington, D.C., EPA/600/R-99/064, 2000.
56. U.S. Environmental Protection Agency, Guidelines for Ecological Risk Assessment, U.S. Environmental Protection Agency, Washington, D.C., EPA/630/R-95/002F, 1998.
57. Stephan, C. E., Methods for calculating an $LC_{50}$, in *Aquatic Toxicology and Hazard Evaluation*, Mayers, F. L., and Hamelink, J. L., Eds., American Society for Testing and Materials, Philadelphia, Special Technical Publication, 634, 1977, 65.
58. Litchfield, J. T., and Wilcoxon, F., A simplified method of evaluating dose-effect experiments, *J. Pharmacol. Exp. Ther.*, 96, 99, 1949.
59. Sebaugh, J. L., Wilson, J. D., Thecker, M. W., and Adams, W. J., A study of the shape of dose-response curves for acute lethality at low response: A megadaphnia study, *Risk Analys.*, 11, 633, 1991.
60. Society of Environmental Toxicology and Chemistry, *Aquatic Dialogue Group: Pesticide Risk Assessment and Mitigation*, SETAC Press, Pensacola, FL, 1994, 65.
61. Posthuma, L., Suter, G. W., II, and Traas, T. P., *Species Sensitivity Distributions in Ecotoxicology*, Lewis Publishers, Boca Raton, FL, 2002, 345.
62. Steel, R. G. and Torrie, J. H., Principles and Procedures for Statistics with Special Reference to Biological Sciences, McGraw-Hill, New York, 1960, 1.
63. U.S. Environmental Protection Agency, National Recommended Water Quality Criteria: Notice, Fed. Reg., 63(234), 67548–67558, Monday, December 7, 1998.
64. U.S. Environmental Protection Agency, Methods for Aquatic Toxicity Identifications: Phase I Toxicity Characterization Procedures, U.S. EPA, Office of Research and Development, Environmental Research Laboratory, Duluth, MN, EPA/600/3–88/034, 1988.
65. U.S. Environmental Protection Agency, Generalized Methodology for Conducting Industrial Toxicity Reduction Evaluations (TREs), U.S. EPA, The Chemicals and Chemical Product Branch, Risk Reduction Engineering Laboratory, Cincinnati, Ohio, EPA/600/2–88/070, 1989.
66. U.S. Environmental Protection Agency, Methods for Aquatic Toxicity Identifications: Phase II Toxicity Identification Procedures, U.S. EPA, Office of Research and Development, Environmental Research Laboratory, Duluth, MN, EPA/600/3–88/035, 1989.
67. U.S. Environmental Protection Agency, Methods for Aquatic Toxicity Identification Evaluations: Phase III Toxicity Confirmation Procedures, U.S. EPA, National Effluent Toxicity Assessment Center, Environmental Research Lab, Duluth, Minnesota. EPA-600/3–88/035, 1989.

68. U.S. Environmental Protection Agency, Toxicity Reduction Evaluation Protocol for Municipal Treatment Plants, U.S. EPA, Risk Reduction Engineering Laboratory, Office of Research and Development, Cincinnati, Ohio, EPA/600/2–88–062, 1989.
69. U.S. Environmental Protection Agency, Methods for Aquatic Toxicity Identifications: Phase I Toxicity Characterization Procedures, 2nd ed., U.S. EPA, Environmental Research Laboratory, Duluth, MN, EPA/600/6–91/003, 1991.
70. U.S. Environmental Protection Agency, Toxicity Identification Evaluation: Characterization of Chronically Toxic Effluents, Phase I, U.S. EPA, Office of Research and Development, Duluth, MN, EPA/600/6–91/005, 1991.
71. U.S. Environmental Protection Agency, Sediment Toxicity Identification Evaluation: Phase I (Characterization), Phase II (Identification), and Phase III (Confirmation) Modifications of Effluent Procedures, Draft, U.S. EPA, Office of Research and Development, Duluth, MN, 1991.
72. U.S. Environmental Protection Agency, Marine Toxicity Identification Evaluation (TIE) Guidance Document: Phase I, Draft, U.S. EPA, Environmental Research Laboratory, Narragansett, RI, 1993.
73. U.S. Environmental Protection Agency, Methods for Aquatic Toxicity Identifications: Phase II Toxicity Identification Procedures for Samples Exhibiting Acute and Chronic Toxicity, Environmental Research Laboratory, Duluth, MN, EPA/600/R-92/080, 1993.
74. U.S. Environmental Protection Agency, Methods for Aquatic Toxicity Identifications: Phase III Toxicity Confirmation Procedures for Samples Exhibiting Acute and Chronic Toxicity, Environmental Research Laboratory, Duluth, MN, EPA/600/R-92/081, 1993.
75. U.S. Environmental Protection Agency, Marine Toxicity Identification Evaluation: Phase I Guidance Document, Office of Research and Development, Washington, D.C., EPA/600/R-96/054, 1996.
76. Smrchek, J., Office of Pollution Prevention and Toxics Overview, Presented to ORD, National Health and Environmental Effects Research Laboratory (NHEERL), Western Ecology Division, Corvallis, OR, August 29, 2000.
77. Smrchek, J. C., personal communication, 2001
78. U.S. Environmental Protection Agency, Data Requirements for Pesticide Registration: Final Rule, 40 CFR part 158, *Federal Register*, 49, 42857, 1984.
79. Report of the Aquatic Effects Dialogue Group, *Improving Aquatic Risk Assessment under FIFRA*, World Wildlife Fund, Resolve, Washington, D.C., 1992.
80. Smrchek, J. C. and Zeeman, M. G., Assessing risks to ecological systems from chemicals, in *Handbook of Environmental Risk Assessment and Management,* Calow, P., Ed., Blackwell Sciences, London, 1998.
81. Code of Federal Regulation, Food and Drug Administration, HHS, CFR 21, Part 25, 1987, 194.
82. U.S. Environmental Protection Agency, The Superfund Program: Ten Years of Progress, Office of Solid Waste and Emergency Response, U.S. Environmental Protection Agency, Washington, D.C., EPA/540/8–91/003, 1991, 1.
83. U.S. Environmental Protection Agency, Ecological Risk Assessment Guidance for Superfund: Process for Designing and Conducting Ecological Risk Assessments, U.S. EPA, Office of Solid Waste and Emergency Response, Washington, D.C., 1997.
84. U.S. Environmental Protection Agency, Risk Assessment Guidance for Superfund, Volume II, Environmental Evaluation Manual, U.S. EPA, Office of Solid Waste and Emergency Response, EPA 540 1–89–001, 1998.
85. Warren-Hicks, W., Parkhurst, B. R., and Baker, S. S., Jr., Ecological Assessment of Hazardous Waste Sites, U.S. Environmental Protection Agency, U.S. Environmental Research Laboratory, Corvallis, OR, EPA 600/3–89/013, 1989, 1.
86. EPA/USACE, Environmental Protection Agency/Corps of Engineers Technical Committee on Criteria for Dredged Material into Ocean Waters, Implementation Manual for Section 103 of Public Law 92–532 (Marine Protection, Research, and Sanctuaries Act of 1972), Environmental Effects Laboratory, U.S. Army Engineer Waterways Experiment Station, Vicksburg, MS, 1977.
87. U.S. Environmental Protection Agency, Evaluation of Dredged Material Proposed for Discharge in Waters of the U.S., Testing Manual, U.S. Environmental Protection Agency, Office of Water, Washington, D.C., EPA 823/B-998/004, 1998.
88. Council Directive 91/414/EEC, EC agrochemical registration directive, *Official Journal of the EC*, L230, 1991, 1.

89. Organization for Economic Co-operation and Development, *OECD Guidelines for Testing of Chemicals*, OECD, Paris, 1981, 1.
90. Smrchek, J. C. and Zeeman, M., Harmonization of Test Methods Between the U.S. EPA (OPPTS) and the Organization for Economic Cooperation and Development (OECD): General Overview of Recent Activities with Emphasis on Aquatic Sediment Methods in Environmental Toxicology and Risk Assessment: Science, Policy and Standardization – Implication for Environmental Decisions: Tenth Volume, Greenberg, B. M., Hull, R. H., Roberts, M. H., Jr. and Gensemer, R. W., Eds., American Society for Testing and Materials, West Conshohocken, PA, 2001, 1.
91. McKim, J. M., Early life stage toxicity tests, in *Fundamentals of Aquatic Toxicology*, Rand, G. M. and Petrocelli, S. R. Eds., Hemisphere Publishing Corporation, New York, 1985, Chap. 3.
92. Mount, D. I. and Norberg, T. J., A seven-day life-cycle cladoceran toxicity test, *Environ. Toxic. Chem.*, 3, 425, 1984.
93. Snell, T. W. and Moffat, B. D., A two-dimensional life cycle test with the rotifer *Brachinus calyeiflorus*, *Environ. Toxic. Chem.*, 11, 1249, 1992.
94. Norberg, T. J. and Mount, D. I., A new fathead minnow (*Pimephales promelas*) subchronic test, *Environ. Toxic. Chem.*, 4, 711, 1985.
95. Bulich, A. A., Use of luminescent bacteria for determining toxicity in aquatic environments, in *Aquatic Toxicology*, Marking, L. L. and Kimerle, R. A., Eds., American Society for Testing and Materials, Philadelphia, 98, 1979.
96. Shugart, L. R., Molecular markers to toxic agents, in *Ecotoxicology: A Hierarchical Treatment*, Newman, M. C. and Jagoe, C. H., Eds., Lewis Publishers, Boca Raton, FL, 1996, Chap. 5, 133–161.
97. Huggett, R. J., Kimerle, R. A., Mehrle, P. M., and Bergman, H. L., Eds., *Biomarkers: Biochemical, Physiological and Histological Markers of Anthropogenic Stress*, Lewis Publishers, Boca Raton, FL, 1992.
98. Shugart, L. R., Adams, S. M., Jimenez, B. D., Talmage, S. S., and McCarthy, J. F., Biological markers to study exposure in animals and bioavailability of environmental contaminants, in *Biological Monitoring for Pesticide Exposure: Measurement, Estimation and Risk Reduction*, Wang, R. G. M., Franklin, C. A., Honeycutt, R. C., and Reinert, J. C., Eds., ACS Symposium Series 382, American Chemical Society, Washington, D.C., 1989, 86–97.
99. Shugart, L. R., Bickham, J., Jackim, G., McMahon, G., Ridley, W., Stein, J., and Steinert, S., DNA alterations, in *Biomarkers: Biochemical, Physiological and Histological Markers of Anthropogenic Stress*, Huggett, R. J., Kimerle, R. A., Mehrle, P. M., and Bergman, H. L., Eds. Lewis Publishers, Boca Raton, FL, 1992, 125–153.
100. McCarthy, J. F., and Shugart, L. R., Eds., *Biomarkers of Environmental Contamination*, Lewis Publishers, Boca Raton, FL, 1990, 3–14.
101. Shugart, L. R., McCarthy, J. F., and Halbrook, R. S., Biological markers of environmental and ecological contamination: An overview, *J. Risk Anal.*, 12, 352–360, 1992.
102. Peakall, D. B. and Shugart, L. R., Eds., *Strategy for Biomarker Research and Application in the Assessment of Environmental Health*, Springer-Verlag, Heidelberg, 1992.
103. Fossi, M. C. and Leonzio, C., Eds., *Nondestructive Biomarkers in Vertebrates*, Lewis Publishers, Boca Raton, FL, 1993.
104. Travis, C. C. Ed., *Use of Biomarkers in Assessing Health and Environmental Impacts of Chemical Pollutants*, NATO ASI Series: Life Sciences, Vol. 250, Plenum Press, New York, 1993.
105. Evans, C. W., Hills, J. M., and Dickson, J. M. J., Heavy metal pollution in Antarctica: A molecular ecotoxicological approach to exposure assessment, *J. Fish Biol.*, 57, 8–19, 2000.
106. Wong, C. K. C., Yeung, H. Y., Cheung, R. Y. H., Yung, K. K. L., and Wong, M. H., Ecotoxicological assessment persistent organic and heavy metal contamination in Hong Kong coastal sediment, *Arch. Environ. Contam. Tox.*, 38, 486–493, 2000.
107. Petrovic, S., Ozretic, B., Krajnovic-Ozretic, M., and Bobinac, C, Lysosomal stability and metallothioneins in the digestive gland of mussels (*Mytilus galloprovincialis Lam.*) as biomarkers in a field study, *Mar. Pollu. Bull.*, 24, 1373–1378, 2001.
108. Schramm, M., Muller, E., and Triebskorn, R., Brown trout (*Salmo trutta f. Fario*) liver ultrastructure as a biomarker for assessment of small stream pollution, in *Biomarkers*, Taylor and Francis LTD, 1998.
109. Haasch, M. L., Prince, R., Wejksnorea, P. J., Cooper, K. R., and Lech, J. J., Caged and wild fish: induction of hepatic cytochrome P-450 (CYP1A1) as an environmental biomonitor, *Environ. Tox. Chem.*, 12, 885–895, 1993.

CHAPTER 3

# Model Aquatic Ecosystems in Ecotoxicological Research: Considerations of Design, Implementation, and Analysis

**James H. Kennedy, Thomas W. LaPoint, Pinar Balci, Jacob K. Stanley, and Zane B. Johnson**

## CONTENTS

## 3.1 INTRODUCTION

A number of research studies have made use of model aquatic ecosystems of varying design and complexity for evaluating the fate and effects of contaminants in aquatic ecosystems. These systems are designed to simulate ecosystems or portions of ecosystems. As research tools, model ecosystems contribute to our understanding of the ways in which contaminants affect natural ecosystems.[1] These systems are a tool for allowing ecologists to address hypotheses on a manageable scale and with control or reference systems. They also provide ecotoxicologists with models of ecosystem functioning, in the absence of perturbation, so that direct and indirect effects might be better separated from natural events such as succession or inherent variation.[2]

Traditionally, model ecosystems have been categorized as either microcosms or mesocosms. The distinction between microcosms and mesocosms has been somewhat subjective, with researchers establishing their own criteria, but has mainly been a function of size.[3] The degrees of organizational complexity and realism will often vary when these systems are established, depending largely on study goals and endpoints selected by the researchers.

Giesy and Odum[4] define microcosms as artificially bounded subsets of naturally occurring environments that are replicable, contain several trophic levels, and exhibit system-level properties. Mesocosms are defined as either physical enclosures of a portion of a natural ecosystem or manmade structures such as ponds or stream channels.[5] Voshell[5] further specifies that the size and complexity of mesocosms are sufficient for them to be self-sustaining, making them suitable for long-term studies. In this regard they differ from microcosms, where smaller size and fewer trophic levels do not allow for long study durations, particularly in laboratory systems. Cairns,[6] however, does not distinguish between microcosms and mesocosms because "both encompass higher levels of biological organization and have high degrees of environmental realism." The lack of a defined distinction between microcosm and mesocosm systems has caused some confusion among researchers around the world. The organizing committee of the European Workshop on Freshwater Field Tests (EWOFFT) operationally described microcosms on the basis of size, defining outdoor lentic microcosms as those surrogate ecosystems whose volume contain less than 15 $m^3$ of water and mesocosms as ponds of 15 $m^3$ or larger. Experimental stream channels were also characterized on the basis of size, defining microcosms as smaller and mesocosms as larger than 15 m in length. Such designations are useful categories for standardizing terminology. These distinctions are often used when comparing studies conducted throughout the world, and this paper will define model systems based on the EWOFFT definitions, when needed.

The use of "model" systems in aquatic research has grown considerably since the use of replicated ponds in community structure analysis by Hall, Cooper, and Werner[7] in the late 1960s and the pesticide studies of Hurlbert et al.[8] Studies prior to or concurrent with these, such as Eisenberg's[9] studies of density regulation in pond snails, used experimentally manipulated natural systems. Aspects such as community composition and spatial heterogeneity can be controlled to a greater extent in model (constructed) systems relative to natural ones. Model ecosystems are logistically more manageable and replicable for statistical analyses. In addition, model systems are effective tools in aquatic research because they act as surrogates for important cause-and-effect pathways in natural systems[6,10] yet retain a high degree of environmental realism relative to laboratory single-species bioassays.[6] These tests should be viewed as part of a tiered testing sequence and not as replacements of single-species bioassays.[11] Single-species tests, however, are inadequate when chemical fate is altered significantly under field conditions, when organismal behavior can affect exposure to a toxicant, or when secondary effects occur due to alterations in competitive or predator-prey relationships.[12]

Model ecosystems in ecotoxicological research are used to study the fate and potential adverse effects of chemicals. The ability to detect and accurately measure these effects can be influenced by both system and experimental design that influence variability. This paper addresses key factors that can influence the ability of model systems to accomplish these tasks.

## 3.2 HISTORICAL PERSPECTIVE

The concept of the microcosm was introduced early in ecological thinking through the writings of Forbes.[13] In his work on lake natural history the basic principles of ecological synergism, variability, and dynamic equilibrium as well as the complex interactions of predator and prey were discussed. Though speaking of the lake itself and not of the surrogate systems routinely employed in aquatic research today, Forbes[13] touched upon the rationale for the use of artificial systems in both toxicological and ecological research: "It forms a little world within itself — a microcosm within which all the elemental forces are at work and the play of life goes on in full, but on so small a scale as to bring it easily within the mental grasp."

This assertion — that artificial systems simulate processes that occur in nature enough to be viable surrogates for natural systems — is central to the underlying basis for using microcosms (and mesocosms) in ecotoxicological research.

The initial applications of artificial aquatic systems, such as laboratory microcosms, artificial ponds, and various *in situ* enclosures, were historically utilized in ecological studies of productivity,[7,14–16] community metabolism,[17–19] and population dynamics.[20] The earliest of these experiments, using laboratory microcosms, were those of Woodruff[21] and Eddy,[22] who examined protozoan species succession in hay infusions, and the slightly later studies of Lotka,[23,24] Volterra,[25,26] and Gause,[20] which formed the basis of the now standard quadratic population models in competition and predator-prey interaction. Gause[20] conducted his classic experiments on protozoan competition in glass dish microcosms from which his mathematical theory of competitive exclusion was derived. Gause[20] sought to address important ecological issues in these systems while being cognizant of their (potential) limitations. In discussing earlier studies conducted in laboratory microcosms Gause writes:

> However, in experiments of this type there exists a great number of different factors not exactly controlled, and a considerable difficulty for the study of the struggle for existence is presented by the continuous and regular changes in the environment. It is often mentioned that one species usually prepares the way for the coming of another species. Recollecting what we have said in Chapter II it is easy to see that in such a complicated environment it is quite impossible to decide how far the supplanting of one species by another depends on the varying conditions of the microcosm which oppress the first species, and in what degree this is due to direct competition between them.

The above-cited research helped lay the groundwork for understanding how biotic processes function in artificially bounded and maintained systems. A fundamental knowledge of the ecology of the systems is necessary if there is to be any understanding of how they may be altered by an introduced perturbation. There has been considerable concern and debate over whether model systems, such as microcosms, simulate natural systems closely enough to be used as ecosystem surrogates. Microcosms tend not to closely simulate natural systems at all levels of ecological organization. Traditionally, this has not been viewed as a problem, as the system selected will vary with the research goals and the endpoints of choice. The presence of higher levels of organization may not be necessary to demonstrate effects with some endpoints.

The use of surrogate systems in toxicological research, particularly those encompassing any appreciable scale and complexity, has been relatively recent (ca. 1960). Concern over the effects of insecticides used to control mosquito populations in California prompted a series of field studies on the consequences of chemical control methods on nontarget species such as mosquito fish and waterfowl. Keith and Mulla[27] and Mulla et al.[28] used replicated artificial outdoor ponds to examine the effects of organophosphate-based larvicides on mallard ducks. Hurlbert et al.[8] conducted subsequent studies in the same systems, examining the impact on a greater number of species within several broad taxa (phytoplankton, zooplankton, aquatic insects, fish, and waterfowl). Essentially, system-level impacts were being assessed, with subsequent evaluation of indirect effects (e.g., changes in prey species densities in the absence of a predator species or the effects of emigration from a system) attributable to the pesticide application.

The broad application of microcosms and mesocosms in toxicological studies largely followed the realization that single-species toxicity tests alone were inadequate for predicting effects at the population and ecosystem levels.[29,30] Multispecies tests have the potential to demonstrate effects not evident in laboratory tests that use a single, presumably (most) sensitive, species.[31] As the goals of environmental protection focus on ecosystem-level organization, testing within more complex systems involves less extrapolation, apparently enhancing the prediction of impacts on natural systems. Model ecosystems in ecotoxicological research are seen primarily as a way of studying potential contaminants in systems that simulate parts of the natural environment but that are amenable to experimental manipulations.[1]

An assessment of the ecological risk of pesticides is required under the United States Federal Insecticide, Fungicide and Rodenticide Act (FIFRA). A tiered data-collecting process that results from a progression of increasingly complex toxicity tests is considered together with an estimation of environmental exposure to make an assessment of whether a chemical may pose an unacceptable risk to the aquatic environment. Following tests conducted for each tier, data are evaluated, and risk to the aquatic environment is determined. Based on the outcomes of testing at each tier, the decision is made whether to stop testing or to continue to the next tier. Initial tiers are in the form of laboratory bioassays. The final tier (Tier IV) involves field testing. A description of the tests required at each tier and criteria for their implementation by the Environmental Protection Agency (EPA) is given in the Report of the Aquatic Effects Dialogue Group (AEDG).[32] Registrants may be required by the EPA to conduct higher-tier tests, or they may opt for this level of testing to refute the presumption of unacceptable environmental risk indicated by a lower-tier test.

Prior to the EPA's adoption of the mesocosm technique as part of the ecological risk assessment of pesticides, Tier IV tests were conducted in natural systems that were exposed to the agricultural chemical during the course of typical farming practices. Although these types of studies provided realism in terms of environmental fate of the compound and exposure to the aquatic ecosystem, they were difficult to evaluate, in part because of insufficient or no replication, a high degree of variability associated with the factors being measured, and influences of uncontrollable events such as weather. In the mid-1980s the EPA adopted the use of mesocosms (experimental ponds) as surrogate natural systems in which ecosystem-level effects of pesticides could be measured (Tier IV tests) and included in the ecological risk-assessment process.[33]

Although no longer part of the regulatory requirements in the United States, mesocosm test requirements have stimulated an increased worldwide interest in the use of surrogate ecosystems for the evaluation of the fate and effects of contaminants in aquatic ecosystems, as evidenced by the number of symposia[5,34–38] and workshops[1,39–41] over the last decade.

## 3.3 BIOMAGNIFICATION

Barron[42] presents an overview of the principles and determinants of biomagnification in aquatic food webs. Environmental contaminants affect organisms that are part of an aquatic food chain. Biomagnification is the increase in contaminant body burden (tissue contaminant) caused by the transfer of contaminant residues from lower to higher trophic levels.[43] Rasmussen et al.[44] showed that PCBs in lake trout increased with the length of benthic-based food web and with the lipid content of tissue. Simon et al.[45] have analyzed the trophic transfers of metals (cadmium and methylmercury) between the Asiatic clam *Corbicula fluminea* and crayfish *Astacus astacus*. Their experimental data suggest a small risk of Cd transfer between the crayfish and predators, humans included. However, methylmercury distribution in muscle and accumulation trends in this tissue represent an obvious risk of transfer.

## 3.4 MODEL ECOSYSTEMS

A wide variety of model ecosystems have been developed and used for fundamental and applied aquatic ecological research. Review articles describing these systems are available for microcosms,[46] freshwater mesocosms,[47,48] marine mesocosms,[49–51] and artificial streams.[35] Kennedy et al.[52] review and summarize in table format representative examples of experimental designs used by researchers to study the fate and effects of xenobiotic chemicals in freshwater and marine aquatic environments.

### 3.4.1 Microcosms

Microcosms have been employed extensively in studies of contaminant effects on community-level structure and function. These systems can be viewed as an intermediate to laboratory tests and larger-scale mesocosms. Microcosms — whether indoor or outdoor — may not accurately parallel natural systems at all levels of organization, but important processes such as primary productivity and community metabolism can be studied in them, even in cases where systems cannot support all of the trophic levels found in larger systems.

Outdoor microcosms have taken a variety of forms including small enclosures in larger ponds[53–56] and free-standing tanks of sizes ranging from small aquaria (12 L)[57] suspended in a natural pond to vessels constructed of fiberglass,[58–63] stainless steel,[64] or concrete[65–67] or excavated from the earth.[68,69] Other researchers have used plastic wading pools[70] and temporary pond microcosms.[71]

### 3.4.2 Mesocosms

Likewise, an assortment of mesocosm ecosystems have been devised. Most mesocosm systems can be categorized into one of several systems based on their construction.

### 3.4.3 Enclosures

Limnocorrals — large enclosure systems in open-water areas of lakes — have been used extensively by Canadian researchers. These are systems designed to partition and encompass natural planktonic populations in order to study their responses to perturbations. Kaushik et al.[72] described limnocorral construction. The impact of pesticides on plankton populations has been a frequent focus of enclosure studies.[72–79]

Littoral enclosures, which border the edge of a pond or lake, have been developed and used by the U.S. EPA Research Laboratory at Duluth, MN. These systems (5 m × 10 m surface area) have been used to study the fate and effects of pesticides on water quality parameters, zooplankton, phytoplankton, macroinvertebrates, and fish.[80] Brazner et al.[81] described littoral enclosure construction and endpoints studied and discussed variability (coefficients of variation) of different indicators.

### 3.4.4 Pond Systems

Replicated pond mesocosms have been used extensively to evaluate pesticide fate and toxicological effect relationships.[82] Most ponds used for this purpose are dug in the earth and range in size from 0.04 to 0.1 hectares in surface area.

### 3.4.5 Artificial Streams

Unlike lentic mesocosms, there have been no attempts to standardize the conduct of lotic experimental systems, even though experimental stream ecosystems have been employed to test chemical effects (Tables 3.1 and 3.2). Invariably, the use of these constructed stream ecosystems

**Table 3.1 Use of Stream Mesocosms: Physical Parameters**

| References | Circulation | Length/Size | Volume/Flow |
|---|---|---|---|
| Austin et al.[93] | FT | 0.245 m | 18.0 L/min |
| Belanger et al.[94] | FT | 20.0 L | 1.6 L/min |
| Clements et al.[95] | FT | 0.76 m | 1.6 L/min |
| Clements[96]<br>Farris et al.[242] | FT | 0.76 m | 1.0 L/min |
| Belanger et al.[243]<br>Guckert et al.[244]<br>Lee et al.[245] | FT | 12.0 m | 166.0 L/min |
| Dorn et al.[97–99]<br>Gillespie et al.[100–102]<br>Harrelson et al.[103]<br>Kline et al.[104] | FT | 4.9 m | 77.0 L/min |
| Haley et al.[105]<br>Hall et al.[106] | FT | 110.0 m | 1241.0 L/min |
| Hermanutz et al.[107] | FT | 520.0 m | 0.57 $m^3$/min winter, 0.76 $m^3$/min |
| Kreutzweiser and Capell[108] | FT | 6.0 m | 14.0 L/min |
| Crossland et al.[109]<br>Maltby[110]<br>Mitchell et al.[111]<br>Pascoe et al.[112] | PRC | 5.0 m | 10.0 L/h |
| Richardson and Kiffney[113] | FT | 2.5 m | 0.1–0.2 L/s |

*Note:* RC = recirculating; FT = flow-through; PRC = partially recirculating. Single-spaced references imply use of the same systems.

has involved studying the responses of macrobenthic communities to multiple chemicals, chosen to be "typical" of what might be expected in natural streams. Response variables have differed among previous lotic studies and depend on the research questions asked and approaches taken (Table 3.2). Presently, there are relatively few such systems in the world. Costs associated with building and operating lotic mesocosm systems often limit the number of experimental units. Thus, most stream mesocosm studies have evaluated single chemicals at multiple concentrations with or without treatment replication. The designs range from small recirculating streams[83] to large, in-ground flow-through streams of 520 m in length.[84] Most constructed streams are 3 or 4 m long and around 50 cm wide. Volume flows range considerably and are usually selected to approximate the regional conditions.

The endpoints selected for study are almost always functional and structural endpoints of algae or benthic invertebrate populations (Table 3.2). The size and scale of the artificial streams preclude the use of predator fish, except for the very large systems. For the short term, studies pools may be constructed downstream to place herbivorous minnows or larval predators such as bluegill or bass.

Regression designs are common and suggested for use in risk assessment when experimental units are scarce.[85,86] Despite problems associated with pseudoreplication,[87] lack of replication may be justified because intraunit variability due to treatments can be substantially more important than interunit variability.[88] Limited experimental stream studies have used factorial designs or addressed issues of multiple stressors.[89,90] Factorial designs use ANOVA (requires replication), are efficient, and allow investigation of multiple-factor interactions (multiple stressors).[91,92] Tables 3.1 and 3.2 present representative examples of experimental designs and endpoints used in outdoor stream mesocosms.

## 3.5 DESIGN CONSIDERATIONS

There are many problems to be considered when designing and implementing studies using model systems. These range from the pragmatic (funding, time constraints, etc.) to the heuristic

**Table 3.2 Use of Stream Mesocosm Chemicals Tested and Response Variables**

| References | Chemical(s) | Structural | Functional |
|---|---|---|---|
| Austin et al.[93] (periphyton) | Herbicide | A, biomass | |
| Belanger et al.[94] (clams) | Cu | | Mortality, growth, & bioaccumulation |
| Belanger et al.[246] (clams) | Surfactant | | Mortality, growth, reproduction, cellulolytic enzyme activity, larval colonization |
| Belanger et al.[247] (invertebrates) | Surfactant | A, biomass, H′, trophic functional feeding group | Drift |
| Belanger et al.[243] (periphyton, protozoa, invertebrates) | Surfactant | A, biomass, H′ | Productivity, biodegradation of test chemical |
| Clements et al.[96] (invertebrates) | Cd, Cu, Zn | A | Invertebrate survival, drift, & predation rate |
| Clements et al.[95] (invertebrates) | Cu | A | |
| Crossland et al.[109] (invertebrates) | Effluent | A | Feeding rates and drift |
| Dorn et al.[97, 98, 99] (fish, inverts, macrophytes, periphyton) | LAE | A, biomass | Drift, mortality, growth, reproduction, chlorophyll & pheophytin |
| Farris et al.[242] (clams and snails) | Zn | | Cellulolytic enzyme activity bioaccumulation |
| Gillespie et al.[106, 108] (invertebrates.) | LAE | A | Drift |
| Gillespie et al.[100] (invertebrates) | LAE | A | Drift, feeding rates |
| Guckert et al.[244] (periphyton, protozoa, invertebrates) | Surfactant | A, functional group composition, H′ | Primary production, drift, recruitment |
| Haley et al.[105] (fish, invertebrates, periphyton) | Effluent | A, biomass, H | Mortality, growth, histopathology, chlorophyll, production |
| Hall et al.[106] (fish, invertebrates, periphyton) | Effluent | a, biomass | Mortality, growth, histopathology, reproduction, chlorophyll, production |
| Harrelson et al.[103] (fish) | LAE | | Mortality, growth, reproduction, behavior |
| Hermanutz et al.[107] (fish) | Se | | Bioaccumulation, mortality, growth, development, reproduction |
| Kline et al.[104] (fish, zooplankton) | Surfactant | A | Mortality, growth, reproduction, swimming performance |
| Kreutzweiser and Capell[108] (Invertebrates) | Pesticides | | Mortality, drift |
| Lee et al.[245] (periphyton) | Surfactants | A, community biovolume | Surfactant biodegradation, heterotrophic respiration |
| Maltby[110] (invertebrate) | Effluent | | Scope for growth |
| Mitchell et al.[111] (invertebrates, periphyton) | Lindane | A | Drift, feeding rates, photosynthesis rate |
| Pascoe et al.[112] (invertebrates, periphyton) | Cu, lindane, 3,4-dichloroaniline (DCA) | A, biomass | Drift, growth, precopula disruption, photosynthesis, chlorophyll |
| Richardson and Kiffney[113] (invertebrates, periphyton) | Cu, Zn, Mn, Pb | A | Emigration (drift), chlorophyll, bacterial respiration |

*Note:* H′ = invertebrate diversity; A = abundance; LAE = linear alkyl ethoxylate surfactant.

(What are the study goals? What levels of realism are desired?). The physicochemical and biotic features of a model system will determine to what extent, if any, the system will represent a natural one. These factors also influence contaminant fate and effects in model ecosystems. System design is therefore important in defining what inferences may be drawn from the results of surrogate systems and extrapolated to natural aquatic ecosystems. Using results from the scientific literature on model ecosystems, the following sections seek to provide a synthesis of some key experimental design considerations.

### 3.5.1 Scaling Effects in Artificial System Research

The question of whether artificial aquatic systems are reliable surrogates for natural ones is strongly linked to system scale. Scale includes not only size and physical dimensions of a microcosm or mesocosm but also its spatial heterogeneity and attendant biotic components. Crucial physical and chemical processes behave differently as both a function of, and contributor to, scale. Thus, scaling effects can have implications for community structure and the resultant functional attributes of the system.

The choice of spatial and temporal scales in an experiment may determine whether changes in selected endpoints can be detected during a study and is, therefore, vital to the research methodology. Frost et al.[114] stated that "typically scale has not been incorporated explicitly into sampling protocols and experimental designs." The choice of appropriate time scales, for example, in model aquatic system research must be considered in the selection of both study duration and sampling frequency intervals between sampling events. Both temporal elements should consider life cycles and periodicities of important system species. Sampling intervals should also consider the temporal behavior of key physicochemical processes (often related to pesticide fates and half-lives) and, ultimately, the longevity of the surrogate system as well.

Microcosms, particularly laboratory ones, require little or no equilibration time prior to their use as test systems. Results can be observed quickly, but the systems are not self-sustaining and tend to become unstable over time. Because laboratory microcosms can sustain only a limited number of trophic levels, usually composed of small organisms with short lifespans (days to weeks) and rapid turnover times, frequent sampling regimes and short study durations are required. Unfortunately, frequent sampling in small systems may be damaging to the system and its biotic contingent.[32]

The size and overall dimensions of systems in ecotoxicological research have idiosyncratic implications in the outcome of the project. Dudzik et al.[115] cite the prevalence of biological and chemical activity on the sides and bottoms of microcosms as one of the most important problems in microcosm research. Edge effects have been noted and discussed in enclosure studies as well,[116,117] but the ecological implications of such scaling ramifications in ecotoxicological and ecological studies have yet to be resolved. These concerns present a unique challenge in the toxicological arena, as scaling effects may ultimately hinder the validation process, which is becoming increasingly critical in decision-making, policy-making, enforcement, and litigation issues.

The cause of some edge effects in ecotoxicological work pertain to materials from which littoral and pelagic enclosures are constructed, since they may serve as sorption sites for some toxins (via adsorption).[118–120] This problem was also linked to physical scale and system dimensions, as the ratio of the wall surface area to water volume is greater in smaller test systems. Smaller enclosures and microcosms may remove disproportionate amounts of pesticide from the water column via absorption to container walls.[32]

A study investigating the role of spatial scale on methoxychlor fate and effects in three sizes of limnocorrals found pesticide dissipation was more rapid than expected in the smallest enclosures.[74] These findings were associated with less severe impacts and quicker recovery of zooplankton populations in the smallest enclosures. Such studies are, in part, contingent upon an understanding of the role of spatial factors in biotic organization.

In an attempt to address such concerns, Stephenson et al.[116] studied the spatial distributions of plankton in limnocorrals of three sizes (equal depths) in the absence of perturbations to assess the viability of such systems in community-level toxicant research. The most predominant edge effects were reported in the largest enclosures, where macrozooplankton occurred in significantly higher numbers than microzooplankton. Perhaps differentially sized edge zones contributed to distributional differences (edge zones constituted 100% of the volume of the smallest enclosures), or circular currents that occurred only in the largest enclosures could have affected zooplankton distributions. The actual cause of the zooplankton distribution differences was not determined, but basic research of this type provides important "background" data to better recognize treatment effects once disturbance has been introduced.

The limited size and accompanying physical homogeneity of many microcosms creates additional problems: (1) they are particularly susceptible to stochastic, often catastrophic, events from which system recovery may be highly variable relative to mesocosms,[32] and (2) the often limited species compositions of microcosms induce overly strong biotic couplings, resulting in drastic population oscillations and competitive exclusion events.[114] This latter problem was established early in ecological study with the research of Lotka[24] and Gause[20] when only a limited number of population cycles could be established in small microcosms.

As discussed previously, large outdoor systems, such as pond mesocosms, require colonization and equilibration times of months to years because they may incorporate many trophic levels, and an extensive number of interactions occur as a function of greater physical scale. Frequent (i.e., daily) sampling for many selected parameters may not be logistically feasible or even necessary to detect effects at the population or community levels. Study durations are by necessity and design much longer, since impacts at higher levels of organization, particularly indirect effects, may not be immediately evident. Such systems are presumably self-sustaining enough to permit the study periods necessary for detecting effects at these higher levels.

A variety of scales are to be considered when designing studies using surrogate systems because the scales discussed herein will affect the outcome of research whether the experimenter acknowledges them or not. Most researchers are aware of the implications of system size in fate and effects research, though indirect results in these studies may not always be perceived or attributed to their actual causes. Temporal aspects are also recognized, though the interaction of timing and spatial factors is still not well understood. The treatment of these scaling considerations in a more integrated fashion will ultimately enhance the predictive value and ecological relevance of the results.

### 3.5.2 Variability

Variability is inherent in any biological system, but the limiting of variability is often critical to the scientific process wherein the ultimate goal is prediction. Variability occurs within and among systems such as microcosms or mesocosms. Replication of treatments and the use of controls are necessary to distinguish natural variation from the effects of treatment.[83] Sampling replication can assess intrasystem heterogeneity resulting from spatiotemporal variation in community structure and physicochemical parameters.

Studies of stream benthos, however, indicate that the number of samples required to obtain adequate representation of the community would be quite high and no doubt impractical.[121–123] There is also the risk that accepted sampling regimes in lentic and marine research may similarly underestimate inherent variability in these more "homogeneous" systems. Assessing variability through such methods as coefficients of variation[123] and determining the number of sampling replicates that would be adequate to ensure representative sampling become critical in ecological research. Green[124] emphasizes the importance of conducting pilot studies in ecological research and having adequate replication, both in treatment and sampling. Unfortunately, even though the number of replicates needed to detect changes of a given magnitude can be determined *a priori*, such estimates do not always match the availability of research resource personnel, space, or time.[125] Therefore, sampling must be focused on those variables that convey scientific meaning and provide investigators with resolving power for detecting differences. At the present time these variables are primarily structural.[126]

An alternative to increased sampling replication has been to employ less diverse systems as a means of reducing intrasystem variation. This approach may result in simplification to the extent that model systems will not resemble the natural systems they are attempting to "mimic," thereby affecting predictability[4] and applicability. However, mimicking natural systems may not provide the best experimental models, according to Maciorowski,[127] who further emphasizes that the challenge in ecotoxicological research is to find those phenomena that can be simplified to several salient interactions. Extrapolating the results of model system research to natural systems remains one of the major areas of contention regarding their use.

In any manipulative experiment the assumption is made that observed effects (i.e., significant differences) are due to the treatment. Often, however, observed differences among treatment levels, or even among replicates within a treatment level, may be influenced by factors other than those being tested.[87] When this occurs, it is impossible to separate the covariates, and the hypotheses being tested at the onset may be invalidated. Variability among systems is a frequent contributor to this phenomenon.

Sources of variability may be structural, physicochemical, or biotic. Biotic variability can occur at a variety of levels within the ecosystem and markedly affect system-level processes such as productivity and respiration. Variability can be due to differences among systems prior to study initiation, or it may result from changes that occur during the study. Hurlbert[87] discusses both initial or inherent variability among systems and the temporal changes that occur within systems.

The confounding influence of system variability in ecotoxicological studies involving microcosms and mesocosms has long been recognized,[115] but no uniform approach to a solution has been reached. Some researchers[128,129] have attempted to assess inherent variability and determine the amount of sampling replication required to detect treatment effects. Other solutions involve establishing more stable communities in the hopes that equilibrium within systems will occur, enhancing both similarity among systems and increasing system realism. Giesy and Odum[4] suggest that higher trophic levels assert a controlling influence on lower trophic levels in microcosms being used for effects studies. Giddings and Eddlemon[130] have attempted to assess microcosm variability for the purpose of determining the validity of using such model systems in toxicological research.

Methods of limiting intersystem variability sometimes employ design features. One routinely applied method in mesocosm — and sometimes in microcosm — research circulates water among the systems prior to study commencement.[60,131–134] Heimbach[135] developed outdoor microcosms in which three interconnected tanks were joined via wide locks (passageways). Water exchange was allowed during an acclimation period, followed by isolation prior to pesticide application. Systematic "seeding" of the systems with biota and sediments from mature ponds may minimize variability resulting from nonuniform distributions of macroinvertebrates and macrophytes.[136]

### 3.5.3 Colonization and Acclimation

Ecological maturity of mesocosms affects the degree of variability of both physicochemical and biological parameters used to investigate the impact of contaminants.[137] The establishment of biological organism communities is a critical part of microcosm and mesocosm experiments. Adequate time is required to establish a number of interacting functional groups.[4] The colonization methods used in microcosm and mesocosm research will vary predominantly as functions of system size, the type of study, whether it is fate- or effect-oriented, and the endpoints of interest.[138] Studies using limnocorrals and littoral enclosures usually have no acclimation period because it is assumed these systems enclose established communities.[73,119,139] In stream mesocosms stabilization periods of 10 days,[140] 4 weeks,[109] or 1 year[141] have been reported.

The duration of the maturation period for pond mesocosms varies from 1 to 2 months[39] to 2 years.[7] Following initial system preparation a period of acclimation is usually required to allow the various biotic components to adjust to the new environment and establish interspecific and abiotic interactions. Duration of acclimation time depends on system size and complexity. Systems with more trophic levels will form more complex interactions that may require much more time to equilibrate than small systems with fewer species. The time needed to equilibrate will increase with initial system complexity, although the use of natural sediments usually shortens the duration of the stabilization period because natural maturation processes are enhanced.[52] During this acclimation period in outdoor systems the initial preparation of the systems is typically controlled, and natural colonization by insects and amphibians will contribute to biotic heterogeneity and system realism. Continuous colonization, however, presents further problems in that each system tends to follow its own trajectory through time. These trends are most apparent in small-scale systems and in systems that have been in operation longer.[35] Circulation of water between the different systems has frequently been proposed as a way to limit intersystem variability during this period.[142,143,68]

### 3.5.4 Macrophytes

Aquatic vascular plants play a key role in system dynamics within natural lakes, and their presence in model ecosystems makes them more representative of littoral zones in natural systems. However, once introduced into model ecosystems, macrophyte growth is difficult to control and may vary greatly among replicates. This is of particular concern in field studies because macrophytes can influence the fate of chemicals, the occurrence and spatial distribution of invertebrates, and, if present, the growth of fish. Thus, variations of plant density and diversity in model ecosystems can be a major contributor to system variability and subsequent inability to detect changes in ecosystem structure and function.

Macrophyte densities can affect chemical fate processes by increasing the surface area available for sorption of hydrophobic compounds. The pyrethroid insecticide deltamethrin accumulated rapidly in aquatic plants and filamentous algae during a freshwater pond chemical fate study.[144] Caquet et al.[137] reported the residues of deltamethrin and lindane in the macrophyte samples for 5 weeks after treatment but never in the sediment. A microcosm study with permethrin demonstrated similar results, with extensive partitioning to macrophytes.[145] Weinberger and others[146] evaluated fenitrothion uptake by macrophytes in freshwater microcosms and found that pesticide accumulation was two- to fivefold greater in the light compared with microcosms in the dark. They concluded that both uptake and degradation of fenitrothion appeared to be photocatalyzed.

Macrophytes can also affect physicochemical composition in surrounding waters, influencing the distribution and community structures of many aquatic organisms.[147] In addition, macrophytes provide three-dimensional structure within constructed ecosystems, which affects organism distribution and interactions. Brock et al.[148] in a study with the insecticide Dursban 4E observed considerable invertebrate taxa differences between Elodea-dominated and macrophyte-free systems. Other workers have shown that macroinvertebrate community diversity is influenced by patchy macrophyte abundance[149] and specific macrophyte types.[150,151] Cladoceran communities are also associated with periphytic algae on aquatic macrophytes.[152]

Impacts of chemicals on macrophytes densities may cause indirect effects on organisms by influencing trophic linkages such as predator-prey interactions between invertebrates and vertebrates. Bluegill utilization of epiphytic prey may be much greater than predation upon benthic organisms.[153] Excessive macrophyte growth may force fish that normally forage in open water to feed on epiphytic macroinvertebrates, where the energy returns may not be as great.[154] Fish foraging success on epiphytic macroinvertebrates depends on macrophyte density[155] and plant growth form (i.e., cylindrical stems vs. leafy stems).[107,156–158] Dewey[159] studied the impacts of atrazine on aquatic insect community structure and emergence. Decreases in the number of insects in this study were correlated with reductions in aquatic macrophytes and associated algae and were not a direct effect of the pesticide.

The wide range of chemical, structural, and biotic interactions dependent upon macrophyte type and density as outlined above emphasizes the central role of this part of the community in lentic systems. It is apparent that the design of surrogate ecosystems needs to consider plant density and diversity as a contributor to system variability and the inability to detect ecosystem changes.

### 3.5.5 Fish

Whether to include fish, what species or complex of species to select, the loading rates, and their potential for reproduction are critical factors to consider in experiment design. Fish populations are known to have direct and indirect effects on ecosystem functioning. Fish predation is known to alter plankton community composition,[160–162] and the presence of fish in limnocorral or microcosm experiments may alter nutrient dynamics and cycling.[163,164] For example, during an outdoor microcosm experiment, Vinyard and others[162] found that filter feeding cichlids altered the "quality" of nitrogen (shifting dominant form) and decreased limnetic phosphorus levels via sedimentation of fecal pellets. Additionally, unequal fish mortality among replicate microcosms may influence nutrient levels independently of any other treatment manipulations.[165]

In separate limnocorral studies Brabrand et al.[166] and Langeland et al.[167] both concluded that fish predation alters planktonic communities in eutrophic lakes and that the very presence of certain fish species may contribute to the eutrophication process. These studies offered a number of interesting hypotheses regarding fish effects in limnetic systems; unfortunately, the experimental designs of these studies lacked treatment replication, limiting their inferential capability.

Many studies completed in the United States from 1986 through 1992 under U.S. EPA guidelines[33] for pesticide studies require that mesocosms include a reproducing population of bluegill sunfish (*Lepomis macrochirus* Rafinesque). Presumably, these fish and their offspring are integrators of system-level processes, and differences in numbers, biomass, and size distribution between pesticide exposure levels provide requisite endpoints for risk-management decisions. Chemical registration studies by Hill et al.,[47,168] Giddings et al.,[169] Johnson et al.,[65] Morris et al.,[170] and Mayasich et al.[171] have determined that the abundance of young bluegill in mesocosm experiments obscured or complicated the evaluation of pesticide impacts on many invertebrate populations. This is consistent with Giesy and Odum's[4] suggestion that higher trophic levels assert a controlling influence on lower trophic levels in microcosms being used for effects studies. Ecological research with freshwater plankton and pelagic fish communities indicates that both "top-down" and "bottom-up" influences affect planktonic community structure and biomass.[172–174] These relationships have not been investigated to the same degree in littoral zone communities, and the role of benthic macroinvertebrates in these trophic relationships requires further study. Along these lines Deutsch et al.[175] stocked largemouth bass in pond mesocosms in order to control unchecked bluegill population growth, thereby potentially limiting intersystem variability and provide a more natural surrogate system. However, the desirability of adding bass to mesocosms must be balanced against possible increases in experimental error variances that may result from differential predation on bluegill if variable bass mortality occurs in the ponds.[176] The only way to control variability in predation of bluegill would be to maintain equal levels of predator mortality in all ponds.

The requirement of using a single test-fish species (bluegill sunfish) in mesocosm experiments may not be sufficiently protective of natural fish communities, for a number of reasons. First, the inherent sensitivity of other fishes compared with bluegill is not known with any degree of certainty. Second, due to a variety of life history adaptations, other fish might experience differential exposure to chemicals. For example, surface-dwelling fish, such as top-minnows, would potentially be exposed to high initial pesticide concentrations found in the surface layer following treatment. Alternatively, contaminants that sorb to sediments (including many pesticides) might be expected to impact bottom-feeding fish selectively. Drenner et al.[177] studied the effects of a pyrethroid insecticide on gizzard shad, *Dorosoma cepedianum*, in outdoor microcosms. These fish are filter feeders and commonly have large amounts of bottom sediments and detritus in their digestive systems. This study[177] is unique in the use of "nonstandard" fish species. Similar field studies utilizing other fish species should be pursued in order to evaluate the influence of feeding behavior and habitat selection on chemical exposure. Following appropriate research it is conceivable that a multispecies assemblage (i.e., surface feeder, water-column planktivore, and bottom feeder) might eventually be used to better represent potential impacts to natural fish communities.

The reader is cautioned, however, that additional research in this area is needed. Scaling is an important consideration, and criteria for fish stocking levels are highly dependent on system size. The fish population should not exceed the "carrying capacity" of the test system.[168] Biomass densities should generally not exceed 2 $g/m^3$.[178]

It may be useful to stock the mesocosm with a low adult density and remove adults and larvae after spawning. However, the life stage, number, and biomass of fish added will depend on the purpose of the test. For example, should the emphasis be on an insecticide, larval fish may be added to monitor their growth in relation to the invertebrate food base.

## 3.6 DOSING CONTAMINANT EXPOSURE

### 3.6.1 Chemical Fate Considerations

The primary assessment of the potential that a chemical has to affect an aquatic ecosystem is the prediction of its environmental fate. This includes how it is transported, its persistence, its distribution or partitioning among various environmental compartments, and an estimation of its bioavailability and potential to bioaccumulate.[179] Various chemical characteristics affecting fate are currently measured in the laboratory such as solubility, octonal/water and soil/water partitioning, and bioaccumulation in different organisms. More comprehensive estimates of the fate of the chemical are manifested in mathematical and physical models of aquatic ecosystems. Boyle[179] provided a list of examples of different representative types of mathematical models from the literature used to determine the fate of a potential contaminant. Rand et al.[63] described the design, specific techniques, and fate of pyridaben in microcosms and discussed the usefulness of microcosms to study the fate of a chemical under environmental conditions that are more representative of the field.

### 3.6.2 Application Method and Dosing

Test chemicals, such as pesticides and other toxicants, are commonly applied to treatment mesocosms, with application method, frequency, and concentration of test chemical used being the major considerations.[137] The method used for application of the test chemical can have considerable effect on its fate and the exposure of organisms.[137]

Because of their scale microcosms usually lend themselves to somewhat less complicated methods of chemical application compared to similar mesocosm experiments. Microcosm experiments have used systems to distribute the test material that range from simply pouring the solution into the test chamber and stirring,[180] to a continuous-flow system.[181] Stay et al.[180] poured in the selected concentration of the chemical and used a magnetic stirring bar to thoroughly mix the contents of the microcosm before any measurements or samples were taken. Staples et al.[181] used a flow-through system in which dilution water and the chemical mixture flowed into a mixing tube and dispensed at three subsurface levels. A stirring paddle was employed to consistently mix the chemical solution in the microcosm. In their edge-of-field runoff study Huckins et al.[182] placed topsoil in a flask, spiked it with pesticide, and mixed it thoroughly. Then water was added, and mixing was achieved using a magnetic stirring bar.

Preparing a stock solution in a diluter or mixing chamber of some type and pumping it into the system is also common in these studies.[181,183,184] For example, experiments by Cairns et al.[183] and Pratt et al.[184] demonstrated this application procedure in which the dilutent flowed to a headbox, into a mixing chamber where the toxicants were added, and delivered to the microcosm test chambers with a peristaltic pump. Koerting-Walker and Buck[185] added their chemical to sediment samples to use in their 135 mm × 15 mm sediment tube microcosms.

Outdoor microcosm experiments also demonstrate a variety of dosing procedures. Pratt et al.[184] added and mixed chlorine in 130-L sediment, and water-filled polyethylene bags that were floating in a lake. Lehtinen et al.[186] used a continuous-flow system with 400-L fiberglass tanks to which effluent was continuously pumped. In his simulated wetland microcosms Johnson[187] prepared and poured a soil/water slurry onto the water surface to simulate field runoff. Similar methods were used by researchers at the University of North Texas Water Research Field Station,[65,170] where a pesticide/water solution and pesticide/soil/water slurry were prepared and poured onto the water surface of concrete microcosms to simulate spray drift and runoff events, respectively.

Complexity of dosing methods for mesocosm studies varies with the purpose of the study. The contaminant may be added to the water surface or subsurface or on the sediments by pouring the active ingredient or a mixture of soil and toxicant surface,[188–191] spraying with hand-held sprayers and spanners that release the solution onto the water surface,[192–198] or pumping via a flow-through system.[199–201] Subsurface dosing can also be achieved by placing the spray nozzle or hand-held sprayer below the water level.[131,202]

Some application methodologies are quite innovative. Wakeham et al.[203] spiked the water column of their fiberglass tank mesocosms with volatile organic compounds (VOCs), using Teflon tubing that released the VOC at about mid-tank depth, while the tank was mixed for several hours to ensure uniform VOC dispersal in the water column. Stephenson and Kane[204] applied their stock solution by allowing it to run out a separating funnel through a diffuser that was raised and lowered within the water column. De Noyelles et al.[195] used a boat to achieve access to multiple portions of a pond and dispensed a herbicide through a fine screen just below the water surface so that undissolved portions would be finely dispersed. Lay et al.[205] soaked strips of polyethylene in p-chloroaniline and placed these in the mesocosm to achieve a slow-release technique type of application. Giddings et al.[169] used a circulating system of reservoirs and tanks to simulate a typical runoff event. A stock-solution reservoir was metered to ensure the desired concentration passed into the mixing tank. The mesocosm water was then circulated into the mixing tank and pumped back into the mesocosm at three different places to ensure that each test system received a similar hydrologic treatment.

Reviewing the literature, one comes to realize that there are nearly as many application methods as there are researchers designing microcosm and mesocosm studies. It should be noted that the method chosen for the application of the test material can have considerable influence on its fate and subsequent exposure to organisms. For example, the size of droplets reaching the water surface from a spray nozzle held near the water surface of an experimental system may differ from that of droplets deposited on a natural body of water following agricultural application to adjacent land.[1] In turn, droplet size may be critical since volatilization from the water-surface microlayer can be a very rapid process and may be a major route of dissipation.[206] Thus, the decision to either spray a chemical on the water surface or inject it underneath can have a major influence on its half-life. Clearly, the method of test material application must be chosen so that realistic exposures are obtained.

## 3.7 EXPERIMENTAL DESIGN AND STATISTICAL CONSIDERATIONS

### 3.7.1 Experimental Design Considerations

Key issues in designing microcosm and mesocosm tests that need attention are replication of treatments, sample size and power, optimization criteria in design selection, choice of number and spacing of dose levels, inference on "safe dose," and defining the dose-response curve.[207] Biological variables measured in field studies have a large amount of variability associated both within and between test systems that can decrease our ability to detect ecosystem effects.[208] One approach to improving designs is by reducing variation. While this may be done by increasing the number of microcosms, this is not always possible due to cost. Information that can be gathered through power analysis can be used to maximize manpower resources and project expenditures to produce the best possible sample design as well as to determine which biological parameters should be included in a study protocol.[208]

### 3.7.2 Endpoint Selection

Toxicological endpoints are values derived from toxicity tests that are the results of specific measurements made during or at the conclusion of the test.[209] Two broad categories of endpoints are widely used: assessment and measurement endpoints. The determination, selection, and measurement of assessment and measurement endpoints are among the most critical factors in conducting an ecological risk assessment.[210] Assessment endpoints refer to the population, community, and ecosystem parameters that are to be protected, such as population growth rate, or something specific and quantifiable, such as eutrophication.[209–211] Measurement endpoints refer to the variables measured, often at the individual level, that are used to evaluate the assessment endpoints.[209] The measurement endpoints describe the variables of interest for a given test. The most common measurement endpoints include descriptions of the effects of toxic agents on survival, growth, and reproduction of a single species. Other measurement endpoints include descriptions of community effects (respiration, photosynthesis, or diversity) or cellular effects. In each case the endpoint is a variable that can be quantitatively measured and used to evaluate the effects of the toxic agent on a given individual, population, or community. Sometimes it is not possible to examine the assessment endpoint directly. In this case measurement endpoints are used to describe the organism or entity of concern.[210] The underlying assumption in making toxicological endpoint measurements is that the endpoints can be used to evaluate or predict the effects of toxic agents in natural environments. Suter[211] discussed the endpoints for the different levels of organization: suborganismal endpoints, organismal endpoints, population endpoints, and ecosystem endpoints. EPA risk assessment guidelines provide information on how endpoints can be used in the environmental risk-assessment process.[212]

### 3.7.3 Level of Taxonomic Analysis

Frost et al.[114] discussed as a scale of concern the taxonomic or functional levels to which organisms are identified or the degree of resolution. In this context organisms may be analyzed in trophic levels or functional groups or at some taxonomic level depending on the research focus and the questions proposed. Theoretically, species-level identifications have the greatest potential for identifying impacts of chemicals on aquatic organisms.[213] However, from a practical and technological standpoint, our ability to identify many organisms to the lowest taxonomic levels is limited. Many endpoints measured in mesocosm and microcosm studies require the identification of invertebrates. Increasing the taxonomic resolution used in a study increases the expertise and time needed to complete a study. As a result, taxonomic resolution used in field monitoring studies has traditionally been determined by budgetary considerations, the familiarity of the researchers with critical taxa, and the availability of reliable identification guides.

Decisions regarding the appropriate level of identification should be made with some knowledge of an organism's habitat and life history, combined with information regarding the fate of the chemical. Coarser identifications may obscure results and failure to identify organismal responses to stressors. Taxonomic sufficiency has been defined as the highest level of identification where toxic response is similar to that occurring at the lower levels of identification.[214,215] In a mesocosm study Kennedy et al.[216] demonstrated that identification to the family level for a dominant macrobenthic group failed to detect statistical differences when compared to the reference populations that were evident if subfamily identifications were determined. However, identification of these invertebrates to the genus and species level failed to detect statistical differences. This observation demands that we consider the influence counts of organisms made at lower taxonomic levels can have on statistical tests. As diverse populations are identified to lower taxonomic levels progressively lower counts and greater variability result at each level. Lower counts and increased variability results in tests of lower statistical power, which compromises our ability to detect statistical differences between populations.[208,217] If preliminary studies indicate that greater taxonomic resolution is needed, then specialized sampling methods and strategies need to be developed to increase sampling effectiveness, thereby increasing counts and reducing variability.

### 3.7.4 Species Richness, Evenness, Abundance, and Indicator Organisms

The presence of species and their relative abundance are used as a measure of the degree of contamination of an aquatic habitat.[218,219] These parameters are often used to calculate diversity indices. Although diversity indices have been shown to be insensitive to slight to moderate perturbations,[220,221] they are still reported in biological monitoring.[222,223] Species richness (the number of different species) and evenness (the distribution of individuals among species present) have been shown to better reflect impacts to aquatic communities than diversity indices.[224] The abundance of species has been a standard measure for "good quality" habitat since early studies of habitat perturbation.[225]

### 3.7.5 Univariate Methods

Univariate techniques, particularly analysis of variance (ANOVA) using parametric or log $(x + 1)$ transformed data, are the most commonly used analysis method, with either Dunnett's or the Student-Newman-Keuls (SNK) being the most common post hoc test.[208] Linear regression and correlation have also been used, but with less frequency.[226] When the assumptions of parametric tests, normality, and homogeneity cannot be met, nonparametric tests, such as Spearman, Wilcoxon rank statistic, and the Kruskal-Wallis tests, have been employed.[208]

However, these univariate methods of hypothesis testing are inappropriate for multispecies toxicity tests. As such, these methods are an attempt to understand a multivariate system by looking at one univariate projection after another, attempting to find statistically significant differences. Often, the powers of the statistical tests are quite low due to the few replicates, the high inherent variance of many of the biotic variables, and the zero counts for some organisms that were eradicated during the experiment.[227] Ammann et al.[217] and Kennedy et al.[208] proposed a statistical program, TAX-ALLN.Q, that overcomes the problems of high variability and zero counts as well as provides a measurement of statistical power that is an important design criterion in experimental studies. Perhaps the greatest danger of the use of ANOVA and related univariate tools is the perpetuation of NOECs, LOECs, and related terms based on univariate hypothesis testing. NOECs and LOECs are so dependent upon the statistical power and the concentrations chosen by the experimenter that they are artifacts of the experimental design rather than reflections of the intrinsic hazard of the toxicant.[228]

### 3.7.6 Multivariate Methods

Highly variable data are common in aquatic mesocosm studies. This can be a problem when univariate statistical procedures are used to analyze these data. The statistical power to detect effect is so low that the usefulness of conducting the analysis is questionable because even if effects exist, they may not be detected.[208,217,228] However, even if the univariate procedures are performed with satisfactory power, the interactions between species, populations, or communities are usually not considered and are therefore inadequate to elucidate ecological effects.[124] Additionally, standard univariate statistical methods can only properly analyze the information on a limited number of taxa (usually the abundant ones).[229,230] Furthermore, with the vast number of potentially confounding variables that can affect population dynamics, such approaches can lead to problems in determining cause and effect.[231]

A variety of multivariate techniques offer potential solutions to these analytical and interpretational problems.[232] Analyzing ecotoxicological field studies with multivariate techniques has some clear advantages. Community-level approaches have more ecological relevance than studies at lower levels of biological organization, and so far no compelling evidence suggests that they are any less sensitive at detecting the biological effects of pollution, especially when multivariate analyses are applied. Multivariate statistics analyzes all available data, and it is more likely to discriminate between treatments than simple univariate summaries of the same data.[210] Consequently, these approaches may be more helpful in determining the ecological significance of toxicant impacts and may help the evaluator of the study reach conclusions based on ecologically important effects, a fundamental responsibility in field studies.[233] Cost effectiveness is important, and costs can be reduced dramatically by considering taxonomic sufficiency and sampling design appropriate for the subsequent statistical analysis.[234] Multivariate techniques are also ideal for handling large amounts of data and endpoints more effectively. Kedwards et al.[67] showed how multivariate techniques can aid in the interpretation of biological monitoring studies, which present difficulties related to the sometimes semiquantitative nature of the data and the unavailability of true control sites, replication, and experimental manipulation.

Ludwig and Reynolds[235] provide an introduction to the assumptions, derivations, and use of several multivariate techniques commonly used for the analysis of ecological communities. Van Wijgaarden et al.[236] compare DCA, PCA, and RDA and their usage in mesocosm research in more detail. Van den Brink et al.[237] proposed a multivariate method based on redundancy analysis (RDA). Clarke[238] showed the use of nonmetric multivariate analysis in community-level ecotoxicology, which does not require the restrictive assumptions of parametric techniques. Multivariate techniques have become more accessible and user-friendly with the availability of software such as the principle response curves method[233] and the routines in the PRIMER software package.[238] Major steps have also been taken to produce outputs readily interpretable by both ecologists and environmental managers and regulators. Multivariate techniques now provide ecotoxicologists with powerful tools to visualize and present impacts at the community and ecosystem level.

## 3.8 SUMMARY

This chapter has focused on key factors that need to be considered in the experimental design of outdoor model ecosystem studies to increase their realism, reduce variability, and ultimately assess the ability of these systems to detect changes. The success in using such systems depends on the establishment of appropriate scales of sampling, both temporal and spatial. As systems need to be sampled with response times for species taken into account, so sampling programs should reflect the variance in activities, life span, and reproductive potential of the species of interest. The failure to observe patterns (predicted or otherwise) or establish equilibrium conditions in experimental plots is often a result of the scales selected.

In performing a model ecosystem study it is important to determine the ecological relevance of effects identified in linked laboratory studies. The studies therefore include several species, functional groups, or habitat types. Interpretation of the field study focuses on effects at the community level, potential indirect effects, and the recovery potential of aquatic populations and communities. A second important reason for conducting a model ecosystem study is to measure consequences of the chemical under environmentally realistic exposure conditions (realistic fate and distribution). Such conditions could include partitioning to sediments and plants, photolysis, and other processes that may influence the fate of the pesticide. Moreover, these studies incorporate natural abiotic conditions (temperature, light, pH, etc.) that may influence the response of certain organisms.[40]

Based on the discussions and examples given in the previous pages it should be evident that there is no single "best" experimental design or test system.[46] There are a number of options that can be chosen depending on available budgets and facilities. The experimental design needs to address the objectives of the study and must consider the characteristics of the contaminant being studied and the ecosystem being impacted. Outdoor meso- or microcosm studies can be performed with artificial tanks or ponds or in parts of existing ecosystems that are enclosed in a way that causes minimal disturbance. The size of the mesocosm depends on the nature of the study and size and habitat of the organisms of interest. Typically, for "pond" studies, volumes of 1 to 20 $m^3$ are usually regarded as appropriate for outdoor meso- or microcosm studies. In situations in which planktonic species are the main concern microcosms of 100 to 1000 L may also be appropriate. The size to be selected for a meso- or microcosm study will depend on the objectives of the study and the type of ecosystem that is to be simulated. In general, studies with smaller tanks (about 1 to 5 $m^3$) are more suitable for shorter studies (up to about 1 month), and larger volumes are more suitable for longer studies (e.g., up to 6 months or longer). Benthic and planktonic invertebrates are often added to mesocosms with sediments and water. Invertebrates typically include rotifers, cladocera, copepods, annelids, benthic crustaceans, gastropods, and insect larvae.

In general, approaches to ecosystem-level testing using surrogate systems have been overly simplistic. Continued development of innovative approaches to data collection and analysis are needed. Historically, mesocosm tests have been viewed as a "series" of single-species tests (the ANOVA statistical approach is currently favored). Ecosystem-level studies often require the prediction of responses of many biological variables given information on the state of environmental or other biological variables.[239]

Methods that evaluate endpoints in a more integrated and holistic fashion should be applied to these studies. Multivariate statistics are one tool for viewing the "big picture." Multivariate techniques, however, are not a panacea for data analysis[240] but should be part of an integrated approach that encompasses both reductionist component analysis and other holistic approaches such as modeling.[127] Ultimately, the value of research using surrogate systems lies in their potential to provide prediction of and probabilities for ecosystem responses to contaminants.

A number of ecotoxicological studies with similar experimental designs have been completed in North America and Europe for purposes of pesticide registration. The existence of such a large number of similarly designed and conducted large-scale model ecosystem studies provides a unique opportunity for further research. If results of these studies were compiled into a common database, analysis of these data could help identify common results, allowing generalization for given classes of stressors. Evaluation of these results in the light of current ecological theory should allow for the formulation of alternative hypotheses, future study designs to test these hypotheses, and subsequent validation (or refutation) of these ideas. A clear-cut, systematic synthesis of the existing information will help enable model ecosystems to reach their full potential as a tool for predicting impacts (as opposed to simply effects assessment). Until this happens, regulators, the manufacturing industries, and researchers will continue to argue over the meaning of community and ecosystem responses measured in these studies. As Cairns[241] succinctly stated: "If environmental toxicology is to come of age, it must begin to ask more searching questions, develop broader hypotheses involving natural systems, and develop models that are validated in landscapes, not laboratories."

## REFERENCES

1. Crossland, N. O., Heimbach, F., Hill, I. R., Boudou, A., Leeuwangh, P., Matthiessen, P., and Persoone, G., Summary and Recommendations of the European Workshop on Freshwater Field Tests (EWOFFT), Potsdam, Germany, 1993, 37.

2. Crow, M. E. and Taub, F. B., Designing a microcosm bioassay to detect ecosystem-level effects, *Int. J. Environ. Stud.*, 13, 141, 1979.
3. Giesy, J. P., Jr. and Allred, P. M., Replicability of aquatic multispecies test systems, in *Multispecies Toxicity Testing*, Cairns, J., Jr., Ed., Pergamon Press, New York, 1985, 187.
4. Giesy, J. P., Jr. and Odum, E. P., Microcosmology: introductory comments, in *Microcosms in Ecological Research*, Giesy, J. P., Jr., Ed., Dept. of Energy Symposium Series 52, Conf. 781101, National Technical Information Service, Springfield, VA, 1, 1980.
5. Voshell, J. R., Jr., Introduction and overview of mesocosms, in *North American Benthological Society*, 1990, in *Experimental Ecosystems: Applications to Ecotoxicology*, Technical Information Workshop, Virginia Polytechnic Institute and State University, Blacksburg, VA, 1990.
6. Cairns, J., Jr., Putting the eco in ecotoxicology, *Reg. Toxicol. Pharmacol.*, 8, 226, 1988.
7. Hall, D. J., Cooper, W. E., and Werner, E. E., An experimental approach to the production dynamics and structure of freshwater animal communities, *Limnol. Oceanogr.*, 15, 839, 1970.
8. Hurlbert, S. H., Mulla, M. S., Keith, J. O., Westlake, W. E., and Düsch, M. E., Biological effects and persistence of Dursban? in freshwater ponds, *J. Econ. Entomol.*, 63, 43, 1970.
9. Eisenberg, R. M., The regulation of density in a natural population of the pond snail Lymnaea elodes, *Ecology*, 47, 889, 1966.
10. Odum, E. P., The mesocosm, *Bioscience*, 34, 558, 1984.
11. Cairns, J., Jr., Applied ecotoxicology and methodology, in *Aquatic Ecotoxicology: Fundamental Concepts and Methodologies,* Vol. I, Boudou, A. and Ribeyre, F., Eds., CRC Press, Boca Raton, FL, 1989, 275.
12. LaPoint, T. W., Fairchild, J. F., Little, E. E., and Finger, S. E., Laboratory and field techniques in ecotoxicological research: Strengths and limitations, in *Aquatic Ecotoxicology: Fundamental Concepts and Methodologies,* Vol. I, Boudou, A., and Ribeye, F., Eds., CRC Press, Boca Raton, FL, 1989, 239.
13. Forbes, S. A., The lake as a microcosm, *Illinois Nat. Hist. Survey Bull.*, 15, 537, 1887.
14. Kevern, N. R. and Ball, R. C., Primary productivity and energy relationships in artificial streams, *Limnol. Oceanogr.*, 10, 74, 1965.
15. McConnell, W. J., Productivity relations in carboy microcosms, *Limnol. Oceanogr.*, 7, 335, 1962.
16. McConnell, W. J., Relationship of herbivore growth to rate of gross photosynthesis in microcosms, *Limnol. Oceanogr.*, 10, 539, 1965.
17. Beyers, R. J., Relationship between temperature and the metabolism of experimental ecosystems, *Science*, 136, 980, 1962.
18. Beyers, R. J., The metabolism of twelve aquatic laboratory microecosystems, *Ecol. Monogr.*, 33, 281, 1963.
19. Copeland, B. J., Evidence for regulation of community metabolism in a marine ecosystem, *Ecology*, 46, 563, 1965.
20. Gause, G. F., *The Struggle for Existence*, Hafner, New York, 1934 (reprinted 1971 by Dover Publishers, New York).
21. Woodruff, L. L., Observations on the origin and sequence of the protozoan fauna of hay infusions, *J. Exp. Zool.*, 12, 205, 1912.
22. Eddy, S., Succession of Protozoa in cultures under controlled conditions, *Trans. Am. Microsc. Soc.*, 47, 283, 1928.
23. Lotka, A. J., *Elements of Physical Biology*, Williams and Wilkins, Baltimore, 1925, 460.
24. Lotka, A. J., The growth of mixed populations: Two species competing for a common food supply, *J. Wash. Acad. Sci.*, 22, 461, 1932.
25. Volterra, V., Fluctuations in the abundance of a species considered mathematically, *Nature*, 118, 558, 1926.
26. Volterra, V., Appendix, in *Animal Ecology*, Chapman, R. N., Ed., McGraw-Hill, New York, 1939, 409.
27. Keith, J. O. and Mulla, M. S., Relative toxicity of five organophosphorus mosquito larvicides to mallard duck, *J. Wildl. Manage.*, 30, 553, 1966.
28. Mulla, M. S., Keith, J. O., and Gunther, F. A., Persistence and biological effects of parathion residues in waterfowl habitats, *J. Econ. Entomol.*, 59, 108, 1966.
29. Cairns, J., Jr., Are single species toxicity tests alone adequate for estimating environmental hazard?, *Hydrobiologia*, 100, 47, 1983.
30. Kimball, K. D., and Levin, S. A., Limitations of laboratory bioassays: The need for ecosystem-level testing, *Bioscience*, 35, 165, 1985.
31. Cairns, J., Jr., The myth of the most sensitive species, *Bioscience*, 36, 670, 1986.

32. A.E.D.G., Improving Aquatic Risk Assessment under FIFRA Report of the Aquatic Effects Dialogue Group, World Wildlife Fund, 1992.
33. Touart, L. W., Hazard Evaluation Division, Technical Guidance Document: Aquatic Mesocosm Tests to Support Pesticide Registrations, EPA-540/09-88-035, Environmental Protection Agency, Office of Pesticide Programs, Ecological Effects Branch, 1988.
34. Voshell, J. R., Jr., Ed., Using mesocosms for assessing the aquatic ecological risk of pesticides: Theory and practice, *Misc. Publ. Entomol. Soc. Am.*, 75, 88, 1989.
35. Cuffney, T. F., Hart, D. D., Wolbach, K. C., Wallace, J. B., Lugthart, G. J., and Smith-Cuffney, F. L., Assessment of community and ecosystem level effects in lotic environments: The role of mesocosm and field studies, in *North American Benthological Society Tech. Info. Workshop, Exp. Ecosystems: Applications to Ecotoxicology*, Virginia Polytechnic Institute and State University, Blacksburg, VA, 1990, 40.
36. Graney, R. L., Giesy, J. P., Jr., DiToro, D., and Hallden, J. A., Mesocosm experimental design strategies: Advantages and disadvantages in ecological risk assessment, in *Using Mesocosms for Assessing the Aquatic Ecological Risk of Pesticides: Theory and Practice*, Voshell, J. R., Jr., Ed., Misc. Publ. Entomol. Soc. Am., 75, 74, 1989.
37. Campbell, P. J., Arnold, D. J. S., Brock, T. C. M., Grandy, N. J., Heger, W., Heimbach, F., Maund, S. J., and Streloke, M., Eds., Guidance Document on Higher-Tier Aquatic Risk Assessment for Pesticides (HARAP), SETAC – Europe Brussels, Belgium, 1999.
38. Graney, R., Rodgers, J. H., and Kennedy, J. H., *Aquatic Mesocosms in Ecological Risk Assessment Studies*, Special Publ., Society of Environmental Toxicology and Chemistry, Michigan, Lewis Publishers, 1994.
39. SETAC Foundation for Environmental Education & RESOLVE 1992, Workshop on Aquatic Microcosms for Ecological Assessment of Pesticides, Workshop Report, Oct. 6–11, 1991, Wintergreen, VA, 1992, 56.
40. Campbell, P. J., Arnold, D. J. S., Brock, T. C. M., Grandy, N. J., Heger, W., Heimbach, F., Maund, S. J., and Streloke, M., Eds., Guidance Document on Higher-Tier Aquatic Risk Assessment for Pesticides (HARAP), SETAC Europe Publication, Brussels, Belgium, 1999, 179.
41. Campbell, P. J., Arnold, D. J. S., Brock, T. C. M., Grandy, N. J., Heger, W., Heimbach, F., Maund, S. J., and Streloke, M., Eds., *Proc. of the CLASSIC (Community level Aquatic System Studies Interpretation Criteria) Workshop*, SETAC Europe Publication, Brussels, Belgium, in press, 2002.
42. Barron, M. G., Bioaccumulation and bioconcentration in aquatic organisms, in *Handbook of Ecotoxicology*, Calow, P., Ed., Blackwell Scientific Publications, Cambridge, MA, 1995, 652.
43. Thomann, R. V., Connolly, J. P., and Parkerton, T. F., An equilibrium model of organic chemical accumulation in aquatic food webs with sediment interaction, *Environ. Toxicol. Chem.*, 11, 615, 1992.
44. Rasmussen, J. B., Rowan, D. J., Lean, D. R. S., and Casey, J. H., Food chain structure in Ontario lakes determines PCB levels in lake trout (*Salvelinus namaycush*) and other pelagic fish, *Can. J. Fish. Aquat. Sci.*, 47, 2020, 1990.
45. Simon, O., Ribeyre, F., and Boudou, A., Comparative experimental study of cadmium and methylmercury trophic transfers between the Asiatic clam Corbicula fluminea and the crayfish *Astacus astacus*, *Arch. Environ. Contam. Toxicol.*, 38, 317, 2000.
46. Giddings, J. M., Types of aquatic mesocosms and their research applications, in *Microcosms in Ecological Research*, Giesy, J. P., Jr., Ed., Dept. of Energy Symposium Series 52, Conf. 781101, National Technical Information Center, Springfield, VA, 1980, 248.
47. Hill, I. R., Hadfield, S. T., Kennedy, J. H., and Ekoniak, P., Assessment of the impact of PP321 on aquatic ecosystems using tenth-acre experimental ponds, *Proc. Brighton Crop Protection Conference — Pest and Diseases*, Brighton, England, 1988, 309.
48. Solomon, K. R. and Liber K., Fate of pesticides in aquatic mesocosm studies — An overview of methodology, *Proc. Brighton Crop Protection Conference — Pests and Diseases*, Brighton, England, 1988, 139.
49. Grice, G. D. and Reeve, R. R., *Marine Mesocosms: Biological and Chemical Research in Experimental Ecosystems*, Springer Verlag, New York, 1982.
50. Lalli, C. M., *Enclosed Experimental Marine Ecosystems: A Review and Recommendations*, Springer Verlag, New York, 1990.

51. Gearing, J. N., The role of aquatic microcosms in ecotoxicologic research as illustrated by large marine systems, in *Ecotoxicology: Problems and Approaches*, Levin, S. A., Harwell, M. A., Kelly, J. R., and Kimbell, K. D., Eds., Springer Verlag, New York, 1989, 411.
52. Kennedy, J. H., Johnson, Z. B., Wise, P. D., and Johnson, P. C., Model aquatic ecosystems in ecotoxicological research: Consideration of design, implementation, and analysis, in *Handbook of Ecotoxicology*, Hoffman, D. J., Rattner, B. A., Burton, G. A., and Cairns, J., Jr., CRC Press, Boca Raton, FL, 117,1995.
53. Schuaerte, W., Lay, J. P., Klein, W., and Korte, F., Influence of 2,4,6-trichlorophenol and pentachlorophenol on the biota of aquatic systems, *Chemosphere*, 11, 71, 1982.
54. Maund, S. J., Peither, A., Taylor, E. J., Juttner, I., Beyerle-Pfnur, R., Lay, J. P., and Pascoe, D., Toxicity of lindane to freshwater insect larvae in compartments of an experimental pond, *Ecotoxicol. Environ. Saf.*, 23, 76, 1992.
55. Yasuno, M., Hanazato, T., Iwakuna, T., Takamura, K., Ueno, R., and Takamura, N., Effects of permethrin on phytoplankton and zooplankton in an enclosure ecosystem in a pond, *Hydrobiologia*, 159, 247, 1988.
56. Zrum, L., Hann, B. J., Goldsborough, L. G., and Stern, G. A., Effects of organophosphorus insecticide and inorganic nutrients on the planktonic microinvertebrates and algae in a prairie wetland, *Arch. Hydrobiol.*, 147, 373, 2000.
57. Lay, J. P., Muller, A., Peichl, L., Lang, R., and Korte, F., Effects of $\gamma$-BHC (lindane) on zooplankton under outdoor conditions, *Chemosphere*, 16, 1527, 1987.
58. Howick, G. L., de Noyelles, F., Jr., Giddings, J. M., and Graney, R. L., Earthen ponds vs. fiberglass tanks as venues for assessing the impact of pesticides on aquatic environments: A parallel study with sulprofos, in *Simulated Field Studies in Aquatic Ecological Risk Assessment,* Graney, R. L., Kennedy, J. H., and Rodgers, R. H., Jr., Eds., Lewis Publishers, Boca Raton, FL, 321, 1994.
59. Drenner, R. W., Hambright, K. D., Vinyard, G. L., Gopher, M., and Pollingher, U., Experimental study of size-selective phytoplankton grazing by a filter-feeding cichlid and the cichlid's effects on plankton community structure, *Limnol. Oceanogr.*, 32, 1138, 1987.
60. Kennedy, J. H., Johnson, P. C., Morris, R. G., Moring, J. B., and Hambleton, F. E., Case history: Microcosm research at the University of North Texas, presented at *Society of Environmental Toxicology and Chemistry Microcosm Workshop*, Wintergreen, VA, 1991, 52.
61. Giddings, J. M., Biever, R. C., Annuniziato, M. F., and Hosmer, A. J., Effects of diazinon on large outdoor microcosms, *Environ. Toxicol. Chem.*, 15, 618, 1996.
62. Shaw, J. L. and Manning, J. P., Evaluating macroinvertebrate populations and community level effects in outdoor microcosms: Use of *in situ* bioassays and multivariate analysis, *Environ. Toxicol. Chem.*, 15, 608, 1996.
63. Rand, G. M., Clark, J. R., and Holmes, C. M., Use of outdoor freshwater pond microcosm: I. Microcosm design and fate of pyridaben, *Environ. Tox. Chem.*, 19, 387, 2000.
64. Heimbach, F., Plfueger, W., and Ratte, H.-T., Use of small artificial ponds for assessment of hazards to aquatic ecosystems, *Environ. Toxicol. Chem.*, 11, 27, 1992.
65. Johnson, P. C., Kennedy, J. H., Morris, R. G., Hambleton, F. E., and Graney, R. L., Fate and effects of cyfluthrin (pyrethroid insecticide) in pond mesocosms and concrete microcosms, in *Simulated Field Studies in Aquatic Ecological Risk Assessment*, Graney, R. L., Kennedy, J. H., and Rodgers, J. H., Jr., Eds., Special Publication of the Society of Environmental Toxicology and Chemistry, Lewis Publishers, Boca Raton, FL, 337, 1994.
66. Hill, I. R., Hadfield, S. T., Kennedy, J. H., and Ekoniak, P., Assessment of the impact of PP321 on aquatic ecosystems using tenth-acre experimental ponds, in *Proc. Brighton Crop Protection Conference — Pests and Diseases*, 309, 1988.
67. Kedwards, T. J., Maund, S. J., and Chapman P. F., Community level analysis of ecotoxicological field studies: II Replicated design studies, *Environ. Toxicol. Chem.*, 18, 158, 1999.
68. Heimbach, F., Pflueger, W., and Ratte, H. T., Two artificial pond ecosystems of differing size, in *Simulated Field Studies in Aquatic Ecological Risk Assessment*, Graney, R. L., Kennedy, J. H., and Rodgers, J. H., Jr., Eds., Special Publication of the Society of Environmental Toxicology and Chemistry, Lewis Publishers, Boca Raton, FL., 303, 1994.

69. Luccassen, W. and Leewangh, P., Response of zooplankton to Dursban 4E insecticide in a pond experiment, in *Simulated Field Studies in Aquatic Ecological Risk Assessment*, Graney, R. L., Kennedy, J. H., and Rodgers, J. H., Jr., Eds., Special Publication of the Society of Environmental Toxicology and Chemistry, Lewis Publishers, Boca Raton, FL, 1994, 517.
70. Scott, I. M. and Kaushik, N. K., The toxicity of a new insecticide to populations of Culicidae and other aquatic invertebrates as assessed in *in situ* microcosms, *Arch. Environ. Cont. Toxicol.,* 39, 329, 2000.
71. Barry, M. J., and Logan, D. C., The use of temporary pond microcosms for aquatic toxicity testing: Direct and indirect effects of endosulfan on community structure, *Aquat. Toxicol.,* 41, 101, 1998.
72. Kaushik, N. K., Solomon, K. R., Stephenson, G. L., and Day, K. E., Use of limnocorrals in evaluating the effects of pesticides on zooplankton communities, in *Community Toxicity Testing*, ASTM STP 920, Cairns, J., Jr., Ed., American Society for Testing and Materials, Philadelphia, 1986, 269.
73. Solomon, K. R., Yoo, J. Y., Lean, D., Kaushik, N. K., Day, K. E., and Stephenson, G. L., Dissipation of permethrin in limnocorrals, *Can. J. Fish. Aquat. Sci.,* 42, 70, 1985.
74. Solomon, K. R., Stephenson, G. L., and Kaushik, N. K., Effects of methoxychlor on zooplankton in freshwater enclosures: Influence of enclosure size and number of applications, *Environ. Toxicol. Chem.,* 8, 659, 1989.
75. Day, K. E., Kaushik, N. K., and Solomon, K. R., Impact of fenvalerate on enclosed freshwater planktonic communities and on *in situ* rates of filtration of zooplankton, *Can. J. Fish. Aquat. Sci.,* 44, 1714, 1987.
76. Solomon, K. R., Yoo, J. Y., Lean, D., Kaushik, N. K., Day, K. E., and Stephenson, G. L., Methoxychlor distribution, dissipation, and effects in freshwater limnocorrals, *Environ. Toxicol. Chem.*, 5, 577, 1986.
77. Stephenson, G. L., Kaushik, N. K., Solomon, K. R., and Day, K., Impact of methoxychlor on freshwater communities of plankton in limnocorrals, *Environ. Toxicol. Chem.*, 5, 587, 1986.
78. Kaushik, N. K., Stephenson, G. L., Solomon, K. R., and Day, K. E., Impact of permethrin on zooplankton communities in limnocorrals, *Can. J. Fish. Aquat. Sci.,* 42, 77, 1985.
79. Hanazato, T., and Yasuno, M., Effects of carbaryl on the spring zooplankton communities in ponds, *Environ. Poll.,* 56, 1, 1989.
80. Siefert, R. E., Lozano, S. J., Brazner, J. C., and Knuth, M. L., Littoral enclosures for aquatic field testing of pesticides: Effects of chlorpyrifos on a natural system, in *Using Mesocosms to Assess the Aquatic Ecological Risk of Pesticides: Theory and Practice*, Voshell, J. R., Jr., Ed., Miscellaneous Publications No. 75, Entomological Society of America, 1989, 57.
81. Brazner, J. C., Heinis, L. J., and Jensen, D. A., A littoral enclosure for replicated field experiments, *Environ. Toxicol. Chem.*, 8, 1209, 1989.
82. Touart, L. W., and Slimak, M. W., Mesocosm approach for assessing the ecological risk of pesticides, in *Using Mesocosms to Assess the Aquatic Ecological Risk of Pesticides: Theory and Practice*, Voshell, J. R., Ed., Miscellaneous Publications No. 75, Entomological Society of America, 1989, 33.
83. Crossland, N. O., and La Point, T. W., The design of mesocosm experiments, *Environ. Toxicol. Chem.*, 11, 1, 1992.
84. Hermanutz, R. O., Allen, K. N., Roush, T. H., and Hedtke, S.F., Effects of elevated selenium concentrations on bluegills (*Lepomis macrochirus*) in outdoor experimental streams, *Environ. Toxicol. Chem.*, 11, 217, 1992.
85. Shaw, J. L. and Manning, P. J., Evaluating macroinvertebrate population and community level effects in outdoor microcosms: Use of *in situ* bioassays and multivariate analysis, *Environ. Toxicol. Chem.,* 15, 508, 1996.
86. Dyer, S. D. and Belanger, S. E., Determination of the sensitivity of macroinvertebrates in stream mesocosms through field-derived assessments, *Environ. Toxicol. Chem.,* 18, 2903, 1999
87. Hurlbert, S. H., Pseudoreplication and the design of ecological field experiments, *Ecol. Monogr.,* 54, 187, 1984.
88. Belanger, S. E., Literature review, An analysis of biological complexity in model stream ecosystems: Influence of size and experimental design, *Ecotoxicol. Environ. Saf.,* 36, 1, 1997.
89. Carder, J. P. and Hoagland, K. D., Combined effects of Alachlor and Atrazine on benthic algal communities in artificial streams, *Environ. Toxicol. Chem.*, 17, 1415, 1998.
90. La Point, T. W. and Perry, J. A., Use of experimental ecosystems in regulatory decision making, *Environ. Manage.,* 13, 539, 1989

91. Groten, J. P., Schoen, E. D., and Feron, V. J., Use of factorial designs in combination toxicity studies, *Food Chem. Toxicol.,* 34, 1083, 1996.
92. Underwood, A. J., *Experiments in Ecology,* Cambridge University Press, Cambridge, UK,1997.
93. Austin, A. P., Harris, G. E., and Lucey, W. P., Impact of an organophosphate herbicide (Glyphosate®) on periphyton communities developed in experimental streams, *Bull. Environ. Contam. Toxicol.,* 47, 29, 1991.
94. Belanger, S. E., Farris, J. L., Cherry, D. S., and Cairns, J., Jr., Validation of *Corbicula fluminea* growth reductions induced by copper in artificial streams and river systems, *Can. J. Fish. Aquat. Sci.,* 47, 904, 1990.
95. Clements, W. H., Cherry, D. S., and Cairns, J., Jr., Macroinvertebrate community responses to copper in laboratory and field experimental streams, *Arch. Environ. Contam. Toxicol.,* 19, 361, 1990.
96. Clements, W. H., Metal tolerance and predator-prey interactions in benthic macroinvertebrate stream communities, *Ecol. Appl.,* 9, 1999.
97. Dorn, P. B., Rodgers, J. H., Jr., Dubey, S. T., Gillespie, W. B., Jr., and Figueroa, A. R., Assessing the effects of a C14-15 linear alcohol ethoxylate surfactant in stream mesocosms, *Ecotoxicol. Environ. Saf.,* 34, 196, 1996.
98. Dorn, P. B., Rodgers, J. H., Jr., Dubey, S. T., Gillespie, W. B., Jr., and Lizotte, R. E., An assessment of the ecological effects of a C9-11 linear alcohol ethoxylate surfactant in stream mesocosm experiments, *Ecotoxicology,* 6, 275, 1997.
99. Dorn, P. B., Rodgers, J. H., Jr., Gillespie, W. B., Jr., Lizotte, R. E., Jr., and Dunn, A. W., The effects of a C12-13 linear alcohol ethoxylate surfactant on periphyton, macrophytes, invertebrates and fish in stream mesocosms, *Environ. Toxicol. Chem.,* 16, 8, 1634–1645, 1997.
100. Gillespie, W. B., Jr., Rodgers, J. H., Jr. and Crossland, N. O., Effects of a nonionic surfactant (C14-15 AE-7) on aquatic invertebrates in outdoor stream mesocosms, *Environ. Toxicol. Chem.,* 15, 1418, 1996.
101. Gillespie, W. B., Jr., Rodgers, J. H., Jr., and Dorn, P. B., Responses of aquatic invertebrates to a C9-11 non-ionic surfactant in outdoor stream mesocosms, *Aquat. Toxicol.,* 37, 221, 1997.
102. Gillespie, W. B., Jr., Rodgers, J. H., Jr., and Dorn, P. B., Responses of aquatic invertebrates to a linear alcohol ethoxylate surfactant in stream mesocosms, *Ecotoxicol. Environ. Saf.,* 41, 215, 1998.
103. Harrelson, R. A., Rodgers, J. H., Jr., Lizotte, R. E., Jr., and Dorn. P. B., Responses of fish exposed to a C9-11 linear alcohol ethoxylate nonionic surfactant in stream mesocosms, *Ecotoxicology,* 6, 321, 1997.
104. Kline, E. R., Figueroa, R. A., Rodgers, J. H., Jr., and Dorn, P. B., Effects of a nonionic surfactant (C14-15 AE-7) on fish survival, growth and reproduction in the laboratory and in outdoor stream mesocosms, *Environ. Toxicol. Chem.,* 15, 997, 1996.
105. Haley, R. K., Hall, T. J., and Bousquet, T. M., Effects of biologically treated bleached-kraft mill effluent before and after mill conversion to increased chlorine dioxide substitution: Results of an experimental streams study, *Environ. Toxicol. Chem.,* 14, 287, 1995.
106. Hall, T. J., Haley, R. K., and LaFleur L. E., Effects of biologically treated bleached kraft mill effluent on cold water stream productivity in experimental stream channels, *Environ. Toxicol. Chem.,* 10, 1051, 1991.
107. Hermanutz, R. O., Allen, K. N., Roush, T. H., and Hedtke, S.F., Effects of elevated selenium concentrations on bluegills (*Lepomis macrochirus*) in outdoor experimental streams, *Environ. Toxicol. Chem.,* 11, 217, 1992.
108. Kreutzweiser, D. P. and Capell, S. S., A simple stream-side test system for determining acute lethal and behavioral effects of pesticides on aquatic insects, *Environ. Toxicol. Chem.,* 11, 993, 1992.
109. Crossland, N. O., Mitchell, G. C., and Dorn, P. B., Use of outdoor artificial streams to determine threshold toxicity concentration for a petrochemical effluent, *Environ. Toxicol. Chem.,* 11, 49, 1992.
110. Maltby, L., The use of the physiological energetics of *Gammarus pulex* to assess toxicity: A study using artificial streams, *Environ. Toxicol. Chem.,* 11, 79, 1991.
111. Mitchell, G. C., Bennett, D., and Pearson, N., Effects of lindane on macroinvertebrates and periphyton in outdoor artificial streams, *Ecotoxicol. Environ. Saf.,* 25, 90, 1993.
112. Pascoe, D., Wenzel. A., Janssen, C., Girling, A. E., Jüttner, I., Fliedner, A., Blockwell, S. J., Maud, S. J., Taylor, E. J., Diedrich, M., Persoone, G., Verhelst, P., Stephenson, R. R., Crossland, N. O., Mitchell, G. C., Pearson, N., Tattersfield, L., Lay, J. P., Peither, A., Neumeier, B., and Velletti, A. R., The development of toxicity tests for freshwater pollutants and their validation in stream and pond mesocosms, *Wat. Res.,* 34, 2323, 2000.

113. Richardson, J. S. and Kiffney, P. M., Responses of a stream macroinvertebrate community from a pristine, Southern British Columbia, Canada, stream to metals in experimental mesocosms, *Environ. Toxic. Chem.,* 19, 736, 2000.
114. Frost, T. M., DeAngelis, D. L., Bartell, S. M., Hall, D. J., and Hurlbert, S. H., Scale in the design and interpretation of aquatic community research, in *Complex Interactions in Lake Communities*, Carpenter, S. R., Ed., Springer-Verlag, New York, 1988, 229.
115. Dudzik, M., Harte, J., Jassby, A., Lapan, E., Levy, D., and Rees, J., Some considerations in the design of aquatic microcosms for plankton studies, *Int. J. Environ. Stud.,* 13, 125, 1979.
116. Stephenson, G. L., Hamilton, P., Kaushik, N. K., Robinson, J. B., and Solomon, K. R., Spatial distribution of plankton in enclosures of three sizes, *Can. J. Fish. Aquat. Sci.,* 41, 1048, 1984.
117. Arumugam, P. T. and Geddes, M. C., An enclosure for experimental field studies with fish and zooplankton communities, *Hydrobiologia,* 135, 215, 1986.
118. Chant, L., and Cornett, R. J., Measuring contaminant transport rates between water and sediments using limnocorrals, *Hydrobiologia,* 159, 237, 1988.
119. Heinis, L. J. and Knuth, M. L., The mixing, distribution and persistence of esfenvalerate within littoral enclosures, *Environ. Toxicol. Chem.,* 11, 11, 1992.
120. Siefert, R. E., Lozano, S. J., Knuth, M. L., Heinis, L. J., Brazner, J. C., and Tanner, D. K., Pesticide testing with littoral enclosures, in *Experimental Ecosystems: Applications to Ecotoxicology*, North American Benthological Society, Technical Information Workshop, Virginia Polytechnic Institute and State University, Blacksburg, VA, 1990, 13.
121. Needham, P. R. and Usinger, R. L., Variability in the macrofauna of a single riffle in Prosser Creek, California, as indicated by the Surber sampler, *Hilgardia,* 24, 383, 1956.
122. Chutter, F. M. and Noble, R. G., The reliability of a method of sampling stream invertebrates, *Arch. Hydrobiol.,* 62, 95, 1966.
123. Dickson, K. L. and Cairns, J., Jr., The relationship of fresh-water macroinvertebrate communities collected by floating artificial substrates to the MacArthur-Wilson equilibrium model, *Am. Midl. Nat.*, 88, 68, 1972.
124. Green, R. H., *Sampling Design and Statistical Methods for Environmental Biologists*, John Wiley and Sons, New York, 1979.
125. Pratt, J. R. and Bowers, N. J., Variability of community metrics: Detecting changes in structure and function, *Environ. Toxicol. Chem.,* 11, 451, 1992.
126. Schindler, D. W., Detecting ecosystem responses to anthropogenic stress, *Can. J. Fish. Aquat. Sci.,* 44, 6, 1987.
127. Maciorowski, A. F., Populations and communities: Linking toxicology and ecology in a new synthesis, *Environ. Toxicol. Chem.,* 7, 677, 1988.
128. O'Neil, P. E., Harris, S. C., and Mettee, M. F., Experimental stream mesocosms as applied in the assessment of produced water effluents associated with the development of coalbed methane, in *Experimental Ecosystems: Applications to Ecotoxicology*, North American Benthological Society, Technical Information Workshop, Virginia Polytechnic Institute and State University, Blacksburg, VA, 1990, 14.
129. Van Christman, D., Voshell, J. R., Jr., Jenkins, D. G., Rosenzweig, M. S., Layton, R. J., and Buikema, A. L., Jr., Ecology development and biometry of untreated pond mesocosms, in *Simulated Field Studies in Aquatic Ecological Risk Assessment*, Graney, R. L., Kennedy, J. H., and Rodgers, J. H., Jr., Eds., Special Publication of the Society of Environmental Toxicology and Chemistry, Lewis Publishers, Boca Raton, FL, 1994, 105.
130. Giddings, J. M. and Eddlemon, G. K., The effects of microcosm size and substrate type on aquatic microcosm behavior and arsenic transport, *Arch. Environ. Contam. Toxicol.,* 6, 491, 1977.
131. Crossland, N. O., and Bennett, D., Fate and biological effects of methyl parathion in outdoor ponds and laboratory aquaria. I. Fate, *Ecotoxicol. Environ. Saf.,* 8, 471, 1984.
132. Crossland, N. O. and Bennett, D., Fate and biological effects of methyl parathion in outdoor ponds and laboratory aquaria. II. Effects, *Ecotoxicol. Environ. Saf.,* 8, 482, 1984.
133. Crossland, N. O., Bennett, D., Wolff, C. J. M., and Swannell, R. P. J., Evaluation of models to assess the fate of chemicals in aquatic systems, *Pest. Sci.,* 17, 297, 1986.
134. Wolff, C. J. M. and Crossland, N. O., Fate and effects of 3,4-Dichloroaniline in the laboratory and in outdoor ponds: I, Fate, *Environ. Toxicol. Chem.,* 4, 481, 1985.

135. Heimbach, F., Pflueger, W., and Ratte, H. T., Use of small artificial ponds for assessment of hazards to aquatic ecosystems, *Environ. Toxicol. Chem.,* 11, 27, 1992.
136. Howick, G. L., Giddings, J. M., de Noyelles, F., Ferrington, L. C., Jr., Kettle, W. D., and Baker, D., Rapid establishment of test conditions and trophic-level interactions in 0.04-hectare earthen pond mesocosms, *Environ. Toxicol. Chem.,* 11, 107, 1992.
137. Caquet, T. H., Lagadic, L., and Sheffield, S. R., Mesocosms in ecotoxicology (1). Outdoor aquatic systems, *Rev. Environ. Contam. Toxicol.,* 165, 1, 2000.
138. Kennedy, J. H., Johnson, Z. B., Wise, P. D., and Johnson, P. C., Model aquatic ecosystems in ecotoxicological research: Considerations of design, implementation, and analysis, in *Handbook of Ecotoxicology,* Hoffman, D. J., Rattner, B. A., Burton, G. A. Jr., Cairns, J. Jr., Eds., Lewis Publishers, Boca Raton, FL, 1995, 117.
139. Lozano, S. L., O' Halloran, S. L., Sargent, K. W., and Brazner, J. C., Effects of esfenvalerate on aquatic organisms in littoral enclosures, *Environ. Toxicol. Chem.,* 4, 399, 1992.
140. Genter, R. B., Cherry, D. S., Smith, E. P., and Cairns, J. Jr., Algal-periphyton population and community changes from zinc stress in stream mesocosms, *Hydrobiologia,* 153, 261, 1987.
141. Lynch, T. R., Johnson, H. E., and Adams, W. J., Impact of atrazine and hexachlorobiphenyl on the structure and function of model stream ecosystems, *Environ. Toxicol. Chem.,* 4, 399, 1985.
142. Crossland, N. O., Fate and biological effects of methyl parathion in outdoor ponds and laboratory aquaria. II: Effects, *Ecotoxicol. Environ. Saf.,* 8, 482, 1984.
143. Crossland, N. O., Bennet, D., Wolff, C. J. M., and Swannell, R. P. J., Evaluation of models to assess fate of chemicals in aquatic systems, *Pest. Sci.,* 17, 297, 1986.
144. Muir, D. C. G., Rawn, G. P., and Grift, N. P., Fate of the pyrethroid insecticide deltamethrin in small ponds: A mass balance study, *J. Agric. Food Chem.,* 33, 603, 1985.
145. Rawn, G. P., Webster, G. R. B., and Muir, D. C. G., Fate of permethrin in model outdoor ponds, *Environ. Sci. Health,* B17, 5, 463, 1982.
146. Weinberger, P., Greenhalgh, R., Sher, D., and Ouellette, M., Persistence of formulated fenitrothion in distilled, estuarine, and lake water microcosms in dynamic and static systems, *Environ. Contam. Toxicol.,* 28, 5, 484, 1982.
147. Barko, J. W., Godshalk, G. L., Carter, and Rybicki, V., Effects of submersed aquatic macrophytes on physical and chemical properties of surrounding water, Tech Rep. A-88-11, U.S. Army Corps of Engineers Waterways Experiment Station, Vicksburg, MS, 1988.
148. Brock, T. C. M., van den Bogaert, M., Bos, A. R., van Breuklen, S. W. F., Reiche, R., Terwoert, J., Suykerbuyk, R. E. M., and Roijackers, R. M. M., Fate and effects of the insecticide Dursban 4E in indoor Elodea dominated and macrophyte-free freshwater model ecosystems. II. Secondary effects on community structure, *Arch. Environ. Contam. Toxicol.,* 23, 391, 1992.
149. Street, M. and Titmus, G., The colonization of experimental ponds by chironomidae (Diptera), *Aquat. Insects,* 1, 233, 1979.
150. Schramm, H. L., Jr., Jirka, K. J., and Hoyer, M. V., Epiphytic macroinvertebrates on dominant macrophytes in tow central Florida lakes, *J. Fresh. Ecol.,* 4, 151, 1987.
151. Learner, M. A., Wiles, P. R., and Pickering, J. G., The influence of aquatic macrophyte identity on the composition of the chironomid fauna in a former gravel pit in Berkshire, England, *Aquat. Insects,* 11, 183, 1989.
152. Campbell, J. M. and Clark, W. J., The periphytic Cladocera of ponds of Brazos County, Texas, *Texas J. Sci.,* 39, 335, 1987.
153. Schramm, H. L., Jr. and Jirka, K. J., Epiphytic macroinvertebrates as a food resource for bluegills in Florida lakes, *Trans. Am. Fish. Soc.,* 118, 416, 1989.
154. Mittlebach, G. G., Foraging efficiency and body size: A study of optimal diet and habitat use by bluegills, *Ecology,* 62, 1370, 1981.
155. Crowder, L. B. and Cooper, W. E., Habitat structural complexity and the interaction between bluegills and their prey, *Ecology,* 63, 1802, 1982.
156. Gilinsky, E., The role of fish predation and spatial heterogeneity in determining benthic community structure, *Ecology,* 65, 455, 1984.
157. Loucks, O. L., Looking for surprise in managing stressed ecosystems, *Bioscience,* 35, 428, 1985.
158. Dionne, M. and Folt, C. L., An experimental analysis of macrophyte growth forms as fish foraging habitat, *Can. J. Fish. Aquat. Sci.,* 48, 123, 1991.

159. Dewey, S. L., Effects of the herbicide atrazine on aquatic insect community structure and emergence, *Ecology*, 67, 148, 1986.
160. Brooks, J. L. and Dodson, S. I., Predation, body size, and composition of plankton, *Science*, 150, 28, 1965.
161. Drenner, R. W., Threlkeld, S. T., and McCracken, M. D., Experimental analysis of the direct and indirect effects of an omnivorous filter-feeding clupeid on plankton community structure, *Can. J. Fish. Aquat. Sci.*, 43, 1935, 1986.
162. Vinyard, G. L., Drenner, R. W., Gophen, M., Pollingher, U., Winkleman, D. L., and Hambright, K. D., An experimental study of the plankton community impacts of two omnivorous filter-feeding cichlids, *Tilapia galilaea* and *Tilapia aurea*, *Can. J. Fish. Aquat. Sci.*, 45, 685, 1988.
163. Mazumder, A., McQueen, D. J., Taylor, W. D., and Lean, D. R. S., Effects of fertilization and planktivorous fish (yellow perch) predation on size distribution of particulate phosphorus and assimilated phosphate: Large enclosure experiments, *Limnol. Oceanogr.*, 33, 421, 1988.
164. Mazumder, A., Taylor, W. D., McQueen, D. J., and Lean, D. R. S., Effects of fertilization and planktivorous fish on epilimnetic phosphorus and phosphorus sedimentation in large enclosure, *Can. J. Fish. Aquat. Sci.*, 46, 1735, 1989.
165. Threlkeld, S. T., Planktivory and planktivore biomass effects on zooplankton, phytoplankton, and the trophic cascade, *Limnol. Oceanogr.*, 33, 1326, 1988.
166. Brabrand, Å., Faafeng, B., and Nilssen J. P. M., Pelagic predators and interfering algae: Stabilizing factors in temperate eutrophic lakes, *Arch. Hydrobiol.*, 110, 533, 1987.
167. Langeland, A., Koksvik, J. I., Olsen, Y., and Reinertsen, H., Limnocorral experiments in a eutrophic lake — effects of fish on the planktonic and chemical conditions, *Pol. Arch. Hydrobiol.*, 34, 51, 1987.
168. Hill, I. R., Sadler, J. K., Kennedy, J. H., and Ekoniak, P., Lambda-cyhalothium: A mesocosm study of its effects on aquatic organisms, in *Simulated Field Studies in Aquatic Ecological Risk Assessment*, Graney, R. L., Kennedy, J. H., and Rodgers, R. H., Jr., Eds., Lewis Publishers, Boca Raton, FL, 403, 1994.
169. Giddings, J. M., Biever, R. C., Helm, R. L., Howick, G. L., and de Noyelles, F. J., Jr., The fate and effects of Guthion (Azinophos Methyl) in mesocosms, in *Simulated Field Studies in Aquatic Ecological Risk Assessment*, Graney R. L., Kennedy, J. H., and Rodgers, J. H., Jr., Eds., Special Publication of the Society of Environmental Toxicology and Chemistry, Lewis Publishers, Boca Raton, FL, 469, 1994.
170. Morris, R. G., Kennedy, J. H., and Johnson, P. C., Comparison of the effects of the pyrethroid insecticide cyfluthrin on bluegill sunfish, in *Simulated Field Studies in Aquatic Ecological Risk Assessment*, Graney, R. L., Kennedy, J. H., and Rodgers, R. H., Jr., Eds., Lewis Publishers, Boca Raton, FL, 303, 1994.
171. Mayasich, J., Kennedy, J. H., and O'Grodnick, J. S., in *Simulated Field Studies in Aquatic Ecological Risk Assessment*, Graney, R. L., Kennedy, J. H., and Rodgers, R. H., Jr., Eds., Lewis Publishers, Boca Raton, FL, 497, 1994.
172. Carpenter, S. R., Kitchell, J. F., and Hodgson, J. R., Cascading trophic interactions and lake productivity, *Bioscience*, 35, 634, 1985.
173. Threlkeld, S. T., Experimental evaluation of trophic-cascade and nutrient-mediated effects of planktivorous fish on plankton community structure, in *Predation: Direct and Indirect Impacts on Aquatic Communities*, Kerfoot, W. C. and Sih, A., Eds., New England University Press, 1987, 161.
174. McQueen, D. J. and Post, J. R., Cascading trophic interactions: Uncoupling at the zooplankton-phytoplankton link, *Hydrobiologia*, 159, 277, 1988.
175. Deutsch, W. G., Webber, E. C., Bayne, D. R., and Reed, C. W., Effects of largemouth bass stocking rate on fish populations in aquatic mesocosms used for pesticide research, *Environ. Toxicol. Chem.*, 11, 5, 1992.
176. Stunkard, C. and Springer, T., Statistical analysis and experimental design, in *Improving Aquatic Risk Assessment Under FIFRA Report of the Aquatic Effects Dialogue Group*, Aquatic Effects Dialogue Group, Eds., World Wildlife Fund, 1992, 65.
177. Drenner, R. W., Hoagland, K. D., Smith, J. D., Barcellona, W. J., Johnson, P. C., Palmieri, M. A., and Hobson, J. F., Effects of sediment-bound bifenthrin on gizzard shad and plankton in experimental tank mesocosms, *Environ. Toxicol. Chem.*, 12, 1297, 1993.
178. Fairchild, J. F., La Point, T. W., Zajicek, J. L., Nelson, M. K., Dwyer, F. J., and Lovely, P. A., Population, community, and ecosystem-level responses of aquatic mesocosm to pulsed doses of a pyrethroid insecticide, *Environ. Toxicol. Chem.*, 11, 115, 1992.

179. Boyle, T. P., Research needs in validating and determining the predictability of laboratory data to the field, *Aquatic Toxicity and Hazard Assessment*, Eighth Symposium, ASTN STP 891, Bahner, R. C. and Hansen, D. J., Eds., American Society for Testing and Materials, Philadelphia, 1985, 61–66.
180. Stay, F. S., Katko, A., Rohm, C. M., Fix, M. A., and Larsen, D. P., The effects of atrazine on microcosms developed from four natural plankton communities, *Arch. Environ. Contam. Toxicol.*, 18, 866, 1989.
181. Staples, C. A., Dickson, K. L., Saleh, F. Y., and Rodgers, J. H., Jr., A microcosm study of lindane and naphthalene for model validation, in *Aquatic Toxicity and Hazard Assessment*, Sixth Symposium, ASTM STP 802, Bishop, W. E., Cardwell, R. D., and Heidolph, B. B., Eds., American Society for Testing and Materials, Philadelphia, 1983, 26.
182. Huckins, J. N., Petty, J. D., and England, D. C., Distribution and impact of trifluralin, atrazine, and fonofos residues in microcosms simulating a northern prairie wetland, *Chemosphere*, 15, 563, 1986.
183. Cairns, J., Jr., Niederlehner, B. R., and Pratt, J. R., Evaluation of joint toxicity of chlorine and ammonia to aquatic communities, *Aquat. Toxicol.,* 16, 87, 1990.
184. Pratt, J. R., Bowers, N. J., Niederlehner, B. R., and Cairns, J., Jr., Effects of chlorine on microbial communities in naturally derived microcosms, *Environ. Toxicol. Chem.*, 7, 9, 679, 1988.
185. Koerting-Walker, C. and Buck, J. D., The effect of bacteria and bioturbation by *Clymenella torquata* on oil removal from sediment, *Water Air Soil Pollut.*, 43, 3–4, 413, 1989.
186. Lehtinen K. J, Kierkegaard, A., Jakobsson, E., and Wandell, A., Physiological effects in fish exposed to effluents from mills with six different bleaching processes, *Ecotoxicol. Environ. Saf.*, 19, 1, 33, 1990.
187. Johnson, B. T., Potential impact of selected agricultural chemical contaminants on a northern prairie wetland: A microcosm evaluation, *Environ. Toxicol. Chem.*, 5, 473, 1986.
188. Cushman, R. M. and Goyert, J. C., Effects of a synthetic crude oil on pond benthic insects, *Environ. Pollut. (Ser. A.)*, 33, 163, 1984.
189. Oviatt, C. A., Quinn, J. G., Maughan, J. T., Ellis, J. T., Sullivan, B. K., Gearing, J. N., Gearing, P. J., Hunt, C. D., Sampou, P. A., and Latimer, J. S., Fate and effects of sewage sludge in the coastal marine environment: A mesocosm experiment, *Mar. Ecol. (Prog. Ser.)*, 41, 2, 187, 1987.
190. Boyle, T. P., Fairchild, J. F., Robinson, W. E. F., Haverland, P. S., and Lebo, J. A., Ecological restructuring in experimental aquatic mesocosms due to the application of diflubenzuron. *Environ. Toxicol. Chem.*, 15, 1806, 1996.
191. Giddings, J. M., Biver, R. C., and Racke, K. D., Fate of chlorpyrifos in outdoor microcosms and effects on growth and survival of bluegill sunfish, *Environ. Toxicol. Chem.*, 16, 2353, 1997.
192. Sugiura, K., Aoki, M., Kaneko, S., Daisaku, I., Komatsu, Y., Shibuya, H., Suzuki, H., and Gogo, M., Fate of 2,4,6-trichlorophenol, pentachlorophenol, p-chlorobiphenyl, and hexachlorobenzene in an outdoor experimental pond: Comparison between observations and predictions based on laboratory data, *Arch. Environ. Contam. Toxicol.*, 13, 6, 745, 1984.
193. Stout, R. J. and Cooper, W. E., Effect of p-cresol on leaf decomposition and invertebrate colonization in experimental outdoor streams, *Can. J. Fish. Aquat. Sci.*, 40, 1647, 1983.
194. De Noyelles, F., Jr., Kettle, W. D., Fromm, C. H., Moffett, M. F., and Dewey, S. L., Use of experimental ponds to assess the effects of a pesticide on the aquatic environment, in *Using Mesocosms for Assessing the Aquatic Ecological Risk of Pesticides: Theory and Practice*, Voshell, J. R., Jr., Ed., Misc. Publ., Entomol. Soc. Am., 75, 1989, 41.
195. Brazner, J. C. and Kline, E. R., Effects of chlorpyrifos on the diet and growth of larval fathead minnows, *Pimephales promelas*, in littoral enclosures, *Can. J. Fish. Aquat. Sci.*, 47, 1157, 1990.
196. Crossland, N. O., Aquatic toxicology of cypermethrin II. Fate and biological effects in pond experiments, *Aquat. Toxicol.*, 2, 205, 1982.
197. Kedwards, T. J., Maund, S. J., and Chapman, P. F., Community level analysis of ecotoxicological field studies. I. Biological monitoring, *Environ. Toxicol. Chem.*, 18, 149, 1999.
198. Ronday, R., Aalderrink, G. H. and Crum, S. J. H., Application methods of pesticides to an aquatic mesocosm in order to simulate effects of spray drift, *Wat. Res.*, 32, 147, 1998.
199. Farke, H., Wonneberger, K., Gunkel, W., and Dahlmann, G., Effects of oil and a dispersant on intertidal organisms in field experiments with a mesocosm, the Bremerhaven Caisson, *Mar. Environ. Res.*, 15, 2, 97, 1985.
200. Zischke, J. A., Arthur, J. W., Hermanutz, R. O., Hedtke, S. F., and Helgen, J. C., Effects of pentachlorophenol on invertebrates and fish in outdoor experimental channels, *Aquat. Toxicol.*, 7, 37, 1985.

201. Bakke, T., Follum, O. A., Moe, K. A., and Soerensen, K., The GEEP workshop: Mesocosm exposures, *Mar. Ecol. (Prog. Ser.)*, 46, 1–3, 13, 1988.
202. Boyle, T. P., Finger, S. E., Paulson, F. L., and Rabeni, C. F., Comparison of laboratory and field assessment of fluorine. Part II: Effects on the ecological structure and function of experimental pond ecosystems, in *Validation and Predictability of Laboratory Methods*, American Society for Testing and Materials, Philadelphia, 1985, 134.
203. Wakeham, S. G., Davis, A. C., and Karas, J. A., Mesocosm experiments to determine the fate and persistence of volatile organic compounds in coastal seawater, *Environ. Sci. Technol.*, 17, 611, 1983.
204. Stephenson, R. R. and Kane, D. F., Persistence and effects of chemicals in small enclosures in ponds, *Arch. Environ. Contam. Toxicol.*, 13, 313, 1984.
205. Lay, J. P., Herrmann, M., Kotzias, D., and Parlar, H., Influence of chemicals upon plankton in freshwater systems, environmental pollution and its impact on life in the Mediterranean region 1985, *Chemosphere*, 581, 1987.
206. Maguire, R. J., Carey, J. H., Hart, J. H., Tkacz, R. J., and Lee, H. B., Persistence and fate of deltamethrin sprayed on a pond, *J. Agric. Food Chem.*, 37, 1153, 1989.
207. Smith, E. P. and Mercante, D., Statistical concerns in the design and analysis of multispecies microcosm and mesocosm experiments, *Toxicity Assessment*, 4, 129, 1989.
208. Kennedy, J. H., Ammann, L. P., Waller, W. T., Warren, J. E., Hosmer, A. J., Cairns, S. H., Johnson, P. C., and Graney, R. L., Using statistical power to optimize sensitivity of analysis of variance designs for microcosms and mesocosms, *Environ. Toxicol. Chem.*, 18, 113–117, 1999.
209. Adams, W. J., Aquatic toxicology testing methods, in *Handbook of Ecotoxicology*, Calow, P., Ed., Blackwell Scientific Publications, Cambridge, MA, Chap. 3, 1995.
210. Landis, W. G., Matthews, G. B., Matthews, R. A., and Sergeant, A., Application of multivariate techniques to endpoint determination, selection and evaluation in ecological risk assessment, *Environ. Toxicol. Chem.*, 13, 1917, 1994.
211. Suter, G. W., Endpoints of interest at different levels of biological organization, in *Ecological Toxicity Testing: Scale, Complexity, and Relevance*, Cairns, J., Jr., Niederlehner, B. R., Eds., Lewis Publishers, Boca Raton, FL, 1995, Chap. 3.
212. U.S. Environment Protection Agency (USEPA), Framework for Ecological Risk Assessment, EPA/630/R-92/001, National technical information service, Springfield, VA, 1992.
213. Resh, V. H. and Unzicker, J. D., Water quality monitoring and aquatic organisms: The importance of species identification, *J. Water Poll. Control Fed.*, 47, 9, 1975.
214. Ferraro, S. P. and Cole, F. A., Taxonomic level and sample size sufficient for assessing pollution impacts on the southern California Bight macrobenthos, *Mar. Ecol. (Prog. Ser.)*, 67, 251, 1990.
215. Ferraro, S. P. and Cole, F. A., Taxonomic level sufficient for assessing a moderate impact on macrobenthic communities in Puget Sound, Washington, USA, *Can. J. Fish. Aquat. Sci.*, 49, 1184, 1992.
216. Kennedy, J. H., Johnson, Z. B. and Johnson, P. C., Sampling and analysis strategy for biological effects in freshwater field tests, in *Freshwater Field Tests for Hazard Assessment of Chemicals*, Lewis, Chelsea, MI, 1993.
217. Ammann, L. P., Waller, W. T., Kennedy, J. H., Dickson, K. L., Mayer, F. L., Lewis, M., Power, sample size and taxonomic sufficiency for measures of impacts in aquatic systems, *Environ. Toxicol. Chem.*, 16, 2421, 1997.
218. Sheehan, P. J., Effects on community and ecosystem structure and dynamics, in *Effects of Pollutants at the Ecosystem Level*, Sheehan, P. J., Miller, D. R., Butler, G. C., and Bourdeau, P., Eds., John Wiley & Sons, New York, 1984, 51–99.
219. Lamberti, G. A. and Resh, V. H., Distribution of benthic algae and macroinvertebrates along a thermal stream gradient, *Hydrobiologia*, 128, 13, 1985.
220. Barton, D. R., A comparison of sampling techniques and summary indices for assessment of water quality in the Yamaska River, Quebec, based on macroinvertebrates, *Environ. Monitoring Assessment*, 21, 225, 1992.
221. Cao, Y., Bark, A. W., and Williams, W. P., Measuring the responses of macroinvertebrate communities to water pollution: A comparison of multivariate approaches, biotic and diversity indices, *Hydrobiologia*, 341, 1, 1996.
222. Camargo, J. A., Macroinvertebrate surveys as a valuable tool for assessing freshwater quality in the Iberian Peninsula, *Environ. Monitoring Assessment*, 24, 71, 1993.

223. Joshi, H., Shishodia, S. K., Kumar, S. N., Saikia, D. K., Nauriyal, B. P., Mathur, R. P., Pande, P. K., Mathur, B. S., and Puri, N., Ecosystem studies on upper region of Ganga River, India, *Environ. Monitoring Assessment*, 35, 181, 1995.
224. Dickson, K. L., Waller, W. T., Kennedy, J. H., and Ammann, L. T., Assessing relationships between effluent toxicity, ambient toxicity and aquatic community responses, *Environ. Toxicol. Chem.*, 11, 1307–1322, 1992.
225. LaPoint, T. W., Signs and measurements of ecotoxicology in the aquatic environment, in *Handbook of Ecotoxicology in Calow*, P., Ed., Graney, R. L., Kennedy, J. H., and Rodgers, J. H., Jr., Eds., Blackwell Scientific Publications, 1995, 337.
226. Liber, K., Kaushik, N. K., Solomon, K. R., and Carey, J. H., Experimental designs for aquatic mesocosm studies: A comparison of the "Anova" and "Regression" design for assessing the impact of tetrachlorophenol on zooplankton populations in limnocorrals. *Environ. Toxicol. Chem.*, 11, 61, 1992.
227. Landis, W. G., Matthews, R. A., and Matthews, G. B., Design and analysis of multispecies toxicity tests for pesticide registration, *Ecol. Appl.*, 7, 1111, 1997.
228. Peterman, R. M., Application of statistical power analysis to the Oregon coho salmon (*Oncorhynhus kisutch*) problem, *Can. J. Fish. Aquat. Sci.*, 46, 1183, 1989.
229. Van den Brink, P. J. and Ter Braak, C. J. F., Multivariate analysis of stress in experimental ecosystems by Principal Response Curves and similarity analysis, *Aquat. Ecol.*, 32, 163, 1998.
230. Van Wijngaarden, R. P. A., van den Brink, P. J., Crum, S. J. H., Oude Voshaar, J. H., Brock, T. C. M., and Leeuwangh, P., Effects of insecticide Dursban 4E (active ingredient chlorpyrifos) in outdoor experimental ditches. I. Comparison of short-term toxicity between the laboratory and field, *Environ. Toxicol. Chem.*, 15, 1133, 1996.
231. Maund, S. J., Chapman, P. K., Edwards, T. J., Tattesfield, L., Matthiessen, P., Warwick, R., and Smith, E., Application of multivariate statistics to ecotoxicological field studies, *Environ. Toxicol. Chem.*, 18, 111, 1999.
232. Sparks, T. H., Scott, W. A., and Clarke, R. T., Traditional multivariate techniques: Potential for use in ecotoxicology, *Environ. Toxicol. Chem.*, 18, 128, 1999.
233. Van den Brink, P. J., and Ter Braak, C. J. F, Principal response curves: Analysis of time-dependent multivariate responses of biological community to stress, *Environ. Toxicol. Chem.*, 18, 138, 1999.
234. Van Breukelen, S. W. F. and Brock, T. C. M., Response of a macroinvertebrate community to insecticide application in replicated freshwater microcosms with emphasis on the use of principal component analysis, *Sci. Total Environ.*, 0(suppl. part 2), 1047, 1993.
235. Ludwig, J. A. and Reynolds, J. F., *Statistical Ecology*, John Wiley and Sons, New York, 1988.
236. Van Wijngaarden, R. P. A., van den Brink, P. J., Oude Voshaar, J. H., and Leeuwangh, P., Ordination techniques for analyzing response of biological communities to toxic stress in experimental ecosystems, *Ecotoxicology*, 4, 61, 1995.
237. Van den Brink, P. J., van Wijgaarden, R. P. A., and Lucassen, W. G. H., Effects of the insecticide Dursban? 4E (active ingredient chlorpyrifos) in outdoor experimental ditches. II. Invertebrate community responses and recovery, *Environ. Toxicol. Chem.*, 15, 1143, 1996.
238. Clarke, K. R., Nonmetric multivariate analysis in community level ecotoxicology, *Environ. Toxicol. Chem.*, 18, 118, 1999.
240. Green, R. H., Multivariate approaches in ecology: The assessment of ecologic similarity, *Annu. Rev. Ecol. Syst.*, 11, 1, 1980.
240. James, F. C. and McCulloch, C. E., Multivariate analysis in ecology and systematics: Panacea or Pandora's box?, *Annu. Rev. Ecol. Syst.*, 21, 129, 1990.
241. Cairns, J., Jr., Paradigms flossed: The coming of age of environmental toxicology, *Environ. Toxicol. Chem.*, 11, 285, 1992.
242. Farris, J. L., Grudzien, J. L., Belanger, S. E., Cherry, D. S., and Cairns, J., Jr., Molluscan cellulolytic activity responses to zinc exposure in laboratory and field stream comparisons, *Hydrobiologia*, 287, 161, 1994.
243. Belanger, S. E., Guckert, J. B., Bowling, J. W., Begley, W. M., Davidson, D. H., LeBlanc, E. M., and Lee, D. M., Responses of aqautic communities to 25-6 alcohol ethoxylate in model stream ecosystems, *Aquat. Toxicol.*, 28, 135, 2000.
244. Guckert, J. B., Belanger, S. E., and Barnum, J. B., Testing single-specieis predictions for a cationic surfactant in a stream mesocosm, *Sci. Total Environmen.*, Supplement 1993, 1011, 1993.

245. Lee, D. M., Guckert, J. B., Belanger, S. E., and Feijtel, T. C. J., Seasonal temperature declines do not decrease periphytic surfactant biodegradation or increase algal species sensitivity, *Chemosphere*, 35, 1143, 1997.
246. Belanger, S. E., Davidson, D.H., Farris, J. L., Reed, D., and Cherry, D. S., Effects of cationic surfactant exposure to a bivalve mollusc in stream mesocosms, *Environ. Toxicol. Chem.*, 12, 1789, 1993.
247. Belanger, S. E., Meiers, E. M., and Bausch, R. G., Direct and indirect ecotoxicological effects of alkyl sulfate and alkyl ehoxysulfate on macroinvertebrates in stream mesocosms, *Aquat. Toxicol.*, 33, 65. 1995.

## CHAPTER 4

# Wildlife Toxicity Testing

**David J. Hoffman**

## CONTENTS

## 4.1 INTRODUCTION AND HISTORICAL BACKGROUND

Wildlife toxicology is the study of potentially harmful effects of toxic agents on wild animals. Fish and aquatic invertebrates are usually not included as part of wildlife toxicology since they fall within the field of aquatic toxicology, but collectively both fields often provide insight into one another and both are integral parts of ecotoxicology. Wildlife toxicology endeavors to predict the effects of toxic agents on nontarget wildlife species and, ultimately, populations in natural environments. Wildlife toxicology has often focused on highly visible species, including certain birds and a few mammals, that are of aesthetic or economic interest to humans.[1] However, during the past decade wildlife toxicology has expanded to include the effects of environmental contaminants on reptiles, amphibians, and terrestrial invertebrates.

Reports of anthropogenic environmental contaminants affecting free-ranging wildlife began to accumulate during the industrial revolution of the 1850s. Early reports included cases of arsenic and lead-shot ingestion and industrial smokestack-emission toxicity. One report described the death of fallow deer (*Dama dama*) due to arsenic emissions from a silver foundry in 1887 in Germany,[2] with subsequent reports of widespread killing of game animals, including deer and foxes, by arsenic emissions from metal smelters.[3] Another report described hydrogen sulfide fumes in the vicinity of a Texas oil field that resulted in a large die-off of many species of wild birds and mammals.[2] Mortality in waterfowl and ring-necked pheasants (*Phasianus colchicus*) from ingestion of spent lead shot was recognized as early as 1874 when lead-poisoned birds were reported in Texas and North Carolina.[4] Waterfowl mortality in the vicinity of mining and smelting operations was first reported in the early 1900s and subsequently linked to metallic or lead poisoning.[5] At about the same time the potentially devastating effects, including the death of seabirds such as puffins, of crude oil spills were noticed.[6]

Prior to World War II most agricultural and household pesticides were relatively simple derivations of naturally occurring minerals and plant products. The advent of synthetic organic insecticides evolved from World War II. Many of these post-war insecticides exhibited vast biological activity and were remarkably persistent in the natural environment. Initially, this persistence seemed desirable, especially from an economical perspective, i.e., long-term pest control. However, within a few years insects began to show resistance to many of these "modern" pesticides, and even trivial amounts of some impaired reproduction in certain wildlife species. For example, dichlorodiphenyltrichloroethane (DDT) was introduced in 1943, and by the end of the decade ecological problems other than incidents of acute mortality began to surface. Farsighted scientists, such as Lucille Stickel and others at the Patuxent Wildlife Research Center, Laurel, Maryland, cautioned users of the potential hazards of application of DDT to wildlife.[7,8] Little attention was paid to ecological hazards of pesticides until Rachel Carson published *Silent Spring*[9] in 1962. This seminal treatise effectively sensationalized many ecologically significant happenings such as a decline in the population of American robins (*Turdus migratorius*) by the early 1950s, which was linked to DDT spraying to fight Dutch Elm disease, and evidence that bald eagles (*Haliaeetus leucocephalus*), osprey (*Pandion haliaetus*), and populations of fish-eating mammals were at risk. Research then revealed that DDT and other chlorinated hydrocarbon insecticides, including dichlorodiphenyldichloroethane (DDD), endrin, aldrin, and dieldrin, when incorporated into the diets of pheasants and quail, impaired reproductive success without necessarily having other adverse effects on adults.[10] Eggshell thinning, related to DDT and ultimately dichlorodiphenyldichloroethylene (DDE), was determined to be an important factor in reproductive failure in European and North American birds of prey[11–13] as well as in brown pelicans (*Pelecanus occidentalis*).[14] This and similar research played a major role in the cancellation of many highly persistent pesticides in the United States.

Even pesticides that are comparatively labile in a natural system may be problematic. For example, many wildlife losses have been documented due to organophosphorus and carbamate insecticides.[15–18] These poisonings of many species of birds and mammals are due to acute lethal toxicity from cholinesterase inhibition. Agricultural practices other than pesticide application that have received focus due to adverse effects on wildlife include subsurface drainage of irrigation water, wherein bioaccumulation of selenium and other trace elements in the aquatic food chain has proven highly embryotoxic and teratogenic to numerous species of waterbirds.[19–22]

Certainly, factors other than agricultural practices may pose toxic hazards to wildlife. For example, concern has arisen over globally increasing concentrations of methylmercury in aquatic biota, even evident at remote and semiremote sites, as a consequence of multiple anthropogenic sources and their emitting mercury into the environment.[23] A case in point — in the marine food web of the North Atlantic Ocean — is the steady long-term increase in concentration of methylmercury in feathers of fish-eating seabirds sampled from 1885 through 1994.[24] This increase has averaged 1.9% per year in Cory's shearwater (*Calonectris diomedea borealis*) and 4.8% per year in Bulwer's petrel (*Bulweria bulwerii*). These increases are attributed to global trends in mercury contamination rather than local or regional sources. Mercury concentrations have also increased over the past century in other species of seabirds.[25]

Another concern is endocrine disruption in wildlife species.[26–28] Many of the endocrine disruptor reports in wildlife are based upon observed adverse reproductive and developmental effects rather than direct evidence of endocrine-modified function or defined endocrine pathways. A wide variety of chemicals have been reported as potential endocrine disruptors and are described by Gross et al. in Chapter 39 of this book. These include polycyclic aromatic hydrocarbons; polychlorinated and polybrominated biphenyls, dibenzo-*p*-dioxins and dibenzo-*p*-furans; organochlorine pesticides and fungicides; some nonorganochlorine pesticides; complex environmental mixtures; and a few metals. Collectively, there is strong evidence of altered reproductive and developmental processes in wildlife exposed to endocrine disruptors. The U.S. Congress has passed legislation (listed in federal register notice, 63 FR 71542) requiring the U.S. Environmental Protection Agency (EPA) to develop, validate, and implement an Endocrine Disruptor Screening Program (EDSP) for identifying potential endocrine-disrupting chemicals.

Wildlife toxicology involves the integration of three principal strategies for understanding effects of toxic agents on wildlife.[1,29] The first strategy is *chemical screening*. A variety of toxicological tests are performed in the laboratory or in outdoor pens. Representative species are tested to help predict potential effects of a given chemical in natural populations of the same or closely related species. The second strategy is the *controlled field* or *mesocosm study*. Wildlife species of interest are exposed to operational chemical applications in a confined environment, simulating a natural system. The third strategy is *field ecology assessment*. Natural populations are studied in environments subjected to high levels of contamination.

With the advent of synthetic insecticides as well as the release of industrial pollutants and consequent wildlife losses, screening of pesticides and other chemicals for adverse effects has become an integral part of wildlife toxicology. A wide variety of wildlife testing protocols have been developed for regulatory use by the U.S. EPA under the Federal Insecticide, Fungicide and Rodenticide Act (FIFRA) and the Toxic Substances Control Act (TSCA). The U.S. EPA has established a unified library of test guidelines issued by the Office of Prevention, Pesticides and Toxic Substances (OPPTS), for the Series 850-Ecological Effects Test Guidelines. This document outlines testing requirements and protocols for review by the U.S. EPA under TSCA and FIFRA. The purpose of harmonizing these guidelines into a single set of OPPTS guidelines is to minimize variations among test procedures that must be performed to meet the U.S. EPA data requirements. These guidelines are a compilation of the testing guidance and requirements of the Office of Pollution Prevention and Toxics (OPPT; Title 40, Chapter I, Subchapter R of the Code of Federal Regulations), the Office of Pesticide Programs (OPP), and the Organization for Economic Cooperation and Development (OECD). Table 4.1 summarizes the guidelines for terrestrial wildlife in this series and the above sources from which they were derived.

This chapter provides a summary of the toxicity tests commonly used in wildlife toxicology. The focus is on avian studies because birds have served as primary models for terrestrial wildlife toxicology since the 1950s. In contrast, though mammalian wildlife species were considered

**Table 4.1 Summary of Ecological Effects Test Guidelines for Terrestrial Wildlife, Soil Microbes, and Environmental Chemistry**

| OPPTS[a] Number | Name of Test | OTS[b] | OPP[c] | OECD[d] | EPA Pub. No. |
|---|---|---|---|---|---|
| | **Terrestrial Wildlife Tests** | | | | |
| 850.2100 | Avian acute oral toxicity test | 797.2175 | 71–1 | none | 712-C-96–139 |
| 850.2200 | Avian dietary toxicity test | 797.2050 | 71–2 | 205 | 712-C-96–140 |
| 850.2300 | Avian reproduction test | 797.2130, .2150 | 71–4 | 206 | 712-C-96–141 |
| 850.2400 | Wild mammal acute toxicity | none | 71–3 | none | 712-C-96–142 |
| 850.2450 | Terrestrial (soil-core) microcosm test | 797.3775 | none | none | 712-C-96–143 |
| 850.2500 | Field testing for terrestrial wildlife | none | 71–5 | none | 712-C-96–144 |
| | **Beneficial Insects and Invertebrates Tests** | | | | |
| 850.3020 | Honey bee acute contact toxicity | none | 141–1 | none | 712-C-96–147 |
| 850.3030 | Honey bee toxicity of residues on foliage | none | 141–2 | none | 712-C-96–148 |
| 850.3040 | Field testing for pollinators | none | 141–5 | none | 712-C-96–150 |
| *Toxicity to Microorganisms Tests* | | | | | |
| 850.5100 | Soil microbial community toxicity test | 797.3700 | none | none | 712-C-96–161 |
| | **Chemical Specific Tests** | | | | |
| 850.6200 | Earthworm subchronic toxicity test | 795.150 | none | 207 | 712-C-96–167 |
| 850.6800 | Modified activated sludge, respiration inhibition test for sparingly soluble chemicals | 795.170 | none | 209 | 712-C-96–168 |
| | **Field Test Data Reporting** | | | | |
| 850.7100 | Data reporting for environmental chemistry methods | none | none | none | 712-C-96–348 |

[a] Office of Prevention, Pesticides and Toxic Substances
[b] Office of Toxic Substances (for TSCA)
[c] Office of Pesticide Programs (for FIFRA)
[d] Organization for Economic Cooperation and Development

important, it was generally accepted that the array of baseline tests routinely conducted with laboratory mammals would usually suffice for at least provisional intertaxa comparisons. The main avian tests described herein are acute, subacute, subchronic, chronic, developmental, field, and behavioral (Figure 4.1). Coverage of toxicity testing for wild mammals, amphibians, and reptiles is also provided but in somewhat less detail since the present body of information on these is more limited in scope and requirement than avian wildlife toxicity testing.

## 4.2 SINGLE-DOSE ACUTE ORAL AND SHORT-TERM SUBACUTE DIETARY AVIAN TOXICITY TESTS

Basic protocols with lethality as the principal endpoint have been used for first-line toxicity testing with birds by the U.S. Fish and Wildlife Service, the U.S. Geological Survey, and the U.S. EPA. These include experiments designed to estimate the acute oral median lethal dosage ($LD_{50}$), the 5-day median lethal dietary concentration ($LC_{50}$), and relevant statistical parameters.[30]

### 4.2.1 Single-Dose Acute Oral

Reports on single-oral-dose avian $LD_{50}$s contain data for adults of nearly 75 species of birds and more than 1000 chemicals tested.[31–33] Full-scale acute oral toxicity tests are required by

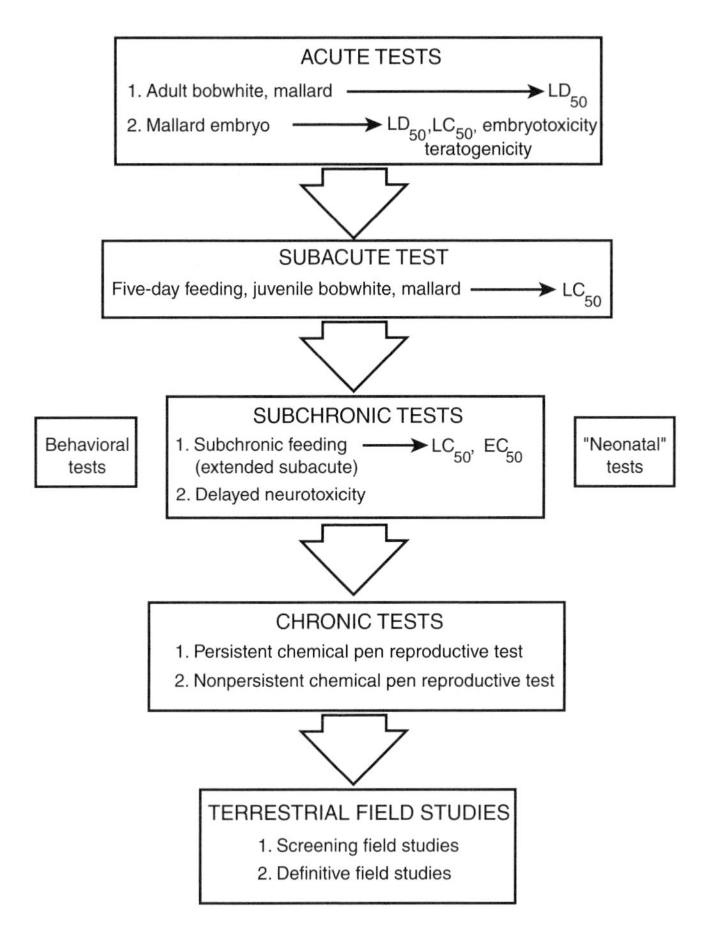


**Figure 4.1** Protocols used in avian toxicity testing.

regulatory guidelines for pesticide registration in the United States (Table 4.1). Testing is with one of two species of birds, usually the mallard (*Anas platyrhynchos*) or the northern bobwhite (*Colinus virginianus*), and provides the $LD_{50}$ and associated statistics including the dose-response curve.[34] Acute oral tests provide a preliminary indication of the lethal hazard of a substance.

These tests are rapid, uncomplicated, inexpensive, statistically reliable within the lethality curve (i.e., ± 1 SD of the $LD_{50}$), and, most importantly, provide insight necessary for further hazard evaluation.[35] Overnight-fasted birds receive a single dose of test substance at midmorning, usually administered by gavage or by capsule, at each of five or six geometrically arranged dosage levels that were predetermined from a preliminary study to span the expected 10 to 90% mortality levels.[30,35] Feed is provided immediately postdosing, and observations for signs of intoxication are continued throughout the day. Special attention is given to the length of time to first evidence of toxicity, death, and recovery. Observations are continued twice daily, or more often as indicated, for 14 days posttreatment, or longer if toxic signs persist. Gross necropsy should be performed on all birds that die and on a subsample of survivors.

Optimal use of the acute test requires statistical estimation of the lethality curve and its midpoint and descriptive information on toxic response. The $LD_{50}$, expressed as mg (active ingredient)/kg of body mass, its 95% confidence interval, and the slope and error of the dose-response curve are derived by probit, logit, or other appropriate analysis. When only the general order of acute toxicity is desired for wide-scale initial comparisons of many species or finished product formulations, then a choice of several approximate tests of lethality may be used instead (e.g., up-and-down, moving averages).[31,32,35]

Approximate tests conserve test animals. For example, as few as three groups of three to five birds are tested against a standardized fixed-dosage arrangement; toxicity statistics are derived from published tables.[31] The up-and-down procedure may use even fewer animals and is based on dosing single subjects at a time and adjusting each subsequent dose up or down, depending on the response of the preceding subject. Approximate tests provide an estimate of the $LD_{50}$ and its 95% confidence interval but do not provide a statistical estimate of the dose-response curve.

The single acute oral test is best suited for substances whose principal action is lethality (e.g., insecticides, certain heavy metals, natural toxins), but (with less statistical confidence) the $LD_{50}$ can be used as a toxicity marker for any substance with substantial bioactivity. Thus, the $LD_{50}$ provides a statistical basis for comparisons among contaminants and species as well as with measured or estimated amounts of pesticide or other contaminant residue in the terrestrial or aquatic environment that could be consumed by wildlife in a single oral exposure.

However, care must be taken in how acute data are evaluated and whether the test is being used generally for regulatory purposes or specifically for risk assessment or research planning. For example, an avian database was compiled for 147 cholinesterase-inhibitor pesticides, with 837 acceptable $LD_{50}$ determinations, and 733 noncholinesterase-inhibitor pesticides with 1601 acceptable $LD_{50}$ values.[33] These authors found that when fitting a distribution to $LD_{50}$s expressed as mg/kg body weight, the $LD_{50}$ values were overestimated for small-body birds. Therefore, the use of scaling factors for body weight has been recommended to improve cross-species comparisons of acute toxicological sensitivity.[36] After the protection level had been arbitrarily fixed at the 5th percentile of the species distribution (termed $HD_5$; Hazardous Dose 5%), it was shown that of all of the above pesticides, 34 had an $HD_5$ less than 1 mg/kg, of which 24 were cholinesterase inhibitors. Of the remaining ten, two were insecticides (including the very new pyrrole insecticide chlorfenapyr), two were fungicides, and the others were rodenticides, including three of the coumarin anticoagulant products.

In another example generally similar acute toxicities between mallards and laboratory rats were reported for insecticides and herbicides.[37] However, subsequent evaluations have revealed important and major differences in acute sensitivity to pesticides between birds and mammals (laboratory rats) and have been reviewed by Walker[38] and by Hill.[35] To illustrate, the organophosphorus pesticides dimethoate, fenitrothion, and temephos have $LD_{50}$s of, respectively, 215, 740, and 8600 mg/kg of body mass in rats, but are extremely toxic to both pheasants and blackbirds with $LD_{50}$s of 7 to 42 mg/kg (Table 4.2.). It was concluded that the laboratory rat is not a good model for the prediction of acute toxicity to birds, even for chemicals that elicit their toxicity in the same manner.

**Table 4.2 Acute Avian Toxicity Testing of Organophosphorus Pesticides of Widely Variable Mammalian Toxicity[a]**

| Pesticide | Rat[b] | Pheasant[c] | Blackbird[d] |
|---|---|---|---|
| | Single-Dose Oral $LD_{50}$ | | |
| Phorate | 2 | 7 | 1 |
| Azinphos-methyl | 13 | 75 | 8 |
| Ethion | 65 | 1297 | 45 |
| Dimethoate | 215 | 20 | 7 |
| Fenitrothion | 740 | 26 | 25 |
| Temephos | 8600 | 35 | 42 |

[a] Derived from reference 35.
[b] Sherman strain male laboratory rats, 3 months old, n = 50–60 per test.
[c] Farm-reared male and female ring-necked pheasants, 3 to 4 months old, n = 8–29 per test.
[d] Wild-captured pen-conditioned male and female red-winged blackbirds, adult, n = 8–28 per test.

Such remarkable differences between birds and rats clearly justify the need for acute avian toxicity testing and confirm that reliance on widely existing rat data is not adequate for prediction of avian hazard. Wide variation in sensitivity to acute exposure to pesticides exists even among avian species. For example, the red-winged blackbird (*Agelaius phoeniceus*) was one of the most sensitive of seven species tested with ten different anticholinesterase pesticides.[35] In another study adult red-winged blackbirds were 137 times more sensitive to terbufos and 65 times more sensitive to diazinon than adult European starlings (*Sturnus vulgaris*).[39] Furthermore, a difference of nearly 70-fold in sensitivity was detected between American kestrels (*Falco sparverius*) and eastern screech owls (*Otus asio*) for the anticholinesterase insecticide EPN; screech owls were relatively tolerant ($LD_{50}$ = 274 mg/kg), but kestrels were not ($LD_{50}$ = 4 mg/kg).[40]

$LD_{50}$ tests have also been used to demonstrate practical differences in acute hazard of technical grade and liquid and granular formulations of pesticides.[35,41–44] While the toxicity of liquid formulations was found to be equal or somewhat more toxic than the technical grade, granulars were generally less toxic than the technical grade. However, the hazard involved in risk assessment of granulars was more dependent upon which avian species (size and feeding behavior) was likely to inhabit a treated area rather than on actual application rate.[45] If ingestion is haphazard, then the application rate becomes more critical, but if ingestion is selective, then even the most stringent attempts to reduce granule availability may fail to reduce the hazard. Recent studies with bobwhite in planted enclosures have suggested that ground-feeding birds are more susceptible to granular insecticides than flowable applications.[46] The color, size, texture, granule base, taste, and application rate are all factors for consideration in reduction of granular hazard to wildlife.[47–54]

### 4.2.2 Subacute Dietary

The subacute test ($LC_{50}$), a 5-day feeding trial, is required for two species, including upland game birds and waterfowl, to support registration of pesticides (Table 4.1). This guideline was modified from 40 CFR 797.2050 and OPP 71–2 to be harmonized with OECD guideline 205, allowing the use of coturnix (Japanese quail, *Coturnix japonica*) as an acceptable test species. This test serves as a composite indicator of vulnerability to a contaminated diet, allowing for metabolic changes that occur over time. The test was developed to quantify the toxicity of dietary residues that were considered an important source of exposure of wildlife to environmental contaminants.[43,44,54] This test was optimized with young precocial birds, including ducks and quail, but almost any species can be tested provided it can be maintained in captivity in good health and cannot survive for 5 days without eating.[35] Mortality and signs of intoxication are monitored at least twice daily, and food consumption is measured at 24-hour intervals. After the fifth day, all feed is replaced with untreated feed, and the study is continued for at least 3 days or until complete remission of overt toxic signs.

The $LC_{50}$ is expressed as mg (active ingredient)/kg of feed (or ppm) with its 95% confidence interval and slope with error of the dose-response curve as done for acute tests. Results of tests on more than 200 pesticides with young northern bobwhite, coturnix, ring-necked pheasants, and mallards have been published.[43,44] When $LC_{50}$ tests are compared with $LD_{50}$ tests, subacute $LC_{50}$ results often describe relationships among species and chemicals that are quite different from those for $LD_{50}$s because $LC_{50}$ tests measure ability of birds to cope with toxic feed for a set duration.[35] $LC_{50}$s must be used carefully when comparing the toxicity of pesticides among species because there is no assurance that the species of interest have been equally challenged by the test protocol.[35] For example, if one species has a greater tendency to refrain from eating the test diet, or if a portion of the population can survive severely reduced nutriment for the duration of the test, then responses may vary considerably.

Age is an important consideration when evaluating $LC_{50}$s. In precocial species there is generally an increase in resistance to chemicals with increasing age during early growth.[55,56] This increase occurs across a given class of chemical or pesticide and appears to be the result of changes in ability to cope with a toxic diet for the exposure duration, where older (larger) chicks that eat less proportional to body mass are better able to survive the 5-day trial by reducing food consumption and hence toxic exposure.[56] During the first 21 days average increases in $LC_{50}$s in coturnix for nine pesticides (three organophosphorus and two each of carbamate, chlorinated hydrocarbon, and methylmercury) were 36% for days 1–7, 43% for days 7–14, and 28% for days 14–21.[56] This corresponded to reductions of normal food consumption in controls of about 35, 23, and 21% per week from hatch to 3 weeks of age. In contrast, acute oral $LD_{50}$s show a dichotomy of change with age. For example, mallard $LD_{50}$s for anticholinesterases that require activation for maximum potency tend to decrease between hatch and 7 days and then increase through adulthood, whereas the opposite pattern occurs for direct-acting organophosphorus and carbamate anticholinesterases.[57]

## 4.3 AVIAN SUBCHRONIC DIETARY TOXICITY TEST

This test was developed as an extension of the subacute $LC_{50}$ test but with greater emphasis on sublethal indicators of toxicity. The test was designed as a precursor to provide a biological indication of the necessity for conducting a full-scale reproductive trial and to provide a possible hazard index based on the ratio of sublethal to lethal toxicity values.[30] The first test of this kind was conducted to compare the effects of organic and inorganic mercury on various physiological parameters, including indicator enzymes and blood chemistries, in coturnix through 9 weeks of age, which is full maturity in this species; calculation of periodic $EC_{50}$s (median effective concentration) for each responding variable was used to develop hazard indices relating the $EC_{50}$ to the oral $LD_{50}$ and 5-day $LC_{50}$.

Other special studies of a subchronic nature (exposure period generally 1 to 3 months) have been conducted on pesticides and other environmental contaminants using farm-reared mallards and quail, as well as wild-captured species of birds. The emphasis of these studies has been on biochemical indicators of toxicity and toxicokinetics. For example, when studying the effects of an organic form of selenium, selenomethionine, implicated in agricultural drainwater toxicity to waterbirds in California, 2-year-old mallard drakes were fed diets containing supplementation of excess selenium from 1 to 32 ppm for 14 weeks.[58] Selenium accumulated readily in the liver in a dose-dependent manner. Mortality (10%) and histopathological effects, including bile duct hyperplasia and hemosiderin pigmentation of the liver and spleen, occurred at the highest level tested. These histopathological effects were accompanied by elevated plasma-alkaline-phosphatase activity, which is indicative of cholestatic liver injury. Other manifestations of hepatotoxicity included significant dose responses for hepatic oxidized glutathione (GSSG) concentrations and increased ratio of GSSG to reduced glutathione (GSH). Mean hepatic malondialdehyde (a measure of lipid peroxidation) concentration was elevated at the two highest levels tested. A number of these subchronic effects were similar to effects reported in wild waterfowl in a seleniferous location.

Subchronic studies have also been used for comparative purposes among species. For example, the potential hazard of ingestion of lead-contaminated sediment was assessed in mallard ducklings and goslings of Canada geese (*Branta canadensis*) from hatching for 6 weeks.[59–61] At similar dietary concentrations of lead, blood and liver lead concentrations increased almost twice as much in mallards as in geese. Yet when species were compared for responses at similar blood and liver lead concentrations, manifestations of toxicity were greater in geese, as reflected by reduced survival and growth, hematological effects, and hepatic oxidative stress. Hepatic GSH-S-transferase activity was nearly three times higher in geese than in mallards and presumably had a role in the binding of lead to GSH and subsequent biliary excretion. In contrast, mallards showed higher hepatic levels of GSH and activities of GSH peroxidase (GPX) and GSSG reductase (GR). An increase of lipid peroxidation with lead exposure was more evident in geese than mallards. Hepatic GSH was inversely related to hepatic lipid peroxidation — only in mallards and in agreement with the differences observed in GPX and GR activities. This apparent lower resistance to lipid peroxidation of Canadian geese compared to mallards may explain why geese found dead in the field by lead ingestion often have lower liver lead concentrations than mallards.

A third example of specialized subchronic testing comes from studying the potential of several organophosphorus insecticides to induce delayed neurotoxicity (OPIDN) in mallards.[62] Here mallard hens received up to 270 ppm technical grade EPN in the diet for 90 days. Muscular incoordination, or ataxia, was first observed with 270 ppm in the diet after 16 days, with 90 ppm after 20 days, and with 30 ppm after 38 days; 10 ppm failed to produce ataxia. Brain neurotoxic esterase activity was inhibited by about 70% or more in groups experiencing ataxia. Brain and blood plasma cholinesterases and plasma alkaline phosphatase were significantly inhibited as well. Distinct histopathological manifestations of OPIDN were seen at concentrations as low as 30 ppm, which included demyelination and degeneration of axons of the spinal cord. Additional ducks exposed in a similar manner to leptophos experienced similar behavioral, biochemical, and histopathological alterations. These findings showed that adult mallards were sensitive to OPIDN but somewhat less so than chickens, which have served as the traditional model for screening substances for OPIDN.

## 4.4 AVIAN CHRONIC TOXICITY TESTS

Most avian chronicity tests are designed with reproduction as the primary endpoint. Such tests permit study of simulated field exposure under controlled conditions with a relevant species and exposure route. The Avian Reproduction Test guideline has been modified from 40 CFR 797.2130 and 797.2150 and OPP 71–4 to be harmonized with OECD guideline 206. The OECD guideline identifies Japanese quail as an acceptable species, but until several technical issues are resolved for this species the United States is listing only mallard and bobwhite as acceptable species for avian reproduction testing. For certain other tests, such as part of the Endocrine Disruptor Screening Program (EDSP, listed in federal register notice, 63 FR 71542), a two-generation test with Japanese quail is allowed. This test includes endocrine-specific endpoints in addition to the conventional fitness endpoints of existing avian reproduction test designs for identifying and characterizing endocrine effects of pesticides, industrial chemicals, and environmental contaminants.

Avian reproductive studies are required by the U.S. EPA[63,64] with both waterfowl (mallards) and an upland game species (northern bobwhite) to support the registration of an end-use product that meets one or more of the following criteria: (1) the end-use product is intended for use where birds may be subjected to repeated or continuous exposure to the pesticide or any of its major metabolites or degradation products, especially preceding or during the breeding season; (2) the pesticide or its products may persist in the food at toxic levels; (3) the pesticide or its products accumulate in plant or animal tissues; and (4) the pesticide or its products cause adverse reproductive effects in mammals. The basic study usually required by the U.S. EPA is designed to elucidate reproductive effects of chemicals that persist in potentially toxic amounts in wildlife habitats throughout the reproductive season. Alternative studies have also been used to evaluate reproductive effects from pesticides that have a short life span in nature but may be applied repeatedly during the reproductive season. The basic reproductive study and an example of an altered reproduction study follow.

### 4.4.1 Reproduction Studies — Basic Protocol

This basic protocol was designed for testing environmentally persistent chemicals and may be required for pesticide registration (Table 4.1). Test concentrations for regulatory purposes are based on measured or calculated residues expected in the diet from the proposed use pattern, including an actual field exposure level and a multiple level, such as ten times the field level. Treatment commences at least 10 weeks in advance of breeding and continues through egg laying. Eggs are collected daily and set at weekly intervals for artificial incubation in the laboratory. Reproductive endpoints in such studies include number of eggs laid, percent fertility, live 3-week embryos, hatchability, 14-day-old posthatch survivors, and eggshell thickness. Two examples of reproductive studies conducted with mallards include: (1) studies with endrin, a chlorinated hydrocarbon pesticide, and (2) studies with selenium, an agricultural drainwater contaminant. Both contaminants have been associated with avian mortality in the field.

In the early 1980s endrin was found to be quite toxic to birds as the cyclodiene insecticide was used in the western United States. Reproductive effects were suspected mostly in waterfowl. Therefore, studies were conducted in which concentrations of 0, 1, and 3 ppm endrin in dry feed were fed to mallards from late fall through the following breeding season; health and reproduction were monitored.[65] Endrin at 3 ppm increased adult mortality, reduced hatching success, and delayed hatching. In a comparative study screech owls were more sensitive than mallards to reproductive effects of endrin yet accumulated less endrin in eggs and tissues than mallards.[66] Eggshell thickness was not affected in either species.

Another example is from reproductive studies with mallards fed selenium. The purpose was to determine the form of selenium, dietary concentration, and concentration in eggs responsible for high embryonic mortality and teratogenicity in aquatic birds exposed to evaporation ponds receiving agricultural drainwater at the Kesterson Reservoir in central California and other western locations. Another facet of the studies was to examine the interaction of selenium and mercury. In summary, the organic form of selenium, selenomethionine, accumulated to levels in eggs comparable to those found in eggs of wild birds in contaminated locations; sodium selenite and selenocystine did not accumulate readily or produce reproductive impairment similar to selenomethionine.[22] Characteristic malformations of the extremities occurred with selenomethionine, and included deformities of the bill, eyes, brain, and feet identical to those found at Kesterson.[67] Threshold concentrations for reproductive impairment from selenomethionine were 4 to 8 ppm in the diet (dry diet) and >3 ppm (wet weight) in the eggs. Studies of selenium and mercury interaction revealed antagonistic effects in adult mallards, but the two substances were synergistic in their effects on the reproductive process.[68] Individually, selenium and mercury lowered hatching success and survival of ducklings; the combination of mercury and selenium further reduced reproduction. Controls produced an average of 7.6 young per female — females fed 10 ppm selenium produced an average of 2.8 young, females fed 10 ppm mercury produced 1.1 young, and females fed both mercury and selenium produced 0.2 young. Furthermore, teratogenic effects were notably increased for the combined treatment; deformities were recorded in 6.1% of the embryos of controls, 16.4% for those fed methylmercury chloride, 36.2% for those fed selenomethionine, and 73.4% for those fed combined methylmercury chloride and selenomethionine.

### 4.4.2 Reproduction Studies: An Alternative Protocol

Several protocols have been used for testing comparative labile chemicals such as anticholinesterase pesticides.[63,69–74] Because of use patterns the initial contact with these pesticides may occur at any time during reproduction but usually not in advance of nesting. Therefore, a shorter exposure period is initiated once the test population is in egg production. Northern bobwhite reproductive tests with organophosphorus insecticides have shown significant effects with treatment periods of 8 days,[70] 10 days,[71] and 3 weeks.[69,72–74] A mallard test of methyl parathion in the diet for 8 days resulted in reduced daily egg production, changes in incubation behavior of hens, and decreased number of hatchlings.[75]

Test methods used in the above studies varied substantially and should be further considered if standardization is desired. There are many potential advantages to using short-term exposure tests including: (1) known layers of fertile eggs can be used, hence reducing variability in test data; (2) pretreatment values for each pen can serve as an additional control; and (3) timing of the test can coincide with maximum egg production. However, the most important advantage may be to mimic actual exposure spikes as they may be encountered in nature.

## 4.5 SINGLE-DOSE AVIAN EMBRYOTOXICITY AND TERATOGENICITY TESTS

Toxic contaminants can enter the natural environment by numerous routes, including air pollution from mobile and stationary sources, pesticide spray applications, or aquatic translocation. All these routes have the potential to contaminate avian eggs, which are well known for susceptibility to direct applications of toxicant.[30,76] Even a parent bird may be the source of contamination. For example, petroleum pollutants carried to the nest on breast feathers, feet, or nesting materials of sea birds caused reduced hatchability of contaminated eggs.[77] Laboratory studies showed that as little as 1 to 10 μL of crude or refined oil topically applied to eggs of various species was embryotoxic or teratogenic. The extent of toxicity was from egg penetration and the aromatic hydrocarbon composition, rather than from blockage of shell pores and interference with oxygen transfer as previously believed.[78,79] Dose-dependent liver necrosis was one of the best indices of embryotoxicity dependent on petroleum composition and stage of development.[80]

A variety of pesticides, including paraquat and parathion, have been shown to be embryotoxic and teratogenic to several avian species. Embryos did not develop normally when eggs were sprayed with or immersed in chemical formulations at concentrations encountered in nature.[76,81]

With this background, routine test methods were developed for embryotoxicity and teratogenicity with mallard eggs. Measurement of embryolethality was modeled after the avian acute toxicity test, that is, the median lethal effect ($LD_{50}$ or $LC_{50}$) is derived by exposure of eggs or embryos to at least three geometrically arranged treatments. Treatment has been by egg immersion ($LC_{50}$) or topical application by spraying or pipetting ($LD_{50}$).

An evaluation of the potential hazard of external exposure of mallard eggs to petroleum, pesticides, and industrial effluents was carried out; tests conducted on more than 70 environmental contaminants indicated that 8 of 30 pesticides tested caused teratogenic effects at exposure levels well below the calculated $LC_{50}$ or $LD_{50}$.[81] Other measurements of embryotoxicity included presence of subcutaneous edema, blisters, hemorrhages, and stunted growth. Studies with different species (e.g., northern bobwhite) but the same classes of pesticides often lead to similar conclusions.[76] Several studies with insecticides and herbicides are summarized below.

The order of toxicity of commercial formulations was determined for 14 insecticides in aqueous emulsion, as reflected by $LC_{50}$ values in mallard eggs.[81] The $LC_{50}$ values for these insecticides ranged from 30 g/L to greater than 600 g/L, and order of toxicity was endrin > sulprofos > parathion > acephate > lindane > temephos > diazinon > dimethoate > toxaphene > malathion > carbaryl, permethrin, phosmet, and methomyl. However, the order of potential hazard in terms of the highest permissible field level of application in the United States was toxaphene > malathion > endrin = dimethoate > lindane > sulprofos > diazinon > parathion > acephate > temephos > carbaryl, phosmet, methomyl, and permethrin. Apparent differences between the ranking based on toxicity and potential hazard were due to differences in maximum permissible levels of application, which, for example, were extremely high for toxaphene and malathion. Insecticides with $LC_{50}$ values that occurred at approximately ten times the highest permissible field level of application or less included dimethoate, endrin, malathion, and toxaphene. Of these, all were teratogenic but only dimethoate and endrin were teratogenic at levels below the $LC_{50}$. Subsequent observations have shown that many of these abnormal embryos are unable to hatch, and this would have lowered the calculated $LC_{50}$ if hatching success had been included.

The order of toxicity of herbicides in aqueous emulsion ($LC_{50}$ values) was paraquat = trifluralin > propanil, bromoxynil with MCPA (Bronate®) > methyl-diclofop > prometon > picloram > 2,4,5,-T > amitrole > glyphosate > 2,4,-D > atrazine, dicamba, and dalapon.[81] However, the order of potential hazard in terms of the highest permissible field level of application in the United States was trifluralin > paraquat > prometon > methyldiclofop = propanil > bromoxynil with MCPA (Bronate®) > picloram > amitrole > 2,4,5,-T > 2,4,-D > glyphosate > atrazine and dicamba. Herbicides with $LC_{50}$ values that occurred at ten times the field level of application or less included bromoxynil with MCPA, methyldiclofop, paraquat, prometon, propanil, and trifluralin.

Embryotoxicity testing has also involved hazard evaluations of aquatic weed control,[82] mosquito control,[83,84] and wildfire control agents.[85] Some examples follow. (1) Diquat dibromide, a bipyridylium compound, is commonly used as an aquatic weed control. Tests with mallard eggs indicated that concentrations of diquat in aqueous solutions as used for aquatic weed control would probably have little impact on mallard embryos. This is, of course, dependent upon the dilution effect of average water depth of the application area. However, concentrations applied above ground to weeds and cattails along ditches could affect the survival and normal development of mallard embryos and presumably other avian species nesting in such habitats.[82]

(2) Golden Bear Oil or GB-1111 (also GB-1313) is a petroleum distillate that is used throughout the United States as a larvicide for mosquito pupae. External application of the product to mallard, bobwhite, and red-winged blackbird eggs reduced hatching success but at different levels of treatment. Hatching was significantly reduced in mallards treated on day 4 or day 11 at three and ten times the maximum field application; the $LD_{50}$ was 1.9 times the maximum field application.[83,84] Hatching success of bobwhite and red-winged blackbirds was only reduced at ten times the maximum field application. Recommended rates of field application of GB-1111 were potentially toxic to mallard embryos, especially under conditions of spray overlap but unlikely to impair the survival or development of bobwhite or red-winged blackbird embryos.

(3) Eggs of nesting birds situated in peripheral areas serving as fire breaks are at risk of being sprayed with fire-control chemicals. Acute toxicity tests were conducted with northern bobwhite quail eggs using different water-based concentrations of Silv-Ex® (S-E), a fire suppressant chemical, and Phos-Chek® G75-F (P-C), a fire-retardant chemical, on day 4 or day 11 of incubation.[85] Mortality appeared higher in most groups exposed on day 11 than on day 4, suggesting that on day 11 the extensive chorioallantoic vascular network permitted greater uptake of chemical. A combination of S-E and P-C was synergistic (202 g/L P-C and 50 g/L S-E) at day 11 of incubation resulting in a large decrease in hatching success. However, lower combined concentrations of S-E (10 g/L or 30 g/L) with 202 g/L of P-C appeared antagonistic. This may be due to S-E, as a surfactant, altering the ability of P-C to penetrate. It was concluded that Phos-Chek G75-F could pose a potential threat to developing galliform species, especially during heavy application in densely wooded areas (the $LC_{50}$ was 220 g/L compared to the standard application concentration of 135 g/L).

Other studies have utilized egg injections as a short cut for initial comparison of relative toxicities of environmental contaminants that could be ingested by birds in nature to an array of avian species.[86–88] For example, the effects of polychlorinated biphenyl (PCB) congeners, PCB 126 and PCB 77, were examined in chicken (*Gallus gallus*), American kestrel, and common tern (*Sterna hirundo*) embryos through hatching following air cell injections on day 4.[87] PCB 126 caused malformations and edema in chickens starting at 0.3 ppb, in kestrels at 2.3 to 23 ppb, but in terns only at levels affecting hatching success (44 ppb). The estimated $LD_{50}$ for PCB 126 in chickens, kestrels, and terns was 0.4, 65 and 104 ppb, respectively, and for PCB 77 it was 2.6 and 316 ppb, respectively, for chickens and kestrels. High concentrations of PCB 126 found in bald eagle eggs in nature are nearly 20-fold higher than the lowest toxic concentration detected in kestrel eggs suggesting potential for toxicity in nature. In contrast, concentrations of PCB 126 causing low-level toxic effects in common tern eggs are comparable to highest levels in common terns and Forster's terns found in nature, suggesting additional involvement of other compounds causing toxicity in areas such as the Great Lakes.

Another example of using egg injections as a short cut for comparing relative embryotoxicities among species involves the evaluation of mercury in the environment, which is especially hazardous to fish-eating birds since it accumulates in the food chain. Therefore, studies were conducted by injecting methylmercury (dissolved in corn oil) into eggs of several species.[88] When mallard eggs were injected, hatching success was 76% for controls, and 56, 62, 53, 44, and 29% for eggs injected with 0.1, 0.2, 0.4, 0.8, and 1.6 ppm mercury, respectively. With white ibis (*Eudocimus albus*) eggs, hatching success was 62% for controls and 10, 25, and 20% for eggs injected with 0.2, 0.4, and 0.8 ppm mercury, respectively. For tricolored herons success was 60% for controls and 10% for eggs injected with 0.4 ppm mercury. For great egrets (*Casmerodius albus*) success was 60% for controls and 0% for eggs injected with either 0.4 or 1.3 ppm mercury. These results indicate that the embryos of some fish-eating birds may be more sensitive to methylmercury than mallard embryos and that estimates of toxic levels of mercury in eggs, based on reproductive trials with mallards, may have to be re-evaluated.

## 4.6 "NEONATAL" TOXICITY TESTING IN ALTRICIAL NESTLING BIRDS

Neonatal mammals are especially susceptible to accumulation and toxicity of many environmental contaminants. With this as a guide, studies with altricial and semi-altricial nestling birds were conducted and have shown that nestlings of different species including American kestrels, European starlings, red-winged blackbirds, herring gulls (*Larus argentatus*), and great egrets are considerably more sensitive to oral ingestion of contaminants, such as heavy metals, herbicides, OP insecticides and PCBs, than are adults of the same species or the young of fully precocial species such as northern bobwhite, mallards, and ring-necked pheasants.[89] Following are examples of the effective use of nestlings of various species in elucidation of contaminant hazard to birds.

### 4.6.1 American Kestrels

Studies have been conducted with American kestrel nestlings that were orally intubated daily for the first 10 days posthatching. Environmental contaminants administered were lead, herbicides (paraquat, nitrofen, bifenox, and oxyfluorfen), and PCB congeners (Table 4.3). Endpoints have included survival; body, organ, and skeletal growth; and blood and organ biochemistry.

Overt toxicity of metallic lead was manifested by 40% nestling mortality at the dose level of 625 mg/kg, with significant impairment of growth occurring at 125 mg/kg.[90] Since nestling kestrels consume approximately their own body weight in food per day, 625 mg/kg of lead daily can be considered equivalent in daily lead intake to approximately 625 ppm lead in the diet, and 125 mg/kg equivalent to 125 ppm in the diet. Lead ingestion was considerably more toxic to nestlings than previously reported for adults and young precocial species, where $LC_{50}$s were greater than 5000 ppm. Biochemical and hematological alterations in nestlings were also more severe than those reported in adult kestrels or precocial young birds exposed to lead.[91]

Paraquat ingestion by nestling American kestrels was also more toxic than previously reported for adults and young precocial species.[92] For northern bobwhite, coturnix, ring-necked pheasants, and mallards, the $LC_{50}$s varied from nearly 1000 ppm to over 4000 ppm after a 5-day feeding trial. In kestrels 60 mg/kg of paraquat resulted in 44% mortality on days 4 to 8 of the experimental treatment period.

Diphenyl ether herbicides, including bifenox, nitrofen, and oxyfluorfen, were examined for developmental toxicity in kestrels. Bifenox was found to be more toxic than reported for precocial species; for ducks and pheasants, the 5-day $LC_{50}$s for bifenox were reported to be greater than 5000 ppm in the diet.[93] Precocial young also appear to be less sensitive than nestling kestrels to nitrofen ingestion. Bifenox at 500 mg/kg caused 66% mortality in nestling kestrels, whereas 500 mg/kg nitrofen caused 100% mortality within 5 days. Levels of bifenox or nitrofen at 250 mg/kg caused 9% mortality within 7 days. In addition to reduced growth and survival, nitrofen caused some

**Table 4.3 "Neonatal" Toxicity Testing of Environmental Contaminants in Nestling Birds**

| Species | Exposure Method | Chemical | Observation Period | Effects | References |
|---|---|---|---|---|---|
| American kestrel | Daily oral | Lead, metallic | Days 1–10 | 525 mg/kg, high mortality; 125 mg/kg, reduced growth; 25 mg/kg, altered physiology | 90, 91 |
| American kestrel | Daily oral | Paraquat | Days 1–10 | 60 mg/kg, high mortality; 10 to 25 mg/kg, reduced growth, altered physiology | 92 |
| American kestrel | Daily oral | Bifenox | Days 1–10 | 500 mg/kg, high mortality, 250 mg/kg, reduced growth, altered physiology | 93 |
| American kestrel | Daily oral | Nitrofen | Days 1–10 | 500 mg/kg, complete mortality; 250 mg/kg, reduced growth; 50 mg/kg, altered physiology | 93 |
| American kestrel | Daily oral | Oxyfluorfen | Days 1–10 | 500 mg/kg, few effects | 93 |
| American kestrel | Daily oral | PCB 126 | Days 1–10 | 50 ug/kg, onset of lymphoid depletion, decreased thyroid content, hepatic necrosis | 94 |
| European starling | Single oral (day 5 or 15) | Dicrotophos | 24 h postdose | Day 5 $LD_{50}$ = 4.9 mg/kg Day 15 $LD_{50}$ = 9.0 mg/kg, reduced growth, brain cholinesterase | 95 |
| European starling | Single oral | Diazinon | 24 h postdose | Day 1 $LD_{50}$ = 13 mg/kg fledgling $LD_{50}$ = 145 mg/kg | 96, 39 |
| European starling | Single oral | Terbufos | 24 h postdose | Day 2 $LD_{50}$ = 2.3 mg/kg fledgling $LD_{50}$ = 61 mg/kg | 39 |
| Red-winged blackbird | Single oral | Diazinon | 24 h postdose | Day 0–3 $LD_{50}$ = 2.4 mg/kg fledgling $LD_{50}$ = 8.3 mg/kg | 39 |
| Red-winged blackbird | Single oral | Terbufos | 24 h postdose | Day 0–3 $LD_{50}$ = 0.4 mg/kg fledgling $LD_{50}$ = 3.3 mg/kg | 39 |
| Herring gull | Single oral (3–4 weeks old) | Crude oil | For 9 days | 0.3 mg/kg reduced growth, altered physiology | 97 |
| Herring gull | Single i.p. injection (day 2 or 6) | Lead nitrate | 19 days of age or older | 100 mg/kg on day 6 was a critical period for disruption of behavior | 99, 100 |
| Herring gull | Single i.p. injection (day 2 or 6) | Chromium nitrate | 50 days of age | 50 mg/kg on day 2 affected growth and 12 of 14 behaviors | 101 |
| Herring gull | Single i.p. injection (day 2 or 6) | Manganese acetate | 50 days of age | 50 mg/kg on day 2 affected growth and 12 of 14 behaviors | 101 |
| Black guillemot | Single oral | Crude oil (weathered) | For 22 days | 0.1–0.2 mL reduced growth, altered physiology | 98 |
| Great egret | Dosed orally from 1 to 14 weeks | Methylmercury chloride | Through 14 weeks of age | 0.5 mg/kg affected immune system and behavior; 5 mg/kg caused severe ataxia and neural lesions | 103–105 |

hepatotoxicity, including an increase in liver weight relative to body weight with significant increases in activities of several plasma enzymes. Hepatic tissue GSH peroxidase activity was significantly higher in all nitrofen-treated groups and in the two treated with the highest doses of bifenox (50 mg/kg and 250 mg/kg). Other alterations included an increase in plasma total thyroxine concentration (T4) by nitrofen, which may indicate thyromimetic activity, as suggested for mammals.

The developmental toxicity of the planar PCB 126 was also studied in the posthatching kestrel as a model for the eagle.[94] Nestlings were dosed with 5 μL/g body weight of corn oil (controls) or PCB 126 at concentrations of 50, 250, or 1000 ng/g body weight. Dosing with 50 ng/g of PCB 126 resulted in a hepatic concentration of 156 ng/g wet weight, liver enlargement and mild coagulative necrosis, over tenfold increases in hepatic microsomal ethoxyresorufin-O-dealkylase and benzyloxyresorufin-O-dealkylase, and a fivefold increase in methoxyresorufin-O-dealkylase. At this dose mild to moderate lymphoid depletion of the spleen was apparent, as were decreased follicle size and content of the thyroid. At 250 ng/g concentration of PCB 126 in the liver was 380 ng/g, with increasing multifocal coagulative necrosis, decreased bone growth, decreased spleen weight with lymphocyte depletion of the spleen and bursa, and degenerative lesions of the thyroid. At 1000 ng/g the liver concentration was 1098 ng/g, accompanied by decreased bursa weight, decreased hepatic thiol concentration, and increased plasma enzyme activities (ALT, AST, and LDH-L), in addition to the previous effects. Highly significant positive correlations were noted between liver concentrations of PCB 126 and the ratio of oxidized to reduced glutathione. These findings indicated that nestling kestrels are more susceptible to PCB 126 toxicity than are adults but less sensitive than embryos.

### 4.6.2 European Starlings and Red-Winged Blackbirds

Acute oral $LD_{50}$s for European starlings were determined for nestlings from hatching until fledging. The $LD_{50}$ of dicrotophos for free-living 5-day-old nestlings was about half that obtained for 15-day-old nestling and adults.[95] Brain cholinesterase activity was depressed by 74 to 94% in all birds that died, but neither the degree of inhibition nor the baseline cholinesterase activity varied with age. In another study of anticholinesterase pesticides to starlings newly hatched young were 20 times more sensitive to a single dose of diazinon than were fledglings of about 21 days of age.[96] Nestling sensitivity was supported by decreased plasma and brain cholinesterase activities. Acute age-dependent toxicity of terbufos and diazinon was evaluated in nestling starlings and red-winged blackbirds.[39] In brief, for starlings, $LD_{50}$s for turbufos increased from 2 days of age to 9 days by nearly ninefold and by 26-fold at 19 days and fledging; whereas for diazinon the increases between 2 days and 9 and 19 days were 7- and 11-fold. In comparison, blackbird nestlings were tested at 0 to 3 days and about halfway to fledging (8 to 11 days). The $LD_{50}$ increased during this period by ninefold for turbufos and threefold for diazinon. At both ages the red-winged blackbird was substantially more sensitive to turbufos and diazinon than was the European starling. Baseline cholinesterase activities in both brain and plasma increased with age in both species and may have contributed to the changes in sensitivity between hatching and fledging. It was also noted that fledgling starlings (19 days) were about three to four times more sensitive than adults to turbufos and diazinon.

### 4.6.3 Herring Gulls and Black Guillemots

Nestling birds are quite sensitive to exposure of petroleum hydrocarbons containing an aromatic fraction. For example, a single small oral dose of Kuwait Crude Oil (KCO) or South Louisiana Crude Oil (SLCO) at approximately 0.3 mL/kg body weight caused reduced growth, osmoregulatory impairment, and hypertrophy of hepatic, adrenal, and nasal gland tissue in herring gull nestlings living in a simulated marine environment.[97] Weathered SLCO caused similar effects in black guillemot (*Cepphus grylle*) nestlings in nature.[98] This suggests that ingestion of oil by nestlings could reduce a fledgling's ability to survive at sea.

Another example is a series of studies with herring gull chicks designed to examine the relationship between dose, tissue level, and response to lead.[99,100] These studies showed that lead affected neurobehavioral development at critical periods. Seventy-two 1-day-old herring gull chicks were randomly assigned to six treatment groups to receive a lead nitrate concentration of 100 µg/g at age 2 days or at age 6 days, a similar cumulative dose evenly divided on days 2, 4, and 6, or matched-volume saline injections on the same days. Behavioral tests were performed (some at 2- and others at 5-day intervals) to examine locomotion, balance, righting response, thermoregulation, and visual cliff. Overall, the data showed that the lead-6 group was more affected by the dose than the other groups, suggesting that 6 days of age may be a more critical period than earlier ages for some behaviors.

The same researchers also examined the effects of chromium and manganese on early neurobehavioral development in herring gulls.[101] Each of 36 two-day-old herring gull chicks was randomly assigned to one of three treatment groups to receive chromium nitrate (50 mg/kg), manganese acetate (50 mg/kg), or a control dose of sterile saline solution. Behavioral tests examined food begging, balance, locomotion, righting response, recognition, thermoregulation, and perception. There were significant differences in begging behavior by 5 days postinjection, and there were significant differences in weight gain throughout development until 50 days of age, when the experiment was terminated. Behavioral tests, administered from 18 to 48 days postinjection, indicated differences between control and the exposed groups for time to right themselves; thermoregulation behavior; and performance on a balance beam, inclined plane, actual cliff, and visual cliff. Of the 14 behavioral measures with significant differences, control birds performed best on 12.

### 4.6.4 Great Egrets

The effects of mercury were studied on captive great egret nestlings, which were either maintained as controls or were dosed from 8 days of age for 13 weeks with 0.5 or 5 mg methylmercury chloride/kg wet weight in fish.[102] Low-dosed birds were given methylmercury at concentrations comparable to exposure of wild birds in the Florida Everglades. Subacute toxicity was indicated for birds dosed with 5 mg/kg after 9 weeks. Growing feather concentrations of mercury were closely correlated with cumulative mercury consumed per unit of body weight. After 8 weeks of exposure, appetite and weight index (weight/bill length) declined significantly in the high-dose group, with the same response noted a week later in the low-dose group.

Other effects indicative of mercury toxicity in low-dosed birds included lower packed cell volumes, dingy feathers, increased lymphocytic cuffing in a skin test, increased bone marrow cellularity, decreased bursal wall thickness, decreased thymic lobule size, fewer lymphoid aggregates in lung, increased perivascular edema in lung, and decreased phagocytized carbon in lung.[103] High-dosed birds became severely ataxic and had severe hematologic, neurologic, and histologic changes. The most severe lesions were in immune- and nervous-system tissues. Manifestations of oxidative stress and elevated plasma enzyme activities were also apparent.[104] Comparison of responses in captive and wild birds indicated that sublethal effects of mercury occurred at lower levels in captive than in wild birds. This may be due to the reduced sources of variation characteristic of the highly controlled laboratory study. Conversely, thresholds for more severe changes (death, disease) occurred at lower concentrations in wild birds than in captive birds, possibly because wild birds were exposed to multiple stressors. Thus, caution should be used in applying lowest observed effect levels between captive and wild studies.

As an integral part of the above study, behavioral effects, including activity levels, maintenance behavior, and hunting behavior, were measured.[105] Mercury affected activity and maintenance behavior. Birds dosed with 5 mg/kg in fish (as above) became severely ataxic (lost muscle coordination) and were euthanized by 12 weeks of age. The low-dosed birds exhibited less tendency to hunt fish. Therefore, even at 0.5 mg/kg mercury concentration in the food there were significant effects of methylmercury on activity, tendency to seek shade, and motivation to hunt prey.

## 4.7 AVIAN TERRESTRIAL FIELD STUDIES

In the United States the EPA no longer requires terrestrial field studies for pesticide registration except under extremely compelling circumstances; however, protocols are available for limited testing or if the requirements are reinstated (Table 4.1). In place of the field study the U.S. EPA's ecological risk-assessment process uses laboratory toxicity data in quotient indices to characterize risks to wildlife. However, these data sometimes fall short of predicting actual field effects.[106,107] For example, laboratory and field results of the toxicity of azinphos-methyl were compared with northern bobwhite.[106] Chick survival, brain cholinesterase activity, and growth in the field exposure study were significantly different from equivalent exposures in the laboratory, and temporal patterns of effects differed between field and laboratory. It was concluded that the effects observed in the field differed from that predicted by risk quotients because the quotient method did not consider alternate routes of exposure, behavioral responses, influence of spatial and temporal environmental variability, or indirect effects.

Avian die-offs due to anticholinesterase insecticide exposure include incidents varying from small-scale poisonings of only a few birds in a barnyard or on a golf course to massive die-offs with at least several hundred colonial breeding birds or migrants.[15] The dimension of this problem is indicated by a review of raptor mortality in the United States, the United Kingdom, and Canada for the period 1985 to 1995; of 520 incidents evaluated, the largest number of poisonings were attributed to anticholinesterase pesticides.[18] Anticholinesterase poisonings have also involved many other species, ranging in size and diversity from American robins and warblers (*Vermivora* spp.) to Canada geese and great blue herons (*Ardea herodias*).[29] Most of these were highly visible situations, and undoubtedly there are a multitude of undetected incidents for every one discovered. In a recent review on factors influencing estimation of pesticide-related wildlife mortality, it was concluded that most effects on wildlife are not observed, much of observed mortality is not reported, and the actual number of affected animals per mortality event typically exceeds the number recovered.[107] Therefore, larger-than-expected net losses could occur because of such cumulative events, but the real impact on bird populations is not known. Secondary poisoning of predatory birds and other carnivorous animals consuming prey poisoned by anticholinesterase pesticides is also an important contributing factor to wildlife mortality that had been generally disregarded prior to 1980.[45,108–110]

### 4.7.1 Prerequisites for Testing

In assessing the need for required terrestrial field studies under FIFRA, the U.S. EPA considers the following prerequisites: (1) environmental concentration of the substance, which must exceed the lowest-effect level eliciting a biological response; (2) chemical properties of the pesticide (e.g., persistence, toxic metabolites and degradates, retention on food); (3) intended use pattern (areas and species likely to be exposed, treatment intervals); (4) margin between lowest-effects level and estimated environmental concentration; and (5) dose-response slopes in laboratory studies.[34,111] Field studies are designed to evaluate survival and reproductive success of nontarget wildlife species under actual conditions of use of the pesticide. Potential outcomes of pesticides in field tests can include (1) direct poisoning by ingestion, dermal, or inhalation exposure; (2) sublethal effects indirectly causing death by reducing resistance to natural environmental stresses such as disease, weather, or predators; (3) altered behavior such as abandonment of nests or young and change in parental care; (4) habitat alteration that results in reduced food resources or greater vulnerability to predators; and (5) reduced productivity.

### 4.7.2 Types of Field Studies

Field studies consist of two types: (1) *screening field studies* to evaluate whether impacts are occurring and (2) *definitive field studies* to estimate the magnitude of the impact.[111] Generally, species representative of areas where pesticide applications occur are utilized. In most instances screening studies monitor for overt signs of toxicity such as mortality or aberrant behavior and changes in biochemical and histological indicators of toxicity. Carcass searches, radiotelemetry, depression of cholinesterase in the event of anticholinesterase pesticide exposure, residue analysis, behavioral observations, and population parameters may all be components of screening field studies.

With definitive field studies, mark-recapture techniques and radiotelemetry are often used to monitor survival and behavior. Extensive monitoring of reproduction and survival of dependent young is often required. Active nests are periodically monitored at the study site for number of eggs laid, hatched, young fledged, and nest abandonment. Sometimes artificial nest structures are provided to increase nest sites. Incubation behavior and parental care are sometimes monitored. A further indicator of reproductive effects may include comparison of young-adult ratios between treated and untreated plots. A number of pesticide–related field studies using different resident avian species are summarized below.

#### 4.7.2.1 Sage Grouse (*Centrocercus urophasianus*)

Avian studies have focused on a series of die-offs of sage grouse associated with increased agricultural applications of organophosphorus insecticides to irrigated meadows and cropland.[112] Sage grouse in Idaho were radio-tagged, and their brain cholinesterase activity was monitored over several seasons; nearly 20% of the birds died or were seriously affected by organophosphorus insecticides. Mortality was higher among juvenile grouse than adults. Since these studies deal with a species that is both vulnerable and nonmigratory (which provides ease of tracking), there is high probability that such studies will further elucidate the impact of pesticide usage on population dynamics. At present, sage grouse in several parts of the western United States are being considered for listing under the endangered Species Act.

#### 4.7.2.2 Prairie Pothole Waterfowl Studies

The prairie potholes region of the northern plains of the United States and Canada provides breeding habitat for more than 50% of North American waterfowl production. Drainage of prairie wetlands for agricultural purposes has been intense; only 35% of the original wetland area remains.[113] The potential for agricultural chemicals to enter the remaining wetlands and impact wildlife is substantial. Many of the most widely used organophosphorus and carbamate insecticides within the region have been found to be highly toxic to aquatic animals. Both direct (mortality and toxicity) and indirect (loss of invertebrate food items) effects were greater than expected when parathion or methyl parathion was applied by aircraft according to label instructions and county agent recommendations.[113–115] The use of tank mixture combinations of insecticides and of herbicides during spraying further contributes to the toxicity. Of 16 widely used insecticides in North Dakota, nine were implicated in wildlife mortality. In addition, 13 of these were either highly toxic to aquatic invertebrates or vertebrates. Reduced availability of aquatic invertebrates as food for ducklings and egg-laying females affected waterfowl productivity.

#### 4.7.2.3 Passerine Studies

Aerial spraying for western budworms (*Choristoneura occidentalis*) with relatively low volume trichlorfon and carbaryl in Montana forests did not alter the success of nests with eggs or young birds of various species.[116] However, application rates for these two insecticides were lower than normal. In another study the effects of pesticides on mourning doves (*Zenaida macroura*) nesting in orchards in southern Illinois were evaluated.[117] The pesticides usually consisted of a combination of an insecticide and a fungicide and were sprayed at intervals of approximately 10 days from March through September. Most adult doves incubating eggs left during the actual spraying application but all returned within 30 min. Nearly 90% of the nests that contained eggs during spraying were unsuccessful, with most failures occurring four or more days postspray, suggesting direct embryotoxicity from contact and penetration of pesticides on eggs.

Reproductive effects were assessed for multiple and varied organophosphorus and carbamate operational exposures on avian productivity.[118] Nest, egg, and nestling daily survival rates (DSRs) were determined for northern mockingbirds (*Mimus polyglottos*), brown thrashers (*Toxostoma rufum*), and northern cardinals (*Cardinalis cardinalis*) nesting along edges of pecan orchards and row crops in southern Georgia (United States). Egg and nestling DSRs for all species combined were negatively correlated with exposure. Nestling growth was reduced with increasing exposure. Brain cholinesterase activities were age-dependent and substantiated adult, but not nestling, exposure. The authors concluded that increasing exposure to operational pesticide use may reduce songbird productivity.

Insecticide hazard to breeding birds in Christmas tree plantations in Quebec were assessed by examining potential deleterious effects of three insecticides (i.e., dimethoate, diazinon, and insecticidal soap) on American robins and song sparrows (*Melospiza melodia*).[119] Cases of complete or partial mortality were recorded in nests. Abandonment of nests and egg infertility were ruled out as possible causes of mortality. No mortality was recorded for broods exposed to the insecticidal soap. The cases of total mortality observed in broods of both species exposed to dimethoate were similar to those recorded for control nests (18 and 25% vs. 14 and 21%, respectively). However, among robin and sparrow nestlings exposed to diazinon, about twice as many cases of total mortality (31 and 38%) were recorded than for the control nests. The authors concluded that American robin eggs are sensitive to diazinon and dimethoate, particularly when spraying is carried out early in the incubation stage. However, for song sparrows, it is mainly the nestlings that succumb after diazinon is sprayed on them or when dimethoate applications are made during incubation.

Organophosphorus insecticides were studied through examination of cholinesterase activity in tree swallows (*Tachycineta bicolor*) and eastern bluebirds (*Sialia sialis*) in apple orchards in Ontario, Canada treated with azinphos-methyl, diazinon, phosalone, or phosmet.[120] In nestlings, brain cholinesterase activities obtained postspray often fell below predicted activities calculated from control siblings. This trend was especially apparent in the younger nestlings, less than 6 days old. However, for bluebirds, the rate of increase of brain cholinesterase with age in nestlings from treated sites was lower than in nestlings from control sites. Results of depressed cholinesterase levels in tree swallows and eastern bluebirds inhabiting apple orchards were consistent with those in avian species in other orchard monitoring studies. However, there was no indication that organophosphorus exposure due to agricultural spraying in apple orchards adversely affected the survival of the birds monitored. Reproduction of tree swallows and eastern bluebirds was further assessed in pesticide-sprayed apple orchards in southern Ontario to evaluate egg fertility, clutch size, egg and chick survival, and pesticide exposure.[121] In this study of cavity-nesting birds reproduction was compared for nest boxes in sprayed and nonsprayed apple orchards from 1988 through 1994. There was a significant increase in unhatched eggs in bluebirds, as organochlorine levels increased in eggs. There were significant associations between toxicity scores of current-use pesticides and at least one avian reproductive parameter in every year of the study, but the reduction in reproductive rates associated with pesticides did not exceed 14%, for either species, in any year. Reduced reproduction occurred in six years for tree swallows, but for bluebirds this occurred in only four years.

Increased spray drift was suspected to be a factor of potential concern experienced by nestling and adult great tits (*Parus major*).[122] Nest boxes were placed in hedgerows bordering fields sprayed with pesticides (pirimicarb or dimethoate). One hedge was sprayed directly with pirimicarb to simulate maximum drift effect. Two hedges were left untreated to serve as control areas. Significant inhibition of blood plasma cholinesterase activity was detected within 24 h in adult birds exposed to drift of dimethoate and in adult birds from the hedge sprayed directly with pirimicarb. Inhibition of nestling plasma cholinesterase activity was found in all treated hedges after 24 h. A tendency toward reductions in weight gain, though not significant, was found in nestlings both between 0 to 24 h and 24 to 48 h after treatment in all the treated hedges compared to nestlings from the control hedge.

Studies with European starlings have provided a valuable avian model for field testing because starlings readily occupy artificial nest boxes installed in test spray fields.[95,123] Nest boxes provide a large synchronous breeding population of a passerine species at treatment sites. Starlings consume soil invertebrates, which come in direct contact with pesticides in the soil. Using this technique, the effects of application of methyl parathion at 1.4 kg active ingredient (a.i.)/ha. to a cultivated field were examined, and it was determined that there was depression of brain cholinesterase in adults and nestlings, and the number of nestlings fledging from treated fields was reduced compared with the control field.[124] In a similar study with the same insecticide control birds had over 50% successful nests, whereas those in a field treated with 2.5 kg a.i./ha. of methylparathion had only a 20% success rate; fledglings were found to be more susceptible to postfledgling mortality when nestlings from each field were radio-collared and followed for 2 weeks after leaving the nest.[96]

## 4.8 AVIAN BEHAVIORAL TOXICITY TESTING

Behavioral aberrations in wildlife can be manifested at one or two orders of magnitude below lethal levels and therefore can be regarded as sensitive toxic-response indicators.[29] In the laboratory subtle alterations in behavior have been associated with exposure to toxic substances. Peakall[125] reviewed behavioral responses of birds to pesticides and other contaminants and concluded that certain operant tests are relatively simple and reproducible, but that other more complex tests, such as breeding behavior and prey capture, are probably more relevant to survival in the wild. In field studies the presence of interacting environmental factors, such as ambient temperature and weather conditions, complicate establishing a cause-and-effect relationship. The best-documented field cases of behavioral aberrations in wildlife include exposure of birds to anticholinesterase insecticides and of fish-eating birds in the lower Great Lakes to organochlorines. Both of these classes of chemicals have affected reproduction by causing decreased nest attentiveness.[126–128]

Laboratory and pen studies have documented changes in mallard hen and brood behaviors in response to anticholinesterase pesticides. Methyl parathion caused broods to mostly preen and loaf on land, while control broods primarily fed and swam in open water.[129] Methyl parathion also affected incubation behavior, causing nest abandonment and decreased nest attentiveness.[70] Several other studies have demonstrated increased vulnerability of birds, including sharp-tailed grouse (*Tympanuchus phasianellus*), northern bobwhite, and European starlings, to predation following organophosphorus insecticide exposure and field release[130,131] or increased susceptibility to experimental predation by a domestic cat.[132] Evidence has also been developed indicating that migratory orientation may be affected in captured migratory white-throated sparrows (*Zonotrichia albicollis*) exposed to dietary acephate for 14 days.[133] Adult sparrows did not establish a preferred orientation, but juveniles displayed a seasonally correct southward migratory orientation. It was hypothesized that acephate produced aberrant migratory behavior by affecting the memory of the migratory route and wintering ground.

Two laboratory tests of natural behavior that may be performed on the same subject consist of response of newly hatched precocial birds to the maternal call (measured time a chick takes to approach a recorded call) and avoidance of a fright stimulus. Both tests are conducted in partitioned runways, permitting several chicks to be tested simultaneously with recorded responses. These tests showed that mallard ducklings from parents fed as little as 0.5 ppm methylmercury were less responsive to maternal calls and more responsive to a fright stimulus than were controls.[134] In another study opposite responses were observed for ducklings whose parents were fed 3 ppm DDE.[135] In a test of a carbamate insecticide, decreased duckling approach-response behavior was decreased following exposure to carbofuran-sprayed vegetation.[136]

Tests of learning ability based on operant conditioning appear to be sensitive indicators of toxicant-induced effects in birds. Recent hatchlings are fed low dietary concentrations of toxicant for several months and are then trained via hunger motivation to peck a lighted key in the conditioning box for a food reward. After the correct pattern is learned, the pattern is reversed or changed, and the ability to adjust is measured. This technique has been applied to test responses at dietary concentrations of less than 1.0 ppm endrin, 10 ppm toxaphene, and up to 100 ppm paraquat in northern bobwhite.[137,138] Learning impairment was caused by endrin and toxaphene, but not by paraquat.

### 4.8.1 Time-Activity Budgets

Time-activity budgets are important aspects of behavioral ecology that aid in understanding habitat utilization and energy consumption. Such measurements were used effectively in evaluation of the hazards to mallard ducklings of toxic drainwater components and lead-contaminated sediments. In one study environmentally reasonable concentrations of either boron or arsenic in the diet affected the activity schedules of developing ducklings, including increased time at rest and selection of supplementary warmth, with less time in alert behavior and in the water compared to controls.[139] In another study the incidence and duration of ten behaviors (resting, standing, moving, drinking, dabbling, feeding, pecking, preening, bathing, and swimming) were recorded. Consumption of diets containing 12 or 24% lead-contaminated sediment (3449 ug/g lead) affected the proportion of time spent swimming but did not affect any of the other recorded behaviors.[140] However, there were signs of impaired balance and mobility and effects on the brain due to lead accumulation (e.g., oxidative stress, altered metabolites, and decreased brain weights).[141]

The behavior of captive great egret nestlings was evaluated in a study of methylmercury.[105] Birds were randomly divided into a control group and groups that received 0.5 or 5 mg methylmercury chloride per kg of food between 12 and 105 days of age. Activity levels, maintenance behavior, and foraging efficiency were studied. During the postfledging period there were no differences between low-dosed and control birds in time required to capture live fish in pools or in efficiency of capture. However, the methylmercury affected their activity, tendency to seek shade, and motivation to hunt prey. Alterations in the allocation of energy in developing ducklings and egret chicks as seen above would have obvious drawbacks in the natural environment including avoidance of predators and foraging strategies. Birds dosed with 5 mg/kg became severely ataxic and were euthanized by 12 weeks of age.

### 4.8.2 Critical Periods of Development

The temporal effects of lead were evaluated on developing herring gull chicks.[99,100] In these studies, 1-day-old gull chicks were placed into treatment groups to receive a lead acetate dose on day 2 (50 or 100 μg/g), on day 6 (100 μg/g), or on day 12 (50 or 100 μg/g); controls received saline injections on the same days. Behavioral tests were performed at 2- to 5-day intervals to examine locomotion, balance, righting response, thermoregulation, and visual cliff. Flight behavior was examined at fledging. Righting response and balance were disrupted immediately after exposure, regardless of the timing of exposure. Thermoregulatory, visual cliff, and individual recognition behavior were more affected by exposure at 2 to 6 days, and there was little effect with exposure at 12 days. Overall, the data showed that treatment at 6 days of age may be a more critical period than earlier ages for some behaviors. However, chicks treated with single doses of lead acetate solution (100 mg/kg) at day 2 experienced disrupted sibling recognition through fledging at 26 days of age. In nature lead-impaired chicks might be unable to use siblings as a cue to find their nests and could experience higher mortality from territorial adults and chicks as well as from cannibalistic adults. Feathers of some roseate terns (*Sterna dougallii*), herring gulls, and black skimmers (*Rynchops niger*) contain lead in concentrations that have been experimentally correlated with behavioral impairment and growth retardation.

### 4.8.3 Food Discrimination and Feeding Behavior

Behavior studies have demonstrated the effects of contaminants on food discrimination.[142–146] Procedures for evaluating the potential ability of birds to avoid chemically contaminated food have been published.[144–146] The discrimination threshold is defined as the dietary concentration above which test animals will decrease the proportion of treated food they consume if alternative untreated foods are available. The field conditions under which wildlife species could utilize the ability to detect and avoid toxic foods are largely unknown. However, when alternative food choices exist, vulnerability to poisoning in northern bobwhite chicks by organophosphorus and carbamate insecticides can be reduced by the number and relative abundance of choices as well as by the bird's ability to detect the chemical.[146]

Low-grade exposure to organophosphorus insecticides may produce long-term changes in bird feeding behavior.[147] This was demonstrated through tests for conditioned taste aversion in a series of field experiments in which independent replicates were large numbers of breeding territories of red-winged blackbirds. Birds freely consumed untreated insect prey offered to them in control territories, but those in treated territories consumed up to three meals of prey tainted with parathion and then avoided offered prey in the treated territories long after parathion tainted prey were no longer present. This long-term change in feeding behavior was produced by organophosphorus in amounts insufficient to induce signs of toxicity or to depress brain cholinesterase activity. The effect was long-term because, unlike noxious repellency, conditioned taste aversion induced by trivial amounts of parathion denied birds the opportunity to discriminate between tainted and untainted prey. Although birds may be spared repeated exposure to the toxic substance, continued avoidance of untainted prey disrupts foraging, endangers breeding efficiency, and reduces predation upon pest insects.

Other behavioral studies have been conducted for assessing and reducing avian risk from granular pesticides. The extant information regarding bird response to grit and granule characteristics (i.e., granule carrier type, color, size, shape, and surface texture) and pesticide load per granule has been summarized.[52] When the efficacy of eight taste repellents for deterring the consumption of granular insecticides in northern bobwhite was evaluated, the most effective were d-pulegone and quinine hydrochloride.[53] The authors concluded that treating pesticide granules with a potent taste repellent, such as d-pulegone, is a promising approach to reduce the risk of their ingestion by birds.

## 4.9 MAMMALIAN WILDLIFE TOXICITY TESTING

Fewer laboratory studies have been conducted with mammalian than avian wildlife, most likely due to the fact that the human health effects literature and agricultural nutrition literature is abundant with studies conducted with laboratory rodents as well as other species of domestic mammals that are viewed by some as surrogate species for mammalian wildlife.

Toxicity testing with mammalian wildlife for regulatory purposes has been limited in scope and requirement compared to avian wildlife toxicity testing (Table 4.1). Much of the U.S. EPA's FIFRA mammalian toxicity data is derived from routine studies with the laboratory rat (*Rattus norvegicus*) for pesticide registration. However, if the margin of safety appears small, or if the likelihood is high that specific mammals of concern will be exposed, then additional and more ecologically relevant tests may be required. This next level of testing requires a dietary $LC_{50}$ study or acute oral $LD_{50}$ study with a nonendangered and representative species that is likely to be exposed (e.g., microtine rodent). Occasionally, other reproductive and secondary toxicity tests are mandated.

The U.S. EPA has been encouraged to expand its required mammalian studies to include a wild herbivore test species (a microtine rodent such as the meadow vole, *Microtus pennsylvanicus*), an omnivore (the deer mouse, *Peromyscus maniculatus*), and a carnivore (the mink, *Mustela vison*, or european ferret, *Mustela putorius furo*), as recommended by Ringer.[148] The mink may be a preferred carnivore because it (1) is indigenous to North America, (2) can be reared in the laboratory, (3) has a large biological database, and (4) is among, if not is, the mammalian species most sensitive to PCBs, PBBs, hexachlorobenzene, TCDD, and aflatoxins.[149–155] However, it has been argued that the mink is no more sensitive than laboratory rodents to many other chemicals including DDT, dieldrin, and *o*-cresol.[148,155–157] The use of wild mammals has also been proposed for setting water quality criteria for those species that consume aquatic life including the mink as a piscivorous species, the northern short-tailed shrew (*Blarina brevicauda*) as an insectivorous consumer of aquatic invertebrates, and the deer mouse as an omnivorous species.[158]

Three published protocols describe guidelines for testing mink and European ferrets for conducting dietary $LC_{50}$ and reproductive toxicity tests and for assessing the primary vs. secondary toxicity of test substances.[156] For the $LC_{50}$ test, the dietary exposure period is 28 days, over which signs of toxicity and mortality are recorded. The reproductive protocol is designed to evaluate dietary exposure prior to and during the breeding period and through gestation and lactation. The main endpoints include adult survival, oogenesis and spermatogenesis, reproductive indices, embryo and fetal development, and offspring growth and survival. The third protocol compares the dietary toxicity and lethality for the parent compound of a test substance (primary toxicity test) with the same substance fed at identical concentrations but contained in animal tissue (prey) contaminated by previous exposure (secondary toxicity test). Secondary toxicity testing of Aroclor 1254 revealed enhanced toxicity of the metabolized form in mink.[159] PCB levels in wild mink on Lake Ontario are similar to those reported to cause reproductive problems in controlled-feeding studies, and correlations between organochlorine levels in fish and levels in mink and otter (*Lutra canadensis*) are apparent.[149–160]

In a comparative toxicity study of potential mammalian models, $LD_{50}$ and 5-day $LC_{50}$ tests were conducted in the same laboratory. The laboratory mouse (*Mus musculus*), the meadow vole, and the white-footed mouse (*Peromyscus leucopus*) were equally sensitive to the organophosphorus insecticide acephate.[161] A database of tests conducted at a single laboratory has been developed for acute oral toxicities of 933 chemicals to deer mice and house mice.[162] These were first-line screening tests for discovery of potential economic poisons for use in invertebrate pest control; the authors did not test for, nor offer any opinion on, which species was the best model for regulatory purposes. In one effort to reconcile the need for wild mammal testing the U.S. EPA sponsored a study comparing oral $LD_{50}$ and 30-day dietary $LC_{50}$ tests with four members of the genus *Microtus*.[163] The results were compared with published values for standard tests of laboratory rats and mice. Though the comparison was limited to ten widely different pesticides, the authors concluded that laboratory rodents were generally more sensitive than voles to the compounds tested.[163]

Factors such as food and habitat preference may affect routes and degree of exposure in the field, thereby rendering some species of wild rodents ecologically more vulnerable to certain contaminants. In studies with microtine rodents aversion to carbofuran-treated feed was associated with delays in the time to first breeding, whereas a female-biased sex ratio in offspring of breeding pairs receiving paraquat in the diet was apparent.[164] White-footed mice exposed to PCBs in the diet at 10 ppm through the second generation exhibited poor reproductive success in comparison with second generation controls and the parental generation. PCB-treated young were significantly smaller at 4, 8, and 12 weeks of age.[165] Other studies have suggested inhibition of reproduction and changes in liver, spleen, adrenal, and testis function at a PCB-contaminated field site for this species.[166] Experimental feeding studies with lead have revealed mortality and impaired postnatal growth in young bank voles (*Clethrionomys glareolus*) when their mothers received lead-contaminated food after giving birth.[167] Though wild mammals have not been used extensively in laboratory studies with environmental contaminants, they have been used as monitors of environmental contaminants in nature, especially in studies of metals, organics, or radionuclides.[168–170]

In other situations studies have been conducted with unique species of wild mammals with varied sensitivities to certain classes of contaminants.[171] A review of the recent book *Ecotoxicology of Wild Mammals* by Shore and Rattner[171] clearly indicates the need for more controlled comparative laboratory studies with unique wild mammals. For example, much of the ecotoxicological research of contaminant effects on insectivorous mammals has focused on terrestrial shrews, but little is known about contaminant effects on water shrews, moles, or hedgehogs.[172] Fish-eating marine mammals, including seals, occupy high trophic levels in the aquatic food chain and accumulate high levels of contaminants including polychlorinated biphenyls, polychlorinated dibenzo-p-dioxins, and polychlorinated dibenzofurans. Such chemicals have been found to be immunotoxic at low doses in studies with laboratory mammals and mink. Recent associations have been established between such contaminants and effects on the immune system, e.g., increased incidence of disease was noted for certain free-ranging seal population; laboratory studies with captive harbor seals have confirmed these findings.[173] On the basis of these and other studies it was concluded that complex mixtures of environmental contaminants may represent a real immunotoxic risk to free-ranging marine mammals in many areas of Europe and North America.[173]

## 4.10 AMPHIBIAN AND REPTILE TOXICITY TESTING

Over the past decade widespread population declines of amphibians have been documented in North America, Europe, Australia, and Central and South America. Population declines in eastern Europe, Asia, and Africa have also been suggested but are not as well documented.[174–177] Contaminants may be involved with amphibian population declines including their possible interaction with other factors, as discussed in Chapter 40 of this book and in a thorough review of the literature as provided in the recent book, *Ecotoxicology of Amphibians and Reptiles,* by Sparling, Linder, and Bishop.[177] In the past it was presumed that tests conducted on fish, birds, and mammals would be sufficiently conservative to protect amphibians and reptiles. This concept can no longer be supported. Comparative toxicities of organic compounds and metals between amphibians and fish have been summarized for a standard embryo larval test (exposure from fertilization through 4 days posthatching).[178]

Fish species commonly used in toxicity tests, such as the rainbow trout (*Oncorhynchus mykiss*), fathead minnow (*Pimephales promelas*), and largemouth bass (*Micropterus salmoides*), were used for comparisons that were standardized as much as possible including comparable life stages, water chemistry, and durations. These tests included 28 species of native amphibians from the families of Ambystomatidae, Microhylidae, Hylidae, Ranidae, and Bufonidae and the African clawed frog (*Xenopus laevis*). Median lethal toxicity values for metals in amphibians varied by 100-fold. In all, 50 metals and inorganics as well as 13 organic compounds were tested, for a total of 694 comparisons of amphibians and fish. In summary, amphibians had lower $LC_{50}$ values than fish in: (1) 64% of all the tests, (2) 74% of the comparisons among the 15 most sensitive amphibian species and fishes, (3) 80% of the comparisons involving amphibians and warm-water fishes, (4) 66% of the 13 most metal-sensitive amphibians and rainbow trout, and (5) 74% of all amphibian species vs. the fathead minnow. The overall conclusions were that there was great variation among amphibian species in their sensitivity to metal and organic contaminants, that amphibians generally were more sensitive than fish, and that water quality criteria established for fish may not be protective of amphibians.

The U.S. EPA drafted a guideline (draft revised FIFRA Guidelines Document-Subdivision E, March 1988) with several acceptable protocols for preregistration testing of chemicals for acute lethal toxicity to amphibians and reptiles.[179,180] The provisional species of choice are frog tadpoles (*Rana spp.*) and adult green anole lizards (*Anolis carolinensis*). There is also a U.S. EPA testing guideline for a "tadpole/sediment subchronic toxicity test" under aquatic testing guidelines of the series "850-Ecological Effects Test Guidelines." This guideline is used to develop data on the subchronic toxicity of chemical substances and mixtures of chemicals sorbed to natural sediments to bullfrog tadpoles. Here, test chambers are filled with appropriate volumes of dilution water (control) or appropriate amounts of contaminated natural sediments and dilution water. If a flow-through test is performed, the flow of dilution water through each chamber is adjusted to the rate desired. This toxicity test may be performed by either of two methods: (1) dosing the tadpoles orally with a sediment/test substance slurry and maintaining tadpoles in test chambers with only clean dilution water for 30 days or (2) maintaining tadpoles in test chambers containing contaminated sediments and allowing tadpoles to ingest the contaminated sediments. Concentration-response curves and $LC_{50}$, $EC_{50}$, LOEC, and NOEC values for the test substance are developed from survival and growth responses. Any abnormal behavior (e.g., erratic swimming, loss of reflex, increased excitability, lethargy) and any changes in appearance or physiology such as discoloration (e.g., reddened leg, excessive mucus production, opaque eyes, curved spine, or hemorrhaging) are also evaluated.

As mentioned in Section 4.1 of this chapter, the U.S. EPA is implementing an Endocrine Disruptor Screening Program that includes several amphibian-based ecotoxicological screens and tests for identifying and characterizing endocrine effects of pesticides, industrial chemicals, and environmental contaminants. These include a "frog metamorphosis assay" and an "amphibian life cycle reproductive toxicity study with endocrine endpoints."

A strategy for protection of herpetofauna has been developed.[181] Rather inclusive recommendations include: (1) research should examine the relative sensitivity of major groups of amphibians and reptiles to major classes of environmental contaminants to detect possible inherent (taxonomic) variability in response; (2) chemicals with selective toxicity should be examined first, permitting comparison of data from aquatic and terrestrial tests; (3) a variety of *in vitro* procedures, such as effects in embryos or larvae, may reduce cost and time; and (4) laboratory investigation should provide a guide, but should not obviate the need, for well-designed field tests and postregistration vigilance by field biologists.

A summary of acute toxicity to amphibians of more than 200 different contaminants and field studies of over 50 contaminants indicated that neither test species nor protocols were standardized.[182] A searchable database — RATL (Reptile and Amphibian Toxicology Literature, http://www.cws-scf.ec.gc.ca/nwrc/ratl/index_e.htm) — for published ecotoxicological data from studies with amphibians and reptiles has been created by the Canadian Wildlife Service. Currently, there are approximately 2000 references in this database for approximately 6200 contaminant-related studies, divided almost equally between reptiles and amphibians. Approximately 650 different species are included in the database.

Deformities in tadpoles have been studied as a possible sensitive indicator of environmental contaminants; multiple causes and types of deformities in amphibians have been reviewed.[183] Caged tadpoles in water receiving runoff or spray drift from agricultural fields has been used to identify potential hazard.[184] A standard test (FETAX, Frog Embryo Teratogenesis Assay: *Xenopus*) has been developed with embryos of the clawed frog (*Xenopus laevis*) as an assay for teratogenicity of chemicals and mixtures of contaminants.[185] Under the auspices of the ASTM, a comprehensive guideline for FETAX was published in 1991 and updated in 1998 ["Standard Guide for Conducting the Frog Embryo Teratogenesis Assay-*Xenopus (FETAX)*," Annual Book of ASTM Standards, Designation E143998]. The similarity of response to dieldrin between *Xenopus laevis* and two ranid species in a study conducted by the U.S. EPA supports the utility of this species.[186] Results with nine combinations of developmental toxicants have indicated that FETAX is useful for hazard assessment of mixture toxicity and of sediment extracts.[187]

Short-term toxicity tests with *Xenopus laevis* and *Rana pipiens* were conducted with a 96-hour modified FETAX to assess paraquat toxicity.[188] The commercial formulation was three times as acutely toxic as the technical-grade chemical, and the 96-hour $LC_{50}$ of either form at least sixfold lower for *Rana pipiens* than *Xenopus laevis*. In another comparative study the embryotoxicity of the nonionic surfactant nonylphenol ethoxylate was determined in *Xenopus laevis* and the Australian frogs *Litoria adelaidensis* and *Crinia insignifera* using the FETAX protocol.[189] Growth inhibition as assessed by embryo length was the most sensitive indicator of effect in all three species. *Xenopus laevis* was the most sensitive of the three species and the only species that displayed indisputable terata. Integrated field and laboratory studies have been used for evaluating amphibian responses in wetlands impacted by mining activities in the western United States, where FETAX was conducted in the lab and *in situ* in the field using the bullfrog.[190]

For reptiles, a considerable portion of the published toxicological research has focused on turtles, especially snapping turtles (*Chelydra serpentina*) and sea turtles.[191] Focus on these groups may be justified in part because snapping turtles are large, long-lived omnivores that live intimately with aquatic sediments and thus are considered excellent bioindicators of wetland conditions, whereas many species of sea turtles are rare or endangered. Snapping turtles have been used as monitors of environmental contaminants including use in tidal wetlands, freshwater ponds, rivers, and lakes.[192–194] Contaminant-related DNA damage was detected in snapping turtles and in sliders (*Pseudemys scripta*).[192,194] Higher rates of deformities and unhatched eggs were related to PCB exposure of snapping turtle in highly contaminated areas of the Great Lakes.[193,195] Reptiles in general, and snakes and lizards in particular, are important although often neglected components of terrestrial and aquatic ecosystems and should be included in studies of environmental contamination.[196]

Accumulation and effects of environmental contaminants on snakes and lizards have been comprehensively reviewed.[197,198] Since all snakes are secondary, tertiary, and top predators, they are especially subject to the bioaccumulation of environmental contaminants. Their unique life histories make their roles in food webs diverse and important, and they are crucial to the proper functioning of many ecological processes. Lizards may also be excellent bioindicators of contamination.[198] Lizards are a significant part of many ecosystems as well as an important link in many food chains. There are large gaps in data for many environmental contaminants on lizards. Ecotoxicological studies on a wide variety of lizard species are needed; both laboratory and field studies would provide useful information. Because the majority of lizards are insectivores, studies of the effects and accumulation of pesticides are essential. Furthermore, many species are listed as threatened or endangered in the United States.

Reptiles, including turtles and alligators, have been studied for evidence of contaminant-related endocrine disruption, as discussed in Chapter 39 of this book and elsewhere.[195,199]

## 4.11 SUMMARY

Wildlife toxicology is the study of potentially harmful effects of toxic agents on wild animals. Wildlife toxicology endeavors to predict the effects of toxic agents on nontarget wildlife species and, ultimately, populations in natural environments. Avian toxicity testing protocols were first utilized by the U.S. Fish and Wildlife Service as a consequence of wildlife losses in the 1950s due to the increased use of DDT and other pesticides. The first testing protocols focused on single-dose acute oral toxicities with lethality as the major endpoint. Further protocol development resulted in subacute 5-day dietary tests. These, along with the single acute oral dose tests, are currently required by the U.S. Environmental Protection Agency in support of pesticide registration. The avian subchronic dietary toxicity test was developed as an extension of the subacute test as a precursor to full-scale reproductive studies, but it is not routinely required for regulatory purposes. Subchronic testing has been applied to compare the sublethal effects of different forms of mercury, to study hepatotoxicity of organic selenium, to comparatively study contaminated sediment ingestion, and to study delayed neurotoxicity of certain organophosphorus insecticides. Avian chronic toxicity tests are designed with reproduction as the primary endpoint and are required for both waterfowl and upland gamebirds during chemical registration. Persistent chemicals, such as chlorinated hydrocarbons, require relatively long-term exposures (at least 10 weeks) in advance of breeding, whereas shorter-term exposures may be utilized for less persistent chemicals such as organophosphorus insecticides. The U.S. EPA is presently implementing an Endocrine Disruptor Screening Program (EDSP), which includes an avian two-generation test with Japanese quail with endocrine-specific endpoints in addition to conventional reproduction endpoints.

Single-dose avian embryotoxicity and teratogenicity tests were developed in part to assess the potential contaminant hazard of external exposure of bird eggs; tests with multiple species and chemicals have revealed differential toxicities of a spectrum of chemicals and sensitivities among species. Recent focus has included hazard evaluations of aquatic weed control, mosquito control, and wildfire control agents. Developmental toxicity testing has also focused on the vulnerability of "neonatal" nestling birds, including kestrels, starlings, red-winged blackbirds, great egrets, and gulls, to oral ingestion of environmental contaminants.

Avian terrestrial field studies are basically of two types: (1) screening studies to ascertain whether impacts are occurring, and (2) definitive studies to estimate the magnitude. These studies often require extensive monitoring of reproductive success and survival of young wherein active nests are periodically examined and mark-recapture techniques as well as radiotelemetry are used. Sometimes, artificial nest structures are provided in the vicinity of test-spray fields to increase nest density, and hence experimental sample size, for species such as American kestrels and European starlings. Avian terrestrial field studies have been successfully undertaken using diverse species, including northern bobwhite, sage grouse, waterfowl, and at least five different passerine species, during and following applications of agricultural pesticides.

Behavioral aberrations in wildlife can be manifested at one or two orders of magnitude below lethal levels of environmental contaminants. In birds, changes in nest attentiveness, brood behavior, and increased vulnerability to predation have been documented in field and pen studies. Response time to maternal call, avoidance of fright stimulus, and tests of operant learning ability as well as time-activity budgets, effects at critical periods of development, and altered food discrimination and feeding behavior have been successfully applied to laboratory studies.

Laboratory studies with environmental contaminants and mammalian wildlife have been limited compared to avian studies. This may be due to the fact that the human health effects and agricultural nutrition literature is abundant with studies conducted with laboratory rodents as well as other species of domestic mammals that are viewed as surrogate species for mammalian wildlife. Mammalian toxicity data of the U.S. EPA FIFRA has consisted largely of laboratory rat data for pesticide registration, whereas wildlife testing requires a dietary $LC_{50}$ or acute oral $LD_{50}$ study with a nonendangered representative species likely to be exposed, quite often a microtine rodent. Previous recommendations have included use of an omnivore, such as the deer mouse, and a carnivore, the mink, for which there is a large biological database for both laboratory and field studies. The use of wild mammals has also been proposed for setting water quality criteria for those species that consume aquatic life, including the mink as a piscivorous species, the northern short-tailed shrew as an insectivorous consumer of aquatic invertebrates, and the deer mouse as an omnivorous species. There are many unique species of wild mammals with varied sensitivities to certain classes of contaminants; controlled comparative laboratory studies are needed for many of these species.

Worldwide concern over declining populations of amphibians and reptiles has revealed that numerous taxa of amphibians and reptiles are endangered or threatened. Comparative toxicities of organic compounds and metals indicate a wide variation in sensitivity among amphibian species. Amphibians are generally more sensitive than fish, suggesting that water quality criteria established for fish may not be protective of amphibians. The U.S. EPA is developing guidelines for preregistration testing under FIFRA for acute lethal toxicity to amphibians and reptiles. Also, the U.S. EPA is implementing an Endocrine Disruptor Screening Program (EDSP), which includes several amphibian-based ecotoxicological screens and tests for identifying and characterizing endocrine effects of pesticides, industrial chemicals, and environmental contaminants. These include a "frog metamorphosis assay" and an "amphibian life cycle reproductive toxicity study with endocrine endpoints."

Reptiles, which are critical components of many food chains, are often neglected in studies of terrestrial and aquatic contamination. A considerable portion of the toxicological research on reptiles has focused on turtles, especially snapping turtles and sea turtles. Snapping turtles live intimately with aquatic sediments and are considered excellent bioindicators of wetland conditions. Turtles and alligators are being studied for evidence of contaminant-related endocrine disruption. All snakes are secondary, tertiary, and top predators and are susceptible to the bioaccumulation of environmental contaminants. Lizards provide important links in many food chains and are perhaps more influenced by contamination than previously believed.

## ACKNOWLEDGMENTS

Reviews of the content of this manuscript provided by E. F. Hill, P. H. Albers, and L. Touart are gratefully acknowledged.

## REFERENCES

1. Hill, E. F., Wildlife toxicology, in *General and Applied Toxicology*, 2nd ed., Ballantyne, B., Marrs, T. C., and Syversen, T., Eds., Macmillan Reference, London, 1999, 1327.
2. Newman, J. R., Effects of industrial air pollution on wildlife, *Biol. Conserv.*, 15, 81, 1979.
3. Eisler, R., *Handbook of Chemical Risk Assessment: Health Hazards to Humans, Plants, and Animals, Vol. 1: Metals*, Lewis Publishers, CRC Press, Boca Raton, FL, 2000, 738.
4. Forbes, R. M. and Sanderson, G. C., Lead toxicity in domestic animals and wildlife, in *The Biogeochemistry of Lead in the Environment, Part B*, Nriagu, J. O., Ed., Elsevier/North Holland Biomedical Press, Amsterdam, 1978, Chap. 16.
5. Chupp, N. R. and Dalke, P. D., Waterfowl mortality in Coeur d'Alene River Valley, Idaho, *J. Wildl. Manage.*, 28, 692, 1964.
6. Bourne, W. R. P., Oil pollution and bird populations, in *The Biological Effects of Oil Pollution on Littoral Communities*, McCarthy, J. D. and Arthur, D. R., Eds., Field Studies 2 (Suppl.), 1968, 99.
7. Stickel, L. F., Field studies of a *Peromyscus* population in an area treated with DDT, *J. Wildl. Manage.*, 10, 216, 1946.
8. Hall, R. J., Impacts of pesticides in bird populations, in *Silent Spring Revisited*, Marcu, G. J., Hollingworth, R. M., and Durham, W., Eds., American Chemical Society, Washington, D.C., 1987, chap. 6.
9. Carson, R., *Silent Spring*, Houghton Mifflin, Boston, 1962, 103.
10. Dewitt, J. B., Effects of chlorinated hydrocarbon insecticides upon quail and pheasants, *J. Agric. Food Chem.*, 3, 672, 1955.
11. Ratcliffe, D. A., Decrease in eggshell weight in certain birds of prey, *Nature*, 215, 208, 1967.
12. Hickey, J. J. and Anderson, D. W., Chlorinated hydrocarbons and eggshell changes in raptorial and fish-eating birds, *Science*, 162, 271, 1968.
13. Wiemeyer, S. N., and Porter, R. D., DDE thins eggshells of captive American kestrels, *Nature*, 227, 737, 1970.
14. Blus, L. J., Gish, C. D., Belisle, A. A., and Prouty, R. M., Logarithmic relationship of DDE residues to eggshell thinning, *Nature*, 235, 376, 1972.
15. Hill, E. F. and Fleming, W. J., Anticholinesterase poisoning of birds; field monitoring and diagnosis of acute poisoning, *Environ. Toxicol. Chem.*, 1, 27, 1982.
16. Grue, C. E., Fleming, W. J., Busby, D. G., and Hill, E. F., Assessing hazards of organophosphate pesticide to wildlife, *Trans. N. Am. Wildl. Nat. Res. Conf.*, 48, 200, 1983.
17. Mendelssohn, H. and Paz, U., Mass Mortality of birds of prey by Azodrin, an organophosphorus insecticide, *Biol. Conserv.*, 11, 163, 1997.
18. Mineau, P., Fletcher, M. R., Glaser, L. C., Thomas, N. J., Brassard, C., Wilson, L. K., Elliott, J. E., Lyon, L. A., Henny, C. J., Bollinger, T., and Porter S. L., Poisoning of raptors with organophosphorus and carbamate pesticides with emphasis on Canada, U.S. and U.K., *J. Raptor Res.*, 33, 1, 1999.
19. Ohlendorf, H. M., Hoffman, D. J., Saiki, M. K., and Aldrich, T. W., Embryonic mortality and abnormality of aquatic birds: Apparent impacts by selenium from irrigation drainwater, *Sci. Total Environ.*, 52, 49, 1986.
20. Hoffman, D. J., Ohlendorf, H. M., and Aldrich, T. W., Selenium teratogenesis in natural populations of aquatic birds in central California, *Arch. Environ. Contam. Toxicol.*, 17, 519, 1988.

21. Skorupa, J. P. and Ohlendorf, H. M., Contaminants in drainage water and avian risk thresholds, in *The Economics and Management of Water and Drainage in Agriculture*, Dinar A. and Zilberman D., Eds., Kluwer Academic Publishing Co., Boston, 1991, 345.
22. Heinz, G. H., Hoffman, D. J., and Gold, L. G., Impaired reproduction of mallards fed an organic form of selenium, *J. Wildl. Manage.,* 53, 418, 1989.
23. Fitzgerald, W. F., Engstrom, D. R., Mason, R. P., and Nater, E. A., The case for atmospheric mercury contamination in remote areas, *Environ. Sci. Technol.*, 32, 1, 1998.
24. Monteiro, L. R. and Furness, R. W., Accelerated increase in mercury contamination in North Atlantic mesopelagic food chains as indicated by time series of seabird feathers, *Environ. Toxicol. Chem.*, 16, 2489, 1997.
25. Thompson, D. R., Furness, R. W., and Walsh, P. M., Historical changes in mercury concentrations in the marine ecosystem of the north and north-east Atlantic ocean as indicated by seabird feathers, *J. Appl. Ecol.*, 29, 79, 1992.
26. Colborn, T., von Saal, F. S., and Soto, A. M., Developmental effects of endocrine-disrupting chemicals in wildlife and humans, *Environ. Health Perspect.,* 101, 378, 1993.
27. Kavlock, R. J. and Ankley, G. T., A perspective on the risk assessment process for endocrine-disruptive effects on wildlife and human health, *Risk Anal.,* 16, 731, 1996.
28. Kendall, R. J., Dickerson, R. L., Geisey, J. P., and Suk, W. P., *Principles and Processes for Evaluating Endocrine Disruption in Wildlife*, SETAC Press, Pensacola, FL, 1998.
29. Hoffman, D. J., Rattner, B. A., and Hall, R. J., Wildlife toxicology, *Environ. Sci. Technol.,* 24, 276, 1990.
30. Hill, E. F. and Hoffman, D. J., Avian models for toxicity testing, *J. Am. Coll. Toxicol.,* 3, 357, 1984.
31. Tucker, R. K. and Crabtree, D. G., Handbook of Toxicity of Pesticides to Wildlife, Res. Publ. 84, Washington, D.C.: U.S. Dept. of Interior, 1970.
32. Schafer, E. W., Jr., Bowles, W. A., Jr., and Hurlbut, J., The acute oral toxicity, repellency, and hazard potential of 998 chemicals to one or more species of wild and domestic birds, *Arch. Environ. Contam. Toxicol.,* 12, 355, 1983.
33. Mineau, P., Baril, A., Collins, B. T., Duffe, J., Joerman, G., and Luttik, R., Pesticide acute toxicity reference values for birds, *Rev. Environ. Contam. Toxicol.,* 170,13, 2001.
34. Fite, E., *The Environmental Protection Agency's Avian Pesticide Assessment Model in Population Ecology and Wildlife Toxicology of Agricultural Pesticide Use: A Modelling Initiative for Avian Species*, Kendall, R. J. and Lacher, T. E., Eds., Lewis Publishers, Chelsea, MI, 1994.
35. Hill, E. F., Acute and subacute toxicology in evaluation of pesticide hazard to wildlife, in *Population Ecology and Wildlife Toxicology of Agricultural Pesticide Use: A Modelling Initiative for Avian Species,* Kendall, R. J. and Lacher, T. E., Eds., Lewis Publishers, Chelsea, MI, 1994.
36. Mineau, P., Collins, B. T., and Baril, A., On the use of scaling factors to improve interspecies extrapolation of acute toxicity in birds, *Regul. Toxicol. Pharmacol.,* 24, 24, 1996.
37. Kenaga, E. E., Test organisms and methods useful for early assessment of acute toxicity of chemicals, *Environ. Sci. Technol.,* 12, 1322, 1978.
38. Walker, C. H., Pesticides and birds: Mechanisms of selective toxicity, *Agric. Ecosyst. Environ.*, 9, 211, 1983.
39. Wolfe, M. F. and Kendall, R. J., Age-dependent toxicity of diazinon and terbufos in European starlings (*Sturnus vulgaris*) and red-winged blackbirds (*Agelaius phoeniceus), Environ. Toxicol. Chem.,* 17, 1300, 1998.
40. Wiemeyer, S. N. and Sparling, D. W., Acute toxicity of four anti-cholinesterase insecticides to American kestrels, eastern screech owls and northern bobwhite, *Environ. Toxicol. Chem.,* 10, 1139, 1991.
41. Balcomb, R., Stevens, R., and Bowen, C., II, Toxicity of 16 granular insecticides to wild-caught song birds, *Bull. Environ. Contam. Toxicol.,* 33, 302, 1984.
42. Hill, E. F. and Camardese, M. B., Toxicity of anticholinesterase insecticides to birds: Technical grade versus granular formulations, *Ecotoxicol. Environ. Saf.,* 8, 551, 1984.
43. Hill, E. F., Heath, R. G., Spann, J. W., and Williams, J. D., Lethal Dietary Toxicities of Environmental Pollutants to Birds, Spec. Sci. Rep. Wildl., 191. Washington, D.C.: U.S. Dept. of Interior, 1975.
44. Hill, E. F. and Camardese, M. B., Lethal Dietary Toxicities of Environmental Contaminants and Pesticides to Coturnix, U.S. Fish Wildl. Serv. Tech. Rep. 2, Washington, D.C., 1987.

45. Stinson, E. R., Hayes, L. E., Bush, P. B., and White, D. H., Carbofuran affects wildlife on Virginia corn fields, *Wildl. Soc. Bull.*, 22, 566, 1994.
46. Wang, G., Edge, W. D., and Wolff, J. O., Response of bobwhite quail and gray-tailed voles to granular and flowable diazinon applications, *Environ. Toxicol. Chem.*, 20, 406, 2001.
47. Best, L. B. and Fischer, D. L., Granular insecticides and birds: Factors to be considered in understanding exposure and reducing risk, *Environ. Toxicol. Chem.*, 11, 1495, 1991.
48. Best, L. B., Stafford, T. R., and Mihaich, E. M., House sparrow preferential consumption of pesticide granules with different surface coatings, *Environ. Toxicol. Chem.*, 15, 1763, 1996.
49. Gionfriddo, J. P. and Best, L. B., Grit color selection by house sparrows and northern bobwhite, *J. Wildl. Manage.*, 60, 836, 1996.
50. Stafford, T. R., Best, L. B., and Fischer, D. L., Effects of different formulations of granular pesticides on birds, *Environ. Toxicol. Chem.*, 15, 1606, 1996.
51. Stafford, T. R. and Best, L. B., Effects of application rate on avian risk from granular pesticides, *Environ. Toxicol. Chem.*, 17, 526, 1998.
52. Stafford, T. R. and Best, L. B., Bird response to grit and pesticide granule characteristics: Implications for risk assessment and risk reduction, *Environ. Toxicol. Chem.*, 18, 722, 1999.
53. Mastrota, F. N. and Mench, J. A., Evaluation of taste repellents with northern bobwhites for deterring ingestion of granular pesticides, *Environ. Toxicol. Chem.*, 14, 631, 1995.
54. Heath, R. G. and Stickel, L. F., Protocol for testing the acute and relative toxicity of pesticides to penned birds, in The Effects of Pesticides on Wildlife, Circ. 226. Washington, D.C.: U.S. Dept. of Interior, 1965, 18.
55. Hill, E. F., Spann, J. W., and Williams, J. D., Responsiveness of 6 to 14 generations of birds to dietary dieldrin toxicity, *Toxicol. Appl. Pharmacol.*, 42, 425, 1977.
56. Hill, E. F. and Camardese, M. B., Subacute toxicity testing with young birds: Response in relation to age and interest variability of $LC_{50}$ estimates, in *Avian and Mammalian Wildlife Toxicology Second Conference*, Lamb, D. W. and Kenaga, E. E., Eds., Philadelphia: American Society for Testing and Materials, STP 757, 1981, 41.
57. Hudson, R. H., Tucker, R. K., and Haegele, M. A., Effect of age on sensitivity: Acute oral toxicity of 14 pesticides to mallard ducks of several ages, *Toxicol. Appl. Pharmacol.*, 22, 556, 1972.
58. Hoffman, D. J., Heinz, G. H., LeCaptain, L. J., Bunck, C. M., and Green, D. E., Subchronic hepatotoxicity of selenomethionine in mallard ducks, *J. Toxicol. Environ. Health*, 32, 449, 1991.
59. Hoffman, D. J., Heinz, G. H., Sileo, L., Audet, D. J., Campbell, J. K., LeCaptain, L. J., and Obrecht, H. H., Developmental toxicity of lead-contaminated sediment in Canada geese (*Branta canadensis*), *J. Toxicol. Environ. Health*, 59, 235, 2000.
60. Hoffman, D. J., Heinz, G. H., Sileo, L., Audet, D. J., and LeCaptain, L. J., Developmental toxicity of lead-contaminated sediment to mallard ducklings, *Arch. Environ. Contam. Toxicol.*, 39, 221, 2000.
61. Mateo, R. and Hoffman, D. J., Differences in oxidative stress between young Canada geese and mallards exposed to lead-contaminated sediment, *J. Toxicol. Environ. Health*, Part A, 64, 531, 2001.
62. Hoffman, D. J., Sileo, L., and Murray, H. C., Subchronic organophosphorus ester-induced delayed neurotoxicity in mallards, *Toxicol. Appl. Pharmacol.*, 75, 128, 1984.
63. Bennett, R. S. and Ganio, L. M., Overview of Methods for Evaluating Effects of Pesticides on Reproduction in Birds, EPA Rep. 600–3–91–048, U.S. Environmental Protection Agency, Washington, D.C., 1991.
64. Akerman, J. W., Environmental Protection Agency's regulatory requirements for wildlife toxicity testing, in *Avian and Mammalian Wildlife Toxicology, STP 693*, Kenaga, E. E., Ed., American Society for Testing and Materials, Philadelphia, 1979, 3.
65. Spann, J. W., Heinz, G. H., and Hulse, C. S., Reproduction and health of mallards fed endrin, *Environ. Toxicol. Chem.*, 5, 755, 1986.
66. Fleming, W. J., McLane, A. R., and Cromartie, E., Endrin decreases screech owl productivity, *J. Wildl. Manage.*, 46, 462, 1982.
67. Hoffman, D. J. and Heinz, G. H., Embryotoxic and teratogenic effects of selenium of selenium in the diet of mallards, *J. Toxicol. Environ. Health*, 24, 477, 1988.
68. Heinz, G. H. and Hoffman, D. J., Methylmercury chloride and selenomethionine interactions on health and reproduction in mallards, *Environ. Toxicol. Chem.*, 17, 139, 1998.

69. Stromborg, K. L., Reproductive tests of diazinon on bobwhite quail, in *Avian and Mammalian Wildlife Toxicology-Second Conference. STP 757.* Kenaga, E. E., Ed., American Society for Testing and Materials, Philadelphia, 1981, 19.
70. Bennett, J. K. and Bennett, R. S., Effects of dietary methyl parathion on northern bobwhite egg production and eggshell quality, *Environ. Toxicol. Chem.,* 9, 1481, 1990.
71. Rattner, B. A., Sileo, L., and Scanes, C. G., Oviposition and the plasma concentrations of LH, progesterone and corticosterone in bobwhite quail (*Colinus virginianus*) fed parathion, *J. Reprod. Fert.,* 66, 147, 1982.
72. Stromborg, K. L., Reproduction of bobwhites fed different dietary concentrations of an organophosphate insecticide, methamidophos, *Arch. Environ. Contam. Toxicol.*, 15, 143, 1986.
73. Stromborg, K. L., Reproductive toxicity of monocrotophos to bobwhite quail, *Poult. Sci.,* 65, 51, 1986.
74. Bennett, R. S., Bentley, R., Shiroyama, T., and Bennett, J. K., Effects of the duration and timing of dietary methyl parathion exposure on bobwhite reproduction, *Environ. Toxicol. Chem.*, 9, 1473, 1990.
75. Bennett, R. S., Williams, B. A., Schmedding, D. W., and Bennett, J. K., Effects of dietary exposure to methyl parathion on egg laying and incubation in mallards, *Environ. Toxicol. Chem.,* 10, 501, 1991.
76. Hoffman, D. J., Embryotoxicity and teratogenicity of environmental contaminants to bird eggs, *Rev. Environ. Contam. Toxicol.,*115, 40, 1990.
77. Birkhead, T.R., Lloyd, C., and Corkhill, P., Oiled seabirds successfully cleaning their plumage, *Br. Birds,* 66, 535, 1973.
78. Albers, P. H., Effects of external application of fuel oil on hatchability of mallard eggs, in *Fate and Effects of Petroleum Hydrocarbons in Marine Ecosystems and Organisms*, Wolfe, D. A., Ed., Pergamon Press, New York, 1977, 158.
79. Hoffman, D. J., Embryotoxic effects of crude oil in mallard ducks and chicks, *Toxicol. Appl. Pharmacol.,* 46, 183, 1978.
80. Couillard, C. M. and Leighton, F. A., Bioassays for the toxicity of petroleum oils in chicken embryos, *Environ. Toxicol. Chem.,* 10, 533, 1991.
81. Hoffman, D. J. and Albers, P. H., Evaluation of potential embryotoxicity and teratogenicity of 42 herbicides, insecticides, and petroleum contaminants to mallard eggs, *Arch. Environ. Contam. Toxicol.,* 13, 15, 1984.
82. Sewalk, C. J., Brewer, G. L., and Hoffman, D. J., The effects of diquat, an aquatic herbicide, on the development of mallard embryos, *J. Toxicol. Environ. Health,* 62, 101, 2001.
83. Albers, P. H., Hoffman, D. J., Buscemi, D. M., and Melancon, M. J., Effects of the mosquito larvicide GB1111 on red-winged blackbird embryos, *The Wildlife Society 7th Annual Conference*, Nashville, September 12–16, 2000.
84. Hoffman, D. J., Albers, P. H., Melancon, M. J., and Miles, K., Effects of the mosquito larvicide GB1111 on mallard and bobwhite embryos, *Society of Environmental Toxicology and Chemistry 21st Annual Meeting,* Nashville, November 12–16, 2000.
85. Buscemi, D. M., Hoffman, D. J., Vyas, N. B., Spann, J. W., and Kuenzel, W. J., Effects of Phos-Chek® G75-F and Silv-Ex® on developing northern bobwhite quail (*Colinus virginianus*), *Arch. Environ. Contam. Toxicol.,* in press, 2002.
86. Brunstrom, B. and Reutergardh, L., Difference in sensitivity of some avian species to the embryotoxicity of a PCB, 3,3',4,4'-tetrachlorobiphenyl, injected into the eggs, *Environ. Pollut.,* (A), 42, 37, 1986.
87. Hoffman, D. J., Melancon, M. J., Klein, P. N., Eisemann, J. D., and Spann, J. W., Comparative developmental toxicity of planar PCB congeners in chickens, american kestrels and common terns, *Environ. Toxicol. Chem.,* 17, 747, 1998.
88. Heinz, G. H., Hoffman, D. J., Murray, D. R., and Erwin, C. A., Using egg injections to measure the toxicity of methylmercury to avian embryos, *Society of Environmental Toxicology and Chemistry 22nd Annual Meeting,* Baltimore, November 11–15, 2001.
89. Hoffman, D. J., Measurements of toxicity and critical stages of development, in *Population Ecology and Wildlife Toxicology of Agricultural Pesticide Use: A Modelling Initiative for Avian Species,* Kendall, R. J. and Lacher, T. E., Eds., Lewis Publishers, Chelsea, MI, 1993.
90. Hoffman, D. J., Franson, J. C., Pattee, O. H., Bunck, C. M., and Anderson, A., Survival, growth and accumulation of ingested lead in nestling American kestrels, *Arch. Environ. Contam. Toxicol.,* 14, 89, 1985.

91. Hoffman, D. J., Franson, J. C., Pattee, O. H., Bunck, C. M., and Murray, H. C., Biochemical and hematological effects of lead ingestion in nestling American kestrels, *Comp. Biochem. Physiol.*, 80c, 431, 1985.
92. Hoffman, D. J., Franson, J. C., Pattee, O. H., and Bunck, C. M., Survival growth, and histopathological effects of paraquat ingestion in nestling American kestrels, *Arch. Environ. Contam. Toxicol.,* 14, 495, 1985.
93. Hoffman, D. J., Spann, J. W., LeCaptain, L. J., Bunck, C. M., and Rattner, B. A., Developmental toxicity of diphenyl ether herbicides in nestling American kestrels, *J. Toxicol. Environ. Health,* 34, 323, 1991.
94. Hoffman, D. J., Melancon, M. J., Klein, P. N., Rice, C. P., Eisemann, J. D., Hines, R. K., Spann, J. W., and Pendleton, G. W., Developmental toxicity of PCB126 (3,3',4,4',5-pentachlorobiphenyl) in nestling american kestrels (*Falco sparverius*), *Fund. Appl. Toxicol.,* 34, 188, 1996.
95. Grue, C. E. and Shipley, B. K., Sensitivity of nestling and adult starlings to dicrotophos, an organophosphate pesticide, *Environ. Res.,* 35, 454, 1984.
96. Hooper, M. J., Brewer, L. W., Cobb, G. P., and Kendall, R. J., An integrated laboratory and field approach for assessing hazards of pesticide exposure to wildlife, in *Pesticide Effects on Terrestrial Wildlife*, Somerville, L. and Walker, C. H., Eds., Taylor and Francis, New York, 1990, 271.
97. Miller, D. S., Peakall, D. B., and Kinter, W. B., Ingestion of crude oil: Sublethal effects on herring gull chicks, *Science,* 199, 315, 1978.
98. Peakall, D. B., Hallett, D., Miller, D. S., Butler, R. G., and Kinter, W. B., Effects of ingested crude oil on black guillemots: A combined field and laboratory study, *AMBIO* 9, 28, 1980.
99. Burger, J. and Gochfeld, M., Lead and neurobehavioral development in gulls: A model for understanding effects in the laboratory and the field, *Neurotoxicology*, 18, 495, 1997.
100. Burger, J. and Gochfeld, M., Effects of varying temporal exposure to lead on behavioral development in herring gull (*Larus argentatus*) chicks, *Pharmacol. Biochem. Behav.*, 52, 601, 1995.
101. Burger, J. and Gochfeld, M., Growth and behavioral effects of early postnatal chromium and manganese exposure in herring gull (*Larus argentatus*) chicks, *Pharmacol. Biochem. Behav.*, 50, 607, 1995.
102. Spalding, M. G., Frederick, P. C., McGill, H. C., Bouton, S. N., and McDowell, L. R., Methylmercury accumulation in tissues and its effects on growth and appetite in captive great egrets, *J. Wildl. Dis.*, 36, 411, 2000.
103. Spalding, M. G., Frederick, P. C., McGill, H. C., Bouton, S. N., Richey, L. J., Schumacher, I. M., Blackmore, C. G. M., and Harrison, J., Histologic, neurologic, and immunologic effects of methylmercury in captive great egrets, *J. Wildl. Dis.*, 36, 423, 2000.
104. Hoffman, D., Spalding, M., and Frederick, P., Subchronic effects of methylmercury in great egret nestlings, *Society of Environmental Toxicology and Chemistry 18th Annual Meeting,* San Francisco, November 16–20, 1997.
105. Bouton, S. N., Frederick, P. C., Spalding, M. G., and Mcgill, H., Effects of chronic, low concentrations of dietary methylmercury on the behavior of juvenile great egrets, *Environ. Toxicol. Chem.,* 18, 1934, 1999.
106. Matz, A. C., Bennett, R. S., and Landis, W. G., Effects of azinphos-methyl on northern bobwhite: A comparison of laboratory and field results, *Environ. Toxicol. Chem.,* 17, 1364, 1998.
107. Vyas, N. B., Factors influencing estimation of pesticide-related wildlife mortality, *Toxicol. Indust. Health,* 15, 186, 1999.
108. Hill, E. F. and Mendenhall, V. M., Secondary poisoning of barn owl with famphur, an organophosphate insecticide, *J. Wildl. Manage.,* 44, 676, 1980.
109. Henny, C. J., Blus, L. J., Kolbe, E. J., and Fitzner, R. E., Organophosphate insecticide (famphur) topically applied to cattle kills magpies and hawks, *J. Wildl. Manage.,* 49, 648, 1985.
110. Elliott, J. E., Wilson, L. K., Langelier, K. M., Mineau, P., and Sinclair, P. H., Secondary poisoning of birds of prey by the organophosphorus insecticide, phorate, *Ecotoxicology*, 6, 219, 997.
111. Fite, E. C., Turner, L. W., Cook, J. J., and Stunkard, C., Guidance Document for Conducting Terrestrial Field Studies, EPA 540–09–88–109, U.S. Environmental Protection Agency, Washington, D.C., 1988.
112. Blus, L. J., Staley, C. S., Henny, C. J., Pendleton, G. W., Craig, T. H., Craig, E. H., and Halford, D. K., Effects of organophosphorus insecticides on sage grouse in southeastern Idaho, *J. Wildl. Manage.,* 53, 1139, 1989.

113. Grue, C. E., DeWeese, L. R., Mineau, P., Swanson, G. A., Foster, J. R., Arnold, P. M., Huckins, J. N., Sheeham, P. J., Marshall, W. K., and Ludden, A. P., Potential impacts of agricultural chemicals on waterfowl and other wildlife inhabiting prairie wetlands: An evaluation of research needs and approaches, *Trans. N. A. Wildl. Nat. Res. Conf.*, 51, 357, 1986.
114. Grue, C. E., Tome, M. W., Swanson, G. A., Borthwick, S. M., and Deweese, L. R., Agricultural chemicals and the quality of prairie-pothole wetlands for adult and juvenile waterfowl — What are the concerns?, in Proc. of the National Symposium on the Protection of Wetlands from Agricultural Impacts, Stuber, P. J., Coord., U.S. Fish Wildl. Serv. Biol. Rep. 88, 16, 1988, 55.
115. Tome, M. W., Grue, C. E., and Henry, M. G., Case studies: Effects of agricultural pesticides on waterfowl and prairie pothole wetlands, in *Handbook of Ecotoxicology*, Hoffman, D. J., Rattner, B. A., Burton, G. A., Jr., and Cairns, J., Jr., Eds., CRC Press, Boca Raton, FL, 1995, 565.
116. DeWeese, L. R., Henny, C. J., Floyd, R. L., Bobal, K. A., and Schultz, A. W., Response of Breeding Birds to Aerial Sprays of Trichlorfon (Dylox) and Carbaryl (Sevin-4-Oil) in Montana Forests, U.S. Dept. of Interior, Fish Wildl. Serv. No. 224, Washington, D.C., 1979.
117. Putera, J. A., Woolf, A., and Klimstra, W. D., Mourning dove use of orchards in southern Illinois, *Wildl. Soc. Bull.*, 13, 496, 1985.
118. Patnode, K. A. and White, D. H., Effects of pesticides on songbird productivity in conjunction with pecan cultivation in southern Georgia: A multiple exposure experimental design, *Environ. Toxicol. Chem.*, 10, 1479, 1991.
119. Rondeau, G. and Desgranges, D., Effects of insecticide use on breeding birds in Christmas tree plantations in Quebec, *Ecotoxicol*ogy, 4, 281, 1995.
120. Burgess, N. M., Hunt, K. A., Bishop, C., and Weseloh, D. V., Cholinesterase inhibition in tree swallows (*Tachycineta bicolor*) and eastern bluebirds (*Sialia sialis*) exposed to organophosphorus insecticides in apple orchards in Ontario, Canada, *Environ. Toxicol. Chem.*, 18, 708, 1999.
121. Bishop, C. A., Collins, B., Mineau, P., Burgess, N. M., Read, W. F., and Risley, C., Reproduction of cavity-nesting birds in pesticide-sprayed apple orchards in southern Ontario, Canada, 1988–1994, *Environ. Toxicol. Chem.*, 19, 588, 2000.
122. Cordi, B. and Fossi, C., Temporal biomarker responses in wild passerine birds exposed to pesticide spray drift, *Environ. Toxicol. Chem.*, 16, 2118, 1997.
123. Kendall, R. J., Farming with agro-chemicals — the response of wildlife, *Environ. Sci. Technol.*, 26, 239, 1992.
124. Robinson, S. C., Kendall, R. J., Robinson, R., Driver, C. J., and Lacher, T. E., Jr., Effects of agricultural spraying of methyl parathion on cholinesterase activity and reproductive success in wild starlings (*Sturnus vulgaris*), *Environ. Toxicol. Chem.*, 7, 343, 1988
125. Peakall, D. B., Behavioral responses of birds to pesticides and other contaminants, *Residue Rev.*, 96, 45, 1985.
126. Fox, G. A., Gilman, A. P., Peakall, D. B., and Anderka, F. W., Behavioral abnormalities of nesting Lake Ontario herring gulls, *J. Wildl. Manage.*, 42, 477, 1978.
127. White, D. H., Mitchell, C. A., and Hill, E. F., Parathion alters incubation behavior of laughing gulls, *Bull. Environ. Contam. Toxicol.*,31, 93, 1983.
128. Brewer, L. W., Driver, C. J., Kendall, R. J., Zenier, C., and Lacher, T. E., Jr., Effects of methyl parathion in ducks and duck broods, *Environ. Toxicol. Chem.*, 7, 375, 1988.
129. Fairbrother, A., Meyers, S. M., and Bennett, R. S., Changes in mallard hen and brood behaviors in response to methyl parathion induced illness of ducklings, *Environ. Toxicol. Chem.*, 7, 499, 1988.
130. McEwen, L. C. and Brown, R. L., Acute toxicity of dieldrin and malathion to wild sharp-tailed grouse, *J. Wildl. Manage.*, 30, 604, 1966.
131. Buerger, T. T., Kendall, R. J., Mueller, B. S., DeVos, T., and Williams, B. A., Effects of methyl parathion on northern bobwhite survivability, *Environ. Toxicol. Chem.*, 10, 527, 1991.
132. Galindo, J. C., Kendall, R. J., Driver, C. J., and Lacher, T. J., Jr., The effect of methyl parathion on susceptibility of bobwhite quail (*Colinus virginianus*) to domestic cat predation, *Behav. Neural Biol.*, 43, 21, 1985.
133. Vyas, N. B., Kuenzel, W. J., Hill, E. F., and Sauer, J. R., Acephate affects migratory orientation of the white-throated sparrow (*Zonotrichia albicollis*), *Environ. Toxicol. Chem.*, 14, 1961, 1995.
134. Heinz, G. H., Methylmercury: Reproductive and behavioral effects on three generations of mallard ducks, *J. Wildl. Manage.*, 43, 394, 1979.

135. Heinz, G. H., Behavior of mallard ducklings from parents fed 3 ppm DDE, *Bull. Environ. Contam. Toxicol.,16, 640, 1976.*
136. Martin, P. A., Solomon, K. R., Forsyth, D. J., Boermans, H. J., and Westcott, N. D., Effects of exposure to carbofuran-sprayed vegetation on the behavior, cholinesterase activity and growth of mallard ducklings (*Anas platyrhynchos*), *Environ. Toxicol. Chem.*, 10, 901, 1991.
137. Kreitzer, J. F., Effects of toxaphene and endrin at very low dietary concentrations on discrimination acquisition and reversal in bobwhite quail, *Colinus virginianus*, *Environ. Pollut. (Ser. A),* 23, 217, 1980.
138. Bunck, C. M., Bunck, T. J., and Sileo, L., Discrimination learning in adult bobwhite quail fed paraquat, *Environ. Toxicol. Chem.,* 5, 295, 1986.
139. Whitworth, M. R., Pendleton, G. W., Hoffman, D. J., and Camardese, M. B., Effects of dietary boron and arsenic on the behavior of mallard ducklings, *Environ. Toxicol. Chem.*, 10, 911, 1991.
140. Douglas-Stroebel, E., Brewer, G. L., and Hoffman, D. J., Effects of lead-contaminated sediment on mallard duckling behavior, in *Proc. SETAC 19th Annual Meeting*, Charlotte, 19, 1998.
141. Douglas-Stroebel, E., Brewer, G. L., and Hoffman, D. J., Effects of lead-contaminated sediment and nutrient level on mallard duckling brain growth and biochemistry, in *Proc. SETAC 22nd Annual Meeting*, Baltimore, 2001.
142. Bussiere, J. L., Kendall, R. J., and Lacher, T. E., Jr., Effect of methyl parathion on food discrimination in northern bobwhite (*Colinus virginianus*), *Environ. Toxicol. Chem.,* 8, 1125, 1989.
143. Bennett, R. S., Role of dietary choices in the ability of bobwhite to discriminate between insecticide-treated and untreated food, *Environ. Toxicol. Chem.,* 8, 731, 1989.
144. Kononen, D. W., Hochstein, J. R., and Ringer, R. K., A quantitative method for evaluating avian food avoidance behavior, *Environ. Toxicol. Chem.,* 5, 823, 1986.
145. Bennett, R. S. and Schafer, D. W., Procedure for evaluating the potential ability of birds to avoid chemically contaminated food, *Environ. Toxicol. Chem.,* 7, 359, 1988.
146. Bennett, R. S., Factors influencing discrimination between insecticide-treated and untreated foods by northern bobwhite, *Arch. Environ. Contam. Toxicol.*,18, 697, 1989.
147. Nicolaus, L. K. and Lee, H., Low acute exposure to organophosphate produces long-term changes, in bird feeding behavior, *Ecol. Appl.*, 9, 1039, 1999.
148. Ringer, R. K., The future of a mammalian wildlife toxicology test: One researcher's opinion, *Environ. Toxicol. Chem.*, 7, 339, 1988.
149. Aulerich, R. J. and Ringer, R. K., Current status of PCB toxicity to mink, and effects on their reproduction, *Arch. Environ. Contam. Toxicol.,* 6, 279, 1977.
150. Aulerich, R. J. and Ringer, R. K., Toxic effects of dietary polybrominated biphenyls on mink, *Arch. Environ. Contam. Toxicol.,* 8, 487, 1979.
151. Bleavins, M. R., Aulerich, R. J., and Ringer, R. K., Effects of chronic dietary hexachlorobenzene exposure on the reproductive performance and survivability of mink and European ferrets, *Arch. Environ. Contam. Toxicol.,* 13, 357, 1984. 154.
152. Hochstein, J. R., Aulerich, R. J., and Bursian, S. J., Acute toxicity of 2,3,7,8-tetrachlorodibenzo-p-dioxin to mink, *Arch. Environ. Contam. Toxicol.,* 17, 33, 1988.
153. Chou, C. C., Marth, E. H., and Shackelford, R. M., Experimental acute aflatoxicosis in mink, *Am. J. Vet. Res.,* 37, 1227, 1976.
154. Tillitt, D. E., Gale, R. W., Meadows, J. C., Zajicek J. L., Peterman, P. H., Heaton S. N., Jones, P. D., Bursian, S. J., Kubiak, T. J., Giesy, J. P., and Aulerich, R. J., Dietary exposure of mink to carp from Saginaw Bay. 3. Characterization of dietary exposure to planar halogenated hydrocarbons, dioxin equivalents, and biomagnification, *Environ. Sci. Technol.*, 30, 283, 1996.
155. Mason, C. F. and Wren, C. D., Carnivora, in *Ecotoxicology of Wild Mammals,* Shore, R. F. and Rattner, B. A., Eds., John Wiley and Sons, Chichester, 2001, 315.
156. Ringer, R. K., Hornshaw, T. C., and Aulerich, R. J., Mammalian Wildlife (Mink and Ferret) Toxicity Test Protocols (LC50, Reproduction, and Secondary Toxicity), U.S. Environmental Protection Agency, Corvallis, OR, EPA/600/3–91/043 (NTIS PB91–216507), 1991.
157. Hornshaw, T. C., Aulerich, R. J., and Ringer, R. K., Toxicity of *o*-cresol to mink and European ferrets, *Environ. Toxicol. Chem.,* 5, 713, 1986.
158. Fairbrother, A. and Ringer, R. K., Criteria for the protection of wildlife, in *Re-evaluation of the State of the Science for Water Quality Criteria Development,* Adams, W., Hodson, P., Reilley, M., and Stubblefield, W., Eds., SETAC Press, Pensacola, FL, in press, 2002.

159. Aulerich, R. J. and Ringer, R. K., Safronoff, J., Assessment of primary versus secondary toxicity of aroclor 1254 to mink, *Arch. Environ. Contam. Toxicol.,* 15, 393, 1986.
160. Foley R. E., Jackling, S. J., Sloan, R. J., and Brown, M. K., Organochlorine and mercury residues in wild mink and otter: Comparison with fish, *Environ. Toxicol. Chem.*, 7, 363, 1988.
161. Rattner, B. A. and Hoffman, D. J., Comparative toxicity of acephate in laboratory mice, white-footed mice, and meadow voles, *Arch. Environ. Contam. Toxicol.,* 13, 483, 1984.
162. Schafer, E. W., Jr. and Bowles, W. A., Jr., Acute oral toxicity and repellency of 933 chemicals to house and deer mice, *Arch. Environ. Contam. Toxicol.,* 14, 11, 1985.
163. Cholakis, J. M., McKee, M. J., Wong, L. C. K., and Gile, J. D., Acute and subacute toxicity of pesticides in microtine rodents, in *Avian and Mammalian Wildlife Toxicology: Second Conference STP 757,* Lamb, D. W. and Kenaga, E. E., Eds., American Society for Testing and Materials, Philadelphia, 1981, 143.
164. Linder, G. and Richmond, M. E., Feed aversion in small mammals as a potential source of hazard reduction for environmental chemicals: Agrochemical case studies, *Environ. Toxicol. Chem.,* 9, 95, 1990.
165. Linzey, A. V., Effects of chronic polychlorinated biphenyls exposure on growth and reproduction of second generation white-footed mice (*Peromyscus leucopus*), *Arch. Environ. Contam. Toxicol.,* 17, 39, 1988.
166. Batty, J., Leavitt, R. A., Biondo, N., and Polin, D., An ecotoxicological study of a population of the white footed mouse (*Peromyscus leucopus*) inhabiting a polychlorinated biphenyls-contaminated area, *Arch. Environ. Contam. Toxicol.,* 19, 283, 1990.
167. Zakrzewska, M., Effect of lead on postnatal development of the bank vole (*Clethrionomys glareolus*), *Arch. Environ. Contam. Toxicol.*, 17, 365, 1988.
168. Wren, C. D., Mammals as biological monitors of environmental metal levels, *Environ. Monitoring Assessment,* 6, 127, 1986.
169. Talmage, S. S. and Walton, B. T., Small mammals as monitors of environmental contaminants, *Rev. Environ. Contam. Toxicol.,* 119, 47, 1991.
170. Sheffield, S. R., Sawicka-Kapusta, K., Cohen, J. B., and Rattner, B. A., Rodentia and lagomorpha, in *Ecotoxicology of Wild Mammals,* Shore, R. F. and Rattner, B. A., Eds., John Wiley and Sons, Chichester, 2001, 215.
171. Shore, R.F. and Rattner, B.A., Eds., *Ecotoxicology of Wild Mammals*, John Wiley and Sons, Chichester, 730, 2001.
172. Ma, W-C. and Talmage, S., Insectivora, in *Ecotoxicology of Wild Mammals,* Shore, R. F. and Rattner, B. A., Eds., John Wiley and Sons, Chichester, 2001, 123.
173. Ross, P. S., de Swart, R. L., van Loveren, H., Osterhaus, A. D. M. E., and Vos, J. G., The immunotoxicity of environmental contaminants to marine wildlife: A review, *Annu. Rev. Fish Dis.*, 6, 151, 1996.
174. Wake, D. B., Declining amphibian populations, *Science,* 253, 860, 1991.
175. Houlahan, J. E., Findlay, C. S., Schmidt, B. R., Meyers, A. H., and Kuzmin, S. L., Quantitative evidence for global amphibian population declines, *Nature,* 404, 752, 2000.
176. Alford, R. A. and Richards, S. J., Global amphibian declines: A problem in applied ecology, *Ann. Rev. Ecol. System,* 30, 133, 1999.
177. Sparling, D. W., Linder, G., and Bishop, C. A., Eds., *Ecotoxicology of Amphibians and Reptiles*, SETAC Press, Pensacola, FL, 2000, 877.
178. Birge, W. J., Westerman, A. G., and Spromberg, J. A., Comparative toxicology and risk assessment of amphibians, in *Ecotoxicology of Amphibians and Reptiles*, Sparling, D. W., Linder, G., and Bishop, C. A., Eds., SETAC Press, Pensacola, FL, 2000, 727.
179. Hall, R. J., and Swineford, D., Effects of endrin and toxaphene on the southern leopard frog *Rana sphenocephala, Environ. Pollut. Ser.* A, 23, 53, 1980.
180. Hall, R. J. and Clark, D. R., Jr., Responses of the iguanid lizard *Anolis carolinensis* to four organophosphorus pesticides, *Environ. Pollut. Ser. A*, 28, 45, 1982.
181. Hall, R. J. and Henry, P. F. P., Assessing effects of pesticides on amphibians and reptiles: Status and needs, *Herpet. J.,* 2, 65, 1992.
182. Power, T., Clark, K. L., Harfenist, A., and Peakall, D. B., *A Review and Evaluation of the Amphibian Toxicological Literature*, Can. Wildl. Serv. Tech. Rep. Ser. 61, 1989.
183. Ouellet, M., Amphibian deformities: Current state of knowledge, in *Ecotoxicology of Amphibians and Reptiles*, Sparling, D. W., Linder, G., and Bishop, C. A., Eds., SETAC Press, Pensacola, FL, 2000, 617.

184. Cooke, A. S., Tadpoles as indicators of harmful levels of pollution in the field, *Environ. Pollut. Ser. A*, 25, 123, 1981.
185. Dumont, J. N., Shultz, T. W., Buchanan, M., and Kao, G., Frog embryo teratogenesis assay- *Xenopus* (FETAX) — A short term assay applicable to comply environmental mixtures, in *Symposium on the Application of Short-Term Bioassays in the Analysis of Complex Environmental Mixtures III.* Waters, M. D., Sandhu, S. S., Lewtas, J., Claxton, L., Chernoff, N., and Nesnow, S., Eds., Plenum Press, New York, 1983, 393.
186. Schuytema, G. S., Nebeker, A. B., Griffis, W. L., and Wilson, K. N., Teratogenesis, toxicity, and bioconcentration in frogs exposed to dieldrin, *Arch. Environ. Contam. Toxicol.,* 21, 332, 1991.
187. Dawson, D. A. and Wilke, T. S., Evaluation of the frog embryo teratogenesis assay: *Xenopus* (FETAX) as a model system for mixture toxicity hazard assessment, *Environ. Toxicol. Chem.,* 10, 941, 1991.
188. Linder, G., Barbitta, J., and Kwaiser, T., Short-term amphibian toxicity tests and paraquat toxicity assessment, in *Aquatic Toxicology and Risk Assessment. Thirteenth Symposium, ASTM STP 1096*, American Society for Testing and Materials, Philadelphia, 1990, 189, 118.
189. Mann, R. M. and Bidwell, J. R., Application of the FETAX protocol to assess the developmental toxicity of nonylphenol ethoxylate to *Xenopus laevis* and two Australian frogs, *Aquat. Toxicol.,* 51, 19, 2000.
190. Linder, G. and Grillitsch, B., Ecotoxicology of metals, in *Ecotoxicology of Amphibians and Reptiles*, Sparling, D. W., Linder, G., and Bishop, C. A., Eds., SETAC Press, Pensacola, FL, 2000, 325.
191. Sparling, D. W., Bishop, C. A., and Linder, G., The current status of amphibian and reptile ecotoxicological research, in *Ecotoxicology of Amphibians and Reptiles*, Sparling, D. W., Linder, G., and Bishop, C. A., Eds., SETAC Press, Pensacola, FL, 2000, 1.
192. Meyers-Shone, L. and Walton, B. T., Comparison of Two Freshwater Turtle Species as Monitors of Environmental Contamination, Environmental Sciences Division Publication No. 3454, Oak Ridge National Laboratory, Oak Ridge, Tennessee, 1990.
193. Bishop, C. A., Brooks, R. J., Carey, J. H., Ng, P., Norstrom, R. J., and Lean, D. R. S., The case for a cause-effect linkage between environmental contamination and development in eggs of the common snapping turtle (*Chelydra s. serpentina*) from Ontario, Canada, *J. Toxicol. Environ. Health,* 33, 521, 1991.
194. Bickham, J. W., Hanks, B. G., Smolen, M. J., Lamb, T., and Gibbons, J. W., Flow cytometric analysis of the effects of low-level radiation exposure on natural populations of slider turtles (*Pseudemys scripta*), *Arch. Environ. Contam. Toxicol.,* 17, 837, 1988.
195. De Solla, S. R., Bishop, C. A., van der Kraak, G., and Brooks, R. J., Impact of organochlorine contamination on levels of sex hormones and external morphology of common snapping turtles (Chelydra serpentina serpentina) in Ontario, Canada, *Environ. Health Perspect.,* 106, 253, 1998.
196. Hall, R. J., Effects of Environmental Contaminants on Reptiles: A review, U.S. Fish Wildl. Serv. Spec. Sci. Rep. Wildl., 228, 1980.
197. Campbell, K. R. and Campbell, T. S., The accumulation and effects of environmental contaminants on snakes: A review, *Environ. Monitoring Assessment*, 70, 253, 2001.
198. Campbell, K. R. and Campbell, T. S., Lizard contaminant data for ecological risk assessment, *Rev. Environ. Contam. Toxicol.*, 165, 39, 2000.
199. Guillette, L. J., Jr. and Crain, D. A., Endocrine-disrupting contaminants and reproductive abnormalities in reptiles, *Comments Toxicol.*, 5, 381, 1996.

CHAPTER 5

# Sediment Toxicity Testing: Issues and Methods

G. Allen Burton, Jr., Debra L. Denton, Kay Ho, and D. Scott Ireland

## CONTENTS

## 5.1 WHY TEST SEDIMENTS?

### 5.1.1 The Source and Sink

The significant role that sediments play in aquatic ecosystems is well known.[1] They serve as both a sink and a source of organic and inorganic materials, where critical cycling processes for organic matter and the critical elements (e.g., C, N, P, and S). The majority of decomposition of allocthonous and autochthonous inputs occur in the sediment. Since most chemicals of anthropogenic origin (e.g., pesticides, polycyclic aromatic hydrocarbons (PAHs), and chlorinated hydrocarbons) tend to sorb to sediments and organic materials, they also end up concentrating in the sediment. Sediment contamination can have many detrimental effects on an ecosystem, some of which are evident and others more discrete or unknown. For example, benthic invertebrate communities can be totally lost or converted from sensitive to pollution-tolerant species. These tolerant species process a variety of materials, and their metabolic products may also be different. These differences mean that ecosystem functions, such as energy flow, productivity, and decomposition processes, may be significantly altered.[2]

Loss of any biological community in the ecosystem can indirectly affect other components of the system. For example, if the benthic community is significantly changed, nitrogen cycling may be altered such that forms of nitrogen necessary for key phytoplankton species are lost and the phytoplankton are replaced with blue-green algae (cyanobacteria) capable of nitrogen fixation. The production of neuro- and hepatotoxins by the blue-green algae may then affect herbivorous fish and consumers of the water such as cattle and humans.[3] Other effects from sediment contamination are direct, as observed in the Great Lakes, where top predator fish have become highly contaminated from consuming bottom-feeding fish and benthic invertebrates that are laden with sediment-associated pollutants such as PAHs, PCB, mercury, and pesticides. Effects on ecosystem processes have been very dramatic in areas affected by both acid precipitation and acid mine drainage. However, in most areas receiving pollutant loadings the effects are difficult to observe and require use of a variety of assessment tools such as benthic macroinvertebrate community analyses, chemical testing, quantification of habitat characteristics, and toxicity testing (Figure 5.1).[4]

### 5.1.2 Regulatory Concerns

In 1998 the U.S. Environmental Protection Agency (EPA) released its Contaminated Sediment Management Strategy. This strategy shows how the EPA has authority under a variety of statutes to manage contaminated sediments (e.g., Comprehensive Environmental Response, Compensation and Liability Act, CERCLA; Clean Water Act, CWA; and Resource Conservation and Recovery Act, RCRA). Currently, the various EPA programs are defining the methods and policies they will utilize to manage this pervasive problem. One of the goals of the Strategy is to "develop scientifically sound sediment management tools for use in pollution prevention, source control, remediation, and dredged material management."[5]

In support of this goal the EPA has developed standardized methods for collecting sediments as well as methods for conducting sediment toxicity tests for freshwater and marine/estuarine sediments. In 1994 the EPA released guidelines on evaluating acute sediment toxicity utilizing the marine and estuarine amphipods *Ampelisca abdita, Leptocheirus plumulosus, Rhepoxynius abronius,* and *Eohaustorius estuarius*.[6] Also in 1994 the EPA released guidelines on evaluating acute

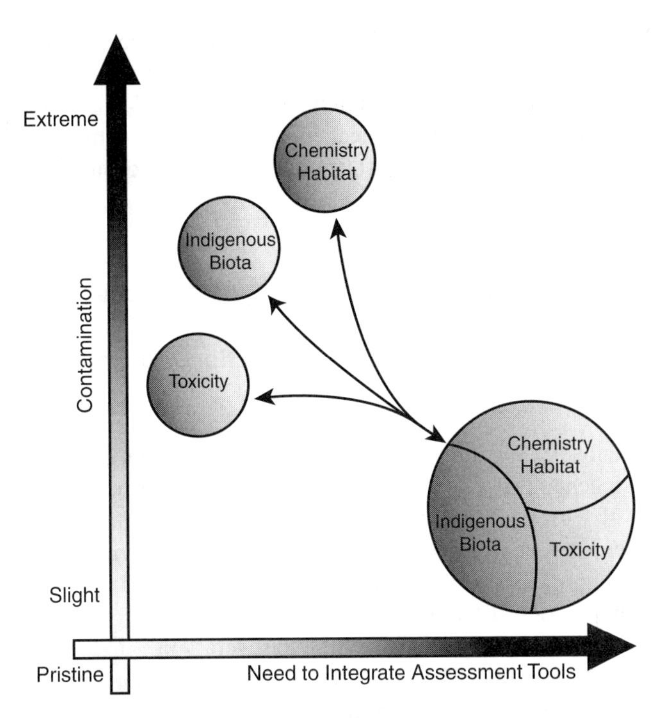


**Figure 5.1** Relationship between need to integrate assessment components and degree of ecosystem contamination. (From Burton, G.A. et al., in *Hazard Assessment of Chemicals — Current Developments,* Taylor & Francis, London, 1993, 171. With permission.)

sediment toxicity utilizing the freshwater amphipod *Hyalella azteca* and the freshwater midge *Chironomus tentans*.[7] In 2000 the EPA released guidelines on evaluating chronic sublethal sediment toxicity utilizing the freshwater amphipod *H. azteca* and the freshwater midge *C. tentans*.[8] Both the EPA 1994 and 2000 freshwater sediment toxicity test methods manuals outline methods for measuring bioaccumulation in freshwater invertebrates using the oligochaete *Lumbriculus variegatus*. The EPA has recently released guidelines on evaluating chronic sublethal sediment toxicity utilizing the marine amphipod *L. plumulosus*. [9]

A variety of theoretical and empirical approaches have been used to create sediment quality guidelines in aquatic ecosystems.[10] The empirical approach derives a sediment concentration of contaminants associated with toxic responses measured in laboratory exposures or in field assessments (i.e., Apparent Effects Threshold (AET), Effects Range Low and Medium (ERL/ERM), and Threshold and Probable Effects Levels (TEL/PEL). The EPA has used an equilibrium partitioning (EqP) approach (theoretical) to develop draft equilibrium partitioning sediment guidelines (ESGs). This approach focuses on predicting the chemical interaction among sediments, interstitial water, and the contaminants. EqP theory holds that a nonionic chemical in sediment partitions between sediment organic carbon, interstitial (pore) water, and benthic organisms.[11] At equilibrium, if the concentration in any one phase is known, then the concentration in the others can be predicted. Because the water quality criterion (WQC) is the concentration of a chemical in water that is protective of the presence of aquatic life and that is appropriate for benthic organisms, the product of the final (or secondary) chronic value (FCV or SCV) from the WQC and the organic carbon partition coefficient ($K_{OC}$) represents the concentration of a nonionic chemical in sediments that, on an organic carbon basis, is protective of benthic organisms.

EqP theory can also be used for a mixture of cationic metals.[12,13] For mixtures of cadmium, copper, lead, nickel, silver, and zinc, if the sum of the molar concentrations of simultaneously extracted metals (SEM, the metal extracted in the AVS extraction procedure) is less than the molar concentration of AVS, the sediments are predicted to be not toxic because of these metals. The use of (SEM-AVS)/$f_{OC}$ reduces variability associated with prediction of when sediments will be toxic.[13] These draft guidelines do not account for interactive (synergistic, antagonistic) effects of all chemicals in sediment, aside from the chemical mixtures addressed by specific draft guidelines (e.g., cationic metals and PAHs). These draft guidelines do not account for bioaccumulation or food chain effects.

## 5.2 ASSESSMENT ISSUES AND CONCERNS

### 5.2.1 Physical, Chemical, and Biological Interactions: Maintaining Sediment Integrity

Sediments are a semisolid medium comprised of minerals, organic material, interstitial water, and a myriad of physicochemical and biological components, all structured into micro- to macro-environments, many of which are interlinked. At the microenvironment level are approximately one billion bacteria per gram of sediment that, over the space of millimeters, occupy critical niches for metabolizing–cycling specific organic compounds, organic acids, nitrogen, sulfur, methane, and hydrogen. These populations lie at different depths in the sediment — from the narrow oxidized layer at the surface where the redox potential is high to several milli- to centimeters in depth where redox potential is strongly negative. The metabolic by-products from one zone are energy sources for the adjacent microbiological zones. Two common by-products produced in sediments containing adequate concentrations of organic matter are ammonia and hydrogen sulfide, both highly toxic and produced by numerous aquatic microorganisms. Many freshwater sediments in areas where urban and industrial inputs or organic materials have occurred contain elevated levels of ammonia. The unionized form of ammonia, which predominates in slightly alkaline conditions, has been shown to be the most toxic form for coldwater fish; however, *H. azteca* is affected equally by total ammonia.[14]

Chemical partitioning to sediment and colloidal particles via adsorption and absorption is controlled to some extent by sediment temperature, interstitial spacing and contact between particles, and dissolved organic and inorganic compounds present in the interstitial water. The surface layers of sediments are normally oxidized over a depth of 1 to several mm. This oxidized layer serves as a barrier to the transport of dissolved and reduced inorganic complexes such as ferrous iron, where reduced species become oxidized and complex with other anions and cations.

Benthic macroinvertebrates, which are infaunal, burrow into the sediment's anoxic zones (down to several centimeters) and create oxidized conditions within their burrows (tubes). This process can alter their exposure to ionic species of contaminants and result in transport of contaminants to and from contaminated sediments (by bioturbation). Most of the sediment environment is anaerobic, where the most toxic and problematic reduced chemical species exist. Though benthic macrofauna are in close proximity to these reduced sediments, they require oxygen and circulate overlying dissolved oxygen into their microenvironment. Some species of chironomids (midges) and oligochaetes (aquatic worms) are very tolerant of low dissolved oxygen levels or short periods of anoxia.

Many northern temperate freshwater lake sediment environments remain at low temperatures throughout the year. This means that the microbiological community is basically a psycrophilic one, which metabolizes best at cold temperatures. Therefore, when sediment samples are stored at 4°C, as is recommended for toxicological testing,[15] sediment metabolism does not stop but continues on at a rate reflective of the nutrient supply and the microenvironmental conditions.

In 2001 the EPA Office of Water released a guidance document on the collection, storage, and manipulation of sediments for toxicity and chemical testing.[15a] The guidance builds on recent guidance from ASTM and Environment Canada.[15,16] The guidance rarely dictates methods that *must* be followed, but rather makes recommendations at to those that *should* be followed. The lack of definitive guidance is because the optimal method to follow often is a function of both the sediment's characteristics and the study objectives. In addition, the lack of comparative studies of various methods dictates that "best professional judgments" have been made to arrive at consensus on optimal methods. Some of the more common and contentious issues of sediment handling are discussed below.

### 5.2.2 Sample Alteration

#### *5.2.2.1 Collection, Transport, and Storage*

Sediment samples cannot be collected in the field, transported to the laboratory, stored, and then tested for toxicity without altering their original structure/characteristics due to disruption of their fragile integrity. Some methods of sample collection and testing are more disruptive than others. For example, dredge (grab) samplers (e.g., Ponar, Ekman, Van Veen, Shipek, or Peterson) are more disruptive than core samplers. A standard core sampler is more disruptive than a box core sampler, and box core samplers are more disruptive than *in situ* sampling, which requires no sediment extraction.[17]

When the sediment is collected, the sampler tends to disrupt and lose the surficial fine particle layer, which is the most biologically active component and may contain the highest concentration of contaminants. Dredge samplers have a greater tendency to lose surficial fines than core samplers. The Ekman is perhaps the least disruptive of the dredge samplers, but it cannot be used in deeper waters and consolidated sediments. The sides of the dredge or core sampler also destroy the stratified gradients that exist vertically in the sediment. In addition, the underlying anoxic layers are exposed to oxygen in the collection processes most commonly used. This exposure results in the conversion of reduced species to oxidized complexes within hours; these complexes are less available for biological uptake/effects.

Sediment transport and storage are often performed under an aerobic atmosphere, which intensifies oxidation at the "new" sediment surfaces in the sampling container. This is readily observed in sediment sample bottles with the reddish-brown coloration of oxidized iron on the bottle sides and sample surface. The underlying sediments will return to the anoxic state. There have been reports that contaminated sediment samples did not show significant changes in chemistry or toxicity after several months of storage at 4°C,[18] while others have shown significant changes within days to weeks.[19,20] It is likely that sediments primarily contaminated with nonpolar, nonvolatile organics will change little during storage, due to their relatively high resistance to biodegradation and sorption to solids. Metals and metalloids, however, are more likely to be affected by changing redox or oxidation reduction and microbial metabolism (such as with arsenic, selenium, mercury, lead, and tin, all of which are methylated by a number of bacteria and fungi). Metal-contaminated sediments should be tested relatively soon after collection with as little manipulation as possible.

#### *5.2.2.2 Manipulation*

When sediments are removed from sample containers for toxicity testing, they are typically either stirred first in the bottle or placed in another container for mixing. Some methods commonly follow the mixing with sieving to remove large debris and indigenous organisms that may interfere with the toxicity assay and to better homogenize the sediment sample. All of these processes result in the introduction of oxygen to anaerobic environments, complete disruption of redox and microbiological gradients/microenvironments, alteration of equilibrium conditions for chemicals associated with the interstitial water and sediment particles, and the introduction of new compounds (such as nitrogen species, oxyhydroxy-metal compounds) into the sample via desorption and oxidation processes.[19] The microenvironments of microorganisms are usually quickly depleted of available nutrients due to the rapid sorption and metabolism of compounds by bacteria (occurring within minutes to hours). The microbial community is an excellent early warning indicator of environmental conditions because of the ability to respond so quickly to the immediate environment. For instance, regulatory testing of fecal coliforms recommends that sample analysis begin within 6 h of collection because of the quickly changing microbial population. The varying levels of sediment disruption described above will result in the introduction of new microenvironments with new sources of nutrients, thereby increasing metabolism and production of by-products (e.g., ammonia). In addition, contaminants may become either more available (from desorption or chelation) or less available for biological uptake/effects (due to sorption, complexation, biodegradation, hydrolysis, or oxidation/reduction).

There are a number of reasons why testing the sediment's interstitial (pore) water is useful. This is perceived as being the primary route of exposure for many infaunal benthic invertebrates in contaminated sediments, although such may not be the case for those species that ingest sediments. The EPA EqP approach[10,21] utilizes the interstitial water as the medium where sediment quality guidelines are applied and threshold effects equated to overlying water quality criteria. This fraction is simpler to test in many toxicity assays since test organism effects are easier to see and recover. It can also be used for assays that cannot be readily tested in solid-phase exposures, such as the algal growth assay with *S. capricornutum.*

There are a number of methods one may utilize to extract interstitial water[15] including filtration, suction, centrifugation, and *in situ* sampling with "peepers" consisting of membrane bags or chambers.[22,23] Of the laboratory methods used, the most commonly recommended is centrifugation (EPA in press).[16] All of the laboratory methods alter interstitial water chemistry, but centrifugation has been shown to alter it the least.[24] With centrifugation, the force used is important, as small-sized clays and colloids that bind toxicants may not be removed easily. This makes it difficult to make interlaboratory comparisons if extraction methods are not similar. The pH of centrifuged interstitial water has been shown to increase slightly, which increases the fraction of unionized ammonia. Some laboratories have successfully used sediment presses whereby the interstitial water is forced out via a Teflon filter.

The most desired method is to extract the pore water *in situ,* since less sample manipulation occurs (EPA in press).[25] However, this is often difficult to accomplish and produces small quantities of water that may be inadequate for toxicity testing. In this approach, dialysis membrane bags of chambers with membrane walls are embedded in the sediment and allowed to equilibrate before the pore water is extracted either by bag or chamber removal or by suction of water through access ports on the chamber. Others have embedded air stones in the sediment attached to tubing and then withdrawn pore water with a suction syringe. The reader should refer to other reviews on interstitial water sampling for additional information (EPA, in press).[15,17,22]

Testing of the elutriate (water-extractable) fraction of the sediment is a commonly used technique. It has the advantages noted for interstitial water, as the test method is similar to water-column testing and is easy to perform. Typically, sediments are mixed for 30 min in a 1:4 sediment:water ratio. They are allowed to settle for 1 h, and then the supernatant is withdrawn for testing (with or without subsequent filtration). This protocol was developed by the U.S. Army Corps of Engineers[26] to simulate dredging conditions and evaluate water quality effects during dredging operations. It is most appropriately used in dredged material evaluations. The elutriate fraction produces biological effects that are procedurally defined; if the method was modified, the responses would be different. It is obvious that the conditions used in the extraction process and assay are never identical to field conditions, so results must be extrapolated to the field with caution. Elutriate responses tend to be slightly less sensitive than exposures to solid-phase or interstitial water fractions.[14,27] The strengths and weaknesses of elutriate, pore water, and solid-phase testing have been discussed previously.[15,19]

Another manipulation of sediments occasionally employed is sediment dilution to obtain effect levels such as $LC_{50}$ or no observable effect concentrations (NOECs). There is no standard protocol, and methods have varied widely,[15,19] including diluting the sediment with water, diluting sediment with sand or reference sediment of a similar quality, or extracting the interstitial water or elutriate fraction and diluting it. Results have been inconclusive, and the optimal method is likely sediment specific. A key issue in sediment dilutions is the alteration of contaminant partitioning from the natural conditions and allowing equilibration to return. Introduction of new binding sites that may or may not be similar to site characteristics will influence bioavailability to varying degrees depending on the toxicant type and sediment characteristics. It would appear that dilutions with a nearby reference sediment with similar characteristics would be optimal. However, in this circumstance one must still allow an appropriate amount of time to pass after dilution for equilibration to occur between the contaminants and the reference sediment. These unknowns place sediment dilutions into the category of research. When used for regulatory purposes sediment dilution results should be used with caution.

A useful manipulation for identifying single chemical effect response patterns is spiking (dosing) of sediments with pure compounds. Here again there are no standard methods and few studies that have evaluated spiking methods.[20,24] There are different approaches used for spiking metals and organics.[15] Metals are usually dissolved in water and added to the sediment, which is mixed by rolling or shaking for minutes to hours. After an equilibration time (which varies widely), testing is initiated and samples are removed for toxicity testing and chemical analyses. With organics, the nonpolar compounds must be dissolved in a carrier (such as acetone) and inoculated into the sediment directly, or added to an empty container and allowed to dry to the wall of the container, with the carrier evaporating. Sediments are then added, followed by mixing. As discussed with sediment dilutions, equilibration time is an important issue for which no standard guidance exists. This may have a profound impact on the results of the toxicity test, as it will determine bioavailability and affect the accuracy of laboratory-to-field extrapolations. It is recommended that spiking studies include routine analysis of the spiked chemical in the overlying water, interstitial water, and bulk sediment to ascertain the degree of partitioning. In bioaccumulation studies tissue residues should be monitored at several points in time to determine when uptake plateaus.

#### 5.2.2.3 Summary of Manipulation Effects

Unfortunately, when we collect and test sediments we know little more than the fact that changes are occurring in the sediment.[28] We have precious few data that quantify any of the multitudinous alterations occurring due to sample collection, handling, and manipulation. Therefore, one cannot predict if the effects will be of significant consequence to the study objectives. It may be that the alterations that occur do not significantly affect measurable toxicity. Given that the contaminants of concern and the influencing sediment characteristics are not always known *a priori,* it is recommended that toxicity testing commence soon after collection, not to exceed 2 weeks' storage at 4°C.[15]

#### 5.2.2.4 Laboratory-to-Field Extrapolations of Toxicity Data

When toxicity assays are conducted in the laboratory one typically exposes organisms in a static, static-renewal, or flow-through system for 48 h to 10 days. The effect of the exposure system design on toxicity can be substantial, with dramatically differing results from the same contaminated sample. In solid-phase sediment assays flow-through tests that replace overlying water quickly may show fewer effects than systems in which overlying water is slowly replaced. "Static" toxicity assays, where waters are not renewed, are often considered a "worst-case" exposure, producing the most detrimental effects possible from the sample. However, this is not true in all situations. For example, if the sediments are calcareous, they will leach carbonates into the overlying water, thereby increasing hardness.[29] If the sediment contaminants are metals, metal toxicity may be reduced by the carbonates in the water column.

Given the above discussions on manipulation effects, it is obvious that laboratory results may not reflect *in situ* conditions, so there should always be some degree of field validation. Studies in the past decade have provided more information on the ecological relevancy of sediment toxicity testing. Ecological relevance can be evaluated in terms of the linkage between response endpoints and the ecological resources to be protected. Survival of amphipods exposed to field-collected sediments in 10- to 28-day toxicity tests has been correlated with abundance of amphipods, species richness, and other measures of community structure in the field. Also, growth of the amphipod *H. azteca* or the midge *C. tentans* has been correlated to measures of community structure.[30–34] Burton et al.[34] compared results from laboratory sediment toxicity tests to colonization of artificial substrates exposed *in situ* to contaminated Great Lakes sediment. Survival and growth of *H. azteca* and *C. tentans* in laboratory exposures were negatively correlated to percent chironomids and percent taxa colonizing artificial substrates in the field. Schlekat et al.[35] also reported general good agreement between sediment toxicity tests with *H. azteca* and benthic community responses in the Anacostia River in Washington, D.C.

There may be concern over the use of surrogate species as indicators of indigenous species effects at the same or differing levels of biological organization. The EPA conducted several field validation studies of the toxicity assay approach in their Complex Effluent Testing Program during the 1980s.[36] The results supported those of many other investigators — that sensitive species and assays (such as the *C. dubia* 7-day survival-reproduction and the *P. promelas* 7-day early-life-stage assays) did show responses indicative of effects in resident fish and macroinvertebrate populations. The EPA[37] provides a comprehensive review of the reliability of single-species toxicity test results for predicting aquatic ecosystem impacts (e.g., ecological relevance of laboratory single-species toxicity tests). Others have shown the utility of toxicity assays as predictors of receiving water quality.[38] There have been other studies,[39] however, that have shown the opposite and point to the need for multiple assays and assessment of resident communities.

Grab samples provide a "snapshot" of *in situ* conditions. Contaminant concentrations, modifying ecosystem factors, and resident communities vary over space and time in aquatic ecosystems. Fortunately, the grab sample approach of monitoring has been effective in detecting contamination, though it is uncertain how frequently false negatives may have been produced in these monitoring studies. The "gray" zones of contamination (where significant effects are more difficult to discern) will require more intensive sampling than has often been utilized in routine assessment studies. Stemmer et al.[40] showed that sediment toxicity could vary from 0 to 100% mortality in samples collected centimeters apart, while particle size and water content of the samples were equal. One approach to dealing with spatial sediment variation is composite sampling. While this is better than collecting one grab, it serves to moderate high levels of contamination and may produce false negatives.

### *5.2.2.5 The Integrated Approach*

For most sites containing contaminated sediments there is a zone of *significant* contamination and a zone of *insignificant* contamination (pollution). Determining the location of the boundary area (also known as the gray zone) requires the use of multiple assessment tools. Contaminant concentrations typically are highest in one localized area and decrease laterally in all directions. This zone can range in size from 1 m to several kilometers and vertically from centimeters in sediment depth to several meters.

Chapman and others[41–44] have promoted the use of an integrated assessment approach that utilizes measures of chemical concentrations, physical habitat, *in situ* biological communities, and sediment toxicity. No one component can delineate significant levels of contamination in the gray zone that exists at every site. Each component provides unique information about the system but results in uncertainty if used in isolation from the other components. Increased use of a variety of assessment components provides a clearer ecosystem description. The chemical approach quantifies levels of specific contaminants and can be related to water quality criteria. However, it provides no information on chemical interactive effects with other chemicals (such as additivity, antagonism, or synergism) and does not provide information either on the quantity of the chemical that is bioavailable or its effect.

The biological-community approach (discussed in other chapters) is a direct measure of *in situ* conditions and has been used effectively for decades in regulatory monitoring and assessment programs. It provides information on both community structure and function and human health risk from consumption of contaminated fish. In areas of questionable contamination the data may be inconclusive, as (1) biological communities are affected by natural disturbances (e.g., high-flow events or drought), habitat alteration (e.g., siltation from erosion or channelization), and natural spatial/temporal variation; (2) the populations being monitored may be relatively resistant to contamination (e.g., some chironomids and oligochaetes); and (3) the populations may not be responsive to short-term pollution events. Toxicity testing shows the integrative effects of multiple contaminants, providing data that are easy to interpret. The assays are relatively simple to perform and inexpensive. However, they may not provide a sensitive measure of long-term exposure to some contaminants, such as PCBs, which have carcinogenic and other chronic effects and which are tightly bound to the sediment. Another shortcoming is the need to extrapolate laboratory data to field conditions and extrapolate surrogate organism responses to indigenous species at various levels of biological organization.

The uncertainty produced by laboratory extrapolations is reduced by exposing test organisms *in situ*.[27] This approach has been used extensively in marine bioaccumulation studies using mussels. It is now being used effectively in sediment and stormwater contamination studies using cladocerans, fish, and amphipods.[23,27,45–48] Uncertainties still exist, however, in the extrapolation of the effect levels of one indicator species to effects in the test ecosystem. In addition, *in situ* exposures reduce the effects of potential stressors such as predation, stream power, and suspended solids.

The integrated approach may be most efficient when conducted in a tiered manner,[41,49] whereby toxicity testing is first conducted to guide sampling and analyses of more expensive chemical tests. If an accurate assessment is not conducted, the spatial extent of the problem may go undetected and "remedial" activities will be ineffective. Or, remedial activities may waste thousands to millions of dollars cleaning up areas that are not causing significant damage to the natural resources. This can be the more likely scenario when contamination is defined using only chemical data, as sediments can accumulate relatively high concentrations of contaminants that may or may not be biologically available.

It is evident that each approach for assessing sediment contamination has its own unique advantages and disadvantages. The magnitude or importance of these strengths and weaknesses will, to a large degree, be site- and study-objective-specific. No one rigid assessment protocol can be recommended for all situations. The sediment environment is extremely complex, and when combined with natural biological and physicochemical variability determining or predicting toxicant effects becomes an involved process requiring multidisciplinary expertise. While this may frustrate the regulator or environmental manager, it is a reality that can be and has been effectively dealt with using tiered integrative assessment strategies.[41–43,49]

A cost-effective approach is to utilize historical data to target critical areas and collect numerous samples for testing with inexpensive tests (e.g., total organic carbon, ammonia, Microtox). Fewer samples from each site can be tested with a minimum of two toxicity assays such as the 10-day solid-phase *H. azteca* survival and *C. dubia* survival-reproduction tests. These assays have been shown in many studies to be relatively sensitive and discriminatory of sediment contamination.[4,30,34] This second tier of testing should also include chronic sediment toxicity testing (e.g., the 28-day solid-phase *H. azteca* assay) and would target sites where toxicity was detected or questionable, using benthic community and habitat analyses. Once sites of significant impact were identified, chemical analyses could be better focused to help establish the impact zone, pollutant sources, and appropriate clean-up criteria.

Supplemental toxicity testing with other species will also help delineate zones of contamination and reduce the uncertainty of the "gray" zone. It has been shown at many sites[4,19] that the need to integrate these various assessment components and to use additional measures within each assessment component (such as several toxicity measures or several benthic indices) is a direct function of the degree of contamination (Figure 5.1).[10] As contamination increases, it becomes more apparent and can be detected with fewer methods.

## 5.3 TOXICITY TESTING

### 5.3.1 Recommended Assays and Test Conditions

Sediment toxicity tests have been performed with a wide variety of freshwater and marine organisms. These tests have been applied to both "water column" and benthic species with measured endpoints including survival, growth, reproduction, larval development, and luminescence. The choice of the appropriate endpoint (response) to measure is an important step in the assessment process. All toxicants do not affect the same metabolic processes and result in the same effects since they have differing modes of action and target receptors. Some toxicants may interfere with a process essential to reproduction or growth. They may also stimulate a process due to interruption of a feedback mechanism or, conversely, be essential nutrients at low concentrations. Stimulation at low concentrations of toxicant exposure has been commonly reported in the literature and is often called *hormesis*. Some responses are much more sensitive than others (e.g., enzyme inhibition vs. lethality) and should not be considered equally in evaluating the significance of ecosystem impact. For assessing the extent of contamination the most sensitive assay response may be the most useful for Tier I assessments.[42,49] Some techniques were developed specifically for testing benthic organisms (e.g., amphipods) with whole sediments. Other tests were developed by modifying methods developed for effluent toxicity testing and applying them to whole sediments, elutriates, pore water, or extracts. Test methods for some of the more commonly used organisms are described in the following pages.

### 5.3.2 General Requirements for Test Organism Care

Test organisms may be obtained from laboratory cultures, commercial suppliers, or by field collection of wild organisms. Laboratory cultures of such organisms as algae, cladocerans, fish, insects, and some amphipods can provide organisms that have a known history, age, size, and quality and that are similar at all times of the year. Adult bivalves, echinoderms, and some amphipod species are usually collected from wild populations. Test species obtained directly from wild populations should be collected only in areas documented to be free of contamination. Permits for collection, transportation, and importation of these organisms may be required by local or state agencies. These requirements, which may also specify organism disposal procedures, should be confirmed with the appropriate authorities before planning any collections.

All organisms used in a test must be from the same source or batch and must be of acceptable quality.[50–52] Damaged, dead, abnormal, and diseased organisms should be disposed of as appropriate. A group of organisms should not be used for a test if the individuals appear to be diseased or otherwise stressed or if more than 5% die during the 48-hour period immediately preceding the test. If a group of organisms fails to meet these criteria, all individuals should be discarded. The procedures for organism collection and holding should be reviewed prior to obtaining a new supply to try to identify the source of the problem. An alternate source of test organisms should always be available. In general, it is preferable to replace unhealthy organisms, since treatment may alter their sensitivity and impose additional handling stress.

The conditions for holding test organisms will vary with each species; holding conditions may be static, static-renewal, or flow-through and may be with or without substrate material. The holding tanks and any areas used for maintaining live organisms should be located in a room or space separate from that in which toxicity tests are conducted, stock solutions or test materials are prepared, or equipment is cleaned. Organisms should be cared for and handled in such a manner that they are not unnecessarily stressed. Any organisms that are accidentally dropped or handled directly must be discarded. Test organisms should not be crowded or subjected to rapid changes in temperature or water quality. Dissolved oxygen concentrations must be provided. Monitoring of ammonia levels is recommended if organisms are held under static conditions. An appropriate photoperiod should also be maintained. Holding containers may need to be covered to minimize disturbance of the organisms.

Whenever aquatic animals are brought into a laboratory, they should be kept separate from other organisms, including those of the same species. In addition to maintaining any necessary quarantine requirements, this ensures that only organisms from a particular batch are used for a given test. Dip nets and other handling equipment should be disinfected before and after each use.

Organisms must be fully acclimated to laboratory test conditions before testing begins. If a different water supply is to be used, then the water in the acclimation tank should be gradually changed from 100% holding water to 100% dilution water over a period of 1 or more days. Similarly, the water temperature should be changed over a period of several hours until the test temperature is reached. They should not be tested at temperatures outside the recommended range. With organisms that can tolerate a range of salinities, acclimation should not usually exceed 3 to 5 ppt every 2 days.[50] All organisms, with the exception of embryonic or larval stages, should be maintained in dilution water at the test temperature for at least the last 48 h before they are placed in test chambers.

### 5.3.3 Freshwater Test Organisms

Examples of test species and their associated methods are presented in the following sections, encompassing trophic levels ranging from decomposers and producers to predators and representing multiple levels of biological organization. The freshwater species that have been used successfully in assessments of sediment toxicity include both water column and benthic species: *Selenastrum capriconutum, Daphnia magna, Ceriodaphnia dubia, Pimephales promelas, Hyalella azteca, Chironomus tentans* and *Chironomus riparius,* and *Hexagenia limbata.* A few "standard" methods exist for freshwater sediment toxicity tests.[8] The EPA's Office of Science and Technology and Office of Research and Development have developed methods for *H. azteca* and *C. tentans* 10-day whole sediment exposures and a 28-day bioaccumulation test with the oligochaete *L. variegatus.* Standard guides for whole-sediment toxicity testing have been published by the American Society for Testing and Materials for invertebrates, including *H. azteca, C. tentans, C. riparius,* the cladocerans *Daphnia magna* and *C. dubia*, the mayfly *Hexagenia,* the Great Lakes amphipod *Diporeia,* and the oligochaete *Tubifex tubifex.*[51] Dredged material evaluations in the United States should follow protocols published by the EPA and U.S. Army Corps of Engineers — in 1991 for dredged material proposed for ocean disposal, and protocols published in 1998 for dredged material proposed for disposal in "inland" waters of the United States.[26,53] Canada's Ontario Ministry of the Environment has also published solid-phase test methods.[54]

#### *5.3.3.1 Algae*

The standard freshwater algal toxicity test has been the 96-hour growth test with the unicellular green alga *Selenastrum capricornutum.* Methods for culture and testing are described by the EPA.[55] This method was originally developed for use with effluents, receiving waters, and pure compounds, but it is also suitable for testing sediment elutriates and pore water. Inasmuch as the cells must be kept in suspension during the exposure period, this procedure is not suitable for use with whole sediments. In fact, it is seldom used in sediment quality assessments for this reason and the tendency for stimulated responses in the presence of nutrients. The test can be used to measure changes in cell density, biomass, chlorophyll content, or absorbance.

Starter cultures of the algae can be obtained from biological supply houses or certain university culture collections (e.g., University of Texas or University of Toronto) and then maintained in continuous culture in the laboratory. Instructions for the preparation of glassware and nutrient stock solutions are detailed by the EPA.[55] The use of appropriate aseptic techniques is critical to prevent contamination. Cultures are normally grown in a nutrient medium that includes EDTA and then tested in a medium without EDTA. The cultures can be grown in any size flask, ranging from 125 mL to 4 L; 250-mL glass Erlenmeyer flasks containing 100 mL culture medium are typical. New cultures should be started each week. The flasks are sealed with foam plugs to prevent contamination and placed in an incubation chamber at 25 ± 1°C under continuous illumination. The flasks should be placed on a shaker table or manually shaken twice daily. Subsamples should be examined regularly for contamination and growth assessment. To initiate a toxicity test the stock culture should have a cell density of $1 \times 10^2$ cells/mL and be entering a log-phase growth stage. Healthy cultures are characterized by rapid growth and a bright green color; older cultures tend to be yellowish-green in color.

The test may be conducted as a screen (100% concentration) or by using a series of dilutions to determine the no-observed effect concentration (NOEC). It may be necessary to filter elutriate or pore water through a 0.45-µm filter to remove particulate material (filtration may also reduce toxicity). The sample is warmed to test temperature, and water quality values are measured. If the dissolved oxygen (DO) level is outside the range of 40 to 100% saturation, a few minutes of aeration may be required to bring the DO to an acceptable level. Nutrients (without EDTA) are added to ensure that reduction in growth is not due to nutrient limitation. The test is generally conducted in 250-mL Erlenmeyer flasks, with a test volume of 100 mL and three replicates per treatment. Methods are also being developed to make use of microplate testing techniques. The negative (clean) control consists of three replicates of culture medium, prepared without EDTA. The inoculum density is adjusted to $1 \times 10^6$ cells/mL and each flask is inoculated with 1 mL stock culture. Each flask is seated with a foam plug and placed randomly in an incubation chamber at 25 ± 1°C with a continuous photoperiod. The flasks are swirled twice daily by hand, if a shaker table is not available, and randomly repositioned in the incubation chamber.

If direct cell counts are being made to measure changes in cell density, duplicate subsamples can be removed from each well-mixed replicate container. The samples can be counted immediately or preserved (e.g., 10% Lugol's solution). Cell counts are made of duplicate samples using a haemocytometer or a particle size counter (e.g., Coulter), and the average count (cells/mL) is calculated for each treatment. Percent inhibition and stimulation of growth relative to the control are also determined. The mean cell counts are used to calculate the no-observed effect level (NOEL) value. Water quality measures should include alkalinity, hardness, pH, and temperature at test initiation and termination.

#### 5.3.3.2 *Daphnia magna*

*Daphnia magna* has been used extensively to determine the toxicity of effluents and water samples and has been demonstrated to be sensitive to many environmental contaminants. Methods for conducting acute and sublethal (partial life cycle) tests with liquid samples have been well established[51,55,56] and may be readily adapted for testing elutriates and pore water. Extensive research has also been performed to evaluate the effects of different culture and testing regimes (e.g., diet, temperature, or culture medium). *D. magna* has largely been replaced in the United States by *Ceriodaphnia dubia*, but it is still often used in other countries. Standardized techniques involving whole sediment exposures, with acute or sublethal endpoints, have been developed.[51]

Daphnids are very sensitive organisms, and their health is easily affected by the diet and culture medium used. Cultures are prone to periodic "crashes." Once suitable culture conditions have been refined for a particular laboratory, maintenance of the cultures is relatively straightforward. Generally, *Daphnia magna* are cultured at 20 to 25°C, under a photoperiod of 16:8 h light:darkness and a light intensity of 10 to 20 $\mu E/m^2/s$ (50 to 100 ft-c). A number of water sources may be suitable for culturing *D. magna,* including the use of natural spring water. One acceptable culture medium is deionized water, reconstituted to a hardness of 160 to 180 mg/L as $CaCO_2$. Mass cultures of *D. magna* can be raised in 2-L glass beakers at a maximum density of 20 to 25 daphnids per liter. The culture water is replaced three times per week, and cultures are thinned at that time. Daphnids used in toxicity tests are usually neonates (<24 h old), although slightly older organisms (5 days old) are used for short-term chronic whole sediment toxicity tests. Brood adults should be 2 to 4 weeks old, and young from the first and second broods should be discarded. Neonates used for a particular test must all be from the same brood. To obtain test organisms brood adults are transferred to fresh culture media the day before the test is initiated. On the following day neonates (<24 h old) may be collected and used in toxicity tests. Neonates collected for a 7-day partial life cycle test can be held in 2-L glass beakers for 5 days and treated as above until the test is initiated. Daphnids can be fed a YCT (yeast-Cerophyl-trout chow) diet, supplemented with algae, on a daily basis. Healthy cultures are characterized by a golden color, absence of males, no ephippia present, and a high ratio of young to adults. Daphnids are very fragile organisms that must be handled carefully; a large-bore glass pipette should be used to transfer daphnids between containers, and they should always be released below the water surface.

A static 48-hour acute test, with an elutriate or pore water sample, may be conducted as a screen (100% concentration) or by using a series of dilutions to determine the 48-hour $LC_{50}$ value. A sufficient number of neonates of acceptable age (<24 h old) and health should be collected from a laboratory culture. The test is conducted in 250-mL glass beakers, with a test volume of 200 mL and two to four replicates per treatment. The negative (clean) control consists of an equal number of replicate containers containing dilution medium. Dissolved oxygen and temperature should be measured daily, while pH, ammonia, conductivity, alkalinity, and hardness should be determined at test initiation and termination. Each beaker is randomly seeded with five neonates, taking care to gently release the organisms below the water surface. The daphnids should be checked after about 2 h to ensure none are floating on the water surface. Water chemistry is measured daily, and each beaker is checked daily for mortalities. Daphnids that become trapped at the air-water interface are gently resubmerged. After 48 h, survival data from replicate beakers are pooled to determine the $LC_{50}$ value.

The procedure for performing static 48-hour toxicity tests using whole sediments is similar to that described above, except that there are no dilutions of sediment. Each test container is a 250-mL glass beaker, and there are normally three to five replicates per treatment. Approximately 50 g sediment (wet weight) is placed in each beaker, and then 200 mL culture water is added, taking care not to disturb the sediments. Smaller test containers can be used, as long as the 1:4 ratio of sediment to water is maintained.[51] After a brief settling period (reported times vary from 1 h to 4 days), ten neonates are added to each container, and care is taken to release them below the water surface. Monitoring of water quality is performed as described above. After 48 h, survival data are pooled and treatments are compared to the negative control.

A 7-day partial life cycle test may be conducted with 5-day-old daphnids and whole sediments. Daphnids of this age are almost ready to produce their first brood of offspring. The test is conducted in 300-mL glass beakers, with 10 replicates for each treatment. Each beaker contains 5 g sediment and 20 mL overlying water, and each is seeded with one daphnid. The overlying water is normally the same as that used for culture, but the experimental design may also require that reconstituted or receiving water be used. During the 7-day exposure the sediments are left undisturbed, but the overlying water is renewed daily or three times per week. This renewal is necessary because of the need to count and remove offspring and to prevent a buildup of food in the text containers. Each day, the adult is carefully transferred to a beaker of culture water, and any offspring are counted and removed. A total of 15 mL overlying water is removed and replaced, and care is taken not to disturb the sediments. At that time food is added to each container. The adult daphnid is then returned to its test container, and the exposure is allowed to continue. The endpoints are survival and reproduction. For toxicity tests with *Daphnia* to be considered valid, mean control survival must be at least 80%. The test is terminated when at least 60% of the control or reference sample have produced their third brood, which should occur by day 7 when tested at 25°C.

#### 5.3.3.3 *Ceriodaphnia dubia*

Methods for culturing *Ceriodaphnia dubia* and for performing 7-day partial life cycle tests with liquid samples are described by the EPA.[55] As described above for *D. magna*, they have been used for whole sediment testing as well as elutriate and pore water testing and are very sensitive to the presence of sediment contaminants.[30,34,51] They are much smaller than *D. magna* and therefore more difficult to see in turbid samples. However, their faster reproduction rate and slightly greater sensitivity than *D. magna* make them a desirable test organism. Test organisms should be obtained from an in-house laboratory culture. *C. dubia* are cultured at 25 ± 1°C, under a photoperiod of 16:8 h light:darkness. The culture medium may be either dilute mineral water (deionized water plus 10% mineral water), or moderately hard reconstituted water, with a hardness of 80 to 100 mg/L as $CaCO_3$.

Mass cultures of *C. dubia* can be reared in 0.6 to 2-L glass beakers at a density of approximately 50 daphnids per liter. The culture water is changed three times per week, and the cultures are thinned at that time. Organisms used in toxicity tests must be neonates (<24 h old). To obtain test organisms daphnids that have not yet produced their first brood are placed in individual 30-mL glass beakers containing 15 to 25 mL culture water and reared until they are 14 days old. The water is changed three times per week, and offspring are removed and counted daily. Neonates from the first and second broods should be discarded. Neonates for testing should only be collected from females that produce a minimum of 15 offspring and at least 5 neonates in their third brood. *C. dubia* may be fed YCT diet, supplemented with green algae or a *S. capricornutum*-Cerophyl mixture, on a daily basis. Healthy cultures are characterized by a golden color, absence of males, and a high ratio of young to adults. Daphnids are very fragile organisms that must be handled carefully; a large-bore glass pipette should be used to transfer daphnids between containers, and they should always be released below the water surface.

The techniques used for performing 48-hour acute and 7-day partial life cycle tests and the water quality measurements with *C. dubia* are the same as those described above for *D. magna.* The tests should be performed with neonates that are less than 24 h old. The endpoint for the acute test is survival, while the endpoints for the 7-day tests are survival and reproduction. As with the *D. magna,* short-term chronic toxicity test is terminated when at least 60% of the controls or reference sample have produced their third brood.

#### 5.3.3.4 Fathead Minnow

Toxicity tests with fathead minnow, *Pimephales promelas,* may be performed with the juvenile stage (1 to 14 days old) or with <24-hour-old larvae. Juvenile fish are typically used for acute exposures, while the larval fish are used for 7-day sublethal tests that have endpoints of survival and growth. Methods for testing fathead minnow with liquid samples are well established.[55,56] These test procedures can be applied when using elutriates, pore water, or whole sediment.[34]

Juvenile fathead minnows can be obtained from a laboratory culture or commercial supplier. Moderately hard reconstituted water should be used for culture and testing. Reagent-grade chemicals are added to Milli-Q deionized water to achieve a hardness of 80 to 100 mg/l as $CaCO_3$. The EPA[55] specifies the use of 1- to 14-day-old fish. The minimum size of test container that is recommended for an acute test is a 250-mL glass beaker, with a 200-mL test-solution volume. Elutriates or pore water may be tested at a single concentration (usually 100%) or by using a series of concentrations to determine the $LC_{50}$ value. Acute tests may also be performed with whole sediment, although without dilutions.

When testing whole sediments, a 1:4 ratio of sediment to water should be maintained in the test containers, and the water should be added carefully to reduce sediment disturbances. Culture water is usually used for the negative (clean) control. For each treatment, there should be a minimum of three replicates, with 10 fish in each. The test solutions or sediments should be distributed into their respective beakers, and each beaker carefully seeded with 10 fish. Test duration may be 24, 48, or 96 h. If the test is longer than 48 h, test solutions or overlying water should be renewed and the larvae fed daily. The tests are conducted at 20 or 25°C with a photoperiod of 16:8 h light:darkness. Water quality measurements are as described above for *D. magna.* The test endpoint is survival; observations are made daily to determine responses in the test containers. At the end of the exposure survival is recorded in each test container, and replicate data are then pooled for each treatment.

The 7-day larval fathead minnow growth and survival test can be conducted as described by the EPA,[55] using either elutriates, pore water, or whole sediments.[34,57] Pore water testing is seldom conducted due to the volume of pore water required. Test conditions and experimental designs are the same as those used for the acute tests with juvenile fish, except that the tests are normally performed at 25°C. Fathead minnow eggs can be obtained from a laboratory culture or commercial supplier. The eggs are laid on ceramic tiles (or PVC-halved pipe) that are placed in plastic basins filled with culture water and supplied with vigorous aeration. Larvae that hatch within 24 h are used for each test. Ten randomly selected larvae are placed in each test container, and test containers should be randomized with respect to position. Each replicate should be fed 0.1 mL (700 to 1000) newly hatched *Artemia nauplii* two to three times daily on days 1 through 6. Replicates are not fed on day 7. Water quality is measured each day. Dissolved oxygen and pH are measured at the start and end of each 24-h period in one test container from each treatment.

For elutriate tests, test chambers are cleaned and test solutions are renewed every 24 h. For whole sediment tests, a portion of the overlying water should be replaced each day and dead brine shrimp should be carefully siphoned from the sediment surface in each container. Daily observations should be made and the number of live larvae recorded each day. Dead larvae should be discarded. At the conclusion of the test, final counts of surviving larvae should be made and live larvae collected and killed by immersion in cold deionized water. The larvae should be rinsed with deionized water, placed in preweighed aluminum pans, and oven dried at 100°C for 2 h to obtain final dry weights. Mean survival and growth (measured as dry weight) of larvae exposed to test samples can then be compared to mean survival and growth of larvae exposed to controls and reference samples.

### *5.3.3.5 Hyalella azteca*

*Hyalella azteca* is a small freshwater amphipod that has been shown to be a sensitive indicator of the presence of contaminants in freshwater sediments. *H. azteca* is an epibenthic detritivore and herbivore and will burrow in the surface sediments in search of food. Its short life cycle, widespread and abundant distribution, ease of culture, and wide tolerance of sediment grain size and salinity make it a very suitable test species. Methods for culture and testing are summarized in the ASTM and EPA.[51,7,8] They have been used extensively for whole sediment toxicity testing in North America. In Europe *Gammarus* is the amphipod of preference.

*H. azteca* can be obtained from a commercial supplier or laboratory culture. The amphipods can be held in 80-L glass aquaria filled with about 50 L moderately hard reconstituted water (80 to 100 mg/L as $CaCO_3$). [29] A flaked food (e.g., Tetrafin®) is added to each culture chamber, receiving daily water renewals to provide about 20 g dry solids/50 L of water twice weekly in an 80-L culture chamber.[8] Each culture chamber has a substrate of maple leaves and artificial substrates (six 20-

cm diameter sections per 80-L aquaria of nylon "coiled-web material"; 3-M, St. Paul, MN). Before use, leaves are soaked in 30% salt water for about 30 days to reduce the occurrence of planaria, snails, or other organisms on the substrate. The leaves are then flushed to remove the salt water and residuals of naturally occurring tannic acid before placement in the cultures.[8]

Cultures should be maintained at 23°C under a 16:8-hour photoperiod at an illuminance of about 100 to 1000 lux. Gentle aeration is provided. Water in culture chambers is changed weekly. Survival of adults and juveniles and production of young should be measured at this time. Mixed-age amphipods may be separated by sieving the amphipods through 250-λm, 425-λm, and 600-λm sieves. Sieves should be held under water to isolate the amphipods. Artificial substrates or leaves are placed in the 600-λm sieve. Culture water is rinsed through the sieves, and small amphipods stopped by the 250-λm sieve are washed into a collecting pan. Larger amphipods in the other two sieves are returned to the culture chamber. The smaller amphipods are then placed in 1-L beakers containing culture water and food (about 200 amphipods per beaker) with gentle aeration. Newborn amphipods should be held for 6 to 13 days to provide 7- to 14-day-old organisms to start a 10-day test or should be held for 7 days to provide 7- to 8-day-old organisms to start a long-term test. [8]

Assessment of whole sediment toxicity involves a 10- to 4 two-dimensional exposure of juvenile amphipods to sediments, using procedures described by the EPA and ASTM.[8,51] Both the short-term (10-day) and long-term (42-day) sediment toxicity tests are conducted in 300-mL high-form lipless beakers. The sediment volume for both tests is 100 mL with 175 mL of overlying water. The recommended number of replicate chambers for routine testing for the 10-day sediment test is 8, and the recommended number of replicate chambers for the 42-day sediment test is 12.[8] Sediments are prepared the day before test initiation and allowed to equilibrate overnight. The following day (day 0), test organisms are added (10 organisms per chamber for both the short- and long-term sediment tests), and the experiment begins. Prior to distribution to the test containers, the overlying water is renewed, with two volume replacements per day, continuous or intermittent (e.g., 1 volume addition every 12 h), thereafter.[8] Other successful test methods have consisted of a 1:4 sediment:water ratio in 30 (1 organism per beaker, 10 replicates) to 250-mL beakers (10 organisms per beaker, 3 replicates), with daily water renewal; however, these methods do not follow the EPA methods.[57] A negative (clean) control, consisting of fine silica sand (culture material) or mesh, is tested concurrently. Juvenile amphipods (as outlined above) are sieved or picked from their holding containers, and 10 amphipods are randomly distributed to each test container. Sieved organisms should be held for 1 to 3 days prior to testing to check for sieve-related mortality.

The short-term sediment test is allowed to proceed for 10 days, and the long-term sediment test is allowed to proceed for 42 days. Both tests are conducted at 23 ± 1°C under a 16:8-hour light:dark photoperiod. Gentle aeration is provided if needed to keep the DO greater than 2.5 mg/L throughout the test. Hardness, alkalinity, conductivity, and ammonia are monitored at the beginning and the end of both sediment tests. For the 10-day sediment test, pH is also monitored at the beginning and end of the test. Also for the 10-day sediment test with *H. azteca*, temperature and DO are monitored daily. For the 42-day sediment test, temperature is monitored daily, conductivity weekly, and DO and pH are three times per week. Amphipods are fed 1.0 mL of YCT daily to each test chamber At test termination of the 10-day sediment test the sediments are sieved and the number of live, dead, and missing amphipods in each test container recorded. The test endpoints from the short-term sediment test with *H. azteca* are survival and growth measured on day 10.

For the long-term sediment test, the amphipods are isolated on day 28 from the sediment and placed in water-only chambers where reproduction is measured on days 35 and 42. Endpoints measured in the long-term amphipod test include survival (days 28, 35, and 42), growth (days 28 and 42), and reproduction (number of young per female from day 28 to 42). For best recovery of the live organisms, the test beakers should be gently swirled several times to resuspend the upper layer of sediment and then quickly poured into the sieve. Since the *H. azteca* do not burrow into the deeper layers, this method allows recovery with minimal sieving. Test acceptability for the 10-

day sediment test is a minimum mean control survival of 80% and measurable growth of test organism in the control sediment. For the 42-day sediment test, acceptability is defined as a minimum mean control survival of 80% on day 28. For both tests, additional performance-based criteria specifications are outlined in EPA.[8]

### 5.3.3.6 *Chironomus tentans* and *Chironomus riparius*

This test provides information on the bioavailability and adverse effects of contaminants associated with whole sediment using the freshwater midges. *C. tentans* and *C. riparius* have been used successfully in freshwater toxicity tests because these species are fairly large midges with a short generation time, are easily cluttered, and the larvae are in direct contact with the sediment. They have been shown to be sensitive to many contaminants associated with sediments. Methods for culture and testing are summarized in ASTM for both *C. tentans* and *C. riparius*, and in EPA for *C. tentans.* [8,51] *C. tentans* has been used extensively for whole sediment testing in the United States, and *C. riparius* has been used often in Canada and western Europe.

Assessment of whole sediment toxicity involves a 10–day to about 60-day exposure of juvenile *C. tentans* to sediments. Both the short-term (10-day) and long-term (about 60-day) sediment toxicity tests are conducted in 300-mL high-form lipless beakers. The sediment volume for both tests is 100 mL with 175 mL of overlying water. The recommended number of replicate chambers for routine testing for the 10-day sediment test is 8, and the recommended number of replicate chambers for the 60-day sediment test is 16. Sediments are prepared the day before test initiation and allowed to equilibrate overnight. The following day (day 0) test organisms are added (10 organisms per chamber for the short-term sediment test, and 12 organisms per chamber for the long-term sediment test), and the experiment begins. Prior to distribution to the test containers the overlying water is renewed, with two volume replacements per day, continuous or intermittent (e.g., 1 volume addition every 12 h), thereafter. [8]

Second- to third-instar larvae (approximately 10 to 12 days postdeposition; all organisms must be third instar or younger with at least 50% of the organisms at third instar) are used in the short-term sediment test. Less than 24-hour-old larvae are used to initiate the long-term *C. tentans* sediment test. *C. tentans* can be cultured in soft water in glass aquaria. A water volume of about 6 to 8 L in these flow-through chambers can be maintained by drilling an overflow hole in one end 11 cm from the bottom. The top of the aquarium is covered with a mesh material to trap emergent adults. Panty hose with the elasticized waist are positioned around the chamber top, and the legs are cut off. About 200 to 300 mL of 40-mesh silica sand is placed in each chamber. The stocking density of the number of *C. tentans* eggs should be about 600 eggs per 6 to 8 L of water. Fish food flakes (i.e., Tetrafin®) are added to each culture chamber to provide a final food concentration of about 0.4 mg dry solids/mL of culture water.[8]

The short-term sediment test is allowed to proceed for 10 days, and the long-term sediment test is allowed to proceed for about 60 days. Both tests are conducted at 23 ± 1°C under a 16:8 h light:dark photoperiod. Gentle aeration is provided if needed to keep the DO greater than 2.5 mg/L throughout the test. For the short-term sediment test, hardness, alkalinity, conductivity, pH, and ammonia are monitored at the beginning and end of the test. Temperature and DO are monitored daily. For the long-term sediment test, hardness, alkalinity, conductivity, and ammonia are monitored at the beginning, on day 20, and at the end of the test. Temperature is monitored daily, DO and pH three times per week, and conductivity weekly. For the short-term test, each test chamber is fed 1.5 mL of Tetrafin® (1.5 mL contains 6.0 mg of dry solids) daily. For the long-term sediment test with *C. tentans*, each chamber is fed 1.5 mL of Tetrafin® (1.0 mL contains 4.0 mg of dry solids) daily starting the day prior to when the organisms are added to the test chambers. The test endpoints for the short-term sediment test is survival, and growth is measured on day 10.

The test endpoints measured in the long-term sediment exposures with *C. tentans* include survival and growth measured on day 20, with emergence and reproduction being measured daily starting on day 23 to the end of the test (approximately 60 days). The number of eggs per female is determined for each egg mass, which is incubated for 6 days to determine hatching success. Test acceptability for the short-term test is a minimum control survival of 70%, with a minimum mean weight/surviving control organism of 0.48 mg ash-free dry weight (AFDW). For the long-term test, test acceptability is defined as the average size of *C. tentans* in the control sediment at 20 days that must be at least 0.6 mg/surviving organism as dry weight or 0.48 mg/surviving organism as AFDW. Emergence should be greater than or equal to 50%. For both tests, additional performance-based criteria specifications are outlined in EPA.[8]

### 5.3.3.7 *Hexagenia limbata*

*Hexagenia* are large-bodied mayflies belonging to the order Ephemeroptera. The most common test species include *H. limbata* (Serville), *H. rigida* (McDunnough), *H. bilineata* (Say), and *H. munda* (Eaton), and they are common to the United States and Canada.[58] Mayfly nymphs live in U-shaped tubes that are formed in freshwater aquatic sediments and are continuously exposed to sediment, pore water, and overlying water.[65,66] Mayfly nymphs have been used in whole sediment toxicity tests [58–63] and pore water exposures[65,67] as well as for examining the bioaccumulation dynamics of sediment-associated contaminants.[68] Their life history consists of four stages: eggs, nymphs, subimagoes, and adults or imagoes; their life cycle is 1 to 2 years, depending on temperature. The emergence of adults occurs over a short period of time, culminating in massive swarms during the summer months. Each female adult can produce an average of 4000 eggs. Edmunds et al. [69] report body length of the nymph between 12 and 32 mm and adult wing length between 10 and 25 mm. They are tolerant of a wide range of grain size and organic content, but growth has been reported to be limited in sediments with a high sand fraction and low total organic carbon. An annex to ASTM,[51] which provides guidance for conducting *Hexagenia* tests, has recently been developed.

Organisms can be obtained from the wild, another laboratory, or a commercial source either in the form of nymphs[61] or eggs.[70] All individuals in a test should be obtained from the same sources and collected from a clean site. Mayfly nymphs used for testing are reared in the laboratory and not continuously cultured due to the length of the mayfly's cycle and the conditions necessary to mature successfully. Therefore, nymphs or eggs are collected as necessary. [51] Mayfly nymphs are found in lakes, rivers, and ponds with soft mud and fine silt/clay bottoms and are found infrequently in areas containing gravel, sand, or peat.[71] The method of collection will vary (e.g., dipnet, grab sampler) depending on water depth, current, and substrate characteristics. Once collected, they are collected from their native sediments and then transported to the laboratory in cool, aerated water. *Hexagenia* spp. nymphs brought back into the laboratory should be acclimated (at a rate not to exceed 2°C within 24 h) to the culture water by gradually changing the water in the holding chamber from the water in which they were transported to 100% culture water. Adult mayflies can be collected during the summer to obtain fertilized eggs.[48] The eggs can be stored at 8°C within polypropylene bags or petri dishes holding clean water and containing an air space. These eggs can be held at 20°C for 6.5 days and gradually cooled to 8°C in 4°C increments every 4 days. Eggs can be stored for 41 weeks, and for successful hatching the eggs may be transferred from 8°C directly to 20°C (room temperature) or warmed in 4°C increments every 7 days (12°C for 7 days, 16°C for 7 days). Hatching begins in 6 to 8 days at 20°C using either approach.[51] Newly hatched nymphs have been held in static, aerated aquaria ranging from 1 to 40 L in size. The photoperiod can be maintained on a 16l:8D or the natural photoperiod of the region (196, 277, 281). Dechlorinated water has been used for rearing *Hexagenia* spp., with a pH of 7.1 to 8.2 and total hardness of 100 to 144 mg/L as $CaCO_3$ (277). Nymph density will vary with organism size.

Fine silt/clay sediment can be obtained from a native area known to support mayfly populations[59,62,70] or can be made by mixing reconstituted potting soil and clay[63] with the addition of silica sand.[72] The field-collected sediment is prepared by initially autoclaving the sediment,[59,70] which may be followed by an exposure to air for 48 h.[62] The sediment is placed into aquaria to a uniform depth (1 to 2 cm[62,70]), overlain with water, and allowed to settle.[70] Newly hatched nymphs are interstitial sediment dwellers and do not require feeding for the first 7 days since the sediment can provide sufficient nourishment for establishment. Young organisms may be fed an algal suspension for the first month of development (that is, 10% *S. capricornutum* and 10% *Chlorella fusca*[62]). The feeding solutions are prepared by blending the appropriate amount of material in water until a fine slurry is achieved.[51]

As stated earlier, toxicity tests can be performed using whole sediments or aqueous fractions, such as pore water or elutriates. If the mayflies are tested without sediment, artificial burrows (e.g., small glass tubes) must be provided as a burrowing substrate to reduce stress.[67] Nymphs used in toxicity tests are usually approximately 3 to 5 months old. Nebeker et al. [59] described 10-day, static, whole sediment toxicity tests using 10 young mayfly nymphs (<10 mm long) placed into 1-L beakers or 10 older nymphs (>10 mm long) placed into 4-L glass beakers. The 1-L jars contained 200 mL of sediment and 800 mL of overlying water, and the 4-L jars contained 800 mL of sediment and 3200 mL of overlying water, both achieving a 4:1 (v:v) water:sediment ratio. Organisms are not fed in either the static or recirculating short-term tests.[59,63,73]

In addition to the standard 10-day exposure, Bedard et al. [74] have also developed a 21-day sediment test that has growth and survival as endpoints. This 21-day, static, whole sediment toxicity tests uses ten early instar mayfly nymphs (<8 mm long, about 5-mg wet weight) placed into 1.8-L (11.5 × 11.5 × 14.5-cm) wide-mouthed glass jars with a minimum of three replicates. The 1.8-L jars held 325 mL of sediment and 1300 mL of water, providing a 4:1 (v:v) water:sediment ratio. Air was bubbled through Pasteur pipettes positioned just below the water surface to maintain a dissolved oxygen concentration of 7 to 10 mg/L in the overlying water. A flowmeter was used to regulate the air supply to every six test chambers. The test chambers were covered with loosely fitting plastic lids.

Organisms might not need to be fed during the 21-day tests depending on the natural food content of the sediment. Previous studies indicated that diet did not influence early instar nymph growth or survival over a 21-day test exposure for a number of sediment types.[74] The endpoint for short-term tests is mortality.[59,61,63] The endpoints for long-term tests are survival and growth. Burrowing behavior[63] and molting frequency[63,73] can also be monitored throughout the test depending on the turbidity of the overlying water. For any toxicity tests performed with *Hexagenia,* mean control survival must be at least 80% for the tests to be considered valid. Additional performance-based criteria specifications are outlined in the ASTM annex. [51]

### *5.3.3.8 Estuarine and Marine Test Species*

The following section presents examples of toxicity tests for marine and estuarine species; these range from fish to bacterial levels of biological organization. The marine and estuarine species frequently and successfully used in studies of sediment contamination include the fish *Cyprinodum variegatus* and *Menidia beryllina,* and amphipods *Rhepoxynius abronius, Ampelisca abdita, Eohaustorius estuarius, Grandidierella japonica* and *Leptocheirus plumulosus,* the bivalves *Haliotis rufescens, Crassostrea gigas,* and *Mytilus* sp., the echinoderms *Strongylocentrotus purpuratus, Dendraster excentricus,* and *Arbacia punctulata,* the polychaetes *Dinophilus gyrociliatus* and *Neanthea arenaceodentata,* the mysid *Americamysis bahia* (formerly *Mysidopsis bahia),* the algal *Champia parvula,* and the bacterial assay Microtox using *Vibrio fischeri.* As with freshwater tests, several "standard" methods for marine and estuarine organisms exist. The EPA has standardized protocols for toxicity testing with the amphipods *Rhepoxynius, Ampelisca, Eohaustorius, and Leptocheirus* using a 10-day survival assay, a chronic 28-day survival, growth reproduction assay with *Leptocheirus,* and 28-day bioaccumulation assays with *Macoma* and *Nereis.*[6,9] The European OECD (Organization of Economic Cooperation and Development) is discussing the possibility of developing standardized sediment tests. Environment Canada has developed methods for sediments, elutriates, and leachate samples using *Vibrio fisheri* luminescence, sea urchins and sand dollars (echinoids), and seven amphipod species. Methods were recently published by the Chesapeake Bay Program for benthic testing using *Lepocheirus plumulosus, Ampelisca abdita, Lepidactylus dytiscus,* and *Monoculodes edwardsi.* Dredged materials in the United States should be tested using EPA and U.S. Army Corps of Engineers protocols.[26,53] The American Society for Testing and Materials has standard procedures for amphipods, fish, mysids, echinoderm and oyster early life stages, polychaetes, and algae.[75–78]

#### *5.3.3.8.1 Larval Fish*

Sediment toxicity tests with estuarine and marine samples can be performed using the juvenile or larval stages of the sheepshead minnow, *Cyprinodum variegatus,* or the inland silverside, *Menidia beryllina.* Juvenile fish are typically used for acute exposures, while the larval fish are used for 7-day sublethal tests that have endpoints of survival and growth.[55,77] These test procedures may also be applied when testing elutriates, pore water, or whole sediment; however, as with the algae tests, volume requirements (200 mL/rep) prohibit the use of porewaters except under the most unusual conditions.

Juvenile fish (1 to 14 days old for sheepshead minnow, 9 to 14 days old for inland silverside) can be obtained from a laboratory culture or commercial supplier. Methods for culturing and testing sheepshead minnow and silverside with liquid samples are well established.[55,56,77] Both species are tolerant of a wide range of salinities 0 to 32 ppt if they are acclimated following hatching, making them suitable for testing estuarine and marine samples. Water used for culture and testing should be either natural seawater from an uncontaminated source or artificial seawater (prepared from hypersaline brine or aquarium salt mix), adjusted to the salinity required for a particular test. The minimum recommended test container size for an acute test is a 250-mL glass beaker, with a 200-mL test solution volume. Elutriates and pore waters may be tested at a single concentration (usually 100%) or using a series of concentrations to determine the $LC_{50}$ value. Acute tests may also be performed with whole sediment.[79] When testing whole sediments a 1:4 ratio of sediment to water should be maintained in the test containers, and the water should be added carefully to reduce sediment disturbance. Culture water is usually used for the negative (clean) control. For each treatment there should be a minimum of three replicates, with 10 fish in each. The test solutions or sediments should be distributed into their respective beakers and each beaker carefully seeded with 10 fish. If the test is longer than 48 h, test solutions or overlying water should be renewed and the larvae fed. The tests are conducted at 20 or 25°C with a photoperiod of 16:8 h light:darkness. Temperature, dissolved oxygen, salinity, and pH should be measured in each treatment every 24 h. The test endpoint is survival; observations are made daily to determine responses in the test containers. At the end of the exposure survival is recorded in each test container, and then replicate data are pooled for each treatment.

The 7-day larval growth and survival test can be conducted with either species for effluents, using pore water, elutriates, or whole sediments.[56,77,79] Test conditions and experimental designs are the same as those used for the acute tests with juvenile fish, except that the tests are normally performed at 25°C. Larval fish (<24-hour-old sheepshead minnow or 7- to 11-day-old inland silversides) can be obtained from a laboratory culture or commercial supplier. Larvae that hatch within 24 h of each other are used for each test. Ten randomly selected larvae are placed in each test container, which should be randomized with respect to position. Each replicate should be fed 0.1 to 0.15 mL newly hatched *Artemia nauplii* suspension daily on days 1 through 6. Replicates are not fed on day 7. Water quality is measured each day. Dissolved oxygen and pH are measured at the start and end of each 24-hour period in one test container from each treatment. For elutriate or pore water tests, chambers are cleaned and test solutions are renewed every 24 h.

For whole sediment tests, a portion of the overlying water should be replaced each day and dead brine shrimp carefully siphoned from the sediment surface in each container. Daily observations should be made and the number of live larvae recorded each day. Dead larvae should be discarded. At the conclusion of the test final counts of surviving larvae should be made and live larvae collected and killed by immersion in cold deionized water. The larvae should be rinsed with deionized water, placed in preweighed aluminum pans, and oven-dried to constant weight to obtain final dry weights. Mean survival and growth (measured as dry weight) of larvae exposed to test samples can then be compared to mean survival and growth of larvae exposed to controls. An acceptable test should have 80% survival in the controls.

#### *5.3.3.8.2 Amphipods*

Static or flow-through acute 10-day toxicity tests can be used to determine the relative toxicity of marine and estuarine sediments using the amphipods *Eohaustorius estuarius, Ampelisca abdita,* and *Leptocheirus plumulosus. Rhepoxynius abronius* and *Grandidierella japonica* can be used to test marine sediments with salinities >25 ppt. These tests can be conducted according to procedures described in several publications.[6,9,75,80–88] Both the EPA and ASTM provide guidance for conducting tests with these species that summarizes ecological and testing conditions for each.[6,75] On the west coast of the United States, *R. abronius* has been widely used in sediment toxicity assessments and has been recommended as a benchmark test species for comparing different data sets and assessing the sensitivity of newly developed test methods.[75,81,82,89] It is suitable for marine sediments having interstitial salinities of at least 25 ppt. These tests are conducted using filtered seawater at a salinity of 28 ± 2 ppt. *Grandidierella japonica* has also been used in studies along the southern west coast of the United States; its ability to survive in varying sediment types makes it useful for a variety of research and regulatory applications. *A. abdita* has been tested in marine sediments at interstitial water salinities of 0 to 35 ppt when overlying water salinity is >28 ppt. The estuarine amphipods *E. estuarius and Leptocheirus plumulosus* are suitable for testing sediments for interstitial salinities ranging from approximately 1 to 33 ppt.

*Rhepoxynius abronius* is found on the west coast of the United States, from Puget Sound, Washington to California. It is a free-burrowing amphipod, typically found in the low intertidal-subtidal areas in the upper 2 cm of clean fine sands.[81] This organism may exhibit mortality when held in sediments composed largely of silt and clay.[82] Test organisms are obtained from field collections at uncontaminated sites, rather than from laboratory cultures. The amphipods can be collected with an epibenthic dredge, using short hauls (20 m) to prevent damaging the animals. Amphipods are sieved from the sediment for enumeration and then placed in holding containers with clean seawater and about 10 cm sieved sediment; up to 1000 amphipods can be held in a 20-L plastic bucket. The buckets should be kept cool and transported to the laboratory as soon as possible. About 5-L sieved collection site sediment should also be retained for use as the negative control. In the laboratory amphipods are held at 15 ± 1°C under continuous light with gentle aeration provided. Swartz et al.[81] recommended sorting the amphipods, transferring them into small containers (20 in each) with clean sediment, and holding them in flowing seawater until used for testing. The amphipods can be held under static conditions if the overlying seawater is replaced every 1 to 2 days. The cultures are not fed, but the seawater (28 ± 2 ppt) is replaced every 1 to 2 days. The amphipods should be used within 2 to 10 days of collection.

*Ampelisca abdita* is found on the east coast, from Maine to northern Florida, and in the eastern Gulf of Mexico; there is also an introduced population in San Francisco Bay.[84] Unlike burrowing *Rhepoxynius abronius* and *Eohaustorius estuarius,* this species is a tube dweller, building soft 2- to 3-cm-long tubes in fine silty surface sediments. There is some evidence that sediments containing >95% sand may elicit mortality (John Scott, SAIC Narragansett, RI, personal communication). *A. abdita* has a shorter life cycle (6 weeks at 20°C) than other amphipods, which makes it a candidate for tests using reproduction as an endpoint.[84] Test organisms can be purchased or obtained from field collections. For field collections, *A. abdita* are collected using a small dredge or grab or by skimming surficial sediments with a hand-held long-handled net. The amphipod tubes are then gently sieved in the field to separate them from surrounding sediments and the amphipods and tubes are immediately transported back to the laboratory in clean buckets with overlying seawater and maintained at temperatures either at or below collection temperature.

In the laboratory the amphipods and tubes are placed on a sieve series consisting of 2.0-, 1.0-, and 0.5-mm sieves. Collection temperature seawater is sprayed over the tubes to separate the amphipods from the tubes. The amphipods fall through the sieve series and are sorted according to size. In the laboratory *A. abdita* are held in clean sediment under flow-through conditions at collection. If need be, they are acclimated at 2 to 3ºC/day until they reach test conditions (20ºC). During holding and acclimation they are fed a diatom algae *(Phaeodactylum tricornutum or Skeletonema)* daily.[6] Organisms should be used within 2 to 10 days of collection.

*Eohaustorius estuarius* is distributed from British Columbia to central California[83] and is found in clean, fine sediments in the intertidal region. Amphipods can be collected near the shore using a shovel and handled and maintained using the same methods as for *R. abronius,* except that the holding containers are placed in aquaria filled with water at the same salinity as the collection site (3 to 10 ppt). *E. estuarius* has been successfully tested in salinity ranges from 0 to 34 ppt and in sediments with a wide range of grain-size characteristics.[91] Tests are generally performed at the salinity of the test sediments, so it may be necessary to acclimate the amphipods to increased salinities. *E. estuarius* can acclimate to changes from 5 to 25 ppt within a matter of hours, but such a short acclimation period should probably be avoided.

*Grandidierella japonica* has been used for environmental studies in southern California and is found throughout California bays and estuaries. Its ability to survive well in a variety of different sediment types makes it useful as a test species. It can be collected from intertidal and subtidal sediments, by collecting the top 2 to 4 cm of sediment in a bucket, stirring the sediment gently, and pouring the supernatant fluid through a 1.0-mm sieve. The material collected on the sieve should then be transported to the laboratory for sorting. Females with embryos in their marsupium should not be used. Tests should be conducted between 15 and 19ºC with overlying water of 30 to 35 ppt.[75]

*Leptocheirus plumulosus* has been reported in the literature as a useful species for testing the toxicity of estuarine sediments.[86,87] It is found along the east coast of the United States from Massachusetts to northern Florida. It inhabits both oligohaline and mesohalineareas in subtidal sediments ranging from fine sand to clay. Organisms may be either field collected or cultured.[6] Field-collected organisms are obtained using a small dredge or grab and should be held no longer than 4 days prior to testing. Organisms should be 2 to 4 mm in length or retained on a 0.5-mm sieve after passing through a 0.71-mm sieve. *L. plumulosus* may be tested in an acute or chronic test.[9] The chronic test method is similar to the acute test method described below, with the following exceptions. The test organisms should not vary more than 2 to 3 mm in size, and 10 organisms per replicate are used. In a 28-day exposure organisms are fed approximately three times per week — 20 mg finely ground TetraMin during days 0 to 13 and 40 mg finely ground TetraMin during days 14 to 28. One third of the water in each jar is replaced two times per week. At test termination survival, young production, and growth (length) are the response measures. A 28-day exposure assay is acceptable with 80% survival in the controls.

Test containers for 10-day whole sediment toxicity tests are 1-L glass jars or beakers, each containing a 2-cm layer of sediment and 800 mL filtered seawater (adjusted to the appropriate salinity as necessary). Test containers for *A. abdita are* 1-L glass jars with a screened hole drilled near the top to allow water overflow in a flow-through arrangement. Negative controls consist of sediment from the amphipod collection site or "clean" sediments proven to be nontoxic and filtered seawater. The sediments are prepared the day before test initiation and allowed to equilibrate overnight, and the amphipods are added the following day (day 0).

Replication of treatments is dependent upon the objectives of the experiment; however, a minimum of five replicates is recommended for each treatment. Each replicate should contain 20 amphipods. Daily monitoring of water quality parameters (temperature, pH, dissolved oxygen, and salinity) should be conducted and may be done in one of the replicates or an additional separate monitoring replicate. The test is conducted at 15 ± 1°C (20 ± 1°C for *A. abdita*, 25 ± 2°C for *L. plumulosus)*, under continuous illumination for 10 days, with gentle aeration provided. Water quality and emergence *(R. abronius and E. estuarius* only) are measured daily. Organisms are not fed during the test. After 10 days the sediments are sieved, and the number of living, dead, and missing amphipods is determined for each replicate.

The level of effort required to recover surviving amphipods varies with the species. *R. abronius* are the easiest to recover because they are more easily distinguished by their pink color and because they tend to float during sieving. *A. abdita* are the most difficult to recover from their tubes. Response criteria include mortality, emergence from sediment, and ability to rebury in clean sediment after a 10-day exposure. Test acceptability in all the amphipod 10-day exposures is 90% survival.

##### *5.3.3.8.3 Bivalves*

Methods for performing toxicity tests with aqueous samples using four species of bivalve embryos are described in the EPA[88] and ASTM[50] and the red abalone larval development in the EPA.[88] Application of the oyster larvae test method to sediment toxicity assessment was developed by Chapman.[89] Species of bivalves are commonly used to conduct 48-hour larval development toxicity tests: the Pacific oyster, *Crassostrea gigas,* and the mussel, *Mytilus sp.* (*M. edulis*, *M. californianus*, *M. galloprovincialis*, or *M. trossulus*). Since the different mussel species have been difficult to distinguish, on the west coast bay mussels are referred to at the genus level for toxicity testing. Although the larvae do remain in the water column during the exposure period, tests can be conducted in the presence of pore water[90–93] or by using elutriates.[26]

Adult oysters may be obtained from wild populations, oyster farms, or commercial hatcheries, depending on the season. The natural spawning season for oysters in the Pacific Northwest is June to September, but this can be extended with some success by artificial conditioning and by obtaining gravid oysters from California. Upon arrival, the oysters should be cleaned of fouling organisms and held in either a flow-through or static-renewal seawater system at 20 ± 1°C, 30 ± 2 ppt salinity, and a photoperiod of 16:8 h light:darkness. The oysters should be kept in small shallow trays (5 to 10 oysters per tray) to minimize losses due to mass spawnings. Care must be taken with the animals as rough handling may cause spontaneous spawning of very mature oysters. Oysters being brought into spawning condition should be held initially at the collection site temperature and acclimated to the conditioning temperature at intervals of 2°C every 2 days. During the summer oysters may require little or no conditioning, but in the winter the process may take 6 weeks or more and may be unsuccessful. The condition of the gonads should be checked regularly by sacrificing a few oysters and examining them by dissection. Healthy, mature oysters are characterized by plump, milky gonads, with females having many round or teardrop-shaped granular eggs and males having active, motile sperm. Conditioned oysters should not be held for more than 6 to 8 weeks, as gamete condition will begin to deteriorate. The oysters should be fed daily, using an algal suspension prepared from a marine diatom paste. Fecal material should be siphoned from the tanks daily when any mortalities are removed and recorded.

Adult mussels can be obtained from wild populations collected from uncontaminated areas or from commercial suppliers. The natural spawning season for local mussels on the U.S. west coast can be extended all year by locating different mussel species up and down the coast. Much of the information provided for oysters also applies to mussels. The mussels are kept in small shallow trays (10 to 20 per tray) to prevent mass spawnings, at 15 to 1°C. Development rates may depend on the species, local population characteristics, or other factors. Healthy, mature mussels are characterized by plump, orange gonads along the insides of the shells.

Toxicity tests are used to provide information about the effects of marine sediments (solid-phase or elutriate) on the embryos and larvae of Pacific oysters and mussels.[94] Mature adult bivalves are stimulated to spawn when they are placed in individual spawning dishes in a warm-water bath (30°C for oysters and 20°C for mussels) and adding a small amount of sperm suspension obtained from a sacrificed male bivalve. Once spawning begins the bivalves are removed from the water bath and allowed to spawn for approximately 30 min. The eggs and sperm should be examined under a microscope to assess their viability, and then all the eggs from at least one female and an aliquot (20 to 50 mL) of sperm suspension from a male should be combined in a stock beaker. While spawning and fertilization are in progress, the treatments can be prepared.

For tests conducted in the presence of sediment, the methods described by the EPA[88] or the ASTM[50] can be used. The test containers are clean 1-L containers. For each treatment, five replicates are prepared: 20 g of material are weighed into each test jar and covered with 1 L of filtered seawater. The sediments are shaken for 10 s and then allowed to settle for 4 h before inoculation with newly fertilized embryos. Sediment elutriates can be prepared according to the EPA/COE.[26] The elutriates are prepared by combining sediment and clean seawater together in a 1:4 ratio, stirring for 30 min, and then allowing the mixture to settle for 60 min. The supernatant is then decanted and used for testing. Test containers are clean 250-mL beakers, with a 200-mL test solution volume. Each elutriate is tested at three concentrations (10, 50, and 100%), with five replicates per treatment. Negative (clean) controls for both tests consist of filtered seawater.

Within 1.6 to 2 h of fertilization the embryo density in the stock beaker is measured by making duplicate counts of a 1:99 dilution of embryo suspension. Each test container is inoculated with a sufficient number of embryos to provide a density of 20 to 30 embryos per milliliter. Additional "zero-time" controls are also prepared. These are used to collect subsamples after inoculation to confirm the embryo density and to monitor the progress of the test without disturbing the test organisms. Counts from the "zero-time" controls are used in subsequent larval mortality calculations. The test containers are allowed to incubate at test temperature (20 ± 1°C for oysters and 15 or 18 ± 1°C for mussels) for 48 ± 4 h. After 48 h, the water in each test container is carefully decanted (if sediment was present) and then mixed to resuspend the larvae. Multiple 10-mL aliquots of test solution are removed by pipette, transferred to screw-cap glass vials, and preserved with 5% buffered formalin. The larvae are examined under a compound microscope at 100× to enumerate normal and abnormal larvae. Larvae that fail to transform to the fully shelled, D-shaped prodissoconch I stage are considered abnormal. The test endpoints are larval abnormality and mortality. Control survival must be = 70% for oyster embryos or = 50% for mussel embryos and = 90% normal shell development in surviving controls.

The red abalone (*Haliotis rufescens*) is a 48-hour embryo-larval development test[88] that has been utilized with pore water samples.[90] Adult male and female abalone are inducted to spawn by exposure to hydrogen peroxide solution in seawater for approximately 2.5 h. Fertilized eggs are distributed to test chambers within 1 h of fertilization. The test is conducted at 15 ± 1°C, 34 ± 2 ppt salinity, and a photoperiod of 16:8 h light:darkness. At the end of the 48-hour period, larvae are preserved and counted to determine the number of normal and abnormal larvae. Control normal larval development must be = 80%.

#### *5.3.3.8.4 Echinoderms*

Several echinoderm species can be used to conduct echinoderm sperm cell fertilization or echinoderm larval development toxicity tests. On the west coast the purple sea urchin, *Strongylocentrotus purpuratus,* and the sand dollar, *Dendraster excentricus,* are commonly used for both test endpoints.[88] A method for performing a sperm cell fertilization test with the east coast, sea urchin, *Arbacia punctulata,* has been published.[95] The sperm cell fertilization test has a very short duration and requires very little sample; echinoderm sperm are exposed to test solutions for 60 min for *Arbacia* and 20 min for the west coast species. After sperm exposure, echinoderm eggs are added for a 20-min fertilization period; then the eggs are preserved for later enumeration. The endpoint is fertilization success, demonstrated by the formation of the fertilization membrane. This test was developed for testing aqueous samples; because the eggs settle after preservation and particles may interfere with fertilization this test is appropriate for pore water and elutriates, but not for whole sediments. In addition, the interstitial water and elutriates need to be relatively clear and particle-free to allow the researcher to view the presence (or absence) of the fertilization membrane and to prevent the sperm from attempting to fertilize sediment particles.

Most recently, the sediment-water interface approach has been adapted for a sea urchin larval development test described by Anderson et al.[96] The test utilizes an intact sediment core with a screen tube inserted to the top of the core with clean overlying seawater. The embryos are inoculated into the screen tube, and the negatively-buoyant embryos settle to the bottom and develop on the screen at the sediment interface. An advantage of this approach is that it more closely represents *in situ* exposure conditions and utilizes intact sediment samples. The larvae remain in the water column during the 48-hour exposure, and the test endpoints are normal development to the echinopluteus stage.

Adult sea urchins can be obtained from wild populations at uncontaminated locations. The natural spawning season for the purple sea urchin, *S. purpuratus,* is approximately September to April, but this may be extended either by conditioning animals in the laboratory through manipulation of temperature and photoperiod[97] or by purchasing animals from California. Some facilities have been able to maintain supplies of gravid sea urchins on a year-round basis[98] and reuse spawned animals after recovery and reconditioning. Locally collected animals should be transported in buckets of seawater with aeration, with care taken not to damage the animals' spines. Purple urchins can be held either in a static aquarium filled with seawater (15 ± 1°C and 34 ± 2 ppt salinity) and equipped with an under-gravel filtration system or preferably in flowing seawater at 15°C. Purple sea urchins should be kept in darkness. A sediment substrate (other than the gravel filtration system) is not required. Urchins can be fed either fresh *Macrocystis pyrifera*, or fresh or dried kelp, *Laminaria* sp., gathered from uncontaminated zones. Animals should be sacrificed periodically, or used in trial spawnings, to assess the condition of the animals.

Adult sand dollars can be obtained from wild populations from uncontaminated locations. The natural spawning season for the sand dollar, *Dendraster excentricus,* is approximately May to October on the west coast, but this may vary from year to year depending on the weather. Animals can be collected at low tide and transported in buckets of sediment and seawater with aeration, with care taken not to damage the animals' exoskeleton. Sand dollars can be held in shallow trays containing a 3- to 5-cm layer of clean sediment with flowing seawater at 15 to 1°C. Sand dollars feed on suspended or benthic materials such as phytoplankton or benthic diatoms. Animals should be sacrificed periodically, or used in trial spawnings, to assess the condition of the animals.

The sperm cell fertilization test can be conducted according to procedures described by the EPA or Environment Canada.[55,88,99] The test species are sand dollars *(Dendraster excentricus)* or sea urchins *(Strongylocentrotus purpuratus)* depending on seasonal availability. High-quality natural seawater, which has been filtered through a 1-μm filter, should be used for the dilution, control water, making hypersaline brine, or all three. The test may be conducted using a single concentration (e.g., 100%) or a series of dilutions, if a definitive dose-response is desired. Spawning is induced by injection of 0.5 M KCl into the coelomic cavity. Female organisms can be wet-spawned into 250-mL beakers filled with seawater, but males should be dry-spawned. Males can be spawned into petri dishes containing a little seawater; the sperm can be collected by pipette before they become activated, transferred to a small vial, and stored on ice. The tests are conducted in small glass test tubes at 15°C for west coast species or 20°C for *Arbacia* with a 20-min sperm exposure for west coast species or 60-min sperm exposure period for *Arbacia* and a 20-min egg fertilization period. Prior to conducting a definitive test, a pretrial fertilization may be performed to determine the sperm:egg ratio that yields the desired control fertilization. Toxicity of the test solutions to echinoderm sperm is based on fertilization success of the eggs. Percentage fertilization is determined by examining subsamples of 100 eggs from each replicate for the presence or absence of a normal fertilization membrane. Control fertilization must be between 70 and 90% for *Arbacia* and = 70% for the west coast species.

Sediment-water interface toxicity tests with the purple urchin echinoderm larvae can be conducted according to methods described by the EPA,[88] as modified from ASTM procedures.[100] Spawning is induced by injection of 0.5 M KC1 into the coelomic cavity. It may be necessary to spawn many animals to obtain a sufficient number of eggs to inoculate a large number of test containers. The eggs and sperm are examined under a microscope to assess their viability and then combined at a density of 500 sperm per egg. At least 95% egg fertilization should be achieved. Sediment samples are placed into the test chambers, as described in Anderson et al.[96] and filled with uncontaminated overlying seawater and equilibrate for 24 h prior to addition of embryos. The test containers are allowed to incubate at 15 ± 1°C. The test duration is between 72 and 96 h to reach the echinopluteus stage. At the end of the test the water in each test container is carefully washed to remove embryos into a funnel, with buffered 5% formalin, capped, and evaluated later. The larvae are examined under a compound microscope at 100× to enumerate normal and abnormal larvae. Larvae that fail to transform to the echinopluteus stage are considered abnormal. The test endpoints are larval normal shell development, with an acceptable test having 80% normal larval development. This test procedure has been utilized in California by Hunt et al.[101] and Phillips et al. [98]

### *5.3.3.8.5 Polychaetes*

Two species of polychaetes used successfully in sediment toxicity tests include *Dinophilus gyrociliatus* and *Neanthea arenaceodentata. D. gyrociliatus* is tested for survival and growth in 4-day pore water exposures.[102,103] Many species of polychaetes have been used in toxicity tests (see Reish for a review).[104] Two types of relatively short-term sediment toxicity tests using *N. arenaceodentata* include a 10-day static, acute test or a 20- to 28-day static-renewal, sublethal growth test, according to Dillon et al.,[105] Standard Methods,[106] or the ASTM.[107] A comparison study was conducted between the amphipod *Rhepoxynius abronius* and the polychaete *N. arenaceodentata*, with 341 samples collected in California.[108] Anderson et al.[108] found that the amphipod was a greater indicator of toxicity; however, the polychaete protocol could be improved by minimizing replicate variability and thereby increasing test sensitivity. Long-term studies (10 weeks) examining the effects of sediment exposure on survival, growth, and reproduction have also been performed with *Neanthes.*[109,110] Juvenile polychaete worms, *Neanthes arenaceodentata,* can be obtained from Dr. Don Reish at California State University, Long Beach, California, or cultured in-house and require very little time to maintain.[110] Laboratory cultures can also be readily maintained.

Worms used in the tests should be 2 to 3 weeks postemergence and of uniform size. If worms are purchased, they are transported in Whirlpak bags (20 each) with seawater and a little *Enteromorpha* algae and shipped by overnight courier. Upon arrival at the laboratory the Whirlpaks should be placed in a small aquarium filled with 20°C seawater. Once the temperature has equilibrated, the bags should be carefully opened and the worms teased out with a fine paintbrush. At this age the worms are small (10 to 15 mm in length) and delicate and must be handled very carefully. A small amount of *Enteromorpha,* presoaked in seawater if available, should be added to the aquarium for the worms to use in tube formation. The worms can be fed *Enteromorpha* or finely ground up TetraMarin as needed. Be careful not to overfeed, especially with TetraMarin since it may foul the water. The worms are held under static conditions at 20 ± 1°C, at 28 to 35 ppt, with gentle aeration provided. The worms should be held for less than 1 week, or they may mature sexually before the 20-day growth test is complete.

The test containers are 1-L glass jars, each containing a 2-cm layer of sediment and 800-mL filtered seawater. The sediments are prepared the day before test initiation and allowed to settle overnight, and then the worms are added the following day (day 0). The negative control consists of sediment from an uncontaminated location and filtered seawater. There are five replicates per treatment.

In the 10-day acute test, each container is seeded with ten worms. There are no water changes, and the worms are not fed. Water quality parameters are measured daily for 10 days. After 10 days the worms are sieved from the sediments, and the number of live, dead, and missing worms is recorded. Replicate data are pooled and compared to responses in the negative control and the reference sediment.

For the 20-day sublethal test, each container is seeded with five worms. Approximately one third of the water volume in each test container is replaced every third day; water quality parameters are measured at that time. The worms are fed ground TetraMarin flakes (8 mg per worm) every second day. On day 0 at least three subsamples of five worms should be set aside, rinsed with distilled water, and dried at 50°C for 24 h to provide an initial estimate of dry weight. At day 20 the worms are carefully removed from the sediments, and the surviving worms are rinsed, dried, and weighed. The average dry weights obtained for each treatment are compared to the initial dry weights to determine the percent change in biomass. Acceptable tests have a minimum of 90% control survival.

Carr et al.[102,103] have described a 4- to 10-day liquid-phase exposure using the archiannelid *Dinophilus gyrociliatus.* The 96-h assay measures survival with the 7- to 10-day assay measure survival and reproductive impairment, that is, egg production, eggs per egg capsule, time until laying and hatching, and juveniles per female. The organisms can be cultured and have been collected from North American coasts as well as Japanese and European waters.[102]

Tests are conducted in 20-mL Stender dishes with a 10-mL test solution. At least four organisms are used per replicate (five per treatment). Tests use 1- to 2-day postemergence females. At time zero 50 μl of 0.5% spinach food suspension is added to each dish. Dishes are examined at days 1, 4, and 7 using a dissecting microscope. At test termination mortality and sublethal reproductive effects are recorded.

#### *5.3.3.8.6 Mysids*

Methods for conducting 96-hour static, acute tests[56] and 7-day static-renewal, sublethal tests[55] with the mysid, *Americamysis bahia* (formerly *Mysidopsis bahia),* are described by the EPA and ASTM. These methods were developed to test aqueous samples such as effluents and receiving waters. *A. bahia* is also listed in the EPA/COE as a possible species for testing elutriates and sediments.[26] Although mysids are usually associated with the water column, they can be used for testing whole sediments or aqueous fractions. Juvenile mysids have been observed to remain in close contact with the aquarium gravel in their culture containers and are generally in contact with sediments in a laboratory experiment. Pore water and elutriates can be tested using standard protocols for effluent tests; whole sediments can be tested using the same type of procedure described for fathead minnow or amphipods.

Juvenile mysids for acute toxicity tests should be 1 to 5 days old within a 24-hour range. Mass cultures of adults can be transmitted in 70-L aquaria filled with seawater and equipped with undergravel filtration systems. If seawater is artificial, it should be aged for several weeks prior to use. New culture tanks should be prepared 1 week before needed and inoculated with a small amount of *Artemia.* A few mysids can be added and observed before making mass transfers. The mysids should be maintained at 25 ± 1°C, 8.0 ± 0.2 pH, and a salinity of 20 to 30 ± 2 ppt. The mysids are fed newly hatched live *Artemia* twice daily. To obtain test organisms gravid females should be removed from the culture tank and placed in a separate tank in an incubation basket fitted with a mesh bottom. After 48 h the incubation basket containing the adults is removed. The juveniles remaining in the tank can be used until they are 5 days old.

Acute toxicity tests can be conducted in 2-L glass beakers equipped with Nitex mesh baskets to hold the mysids. Use of the Nitex baskets facilitates daily mortality observations in turbid treatments, without unduly stressing the organisms. A test volume of 1 L can be used, with at least two replicates per treatment, each containing 10 juvenile mysids. The test is conducted at either 20 (acute) or 25 (chronic) ± 1°C under a 16:8-hour light:dark photoperiod with gentle aeration. The test duration of 96 h and the mysids are fed newly hatched *Artemia* each day. The negative control consists of an equal number of replicate containers containing natural seawater (20 to 30 ± 2 ppt salinity). Successful assays have at least 90% survival in acute exposures and 80% in 7-day exposures.

### *5.3.3.8.7 Algae*

Methods for conducting toxicity using the red alga, *Champia parvula,* are described by the EPA and ASTM.[55,78] The procedure was designed to measure the effects of toxic materials in effluents and receiving waters on sexual reproduction of this marine alga. This test is not suitable for testing whole sediments because test containers are shaken during the exposure, and this would result in disturbance of the sediments. In addition, the large sample volume necessary for this toxicity test (100 mL/replicate) makes this organism unacceptable for interstitial water testing except under the most unusual conditions. The only sediment fraction that this test may be used for are elutriates; however, particulates in elutriates must not interfere with light transmission that may affect test results.

The *Champia* toxicity test involves a two-dimensional exposure of sexually mature male and female plants to test solutions, followed by a 6- to 7-day recovery period in culture medium to allow maturation of cystocarps. Fertilization occurs during the two-dimensional exposure period. Cultures of male and female plants may be obtained from the University of Texas Culture Collection (http://www.bio.utexas.edu/research/utex/). Once established, new cultures can be propagated asexually from the algal branches, making it possible to maintain the same clone continuously. Male and female plants are reared separately in 800 mL aerated culture medium in 1-L glass Erlenmeyer flasks at 23 ± 1°C, using a 16:8-hour light:dark photoperiod and a light intensity of 100 $\mu E/m^2/s$ (500 ft-c). Details for preparation of nutrient stock solutions and culture and testing media are provided by the EPA.[55] One limitation to using this test is that natural seawater is the preferred water source for culture and testing, and suitable supplies may not always be available.

Plants should be checked for maturity prior to initiating a test. Female plants are characterized by the presence of trichogynes (reproductive hairs) at the apex of the branches, while males should have mature spermatial sori (clusters of reproductive bodies that protrude from the edge of the thallus surface). If a female plant is placed in water from a male culture for a few seconds, spermatia should be visible on the trichogynes. Toxicity tests are conducted in 250-mL Erlenmeyer flasks, with a 100-mL test solution (elutriate or interstitial water) volume and a minimum of three replicates per treatment. Test media are prepared without addition of metals or EDTA. One male plant cutting and five female plant cuttings are added to each test container.

Test containers are incubated under the same conditions used for culture and are either swirled by hand twice daily or placed on a rotary shaker at 100 rpm. After 48 h the female plants are transferred to a new set of containers with culture medium, provided with gentle aeration, and incubated for an additional 5 to 7 days. Following this recovery period the plants are examined under a microscope and the number of cystocarps per female enumerated. Cystocarps are characterized by an apical opening for spore release and the presence of dark spores. If cystocarps are not fully developed, the test can be extended a short period (24 to 72 h); no new cystocarps will form without the presence of male plants. Toxic effects are expressed as a reduction in cystocarp formation relative to the negative control. For a test to be considered valid, average control mortality must be = 20%, and the average number of cystocarps must be at least 10.

### *5.3.3.8.8 Microtox*

The Microtox test uses the photoluminescent marine bacterium *Vibrio fischeri* and may be used in freshwater, estuarine, or marine studies.[30,110,112] Changes in bioluminescence in response to exposure to test solutions are detected using a Microtox analyzer. Methods for preparing saline and organic extracts from whole sediments are described by the EPA.[110] In addition, a solid-phase test protocol was recently developed. Initial testing of this method shows it to be at least as sensitive as the conventional elutriate fraction assay.[113] In addition, interlaboratory comparison tests performed with the solid-phase test indicates test variability to be within normal biological testing parameters.[114] The solid-phase protocol is similar to the liquid phase; however, there is a sediment exposure/filtration step added.

Freeze-dried bacteria are rehydrated with 1-mL reconstitution solution and stored in the Microtox analyzer at 4°C. Serial dilutions of 0, 12.5, 25, 50, and 100% are prepared with the elutriate or pore and Microtox diluent (2% NaCl in sterile water). The 0% dilution is a control blank containing only diluent. Each of 10 cuvettes receives 10 μl of the bacterial suspension and 350 μl of diluent. These are allowed to incubate at 15°C for 5 or 15 min; then, an initial light reading is taken. Aliquots of 500 μl are added to two of the cuvettes from each extract dilution (two replicates per dilution). At 5 or 15 min after addition of the extract a final light reading is taken.

Measured values are relative to standard light levels set at the beginning of the test. The light given off by the bacteria diminishes naturally over time. Light loss in the blank controls is used as a ratio to normalize results from the cuvettes containing sample supernatant. The blank response is the ratio of light levels read at time zero and after 15 min. Light-level results of the test chambers are normalized against the blank response to yield gamma values. The gamma value is a percent decrease (represented as a positive value) in light that has been adjusted to take into account the natural loss in luminescence over time of the bacteria. Negative gamma values represent an increase in light output in the system. The 15-min $EC_{50}$ (effective concentration yielding a 50% response in the test system) values for each sample can be calculated.

With some samples it is necessary to alter the method to optimize sensitivity. Some samples have turbidity or color associated with them that can be determined and removed from the effect determination. For samples containing ionic compounds, such as ammonia, the response can be optimized by replacing the NaCl osmotic adjusting solution with 20.4% sucrose.[115] Because turbidity, grain size, and color play a large role in the interpretation of the solid-phase test, factoring in these effects should be considered in the assessment of toxicity.[116]

## 5.4 QUALITY ASSURANCE AND QUALITY CONTROL (QA/QC)

Quality assurance (QA) is a total integrated program organized and designed to provide accurate and precise results to assure data reliability. Included in this program are selection of proper technical methods or laboratory procedures, sample collection and handling, data quality objectives, evaluation of data, quality control, and qualifications and training of personnel. Quality control (QC) includes specific actions for obtaining prescribed standards of performance as part of a quality assurance program. Included are standardizations, calibration, replicates, and control and reference samples suitable for providing statistical estimates of data confidence. The following general QA/QC guidelines[8,9,55,56,77,110] apply to all toxicity tests.

### 5.4.1 Negative Controls

All toxicity assays must be conducted using well-established negative (clean) controls. For every test series with a particular organism or biological system, one test chamber (or series of chambers) must contain diluent water (or clean sediment plus diluent water) only. If solvent carriers are used to introduce organic compounds (e.g., pesticides), then the concentration of the solvent used in the highest treatment must be added to clean dilution water and used as an additional set of controls. In the case of tests involving chemical extraction procedures, solvent controls and solvent extracts of clean, inert sediment comprise the negative controls. The complete test series must be repeated if more than 10 to 20% (depending on the test method's test acceptability criteria) of the control organisms die or show evidence of sublethal effects.

Sediments should be tested as soon as possible after collection. It is recommended that the storage time not exceed 2 weeks.[15] If storage is to exceed a 6-week period, then an initial chemical or toxicity analysis (e.g., ammonia, Microtox) should be conducted when the samples are brought into the laboratory and immediately prior to toxicity testing to verify stability.

### 5.4.2 Positive Controls (Reference Toxicants)

All toxicity assays should include positive (toxic) controls, conducted with well-established standard reference toxicants. For sediment tests, these are conducted in the absence of sediment. Reference toxicants (e.g., KCl, Cu, Cd, Zn) are used to provide insight into mortalities or changes in organism sensitivity that may occur as a result of acclimation, disease, loading density, or handling stress. Concurrent tests using a reference toxicant should be implemented at regular intervals (e.g., monthly for cultures maintained in-house) and for each new batch of test organisms obtained. Control charts as described by the EPA[55,56,88] are constructed for each species, and reference toxicant is used. The cumulative mean value (recalculated with successive data points until the results are stable) and upper and lower control limits (mean ± 2 S.D.) are plotted on each chart. These charts should be available in the laboratory so that the results of each reference toxicant test can be compared to the mean.

Reference toxicant testing can be conducted using either water-only exposures of 48 to 96 h or spiked sediments of 7- to 10-day exposures. The water-only exposure is preferred as there are fewer variables involved; thus, results can be more consistent than when spiking sediments. Standard spiking protocols currently do not exist, and the toxicity of the sample will be influenced by procedural differences and sediment characteristics. For reference toxicant testing in water-only exposures using benthic species, it is advantageous to provide a substrate to reduce organism stress. For *H. azteca, C. tentans,* and *H. limbata,* this involves use of a plastic mesh, sand monolayer, and glass tube, respectively. Organisms may be fed at time zero and at 48 h. There has been limited testing of spiked sediments.[15,19,2,0,24] Environment Canada has developed a standard guidance for reference toxicant testing with spike sediments.[117] At present, however, there has been little research into this area, and issues such as proper spiking methods, equilibration times of spiked sediments, and recipes for formulated sediments have not been adequately resolved.

### 5.4.3 Reference Sediments

Reference sediments are most commonly required for sediment tests but may also be appropriate for water-column tests. Physical and chemical characteristics of control sediments may be very different from test sediments. Reference sediments are collected from an area documented to be free of chemical contamination and should represent the range of important physical and chemical variables found in the test sediments. The ASTM defines a reference sediment as a whole sediment near an area of concern used to assess sediment conditions, exclusive of materials of interest. A control sediment is defined as a sediment essentially free of contaminants and is used routinely to assess the acceptability of a test. Data derived from such samples can be used to partition toxicant effects from unrelated effects such as sediment grain size.

### 5.4.4 Test Organisms

Only healthy organisms of similar size, age, and life history stage are used for toxicity assays. Taxonomic identification should be confirmed by a qualified taxonomist. Records of collection, shipping, and holding should be maintained for species obtained outside the laboratory. Records of culture history, culture conditions, food, and water quality should be maintained for laboratory cultures.

### 5.4.5 Blind Testing

All treatment and test containers should be randomized, and testing should be conducted without laboratory personnel knowing the identities of the samples. Replicates of each treatment should be assigned a sample code during testing and randomized in the test sequence.

### 5.4.6 Replication

The number of replicates required varies from one protocol to another but should be sufficient to account for variability in test-organism response. Some tests may require two to five replicates. Sediment bioassays typically require five replicates per treatment. Unless otherwise specified in the experimental design, each treatment in a test series must begin with the same number of replicates. Statistical significance of toxicity test results has been evaluated for certain marine sediment tests by determining the ninetieth-percentile minimum significant difference.[118] The examination of detectable difference is important for assessing whether an individual test result had either low sample variance (e.g., able to discern differences) or high sample variance (e.g., differences cannot be found).

### 5.4.7 Water Quality Measurement/Maintenance

Toxicity assays involving exposure of organisms in aqueous media require that the media be uncontaminated and that proper water quality conditions be maintained to ensure the survival of the organisms and to ensure that undue stress is not exerted on the organisms, irrespective of the test materials. Appropriate water quality parameters must be measured at least at the start and end of a test, and preferably every 24 h. Typically, alkalinity, hardness, pH, ammonia, and conductivity are measured at test initiation and termination. Dissolved oxygen, pH, salinity, and temperature are measured as specified in each test protocol.

### 5.4.8 Standard Laboratory Procedures

Standard laboratory procedures must be followed in all testing. These include proper documentation, proper cleaning, avoidance of contamination, and maintenance of appropriate test conditions. All unusual observations or deviations from established procedures must be recorded and reported. Records should be recorded and signed in ink.

## 5.5 SUMMARY

Sediment toxicity testing has only been a recent activity in ecological assessments. The first testing was spawned by concerns over dredged material contamination and its suitability for open-water disposal by the U.S. Army Corps of Engineers in the late 1960s and early 1970s. There was relatively little testing until the 1980s, with a dramatic increase in recent years.[19] The science has been able to progress at a relatively rapid rate because of the similarities to, and state of, the more traditional water-column and effluent toxicity testing. As discussed in a preceding section, the EPA has initiated new efforts in managing contaminated sediments and method standardization that will undoubtedly result in an even greater degree of sediment toxicity testing, regulation, and research in the near future.[5,53]

**Table 5.1 Desirable Traits of Sediment Toxicity Assays**

| |
|---|
| Sensitivity (responsiveness and protective capacity) |
| Discriminatory (discerns degrees of contamination) |
| Relevance (to ecosystem or study objectives) |
| Validity (field verification, lack of false positives or negatives) |
| Comprehensiveness (surrogate for wide range of species) |
| Reliability (consistent cultures and successfully completed test assays) |
| Reasonable resource requirements (expertise, time, cost, availability of organisms) |
| Standardization (accepted methods, large database, QA/QC criteria) |
| Uniqueness (produces nonredundant ecosystem information) |
| Confirmatory (opposite of uniqueness, provides weight-of-evidence) |
| Significance (critical ecosystem link or species) |

The process of selecting the optimal tests for use in an ecosystem assessment is not as simple and straightforward as many would like. The optimal assay can only be selected when the study's objectives and associated data quality objectives have been defined and there is a reasonable understanding of the physical, chemical, and biological characteristics of the study site. This information must be combined with an understanding of the strengths and weaknesses of the various sediment toxicity assays available. Once standardized methods are recognized by the EPA, the process will be dictated for many studies. These methods provide information on successfully used procedures with advice on which conditions are the most critical; however, they are not definitive test methods and allow some investigator flexibility, which may be inappropriate for use by the inexperienced or for regulatory programs.

No one assay is superior to all others. A number of useful tests have been evaluated in freshwater and marine studies.[4,19,119–124] The better tests have been described in the previous sections. To reduce both uncertainty and the likelihood of obtaining false positive or false negative results it is important to test more than one species. The importance of testing multiple species increases with the importance of protecting the ecosystem and the need to define "significant" contamination in the "gray" zone. The methods described below are for assays that represent critical levels of biological organization, have been shown to be sensitive to a wide variety of contaminants, are relatively discriminatory of differing levels of contamination, use relevant species for many desirable ecosystems, and have been successfully used in sediment contamination studies.[5] The tests include optimal characteristics in that they are relatively standardized, the species can be obtained and cultured in a number of locations throughout North America, have a high degree of reliability (successful completion), are relatively inexpensive and short-term in nature, and have associated performance criteria to evaluate testing quality assurance and quality control (Table 5.1).

For most applications, a test battery consisting of two to three species can be recommended. These recommendations are for waters in temperate regions of North America, but they may be useful elsewhere and are based on the above characteristics and on comparison studies where multiple species have been used simultaneously in sediment contamination investigations.[4,19,30,34,41,119–124] For both freshwater and marine systems, Microtox is recommended as a screening tool; it is useful in assessing a large number of samples and directing more intensive sampling. The bioluminescence response has been shown to be relatively sensitive to many contaminants and often correlated significantly with responses of other test species. It should never be used as the sole toxicity test method. The resource efficiency of the method; its degree of standardization and simplicity; and its ability to test water, pore water, elutriates, and a new solid-phase procedure make it versatile and relevant.

Both freshwater and marine studies should also use a water column and benthic species in solid-phase exposures. For freshwater studies in North America, the useful water-column tests are the early life stage *P. promelas* assays (larval growth or embryo-larval growth and deformities) or the 3-brood, survival, and reproduction assays with *C. dubia* or *D. magna.* Both of these cladoceran assays recently became ASTM methods.[51] The freshwater benthic assays that have been most

**Table 5.2 Toxicity Test Batteries: Recommended Assays for the EPA Assessment and Remediation of Contamination Sediments (ARCS) Program**

Minimal size recommended: two test species and four measurement endpoints from at least three groupings.

Group A: *Chironomus riparius* 14-day survival; *Chironomus tentans* 10-day length; *Hyalella azteca* 28-day survival, length, sexual maturation, and antennae length; *H. azteca* 14-day survival; *Daphnia magna* 7-day reproduction; *Lemna minor* 4-day frond growth; and *Ceriodaphnia dubia* 7-day reproduction.

Group S: *C. dubia* 7-day survival; *C. riparius* 14-day length; *H. azteca* 14-day sexual maturation; and *Pimephales promelas* 7-day larval weight.

Group C: *Hydrilla verticillata* 10-day root growth; *Diporeia* 5-day avoidance/preference; and *H. azteca 14-day* survival.

Group D: *Hexagenia bilineata* 10-day survival and molting frequency.

Minimal Groupings with the Recommended Size Limits:[a]

Option 1: a. *H. azteca* (14 days)
b. *C. dubia, C. riparius, D. magna, P. promelas, Diporeia* or *H. bilineata*

Option 2: a. *C. dubia* or *C. riparius*
b. *Diporeia* or *H. bilineata*

Option 3: a. *D. magna*
b. *P. promelas*
c. *Diporeia* or *H. bilineata*

[a] All endpoints associated with each of the assays should be measured. Microtox should be used in reconnaissance surveys due to its high degree of correlation with the above assay response[30,34] patterns. A predominant amphipod in the Great Lakes, *Diporeia* sp. (formerly *Pontoporeia hoyi*), survival and avoidance assay (5 days) showed behavior to be the most sensitive measurement endpoint in a U.S. EPA study that compared 25 toxicity assays comprised of over 50 endpoints.[30] For marine studies in North America, the choices are more varied with some geographic specificity. For water column species, the sea urchin *A. punctulata* and the red macroalga *C. parvula* are useful indicators of sediment toxicity. For benthic species testing, 10-day static sediment survival tests are used, with the Pacific coast amphipod species including *R. abronius* and *E. estuarius*, and on the Atlantic coast *A. abdita and L. plumulosus.*

successful are the 7- to 10-day survival test with *H. azteca,* the 10-day survival and growth assays in *C. tentans* or *C. riparius,* or the 10-day survival and molting frequency assay with *Hexagenia* sp. In addition, Ingersoll et al.[29] observed *H. azteca* growth, including length, antenna length, and sexual maturation, to be very sensitive indicators in 14- and 28-day exposures.

Also, with the recent publication by U.S. EPA and ASTM long-term sublethal sediment toxicity tests using *H. azteca* and *C. tentans*, various investigators are reporting sublethal endpoints (e.g., amphipod growth) that correlate with benthic response in the field.[30–34] Burton and Ingersoll[30] have recommended three test battery options based on studies of "Areas of Concern" in the Great Lakes basin (Table 5.2). These recommendations were based on assay comparisons of toxicity sensitivity, ability to discriminate differing degrees of sample toxicity, response relationships (correlations) to other endpoint response patterns, and factor analysis that showed four unique response patterns. These test batteries will monitor at least two species, four measurement endpoints, and represent three response patterns.

These test methods have proven useful for assessing sediment toxicity in North America on a routine basis and can be conducted at a reasonable cost and with a reasonable level of training. The assays reported here are only those that have become "relatively" standardized or used extensively in assessments of sediment contamination. There are a number of other assays that show promise and may be standardized in the future. The recent and near future standardization of the methods reported in this chapter allows them to be incorporated into regulatory programs to provide more comprehensive assessments of protection of our aquatic ecosystems.

## REFERENCES

1. Wetzel, R. G., *Limnology,* W. B. Saunders Co., Philadelphia, 1975.
2. Griffiths, R. P., The importance of measuring microbial enzymatic functions while assessing and predicting long-term anthropogenic perturbations, *Mar. Pollut. Bull.,* 14,162, 1983.

3. Carmichael, W. W., Freshwater blue-green algae (cyanobacteria) toxins: A review, in *The Water Environment: Algal Toxins and Health,* Carmichael, W. W., Ed., Plenum Press, New York, 1981, 1.
4. Burton, G. A., Jr. and Scott, K. J., Sediment toxicity evaluations: Their niche in ecological assessments, *Environ. Sci. Technol.,* 26, 2068, 1992.
5. U.S. EPA., Contaminated Sediment Management Strategy, EPA 823-R-98–001, Office of Water, Washington, D.C., 1998.
6. U.S. EPA, Methods for Assessing the Toxicity of Sediment-Associated Contaminants with Estuarine and Marine Amphipods, EPA/600/R-94/025, Office of Research and Development, Washington, D.C., 1994a.
7. U.S. EPA, Methods for Assessing the Toxicity and Bioaccumulation of Sediment-Associated Contaminants with Freshwater Invertebrates, 1st ed., EPA/600/R-94/024, Duluth, MN, 1994b.
8. U.S. EPA, Methods for Assessing the Toxicity and Bioaccumulation of Sediment-Associated Contaminants with Freshwater Invertebrates, 2nd ed., EPA/600/R-99/064, Duluth, MN, 2000.
9. U.S. EPA, Method for Assessing the Chronic Toxicity of Marine and Estuarine Sediment-Associated Contaminants with the Amphipod *Leptocheirus plumulosus,* 1st ed., EPA/600/R-01/020, Washington, D.C., 2001.
10. Burton, G. A., Jr., La Point, T., and Zarba, C., Contamination assessment of sediments in freshwater ecosystems, in *Hazard Assessment of Chemicals: Current Developments,* Saxena, J., Ed., Taylor & Francis, Washington, D.C., 1993, 171.
11. Di Toro, D. M., Zarba, C. S., Hansen, D. J., Berry, W. J., Swartz, R. C., Cowen, C. E., Pavlou, S. P., Allen, H. E., Thomas, N. A., and Paquin, P. R., Technical basis for establishing sediment quality criteria for nonionic organic chemicals using equilibrium partitioning, *Environ. Toxicol. Chem.,* 10, 1541–1583, 1991.
12. Ankley, G. T., Di Toro, D. M., Hansen, D. J., Berry, and W. J., Technical basis and proposal for deriving sediment quality criteria for metals, *Environ. Toxicol. Chem.,* 15, 2056–2066, 1996.
13. Di Toro, D. M., Hansen, D. J., McGrath, J. A., and Berry, W. J., Predicting the toxicity of metals in sediments using SEM and AVS, (Draft in preparation), 2001.
14. Ankley, G. T., Schubauer-Berigan, M. K., and Dierkes, J. R., Predicting the toxicity of bulk sediments to aquatic organisms with aqueous test fractions: Pore water vs. elutriate, *Environ. Toxicol. Chem.,* 10, 1359, 1991.
15. ASTM, Standard guide for collection, storage, characterization, and manipulation of sediment for toxicological testing, Method E1391–94, *Annual Book of ASTM Standards, Water and Environmental Technology,* Vol. 11.04, American Society for Testing and Materials, Philadelphia, 1994.
15a. U.S. EPA, Methods for Collection, Storage and Manipulation of Sediments for Chemical and Toxicological Analyses: Technical Manual, Office of Water, EPA/823/B/01/002, Washington, D.C., 2001.
16. Environment Canada, Guidance Document on Collection and Preparation of Sediment for Physicochemical Characterization and Biological Testing. Report EPS/1/RM/29, Technology Development Directorate, Ottawa, 1994.
17. Burton, G. A, Jr., Sediment collection and processing: Factors affecting realism, in *Sediment Toxicity Assessment,* Lewis Publishers, Boca Raton, FL, 37, 1992.
18. Othoudt, R. A., Giesy, J. P., Grzyb, K. R., Verbrugge, D. A., Hoke, R. A., Drake, J. B., and Anderson, D., Evaluation of the effects of storage time on the toxicity of sediments, *Chemosphere,* 22, 801, 1991.
19. Burton, G. A., Jr., Assessing the toxicity of freshwater sediments, *Environ. Toxicol. Chem.,* 10, 1585, 1991.
20. Stemmer, B. L., Burton, G. A., Jr., and Leibfritz-Frederick, S., Effect of sediment test variables on selenium toxicity to *Daphnia magna, Environ. Toxicol. Chem.,* 9, 381, 1990.
21. Zarba, C. S., Equilibrium Partitioning Approach, in Sediment Classification Methods Compendium, EPA 823-R-92–006, Office of Water (WH-556), U.S. Environmental Protection Agency, Washington, D.C., 1992.
22. Adams, D. D., Sediment pore water sampling, *CRC Handbook of Techniques for Aquatic Sediments Sampling,* Mudroch, A. and MacKnight, S. D., Eds., CRC Press, Boca Raton, FL, 1991, 131.
23. Skalski, C., Fisher, R., and Burton, G. A., Jr., An *in situ* interstitial water toxicity test chamber, Abstr. Annu. Meet. Soc., *Environ. Toxicol. Chem.*, 132, 58, 1990.
24. Ditsworth, G. R., Schults, D. W., and Jones, J. K. P., Preparation of benthic substrates for sediment toxicity testing, *Environ. Toxicol. Chem.,* 9, 1523, 1990.
25. Sarda, N. and Burton, G. A., Jr., Ammonia variation in sediments: Spatial, temporal and method-related effects, *Environ. Toxicol. Chem.* 14, 1499, 1995.

26. U.S. EPA-USACE, Evaluation of Dredged Material Proposed for Ocean Disposal (Testing Manual), EPA-68-C8–0105, USEPA, Office of Marine and Estuarine Protection, Washington, D.C., 1991.
27. Sasson-Brickson, G. and Burton, G. A., Jr., *In situ* and laboratory sediment toxicity testing with *Ceriodaphnia dubia, Environ. Toxicol. Chem., 10,* 201, 1991.
28. Burton, G. A., Jr. Realistic assessments of ecotoxicity using traditional and novel approaches, *J. Aquatic Ecosys. Health Manage.*, 2, 1, 1999.
29. Ingersoll, C. G. and Nelson, M. K., Testing sediment toxicity with *Hyalella azteca* (Amphipoda) and *Chironomus riparius* (Diptera), in *Aquatic Toxicology and Risk Assessment*, Landis, W. G. and van der Schalie, W. H., Eds., ASTM STP 1096, American Society for Testing and Materials, Philadelphia, 1990.
30. Burton, G. A, Jr. and Ingersoll, C. G., Evaluating the Toxicity of Sediments, The ARCS Assessment Guidance Document, EPA/905-B94/002, Chicago, IL., 1994.
31. Canfield, T. J., Kemble, N. E., Brumbaugh, W. G., Dwyer, F. J., Ingersoll, C. G., and Fairchild, J. F., Use of benthic invertebrate community structure and the sediment quality triad to evaluate metal-contaminated sediment in the Upper Clark Fork River, Montana, *Environ. Toxicol. Chem.,* 13,1999, 1994.
32. Canfield, T. J., Dwyer, F. J., Fairchild, J. F., Haverland, P. S., Ingersoll, C. G., Kemble, N. E., Mount, D. R., La Point, T. W., Burton, G. A., and Swift, M. C., Assessing contamination in Great Lakes sediments using benthic invertebrate communities and the sediment quality triad approach. *J. Great Lakes Res.,* 22, 565, 1996.
33. Canfield, T. J., Brunson, E. L., Dwyer, F. J., Ingersoll, C. G., and Kemble, N. E., Assessing sediments from the upper Mississippi river navigational pools using benthic community invertebrate evaluation and the sediment quality triad approach, *Arch. Environ. Contam. Toxicol.*, 35, 202, 1998.
34. Burton, G. A., Jr., Ingersoll, C., Burnett, L., Henry, M., Hinman, M., Klaine, S., Landrum, P., Ross, P., and Tuchman, M., A comparison of sediment toxicity test methods at three Great Lake Areas of concern, *J. Great Lakes Res.*, 22, 495, 1996.
35. Schlekat, C. E., McGee, B. L., Boward, E., Reinharz, E., Velinsky, D. J., and Wade, T. L., Tidal river sediments in the Washington, D.C. area. III. Biological effects associated with sediment contamination, *Estuaries,* 17, 334, 1994.
36. U.S. EPA, Biomonitoring to Achieve Control of Toxic Effluents, Office of Water, EPA/625/8–87/013, Washington, D.C., 1987.
37. U.S. EPA, A Review of Single Species Toxicity Tests: Are the Tests Reliable Predictors of Aquatic Ecosystem Community Responses? deVlaming, V., Norberg-King, T.J., Eds., EPA/600/R-97/114, Office of Research and Development. Duluth, MN, 1999.
38. Grothe, D., Dickson, K., and Reed-Judkins, D., Eds., *Whole Effluent Toxicity Testing: An Evaluation of Methods and Prediction of Receiving System Impacts*, SETAC Press, Pensacola, FL, 1996.
39. Pontasch, K. W., Niederiehner, B. R., and Cairns, J., Jr., Comparison of single species, microcosm and field responses to a complex effluent, *Environ. Toxicol. Chem.,* 8, 521, 1989.
40. Stemmer, B. L., Burton, G. A., Jr., and Sasson-Brickson, G., Effect of sediment spatial variance and collection method on cladoceran toxicity and indigenous microbial activity determinations, *Environ. Toxicol. Chem.,* 9, 1035, 1990.
41. Chapman, P., Power, E. A., and Burton, G. A., Jr., Integrative assessments in aquatic ecosystems, *Sediment Toxicity Assessment,* Burton, G. A., Jr., Ed., Lewis Publishers, Boca Raton, FL, 1992, 313.
42. Adams, W. J., Kimerle, R. A., and Barnett, J. R., Jr., Sediment quality and aquatic life assessment, *Environ. Sci. Technol.,* 26, 1865, 1992.
43. U.S. EPA, Assessment and Remediation of Contaminated Sediments (ARCS), 1992 Work Plan, Great Lakes National Program Office, U.S. Environmental Protection Agency, Chicago, 1992.
44. Long, E. R. and Chapman, P. M., A sediment quality triad: Measures of sediment contamination, toxicity and infaunal community composition in Puget Sound, Mar. *Pollut. Bull.,* 16, 405, 1985.
45. Bascombe, A. D., Ellis, J. B., Revitt, M., and Shutes, R. B. E., The development of eco-toxicological criteria in urban catchments, *Water Sci. Technol.,* 22, 173, 1990.
46. Ellis, J. B., Shutes, R. B., and Revitt, D., Ecotoxicological approaches and criteria for the assessment of urban runoff impacts on receiving waters, *Proc. Effects of Urban and Receiving,* Herricks, E. E., Jones, J. E., and Urbonas, B., Eds., Systems Symposium, American Soc. of Civil Engineers, Mt. Crested Butte, CO, 1993.

47. McCahon, C. P., Poulton, M. J., Thomas, P. C., Xu, Q., Pascoe, D., and Turner, C., Lethal and sublethal toxicity off field simulated farm waste episodes to several freshwater invertebrate species, *Water Res.,* 25, 661, 1991.
48. Ireland, D. S., Burton, G. A., Jr., and Hess, G. G., *In Situ* toxicity evaluations of turbidity and photoinduction of polycyclic aromatic hydrocarbons. *Environ. Toxicol. Chem.*, 15, 574, 1996.
49. Burton G. A., Jr., Rowland C. D., Greenberg M. S., Lavoie D. R., Nordstrom J. F., and Eggert L. M., A tiered, weight-of-evidence approach for evaluating aquatic ecosystems, in *Sediment Quality Assessment: Approach, Insights and Technology for the 21st Century,* Munawar, M., Ed., Ecovision World Monograph Series. Backhuys Publ., Leiden, The Netherlands, (in press), 2002.
50. ASTM, Standard guide for conducting static acute toxicity tests starting with embryos of four species of saltwater bivalve molluscs. Method E724–89, *Annual Book of ASTM Standards, Water and Environmental Technology,* Vol. 11.04, American Society for Testing and Materials, Philadelphia, 1991.
51. ASTM, Test Method for Measuring the Toxicity of Sediment-Associated Contaminants with Freshwater Invertebrates. E1706–00. In *Annual Book of ASTM Standards*, Vol. 11.05, West Conshohocken, PA, 2000.
52. American Public Health Association, Water Pollution Control Federation, and American Water Works Association, *Standard Methods for the Examination of Water and Wastewater,* 18th ed., APHA, Washington, D.C., 1989.
53. USEPA-USACE, Evaluation of Dredged Material for Discharge in Waters of the U.S., Testing Manual, EPA-823-B-98–005, Washington, D.C., 1998.
54. Bedard, D. C, Hayton, A., and Persaud, D, Ontario Ministry of the Environment Laboratory Sediment Biological Testing Protocol, Ontario Ministry of the Environment, Toronto, Ontario, Canada, draft report, January 1992.
55. U.S. EPA, Short-Term Methods for Estimating the Chronic Toxicity of Effluents and Receiving Waters to Freshwater Organisms, 3rd ed., Klemm, D. J., Lazorchak, J. M., Norberg-King, T. J., Peltier, W. H., and Heber, M. A., Eds., Office of Research and Development, EPA/600/4–91/002, Cincinnati, 1994.
56. U.S. EPA, Methods for Measuring the Acute Toxicity of Effluents and Receiving Waters to Freshwater and Marine Organisms, 4th ed., Office of Research and Development, EPA/600/4–90/027F, Washington, D.C., 1991.
57. Burton, G. A., Jr., Stemmer, B. L, Winks, K. L., Ross, P. E., and Burnett, L. C., A multitrophic level evaluation of sediment toxicity in Waukegan and Indiana harbors, *Environ. Toxicol. Chem.,* 8,1057,1989.
58. McCafferty, W. P., The burrowing mayflies (*Ephemeroptera*: Ephemeridea) of the United States, *Trans. Am. Entomolog. Soc.,* 101, 447, 1975.
59. Nebeker, A. V., Cairns, M. A., Gakstatter, J. H., Malueg, K. W., Schuytema, G. S., and Krawczyk, D. F., Biological methods for determining toxicity of contaminated freshwater sediments to invertebrates, *Environ. Toxicol. Chem.*, 3, 617, 1984.
60. Prater, B. L., and Anderson, M. A., A 96-hour sediment bioassay of Duluth and Superior Harbor basins (Minnesota) using *Hexagenia limbata, Asellus communis, Daphnia magna,* and *Pimephales promelas* as test organisms, *Bull. Environ. Contamination Toxico.,* 18, 159, 1977.
61. Giesy, J. P., Rosiu, C. R., and Graney, R. L., Benthic invertebrate bioassays with toxic sediment and pore water, *Environ. Toxicol. Chem.,* 9, 233, 1990.
62. Bedard, D., Hayton, A., and Persaud, D., Ontario Ministry of the Environment Laboratory Sediment Biological Testing Protocol, Ontario Ministry of the Environment, Toronto, Ontario, 26, 1992.
63. Giesy, J. P., Rosiu, C. J., Graney, R. L., and Henry, M. G., Benthic invertebrate bioassays with toxic sediment and pore water, *Environ. Toxicol. Chem.* 9, 233, 1990.
64. Fremling, C. R. and Mauck, W. L., Methods for using nymphs of burrowing mayflies (*Ephemeroptera, Hexagenia*) as toxicity test organisms, *Aquatic Invertebrate Bioassays, ASTM STP 715,* Buikema, A. L., Jr., and Cairns, J., Jr., Eds., ASTM, Philadelphia, 81, 1980.
65. Fremling, C. R. and Mauck, W. L., Methods for using nymphs of burrowing mayflies (*Ephemeroptera, Hexagenia*) as toxicity test organisms, *Aquatic Invertebrate Bioassays, ASTM STP 715,* Buikema, A. L., Jr., and Cairns, J., Jr., Eds., ASTM, Philadelphia, 81,1980.
66. Giesy, J. P. and Hoke, R. A., freshwater sediment toxicity bioassessment: Rationale for species selection and test design, *J. Great Lakes Res.,* 15, 539, 1989.

67. Henry, M. G., Chester, D. N., and Mauck, W. L., Role of artificial burrows in *hexagenia* toxicity tests: Recommendations for protocol development, *Environ. Toxicol. Chem.*, 5, 553, 1986.
68. Landrum, P. F. and Poore, R., Toxicokinetics of selected xenobiotics in *Hexagenia limbata, J. Great Lakes Res.*, 14, 427, 1988.
69. Edmunds, G. F., Jr., Jensen, S. L., and Berner, L., *The Mayflies of North and Central America,* University of Minnesota Press, St. Paul, 1976.
70. Friesen, M. K., *Hexagenia rigida* (McDunnough), in Manual for the Culture of Selected Freshwater Invertebrates, Lawrence, S. G., Ed., *Canadian Special Publication of Fisheries Aquatic Science,* 54, 127, 1981.
71. Wright, L. L., and Mattice, J. J., Substrate selection as a factor in *Hexagenia* distribution, *Aquatic Insects,* 3, 13, 1981.
72. Hanes, E. C. and Ciborowski, J. J. H., The effects of larval density and food limitation on growth of *Hexagenia* (Ephemeroptera:Ephemeridae) under laboratory conditions, *Proc. Bull. North Am. Benthol. Soc.,* 6, 183, 1989.
73. Malueg, K. W., Schuytema, G. S., Krawczyk, D. F., and Gakstatter, J. H., Laboratory sediment toxicity tests, sediment chemistry and distribution of benthic macroinvertebrates in sediment from the Keweenaw Waterway, Michigan, *Environ, Toxicol. Chem.*, 3, 233, 1984.
74. Bedard, D., Sediment bioassay development: Addressing the pros and cons, *Proc. Technol. Transfer Conf.,* 1, No. 10, 471, 1989.
75. ASTM, Annual Book of ASTM Standards. Section 11: Water and Environmental Technology E1367, Biological Effects and Environmental Fate; Biotechnology; Pesticides: Standard guide for conducting Standard guide for conducting 10-day static sediment toxicity tests with marine and estuarine amphipods, Section 11.05 American Society for Testing and Materials, West Conshohocken, PA. 2000.
76. ASTM, *Annual Book of ASTM Standards*, Section 11: Water and environmental technology E1191, biological effects and environmental fate; biotechnology; pesticides: Standard guide for conducting life-cycle toxicity tests with saltwater mysids, Section 11.05, American Society for Testing and Materials, West Conshohocken, PA. 2000.
77. ASTM, *Annual Book of ASTM Standards*, Section 11: Water and environmental technology E729, biological effects and environmental fate; biotechnology; pesticides: Standard guide for conducting acute toxicity tests on test materials with fishes, macroivertebrates and amphipods, Section 11.05, American Society for Testing and Materials. West Conshohocken, PA. 2000.
78. ASTM, *Annual Book of ASTM Standards*, Section 11: Water and environmental technology E1498. biological effects and environmental fate; biotechnology; pesticides: Standard guide for conducting sexual reproduction tests with seaweeds, Section 11.05, American Society of Testing Materials, West Conshohocken, PA. 2000.
79. ASTM, *Annual Book of ASTM Standards*. Section 11: Water and environmental technology E1525, biological effects and environmental fate; biotechnology; pesticides: Standard guide for designing biological tests with sediments, Section 11.05, American Society of Testing Materials, West Conshohocken, PA. 2000
80. Swartz, R. C., Cole, F. A., Lamberson, J. O., Ferraro, S. P., Schultz, D. W., DeBen, W. A., Lee, H., II, and Ozretich, R. J., Sediment toxicity, contamination and amphipod abundance at a DDT and dieldrin contaminated site in San Francisco Bay, *Environ. Toxicol. Chem*, 13, 6, 1994.
81. Swartz, R. C., DeBen, W. A., Phillips, J. K., Lamberson, J. O., and Cole, F. O., Phoxocephalid amphipod bioassay for marine sediment toxicity, in *Aquatic Toxicology and Hazard Assessment: Proc. 7th Annu. Symposium,* STP 854, Cardwell, R. D., Purdy, R., and Bahner, R. C., Eds., American Society for Testing and Materials, Philadelphia, 1985.
82. DeWitt, T. H., Ditsworth, G. R., and Swartz, R. C., Effects of natural sediment features on survival of the Phoxocephalid amphipod, *Rhepocytiius abronitis, Mar. Environ. Res.,* 25, 99, 1988.
83. DeWitt, T. H., Swartz, R. C., and Lamberson, J. O., Measuring the acute toxicity of estuarine sediments, *Environ. Toxicol. Chem.,* 8, 1035, 1989.
84. Scott, K. J. and Redmond, M. S., The effects of a contaminated dredged material on laboratory populations of the tubicolous amphipod *Ampeliaca abdita, Aquatic Toxicology and Hazard Assessment,* Vol. 12, STP 1027; Cowgill, U. M. and Williams, L. R., Eds., American Society for Testing and Materials, Philadelphia, 1989, p. 289.

85. Williams, L. G., Chapman, P. M., and Ginn, T. C., A comparative evaluation of marine sediment toxicity using bacterial luminescence, oyster embryo, and amphipod sediment bioassays, *Mar. Environ. Res.,* 19, 75, 1986.
86. Schlekat, C. E., McGee, B. L., and Reinharz, E., Testing sediment toxicity in Chesapeake Bay with the amphipod *Leptocheirus plumulosus:* An evaluation, *Environ. Toxicol. Chem.,* 11, 225,1992.
87. DeWitt, T. H., Redmond, M. S., Sewall, J. E., and Swartz, R. C., Development of a chronic sediment toxicity test for marine benthic amphipods, Chesapeake Bay Program, CVP/TRS, 89, 93, 1992.
88. U.S. EPA, Short-Term Methods for Estimating the Chronic Toxicity of Effluents and Receiving Waters to West Coast Marine and Estuarine Organisms, Office of Research and Development, Chapman, G. A., Denton, D. L., and Lazorchak, J. M., Eds., Cincinnati, EPA/600/R-95–136, 1995.
89. Chapman, P. M., Marine sediment toxicity tests, in *Chemical and Biological Characterization of Sludges, Sediments, Dredge Spoils and Drilling Muds,* Lichtenberg, J. J., Winter, F. A., Weber, C. I., and Fredkin, L., Eds., STP 976, American Society for Testing and Materials, Philadelphia, 1988, 391.
90. Anderson, B. S, Hunt, J. W., Phillips, B. M., Fairey, R., Roberts, C. A., Oakden J. M., Puckett, H. M., Stephenson, M., Tjeerdema, R. S., Long, E. R., Wilson, C. J., and Lyons, J. M., Sediment quality in Los Angeles Harbor, USA: A triad assessment, *Environ. Toxicol. Chem.*, 20, 359, 2001.
91. Environment Canada, Biological Test Method: Acute Test for Sediment Toxicity Using Marine or Estuarine Amphipods, Report EPS I/RM/26, Method Development & Application Division, Ottawa, Ontario, Canada, 1992.
92. Hunt, J. W., Anderson, B. S., and Phillips, B. M., Recent advances in microscale toxicity testing with marine mollusks, in *Microscale Testing in Aquatic Toxicology: Advances, Techniques, and Practice*, Well, P. G., Lee, K., and Blaise, C., Eds., CRC Press, Boca Raton, FL, 1998.
93. Anderson, S. L., Knezovich, J. P., Jelinski, J., and Steichen, D. J., The Utility of Pore Water Toxicity Testing for Development of Site-Specific Marine Sediment Quality Objectives for Metals, Report LBL-37615 UC-000, Lawerence Berkeley National Laboratory, University of California, Berkeley, CA, 1995.
94. Phillips, B. M., Anderson, B. S., Hunt, B. S., Investigations of Sediment Elutriate Toxicity at Three Estuarine Stations in San Francisco Bay, California, San Francisco Estuary Regional Monitoring Program, Regional Monitoring Program Report #43, 2000.
95. U.S. EPA, Short Term Methods for Estimating the Chronic Toxicity of Effluents and Surface Waters to Marine and Estuarine Organisms, 2nd ed., Cincinnati, EPA/600/4–91–003, 1994.
96. Anderson, B. S., Hunt, J. W., Hester, M. N., and Phillips, B. M., Assessment of sediment toxicity at the sediment-water interface, in *Techniques in Aquatic Toxicology*, Ostrander, G. K., Ed., Lewis Publishers, Boca Raton, FL, 1996, 609.
97. Leahy, P. S., Tutschulte, T. C., Britten, R. J., and Davidson, E. H., Large-scale laboratory maintenance system for gravid purple sea urchins (*Strongylocentrotus purpuratus*), *J. Exp. Zool.* 204, 369, 1978.
98. Phillips, B. M., Anderson, B. S., and Hunt, J. W., Spatial and temporal variation in results of sea urchin (*Strongylocentrotus purpuratus*) toxicity with zinc, *Environ. Toxicol. Chem.*, 17, 453, 1998.
99. Environment Canada, Biological Test Method: Fertilization Assay Using Echinoids (Sea Urchins and Sand Dollars), Report EPS I/RM/27, Method Development & Application Division, Ottawa, Ontario, Canada, 1992.
100. ASTM, Standard Guide for Conducting Static Acute Toxicity Tests with Echinoid Species, ASTM, Philadelphia, 1029, 1995.
101. Hunt, J. W., Anderson, B. S., Phillips, B. M., Tjeerdema, R. S., Taberski, K. M., Wilson, C. J., Puckett, M., Stephenson, M., Fairey, R., and Oakden, J., A large-scale categorization of sites in San Francisco Bay, USA, based on the sediment quality triad, toxicity identification evaluations, and gradient studies. *Environ. Toxicol. Chem.*, 20, 1252, 2001.
102. Carr, R. S., Curran, M. D., and Mazurkiewicz, M., Evaluation of the archiannelid *Dinophiltis gyrociliatits* for use in short-term life-cycle toxicity tests, *Environ. Toxicol. Chem.,* 5, 703, 1986.
103. Carr, R. S., Williams, J. W., and Fragata, C. T. B., Development and evaluation of a novel marine sediment pore water toxicity test with the polychaete *Dinophilus gyrociliatus, Environ. Toxicol. Chem.,* 8, 533, 1989.
104. Reish, D. J. and Gerlinger, T. V., A review of the toxicological studies with polychaetes annelids, *Bull. Mar. Sci.*, 60, 584, 1997.

105. Dillon, T. M., Moore, D. W., and Reish, D. J., A 28-day sediment bioassay with marine polychaete, *Nereis* (*Neanthes*) *arenaceodentata*, in *Environmental Toxicology and Risk Assessment: Third Volume*, Hughes, J. S., Biddenger, G. R., and Homes, E., Eds., *American Society of Testing Materials*, STP 1218, Philadelphia, 1995.
106. American Public Health Association, Water Environment Federation, 8510 D, Sediment test procedures using the marine polychaete *Neanthes arenaceodentata*, in *Standard Methods for Examination of Water and Wastewater*, Washington, D.C., 1998.
107. ASTM, Guide for Conducting Sediment Toxicity Tests with Marine and Estuarine Polychaetous Annelids, ASTM, Philadelphia, 1992.
108. Anderson, B. S., Hunt, J. W., Phillips, B. M., Tudor, S., Fairey, R., Newman, J., Puckett, H. M., Stephenson, M., Long, E. R., Wilson, C. J., and Tjeerdema, R. S., Comparison of marine sediment toxicity test protocols for the amphipod *Rhepoxynius abronius* and the polychaete worm *Nereis* (Neanthes) *arenaceodentata*, *Environ. Toxicol. Chem.,* 17, 859, 1998.
109. Murdoch, M. H., Chapman, P. M., Johns, D. M., and Paine, H. D., Chronic effects of organochlorine exposure in sediment to the marine polychaete *Neanthes arenaceodentata*, *Environ. Toxicol. Chem.*, 16, 1494, 1997.
110. U.S. EPA, Recommended Guidelines for Conducting Laboratory Bioassays on Puget Sound Sediments, prepared by PTI Environmental Services for U.S. Environmental Protection Agency, Region 10, Seattle, WA, 1990.
111. Reish, D. J., The use of larvae and small species of polychaetes in marine toxicological testing, in *Microscale Testing in Aquatic Toxicology: Advances, Techniques, and Practice*, Well, P. G., Lee, K., Blaise, C., Eds., CRC Press, Boca Raton, FL, 1998.
112. Environment Canada, Biological Test Method: Toxicity Test Using Luminescent Bacteria *(Photobacterium phosphoreum),* Report EPS I/RM/24, Method Development & Application Division, Ottawa, Ontario, Canada, 1992.
113. Tay, K. L., Doe, K. G., Wade, S. J., Vaughan, D. A., Berrigan, R. E., and Moore, M., Sediment bioassessment in Halifax harbour, *Environ. Toxicol. Chem., 11,* 1567, 1992.
114. Ross, P., Burton, G. A., Greene, M., Ho, K., Meier, P, Sweet, L., Auwarter, A., Bispo, A., Doe, K., Erstfeld, K., Goudey, S., Goyvaerts, M., Jourdain, H. D., and Jourdain, M., Interlaboratory precision study of a whole sediment toxicity test with the bioluminescent bacterium *Vibrio fischeri, Environ. Toxicol.,* 14, 339. 1999.
115. Hinwood, A. L. and McCormick, M. J., The effect of ionic solutes on EC 50 values measured using the Microtox test, *Tox. Assessment,* 2, 499, 1987.
116. Benton, M. J., Malott, M. L., Knight, S. S., Cooper, C. M. and Benson, W. H., Influence of sediment composition on apparent toxicity in a solid-phase test using bioluminescent bacteria, *Environ. Toxicol. Chem.,* 14, 411, 1995.
117. Environment Canada, Guidance Document on Measurement of Toxicity Test Precision Using Control Sediments Spiked with a Reference Toxicant, EPS 1/RM/30, Environmental Technology Advancement Directorate, Ottawa, 1995.
118. Phillips, B. M., Hunt J. W., Anderson B. S., Puckett, H. M., Fairey, R., Wilson, C. J., and Tjeerdema, R., Statistical significance of sediment toxicity test results: Threshold values derived by the detectable significance approach, *Environ. Toxicol. Chem.,* 20, 371, 2001.
118. Burgess, R. M. and Scott, K. J., The significance of in-place contaminated marine sediments on the water column: Processes and effects, *Sediment Toxicity Assessment,* Burton, G. A., Jr., Ed., Lewis Publishers, Boca Raton, FL, 129, 1992.
119. Lamberson, J. O., DeWitt, T. H., and Swartz, I. L. C., Assessment of sediment toxicity to marine benthos, *Sediment Toxicity Assessment,* Burton, G. A., Jr., Ed., Lewis Publishers, Boca Raton, FL, 183, 1992.
120. Giesy, J. P. and Hoke, R. A., Freshwater sediment toxicity bioassessment: Rationale for species selection and test design, 1. *Great Lakes Res., 15,* 539, 1989.
121. Giesy, J. P., Graney, R. L., Newstad, J. L, Rosiu, C., Benda, A., Kreis, R. G., and Horvath, F. J., A comparison of three sediment, bioassay methods for Detroit River sediments, *Environ. Toxicol. Chem.,* 7, 483, 1988.

122. Hoke, R. A., Giesy, J. P., Ankley, G. T., Newsted, J. L., and Adams, J. R., Toxicity of sediments from western Lake Erie and the Maumee River at Toledo, Ohio, 1987: Implications for current dredged material disposal practices, 1. *Great Lakes Res.,* 16, 457, 1990.
123. Long, E. R. and Buchman, M. F., An Evaluation of Candidate Measures of Biological Effects for the National Status and Trends Program, NOAA Technical Memorandum NOS OMA 45, National Oceanic and Atmospheric Administration, Seattle, WA, 1989.

CHAPTER 6

# Toxicological Significance of Soil Ingestion by Wild and Domestic Animals

W. Nelson Beyer and George F. Fries

## CONTENTS



## 6.1 INTRODUCTION

Most wild and domestic animals ingest some soil or sediment, and some species may routinely, or under special circumstances, ingest considerable amounts. Ingested soil supplies nutrients, exposes animals to parasites and pathogens, and may play a role in developing immune systems.[1] Soil ingestion is also sometimes the principal route of exposure to various environmental contaminants.[2–7] Ingestion of soil and earthy material is defined as geophagy and may be either intentional or unintentional, occurring as an animal eats or grooms.

The importance of soil ingestion for estimating exposure to environmental contaminants has been best documented in assessments of pesticides or wastes applied to land supporting farm animals. Soil ingestion tends to be most important for those environmental contaminants that are found at concentrations in soil that are high relative to concentrations in a soil-free diet. Lead is an excellent example. Reviews of soil ingestion by domestic animals that emphasize effects on mineral nutrition and the intake of chemical pollutants have been prepared by Healy,[8] Fries and Paustenbach,[9] and Fries.[10] Evaluation of the soil-animal-product pathway is now a part of all risk assessments that involve the presence or potential introduction of pollutants into agricultural environments.[11,12]

The next section of this review examines the methods for determining soil ingestion and identifies the limitations and uncertainties associated with these methods. Intentional geophagy, discussed in the third section, has in general not been considered in risk assessments, although it is potentially an important route of exposure. The fourth and fifth sections present general conclusions from studies of soil ingestion by domestic and wild animals and include extensive tables summarizing rates of ingestion. The sixth section presents laboratory and field studies demonstrating injury through ingestion of contaminated soils or sediments. The seventh section addresses the texture of ingested soil, which is relevant to estimating exposure since environmental contaminants tend to be associated with particular soil fractions. The eighth section considers the bioavailability of environmental contaminants in soil. This topic is widely recognized as important, although risk assessors generally lack reliable measures of bioavailability that can be incorporated into risk assessments. The final section discusses default values in regulatory guidance documents of the U.S. Environmental Protection Agency (EPA).

## 6.2 METHODS FOR DETERMINING SOIL INTAKE

Studies of soil ingestion by domestic animals have employed a marker that was present in soil, taken up poorly by plants, and absorbed poorly by animals. The concentration of the marker in feces is thus a measure of the concentration of soil in feces, which with appropriate assumptions provides a means to estimate soil intake. Acid-insoluble ash and titanium oxide were the usual markers used in studies of soil ingestion by farm animals. At times, other markers were used for measuring soil ingestion by humans because titanium oxide is present in some consumer products.[13,14] No one marker has been found to be consistently superior to the others. Acid-insoluble ash provides the advantage of ease of analysis, whereas analyses of titanium oxide, a single chemical entity, should be more precise and reproducible. The concentrations of a marker in the soil and fecal dry matter are the only analytical measurements required for estimation of soil ingestion. Soil calculated as a fraction of the feces is

$$F_{soil} = C_{feces}/C_{soil} \quad (6.1)$$

where $C_{feces}$ is the concentration of marker in feces and $C_{soil}$ is the concentration of marker in the soil. The amount of soil ingested is estimated by the equation

$$I_{soil} = I_{DM} * (1\text{-}D) * F_{soil}/(1\text{-}D * F_{soil}) \quad (6.2)$$

and soil as a fraction of dry matter intake is

$$I_{soil}/I_{DM} = (1\text{-}D) * F_{soil}/(1\text{-}D * F_{soil}) \quad (6.3)$$

where $I_{soil}$ is the amount of soil ingested (kg/day), $I_{DM}$ is the dry matter consumed (kg/day), D is the fraction of dry matter that disappears in the digestive tract, and $F_{soil}$ is the fraction of soil in feces.[15] The soil content of feces is measured, but values for digestibility and feed intake are estimated from the literature because these parameters are difficult to measure directly in grazing animals.

Several types of errors and uncertainties are associated with the estimation of soil ingestion. The least serious of these are the analytical errors in the determination of acid-insoluble ash or titanium oxide. Certain plants may accumulate silicates, and with some exceptions like McGrath et al.,[39] this potential contribution to the intake of acid-insoluble ash is not always corrected for in published studies. Also, the soil samples collected for determining marker concentrations may differ from the soil ingested by the animals studied.

By far, the greatest sources of uncertainty are associated with (1) the necessity to use estimated values for the amount of dry matter consumed by domestic animals and (2) digestibility, which is the fraction of the dry matter that disappears in the gastrointestinal tract. Typically, digestibility decreases as plants mature. If a single value for digestibility is used for the entire grazing season, soil ingestion will be underestimated early in the season when digestibility is high and overestimated later in the season as digestibility decreases. Feed intake in most studies was assumed to be constant throughout the grazing season. However, dry matter intake will decline if the amount of forage available per animal is low.[16] If dry matter intake is reduced, the same amount of ingested soil would yield a higher concentration in the feces than would occur with normal intake. Thus, when calculating intake, soil as a fraction of intake will be overestimated. It has been shown that estimates of soil ingestion can vary as much as twofold, depending on the assumed values for feed intake and digestibility.[17]

Methods for estimating soil and sediment ingestion by wildlife are slightly different from those used for domestic animals.[6] Because the amounts of food eaten by wild animals are not well documented, soil and sediment ingestion is generally expressed as a fraction of the diet, rather than as an amount per day. Also, it may be difficult to estimate the concentration of a soil marker, since that concentration may vary over the feeding range of the animal. The use of acid-insoluble ash may be a more reliable marker when little is known about the particular soil that was ingested. However, acid-insoluble ash is not as accurate as some other markers for estimating very low rates of soil or sediment ingestion because the diet contains some acid-insoluble ash.

Analyzing prey or plant material for soil provides an alternative approach to estimating ingestion rates, although this approach may sometimes be misleading. For example, consider the hypothetical food chain from soil to earthworms to predators, which is frequently used in risk assessments of wildlife on contaminated soils. The soil content of earthworms is variable but is typically 20 to 30% of the dry weight of the whole body. In general, when an animal eats an earthworm, it ingests all of the soil inside the earthworm and additional soil on the outside of the earthworm. Predators of earthworms would be expected to ingest 20 or 30% soil in their diets. In some instances, however, predators are exposed to less soil than would be predicted. Moles, for example, are known to squeeze out some of the soil from an earthworm as it is swallowed. Gorman and Stone[18] describe moles as eating with finesse, squeezing out soil like toothpaste from a tube, possibly to reduce tooth wear associated with ingesting soil. Estimating soil ingestion associated with plant material may also be inexact because the way in which a biologist collects plant material and prepares it for analysis may be different from the ways in which animals ingest plant material. Grazers might crop off vegetation above the soil surface, ingesting little soil, or they might pull up vegetation by its roots, ingesting much soil. Analyzing dietary items of waterfowl for sediment may also be misleading, since some waterfowl clean dietary items by shaking them and by filtering them in their mouths.[19]

## 6.3 INTENTIONAL GEOPHAGY IN ANIMALS

Most studies on intentional geophagy have emphasized nutritional implications, although at contaminated sites intentional geophagy may also lead to exposure to environmental contaminants. In reviewing the literature, Setz et al.[20] suggested that animals ingest soil for supplemental minerals, to reduce gastric disturbances, to reduce acidity in the gut, to adsorb toxic chemicals, and for the tactile sensations created in the mouth. At least fifty species of animals, particularly ruminants, are known to ingest soil purposely at licks, possibly for sodium.[21–24] Rhesus macaques (*Macaca mulatta*) and mountain gorillas (*Gorilla gorilla*) mine and ingest tropical soil, possibly to reduce intestinal ailments.[25,26] African elephants (*Loxodonta africana*) excavate termite mounds and eat the associated soil, which tends to have elevated concentrations of bases, pH, and clay content.[27] Diamond et al.[28] suggested that the geophagy they observed in about ten species of frugivorous birds in New Guinea was a means to reduce the toxicity or bitterness of ingested seeds and fruits. Setz et al.[20] examined several explanations of ingestion of soil from arboreal termite nests by golden-faced saki monkeys (*Pithecia pithecia*) but failed to find any of them convincing. Lizards, crocodilians, tortoises,[29] and box turtles[30] also intentionally ingest soil. Many species of birds that eat seeds ingest grit, which is coarse material used to grind food in the gizzard.[31] Some grit supplies calcium and other minerals,[32] whereas grit composed of sand and gravel from quartz have little nutritional value beyond the grinding of the diet. The search for grit may endanger birds when they mistakenly ingest pesticide granules or lead shot.[32]

## 6.4 SOIL INGESTION BY DOMESTIC ANIMALS

Soil ingestion has been measured in most domestic animal species and in production classes that are important in carrying out exposure and risk assessments involving persistent environmental pollutants. The original impetus for research on soil ingestion was to evaluate the nutritional significance of soil ingestion as a source of trace elements.[33] A second impetus was the discovery of an association of soil ingestion with excessive wear of teeth of sheep grazing on poor pastures.[34] It was soon recognized that soil ingestion could be an important pathway of animal exposure to persistent organic pollutants, like DDT, when present in soil.[35]

Studies on sheep provide the most comprehensive database on the effects of management and climatic conditions on soil ingestion.[33,34,36–39] Dairy and beef cattle have been studied less extensively, but data are available for temperate pastures, semiarid rangeland, and unpaved confinement lots.[3,15,40] Data on swine are limited to a single study.[41]

Results of representative studies with sheep, cattle, and swine are summarized in Table 6.1. The extensive data on sheep were reasonably consistent, and several broad conclusions may be drawn. When pasture was the sole animal feed source, soil ingestion generally was related inversely to the availability of forage. The lowest soil ingestion values occurred during spring, when grass growth was the greatest (Table 6.1). The lowest soil-ingestion rates of sheep were as little as 1 or 2% of dry matter intake. When forage was sparse, during fall or winter, intake of soil was estimated to be as great as 18% of the diet. The high values of soil ingested when forage is sparse can be attributed mainly to the need of the animals to graze close to the ground. The calculated yearly average soil intake was about 4.5% of dry matter intake for sheep when animals in New Zealand grazed 365 days a year and pasture was the only feed source.[34] This average is equivalent to intake of 45 g/day soil by a 50-kg sheep consuming 1 kg/day dry matter.

Two management factors had a significant effect on the amount of soil consumed by sheep. Soil ingestion was reduced about 50% when sheep were offered supplemental feed during the winter months.[37,38] This reduction would be expected because the sheep would depend less on pasture for adequate feed intake. It can be inferred from the increased soil ingestion during periods of spare forage availability that increasing the number of animals per unit area would increase soil intake because less forage would be available per animal. The best example of the effect of stocking rate is given by McGrath et al.,[39] in which an increase in stocking rate from 11 to 15 animals/hectare resulted in a sixfold increase in soil ingestion.

Studies with cattle are less numerous than with sheep, but the qualitative findings are similar. The three cattle studies involving pasture have similar high and low soil intake values related to the availability of forage. The average intake for the English study in Table 6.1 was somewhat lower because the average values included dairy herds, which under English practice would have received supplemental feed. The soil ingested by a 500-kg dairy cow in the New Zealand study was estimated to average 900 g/day when it was assumed that dry matter intake was 15 kg/day, that cows grazed 365 days per year, and that no supplemental feed was offered.

**Table 6.1 Results of Representative Studies of Soil Ingestion by Farm Animals**

| | | | Soil Intake, g/day | | | Intake, g | | |
|---|---|---|---|---|---|---|---|---|
| Location | Local Season | Other Feed | Mean | High | Low | Body Wt., kg | Feed, g | Reference |
| | | | **Sheep** | | | | | |
| New Zealand | Winter | No | 60 | 150 | 5 | 1.2 | 0.06 | 34 |
| New Zealand | Spring–Fall | No | 4 | 10 | 0 | 0.1 | 0.005 | 34 |
| New Zealand | Apr.–Oct. | No | 63 | 108 | 1 | 1.2 | 0.06 | 37 |
| New Zealand | Jul.–Aug. | Yes | >1 | >1 | — | — | — | 37 |
| New Zealand | Jul.–Oct. | No | 90 | — | — | 1.8 | 0.09 | 37 |
| New Zealand | Jul.–Oct. | Yes | 35 | — | — | 0.7 | 0.035 | 37 |
| New Zealand | Winter | No | 83 | 125 | 43 | 1.7 | 0.085 | 38 |
| New Zealand | Winter | Yes | 48 | 68 | 26 | 1.0 | 0.05 | 38 |
| New Zealand | Winter | No | 30 | 41 | 21 | 0.6 | 0.03 | 38 |
| Scotland | May–Dec. | No | 70 | 198 | 9 | 0.9 | 0.046 | 33 |
| Ireland | Apr.–Oct. | No | 40 | 159 | 7 | 0.7 | 0.036 | 39 |
| | | | **Cattle** | | | | | |
| New Zealand[1] | All Year | No? | 770 | 2070 | 260 | 1.9 | 0.063 | 36 |
| England[2] | Apr.–Aug. | ? | 310 | 2400 | 27 | 0.7 | 0.022 | 3 |
| United States[3] | Jun.–Nov. | No | 400 | 1500 | 100 | 1.1 | 0.055 | 40 |
| United States[4] | May–Nov. | Yes | 113 | 146 | 83 | 0.4 | 0.019 | 15 |
| | | | **Pigs** | | | | | |
| United States | Jun.–Aug. | Yes | 197 | 392 | 37 | 2.0 | 0.061 | 15 |

[1] Dairy cattle.
[2] Stated to be both dairy and beef cattle. Authors did not separate the results by production class.
[3] Beef cattle grazing semiarid range.
[4] Dairy cattle on unpaved confinement lots.

There are several situations other than grazing in which soil ingestion can be the source of animal exposure and residues in products. Cattle confined in unpaved lots ingest a small amount of soil that may amount to 1 to 2% dry matter intake.[15] Although most commercial swine and poultry production is conducted in confined operations, soil ingestion can be quite significant when these species have access to soil. Pigs may ingest as much as 8% of their dry matter intake as soil because of their rooting habits.[41] Intuitively, it might be expected that free-range poultry would have high rates of soil consumption, but no data could be found in the literature.

## 6.5 SOIL AND SEDIMENT INGESTION BY WILDLIFE

Table 6.2 provides estimates of soil and sediment ingestion rates for various wildlife species. It should be emphasized that some of the sample sizes are small and that a mean from a single collection does not adequately represent variation throughout the year or in different habitats. In general, animals that browse (deer, elk, moose) ingest scant soil with their diet. Browsers also tend to be the species commonly associated with deliberate ingestion of soil at salt licks. Mice, voles, woodchuck, and other herbivores also tend to ingest only small amounts of soil. Bison ingest somewhat more soil. The inverse relation between forage availability and soil ingestion rates observed in domestic grazers presumably holds for wild grazers.

Soil ingestion rates of herbivores tend to be high in dry climates, possibly because soil is pulled up with roots of sparsely available forbs and grasses. Jackrabbits and pronghorn in sagebrush and black-tailed prairie dogs in a short-grass prairie[4] ingest more soil than do most other herbivores. Arthur and Gates[4] suggested that some of the soil ingestion by jackrabbits and pronghorn is related

**Table 6.2 Mean Percent Soil or Sediment Ingested by Wildlife (Dry Weight)**

| Species | Mean % Soil or Sediment Ingested | State (No.) Comments | Reference |
|---|---|---|---|
| **Turtles** | | | |
| Box turtle *Terrapene carolina* | 4.5 | MD (8) summer | 6 |
| Eastern painted turtle *Chrysemys picta* | 5.9 | MD (9) summer | 6 |
| **Ungulates** | | | |
| Pronghorn *Antilopa capra* | 5.4 | ID (177) all year | 4 |
| White-tailed deer *Odocoileus virginianus* | <2 | MD (6) summer | 6 |
| Mule deer *Odocoileus hemionus* | <2 | WY (5) summer | 6 |
| Mule deer | 0.6–2.1 | CO (78) seasonal range | 2 |
| Elk *Cerrus elaphus* | <2 | WY (4) summer | 6 |
| Moose *Alces alces* | <2 | WY (3) summer | 6 |
| Bison *Bison bison* | 6.8 | WY (4) summer | 6 |
| Feral hog *Sus scrofa* | 2.3 | SC (15) all year | 6 |
| **Rodents** | | | |
| White-footed mouse *Peromyscus leucopus* | <2 | MD (9) summer | 6 |
| Meadow vole *Microtus pennsylvanicus* | 2.4 | MD (7) summer | 6 |
| Black-tailed prairie dog *Cynomys ludovicianus* | 7.7 | KS (12) summer | 6 |
| White-tailed prairie dog *Cynomys leucurus* | 2.7 | CO (5) summer | 6 |
| Hispid cotton rat *Sigmodon hispidus* | 2.8 | TN (18) | 42 |
| Woodchuck *Marmota monax* | <2 | MD (6) summer | 6 |
| **Other mammals** | | | |
| Opossum *Didelphis virginiana* | 9.4 | MD, SC (16) spring, summer | 6 |
| Nine-banded armadillo *Dasypus novemcinctus* | 17 | LA (5) fall | 6 |
| Jackrabbit *Lepus californicus* | 6.3 | ID (118) all year | 4 |
| Raccoon *Procyon lotor* | 9.4 | MD (4) summer | 6 |
| Red fox *Vulpes vulpes* | 2.8 | MD (7) summer | 6 |
| **Waterfowl** | | | |
| Blue-winged teal *Anas discors* | <2 | MN (12) summer | 6 |
| Ring-necked duck *Aythya collaris* | <2 | MN (6) summer | 6 |
| Wood duck *Aix sponsa* | 11 | MN (7) summer | 43 |
| Wood duck | <2 | ID (41) summer | 43 |
| Mallard *Anas platyrhynchos* | 3.3 | MN (88) summer | 6 |
| Mallard | 4.9[1] | ID (118) all year | 44 |
| | 12 | VA (30) summer | 45 |

**Table 6.2 Mean Percent Soil or Sediment Ingested by Wildlife (Dry Weight) *(Continued)***

| Species | Mean % Soil or Sediment Ingested | State (No.) Comments | Reference |
|---|---|---|---|
| American black duck *Anas rubripes* | <1 | DE (9) fall | 46 |
| Green-winged teal *Anas crecca* | <1 | DE (8) fall | 46 |
| Northern pintail *Anas acuta* | <1 | DE (13) fall | 46 |
| American widgeon *Anas americana* | 3 | DE (5) fall | 46 |
| Canada goose *Branta canadensis* | 8.2 | MD (23) feeding on grass, spring | 6 |
| Canada goose | 9.0 | ID (402) in wetlands, all year | 44 |
| | 5 | VA (20) summer | 45 |
| Tundra swan *Cygnus columbianus* | 9.0 | ID (93) all year | 44 |
| Mute swan *Cygnus olor* | 3.2 | MD (40) spring | 47 |
| **Shorebirds** | | | |
| Stilt sandpiper *Micropalama himantopus* | 17 | KS (1) summer | 6 |
| Semipalmated sandpiper *Calidris pusilla* | 30 | KS (1) summer | 6 |
| Least sandpiper *Calidris minutilla* | 7.3 | KS (1) summer | 6 |
| Western sandpiper *Calidris mauri* | 18 | KS (1) summer | 6 |
| Black-bellied plover *Pluvialis squatarola* | 29 | CA (8) fall, winter | 48 |
| Willet *Catoptrophorus semipalmatus* | 3 | CA(16) fall, winter | 48 |
| 8 species | 10–60 | CA (27) Sand in alimentary tract, spring | 49 |
| **Other birds** | | | |
| American woodcock *Scolopax miner* | 10 | ME (7) | 6 |
| Wild turkey *Meleagris gallopavo* | 9.3 | SD (12) summer | 6 |

[1] Calculated assuming a digestibility of 50%.

to the animals' needs to ingest adequate sodium relative to potassium. Armadillos and opossums, which prey on soil organisms, tend to ingest more soil than do herbivores. Among waterfowl, dabblers tend to ingest little sediment, generally less than 2%,[6,46] although occasionally individuals are found that have ingested large amounts of sediment. Sediment ingestion by dabblers may increase in winter. Canada geese and tundra swans, which feed closer to the bottom than do dabblers, ingested about 9% sediment in their diets. Mute swans were observed to ingest sediment at a lower rate (3%) than tundra swans (9%). We are uncertain whether the difference is related to the feeding habits of the two species or to a difference in the collection times. Samples of digesta from the mute swans were collected when submerged aquatic vegetation was plentiful, rather than at a time when the swans would have been feeding on plant material in the sediment. As a group, shorebirds tend to ingest the most sediment, although willets, feeding on mollusks, ingested only 3% sediment in one study.

At least some species of animals can tolerate surprisingly large amounts of ingested soil if it is not contaminated. For example, mallards were able to maintain their weights for two weeks when fed diets containing 50, 60, or 70% sediment in a mash diet (G. Heinz, personal communication, Patuxent Wildlife Research Center).

## 6.6 TOXICITY OF ENVIRONMENTAL CONTAMINANTS IN SOIL OR SEDIMENT TO ANIMALS

Laboratory studies have demonstrated both the bioavailability and the toxicity of environmental contaminants in soil and sediment ingested by wildlife. When Connor et al.[45] fed to bobwhite a diet containing contaminated sediment from the Coeur d'Alene River Basin, their blood lead increased to concentrations that the authors considered clinically significant. When mute swans were fed a high concentration (24%) of sediment from the Anacostia River in their diet for six weeks,[50] the dietary lead concentration (85 mg/kg) was high enough to reduce the activity of red blood cell ALAD (amino-levulinic acid dehydratase) activity. The presence of polynuclear aromatic hydrocarbons in this sediment was the probable cause of an observed increase in hepatic microsomal monooxygenase activity in the livers of these swans.

A series of laboratory studies was conducted on the toxicity to waterfowl of ingested sediments from the Coeur d'Alene River Basin, known to be severely contaminated with lead and other metals. A survey of scat from tundra swans from the site had found that swans on average were ingesting about 9% sediment in their diets and that 10% of the swans ingested 22% or more sediment in their diets.[44] In a controlled study, Heinz et al.[51] demonstrated a linear dose response in mallards between ingestion (0, 3, 6, 12, and 24% of diet) of that sediment and blood lead. One of the mallards fed the 24% sediment diet died, and the others in that group showed signs of lead poisoning. Hoffman et al.[52,53] demonstrated the toxicity of this sediment to goslings of Canada geese and to mallard ducklings, and Day[54] demonstrated its toxicity to mute swans.

The poisoning of wild animals from the ingestion of contaminated soils or sediments has been documented at a few sites. Hundreds of waterfowl feeding in the severely polluted Coeur d'Alene River Basin have been diagnosed as lead-poisoned.[55–57] These waterfowl rarely contained the lead shot or sinkers usually associated with lead poisoning. Although some of the lead ingested by these waterfowl was associated with their diets, the great majority was associated with ingested sediment, as shown by analyses of the feces.[44] Similar waterfowl mortality from lead poisoning has been recorded previously at other mining sites.[58,59] Ingestion of white phosphorus in sediment has killed waterfowl in Alaska.[60] Soil ingestion was suggested as an important route of exposure of small mammals to lead in a skeet range contaminated with lead shot.[61] Pain et al.[62] suggested that the contamination of the Coto Doñana avian overwintering site in Spain with mining wastes presents hazards to European birds.

## 6.7 RELATION OF PARTICLE SIZE OF INGESTED SOIL TO EXPOSURE TO CONTAMINANTS

Ingested soil or sediment may have a different texture from the soil or sediment in an animal's habitat as a whole. When birds seek grit, for example, they selectively ingest coarse particles. Geophagous birds, in contrast, commonly select clay or silt.[28] Sheppard[7] concluded that the soil associated with plant surfaces is especially fine. Whereas clay may become incorporated into the wax on leaf surfaces, coarse particles are removed when soil on plants is disturbed or washed by rainfall. Soil found adhering to human fingers has been found to be greatly enriched in clay, increasing lead concentrations up to twentyfold.[7] McGrath et al.[39] found that soil ingested by sheep was enriched in silt and clay. Enrichment of the fine particles is important because metals tend to be associated with the fines and because the fines are the most chemically reactive portion of the mineral soil.

Digesta of mute swans feeding on aquatic vegetation in Chesapeake Bay contained finer sediment than that present in the habitat as a whole.[47] These swans were ingesting about 3% sediment in their diet. Presumably the particles that settled onto vegetation eaten were representative of the fine particles suspended in the water column rather than the coarser material found on the bottom. The lead content observed in the digesta of these swans was two or three times that expected from the lead content of the sampled sediment and from the acid-insoluble ash content of the digesta, a measure of total sediment ingested. Lead concentrations in their digesta were closely correlated with concentrations of aluminum, which is associated with clay. In contrast, lead concentrations of waterfowl digesta samples collected from the Coeur d'Alene River Basin were precisely those expected if the waterfowl had been ingesting unsorted sediment from the bottom.[44] The difference between the studies is probably related to the rates of sediment ingestion. When only small amounts of soil or sediment are ingested, the material may be sorted by texture, but when waterfowl ingest an average of 9% sediment in their diet, they are presumably eating sediment from the bottom rather than just a thin film of clay on vegetation. When sorting by particle size has occurred, uncorrected bulk soil or sediment samples may provide inaccurate estimates of exposure.

## 6.8 BIOAVAILABILITY OF ORGANIC CONTAMINANTS IN SOIL

Hazards to animals from contaminants ingested with soil or sediment depend on the bioavailability of the contaminants as well as on the exposure and the toxicity of the contaminants. Risk assessors recognize the importance of including a measure of bioavailability in their analyses.[63]

The term *bioavailability* refers to the fraction of the compound that is absorbed, which is then stored, metabolized, or excreted from the animal. True bioavailability can be evaluated only in mass balance experiments, which rarely have been conducted with environmental contaminants. But relationships between dietary intakes and concentrations in tissues or products like milk are established for many compounds. Measurement of the residue concentrations produced by feeding a soil-borne contaminant will provide a measure of the bioavailability from soil relative to the bioavailability from the normal diet.

An environmental contaminant associated with soil may have lower bioavailability than the same contaminant incorporated into other matrices. The effect of matrices on uptake of lipophilic compounds can be illustrated with 2,3,7,8-tetrachloro-*p*-dioxin (TCDD), which has been studied in a variety of matrices. Net absorption of TCDD by rats was 75 to 80% when administered to rats in corn oil[64] but only 50 to 55% when TCDD was contained in normal rat and cow diets.[65,66] Reviews of the available literature indicated uptake of TCDD from soils is typically about 40% and is less than 20% from soils at some industrial sites.[67–69] Stephens et al.[70] found that about 70 to 80% of a dose of tetrachlorinated dioxin congeners in soil was absorbed by chickens, which suggests that soil had very little effect on the bioavailability of dioxin to chickens. When TCDD was administered with activated carbon, a strong adsorbent, uptake by rats was negligible.[71] Data are not available for such comprehensive comparisons for other lipophilic compounds, but comparable results can be anticipated. Although the uptake from soil is consistently lower than from normal diets, the reduction may not be great enough to merit inclusion in risk assessments.

Lipophilic compounds in soil tend to be adsorbed to the organic matter fraction.[72] Organic matter content varies greatly among soils and can be altered by manure, sludge, and other amendments. Differences in bioavailability due to organic matter content might be expected, but the uptake of PCBs by rats was unaffected by the organic matter content of the soil to which the soil had been added.[73] Similarly, animal uptake of PBBs from soil was not affected significantly by amending soil with activated carbon.[5] The failure to detect differences due to organic matter content and amendments suggests that conditions of the gastrointestinal tract favor dissociation from soil organic matter to dietary lipids.

Kadry et al.[74] found that when rats were exposed to phenanthrene, the residues in the plasma accounted for about 60% of the dose. There was little difference whether the phenanthrene was administered in the pure form or adsorbed to soil, which suggests that soil does not significantly reduce its bioavailability.

The absorption of polar compounds to soil involves an electrostatic attraction to negatively charged clay and mineral particles and may differ by orders of magnitude depending on the properties of the soils and the compounds.[75] The acidic conditions of parts of the gastrointestinal tract may lead to the dissociation of chemicals that are adsorbed at the pH range of normal soils, thereby making the compounds available for absorption. However, because polar compounds are readily metabolized and excreted in urine by mammals, they tend not to be an important concern when conducting human risk assessments.

## 6.9 BIOAVAILABILITY OF INORGANIC CONTAMINANTS IN SOIL

The ingestion of soil can have both beneficial and deleterious effects on the mineral status of the animal. Ingested soil may provide a number of necessary trace elements to an animal that might otherwise be deficient.[8,76] On the other hand, excessive soil intake can inhibit the absorption of some minerals. The production of iodine deficiency is a prominent example.[77] Concerns have been raised about the uptake of heavy metals by domestic animals from such activities as deposition of emissions from smelting activities and addition of sewage sludge to land, although Thornton and Abrahams[3] and Hogue et al.[78] failed to find adverse effects in their studies. Although soil ingestion can be a significant source of heavy metal intake by grazing animals, the heavy metals are generally not taken in at toxic concentrations by human consumers if these metals are stored in animal tissues, such as bone, that are not used for food.

Rates of intestinal absorption depend on the species and experimental conditions. Ingested soil generally reduces absorption of metals from digestive tracts. Based on the cadmium concentration in the blood of rats fed cadmium chloride in soil compared to the cadmium concentration in the blood of rats fed cadmium chloride at the same dose in a saline solution, Schilderman et al.[79] concluded that incorporation into soil reduced the bioavailability of cadmium to 43% of its reference value. Diamond et al.[28] and Mahaney et al.[25] suggested that animals deliberately ingest soil to bind to toxic substances in their diet. Soil was deliberately added to the diets of domestic animals to reduce their uptake of cesium released during the Chernobyl accident.[7]

Chaney and Ryan[80] concluded that as a laboratory mammal is fed greater amounts of soil contaminated with lead, lead absorption by the animal does not increase linearly with dose but reaches a plateau. The authors' model predicted that the concentrations of free lead ions in solution would level off as additional soil in the gut adsorbed additional lead. These studies were designed to be relevant to a child ingesting soil on an empty stomach. In contrast, when Canada geese, mute swans, and mallards were fed increasing amounts of lead-contaminated sediment, up to 48% of the diet, blood lead concentrations increased linearly with dose.[81] We presume that the difference in absorption curves between the laboratory mammals and the waterfowl results from the experimental conditions; the laboratory mammals were fed doses of soil without food, whereas the waterfowl were chronically fed large amounts of sediment incorporated into a diet for waterfowl. The presence of food in an animal's digestive system, as well as soil, affects the bioavailability. Maddaloni et al.[82] estimated that human volunteers who had fasted absorbed 26% of a dose of lead in soil from a mining site, but that volunteers who had eaten absorbed about 2.5% of that same dose.

It is widely accepted that the chemical form of an ingested metal affects its absorption and potential toxicity. When the National Research Council[83] established "maximum tolerable levels of dietary minerals for domestic animals," it noted that the elements used in the supporting studies were generally soluble salts and that the recommended values might not apply to elements naturally found in diets. When data are available, a "relative absorption factor" may be incorporated into risk assessments to adjust for the differences in chemical form.[84] Absorption of lead from soil has been widely studied because soil ingestion is the most important route of exposure of lead to children. Mosby[85] measured the uptake of lead by juvenile swine fed lead-contaminated soils from a mining site in Missouri. (Juvenile swine are considered appropriate models for children in their absorption of lead.) Compared to a control diet containing lead acetate, untreated mine soils had relative bioavailabilities of 0.67 and 0.63. Amending the contaminated soils with phosphoric acid reduced bioavailability by 27% to 38% (compared to the untreated contaminated soil). In a similar study on juvenile swine, Casteel et al.[86] concluded that lead in two severely contaminated soils from a mining site in Colorado had bioavailabilities of 58% and 60% compared to lead acetate in a diet fed without soil. Summarizing the relative absorption factors reported for soil lead primarily from mining wastes, Ruby et al.[84] found that relative absorption values were generally between 0.2 and 0.6, but were as low as 0.01. Values for arsenic were generally between 0.2 and 0.4, but were as low as 0.03. Because the experiments used to derive the relative absorption factors are expensive, Ruby et al.[84] have developed a quick *in vitro* procedure, designed to measure solubility of metals under conditions that mimic those in a human gut.

Hoffman et al.[53] measured the bioavailability of lead from Coeur d'Alene River Basin sediments in a feeding study on mallards based on blood lead concentrations. Blood lead concentrations were about half as high in ducks fed the sediment as in ducks fed lead acetate at a comparable rate of exposure in the diet. Sheppard et al.[87] concluded that adding soil to a diet fed to mice reduced bioavailability of isotopes of four cations by a factor of 1.1 for lead, 2.1 for cadmium, 2.3 for mercury, and 4.6 for cesium.

Some elements may be sparingly bioavailable in soil and are rarely, if ever, toxic threats to terrestrial animals. Examples include aluminum, barium, and iron, which are ingested by many animals primarily with soil, sometimes at high concentrations. Earthworms from confined disposal facilities for dredged materials were found to contain medians of 3400 mg/kg of aluminum, 31 mg/kg of barium, and 8000 mg/kg of iron, mainly associated with ingested soil in the earthworms.[88] All three of these concentrations are well above the tolerances of these minerals established for domestic animals (200 to 1000 ppm aluminum, 20 ppm barium, 500 to 3000 ppm iron) by the National Research Council.[83] Even higher concentrations of these three metals have been found in tadpoles,[89] probably associated with sediment in the tadpole gut. From a risk assessment based on feeding studies of soluble salts of the metals an assessor might conclude that these metals are toxic to predators of earthworms or tadpoles. However, it seems unlikely that these metals would be toxic under normal circumstances, and the examples illustrate the potential error introduced when the form of a metal in a feeding study does not match the form of that metal in an environmental sample being evaluated.

A science panel convened by the Agency for Toxic Substances and Disease Registry concluded that assuming 100% bioavailability of mercury in contaminated soils is excessively conservative; however, the panel was unable at that time to suggest default assumptions or to develop guidelines for determining relative bioavailability in soils.[90] Many risk assessors recognize that ignoring the bioavailability of elements in soil and sediments may lead to erroneous conclusions about hazards from ingestion. However, they lack adequate studies for most elements to account for bioavailability in risk assessments.

## 6.10 APPLICATIONS TO RISK ASSESSMENTS

The soil-animal-product pathway to humans is important in many risk assessments of persistent organic pollutants in agriculture. It may be the most important pathway when the pollutant is introduced directly to the soil, as in the application of sewage sludge to agricultural land. For a given measured or projected concentration of a contaminant in soil, the two most important factors that determine the amount of contaminant transported to human foods are the amount of soil consumed and the bioavailability of the contaminant. Very conservative (i.e., protective) default values are generally suggested in risk assessments.[11,12] We will identify areas of uncertainty and conservatism, and site-specific circumstances that would allow the use of less conservative parameters will be identified.

Although the extensive data available for sheep are useful for establishing general principles, agricultural risk assessments are generally based on beef and dairy cattle. The cattle studies

(Table 6.1) provided the basis for deriving the default values for cattle soil ingestion included in regulatory agency guidance documents.[11,12] In general, the default values will overestimate the amount of soil consumed in most circumstances.

The expression of soil ingestion as a quantity rather than a fraction of intake has been the normal practice in regulatory guidance. Estimation of the quantity of soil ingested requires assumed values for digestibility and dry matter intake (Equation 6.2), whereas estimation of soil as a fraction of intake requires only an assumption for digestibility (Equation 6.3). Generally, digestibility can be estimated reasonably well, but the assumption of single values for dry matter intake may be erroneous in periods of low forage availability. Thus, the average values for the amounts of soil ingested that are listed in Table 6.1 may be too high. The fraction of soil in the diet does not require assumptions about dry matter intake. The use of the bioconcentration factor approach (concentration in product/concentration in diet) has been adopted for dioxins and related compounds in some guidance documents.[11,12] This approach does not require assumptions concerning the amount of soil and dry matter consumed, which should reduce uncertainty.

The recommended default intake of 0.4 kg/day for dairy cattle[11] is somewhat less than the yearly mean for New Zealand cattle offered no supplemental feed (Table 6.1). This recommendation recognizes that standard dairy feeding practices in the United States include supplemental feeds in the form of concentrates (grain, etc.) and hay or silage in periods of sparse pasture. The default value is equal to approximately 2% of dry matter intake if the cow is consuming 20 kg of dry matter per day. This fairly conservative value is reasonable for the subsistence farmer scenarios commonly included in risk assessments. However, the value greatly overestimates soil intake under commercial dairy conditions because use of pasture for lactating dairy cows is not a common practice in the United States.

The suggested value for beef cattle, 0.5 kg/day,[11] is greater than the average intake for beef cattle in Table 6.1 because it includes consideration of the dairy cattle results. The value is probably a reasonable representation of beef cattle being maintained for breeding purposes on pasture only. The value, however, was derived assuming a dry matter intake typical of mature cattle, and the value would overestimate soil intake by the younger animals that had not attained their mature weight. This younger class of animals would represent the largest portion of the meat supply, and use of the high default value would overestimate exposures to large populations. On the other hand, it would probably provide a reasonable worst case for the subsistence farmer scenario.

Most commercial swine and poultry do not have access to soil, but soil consumption can be quite high when these species do have access. The suggested default intake for swine in guidance documents was 0.37 kg/day,[11] which is approximately the same as the highest measured value in the only published study on swine (Table 6.1). There have been no direct measurements of soil ingestion by free-range chickens. Soil ingestion by free-range chickens has been inferred in several cases,[70,90] and a default value of 10% of dry matter intake has been suggested.[11] This value seems reasonable because it agrees with the 9% intake measured in wild turkeys, which are birds with similar eating habits.[6]

Ingestion rates of soil and sediment are not as well documented for wildlife as for farm animals, and less guidance for conducting risk assessments is available. Wildlife risk assessors may benefit from generalizations about soil ingestion by farm animals but must consider the range of wildlife species present at a site and may prefer not using a mean for a species when diets may vary greatly with seasons and habitats. At this time there are no default values recommended for use with wildlife as there are for farm animals. Although the importance of estimating bioavailability of soil contaminants is widely accepted, risk assessors are likely to have difficulty in incorporating a measure of bioavailability without conducting feeding experiments.

Inclusion of estimates of soil and sediment ingestion may change the emphasis of risk assessments. Whereas wildlife risk assessments have often emphasized bioaccumulation of persistent contaminants in top predators, the soil and sediment ingestion route of exposure is most important to those species lower in the food chain. Raptors, for example, might be poisoned by organochlorines or through ingestion of lead shot, but they have low exposure to lead through the soil-ingestion route. Predators of earthworms and other soil organisms, shorebirds, and waterfowl feeding in sediments may be key species in risk assessments where the contaminant is found in the soil and sediment. Soil and sediment ingestion is most important for contaminants that are poorly absorbed and do not biomagnify. Risk assessors should also bear in mind that soil and sediment are not just vehicles for contaminants, but they supply nutrients and modify the bioavailability of contaminants. Research on contaminants to date has considered only the unintentional ingestion of soil and sediment, but it is possible that wildlife will be found sometimes to intentionally ingest contaminated soil because of its taste.

## REFERENCES


1. Hamilton, G., Let them eat dirt, *New Scientist*, 2143, 26,1998.
2. Arthur, W. J., III and Alldredge, A. W., Soil ingestion by mule deer in northcentral Colorado, *J. Range Manage.,* 32, 67, 1979.
3. Thornton, I. and Abrahams, P., Soil ingestion — A major pathway of heavy metals into livestock grazing contaminated land, *Sci. Total Environ.*, 28, 287, 1983.
4. Arthur, W. J., III and Gates R. J., Trace element intake via soil ingestion in pronghorns and in black-tailed jackrabbits, *J. Range Manage.*, 41, 162, 1988.
5. Fries, G. F., Bioavailability of soil-borne polybrominated biphenyls ingested by farm animals, *J. Toxicol. Environ. Health,* 16, 565, 1985.
6. Beyer, W. N., Connor, E. E., and Gerould, S., Estimates of soil ingestion by wildlife, *J. Wildl. Manage.,* 58, 375, 1994.
7. Sheppard, S. C., Geophagy: Who eats soil and where do possible contaminants go?, *Environ. Geol.*, 33, 109, 1998.
8. Healy, W. B., Nutritional aspects of soil ingestion by grazing animals, in *Chemistry and Biochemistry of Herbage*, Butler, G. W. and Bailey, R. W., Eds., Academic Press, New York, Vol. 1, 1973, 567.
9. Fries, G. F. and Paustenbach, D. J., Evaluation of potential transmission of 2,3,7,8-tetrachlorodibenzo-*p*-dioxin contaminated incinerator emissions to humans via foods, *J. Toxicol. Environ. Health,* 29, 1, 1990.
10. Fries, G. F., Organic contaminants in terrestrial food chains, in *Organic Contaminants in the Environment,* Jones, K. C., Ed., Elsevier Appl. Sci., New York, 1991, 207.
11. USEPA, Human Health Risk Assessment Protocol for Hazardous Waste Combustion Facilities, Report No. EPA530-D-98/001a, Office of Solid Waste and Emergency Response, Environmental Protection Agency, Washington, D.C., 1998.
12. USEPA, Estimating Exposure to Dioxin-like Compounds, Report No. EPA/600/P-00/001Bb, Office of Research and Development, Environmental Protection Agency, Washington, D.C., 2000.
13. Calabrese, E. J. and Stanek, E. J., Resolving intertracer inconsistencies in soil ingestion estimation, *Environ. Health Perspect.,* 103, 454, 1995.
14. Stanek, E. J. and Calabrese, E. J., Daily estimates of soil ingestion by children, *Environ. Health Perspect.,* 103, 276, 1995.
15. Fries, G. F., Marrow, G. S., and Snow, P. A., Soil ingestion by dairy cattle, *J. Dairy Sci.*, 65, 611, 1982.
16. Subcommittee on Feed Intake, Predicting Feed Intake of Food-Producing Animals, National Academy Press, Washington, D.C., 1987.
17. Fries, G. F., Ingestion of sludge applied organic chemicals by animals, *Sci. Total Environ.,* 185, 93, 1996.
18. Gorman, M. L. and Stone, R. D., *The Natural History of Moles*, Cornell University Press, Ithaca, New York, 1990.
19. Goodman, D. C. and Fisher, H. I., *Functional Anatomy of the Feeding Apparatus in Waterfowl (Aves: Anatidae)*, Southern Illinois Univ. Press, Carbondale, 1962.
20. Setz, E. Z. F., Enzweiler, J., Solferini, V. N., Amendola, M. P., and Berton, R. S., Geophagy in the golden-faced saki monkey (*Pithecia pithecia chrysocephala*) in the Central Amazon, *J. Zool. Lond.*, 247, 91, 1999.

21. Cowan, I. M. and Brink, V. C., Natural game licks in the Rocky Mountain national parks of *Can. J. Mammal.*, 30, 379, 1949.
22. Hebert, D. and Cowan, I. M., Natural salt licks as a part of the ecology of the mountain goat. *Can. J. Zool.*, 49, 605, 1971.
23. Weeks, H. P., Jr. and Kirkpatrick, C. M., Adaptations of white-tailed deer to naturally occurring sodium deficiencies, *J. Wildl. Manage.*, 40, 610, 1976.
24. Kreulen, D. A. and Jager, T., The significance of soil ingestion in the utilization of arid rangelands by large herbivores, with special reference to natural licks on the Kalahari pans, in *Int. Symposium on Herbivore Nutrition in the Subtropics and Tropics* (1983: Pretoria, S. Africa), Gilchrist, F. M. C. and Mackie, R. I., Eds., Science Press, Chraignall, S. Africa, 1984, 204.
25. Mahaney, W. C., Aufreiter, S., and Hancock, R. G. V., Mountain gorilla geophagy: A possible seasonal behavior for dealing with the effects of dietary changes, *Int. J. Primatol.*, 16, 475, 1995.
26. Mahaney, W. C., Stambolic, A., Knezevich, M., Hancock, R. G. V., Aufreiter, S., Sanmugadas, K., Kessler, M. J., and Grynpas, M. D., Geophagy amongst rhesus macaques on Cayo Santiago, Puerto Rico, *Primates*, 36, 323, 1995.
27. Ruggiero, R. G. and Fay, M., Utilization of termitarium soils by elephants and its ecological implications, *Afr. J. Ecol.*, 32, 222, 1994.
28. Diamond, J., Bishop, K. D., and Gilardi, J. D., Geophagy in New Guinea birds, *Ibis*, 141, 181, 1999.
29. Sokal, O. M., Lithophagy and geophagy in reptiles, *J. Herpetol.*, 5, 69, 1971.
30. Kramer, D. C., Geophagy in *Terrepene ornata ornata* Agassiz, *J. Herpetol.*, 7, 138, 1973.
31. Meinertzhagen, Col. R., Grit, *Bull. Br. Ornithologists' Club*, 74, 97, 1954.
32. Gionfriddo, J. P. and Best, L. B., Grit use by birds, a review, *Curr. Ornithol.*, 15, 89, 2000.
33. Field, A. C. and Purves, D., The intake of soil by grazing sheep, *Proc. Nutr. Soc.*, 23, xxiv, 1964.
34. Healy, W. B. and Ludwig, T. G., Wear of sheep's teeth. I. The role of ingested soil, *N. Z. J. Agric. Res.*, 8, 737, 1965.
35. Harrison, D. L., Mol, J. C. M., and Healy, W. B., DDT residues in sheep from the ingestion of soil, *N. Z. J. Agric. Res.*, 13, 664, 1970.
36. Healy, W. B., Ingestion of soil by dairy cows, *N. Z. J. Agric. Res.*, 11, 487, 1968.
37. Healy, W. B., Cutress, T. W., and Michie, C., Wear of sheep's teeth. IV. Reduction of soil ingestion and tooth wear by supplementary feeding, *N. Z. J. Agric. Res.,* 10, 201, 1967.
38. Healy, W. B. and Drew, K. R., Ingestion of soil by hoggets grazing swedes, *N. Z. J. Agric. Res.,* 13, 940, 1970.
39. McGrath, D., Poole, D. B. R., Fleming, G. A., and Sinnott, J., Soil ingestion by grazing sheep. *Irish J. Agric. Res.*, 21, 135, 1982.
40. Mayland, H. F., Florence, A. R., Rosenau, R. C., Lazar, V. A., and Turner, H. A., Soil ingestion by cattle on semiarid range as reflected by titanium analysis of feces, *J. Range Manage.*, 28, 448, 1975.
41. Fries, G. F., Marrow, G. S., and Snow, P. A., Soil ingestion by swine as a route of contaminant exposure, *Environ. Toxicol. Chem.*, 1, 201, 1982.
42. Garten, C. T., Jr., Ingestion of soil by hispid cotton rats, white-footed mice and eastern chipmunks, *J. Mamm.*, 61, 136, 1980.
43. Beyer, W. N., Blus, L. J., Henny, C. J., and Audet, D., The role of sediment ingestion in exposing wood ducks to lead, *Ecotoxicology,* 6, 181, 1997.
44. Beyer, W. N., Audet, D. J., Morton, A., Campbell, J. K., and LeCaptain, L., Lead exposure of waterfowl ingesting Coeur d'Alene River Basin sediments, *J. Environ. Qual.*, 27, 1533, 1998.
45. Connor, E. E., Scanlon, P. F., and Kirkpatrick, R. L., Bioavailability of lead from contaminated sediment in northern bobwhites, *Colinus viginianus, Arch. Environ. Contam. Toxicol.*, 27, 60, 1994.
46. Beyer, W. N., Spann, J., and Day, D., Metal and sediment ingestion by dabbling ducks, *Sci. Total Environ.*, 231, 235, 1999.
47. Beyer, W. N., Day, D., Morton A., and Pachepsky, Y., Relation of lead exposure to sediment ingestion in mute swans on the Chesapeake Bay, *Environ. Toxicol. Chem.*, 17, 2298, 1998.
48. Hui, C. A. and Beyer, W. N., Sediment ingestion of two sympatric shorebird species, *Sci. Total Environ.*, 224, 227, 1998.
49. Reeder, W. G., Stomach analyses of a group of shorebirds, *Condor,* 53, 43,1951.
50. Beyer, W. N., Day, D., Melancon, M. J., and Sileo, L., Toxicity of Anacostia River, Washington, D.C., USA, sediment fed to mute swans (Cygnus olor). *Environ. Toxicol. Chem.*, 19, 731, 2000.

51. Heinz, G. H., Hoffman, D. J., Sileo, L, Audet, D. J., and LeCaptain, L. J., Toxicity of lead-contaminated sediment to mallards, *Arch. Environ. Contam. Toxicol.*, 36, 323, 1999.
52. Hoffman, D. J., Heinz, G. H., Sileo, L., Audet, D. J., Campbell, J. K., LeCaptain, L. J., and Obrecht, H. H., III, Developmental toxicity of lead-contaminated sediment in Canada geese (*Branta canadensis*), *J. Toxicol. Environ Health (Part A)*, 59, 235, 2000.
53. Hoffman, D. J., Heinz, G. H., Sileo, L., Audet, D. J., Campbell, J. K., and LeCaptain, L. J., Developmental toxicity of lead-contaminated sediment to mallard ducklings, *Arch. Environ. Contam. Toxicol.*, 39, 221, 2000.
54. Day, D. D., The toxicity of lead-contaminated sediment to mute swans, MS thesis, University of Maryland, College Park, MD, 1998.
55. Chupp, N. R. and Dalke, P. D., Waterfowl mortality in the Coeur d'Alene River Valley, Idaho, *J. Wildl. Manage.*, 28, 692,1964.
56. Blus, L. J., Henny, C. J., Hoffman, D. J., Sileo, L., and Audet, D. J., Persistence of high lead concentrations and associated effects in tundra swans captured near a mining and smelting complex in northern Idaho, *Ecotoxicology*, 8,125, 1999.
57. Sileo, L., Creekmore, L. H., Audet, D. J., Snyder, M. R., Meteyer, C. U., Franson, J. C., Locke, L. N., Smith, M. R., and Finley, D. L., Lead poisoning of waterfowl by contaminated sediment in the Coeur d'Alene River, *Arch. Environ. Contam, Toxicol.*, 41, 364, 2001.
58. Phillips, J. C. and Lincoln, F. C., *American Waterfowl, Their Present Situation and the Outlook for Their Future*, Houghton Mifflin Co., Boston, 1930, 162.
59. Sanderson, G. C. and Bellrose, F. C., A review of the problem of lead poisoning of waterfowl. Special Publication No.4, Ill. Nat. Hist. Sur., Champaign, Illinois, 1986.
60. Sparling, D. W., Vann, S., and Grove, R. A., Blood changes in mallards exposed to white phosphorus, *Environ. Toxicol. Chem.*, 17, 2521, 1998.
61. Stansley, W. and Roscoe, D. E., The uptake and effects of lead in small mammals and frogs at a trap and skeet range, *Arch. Environ. Contam. Toxicol.*, 30, 220, 1996.
62. Pain, D. J., Sánchez, A., Meharg, A. A., The Doñana ecological disaster: Contamination of a world heritage estuarine marsh ecosystem with acidified pyrite mine waste, *Sci. Total Environ.*, 222, 45, 1998.
63. Anderson, W. C., Loehr, R. C., and Smith, B. P., Eds., *Environmental Availability of Chlorinated Organics, Explosives, and Metals in Soils*, American Academy of Environmental Engineers, Annapolis, MD, 1999.
64. Rose, J. Q., Ramsey, J. C., Wentzler, T. H., Hummel, R. A., and Gehring, P. J., The fate of 2,3,7,8-tetrachloro-*p*-dioxin following single and repeated doses to the rat, *Toxicol. Appl. Pharmacol.*, 36, 209, 1976.
65. Fries, G. F. and Marrow, G. S., Retention and excretion of 2,3,7,8-tetrachloro-*p*-dioxin by rats, *J. Agric. Food Chem.*, 23, 265, 1975.
66. Jones, D., Safe, S., Morcom, E., Holcomb, M., Coppock, C., and Ivie, W., Bioavailability of grain and soil-borne tritiated 2,3,7,8-tetrachlorodibenzo-*p*-dioxin (TCDD) administered to lactating Holstein cows, *Chemosphere,* 18, 1257, 1989.
67. McConnell, E., Lucier, G., Rumbaugh, R., Albro, P., Harvan, D. Hass, J., and Harris, M., Dioxin in soil: Bioavailability after ingestion by rats and guinea pigs, *Science,* 223, 1077, 1984.
68. Shu, H., Paustenbach, D., Murray, F. J., Marple, L., Brunck, B., Dei Rossi, D., and Teitelbaum, P., Bioavailability of soil-bound TCDD: Oral bioavailability in the rat, *Fundam. Appl. Toxicol.*, 10, 335, 1988.
69. Umbreit, T. H., Hesse, E. J., and Gallo, M. A., Bioavailability of dioxin in soil from a 2,4,5-T manufacturing site, *Science,* 232, 497, 1986.
70. Stephens, R. D., Petreas, M. X., and Hayward, D. G., Biotransfer and bioaccumulation of dioxins and furans from soil: Chickens as a model for foraging animals, *Sci. Total Environ.,* 175, 253, 1995.
71. Poiger, H. and Schlatter, C., Influence of solvents and adsorbents on dermal and intestinal absorption of TCDD, *Food Cosmet. Toxicol.*, 18, 477, 1980.
72. Burchill, S., Greenland, D. J., and Hayes, M. H. B., Adsorption of organic molecules, in *The Chemistry of Soil Processes*, Greenland, D. J. and Hayes, M. H. B., Eds., John Wiley & Sons, Chichester, 1981, 621.
74. Fries, G. F. and Marrow, G. S., Influence of soil properties on the uptake of hexachlorobiphenyls by rats, *Chemosphere,* 24, 109, 1992.

74. Kadry, A. M., Skowronski, G. A., Turkall, R. M., and Abdel-Rahman, M. S., Comparison between oral and dermal bioavailability of soil-adsorbed phenanathrene in female rats, *Toxicol. Lett.,* 78, 153, 1995.
75. Weber, J. B., Interaction of organic pesticides with particulate matter in aquatic and soil systems, *Advan. Chem. Series,* 111, 55, 1972.
76. Healy, W. B., McCabe, W. J., and Wilson, G. F., Ingested soil as a source of microelements for grazing animals, *N. Z. J. Agric. Res.,* 13, 503, 1970.
77. Healy, W. B., Crouchley, G., Gillett, R. L., Rankin, P. C., and Watts, H. M., Ingested soil and iodine deficiency in lambs, *N. Z. J. Agric. Res.,* 15, 778, 1972.
78. Hogue, D. E., Parrish, J. J., Foote, R. H., Stouffer, J. R., Anderson, J. L., Stoewsand, G. S., Telford, J. N., Bache, C. A., Gutenmann, W. H., and Lisk, D. J., Toxicological studies with male sheep grazing on municipal sludge-amended soil, *J. Toxicol. Environ. Health,* 14, 153, 1984.
79. Schilderman, P. A. E. L., Moonen, E. J. C., Kempkers, P., and Kleinjans, J. C. S., Bioavailability of soil-adsorbed cadmium in orally exposed male rats, *Environ. Health Perspect.,* 105(2), 234, 1997.
80. Chaney, R. L. and Ryan, J. A., Risk based standards for As, Pb, and Cd in urban soils, in *State-of-the-Art in Evaluating the Risks of As, Cd, and Pb in Urban Soils for Plants, Animals, and Humans,.* Proc. Conf. Criteria for Decision Finding in Soil Protection: Evaluation of Arsenic, Lead, and Cadmium in Contaminated Urban Soils (Oct. 9–11, 1991; Braunschweig, FRG), DECHEMA, Frankfurt, Germany, 1994, pp. 59–88.
81. Beyer, W. N., Audet, D. J., Heinz, G. H., Hoffman, D. J., and Day, D., Relation of waterfowl poisoning to sediment lead concentrations in the Coeur d'Alene River Basin, *Ecotoxicology,* 9, 207, 2000.
82. Maddaloni, M., Lolacono, N., Manton, W., Blum, C., Drexler, J., and Graziano, J., Bioavailability of soilborne lead in adults, by stable isotope dilution, *Environ. Health Perspect.*, 106, 1589, 1998.
83. National Research Council, Mineral Tolerance of Domestic Animals, National Academy of Sciences, Washington, D.C., 1980.
84. Ruby, M. V., Schoof, R., Brattin, W., Goldade, M., Post, G., Harnois, M., Mosby, D. E., Casteel, S. W., Berti, W., Carpenter, M., Edwards, D., Cragin, D., and Chappell, W., Advances in evaluating the oral bioavailability of inorganics in soil for use in human health risk assessment, *Environ. Sci. Technol.*, 33, 3697, 1999.
85. Mosby, D. E., Reductions of Lead Bioavailability in Soil to Swine and Fescue by Addition of Phosphate, Masters Thesis, University of Missouri-Columbia, May 2000.
86. Casteel, S. W., Cowart, R. P., Weis, C. P., Henningsen, G. M., Hoffman, E., Brattin, W. J., Guzman, R. E., Starost, M. F., Payne, J. T., Stockham, S. L., Becker, S. V., Drexler, J. W., and Turk, J. R., Bioavailability of lead to juvenile swine dosed with soil from the smuggler mountain NPL site of Aspen, Colorado, *Fundam. Appl. Toxicol.*, 36,177, 1997.
87. Sheppard, S. C., Evenden, W. G., and Schwartz, W. J., Ingested soil: Bioavailability of sorbed lead, cadmium, cesium, iodine, and mercury, *J. Environ. Qual.*, 24, 498, 1995.
88. Beyer, W. N. and Stafford, C., Survey and evaluation of contaminants in earthworms and in soils derived from dredged materials at confined disposal facilities in the Great Lakes Region, *Environ. Monit. Assess.*, 24, 151, 1993.
89. Sparling, D. W. and Lowe, T. P., Metal concentrations of tadpoles in experimental ponds, *Environ. Pollut.*, 91, 149, 1996.
90. Canady, R. A., Hanley, J. E., and Susten, A. S., ATSDR Science Panel on the bioavailability of mercury in soils: Lessons learned, *Risk Anal.*, 17, 527, 1997.
91. Schuler, F., Schmid, P., and Schlatter, C., The transfer of polychlorinated dibenzo-*p*-dioxins and dibenzofurans from soil into eggs of foraging chicken, *Chemosphere,* 34, 711, 1997.

CHAPTER 7

# Wildlife and the Remediation of Contaminated Soils: Extending the Analysis of Ecological Risks to Habitat Restoration

**Greg Linder, Gray Henderson, and Elaine Ingham**

## CONTENTS

*They paved paradise, put up a parking lot.*

**Joni Mitchell, 1970**

## 7.1 INTRODUCTION

As a process, ecological risk assessment (ERA) may benefit efforts to remediate soils contaminated with chemicals intentionally or unintentionally released to the environment. The ERA process has gained increased application across a wide range of environmental issues because it attempts to integrate and synthesize available technical information that may contribute to environmental decisions and to help forge environmental policy developed by environmental decision-makers. By applying the process and technical tools available in ERA to a wide range of land-use and water-use issues, fish and wildlife scientists and resource managers can

- Identify resources that are "at risk" from a variety of environmental stressors (chemical, physical, and biological) that may be characteristic of habitats variously impacted by a range of anthropogenic activities,
- Prioritize data collection, including the design of field investigations for assessment and monitoring when the current status of "at-risk" resources is adequately characterized, and
- Link human activities — both causal and remedial in character — or naturally occurring hazards with their potential ecological effects during the restoration or rehabilitation of habitats of concern.

This chapter will focus on applications of the ERA process beyond an evaluation of risks associated with chemical release. Here, we will briefly summarize (1) ERA tools used to characterize a lower bound on the concentration of residual chemicals remaining in soil following remediation (that is, contribute to establishing preliminary remedial goals, or PRGs, for habitat restoration), (2) the role of bioavailability in critically evaluating these lower-bound PRGs and establishing residual chemical concentrations that reflect field conditions, and (3) remediation measures that are intended to address field conditions and modify soil in order to decrease a chemical's immediate bioavailability, while increasing the likelihood of recovery to habitats suitable to future use by fish and wildlife.

## 7.2 TERMINOLOGY AND APPLICATION

Because words often acquire meaning through their regulatory context, the terms restoration, remediation, reclamation, and rehabilitation will be defined as used in this chapter, so their definitions will not confound the discussion that follows. Conceptually, ecological restoration is "the process of assisting the recovery and management of ecological integrity, where ecological integrity is specified by a critical range in the variability of biological diversity, ecological processes and structures, regional and historical contexts, and sustainable cultural practices."[1] Given the scope of that definition, however, a relatively simple dictionary definition seems more appropriate to our present discussion. Here, restoration will be considered as the act or process involved with bringing habitats back to a former condition, given a past perturbation that has yielded an "impacted," "disturbed," or "adversely affected" condition. Adverse effects may be associated with chemical releases (e.g., oil spill or other chemical release), physical perturbations (e.g., habitat alteration resulting from vegetation removal consequent to construction or remedial activities), or release of biological agents (e.g., nonnative species). A common observation finds chemical and physical stressors occurring in tandem when a habitat is degraded or altered from its initial preimpact condition.

Restoration and remediation are closely related, since the latter act or process reflects the implementation of a remedy for a past action that has yielded an impacted system. For our purposes, restoration may be one possible goal of remediation focused on an impacted habitat; that is, the remedial activity is intended to return a habitat to its preimpact condition. The goal of "return to preimpact condition" may be difficult, if not impossible, to achieve depending on both qualitative attributes of the release — "what was released" — and the magnitude of the release — "how much was released." Like restoration in a dictionary definition, rehabilitation may also be defined as the act or process of returning an affected area to a former condition or capacity, which appears largely synonymous with restoration. However, from a practical perspective, shades of difference between restoration and rehabilitation are often reflected by differences in likelihood of success in attaining preimpact or initial condition. For a restoration project, a return to preimpact condition is generally more likely to be achieved, while rehabilitation entails the reestablishment of a specified target condition that, from necessity, is not truly a return to preimpact condition. Differences between restoration and rehabilitation, then, reflect the degree of success in recovery of a habitat's preimpact or initial condition.

In contrast to restoration or rehabilitation, reclamation is generally considered the act or process of recovering a given habitat from some "undesirable" condition. Reclamation generally reflects the intentional recovery of lands for subsequent human use, whether that initial condition for reclamation is an unwanted natural state (e.g., wetland conversion to farmland or to urban development through drain-and-fill programs) or an impacted managed landscape (e.g., brownfield developments targeted for continued human use or otherwise intensively managed condition).

Application of the tools and process of ERA under the auspices of restoration or rehabilitation provides a foundation for comparing different management options and potentially helps policymakers and the public to make better-informed decisions about the management of natural resources that have variously been impacted by anthropogenic activities. These natural-resource-management decisions, however, must occur against a backdrop of naturally occurring hazards, ranging from relatively short-term (but potentially high-consequence) events such as floods, volcanic activity, earthquakes, and extreme atmospheric events, such as hurricanes, or relatively long-term events such as cumulative impacts associated with human use of a particular geographic area.[2]

### 7.2.1 Tools Available for Evaluating Habitat Remediation

Fish and wildlife scientists and resource managers must recognize that many of the ecosystems in which they are interested have been variously impacted by anthropogenic activities, ranging from those systems having relatively limited impacts to those highly disturbed systems having marked physical and chemical perturbations from some initial condition.[3–6] Chemicals released into the environment are only one source of environmental perturbation likely to drive the restoration or rehabilitation of habitats within a landscape. From an ecotoxicological perspective the evaluation of adverse effects associated with chemicals released into terrestrial and wetland habitats may directly support efforts beyond retrospective risk assessment focused on current conditions at recovering contaminated terrestrial and wetland habitats. A retrospective risk assessment for a terrestrial or wetland location often considers remediation of soils as one of the outcomes of the ERA process. Although this brief overview of the ERA process applied to restoration and rehabilitation is focused on chemically degraded terrestrial and wetland habitats, from an ecological perspective the interrelationship of habitats (especially for wetlands) clearly suggests that remediation of soils in these habitats could also impact sediments, surface water, and groundwater.[7,8]

In borrowing tools from ERA, the techniques and methods applicable to restoration and rehabilitation efforts include diagnostic methods focused on biochemical and molecular-, cellular-, tissue-, and organ-level measures as well as physiological measures of adverse effects associated with exposure (e.g., biomarkers).[8–13] Other tools may be focused on population and community measures that potentially reflect adverse changes in higher levels of biological organization associated with stressors in the environment.[10,11] Across these levels of biological organization, these tools may be applied to the (1) evaluation of current conditions, (2) characterization of reference condition, or (3) monitoring of recovery during the restoration or rehabilitation process. Regardless of their application, current practice largely depends on species-specific tools (e.g., individuals as surrogates for guilds), with subsequent extrapolation to potential ecological effects associated with these measures of exposure and effects. Population- and community-level analysis may be used to confirm whether these extrapolations are on target or whether the calibration of the lower-level analysis is required. Ideally, population- or community-level metrics are included in monitoring plans instituted to evaluate recovery in restored or rehabilitated habitats.[11]

### 7.2.2 Setting Targets for Soil Preliminary Remediation Goals (PRGs) Focused on Habitat Remediation

From an ecotoxicity and bioaccumulation perspective the development of target preliminary remediation goals (PRGs) for contaminated upland and wetland soils can be considered through a process that answers a simple question: "How much residual material can be left *in situ* yet not be associated with risks or injury related to long-term exposures to chemicals that occur at low levels in the environment?" Such a simple question assumes that low-level exposures would not be characterized by acute toxicity since risk assessments reliant on available guidance should have clearly identified those levels as part of the risk-assessment process.

For example, while benchmark values for soils are presently limited (few, if any) or under development (e.g., Ecological Soil Screening Levels, or EcoSSLs[12]), biological assessments conducted using standard test methods may directly evaluate the short-term effects of soil exposures, for example, to plants and soil invertebrates.[13] Similarly, guidance values for sediments, such as probable effect concentrations (PECs) and threshold effect concentrations (TECs) as characterized by McDonald et al.[14] for freshwater sediments, or effects range-medians (ERMs) and effects range-lows (ERLs) for marine and estuarine sediments,[15,16] may provide preliminary benchmarks for hydric soils indicative of wetland habitats. These latter tools from sediment risk assessment may have been considered within a risk context to characterize concentrations that directly relate to toxicity of wetland soils (e.g., emergent zone habitats or tidal mudflats), and their extension to restoration or rehabilitation activities may be pertinent to restoration or rehabilitation issues.

However, for remedial application, the concern associated with low-level chemical exposure should primarily be focused on chronic effects (e.g., altered reproductive capacity, altered immune function) and bioaccumulation.[13,17] These endpoints may be incompletely captured by available risk-based guidance values, and additional efforts may be required to assure restoration or rehabilitation success. For example, with a focus on chemicals having bioaccumulative potential, trophic transfers resulting from food-web interactions among exposed biota and their likely "predators" (that is, any consumer dependent upon exposed organisms as a food source) must figure prominently in the analysis and derivation of soil PRGs. Existing management practices (e.g., land and water use, especially as they relate to potential habitat interactions) and environmental fate issues (e.g., related to sediment transport subsequent to soil erosion) should also be sufficiently characterized to assure that pathways potentially linking sources with receptors account for future events such as soil erosion and increased sedimentation of surface waters adjacent to remediated habitats.[17]

Even with a technical analysis sufficient to the issues related to trophic transfer of *in situ* chemicals, this focus on biological interactions related to food-chain contamination may not be directly associated with ecological effects linked to residual chemicals remaining *in situ* at low levels. The available information regarding bioaccumulation and adverse ecological effects (e.g., change in populations and community structure) is very poorly developed, and depending on the chemicals of ecological concern, an analysis of bioaccumulation potential may not be the only aspect of exposure critical to management decisions regarding cleanup.[18] For example, an evaluation of bioaccumulation potential is not particularly pertinent to chemicals that are metabolized or otherwise not accumulated in the tissues of receptors of concern, yet long-term exposures to residual chemicals or their degradation products in remediated soils must be considered in the restoration or rehabilitation process. The following discussion reflects a primary focus on bioaccumulative chemicals of concern (BCCs) and the analysis of *in situ* BCCs that may remain following a remedial action.

#### *7.2.2.1 Bioaccumulative Chemicals of Concern: Developing Soil PRGs Focused on Habitat Remediation*

For evaluating risks of residual chemicals remaining in soil following remediation, chemicals with bioaccumulation potential may be a primary concern, and a two-step analysis of residual risks may be followed: (1) to estimate the exposed dose of BCCs that organisms encounter in chemical transfers from soil-to-soil biota and (2) to estimate exposed dose of BCCs to higher trophic levels exposed to chemicals in soil either directly through coincidental ingestion of soil or indirectly through ingestion of soil-exposed food sources.[19] For remediation purposes, developing soil PRGs should initially be focused on chemical risks and adverse biological effects associated with long-term exposure to low-level chemicals, with their derivation being based in part on management practices applicable to a given remedial activity.[2,17]

Technically, three general approaches to the evaluation of bioaccumulation are applicable to the restoration and rehabilitation process. These tools may be used singly or in combination in the development of soil PRGs:

- A laboratory approach exposing organisms to soil under controlled conditions
- A field approach collecting species from a study area
- Models to estimate bioaccumulation, including the assessment of food-web transfer (generally through food-chain analysis as the primary tool to evaluate exposure)

*Laboratory approach*: Individuals of a single species are exposed under controlled laboratory conditions to soils collected from the study area being assessed. After a defined exposure period, tissues of test organisms are analyzed for the chemicals of concern. Bioaccumulation is generally considered to have occurred if the final concentration in tissues exceeds concentrations that were present in tissues before the exposure was started (i.e., tissue residues characteristic of organism's baseline condition at test initiation).[13,18,19]

*Field approach*: Tissue concentrations of contaminants are determined by collecting one or more species exposed to soils at the study area being assessed. These concentrations are compared to some reference condition (e.g., local or population of reference areas) for interpretation.[20,21] Two methods have been used to determine bioaccumulation in the field: (1) organisms resident at the area are collected *in situ* for analysis or (2) organisms are transplanted from another location (presumably with a history of little contaminant exposure) to the area of concern and then re-collected, and tissues are analyzed after an established period of exposure.

*Models for evaluation of bioaccumulation and food-chain transfer*: Although not validated, models that describe bioaccumulation are widely applied to ERA and, by extension, to evaluations focused on residual chemicals present in remediated soils.[12,13,20–22] For example, site-specific models may be developed to estimate bioaccumulation on the basis of laboratory-derived methods (using standard test species or, alternatively, more habitat-specific species)[13] and field-determined chemical measurements at the area of concern. Field studies should also be considered to characterize bioaccumulation potentials of chemicals *in situ*, although exposure may confound linkages directly with soils, given the multiple chemical inputs captured in most field exposures.[13,23] Models that link (1) bioaccumulation of chemicals from soil to (2) transfer of chemical from soil biota to "predators" through a food-chain analysis have been technically developed but have not been validated with respect to their predictive capabilities for estimating field-measured tissue residues in consumers (primary or secondary).[12,20,21]

Regardless of the tools used, variability is a common problem in bioaccumulation studies and can lead to imprecise estimates of exposure. For example, laboratory bioaccumulation tests of soils (upland or wetland) potentially yield precise estimates of bioaccumulation; however, their precision is directly dependent upon controlling biological factors through test standardization, e.g., specification of appropriate test organisms or use of organisms of similar size, age, and life-history stage.[13,24] Under field conditions bioaccumulation may be highly variable, since organisms variously alter exposure through, for example, avoidance, or multiple stressors and life-stage differences apparent in the exposed population serve to confound the answer to a relatively simple couplet interrogative focused on bioaccumulation ("Does bioaccumulation occur and, if so, to what extent?"). Also, under field conditions individuals may respond differently to residual chemicals in soil, and if bioaccumulation is sufficient to alter biological response, the kinetics of bioaccumulation may be affected.[25] Depending on exposure in the field, organisms exposed to lower doses in remediated soils may accumulate greater tissue residues than estimated.

Bioaccumulation models are generally imprecise because of the knowledge gaps that characterize model parameters, such as ingestion rates and body-size-related functions, that influence exposure.[18] Key parameter values, such as bioaccumulation factors (BAFs), may differ by orders of magnitude among studies. Hence, predictions differ and imprecision of model outputs must be considered when values for soil PRGs are evaluated within the context of habitat restoration.

Whether framed about soil-to-soil biota transfers or on food chains dependent on soil-dwelling invertebrates, a limitation to food-chain evaluations applied to the restoration of soil is the poorly characterized relationships among model outputs (e.g., a soil concentration term), ecological effects, and unacceptable change[7,18,22] potentially linked to that soil PRG. Bioaccumulation does not necessarily mean an adverse biological or ecological effect is occurring.[13] Correlations between bioaccumulated contaminants and effects of contaminants are frequently not well established, and the relevance of bioaccumulation stems mainly from its value in characterizing exposure, which in turn reinforces the need to characterize dose. Understanding dose is essential to understanding effects, whether the dose is from the chemical itself or from its metabolites.[25] The concentration of contaminants assimilated into tissues may be one of the most sensitive measures of the dose an organism can realize, given the uncertainties in defining bioavailability from chemical measures alone.[12,13]

Exposures under field conditions are complex, and the temporal and spatial complexities of exposures are likely reasons why effects of contaminants may be difficult to demonstrate or to interpret from field data alone.[26] Collection of organisms exposed in the field, food web bioaccumulation estimates, and both empirical and site-specific models provide direct measures of contaminant concentrations in wildlife foods, which may suggest alternatives applicable to the characterization of soil PRGs.[20,21,27] For example, some soil PRGs might be developed given available regulatory values that are triggered by bioaccumulation concerns (e.g., FDA action levels), while others would necessarily reflect case-specific studies (or equally accepted modeling efforts) that would follow a stepwise process similar to that detailed for evaluating ecological risks.

Rather than rely solely on a single tool for their derivation, soil PRGs could be technically developed as a stepwise process based on laboratory-derived or field-derived estimates of bioaccumulation potential.[12,13] Such information should be in hand for soils targeted for remediation, as should estimates of reference conditions as baseline or reference areas. These empirical values should reflect some specified "co-location" of soil and tissue samples so the spatial context of exposure can be identified.[20,21,22] Ideally, a conceptual site model or linked conceptual site models for soil remediation are consistent with those from the ERA,[7,22] where critical receptors (most often mammalian or avian wildlife) are specified such that derivation of soil PRGs associated with soil concentrations or trigger values for tissue residues can be estimated using available empirical data collected during ERA activities. Regardless of the tools used to derive soil PRGs, the role of post-remediation monitoring should not be ignored, especially given the relatively large uncertainties associated with even the "best" practice available for developing soil PRGs and implementing restoration or rehabilitation plans.[3–6]

### *7.2.2.2 Derivation and Calculation of Soil PRGs*

Lower bound targets for soil PRGs may be derived from a "back calculation" of a species-specific exposed dose calculation routinely used in evaluation of chemical risks to vertebrate wildlife (birds and mammals).[12,13,18] Here, our starting point is the basic exposure model derived from food-chain analysis,

$$\frac{(\mathrm{Cs} \times \mathrm{IRs}) + (\mathrm{Cp} \times \mathrm{IRp})}{\mathrm{BW}} = \mathrm{Edose} \tag{7.1}$$

where:
Cp = concentration of chemical x in prey, mg *X*/kg prey tissue;
IRp = ingestion rate of prey, kg prey tissue/day;
Cs = concentration of chemical *X* in soil, mg *X*/kg soil;
IRs = ingestion rate of soil, kg soil/day;
BW = body weight of receptor of concern, kg; and
Edose = exposed dose, mg/kg/day.

For an analysis focused on soil, inhalation routes of exposure, dermal uptake, and water consumption are considered negligible and are not considered a significant contribution to exposed dose (Ed). Additionally, with the role that BAF (bioaccumulation factor) plays in the derivation of soil PRGs (especially in the absence of empirical bioaccumulation data),

$$(\mathrm{Cs} \times \mathrm{BAF}) = \mathrm{Cp} \tag{7.2}$$

may be substituted for Cp, the concentration of chemical in prey tissue (in mg *X*/kg prey tissue where *X* is a chemical of concern) with

BAF = bioaccumulation factor, [(mg *X*/kg tissue)/(mg *X*/kg soil)]

Upon substitution, this yields the basic equation for evaluating a dietary exposure to a chemical as:

$$\frac{(\mathrm{Cs} \times \mathrm{IRs}) + (\mathrm{Cs} \times \mathrm{BAF} \times \mathrm{IRp})}{\mathrm{BW}} = \mathrm{Edose} \tag{7.3}$$

The equation can then be rearranged to solve for Cs, the concentration of chemical in soil (in mg *X*/kg soil where *X* is a chemical of concern).

Then, setting hazard quotient (HQ) to unity[12,13,18]

$$\mathrm{HQ} = \frac{\mathrm{Edose}}{\mathrm{TRV}} \equiv \mathrm{TRV} \times \mathrm{HQ} = \mathrm{Edose} \tag{7.4}$$

where:

TRV = toxicity reference value, mg *X*/kg tissue/day

subsequently yields

$$\frac{(\mathrm{Cs} \times \mathrm{IRs}) + (\mathrm{Cs} \times \mathrm{BAF} \times \mathrm{IRp})}{\mathrm{BW}} = \mathrm{TRV} \tag{7.5}$$

on substitution. Rearrangement then yields,

$$Cs(IRs + [BAF \times IRp]) = TRV \times BW \tag{7.6}$$

$$Cs = \frac{TRV \times BW}{IRs + [BAF \times IRp]} \tag{7.7}$$

The simplest, screening-level derivation of Cs views BAF as a simple ratio estimator, ideally developed from study-specific data (either literature-derived or site-specific).[12,13,18] Regression analysis of empirical data capturing a range of exposure concentrations and associated tissue residue values, however, will likely improve estimates of Cp, which would subsequently improve derived values of Cs, which could be considered a lower bound estimate of a soil PRG for a particular chemical of concern.[28,29]

### 7.2.3 Bioavailability: Its Evaluation and Role in Designing and Implementing Remediation Plans Focused on Wildlife

The derivation of soil PRGs may yield target values that are not attainable given the remedial technology available for cleanup.[27] For some chemicals of concern, target estimates for soil PRGs are frequently less than reference condition values. Risk-management tools must be applied when targeted values for soil PRGs cannot be attained owing to the limitations of remedial technology (e.g., institution controls such as capping are likely insufficient for remediation goals) or to a poor characterization of a chemical's bioavailability in the particular soils. When values for soil PRGs are at odds with background values or impractical to achieve given the available cleanup technology, the bioavailability of a chemical of concern must be considered in a critical evaluation of the target soil PRG.[13]

Bioavailability refers to a measure of actual entry of chemicals into receptors and is specific to the receptor, route of entry, time of exposure, type of chemical (e.g., metal, polycyclic aromatic hydrocarbon, etc.), chemical concentration, and the matrix containing the chemical.[12,13,18] As such, bioavailability is not necessarily measured as an adverse effect but is simply measured as uptake of chemical by an organism. As a biological process common to life on earth, bioaccumulation was critical to biological systems long before its association with ecotoxicology (for example, in the accumulation of nutrients from the environment). As with numerous terms in environmental science, the definition of bioavailability may vary from user to user, which in part reflects the context supporting the term's use. For terrestrial vertebrates, bioavailability of a chemical in soil is generally considered the fraction of exposed dose that enters into the blood stream.[12] Here, the classical method for quantifying bioavailability commonly relies on the toxicological concept of "area under the time-vs.-blood concentration curve" (AUC) where the toxicologist follows the time-course change in concentration of chemical in circulation following an intravenous injection against that time-course described by AUC following oral ingestion of an equivalent dose of the same chemical.[25]

In contrast, bioavailability for terrestrial and wetland plants, soil invertebrates, and soil microorganisms is that fraction of chemical in soil that is available for uptake.[31,32] Terms from soil science have been adopted to characterize bioavailability for soil biota, since the vast majority often lack physiological systems amenable to the approach applied to birds and mammals. Here, we are concerned with other descriptors that may estimate the bioavailable fraction of chemical. For example, exchangeable ions occur at the interface of soil-soil interstitial water surrounding soil particles. Exchangeable ions may be removed from the soil by solutions of neutral salts and may be referred to as the fraction of total chemical concentration adsorbed on soil particle surfaces, excluding irreversibly adsorbed and precipitated fractions. Similarly, adsorbed exchangeable ions are generally assumed to occur in dynamic equilibrium with that fraction that can be exchanged back to interstitial water. Overall, exchangeability of any chemical refers to the capacity of a soil to exchange adsorbed ions with those occurring in the soil interstitial water. Cation exchange capacity (CEC) and related measures of a soil's exchangeability are frequent measures of a chemical constituent's bioavailability. Generally, chemicals occurring in soil interstitial water are more directly available to soil biota, while exchangeable forms are potentially bioavailable.[31,32]

For our purposes, absolute bioavailability (also referred to as realized bioavailability) refers to the fraction of a chemical (as elemental or compound) that is ingested and actually absorbed and reaches the systemic circulation (e.g., of a vertebrate receptor) or the internal milieu of the soil organisms (e.g., uptake by plant root or soil bacteria).[12,13] On the other hand, relative bioavailability refers to a comparison of the absolute bioavailability of two or more forms of a specific chemical (e.g., elemental lead vs. oxides of lead), different vehicles (e.g., lead in food, soil, or water), or different doses (e.g., lead at 1 μg/kg vs. lead at 100 μg/kg).[12,13] Most often, relative bioavailability is expressed as the ratio of contrasts, e.g., the realized dose or absorbed fraction achieved from exposure to soil relative to the absorbed fraction from the exposure medium used in a toxicity study used to characterize a benchmark value such as a TRV. Then, given this simple ratio estimator, simple arithmetic calculations may be used to calibrate the models specified in Equations 7.1 through 7.7, and target soil PRGs derived from such simple models may be revised toward achievable soil PRGs that incorporate bioavailability in the derivation process.

Similarly, in an effort to bring physiological processes to food-chain analysis and the derivation of soil PRGs, the incorporation of a "relative absorption fraction" (RAF) or "gut absorption factor" (GAF) has frequently been incorporated into the basic exposure equation.[33] Early implementations of food-chain analysis developed "absorption-transfer factors" (ATFs) as macrorate coefficients or factors that accounted for various transfers of chemical from one compartment to another in a food web.[20,21] Regardless of the nomenclature, in each of these instances the fraction of exposed dose that is actually incorporated into tissues is estimated by empirically characterizing the relationships between exposed dose and the fraction of that dose bioaccumulated in tissues. Food-chain multipliers have also been applied to the evaluation of BCCs that are highly lipophilic and likely to be biomagnified.[18] Such modifying factors may be developed largely from a modeling effort, or they may be empirically developed depending on the case-specific data required in the evaluation process. Either a modeling or empirical characterization of these factors may benefit the estimation of bioaccumulation as a refinement beyond a simple calculation of exposed dose, since these relatively simple fractional estimators of bioavailability reflect biological and ecological processes (e.g., biotic and abiotic factors) that determine a chemical's bioaccumulation in terrestrial wildlife.

## 7.3 WORKING WITH SOIL PRGs AND DIFFERENTIAL BIOAVAILABILITY: REMEDIATION OF TERRESTRIAL AND WETLAND HABITATS

Whether focused on relatively small parcels of land characteristic of brownfield developments or more widely disturbed landscapes characteristic of mining districts in the western United States,[34,35] successful restoration or rehabilitation projects all require the design and implementation of soil remediation and revegetation plans in order to achieve habitat restoration or rehabilitation.[36] For the most part, the design and remediation of chemically contaminated and physically disturbed soils depend on management plans supporting future land-use goals identified for the habitats of concern. The development and application of technical methods required to achieve these goals are highly interdependent. While land-use decisions and management practices are critical to landscape restoration and must be clearly developed to assure technical success,[36] our focus for the balance of this chapter is on soils and plant interactions that influence the success of a revegetation effort.

Whether robust with respect to reestablishing habitats suited to all vertebrate wildlife or just a limited population, the ultimate goal of many restoration and rehabilitation plans frequently centers on birds and mammals; however, here we will center our discussions on soil characteristics —

abiotic and biotic — that must be considered to ensure successful revegetation. Without habitat, successful restoration or rehabilitation focused on wildlife resources would be frustrated because soils and their interrelationships with vegetation are a core requirement in establishing a sustainable system supporting the desired wildlife species.[3–6] We will also focus on potential remedies that alter the bioavailability of residual chemicals, while attending to problems associated with diminished soil tilth and its impact on achieving long-term revegetation success.

### 7.3.1 Soils: Abiotic and Biotic Factors Influencing Revegetation Success

Two highly interrelated components of the revegetation process should gain the primary focus in restoration or rehabilitation activities focused on wildlife, especially when remedial activities have resulted in a highly disturbed soil bearing residual chemicals. First, abiotic factors critical to successful restoration or rehabilitation will be identified and reviewed within the context of developing and implementing a revegetation plan.[26,31,32] These soil characteristics must be evaluated and are the most likely to limit the success of revegetation, especially those associated with physicochemical characteristics of the soil matrix. Second, we will review biotic factors characteristic of soils that are critical to the success of revegetation projects. Here, a brief summary of the soil processes that are critical to revegetation success will be reviewed.

#### *7.3.1.1 Soil Classification and Evaluation of Soil Matrix and Site for Metal Impacts*

In disturbed habitats identified for restoration or rehabilitation the existing physical, chemical, and biological properties of the soil will determine the soil's productivity and influence the success of any revegetation project.[3–6,26,36] Here, we will briefly discuss methods for evaluating the effects of disturbance and metals on the soil's productivity as well as soil tilth in general. Most habitats disturbed by human activities result in altered soil physical and chemical properties that may also influence — directly or indirectly — biological properties of the soil.[26,37–39] These existing, or ambient, soil properties will predispose a site to a certain productivity level; therefore, it is necessary to evaluate basic soil properties to assess this ambient condition and any deviation attributable to residual chemicals in the soils. In addition, a quantitative knowledge of certain soil properties is often necessary for calculations involving the design of the restoration or rehabilitation project. For example, some designs may specify the addition of soil amendments that require the conversion of residual chemical concentrations to loading rates on a spatial basis. These calculations require knowledge of the bulk density and depth of the soil horizon bearing the residual chemicals of concern.[26]

Adequate description of soil properties involves both field observations and laboratory measurements. Field observations include variables that are commonly recognized as site characteristics, while laboratory measurements are those commonly associated with traditional soil testing. Both are important in defining soil productivity. This discussion of soil properties is separated into a field and a laboratory section and is summarized in Table 7.1.[40–43]

##### *7.3.1.1.1 Field Observations*

Field observations can be subdivided into those that describe general site conditions and those that specifically describe the soil or soils that are found at a given site. General site parameters are type of overstory and understory vegetation, aspect, slope, and slope position. Aspect and slope characteristics are important parameters related to the water relations of a site. Vegetation type and distribution are often indicators of spatial variability of soils or metal distribution in soils.

Specific soil parameters are those integral to a complete description of the soil profile.[44,45] This profile description should be conducted to a depth sufficient to adequately represent the rooting volume of the soil and for all major soil types at a site. The description should delineate the depth

**Table 7.1 Summary of Applicable Methods for Measurement of Soil Properties in the Field and Laboratory**

| Soil Property | Determination in Field or Laboratory | Electrical Conductivity | References |
|---|---|---|---|
| Texture | Laboratory | Percent by weight in various size classes (i.e., % sand, silt, clay) | 40 |
| Organic matter (carbon) content | Laboratory | Percent of dry weight (% C or % OM) | 43 |
| pH | Laboratory or field | pH units | 43 |
| Cation exchange capacity (CEC) | Laboratory | cmol(+)/kg meq/100 g | 43 |
| Base saturation | Laboratory | Percent of CEC | 43 |
| Extractable acidity | Laboratory | cmol/kg mEq/100 g | 43 |
| Extractable nutrients (Bases: Ca, Mg, K, Na) | Laboratory | cmol/kg meq/100 g | 40, 43 |
| Nitrogen (Various forms) | Laboratory | Percent of dry weight or mg/kg | 43 |
| Phosphorous | Laboratory | mg/kg | 43 |
| Electrical conductivity | Laboratory | mmhos/cm | 43 |
| Water retention (available water capacity) | Laboratory | Volume percent (cm/cm) | 40 |
| Bulk density | Field and laboratory | $g/cm^3$ $mg/m^3$ | 40 |
| Infiltration capacity | Field | cm/hour | 40 |
| Hydraulic conductivity | Field or laboratory | cm/day or cm/hour | 40 |

*Source:* Modified from Linder et al.[26]

and thickness of the horizons present in the profile including the litter *O* horizon, if present. For each horizon, the following information should be collected: color, structure, rock content, bulk density, and root distribution. Homogeneous samples should be collected from each horizon for laboratory testing. In addition, it may be desirable to measure infiltration capacities of water to the soil surface and hydraulic conductivities of water within the profile. These would be important if metals in the soil are susceptible to being carried by water as it moves over or through the soil profile. These soil parameters are briefly summarized in Table 7.1.

*Horizon description.* Delineating the types and sequences of horizons present accomplishes two major tasks. When compared to undisturbed sites, a complete description documents the absence (removal), enrichment (augmentation or deposition), and mixing of soil horizons and thus indicates the degree of disturbance. Second, the description forms the basis for field observation and sampling for laboratory analyses. In this context the depth of various horizons can be used to calculate volumetric quantities of various elements, including metals or other chemical constituents.

*Color.* A Munsell color chart (book) is used to evaluate soil color. Color is used primarily to indicate organic matter status of a soil horizon or the water relations in a particular horizon or profile. Dark colors indicate high content of organic matter, while light colors generally point to lower content. Bright colors (generally reds and yellows) indicate soils that have greater internal aeration; that is, they are better drained. Gleyed soils (blue-gray colors) indicate anaerobic conditions (poor internal drainage) in which a soil is commonly saturated for a major portion of the year. Variable coloring with interspersed gley and bright colors (mottles) indicates a horizon or profile that is subject to a fluctuating water table and therefore experiences alternating aerobic and anaerobic conditions. The internal oxygen/water relationship is important because it determines the reduction-oxidation potential of a soil and hence influences chemical forms and some chemical reactions that can occur in the soil matrix.

*Structure.* Structure refers to the way and degree to which individual soil particles bind together in larger units. The type of structure in a soil influences the paths of water movement (in reality, is partially dependent on water movement) and paths of root penetration and expansion. Lack of structure can be an indication of intense physical disturbance.

*Bulk density.* Bulk density refers to the weight per unit volume of solid particles in a soil. It is a measurement that must be taken in the field; it cannot be successfully reconstructed from laboratory data. Bulk density indicates the degree of natural or induced compaction in a given horizon. It is inversely related to porosity, or the amount of air space in a soil. Compacted (bulk densities with greater values) horizons are barriers to both root growth and vertical (downward) water movement and may induce lateral pathways of water flow. Bulk density is also a parameter that is necessary in order to convert concentrations of a chemical constituent, such as a metal (usually reported on a weight basis), to an area or volume basis. It is used in conjunction with horizon thickness to calculate the content of a constituent (element) in that horizon.

*Rock content.* The rock or stone content of a soil horizon can be an important consideration because of its influence on water relations, the volume of finer soil particles important in soil chemical reactions, and the soil volume available for root growth. With respect to water relations, greater stone content reduces the soil volume active in storing soil water for subsequent plant use (lowers soil water-holding capacity). In addition, the paths of water movement within a soil are determined by the amount and arrangement of any rock that is present. In general, the greater the rock content the greater the volume of water interacting with the fine-earth portion of a soil horizon and the greater the potential for chemical alteration or leaching.

*Root distribution.* Root growth and the patterns of root distribution found in a soil profile provide an integration of most of the soil properties discussed above as well as the soil's chemical properties. Lack of roots (or reduced rooting) in a particular horizon or soil serve as an indication that a chemical, physical, or biological barrier is inhibiting root growth. However, it is necessary to recognize there is a "natural" root distribution characterized by decreasing root distribution with depth. This natural distribution is a function of vegetation type and regional climate and can be characterized to provide a standard for comparison of subject sites.

*Infiltration capacity and hydraulic conductivity.* Infiltration capacity is a measurement of the ability of the soil surface to absorb water. The greater the value the greater the rate at which water is taken into a soil and the lesser the chance of surface runoff and, hence, erosion. Infiltration capacity is strongly related to texture and organic matter content of surface horizons. Hydraulic conductivity refers to the rates of movement within a soil profile. It is also referred to as permeability. It is important because it characterizes the "residence time" of water within a soil profile and hence its chances of interacting with soil chemical constituents. Hydraulic conductivity is directly related to infiltration capacities. The slower the hydraulic conductivity the lesser the infiltration capacity.

#### *7.3.1.1.2 Laboratory Analysis*

Various laboratory analyses are typically conducted on samples obtained from field sites. These analyses are important for understanding the chemical, physical, and biological reactions that might take place in the soil. They are also important for assessing the mobility and potential pathways of movement of a contaminant, especially with regard to water movement and water-mediated biological decomposition. However, the results of these analyses only represent the properties of the sample itself. Adequate interpretation must consider the entire site setting and the horizon the sample came from in the soil profile. As such, the thoroughness of field observations is especially critical.

Laboratory analyses suggested for routine characterization and interpretation of soils include those for texture, organic matter, pH, cation exchange capacity, base saturation, extractable acidity, electrical conductivity, water retention, and concentrations of the elements nitrogen, phosphorus, potassium, calcium, magnesium, and sodium. The importance of these soil properties is discussed below.

*Texture.* Texture refers to the proportion of particles of different size classes (sand, silt, clay) that make up a sample. Knowledge of texture is important for assessing water relations, erodibility, and retention of cations, anions, and toxicants.

*Organic matter content.* The organic matter content of a soil is an indicator of its fertility, ability to support microbial populations, retention of elements, and water relations. In general, the greater the organic matter content the better the soil serves as a medium for plant growth.

*pH.* The pH of a soil sample is a measure of its hydrogen ion concentration (or, more accurately, its activity) and thus its acidity or alkalinity. This is an important property because it influences (or regulates) many chemical and biological processes occurring in a soil including the availability of soil nutrients and metals.

*Cation exchange capacity.* Cation exchange capacity (CEC) is a measure of a soil's ability to retain positively charged ions. The greater the CEC the greater the capacity of a soil to retain cations, and the greater its "storage" or "buffering" capacity. CEC is especially important for assessment of the mobility of positively charged metallic elements.

*Base saturation and extractable acidity.* These two measurements characterize opposite ends of the spectrum relative to the composition of cations occupying the cation exchange complex of a soil. They are strongly related to the pH of a soil. Base saturation represents the proportion of the CEC electrically satisfied by base cations such as calcium, magnesium, potassium, and sodium, whereas extractable acidity measures the proportion satisfied by the acidity cations — hydrogen and aluminum.

*Electrical conductivity.* Electrical conductivity is an indirect measurement of the ionic content of a soil sample. The greater the conductivity the greater the ion content of a soil. It is an especially important measurement in soils that have "free" salts, such as saline soils, and serves as an indicator of adverse conditions for growth of many plants in such soils.

*Water retention.* Water retention refers to the amount of water that can be retained by soil when it is subjected to different pressures. Characteristically a "water retention curve" is established for a soil sample by measuring its water content (on either a volumetric or weight basis) at 1/10-, 1/3-, 1-, 5-, and 15-bar pressures. The difference in water contents between the 1/3-bar (also termed field capacity) and 15-bar (wilting point) pressures is termed the available water content of a soil. It represents the amount of water storage the soil can provide against the force of gravity and the amount available for plant uptake.

*Concentrations of various elements.* Knowing the concentrations of the complete spectrum of nutrient elements in a soil allows interpretation of whether a specific nutrient deficiency is possible or whether there is an imbalance in the distribution of elements present in a soil or soil sample. Specifically, these analyses of extractable element concentrations provide a measure of their availability for plant or microbial uptake.

References such as the American Society of Agronomy (ASA) Monograph Series[40–43] provide information valuable in the interpretation of analytical results and are the source of instructions for performing specific analyses given in Table 7.1.

### 7.3.2 Soil Factors Commonly Limiting Revegetation Success on Disturbed Soils

While each of the soil properties and characteristics briefly outlined above may contribute to problems encountered in revegetation efforts of disturbed soils, the combination and interaction of these factors (e.g., interactions of residual metals with soils or overburden at physically disturbed mining sites) frequently contributes to the relatively poor revegetation of some remediated soils bearing residual chemicals. In general, the factors that recur as prominent causes of failure of revegetation projects are associated with pH (or hydrogen ion concentration), salinity, sodicity (elevated sodium concentrations), cation exchange capacity, organic matter, soil texture and structure, and bulk density and porosity.[46] Each of these soil factors, their critical values for enhancing revegetation success, and soil manipulations that may promote remediation efforts when these critical values are not present are briefly summarized in Table 7.2 and in the following section. Although each restoration or rehabilitation activity will require independent analyses to determine a course to assure successful habitat recovery, Table 7.2 lists critical values for some of these soil

**Table 7.2 Critical Values for Selected Soil Factors (Physical and Chemical) that May Yield Poor Revegetation**

| Soil Factor | Critical Value |
|---|---|
| **Physical Characteristic** | |
| pH | Less than pH 5.5 or greater than pH 8.5 |
| Conductivity | 4–6 mmhos/cm or greater |
| Sodium Absorption Ratio (SAR) | Greater than or equal to 12 |
| Texture | 40% clay, or loamy sand to sand |
| **Chemical Characteristic (μg/g)** | |
| Boron (B) | 8 or greater |
| Cadmium (Cd) | 0.1–1.0 or greater |
| Copper (Cu) | 40 or greater |
| Lead (Pb) | 10–15 (at pH less than 6.0)<br>15–20 (at pH greater than 6.0) |
| Manganese (Mn) | 60 or greater |
| Selenium (Se) | 2.0 or greater |
| Zinc (Zn) | 40 or greater |

*Source:* Modified from Munshower.[46]

factors that should be viewed with caution when soils are being considered for revegetation, especially in soils of the western United States.

### 7.3.3 Evaluation and Reestablishment of a Soil Biota for Long-Term Revegetation Success

The development of a habitat-restoration or rehabilitation plan requires, in addition to the physicochemical evaluations outlined above, a characterization of the existing biological integrity of the soils.[26] Many revegetation efforts fail, or do not achieve long-term restoration goals, when soils are not critically evaluated during baseline activities of the revegetation phase of the project. The level of effort devoted to these biological characterizations will depend on the restoration goals identified in the management plans for future use as well as the time line developed for the activities to accomplish those goals.

#### *7.3.3.1 Overview of Soil Biota and Its Role in Successful Revegetation*

Nutrient limitations frequently diminish the success of long-term revegetation or restoration efforts. For example, of the major nitrogen transformations mediated by microorganisms, nitrogen cycling is one of the most important and directly related to plant productivity and successful revegetation.[47–50] Nutrient cycling, however, cannot occur without soil organisms to perform the majority of the processes critical to the soil-rhizosphere interactions required for nutrient uptake. While abiotic factors like those physicochemical properties described previously may influence the presence and activity of soil nutrients, an understanding of the biological factors along with these physicochemical properties improves the site-specific characterization of processes like nitrification and respiration that will determine the overall success of any revegetation effort. Soil organisms, especially in concert with plant roots, are the filter through which all nutrients pass, and an understanding of their ability to cycle nutrients is critical to ensuring a successful revegetation or restoration project (Figure 7.1).

Soil organisms perform many processes, often with varying degrees of redundancy. This redundancy is important to resiliency of soil following disturbance. In healthy soil, there are usually (but

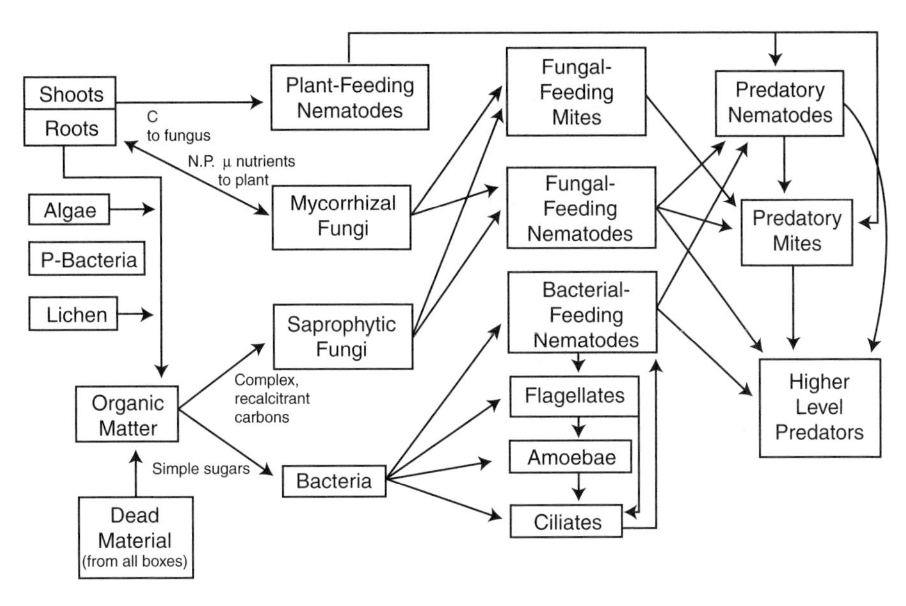


**Figure 7.1** Generalized soil foodweb illustrating interrelationships of soil biota.

not always) several organism groups that perform any particular process, and in highly disturbed soils these organisms may be lacking or sufficiently impacted to dramatically limit the success of revegetation. For example, in soils of good tilth, some organisms function in colder conditions, some in warmer, wetter, or drier conditions, higher or lower organic matter, high or low pH, and so forth. Thus, nutrient cycling in soils of good tilth, as well as disturbed soils like those in the mining districts of the western United States, depend on soil microorganisms performing these functions. Regardless of the disturbance, functional redundancy can be diminished or destroyed in soils. For example, in many soils of the western United States, successful revegetation will depend on the presence of mycorrhizal fungi and on the functional soil organism-nutrient cycling system characteristic of undisturbed soils. At least some plants preferred for revegetation and restoration are obligately dependent on symbiotic organisms for establishment or survival.[26,47]

In healthy soils, soil organisms are functionally interrelated via foodwebs. A foodweb consists of primary producers, primary consumers, secondary consumers, consumer generalists, and higher-level consumers.[26] One can consider the first "trophic" step as organic matter that drives the system, derived from leaf litter, roots, algae, moss, small mammal and bird fecal material, and secondary metabolites from microbial decomposition. These materials are used by the two primary decomposer groups, saprophytic fungi and bacteria, corresponding to the primary consumer trophic level. Nutrient and energy from living plant material also moves into nondecomposer, root-mutualist fungi and bacteria, such as mycorrhizae and nitrogen-fixing bacteria, and root-feeding organisms (parasites, pathogens) such as plant root-feeding nematodes. These organisms should be considered primary consumers as well.

The next trophic level is comprised of secondary consumers, which feed on both the decomposers and root-associates, i.e., bacterial-feeding (nematodes and protozoa) and fungal-feeding organisms (nematodes, certain amoebae, and a number of microarthropods). Secondary consumers are fed upon by higher-level predatory fauna such as mesostigmatid mites, predatory nematodes, insects, insect larvae, small mammals, and birds. There are a variety of generalist consumers within the soil such as enchytraeids, earthworms, rotifers, and microarthropods. Thus, a typical soil foodweb is complex, redundant, highly diverse, and quite stable. Disturbed soils, and especially soils that have been mined, generally have low diversity and, depending on the site, may be relatively depauperate and poorly colonized with indigenous microbial and fungal populations.

By characterizing soil foodwebs in part or in total, short-term and long-term revegetation success can be critically evaluated and remedial activities developed, if necessary. For example, reductions in foodweb organism numbers in disturbed mining soils as compared to undisturbed soils may indicate a negative impact on long-term revegetation success unless remedial options intervene. Significant decreases in soil organism populations may suggest a significant impact on revegetation success, and with time the ecological effect of reductions in numbers of organisms may be more clearly interpreted with respect to whether an impact improves or degrades soil fertility and ecosystem productivity. While the wildlife manager likely does not present sufficient knowledge to actively participate in the remediation of soil, if the biological components are critical to the long-term sustainability of habitat, their perspective on habitat requirements of vertebrate wildlife must be considered when the soil scientist's focus is drawn toward remedial measures channeled toward recovery of a soil's biological integrity (e.g., bacterial and fungal communities as well as soil protozoa and nematodes communities) and nominal processes (e.g., microbial nitrification and soil respiration).

### 7.3.4 Soil Tilth and Vegetation: Restoring Remediated Soils to Assure Long-Term Sustainability

While not all restoration or rehabilitation projects will require an exhaustive evaluation of the disturbed soil's baseline properties, when the soil physicochemical and biological characteristics outlined above are adequately characterized, the likelihood of successful revegetation will be enhanced. If soil conditions permit, successful revegetation will then depend on plant species selection and assurance that plant nutrients are adequate to support both initial and long-term success for the revegetation effort. Depending on the short-term as well as the long-term restoration or rehabilitation goals established for the project, various seed mixes, nursery material (bareroot or container-stock), and wilding transplants may be used to achieve site management goals.

While beyond the scope of the present overview, species lists should reflect a selection of plant materials consistent with the native flora of the region, especially at those locations identified as self-sustaining and likely to have little, if any, long-term landscape management. Indeed, within the site or landscape setting, an exhaustive list of grasses and grass-like species, forbs, shrubs, and trees may not necessarily satisfy the needs of site managers or revegetation specialists; rather, lists of plant species conducive to a variety of revegetation and restoration goals should be developed within the context of the ecoregion where the disturbance site is located. With such lists, plants species may be selected that are physiologically adapted to the soils and climatological setting of the site. Some locations, especially if they cover large areas of space or contain numerous microsites, may require numerous different lists of plant species or different seed mixes to assure long-term revegetation or restoration success in very heterogeneous sites or landscapes.

The success of any restoration effort depends on many interrelated technical factors including the characterization of the physicochemical and biological properties of soil and selection of plant species that have successfully been used in revegetation projects in the surrounding areas. While site-specific variation in soils may require that expert judgment and experience be used in the selection of methods to characterize the site-soils identified for revegetation, these soil properties should be considered early in the design and implementation of the site-management plan, especially if long-term restoration is driving the revegetation effort. When the evaluation of the existing physicochemical and biological properties of the disturbed soils suggests that no critical values (e.g., pH, salinity, plant nutrients) are deficient or in excess and that soil tilth should adequately support the plant species identified in the revegetation plan, the likelihood for a successful revegetation of the site will be increased. But if the baseline characterization of the disturbed soils is incomplete or indicates that soil tilth is insufficient to achieve revegetation or restoration goals, then soil remediation activities will have to be pursued prior to the revegetation effort.

**Table 7.3 Trace Elements in Western Soils and Their Relationship to Plant Nutrition[46,51,55,56]**

| | Ranges | | Extractable | |
|---|---|---|---|---|
| Element[a] | Total | Extractable or Soluble | Low or Deficient | High or Toxic |
| Al | 0.5%–10% | Varies with pH | NA[b] | 1–5 |
| B | 20–70 | < 1–8 | 0.5 | 8–10 |
| Cd | 0.1–2 | 0.01–0.05 | NA | Not known (varies with soil type) |
| Cu | 2–90 | 0.1–0.7 | 0.2 | Not known (varies with soil type) |
| Fe | 0.3%–6.8% | 1.5–200 | 2.0 | Not known under natural conditions |
| Mn | 65–1300 | 1–40 | 1.0 | Not known (varies with soil type) |
| Mo | 0.5–20 | 0.05–0.3 | Varies with Cu concentration | Varies with Cu concentration |
| Pb | 3–25 | 0.1–3 | NA | Not known (varies with soil type) |
| Se | 2–20 | 0.01–10 | Vegetation levels < 0.05 | Vegetation levels > 0.05 |

[a] Concentrations in μg/g unless otherwise noted.
[b] NA = not applicable (element not required for plant growth).

Disturbed habitats frequently are characterized by poor soil tilth, where the relationship between the plant and the physical condition of the soil is poorly developed.[51–53] While the soil analyses previously identified are critical in designing and implementing any revegetation project, they enable us to regard soil as only a single "snapshot" in time. The successful revegetation project must consider numerous snapshots collected throughout the project's lifetime if long-term sustainability and revegetation success are to be achieved.[54] Soils, and especially disturbed soils, are continually changing, albeit at a relatively slow rate, throughout the restoration process. And depending on the initial conditions of the soils and the extent of intervention planned as part of the restoration effort, changes in soil will vary from being quite rapid to very slow. These changes, whether relatively slow with little intervention or rapid with much design and intervention, may improve soil tilth, or they may contribute to the continued degradation of the soil resource. Similarly, plant nutrients, such as nitrogen, phosphorus, potassium, calcium, magnesium, and sulfur (i.e., macronutrients), should be characterized throughout the restoration or rehabilitation project to assure their sufficiency, especially within the management context for the project as a whole (e.g., planned periodic fertilization programs).[3–6,46]

Under baseline conditions trace elements and micronutrients (e.g., boron, copper, manganese, molybdenum, zinc) generally occur at highly variable concentrations in soils. For example, Table 7.3 briefly lists some frequently considered trace elements and plant micronutrients that recur as critical components in revegetation projects in the western United States.[46] The role of these micronutrients is frequently poorly characterized or varies from plant species to plant species, which reinforces the role that soil physicochemical characterization plays in the restoration or rehabilitation process, since some revegetation projects may have to consider their role in plant-soil interactions to assure long-term success.[53]

#### *7.3.4.1 Physical Manipulations*

Successful revegetation of any disturbed soil, whether related to chemical releases or not, is only assured if sufficient planning has preceded the implementation of the project. Depending on the site and long-term land use plans associated with its restoration, some restoration strategies may have to include short-term as well as long-term plans, for example, to assure an initial stabilization of the disturbed soils as a developmental step in the long-term restoration of the site with a highly diverse plant community. Such short-term practices may include liming amendments that, initially designed to alleviate soil pH problems, will be discontinued once soils have recovered their normal biogeochemical processes for nutrient cycling.[46]

While various activities will be required in the restoration of vegetation to disturbed soils, the physical manipulations required to assure success will likely alter residual chemicals' bioavailability while improving soil tilth. Methods commonly available to the remediation process include recontouring and slope reconstruction, topsoiling, surface manipulation,[46] and the addition of various soil amendments, fertilizers, and mulches.[46,56]

*Recontouring and slope reconstruction.* When water is a major factor limiting the success of revegetation efforts (for example, in the western United States), the control of erosion by wind and water runoff requires interrelated physical manipulations that are closely related to the maintenance of soil surfaces to enhance the capture of water when it is occurs as rain and snow. Land recontouring should establish stable slopes and the reconstruction of the original topography to assure satisfactory surfaces for reestablishing plants and maintaining a drainage basin to capture water. Recontouring should reestablish linkages to the surrounding undisturbed landscape; ideally, long-term revegetation success will depend on the successful germination and establishment of native vegetation to obscure the boundary between the restored and undisturbed plant communities.[3–6]

*Topsoiling.* Throughout the United States topsoil is frequently preferred as a growth medium for plants being used in a revegetation project, primarily because spoil material (e.g., from mining or other excavation activity) stored for restoration purposes may be nutrient deplete and physically unsuited for establishing sustainable plant-soil relationships.[46,55] As a general term, cover soil may be more appropriate. Nonetheless, topsoil often times refers to the plow layer, or that layer of surface soil that is a minimum depth removed through operation of an earth scraper, and still possesses good plant-growing characteristics. In disturbed arid lands like those of the western United States, there is little, if any, *A* horizon in topsoil. More frequently, topsoil at disturbed mine sites is a mixture of soil horizons.[35,36,46]

Salvaged surface soils at disturbed sites are preferred for restoration because of its more favorable physical and chemical properties relative to other materials. Ideally, cover soil for revegetation projects should have sufficient organic matter, as well as an indigenous soil flora and fauna to enhance and promote plant growth. The long-term sustainability of any revegetation project will depend on the development of a soil community capable of cycling plant nutrients and maintaining a favorable microenvironment in the rhizosphere. Even a shallow layer of a few centimeters of surficial soil improves infiltration, and seed germination and early seedling survival are improved in topsoil materials.[46,55]

*Surface manipulation.* The preparation of a seedbed is critical to the germination and establishment of seeds, and to a certain extent the successful propagation of transplanted plant materials also depends on adequate preparation of the cover soil's surface. These surface manipulations most generally include plowing, chiselling, disking, and harrowing and should be practiced on the contour. Special manipulations may be required on some sites to minimize the long-term, as well as short-term, problems associated with compaction and the development of slippage planes between subsoil and overburden and cover soil. Gouge depressions and "dozer" basins may also be used when water infiltration is considered a problem for a specific site. Often plants may be initially established in these surface depressions to promote growth in sheltered, well-watered areas; then, these "islands" of growth may fill in the bare ground between surface depressions in subsequent seasons, especially when rhizomatous species are used to establish the initial vegetation cover.[46]

*Soil-incorporated amendments, biosolids, and commercial fertilizers.* Amendments are additions to surface soils and overburden that occur immediately prior to or following seeding operations; in general, these amendments are intended to provide a better medium for growth, especially with respect to the rooting medium. Overall, amendments should provide supplemental nutrients and increase soil organic matter, adjust soil pH to enhance plant growth, and improve physical properties to enhance soil moisture. On a larger scale amendments may also be used to reduce erosion and modify temperature.[46,56]

To achieve these objectives related to the promotion of plant growth several types of amendments are available to the restorationist, including commercial fertilizers (as sources of nitrogen, phosphorus, and potassium) and various sources of organic material (such as composted manures, wood chips, sludge, and straw). For acid soils characteristic of disturbed mining lands, lime amendments may be used to alter pH regimes in the rhizosphere, and salts may be added to soils to promote the leaching of growth-inhibiting anions and cations.[31,46,55]

*Mulches and surface-applied organic amendments.* Closely related to soil amendments are mulches of various types, including surface-applied organic amendments. In general, mulches are organic materials applied to the soil surface after seeding or transplanting of root stock. Like soil-incorporated amendments, mulches are diverse in their character and include wood chips and other wood residues, straw, and native hay. While not strictly a mulch, composted manures and sludges may also be used as a surface-applied material after planting. Regardless of the type, mulches reduce erosion and evaporative water loss from soil, help prevent soil crusting, and reduce soil temperatures, which may be critical to seed germination and early seedling survival. Surface-applied mulches may be subsequently incorporated into the soil through crimping or rototilling (or any similar practice) and as such become an organic amendment that will affect soil physical structure and chemical characteristics.[31,46,55]

### 7.3.5 Interactions between Remediated Soils and Surface Water Habitats

Soil erosion includes the processes of detachment of soil particles from the soil mass and the subsequent transport and deposition of those soil particles on land surfaces or as bed sediments in surface waters. Erosion is the source of 99% of sediment loads in surface waters in the United States.[57,58] Soil erosion and sedimentation contribute to the degradation of streams, lakes, and reservoirs as well as disturbance to wildlife habitat. Additionally, financial costs (e.g., costs associated with dredging) are also directly related to soil erosion and sedimentation processes. For residual chemicals in soil, this cost could extend to issues associated with what may become "contaminated sediments" when erosion occurs, releasing soil-bound chemicals to surface waters.

Erosion of soil is an important process involved in the transport of soil-bound chemicals (e.g., agricultural chemicals, waste chemicals derived from mining operations, and chemicals associated with various land-use practices) from nonpoint source areas to waterways, where the resulting soils may create water- and sediment-quality problems.[57,58,59] From a water-quality perspective, soil erosion is the source of 80% of the total phosphorus and 73% of the total Kjeldahl nitrogen in the waterways of the United States.[58,59] In addition to providing a source of nutrient enrichment, unless adequate assurance is available the likelihood that residual chemicals in soil will be lost to surface waters through erosion will diminish the success of any soil-restoration or rehabilitation plan. A characterization of erosion potential can be achieved through various sources including the Revised Universal Soil Loss Equation (RUSLE),[58,59] which may be beneficial to evaluating the overall restoration or rehabilitation plan, particularly if residual chemicals in soil present a potential risk if transported to the sediment environments of adjacent surface waters.

## 7.4 SUMMARY

Keys to successful remediation targeted at restoring terrestrial wildlife populations center on estimating remediation goals in terrestrial and wetland habitats, identifying functional goals for ecosystems to be remediated and the spatiotemporal scales for the remediation activity, identifying locations inappropriate for future wildlife use, and designing management plans to minimize adverse effects associated with such areas. Currently, there is little or no federal, state, or other guidance to derive soil cleanup values or ecological-based remedial goal options, and derivation of "cleanup levels" is usually subject to negotiation among regulators and others having an interest in the habitat of concern.[60] The derivation process for soil PRGs focused on habitat remediation consists of:

- Developing target PRGs for soil that reflect lower bounds for residual chemicals likely to present risk (negligible or no risk) to wildlife exposed to low levels over long periods of time, and
- When target PRGs are not attainable, data on a chemical's bioavailability are considered to modify these target PRGs to levels and characterize soil PRGs that can be achieved with the available technology and will be allowed to remain at the site without further remedial action.

Within a restoration and rehabilitation context, when a target value for a soil PRG cannot be reached, a chemical concentration should be characterized by its differential bioavailability in order to reach this attainable soil PRG.

As was detailed earlier in this chapter, bioavailability varies as a function of (1) the receptor, (2) the route of entry or mode of exposure, (3) time of exposure (i.e., biotic-abiotic frame of reference, largely spatiotemporal), (4) type of chemical and its concentration, and (5) the matrix containing the chemical. The biological components characteristic of disturbed habitats that should be considered when chemical releases and remedial activities are concerned include soil biota (soil-dwelling flora and fauna, including rhizosphere biota), soil surface-dwelling biota, including macro-invertebrates, plants, and associated fauna (e.g., plant-dwelling invertebrates and mycorhizzal fungi), invertebrate wildlife, and vertebrate wildlife. Ultimately, in view of our relatively limited set of truly predictive tools, the development of monitoring programs to evaluate remediation and restoration of terrestrial and wetland habitats is essential to any restoration or rehabilitation plan likely to be successful. The costs both monetary and aesthetic may be unacceptable if our best estimate of soil PRG proves to be associated with subsequent risk or injury.

## REFERENCES


1. Jackson, L. L., Lopoukhine, N., and Hillyard, D., Ecological restoration: A definition and comments, *Restoration Ecol.,* 3, 71, 1995.
2. U.S. Environmental Protection Agency (U.S. EPA), Guidance on Cumulative Risk Assessment, Part 1, Planning and Scoping, Science Policy Council, U.S. Environmental Protection Agency, Washington, D.C., 1997.
3. Urbanska, K. M., Webb, N. R., and Edwards P. J., Eds., *Restoration Ecology and Sustainable Development*, Cambridge University Press, Cambridge, England, 1997.
4. Harris, J. A., Burch, P., and Palmer, J. P., *Land Restoration and Reclamation: Principles and Practice*, Addison Wesley Longman, Essex, England, 1996.
5. Whisenant, S. G., *Repairing Damaged Wildlands*, Cambridge University Press, Cambridge, England, 1999.
6. Reith, C. C. and Potter, L. D., Eds., *Principles and Methods of Reclamation Science*, University of New Mexico Press, Albuquerque, NM, 1986.
7. U.S. EPA, Guidelines for Ecological Risk Assessment, EPA/630/R-95/002F, Risk Assessment Forum, U.S. Environmental Protection Agency, Washington, D.C., 1998.
8. Lemly, A. D., Best, G. R., Crumpton, W. G., Henry, M. G., Hook, D. D., Linder, G., Masscheleyn, P. H., Peterson, H. G., Salt, T., and Stahl, R.G., Jr., Contaminant fate and effects in freshwater wetlands, in *Ecotoxicology and Risk Assessment for Wetlands*, Lewis, M. A., Mayer, F. L., Powell, R. L., Nelson, M. K., Klaine, S. J., Henry, M. G., and Dickson, G. W., Eds., SETAC Press, Pensacola, FL., 1999, chap. 4.
9. Stephenson, G., Kuperman, R., Linder, G., and Visser, S., Toxicity tests for assessing contaminated soils and ground water, in *Environmental Analysis of Contaminated Sites: Tools to Measure Success or Failure*, Sunahara, G. I., Renoux, A. Y., Thellen, C., and Gaudet, C. L., Eds., John Wiley & Sons, Ltd. Chichester, Sussex, England, 2002, chap. 3.

10. Pastorok, R. A. and Linder, G., A tiered ecological assessment strategy for contaminated soil cleanup, in *Hydrocarbon Contaminated Soils and Groundwater: Analysis, Fate, Environmental and Public Health Effects, and Remediation*, Vol. 2, Calabrese, E. J. and Kostecki, P. T., Eds., Lewis Publishers, CRC Press, Boca Raton, FL, 1993, Chap 33.
11. Suter, G., Antcliff, B. L., Davis, W., Dyer, S., Gerritsen, J., Linder, G., Munkittrick, K., and Rankin, E., Conceptual approaches to identify and assess multiple stressors: Biologically directed approach, in *Multiple Stressors in Ecological Risk and Impact Assessment*, Foran, J. A. and Ferenc, S. A., Eds., SETAC Press, Pensacola, FL, 1999, Chap. 1.
12. U.S. EPA, Ecological Soil Screening Levels, World-Wide Web page accessed 12.18.2001, http://www.epa.gov/superfund/programs/risk/ecorisk/ecossl.htm.
13. Lanno, R. P., Ed., *Contaminated Soils: From Soil-Chemical Interactions to Ecosystem Management*, SETAC Press, Pensacola, FL, 2002.
14. MacDonald, D. D., Ingersoll, C. G., and Berger, T. A., Development and evaluation of consensus-based sediment quality guidelines for freshwater ecosystems, *Arch. Environ. Contam. Toxicol.*, 39, 20, 2000.
15. Long, E. R. and Morgan, L. G., *The potential for biological effects of sediment-sorbed contaminants tested in the National Status and Trends Program*, NOAA Technical Memorandum NOS OMA 52, Seattle, WA, 1991.
16. Long, E. R., MacDonald, D. D., Smith, S. L., and Calder, F. D., Incidence of adverse biological effects within ranges of chemical concentrations in marine and estuarine sediments, *Environ. Manage.*, 19, 81, 1995.
17. U.S. EPA, Issuance of Final Guidance: Ecological Risk Assessment and Risk Management Principles for Superfund Sites, OSWER Directive 9285.7–28 P, U.S. EPA, Office of Solid Waste and Emergency Response, Washington, D.C., 1999.
18. Linder, G. and Joermann, G., Hazard and risk assessment for wild mammals, in *Ecotoxicology of Wild Mammals*, Shore, R. and Rattner, B., Eds., John Wiley & Sons, New York, chap. 13.
19. Linder, G., Bollman, M., Callahan, C., Gillette, C., Nebeker, A., and Wilborn, D., Bioaccumulation and food-chain analysis for evaluating ecological risks in terrestrial and wetland habitats. I. Availability-transfer factors (ATFs) in "soil → soil macroinvertebrate → amphibian" food chains, in *Superfund Risk Assessment in Soil Contamination Studies: Third Volume*, ASTM STP 1338, Keith Hoddinott, Ed., American Society for Testing and Materials, 51, 1998.
20. Pascoe, G. A., Blanchet, R. J., and Linder, G., Bioavailability of metals and arsenic to small mammals at a mining waste-contaminated wetland, *Arch. Environ. Contam. Toxicol.*, 27, 44, 1994.
21. Pascoe, G. A., Blanchet, R. J., and Linder, G., Food chain analysis of exposures and risks to wildlife at a metals-contaminated wetland, *Arch. Environ. Contam. Toxicol.*, 30, 306, 1996.
22. U.S. EPA, Ecological Risk Assessment Guidance for Superfund: Process for Designing and Conducting Ecological Risk Assessments, Interim Final, EPA540-R97–006, U.S. EPA, Environmental Response Team, Edison, NJ, 1997.
23. U.S. EPA, Ecological Assessment of Hazardous Waste Sites: A Field and Laboratory Reference, Warren-Hicks, W., Parkhurst, B. R. and Baker, S. S, Eds., EPA 600–3–89–013, U.S. Environmental Protection Agency, Environmental Research Laboratory, Corvallis, OR, 1989.
24. American Society for Testing and Materials (ASTM*), Annual Book of Standards, Volume 11.05*, Committee E47 on Biological Effects and Environmental Fate, ASTM, West Conshohocken, PA, 2001.
25. Klaassen, C., Amdur, M., and Doull, J, Eds*., Casarett and Doull's Toxicology — The Basic Science of Poisons*, 5th ed., McGraw-Hill, New York, 1996.
26. Linder, G., Ingham, E., Brandt, J., and Henderson, G., Evaluation of Terrestrial Indicators for Use in Ecological Assessments at Hazardous Waste Sites, 600/R-92/183, U.S. EPA, Environmental Research Laboratory, Corvallis, OR, 1992.
27. Pascoe, G. A. and Linder, G., Metals bioavailability as a determining factor in natural remediation at Milltown reservoir wetlands in western Montana, in *Natural Remediation of Environmental Contaminants: Its Role in Ecological Risk Assessment and Risk Management*, Swindoll, M., Stahl, R. G., Jr., and Ells, S. J., Eds., SETAC Press, Pensacola, FL, 2001, chap. 15.
28. Sample, B. E., Beauchamp, J. J., Efroymson, R., and Suter, G. W., II, Literature-derived bioaccumulation models for earthworms: Development and validation. *Environ. Toxicol. Chem.*, 18, 2110, 1999.
29. Sample, B. E, Beauchamp, J. J., Efroymson, R., and Suter, G. W., II, Development and Validation of Bioaccumulation Models for Small Mammals, ES/ER/TM-219, Oak Ridge National Laboratory, Lockheed Martin Energy Systems Environmental Restoration Program, 1998.

30. U.S. EPA, Soil Screening Guidance: Technical Background Document, EPA/540/R-95/128. Office of Emergency and Remedial Response, Washington, D.C., May; U.S. EPA; Soil Screening User's Guide, Office of Emergency and Remedial Response, Washington, D.C., EPA/540/R-96/018, 1996.
31. Bohn, H. L., McNeal, B., and O'Connor, G., *Soil Chemistry*, 2nd ed., John Wiley and Sons, New York, 1985.
32. Barber, S. A., *Soil Nutrient Bioavailability — A Mechanistic Approach*, 2nd ed., John Wiley and Sons, New York, 1995.
33. Bischoff, K. B., Physiologically based pharmacokinetic modeling, in *Pharmacokinetics in Risk Assessment*, National Research Council, Subcommittee on Pharmacokinetics, Gillette, J. R. and Jollow, D. J., co-chairs, National Academy Press, Washington, D.C., 36, 1987.
34. Fishman, B. E. and Reinert, K. H., Ecological considerations in brownfields redevelopment, *Environ. Toxicol. Chem.* 19, 257, 2000.
35. Potter, L. D., Pre-mining assessments of reclamation potential, in *Principles and Methods of Reclamation Science*, Reith, C. C. and Potter, L. D., Eds., University of New Mexico Press, Albuquerque, 41, 1986.
36. Vogel, W. G., A Manual for Training Reclamation Inspectors in the Fundamentals of Soils and Revegetation, Prepared by U.S. Department of Agriculture, Northeastern Forest Experiment Station, Berea, KY and Soil and Water Conservation Society, Ankeny, IA for U.S. Dept. of Interior, Office of Surface Mining and Enforcement, Washington, D.C., 1987.
37. Alexander, M., *Introduction to Soil Microbiology*, 3rd ed., John Wiley and Sons, New York, 1986.
38. Richards, B.N., *The Microbiology of Terrestrial Ecosystems*, Longman Scientific and Technical, UK, 1987.
39. Dindal, D., *Soil Biology Guide*, John Wiley and Sons, New York, 1990.
40. Klute, A., Ed., *Methods of Soil Analysis, Part 1: Physical and Mineralogical Methods*, Second Edition, American Society of Agronomy Monograph 9, Madison, WI, 1986.
41. Page, A. L, Miller, R. H., and Keeney, D. R., Eds., *Methods of Soil Analysis, Part 2: Chemical and Microbiological Properties*, 2nd ed., American Society of Agronomy Monograph 9, Madison, WI, 1994.
42. Weaver, R. W., Angle, S., Bottomly, P., Bezdicek, D., Smith, S., Tabatabai, A., and Wollum, A., Eds., *Methods of Soil Analysis, Part 2, Microbiological and Biochemical Properties*, 2nd ed., Number 5, Soil Science Society of America, Inc., Madison, WI, 1994.
43. Sparks, D., Page, A. L., Helmke, P. A., Loeppert, R. H., Soltanpour, P. N., Tabatabai, M. A., Johnston, C. T., and Sumner, M. E., Eds., *Methods of Soil Analysis, Part 3, Chemical Methods*, Number 5, Soil Science Society of America, Inc., Madison, WI, 1996.
44. Soil Conservation Service, Soil Survey Manual, U.S. Department of Agriculture Handbook No. 18, Washington, D.C., 1951 (online revision, 1993).
45. Soil Conservation Service, Procedures for Collecting Soil Samples and Methods of Analysis for Soil Survey, Soil Survey Investigation Report No. 1, U.S. Department of Agriculture, Washington, D.C., 1982.
46. Munshower, F. F., *Practical Handbook of Disturbed Land Revegetation*, Lewis Publishers, CRC Press, Boca Raton, FL, 1994.
47. Paul, E. A. and Clarke, F. E., *Soil Microbiology and Biochemistry*, Academic Press, San Diego, 1990.
48. Miller, R. M., Mycorrhizae and succession, in *Restoration Ecology: A Synthetic Approach To Ecological Research*, Jordan, W.R., III, Gilpin, M. E., and Aber, J. D., Eds., Cambridge University Press, Cambridge, UK, 205, 1987.
49. Ingham, R. E., Trofymow, J. A., Ingham, E. R., and Coleman, D. C., Interactions of bacteria, fungi, and their nematode grazers: Effects on nutrient cycling and plant growth, *Ecol. Monogr.*, 55, 119, 1985.
50. Ingham, E. R. and Molina, R., Interactions of mycorrhizal fungi, rhizosphere organisms and plants, in *Microorganisms, Plants and Herbivores,* Barbosa, P., Ed., John Wiley and Sons, New York, 1991.
51. Gershuny, G. and Smillie, J., *The Soul of Soil: A Guide to Ecological Soil Management,* 3rd ed., AgAccess, Davis, CA, 1995.
52. Coleman, D. C., Through a ped darkly: An ecological assessment of root-soil microbial-faunal interactions, in *Ecological Interactions in Soil*, Fitter, A. H., Atkinson, D., Read, D. J., and Usher, M. B., Eds., Blackwell Scientific Publications, Cambridge, England, 1, 1985.
53. Nannipieri, P., Grego, S., and Ceccanti, B., Ecological significance of the biological activity in soil, *Soil Biochem.*, 6, 293, 1990.

54. Moore, J. C. and De Ruiter, P. C., Temporal and spatial heterogeneity of trophic interactions within belowground food webs, in *Modern Techniques in Soil Ecology*, Crossley, D. A., Jr., Ed., Elsevier, Amsterdam, 1990, 371.
55. Marschner, H., *Mineral Nutrition of Higher Plants,* Academic Press, London, 1985.
56. Sopper, W. E., *Municipal Sludge Use in Land Reclamation,* Lewis Publishers, Boca Raton, FL, 1993.
57. Julien, P. Y., *Erosion and Sedimentation,* Cambridge University Press, Cambridge, England, 1998.
58. Renard, K. G., Foster, G. R., Weesies, G. A., McCool, D. K., and Yoder, D. C., Predicting Soil Erosion by Water: A Guide to Conservation Planning with the Revised Universal Soil Loss Equation (RUSLE), U.S. Department of Agriculture, Agriculture Handbook No. 703, 1997.
59. Toy, T. J and Foster, G. R., Eds., Guidelines for the Use of Revised Universal Soil Loss Equation (RUSLE) Version 1.06 on Mined Lands, Construction Sites, and Reclaimed Lands, Office of Surface Mining, Western Regional Coordinating Center, Denver, 1998.
60. Glanders, G. A., Standardizing the standards, *Environ. Protec.,* 10, 38, 1999.

# CHAPTER 8

# Phytotoxicity

**Stephen J. Klaine, Michael A. Lewis, and Sandra L. Knuteson**

## CONTENTS



## 8.1 INTRODUCTION

The evaluation of the phytotoxicity of a chemical is an essential component of the ecological risk assessment of that compound. Primary producers form an essential trophic level of any ecosystem. Further, since all chemicals introduced into the environment ultimately find their way into aquatic ecosystems, aquatic algal and plant toxicity evaluations are particularly critical. This

**Table 8.1 Examples of Contaminants Reported to be More Toxic to Aquatic Plants than to Faunal Species**

| | |
|---|---|
| • 2,4-D | • Hydrazine hydroxide |
| • Acridine | • Industrial effluents |
| • Acrylates | • Monobromoacetic acid |
| • Aldrin | • Monochloroacetic acid |
| • Arsenate | • Nickel |
| • Atrazine | • Organotin |
| • Cadmium | • Phenol |
| • Cationic surfactants | • Phosphoric acid tributyl ester |
| • Chloramine | • Phthalic acid dialyl ester |
| • Chlordane | • Polycationic polymers |
| • Chloroacetaldehyde | • Potassium chlorate |
| • Chromium | • Potassium dichromate |
| • Copper | • Refinery effluents |
| • Dinitrotoluene | • Sodium salts |
| • Diquat | • Sodium tetraborate |
| • Disodium hydrogen | • Soil elutriates |
| • Endrin | • Textile effluents |
| • Glyphosate | • Zinc |
| • Herbicide effluent | |

*Note:* For more detail, see, among others, References 23, 51, and 72–74.

chapter presents the current state of phytotoxicity testing with particular attention to algal and vascular plant bioassays. The algal bioassay section focuses on test methods due to the relatively long history of algal toxicity testing. The vascular plant section focuses on the different plants used for bioassays and the various endpoints used in these bioassays.

## 8.2 ALGAE

### 8.2.1 Introduction

Phytoplankton, benthic and epiphytic microalgae, and macroalgae are energy sources critical to most aquatic ecosystems. Changes in their density and composition can affect the chemical and biological quality of the habitat. Therefore, the evaluation of the phytotoxicity of a contaminant is an essential component of any ecological risk assessment. Freshwater microalgae are used more frequently in phytotoxicity tests than any other type of freshwater or marine plant. However, algae are used less often than animal species. This is attributable to several factors. First, technical expertise is needed to conduct phytotoxicity tests. In addition, some consider plants to be less sensitive than animals to toxicants. However, algae have been found to be more sensitive than animal species to a variety of potential contaminants (Table 8.1). A review of the toxicity database for PMN submissions for the Toxic Substances Control Act (TSCA) indicated that algae were more sensitive than animal species in 50% of the submissions and less sensitive in 30%.[1] It is clear that the sensitivities of plants and animals to toxicants is unpredictable and that no assumptions of equivalency should be made. Therefore, the use of algae is necessary to estimate the environmental hazard of chemicals. This use is most frequent in the United States for commercial chemicals to comply with requirements for TSCA and the Federal Insecticide, Fungicide and Rodenticide Act (FIFRA). In contrast, the phytotoxic effects of most municipal and industrial effluents, contaminated sediments, and hazardous wastes are unknown.

Most of the information available to the scientific community concerning the phytotoxic effects of chemicals and other potential toxicants is based on the results for a few green algal species, such as *Selenastrum capricornutum*, and several *Scenedesmus* species. The use of *Selenastrum* in

**Table 8.2 Toxicity Test Methods for Freshwater Algae and Their Intended Use**

| Use | Reference |
|---|---|
| • Chemical compounds | 5, 47,[a] 48, 53,[b] 54,[c] 75,[d] 76,[e] 77[f] |
| • Industrial and municipal effluents | 55, 78 |
| • Human and animal drugs | 32 |
| • Freshwater dredge material | 79, 80 |
| • Contaminated sediments/elutriates | 68, 81–84 |
| • Hazardous chemical wastes | 85 |
| • Groundwater contamination | 86, 87 |

[a] APHA — American Public Health Association
[b] OECD — Organization for Economic Cooperation and Development
[c] ISO — International Organization for Standardization
[d] Method for Federal Insecticide, Fungicide, Rodenticide Act
[e] Method for Toxic Substances Control Act
[f] EEC — European Economic Community

various studies conducted during the 1960s and the 1970s has been reviewed by Leischman et al.[2] The majority of the standard test methods and culture techniques developed in the U.S. have been designed for its use. It is important to note that toxicity data for these and other algal species have been used as surrogates for marine algae, macrophytes, and even terrestrial plants. However, this type of data extrapolation is decreasing as methodologies for marine algae and vascular plants become more available.

### 8.2.2 Test Methodologies

Algicidal or acute toxicity tests are conducted less frequently than chronic tests with algae. Numerous chronic methods are available that can be used to evaluate the phytotoxicities of toxicants (Table 8.2). These chronic methods are based on the Algal Assay Procedure Bottle Test[3] or a modification of it.[4] Other test methods have been described that are either variations of those based on population growth, such as that by Payne and Hall,[5] or others that are more different such as those using microplate methods,[6] chemostat culture,[7, 8] flow cytometry,[9] and immobilized algae.[10]

The freshwater algal test methods currently used (Table 8.2) were developed during the past 20 years by a variety of organizations. These tests were designed to be conducted with an easily cultured alga, such as *Selenastrum capricornutum*, that rapidly grows and can be easily enumerated. The basic design of most methods is similar; the effect of a toxicant, usually five concentrations, is determined in the algal exponential-growth phase that occurs during the normal 3- to 4-day exposure period. This static exposure occurs in a nutrient-enriched medium that is diluted with well, reconstituted, dechlorinated tap, or river water under conditions of controlled pH, light, and temperature. See Figure 8.1 for an example of a typical 96-hour laboratory algal flask toxicity test conducted on an oscillation shaker. The effects of the toxicant are based on changes in biomass and growth rate that are monitored daily, in most cases by electronic or manual cell enumeration techniques. The response of the growing algae to the usual five test concentrations can be stimulatory, but more often it is inhibitory (Figure 8.2). Examples of the diversity of toxicants used in algal toxicity tests appear in Table 8.3.

The experimental conditions in the various types of algal toxicity tests available to the scientific community differ in some aspects (Table 8.4). The test results can be affected by several factors including the choice of test species, light intensity, temperature, and test duration. The effects of these and other variables on the test results have been reported on numerous occasions (Table 8.5) and have often exceeded several orders of magnitude. Another factor to consider is the use of a reference toxicant. There is no preferred compound, and cadmium chloride, sodium chloride, and sodium dodecyl sulfate have been used. Analytical verification of the test concentration is recommended in a few methods, but it is seldom performed. Nevertheless, this is an important issue that

**Figure 8.1** Example of a flask algal toxicity test conducted on an oscillator shaker.

**Table 8.3 Examples of Contaminants for Which Algal Toxicity Has Been Recently Determined**

| Test Compound | Reference |
|---|---|
| • Dyes | 88 |
| • Fire control chemicals | 89 |
| • Fuel oil | 90 |
| • Oil refinery chemicals | 91 |
| • Manure runoff | 92 |
| • Metals | 93, 94 |
| • Herbicides/pesticides | 95–98 |
| • PCBs | 99 |
| • Waste dump leachates | 100 |
| • Sediment | 101, 102 |
| • Wastewaters | 103–105 |
| • Landfill leachate | 106 |

should be considered since the toxicant solutions are not renewed in algal toxicity tests and the exposure concentration can be considerably different than nominal concentrations. A brief discussion of several important experimental factors in conducting algal toxicity tests follows. See the publications by Sirois[11] and Nyholm and Kallqvist[12] for additional details and evaluation.

### *8.2.2.1 Nutrient Medium*

An important component of algal cultures and algal phytotoxicity tests is the nutrient medium. The concentrations of the micro- and macronutrients comprising the medium exceed those in natural waters, and therefore its use affects the environmental relevance of the results. Algal growth is rapid in the medium, and the population can double or triple within 24 h (Figure 8.2). A variety of suitable culture media have been reported[13–15] in addition to those described in the published test methods. Test results can differ by as much as an order of magnitude depending on the type of medium.[16–19] Media components of greater research interest have included the chelator, ethylenediamine tetracetic acid (EDTA), and nitrogen and phosphorus.[7,20–22]

**Table 8.4 Experimental Conditions in Several Toxicity Tests Conducted with Freshwater Microalgae**

| Test Species | Duration (Days) | Test Chambers | Replicates | Light Intensity ($\mu Em^{-2}s^{-1}$) | Temperatures (°C) | Calculations | Reference |
|---|---|---|---|---|---|---|---|
| *Selenastrum capricornutum* | 5 (exposure)<br>9 (recovery) | Erlenmeyer flasks | 3 | 200–400 fc[a] | 19–24 | Algistatic, NOEC | 5 |
| *S. capricornutum* Printz<br>*Anabaena flos-aquae* (Lyng) Breb.<br>*Navicula seminulum* | 5 | Erlenmeyer flasks | 3 | 40–85 | 24 ± 2 | $EC_{10}$, $EC_{50}$, EC90 | 75 |
| *S. capricornutum* Printz<br>*Scenedesmus subspicatus* Chodat | 3 | Conical flasks | 3 | 120 (± 20%) | 21–25 | $EC_{50}$, NOEC | 53 |
| *S. capricornutum* Printz<br>*Scenedesmus quadricauda* (Turp.) Breb.<br>*Chlorella vulgaris* Beij | 4 | Erlenmeyer flasks | 3 | 300 ± 25 | 20–24 | $EC_{10}$, $EC_{50}$, EC90<br>Algistatic and algicidal effects | 50 |
| *S. capricornutum* Printz<br>*Microcystis aeruginosa* Kutz | 14, 21 | Erlenmeyer flasks | 3–5 | 175–300 | 24 ± 2 | MIC, NOEC | 32 |
| *S. capricornutum* Printz<br>*S. subspicatus* Chodat | 3–4 | Erlenmeyer flasks | 3–6 | 120 (±20%) | 23 ± 2 | $EC_{50}$, NOEC | 54 |
| *C. vulgaris* Beij. | 4 | Glass bottles | Not stated | Not stated | 4 | 5–20% | 81 |
| *S. capricornutum* Printz<br>*S. capricornutum* Printz | 4<br>4 | Erlenmeyer flasks<br>Erlenmeyer flasks | 3<br>2 | 86 ± 8.6<br>400 | 25 ± 1<br>24 | LOEC, NOEC<br>$EC_{50}$ | 107<br>84 |
| *S. capricornutum* Printz<br>*S. subspicatus* Chodat<br>*C. vulgaris* Beij.<br>*M. aeruginosa* Kutz<br>*A. flos-aquae* (lyng) Breb.<br>*Navicula pelliculosa* Grun. | 4 | Erlenmeyer flasks | 3 | 30–90 | 20–24 | $IC_{50}$, NOEC | 48 |
| *Chlorella vulgaris* Beij.<br>*Chlorella pyrenoidosa*<br>*Dunaliella tertiolecta* Butcher<br>*Scenedesmums* sp. | 12 | Culture tubes | 3 | 25 | 25 | $EC_{50}$ | 106 |

*Source:* From Lewis (1995).

**Table 8.5 Experimental Factors that Affect the Results and Data Interpretation of Toxicity Tests Conducted with Freshwater Algae**

| Experimental Factor | References |
|---|---|
| • Statistical analyses | 52, 56, 57, 108 |
| • Nutrient medium composition | 7,16, 17, 18, 19, 21, 22,109–112 |
| • Effect parameters/calculations | 5, 49, 113 |
| • Test solution volume | 114 |
| • Test duration | 78, 115–117 |
| • Solvent use | 108,118, 119, |
| • pH, light, and temperature | 8, 19, 120–124 |
| • Interspecific variation in response | 23, 24, 35, 37 |
| • Test species selection | 26, 29, 31, 125, 126 |

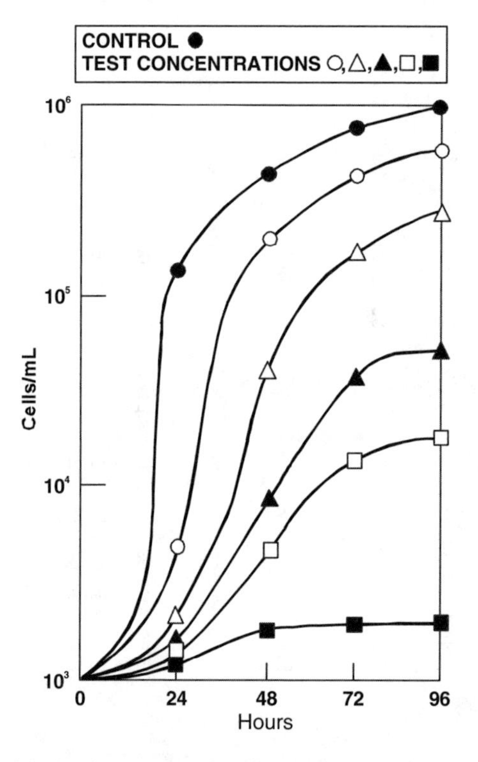


**Figure 8.2** Response of a unialgal culture exposed to five test concentrations.

### *8.2.2.2 Test Species*

The selection of the test species is an important consideration in phytotoxicity tests with algae, particularly if environmental realism is desired. The choice of a species has been usually dependent upon its availability, culture requirements, and case of use. Typically, only one species is used to determine the phytotoxicity of a commercial chemical, and usually a green alga has been the species of choice. Blue-green algae and diatoms are used infrequently because of their slow growth and difficulty in culture. However, there are exceptions, as shown for pesticides (Table 8.6).

Interspecific variation has been demonstrated for a variety of algae, including *S. capricornutum* and several compounds (Table 8.7). The difference in sensitivity can be less than ten or several orders of magnitude.[23–26] In addition to interspecific variation, the sensitivities of different clones and geographical races of algae can differ,[24,27,28] and the presence of other algae can affect the test results.[29–31] The overall effect of the considerable interspecific response is the inability to predict the phytotoxic effects of chemicals from the results for only one or two algal test species, which

**Table 8.6 Diversity of Algal Test Species Used to Determine the Toxicity of Pesticides in Laboratory Tests**

| Microalgae | No. of Citations | No. of Pesticides |
|---|---|---|
| ***Chlorophyceae*** | | |
| *Ankistrodesmus falcatus* | 9 | 9 |
| *Chlamydomonas reinhardtii* or sp. | 8 | 8 |
| *Chlorella vulgaris* | 11 | 7 |
| *Chlorella pyrenoidosa* | 9 | 17 |
| *Oedogonium cardiacum* | 9 | 8 |
| *Scenedesmus quadricauda* | 14 | 15 |
| *Scenedesmus obliquus* | 7 | 7 |
| *Selenastrum capricornutum* | 5 | 16 |
| ***Cyanophyceae*** | | |
| *Anabaena flos-aquae* or *cylindrica* | 8 | 6 |
| *Microcystis aeruginosa* | 8 | 6 |
| ***Bacillariophyceae*** | | |
| *Navicula pelliculosa* or *Navicula* sp. | 5 | 6 |

From Swanson et al.[26]

**Table 8.7 Several Chemicals for Which Interspecific Variation in Toxicity have been Observed**

| 1–9[a] | 10–99 | 100–499 | 499–1000 | > 1000 |
|---|---|---|---|---|
| 4-Nitrophenol | 2-Ethylhexyl acrylate | 2, 4-D | Ceresan | Disodium hydrogen |
| Aliquat | Acetic acid | Acrylic acid | Arsenic | Linuron |
| Amitrole | Acetone | Bromacil | Selenium | Propazine |
| Anthracene | Aniline | Diuron | Streptomycin arsenate | Sodium nitrate |
| Biphenyl | Atrazine | Hydroquinone | | |
| Cadmium | Benzoic acid | Nickel chloride | | |
| Chlordane | Benzonitrile | Paraquat | | |
| Industrial effluent | Cadmium | Phenylmercury acetate | | |
| Lindane | Carbaryl | Slimecide | | |
| Mercury | Chlorophenols | Sodium lauryl sulfate | | |
| Mercury | Chromium | Sodium dichromate | | |
| Methyl acrylate | Copper sulfate | | | |
| Naphthalene | Dithane | | | |
| Nickel | Mercuric chloride | | | |
| NTA | Potassium dichromate | | | |
| Organotin | Simazine | | | |
| Paper mill effluent | Surfactants | | | |
| PCP | Trichloroethylene | | | |
| Phenol | | | | |
| Pyridine | | | | |

[a] Differences in effect concentrations.

Values taken from several sources, including References 23 and 24.

is the usual practice. Therefore, a species battery approach is needed where several taxonomically different species are used. This battery may include "nontraditional" species. Swanson et al.[26] reported that algal species of 35 genera have been used to evaluate the toxicity of atrazine.

#### 8.2.2.3 Test Duration

The test duration is an important experimental consideration since it can influence the test result (Table 8.8). Most toxicity tests with algae are conducted for 3 to 4 days. Shorter durations have been recommended to reduce the possibility of a decrease in the test concentration due to sorption of the toxicant to the rapidly increasing algal biomass and test container walls as well as to pH changes in the test water. In contrast, tests of longer duration (5 to 21 days) allow for the determination of algistatic concentrations,[5] the toxicities of intermediate biodegradation products, and stimulatory effects.[32]

**Table 8.8 Effects of the Test Duration on the Toxicities of Several Pesticides**

| Test Compound | Test Duration | Concentration (μg/L) |
|---|---|---|
| DDT[a] | 48 h | 100 |
| | 96 h | 100 |
| | 6 days | 1000 |
| | 8 days | 1000 |
| Carbaryl[a] | 2 days | 100 |
| | 96 h | 1000 |
| | 6 days | 1000 |
| 2,4-D[a] | 48 h | 1000 |
| | 96 h | 1000 |
| | 6 days | 100 |
| | 8 days | 100 |
| | 10 days | 1000 |
| Carbaryl[b] | 4 days | 100 |
| | 7 days | 100,000 |

[a] For *Scenedesmus quadricauda.*
[b] For *Chlorella pyrenoidosa.*
Data from Swanson et al.[127]

**Table 8.9 Experimental Conditions in Two Short-Term Phytotoxicity Tests**

| Parameter | Short-Term Photosynthesis[a] | Microplate Technique[b] |
|---|---|---|
| Test species | *S. capricornutum* | *S. capricornutum* |
| Duration | 2 h | 4 h to 8 days |
| No. test concentrations | 8 | 8 |
| No. replicates | 2 | 6–8 |
| Temperature (°C) | 24 | 24 ± 2 |
| Light ($\mu Em^{-2}s^{-1}$) | 84 | 95 |
| Test chambers | BOD bottles | 96 well plate |
| Calculations | $EC_{50}$ | $EC_{50}$ |
| Effect | Photosynthesis ($^{14}C$ uptake) | Growth (cell counts) |

[a] Data from Delistraty.[40]
[b] Data from Blaise et al.[6]

Algal toxicity tests can be 1 h or less if photosynthetic activity is the effect parameter. The effects of several herbicides,[33–36] surfactants,[37] metals,[38] effluents,[39] and oil products[40,41] have been determined based on the photosynthetic response of algae as measured by carbon uptake or oxygen evolution. These studies have been conducted with single cultured species of algae as well as with natural periphyton and phytoplankton communities. Although the short duration of these studies is desirable, short-term photosynthesis has been a less sensitive effect parameter than population growth monitored during the 3- to 4-day traditional studies.[36,42]

There has been an effort recently to develop and validate tests of shorter duration that provide relatively rapid estimates of phytotoxicity; i.e., the microplate technique.[20] Microplate techniques are being used more frequently, and they have been shown in some cases to provide results comparable to those from flask bioassay procedures[43] such as shown in Figure 8.1. Examples of the experimental techniques used in the microplate method as well as in a short-term photosynthesis test appear in Table 8.9.

### *8.2.2.4 Light, Temperature, and pH*

Light intensity, temperature, and pH may impact the results of phytotoxicity tests. This should not be unexpected since they affect the toxicities of chemicals to aquatic animals. Increases in light intensity and temperature generally increase toxicity.[44] The effect of pH is less predictable; toxicity may decrease with increasing pH,[45] or it may be greatest at some intermediate level.[46]

#### *8.2.2.5 Response Parameters and Calculations*

A variety of effects have been monitored in algal toxicity tests, including physiological and morphological changes. However, the inhibitory and, to a lesser extent, stimulatory effects of the toxicant on biomass are determined in most cases. Effects on biomass are directly or indirectly determined using measurements such as dry weight, cell counting, and chlorophyll *a*. Counting chambers (hemacytomer), spectrophotometers, and fluorometers are commonly used for this purpose.

The test results are usually reported as a change in standing crop, growth rate, or area under the growth curve. These measurements, as well as others, are described in the various published test methods reported by APHA et al.,[47] and ASTM.[48] Typically, an $IC_{50}$ value and the no-observed effect concentration (NOEC) are calculated. However, these calculations are considered by some not to be as environmentally meaningful as algistatic and algicidal concentrations,[5,49] which are seldom reported. Methodologies for determining algistatic and algicidal effects are provided by Payne and Hall[5] and the U.S. EPA,[50] and they include the use of extended exposure and recovery periods as well as vital staining techniques.

Stimulation is not commonly reported in phytotoxicity tests, although it is occasionally observed. The calculation used most often to report stimulatory effects is the $SC_{20}$ value, which was first described by Walsh et al.[51] It represents the test concentration that results in a 20% increase in biomass relative to that for the control algal population. The importance of stimulation in the data interpretation and the definition of biologically relevant levels are issues that need to be resolved.

Appropriate statistical methods to analyze algal toxicity data have been summarized and discussed previously.[47,52–56] Graphical interpolation is a simple and useful method to determine the $EC_{50}$ and $IC_{50}$ value,[57] and 95% confidence intervals can be provided using linear and nonlinear regression methods.[55] Parametric and nonparametric analyses have been used to determine the NOEC and the first-effect values, which is recommended in some methods. An inverse linear regression technique is used to calculate the algistatic concentration.[5]

### 8.2.3 Environmental Significance of Results

The environmental relevance of laboratory-derived phytotoxic concentrations for single species of microalgae is not known for most chemicals. The large difference in interspecific variation, the usual controlled experimental conditions, and the unknown impact of environmental modification such as biological adaptation[58] are several factors that reduce the relevance of these data as they are currently derived. Consequently, results from laboratory phytotoxicity tests are considered generally to be unrealistic estimates of concentrations expected to cause detrimental ecosystem effects.

Several experimental designs have been used to provide more realistic phytotoxicity estimates. These include adaptations to the traditional laboratory methods such as using river water for dilution,[59] the use of chemical mixtures,[60] and the simultaneous exposure of multiple algal species.[61] In addition, more complex multispecies designs have been used in the laboratory and field, and these include microcosms, experimental ponds and streams, enclosures, and *in situ* dosing. The effects of various contaminants have been determined on natural phytoplankton and periphyton communities using these techniques (Table 8.10).

The results from the multispecies tests have been compared in several cases to those from the laboratory single-species tests conducted with the same effluents,[39] pesticides,[35,62] metals, and surfactants.[63–65] The outcomes of these data comparisons, which are relatively few, have been

**Table 8.10 Methodologies Used to Determine the Effects of Contaminants on Aquatic Plant Species, Populations, and Communities**

| Test Compound | Biota | Experimental Design | Reference |
|---|---|---|---|
| Coal-derived oil | Algae, macrophytes | Laboratory microcosms | 128–130 |
| Atrazine | Phytoplankton | | |
| Chlorpyrifos | Phytoplankton, periphytic algae | | |
| Pesticides | macrophytes | | |
| Diflubenzuron | Periphytic algae | Experimental streams | 131–135 |
| Atrazine, trifluralin, paraquat | Spring algae | | |
| Ammonia | Periphytic algae | | |
| Lindane | Periphytic algae | | |
| Nonionic surfactant | Periphytic algae | | |
| Coal-derived oil | Phytoplankton, macrophytes | Experimental ponds | 136–139 |
| Arazine | Phytoplankton | | |
| Flourene | Phytoplankton, macrophytes | | |
| Lindane, deltamethrin | Macrophytes, phytoplankton, periphytic algae | | |
| Cadmium | Lake phytoplankton | Enclosures | 140–145 |
| Surfactants | Lake phytoplankton | | |
| Chlorpyrifos | Periphytic algae, phytoplankton | | |
| Triazine herbicides | Periphytic algae | | |
| Copper | Pond phytoplankton | | |
| Esfenvalerate | Lake macrophytes, algae | | |

Adapted from Lewis.[127]

species- and compound-specific. For example, cultured algae were less sensitive to anionic surfactants than a phytoplankton community but not to nonionic and cationic surfactants.[37] This inconsistent trend in sensitivity has been observed for different metals as well. Based on these and other reported results, the inability to generalize or predict ecosystem impacts from laboratory-derived toxicity results for microalgae and most chemicals is obvious.

### 8.2.4 Usefulness of Algal Phytotoxicity Tests

The importance of phytotoxicity tests in ecotoxicology and in the risk assessment process has not always been obvious. As stated earlier, these tests were once thought to be less worthwhile than those conducted with animals.[66] This opinion has contributed to the current minimal phytotoxicity database.[67] However, comparative toxicity evaluations have shown phytotoxicity tests to be more useful than some animal toxicity tests,[68] and they are conducted more frequently than animal chronic toxicity tests in support of the TSCA process.[69, 70]

Current scientific knowledge concerning the phytotoxic effects of most chemicals is based primarily on the laboratory-derived results for freshwater green algal species. As stated earlier, these results are used, sometimes with little scientific justification, as surrogates for other types of freshwater algae, marine algae, vascular aquatic plants, and, in some cases, terrestrial plants.[71] Furthermore, laboratory-derived results for single species of cultured microalgae are also used to predict ecosystem effects. The validity of this latter extrapolation has not been proven and should not be attempted unless supporting data are available.

In summary, laboratory-derived algal toxicity tests are worth the effort, despite current limitations. To improve their usefulness several taxonomically distinct test species need to be used, and the biological significance of the inhibitory and stimulatory effect levels need determination. The derivation of these data should be of high priority to the scientific, regulatory, and regulated communities since their availability will increase the relevance of the current risk assessment process for potential phytotoxic substances.

## 8.3 AQUATIC VASCULAR PLANTS

### 8.3.1 Introduction

Traditionally, algae have been the primary test organisms for aquatic phytotoxicity evaluations due, primarily, to the simplicity of algal culture systems. In fact, toxicity data for algae are often used as surrogates for vascular and terrestrial plants. Recently, this practice has been challenged since algae were less sensitive than vascular plants 20% of the time.[71]

Aquatic vascular plants, or macrophytes, are an extremely important component of the primary producers in many ecosystems. They include mosses, liverworts, ferns, and angiosperms and are usually classified as floating, submersed, or emergent. Taxonomical, ecological, and physiological characteristics of aquatic macrophytes are presented in detail elsewhere.[128–130] Both freshwater and saltwater vascular plants serve as food sources for waterfowl, fish, and invertebrates; they capture solar energy, produce oxygen, participate in nutrient cycling, assimilate carbon, and provide habitat for a variety of aquatic life. For these reasons, this trophic level is critical to the stability of aquatic ecosystems, and pollutant burden on these organisms may have indirect repercussions on other trophic levels.

During the past 30 years the major focus of aquatic vascular plant research was directed toward controlling and eradicating troublesome vegetation. Literature is available on the control of the Haloragaceae (water-milfoil),[131] Potamogetonaceae (Pondweeds),[132] Najadaceae (Water nymphs),[133] Pontederiaceae (hyacinths),[134] Hydrocharitaceae (Frog's bit),[135] and several others.

In the last decade, however, the importance of aquatic vascular plants in the environmental risk assessment process has been recognized. These organisms are not only important as indicators of contaminant stress in the aquatic ecosystem but also as significant routes of chemical disposition, movement, and bioavailability.[136] Problems associated with the use of aquatic vascular plants in toxicity and fate studies include inadequate knowledge of typical growth conditions and the inability to obtain optimum and reproducible growth with the test species. In addition, these organisms do not lend themselves to nondestructive quantitative measurements of growth with time. Rooted species present an added difficulty if a complex rooting substrate is needed. Development of culture techniques, test methods, and sensitive endpoints for aquatic vascular plant bioassays has been limited. A few species of floating and submerged plants, however, have been used to successfully assess the toxicity and fate of aquatic contaminants.

### 8.3.2 Duckweeds

By far, the most commonly used vascular plants in toxicity tests are duckweeds, members of the Lemnaceae. Advantages of these floating macrophytes include their small size, their ability to be grown in axenic culture in a defined inorganic medium, the fact that they are not rooted in the hydrosoil, and the fact that both their growth rate and final biomass yield can be determined visually and nondestructively. A review of the taxonomic literature has been presented by Hillman.[137] The use of duckweeds has been suggested or required by the Toxic Substances Control Act and the Federal Insecticide, Fungicide and Rodenticide Act.[138,139] While there are four genera in the Lemnaceae family, *Lemna* are most often used in toxicity testing. The four species that are globally distributed are *Lemna minor*, *Lemna trisulca*, *Spirodela polyrhiza*, and *Spirodela oligorhiza*.[137] *L. minor* and *L. gibba* are the more commonly used bioassay species.

Duckweed toxicity bioassays can be static, static-renewal, or flow-through design.[48,138–142] A variety of test lengths (4 to 10 days), light intensities (30 to 300 mE $m^{-2}$ $sec^{-1}$), and growth media (nutrient-enriched algal media[50] and Hoagland's media[143]) have been used in these assays. Grant[144] optimized the growth of *L. minor* and *S. polyrhiza* in static toxicity bioassays. Experimental conditions that provided maximum growth rate included 33% Hutner's Medium[145] buffered to pH 6.5 with 0.015

M phosphate buffer, light intensity equal to 151 mE $m^{-2}$ $sec^{-1}$ with cool white fluorescent tubes and a temperature of 25°C. Hutner's reported sensitivity of *L. minor* to the anionic surfactant linear alkylbenzene sulfonate during a 10-day test was greater than that reported by Bishop and Perry.[146]

Wang[147] reviewed the use of duckweed in toxicity tests and reported that the doubling time for *Lemna* species ranged from 0.35 to 2.8 days. In Grant's growth optimized system doubling times for *L. minor* and *S. polyrhiza* were 1.43 and 2.24 days, respectively.[144] To be a valid test the biomass of duckweed in the control needs to increase at least five-fold during one 7-day test.[48] The most commonly used calculations are the values that are based on a variety of effects such as the number of fronds and on biomass, root number, plant number, root length, and pigment content. Other calculations, such as the phytostatic, phytocidal, EC10, and EC90 values, have also been used.[138,148]

Little information on the comparative sensitivities of various species of duckweed exists. Grant[144] reported that *L. minor* was more sensitive than *S. polyrhiza* to the herbicide Hydrothal-191 (Endothal) and to linear alkylbenzene sulfonate: the two were equally sensitive to ammonium carbamate. King and Coley[149] reported that *L. gibba* was less sensitive than *L. minor* and *L. perpursilla* to extracts of oils and coal distillates.

In some studies duckweed has been reported to be more sensitive to some chemicals than other macrophytes and algae.[146,150] Wang[147] reported that duckweed was more sensitive than fish to Cd, Cr, Pb, Ni, and Se. Taraldsen and Norberg-King[151] developed a test method for effluents and reported that duckweed was more sensitive than the fathead minnow or *Ceriodaphnia dubia* to two of three effluents tested. Other studies, however, have illustrated that duckweed is not as sensitive as some algae to chromium (VI), solid waste leachates, chlorinated phenolic compounds, and coal ash waste.[152–155]

There is great need for additional data concerning the relative sensitivity of duckweeds to a wide variety of toxic compounds. Only then will they be more widely incorporated into chemical registration and NPDES permit testing. In addition, the ecological significance of duckweed bioassay results and the utility of these organisms as surrogates for other aquatic vascular plants need to be examined more thoroughly.

### 8.3.3 Submersed and Emergent Vegetation

Submersed and emergent aquatic vegetation has been used even less in pollutant toxicity and fate bioassays than duckweeds. Most of these organisms are rooted, and culture and standard growth measurements are more difficult than with duckweeds. Issues as basic as suitable indicator species, method of contaminant exposure, duration of tests, and exposure-effect endpoints are yet to be resolved. One attribute that makes these organisms particularly desirable is that they are exposed to contaminants in both the sediment and the water column. While there exists an abundance of water-column bioassays, few organisms sample the sediment interstitial water like the roots of vascular plants. These organisms are naturally adapted to alter the physical and chemical parameters of the rhizosphere in order to make nutrients more bioavailable. Thus, sediment-incorporated contaminants may be more or less bioavailable in an active rhizosphere than in sediments without rooted macrophytes. Recent efforts by USEPA and others to establish sediment quality criteria may increase the use of these plants in sediment bioassays.

Numerous efforts to quantify the relationship between contaminant exposure and rooted aquatic vascular plant response have been reported. Test conditions vary widely, and as yet no attempt has been made to standardize experimental parameters or determine the most appropriate test species.

The majority of the research reported in the 1970s concerning inorganic contaminants centered on uptake, accumulation, and elimination of these elements.[156–162] Many of these studies demonstrated that heavy metals were accumulated by vascular plants to very high levels without serious reductions in the growth of the organisms. Some studies, however, have demonstrated a stress response by some rooted macrophytes to heavy metal burden.

*Elodea canadensis* has been used to assess copper toxicity.[163] A static bioassay was developed in which the growth of the organism was optimized with respect to illumination, temperature, and dissolved carbon dioxide. Relative toxicity was judged by reduction of photosynthetic oxygen evolution and visual evaluation of plant damage.

Another member of the Hydrocharitaceae family, *Hydrilla verticillata*, was used to assess copper toxicity under static conditions.[164] Whole plant and root exposures were conducted for 14 days in Hoagland's medium augmented with 200 mg/L $Na_2HCO_3$.[135] Shoot growth, root growth, changes in chlorophyll *a,* and changes in whole plant dehydrogenase were determined. The most sensitive endpoint was root growth.

Both *Najas quadulepensis* (Southern Naiad) and *Myriophyllum spicatum* (Eurasian Watermilfoil) have also been used in static bioassays with heavy metals. In one study visual evidence of cadmium toxicity to *N. quadulepensis* was present at 0.007 mg/L.[165] In another study the toxicity of several heavy metals to *M. spicatum* was determined using reduction in dry weight as the endpoint.[166] Methylmercury was concentrated in the younger tissues of *Elodea densa,* while inorganic mercury was concentrated in the older tissues.[167,168] Methylmercury was better absorbed and was more toxic than the inorganic form, which was released more readily from plants containing both forms.

Several different macrophytes have been used to assess the toxicity of surfactants, chlorine, and pesticides.[166,169–173] These efforts have utilized not only different species and methods but also different endpoints.

One endpoint of interest is the peroxidase response reported by Byl and Klaine.[172] They report a quantitative dose-response relationship with several different contaminants in the water column. Byl et al.[174] reported further on the peroxidase response to a range of chemicals including anthracene, the herbicide sulfometuron methyl, $Cd^{2+}$, $Cr^{6+}$, $Cu^{2+}$, $Mn^{2+}$ and $Se^{4+}$. In many instances a statistically significant increase in peroxidase was observed at toxicant levels two or three orders of magnitude less than that required to decrease growth. They suggest that peroxidase may be a valuable measure of contaminant exposure but not necessarily stress.

Further studies indicated that peroxidase was also a good indicator of the bioavailability of sediment-incorporated metals.[175,176] A strong relationship between plant metal content and peroxidase was demonstrated. Byl et al.[176] also demonstrated that rooted aquatic plants could oxidize sediment-incorporated cadmium-sulfide complexes. This made the cadmium more bioavailable and resulted in significant incorporation of cadmium into the plant material, even under anoxic sediment conditions.

Sutton[177] examined the peroxidase response of several aquatic and wetland plants to exposure to a variety of metals and pesticides. In general, the response of these organisms to contaminant stress was more variable and less sensitive than agronomic species typically used to evaluate phytotoxicity.

The development of biochemical endpoints of stress and exposure in aquatic vascular plants may result in increased sensitivities of these organisms to contaminant burden. Research is needed to determine when a change in a biochemical endpoint, such as peroxidase activity, represents stress or a decreased fitness of the organism in particular and the population in general.

Biochemical indices of stress have been used in terrestrial vascular plants to monitor air-pollution stress. Richardson et al.[178] demonstrated significant increases in the enzymes superoxide dismutase and peroxidase in loblolly pines (*Pinus taeda*) subjected to five ozone and two acid treatments. Concurrent with these increases was a decrease in net photosynthesis. They suggest that antioxidants may be useful as early-warning biomarkers of oxidative stress in trees.

It is important to note that two standardized rooted aquatic macrophyte bioassays have been developed. The American Society of Testing Methods (ASTM) issued two protocols. The first is a general protocol for the use of freshwater emergent macrophytes,[179] while the second is more specific to the use of an axenic culture of *Myriophyllum sibiricum* Komarov.[180] These procedures, while straight-forward, are much more complex than the duckweed bioassays.

**Table 8.11 Bioassays Used with Terrestrial Plants**

| Bioassay | References |
|---|---|
| Soil | 186, 192–198, 209, 254, 255 |
| Foliar | 186, 193, 197, 204–206, 235, 256, 257 |
| Petri dish | 186, 208 |
| Hydroponics | 207, 215 |
| Cell culture | 186, 217, 244, 245 |

## 8.4 TERRESTRIAL VASCULAR PLANTS

### 8.4.1 Introduction

Terrestrial plant bioassays have been used for several decades for determining herbicidal effectiveness in the agricultural and horticultural industries. Studies were used to identify agricultural crop and weed species sensitivity, resistance, and herbicidal mode of action/biochemical pathways.[181–184] Several different types of bioassays (Table 8.11) exist such as soil or hydroponic, foliar, petri dish, and tissue culture assays;[185,186] however, standardized phytotoxicity tests have centered mainly around algal and aquatic vascular plants. Fairchild et al.[187] performed comparative phytotoxicity tests with several algal and vascular plant species using atrazine, metribuzin, alachlor, and metolachlor. Two of the commonly tested species, the green alga (*Selanastrom capricornutum*) and duckweed (*Lemna minor*), were shown to have $EC_{50}$s over three times that of the macrophyte *Ceratophyllum demersum.* Jones and Winchell[33] exposed four macrophytes to atrazine and found a narrow range of $EC_{50}$s (77–104 μg/L); while Forney and Davis[150] exposed five macrophytes to atrazine and found a much greater range of $EC_{50}$s (80–1104 μg/L). Each of these studies measured a different endpoint including plant growth, chlorophyll fluorescence, and frond count. The results of these studies illustrate the large variation among existing toxicity tests. Lewis[127,188] compared the sensitivities of plants and animals to several environmental contaminants and showed that sensitivity to chemicals is not only chemical-specific but also species-specific. Fletcher[71] compared the alga *Chlorella* with several terrestrial plants using the PHYTOTOX database and came to similar conclusions. This suggests that a wide selection of both fauna and flora species should be examined for their sensitivity when characterizing the effects of a toxicant on nontarget species.

### 8.4.2 Test Methods

Recently, the Office of Prevention, Pesticides and Toxic Substances (OPPTS) of the U.S. Environmental Protection Agency (EPA) has developed a series of Ecological Effects Test Guidelines.[189–203] These guidelines combine current standardized bioassays used under the Toxic Substances Control Act (TSCA) and the Federal, Insecticide, Fungicide, and Rodenticide Act (FIFRA). Group D of these guidelines is designed to estimate the effects of toxicants on nontarget plants and includes tests for terrestrial plant species (Table 8.12).

Several different types of bioassays exist including soil or hydroponics, foliar, petri dish, and tissue culture assays.[185,186] Soil bioassays incorporate the toxicant into the soil, exposing the plant through the roots or as a seed.[186] In foliar bioassays the plant is exposed by application of the toxicant directly onto the leaves.[204–206] Plants can be grown in either soil or hydroponic systems for foliar assays. Petri-dish bioassays are generally used for seed-germination studies, where seeds of the test species are directly exposed to a substrate (moist filter paper, sand, or glass beads) containing the toxicant.[207,208] However, plant tissues may also be exposed in petri dishes.[185] Tissue culture assays use *in vitro* cultivation of plant cells or tissues for phytotoxicity, biochemical, and mode-of-action studies.[185]

**Table 8.12 Ecological Effects Test Guidelines, Group D — Nontarget Plants Test Guidelines, Developed by the Office of Prevention, Pesticides, and Toxic Substances (OPPTS)[189–203]**

| OPPTS # | EPA # | Name |
|---|---|---|
| 850.4000 | 712-C-96-151 | Background — nontarget plant testing |
| 850.4025 | 712-C-96–152 | Target area phytotoxicity |
| 850.4100 | 712-C-96–153 | Terrestrial plant toxicity, Tier I (seedling emergence) |
| 850.4150 | 712-C-96–163 | Terrestrial plant toxicity, Tier I (vegetative vigor) |
| 850.4200 | 712-C-96–154 | Seed germination/root elongation toxicity test |
| 850.4225 | 712-C-96–363 | Seedling emergence, Tier II |
| 850.4230 | 712-C-96–347 | Early seedling growth toxicity test |
| 850.4250 | 712-C-96–364 | Vegetative vigor, Tier II |
| 850.4300 | 712-C-96–155 | Terrestrial plants field study, Tier III |
| 850.4400 | 712-C-96–156 | Aquatic plant toxicity test using *Lemna* spp., Tiers I & II |
| 850.4450 | 712-C-96–157 | Aquatic plants field study, Tier III |
| 850.4600 | 712-C-96–158 | Rhizobium-legume toxicity |
| 850.4800 | 712-C-96–159 | Plant uptake and translocation test |

### 8.4.3 Plant Species

For phytotoxicity tests with terrestrial plants, agricultural crop species have generally been chosen. The OPPTS guidelines state that ten plant species must be used for each test. Soybean (*Glycine max* L. [Fabaceae]), corn (*Zea mays* L. [Poaceae]), and a root crop such as carrot (*Daucas carrotta* L. [Umbellifereae]) or onion (*Allium cepa* L. [Amaryllidaceae] must be used. The investigator can choose the other seven species. Although native species may be used, crops such as beet (*Beta vulgaris* L. [Chenopodiaceae], tomato (*Lycopersicon esculentum* Mill. [Solanaceae], cucumber (*Cucumis sativa* L. [Cucurbitaceae], cabbage (*Brassica oleracea* L. [Brassicaceae], lettuce (*Lactuca sativa* L. [Asteraceae]) or relevant weedy species are the most commonly used.[185,192–197,209,210]

### 8.4.4 Support Media and Exposure

Various types of support media can be used depending on the growth characteristics or needs of the plant species and the experimental conditions. When mixing the toxicant into a soil substrate, common greenhouse potting soils may skew results of the assay due to a higher binding capacity than with field soils. Generally, a sterilized standard sandy, sandy loam, or clay loam soil with up to 3% organic matter is used. The toxicant is incorporated into the dry substrate by dissolving the toxicant in a solvent and mixing it thoroughly into the soil. The solvent must be allowed to completely evaporate overnight before use in phytotoxicity testing.[186,192–197,209,211] Soil characteristics, such as pH, composition (sand/silt/clay content) organic content, and binding capacities, may affect the bioavailability of the toxicant and should be determined.[186]

When plants are exposed via a foliar or nutrient solution, tests may be designed for hydroponics. The classic paper by Hoagland and Arnon[143,212] describes a hydroponics system for growing terrestrial plants in nutrient solution. This system can be adapted to various phytotoxicity bioassays.[213–215]

### 8.4.5 Measured Responses

Various plant responses (Table 8.13) can be measured in terrestrial phytotoxicity bioassays depending on the design of the experiment. Over all, effects of toxicants on the plants may be analyzed using seedling germination, seedling emergence, plant growth measured by fresh weight, shoot or root length change over the duration of the test, or physical observations such as chlorosis, yellowing of the leaves, cupping of the leaves, etc.[186,192–197,209] Biochemical endpoints, such as

**Table 8.13 Measured Responses to Toxicants in Terrestrial Phytotoxicity Tests**

| Response | References |
|---|---|
| Seedling germination | 186, 194, 209 |
| Seedling emergence | 192, 195, 209 |
| Plant growth | 186, 192–199, 209, 235, 236 |
| Chlorosis | 186 |
| Cupping of the leaves | 204 |
| Photosynthesis | 216, 217, 236–238 |
| Respiration | 217, 239 |
| $CO_2$ assimilation | 216, 217 |
| Ethylene and ACC | 218, 220, 221, 240–242 |
| Protein content | 231 |
| Fatty acid composition | 222 |
| Glutathione (GSH) | 230 |
| Enzyme activities | 243–245, 259 |
| Glutathione S-transferase activity | 219, 231, 243, 246–248, 258 |
| Phtyochelatins | 230, 249–251 |
| $^{15}N$ metabolism | 252 |
| Transpiration | 215, 253 |

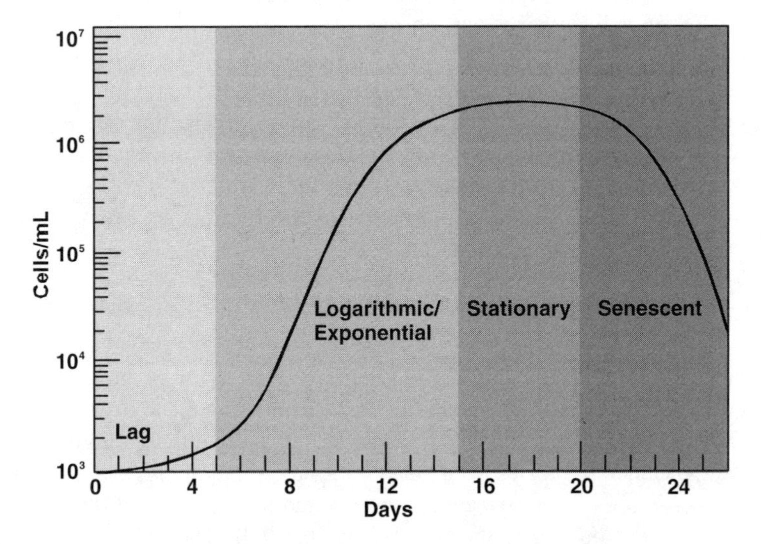


**Figure 8.3** Typical growth curve for an algal culture in a nutrient-enriched medium.

photosynthesis, respiration, $CO_2$ assimilation, protein synthesis, enzyme activities, and lipid composition, may be used.[216–222] Several review papers can be found on photosynthesis,[217,223–226] ethylene synthesis,[220,227,228] phytochelatins,[229,230] glutathione and glutathione-S-transferases,[231–233,258] and other enzymes.[234]

## 8.5 SUMMARY

Phytotoxicity assessment remains an important aspect of hazard assessment. In fact, the importance of phytoxicity evaluations is better recognized today than anytime in the past. Currently, most pesticides are registered and regulated on the basis of toxicity bioassays using aquatic and terrestrial fauna, with relatively little contribution from data on flora. Pesticide regulators must rely on a restricted historical phytotoxicity database and a series of standardized phytotoxicity tests. Risk assessments are routinely performed using laboratory toxicity tests to characterize ecological effects of contaminants. The endpoints of these tests are generally the mean lethal concentration or dose ($LC_{50}$ or $LD_{50}$) for long-term faunal studies (Lytle and Lytle[210]). A limited macrophyte toxicity database results in the extrapolation of toxicity data from algae to macrophytes. For terrestrial plants, the situation is worse. The U.S. Environmental Protection Agency requires terrestrial phytoxicity data for pesticide registration. This data set, however, is composed entirely of results from agronomic species, which may have little relationship with nontarget native plants. In fact, very little information is available to characterize the uncertainty in the extrapolation of these data to nontarget native plants. This is a huge research need in comparative phytotoxicity. Other research needs include the development and implementation of physiological, biochemical, and molecular endpoints to evaluate phytotoxicity and the characterization of factors controlling contaminant bioavailability in the soil environment.

## REFERENCES


1. Benenati, F., Keynote Address: Plants — keystone to risk assessment, in *Plants for Toxicity Assessment*, ASTM STP 1091, Wang, W., Gorsuch, J. W., and Lower, W. R., American Society for Testing and Materials, Philadelphia, 5, 1991.
2. Leischman, A. A., Greene, J. C., and Miller, W. E., Bibliography of literature pertaining to the genus *Selenastrum*, U.S. EPA, Corvallis, OR, EPA-600/9–79–021, 1979.
3. U.S. Environmental Protection Agency, Algal Assay Procedure Bottle Test, National Eutrophication Research Program, Pacific Northwest Environmental Research Laboratory, Corvallis, OR, 1971.
4. U.S. Environmental Protection Agency, The *Selenastrum capricornutum* Printz Algal Assay Bottle Test: Experimental Design, Application, and Data Interpretation Protocol, U.S. EPA-600/9–78–018, Corvallis, OR, 126, 1978.
5. Payne, A. G. and Hall, R. H., A method for measuring algal toxicity and its application to the safety assessment of new chemicals, in *Aquatic Toxicology*, STP 667, Marking, L. L., and Kimerle, R. A., Eds., American Society for Testing and Materials, Philadelphia, 171, 1979.
6. Blaise, C., Legault, R., Bermingham, N., van Coillie, R., and Vasseur, P., A simple microplate algal assay technique for aquatic toxicity assessment, *Toxicity Assessment*, 1, 261, 1986.
7. Hall, J., Healey, F. P., and Robinson, G. G. C., The interaction of chronic copper toxicity with nutrient limitation in chemostat cultures of *Chlorella*, *Aquat. Toxicol.*, 14, 15, 1989.
8. Peterson, H. G., Toxicity testing using a chemostat-grown green alga, *Selenastrum capricornutum,* in *Plants for Toxicity Assessment: 2nd Volume*, ASTM STP 1115, Gorsuch, J. W., Lower, W. R., Wang, W., and Lewis, M. A., Eds., American Society for Testing and Materials, Philadelphia, 107, 1991.
9. Gala, W. R. and Giesy, J. P., Flow cytometric techniques to assess toxicity to algae, in *Aquatic Toxicology and Risk Assessment:* Vol. 13, ASTM STP 1096, W. Landis and W.H. Vander Schalie, Eds, American Society for Testing and Materials, Philadelphia, 237, 1990.
10. Bozeman, J., Koopman, B., and Bitton, G., Toxicity testing using immobilized algae, *Aquat. Toxicol.*, 14, 345, 1989.
11. Sirois, D. L., Evaluation of protocols for the assessment of phytotoxicity, in *Plants for Toxicity Assessment*, ASTM STP 1091, Wang, W., Gorsuch, J. W., and Lower, W. R., Eds., American Society for Testing and Materials, Philadelphia, 225, 1990.
12. Nyholm, N. and Källqvist, T., Methods for growth inhibition toxicity tests with freshwater algae, *Environ. Toxicol. Chem.,* 8, 689, 1989.
13. Droop, M. R., *Methods in Microbiology* 3B, Norris, J. R., and Ribbons, W. D., Eds., Academic Press, New York, 269, 1969.
14. Stein, J. R., Ed., *Handbook of Phycological Methods*, Cambridge University Press, Cambridge, England, 1973.
15. Fogg, G. E., *Algal Cultures and Phytoplankton Ecology*, 2, University of Wisconsin Press, Madison, 1975.
16. Adams, N. and Dobbs, A. J., A comparison of results from two test methods for assessing the toxicity of aminotriazole to *Selenastrum capricornutum, Chemosphere,* 13, 965, 1984.

17. Millington, L. A., Goulding, K. H., and Adams, N., The influence of growth medium composition on the toxicity of chemicals to algae, *Water Res.,* 22, 1593, 1988.
18. Eloranta, V., Effect of the slimecide Fennosan F 50 on algal growth in different test media, *Paperi ja Puu,* 64, 129, 1982.
19. Smith, P. D., Brockway, D. L., and Stancil, F. E., Effects of hardness, alkalinity and pH on the toxicity of pentachlorophenol to *Selenastrum capricornutum* (Printz), *Environ. Toxicol. Chem.,* 6, 891, 1987.
20. Blaise, C., Microbiotests in aquatic ecotoxicology: Characteristics, utility and prospects, *Environ. Toxicol. Water Qual.*, 6, 145, 1991.
21. Green, J., Peterson, S. A., Parrish, L. and Nimmo, D., Zinc Sensitivity of *Selenastrum Capricornutum* in Algal Assay Medium with Various EDTA Concentrations, Canadian Technical Report of Fisheries and Aquatic Sciences, 1991.
22. Riseng, C. M., Gensemer, R. W. and Kilham, S. S., The effect of pH, aluminum and chelator manipulations on the growth of acidic and circumneutral species of *Asterionella, Water Air Soil Pollut.*, 3, 231, 1991.
23. Bringmann, G. and Kuhn, R., Comparison of toxicity thresholds of water pollutants to bacteria, algae, and protozoa in the cell multiplication test, *Water Res.*, 14, 23, 1980.
24. Blanck, H., Wallin, G., and Wangberg, S., Species-dependent variation in algal sensitivity to compounds, *Ecotoxicol. Environ. Saf.*, 8, 339, 1984.
25. Wangberg, S. and Blanck, H., Multivariate patterns of algal sensitivity to chemicals in relation to phylogeny, *Ecotoxicol. Environ. Saf.*, 16, 72, 1988.
26. Swanson, S. M., Rickard, C. P., Freemark, K. E., and MacQuarrie, P., Testing for pesticide toxicity to aquatic plants: Recommendations for test species, in *Plants for Toxicity Assessment*, ASTM STP 1115, Gorsuch, J. W. and Lower, W. R., Eds., American Society for Testing and Materials, Philadelphia, 1991.
27. Millie, D. F. and Hersh, C. M., Statistical characterizations of the atrazine induced photosynthetic inhibition of *Cyclotella menghiniana* (Bacillariophyta), *Aquat. Toxicol.*, 10, 239, 1987.
28. Riedel, G. F., Interspecific and geographical variation of the chromium sensitivity of algae, in *Aquatic Toxicology and Environmental Fate: Eleventh Volume*, ASTM STP 1007, Suter, G. W., II and Lewis, M. A., Eds., American Society for Testing and Materials, Philadelphia, 537, 1989.
29. Walsh, G. E. and Alexander, S. V., A marine algal bioassay method: Results with pesticides and industrial waters, *Water Air Soil Pollut.*13, 45, 1980.
30. Harrass, M. C., Kindig, A. C. and Taub, F. B., Responses of blue-green and green algae to streptomycin in unialgal and paired culture, *Aquat. Toxicol.* 6, 1, 1985.
31. Outridge, P. M., Comparing Cd toxicity tests with plants in monocultures and species mixtures, *Bull. Environ. Contam. Toxicol.,* 48, 344, 1992.
32. Erikson, C., Harrass, M. C., Osborne, C. M., Sayre, P. G. and Zeeman, M., Environmental Technical Assessment Assistance Handbook, Food and Drug Administration PB 87–175345, NTIS, Springfield, VA, 1987.
33. Jones, T. W. and Winchell, L., Uptake and photosynthetic inhibition by atrazine and its degradation products on four species of submerged vascular plants, *J. Environ. Qual.*, 13, 243, 1984.
34. Jones, T. W., Kemp, W. M., Estes, P. S. and Stevenson, J. C., Atrazine uptake, photosynthetic inhibition and short-term recovery for the submersed vascular plant, *Potamogeton perfoliatus L., Arch. Environ. Contam. Toxicol.*, 15, 277, 1986.
35. Larsen, D. P., de Noyelles, F., Stay, F., and Shiroyama, T., Comparisons of single species, microcosm and experimental pond responses to atrazine exposure, *Environ. Toxicol. Chem.,* 5, 179, 1986.
36. Turbak, S. C., Olson, S. B. and McFeters, G. A., Comparison of algal systems for detecting waterborne herbicides and metals, *Water Res.,* 20, 91, 1986.
37. Lewis, M. A. and Hamm, B. G., Environmental modification of the photosynthetic response of lake plankton to surfactants and significance to a laboratory-field comparison, *Water Res.,* 12, 1575, 1986.
38. Spencer, D. F. and Greene, R. W., Effects of nickel on seven species of freshwater algae, *Environ. Pollut.* (Series A), 25, 241, 1981.
39. Ludyanskiy, M. L. and Pasichny, A. P., A system for water toxicity estimation, *Water Res.*, 5, 689, 1992.
40. Delistraty, D., Growth and photosynthetic response of a freshwater alga, *Selenastrum capricornutum* to an oil shale by-product water, *Bull. Environ. Contam. Toxicol.,* 36, 114, 1986.
41. Gaur, J. P. and Singh, A. K., Growth, photosynthesis and nitrogen fixation of *Anabaena doliolum* to Assam crude extract, *Bull. Environ. Contam. Toxicol.*, 44, 494, 1990.

42. Versteeg, D. J., Comparison of short- and long-term toxicity tests results for the green alga, *Selenastrum capricornutum*, in *Plants for Toxicity Assessment*, ASTM STP 1091, Wang, W., Gorsuch, J. W., and Lower, W. R., Eds., American Society for Testing and Materials, Philadelphia, 40, 1990.
43. St. Laurent, Blaise, C., MacQuairie, P., Scroggins, R., and Troffier, B., Comparative assessment of herbicide phytotoxicity to *Selenastrum capricornutum* using microplate and flask bioassay procedures, *Environ. Toxicol. Water Qual.*, 7, 35, 1992.
44. Gaur, J. P. and Singh, A. K., Regulatory influence of light and temperature on petroleum toxicity to *Anabaena doliolum*, *Environ. Toxicol. Water Chem.*, 6, 341, 1991.
45. Smith, P. D., Brockway, D. L., and Stancil, F. F., Effects of hardness, alkalinity and pH on the toxicity of pentachlorophenol to *Selenastrum capricornutum* (Printz), *Environ. Toxicol. Chem.*, 6, 891–900, 1987.
46. Skowronski, T., Szubinska, S., Pawlik, B., Jakubowski, M., Bilewicz, R., and Cukrowska, E., The influence of pH on cadmium toxicity to the green alga *Stichococcus bacillaris* and on the cadmium forms present in the culture medium, *Environ. Pollut.*, 74, 89, 1991.
47. American Public Health Association, American Water Works Association and Water Pollution Control Federation, Toxicity testing with phytoplankton, in *Standard Methods for the Examination of Water and Wastewater*, 20th ed., American Public Health Association, Washington, D.C., 1998.
48. American Society for Testing and Materials, Standard Guide for Conducting Static 96 h Toxicity Tests with Microalgae, E 1218–90, American Society for Testing and Materials, Philadelphia, 1990.
49. Hughes, J. S., Alexander, M. M. and Balu, K., An evaluation of appropriate expressions of toxicity in aquatic plant bioassays as demonstrated by the effects of atrazine on algae and duckweed, in *Aquatic Toxicology and Hazard Assessment*, 10, ASTM STP 971, Adams, W. J., Chapman, G. A., and Landis, W. G., Eds., American Society for Testing and Materials, Philadelphia, 531, 1988.
50. U.S. Environmental Protection Agency, Environmental Effects Test Guidelines, Federal Register, 50, 39321, 1985.
51. Walsh, G. E., Bahner, L. H. and Horning, W. B., Toxicity of textile mill effluents to freshwater and estuarine algae, crustaceans, and fishes, *Environ. Pollut.* (Series A), 21, 169, 1980.
52. Kooijman, S. A. L. M., Hanstveit, A. O., and Oldersma, H., Parametric analyses of population growth in bioassays, *Water. Res.*, 17, 527, 1983.
53. Organization for Economic Cooperation and Development (OECD), Alga growth inhibition test, Test Guideline No. 201, OECD Guidelines for Testing of Chemicals, Paris, 1984.
54. International Organization for Standardization (ISO), Water quality — Algal growth inhibition test, No. 8692, Paris, 1987.
55. U.S. Environmental Protection Agency, Short-Term Methods for Estimating the Chronic Toxicity of Effluents and Receiving Waters to Freshwater Organisms, U.S. Environmental Protection Agency 600/4–89/001, Environmental Monitoring Systems Laboratory, Cincinnati, 1989.
56. Bruce, R. D. and Versteeg, D. J., A statistical procedure for modeling continuous toxicity data, *Environ. Toxicol. Chem.*, 11, 1485–1494, 1992.
57. Walsh, G. E., Deans, C. H. and McLaughlin, L. L., Comparison of the $EC_{50}$'s of algal toxicity tests calculated by four methods, *Environ. Toxicol. Chem.*, 6, 767, 1987.
58. Stockner, J. G. and Antia, N. J., Phytoplankton adaptation to environmental stresses from toxicants, nutrients and pollutants - a warning, *J. Fish. Res. Board Can.*, 33, 2089, 1976.
59. Wang, W., The effect of river water on phytotoxicity of Ba, Cd, and Cr, *Environ. Pollut.* (Series B), 11, 193, 1987.
60. Aoyama, I., Okamura, H., and Yagi, M., The interaction effects of toxic chemical combinations on *Chlorella ellipsoidea*, *Toxicity Assessment*, 2, 341, 1987.
61. Claesson, A., Use of a mixed algal culture to characterize industrial wastewaters, *Ecotoxicol. Environ. Saf.*, 8, 80, 1984.
62. Plumley, F. G. and Davis, D. E., The effects of a photosynthetic inhibitor atrazine on salt marsh edaphic algae, in culture, microecosystems and in the field, *Estuaries*, 3, 271, 1980.
63. Lewis, M. A., Taylor, M. J., and Larson, R. J., Structural and functional response of natural phytoplankton and periphyton communities to a cationic surfactant with considerations on environmental fate, in *Community Toxicity Testing* STP 920, Cairns, J., Ed., American Society for Testing and Materials, Philadelphia, 241, 1986.

64. Lewis, M. A., Pittinger, C. A., Davidson, D. H., and Ritchie, C. J., *In-situ* response of natural periphyton to an anionic surfactant and an environmental safety assessment for phytotoxic effects, *Environ. Toxicol. Chem.*, 12, 1803–1812, 1993.
65. Huber, W., Zieris, F. J., Feind, D., and Neugebaur, K., Ecotoxicological evaluation of environmental chemicals by means of aquatic model, Bundesministerium für Forschung und Technologie, Research Report 03–7314–0, Bonn, 1987.
66. Macek, K., Birge, W., Mayer, F., Buikema, A., and Maki, A., Discussion session synopsis, in *Estimating the Hazard of Chemical Substances to Aquatic Life* STP 657, Cairns, J., Dickson, K. L., and Maki, A. W., Eds., American Society for Testing and Materials, Philadelphia, 1978.
67. Benenati, F. and Smith, B. M., An inter- and intra-agency survey of the use of plants for toxicity assessment, in *Plants for Toxicity Assessment, Vol. 2,* ASTM STP 1115, Gorsuch, J. W., Lower, W. R., Wang, W., and Lewis, M. A., Eds., American Society for Testing and Materials, Philadelphia, 41, 1991.
68. Giesy, J. P. and Hoke, R. A., Freshwater sediment quality criteria: Toxicity bioassessment, in *Sediments: Chemistry and Toxicity of In-Place Pollutants*, Baudo, R., Giesy, J., Muntau, H., Eds., Lewis Publishers, Chelsea, MI, 265, 1989.
69. Nabholz, J. V., Miller, P., and Zeeman, M., Environmental risk assessment of new chemicals under the Toxic Substances Control Act, in *Environmental Toxicology and Risk Assessment*, ASTM STP 1179, Landis, W. G., Hughes, J. S., and Lewis, M. A., Eds., American Society for Testing and Materials, Philadelphia, 40–53, 1993.
70. Walker, J. D. and Brink, R. H., New cost-effective computerized approaches to selecting chemicals for priority testing consideration, in *Aquatic Toxicology and Environmental Fate: Eleventh Vol.*, STP 1007, Suter, G. W. and Lewis, M. A., Eds., American Society for Testing and Materials, Philadelphia, 507–536, 1989.
71. Fletcher, J. S., Use of algae versus vascular plants to test for chemical toxicity, in *Plants for Toxicity Assessment*, ASTM STP 1091, Wang, W., Corsuch, J. W., and Lower, W. R., Eds., American Society for Testing and Materials, Philadelphia, 33, 1990.
72. Meyerhoff, R. D., Grothe, D. W., Sauter, S., and Dorulla, G., Chronic toxicity of tebuthiuron to an alga (*Selenastrum capricornutum*), a cladoceran (*Daphnia magna*) and the fathead minnow (*Pimephales promelas*), *Environ. Toxicol. Chem.,* 4, 695, 1985.
73. Blaylock, B. G., Frank, M. L., and McCarthy, J. M., Comparative toxicity of copper and acridine to fish, Daphnia and algae, *Environ. Toxicol. Chem.,* 4, 63, 1985.
74. Kuhn, R. and Pattard, M., Results of the harmful effects of water pollutants to green algae (*Scenedesmus subspicatus*) in the cell multiplication inhibition test, *Water Res.,* 24, 31, 1990.
75. Holst, R. W. and Ellwanger, T. C., Pesticide Assessment Guidelines, Subdivision J. Hazard Evaluation: Nontarget Plants, U.S. Environmental Protection Agency, Washington, D.C., EPA-54019–82–020, 1982.
76. U.S. Environmental Protection Agency, Toxic Substances Control Act Test Guidelines; Final Rules, Federal Register, 50, 797.1050, 797.1075, and 797.1060 (see also Technical Support Document), 1985.
77. European Economic Community (EEC), *Methods for the Determination of Ecotoxicity: Algal Inhibition Test,* EEC Directive, 79/831, Annex V, Part C, 1987.
78. Walsh, G. E., Algal Bioassay of Industrial and Energy Process Effluents, U.S. Environmental Protection Agency, Gulf Breeze, FL, EPA-600/D-82–141, 1982.
79. U.S. Army Corps of Engineers, A Plant Bioassay for Assessing Plant Uptake of Heavy Metals from Contaminated Freshwater Dredged Material, EEDP-04–11, U.S. Army Engineer Waterways Experiment Station, Vicksburg, MS, 1989.
80. Folsom, B. L., Jr. and Price, R. A., A plant bioassay for assessing plant uptake of contaminants from freshwater soils or dredged material, in *Plants for Toxicity Assessment: 2nd Volume*, ASTM STP 1115, Gorsuch, J. W., Lower, W. R., Wang, W., and Lewis, M. A., Eds, American Society for Testing and Materials, Philadelphia, 172, 1991.
81. Munawar, M. and Munawar, I. F., Phytoplankton bioassay for evaluating toxicity of *in situ* sediment contaminants, *Hydrobiologia,* 149, 87, 1987.
82. Ross, P., Jarry, V., and Sloterdijk, H., A rapid bioassay using the green alga *Selenastrum capricornutum* to screen for toxicity in St. Lawrence River sediment elutriates, in *Functional Testing of Aquatic Biota for Estimating Hazards of Chemicals,* ASTM STP 988, Cairns, J. and Pratt, J. R., Eds., American Society for Testing and Materials, Philadelphia, 68, 1988.

83. Burton, G. A., Stemmer, B. L., Winks, K. L., Ross, P. E., and Burnett, L. C., A multitrophic level evaluation of sediment toxicity in Waukegan and Indiana Harbors, *Environ. Toxicol. Chem.*, 8, 1, 1989.
84. Ireland, F. A., Judy, B. M., Lower, W. R., Thomas, M. W., Krause, G. F., Asfaw, A., and Sutton, W. W., Characterization of eight soil types using the *Selenastrum capricornutum* bioassay, in *Plants for Toxicity Assessment: 2nd Volume*, ASTM STP 1115, Gorsuch, J. W., Lower, W. R., Wang, W., and Lewis, M. A., Eds., American Society for Testing and Materials, Philadelphia, 217, 1991.
85. Thomas, J. M., Skalski, J. R., Cline, J. F., McShane, M. C., Miller, W. E., Simpson, J. C, Peterson, S. A., Callahan, C. A., and Greene, J. C., Characterization of chemical waste site contamination and determination of its extent using bioassays, *Environ. Toxicol. Chem.*, 5, 487, 1986.
86. Greene, J. C., Miller, W. E., Debacon, M., Long, M. A. and Bartels, C. L., Use of *Selenastrum capricornutum* to assess the toxicity potential of surface and ground water contamination toxicity caused by chromium waste, *Environ. Toxicol. Chem.*, 7, 35, 1988.
87. American Society for Testing and Materials, Standard practice for conducting static chronic 96-hour toxicity tests on hazardous chemicals wastes using the freshwater green alga, *Selenastrum capricornutum*, ASTM, Philadelphia, 1991.
88. Comber, M. H. I., Smyth, D. V., and Thompson, R. S., Assessment of the toxicity to algae of colored substances, *Bull. Environ. Contam. Toxicol.*, 55, 922–928, 1995.
89. McDonald, S. F., Hamilton, S. J., Buhl, K. J., and Heisinger, J. F., Acute toxicity of fire control chemicals to *Daphnia magna* and *Selenastrum capricornutum*, *Ecotoxicol. Environ. Saf.*, 33, 62–72, 1996.
90. El-Dib, M. A., Abou-Waly, H. F., and El-Naby, A. M. H., Impact of fuel oil on the freshwater alga *Selenastrum capricornutum*, *Bull. Environ. Contam. Toxicol.*, 59, 438–444, 1997.
91. Roseth, S., Edvardsson, T., Botten, T. M., Fuglestad, J., Fonnum, F., and Stenersen, J., Comparison of acute toxicity of precess chemicals used in oil refinery industry, tested with the diatom, *Chaetoceros gracilis*, the flagellate, *Isochrysis galbana* and the zebra fish, *Brachydanio rerio*, *Environ. Toxicol. Chem.*, 15, 1214–1217, 1996.
92. Couillard, D. and Li, J. F., Assessment of manure-application effects upon the runoff water quality by algal assays and chemical analyses, *Environ. Pollut.*, 80, 922–928, 1993.
93. Fargasova, A., Effects of five toxic metals on the algal *Scenedesmus-quadricauda, Biologia*, 48, 301–304, 1993.
94. Issa, A. A. and Adam, M. S., Influence of selenium on toxicity on some heavy metals in the green alga *Scenedesmus obliquus*, *Folia Microbiologica*, 44, 406–410. Publisher, Folia Microbiologica Inst. Microbiology, Videnska 1983, Prague, 4, 142, 20, Czech Republic, 1999.
95. Fairchild, J. F., Ruessler, D. S., Haverland, P. S., and Carlson, A. R., Comparative sensitivity of *Selenastrum capricornutum* and *Lemna minor* to sixteen herbicides, *Arch. Environ. Cont. Toxicol.*, 32, 353–357, 1997.
96. Simmons, N. K., The effects of atrazine on the competitive interactions of three green algae (herbicide, *Scenedesmus quadricauda, Chlamydomonas musicola, Chlorella vulgaris*, community), The University of Manitoba, Canada, Dissertation Abstracts International, 58, 5761, 1997.
97. Carrasco, J. M. and Sabater, C., Toxicity of atrazine and chlorsulfuron to algae, *Toxicolog, Environ. Chem.*, 59, 89, 1997.
98. Piri, M. and Oerdog, V., Herbicides and insecticide effects on green algae and cyanobacteria strain, *Iran J. Fish. Sci.*, 1, 48–58, 1999.
99. Mayer, P., Toxic cell concentrations of three polychlorinated biphenyl congeners in the green alga, *Selenastrum capricornutum*, *Environ. Toxicol. Chem.*, 17, 1848–1851, 1998.
100. Magdaleno, A. and DeRosa, E., Chemical composition and toxicity of waste dump leachates using *Selenastrum capricornutum*, Printz (*Chlorococcales, Chlorophyta*), *Environ. Toxicol.*, 15, 76–80, 2000.
101. Hall, N. E., Fairchild, J. F., LaPoint, T. W., Heine, P. R., Ruessler, D. S., and Ingersoll, C. G., Problems and recommendations in using algal toxicity testing to evaluate contaminated sediments, *J. Great Lakes Res.*, 22, 545–556, 1996.
102. Blaise, C. and Menard, L., A micro-algal solid-phase tests to assess the toxic potential of freshwater sediments, *Water Qual. Res. J. Can.*, 33, 133, 1998.
103. Gamila, H. A. and Naglaa, F. A., *Estimation of the Hazard Concentration of Industrial Wastewaters Using Algal Bioassay*, Springer-Verlag, New York, 1999.
104. Okay, O. S., Morkoc, E., and Gaines, A., Effects of two herbicidal wastewaters on *Chlorella* sp. and *Phaeodactylum tricornutum*, *Environ. Pollut.*, 84, 1–6, 1994.

105. Danilov, R. and Ekelund, N. G. A., Influence of wastewater from the paper industry and UV-B radiation on the photosynthetic efficiency of *Euglena gracilis*, *J. Appl. Phycol.,* 2, 157–163, 1999.
106. Cheung, K. C., Chu, L. M., and Wong, M. H., Toxic effect of landfill leachate on microalgae, *Water Air Soil Pollut.*, 69, 337–49, 1993.
107. Weber, C. I., Peltier, W. H., Norberg-King, T. J., Horning, W. B., Kessler, F. A., Menkedick, J. R., Neiheisel, T. W., Lewis, P. A., Klemm, D. J., Pickering, W. H., Robinson, E. L., Lazorchak, J. M., Wymer, L. J., and Freyberg, R. W., Short-Term Methods for Estimating the Chronic Toxicity of Effluents and Receiving Waters to Freshwater Organisms, U.S. Environmental Protection Agency, 600/4–89/001, Environmental Monitoring Systems Laboratory, Cincinnati, 1989.
108. Adams, N., Goulding, K. H., and Dobbs, A. J., Effect of acetone on the toxicity of four chemicals to *Selenastrum capricornutum*, *Bull. Environ. Contam. Toxicol.*, 35, 254, 1986.
109. Lukavsky, J., The evaluation of algal growth potential by cultivation on solid media, *Water Res.*, 17, 549, 1983.
110. Vasseur, P. and Pandard, P., Influence of some experimental factors on metal toxicity to *Selenastrum capricornutum*, *Toxicity Assessment,* 3, 331, 1988.
111. Fernandez-Pinas, F., Mateo, P., and Bonilla, I., Binding of cadmium by cyanobacterial growth media, free on concentration as a toxicity index to the cyanobacterium NOSTOC VAM 208, *Arch. Environ. Contam. Toxicol.,* 21, 425, 1991.
112. Huebert, D. B. and Shay, J. M., The effect of EDTA on cadmium and zinc uptake and toxicity in *Lemna trisulca* L., *Arch. Environ. Contam. Toxicol.,* 22, 313, 1992.
113. Nyholm, N., Response variable in algal growth inhibition tests-biomass or growth rate, *Water Res.,* 19, 273, 1985.
114. Stratton, G. W. and Giles, J., Importance of bioassay volume in toxicity tests using algae and aquatic invertebrates, *Bull. Environ. Contam. Toxicol.,* 44, 420, 1990.
115. Eloranta, V. and Laitinen, O., Evaluation of sample preparation for algal assays on water receiving cellulose effluents, *Verh. Inst. Verein. Limnol.,* 21, 738, 1981.
116. Larson, L. J., The influence of test length and bacteria on the results of algal bioassays with mono-phenolic acids, in *Plants for Toxicity Assessment*: *2nd Volume*, ASTM STP 1115, Gorsuch, J. W., Lower, W. R., Wang, W., and Lewis, M. A., Eds., American Society for Testing and Materials, Philadelphia, 230, 1991.
117. Thompson, P. A. and Couture, P., Short- and long-term changes in growth and biochemical composition of *Selenastrum capricornutum* populations exposed to cadmium, *Aquat. Toxicol.,* 21, 135, 1991.
118. Hughes, J. S. and Vilkas, A. G., Toxicity of N,N-dimethylformamide used as a solvent in toxicity tests with the green alga, *Selenastrum capricornutum*, *Bull. Environ. Contam. Toxicol.,* 31, 98, 1983.
119. Stratton, G. W. and Smith, T. M., Interaction of organic solvents with the green alga *Chlorella pyrenoidosa*, *Bull. Environ. Contam. Toxicol.,* 40, 736, 1988.
120. Nyberg, H., The effects of some detergents on the growth of *Nitzchia holsatica* Hust. (Diatomeve), *Ann. Bot. Fenn.,* 13, 65, 1976.
121. Vocke, R. W., Sears, K. L., O'Toole, J. J., and Wildman, R. B., Growth responses of selected freshwater algae to trace elements and scrubber ash slurry generated by coalfired power plants, *Water Res.,* 14, 141, 1980.
122. Spencer, D. F. and Greene, R. W., Effects of nickel on seven species of freshwater algae, *Environ. Pollut.* (Ser. A), 25, 241, 1981.
123. Steeman-Nielsen, E. and Willemoes, M., How to measure the illumination rate when investigating the rate of photosynthesis of unicellular algae under various light conditions, *Int. Rev. Gesamier Hydrobiol.*, 56, 541–556, 1971.
124. McCormick, P. V. and Cairns, J., Jr., Algae as indicators of environmental change, *J. Appl. Phycol.*, 6, 509–526, 1994.
125. Mosser, J. L., Fisher, N. S., and Wurster, C. F., Polychlorinated biphenyls and DDT alter species composition in mixed cultures of algae, *Science*, 176, 633, 1972.
126. Lundy, P., Wurster, C. F., and Rowland, R. F., A two-species marine algal bioassay for detecting aquatic toxicity of chemical pollutants, *Water Res.,* 18, 187, 1984.
127. Lewis, M. A., Use of freshwater plants in the environmental risk assessment process, *Environ. Pollut.,* 87, 318–336, 1995.
128. Sculthorpe, C. D., *The Biology of Aquatic Vascular Plants,* St. Martin's Press, New York, 1967.

129. Correll, D. S. and Correll, H. B., *Aquatic and Wetland Plants of Southwestern United States*, Stanford University Press, Palo Alto, 1975.
130. Hutchinson, G. E., *A Treatise on Limnology,* Volume III: Limnological Botany, Wiley, New York, 1975.
131. Sutton, D. L., Blackburn, R. D. and Barlowe, W. C., Response of aquatic plants to combinations of endothall and copper, *Weed Sci.* 19, 643, 1971.
132. Dutta, T. R., Prasad, J., and Singh, R. P., Evaluations of herbicides for submerged weeds in Chambal and Bhakra-Nangal canal systems, *Ind. J. Agric. Sci.,* 42, 70, 1972.
133. Sutton, D. L., Weldon, L. W., and Blackburn, R. D., Effect of diquat on uptake of copper in aquatic plants, *Weed Sci.,* 18, 703, 1970.
134. Stewart, K. K., Improving Technology for Chemical Control of Aquatic Weeds, Technical report A-81–2, U.S. Army Corps of Engineers, Environmental Laboratory, Waterways Experimental Station, Vicksburg, MS, 1981.
135. Klaine, S. J. and Ward, C. H., Environmental and chemical control of bud production in *Hydrilla verticillata*, *Ann. Bot.,* 53, 503, 1984.
136. Hinman, M. L. and Klaine, S. J., Uptake and translocation of selected organic pesticides by the rooted aquatic plant *Hydrilla verticillata* Royle, *Environ. Sci. Technol.,* 26, 609, 1992.
137. Hillman, W. S., The Lemnaceae or duckweeds: A review of the descriptive and experimental literature, *Botanical Rev.*, 27, 221, 1961.
138. Holst, R. W. and Ellwanger, T. C., Pesticide Assessment Guidelines, Subdivision J. Hazard Evaluation: Nontarget Plants, U.S. EPA, Washington, D.C., EPA-54019–82–020, 1982.
139. U.S. Environmental Protection Agency, *Lemna* Acute Toxicity Test, Federal Register 50, 39331, 1985.
140. Wallbridge, C. T., A Flow-Through Testing Procedure with Duckweed (*Lemna minor*), U.S. Environmental Protection Agency, Duluth, MN, EPA-600/3–77–108, 1977.
141. Davis, J. A., Comparison of Static-Replacement and Flow-Through Bioassays Using Duckweed, *Lemna gibba* G-3, U.S. EPA 560/6–81–003, Washington, D.C., 1981.
142. Cowgill, U. M. and Midazzo, D. P., The culturing and testing of two species of duckweed, *Aquatic Toxicology and Hazard Assessment: 12th Volume, ASTM STP 1027*, Cowgill, U. M. and Williams, L. R., Eds., American Society for Testing and Materials, Philadelphia, 379, 1989.
143. Hoagland, D. R. and Arnon, D. I., The water culture method for growing plants without soil, California Agricultural Experiment Station, Circular 347 (Revised) Berkeley, CA, 1950.
144. Grant, J. G., An Aquatic Bioassay Utilizing the Lemnaceae (Duckweeds) in a Static System, Master of Science thesis, Rice University, Houston, 1982.
145. Hutner, S. H., Comparative physiology of heterotrophic growth in plants, *Growth and Differentiation in Plants*, Loomis, W. E., Ed., Iowa State College Press, Ames, 1953.
146. Bishop, W. H. and Perry, R. L., The development and evaluation of a flow-through growth inhibition test with duckweed (*Lemna minor*), in *Aquatic Toxicology and Hazard Assessment: Fourth Conference, ASTM STP 737*, Branson, D. R. and Dickson, K. L., Eds., American Society for Testing and Materials, Philadelphia, 421, 1981.
147. Wang, W., Review: Literature on duckweed toxicity testing, *Environ. Res.*, 52, 7, 1990.
148. Hughes, J. S., Alexander, M. M., and Balu, K., An evaluation of appropriate expresssions of toxicity in aquatic plant bioassays as demonstrated by the effects of atrazine on algae and duckweed, in *Aquatic Toxicology and Hazard Assessment: 10th Volume, ASTM STP 971*, Adams, W. J., Chapman, G. A. and Landis, W. G., Eds., American Society for Testing and Materials, Philadelphia, 531, 1988.
149. King, J. M. and Coley, K. S., Toxicity of aqueous extracts of natural and synthetic oils to three species of *Lemna*, in *Aquatic Toxicology and Hazard Assessment: 8th Volume, ASTM STP 891*, American Society for Testing and Materials, Philadelphia, 302, 1985.
150. Forney, D. R. and Davis, D. E., Effects of low concentrations of herbicides on submersed aquatic plants, *Weed Sci.,* 29, 677, 1981.
151. Taraldsen, J. E. and Norberg-King, T. J., New method for determining effluent toxicity using duckweed (*Lemna minor*), *Environ. Toxicol. Chem.,* 9, 761, 1990.
152. Mangi, J., Schmight, K., Pankow, J., Gaines, L., and Turner, P., Effects of chromium on some aquatic plants, *Environ. Pollut.,* 16, 285, 1978.
153. Rodgers, J. H., Cherry, D. S., and Guthrie, R. K., Cycling of elements in duckweed (*Lemna perpusilla*) in an ash settling basin and swamp drainage system, *Water Res.* 12, 765, 1978.

154. Rowe, E. L., Ziobro, R. J., Wang, C. J., and Dence, C. W., The use of an alga *Chlorella pyrenoidesa* and a duckweed *Lemna perpusilla* as test organisms for toxicity bioassays of spent bleaching liquors and their compounds. *Environ. Pollut. (Series A),* 27, 289, 1982.
155. Klaine, S.J., Toxicity of coal gasifier solid waste to the aquatic plants *Selenastrum capricornutum* and *Spirodela oligorhiza, Bull. Environ. Contam. Toxicol.,* 35, 551, 1985.
156. Harding, J. P. C. and Whitton, B. A., Zinc, cadmium, and lead in water, sediments and submerged plants of the Derwent reservoir, Northern England, *Water Res.,* 12, 307, 1978.
157. Nakada, M., Fukaya, K., Takeshita, S., and Wada, Y., The accumulation of heavy metals in the submerged plant *Elodea nuttallii, Bull. Environ. Contam. Toxicol.* 22, 21, 1979.
158. Burton, M. A. S. and Peterson, P. J., Metal accumulation by aquatic bryophytes from polluted mine streams, *Environ. Pollut.,* 19, 39, 1979.
159. Behan, M. J., Kinraide, T. B. and Selser, W. I., Lead accumulation in aquatic plants from metallic sources including shot, *J. Wildl. Manage.,* 43, 240, 1979.
160. Bergamini, P. G., Palmas, G., Piantelli, F., Sani, M., Banditelli, P., Previtera, M., and Sodi, F., Study of $^{137}$Cs absorption by *Lemna minor, Health Phys.* 37, 315, 1979.
161. Kozuchowski, J. and Johnson, D. L., Gaseous emissions of mercury from an aquatic vascular plant, *Nature,* 274, 468, 1978.
162. Suckcharven, S., Mercury contamination of terrestrial vegetation near a caustic soda factory in Thailand, *Bull. Environ. Contam. Toxicol.,* 24, 463, 1980.
163. Brown, B. T. and Rattigan, B. M., Toxicity of soluble copper and other metal ions to *Elodea canadensis, Environ. Pollut.*, 20, 303, 1979.
164. Hinman, M. L., Utility of Rooted Aquatic Vascular Plants for Aquatic Sediment Hazard Evaluation, Doctoral dissertation, Memphis State University, Memphis, TN, 1989.
165. Clearley, J. E. and Coleman, R. L., Cadmium toxicity and accumulation in southern naiad. *Bull. Environ. Contam. Toxicol.,* 9, 100, 1973.
166. Stanley, R. A., Toxicity of heavy metals and salts to eurasian watermilfoil (*Myriophyllum spicatum* L.), *Arch. Environ. Contam. Toxicol.* 2, 331, 1974.
167. Czuba, M. and Mortimer, D.C., Stability of methylmercury and inorganic mercury in aquatic plants (*Elodea densa*), *Can. J. Bot.,* 58, 316, 1980.
168. Mortimer, D. C. and Kudo, A., Interaction between aquatic plants and bed sediments in mercury uptake from flowing water, *J. Environ. Qual.,* 4, 491, 1975.
169. Correll, D.L. and Wu, T.L., Atrazine toxicity to submersed vascular plants in simulated estuarine microcosms, *Aquat. Bot.,* 12, 151, 1982.
170. Watkins, C. H. and Hammerschlag, R. S., The toxicity of chlorine to a common vascular aquatic plant, *Wat. Res.,* 18, 1037, 1984.
171. Mitchell, C., Growth of *Halodule wrightii* in culture and the effects of cropping, light, salinity and atrazine, *Aquat. Bot.,* 28, 25, 1987.
172. Byl, T. D. and Klaine, S. J., Peroxidase activity as an indicator of chemical stress in *Hydrilla verticillata* Royle, in *Plants for Toxicity Assessment, Vol. 2 STP 1115,* Gorsuch, J. W., Lower, W. R., Lewis, M. A., and Wang, W., Eds., ASTM Publication, Philadelphia, 101, 1990.
173. Garg, P. and Chandra, P., Toxicity and accumulation of chromium in *Cerataphyllum demersum* L., *Bull. Environ. Contam. Toxicol.,* 44, 473, 1990.
174. Byl, T. D., Sutton, H. D. and Klaine, S. J., Evaluation of peroxidase as a biochemical indicator of toxic chemical exposure in the aquatic plant *Hydrilla verticillata* Royle, *Environ. Toxicol. Chem.,* 13, 5509, 1994.
175. Byl, T. D., The Influence of Sediment Physicochemical Parameters on Metal Bioavailablity and the Peroxidase Response in the Aquatic Macrophyte *Hydrilla verticillata* Royle, Ph.D. dissertation, Memphis State University, Memphis, TN, 1992.
176. Byl, T. D., Warren, J. E., Bailey, F. C., and Klaine, S. J., Oxidation of the rhizosphere by aquatic plant roots: Does acid volatile sulfide adequately predict metal availability?, *Proc. Soc. Environ. Toxicol. Chem.*, 13, 208, 1992.
177. Sutton, H. D., Contaminant-Induced Peroxidase Response in Submerged and Wetland Plants, Ph.D. dissertation, Clemson University, Clemson, SC.
178. Richardson, C. J., Di Giulio, R. T., and Tandy, N. E., Free-radical mediated processes as markers of air pollution stress in trees, in *Biologic Markers of Air-Pollution Stress and Damage in Forests,* National Research Council Commission on Life Sciences, National Academy Press, Washington, D.C., 251, 1989.

179. American Society for Testing and Materials (ASTM), *Standard Guide for Conducting Renewal Phytotoxicity Tests with Freshwater Emergent Macrophytes*, E 1841–96, ASTM, West Conshohocken, PA, 1997.
180. American Society for Testing and Materials (ASTM), *Standard Guide for Conducting Static, Axenic, 14-Day Phytotoxicity Tests in Test Tubes with the Submersed Aquatic Macrophyte, Myriophyllum sibiricum,* Komarov, E 1913–97, ASTM, West Conshohocken, PA, 1998.
181. Royce, C., Fletcher, J., Risser, P., McFarlane, J., and Benenati, F., PHYTOTOX: A database dealing with the effects of organic chemicals on terrestrial vascular plants, *J. Chem. Inf. Comput. Sci.,* 24, 7, 1984.
182. Fletcher, J. S., Muhitch, M. J., Vann, D. R., McFarlane, J. C., and Benenati, F. E., PHYTOTOX database evaluation of surrogate plant species recommended by the US Environmental Protection Agency and the Organization for Economic Cooperation and Development, *Environ. Toxicol. Chem.,* 4, 523, 1985.
183. Fletcher, J. S., Johnson, F. L., and McFarlane, J. C., Database assessment of phytotoxicity data published on terrestrial vascular plants, *Environ. Toxicol. Chem.,* 7, 615, 1988.
184. Smith, B. N., An inter- and intra-agency survey of the use of plants for toxicity assessment, in *Plants for Toxicity Assessment*, Gorsuch, J. W. et al., Eds., American Society for Testing and Materials, Philadelphia, 1991, 41.
185. Camper, N. D., Herbicide studies with plant tissue and cell cultures, in *Research Methods in Weed Science*, Camper, N. D., Ed., Southern Weed Science Society, Champaign, IL, 1986, 385.
186. Lavy, T. L. and Sentelmann, P. W., Herbicide bioassay as a research tool, in *Research Methods in Weed Science*, Camper, N. D., Ed., Southern Weed Science Society, Champaign, IL, 1986, 201.
187. Fairchild, J. F., Ruessler, D. S., and Carlson, A. R., Comparative sensitivity of five species of macrophytes and six species of algae to atrazine, metribuzin, alachlor, and metolachlor, *Environ. Toxicol. Chem.,* 17, 1830, 1998.
188. Lewis, M. A., Periphyton photosynthesis as an indicator of effluent toxicity: Relationship to effects on animal test species, *Aquat. Toxicol.,* 23, 279, 1992.
189. OPPTS, Ecological Effects Test Guidelines: Special Considerations for Conduction Aquatic Laboratory Studies, Office of Prevention, Pesticides and Toxic Substances (OPPTS), U.S. Environmental Protection Agency (EPA), EPA 712-C-96–113, 1996.
190. OPPTS, Ecological Effects Test Guidelines: Background — Nontarget Plant Testing, Office of Prevention, Pesticides and Toxic Substances (OPPTS), U.S. Environmental Protection Agency (EPA), EPA 712-C-96–151, 1996.
191. OPPTS, Ecological Effects Test Guidelines: Target Area Phytotoxicity, Office of Prevention, Pesticides and Toxic Substances (OPPTS), U.S. Environmental Protection Agency (EPA), EPA 712-C-96–152, 1996.
192. OPPTS, Ecological Effects Test Guidelines: Terrestrial Plant Toxicity, Tier 1 (seedling emergence), Office of Prevention, Pesticides and Toxic Substances (OPPTS), U.S. Environmental Protection Agency (EPA), EPA 712-C-96–153, 1996.
193. OPPTS, Ecological Effects Test Guidelines: Terrestrial Plant Toxicity, Tier 1 (vegetative vigor), Office of Prevention, Pesticides and Toxic Substances (OPPTS), U.S. Environmental Protection Agency (EPA), EPA-712-C-96–163, 1996.
194. OPPTS, Ecological Effects Test Guidelines: Seed Germination/Root Elongation Test, Office of Prevention, Pesticides and Toxic Substances (OPPTS), U.S. Environmental Protection Agency (EPA), EPA-712-C-96–154, 1996.
195. OPPTS, Ecological Effects Test Guidelines: Seedling Emergence, Tier II, Office of Prevention, Pesticides and Toxic Substances (OPPTS), U.S. Environmental Protection Agency (EPA), EPA-712-C-96–363, 1996.
196. OPPTS, Ecological Effects Test Guidelines: Early Seedling Growth Toxicity Test, Office of Prevention, Pesticides and Toxic Substances (OPPTS), U.S. Environmental Protection Agency (EPA), EPA-712-C-96–347, 1996.
197. OPPTS, Ecological Effects Test Guidelines: Vegetative Vigor, Tier II, Office of Prevention, Pesticides and Toxic Substances (OPPTS), U.S. Environmental Protection Agency (EPA), EPA-712-C-96–364, 1996.
198. OPPTS, Ecological Effects Test Guidelines: Terrestrial Plants Field Study, Tier III, Office of Prevention, Pesticides and Toxic Substances (OPPTS), U.S. Environmental Protection Agency (EPA), EPA-712-C-96–155, 1996.

199. OPPTS, Ecological Effects Test Guidelines: Aquatic Plant Toxicity Test Using *Lemna* spp. Tiers I and II, Office of Prevention, Pesticides and Toxic Substances (OPPTS), U.S. Environmental Protection Agency (EPA), EPA-712-C-96–156, 1996.
200. OPPTS, Ecological Effects Test Guidelines: Aquatic Plants Field Study, Tier III, Office of Prevention, Pesticides and Toxic Substances (OPPTS), U.S. Environmental Protection Agency (EPA), EPA-712-C-96–157, 1996.
201. OPPTS, Ecological Effects Test Guidelines: Rhizobium-legume Toxicity, Office of Prevention, Pesticides and Toxic Substances (OPPTS), U.S. Environmental Protection Agency (EPA), EPA-712-C-96–158, 1996.
202. OPPTS, Ecological Effects Test Guidelines: Plant Uptake and Translocation Test, Office of Prevention, Pesticides and Toxic Substances (OPPTS), U.S. Environmental Protection Agency (EPA), EPA-712-C-96–159, 1996.
203. OPPTS, Ecological Effects Test Guidelines: Algal Toxicity, Tiers I and II, Office of Prevention, Pesticides and Toxic Substances (OPPTS), U.S. Environmental Protection Agency (EPA), EPA-712-C-96–164, 1996.
204. Leonard, O. A., Weaver, R. J., and Kay, B. L., Bioassay method for determining 2,4-D in plant tissues, *Weeds,* 10, 20, 1962.
205. Muzik, T. J. and Whitworth, J. W., Growth-regulating chemicals persist in plants: Qualitative bioassay, *Science,* 140, 1212, 1963.
206. Leasure, J. K., Bioassay methods for 4-amino-3,5,6-trichloropicolinic acid, *Weeds,* 12, 232, 1964.
207. Swanson, C. P., A simple bioassay method for the determination of low concentrations of 2,4-D in aqueous solutions, *Bot. Gaz.,* 107, 507, 1946.
208. Ready, D. and Grant, V. Q., A rapid bioassay for simultaneous identification and quantitation of picloram in aqueous solution, *Bot. Gaz.,* 109, 39, 1947.
209. American Society for Testing and Materials (ASTM), Standard Guide for Conducting Terrestrial Plant Toxicity Tests, E 1963–98, ASTM, West Conshohocken, PA, 1999.
210. Lytle, J.S. and Lytle, T. F., Use of plants for toxicity assessment of estuarine ecosystems, *Environ. Toxicol. Chem.,* 20, 68, 2001.
211. Frans, R., Talbert, R., Marx, D., and Crowley, H., Experimental design and techniques for measuring and analyzing plant responses to weed control practices, in *Research Methods in Weed Science*, Camper, N. D., Ed., Southern Weed Science Society, Champaign, IL, 1986, 29.
212. Hoagland, D. R. and Arnon, D. I., The Water Culture Method for Growing Plants without Soil, University of California, College of Agriculture, Agricultural Experiment Station, Circular 347, 1938.
213. Burnet, M. W. M., Loveys, B. R., Holtum, J. A. M., and Powles, S. B., Increased detoxification is a mechanism of simazine resistance in *Lolium rigidum, Pest. Biochem. Physiol.,* 46, 207, 1993.
214. Etzion, O. and Neumann, P. M., Screening for phyto-toxic contaminants in an industrial source of nutrients intended for foliar and root fertilization, *J. Plant Nut.,* 16, 1385, 1993.
215. Wilson, P. C., Whitwell, T., and Klaine, S. J., Phytotoxicity, uptake, and distribution of [14C]simazine in *Canna hybrida* "Yellow King Humbert," *Environ. Toxicol. Chem.,* 18, 1462, 1999.
216. Cseplo, A. and Medgyesy, P., Characteristic symptoms of photosynthesis inhibition by herbicides are expressed in photomixotrophic tissue cultures of *Nicotiana, Planta,* 168, 24, 1986.
217. Truelove, B. and Davis, D. E., The measurement of photosynthesis and respiration using whole plants or plant organs, in *Research Methods in Weed Science*, Camper, N. D., Ed., Southern Weed Science Society, Champaign, IL, 1986, 325.
218. Tittle, F. L., Goudey, J. S., and Spencer, M. S., Effect of 2,4-dichlorophenoxyacetic acid on endogenous cyanide, beta-cyanoalanine synthase activity, and ethylene evolution in seedlings of soybean and barley, *Plant Physiol.,* 94, 1143, 1990.
219. Irzyk, G. P. and Fuerst, E. P., Purification and characterization of a glutathione S-transferase from Benoxacor-treated maize (*Zea mays*), *Plant Physiol.,* 102, 803, 1993.
220. Grossmann, K., Quinclorac belongs to a new class of highly selective auxin herbicides, *Weed Sci.,* 46, 707, 1998.
221. Lidon, F. C., Barreiro, M. G., Ramalho, J. C., and Lauriano, J. A., Effects of aluminum toxicity on nutrient accumulation in maize shoots: Implications on photosynthesis, *J. Plant Nut.,* 22, 397, 1999.
222. Verdoni, N., Mench, M., Cassagne, C., and Bessoule, J. J., Fatty acid composition of tomato leaves as biomarkers of metal-contaminated soil, *Environ. Toxicol. Chem.,* 20, 382, 2001.

223. Krause, G. H. and Engelbert, W., Chlorophyll fluorescence as a tool in plant physiology. II. Interpretation of fluorescence signals, *Photo. Res.*, 5, 139, 1984.
224. Karukstis, K.K., Chlorophyll fluorescence as a physiological probe of the photosynthetic apparatus, in *Chlorophylls*, Scheer, H., Ed., CRC Press, Boca Raton, FL, 1991, 769.
225. Maxwell, K. and Johnson, G. N., Chlorophyll fluorescence — A practical guide, *J. Exp. Bot.*, 51, 659, 2000.
226. Jenkins, C. D., Furbank, R. T., and Hatch, M. D., Mechanism of C4 photosynthesis: A model describing the inorganic carbon pool in bundle sheath cells, *Plant Physiol.*, 91, 1372, 1989.
227. Yang, S. F., Metabolism of 1-aminocyclopropane-1-carboxylic acid in relation to ethylene biosynthesis, in *Recent Advances in Phytochemistry: Plant Nitrogen Metabolism*, Poulton, J. E., Romeo, J. T., and Conn, E. E., Eds., Plenum Press, New York, 1989, 263.
228. Kende, H., Ethylene biosynthesis, *Annu. Rev. Plant Physiol. Plant Mol. Biol.*, 44, 283, 1993.
229. Cobbett, C. S., Phytochelatins and their roles in heavy metal detoxification, *Plant Physiol.*, 123, 825, 2000.
230. Mehra, R. K. and Tripathi, R. D., Phytochelatins and metal tolerance, in *Environmental Pollution and Plant Responses*, S.B. Agrawal and M. Agrawal, Eds., CRC Press, Boca Raton, FL, 2000, 367.
231. Cummins, I., Cole, D. J., and Edwards, R., Purification of multiple glutathione transferases involved in herbicide detoxification from wheat (*Triticum aestivum* L.) treated with the safener fenchlorazole-ethyl, *Pest. Biochem. Physiol.*, 59, 35, 1997.
232. Neuefeind, T., Reinemer, P., and Bieseler, B., Plant glutathione S-transferases and herbicide detoxification, *Biol. Chem.*, 378, 199, 1997.
233. Pflugmacher, S., Schroder, P., and Sandermann, H., Taxonomic distribution of plant glutathione S-transferases acting on xenobiotics, *Phytochem.*, 54, 267, 2000.
234. McNally, S. F., Hirel, B., Gadal, P., Mann, F. A., and Stewart, G. R., Glutamine synthetase of higher plants. Evidence for a specific isoform content related to their possible physiological role and their compartmentation within the leaf, *Plant Physiol.*, 72, 22, 1983.
235. Wilcut, J. W., Wehtje, G. R., Patterson, M. G., Cole, T. A., and Hicks, T. V., Absorption, translocation, and metabolism of foliar-applied chlorimuron in soybeans (*Glycine max*), peanuts (*Arachis hypogaea*), and selected weeds, *Weed Sci.*, 37, 175, 1989.
236. De Prado, R., Romera, E., and Menendez, J., Atrazine detoxification in *Panicum dichotomiflorum* and target site *Polygonum lapathifolium*, *Pest. Biochem. Physiol.*, 52, 1, 1995.
237. Hill, J. M., The changes with age in the distribution of copper and some copper containing oxidases in red clover (*Trifolium pratense* L cv. Dorset Marlgrass), *J. Exp. Bot.*, 45, 621, 1974.
238. Droppa, M., Masojidek, J., Rozsa, Z., Wolak, A., Horvath, L. I., Farkas, L., and Hortath, G., Characteristics of Cu deficiency-inhibition of photosynthetic electron transport in spinach chloroplasts, *Biochim. Biophys. Acta*, 891, 75, 1987.
239. Funderburk, H. H. and Davis, D. E., The metabolism of C14 chain- and ring-labeled simazine by corn and the effect of atrazine on plant respiratory systems, *Weeds*, 11, 101, 1963.
240. Meigh, D. F., Norris, K. H., Craft, C. C., and Lieberman, M., Ethylene production by tomato and apple fruits, *Nature*, 186, 902, 1960.
241. Grossmann, K. and Kwiatkowsk, J., Selective induction of ethylene and cyanide biosynthesis appears to be involved in the selectivity of the herbicide quinclorac between rice and barnyard grass, *J. Plant Physiol.*, 142, 457, 1993.
242. Mueller, L. A., Goodman, C. D., Silady, R. A., and Walbot, V., AN9, a petunia glutathione S-transferase required for anthocyanin sequestration, is a flavonoid-binding protein, *Plant Physiol.*, 123, 1561, 2000.
243. Forlani, G., Lejczak, B., and Kafarski, P., The herbicidally active compound *N*-2-(5-chloro-pyridyl)aminomethylene bisphosphonic acid acts by inhibiting both glutamine and aromatic amino acid biosynthesis, *Aust. J. Plant Physiol.*, 27, 677, 2000.
244. Forlani, G., Purification and properties of a cytosolic glutamine synthetase expressed in *Nicotiana plumbaginifolia* cultured cells, *Plant Physiol. Biochem.*, 38, 201, 2000.
245. Gronwald, J. W., Fuerst, E. P., Eberlein, C. V., and Egli, M. A., Effect of herbicide antidotes on glutathione content and glutathione S-transferase activity of sorghum shoots, *Pest. Biochem. Physiol.*, 29, 66, 1987.
246. Viger, P. R., Eberlein, C. V., Fuerst, E. P., and Gronwald, J. W., Effects of CGA-154281 and temperature on metolachlor absorption and metabolism, glutathione content and glutathione-S-transferase activity in corn (*Zea mays*), *Weed Sci.*, 39, 324, 1991.

247. Fuerst, E. P., Iryk, G. P., and Miller, K. D., Partial characterization of glutathione S-transferase isozymes induced by the herbicide safener benoxacor in maize, *Plant Physiol.,* 102, 795, 1993.
248. Chen, J. and Goldsbrough, P. B., Increased activity of gamma-glutamylcysteine synthetase in tomato cells selected for cadmium tolerance, *Plant Physiol.,* 106, 233, 1994.
249. Howden, R., Goldsbrough, P. B., Andersen, C. R., and Cobbett, C. S., Cadmium-sensitive, *cad1*, mutants of *Arabidopsis thaliana* are phytochelatin deficient, *Plant Physiol.,* 107, 1059, 1995.
250. Klapheck, S., Schlunz, S., and Bergmann, L., Synthesis of phytochelatins and homo-phytochelatins in *Pisum sativum* L., *Plant Physiol.,* 107, 515, 1995.
251. Jung, K., Kaletta, K., Segner, H., and Schuurmann, G., 15N Metabolic test for the determination of phytotoxic effects of chemicals and contaminated environmental samples, *ESPR,* 6, 72, 1999.
252. Dietz, A. C. and Schnoor, J. L., Phytotoxicity of chlorinated aliphatics to hybrid poplar (*Populus destoides X nigra* DN34), *Environ. Toxicol. Chem.,* 20, 389, 2001.
253. Akinyemiju, O. A., Dickmann, D. I., and Leavitt, R. A., Distribution and metabolism of simazine in simazine-tolerant and -intolerant poplar (*Populus* sp.) clones, *Weed Sci.,* 31, 775, 1983.
254. Nair, D. R., Burken, J. G., Licht, L. A., and Schnoor, J. L., Mineralization and uptake of triazine pesticide in soil-plant systems, *J. Environ. Eng.,* 119, 842, 1993.
255. Mehta, N., Saharan, G. S., and Kathpal, T. S., Absorption and degradation of metalaxyl in mustard plant (*Brassica juncea*), *Ecotoxicol. Environ. Saf.,* 37, 119, 1997.
256. Pline, W. A., Wu, J., and Hatzios, K. K., Absorption, translocation, and metabolism of glufosinate in five weed species as influenced by ammonium sulfate and pelargonic acid, *Weed Sci.,* 47, 636, 1999.
257. Hatton, P. J., Cummins, I., Price, L. J., Cole, D. J., and Edwards, R., Glutathione transferases and herbicide detoxification in suspension-cultured cells of giant foxtail (*Setaria faberi*), *Pestic. Sci.,* 53, 209, 1998.
258. Casano, L. M., Martin, M., and Sabater, B., Sensitivity of superoxide dismutase transcript levels and activities to oxidative stress is lower in mature-senescent than in young barley leaves, *Plant Physiol.,* 106, 1033, 1994.

# CHAPTER 9

# Landscape Ecotoxicology

**Karen Holl and John Cairns, Jr.**

## CONTENTS



## 9.1 INTRODUCTION

In recent years ecologists have increasingly tended toward studying natural processes not only at the individual, community, or ecosystem level, but over the entire landscape.[1–6] One of the primary reasons for this trend is the increasing recognition that most ecological studies are too small, both spatially and temporally, to detect many important natural processes. Landscape ecology focuses on (a) landscape structure, or the spatial arrangement of ecosystems within landscapes, (b) landscape function, or the interaction among these ecosystems through flow of energy, materials, and organisms, and (c) alterations of this structure and function.[1,7] The growth of landscape ecology has resulted in increasing recognition that ecosystem processes occur within a hierarchy of different spatial and temporal scales.[8,9] Ecological systems are constrained both by the range of potential behavior of lower scales and the environmental limits of higher scales.[10] Measurements of ecological structure and function must be assessed at a scale appropriate to the process being observed.[1]

Most ecotoxicologists have studied the effects of pollutants on single species, and only recently has a considerable amount of research been undertaken on the impacts of toxicants on entire ecosystems.[3,11] However, for ecotoxicology to be a science that accurately assesses certain human impacts on natural processes, it must necessarily consider the effects of pollutants over the entire landscape since ecosystem boundaries are often not clearly defined, and movement of pollutants between ecosystems is commonplace. Since the fate of pollutants is widespread, the impacts of these pollutants must be measured over more encompassing scales.

Cairns and Niederlehner[3] define landscape ecotoxicology as examining "chemicals dispersed over a large spatial scale and their potential for adverse effects on biological systems, including humans." The term *landscape* has been variably used to refer to areas ranging from a few square kilometers to entire continents.[3] A landscape is usually composed of heterogeneous ecosystems; however, in certain regions a single type of ecosystem may dominate a sufficiently large area to cause the ecosystem/landscape distinction to be blurred. The examples cited in this chapter cover the range of definitions of *landscape* used in the literature.

## 9.2 THE NEED FOR A LANDSCAPE VIEW IN ECOTOXICOLOGY

Stated simply, pollutants impact natural systems over broad areas. Chemicals can cause ecological stress throughout large, complex landscapes in a variety of ways. First, many chemical stresses, such as air pollution and acid rain, are either emitted or spread soon after release over immense areas. For example, Hirao and Patterson[12] note that damage from car exhaust in coastal California cities is visible in the Sierra Nevada Mountains located in the eastern part of the state, and Freedman and Hutchinson[13] found that the majority of sulfur dioxide, nickel, copper, and iron from a heavy-metal smelter was carried beyond the 60-km radius of their study area. In a number of cases, toxicants that appear to have adverse effects on a relatively small area may impact areas far from the release through chemical transformation and transport. Volatilization and atmospheric transport of hexachlorocyclohexanes, a common type of pesticide, have resulted in the contamination of surface water throughout the Pacific Ocean.[14]

Second, what appear to be small-scale environmental impacts can become a landscape stress when similar or interactive events occur over a large area.[15,16] Odum[17] describes this phenomenon of cumulative or interactive events, which are individually minor, becoming collectively significant as "the tyranny of small decisions." Between 1973 and 1976, for example, 274,000 hectares of new farmland were brought into use in Georgia because of favorable crop prices.[18] Each hectare of farmland caused runoff of an estimated 9.8 metric tons of sediment, 8.4 kg of nitrogen, and 1.5 kg of phosphorus. Although the impact of any given farm on local rivers is probably not significant, the sum over the entire state is staggering. The disastrous state of the Chesapeake Bay affords another example of the cumulative impacts of both urban and agricultural runoff.[19]

Third, although a toxicant may only directly affect one ecosystem, it may indirectly affect another, resulting in a landscape-scale effect. There are a number of possible ways in which such a process can occur.[20] Organisms that are capable of concentrating heavy metals may move from a contaminated to an uncontaminated ecosystem, thereby introducing the contaminant into the food web. Alternatively, decreased primary production in a stressed ecosystem could cause migration of mobile organisms to healthier systems, resulting in increased competition for resources.

Many chemical stresses impact natural systems over large areas, necessitating the evaluation of response to these stresses within the framework of the natural landscape. Natural landforms affect the spread of disturbance through the landscape.[21] Slope, gradient, elevation, and aspect influence the movement of pollutants, materials, and energy across sites. Physical and climatic factors at a range of spatial scales affect the impact of pollutants. For example, in an extensive study of acidification of lakes in the eastern United States, certain regional trends were observed due to sources of pollution, climate, and soil type. However, a great deal of intra-region variation occurs due to hydrology, weathering rates, and nearby land uses.[22]

In addition to considering broad spatial scales in landscape ecotoxicology, natural processes must also be observed over extended time periods to assess fully the impact of a chemical stress. The PCBs released into the Hudson River were previously believed to present a minimal environmental and health problem because they were trapped in the sediments; however, over time they have slowly been remobilized and moved to other environmental compartments where they were "unexpected."[23] Certain systems may show no-observable effects to a chemical stress initially, but if the same stress is repeated over a period of time, the system may lose its resistance.[24]

Finally, the need for a larger scale view in ecotoxicology is becoming more acute since humans are altering the climate at a global scale, which will affect the large-scale transport of toxicants. Anthropogenic increases in carbon dioxide and other greenhouse gases will likely result in a 1.5 to 5°C increase in global air temperatures near the earth's surface by 2050. This warming will, in turn, alter weather patterns, result in sea-level rise from the melting of the polar ice caps, affect microbial processes, and alter plant and animal species composition,[25,26] all of which will affect the mobility of chemicals on a large scale.

Moreover, temperature and moisture stress will affect the susceptibility of organisms to chemical stress. Changes in precipitation associated with global change are likely to affect mobility of chemicals across the landscape. For example, Steding et al.[27] found that lead from gasoline emissions and hydraulic mining in the 1980s and earlier in California was mobilized during periods of high rainfall in the 1990s. Changes in weather patterns may affect movement of air pollutants. Reid et al.[28] suggest a number of mechanisms by which increased ocean temperatures will influence the effect of chemicals on fish; these include direct effects of temperature on chemical concentrations such as $NH_3$, changes in rates of fish metabolic processes such as heavy metal sequestration, and increased susceptibility to chemical stress because of simultaneous exposure to temperature stress. It is impossible to predict specific effects of climate change on toxicants, but it is certain that their large-scale transport will be altered, necessitating regional and global monitoring.

## 9.3 OBSTACLES TO LANDSCAPE ECOTOXICOLOGY

If the fate of pollutants is so widespread, the question arises as to why the effects of these pollutants at the landscape level have only recently been considered and why the vast majority of the toxicological literature remains focused on laboratory testing of single species. The reasons for this lack of information have been discussed previously and are only briefly reviewed here.[29,30]

First, until recently there has been a paucity of large-scale data. In the field of toxicity testing, the move from the single-species level, an area that is still in need of further research, to the measurement of effects of chemicals on communities and whole ecosystems is still fairly recent.[11] At the next organizational level, the landscape, few precedents are available for indicators of toxicity. Fortunately, recent technological developments, such as Geographic Information Systems (GIS) and remote sensing (discussed later), necessary to measure landscape variables, have emerged at a rapid rate. Unfortunately, other obstacles to landscape ecotoxicology are not as easily overcome.

Second, the effects of pollutants can be detected at different organizational scales.[11,31] It is difficult to distinguish the level or levels at which the chemical stress is acting or how the effects at different levels may be interacting, as effects at one level will be propagated to lower and higher levels.[32] For example, the decline in a certain species could be caused by (a) a substance reaching levels toxic to that species, (b) the decline of a prey species affected by the toxicant, (c) the invasion of a predator species from another ecosystem where the prey population has declined due to a toxicant, (d) a combination of the above, or (e) any number of other reasons. As one tries to determine the effect of a chemical stress on increasing levels of biological organization, a concomitant decrease occurs in the precision of data that is possible[11,15] if the attributes selected are inappropriate for that level of organization.[31]

Third, the majority of ecosystems on the planet are sufficiently disturbed that it is difficult to determine which specific toxicant or interaction of toxicants is causing a given ecosystem response and the source of that toxicant.[30,33] An analogy is the common cold that could be caused by a number of viruses transmitted from any number of sources. As scales of organization become increasingly complex, the different chemical stresses acting and interacting on a given system increase rapidly. A study in central Illinois found that increased phosphorus levels in aquatic systems resulted not from nonpoint source agricultural runoff, as commonly assumed, but rather resulted from urban runoff.[34] It appears that extensive areas of chlorotic dwarf eastern white pine (*Pinus strobus*) were caused by an interaction between ozone and sulfur dioxide stress.[35] One reason for difficulty in identifying the source of a landscape stress is that the spatial and temporal boundaries of a chemical stress are rarely clear.[36]

A fourth obstacle to landscape ecotoxicology is that it becomes increasingly difficult on larger scales to determine direct reactions to toxicants as opposed to natural environmental stochasticity.[15,30] In other words, the question of what to use as a reference system, which is always difficult in ecotoxicology, is an enormous challenge when working on large scales. To some extent, this problem cannot be combatted; the increasing availability of long-term data sets over large areas will help to provide a comparison for natural environmental variation. Standardization of data-collection procedures is essential to make such comparisons.

Finally, traditional research techniques are difficult to use on the landscape scale. Microcosm and mesocosm models have limited application in extrapolating to ecosystems[11,37] and become nearly useless at the landscape level. Large-scale manipulations are possible at the ecosystem level and have been extremely informative,[38–40] but they are minimally used because of cost, ethics, and problems with replication. Comparisons between disturbed and undisturbed systems are always approximations because of lack of control systems; undisturbed reference systems simply do not exist.[15]

## 9.4 LANDSCAPE INDICATORS IN ECOTOXICOLOGY

As stated previously, only recently has it been possible to assess the effects of pollutants at large scales. Therefore, there is only a small, though growing, number of studies of landscape indicators in ecotoxicology. One of the main focuses of the Environmental Monitoring Assessment Program of the U.S. Environmental Protection Agency is to develop landscape indicators, and research in this area is increasing worldwide. Here we discuss the accuracy and utility of various landscape indicators of ecotoxicology within the following framework: (a) the landscape itself with its component ecosystems and their spatial relationships, (b) structural changes such as species spread or decline over entire landscapes, and (c) functional changes such as pollutant flows between systems and recovery from disturbance. We primarily focus on indicators appropriate to the larger scales and measures of the relationships between different ecosystems.

Some stress indicators used at the ecosystem level, such as primary production and species diversity or richness,[41] can be measured over the entire landscape; however, these indicators can become problematic when aggregated over large scales. For example, agricultural runoff could be decreasing production in agricultural lands while increasing production in lakes, thereby masking any landscape-level change. Therefore, such measurements must be compared to baselines appropriate to each ecosystem type. This requires using standardized monitoring procedures to allow comparisons over large spatial and temporal scales in order to be able to separate out natural environmental variation.[6,42,43] Odjsö et al.[44] used samples of guillemot (*Uria algae*) eggs collected annually at specified locations in the Baltic Sea and stored at the Environmental Specimen Bank at the Swedish Museum of Natural History to ascertain changes in PCB concentrations over three decades. This collection was only possible because of systematic sampling and careful cataloging of specimens.

## 9.5 TECHNOLOGICAL ADVANCEMENTS FACILITATING LARGE-SCALE ECOTOXICOLOGY

The ability to assess landscape-scale measurements and to scale up more localized structural and functional measurements relies heavily on increasingly sophisticated technologies. As mentioned in the previous section, rapid technological improvements present many new opportunities to assess effects of chemicals on large scales. The number of large-scale data sets available has increased exponentially in the past decade due to improvements in remotely sensed databases and the increasing availability of data on the World Wide Web (www). Enhanced computer capacities allow for the storage and analysis of such data. Moreover, GISs are becoming commonly used to link large-scale, spatially referenced data with computer simulation models to predict the effects of toxicants on ecosystems and prioritize management decisions. These technologies are critical to the development of the full suite of possible landscape ecotoxicology indicators.

An enormous amount of data that once would have been stored in paper form in one location is becoming easily accessible via the www or by request on CD-ROM. Increasing efforts are being made to provide clearinghouses for such data and to coordinate data-collection protocols across large areas.[6,42,45] The U.S. National Science Foundation has made one of its highest priorities to "increase training in the use of large-scale research and monitoring databases, and increase access to those databases."[46]

With the declassification of military satellites, improving sensors, and improved computing capacities, remotely sensed data are being increasingly used to measure environmental stress on large scales. These technologies are the subject of entire books,[47–49] and we only briefly review them here. Remote sensors can measure radiation in the visible, near-infrared, thermal-infrared, microwave, or radiowave part of the electromagnetic spectrum. They can be mounted on satellites or aircraft, which greatly affects the area measured and level of precision. For example, the two types of satellite imagery most widely used are Landsat Thematic Mapper (resolution of 30 m) and Advanced Very High Resolution Radiometer (AVHRR; resolution of 1 km), both of which cover most of the earth's surface, with five to seven spectral bands in the visible near-infrared, and thermal-infrared parts of the spectrum.[6] Hyperspectral imaging, which has become more common in the past few years, is currently usually done from aircraft and may collect up to 200 or more spectra.[49] Remotely sensed data have numerous applications in landscape ecotoxicology, ranging from measuring weather patterns to soil erosion, spread of oil pollution, and vegetation productivity.[47–49] Herut et al.[50] used airborne hyperspectral imagery to remotely sense chlorophyll-*a* (an indicator of high nutrient levels from sewage) and suspended particulate matter (which was highly correlated with some heavy metals) in Haifa Bay, Israel. The AVHRR data were used to map the ash cloud trajectory from the 1983 eruption of Mt. Etna pollution.[47]

Remotely sensed data have an enormous potential to provide detailed information on stress due to toxicants at temporal and spatial scales not possible for field-based sampling, but they cannot replace field measurements. Ground truthing of measurements is critical. And the type of remote sensing must be carefully applied to the particular question. For example, Allum and Dreisinger[51] used Landsat data to assess vegetation recovery in areas disturbed by nickel and copper mining near Sudbury, Ontario, Canada but did not find spectral signatures that could be used to distinguish nine vegetation types distinguished in the field. They could only distinguish whether an area was totally, partially, or not vegetated and then calculate drastic changes in vegetation over time. Remote sensing can distinguish areas for field surveying, but field surveys are needed to detect subtle changes in chemical concentrations or species composition.[51–53]

GISs are becoming ubiquitous tools for managing complex, spatially referenced databases. GISs can be used to overlay various environmental stresses on climatic, geographic, or biological variables in order to locate points of high risk[54] or to provide hypotheses for the cause of ecosystem deterioration. GISs can be used to better quantify the cumulative amount of nonpoint source pollution.[55,56] Another benefit of GISs is their ability to display spatial arrangements of various ecosystems, thereby allowing for the inclusion of the flow of materials or organisms between systems in large-scale models.[57] The possible uses in landscape ecotoxicology are endless. In the highly polluted "Black Triangle" area of Poland, Germany, and the Czech Republic, a GIS was used to overlay critical load maps, land-use maps, and population data in order to prioritize strategies for reducing $SO_2$ emissions.[58] Zhang et al.[59] used a GIS model to identify factors that caused pesticide leaching in an agricultural county in California to identify areas susceptible to leaching that might be removed from production. Clearly, GISs themselves do not solve the problem of identifying landscape indicators of ecotoxicology, but they are an invaluable tool in the process and are necessary for the evaluation of a number of the landscape indicators discussed.

Computer models can be used in combinations with GISs to predict effects of toxicants over large scales. Schindler et al.[60] and Minn et al.[61] combined chemical and biotic models to predict the effects of regional acidification on the aquatic ecosystems of the eastern United States and eastern Canada, respectively. Feijtel et al.[62] used a GIS and flow-path models to predict areas of high concentrations of chemicals in various European countries and then used this information to focus sampling and decide where to site treatment plants. The limitations of these models have been discussed extensively. Models should rely on reliable databases that often do not exist,[30] parameters are often simplified in order to achieve linear solutions,[30] the scale at which ecological processes occur in models is fixed,[57] and large-scale computer models can seldom be verified.[15] Despite these weaknesses, the use of models is a necessity as it is impossible to take measurements in all ecosystems across an entire landscape or to perform large-scale manipulative experiments to test the effects of all chemical stresses.[61] Although models, no matter how complex, rarely predict surprises,[33] all these tools should be applied carefully and assumptions and data quality evaluated. They are all critical to measuring any of the indicators discussed below.

## 9.6 THE LANDSCAPE COVER AND HETEROGENEITY INDICES

Changes in the amount and spatial distribution of various ecosystems within the landscape may be indicators of landscape stress or increased landscape susceptibility to stress. In the late 1980s a number of mathematical measures to quantify landscape pattern were proposed.[1,63] However, translating these measures of landscape pattern into indicators of alteration of ecosystem processes has proven to be difficult.[54,64] Examples of such indicators include the dominance index, the contagion index, and fractal dimensions.[63] The dominance index indicates the extent to which the landscape is dominated by one type of ecosystem. The contagion index measures the probability of the same type of ecosystem neighboring one another, or in other words, the clumping of ecosystems. Fractal dimensions quantify the complexity of the shapes of the different landscape components. These measures of landscape pattern distinguish differences between types of landscape such as forest and agricultural,[63] but it is debatable whether these measures can be interpreted to indicate ecosystem stress or susceptibility to disturbance.

Increased chemical stress could lead to increased homogeneity or increased heterogeneity of the landscape. The effects of some large-scale natural and human-caused disturbances, such as the Yellowstone Fires of 1988[65] and acid precipitation,[22] may have a mosaic of effects on the landscape as variations of geology and topography within the landscape influence the response to stress. Other disturbances, such as the eruption of Mount St. Helens[66] and heavy-metal smelters,[67] result in large homogenous landscapes. These examples illustrate that, at the present state of knowledge, change in landscape pattern with stress is unpredictable and depends on the ecosystem and stress type. Moreover, in many cases, intentional conversion of land cover types (e.g., clearing forest for agricultural or urbanization) is likely to override changes in land cover due to chemical stress. Therefore, measures of landscape pattern are difficult to interpret as indicators of ecotoxicological disturbances.

However, these measurements, in particular percentage cover of certain land cover types and change in percent cover, may be useful in predicting the susceptibility of ecosystems to chemical stress. Measurements must be defined and selected with specific goals in mind rather than as abstract indices. For example, measures of dominance by land uses that are chemical intensive, such as agriculture, and whether such land uses are adjacent to areas that are particularly sensitive may help to assist with identifying areas of risk and prioritizing areas to take out of agricultural production. Increase or loss of cover of wetlands and riparian buffers may be an indication of the ability of a watershed to uptake nutrients before they enter aquatic ecosystems.[68,69] Basnyat et al.[70] used remotely sensed land-use/land-form data to test the effect of land use (forest, agriculture, urban, orchards) on in-stream nitrate concentrations in Alabama. Their results showed that forest cover in the vicinity of sampling points strongly influenced nitrate concentrations, but the amount of forest in the entire watershed did not have a significant influence, suggesting an appropriate scale for management.

As discussed in the preceding section, improved remotely sensed data will facilitate calculating changes in land use cover over time. But with increasing human impacts, it will undoubtedly be difficult to distinguish between landscape change due to a variety of chemical stresses, landscape change caused by other human impacts, and landscape change resulting from natural disturbance and succession. In summary, landscape spatial structure and change will both be affected by pollutants and affect the spread of pollutants. However, the utility of various measures to quantify spatial structure and change is limited, and landscape indicators should be selected carefully because no general trends of changes in landscape pattern in response to stress are apparent.

## 9.7 STRUCTURAL INDICATORS AT THE LANDSCAPE LEVEL

Landscape-scale stresses caused by pollutants may not cause a change in landscape type but may affect individual components or functions within the system. Traditionally, ecosystem-level indicators of stress have been separated as structural (i.e., species diversity, presence of sensitive indicator species, proportion of r-strategists) or functional (i.e., community respiration, nutrient turnover rates, nutrient loss).[41] Debate over the relative sensitivity and utility of the two types of indicators has been extensive,[71] and clearly the best approach is to consider both types.

Certain ecosystem-level structural measurements, such as species diversity or length of food chains, have less utility at the landscape level because of differences between various landscape components. Comparing species diversity between a forest and a stream in a watershed is meaningless. It is more useful to look at the distribution of certain species in similar ecosystem types throughout the landscape. The decline or spread of certain species, when summed over large areas, can be indicative of large-scale ecotoxicological stress. Also, disturbance may favor certain life history traits.[72]

Numerous criteria for selecting indicator species have been offered: (a) their life histories are well known, (b) they are easily surveyed and identified, (c) their populations fluctuate little in the absence of anthropogenic stress, and (d) they are sensitive to anthropogenic changes to the environment.[73–75] However, few species fulfill all these requirements. At the landscape level, it is also useful if an indicator species utilizes more than one ecosystem type within the landscape. For example, the bald eagle has been suggested as an indicator of organochlorines throughout the Great Lakes Region.[76]

Certain taxonomic groups have been suggested as regional indicators of environmental health because of their large-scale distribution and sensitivity to stress. For example, air-pollution-indicator systems for a number of regions have been developed using lichens,[77,78] as lichens are quite sensitive to air pollution, particularly $SO_2$, that interferes with photosynthetic pigments. Birds are another taxa that have been used in regional monitoring systems[79] because of their widespread distribution; their sensitivity to various toxicants directly,[80,81] through the food chain,[82] or through habitat alteration;[83] and the large database of information that already exists. Karr and Chu[43] and Hall et al.[84] discuss a number of multimetric aquatic indicator systems using either fish or benthic macroinvertebrates to assess the health of large areas of aquatic ecosystems. These indicator systems measure not just the number of species present but also categorize species by habitat type used, trophic level, and sensitivity to chemical stresses, which allows the integration of structural and functional measures. Determining the change in distribution of indicator species is greatly facilitated by the use of GISs to compile and map databases.

Ecotoxicological stress often results in a decline of sensitive species and a concomitant increase in the distribution of resistant species through the landscape. Walker and Everett[85] studied the effect of road dust on moss species along a 577-km stretch of highway in Alaska. *Sphagnum* sp. declined from 37 to 0% within 25 m of roads, while minerotrophic species, such as *Auracomnium turgidum,* increased. Stokes et al.[86] recorded a shift to heavy-metal-tolerant algal species in lakes near smelters in Ontario, Canada. Often, human disturbance favors the spread of nonnative species,[87,88] a phenomenon that has been proposed as one of the EMAP indicators.[54] For example, *Agrostis perenaris* and *Arenaria patula*, two introduced grass species rarely seen in Pennsylvania, were common in areas with high soil zinc concentrations derived from nearby smelters.[67] Exotic trout species appear better adapted than native trout to the highly polluted Great Lakes; because of their shorter life spans, the exotic species do not accumulate toxicants to the levels of their native counterparts.[89]

It is important that the spread or decline of species is studied at the landscape scale as, ultimately, species distribution at the larger scale regulates the presence or absence of populations within a given ecosystem. The loss of habitat for a species due to a single ecosystem being perturbed can become extirpation or even extinction of this species if the stress is repeated or spread over a large area. As more and more components within a landscape are disturbed, sources of recolonizing populations become increasingly rare and the ability of an ecosystem within a landscape to recover decreases.

Although certain structural components of the landscape are sensitive to and may be useful as indicators of large-scale ecotoxicological stress, the choice of suitable indicator species and interpretation of results can be difficult. Ideally, indicator species are those that, in addition to the previously discussed characteristics, are the most sensitive to all toxicants, are widespread throughout the landscape, and are important to ecosystem function. However, fulfilling all these requirements is nearly impossible for a number of reasons that have been discussed at length elsewhere.[90]

First, a "most sensitive species" to all chemical stresses does not exist.[91] Second, species may exist in ecosystem types that are widespread throughout a landscape, but only a few species are present throughout all ecosystems in a given landscape. This presents a particular problem at terrestrial/aquatic interfaces, as there is little overlap in the types of biota. Most of the examples of indicator species and species decline or spread mentioned previously referred to species that were present in one type of ecosystem common in a given landscape, such as mosses on tundra in Alaska or lake algal species.

Finally, determining the specific cause of a structural change in an ecosystem is difficult,[92] even if one is able to select appropriate indicator species. The decline of a given species could be due to a single chemical stress acting either directly or indirectly on the organism, the interaction of a number of chemical stresses, the spread of a disease, habitat fragmentation, natural stochastic events, or the interaction of any or all of the above. It becomes even more complicated to determine the cause of decline or spread of a species that utilizes more than one type of ecosystem since either habitat could be degraded. Therefore, as with landscape indicators, specific guilds of species to be used as indicators should be chosen with a focused question in mind.

## 9.8 FUNCTIONAL LANDSCAPE INDICATORS

As mentioned previously, functional, in addition to structural, measurements are often used to evaluate stress at the ecosystem level. Ecosystem functions that are commonly measured include the ratio of production to total biomass, community respiration, and rate of nutrient cycling.[41] These measures become obsolete when aggregated across the landscape because of the immense variation between different ecosystems within the landscape; however, some, such as primary productivity, may be useful indicators if they can be compared to baseline values for specific systems.[6] Stress from many chemicals causes decreased productivity, whereas increased nutrients may cause increased productivity and eutrophication in aquatic systems. Primary productivity is relatively easily measured using remotely sensed data.

Few precedents exist of landscape functions to measure. We suggest considering the effect of pollutants on the flow of nutrients between ecosystems within a landscape,[6] the rate of spread of disturbance through the landscape, and the ability of the landscape to recover from disturbance as possible functional landscape indicators of stress. As discussed previously, most ecosystems can tolerate some structural changes before functional attributes are altered due to ecosystem redundancy. However, as stress levels increase, a point is eventually reached where nutrient regulatory mechanisms are disrupted.[1,88,92] At the ecosystem level, chemical stress often results in an increased export of nutrients and primary production.[41] On a landscape level, this sums to an increased flow of materials between ecosystems. Jackson and Watson[93] found that lead from a heavy-metal smelter disrupted microbial communities in a forest, thereby increasing leaching of nutrients from the soil into nearby aquatic systems. Acid rain generally has been observed as having a similar effect.[94,95]

Another measure of landscape integrity is the ability to contain the spread of a chemical stress through the landscape. The spatial structure of the different ecosystems within a landscape affects the spread of toxic chemicals through the landscape,[29] as ecosystems retain and detoxify chemicals to differing degrees.[96] Often, different ecosystems within the landscape may serve as sources or sinks for pollutants. For example, forests within agricultural landscapes may serve as sinks for excess nutrients that act as pollutants in aquatic ecosystems,[68,69] and wetlands often help control the spread of storm water pollution.[97] If the amount of forests or wetlands, in the previous examples, is decreased or the amount of pollutant increases, the ability of the ecosystem to resist the spread of these pollutants will be altered. A dominance index, discussed previously, could be useful in quantifying changes in the amount of different ecosystem types within a landscape.

In addition to spatial structure, the spread of a chemical stress through a landscape is dependent on the health of the ecosystem components. The emissions from an iron-sintering plant in Canada caused sufficient acidification of the precipitation and runoff to exhaust the buffering capacity of the lime deposits, which resulted in acidification of nearby lakes.[98] Acidification of peatlands due to acid rain resulted in the release of toxic metals from the peatlands into downstream waters.[99] A study by Giblin[100] found that increased oxidation in the sediments, caused by excess nutrient inputs, resulted in the leaching from wetlands of lead and iron, two metals normally retained in wetlands. In each of these cases, the spread of a chemical stress would have been prevented if the ecosystem were not unhealthy and unable to perform this function because of previous disturbances.

In addition to disrupting the ability of ecosystems within the landscape to resist the spread of disturbances, chemical stress may inhibit the ability of a landscape to recover from disturbance. For example, a study on forests near a zinc smelter showed that the forest was less able to recover from the natural disturbance of fire than similar forests not subjected to intensive air and soil pollution.[67]

As with the other indicators previously discussed, examples of chemical stress disrupting functional attributes within a landscape are numerous, but methods of quantifying such disturbances are limited, though improving with the increased availability of data. Measurement of nutrient and pollutant flows throughout a landscape are difficult to measure,[101] and it is nearly impossible to quantify ecosystem resistance or resilience until these attributes are completely exhausted. Once again, no clear way exists to determine the cause of a functional alteration.

## 9.9 CONCLUSIONS

We have discussed a number of different types of landscape indicators in ecotoxicology, while recognizing that this area of research is rapidly developing. Although chemical stress causes changes in spatial, structural, and functional components of landscapes, quantifying and interpreting these changes is a formidable challenge. With increasing research devoted to investigating landscape-scale processes and rapidly improving technology to analyze large spatial databases, understanding of the structure, function, and natural changes of landscapes is increasing at an accelerating pace. This knowledge should facilitate more accurate selection and measurement of landscape-level indicators of stress. These technologies alone will not answer all the many questions, but they help to organize existing information and prioritize areas for future research. It is important to balance a need for developing standardized measurement procedures to allow for comparison across large scales with the need to fine-tune measurements to specific, localized questions.

As has been stated throughout, numerous obstacles to developing accurate indicators of ecotoxicological stress must be acknowledged. As increasingly large scales of biological organization are considered, processes occurring at different levels and between different components of the same level must be recognized. Just as the ecosystem is more than the sum of its parts,[11] the landscape is more than the sum of its ecosystems. The interaction of all these processes is mind-boggling and challenging to interpret.

Disturbance of the planet has progressed to the extent that separating the effects of the various landscape stresses, such as chemical pollutants, habitat fragmentation, introduction of exotics, and influence of processes occurring at larger scales, is difficult. Just as some landscape processes limit, and are limited by, intraecosystem processes and vice versa, global disturbances, such as global warming, ozone depletion, and land-use change, affect landscape processes. Furthermore, no control systems exist with which to compare these stresses. Despite formidable obstacles, efforts must be made toward better understanding landscape-scale processes. We must work on larger scales, not only because of the large scale of our alterations, but also because policies to regulate chemical emission and treatment are often made at regional or national scales.

## ACKNOWLEDGMENTS

We are grateful to Darla Donald for skilled editorial assistance.

## REFERENCES

1. Forman, R. T. T. and Godron, M., *Landscape Ecology*, John Wiley and Sons, New York, 1986.
2. Gosz, J. R., Biogeochemistry research needs: Observations from the ecosystem studies program of the National Science Foundation, *Biogeochemistry*, 2, 101, 1986.
3. Cairns, J., Jr. and Niederlehner, B. R., Developing a field of landscape ecotoxicology, *Ecol. Appl.*, 6, 3, 790, 1996.
4. O'Neill, R. V. et al., Monitoring environmental quality at the landscape scale, *Bioscience*, 47, 8, 513, 1997.
5. U.S. Environmental Protection Agency, Environmental Monitoring and Assessment Program Research Strategy, October 1997, http://www.epa.gov/emap. Accessed December 27, 2000.
6. National Research Council, *Ecological Indicators for the Nation*, National Academy Press, Washington, D.C., 2000.
7. Risser, P. G., Landscape ecology: State of the art, in *Landscape Heterogeneity and Disturbance*, Turner, M. G., Ed., Springer-Verlag, New York, 1987, 1.
8. Allen, T. F. H. and Starr, T. B., *Hierarchy: Perspectives in Ecological Complexity*, University of Chicago Press, Chicago, 1982.
9. O'Neill, R. V. et al., *A Hierarchical Concept of Ecosystems*, Princeton University Press, Princeton, NJ, 1986.
10. O'Neill, R. V., Johnson, A. K., and King, A. W., A hierarchical framework for the analysis of scale, *Landscape Ecol.*, 3, 193, 1989.
11. Levin, S. A. et al., New perspectives in ecotoxicology, *Environ. Manage.*, 8, 375, 1984.

12. Hirao, Y. and Patterson, C. C., Lead aerosol pollution in the high Sierra overrides: Natural mechanisms which exclude lead from food chains, *Science*, 184, 989, 1974.
13. Freedman, B. and Hutchinson, T. C., Smelter pollution near Sudbury, Ontario, Canada and effects on forest litter decomposition, in *Effect of Acid Precipitation on Temperate Ecosystems*, Hutchinson, T. V. and Havas, M., Eds., Plenum Press, New York, 1978, 395.
14. Kurtz, D. A. and Atlas, E. L., Distribution of hexachlorocyclohexanes in the Pacific Ocean basin, air and water, 1987, in *Long Range Transport of Pesticides,* Kurtz, D. A., Ed., Lewis Publishers, Chelsea, MI, 1990, 143.
15. Hunsaker, C. T. et al., Assessing ecological risk on a regional scale, *Environ. Manage.*, 14, 325, 1990.
16. Suter, G. W., II, Endpoints for regional ecological risk assessment, *Environ. Manage.*, 14, 9, 1990.
17. Odum, W. E., Environmental degradation and the tyranny of small decisions, *Bioscience*, 32, 728, 1982.
18. White, F. C. et al., Relationship between increased crop acreage and nonpoint-source pollution: A Georgia case study, *J. Soil Water Cons.*, 36, 172, 1981.
19. Horton, T. and Eichbaum, W. M., *Turning the Tide: Saving the Chesapeake Bay*, Island Press, Washington, D.C., 1991.
20. Ewel, K. C., Responses of wetlands and neighboring ecosystems to wastewater, in *Ecological Considerations in Wetland Treatment of Municipal Wastewaters*, Godfrey, P. J., Kaynor, E. R., Pelczarski, S., and Benforado, J., Eds., Van Nostrand Reinhold Co., New York, 1985, 67.
21. Swanson, F. J. et al., Landform effects on ecosystem pattern and process, *Bioscience*, 38, 92, 1988.
22. United States Environmental Protection Agency, Characteristics of Lakes in the Eastern United States Vol. 1: Population Descriptions and Physico-Chemical Relationships, EPA/600/4–861007A, Office of Research and Development, Washington, D.C., 1986.
23. Sanders, J. E., The PCB-pollution problem of the Upper Hudson River from the perspective of the Hudson River PCB Settlement Advisory Committee, *Northeast. Environ. Sci.*, 1, 7, 1982.
24. Baker, J. M., Recovery of salt marsh vegetation from successive oil spillages, *Environ. Pollut.*, 4, 223, 1973.
25. Houghton, J. T., *Global Warming: The Complete Briefing*, Cambridge University Press, New York, 1997.
26. Houghton, R. A., Global effects of deforestation, in *Handbook of Ecotoxicology*, Hoffman, D. J., Rattner, B. A., Burton, G. A., Jr., and Cairns, J., Jr., Eds., Lewis Publishers, Boca Raton, FL, 1995, 492.
27. Steding, D. J., Dunlap, C. E., and Flegal, A. R., New isotopic evidence for chronic lead contamination in the San Francisco Bay estuary system: Implications for the persistence of past industrial lead emissions in the biosphere, *Proc. Nat. Acad. Sci.*, 97, 21, 11181, 2000.
28. Reid, S. D., McDonald, D. G., and Wood, C. M., Interactive effects of temperature and pollutant stress, in *Society for Experimental Biology Seminar Series 61: Global Warming: Implications for Freshwater and Marine Fish*, Wood, C. M. and McDonald, D. G., Eds., Cambridge University Press, Cambridge, MA, 1996, 325.
29. Kelly, J. R. and Harwell, M. A., Indicators of ecosystem response and recovery, in *Ecotoxicology: Problems and Approaches*, Levin, S. A., Harwell, M. A., Kelly, J. R., and Kimball, K. D., Eds., Springer-Verlag, New York, 1989, 9.
30. Harwell, M. A. and Harwell, C. C., Environmental decision making in the presence of uncertainty, in *Ecotoxicology: Problems and Approaches*, Levin, S. A., Harwell, M. A., Kelly, J. R., and Kimball, K. D. Eds., Springer-Verlag, New York, 1989.
31. Cairns, J., Jr., The case for testing at different levels of biological organization, in *Aquatic Toxicology and Hazard Assessment*: Sixth Symposium, STP802, Bishop, W. E., Cardwell, R. D., and Heidolph B. B., Eds., American Society for Testing and Materials, Philadelphia, 1983, 111.
32. Patten, B. C. et al., Propagation of cause in ecosystems, in *Systems Analysis and Ecosystem Ecology* Vol. IV, Patten, B. C., Ed., Academic Press, New York, 1976, 457.
33. Messer, J., EMAP indicator concepts, in Ecological Indicators for the Environmental Monitoring and Assessment Program, EPA 600/3–90/060, Hunsaker, C. T. and Carpenter, D. E., Eds., U.S. Environmental Protection Agency, Office of Research and Development, Research Triangle Park, NC, 1990, chap. 2.
34. Osborne, L. L. and Wiley, M. J., Empirical relationships between land use/cover and stream water quality in an agricultural watershed, *J. Environ. Manage.*, 26, 9, 1988.

35. Dochinger, L. S. et al., Chlorotic dwarfs of eastern white pine caused by an ozone and sulphur dioxide interaction, *Nature*, 225, 476, 1970.
36. Gosselink, J. G. and Lee, L. C., Cumulative impact assessment in bottomland hardwood forests, *Wetlands*, 9, 116, 1989.
37. Cairns, J., Jr., Are single species toxicity tests alone adequate for estimating hazard?, *Hydrobiologia*, 100, 47, 1983.
38. Bormann, F. H. and Likens, G. E., *Pattern and Process in a Forested Ecosystem*, Springer-Verlag, New York, 1979.
39. Watras, C. J. and Frost, T. M., Little Rock Lake (Wisconsin): Perspectives on an experimental ecosystem approach to seepage lake acidification, *Arch. Environ. Contam. Toxicol.*, 18, 157, 1989.
40. Schindler, D. W., Experimental perturbations of whole lakes as tests of hypotheses concerning ecosystem structure and function, *Oikos*, 57, 25, 1990.
41. Odum, E. P., Trends expected in stressed ecosystems, *Bioscience*, 35, 419, 1985.
42. Hale, S. S. et al., Managing scientific data: The EMAP approach, *Environ. Monitor. Assess.*, 51, 1–2, 429, 1998.
43. Karr, J. R. and Chu, E. W., *Restoring Life in Running Water*, Island Press, Washington, D.C., 1999.
44. Odsjö, T. et al., The Swedish Environmental Specimen Bank — Application in trend monitoring of mercury and some organohalogenated compounds, *Chemosphere*, 34, 9–10, 2059, 1997.
45. Hale, S. S. and Buffum, H. W., Designing environmental databases for statistical analyses, *Environ. Monitor. Assess.*, 64, 1, 55, 1998.
46. Thompson, J. N. et al., Frontiers in ecology, *Bioscience*, in press, 2001.
47. Barrett, E. C. and Curtis, L. F., *Introduction to Environmental Remote Sensing*, Chapman and Hall, London, 1992.
48. Cracknell, A. and Hayes, L., *Introduction to Remote Sensing*, Taylor & Francis, London, 1991.
49. Campbell, J. B., *Introduction to Remote Sensing*, Guilford Press, New York, 1996.
50. Herut, B. et al., Synoptic measurements of chlorophyll-a and suspended particulate matter in a transitional zone from polluted to clean seawater utilizing airborne remote sensing and ground measurements, Haifa Bay (SE Mediterranean), *Marine Pollut. Bull.*, 38, 9, 762, 1999.
51. Allum, J. A. E. and Dreisinger, B. R., Remote sensing of vegetation change near Inco's Sudbury mining complexes, *Int. J. Remote Sensing*, 8, 399, 1987.
52. Phinn, S. R., Stow, D. A., and Zedler, J. B., Monitoring wetland habitat restoration in southern California using airborne multispectral video data, *Restor. Ecol.*, 4, 4, 412, 1996.
53. Stehman, S. V. et al., Combining accuracy assessment of land-cover maps with environmental monitoring programs, *Environ. Monitor. Assess.*, 64, 1, 115, 2000.
54. Hunsaker, C. T. et al., Indicators relevant to multiple resource categories, in Ecological Indicators for the Environmental Monitoring and Assessment Program, EPA 600/3–90/060, Hunsaker, C. T. and Carpenter, D. E., Eds., U.S. Environmental Protection Agency, Office of Research and Development, Research Triangle Park, NC, 1990, chap. 9.
55. Halliday, S. L. and Wolfe, M. L., Assessing ground water pollution potential from nitrogen fertilizer using a geographic information system, *Water Resour. Bull.*, 27, 237, 1991.
56. Petersen, G. W. et al., Evaluation of Agricultural Nonpoint Pollution Potential in Pennsylvania Using a Geographic Information System, Environmental Resources Unit, University Park, PA, 1991.
57. Baker, W. L., A review of models of landscape change, *Landscape Ecol.*, 2, 111, 1989.
58. Lowles, I. et al., Integrated assessment models: Tools for developing emission abatement strategies for the Black Triangle region, *J. Hazard. Materials*, 61, 1–3, 229, 1998.
59. Zhang, M. et al., Pesticide occurrence in groundwater in Tulare County, California, *Environ. Monitor. Assess.*, 45, 2, 101, 1997.
60. Schindler, D. W., Kasian, S. E. M., and Hesslein, R. H., Biological impoverishment in lakes of the midwestern and northeastern United States from acid rain, *Environ. Sci. Technol.*, 23, 573, 1989.
61. Minns, C. K., et al., Assessing the potential extent of damage to inland lakes in eastern Canada due to acidic deposition. III. Predicted impacts on species richness in seven groups of aquatic biota, *Can. J. Fish. Aquat. Sci.*, 47, 821, 1990.
62. Feijtel, T. et al., Development of a geography-referenced regional exposure assessment tool for European rivers-GREAT-ER Contribution to GREAT-ER number 1, *Chemosphere*, 34, 11, 2351, 1997.
63. O'Neill, R. V. et al., Indices of landscape pattern, *Landscape Ecol.*, 1, 153, 1988.

64. Turner, M. G., Landscape ecology: The effect of pattern on process, *Ann. Rev. Ecol. Syst.*, 20, 171, 1989.
65. Christensen, N. L. et al., Interpreting the Yellowstone fires of 1988, *Bioscience*, 39, 678, 1989.
66. Dale, V. H., Revegetation of Mount St. Helens debris avalanche 10 years posteruptive, *Nat. Geogr. Res. Explor.*, 7, 328, 1991.
67. Jordan, M. J., Effects of zinc smelter emissions and fire on a chestnut-oak woodland, *Ecology*, 56, 78, 1975.
68. Peterjohn, W. T. and Correll, D. L., Nutrient dynamics in an agricultural watershed: Observations on the role of a riparian forest, *Ecology*, 65, 1466, 1984.
69. Lowrance, R. et al., Riparian forests as nutrient filters in agricultural watersheds, *Bioscience*, 34, 374, 1984.
70. Basnyat, P. et al., The use of remote sensing and GIS in watershed level analyses of non-point source pollution problems, *Forest Ecol. Manage.*, 128, 1–2, 65, 2000.
71. Cairns, J., Jr. and Pratt, J. R., Developing a sampling strategy, in *Rationale for Sampling and Interpretation of Ecological Data in the Assessment of Freshwater Ecosystems*, STP894, American Society for Testing and Materials, Philadelphia, 1986, 168.
72. Novotny, V., Effect of habitat persistence on the relationship between geographic distribution and local abundance, *Oikos*, 61, 431, 1991.
73. Ryder, P. A. and Edwards, C. J., Eds., *A Conceptual Approach for the Application of Biological Indicators of Ecosystem Quality in the Great Lakes Basin*, International Joint Commission and the Great Lakes Fisheries Commission, Windsor, Ontario, 1985.
74. Noss, R. F., Indicators for monitoring biodiversity: A hierarchical approach, *Conserv. Biol.*, 4, 355, 1990.
75. McGeoch, M. A., The selection, testing and application of terrestrial insects as bioindicators, *Biol. Rev.*, 73, 2, 181, 1998.
76. Strachan, W. M. J. and Henry, M. G., Eds., *Final Report of the Ecosystem Objectives Committee*, International Joint Commission and the Great Lakes Fisheries Commission, Windsor, Ontario, 1990.
77. LeBlanc, F. and DeSlover, J., Relation between industrialization and the distribution and growth of epiphytic lichens and mosses in Montreal, *Can. J. Bot.*, 48, 1485, 1970.
78. Hawsksworth, D. L., Lichen as litmus for air pollution: A historical review, *Int. J. Environ. Stud.*, 1, 281, 1971.
79. Graber, J. W. and Graber, J. R., Environmental evaluations using birds and their habitats, *Biological Notes, No. 97*, Ill. Nat. Hist. Survey, 1976.
80. Newman, J. R., Effects of industrial air pollution on wildlife, *Biol. Conserv.*, 15, 1979, 181.
81. Stanley, P. J. and Bunyan, P. J., Hazards to wintering geese and other wildlife from the use of dieldrin, chlorfernvinfos, and carbophenothion as wheat seed treatments, *Proc. R. Soc. London (B)*, 205, 31, 1979.
82. Blancher, P. J., Modelling waterfowl response to wetland acidification, *Can. Comm. Ecol. Land Classification News*, 18, 12, 1989.
83. Morrison, M. L. and Menslow, E. C., Impacts of forest herbicides on wildlife toxicity and habitat alteration, *Trans. North Am. Wildl. Nat. Res. Conf.*, 175, 1983.
84. Hall, R.K. et al., Status of aquatic bioassessment in U.S. EPA Region IX, *Environ. Monitor. Assess.*, 64, 1, 17, 2000.
85. Walker, D. A. and Everett, K. R., Road dust and its environmental impact on Alaskan taiga and tundra, *Arctic Alpine Res.*, 19, 479, 1987.
86. Stokes, P. M., Hutchinson, T. C., and Kranter, K., Heavy metal tolerance in algae isolated from polluted lakes near the Sudbury, Ontario Smelters, *Water Pollut. Res. J. Can.*, 8, 178, 1973.
87. Elton, C. S., *The Ecology of Invasions by Animals and Plants*, Methuen, London, 1958.
88. Rapport, D. J., Regier, H. A., and Hutchinson, T. C., Ecosystem behavior under stress, *Am. Nat.*, 125, 617, 1985.
89. Bird, P. M. and Rapport, D. J., *State of the Environment Report of Canada*, Minister of the Environment, Ottowa, 1986.
90. Holl, K. D. and Cairns, J., Jr., Monitoring and appraisal, in *Handbook of Restoration Ecology*, Davy, A. J. and Perrow, M., Eds., Blackwell Science, Oxford, 2001.
91. Cairns, J., Jr., The myth of the most sensitive species, *Bioscience*, 36, 66, 1986.

92. Borman, F. H., Landscape ecology and air pollution, in *Landscape Heterogeneity and Disturbance*, Turner, M. G., Ed., Springer-Verlag, New York, 1987, 37.
93. Jackson, D. R. and Watson, A. P., Disruption of nutrient pools and transport of heavy metals in a forested watershed near a lead smelter, *J. Env. Qual.*, 6, 331, 1977.
94. Gorham, E. and McFee, W. W., Effects of acid deposition upon outputs from terrestrial to aquatic systems, in *Effect of Acid Precipitation on Temperate Ecosystems*, Hutchinson, T. V. and Havas, M., Eds., Plenum Press, New York, 1978, 465.
95. Roberts, T. M. et al., Effects of sulfur loading deposition on litter decomposition and nutrient leaching in coniferous forest soils, in *Effect of Acid Precipitation on Temperate Ecosystems*, Hutchinson, T. V. and Havas, M., Eds., Plenum Press, New York, 1978, 381.
96. Risser, P. G., Landscape pattern and its effect on energy and nutrient redistribution, in *Changing Landscapes: An Ecological Perspective*, Zonneveld, I. S. and Forman, R. T. T., Eds., Springer-Verlag, New York, 1990, 45.
97. Strecker, E. et al., *The Use of Wetlands for Controlling Stormwater Pollution*, Woodward-Clyde Consultants, Seattle, WA, 1990.
98. Gordon, A. G. and Gorham, E., Ecological aspects of air pollution from an iron-sintering plant at Wawa, Ontario, *Can. J. Bot.*, 41, 1063, 1963.
99. Gorham, E., Bayley, S. E., and Schindler, D. W., Ecological effects of acid deposition upon peatlands: A neglected field in "acid rain" research, *Can. J. Fish. Aquat. Sci.*, 41, 1256, 1984.
100. Giblin, A. E., Comparison of the processing of elements in ecosystems, II. Metals, in *Ecological Considerations in Wetland Treatment of Municipal Wastewaters*, Godfrey, P. J., Kaynor, E. R. M., Pelczarski, S., and Benforado, J., Eds., Van Nostrand Reinhold Co., New York, 1985, chap. 5.
101. Riiters, K. H. et al., Indicator strategies for forests, in Ecological Indicators for the Environmental Monitoring and Assessment Program, EPA 600/3–90/060, Hunsaker, C. T. and Carpenter, D. E., Eds., U.S. Environmental Protection Agency, Office of Research and Development, Research Triangle Park, NC, 1990, chap. 6.

CHAPTER 10

# Using Biomonitoring Data for Stewardship of Natural Resources

Robert P. Breckenridge, Marilynne Manguba, Patrick J. Anderson, and Timothy M. Bartish

## CONTENTS



## 10.1 INTRODUCTION

Natural resource management decisions are being made that cost the nation millions of dollars every year. These decisions should be supported by scientifically sound data. This chapter discusses why monitoring programs are needed and how to design a program that is based on a good scientific approach with specific objectives. The purpose of this chapter is to identify many of the large-scale monitoring programs in the United States, how to access the information from these programs, and how this information can be used to improve long-term management of natural resources. Biomonitoring data form the basis upon which most long-term stewardship decisions are made. Monitoring data that supports stewardship decisions provide the critical link between field personnel and decision makers.

## 10.2 IMPORTANCE OF MONITORING IN RESOURCES MANAGEMENT

Monitoring and stewardship share an interesting connection. Both take an active role in the understanding of the condition of a resource to improve long-term sustainability. Over the past decade much change has occurred in the field of monitoring and management of data sources. Some of the change is due to advances in technology that allow easy access to large datasets and enhanced capabilities to evaluate, analyze, and visualize data. Developments in remote sensing, geographic information systems (GISs), the Internet, and personal computers and personal data recorders have provided scientists, managers, and the general public the ability to access and evaluate data. In the past paper maps and overlays were used to identify alternatives for resource management. Today, maps and charts can be generated on the spot and served over network connections, allowing the audience to evaluate options immediately.

Stewardship is a term that has become much more common in natural resource management. A three-volume set titled *Ecological Stewardship*[1] is a good reference for scientists and managers involved in or responsible for ensuring natural resources are maintained so that future generations have options for their management. These three volumes summarize hundreds of years of experience of scientists and managers representing the major land-management agencies in the United States. To be good stewards it is critical that we understand where we have come from, where we are, and what is the desired future condition.

The Federal Government alone spends over $650 million per year on environmental monitoring[2] to support regulatory compliance, evaluate the presence of contaminants, and improve our understanding of environmental systems. As stewards of the environment, scientists, engineers, and managers have critical responsibilities to ensure that monitoring programs are established with logical objectives, are designed to collect monitoring data following a well-established process, and take advantage of existing data. A recent article in *Scientific American*[3] responding to the book, *The Skeptical Environmentalist,*[4] clearly points out that there are multiple views to the environmental issues facing the nation and the world. The responsibility of natural resource stewards is to have as complete an understanding of the resource they are managing, the monitoring process, and a thorough understanding of how to acquire and use data to make decisions. As an analogy, it would be bad money management to invest without doing a thorough evaluation of what data exists on the status and trends of various investment programs.

To be good stewards of our investments, it is important to monitor performance data on a regular basis and adjust our management strategy periodically to ensure our goals are being met. In a similar manner many of us are tasked to design biomonitoring programs for systems we have limited knowledge about. A further complication is that many who are responsible for management of natural resources change work locations frequently. Those who are working on existing monitoring programs or who are designing new ones can often gain great insight by consulting the literature and reviewing existing information on local and regional biomonitoring programs. This review process is an important step to help ensure the program is best designed to meet management objectives. The sources identified in this chapter provide insight to those faced with determining how to use biomonitoring data to support long-term stewardship of natural resources.

Resource management requires accurate information that can be used in the decision-making process. We must take advantage of existing data where available and access monitoring networks to identify what information can help us better interpret how humanity's activities are affecting the biological systems we are tasked to manage.

The National Research Council (NRC)[5] recently published a book on the grand challenges facing the nation in the environmental sciences. The NRC, at the request of the National Science Foundation, identified eight grand challenges: (1) biogeochemical cycles, (2) biological diversity and ecosystem function, (3) climate variability, (4) hydrologic forecasting, (5) infectious disease and the environment, (6) institutional and resource use, (7) land use dynamics, and (8) reinventing the use of materials. In their recommendations section the authors state that for most of these challenges, there needs to be a comprehensive understanding of relationships that govern processes (e.g., the relationship between ecosystem structure and function and biological diversity) and forecasting information to improve our comprehension of consequences.

To address the challenges put forth by the NRC, we must delve into the literature and identify what information exists from previous monitoring programs and how this can be used to improve our understanding. Many of us facing challenges in the biological sciences feel that we are pioneers in addressing stewardship of the resources we are tasked to manage. However, in reality humans have been stewards of natural resources for many years. One challenge we face is to state a meaningful question. Once the question is identified, we can set out to determine how and if existing information can be used to help establish baselines and identify critical data gaps and how these gaps can be filled with new data.

Monitoring data, both existing and new, can play an important role in the design and evaluation of ecotoxicological studies. Anthropogenic impacts provide numerous stresses to environmental systems. The advent of low-level chemical analysis and techniques to evaluate genetic effects from environmental stress provide scientists new tools to better understand how biological systems can be impacted from chemical stresses. However, it is important that scientists use existing information from the sources identified in this chapter as well as others to identify the degree and type of management action required and understand how this information should be used in context.

Volumes of data are available through environmental monitoring program offices of the various agencies identified in this chapter. Even though much of these data may be obtained over the Web or with a phone call, these resources are underutilized. When designing new biomonitoring programs or ecotoxicology studies, it is important to avoid the trap of developing hypotheses and setting out to prove them without consulting existing data and benefiting from past experience. The hesitancy to use existing data can be traced to a lack of familiarity with the types of monitoring data available and to uncertainty about data quality, compatibility, and the means by which different program designs may be integrated. To make biomonitoring cost-effective and justify its benefits, scientists need to develop clear goals and objectives that build upon existing datasets and learn from previous studies. It is often more cost effective to invest in early detection than pay later for remediation.

Evaluating existing data and information from monitoring programs and previous ecotoxicology studies has benefits analogous to conducting a market survey before releasing a new product to the public. Reviewing and evaluating existing data sources can be tedious; however, the effort allows the study or program to

- Enhance understanding of the mechanisms behind observed trends
- Fill critical data gaps
- Define ecosystem interactions
- Develop a conceptual model of how the ecosystem operates
- Identify the best time to monitor for parameters of interest and variability
- Identify quality assurance/quality control issues
- Avoid problems encountered by previous biomonitoring efforts

*Data* in this chapter are defined as the raw values collected in the field or laboratory analysis, while *information* is the interpreted data that provides conclusions for making decisions. The chapter focuses on long-term, large-scale monitoring programs in the United States that can help assess the condition of the nation's natural resources. The term *existing data* is used here to refer to data in the literature or in electronic form on both large-scale and site-specific efforts. A distinction between the two will be made in cases where limitations in the use of the data exist. Other data sources that focus on human health or international monitoring efforts are not addressed in this chapter. Some reference is made to existing ecotoxicology data; however, many other sources like those listed in other chapters of this book are also available.

Monitoring data are available on site-specific, local, regional, and national scales. Data from one scale may be used to support a study on another scale. For example, a scientist interested in studying the site-specific impact of mercury from incineration may use data from a regional biomonitoring program to help identify important ecological endpoints to monitor (e.g., fish, wetland aquatic plants). Evaluating water and air pathways on a large scale can provide significant information on how distant contaminant sources can impact a study or management area.

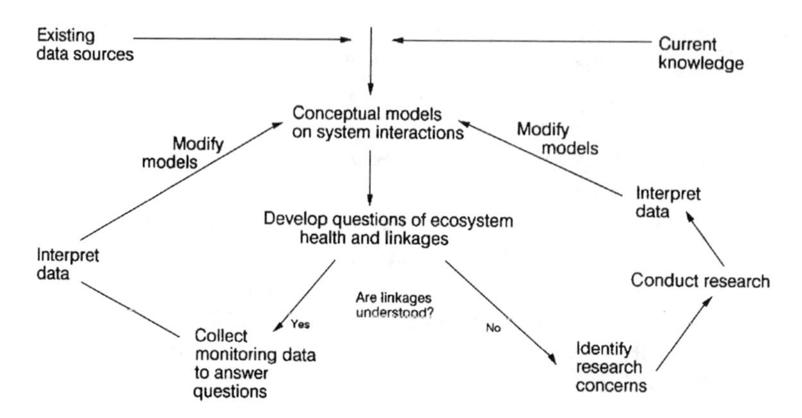


**Figure 10.1** Relationship between existing data and its use developing monitoring questions and research options.

Research at the local scale can also help in the interpretation of results from a large-scale monitoring program. Local- or small-scale ecosystem studies may help address issues associated with sampling design, interactions within and between ecosystem processes (e.g., biogeochemical cycling, productivity, and population dynamics), and partitioning of chemicals in the ecosystem. Knowledge of how local- or small-scale studies can support or improve data interpretation for large-scale programs can be important to resource stewardship. Figure 10.1 presents this relationship and shows the importance of research to any monitoring program. Often questions that require new research to address are raised by a monitoring program.

## 10.3 REVIEW OF BIOMONITORING PROGRAMS

In 1990 scientists from the Idaho National Engineering and Environmental Laboratory and the U.S. Fish and Wildlife Service identified long-term, large-scale monitoring programs that could be useful for assessing the status of contaminants in U.S. fish and wildlife resources and their habitats.[6] This material was published in the original edition of this handbook.[7] Since then many of the data sources have been modified due to changes among agencies, access via the Internet, and addition of new datasets. The material in this chapter builds upon the original findings but focuses on web-accessible databases that provide contaminant information related to biota, sediment, and water. Other databases describing ecological and biological survey and monitoring results were also evaluated. A search was conducted for the existence of national (or large regional) monitoring programs and databases. These searches included the use of the Internet, literature, and phone contact with federal resource agencies, nongovernmental organizations, industrial organizations, and others.

The programs/databases identified were evaluated as to their status (e.g., ongoing, eliminated, modified), their content (e.g., type of data, frequency of collection, locations), and their accessibility. Long-term programs were of particular interest because of their spatial and temporal datasets for evaluating variability issues and their ability to somewhat transcend short-term political changes that occur with changes in administrations. Long-term was defined as 5 years or more; large-scale focused on regional or national scales, but some large site-specific programs were included. The effort identified a number of Federal agencies that are involved in 23 programs (Table 10.1) that provide potentially useful data for ecotoxicological assessments and in the stewardship of natural resources.

A review of existing data sources (Table 10.1) indicated that a number of different media were being monitored, but collectively they provide only a limited picture of toxic threats to ecological

**Table 10.1 National and Regional Contaminant Monitoring and Assessment Programs**

| | Agency | Program Description | Extent | Contaminant Focus | Data/Databases |
|---|---|---|---|---|---|
| | **U.S. Environmental Protection Agency (www.epa.gov)** | | | | |
| 1 | National Fish and Wildlife Program (http://www.epa.gov/waterscience/fish) Contact: Jeffrey Bigler U.S. EPA, 1200 Pennsylvania Ave., NW (4305) Washington, D.C. 20460; email: bigler.jeff@epa.gov | This program provides guidance for assessing health risks associated with the consumption of chemically contaminated noncommercial fish and wildlife to state and tribal officials and provides outreach and current information to advisory programs. | Rivers, streams, and coastal waters of United States and Canada | Bioaccumulative contaminants in fish and wildlife | http://map1.epa.govname = Listing&Cmd = Map http://www.epa.gov/ost/fish/mercurydata.html |
| 2 | National Estuary Program (http://www.epa.gov/owow/estuaries/) Contact: Coastal Management Branch U.S. EPA (4504F) Ariel Rios Bldg. 1200 Pennsylvania Ave., NW Washington, D.C. 20460 Fax: (202) 260–9960 | The National Estuary Program (NEP) identifies, restores, and protects nationally significant estuaries of the United States. There are currently 28 individual NEP programs designed to address a broad range of environmental problems including toxic chemicals and nutrient loadings. | Estuaries of United States | Toxic chemicals (including trace metals and organochlorines) and nutrients in water, sediments, fish, invertebrates, shellfish, and vegetation | Data are accessible through the web pages of many of the individual estuary programs http://www.epa.gov/owow/estuaries/list.htm |
| 3 | National Dioxin Air Monitoring Network (NDAMN) (http://cfpub.epa.gov/ncea/cfm/1pagecfm#air = 22423&partner = ORD-NCEA) Contact: David Cleverly email: cleverly.david@epa.gov | The National Dioxin Air Monitoring Network (NDAMN) was established to determine the temporal and geographical variability of atmospheric CDDs, CDFs, and dioxin-like PCBs at rural locations throughout the United States. | Nationwide | Atmospheric CDDs, CDFs, and PCBs | Data are available through formal reports. The first year's results (1998–1999) are available at: www.epa.gov/ncea/pdfs/dioxin/dei/ NDAMN_PAPER3a.pdf |
| 4 | The Acid Rain Program (http://www.epa.gov/airmarkets/arp/index.html) Contact: U.S. EPA Clean Air Markets Division 1200 Pennsylvania Ave., NW Mail Code 6204N Washington, D.C. 20460 | The overall goal of the Acid Rain Program is to achieve significant environmental and public health benefits through reductions in emissions of sulfur dioxide ($SO_2$) and nitrogen oxides (NOx). This program is part of the EPA's Clean Air Programs. | Nationwide | $SO_2$, $NO_x$, and $CO_2$, emissions from electric power plants and other generation facilities as well as volumetric flow and opacity and criteria pollutants | Data are accessible at: www.epa.gov/airmarkets/data.html#emissions |
| 5 | Great Lakes National Program Contact: U.S. EPA Great Lakes National Program Office 77 West Jackson Boulevard Chicago, Illinois 60604 Phone: (312) 886–4040; Fax: (312) 353–2018 | The goals of the GLNP are to reduce levels of toxic contaminants in the environment, protect and restore vital habitats, and protect the health of the Great Lakes ecosystem's living resources. This program oversees the collection of environmental data on a wide variety of constituents in water, biota, sediment, and air. | Great Lakes ecosystem in United States and Canada | Numerous contaminants in the air, water, sediments, and fish tissue | Great Lakes Environmental Database (GLENDA) (http://www.epa.gov/glnpo/) |
| 6 | Environmental Monitoring and Assessment Program (EMAP) Contact: Mike McDonald (Director) email: emap@epa.gov Contact Page: www.epa.gov/emap/html/emapgrps.html | The EMAP program conducts environmental assessments and develops methods to quantify ecosystem health at the regional and national scales. | National and regional | Numerous contaminants in the air, water, sediments, and biota | Access to data is available below and at the following link: http://www.epa.gov/emap/html/data/index.html National Coastal Assessment Database Mid-Atlantic Integrated Assessment Datasets Regional Environmental Monitoring and Assessment Program Data |

**Table 10.1 National and Regional Contaminant Monitoring and Assessment Programs** ***(Continued)***

| | Agency | Program Description | Extent | Contaminant Focus | Data/Databases |
|---|---|---|---|---|---|
| 7 | Total Maximum Daily Load Program (http://www.epa.gov/OWOW/tmdl/index.html) email: OW-General@epamail.epa.gov | The TMDL Program is designed to work with states and tribes to establish individual TMDL programs and set water quality standards. States and tribes may monitor contaminants as part of their individual program. | States and tribes | Numerous contaminants in surface waters and biota | Data from TDML projects are available from the EPA's STORET System and may also be accessible from individual state programs |
| | **National Oceanic and Atmospheric Administration (NOAA) (http://www.noaa.gov/)** | | | | |
| 8 | National Status and Trends Program (http://www.noaa.gov) Contacts: NS&T Program NOAA/NOS/NSI1 1305. East West Hwy. Silver Spring, MD 20910 Phone: (301) 713-3028 x. 151 Fax: (301) 713-4388 | This program was initiated to determine the current status of and detect changes in the environmental quality of estuarine and coastal waters. The NS&T Program is comprised of several projects. These are: Benthic Surveillance Project, the Mussel Watch Program, the Quality Assurance Project, Historic Trends, the Sediment Coring Project, the Specimen Banking Project, and numerous other projects. | Estuaries and coastal waters of the United States | Contaminants (trace metals, DDE, PCBs, aromatic hydrocarbons, and radionuclides) in sediments, fish, and mussels and oysters and the effects of contaminants in fish and shellfish. | Benthic Surveillance http://ccmaserver.nos.noaa.gov/nsandtdata/NSandTdatasets/benthicsurveillance/welcome.html Mussel http://ccmaserver.nos.noaa.gov/nsandtdata/NSandTdatasets/musselwatch/welcome.html Benthic Community http://140.90.163.135 |
| | **U.S. Geological Survey (http://www.usgs.gov/)** | | | | |
| 9 | National Water Quality Assessment Program Contact: Chief, NAWQA Program U.S. Geological Survey 413 National Center Reston, Virginia 20192 Phone: (703) 648–5716 Fax: (703) 648–6693 email: nawqa_whq@usgs.gov | The NAWQA program has been collecting and analyzing data on over 50 major river basins and aquifers across the nation since 1991 with the goal of developing long-term consistent and comparable information on streams, ground water, and aquatic ecosystems to support sound management and policy decisions. | Nationwide network of streams and rivers | Measurements include numerous chemical concentrations (nutrients, trace elements, and organic chemicals) in water, sediment, and aquatic organisms from a national network of streams | Data are available at the NAWQA Data Warehouse and through NAWQA publications and National Synthesis reports and http://water.usgs.gov/nawqa/natsyn.html |
| 10 | Biomonitoring of Environmental Status and Trends Contact: Tim Bartish (Program Coordinator) 4512. McMurry Ave. Fort Collins, CO 80525 Fax (970) 226–9230 (970) 226–9483 email: tim_bartish@usgs.gov | The BEST program is designed to identify and understand the effects of environmental contaminants on biological resources, particularly those resources under the stewardship of the Department of the Interior. BEST is unique among national monitoring programs with its emphasis on characterizing the effects of environmental contaminants on the health of the biota and their supporting habitat. | Nationwide network of the nation's large rivers | Chemical concentration measurements and biological responses or effect measurements in biota | Data are available at: http://www.best.usgs.gov/data.htm Mississippi River – 1995 at: http://www.cerc.cr.usgs.gov/data/best/index.htm |
| 11 | Toxic Substances Hydrology Program Contact: USGS Toxics Program MS 412 12201 Sunrise Valley Drive Reston, VA 20192 | The Toxics Program conducts investigations of subsurface contamination and watershed- and regional-scale investigations of contamination affecting aquatic ecosystems from nonpoint and point sources. The program has also implemented (1999) a national reconnaissance to provide baseline information on the environmental occurrence of emerging contaminants in the Nations water resources, The Toxics Program also includes the Abandoned Mine Lands Initiative, and represents USGS Assistance for TMDLs. | Nationwide (watershed to regional level investi-gations) | Nutrients, organic chemicals, metals, pharmaceuticals, and pathogens in surface water, groundwater, sediments | Data are available in reports, by contacting investigators, and through the USGS National Water Information System (http://water.usgs.gov/nwis) |

| | | | | | |
|---|---|---|---|---|---|
| 12 | National Stream Quality Accounting Network<br>Contact: Rick Hooper (National Program Coordinator)<br>Phone: (508/490–5065)<br>email: mailto: rphooper@usgs.gov | The NASQAN Program monitors the water quality of four of the nation's largest river systems — the Mississippi (including the Missouri and Ohio), the Columbia, the Colorado, and the Rio Grande. NASQAN uses a flux-based approach to water-quality monitoring to describe and compare yields of contaminants among large regional basins, describe long-term trends in the mass flux and concentration of constituents at key locations, and calculating loads to receiving waters. | Nationwide network of the nation's largest river systems | Constituents include nutrients, major ions, sediment, dissolved and sediment associated trace elements, and hydrolphylic pesticides | Data are available at: http://water.usgs.gov/nasqan/data/index.html |
| 13 | The Hydrologic Benchmark Network<br>Contact: Rick Hooper<br>Phone: (508/490–5065)<br>email: mailto: rphooper@usgs.gov | The HBN which operated between 1963 to 1995 was established to provide long-term measurements of streamflow and water quality in areas that were minimally affected by human activities. | Nationwide network of streams from 63 drainage basins in 39 states | Measurements included major dissolved constituents, trace elements, radiochemical constituents, nutrients, and biological constituents | Data collected by the HBN program between 1962 and 1995 are available from USGS (Data from Selected U.S. Geological Survey National Stream Water-Quality Monitoring Networks [WQN] are also available on the Web.) http://water.usgs.gov/pubs/dds/wqn96 |
| | **U.S. Fish and Wildlife Service (www.fws.gov)** | | | | |
| 14 | Environmental Contaminants Program – Division of Environmental Quality (www.contaminants.fws.gov/)<br>Contact: Division of Environmental Quality<br>4401 North Fairfax Dr., Suite 322<br>Arlington, Virginia 22203<br>(703)358–2148<br>email: contaminants@fws.gov | The Contaminants Program focuses on toxic chemicals and their effects to fish, wildlife and their habitats and removing toxic chemicals and restoring habitat. The Contaminant Program's operations are integrated into all other service activities. | Nationwide | Investigates a broad range of contaminants and contaminant effects on trust lands, waters, and species | Access to data about contaminants, refuges, species, studies and reports is available through http://ecos.fws.gov/cimas_98 |
| 15 | National Contaminant Biomonitoring Program (NCBP) (http://www.cerc.usgs.gov/data/ncbp/ncbp.html)<br>Contact: Chris Schmitt<br>Columbia Environmental Research Center<br>4200. E. New Haven Rd.<br>Columbia, MO 65201–9634<br>(573) 875–5399,<br>Fax: (573) 876–1896<br>email: christopher_schmitt@usgs.gov | The NCBP which was established to document trends in the occurrence of persistent toxic chemicals that may threaten fish and wildlife resources. Monitoring efforts were focused on streams and associated habitats throughout the Nation. The FWS maintained this program from 1967 through 1987. | Nationwide network of streams | The NCBP measured contaminant concentrations (pesticides, metals, and other long-lived toxic contaminants) in fish and birds | Access to fish data. http://www.cerc.usgs.gov/data/ncbp/fish.htm<br>Access to starling residue data http://www.cerc.usgs.gov/data/ncbp/starling/starling.htm |
| | **National Park Service (www.nps.gov)** | | | | |
| 16 | Gaseous Pollutant Monitoring Program (GPMP)<br>Program Manager - John Ray (NPS-ARD) (303) 969–2820<br>email: john_d_ray@nps.gov | The GPMP identifies and monitors air pollutants with the potential to injure or damage park natural resources. The program is currently concentrating on the levels of ozone and sulfur dioxide since they are particularly toxic to native vegetation. | Nationwide in National Parks | Ozone, sulfur dioxide | http://www.epa.gov/air/data/index.html |

**Table 10.1 National and Regional Contaminant Monitoring and Assessment Programs *(Continued)***

| | Agency | Program Description | Extent | Contaminant Focus | Data/Databases |
|---|---|---|---|---|---|
| 17 | The Interagency Monitoring of Protected Visual Environments (IMPROVE)<br>Contact: Marc Pitchford<br>Desert Research Institute<br>755 East Flamingo Road, Las Vegas, NV 89119 | IMPROVE is designed to establish trends in visibility levels and identify sources of anthropogenic impairments to national parks and wilderness areas (Class I Areas). Agencies involved in this program include the National Park Service (lead agency), EPA, Fish and Wildlife Service, Bureau of Land Management, and the U.S. Forest Service. There are also several nongovernmental organizations that make up the program. | Nationwide in National Parks and wilderness areas | Measures aerosols, particulate matter, and optical data | Data are available at: http://vista.cira.colostate.edu/improve/Data/data.htm and at: http://improve.cnl.ucdavis.edu/. |
| | **Interagency Programs** | | | | |
| 18 | Park Research and Intensive Monitoring Network (PRIMENet) Program<br>Contact: NPS PrimeNet Coordinator: Kathy Tonnessen,<br>NPS Air Resources Division<br>(303) 969–2738<br>email: Kathy_Tonnessen@nps.gov | PRIMENet is an EPA-NPS program designed to assess the effects of environmental stressors on ecological systems nationwide. This network, formerly known as DISPro, consist of 14 monitoring and research sites in national parks which are representative of major ecosystem types and were chosen because of their status as Class 1 air quality parks. At these sites the NPS is sponsoring air quality monitoring, including ozone, wet and dry deposition, visibility, and meteorology. The EPA has added UV-B monitors at these parks to determine changes in irradiance that may be affecting human health and ecosystem processes. | National Parks located throughout the nation | Measures ozone, wet and dry deposition, visibility, UV-B, and meteorology | http://www2.nature.nps.gov/ard/gas/netdata1.htm |
| 19 | The National Atmospheric Deposition Program (NADP)<br>Contact: Van C. Bowersox, Coordinator<br>NADP Program Office<br>Illinois State Water Survey<br>2204 Griffith Drive<br>Champaign, IL 61820–7495<br>email: nadp@sws.uiuc.edu | The NADP monitors wet atmospheric deposition at over 220 National Trends Network sites throughout the United States. The network is a cooperative effort between many different groups including the State Agricultural Experiment Stations, U.S. Geological Survey, U.S. Department of Agriculture, and numerous other governmental and private entities. The NADP also administers two national networks. These are the Mercury Deposition Network (MDN), which measures weekly concentrations of total mercury in precipitation and the seasonal and annual flux of total mercury in wet deposition, and the Atmospheric Integrated Research Monitoring Network (AIRMoN), which monitors wet and dry deposition. This network is sponsored from the National Oceanic and Atmospheric Administration. | Nationwide network of sites | Precipitation chemistry with a focus on atmospheric deposition constituents (SO4, NO3-N, NH4-N, Ca, Mg, Na, K, pH, specific conductance, and PO4) and mercury. | NADP Data at: http://nadp.sws.uiuc.edu/nadpdata/<br>AIRMon Data at: http://nadp.sws.uiuc.edu/airmon/GetAMdata.asp<br>MDN Data at: http://nadp.sws.uiuc.edu/nadpdata/mdnsites.asp |

| | | | | | |
|---|---|---|---|---|---|
| 20 | The Clean Air Status and Trends Network (CASTNET)<br>Contact: Gary Lear<br>(202)564–9159<br>email: lear.gary@epa.gov | CASTNET provides atmospheric data on the dry deposition component of total acid deposition, ground-level ozone, and other forms of atmospheric pollution. CASTNET is considered the nation's primary source for atmospheric data for estimating dry acidic deposition and to provide data on rural ozone levels. CASTNET, established in 1987, is comprised of over 70 monitoring stations across the United States and is cooperatively funded and operated between the EPA and the National Park Service. | Nationwide | Acid deposition, ozone | CASTNET data are available at: http://www.epa.gov/castnet/data .html |
| 21 | State and Local Air Monitoring Stations/National Air Monitoring Stations (SLAMS/NAMS)<br>(http://www.epa.gov/oar/oaqps/qa/monprog.html)<br>Contact: David Lutz<br>(919)541–5476 Fax: (919) 541–1903<br>email: lutz.david@epa.gov | NAMS and SLAMS are designed to measure air criteria pollutants, visibility, fine particulates, toxic chemicals, and ozone in urban areas throughout the United States. The network now includes over 4000 sampling stations. This network is part of the EPA National Air Ambient Monitoring Program and includes other federal and state agencies. | Nationwide | Air criteria pollutants, toxic chemicals, ozone, visibility | Data is accessible through the AIRData System<br>http://www.epa.gov/air/data/ |
| 22 | Photochemical Assessment Monitoring Stations (PAMS)<br>(http://www.epa.gov/oar/oaqps/pams/)<br>Contact: Rich Scheffe (Program Mgt.)<br>email: scheffe.rich@epa.gov | PAMS includes measures of ozone and ozone precursors in areas known to have persistently high ozone levels (nonattainment areas). This network is part of the EPA National Air Ambient Monitoring Program and includes other federal and state agencies. | Nationwide | Ozone measurements (approximately 60 volatile hydrocarbon and carbonyl compounds) and nitrogen in non-attainment areas | Data from PAMS are currently in an interim database available through AIRData with plans for access to be through the AIRS system<br>http://www.epa.gov/air/data/ |
| 23 | The USDA UVB Radiation Monitoring Program<br>Dr. James R. Slusser (Program Director)<br>USDA UVB Radiation Monitoring Program<br>Natural Resource Ecology Laboratory<br>Colorado State University<br>Fort Collins, CO 80523<br>Fax: (970) 491–3601<br>sluss@uvb.nrel.colostate.edu | The UVB Radiation monitoring Program is a program of the U.S. Department of Agriculture's Cooperative State Research, Education and Extension Service (CSREES). The program was initiated in 1992 (through a grant to Colorado State University) to provide information on the geographical distribution and temporal trends of UVB (ultraviolet -B) radiation in the U.S. Information from these efforts are used to evaluate the potential damage effects of UVB to agricultural crops and forests. | Nationwide | UVB radiation | Data is available at: http://uvb.nrel.colostate.edu/UV B/uvb_program_overview.html |
| | **State and/or Regional Monitoring Programs** | | | | |
| 24 | The Southern Oxidants Study<br>Contact: Office of the Director<br>1509 Varsity Drive<br>Raleigh, NC 27606<br>919–515–4260 or Fax 919–515–1700<br>sos@ncsu.edu | The SOS is an alliance of research scientists, engineers, and air quality managers from universities, federal and state governments, industry, and public interest groups working together to design and execute scientific research and modeling programs to increase the present understanding of ozone accumulation in the atmosphere. | Southern United States | Continuous monitoring of regional ozone concentrations, ozone precursor concentrations and other oxidants, and particulate matter | Data access for SOS is at http://vortex.atmos.uah.edu/essl /sos |

**Table 10.1 National and Regional Contaminant Monitoring and Assessment Programs *(Continued)***

| | Agency | Program Description | Extent | Contaminant Focus | Data/Databases |
|---|---|---|---|---|---|
| 25 | Great Lakes National Program<br>Contact: Great Lakes National Program<br>77 West Jackson Boulevard<br>Chicago, Illinois 60604<br>phone: (312) 886–4040 fax: (312) 353–2018 | The Great Lakes National Program is a collaboration of the EPA and Great Lakes states and private groups involved with monitoring and assessment activities related to the health of the Great Lakes. | Great Lakes ecosystem | Collects environmental data on a wide variety of constituents in water, biota, sediment, and air. | (http://www.epa.gov/glnpo/glenda/aboutglenda.html) |
| 26 | National Irrigation Water Quality Program (NIWQP)<br>Contact: N. John Harb Manager, National Irrigation Water Quality Program<br>Bureau of Reclamation<br>P.O. Box 25007 (D-6200)<br>Denver CO 80225–0007<br>Office: (303) 445–2789<br>Fax: (303) 445–6693<br>email: nharb@do.usbr.gov | The NIWQP is an inter-bureau program managed by the Department of the Interior. Participating bureaus within the Department include Bureau of Indian Affairs, Bureau of Reclamation, U.S. Fish and Wildlife Service, and U.S. Geological Survey. The scope of the program is to identify and address irrigation-induced water quality and contamination problems related to DOI water projects in the west. Reconnaissance investigations have been completed for 26 study areas. | Irrigation waters throughout the western United States | Measures contaminants in water, sediments, vegetation, invertebrates, fish and birds with an emphasis on arsenic, boron, copper, DDT, mercury, molybdenum, salinity, selenium, and zinc | Chemical data from water, bottom-material, inorganic and organic biological samples are available for 26 study areas |
| 27 | The Long-Term Resource Monitoring Program (LTRMP)<br>(http://www.umesc.usgs.gov/ltrmp.html)<br>Contact: Upper Midwest Environmental Sciences Center<br>2630 Fanta Reed Road<br>La Crosse, WI 54602–0818<br>Phone: (608) 783–6451 | The LTRMP is being implemented by the U.S. Geological Survey (USGS) in cooperation with the five Upper Mississippi River System states (Illinois, Iowa, Minnesota, Missouri, and Wisconsin), with guidance and overall program responsibility provided by the U.S. Army Corps of Engineers. The LTRMP monitors nutrients in water and sediments and environmental trends of invertebrates, birds, and fish. | Upper Mississippi River Basin | Water quality and nutrients in the Upper Mississippi River System | http://www.umesc.usgs.gov/data_library/sediment_nutrients/sediment_nutrient_page.html<br>http://www.umesc.usgs.gov/data_library/sediment_contaminants/sediment_contaminant_page.html |
| 28 | The Chesapeake Bay Program<br>Contact: Chesapeake Bay Program Office:<br>410 Severn Avenue, Suite 109, Annapolis, MD 21403/Tel: (800) YOUR-BAY/Fax: (410) 267–5777 | Chesapeake Bay Program resulted from the Chesapeake Bay Agreement of 1983 which was designed to restore the Bay. Partners of this Program include the states of Maryland, Pennsylvania and Virginia; the District of Columbia; the Chesapeake Bay Commission (a tri-state legislative body); the US Environmental Protection Agency (EPA), which represents the federal government; and participating citizen advisory groups. | Chesapeake Bay | Monitoring efforts include measures of nutrients, metals, PAHs, PCBs, pesticides, etc. in water, sediment, and fish tissue | Databases from the CBP and other sources are available at: http://www.chesapeakebay.net/data/index.htm |
| 29 | The Chesapeake Bay River Monitoring (RIM) Program<br>Contact: Doug Moyer<br>(dlmoyer@usgs.gov)<br>Mick Senus<br>(mpsenus@usgs.gov) | The RIM Program was established to quantify loads and long-term trends in concentrations of nutrients and suspended material entering the tidal part of the Chesapeake Bay Basin from its nine major tributaries. | Chesapeake Bay and major tributaries | Nutrient concentrations and loads (nitrogen, phosphorus, ammonia), and suspended sediments | Concentrations, loads, and streamflow are available at: http://va.water.usgs.gov/chesbay/RIMP/dataretrieval.html |

| | | | | | |
|---|---|---|---|---|---|
| 30 | Alaska Marine Mammal Tissue Archival Project (AMMTAP)<br>Lyman K. Thorsteinson (Program Manager)<br>USGS-BRD, Western Regional Biologist<br>909 First Avenue, Suite 800<br>Seattle, WA 98104<br>Telephone: (206) 220–4614<br>email: lyman_thorsteinson@usgs.gov | AMMTAP collects tissue samples from marine mammals for future retrospective analysis. The project is a joint effort by the U.S. Geological Survey Biological Resources Division (USGS\BRD), the National Oceanic and Atmospheric Administration's National Marine Fisheries Service (NOAA Fisheries), and the National Institute of Standards and Technology (NIST). | Alaska, arctic mammals | Collects baseline levels of chlorinated hydrocarbons and trace elements, including heavy metals, in arctic marine mammal species | Data is available through reports and requests (http://www.absc.usgs.gov/research/ammtap/index.htm) |
| 31 | The Seabird Tissue Archival and Monitoring Project (STAMP)<br>Contact: Geoff York<br>USGS, Alaska Science Center<br>1011. East Tudor Road<br>Anchorage, AK 99503–6199<br>Telephone: (907) 786–3928<br>email: geoff_york@usgs.gov | The STAMP program mirrors the AMMTAP Program in purpose and procedure except the focus is on contaminants in eggs of arctic bird species. This project is a joint effort between USGS, NIST, and U.S. FWS. | Arctic United States | Collects levels of chlorinated hydrocarbons and trace elements, including heavy metals, in eggs of arctic bird species | Data are available through reports and requests (http://www.absc.usgs.gov/research/ammtap/index.htm) |
| 32 | The Mid-Atlantic Integrated Assessment (MAIA)<br>Contact: Tom DeMoss, EPA/Region 3<br>Email: demoss.tom@epamail.epa.gov | MAIA is an *interagency,* multidisciplinary program research, monitoring, and assessment initiative. Its main goal is to develop high-quality scientific information on the condition of the natural resources of the Mid-Atlantic region of the eastern United States. Partners include numerous Federal agencies (headed by the US EPA), states, and public and private organizations. | Mid-Atlantic region of United States | Measurement of numerous contaminants in streams, sediments, groundwater, estuaries, and biota | Data are available at: http://www.epa.gov/emap/maia/html/data.html |
| 33 | The Gulf of Mexico Program<br>Contact: James Giattina (Program Director)<br>Email: giattina.jim@epa.gov<br>Bldg. 1103, Room 202, Stennis Space Center, MS 39529–6000<br>Main Telephone: (228) 688–3726<br>Fax: (228) 688–2709/2306 | The Gulf of Mexico Program was formed in 1988 to develop and implement voluntary, incentive-based management strategies to protect, restore, and maintain the health and productivity of the Gulf of Mexico ecosystem. Partners of the Gulf of Mexico Program include numerous Federal agencies (headed by the U.S. EPA), states, and public and private organizations. | Gulf of Mexico | Numerous contaminants and environmental health measurements in water and biota | Data is available at: http://pelican.gmpo.gov/datapage.html<br>http://pelican.gmpo.gov/hgreport.pdf |

**Table 10.1 National and Regional Contaminant Monitoring and Assessment Programs** ***(Continued)***

| | Agency | Program Description | Extent | Contaminant Focus | Data/Databases |
|---|---|---|---|---|---|
| 34 | The Gulf of Maine<br>Contacts: Laura A. Marron<br>Gulf of Maine Council Coordinator<br>New Hampshire Department of Environmental Services<br>P.O. Box 95<br>Concord, NH 03302–0095<br>Phone: 603–271–8866<br>Fax: 603–271–0656<br>email: info@gulfofmaine.org | The Gulf of Maine (GOM) is a collaborative effort by federal agencies of the United States and Canada, states, and public and private organizations. The goals of GOM are to: protect and restore coastal habitats, restore shellfish habitats, and protect human health and ecosystem integrity from toxic contaminants in marine habitats. Nested under these efforts is the Gulf Watch Program which uses mussels as a sentinel species for habitat exposure to organic and inorganic contaminants. | Gulf of Maine | Contaminant measures include mercury deposition, and organic and inorganic contaminants. Mussel tissue are analyzed for trace metals (Ag, Al, Cd, Cr, Cu, Fe, Hg, Ni, Pb, Zn), PAHs, PCB congeners, and chlorinated pesticides | Mussel data are available at:<br>http://www.gulfofmaine.org/library/gulfwatch/#data<br>http://ccmaserver.nos.noaa.gov/NSandT/NewNSandT.html |
| 35 | Arctic Monitoring and Assessment Program (http://www.amap.no)<br>Contact: Arctic Monitoring and Assessment Programme Secretariat (AMAP)<br>P.O. Box 8100 Dep<br>N-0032 Oslo, Norway<br>email: amap@amap.no | Efforts of AMAP include monitoring levels of, and assess the effects of, anthropogenic pollutants in the Arctic environment. This program is an effort by eight Arctic Rim countries (Canada, Denmark/Greenland, Iceland, Norway, Sweden, Soviet Union, and United States). | Arctic regions | Numerous contaminants (including persistent organic pollutants and heavy metals) in the arctic ecosystem and biota | http://www.amap.no/data-gis/data-gis.htm |

resources in the United States. Numerous programs monitor the media that support fish and wildlife resources (e.g., water, soil, sediment, air), but data are limited to interactions of contaminants across media.

Existing data can help researchers seeking insight on the partitioning of contaminants in an ecosystem. Water-quality and sediment-monitoring data from a river can help identify areas where contaminants concentrate (usually fine sediments) or those areas that have elevated values because of human-induced contaminants or enriched natural-metal deposits. Information that can point to unique features of an area provides a new context for the study and can be used to evaluate the need for additional research.

Often large-scale programs select their monitoring stations randomly to avoid being biased by local contaminant sources. Theoretically, this should allow examination of trends in contaminants in different media to make associations. These types of associations are possible at the qualitative level, but quantitative statements are difficult because of inconsistencies in sample timing, collection, analytical methods, and level of quality control.

Presently, due to financial constraints many studies have to focus on only a part of an ecosystem (e.g., the aquatic or avian portion). There are often interactions between the aquatic and terrestrial portions that need to be considered. A study that considers the interaction between the terrestrial and aquatic systems at the species, population, community, or ecosystem level can benefit greatly from data collected in complementary studies and determine if they can be combined in a manner that conceptually describes associations between systems.

Scientists addressing contaminant impact issues need to consider several factors when evaluating the use of existing data: (a) Is the chemical of interest new or existing? (b) Is the projected impact local or regional? (c) Will the data be used to make a specific management decision (e.g., license a chemical) or as part of an ecological risk assessment? and (d) Is the potential impact associated with a common or limited natural resource?

Biomonitoring programs often provide data on selected chemicals. Table 10.1 lists the chemicals that most of the large-scale programs have focused on. If a monitoring program is to be manageable, it must focus on a select subset of chemicals or responses of a biological system and track these over time. Existing programs have limitations because they cannot track all chemicals but must select those that are representative of classes of chemicals.

Most biomonitoring programs do not collect data to determine toxicity values for a chemical, but this information is available from other sources. There are several good references available in supporting chapters of this book to provide data on toxicity values of existing chemicals.

Existing biomonitoring data can be used to provide long-term averages for evaluating site-specific data. For example, if there was concern about mercury causing toxicity problems in a specific area, water and fish samples could be collected and evaluated against regional values from large-scale programs, such as the USGS National Water Quality Assessment (NAWQA) (Table 10.1, item 9) program or National Contaminant Biomonitoring Program fish data collected from 1967 to 1987 (Table 10.1, item 15). Since mercury is often transported via the air pathway, scientists should consult the Mercury Deposition Network nested under the National Atmospheric Deposition Program (Table 10.1, item 20) to see if monitoring sites are close to their area. These data can also be compared with published toxicology data to assess threat to a resource or the need for corrective action.[8] If, however, the chemical of concern is new, existing data may not be available.

The sciences of ecological risk assessment and ecotoxicology are developing quickly, but we are only starting to understand dose-response relationships at the species level. Toxicology needs to have a direct relationship with ecology if ecotoxicology is to emerge and face the challenge of assessing risk at community and population levels across complex natural ecosystems. The programs identified in Table 10.1 can help provide data needed to support risk assessment and toxicology studies. In conducting ecotoxicology studies conceptual models should be developed to link sources of contaminants to receptors. Van Horn[9] provides a brief overview of risk assessment and toxicology and explains in detail how contaminant sources, exposure pathways, and receptors are linked using conceptual models.

Some data are available to help scientists make habitat assessments of marine, upland, inland, and coastal areas. Better information is available for coastal areas than upland and inland areas, possibly because these coastal areas are higher in population and the resources are targeted for use, exploitation, and protection by the public and many different agencies and organizations. These coastal areas also serve as some of the major migratory routes for fish, waterfowl, and nongame birds.[10] NOAA strives to make multimedia assessments in coastal areas, and interagency programs are being established to study larger U.S. estuaries. The U.S. Environmental Protection Agency's (EPA's) Environmental Monitoring and Assessment Program (EMAP) is a research program designed to develop the tools necessary to monitor and assess the status and trends of national ecological resources (see EMAP Research Strategy). EMAP's goal is to develop the scientific understanding for translating environmental monitoring data from multiple spatial and temporal scales into assessments of ecological condition and forecasts of the future risks to the sustainability of our natural resources. Data collected by EMAP are available at http://www.epa.gov/emap/html/dataI/index.html.

## 10.4 USE OF EXISTING DATABASES

Data from existing monitoring programs can be very useful in designing ecotoxicology studies and monitoring projects. However, existing data are only tools with both applications and limitations. The most difficult problem is determining how, when, and where to use them in designing a new project. This discussion focuses on some general issues regarding the use of existing data and how those data can be helpful in conducting natural resource damage assessments, ecological risk assessments, and designing environmental monitoring projects.

Datasets from various federal agencies have limitations and cannot be combined to produce meaningful assessments unless the data are comparable, applicable, and transferable. There are three types of constraints that scientists must check when attempting to integrate existing datasets into new-program design: data, program, and institutional. Data constraints are usually related to whether the data can be obtained in a cost-effective manner, the format in which they are available, whether they are compatible and at the appropriate scale, their geographic and temporal coverage, and the types of analytes. Program constraints include compatibility of objectives, differences in sampling periods, intensity, statistical design, resource of interest, and level of quality assurance and quality control. Institutional constraints are mainly an issue when crossing over between agencies and include different agency missions, requirements for upward reporting, and interagency reporting.

The development of geographic information systems (GISs) and global positioning systems (GPSs) and efforts by the Federal Geographic Data Committee (FGDC) to coordinate the development of the National Spatial Data Infrastructure (NSDI) to develop policies, standards, and procedures for organizations to cooperatively produce and share geographic data are improving the accessibility and utility of new and existing data. The FGDC is comprised of 17 federal agencies along with organizations from state, local, and tribal governments, the academic community, and the private sector. Part of this effort includes establishing regional and state data clearinghouses (see http://www.fgdc.gov) and developing a standard for metadata, (i.e.,"data about data" that describe the content, quality, condition, and other characteristics of data). It is important that scientists check the type and amount of supporting documentation that exists to determine whether datasets can be integrated. Topics to evaluate include sampled media, sampling and analytical methods, statistical design, field collection methods, and data reporting.

Existing data that has a geographic component can now be converted, digitized, or formatted to be used with a GIS, allowing managers to assess data in ways that would have been impossible in the past. Application of spatial analysis and modeling techniques to existing data that have been digitized can provide information on relationships of ecosystem components, long-term trends, regional- or landscape-level influences, location of contaminant sources, with the added possibility of adding new georeferenced data.

Existing data can be a valuable asset in the Natural Resource Damage Assessment process. For example, monitoring data collected for a river can be used as baseline data. Similarly, data from existing long-term, regional monitoring programs can help interpret data from a new program by providing insight on natural variability and trends for the region. The existing data provides a link to past conditions, thereby providing scientists and managers with the opportunity to assess trends or, if needed, to make informed decisions based on limited new data. This aspect is becoming more important given the global transport of many contaminants.

The variability of data is directly related to its quality. Scientists should consider this when investing in monitoring resources. Good quality monitoring data can form the basis by which scientists and managers assess the degree of change that has occurred over time.

If data quality is sufficient, the data may be allowable in legal defenses. There are several important factors that scientists need to consider when assessing the quality requirement of existing or new data. Items that should be checked if the data are likely to be used in a legal defense include, but are not limited to:

- Established and documented quality assurance/quality control plan that governs data collection, handling, and reporting
- Rigorous and defensible statistical study design and reporting requirements
- Chain-of-custody procedures for all environmental samples submitted for analysis
- Established shipping, holding time, and analytical procedures
- Appropriateness of methods for handling outlier and less than detectable values.

The following is an example of how existing ecotoxicology data can be (and are) used in legal defenses. At numerous hazardous waste sites across the country cleanup is occurring to reduce the hazard to humans and the environment to an acceptable level. It is often difficult to establish background levels for the soils, sediment, and water and determine what is the result of site history and regional or global contaminant input. Existing data can often be used to help determine background and guide remediation discussions. In a similar manner data on species diversity, productivity, and composition for an area located near a spill site are useful when establishing the worth of a resource in a natural resource damage assessment.

Data may also be used in the four main areas of an ecological risk assessment:[11,12] (a) conceptual framework (stressor and environmental characterization, endpoint identification, and conceptual model formulation), (b) hazard assessment, (c) exposure assessment, and (d) risk assessment. These components address the fate, transport, and effect of chemicals in the ecosystem.[13] During the early years of ecotoxicology studies it was acceptable to characterize the risk of using a chemical by evaluating the response of an individual organism to various doses. This approach was borrowed from the human health risk assessment approach that focuses on dose to an individual. However, our knowledge has improved, and we now understand that exposure to an organism from a chemical or contaminant only partially characterizes the risk to an ecosystem. Exposure information addresses the quantity of a contaminant present, how much it had degraded, and how long it was exposed, but it fails to address the hazard of the contaminant (how toxic it is to different species and whether the toxicity is acute or chronic) and the condition of the receptor (species threatened, their life stage, and part of their habitat affected). All this information needs to be compiled in a manner that evaluates the risk across different levels within an ecological system.[14] Existing information from documented studies and ongoing monitoring programs can provide the data required to make these assessments.

Because ecological risk assessment requires an extensive amount of information to establish ecotoxicology relationships, scientists will be forced to rely more on existing or ancillary data to conduct cost-effective studies. Classical toxicology work considers impacts of a chemical to molecules, cells, tissues, organs, and, in some cases, individuals. Most of the literature and data on ecotoxicology focuses on aquatic-related impacts to individuals and populations (e.g., benthic, fish, and birds).[13,15,16] Limited studies have attempted to consider the true ecotoxicology effects on communities or to an ecosystem. Most are conducted on the local scale, which makes it difficult to extend results much past the community level. However, the results and data from these studies do help form the basis for conducting regional ecosystem and global biosphere studies. For example, several ecotoxicology modeling[17] efforts conducted on heavy metals in aquatic and terrestrial systems lead to the conclusion that because mercury has a higher degree of complex biogeochemistry (in the way it reacts with organic matter and bioaccumulates) compared to cadmium in the same environment, different data are needed to assess their impact to an ecosystem. Data from existing literature[18] and existing monitoring programs (e.g., USGS's NAWQA)[19] can help identify relationships useful in establishing models.

Existing information from the literature and indicators used in monitoring programs are useful in characterizing the pathways and receptor components when conducting an ecological risk assessment. For example, if there is a proposal to construct a new hazardous waste incinerator in a rural area, an ecological risk assessment could be used to evaluate different sites. This approach could help select the site that would have the least likelihood of causing adverse ecological effects. These results can be overlaid (using a GIS) with the results from a human health risk assessment to select the best option with the lowest overall risk. Data from existing monitoring programs (Table 10.1) can help guide selection of important ecosystem parameters that should be included in the risk assessment. Programs like NAMS, SLAMS, IMPROVE, and NADP (see Table 10.1) identify critical or sensitive airsheds to avoid; the USGS's NAWQA program[19] is a source for identifying important surface and ground water resources to avoid; and data from the FWS, BLM, USGS, and from NatureServe (http://www.natureserve.org/) can provide information on sensitive areas for plant and animals that should be avoided.

If scientists need to conduct both ecological and human health risk assessments for the incinerator example, they would benefit by looking at several of the existing data sources from the atmospheric monitoring programs listed in Table 10.1 (e.g., programs 11, 12, 16, 19, and 21). Depending on the location these data sources can provide status and trend data to help establish the following:

- Average annual wind patterns for the area
- Importance and extent of atmospheric pathway
- Critical aquatic and terrestrial receptors or indicators of ecological change for selected contaminants
- Data that would assist in developing a conceptual model depicting the interaction of the atmospheric pathway with ecological receptors like those detailed in Figure 10.2
- Information about sensitive species and habitats

Existing atmospheric data, in combination with other data sources from other media, are a powerful tool when conducting ecological risk assessments. However, some level of independent verification (i.e., collection of field data) must be done on any assessment that relies solely on existing data.

Collecting and analyzing samples and interpreting data to evaluate all components of an ecotoxicology study can be cost-prohibitive. But, depending on the scale of the study, there may be several data sources available to provide information about the resource or topic. For example, if the objective of the study is to assess ecological risk from a new pesticide on wildlife, existing data would be most useful in assessing information about pathways and critical receptors from an existing chemical with similar properties. Existing data can help provide baseline values and identify pathways for movement of contaminants in the ecosystem of concern — areas where contaminants

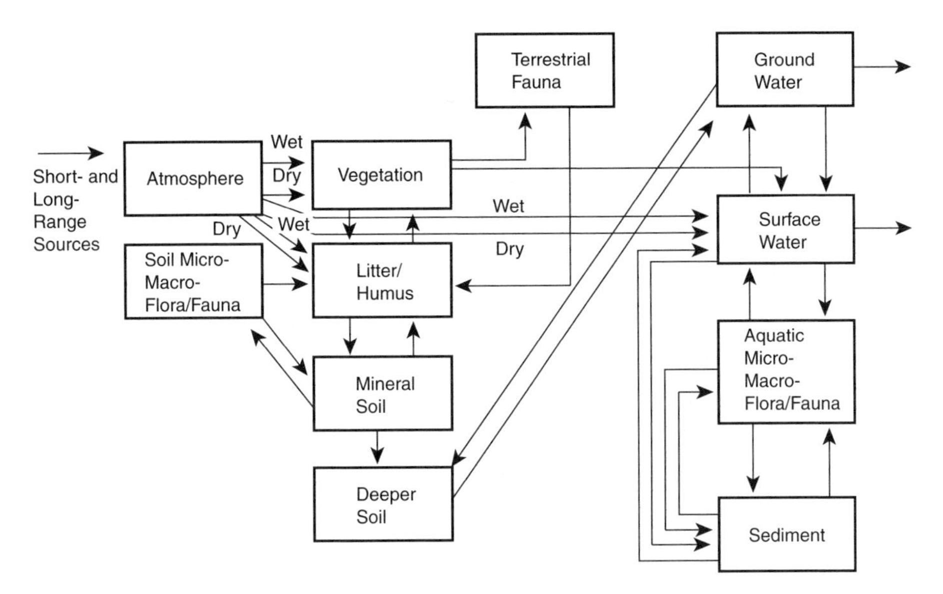


**Figure 10.2** Existing data can be used to help identify ecosystem dynamics and formulate these into a conceptual model. This model was used to design a monitoring program to assess the impact of atmospheric deposition of contaminants on a high alpine ecosystem. (From Bruns, D. A. et al., *Environ. Monitor. Assess.*, 17, 3, 1991. With permission.)

partition — and important indicator species; however, specific parameters for the chemical in question will still need to be provided to meet regulatory requirements.

In the past monitoring systems often concentrated on a single medium for sampling (e.g., air, water). This is clearly demonstrated in the types of programs identified in Table 10.1 at an annual cost of billions of dollars.[20] This single-medium focus has been fostered by the way regulations have been promulgated. Because of regulations like the Clean Air Act and Clean Water Act and the organization of Federal agencies, the interactions within and among ecosystems have not been appreciated or studied. Ecotoxicology is pushing regulators to reconsider how and why the regulations were developed. If the purpose is to protect natural resources, then regulations will change to reflect the enhanced interest in protecting the complex nature of natural systems.

Existing data are very useful in the design and interpretation of results of new ecotoxicology and monitoring studies. Existing data can

1. Provide information on media to monitor, identify optimal collection time (index period), assess variability, guide statistical design, and develop conceptual models of system- and process-level interactions (e.g., EMAP, NAWQA) (see Figures 10.1 and 10.2).
2. Provide data on reaction rates or partitioning coefficients between components when designing integrated monitoring networks.[17,21]
3. Identify problems and limitations of previous research.
4. Suggest reporting formats and appropriate distribution networks for various users (e.g., management, research scientist, and public).
5. Provide approximate estimates for scheduling time and costs.

Study design can be improved by using existing data sources in combination with current knowledge to develop a conceptual model of how the system of interest interacts. Figures 10.1 and 10.2 illustrate this concept of how models can be used to help identify questions and level of understanding about the interactions between various components in the ecosystem (e.g., influence of atmospheric deposition on biogeochemical cycling). An exercise evaluating existing data often reveals the most sensitive part of the system to monitor or identifies data gaps related to the

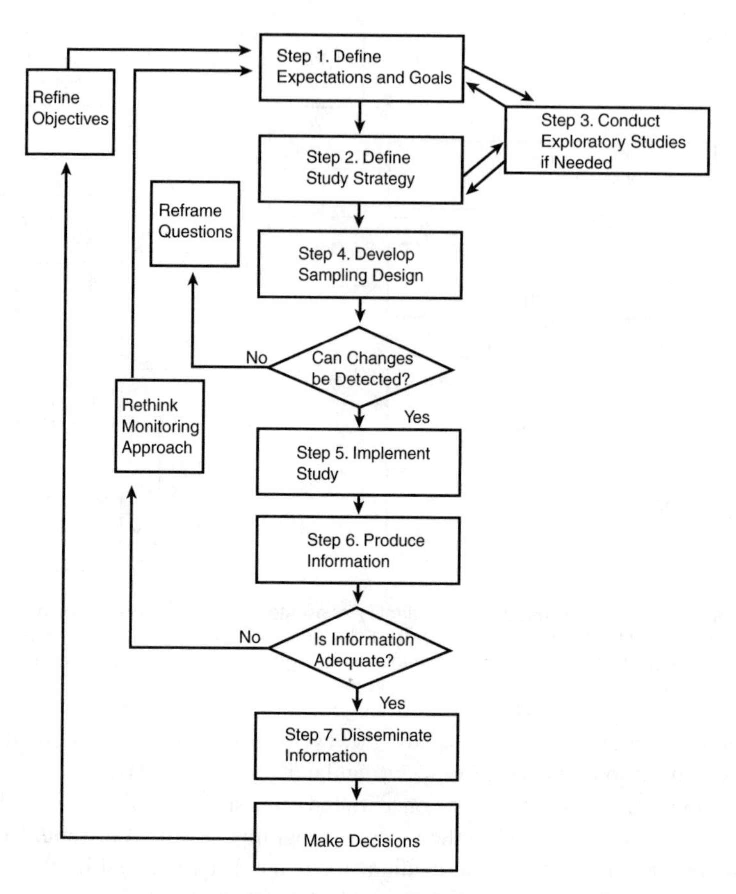


**Figure 10.3** Key elements of designing and implementing a monitoring program. (From National Research Council, *Managing Troubled Waters: The Role of Marine Environmental Monitoring,* National Academy Press, Washington, D.C., 1990. With permission.)

interactions and linkages being studied. The feedback loops presented in Figure 10.1 help ensure that objectives are properly formulated and addressed. Once the generic linkages are established, they can be focused to meet the need of the individual study. Figure 10.2 provides an example of how this process was applied to establish the relationships important in addressing questions about the impacts of wet and dry deposition on a high alpine ecosystem.[22]

The NRC published a seven-step conceptual approach to designing and implementing monitoring programs.[23] Developed by the Committee on a Systems Assessment of Marine Environmental Monitoring, the approach also has application to designing research projects and large-scale monitoring programs. Existing information plays an important role in providing answers to many of the steps. Existing data can be accessed at several steps in the process to make it a more powerful tool in designing useful research projects and monitoring programs. The general elements of the seven-step approach are presented in Figure 10.3. Each of the steps in the process is presented in Table 10.2 along with discussion and examples on how to include existing data and recognize data limitations.

Once the study objectives are selected, the next step should be to seek out existing information and data that can help address study objectives. A question often asked by management and Congress is, why do we need another monitoring program? To answer this question scientists need to spend time identifying which environmental media are being monitored, by whom, and how often. The result of this effort can provide the convincing argument as to what is the unique need for existing and new programs.

**Table 10.2 Benefits and Limitations on Using Existing Data to Help Design and Implement Ecotoxicology Studies and Monitoring Programs**

| Steps for Implementing a Monitoring Program[a] | Benefits of Considering Existing Data and Information | Limitations with Existing Data |
|---|---|---|
| 1 — Define expectations and goals for study | Existing data identify compliance requirements for the area (e.g., water contaminants, NPDES permit, Class I airshed, etc.).<br>Identify past and current stressors[24] and metals,[25] areas sensitive to acid deposition (NADP).<br>Identify gaps in existing monitoring efforts (i.e., currently little known about numbers or contaminant effects on terrestrial or amphibian species).<br>Identify data quality needs (e.g., previous data not legally defensible). | Access and time required to collect and evaluate data is time-intensive; data referral services are very useful.<br>Quality of existing data may be too poor to be useful |
| 2 — Define study strategy | Identify resources at risk from different stresses; large-scale effort can identify regional stresses (e.g.,acid deposition, mercury deposition). Local stresses identified by regional, state, and private efforts.<br>Help develop conceptual models of interaction occurring or expected in system. Data from literature important for establishing exchange rates (first order kinetics) between different part of system of interest.[21]<br>Provide data to assess natural variability of system or components of interest, thus, allows investigator to assess signal to noise-ratio.<br>Identify sources of contaminants coming into study area; can help establish rough mass balance of contaminant load (e.g., $SO_2$ or $NO_x$ deposition rate).<br>Identify baseline data for study area. | Possible to overlook unidentified stress<br>Existing data helpful in establishing initial model, but data need to be verified with field or laboratory-specific data.<br>Statistical value must be checked<br>Quality of data and collection methods may not be comparable |
| 3 — Conduct exploratory studies | Identify data-rich area useful to help plan study and improve data interpretation potential (e.g., with good existing data set could make trend assessment from 1 year of data).<br>Identify previous problems with access, logistics, or regulatory issues (e.g., does National Environmental Policy Act (NEPA) apply?). | Quality of data and access need to be checked carefully |

**Table 10.2 Benefits and Limitations on Using Existing Data to Help Design and Implement Ecotoxicology Studies and Monitoring Programs** *(Continued)*

| Steps for Implementing a Monitoring Program[a] | Benefits of Considering Existing Data and Information | Limitations with Existing Data |
|---|---|---|
| 4 — Develop sampling design | Quantify variability levels to establish data quality objective before starting work; data most useful for aquatic and near coastal resources.<br>Identify what has been previously monitored in similar studies and success rates (e.g., Pb levels in blood of birds have been used for assessing metal smelter.[29] Metals in different species of fish can be compared, but not directly because many species have different accumulation factors.[28]).<br>Enhance statistical base for making comparisons or statements from short-term study. | Chemical-specific data are limited<br>Limited data for assessing contaminants in terrestrial species or systems.[4] |
| 5 — Implement study | Identify proven methods and assess usefulness<br>Identify logistical problems (e.g., contact point for access, environmental and safety hazards). | Must be careful data and methods are comparable<br>Existing data and information are always subject to change; thus, they must be verified, especially if regulations change |
| 6 — Produce information | Provide good examples on how to conduct assessments, statistical tests, and format report.<br>Provide basis for comparison of study results with previous studies. | Quality of data must be checked |
| 7 — Disseminate information | Help identify different types of audiences and level of report needed or expected.<br>Identify appropriate means and places to distribute information (e.g., journals that previously accepted similar results, use of different media for internal vs. external audiences).<br>Identify limitations or requirements associated with dissemination of results (e.g., path clearance, peer review, and transmittal process). | Previous studies can set precedent for type and level of detail expected by management or peers |

[a] Steps from NRC[23] (see Figure 10.3).

## 10.5 ACCESSING EXISTING DATABASES

Many of the Federal datasets have been organized and made available via the Internet or can be ordered on CD or to be downloaded (see Table 10.1). Also available are data compilation or reference services that sell or provide computer access to large datasets. Most of these services place advertisements in trade and scientific journals. This information can generally be accessed either by using standard personal computer systems with CD-ROM readers or via the Internet. The National Technical Information Service is also a good source of data that can be accessed through most technical libraries.

Because of the large quantity of and increasing interest in accessing existing databases, most agencies have established an office of information management. Many of these information management systems are accessible through computer and phone line access (see Table 10.1). For example, the EPA has developed an Environmental Information Management System (EIMS) (http://www.epa.gov/eims/emshome.html) that stores, manages, and delivers descriptive information (metadata) for datasets, databases, documents, models, multimedia, projects, and spatial information. Users including scientists, resources managers, and others searching for information sources of interest based on topic or criteria related to types of environmental resources, geographical extent, data, or content origin.

## 10.6 CONCLUSION

Capitalizing on the millions of dollars invested in monitoring of the United States' environmental systems requires an understanding of opportunities and limitations. Managers are becoming increasingly aware of the high cost of conducting ecotoxicology and monitoring studies. Existing information can play an important role in reducing costs of performing an ecological risk assessment or in the design of new ecotoxicology or monitoring studies. Scientists can use existing information to obtain support for a new program or hypothesis by identifying both gaps in existing information and a need for new programs.

In order for scientists to produce high-quality studies or projects in the most cost-effective manner one of the initial steps should be to seek out applicable information and assess its usefulness. Many scientists do not take full advantage of existing data because this can be a tedious effort. However, by developing a set of selection criteria tailored to the special interest of the new effort, the new project can often be given a boost during its initial stages by taking advantage of lessons learned from previous efforts, by identifying important components to monitor, and by improving interpretation of results even with only limited data.

Existing data are useful when they can be exchanged between users in a comparable manner. For this to occur, the existing data must have good documentation as to quality, purpose, place and time period of collection, and limitations. Use of GISs and spatial analysis and modeling techniques enhance the use of existing and new data, allowing scientists and managers to assess data in ways that were impossible in the past.

Environmental science is at an exciting stage in its development. With each year come new and better ways of conducting ecotoxicology and ecological risk assessment studies. Scientists have an obligation to consider future activities and efforts when designing their studies. The data and information collected from the program must be documented so they are useful to future investigators. This requires that the data be of known quality and that future users are able to identify why, when, where, and how they were collected. Teaching new scientists about the benefits and utility of existing data is the responsibility of all established scientists. Users must be aware of the limitations and biases of the existing data and use those data appropriately. The right fit of the data can enhance the development of a new study; the wrong application (because it does not identify limitations) can cause serious design or interpretation problems. By establishing acceptance criteria tailored to their needs, scientists can identify and benefit from existing information.

## ACKNOWLEDGMENTS

Review of this manuscript by Gail L. Olson is gratefully acknowledged.

## REFERENCES

1. Johnson, N. C., Malk, A. J., Szaro, R. C., and Sexton, W. T., Eds., *Ecological Stewardship: A Common Reference for Ecosystem Management*, Elsevier Science, Oxford, UK, 1999.
2. U.S. Geological Survey, Biological Resources Division, The Status and Trends of the Nation's Biological Resources, U.S. Dept. of Interior, Washington, D.C., 2000.
3. Rennie, J., Ed., Misleading math about the earth, *Scientific American*, 62–72, January 2002.
4. Lomborg, B., *The Skeptical Environmentalist*, Cambridge University Press, Cambridge, UK, 2001.
5. National Research Council, *Grand Challenges in Environmental Sciences*, National Academy of Sciences, Washington, D.C., 2001.
6. Olson, G. L. and Breckenridge, R. P., *Federal Contaminant Monitoring Programs and Databases: A Fish and Wildlife Perspective*, EG&G Idaho, Informal Report No. EGG-EST-9341, Idaho Falls, Idaho, November 1990.
7. Breckenridge, R. P. and Olson, G. L., Identification and use of biomonitoring data, in *Handbook of Ecotoxicology*, Hoffman, D. J., Barnett, A., Rattner, Burton, G. A., Jr., and Cairns, J., Jr., Eds., CRC Press, Boca Raton, FL, 1995.
8. Sine, C., Ed., *Farm Chemicals Handbook*, Meister Publishing Company, Willoughby, OH, 2002.
9. Van Horn, R. L, Risk analysis and toxicology, in *Hazardous and Radioactive Waste Treatment Technologies Handbook*, CRC Press, Boca Raton, FL, 2001.
10. U.S. Dept. of Interior, Fish and Wildlife Service, Flyways: Pioneering Waterfowl Management in North America, Hawkins, A. S., Hanson, R. C., Nelson, H. K., and Reeves, H. M., Eds., U.S. Government Printing Office, Washington, D.C., 1987.
11. Norton, S. B., Rodier, D. J., Gentile, J. H., van der Schalie, W. H., Wood, W. P., and Slimak, M. W., A framework for ecological risk assessment at the EPA, *Environ. Toxicol. Chem.*, 11, 1663, 1992.
12. U.S. Environmental Protection Agency, Peer Review Workshop Report on a Framework for Ecological Risk Assessment, Risk Assessment Forum, U.S. Environmental Protection Agency, Washington, D.C., EPA Report EPA/625/3–91–022, 1992.
13. Levin, S. A., Harwell, M. A., Kelly, J. R., and Kimball, K. D., Eds., *Ecotoxicology: Problems and Approaches*, Springer-Verlag, New York, 1989, chaps. 9 and 17.
14. Hope, B. K., Ecological considerations in the practice of ecotoxicology, Letter to the editor, *Environ. Toxicol. Chem.*, 12, 205, 1993.
15. Bartell, S. M., Gardner, R. H., and O'Neill, R. V., *Ecological Risk Estimation*, Lewis Publishers, Chelsea, MI, 1992.
16. Suter, G. W., Barnthouse, L. W., Bartell, S. M., Mill, T., Mackay, D., and Paterson, S., *Ecological Risk Assessment*, Lewis Publishers, Chelsea, MI, 1993.
17. Jorgensen, S. E., Ed., *Modelling in Ecotoxicology*, Elsevier Publishers, New York, 1992.
18. Kokoszka, L. C., Guide to federal environmental databases, *Pollut. Eng.*, 83, 1992.
19. Hirsch, R. M., Alley, W. M., and Wilber, W. G., Concepts for a National Water Quality Assessment Program, U.S. Geological Survey Circular #1021, 1988.
20. Bureau of the Census, Statistical Abstract of the United States 2000, U.S. Government Printing Office, Washington, D.C., 2000.
21. Wiersma, G. B. and Otis, M. D., Multimedia design principles applied to the development of the global baseline integrated monitoring network, in *Pollutants in a Multimedia Environment*, Wolfe, D. A., and O'Connor, T. P., Eds., Krieger Publishing, Malabar, FL, 1986, 3–20.

22. Bruns, D. A., Wiersma, G. B., and Rykiel, E. J., Jr., Ecosystem monitoring at global baseline sites, *Environ. Monit. Assess.,* 17, 3, 1991.
23. National Research Council, *Managing Troubled Waters: The Role of Marine Environmental Monitoring*, Committee on a Systems Assessment of Marine Environmental Monitoring, Marine Board, Commission on Engineering and Technical Systems, National Academy Press, Washington, D.C., 1990.
24. Schmitt, C. J., Ricbick, M. A., Ludke, J. L., and May, T. W., National Pesticide Monitoring Program: Organiochlorine Residues in Freshwater Fish, 1976–79. U.S. Dept. of Interior, Fish and Wildlife Service, Resource Publication 152, Washington, D.C., 1983.
25. Eisler, R., Lead Hazards to Fish, Wildlife, and Invertebrates: A Synoptic Review, U.S. Department of the Interior Fish and Wildlife Service, Biological Report 85(1.14) Contaminant Hazard Reviews Report No. 14.

CHAPTER 11

# Bioindicators of Contaminant Exposure and Effect in Aquatic and Terrestrial Monitoring

**Mark J. Melancon**

## CONTENTS

## 11.1 INTRODUCTION

There are many different types of contaminants present in the environment, ranging from synthetic chemicals, which would not be present in the environment without human intervention, to trace metals that are required for life. Concerns range from possible harmful effects on flora and fauna to possible harm to humans from consuming contaminated organisms. The complete chemical analysis for all possible environmental contaminants in sediments, water, air, and every species and sample of interest would not only be excessively costly, but the facilities to handle so many samples do not exist. Furthermore, such chemical analysis would only establish the presence of contaminants, without revealing how available or active they were within the organism.

Bioindicators, sometimes called biomarkers, are responses of living organisms that may simply signify exposure to contaminants, may predict future harm due to the exposure, or may themselves be harmful effects. In this chapter *bioindicator* will be interpreted to mean the same as a biomarker (a biochemical physiological or morphological response), but not a population or ecosystem bioindicator such as a sentinel species, genetic variability, or species richness. Thus, a bioindicator (biomarker) may be studied in a particular species to assess the status of that species or in a sentinel species to assess habitat status. Individuals of the sentinel species might be those already residing at the site of interest or might be pristine organisms brought to the site and maintained in cages (fish) or released and recovered (birds). Although most bioindicators are measured in tissue samples taken from the study organisms, the exposure itself must have occurred in a living organism.

Systematically, the responses can range from minor biochemical or physiological responses (homeostatic responses) in individual organisms to major toxicity (in an individual, a species, a community, or an ecosystem).

At the least, a bioindicator response shows that a contaminant (or contaminants) is present in the environment, that it is available to the organism, and that it has reached the affected tissue or organ in sufficient amounts for a period of time long enough to cause the observed response. The relative value of such a response to indicate harm ranges from a simple homeostatic response to a progression of levels of response indicating possible damage. The progression includes responses that indicate that a process that may lead to harm has been initiated, that a process that definitely leads to harm has been initiated, that the indicator itself is a harmful effect, that this harmful effect will lead to death or decreased reproduction for the individual, or that this harmful effect will lead to significant population effects.

For a response to a contaminant to be a useful biomarker, there are a number of properties of the response that are desirable, some of them even necessary. There must be a known specificity of the response to a contaminant presence or exposure, some correlation of the presence or magnitude of the response to the amount of the contaminant present, and a known temporal relationship between exposure and effect. The biomarker response should precede harm, but ideally it would be a predictor of future harm. Samples should be readily obtainable in quantities needed for measurement. Ideally, the sampling would be nonlethal, which would greatly increase its applicability. The assay method must be accurate and reproducible. The overall costs should be reasonable in regard to equipment, supplies, and technician time when compared to the cost of nonbiomarker approaches.

Although the biomarkers are discussed individually in this chapter, their most effective application may be as a suite of biomarkers, unless information is needed only on a particular contaminant or class of contaminants for which there is a highly specific biomarker. Because in most cases a variety of contaminants may be present, and most biomarkers have limited specificity, a battery of biomarkers is required to assess multiple classes of contaminants. An added advantage of using a suite of biomarkers is that it might also distinguish types of contaminants that can affect more than one element in a suite of biomarkers. This chapter will present an overview of selected biomarkers that are presently in use or ready to use in the field and present suggestions for future directions in the area of bioindicators. Four books, *Biomarkers of Environmental Contamination*;[1] *Biomarkers:*

*Biochemical, Physiological and Histological Markers of Anthropogenic Stress*;[2] *Animal Biomarkers as Pollution Indicators*;[3] and *Nondestructive Biomarkers in Vertebrates*[4] contain details on a wide variety of biomarkers and provide more extensive information on the biomarkers covered in this review. More recently, there have been manuscripts and chapters devoted to the biomarker concept and biomarker descriptions.[5–9]

## 11.2 INDIVIDUAL BIOINDICATORS

### 11.2.1 Cholinesterase Inhibition

In most cases a bioindicator response to a particular contaminant is fortuitous because contaminants generally arise from anthropogenic processes independent of specific processes in fish and wildlife. However, the anticholinesterase pesticides, such as the carbamate and organophosphorus compounds, were designed to cause death by inhibition of cholinesterase activity. Even though designed as insecticides, they are well known as agents causing kills of wild vertebrates and as such present an unusual opportunity for assessing exposure of fish and wildlife. Thus, it is possible to monitor terrestrial and aquatic organisms for inhibition of a system that a contaminant was designed to affect. Although cholinesterase activity is present in many tissues, such inhibition in environmental samples is generally determined from samples of brain tissue or blood (serum or plasma). Brain contains mainly acetylcholinesterase activity, while blood generally contains both acetylcholinesterase and butrylcholinesterase activities.

The use of cholinesterase inhibition in the determination of cause of death of an organism is more definitive than the use of such inhibition in a living organism. A finding of ≥50% inhibition of brain cholinesterase activity, along with the residual presence of a carbamate or organophosphorus pesticide in the body (GI tract), is considered diagnostic for the cause of death in birds.[10] Fish in general present a more variable relation between death and brain cholinesterase inhibition; lethal inhibitions reported range from 40 to 80%.[11–13]

It has been demonstrated that normal cholinesterase activity varies between species[14–16] and that age affects cholinesterase activity significantly in species such as altricial birds[17] and mammals.[16] There are several approaches for determining the control or normal levels to which to compare the possibly inhibited levels. The best approach is to collect concurrent reference samples of comparable individuals of the species of interest from an uncontaminated area. Other approaches are to use previously collected data on comparable uncontaminated individuals of the same species or to treat the sample so as to reverse or remove the inhibition (reactivation assay[18]) so that the sample can serve as its own control. Although 50% inhibition of cholinesterase activity may be found in some dead birds, other individuals may survive with even greater inhibition.

The effectiveness of using cholinesterase inhibition as a biomarker in living, field-collected animals depends on the quality of the control values and whether possibly inhibited samples can be demonstrated to be significantly different from control values. Johnson et al.[19] have reported on the stability and correlation of tissue pesticide residues and cholinesterase activity in attributing death to such pesticides. Depressed brain cholinesterase activity is indicative of significant exposure to anticholinesterase pesticides, and the inhibition of activity may remain for several weeks following exposure and correlate well with effect. Depressed blood or plasma cholinesterase activity is indicative of exposure to anticholinesterase pesticides rather than harm, and the inhibition of activity is short-lived. Despite these limitations blood sampling has the advantages that animals do not have to be sacrificed for sampling, serial samples may be collected from the same animal, and continuing exposure can be detected. Although more definitive results are usually obtained with brain cholinesterase inhibition than with blood or plasma cholinesterase inhibition,[20] for biomonitoring the advantage of obtaining blood samples without sacrificing an animal may be the more important consideration.

The use of cholinesterase as a biomarker has been studied, including field validation, in many species[14–16] and is widely accepted as a biomarker in avian, mammalian, and aquatic species. The number of species to which this technique has been applied continues to increase. The use of cholinesterase inhibition in additional species requires sufficient testing to ensure adequate baseline data for the analytical technique selected.

Pertinent reports include that of Hill on brain cholinesterase activity of apparently normal wild birds,[16] two letters to the editor expanding upon this report,[21,22] the book, *Cholinesterase-Inhibiting Insecticides: Their Impact on Wildlife and the Environment*,[23] a brief review of esterases as biomarkers, including cholinesterase inhibition,[24] and a Canadian Wildlife Service methods report.[25]

### 11.2.2 Cytochromes P450

The cytochromes P450 are a class of hemoproteins present in all tissues but found in especially high concentrations in the liver. They are inducible, that is, increased amounts of specific cytochromes P450 are found after treatment of organisms with a variety of organic chemicals. Of greatest interest for biomonitoring is that several major classes of environmental contaminants induce cytochromes P450. The two major types of cytochromes P450 that are induced by contaminants are those classically described as the 3-methylcholanthrene-inducible and the phenobarbital-inducible types. Currently, these inducible cytochrome P450 classes are referred to as CYP1A and CYP2B, although in some species the phenobarbital-inducible cytochromes P450 have been designated differently within the CYP2 family.[26] Although in laboratory studies, administration of large quantities of an inducing agent can increase the amount of total cytochromes P450 several-fold, such magnitude of induction usually is not the case in environmental exposure. Environmental exposures to contaminants that are inducing agents usually occur at concentrations much below those for maximal induction, so the total mass of cytochromes P450 may not increase significantly. Thus, more sensitive methods must be used to discriminate between constitutive cytochromes P450 and induced cytochromes P450.

#### *11.2.2.1 Monooxygenase Activity of Cytochromes P450 in Microsomes or S9 Fraction*

Associated with the cytochromes P450 is a wide range of enzyme activities referred to as monooxygenase activities. In an organism these enzyme activities increase the polarity of lipophilic xenobiotics, thereby speeding detoxication and elimination, but they can also metabolize some xenobiotics to more toxic forms. A number of highly sensitive monooxygenase assays, mainly oxidases, hydroxylases, and dealkylases, have been developed for evaluating the amounts of induced cytochromes P450 present.[27–30] Appropriate assay conditions and specificity of response (CYP1A- or CYP2B-linked) must be ascertained for each species. Ethoxyresorufin-*O*-dealkylase (EROD) is the monooxygenase assay usually used in biomonitoring because it is a highly sensitive fluorometric assay, and because in most species, it is CYP1A-linked rather than CYP2B-linked, and it is effective in species such as fish where CYP1A, but not CYP2B, is induced by contaminants.[31]

#### *11.2.2.2 Immunological Assessment of Cytochromes P450*

Cytochromes P450 are frequently evaluated by Western blot. Both monoclonal and polyclonal antibodies have been used successfully for avian, mammalian, and aquatic species.[32–35] A major advantage of Western blot over monooxygenase assays is that the constitutive cytochromes P450 may possess some activity in a particular monooxygenase assay, but the antibody method is specific for an individual or a family of induced cytochromes P450. The Western blot technique might also permit assessing cytochromes P450 in samples that have not been stored adequately for assaying monooxygenase activity. The use of antibody to fish CYP1A has also been applied to a semiquantitative ELISA.[36,37]

#### 11.2.2.3 Cytochromes P450 mRNA

Increase in the hepatic mRNA for CYP1A has been demonstrated in a number of fish species following exposure to contaminants by exposure routes ranging from manual dosing to environmental exposure.[38–40]

#### 11.2.2.4 Nonlethal Methods of Assessing Cytochromes P450

The use of *in vivo* metabolism of xenobiotics and tissue biopsy samples for P450 quantitation in wildlife is still in development. Such nonlethal methods are important because they provide the opportunity (1) to use cytochromes P450 as a biomarker in species for which sacrificing individuals is not possible and (2) to monitor the same individuals over time. The sleeping-time method is an indirect method for evaluating the cytochrome P450 status of an individual by comparing sleep duration after administration of phenobarbital to animals with suspected contaminant exposure and control animals. If the contaminant caused increased metabolism of phenobarbital, then sleeping time for the contaminant-exposed group will be shorter than that of the control group. This has been observed in wild-trapped muskrats, *Ondatra zibethicus*.[41] Also, the metabolism of an administered chemical like caffeine may be determined by several methods, including disappearance of the caffeine from the blood, appearance of metabolites in excreta, or appearance of tagged $CO_2$ from $^{13}C$- or $^{14}C$-labeled caffeine in exhaled air.[42,43] It also is possible to apply monooxygenase assays or Western blot to tissue obtained by biopsy from species with sufficient size and hardiness to exhibit rapid and complete recovery after the removal of approximately 0.05 to 0.25 g of liver tissue. The observation of *in vivo* ethoxyresorufin dealkylation in killifish embryos has also been reported.[44]

Cytochromes P450 have been studied in too many fish species to list them all in this brief review, and the number of wildlife species studied, including mammals, birds, reptiles, amphibians, and aquatic organisms is increasing rapidly. Numerous review articles have covered induction of cytochromes P450 as a biomarker for exposure to environmental contaminants.[31,45–49]

### 11.2.3 Reproductive Problems

#### 11.2.3.1 Eggshell Thinning

Severe eggshell thinning can result in breakage of an egg before hatching, effectively preventing reproduction. In such cases determination of the shell thickness of the broken eggs would be diagnostic. Less severe, but significant, eggshell thinning provides an indicator for possible reduced reproductive success. DDE, the metabolite of DDT, is the major, if not the only, cause of contaminant-induced shell thinning in many species of fish-eating birds.[55–56] Eggshell thickness can be measured directly with a micrometer or the thickness index calculated as the weight of the shell (mg) divided by the product of shell length and breadth ($mm^2$).[57] Recent studies show a better correlation for shell strength than for shell thickness to contaminant-associated abnormalities.[58–60] Because the shell is partially used up by the developing embryo, these measurements must be done early in incubation. Because only one egg is taken early in incubation, for many species the use of this biomarker would have minimal effect on reproduction. Though one of the earliest biomarkers, it is still being used.[61]

#### 11.2.3.2 Terata

Environmental contaminants can cause a variety of developmental abnormalities in aquatic and terrestrial organisms. In the environment, exposure of the embryo could occur via the mother (placental transfer or egg formation) and also (in cases such as birds and fish) via direct exposure of the eggs after they are released. Laboratory studies have utilized exposure by injection of contaminants into the egg air sac or yolk. The embryotoxicity and teratogenicity of externally applied petroleum and related materials and pesticides to avian embryos have been summarized by Hoffman.[62] Some of the most clearly established relations between terata and contaminants include the crossed bills caused by chlorinated hydrocarbons in water-birds in the Great Lakes and beak and other abnormalities caused by selenium in waterfowl and shorebirds using irrigation drainwater ponds. The most extensive study of bill malformations caused by chlorinated hydrocarbons included the examination of over 31,000 cormorant chicks from colonies in the Great Lakes and almost 21,000 chicks in reference areas.[63] Only two chicks in reference areas had bill deformities, while 70 of those in the Great Lakes had bill deformities. Selenium exposure of American coot (*Fulica americana*), mallard (*Anas platyrhynchos*), eared grebe (*Podiceps nigricollis*), and black-necked stilt (*Himantopus mexicanus*) resulted in embryo death[64] and terata in embryonic livers.[65] Embryo deaths were found in 17 to 60% of the nests of these species involving 2.5 to 31.7% of the eggs. Based on observations of at least two viable embryos per nest, abnormalities of eyes, brain, feet, or beak were found in 13 to 53.8% of nests and 8.6 to 16.4% of embryos.

#### *11.2.3.3 Vitellogenin as a Biomarker for Estrogenic Activity in Males*

Since the discovery of male alligators at Lake Apopka, Florida, with smaller than normal external genitalia, there has been concern about environmental chemicals that modify androgenic and estrogenic activity. In the mid-1990s blood level of vitellogenin was suggested as a biomarker for estrogenic activity in oviparous species.[66,67] This is a very sensitive biomarker in male and immature organisms that should not have vitellogenin present and is less sensitive in mature females, where it is normally present. This biomarker utilizes antibodies to vitellogenin in Western blot or ELISA, and it quickly became established as a biomarker in fish.[68–71] Detailed directions for preparing the necessary reagents and performing the antibody-based vitellogenin assays were recently published.[72] An assay for vitellogenin mRNA in largemouth bass has also been reported.[73] This and other hormone-related effects are covered in depth in Chapter 38.

### 11.2.4 Aberrations of Hemoglobin Synthesis

#### *11.2.4.1 Red Blood Cell-δAminolevulinic Acid Dehydratase (ALAD)*

Activity of ALAD, the enzyme responsible for condensing two molecules of aminolevulinic acid to form porphobilinogen in the synthesis of hemoglobin in red blood cells, is very sensitive to lead. Inhibition of red blood cell ALAD activity as a direct effect of lead has been clearly demonstrated in numerous human studies and in fish[74] and wildlife.[75] The negative correlation between red blood cell ALAD activity and blood lead concentration was studied in fish[76,77] and birds.[78,79] Examples of species in which the effects of environmental exposure to lead on ALAD activity have been monitored are canvasbacks (*Aythya valisineria*),[80,81] European starlings (*Sturnus vulgaris*),[82] suckers (catostomid fishes),[83] and snapping turtles (*Chelydra serpentina*).[84] The sensitivity of response must be determined in each species to be monitored. Two approaches are available for assessing the inhibition. Reference values or concurrent samples from relatively unexposed individuals may be used, or the inhibition can be reversed with zinc and the ratio of inhibited to restored activities can be used.[85] Among the advantages of this assay are that it does not require sacrifice of the animal to be monitored, the sample may be stored in an ultracold freezer for up to two months before assay, and only readily available laboratory equipment is required for the assay.

#### *11.2.4.2 Porphyria*

A number of chlorinated organics affect hemoglobin synthesis at points closer to hemoglobin formation and result in the accumulation of highly carboxylated porphyrins that may be found in liver, blood, urine, or feces.[86–88] Although an early approach to such measurements gave a total porphyrin value, the use of high-performance liquid chromatography (HPLC) makes possible the determination of individual porphyrins, which gives greater sensitivity and specificity to this method.[89,90] This effect was clearly shown in herring gulls (*Larus argentatus*) from the Great Lakes, which were contaminated with chlorinated aromatics, as compared with herring gulls from control locations[88] and in river otters (*Lontra canadensis*) following a petroleum spill.[91] Lead has also been demonstrated to cause increased concentration of protoporphyrin in erythrocytes in mallard ducks[92,93] and black ducks.[94,95] As a biomarker, porphyria has the advantage that it is a negative effect on a necessary pathway, not merely a homeostatic response. Depending on the quantity of porphyrins found, this response may be representative of minor effects, probably without implications for survival, through a gradation of responses, to severe liver damage. Because of the requirement for HPLC, quantitation of porphyrins is not a method that is amenable to occasional use but is not overly complex if set up and run on a regular basis.

### 11.2.5 DNA Damage

After uptake by an organism, some chemicals or their metabolites or both are capable of interacting with DNA to modify or damage it. The common methods for evaluating such interactions are based on direct measurement of DNA structural damage, assessment of DNA repair, or determination of mutations present in genomes.

The activated forms of a variety of polycyclic aromatic hydrocarbons can become covalently bound to DNA. Tests for assessing the presence of such adducts may be general or specific. The $^{32}P$-postlabeling technique[96] is a sensitive method for determining the quantity of DNA adducts present, regardless of the adduct, and has been applied to aquatic species from many contaminated areas.[97–101] Researchers at the National Marine Fisheries Service laboratory in Seattle have described extensive studies on the presence of DNA adducts in liver, some of which relate their presence to PAH-contaminated sediments and the occurrence of tumors in several species of marine fish.[97,102–104] The sensitivity is approximately one adduct per $10^{10}$ bases. It is also possible to hydrolyze the DNA and determine the specific adduct present by gas chromatography/mass spectrometry. This approach is much less sensitive, however. Attempts are being made to quantify such adducts via antibodies. Modification of DNA, such as by adduct formation, is the first step in contaminant effects on DNA and can lead to greater changes in the DNA, which can be detected or measured by other methods.

Strand breakage can be quantified by an alkaline unwinding assay in which the rate of unwinding, measured in an alkaline medium, is facilitated by strand breaks. This approach has been applied successfully to demonstrate increased strand breakage in freshwater fish exposed to benzo(a)pyrene[105,106] and in turtles from contaminated, as compared with clean, sites.[107]

Standard chromosome analysis and sister chromatid exchange are sensitive methods for detecting mutagenic damage. Flow cytometry, which has been used to detect genetic damage in small mammals[108] and birds,[109,110] is a good approach to assess such damage because of its sensitivity, convenience, and low cost. Several reviews[111,112] discuss DNA alterations and their role as a biomarker for contaminant exposure. See also Chapter 40.

### 11.2.6 Histopathological Effects

Many contaminants are capable of damaging specific cell types or organ systems. Unless severe and extensive, and therefore readily observable at necropsy, such damage may best be evaluated by histopathological techniques. Because of the variety of cell types and the many organs and organ systems present in any individual, there are an overwhelming number of histological approaches to detecting the effects of contaminants.[113] The main limitations are field validation studies in the species in question, assessment of whether convenience and cost are better for this approach, and whether this would serve as part of a suite of biomarkers.

Tumors are probably the most widely occurring aberration studied by histological methods. Tumors have been observed in a variety of fish species from contaminated areas, with the most common occurrences in bottom-dwelling fish in areas of PAH contamination, flatfish (*pleuronectitums*) in the marine environment, and bullheads (*Amerinons* spp.) in freshwater.[114–118] A notable exception is the finding of tumors in saugers (*Stizostedion canadense*) and walleyes (*S. vitreum*) in Torch Lake in Michigan.[119] There have been good correlations between sediment PAH levels, DNA adduct formation, and presence of tumors in flatfish.[117–119] A recent study with brown bullheads found tumor occurrences as high as 60% in liver and 37% in skin.[120]

This approach is well established and may be applied whenever a species is known to produce tumors in response to the contaminants of interest, reference values are available, and the frequency of tumor formation is high enough for likely detection.

### 11.2.7 Metabolites of Contaminants in Bile

Some contaminants, such as polycyclic aromatic hydrocarbons (PAHs) and substituted phenols, are metabolized so rapidly that concentrations of parent contaminant in an organism might not be indicative of the extent of exposure. In such cases analysis for the metabolites produced by the organism would be an advantageous approach. Although a variety of tissues could be selected for this analysis, bile is a good choice because of the relatively high concentration of metabolites in this fluid[121–123] and its relatively clean state (i.e., free of tissue lipids, etc.). Krahn and co-workers[124] developed a protocol for measurement of metabolites of PAHs in bile that involves separation of such metabolites into three fractions and quantification of these classes via fluorescence measurements at three wavelength pairs. Using this methodology they have demonstrated the presence of metabolites of a number of PAHs in bile from flatfish from areas thought to be contaminated.[125,126] In subsequent studies other researchers at the same laboratory have related such measurements to induction of cytochrome P450, DNA adduct formation, and the presence of tumors in several species of marine bottom fish.[102–104] Assessment of PAH metabolites in bile continues to be utilized in both freshwater and saltwater fish.[39,127–129] Metabolites of chlorinated phenols and resin acids have been found in bile of fish collected downstream from pulp paper mills.[130] A review of metabolites as biomarkers of contaminant exposure has been published.[131]

### 11.2.8 Oxidative Damage

Contaminants such as polycyclic aromatic hydrocarbons, halogenated aromatic hydrocarbons, heavy metals, selenium, industrial solvents, and many of their metabolites are capable of causing oxidative damage within an organism. The role of free radicals in such oxidative stress has been reviewed by Kehrer et al.[132] In response to oxidative stress, there may be adaptive responses of the protective (antioxidant) systems, modification of cellular macromolecules, and ultimately tissue damage. Changes in the antioxidant systems and modified macromolecules can serve as bioindicators for a variety of contaminant exposures.[133] The protective systems include reduced glutathione (GSH)/oxidized glutathione (GSSG)/glutathione reductase, catalase, superoxide dismutase, peroxidase, ascorbate, and alpha-tocopherol. Macromolecules that may be affected include lipids, proteins, and nucleic acids. Although most of the early studies in this area utilized typical laboratory research species, some recent studies have utilized wildlife species.

Excess dietary selenium caused increased glutathione peroxidase and glutathione reductase activities, GSSG/GSH ratios, and hepatic lipid peroxidation in mallard ducks.[134–136] Livers of adult coots with high levels of selenium collected from the Kesterson National Wildlife Refuge had elevated glutathione peroxidase activity and decreased thiol concentrations.[64] Spot (*Leiostomus xanthurus*) from a polluted site had elevated superoxide dismutase as compared to a clean site,[137] bivalves from contaminated areas had antioxidant parameters that related to heavy-metal concentrations,[138] and scaup from a contaminated area had elevated hepatic lipid peroxidation that correlated with several heavy metals and with PCBs.[110]

The potential of both adaptive and damaging responses to oxidative stress to serve as biomarkers of contaminant exposure appears excellent. Because of the wide variety of contaminants that can cause oxidative stress, these biomarkers can serve as indicators for exposure for a wider range of contaminants than can most other biomarkers. This lack of specificity also requires that this be used as part of a suite of biomarkers or that follow-up work would be required to determine more precisely the types of contaminants causing the response. Field validation studies are needed, both for specific oxidative stress-related responses and for concurrent exposures to multiple classes of contaminants.

### 11.2.9 Immune Competence

A variety of contaminants are known to affect the immune system. Phylogenetically, the presence of an immune system is widespread, and effects of contaminants on immune systems have been reported from earthworms (*Lumbricus terrestris*) to mammals.[139,140] Because compromise of the immune system by contaminants may lead to serious effects in aquatic and terrestrial organisms, these responses are not only indicators of exposure but also of potential or actual harm.[141–142] Numerous assays for immune function are available because of the complexity of this function and the need to assess its many parts.

An early study with a wildlife species examined the effects of organochlorine residues on susceptibility to infection in mallard ducks.[143,144] More recently, there have been demonstrations of impaired immune response in free-ranging wildlife. Macrophages isolated from kidneys of spot and hogchoker (*Trinectes maculatus*) from the Elizabeth River were less than those isolated from those species from a control location.[145] In dolphins immune response correlated inversely with blood PCBs and DDT.[146] In nestlings of Great Lakes terns and gulls T-lymphocyte function, as assessed by the phytohemagglutinin skin test, was inversely correlated with egg organochlorine contamination.[147] In a more controlled study, seals fed fish from a contaminated site had reduced immune function as compared to those fed fish from a cleaner site.[148]

The impact of toxic metals on immune function in earthworms has been evaluated in a number of studies.[149–154] Effects of contaminants on immune function have been demonstrated in controlled exposure studies of polychaetes,[155] bivalves,[156,157] starling,[158] and white-footed mouse.[159] Many of these assays are suitable for environmental application, needing only background information on responses in particular species and field validation.

### 11.2.10 Metallothionein

Metallothioneins are a class of small proteins, rich in cysteine, capable of binding metal ions and inducible by cadmium, copper, mercury, zinc, cobalt, bismuth, nickel, and silver ions[160] and by a variety of nonmetal factors.[161,162] Because they are induced by the metal ions enumerated above, they have been viewed as possible biomarkers for exposure to those metal ions. The methods currently used evaluate individual metallothioneins via the use of antibodies specific to those metallothioneins[163–167] or quantify the mRNAs for metallothioneins by the use of cDNAs specific to those mRNAs.[168–170] Evaluating metallothionein is slower and more expensive than analysis for the causative metals themselves. It must be determined on an individual basis whether the additional information on the route and time of exposure that are provided by this approach warrants the increased costs.

### 11.2.11 Heat Stress Proteins

Heat stress proteins (HSPs) are one group of a larger class of proteins referred to as stress proteins.[171] Previously referred to as heat shock proteins because of their rapid appearance following heat stress, they have since been found to increase in response to a variety of chemical and nonchemical stressors.[172–175] Initially, this broad responsiveness suggested that HSPs lacked adequate specificity to be of value as a biomarker. Subsequent studies demonstrated that there are many stress proteins, falling into major groupings by weight. These include HSP90 (80 to 100 kda), HSP70 (65 to 75 kda), HSP60 (chaperonin, approximately 60 kda), low molecular weight (20 to 50 kda), and ubiquitin (8 kda).[176–178]

Although all of these heat stress proteins deal with the structural integrity of other proteins, they differ in their functions in stressed and nonstressed individuals, baseline levels in nonstressed individuals, and magnitudes of increases following stress.[179] Although HSPs are referred to as constitutive and inducible, constitutive HSPs are also inducible. The tissues in which induction of heat stress proteins occurs and the patterns of individual heat stress proteins induced appear to be specific to species and to different types of stressors.

With the development of antibodies to some heat stress proteins,[180–184] Western blotting is the best of the already developed analytical methods for evaluating heat stress proteins.

In general, heat stress protein responses have been studied in laboratory experiments.[185–187] Although there have been some field studies,[188–190] investigations in additional species, including more field studies, are required before this biomarker can be widely applied to environmental samples. Much more information is needed on the response of particular stress proteins to particular contaminant and noncontaminant stressors. It also must be determined whether the information obtained by heat stress protein analysis is equal to or more valuable than results of chemical analysis for contaminants or other biochemical or physiological responses, whether heat stress proteins serve better as integrative or specific biomarkers, and whether such approaches are cost effective.

### 11.2.12 Other

A wide variety of biochemical and physiological responses to contaminants have been suggested as biomarkers, and it seems that new candidates are proposed in almost every environmental journal issue. It is beyond the scope of this chapter to discuss all of these candidate biomarkers, but a number of them have been covered in the publications cited previously.[1–4,6,9] Many of them are included in Table 11.1 along with those covered in this chapter. Most of these approaches are not as well established as those that are covered in this chapter, and interested readers are referred to these publications.

There are two emerging areas of study that must be mentioned. One area that is developing rapidly in the human health area is genomics and the use of cDNA microarrays to detect the presence of mRNAs in response to disease or other challenges. Some mRNAs are now studied individually as biomarkers in wildlife. Examples are those for cytochromes P450, stress proteins, and vitellogenin. What is lacking is background information on DNA for wildlife species to permit development of gene arrays. To date, most work in genomics has pertained to humans and to human-health-related laboratory species. When genomic information is available for wildlife species, or closely related species, it will be feasible to make gene arrays to use in broad screens of wildlife status or response to environmental challenges of many kinds. Similarly, broad-based approaches to the synthesis of specific proteins (proteomics) in wildlife species have yet to be developed. Regardless of the quantitation method, enzyme activity, Western blot, etc., many of the current biomarkers evaluate the presence of a single protein or a class of related proteins. Current separation techniques and analytical software present the potential for examining multiple proteins simultaneously. Basically, this would be the simultaneous assessment of multiple gene products that would arise from the mRNA responses described in the preceding paragraph, but would not require the genomic information. Early work of this type utilized two-dimensional electrophoresis followed by visualization and computer-based pattern comparisons, but arrays of antibodies for specific proteins are currently under development.

**Table 11.1 Presently Used and Possible Bioindicators[a]**

| Bioindicator | Description | Stressor |
|---|---|---|
| Protein-related | | |
| Proteomics | Broadly encompasses quantity and functioning of any protein; approaches being developed that may be protein equivalency of DNA arrays | Specificity will depend upon contaminant and technique |
| Enzymes | Many proteins are enzymes, and the quantities and activities and many of them are impacted by contaminants; some of the major ones follow | Specificity as indicated |
| ALAD | Change in the measurable activity of δ-aminolevulinic acid dehydratase | Lead |
| Cholinesterase | Assess acetylcholinesterase inhibition in brain; acetylcholinesterase and butrylcholinesterase inhibition in serum or plasma | Activity inhibited by carbamate and organophosphorus pesticides |
| Conjugating enzymes | Phase 2 metabolizing enzymes such as glutathione-S-transferase. | Increases may be caused by (PAHs), halogenated PAHs, and heavy metals |
| Cytochromes P450 | Changes, generally increases, in various families of cytochromes P450 measured in any tissue by a variety of methods | Increases generally caused by polycyclic aromatic hydrocarbons (PAHs), halogenated PAHs; may be inhibited by heavy metals |
| Metallothionein | Increase in class of -SH rich polypeptides in tissues such as liver, kidney, and testes | Caused by heavy metals |
| Stress proteins glucose-regulated proteins heat stress proteins | There are several families of stress proteins that differ in molecular weight and effectors | Relatively low specificity, differs with family |
| DNA-related | A wide variety of factors impact DNA transcription and DNA structure | Specificity dependent upon contaminant and technique |
| Genomics | DNA transcription can be affected by most contaminants; measurements can be from one particular mRNA or from an array of dozens, hundreds, or thousands | Specificity dependent upon contaminant and mRNAs that can be assessed; the use of arrays in contaminant studies is under development |
| DNA modifications | Contaminant-induced changes in DNA include adduct formation, sister chromatid exchange, flow cytometry | Can have many causes; specificity varies with contaminant and evaluation technique |
| Reproduction- and growth-related | There are many possible effects ranging from slight changes to lethal effects. Depending upon species and nature of reproduction, may be due to direct impact on developing organism or to impact on parent organism | Specificity depends upon the species and the life stage where exposure to contaminants occurs and the contaminant itself |
| Eggshell changes | Thinning or weakening of eggshells | Caused by DDT and other halogenated organics |

**Table 11.1 Presently Used and Possible Bioindicators[a] *(Continued)***

| Bioindicator | Description | Stressor |
|---|---|---|
| Developmental problems — teratogenicity | Pertains specifically to failure of one or more organs or systems to develop properly. | Specificity depends upon the mechanism of action of the contaminant, time at which exposure occurs, and exposure level |
| Skeletal abnormalities | Greater than normal bilateral asymmetry | May be caused by halogenated polycyclic aromatic hydrocarbons |
| Growth rate | Reduced growth rate of young organisms | Nonspecific; can be due to direct effect on organism, effect on habitat, or effect on parent |
| Clinical-type measurements | | |
| Blood chemistry | Change in levels of blood constituents that would be measured in typical human clinical testing; these include enzymes released by tissue damage, hormones (particularly corticosteroids, growth hormone, insulin and glucagon, reproductive hormones, thyroid hormones), vitamins (particularly vitamin A and its metabolites and vitamin C) | Many possibilities with varying specificities |
| Histopathologic | Impacts on a variety of tissues; can be detected by histochemistry | Specificity depends on contaminant, species, and tissue |
| Oxidant-mediated responses | Changes in levels of compounds such as glutathione (both oxidized and reduced and their ratios), ascorbate, and α-tocopherol, and in a variety of enzymes involved in radical scavenging, etc. | Caused by a wide variety of organic and inorganic contaminants |
| Porphyria | Disruption of normal porphyrin metabolism | Caused by halogenated PAHs and heavy metals |
| Bile metabolites | Metabolites of contaminants that are metabolized rapidly and may not be found as the parent chemicals in amounts that are readily measurable, may be found in measurable concentrations in bile | Variety of organic contaminants, but commonly used for rapidly metabolized PAHs |
| Immune competence | Impaired function of parts of immune process | Can be caused by chlorinated organics and heavy metals |

[a] Assignments to categories are arbitrary, some of the bioindicators could fit under several categories.

## 11.3 SUMMARY

Bioindicators of contaminant exposure presently used in environmental monitoring are discussed. Some have been extensively field-validated and are already in routine application. Included are (1) inhibition of brain or blood cholinesterase by anticholinesterase pesticides, (2) induction of hepatic microsomal cytochromes P450 by chemicals such PAHs and PCBs, (3) reproductive problems such as terata and eggshell thinning, and (4) aberrations of hemoglobin synthesis, including the effects of lead and of certain chlorinated hydrocarbons. Many studies on DNA damage and of histopathological effects, particularly in the form of tumors, have already been completed. There are presently numerous other opportunities for field validation. Bile metabolites of contaminants in fish reveal exposure to contaminants that might otherwise be difficult to detect or quantify. Bile analysis is beginning to be extended to species other than fishes. Assessment of oxidative damage and immune competence appear to be valuable biomarkers, needing only additional field validation for wider use.

The use of metallothioneins as biomarkers depends on the development of convenient, inexpensive methodology that provides information not available from measurements of metal ions. The use of stress proteins as biomarkers depends on development of convenient, inexpensive methodology and field validation. Gene arrays and proteomics hold promise as bioindicators for contaminant exposure or effect, particularly because of the large amount of data that could be generated, but they still need extensive development and testing.

## REFERENCES


1. Huggett, R. J., Kimerle, R. A., Mehrle, P. M., Jr., and Bergman, H. L., *Biomarkers: Biochemical, Physiological, and Histological Markers of Anthropogenic Stress*, Lewis Publishers, Chelsea, MI, 1992.
2. McCarthy, J. F. and Shugart, L. R., *Biomarkers of Environmental Contamination*, Lewis Publishers, Boca Raton, FL, 1990.
3. Peakall, D., *Animal Biomarkers as Pollution Indicators*, Chapman and Hall, London, 1992.
4. Fossi, M. C. and Leonzio, C., *Nondestructive Biomarkers in Vertebrates*, Lewis Publishers, Boca Raton, FL, 1994.
5. Van Gestel, C. A. M. and van Brummelen, T. C., Incorporation of the biomarker concept in ecotoxicology calls for a redefinition of terms, *Ecotoxicology*, 5, 217, 1996.
6. Peakall, D. B. and Fairbrother, A., Biomarkers for monitoring and measuring effects, in *Pollution Risk Assessment and Management*, Douben, P. E. T., Ed., John Wiley and Sons, Chichester, UK, 1998, 351.
7. Handy, R. D. and DePledge, M. H., Physiological responses: Their measurement and use as environmental biomarkers in ecotoxicology, *Ecotoxicology*, 8, 329, 1999.
8. Schlenk, D., Necessity of defining biomarkers for use in ecological risk assessments, *Mar. Poll. Bull.*, 39, 48, 1999.
9. Peakall, D. B. and McBee, K., Biomarkers for contaminant exposure and effects in mammals, in *Ecotoxicology of Wild Mammals*, Shore, R. F. and Rattner, B. A., Eds., John Wiley and Sons, Chichester, UK, 2001, 551.
10. Ludke, J. L., Hill, E. F., and Dieter, M. P., Cholinesterase (ChE) response and related mortality among birds fed ChE inhibitors, *Arch. Environ. Contam. Toxicol.*, 3, 1, 1975.
11. Weiss, C. M., The determination of cholinesterase in the brain tissue of three species of fresh water fish and its inactivation *in vivo*, *Ecology*, 39, 194, 1958.
12. Weiss, C. M., Physiological effect of organic phosphorus insecticides on several species of fish, *Trans. Am. Fish. Soc.*, 90, 143, 1961.
13. Gibson, R. F., Ludke, J. L., and Ferguson, D. E., Sources of error in the use of fish-brain acetylcholinesterase as a monitor for pollution, *Bull. Environ. Contam. Toxicol.*, 4, 17, 1969.
14. Westlake, G. E., Martin, A. D., Stanley, P. I. and Walker, C. H., Control enzyme levels in the plasma, brain and liver from wild birds and mammals in Britain, *Comp. Biochem. Physiol.*, 76C, 15, 1983.

15. Smith, G. J., Pesticide Use and Toxicology in Relation to Wildlife: Organophosphorus and Carbamate Compounds, U.S. Fish Wildl. Serv., Resour. Publ. 170, 1987.
16. Hill, E. F., Brain cholinesterase activity of apparently normal wild birds, *J. Wildl. Dis.*, 24, 51, 1988.
17. Grue, C. E. and Hunter, C. C., Brain cholinesterase activity in fledgling starlings: Implications for monitoring exposure of songbirds to ChE inhibitors, *Bull. Environ. Contam. Toxicol.*, 32, 282, 1984.
18. Martin, A. D., Norman, G., Stanley, P. I., and Westlake, G. E., Use of reactivation techniques for the differential diagnosis of organophosphorus and carbamate pesticide poisoning in birds, *Bull. Environ. Contam. Toxicol.*, 26, 775, 1981.
19. Johnson, G. D., Palmer, D. A., Krueger, H. O. and Fischer, D. L., Stability of residues and brain cholinesterase activity in carcasses of northern bobwhite exposed to three organophosphate insecticides, *Environ. Toxicol. Chem.*, 12, 673, 1993.
20. Fairbrother, A., Bennett, R. S. and Bennett, J. K., Sequential sampling of plasma cholinesterase in mallards (*Anas platyrhynchos*) as an indicator of exposure to cholinesterase inhibitors, *Environ. Toxicol. Chem.*, 8, 117, 1989.
21. Fairbrother, A. and Bennett, J. K., The usefulness of cholinesterase measurement, *J. Wildl. Dis.,* 24, 587, 1988.
22. Hill, E. F., The usefulness of cholinesterase measurement: A response, *J. Wildl. Dis.*, 24, 591, 1988.
23. Mineau, P., *Cholinesterase-Inhibiting Insecticides: Their Impact on Wildlife and the Environment*, Elsevier Science Publishers, Amsterdam, 1991.
24. Thompson, H. M., Esterases as markers of exposure to organophosphates and carbamates, *Ecotoxicol*ogy, 8, 369, 1999.
25. Trudeau, S. and Sans Cartier, G., Biochemical Methods to Determine Cholinesterase Activity in Wildlife Exposed to Pesticides, Technical Report Series No. 338, Canadian Wildlife Service, Headquarters, Hull, Quebec, Canada, 2000.
26. Nebert, D. W., Nelson, D. R., Adesnik, M., Coon, M. J., Estabrook, R. W., Gonzalez, F. J., Guengerich, F. P., Gunsalus, I. C., Johnson, E. F., Kemper, B., Levin, W., Phillips, I. R., Sato, R., and Waterman, M. R., The P450 superfamily: Updated listing of all genes and recommended nomenclature for the chromosomal loci, *DNA,* 8, 1, 1989.
27. Van Cantfort, J., De Graeve, J., and Gielen, J. E., Radioactive assay for aryl hydrocarbon hydroxylase. Improved method and biological importance, *Biochem. Biophys. Res. Commun.*, 79, 505, 1977.
28. Burke, M. D. and Mayer, R. T., Ethoxyresorufin: Direct fluorometric assay of a microsomal O-dealkylation which is preferentially inducible by 3-methylcholanthrene, *Drug Metab. Disp.*, 2, 583,1974.
29. Burke, M. D., Thompson, S., Elcombe, C. R., Halpert, J., Haaparanta, T., and Mayer, R. T., Ethoxy, pentoxy- and benzyloxyphenoxazones and homologues: A series of substrates to distinguish between different induced cytochromes P-450, *Biochem. Pharmacol.*, 34, 3337, 1985.
30. Ullrich, V. and Weber, P., The O-dealkylation of 7-ethoxycoumarin by liver microsomes, *Hoppe-Seyler's Z. Physiol. Chem.*, 353, 1171, 1972.
31. Melancon, M. J., Binder, R. L., and Lech, J. J., Environmental induction of monooxygenase activity in fish, in *Toxic Contaminants and Ecosystem Health; A Great Lakes Focus*, Evans, M. S., Ed., John Wiley and Sons, New York, 1988, 215.
32. Kloepper-Sams, P. J., Park, S. S., Gelboin, H. V. and Stegeman, J. J., Specificity and cross-reactivity of monoclonal and polyclonal antibodies against cytochrome P-450E of the marine fish scup, *Arch. Biochem. Biophys.*, 253, 268, 1987.
33. Stegeman, J. J., Teng, F. Y., and Snowberger, E. A., Induced cytochrome P450 in winter flounder (*Pseudopleuronectes americanus*) from coastal Massachusetts evaluated by catalytic assay and monoclonal antibody probes, *Can. J. Fish. Aquat. Sci.*, 44, 1270, 1987.
34. Van Veld, P. A., Westbrook, D. J., Woodin, B. R., Hale, R. C., Smith, C. L., Huggett, R. J., and Stegeman, J. J., Induced cytochrome P-450 in intestine and liver of spot (*Leiostomus xanthurus*) from a polycyclic aromatic hydrocarbon contaminated environment, *Aquat. Toxicol.*, 17, 119, 1990.
35. Ronis, M. J. J., Hansson, T., Borlakoglu, J., and Walker, C. H., Cytochromes P-450 of sea birds: Cross-reactivity studies with purified rat cytochromes, *Xenobiotica*, 19, 1167, 1989.
36. Goksøyr, A., A semi-quantitative cytochrome P4501A1 ELISA: A simple method for studying the monooxygenase induction response in environmental monitoring and ecotoxicological testing of fish, *Sci. Total Environ.*, 101, 255, 1991.

37. Agradi, E., Baga, R., Cillo, F., Ceradini, S., and Heltai, D., Environmental contaminants and biochemical response in eel exposed to Po river water, *Chemosphere,* 41, 1555, 2000.
38. Campbell, P. M., Kruzynski, G. M., Birtwell, I. K., and Devlin, R. H., Quantitation of dose-dependent increases in CYP1A1 messenger RNA levels in juvenile chinook salmon exposed to treated bleached-kraft mill effluent using two field sampling techniques, *Environ. Toxicol. Chem.,* 15, 1123, 1996.
39. Willett, K. L., McDonald, S. J., Steinberg, M. A., Beatty, K. B., Kennicutt, M. C., and Safe, S. H., Biomarker sensitivity for polynuclear aromatic hydrocarbon contamination in two marine fish species collected in Galveston Bay, Texas, *Environ. Toxicol. Chem.*, 16, 1472, 1997.
40. Nagler, J. J. and Cyr, D. G., Exposure of male American Plaice (*Hippoglossoides platessoides*) to contaminated marine sediments decreases the hatching success of their progeny, *Environ. Toxicol. Chem.*, 16, 1733, 1997.
41. Halbrook, R. S. and Kirkpatrick, R. L., Use of barbiturate-induced sleeping time as an indicator of exposure to environmental contaminants in the wild, in *Biomarkers of Environmental Contamination*, McCarthy, J. F. and Shugart, L. R., Eds., Lewis Publishers, Boca Raton, FL, 1990, 151.
42. Feyk, L. A., Giesy, J. P., and Lambert, G. H., Relationship between polychlorinated biphenyl 126 treatment and cytochrome P4501A activity in chickens, as measured by *in vivo* caffeine and *ex vivo* ethoxyresorufin metabolism, *Environ. Toxicol. Chem.*, 18, 2013, 1999.
43. Feyk, L. A., Giesy, J. P., Bosveld, A. B. T., and van den Berg, M., Changes in cytochrome P4501A activity during development in common tern chicks fed polychlorinated biphenyls, as measured by the caffeine breath test, *Environ. Toxicol. Chem.*, 19, 712, 2000.
44. Nacci, D., Coiro, L., Kuhn, A., Champlin, D., Munns, W., Jr., Specker, J., and Cooper, K., Nondestructive indicator of ethoxyresorufin-*O*-deethylase activity in embryonic fish, *Environ. Toxicol. Chem.*, 17, 2481, 1998.
45. Ronis, M. J. J. and Walker, C. H., The microsomal monooxygenases of birds, *Rev. Biochem. Toxicol.,* 10, 301, 1989.
46. Payne, J. F., Mixed-function oxygenases in biological monitoring programs: Review of potential usage in different phyla of aquatic animals, in *Ecological Testing for the Marine Environment, Vol. 1*, Persoone, G., Jaspers, E., and Claus, C., Eds., Bredene, Belgium, 1984, 798.
47. Payne, J. F., Fancey, L. L., Rahimtula, A. D., and Porter, E. L., Review and perspective on the use of mixed-function oxygenase enzymes in biological monitoring, *Comp. Biochem. Physiol.*, 86C, 233, 1987.
48. Rattner, B. A., Hoffman, D. J., and Marn, C. M., Use of mixed-function oxygenases to monitor contaminant exposure in wildlife, *Environ. Toxicol. Chem.*, 8, 1093, 1989.
49. Peakall, D., *Animal Biomarkers as Pollution Indicators*, Chapman and Hall, London, 1992, 86.
50. Blus, L. J., Heath, R. G., Gish, C. D., Belisle, A. A., and Prouty, R. M., Eggshell thinning in the brown pelican: Implication of DDE, *Bioscience*, 21, 1213, 1971.
51. Cooke, A. S., Shell thinning in avian eggs by environmental pollutants, *Environ. Poll.*, 4, 85, 1973.
52. Lincer, J. L., DDE-induced eggshell-thinning in the American kestrel: A comparison of the field situation and laboratory results, *J. Appl. Ecol.*, 12, 781, 1975.
53. Cooke, A. S., Changes in eggshell characteristics of the sparrowhawk (*Accipiter nisus*) and peregrine (*Falco peregrinus*) associated with exposure to environmental pollutants during recent decades, *J. Zool.*, 187, 245, 1979.
54. Wiemeyer, S. N., Lamont, T. G., Bunck, C. M., Sindelar, C. R., Gramlich, F. J., Fraser, J. D., and Byrd, M. A., Organochlorine pesticide, polychlorobiphenyl, and mercury residues in bald eagle eggs — 1969–79 — and their relationships to shell thinning and reproduction, *Arch. Environ. Contam. Toxicol.*, 13, 529, 1984.
55. Lundholm, E., Thinning of eggshells in birds by DDE: Mode of action on the eggshell gland, *Comp. Biochem. Physiol.*, 88C, 1, 1987.
56. Wiemeyer, S. N., Bunck, C. M., and Krynitsky, A. J., Organochlorine pesticides, polychlorinated biphenyls, and mercury in osprey eggs — 1970–79 — and their relationship to shell thinning and productivity, *Arch. Environ. Contam. Toxicol.,* 17, 767, 1988.
57. Ratcliffe, D. A., Decrease in eggshell weight in certain birds of prey, *Nature*, 215, 208, 1967.
58. Carlisle, J. C., Lamb, D. W., and Toll, P. A., Breaking strength: An alternative indicator of toxic effects on avian eggshell quality, *Environ. Toxicol. Chem.*, 5, 887, 1986.

59. Bennett, J. K., Ringer, R. S., Bennett, R. S., Williams, B. A., and Humphrey, P. E., Comparison of breaking strength and shell thickness as evaluators of eggshell quality, *Environ. Toxicol. Chem.,* 7, 351, 1988.
60. Henny, C. J. and Bennett, J. K., Comparison of breaking strength and shell thickness as evaluators of white-faced ibis eggshell quality, *Environ. Toxicol. Chem.*, 9, 797, 1990.
61. Custer, T. W., Custer, C. M., Hines, R. K., Stromber, K. L., Allen, P. D., Melancon, M. J., and Henshel, D. S., Organochlorine contaminants and biomarker response in double-crested cormorants nesting in Green Bay and Lake Michigan, WI, *Environ. Toxicol. Chem.*, 40, 89, 2001.
62. Hoffman, D. E., Embryotoxicity and teratogenicity of environmental contaminants to bird eggs, *Rev. Environ. Contam. Toxicol.*, 115, 39, 1990.
63. Fox, G. A., Collins, B., Hayakawa, E., Weseloh, D. V., Ludwig, J. P., Kubiak, T. J., and Erdman, T. C., Reproductive outcomes in colonial fish-eating birds: A biomarker for developmental toxicants in Great Lakes food chains. II. Spatial variation in the occurrence and prevalence of bill defects in young double-crested cormorants in the Great Lakes, 1979–1987, *J. Great Lakes Res.,* 17, 158, 1991.
64. Ohlendorf, H. M., Kilness, A. W., Simmons, J. L., Stroud, R. K., Hoffman, D. J., and Moore, J. F., Selenium toxicosis in wild aquatic birds, *J. Toxicol. Environ. Health*, 24, 67, 1988.
65. Hoffman, D. J., Ohlendorf, H. M., and Aldrich, T. W., Selenium toxicosis in natural populations of aquatic birds in central California, *Arch. Environ. Contam. Toxicol.*, 17, 519, 1988.
66. Heppell, S. A., Denslow, N. D., Folmar, L. C., and Sullivan, C. V., A universal assay of vitellogenin as a biomarker for environmental estrogens, *Environ. Health Perspect.*, 103, 9, 1995.
67. Folmar, L. C., Denslow, N. D., Rao, V., Chow, M., Crain, D. A., Enblom, J., Marcino, J., and Guillette, L. J., Jr., Vitellogenin induction and reduced serum testosterone concentrations in feral male carp (*Cyprinus carpio*) captured near a major metropolitan sewage treatment plant, *Environ. Health Persp.*, 104, 1096, 1996.
68. Denslow, N. D., Chow, M. M., Folmar, L. C., Bonomelli, S., Heppell, S. A., and Sullivan, C. V., Development of antibodies to teleost vitellogenins: Potential biomarkers for environmental estrogens, in *Environmental Toxicology and Risk Assessment: 5$^{th}$ Volume*, ASTM STP 1306, Bengston, D. A. and Henshel, D. A., Eds., American Society for Testing and Materials, Philadelphia, 1997, 23.
69. Harries, J. E., Sheahan, D. A., Jobling, S., Matthiessen, P., Neall, P., Sumpter, J. P., Tylor, T., and Zaman, N., Estrogenic activity in five United Kingdom rivers detected by measurement of vitellogenesis in caged male trout, *Environ. Toxicol. Chem.*, 16, 534, 1997.
70. Carlson, D. B. and Williams, D. E., Sex-specific vitellogenin production in immature rainbow trout, *Environ. Toxicol. Chem.*, 18, 2361, 1999.
71. Parks, L. G., Cheek, A. O., Denslow, N. B., Heppell, S. A., McLachlan, J. A., LeBlanc, G. A., and Sullivan, C. V., Fathead minnow (*Pimephales promelas*) vitellogenin: Purification, characterization and quantitative immunoassay for the detection of estrogenic compounds, *Comp. Biochem. Physiol.*, 123, 113, 1999.
72. Denslow, N. D., Chow, M. C., Kroll, K. J., and Green, L., Vitellogenin as a biomarker of exposure for estrogen or estrogen mimics, *Ecotoxicol*ogy, 8, 385, 1999.
73. Bowman, C. J. and Denslow, N. B., Development and validation of a species- and gene-specific molecular biomarker: Vitellogenin mRNA in largemouth bass (*Micropterus salmoides*), *Ecotoxicology*, 8, 399, 1999.
74. Johansson-Sjobeck, M.-L. and Larsson, A., Effects of inorganic lead on delta-aminolevulinic acid dehydratase activity in the rainbow trout, *Salmo gairdneri*, *Arch. Environ. Contam. Toxicol.*, 8, 419, 1979.
75. Scheuhammer, A. M., Erythrocyte delta-aminolevulinic acid dehydratase in birds. I. The effects of lead and other metals *in vitro*, *Toxicology*, 45, 155, 1987.
76. Hodson, P. V., Blunt, B. R., Spry, D. J., and Austen, K., Evaluation of erythrocyte-amino levulinic acid dehydratase activity as a short-term indicator in fish of a harmful exposure to lead, *J. Fish. Res. Bd. Can.*, 34, 501, 1977.
77. Hodson, P. V., Blunt, B. R., and Whittle, D. M., Biochemical monitoring of fish blood as an indicator of biologically available lead, *Thalassia Jugosl.*, 1, 389, 1980.
78. Finley, M. T., Dieter, M. P., and Locke, L. N., Delta-aminolevulinic acid dehydratase: Inhibition in ducks dosed with lead shot, *Environ. Res.*, 12, 243, 1976.

79. Beyer, W. N., Spann, J. W., Sileo, L., and Franson, J. C., Lead poisoning in six captive avian species, *Arch. Environ. Contam. Toxicol.*, 17, 121, 1988.
80. Dieter, M. P., Perry, M. C., and Mulhern, B. M., Lead and PCB's in canvasback ducks: Relationship between enzyme levels and residues in blood, *Arch. Environ. Contam. Toxicol.*, 5, 1, 1976.
81. Dieter, M. P., Blood delta-aminolevulinic acid dehydratase (ALAD) to monitor lead contamination in canvasback ducks (*Aythya valisineria*), in *Animals as Monitors of Environmental Pollutants*, National Academy of Sciences, Washington, D.C., 1979, 177.
82. Grue, C. E., Hoffman, D. J., Beyer, W. N., and Franson, L. P., Lead concentrations and reproductive success in European starlings *Sturnus vulgaris* nesting within highway roadside verges, *Environ. Pollut.*, 42, 157, 1986.
83. Schmitt, C. J., Dwyer, F. J., and Finger, S. E., Bioavailability of Pb and Zn from mine tailings as indicated by erythrocyte delta-aminolevulinic acid dehydratase (ALA-D) activity in suckers (Pisces: Catostomidae), *Can. J. Fish. Aquat. Sci.*, 41, 1030, 1984.
84. Overmann, S. R. and Krajicek, J. J., Snapping turtles (*Chelydra serpentina*) as biomonitors of lead contamination of the Big River in Missouri's old lead belt, *Environ. Toxicol. Chem.*, 14, 689, 1995.
85. Scheuhammer, A. M., Erythrocyte-δ–aminolevulinic acid dehydratase in birds. II. The effects of lead exposure *in vivo*, *Toxicology,* 45, 165, 1987.
86. Goldstein, J. A., Hickman, P., Bergman, H., and Vos, J. G., Hepatic porphyria induced by 2,3,7,8-tetrachlorodibenzo-*p*-dioxin in the mouse, *Res. Commun. Chem. Pathol. Pharmacol.*, 6, 919, 1973.
87. Strik, J. J. T. W. A., Porphyrins in urine as an indication of exposure to chlorinated hydrocarbons, *Ann. N.Y. Acad. Sci.* 320, 308, 1979.
88. Fox, G. A., Kennedy, S. W., Norstrom, R. J., and Wigfield, D. C., Porphyria in herring gulls: A biochemical response to chemical contamination of Great Lakes food chains, *Environ. Toxicol. Chem.*, 7, 831, 1988.
89. Kennedy, S. W., Wigfield, D. C., and Fox, G. A., Tissue porphyrin pattern determination by high-speed high-performance liquid chromatography, *Anal. Biochem.*, 157, 1, 1986.
90. Reddy, V. R., Christenson, W. R., and Piper, W. N., Extraction and isolation by high performance liquid chromatography of uroporphyrin and coproporphyrin isomers from biological tissues, *J. Pharmacol. Methods*, 17, 51, 1987.
91. Taylor, C., Durry, L. K., Bowyer, R. T., and Blundell, G. M., Profiles of fecal poryphrins in River Otters following the *Exxon Valdez* oil spill, *Mar. Poll. Bull.,* 40, 1132, 2000.
92. Roscoe, D. E., Nielson, S. W., Lamola, A. A., and Zuckerman, D., A simple, quantitative test for erythrocytic protoporphyrin in lead-poisoned ducks, *J. Wildl. Dis.,* 15, 127, 1979.
93. Srebocan, E. and Rattner, B. A., Heat exposure and the toxicity of one number four lead shot in mallards, *Anas platyrhynchos*, *Bull. Environ. Contam. Toxicol.,* 40, 165, 1988.
94. Pain, D. J. and Rattner, B. A., Mortality and hematology associated with the ingestion of one number four lead shot in black ducks, *Anas rubripes*, *Bull. Environ. Contam. Toxicol.*, 40, 159, 1988.
95. Pain, D. J., Haematological parameters as predictors of blood lead and indicators of lead poisoning in the black duck (*Anas rubripes*), *Environ. Pollut.,* 60, 67, 1989.
96. Randerath, K., Reddy, M., and Gupta, R. C., $^{32}$P-postlabeling analysis for DNA damage, *Proc. Natl. Acad. Sci. U.S.A.*, 78, 6126, 1981.
97. Collier, T. K. and Varanasi, U., Biochemical indicators of contaminant exposure in flatfish from Puget Sound, WA, *Proceedings Oceans 87*, Vol. 5., IEEE, Washington, D.C., 1987, 1544.
98. Dunn, B., Black, J., and Maccubbin, A., $^{32}$P-postlabeling analysis of aromatic DNA adducts in fish from polluted areas, *Cancer Res.*, 47, 6543, 1987.
99. Martineau, D., Legace, A., Beland, P., Higgins, R., Armstrong, D., and Shugart, L. R., Pathology of stranded beluga whales (*Delphinapterus leucas*) from the St. Lawrence estuary, Quebec, Canada, *J. Comp. Pathol.*, 98, 287, 1988.
100. Dunn, B. P., Fitzsimmons, J., Stalling, D., McCubbin, A. E., and Black, J. J., Pollution-related aromatic DNA adducts in liver from populations of wild fish, *Proc. Am. Assoc. Cancer Res.*, 31, 1990, 96.
101. Poginsky, B., Blomeke, B., Hewer, A., Phillips, D. H., Karbe, L., and Marquardt, H., $^{32}$P-postlabeling analysis of hepatic DNA of benthic fish from European waters, *Proc. Am. Assoc. Cancer Res.*, 31, 1990, 96.
102. Stein, J. E., Reichert, W. L., Nishimote, M., and Varanasi, U., $^{32}$P-postlabeling of DNA: A sensitive method for assessing environmentally induced genotoxicity, *Proc. Oceans 89*, Vol. 7, 1989, 385.

103. Varanasi, U., Reichert, W. L., Eberhart, B.-T., Stein, J., Formation and persistence of benzo(a)pyrene-diolepoxide-DNA adducts in liver of English sole (*Parophrys vetulus*), *Chem.-Biol. Interact.*, 69, 203, 1989.
104. Varanasi, U., Reichert, W. L., and Stein, J., $^{32}P$-postlabeling analysis of DNA adducts in liver of wild English sole (*Parophrys vetulus*) and winter flounder (*Pseudopleuronectes americanus*), *Cancer Res.*, 49, 1171, 1989.
105. Shugart, L. R., An alkaline unwinding assay for the detection of DNA damage in aquatic organisms, *Marine Environ. Res.*, 24, 321, 1988.
106. Shugart, L.R., Quantitation of chemically induced damage to DNA of aquatic organisms by alkaline unwinding assay, *Aquat. Toxicol.*, 13, 43, 1988.
107. Meyers-Schone, L., Shugart, L. R., Beauchamp, J. J., and Walton, B. T., Comparison of two freshwater turtle species as monitors of radionuclide and chemical contamination: DNA damage and residue analysis, *Environ. Toxicol. Chem.*, 12, 1487, 1993.
108. Bickham, J. W., Flow cytometry as a technique to monitor the effects of environmental genotoxins on wildlife populations, in *In situ Evaluation of Biological Hazards of Environmental Pollutants*, Sandhu, S., Lower, W. R., DeSerres, F. J., Suk, W. A., and Tice, R. R., Eds., Environmental Research Series, Vol. 38, Plenum Press, New York, 1990, 97.
109. Custer, T. W., Bickham, J. W., Lyne, T. B., Lewis, T., Ruedas, L. A., Custer, C. M., and Melancon, M. J., Flow cytometry for monitoring contaminant exposure in black-crowned night-herons, *Arch. Environ. Contam. Toxicol.*, 27, 176, 1994.
110. Custer, T. W., Custer, C. M., Hines, R. K., Sparks, D. W., Melancon, M. J., Hoffman, D. J., Bickham, J. W., and Wickliffe, J. K., Mixed-function oxygenases, oxidative stress, and chromosomal damage measured in lesser scaup wintering on the Indiana Harbor Canal, *Arch. Environ. Contam. Toxicol.*, 38, 522, 2000.
111. Peakall, D., Studies on genetic material, in *Animal Biomarkers as Pollution Indicators*, Peakall, D., Ed., Chapman and Hall, London, 1992, 70.
112. Shugart, L. R., DNA damage as a biomarker of exposure, *Ecotoxicology*, 9, 329, 2000.
113. Hinton, D. E., Baumann, P. C., Gardner, G. R., Hawkins, W. E., Hendricks, J. D., Murchelano, R. A., and Okihiro, M. S., Histopathologic biomarkers, in *Biomarkers: Biochemical, Physiological, and Histological Markers of Anthropogenic Stress*, Huggett, R. J., Kimerle, R. A., Mehrle, P. M., Jr., and Bergman, H. L., Eds., Lewis Publishers, Chelsea, MI, 1992, 155.
114. Stich, H. F. and Acton, A. B., The possible use of fish tumors in monitoring for carcinogens in the marine environment, *Prog. Exp. Tumor Res.*, 20, 44, 1976.
115. Brown, E. R., Hazdra, J. J., Keith, L., Greenspan, I., and Kwapinski, J. B. G., Frequency of fish tumors found in a polluted watershed as compared to nonpolluted Canadian waters, *Cancer Res.*, 33, 189, 1973.
116. McCain, B. B., Pierce, K. V., Wellings, S. R., and Miller, B. S., Hepatomas in marine fish from an urban estuary, *Bull. Environ. Contam. Toxicol.*, 18, 1, 1977.
117. Grizzle, J. M., Melius, P., and Strength, D. R., Papillomas on fish exposed to chlorinated wastewater effluent, *J. Natl. Cancer Inst.*, 73, 1133, 1984.
118. Fabacher, D. L. and Baumann, P. C., Enlarged livers and hepatic microsomal mixed-function oxidase components in tumor-bearing brown bullheads from a chemically contaminated river, *Environ. Toxicol. Chem.*, 4, 703, 1985.
119. Black, J. J., Evans, E. D., Harshbarger, J. C., and Ziegel, R. F., Epizootic neoplasms in fishes from a lake polluted by copper mining wastes, *J. Natl. Cancer Inst.*, 69, 915, 1982.
120. Pinkney, A. E., Harshbarger, J. C., May, E. B., and Melancon, M. J., Tumor prevalence and biomarkers of exposure in brown bullheads (*Ameiurus nebulosus*) from the tidal Potomac River, USA, watershed, *Environ. Toxicol. Chem.*, 20, 1196, 2001.
121. Lee, R. F., Sauerheber, R., and Dobbs, G. H., Uptake, metabolism and discharge of polycyclic aromatic hydrocarbons in marine fish, *Mar. Biol.*, 17, 201, 1972.
122. Statham, C. N., Melancon, M. J., and Lech, J. J., Bioconcentration of xenobiotics in trout bile: A proposed monitoring aid for some waterborne chemicals, *Science*, 193, 680, 1976.
123. Roubal, W. T., Collier, T. K., and Malins, D. C., Accumulation and metabolism of carbon-14 labeled benzene, naphthalene, and anthracene by young coho salmon (*Oncorhynchus kisutch*), *Arch. Environ. Contam. Toxicol.*, 5, 513, 1977.

124. Krahn, M. M., Myers, M. S., Burrows, D. G., and Malins, D. C., Determination of metabolites of xenobiotics in the bile of fish from polluted waterways, *Xenobiotica*, 14, 633, 1984.
125. Krahn, M. M., Rhodes, L. D., Myers, M. S., Moore, L. K., MacLeod, W. D., Jr., and Malins, D. C., Associations between metabolites of aromatic compounds in bile and the occurrence of hepatic lesions in English sole (*Parophrys vetulus*) from Puget Sound, Washington, *Arch. Environ. Contam. Toxicol.*, 15, 61, 1986.
126. Krahn, M. M., Burrows, D. G., MacLeod, W. D., Jr., and Malins, D. C., Determination of individual metabolites of aromatic compounds in hydrolyzed bile of English sole (*Parophrys vetulus*) from polluted sites in Puget Sound, Washington, *Arch. Environ. Contam. Toxicol.*, 16, 511, 1987.
127. Britvic, S., Lucic, D., and Kurelec, B., Bile fluorescence and some early biological effects in fish as indicators of pollution by xenobiotics, *Environ. Toxicol. Chem.*, 12, 765, 1993.
128. Leadly, T. A., Arcand-Hoy, L. D., Haffner, G. D., and Metcalfe, C. D., Fluorescent aromatic hydrocarbons in bile as a biomarker of exposure of brown bullheads (*Ameiurus nebulosus*) to contaminated sediments, *Environ. Toxicol. Chem.*, 18, 750, 1999.
129. Pointet, K. and Milliet, A., PAHs analysis of fish whole gall bladders and livers from the natural reserve of Camargue by GC/MS, *Chemosphere,* 40, 293, 2000.
130. Okari, A. O. J., Metabolites of xenobiotics in the bile of fish in waterways polluted by pulpmill effluents, *Bull. Environ. Contam. Toxicol.*, 36, 429, 1986.
131. Melancon, M. J., Alscher, R., Benson, W., Kruzynski, G., Lee, R. F., Sikka, H. C., and Spies, R. B., Metabolic products as biomarkers, in *Biomarkers: Biochemical, Physiological, and Histological Markers of Anthropogenic Stress*, Huggett, R. J., Kimerle, R. A., Mehrle, P. M., Jr., and Bergman, H. L., Eds., Lewis Publishers, Chelsea, MI, 1992, 87.
132. Kehrer, J. P., Mossman, B. T., Sevanian, A., Trush, M. A., and Smith, M. T., Free radical mechanisms in chemical pathogenesis, *Toxicol. Appl. Pharmacol.*, 95, 349, 1988.
133. Di Guilio, R. T., Washburn, P. C., Wenning, R. J., Winston, G. W., and Jewell, C. S., Biochemical responses in aquatic animals: A review of determinants of oxidative stress, *Environ. Toxicol. Chem.*, 8, 1103, 1989.
134. Hoffman, D. J. and Heinz, G. H., Embryotoxic and teratogenic effects of selenium in the diet of mallards, *J. Toxicol. Environ. Health,* 24, 477, 1988.
135. Hoffman, D. J., Heinz, G. H., and Krynitsky, A. J., Hepatic glutathione metabolism and lipid peroxidation in response to excess dietary selenomethionine and selenite in mallard ducklings, *J. Toxicol. Environ. Health*, 27, 263, 1989.
136. Hoffman, D. J., Heinz, G. H., LeCaptain, L. J., Bunck, C. M., and Green, D. E., Subchronic hepatoxicity of selenomethionine ingestion in mallard ducks, *J. Toxicol. Environ. Health*, 32, 449, 1991.
137. Roberts, M. H., Sved, D. W., and Felton, S. P., Temporal changes in AHH and SOD activities in feral spot from the Elizabeth River, a polluted subestuary, *Mar. Environ. Res.*, 23, 89, 1987.
138. Regoli, F., Hummel, H., Amiard-Triquet, C., Larroux, C., and Sukhotin, A., Trace metals and variations of antioxidant enzymes in arctic bivalve populations, *Arch. Environ. Contam. Toxicol.*, 35, 594, 1998.
139. Rodriguez-Grau, J., Venables, B. J., Fitzpatrick, L. C., Goven, A. J., and Cooper, E. L., Suppression of secretory rosette formation by PCBs in *Lumbricus terrestris*: An earthworm assay for humoral immunotoxicity of xenobiotics, *Environ. Toxicol. Chem.*, 8, 1201, 1989.
140. Sharma, R. P., *Immunologic Considerations in Toxicology,* Vols. I and II, CRC Press, Boca Raton, FL, 1981.
141. Weeks, B. A., Anderson, D. P., DuFour, A. P., Fairbrother, A., Goven, A. J., Lahvis, G. P., and Peters, G., *Biomarkers: Biochemical, Physiological, and Histological Markers of Anthropogenic Stress*, Huggett, R. J., Kimerle, R. A., Mehrle, P. M., Jr., and Bergman, H. L., Eds., Lewis Publishers, Chelsea, MI, 1992, 211.
142. Wong, S., Fournier, M., Corderre, D., Banska, W., and Krzystyniak, K., Environmental immunotoxicology, in *Animal Biomarkers as Pollution Indicators*, Peakall, D., Ed., Chapman and Hall, London, 1992, 167.
143. Friend, M. and Trainer, D. O., Experimental DDT-duck hepatitis virus interaction studies, *J. Wildl. Manage.*, 38, 887, 1974.
144. Friend, M. and Trainer, D. O., Experimental dieldrin-duck hepatitis virus interaction studies, *J. Wildl. Manage.*, 38, 896, 1974.

145. Weeks, B. A., Huggett, J. E., Warinner, J. E., and Mathews, E. S., Macrophage responses of estuarine fish as bioindicators of toxic contamination, in *Biomarkers of Environmental Contamination*, McCarthy, J. F. and Shugart, L. R., Eds., Lewis Publishers, Boca Raton, FL, 1990, 193.
146. Lahvis, G. P., Wells, R. S., Kuehl, D. W., Stewart, J. L., Rhinehart, H. L., and Via, C. S., Decreased lymphocyte responses in free-ranging Bottlenose Dolphins (*Tursiops truncatus*) are associated with increased concentrations of PCBs and DDT in peripheral blood, *Environ. Health Persp.*, 103, 67, 1995.
147. Grasman, K. A., Fox, G. A., Scanlon, P. F., and Ludwig, J. P., Organochlorine-associated immunosuppression in prefledgling Caspian Terns and Herring Gulls from the Great Lakes: An ecoepidemiological study, *Environ. Health Persp.*, 104, 829, 1996.
148. Ross, P. S., DeSwart, R. L., Rejinders, P. J. H., van Loveren, H., Vos, J. G., and Osterhaus, A. D. M. E., Contaminant-related suppression of delayed-type hypersensitivity and antibody responses in Harbor Seals fed herring from the Baltic Sea, *Environ. Health Persp.,* 103, 162, 1995.
149. Chen, S. C., Fitzpatrick, L. C., Goven, A. J., Venables, B. J., and Cooper, E., Nitroblue tetrazolium dye reduction by earthworm (*Lumbricus terrestris*) coelomocytes: An enzyme assay for nonspecific immunotoxicity of xenobiotics, *Environ. Toxicol. Chem.*, 10, 1037, 1991.
150. Venables, B. J., Fitzpatrick, L. C., and Goven, A. J., Earthworms as indicators of ecotoxicity, in *Ecotoxicology of Earthworms*, Grieg-Smith, P. W., Becker, H., Edwards, P. J., and Heimbach, F., Eds., Intercept Ltd, Andover, U.K., 1992, 197.
151. Ville, P., Roch, P., Cooper, E. L., and Narbonne, J.-F., Immuno-modulator effects of carbaryl and 2,4 D in the earthworm *Eisenia fetida andrei*, *Arch. Environ. Contam. Toxicol.*, 32, 291, 1997.
152. Svendsen, C. and Weeks, J. M., Relevance and applicability of a simple earthworm biomarker of copper exposure, 1. Links to ecological effects in a laboratory study with *Eisenia andrei*, *Ecotoxicol. Environ. Saf.*, 36, 72, 1997.
153. Nusetti, O., Parejo, E., Esclapès, M. M., Rodiguez-Grau, J., and Marcano, L., Acute-sublethal copper effects on phagocytosis and lysosome activity in the earthworm *Amynthas hawayanus*, *Bull. Environ. Contam. Toxicol.*, 63, 350, 1999.
154. Spurgeon, D. J., Svendsen, C., Rimmer, V. R., Hopkin, S. P., and Weeks, J. M., Relative sensitivity of life-cycle and biomarker responses in four earthworm species exposed to zinc, *Environ. Toxicol. Chem.*, 19, 1800, 2000.
155. Marcano, L., Nusetti, O., Rodiguez-Grau, J, Briceno, J., and Vilas, J., Coelomic fluid lysozyme activity induction in the Polychaete *Eurythoe complanata* as a biomarker of heavy metal toxicity, *Bull. Environ. Contam. Toxicol.*, 59, 22, 1997.
156. Cima, F., Marin, M. G., Matozzo, V., Da Ros, L., and Ballarin, L., Immunotoxic effects of organotin compounds in *Tapes philippinarum*, *Chemosphere,* 37, 3035, 1998.
157. Anderson, R. S., Mora, L. M., and Thomsom, S. A., Modulation of oyster (*Crassostrea virginica*) hemocyte immune function by copper, as measured by luminol-enhanced chemiluminescence, *Comp. Biochem. Physiol.*, 108C, 215, 1994.
158. Trust, K. A., Fairbrother, A., and Hooper, M. J., Effects of 7,1 2-dimethylbenz[a]anthracene on immune function and mixed-function oxygenase activity in the European Starling, *Environ. Toxicol. Chem.*, 13, 821, 1994.
159. Wu, P. J., Greeley, E. H., Hansen, L. G., and Segre, M., Immunological, hematological, and biochemical responses in immature white-footed mice following maternal Aroclor 1254 exposure: A possible bioindicator, *Arch. Environ. Contam. Toxicol.*, 36, 469, 1999.
160. Bracken, W. M. and Klassen, C. D., Induction of metallothionein in rat primary hepatocytes culture: Evidence for direct and indirect induction, *J. Toxicol. Environ. Health*, 22, 163, 1987.
161. Petering, D. H., Goodrich, M., Hodgman, W., Krezowski, S., Weber, D., Shaw, C. F., III, Spieler, R., and Zettergren, L., Metal-binding proteins and peptides for the detection of heavy metals in aquatic organisms, in *Biomarkers of Environmental Contamination*, McCarthy, J. F. and Shugart, L. R., Eds., Lewis Publishers, Boca Raton, FL, 1990, 239.
162. Benson, W. H., Baer, K. N., and Watson, C. F., Metallothionein as a biomarker of environmental metal contamination, in *Biomarkers of Environmental Contamination*, McCarthy, J. F. and Shugart, L. R., Eds., Lewis Publishers, Boca Raton, FL, 1990, 255.
163. Garvey, J. S., Metallothionein: A potential biomonitor of exposure to environmental toxins, in *Biomarkers of Environmental Contamination*, McCarthy, J. F. and Shugart, L. R., Eds., Lewis Publishers, Boca Raton, FL, 1990, 267.

164. Hogstrand, C., Galvez, F., and Wood, C. M., Toxicity, silver accumulation and metallothionein induction in freshwater rainbow trout during exposure to different silver salts, *Environ. Toxicol. Chem.*, 15, 1102, 1996.
165. Teigen, S. W., Anderson, R. A., Daae, H. L., and Skaare, J. U., Heavy metal content in liver and kidneys of grey seals (*Halichoerus grypus*) in various life stages correlated with metallothionein levels: Some metal-binding characteristics of this protein, *Environ. Toxicol. Chem.*, 18, 2364, 1999.
166. Wall, K. L., Jessen-Eller, K., and Crivello, J. F., Assessment of various biomarkers in winter flounder from coastal Massachusetts, USA, *Environ. Toxicol. Chem.*, 17, 2504, 1998.
167. Dethloff, G. M., Schlenk, D., Khan, S., and Bailey, H. C., The effects of copper on blood and biochemical parameters of rainbow trout (*Oncorhynchus mykiss*), *Arch. Environ. Contam. Toxicol.*, 36, 415, 1999.
168. Baršytė, D., White, K. N., and Lovejoy, D. A., Cloning and characterization of metallothionein cDNAs in the mussel *Mytilus edulis L.* digestive gland, *Comp. Biochem. Physiol.*, 122, 287, 1999.
169. Savva, D. and Li, B., Characterization of two metallothionein cDNAs from the shore crab for use as biomarkers of heavy metal pollution, *Ecotoxicology*, 8, 485, 1999.
170. Gerpe, M., Kling, P., Berg, A. H., and Olsson, P.-E., Arctic char (*Salvelinus alpinus*) metallothionein: cDNA sequence, expression, and tissue-specific inhibition of cadmium-mediated metallothionein induction by 17β-estradiol, 4-OH-PCB 30 and PCB 104, *Environ. Toxicol. Chem.*, 19, 638, 2000.
171. Lindquist, S., The heat shock response, *Ann. Rev. Biochem.*, 55, 1151, 1986.
172. Caltabiano, M. M., Koestler, T. P., Poste, G., and Grieg, R. G., Induction of 32 and 34-kDa stress proteins by sodium arsenite, heavy metals, and thiol-reactive agents, *J. Biol. Chem.*, 261, 13381, 1986.
173. Sanders, B. M., The role of the stress proteins response in physiological adaptation of marine molluscs, *Mar. Environ. Res.*, 24, 207, 1988.
174. Hassanein, H. M. A., Banhawy, M. A., Soliman, F. M., Abdel-Rehim, S. A., Müller, W. E. G., and Schröder, H. C., Induction of Hsp70 by the herbicide Oxyfluorfen (Goal) in the Egyptian Nile fish, *Oreochromis niloticus*, *Arch. Environ. Contam. Toxicol.*, 37, 78, 1999.
175. Köhler, H.-R., Knödler, C., and Zanger, M., Divergent kinetics of hsp70 induction in *Oniscus asellus* (Isopoda) in response to four environmentally relevant organic chemicals (B[a]P, PCB52, γ-HCH, PCP): Suitability and limits of a biomarker, *Arch. Environ. Contam. Toxicol.*, 36, 179, 1999.
176. Sanders, B., Stress proteins: Potential as multitiered biomarkers, in *Biomarkers of Environmental Contamination*, McCarthy, J. F. and Shugart, L. R., Eds., Lewis Publishers, Boca Raton, FL, 1990, 165.
177. Sanders, B. M., Stress proteins in aquatic organisms; An environmental perspective, *Crit. Rev. Toxicol.*, 23, 49, 1993.
178. Lewis, S., Handy, R. D., Cordi, B., Billinghurst, Z., and Depledge, M. H., Stress proteins (HSPs): Methods of detection and their use as an environmental biomarker, *Ecotoxicology*, 8, 351, 1999.
179. Welch, W. J., Mizzen, L. A., and Arrigo, A.-P., Structure and function of mammalian stress proteins, in *Stress-Induced Proteins*, Pardue, M. L., Feramisco, J. R., and Lindquist, S., Eds., Alan R. Liss, New York, 1989, 187.
180. Bradley, B. P. and Ward, J. B., Detection of a major stress protein using a peptide antibody, *Mar. Environ. Res.*, 28, 471, 1989.
181. Gutierrez, J. A. and Guerriero, V., Jr., Quantitation of hsp70 in tissues using a competitive enzyme-linked immunosorbant assay, *J. Immunol. Methods,* 143, 8, 1991.
182. Hara, I., Sato, N., Matsumura, A., Cho, J.-M., Weimin, Q., Torigoe, T., Shinnick, T. M., Kamidono, S., and Kikuchi, K., Development of monoclonal antibodies reacting against mycobacterial 65 kDa heat shock protein by using recombinant truncated products, *Microbiol. Immunol.*, 35, 995, 1991.
183. Heine, L., Drabent, B., Benecke, B.-J., and Gunther, E., A novel monoclonal antibody directed against the heat-inducible 68kDa heat shock protein, *Hybridoma*, 10, 721, 1991.
184. Margulis, B. A., Nacharov, P. V., Tsvetkova, O. I. Welsh, M., and Kinev, A. V., The characterization and use of different antibodies against the hsp70 major heat shock protein family for the development of an immunoassay, *Electrophoresis,* 12, 670, 1991.
185. Williams, J. H., Farag, A. M., Stansbury, M. A., Young, P. A., Bergman, H. L., and Peterson, N. S., Accumulation of HSP70 in juvenile and adult rainbow trout gill exposed to metal-contaminated water and/or diet, *Environ. Toxicol. Chem.*, 15, 1324, 1996.
186. Werner, I. and Nagel, R., Stress proteins HSP60 and HSP70 in three species of amphipods exposed to cadmium, diazinon, dieldrin and fluoranthene, *Environ. Toxicol. Chem.,* 16, 2393, 1997.

187. Bradley, B. P., Are the stress proteins indicators of exposure or effect?, *Mar. Environ. Res.*, 35, 85, 1993.
188. Köhler, H.-R., Belitz, B., Eckwert, H., Adam, R., Rahman, B., and Trontelj, P., Validation of hsp70 stress gene expression as a marker of metal effects in *Deroceras reticulatum (pulmonata)*: Correlation with demographic parameters, *Environ. Toxicol. Chem.*, 17, 2246, 1998.
189. Oberdörster, E., Martin, M., Ide, C. F., and McLachlan, J. A., Benthic community structure and biomarker induction in grass shrimp in an estuarine system, *Arch. Environ. Contam. Toxicol.*, 37, 512, 1999.
190. Schröder, H. C., Batel, R., Hassanein, H. M. A., Lauenroth, S., Jenke, H.-St., Simat, T., Steinhart, H., and Müller, W.E.G., Correlation between the level of the potential biomarker, heat-shock protein, and the occurrence of DNA damage in thc dab *Limanda limanda*. A field study in the North Sea and the English Channel, *Mar. Environ. Res.*, 49, 201, 1999.

SECTION II

# Contaminant Sources and Effects

CHAPTER 12

# Wildlife Toxicology of Organophosphorus and Carbamate Pesticides

**Elwood F. Hill**

## CONTENTS

## 12.1 INTRODUCTION

Organophosphorus (OP) and carbamate (CB) pesticides are used in domestic and natural environments for control of a wide variety of insect pests and disease vectors, other invertebrates, fungi, birds, and mammals; some CBs are used as avian repellents, and others have herbicidal properties. These two pesticidal classes, numbering about 200 OPs and 50 CBs, have been formulated into thousands of products that are available in the world's marketplace for varied applications to wetlands, rangelands, cultivated crops, forests, and rural and urban environs.[1–5] Except for mosquito control, these products are applied mostly on terrestrial landscapes. When applied according to the label, the active ingredient should be reasonably well contained within the intended treatment area. However, due to drift, runoff, or applicator error, the pesticide or toxic degradates are inevitably detected in water, soils, and vegetation outside the treated area — sometimes in toxic concentrations and for durations well beyond the expected residual life of the product. Off-site contamination of OPs and CBs has resulted in massive episodes of mortality of aquatic organisms,[6] but the long-term implications of periodic transient mortality on ecosystem productivity has not been thoroughly evaluated. Free-ranging vertebrates have also suffered large-scale mortality from acute exposure to OP and CB insecticides within and peripheral to the treated area.[7–10] OPs and CBs are acutely toxic to most animals with the potency of a chemical often widely variable among species.[11–20]

Most widely used OP and CB insecticides are highly toxic but relatively short-lived in nature (e.g., 2 to 4 weeks) and are readily metabolized and excreted by homoiothermic animals.[2,3] These factors and broad-spectrum insecticidal efficacy favored OP and CB pesticides as replacements for the persistent and problematical mercurial and organochlorine compounds.[21–23] For example, certain organochlorine pesticides and metabolites bioaccumulate in food chains, inhibit proper eggshell formation, and severely jeopardize populations of fish-eating birds such as brown pelicans (*Pelecanus occidentalis*), bald eagles (*Haliaeetus leucocephalus*) and ospreys (*Pandion haliaetus*).[24] This type of chronic manifestation has not been demonstrated for OP or CB exposure, and it is not certain that OP- and CB-induced mortalities of nontarget vertebrates have critical effects at the population level. However, there is increasing evidence that mortality of raptorial birds from OP and CB poisoning may be affecting some species at the regional level.[10,25]

The ecological hazard of OP and CB pesticides to wildlife is primarily from acute anticholinesterase toxicity but also includes species habitat association and foraging preference. Exposure may be directly from the pesticidal application, contact with or ingestion of contaminated water, soil or vegetation, or ingestion of contaminated prey or pesticide impregnated seeds or granules. Other factors also bear on wildlife tolerance of an OP- or CB-contaminated environment. For example, the prey base may be altered and affect foraging success; sublethal exposure may affect critical behaviors such as reproduction and migration; and proper balance between producer and consumer organisms in soil and aquatic systems may be disrupted.[26] Fish and other aquatic organisms also vary widely in tolerance of OP and CB exposure depending on inherent sensitivity and factors of water quality, chemistry, and temperature.[27]

This chapter provides an overview of the hazard of organophosphorus and carbamate pesticides to avian and mammalian wildlife. Attention is given to practical environmental considerations rather than interpretation of laboratory studies that were detailed in the first edition of this book.[28] Invertebrates, fish, amphibians, and reptiles are exemplified as ecosystem components or for comparison with birds and mammals, but the toxicology of OP and CB pesticides to these taxa is presented in other chapters. The focus herein is on concepts of ecological toxicology of birds and mammals related to natural systems as affected by pesticidal application in agriculture and public health. The environmental fate of representative OP and CB pesticides, the availability of these pesticides to wildlife, and toxicology as related to ambient factors, physiological cycles and status, product formulations, and sources of exposure are discussed.

## 12.2 GENERAL TOXICOLOGY

### 12.2.1 Organophosphorus Pesticides

Organophosphorus chemicals comprise more than one third of the registered pesticides on the world market. Most registrations are for control of a large array of insect pests and disease vectors, but OPs effectively control many animal pests including other invertebrates and terrestrial vertebrates. Over 95% of the OP products presently in production are used in agriculture and mosquito control.[29] In the United States about 70 OPs are registered as the active ingredient (AI) in thousands of products, but only 10 to 15 of the chemicals account for over three fourths of the use. Examples of OPs that have been widely used in U.S. agriculture include azinphos-methyl, chlopyrifos, fonophos, malathion, methyl parathion, parathion, phorate, phosalone, and terbufos; OPs used extensively for mosquito control are fenthion, malathion, naled, and temephos. Many of these chemicals have been reviewed by regulatory agencies for environmental and public health concerns and are now classified as restricted-use pesticides in the United States; some uses have been cancelled. OPs registered for outdoor use have label warnings about toxicity to wildlife and application to wetlands.

The main concern about OPs is their acute lethal toxicity. Based on single-dose oral $LD_{50}$ tests, the above chemicals are among the most toxic OP pesticides to laboratory rodents and wildlife.[11,13,15,16,30] Of the 12 chemicals, ten are classed highly toxic to birds ($LD_{50}$, <50 mg AI/kg body mass[31]), and seven are highly toxic to mammals. Only malathion is considered to be of a low order of acute toxicity to both birds and mammals ($LD_{50}$, >500 mg/kg). Chlorpyrifos and fenthion are moderately toxic ($LD_{50}$, 50 to 500 mg/kg) to mammals, and naled is moderately toxic to birds and mammals. Some OPs that have had most uses withdrawn or cancelled in the United States (e.g., dimethoate, EPN, monocrotophos, parathion, TEPP)[5] remain available on the international market in spite of their demonstrated environmental hazard to human health and wildlife. In the U.S.S.R., pesticides with a single-dose oral mammalian $LD_{50}$ less than 50 mg/kg were banned in the 1960s; a few exceptions were permitted for use of granular formulations in agriculture.[32]

The principal toxicity of OP pesticides is based on disruption of the nervous system by inhibition of cholinesterase (ChE; acetylcholinesterase, EC 3.1.1.7, and a mixture of nonspecific esterases) activity in the central nervous system and at neuromuscular junctions with death generally attributed to acute respiratory failure.[33] When OP binds to ChE, a relatively stable bond is formed and prevents the ChE from deactivating the neurotransmitter acetylcholine. This permits buildup of acetylcholine and overstimulation of the cholinergic nervous system. Some of the nonspecific signs following acute OP exposure of small mammals and birds include lethargy, labored breathing, excessive bronchial secretion, vomiting, diarrhea, tremors, convulsions, and death. These toxic indicators are useful when sick animals are found near an area of recent OP application, but the signs are not uniquely different from poisoning by other neurotoxic agents.[1,16,33] In nature, notation of toxic signs is important in the investigation of a wildlife incident, while conclusive diagnosis depends on biochemical and chemical analyses for brain ChE inhibition and OP residues in the carcass.[34–36] Biochemical diagnosis of OP and CB exposure is described in Section 12.4.7.

Two additional syndromes of single or very short-term OP exposure have been demonstrated in the laboratory. The first, referred to as an "intermediate syndrome," is a potentially lethal paralytic condition of the neck, limbs, and respiratory muscles.[37,38] The paralysis from muscle necrosis follows an acute OP exposure by 2 to 3 days and is apparently initiated by depressed ChE activity and calcium accumulation in the region of the motor end-plate.[38,39] This syndrome has not been identified as such in either laboratory or field studies of wildlife, but it could be an important factor in OP hazard in nature. For example, when flocking birds enter a hazardous OP-treated area, onset of acute toxicity often occurs within a few minutes in some birds, while others appear unaffected. If the intermediate syndrome plays a significant role, mortality away from the treated area could be much higher than generally believed. Examples of OP pesticides demonstrated to induce the intermediate syndrome in mammals include fenthion, malathion, and monocrotophos.[38] Fenthion and monocrotophos have caused large-scale episodes of wildlife mortality; malathion has not. The possibility of intermediate syndrome from malathion has not been investigated.

The second syndrome of single-dose OP exposure is OP-induced delayed neurotoxicity (OPIDN) in which a relatively small dosage (e.g., 1/25 $LD_{50}$) of OP causes degeneration in the myelin sheath of long peripheral nerves and the spinal cord.[40] This debilitating malady develops in 1 to 3 weeks, causing a stumbling gait and incoordination. OPIDN has been demonstrated in a variety of laboratory animals including rodents, chickens, and mallards (*Anas platyrhynchos*); mallards do not appear to be as susceptible as chickens to OPIDN.[41] Apparently, OPIDN is not related to anticholinesterase action of OPs.[42] Most OPIDN-inducing pesticides are no longer on the market (e.g., leptophos and mipafox), but a few remain in use in some countries (e.g., cyanofenphos, EPN, methamidophos, trichlorfon, and trichloronate).[38,42]

Subacute or subchronic exposure to repeated sublethal doses of OP have been demonstrated to affect birds and mammals in captivity[28,43,44] and undoubtedly influences critical physiological functions in nature. However, validation of low-grade OP hazard to natural wildlife populations remains elusive. Some of the more likely sublethal effects on wildlife that live in or depend regularly on OP-contaminated forage and water include changes in response to ambient stressors, changes in foraging and reproductive behavior, and possible alteration in migration orientation. Whereas these effects appear related to anticholinesterase insult, other effects may be entirely independent of ChE inhibition. Examples include mutagenicity, carcinogenicity, and organ-specific toxicity to the heart and kidneys.[42] Putative sublethal effects of OP exposure are discussed in Section 12.5.

### 12.2.2 Carbamate Pesticides

Fewer than one fourth of the registered carbamates in the world market are insecticides with significant anticholinesterase activity; the others are fungicides and herbicides with little acute hazard to birds and mammals. Of approximately 50 registered CB pesticides, only about eight (aldicarb, carbaryl, carbofuran, formetanate, methiocarb, methomyl, oxamyl, and propoxur) are used for insect control on crops, forests, and rangelands; methiocarb and methomyl are also used as avian repellents. Of these eight, carbofuran, methomyl, and carbaryl account for more than 90% of the use. As with OPs, CB insecticides have label warnings about toxicity to wildlife and application to wetlands.

CB insecticides exert their toxicity through acute ChE inhibition, and all of the above-named except carbaryl are classed highly toxic to birds and mammals.[12,16] $LD_{50}$s are generally less than 20 mg/kg for both taxa and as low as 0.8 mg/kg for aldicarb with male laboratory rats and 0.5 mg/kg for carbofuran with male mallards. In contrast, the $LD_{50}$s for carbaryl are reported as 850 and >2,500 mg/kg for male rats and mallards.

Acute toxicity of CB insecticides including toxic signs are similar to that of OPs, except onset of and recovery from CB is faster than for equipotent exposure to OP insecticides.[45] The rapid reaction is partly because CB insecticides are direct ChE inhibitors that do not require metabolic activation for full potency as do most OPs. Rapid recovery is a product of near spontaneous reactivation of carbamylated ChE. Thus, an equipotent sublethal exposure of CB is generally less severe than OP exposure, and factors of delayed toxicity described for OPs do not occur with CBs.[1,42]

### 12.2.3 Some Considerations

The toxicity of organophosphorus and carbamate pesticides varies considerably among vertebrates (Table 12.1). OPs and CBs do not bioaccumulate in food chains. Cumulative depression of ChE enzyme may occur and persist from repeated exposure to some OPs but generally not from CBs. CB and a variety of OP esters (e.g., acephate, monocrotophos, trichlorfon) are direct ChE inhibitors, but most OP pesticides (e.g., diazinon, malathion, parathion) must first undergo an

**Table 12.1 Acute Response of Fish, Laboratory Rats, and Birds for Anticholinesterase Pesticides of Widely Variable Mammalian Toxicity**

| Pesticide | Rainbow Trout[a] | Bluegill[a] | Laboratory Rat[b,c] | Ring-Necked Pheasant[b,d] | Red-Winged Blackbird[b,e] |
|---|---|---|---|---|---|
| Aldicarb | 560 | 50 | 0.8 | 5.3 | 1.8 |
| Phorate | 13 | 2 | 2.3 | 7.1 | 1.0 |
| Carbofuran | 380 | 240 | 11 | 4.1 | 0.4 |
| Azinphos-Methyl | 4 | 22 | 13 | 75 | 8.5 |
| Mexacarbate | 12,000 | 22,900 | 37 | 4.6 | 10 |
| Ethion | 500 | 210 | 65 | 1297 | 45 |
| Methiocarb | 800 | 210 | 70 | 270 | 4.6 |
| Dimethoate | 6200 | 6000 | 215 | 20 | 6.6 |
| Carbaryl | 1950 | 6760 | 859 | 707 | 56.0 |
| Temephos | 3490 | 21,800 | 8600 | 35 | 42 |

[a] $LC_{50}$ = µg of active ingredient per liter of water calculated to kill 50% of test population during a standard 96-h exposure. Tests were conducted under static conditions (pH 7.2-7.5 at 10-13°C for trout or 20-22°C for bluegills), *n* = 50-60 per test.[14]

[b] $LD_{50}$ = mg of active ingredient per kilogram of body mass calculated to kill 50% of test population.

[c] Sherman strain male laboratory rats, 3 months old, *n* = 50 to 60 per test; dosage by gavage in peanut oil.[11,12]

[d] Farm-reared male and female ring-necked pheasants, 3 to 4 months old, *n* = 8 to 28 per test; dosage by gelatin capsule.[16]

[e] Wild-captured pen-conditioned male and female red-winged blackbirds, adult, *n* = 8 to 28 per test; dosage by gavage in propylene glycol.[15]

oxidative desulfuration step for maximum anticholinesterase potency.[2,33] This toxicating step is mediated by mixed-function oxidases (MFO) in the liver of vertebrates and in the fat body, malpighian tubules, and digestive tract of invertebrates. MFO activity varies widely among vertebrates in the following order: mammals > birds > fish.[46,47] (The place of amphibians and reptiles in this ranking has not been determined.)[48,49] The same physiologic system responsible for toxication of most OPs also has a primary role in their detoxication. Since OP and CB metabolism occurs primarily in the liver of vertebrates, the portal of entry into the circulatory system is important to acute toxicity. CBs and direct-acting OPs may be more hazardous through inhalation than through ingestion where substances are routed through the liver and first-phase metabolism. The environmental hazard of inhalation toxicity has not been thoroughly evaluated but is generally believed less important than ingestion. Likewise, percutaneous exposure to OP and CB pesticides has not been adequately studied in wildlife, but in nature the composite of inhalation, percutaneous exposure, and ingestion must all be considered in hazard prediction and understanding differences of species' response to OP and CB applications. It is important that OP toxication via oxidative desulfuration also occurs in nature as mediated by microbial metabolism in soils and by phytometabolism, but this process is slow compared to MFO metabolism in vertebrates.

The toxic consequences of OP or CB application to aquatic and terrestrial wildlife usually last only a few days. However, multiple applications of pesticide are the rule during the growing season; this extends the potential hazard to wildlife. In areas where a variety of crops are cultivated, exposure to a variety of pesticides and unexpected hazard may occur, especially to highly mobile wildlife. Other factors of variability must also be considered in hazard assessment of OP and CB pesticides. Predator-prey or competitor balance among invertebrates and aquatic vertebrates may be disrupted. Daily activity patterns, energy budgets, and various behaviors of terrestrial vertebrates may be affected. Repeated application of these biologically labile chemicals may cause cumulative physiological effects without a corresponding accumulation of chemical residues such as occurs for heavy metals and organochlorine pesticides. Recovery from anticholinesterase exposure may differ among pesticidal classes. For example, 1 to 3 weeks may be required for brain ChE activity to recover in vertebrates receiving a single exposure to OPs,[50,51] but only a few hours may be necessary to recover from CB exposure.[52–53] Many of the confounding variables will become evident in the following section on environmental fate and hazard of several representative OP and CB pesticides.

## 12.3 ENVIRONMENTAL FATE AND HAZARD

The first considerations of environmental fate and hazard of organophosphorus and carbamate pesticides are agricultural crop, products formulation, rate and method of application, likelihood of wildlife exposure, and the biological and physical characteristics of any wetlands likely to receive runoff. OP and CB pesticides are comparatively labile in circumneutral environments and readily degrade under conditions of alkalinity, rain, sunlight, and temperature.[2,3,54] Neither bioaccumulation nor biomagnification occurs to any important degree in aquatic or terrestrial food chains. The main hazard of OP and CB exposure to wildlife is from short-term, potentially lethal exposure to pesticidal treatment. This may include direct contact via inhalation and percutaneous routes or from ingestion of contaminated food and water. Classic secondary poisoning through biomagnification of highly lipid soluble pesticides, such as chlorinated hydrocarbons, does not happen with OP and CB compounds. Instead, predators are often poisoned by feeding on prey that had been contaminated by pesticidal application. For example, invertebrates and aquatic vertebrates adsorb OP and CB pesticide on their cutaneous surface or mucosal coating, which is then readily dissociated and absorbed when eaten by another animal.[28] Predatory birds have also been poisoned from eating insects that had ingested systemic OP and CB while foraging on vegetation.[10] The stomach contents of poisoned birds and mammals may contain large concentrations of OP and CB that have proven toxic to predators and scavengers.[52]

OP pesticides and anticholinesterase CBs have broad-spectrum toxicity to most animals with a cholinergic-dependent nervous system. OP and CB products are used as stomach and contact poisons for nearly any type of insect control; about 20% of OP and 40% of CB insecticides have systemic activity. They are also used as acaricides, nematicides, rodenticides, avicides, and bird repellents. As mentioned, a few OPs (e.g., butylate, EPTC, molinate) and many CBs have herbicidal properties, and some CBs receive wide use as fungicides (e.g., benomyl, maneb, mancozeb). In spite of their broad toxicity, OPs such as fenthion, naled, and malathion are widely used over wetlands and areas of human habitation for mosquito control.[29] ChE-inhibiting CBs are not as extensively applied on natural wetlands as are OPs, and CBs other than carbaryl are not applied extensively near human habitation.

Except for mosquito control, nearly all OP and CB application is on terrestrial landscapes. Nonetheless, pesticidal treatment of farmlands, forests, and cities results in the contamination of adjacent aquatic systems by some of the pesticide and its degradates. It has been theorized that all pesticidal chemicals will eventually enter an aquatic environment and affect a much larger number of species than originally intended.[19] The hazard of OP and CB pesticides to nontarget life is a product of the amount, rate, and form of residual entering the aquatic system as well as the dynamics and chemistry of the system. For example, if runoff from a pesticidal application enters a stream, rapid dilution and dispersal may soon render the contamination ecologically innocuous. In contrast, if the same runoff enters an aquatic sink, such as a farm pond rich with detritus, the ecological consequences may be considerable to organisms at all levels including terrestrial species. In general, the residue levels of OP and CB compounds in natural waters are as follows: closed pond > free-flowing waters > lakes > estuaries > open sea.[19]

Contamination of small closed ponds has resulted in large incidents of mortality of waterfowl and predatory birds including northern harriers (*Circus cyaneus*) and golden eagles (*Aquila chrysaetos*) in the northern plains of the United States. In this example a granular formulation of phorate was incorporated into the soil for nematode control. Heavy rains occurred several months after application and facilitated the transport of phorate and degradates into small ponds, resulting in mortality of aquatic and terrestrial animals for 2 to 3 weeks following the storm event. Similar scenarios leading to mortality from OP and CB leaching from granular formulations have been documented for agricultural application of pesticides such as parathion, diazinon, and carbofuran.[6] Wildlife have been acutely poisoned by ingesting OP- and CB-impregnated granules and by foraging in areas treated with flowable and emulsifiable concentrate formulations and technical grade pesticide.[4,6,10,52]

### 12.3.1 Organophosphorus Pesticides

Phosphorothioic acids contribute more than 90% of the OP pesticide use in the United States. These pesticides require metabolic activation for maximum anticholinesterase effect and vary widely in acute toxicity and chemical stability in the environment. Phosphoric acids, representing nearly all of the remaining OP pesticides, do not require metabolic activation for toxicity and, except for acephate and ethoprop, are presently of little importance in the United States. This is because voluntary withdrawal and regulatory restrictions have reduced the market for some previously widely-used and extremely toxic products such as dicrotophos, mevinphos, monocrotophos, and phosphamidon.[5] The major phosphoric acid products on the U.S. market are acephate and ethoprop, while the others remain available on other world markets.

As discussed, OP pesticides are labile in natural environments, which generally limits their hazard to only a few days for vertebrates depending on a variety of ambient factors related to method of application and formulation of product. In contrast, a few OPs, such as phorate in the above example, have hazard potential to wildlife for a longer period due to systemic action and toxic degradates.

#### *12.3.1.1 Case Study: Phorate*

Phorate is one of a small but widely used group of highly efficacious insecticide-acaricide compounds whose environmental degradates are more toxic and stable than the parent chemical. This feature increases their potential hazard to wildlife but also increases their marketability as broad-spectrum systemic pesticides for foliar application and incorporation into the soil. Examples of these structurally similar OPs include phorate, (EtO)2 PS.SCH2 SEt; disulfoton, (EtO)2 PS.SCH2CH2Set, and terbufos, (EtO)2 PS.SCH2 SBu.t55, are extremely toxic to mammals and birds ($LD_{50}$, <10 mg/kg).[12,16] For comparison, the acute toxicity of methyl parathion and carbofuran has been reported as 14 and 11 mg/kg for rats[12] and 10 and 0.5 mg/kg for mallards.[16] These pesticides are also highly toxic by the dermal route of exposure.[5,12]

This group of pesticides has an additional oxidative pathway of toxication in the environment that is mediated primarily by UV irradiation in water and soil, to a lesser degree by microbial metabolism,[56] and by phytometabolism in plants.[57] As with the parent compound, the sulfoxide and sulfone degradates also require the oxidative desulfuration step to increase anticholinesterase potency.

With phorate as the model, several observations on environmental fate are important to the ecological hazard of these unique phosphorothioic acid pesticides. The fate of phorate in water is determined by pH, temperature, and photolysis.[2] Phorate is stable in water at about pH 5; its half-life in pH 6 water at 25°C is about 7 days.[58] The rate of hydrolysis increases about tenfold per pH unit under alkaline conditions. Irradiation by UV light oxidizes phorate within minutes to its more potent ChE-inhibiting sulfoxide and sulfone degradates.[56] Neither aquatic invertebrates nor fish tend to bioaccumulate phorate in model ecosystems,[59] but phorate is highly toxic to fish. For example, in comparable standardized 96-hour $LC_{50}$ tests, phorate is more than 100 times as toxic as methyl parathion and carbofuran to an array of both warm- and cold-water game fish.[14]

The fate of phorate in soil is affected by pesticide formulation, the method and rate of application, soil type, pH, temperature, moisture content, irrigation and water percolation, vegetation type and abundance, and microbial populations.[60,61] Surface application of either granular or emulsifiable phorate results in 15 to 20% loss due to volatization within 1 h; thereafter, nearly all the residual phorate remains bound to the soil particles but undergoes UV irradiation and oxidation to sulfoxide and sulfone degradates within a few days. Up to 80% of these highly toxic degradates persist in various soil types for 1 to 2 months.[60] The terrestrial dissipation half-life of phorate in irrigated soils is 2 days in sandy loam and 9 to 15 days in silty loam. For sulfoxide and sulfone degradates, the half-life in sandy loam is 12 to 18 weeks.[62] Phorate movement is more rapid in summer than in winter, but it is more stable in winter, probably due to lower microbial activity.[61] Phorate is poorly soluble in water (~50 mg/L), and residues do not migrate extensively from the treated area. However, residues may be transported by erosion and runoff from agricultural fields into aquatic systems, where they have caused major episodes of fish and bird mortality.[6]

Soil-incorporated phorate is readily translocated through the roots and stems of plants and provides insecticidal protection to plants for a relatively long time because of the greater persistence of the sulfoxide metabolite.[2] The initial oxidation to sulfoxide proceeds rapidly in plants, whereas further oxidation to sulfone and desulfuration goes more slowly. Anticholinesterase activity increases as phorate oxidation proceeds to the most toxic sulfone. The oxon analog of each form is even more potent. Therefore, the anticholinesterase activity of phorate assimilated into plants increases for several days and only then gradually loses activity over 2 to 5 weeks.[2] Because of this systemic activity, phorate poses a potential hazard to herbivorous animals including wildlife and livestock.

In summary, phorate, like most widely used OP pesticides, is highly toxic to vertebrates. However, where most OPs rapidly degrade to biologically inert products, phorate degrades to more stable and potentially more potent anti-ChE products. This contributes to the systemic efficacy of phorate but also extends its hazard compared to other systemic OP insecticides. Examples of other systemic OPs include acephate, dicrotophos, oxydemeton-methyl, and phosphamidon; all are direct-acting ChE inhibitors.

### 12.3.2 Carbamate Pesticides

Carbamates with significant anticholinesterase activity were developed as analogs of alkaloid extracts of the calabar bean (*Physostigma venenosum*). Elimination of the polar moiety of the natural drug enhanced lipid solubility and penetration of the insect cuticle and nerve sheath, leading to broad-spectrum insecticidal activity.[3,54] Most insecticidal products are phenyl N-methylcarbamates (e.g., carbaryl, carbofuran, propoxur); the others are oxime carbamates (e.g., aldicarb, methomyl, oxamyl).[55] Carbofuran and most oxime carbamates have significant systemic activity in plants. All CB insecticides are potent anticholinesterases and highly toxic to most invertebrates, and many are highly toxic to vertebrates. Birds and mammals receiving sublethal exposure to CB readily detoxicate the molecule and appear to fully recover within a few hours. There is little evidence that herbicidal and fungicidal CBs pose important hazard to terrestrial wildlife.[4] For example, none of these CBs have significant anticholinesterase activity, and only one of the widely used products has a mammalian $LD_{50}$ of less than 1000 mg/kg (molinate, 720 mg/kg for laboratory rats), while most yield $LD_{50}$s above 2000 mg/kg for birds and mammals. However, some products are highly toxic to fish.[14]

Carbamate insecticides are generally short-lived as foliar applications (e.g., 1 to 4 weeks) and require several applications during growing seasons; when applied to soil, significant residues may persist 2 to 16 months.[3] Little information is available on the biological hazard of multiple applications or on the importance of comparatively more persistent but less toxic degradates of CBs such as aldicarb and carbofuran. Both of these compounds are highly toxic to terrestrial animals[3,4,6,] and aldicarb is toxic to fish;[14] carbofuran is not nearly as toxic as aldicarb in 96-hour $LC_{50}$ tests with freshwater fish.[14] Toxic residues of aldicarb and its degradates (mostly aldicarb-sulfoxide and sulfone) have been detected in ground water and farm produce as long as a year after application.[63,64] Due to extreme mammalian toxicity and hazard to farm workers, aldicarb is a restricted-use pesticide in the United States. Carbofuran is also a restricted-use pesticide, but it is still widely used in spite of regulatory concerns about its environmental hazard.[6,65]

#### 12.3.2.1 Case Study: Carbofuran

Carbofuran is used mostly in granular formulations for control of soil and foliage insects and agricultural nematodes. Flowable formulations were used extensively in the United States through the early 1970s and resulted in large-scale episodes of wildlife, mostly waterfowl, mortality. Thereafter, granular formulations were favored, but similar incidents of mortality continued.[65,66] These incidents usually occurred within the first 2 to 3 days following approved pesticidal application to field crops. Additional incidents of mortality of both terrestrial and aquatic wildlife have also been documented subsequent to heavy rainfall for more than 6 months following application.

Carbofuran is the epitome of an acutely toxic environmental poison. That is, a trivial exposure (e.g., 0.1-5 mg/kg) may be lethal in less than 5 min, especially to some waterfowl and passerine birds.[13,16] Other individuals exposed to the same and even higher dosages (e.g., > $LD_{50}$) may exhibit intense signs of poisoning within the same time frame and then abruptly recover and become fully alert and ambulatory within as little as 30 min posttreatment.[45,67] In captivity, survivors of this initial toxicity appeared to recover fully; whether similar recovery would occur in nature is questionable. For example, on the day after an application of carbofuran on alfalfa both dead and living American wigeon (*Anas americana*) were in the field.[36] After a brief period of observation, the field was entered, and many of the birds flew to a nearby irrigation canal only to die soon thereafter. It was not known how long the wigeon had been in the field when approached or whether they were experiencing the onset of toxicity or were recovering from toxicity when flushed. However, during initial observations many normal appearing birds were noted feeding on the alfalfa in the same location where others had died.

It has been determined experimentally that birds may continue foraging on potentially lethal carbofuran-treated feed immediately upon recovery from a toxic episode, and without apparent effect.[68] In 5-day feeding studies with 1- to 21-day-old Japanese quail (*Coturnix japonica*), nearly all mortality and overt toxic signs from all graded exposures subsided within about 6 h of initial presentation of carbofuran. Both recovered, and apparently unaffected birds continued feeding on their carbofuran diets at the same rate and with comparable growth as controls on untreated diet. A few birds died later in the studies at most levels of exposures. Because these birds were tested in groups and not handled during treatment, it was not determined which individuals had previously shown overt toxicity. However, once toxicity occurred on day 3 or 4, the birds showed signs characteristic of acute poisoning and promptly died. In reference to the above wigeon example: (1) wild birds may initially recover from acute carbofuran exposure and then continue foraging on the same diet; (2) it is not certain that such recovered wild birds can also tolerate natural stressors such as fright response, extended flight, or adverse weather in the short term.

As mentioned, the most common hazard of carbofuran to wildlife is foraging in treated fields shortly after pesticidal application to foliage. Ambient factors of wind, rain, and UV irradiation rapidly reduce the hazard of readily available pesticide in its most toxic form within 2 to 3 days, even though systemic insecticidal activity may continue for more than a month.[3] The availability of contaminated invertebrate prey diminishes precipitously immediately following a carbofuran treatment. Even for grazing waterfowl, such as the American wigeon and Canada goose (*Branta canadensis*), the hazard of carbofuran poisoning is greater immediately after treatment than after its incorporation into plant tissue. Another hazard that has been substantial to both aquatic and terrestrial life is periodic surges of pesticide into wetlands from runoff of heavy rainfall.[6] In this situation the source of the carbofuran has often been granular product that had been incorporated in the soil. Though soil incorporation of granular would seem safe, wildlife mortality has occurred within a day following careful application of carbofuran according to the best available technology.[69] A form of secondary poisoning other than eating contaminated insects has also been documented for carbofuran. Predatory and scavenging birds and mammals have been poisoned by eating the entrails, including unabsorbed carbofuran in the stomachs of dead animals,[69,70] sometimes as long as 6 months after application of pesticides for control of soil pests.[71]

General factors affecting the hazard of carbofuran to wildlife include formulation and application, its water solubility and potential for leaching and transport out of the treated area, and the likelihood of its entrance into wetlands. Carbofuran is soluble in water to about 700 mg/L and essentially stable in acidic medium.[3] The fate of carbofuran in water is influenced by pH, photolysis, temperature, and trace impurities.[72] The half-life of carbofuran in distilled water at 25°C and pH 5.5, 7, 8, and 9 is about 16 years, 1 month, 6 days, and 6 h, respectively.[73] The rate of hydrolysis is positively correlated with ambient temperature. Neither aquatic invertebrates nor fish tend to bioaccumulate carbofuran in slightly alkaline model ecosystems.[73–75]

The fate of carbofuran in soil is affected by pesticide formulation, rate and method of application, soil type, pH, rainfall and irrigation, temperature, moisture content, and microbial populations.[3,76] Carbofuran decomposes rapidly in alkaline soil and is stable at pH 5.5; the hydrolytic half-life in soil at pH 7 is about 35 days.[73] Temperature and moisture content are positively correlated with degradation; maximum degradation to hydrolytic metabolites occurs at 27°C.[77] The terrestrial dissipation half-life of carbofuran in irrigated soils is 4 to 11 days in sandy loam, 1 month in loam, and less than 5 months in silty loam.[6] Carbofuran is mobile and has been found in streams, surface water, and runoff sediments from treated watersheds.[6] This suggests carbofuran may be quite stable in association with acid precipitation.[28] It is also suggested that carbofuran could leach into ground water, where it could persist at characteristically acidic pH, like aldicarb.[64,65]

Carbofuran degrades within about 7 to 10 days on crops such as alfalfa sprayed with flowable formulations.[78] In-furrow application of granular formulation is readily translocated through the roots and stems with insecticidal activity continuing in foliage for 2 to 4 months.[3,79] Even when granules are covered by soil, toxic carbofuran residues may be available to herbivorous animals.

Carbofuran is readily metabolized in living animals and is rapidly excreted. Toxicologically important bioaccumulation of either parent compound or degradates does not occur in either invertebrates or vertebrates.[73] Nonetheless, secondary hazard from carbofuran occurs in predatory vertebrates that feed on dead and struggling insects, earthworms, and small birds and mammals. As mentioned above and in the case study of phorate, this source of poisoning is most likely from unabsorbed carbofuran in the gut of the primary subject.

### 12.3.3 Wetlands

Organophosphorus and carbamate insecticides are generally not permitted for use on wetlands. An exception is a few OP products used for mosquito control, most notably malathion, but even this use is decreasing in favor of synthetic pyrethroids and biological control agents. However, wetlands are routinely contaminated during agricultural use of a wide array of OP and CB products. Drift, direct overspray, and runoff are sources of contamination when farmlands border natural wetlands, irrigation canals, and drainage ditches.[6] Ground application of pesticide contaminates wetlands through volatilization, migration, and runoff, but drift from this source is not equal to that of aerial treatment. It has been estimated that downwind drift from aerial application is conservatively four times higher than produced by higher clearance ground sprayers.[80]

Small streams and ponds are especially prone to OP and CB contamination from overspray, drift, and runoff. This is partly because they are difficult to avoid during aerial application when crops abut the wetlands. Of more importance, however, is the fact that these small wetlands, often less than a hectare in size and only 1 to 2 m deep, are not fully appreciated for their intrinsic value in regions dominated by expansive farmlands and monocultures. For example, in North America's prairie pothole region of south-central Canada and the north-central United States, about 65% of the original wetlands have been drained and put in cultivation.[81] The remaining wetlands contribute more than 50% of North American's duck production.[82] A very large proportion of these wetlands are near or surrounded by agricultural fields and are considered of high potential for excessive agricultural pesticide contamination.[83,84] OP products constitute the majority of the pesticide use in the region.

Agricultural contamination of small and sometimes only seasonal wetlands is probably the most ecologically abusive use of OP and CB pesticides. Many of these wetlands are extremely productive in terms of biomass and wildlife dependence during critical spring and summer periods. These small ponds contain proportionally little volume to surface area for dilution and buffering of toxicity compared to larger ponds and lakes. OP and CB insecticides may seriously offset invertebrate populations for the short-term. However, most OP and CB products are inactivated within a few hours in circumneutral aquatic environments and within a few days to 1 to 2 weeks in more acidic systems. This rapid degradation is ecologically advantageous because most knocked-down invertebrate populations seem to recover rapidly from a single operational OP or CB treatment. The cumulative effects of multiple applications on species diversity, balance, abundance, and total biomass are not well understood for different product formulations or aquatic systems.

The U.S. Fish and Wildlife Service developed a comprehensive research plan in the mid-1980s to evaluate hazard of pesticides to aquatic invertebrates and waterfowl in the prairie pothole region of North America.[85] The studies focused on an array of "average" potholes of similar size and depth to determine the response of mallard and blue-winged teal (*Anas discors*) to "typical" treatments of parathion and methyl parathion on wheat and sunflowers. In both studies each of five broods was confined to a fenced pothole and permitted to roam throughout the pond and riparian zone to the edge of the treated crop that surrounded the pothole. Each pothole was believed to be of adequate size and quality to provide necessary cover, spatial, and nutritional requirements for simultaneous fledging of several mallard broods. (Details of this research initiative have been summarized in several case studies.[86])

In the study of spring wheat the first step was to treat all of the fields, including the ten potholes, with an aerial application of the herbicide 2,4-D (0.3 kg AI/ha), which is a standard practice in the region. One month later one half of the fields and ponds were treated by standard operational application of parathion (0.6 kg AI/ha), and other fields and ponds were treated with methyl parathion (0.3 kg AI/kg). All treatments were made from 1.5 to 2 m above the crop, which resulted in a spray swath of about 18 m. Three days prior to the insecticide treatments 10 to 12 3-week-old mallards and 10 to 12 3-week-old teal were released on each wetland. The ducklings were sampled at 2 and 7 days after the treatment, and then at about weekly intervals through 30 days posttreatment. Body mass, tissue residues of OP, and brain ChE activity were evaluated at each sample period. None of the ducklings died from OP poisoning, but within 2 days posttreatment, 27% of the ducklings on parathion-treated ponds and 29% of those on the methyl parathion treatment had significantly depressed ChE activity. Body mass was not affected. Behavioral and population studies did not indicate any differences in the number of wild broods or adults of either species on treated and control fields.

In the study of sunflowers an operational aerial treatment of parathion was evaluated.[87] The main differences from spring wheat studies were the use of younger mallards (1 to 3 days of age), the release of two broods of 10 to 12 ducklings with their hens in each pond, and near doubling of the rate of parathion treatment (1.12 kg AI/ha). This was the maximum recommended treatment for sunflowers in North Dakota. One half of the study wetlands were oversprayed to test the worst-case scenario. For the other half of the wetlands, the operator attempted to achieve complete crop coverage, while avoiding overspray of the ponds. Overall, more than 90% of the ducklings in the oversprayed ponds were dead within 3 days posttreatment. Brain ChE activity was inhibited an average of 76% compared with ducklings collected on control wetlands. In this study parathion residues were determined in several matrices (water, vegetation, insects, and snails) 1 days prior to spray and 1, 7, 14, and 29 days posttreatment. All prespray samples were clear of parathion. All samples contained parathion (mean residues: water, 0.12 μg/mL; insects, 0.29 μg/g; snails, 9.04 μg/g; and vegetation, 0.42 μg/g) 1 day posttreatment and by 7 days all samples were clear of detectable residues. Ground searching of thc oversprayed ponds and their riparian zones yielded more than 30 dead birds with inhibited brain ChE activity. Over two thirds of the dead birds were juveniles of blue-winged teal, American coot (*Fulica americana*), red-winged blackbird (*Agelaius phoeniceus*), and yellow-headed blackbird (*Xanthocephalus xanthocephalus*). For the wetlands where reasonable effort was made to avoid direct pond overspray while providing complete crop coverage, parathion was not attributed to the death of any of the ducklings. Likewise, there was not evidence of other birds having been acutely poisoned by the treatment. Wetland residues of parathion were also much less than for the oversprayed ponds. The mean residues for key wetland matrices 1 day posttreatment were also much less than for the oversprayed ponds. The mean residues for key wetland matrices 1 day posttreatment were: water, 0.01 μg/mL; insects, 0.13 μg/g; snails, 3.04 μg/g; and vegetation, 0.35 μg/g. The effort to avoid direct overspray of small wetlands eliminated OP-induced avian mortality and reduced parathion in water (> 90% and invertebrates 34 to 45%).

These studies were invaluable in demonstrating that: (1) standard commercial aerial application of flowable OP may be directly lethal to avian wildlife on small wetlands and (2) adjustment in application to avoid direct overspray reduced the inherent hazard of highly toxic OP to avian wildlife and critical components of the pothole ecosystem. Though an inconvenience to the applicator, the benefits of improved precision in broad-spectrum pesticidal application are clearly warranted in protection of North American waterfowl production. Similar studies, generally absent from the literature, are needed to identify practical problems and provide solutions. Even though OP and CB pesticides are not considered persistent in the same terms as organochlorine pesticides, there is ample evidence of wildlife mortality in wetlands as long as 6 months after application of granular formulations, e.g., phorate, fensulfothion, carbofuran.[4,6,66,71] These mortalities were mostly delayed effects of leaching and pesticide availability in water killing waterfowl, which were then eaten by scavenging predatory birds and mammals. It has not been demonstrated that flowable formulations of either OP or CB insecticides produce the same long-term hazard as granular formulations.

### 12.3.4 Mosquito Control

Mosquito control is the only major use of organophosphorus pesticides on wetlands in the United States, and carbamates are not used in mosquito control. However, OPs and CBs are used extensively in mosquito control outside the United States. OP use is being supplanted by biological control agents such as BTI (*Bacillus thruingiensis israelensis*) and other subspecies (e.g., *B.t. darmstadiensis*, *B.t. kyushuiensis*, *B.t. galleriae*), fungus (e.g., *Lagenidium giganteum*) and anti-juvenile hormone enzymes.[88] This transition is in response to pesticides needed for wide-scale application to wetlands and areas of human habitation and is becoming necessary to overcome developing resistance of some mosquito species to chemical control agents.

Mosquito control routinely uses OPs throughout the world for suppression of nuisance mosquitoes and disease vectors. For example, under the threat of encephalitis epidemics following hurricanes and floods, thousands of hectares of contiguous bottomlands and major populated areas of the United States have been aerially sprayed with undiluted OPs, such as malathion, at rates as small as 219 mL/ha.[29] These ultra-low-volume treatments are efficient and comparatively safe for use wherever there is a need for rapid application of OP to large or remote lands.[89] Only a few OPs, such as malathion, naled, and fenthion, have been used extensively for mosquito control as ULV sprays, and none has been implicated in serious human health effects. This does not imply such treatments are without environmental hazard. In May 1969 undiluted fenthion was applied by helicopter at the recommended rate of 95 mL/ha to approximately 600 ha of a portion of the Red River and residential and park areas of Grand Forks, North Dakota.[90] Over the course of 2 days more than 450 birds representing 37 species were found dead or moribund within the fenthion-treated area. Most of the dead birds were warblers and other species that feed mainly on invertebrates. The mortality was finally estimated at more than 5000 birds in and around the city. In part because of this extreme incident of wildlife mortality from a standard mosquito-control application, many other studies have been conducted on fenthion mosquito control.[29] Though most studies resulted in avian mortality, fenthion remains a favored OP for use in mosquito control throughout much of the world.

Mosquito-control agents are normally used in areas of human activity and habitation and should not be either acutely toxic to mammals or induce serious side effects. The oral $LD_{50}$s for laboratory rats of favored mosquito-control agents in the United States, such as fenthion, malathion, temephos, chlorpyrifos, and naled, vary from about 150 (chlorpyrifos)16 to 8600 mg/kg (temephos).[12] Thus, $LD_{50}$s for these OPs are considered only moderately (chlorpyrifos, fenthion, and naled) to slightly (malathion and temephos) toxic to mammals, according to the acute toxicity ranking of Loomis and Hayes.[31] However, all except malathion are classed highly toxic to birds. Fenthion may also induce potentially lethal "intermediate syndrome" in mammals.[37,38] This syndrome has not yet been diagnosed in birds or wild mammals.

All of the above mosquito-control agents except temphos are broad-spectrum insecticides that are also widely used in agriculture. Temephos is used almost exclusively for mosquito, midge, and black-fly larvae control in lakes, ponds, and wetland habitats.[4] All of the OPs are acutely toxic to susceptible invertebrates and fish within a few minutes of exposure but are poorly soluble and rapidly degraded in water (e.g., 2 to 72 h in circumneutral water) and have little potential for bioaccumulation in nature.[91] For reference, except for temephos, all of these mosquito-control agents are generally more toxic to fish and aquatic invertebrates than are parathion or methyl parathion, are restricted-use pesticides in the United States, and are extremely toxic to natural pollinators such as honey bees (*Apis mellifera*).[5] The potential hazard of these mosquito-control agents to bee pollinators is greater during general aerial application for nuisance and disease vector control than when applied to wetlands for larvae control.

In conclusion, malathion remains the primary OP adulticide for use in event of mosquito-transmitted disease epidemics in the United States, but with developing evidence of OP resistance in many mosquito populations its continued efficacy and large-scale use is being reconsidered. Development of alternative methods is in progress, and these alternative methods are generally believed to be less hazardous when used in wetlands. OP larvicides are likely to soon be replaced by alternative larvicides such as BTI and other biologicals. Such alternatives are expected to be safer in wetland applications.

## 12.4 FACTORS OF ACUTE HAZARD

Short-term life-threatening exposure is the principal hazard of organophosphorus and carbamate insecticides to aquatic and terrestrial wildlife. Nearly all applications of the more toxic products are on cultivated crops and, depending on the formulation and treatment, may remain potentially lethal to nontarget vertebrates for several days. These same treatments have sometimes contaminated nontarget wetlands. Such aquatic incidents usually involve large numbers and kinds of wildlife, often including birds and mammals. Sometimes aquatic toxicity lasts only a few hours depending on the system's capacity to dilute and degrade the contamination. In contrast, granular formulations, even when incorporated into the soil, have proved toxic to birds and mammals immediately posttreatment and again later when either granules or leachate become available to wildlife.[71]

Considering different scenarios of exposure and that multiple applications of insecticides are routine during a growing season, many forms of toxicity could manifest from OP and CB exposure. Perhaps they do, as will be discussed later, but the overwhelming effect is lethality. Most factors contributing to the acute toxicity of OP and CB insecticides in nature have been reasonably understood for at least two decades; therefore, only basic concepts are summarized here.

### 12.4.1 Comparative Toxicology

Organophosphorus and carbamate insecticides and product formulations vary widely in toxicity to different aquatic and terrestrial vertebrates.[4,14,48,49,92] Those pesticides that are esters of carbamic acid and phosphoric acid are direct ChE inhibitors that invoke toxicity immediately upon absorption into the circulatory system. Most OPs are esters of phosphorothioic acid and must undergo an oxidative desulfuration step for maximum anticholinesterase potency, as described earlier under general toxicology. Direct-acting ChE inhibitors may be more hazardous through inhalation and percutaneous exposure than are phosphorothioic acids.

Metabolic responses to OP and CB insecticides are similar for birds and mammals with any difference more quantitative than qualitative.[93] Several quantitative enzymatic differences between birds and mammals are important to differential response to acute toxic exposure. Birds generally have lower levels of hepatic metabolizing enzymes and A-esterase activity than do mammals, which tends to make birds more sensitive to acute anticholinesterase poisoning.[46,47,94] However, the majority of widely used anticholinesterase pesticides are phosphorothioic acids that must be metabolically activated to their most potent analog. This step is mediated by the same hepatic enzymes as detoxication in birds and mammals.[2] This would suggest that mammals should toxicate phosphorothioic acids more efficiently than birds and be the more sensitive taxa. But this is not so, because activated OP analogs are substrates for A-esterase hydrolysis and are rapidly detoxified in the liver and blood. In a study of 14 species of birds, three species of laboratory mammals, domestic sheep, and humans, plasma A-esterase was at least 13 times higher in all mammals than in any of the avian species.[94] The importance of these enzymatic differences to birds was demonstrated with dimethoate, a phosphosothioic acid, in which adult ring-necked pheasants (*Phasianus colchicus*) and laboratory rats were compared. The toxic oxygen analog was rapidly formed and accumulated in pheasants but was rapidly degraded to inactive metabolites in the rats.[95] This undoubtedly contributed to the more than tenfold difference in the acute oral sensitivity of pheasants ($LD_{50}$, 20 mg/kg)[16] and rats ($LD_{50}$, 215 mg/kg).[12]

In general, the array of domestic and wild bird species commonly studied in the laboratory are consistently more sensitive to acute oral toxicity of OP and CB insecticides than are laboratory mammals and wild mice (*Peromyscus* spp.) and voles (*Microtus* spp.).[11–16,30] Though this generalization of differential acute sensitivity is convenient for preliminary risk assessments, it is not reliable for all species and pesticides.[26] For example, the acute toxicity of a series of phosphorothioic acid pesticides was compared for standardized $LD_{50}$ tests of adult male laboratory rats,[12] ring-necked pheasants,[16] and red-winged blackbirds[16] (Table 12.1). Pheasants and blackbirds were used because both species have general feeding habits but represent extreme body mass compared with rats (e.g., blackbirds, ~65 g; rats, ~200 g; pheasant, ~1000 g). The rat $LD_{50}$s were graded from 2.3 (phorate) to 8600 mg/kg (temephos). By most criteria for ranking acute mammalian toxicity, phorate is classed extremely toxic, and temephos is practically nontoxic. At the one extreme, though phorate is also highly toxic to pheasants ($LD_{50}$, 7.1 mg/kg), it is about three times more toxic to rats. At the other extreme, temephos is nearly 250 times more toxic to pheasants ($LD_{50}$, 35 mg/kg) than to rats. The original premise that birds are more sensitive to anticholinesterase insecticides was borne out for red-winged blackbirds, which were 2 to 200 times as sensitive as the laboratory rat.

The limited comparison of phosphorothioic acids has important implications for ecological risk assessment. First, avian mixed-function oxidase activity tends to be inversely related to body mass,[46,47] and therefore red-winged blackbirds should be more tolerant of OP poisoning than is the ring-necked pheasant, but they are not. This may be partially explained by the fact that red-winged blackbirds (and possibly other small passerines) are particularly deficient in liver-detoxication enzymes, but it also may have been influenced by small birds having a much higher metabolic rate. Second, it was noted that OPs with rat $LD_{50}$s above 200 mg/kg were usually more than ten times as toxic to pheasants as to rats; but for OP $LD_{50}$s less than 200 mg/kg, rats were 2 to 20 times more sensitive than the pheasants (Table 12.1). This is especially important because rat $LD_{50}$s are often used as a primary source for preliminary risk assessments, and $LD_{50}$s above 200 mg/kg are considered only moderately toxic for wildlife.[4] This conclusion would be erroneous for OPs such as dimethoate, fenitrothion, and temephos, all of which are classed highly toxic, with $LD_{50}$s of less than 50 mg/kg for many birds. In nature, fenitrothion applied at recommended rates on rangelands for grasshopper control resulted in bird mortality and decreases in breeding populations.[4]

Based on the foregoing, the laboratory rat is not a good model for prediction of acute anticholinesterase toxicity in birds. In contrast, laboratory rats and mice are conservative predictors of acute toxicity to wild rodents. In a study of the acute oral toxicity of a spectrum of pesticides including OPs to four species of voles and laboratory rats and mice it was determined that laboratory rodents are generally more sensitive to acute exposure than the most sensitive of the voles, (*Microtus canicaudus*).[96] Laboratory mice were also found to be more sensitive than deer mice (*Permyscus maniculatus*), but the relationship was not consistent.[30]

#### *12.4.1.1 Acute Toxicity Testing*

Acute exposure is the main hazard of organophosphorus and carbamate insecticides to wildlife, and of all standard toxicological tests, it is best represented by single-dose $LD_{50}$ study. Response to the dose of anticholinesterase is usually rapid, and, if not fatal, recovery normally occurs within a few hours. This test provides a sound method for quantification of naive sensitivity to OPs and CBs and comparisons such as differences between sexes, age classes, and formulated products. The test also provides characterization of the dose-response curve, which is essential for proper hazard evaluation and risk assessment. However, the acute test provides only an estimate of relative sensitivity based on a single potentially lethal exposure; whereas in the field exposure varies widely from repeated small doses during feeding (the degree of response depends on the level of contamination and susceptibility of the individual) to massive overdose from rapid ingestion of highly contaminated water or bolus of food. Therefore, a second "acute" test has also been developed to check the short-term response of birds and mammals to OP and CB-contaminated forage. The feeding trial is designed, replicated, and analyzed statistically the same as the single-dose acute test but provides graded levels of contaminant for 5 days. The critical statistic is the $LC_{50}$ (median lethal concentration). This feeding trial was developed in the 1960's as an alternative to the acute test for highly persistent pesticides that were not acutely toxic.[97] As with the acute test, this feeding trial does not adequately represent field exposure of wildlife to OPs and CBs, but in combination the two tests provide insight into how wildlife may respond. Where the acute test provides information on inherent sensitivity to OP and CBs, the feeding trial provides information on response to repeated chemical exposures as may be encountered in nature. The use of these tests and their value in evaluation of ecological hazard has been critiqued in detail.[97–100]

### 12.4.2 Acute Environmental Hazard

Low $LD_{50}$s indicate that most common organophosphorus and carbamate insecticides are extremely toxic to birds and mammals. However, only a small number of these pesticides are responsible for the majority of large-scale incidents of wildlife mortality. This is probably because certain uses are more likely to bring large numbers of wildlife in contact with a few of the more commonly used products. It is also likely that many incidents of wildlife mortality are not detected or reported. Most reported incidents are of three types: (1) a few dead songbirds in a neighborhood park or backyard, (2) an unusual concentration of dead flocking birds, or (3) large conspicuous water birds or special interest species such as raptors. In contrast, sometimes large flocks of birds were found dead and not reported because the discovery was believed to be part of an avian-control program. (Note: OPs and CBs have been used in wildlife damage and hazard control.)[101] The prevalence of small mammals may range from common to abundant in many habitats routinely treated with OP and CB pesticides, but these mammals are rarely listed in reports of even large-scale terrestrial mortality. This omission is more likely due to difficulty in detection of small secretive species (e.g., shrews, mice, voles) than to any special tolerance. There has been comparatively little documentation of reptile and amphibian mortality from OP or CB pesticide treatments.[48,49]

Prediction of acute OP or CB hazard to wildlife is confounded by many factors of anthropogenic and natural origin. A primary factor, and perhaps most easily addressed through limited additional testing, is the influence of pesticidal formulations on availability and toxicity to wildlife. Regulatory agencies have only recently begun to consider the differential ecological hazard of finished product formulations, and then only after years of reported episodes of wildlife mortality from certain products.[26] Some of the other factors that may affect acute OP and CB hazard to wildlife to different degrees include the route, source, and timing of exposure and possible interactions with other chemicals, infectious diseases, and weather stressors.

#### *12.4.2.1 Confounding Variables*

Tolerance of anticholinesterase exposure may be affected by growth, maturation, gender, reproductive status, lineage, nutritional status, and exogenous and endogenous stressors. Few rigorous studies have been conducted on the influence of normal physiological factors on the sensitivity of wildlife to pesticides, or whether captive wildlife adequately represents nature. This is a critical uncertainty that must be attended to for reliable hazard and risk assessment. Even species raised commercially, such as quail and ducks, are not well understood, and they serve as the primary models for regulatory hazard assessment in North America and Europe.[26] Another uncertainty is the suitability of wild-captured animals for routine toxicity testing. Simple survival and weight maintenance for a few weeks in captivity may not reflect variables, such as nutritional imbalance or stress response, to confinement, isolation, or crowding.[28]

Maturation alone does not always increase avian tolerance of anticholinesterase pesticides. For example, in a seminal study of acute toxicity with mallards at 1.5 days, 1 week, 1 month, and 6 months of age, $LD_{50}$s tended to follow different patterns for esters of phosphoric and phosphorothioic acid insecticides.[102] $LD_{50}$s between 1 week and 6 months decreased by 45 and 53% for demeton and monocrotophos, while they increased by 60 and 190% for parathion and chlorpyriphos. Tolerance patterns for the two CBs tested — carbofuran and aldicarb — were similar to dementon and dicrotophos, both direct-acting OP insecticides. The apparently higher tolerance of young ducklings compared to adults was attributed to an immature nervous system being less sensitive to certain neurotoxic agents. The authors offered no explanation for the differences between direct-acting ChE inhibitors and those requiring metabolic activation for potency.

In contrast to single-dose $LD_{50}$ tests, dietary $LC_{50}$s typically increased with growth of precocial birds.[68,103] The increase is probably due to larger (older) chicks eating less proportional to body mass, which reduces practical exposure to chemicals over the duration of the experiment. Physiological tolerance undoubtedly plays a role, but it was demonstrated that relative food consumption normally decreases about 60% for Japanese quail from hatch through 3 weeks of age.[68] Food consumption of controls in proportion to body mass averaged 48 g/100 g at 3 days of age, 31 g at 10 days, 24 g at 17 days, and 19 g at 24 days. This decreased energy requirement was correlated with an overall three- to fourfold increase of OP and CB $LC_{50}$s from hatch to 3 weeks of age. This study included chlorinated hydrocarbon and mercurial pesticides, which also yielded, increased $LC_{50}$s over age, but the differences between hatch and 3 weeks were consistently less than twofold. These differences indicate the more biologically labile and less hazardous nature of OP and CB pesticides compared to highly lipid soluble and cumulative products.

Historically, the sex of a bird or mammal was not considered important in studies of acute OP and CB exposure, unless the animal was in reproductive condition.[104] Therefore, to conserve resources for testing as many species and pesticides as possible $LD_{50}$ tests were designed for use of approximate statistical methods[105,106] and reproductively quiescent animals.[16,107] It was felt that routine tests reported for laboratory rats provided sufficient insight on sex effects for most hazard assessments.[104] For example, it was known that females in estrus were significantly more sensitive than males to about 30% of technical-grade OP and CB insecticides tested; males were more sensitive about 10% of the time.[11,12] When sex differences were detected, females were usually 1.5 to 5 times as sensitive as males; when males were most sensitive, the difference was usually less than 50%. These relationships have not been verified for commercially reared mallards and quail that are easily brought into reproductive status and are routinely used in regulatory decisions on ecological hazard. (Note: In the primary source on acute pesticide toxicity to wildlife, sexes are indicated, but most animals were reproductively quiescent.[16]) The need for such testing is critical because OP and CB pesticides are intensively applied throughout breeding seasons of many species. The importance of this variable was indicated by an acute test of fenthion in which ovulating northern bobwhites were over twice as sensitive as males.[108] This disproportional sensitivity to OP and CBs is potentially critical at the population level because females of most species are more important than males in population recruitment.

### 12.4.3 Routes of Exposure

Wildlife are most often exposed to organophosphorus and carbamate pesticides through ingestion of contaminated food and water. Inhalation and dermal exposure also occur but are believed to be less common and less hazardous to wildlife than is oral ingestion.[109] Accordingly, nearly all studies of anticholinesterase toxicity to wildlife involve oral administration of chemical, but such studies are usually based on technical grade chemical via gelatin capsule, gavage, or dry mash. Little is known about the toxicity of the finished formulated pesticide complexed with natural matrices or alternative routes and conditions of exposure.

As discussed earlier, the toxicity of OP and CB insecticides is primarily a function of hepatic metabolism regardless of the route of entry.[110] Thus, esters of phosphoric and carbamic acids are potentially more potent if they enter the circulatory system by routes other than the alimentary canal, which shunts directly to the liver for first-pass metabolism and detoxication. Esters of phosphorothioic acids that require hepatic metabolism for activation (which includes most OP pesticides) are theoretically more toxic when ingested. Usually, however, pesticides are not rapidly absorbed through the skin and scales, the main alternative routes to ingestion. Percutaneous hazard depends on the rate (dose and time) at which chemical penetrates the skin and varies widely among chemical formulations, sites of application, and species.[111] An array of anticholinesterase pesticides have been tested by application to the wrapped feet (tarsometatarsis, phalanges, and webbing; ~12% of body surface) of adult mallards for 24 h,[112] and to a 1 $cm^2$ of a featherless skin under the wing joint of small passerine birds.[113] At the same time, companion birds were dosed orally. Percutaneous and oral $LD_{50}$s were positively correlated in both studies (mallard, $n = 19$, $r^2 = 0.42$, $p < 0.01$; passerine, $n = 17$, $r^2 = 0.72$, $p < 0.01$). The $LD_{50}$s were consistently highest for percutaneous routes.

Though potentially more important because OP and CB insecticides readily cross the mucus membrane of the lung, inhalation toxicity is considered minor compared with deposition of chemical on the skin.[109,114] This concept has not been rigorously tested on wildlife species. For example, it is not known to what degree bats are at risk when returning to their roost during morning twilight, the time considered optimal for aerial application of pesticides because wind is usually less of a factor.

### 12.4.4 Sources of Exposure

Wildlife are exposed to organophosphorus and carbamate pesticides by ingestion of contaminated water, soil, seeds, foliage, invertebrates, vertebrates, and formulated granular particles. All of these sources have killed large numbers of wild birds and mammals under varied environmental circumstances. Water is a common source of exposure that is poorly documented for terrestrial vertebrates. Potential OP and CB hazard is dependent on widely variable factors of ambient water quality, movement, and the solubility and stability of the product. Water soluble formulations usually remain available longest and tend to be the most acutely toxic.[2,3] How these studies relate to waterborne exposure in nature is not clear because rates of feed and water consumption vary widely among wildlife species at different ages and seasons of the year. Smaller birds and mammals have a much higher water requirement relative to body mass than do larger species under similar ambient conditions.[115] However, even closely related birds of similar sizes and feeding habits vary their rates of free-water consumption from about 15 to 40% of their body mass per day at the ambient temperature of 25°C.[116] Though aquatic wildlife undoubtedly contact and ingest more water through feeding, swimming, and wading than do terrestrial species that use water primarily for hydration, terrestrial species may opportunistically ingest large amounts of highly contaminated water, especially in arid regions.

Few studies of OP and CB pesticides in water have been reported for wildlife species. However, as previously mentioned, field applications of technical-grade fenthion in various formulations of 47 to 100 g of active ingredient per hectare over wetlands of various water depths killed a variety of passerine and water birds.[90,117,118] The authors concluded that contaminated insects were an important source of fenthion in the avian mortalities, but the importance of contaminated water was not dismissed. Mortality of wildlife from puddling and run-off from agricultural fields has been documented for some of the more acutely toxic OP and CB pesticides, as discussed in Section 12.3.3.

OP and CB residues in foliage, whether by topical or systemic soil application, may be hazardous to wildlife in or near the treated area. Treatments to control insects in forests and orchards and on cultivated crops, such as small grains, alfalfa, and turf grasses, have all resulted in excessive mortality over the years. Foliar treatments may be especially hazardous when animals are in the spray zone during treatment and subject to multiple routes of exposure. Turf grass treatment with OPs (e.g., diazinon) has proven particularly hazardous to waterfowl such as American wigeon and the Canada goose.[119] This unique hazard to grazing waterfowl resulted in the U.S. Environmental Protection Agency's issuing a cancellation notice for the use of diazinon products on golf courses and sod (turf) farms.[120]

Seeds, like granular pesticides, are an important source of exposure to wildlife when the seeds are treated with pesticides for soil insect control.[6,121,122] OP- and CB-treated seeds are readily eaten by small granivorous animals in spite of being brightly colored for safeguard of human health. This hazard is not limited to seeds on the surface of the soil or to small animals; large-scale mortality of greylag geese (*Anser anser*) was attributed to the uprooting and ingestion of germinating OP-treated seeds.[123,124] It is not known whether phytometabolism of systemic OP and CB has contributed to wildlife mortality, but increased hazard to herbivores is plausible because metabolites are more persistent and may be as potent as the parent compound.

Contaminated arthropods have been proven lethal to wildlife after application of OP and CB insecticides such as terbufos, monocrotophos, dimethoate, fenthion, trichlorfon, and carbofuran.[4,9,10,90,125,126] Such poisonings are most likely from insecticides adsorbed on the cuticle of the anthropods. An adsorbed chemical may be rapidly dissociated in the stomach of the consumer. However, pesticide ingested by arthropods may also contribute to a consumer's exposure. Recently, there was a documented incident of disulfoton-treated cotton seeds passing sufficient insecticide through the plant to grazing insects to be lethal to Swainson's hawks (*Buteo swainsoni*).[10] In another incident as many as 3000 Swainson's hawks died from eating freshly-sprayed insects following monocrotophos treatment for grasshopper control.[10] Monocrotophos is extremely toxic to birds with acute oral $LD_{50}$s consistently less than 5 mg/kg and as low as 0.8 and 0.2 mg/kg for California quail (*Callipepla californica*) and the golden eagle.[16] Monocrotophos has been cancelled in the United States.[5]

Predatory and carrion-eating birds and mammals are known to have died from eating prey and carcasses contaminated with anticholinesterase pesticides such as monocrotophos, fenthion, mevinphos, phorate, famphur, and carbofuran.[6,10,127] Most of these secondary poisonings were probably from eating unaltered chemical in the alimentary canal of prey or carcasses. The liver and kidneys may contain some biologically available anticholinesterase residues (e.g., oxons, sulfoxides, sulfones), but other postabsorptive tissues and fluids are not as hazardous as a source of secondary poisoning. For example, in an experiment with barn owls (*Tyto alba*) fed quail that had been killed with oral dosage of famphur, intact carcasses caused significant ChE inhibition in the owls, whereas owls fed quail with the entrails removed were unaffected.[128] Famphur presented a unique form of secondary poisoning because the product is applied directly on livestock for vermin and parasite control. American magpies (*Pica hudsonia*) were poisoned by eating contaminated hair of cattle and possibly invertebrates in stockyards.[127]

The potential for secondary poisoning from aquatic vertebrates has also been demonstrated.[129] Tadpoles exposed to parathion at 1 mg/L of water for 96 h were force-fed to 14-day-old mallards at the rate of 5% of body mass. A single meal was lethal to ducklings within 30 min. Because only parathion, and not its oxygen degradate, was found in the tadpoles and stomachs of the dead ducklings, it is likely that parathion concentrated in the protective outer mucus layer of the tadpoles and was almost immediately available for absorption upon ingestion by the ducklings. The treated tadpoles appeared healthy when fed to the ducklings.

Percutaneous, ocular, and inhalation exposure undoubtedly occur when wildlife are directly oversprayed by OP and CB pesticides, or when entering a freshly-sprayed area. The few studies conducted on these alternative routes of exposure have demonstrated that they may be important to toxicity in some circumstances, but most often ingestion is the primary source of poisoning.[109]

### 12.4.5 Pesticide Formulations

Organophosphorus and carbamate residues on seed grains, vegetation, and formulated pesticide granules have killed large numbers of wildlife under varied environmental circumstances. Some of the kills were due to misuse, but some of the problem was due to general lack of information on the comparative toxicology of pesticidal formulations and hazard associated with various application techniques. Most often, potential hazard or risk to wildlife is estimated by comparison of the theoretical concentration of the active ingredient in a food item to results of standard acute tests of technical-grade chemical with northern bobwhites and mallards but without regard to the differential effects of finished product on absorption, fate, and toxicity.[28,130]

Only rarely is technical-grade chemical applied in the field, and then only in very low volume, as described earlier for mosquito control. Instead, pesticides are normally applied as a formulated product that may differ substantially in acute toxicity compared to the technical-grade material tested. As a general rule, it has been determined that granular formulations are most often less toxic than technical-grade materials, whereas liquid formulations are usually more toxic than technical-grade products. This relationship was demonstrated with acute tests of northern bobwhites in which emulsifiable concentrates (48% AI) of diazinon and carbofuran were significantly more toxic than either technical-grade (99% AI) or granular formulation (14 to 15% AI.).[52] Sometimes the $LD_{50}$s among anticholinesterase formulations varied as much as fourfold. Aqueous solutions were consistently more toxic than oil-based ones. In contrast, pesticides were more toxic to mallard embryos when applied to eggs in an oil vehicle than when applied in an aqueous emulsifiable concentrate.[131,132] Apparently, the oil medium retarded volitalization, increased the time of contact, and facilitated pesticide transport through the shell and membranes.

Though granular OP and CB insecticides may be less toxic in dosing studies than other formulations and are safer to handle during application, their potential environmental hazard is excessive.[45,133,134] The hazard of granulars in nature depends on whether they are haphazardly or selectively ingested. If ingestion is haphazard, then the application rate is the more critical variable, but if ingestion is selective, then even the most stringent attempts to reduce granule availability may fail to reduce the hazard.[69] The color, size, texture, granule base, and application rate are all factors for consideration in reduction of granular hazard to wildlife.[135–139] The substantial hazard to wildlife of granules leaching into wetlands is discussed in Section 12.3.3.

### 12.4.6 Toxic Interactions: Chemical and Environmental

Interaction among organophosphorus and carbamate pesticides and between anticholinesterase and other common environmental xenobiotics has not been thoroughly studied for wildlife species.

A few studies have been conducted on subchronic exposure of birds to expected field concentrations of persistent pesticides followed by acute challenge with OP or CB, or studies of simultaneous feeding on combinations of OPs for 5 days. Results of these studies are generally consistent with similar studies of laboratory mammals, but there are some differences that may affect ecological hazard assessment. For example, when laboratory rodents were pretreated with chlorinated hydrocarbon pesticides that increased hepatic mixed-function oxidase activity, their sensitivity to OP insecticides was reduced.[140–142] In contrast, when the chlorinated hydrocarbon DDE was fed to Japanese quail for 3 months, the sensitivity to a single dose of parathion increased significantly.[143] When the quail were pretreated with chlordane, another chlorinated hydrocarbon, acute sensitivity to parathion was decreased. This latter relationship was also observed in similar studies with mice.[141]

In general, response of naive birds and rodents is additive when acutely challenged with paired anticholinesterases or anticholinesterases with chlorinated hydrocarbon pesticides.[52] When more than additive effects are detected, the level of synergy is usually less than twofold, whether exposure is acute by oral dosage[144–146] or short-term dietary presentation.[147] Little information is available on the effects of sequential exposures of wildlife, but the potential hazard may depend on the order in which the chemicals are encountered. When the initial exposure is to CB, some protection from OP may occur.[148] In contrast, when the initial exposure is to an OP, toxicity of subsequent exposure to either OP or CB may be increased.[149] Sequential exposure to different pesticides is a reasonable possibility in nature, especially for birds that are highly mobile and may forage among several crops.

Temperature extremes and season of the year are natural stressors that can affect the acute toxicity of anticholinesterase insecticides to wildlife. Abrupt changes in climate, particularly late spring cold fronts with heavy rains, may profoundly influence avian nesting success. Whether effects of OP exposure on thermoregulation so far demonstrated in the laboratory would exacerbate an already dramatic physiological challenge is not known. At the other extreme, heat stress was suggested as a contributor to dimethoate toxicity in an episode of sage grouse (*Centrocercus urophasianus*) poisoning in Idaho.[150] Environmental factors affecting OP and other xenobiotic toxicity in wildlife have been reviewed.[44,151,152] Effects of seasonal changes on OP and CB toxicity are difficult to elucidate because many confounding endocrine changes are cued to photoperiod and temperature changes, with the more profound differences usually in females.

### 12.4.7 Diagnosis of Anticholinesterase Exposure

Organophosphorus and carbamate pesticides have resulted in hundreds of incidents of wildlife mortality from disease vector control and agriculture (including forest and range management) throughout the world. When many dead and moribund animals of mixed species are found in an area of known OP or CB treatment, the casual association may be evident but is not conclusive without biochemical and chemical confirmation. Proper diagnosis is then contingent upon demonstration of brain ChE inhibition consistent with levels indicative of toxicity or exposure and chemical detection of residues of the causative agent. This last step is sometimes difficult because neither OP nor CB residues tend to accumulate in postabsorptive tissues. However, a strong inferential diagnosis is possible by demonstrating depressed brain ChE activity and detection of a known anticholinesterase in either ingesta or tissues.[34,36,123]

A conservative threshold of about 50% depression in whole brain ChE activity is generally considered diagnostic of death from anticholinesterase poisoning,[35] though depression of 70 to 95% is commonly reported for birds and mammals killed in nature by OP pesticides.[36] In contrast, when animals are killed in the field by CB pesticides, whole brain ChE activity may vary from near normal to depressions of only 60 to 70%.[36,153] Apparently high levels of CB exposure kill by systemic neuromuscular blocking before significant penetration of the central nervous system has occurred.[53,154] Also, lesser degrees of ChE inhibition may reflect spontaneous postmortem reactivation of carbamylated enzyme.[67] Brain ChE can be determined by many techniques,[155,156] but either a laboratory norm must be developed for each species or a suitable enzyme reactivation technique must be used to determine the degree of inhibition. In cases of poisoning by OP compounds ChE activity can be reactivated *in vitro* by the oxime 2-PAM (pyridine-2-aldoxime methyl chloride).[34] Carbamylated ChE is much less stable than phosphorylated ChE; therefore, simple *in vitro* heat reactivation will serve as a rapid indicator of carbamate exposure,[36] as will dilution techniques.[157] OP and CB reactivation techniques provide important guidance for analytical chemistry.

Blood plasma or serum ChE may be used as a nondestructive technique for detection of OP or CB exposure. As for brain ChE, the species-specific norm must be developed for diagnostic reference. If ChE depression exceeds the lower to end of normality, i.e., more than two standard deviations below the baseline mean, the subject is considered to have received significant exposure to anticholinesterase compound. Again, heat and 2-PAM reactivation may be used as provisional indicators of CB and OP ChE inhibition. These concepts have been reviewed in detail for wildlife including fish.[158]

## 12.5 SUBLETHAL ENVIRONMENTAL HAZARD

### 12.5.1 Subchronic and Behavioral Effects

Wild birds and mammals are relatively tolerant of low-level exposure to organophosphorus and carbamate pesticides.[43,44] This is partly because the chemicals are labile and readily excreted by warm-blooded animals. Also, low-grade exposure to anticholinesterase may stimulate changes in synaptic physiology, which may include reduction of both axonal release of acetylcholine transmitter and the density of postsynaptic acetylcholine receptors. This apparent tolerance was determined from laboratory studies that do not involve the rigors of nature that affect free-living animals, such as extreme weather and the ability to capture prey or avoid predation. For example, a single OP or CB dose of about 5% of its $LD_{50}$ (essentially nonlethal but induces significant brain ChE inhibition) may reduce the core temperature as much as 2°C in homoiothermic animals acclimatized to moderate ambient conditions of 25 to 30°C. If the ambient temperature is abruptly shifted to mimic a cold front, the core temperature may drop as much as 3 to 6°C from the same dose.[159–161] However, it has been demonstrated that even under these chilled ambient conditions the core temperature of OP-induced hypothermic animals returns to normal within about a day.[44] Generally, hypothermia occurs with brain ChE inhibition of about 50%, which is the degree of inhibition often associated with general physiologic deficit and sometimes death. Hypothermia is also associated with reduced metabolic efficiency, which in turn may slow metabolic degradation and excretion of OP and CB, thereby extending contact at receptor sites and enhancing toxicity.

Hypothermic animals are more sensitive to OP poisoning than normal.[161,162] Young birds may be especially susceptible to anticholinesterase interference with thermoregulation because many species are not fully homoiothermous for 1 to 3 weeks after hatching.[163] Interaction between OP and ambient chilling was studied with 14-day-old northern bobwhites acclimatized to 35°C and then subjected to 27.5°C for 4 h.[164] Brain ChE inhibition from a single dose of chlorpyrifos in the chilled chicks was depressed about twice as much as in chicks continued at 35°C. Both nestling and precocial chicks could be further compromised because parental care may be affected when the female is exposed to anticholinesterase.[165,166] Temperatures below 15°C are common throughout the breeding season of birds and small mammals in temperate climates. Mechanisms of thermoregulation and interaction with anticholinesterase have been thoroughly reviewed.[43,151]

The ability of birds and mammals to capture prey or avoid predation after subacute exposure to OP and CB pesticides has not been properly evaluated. Controlled experiments are difficult to interpret. Treatment levels were usually based on some level of acute dosage that at least initially rendered the subject, predator or prey, critically ill and often immobile. The results tend to be predictable, i.e., predators will not hunt, and prey cannot escape. When levels are tested that do not indicate overt toxicity, predators and prey seem to respond normally. Field studies are hindered by uncertainty of the subjects' exposure history and inability to follow movements and observe behavior of highly mobile or reclusive species. Radio tracking has provided some insight into effects of OP on northern bobwhite survival but not necessarily cause of death. One hundred ninety-seven wild quail were captured, equipped with a small radio transmitter, dosed once with either 0, 2, 4, or 6 mg/kg of methyl parathion, and their movements were monitored for 14 days.[167] Quail receiving the highest dosage had lower survival than did the controls; otherwise, there were no apparent treatment effects. The authors concluded that reduced survivability was due to increased OP-induced vulnerability to predation, perhaps due to reduced covey integrity. A similar study was conducted with wild northern bobwhites dosed with 0, 3, 5.6, or 21 mg/kg of terbufos and realized essentially the same results as for methyl parathion.[168]

Subacute exposure of birds and mammals to various anticholinesterases has been shown to affect an array of behaviors, such as activity level, alertness, aggression, foraging and drinking, learning and memory, navigation, and reproduction.[43,44,169,170] Though most of these behavioral studies were well planned and have important theoretical implications for survivability, laboratory studies are highly restrictive, and their projection to natural populations is speculative.[170] Or, as aptly stated, whereas behavioral tests seem desirable, there are two fundamental difficulties in the use of such tests: (1) the best-studied and most easily performed and quantified have the least environmental relevance and (2) the most relevant behaviors are the most strongly conserved against change.[171]

### 12.5.2 Chronic and Reproductive Effects

The chronic toxicity of organophosphorus and carbamate pesticides has not been extensively studied with wildlife because such chemicals were believed to be too labile in nature to pose a serious sublethal hazard. However, as discussed earlier, some OPs and CBs may remain available in the soil and vegetation for several months, and they may be applied to wildlife habitat several times during the growing season, which may coincide with critical periods of reproduction. Specially designed research is needed to evaluate: (1) intermittent exposure from repeated pesticide application, (2) exposure of naive animals to pesticide application at different stages in the reproductive cycle (e.g., courtship, onset of lay, incubation), (3) exposure to systemic pesticide in plant tissue, and (4) effects of potential interaction of tank mixtures and different anticholinesterase application on nearby crops.

Chronic studies of OP and CB pesticides with wildlife were usually some modification of the standard reproduction trial developed for pesticide registration. These tests were developed for evaluation of more persistent chlorinated hydrocarbons and heavy metals. The studies, usually of mallards or northern bobwhites, expose first-time breeders to constant rate of chemical from several weeks prior to lay through chick hatchability. Chicks were then observed for 2 weeks for evidence of mortality. Main effects noted in these studies included reduced feeding and corresponding weight loss in adults followed by predictable reduction in rate of lay and egg hatchability. Pharmacologic action on the endocrine system has also been demonstrated.[43,44]

Some of the most acutely toxic OP and CB pesticides may also pose hazards to wildlife that have gone unnoticed. Monocrotophos has been implicated in some of the largest wildlife kills, but trivial residues have also been shown to depress egg production and hatchability in northern bobwhites.[172,173] In another study with northern bobwhites the objective was to determine how reproductively active birds responded to decreasing concentrations of OP, as would be expected from a single application in nature.[174] Monocrotophos at concentrations of 0.1 to 1.0 mg/kg in diet were provided to breeding pairs. Then, at 3-day intervals the basic concentration was either continued or reduced so that at the midpoint and end of the 15-day study the concentrations were reduced by 50 and 75%. Finally, all birds were fed untreated diet for 2 weeks. Food consumption and egg production was negatively correlated with increased concentrations of OP. Inhibition of oviposition was not permanent, and time to recovery was dose-related. There was no evidence of monocrotophos effect that could not be attributed to OP-induced anorexia. In studies of methyl parathion on avian reproduction, effects were also mediated by anorexia.[175,176]

The most important effect of OP and CB pesticides on avian reproduction in nature, other than killing or incapacitating the parents, is the removal of the prey base.[122] When prey is depleted, birds may abandon nests and leave the treatment area, or at least have more difficulty in caring for their young. Abandonment of the first nesting attempt is especially critical to population success because subsequent attempts are usually less successful.[176] Some of the subtle effects of sublethal parental poisoning have been studied for a variety of free-living birds. Female red-winged blackbirds were captured on their nests and given a single dose of methyl parathion (0, 2.4, or 4.2 mg/kg) and then observed for 5 h to document acute or gross behavioral responses including times spent incubating.[177] Each nest was then monitored through fledging. Females at the highest dose showed classic signs of acute anticholinesterase poisoning, but they all recovered, and there was no apparent effect on nest success. In a similar study[178] European starlings (*Sturnus vulgaris*) were induced to nest in artificial nest boxes. When nestlings were 10 days old, the male parent was eliminated, and the female parent was dosed once with dicrotophos at 2.5 mg/kg, and her activities were monitored at 2-hour intervals for the next 3 days. OP-dosed females made fewer trips in search of food for their young and stayed away from their nests longer than did the controls. Nestlings of treated females lost significant amounts of weight, which could have affected their postfledgling success.

Though the potential for reduced prey availability causing a negative effect on avian reproduction is evident, neither decreased nestling growth nor fledgling success was detected for free-living passerines in spite of 50 to 70% depletion of primary insect prey due to aerial application of fenthion[179] or trichlorfon.[125] The importance of relative depletion of insect prey probably varies widely depending on prey abundance at the time of pesticide application and the size, mobility, and energy demands of the insectivore.

Studies of nest attentiveness or other breeding behavior in wild birds gave mixed results when subjects were dosed with OPs at rates producing 10 to 50% brain ChE inhibition, but without causing observable signs of toxicity. Sharp-tailed grouse (*Tympanuchus phasianellus*) given a single dose of malathion at 200 mg/kg were less effective in defending breeding territories on leks.[180] One member per pair of incubating laughing gulls (*Larus atricilla*) was dosed once with parathion at 6 mg/kg, and incubation behavior was observed for 10-minute intervals throughout the day for 3 days. No effects on incubation were detected on the day of dosing, but parents dosed with parathion spent less time incubating on day 2 and the morning of day 3; activities appeared normal thereafter.[181] This study was motivated by a natural event in which adult laughing gulls gathered parathion-poisoned insects in nearby cotton fields and either died leaving chicks to starve or returned and poisoned their chicks through presentation of parathion-contaminated insects.[182]

OP and CB pesticides are not believed to pass through the parent to the egg in biologically important amounts, but such pesticides may be deposited on the egg from parents' feathers during incubation or from direct contamination by pesticide application. Effects of such topical exposures have been studied extensively with northern bobwhite and mallard eggs at day 3 of incubation. Eggs were immersed for 30 sec in aqueous emulsion or were dosed with a single topical application of pesticide in nontoxic oil. OPs were shown to be as much as 18 times more toxic when applied to the shell in oil than when immersed in water.[131] Embryotoxicity and teratogenicity of OPs and CBs applied to bird eggs have been thoroughly reviewed.[132]

## 12.6 CONCLUSIONS AND RECOMMENDATIONS

About 200 organophosphorus and 50 carbamate pesticides have been formulated into thousands of products that are available in the world's marketplace for application to forests, rangelands, wetlands, cultivated crops, cities, and towns. Though most applications are on field crops and other terrestrial habitats, the chemicals often drift or otherwise translocate into nontarget aquatic systems and affect a much larger number of species than originally intended.

OP and CB insecticides, though acutely toxic to most animals, are readily metabolized by vertebrates and are quite labile in nature. Therefore, OPs and CBs have generally been underestimated as important environmental hazards beyond periodic incidents of short-lived mortality of fish and birds. These incidents, sometimes involving hundreds and even thousands of nontarget animals, were often justified philosophically by resource managers and regulators by equating them with mortality from natural phenomena such as storm fronts and infectious disease. This justification is inappropriate because wildlife populations evolved in coexistence with nature, but not with the extra burden of anthropogenic factors of habitat modification and reduction or application of potentially incapacitating or lethal chemicals during critical periods of reproduction.

OP and CB pesticides have not caused the type of population effects attributed to highly persistent chlorinated hydrocarbon pesticides, but there is increasing evidence that populations of some raptorial birds have been affected at the regional level. This is an important concept because prior to the mid-1980s predatory birds and mammals other than those foraging regularly on large insects were not generally believed to be subject to OP and CB hazard. It has since been determined that predators and carrion feeders are regularly poisoned by foraging on the OP- and CB-contaminated entrails of small birds and mammals, and small insectivorous species are often found dead in OP- and CB- treated habitats.

Granular formulations are another aspect of OP and CB hazard that has been increasingly documented for wildlife since the mid-1980s. Granulars were generally considered safer than flowable products, and controlled release granular formulations were highly efficacious. However, recurring incidents of toxicity were noted in association with rainfall and when granulars used in soil treatment were not properly buried. Sometimes, the hazard persisted as long as 6 months posttreatment.

Despite evidence that OP and CB toxicity to wildlife may persist well beyond the first few days posttreatment and that there are many uncertainties about their environmental hazard, regulatory controls of these widely used pesticides are almost exclusively based on incidents of aquatic and terrestrial wildlife mortality. Little information has been acquired to properly account for sublethal effects of multiple applications during the growing season or effects on ecosystem structure and productivity. Some areas in which aggressive research is needed are listed below:

1. Pesticidal formulations need to be tested for comparison to existing data on technical grade active ingredient. Technical grade pesticide is rarely used other than for emergency mosquito control. Finished-product and technical-grade chemicals are not equally absorbed by vertebrates, and differences of acute toxicity among formulations may exceed threefold.
2. The rate and method of pesticidal applications affect hazards to nontarget animals. Additional field studies of natural populations are essential and are critical to advance concepts of probabilistic risk assessment.
3. The stability and fate of finished product in nontarget habitats is critical. Studies are in progress on aquatic macrocosms, but little information is available for aquatic or terrestrial systems adjoining croplands.
4. OP and CB pesticides are applied at different intervals during the growing season depending on the crop and target pest. These treatments coincide with the reproductive cycle for many species. Information is generally lacking on wildlife behavior and productivity during this critical period.
5. Studies are needed to understand population effects relating to OP- and CB-induced mortality when coupled with natural wildlife mortality from climatic events and infectious diseases.

## REFERENCES

1. Ecobichon, D. J., Toxic effects of pesticides, in *Casarett and Doull's Toxicology: The Basic Science of Poisons*, 5th ed., Klaassen, C. D., Amdur, M. O., and Doull, J., Eds., McGraw-Hill, New York, 1996, 643.
2. Eto, M., *Organophosphorus Pesticides: Organic and Biological Chemistry*, CRC Press, Cleveland, OH, 1974.
3. Kuhr, R. J. and Dorough, H. W., *Carbamate Insecticides: Chemistry, Biochemistry, and Toxicology*, CRC Press, Cleveland, OH, 1976.
4. Smith, G. J., Pesticide Use and Toxicology in Relation to Wildlife: Organophosphorus and Carbamate Compounds, U.S. Fish and Wildlife Service, Resour. Pub. 170, Washington, D.C., 1987.
5. Briggs, S. A., *Basic Guide to Pesticides: Their Characteristics and Hazards*, Taylor and Francis, Washington, D.C. 1992.
6. U.S. Environmental Protection Agency, Formal Request for Endangered Species Act, Section 7, Consultation on 31 Pesticides for All Uses, on All Listed Species, U.S. Environmental Protection Agency, Office of Pesticides and Toxic Substances, Washington, D.C., 1991.
7. Mendelssohn, H. and Paz, U., Mass mortality of birds of prey by azodrin, an organophosphorus insecticide, *Biol. Conserv.*, 11, 163, 1997.
8. Greig-Smith, P. W., Understanding the impact of pesticides on wild birds by monitoring incidents of poisoning, in *Wildlife Toxicology and Population Modeling: Integrated Studies of Agroecosystems*, Kendall, R. J. and Lacher, T. E., Jr., Eds., CRC Press, Boca Raton, FL, 1994, 301.
9. Goldstein, M. I. et al., An assessment of mortality of Swainson's hawks on wintering grounds in Argentina, *J. Raptor Res.*, 30, 106, 1996.
10. Mineau, P. et al., Poisoning of raptors with organophosphorus and carbamate pesticides with emphasis on Canada, U.S., and U.K., *J. Raptor Res.*, 33, 1, 1999.
11. Gaines, T. B., The acute toxicity of pesticides to rats, *Toxicol. Appl. Pharmacol.*, 2, 88, 1960.
12. Gaines, T. B., Acute toxicity of pesticides, *Toxicol. Appl. Pharmacol.*, 14, 515, 1969.
13. Schafer, E. W., The acute toxicity of 369 pesticidal, pharmaceutical, and other chemicals to wild birds, *Toxicol. Appl. Pharmacol.*, 21, 315, 1972.
14. Johnson, W. W. and Finley, M. T., Handbook of Acute Toxicity of Chemicals to Fish and Aquatic Invertebrates, U.S. Fish and Wildlife Service, Resour. Publ. 137, Washington, D.C., 1980.
15. Schafer, E. W., Jr., Bowles, W. A., Jr., and Hurlbut, J., The acute oral toxicity, repellency, and hazard potential of 998 chemicals to one or more species of wild and domestic birds, *Arch. Environ. Contam. Toxicol.*, 12, 355, 1983.
16. Hudson, R. H., Tucker, R. K., and Haegele, M. A., Handbook of Toxicity of Pesticides to Wildlife, U.S. Fish and Wildlife Service, Resour. Publ. 153, Washington, D.C., 1984.
17. Mayer, F. L., Jr. and Ellersieck, M. R., Manual of Acute Toxicity: Interpretation and Data Base for 410 Chemicals and 66 Species of Freshwater Animals, U.S. Fish Wildlife Service, Resour. Publ. 160. Washington, D.C., 1986.
18. Edwards, C. A. and Fisher, S. W., The use of cholinesterase measurements in assessing the impacts of pesticides on terrestrial and aquatic invertebrates, in *Cholinesterase-Inhibiting Insecticides: Their Impact on Wildlife and the Environment*, Mineau, P., Ed., Elsevier Publishers, Amsterdam, 1991, 255.
19. Murty, A. S. and Ramani, A. V., Toxicity of anticholinesterases to aquatic organisms, in *Clinical and Experimental Toxicology of Organophosphates and Carbamates*, Ballantyne, B. and Marrs, T. C., Eds., Butterworth-Heinemann, Oxford, 1992, 305.
20. Devillers, J. and Exbrayat, J. M., Eds., *Ecotoxicology of Chemicals to Amphibians*, Gordon and Breach, Philadelphia, 1992.
21. Stickel, W. H., Some effects of pollutants in terrestrial ecosystems, in *Ecological Toxicology Research*, McIntyre, A. D. and Mills, C. F., Eds., Plenum Publishing Company, New York, 1975, 25.
22. Edwards, C. A., The impact of pesticides on the environment, in *The Pesticide Question: Environment, Economies, and Ethics*, Pimentel, D. and Lehman, H., Eds., Chapman & Hall, New York, 1993, 13.
23. Osteen, C., Pesticides use, trends, and issues in the United States, in *The Pesticide Question: Environment, Economics, and Ethics*, Pimentel, D. and Lehman, H., Eds., Chapman and Hall, New York, 1993, 307.

24. Blus, L. J., Organochlorine pesticides, in *Handbook of Ecotoxicology*, Hoffman, D. J. et al., Eds., CRC Press, Boca Raton, FL, 1995, 275.
25. Vyas, N., Factors influencing estimation of pesticide-related wildlife mortality, *Toxicol. Indust. Health*, 15, 186, 1999.
26. Hill, E. F., Wildlife toxicology, in *General and Applied Toxicology*, 2nd ed., Ballantyne, B., Marrs, T. C., and Syversen, T., Eds., Macmillan, London, 1999, 1327.
27. Rattner, B. A. and Heath, A. G., Environmental factors affecting contaminate toxicity in aquatic and terrestrial vertebrates, in *Handbook of Ecotoxicology*, Hoffman, D. J. et al., Eds., CRC Press, Boca Raton, FL, 1995, 519.
28. Hill, E. F., Organophosphorus and carbamate pesticides, in *Handbook of Ecotoxicology*, Hoffman, D. J. et al., Eds., CRC Press, Boca Raton, FL, 1995, 243.
29. Hill, E.F., Organophosphorus agents and the environment, in *Organophosphates and Health*, Karalliedde, L. et al., Eds., Imperial College Press, London, 2001, 317.
30. Schafer, E. W. and Bowles, W. A., Jr., Acute oral toxicity and repellency of 933 chemicals to house and deer mice, *Arch. Environ. Contam. Toxicol.*, 14, 11, 1985.
31. Loomis, T. A. and Hayes, A. W., *Essentials of Toxicology*, 4th ed., Academic Press, San Diego, 1996.
32. Kundiev, Y. I. and Kagan, Y. S., Anticholinesterases used in the USSR: Poisoning, treatment and preventable measures, in *Clinical and Experimental Toxicology of Organophosphates and Carbamates*, Ballantyne, B. and Marrs, T. C., Eds., Butterworth-Heinemann Ltd., Oxford, 1992, 494.
33. O'Brien, R. D., *Insecticides Action and Metabolism*, Academic Press, New York, 1967.
34. Fairbrother, A., Cholinesterase-inhibiting pesticides, in *Noninfectious Diseases of Wildlife*, 2nd ed., Fairbrother, A., Locke, L. N., and Hoff, G. L., Eds., University Press, Ames, IA, 1996, 52.
35. Ludke, J. L., Hill, E. F., and Dieter, M. P., Cholinesterase (ChE) response and related mortality among birds fed ChE inhibitors, *Arch. Environ. Contam. Toxicol.*, 3, 1, 1975.
36. Hill, E. F. and Fleming, W. J., Anticholinesterase poisoning of birds: Field monitoring and diagnosis of acute poisoning, *Environ. Toxicol. Chem.*, 1, 27, 1982.
37. Senanayaka, N. and Karalliedde, L., Neurotoxic effects of organophosphorus insecticides: An intermediate syndrome, *N. Engl. J. Med.*, 316, 761, 1987.
38. DeBleecker, J. L., The intermediate syndrome, in *Organophosphates and Health*, Karalliedde, L. et al., Eds., Imperial College Press, London, 2001, 141.
39. Inns, R. H. et al., Histochemical demonstration of calcium accumulation in muscle fibres after experimental organophosphate poisoning, *Hum. Exp. Toxicol.*, 9, 245, 1990.
40. Abou-Donia, M. B., Organophosphorus ester-induced delayed neurotoxicity, *Ann. Rev. Pharmacol. Toxicol.*, 21, 511, 1981.
41. Hoffman, D. J., Sileo, L., and Murray, H. C., Subacute organophosphorus ester-induced delayed neurotoxicity in mallards, *Toxicol. Appl. Pharmacol.*, 75, 128, 1984.
42. Marrs, T. C. and Dewhurst, I., Toxicology of pesticides, in *General and Applied Toxicology*, 2nd ed., Ballantyne, B., Marrs, T. C., and Syversen, T., Eds., Macmillan Reference, London, 1999, 1993.
43. Grue, C. E., Hart, A. D. M., and Mineau, P., Biological consequences of depressed brain cholinesterase in wildlife, in *Cholinesterase-Inhibiting Insecticides: Their Impact on Wildlife and the Environment*, Mineau, P., Ed., Elsevier Publishers, Amsterdam, 1991, 151.
44. Grue, C. E., Gilbert, P. L., and Seeley, M. E., Neurophysiological and behavioral changes in non-target wildlife exposed to organophosphate and carbamate pesticides: Thermoregulation, food consumption, and reproduction, *Am. Zool.*, 37, 369, 1997.
45. Hill, E. F. and Camardese, M. B., Toxicity of anticholinesterase insecticides to birds: Technical grade versus granular formulations, *Ecotoxicol. Environ. Saf.*, 8, 551, 1984.
46. Walker, C. H., Species differences in microsomal monooxygenase activities and their relationship to biological half lives, *Drug Metab. Rev.*, 7, 295, 1978.
47. Walker, C. H., Species variations in some hepatic microsomal enzymes that metabolize xenobiotics, *Prog. Drug Metab.*, 5, 118, 1980.
48. Cowman, D. F. and Mazanti, L. E., Ecotoxicology of "new generation" pesticides to amphibians, in *Ecotoxicology of Amphibians and Reptiles*, Sparling, D. W., Linder, G., and Bishop, C. A., Eds., SETAC Press, Pensacola, FL, 2000, 233.

49. Pauli, B. D. and Money, S., Ecotoxicology of pesticides in reptiles, in *Ecotoxicology of Amphibians and Reptiles*, Sparling, D. W., Linder, G., and Bishop, C. A., Eds., SETAC Press, Pensacola, FL, 2000, 269.
50. Fleming, W. J., Recovery of brain and plasma cholinesterase activities in ducklings exposed to organophosphorus pesticides, *Arch. Environ. Contam. Toxicol.*, 10, 215, 1981.
51. Fleming, W. J. and Grue, C. E., Recovery of cholinesterase activity in five avian species exposed to dicrotophos, an organophosphorus pesticide, *Pest. Biochem. Physiol.*, 16, 129, 1981.
52. Hill, E. F., Avian toxicology of anticholinesterases, in *Clinical and Experimental Toxicology of Organophosphates and Carbamates*, Ballantyne, B. and Marrs, T. C., Eds., Butterworth-Heinemann, Oxford, 1992, 272.
53. Westlake, G. E. et al., Carbamate poisoning. Effects of selected carbamate pesticides on plasma enzymes and brain esterases of Japanese quail (*Coturnix coturnix japonica*), *J. Agric. Food Chem.*, 29, 779, 1981.
54. Matsumura, F., *Toxicology of Insecticides*, 2nd ed., Plenum, New York, 1985.
55. Corbett, J. R., Wright, K., and Baillie, A. C., *The Biochemical Mode of Action of Pesticides*, 2nd ed., Academic Press, London, 1984.
56. Mitchell, T. H. et al., The chromatographic determination of organophosphorus pesticides. III. The effect of irradiation on the parent compounds, *J. Chromatogr.*, 32, 17, 1968.
57. Bowman, J. S. and Casida, J. E., Further studies on the metabolism of Thimet by plants, insects, and mammals, *J. Econ. Entomol.*, 51, 838, 1958.
58. Ruzicka, J. H., Thomson, J., and Wheals, B. B., The gas chromatographic determination of organophosphorus pesticides. II. A comparative study of hydrolysis rates, *J. Chromatogr.*, 31, 37, 1967.
59. Lichtenstein, E. P., Liang, T. T., and Fuhrman, T. W., A compartmentalized microcosm for studying the fate of chemicals in the environment, *J. Agric. Food Chem.*, 26, 948, 1978.
60. Getzin, L. W. and Chapman, R. K., The fate of phorate in soils, *J. Econ. Entomol.*, 53, 47, 1960.
61. Singh, G. and Singh, Z., Persistence and movement of phorate at high concentrations in soil, *Ecotoxicol. Environ. Saf.*, 8, 540, 1984.
62. Getzin, L. W. and Shanks, C. H., Persistence, degradation, and bioactivity of phorate and its oxidative analogs in soil, *J. Econ. Entomol.*, 63, 52, 1970.
63. Zaki, M. H., Moran, D., and Harris, D., Pesticides in ground water: The aldicarb story in Suffolk County, NY, *Am. J. Public Health*, 72, 1391, 1982.
64. Hall, A. H. and Rumack, B. H., Incidence, presentation and therapeutic attitudes to anticholinesterase poisoning in the USA, in *Clinical and Experimental Toxicology of Organophosphates and Carbamates*, Ballantyne, B. and Marrs, T. C., Eds., Butterworth-Heinemann, Oxford, 1992, 471.
65. U.S. EPA, Carbofuran Special Review Technical Support Document, U.S. Environmental Protection Agency, Office of Pesticides and Toxic Substances, Washington, D.C. 1989.
66. Mineau, P., The Hazard of Carbofuran to Birds and other Vertebrate Wildlife, Canadian Wildlife Service, Tech. Rep. 177, Ottawa, 1993.
67. Hill, E. F., Divergent effects of postmortem ambient temperature on organophosphorus- and carbamate-inhibited brain cholinesterase activity in birds, *Pest. Biochem. Physiol.*, 33, 264, 1989.
68. Hill, E. F. and Camardese, M. B., Subacute toxicity testing with young birds: Response in relation to age and intertest variability of $LC_{50}$ estimates, in *Avian and Mammalian Wildlife Toxicology: Second Conference*, Lamb, D. W. and Kenega, E. E., Eds., ASTM STP 757, American Society for Testing and Materials, Philadelphia, 1981, 41.
69. Stinson, E. R. et al., Carbofuran affects wildlife on Virginia corn fields, *Wildl. Soc. Bull.*, 22, 566, 1994.
70. Stinson, E. R. and Bromley, P. T., Pesticides and Wildlife: A Guide to Reducing Impacts on Animals and Their Habitat, Virginia Department of Game and Inland Fisheries, Publ. 420-004, Richmond, VA, 1991.
71. Elliott, J. E. et al., Poisoning of bald eagles and red-tailed hawks by carbofuran and fensulfothion in the Fraser Delta of British Columbia, Canada, *J. Wildl. Dis.*, 32, 486, 1996.
72. Seiber, J. N., Catahan, M. P., and Barril, C. R., Loss of carbofuran from rice paddy water: Chemical and physical factors, *J. Environ. Sci. Health*, B, 13, 131, 1978.
73. Finlayson, D. G. et al., Carbofuran: Criteria for Interpreting the Effects of Its Use on Environmental Quality, National Resources Council of Canada, Publ. RRCC 16740, Ottawa, 1979.

74. Metcalf, R. L., Saugha, G. K., and Kapoor, I. P., Model ecosystems for the evaluation of pesticide biodegradability and ecological magnification, *Environ. Sci. Tech.*, 5, 709, 1971.
75. Neely, W. B., Branson, D. R., and Blau, G. E., Partition coefficient to measure bioconcentration potential of organic chemicals in fish, *Environ. Sci. Tech.*, 8, 1113, 1974.
76. Eisler, R., Carbofuran, in *Handbook of Chemical Risk Assessment: Health Hazards to Humans, Plants, and Animals, Vol. 2, Organics*, CRC Press, Boca Raton, FL, 2000, 799.
77. Ou, L. et al., Influence of soil temperature and soil moisture on degradation and metabolism of carbofuran in soil, *J. Environ. Qual.*, 11, 293, 1982.
78. Fahey, J. E., Wilson, M. C., and Armbrust, E. J., Residues of supracide and carbofuran in green and dehydrated alfalfa, *J. Econ. Entomol.*, 63, 589, 1970.
79. Turner, B. C. and Caro, J. H., Uptake and distribution of carbofuran and its metabolites in field grown corn plants, *J. Environ. Qual.*, 2, 245, 1973.
80. Ware, G. W. et al., Pesticide drift: 1. High clearance versus aerial application of sprays, *J. Econ. Entomol.*, 62, 840, 1969.
81. Dahl, T., Wetland Losses in the United States — 1780's to 1980's, U.S. Fish and Wildlife Service, Washington, D.C. 1990.
82. VanderValk, A., *Northern Prairie Wetlands*, Iowa State University Press, Ames, 1989.
83. Sheehan, P. J. et al., The Impact of Pesticides on the Ecology of Prairie Nesting Ducks, Canadian Wildlife Service, Tech. Rep. 19, Ottawa, 1987.
84. Grue, C. E. et al., Agricultural chemicals and the quality of prairie pothole wetlands for adult and juvenile waterfowl — What are the concerns? in *Proc. Natl. Symp. Prot. of Wetlands from Agricultural Impacts*, Stuber, P. J., Coord., U.S. Fish and Wildlife Service, Biol. Rep. 88(16), Washington, D.C., 1988.
85. Grue, C. E. et al., Potential impacts of agricultural chemicals on waterfowl and other wildlife inhabiting prairie wetlands: An evaluation of research needs and approaches, *Trans. N. Am. Wildl. Nat. Resour. Conf.*, 51, 357, 1986.
86. Tome, M. W., Grue, C. E., and Henry, M. G., Case studies: Effects of agricultural pesticides on waterfowl and prairie pothole wetlands, in *Handbook of Ecotoxicology*, Hoffman, D. J. et al., Eds., CRC Press, Boca Raton, FL, 1995, 565.
87. Tome, M. W., Grue, C. E., and DeWeese, L. R., Ethyl parathion in wetlands following aerial application to sunflowers in North Dakota, *Wildl. Soc. Bull.*, 19, 450, 1991.
88. Elder, B. F., Personal communication, 1996.
89. Richter, E. D., Aerial application and spray drift of anticholinesterase: Protective measures, in *Clinical and Experimental Toxicology of Organophosphates and Carbamates*, Ballantyne, B. and Marrs, T. C., Eds., Butterworth-Heinemann, Oxford, 1992, 623.
90. Seabloom, R. W. et al., An incident of fenthion mosquito control and subsequent avian mortality, *J. Wildl. Dis.*, 9, 18, 1973.
91. Kenaga, E. E., Predicted bioconcentration factors and soil sorption coefficients of pesticides and other chemicals, *Ecotoxicol. Environ. Saf.*, 4, 26, 1980.
92. Hill, E. F. et al., Lethal Dietary Toxicity of Environmental Pollutants to Birds, U.S. Fish Wildlife Service, Spec. Sci. Rep. Wildl. 191, Washington, D.C., 1975.
93. Pan, H. P. and Fouts, J. R., Drug metabolism in birds, *Drug Metab. Rev.*, 7, 1, 1978.
94. Brealey, C. J., Walker, C. H., and Baldwin, B. C., A-esterase activities in relation to the differential toxicity of pirimiphos-methyl to birds and mammals, *Pest. Sci.*, 11, 546, 1980.
95. Sanderson, D. M. and Edson, E. F., Toxicological properties of the organophosphorus insecticide dimethoate, *Brit. J. Indust. Med.*, 21, 52, 1964.
96. Cholakis, J. M. et al., Acute and subacute toxicity of pesticides in microtine rodents, in *Avian and Mammalian Wildlife Toxicology: 2nd Conference*, Lamb, D. W. and Kenaga, E. E., Eds., ASTM STP 757, American Society for Testing and Materials, Philadelphia, 1981, 143.
97. Heath, R. G. and Stickel, L. F., Protocol for testing the acute and relative toxicity of pesticides to penned birds, in The Effects of Pesticides on Wildlife, U.S. Department of the Interior, Circ. 226, Washington, D.C., 1965, 18.
98. Baril, A. et al., A Consideration of Inter-Species Variability in the Use of the Median Lethal Dose (LD50) in Avian Risk Assessment, Canadian Wildlife Service, Tech. Rep. 216, Hull, Quebec, 1994.

99. Hill, E. F., Acute and subacute toxicology in evaluation of pesticide hazard to avian wildlife, in *Wildlife Toxicology and Population Modeling: Integrated Studies of Agroecosystems*, Kendall, R. J. and Lacher, T. E., Jr., Eds., CRC Press, Boca Raton, FL, 1994, 207.
100. Mineau, P. and Baril, A., A Critique of the Avian 5-Day Dietary Test (LC50) as the Basis of Avian Risk Assessment, Canadian Wildlife Service, Tech. Rep. Ser. 215, Hull, Quebec, 1994.
101. Mason, J. R., Repellents in Wildlife Management: Proceedings of a Symposium, U.S. Department of Agriculture, Fort Collins, CO, 1997.
102. Hudson, R. H., Tucker, R. K., and Haegele, M. A., Effect of age on sensitivity: Acute oral toxicity of 14 pesticides to mallard ducks of several ages, *Toxicol. Appl. Pharmacol.*, 22, 556, 1972.
103. Hill, E. F., unpublished data.
104. Tucker, R. K. and Leitzke, J. S., Comparative toxicology of insecticides for vertebrate wildlife and fish, *Pharmacol. Ther.*, 6, 167, 1979.
105. Thompson, W. R., Use of moving averages and interpolation to estimate median effective dose, I. Fundamental formulas, estimation of error, and relation to other methods, *Bacteriol. Rev.*,11, 115, 1947.
106. Weil, C. S., Tables for convenient calculation of median effective dose (LD50 or ED50) and instructions in their use, *Biometrics*, 8, 249, 1952.
107. Tucker, R. K. and Crabtree, D. G., Handbook of Toxicity of Pesticides to Wildlife, U.S. Dept. of Interior, Resour. Pub. 84, Washington, D.C., 1970.
108. Wiemeyer, S. N. and Sparling, D. W., Acute toxicity of four anticholinesterase insecticides to American kestrels, eastern screech-owls, and northern bobwhite, *Environ. Toxicol. Chem.*, 10, 1139, 1991.
109. Driver, C. J. et al., Routes of uptake and their relative contribution to the toxicological response to northern bobwhite (*Colinus virginianus*) to an organophosphate pesticide, *Environ. Toxicol. Chem.*, 10, 21, 1991.
110. Natoff, I. L., Influence of route of administration on the toxicity of some cholinesterase inhibitors, *J. Pharm. Pharmacol.*, 19, 612, 1967.
111. McCreesh, A. H., Percutaneous toxicity, *Toxicol. Appl. Pharmacol.*, 7, 20, 1965.
112. Hudson, R. H., Haegele, M. A., and Tucker, R. K., Acute oral and percutaneous toxicity of pesticides to mallards: Correlations with mammalian toxicity data, *Toxicol. Appl. Pharmacol.*, 47, 451, 1979.
113. Schafer, E. W., Jr. et al., Comparative toxicity of seventeen pesticides to the Quelea, house-sparrow, and red-winged blackbird, *Toxicol. Appl. Pharmacol.*, 26, 154, 1973.
114. Vinson, L. J. et al., The nature of epidermal barrier and some factors influencing skin permeability, *Toxicol. Appl. Pharmacol.*, 7 (Suppl.), 7, 1965.
115. Robbins, C. T., *Wildlife Feeding and Nutrition*, Academic Press, New York, 1983.
116. Bartholomew, G. A. and Cade, T. J., The water economy of land birds, *Auk*, 80, 504, 1963.
117. Zinkl, J. G. et al., Fenthion poisoning of wading birds, *J. Wildl. Dis.*, 17, 117, 1981.
118. DeWeese, L. R. et al., Effects on birds of fenthion aerial application for mosquito control, *J. Econ. Entomol.*, 76, 906, 1983.
119. Stone, W. B. and Gradoni, P. B., Wildlife mortality related to the use of the pesticide diazinon, *Northeast Environ. Sci.*, 4, 30, 1985.
120. U.S. EPA, Intent to Cancel Registrations on Denial of Applications for Registration of Pesticide Products Containing Diazinon; Conclusion of Special Review, *Fed. Regis.*, 51, 35034, 1986.
121. Stromborg, K. L., Seed treatment pesticide effects on pheasant reproduction at sublethal doses, *J. Wildl. Manage.*, 41, 632, 1977.
122. Grue, C. E. et al., Assessing hazards of organophosphate pesticides to wildlife, *Trans. N. Am. Wildl. Nat. Resour. Conf.*, 48, 200, 1983.
123. Hamilton, G. A. et al., Poisoning of wild geese by carbophenothion-treated winter wheat, *Pest. Sci.*, 7, 175, 1976.
124. Hamilton, G. A. et al., Wildlife deaths in Scotland resulting from misuse of agricultural chemicals, *Biol. Cons.*, 21, 315, 1981.
125. DeWeese, L. R. et al., Response of Breeding Birds to Aerial Sprays of Trichlorfon (Dylox) and Carbaryl (Sevin-4-Oil) in Montana Forests, U.S. Fish and Wildlife Service, Spec. Sci. Rep. Wildl. 224,Washington, D.C., 1979.
126. Fox, G. A. et al., The Impact of the Insecticide Carbofuran (Furadan 480F) on the Burrowing Owl in Canada, Canadian Wildlife Service, Tech. Rep. 72, Hull, Quebec, 1989.

127. Henny, C. J., Kolbe, E. J., and Fitzner, R. E., Organophosphate insecticide (Famphur) topically applied to cattle kills magpies and hawks, *J. Wildl. Manage.*, 49, 648, 1985.
128. Hill, E. F. and Mendenhall, V. M., Secondary poisoning of barn owls with famphur, an organophosphate insecticide, *J. Wildl. Manage.*, 44, 676, 1980.
129. Hall, R. J. and Kolbe, E. J., Bioconcentration of organophosphorus pesticides to hazardous levels by amphibians, *J. Toxicol. Environ. Health*, 6, 853, 1980.
130. U.S. EPA, Ecological Effects Test Guidelines, U.S. Environmental Protection Agency, EPA-712-C-96-139, Washington, D.C., 1996.
131. Hoffman, D. J. and Albers, P. H., Evaluation of potential embryotoxicity and teratogenicity of 42 herbicides, insecticides, and petroleum contaminants to mallard eggs, *Arch. Environ. Contam. Toxicol.*, 13, 15, 1984.
132. Hoffman, D. J., Embryotoxicity and teratogenicity of environmental contaminants to bird eggs, *Rev. Environ. Contam. Toxicol.*, 115, 40, 1990.
133. Balcomb, R., Stevens, R., and Bowen, C., II, Toxicity of granular insecticides to wild-caught songbirds, *Bull. Environ. Contam. Toxicol.*, 33, 302, 1984.
134. Mineau, P., Avian mortality in agro-ecosystems. 1. The case against granular insecticides in Canada, in *Field Methods for the Study of Environmental Effects of Pesticides*, Greaves, M. P., Smith, B. D., and Greig-Smith, P. W., Eds., Brit. Crop. Prot. Monog., 40, 3, 1988.
135. Best, L. B. and Fischer, D. L., Granular insecticides and birds: Factors to be considered in understanding exposure and reducing risk, *Environ. Toxicol. Chem.*, 11, 1495, 1991.
136. Best, L. B., Stafford, T. R., and Mihaich, E. M., House sparrow preferential consumption of pesticide granules with different surface coatings, *Environ. Toxicol. Chem.*, 15, 1763, 1996.
137. Gionfriddo, J. P. and Best, L. B., Grit color selection by house sparrows and northern bobwhite, *J. Wildl. Manage.*, 60, 836, 1996.
138. Stafford, T. R., Best, L. B., and Fischer, D. L., Effects of different formulations of granular pesticides on birds, *Environ. Toxicol. Chem.*, 15, 1606, 1996.
139. Stafford, T. R. and Best, L. B., Effects of application rate on avian risk from granular pesticides, *Environ. Toxicol. Chem.*, 17, 526, 1998.
140. Ball, W. I. et al., Modification of parathion's toxicity for rats by pretreatment with chlorinated hydrocarbon insecticides, *Can. J. Biochem. Physiol.*, 32, 440, 1954.
141. Triolo, A. J. and Coon, J. M., Toxicologic interactions of chlorinated hydrocarbon and organophosphate insecticides, *J. Agric. Food Chem.*, 14, 549, 1966.
142. Menzer, R. E., Effect of chlorinated hydrocarbons in the diet on the toxicity of several organophosphorus insecticides, *Toxicol. Appl. Pharmacol.*, 16, 446, 1970.
143. Ludke, J. L., DDE increases the toxicity of parathion to coturnix quail, *Pest. Biochem. Physiol.*, 7, 28, 1977.
144. DuBois, K. P., Potentiation of the toxicity of organophosphorus compounds, *Adv. Pest Cont. Res.*, 4, 117, 1961.
145. Durham, W. F., The interaction of pesticides with other factors, *Residue Rev.*, 18, 21, 1967.
146. Murphy, S. D., Mechanisms of pesticide interactions in vertebrates, *Residue Rev.*, 25, 201, 1969.
147. Kreitzer, J. F. and Spann, J. W., Tests of pesticidal synergism with young pheasants and Japanese quail, *Bull. Environ. Contam. Toxicol.*, 9, 250, 1973.
148. Gordon, J. J., Leadbeater, L., and Miadment, M. P., The protection of animals against organophosphate poisoning by pretreatment with a carbamate, *Toxicol. Appl. Pharmacol.*, 43, 207, 1978.
149. Takahashi, H. et al., Potentiations of N-methylcarbamate toxicities by organophosphorus insecticides in male mice, *Fundam. Appl. Toxicol.*, 8, 139, 1987.
150. Blus, L. J. et al., Effects of organophosphorus insecticides on sage grouse in southeastern Idaho, *J. Wildl. Manage.*, 53, 1139, 1989.
151. Rattner, B. A. and Fairbrother, A., Biological variability and the influence of stress on cholinesterase activity, in *Cholinesterase-Inhibiting Insecticides: Their Impact on Wildlife and the Environment*, Mineau, P., Ed., Elsevier Publishers, Amsterdam, 1991, 89.
152. Ratter, B. A. and Heath, A. G., Environmental factors affecting contaminant toxicity in aquatic and terrestrial vertebrates, in *Handbook of Ecotoxicology*, Hoffman, D. J. et al., Eds., CRC Press, Boca Raton, FL, 1995, 519.

153. Flickinger, E. L. et al., Bird poisoning from misuse of the carbamate Furadan in a Texas rice field, *Wildl. Soc. Bull.*, 14, 59, 1986.
154. Westlake, G. E. et al., Organophosphate poisoning. Effects of selected organophosphate pesticides on plasma enzymes and brain esterases of Japanese quail (*Coturnix coturnix japonica*), *J. Agric. Food Chem.*, 29, 772, 1981.
155. Thompson, H. M. and Walker, C. H., Serum B esterases as indicators of exposure to organophosphorus and carbamate insecticides, in *Nondestructive Biomarkers in Vertebrates*, Fosi, C. and Leonzio, C., Eds., CRC Press, Boca Raton, FL, 1994, 35.
156. Fairbrother, A. et al., Methods used in determination of cholinesterase activity, in *Cholinesterase-inhibiting Insecticides: Their Impact on Wildlife and the Environment*, Mineau, P., Ed., Elsevier Publishers, Amsterdam, 1991, 35.
157. Hunt, K. A. and Hooper, M. J., Development and optimization of reactivation techniques for carbamate-inhibited brain and plasma cholinesterases in birds and mammals, *Anal. Biochem.*, 212, 335, 1993.
158. Mineau, P., Ed., *Cholinesterase-Inhibiting Insecticides: Their Impact on Wildlife and the Environment*, Elsevier Publishers, Amsterdam, 1991.
159. Ahdaya, S. M., Shar, P. V., and Guthrie, F. E., Thermoregulation in mice treated with parathion, carbaryl, or DDT, *Toxicol. Appl. Pharmacol.*, 35, 575, 1976.
160. Chattopadhyay, D. P. et al., Changes in toxicity of DDVP, DFP, and parathion in rats under cold environment, *Bull. Environ. Contam. Toxicol.*, 29, 605, 1982.
161. Rattner, B. A., and Franson, J. C., Methyl parathion and fenvalerate toxicity in American kestrels: Acute physiological responses and effects of cold, *Can. J. Physiol. Pharmacol.*, 62, 787, 1984.
162. Rattner, B. A., Becker, J. M., and Nakatsugawa, T., Enhancement of parathion toxicity to quail by heat and cold exposure, *Pest. Biochem. Physiol.*, 27, 330, 1987.
163. Shilov, I. A., *Heat Regulation in Birds*, Amerind Publishing Co., New Delhi, 1973.
164. Maguire, C. C. and Williams, B. A., Cold stress and acute organophosphorus exposure: Interaction effects on juvenile northern bobwhite, *Arch. Environ. Contam. Toxicol.*, 16, 477, 1987.
165. Grue, C. E., Powell, G. V. N., and McChesney, M. J., Care of nestlings by wild female starlings exposed to an organophosphate pesticide, *J. Appl. Ecol.*, 19, 327, 1982.
166. Brewer, L. W. et al., Effects of methyl parathion in ducks and duck broods, *Environ. Toxicol. Chem.*, 7, 375, 1988.
167. Buerger, T. T. et al., Effects of methyl parathion on northern bobwhite survivability, *Environ. Toxicol. Chem.*, 10, 527, 1991.
168. Brewer, R. A. et al.,Toxicity, survivability, and activity patterns of northern bobwhite quail dosed with the insecticide terbufos, *Environ. Toxicol. Chem.*, 15, 750, 1996.
169. Peakall, D. B., Behavioral responses of birds to pesticides and other contaminants, *Residue Rev.*, 96, 45, 1985.
170. Bennett, R. S., Do behavioral responses to pesticide exposure affect wildlife population parameters?, in *Wildlife Toxicology and Population Modeling: Integrated Studies of Agroecosystems*, Kendall, R. J. and Lacher, T. E., Jr., Eds., CRC Press, Boca Raton, FL, 1994, 241.
171. Walker, C. H. et al., *Principles of Ecotoxicology*, Taylor and Francis, London, 1996.
172. Schom, C. B., Abbott, U. K., and Walker, N. E., Adult and embryo responses to organophosphate pesticide: Azodrin, *Poultry Sci.*, 58, 60, 1979.
173. Schom, C. B. and Abbott, U. K., Studies with bobwhite quail: Reproductive characteristics, *Poultry Sci.*, 53, 1860, 1974.
174. Stromborg, K. L., Reproductive toxicity of monocrotophos to bobwhite quail, *Poultry Sci.*, 65, 51, 1986.
175. Bennett, J. K. and Bennett, R. S., Effects of dietary methyl parathion on northern bobwhite egg production and eggshell quality, *Environ. Toxicol. Chem.*, 9, 1481, 1990.
176. Bennett, R. S. et al., Effects of dietary exposure to methyl parathion on egg laying and incubation in mallards, *Environ. Toxicol. Chem.*, 10, 501, 1991.
177. Meyers, S. M., Cummings, J. L., and Bennett, R. S., Effects of methyl parathion on red-winged blackbirds (*Agelaius phoeniceus*) incubation behavior and nesting success, *Environ. Toxicol. Chem.*, 9, 808, 1990.
178. Stromborg, K. L. et al., Postfledging survival of European starlings exposed to an organophosphorus insecticide, *Ecology*, 69, 590, 1988.

179. Powell, G. V. N., Reproduction by an altricial songbird, the red-winged blackbird, in fields treated with the organophosphate insecticide fenthion, *J. Appl. Ecol.*, 21, 83, 1984.
180. McEwen, L. C. and Brown, R. L., Acute toxicity of dieldrin and malathion to wild sharp-tailed grouse, *J. Wildl. Manage.*, 30, 604, 1966.
181. White, D. H. et al., Parathion alters incubation behavior of laughing gulls, *Bull. Environ. Contam. Toxicol.*, 31, 93, 1983.
182. White, D. H. et al., Parathion causes secondary poisoning in a laughing gull breeding colony, *Bull. Environ. Contam. Toxicol.*, 23, 281, 1979.

# CHAPTER 13

# Organochlorine Pesticides

**Lawrence J. Blus**

## CONTENTS

## 13.1 INTRODUCTION

### 13.1.1 Background

The discovery of the insecticidal properties of DDT that led to its subsequent use in pest control was hailed as a tremendous scientific achievement. Initial success with DDT in controlling human health pests during World War II and subsequent success in agricultural pest control stimulated synthesis and development of related organochlorine pesticides; their use increased exponentially following the war.[1] At first, evidence slowly accumulated that nearly all of these compounds were having widespread adverse effects on nontarget organisms. Later, a veritable mountain of evidence was amassed relating to their toxicity, persistence, and lipophilic characteristics that resulted in accumulation of residues, mortality, lowered reproductive success, and decline — even extirpation — of certain populations of wildlife.[2–3] Ecotoxicological data for organochlorine pesticides are limited to relatively few countries where most research has been conducted. Despite documented damage of organochlorine pesticides to the environment a number of these compounds, particularly DDT, currently are used in a number of countries.[4] It is likely that no other group of contaminants of anthropogenic origin has exacted such a heavy toll on the environment as have the organochlorine pesticides.

### 13.1.2 Objectives

The objectives of this review are to document the nature and extent of ecotoxicological problems resulting from use of organochlorine pesticides for over a half century and to discuss future relevance of these problems.

## 13.2 CHARACTERISTICS

### 13.2.1 Grouping

The five major groups of organochlorine pesticides are: DDT and its analogs, hexachlorocyclohexane (HCH), cyclodienes and similar compounds, toxaphene and related chemicals, and the caged structures mirex and chlordecone (Kepone®).[5] This grouping is based on chemical structure; therefore, it is difficult to generalize group characteristics and their resultant toxic effects.

DDT and ten of its analogs include the commercial pesticides dicofol (Kelthane®), DDD (TDE), Bulan®, chlorfenethol (DMC), chlorobenzilate, chloropropylate, DFDT, ethylan (Perthane®), methoxychlor, and Prolan®. Some of these compounds have strikingly different properties; for example, the slow metabolism and marked storage of DDT and its metabolite DDE is in contrast to the rapid metabolism and negligible storage of methoxychlor.[5] Most of these compounds will not be discussed in this paper because there are limited ecotoxicological data available, primarily because they were used sparingly.

Hexachlorocyclohexane (HCH) consists of eight steric isomers, including the well-known $\gamma$ isomer — lindane. Like DDT and its analogs, the isomers have remarkably different properties.

Cyclodienes and related compounds include aldrin, isodrin, dieldrin, endrin, telodrin, heptachlor, isobenzam, chlordane, and endosulfan. This group contains the most toxic organochlorine pesticides (OCs), particularly in relation to acute poisoning.

The pesticide toxaphene is made up of a complex mixture of chemicals, many of which are still unidentified.

Mirex and chlordecone are slowly metabolized and are readily stored in the body.

### 13.2.2 Lipophilicity

One common characteristic of OCs is a generally high solubility in lipids or lipophilicity; this is expressed as the n-octanol/water partition coefficient.[6] As a result, the highest concentrations of OCs are found in lipids; residues (presented on a wet weight basis unless otherwise indicated) may be stored in fat depots throughout the organism.[2] With decreases in body weight, as during times of stress, lipids and residues of organochlorines are mobilized into the blood.[7] The residues may be transported to the brain or other compartments, where they may induce mortality or serious sublethal effects.[8] Residues of organochlorines occur in lipids of biota throughout the world. Residues have decreased drastically in some areas, particularly in developed Western countries, due to banning of most problem organochlorines.[3] In many undeveloped countries use persists, and there are few data available on resides or effects on the biota.[9]

### 13.2.3 Persistence

One of the best-known characteristics of organochlorine pesticides is their persistence in the environment. Persistence is based on the half-life both in the physical environment and in the organism.[10] Half-lives range from months to years. Residues of some organochlorines may persist for decades — possibly centuries — in the environment;[11] but there are many physical factors that influence this including temperature, light, pH, and moisture.[12] Also, some microorganisms have the ability to break down organochlorine pesticides in the environment[13] as well as in the living organism.[14] Hibernation,[15] reproductive activity,[16–17] and other factors can influence the rate of loss of residues. Therefore, the range in half-life of an OC may vary dramatically according to existing physical and biological conditions. Persistence varies greatly among compounds. Several pesticides, including heptachlor and aldrin, are rapidly metabolized once inside the organism; however, their major metabolites, heptachlor epoxide and dieldrin, are indeed persistent and as toxic as the parent compound.[18] Endrin and dicofol are relatively short-lived in organisms and the physical environment as well. Dicofol has no toxic metabolites of major significance. The highly toxic metabolite of endrin, 12-ketoendrin, is rarely found in birds but is important in a few mammalian species, including the laboratory rat.[18]

### 13.2.4 Bioaccumulation

The terminology related to buildup of OC residues in organisms is not standardized, and confusion exists as to the definition of the various terms.[19] Definitions of the mechanisms and routes listed here follow Moriarty;[2] they are relatively straightforward and adequately define the processes.

Bioaccumulation is the characteristic that results in concentration of organochlorine pesticides in the biota, whether it be from water directly into an organism through gills, epidermis, and similar media (bioconcentration) or from food only (the food-chain effect or biomagnification). Bioaccumulation includes both biomagnification and bioconcentration; mirex residues bioaccumulated at a rate of $25 \times 10^6$ from water to eggs of birds.[20] Bioconcentration is the process whereby residues of certain organochlorines may be increased as much as $10^6$ times from water to aquatic or marine organisms.[21] American oysters (*Crassostrea virginica*) can bioconcentrate DDT from surrounding water up to 700,000 times.[22] Matsumura[23] indicated that astronomical bioconcentration figures, such as those cited above, were misleading, primarily because of the method of calculation of biomagnification with respect to water. Moriarty[2] discussed problems and pitfalls in calculating and interpreting bioaccumulation; one major problem is comparing levels in whole bodies to individual tissues or eggs. Biomagnification is related to increases in various organochlorines to about 30- to 100-fold from prey to tissues and eggs of fish-eating birds including the herring gull (*Larus argentatus*), and brown pelican (*Pelecanus occidentalis*),[16,24] and up to tenfold in terrestrial organisms.[25] Harris et al.[26] calculated mean increases in residues of 42-fold (DDE) and 60-fold (DDT) from earthworms to eggs of American robins in orchards in Canada.

Evidence suggests that intake from food is not the major source of residues from persistent organic pollutants in aquatic species such as fish, oysters, and related organisms.[2,21] For example, there is a linear correlation between concentrations in biota, particularly invertebrates, and those in water.[21] Food is the principal source of persistent organochlorines for most trophic levels in terrestrial systems[2] and in certain aquatic species such as the Nile crocodile (*Crocodylus niloticus*) whose eggs contained total DDT residues that were tenfold higher than in fish.[27] There was a 37-fold increase in DDT and metabolites from whole bodies of fish to blubber of gray (*Halichoerus grypus*) and harbor seals (*Phoca vitulina*) in Norway.[28] There are also special cases that do not fit the usual bioaccumulation schemes in that the organisms are exposed to very high concentrations through ingestion, dermal absorption, or other routes of exposure that may lead to acute poisoning. Examples include control programs that use treated substrates such as bird perches,[29] spraying of bat roosts[30] or bird nesting areas,[31] intentional use of treated seed[32] or laced carcasses[33] for the specific purpose of killing nuisance wildlife, unintentional spillage of treated seed during transporting or planting operations, dumping of excess treated grain, and spills into bodies of water.[34–36]

### 13.2.5 Toxicity

Molecules of organochlorine pesticides and other pollutants attach to and act upon receptors — molecules or parts of molecules — within the affected organism.[2] This combination of pollutant with receptor, termed the biochemical lesion,[37] is the first step before effects are manifested.[38] As an example, the biochemical lesion of DDT occurs in the axonic membrane of the nerve fiber,[39] but some of the details are still uncertain.[40] Matsumura[23] discussed modes of action of pesticides in some detail. Basically, OCs are neuroactive agents whose modes of action include effects on ion permeability (DDT and its analogs) or as agents for nerve receptors (cyclodienes and BHC).

Toxicity of organochlorines is determined by many factors, not the least of which is the chemical structure of the various compounds. All of these compounds can be absorbed through respiratory and dermal routes; the importance of the dermal route varies greatly among the toxicity of different pesticides.[5] Toxicity of OC pesticides, as expressed by the $LD_{50}$ (calculated single oral dose that will kill 50% of exposed organisms), ranged from 1 mg/kg to > 2,080 mg/kg in birds[41] and 8 to 6000 mg/kg in laboratory rats.[42] The $LC_{50}$ (estimated dietary concentration that is lethal to 50% of the exposed animals) ranged from 14 to > 5000 μg/g in birds.[43] The 96-hour $LC_{50}$ (estimated concentration in water that is lethal to 50% of exposed fish or other aquatic organisms within 96 hours) ranged from <1 μg/L to 4300 mg/L.44

In the environment mortality from organochlorines can occur within hours after exposure,[45] but usually death occurs after a buildup in residues after weeks or months. Determination of lethal brain residues of DDT in laboratory rats[46] and birds[47] and subsequently with the parent compound and the major metabolites DDE and DDD in experimental birds[48–51] provided the mechanism whereby die-offs in the field were linked with the DDT group. Subsequently, diagnostic lethal residues in brains of experimental birds were determined for heptachlor epoxide, dieldrin, endrin, mirex, several chlordane-related compounds, and telodrin.[52–58] It should be noted that some investigators have used residues in livers of birds and mammals to document mortality from OCs,[2] despite the fact that extensive research with lethal residues in brains and livers of experimental animals indicates that use of the brain for this purpose provides a more accurate diagnostic tool.[47–57] Lethal residues are a diagnostic indicator of mortality — but not necessarily a direct measure of their effects on the action of the biochemical lesion that results in death.

Sex and age of vertebrates, repellency, type of exposure, formulation of the pesticide, interaction with other contaminants, and numerous other factors affect toxicity of pesticides.[59] Stresses such as shortage of food, weather extremes, migration, reproductive activities, hibernation, molting, and disturbance also enhance toxicity.[15,49,60–65] Species sensitivity is an important factor influencing toxicity of pesticides.[43,66]

Resistance to DDT was reported as early as 1947. Genetic resistance to OCs has been reported in some frogs[67] and fish[68] and is commonly described in invertebrates,[69] but aside from documentation of apparent resistance to endrin in the woodland vole (*Microtus pinetorum*),[70] there are no other examples for mammals or birds in the field. Even in the woodland vole, genetic resistance was not verified because there were no consistent differences in population genetic indices in individuals from orchards with or without a history of endrin treatment.[71] Deer mice (Peromyscus maniculatus) occupying a dieldrin "hotspot" were thought to have developed possible resistance to dieldrin,[72] but there was no real evidence (see section on dieldrin). Cross-resistance may occur between and within classes of compounds; for example, DDT provides some cross-resistance to methoxychlor but not to lindane or dieldrin.[73]

### 13.2.6 Physiological Responses

Metabolism is the effect of an organism, through its enzymes, on the chemical structure of foreign compounds or xenobiotics.[18] Induction of microsomal enzyme activity by organochlorine pesticides was first discovered in laboratory mammals exposed to DDT or chlordane.[74–75] Enzyme induction is, within normal physiological limits, not deleterious;[2] it may be thought of as a preparatory step for the liver to metabolize the xenobiotics. Wide interspecific differences occur in efficiency of the liver microsomal enzymes to metabolize organochlorine substrates. These affect persistence of lipid-soluble compounds and the rate of excretion of their water-soluble metabolites.[6] For example, this enzyme system in the shag (*Phalacrocorax aristotelis*) is relatively inefficient in comparison to those of other species tested.[5]

DDT was the first OC demonstrated to exhibit estrogenic activity;[76] other OCs, including methoxychlor, chlordecone, and DDE, also exhibit estrogenic activity.[3] Organochlorine pesticides stimulated production of liver microsomal enzymes that led to hydroxylation of steroids in experimental animals that may have adversely affected reproductive success.[77,78] Peakall[78] demonstrated that DDT in experimental birds induced breakdown of sex hormones that regulate the mobilization of calcium. Hatchling and juvenile American alligators (*Alligator mississipensis*) in Lake Apopka, Florida, exhibited a number of abnormalities in gonadal morphology and concentrations of sex steroids in both sexes.[36] Although these changes may have been related to residues of DDE and other contaminants, it was not possible to pinpoint the responsible pollutants. There are also adverse effects of OCs on thyroid secretion,[79] adrenal function,[80] migratory condition,[81] biogenic amines,[82] the immune system,[63] and many other physiological responses.[1–3,23,29,38] Few of these physiological responses have received extensive study in the field; outstanding success has been achieved in quantifying effects of residues on endpoints like mortality, reproductive success, and eggshell thinning that are used in estimating degree of risk of OCs to the individual and the population.

### 13.2.7 Interactions

Organisms in the environment are frequently exposed to many OCs and other contaminants at the same time; residues of most OCs in animal tissues are positively intercorrelated.[84] As a result, synergism, potentiation, antagonism, or additivity of toxic effects may result.[59] Experimental studies have documented interactions of OCs and other stressors, but these are inadequate to make predictions to the field, where many complex factors are interacting.[3] Prior treatment of birds with DDE enhanced the toxic effects of parathion to Japanese quail (*Coturnix japonica*), whereas prior treatment with chlordane was antagonistic to parathion toxicity but not to its metabolite paraoxon.[85] Northern bobwhites (*Colinus virginianus*) pretreated with chlordane for 10 weeks, followed by endrin, had greater accumulations of chlordane in their brains than did birds treated only with chlordane.[86] Aldrin markedly increased storage of DDT in beagle dogs.[87] In an intensive study of over 100 combinations of eight OCs with organophosphorus and carbamate pesticides given to laboratory rodents, small or questionable potentiation was found between chlordane and methoxychlor, and a minor level of antagonism was noted between aldrin and trithion.[88] For industrial chemicals and pesticides, additivity seems the most common joint action.[89] Moriarty[2] concluded that for combinations of pollutants, the toxic unit (concentration of a pollutant producing a standard response) is considered to be additive. Experimental birds, dying while receiving DDT, DDD, or DDE in their diets, exhibited an approximately additive lethal ratio in their brains of 1:5:15 of DDT, DDD, and DDE, respectively.[50]

## 13.3 SOURCES, USE, AND EFFECTS

### 13.3.1 DDT and Metabolites

DDT is the best-known organochlorine pesticide. It was synthesized in 1874, and discovery of its insecticidal activity by Paul Müller in 1939 subsequently led to his receiving the Nobel Prize.[90] Early work with heavy applications of DDT resulted in mortality of nontarget organisms,[91–93] but it was not until the program to control Dutch elm disease was initiated that the lethal risk caused widespread concern. More than 90 species of birds, particularly American robins (*Turdus migratorius*), were found dead in neighborhoods where DDT was applied to trees.[47,94–95] Subsequently, a large-scale mortality of insects and other invertebrates was discovered as well; earthworms accumulated high concentrations of DDT and metabolites that subsequently proved lethal to robins and other birds that ingested contaminated prey.[25]

Determination of lethal brain residues of DDT and its metabolites in experimental animals[46–51] provided the scientific evidence whereby these die-offs were linked with the DDT group.

It took several additional years to document the major sublethal risk of DDT; that is, the effects of its metabolite DDE on embryotoxicity, eggshell thinning, and related effects that adversely influence avian reproductive success. In Great Britain, Ratcliffe[96] initially documented eggshell thinning, depressed reproduction, and population declines of peregrine falcons (*Falco peregrinus*) and other birds. Experimental studies with birds verified that DDE was primarily responsible for thinning of eggshells and adverse affects on reproduction.[97–99] Of the DDT group, DDE residues are usually highest and are found most frequently in environmental samples, particularly at higher trophic levels.[100] After application of DDT, DDE residues increase because of breakdown of the parent compound.[11,100] The mode of action of DDE in eggshell thinning has been the subject of studies that should be consulted for further information on physiological relationships and structural changes.[66,100,101] The latest information indicates that thinning is related to inhibition of prostaglandin synthetase in the eggshell gland mucosa.[102]

Shortly after the initial work, many studies documented eggshell thinning in 18 families of wild birds in North America[104–106] and throughout the world.[101–107] In field studies, statistical analyses revealed that DDE was the contaminant primarily responsible for eggshell thinning[108–110] and associated lowered productivity.[111–112] Generally, thinning of 18 to 20% or more over several years is related to population decline.[104] Eggshell thinning and a number of other factors are related to adverse effects of DDE on productivity.[112] Despite the devastating effects of DDE on avian productivity and resultant population decreases described here, this metabolite has caused few documented mortalities in wild adult birds.[113–114] Additionally, some segments of the avian fauna, including a raptorial species like the great horned owl (*Bubo virginianus*), a fish-eating species like the great blue heron (*Ardea herodias*), and the gray heron (*A. cinera*), showed minimal effects and no population declines even at the peak of usage of OCs.[115–116]

The brown pelican is the most sensitive avian species to effects of DDE on reproduction; 3 µg/g in the egg is associated with near total reproductive failure.[112] Based on individual eggs, residues are a more accurate indicator of nest success than eggshell thickness.[117] The most extreme incidence of avian eggshell thinning was noted in 1969 on Anacapa Island in southern California; nearly all eggs of brown pelicans in the colony collapsed under incubating birds.[118–120] Eggshell thinning of collapsed eggs was 53% compared with 34% in intact eggs. Only five young pelicans hatched from nearly 1300 nesting attempts in 1969.[119] Eggshell thickness and reproductive success were only slightly higher at three colonies of brown pelican in northwestern Mexico in 1969.[121] In comparison with other colonies, reproductive success and eggshell thickness were highest at three pelican colonies in the Gulf of California, but productivity was less than normal.[118] Residues (lipid basis) in 1969 were extremely high and means ranged from 58 μg/g at Isla Piojo in the Gulf of California, 1310 μg/g at Coronados (northwestern Mexico) to 1818 μg/g at Anacapa Island.[118,121] Residues largely originated from a DDT manufacturing plant near Los Angeles; residues remained high and pelicans experienced serious reproductive problems for many years.[122] Except in Louisiana (see the endrin section), DDE was primarily responsible for declines of brown pelican populations in the United States and part of Mexico.[112,121–123]

The peregrine falcon is considered sensitive to reproductive effects from DDE (but also see the section on dieldrin) because there were population declines associated with eggshell thinning and decreased productivity throughout much of its range.[107,124–127] The sensitivity of the peregrine falcon to DDE depend not so much on the levels that adversely affect reproductive success as on its very high potential for bioaccumulation. The lowest level of DDE in peregrine eggs that adversely affects reproductive success was estimated at 15 to 20 μg/g.[126] However, a recent study more clearly defined this level at near 30 μg/g.[128] The peregrine is at the top of a complex food chain and has a far greater capacity to bioaccumulate residues of DDE and other OCs than a more sensitive species such as the brown pelican that is situated at a lower trophic level atop a less complicated food chain. Therefore, quantifying differences in species is complex; the above example with falcons and pelicans demonstrates that many factors affect toxicity of OCs to various segments of the biota.

Other examples of species that experienced widespread declines, primarily related to DDE, included the double-crested cormorant (*Phalacrocorax auritus*),[129,130] osprey (*Pandion haliaetus*),[132–133] bald eagles (*Haliaeetus leucocephalus*),[134–135] merlins (*Falco columbarius*),[136] great cormorants (*Phalacrocorax carbo*),[137] Mexican free-tailed bats (*Tadarida brasiliensis*),[138] and other species.

The bald eagle seemed moderately sensitive to DDE with normal reproductive success when sample eggs contained ≤ 3 μg/g and nearly complete failure when eggs contained ≥ 16 μg/g.[135] The critical level of DDE in eggs of osprey that resulted in reproductive impairment was about 10 μg/g.[139] Although the evidence indicated that DDE was primarily related to these declines, some interaction with other contaminants is possible (see the section on dieldrin).

American kestrels (*Falco sparverius*) seemed tolerant of DDE in eggs; success was apparently unaffected when residues ranged from 5 to 100 μg/g.[140] Great blue herons produced fledglings from nests with sample eggs that contained up to 16 μg/g of DDE.[141] Pipped great blue heron eggs with viable embryos contained as much as 78 μg/g.[142] Viable herring gull eggs collected from an island in the Great Lakes in 1964 contained very high mean residues of DDT (19 μg/g), DDD (6 μg/g), and DDE (202 μg/g); however, productivity was reduced to only about 50% of the normal (pre-DDT) levels[143] and returned to normal in 1965, despite persistence of high residues.[144] Domestic chickens experienced few effects on eggshell thickness or reproductive success even at high dietary levels of DDT or DDE.[145–146] Lundholm[66] characterized birds as very sensitive, moderately sensitive, or insensitive to effects of DDE on eggshell thinning that may persist for up to 1 year after dosage ceases.[147–148]

Although ecotoxicological effects of DDT were demonstrated most often in birds, and to a lesser extent in mammals,[149] particularly bats,[30,138] other segments of the biota have been affected. Lake trout (*Salvelinus namaycush*) fry that absorbed the yolk sac and died contained ≥ 2.9 μg/g of DDT;[150] this apparently had some adverse effect on the population.

Usage of DDT in the United States peaked in 1959 when $78 \times 10^6$ pounds were applied, but only one third that amount was used in 1970.[1] The primary reason for decline in use, before the ban of nearly all uses of DDT in the United States in 1972,[151] was the development of resistance in major invertebrate pests. With the ban in the United States and many other countries, adverse effects of DDT on wildlife decreased, and depressed populations of brown pelicans,[112,122] bald eagles,[152,153] osprey,[154,155] merlins (*Falco columbarius*),[136] double-crested cormorants,[156] great cormorants,[137] Mexican free-tailed bats,[138] and other species experienced full or partial recovery in much of North America, Europe, and certain other areas. Recent changes in peregrine falcon populations have ranged from small increases to essentially full recovery.[157]

DDT proved more persistent than first estimated, particularly around sites of heavy contamination such as fruit orchards[158] other agricultural lands[159,160] and manufacturing and formulating plants.[122,161–163] DDT use continues in Brazil, Mexico, South Africa, and other countries,[164] where it is used primarily for mosquito control[3] and in Zimbabwe and possibly other African countries, where it was used for tsetse fly (*Glossina* spp.) control.[165] Residues sufficient to induce eggshell thinning, reproductive effects, and mortality were documented recently in some avian populations including the white-headed black chat (*Thamnolaea arnoti*) in Zimbabwe.[165]

Bald eagles were extirpated on Santa Catalina Island in California about 50 years ago. The current breeding population originated from a transplanting program initiated in the 1980s (first egg laid in 1987) and extending into the 1990s. Eggs of eagles on Santa Catalina contained high residues of DDE (25 to 67 μg/g) in 1990; all nests on the island failed.[166] The bald eagle transplanting program was initiated because brown pelicans in the Channel Islands were recovering from effects of DDT. In retrospect, it seems unreasonable to continue the transplanting program as long as the serious problems from DDE persist in bald eagles. This tale of response of two species to a pollutant in a common area further emphasizes the complexity of ecotoxicological research.

In Nevada in 1996 mean residues of 2.7 μg/g of DDE in eggs of white-faced ibis (*Plegadis chihi*) were associated with decreases of 20% in productivity and 18% in eggshell thickness.[167] Residues in eggs of ibis were similar to those reported 10 years earlier; residues were apparently accumulated on the wintering ground. In 2000 another study of the white-faced ibis along the Colorado River in Arizona also revealed continuing problems from DDE. Average eggshell thinning of 15% were associated with DDE a mean residue of 2.2 μg/g in eggs apparently was related to poor reproductive success.[168]

In Canada from 1993 to 1995 American robins occupying fruit orchards were accumulating very high levels ($\bar{x} = 85$ μg/g) of DDE several decades after the last use of DDT in the orchards.[26] Although these high DDE levels had no demonstrable effect on reproductive success of the robins, predators consuming robins were probably at risk.[169] Another study (1988–1994) of DDE persistence in Canadian apple orchards revealed that eastern bluebirds (*Sialia sialis*) in Ontario were accumulating residues as high as 105 μg/g in pooled egg samples.[170] There was some indication that organochlorine pesticides were associated with a reduction in reproduction of ≤ 14% in 4 of 6 years; however, pooling of eggs for analysis probably confounded the conclusions.

From 1994 to 1995 some eggs of double-crested cormorants in Green Bay, Wisconsin contained DDE residues sufficient to reduce hatching success of some eggs; however, there was no apparent effect on cormorant population trends.[171]

A study of burrowing owls (*Athene cunicularia*) in 1996 in four areas in southern California revealed no effects of contaminants in eggs on reproductive success or the owls, even though DDE residues in one area averaged 7.5 μg/g and eggshell thinning averaged 42%.[172] The extreme thinning of burrowing owl eggshells appears anomalous and was apparently too high. Use of the pre-1947 mean thickness listed by Henny et al.,[173] compared to the data from California, indicated 12% thinning in the most contaminated area and 4% in the least contaminated areas.

The only major problem with DDD occurred in California when it was applied three times from 1949 to 1957 to Clear Lake to control gnats (*Haoporus astictopus*).[174,175] Residues of DDD reached high levels in the biota, especially in western grebes (*Aeschmorphorus occidentalis*), where levels (lipid basis) were as high as 1200 μg/g in body fat and 1007 μg/g in eggs. The Clear Lake study was probably the first to document trophic level accumulation that proved so important in explaining the persistence of residues and residual effects of the OCs.[177] After the first application of DDD, the number of breeding pairs of western grebes decreased from 1000 in 1949 to none the following year. The western grebe breeding population had increased to only 165 pairs in 1967, and their reproductive success was still poor. The only other known environmental problem from DDD was mortality of a common loon (*Gavia immer*).[178]

### 13.3.2 Aldrin/Dieldrin

These two closely related OCs are very toxic and were involved in numerous incidents of wildlife mortality. When aldrin is applied in the field, it is rapidly broken down to dieldrin; aldrin is found in biological samples only near application sites.[179]

In 1960 and 1961 aldrin was applied in April to agricultural land in Illinois to control an infestation of Japanese beetles (*Popillia japonica*). The solid-block application of 2.2 kg/ha resulted in an estimated mortality of 25 to 50% of the adult ring-necked pheasants (*Phasianus colchicus*) within 1 month.[180] Reproduction of pheasants the year of application was severely depressed, as evidenced by the low numbers of chicks and the abnormally high proportion (52 to 56%) of broodless hens, the low ratio of young per adult in autumn, and reduced numbers of cock-calls the following spring. Reproduction on the treated area returned to normal the following year. Dieldrin residues ranged from 2.9 to 15 μg/g in breast muscle of pheasants found dead and from 4.6 to 7.9 μg/g in eggs.[180]

In experimental studies as little as 1 μg/g of aldrin in the diet caused mortality of 20% of adult ring-necked pheasants.[181] Egg production ceased by 10 weeks in hens receiving only 1 or 2 μg/g of aldrin in their diet.[184] Fulvous whistling-ducks (*Dendrocygna bicolor*), snow geese (*Chen caerulescens*), and a number of other birds and invertebrates died after ingesting aldrin-treated rice seed in Texas from the late 1960s until 1974, when this use of aldrin was banned.[183,184] Brains of most waterfowl and other birds found dead contained lethal levels of dieldrin as determined by the lower diagnostic lethal level of 5 μg/g in brains of experimental birds receiving dieldrin-contaminated diets.[52] The fulvous whistling-duck population decrease was apparently related to aldrin use.[183] The whistling-duck population started to increase after the 1974 ban,[185] but the population estimates in the mid-1980s were still less than those of the early 1950s.[186]

In one of the most vivid examples of the toxicity of dieldrin, Scott et al.[187] found virtual elimination or extreme mortality of most species of terrestrial mammals when dieldrin was applied in Illinois to control the Japanese beetle. Dieldrin was also used for a short time in the attempted eradication of the imported fire ant (*Solenopsis saevissima*) in the southeastern United States.[188]

Dieldrin induced adverse effects on reproduction of birds in experimental studies,[189,190] but there are few data from the field. Shags had low productivity when dieldrin residues reached 2 to 3 μg/g in their eggs.[191] Golden eagles (*Aquila chrysaetos*) in Scotland and brown pelicans in the southeastern United States seemed to have lowered productivity when dieldrin residues in sample eggs were ≥1 μg/g.[111,192] However, further investigation seemed to indicate that at least the pelican relationship was spurious because brown pelicans in Louisiana had good productivity when dieldrin residues averaged nearly 1 μg/g in their eggs.[112,193,194] Such interpretive problems arise because many of the OC residues are positively intercorrelated; this may confound identification of effect levels of individual compounds because effects of multiple contaminants may or may not be additive. Dieldrin only induced up to 6% shell thinning in eggs of experimental birds;[195,196] evidence from field studies also indicated that this OC is unimportant in eggshell thinning.[109,132]

In the United States dieldrin killed 28 of 444 bald eagles analyzed from 1964 to 1977; however, it was concluded that DDE was the primary factor in the widespread population declines.[135] In contrast, several investigators have postulated that dieldrin, in concert with DDT, was responsible for population declines of certain birds including the Eurasian sparrowhawk (*Accipiter nisus*), the merlin, and the peregrine falcon in Great Britain,[17,197–199] and the peregrine falcon in the eastern United States.[200] It was postulated that DDE adversely affected eggshell thickness and reproduction and that dieldrin and related cyclodienes (heptachlor and aldrin) induced mortality of adults in these populations. In the Eurasian sparrowhawk studies the role of dieldrin was emphasized because the population apparently crashed more quickly than could be accounted for by reproductive failure alone.[201] Secondary exposure of Eurasian sparrowhawks to dieldrin was mostly through ingestion of birds that had consumed treated seed.

A recent study[202] analyzed population trends of Eurasian sparrowhawks from 1963 to 1986; dieldrin accounted for 29% of sparrowhawks found dead. In conjunction with adverse reproductive problems related to DDE, it was postulated to cause a population decline of 20% per year. The population recovery of Eurasian sparrowhawks and merlins in Great Britain proceeded as residues of dieldrin and DDE declined.[203,204]

Ratcliffe[198] concluded that the dramatic decline of the peregrine falcon population in Great Britain was related to the combined effects of dieldrin, aldrin, and heptachlor (cyclodienes) that were used as seed treatments, soil treatments, sheep dips, and other purposes. DDT was apparently of secondary importance because it was used years before onset of the decline that coincided more closely with initial use of the other three insecticides. Peregrines in the Scottish Highlands during 1962 to 1966 were reproducing well despite a high mean level (1.2 μg/g) of dieldrin in their eggs.[205] This seems to reinforce the conclusion that if population declines were attributable to dieldrin, the critical aspect was mortality of adults. In addition, the population in Great Britain increased shortly after restrictions on the cyclodienes were imposed in 1962 and apparently reached the pre-1940 level by 1985.[198]

Declines of peregrine populations occurred throughout Europe; lowest levels were found in the 1960s and early 1970s.[206] Populations increased dramatically in most countries by 1985, but few populations were near pre-1940 levels. The pre-1940 population of 2000 to 3500 pairs in Fennoscandia decreased to a low of 65 pairs in 1975 and had increased to just 120 to 150 pairs by 1985.[207] Newton[206] concluded that the cyclodienes were primarily responsible for declines of peregrine populations throughout Europe; DDT played a secondary role in spite of adverse effects on eggshell thickness and reproductive success. In Fennoscandia, however, Lindberg et at.[207,208] stressed the role of DDE and polychlorinated biphenyls (OC contaminants of industrial origin) in peregrine declines in that region.

Nisbet[200] concluded that declines in peregrine falcon populations in some regions of North America were too rapid to have been caused by reproductive failure alone. Actual and potential exposure of peregrines to dieldrin was considered sufficiently high and widespread by the mid-1950s to have caused excessive adult mortality. Nevertheless, Nisbet[200] concluded that the role of dieldrin in declines of the North American and British populations was based largely on circumstantial evidence and that other OCs may have been involved. Peakall and Kiff[127] concluded that DDE has been and continues to be the pollutant of primary importance adversely affecting peregrine falcon populations.

The gray bat (*Myotis grisescens*) population occupying a cave in Central Missouri experienced mortality from dieldrin (originating from agricultural applications of aldrin). Lethal levels of dieldrin in brains of gray bats were first noted in 1976, when the population contained 1800 individuals;[209] the population was extirpated by 1979. Residues of dieldrin reached 89 μg/g in milk of lactating bats. Beginning in 1977 residues of heptachlor epoxide (heptachlor was intended as a substitute for aldrin) increased and also reached lethal levels in brains of some gray bats. After use of dieldrin and heptachlor was suspended, residues in the bats and their food decreased. Bats then returned to the colony, and the population increased from 8000 to 38,000 by 1994.[138] Gray bats at several other caves in central Missouri also died with lethal levels of dieldrin and heptachlor epoxide in their brains.

Except for limited use, aldrin and dieldrin were canceled in the United States in 1974;[151] several other countries also banned these pesticides in the 1970s. As a result, problems from these pesticides largely dissipated in these areas. Nevertheless, wildlife mortalities from dieldrin persisted long after the ban in the United States.[210] A dieldrin "hotspot" still persists at the Rocky Mountain Arsenal National Wildlife Refuge (RMANWR) near Denver, Colorado. Several pesticides and other toxic chemicals were manufactured at the site until 1987. A study of great horned owls from 1994 to 1996 revealed dieldrin-related deaths of a few adult owls and a dispersing juvenile. Although no adverse effects were noted during the nesting period, plasma samples from 9- to 10-week-old young that later died during the post-fledging period tended to have the highest residues of dieldrin.[211]

Carcasses and brains of big brown bats captured at the RMANWR in 1997 and 1998 contained elevated residues of dieldrin; however, the bats experienced no obvious adverse effects.[212] High residues of dieldrin in soil apparently had no effect on deer mouse (*Peromyscus maniculatus*) populations at the RMANWR. Paradoxically, populations tended to be higher and greater proportions of females were reproductively active in the highly contaminated sites compared to a reference area.[72] While genetic resistance of deer mice to dieldrin was mentioned as a possible mechanism for their success, absence of residue analysis in deer mouse tissues precludes any factual basis for resistance.

### 13.3.3 Endrin

Unlike most other OCs, endrin has a relatively short half-life in tissues of mammals[213,214] and birds.[215,216] Endrin rivals telodrin as the most acutely toxic OC pesticide.

In the United States the most extreme effect of endrin resulted from contamination of the Mississippi River from a plant at Memphis, Tennessee. The endrin moved downstream and killed millions of fish, especially the Gulf menhaden (*Brevoortia patronus*), near the mouth of the river.[217] Coincidentally, and without evidence of involvement by endrin, the large population of brown pelicans (estimated at 12,000 to 85,000 birds from 1918 to 1933[123]) in the Louisiana estuaries was first noted undergoing a decline in 1958[218] and was extirpated by 1963.[219] The last breeding record was 1961 when a study of behavior was conducted at a small colony where the investigator was unaware of the plight of the population.[220] From 1968 to 1980 a program was initiated to restore the brown pelican in Louisiana by transplanting about 1300 birds from Florida.[221,222] The initial population growth was slow, and the short-term increase in use of endrin after the DDT ban, primarily on sugar cane, caused a die-off of about 35 to 45% of the total population of 400 to 459 pelicans in the mid-1970s.[193,221] Brains of pelicans found dead in the mid-1970s contained endrin residues[193] similar to the lethal diagnostic residues ($\geq 0.8$ µg/g) established in experimental birds.[55] Their eggs contained elevated levels suggestive of possible effects[193] based on studies with experimental eastern screech owls (*Otus asio*).[223] Once endrin use was discontinued in the area, endrin residues dropped dramatically and problems quickly dissipated,[193] and the breeding population of brown pelicans in Louisiana increased to about 700 pairs by 1984.[222]

In field experiments endrin was shown to cause reductions in populations of meadow voles (*Microtus pennsylvanicus*) over the short term and in deer mice (*Peromyscus maniculatus*) over the long term.[224] Endrin use as a rodenticide in Washington State orchards killed many birds.[45,225] Circumstantial evidence indicated that avian mortality may occur within hours after heavy applications in orchards despite a minimal time to death of several days in experimental studies.[45] Coveys of California quail (*Callipepla californica*) and chukars (*Alectoris chukar*) died after exposure to endrin spray. Use of endrin was restricted in 1979.[151] Mortality of wildlife in Washington orchards dramatically declined after severe restrictions on endrin use were implemented in 1982 and other rodenticides, such as anticoagulants, were substituted.[45] In the Washington orchards endrin was applied postharvest in late fall or early winter, so there were no effects on reproductive success. Residues were present in some birds' eggs, but they were not high enough to impair reproduction.[45]

In Montana relatively small quantities of endrin were applied to wheat fields to control insect pests.[226] Residues of endrin were found in birds and mammals in Montana, but few detrimental effects were substantiated. In fruit orchards in New York, use of endrin as a rodenticide resulted in buildup of residues of endrin, but few adverse effects were noted.[227]

Endrin has one of the most interesting metabolic pathways of the OCs. In some mammals the pesticide is broken down into a metabolite, 12-ketoendrin, which is much more toxic than the parent compound. In laboratory rodents dying on endrin-contaminated diets residue analyses of rat brains revealed low levels of endrin but apparently lethal quantities of 12-ketoendrin (means ranged from 0.13 μg/g[228] to 0.28 μg/g),[229] whereas brains of mice contained lethal residues of endrin and little 12-ketoendrin.[228] Experimental birds dying from dietary endrin had lethal levels of endrin in their brains, but only one of these contained 12-ketoendrin.[228,230] There are no records of 12-ketoendrin residues in wild birds, even those dying from endrin poisoning.[45] In the field apparent deaths from 12-ketoendrin have occurred in an opossum (*Didelphis virginiana*) in New York[227] and two deer mice in Montana.[226] A montane vole (*Microtus montanus*) found dead after an endrin application in a Washington orchard contained a lethal level of endrin in its brain, but no 12-ketoendrin was detected.[45]

### 13.3.4 Chlordane

Technical chlordane is a mixture of organochlorines that has been used as an insecticide since the late 1940s.[231] Chlordanes and their metabolites are ubiquitous in the environment; low concentrations are commonly found in environmental samples. Despite widespread use, persistence, and lipophilic properties, there was no firm evidence of either lethal or sublethal effects of chlordane on wild vertebrates when used in terrestrial systems[54,231] until diagnostic lethal residues of heptachlor epoxide and oxychlordane were established in experimental birds given technical chlordane in the diet.[54] Residues of the two critical compounds in brains of experimental birds dying from chlordane ranged from 3.4 to 8.3 μg/g for heptachlor epoxide and 1.1 to 5.0 μg/g for oxychlordane. The lower ends of these ranges are imprecise. The residues of each compound should be weighed according to their individual lethal residues. For example, the lower limit of the lethal range is 8 μg/g of heptachlor epoxide alone or 5 μg/g oxychlordane alone in brains of birds.[54] Therefore, assuming straightforward additivity, examples of lethal combinations are 4 μg/g of heptachlor epoxide and 2.5 μg/g of oxychlordane or 2 μg/g of heptachlor epoxide and 3.8 μg/g of oxychlordane.

The first documented problems in the terrestrial environment occurred when three raptorial birds died with lethal combinations of the two toxic components of technical chlordane. One of the raptors also had a near-lethal level of dieldrin in its brain.[232] Other birds also died from chlordane poisoning.[233] In New York State and New Jersey mortalities related to technical chlordane are recurring,[210,234] even though the only registered use for chlordane from the late 1970s to 1988 was for termite control through subterranean applications.[235] The persistence of chlordane problems in the United States seems unique to the New York-New Jersey area;[210,234] however, routine analysis for residues of organochlorine pesticides in other areas probably have declined. Previously, chlordane was heavily used on lawns, golf courses, and certain crops. Continuing problems are related to these past uses. To support this a recent study in New Jersey in 1998[236] verified that beetles were accumulating high residues of chlordane-related residues in areas formerly treated with chlordane; these residues were sufficient to account for mortality of over 400 birds in 1996 to 1997, including common grackles, (*Quiscalus quiscula*), European starlings (*Sturnus vulgaris*), American robins, and Cooper's hawks (*Accipiter cooperi*). Big brown bats (*Eptesicus fuscus*) and little brown bats (*Myotis lucifugus*) also died from chlordane.[236]

### 13.3.5 Heptachlor

Heptachlor was formerly a widely used insecticide; it was gradually phased, out and most of its uses were withdrawn by 1983.[151] One of its early uses that pointed toward its extreme toxicity to wildlife was an attempt to eradicate the imported fire ant in the southeastern United States. Excessive mortality of passerine birds and nontarget invertebrates on small plots[237] resulted in its withdrawal for fire-ant control — but not until after several years of use.[238]

Treatment of wheat seeds with cyclodiene insecticides, including heptachlor, aldrin, or dieldrin, resulted in mortality of a large number of wood pigeons (*Columba palumbus*) and other birds in England;[239] nearly 6000 wood pigeons died on 600 ha in 1961.[240] Experiments verified that cyclodiene seed treatments were lethal to birds.[241]

Heptachlor seed treatments resulted in the death of about 1300 foxes (*Vulpes vulpes*) in England during the winter of 1959–1960.[242] Livers of dead foxes contained 10 to 90 μg/g heptachlor epoxide and 3 to 13 μg/g dieldrin. The foxes were feeding on wood pigeons that had ingested the treated wheat seed; captive foxes died within 1 to 2 weeks after ingesting three to six contaminated wild wood pigeons.[242] Heptachlor also induced mortality of gray bats in central Missouri and, along with dieldrin, may have caused extirpation of one colony (see section on dieldrin).[138,209]

In the late 1970s in Oregon and Washington, heptachlor seed treatments were associated with mortality, lowered reproductive success, and decline of a local population of Canada geese (*Branta canadensis*).[34–35] American kestrels that fed on prey that had ingested treated wheat seed also experienced lowered reproductive success and some adult mortality.[140] Kestrels seemed more sensitive to heptachlor epoxide than geese; nest success of geese was adversely affected when sample eggs contained > 10 μg/g compared with < 3 μg/g for kestrels.[35,140] Once heptachlor seed treatments were banned in the area and another OC, lindane, was used as a replacement, die-offs of birds ceased. Following the ban, reproductive success improved; and the local Canada goose breeding population doubled in size in less than a decade.[35] Most of the treated grain was made available to geese from spillage during planting or transporting operations and from intentional dumping of excess grain.[35]

### 13.3.6 Hexachlorocyclohexane (HCH)

This pesticide, also known by the misnomer benzene hexachloride (BHC), consists of a mixture of five isomers; the gamma isomer (lindane) is the main insecticidal component.[235]

Several isomers of HCH impart an unpleasant flavor to food, crops, and poultry products consumed by humans;[1] therefore, its use in the United States was voluntarily canceled by the principal manufacturer in 1978. The alpha and beta isomers also had some repellency to wildlife when used as a seed treatment, particularly to ring-necked pheasants.[243] Lindane is odorless and tasteless, but the technical formulation contains other isomers that impart repellency.[244] After ingestion, lindane is rapidly metabolized to water-soluble chlorophenols and chlorobenzenes that are readily excreted. Other characteristics of lindane that reduce its hazard to wildlife are a much lower half-life in tissues and eggs of birds than most OCs[215,245] and rapid degradation after application in the field.[246] Although high levels of lindane in the diet inadvertently given to domestic pigeons (*Columba livia*) resulted in some mortalities[247] and lindane seed treatment was suspected of being hazardous to wild birds,[248] lindane has not been implicated in any problem with populations in the field,[249] except for a population of greater horseshoe bats (*Rhinolophus ferrumequinum*) in Dorset in Britain that occupied 3200 $km^2$ of habitat. This population was thought extirpated by lindane, but there were no residue data available. After roosts in buildings were sprayed in 1953 with a 1% solution of lindane, 15,000 bats died; mortality and reproductive failure were noted for 26 years following the single application (summary by Clark and Shore[138]). Die-offs of bats in Britain also occurred in Cornwall; high residues of dieldrin were noted in timber samples after this material was applied as a chiropteracide.

When lindane was substituted for heptachlor as a seed treatment in the northwestern United States (see heptachlor section), problems dissipated and residues of lindane in tissues or eggs were rarely found in seed eaters[35,250] and were never found in predators of seed eaters.[140,251] For these reasons, lindane is one of the few OCs that is still widely used. Nevertheless, lindane is sufficiently hazardous under confined conditions that its uses for control of insect pests in poultry facilities or human habitation were banned. Also, experimental studies indicate that lindane in the diet of domestic chickens reduced hatchability and egg production, increased embryonic mortality, and induced eggshell thinning of 8 to 18%.[252] Remarkably, Sauter and Steele[252] reported that effects at 0.1 μg/g in the diet were essentially the same as those at 1 or 10 μg/g. Ware and Naber[253] found no effect of 10 μg/g dietary lindane on reproduction of domestic chickens, but only egg production and weight gain of hens were measured. Also, domestic chickens receiving 200 μg/g of lindane in the diet for 1 month laid eggs with normal eggshell thickness.[254]

### 13.3.7 Toxaphene

Toxaphene is a broad-spectrum pesticide that is produced by reacting camphene with chlorine in the presence of ultraviolet radiation and certain catalysts to yield a mixture of 177 chlorinated camphenes with a chlorine content of 67 to 69%.[235] Toxaphene was first used in the United States in 1948; it was the most heavily used insecticide in 1980. Annual production exceeded 45 million kg, with primary use in agricultural crops, particularly cotton. Toxaphene use was canceled in the United States in 1982, but existing stocks could be used after that time.[151]

Residues of toxaphene were as high as 50 μg/g in livers of colonial waterbirds found dead at the Klamath Basin National Wildlife Refuges.[175–255] The deaths could not definitively be attributed to toxaphene because high residues of DDT and its metabolites were present. A more convincing case against toxaphene occurred when Big Bear Lake in southern California was treated with toxaphene to control rough fish populations. Besides the large die-off of fish, over 200 birds were found dead.[175] Tissues of dead birds were not analyzed, but toxaphene residues in fat of ducks and water birds that were shot ranged from 12 to 1700 μg/g.[175] The lake was toxic to fish for over 1 year. Another die-off of birds also occurred in California after toxaphene was used to control grasshoppers on a shortgrass range; carcasses of birds contained from 0.1 to 9.6 μg/g.[256] There are no experimental studies with lethal diagnostic residues of toxaphene to evaluate amounts in birds found dead, but circumstantial evidence indicated that it was the major mortality factor in the last two examples.

Experimental studies to determine effects of dietary levels of toxaphene on reproduction of birds indicated no effects at 50 ppm for black ducks (*Anas rubripes*)[257] or 100 ppm for ring-necked pheasants.[182] Some reproductive problems with the pheasants occurred at a dietary level of 300 ppm.[182] Keith[255] reported that some young American white pelicans (*Pelecanus erythrorhynchos*) died after 4 to 5 weeks on a diet that contained 50 ppm toxaphene, but there was also some mortality of young on clean food. Domestic chickens given dietary levels of 0.5 to 100 ppm toxaphene exhibited no significant effects on productivity; however, sternal or keel deformation was noted in birds fed Ï5 ppm, and renal lesions were detected in those fed Ï50 ppm.[258]

Considering the widespread use of toxaphene, there are few serious known effects on the biota other than problems associated with its use as a piscicide and effects on experimental birds — primarily at high levels.

### 13.3.8 Dicofol

Several million pounds of dicofol are used annually in the United States for control of phytophagous mites, with most being applied to cotton and citrus.[259] Dicofol is produced from DDT as the starting material; therefore, there are a number of DDT-related contaminants in technical dicofol.[260,261] In ring-neck doves (*Streptopelia risoria*) and laboratory mice, there are three fat-soluble metabolites of dicofol including dichlorobenzophenone, dichlorobenzhydrol, and mono-dechlorin-ated dicofol.[261,262] These metabolites are generally less toxic than the parent compound. Some of the experimental birds given diets containing pure or nearly pure dicofol at concentrations ranging from about 5 to 93 mg/kg (dry weight) experienced eggshell thinning and adverse effects on reproductive success including feminization of male embryos[263–266] (MacLellan et al. 1996). Reproduction of mammals and fish was also adversely affected by dicofol in experimental studies.[259]

Experimental studies indicate that dicofol has the potential to induce serious adverse effects when applied in the field because it and its metabolites accumulate in tissues and eggs of birds given dietary dosages.[262,265] However, in the best definition of the value of ecotoxicological studies there were no adverse effects from dicofol documented in field studies; moreover, residues are relatively rare in environmental samples. Residues of dicofol and metabolites were low in fish and eggs of birds, with maximum residues of about 5 μg/g (summarized by Clark[259]). In a later study in California and Texas, very low residues were detected in only one of 24 eggs of birds sampled in areas of high dicofol use. Several species of lizards in Texas contained low residues of dicofol and its metabolites, whereas lizards from California contained no residues.[267] The legal sale and use of dicofol products that contained > 0.1% of DDT and metabolites ceased in early 1989.[268]

### 13.3.9 Telodrin

Telodrin is probably the most acutely toxic OC pesticide,[269] but it was apparently never used in the United States and was used for only a short time in a relatively isolated area in Europe. This insecticide was involved in a decline in numbers of breeding pairs of Sandwich terns (*Sterna sandvicensis*) in The Netherlands from 40,000 in 1954 to 650 in 1965.[270–272] After the telodrin factory at Rotterdam was closed in 1965 and subsequent improvements in the purification plant reduced output of the other pesticides,[272] the number of pairs of terns increased to > 4000 by 1974.[273] Apparently, the Sandwich tern decline was unique to The Netherlands because there were no decided declines in breeding pairs in other western European countries.[272] Adverse effects from telodrin were not quantified; rather, the evidence suggested joint involvement with several OCs, including endrin and dieldrin.[271] Brains of several seabirds found dead contained lethal levels of endrin.[274]

### 13.3.10 Methoxychlor

This insecticide more or less replaced DDT in control of Dutch elm disease in American elms.[275] American robins experienced a low rate of mortality when exposed to methoxychlor in either laboratory or field.[276] Methoxychlor is rapidly broken down in warm-blooded animals, so that residues are rarely found in either wild birds in areas sprayed for Dutch elm disease[276] or experimental birds given diets containing this insecticide.[277] In American robin populations decimated by DDT numbers increased dramatically with use of methoxychlor.[276]

### 13.3.11 Mirex

Mirex was used as a replacement for dieldrin and heptachlor in attempts to control the imported fire ant in the southeastern United States[278] and as a fire retardant in electronic components, fabrics, and plastics.[279] Mirex accumulated in wildlife near the sites of application.[280] Residues also were detected in migratory birds, mammals, and other biota throughout the United States and Canada.[279] Mirex has potential for chronic toxicity because it is only partly metabolized and is eliminated slowly; there also may be a high rate of bioaccumulation of residues in fat, brain, eggs, and other tissues.[20] Although mirex was said to cause severe damage to fish and wildlife in the southeastern United States and the Great Lakes,[279] the evidence supporting this claim is lacking. Adverse effects in experimental birds occurred only at very high dietary levels,[53,281] and the field data only indicated possible effects.[278] All uses of mirex in the United States were banned in 1978.[279]

### 13.3.12 Chlordecone

The major environmental problem from chlordecone resulted from illegal dumping from a manufacturing plant in Virginia.[282] Residues of chlordecone were found throughout the James River and environs; contamination also was found in Chesapeake Bay.[252] Chlordecone was also used to control fire ants in the southeastern United States.[3] Chlordecone induced reversal of external sex characteristics in experimental male ring-necked pheasants.[181] Similar to mirex, chlordecone effects in experimental animals occur only at high dietary levels, but both closely related compounds are relatively persistent. Use of chlordecone in the United States was canceled in 1978.[151]

## 13.4 PROJECTIONS FOR THE FUTURE

Although most of the problem OCs have been banned in a number of countries, exposure, bioaccumulation, and ecotoxicological effects will be found far into the future because of the environmental persistence of many compounds and continued use in a fairly large area. The recent findings in the United States and Canada that DDE, dieldrin, and chlordane are still exerting important adverse effects on birds and other animals 20 to 30 years after use was discontinued reinforces concerns for the future. Worldwide sales of OCs dropped 20% from 1983 to 1986,[164] but the decline since then has probably leveled off faster than anticipated.

Recent developments include a proposed treaty that would instigate a global ban on persistent organic pollutants including DDT and other OCs.[283] However, there is a debate about the retention of DDT for control of mosquitoes that carry the malaria parasite.[9] In the Philippines in 1993 DDT was banned for all uses including mosquito control.[4] Several replacement insecticides, primarily organophosphorus compounds, seemed to control the mosquitoes, but there are plans to resort to one insecticide — the synthetic pyrethroid ethofenprox.

The recent settlement of the lawsuit regarding DDT contamination in the California Bight for $73 million[284] provided the legal foundation for assessing past and continuing problems to the environment; the ecotoxicological foundation was established over 50 years ago.

## 13.5 CONCLUSIONS AND PERSPECTIVES

Organochlorine pesticides, although frequently regarded as uniformly highly toxic and persistent to the organism and the ecosystem as a whole, are a diverse group whose toxicity and persistence are highly variable. Toxicity of OCs is influenced by species, sex, age, stress of various kinds, formulations used, and numerous other factors. Sensitivity by species is highly variable among all taxa. Most OCs, but particularly DDT, telodrin, aldrin, endrin, dieldrin, and heptachlor, are too toxic for use in the environment; this is particularly true for the most sensitive species that are constantly at risk from these compounds. A few OCs, such as lindane and dicofol, that are still in widespread use present a low risk to the environment.

The eggshell-thinning phenomenon, depressed productivity, and mortality of birds in the field led to experimental studies with DDE and DDT and other OCs that are pertinent to actual environmental problems. In turn, experimental studies under controlled conditions resolve complex problems, such as the metabolism of endrin to 12-ketoendrin, that are impractical to study in the field. Results from controlled experiments are essential in interpreting field data that are correlative in nature. Such studies point the way to more productive work in the field; however, it is evident that experimental studies and more recent stringent registration and reregistration testing requirements have limitations in duplicating environmental conditions. In dealing with the OCs as well as other contaminants the ecotoxicologist must use a mix of data obtained from controlled experiments as well as those obtained from the field. Field experimentation, where the investigator controls certain variables and where check or reference areas are used as a comparison with the contaminated area,[35] present the best experimental design for obtaining accurate and pertinent results. It seems obvious that too much emphasis has been placed on obtaining large numbers of samples for residue analysis rather than using the best design to tie OC residues to important effects. As such, much of the residue data base is of doubtful scientific value, insofar as it can be used in defining ecotoxicological effects.[2] Controversies over whether DDE or dieldrin were more important in causing declines of peregrine falcons and other raptors in Great Britain or whether endrin was responsible for declines of brown pelicans in Louisiana probably could have been resolved through use of the sample egg technique[117] and other innovations.[35]

## REFERENCES


1. Hayes, W. J., Jr., Introduction, in *Handbook of Pesticide Toxicology*, Vol. 1, Hayes, W. J., Jr. and Laws, E. R., Jr., Eds., Academic Press, San Diego, 1991, Chap.1.
2. Moriarty, F., *Ecotoxicology*, Academic Press, London, 1990, 289.
3. Peterle, T. J., *Wildlife Toxicology*, Van Nostrand Reinhold, New York, 1991, 322.
4. World Wildlife Fund, Resolving the DDT dilemma: Protecting diversity and human health, WWF Canada (Toronto) and WWF United States (Washington, D.C.), 1998, 52.
5. Smith, A. G., Chlorinated hydrocarbon insecticides, in *Handbook of Pesticide Toxicology*, Vol. 3, Hayes, W. J., Jr., and Laws, E. R., Jr., Eds., Academic Press, San Diego, 1991, Chap. 15.
6. Walker, C. H., Variations in the intake and elimination of pollutants, in *Organochlorine Insecticides: Persistent Organic Pollutants*, Moriarty, F., Ed., Academic Press, London, 1975, 73.
7. Sodergren, A. and Ulfstrand, S., DDT and PCB relocate when caged robins use fat reserves, *Ambio* 1, 36, 1972.
8. Van Velzen, A. C., Stiles, W. B., and Stickel, L. F., Lethal mobilization of DDT by cowbirds, *J. Wildl. Manage.*, 36, 733, 1972.
9. Curtis, C. F., and Lines, J. D., Should DDT be banned by international treaty?, *Parasitol. Today*, 16, 119, 2000.
10. Forsyth, D. J., Peterle, T. J., and Bandy, L. W., Persistence and transfer of $^{36}$Cl DDT in the soil and biota of an old-field ecosystem: A six-year balance study, *Ecology*, 64, 1620, 1983.
11. Beyer, W. N. and Gish, C. D., Persistence in earthworms and potential hazards to birds of soil applied DDT, dieldrin and heptachlor, *J. Appl. Ecol.*, 17, 295, 1980.
12. Brown, N. J. and Brown, A. W. A., Biological fate of DDE in a sub-arctic environment, *J. Wildl. Manage.*, 34, 929, 1970.
13. Lichtenstein, E. P., Anderson, J. P., Fuhremann, T. W., and Schulz, K. R., Aldrin and dieldrin: Loss under sterile conditions, *Science*, 159, 1110, 1968.
14. Wedemeyer, G., Role of intestinal microflora in the degradation of DDT by rainbow trout, *Life Sci.*, 7, 219, 1968.
15. Jefferies, D. J., Organochlorine insecticide residues in British bats and their significance, *J. Zool.*, 166, 245, 1972.
16. Anderson, D. W. and Hickey, J. J., Dynamics of storage of organochlorine pollutants in herring gulls, *Environ. Pollut.*, 10, 183, 1976.
17. Newton, I. and Bogan, J., The role of different organochlorine compounds in the breeding of British sparrowhawks, *J. Appl. Ecol.*, 15, 105, 1978.
18. Hodgson, E., Silver, I. S., Butler, L. E., Lawton, M. P., and Levi, P. E., Metabolism, in *Handbook of Pesticide Toxicology*, Vol. 1, Hayes, W. J., Jr. and Laws, E. R., Jr., Eds., Academic Press, San Diego, 1991, Chap.3.
19. Bevenue, A., The "bioconcentration" aspects of DDT in the environment, *Residue Rev.*, 61, 37, 1976.
20. Norstrom, R. J., Hallett, D. J., and Sonstegard, R. A., Coho salmon (*Oncorhynchus kisutch*) and herring gulls (*Larus argentatus*) as indicators of organochlorine contamination in Lake Ontario, *J. Fish. Res. Board Can.*, 35, 1401, 1978.
21. Hamelink, J. L., Waybrant, R. C., and Ball, R. C., A proposal: Exchange equilibria control the degree chlorinated hydrocarbons are biologically magnified in lentic environments, *Trans. Am. Fish. Soc.*, 100, 207, 1971.
22. Butler, P. A., Pesticides in the marine environment, *J. Appl. Ecol.*, 3 (Suppl.), 253, 1966.
23. Matsumura, F., *Toxicology of Insecticides*, 2nd ed., Plenum Press, New York, 1985, 598.
24. Blus, L. J., Neely, B. S., Jr., Lamont, T. G., and Mulhern, B. M., Residues of organochlorines and heavy metals in tissues and eggs of brown pelicans, 1969–1973, *Pestic. Monit. J.*, 11, 40, 1977.
25. Barker, R. J., Notes on some ecological effects of DDT sprayed on elms, *J. Wildl. Manage.*, 22, 269, 1958.

26. Harris, M. L., Wilson, L. K., Elliott, J. E., Bishop, C. A., Tomlin, A. D., and Henning, K. V., Transfer of DDT and metabolites from fruit orchard soils to American robins (*Turdus migratorius*) twenty years after agricultural use of DDT in Canada, *Arch. Environ. Contam. Toxicol.*, 39, 205, 2000.
27. Skaare, J. U., Ingebrigtsen, K., Aulie, A., and Kanui, T. I., Organochlorines in crocodile eggs from Kenya, *Bull. Environ. Contam. Toxicol.*, 4, 126, 1991.
28. Ruus, A., Ugland, K. I., Espeland, O., and Skaare, J. U., Organochlorine contaminants in a local marine food chain from Jarfjord, northern Norway, *Mar. Environ. Res.*, 48,131,1999.
29. Brown, A. W. A., *Ecology of Pesticides*, John Wiley and Sons, New York, 1978, 525.
30. Clark, D. R., Jr., Bats and environmental contaminants: A review, *Spec. Sci. Rep. Wildl.*, 235. Washington, D.C.: U.S. Dept. of Interior, 1981.
31. McKay, W. D., Notes on purple gallinules in Columbia ricefields, *Wilson Bull.*, 93, 267, 1981.
32. Mendelssohn, H., The impact of pesticides on bird life in Israel, *Bull. Int. Counc. Bird Preserv.*, 11, 75, 1972.
33. Spillett, J., Pesticide poisoning of tigers, *Oryx*, 9, 1983, 1967.
34. Blus, L. J., Henny, C. J., Lenhart, D. J., and Grove R. A., Effects of heptachlor-treated cereal grains on Canada geese in the Columbia Basin, in *Management and Biology of Pacific Flyway Geese: A Symposium*, Jarvis, R. L. and Bartonek, J. C., Eds., Oregon State University Bookstores, Corvallis, 1979, 105.
35. Blus, L. J., Henny, C. J., Lenhart, D. J., and Kaiser, T. E., Effects of heptachlor- and lindane-treated seed on Canada geese, *J. Wildl. Manage.*, 48, 1097, 1984.
36. Guillette, L. J., Jr., Brock, J. W., Rooney, A. A., and Woodward, A. R., Serum concentrations of various environmental contaminants and therir relationship to sex steroid concentrations and phallus size in American alligators, *Arch. Environ. Contam. Toxicol.*, 36, 447, 1999.
37. Peters, R. A., The biochemical lesion and its historical development, *Br. Med. J.*, 25, 223, 1969.
38. Ariëns, E. J., Simonis, A. M., and Turner, R. G., *Introduction to General Toxicology*, Academic Press, New York, 1976.
39. Narahashi, T., Effects of insecticides on excitable tissues, *Adv. Insect Physiol.*, 8, 1, 1971.
40. Beeman, R. W., Recent advances in mode of action of insecticides, *Ann. Rev. Entomol.*, 27, 253, 1982.
41. Hudson, R. R., Tucker, R. K., and Haegele, M. A., Handbook of Toxicity of Pesticides to Wildlife, U.S. Fish and Wildlife Service Resour. Publ., 153., Washington, D.C., 1984.
42. Pimentel, D., Ecological Effects of Pesticides on Non-Target Species, Office of Sci. Technol., Supt. Doc., Washington, D.C., 220.
43. Hill, E. F., Heath, R. G., Spann, J. W., and Williams, J. D., Lethal dietary toxicities of environmental pollutants to birds, *Spec. Sci. Rep. Wildl.*, 191, Washington, D.C.: U.S. Dept. of Interior, 1975.
44. Mayer, F. L. and Ellersleck, M. R., Manual of Acute Toxicity: Interpretation and Data Base for 410 Chemicals and 66 Species of Freshwater Animals, U.S. Fish and Wildlife Service Resource Publ., 160. Washington, D.C., 1986.
45. Blus, L. J. Henny, C. J., and Grove, R. A., Rise and fall of endrin usage in Washington state fruit orchards: Effects on wildlife, *Environ. Pollut.*, 60, 331, 1989.
46. Dale, W. E., Gaines, T. B., and Hayes, W. J., Jr., Poisoning by DDT: Relation between clinical signs and concentration in rat brain, *Science*, 142, 1474, 1963.
47. Bernard, R. F., *Studies of the Effects of DDT on Birds*, Michigan State Univ. Mus. Pub. Biol. Ser. 2, 155, 1963.
48. Stickel, L. F., Stickel, W. H., and Christensen, R., Residues of DDT in brains and bodies of birds that died on dosage and in survivors, *Science*, 151, 1549, 1966.
49. Stickel, L. F. and Stickel, W. H., Distribution of DDT residues in tissues of birds in relation to mortality, body condition and time, *Ind. Med. Surg.*, 38, 44, 1969.
50. Stickel, W. H., Stickel, L. F., and Coon, F. B., DDE and DDD residues correlated with mortality of experimental birds, in *Pesticides Symposia, 7th Int. Am. Conf. Toxicol. Occupational Med.*, Deichmann, W. B., Ed., Helios and Assoc., Miami, Florida, 1970, 287.
51. Stickel, W. H., Stickel, L. F., Dyrland, R. A., and Hughes, D. L., DDE in birds: Lethal residues and loss rates, *Arch. Environ. Contam. Toxicol.*, 13, 1, 1984.
52. Stickel, W. H., Stickel, L. F., and Spann, J. S., Tissue residues of dieldrin in relation to mortality in birds and mammals, in *Chemical Fallout: Current Research on Persistent Pesticides*, Miller M. W., and Berg, G. G., Eds., Charles C Thomas, Springfield, IL, 1969, 174.

53. Stickel, W. H., Galyen, J. A., Dyrland, R. A., and Hughes, D. L., Toxicity and persistence of mirex in birds, in *Pesticides and the Environment: A Continuing Controversy*, Deichmann, W. B., Ed., Symposia Specialists, North Miami, FL, 1973, 437.
54. Stickel, L. F., Stickel, W. H., McArthur, R. D., and Hughes, D. L., Chlordane in birds: A study of lethal residues and loss rates, in *Toxicology and Occupational Medicine*, Deichmann, W. B., Ed., Elsevier, North Holland, New York, 1979, 387.
55. Stickel, W. H., Reichel, W. L., and Hughes, D. L., Endrin in birds: Lethal residues and secondary poisoning, in *Toxicology and Occupational Medicine*, Deichmann, W. B., Ed., Elsevier, North Holland, New York, 1979, 397.
56. Stickel, L. F., Stickel, W. H., Dyrland, R. A., and Hughes, D. L., Oxychlordane, HCS-3260, and nonachlor in birds: Lethal residues and loss rates, *J. Toxicol. Environ. Health*, 12, 611, 1983.
57. Robinson, J., Brown, V. K. H., Richardson, A., and Roberts, M., Residues of dieldrin (HEOD) in the tissues of experimentally poisoned birds, *Life Sci.*, 6, 1207, 1967.
58. Koeman, J. H., The occurrence and toxicological implications of some chlorinated hydrocarbons in the Dutch coastal area in the period from 1965 to 1970, Doctoral dissertation, Univ. Utrecht, The Netherlands, 1971, 139.
59. Hayes, W. J., Jr., Dosage and other factors influencing toxicity, in *Handbook of Pesticide Toxicology*, Vol. 1, Hayes, W. J., Jr., and Laws, E. R., Jr., Eds., Academic Press, San Diego, 1991, Chap. 2.
60. Stickel, W. H., Delayed mortality of DDT-dosed cowbirds in relation to disturbance, in The Effects of Pesticides on Fish and Wildlife, Circ. 226, Washington, D.C., U.S. Dept. of Interior, 1965, 17.
61. De Freitas, A. S. W., Hart, J. S., and Morley, H. V., Chronic cold exposure and DDT toxicity, in *Chemical Fallout: Current Research on Persistent Pesticides*, Miller, M. W. and Berg, G. G., Eds., Charles C Thomas, Springfield, IL, 1969, 361.
62. Gish, C. D. and Chura, N. J., Toxicity of DDT to Japanese quail as influenced by body weight, breeding condition, and sex, *Toxicol. Appl. Pharmacol.*, 17, 740, 1970.
63. Stickel, L. F. and Rhodes, L. I., The thin eggshell problem, in *Biological Impact of Pesticides in the Environment*, Gillett, F. W., Ed., Oregon State Univ. Environ. Health Ser. No. 1, 1970, 31.
64. Van Velzen, A. C., Stiles, W. B., and Stickel, L. F., Lethal mobilization of DDT by cowbirds, *J. Wildl. Manage.*, 36, 733, 1972.
65. Keith, J. O., Synergistic effects of DDE and food stress on reproduction in brown pelicans and ring doves, Ph.D. Disser., Ohio State Univ., Columbus, 1980.
66. Lundholm, E., Thinning of eggshells in birds by DDE: Mode of action on the eggshell gland, *Comp. Biochem. Physiol.*, 88C, 1, 1987.
67. Boyd, C. E., Vinson, S. B., and Ferguson, D. E., Possible DDT resistance in two species of frogs, *Copeia*, 1963, 426, 1963.
68. Ferguson, D. E., The ecological consequences of pesticide resistance in fishes, *Trans. N. Am. Nat. Resour. Conf.*, 32, 103, 1967.
69. Dover, M. J. and Croft, B. A., Pesticide resistance and public policy, *Bioscience*, 36, 78, 1986.
70. Webb, R. E. and Horsfall, F., Jr., Endrin resistance in the pine mouse, *Science,* 156, 1762, 1967.
71. McBride, J. E., Kirkpatrick, R. L., Ney, J. J., and Tipson, A. R., Variation of genetic indices in the pine vole before and after endrin treatment, *Am. Midl. Nat.*, 111, 198, 1984.
72. Allen, D. L. and Otis, D. L., Relationship between deer mouse population parameters and dieldrin contamination in the Rocky Mountain Arsenal National Wildlife Refuge, *Can. J. Zool.*, 76, 243, 1998.
73. O'Brien, R. D., *Insecticides: Action and Metabolism*, Academic Press, New York, 1967, 332.
74. Hart, L. G. and Fouts, J. R., Effects of acute and chronic DDT administration on hepatic microsomal drug metabolism in the rat, *Proc. Soc. Exp. Biol. Med.*, 114, 388, 1963.
75. Hart, L. G., Shultice, R. W., and Fouts, J. R., Stimulatory effects of chlordane on hepatic microsomal drug metabolism in the rat, *Toxicol. Appl. Pharmacol.*, 5, 371, 1963.
76. Fisher, A. L., Keasling, H. H., and Schueler, F. W., Estrogenic action of DDT and some analogues, *Fed. Proc.*, 11, 345, 1952.
77. Conney, A. H., Pharmacological implications of microsomal enzyme induction, *Pharmacol. Rev.*, 19, 317, 1967.
78. Peakall, D. B., Pesticide-induced enzyme breakdown of steroids in birds, *Nature*, 216, 505, 1967.
79. Jefferies, D. J. and French, M. C., Changes induced in the pigeon thyroid by *p,p'*-DDE and dieldrin, *J. Wildl. Manage.*, 36, 24, 1972.

80. Lehman, J. W., Peterle, T. J., and Mills, C. M., Effects of DDT on bobwhite quail (*Colinus virginianus*) adrenal gland, *Bull. Environ. Contam. Toxicol.*, 11, 407, 1974.
81. Mahoney, J. J., Jr., DDT and DDE effects on migratory condition in white-throated sparrows, *J. Wildl. Manage.*, 39, 520, 1975.
82. Heinz, G. H., Hill, W. F., and Contrera, J. F., Dopamine and norepinephrine depletion in ring doves fed DDE, dieldrin, and arochlor 1254, *Toxicol. Appl. Pharmacol.*, 53, 75, 1980.
83. Friend, M. and Trainer, D. O., Experimental DDT-duck hepatitis virus interaction studies, *J. Wildl. Manage.*, 38, 887, 1974.
84. Blus, L. J., Belisle, A. A., and Prouty, R. M., Relations of the brown pelican to certain environmental pollutants, *Pest. Monit. J.*, 7, 181, 1974.
85. Ludke, J. L., DDE increases the toxicity of parathion to coturnix quail, *Pest. Biochem. Physiol.*, 7, 28, 1977.
86. Ludke, J. L., Organochlorine pesticide residues associated with mortality: Additivity of chlordane and endrin, *Bull. Environ. Contam. Toxicol.*, 16, 253, 1976.
87. Deichmann, W. B., MacDonald, W. E., and Cubit, D. A., DDT tissue retention: Sudden rise induced by the addition of aldrin to a fixed DDT intake, *Science*, 172, 275, 1971.
88. Keplinger, M. L. and Deichmann, W. B., Acute toxicity of combinations of pesticides, *Toxicol. Appl. Pharmacol.*, 10, 586, 1967.
89. Smyth, H. F., Jr., Weil, C. S., West, J. S., and Carpenter, C. P., An exploration of joint toxic action; twenty-seven industrial chemicals intubated in rats in all possible pairs, *Toxicol. Appl. Pharmacol.*, 14, 340, 1969.
90. Carson, R., *Silent Spring*, Houghton Mifflin, Boston, 1962, 368.
91. Cottam, C. and Higgins, E., DDT: Its Effects on Fish and Wildlife, U.S. Fish and Wildlife Service Circ. 11, Washington, D.C., 1946.
92. Barnett, D. C., The effect of some insecticide sprays on wildlife, *Proc. Ann. Conf. West. Assoc. State Game and Fish Comm.*, 30, 125, 1950.
93. Mohr, R. W., Telford, H. S., Peterson, E. H., and Walker, K. C., Toxicity of orchard insecticides to game birds in eastern Washington, *Wash. Agric. Exp. Sta. Circ.*, 170, 1951, 22.
94. Hickey, J. J. and Hunt, I. B., Initial song bird mortality following Dutch elm disease control program, *J. Wildl. Manage.*, 24, 259, 1960.
95. Wurster, C. F., Wurster, D. H., and Stickland, W. N., Bird mortality after spraying for Dutch elm disease with DDT, *Science*, 148, 90, 1965.
96. Ratcliffe, D. A., Decrease in eggshell weight in certain birds of prey, *Nature*, 215, 208, 1967.
97. Heath, R. G., Spann, J. W., and Kreitzer, J. F., Marked DDE impairment of mallard reproduction in controlled studies, *Nature*, 224, 47, 1969.
98. Wiemeyer, S. N. and Porter, R. D., DDE thins eggshells of captive American kestrels, *Nature*, 227, 737, 1970.
99. Davison, K. L. and Sell, J. L., DDT thins shells of eggs from mallard ducks maintained on ad libitum or controlled feeding regimens, *Arch. Environ. Contam. Toxicol.*, 2, 222, 1974.
100. Stickel, L. F., Pesticide residues in birds and mammals, in *Environmental Pollution by Pesticides*, Edwards, C. A., Ed., Plenum, London, 1973, 254.
101. Cooke, A. S., Shell thinning in avian eggs by environmental pollutants, *Environ. Pollut.*, 4, 85, 1973.
102. Peakall, D. B., Lincer, J. L., Risebrough, R. W., Pritchard, J. B., and Kinter, W. B., DDE-induced eggshell thinning: Structural and physiological effects in three species, *Comp. Gen. Pharmacol.*, 4, 305, 1973.
103. Lundholm, C. E., DDE-induced eggshell thinning in birds: Effects of *p,p*'-DDE on the calcium and prostaglandin metabolism of the eggshell gland, *Comp. Bio. Physiol.*, 118, 113, 1997.
104. Hickey, J. J. and Anderson, D. W., Chlorinated hydrocarbons and eggshell changes in raptorial and fish-eating birds, *Science*, 162, 271, 1968.
105. Risebrough, R. W., Pesticides and bird populations, *Curr. Ornithol.*, 3, 397, 1986.
106. Anderson, D. W. and Hickey, J. J., Eggshell changes in certain North American birds, *Proc. Int. Ornithol. Congr.*, 15, 514, 1972.
107. Ratcliffe, D. A., Changes attributable to pesticides in egg breakage frequency and eggshell thickness in some British birds, *J. Appl. Ecol.*, 7, 67, 1970.

109. Blus, L. J., Heath, R. G., Gish, C. D., Belisle, A. A., and Prouty, R. M., Eggshell thinning in the brown pelican: Implication of DDE, *Bioscience*, 21, 1213, 1971.
109. Blus, L. J., Gish, C. D., Belisle, A. A., and Prouty, R. M., Logarithmic relationship of DDE residues to eggshell thinning, *Nature*, 235, 376, 1972.
110. Blus, L. J., Gish, C. D., Belisle, A. A., and Prouty, R. M., Further analysis of the logarithmic relationship of DDE residues to nest success, *Nature*, 240, 164, 1972.
111. Blus, L. J., Neeley, B. S., Belisle, A. A., and Prouty, R. M., Organochlorine residues in brown pelican eggs: Relation to reproductive success, *Environ. Pollut.*, 7, 81, 1974.
112. Blus, L. J., Further interpretation of the relation of organochlorine residues in brown pelican eggs to reproductive success, *Environ. Pollut.*, A28, 15, 1982.
113. Belisle, A. A., Reichel, W. L., Locke, L. N., Lamont, T. G., Mulhern, B. M., Prouty, R. M., DeWolf, R. B., and Cromartie, E., Residues of organochlorine pesticides, polychlorinated biphenyls, and mercury and autopsy data for bald eagles, 1969 and 1970, *Pest. Monit. J.*, 6, 133, 1972.
114. Call, D. J., Shave, H. J., Binger, H. C., Bergeland, M. E., Ammann, B. D., and Worman, J. J., DDE poisoning in wild great blue heron, *Bull. Environ. Contam. Toxicol.*, 16, 310, 1976.
115. Prestt, I., The heron *Andea cinerea* and pollution, *Ibis*, 112, 147, 1970.
116. Henny, C. J., An Analysis of the Population Dynamics of Selected Avian Species with Special Reference to Changes during the Modern Pesticide Era, U.S. Fish and Wildlife Service Wildlife Research Report, 1. Washington, D.C., 1972.
117. Blus, L. J., DDE in birds' eggs: Comparison of two methods for estimating critical levels, *Wilson Bull.*, 96, 268, 1984.
118. Keith, J. O., Woods, L. A., Jr., and Hunt, E.G., Reproductive failure in brown pelicans on the Pacific Coast, *Trans. North Am. Wildl. Nat. Resour. Conf.*, 35, 56, 1970.
119. Risebrough, R. W., Davis, J., and Anderson, D. W., Effects of various chlorinated hydrocarbons, in *Biological Impact of Pesticides in the Environment*, Gillett, F. W., Ed., Oregon State Univ. Environ. Health Ser. No. 1, 1970, 40.
120. Risebrough, R. W., Sibley, F. C., and Kirven, M. N., Reproductive failure of the brown pelican on Anacapa Island in 1969, *Am. Birds*, 25, 8, 1971.
121. Jehl, J. R., Jr., Studies of a declining population of brown pelicans in northwestern Baja California, *Condor*, 75, 69, 1973.
122. Anderson, D. W. and Gress, F., Status of a northern population of California brown pelicans, *Condor*, 85, 79, 1983.
123. King, K. A., Flickinger, E. L., and Hildebrand, H. H., The decline of brown pelicans on the Louisiana and Texas Gulf Coast, *Southwest. Nat.*, 21, 417, 1977.
124. Hickey, J. J., Ed., *Peregrine Falcon Populations: Their Biology and Decline*, University of Wisconsin Press, Madison, 1969, 596.
125. Cade, T. J., Lincer, J. L., White, C. M., Roseneau, D. G., and Schwartz, L. G., DDE residues and eggshell changes in Alaska falcons and hawks, *Science*, 172, 955, 1971.
126. Peakall, D. B., Cade, T. J., White, C. M., and Haugh, J. R., Organochlorine residues in Alaskan peregrines, *Pest. Monit. J.*, 8, 255, 1975.
127. Peakall, D. B. and Kiff, L. F., DDE contamination in peregrines and American kestrels and its effect on reproduction, in *Peregrine Falcon Populations: Their Management and Recovery*, Cade, T. J., Enderson, J. H., Thelander, C. G., and White, C. M., Eds., The Peregrine Fund, Boise, 1988, 337.
128. Ambrose, R. E., Henny, C. J., Hunter, R. E., and Crawford, J. A., Organochlorines in Alaskan peregrine falcon eggs and their current impact on productivity, in *Peregrine Falcon Populations: Their Management and Recovery*, Cade, T. J., Enderson, J. H., Thelander, C. G., and White, C. M., Eds., The Peregrine Fund, Boise, 1988, 385.
129. Anderson, D. W., Hickey, J. J., Risebrough, R. W., Hughes, D. F., and Christensen, R. E., Significance of chlorinated hydrocarbon residues to breeding pelicans and cormorants, *Can. Field-Nat.*, 83, 91, 1969.
130. Gress, F., Risebrough, R. W., Anderson, D. W., Kiff, L. F., and Jehl, J. R., Jr., Reproductive failures of double-crested cormorants in southern California and Baja California, *Wilson Bull.*, 85, 197, 1973.
131. Ames, P. L., DDT residues in the eggs of the osprey in the north-eastern United States and their relation to nesting success, *J. Appl. Ecol.*, 3 (Suppl.), 87, 1966.

132. Henny, C. J., Byrd, M. A., Jacobs, J. A., McLain, P. D., Todd, M. R., and Halla, B. F., Mid-Atlantic coast osprey population: Present numbers, productivity, pollutant contamination, and status, *J. Wildl. Manage.*, 41, 254, 1977.
133. Wiemeyer, S. N., Bunck, C. M., and Krynitsky, A. J., Organochlorine pesticides, polychlorinated biphenyls, and mercury in osprey eggs — 1970–1979 — and their relationships to shell thinning and productivity, *Arch. Environ. Contam. Toxicol.*, 17, 767, 1988.
134. Sprunt, A., IV, Robertson, W. B., Jr., Postupalsky, S., Hensel, R. J., Knoder, C. E., and Ligas, F. J., Comparative productivity of six bald eagle populations, *Trans. N. Am. Wildl. Nat. Resour. Conf.*, 38, 96, 1973.
135. Wiemeyer, S. N., Lamont, T. G., Bunck, C. M., Sindelar, C. R., Grandich, F. J., Fraser, J. D., and Byrd, M. A., Organochlorine pesticide, polychlorobiphenyl, and mercury residues in bald eagle eggs, 1969–79, and their relationships to shell thinning and reproduction, *Arch. Environ. Contam. Toxicol.*, 13, 529, 1984.
136. Nygard, T., Long term trends in pollutant levels and shell thickness in merlin in Norway, in relation to its migration pattern and numbers, *Ecotoxicology*, 8, 23, 1999.
137. Mason, C. F., Ekins, C., and Ratford, J. R., PCB congeners, DDE, dieldrin and mercury in eggs from an expanding colony of cormorants (*Phalacrocorax carbo*), *Chemosphere*, 34, 1845, 1997.
138. Clark, D. R., Jr. and Shore, R. F., *Chiroptera*, in *Ecotoxicology of Wild Mammals*, Shore, R.F., and Rattner, B.A., Eds, John Wiley and Sons, Chichester, U.K., 2001, 159.
139. Johnson, D. R., Melquist, W. E., and Schroeder, G. J., DDT and PCB levels in Lake Coeur d'Alene, Idaho osprey eggs, *Bull. Environ. Contam. Toxicol.*, 13, 401, 1975.
140. Henny, C. J., Blus, L. J., and Stafford, C. J., Effects of heptachlor on American kestrels in the Columbia Basin, Oregon, *J. Wildl. Manage.*, 47, 1080, 1983.
141. Blus, L. J., Henny, C. J., and Kaiser, T. E., Pollution ecology of breeding great blue herons in the Columbia Basin, Oregon and Washington, *Murrelet*, 61, 63, 1980.
142. Vermeer, K. and Reynolds, L. M., Organochlorine residues in aquatic birds in the Canadian Prairie Provinces, *Can. Field-Nat.*, 84, 117, 1970.
143. Keith, J. A., Reproduction in a population of herring gulls (*Larus argentatus*) contaminated by DDT, *J. Appl. Ecol.*, 3 (Suppl.), 57, 1966.
144. Buckley, J. L., Hickey J. J, Prestt, I., Stickel, L. F., and Stickel, W. H., Pesticides as possible factors affecting raptor populations, in *Peregrine Falcon Populations — Their Biology and Decline*, Hickey, J. J., Ed., Univ. Wisconsin Press, Madison, 1969, 461.
145. Cecil, H. C., Fries, G. F., Bitman, J., Harris, S. J., Lillie, R. J., and Denton, C. A., Dietary *p,p*'-DDT, *o,p*'-DDT or *p,p*'-DDE and changes in egg shell characteristics and pesticide accumulation in egg contents and body fat of caged white leghorns, *Poult. Sci.*, 51, 130, 1972.
146. Davison, K. L. and Sell, J. L., Dieldrin and *p,p*'-DDT effects on egg production and eggshell thickness of chickens, *Bull. Environ. Contam. Toxicol.*, 7, 9, 1972.
147. Haegele, M. A. and Hudson, R. H., Eggshell thinning and residues in mallards one year after DDE exposure, *Arch. Environ. Contam. Toxicol.*, 2, 356, 1974.
148. Longcore, J. F. and Stendell, R. C., Shell thinning and reproductive impairment in black ducks after cessation of DDE dosage, *Arch. Environ. Contam. Toxicol.*, 6, 293, 1977.
149. Shore, R. F. and Rattner, B. A., Eds., *Ecotoxicology of Wild Mammals*, John Wiley and Sons, Chichester, U.K., 2001, 752.
150. Burdick, G. E., Harris, E. J., Dean, H. J., Walker, T. M., Skea, J., and Colby D., The accumulation of DDT in lake trout and the effect on reproduction, *Trans. Am. Fisher. Soc.*, 93, 127, 1964.
151. Fleming, W. J., Clark, D. R., and Henny, C. J., Organochlorine pesticides and PCB's: A continuing problem in the 1980's, *Trans. N. Am. Wildl. Nat. Resour. Conf.*, 48, 186, 1983.
152. Grier, J. W., Ban of DDT and subsequent recovery of reproduction in bald eagles, *Science*, 218, 1232, 1982.
153. Bowerman, W. W., Giesy, J. P., Best, D. A., and Kramer, V. J., A review of factors affecting productivity of bald eagles in the Great Lakes region: Implications for recovery, *Environ. Health Perspect.*, 103, 51, 1995.
154. Spitzer, P. R., Risebrough, R. W., Walker, W., II, Hernandez, R., Poole, A., Puleston, D., and Nisbet, I. C. T., Productivity of ospreys in Connecticut-Long Island increases as DDE residues decline, *Science*, 202, 333, 1978.

155. Elliott, J. E., Wilson, L. K., Henny, C. J., Trudeau, S. F., Leighton, F. A., Kennedy, S. W., and Cheng, K. M., Assessment of biological effects of chlorinated hydrocarbons in osprey chicks, *Environ. Toxicol. Chem.*, 20, 866, 2001.
156. Fox, G. A., Collins, B., Hayakawa, E., Weseloh, D. V., Ludwig, J. P., Kubiak, T. J., and Erdman, T. C., Reproductive outcomes in colonial fish-eating birds: A biomarker for developmental toxicants in Great Lakes food chains, *J. Great Lakes Res.*, 17, 158, 1991.
157. Cade, T. J., Enderson, J. H., Thelander, C. G., and White, C. M., Eds., *Peregrine Falcon Populations: Their Management and Recovery*, Peregrine Fund, Boise, 1988, 49.
158. Blus, L. J., Henny, C. J., Stafford, C. J., and Grove, R. A., Persistence of DDT and metabolites in wildlife from Washington state orchards, *Arch. Environ. Contam. Toxicol.*, 16, 467, 1987.
159. Rudd, R. L., Craig, R. B., and Williams, W. S., Trophic accumulation of DDT in a terrestrial food web, *Environ. Pollut.*, 25, 219, 1981.
160. Hitch, R. K., and Day, H. R., Unusual persistence of DDT in some western USA soils, *Bull. Environ. Contam. Toxicol.*, 48, 259, 1992.
161. Young, D. R., Heesen, T. C., Esra, G. N., and Howard, E. B., DDE-contaminated fish off Los Angeles are suspected cause in deaths of captive marine birds, *Bull. Environ. Contam. Toxicol.*, 21, 584, 1979.
162. Fleming, W. J. and O'Shea, T. J., Influence of a local source of DDT pollution on statewide DDT residues in waterfowl wings, northern Alabama, 1978–79, *Pest. Monit. J.*, 14, 86, 1980.
163. O'Shea, T. J., Fleming, W. J., and Cromartie, E., DDT contamination at Wheeler National Wildlife Refuge, *Science*, 209, 509, 1980.
164. Hirano, M., Characteristics of pyrethroids for insect pest control in agriculture, *Pest. Sci.*, 27, 353, 1989.
165. Douthwaite, R. J., Effects of DDT treatments applied for tsetse fly control on white-headed black chat (*Thamnolaea arnoti*) populations in Zimbabwe, Part I: Population changes, *Ecotoxicology*, 1, 17, 1992.
166. Jenkins, J. M., Jurek, R. M., Garcelon, D. K., Mesta, R., Hunt, W. G., Jackman, R. E., Driscoll, R. E., and Risebrough, R. W., DDE contamination and population parameters of bald eagles (*Haliaeetus leucocephanus*) in California and Arizona, USA, in *Raptor Conservation Today*, Meyburg, B.-U., and Chancellor, R. D., Eds., The Pica Press, Robertsbridge, Sussex, U.K., 1994, 751.
167. Henny, C. J., DDE still high in white-faced ibis eggs from Carson Lake, Nevada, *Colonial Waterbirds*, 20, 478, 1997.
168. King, K. A., Zaun, B. J., Schotborgh, H. M., and Hurt, C., DDE-induced eggshell thinning in white-faced ibis: A continuing problem in western United States, *Southwest. Nat.*, in press, 2002.
169. Elliott, J. E., Martin, P. A., Arnold, T. W., and Sinclair, P. H., Organochlorines and reproductive success of birds in orchard and non-orchard areas of central British Columbia, Canada, 1990–91, *Arch. Environ. Contam. Toxicol.*, 26, 435, 1994.
170. Bishop, C. A., Collins, B., Mineau, P., Burgess, N. M., Read, W. F., and Risley, C., Reproduction of cavity nesting birds in pesticide-sprayed apple orchards in southern Ontario, Canada, 1998–1994, *Environ. Toxicol. Chem.*, 19, 588, 2000.
171. Custer, T. W., Custer, C. M., Hines, R. K., Gutreuter, S., Stromborg, K. L., Allen, P. D., and Melancon, M. J., Organochlorine contaminants and reproductive success of double-crested cormorants from Green Bay, Wisconsin, USA, *Environ. Toxicol. Chem.*, 18, 1209, 1999.
172. Gervais, J. A., Rosenberg, D. K., Fry, D. M., Trulio, L., and Sturm, K. K., Burrowing owls and agricultural pesticides: Evaluation of residues and risks for three populations in California, USA, *Environ. Contam. Toxicol.*, 19, 337, 2000.
173. Henny, C. J., Blus, L. J., and Kaiser, T. E., Heptachlor seed treatment contaminates hawks, owls, and eagles of Columbia Basin, Oregon, *Raptor Res.*, 18, 41, 1984.
174. Hunt, E. G., and Bischoff, A. I., Inimical effects on wildlife of periodic DDD applications to Clear Lake, *Calif. Fish and Game*, 46, 91, 1960.
175. Rudd, R. L., *Pesticides and the Living Landscape*, Univ. Wisconsin Press, Madison, 1964, 320.
176. Herman, S. G., Garrett, R. L., and Rudd, R. L., Pesticides and the western grebe, in *Chemical Fallout*, Miller, M. W. and Berg, G. G., Eds., Charles C Thomas, Springfield, 1969, 24.
177. Keith, J. O., Historical perspectives, in *Wildlife Toxicology*, Van Nostrand Reinhold, New York, 1991, Chap.1.
178. Prouty, R. M., Peterson, J. E., Locke, L. N., and Mulhern, B. M., DDD poisoning in a loon and the identification of the hydroxylated form of DDD, *Bull. Environ. Contam. Toxicol.*, 14, 385, 1975.

179. Korschgen, L. J., Soil-food-chain-pesticide wildlife relationships in aldrin-treated fields, *J. Wildl. Manage.*, 34, 186, 1970.
180. Labisky, R. F. and Lutz, R. W., Responses of wild pheasants to solid-block applications of aldrin, *J. Wildl. Manage.*, 31, 13, 1967.
181. DeWitt, J. B., Chronic toxicity to quail and pheasants of some chlorinated insecticides, *J. Agric. Food Chem.*, 4, 863, 1956.
182. Genelly, R. E. and Rudd, R. L., Effects of DDT, toxaphene and dieldrin on pheasant reproduction, *Auk*, 73, 529, 1956.
183. Flickinger, E. L. and King, K. A., Some effects of aldrin-treated rice on Gulf Coast wildlife, *J. Wildl. Manage.*, 36, 706, 1972.
184. Flickinger, E. L., Effects of aldrin exposure on snow geese in Texas rice fields, *J. Wildl. Manage.*, 43, 94, 1979.
185. Flickinger, E. L., Lobpries, D. S., and Bateman, H. A., Fulvous whistling duck populations in Texas and Louisiana, *Wilson Bull.*, 89, 329, 1977.
186. Flickinger, E. L., Mitchell, C. A., and Krynitsky, A. J., Dieldrin and endrin residues in fulvous whistling-ducks in Texas in 1983, *J. Field Ornithol.*, 57, 85, 1986.
187. Scott, T. G., Willis, Y. L., and Ellis, J. A., Some effects of a field application of dieldrin on wildlife, *J. Wildl. Manage.*, 23, 409, 1959.
188. Clawson, S. G. and Baker, M. F., Immediate effects of dieldrin and heptachlor on bobwhites, *J. Wildl. Manage.*, 23, 215, 1959.
189. Baxter, W. L., Linder, R. L., and Dahlgren, R. B., Dieldrin effects in two generations of penned hen pheasants, *J. Wildl. Manage.*, 33, 96, 1969.
190. Dahlgren, R. B. and Linder, R. L., Effects of dieldrin in penned pheasants through the third generation, *J. Wildl. Manage.*, 38, 320, 1974.
191. Potts, G. R., Success of eggs of the shag on the Farne Islands, Northumberland, in relation to their content of dieldrin and *p,p'*-DDE, *Nature*, 217, 1282, 1968.
192. Lockie, J. D. and Ratcliffe, D. A., Insecticides and Scottish golden eagles, *Br. Birds*, 57, 89, 1964.
193. Blus, L., Cromartie, E., McNease, L., and Joanen, T., Brown pelican: Population status, reproductive success, and organochlorine residues in Louisiana, 1971–1976, *Bull. Environ. Contam. Toxicol.*, 22, 128, 1979.
194. Blus, L. J., Lamont, T. G., and Neely, B. S., Jr., Organochlorine residues, eggshell thickness, reproduction, and population status of brown pelicans in South Carolina and Florida, 1969–76, *Pest. Monit. J.*, 12, 172, 1979.
195. Lehner, P. N. and Egbert, A., Dieldrin and eggshell thickness in ducks, *Nature*, 224, 1218, 1969.
196. Mendenhall, V. M., Klaas, E. E., and McLane, M. A. R., Breeding success of barn owls (*Tyto alba*) fed low levels of DDE and dieldrin, *Arch. Environ. Contam. Toxicol.*, 12, 235, 1983.
197. Newton, I. and Bogan, J., Organochlorine residues, eggshell thinning and hatching success in British sparrowhawks, *Nature*, 249, 582, 1974.
198. Ratcliffe, D. A., The peregrine falcon population of Great Britain and Ireland, 1969-1985, in *Peregrine Falcon Populations: Their Management and Recovery*, Cade, T. J., Enderson, J. H., Thelander, C. G., and White, C. M., Eds., The Peregrine Fund, Boise, 1988, 147.
199. Newton, I., Dale, L., and Little, B., Trends in organochlorine and mercurial compounds in the eggs of British merlins Falco columbarius, *Bird Study*, 20, 241, 1973.
200. Nisbet, I. C. T., The relative importance of DDE and dieldrin in the decline of peregrine falcon populations, in *Peregrine Falcon Populations: Their Management and Recovery*, Cade, T. J., Enderson, J. H., Thelander, C. G., and White, C. M., Eds., The Peregrine Fund, Boise, 351.
201. Newton, I., *The Sparrowhawk*, T and A. D. Poyser, Calton, U.K., 1986.
202. Sibly, R. M., Newton, I., and Walker, C. H., Effects of dieldrin on population growth rates of sparrowhawks 1963–1986, *J. Appl. Ecol.*, 37, 540, 2000.
203. Newton, I. and Haas, M. B., The return of the sparrowhawk, *Br. Birds*, 77, 47, 1984.
204. Newton, I., Dale, L., and Little, B., Trends in organochlorine and mercurial compounds in the eggs of British merlins (*Falco columbarius*), *Bird Study*, 46, 356, 1999.
205. Ratcliffe, D. A., Peregrine population in Great Britain in 1971, *Bird Study*, 19, 117, 1972.

206. Newton, I., Changes in the status of the peregrine falcon in Europe: An overview, in *Peregrine Falcon Populations: Their Management and Recovery*, Cade, T. J., Enderson, J. H., Thelander, C. G., and White, C. M., Eds., The Peregrine Fund, Boise, 227.
207. Lindberg, P., Schei, P. J., and Wikman, M., The peregrine falcon in Fennoscandia, in *Peregrine Falcon Populations: Their Management and Recovery*, Cade, T. J., Enderson, J. H., Thelander, C. G. and White, C. M., Eds., The Peregrine Fund, Boise, 159.
208. Lindberg, P., Population status, pesticide impact and conservation efforts for the peregrine (*Falco peregrinus*) in Sweden, with some comparative data from Norway and Finland, in Conservation Studies on Raptors, ICBP Tech Pub. No. 5, Newton, I., and Chancellor, R.D., Eds., Int. Council for Bird Preservation, London, 1985, 343.
209. Clark, D. R., Jr., Bunck, C. M., and Cromartie, E., Year and age effects on residues of dieldrin and heptachlor in dead gray bats, Franklin County, Missouri — 1976, 1977, and 1978, *Environ. Toxicol. Chem.*, 2, 387, 1983.
210. Stone, W. B. and Okoniewski, J. C., Organochlorine pesticide-related mortalities of raptors and other birds on New York, 1982–1986, in *Peregrine Falcon Populations: Their Management and Recovery*, Cade, T. J., Enderson, J. H., Thelander, C. G., and White, C. M., Eds., The Peregrine Fund, Boise, 429.
211. Frank, R. A. and Lutz, R. S., Productivity and survival of great horned owls exposed to dieldrin, *Condor*, 101, 331, 1999.
212. O'Shea, T. J., Everette, A. L., and Ellison, L. E., Cyclodiene insecticide, DDE, DDT, arsenic, and mercury contamination of big brown bats (*Eptesicus fuscus*) foraging at a Colorado Superfund site, *Arch. Environ. Contam. Toxicol.*, 40, 112, 2001.
213. Cole, J. F., Klevay, L. M., and Zavon, M. R., Endrin and dieldrin: A comparison of hepatic excretion in the rat, *Toxicol. Appl. Pharmacol.*, 16, 547, 1970.
214. Brooks, G. T., Chlorinated Insecticides, Vol. 1, *Biological and Environmental Aspects*, CRC Press, Cleveland, 1974.
215. Cummings, J. G., Eidelman, M., Turner, V., Reed, D., Zee, K. T., and Cook, R. E., Residues in poultry tissues from low level feeding of five chlorinated hydrocarbon insecticides to hens, *J. Assoc. Off. Agric. Chem.*, 50, 418, 1967.
216. Heinz, G. H. and Johnson, R. W., Elimination of endrin by mallard ducks, *Toxicology*, 12, 189, 1979.
217. Mount, D. I. and Putnicki, G. J., Summary report of the 1963 Mississippi fish kill, *Trans. N. Am. Wildl. Nat. Resour. Conf.*, 31, 177, 1966.
218. Newman, R. J., Central southern region, *Audabon Field Notes*, 12, 284, 1958.
219. James, F. C., Central southern region, *Audabon Field Notes*, 17, 329, 1963.
220. Van Tets, G. F., A comparative study of some social communication patterns in the Pelecaniformes, *Am. Ornithol. Union, Ornithol. Monogr.*, 2, 1965.
221. Nesbitt, S. A., Williams, L. E., McNease, L., and Joanen, T., Brown pelican restocking efforts in Louisiana, *Wilson Bull.*, 90, 443, 1978.
222. McNease, L., Joanen, T., Richard, D., Shepard, J., and Nesbitt, S. A., The brown pelican restocking program in Louisiana, *Proc. Ann. Conf. Southeast. Assoc. Fish Wildl. Agencies*, 38, 165, 1984.
223. Fleming, W. J., McLane, M. A. R., and Cromartie, E., Endrin decreases screech owl productivity, *J. Wildl. Manage.*, 46, 462, 1982.
224. Morris, R.D., The effects of endrin on Microtus and Peromyscus. I. Unenclosed field populations, *Can. J. Zool.*, 48, 695, 1970.
225. Blus, L. J., Henny, C. J., Kaiser, T. E., and Grove, R. A., Effects on wildlife from the use of endrin in Washington state orchards, *Trans. N. Am. Wildl. Nat. Resour. Conf.*, 48, 159, 1983.
226. Schladweiler, P. and Weigand, J. P., Relationships of Endrin and Other Chlorinated Hydrocarbon Compounds to Wildlife in Montana, 1981–1982, Montana Dept. Fish Wildl. and Parks, Helena, 1983.
227. Mungari, R. J. (Compiler), A Report on the Monitoring Activities Associated with the Emergency Release of Endrin for Pine Vole Control, New York State Dept. Agric. Markets, Albany, 1978.
228. Stickel, W. H., Kaiser, T. E., and Reichel, W. L., Endrin versus 12-ketoendrin in birds and rodents, in *Avian and Mammalian Wildlife Toxicology*, Kenaga, E. E., Ed., Spec. Tech Publ. 693, Am. Soc. Test. Mater., Philadelphia, 1979, 61.
229. Bedford, C. T., Hutson, D. H., and Natoff, I. L, The acute toxicity of endrin and its metabolites to rats, *Toxicol. Appl. Pharmacol.*, 33, 115, 1975.

230. Baldwin, M. K., Crayford, J. V., Hutson, D. H., and Street, D. L., The metabolism and residues of [$^{14}$C] endrin in lactating cows and laying hens, *Pest. Sci.*, 7, 575, 1976.
231. National Research Council of Canada, Chlordane: Its effects on Canadian Ecosystems and Its Chemistry, Publ. No. NRCC 14094, 1974, 189.
232. Blus, L. J., Pattee, O. H., Henny, C. J., and Prouty, R. M., First records of chlordane-related mortality in wild birds, *J. Wildl. Manage.*, 47, 196, 1983.
233. Blus, L. J., Henny, C. J., and Krynitsky, A. J., Organochlorine-induced mortality and residues in long-billed curlews from Oregon, *Condor*, 87, 563, 1985.
234. Stansley, W. and Roscoe, D. E., Chlordane poisoning of birds in New Jersey, USA, *Environ. Toxicol. Chem.*, 18, 2095, 1999.
235. Sittig, M., Ed., *Priority Toxic Pollutants*, Noyes Data Corp., Park Ridge, NJ, 1980, 368.
236. Stansley, W., Roscoe, D. E., Hawthorne, E., and Meyer, R., Food chain aspects of chlordane poisoning in birds and bats, *Arch. Environ. Contam. Toxicol.*, 40, 285, 2001.
237. Ferguson, D. E., Some ecological effects of heptachlor on birds, *J. Wildl. Manage.*, 28, 158, 1964.
238. Rosene, W., Jr., Effects of field applications of heptachlor on bobwhite quail and other wild animals, *J. Wildl. Manage.*, 29, 554, 1965.
239. Murton, R. K. and Vizoso, M., Dressed cereal seed as a hazard to wood pigeons, *Ann. Appl. Biol.*, 52, 503, 1963.
240. Cramp, S., The effects of pesticides on British wildlife, *Br. Vet. J.*, 129, 315, 1973.
241. Turtle, E. E., Taylor, A., Wright, E. N., Thearle, J. P., Egan, H., Evans, W. H., and Soutar, N. M., The effects on birds of certain chlorinated insecticides used as seed dressings, *J. Sci. Food Agric.*, 14, 567, 1963.
242. Blackmore, D. K., The toxicity of some chlorinated hydrocarbon insecticides to British wild foxes (*Vulpes vulpes*), *J. Compar. Pathol. Ther.*, 73, 391, 1963.
243. Burrage, R. H. and Saha, J. G., Insecticide residues in pheasants after being fed on wheat seed treated with heptachlor and 14C-lindane, *J. Econ. Entomol.*, 65, 1013, 1972.
244. Rudd, R. L. and Genelly, R. E., Lindane repellent to pheasants, *Calif. Ag.*, 8, 13, 1954.
245. Cummings, J. G., Zee, K. T., Turner, V., and Quinn, F., Residues in eggs from low level feeding of five chlorinated hydrocarbon insecticides to hens, *J. Assoc. Off. Agric. Chem.*, 49, 354, 1966.
246. Copeland, M. F. and Chadwick, R. W., Bioisomerization of lindane in rats, *J. Environ. Pathol. Toxicol.*, 2, 737, 1979.
247. Blakely, B. R., Lindane toxicity in pigeons, *Can. Vet. J.*, 23, 267, 1982.
248. French, M. C. and Jefferies, D. J., Disappearance of BHC from avian liver after death, *Nature*, 219, 164, 1968.
249. Koeman, J. H., Chemicals in the environment and their effects on ecosystems, in *Advances in Pesticide Science*, Geissbuhler, H., Ed., Pergamon Press, Oxford, 1979, 25.
250. Blus, L. J., Henny, C. J., and Krynitsky, A. J., The effects of heptachlor and lindane on birds, Columbia Basin, Oregon and Washington 1976–1981, *Sci. Total Environ.*, 46, 73, 1985.
251. Henny, C. J., Blus, L. J., and Kaiser, T. E., Heptachlor seed treatment contaminates hawks, owls, and eagles of Columbia Basin, Oregon, *Raptor Res.*, 18, 41, 1984.
252. Sauter, E. A. and Steele, E. E., The effect of low level pesticide feeding on the fertility and hatchability of chicken eggs, *Poult. Sci.*, 51, 71, 1972.
253. Ware, G. W. and Naber, E. C., Lindane in eggs and chicken tissues, *J. Econ. Entomol.*, 54, 675, 1961.
254. Whitehead, C. C., Downing, A. G., and Pettigrew, R. J., The effects of lindane on laying hens, *Br. Poult. Sci.*, 13, 293, 1972.
255. Keith, J. O., Insecticide contaminations in wetland habitats and their effects on fish-eating birds, *J. Appl. Ecol.*, 3 (Suppl.), 71, 1966.
256. Pollock, G. A. and Kilgore, W. W., Toxaphene, *Residue Rev.*, 69, 87, 1978.
257. Haseltine, S. D., Finley, M. T., and Cromartie, E., Reproduction and residue accumulation in black ducks fed toxaphene, *Arch. Environ. Contam. Toxicol.*, 9, 461, 1980.
258. Bush, P. B., Kiker, J. T., Page, R. K., Booth, N. H., and Fletcher, O. J., Effects of graded levels of toxaphene on poultry residue accumulation, egg production, shell quality, and hatchability in white leghorns, *J. Agr. Food Chem.*, 25, 928, 1977.
259. Clark, D. R., Jr., Dicofol (Kelthane) as an Environmental Contaminant: A Review, U.S. Fish and Wildl. Serv. Tech. Rep., 29, Washington, D.C., 1990.

260. Clark, D. R., Jr. and Krynitsky, A. J., DDT: Recent contamination in New Mexico and Arizona?, *Environment*, 25, 27, 1983.
261. Brown, M. A. and Casida, J. E., Metabolism of a dicofol impurity, alpha-chloro-DDT, but not dicofol or dechlorodicofol, to DDE, *Xenobiotica*, 17, 1169, 1987.
262. Schwarzbach, S. E., Fry, D. M., Rosson, B. E., and Bird, D. M., Metabolism and storage of *p,p*'-dicofol in American kestrels (*Falco sparverius*) with comparisons to ring neck doves (*Streptopelia risoria*), *Arch. Environ. Contam. Toxicol.*, 20, 206, 1991.
263. Wiemeyer, S. N., Spann, J. W., Bunck, C. M., and Krynitsky, A. J., Effects of kelthane on reproduction of captive eastern screech-owls, *Environ. Toxicol. Chem.*, 8, 903, 1989.
264. Clark, D. R., Jr., Spann, J. W., and Bunck, C. M., Dicofol (Kelthane)-induced eggshell thinning in captive American kestrels, *Environ. Toxicol. Chem.*, 9, 1063, 1990.
265. Schwarzbach, S. E., Shull, L., and Grau, C. R., Eggshell thinning in ring doves exposed to *p,p*'-dicofol, *Arch. Environ. Contam. Toxicol.*, 17, 219, 1988.
266. MacLellan, K. N., Bird, D. M., Fry, D. M., and Cowles, J. L., Reproductive and morphological effects of *o,p*'-dicofol on two generations of captive American kestrels, *Arch. Environ. Contam. Toxicol.*, 30, 364, 1996.
267. Clark, D. R., Jr., Flickinger, E. L., White, D. H., Hothem, R. L., and Belisle, A. A., Dicofol and DDT residues in lizard carcasses and bird eggs from Texas, Florida, and California, *Bull. Environ. Contam. Toxicol.*, 54, 817, 1995.
268. Moore, J. A., Dicofol: Intent to Cancel Registrations of Pesticide Products Containing Dicofol: Denial of Applications for Registration of Pesticide Products Containing Dicofol: Conclusion of Special Review: Notice of Final Determination, *Fed. Regis.* 51, 19508, 1986.
269. Jager, K. W., *Aldrin, Dieldrin, Endrin, and Telodrin*, Elsevier Publ., Amsterdam, The Netherlands, 1970, 234.
270. Rooth, J. and Mörzer Bruyns, M. F., *De Grote Stern als Broedvogel in Nederland*, Limosa 43, 31, 1959.
271. Koeman, J. H. Oscamp, A. A. G., van den Broek, E., and van Genderen, H., Insecticides as a factor in the mortality of the Sandwich tern (*Sterna sandvicensis*), *Meded. Rijksfac. Landbouwwetensch*, 32, 841, 1967.
272. Rooth, J. and Jonkers, D. A., The status of some piscivorous birds in The Netherlands, *TNO-Nieuws*, 27, 551, 1972.
273. Koeman, J. H., The toxicological importance of chemical pollution for marine birds in The Netherlands, *Die Vogelwarte*, 28, 145, 1975.
274. Koeman, J. H. and van Genderen, H., Some preliminary notes on residues of chlorinated hydrocarbon insecticides in birds and mammals in The Netherlands, *J. Appl. Ecol.*, 3 (Suppl.), 99, 1966.
275. Wootten, J. F., Methoxychlor: Safe and Effective Substitute for DDT in Controlling Dutch Elm Disease, Central States Forest Expt. Sta. Note 156, 1962, 1.
276. Hunt, L. B. and Sacho, R. J., Response of robins to DDT and methoxychlor, *J. Wildl. Manage.*, 33, 336, 1969.
277. McCaskey, T. A., Stemp, A. R., Liska, B. J., and Stadelmann, W. J., Residues in egg yolks and tissues from laying hens administered selected chlorinated hydrocarbon insecticides, *Poult. Sci.*, 47, 564, 1968.
278. Baker, M. F., Studies on possible effects of mirex bait on the bobwhite quail and other birds, *Proc. Annu. Conf. S. E. Assoc. Fish and Game Comm.*, 18, 1, 1964.
279. Eisler, R., Mirex Hazards to Fish, Wildlife, and Invertebrates: A Synoptic Review, U.S. Fish and Wildl. Serv. Biol. Rep., 85, Washington, D.C., 1990, 42.
280. Wolfe, J. L., and Norment, B.R., Accumulation of mirex residues in selected organisms after an aerial treatment, Mississippi, 1971–1972, *Pest. Monit. J.*, 7, 112, 1973.
281. Heath, R. G. and Spann, J. W., Reproduction and related residues in birds fed mirex, in *Pesticides and the Environment: A Continuing Controversy*, Deichmann, W.B., Ed., Symposia Specialists, North Miami, 1973, 421.
282. National Academy of Sciences, Kepone/Mirex/Hexachlorocyclopentadiene: An Environmental Assessment, National Academy of Sciences, Washington, D.C., 1978, 71.
283. Anonymous, Countries talk about banning DDT, *World Scientist*, press release, March 2000.
284. Cable News Network, Companies agree to settle DDT pollution case for $73 million, press release, December 2000.

CHAPTER 14

# Petroleum and Individual Polycyclic Aromatic Hydrocarbons

**Peter H. Albers**

## CONTENTS

## 14.1 INTRODUCTION

Crude petroleum, refined petroleum products, and individual polycyclic aromatic hydrocarbons (PAHs) contained within petroleum are found throughout the world. Their presence has been detected in living and nonliving components of ecosystems. Petroleum can be an environmental hazard for all organisms. Individual PAHs can be toxic to organisms, but they are most commonly associated with illnesses in humans. Because petroleum is a major environmental source of these PAHs, petroleum and PAHs are jointly presented in this chapter. Composition, sources, environmental fate, and toxic effects on all organisms of aquatic and terrestrial environments are addressed.

Petroleum spills raised some environmental concern during the early twentieth century when ocean transport of large volumes of crude oil began.[1] World War I caused a large number of oil spills that had a noticeably adverse effect on marine birds. The subsequent conversion of the economy of the world from coal to oil, followed by World War II, greatly increased the petroleum threat to marine life. Efforts to deal with a growing number of oil spills and intentional oil discharges at sea continued during the 1950s and 1960s.[1] The wreck of the *Torrey Canyon* off the coast of England in 1967 produced worldwide concern about the consequences of massive oil spills in the marine environment. Research on the environmental fate and biological effects of spilled petroleum increased dramatically during the 1970s. The *Exxon Valdez* oil spill in Prince William Sound, Alaska, in 1989, and the massive releases of crude oil into the Arabian Gulf during the 1991 Gulf War again captured international attention and resulted in an increase in environmental research. Despite considerable progress in developing methods to clean up spills, the adoption of numerous national and international controls on shipping practices, and high public concern (e.g., passage of the Oil Pollution Act of 1990 [33 USCA Sec. 2701-2761] in the United States), petroleum continues to be a widespread environmental hazard.

The association between skin cancer in chimney sweeps and exposure to contaminants in soot was made in England during the late eighteenth century. By the early twentieth century, soot, coal tar, and pitch were all known to be carcinogenic to humans. In 1918 benzo(a)pyrene was identified as a major carcinogenic agent; other PAHs have since been identified as carcinogenic or tumorigenic. Many toxic and carcinogenic effects of PAHs on humans, laboratory animals, and wildlife have been described in numerous reviews.[2]

## 14.2 COMPOSITION AND CHARACTERISTICS

### 14.2.1 Petroleum

Petroleum consists of crude oils and a wide variety of refined oil products. Crude oils vary in chemical composition, color, viscosity, specific gravity, and other physical properties. Color ranges from light yellow-brown to black. Viscosity varies from free flowing to a substance that will barely pour. Specific gravity of most crude oils varies between 0.73 and 0.95.[3] Refined oil products most often spilled in large quantities are aviation fuel, gasoline, No. 2 fuel oil (diesel fuel), and No. 6 fuel oil (bunker C). Fuel oils (Nos. 1 to 6) become increasingly dense and viscous and contain increasingly fewer volatile compounds as their numeric classification proceeds from one to six.

Crude oil is a complex mixture of thousands of hydrocarbon and nonhydrocarbon compounds. Hydrocarbons comprise more than 75% of most crude and refined oils; heavy crude oils can contain more than 50% nonhydrocarbons.[3,4] Hydrocarbons in petroleum are divided into four major classes: (1) straight-chain alkanes (n-alkanes or n-paraffins), (2) branched alkanes (isoalkanes or isoparaffins), (3) cycloalkanes (cycloparaffins), and (4) aromatics (Figure 14.1). Alkenes occur in crude oil, but they are rare. A variety of combinations of the different types of compounds occur. Low-molecular-weight members of each class predominate in crude oils. Aliphatic hydro-

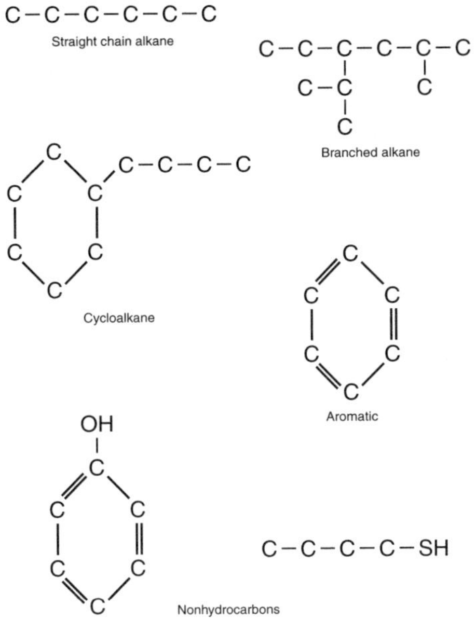

C–C–C–C–C–C

Straight chain alkane

Branched alkane

Cycloalkane

Aromatic

C–C–C–C–SH

Nonhydrocarbons

**Figure 14.1** Types of molecular structures found in petroleum.[3] Hydrogen atoms bonded to carbon atoms are omitted.

carbons have the carbon atoms arranged in an open chain (straight or branched). Aromatic hydrocarbons have the carbon atoms arranged in ring structures (up to six), with each ring containing six carbon atoms with alternating single and double bonds. The aliphatic hydrocarbons (except alkenes) have maximum hydrogen content (saturated), whereas the aromatic hydrocarbons do not have maximum hydrogen content (unsaturated) because of the alternating double bonding between carbon atoms. Nonhydrocarbons in petroleum are compounds containing oxygen (O), sulfur (S), nitrogen (N) (Figures 14.1 and 14.3), or metals, in addition to hydrogen and carbon, and can range from simple open-chain molecules to molecules containing 10 to 20 fused aromatic and cycloalkane carbon rings with aliphatic side chains. The largest and most complex nonhydrocarbons are the resins and asphaltenes.

Crude oils are classified according to physical properties or chemical composition.[3,5] Specific gravity determines whether oil is classified as light, medium, or heavy. Crude oils also can be partitioned into chemical fractions according to boiling point. The relative amounts of compounds in various categories are sometimes used to classify oil (e.g., paraffinic, napthenic, high or low sulfur).

### 14.2.2 PAHs

Polycyclic aromatic hydrocarbons are aromatic hydrocarbons with two or more fused carbon rings that have hydrogen or an alkyl (Cn H2n+1) group attached to each carbon. Compounds range from naphthalene (C10 H8, two rings) to coronene (C24 H12, seven rings) (Figure 14.2). Crude oils contain 0.2 to 7% PAHs, with configurations ranging from two to six rings; PAH content increases with the specific gravity of the oil.[6–8] In general, PAHs have low solubility in water, high melting and boiling points, and low vapor pressure. Solubility decreases, melting and boiling points increase, and vapor pressure decreases with increasing molecular volume. Investigators assessing biological effects sometimes group true PAHs with compounds consisting of aromatic and nonaromatic rings, or compounds with N, S, or O within the ring (heterocycle) or substituted for attached hydrogen (Figure 14.3).

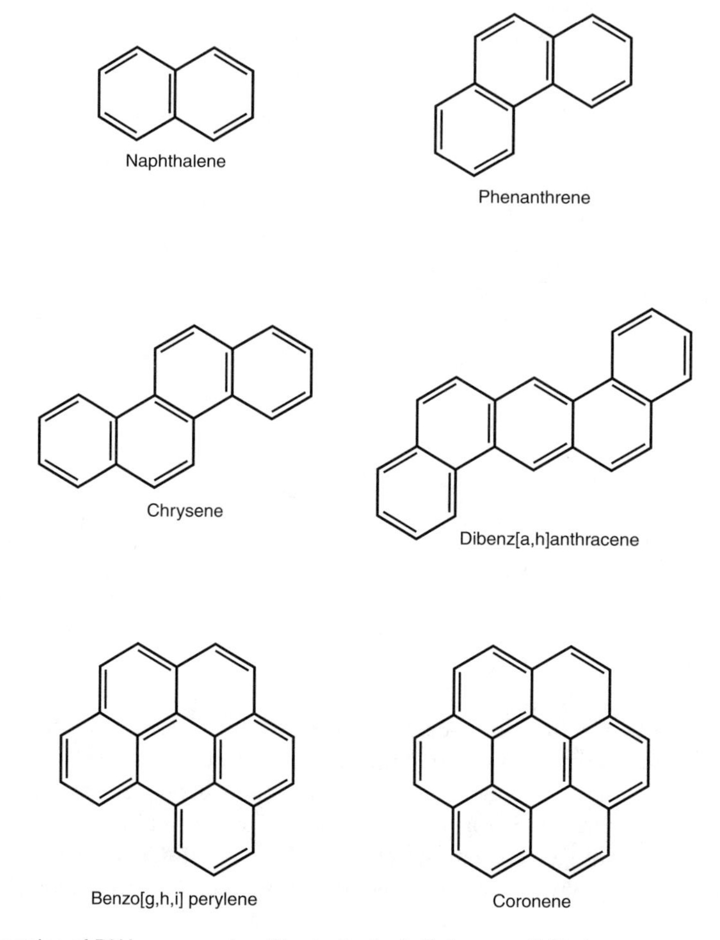


**Figure 14.2** Examples of PAH compounds without attached alkyl groups.[16] Hydrogen atoms bonded to carbon atoms are omitted.

## 14.3 SOURCES

### 14.3.1 Petroleum

During the early 1970s about 35% of the petroleum hydrocarbons in the marine environment came from spills and discharges related to marine transportation; the remainder came from offshore oil and gas production, industrial and municipal discharges, stormwater discharges, river runoff, atmospheric deposition, and natural seeps[9,10] (Figure 14.4). Transportation spills and discharges probably accounted for less than 35% of the total oil discharged onto land and freshwater environments.[10] Estimates for the late 1970s indicated that about 45% of the petroleum hydrocarbons in the marine environment came from spills and discharges related to marine transportation.[11] In heavily used urban estuaries, the contribution of transportation spills and discharges to total petroleum hydrocarbon input can be 10% or less.[12,13] By contrast, the largest source of petroleum in coastal or inland areas removed from urban or industrial centers is petroleum transportation. In the 1980s and 1990s, war, terrorism, vandalism, and theft became additional, and sometimes major, causes of petroleum discharges into water and land environments.[14,15]

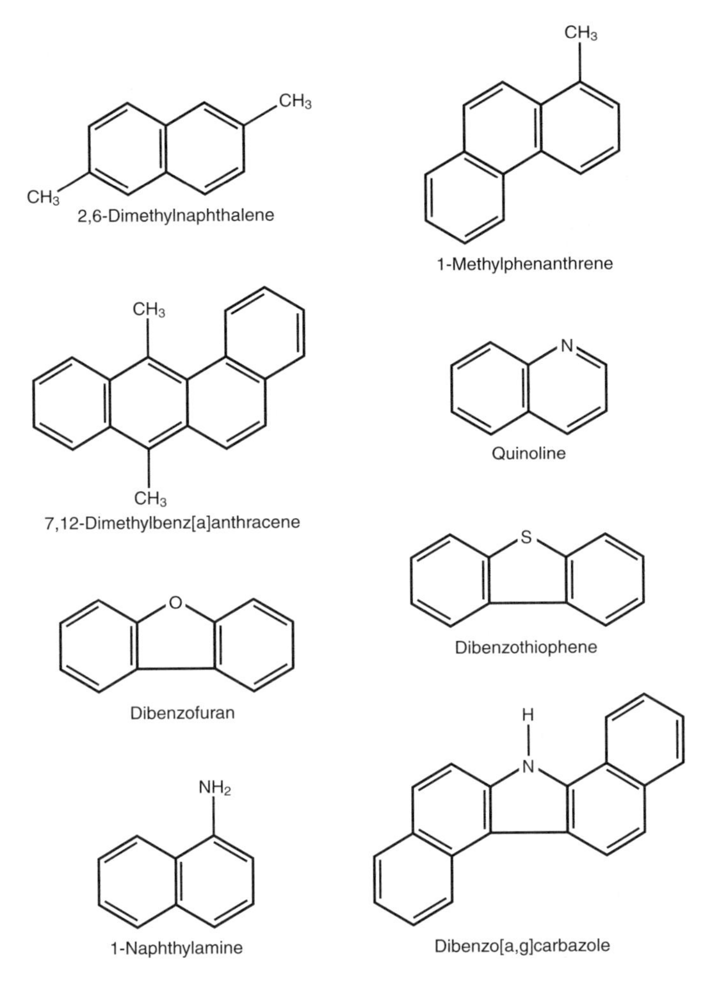


**Figure 14.3** Examples of PAH compounds with attached alkyl groups and aromatic compounds with nitrogen, sulfur, or oxygen.[8] Except for the alkyl groups, hydrogen atoms bonded to carbon atoms are omitted.

## 14.3.2 PAHs

Most PAHs are formed by a process of thermal decomposition (pyrolysis) and subsequent recombination (pyrosynthesis) of organic molecules.[16] Incomplete combustion of organic matter produces PAHs in a high-temperature (500 to 800°C) environment. All forms of combustion, except flammable gases well mixed with air, produce some PAHs. Subjection of organic material in sediments to a low-temperature (100 to 300°C) environment for long periods of time produces PAHs as coal and oil deposits within sedimentary rock formations (a.k.a. diagenesis).[7,17] Although the PAH compounds formed by high- and low-temperature processes are similar, the occurrence of alkylated PAHs is greater in the low-temperature process.[14] Biological formation of PAHs occurs in chlorophyllous plants, fungi, and bacteria.[2,8]

Natural sources of PAHs include forest and grass fires, oil seeps, volcanoes, plants, fungi, and bacteria. Anthropogenic sources of PAHs include petroleum, electric power generation, refuse incineration, and home heating; production of coke, carbon black, coal tar, and asphalt; and internal combustion engines.[2,7] The primary mechanism for atmospheric contamination by PAHs is incom-

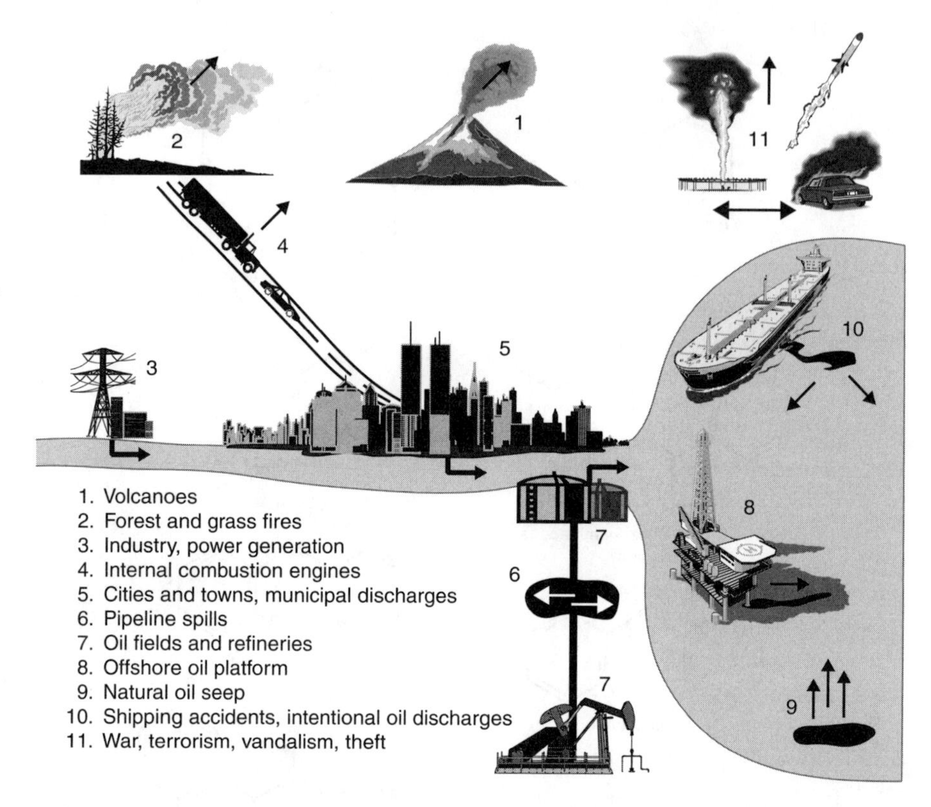


**Figure 14.4** Sources of petroleum and PAHs in the environment. Arrows indicate the initial movement of PAH and petroleum into the air, water, and soil.

plete combustion of organic matter from previously mentioned sources.[15] Aquatic contamination by PAHs is caused by petroleum spills, discharges, and seepages; industrial and municipal wastewater; urban and suburban surface runoff; and atmospheric deposition.[2] Land contamination by PAHs is caused by petroleum spills and discharges, forest and grass fires, volcanoes, industrial and municipal solid waste, and atmospheric deposition.

## 14.4 ENVIRONMENTAL FATE

### 14.4.1 General Considerations

Petroleum is monitored as a liquid composed of a diverse array of thousands of compounds, but PAHs are monitored as a group of individual compounds with similar molecular structures. Polycyclic aromatic hydrocarbons from low- or high-temperature pyrolysis and pyrosynthesis of organic molecules have similar fates in the environment. Whereas PAHs from crude and refined oils and coal originate from a concentrated hydrocarbon source, PAHs produced by high temperature (combustion or industrial processes) are dispersed in the air, scattered on the ground, or included as a component of liquid waste and municipal sewage discharges.

### 14.4.2 Physical and Chemical

Petroleum discharged on water spreads quickly to cover large areas with a layer of oil varying from micrometers to centimeters thick. Some oils, especially heavy crudes and refined products,

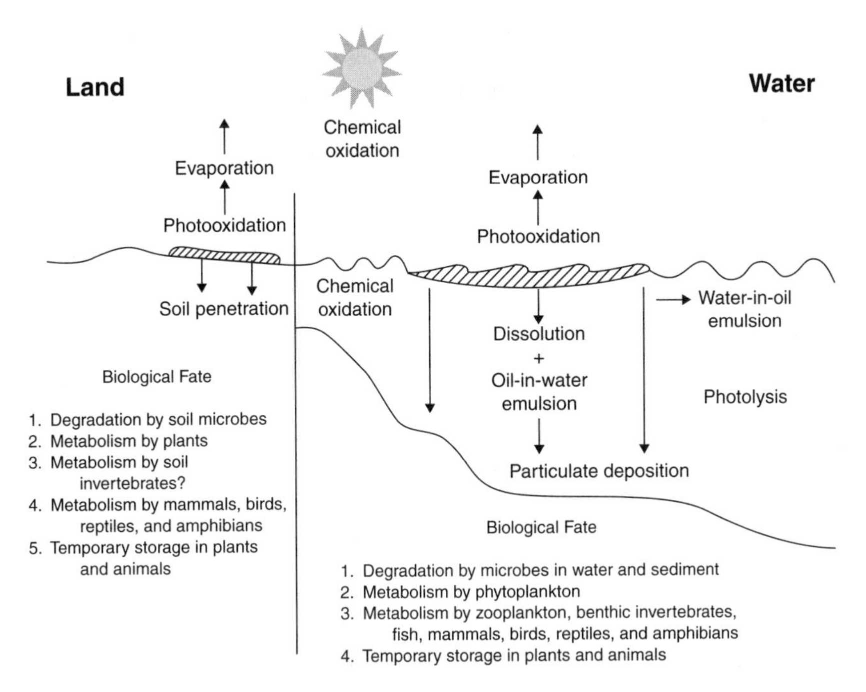


**Figure 14.5** Chemical and biological fate of petroleum and PAHs in water and on land.

sink and move below the surface or along the bottom of the water body. Wave action and water currents mix the oil and water and produce either an oil-in-water emulsion or a water-in-oil emulsion. The former increasingly disperses with time, but the latter resists dispersion. Water-in-oil emulsions have 10 to 80% water content; 50 to 80% of emulsions are often described as "chocolate mousse" because of the thick, viscous, brown appearance. Oil remaining on the water eventually forms tar balls or pancake-shaped patches of surface oil that drift ashore or break up into small pieces and sink to the bottom.

Polycyclic aromatic hydrocarbons released into the atmosphere have a strong affinity for airborne organic particles and can be moved great distances by air currents. The molecules are eventually transported to earth as wet or dry particulate deposition.[18]

Crude and refined oil products begin to change composition on exposure to air, water, or sunlight[3] (Figure 14.5). Low-molecular-weight components evaporate readily; the amount of evaporation varies from about 10% of the spilled oil for heavy crudes and refined products (No. 6 fuel oil) to as much as 75% for light crudes and refined products (No. 2 fuel oil, gasoline). Less than 5% of a crude oil or refined product (primarily low-molecular-weight aromatics and polar nonhydrocarbons) dissolves in water. Hydrocarbons exposed to sunlight, in air or water, can be converted to polar oxidized compounds (photooxidation). Degradation of hydrocarbons in water by photolysis occurs when oxygen is insufficient for photooxidation; high-molecular-weight aromatic hydrocarbons are particularly likely to be altered by this mechanism.[7] Chemical oxidation of aromatic hydrocarbons can result from water and wastewater treatment operations[7] and chemical reactions in the atmosphere.[18]

### 14.4.3 Biological

The movement of oil from the water surface into the water column by dissolution and emulsion exposes the molecules and particles of oil to degradation and transport by organisms. Microbes (bacteria, yeast, filamentous fungi) in the water metabolize the light and structurally simple hydrocarbons and nonhydrocarbons.[3,19] Heavy and complex compounds are more resistant to microbial degradation and eventually move into bottom sediments. Oil particles and individual hydrocarbons (petroleum or recent pyrosynthesis) also adhere to particles (detritus, clay, microbes, phytoplankton) in the water and settle to the bottom, where a variety of microbes metabolize the light and structurally simple compounds. About 40 to 80% of a crude oil can be degraded by microbial action.[3]

Oil and anthropogenic PAHs are ingested by a variety of invertebrate and vertebrate organisms, in addition to microbes. Mammals, birds, fish, and many invertebrates (crustaceans, polychaetes, echinoderms, insects) metabolize and excrete some of the hydrocarbons ingested during feeding, grooming, and respiration.[2,20–24] Although bivalve mollusks and some zooplankton are either unable or marginally able to metabolize oil, they can transport and temporarily store it. Terrestrial plants and aquatic algae can assimilate and metabolize hydrocarbons.[2,25,26] Some soil invertebrates could be capable of metabolizing oil and anthropogenic PAHs, but evidence is absent from the literature.

Accumulation of hydrocarbons is usually inversely related to the ability of the organism to metabolize hydrocarbons.[2,6] For example, bivalve mollusks have a poorly developed mixed function oxygenase (MFO) capability and accumulate hydrocarbons rapidly. After acute exposure, depuration of light for structurally simple hydrocarbons, especially aliphatic hydrocarbons, is more rapid than for heavy hydrocarbons, especially aromatic hydrocarbons.[2,8,27,28] Hydrocarbons accumulated by bivalves through chronic exposure are depurated slowly.[8,29] Organisms, such as fish and some crustaceans, that possess a well-developed MFO system (microsomal monooxygenases) are capable of metabolizing hydrocarbons and only accumulate hydrocarbons in heavily polluted areas.[2] Aquatic environmental factors that reduce the potential for hydrocarbon uptake and retention include high levels of dissolved or suspended organic material and warm water temperatures. Increases in PAH accumulation have not been observed in the trophic levels of aquatic ecosystems.[8,30] Aquatic and terrestrial mammals, birds, reptiles, and amphibians have well-developed MFO systems, but enzyme induction by hydrocarbons in these species is not as well described as in fish and some aquatic invertebrates.[31] Phytoplankton can accumulate aromatic hydrocarbons from water.[2] Terrestrial plants are poor accumulators of soil PAHs, presumably because PAHs strongly adsorb to soil organic material. Most of the PAHs detected in terrestrial plants appear to be derived from the atmosphere; PAHs adhere to or are absorbed by plant tissue. Following the death of vegetation, PAHs are deposited into the surface litter of soil.[2,27,32]

Organisms with high lipid content, a poor MFO system, and that exhibit activity patterns or distributions that coincide with the location of hydrocarbon pollution are most likely to accumulate hydrocarbons.[2,8] Once assimilated, heavy aromatic hydrocarbons (four or more benzene rings) are the most difficult group of hydrocarbons to excrete, regardless of MFO capability.[2,33]

### 14.4.4 Residence Time

Residence time for petroleum in water is usually less than 6 months. Residence time for petroleum deposited on nearshore sediments and intertidal substrate is determined by the characteristics of the sediment and substrate. Oil-retention times for coastal environments range from a few days for rock cliffs to as much as 20 years for cobble beaches, sheltered tidal flats, and wetlands.[34–38] Microbes in the sediment and on the shoreline metabolize petroleum compounds; microbial degradation is reduced considerably by anaerobic conditions.[39,40] In cold climates ice, low wave energy, and decreased chemical and biological activity cause oil to remain in sediments or on the shore longer than in temperate or tropical climates; sheltered tidal flats and wetlands can be expected to retain oil for long periods of time.[41,42] Total petroleum hydrocarbons or individual PAHs in undisturbed estuarine subtidal sediments can remain as identifiable deposits for many decades.[43]

Oil spilled on land has little time to change before it penetrates into the soil. Oil spilled on lakes and streams usually has less opportunity to change before drifting ashore than oil spilled on the ocean because the water bodies are smaller. Petroleum spilled directly on land is degraded by evaporation, photooxidation, and microbial action. Residence times for hydrocarbons in terrestrial soils are less well documented than for sediments but are determined by the same conditions (substrate type, oxygen availability, temperature, surface disruption) that determine residence times in intertidal sediments. Total petroleum hydrocarbons and individual PAHs can persist in soils of cold or temperate regions for at least 20 to 40 years; persistence of PAHs increases with increase in the number of benzene rings.[32,44–46]

### 14.4.5 Spill Response

Petroleum spills are sometimes left to disperse and degrade naturally,[47] but cleanup efforts are initiated when economically or biologically important areas are threatened. Oil-response actions include mechanical removal, chemical treatment, enhanced biodegradation, and restoration.[48,49] The addition of chemicals to floating oil for purposes of gelling, herding, or dispersing the oil can modify the expected effects of the oil. The chemicals themselves often have some toxicity, and in the case of dispersants, movement of oil into the water column is greatly accelerated. Enhanced biodegradation practices are evolving through experimentation, but the most common practice is the stimulation of indigenous microbial populations through application of supplemental nitrogen or phosphorus to oil on land or water. The addition of nitrogen and phosphorus to oiled shorelines in Prince William Sound, Alaska, accelerated biodegradation and did not increase toxicity or environmental effects of the oil.[50] The literature on oil-spill response procedures, cleanup methods, and restoration methods is extensive; the interested reader is encouraged to consult literature cited in references[47–50] and other sources for a more comprehensive treatment of this topic.

## 14.5 EFFECTS ON ORGANISMS

### 14.5.1 General

Petroleum can adversely affect organisms by physical action (smothering, reduced light), habitat modification (altered pH, decreased dissolved oxygen, decreased food availability), and toxic action. Large discharges of petroleum are most likely to produce notable effects from physical action and habitat modification. The aromatic fraction of petroleum is considered to be responsible for most of the toxic effects.

Polycyclic aromatic hydrocarbons affect organisms through toxic action. The mechanism of toxicity is reported to be interference with cellular membrane function and enzyme systems associated with the membrane.[7] Although unmetabolized PAHs can have toxic effects, a major concern in animals is the ability of the reactive metabolites, such as epoxides and dihydrodiols, of some PAHs to bind to cellular proteins and DNA. The resulting biochemical disruptions and cell damage lead to mutations, developmental malformations, tumors, and cancer.[2,6,51] Four-, five-, and six-ring PAHs have greater carcinogenic potential than the two-, three-, and seven-ring PAHs.[2,7] The addition of alkyl groups to the base PAH structure often produces carcinogenicity or enhances existing carcinogenic activity (e.g., 7,12-dimethylbenz[a]anthracene). Some halogenated PAHs are mutagenic without metabolic activation,[52] and the toxicity and possibly the carcinogenicity of PAHs can be increased by exposure to solar ultraviolet radiation.[53,54] Cancerous and precancerous neoplasms have been induced in aquatic organisms in laboratory studies, and cancerous and noncancerous neoplasms have been found in feral fish from heavily polluted sites.[2,7,55,56] However, sensitivity to PAH-induced carcinogenesis differs substantially among animals.[2,7]

In water the toxicity of individual PAHs to plants and animals increases as molecular weight (MW) increases up to MW 202 (fluoranthene, pyrene). Beyond MW 202 a rapid decline in solubility reduces PAH concentrations to less than lethal levels, regardless of their intrinsic toxicity. However, sublethal effects can result from exposure to these very low concentrations of high MW compounds.[7] Except for the vicinity of chemical or petroleum spills, environmental concentrations of PAHs in water are usually several orders of magnitude below levels that are acutely toxic to aquatic organisms. Sediment PAH concentrations can be much higher than water concentrations, but the limited bioavailability of these PAHs greatly reduces their toxic potential.[2] In general, caution should be employed when assessing the aquatic "toxicity" of biogenic or anthropogenic PAHs because bioavailability (solubility, sediment sequestration, mechanism of exposure) and chemical modification determine how much toxicity is realized.

### 14.5.2 Plants and Microbes

Reports of the effects of petroleum spills or discharges on plants and microbes contain accounts of injury or death to mangroves,[57,58] seagrasses,[59] and large intertidal algae;[60,61] severe and long-lasting (> 2 years) destruction of salt marsh vegetation[38,62–64] and freshwater wetland vegetation;[65,66] enhanced or reduced biomass and photosynthetic activity of phytoplankton communities;[67,68] genetic effects in mangroves and terrestrial plants;[69,70] and microbial community changes and increases in numbers of microbes (Table 14.1).[71–73] Differences in species sensitivity to petroleum are responsible for the wide variation in community response for phytoplankton and microbes.

Recovery from the effects of oil spills on most local populations of nonwoody plants can require from several weeks to 5 years, depending on the type of oil, circumstances of the spill, and species affected. Mechanical removal of petroleum in wetlands can markedly increase the recovery time.[74,75] Complete recovery by mangrove forests could require up to 20 years.[76] Phytoplankton and microbes in the water column of large bodies of water return to prespill conditions faster than phytoplankton and microbes in small bodies of water because of greater pollutant dilution and greater availability of colonizers in nearby waters.[77] Lethal and sublethal effects are caused by contact with oil or dissolved oil, systemic uptake of oil compounds, blockage of air exchange through surface pores, and possibly by chemical and physical alteration of soil and water, such as depletion of oxygen and nitrogen, pH change, and decreased light penetration.

The effects of petroleum on marine plants, such as mangroves, sea grasses, saltmarsh grasses, and micro- and macroalgae, and microbes have been studied with laboratory bioassays, experimental ecosystems, and field experiments and surveys (Table 14.1).[78–87] Petroleum caused death, reduced growth, and impaired reproduction in the large plants. Microalgae were either stimulated or inhibited, depending on the species and the type and amount of oil; response was expressed as changes in biomass, photosynthetic activity, and community structure. In response to petroleum exposure, community composition of indigenous bacteria was altered and total biomass increased.

The effects of petroleum on freshwater phytoplankton, periphyton, and microbes have been studied with laboratory bioassays and field experiments.[88–93] Petroleum induced effects similar to those described for marine microalgae and bacteria. Domestic and wild plants have been exposed to oiled soil in laboratory experiments[94–96] and tundra vegetation has been subjected to an experimental spill of crude oil.[97] Inhibition of seed germination, plant growth, and fungal colonization of roots was demonstrated in the laboratory bioassays. Death of herbaceous and woody plants and long-term community alteration were caused by the experimental tundra spill.

Individual PAHs, mostly two- and three-ring compounds, at low concentrations (5 to 100 ppb) can stimulate or inhibit growth and cell division in aquatic bacteria and algae. At high concentrations (0.2 to 10 ppm) the same PAHs interfere with cell division of bacteria and cell division and photosynthesis of algae and macrophytes; they can also cause death.[2,7]

**Table 14.1 Effects of Petroleum or Individual PAHs on Organisms**

| | Type of Organism | | | | | |
|---|---|---|---|---|---|---|
| Effect[a] | Plant or Microbe | Invertebrate | Fish | Reptile or Amphibian | Bird | Mammal[b] |
| **Individual Organisms** | | | | | | |
| Death | X | X | X | X | X | X |
| Impaired reproduction | X | X | X | X | X | X |
| Reduced growth and development | X | X | X | X | X | |
| Altered rate of photosynthesis | X | | | | | |
| Altered DNA | X | X | X | X | X | X |
| Malformations | | | X | | X | |
| Tumors or lesions | | X | X | X | | X |
| Cancer | | | X | X | | X |
| Impaired immune system | | | X | | X | X |
| Altered endocrine function | | | X | | X | |
| Altered behavior | | X | X | X | X | X |
| Blood disorders | | X | X | X | X | X |
| Liver and kidney disorders | | | X | | X | X |
| Hypothermia | | | | | X | X |
| Inflammation of epithelial tissue | | | | X | X | X |
| Altered respiration or heart rate | | X | X | X | | |
| Impaired salt gland function | | | | X | X | |
| Gill hyperplasia | | | X | | | |
| Fin erosion | | | X | | | |
| **Groups of Organisms[c]** | | | | | | |
| Local population change | X | X | X | | X | X |
| Altered community structure | X | X | X | | X | X |
| Biomass change | X | X | X | | | |

[a] Some effects have been observed in the wild and in the laboratory, whereas others have only been induced in laboratory experiments or are population changes estimated from measures of reproduction and survival.
[b] Includes a sampling of literature involving laboratory and domestic animals.
[c] Populations of microalgae, microbes, soil invertebrates, and parasitic invertebrates can increase or decrease in the presence of petroleum, whereas populations of other plants and invertebrates and populations of vertebrates decrease.

### 14.5.3 Invertebrates

Reports of the effects of aquatic oil spills or discharges and oil-based drill cuttings often contain accounts of temporary debilitation, death, population change, or invertebrate community change for marine water column,[98,99] deepwater benthic,[100,101] nearshore subtidal,[102–105] intertidal,[64,106,107] coastal mangrove organisms,[102,108] and stream[109,110] and lake[111] organisms. For a review of the effects of petroleum on marine invertebrates, see Suchanek.[112] Observed effects are caused by smothering; contact by adults, juveniles, larvae, eggs, and sperm with whole or dissolved oil; ingestion of whole oil or individual compounds; and possibly by chemical changes in the water, including oxygen depletion and pH change. Accounts of the effects of petroleum spills or discharges on terrestrial invertebrates have not been published.

Recovery from the effects of oil spills on local populations of invertebrates can require as little as a week for pelagic zooplankton or as much as 10 years, or more, for nearshore subtidal meiofauna. Uncertainty of recovery time is particularly evident for intertidal, nearshore subtidal, and coastal mangrove biota; the prognosis for recovery is a function of the size of the spill, type of oil, season of year, weather, characteristics of affected habitat, and species affected. Chronic input of petroleum from natural seeps or anthropogenic sources produces communities of biota that are adapted to the hydrocarbon challenge; preexisting or continuing chronic input can complicate estimates of invertebrate recovery from ephemeral input. Aggressive cleaning of shorelines can retard recovery.[113] Zooplankton in large bodies of water return to prespill conditions faster than zooplankton in small bodies of water (isolated estuaries, lakes, stream headwaters) because of greater pollutant dilution and greater availability of colonizers in nearby waters.[77]

Much work has been done on the effects of petroleum on marine invertebrates with laboratory bioassays, mesocosms, enclosed ecosystems, and field experiments and surveys.[82,114–124] Less work has been conducted on freshwater invertebrates with laboratory bioassays and field experiments.[88,125–129] Among the effects reported are reduced survival, altered physiological function, soft-tissue abnormalities, inhibited reproduction, altered behavior, and changes in species populations and community composition.

Soil contaminated with petrochemicals has been associated with a reduced number of species and reduced species abundance of rodent parasites.[130] In contrast, petrochemical contamination of soil increased the abundance of isopods,[131] and chronic exposure to low concentrations of PAHs appeared to stimulate populations of several groups of soil invertebrates.[132] In soil remediation studies earthworms have been found to be a sensitive indicator of soil quality.[133]

In short-term exposure trials (24 to 96 h) on selected aquatic invertebrates, individual PAH compounds had $LC_{50}$ values in water ranging from 0.1 to 5.6 ppm.[2,7] Eggs and larvae are more sensitive than juveniles or adults to dissolved PAHs. Larvae with a high tissue burden of PAHs from maternal transfer are likely to be at risk for increased toxicity from ultraviolet radiation.[134] Sublethal effects include reduced reproduction, inhibited development of embryos and larvae, delayed larval emergence, decreased respiration and heart rate, abnormal blood chemistry, and lesions.[2,7]

### 14.5.4 Fish

Adult and juvenile fish, larvae, and eggs are exposed to petroleum through contact with whole oil, dissolved hydrocarbons, particles of oil dispersed in the water column, or ingestion of contaminated food and water.[135,136]

Death of fish in natural habitat usually requires a heavy exposure to petroleum. Consequently, it is unlikely that large numbers of adult fish inhabiting large bodies of water would be killed by the toxic effects of petroleum. Fish kills usually are caused by large amounts of oil moving rapidly into shallow waters.[137,138] However, fresh and weathered crude oils and refined products vary considerably in their composition and toxicity, and the sensitivity of fish to petroleum differs among species. Petroleum concentrations (total petroleum hydrocarbons) in water of less than 0.5 ppm during long-term exposure[139] or higher concentrations (several to more than 100 ppm) in moderate- or short-term exposures can be lethal.[140–143] Sublethal effects begin at concentrations of less than 0.5 ppm and include changes in heart and respiratory rates, gill structural damage, enlarged liver, reduced growth, fin erosion, corticosteriod stress response, immunosuppression, impaired reproduction, increased external and decreased internal parasite burdens, behavioral responses, and a variety of biochemical, blood, and cellular changes.[136,142,144–154]

Eggs and larvae can suffer adverse effects, including death, when exposed to concentrations of petroleum in water ranging from less than 1 ppb (total PAHs) up to 500 ppb (total PAHs or total petroleum hydrocarbons).[155–158] Eggs, larvae, and early juveniles are generally more vulnerable than adults to oil spills and discharges because they have limited or no ability to avoid the oil, and they reside in locations that receive the most severe exposures (near the water surface or in shallow nearshore areas and streams). Effects of oil on eggs and larvae include death of embryos and larvae, abnormal development, reduced growth, premature and delayed hatching of eggs, DNA alterations, and other cellular abnormalities.[136,156–161] Evidence of continued adverse effects on the viability of pink salmon (*Oncorhynchus gorbuscha*) embryos for 4 years after the *Exxon Valdez* oil spill in Prince William Sound, Alaska, presented the possibility of gametic damage to exposed fish.[162]

Assessing the effect of petroleum spills and discharges on fish populations has been attempted with a variety of approaches, including taking fish samples from oiled and control areas after a spill,[163] monitoring postspill fish harvest and recruitment,[138] aerial surveys of fish stocks before and after spill events,[164] translating abundance of benthic prey into estimates of demersal fish biomass,[104] using a life-history model to estimate effects on a species,[165] measuring differences in fish community composition above and below a discharge site on a stream,[149] and evaluating the potential of deleterious heritable mutations induced by a one-time spill event to produce measurable population change.[166] Several of these attempts revealed likely effects of petroleum on fish populations.

Also, the potential effects of oil spills on regional pelagic fish populations were evaluated with an extensive modeling effort of the Georges Bank fishery off the northeastern coast of the United States.[167,168] Unfortunately, normal variation in natural mortality of eggs and larvae for such species as Atlantic cod (*Gadus morhua*), haddock (*Melanogrammus aeglefinus*), and Atlantic herring (*Clupea harengus*) is often larger than mortality estimates for the largest of spills. Consequently, direct field observations would not be able to distinguish the effect of a major oil spill on a single year class from natural variation in recruitment. The authors concluded that a comprehensive recruitment model was needed to separate the effect of spilled oil from expected natural mortality.

In general, it is difficult to determine the effect of an ephemeral petroleum discharge on fish populations in large bodies of water. It is beneficial to have before-and-after data, and some modeling appears necessary to identify effects that are difficult to measure. Also, time and geographic scale are important considerations in any population assessment.

In short-term exposure trials (24 to 96 h) on selected species of fish, individual PAH compounds had $LC_{50}$ values in water ranging from 1.3 to 3400 ppb.[2] The primary target organ for toxic action is the liver. Sublethal effects on eggs, larvae, juveniles, and adult fish[2] are generally similar to those previously described for exposure to fresh or weathered petroleum and separate aromatic fractions but with greater emphasis on neoplasm induction and DNA alteration.

Induction of precancerous cellular changes in laboratory studies with PAHs and high frequencies of lesions and cancerous and noncancerous neoplasms in bottom-dwelling fish from areas contaminated with PAHs provide support for a causal relation between PAHs in sediment and the presence of cancer in several species of bottom-dwelling marine fish.[7,33,55,56,169–171] Fish from Puget Sound, Washington,[172] tributaries of southern Lake Erie, Ohio,[55,173,174] and the Elizabeth River, Virginia,[175] had cancerous and noncancerous skin and liver neoplasms, fin erosion, and a variety of other external abnormalities. Concentrations of total PAHs in the sediment were sometimes > 100 ppm and 50 to 10,000 times greater than in reference areas.[2,55,173,175]

There is evidence that exposure to high concentrations of PAHs can affect fish populations and communities. Lesion frequency, overall health assessment, and population age structure were useful biological measures for differentiating a population of brown bullheads (*Ameiurus nebulosus*) in an industrialized urban river (Schuylkill River, Philadelphia, Pennsylvania) from a population in a nonindustrialized suburban pond (Haddonfield, New Jersey).[176] Analysis of fish community structure revealed lower species diversity for contaminated Lake Erie tributaries (Black and Cuyahoga Rivers, Ohio) than for a reference stream (Huron River, Ohio).[174]

Metabolism of PAHs by feral fish in areas chronically contaminated with multiple pollutants is poorly studied; studies such as van der Oost[177] are rare. Also lacking is information on dose-effect and temporal aspects of *in situ* exposure to carcinogens.[33,55] Species differences in PAH metabolism and incidence of neoplasms, even among closely related species,[33] further complicate efforts to generalize about findings from individual studies.

### 14.5.5 Reptiles and Amphibians

The response of reptiles and amphibians to petroleum exposure is not well characterized. Various species of reptiles and amphibians were killed by a spill of bunker C fuel oil in the St. Lawrence River (E.S. Smith, New York Department of Environmental Conservation, Albany, NY, unpublished report). Sea snakes were presumed to have been killed by crude oil in the Arabian Gulf.[178] Petroleum could have been a contributing factor in the deaths of sea turtles off the coast of Florida,[179] in the

Gulf of Mexico following the *Ixtoc I* oil spill,[180] and in the Arabian Gulf after the Gulf War oil spills,[181] but the cause of death was not determined. Ingestion of oil and plastic objects has been reported for green (*Chelonia mydas*), loggerhead (*Caretta caretta*), and Atlantic Ridley (*Lepidochelys kempi*) turtles.[179,180,182]

Experimental exposure of juvenile loggerhead turtles to crude oil slicks revealed effects on respiration, skin, blood characteristics and chemistry, and salt gland function.[183] Juvenile loggerhead turtles were exposed to artificially weathered crude oil for 4 days followed by an 18-day recovery period; blood abnormalities and severe skin and mucosal changes from exposure were reversed during the recovery period.[184] Atlantic Ridley and loggerhead embryos died or developed abnormally when the eggs were exposed to oiled sand; weathered oil was less harmful to the embryos than fresh oil (T.H. Fritts and M.A. McGehee, Fish and Wildlife Service, Denver Wildlife Research Center, Denver, CO, unpublished report).

Bullfrog tadpoles (*Rana catesbeiana*) were exposed to amounts of No. 6 fuel oil that could be expected in shallow waters following oil spills; death was most common in tadpoles that were in the late stages of development, and sublethal effects included grossly inflated lungs, fatty livers, and abnormal behavior.[185] Larvae of the wood frog (*Rana sylvatica*), the spotted salamander (*Ambystoma maculatum*), and two species of fish were exposed to several fuel oils and crude oils in static and flow-through tests; sensitivity of the amphibian larvae to oil was slightly less than that of the two species of fish.[140] Exposure of green treefrog (*Hyla cinerea*) embryos and larvae and larval mole salamanders (*Ambystoma opacum* and *A. tigrinum*) to used motor oil in natural ponds, field enclosures, or laboratory containers caused reduced growth or reduced food (algae) densities and prevented metamorphosis of green frogs at high exposure.[186,187]

The injection or implantation in amphibians of perylene or crystals of benzo(a)pyrene and 3-methylcholanthrene induced cancerous and noncancerous tissue changes.[2] It has been suggested that amphibians are more resistant to PAH carcinogenesis than mammals because of the demonstrated inability of the hepatic microsomes of the tiger salamander (*A. tigrinum*) to produce mutagenic metabolites.[188] The toxicity and genotoxicity of benzo(a)pyrene and refinery effluent to larval newts (*Pleurodeles waltlii*), in the presence and absence of ultraviolet radiation, was established in a series of experiments by Fernandez and l'Haridon.[189]

### 14.5.6 Birds

Birds can be affected by petroleum through external oiling, ingestion, egg oiling, and habitat changes. External oiling disrupts feather structure, causes matting of feathers, and produces eye and skin irritation. Death often results from hypothermia and drowning.[1,190–192] Bird losses in excess of 5000 individuals are common for moderate to large petroleum spills. Birds that spend much of their time in the water, such as alcids (*Alcidae*), waterfowl (*Anatidae*), and penguins (*Spheniscidae*), are the most vulnerable to surface oil.

Petroleum can be ingested through feather preening, consumption of contaminated food or water, and inhalation of fumes from evaporating oil. Ingestion of oil is seldom lethal, but it can cause many debilitating sublethal effects that promote death from other causes, including starvation, disease, and predation. Effects include gastrointestinal irritation, pneumonia, dehydration, red blood cell damage, impaired osmoregulation, immune system suppression, hormonal imbalance, inhibited reproduction, retarded growth, and abnormal parental behavior.[192–203]

Bird embryos are highly sensitive to petroleum. Contaminated nest material and oiled plumage are mechanisms for transferring oil to the shell surface. Small quantities (1 to 20 μL) of some types of oil (light fuel oils, certain crude oils) are sufficient to cause death, particularly during the early stages of incubation.[204,205] Eggshell applications of petroleum weathered for several weeks or longer are less toxic to bird embryos than fresh or slightly weathered petroleum.[206,207]

Petroleum spilled in avian habitats can have immediate and long-term effects on birds. Fumes from evaporating oil, a shortage of food, and cleanup activities can reduce use of an affected area.[208,209]

Long-term effects are more difficult to document, but severely oiled wetlands and tidal mud flats are likely to have altered plant and animal communities for many years after a major spill.[62,82]

The direct and indirect effects of oil spills are difficult to quantify at the regional or species level. Death from natural causes and activities of humans (e.g., commercial fishing), weather, food availability, and movement of birds within the region can obscure the effects of a single or periodic catastrophic event. Regional population assessments in the northern Gulf of Alaska for bald eagles (*Haliaeetus leucocephalus*) and common murres (*Uria aalge*) after the *Exxon Valdez* oil spill in March 1989 failed to identify population changes attributable to the spill.[210–212] During the last 50 years seabird populations of western Europe have increased or decreased without apparent relation to losses from oil spills.[213,214]

Effects of oil spills are most likely to be detected at the level of local populations. Also, measures of survival, reproduction, and habitat use of numerous individual birds can be extrapolated to local or regional populations. The *Exxon Valdez* oil spill in Prince William Sound, Alaska caused the loss of an estimated 250,000 to 375,000 birds.[215,216] Adverse effects of this spill have been described for individuals of several species of seabirds or wintering waterfowl in Prince William Sound and extrapolated to populations.[217–220] Wiens et al.[218] also combined the species effects into assessments of avian community composition. Marine bird population declines have also been reported after the 1996 *Sea Empress* spill off the coast of England[221] and the 1991 Arabian Gulf oil spill.[222] Most of the changes in performance measures or local population size were no longer detectable by 2 years after the spill; exceptions were described by Wiens et al.,[218] Esler et al.,[221] and Symens and Suhaibani.[222]

The consequences of direct and indirect effects of oil spills on seabird populations have also been estimated with simulation models.[223–226] Models have shown that (1) an occasional decrease in survival of breeding adults will have a greater effect on seabird populations than an occasional decrease in reproductive success, (2) long-lived seabirds with low reproductive potential will have the most difficulty recovering from a catastrophic oil spill, and (3) recovery of a seriously depleted population of long-lived seabirds will be greatly hindered if adult survival and reproduction show small, but sustained, decreases. The overall recovery potential for a species depends on the reproductive potential of the survivors and the immigration potential from surrounding areas.[227,228]

Much of the available information on effects of PAHs on birds was produced by studies of the effects of petroleum on eggs. Experiments have shown that the PAH fraction of crude and refined oils is responsible for the lethal and sublethal effects on bird embryos caused by eggshell oiling.[205] Further, the toxicity of PAHs to bird embryos is a function of the quantity and molecular structure of the PAHs.[229] Brunstrom et al.[230] injected a mixture of 18 PAHs (2.0 mg/kg of egg) into eggs of the chicken (*Gallus domesticus*), turkey (*Meleagris gallopavo*), domestic mallard (*Anas platyrhynchos*), and common eider (*Somateria mollissima*) and found the mallard to be the most sensitive species and benzo[k]fluoranthene (four rings) and indeno[1,2,3-cd]pyrene (five rings) to be the most toxic of the PAHs tested. The most toxic PAHs were found to have additive effects on death of embryos, and the cause of toxicity was proposed to be a mechanism controlled by the Ah receptor.[231] Naf et al.[232] injected a mixture of 16 PAHs (0.2 mg/kg of egg) into chicken and common eider eggs and reported that > 90% was metabolized by day 18 of incubation (chicken), with the greatest concentration of PAHs in the gall bladder of both species. Mayura et al.[233] injected fractionated PAH mixtures from coal tar (0.0625 to 2.0 mg/kg of egg) into the yolk of chicken eggs and found that death, liver lesions and discoloration, and edema increased as the proportion of five- and > five-ring aromatics, compared to the proportion of two- to four-ring aromatics, increased in the mixture.

Studies with adults and nestlings revealed a variety of sublethal toxic effects induced by PAHs. Male mallard ducks fed 6000 ppm of a mixture of 10 PAHs combined with 4000 ppm of a mixture of 10 alkanes for 7 months in a chronic ingestion study had greater hepatic stress responses and higher testes weights than male mallards fed 10,000 ppm of the alkane mixture.[234] Nestling herring gulls (*Larus argentatus*) were administered single doses (0.2 or 1.0 mL) of crude oils, their aromatic or aliphatic fractions, or a mixture of crude oil and dispersant.[235,236] Retardation of nestling weight gain and increased adrenal and nasal gland weights was attributed to the PAHs with four or more rings. Immune function and mixed-function oxidase activity of adult European starlings (*Sturnus vulgaris*) were altered by subcutaneous injections (25 mg/kg body weight) of 7,12-dimethylbenz[a]anthracene, a four-ring PAH, every other day for 10 days.[237] Oral doses in adult birds (25 mg/kg body weight) and nestlings (20 mg/kg body weight) also altered immune function. The coefficient of variation of nuclear DNA volume of red blood cells from wild lesser scaup (*Athya affinis*) was positively correlated with the concentration of total PAHs in scaup carcasses.[238]

### 14.5.7 Mammals

Marine mammals that rely primarily on fur for insulation, such as the sea otter (*Enhydra lutris*), polar bear (*Ursus maritimus*), Alaska fur seal (*Callorhinus ursinus*), and newborn hair seal pups (*Phocidae*), are the most likely to die after contact with spilled oil.[22,239,240] Oiled fur becomes matted and loses its ability to trap air or water, resulting in hypothermia. Adult hair seals, sea lions (*Eumetopias jubatus, Zalophus californianus*), and cetaceans (whales, porpoises, dolphins) depend primarily on layers of fat for insulation; thus, oiling causes much less heat loss. However, skin and eye irritation and interference with normal swimming can occur. Skin absorption of oil has been reported for seals and polar bears.

Oil ingested in large quantities can have acute effects on marine mammals. However, marine mammals, such as seals and cetaceans, are capable of rapid hydrocarbon metabolism and renal clearance.[239] Ingested oil can cause gastrointestinal tract hemorrhaging in the European otter (*Lutra lutra*);[241] renal failure, anemia, and dehydration in the polar bear;[242] pulmonary emphysema, centrilobular hepatic necrosis, hepatic and renal lipidosis, increased nuclear DNA mass, altered blood chemistry, and possible gastric erosion and hemorrhage in sea otters,[243–245] and altered blood chemistry and reduced body weight in river otters (*Lontra canadensis*).[246] Inhalation of evaporating oil is a potential respiratory problem for mammals near or in contact with large quantities of unweathered oil.[239,247] Some of the previously described disorders are thought to be caused by hypothermia, shock, and stress rather than direct toxic action; distinguishing between the two types of causes can be difficult.[248]

Effects of the *Exxon Valdez* oil spill on populations of sea otter, harbor seals (*Phoca vitulina*), Stellar sea lions, killer whales (*Orcinus orca*), and humpback whales (*Megaptera novaeangliae*) in Prince William Sound, Alaska, are summarized in the overview provided by Loughlin et al.[245] Sea otters and harbor seals were the most affected, with loss estimates in the thousands for otters and in the hundreds for seals. Age distributions of dead sea otters systematically collected from oiled areas of western Prince William Sound between 1976 and 1998 revealed a reduction in the survival rate of otters during the 9 years after the spill (1990–1998).[249] Aerial counts of harbor seals at seven sites affected by the oil spill and 18 unaffected sites in central and eastern Prince William Sound revealed a 28% population reduction during the period 1990–1997; however, harbor seal populations on the survey route were declining prior to the 1989 spill.[250] Effects of the Gulf War oil spills on cetaceans in the Arabian Gulf were thought to be minimal.[251] Overall, the consequences of local effects (acute, chronic, and indirect) of catastrophic oil spills on regional populations of marine mammals have proven difficult to determine because of a lack of prespill population information, movement of animals within the region, and natural fluctuations in survival and reproduction.

Documentation of the effects of oil spills on wild nonmarine mammals is less than for marine mammals. Large numbers of muskrats (*Ondatra zibethica*) were killed by a spill of bunker C fuel oil in the St. Lawrence River (E.S. Smith, New York Department of Environmental Conservation, Albany, NY, unpublished report). Giant kangaroo rats (*Dopodomys ingens*) in California were found dead after being oiled,[252] beaver (*Castor canadensis*) and muskrats were killed by an aviation kerosine spill in a Virginia river,[253] and rice rats (*Oryzomys palustris*) in a laboratory experiment died after swimming through oil-covered water.[254] Oil field pollution in wooded areas in Russia affected blood characteristics, organ indices, species composition, relative abundance, and population age and sex structure of small mammals.[255] Cotton rats occupying old petrochemical sites were characterized by increased cell apoptosis in the ovary and thymus,[256] and rodent populations and assemblages on the same types of sites were altered compared to nearby reference sites.[257] The literature on effects of crude or refined petroleum on laboratory and domestic animals is substantial; recent examples are Lee and Talaska (DNA adduct formation in mouse skin),[258] Feuston et al. (systemic effects of dermal application in rats),[259] Khan et al. (clinical and metabolic effects in dosed cattle),[260] and Mattie et al. (pathology and biochemistry of dosed rats).[261]

The effects of major oil spills on the environment of mammals could include food reduction, an altered diet, and changed use of habitat.[262,263] These effects could be short- or long-term and would be most serious during the breeding season, when movement of females and young is restricted.

The metabolism and effects of some PAHs have been well documented in laboratory rodents and domestic mammals but poorly documented in wild mammals. Acute oral $LD_{50}$ values for selected PAHs in laboratory rodents range from 50 to 2000 mg/kg body weight.[2] Target organs for PAH toxic action are skin, small intestine, kidney, and mammary gland; tissues of the hematopoietic, lymphoid, and immune systems; and gametic tissue. Nonalkylated PAHs are rapidly metabolized; hence, accumulation is less likely than for alkylated PAHs.[2] Partially aromatic PAHs, alkylated fully and partially aromatic PAHs, and metabolites of nonalkylated, fully aromatic PAHs have the greatest potential to alter DNA and induce cancerous and noncancerous neoplasms in epithelial tissues of laboratory and domestic mammals.[2] Species differences in sensitivity to carcinogenesis appears to be a function of differences in levels of mixed-function oxidase activities. Background exposure concentrations of PAHs and studies on mixtures of PAHs in mammals are needed to accurately assess the potential hazard of exposure to multiple PAHs in polluted environments.[2]

Although PAHs can produce systemic effects, DNA alterations, and cancer, concerns about the carcinogenic and mutagenic potential predominate, especially for human health. Consequently, PAHs that are carcinogenic or mutagenic are most studied.[2] With regard to wild mammals, a few investigations have documented the presence of PAH adducts on DNA of marine mammals,[264–266] and participants of a recent workshop on marine mammals and persistent ocean contaminants[267] described PAHs as a "less widely recognized" contaminant that should be further studied because of potential mutagenic and genotoxic effects.

### 14.5.8 The *Exxon Valdez* and Arabian Gulf Oil Spills

The 1989 *Exxon Valdez* oil spill (EVOS) in Prince William Sound, Alaska and the Gulf War oil spills (GWOS) in the Arabian Gulf in 1991 are worthy of special comment. Each was significant for different reasons, and reports of the fate and effects of these spills dominated the petroleum pollution literature during the 1990s.

The EVOS consisted of 36,000 metric tons of crude oil released in just a few days into a high-latitude, relatively pristine marine ecosystem. Although many times smaller than the Gulf War oil discharges, the scientific and societal response was immediate and intense. The spill resulted in a massive oil-removal effort,[268] a thorough crude oil mass balance determination spanning the period 3/24/89 to 10/1/92[269] and the largest wildlife rescue and rehabilitation effort ever attempted (at a cost of $45 million).[270]

In the United States, the Clean Water Act and the Comprehensive Environmental Response, Compensation, and Liability Act determined the response requirements for trustees (federal, state, tribal) of affected natural resources. Trustees initiated research that sought evidence of "injury" to natural resources, and Exxon Corporation initiated research that sought to demonstrate minimal injury and subsequent restoration of affected resources. This adversarial legal process discouraged cooperation among scientists and delayed public access to the results of the studies for several years.[268] Two books containing studies sponsored mostly by Exxon[271] and Trustees[272] provide many examples of conflicting interpretations of the effects of the spill; Wiens[273] provides a discussion of the effects of advocacy on investigations of the effects of the spill on birds. The EVOS is rapidly becoming the most studied oil spill in history; a stream of scientific reports continues to be generated by scientists assessing the biological consequences.

The GWOS consisted of 240 million metric tons of crude oil released into the northern Arabian Gulf over a 6-month period; the Gulf had a previous history of petroleum pollution from oil production activities and warfare. The spillage was caused by acts of war, and scientific coverage of the progression of the spill and its biological effects was delayed until hostilities ended and munitions were cleared from areas affected by discharged oil. Investigators from a number of countries performed assessments of the effects of the oil, which appear to have been less severe on plants and animals of aquatic and coastal environments than the effects reported for the EVOS. In contrast, the contamination of terrestrial environments caused by destruction of oil wells and pipelines was severe and is likely to affect terrestrial plants and animals for many years to come.[274,275]

### 14.5.9 Conclusions

The effects of petroleum on organisms are as varied as the composition of petroleum and the environmental conditions accompanying its appearance. Petroleum can cause environmental harm by toxic action, physical contact, chemical and physical changes within the soil or water medium, and habitat alteration. Oil spills have caused major changes in local plant and invertebrate populations lasting from several weeks to many years. Effects of oil spills on populations of mobile vertebrate species, such as fish, birds, and mammals, have been difficult to determine beyond an accounting of immediate losses and short-term changes in local populations. Reptiles and amphibians need further study. Knowledge of the biological effects of petroleum in freshwater environments continues to increase but still lags behind comparable information for saltwater environments.

Concentrations of individual PAHs in air, soil, and water are usually insufficient to be acutely toxic, but numerous sublethal effects can be produced. The induction of lesions and neoplasms in laboratory animals by metabolites of PAHs and observations of lesions and neoplasms in fish from PAH-contaminated sites indicate potential health problems for animals with a strong MFO system capable of metabolizing PAHs. Although evidence linking environmental PAHs to the incidence of cancerous neoplasms in wild vertebrates is primarily limited to fish, the growing quantities of PAHs entering our environment are a cause for concern.

## 14.6 SUMMARY

Petroleum and individual PAHs from anthropogenic sources are found throughout the world in all components of ecosystems. Crude and refined oils are highly variable in composition and physical characteristics and consist of thousands of hydrocarbon and nonhydrocarbon compounds. Polycyclic aromatic hydrocarbons are aromatic hydrocarbons with two to seven fused benzene rings that can have alkyl groups attached to the rings. Less than half of the petroleum in the environment comes from spills and discharges associated with petroleum transportation. Most of the petroleum comes from industrial, municipal, and household discharges; motorized vehicles; natural oil seeps; and acts of war, terrorism, vandalism, and theft. Most PAHs are formed by a process of thermal decomposition and subsequent recombination of organic molecules (pyrolysis and pyrosynthesis). Low-temperature processes produce the PAHs in coal and oil. High-temperature processes can occur naturally (forest and grass fires, volcanoes) or can be caused by anthropogenic activities (oil, coal, and wood combustion; refuse incineration; industrial activity). Most high-temperature PAHs enter the environment as combustion emissions or components of liquid waste effluents from industrial sites and municipal sewage plants.

Crude and refined petroleum spreads rapidly on water and begins to change composition upon exposure to air, water, or sunlight. Anthropogenic aromatic hydrocarbons are dispersed by air currents and movement of waters receiving wastewater effluents. Hydrocarbons are primarily degraded by microbial metabolism; mammals, birds, fish, and many macroinvertebrates can also metabolize ingested hydrocarbons. Hydrocarbons are also degraded by photooxidation, photolysis, and chemical oxidation.

Organisms with high lipid content, activity patterns or distributions that coincide with the location of the hydrocarbon source, and a poor mixed-function oxygenase system are most likely to accumulate hydrocarbons. Trophic-level increases in accumulation have not been observed. Residence time for petroleum in the water column is usually less than 6 months. Coastal environments can retain oil from several days to 20 years, depending on the configuration of the shoreline, type of substrate, and climate. High-molecular-weight hydrocarbons, particularly aromatics, can persist for long periods of time (> 20 years) in sediments and terrestrial soils.

Petroleum can have lethal or sublethal effects in plants and microbes. Recovery from the effects of oil spills requires as little as a few weeks for water column microalgae up to 5 years for most wetland plants; mangroves could require up to 20 years. Individual PAHs at low concentrations can induce positive or negative sublethal effects in aquatic bacteria and algae; high concentrations can lead to death.

Oil spills often have pronounced effects on local populations of invertebrates; recovery could require a week for zooplankton or 10 years for intertidal populations of mollusks. A large and diverse amount of experimental and survey research, performed with saltwater and freshwater invertebrates, has demonstrated lethal and many sublethal effects as well as population and community effects. Individual PAHs can be lethal at high concentrations and cause sublethal effects at low concentrations.

Eggs, larvae, and early juvenile stages of fish are more vulnerable to oil because they have limited or no ability to avoid it. Large losses of adult fish are usually limited to situations where a large quantity of oil rapidly moves into shallow water. Many sublethal effects of oil on fish have been documented. Effects of oil spills on fish populations in large bodies of water are difficult to determine because of large natural variation in annual recruitment. Laboratory studies of PAH metabolism and lesions and tumors in fish collected from areas heavily contaminated with PAHs imply that PAH exposure can cause cancerous and noncancerous tissue changes in feral fish. Uncertainties about the interactive effects of multiple pollutants at heavily contaminated sites and inadequate knowledge of dose-effect responses and temporal aspects of *in situ* exposure to carcinogens complicate efforts to link environmental PAHs to neoplasms or local population changes.

Adult reptiles and amphibians can be killed and their eggs and amphibian larvae killed or sublethally affected by petroleum. However, available information is inadequate to compare their sensitivity to petroleum or individual PAHs to that of other vertebrates.

Birds are often killed by oil spills, primarily because of plumage oiling and oil ingestion. Birds that spend much of their time on the water surface are the most vulnerable to spilled oil. Ingested oil can cause many sublethal effects. Effects of oil spills are more likely to be detected at the level of local populations than at the regional or species level. When the quantity of data is large, investigators often extrapolate responses of individual birds to local and regional populations. Population modeling studies have shown that long-lived birds with low reproductive success will have the most difficulty recovering from a major oil spill. Recovery potential for a species is a function of the reproductive potential of the survivors and the immigration potential at the spill site. Experiments with individual or groups of PAHs and bird eggs, nestlings, and adults have revealed a variety of toxic responses and shown that PAHs are responsible for most of the toxic effects attributed to petroleum exposure.

Mammals that rely on fur for insulation (polar bear, otters, fur seals, muskrat) are the most likely to die from oiling. Mammals that rely on layers of fat for insulation (seals, cetaceans) are infrequently killed by oil. Ingested oil is rapidly metabolized and cleared, but it can cause many sublethal effects. Effects of spilled oil on local and regional populations of marine mammals have proven difficult to determine because of a lack of prespill population information, movement of animals within the region, and natural fluctuations in survival and reproduction. Laboratory mammals, but not wild mammals, have been extensively used to study the toxic and carcinogenic potential of individual PAHs. Partially aromatic PAHs, alkylated, fully and partially aromatic PAHs, and metabolites of nonalkylated, fully aromatic PAHs have the greatest potential to cause neoplasms in tissues of laboratory and domestic mammals. Investigations involving mixtures of PAHs are especially needed.

In general, petroleum negatively affects living organisms through physical contact, toxic action, and habitat modification, whereas individual PAHs have toxic effects. Partially metabolized and alkylated PAHs can induce genetic damage, developmental abnormalities, and cancerous and noncancerous tissue changes. Evidence linking environmental concentrations of PAHs to induction of cancer in wild animals is strongest for fish. Although concentrations of individual PAHs in aquatic environments are usually much lower than concentrations that are acutely toxic to aquatic organisms, sublethal effects can be produced. Effects of spills on populations of mobile species have been difficult to determine beyond an accounting of immediate losses and, sometimes, short-term changes in local populations.

## REFERENCES


1. Bourne, W. R. P., Oil pollution and bird populations, in *The Biological Effects of Oil Pollution on Littoral Communities*, McCarthy, J. D. and Arthur, D. R., Eds., Field Studies 2 (Suppl.), 1968, 99.
2. Eisler, R., Polycyclic aromatic hydrocarbons, in *Handbook of Chemical Risk Assessment*, Vol. 2, Lewis Publishers, Boca Raton, FL, 2000, Chap. 25.
3. Atlas, R. M. and Bartha R., Fate and effects of polluting petroleum in the marine environment, *Residue Rev.*, 49, 49, 1973.
4. Posthuma, J., The composition of petroleum, Rapp. P.-V., *Reun. Cons. Int. Explor. Mer.*, 171, 7, 1977.
5. Farrington, J. W., Analytical techniques for the determination of petroleum contamination in marine organisms, in *Background Information: Workshop on Inputs, Fates, and Effects of Petroleum in the Marine Environment*, Ocean Affairs Board, National Academy of Sciences, Washington, D.C., 1973, 157.
6. Santodonato, S., Howard, P., and Basu, D., Health and ecological assessment of polynuclear aromatic hydrocarbons, *J. Environ. Pathol. Toxicol.*, Special Issue, 5, 1981.
7. Neff, J. M., Polycyclic aromatic hydrocarbons, in *Fundamentals of Aquatic Toxicology, Rand*, G. M. and Petrocilli, S. R., Eds., Hemisphere, New York, 1985, Chap. 14.
8. McElroy, A. E., Farrington, J. W., and Teal, J. M., Bioavailability of polycyclic aromatic hydrocarbons in the aquatic environment, in *Metabolism of Polycyclic Aromatic Hydrocarbons in the Aquatic Environment*, Varanasi, U., Ed., CRC Press, Boca Raton, FL, 1989, Chap. 1.
9. National Academy of Sciences, Petroleum in the Marine Environment. Workshop on Inputs, Fates and the Effects of Petroleum in the Marine Environment, Ocean Affairs Board, Washington, D.C., 1975, Chap. 1.
10. Grossling, B. F., An estimate of the amounts of oil entering the oceans, in *Sources, Effects and Sinks of Hydrocarbons in the Aquatic Environment*, American Institute of Biological Sciences, Arlington, VA, 1976, 6.
11. National Research Council, *Oil in the Sea. Inputs, Fates, and Effects*, Board on Ocean Science and Policy, Washington, D.C., 1985, Chap. 1.
12. Whipple, W., Jr. and Hunter, J. V., Petroleum hydrocarbons in urban runoff, *Water Resour. Bull.*, 15, 1096, 1979.
13. Connell, D. W., An approximate petroleum hydrocarbon budget for the Hudson Raritan estuary — New York, *Mar. Pollut. Bull.*, 13, 89, 1982.
14. Readman, J. W. et al., Oil and combustion-product contamination of the Gulf marine environment following the war, *Nature*, 358, 662, 1992.

15. Hosmer, A. W., Stanton, C. E., and Beane, J. L., Intent to spill: Environmental effects of oil spills caused by war, terrorism, vandalism, and theft, in *Proc. 1997 Int. Oil Spill Conf.*, Publ. 4651, American Petroleum Institute, Washington, D.C., 1997, 157.
16. Mastral, A. M. and Callen, M. S., A review on polycyclic aromatic hydrocarbon (PAH) emissions from energy generation, *Environ. Sci. Technol.*, 34, 3051, 2000.
17. Grimmer, G., in *Environmental Carcinogens: Polycyclic Aromatic Hydrocarbons*, Grimmer, G., Ed., CRC Press, Boca Raton, FL, 1983, Chap. 2.
18. Baek, S. O. et al., A review of atmospheric polycyclic aromatic hydrocarbons, sources, fate and behavior, *Water Air Soil Pollut.*, 60, 279, 1991.
19. Colwell, R. R. and Walker, J. D., Ecological aspects of microbial degradation of petroleum in the marine environment, *Crit. Rev. Microbiol.*, 5, 423, 1977.
20. Lee, R. F., Fate of oil in the sea, in *Proc. 1977 Oil Spill Response Workshop*, Four, P. L., Ed., U.S. Fish and Wildlife Service FWS/OBS/77-24, 1977, 43.
21. McEwan, E. H. and Whitehead, P. M., Uptake and clearance of petroleum hydrocarbons by the glaucous-winged gull (*Larus glaucescens*) and the mallard duck (*Anas platyrhynchos*), *Can. J. Zool.*, 58, 723, 1980.
22. Engelhardt, F. R., Petroleum effects on marine mammals, *Aquat. Toxicol.*, 4, 199, 1983.
23. James, M. O., Microbial degradation of PAH in the aquatic environment, in *Metabolism of Polycyclic Aromatic Hydrocarbons in the Aquatic Environment*, Varanasi, U., Ed., CRC Press, Boca Raton, FL, 1989, Chap. 3.
24. Jenssen, B. M., Ekker, M., and Zahlsen, K., Effects of ingested crude oil on thyroid hormones and on the mixed function oxidase system in ducks, *Comp. Biochem. Physiol.*, 95C, 213, 1990.
25. Lytle, J. S. and Lytle, T. F., The role of *Juncus roemerianus* in cleanup of oil-polluted sediments, in *Proc. 1987 Oil Spill Conf.*, Publ. 4452, American Petroleum Institute, Washington, D.C., 1987, 495.
26. Simonich, S. L. and Hites, R. A., Importance of vegetation in removing polycyclic aromatic hydrocarbons from the atmosphere, *Nature*, 370, 49, 1994.
27. Clement, L. E., Stekoll, M. S., and Shaw, D. G. Accumulation, fractionation and release of oil by the intertidal clam *Macoma balthica*, *Mar. Biol.*, 57, 41, 1980.
28. Soler, M., Grimalt, J. O., and Albaiges, J., Vertical distribution of aliphatic and aromatic hydrocarbons in mussels from the Amposta offshore oil production platform (western Mediterrean), *Chemosphere*, 18, 1809, 1989.
29. Boehm, P. D. and Quinn, J. G., The persistence of chronically accumulated hydrocarbons in the hard shell clam *Mercenaria mercenaria*, *Mar. Biol.*, 44, 227, 1977.
30. Broman, D., Naf, C., Lundbergh, I., and Zebuhr, Y., An *in situ* study on the distribution, biotransformation and flux of polycyclic aromatic hydrocarbons (PAHs) in an aquatic food chain (seston — *Mytilus edulis L. — Somateria mollissima L.*) from the Baltic: An ecotoxicological perspective, *Environ. Toxicol. Chem.*, 9, 429, 1990.
31. Rattner, B. A., Hoffman, D. J., and Marn, C. N., Use of mixed-function oxygenases to monitor contaminant exposure, *Environ. Toxicol. Chem.*, 8, 1093, 1989.
32. Wild, S. R. et al., Polynuclear aromatic hydrocarbons in crops from long-term field experiments amended with sewage sludge, *Environ. Pollut.*, 76, 25, 1992.
33. Varanasi, U., Stein, J. E., and Nishimoto, M., Biotransformation and disposition of PAH in fish, in *Metabolism of Polycyclic Aromatic Hydrocarbons in the Aquatic Environment*, Varanasi, U., Ed., CRC Press, Boca Raton, FL, 1989, Chap. 4.
34. Gundlach, E. R. and Hayes, M. O., Classification of coastal environments in terms of potential vulnerability to oil spill damage, *Mar. Technol. Soc. J.*, 12, 18, 1978.
35. Page, D. S. et al., Long-term weathering of *Amoco Cadiz* oil in soft intertidal sediments, in *Proc. 1989 Oil Spill Conf.,* American Petroleum Institute Publ. 4479, Washington, D.C., 1989, 401.
36. Corredor, J. E., Morell, J. M., and Del Castillo, C. E., Persistence of spilled crude oil in a tropical intertidal environment, *Mar. Pollut. Bull.*, 21, 385, 1990.
37. Vandermeulen, J. H. and Singh, J. G., ARROW oil spill, 1970–90; persistence of 20-yr weathered bunker C fuel oil, *Can. J. Fish. Aquat. Sci.*, 51, 845, 1994.
38. Baker, J. M. et al., Long-term fate and effects of untreated thick oil deposits on salt marshes, in *Proc. 1993 Oil Spill Conf.,* Publ. 4580, American Petroleum Institute, 1993, 395.
39. Gundlach, E. R. et al., The fate of *Amoco Cadiz* oil, *Science*, 221, 122, 1983.

40. DeLaune, R. D. et al., Degradation of petroleum hydrocarbons in sediment receiving produced water discharge, *J. Environ. Sci. Health*, A35, 1, 2000.
41. Gundlach, E. R., Domeracki, D. D., and Thebeau, L. C., Persistence of Metula oil in the Strait of Magellan six and one-half years after the incident, *Oil Petrochem. Pollut.*, 1, 37, 1982.
42. Haines, J. R. and Atlas, R. M., *In situ* microbial degradation of Prudhoe Bay crude oil in Beaufort Sea sediments, *Mar. Environ. Res.*, 7, 91, 1982.
43. Huntley, S. L., Bonnevie, N. L., and Wenning, R. J., Polycyclic aromatic hydrocarbons and petroleum hydrocarbon contamination in sediment from the Newark Bay Estuary, New Jersey, *Arch. Environ. Contam. Toxicol.*, 28, 93, 1995.
44. Olivera, F. L. et al., Prepared bed land treatment of soils containing diesel and crude oil hydrocarbons, *J. Soil Contam.*, 7, 657, 1998.
45. Wang, Z. et al., Study of the 25-year-old Nipisi oil spill: Persistence of oil residues and comparisons between surface and subsurface sediments, *Environ. Sci. Technol.*, 32, 2222, 1998.
46. Aislabie, J. et al., Polycyclic aromatic hydrocarbons in fuel-oil contaminated soils, Antarctica, *Chemosphere*, 39, 2201, 1999.
47. Jahn, A. E. and Robilliard, G. A., Natural recovery: A practical natural resource restoration option following oil spills, in *Proc. 1997 Oil Spill Conf.*, Publ. 4651, American Petroleum Institute, Washington, D.C., 1997, 665.
48. American Petroleum Institute, Oil Spill Cleanup: Options for Minimizing Adverse Ecological Impacts, Publ. 4435, American Petroleum Institute, Washington, D.C., 1985, Chap. 2.
49. Vandermeulen, J. H. and Ross, C. W., Oil spill response in freshwater: Assessment of the impact of cleanup as a management tool, *J. Environ. Manage.*, 44, 297, 1995.
50. Atlas, R.M., Petroleum biodegradation and oil spill bioremediation, *Mar. Pollut. Bull.*, 31, 178, 1995.
51. Varanasi, U., Metabolic activation of PAH in subcellular fractions and cell cultures from aquatic and terrestrial species, in *Metabolism of Polycyclic Aromatic Hydrocarbons in the Aquatic Environment*, Varanasi, U., Ed., CRC Press, Boca Raton, FL, 1989, Chap. 6.
52. Fu, P. P. et al., Halogenated-polycyclic aromatic hydrocarbons: A class of genotoxic environmental pollutants, *Environ. Carcino. Ecotox. Revs.*, C17, 71, 1999.
53. Ren, L. et al., Photoinduced toxicity of three polycyclic aromatic hydrocarbons (fluoranthene, pyrene, and naphthalene) to the duckweed *Lemna gibba L.* G-3, *Ecotox. Environ. Saf.*, 28, 160, 1994.
54. Arfsten, D. P., Schaeffer, D. J., and Mulveny, D. C., The effects of near ultraviolet radiation on the toxic effects of polycyclic aromatic hydrocarbons in animals and plants: A review, *Ecotox. Environ. Saf.*, 33, 1, 1996.
55. Baumann, P. C., PAH, metabolites, and neoplasia in feral fish populations, in *Metabolism of Polycyclic Aromatic Hydrocarbons in the Aquatic Environment*, Varanasi, U., Ed., CRC Press, Boca Raton, FL, 1989, Chap. 8.
56. Chang, S., Zdanowicz, V. S., and Murchelano, R. A., Associations between liver lesions in winter flounder (*Pleuronectes americanus*) and sediment chemical contaminants from north-east United States estuaries, *ICES J. Mar. Sci.*, 55, 954, 1998.
57. Baker, J. M. et al., Tropical marine ecosystems and the oil industry; with a description of a post-oil survey in Indonesian mangroves, in *Proc. PETROMAR '80*, Graham and Trotman, London, 1980, 617.
58. Duke, N. C., Pinzon, Z. S., and Prada, T. M. C., Large-scale damage to mangrove forests following two large oil spills in Panama, *Biotropica*, 29, 2, 1997.
59. Jackson, J. B. C. et al., Ecological effects of a major oil spill on Panamanian coastal marine communities, *Science*, 243, 37, 1989.
60. Floc'h, J.-Y. and Diouris, M., Initial effects of *Amoco Cadiz* oil on intertidal algae, *Ambio*, 9, 284, 1980.
61. Thomas, M. L. H., Long term biological effects of Bunker C oil in the intertidal zone, in *Fate and Effects of Petroleum Hydrocarbons in Marine Organisms and Ecosystems*, Wolfe, D. A., Ed., Pergamon Press, New York, 1977, 238.
62. Baca, B. J., Lankford, T. E., and Gundlach, E. R., Recovery of Brittany coastal marshes in the eight years following the *Amoco Cadiz* incident, in *Proc.1987 Oil Spill Conf.*, Publ. 4452, American Petroleum Institute, Washington, D.C., 1987, 459.
63. Mendelssohn, I. A., Hester, M. W., and Hill, J. M., Assessing the recovery of coastal wetlands from oil spills, in *Proc. 1993 Int. Oil Spill Conf.*, Publ. 4580, American Petroleum Institute, Washington, D.C., 1993, 141.

64. Jones, D. A. et al., Long-term (1991–1995) monitoring of the intertidal biota of Saudi Arabia after the 1991 Gulf War oil spill, *Mar. Pollut. Bull.*, 36, 472, 1998.
65. Burk, C. J., A four year analysis of vegetation following an oil spill in a freshwater marsh, *J. Appl. Ecol.*, 14, 515, 1977.
66. Baca, B. J., Getter, C. D., and Lindstedt-Siva, J., Freshwater oil spill considerations: Protection and cleanup, in *Proc. 1985 Oil Spill Conf.*, Publ. 4385, American Petroleum Institute, Washington, D.C., 1985, 385.
67. Johansson, S., Larsson, U., and Boehm, P., The *Tsesis* oil spill. Impact on the pelagic ecosystem, *Mar. Pollut. Bull.*, 11, 284, 1980.
68. Shailaja, M. S., The influence of dissolved petroleum hydrocarbon residues on natural phytoplankton biomass, *Mar. Environ. Res.*, 25, 315, 1988.
69. Klekowski, E. J., Jr. et al., Petroleum pollution and mutation in mangroves, *Mar. Pollut. Bull.*, 28, 166, 1994.
70. Malallah, G. et al., Genotoxicity of oil pollution on some species of Kuwaiti flora, *Biologia*, Bratislava, 52, 61, 1997.
71. Heitkamp, M. A. and Johnson, B. T., Impact of an oil field effluent on microbial activities in a Wyoming river, *Can. J. Microbiol.*, 30, 786, 1984.
72. Braddock, J. F., Lindstrom, J. E., and Brown, E. J., Distribution of hydrocarbon-degrading microorganisms in sediments from Prince William Sound, Alaska, following the *Exxon Valdez* oil spill, *Mar. Pollut. Bull.*, 30, 125, 1995.
73. Megharaj, M. et al., Influence of petroleum hydrocarbon contamination on microalgae and microbial activities in a long-term contaminated soil, *Arch. Environ. Contam. Toxicol.*, 38, 439, 2000.
74. Gilfillan, E. S. et al., Use of remote sensing to document changes in marsh vegetation following the *Amoco Cadiz* oil spill (Brittany, France, 1978), *Mar. Pollut. Bull.*, 30, 780, 1995.
75. Houghton, J. P. et al., Prince William Sound intertidal biota seven years later: Has it recovered?, in *Proc. 1997 Oil Spill Conf.,* Publ. 4651, American Petroleum Institute, Washington, D.C., 1997, 679.
76. Burns, K. A., Garrity, S. D., and Levings, S. C., How many years until mangrove ecosystems recover from catastrophic oil spills?, *Mar. Pollut. Bull.*, 26, 239, 1993.
77. Davenport, J., Oil and planktonic ecosystems, *Phil. Trans. R. Soc. Lond.*, B297, 369, 1982.
78. Griffiths, R. P. et al., The long-term effects of crude oil on microbial processes in subarctic marine sediments, *Estuar. Coastal Shelf Sci.*, 15, 183, 1982.
79. Vargo G. A., Hutchins, M., and Almquist, G., The effect of low, chronic levels of No. 2 fuel oil on natural phytoplankton assemblages in microcosms: 1. Species composition and seasonal succession, *Mar. Environ. Res.*, 6, 245, 1982.
80. Getter, C. D., Ballou, T. G., and Koons, C. B., Effects of dispersed oil on mangroves. Synthesis of a seven-year study, *Mar. Pollut. Bull.*, 16, 318, 1985.
81. Parra-Pardi, G., Sutton, E. A., and Rincon, N. E., Effects of petroleum on algal blooms in Lake Maracaibo, in *Proc. 1985 Oil Spill Conf.*, Publ. 4385, American Petroleum Institute, Washington, D.C., 1985, 373.
82. Ballou, T. T. et al., Tropical oil pollution investigations in coastal systems (tropics): The effects of untreated and chemically dispersed Prudhoe Bay crude oil on mangroves, seagrasses, and corals in Panama, in *Oil Dispersants: New Ecological Approaches*, Flaherty, L. M., Ed., American Society for Testing and Materials STP 1018, Philadelphia, 1989, 229.
83. Siron, R. et al., Water-soluble petroleum compounds: Chemical aspects and effects on the growth of microalgae, *Sci. Total Environ.*, 104, 211, 1991.
84. Sjotun, K. and Lein, T. E., Experimental oil exposure of *Ascophyllum nodosum* (*L.*) Le Jolis, *J. Exp. Mar. Biol. Ecol.*, 170, 197, 1993.
85. Antrim, L. D. et al., Effects of petroleum products on bull kelp (*Nereocystis luetkeana*), *Mar. Biol.*, 122, 23, 1995.
86. Braddock, J. F. et al., Patterns of microbial activity in oiled and unoiled sediments in Prince William Sound, in *Proc. Exxon Valdez Oil Spill Symp.*, Rice, S. D. et al., Eds., Symposium 18, American Fisheries Society, Bethesda, 1996, 94.
87. Pezeshki, S. R. et al., The effects of oil spill and clean-up on dominant US Gulf coast marsh macrophytes: A review, *Environ. Pollut.*, 108, 129, 2000.

88. Snow, N. B. and Scott, B. F., The effect and fate of crude oil spilt on two Arctic lakes, in *Proc. 1975 Conf. Prevention Control Oil Pollut.*, American Petroleum Institute, Washington, D.C., 1975, 527.
89. Bott, T. L., Rogenmuser, K., and Thorne, P., Effects of No. 2 fuel oil, Nigerian crude oil, and used crankcase oil on benthic algal communities, *J. Environ. Sci. Health*, A13, 751, 1978.
90. Bott, T. L. and Rogenmuser, K., Effects of No. 2 fuel oil, Nigerian crude oil, and used crankcase oil on attached algal communities: Acute and chronic toxicity of water-soluble constituents, *Appl. Environ. Microbiol.*, 36, 673, 1978.
91. Federle, T. W. et al., Effects of Prudhoe Bay crude oil on primary production and zooplankton in Arctic tundra thaw ponds, *Mar. Environ. Res.*, 2, 3, 1979.
92. Baker, J. H. and Morita, R. Y., A note on the effects of crude oil on microbial activities in a stream sediment, *Environ. Pollut. (Ser. A)*, 31, 149, 1983.
93. El-Dib, M. A., Abou-Waly, H. F., and El-Naby, A. M. H., Impact of fuel oil on the freshwater alga *Selenastrum capricornutum*, *Bull. Environ. Contam. Toxicol.*, 59, 438, 1997.
94. Anoliefo, G. O. and Vwioko, D. E., Effects of spent lubricating oil on the growth of *Capsicum annum L.* and *Lycopersicon esculentum* Miller, *Environ. Pollut.*, 88, 361, 1995.
95. Chaineau, C. H., Morel, J. L., and Oudot, J., Phytotoxicity and plant uptake of fuel oil hydrocarbons, *J. Environ. Qual.*, 26, 1478, 1997.
96. Nicolotti, G. and Egli, S., Soil contamination by crude oil: Impact on the mycorrhizosphere and on the revegetation potential of forest trees, *Environ. Pollut.*, 99, 37, 1998.
97. Collins, C. M., Racine, C. H., and Walsh, M. E., The physical, chemical, and biological effects of crude oil spills after 15 years on a black spruce forest, interior Alaska, *Arctic*, 47, 164, 1994.
98. Johansson, S., Larsson, U., and Boehm, P., The Tsesis oil spill, *Mar. Pollut. Bull.*, 11, 284, 1980.
99. Batten, S. D., Allen, R. J. S., and Wotton, C. O. M., The effects of the *Sea Empress* oil spill on the plankton of the southern Irish Sea, *Mar. Pollut. Bull.*, 36, 764, 1998.
100. Kingston, P. F., Impact of offshore oil production installations on the benthos of the North Sea, *ICES J. Mar. Sci.*, 49, 45, 1992.
101. Daan, R. and Mulder, M., On the short-term and long-term impact of drilling activities in the Dutch sector of the North Sea, *ICES J. Mar. Sci.*, 53, 1036, 1996.
102. Jackson, J. B. C. et al., Ecological effects of a major oil spill on Panamanian coastal marine communities, *Science*, 243, 37, 1989.
103. Lee, R. F. and Page, D. S., Petroleum hydrocarbons and their effects in subtidal regions after major oil spills, *Mar. Pollut. Bull.*, 34, 928, 1997.
104. Dauvin, J.-C., The fine sand *Abra alba* community of the Bay of Morlaix twenty years after the *Amoco Cadiz* oil spill, *Mar. Pollut. Bull.*, 36, 669, 1998.
105. Jewett, S. C. et al., "*Exxon Valdez*" oil spill: Impacts and recovery in the soft-bottom benthic community in and adjacent to eelgrass beds, *Mar. Ecol. Prog. Ser.*, 185, 59, 1999.
106. Sanders, H. L. et al., Anatomy of an oil spill: Long-term effects from the grounding of the barge *Florida* off West Falmouth, Massachusetts, *J. Mar. Res.*, 38, 265, 1980.
107. Highsmith, R. C. et al., Impact of the *Exxon Valdez* oil spill on intertidal biota, in *Proc. Exxon Valdez Oil Spill Symp.*, Rice, S. D. et al., Eds., Symposium 18, American Fisheries Society, Bethesda, 1996, 212.
108. Levings, S. C., Garrity, S. D., and Burns, K. A., The *Galeta* oil spill. III. Chronic reoiling, long-term toxicity of hydrocarbon residues and effects on epibiota in the mangrove fringe, *Estuar. Coastal Shelf Sci.*, 38, 365, 1994.
109. Crunkilton, R. L. and Duchrow, R. M., Impact of a massive crude oil spill on the invertebrate fauna of a Missouri Ozark stream, *Environ. Pollut.*, 63, 13, 1990.
110. Poulton, B. C. et al., Effects of an oil spill on leafpack-inhabiting macroinvertebrates in the Chariton River, Missouri, *Environ. Pollut.*, 99, 115, 1998.
111. Neff, J. et al., An oil spill in an Illinois lake: Ecological and human health assessment, in *Proc. 1995 Int. Oil Spill Conf.*, Publ. 4620, American Petroleum Institute, Washington, D.C., 1995, 415.
112. Suchanek, T. H., Oil impacts on marine invertebrate populations and communities, *Am. Zool.*, 33, 510, 1993.
113. Driskell, W. B. et al., Recovery of Prince William Sound intertidal in fauna from *Exxon Valdez* oiling and shoreline treatments, 1989 through 1992, in *Proc. Exxon Valdez Oil Spill Symp.*, Rice, S. D. et al., Eds., Symposium 18, American Fisheries Society, Bethesda, 1996, 362.

114. Straughan, D. and Hadley, D., Experiments with *Littorina* species to determine the relevancy of oil spill data from southern California to the Gulf of Alaska, *Mar. Environ. Res.*, 1, 135, 1978.
115. Davies, J. M. et al., Some effects of oil-derived hydrocarbons on a pelagic food web from observations in an enclosed ecosystem and a consideration of their implications for monitoring, *Rapp. P.-v. Reun. Cons. Int. Explor. Mer.*, 179, 201, 1980.
116. Carr, R. S. and Linden, O., Bioenergetic responses of *Gammarus salinus* and *Mytilus edulis* to oil and oil dispersants in a model ecosystem, *Mar. Ecol.Prog. Ser.*, 19, 285, 1984.
117. Frithsen, J. B., Elmgren, R., and Rudnick, D. T., Responses of benthic meiofauna to long-term, low-level additions of No. 2 fuel oil, *Mar. Ecol. Prog. Ser.*, 23, 1, 1985.
118. Cross, W. E., Martin, C. M., and Thomson, D. H., Effects of experimental releases of oil and dispersed oil on arctic nearshore macrobenthos. II. Epibenthos, *Arctic*, 40 (Suppl. 1), 201, 1987.
119. Moore, M. N., Livingstone, D. R., and Widdows, J., Hydrocarbons in marine mollusks: Biological effects and ecological consequences, in *Metabolism of Polycyclic Aromatic Hydrocarbons in the Aquatic Environment*, Varanasi, U., Ed., CRC Press, Boca Raton, FL, 1989, Chap. 9.
120. Nance, J. M., Effects of oil/gas field produced water on the macrobenthic community in a small gradient estuary, *Hydrobiologia*, 220, 189, 1991.
121. Burger, J., Brzorad, J., and Gochfeld, M., Immediate effects of an oil spill on behavior of fiddler crabs (*Uca pugnax*), *Arch. Environ. Contam. Toxicol.*, 20, 404, 1991.
122. Brown, R. P., Cristini, A., and Cooper, K. R., Histopathological alterations in *Mya arenaria* following a #2 fuel oil spill in the Arthur Kill, Elizabeth, New Jersey, *Mar. Environ. Res.*, 34, 65, 1992.
123. Shelton, M. E. et al., Degradation of weathered oil by mixed marine bacteria and the toxicity of accumulated water-soluble material to two marine crustacea, *Arch. Environ. Contam. Toxicol.*, 36, 13, 1999.
124. Carman, K. R., Fleeger, J. W., and Pomarico, S. M., Does historical exposure to hydrocarbon contamination alter the response of benthic communities to diesel contamination?, *Mar. Environ. Res.*, 49, 255, 2000.
125. Rogerson, A., Berger, J., and Grosso, C. M., Acute toxicity of ten crude oils on the survival of the rotifer *Asplanchna sieboldi* and sublethal effects on rates of prey consumption and neonate production, *Environ. Pollut. (Ser. A)*, 29, 179, 1982.
126. Woodward, D. F. and Riley, R. G., Petroleum hydrocarbon concentrations in a salmonid stream contaminated by oil field discharge water and effects on macrobenthos, *Arch. Environ. Contam. Toxicol.*, 12, 327, 1983.
127. Miller, M. C. and Stout, J. R., Effects of a controlled under-ice oil spill on invertebrates of an arctic and a subarctic stream, *Environ. Pollut. (Ser. A)*, 42, 99, 1986.
128. Bobra, A. M. et al., Acute toxicity of dispersed fresh and weathered crude oil and dispersants to *Daphnia magna*, *Chemosphere*, 19, 1199, 1989.
129. Calfee, R. D. et al., Photoenhanced toxicity of a weathered oil on *Ceriodaphnia dubia* reproduction, *Environ. Sci. Pollut. Res.*, 6, 207, 1999.
130. Faulkner, B. C. and Lochmiller, R. L., Ecotoxicity revealed in parasite communities of *Sigmodon hispidus* in terrestrial environments contaminated with petrochemicals, *Environ. Pollut.*, 110, 135, 2000.
131. Faulkner, B. C. and Lochmiller, R. L., Increased abundance of terrestrial isopod populations in terrestrial ecosystems contaminated with petrochemical wastes, *Arch. Environ. Contam. Toxicol.*, 39, 86, 2000.
132. Erstfeld, K. M. and Snow-Ashbrook, J., Effects of chronic low-level PAH contamination on soil invertebrate communities, *Chemosphere*, 39, 2117, 1999.
133. Dorn, P. B. et al., Assessment of the acute toxicity of crude oils in soils using earthworms, Microtox, and plants, *Chemosphere*, 37, 845, 1998.
134. Pelletier, M. C. et al., Importance of maternal transfer of the photoreactive polycyclic aromatic hydrocarbon fluoranthrene from benthic adult bivalves to their pelagic larvae, *Environ. Toxicol. Chem.*, 19, 2691, 2000.
135. Bowman, R. E. and Langton, R. W., Fish predation on oil-contaminated prey from the region of the Argo Merchant oil spill, in *In the Wake of the Argo Merchant*, University of Rhode Island, Kingston, 1978, 137.
136. Malins, D. C. and Hodgins, H. O., Petroleum and marine fishes: A review of uptake, disposition, and effects, *Environ. Sci. Technol.*, 15, 1272, 1981.

137. Hampson, G. R. and Sanders, H. L., Local oil spill, *Oceanus*, 15,8, 1969.
138. Teal, J. M. and Howarth, R. W., Oil spill studies: A review of ecological effects, *Environ. Manage.*, 8, 27, 1984.
139. Woodward, D. F., Riley, R. G., and Smith, C. E., Accumulation, sublethal effects, and safe concentration of a refined oil as evaluated with cutthroat trout, *Arch. Environ. Contam. Toxicol.*, 12, 455, 1983.
140. Hedtke, S. F. and Puglisi, F. A., Short-term toxicity of five oils to four freshwater species, *Arch. Environ. Contam. Toxicol.*, 11, 425, 1982.
141. Anderson, J. W. et al., Toxicity of dispersed and undispersed Prudhoe Bay crude oil fractions to shrimp and fish, in *Proc. 1987 Oil Spill Conf.*, Publ. 4452, American Petroleum Institute, Washington, D.C., 1987, 235.
142. Barnett, J. and Toews, D., The effects of crude oil and the dispersant, Oilsperse 43, on respiration and coughing rates in Atlantic salmon (*Salmo salar*), *Can. J. Zool.*, 56, 307, 1978.
143. Little, E. E. et al., Assessment of the photoenhanced toxicity of a weathered oil to the tidewater silverside, *Environ. Toxicol. Chem.*, 19, 926, 2000.
144. Chambers, J. E. et al., Enzyme activities following chronic exposure to crude oil in a simulated ecosystem, *Environ. Res.*, 20, 140, 1979.
145. Thomas, P., Woodin, B. R., and Neff, J. M., Biochemical responses of the striped mullet *Mugil cephalus* to oil exposure, I. Acute responses — interrenal activations and secondary stress responses, *Mar. Biol.*, 59, 141, 1980.
146. Fletcher, G. L., Kiceniuk, J. W., and Williams, U. P., Effects of oiled sediments on mortality, feeding and growth of winter flounder *Pseudopleuronectes americanus*, *Mar. Ecol. Prog. Ser.*, 4, 91, 1981.
147. Weber, D. D. et al., Avoidance reactions of migrating adult salmon to petroleum hydrocarbons, *Can. J. Fish. Aquat. Sci.*, 38, 779, 1981.
148. Thomas, P. and Budiantara, L., Reproductive life history stages sensitive to oil and naphthalene in Atlantic croaker, *Mar. Environ. Res.*, 39, 147, 1995.
149. Kuehn, R. L. et al., Relationships among petroleum refining, water and sediment contamination, and fish health, *J. Toxicol. Environ. Health*, 46, 101, 1995.
150. Willette, M., Impacts of the *Exxon Valdez* oil spill on the migration, growth, and survival of juvenile pink salmon in Prince William Sound, in *Proc. Exxon Valdez Oil Spill Symp.*, Symposium 18, American Fisheries Society, Bethesda, 1996, 533.
151. Gregg, J. C, Fleeger, J. W., and Carman, K. R., Effects of suspended, diesel-contaminated sediment on feeding rate in the darter goby, *Gobionellus boleosoma* (*Teleostei: Gobiidae*), *Mar. Pollut. Bull.*, 34, 269, 1997.
152. Moles, A. and Norcross, B. L., Effects of oil-laden sediments on growth and health of juvenile flatfishes, *Can. J. Fish. Aquat. Sci.*, 55, 605, 1998.
153. Carls, M. G. et al., Expression of viral hemorrhagic septicemia virus in prespawning Pacific herring (*Clupea pallasi*) exposed to weathered crude oil, *Can. J. Fish. Aquat. Sci.*, 55, 2300, 1998.
154. Khan, R. A., Study of pearl dace (*Margariscus margarita*) inhabiting a stillwater pond contaminated with diesel fuel, *Bull. Environ. Contam. Toxicol.*, 62, 638, 1999.
155. Tilseth, S., Soldberg, T. S., and Westheim, K., Sublethal effects of the water-soluble fraction of Ekofisk crude oil on the early larval stages of cod (*Gadus morhua L.*), *Mar. Environ. Res.*, 11, 1, 1984.
156. Marty, G. D. et al., Histopathology and cytogenetic evaluation of Pacific herring larvae exposed to petroleum hydrocarbons in the laboratory or in Prince William Sound, Alaska, after the *Exxon Valdez* oil spill, *Can. J. Fish. Aquat. Sci.*, 54, 1846, 1997.
157. Heintz, R. A., Short, J. W., and Rice, S. D., Sensitivity of fish embryos to weathered crude oil: Part II. Increased mortality of pink salmon (*Oncorhynchus gorbuscha*) embryos incubating downstream from weathered *Exxon Valdez* crude oil, *Environ. Toxicol. Chem.*, 18, 494, 1999.
158. Carls, M. G., Rice, S. D., and Hose, J. E., Sensitivity of fish embryos to weathered crude oil: Part I. Low-level exposure during incubation causes malformations, genetic damage, and mortality in larval Pacific herring (*Clupea pallasi*), *Environ. Toxicol. Chem.*, 18, 481, 1999.
159. Falk-Petersen, I. and Kjorsvik, E., Acute toxicity tests of the effects of oils and dispersants on marine fish embryos and larvae — A review, *Sarsia*, 72, 411, 1987.
160. Hughes, J. B., Cytological-cytogenetic analyses of winter flounder embryos collected from the benthos at the barge North Cape oil spill, *Mar. Pollut. Bull.*, 38, 30, 1999.

161. Brown, E. D. et al., Injury to the early life history stages of Pacific herring in Prince William Sound after the *Exxon Valdez* oil spill, in *Proc. Exxon Valdez Oil Spill Symp.*, Symposium 18, American Fisheries Society, Bethesda, 1996, 448.
162. Bue, B. G., Sharr, S., and Seeb, J. E., Evidence of damage to pink salmon populations inhabiting Prince William Sound, Alaska, two generations after the *Exxon Valdez* oil spill, *Trans. Am. Fish. Soc.*, 127, 35, 1998.
163. Mankki, J. and Vauras, J., Littoral fish populations after an oil tanker disaster in the Finnish SW archipelago, *Ann. Zool. Fennici*, 11, 120, 1974.
164. Squire, J., Jr., Effects of the Santa Barbara, Calif., oil spill on the apparent abundance of pelagic fishery resources, *Mar. Fish. Rev.*, 54, 7, 1992.
165. Geiger, H. J. et al., A life history approach to estimating damage to Prince William Sound pink salmon caused by the *Exxon Valdez* oil spill, in *Proc. Exxon Valdez Oil Spill Symp.*, Symposium 18, American Fisheries Society, Bethesda, 1996, 487.
166. Cronin, M. A. and Bickham, J. W., A population genetic analysis of the potential for a crude oil spill to induce heritable mutations and impact natural populations, *Ecotoxicology*, 7, 259, 1998.
167. Spaulding, M. L. et al., Oil-spill fishery impact assessment model: Application to selected Georges Bank fish species, *Estuar. Coastal Shelf Sci.*, 16, 511, 1983.
168. Reed, M. et al., Oil spill fishery impact assessment modeling: The fisheries recruitment problem, *Estuar. Coastal Shelf Sci.*, 19, 591, 1984.
169. Williams, D. E., Lech, J. J., and Buhler, D. R., Xenobiotics and xenoestrogens in fish: Modulation of cytochrome P450 and carcinogenesis, *Mutation Res.*, 399, 179, 1998.
170. French, B. L. et al., Accumulation and dose-response of hepatic DNA adducts in English sole (*Pleuronectes vetulus*) exposed to a gradient of contaminated sediments, *Aquat. Toxicol.*, 36, 1, 1996.
171. Myers, M. S. et al., Toxicopathic hepatic lesions as biomarkers of chemical contaminant exposure and effects in marine bottomfish species from the Northeast and Pacific Coasts, USA, *Mar. Pollut. Bull.*, 37, 92, 1998.
172. Myers, M. S. et al., Relationships between hepatic neoplasms and related lesions and exposure to toxic chemicals in marine fish from the U.S. west coast, *Environ. Health Perspect.*, 90, 7, 1991.
173. Baumann, P. C., Smith, W. D., and Ribick, M., Hepatic tumor rates and polynuclear aromatic hydrocarbon levels in two populations of brown bullhead (*Ictalurus nebulosus*), in *Polynuclear Aromatic Hydrocarbons: Physical and Biological Chemistry*, Cooke, M. W., Dennis, A. J., and Fisher, G. L., Eds., Battelle Press, Columbus, Ohio, 1982, 93.
174. Smith, S. B., Blouin, M. A., and Mac, M. J., Ecological comparisons of Lake Erie tributaries with elevated incidence of fish tumors, *J. Great Lakes Res.*, 20, 701, 1994.
175. Roberts, M. H., Jr. et al., Acute toxicity of PAH contaminated sediments to the estuarine fish, *Leiostomus xanthurus*, *Bull. Environ. Contam. Toxicol.*, 42, 142, 1989.
176. Steyermark, A. C. et al., Biomarkers indicate health problems in brown bullheads from the industrialized Schuylkill River, Philadelphia, *Trans. Am. Fish. Soc.*, 128, 328, 1999.
177. Van der Oost, R. et al., Bioaccumulation, biotransformation and DNA binding of PAHs in feral eel (*Anguilla anguilla*) exposed to polluted sediments: A field survey, *Environ. Toxicol. Chem.*, 13, 859, 1994.
178. Pierce, V., The effects of the Arabian Gulf oil spill on wildlife, in *1991 Proc. Am. Assoc. Zoo Veterin.*, Junge, R. E., Ed., Calgary, Canada, 1991, 370.
179. Witham, R., Does a problem exist relative to small sea turtles and oil spills?, in *Proc. Conf. Assess. Ecolog. Impacts Oil Spills*, American Institute of Biological Sciences, Keystone, Colorado, 1978, 630.
180. Hall, R. J., Belisle, A. A., and Sileo, L., Residues of petroleum hydrocarbons in tissues of sea turtles exposed to the *Ixtoc I* oil spill, *J. Wildl. Dis.*, 19, 106, 1983.
181. Symens, P. and Al Salamah, M. I., The impact of the Gulf War oil spills on wetlands and waterfowl in the Arabian Gulf, in *Wetland and Waterfowl Conservation in South and West Asia*, Moser, M. and van Vessem, J., Eds., Special Publ. No. 25, The International Waterfowl and Wetlands Research Bureau, Slimbridge, U.K., 1993, 24.
182. Gramentz, D., Involvement of loggerhead turtle with the plastic, metal, and hydrocarbon pollution in the central Mediterranean, *Mar. Pollut. Bull.*, 19, 11, 1988.
183. Vargo, S. et al., Effects of Oil on Marine Turtles, Minerals Management Service Outer Continental Shelf Rep. MMS 86-0070, Vienna, VA, 1986.

184. Lutcavage, M. E. et al., Physiologic and clinicopathologic effects of crude oil on loggerhead sea turtles, *Arch. Environ. Contam. Toxicol.*, 28, 417, 1995.
185. McGrath, E. A. and Alexander, M. M., Observations on the exposure of larval bullfrogs to fuel oil, *Trans. N.E. Sect. Wildl. Soc.*, 36, 45, 1979.
186. Mahaney, P. A., Effects of freshwater petroleum contamination on amphibian hatching and metamorphosis, *Environ. Toxicol. Chem.*, 13, 259, 1994.
187. Lefcort, H. et al., The effects of used motor oil, silt, and the water mold *Saprolegnia parasitica* on the growth and survival of mole salamanders (Genus *Ambystoma*), *Arch. Environ. Contam. Toxicol.*, 32, 383, 1997.
188. Anderson, R. S., Doos, J. E., and Rose, F. L., Differential ability of *Ambystoma tigrinum* hepatic microsomes to produce mutagenic metabolites from polycyclic aromatic hydrocarbons and aromatic amines, *Cancer Lett.*, 16, 33, 1982.
189. Fernandez, M. and l'Haridon, J., Effects of light on the cytotoxicity and genotoxicity of benzo(a)pyrene and an oil refinery effluent in the newt, *Environ. Molec. Mutagen.*, 24, 124, 1994.
190. Vermeer, K. and Vermeer, R., Oil threat to birds on the Canadian coast, *Can. Field-Nat.*, 89, 278, 1975.
191. Tseng, F. S., Care of oiled seabirds: A veterinary perspective, in *Proc. 1993 Int. Oil Spill Conf.*, Publ. 4580, American Petroleum Institute, Washington, D.C., 1993, 421.
192. Jenssen, B. M., Review article: Effects of oil pollution, chemically treated oil, and cleaning on the thermal balance of birds, *Environ. Pollut.*, 86, 207, 1994.
193. Hartung, R. and Hunt, G. S., Toxicity of some oils to waterfowl, *J. Wildl. Manage.*, 30, 564, 1966.
194. Miller, D. S., Peakall, D. B., and Kinter, W. B., Ingestion of crude oil: Sublethal effects in herring gull chicks, *Science*, 199, 15, 1978.
195. Szaro. R. C., Hensler, G., and Heinz, G. H., Effects of chronic ingestion of No. 2 fuel oil on mallard ducklings, *J. Toxicol. Environ. Health*, 7, 789, 1981.
196. Albers, P. H., Effects of oil on avian reproduction: A review and discussion, in *The Effects of Oil on Birds. A Multi-Discipline Symposium*, Tri-State Bird Rescue and Research, Inc., Wilmington, DE, 1983, 78.
197. Leighton, F. A., The toxicity of petroleum oils to birds: An overview, in *The Effects of Oil on Wildlife*, White, J. et al., Eds., The Sheridan Press, Hanover, PA, 1991, 43.
198. Burger, A. E. and Fry, D. M, Effects of oil pollution on seabirds in the northeast Pacific, in *The Status, Ecology, and Conservation of Marine Birds of the North Pacific*, Vermeer, K. et al., Eds., Canadian Wildlife Service Special Publication, Ottawa, 1993, 254.
199. Fry, D. M. et al., Reduced reproduction of wedge-tailed shearwaters exposed to weathered Santa Barbara crude oil, *Arch. Environ. Contam. Toxicol.*, 15, 453, 1986.
200. Eppley, Z. A., Assessing indirect effects of oil in the presence of natural variation: The problem of reproductive failure in south polar skuas during the Bahia Paraiso oil spill, *Mar. Pollut. Bull.*, 25, 307, 1992.
201. Fowler, G. S., Wingfield, J. C., and Goersma, P. D., Hormonal and reproductive effects of low levels of petroleum fouling in Magellanic penguins (*Spheniscus magellanicus*), *The Auk*, 112, 382, 1995.
202. Walton, P. et al., Sub-lethal effects of an oil pollution incident on breeding kittiwakes, *Rissa tridactyla*, *Mar. Ecol. Prog. Ser.*, 155, 261, 1997.
203. Briggs, K. T., Gershwin, M. E., and Anderson, D. W., Consequences of petrochemical ingestion and stress on the immune system of seabirds, *ICES J. Mar. Sci.*, 54, 718, 1997.
204. Parnell, J. F., Shields, M. A., and Frierson, D., Jr., Hatching success of brown pelican eggs after contamination with oil, *Colonial Waterbirds*, 7, 22, 1984.
205. Hoffman, D. J., Embryotoxicity and teratogenicity of environmental contaminants to bird eggs, *Rev. Environ. Contam. Toxicol.*, 115, 39, 1990.
206. Szaro, R. C., Coon, N. C., and Stout, W., Weathered petroleum: Effects on mallard egg hatchability, *J. Wildl. Manage.*, 44, 709, 1980.
207. Stubblefield, W. A. et al., Evaluation of the toxic properties of naturally weathered *Exxon Valdez* crude oil to surrogate wildlife species, in *Exxon Valdez Oil Spill: Fate and Effects in Alaskan Waters*, Wells, P. G., Butler, J. N., and Hughes, J. S., Eds., ASTM STP 1219, American Society for Testing and Materials, Philadelphia, 1995, 665.
208. Parsons, K. C., The Arthur Kill oil spills: Biological effects in birds, in *Before and After an Oil Spill: The Arthur Kill*, Burger, J., Ed., Rutgers University Press, Rutgers, NJ, 1994, 305.
209. Day, R. H. et al., Effects of the *Exxon Valdez* oil spill on habitat use by birds in Prince William Sound, Alaska, *Ecol. Appl.*, 7, 593, 1997.

210. Piatt, J. F. and Anderson, P., Response of common murres to the *Exxon Valdez* oil spill and long-term changes in the Gulf of Alaska marine ecosystem, in *Proc. Exxon Valdez Oil Spill Symp.*, Symposium 18, American Fisheries Society, Bethesda, Maryland, 1996, 720.
211. Bowman, T. D., Schempf, P. F., and Hodges, J. I., Bald eagle population in Prince William Sound after the *Exxon Valdez* oil spill, *J. Wildl. Manage.*, 61, 962, 1997.
212. Piatt, J. F. and van Pelt, T. I., Mass-mortality of guillemots (*Uria aalge*) in the Gulf of Alaska in 1993, *Mar. Pollut. Bull.*, 34, 656, 1997.
213. Dunnet, G. M., Oil pollution and seabird populations, *Phil. Trans. R. Soc. Lond.*, B297, 413, 1982.
214. Piatt, J. F., Carter, H. R., and Nettleship, D. N., Effects of oil pollution on marine bird populations, in *The Effects of Oil on Wildlife*, White, J. et al., Eds., The Sheridan Press, Hanover, PA, 1991, 125.
215. Piatt, J. F. and Gord R. G., How many seabirds were killed by the *Exxon Valdez* oil spill?, in *Proc. Exxon Valdez Oil Spill Symp.*, Symposium 18, American Fisheries Society, Bethesda, 1996, 712.
216. Ford, R. G. et al., Total direct mortality of seabirds from the *Exxon Valdez* oil spill, in *Proc. Exxon Valdez Oil Spill Symp.*, Symposium 18, American Fisheries Society, Bethesda, 1996, 684.
217. Bernatowicz, J. A., Schempf, P. F., and Bowman, T. D., Bald eagle productivity in south-central Alaska in 1989 and 1990 after the *Exxon Valdez* oil spill, in *Proc. Exxon Valdez Oil Spill Symp.*, Symposium 18, American Fisheries Society, Bethesda, 1996, 785.
218. Wiens, J. A. et al., Effects of the *Exxon Valdez* oil spill on marine bird communities in Prince William Sound, Alaska, *Ecol. Appl.*, 6, 828, 1996.
219. Andres, B. A., The *Exxon Valdez* oil spill disrupted the breeding of black oystercatchers, *J. Wildl. Manage.*, 61, 1322, 1997.
220. Esler, D., Winter survival of adult female harlequin ducks in relation to history of contamination by the *Exxon Valdez* oil spill, *J. Wildl. Manage.*, 64, 839, 2000.
221. Parr, S. J., Haycock, R. J., and Smith, M. E., The impact of the *Sea Empress* oil spill on birds of the Pembrokeshire coast and islands, in *Proc. 1997 Int. Oil Spill Conf.*, Publ. 4651, American Petroleum Institute, Washington, D.C., 1997, 217.
222. Symens, P. and Suhaibani, A., The impact of the 1991 Gulf War oil spill on bird populations in the northern Arabian Gulf — A review, *Courier Forsch.-Inst. Senckenberg*, 166, 47, 1994.
223. Samuels, W. B. and Lanfear, K. J., Simulations of seabird damage and recovery from oilspills in the northern Gulf of Alaska, *J. Environ. Manage.*, 15, 169, 1982.
224. Samuels, W. B. and Ladino, A., Calculations of seabird population recovery from potential oilspills in the mid-Atlantic region of the United States, *Ecol. Model.*, 21, 63, 1984.
225. Wiens, J. A. et al., Simulation modelling of marine bird population energetics, food consumption, and sensitivity to perturbation, in *Environmental Assessment of the Alaskan Continental Shelf*, Annual Report Vol. 1, U.S. Department of Commerce and U.S. Dept. of Interior, Boulder, 1979, 217.
226. Ford, R. G. et al., Modelling the sensitivity of colonially breeding marine birds to oil spills: Guillemot and kittiwake populations on the Pribilof Islands, Bering Sea, *J. Appl. Ecol.*, 19, 1, 1982.
227. Albers, P. H., Effects of oil and dispersants on birds, in *Proc. 1984 Region 9 Oil Dispersants Workshop*, U.S. Coast Guard, Santa Barbara, California, 1984, 101.
228. Cairns, D. K. and Elliot, R. D., Oil spill impact assessment for seabirds: The role of refugia and growth centres, *Biol. Conserv.*, 40, 1, 1987.
229. Matsumoto, H. and Kashimoto, T., Embryotoxicity of organic extracts from airborne particulates in ambient air in the chicken embryo, *Arch. Environ. Contam. Toxicol.*, 15, 447, 1986.
230. Brunstrom, B., Broman, D., and Naf, C., Embryotoxicity of polycyclic aromatic hydrocarbons (PAHs) in three domestic avian species, and of PAHs and coplanar polychlorinated biphenyls (PCBs) in the common eider, *Environ. Pollut.*, 67, 133, 1990.
231. Brunstrom, B., Embryolethality and induction of 7-ethoxyresorufin *O*-deethylase in chick embryos by polychlorinated biphenyls and polycyclic aromatic hydrocarbons having Ah receptor affinity, *Chem.-Biol. Interactions*, 81, 69, 1991.
232. Naf, C., Broman, D., and Brunstrom, B., Distribution and metabolism of polycyclic aromatic hydrocarbons (PAHs) injected into eggs of chicken (*Gallus domesticus*) and common eider duck (*Somateria mollissima*), *Environ. Toxicol. Chem.*, 11, 1653, 1992.
233. Mayura, K. et al., Multi-bioassay approach for assessing the potency of complex mixtures of polycyclic aromatic hydrocarbons, *Chemosphere*, 38, 1721, 1999.

234. Patton, J. F. and Dieter, M. P., Effects of petroleum hydrocarbons on hepatic function in the duck, *Comp. Biochem. Physiol.*, 65C, 33, 1980.
235. Miller, D. S., Hallett, D. J., and Peakall, D. B., Which components of crude oil are toxic to young seabirds?, *Environ. Toxicol. Chem.*, 1, 39, 1982.
236. Peakall, D. B. et al., Toxicity of Prudhoe Bay crude oil and its aromatic fractions to nestling herring gulls, *Environ. Res.*, 27, 206, 1982.
237. Trust, K. A., Fairbrother, A., and Hooper, M. J., Effects of 7,12-dimethylbenz[a]anthracene on immune function and mixed-function oxygenase activity in the European starling, *Environ. Toxicol. Chem.*, 13, 821, 1994.
238. Custer, T. W. et al., Trace elements, organochlorines, polycyclic aromatic hydrocarbons, dioxins, and furans in lesser scaup wintering on the Indiana Harbor Canal, *Environ. Pollut.*, 110, 469, 2000.
239. Hansen, D. J., The Potential Effects of Oil Spills and Other Chemical Pollutants on Marine Mammals Occurring in Alaskan Waters, Minerals Management Service Outer Continental Shelf Rep. MMS 85-0031, Anchorage, Alaska, 1985.
240. Waldichuk, M., Sea otters and oil pollution, *Mar. Pollut. Bull.*, 21, 10, 1990.
241. Baker, J. R. et al., *Otter Lutra lutra* L. mortality and marine oil pollution, *Biol. Conserv.*, 20, 311, 1981.
242. Oritsland, N. A. et al., Effect of Crude Oil on Polar Bears, Environmental Studies No. 24, Northern Affairs Program, Ottawa, 1981.
243. Lipscomb, T. P. et al., Histopathologic lesions in sea otters exposed to crude oil, *Vet. Pathol.*, 30, 1, 1993.
244. Bickham, J. W. et al., Flow cytometric determination of genotoxic effects of exposure to petroleum in mink and sea otters, *Ecotoxicology*, 7, 191, 1998.
245. Loughlin, T. R., Ballachey, B. E., and Wright, B. A., Overview of studies to determine injury caused by the *Exxon Valdez* oil spill to marine mammals, in *Proc. Exxon Valdez Oil Spill Symp.*, Symposium 18, American Fisheries Society, Bethesda, 1996, 798.
246. Duffy, L. K. et al., Evidence for recovery of body mass and haptoglobin values of river otters following the *Exxon Valdez* oil spill, *J. Wildl. Dis.*, 30, 421, 1994.
247. Jenssen, B. M., An overview of exposure to, and effects of, petroleum oil and organochlorine pollution in Grey seals (*Halichoerus grypus*), *Sci. Total Environ.*, 186, 109, 1996.
248. Williams, T. M., O'Connor, D. J., and Nielsen, S. W., The effects of oil on sea otters: Histopathology, toxicology, and clinical history, in *Emergency Care and Rehabilitation of Oiled Sea Otters: A Guide for Oil Spills Involving Fur-Bearing Marine Mammals*, Williams, T. M. and Davis, R. W., Eds., University of Alaska Press, Fairbanks, 1995, Chap. 1.
249. Monson, D. H. et al., Long-term impacts of the *Exxon Valdez* oil spill on sea otters, assessed through age-dependent mortality patterns, *Proc. Natl. Acad. Sci.*, 97, 6562, 2000.
250. Frost, K. J., Lowry, L. F., and Ver Hoef, J. M., Monitoring the trend of harbor seals in Prince William Sound, Alaska, after the *Exxon Valdez* oil spill, *Mar. Mammal Sci.*, 15, 494, 1999.
251. Robineau, D. and Fiquet, P., Cetaceans of Dawhat ad-Dafi and Dawhat al-Musallamiya (Saudi Arabia) one year after the Gulf War oil spill, *Courier Forsch.-Inst. Senckenberg*, 166, 76, 1994.
252. Simons, E.A. and Akin, M., Dead endangered species in a California oil spill, in *Proc. 1987 Oil Spill Conf.*, American Petroleum Institute, Publ. 4452, Washington, D.C., 1987, 417.
253. Albers, P. H. and Gay, M. L., Unweathered and weathered aviation kerosine: Chemical characterization and effects on hatching success of duck eggs, *Bull. Environ. Contam. Toxicol.*, 28, 430, 1982.
254. Wolfe, J. L. and Esher, R. J., Effects of crude oil on swimming behavior and survival in the rice rat, *Environ. Res.*, 26, 486, 1981.
255. Gashev, S. N., Effect of oil spills on the fauna and ecology of small mammals from the Central Ob' region, *Soviet J. Ecol.*, 23, 99, 1992.
256. Savabieasfahani, M., Lochmiller, R. L., and Janz, D. M., Elevated ovarian and thymic cell apoptosis in wild cotton rats inhabiting petrochemical-contaminated terrestrial ecosystems, *J. Toxicol. Environ. Health*, Part A, 57, 521, 1999.
257. Lochmiller, R. L. et al., Disruption of rodent assemblages in disturbed tallgrass prairie ecosystems contaminated with petroleum wastes, in *Environmental Contaminants and Terrestrial Vertebrates: Effects on Populations, Communities, and Ecosystems*, Albers, P. H., Heinz, G. H., and Ohlendorf, H. M., Eds., SETAC Press, Pensacola, FL, 2000, Chap. 13.

258. Lee, J. H. and Talaska, G., Effects of kerosene cleaning on the formation of DNA adducts in the skin and lung tissues of mice dermally exposed to used gasoline engine oil, *J. Toxicol. Environ. Health*, Part A, 56, 463, 1999.
259. Feuston, M. H. et al., Systemic toxicity of dermally applied crude oils in rats, *J. Toxicol. Environ. Health*, 51, 387, 1997.
260. Khan, A. A. et al., Biochemical effects of *Pembina Cardium* crude oil exposure in cattle, *Arch. Environ. Contam. Toxicol.*, 30, 349, 1996.
261. Mattie, D. R. et al., The effects of JP-8 jet fuel on male Sprague-Dawley rats after a 90-day exposure by oral gavage, *Toxicol. Industrial Health*, 11, 423, 1995.
262. Bowyer, R. T. et al., Changes in diets of river otters in Prince William Sound, Alaska: Effects of the *Exxon Valdez* oil spill, *Can. J. Zool.*, 72, 970, 1994.
263. Bowyer, R. T., Testa, J. W., and Faro, J. B., Habitat selection and home ranges of river otters in a marine environment: Effects of the *Exxon Valdez* oil spill, *J. Mammal.*, 76, 1, 1995.
264. Ray, S. et al., Aromatic DNA-carcinogen adducts in beluga whales from the Canadian Arctic and Gulf of St. Lawrence, *Mar. Pollut. Bull.*, 22, 392, 1991.
265. Martineau, D. et al., Pathology and toxicology of beluga whales from the St. Lawrence Estuary, Quebec, Canada. Past, present and future, *Sci. Total Environ.*, 154, 201, 1994.
266. Mathieu, A. et al., Polycyclic aromatic hydrocarbon-DNA adducts in beluga whales from the Arctic, *J. Toxicol. Environ. Health*, 51, 1, 1997.
267. Marine Mammal Commission, Marine mammals and persistent ocean contaminants, in *Proc. Marine Mammal Comm. Workshop,* Keystone, CO, 12–15 October 1998, Bethesda, 1999.
268. Paine, R. T. et al., Trouble on oiled waters: Lessons from the *Exxon Valdez* oil spill, *Annu. Rev. Ecol. Syst.*, 27, 197, 1996.
269. Wolfe, D. A. et al., The fate of the oil spilled from the *Exxon Valdez*, *Environ. Sci. Technol.*, 28, 561A, 1994.
270. Monahan, T. P. and Maki, A. W., The *Exxon Valdez* 1989 wildlife rescue and rehabilitation program, in *Proc. 1991 Int. Oil Spill Conf.*, Publ. 4529, American Petroleum Institute, Washington, D.C., 1991, 131.
271. American Society for Testing and Materials, *Exxon Valdez Oil Spill: Fate and Effects in Alaskan Waters*, Wells, P. G., Butler, J. N., and Hughes, J. S., Eds., STP 1219, Philadelphia, 1995.
272. American Fisheries Society, *Proc. Exxon Valdez Oil Spill Symp.*, Symposium 18, Bethesda, 1996.
273. Wiens, J. A., Oil, seabirds, and science, *Bioscience*, 46, 587, 1996.
274. Saeed, T., Al-Hashash, H., and Al-Matrouk, K., Assessment of the changes in the chemical composition of the crude oil spilled in the Kuwait desert after weathering for five years, *Environ. Int.*, 24, 141, 1998.
275. Al-Senafy, M. N. et al., Soil contamination from oil lakes in northern Kuwait, *J. Soil Contam.*, 6, 481, 1997.

CHAPTER 15

# Lead in the Environment

**Oliver H. Pattee and Deborah J. Pain**

## CONTENTS

## 15.1 INTRODUCTION

Lead (Pb) is a nonessential, highly toxic heavy metal, whose known effects on biological systems are always deleterious. Lead is very soft (< 35 diamond pyramid hardness) and dense (11.34 g/cm$^3$) and occurs as an important constituent of more than 200 minerals. The average concentration in the earth's crust is 0.016g Pb/kg soil,[1] making it a relatively rare metal. Small amounts of lead are released into the environment by natural processes, including the weathering of rocks, igneous activity, and radioactive decay. Present anthropogenic lead emissions have resulted in soil and water lead concentrations up to several orders of magnitude higher than naturally occurring concentrations.[2]

Lead exists in both inorganic forms (Pb[II] and less frequently Pb[IV]), in organic forms. Significantly more inorganic lead is cycled globally than organic lead. However, organolead compounds may contribute significantly on a local scale.[3] Both the chemical and physical forms of lead influence its distribution and behavior within the environment and its potential for absorption by and toxicity to living organisms. Solubility and particle size are particularly important, as they affect the intercompartmental and geographical distributions of lead, its adsorption onto sediments, and absorption by living organisms.

Lead is one of the easiest metals to mine, and smelting requires only moderate temperatures. Consequently, lead has been exploited by humans for millennia. Over 7000 years ago the Egyptians used lead for weights and anchors, cooking utensils and piping, solder, and pottery glaze. The Romans also used lead extensively. The solubilization of lead from cooking pots used for preparing grape syrup for wine and preserved fruit by the aristocracy has been suggested as a major reason behind the fall of the Roman Empire. The high lead content of exhumed bones of Roman aristocrats has lent credibility to this hypothesis.[4]

Although the use of lead has a long history, demand increased greatly during the industrial revolution of the 18th and 19th centuries, and again following the adoption of organic lead compounds as antiknock agents in gasoline (initially in 1923,[5] then proliferating following World War II). Anthropogenic uses of lead, particularly over the last century, have altered the availability and distribution of lead, locally and globally, more than for any other toxic element. High lead emissions from urban centers have resulted in urban lead concentrations several thousand times higher than in pretechnological times, and the small lead particles present in most atmospheric emissions may be transported globally by trade winds. Consequently, rural environments also have elevated lead levels. Even the most recently deposited ice layers at the North Pole have lead concentrations one to two orders of magnitude higher than in prehistoric times. At the South Pole the deposition rate of lead is 3 to 5 times higher than in pretechnological times.[2] World lead production in 1987 was 2,110,000 metric tons: USSR 506,000 metric tons, Australia 485,000 metric tons, Canada 422,000 metric tons, United States 317,000 metric tons, Mexico 190,000 metric tons, and Peru 110,000 metric tons.[6] In 1994 world production was 2,800,000 metric tons primary production and 5,400,000 metric tons from secondary production (recycling).[7] The principal uses of lead are for storage batteries, tetraalkyl lead for gasoline, cable coating, pigments and chemicals, building and construction, and caulking.[8] Principal use patterns for lead in Europe, the United

**Table 15.1 Use Patterns for Lead in Selected Countries**

| Use | Thousands of Metric Tons (Percent) | | |
|---|---|---|---|
| | United States | Europe[a] | Japan |
| Storage batteries | 613 (47) | 392 (34) | 93 (40) |
| Cable sheathing | 14 (1) | 145 (13) | 16 (7) |
| Pigments/chemicals | 303 (23) | 294 (26) | 62 (27) |
| Alloys | 75 (6) | 50 (4) | 15 (7) |
| Ammunition | 66 (5) | (–)[b] | (–)[c] |
| Other | 226 (18) | 267 (23) | 44 (19) |
| Total | 1297 | 1148 | 230 |

[a] France, West Germany, Italy, United Kingdom.
[b] Not reported.

*Source:* Eisler, R., *Handbook of Chemical Risk Assessment: Health Hazards to Humans, Plants, and Animals,* Vol. 1, CRC Press, Boca Raton, FL, 2000.

States, and Japan are illustrated in Table 15.1. Use patterns are changing along with national and international restrictions on many of the uses of lead (e.g., as an additive in gasoline). By 1992 the shift in lead usage was obvious; batteries accounted for 63% of the total use and paint pigments (12.8%) and gas additives (2.2%) declined precipitously.[7] Most anthropogenic lead emissions result from the mining, smelting, and refining of lead and other metal ores; vehicle emissions; and industrial emissions resulting from the production, use, recycling, and disposal of lead-containing products. The United States recycles an estimated 50,000 tons annually and discards an additional 25,000 tons into landfills nationwide.[9]

Lead poisoning has occurred for at least 2500 years,[10] and the adverse health effects of lead have been widely recognized for centuries. Benjamin Franklin, in a letter to a friend in 1786, mentioned the adverse human health effects of lead and stated that "the Opinion of this mischievous Effort from Lead is at least above Sixty Years old; and you will observe with Concern how long a useful Truth may be known and exist, before it is generally received and practiced on."[11] However, today, due to stricter industrial health control measures, the acute effects of lead poisoning to which Franklin was referring are rare and are generally only associated with specific localized sources of exposure. For many years, the major concern has been over health effects associated with continuous exposure to low levels of lead (i.e., chronic exposure, which does not always produce overt signs of poisoning). Particular concern has focused on the effects of lead upon intelligence quotient and behavior in children. A wide range of studies has been conducted since the 1940s,[12–15] and it is now understood that juveniles, including human children, are more susceptible than adults to lead poisoning. This is a consequence of higher lead uptake, incomplete development of detoxifying metabolic pathways, and age-related differences in the permeability of the blood-brain barrier.[16] In the United States, blood lead levels in 9% of the children aged 1 to 5 years exceeded 100 μg/L; blood lead levels exceeding 1000 μg/L have been reported in children living near lead smelters in undeveloped Asiatic and eastern European countries.[17]

As the technological capability to measure ever-lower concentrations of lead has increased, levels at which physiological effects have been measured have continued to decrease. Langford[18] noted that "the concept of differentiating lead absorption from lead poisoning is no longer tenable, and it has been suggested that it may be more useful to speak in terms of its chemical toxicity (without symptoms) and clinical toxicity (with symptoms), recognizing that one merges into the other."

It is significant to note that lead appears to be the only toxic chemical pollutant to have accumulated in humans to average concentrations that approach the threshold for potential clinical poisoning.[19] Although there are increasingly strict controls on the uses of lead and lead emissions, general environmental exposure still induces a wide range of health effects in animals, and specific sources of lead still result in animal mortality. Sources of lead, environmental concentrations, uptake by living organisms, and health effects in animals will be discussed in this chapter.

## 15.2 SOURCES OF LEAD IN THE ENVIRONMENT

Anthropogenic emissions of lead into the environment may be directly into air, water, and soil. Although emissions into these media can be easily quantified, there is a continuous intercompartmental exchange of lead among them. The geographical distribution of atmospheric lead, relative to the point of emission, is largely dependent upon particle size. Most airborne lead is eventually deposited onto some surface, including plants, soil, bodies of water, artificial surfaces, and the respiratory tracts of animals, by dry or wet deposition processes. Dry deposition is either by gravitational settling of larger particles (> 10 μm)[20] or impaction of particles of all sizes (especially smaller particles). Wet deposition results either from the incorporation of particles into water droplets within clouds or from the accumulation of particles by falling precipitation. Industrial discharges, highway runoff, and sewage effluent are responsible for much of the lead in water, with some wet deposition of atmospheric lead, and direct dry deposition, which is more significant for large bodies of water. The distribution of lead in water depends upon the chemical form of the lead. Lead in soils results mainly from dry and wet deposition of atmospheric lead, particularly close to emission sources, and the disposal of sewage sludge, often onto agricultural land. Lead in soils is relatively immobile compared to that in other environmental media, and soils and, ultimately, ocean bed sediments are the "sinks" for anthropogenically emitted lead.

Metallic lead also enters the environment from human activities. Hunting, fishing, and target shooting produces "hot spots" of metallic lead that have been responsible for the deaths of a number of species, ranging from ducks to eagles.[21,22] Using stable lead isotope ratios, it is possible to determine the source of the lead in a variety of matrices, including avian bone. Juvenile herring gulls (*Larus argentatus)* from the Canadian Great Lakes had lead isotope ratios similar to that of lead originating from combustion of Canadian gasoline, whereas waterfowl and eagles had lead ratios most similar to the ratios found in lead shot and sinkers.[23] Additionally, the isotope ratios for all species were higher than those derived from lead from mining or smelting.

### 15.2.1 Atmospheric Emissions

Estimated total lead emissions into the atmosphere from natural sources (1979 to 1984) ranged from (Table 15.2) 8.6 to 54.1 x $10^9$ gm/year.[24] Atmospheric lead emissions resulting from anthropogenic activities are estimated to be at least one to two orders of magnitude greater than from natural sources (Table 15.2). The most important source of atmospheric lead is the combustion of fuels containing tetraalkyl lead antiknock agents. This category by itself exceeds all of the natural emissions of lead. The lead from automobile exhaust is typically in the form of fine particulates. However, lead profiles in dated sediment cores and atmospheric aerosols show a decline since 1970, presumably reflecting the reduction in the use of lead in gasoline.[25] Other than gasoline-derived lead, most sources are point sources, resulting in locally elevated atmospheric lead concentrations.

#### *15.2.1.1 Vehicular Emissions*

The useful properties of tetraalkyl leads as antiknock agents (i.e., agents that allow for high engine compression ratios without spontaneous air/gasoline ignition in the cylinder) were first discovered in 1923.[26] Lead is added to gasoline in organic tetraalkyl forms, as tetraethyl lead, tertramethyl lead, and mixed alkyls of triethylmethyl lead, diethyldimethyl lead, and ethyltrimethyl lead. During fuel combustion in the cylinder lead alkyls are converted to oxides, which foul spark plugs and remain in the cylinder. To prevent this, ethylene dichloride and ethylene dibromide are added to scavenge the lead. The lead chlorides and bromides consequently formed are relatively volatile and allow for approximately 70 to 75% of gasoline lead to be emitted into the atmosphere with exhaust fumes. These inorganic lead salts are emitted as very small particles, of approximately 0.015 μm aerodynamic diameter, that are considered to increase in size rapidly in the atmosphere by coagulation with other particles.[27] A small amount of lead may also be emitted as larger particles

**Table 15.2 Global Estimates of Anthropogenic Emissions of Lead to the Atmosphere from Production and/or Combustion in 1979/1980 ($10^9$gm/year)[24]**

| Source | Nriagu | United States × 5[a] | Europe × 5[a] |
|---|---|---|---|
| Gasoline/waste oil | 177[b] | 176 | 176 (372)[c] |
| Waste incineration | 8.9 | 4.2 | 4.0 |
| Coal combustion | 14 | 4.8 | 4.0 |
| Primary nonferrous metal | 85 | 9.8 | 143.0 |
| Secondary nonferrous metal | 0.8 | — | 2.2 |
| Iron/steel | 50 | 4.1 | 73 |
| Industrial applications | 7.4 | 1.6 | — |
| Wood combustion | 4.5 | — | 2.8 |
| Phosphate fertilizers | 0.1 | 0.4 | 0.03 |
| Miscellaneous | 5.9 | 1.2 | 3.8 |
| Total | 354 | 202 | 419 |

[a] Figures for the United States and for Europe multiplied by a factor of 5 to give a global estimate.
[b] Assumes a 35% reduction in leaded gasoline use between 1975 (the base year used in the original calculations) and 1979/1980.
[c] The figure in brackets, which corresponds to the reported estimate for Europe, seems to be too high.

*Source:* From Hutchinson, T.C. and Meemia, K.M., Eds., *SCOPE31, Lead Mercury, Cadmium, and Arsenic in the Environment,* John Wiley and Sons, Chichester, U.K., 1987. With permission.

(5 to 50 μm) resuspended from the exhaust system.[28] The atmospheric lifetime of emitted lead depends upon particle size, topography, and weather conditions. Particles > 10 μm are usually deposited rapidly close to the emission source (e.g., within 30 m of roads), either through gravitational settling or turbulent deposition. Due to their longer atmospheric residence time smaller particles can be carried by winds and deposited over very large areas.

Although most organic leads in gasoline are converted to inorganic forms during combustion, evaporation and incomplete combustion can release organic leads in the vapor phase into the atmosphere. Tetraalkyllead is relatively stable in the atmosphere, since it is not readily attacked by oxygen or light, and probably accounts for 1 to 6% of total lead at urban sites and possibly a similar percentage at inland rural sites.[28] Most organic lead in the environment results from gasoline additives. Leaded gasoline usually contains 0.15 to 0.40 g Pb/L, and unleaded < 0.01 g Pb/L. Exhaust fumes may contain 2000 to 10,000 μg Pb/m$^3$ when leaded gasoline is used.[28] These fine particles tend to persist in the atmosphere and can be transported long distances; their small diameter further enhances their absorption into the bloodstream.[29]

### *15.2.1.2 Fossil Fuel Combustion*

The combustion of fossil fuels contributes only a small percentage of total lead emissions to the atmosphere, even relative to natural emissions (Table 15.2). However, this source of lead may be important locally. Coal contains, on average, approximately 25 ppm lead, with 17 ppm lead in British coal compared with 35 to 44 ppm in the United States.[30,31] During combustion most lead in coal becomes deposited in fly ash, which may have lead concentrations an order of magnitude higher than the initial fuel. Little fly ash reaches the atmosphere during domestic coal combustion, although high chimney gas velocities may result in some loss of fly ash during industrial combustion. The amount lost depends upon the efficiency of control devices used. Modern electrostatic precipitators have greatly reduced atmospheric lead emissions from this source. However, other environmental media may become contaminated through the disposal of fly ash. Crude petroleum oil contains relatively less lead than coal, generally 0.001 to 0.2 ppm, with maximum concentrations reaching 2 ppm.[31] During fractionation, most of the lead becomes concentrated in heavy fuel oils or residues, and total lead emissions are low.

#### 15.2.1.3 Industrial Emissions

Elevated lead emissions may be recorded in the vicinity of lead and other metal mines, primary and secondary smelters, and many industries that produce, use, recycle, or dispose of lead-containing products. These include the pigment and dye industry,[32] car battery manufacture,[33] the manufacture of tetraalkyl additives for gasoline,[34] and municipal waste incinerators.

#### 15.2.1.4 Dusts

Many sources contribute to lead in dusts, and concentrations may become elevated in urban environments, including within dwellings. A large proportion of lead in dusts probably results from aerial deposition, although paint (sanding and scraping of buildings) or other sources may be the major contributors. Household dusts usually contain < 1000 ppm dry weight (d.w.[48]) Pb[35] and urban and street dusts 500 to 5000 ppm d.w.[36]

### 15.2.2 Lead Emissions into Water

Present-day lead concentrations in marine water are probably up to an order of magnitude greater than natural concentrations. Table 15.3 illustrates industrial inputs of lead annually into all oceans. These inputs are some 2.6 times higher than estimated preindustrial inputs. However, most industrial and domestic lead inputs to waterways occur in urban or coastal areas, and in rural areas the natural mineralization of parent rock may provide the most significant contribution of lead into the aquatic environment.

Using isotopic lead ratios and lead/calcium atomic ratios, the lead in teeth of California sea otters (*Enhydra lutis*) from Alaska have increased 2- to 15-fold over otters from 2000 years ago. The lead in preindustrial otters was derived from natural sources, whereas the lead in recent animals is derived from aerosols or industrial lead waste deposits.[37,38] Likewise in birds, lead burdens are derived from recent anthropogenic sources, primarily gasoline combustion or recently manufactured lead shot and lead sinkers.[23]

Anthropogenic inputs of lead into water include dry and wet deposition of atmospheric lead, which act globally, or at least hemispherically, and point sources of emission, which influence the environment more locally and often involve effluents with high lead concentrations. Point sources include discharges from sewage works and direct industrial and mine works discharges. In addition to such point sources, lead may be leached from soils and mine tailings, and highway runoff frequently reaches local waterways. Lead in freshwater often becomes adsorbed onto sediment and ultimately reaches oceans.

**Table 15.3 Approximate Lead Input to All Oceans**

| Input | Tons/Year |
|---|---|
| Industrial inputs | |
| Aerosols (gasoline) | 37,000 |
| Aerosols (smelters, forest fires) | 3,000 |
| Rivers/sewers (soluble, mainly aerosols) | 60,000 |
| Rivers/sewers (solids) | 200,000 |
| Preindustrial inputs | |
| Aerosols | 1,000 |
| Rivers (soluble) | 13,000 |
| Rivers (solids) | 100,000 |

*Source:* From Harrison, R.M. and Laxen, D. P. H., *Lead Pollution, Causes and Control,* Chapman and Hall, London, 1981. With permission.

#### *15.2.2.1 Vehicular Emissions*

Although the contribution of gasoline lead in water is mainly indirect, through dry and wet deposition of atmospheric lead and highway runoff, it remains very important. Trefry et al.[39] found that lead inputs into the Gulf of Mexico from the Mississippi River in the United States had decreased by 40% 10 years after the introduction of regulations reducing the lead content of gasoline. The Mississippi carries over half the total water and sediment carried by all U.S. rivers, and the 40% decrease in lead transported was comparable to the decrease in the use of leaded gasoline over that period.

Gasoline lead deposited onto roads constitutes the most important source of lead in urban runoff.[40,41] The amounts of lead deposited on road surfaces from vehicles in urban areas depend upon a wide range of factors including traffic intensity, driving mode, weather and climatic factors, road surface area, etc. Runoff water is usually either directed to sewage treatment plants, in which case only a small proportion will reach waterways in the short term, or transported directly to local water courses. Small amounts of gasoline lead may also reach water directly, through combustion by boat engines, although only 9 to 31% of gasoline lead is found in boat exhaust water.[42]

#### *15.2.2.2 Lead in Sewage Effluent*

The amount of lead released into water from sewage treatment plants depends upon the range and nature of effluents received by the plant and the treatment process. Sewage treatment plants may receive, in addition to domestic wastewater, certain industrial discharges and highway runoff, particularly following storms. A large proportion (80 to 100%) of lead present in the effluent is incorporated into sewage sludge during treatment,[43] and the remaining effluent is discharged into waterways.

#### *15.2.2.3 Leaching and Seepage*

Lead may reach waterways by leaching from fly ash and from soils that have been subject to additions of lead (e.g., sewage sludge) and by leaching and seepage from wastedumps, mine tailings, and tailing pond decant. Lead is relatively immobile in soils; therefore, leaching is only likely to contribute a small amount of lead in water. A certain amount of lead may be leached from the fly ash resulting from fossil fuel combustion. It has been reported that distilled water can leach 3.4% of the total lead in fly ash in 48 h.[44]

#### *15.2.2.4 Other Sources*

Certain industrial discharges directly entering waterways result in increased lead concentrations and local contamination, and accidental releases of lead from industrial and other sources can cause considerable water contamination. Lead also reaches water in less obvious ways such as through the use of lead plumbing systems. In areas of soft water, where no calcium layer coats the inside of plumbing systems, elevated lead concentrations can accumulate in static water, for example, overnight.

### 15.2.3 Lead in Soils and Sediments

The major natural source of lead in soils is the weathering and mineralization of parent material, and in rural areas receiving little lead pollution, there is often a strong relationship between lead concentrations in soil and parent material. Naturally high soil lead concentrations can be found in local areas of high mineralization, but even these rarely approach those resulting from anthropogenic activities around urban areas. Emitted lead reaches soils via a wide range of pathways. This may occur on a local scale from point sources, such as the application of sewage sludge to land or land disposal of mine wastes and fly ash, or universally through the deposition of atmospheric lead. By one pathway or another, the vast majority of atmospheric lead, and much lead in water, eventually becomes associated with soils or sediments, due to the strong binding capacities of many soil components for metals.

Lead mine tailings comprise a localized but significant source of lead-bearing sediments. These sediments wash far down exposed waterways, where they are ingested by wildlife, especially waterfowl.[45] This venue constitutes one of the few instances where wild birds have died of lead poisoning without having ingested lead shot.[46]

#### 15.2.3.1 Intercompartmental Lead Transfer to Soils

The extent of dry deposition of airborne lead onto soils from point emission sources is related to the atmospheric lead concentration, particle sizes, weather, wind direction, and soil surface characteristics. Gravitational and turbulent settling of large particles is high near the emission source, decreasing with distance. Deposition of smaller particles by impaction, or diffusion mechanisms (e.g., Brownian diffusion), takes place over a much larger geographical area and is related to surface characteristics. Deposition tends to be lowest on smooth surfaces, with concentrations of lead up to eight times higher on rough or hairy leaf surfaces than on smooth leaf surfaces.[47] Lead deposited onto plant surfaces may finally reach the soil when plants die, if they are not grazed or cropped. In addition, lead reaching plant and other natural and artificial surfaces via dry deposition may be washed off and reach the soil in water.

Water plays a very important role in the transport of lead to soils and sediments. Dry deposition of atmospheric lead gives rise to contamination close to emission sources, whereas wet deposition through precipitation is universal. Both are thought to deposit approximately similar amounts of lead in rural areas.[28] Rain and dew also transport lead to soils by washing deposited lead off plant surfaces, roads, and other artificial surfaces. The movement of lead-contaminated leachates and seepage water contributes locally to soil contamination, as does accidental flooding with water contaminated by industrial processes, sewage processing plants, and waste dumps.

#### 15.2.3.2 Sewage Sludge

As the sludge from sewage treatment facilities is rich in plant nutrients, particularly nitrogen and phosphorus, its addition to agricultural land as fertilizer is considered a convenient and useful means of disposal. Sewage sludge contains variable amounts of lead, depending upon the effluent types received by the treatment plants. Most sewage sludges contain within the range of 120 to 3000 ppm d.w., and surveys in the United Kingdom have found sludges to contain from as low as 19 to as high as 45,400 ppm d.w. of lead.[49,50] Although heavy-metal concentrations are often high in sludge, agricultural use has been permitted under the premise that most lead becomes immobilized within soils and sediments and is unavailable for uptake by plants. However, in cases of surface sludge application, grazing animals may ingest sludge deposited on the surface of, or along with, plants. The U.K. government has regulated the maximum permissible concentrations of lead and other metals in sludge to be applied in this way (2000 ppm d.w. for Pb), along with maximum concentrations to be disposed over a given period of time.[19]

#### 15.2.3.3 Other Sources

Local soil contamination may occur around storage battery reclamation plants, near mines and waste, dumps, and around mainly older buildings where leaded paint has been used and is flaking. An additional source of lead in soils is lead ammunition.[51,52] The production of lead ammunition, both for military purposes and for hunting, has continued to increase throughout the last three centuries. Today, lead shot is regularly and widely distributed in the environment, mainly by hunters. Although lead shot is relatively stable in most soils, it has resulted in significant soil-pollution problems at specific sites, and the ingestion of spent ammunition causes large-scale avian mortality. However, the use of lead in ammunition has declined or, in the case of lead shot for waterfowl, been banned. After 6 years of the ban, an estimated 1.4 million ducks were spared from lead poisoning in 1997.[53]

## 15.3 INTAKE AND UPTAKE OF LEAD

Amounts of lead entering and being assimilated by plants and animals from the atmosphere, soil, and water are not always directly related to environmental concentrations. Lead intake and uptake depends upon many factors including its chemical and physical form, exposure route, and the biology of the exposed organism. Lead reaches plants via wet and dry deposition onto plant surfaces, including road and soil splash, and by uptake from roots. Inorganic lead reaches terrestrial animals mainly via inhalation and ingestion, and absorption through the skin or gills is important for aquatic animals. Organic lead may be absorbed through the skin of all animals, although such uptake is generally insignificant.

### 15.3.1 Plants

#### *15.3.1.1 Aquatic Plants*

The availability of lead (and other metals) for uptake by aquatic plants is related to a wide range of chemical and physical variables, including the chemical form of lead, the pH of water, the presence and quantity of calcium and magnesium ions (water hardness) and nutrients, and the quantity and nature of suspended material. Lead from water usually enters plants in an ionic form, and uptake by aquatic plants can be from water, soil, air, or a combination of the three depending upon the type of plant (rooted/submerged, etc.). Crowder[54] reviewed lead uptake by plants. Of wetland plants, the submergent macrophytes generally tend to have the highest lead concentrations, although uptake is species-specific. Subterranean parts tend to accumulate the highest lead concentrations, and senescent foliage can contain higher lead concentrations than live foliage. Under certain conditions lead concentrations in aquatic plants have been correlated with sediment lead concentrations, and plant lead concentration may be negatively correlated with water pH. Under experimental conditions, pH and electrode potential (Eh) interact to influence lead uptake.[54]

In experimental studies with rooted macrophytes, lead was rapidly taken up from solution containing 1.0 mg $Pb^{2+}$/L via passive mechanisms.[55] A large proportion of lead taken up within one hour by shoots of *Elodea canadensis* was released within 2 weeks of transfer to clean water, although 10% appeared to remain irreversibly bound.

There has been concern over the uptake of lead deposited as gunshot by aquatic and terrestrial plants. Plants can take up large amounts of lead from spent gunshot deposited in acidic soils and sediments (maximum of 24,892 μg/g d.w. — T. Elkington, pers. comm.). However, most wetland sites are not particularly acidic, and Behan et al.[56] found similar lead concentrations in plants collected from heavily hunted wetlands and from refuges. In spite of this some lead from deposited shot may become available for uptake, and these authors found that 3 years after seeding a 0.17-ha pond with 227 kg lead shot, the roots and shoots of aquatic plants had significantly higher lead concentrations than controls. However, this amount of shot is at least an order of magnitude higher than that found in most heavily shot-over wetlands. The exception to this is the fall zone from shotgun target ranges, where lead shot density ranged from 4.15 x $10^6$ to 3.70 x $10^9$ pellets/ha.[57]

#### *15.3.1.2 Terrestrial Plants*

Quantities of lead accumulated on plant surfaces are related to atmospheric concentrations, particle size, topography, climate, and surface characteristics of plants. Little is known of the movement of aerosol lead following deposition on plant surfaces or in stomata, although it appears that most deposited lead is immobile, perhaps being embedded in the cuticle. Although foliar transport of aerosol lead appears limited in most species, Rule et al.,[58] using a solution of $^{210}Pb$, illustrated foliar absorption and translocation of lead to the youngest leaves of lettuce and the storage roots of white icicle radish plants. Translocation of atmospherically derived lead within some species has been confirmed using similar $^{210}Pb$ tracer studies.[59]

The contribution of atmospheric lead to total plant lead may be high both in areas of low soil and low atmospheric lead concentrations as well as in areas with high atmospheric lead concentrations. Tjell et al.[60] found that > 90% of total lead in grass in rural areas of Denmark resulted from atmospheric deposition. Plants with rough foliar surfaces tend to retain more lead than those with smooth or waxy surfaces,[61] and surface characteristics may also influence the amount of lead removed by precipitation. Large seasonal fluctuations in the lead concentrations of foliar parts of plants have been recorded, with values sometimes an order of magnitude higher in winter than in summer months. Increased concentrations in winter are thought to be related to higher deposition rates resulting from increased thermal stability in winter,[62] possible higher lead uptake by dead and decaying material due to cuticular breakdown increasing permeability and facilitating lead access,[63] and slower winter pasture turnover rates.

The availability of soil lead for uptake by plant roots depends upon physical and chemical soil characteristics. Lead is generally strongly adsorbed onto solid soil components with high organic matter cation exchange capacity and high percentage silt or clay. Under such conditions little ionic lead is available in the aqueous phase for uptake by plant roots. When soils are low in the components described and under conditions of poor drainage, low soil pH, and low concentrations of iron oxides and phosphorus, the availability of lead to plant roots increases. The application of calcium or phosphate to soils can reduce lead availability to plants through the formation of lead hydroxide, carbonate, or phosphate compounds of low solubility.[64] Lead uptake by roots can be passive[65] or active,[66] and uptake is influenced by species, cultivar, age, and growth rate. Metal uptake by roots can be accelerated in species in which roots release solubilizing agents or carriers into the rhizosphere, as some of these may act as ligands or chelate metals.[54] Transport of lead between roots and foliar parts of plants tends to be limited, and it has been suggested that lead may be precipitated in vesicles that fuse with cell walls in roots, immobilizing and inactivating it.[66] However, although most lead taken up from soil appears to accumulate in roots, Rolfe[67] found that significant concentrations of lead also accumulated in the leaves and stems of eight tree species within one growing season in soils of high lead content.

The high binding capacity of soil for lead, its limited transport from roots, and the tendency for lead to accumulate at xylem sites within the plant all act as natural barriers against the movement of lead to foliar parts.[68] Although total plant lead concentrations are generally higher with high soil lead concentrations, the increase in plant lead is generally small proportional to that in soils. The Ministry of Agriculture, Food and Fisheries (MAFF)[69] investigated lead concentrations in plants grown in a variety of soils and found that, in soils with up to 25 times background lead concentrations, plant lead concentrations did not more than double.

### 15.3.2 Animals

#### *15.3.2.1 Aquatic Animals*

Lead uptake by aquatic organisms is via water (absorption through skin, gills, intestine, etc.) and food. The chemical form of lead and other elements present both in water and diet, along with species and other biological and environmental factors, influences lead uptake. Lead uptake by certain freshwater fish, such as the pumpkinseed sunfish (*Lepomis gibbosus*) and rainbow trout (*Salmo gairdneri*), has been shown to increase as water pH decreases.[70,71] As in many terrestrial animals, calcium appears to have an important influence on lead transfer. Lead uptake and retention in the skin and skeleton of coho salmon (*Oncorhynchus kisutch*) was reduced when waterborne or dietary Ca was increased.[72]

Organic lead compounds, which accumulate in lipids, tend to be taken up and accumulated by freshwater teleosts more readily than inorganic lead compounds. Organic lead compounds are generally more toxic than inorganic compounds to aquatic organisms, and toxicity increases with the degree of alkylation.[73] A wide range of aquatic organisms can absorb and accumulate very high lead concentrations, and the residence time of lead appears to be related to the route of administration.[74]

#### 15.3.2.2 Terrestrial Animals

Ingestion and inhalation are the most important exposure routes of lead to terrestrial animals. The relative contributions of those routes to total body burden will vary according to environmental concentrations, other environmental factors, and biological considerations such as species, age, sex, and diet.

##### 15.3.2.2.1 Inhaled Lead

Most atmospheric lead is present as particulate inorganic salts, which, following inhalation, may be exhaled, deposited in the lungs and lower respiratory tract, or deposited in the windpipe and subsequently swallowed. Maximum lung retention occurs when particle sizes are 1.5 to 2.5 μm aerodynamic diameter.[75] Although deposition rates vary in relation to air lead concentration, particle size, and respiration rates, approximately 50% of inhaled lead is deposited in the lungs, even when atmospheric lead concentrations are relatively high.[76] Practically all lead deposited in the lungs is absorbed into the bloodstream. Half-lives of removal of lead from the lungs have been estimated at 6 h for humans[77] and 12 h in dogs.[78] Very little lead is thought to remain in the lungs after 3 days. Although most atmospheric lead is inorganic, a similar proportion of organic lead may be absorbed.[79]

##### 15.3.2.2.2 Ingested Lead

Lead is usually ingested along with food and water but may be ingested independently. Examples of nonfood ingestion include the ingestion of spent lead gunshot and anglers weights by waterfowl and ingestion of paint chips and soil by children and other young animals, known as "pica" (the compulsive active ingestion of nonfood objects). Contaminated sediments adhering to food items not only contribute to the body burden but may reach toxic levels when sediment distribution is restricted but lead content is high. Sediments in the Coeur d'Alene river system have been implicated in numerous lead poisoning instances in tundra swans.[45]

Following the ingestion of lead in the diet, a large proportion is eliminated directly with the feces. The proportion of ingested lead absorbed from the intestine is small relative to the proportion of inhaled lead absorbed and varies according to a wide range of factors, primarily dietary. Absorption of ingested lead is reduced when ingestion occurs along with food or if food is already present in the intestine.[80] It is believed that certain dietary constituents may either inhibit lead absorption through the intestinal wall or stimulate the excretion of lead into feces, or both.[81] The strong influence of dietary and other factors on intestinal lead absorption is such that values given in the literature for intestinal lead absorption in humans vary by up to an order of magnitude.

However, the majority of studies have found an 8 to 18% absorption of ingested lead, and 10% is most frequently quoted.[16,82,83] In other animals gastrointestinal absorption efficiencies are different, and even within species large variations are recorded.

*15.3.2.2.2.1 Factors Influencing Lead Absorption in the Intestine* — The amount of lead absorbed from the intestine depends upon a range of factors including the chemical form of lead ingested. As early as 1923 Hanzlik and Presho[84] illustrated experimentally that metallic lead was more toxic to pigeons than similar amounts of lead as lead chloride, iodide, sulfide, carbamate, or acetate. They suggested that this might result from reduced lead solubility and, consequently, reduced intestinal uptake when other elements were present along with the lead. The physical as well as chemical form of lead also influences absorption, and smaller lead particles (<180 μm) may be absorbed from the intestine more rapidly than larger ones.[1]

As discussed, only a small proportion of inorganic lead ingested is absorbed into the body. This is considered to account for approximately 10% of ingested lead in humans[83] and may be as low as 1 to 2% in cattle and sheep.[85] Studies in which animals have been fed various forms of lead have shown widely different tolerances between species. Absorbed lead normally bypasses the soft tissues and is sequestered in the bones; nutritional status also plays an important role in this process.[86] Differences in response are probably attributable to interspecific differences in abilities to absorb, store, detoxify, and excrete lead. These factors appear to be largely related to diet as well as physiological and environmental factors. Age differences in susceptibility to lead poisoning have been widely studied. Although uptake and storage of lead does vary in many species with age, increased toxicity of lead to young animals is related more to an increased lead sensitivity of certain body systems during early stages of development. In some young birds increased storage of lead by juveniles in the developing bones may serve to reduce the toxic effects of a given lead exposure over those of adults. In addition, the bones of female birds exposed to lead frequently have more elevated concentrations than those of males.[87,88] This is related to an increased calcium metabolism and turnover in bones of females, necessitated by eggshell production.

In a nationwide survey of lead-poisoned waterfowl, hepatic lead levels were independent of age and sex; lead-poisoned ducks tended to have higher hepatic lead levels than geese or swans, but the difference could be attributed to differences in body weight rather than kinetic differences between species.[89] Differential absorption and sensitivity to lead may be genetically based as well. Black ducks (*Anas rubripes*) and mallards (*A. platyrhynchos*) were not different in their response to dosing with lead shot, and there was no seasonal effect.[90] However, there was a significant difference in vulnerability; the wild-caught birds exhibited greater mortality and weight loss, probably due to captive-related stress. Lead accumulation in bone and blood of smelter workers was varied and apparently related to the allele they carried for delta-aminolevulinate dehydratase (ALAD).[91] Further analysis has linked response and accumulation to alleles not only for ALAD but also the vitamin D receptor gene and the enzyme aminolevulinic acid.[92]

A considerable amount of research related to lead absorption has been carried out in laboratory studies using the rat. As early as 1939 a component of apples — probably pectin — inhibited lead assimilation in growing rats.[93] Researchers also found higher lead retention in growing than adult rats. Milk in the diet has been found to influence lead absorption in rats. Dietary calcium appears to be one of the most important factors influencing lead absorption and toxicity. Lead uptake, toxicity, and soft tissue storage are increased in lead-exposed animals fed low-calcium diets. This has been illustrated in a wide range of birds and mammals including waterfowl, rats, dogs, horses, and sheep.[94–97] In sheep low-calcium diets adequate in other minerals fed to lambs along with 400 mg Pb/kg resulted in death within 5 weeks. With adequate dietary calcium, lambs survived for up to 10 months.[98] Other dietary factors that have been shown to influence lead uptake or toxicity are phosphorus, zinc, iron, ascorbic acid, vitamin D, and protein. Iron-, calcium-, phosphorus-, zinc-, and protein-deficient diets tend to result in increased lead uptake and toxicity, and dietary supplements of these constituents tend to decrease lead uptake or alleviate signs of lead poisoning in many species.[8,95–99] However, the interactions between lead and other dietary constituents are very complex, and under certain circumstances, and in some species, dietary excess of many elements, including those listed above, may exacerbate lead toxicity.

In some animals there appears to be a seasonal peak in lead poisoning. In children, cattle, and dogs, lead-poisoning cases are most common in spring or summer months.[97,100,101] Although this remains largely unexplained, it is thought that this might be related to increases in vitamin D levels, stimulated by increased sunlight. Vitamin D appears to be associated with intestinal lead absorption, and increased levels of dietary vitamin D tend to increase lead absorption and retention in some animals.[102] The seasonal incidence of lead poisoning in some animals may also be related to other physiological factors such as dehydration[102] or differential access to lead-contaminated dietary material.

When nonfood ingestion of lead occurs, similar factors will influence absorption and retention, although most animals usually rapidly egest large lead objects. However, this is not the case with waterfowl and other birds ingesting lead shot or anglers weights. Waterfowl actively ingest grit, which is retained in their muscular gizzard to help grind up and breakdown ingested material. Small ingested lead objects may be similarly retained and ground down within the gizzard. The retention and absorption of ingested shot by waterfowl is related to the quantity of food ingested and its physical characteristics as well as its chemical characteristics, as described above. Smaller, softer foods, especially those low in fiber, generally pass more rapidly through the intestine,[103] possibly increasing the expulsion rate of ingested shot and any lead particles eroded from the shot's surface. However, an increased shot expulsion rate does not always imply decreased absorption. Mourning doves (*Zenaida macroura*) fed a seed diet had lower shot retention rates but higher tissue lead concentrations than those on a pelleted diet, probably due to differences in lead absorption related to the chemical constituents of the diets.[104] Lead absorption rates from ingested gunshot are lowest when physical characteristics of diet facilitate the rapid passage of lead through the intestine and chemical components reduce lead absorption. Large-scale avian mortality has occurred due to shot ingestion.[105,106]

## 15.4 LEAD CONCENTRATIONS

### 15.4.1 Plants

Lead concentrations in plants depend upon a wide range of factors and may very between different parts of the plant.[62,107] Plants grown in uncontaminated environments generally contain < 1 ppm fresh weight lead (dry-weight values are commonly 2 to 20 times fresh-weight values depending upon plant/part of plant sampled). Some plants have been found to accumulate very large quantities of lead. Extreme examples are *Potamogeton* sp. growing in a tailings pond in Missouri recorded to have 11,300 ppm d.w. lead[108] and certain aquatic bryophytes recorded with maximum lead concentrations of 14,825 ppm d.w.[54] Lead concentrations in a wide range of aquatic and terrestrial plants have been related to their distance from a lead-emission source, suggesting their use as biological indicators of lead pollution. The analyses of mosses and lichens for lead has been used for decades to monitor airborne pollution.[109] Concentrations of lead in plants are illustrated in Table 15.4 and have been reviewed by Jenkins[108] and Eisler.[8]

### 15.4.2 Animals

#### *15.4.2.1 Distribution and Concentrations of Lead within the Body*

Following absorption, lead moves into the bloodstream. Most inorganic lead (> 90%) is transported by the circulatory system attached to the surfaces of red blood cells.[110] Only part of the

**Table 15.4 Lead Concentrations in Representative Plants (ppm, dry weight)**

| Variables | Lead Concentration | Reference[a] |
|---|---|---|
| Mean concentration in 8 aquatic plants | | 56 |
| Lead shot seeded area | | |
| Roots | 19.2 + 7.0 | |
| Shoots | 5.0 + 1.5 | |
| Control area | | |
| Roots | 2.9 + 0.7 | |
| Shoots | 1.2 + 0.3 | |
| Mean concentration in parts of bean grown in nutrient solution (ppm d.w.) | | 240 |
| Control | | |
| Top | 1.04 | |
| Root | 11.00 | |
| Blade | 0.83 | |
| Petiole | 1.04 | |
| 10 ppm Pb (solution) | | |
| Top | 72.28 | |
| Root | 1103.84 | |
| Blade | 81.39 | |
| Petiole | 88.22 | |
| Pondweed, *Potamogeton* sp. | | 108 |
| Missouri | | |
| Tailings pond | 11,300 | |
| 1.6 km Downstream | 3500 | |
| 8.1 km Downstream | 100 | |
| Ryegrass *Lolium perenne* leaves, 10 m from road edge (whole pasture) | | 62 |
| October 1979 | 75 | |
| December 1979 | 164 | |
| January 1980 | 174 | |
| February 1980 | 254 | |
| March 1980 | 132 | |
| May 1980 | 21 | |

[a] See also Jenkins[108] and Eisler.[8]

absorbed lead remains in the bloodstream, the rest being deposited within minutes in soft tissues, primarily the liver and kidney, and in bone.

In the kidney, lead tends to be deposited in the nuclei of cells, particularly in proximal convoluted tubule cells of the renal cortex. This lead is bound to dense proteinaceous structures forming acid-fast intranuclear inclusion bodies.[111] It has been suggested that these bodies may act as an immobile storage system for lead, reducing its cellular toxicity. Intranuclear inclusion bodies have been reported in a wide range of bird species[110,112,113] as well as in bats, rats, rabbits, dogs, swine, cattle, and primates.[111,114–117] Intranuclear inclusion bodies have also been reported in plant cells.[111]

In humans, lead has a half-life of 20 days in blood and 600 to 3000 days in bone.[28] Lead-retention time in bones is related to the medullary content, retention times being greater in bones with a high medullary content.[118,119] Sex and physiological condition influence lead kinetics in avian bone, as medullary bone undergoes sequences of formation and destruction in relation to calcium storage and liberation for eggshell formation.[88] Residence times of lead in soft tissues vary from several weeks to 6 or 7 months. Half-lives vary interspecifically but are of a similar order of magnitude. Blood lead remains in a fairly mobile equilibrium with that in soft tissues. In most vertebrates blood, liver, or kidney lead concentrations indicate recent absorption of inorganic lead, whereas bone lead concentrations indicate both recent and anterior absorption. Bone lead concentrations, therefore, generally tend to be higher in older individuals.

**Table 15.5 Lead Concentrations in the Tissues of Exposed Birds (ppm, wet weight)**

| | Lead Concentration (ppm w.w.) | | | | |
|---|---|---|---|---|---|
| | Trimethyl Lead | | Trialkyl Lead | Total Lead (Inorganic Lead) | |
| | Starlings[a] | | | | |
| Tissue | AI | AII | Dunlin[b] | Bald Eagle[c] | Canada Geese[d] |
| Muscle | 3.07 | 11.0 | 4.05 | 0.90 | — |
| Brain | 3.50 | 16.7 | 7.69 | 1.40 | 3.0 |
| Bone | 0.39 | 4.3 | 1.04 | 10.40 | 26.75 |
| Liver | 3.70 | 32.4 | 10.10 | 16.60 | 7.80 |
| Kidney | 5.38 | 30.2 | 13.80 | 6.00 | — |

[a] Starlings *Sturnus vulgaris*. Experimental study in which birds were given a low (AI = 200 ug/day) and high (AII = 2mg/day) dose of trimethyl lead. 6 birds in each group (Osborne et al.[241]).
[b] Dunlin *Calidris alpina*. Wild birds from the Mersey estuary, U.K., 1979, after being poisoned with alkyl lead from a petrochemical works. 10 birds (Bull et al.[221]).
[c] Bald eagles *Haliaeetus leucocephalus*. Four captive birds that died following dosage with 10+ lead shot (Pattee et al.[242]).
[d] Canada geese *Branta canadensis*. Four wild birds that died following shot ingestion (Bagley et al.[243]).

Exposure regime (acute or chronic) influences the distribution of lead within an animal. For example, an animal suffering acute poisoning may have higher lead concentrations in soft tissue than bone, whereas chronic poisoning is likely to result in higher bone lead concentrations. Wild urban pigeons (*Columbia livia*) accumulated very high tissue lead concentrations without several of the signs usually associated with acute lead toxicity in experimental studies.[120] The results of this study similarly suggest that exposure regime may influence the subcellular distribution and thus the toxic effects of lead.

Organic lead in the bloodstream is more evenly distributed than inorganic lead between the plasma and red blood cells. Organic lead disappears rapidly from the bloodstream and is converted in the liver to trialkyl lead, some of which may reappear in the blood 5 to 10 h after initial absorption.[28] The high liposolubility of organoleads results in their accumulation in nonbony tissues such as the liver, kidney, and brain. Example lead concentrations in different tissues of wild and captive avian species following exposure to inorganic and organic lead are provided in Table 15.5.

The high liposolubility of organolead compounds may partly explain the generally higher toxicity of organic than inorganic forms of lead to animals. However, regardless of the higher toxicity of organoleads, most lead accumulation and toxicity events in animals results from inorganic lead. For example, American kestrels (*Falco sparvarius)* fed for 60 days on a diet containing 48 ppm d.w. of biologically incorporated lead (cockerels fed lead acetate for 4 weeks) exhibited only minor effects besides increasing tissue lead burdens; there were no signs of lead intoxication, significant changes in weight, or development of anemia.[121]

### *15.4.2.2 Lead Concentrations in Relation to Exposure*

In most species of wild vertebrates from uncontaminated areas (both aquatic and terrestrial), blood lead concentrations are generally $< 30$ μg/dL, liver lead concentrations are usually $< 10$ ppm d.w. (often $< 5$ ppm d.w.), and bone lead concentrations are usually $< 20$ ppm d.w. In exposed individuals these values may increase by an order of magnitude, although many species will not tolerate such burdens and may die before they are achieved. Elevated bone lead concentrations are more frequently observed, as bone lead reflects lifetime absorption and is relatively immobile.

Earthworms are widely studied invertebrates in relation to lead accumulation as they are ubiquitous and a food source for a wide range of animals, especially birds. Beyer and Cromartie[122] investigated total body levels in different species of earthworms from natural and lead-contaminated sites. They found that from nine natural sites with low soil lead concentrations (5 to 14 ppm d.w.), lead levels in earthworms ranged from 1.4 to 2100 ppm d.w., and concentrations in different species varied considerably within the same site. Earthworms of *Noides* accumulated large amounts of lead, even at uncontaminated sites (2100 ppm d.w.). They suggested that if earthworms are used as indicators of environmental contamination, it is important to identify the species and collect similar earthworms from uncontaminated soil with similar characteristics for comparison. Gish and Christensen[123] sampled earthworms (mainly *Allolobophora trapezoids* and *A. turgida*) from soil near two major highways in the United States, and found good correlation between whole body lead concentrations, lead concentrations in soil, and distance from highway.

Toads were fed a diet of earthworms containing 10 ppm w.w. total lead (5.6 ppm available lead); a second group was maintained on a diet containing 308 ppm lead w.w. (93 ppm available lead) for 4 weeks followed by 816 ppm w.w. (65 ppm available lead) for 4 weeks. Feeding rates were controlled and were not significantly different between the two groups.[124] Table 15.6 illustrates lead concentrations in different tissues of the two groups. Toads on the high lead diet accumulated lead in all tissues examined, with particularly high levels in bone, kidney, and liver. Toads on the high lead diet also had significantly reduced activity of the heme-biosynthetic enzyme ALAD in erythrocytes. The soft tissue lead levels were sufficient to pose a hazard to predators. However, these concentrations were far lower than those recorded in the earthworms being eaten by the toads.

Lead concentrations in a range of small mammals living adjacent to roads with high and low traffic density were highest in those mammals living close to highways of high traffic density (Table 15.6). There was some evidence of a relationship between feeding habits and tissue lead in those species restricted to roadsides by habitat requirements.[125] However, the elevated tissue concentrations found in the most contaminated environments were not as high as those considered to result in toxic effects in mammals, and the biological significance of the levels recorded is uncertain.

Animal tissue lead levels have been shown to be derived from a variety of lead polluting sources. Table 15.6 illustrates the breadth of the sources and of the animals impacted. Examples include box turtles (*Terrapene carolina*) living close to and far from primary lead smelters, songbirds living in urban and rural environments, and flounder (*Platichthyes flesus*) from two different areas. Hardisty et al.[126] also demonstrated an age-related increase in total body lead of flounder from the two sites.

Several species have been suggested as indicators of lead contamination in human environments. The pigeon (or rock dove), which is ubiquitous and typically feeds on the ground in urban environments where it encounters substrates likely to be contaminated with lead-rich dust, has often been used as an indicator of urban lead contamination. Significantly elevated lead concentrations have been found in the tissues of urban pigeons in many countries.[87,127–129] Tissue lead concentrations in Tokyo pigeons decreased following a reduction in the use of leaded gasoline.[129] Whole body burdens of lead in fish collected throughout the United States declined steadily between 1976 and 1984, suggesting a reduced influx of lead to the aquatic environment following stricter regulatory measures.[130] The level of lead exposure and tissue lead concentrations are not always correlated. This also applies to the relationship between tissue lead concentrations and effects, as mechanisms of storage and detoxification render some individuals or species more susceptible to lead toxicity than others.

#### *15.4.2.2.1 The Passage of Lead through the Food Chain*

Biological magnification (i.e., a progressive increase in concentrations from the source of exposure through the trophic levels of living organisms), with primary producers having the lowest concentrations and predatory consumers the highest, does not appear to occur with lead. Namminga et al.[70] studied the distribution of lead in a pond ecosystem and found concentrations of 0.0013 mg/L, 281 ppm d.w., 37 ppm d.w. and 11.5 ppm d.w. in water, plankton, benthos, and fish (*Gambusia*

**Table 15.6 Lead Concentrations in a Range of Vertebrates from Lead-Contaminated and Uncontaminated Environments (ppm d.w. unless otherwise stated)**

| Variables | Mean Lead Concentrations | | Reference |
|---|---|---|---|
| Toads *Xenopus laevis* fed lead-contaminated earthworms[a] | | | 124 |
| | **I** | **II** | |
| Kidney | 81.3 + 10.4 | 19.1 + 2.6 | |
| Bone | 56.0 + 19.8 | 8.9 + 0.4 | |
| Liver | 30.8 + 7.8 | 4.3 + 0.3 | |
| Muscle | 5.7 + 0.4 | 2.9 + 0.2 | |
| Small mammals trapped near highways[b] | | | |
| *Blarina brevicauda*[c] | | | 125 |
| | **III** | **IV** | |
| Bone | 67.1 + 53.0 | 12.2 + 8.3 | |
| Kidney | 12.4 + 5.7 | 3.9 + 4.1 | |
| Liver | 4.6 + 2.5 | 1.0 + 0.5 | |
| Muscle | 9.7 + 9.8 | 5.4 + 4.7 | |
| Total body | 18.4 + 11.0 | 5.7 + 5.6 | |
| *Peromyscus maniculatus*[d] | | | 125 |
| Bone | 24.6 + 26.3 | 5.7 + 11.1 | |
| Kidney | 7.9 + 2.7 | 1.8 + 3.9 | |
| Liver | 3.5 + 2.1 | 1.1 + 0.6 | |
| Muscle | 6.8 + 5.0 | 2.1 + 1.5 | |
| Total body | 6.3 + 5.3 | 3.1 + 3.3 | |
| Turtles, *Terrapene carolina*[e] (ppm, w.w.) | | | 244 |
| | **V** | **VI** | |
| Bone (femur) | 65.5 + 49.5 | 3.8 + 1.5 | |
| Liver | 21.6 + 17.2 | 1.2 + 0.7 | |
| Kidney | 24.3 + 19.7 | 1.8 + 2.0 | |
| Blood | 6.0 + 4.2 | 0.1 + 0.1 | |
| House sparrow, *Passer domesticus*[f] | | | 125 |
| | **VII** | **VIII** | |
| Liver | 12.0 | 0.6 | |
| Kidney | 33.9 | 3.5 | |
| Bone (femur) | 130.4 | 16.9 | |
| Muscle | 2.1 | 0.9 | |
| Robin, *Turdus migratorius*[g] | | | 125 |
| Liver | 10.5 | 2.4 | |
| Kidney | 25.0 | 7.3 | |
| Bone (femur) | 133.7 | 41.3 | |
| Muscle | 1.2 | 1.0 | |
| Flounders, *Platichthyes flesus*[h] | | | 126 |
| | **IX** | **X** | |
| 2+ | 14.1 + 1.24 | 20.5 + 1.45 | |
| 3+ | 16.0 + 1.24 | 24.0 + 1.58 | |
| 4+ | 18.0 + 0.71 | 26.2 + 1.00 | |
| 5+ | 19.1 + 1.21 | 28.2 + 1.21 | |

[a] I = high-lead diet. II = low-lead diet. Means + standard error, n = 6.
[b] Means + standard deviation. III = high traffic density (19,600 vehicles per day). IV = low traffic density (340 vehicles per day or controls).
[c] N = 49, 32, respectively.
[d] N = 18, 40, respectively
[e] V = near rural sites of primary lead smelters. VI = less contaminated rural sites. Means + standard deviation, n = 4.
[f] VII = urban area; n = 11. VIII = rural area; n = 16.
[g] N = 10, VII. N = 10, VIII.
[h] IX = Barnstable Bay; X = Severn Estuary. Means + standard deviation.

*affinis)*, respectively. Using a laboratory continuous-flow system, Vighi[74] studied the transfer of lead between a primary producer, the algae *Selenastrum capricornutum*; a primary consumer, the zooplanktonic crustacea *Daphnia magna*; and a secondary consumer, the fish *Poecilia reticulata*. Vighi found that lead accumulated within the trophic chain with a decreasing concentration factor from the lowest to highest trophic levels. Similar results were reported by Leland and McNurney[71] in a river ecosystem. Increasing concentration factors were reported by Parslow et al.,[131] with concentrations of 0.03 ppb w.w. lead in unpolluted seawater, < 0.002 ppm w.w. in fish, (sand-eels and herrings *Ammodytes* spp. and *Clupea* spp.), and 0.036 ppm w.w. in the puffin *Fratercula arctica*, a fish predator. However, because many different sources of lead may have contributed to the lead body burden of the puffin, no conclusions can be drawn as to the part that biomagnification may have played.

## 15.5 EFFECTS OF LEAD

Most organic lead compounds are found as contaminants of aquatic environments, and there is some evidence that certain microorganisms in soils and lake sediments can biotransform inorganic lead to organic leads.[132,133] Although organic lead is much less frequent than inorganic lead contamination, organic lead is generally more readily absorbed and more toxic. Lead also influences community structure and biological diversity. For example, although certain aquatic organisms can develop a tolerance to long-term chronic lead exposure, macroinvertebrate and floral diversity is reduced in rivers with elevated lead concentrations.[134–136]

### 15.5.1 Plants

The toxicity of lead to plants is related to its chemical form, availability for and mechanisms of uptake, translocation and storage/detoxification, and physiochemical characteristics of the plant's direct environment. Although lead is toxic to plants under experimental conditions, little lead is generally available for uptake by plants in soils and sediments, and very high soil/sediment lead concentrations are often required before toxic effects upon plants are observed. In addition, many plants appear to have mechanisms of storing lead in an insoluble form, or for limiting its translocation within the plant. Toxic effects of lead in plants are most likely to occur near point sources of lead emission, where environmental concentrations are very high, or when the availability of lead for uptake by plants is high (e.g., under conditions of low soil or water pH, low soil calcium and phosphate, etc.).

Interspecific differences in lead toxicity to plants may be related to lack of translocation of lead within the plant due to physical or chemical isolation of the lead. In plant tissues lead has been reported to precipitate with intracellular orthophosphate.[137] The chemical isolation of lead within insoluble phosphate granules may reduce lead toxicity and its penetration into cells and cellular organelles.[66,138] The immobilization of lead by binding to cell walls, particularly in roots, has also been reported.[139] Silverberg[140] found that in the cells of *Stigeoclonium tenue* (Chlorophyceae), lead was present in an insoluble form in the cytoplasmic vacuoles and was not observed in the lead sensitive sites of cell metabolism (mitochondria, plastids, nuclei). Lead tolerance by lichens, which may accumulate and tolerate very high concentrations, may be related to an ability to precipitate extracellular lead, possibly as lead hydrocerussite.[141]

The existence of lead-tolerant ecotypes of a common grass species was demonstrated as early as 1952 by Bradshaw,[142] and lead tolerance has since been demonstrated in many species growing in areas of high lead concentration. In roadside plant populations lead tolerance decreases with distance from the road. Such tolerance is determined genetically and may be related to morphological changes and tolerance to other metals.[19] Of all aquatic biota, algae appear the most resistant to lead toxicity, although this observation may be influenced by the methods of toxicity testing used.[143]

Lead has been reported to have a wide range of effects in plants including disruption of cell membranes and mitosis; inhibition of plant growth; ATP (adenosine triphosphate) synthesis and structural protein formation; reduction of photosynthesis, water absorption, and transpiration rates; increase in generation time; and decrease in pollen germination and seed viability.[139,143–147]

Koepe and Miller[145] studied the effects of lead upon mitochondrial respiration in corn *Zea mays*. They concluded that if sufficient phosphate is present, lead will precipitate, and enzymatic reactions in corn mitochondria are not likely to be affected. However, it was suggested that plant growth could be dramatically reduced by lead under conditions of phosphate deficiency. Lead has been shown to inhibit respiration and photosynthesis in phytoplankton (*Phaeodactylum tricornutum*), with the magnitude of effect directly related to lead concentration and duration of exposure.[148] Dayton and Lewin[146] found that lead exposure increased the generation time of *P. tricornutum* and *Phaeodactylum tricornutum*, with the magnitude of effect directly related to lead concentration and duration of exposure.[142] Reduced photosynthesis and transpiration (by 10 to 25%) was recorded in 7-week-old seedlings of loblolly pine (*Pinus taeda*) and autumn olive (*Elaegnus umbellata*) 4 weeks after exposure to 320 ppm lead ($PbCl_2$) in potting medium.[147] The simultaneous decline in transpiration and photosynthesis suggests that the leaf stomatal resistance to $CO_2$ and water vapor diffusion had changed.

Krishna and Bedi[144] found a decrease in the percentage pollen germination and seed viability in *Cassia* spp. growing near a highway. These improved as distance from the highway increased. The life-cycle period was also reduced in relation to lead accumulation in plants growing near the highway. It is possible that atmospheric lead may have played a part in the decline of European spruce forests, as the lead concentration in needles and litter was significantly higher where tree decline was most pronounced.[149] The effects of lead in plants are reviewed in Jaworski.[150]

### 15.5.2 Animals

Lead is a nonspecific toxicant at the molecular level and inhibits the activities of many enzymes necessary for normal biological functioning. The most widely studied effects of lead are those upon the hematological system, the brain and nervous system, learning and behavior, and reproduction and survival. Research into the effects of lead upon animals was considerably stimulated in the 1950s when it was suggested that learning ability and coordination in children might be affected by lead. Subsequent research has suggested that young animals are more susceptible than adults to certain of the neurological and behavioral effects, as lead may affect the developing brain and nervous system.[150] Lead may act upon both nervous system structure and function, resulting in impairment of intellectual, sensory, neuromuscular, and psychological functions.[4,8]

Large inter- and intraspecific differences are observed in lead toxicity in relation to mechanisms and factors that influence lead absorption, retention, detoxification, and elimination. Diet is the most important variable influencing lead absorption; age, sex, physiological condition, and environmental factors also influence lead toxicity. Increases in tissue lead concentrations are normally associated with the toxic effects of lead, but it is difficult to determine tissue threshold concentrations for effects, as many factors, such as exposure regime and subcellular distribution, influence toxicity (Table 15.7).[4,8,95,150]

Table 15.7 documents effects of lead exposure upon animals. Although such experiments form an essential part of our understanding of metal toxicity, they should not be directly extrapolated as predictors of effect upon wild animals, as many biological and environmental factors may influence the absorption and distribution — and thus toxic effects — of lead under natural conditions. In addition, under experimental conditions animals are often only exposed to the contaminant under investigation, whereas in the natural environment they will be exposed to multiple contaminants, which may act antagonistically or synergistically.

Before discussing selected aspects of lead toxicity in different groups of animals, the hematological effects of lead will be considered in some detail, as they are common to most vertebrates.

**Table 15.7 Effects of Lead upon Selected Animals in Experimental Studies**

| Dose | Response | Reference |
|---|---|---|
| | **Fish** | |
| Plaice, *Pleuronectes platessa* | $LC_{50}$, 96-hour exposure (μg of Pb per liter) | 245 |
| Tetramethyl Pb | 50 | |
| Tetraethyl Pb | 230 | |
| Triethyl Pb | 1700 | |
| Trimethyl Pb | 24,600 | |
| Diethyl Pb | 75,000 | |
| $Pb^{2+}$ | 180,000 | |
| Dimethyl Pb | 300,000 | |
| | **Birds** | |
| 10 Mallards in each of three groups with different diets fed 4 No. 4 lead shot (721 mg Pb/kg). | | 246 |
| Pelleted calcium supplemented corn diet | 50% mortality | |
| Pelleted commercial duck ration | No mortality | |
| Cracked corn | 100% mortality | |
| 10 Mallard drakes in each of 4 groups with different diets fed 1 No. 6 lead shot. No mortality occurred in controls fed no shot. | | 247 |
| Corn diet | 60% mortality | |
| Mixed grain diet | 60% mortality | |
| Small grains, tame rice and smartweed seed | 10% mortality | |
| Coontail and mixed grains | No mortality | |
| Mallards (7M, 7F) fed 1 No. 4 | One month after dosing, lead shot (ca. 200 mg Pb) ALAD activity reduced 75% in blood, 42% in liver, 50% in cerebellum, and 35% in cerebral hemisphere | 151 |
| 200 μg trialkyl Pb chloride/days (ca. 28 mg/kg/days) given to Starlings, *Sturnus vulgaris* | | 241 |
| Triethyl lead | Reduced respiratory rate and labored breathing; abnormal head and body posture; fluffed feathers; altered feeding activity; reduced body weight, liver weight, and pectoral muscle. | |
| Trimethyl lead | Lack of coordination; impaired balance; tremors; fluffed feathers; inability to fly; bright green watery droppings; altered feeding activity; reduced body weight, liver and kidney weight, and pectoral muscle. 100% mortality within 6 days for both groups. However, pre-death symptoms more severe with trimethyl Pb. | |
| | **Mammals** | |
| ***Rats*** | | |
| 4% lead carbonate fed to lactating females resulting in 46 ppm Pb in milk to young | Growth retardation and paraplegia 26 to 30 days after birth. 85 to 90% mortality in 2 weeks. | 95 |
| Young rats weaned at 25 days on milk as above, and fed 4% Pb carbonate for 5 more days | Urinary incontinence at 22 to 24 days, caudal paraplegia 2 to 4 days later. Restricted growth and delayed maturation of brain cells. | 95 |
| Lead acetate administered to adult (100 days) and weanling (22 days) rats via intraperitoneal injections (0 to 1.2 mg/100 g) for 21 days pretest and 16 days testing. | Urinary ALA excretion increased in all exposed rats. At highest dose some clinical signs of poisoning and mortality. No behavioral effects measured. | 95 |

**Table 15.7 Effects of Lead upon Selected Animals in Experimental Studies** ***(Continued)***

| Dose | Response | Reference |
|---|---|---|
| Lead acetate (0.8 mg/100 g) administered to pregnant rats for 21 days intraperitoneally. | 100% reproductive failure | 95 |
| 3.5 mg Pb/100 g given by gavage to lactating dam. | Impaired T-maze learning in weanlings nursed by dams dosed 1–10, but not 11–20 days postpartum | 95 |
| Lactating rats fed 35 mg/kg $Pb^{2+}$/kg for 1–10 days postparturation. | Pups receiving lead from mothers milk exhibited decreased learning ability that persisted into adulthood. | 248 |
| ***Sheep*** | | |
| Adult sheep fed powdered elemental Pb (4.5 mg/kg/days) for 6 months prior to birth of lambs. Blood lead was 34 μg/dl in adults and 24 μg/dl in 2-week-old lambs. Postnatally, lambs were exposed only to Pb in ewe's milk. | Between 5 and 15 months, lambs trained on a simultaneous, 2-choice, nonspatial visual discrimination task. Significant increase in days required by dosed group to master visual discrimination test. | 95 |
| ***Primates*** | | |
| Lead acetate given orally to infant rhesus monkeys, 0.5–9.0 mg/kg. | Seizures occurred with blood Pb of 300 ug/dl, but stopped when exposure ceased. With blood Pb 60–100 ug/dl monkeys were hyperactive, insomniac, and hemoglobin and hematocrit gradually declined. | 95 |
| Lead acetate in drinking water 20 mg/kg resulting in blood Pb of 135 μg/dl in juvenile monkeys. | No behavioral changes. | 95 |
| Lead acetate 250–500 mg/kg three times weekly to 5- to 6-month-old rhesus monkeys resulting in blood Pb 30–1000 ug/dl. | Overt toxicosis after 7–16 weeks: anorexia; lethargy; muscular weakness; convulsions. | 95 |
| Lead chloride intravenously 1.61 mg/kg/day to baboons. | Two individuals died, at 68 and 85 days with blood Pb 1500 and 950 μg/dl. | 88 |
| Lead fed to infant rhesus monkeys at 3–4 mg/kg/day for 5 weeks or 9–10 mg/kg/day for 3 weeks. | Behavior of experimental animals differed markedly from controls. Infant monkeys more susceptible to Pb-induced behavioral effects than adolescents and adults. | 249 |
| Rhesus monkey infants fed 0.5 mg/kg diet for 4 weeks. Adults given 20 mg/L Pb in drinking water for 4 weeks. | Hyperactivity, insomnia and abnormal social behavior in infants, no effect in adults. | 250 |

Unlike many of the other effects of lead, hematological effects can be quantified in wild animals as precisely as those used in controlled experiments. Several parameters are useful as indicators of lead contamination in wild populations.

#### *15.5.2.1 Effects of Lead in the Bloodstream*

The first measurable effects of lead following absorption occur in the bloodstream. Lead inhibits several heme-biosynthetic enzymes, notably ALAD and heme synthetase. ALAD is the most lead-sensitive erythrocyte enzyme, and lead is a relatively specific inhibitor of ALAD activity. ALAD activity becomes inhibited within hours of lead absorption and, depending upon the level and duration of exposure, may remain inhibited for several weeks or several months after cessation of exposure. Inhibition of erythrocyte ALAD activity following lead exposure has been reported in many species of freshwater and marine fish, in birds, and many mammals including mice, rats, rabbits, sheep, cattle, swine, and humans.[8,143] The sensitivity of ALAD to low concentrations of lead has been demonstrated in a wide range of avian species including mallard, black duck (*Anas rubripres*), canvasback (*Aythya valisineria*), American kestrels, bald eagles (*Haliaeetus leucocephalus*), barn swallows (*Hirundo rustica*), pheasants (*Phasianus colchicus*), Japanese quail (*Cortunix japonica*), and other species.[112,151–158] Inhibition of ALAD activity has been recorded at blood lead concentrations of < 5 μg/dL in some avian species,[152] and in most species > 90% inhibition occurs when blood lead concentrations exceed 120 to 150 μg/dL. Interspecific differences in the sensitivity of ALAD to lead have been recorded in birds.[159] Lead affects fish in a similar way to birds and mammals, interfering with the activities of enzymes, notably those of the heme-biosynthetic pathway, and affecting a wide range of body systems resulting in impaired growth, reproduction, and survival. ALAD activity in cattle and humans appears to be more sensitive to *in vitro* lead inhibition than that of pigs or dogs.[160] Lead inhibition of erythrocyte ALAD activity may be accompanied by a corresponding inhibition of ALAD activity in the brain, liver, and kidney of birds and mammals.[151,154,161,162]

Inhibition of ALAD activity results in the accumulation of amino levulinic acid (ALA), which is excreted in the urine in abnormal concentrations when blood lead is > 40 μg/dl in humans,[163] and high urinary ALA concentrations are characteristic of lead poisoning.[143] In many animals significant inhibition of ALAD activity can occur without a corresponding decrease in heme production, possibly related to a compensatory increase in ALAD production. Many animals can compensate for the inhibition of ALAD activity, and dogs fed sufficient lead to inhibit most ALAD activity may remain apparently healthy.[164] In some avian species ALAD activity may be inhibited by up to 80% by lead, and this inhibition can be sustained over several months without the occurrence of anemia.[165]

Another enzyme in the heme-biosynthetic pathway influenced by lead is heme synthetase, responsible for the incorporation of ferrous iron $Fe_{2+}$ into protoporphyrin 9 (PP9) for heme production. Inhibition of heme-synthetase activity results in elevated levels of metal-free erythrocyte porphyrin (FEP) and zinc protoporphyrin (ZPP), in which zinc becomes incorporated into PP9 instead of iron. Elevated concentrations of PP9 (FEP or ZPP) in the blood following lead exposure have been reported in many animals including humans, rabbits, small mammals, and birds.[152, 166–170] Both FEP and ZPP fluoresce when actuated by specific wavelengths of light, and this fluorescence is used for the quantitative estimation of protoporphyrin concentration. Both FEP/ZPP concentrations and ALAD activity have been used as indicators of lead exposure in many species of fish, birds, and mammals.[8]

In cases of acute or long-term chronic lead exposure, the inhibition of heme-biosynthetic enzymes may result in a reduction in total blood hemoglobin concentration and anemia. Lead may also cause anemia by impairing red blood cell production or initiating premature cell degradation, particularly in cases of acute exposure. Pain and Rattner[171] reported significant decreases in both hemoglobin concentration and hematocrit of black duck within 3 days of ingesting one No. 4 lead shot. Anemia is a classic sign of lead poisoning and has been reported in fish, birds, and mammals.[8,143,172]

Other hematological features associated with lead poisoning include reticulocytosis,[173,174] basophilic stippling of erythrocytes,[175,176] and a high degree of variability in erythrocyte morphology, including characteristic "tear-drop" shaped cells in waterfowl.[103,177] Morphological alterations in red blood cells may result in increased hemolysis and, consequently, contribute to anemia.

#### *15.5.2.2 Aquatic Animals*

The effects of lead upon aquatic organisms vary according to species, duration of exposure, development of tolerance, lead concentration, and environmental factors, such as water hardness and pH, that influence lead solubility. For example, the $LC_{50}$ (concentration of total lead required to kill 50% of individuals) for *Daphnia magna* exposed for 96 h varies from 612 μg Pb/LL with a water hardness of 54 mg $CaCO_3$/L to 1910 μg Pb/L with water hardness 152 mg $CaCO_3$/L.[178] A similar increase in $LC_{50}$ with increasing water hardness has been reported for minnows (*Pimephales promelas*), rainbow trout, and bluegills (*L. macrochirus*), although the relationship between water hardness and lead toxicity was not constant across species.[127]

Lead-tolerant strains of the isopod *Asellus meridianus* have been reported with an $LC_{50}$ of 280 μg Pb/L (48-h exposure) for a nontolerant strain compared with 3500 μg Pb/L for a strain from a lead-contaminated river.[143] Lead tolerance may be either genetically controlled or achieved by acclimation during the animal's lifetime. Evidence of the former has been provided by Brown[179] in *A. meridianus* and the latter by Fraser[180] in *A. aquaticus*. Although tolerance to the lethal effects of relatively high concentrations of lead can and does develop in certain aquatic organisms, rivers affected by lead mine waste support few macroinvertebrates and flora. Demayo et al.[143] suggested that the observed sublethal effects of lead at concentrations > 20 μg Pb/L on invertebrates may be sufficient to extinguish some populations.

The chemical form of lead influences its toxicity, with organic forms tending to produce toxic effects at lower concentrations than $Pb^{2+}$, as is illustrated by the $LC_{50}$ concentrations in plaice (*Pleuronectes platessa*) (Table 15.7). Although the concentration of lead resulting in lethality varies considerably with the chemical form of lead and composition of the water, a variety of sublethal effects can occur at very low water lead concentrations (7 μg/L[143]), and the activity of certain hematopoietic enzymes can be inhibited by water Pb concentrations of 10 μg/L.[181] Elevated water lead concentrations may result in anemia, reduced egg hatching, darkening of the dorsal tail region and degeneration of the caudal fin, and scoliosis (lateral curvature of the spine).[143] Lethal solutions of lead cause increased mucus formation, which coagulates over the gills and body resulting in death by anoxia.[182,183] Certain fish appear to be more sensitive to the effects of lead when exposed at the egg or young hatchling stage, possibly due to an increased sensitivity of the developing nervous system or an increased ability to take up lead relative to older fish. Reduced community diversity and general absence of fish in parts of a river with higher lead content in suspended and bed sediments were reported by Leland and McNurney,[71] although lead was not the only factor influencing community structure, as oxygen levels also affected the biota.

#### 15.5.2.3 Amphibians

Lead exposure has resulted in a range of effects in amphibians including decreased red and white blood cells; neutrophils and monocytes; sloughing of the skin; excessive bile excretion; hypertrophy of liver, spleen, and stomach; decreased muscle tone and loss of normal semierect posture; salivation, excitement, and muscular twitching; and delayed metamorphosis.[8,184] $LC_{50}$ values vary between species and according to water chemistry, but in soft water (99 mg $CaCO^3$/L), an 8-day exposure to 1.4 mg Pb/L was sufficient to cause mortality in some salamanders (*Ambystoma opacum*).[8] Surface water runoff from the fall zone of a shotgun range (total lead ranged from 840 to 3150 μg/L) was toxic pickerel frog (*Rana palustris*) tadpoles but not to bullfrog (*Rana catesbeiana*) tadpoles.[180] Green frogs (*Rana clamitans*) on a shotgun target range had femur lead levels of 1728 μg/g, nearly 1000-fold higher than a nearby neighbor.[181]

#### 15.5.2.4 Birds

##### 15.5.2.4.1 The Ingestion of Lead Gunshot and Bullets

The effects of lead upon wild birds have perhaps been observed more frequently than for any other group of wild animals. This results primarily from the widespread exposure of avian species to spent lead gunshot. The two main groups of birds that become poisoned from this source are those feeding in substrates in shot-over areas, such as wetlands, clay-pigeon shoots, firearms training facilities,[185] and upland game areas; and those species that ingest shot while feeding on animals carrying shot in their flesh. Waterfowl, waders, rails, coots, pigeons, and doves fall into the first category, and raptors (including eagles, buzzards, harriers, and vultures) into the second.[186]

Lead poisoning from gunshot ingestion is considered to be directly responsible for the deaths of millions of wild birds each year.[105,251] In the United States, before the widespread use of nontoxic gunshot, 1.6 to 2.4 million waterfowl were estimated to die annually as a direct result of gunshot ingestion.[252] Lead poisoning threatens waterfowl worldwide,[187] with particularly high levels of exposure in some parts of Europe, such as Mediterranean wetlands,[188–190] where other waterbirds, such as the greater flamingo (*Phoenicopterus rubber roseus*), have also suffered lead-poisoning mortality.[191] Several waterfowl species classified as globally threatened,[192] including the marbled teal (*Marmaronetta angustirostris*) and the white-headed duck (*Oxyura leucocephala*), are susceptible to lead poisoning in Spain.[193] Historical analyses of ringing recoveries from x-rayed mallards has shown reduced survival in birds carrying ingested shot.[194]

Both predatory and scavenging raptors are vulnerable to lead poisoning.[195–201] Elevated tissue lead levels and mortality have been reported, for example, in bald (*Haliaeetus leucocephalus*) and golden eagles (*Aquila chrysaetos*) from Canada;[195] in Steller's sea eagles (*Heliaeetus pelagicus*) from Japan;[196] in white-tailed sea eagles (*Heliaeetus albicilla*) from Japan, Austria, and Germany;[196,202,203] in golden eagles from California;[197] in bald eagles from the great plains of North America; in golden and bald eagles from the upper Midwest[198] (United States); in bald and golden eagles from Idaho;[199–201] in marsh harriers (*Circus aeruginosus*) from France and Spain;[204, 205] in Griffon vultures (*Gyps fulvus*) from Spain;[206] and California condors (*Gymnogyps californianus*).[207]

For example, the Raptor Center at the University of Minnesota treated 654 golden and bald eagles from 1980 to 1995; 138 were lead poisoning cases.[199] More than 170 bald eagle deaths from lead poisoning have been documented by the National Wildlife Health Research Center (United States, pers. comm.). Several globally threatened raptor species are exposed to lead gunshot or bullet fragments in their prey including the Spanish imperial eagle[208] and the California condor.[207] The remnant wild population of California condors exhibited a rapid population decline in the 1980s, largely as a result of lead poisoning through ingesting bullet fragments in carrion, and this remains a problem for birds released in the 1990s subsequent to a successful captive breeding program.[209]

External signs of lead poisoning in birds following shot ingestion were first recognized in 1876 in pheasants (*Phasianus colchicus*), emaciated and with paralyzed legs, the feet being held in a similar manner to the "drop-hand" condition typical of lead-poisoned humans.[210] Lead-poisoned birds are frequently emaciated, have a prominent keel, and are weak and unable to walk or fly. The muscular paralysis caused by lead poisoning results in characteristic signs such as abnormal carriage of wings by waterfowl (either wing-droop or carriage in a hat-shaped position) and "kinky" necks in swans following the ingestion of lead fishing sinkers. Wing-droop has also been reported in Laysan albatross (*Diomeda immutabilis)* chicks after ingesting paint chips containing lead.[211] Jacobsen et al.[212] noted that bald eagles were unable to maintain an upright position following shot ingestion.

Not all birds exposed to shot die, as shot retention and lead absorption and distribution within the body vary according to a wide range of factors, including age, sex, environmental factors, and diet. The importance of diet upon lead toxicity is illustrated in mallards in Table 15.7. Waterfowl that die following shot ingestion may die within several days of ingesting a large number of shot (e.g., ten), with little loss of weight and few external signs of poisoning. However, more frequently, waterfowl die several weeks after the ingestion of a smaller number of shot. At necropsy, birds that have died from lead poisoning are often underweight, lacking subcutaneous fat reserves. Substantial weight losses, occasionally of > 50% of body weight, have been recorded in many species of waterfowl, game birds, raptors, and passerines.[105,112,174,213,214] Tissues of dead birds are generally anemic, the gall bladder is often distended and bile-filled, and the gastrointestinal tract may be impacted with food. Intestinal impaction due to muscular paralysis may result in death from starvation.

#### *15.5.2.4.2 Nonammunition Sources*

Lead poisoning of mute swans (*Cygnus olor)* and other species has occurred due to the ingestion of lead fishing sinkers.[186,215] Concern over Mute swan mortality resulted in the banning of most lead fishing sinkers in the United Kingdom in 1986. Lead poisoning from lead fishing weights has been identified as a source of mortality in common loons (*Gavia immer*); 16 of 31 dead loons examined died of lead poisoning,[216] and 21% of 105 loons found dead in New York from 1972 to 1979 had ingested lead fishing weights.[217]

An additional source of lead poisoning, although rarely reported in wild birds, is leaded paint. Ingestion of chips of flaking paint from a deserted military base on Midway Atoll resulted in mortality of Laysan albatross chicks in 1983.[211] Cases of lead poisoning from leaded paint are more often reported in caged birds.[186] Birds are exposed to many sources of environmental lead contamination, and elevated tissue lead concentrations and associated physiological effects are frequently found in birds occupying or feeding in urban areas, industrial areas, or waste dumps.[87,218] These sources of lead contamination usually result in sublethal effects in birds and other animals, the biological significance of which is difficult to quantify. However, recent work in the United States has shown that in areas where sediment has become lead contaminated (e.g., through mining or smelting) sediment ingestion is an important route of lead contamination in waterfowl.[219,220]

Although most avian mortality from lead poisoning results from the ingestion of lead objects, sporadic cases of mass avian mortality from other sources do occur. One example is the deaths of several thousand shorebirds and gulls in the Mersey Estuary, United Kingdom, in 1979 and the early 1980s following the ingestion of prey contaminated with organic lead from a petrochemical works.[221] Signs of lead poisoning following exposure to organic lead compounds are presented in Table 15.7.

#### *15.5.2.5 Mammals*

From experimental studies, the toxic effects of lead in mammals have included blindness, histopathology of liver and kidney, hemorrhaging, depressed food intake and anorexia, anemia, inhibition of ALAD activity and increased proptoporphyrin concentrations, reduced brain weight and cerebral pathology, lack of coordination, convulsions, impaired motor skills, impaired visual discrimination and learning behavior, abnormal social behavior, increase in aggression, hyperactivity, disturbed sleep patterns and insomnia, reproductive impairment, increased fetal deaths and abortions, and reduced survival and longevity.[8,143,222]

The minimum level of lead intake that results in sublethal lead poisoning or death in cattle has been shown to be 5 to 7 mg ingested Pb/kg body weight/day.[8,223] However, lower lead doses have resulted in toxic effects in some experiments, and the form of lead administered, age, and physiological state of test animals and diet all appear to influence the minimum toxic dose. A dose of 5 mg Pb/kg/day over a 10- to 20-day period resulted in some clinical lead poisoning and mortality in calves, whereas 2.7 mg/kg body weight given as lead acetate to calves for 20 days on a milk diet resulted in mortality.[224] Mortality in horses has resulted from the ingestion of 2.4 mg Pb/kg body weight/day.[143] Only a percentage of lead administered orally is absorbed, and toxic effects of lead are observed at much lower doses when lead is administered via intraperitoneal or intramuscular injection. Even when no clinical signs of lead poisoning are apparent, lead may increase the susceptibility of animals to disease. Hemphill et al.[225] illustrated that by exposing mice to lead nitrate at concentrations that provoked no clinical signs of poisoning. Their susceptibility to infection by the bacteria *Salmonelle typhimurium* increased.

Controlled studies in a wide range of mammals, including rats, sheep, and primates, have illustrated various effects of lead on learning and behavior (Table 15.7). A common finding of these studies is that the age at exposure to lead is critical; prenatal and very early postnatal lead exposure is more likely to affect animals than exposure at later stages.[150] Maternal or dual-parent exposure to lead has been shown to influence learning and behavior in offspring of rats[226,227] and sheep.[228] Transfer of lead from mother to young is both via the placenta and milk. Lambs from pregnant ewes fed 4.5 mg Pb/kg/days for 6 months prenatally required more days to learn visual discrimination processes than controls (Table 15.7).[228,229] No data are available concerning the effects of lead upon learning and altered social behavior in wild animals.

Cases of lead poisoning and mortality in domestic and wild mammals usually result from feeding in areas contaminated by industrial lead operations, near disposal sites of lead containing waste, or in areas where leaded paints have been used.[230] Mortality from lead poisoning has been reported in horses and cattle in the vicinity of a California lead smelter,[231] and lead poisoning has been reported in buffalo and cattle in the vicinity of a plant that recycles lead from old batteries.[232] Lead toxicity has also been reported in cattle consuming silage cut from fields that had been used for clay-pigeon shoots.[233,234] Frape and Pringle[233] reported soluble lead concentrations in shot-contaminated silage of 3800 ppm d.w. soluble lead (pH was 3.9 to 4.4). Affected cows showed lack of appetite and coordination as well as stiff and swollen joints; in addition, a large number of stillbirths or abortions occurred. Many zoo animals have suffered lead poisoning including primates, bats, foxes, ferrets, panthers, and a bear.[235] In almost every case the source of lead poisoning was lead-containing paint.

## 15.6 CONCLUSIONS

Anthropogenic uses of lead have probably altered its availability and environmental distribution more than any other toxic element. Consequently, lead concentrations in many living organisms may be approaching thresholds of toxicity for the adverse effects of lead. Such thresholds are difficult to define, as they vary with the chemical and physical form of lead, exposure regime, other elements present and also vary both within and between species. The technological capability to accurately quantify low lead concentrations has increased over the last decade, and physiological and behavioral effects have been measured in wildlife with tissue lead concentrations below those previously considered safe for humans.[8,236] Consequently, lead criteria for the protection of wildlife and human health are frequently under review, and "thresholds" of lead toxicity are being reconsidered. Proposed lead criteria for the protection of natural resources have been reviewed by Eisler.[8]

Uptake of lead by plants is limited by its generally low availability in soils and sediments, and toxicity may be limited by storage mechanisms and its apparently limited translocation within most plants. Lead does not generally accumulate within the foliar parts of plants, which limits its transfer to higher trophic levels. Although lead may concentrate in plant and animal tissues, no evidence of biomagnification exists.

Acid deposition onto surface waters and soils with low buffering capacity may influence the availability of lead for uptake by plants and animals, and this may merit investigation at susceptible sites. The biological significance of chronic low-level lead exposure to wildlife is sometimes difficult to quantify. Animals living in urban environments or near point sources of lead emission are inevitably subject to greater exposure to lead and enhanced risk of lead poisoning.

Increasingly strict controls on lead emissions in many countries have reduced exposure to lead from some sources, and the reduction of lead in gasoline has resulted in lower tissue lead concentrations in humans and wildlife from many, particularly urban, locations.[237] However, it has been suggested that increasing use of organic lead compounds as catalysts for the production of plastics and as wood preservatives and biocides could adversely affect wildlife.[8]

The most significant source of direct wildlife mortality from lead is spent gunshot and fishing sinkers. Elevated mortality from shot ingestion in avian species resulted in the introduction of nontoxic (steel) shot zones along certain flyways in the United States in the mid-1970s and a total ban on the use of lead for waterfowl and coot hunting nationwide by 1992.[251,252] Several other countries are now following suit and have either banned or are in the process of restricting the use of lead shot for waterfowl hunting.[238] In the United States it has been estimated that since the 1986 hunting season, when the use of nontoxic shot became widespread, over 6 million ducks have not been lost to lead poisoning.[239] Raptors, especially eagles, have also apparently benefited, although lead poisoning from ingestion of bullet fragments remains a problem for the critically threatened California condor.[209] Quantifying reductions in lead mortality rates would be difficult since eagle populations throughout North America are rapidly recovering from other anthropogenic perturbations, especially organochlorine pesticides.

## REFERENCES


1. U.S. Environmental Protection Agency, Ambient Water Quality Criteria for Lead, U.S. Environmental Protection Agency Rep. 440/5–80–057, Available from Natl. Tech. Infor. Serv., 5285 Port Royal Road, Springfield, Virginia 22161, 1980.
2. Boultron, C. F. and Patterson, C. C., The occurrence of lead in Antarctic recent snow firn deposited over the last two centuries and prehistoric ice, *Geochim. Cosmochim. Acta*, 47, 1355, 1983.
3. Grandjean, P., Ed., *Biological Effects of Organolead Compounds*, CRC Press, Boca Raton, FL, 1984.
4. Nriagu, J. O., Ed., *The Biogeochemistry of Lead in the Environment. Part A. Ecological Cycles,* Elsevier/North Holland Biomedical Press, Amsterdam, 1978.
5. Waldron, H. A. and Stoffen, D., *Subclinical Lead Poisoning*, Academic Press, London, New York, 1974.
6. Albert, L. A. and Badillo, F., Environmental lead in Mexico, *Rev. Environ. Contam. Toxicol.,* 117, 1, 1991.
7. Scheuhammer, A. M. and Norris, S. L., A Review of the Environmental Impacts of Lead Shotshell Ammunition and Lead Fishing Weights in Canada, Occasional Paper Number 88, Canadian Wildlife Service, 1999.
8. Eisler, R., *Handbook of Chemical Risk Assessment: Health Hazards to Humans, Plants, and Animals. Volume 1*, Lewis Publishers, CRC Press, Boca Raton, FL, 2000.
9. U.S. Public Health Service (USPHS), Toxicological Profile for Lead, Update, U.S. Dept. Health Human Serv., Publ. Health. Serv., Agen. Toxic Subst. Dis. Regis., TP-92/12, 1993.
10. Barth, D., Berlin, A., Engel, R., Recht, P., and Smeets, J. Eds., *Proc. Int. Symp. Environmental Health Aspects of Lead,* Commis. Eur. Commun., Luxembourg, 1, 1973.
11. Tackett, S. L., The Franklin letter on lead poisoning, *J. Chem. Ed.*, 58, 274, 1981.
12. Byers, R. K. and Lord, E. E., Late effects of lead poisoning on mental development, *Am. J. Dis. Child.,* 66, 471, 1943.
13. Needleman, H. L., Gunnoe, C., Leviton, A., Reed, R. R., Peresie, H., Maher, C., and Barrett, P., Deficits in psychologic and classroom performance in children with elevated dentine lead levels, *N. Engl. J. Med.,* 300, 698, 1979.
14. Needleman, H. L., Leviton, A., and Bellinger, D., Lead-associated intellectual deficit, *N. Engl. J. Med.,* 306, 367, 1982.
15. Landrigan, P., Baloh, R., Whitworth, R., Staehling, N., and Rosenblum, B. F., Neuropsychological dysfunction in children with chronic low level lead absorption, *Lancet,* 1, 708, 1975.
16. U.S. Environmental Protection Agency, Air Quality Criteria for Lead, EPA Report, EPA–600/8–77–017. 1977.
17. Ronis, M. J. J., Gandy, J., and Badger, T., Endocrine mechanisms underlying the growth effects of developmental lead exposure in the rat, *J. Toxicol. Environ. Health,* 54A, 77, 1998.
18. Lankford, D., Lead determinations in underground parking garage, Ph.D. thesis, University of Oklahoma, 1975.
19. Southwood, T. R. E. (Chairman), *Lead in the Environment,* Royal Commission on Environmental Pollution, Cmnd. 8852. HMSO, London, 1983.
20. Harrison, R. M. and Parker, J., Analysis of particulate pollutants, in *Handbook of Air Pollution Analysis,* Perry, R. and Young, R. J., Eds., Chapman and Hall, London, 1977.
21. Feierabend, J. S., Steel Shot and Lead Poisoning in Waterfowl, National Wildl. Fed. Sci. Tech. Series, No. 8, 1983.
22. Duerr, A. E. and DeStefano, S., Using a metal detector to determine lead sinker abundance in waterbird habitat, *Wildl. Soc. Bull.,* 27, 952, 1999.
23. Scheuhammer, A. M. and Templeton, D. M., Use of stable isotope ratios to distinguish sources of lead exposure in wild birds, *Ecotoxicology*, 7, 37, 1998.
24. Hutchinson, T. C. and Meemia, K. M., Eds., *SCOPE 31, Lead, Mercury, Cadmium, and Arsenic in the Environment*, John Wiley and Sons, Chichester, U.K., 1987.
25. Eisenreich, S. J., Metzer, N. A., Urban, N. R., and Robbins, J. A., Response of atmospheric lead to decreased use of lead in gasoline, *Environ. Sci. Tech.,* 20, 171, 1986.

26. Anon., The Use of Tetraethyl Lead Gasoline in its Relation to Public Health, Public Health Bull. No. 163, Washington, D.C., 1926.
27. Chamberlain, A. C., Heard, M. J., Little, P., and Wiffen, R. D., The Dispersion of Lead from Motor Exhausts, *Proc. of Royal Soc. Discussion Meeting Pathways of Pollutants in the Atmosphere*, London, 1977; *Phil. Trans. Roy. Soc. Lond. A.*, 290, 577–89, 1979.
28. Harrison, R. M. and Laxen, D. P. H., *Lead Pollution, Causes and Control*, Chapman and Hall, London and New York, 1981.
29. Gaghate, D. G. and Hasan, M. Z., Ambient lead levels in urban areas, *Bull. Environ. Contam. Toxicol.,* 62, 403, 1999.
30. Bertine, K. K. and Goldberg, E. D., Fossil fuel combustion and the major sedimentary cycle, *Science,* 173, 233, 1971.
31. Central Unit of Environmental Pollution, Department of the Environment, *Lead in the Environment and its Significance to Man*, HMSO, London, 1974.
32. Olivio, R., Vivoli, G., Ferrari, L. R., and Vecchi, G., Atmospheric pollution caused by ceramics factories in the area of Sassuolo and Fionano and its effect on health, *Ann. Sanita Publica*, 35, 4, 265, 1974.
33. Crandall, C. J. and Rodenberg, J. R., Waste lead oxide treatment of lead acid battery manufacturing wastewater, *Eng. Bull. Purdue Univ. Eng. Ext. Ser.*, 145, 194, 1974.
34. Nozaki, K. and Hatotani, H., Treatment of tetraethyl lead manufacturing wastes, *Water Res.*, 1, 167, 1967.
35. Moorcroft, S., Watt, J., Thornton, I., Wells, J., Strehlow, C. D., and Barltrop, D., The chemical composition of house and road dusts and garden soils in an apparently uncontaminated rural village in Southwest England-implications to human health, in *Trace Substances in Environmental Health XVI*, Hemphill, D. D., Ed., University of Missouri, Columbia, 1982.
36. Millar, I. B. and Cooney, P. A., Urban lead — A study of environmental health and its significance to school children in the vicinity of a major trunk road, *Atmosph. Environ.*, 16, 615, 1982.
37. Smith, D. R., Niemmeyer, S., Estes, J. A., and Flegal, A. R., Stable lead isotopes evidence anthropogenic contamination in Alaskan waters, *Environ. Sci. Tech.,* 24, 1517, 1990.
38. Smith, D. R., Niemeyer, S., and Flegal, A. R., Lead sources to California sea otters: industrial input circumvents natural biodepletion mechanisms, *Environ. Res.,* 57, 163, 1992.
39. Trefry, J. H., Metz, S., and Trocine, R. P., A decline in lead transport by the Mississippi River. *Science,* 230, 14, 1985.
40. Laxen, D. P. H. and Harrison, R. M., The highway as a source of water pollution: An appraisal with the heavy metal lead, *Water Res.*, 1, 1, 1977.
41. Water Research Centre, Pollution from Urban Run-off, *Notes on Water Res.*, 12, Stevenage, Herts, 1977.
42. Byrd, J. E. and Perona, M. J., The temporal variations of lead concentrations in a freshwater lake, *Water Air Soil Pollut.*, 13, 207, 1980.
43. Lester, J. N., Harrison, R. M. and Perry, R., The balance of heavy metals through a sewage treatment works. 1. Lead, cadmium and copper, *Sci. Tot. Environ.*, 12, 13, 1979.
44. Natusch, D. F. S., Bauer, C. F., Matusiewicz, H., Evans, C. A., Baker, J., Loh, A., and Linton, R. W., Characterization of trace elements in fly ash, in *Proc. Int. Conf. on Heavy Metals in the Environ.*, Toronto, Ontario, Oct. 27–31, 1975.
45. Blus, L. J., Henny, C. J., Hoffman, D. J., Sileo, L., and Audet, D. J., Persistence of high lead concentrations and associated effects in tundra swans captured near a mining and smelting complex in Northern Idaho, *Ecotoxicology*, 8, 125, 1999.
46. Henny, C. J., Blus, L. J., Hoffman, D. J., Grove, R. A., and Hatfield, J. S., Lead accumulation and osprey production near a mining site on the Coeur d'Alene River, Idaho, *Arch. Environ. Contam. Toxicol.,* 21, 415, 1991.
47. Little, P. and Wiffen, R. D., Emission and deposition of petrol engine exhaust Pb, I. Deposition of exhaust Pb to plant and soil surfaces, *Atmos. Environ.,* 11, 437, 1977.
48. Berrow, M. L. and Webber, J., Trace elements in sewage sludges, *J. Sci. Food Agric.,* 23, 13, 1972.
49. Department of the Environment and National Water Council, Report of the Sub-Committee on the Disposal of Sewage Sludge to Land, DOE Standing Technical Committees: Report No. 20. HMSO, London, 1981.
50. Sterritt, R. M. and Lester, J. N. Concentration of heavy metals in forty sewage sludges in England, *Water Air Soil Pollut.*, 14, 125, 1981.

51. Stansley, W. and Roscoe, D. E., The uptake and effects of lead in small mammals and frogs at a trap and skeet range, *Arch. Environ. Contam. Toxicol.*, 30, 220, 1996.
52. Stansley, W., Kosenak, M. A., Huffman, J. E., and Roscoe, D. E., Effects of lead-contaminated surface water from a trap and skeet range on frog hatching and development, *Environ. Pollut.*, 96, 69, 1997.
53. Anderson, W. L., Havera, S. P., and Zercher, B. W., Ingestion of lead and nontoxic shotgun pellets by ducks in the Mississippi flyway, *J. Wild. Manage.*, 64, 848, 2000.
54. Crowder, A., Acidification, metals and macrophytes, *Environ. Pollut.*, 71, 171, 1991.
55. Everard, M. and Denny, P., Flux of lead in submerged plants and its relevance to a freshwater system, *Aquat. Bot.*, 21, 181, 1985.
56. Behan, M. J., Kinraide, T. B., and Selser W. I., Lead accumulation in aquatic plants from metallic sources including shot, *J. Wildl. Manage.*, 43, 240, 1979.
57. Stansley, W., Widjeskog, L., and Roscoe, D. E., Lead contamination and mobility in surface water at trap and skeet ranges, *Bull. Environ. Contam. Toxicol.,* 49, 640, 1992.
58. Rule, J., Hemphill, D., and O'Pierce, J. O., The use of $^{210}$Pb and $^{109}$Cd isotopes in a preliminary study of their uptake and translocation in plants, *Trace Contam. Environ. Proc. 2$^{nd}$ Annu. NSF-RANN Trace Contam. Conf.*, 1974.
59. Harrison, R. M. and Johnston, W. R., Experimental investigations on the relative contribution of atmosphere and soils to the lead content of crops, in *Pollutant Transport and Fate in Ecosystems*, Coughtrey, R. J., Martin, M. H., and Unsworth, M., Eds., Blackwell's Scientific Publications, London, 1987.
60. Tjell, J. C., Houmand, M. F., and Mosboek, H., Atmospheric lead pollution of grass grain in a background area in Denmark, *Nature*, 280, 1979.
61. Wedding, J. B., Carlson, R. W., Stukel, J. J., and Buzaz, F. A., Aerosol deposition on plant leaves, *Environ. Sci. Technol.*, 9, 151, 1975.
62. Crump, D. R. and Barlow, P. J., Factors controlling the lead content of pasture grass, *Environ. Pollut.*, B, 181, 1982.
63. Ratcliffe, D. and Beeby, A., Differential accumulation of lead in living and decaying grass on roadside verges, *Environ. Pollut. A*, 279, 1980.
64. Walsh, L. M., Sumner, M. E., and Corey, P. B., Consideration of soils for accepting plant nutrients and potentially toxic non-essential elements, in *Land Application of Waste Materials*, Soil Conservation Society of America, Ankeny, Iowa, 22, 1976.
65. Brown, O. H. and Slingsby, D. R., The cellular location of lead and potassium in the lichen *Cladonia rangiformis (L.)* Hoffm., *New Phytology*, 71, 297, 1972.
66. Malone, C., Koeppe, D. E., and Miller, R. J., Localization of lead accumulated by corn plants, *Plant Physiol.*, 53, 388, 1974.
67. Rolfe, G. L., Lead uptake by selected tree seedlings, *J. Environ. Qual.*, 2, 1, 153, 1973.
68. Rutter, M. and Russell-Jones, R., Eds., *Lead versus Health: Source and Effects of Low Level Exposure,* John Wiley, Chichester, 1983
69. M. A. F. F., Survey of Lead in Food: Second Supplementary Report, Food Surveillance Paper No.10. HMSO, London, 1982.
70. Namminga, H. E., Scott, J. E., and Burks, S. L., Distribution of copper, lead and zinc in selected components of a pond ecosystem, *Proc. Okla. Acad. Sci.*, 54, 62, 1974.
71. Leland, H. V. and McNurney, J. M., Lead transport in a river ecosystem, *Proc. Int. Conf. on Transport of Persistent Chemicals in Aquatic Ecosystems*, Ottawa, 1974, 111–117.
72. Varansi, U. and Gmur, D. J., Influence of water-borne and dietary calcium on uptake and retention of lead by coho salmon *Oncorhynchus kitutch*, *Toxicol. Appl. Pharmacol.*, 46, 65, 1978.
73. Chau, Y. K., Wong, P. T. S., Kramer, O., Bengert, G. A., Cruz, R. B., Kinrade, J. O., Lye, J., and van Loon, J. C., Occurrence of tetraalkyllead compounds in the aquatic environment, *Bull. Environ. Contam. Toxicol.*, 24, 265, 1980.
74. Vighi, M., Lead uptake and release in an experimental tropical chain, *Ecotoxicol. Environ. Safety*, 5, 177, 1981.
75. Tomashefski, J. F. and Mitchell, R. I., Under What Circumstances is Inhalation of Lead Dangerous?, in Symp. on Environ. Lead Contam., December 13–15, 1965, U.S. Dept. of Health, Education and Welfare, Public Health Service Pub. No. 1440, 1966, 39–49.
76. Danielson, L., Gasoline containing lead, Swedish Natural Science Research Council, *Ecological Research Committee Bull. No. 6*, 1970.

77. Chamberlain, A. C., Clough, W. S., Heard, M. J., Newton, D., Stott, A. N. B., and Wells, A. C., Uptake of lead by inhalation of motor exhaust, *Proc. Royal Soc. Lond. B.*, 192, 77, 1975.
78. Bianco, A., Gibb, F. R., and Marrow, T. T., Inhalation study of a submicron size lead-212 aerosol, *Third Int. Congr. I.R.P.A.*, Conf. 730907, Part II, 1214, 1973.
79. Heard, M. J., Wells, A. C., Newton, D., and Chamberlain, A. C., Human uptake and metabolism of tetra ethyl and tetra methyl lead vapour labelled with $^{203}$Pb, *Proc. of the Int. Conf. of Heavy Metals in the Environment*, London, Sept. 1979, C.E.P. Consultants, Edinburgh, 1979, 103–108.
80. Chamberlain, A. C., Prediction of response of blood lead to airborne and dietary lead from volunteer experiments with lead isotopes, *Proc. Royal Soc. Lond. B.*, 224, 149, 1985.
81. Ito, Y., Niiya, Y., Otami, M., Sarai, S., and Shima, S., Effects of Food Intake on Blood Lead Concentration in Workers Occupationally Exposed to Lead, *Toxicol. Lett.*, 37, 105, 1987.
82. Chamberlain, A. C., Heard, M. J., Little, P., Newton, D., Wells, A. C., and Wiffen, R. D., *Investigations into Lead from Motor Vehicles,* HMSO, London, 1978.
83. Kehoe, R. A., The metabolism of lead in man in health and disease, The Harben Lectures 1960, Lecture No.1, The normal metabolism of lead, *J. Royal Inst. Public Health*, 24, 1961.
84. Hanzlik, P. J. and Presho, E. J., *Pharmacol. Exp. Ther.*, 21, 145, cited in ref. 167, 1978.
85. Blaxter, K. L., Lead as a nutritional hazard to farm livestock II. The absorption and excretion of lead by sheep and rabbits, *J. Comp. Pathol.*, 60, 140, 1950.
86. DeMichele, S. J., Nutrition of lead, *Comp. Biochem. Physiol.*, 78A, 401, 1984.
87. Hutton, M. and Goodman, G. T., Metal contamination of feral pigeons *Columbia livia* from the London, area: 1. Tissue accumulation of lead, cadmium and zinc, *Environ. Pollut. A.*, 22, 3, 207, 1980.
88. Finley, M. T. and Dieter, M. P., Influence of laying on lead accumulation in bone of Mallard ducks, *J. Toxicol. Environ. Health*, 4, 1, 123, 1978.
89. Beyer, W. N., Franson, J. C., Locke, L. N., Stroud, R. K., and Sileo, L., Retrospective study of the diagnostic criteria in a lead-poisoning survey of waterfowl, *Arch. Environ. Contam. Toxicol.*, 35, 506, 1998.
90. Rattner, B. A., Fleming, W. J., and Bunck, C. M., Comparative toxicity of lead shot in black ducks (*Anas rubripes*) and mallards (*Anas platyrhynchos*), *J. Wild. Dis.*, 25, 175, 1989.
91. Fleming, D. E. B., Chattle, D. R., Wetmur, J. G., Desnick, R. J., Robin, J. P., Boulay, D., Rochard, N. S., Gordon, C. L., and Webber, C. E., Effect of the delta-aminolevulinate dehydratase polymorphism on the accumulation of lead in bone and blood in lead smelter workers, *Environ. Res. Sec. A.*, 77, 49, 1998.
92. Onalaja, A. O. and Claudio, L., Genetic susceptibility to lead poisoning, *Environ. Health Perspec.*, 108, 23, 2000.
93. Shields, J. B., Mitchell, H. H., and Ruth, W. A., *J. Nutr.,* 18, 87, cited in ref. 152, 1978.
94. Mikula, E. J., *Wasted Waterfowl*, Rep. Mississippi Flyway Counc. Planning Comm. 1965, pp. 61–65.
95. Van Gelder, G. A., Lead and the nervous system, in *Toxicity of Heavy Metals in the Environment*, Part I. Oehme, F. W., Ed., Marcel Dekker, New York, 1978, pp. 101–121.
96. Koranda, J., Moore, K., Stuart, M., and Conrado, C., Dietary effects on lead uptake and trace element distribution in Mallard ducks dosed with lead shot. Report: ISS UCID-18044, 1979.
97. Osweiler, G. D., van Gelder, G. A., and Buck, W. B., Epidemiology of lead poisoning in animals, in *Toxicity of Heavy Metals in the Environment*, Part I., Oehme, F. W., Ed., Marcel Dekker, New York and Basel, 1978, pp. 143–171.
98. Morrison, J. N., Quarterman, J., and Humphries, W. R., The effect of dietary calcium and phosphate on lead poisoning in lambs, *J. Comp. Path.*, 87, 3, 417, 1977.
99. Shields, J. B. and Mitchell, H. H., The effect of cadmium and phosphate on the metabolism of lead, *J. Nutr.*, 21, 541, 1941.
100. Zook, B. C., Lead intoxication in urban dogs, in *Toxicity of Heavy Metals in the Environment*, Part I., Oehme, F. W., Ed., Marcel Dekker, New York and Basel, 1978, pp. 179–190.
101. Jacobziner, H., Lead poisoning in childhood: Epidemiology, manifestations and prevention, *Clin. Pediatr.*, 5, 277, 1966.
102. Kehoe, R. A., Cholak, J., Hubbard, D. M., Burnbach, K., McNary, R. R., and Story, D. V., Experimental studies on ingestion of lead compounds, *J. Ind. Hyg.*, 27, 381, 1940.
103. Clemens, E. T., Krook, L., Aronsen, A. L., and Stevens, C. E., Pathenogenesis of lead shot poisoning in the Mallard duck, *Cornell Vet.*, 65, 248, 1975.

104. Marn, C. M., Mirarchi, R. E., and Lisane, M. E., Effects of diet and cold exposure on captive female Mourning Doves dosed with lead shot, *Arch. Environ. Contam. Toxicol.* 17, 589, 1988.
105. Sanderson, G. C. and Bellrose, F. C., A review of the problem of lead poisoning in waterfowl, *Ill. Nat. Hist. Surv. Spec.*, Publ. No. 4, 1986.
106. Pain, D. J., Ed., Lead poisoning in waterfowl, *Proc. IWRB Workshop*, Brussels, Belgium, 13–15 June 1991, IWRB Spec. Pub. 16. Slimbridge, U.K., 1992.
107. Mitchell, R. L. and Reith, J. W. S., The lead content of pasture herbage, *J. Sci. Fd. Agric.*, 17, 437, 1966.
108. Jenkins, D. W., Biological Monitoring of Trace Metals, Toxic Trace Metals in Plants and Animals of the World. Part II., Vol. 2, U.S. Environmental Protection Agency Rep. 600/3–80–091:619, 1980.
109. Ruhling, A. and Tyler, G., An ecological approach to the lead problem, *Bot. Notiser.*, 121, 321, 1968.
110. Anders, E., Dietz, D. D., Bagnell, C. R., Gaynor, J., Krigman, M. R., Ross, D. W., Leander, J. D., and Mushak, P., Morphological, pharmokinetic and hematological studies of lead-exposed pigeons, *Environ. Res.*, 28, 344, 1982.
111. Goyer, R. A. and Moore, J. F., Cellular effects of lead, *Adv. Exp. Med. Biol.*, 48, 447, 1974.
112. Beyer, W. N., Spann, J. W., Sileo, L., and Franson, J. C., Lead poisoning in six captive avian species, *Arch. Environ. Contam. Toxicol.*, 17, 121, 1988.
113. Birkhead, M., Luke, B., and Mann, S., Intracellular localization of lead in tissues of the Mute swan, *Tissue Cell*, 14, 4, 691, 1982.
114. Zook, B. C., Sauer, R. M., and Garner, F. M., Lead poisoning in captive wild animals, *J. Wildl. Dis.*, 8, 264, 1972.
115. Osweiler, G. D. and van Gelder, G. A., Epidemiology of lead poisoning in animals, in *Toxicity of Heavy Metals in the Environment*, Part I., Oehme, F. W., Ed., Marcel Dekker, New York and Basel, 1978, pp. 143–177.
116. Colle, A., Grimaud, J. A., Boucherat, M., and Manvel, Y., Lead poisoning in monkeys: Functional and histopathological alterations of the kidneys, *Toxicology*, 18, 145, 1980.
117. Tachon, P., Laschi, A., Briffaux, J. P., and Brain, G., Lead poisoning in monkeys during pregnancy and lactation, *Sci. Total Environ.*, 30, 221, 1983.
118. Tsuchiya, K., Lead, in *Handbook on the Toxicology of Metals,* Friberg, L., Nordberg, G. E., and Vouk, V. B., Eds., Elsevier/North Holland Biomedical Press, Amsterdam, 1979, 451–458.
119. Marcus, A. H., Multicompartment kinetic models for lead. I. Bone diffusion models for long-term retention, *Environ. Res.*, 36, 441, 1985.
120. Hutton, M., Metal contamination of feral pigeons, *Columbia livia*, from the London area (England, U.K.): 2. Biological effects of lead exposure, *Environ. Pollut. Ser. A.*, 22, 4, 281, 1980.
121. Custer, T. W., Franson, J. C., and Pattee, O. H., Tissue lead distribution and hematologic effects in American kestrels (*Falco sparverius*) fed biologically incorporated lead, *J. Wildl. Dis.,* 20, 39, 1984.
122. Beyer, W. N. and Cromartie, E. J., A survey of Pb, Cu, Zn, Cd, Cr, As, and Se in earthworms and soil from diverse sites, *Environ. Monitor. Assess.*, 8, 27, 1987.
123. Gish, C. D. and Christensen, R. E., Cadmium, nickel, lead and zinc in earthworms from roadside soil, *Environ. Sci. Technol.*, 7, 1060, 1973.
124. Ireland, M. P., Lead retention in toads *Xenopus laevis* fed increasing levels of lead contaminated earthworms, *Environ. Pollut.*, 12, 85, 1977.
125. Getz, L. L., Best, L. B., and Prather, M., Lead in urban and rural song birds, *Environ. Pollut.*, 12, 3, 235, 1977.
126. Hardisty, M. W., Kartar, S., and Sainsbury, M., Dietary habits and heavy metal concentrations in fish from the Severn Estuary and Bristol Channel, *Mar. Pollut. Bull.*, 5, 61, 1974.
127. Jenkins, C., Utilisation du pigeon biset (*Columba livia Gm.*) comme temoin de la pollution atmospherique par le plomb, *Comptes rendus-Academie de Sciences, Paris, Serie D.*, 281, 1187, 1975.
128. Ohi, G., Seki, H., Akiyama, K., and Yagyu, H., The pigeon, a sensor of lead pollution, *Bull. Environ. Contam. Toxicol.*, 12, 1, 92, 1974.
129. Ohi, G., Seki, H., Minowa, K., Ohsawa, M., Mizoguchi, I., and Sugimori, F., Lead pollution in Tokyo — The pigeon reflects its amelioration, *Environ. Res.*, 26, 1, 125, 1981.
130. Schmitt, C. J. and Brumbaugh, W. G., National contamination biomonitoring program: Concentrations of arsenic, cadmium, copper, lead, mercury, selenium, and zinc in U.S. freshwater fish, 1976–1984, *Arch. Environ. Contam. Toxicol.*, 19, 731–747, 1990.

131. Parslow, J. L. F., Jeffries, D. J., and French, M. C., Ingested pollutants in puffins and their eggs, *Bird Study*, 19, 18, 1972.
132. Wong, P. T. S., Chau, Y. K., and Luxon, P. L., Methylation of lead in the environment, *Nature*, 253, 263, 1975.
133. O'Hare, J. P., Cheng, C. N., and Focht, D. D., Volatile methyl lead compounds from soils and lakeshore deposits amended with lead salts, *Abstr. ASM Annu. Meeting*, 80, 1977, p. 242.
134. Carpenter, K. E., *Ann. Appl. Biol.*, 11, 1, 1924, cited in Wong, P. T. S., Silverberg, B. A., Chau, Y. K., and Hodson, P. V., Lead and the Aquatic Biota, Canadian Center for Inland Waters, Department of the Environment, Burlington, Ontario, 1976.
135. Jones, J. R. E., *Ann. Appl. Biol.*, 27, 367, 1940, cited in Wong, P. T. S., Silverberg, B. A., Chau, Y. K., and Hodson, P. V., Lead and the Aquatic biota, Canadian Center for Inland Waters, Department of the Environment, Burlington, Ontario, 1976.
136. Jones, J. R. E. J., *Amin. Ecol.*, 27, 1, 1958, Wong, P. T. S., Silverberg, B. A., Chau, Y. K., and Hodson, P. V., Lead and the Aquatic Biota, Canadian Center for Inland Waters, Department of the Environment, Burlington, Ontario, 1976.
137. Libanti, C. M. and Tandler, C. J., The distribution of water-soluble inorganic phosphate ions within the cell: Accumulation within the nucleus, *J. Cell Biol.*, 42, 754, 1969.
138. Sharp, V. and Denny, P., Electron microscope studies on the absorption and localization of lead in the leaf tissue of *Potamogeton pectinatus L.*, *J. Exp. Bot.*, 27, 1155, 1976.
139. Ormerod, D. P., Impact of trace element pollution on plants, in *Air Pollution and Plant Life*, Treshow, M., Ed., John Wiley and Sons, New York, 1984.
140. Silverberg, B. A., Ultrastructural localization of lead in *Stigeoclonium tenue (Chlorophyceae Ulotrichales)* as demonstrated by cytochemical and x-ray microanalysis. *Phycologia*, 14, 4, 265, 1975.
141. Jones, D., Wilson, M. J., and Laundon, J. R., Observation on the location and form of lead in *Stereocaulon vesuianum*, *Lichenologist*, 14, 281, 1982.
142. Bradshaw, A. D., Populations of *Agrostis tenuis* resistant to lead and zinc poisoning. *Nature*, 169, 1098, 1952.
143. Demayo, A., Taylor, M. C., Taylor, K. W., and Hodson, P. V., Toxic effects of lead and lead compounds on human health, aquatic life, wildlife plants and livestock, in *Guidelines for Surface Water Quality*, Vol. 12, 4, 257–305, Inorganic Chemical Substances, Vol. I, 1982.
144. Krishnayya, N. S. R. and Bedi, S. J., Effect of automobile lead pollution in *Cassia tora L.* and *Cassia occidentalis L.*, *Environ. Pollut.*, 40A, 221, 1986.
145. Koeppe, D. E. and Miller, R. J., Lead effects on corn mitochondrial respiration, *Science*, 167, 1376, 1970.
146. Dayton, L. and Lewin, R., The effect of lead on algae. III. Effects of lead on population growth curves in two-membered cultures of phytoplankton, *Arch. Hydrobiol. Suppl.*, 49, 25, 1975.
147. Rolfe, G. L. and Bazzaz, F. A., Effect of lead contamination on transpiration and photosynthesis of Lobolly pine and Autumn olive, *Forest Sci.*, 21, 1, 33, 1975.
148. Woolery, M. L. and Lewin, R. A., The effects of lead on algae. IV. Effects of lead on respiration and photosynthesis of *Phaeodactylum tricornutum (Bacillariophyceae)*, *Water Air Soil Pollut.*, 6, 25, 1976.
149. Backhaus, B. and Backhaus, R., Is atmospheric lead contributing to mid-European forest decline?, *Sci. Total Environ.*, 50, 223, 1986.
150. Jaworski, J. F., Effects of Lead in the Environment, Quantitative Aspects, Natl. Res. Coun. Canada, NRCC No. 16763, 1978.
151. Dieter, M. P. and Finley, M. T., Erythrocyte delta-aminolevulinic acid dehydratase activity in mallard ducks: duration of inhibition after lead shot dosage, *J. Wildl. Manage.*, 42, 3, 621, 1979.
152. Pain, D. J., Haematological parameters as predictors of blood lead and indicators of lead poisoning in the Black Duck (*Anas rubripes*), *Environ. Pollut.*, 60, 67, 1989.
153. Dieter, M. P., Blood delta-aminolevulinc acid dehydratase (ALAD) to monitor lead contamination in Canvasback ducks (*Aythya valisineria*), *Anim. Monitor. Environ. Pollut.*, (*Symp. Pathobiol. Environ. Pollut.: Anim. Models Wildl. Monit.*), 1979, pp. 177–91.
154. Hoffman, D. J., Eastin, W. C., Jr., and Gay, M. L., Embryotoxic and biochemical effects of waste crankcase oil on birds eggs, *Toxicol. Appl. Pharmacol.*, 63, 2, 230, 1982.
155. Hoffman, D. J., Franson, J. C., Pattee, O. H., Bunck, C. M., and Murray, H. C., Biochemical and hematological effects of lead ingestion in nestling American kestrels (*Falco sparvensis*), *Comp. Biochem. Physiol.(c)*, 80, 2, 431, 1985.

156. Grue, C. E., O'Shea, T. J., and Hoffman, D. J., Lead concentrations and reproduction in highway-nesting barn swallows (*Hirundo rustica*), *Condor*, 86, 4, 383, 1985.
157. Eastin, W. C., Hoffman, D. J., and O'Leary, T., Lead accumulation and depression of delta aminolaevulinic acid dehydratase (ALAD) in young birds fed automotive waste oil, *Arch. Environ. Contam. Toxicol.*, 12, 31, 1983.
158. Stone, C. L. and Fox, M. R. S., Effects of low levels of dietary lead and iron on hepatic RNA, protein and minerals in young Japanese quail, *Environ. Res.*, 33, 2, 322, 1984.
159. Pain, D. J., Lead poisoning of waterfowl: An investigation of sources and screening techniques, D. Phil. thesis, Oxford University, 1987.
160. Hapke, H. J. and Prigge, E., *Berl. Muench. Tieraerztl. Wochenschr.* 86, 410, 1973, cited in ref. 152, 1978.
161. Dieter, M. P. and Finley, M. T., Delta-aminolevulinic acid dehydratase enzyme activity in blood, brain and liver of lead-dosed ducks, *Environ. Res.*, 19, 1, 127, 1979.
162. Mouw, D., Kalitis, K., Anver, M., Schwartz, J., Constan, A., Harting, R., Cohen, B., and Ringler, D., Possible toxicity in urban vs. rural rats, *Arch. Environ. Health*, 30, 276, 1976.
163. Hammond, P. M., Metabolism and metabolic action of lead and other heavy metals, in *Toxicity of Heavy Metals in the Environment*, Part I, Oehme, F. W., Ed., Marcel Dekker, New York and Basel, 1978, pp. 87–99.
164. Tepper, L. B. and Pfitzer, E. A., Report of a symposium, Department of Environmental Health, College of Medicine, University of Cincinnati, 1970.
165. Franson, J. C., Sileo, L., Pattee, O. H., and Moore, J. F., Effects of chronic dietary lead in American kestrels (*Falco sparverius*), *J. Wildl. Dis.*, 19, 2, 110, 1983.
166. Secchi, G. C., Alessio, L., Cambiaghi, G., and Andreoletti, F., ALA-dehydratase activity in erythrocytes and blood levels in "critical" population groups, in *Proc. Int Symp. on Environ. Health Aspects of Lead*, Amsterdam, Oct. 2–6, 1972, European Atomic Energy Community, Rep. EUR 500d-e-f. pp. 595–602.
167. Roscoe, D. E., Nielsen, S. W., Eaton, H. D., and Rousseau, J. E., Jr., Chronic plumbism in rabbits: A comparison of 3 diagnostic tests, *Am. J. Vet. Res.*, 26, 1225, 1975.
168. Peter, F. and Strunc, G., Effects of ingested lead on concentration of blood and tissue lead in rabbits, *Clin. Biochem.*, 16, 202, 1983.
169. Roscoe, D. E., Nielsen, S. W., Lamola, A. A., and Zuckerman, D., A simple, quantitative test for erythrocytic protoporphyrin in lead-poisoned ducks, *J. Wildl. Dis.*, 15, 1, 127, 1979.
170. Birkhead, M., Lead levels in the blood of Mute Swans *Cygnus olor* on the river Thames, *J. Zool. London*, 199, 59, 1983.
171. Pain, D. J. and Rattner, B. A., Mortality and hematology associated with the ingestion of one number four lead shot in Black Ducks, *Anas rubripes, Bull. Environ. Contam. Toxicol.*, 40, 159, 1988.
172. Forbes, R. M. and Sanderson, G. C., Lead toxicity in domestic animals and wildlife, in *The Biogeochemistry of Lead in the Environment,* Part B, Nriagu, J. O., Ed., Elsevier/North Holland Biomedical Press, Amsterdam, 1978, 226–276.
173. Kulish, O. P., Erythropoiesis and erythrocyte activity of the blood plasma in lead anemia, *Bull. Expl. Biol. Med.*, 74, 631, 1973.
174. Roscoe, D. E. and Nielsen, S. W., Lead poisoning in Mallard ducks (*Anas platyrhynchos*), *Animal Monit. Environ. Pollut., Symp., Pathobiol. Environ. Pollut., Animal Models Wildl. Monit.*, 165, 1979.
175. Johns, F. M., A study of punctate stippling as found in lead poisoning of wild ducks, *J. Lab. Clin. Med.*, 19, 514, 1934.
176. Link, R. P. and Pensinger, R. R., Lead toxicosis in swine, *Am. Vet. Res.*, 27, 759, 1966.
177. Bates, F. Y., Barnes, D. M., and Higbee, J. M., Lead toxicosis in Mallard ducks, *Bull. Wildl. Dis. Assoc.*, 4, 116, 1968.
178. Environmental Protection Agency, Ambient Water Quality Criteria for Lead — 1984, U.S. Environmental Protection Agency Rep. 440/5–84–027, Natl. Tech. Infor. Serv., 5285 Port Royal Road, Springfield, Virginia 22161, 1985.
179. Brown, B. E., Uptake of copper and lead by a metal-tolerant isopod *Asellus meridianus, Freshwater Biol.*, 7, 3, 235, 1977.
180. Fraser, J., Acclimation to lead in the freshwater isopod *Asellus aquaticus, Oceologia*, 45, 419, 1980.
181. Hodson, P. V., Blunt, B. R., Spry, D. J., and Austen, K., Evaluation of erythrocyte d-aminolevulinic acid dehydratase activity as a short term indicator in fish of harmful exposure to lead, *J. Fish Res. Board Can.*, 34, 501, 1977.

182. Aronsen, A. L., Biologic effects of lead in fish, *J. Wash. Acad. Sci.*, 61, 124, 1971.
183. NRCC, Lead in the Canadian Environment. Natl. Res. Coun. Canada Publ. BY73–7 (ES). Publications NRCC/CNRC, Ottawa, Canada K1A OR6, 1973.
184. Sparling, D. W., Linder, L., and Bishop, C. A., Eds., *Ecotoxicology of Amphibians and reptiles*, Society of Environmental Toxicology and Chemistry, SETAC Press, Pensacola, FL, 2000
185. Lewis, L. A., Poppenga, R. J., Davidson, W. T., Fischer, J. R., and Morgan, K. A., Lead toxicosis and trace element levels in wild birds and mammals at a firearms training facility, *Arch. Environ. Contam. Toxicol.*, 41, 208, 2001.
186. Pain, D. J., Lead poisoning in birds: An international perspective, *Acta XX Congressus Internationalis Ornithologici,* 2343, 1991.
187. Ochiai,K., Kimura, T., Uematsu, K., Umemura, T., and Itakura, C., Lead poisoning in waterfowl in Japan, *J. Wildl. Dis.*, 35, 766, 1999.
188. Mateo, R., Belliure, J., Dolz, J. C., Serrano, J. M. A., and Guitart, R., High prevalences of lead poisoning in wintering waterfowl in Spain, *Arch. Environ. Contam. Toxicol.*, 35, 342, 1998.
189. Pain, D. J., Lead shot ingestion by waterbirds in the Camargue, France: An investigation of levels and interspecific differences, *Environ. Pollut.*, 66, 273, 1990.
190. Pain, D. J. and Handrinos, G. I., The incidence of ingested lead shot in ducks of the Evros delta, Greece, *Wildfowl*, 41, 167, 1990.
191. Mateo, R., Dolz, J. C., Aguilar Serrano, J. M., Belliure, J., and Guitart, R., An epizootic of lead poisoning in greater flamingos (*Phoenicopterus rubber roseus*) in Spain, *J. Wildl. Dis.*, 33, 131, 1997.
192. BirdLife International, *Threatened Birds of the World*, Lynx Ediciones and BirdLife International, Barcelona and Cambridge, U.K., 2000
193. Mateo, R., Green, A. J., Jeske, C. W., Urios, V., and Gerique, C., Lead poisoning in the globally threatened marbled teal and white-headed duck in Spain, *Environ. Toxicol. Chem.*, in press.
194. Tavecchia, G., Pradel, R., Lebreton, J. D., Johnson, A. R., and Mondain-Monval, J.Y., The effect of lead exposure on survival of adult mallards in the Camargue, southern France, *J. Appl. Ecol.*, 38, 1197, 2001.
195. Wayland, M. and Bollinger, T., Lead exposure and poisoning in bald eagles and golden eagles in the Canadian prairie provinces, *Environ. Pollut.,* 104, 341, 1999.
196. Kim, E-Y., Goto, R., Iwata, H., Masuda, Y., Tanabe, S., and Fujita, S., Preliminary survey of lead poisoning of Steller's sea eagles (*Haliaeetus pelagicus*) and white-tailed sea eagle (*Haliaeetus albicilla*) in Hokkaido, Japan, *Environ. Toxicol. Chem.,* 18, 448, 1999.
197. Pattee, O. H., Bloom, P. H., Scott, J. M., and Smith, M. M., Lead hazards within the range of the California condor, *Condor,* 92, 931, 1990.
198. Miller, M. J., Restani, M., Harmata, A. R., Bartolotti, C. R., and Wayland, M. E., A comparison of blood lead levels in bald eagles from two regions of the great plains of North America, *J. Wildl. Dis.,* 34, 704, 1998.
199. Kramer, J. L. and Redig, P. T., Sixteen years of lead poisoning in eagles, 1980–95: An epizootiologic view, *J. Raptor Res.,* 31, 327, 1997.
200. Craig, T. H., Connelly, J. W., Craig, E. H., and Parker, T. L., Lead concentrations in golden and bald eagles, *Wilson Bull.,* 102, 170, 1990.
201. Harmata, A. R. and Restani, M., Environmental contaminants and cholinesterase in blood of vernal migrant bald and golden eagles in Montana, *Intermount. J. Sci.,* 1, 1995.
202. Muller, K., Krone, O., Gobel, T., and Brunnberg, L., Acute lead intoxication in two white-tailed sea eagles (*Haliaeetus albicilla*), *Tierarztliche Praxis Ausgabe Kleintiere Heimtiere,* 29, 209, 2001.
203. Kenntner, N., Tataruch, F., and Krone, O., Heavy metals in soft tissue of white-tailed eagles found dead or moribund in Germany and Austria from 1993 to 2000, *Environ. Toxicol. Chem.*, 20, 1831, 2001.
204. Pain, D. J., Bavoux, C., and Burneleau, G., Seasonal blood lead concentrations in marsh harriers *Circus aeruginosus* from Charente-Maritime, France: Relationship with the hunting season, *Biol. Conserv.*, 81, 1, 1997.
205. Mateo, R., Estrada, J., Paquet, J., Riera, X., Dominguez, L., Guitart, R., and Martinez-Vilalta, A., Lead shot ingestion by marsh harriers *Circus aeruginosus* from the Ebro delta, Spain, *Environ. Pollut.*, 104, 435, 1999.
206. Mateo, R., Molina, R., Grifols, J., and Guitart, R., Lead poisoning in a free ranging griffon vulture (*Gyps fulvus*), *The Vet. Record*, 140, 47, 1997.

207. Wiemeyer, S. N., Scott, J. M., Andersen, M. P., Bloom, P. H., and Stafford, C. J., Environmental Contaminants in Californian Condors, *J. Wildl. Manage.*, 52, 2, 238, 1988.
208. Mateo, R., Cadenas, R., Manez, M., and Guitart, R., Lead shot ingestion in two raptor species from Donana, Spain, *Ecotoxicol. Environ. Saf.*, 48, 6, 2001.
209. Meretsky, V. J., Snyder, N. F. R., Beissinger, S. R., Clenenden, D. A., and Wiley, J. W., Demography of the California condor: Implications for reestablishment, *Conserv. Biol.*, 14, 957, 2000.
210. Calvert, H. J., Pheasants poisoned by swallowing shots, *The Field*, 47, 189, 1876.
211. Sileo, L. and Fefer, S. I., Paint chip poisoning of Laysan albatross at Midway Atoll, *J. Wildl. Dis.*, 23, 3, 432, 1987.
212. Jacobson, E., Carpenter, J. W., and Novilla, M., Suspected lead toxicosis in a Bald eagle, *J. Am. Vet. Med. Assoc.*, 171, 9, 952, 1977.
213. Sileo, L., Jones, R. N., and Hatch, R. C., Effect of ingested lead shot on the electrocardiogram of Canada geese, *Avian Dis.*, 17, 2, 308, 1973.
214. Fimreite, N., Effects of lead shot ingestion in willow grouse, *Bull. Environ. Contam. Toxicol.*, 33, 1, 121, 1984.
215. Sears, J., Regional and seasonal variations in lead poisoning in the Mute swan *Cygnus olor* in relation to the distribution of lead and lead weights in the Thames area England, *Biol. Conserv.*, 46, 115, 1988.
216. Pokas, M. A. and Chafel, R., Lead toxicosis from ingested fishing sinkers in adult common loons (*Gavia immer*) in New England, *J. Zool. Wildl. Med.*, 23, 92, 1992.
217. Stone, W. B. and Okoniewski, J. C., Necropsy findings and environmental contaminants in common loons from New York, *J. Wildl. Dis.*, 37, 178, 2001.
218. Leonzio, C., Fossi, C., and Focardi, S., Lead, mercury, cadmium and selenium in two species of gull feeding on inland dumps, and in marine areas, *Sci. Total Environ.*, 57, 121, 1986.
219. Beyer, W. N., Audet, D. J., Morton, A., Campbell, J., and LeCaptain, L., Lead exposure of waterfowl ingesting Coeur d'Alene river basin sediments, *J. Environ. Qual.*, 27, 1533, 1998.
220. Beyer, W. N., Audet, D. J., Heinz, G. H., Hoffman, D. J., and Day, D., Relation of waterfowl poisoning to sediment lead concentrations in the Coeur d'Alene river basin, *Ecotoxicology*, 9, 207, 2000.
221. Bull, K. R., Every, W. J., Freestone, P., Hall, J. R., and Osborne, D., Alkyl lead pollution and bird mortalities on the Mersey estuary, U.K., 1979–1981, *Environ. Pollut.*, 31, 239, 1983.
222. Oehme, F. W., Ed., *Toxicity of Heavy Metals in the Environment*, Part I, Marcel Dekker, New York and Basel, 1978.
223. Hammond, P. B. and Aronsen, A. L., Lead poisoning in cattle and horses in the vicinity of a smelter, *Ann. N.Y. Acad. Sci.*, 111, 595, 1964.
224. Zmudzki, J., Bratton, G. R., Womac, C., and Rowe, L., Lead poisoning in cattle: Reassessment of the minimum toxic oral dose, *Bull. Environ. Contam. Toxicol.*, 30, 435, 1983.
225. Hemphill, F. E., Kaeberle, M. L., and Buck, W. B., *Science*, 1031, 1972, cited in ref. 152, 1978.
226. Brown, D. R., Long-term effects of lead on learning and organ development in the growing rat, *Toxicol. Appl. Pharmacol.*, 25, 466, 1973.
227. Brady, K., Herrera, Y., and Zenick, H., Influence of parental lead exposure on subsequent learning ability of offspring, *Pharmacol. Biochem. Behav.*, 3, 4, 561, 1975.
228. Carson, T. L., van Gelder, G. A., Karas, G. G., and Buck, W. B., Slowed learning in lambs prenatally exposed to lead, *Arch. Environ. Health*, 29, 3, 154, 1974.
229. Carson, T. L., van Gelder, G. A., Karas, G. G., and Buck, W. B., Development of behavioral tests for the assessment of neurological effects of lead in sheep, *Environ. Health Perspect.*, 7, 233, 1974.
230. Shore, R. F. and Rattner, B. A., Eds., *Ecotoxicology of Wild Mammals*, John Wiley and Sons, New York, 2000.
231. Burrows, G. E., Lead toxicosis in domestic animals: A review of the role of lead mining and primary lead smelters in the United States, *Vet. Human Toxicol.*, 23, 337, 1981.
232. Kwatra, M. S., Gill, B. S., Singh, R., and Singh, M., Lead toxicosis in buffaloes and cattle in Punjab, *Indian J. Anim. Sci.*, 56, 412, 1986.
233. Frape, D. L. and Pringle, J. D., Toxic manifestations in a dairy herd consuming haylage contaminated by lead, *Vet. Record,* 114, 25, 614, 1984.
234. Clausen, B., Lead poisoning control measures in Denmark, in *Lead Poisoning in Waterfowl,* Proc. IWRB workshop, Pain, D. J., Ed., Brussels, 13–15 June 1991, 68–70, IWRB Spec. Pub. 16., Slimbridge, U.K., 1992.

235. Zook, B. C., Lead poisoning in zoo animals, *Zool. Garten N.F. Jena*, 45, 2, 143, 1975.
236. Rice, D. C., Chronic low-lead exposure from birth products deficits in discrimination reversal in monkeys, *Toxicol. Appl. Pharmacol.*, 77, 201, 1985.
237. White, D. H., Bean, J. R., and Longcore, J. R., Nationwide residues of mercury, lead, cadmium, arsenic, and selenium in starlings, 1973, *Pest. Monitor. J.*, 11, 35, 1977.
238. Pain, D. J., Ed., *Lead Poisoning in Waterfowl,* Proc. IWRB workshop, Brussels, 13–15 June 1991, IWRB Spec. Pub. 16., Slimbridge, U.K., 1992.
239. Morehouse, K., Crippling loss and shot type: The United States experience, in Lead Poisoning in Waterfowl, *Proc. IWRB Workshop*, Pain, D. J., Ed., Brussels, 1991, 32–71, IWRB Spec. Pub. 16. Slimbridge, UK, 1992.
240. Broyer, T. C., Johnson, C. M., and Paull, R. E., Some aspects of lead in plant nutrition, *Plant Soil*, 36, 301, 1972.
241. Osborne, D., Every, W. J., and Bull, K. R., The toxicity of trialkyl lead compounds to birds, *Environ. Pollut. A*, 31, 261, 1983.
242. Pattee, O. H., Wiemeyer, S. N., Mulhern, B. M., Sileo, L., and Carpenter, J. W., Experimental lead-shot poisoning in Bald Eagles, *J. Wildl. Manage.*, 45, 3, 806, 1981.
243. Bagley, G. E., Locke, L. N., and Nightingale, G. T., Lead poisoning in Canada geese in Delaware, *Avian Dis.*, 11, 601, 1967.
244. Beresford, W. A., Donovan, M. P., Henninger, J. M., and Waalkes, M. P., Lead in the bone and soft tissues of box turtles caught near smelters, *Bull. Environ. Contam. Toxicol.*, 27, 349, 1981.
245. Maddock, B. G. and Taylor, D., The acute toxicity and bioaccumulation of some lead alkyl compounds in marine animals, in *Lead in the Marine Environment*, Branica, M. and Konrad, Z., Eds., Pergamon Press, Oxford, 1980.
246. Carlson, B. L. and Nielsen, S. W., Influence of dietary calcium on lead poisoning in Mallard ducks (*Anas platyrhynchos*), *Am. J. Vet. Res.*, 46, 1, 277, 1985.
247. Jordan, J. S., Influence of diet in lead poisoning in waterfowl, *Trans. NE Section of the Wildl. Soc. 25th NE Fish & Wildl. Conf.,* Jan. 14–17, 1968. pp. 143–170.
248. Brown, D. R., Long-term effects of lead on learning and organ development in the growing rat, *Toxicol. Appl. Pharmacol.*, 25, 466, 1973.
249. Allen, J. R., McWey, P. J., and Suomi, S. J., Pathobiological and behavioural effects of lead intoxication in the infant rhesus monkey, *Environ. Health Perspect.*, 7, 239, 1974.
250. Nriagu, J. O., Ed., *The Biogeochemistry of Lead in the Environment, Part B Biological Effects*, Elsevier/North Holland Biomedical Press, Amsterdam, 1978.
251. U.S.F.W.S., Use of Lead Shot for Hunting Migratory Birds in the United States, Final Supplement Environmental Impact Statement, Dept. of Interior, Fish and Wildlife Service, 1986.
252. U.S. Fish and Wildlife Service, Steel, Final Environmental Statement. Proposed Use of Steel Shot for Hunting Waterfowl in the United States, U.S. Government Printing Office, Washington, D.C., 1985.

# CHAPTER 16

# Ecotoxicology of Mercury

**James G. Wiener, David P. Krabbenhoft, Gary H. Heinz, and Anton M. Scheuhammer**

## CONTENTS



## 16.1 INTRODUCTION

This chapter describes selected aspects of the behavior of mercury in the environment and examines the ecotoxicology of this highly toxic metal. The widespread geographic extent and adverse consequences of mercury pollution continue to prompt considerable scientific investigation. Furthermore, the environmental sources, biogeochemistry, transformations, transport, fate, and effects of mercury in the environment are subjects of frequent symposia, workshops and a large, steadily expanding body of scientific literature. We characterize the environmental mercury problem, critically review the ecotoxicology of mercury, and describe the consequences of methylmercury contamination of food webs. We discuss processes and factors that influence exposure to methylmercury, the highly neurotoxic form that readily accumulates in exposed organisms and can biomagnify in aquatic and terrestrial food webs to concentrations that can adversely affect organisms in upper trophic levels, including humans. Emphasis is given to aquatic food webs, where the problem of methylmercury contamination is greatest.[1] When available, recent reviews have been cited for readers interested in more detailed coverage.

Concerns about environmental mercury pollution and contamination of aquatic food webs stem largely from the human health risks of dietary exposure to methylmercury, the dominant form of mercury in the edible flesh of fish and aquatic mammals.[2–4] The human health risks associated with mercury in surface waters and aquatic biological resources are not reviewed here but have been critically examined in several case studies and recent reviews.[5–16] Nonetheless, our discussion of processes and factors affecting exposure of fish and wildlife to methylmercury is directly relevant to the issue of human exposure to methylmercury, which results largely from consumption of fish, shellfish, and aquatic mammals and birds.[10,12,17]

## 16.2 EVOLUTION OF THE ENVIRONMENTAL MERCURY PROBLEM

Humans have been using mercury for more than 2000 years for a wide variety of applications,[18,19] and centuries of emissions and reemissions of anthropogenic mercury have caused widespread environmental contamination over large regions of the globe.[20,21] Cinnabar, $HgS_{(S)}$, the principal mercury ore, was used as a red pigment long before the process for refining mercury ore to recover elemental mercury, $Hg^0$, was discovered. Since the advent of refining cinnabar, five mining areas have dominated the historical global production of elemental mercury: the Almadén district in Spain, the Idrija district in Slovenia, the Monte Amiata district in Italy, the Huancavelica district in Peru, and the state of California in the United States.[18,22] At Almadén, Spain, mercury was first mined about 430 B.C.,[23] and during the next 25 centuries the Almadén mines produced more than 280,000 metric tons of the estimated total global production of about 800,000 tons.[22] The mining and smelting of cinnabar and other mercury ores have caused substantial contamination of air, soil, water, biota, and sediment in the vicinity of such operations, and mercury-containing wastes at mining and smelting sites continue to emit mercury, including methylmercury, to the environment for decades or centuries after operations cease.[22,24–30]

From 1550 to 1930 an estimated 260,000 tons or more of mercury were released globally from mining operations that used the mercury-amalgamation process to recover gold and silver.[31] In the United States, gold mining was the primary use of mercury during the latter half of the 1800s, and the demand created by gold and silver mining stimulated mining for mercury as well.[18,32] The mining of mercury deposits (primarily cinnabar) along 400 km of the Coast Range of California, for example, was stimulated by the California gold rush in the mid-1800s.[28,33,34]

Gold or silver was mined throughout much of North America, and large quantities of mercury were used for precious-metal mining in California, Nevada, and South Dakota.[31,34] Contaminated tailings and alluvium originating from mining sites are consequently widespread in North America and elsewhere.[29,31,35–37] Emissions of mercury from contaminated mine tailings and lands include volatilization of $Hg^0$ to the atmosphere, aqueous dissolution by infiltrating water and entrainment with stream flow, and physical erosion and downstream transport of mercury-enriched geologic materials.[37–42] Contaminated tailings can remain a source of mercury emissions for decades or centuries after mining operations have ceased.[31,37] In some drainage basins, exemplified by the

Carson River (Nevada), contaminated sediment originating from historic mining sites has been transported, deposited, and redistributed far downstream, causing persistent contamination of stream and river channels, river banks, floodplains, and reservoirs along extensive reaches of the watershed.[33,35–38,40,41,43,44] The natural burial of such mercury-contaminated deposits by more recent, "clean" sediments may mitigate these settings only temporarily, given that large floods can reexpose the underlying, contaminated deposits.[36]

Since the early 1970s there has been a resurgence of gold-mining operations that use the mercury-amalgamation process, particularly in South America, Southeast Asia, China, and parts of Africa.[31,37,45–47] These ongoing mining activities, which seem to be stimulated partly by economic recession,[46] are widely dispersed in hundreds to thousands of operations — often small and in remote areas — involving millions of people worldwide.[31,45,46] Total emissions from these operations are now and could remain a globally significant source of new anthropogenic mercury for decades.[31,37,39,45] Recent emissions to the global environment from this "new gold rush" may total as much as 460 metric tons per year (about 10% of annual, anthropogenic global emissions),[48] with roughly two thirds of this total emitted to the atmosphere and one third emitted to land or water.[31] In Brazil, gold mining has become the major source of anthropogenic mercury emissions.[45]

Mercury also has a long history of usage in industrial applications, particularly in chlor-alkali plants and pulp and paper mills, and pollution from these sources has been well documented in recent decades.[21,25,49,50] The most publicized industrial releases occurred in Minamata and Niigata, Japan, in the 1950s and 1960s, when many humans were poisoned by methylmercury after eating fish that were highly contaminated by mercury from direct industrial sources.[5,8] These tragedies focused global attention on environmental mercury pollution[51] and prompted efforts, beginning around 1970 in the United States, Canada, and many other industrialized countries, to identify industrial sources of mercury pollution and to reduce intentional discharges of mercury into surface waters.[25,45] As a result, mercury levels in fish and sediments in such industrially affected waters typically declined in subsequent years and decades.[25,52–60] In many cases, the concentrations of mercury in fish decreased by 50% or more during the first decade after discharges were reduced, and the rate of decrease in concentration then slowed considerably, or concentrations leveled off, to values that were elevated relative to lesser contaminated waters nearby.[25,52,53,55,58,61] At some mercury-contaminated sites, however, the decline in concentrations of mercury in fish has been slow or delayed in the affected aquatic ecosystem.[25,62,63]

In industrially polluted Clay Lake, Ontario, mercury concentrations in gamefish have declined from peak levels but remained substantially above the Canadian mercury limit of 0.5 μg/g wet weight nearly three decades after operations ceased at the industrial source, a chlor-alkali plant near Dryden that operated from 1962 to 1970.[52,64] Mercury concentrations in axial muscle of 50-cm walleye (*Stizostedion vitreum*) from Clay Lake decreased rapidly after operations ceased at the chlor-alkali plant — from about 15 μg/g wet weight in 1970 to about 7.5 μg/g in 1972 — and then declined gradually to about 3.5 μg/g in 1983.[52] Concentrations apparently declined little during the next 15 years, given that total mercury averaged 2.7 μg/g in a sample of 14 walleyes (mean fork length, 53 cm) taken from Clay Lake in 1997 and 1998.[64] Persistent problems with methylmercury contamination of aquatic biota at historically contaminated sites may result from continuing, unintended emissions of mercury from the source area, from recycling and methylation of the mercury present in contaminated sediments, from temporal increases in the bioavailability of mercury or in the habitability of highly contaminated zones within the ecosystem, from changes in food-web structure, from atmospheric deposition of mercury from other sources, or from a combination of these and other factors.[25,50,62,65–69] Indeed, the physical and chemical properties that made mercury so useful in industrial applications (e.g., liquid state at ambient temperature, high volatility, and ease of reduction) also make this metal very difficult to contain and recover from the environment.[25]

The growing awareness of the hazards of mercury exposure led to widespread discontinuation or phased reductions in usage of the metal in a variety of applications and consumer goods beginning in the late 1960s.[18,19,29] For example, the use of mercurial fungicides in seed grain, which began in the 1940s, had severe consequences for humans and wildlife. Thousands of humans were poisoned, and hundreds died, when methylmercury-treated grains were eaten (rather than planted) by Iraqi farmers and their families.[6,7,9] Incidents of high mortality of wild birds were reported after planting of seeds treated with alkylmercury compounds,[70] and both seed-eating birds and their predators were poisoned.[71] The use of mercury compounds as seed dressings was decreased or banned in Sweden, Canada, and the United States in the 1960s and 1970s.

Mining of mercury decreased abruptly in response to rapidly declining demand and prices. Mercury production in the United States, for example, had peaked in 1877 at more than 2700 metric tons per year, and as recently as 1969 there were more than 100 active mercury mines in the country.[18] The mercury-mining industry in the United States collapsed in the early 1970s. Fewer than ten mines remained in production in late 1976, and the last mine in the country to produce mercury as its principal product closed in November 1990.[18]

In the late 1970s and 1980s, concentrations of mercury exceeding 0.5 or 1.0 μg/g wet weight — sufficient to prompt fish-consumption advisories — were reported in predatory fishes from aquatic ecosystems lacking substantive, on-site anthropogenic or geologic sources of mercury.[72–76] Subsequent investigations have shown that in certain aquatic systems concentrations of methylmercury in aquatic invertebrates, fish, and piscivorous wildlife are commonly elevated — a situation frequently reported for humic and low-alkalinity lakes (including low-pH lakes),[77–83] newly flooded reservoirs,[84–88] and wetlands or wetland-influenced ecosystems.[67,89–91] Many such environments can be characterized as lightly contaminated systems in which the amount of inorganic Hg(II) being converted to methylmercury is sufficient to contaminate food webs supporting production of fish and wildlife.[92–100]

Reliable records of temporal trends in mercury deposition can be obtained by analyses of dated cores of depositional sediments from lakes or reservoirs, of peat from ombrotrophic bogs, and, in some cases, of glacial ice.[20,59,101–106] At a site in northwestern Spain about 600 km northwest of the Almadén mines, substantive anthropogenic emissions of mercury to the atmosphere are reflected in peat deposited more than 1000 years ago in a core from an ombrotrophic bog.[23] The oldest anthropogenic mercury in this core was deposited about 2500 years ago, coinciding with the start of mining at Almadén and accounting for about 10 to 15% of the total mercury deposited in peat at that time.[23] In remote and semiremote areas of North America, Greenland, and Scotland that lack on-site sources of anthropogenic mercury, the rate of mercury accumulation in many lacustrine sediments has increased by a factor of 2 to 4 since the mid-1800s or early 1900s, based on analyses of sediment and peat cores.[101,102,104,107–110] Moreover, some cores from semiremote sites show evidence of recent declines in atmospheric mercury deposition associated with decreasing regional emissions of anthropogenic mercury into the environment.[102,103,106,109] Much of the mercury deposited onto terrestrial catchments is stored in soils, and the sediments in lakes that receive substantial inputs of mercury from their catchments may be slow to reflect declines in rates of atmospheric deposition of mercury.[110]

Many remote and semiremote ecosystems are contaminated with anthropogenic mercury deposited after long-range atmospheric transport from source areas.[20,109,111,112] Qualitatively, it can be reasonably inferred that a significant fraction of the methylmercury in the aquatic biota of remote or semiremote regions, including marine systems, is derived from anthropogenic mercury entering the aquatic ecosystem or its watershed in atmospheric deposition.[20,94,101,111,113–118] In northern Wisconsin, for example, the total annual atmospheric deposition of mercury to an intensively studied, semiremote seepage lake with no surface inflow and very little groundwater inflow averaged about 0.1 g/ha during 1988 to 1990, an input sufficient to account for the mass of mercury in water, fish, and depositing sediment.[94,114,119]

Concentrations of methylmercury in aquatic biota at remote and semiremote sites have probably increased globally during the past 150 years in response to anthropogenic releases of mercury into the environment. Substantial increases in methylmercury contamination of marine food webs in the North Atlantic Ocean, for example, were revealed by analyses of feathers from two fish-eating

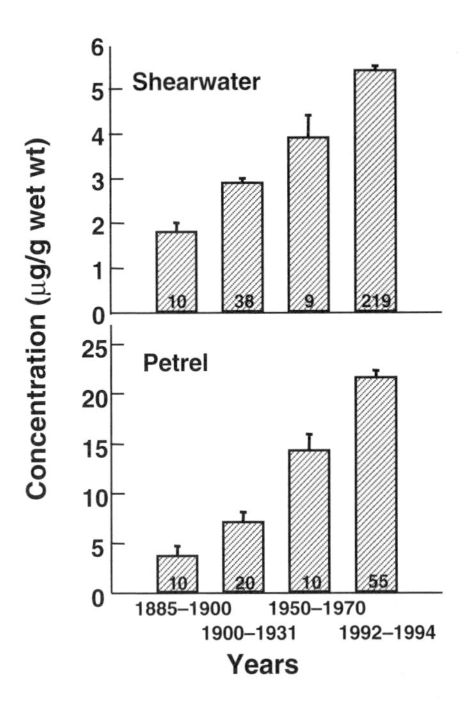


**Figure 16.1** Historical trends of increasing methylmercury concentrations (mean ± 1 standard error, with sample size denoted near the bottom of each bar) in feathers of two species of fish-eating seabirds obtained from the North Atlantic Ocean during 1885 to 1994.[118] Monteiro and Furness[118] determined organic mercury in feathers from museum specimens of the birds to avoid potential errors associated with postmortem contamination of the museum samples with inorganic mercury.

seabirds sampled from 1885 through 1994 (Figure 16.1).[118] The long-term increase in concentration of methylmercury averaged 1.9% per year in Cory's shearwater (*Calonectris diomedea borealis*) and 4.8% per year in Bulwer's petrel (*Bulweria bulwerii*).[118] Monteiro and Furness[118] attributed these increases to global trends in mercury contamination, rather than local or regional sources. Mercury concentrations have also increased during the past century in other species of seabirds.[116]

Quantitatively assessing the relative contributions of anthropogenic and natural emissions to the methylmercury burdens accumulated in biota at remote and semiremote sites is an enormous scientific challenge, partly because of spatial variation in (1) the contribution of natural sources and (2) the biogeochemical transformations and transport of mercury on the landscape.[42] The drainage basins onto which anthropogenic mercury is deposited can vary spatially in many respects. First, there is variation in the natural geologic abundances of mercury in bedrock, soils, sediments, and surface waters.[120–122] Second, surface waters within a region can differ spatially and temporally in the extent to which they receive total mercury and methylmercury exported from the drainage basin.[92,93,101,107,108,123–125] Third, the extent to which inorganic mercury present in aquatic ecosystems is converted to methylmercury can vary considerably, even on spatial scales of a few kilometers to tens of kilometers.[68,92,97,126] To overcome such complexities, new investigations involving the application of stable isotopes of mercury[127,128] are being employed to examine the biogeochemical cycling, bioaccumulation, and food-web transfer of "old" vs. newly deposited mercury in ecosystems.[129]

## 16.3 GLOBAL-SCALE ENVIRONMENTAL CYCLING AND FATE

Our understanding of the biogeochemical cycling of mercury (sources, pathways, and pools) in the environment has increased markedly during the past 10 to 15 years, whether considered in the context of mass balances,[48,130,131] concentrations in environmental media,[20,132,133] or important

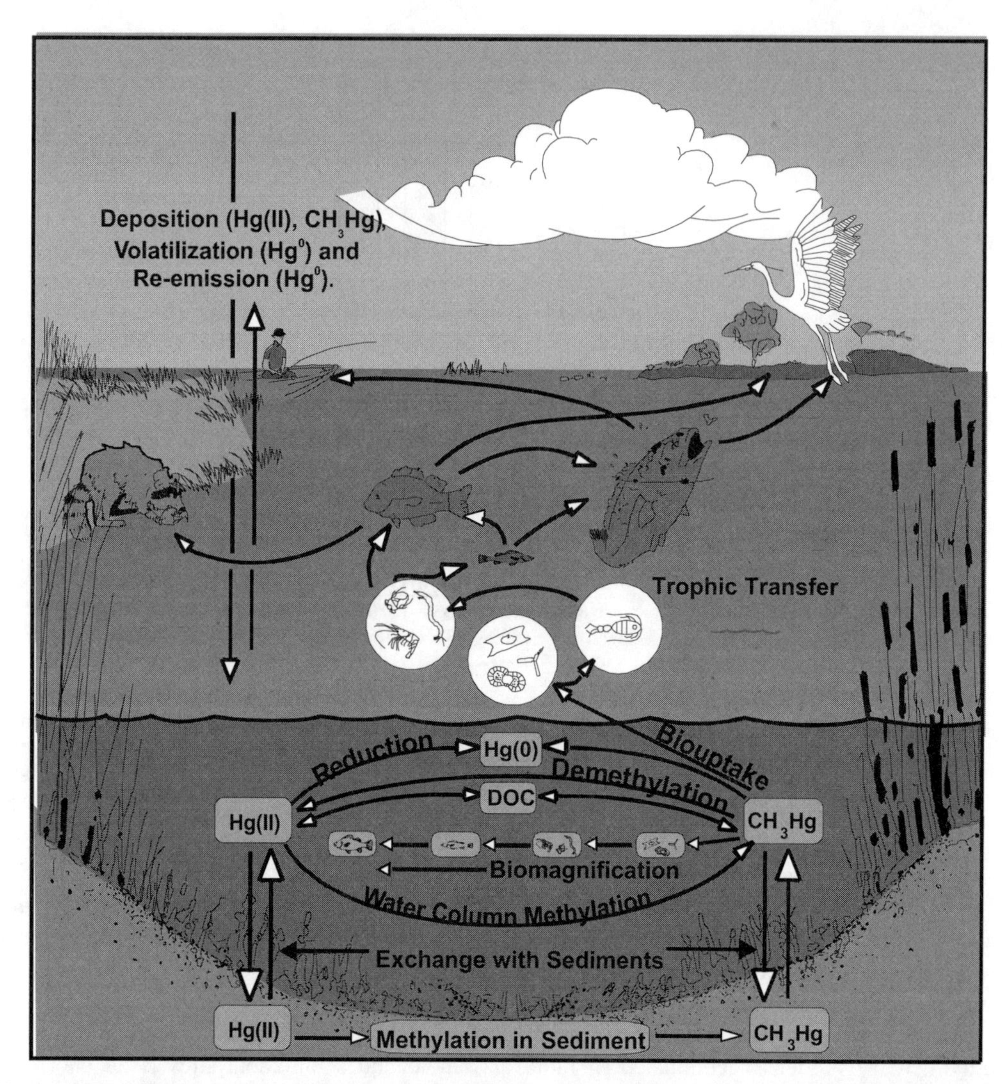


**Figure 16.2** A simplified view of the biogeochemical cycling of mercury in an aquatic ecosystem, depicting pathways and processes that influence exposure of biota to methylmercury. In this illustration, mercury enters the ecosystem largely as inorganic Hg(II) in atmospheric deposition. The mercury cycle includes a complex set of biogeochemical processes, of which methylation is most important from an ecotoxicological perspective. Methylmercury is readily bioaccumulated and transferred in food webs and can biomagnify to high concentrations in predatory fish and wildlife. Biotic exposure to methylmercury in the ecosystem is strongly influenced by the net balance between processes that yield methylmercury and make it available to aquatic biota vs. processes that degrade methylmercury or decrease its bioavailability for uptake.

chemical reactions and rates.[134–140] The environmental mercury cycle (Figure 16.2) has four strongly interconnected compartments: atmospheric, terrestrial, aquatic, and biotic. The atmospheric compartment is dominated by gaseous elemental mercury ($Hg^0$), although Hg(II) dominates the fluxes to the aquatic and terrestrial compartments. The terrestrial compartment is dominated by Hg(II) sorbed to organic matter in soils. The aquatic compartment is dominated by Hg(II)-ligand pairs in water and Hg(II) in sediments, and the biotic compartment is dominated by methylmercury. Mercury is quite reactive in the environment and cycles readily among compartments.

At the global scale, atmospheric processes and pathways dominate the transport of mercury from sources to receptors. The global mercury cycle can be envisioned as a two-way exchange process, in which sources emit elemental mercury ($Hg^0$) in the gas phase and various species of Hg(II) to the atmosphere and the atmosphere loses mercury via oxidation of $Hg^0$ to Hg(II) and the rapid removal of gaseous and particulate species of Hg(II) by wet and dry deposition.[48,139–142] This simple conceptual model reflects the following understanding of atmospheric pathways and processes: (1) that many important sources affecting global mercury cycles (including oceans, fossil fuel combustion, and municipal and medical waste incinerators) emit mostly gaseous $Hg^0$ and, to a lesser extent, gaseous and particulate species of Hg(II);[10,143,144] (2) that gaseous and particulate forms of emitted Hg(II) are subjected to local and regional removal in dry and wet deposition,[139–141,145] limiting their long-range transport; (3) that divalent mercury can be readily reduced to $Hg^0$ by natural processes in both terrestrial and aquatic ecosystems;[135,146] and (4) that $Hg^0$ can be oxidized in the atmosphere to Hg(II), which is efficiently removed in wet and dry atmospheric deposition.[139–142,145]

Mercury deposited onto the land surface in atmospheric deposition is sequestered in terrestrial soils, largely as species of Hg(II) sorbed to organic matter in the humus layer.[131,147,148] Globally, the inventory of mercury in surface soils far exceeds that in the aquatic and atmospheric compartments. The vast majority (947 Mmol) of the estimated total mass of mercury released into the environment during the past century resides in surface soils, compared to 17 Mmol in the atmosphere and 36 Mmol in the oceans.[48] The residence times of mercury in the atmosphere and the oceans are considerably shorter (a year to a few years) than the residence time of mercury in soils. Yet soils should be considered a potential long-term source — as well as a sink — for mercury in the environment, given that the Hg(II) in soils can be reduced and emitted to the atmosphere as $Hg^0$ or slowly exported to aquatic systems located down gradient.[110,147]

Recent advances in our understanding of mercury in the environment have highlighted the dominant influence of human activities — particularly since the industrial revolution — on the redistribution of global mercury pools, the size of actively cycling pools, and the importance of atmospheric pathways to a global pollution problem.[20,48,106] Anthropogenic emissions have greatly increased the mass of mercury now cycling at the earth's surface and in the atmosphere,[48,106,107,109] causing widespread contamination of terrestrial soils and aquatic sediments.[48,102,133,148,149] Mason et al.[48] estimate that two thirds of the mercury in modern global fluxes is from anthropogenic sources, and the remaining one third is from natural emissions. Soil and sediment are considered to be the dominant sinks for atmospherically derived mercury; however, detailed studies have shown that these enriched pools are susceptible to remobilization via volatilization, leaching, or erosion.[21,50,69,100,110,141] Investigators often find that the more closely they look, the more reactive the existing mercury pools appear to be; for example, St. Louis et al.[150] found that soil-canopy-atmosphere transfer rates were up to three times greater than prior estimates, Friedli et al.[151] showed that forest fires can release mercury from burned areas, and Lalonde et al.[152] showed that recently deposited snow can rapidly lose mercury via reevasion to the atmosphere.

The importance of apportioning between natural and anthropogenic mercury emissions is recognized but very difficult to achieve.[20,48,131,141,153] Initially, the estimated ratio of natural to anthropogenic emissions of mercury may have been underestimated.[141] In Europe, for example, estimated natural emissions of mercury during 1995 were in the range of 250 to 300 tons, only slightly less than the estimate of 342 tons for anthropogenic emissions.[144] In addition, emissions of $Hg^0$ from areas that are geologically enriched with mercury or affected by mining activities can be substantial.[42,153] The estimation of natural emissions from diffuse sources is greatly complicated by the fact that original sources — whether natural or anthropogenic — cannot presently be distinguished after mercury has been released into the atmosphere and has entered the global biogeochemical cycle.[21,141]

## 16.4 MERCURY SPECIATION AND ENVIRONMENTAL CONCENTRATIONS

Scientific understanding of mercury speciation in the environment, although far from complete, has increased considerably because of steadily improving analytical and field methods during the past two decades. Mercury exists in the environment in three oxidation states — Hg(0), Hg(I), and Hg(II) — and for each valence many chemical forms can occur in the solid, aqueous, and gaseous phases. The environmental chemistry of mercury is very complex, and subtle changes in chemical, physical, biological, and hydrologic conditions can cause substantial shifts in its physical form and valence state over time scales ranging from hourly to seasonal.[136,138,152] Here we briefly summarize selective aspects of mercury speciation in the atmospheric, aquatic, and terrestrial environments, focusing on aspects most pertinent to the ecotoxicology of mercury, such as the formation and abundance of methylmercury.

### 16.4.1 Atmosphere

In most locations, mercury in the atmosphere is mostly (> 95%) gaseous elemental $Hg^0$, with the remainder composed largely of particulate ionic Hg(II), gaseous divalent mercury (commonly termed "reactive gaseous mercury"), and, on occasion, trace amounts of methylmercury.[141,145] Particulate and reactive gaseous mercury have relatively short travel distances (up to tens of kilometers) and residence times in the atmosphere, whereas gaseous elemental mercury has global-scale transport and an average atmospheric residence time of about 1 year.[48] Recent analyses of air in northern Europe showed that total gaseous mercury averaged 1.98 ng/m$^3$, whereas particulate mercury and reactive gaseous mercury averaged 56 and 22 pg/m$^3$, respectively.[154,155] Monovalent mercury is stable only as the dimer ($Hg_2^{2+}$), which rapidly disproportionates to $Hg^0$ and $Hg,^{2+}$ and is probably only detectable in atmospheric samples at extremely low levels.[135] Over the open oceans concentrations of gaseous elemental mercury increase from the southern hemisphere (~1 ng/m$^3$ at 60° south) to the northern hemisphere (~3 ng/m$^3$ at 60° north),[156,157] reflecting the stronger sources of mercury in the northern hemisphere, which is more industrialized and heavily populated than the southern hemisphere.

Reactive gaseous mercury is generally assumed to be $HgCl_2$, although recent research has shown the existence of $Hg(NO_3)_2{\cdot}H_2O$ in the gas phase.[158] After polar sunrise gaseous $Hg^0$ in the Arctic and Antarctic atmospheres is rapidly depleted via oxidation to reactive gaseous mercury, which increases rapidly in abundance as $Hg^0$ is depleted.[139,140,159,160] During April and May 2000, reactive gaseous mercury often comprised more than 60% of the total gaseous mercury measured in air over Barrow, Alaska.[160] Reactive gaseous mercury is rapidly removed from the atmosphere via both wet and dry deposition[140,155,160] and is considered to be available for methylation once deposited.[140]

### 16.4.2 Aquatic Environments

The methylation of mercury and subsequent exposure of biota to methylmercury are greater in aquatic environments than in terrestrial environments. Many recent investigations of mercury in surface waters have determined methylmercury, gaseous elemental mercury ($Hg^0$), and total mercury (defined as the sum of all mercury species recovered from a strongly oxidized sample).[161] A fraction termed "reactive mercury," which is generally equivalent to mercury reducible by stannous chloride, has also been measured; however, such fractions are often poorly defined and difficult to relate to other environmental factors or processes. The recent development of methods for separating colloidal and truly dissolved fractions of inorganic mercury and methylmercury should advance understanding of aqueous-solid phase partitioning of mercury species and possibly bioavailable fractions.[162] Dimethylmercury has been observed in the marine environment, but only at extremely small concentrations (averaging 0.016 ng/L in the North Atlantic).[163] Dimethylmercury has not been confirmed in fresh waters, however, and its overall importance in the mercury cycle is unknown. We limit this discussion to the three fractions — total mercury, $Hg^0$, and methylmercury — that are most commonly reported for water.

Except under rare geochemical conditions, or in the vicinity of strong geologic or anthropogenic mercury sources, the concentrations of all forms of mercury in most natural waters are very low (picograms to nanograms per liter). Most naturally occurring mercury compounds have very low solubility, although mercury complexes with dissolved organic matter are much more soluble.[164] Among surface waters or within a given lake or stream, the abundances of methylmercury and total mercury can vary widely, and the accurate quantification of their aqueous concentrations requires the steadfast application of trace-metal clean techniques to minimize sample contamination during collection, handling, and analysis, coupled with the application of highly sensitive analytical methods.[132,165] When proper sample collection and preservation protocols are followed, inter-comparisons among laboratories that use accepted analytical methods for total mercury and methylmercury yield similar results.[166]

The speciation of mercury in water is most strongly influenced by the aqueous chemical conditions — most notably redox, pH, organic ligands, and inorganic ligands.[165] Inorganic divalent mercury, Hg(II), and methylmercury are strongly influenced by the chemical makeup of the host water and almost entirely form ion pairs with ligands in the aquatic environment.[167] In most oxic, circumneutral surface waters, ion-pair formation for Hg(II) and methylmercury is dominated by dissolved organic matter and chloride.[168,169] Under anoxic conditions, which can occur in sediment porewaters and in the hypoliminia of certain lakes, or anywhere reduced sulfur species are appreciable, inorganic Hg(II) and methylmercury will dominantly be present as sulfide or sulfhydryl complex ion pairs.[50,168,170] The complexation of Hg(II) with sulfide can substantially affect the availability of mercury for methylation by microbes.[171]

Concentrations of total mercury in unfiltered water samples from lakes and streams lacking substantive, on-site anthropogenic or geologic sources are usually in the range of 0.3 to 8 ng/L.[133,172,173] In waters influenced by mercury mining or industrial pollution, concentrations of total mercury are greater, often in the range of 10 to 40 ng/L.[33,133,173–175] Surface waters with high concentrations of humic substances can also have high concentrations of total mercury, demonstrating the importance of natural organic material on solubility and aqueous transport of the metal.[115,123,127,176,177] Surface waters draining areas with high geologic abundances of mercury or with contaminated tailings from mercury or gold mining can exceed 100 or even 1000 ng/L in total mercury.[3,26–28,33,38,41,165,178]

In oxic waters, concentrations of methylmercury are typically within the range of 0.04 to 0.8 ng Hg/L.[33,41,68,93,97,123,124,133,173,179,180] However, concentrations of 1 to 2 ng Hg/L can occur in surface waters affected by either industrial pollution (e.g., chlor-alkali plants)[174,181] or mercury mine drainage.[28,133] The fraction of total mercury present as methylmercury is generally higher in fresh waters than in estuarine or marine systems,[182] which may result from inhibition of methylation by the abundant sulfide in pore waters of brackish water systems[171] or from the generally low level of dissolved organic matter in marine settings.[127] Within a given drainage basin or geographic area the concentrations and yields of methylmercury, as well as the fraction of total mercury present as methylmercury, are typically highest in surface waters that drain wetlands.[68,93,123–125,133,183] The biogeochemical processes contributing to the methylmercury-wetland association are under investigation; however, it is evident that biogeochemical conditions in wetlands are favorable for methylation and that complexation of methylmercury with the abundant natural organic matter in wetlands can facilitate its export to waters downstream. Methylmercury generally accounts for about 0.1 to 5% and seldom exceeds 10% of the total mercury present in oxic surface water.[123,133,179] Under anoxic conditions, however, methylmercury can be one of the dominant species of mercury present, and concentrations can exceed 5 ng Hg/L.[96,100,179,184]

Early measurements of $Hg^0$ in fresh waters showed concentrations ranging from about 0.01 to 0.10 ng/L, which led to a conclusion of pronounced super saturation of $Hg^0$ in the water column, usually by a factor of 100 to 500, relative to the overlying air,[185] yielding high estimated rates of $Hg^0$ volatilization to the atmosphere. More recent investigations involving diel measurements have generally shown strong correlations between, on the one hand, instantaneous $Hg^0$ in the water column and solar intensity and, on the other, a reequilibration with the atmosphere after sundown, with much lower concentrations of $Hg^0$ in water at night (about 0.005 ng/L).[138,186] Moreover, the rapid reoxidation of $Hg^0$ in surface water has also been demonstrated[187] and, when taken into account, greatly decreases estimated volatilization rates of $Hg^0$ from surface waters.[188] In marine ecosystems, the evasion of $Hg^0$ appears to be a geochemically significant efflux of mercury.[48,189]

### 16.4.3 Terrestrial Environments

Comparatively few data are available on the abundances of total mercury and methylmercury in soils and groundwater in upland settings relative to the substantive information available for surface water, sediment, and peat in aquatic environments. Yet recent estimates indicate that terrestrial soils contain the largest inventories of mercury from natural and anthropogenic emissions.[48,131] In addition, the toxicity, solubility, and volatility of mercury depend highly on its speciation, and such information for soils is scant. Various reductive processes can yield appreciable emissions of $Hg^0$ from contaminated soils, and the mercury in soils may be cycling more actively than previously thought.[147,153,190,191]

The speciation of mercury in most upland soils is probably dominated by divalent mercury species that are sorbed primarily to organic matter in the humus layer and secondarily to mineral constituents in soil.[131,147] Nater and Grigal,[148] who studied forest soils across the upper Midwest of the United States, found that concentrations of total mercury in humus ranged from about 100 to 250 ng/g dry weight, whereas the mineral horizon just below the humus layer contained about 15 to 30 ng/g. In locations near point sources, especially cinnabar ($HgS_s$) deposits or abandoned placer mines, mercury concentrations can be considerably higher — generally in the μg/g range.[26,153,192,193] The speciation of mercury in such highly contaminated soils depends on the origin of the mercury itself (most likely $Hg^0$ used for placer mining and chlor-alkali plants, or $HgS_s$) as well as the chemistry and texture of the soil. Barnett et al.,[194] for example, observed that liquid $Hg^0$ released to anaerobic, hydric soils resulted in the formation and long-term stabilization of mercuric sulfide. Cinnabar, on the other hand, seems to be more stable when exposed to the surface as mine tailings or transported down gradient from mining operations, generally maintaining its $HgS_s$ stoichiometry, although surface coatings of secondary mercury compounds have also been observed on weathered cinnabar.[195] Little is known about the relative stability and reactivity of mercury amalgam; however, it probably behaves similarly to elemental mercury in the environment.

Published information on concentrations and speciation of mercury in upland soils is sparse, especially for methylmercury. Forest soils have been rarely analyzed for methylmercury, and reported concentrations are generally low — about 0.2 to 0.5 ng/g in the humus layer and $< 0.05$ ng/g in the mineral-trophic layer.[150,196,197] Although data are few, the very low concentrations of methylmercury in soils, runoff, and groundwater in upland environments suggest that little methylmercury is produced in upland landscapes.[125,131,198]

## 16.5 MERCURY METHYLATION IN THE ENVIRONMENT

The methylation of inorganic Hg(II) is the most toxicologically significant transformation in the environmental mercury cycle because it greatly increases the bioavailability and toxicity of mercury and increases the exposure of wildlife and humans to methylmercury. It is not surprising that variation in mercury concentrations in fish of a given size (or age) and trophic level among surface waters lacking direct, on-site sources of mercury can be attributed largely to processes and factors that affect the net production and abundance of methylmercury.[92,96,97,126] Mercury *methylation* is the conversion of inorganic Hg(II) to methylmercury by a methyl-group donor. The conversion of methylmercury to inorganic mercury, regardless of the mechanism, is termed *demethylation*. In general, both of these processes (methylation and demethylation) operate simultaneously in aquatic systems. The detection of methylmercury in sediment samples generally indicates a positive net rate of methylation, i.e., the rate of methylation exceeds that of demethylation; however, the abundance of methylmercury is not necessarily a good predictor of *in situ* methylation rate, given that influxes of methylmercury from external sources can be significant in some settings. In this section, we review the current understanding of methylating and demethylating agents as well as the locations where these processes operate in the environment.

Mercury can be methylated through biotic and abiotic pathways, although microbial methylation is generally regarded as the dominant pathway in the environment.[134,199,200] More specifically, sulfate-reducing bacteria are considered to be the most important methylating agents in aquatic environments,[97,134,199–202] and the most important sites of methylation by sulfate-reducing bacteria are thought to be oxic-anoxic interfaces in sediments[203,204] and wetlands.[93,97,123,198,205] Methylation also occurs in aerobic marine and freshwaters,[206,207] floating periphyton mats and the roots of some floating aquatic plants,[208,209] the intestines of fish,[210] and the mucosal slime layer of fish;[211] however, these sites are considered to be much less important quantitatively than are anaerobic sediments and wetlands.[212] In sediments, the microbial methylation of mercury is most rapid in the uppermost 5 cm of the sediment profile, where the rate of sulfate reduction is typically greatest; comparatively little methylmercury is produced in deeper sediments.[27,65,97,202,213,214]

To be methylated by sulfate-reducing bacteria Hg(II) must first cross the cell membrane of a methylating bacterium, presumably as a neutral dissolved species.[170,171] Thus, the speciation of inorganic mercury in aqueous and solid phases controls the fraction of the total mercury pool that is available for microbial methylation.[170,171] At certain concentrations, for example, chloride and sulfide seem to increase bioavailability because they bind $Hg^{2+}$ as the neutrally charged species $HgCl_2$ or HgS;[170,215] however, at higher ligand concentrations these ion pairs become charged (e.g., $HgCl_3^-$), and availability for methylation is decreased. Likewise, when $Hg^{2+}$ is bound to large molecules of dissolved organic matter or to particulates (either organic matter or clay), it is considered unavailable for biotic methylation.[65] Within microbial cells methylation can be facilitated through enzymatic and nonenzymatic pathways, which are distinguished by the presence or absence of active microbial metabolism, although both pathways call upon methylcobalamine as the active methyl donor to the $Hg^{2+}$ ion.[216,217] Methylcobalamine is produced by many microbes in the environment and reacts with Hg(II) to form methylmercury outside of cells in anaerobic or aerobic conditions,[218] although Choi et al.[219] found that the process is catalyzed enzymatically and that production rates are much higher within cells.

Comparatively little is known about abiotic methylation, which in simple chemical terms implies the existence in the environment of a methyl donor. Several methyl-donating compounds that are attributed largely to industrial sources can methylate mercury abiotically,[220,221] but anthropogenic methyl donors have been documented for few environmental settings. At high concentrations methylated tin and lead compounds can transfer a methyl group to Hg(II) to produce methylmercury,[222] but these situations are very limited in extent. Most of the literature on abiotic methylation consequently suggests that the most important methyl donors in the environment are humic acids,[223,224] although this topic has been studied little. In most cases, abiotic formation of methylmercury has been strongly linked to temperature, and at ambient conditions the methylation rates reported for most abiotic pathways are small. Falter and Wilken[225] have shown that small amounts of methylmercury can be formed abiotically in sediments at ambient temperatures. Their results have had implications for the analytical procedures being used in many laboratories to determine methylmercury because methylmercury can be formed as an artifact while processing samples with high concentrations (generally in the μg/g range) of inorganic Hg(II).[226,227]

Although two processes — methylation and demethylation — ultimately control the abundance of methylmercury, the process of demethylation has received comparatively little study.[228] Demethylation, or methylmercury degradation, can occur via a number of abiotic and biotic pathways in the environment. Like methylation, demethylation can occur in near-surface sediments via microbial pathways[229] and in the water column via microbial and abiotic pathways.[137,199]

Much of the early literature suggested that the microbial degradation of methylmercury involved a two-step, enzyme-catalyzed process by microbes encoded with the *mer*-operon gene sequence, also referred to as the *mer* detoxification pathway. The *mer*-operon is widespread in nature and has been found for both Gram-negative and Gram-positive bacteria and under both aerobic and anaerobic conditions,[230] although most investigators have found that this process operates in aerobic conditions.[228] The first step of the *mer* pathway involves cleavage of the carbon-mercury bond by the organomercurial lyase enzyme (encoded by the *mer* B gene) to yield methane and $Hg^{2+}$. A second step involving the mercury-reductase enzyme (*mer* B encoded) reduces $Hg^{2+}$ to $Hg^0$, yielding a mercury species that can evade from the system.[228,231]

More recently, an oxidative demethylation pathway has been proposed and confirmed by the presence of $CO_2$ as the end product of the methyl-group breakdown and does not appear to involve the secondary mercury-reductive step.[229,232] These authors[229,232] have proposed that this pathway is similar to the degradation of methanol or monomethylamine by methanogens. The oxidative pathway has been demonstrated in a wide range of environments including freshwater, estuarine, and alkaline-hypersaline sediments and in both aerobic and anaerobic conditions.[233] In anaerobic sediments of the Everglades, methanogens and sulfate reducers have been identified as the principal anaerobes in the oxidative-demethylation process, and maximal rates are observed in near-surficial sediments, where maximal methylation rates are colocated.[233] The precise mechanisms and triggers that induce the oxidative-demethylation pathway remain unclear because demonstration of the process in pure culture remains elusive. However, examination of both the *mer*-detoxification and the oxidative demethylation pathways across a wide range of mercury-contamination gradients suggest that the *mer*-detoxification pathway predominates in severely contaminated systems, whereas the oxidative pathway is more important in lightly contaminated environments.

The abiotic process of photodegradation of methylmercury in surface waters has been recently examined at a few sites.[125,137,234] It has long been known that methylmercury ion pairs are capable of adsorbing light at appreciable levels and are subject to photolytic breakdown;[235] however, the significance of the photodemethylation process was not recognized until photodemethylation was shown to be quantitatively important in the methylmercury budgets of lakes.[125,234,236] The specific mechanisms causing the degradation of methylmercury in surface waters as well as the factors limiting this process, are unknown, but experimental work suggests that singlet oxygen and peroxide radicals are responsible for the reaction.[237] The mercury end products of photodemethylation have not been determined, and theoretical considerations indicate that any of the three oxidation states are possible.[135,138] Identification of mercury end products is needed to assess the overall effect of photodegradation, given that Hg(II) could be methylated again, whereas $Hg^0$ could evade from the lake to the atmosphere.

## 16.6 MERCURY-SENSITIVE ECOSYSTEMS

Some aquatic ecosystems can be classified as mercury-sensitive because seemingly small inputs or inventories of total mercury (e.g., in the range of $< 1$ to 10 g Hg/ha) can cause significant contamination of fish and wildlife in upper trophic levels with methylmercury. Known mercury-sensitive ecosystems include most wetlands,[93,97,124] low-alkalinity or low-pH lakes,[79,92,94,238] surface waters with upstream or adjoining wetlands,[93,124,239] waters with adjoining or upstream terrestrial areas subjected to flooding,[84,85,96,240] and dark-water lakes and streams.[92,93] One common attribute of mercury-sensitive systems is the efficient conversion of inorganic Hg(II) to methylmercury.[85,92,96,97,133,241] In some cases, concentrations of mercury in game fish inhabiting such ecosystems can equal or exceed concentrations observed in fish from waters heavily contaminated by wastes from industrial point sources such as chlor-alkali plants (Table 16.1).

A recent, but growing, body of evidence indicates that wetlands are mercury-sensitive ecosystems. Wetlands can be important sources of methylmercury on the landscape, given that production and yields of methylmercury in wetland areas can greatly exceed that in other aquatic and terrestrial habitats.[123,124] The production of methylmercury in wetlands can increase greatly during flooding,[96] a periodic event in many wetland systems. A number of ecosystem characteristics probably enhance

**Table 16.1 Elevated Mercury Concentrations in Axial Muscle Tissue of Selected Freshwater Game Fishes in North American Waters**

| Aquatic Environment | Mercury Concentration (μg/g wet weight) Range in Means | Range in Maxima |
|---|---|---|
| Waters polluted by chlor-alkali plants[a] | 1–5 | 2–15 |
| Newly flooded reservoirs[a] | 0.7–3 | 2–6 |
| South Florida wetlands[b] | 0.4–1.4 | 2–4 |
| Low-alkalinity lakes[a] | 0.5–0.9 | 1–3 |

*Note:* Values shown are based on data reported for northern pike (*Esox lucius*), walleye (*Stizostedion vitreum*), largemouth bass (*Micropterus salmoides*), and smallmouth bass (*Micropterus dolomieu*).

[a] From values summarized by Wiener and Spry.[173]

[b] Data for largemouth bass from T. R. Lange (Florida Fish and Wildlife Conservation Commission, Eustis, Florida, USA, personal communication).

the microbial methylation of inorganic Hg(II) in wetlands, including an abundance of labile carbon substrates and dissolved organic matter, anaerobic sediments, high microbial activity, and seasonal water-level fluctuations that can cause oscillating redox cycles.[85,92,96,123,126,133,179,242,243] Yet the quantitative effect of methylmercury production and export from wetland areas on contamination of aquatic food webs in downstream waters supporting fish production, piscivorous wildlife, and recreational fisheries has received little study. Wetlands may differ considerably in their methylmercury-producing potential, and the influence of wetland type on methylmercury yield and the identification of associated controlling mechanisms are areas of needed investigation.

Variation in ecosystem sensitivity to mercury inputs is exemplified in Figure 16.3, which depicts tenfold variation in the mercury content of whole yellow perch (*Perca flavescens*) sampled in 1989 from three nearby lakes in northcentral Wisconsin. The three lakes differed chemically, spanning a spatial gradient in mean pH from about 5 to 7, reflecting the variation in the chemistry of lakes in northcentral Wisconsin, an area with hundreds of lakes with low acid-neutralizing capacity.[244] The three lakes were small seepage basins (no surface inlets or outlets) in rural, mostly forest-covered watersheds having no identifiable on-site anthropogenic or enriched geologic sources of mercury. The low-pH seepage lakes in this area receive very little groundwater inflow; rather, nearly all (> 95%) of their hydrologic inflow is from precipitation falling directly onto the lake surface.[245] Surficial sediments in lakes of the area are enriched with mercury, relative to deeper preindustrial sediments,[246] and the mercury-accumulation rate in sediments deposited about 1990 was three to four times that in the mid-1800s.[107] The three lakes can be regarded as lightly contaminated, with inventories of total mercury in surficial sediments (uppermost 5 cm) ranging from 1.6 to 5.8 g/ha[247] and annual atmospheric inputs of mercury (wet plus dry deposition) averaging about 0.1 g/ha from 1988 to 1990.[114] The yellow perch is one of the more widespread and abundant fishes in area lakes and is an important link in the trophic transfer of mercury in lakes of the region.[248,249] Mercury concentrations in walleyes (a regionally important gamefish) and chicks of common loons (*Gavia immer*, a fish-eating bird) are elevated in area lakes with low pH or low acid-neutralizing capacity,[78,238,250] reflecting variation among lakes in contamination of yellow perch, their preferred prey.[251,252]

## 16.7 BIOACCUMULATION, BIOMAGNIFICATION, AND BIOLOGICAL EFFECTS

### 16.7.1 Biomagnification in Food Webs

Aquatic organisms can obtain methylmercury from food, water, and sediment, and they bioaccumulate methylmercury with continued exposure because elimination is very slow relative to the rate of uptake.[253] Methylmercury readily crosses biological membranes and can biomagnify to high

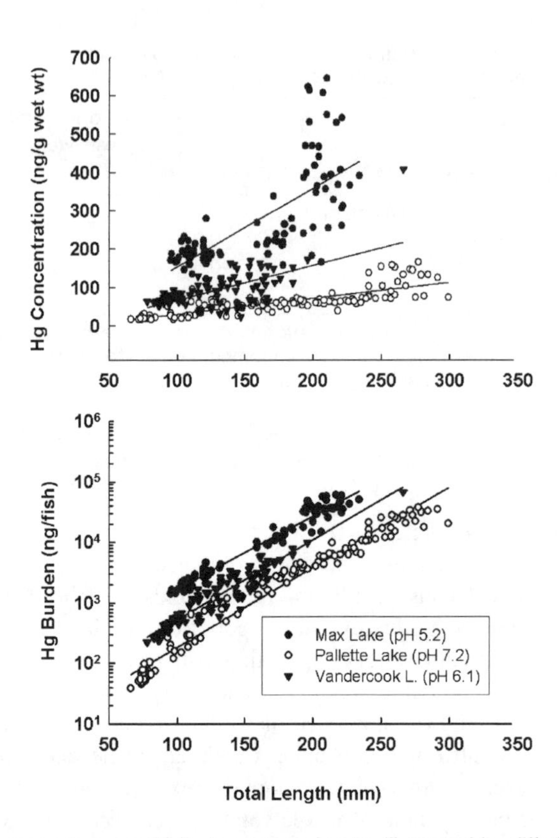


**Figure 16.3** Variation in ecosystem sensitivity to mercury inputs, illustrated by differing concentrations (upper panel) and burdens (lower panel, in logarithmic scale) of mercury in whole yellow perch from three nearby seepage lakes in forest-covered watersheds in Vilas County, Wisconsin. Fish in low-pH lakes in this and many other geographic areas have much higher concentrations of mercury than fish in nearby circumneutral-pH lakes that receive similar inputs of mercury in atmospheric deposition. (From: unpublished data for fish sampled in May 1989 by J. G. Wiener, R. G. Rada, and D. E. Powell, University of Wisconsin-La Crosse, River Studies Center, La Crosse, Wisconsin.)

concentrations in aquatic food webs, despite its seemingly low concentrations (< 1 ng Hg/L) in most surface waters.[57,173,254–256] Concentrations of methylmercury in fish, for example, commonly exceed those in ambient surface water by a factor of $10^6$ to $10^7$.[94,254,256,257]

Most of the mercury in surface waters and sediments is typically inorganic Hg(II), yet the mercury accumulated in fish and higher trophic levels of aquatic food webs is almost entirely methylmercury.[2–4,258–260] Some species of fish-eating seabirds and aquatic mammals exhibit highly variable amounts of inorganic mercury in their internal organs, particularly the liver and kidneys.[258,260–263] The presence of inorganic mercury in piscivorous birds and mammals is generally attributed to the ability of these species to convert methylmercury to the less toxic inorganic mercury via demethylation.[260–263] The presence of inorganic mercury in internal organs and tissues does not indicate dietary uptake or bioaccumulation of inorganic mercury by these organisms.

*Biomagnification*, defined as the increasing concentration of a contaminant with increasing trophic level in a food web, has been widely documented for methylmercury in aquatic ecosystems.[57,257,264] Patterns of methylmercury biomagnification in aquatic food webs are similar, even among aquatic systems that differ in ecosystem type, mercury source, and pollution intensity. This is illustrated in Table 16.2, which summarizes information from three geographically distant, markedly different aquatic environments: one a coastal marine embayment in Western Australia that was contaminated with mercury from a point source, the second a small seepage lake in northern Wisconsin that received mercury almost entirely via atmospheric deposition, and the third a tropical lake in a remote area of New Guinea. Two patterns characteristic of mercury concentrations in food

**Table 16.2 Biomagnification of Methylmercury (MeHg) in Food Webs in Three Substantially Different Aquatic Environments: Princess Royal Harbor, a marine embayment on the south coast of Western Australia that was contaminated with mercury from a super-phosphate plant over a 30-year period;[57,61] Little Rock Lake, a small, temperate seepage lake (no inflowing or outflowing streams) in northern Wisconsin (USA) that received mercury largely from atmospheric deposition directly onto the lake surface;[94] and Lake Murray, a tropical lake in the remote Western Province of Papau New Guinea[257]**

| | Australian Marine Embayment[a] | | Wisconsin Seepage Lake[b] | | Tropical Lake[c] | |
|---|---|---|---|---|---|---|
| Food-Web Component | MeHg (ng/g wet wt) | Total Hg present as MeHg (%) | MeHg (ng/g wet wt) | Total Hg present as MeHg (%) | MeHg (ng/g wet wt) | Total Hg present as MeHg (%) |
| Piscivorous fish | 2300 | >95 | 650 | >95 | 392 | 87 |
| Prey fish | 450 | 93 | 100 | >90 | 26 | 55 |
| Invertebrates | 150 | 45 | 20 | 29 | — | — |
| Algae | 7 | 10 | 4 | 13 | <0.3 | <1 |
| Water | nd | nd | 0.00005 | 5 | 0.000067 | 5 |

[a] Based on data from Francesconi and Lenanton.[57]

[b] Data for the planktonic food web in the untreated reference basin.[286] The MeHg value for piscivorous fish was the estimated concentration in a hypothetical 5-year-old walleye feeding on yellow perch (the prey fish), based on a regression for nearby lakes.[276]

[c] Data for the planktonic food web, from Bowles et al.[257]

webs are evident in Table 16.2. First, the concentration of methylmercury increases up the food web from water and lower trophic levels to fish. Second, the fraction of mercury present as methylmercury increases with increasing trophic level.

In aquatic invertebrates, methylmercury is much more readily assimilated and bioaccumulated than is inorganic mercury.[265–269] In aquatic organisms in trophic levels below fish, the fraction of total mercury present as methylmercury can vary considerably.[67,87,264,270,271] In 15 northern Wisconsin lakes, for example, percent methylmercury ranged from 9 to 82% in aquatic insects and from 46 to 97% in five taxa of crustacean zooplankton.[264] Mean percent methylmercury varied seasonally in net plankton from 12 lakes in northeastern Minnesota (U.S.), increasing from 20% in spring to 52% in autumn.[272] Percent methylmercury in benthic aquatic insects in two hydroelectric reservoirs in northern Quebec, classified by diet, ranged from 20–25% in detritivores, 30–40% in grazers, 60–85% in grazers–predators, and 95% in predatory dragonflies.[270]

The greatest increase in methylmercury concentration in pelagic food webs, relative to concentrations in water, occurs in the phytoplankton or seston (small living plankton and nonliving particulate matter).[240,257,264,273] Bioaccumulation factors for methylmercury between seston and water, for example, ranged from $10^{4.9}$ to $10^{5.6}$ in tropical Lake Murray, New Guinea,[257] and from $10^{4.8}$ to $10^{6.2}$ in 15 northern Wisconsin lakes.[264] Bioaccumulation factors for methylmercury between herbivorous zooplankton and seston are much smaller, averaging 2.5 ($10^{0.4}$) in 12 northern Wisconsin lakes[266] and about 9 ($10^{0.95}$) in natural lakes and the La Grande 2 reservoir in northern Quebec.[240]

Bioaccumulation factors between concentrations of mercury (largely methylmercury) between piscivorous fish and their prey are also small relative to the orders-of-magnitude increases in methylmercury concentration between seston and water. Ratios of mercury concentrations in piscivorous fish (axial muscle) to those in coexisting prey (whole fish) are typically less than 10, with values ranging from ~4 to 9 in freshwater lakes.[256,274–276] Bioaccumulation factors for mercury reported for seabirds (from analysis of contour feathers) and their prey (from analysis of regurgitated food) are considerably greater, ranging from 125 to 225 and averaging more than 150 in six species of seabirds from the Azores archipelago.[277]

The entry of methylmercury into the base of the food web and its subsequent trophic transfer in the lowest levels are poorly understood. The uptake of inorganic mercury by sulfate-reducing bacteria in the ecosystem is an essential step in the methylation of mercury and a prerequisite to the bioaccumulation and trophic transfer of methylmercury.[255,278] The abundance of methylmercury in the lower trophic levels appears to be strongly linked to the net production or supply of methylmercury (i.e., methylation minus demethylation), with the production of methylmercury being tightly coupled to the rate of sulfate reduction by sulfate-reducing bacteria.[97,202,279]

The dominating influence of methylmercury supply on the contamination of an aquatic food web was illustrated by a flooding experiment in the Experimental Lakes Area of Ontario, Canada.[96,98,100] Experimental flooding of the wetland surrounding Lake 979 was followed by decomposition of inundated vegetation,[280] which rapidly depleted dissolved oxygen and imposed anoxic conditions over the inundated wetland surface, stimulating microbial sulfate reduction and mercury methylation.[96,100] The abundance of methylmercury in surface water, seston, and zooplankton in the lake increased rapidly and markedly (tenfold or greater) in response to increased methylmercury production.[96,98] Concentrations of methylmercury in seston and zooplankton were strongly correlated with those in water, and concentrations in zooplankton were strongly correlated with those in seston as well (all $r^2 \geq 0.85$).[98] Bioaccumulation factors for methylmercury in zooplankton (ratio to concentrations in water) were similar before and after flooding, despite the large changes that occurred in water chemistry, waterborne methylmercury concentration, and zooplankton community structure,[98] indicating that the increased contamination of the planktonic food web resulted directly from the increased supply of methylmercury. This conclusion is further supported by a comparison of the biomagnification of methylmercury in planktonic food webs in the La Grande 2 reservoir to that in natural lakes in northern Quebec.[240] In Lake 979, concentrations of methylmercury also increased after flooding in benthic insects, caged fish (finescale dace, *Phoxinus neogaeus*, which fed primarily on benthic invertebrates), and nestling tree swallows (*Tachycineta bicolor*, which fed primarily on emergent dipterans).[87,99,281]

Within a given fish population or community, variation in trophic position accounts for much of the variation in methylmercury concentration, both within and among species.[257,282,283] In species with omnivorous feeding habits, such as adult lake trout (*Salvelinus namaycush*), trophic position can vary substantially — even within a single life stage.[284] Thus, concentrations of methylmercury in adult lake trout are strongly correlated with the trophic position in the pelagic food web.[282,283] Moreover, concentrations of methylmercury in organisms at the top of aquatic food webs increase concomitantly with increasing length (number of trophic levels) of the food chain below.

Unlike methylmercury, inorganic mercury is not readily transferred through successive trophic levels and does not biomagnify in aquatic or terrestrial food webs.[254,264,266,285] Consequently, reliance on data from total-mercury determinations for trophic levels below fish (including water, seston, plants, and invertebrates) can produce misleading assessments of food-web contamination and erroneous estimates of potential methylmercury transfer to fish and higher trophic levels.[57,265,286]

### 16.7.2 Fish

The bioaccumulation of mercury has been more intensively studied in fish than in other aquatic organisms, probably because fish are the primary source of methylmercury in the human diet.[6,12] Nearly all of the mercury in fish muscle and in whole fish is methylmercury.[2,3,57,77,287] There is very little inorganic mercury in either freshwater or marine fish,[2,3,57,77,287] even in aquatic ecosystems with high concentrations of dissolved inorganic mercury.[288] Fish assimilate inorganic mercury much less efficiently than methylmercury from both food and water, and if absorbed, inorganic mercury is eliminated much more rapidly than is methylmercury.[253,289–294]

Dietary uptake probably accounts for more than 90% of the total uptake of methylmercury in wild fishes,[295–298] and fish probably assimilate from 65 to 80% or more of the methylmercury present in the food they eat.[291,294,296] In the laboratory, fish can accumulate high concentrations of methylmercury directly across the gills when exposed to abnormally high concentrations of waterborne methylmercury.[292,299,300] Many of the published laboratory studies on bioaccumulation have exposed test fish to methylmercury concentrations that greatly exceed concentrations of methylmercury in surface waters.[79,173] The mode of uptake (food vs. water) in bioaccumulation experiments, however, seems to have little influence on the distribution of methylmercury among most internal organs and tissues, except that concentrations in the gills are much higher after waterborne (than dietary) exposure and concentrations in the intestines are higher after dietary exposure.[253,300–302]

After crossing the fish gut, methylmercury binds to red blood cells and is transported via the circulatory system to all organs and tissues, readily crossing internal membranes.[294,300,301,303] There is a dynamic internal redistribution of assimilated methylmercury among the tissues and organs of fish exposed to methylmercury in both laboratory and field studies. The masses in the blood, spleen, kidney, liver, and brain decline after exposure to either waterborne or dietary methylmercury ceases, and much of the methylmercury in the body eventually relocates to skeletal muscle, where it accumulates bound to sulfhydryl groups in protein.[290,301–303] Wiener and Spry[173] hypothesized that storage of methylmercury in skeletal muscle serves as a protective mechanism in fishes, given that sequestration in muscle reduces the exposure of the central nervous system to methylmercury.

Within a given fish population, concentrations of methylmercury in muscle tissue or whole fish typically increase with increasing age or body size, a pattern that has been observed repeatedly in surveys of mercury in fishes.[173,253] The increasing concentration with size or age results from the very slow rate of elimination of methylmercury by fish relative to its rapid rate of uptake.[253,293] In a critical analysis of experimental data on methylmercury elimination by fish, Trudel and Rasmussen[293] showed (1) that short-term experiments (< 90 days) substantially overestimate elimination rate, (2) that elimination rate is negatively correlated with body size ($r = -0.65$), (3) that elimination rate is positively correlated with water temperature (r = 0.77) in long-term (> 90 days) experiments, with a $Q_{10}$ of 1.9, and (4) that the concentration or burden of methylmercury in the fish does not influence the rate of elimination.

The bioaccumulation of methylmercury in fish is influenced by an array of biotic, ecological, and environmental variables. Much of the modern spatial variation in fish mercury levels (within a given trophic level) is attributed to differences among lakes and their watersheds in biogeochemical processes and transformations that control the abundance of methylmercury. The production of methylmercury via the microbial methylation of inorganic Hg(II) in the environment is a key process affecting mercury concentrations in fish.[86,92,97,126,242] It follows logically that factors and processes affecting the microbial production of methylmercury on the landscape will also influence the methylmercury content of fish residing in the ecosystem. Some of the variation in mercury concentrations in fish among northwestern Ontario lakes, for example, is caused by the effect of temperature, or lake size, on the microbial net production of methylmercury in the epilimnia.[126] Mean concentrations in axial muscle of walleye and northern pike (*Esox lucius*) ranged from about 0.7 to 1.1 μg Hg/g wet weight in small (89–706 ha) lakes but were less than 0.4 μg/g in nearby lakes that were larger (2219–34,690 ha) and colder.[126] Specific rates of mercury methylation in the lakes were positively correlated with water temperature, whereas specific rates of methylmercury demethylation (microbial destruction of methylmercury) were negatively correlated with temperature.[304]

The dietary uptake of methylmercury in fish is influenced by their size, diet, and trophic position.[274,283,305–307] In piscivorous species, such as the walleye and lake trout, the methylmercury content of the diet and associated rate of mercury accumulation can increase with age, accelerating abruptly when the fish become large enough to switch from a diet of invertebrates to prey fish.[274,306] In adult fish, females often contain higher mercury concentrations than males because they must consume more food than males to support the energy requirements of egg production.[308,309] The increased feeding rates in females cause greater dietary uptake of methylmercury, and only a small fraction of the accumulated methylmercury is transferred to the egg mass and eliminated during spawning.[287,310,311]

The relative contamination of aquatic food webs with methylmercury can be assessed with information on mercury concentrations in fish of a given species and age,[77,80,81,312–314] particularly if the trophic position of the fish analyzed varies little among the water bodies studied. Within a number of geographic areas of North America, mean concentrations of mercury in same-age yellow perch, for example, vary several fold among midcontinental lakes (and presumably receiving similar rates of mercury deposition). Moreover, mean concentrations of mercury in yellow perch are inversely correlated with lake pH and related chemical variables.[77,81,313,315] Similarly, estimated mercury concentrations in 3-year-old largemouth bass (*Micropterus salmoides*) from 53 Florida lakes varied from 0.04 to 1.53 μg/g wet weight and were correlated with lake pH and related chemical factors.[80]

Methylmercury is neurotoxic and can be very harmful to the central nervous system. In the laboratory, long-term dietary exposure of fishes to methylmercury has caused incoordination, diminished appetite or inability to feed, diminished responsiveness and swimming activity, starvation, and mortality.[173,316,317] Adult fishes were adversely affected in at least two cases of extreme industrial mercury pollution during the past century — Minamata Bay (Japan) and Clay Lake (Ontario). In Minamata Bay, coincident with the poisonings of humans and other organisms, resident fish exhibited symptoms of methylmercury intoxication[318,319] that have been subsequently reported in laboratory experiments.[300,316,317] These symptoms included mortality, severely diminished locomotor activity, impaired escape behavior, emaciated condition, and lesions in the brain.[319] Mercury concentrations in axial muscle of "enfeebled" fishes found floating in seawater in the bay averaged 15 μg/g wet weight and ranged from 8.4 to 24 μg/g in six species.[318]

In grossly polluted Clay Lake in the English-Wabigoon River system (northwestern Ontario), northern pike varying in age from 3 to 8 years had mercury concentrations ranging from 6 to 16 μg/g wet weight in axial muscle tissue.[320] Compared with northern pike from a relatively uncontaminated reference lake, fish from Clay Lake were emaciated, had low hepatic fat stores, exhibited symptoms of starvation (low levels of total protein, glucose, and alkaline phosphatase in blood serum), and had low serum cortisol levels.[320] After 1 year, contaminated northern pike transplanted from Clay Lake into the reference lake had serum concentrations of total protein, alkaline phosphatase, and cortisol intermediate to those in Clay Lake and reference-lake fish, suggesting partial recovery. Lockhart et al.[320] did not attribute the poor condition of Clay Lake fish directly to effects of methylmercury; however, similar symptoms have subsequently been observed in fish exposed to dietary methylmercury in a laboratory experiment.[317]

Wiener and Spry[173] derived the following critical tissue concentrations in *adult* fish, based on a review of mercury concentrations associated with toxic effects in freshwater fish. In the brain, concentrations of 7 μg/g wet weight or greater probably cause severe, potentially lethal, effects. In mercury-sensitive species, brain-tissue concentrations of 3 μg/g wet weight or greater probably indicate significant toxic effects. For axial muscle tissue, field studies indicate that concentrations of 6 to 20 μg/g wet weight are associated with toxicity. The range for laboratory studies is similar, with sublethal effects or death associated with concentrations in muscle of 5–8 μg/g in walleyes and 10–20 μg/g in salmonid species. Whole-body concentrations associated with sublethal or lethal effects are about 5 μg/g wet weight for brook trout (*Salvelinus fontinalis*) and 10 μg/g for rainbow trout (*Oncorhynchus mykiss*), whereas estimated no-observed-effect concentrations in salmonid species are 3 μg/g for the whole body and 5 μg/g for brain or axial muscle tissue. However, Wiener and Spry[173] cautioned that the toxicity of methylmercury to fish is influenced by factors such as interspecific and intraspecific variation in sensitivity to methylmercury that contribute uncertainty to estimates of critical tissue concentrations. Moreover, the rate of accumulation seems to affect the toxicity of methylmercury in fish.[292] If methylmercury is accumulated slowly, fish can clearly tolerate higher tissue concentrations of mercury, presumably due to the internal transfer and binding of methylmercury to proteins in skeletal muscle (the primary storage site), which decreases exposure of the central nervous system.[173]

Given the high neurotoxicity of methylmercury, the exposure levels causing adverse behavioral effects are probably much lower than exposure levels associated with overt toxicity.[173] Many fish behaviors are sensitive and ecologically relevant indicators of contaminant toxicity, affected at exposure levels much lower than those causing direct mortality.[321–323] The ability of mosquitofish (*Gambusia affinis*) to avoid predation by largemouth bass, for example, was greatly diminished by aqueous exposure to 10, 50, and 100 μg Hg/L (administered as mercuric chloride), concentrations that otherwise did not influence mortality.[321] The neurotoxic effects of exposure to sublethal concentrations of methylmercury can impair the ability of fish to locate, capture, and ingest prey and to avoid predators.[324–326] For example, Fjeld et al.[325] showed that the feeding efficiency and competitive ability of grayling (*Thymallus thymallus*), exposed as eggs to waterborne methylmercuric chloride for 10 days and having yolk-fry with mercury concentrations of 0.27 μg/g wet weight or greater, were impaired when fish were tested 3 years later.

In a critical review, Wiener and Spry[173] concluded that reduced reproductive success was the most plausible effect of mercury on wild fish populations at contemporary exposure levels in aquatic ecosystems with methylmercury-contaminated food webs. They also suggested, based on the limited information available, that the margin of safety between existing and harmful exposure levels may be small for some fish populations. Methylmercury can impair reproduction of fishes by affecting gonadal development or spawning success in the adults or by reducing the hatching success of eggs and the health and survival of embryolarval stages.[64,300,327,328] The embryolarval and early juvenile life stages of fish are typically most sensitive to toxic contaminants,[300] and exposure of embryos to methylmercury can impair competitive ability and foraging efficiency throughout the lifetime of the fish.[325]

Nearly all of the mercury in the developing eggs of fish is methylmercury derived from maternal transfer.[64,287,311] The amount of methylmercury transferred from the female to the developing egg is small relative to the burden of the metal in the adult, yet the methylmercury content of eggs is strongly related to that of the maternal fish.[64,287,311]

Recent experiments showing diminished reproductive success or fitness of fish exposed to environmentally realistic concentrations of methylmercury indicate that some fish populations may be adversely affected by methylmercury.[64,328] Latif et al.[64] examined the effects of both maternally transferred and waterborne methylmercury on embryos and larvae of walleyes from industrially polluted Clay Lake and two atmospherically contaminated lakes in Manitoba. In their study, the hatching success of eggs and the heart rate of embryos decreased with increasing environmentally realistic concentrations of waterborne methylmercury (range, 0.1–7.8 ng/L), whereas methylmercury concentrations in eggs from maternal transfer did not significantly affect egg-fertilization success, egg-hatching success, or the heart rate of embryos. The growth of larval walleyes (measured 8 days after hatching) and the incidence of larval deformities were unrelated to either maternal or waterborne methylmercury in their study.[64]

Hammerschmidt et al.[328] fed fathead minnows (*Pimephales promelas*) diets containing concentrations of methylmercury present in contaminated food webs, maintained the fish through sexual maturity, and examined the effects of dietary and maternally transferred methylmercury on several reproductive variables. In their study, dietary methylmercury affected the overall reproductive performance of adult fathead minnows, whereas maternally transferred methylmercury did not measurably affect the embryos and larvae produced. In the adults, exposure to dietary methylmercury reduced spawning success, delayed spawning, decreased the instantaneous rate of reproduction, and reduced gonadal development (as reflected by the gonadosomatic index) and reproductive effort of females. These responses were caused by dietary concentrations of methylmercury that are equaled or exceeded in the prey of some piscivorous and invertivorous fish inhabiting low-alkalinity lakes and flooded reservoirs.[328] In contrast, the growth and survival of adult fathead minnows in this study were unrelated to dietary methylmercury. Fertilization success, hatching success, 7-day survival, and 7-day weight of larval fathead minnows varied considerably, but these biological endpoints were not correlated with concentrations of mercury in either the diets or carcasses of parental fish.[328]

### 16.7.3 Birds

The threat of mercury to birds is now largely an aquatic one, given that the probability of exposure of terrestrial birds to high concentrations of organomercurials has diminished substantially since the use of mercury compounds in seed dressings was discontinued. Fimreite[71] reviewed information on bird poisonings caused by mercurial seed dressings. The bioaccumulation and toxic effects of mercury in birds have been reviewed more recently.[329–331] The biomagnification of mercury in aquatic food webs often leads to high concentrations in fish-eating birds.[263,332–334] Moreover, there is evidence that concentrations of methylmercury have increased in some seabirds during the past century and in the past few decades (Figure 16.1).[116,118]

Consumption of fish is the main pathway of methylmercury exposure for birds.[335] Methylmercury, one of the most harmful of contaminants to birds, can adversely affect adult survival, reproductive success, behavior, and cell development.[336] It can also cause teratogenic effects.[337,338] Methylmercury readily crosses the blood-brain barrier[339,340] and is passed from the mother to the eggs.[341–343] When transferred to eggs, nearly 100% of the mercury remains in the methylmercury form and about 85 to 95% is deposited into the albumen.[341]

Incorporation of methylmercury into growing feathers and excretion in the feces are the major routes of mercury elimination in birds,[344] although deposition of mercury into eggs is also an important route for reproducing adult females.[341,343,345] Stickel et al.[346] reported a half-life of mercury in whole bodies of adult male mallards (*Anas platyrhynchos*) of about 12 weeks, and the growth of new feathers toward the end of their 12-week period of study seemed to be responsible for most of the loss. Young birds also excrete methylmercury into their developing feathers.[347]

The concentration of mercury in feathers has been advocated as an indicator of mercury in other avian tissues.[348] Scheuhammer et al.[349] reported significant correlations between mercury in the feathers and blood of chicks of common loons. Caldwell et al.,[335] who failed to find good correlations between mercury concentrations in chick feathers with those in blood, other tissues, and eggs of double-crested cormorants (*Phalacrocorax auritus*), suggested that more information is needed to establish the dynamics of mercury in feathers vs. other tissues. Unlike fish, in which older, larger individuals tend to have higher concentrations of mercury,[173] birds can annually eliminate much of their body burden of methylmercury through the formation of new feathers.[350]

Fish-eating seabirds seem to be able to demethylate methylmercury, mainly in the liver.[262] The rank order of concentrations of methylmercury in various tissues, averaged over several species of seabirds, was liver > kidney > muscle, and the mean percentages of total mercury present as methylmercury were 35% in liver, 36% in kidney, and 66% in muscle. Furthermore, the percentage of total mercury present as methylmercury decreases as the concentration of total mercury increases.

In double-crested cormorants from Caballo reservoir in New Mexico, where prey fish contained from 0.05 to 0.21 μg Hg/g wet weight, cormorant eggs contained a mean of 0.30 μg Hg/g wet weight, and nestlings that were about 7 to 10 days old had concentrations of 0.36, 0.40, 0.18, and 3.54 μg/g wet weight in blood, liver, muscle, and primary feathers, respectively.[335] Kim et al.,[262] who studied nine species of seabirds, found that both total mercury and methylmercury in nearly all species were higher in liver than in feathers, kidney, and muscle; among these latter three tissues there were no consistent relations. With feathers, it is desirable to know when new feathers were formed because mercury in the diet and body burdens at that time seem to control the deposition of methylmercury into developing feathers.[350] In controlled laboratory studies, where continuous methylmercury diets have been fed, species such as chickens (*Gallus gallus*), ring-necked pheasants (*Phasianus colchicus*), and mallards tend to accumulate the highest concentrations of mercury in liver and kidney, with muscle and brain containing lesser concentrations.[343,351,352]

#### 16.7.3.1 Field Studies on Birds

Observations of high concentrations of mercury in fish-eating birds have prompted field studies to assess effects of methylmercury exposure in wild birds. Mortality and impaired reproduction are two effects observed in controlled laboratory experiments with methylmercury that could decrease exposed populations of wild birds. A number of field studies have shown associations between high mercury levels in the diets or tissues of fish-eating birds and suspected harm. In southern Florida, for example, methylmercury exposure may have contributed to deaths from chronic diseases in great white herons (*Ardea herodias occidentalis*).[353] The livers of herons that died of acute causes, such as collisions with power lines or vehicles, had a mean mercury concentration of 1.77 μg/g wet weight, whereas birds that died showing signs of chronic disease had livers with a mean of 9.76 μg/g mercury. Spalding et al.[353] cautioned that little was known about the history of methylmercury exposure in the dead birds examined in their study. In such cases, the wasting of muscle in sick birds could result in the release of mercury from muscle and its further accumulation in liver.[263]

In the Netherlands, many grey herons (*Ardea cinerea*) died during the winter of 1976.[354] The mercury concentration in the livers of 41 of these dead herons averaged 95.5 μg/g dry weight (~27 μg/g wet weight), with a maximum of 773 μg/g dry weight. Necropsies were performed on 26 of the dead herons, and most were severely emaciated. Van der Molen et al.[354] experimentally exposed herons to methylmercury, and lethality was associated with mercury concentrations in the liver that averaged 500 μg/g and ranged from 415 to 752 μg/g dry weight. Only two of the analyzed dead herons from the field had hepatic mercury levels within this lethal range; however, van der Molen et al.[354] postulated that the observed mortality was caused by the sublethal effects of mercury combined with the stress of cold weather and undernourishment. The authors estimated that mercury levels in livers of 20% of the 26 herons examined were sufficiently high to have caused either lethal or serious sublethal effects. In eastern Canada, Scheuhammer et al.[263] reported that the livers and kidneys of common loons found dead or in a weakened, emaciated condition contained levels of mercury that were high enough to have contributed to their ill health, although the authors noted that the wasting of muscle and other tissue in the sick and dead loons could have increased the concentrations of mercury in the remaining tissue. The authors also pointed out that, because only low levels of the mercury in the liver were in the methylmercury form, it was questionable whether the loons were affected by methylmercury toxicity.

The embryos of birds and other vertebrate organisms are much more sensitive than the adult to methylmercury exposure.[6,355] The dietary concentrations of methylmercury that significantly impair avian reproduction are only one-fifth of those that produce overt toxicity in the adult,[355] and possible reproductive impairment of wild birds has been reported in a number of field studies.

Newton and Haas[356] examined the levels of several pollutants, including mercury, in eggs of the merlin (*Falco columbarius*). Concentrations of mercury in eggs from wild merlins were related to the number of young raised by the adults, and higher concentrations were associated with fewer young. Nearly all of the mercury in eggs is methylmercury.[341,357,358] The relation between production of young merlins, and mercury exposure was statistically significant but not clear-cut; for example, some of the most contaminated clutches produced 3 or 4 young, whereas some of the nests with low concentrations of mercury in eggs failed completely. Newton and Haas[356] attributed such variable results to variation in individual sensitivity to methylmercury and to the influences of other environmental factors on reproductive success.

Reproductive failure in common loons, a piscivorous aquatic bird, was attributed to dietary mercury exposure linked to contamination of food webs from industrial mercury pollution of the English-Wabigoon River system. Fimreite[332] observed that young common loons were absent along highly contaminated reaches where mercury concentrations were high in adult loons and other aquatic birds. In a comprehensive field study in the same river system, Barr[357] showed a strong negative correlation between the successful use of breeding territories by common loons and mercury concentrations in lakes in a 160-km reach downstream from a chlor-alkali plant. Barr[357] observed that reductions in egg laying and territorial fidelity were associated with mean mercury concentrations of 0.3 to 0.4 µg/g wet weight in prey organisms and with mean concentrations of 2 to 3 µg/g wet weight in loon eggs and the adult brain. Reproductive effects were more severe when concentrations of mercury in prey fish exceeded 0.4 µg/g wet weight.[357]

The North American breeding range of the common loon includes many semiremote and remote lakes in regions where reported mercury concentrations in fish commonly exceed 0.3 to 0.4 µg/g wet weight, the dietary threshold values for reproductive effects estimated by Barr.[357] These regions include the northcentral and northeastern United States and the eastern Canadian provinces of Ontario, Quebec, New Brunswick, and Nova Scotia. Scheuhammer and Blancher,[359] for example, estimated that as many as 30% of the lakes in central Ontario contained prey fish with mercury levels high enough to impair reproduction in common loons, based on the dietary threshold of 0.3 to 0.4 µg/g estimated by Barr.[357]

In northern Wisconsin, another area with many methylmercury-contaminated fish populations, Meyer et al.[82] examined reproductive success of common loons in relation to mercury levels in blood and feathers. Adults and chicks were studied on 45 lakes (mostly seepage lakes) in an area where atmospheric deposition is the dominant mercury source.[360] The mean concentration of mercury in eggs sampled from nests at these lakes was 0.9 µg/g wet weight,[82] an exposure level that has been associated with reproductive impairment in laboratory experiments with mallards.[343] Production of loon chicks was lowest at the lakes where mercury in the blood of chicks was highest. The concentration in chick blood was negatively correlated ($r^2 = 0.56$) with lake pH,[82] a pattern also observed in the methylmercury content of prey fish (Figure 16.3) and game fish in small lakes of the study area.[78,250]

Descriptive field studies typically yield correlational results, which alone are generally insufficient for establishing a causal linkage between toxicant exposure and a biological response, such as reproductive success, because of the potential confounding influence of other, covarying factors.[361–363] For this reason, Meyer et al.[82] did not conclude a cause-and-effect relation between high methylmercury exposure and low production of loon chicks on northern Wisconsin lakes. They indicated the need to first critically test an alternative hypothesis concerning reproductive success of common loons in northern Wisconsin lakes: that the lower production of loon chicks on low-pH lakes resulted from lesser prey abundance in the low-pH lakes.

Fish assemblages of small seepage lakes used by nesting loons in northern Wisconsin are characterized by low species richness and numerical dominance by a few species, particularly sunfishes (*Lepomis* spp.) and yellow perch.[74,248,249,364,365] The yellow perch, the preferred prey of the common loon,[252] is ubiquitous and typically abundant in these seepage lakes, ranking second in relative numerical abundance (based on catch per unit of effort) only to the bluegill (*Lepomis macrochirus*).[74,248,249,365] Moreover, the yellow perch is an acid-tolerant species, and self-sustaining populations occur in Wisconsin lakes with pH as low as 4.4 standard units.[249,364] Intensive, standardized fish surveys in 12 small lakes in northcentral Wisconsin (fixed sampling effort per lake with four gear types) yielded catches of yellow perch ranging from 285 to 969 fish (mean, 597) in six lakes with low pH (range, 5.1–6.0) and catches ranging from 38–1030 fish (mean, 370) in six lakes with circumneutral pH (range, 6.7–7.5).[74] Two other potential prey-fish species, bluegill and pumpkinseed (*Lepomis gibbosus*), are also abundant in the low-pH lakes.[74] Thus, it is highly improbable that the low production of loon chicks observed by Meyer et al.[82] on low-pH lakes in northern Wisconsin resulted from lesser prey-fish abundance in such lakes. We infer that high methylmercury exposure is a more defensible explanation for the low production of loon chicks on these low-pH lakes.

In Georgia (U.S.), Gariboldi et al.[366] measured mercury levels in prey items regurgitated by nestling wood storks (*Mycteria americana*) at four colonies. The estimated mean concentration in the diet of nestlings at individual colonies ranged from 0.10 to 0.28 μg/g wet weight, equaling or exceeding a lowest observed adverse effect concentration (LOAEC) of 0.1 μg/g wet weight, a value recommended by Eisler[336] as a maximum tolerable dietary concentration for sensitive avian species. The LOAEC of 0.1 μg/g wet weight was derived from laboratory experiments showing that the reproduction and behavior of mallards were affected by a diet containing 0.5 μg Hg/g dry weight (~0.1 μg/g wet weight), administered as methylmercury dicyandiamide. The mean concentration in prey of nestling wood storks was highest (0.28 μg/g wet weight) in an inland colony, where an average of 1.9 young wood storks were fledged per nest — lower than the averages of 2.6 and 2.5 birds fledged per nest at the two coastal colonies, where mean dietary mercury concentrations were 0.10 and 0.19 μg/g wet weight, respectively.[366] Gariboldi et al.[366] stated that it was difficult to separate the effects of dietary mercury exposure from other potential stressors on wood storks such as differences in the abundance of prey among colonies. These authors also noted the uncertainty associated with extrapolating a LOAEC derived from laboratory studies with mallards to wild, fish-eating wood storks. Mercury levels in eggs of wood storks were not measured in this study.

In the English-Wabigoon River system, Fimreite[332] reported a mercury concentration of 3.65 μg/g wet weight in eggs of common terns (*Sterna hirundo*) nesting in Ball Lake, Ontario, where estimated hatching success was less than 27%. In lesser contaminated Wabigoon Lake, mercury averaged 1.0 μg/g in eggs of terns, only seven unhatched eggs were found, and a "large number of fledged young" were observed. The reproductive measurements in this study were made during two visits to each colony and were not systematically collected; however, Fimreite[332] associated the impaired reproduction in terns from Ball Lake with high methylmercury exposure.

The embryos of herring gulls (*Larus argentatus*) seem to be much less sensitive than embryos of the common tern to methylmercury, based on results of Vermeer et al.,[367] who measured mercury in a single egg taken from each of 18 herring gull nests and monitored the subsequent hatching success of the remaining eggs. Mercury levels in whole eggs varied from 2.3 to 15.8 μg/g wet weight. In four eggs, yolk and albumen were analyzed separately, yielding concentrations ranging from 0.9 to 3.5 μg/g in the yolk and from 3.5 to 22.7 μg/g in the albumen. All but two of the remaining eggs in the 18 nests hatched, and those two eggs were in nests where the sampled eggs contained 7.9 and 8.1 μg/g of mercury.[367]

Henny et al.[368] assessed the influence of high concentrations of mercury in eggs on bird reproduction at five national wildlife refuges in the western United States. Concentrations in some eggs of ducks exceeded 3 μg/g dry weight (~0.6 to 0.8 μg/g wet weight), a concentration near that causing reproductive impairment in laboratory studies with mallards.[343] It was not feasible to relate mercury residues measured in a single egg taken from each duck nest to hatching success of the remaining eggs because of heavy predation on the nests. A small sample of eggs from the nests was, therefore, incubated in the laboratory; eggs with more than 3 μg Hg/g (dry weight) hatched as well as eggs with concentrations less than 3 μg/g.[368]

#### 16.7.3.2 Laboratory Experiments on Birds

An inherent limitation of ecotoxicological field studies stems from the difficulty in isolating the biological effects of methylmercury exposure from the effects of other variables. The presence of other environmental contaminants, for example, can complicate the identification of biological responses to methylmercury. Controlled laboratory experiments are useful for addressing the uncertainties inherent in even the best field studies. A principal objective of many laboratory experiments has been to determine the concentrations of methylmercury in the avian diet or in avian tissues and eggs that are associated with mortality or reproductive failure.

Koeman et al.[369] dosed mice with methylmercury dicyandiamide and fed the mice, which contained about 13 μg Hg/g wet weight, to Eurasian kestrels (*Falco tinnunculus*). Methylmercury poisoning in the kestrels became evident after about 15 days, and mortality began after 21 days. The kestrels suffered demyelination of the spinal cord, a symptom of methylmercury poisoning. Concentrations of mercury in the kestrels that died or were sacrificed after showing signs of mercury poisoning ranged from 49 to 122 μg/g wet weight in the liver and from 20 to 33 μg/g in the brain.[369] Methylmercury poisoning and mortality also occurred in goshawks (*Accipiter gentilis*) and red-tailed hawks (*Buteo jamaicencis*) fed chicken flesh containing about 4 to 13 μg Hg/g on a wet-weight basis, as methylmercury.[370,371] Livers of the dead goshawks contained from 103 to 144 μg Hg/g wet weight, and brains contained 36 to 51 μg/g.[370] Livers of poisoned red-tailed hawks contained about 19 to 20 μg/g of mercury.[371]

Finley et al.[372] estimated the concentrations of mercury in tissues associated with the death of birds by feeding 40 μg Hg/g, as methylmercuric dicyandiamide, to European starlings (*Sturnus vulgaris*), common grackles (*Quiscalus quiscula*), red-winged blackbirds (*Agelaius phoeniceus*), and brown-headed cowbirds (*Molothrus ater*). After 5 of the 14 birds of each species had died from methylmercury poisoning, 5 survivors were sacrificed, and mercury concentrations in tissues of the dead and surviving birds were compared. The sacrificed birds showed no overt symptoms of methylmercury intoxication. Concentrations in tissues of most dead birds exceeded, but did not differ statistically from, concentrations in the survivors. This study suggested that there is no specific, single concentration of methylmercury in tissues associated with death of the organism. Although no such threshold concentration was evident in their study, Finley et al.[372] considered 20 μg Hg/g wet weight in the tissues as a hazardous concentration.

In another study to determine harmful tissue levels of mercury, Scheuhammer[373] fed diets containing 5 μg Hg/g dry weight, as methylmercuric chloride, to zebra finches (*Poephila guttata*) for 76 days. One fourth of the birds died, and 40% of the survivors exhibited overt neurological signs of methylmercury poisoning including lethargy and difficulty in balancing on their perches. Mercury levels in the brains of finches that died were no higher than levels in birds that showed overt signs of poisoning but did not die. Finches that survived the 76-day exposure period without exhibiting overt symptoms of methylmercury poisoning typically had less than 15 μg Hg/g wet weight in the brain, whereas birds with symptoms had at least 15 μg/g in the brain. Scheuhammer[373] concluded from these and other data that tissue levels of methylmercury associated with neurological effects are similar in birds of different species, size, and dietary mercury level.

In Pekin ducks (*Anas platyrhynchos*) fed 15 μg Hg/g (as methylmercuric chloride), overt signs of mercury poisoning (loss of appetite, decreased mobility, and leg paralysis) appeared in males after 5 weeks and in females after 8 weeks.[374] Mercury concentrations did not differ between the sexes at death or at the end of the 12-week exposure period; livers of males and females contained 88 and 92 μg/g wet weight, respectively, and brains contained 20 and 23 μg/g.[374] In mallards fed methylmercuric chloride, Pass et al.[375] saw no clear delineation between mercury concentrations in the brains of birds that had developed microscopic lesions in the brain (range, 3.2 to 27.2 μg/g wet weight) and those without detectable lesions (1.8 to 22 μg/g).

Heinz[376] tabulated published concentrations of mercury in the internal tissues of birds that were poisoned, dead, or asymptomatic. He estimated that wet-weight concentrations of mercury associated with harmful methylmercury exposure in adult birds were 15 to 20 μg/g in the brain, 20 to 60 μg/g in the liver, 20 to 60 μg/g in the kidney, and 15 to 30 μg/g in muscle tissue.

The embryos and young of birds are more sensitive to methylmercury than are the adults. Heinz and Locke[340] fed breeding mallards 3 μg Hg/g as methylmercury dicyandiamide, and mercury was accumulated in eggs to mean concentrations between ~5.5 and 7.2 μg/g wet weight in 2 consecutive years. Reproductive success was impaired, and some ducklings died from methylmercury poisoning after hatching. The brains of dead ducklings had mercury concentrations ranging from ~4.9 to 8.7 μg/g wet weight and exhibited demyelination and necrosis characteristic of methylmercury poisoning.[340] Reproduction of black ducks (*Anas rubripes*) fed 3 μg Hg/g as methylmercury dicyandiamide was similarly impaired, with post-hatching mortality of ducklings associated with mercury concentrations between 3.2 and 7.0 μg/g wet weight in the brain.[377]

Concentrations of methylmercury in the maternal diet and in eggs associated with adverse reproductive effects in birds have also been estimated in laboratory studies. Tejning[341] fed chickens a diet containing about 9.2 μg Hg/g dry weight as methylmercury dicyandiamide. Within 3 weeks mercury concentrations in eggs increased to about 25 μg/g wet weight in egg whites and 2 μg/g in egg yolk, and hatching success of exposed eggs decreased to about 10%, relative to about 60% in unexposed controls. When the maternal diet contained about 4.8 μg/g of mercury, administered as methylmercury dicyandiamide, egg whites and yolks contained about 17 and 2 μg/g, and hatching success was 17%.[341] Fimreite[342] fed ring-necked pheasants a diet containing about 3.7 μg Hg/g as methylmercury dicyandiamide. After 12 weeks hatching success was about 10%, compared to about 50 to 55% for controls. In eggs, decreased hatchability was associated with mercury concentrations between 0.5 and 1.5 μg/g wet weight.[342] Borg et al.,[70] who fed breeding pheasants a diet with 15 to 20 μg Hg/g dry weight as methylmercury for 9 days, observed a significant decline in hatching success (55%) relative to controls (74%). Mercury residues in whole eggs associated with this decline ranged from 1.3 to 2.0 μg/g wet weight,[70] in close agreement with the findings of Fimreite.[342]

The diagnosis of methylmercury poisoning in birds based on measured residues of mercury in tissues may be complicated by the influence of other, co-occurring elements that alter the toxicity of methylmercury.[378] In particular, it has been generally believed that selenium protects vertebrate organisms against methylmercury poisoning, even though mercury accumulation in tissues may be increased by selenium.[379]

A notable exception to the presumed protective action of selenium against methylmercury poisoning in wildlife was observed in mallards fed a combination of selenomethionine and methylmercuric chloride.[338] In this experiment, Heinz and Hoffman[338] showed that the toxic effects on the developing bird embryo were much greater when selenium and methylmercury were added jointly to the maternal diet than when methylmercury was added without selenium. In the same experiment, dietary selenium decreased methylmercury toxicity in the adult mallard.[338] Thus, selenium does not seem to protect against reproductive effects of methylmercury.

Few controlled laboratory experiments have been done on the fish-eating birds that are at greatest risk due to methylmercury exposure in the wild. Reproductive experiments with fish-eating birds exposed to dietary methylmercury are urgently needed, given the uncertainties in extending experimental results for laboratory test species to wild, fish-eating birds. Even with laboratory species (such as the mallard, pheasant, and chicken), there is much uncertainty in the threshold concentrations of methylmercury in the maternal diet and in the eggs that elicit reproductive problems. Obtaining information on the reproductive sensitivity of wild, fish-eating birds to methylmercury exposure is perhaps the most pressing research need concerning the avian ecotoxicology of mercury. The combined effects of exposure to methylmercury and other contaminants (particularly selenium and organochlorines), as well as other environmental stressors encountered by wild birds, also merit critical study.

### 16.7.4 Mammals

Mercury intoxication in wild mammals was first reported in association with the widespread use of organomercurial fungicides as seed dressings during the 1950s and 1960s, when individuals of various wild avian and mammalian species, particularly granivores and their predators, were killed from dietary exposure to high concentrations of mercury.[70] Later, methylmercury intoxication was reported as the cause of death of a wild mink (*Mustela vison*) near a mercury-contaminated river[380] and of a wild otter (*Lutra canadensis*) near a lake contaminated with mercury from a chlor-alkali plant.[381] These cases involved outright mortality of adult mammals, probably in response to high dietary exposure to methylmercury. The sources of mercury that caused these past exposures (i.e., emissions from pulp and paper mills and chlor-alkali plants and usage as seed dressings) have largely been discontinued or greatly reduced. As described earlier in this chapter, however, other regionally and globally significant sources of anthropogenic mercury remain, and the methylation of mercury in the environment and biomagnification of methylmercury to high concentrations in food webs continue. Consequently, piscivorous and other top predatory mammals still risk elevated methylmercury exposure in some aquatic environments.

In this section, we review the effects of dietary methylmercury exposure in wild mammals, the *in vivo* demethylation of methylmercury, and interactions between mercury and selenium in mammals. We discuss evidence linking recent methylmercury exposure to toxic effects in certain wild mammal species and in certain environments of North America. For earlier reviews of mercury toxicology in wild mammals, the reader is referred to Wren,[382] Heinz,[376] Thompson,[329] and Wolf et al.[330]

#### *16.7.4.1 Effects of Methylmercury in Mammals, and Critical Concentrations in Tissues and Diets*

Data on tissue concentrations of mercury and methylmercury toxicity are more plentiful for otter and mink than other wild mammals. Combined evidence from wild otter and mink that died after exhibiting signs of methylmercury poisoning[380,381] and from controlled, dietary-dosing experiments on these two species[383–385] indicate that total mercury concentrations in the range of 20 to 100 μg/g wet weight in liver, or > 10 μg/g wet weight in brain, indicate potentially lethal exposure to methylmercury. Reported values for other predatory mammals also fall within these ranges. For example, a fox (*Vulpes vulpes*) that was found staggering and running in circles and that later died had 30 μg/g wet weight of total mercury in its liver and kidneys, and a marten (*Martes martes*) with similar symptoms of methylmercury intoxication had 40 μg Hg/g wet weight in its liver and kidneys.[70] Liver tissue from a Florida panther (*Felis concolor coryi*) suspected of dying of methylmercury poisoning had 110 μg/g wet weight of total mercury.[386] Total mercury concentrations ranging from 37 to 145 μg/g wet weight were found in the livers of feral, domestic cats that died from methylmercury toxicosis.[387] Generally, in the studies cited above only total mercury in tissue was measured, and it was assumed that all or most of the mercury was present as methylmercury, the dominant form to which the animals were exposed. Mammals that die from methylmercury intoxication first exhibit characteristic neurological signs, including some combination of lethargy, weakness, ataxia, paralysis of limbs, tremors, convulsions, and visual impairment.

Chronic exposure to dietary methylmercury concentrations of 1 μg/g wet weight or greater causes neurotoxicity and mortality in adult mink[384,385,388] and otter.[389] Mink die after 3 to 11 months of exposure to 1 μg Hg/g wet weight of methylmercury in the diet.[384–385] Higher dietary concentrations (> 2 μg/g wet weight) hasten the appearance of toxic signs and mortality; however, tissue concentrations of individual mink and otter dying of methylmercury exposure were similar regardless of the methylmercury concentration in the diet, and higher dietary methylmercury concentrations mainly influenced the time required to accumulate toxic tissue concentrations.[388,389] Dietary methylmercury levels of ≤ 0.5 μg Hg/g wet weight are generally not lethal to mink, and consumption of such diets has not caused obvious neurological signs of methylmercury intoxication in mink in controlled feeding experiments.[384,385,390,391]

Comparatively few studies have examined more subtle, sublethal effects of methylmercury in wild mammals. Two studies that examined the effects of dietary methylmercury on reproduction in mink concluded that sublethal exposures (dietary concentrations ≤ 0.5 μg Hg/g) did not adversely affect reproductive variables such as fertility, number of kits born per female, and the survival and growth rates of kits.[385,392] We are unaware of any studies of subtle neurological or neurobehavioral effects of low-level methylmercury exposure in wild mammals such as mink or otter; however, such studies have been done on small mammals used in medical research. Burbacher et al.,[393] who reviewed the medical toxicological literature, concluded that brain-mercury concentrations of 12 to 20 μg/g wet weight during postnatal development are associated with blindness, spasticity, and seizures in small mammals (e.g., rats, mice, and guinea pigs) experimentally exposed to methylmercury; these effects have also been reported in methylmercury-intoxicated mink and otter with similar concentrations in the brain. Lower mercury concentrations (3 to 11 μg/g wet weight) in the brains of small experimental mammals cause behavioral deficits during postnatal development, such as increased activity, poorer maze performance, abnormal auditory startle reflex, impaired escape and avoidance behavior, abnormal visual evoked potentials, and abnormal performance on learning tasks.[393] Similarly, Wobeser et al.[388] concluded that mercury concentrations exceeding 5 μg/g wet weight in the brain, when combined with neurological signs, were consistent with methylmercury toxicity in mink. Mercury concentrations in the brains of free-living otter and mink trapped in Wisconsin and in Manitoba, Ontario, and Quebec were generally in the range of 0.1 to 1.0 μg/g wet weight, although some individuals had 5 to 10 μg/g wet weight.[260,394–397] Results from medical toxicological studies with small mammals indicate that such concentrations in the brain may cause subtle visual, cognitive, or neurobehavioral deficits. Impaired vision and learning ability could be life-threatening to wild, visual predators, given that such dysfunctions could significantly impair ability to catch prey, causing malnutrition, increased susceptibility to disease, or reduced reproductive success.

#### *16.7.4.2 Demethylation and Relationship with Selenium*

Information on concentrations of total mercury in certain commonly analyzed tissues, such as the liver, is not sufficient for diagnosing methylmercury toxicity in wild mammals. Methylmercury is the primary and most toxic form of mercury in the diets of piscivores and other top mammalian predators that are associated with aquatic food webs. However, some wild mammals can demethylate methylmercury to varying degrees, and inorganic mercury often accounts for a significant and highly variable fraction of the total mercury present in the liver, kidney, and brain in such species.[260,389,396] Mammals have also been shown to demethylate methylmercury in controlled experiments. In guinea pigs (*Cavia porcellus*), for example, inorganic mercury accounted for 30 and 60% of total mercury in the liver and kidneys, respectively, after 3 weeks of administration of methylmercury.[398]

The inorganic mercury produced by *in vivo* demethylation of methylmercury can gradually accumulate to very high concentrations in association with selenium in certain tissues without causing any apparent toxicity. Thus, liver-mercury concentrations that would probably be toxic if composed mainly of methylmercury (> 20 μg/g wet weight) may not be toxic if present primarily as Hg-Se complexes. Methylmercury generally predominates when total mercury concentrations in livers of predatory mammals are less than ~9 μg/g wet weight; however, with greater mercury accumulation, an increasingly high proportion of total mercury is often present as inorganic mercury. In marine mammals and aquatic birds that have accumulated high concentrations of mercury in the liver (10 to > 30 μg/g wet weight), more than 85% of total mercury is typically present as inorganic mercury associated with selenium in a molar ratio approximating 1:1.[263,399] Thus, observations of elevated concentrations of total mercury in the liver or kidneys of dead mammals are not sufficient for diagnosing methylmercury intoxication because of demethylation and the subsequent formation of Hg-Se complexes that have relatively low toxicity. Ideally, such diagnoses should be based on information on total mercury, methylmercury, and selenium in the liver, kidneys, and brain. Determination of total mercury in skeletal muscle would also be useful in such assessments because very little demethylation occurs in muscle tissue. In the absence of clinical signs of methylmercury intoxication (e.g., for animals found dead), a mercury concentration exceeding 20 μg/g wet weight in the liver, combined with a concentration exceeding 12 μg/g wet weight in muscle, indicates probable methylmercury intoxication.[383,388,389] Conversely, the same or higher mercury concentration in the liver, in conjunction with low mercury concentrations in muscle, would not be indicative of methylmercury intoxication.

The antagonistic relation between mercury and selenium in biological systems is well known;[379] however, the biochemical mechanisms underlying this antagonism are poorly understood. Studies with rats have demonstrated that selenium protects against or delays the toxicity of methylmercury. Animals that received co-administration of selenium salts with methylmercury had lower mortality, fewer neurological signs of methylmercury intoxication, and better growth rate and weight gain than animals given only methylmercury.[400–404]

In wild mammals, the association between mercury and selenium has been most intensively studied in dolphins. Rawson et al.[405] described pigment granules containing high concentrations of mercury in lysosomes of liver cells of Atlantic bottlenose dolphins (*Tursiops truncates*). All animals with mercury-containing granules had concentrations of total mercury in the liver exceeding 61 μg/g wet weight, whereas animals without pigment had concentrations of less than 50 μg/g. Rawson et al.[405] did not suggest that these mercury-containing pigments were composed of Hg-Se-protein complexes; however, other studies have demonstrated the presence of such compounds, both as Hg-Se-protein complexes and as insoluble HgSe (tiemannite) granules, in liver cells of dolphins.[406–409] As reported for other mammals with high mercury accumulation, dolphins with the highest concentrations of total mercury in the liver (> 100 μg/g wet weight) typically have the lowest percentages of methylmercury (<10 percent of total mercury) and also have high selenium concentrations in the liver.[410] These Hg-Se compounds, which are much less toxic than methylmercury, are very stable and have a long biological half-life, accumulating to high concentrations in older individuals.[405,411] Low concentrations of mercury in the liver of dolphins, and perhaps other marine mammals, are present primarily as methylmercury; however, there is an apparent threshold above which the speciation of mercury is altered by *in vivo* demethylation and concurrent accumulation of selenium with mercury in stable, insoluble complexes.[409] Such Hg-Se complexes have apparently not been reported or studied in otter or mink.

Evans et al.[260] measured concentrations of total mercury and methylmercury in the brain, kidney, liver, and fur of apparently healthy wild otter and mink from Ontario, Canada and reported a greater percent of inorganic mercury in otter than in the same tissues in mink. Mink and otter had comparable levels of total mercury (e.g., in liver, 0.85 to 3.5 μg/g wet weight in mink and 0.87 to 2.3 μg/g in otter), but the mercury in soft tissues of mink was from 80 to 90% methylmercury, whereas the methylmercury fraction ranged from 56 to 81% in otter, leading Evans et al.[260] to suggest that otter are more able than mink to demethylate methylmercury. Wren et al.[396] also reported higher proportions of methylmercury in liver tissue of otter, relative to mink, in animals collected throughout Ontario. Concentrations of mercury and selenium were not correlated in the livers of Ontario otter.[396]

#### 16.7.4.3 Hazard Assessment Studies

Environmental mercury exposure has occasionally been implicated as a possible contributing factor to population declines in mink and otter. Osowski et al.,[412] who noted that mink were completely absent in parts of the Atlantic coastal plain of Georgia, North Carolina, and South Carolina (southeastern United States), where they were historically abundant and where high-quality habitat remained available, assessed the potential role of 17 environmental contaminants in population declines of mink. Mercury (along with PCBs, DDE, and dieldrin) emerged as a potentially important contaminant influencing mink populations. Concentrations of total mercury in kidneys of mink from areas of concern were as high as 25 μg/g wet weight, compared to < 4 μg/g wet weight in reference areas.[412] Notably, the higher mercury levels reported in this study were toxic to mink in controlled feeding experiments.[388]

Giesy et al.[413] conducted a hazard assessment of mercury and other contaminants in mink above and below hydroelectric dams on three rivers flowing into the North American Great Lakes. Based on an assumed dietary no-observed-adverse-effect concentration of 0.05 μg Hg/g wet weight for mink, the hazard assessment concluded that the calculated hazard indices were not high, although mercury levels in fish upstream from the dams were higher than those downstream. Mercury was deemed to be less important than PCBs as a factor potentially affecting Great Lakes mink populations, and Giesy et al.[413] concluded that concentrations of mercury in fish in the rivers examined were probably too low to cause any population-level effect on mink.

Two studies have attempted to assess the risk of adverse effects in otter and mink from environmental mercury and PCB contamination in the Clinch River and Poplar Creek watersheds in Tennessee.[414,415] Mercury in small fish collected in areas of concern averaged about 0.2 to 0.4 μg/g wet weight.[415] Results of Monte Carlo simulations estimating total daily intakes of mercury, integrated with dose-response curves to estimate risks, led Moore et al.[415] to conclude that dietary methylmercury exposure posed a moderate risk to female mink (a 24% probability of at least 15% mortality) within the areas of concern. Similarly, Sample and Suter[414] concluded that there was an 85% probability that mercury exposure in otter in some of the affected sites exceeded the estimated lowest observed adverse effect level for methylmercury.

## 16.8 DEGRADATION OF ECOSYSTEM GOODS AND SERVICES

As a "good" produced by aquatic ecosystems, fish are a high-quality food resource — high in protein, low in saturated fat, and a source of beneficial omega-3 polyunsaturated fatty acids and antioxidants such as vitamin E.[416] Mercury contamination has clearly diminished the economic, nutritional, and cultural values of the fishery resources produced by many freshwater and marine ecosystems.[417–420] The growing awareness of the mercury problem has prompted increasing efforts to survey mercury contamination of fish, producing information that has, in turn, prompted issuance of additional advisories concerning the consumption of sport fish. In Canada, mercury contamination accounted for more than 97% (2572) of all fish-consumption advisories listed in 1997.[421] Most of the Canadian advisories pertained to surface waters in Quebec and Ontario, whereas New Brunswick and Nova Scotia had province-wide advisories in effect for mercury. In the United States, methylmercury contamination accounted for 79% of all fish- and wildlife-consumption advisories in 2000.[421] Out of 50 states, 41 had advisories attributed to mercury, and the number of statewide fish-consumption advisories issued for lakes, rivers, or coastal waters increased substantially during 1993–2000 (Figure 16.4). In Sweden, an estimated 40,000 of the country's 83,000 lakes contained 1-kg northern pike with mercury concentrations higher than the national guideline of 0.5 μg/g wet weight, and an

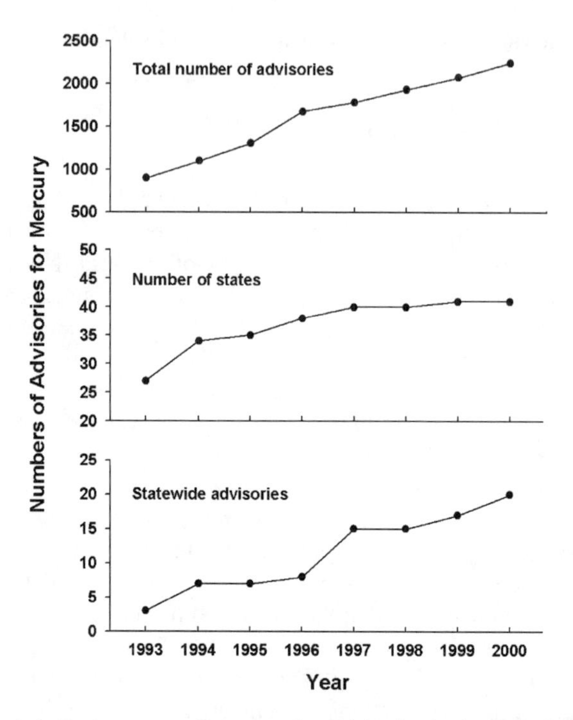


**Figure 16.4** Recent trends in the issuance of consumption advisories in the United States caused by mercury in fish and wildlife. Shown are the total number of advisories for mercury (top panel), the number of states issuing advisories (center), and the total number of statewide advisories for surface waters of a given type (bottom), such as coastal waters, lakes, and rivers. Data reflect the status of advisories through the year 2000.[421] In 2001, the states of Pennsylvania and Maryland each issued a statewide advisory (not included above) for mercury in fish.

estimated 10,000 of these lakes contained 1-kg pike with mercury concentrations exceeding the "blacklisting" limit of 1.0 µg/g.[418] In Uppsala County, Sweden, pollution by toxic substances (specifically mercury and cesium-137) ranked as the second greatest anthropogenic threat to lakes.[422]

The economic losses caused by mercury contamination of fishery resources are largely unknown, having been estimated for only a few cases.[417] The adverse impacts of mercury pollution on some indigenous, or aboriginal, peoples who relied on aquatic ecosystems for subsistence via fishing and hunting are multidimensional, encompassing cultural, social, and health effects as well as economic consequences.[419,420,423] For some of these peoples, the consequences of abandoning subsistence fishing and switching to alternative diets, combined with the social and cultural effects of a disrupted lifestyle, have presented a much more severe overall problem than the direct, clinical effects of exposure to methylmercury via consumption of contaminated fish or wildlife.[419,420,423]

The presence of inorganic mercury on the landscape, whether from anthropogenic or natural sources, can devalue some of the services performed by aquatic ecosystems. Much of the inorganic Hg(II) present on vegetated terrestrial areas inundated to create reservoirs for hydropower development, for example, is methylated after flooding and enters the food chain, accumulating to high concentrations in fish in the reservoirs and riverine reaches downstream.[84,85,100,240,242,424–427] In a new reservoir, the inundation of vegetated landscapes is rapidly followed by a rapid transition to anaerobic conditions caused by rapid decomposition of inundated organic matter and the associated depletion of dissolved oxygen near the soil-water interface. After this redox shift, anaerobic microbial communities (including sulfate reducers) proliferate, and the net rate of methylmercury production increases rapidly.[96] In the first 9 years after creation of the La Grande 2 reservoir (part of the La Grande hydroelectric complex in the Canadian province of Québec), concentrations of mercury increased from ~0.6 to 3.0 µg/g wet weight in 70-cm northern pike and from ~0.7 to

2.8 µg/g in 40-cm walleye.[84] In some cases, mercury levels in fish from recently flooded reservoirs have equaled or exceeded levels in fish from surface waters that were heavily contaminated by direct industrial discharges.[242,428] Moreover, the mercury concentrations in the fishery resources of new impoundments may remain substantially elevated for decades after flooding.[84,242,426,429]

The ecosystem services performed by natural and constructed wetlands can also be devalued by mercury. Freshwater wetlands improve water quality by retaining or partially removing many of the constituents from water passing through the wetland system, including suspended solids, bacteria, biological oxygen demand, chemical oxygen demand, certain heavy metals, phosphorus, and nitrogen.[430–433] Constructed wetlands have been widely used for water treatment and are effective at improving water quality in a variety of situations.[432,433] However, wetlands are also quantitatively important on the landscape as sites of methylmercury production, and they can be important sources of methylmercury for downstream surface waters.[68,100,123–125,133,179,183,196,205,434]

The production and export of methylmercury, particularly in wetland systems where the abundance of mercury has been significantly increased by local anthropogenic sources or atmospheric deposition, can impair the water-treatment function and degrade the biological resources of wetlands and downstream waters. In riparian wetlands adjoining the Sudbury River in Massachusetts, for example, contaminated overbank sediments on the floodplain remained an active source of methylmercury for aquatic biota and downstream riverine reaches long after the primary anthropogenic source of mercury was controlled.[60,67–69] In parts of the Florida Everglades, a nationally renowned wetland ecosystem receiving anthropogenic mercury in atmospheric deposition,[435–437] the rates of methylmercury production are very rapid.[97] The fish and wildlife resources in areas of the Everglades having high methylation rates contain high levels of methylmercury,[89,438–441] diminishing the value of the fishery and posing a threat to wildlife in upper trophic levels.[438,442–444]

## 16.9 MERCURY POLLUTION — A CONTINUING SCIENTIFIC CHALLENGE

Environmental mercury research remains an area of substantive scientific progress and discovery. Virtually hundreds of studies have examined sources, environmental transport, biogeochemical transformations and cycling, bioaccumulation, and biological effects of the metal since global attention first focused on environmental mercury pollution. Amazingly, a variety of recent landmark discoveries indicate that environmental mercury research can be characterized as a relatively "young" field of scientific endeavor.

Examples of recent prominent advances in mercury research include, but certainly are not limited to, the following: (1) the discovery of mercury-sensitive ecosystems characterized by small inventories of total mercury, high rates of mercury methylation, and high concentrations of methylmercury in piscivorous fish and wildlife, (2) the observation that perturbations of the landscape by humans or natural processes (e.g., reservoir creation and prolonged flooding) can markedly increase methylmercury production and contamination of aquatic food webs, (3) the discovery of wetlands as important sites of mercury methylation and export on the landscape, (4) the observation of significant photodegradation of methylmercury in some surface waters, (5) the observation that selenium can worsen — rather than protect against — effects of methylmercury on avian reproduction and developing young, (6) the discovery of highly reactive, gaseous forms of Hg(II) that are rapidly removed from the atmosphere in wet and dry deposition, and (7) observation of the rapid conversion after polar sunrise of gaseous $Hg^0$ in the polar atmospheres to reactive gaseous Hg(II).

Despite this impressive progress, many significant questions remain concerning the exposure and ecotoxicological effects of mercury in the environment. Our collective appraisal is that progress in assessing the biological and ecotoxicological *effects* of methylmercury exposure is lagging far behind progress in understanding biogeochemical processes and environmental factors that influence biological *exposure* to methylmercury. In particular, the following topics pertaining to the ecotoxicology of mercury in biota atop aquatic food webs merit intensive scientific study.

- *Critical examination of the reproductive effects of methylmercury on fish, birds, and mammals:* Which, if any, species and populations are being affected by dietary exposure to methylmercury? Is the vertebrate embryo the weak link in reproductive effects in exposed populations, or are other reproductive endpoints (such as territorial fidelity, courtship and other reproductive behaviors, gonadal development, and spawning success) more sensitive?
- *Variation among taxa of piscivorous fish, birds, and mammals in reproductive and early-life sensitivity to methylmercury:* Are fish-eating marine mammals and seabirds protected against the adverse effects of methylmercury exposure, given their apparent abilities to demethylate methylmercury and subsequently store or eliminate the resulting inorganic mercury? Are fish-eating mammals less vulnerable to methylmercury than fish-eating birds?
- *The combined effects of methylmercury and other co-occurring environmental stressors:* How does mercury interact, in an ecotoxicological sense, with other persistent toxic contaminants that bioaccumulate and biomagnify in food webs?

The modern environmental mercury problem, characterized by large geographic scale and enormous complexity, remains a serious challenge to environmental managers and scientists alike. In an ecotoxicological sense, the modern mercury problem can be viewed rather simply as *biotic exposure to methylmercury.* Given this view, the extraordinary challenge facing scientists and environmental managers is to identify approaches that can decrease biotic exposure to methylmercury. Clearly, a sustained, interdisciplinary effort will be needed to address critical questions concerning methylmercury exposure and the associated risks to wildlife and humans dependent on aquatic food webs.

## 16.10 SUMMARY

Growing awareness of the hazards of mercury exposure prompted widespread reductions in usage and discharges of the metal to many surface waters beginning in the late 1960s. Mercury concentrations in fish and other aquatic biota typically declined in the years and decades following reductions in discharges to industrially polluted waters, although contamination at some sites has decreased slowly and high concentrations persist in fish. Contaminated tailings and alluvium from mining operations are widespread and can remain a source of mercury emissions for decades or centuries. In some basins, contaminated sediment from historic mining sites has been transported to aquatic and floodplain habitats far downstream. Gold mining with the mercury-amalgamation process has resurged in South America, Southeast Asia, China, and Africa in widely dispersed operations that may contribute 10% of modern anthropogenic emissions worldwide.

Since the late 1970s, unexpectedly high concentrations of mercury have been observed in fish from waters lacking on-site anthropogenic or geologic sources of mercury, including low-alkalinity and humic lakes, wetlands, surface waters with adjoining wetlands, waters with adjoining or upstream areas subjected to inundation, and dark-water streams. We classify such systems as mercury-sensitive because seemingly small inputs or inventories of total mercury can cause significant methylmercury contamination in fish and wildlife in upper trophic levels. A common attribute of mercury-sensitive ecosystems is the efficient conversion of inorganic mercury to methylmercury. Concentrations of methylmercury in fish in mercury-sensitive ecosystems can equal or exceed those in fish from industrially contaminated waters.

The modern mercury problem is greatest in aquatic environments, where inorganic Hg(II) can be methylated to methylmercury, the highly toxic form that readily bioaccumulates in exposed organisms and biomagnifies to high concentrations in food webs. Methylation by sulfate-reducing bacteria at oxic-anoxic interfaces in sediments and wetlands is probably the dominant methylation pathway in the environment. Demethylation, or degradation of methylmercury, can occur via a number of abiotic and biotic pathways in the environment, and photodemethylation can significantly affect methylmercury budgets of lakes.

Emissions from anthropogenic sources and long-range atmospheric transport of mercury have contaminated terrestrial and aquatic environments on a global scale. Analyses of sediment, peat, and glacial ice show that the rate of mercury accumulation at semiremote and remote sites has increased substantially (often by two- to fourfold) since the mid-1800s or early 1900s. It can be reasonably inferred that a significant fraction of the methylmercury in aquatic biota in remote or semiremote waters, including marine systems, is derived from anthropogenic mercury in deposition. In the North Atlantic, for example, concentrations of methylmercury in fish-eating seabirds (and their supporting food webs) increased substantially from 1885 through 1994.

Four compartments — atmospheric, terrestrial, aquatic, and biotic — are interconnected in the global mercury cycle. The atmosphere is dominated by gaseous elemental mercury ($Hg^0$), but the fluxes between the atmosphere and both aquatic and terrestrial compartments are dominated by Hg(II). The terrestrial compartment is dominated by Hg(II) in soils, the aquatic compartment by Hg(II)-ligand pairs in water and Hg(II) in sediments, and the biotic compartment by methylmercury. Mercury is reactive and moves readily between compartments. Atmospheric processes and pathways dominate global-scale transport from sources to receptors. The global cycle can be envisioned as a two-way exchange, in which sources emit $Hg^0$ and various species of Hg(II) to the atmosphere and the atmosphere loses mercury via oxidation of $Hg^0$ to Hg(II) and the rapid removal of gaseous and particulate species of Hg(II) in wet and dry deposition. Mercury deposited onto the land is sequestered in soils, largely as Hg(II) sorbed to organic matter in the humus layer. The global inventory of mercury in soils greatly exceeds that in the aquatic and atmospheric compartments. The Hg(II) in soils can be reduced and emitted to the atmosphere as $Hg^0$ or slowly transported down gradient; thus, soils are both a sink and a potential long-term source of mercury. Anthropogenic emissions, particularly since the industrial revolution, have greatly increased the size of cycling mercury pools and the importance of atmospheric pathways to a global pollution problem, increasing the abundance of mercury in the atmosphere, soil, sediment, and biota. Natural emissions of $Hg^0$ can be substantial in areas that are geologically enriched with mercury.

The aqueous abundances of methylmercury and total mercury vary widely. Total mercury in water (unfiltered) ranges 0.3 to 8 ng/L in aquatic systems lacking substantive, on-site anthropogenic or geologic sources, 10 to 40 ng/L in systems influenced by mercury mining or industrial pollution, and can exceed 100 or even 1000 ng/L in systems draining areas with substantive geologic sources or contaminated mine tailings. Methylmercury generally accounts for about 0.1 to 5% of the total mercury in oxic surface water, seldom exceeding 10%. In oxic waters, concentrations of methylmercury are typically in the range of 0.04 to 0.8 ng Hg/L, but can be 1 to 2 ng Hg/L in systems affected by industrial pollution or mercury mine drainage. Methylmercury can be the dominant species under anoxic conditions, and concentrations can exceed 5 ng Hg/L.

Methylmercury readily crosses biological membranes and accumulates in aquatic organisms to concentrations that vastly exceed those in water. Patterns of biomagnification of methylmercury in food webs are similar, even among aquatic systems that differ in ecosystem type, mercury source, and intensity of pollution. The entry of methylmercury into the food web and its concentrations in lower trophic levels are influenced strongly by the supply from methylating environments. Methylmercury concentration increases up the food web from water and lower trophic levels to fish and piscivores. The ratio of methylmercury to total mercury increases with ascending trophic level through fish and can vary greatly in trophic levels below fish. In pelagic food webs, the greatest increase in methylmercury concentration, relative to that in water, occurs between seston and water. Fish accumulate methylmercury mostly via the diet, to concentrations that commonly exceed those in water by $10^6$- to $10^7$-fold. Inorganic mercury, in contrast to methylmercury, is not readily transferred in food webs and does not biomagnify.

In fish, concentrations of methylmercury increase with increasing trophic position and with increasing age or size, because the rate of elimination is very slow relative to uptake. Much of the methylmercury in fish is eventually stored in skeletal muscle, bound to sulfhydryl groups in protein; this may serve as a protective mechanism, given that sequestration in muscle reduces exposure of the central nervous system to methylmercury. In the laboratory, long-term dietary exposure of fish to methylmercury causes incoordination, diminished appetite or inability to feed, diminished responsiveness and swimming activity, starvation, and mortality. Fish inhabiting Minamata Bay (Japan) and Clay Lake (Ontario) — both extreme cases of industrial pollution — had very high mercury concentrations and exhibited multiple symptoms of methylmercury intoxication. In laboratory studies, sublethal exposure to waterborne methylmercury can impair the ability of test fish to locate, capture, and ingest prey and to avoid predators. Methylmercury can impair reproduction by affecting gonadal development or spawning success of adult fish, or by reducing the hatching success of eggs and the health and survival of embryolarval stages. Recent experiments have shown diminished foraging efficiency, reproductive success, health, and fitness in fish exposed to environmentally realistic concentrations of methylmercury, indicating that some fish populations may be adversely affected by existing exposure levels.

The sources of mercury responsible for reported deaths of wild birds and mammals (usage in seed grain and emissions from pulp and paper mills and chlor-alkali plants) have been greatly reduced or largely discontinued, yet methylmercury remains a threat to wildlife in upper trophic levels in many aquatic ecosystems. The present pathway of exposure to methylmercury is largely an aquatic one, and reptiles, birds, and mammals atop aquatic food webs often bioaccumulate high concentrations of methylmercury — even at semiremote or remote sites. The principal routes of elimination in birds and mammals are incorporation of methylmercury into growing feathers or hair and excretion in the feces. Some fish-eating birds (particularly seabirds) and mammals can demethylate methylmercury, and inorganic mercury can account for a significant fraction of the total mercury present in the liver, kidney, and brain in such species.

In laboratory experiments with birds and mammals, methylmercury adversely affects survival, reproduction, behavior, and cellular development and causes teratogenic effects. Animals dying from methylmercury intoxication exhibit characteristic neurological signs, including some combination of lethargy, weakness, ataxia, paralysis of limbs, tremors, convulsions, and impaired vision. In adult mink and otter, chronic exposure to dietary methylmercury of 1 μg/g wet weight or greater causes neurotoxicity and mortality. Mortality and impaired reproduction, two endpoints observed in laboratory experiments, could affect populations of birds and mammals exposed to high levels of methylmercury. In field studies, exposure to high levels of methylmercury was a suspected cause of ill health, emaciation, and mortality in studies of wild white herons, grey herons, and common loons. In these cases, the observed mortality may have resulted from the combined effects of sublethal methylmercury exposure and other stressors. In field studies, it is often difficult to isolate the biological effects of methylmercury exposure from those of co-occurring toxic contaminants or other stressors.

In birds and mammals, early life stages are much more sensitive than the adult to methylmercury. Avian reproduction, for example, is significantly impaired at (maternal) dietary concentrations that are only one fifth of those that produce overt toxicity in the adult. In field studies, impaired reproduction has been associated with high mercury exposure in piscivorous birds, including wild merlins, common loons, wood storks, and common terns. In common loons, reductions in egg laying and territorial fidelity have been associated with mercury concentrations averaging 0.3–0.4 μg/g wet weight in prey organisms, and many lakes in the breeding range of this species contain prey-size fish with concentrations equaling or exceeding this estimated threshold value. Reproductive experiments with fish-eating birds exposed to dietary methylmercury are urgently needed, given the uncertainties in extrapolating results from laboratory test species, such as the mallard, to wild piscivorous birds.

Few studies have examined subtle, sublethal effects of methylmercury in wild, piscivorous mammals. Yet medical studies with small mammals have shown that sublethal exposure to methylmercury can cause subtle visual, cognitive, or neurobehavioral deficits that could indirectly affect the survival and reproductive success of visual predators in the wild.

The antagonistic relation between mercury and selenium is well known in toxicology, but the biochemical mechanisms are poorly understood. In dolphins, there is an apparent threshold concentration in the liver above which the speciation of mercury is altered by *in vivo* demethylation followed by storage with selenium in insoluble, stable Hg-Se complexes. Dietary exposure of mallards to selenomethionine and methylmercuric chloride, separately and in combination, showed that selenium decreased methylmercury toxicity in adults. Yet in the same experiment, the adverse effects on the developing mallard embryo were much greater when selenomethionine and methylmercury were added jointly to the maternal diet than when only methylmercury was added.

Mercury contamination is adversely affecting a number of ecosystem goods and services. The most notable of these is the degradation of fishery resources that have substantial economic, nutritional, and cultural value. The economic losses caused by the widespread contamination of fishery resources have apparently not been estimated. For some aboriginal communities in Canada that once relied on subsistence fishing, the impacts of mercury pollution have encompassed adverse cultural, social, and health effects, as well as economic impacts. For some communities, abandonment of subsistence fishing and a change to less healthy diets, combined with the social and cultural effects of disrupted lifestyle, have presented a more severe overall problem than the direct, clinical effects of exposure to methylmercury via consumption of contaminated fish.

Mercury on the landscape can also devalue services performed by aquatic ecosystems. Much of the inorganic Hg(II) on land inundated to create reservoirs for hydropower is methylated after flooding to methylmercury, which rapidly enters the food web and substantively contaminates fish in the reservoirs and reaches downstream for decades. Mercury also devalues the ecosystem services performed by wetlands, which improve the quality of water passing through the wetland system by partial removal of nutrients and certain other constituents. Wetlands are also sites of mercury methylation, and the production and export of methylmercury can impair their water-treatment function and degrade the biological resources in wetlands and downstream waters.

Environmental mercury research is an area of substantive scientific discovery that can be characterized as a relatively young field of scientific endeavor. Our collective appraisal is that progress in assessing the biological and ecotoxicological *effects* of methylmercury exposure is lagging far behind the recent, impressive advances in our understanding of biogeochemical processes and environmental factors that influence biological *exposure* to methylmercury.

In particular, many questions concerning the ecotoxicology of mercury need to be addressed before the population-level effects of methylmercury exposure in fish and wildlife species atop aquatic food webs are fully understood. The modern environmental mercury problem, large in geographic scale and enormous in complexity, remains a daunting challenge for environmental scientists and managers alike.

## ACKNOWLEDGMENTS

The lead author (JGW) gratefully acknowledges the support provided by the College of Science and Allied Health and the UWL Foundation at the University of Wisconsin-La Crosse during preparation of this chapter. We thank Jeffrey Ziegeweid for assistance with preparation of the figures and the bibliography. Dr. Kristofer Rolfhus and Dr. Mark Sandheinrich provided constructive comments on a draft of the manuscript.

## REFERENCES

1. Wren, C. D., Harris, S., and Harttrup, N., Ecotoxicology of mercury and cadmium, in *Handbook of Ecotoxicology*, Hoffman, D. J., Rattner, B. A., Burton, G. A., Jr., and Cairns, J., Jr., Eds., CRC Press, Boca Raton, FL, 1995, 392–423.

2. Bloom, N. S., On the chemical form of mercury in edible fish and marine invertebrate tissue, *Can. J. Fish. Aquat. Sci.*, 49, 1010–1017, 1992.
3. Kim, J. P., Methylmercury in rainbow trout (*Oncorhynchus mykiss*) from lakes Okareka, Okaro, Rotomahana, Rotorua, and Tarawera, North Island, New Zealand, *Sci. Total Environ.*, 164, 209–219, 1995.
4. Wagemann, R., Trebacz, E., Boila, G., and Lockhart, W. L., Methylmercury and total mercury in tissues of arctic marine mammals, *Sci. Total Environ.*, 218, 19–31, 1998.
5. Tsubaki, T. and Irukayama, K., Eds., *Minamata Disease: Methylmercury Poisoning in Minamata and Niigata, Japan*, Elsevier, Amsterdam, 1977.
6. Clarkson, T. W., Mercury: Major issues in environmental health, *Environ. Health Perspect.*, 100, 31–38, 1992.
7. Gilbert, S. G. and Grant-Webster, K. S., Neurobehavioral effects of developmental methylmercury exposure, *Environ. Health Perspect.*, 103 (Suppl. 6), 135–142, 1995.
8. Hamada, R. and Osame, M., Minamata disease and other mercury syndromes, in *Toxicology of Metals*, Chang, L. W., Magos, L., and Suzuki, T., Eds., Lewis Publishers, Boca Raton, FL, 1996, 337–351.
9. Watanabe, C. and Satoh, H., Evolution of our understanding of methylmercury as a health threat, *Environ. Health Perspect.*, 104 (Suppl. 2), 367–379, 1996.
10. United States Environmental Protection Agency, Mercury Study Report to Congress, USEPA Publ. 452R-97–004, Washington, D.C., 1997.
11. Myers, G. J. and Davidson, P. W., Prenatal methylmercury exposure and children: Neurologic, developmental, and behavioral research, *Environ. Health Perspect.*, 106 (Suppl. 3), 841–847, 1998.
12. National Research Council Committee on the Toxicological Effects of Methylmercury, *Toxicological Effects of Methylmercury*, Nat. Acad. Press, Washington, D.C., 2000.
13. Mahaffey, K. R., Recent advances in recognition of low-level methylmercury poisoning, *Curr. Opinion Neurol.*, 13, 699–707, 2000.
14. Takizawa, Y., Minamata disease in retrospect, *World Resour. Rev.*, 12, 211–223, 2000.
15. Harada, M., Nakanishi, J., Yasoda, E., Pinheiro, M. D. N., Oikawa, T., Guimaraes, G. D., Cardoso, B. D., Kizaki, T., and Ohno, H., Mercury pollution in the Tapajos River basin, Amazon: Mercury level of head hair and health effects, *Environ. Int.*, 27, 285–290, 2001.
16. Clarkson, T. W., The three modern faces of mercury, *Environ. Health Perspect.*, 110 (Suppl. 1), 11–23, 2002.
17. Richardson, M., Mitchell, M., Coad, S., and Raphael, R., Exposure to mercury in Canada: A multimedia analysis, *Water Air Soil Pollut.*, 80, 21–30, 1995.
18. Jasinski, S. M., The materials flow of mercury in the United States, *Resour. Conserv. Recycling*, 15, 145–179, 1995.
19. Sunderland, E. M. and Chmura, G. L., An inventory of historical mercury emissions in Maritime Canada: Implications for present and future contamination, *Sci. Total Environ.*, 256, 39–57, 2000.
20. Fitzgerald, W. F., Engstrom, D. R., Mason, R. P., and Nater, E. A., The case for atmospheric mercury contamination in remote areas, *Environ. Sci. Technol.*, 32, 1–7, 1998.
21. Ebinghaus, R., Tripathi, R. M., Wallschläger, D., and Lindberg, S. E., Natural and anthropogenic mercury sources and their impact on the air-surface exchange of mercury on regional and global scales, in *Mercury Contaminated Sites*, Ebinghaus, R., Turner, R. R., Lacerda, L. D., Vasiliev, O., and Salomons, W., Eds., Springer, Berlin, 1999, 3–50.
22. Ferrara, R., Mercury mines in Europe: Assessment of emissions and environmental contamination, in *Mercury Contaminated Sites*, Ebinghaus, R., Turner, R.R., Lacerda, L.D., Vasiliev, O., and Salomons, W., Eds., Springer, Berlin, 1999, 51–72.
23. Martínez-Cortizas, A., Pontevedra-Pombal, X., Garcia-Rodeja, E., Nóvoa-Muñoz, J. C., and Shotyk, W., Mercury in a Spanish peat bog: Archive of climate change and atmospheric metal deposition, *Science*, 284, 939–942, 1999.
24. Gosar, M., Pirc, S., and Bidovec, M., Mercury in the Idrijca River sediments as a reflection of mining and smelting activities of the Idrija mercury mine, *J. Geochem. Explor.*, 58, 125–131, 1997.
25. Turner, R. R. and Southworth, G. W., Mercury-contaminated industrial and mining sites in North America: An overview with selected case studies, in *Mercury Contaminated Sites*, Ebinghaus, R., Turner, R. R., Lacerda, L. D., Vasiliev, O., and Salomons, W., Eds., Springer, Berlin, 1999, 89–112.
26. Ganguli, P. M., Mason, R. P., Abu-Saba, K. E., Anderson, R. S., and Flegal, A. R., Mercury speciation in drainage from the New Idria mercury mine, California, *Environ. Sci. Technol.*, 34, 4773–4779, 2000.

27. Hines, M. E., Horvat, M., Faganeli, J., Bonzongo, J. C. J., Barkay, T., Major, E. B., Scott, K. J., Bailey, E. A., Warwick, J. J., and Lyons, W. B., Mercury biogeochemistry in the Idrija River, Slovenia, from above the mine into the Gulf of Trieste, *Environ. Res.* (*Sect. A*), 83, 129–139, 2000.
28. Rytuba, J. J., Mercury mine drainage and processes that control its environmental impact, *Sci. Total Environ.*, 260, 57–71, 2000.
29. Trip, L. and Allan, R. J., Sources, trends, implications and remediation of mercury contamination of lakes in remote areas of Canada, *Water Sci. Technol.*, 42, 171–176, 2000.
30. Covelli, S., Faganeli, J., Horvat, M., and Brambati, A., Mercury contamination of coastal sediments as the result of long-term cinnabar mining activity (Gulf of Trieste, northern Adriatic Sea), *Appl. Geochem.*, 16, 541–558, 2001.
31. Lacerda, L. D., Global mercury emissions from gold and silver mining, *Water Air Soil Pollut.*, 97, 209–221, 1997.
32. Averill, C. V., Placer Mining for Gold in California, *Calif. State Div. Mines Geol. Bull.*, 135, 1946.
33. Domagalski, J., Occurrence and transport of total mercury and methyl mercury in the Sacramento River Basin, California, *J. Geochem. Explor.*, 64, 277–291, 1998.
34. Alpers, C. N. and Hunerlach, M. P., Mercury contamination from historic gold mining in California, U.S. Geol. Surv., Fact Sheet FS-061–00, Sacramento, CA, 2000.
35. Leigh, D. S., Mercury-tainted overbank sediment from past gold mining in north Georgia, USA, *Environ. Geol.*, 30, 244–251, 1997.
36. Miller, J. R., Lechler, P. J., and Desilets, M., The role of geomorphic processes in the transport and fate of mercury in the Carson River basin, west-central Nevada, *Environ. Geol.*, 33, 249–262, 1998.
37. Lacerda, L. D. and Salomons, W., Mercury contamination from New World gold and silver mine tailings, in *Mercury Contaminated Sites*, Ebinghaus, R., Turner, R. R., Lacerda, L. D., Vasiliev, O., and Salomons, W., Eds., Springer, Berlin, 1999, 73–87.
38. Wayne, D. M., Warwick, J. J., Lechler, P. J., Gill, G. A., and Lyons, W. B., Mercury contamination in the Carson River, Nevada: A preliminary study of the impact of mining wastes, *Water Air Soil Pollut.*, 92, 391–408, 1996.
39. Artaxo, P., Campos, R. C., Fernandes, E. T., Martins, J. V., Xiao, Z., Lindqvist, O., Fernandez-Jimenez, M. T., and Maenhaut, W., Large scale mercury and trace element measurements in the Amazon basin, *Atmos. Environ.*, 34, 4085–4096, 2000.
40. Blum, M., Gustin, M. S., Swanson, S., and Donaldson, S. G., Mercury in water and sediment of Steamboat Creek, Nevada: Implications for stream restoration, *J. Am. Water Resour. Assoc.*, 37, 795–804, 2001.
41. Domagalski, J., Mercury and methylmercury in water and sediment of the Sacramento River Basin, California, *Appl. Geochem.*, 16, 1677–1691, 2001.
42. Engle, M. A., Gustin, M. S., and Hong Zhang, H., Quantifying natural source mercury emissions from the Ivanhoe Mining District, north-central Nevada, USA, *Atmos. Environ.*, 35, 3987–3997, 2001.
43. Bonzongo, J. C., Heim, K. J., Warwick, J. J., and Lyons, W. B., Mercury levels in surface waters of the Carson River-Lahontan Reservoir System, Nevada: Influence of historic mining activities, *Environ. Pollut.*, 92, 193–201, 1996.
44. Chen, Y., Bonzongo, J. C., and Miller, G. C., Levels of methylmercury and controlling factors in surface sediments of the Carson River system, Nevada, *Environ. Pollut.*, 92, 281–287, 1996.
45. Lacerda, L. D., Evolution of mercury contamination in Brazil, *Water Air Soil Pollut.*, 97, 247–255, 1997.
46. Heemskerk, M., Do international commodity prices drive natural resource booms? An empirical analysis of small-scale gold mining in Suriname, *Ecol. Econ.*, 39, 295–308, 2001.
47. Kambey, J. L., Farrell, A. P., and Bendell-Young, L. I., Influence of illegal gold mining on mercury levels in fish of North Sulawesi's Minahasa Peninsula (Indonesia), *Environ. Pollut.*, 114, 299–302, 2001.
48. Mason, R. P., Fitzgerald, W. F., and Morel, F. M. M., The biogeochemical cycling of elemental mercury: Anthropogenic influences, *Geochim. Cosmochim. Acta,* 58, 3191–3198, 1994.
49. Lodenius, M., Dry and wet deposition of mercury near a chlor-alkali plant, *Sci. Total Environ.*, 213, 53–56, 1998.
50. Gill, G. A., Bloom, N. S., Cappellino, S., Driscoll, C. T., Dobbs, C., McShea, L., Mason, R., and Rudd, J. W. M., Sediment-water fluxes of mercury in Lavaca Bay, Texas, *Environ. Sci. Technol.*, 33, 663–669, 1999.

51. Kudo, A. and Turner, R. R., Mercury contamination of Minamata Bay: Historical overview and progress towards recovery, in *Mercury Contaminated Sites*, Ebinghaus, R., Turner, R. R., Lacerda, L. D., Vasiliev, O., and Salomons, W., Eds., Springer, Berlin, 1999, 143–158.
52. Parks, J. W. and Hamilton, A. L., Accelerating recovery of the mercury-contaminated Wabigoon/English River system, *Hydrobiologia*, 149, 159–188, 1987.
53. Borgmann, U. and Whittle, D. M., Contaminant concentration trends in Lake Ontario lake trout (*Salvelinus namaycush*): 1977 to 1988, *J. Great Lakes Res.*, 17, 368–381, 1991.
54. Borgmann, U. and Whittle, D. M., DDE, PCB, and mercury concentration trends in Lake Ontario rainbow smelt (*Osmerus mordax*) and slimy sculpin (*Cottus cognatus*): 1977 to 1988, *J. Great Lakes Res.*, 18, 298–308, 1992.
55. Lodenius, M., Mercury concentrations in an aquatic ecosystem during twenty years following abatement (of) the pollution source, *Water Air Soil Pollut.*, 56, 323–332, 1991.
56. Becker, D. S. and Bigham, G. N., Distribution of mercury in the aquatic food web of Onondaga Lake, New York, *Water Air Soil Pollut.*, 80, 563–571, 1995.
57. Francesconi, K. A. and Lenanton, R. C. J., Mercury contamination in a semi-enclosed marine embayment: Organic and inorganic mercury content of biota, and factors influencing mercury levels in fish, *Mar. Environ. Res.*, 33, 189–212, 1992.
58. Scheider, W. A., Cox, C., Hayton, A., Hitchin, G., Vaillancourt, A., Current status and temporal trends in concentrations of persistent toxic substances in sport fish and juvenile forage fish in the Canadian waters of the Great Lakes, *Environ. Monit. Assess.*, 53, 57–76, 1998.
59. Balogh, S. J., Engstrom, D. R., Almendinger, J. E., Meyer, M. L., and Johnson, D. K., History of mercury loading in the Upper Mississippi River reconstructed from the sediments of Lake Pepin, *Environ. Sci. Technol.*, 33, 3297–3302, 1999.
60. Frazier, B. E., Wiener, J. G., Rada, R. G., and Engstrom, D. R., Stratigraphy and historic accumulation of mercury in recent depositional sediments in the Sudbury River, Massachusetts, U.S.A., *Can. J. Fish. Aquat. Sci.*, 57, 1062–1072, 2000.
61. Francesconi, K. A., Lenanton, R. C. J., Caputi, N., and Jones, S., Long-term study of mercury concentrations in fish following cessation of a mercury-containing discharge, *Mar. Environ. Res.*, 43, 27–40, 1997.
62. Southworth, G. R., Turner, R. R., Peterson, M. J., Bogle, M. A., and Ryon, M. G., Response of mercury contamination in fish to decreased aqueous concentrations and loading of inorganic mercury in a small stream, *Environ. Monit. Assess.*, 63, 481–494, 2000.
63. Lindeström, L., Mercury in sediment and fish communities of Lake Vänern, Sweden: Recovery from contamination, *Ambio*, 30, 538–544, 2001.
64. Latif, M. A., Bodaly, R. A., Johnston, T. A., and Fudge, R. J. P., Effects of environmental and maternally derived methylmercury on the embryonic and larval stages of walleye (*Stizostedion vitreum*), *Environ. Pollut.*, 111, 139–148, 2001.
65. Rudd, J. W. M., Turner, M. A., Furutani, A., Swick, A. L., and Townsend, B. E., The English-Wabigoon River system: I. A synthesis of recent research with a view towards mercury amelioration, *Can. J. Fish. Aquat. Sci.*, 40, 2206–2217, 1983.
66. Rule, J. H. and Iwashchenko, M. S., Mercury concentrations in soils adjacent to a former chlor-alkali plant, *J. Environ. Qual.*, 27, 31–37, 1998.
67. Naimo, T. J., Wiener, J. G., Cope, W. G., and Bloom, N. S., Bioavailability of sediment-associated mercury to *Hexagenia* mayflies in a contaminated floodplain river, *Can. J. Fish. Aquat. Sci.*, 57, 1092–1102, 2000.
68. Waldron, M. C., Colman, J. A., and Breault, R. F., Distribution, hydrologic transport, and cycling of total mercury and methyl mercury in a contaminated river-reservoir-wetland system (Sudbury River, eastern Massachusetts), *Can. J. Fish. Aquat. Sci.*, 57, 1080–1091, 2000.
69. Wiener, J. G. and Shields, P. J., Mercury in the Sudbury River (Massachusetts, USA): Pollution history and a synthesis of recent research, *Can. J. Fish. Aquat. Sci.*, 57, 1053–1061, 2000.
70. Borg, K., Wanntorp, H., Erne, K., and Hanko, E., Alkyl mercury poisoning in terrestrial Swedish wildlife, *Viltrevy*, 6, 301–379, 1969.
71. Fimreite, N., Accumulation and effects of mercury on birds, in *The Biogeochemistry of Mercury in the Environment,* Nriagu, J. O., Ed., Elsevier, Amsterdam, 1979, 601–627.

72. Abernathy, A. R. and Cumbie, P. M., Mercury accumulation by largemouth bass (*Micropterus salmoides*) in recently impounded reservoirs, *Bull. Environ. Contam. Toxicol.*, 17, 595–602, 1977.
73. Scheider, W. A., Jeffries, D. S., and Dillon, P. J., Effects of acidic precipitation on Precambian freshwaters in southern Ontario, *J. Great Lakes Res.*, 5, 45–51, 1979.
74. Wiener, J. G., Comparative analyses of fish populations in naturally acidic and circumneutral lakes in northern Wisconsin, U.S. Fish Wildl. Serv. Rep. FWS/OBS-80/40.16, La Crosse, WI, 41–56, 1983.
75. Björklund, I., Borg, H., and Johansson, K., Mercury in Swedish lakes — Its regional distribution and causes, *Ambio*, 13, 118–121, 1984.
76. Bodaly, R. A., Hecky, R. E., and Fudge, R. J. P., Increases in fish mercury levels in lakes flooded by the Churchill River Diversion, northern Manitoba, *Can. J. Fish. Aquat. Sci.*, 41, 682–691, 1984.
77. Grieb, T. M., Driscoll, C. T., Gloss, S. P., Schofield, C. L., Bowie, G. L., and Porcella, D. B., Factors affecting mercury accumulation in fish in the upper Michigan peninsula, *Environ. Toxicol. Chem.*, 9, 919–930, 1990.
78. Lathrop, R. C., Rasmussen, P. W., and Knauer, D. R., Mercury concentrations in walleyes from Wisconsin (USA) lakes, *Water Air Soil Pollut.*, 56, 295–307, 1991.
79. Spry, D. J. and Wiener, J. G., Metal bioavailability and toxicity to fish in low-alkalinity lakes: A critical review, *Environ. Pollut.*, 71, 243–304, 1991.
80. Lange, T. R., Royals, H. E., and Connor, L. L., Influence of water chemistry on mercury concentration in largemouth bass from Florida lakes, *Trans. Am. Fish. Soc.*, 122, 74–84, 1993.
81. Simonin, H. A., Gloss, S. P., Driscoll, C. T., Schofield, C. L., Kretser, W. A., Karcher, R. W., and Symula, J., Mercury in yellow perch from Adirondack drainage lakes (New York, U.S.), in *Mercury Pollution: Integration and Synthesis*, Watras, C. J. and Huckabee, J. W., Eds., Lewis Publisher, Boca Raton, FL, 1994, 457–469.
82. Meyer, M. W., Evers, D. C., Hartigan, J. J., and Rasmussen, P. S., Patterns of common loon (*Gavia immer*) mercury exposure, reproduction, and survival in Wisconsin, USA, *Environ. Toxicol. Chem.*, 17, 184–190, 1998.
83. Mierle, G., Addison, E. M., MacDonald, K. S., and Joachim, D. G., Mercury levels in tissues of otters from Ontario, Canada: Variation with age, sex, and location, *Environ. Toxicol. Chem.*, 19, 3044–3051, 2000.
84. Verdon, R., Brouard, D., Demers, C., Lalumiere, R., Laperle, M., and Schetagne, R., Mercury evolution (1978–1988) in fishes of the La Grande hydroelectric complex, Quebec, Canada, *Water Air Soil Pollut.*, 56, 405–417, 1991.
85. Bodaly, R. A., St. Louis, V. L., Paterson, M. J., Fudge, R. J. P., Hall, B. D., Rosenberg, D. M., and Rudd, J. W. M., Bioaccumulation of mercury in the aquatic food chain in newly flooded areas, in *Metal Ions in Biological Systems, Vol. 34, Mercury and Its Effects on Environment and Biology*, Sigel, A. and Sigel, H., Eds., Marcel Dekker, New York, 1997, 259–287.
86. Rosenberg, D. M., Berkes, F., Bodaly, R. A., Hecky, R. E., Kelly, C. A., and Rudd, J. W. M., Large-scale impacts of hydroelectric development, *Environ. Rev.*, 5, 27–54, 1997.
87. Hall, B. D., Rosenberg, D. M., and Wiens, A. P., Methyl mercury in aquatic insects from an experimental reservoir, *Can. J. Fish. Aquat. Sci.*, 55, 2036–2047, 1998.
88. Porvari, P., Development of fish mercury concentrations in Finnish reservoirs from 1979 to 1994, *Sci. Total Environ.*, 213, 279–290, 1998.
89. Ware, F. J., Royals, H., and Lange, T., Mercury contamination in Florida largemouth bass, *Proc. Annu. Conf. Southeast. Assoc. Fish Wildl. Agencies*, 44, 5–12, 1991.
90. Khan, B. and Tansel, B., Mercury bioconcentration factors in American alligators (*Alligator mississippiensis*) in the Florida Everglades, *Ecotoxicol. Environ. Saf.*, 47, 54–58, 2000.
91. Brumbaugh, W. G., Krabbenhoft, D. P., Helsel, D. R., Wiener, J. G., and Echols, K. R., A National Pilot Study of Mercury Contamination of Aquatic Ecosystems along Multiple Gradients: Bioaccumulation in Fish, Biol. Sci. Rep. USGS/BRD/BSR-2001–0009, U.S. Geol. Surv., Reston, VA, 2001.
92. Miskimmin, B. M., Rudd, J. W. M., and Kelly, C. A., Influence of dissolved organic carbon, pH, and microbial respiration rates on mercury methylation and demethylation in lake water, *Can. J. Fish. Aquat. Sci.*, 49, 17–22, 1992.
93. St. Louis, V. L., Rudd, J. W. M., Kelly, C. A., Beaty, K. G., Bloom, N. S., and Flett, R. J., Importance of wetlands as sources of methyl mercury to boreal forest ecosystems, *Can. J. Fish. Aquat. Sci.*, 51, 1065–1076, 1994.

94. Watras, C. J., Bloom, N. S., Hudson, R. J. M., Gherini, S., Munson, R., Claas, S. A., Morrison, K. A., Hurley, J., Wiener, J. G., Fitzgerald, W. F., Mason, R., Vandal, G., Powell, D., Rada, R., Rislove, L., Winfrey, M., Elder, J., Krabbenhoft, D., Andren, A. W., Babiarz, C., Porcella, D. B., and Huckabee, J. W., Sources and fates of mercury and methylmercury in Wisconsin lakes, in *Mercury Pollution: Integration and Synthesis*, Watras, C. J. and Huckabee, J. W., Eds., Lewis Publishers, Boca Raton, FL, 1994, 153–177.
95. Porvari, P. and Verta, M., Methylmercury production in flooded soils: A laboratory study, *Water Air Soil Pollut.*, 80, 765–773, 1995.
96. Kelly, C. A., Rudd, J. W. M., Bodaly, R. A., Roulet, N. P., St. Louis, V. L., Heyes, A., Moore, T. R., Schiff, S., Aravena, R., Scott, K. J., Dyck, B., Harris, R., Warner, B., and Edwards, G., Increases in fluxes of greenhouse gases and methyl mercury following flooding of an experimental reservoir, *Environ. Sci. Technol.*, 31, 1334–1344, 1997.
97. Gilmour, C. C., Riedel, G. S., Ederington, M. C., Bell, J. T., Benoit, J. M., Gill, G. A., and Stordal, M. C., Methylmercury concentrations and production rates across a trophic gradient in the northern Everglades, *Biogeochemistry*, 40, 327–345, 1998.
98. Paterson, M. J., Rudd, J. W. M., and St. Louis, V., Increases in total and methylmercury in zooplankton following flooding of a peatland reservoir, *Environ. Sci. Technol.*, 32, 3868–3874, 1998.
99. Bodaly, R. A. and Fudge, R. J. P., Uptake of mercury by fish in an experimental boreal reservoir, *Arch. Environ. Contam. Toxicol.*, 37,103–109, 1999.
100. Heyes, A., Moore, T. R., Rudd, J. W. M., and Dugoua, J. J., Methyl mercury in pristine and impounded boreal peatlands, Experimental Lakes Area, Ontario, *Can. J. Fish. Aquat. Sci.*, 57, 2211–2222, 2000.
101. Lucotte, M., Mucci, A., Hillaire-Marcel, C., Pichet, P., and Grondin, A., Anthropogenic mercury enrichment in remote lakes of northern Québec (Canada), *Water Air Soil Pollut.*, 80, 467–476, 1995.
102. Engstrom, D. R. and Swain, E. B., Recent declines in atmospheric mercury deposition in the Upper Midwest, *Environ. Sci. Technol.*, 31, 960–967, 1997.
103. Benoit, J. M., Fitzgerald, W. F., and Damman, A. W. H., The biogeochemistry of an ombrotrophic bog: Evaluation of use as an archive of atmospheric mercury deposition, *Environ. Res. (Sect. A)*, 78, 118–133, 1998.
104. Lockhart, W. L., Wilkinson, P., Billeck, B. N., Danell, R. A., Hunt, R. V., Brunskill, G. J., DeLaronde, J., and St. Louis, V., Fluxes of mercury to lake sediments in central and northern Canada inferred from dated sediment cores, *Biogeochemistry*, 40, 163–173, 1998.
105. Lockhart, W. L., Macdonald, R. W., Outridge, P. M., Wilkinson, P., DeLaronde, J. B., and Rudd, J. W. M., Tests of the fidelity of lake sediment core records of mercury deposition to known histories of mercury contamination, *Sci. Total Environ.*, 260, 171–180, 2000.
106. Schuster, P. F., Krabbenhoft, D. P., Naftz, D. L., Cecil, L. D., Olson, M. L., DeWild, J. F., Susong, D. D., Green, J. R., and Abbott, M. L., Atmospheric mercury deposition during the last 270 years: A glacial ice core record of natural and anthropogenic sources, *Environ. Sci. Technol.*, 36, 2303–2310, 2002.
107. Swain, E. B., Engstrom, D. R., Brigham, M. E., Henning, T. A., and Brezonik, P. L., Increasing rates of atmospheric mercury deposition in midcontinental North America, *Science*, 257, 784–787, 1992.
108. Lorey, P. and Driscoll, C. T., Historical trends of mercury deposition in Adirondack lakes, *Environ. Sci. Technol.*, 33, 718–722, 1999.
109. Bindler, R., Renberg, I., Appleby, P. G., Anderson, N. J., and Rose, N. L., Mercury accumulation rates and spatial patterns in lake sediments from West Greenland: A coast to ice margin transect, *Environ. Sci. Technol.*, 35, 1736–1741, 2001.
110. Yang, H., Rose, N. L., Battarbee, R. W., and Boyle, J. F., Mercury and lead budgets for Lochnagar, a Scottish mountain lake and its catchment, *Environ. Sci. Technol.*, 36, 1383–1388, 2002.
111. Downs, S. G., Macleod, C. L., and Lester, J. N., Mercury in precipitation and its relation to bioaccumulation in fish: A literature review, *Water Air Soil Pollut.*, 108, 149–187, 1998.
112. Lin, C. J., Cheng, M. D., and Schroeder, W. H., Transport patterns and potential sources of total gaseous mercury measured in Canadian high Arctic in 1995, *Atmos. Environ.*, 35, 1141–1154, 2001.
113. Mierle, G., Aqueous inputs of mercury to Precambrian Shield lakes in Ontario, *Environ. Toxicol. Chem.*, 9, 843–851, 1990.
114. Fitzgerald, W. F., Mason, R. P., and Vandal, G. M., Atmospheric cycling and air-water exchange of mercury over mid-continental lacustrine regions, *Water Air Soil Pollut.*, 56, 745–767, 1991.

115. Johansson, K., Aastrup, M., Andersson, A., Bringmark, L., and Iverfeldt, Å., Mercury in Swedish forest soils and waters — assessment of critical load, *Water Air Soil Pollut.*, 56, 267–281, 1991.
116. Thompson, D. R., Furness, R. W., and Walsh, P. M., Historical changes in mercury concentrations in the marine ecosystem of the north and north-east Atlantic ocean as indicated by seabird feathers, *J. Appl. Ecol.*, 29, 79–84, 1992.
117. Rolfhus, K. R. and Fitzgerald, W. F., Linkages between atmospheric mercury deposition and the methylmercury content of marine fish, *Water Air Soil Pollut.*, 80, 291–297, 1995.
118. Monteiro, L. R. and Furness, R. W., Accelerated increase in mercury contamination in North Atlantic mesopelagic food chains as indicated by time series of seabird feathers, *Environ. Toxicol. Chem.*, 16, 2489–2493, 1997.
119. Wiener, J. G., Fitzgerald, W. F., Watras, C. J., and Rada, R. G., Partitioning and bioavailability of mercury in an experimentally acidified Wisconsin lake, *Environ. Toxicol. Chem.,* 9, 909–918, 1990.
120. United States Geological Survey, Mercury in the environment, U.S. Geol. Surv. Professional Pap. 713, U.S. Gov. Printing Office, Washington, D.C., 1970.
121. Friske, P. W. B. and Coker, W. B., The importance of geological controls on the natural distribution of mercury in lake and stream sediments across Canada, *Water Air Soil Pollut.*, 80, 1047–1051, 1995.
122. Rasmussen, P. E., Friske, P. W. B., Azzaria, L. M., and Garrett, R. G., Mercury in the Canadian environment: Current research challenges, *Geosci. Can.*, 25, 1–13, 1998.
123. Hurley, J. P., Benoit, J. M., Babiarz, C. L., Shafer, M. M., Andren, A. W., Sullivan, J. R., Hammond, R., and Webb, D. A., Influences of watershed characteristics on mercury levels in Wisconsin rivers, *Environ. Sci. Technol.*, 29, 1867–1875, 1995.
124. St. Louis, V. L., Rudd, J. W. M., Kelly, C. A., Beaty, K. G., Flett, R. J., and Roulet, N. T., Production and loss of methylmercury and loss of total mercury from boreal forest catchments containing different types of wetlands, *Environ. Sci. Technol.*, 30, 2719–2729, 1996.
125. Sellers, P., Kelly, C. A., and Rudd, J. W. M., Fluxes of methylmercury to the water column of a drainage lake: The relative importance of internal and external sources, *Limnol. Oceanogr.*, 46, 623–631, 2001.
126. Bodaly, R. A., Rudd, J. W. M., Fudge, R. J. P., and Kelly, C. A., Mercury concentrations in fish related to size of remote Canadian Shield lakes, *Can. J. Fish. Aquat. Sci.*, 50, 980–987, 1993.
127. Hintelmann, H. and Evans, R. D., Application of stable isotopes in environmental tracer studies: Measurement of monomethylmercury ($CH_3Hg^+$) by isotope dilution ICP-MS and detection of species transformation, *Fresenius J. Anal. Chem.*, 358, 378–385, 1997.
128. Hintelmann, H., Keppel-Jones, K., and Evans, R. D., Constants of mercury methylation and demethylation rates in sediments and comparison of tracer and ambient mercury availability, *Environ. Toxicol. Chem.*, 19, 2204–2211, 2000.
129. Renner, R., Follow the mercury, *Environ. Sci. Technol.*, 35(11), 229A-230A, 2001.
130. Nriagu, J. O., A global assessment of natural sources of atmospheric metals, *Nature*, 338, 47–49, 1989.
131. Lindqvist, O., Mercury in the Swedish environment, *Water Air Soil Pollut.*, 55, 1–261, 1991.
132. Gill, G. A. and Fitzgerald, W. F., Mercury in surface water of the open ocean, *Global Biogeochem. Cycles*, 3, 199–212, 1987.
133. Krabbenhoft, D. P., Wiener, J. G., Brumbaugh, W. G., Olson, M. L., DeWild, J. F., and Sabin, T. J., A national pilot study of mercury contamination of aquatic ecosystems along multiple gradients, in U.S. Geol. Surv. Toxic Substances Hydrol. Program — Proc. Tech. Meeting, Vol. 2, Contamination of Hydrologic Systems and Related Ecosystems, Morganwalp, D. W. and Buxton, H. T., Eds., U.S. Geol. Surv. Water-Resour. Invest. Rep. 99–4018B, 1999, 147–160.
134. Compeau, G. C. and Bartha, R., Sulfate-reducing bacteria: Principal methylators of mercury in anoxic estuarine sediment, *Appl. Environ. Microbiol.*, 50, 498–502, 1985.
135. Schroeder, W. H., Yarwood, G., and Niki, H., Transformation processes involving mercury species in the atmosphere: Results from a literature survey, *Water Air Soil Pollut.*, 56, 653–666, 1991.
136. Amyot, M., Mierle, G., Lean, D. R. S., and McQueen, D. J., Sunlight-induced formation of dissolved gaseous mercury in lake waters, *Environ. Sci. Technol.*, 28, 2366–2371, 1994.
137. Sellers, P., Kelly, C. A., Rudd, J. W. M., and MacHutchon, A. R., Photodegradation of methylmercury in lakes, *Nature*, 380, 694–697, 1996.
138. Krabbenhoft, D. P., Hurley, J. P., Olson, M. L., and Cleckner, L. B., Diel variability of mercury phase and species distributions in the Florida Everglades, *Biogeochemistry*, 40, 311–325, 1998.

139. Ebinghaus, R., Kock, H. H., Temme, C., Einax, J. W., Löwe, A. G., Richter, A., Burrows, J. P., and Schroeder, W. H., Antarctic springtime depletion of atmospheric mercury, *Environ. Sci. Technol.*, 36, 1238–1244, 2002.
140. Lindberg, S. E., Brooks, S., Lin, C. J., Scott, K. J., Landis, M. S., Stevens, R. K., Goodsite, M., and Richter, A., Dynamic oxidation of gaseous mercury in the Arctic troposphere at polar sunrise, *Environ. Sci. Technol.*, 36, 1245–1256, 2002.
141. Schroeder, W. H. and Munthe, J., Atmospheric mercury–an overview, *Atmos. Environ.*, 32, 809–822, 1998.
142. Sommar, J., Feng, X., Gårdfeldt, K., and Lindqvist, O., Measurements of fractionated gaseous mercury concentrations over northwestern and central Europe, 1995–99, *J. Environ. Monit.*, 1, 435–439, 1999.
143. Mason, R. P. and Fitzgerald, W. F., The distribution and biogeochemical cycling of mercury in the equatorial Pacific Ocean, *Deep-Sea Res.*, 40, 1897–1924, 1993.
144. Pacyna, E. G., Pacyna J. M., and Pirrone, N., European emissions of atmospheric mercury from anthropogenic sources in 1995, *Atmos. Environ.*, 35, 2987–2996, 2001.
145. Lindberg, S. E. and Stratton, W. J., Atmospheric mercury speciation: Concentrations and behavior of reactive gaseous mercury in ambient air, *Environ. Sci. Technol.*, 32, 49–57, 1998.
146. Zhang, H. and Lindberg, S. E., Sunlight and iron(III)-induced photochemical production of dissolved gaseous mercury in freshwater, *Environ. Sci. Technol.*, 35, 928–935, 2001.
147. Kim, K. H., Hanson, P. J., Barnett, M. O., and Lindberg, S. E., Biogeochemistry of mercury in the air-soil-plant system, in *Metal Ions in Biological Systems, Vol. 34, Mercury and Its Effects on Environment and Biology*, Sigel, A. and Sigel, H., Eds., Marcel Dekker, New York, 1997, 185–212.
148. Nater, E. A. and Grigal, D. F., Regional trends in mercury distribution across the Great Lakes states, north central USA, *Nature*, 358, 139–141, 1992.
149. Rice, K., Trace element concentrations in streambed sediment across the conterminous United States, *Environ. Sci. Technol.*, 33, 2499–2504, 1999.
150. St. Louis, V. L., Rudd, J. W. M., Kelly, C. A, Hall, B. D., Rolfhus, K. R., Scott, K. J., Lindberg, S. E., and Dong, W., Importance of the forest canopy to fluxes of methyl mercury and total mercury to boreal ecosystems, *Environ. Sci. Technol.*, 35, 3089–3098, 2001.
151. Friedli, H. R., Radke, L. F., and Lu, J. Y., Mercury in smoke from biomass fires, *Geophys. Res. Lett.*, 28, 3223–3226, 2001.
152. Lalonde, J. D., Poulain, A. J., and Amyot, M., The role of mercury redox reactions in snow on snow-to-air mercury transfer, *Environ. Sci. Technol.*, 36, 174–178, 2002.
153. Gustin, M. S., Lindberg, S., Austin, K., Coolbaugh, M., Vetter, A., and Shang, Z., Assessing the contribution of natural sources to regional atmospheric mercury budgets, *Sci. Total Environ.,* 25, 961–971, 2000.
154. Hedgecock, I. M. and Pirrone, N., Mercury and photochemistry in the marine boundary layer — Modelling studies suggest the *in situ* production of reactive gas phase mercury, *Atmos. Environ.*, 35, 3055–3062, 2001.
155. Munthe, J., Wangberg, I., Pirrone, N., Iverfeld, A., Ferrara, R., Ebinghaus, R., Feng, X., Gardfeldt, K., Keeler, G., Lanzillotta, E., Lindberg, S., Lu, J., Mamane, Y., Prestbo, E., Schmolke, S., Schroeder, W. H., Sommar, J., Sprovieri, F., Stevens, R. K., Stratton, W., Tuncel, G., and Urba, A., Intercomparison of methods for sampling and analysis of atmospheric mercury species, *Atmos. Environ.*, 35, 3007–3017, 2001.
156. Fitzgerald, W. F., Is mercury increasing in the atmosphere? The need for an atmospheric mercury network (AMNET), *Water Air Soil Pollut.*, 80, 245–254, 1995.
157. Lamborg, C. H., Rolfhus, K. R., Fitzgerald, W. F., and Kim, G., The atmospheric cycling and air-sea exchange of mercury species in the South and equatorial Atlantic Ocean, *Deep-Sea Res.* II, 46, 957–977, 1999.
158. Stratton, W. J., Lindberg, S., and Perry, C. J., Atmospheric mercury speciation: Laboratory and field evaluation of a mist chamber method for measuring reactive gaseous mercury, *Environ. Toxicol. Chem.*, 35, 170–177, 2001.
159. Schroeder, W. H., Anlauf, K. G., Barrie, L. A., Lu, J. Y., Steffen, A., Schneeberger, D. R., and Berg, T., Arctic springtime depletion of mercury, *Nature*, 394, 331–332, 1998.
160. Lindberg, S., Landis, M. S., Stevens, R. K., and Brooks, S., Comments on "Atmospheric mercury species in the European Arctic: Measurements and modeling" by Berg et al., *Atmos. Environ.*, 14 (2001), 2569–2582, *Atmos. Environ.*, 35, 5377–5378, 2001.

161. Bloom, N. S. and Crecelius, E. A., Determination of mercury in seawater at sub nanogram per liter levels, *Mar. Chem.*, 14, 49–59, 1983.
162. Babiarz, C. L., Hurley, J. P., Hoffmann, S. R., Andren, A. W., Shafer, M. M., and Armstrong, D. E., Partitioning of total mercury and methylmercury to the colloidal phase in freshwaters, *Environ. Sci. Technol.*, 35, 4773–4782, 2001.
163. Mason, R. P., Rolfhus, K. R., and Fitzgerald, W. F., Mercury in the North Atlantic, *Mar. Chem.*, 61, 37–53, 1998.
164. Ravichandran, M., Aiken, G. R., Reddy, M. M., and Ryan, J. N., Enhanced dissolution of cinnabar (mercuric sulfide) by organic matter from the Florida Everglades, *Environ. Sci. Technol.*, 32, 3305–3311, 1998.
165. Gill, G. A. and Bruland, K. W., Mercury speciation in surface freshwater systems in California and other areas, *Environ. Sci. Technol.*, 24, 1392–1400, 1990.
166. Bloom, N. S., Horvat, M., and Watras, C. J., Results of the international aqueous mercury speciation intercomparison exercise, *Water Air Soil Pollut.*, 80, 1257–1268, 1995.
167. Dyrssen, D. and Wedborg, M., The sulphur-mercury (II) system in natural waters, *Water Air Soil Pollut.*, 56, 507–517, 1991.
168. Paquette, K. E. and Helz, G. R., Inorganic speciation of mercury in sulfidic waters: The importance of zero-valent sulfur, *Environ. Sci. Technol.*, 31, 2148–2153, 1997.
169. Ravichandran, M., Aiken, G. R., Ryan, J. N., and Reddy, M. M., Inhibition of precipitation and aggregation of metacinnabar (mercury sulfide) by dissolved organic matter isolated from the Florida Everglades, *Environ. Sci. Technol.*, 33, 1418–1423, 1999.
170. Benoit, J. M., Gilmour, C. C., Mason, R. P., and Heyes, A., Sulfide controls on mercury speciation and bioavailability to methylating bacteria in sediment and pore waters, *Environ. Sci. Technol.*, 33, 951–957, 1999.
171. Benoit, J. M., Mason, R. P., and Gilmour, C. C., Estimation of mercury-sulfide speciation and bioavailability in sediment pore waters using octanol-water partitioning, *Environ. Toxicol. Chem.*, 18, 2138–2141, 1999.
172. Babiarz, C. L. and Andren, A. W., Total concentrations of mercury in Wisconsin (USA) lakes and rivers, *Water Air Soil Pollut.*, 83, 173–183, 1995.
173. Wiener, J. G. and Spry, D. J., Toxicological significance of mercury in freshwater fish, in *Environmental Contaminants in Wildlife: Interpreting Tissue Concentrations*, Beyer, W. N., Heinz, G. H., and Redmon-Norwood, A. W., Eds., Lewis Publishers, Boca Raton, FL, 1996, 297–339.
174. Bigham, G. N. and Vandal, G. M., A drainage basin perspective of mercury transport and bioaccumulation: Onondaga Lake, New York, *NeuroToxicology*, 17, 279–290, 1996.
175. Wang, W. and Driscoll, C. T., Patterns of total mercury concentrations in Onondaga Lake, New York, *Environ. Sci. Technol.*, 29, 2261–2266, 1995.
176. Mierle, G. and Ingram, R., The role of humic substances in the mobilization of mercury from watersheds, *Water Air Soil Pollut.*, 56, 349–357, 1991.
177. Kolka, R. K., Grigal, D. F., Verry, E. S., and Nater, E. A., Mercury and organic carbon relationships in streams draining forested upland/peatland watersheds, *J. Environ. Qual.*, 28, 766–775, 1999.
178. Gray, J. E., Theodorakos, P. M, Bailey, E. A., and Turner, R. R., Distribution, speciation, and transport of mercury in stream-sediment, stream-water, and fish collected near abandoned mercury mines in southwestern Alaska, USA, *Sci. Total Environ.*, 260, 21–33, 2000.
179. Babiarz, C. L., Hurley, J. P., Benoit, J. M., Shafer, M. M., Andren, A. W., and Webb, D. A., Seasonal influences on partitioning and transport of total and methylmercury in rivers from contrasting watersheds, *Biogeochemistry*, 41, 237–257, 1998.
180. Bodaly, R. A., Rudd, J. W. M., and Flett, R. J., Effect of urban sewage treatment on total and methyl mercury concentrations in effluents, *Biogeochemistry*, 40, 279–291, 1998.
181. Bloom, N. S. and Effler, S. W., Seasonal variability in the mercury speciation of Onondaga Lake (New York), *Water Air Soil Pollut.*, 53, 251–265, 1990.
182. Coquery, M., Cossa, D., and Martin, J. M., The distribution of dissolved and particulate mercury in three Siberian estuaries and adjacent arctic coastal waters, *Water Air Soil Pollut.*, 80, 653–664, 1995.
183. Lee, Y. H., Bishop, K. H., Munthe, J., Iverfeldt, Å., Verta, M., Parkman, H., and Hultberg, H., An examination of current Hg deposition and export in Fenno-Scandian catchments, *Biogeochemistry*, 40, 125–135, 1998.

184. Krabbenhoft, D. P., Gilmour, C. C., Benoit, J. M., Babiarz, C. L., Andren, A. W., and Hurley, J. P., Methylmercury dynamics in littoral sediments of a temperate seepage lake, *Can. J. Fish. Aquat. Sci.*, 55, 835–844, 1998.
185. Vandal, G. M., Mason, R. P., and Fitzgerald, W. F., Cycling of volatile mercury in temperate lakes, *Water Air Soil Pollut.*, 56, 791–803, 1991.
186. Poissant, L., Amyot, M., Pilote, M., and Lean, D., Mercury water-air exchange over the upper St. Lawrence River and Lake Ontario, *Sci. Total Environ.*, 34, 3069–3078, 2001.
187. Amyot, M., Gill, G. A., and Morel, F. M. M., Production and loss of dissolved gaseous mercury in coastal seawater, *Environ. Sci. Technol.*, 31, 3606–2611, 1997.
188. Lalonde, J. D., Amyot, M., Kraepiel, A. M. L., and Morel, F. M. M., Photo-oxidation of Hg(0) in artificial and natural waters, *Environ. Sci. Technol.*, 35, 1367–1372, 2001.
189. Rolfhus, K. R. and Fitzgerald, W. F., The evasion and spatial/temporal distribution of mercury species in Long Island Sound, CT-NY, *Geochim. Cosmochim. Acta*, 65, 407–418, 2001.
190. Lindberg, S. E., Kim, K. H., Meyers, T. P., and Owens, J. G., A micrometeorological gradient approach for quantifying air/surface exchange of mercury vapor: Tests over contaminated soils, *Environ. Sci. Technol.*, 29, 126–135, 1995.
191. Hong, Z. and Lindberg, S. L., Processes influencing the emission of mercury from soils: A conceptual model, *J. Geophys. Res.*, 104, 889–896, 1999.
192. Biester, H., Gosar, M., and Covelli, S., Mercury speciation in sediment affected by dumped mining residues in the drainage area of the Idrija Mercury Mine, Slovenia, *Environ. Sci. Technol.*, 34, 3330–3336, 2000.
193. Van Straaten, P., Mercury contamination associated with small-scale gold mining in Tanzania and Zimbabwe, *Sci. Total Environ.*, 259, 105–113, 2000.
194. Barnett, M. O., Harris, L. A., Turner, R. R., Stevenson, R. J., Henson, T. J., Melton, R. C., and Hoffman, D. P., Formation of mercuric sulfide in soil, *Environ. Sci. Technol.*, 31, 3037–3043, 1997.
195. Baily, E. H., Hildebrand, F. A., Christ, C. L., and Fahey, J. J., Schuetteite, a new supergene mercury mineral, *Am. Mineral.*, 44, 1026–1038, 1959.
196. Schwesig, D., Ilgen, G., and Matzner, E., Mercury and methylmercury in upland and wetland acid forest soils of a watershed in NE-Bavaria, Germany, *Water Air Soil Pollut.*, 113, 141–154, 1999.
197. Branfireun, B. A. and Roulet, N. T., Controls on the fate and transport of methylmercury in a boreal headwater catchment, northwestern Ontario, Canada, *Hydrol. Earth Systems Sci.*, 2002, in press.
198. Krabbenhoft, D. P., Benoit, J. M., Babiarz, C. L., Hurley, J. P., and Andren, A. W., Mercury cycling in the Allequash Creek watershed, *Water Air Soil Pollut.*, 80, 425–433, 1995.
199. Winfrey, M. R. and Rudd, J. W. M., Environmental factors affecting the formation of methylmercury in low pH lakes, *Environ. Toxicol. Chem.*, 9, 853–869, 1990.
200. Gilmour, C. C., Henry, E. A, and Mitchell, R., Sulfate stimulation of mercury methylation in freshwater sediments, *Environ. Sci. Technol.*, 26, 2281–2287, 1992.
201. Holmer, M. and Storkholm, P., Sulphate reduction and sulphur cycling in lake sediments: A review, *Freshwat. Biol.*, 46, 431–451, 2001.
202. King, J. K., Kostka, J. E., Frischer, M. E., Saunders, F. M., and Jahnke, R. A., A quantitative relationship that demonstrates mercury methylation rates in marine sediments are based on the community composition and activity of sulfate-reducing bacteria, *Environ. Sci. Technol.*, 35, 2491–2496, 2001.
203. Jensen, S. and Jernelov, A., Biological methylation of mercury in aquatic organisms, *Nature*, 223, 753–754, 1967.
204. Pak, K. R. and Bartha, R., Mercury methylation and demethylation in anoxic lake sediments and by strictly anaerobic bacteria, *Appl. Environ. Microbiol.*, 64, 1013–1017, 1998.
205. Branfireun, B. A., Heyes, A., and Roulet, N. T., The hydrology and methylmercury dynamics of a Precambrian shield headwater peatland, *Water Resour. Res.*, 32, 1785–1794, 1996.
206. Furutani, A. and Rudd, J. W. M., Measurement of mercury methylation in lake water and sediment samples, *Appl. Environ. Microbiol.*, 40, 770–776, 1980.
207. Topping, G. and Davis, I. M., Methylmercury production in the marine water column, *Nature*, 290, 243–244, 1981.
208. Cleckner, L. B., Gilmour, C. C., Hurley, J. P., and Krabbenhoft, D. P., Mercury methylation by periphyton in the Florida Everglades, *Limnol. Oceanogr.*, 44, 1815–1825, 1999.

209. Guimarães, J. R. D., Meili, M., Hylander, L. D., Silva, E. D. E., Roulet, M., Mauro, J. B. N., and de Lemos, R. A., Mercury net methylation in five tropical flood plain regions of Brazil: High in the root zone of floating macrophyte mats but low in surface sediments and flooded soils, *Sci. Total Environ.*, 261, 99–108, 2000.
210. Rudd, J. W. M., Furutani, A., and Turner, M. A., Mercury methylation by fish intestinal contents, *Appl. Environ. Microbiol.*, 40, 777–782, 1980.
211. Jernelov, A., Mercury and food chains, in *Environmental Mercury Contamination*, Hartung, R. and Dinman, B. D., Eds., Ann Arbor Sci. Publishers, Ann Arbor, MI, 174–177, 1972.
212. Ullrich, S. M., Tanton, T. W., and Abdrashitova, S. A., Mercury in the aquatic environment: A review of factors affecting methylation, *Crit. Rev. Environ. Sci. Technol.*, 31, 241–293, 2001.
213. Korthals, E. T. and Winfrey, M. R., Seasonal and spatial variations in mercury methylation and demethylation in an oligotrophic lake, *Appl. Environ. Microbiol.*, 53, 2397–2404, 1987.
214. Bloom, N. S., Gill, G. A., Cappellino, S., Dobbs, C., McShea, L., Driscoll, C., Mason, R., and Rudd, J., Speciation and cycling of mercury in Lavaca Bay, Texas, sediments, *Environ. Sci. Technol.*, 33, 7–13, 1999.
215. Barkay, T., Turner, R. R., Rasmussen, L. D., Kelly, C. A., and Rudd, J. W. M., Luminescence facilitated detection of bioavailable mercury in natural waters, in *Bioluminescence Methods and Protocols,* LaRossa, R. A., Ed., *Methods in Microbiology*, 102, 231–246, Humana Press, Totowa, NJ, 1998.
216. Ridley, W. P., Dizikes, L. J., and Wood, J. M., Biomethylation of toxic elements in the environment, *Science*, 197, 329–330, 1977.
217. Wood, J. M., Alkylation of metals and the activity of metal-alkyls, *Toxicol. Environ. Chem.*, 7, 229–240, 1984.
218. Imura, N., Sukegawa, E., Pan, S., Nagao, K., Kim, J., Kwan, T., and Ukita, T., Chemical methylation of inorganic mercury with methylcobalamin, a vitamin B12 analog, *Science*, 172, 1248–1249, 1971.
219. Choi, S. C., Chase, T., and Bartha, R., Enzymatic catalysis of mercury methylation by *Desulfovibrio desulfuricans LS, Appl. Environ. Microbiol.*, 60, 1342–1346, 1994.
220. Akagi, H., Miller, D. R., and Kudo, A., Photochemical transformation of mercury, in *Distribution and Transport of Pollutants in Flowing Water Ecosystems*, Final Rep., Ottawa River Project, Univ. Ottawa, Nat. Res. Council Can., 1977.
221. Nagase, H., Ose, Y., and Sato, T., Possible methylation of inorganic mercury by silicones in the environment, *Sci. Total Environ.*, 73, 29–36, 1988.
222. Jewett, K. L., Brinckman, F. E., and Bellama, J. M., Chemical factors influencing metal alkylation in water, in *Marine Chemistry in the Coastal Environment*, Church, T. M., Ed., Am. Chem. Soc., Washington, D.C., 1975, 304–318.
223. Nagase, H., Ose, Y., Sato, T., and Ishikawa, T., Methylation of mercury by humic substances in an aquatic environment, *Sci. Total Environ.*, 24, 133–142, 1982.
224. Weber, J. H., Review of possible paths for abiotic methylation of mercury(II) in the aquatic environment, *Chemosphere*, 26, 2063–2070, 1993.
225. Falter, R. and Wilken, R. D., Isotope experiments for the determination of abiotic mercury methylation potential of a River Rhine sediment, *Vom Wasser*, 90, 217–232, 1998.
226. Horvat, M., Bloom, N. S., and Liang, L., Comparison of distillation with other current isolation methods for the determination of MeHg compounds in low level environmental samples. Part I. Sediment, *Anal. Chim. Acta*, 282, 135–152, 1993.
227. Hammerschmidt, C. R. and Fitzgerald, W. F., Formation of artifact methylmercury during extraction from a sediment reference material, *Anal. Chem.*, 73, 5930–5936, 2001.
228. Compeau, G. and Bartha, R., Methylation and demethylation of mercury under controlled redox, pH and salinity conditions, *Appl. Environ. Microbiol.*, 48, 1203–1214, 1984.
229. Oremland, R. S., Culbertson, C. W., and Winfrey, M. R., Methylmercury decomposition in sediments and bacterial cultures — involvement of methanogens and sulfate reducers in oxidative demethylation, *Appl. Environ. Microbiol.*, 57, 130–140, 1991.
230. Robinson, J. B. and Tuovinen, O. H., Mechanisms of microbial resistance and detoxification of mercury and organomercury compounds — Physiological, biochemical and genetic analyses, *Microbiol. Rev.*, 48, 95–125, 1984.
231. Summers, A. O., Organization, expression and evolution of genes for mercury resistance, *Annu. Rev. Microbiol.*, 40, 607–612, 1986.

232. Marvin-DiPasquale, M. and Oremland, R. S., Bacterial methylmercury degradation in Florida Everglades sediment and periphyton, *Environ. Sci. Technol.*, 32, 2556–2563, 1998.
233. Marvin-DiPasquale, M., Agee, J., McGowan, C., Oremland, R. S., Thomas, M., Krabbenhoft, D. P., and Gilmour, C. C., Methyl-mercury degradation pathways — A comparison among three mercury impacted ecosystems, *Environ. Sci. Technol.*, 34, 4908–4916, 2000.
234. Krabbenhoft, D. P., Olson, M. L., Dewild, J. F., Clow, D. W., Striegl, R. S., Dornblaser, M. M., and Van Metre, P., Mercury loading and methylmercury production and cycling in high-altitude lakes from the western United States, *Water Air Soil Pollut. Focus*, 2(2), 233–249, 2002.
235. Baughman, G. L., Gordon, J. A., Wolfe, N. L., and Zepp, R. G., Chemistry of organomercurials in aquatic ecosystems, Rep. EPA-660/3–73–012, U.S. Environ. Protect. Agency, Office of Research and Development, Washington, D.C., 1973.
236. Branfireun, B. A., Hilbert, D., and Roulet, N. T., Sinks and sources of methylmercury in a boreal catchment, *Biogeochemistry*, 41, 277–291, 1998.
237. Suda, I., Suda, M., and Hirayama, K., Degradation of methyl and ethyl mercury by singlet oxygen generated from sea water exposed to sunlight or ultraviolet light, *Arch. Toxicol.,* 67, 365–371, 1993.
238. Meyer, M. W., Evers, D. C., Daulton, T., and Braselton, W.E., Common loons (*Gavia immer*) nesting on low pH lakes in northern Wisconsin have elevated blood mercury content, *Water Air Soil Pollut.*, 80, 871–880, 1995.
239. Westcott, K. and Kalff, J., Environmental factors affecting methyl mercury accumulation in zooplankton, *Can. J. Fish. Aquat. Sci.*, 53, 2221–2228, 1996.
240. Plourde, Y., Lucotte, M., and Pichet, P., Contribution of suspended particulate matter and zooplankton to MeHg contamination of the food chain in midnorthern Quebec (Canada) reservoirs, *Can. J. Fish. Aquat. Sci.*, 54, 821–831, 1997.
241. Xun, L., Campbell, N. E. R., and Rudd, J. W. M., Measurements of specific rates of net methyl mercury production in the water column and surface sediments of acidified and circumneutral lakes, *Can. J. Fish. Aquat. Sci.*, 44, 750–757, 1987.
242. Hecky, R. E., Ramsey, D. J., Bodaly, R. A., and Strange, N. E., Increased methylmercury contamination in fish in newly formed freshwater reservoirs, in *Advances in Mercury Toxicology*, Suzuki, T. et al., Eds., Plenum Press, New York, 1991, 33–52.
243. Hurley, J. P., Krabbenhoft, D. P., Cleckner, L. B., Olson, M. L., Aiken, G. R., and Rawlik, P. S., System controls on the aqueous distribution of mercury in the northern Florida Everglades, *Biogeochemistry*, 40, 293–311, 1998.
244. Eilers, J. M., Brakke, D. F., and Landers, D. H., Chemical and physical characteristics of lakes in the Upper Midwest, United States, *Environ. Sci. Technol.*, 22, 164–172, 1988.
245. Krabbenhoft, D. P. and Babiarz, C. L., The role of groundwater transport in aquatic mercury cycling, *Water Resour. Res.*, 28, 3119–3128, 1992.
246. Rada, R. G., Wiener, J. G., Winfrey, M. R., and Powell, D. E., Recent increases in atmospheric deposition of mercury to north-central Wisconsin lakes inferred from sediment analyses, *Arch. Environ. Contam. Toxicol.*, 18, 175–181, 1989.
247. Rada, R. G., Powell, D. E., and Wiener, J. G., Whole-lake burdens and spatial distribution of mercury in surficial sediments in Wisconsin seepage lakes, *Can. J. Fish. Aquat. Sci.*, 50, 865–873, 1993.
248. Wiener, J. G., Rago, P. J., and Eilers, J. M., Species composition of fish communities in northern Wisconsin lakes: Relation to pH, in *Early Biotic Responses to Advancing Lake Acidification*, Hendrey, G. R., Ed., Butterworth Publishers, Boston, 1984, 133–146.
249. Wiener, J. G. and Eilers, J. M., Chemical and biological status of lakes and streams in the Upper Midwest: Assessment of acidic deposition effects, *Lake Reservoir Manage.*, 3, 365–378, 1987.
250. Wiener, J. G., Martini, R. E., Sheffy, T. B., and Glass, G. E., Factors influencing mercury concentrations in walleyes in northern Wisconsin lakes, *Trans. Am. Fish. Soc.*, 119, 862–870, 1990.
251. Colby, P. J., McNicol, R. E., and Ryder, R. A., Synopsis of biological data on the walleye *Stizostedion v. vitreum* (Mitchill 1818). FAO Fish. Synopsis No. 119, Food Agric. Org. United Nations, Rome, Italy, 1979.
252. Barr, J. F., Aspects of common loon (*Gavia immer*) feeding biology on its breeding ground, *Hydrobiologia*, 321, 119–144, 1996.
253. Huckabee, J. W., Elwood, J. W., and Hildebrand, S. G., Accumulation of mercury in freshwater biota, in *Biogeochemistry of Mercury in the Environment*, Nriagu, J. O., Ed., Elsevier/North-Holland Biomedical Press, New York, 1979, 277–302.

254. Boudou, A. and Ribeyre, F., Mercury in the food web: Accumulation and transfer mechanisms, in *Metal Ions in Biological Systems, Vol. 34, Mercury and Its Effects on Environment and Biology*, Sigel, A. and Sigel, H., Eds., Marcel Dekker, New York, 1997, 289–319.
255. Morel, F. M. M., Kraepiel, A. M. L., and Amyot, M., The chemical cycle and bioaccumulation of mercury, *Annu. Rev. Ecol. Systemat.*, 29, 543–566, 1998.
256. Kim, J. P. and Burggraaf, S., Mercury bioaccumulation in rainbow trout (*Oncorhynchus mykiss*) and the trout food web in lakes Okareka, Okaro, Tarawera, Rotomahana and Rotorua, New Zealand, *Water Air Soil Pollut.*, 115, 535–546, 1999.
257. Bowles, K. C., Apte, S. C., Maher, W. A., Kawei, M., and Smith, R., Bioaccumulation and biomagnification of mercury in Lake Murray, Papua New Guinea, *Can. J. Fish. Aquat. Sci.*, 58, 888–897, 2001.
258. Thompson, D. R., Hamer, K. C., and Furness, R. W., Mercury accumulation in Great Skuas *Catharacta skua* of known age and sex, and its effects upon breeding and survival, *J. Appl. Ecol.*, 28, 672–684, 1991.
259. Furness, R. W., Thompson, D. R., and Becker, P. H., Spatial and temporal variation in mercury contamination of seabirds in the North Sea, *Helgoländer Meeresunters*, 49, 605–615, 1995.
260. Evans, R. D., Addison, E. M., Villeneuve, J. Y., MacDonald, K. S., and Joachim, D. G., Distribution of inorganic and methylmercury among tissues in mink (*Mustela vison*) and otter (*Lutra canadensis*), *Environ. Res. (Sect. A)*, 84,133–139, 2000.
261. Thompson, D. R. and Furness, R. W., The chemical form of mercury stored in South Atlantic seabirds, *Environ. Pollut.*, 60, 305–317, 1989.
262. Kim, E. Y., Murakami, T., Saeki, K., and Tatsukawa, R., Mercury levels and its chemical form in tissues and organs of seabirds, *Arch. Environ. Contam. Toxicol.*, 30, 259–266, 1996.
263. Scheuhammer, A. M., Wong, A. H. K., and Bond, D., Mercury and selenium accumulation in common loons (*Gavia immer*) and common mergansers (*Mergus merganser*) from eastern Canada, *Environ. Toxicol. Chem.*, 17, 197–201, 1998.
264. Watras, C. J., Back, R. C., Halvorsen, S., Hudson, R. J. M., Morrison, K. A., and Wente, S. P., Bioaccumulation of mercury in pelagic freshwater food webs, *Sci. Total Environ.*, 219, 183–208, 1998.
265. Riisgård, H. U. and Famme, P., Accumulation of inorganic and organic mercury in shrimp, *Crangon crangon*, *Mar. Pollut. Bull.*, 17, 255–257, 1986.
266. Back, R. C. and Watras, C. J., Mercury in zooplankton of northern Wisconsin lakes: Taxonomic and site-specific trends, *Water Air Soil Pollut.*, 80, 931–938, 1995.
267. Lawson, N. M. and Mason, R. P., Accumulation of mercury in estuarine food chains, *Biogeochemistry*, 40, 235–247, 1998.
268. Lawrence, A. L. and Mason, R. P., Factors controlling the bioaccumulation of mercury and methylmercury by the estuarine amphipod *Leptocheirus plumulosus*, *Environ. Pollut.*, 111, 217–231, 2001.
269. Simon, O. and Boudou, A., Simultaneous experimental study of direct and direct plus trophic contamination of the crayfish *Astacus astacus* by inorganic mercury and methylmercury, *Environ. Toxicol. Chem.*, 20, 1206–1215, 2001.
270. Tremblay, A., Lucotte, M., and Rheault, I., Methylmercury in a benthic food web of two hydroelectric reservoirs and a natural lake of northern Quebec (Canada), *Water Air Soil Pollut.*, 91, 255–269, 1996.
271. Claisse, D., Cossa, D., Bretaudeau-Sanjuan, J., Touchard, G., and Bobled, B., Methylmercury in molluscs along the French coast, *Mar. Pollut. Bull.*, 42, 329–332, 2001.
272. Monson, B. A. and Brezonik, P. L., Seasonal patterns of mercury species in water and plankton from softwater lakes in northeastern Minnesota, *Biogeochemistry*, 40, 147–162, 1998.
273. Miles, C. J., Moye, H. A., Phlips, E. J., and Sargent, B., Partitioning of monomethylmercury between freshwater algae and water, *Environ. Sci. Technol.*, 35, 4277–4282, 2001.
274. MacCrimmon, H. R., Wren, C. D., and Gots, B. L., Mercury uptake by lake trout, *Salvelinus namaycush*, relative to age, growth, and diet in Tadenac Lake with comparative data from other Precambrian Shield lakes, *Can. J. Fish. Aquat. Sci.*, 40, 114–120, 1983.
275. Suns, K., Hitchin, G., Loescher, B., Pastorek, E., and Pearce, R., Metal accumulations in fishes from Muskoka-Haliburton lakes in Ontario (1978–1984), Ontario Ministry of the Environment, Rexdale, Ontario, 1987.
276. Cope, W. G., Wiener, J. G., and Rada, R. G., Mercury accumulation in yellow perch in Wisconsin seepage lakes: Relation to lake characteristics, *Environ. Toxicol. Chem.*, 9, 931–940, 1990.
277. Monteiro, L. R., Granadeiro, J. P., and Furness, R. W., Relationship between mercury levels and diet in Azores seabirds, *Mar. Ecol. Prog. Ser.*, 166, 259–265, 1998.

278. Benoit, J. M., Gilmour, C. C., and Mason, R. P., The influence of sulfide on solid-phase mercury bioavailability for methylation by pure cultures of *Desulfobulbus propionicus* (1pr3), *Environ. Sci. Technol.*, 35, 127–132, 2001.
279. King, J. K., Kostka, J. E., Frischer, M. E., and Saunders, F. M., Sulfate-reducing bacteria methylate mercury at variable rates in pure culture and in marine sediments, *Appl. Environ. Microbiol.*, 66, 2430–2437, 2000.
280. Heyes, A., Moore, T. R., and Rudd, J. W. M., Mercury and methylmercury in decomposing vegetation of a pristine and impounded wetland, *J. Environ. Qual.*, 27, 591–599, 1998.
281. Gerrard, P. M. and St. Louis, V. L., The effects of experimental reservoir creation on the bioaccumulation of methylmercury and reproductive success of tree swallows (*Tachycineta bicolor*), *Environ. Sci. Technol.*, 35, 1329–1338, 2001.
282. Cabana, G. and Rasmussen, J. B., Modelling food chain structure and contaminant bioaccumulation using stable nitrogen isotopes, *Nature*, 372, 255–257, 1994.
283. Cabana, G., Tremblay, A., Kalff, J., and Rasmussen, J. B., Pelagic food chain structure in Ontario lakes: A determinant of mercury levels in lake trout (*Salvelinus namaycush*), *Can. J. Fish. Aquat. Sci.*, 51, 381–389, 1994.
284. Vander Zanden, M. J. and Rasmussen, J. B., A trophic position model of pelagic food webs: Impact on contaminant bioaccumulation in lake trout, *Ecol. Monogr.*, 66, 451–477, 1996.
285. Gnamus, A., Byrne, A. R., and Horvat, M., Mercury in the soil-plant-deer-predator food chain of a temperate forest in Slovenia, *Environ. Sci. Technol.*, 34, 3337–3345, 2000.
286. Watras, C. J. and Bloom, N. S., Mercury and methylmercury in individual zooplankton: Implications for bioaccumulation, *Limnol. Oceanogr.*, 37, 1313–1318, 1992.
287. Hammerschmidt, C. R., Wiener, J. G., Frazier, B. E., and Rada, R. G., Methylmercury content of eggs in yellow perch related to maternal exposure in four Wisconsin lakes, *Environ. Sci. Technol.*, 33, 999–1003, 1999.
288. Southworth, G. R., Turner, R. R., Peterson, M. J., and Bogle, M. A., Form of mercury in stream fish exposed to high concentrations of dissolved inorganic mercury, *Chemosphere*, 30, 779–787, 1995.
289. Ribeyre, F. and Boudou, A., Bioaccumulation et repartition tissulaire du mercure — $HgCl_2$ et $CH_3HgCl$ — Chez *Salmo gairdneri* apres contamination par voie directe, *Water Air Soil Pollut.*, 23, 169–186, 1984.
290. Ribeyre, F. and Boudou, A., Etude experimentale des processus de decontamination chez *Salmo gairdneri*, apres contamination par voie directe avec deux derives du mercure ($HgCl_2$ et $CH_3HgCl$) — Analyse des transferts aux niveaux "organisme" et "organes," *Environ. Pollut. (Series A)*, 35, 203–228, 1984.
291. Boudou, A. and Ribeyre, F., Experimental study of trophic contamination of *Salmo gairdneri* by two mercury compounds — $HgCl_2$ and $CH_3HgCl$ — Analysis at the organism and organ levels, *Water Air Soil Pollut.*, 26, 137–148, 1985.
292. Niimi, A. J. and Kissoon, G. P., Evaluation of the critical body burden concept based on inorganic and organic mercury toxicity to rainbow trout (*Oncorhynchus mykiss*), *Arch. Environ. Contam. Toxicol.*, 26, 169–178, 1994.
293. Trudel, M. and Rasmussen, J. B., Modeling the elimination of mercury by fish, *Environ. Sci. Technol.*, 31, 1716–1722, 1997.
294. Ribeiro, C. A. O., Rouleau, C., Pelletier, É., Audet, C., and Tjälve, H., Distribution kinetics of dietary methylmercury in the Arctic charr (*Salvelinus alpinus*), *Environ. Sci. Technol.*, 33, 902–907, 1999.
295. Harris, R. C. and Snodgrass, W. J., Bioenergetic simulations of mercury uptake and retention in walleye (*Stizostedion vitreum*) and yellow perch (*Perca flavescens*), *Water Pollut. Res. J. Can.*, 28, 217–236, 1993.
296. Rodgers, D. W., You are what you eat and a little bit more: Bioenergetics-based models of methylmercury accumulation in fish revisited, in *Mercury Pollution: Integration and Synthesis*, Watras, C. J. and Huckabee, J. W., Eds., Lewis Publishers, Boca Raton, FL, 1994, 427–439.
297. Hall, B. D., Bodaly, R. A., Fudge, R. J. P., Rudd, J. W. M., and Rosenberg, D. M., Food as the dominant pathway of methylmercury uptake by fish, *Water Air Soil Pollut.*, 100, 13–24, 1997.
298. Harris, R. C. and Bodaly, R. A., Temperature, growth and dietary effects on fish mercury dynamics in two Ontario lakes, *Biogeochemistry*, 40, 175–187, 1998.

299. Olson, G. F., Mount, D. I., Snarski, V. M., and Thorslund, T. W., Mercury residues in fathead minnows, *Pimephales promelas* Rafinesque, chronically exposed to methylmercury in water, *Bull. Environ. Contam. Toxicol.*, 14, 129–134, 1975.
300. McKim, J. M., Olson, G. F., Holcombe, G. W., and Hunt, E. P., Long-term effects of methylmercuric chloride on three generations of brook trout (*Salvelinus fontinalis*): Toxicity, accumulation, distribution, and elimination, *J. Fish. Res. Board Can.*, 33, 2726–2739, 1976.
301. Boudou, A. and Ribeyre, F., Contamination of aquatic biocenoses by mercury compounds: An experimental ecotoxicological approach, in *Aquatic Toxicology*, Nriagu, J.O., Ed., John Wiley and Sons, New York, 1983, 73–116.
302. Harrison, S. E., Klaverkamp, J. F., and Hesslein, R. H., Fates of metal radiotracers added to a whole lake: Accumulation in fathead minnow (*Pimephales promelas*) and lake trout (*Salvelinus namaycush*), *Water Air Soil Pollut.*, 52, 277–293, 1990.
303. Giblin· F. J. and Massaro, E. J., Pharmacodynamics of methyl mercury in the rainbow trout (*Salmo gairdneri*): Tissue uptake, distribution and excretion, *Toxicol. Appl. Pharmacol.*, 24, 81–91, 1973.
304. Ramlal, P. S., Kelly, C. A., Rudd, J. W. M., and Furutani, A., Sites of methyl mercury production in remote Canadian Shield lakes, *Can. J. Fish. Aquat. Sci.*, 50, 972–979, 1993.
305. Rask, M. and Metsälä, T. R., Mercury concentrations in northern pike, *Esox lucius* L., in small lakes of Evo area, southern Finland, *Water Air Soil Pollut.*, 56, 369–378, 1991.
306. Mathers, R. A. and Johansen, P. H., The effects of feeding ecology on mercury accumulation in walleye (*Stizostedion vitreum*) and pike (*Esox lucius*) in Lake Simcoe, *Can. J. Zool.*, 63, 2006–2012, 1985.
307. Kidd, K. A., Hesslein, R. H., Fudge, R. J. P., and Hallard, K. A., The influence of trophic level as measured by $\delta^{15}N$ on mercury concentrations in freshwater organisms, *Water Air Soil Pollut.*, 80, 1011–1015, 1995.
308. Nicoletto, P. F. and Hendricks, A. C., Sexual differences in accumulation of mercury in four species of centrarchid fishes, *Can. J. Zool.*, 66, 944–949, 1988.
309. Trudel, M., Tremblay, A., Schetagne, R., and Rasmussen, J. B., Estimating food consumption rates of fish using a mercury mass balance model, *Can. J. Fish. Aquat. Sci.*, 57, 414–428, 2000.
310. Niimi, A. J., Biological and toxicological effects of environmental contaminants in fish and their eggs, *Can. J. Fish. Aquat. Sci.*, 40, 306–312, 1983.
311. Johnston, T. A., Bodaly, R. A., Latif, M. A., Fudge, R. J. P., and Strange, N. E., Intra- and inter-population variability in maternal transfer of mercury to eggs of walleye (*Stizostedion vitreum*), *Aquat. Toxicol.*, 52, 73–85, 2001.
312. Wren, C. D. and MacCrimmon, H. R., Mercury levels in the sunfish, *Lepomis gibbosus*, relative to pH and other environmental variables of Precambrian Shield lakes, *Can. J. Fish. Aquat. Sci.*, 40, 1737–1744, 1983.
313. Suns, K. and Hitchin, G., Interrelationships between mercury levels in yearling yellow perch, fish condition and water quality, *Water Air Soil Pollut.*, 50, 255–265, 1990.
314. Frost, T. M., Montz, P. K., Kratz, T. K., Badillo, T., Brezonik, P. L., Gonzalez, M. J., Rada, R. G., Watras, C. J., Webster, K. E., Wiener, J. G., Williamson, C. E., and Morris, D. P., Multiple stresses from a single agent: Diverse responses to the experimental acidification of Little Rock Lake, Wisconsin, *Limnol. Oceanogr.*· 44(3, part 2), 784–794, 1999.
315. Greenfield, B. K., Hrabik, T. R., Harvey, C. J., and Carpenter, S. R., Predicting mercury levels in yellow perch: Use of water chemistry, trophic ecology, and spatial traits, *Can. J. Fish. Aquat. Sci.*, 58, 1419–1429, 2001.
316. Rodgers, D. W. and Beamish, F. W. H., Dynamics of dietary methylmercury in rainbow trout, *Salmo gairdneri*, *Aquat. Toxicol.*, 2, 271–290, 1982.
317. Scherer, E., Armstrong, F. A. J., and Nowak, S. H., Effects of mercury-contaminated diet upon walleyes, *Stizostedion vitreum vitreum* (Mitchill), Fish. Marine Serv. Res. Development Tech. Rep. No. 597, Winnipeg, Manitoba, 1975.
318. Kitamura, S., Determination on mercury content in bodies of inhabitants, cats, fishes and shells in Minamata District and in the mud of Minamata Bay, Chapter 7 in *Minamata Disease*, Study Group of Minamata Disease, Kumamoto Univ., Japan, 1968, 257–266.
319. Takeuchi, T., Pathology of Minamata Disease, in *Minamata Disease*, Study group of Minamata Disease, Kumamoto Univ., Japan, 1968, 211–216.

320. Lockhart, W. L., Uthe, J. F., Kenney, A. R., and Mehrle, P. M., Methylmercury in northern pike (*Esox lucius*): Distribution, elimination, and some biochemical characteristics of contaminated fish, *J. Fish. Res. Board Can.*, 29, 1519–1523, 1972.
321. Kania, H. J. and O'Hara, J., Behavioral alterations in a simple predator-prey system due to sublethal exposure to mercury, *Trans. Am. Fish. Soc.*, 103, 134–136, 1974.
322. Little, E. E. and Finger, S. E., Swimming behavior as an indicator of sublethal toxicity in fish, *Environ. Toxicol. Chem.*, 9, 13–19, 1990.
323. Sandheinrich, M. B. and Atchison, G. J., Sublethal toxicant effects on fish foraging behavior: Empirical vs. mechanistic approaches, *Environ. Toxicol. Chem.*, 9, 107–119, 1990.
324. Weis, J. S. and Weis, P., Swimming performance and predator avoidance by mummichog (*Fundulus heteroclitus*) larvae after embryonic or larval exposure to methylmercury, *Can. J. Fish. Aquat. Sci.*, 52, 2168–2173, 1995.
325. Fjeld, E., Haugen, T. O., and Vøllestad, L. A., Permanent impairment in the feeding behavior of grayling (*Thymallus thymallus*) exposed to methylmercury during embryogenesis, *Sci. Total Environ.*, 213, 247–254, 1998.
326. Samson, J. C., Goodridge, R., Olobatuyi, F., and Weis, J. S., Delayed effects of embryonic exposure of zebrafish (*Danio rerio*) to methylmercury (MeHg), *Aquat. Toxicol.*, 51, 369–376, 2001.
327. Friedmann, A. S., Watzin, M. C., Brinck-Johnsen, T., and Leiter, J. C., Low levels of dietary methylmercury inhibit growth and gonadal development in juvenile walleye (*Stizostedion vitreum*), *Aquat. Toxicol.*, 35, 265–278, 1996.
328. Hammerschmidt, C. R., Sandheinrich, M. B., Wiener, J. G., and Rada, R. G., Effects of dietary methylmercury on reproduction of fathead minnows, *Environ. Sci. Technol.*, 36, 877–883, 2002.
329. Thompson, D. R., Mercury in birds and terrestrial mammals, in *Environmental Contaminants in Wildlife: Interpreting Tissue Concentrations,* Beyer, W. N., Heinz, G. H., and Redmond-Norwood, A. W., Eds., Lewis Publishers, Boca Raton, FL, 1996, 341–356.
330. Wolfe, M. F., Schwarzbach, S., and Sulaiman, R. A., Effects of mercury on wildlife: A comprehensive review, *Environ. Toxicol. Chem.*, 17, 146–160, 1998.
331. Eisler, R., *Handbook of Chemical Risk Assessment: Health Hazards to Humans, Plants, and Animals, Vol. 1, Metals,* Lewis Publishers, Boca Raton, FL, 2000.
332. Fimreite, N., Mercury contamination of aquatic birds in northwestern Ontario, *J. Wildl. Manage.*, 38, 120–131, 1974.
333. Hesse, L. W., Brown, R. L., and Heisinger, J. F., Mercury contamination of birds from a polluted watershed, *J. Wildl. Manage.*, 39, 299–304, 1975.
334. Sepulveda, M. S., Frederick, P. C., Spalding, M. G., and Williams, G. E., Jr., Mercury contamination in free-ranging great egret nestlings (*Ardea albus*) from southern Florida, USA, *Environ. Toxicol. Chem.*, 18, 985–992, 1999.
335. Caldwell, C. A., Arnold, M. A., and Gould, W. R., Mercury distribution in blood, tissues, and feathers of double-crested cormorant nestlings from arid-lands reservoirs in south central New Mexico, *Arch. Environ. Contam. Toxicol.*, 36, 456–461, 1999.
336. Eisler, R., Mercury hazards to fish, wildlife, and invertebrates: A synoptic review, U.S. Fish Wildl. Serv. Biol. Rep. 85 (1.10), 1987.
337. Hoffman, D. J. and Moore, J. M., Teratogenic effects of external egg applications of methyl mercury, *Teratology*, 20, 453–462, 1979.
338. Heinz, G. H. and Hoffman, D. J., Methylmercury chloride and selenomethionine interactions on health and reproduction in mallards, *Environ. Toxicol. Chem.*, 17, 139–145, 1998.
339. Bäckström, J., Distribution studies of mercuric pesticides in quail and some freshwater fishes, *Acta Pharmacol. Toxicol.*, 27 (Suppl. 3), 103 pp., 1969.
340. Heinz, G. H. and Locke, L. N., Brain lesions in mallard ducklings from parents fed methylmercury, *Avian Dis.*, 20, 9–17, 1976.
341. Tejning, S., Biological effects of methyl mercury dicyandiamide-treated grain in the domestic fowl *Gallus gallus* L., *Oikos Suppl.* 8, 116 pp., 1967.
342. Fimreite, N., Effects of dietary methylmercury on ring-necked pheasants, *Occas. Pap. No. 9,* Can. Wildl. Serv., Ottawa, 39 pp., 1971.
343. Heinz, G. H., Methylmercury: Reproductive and behavioral effects on three generations of mallard ducks, *J. Wildl. Manage.*, 43, 394–401, 1979.

344. Lewis, S. A. and Furness, R. W., Mercury accumulation and excretion in laboratory-reared black-headed gull *Larus ridibundus* chicks, *Arch. Environ. Contam. Toxicol.*, 21, 316–320, 1991.
345. Monteiro, L. R. and Furness, R. W., Kinetics, dose-response, and excretion of methylmercury in free-living adult Cory's shearwaters, *Environ. Sci. Technol.*, 35, 739–746, 2001.
346. Stickel, L. F., Stickel, W. H., McLane, M. A. R., and Bruns, M., Prolonged retention of methyl mercury by mallard drakes, *Bull. Environ. Contam. Toxicol.*, 18, 393–400, 1977.
347. Monteiro, L. R. and Furness, R. W., Kinetics, dose-response, excretion, and toxicity of methylmercury in free-living Cory's shearwater chicks, *Environ. Toxicol. Chem.*, 20, 1816–1823, 2001.
348. Burger, J., Metals in avian feathers: Bioindicators of environmental pollution, *Rev. Environ. Contam. Toxicol.*, 5, 203–311, 1993.
349. Scheuhammer, A. M., Atchison, C. M., Wong, A. H. K., and Evers, D. C., Mercury exposure in breeding common loons (*Gavia immer*) in central Ontario, Canada, *Environ. Toxicol. Chem.*, 17, 191–196, 1998.
350. Braune, B. M. and Gaskin, D. E., Mercury levels in Bonaparte's gulls (*Larus philadelphia*) during autumn molt in the Quoddy region, New Brunswick, Canada, *Arch. Environ. Contam. Toxicol.*, 16, 539–549, 1987.
351. Wright, F. C., Younger, R. L., and Riner, J. C., Residues of mercury in tissues and eggs of chickens given oral doses of Panogen 15, *Bull. Environ. Contam. Toxicol.*, 12, 366–372, 1974.
352. Adams, W. J. and Prince, H. H., Mercury levels in the tissues of ring-necked pheasants fed two mercurial fungicides, *Bull. Environ. Contam. Toxicol.*, 15, 316–323, 1976.
353. Spalding, M. G., Bjork, R. D., Powell, G. V. N., and Sundlof, S. F., Mercury and cause of death in great white herons, *J. Wildl. Manage.*, 58, 735–739, 1994.
354. Van der Molen, E. J., Blok, A. A., and de Graaf, G. J., Winter starvation and mercury intoxication in grey herons (*Ardea cinerea*) in the Netherlands, *Ardea*, 70, 173–184, 1982.
355. Scheuhammer, A. M., Effects of acidification on the availability of toxic metals and calcium to wild birds and mammals, *Environ. Pollut.*, 71, 329–375, 1991.
356. Newton, I. and Haas, M. B., Pollutants in merlin eggs and their effects on breeding, *Br. Birds*, 81, 258–269, 1988.
357. Barr, J. F., Population dynamics of the common loon (*Gavia immer*) associated with mercury-contaminated waters in northwestern Ontario, Occas. Pap. 56, Can. Wildl. Serv., Ottawa, Ontario, 1986.
358. Scheuhammer, A. M., Perrault, J. A., and Bond, D. E., Mercury, methylmercury, and selenium concentrations in eggs of common loons (*Gavia immer*) from Canada, *Environ. Monit. Assess.*, 72, 79–94, 2001.
359. Scheuhammer, A. M. and Blancher, P. J., Potential risk to common loons (*Gavia immer*) from methylmercury exposure in acidified lakes, *Hydrobiologia*, 278/280, 445–455, 1994.
360. Lamborg, C. H., Fitzgerald, W. F., Vandal, G. M., and Rolfhus, K. R., Atmospheric mercury in northern Wisconsin: Sources and species, *Water Air Soil Pollut.*, 80, 189–198, 1995.
361. Rago, P. J. and Wiener, J. G., Does pH affect fish species richness when lake area is considered, *Trans. Am. Fish. Soc.*, 115, 438–447, 1986.
362. Custer, T. W., Custer, C. M., Hines, R. K., Gutreuter, S., Stromborg, K. L., Allen, P. D., and Melancon, M. J., Organochlorine contaminants and reproductive success of double-crested cormorants from Green Bay, Wisconsin, USA, *Environ. Toxicol. Chem.*, 18, 1209–1217, 1999.
363. Karasov, W. H. and Meyer, M. W., Testing the role of contaminants in depressing avian numbers, *Revista Chilena Hist. Nat.*, 73, 461–471, 2000.
364. Rahel, F. J. and Magnuson, J. J., Low pH and the absence of fish species in naturally acidic Wisconsin lakes: Inferences for cultural acidification, *Can. J. Fish. Aquat. Sci.*, 40, 3–9, 1983.
365. Rahel, F. J., Biogeographic influences on fish species composition of northern Wisconsin lakes with applications for lake acidification studies, *Can. J. Fish. Aquat. Sci.*, 43, 124–134, 1986.
366. Gariboldi, J. C., Jagoe, C. H., and Bryan, A. L., Jr., Dietary exposure to mercury in nestling wood storks (*Mycteria americana*) in Georgia, *Arch. Environ. Contam. Toxicol.*, 34, 398–405, 1998.
367. Vermeer, K., Armstrong, F. A. J., and Hatch, D. R. M., Mercury in aquatic birds at Clay Lake, western Ontario, *J. Wildl. Manage.*, 37, 58–61, 1973.
368. Henny, C. J., Grove, R. A., and Bentley, V. R., Effects of selenium, mercury, and boron on waterbird egg hatchability at Stillwater, Malheur, Seedskadee, Ouray, and Benton Lake National Wildlife Refuges and surrounding vicinities, Bureau of Reclamation, Nat. Irrigation Water Qual. Program Information Rep. No. 5, 2000.

369. Koeman, J. H., Garssen-Hoekstra, J., Pels, E., and de Goeij, J. J. M., Poisoning of birds of prey by methyl mercury compounds, *Mededelingen Fakulteit Landbouw-wetenschappen Gent*, 36, 43–49, 1971.
370. Borg, K., Erne, K., Hanko, E., and Wanntorp, H., Experimental secondary methyl mercury poisoning in the goshawk (*Accipiter g. gentilis L.*), *Environ. Pollut.*, 1, 91–104, 1970.
371. Fimreite, N. and Karstad, L., Effects of dietary methyl mercury on red-tailed hawks, *J. Wildl. Manage.*, 35, 293–300, 1971.
372. Finley, M. T., Stickel, W. H., and Christensen, R.E., Mercury residues in tissues of dead and surviving birds fed methylmercury, *Bull. Environ. Contam. Toxicol.*, 21, 105–110, 1979.
373. Scheuhammer, A. M., Chronic dietary toxicity of methylmercury in the zebra finch, *Poephila guttata*, *Bull. Environ. Contam. Toxicol.*, 40, 123–130, 1988.
374. Bhatnagar, M. K., Vrablic, O. E., and Yamashiro, S., Ultrastructural alterations of the liver of Pekin ducks fed methyl mercury-containing diets, *J. Toxicol. Environ. Health*, 10, 981–1003, 1982.
375. Pass, D. A., Little, P. B., and Karstad, L. H., The pathology of subacute and chronic methyl mercury poisoning of the mallard duck (*Anas platyrhynchos*), *J. Comp. Pathol.*, 85, 7–21, 1975.
376. Heinz, G. H., Mercury poisoning in wildlife, in *Noninfectious Diseases of Wildlife, 2nd ed.,* Fairbrother, A., Locke, L. N., and Hoff, G. L., Eds., Iowa State Univ. Press, Ames, 1996, 118–127.
377. Finley, M. T. and Stendell, R. C., Survival and reproductive success of black ducks fed methyl mercury, *Environ. Pollut.*, 16, 51–64, 1978.
378. Norheim, G., Levels and interactions of heavy metals in sea birds from Svalbard and the Antarctic, *Environ. Pollut.*, 47, 83–94, 1987.
379. Cuvin-Aralar, M. L. A. and Furness, R. W., Mercury and selenium interaction: A review, *Ecotoxicol. Environ. Saf.*, 21, 348–364, 1991.
380. Wobeser, G. and Swift, M., Mercury poisoning in a wild mink, *J. Wildl. Dis.*, 12, 335–340, 1976.
381. Wren, C. D., Probable case of mercury poisoning in a wild otter, *Lutra canadensis*, in northwestern Ontario, *Can. Field-Nat.*, 99, 112–114, 1985.
382. Wren, C. D., A review of metal accumulation and toxicity in wild mammals. I. Mercury, *Environ. Res.*, 40, 210–244, 1986.
383. Aulerich, R. J., Ringer, R. K., and Iwamota, J., Effects of dietary mercury on mink, *Arch. Environ. Contam. Toxicol.*, 2, 43–51, 1974.
384. Wren, C. D., Hunter, D. B., Leatherland, J. F., and Stokes, P. M., The effects of polychlorinated biphenyls and methylmercury, singly and in combination, on mink. I: Uptake and toxic responses, *Arch. Environ. Contam. Toxicol.*, 16, 441–447, 1987.
385. Dansereau, M., Lariviere, N., Tremblay, D. D., and Belanger, D., Reproductive performance of two generations of female semidomesticated mink fed diets containing organic mercury contaminated freshwater fish, *Arch. Environ. Contam. Toxicol.*, 36, 221–226. 1999.
386. Roelke, M. E., Schultz, D. P., Facemire, C. F., and Sundlof, S. F., Mercury contamination in the free-ranging endangered Florida panther (*Felis concolor coryi*), *Proc. Am. Assoc. Zoo Vet.*, 20, 277–283, 1991.
387. Harada, M. and Smith, A., Minamata disease: A medical report, in *Minimata*, Smith, W. E. and Smith, A. M., Eds., Holt, Rinehart and Winston, New York, 1975, 180–192.
388. Wobeser, G. A., Nielsen, N. O., and Scheifer, B., Mercury and mink. II. Experimental methyl mercury intoxication, *Can. J. Comp. Med.*, 40, 34–45, 1976.
389. O'Connor, D. J. and Nielson, S. W., Environmental survey of methylmercury levels in the wild mink and otter from the north-eastern United States and experimental pathology of methylmercurialism in the otter, in *Worldwide Furbearer Conf. Proc.*, Chapman, J. A. and Pursley, D., Eds., Frostburg, MD, 1981, 1725–1745.
390. Wobeser, G. A., Nielsen, N. O., and Scheifer, B., Mercury and mink. I. The use of mercury contaminated fish as a food for ranch mink, *Can. J. Comp. Med.*, 40, 30–33, 1976.
391. Halbrook, R. S., Lewis, L. A., Aulerich, R. I., and Bursian, S. J., Mercury accumulation in mink fed fish collected from streams on the Oak Ridge Reservation, *Arch. Environ. Contam. Toxicol.*, 33, 312–316, 1997.
392. Wren, C. D., Hunter, D. B., Leatherland, J. F., and Stokes, P. M., The effects of polychlorinated biphenyls and methylmercury, singly and in combination on mink. II. Reproduction and kit development, *Arch. Environ. Contam. Toxicol.*, 16, 449–454, 1987.

393. Burbacher, T. M., Rodier, P. M., and Weiss, B., Methylmercury developmental neurotoxicity: A comparison of effects in humans and animals, *Neurotoxicol. Teratol.*, 12, 191–202, 1990.
394. Sheffy, T. B. and St. Amant, J. R., Mercury burdens in furbearers in Wisconsin, *J. Wildl. Manage.*, 46, 1117–1120, 1982.
395. Kucera, E., Mink and otter as indicators of mercury in Manitoba waters, *Can. J. Zool.*, 61, 2250–2256, 1983.
396. Wren, C. D., Stokes, P. M., and Fischer, K. L., Mercury levels in Ontario mink and otter relative to food levels and environmental acidification, *Can. J. Zool.*, 64, 2854–2859, 1986.
397. Fortin, C., Beauchamp, G., Dansereau, M., Lariviere, N., and Belanger, D., Spatial variation in mercury concentrations in wild mink and river otter carcasses from the James Bay Territory, Quebec, Canada, *Arch. Environ. Contam. Toxicol.*, 40, 121–127, 2001.
398. Komsta-Szumska, E., Czuba, M., Reuhl, K. R., and Miller, D. R., Demethylation and excretion of methyl mercury by the guinea pig, *Environ. Res.*, 32, 247–257, 1983.
399. Dietz, R., Nielsen, C. O., Hansen, M. M., and Hansen, C. T., Organic mercury in Greenland birds and mammals, *Sci. Total Environ.*, 95, 41–51, 1990.
400. Ganther, H. E., Goudie, C., Sunde, M. L., Kopecky, M. J., Wagner, P., Oh, S. H., and Hoekstra, W. G., Selenium: Relation to decreased toxicity of methylmercury added to diets containing tuna, *Science*, 175, 1122–1124, 1972.
401. Stillings, B., Lagally, H., Baurersfeld, P., and Soares, J., Effect of cystine, selenium and fish protein on the toxicity and metabolism of methyl mercury in rats, *Toxicol. Appl. Pharmacol.*, 30, 243–254, 1974.
402. Potter, S. and Matrone, G., Effect of selenite on the toxicity of dietary methyl mercury and mercuric chloride in the rat, *J. Nutr.*, 104, 638–647, 1974.
403. Chang, L. W. and Suber, R., Protective effect of selenium on methylmercury toxicity: A possible mechanism, *Bull. Environ. Contam. Toxicol.*, 29, 285–289, 1982.
404. Miyama, T., Minowa, K., Seki, H., Tamura, Y., Mizoguchi, J., Ohi, G., and Suzuki, T., Chronological relationship between neurological signs and electrophysiological changes in rats with methylmercury poisoning: Special reference to selenium protection, *Arch. Toxicol.*, 52, 173–181, 1983.
405. Rawson, A. J., Patton, G. W., Hofmann, S., Pietra, G. G., and Johns, L., Liver abnormalities associated with chronic mercury accumulation in stranded Atlantic bottlenose dolphins, *Ecotoxicol. Environ. Saf.*, 25, 41–47, 1993.
406. Martoja, R. and Berry, J. P., Identification of tiemannite as a probable product of demethylation of mercury by selenium in cetaceans, *Vie Milieu*, 30, 7–10, 1980.
407. Nigro, M., Mercury and selenium localization in macrophages of the striped dolphin, *Stenella coeruleoalba*, *J. Mar. Biol. Assoc. U.K.*, 74, 975–978, 1994.
408. Cavalli, S. and Cardellicchio, N., Direct determination of seleno-amino acids in biological tissues by anion-exchange separation and electrochemical detection, *J. Chromatogr.*, 706(A), 429–436, 1995.
409. Palmisano, F., Cardellicchio, N., and Zambonin, P. G., Speciation of mercury in dolphin liver: A two-stage mechanism for the demethylation accumulation process and role of selenium, *Mar. Environ. Res.*, 40, 109–121, 1995.
410. Cardellicchio, N., Decataldo, A., Di Leo, A., and Misino, A., Accumulation and tissue distribution of mercury and selenium in striped dolphins (*Stenella coeruleoalba*) from the Mediterranean Sea (southern Italy), *Environ. Pollut.*, 116, 265–271, 2002.
411. Honda, K., Tatsukava, R., Itano, K., Miyazaki, N., and Fujiyama, T., Heavy metals concentration in muscle, liver and kidney tissue of striped dolphin *Stenella coeruleoabla* and their variation with body length, weight, age and sex, *Agric. Biol. Chem.*, 47, 1219–1228, 1983.
412. Osowski, S. L., Brewer, L. W., Baker, O. E., and Cobb, G. P., The decline of mink in Georgia, North Carolina, and South Carolina: The role of contaminants, *Arch. Environ. Contam. Toxicol.*, 29, 418–423, 1995.
413. Giesy, J. P., Verbrugge, D. A., Othout, R. A., Bowerman, W. W., Mora, M. A., Jones, P. D., Newsted, J. L., Vandervoort, C., Heaton, S. N., Aulerich, R. J., Bursian, S. J., Ludwig, J. P., Dawson, G. A., Kubiak, T. J., Best, D. A., and Tillitt, D. E., Contaminants in fishes from Great Lakes-influenced sections and above dams of three Michigan rivers. II: Implications for health of mink, *Arch. Environ. Contam. Toxicol.*, 27, 213–223, 1994.
414. Sample, B. E. and Suter, G. W., Ecological risk assessment in a large river-reservoir: 4. piscivorous wildlife, *Environ. Toxicol. Chem.*, 18, 610–620, 1999.

415. Moore, D. R. J., Sample, B. E., Suter, G. W., Parkhurst, B. R., and Teed, R. S., A probabilistic risk assessment of the effects of methylmercury and PCBs on mink and kingfishers along East Fork Poplar Creek, Oak Ridge, Tennessee, USA, *Environ. Toxicol. Chem.*, 18, 2941–2953, 1999.
416. Egeland, G. M. and Middaugh, J. P., Balancing fish consumption benefits with mercury exposure, *Science*, 278, 1904–1905, 1997.
417. Lodenius, M., The mercury problem and fishing in Finland, in *Economics of Ecosystems Management*, Hall, D. O., Myers, N., and Margaris, N. S., Eds., W. Junk Publishers, Dordrecht, 1985, 99–103.
418. Håkanson, L., A simple model to predict the duration of the mercury problem in Sweden, *Ecol. Modeling*, 93, 251–262, 1996.
419. Wheatley, M. A., Social and cultural impacts of mercury pollution on Aboriginal peoples of Canada, *Water Air Soil Pollut.*, 97, 85–90, 1997.
420. Wheatley, B., Paradis, S., Lassonde, M., Giguere, M. F., and Tanguay, S., Exposure patterns and long term sequelae on adults and children in two Canadian indigenous communities exposed to methylmercury, *Water Air Soil Pollut.*, 97, 63–73, 1997.
421. United States Environmental Protection Agency, Update: National Listing of Fish and Wildlife Advisories, Fact Sheet EPA-823-F-01–010, Office of Water, Washington, D.C., 2001.
422. Brunberg, A. K. and Blomqvist, P., Quantification of anthropogenic threats to lakes in a lowland county of central Sweden, *Ambio*, 30, 127–134, 2001.
423. Chan, H. M. and Receveur, O., Mercury in the traditional diet of indigenous peoples in Canada, *Environ. Pollut.*, 110, 1–2, 2000.
424. Verta, M., Rekolainen, S., and Kinnunen, K., Causes of increased fish mercury levels in Finnish reservoirs, Publ. No. 65, Water Res. Inst., Nat. Board Waters, Helsinki, Finland, 1986, 44–58.
425. Jackson, T. A., The mercury problem in recently formed reservoirs of northern Manitoba (Canada): Effects of impoundment and other factors on the production of methyl mercury by microorganisms in sediments, *Can. J. Fish. Aquat. Sci.*, 45, 97–121, 1988.
426. Ramsey, D. J., Experimental studies of mercury dynamics in the Churchill River diversion, Manitoba, *Collect. Environ. Géol.*, 9, 147–173, 1990.
427. Yingcharoen, D. and Bodaly, R. A., Elevated mercury levels in fish resulting from reservoir flooding in Thailand, *Asian Fish. Sci.*, 6, 73–80, 1993.
428. Surma-Aho, K., Paasivirta, J., Rekolainen, S., and Verta, M., Organic and inorganic mercury in the food chain of some lakes and reservoirs in Finland, *Chemosphere*, 15, 353–372, 1986.
429. Bodaly, R. A. and Johnston, T. A., The mercury problem in hydro-electric reservoirs with predictions of mercury burdens in fish in the proposed Grande Baleine Complex, Québec, James Bay Publ. Series, Hydro-Electric Development Environmental Impacts, Paper No. 3, North Wind Information Serv., Inc., Montreal, Québec, 1992.
430. Carter, V., Wetland hydrology, water quality, and associated functions, in National Water Summary on Wetland Resources, U.S. Geol. Surv. Water-Supply Pap. 2425, 1996, 35–48.
431. Novitzki, R. P., Smith, R. D., and Fretwell, J. D., Wetland functions, values, and assessment, in National Water Summary on Wetland Resources, U.S. Geol. Surv. Water-Supply Pap. 2425, 1996, 79–86.
432. Verhoeven, J. T. A. and Mueleman, A. F. M., Wetlands for wastewater treatment: Opportunities and limitations, *Ecol. Eng.*, 12, 5–12, 1999.
433. Kivaisi, A. K., The potential for constructed wetlands for wastewater treatment and reuse in developing countries: A review, *Ecol. Eng.*, 16, 545–560, 2001.
434. Driscoll, C. T., Holsapple, J., Schofield, C. L., and Munson, R., The chemistry and transport of mercury in a small wetland in the Adirondack region of New York, USA, *Biogeochemistry*, 40, 137–146, 1998.
435. Guentzel, J. L., Landing, W. M., Gill, G. A., and Pollman, C. D., Atmospheric deposition of mercury in Florida: The FAMS Project (1992–1994), *Water Air Soil Pollut.*, 80, 393–402, 1995.
436. Dvonch, J. T., Graney, J. R., Keeler, G. J., and Stevens, R. K., Use of elemental tracers to source apportion mercury in south Florida precipitation, *Environ. Sci. Technol.*, 33, 4522–4527, 1999.
437. Guentzel, J. L., Landing, W. M., Gill, G. A., and Pollman, C. D., Processes influencing rainfall deposition of mercury in Florida, *Environ. Sci. Technol.*, 35, 863–873, 2001.
438. Facemire, C. F., Gross, T. S., and Guillette, L. J., Jr., Reproductive impairment in the Florida panther: Nature or nurture, *Environ. Health Perspect.*, 103, 79–86, 1995.

439. Beyer, W. N., Spalding, M., and Morrison, D., Mercury concentrations in feathers of wading birds from Florida, *Ambio*, 26, 97–100, 1997.
440. Heaton-Jones, T. G., Homer, B. L., Heaton-Jones, D. L., and Sundlof, S. F., Mercury distribution in American alligators (*Alligator mississippiensis*) in Florida, *J. Zoo Wildl. Med.*, 28, 62–70, 1997.
441. Yanochko, G. M., Jagoe, C. H., and Brisbin, I. L., Tissue mercury concentrations in alligators (*Alligator mississippiensis*) from the Florida Everglades and the Savannah River Site, South Carolina, *Arch. Environ. Contam. Toxicol.*, 32, 323–328, 1997.
442. Frederick, P. C., Spalding, M. G., Sepulveda, M. S., Williams, G. E., Nico, L., and Robins, R., Exposure of great egret (*Ardea albus*) nestlings to mercury through diet in the Everglades ecosystem, *Environ. Toxicol. Chem.*, 18, 1940–1947, 1999.
443. Duvall, S. E. and Barron, M. G., A screening level probabilistic risk assessment of mercury in Florida Everglades food webs, *Ecotoxicol. Environ. Saf.*, 47, 298–305, 2000.
444. Spalding, M. G., Frederick, P. C., McGill, H. C., Bouton, S. N., Richey, L. J., Schumacher, I. M., Blackmore, C. G. M., and Harrison, J., Histologic, neurologic, and immunologic effects of methylmercury in captive great egrets, *J. Wildl. Dis.*, 36, 423–435, 2000.

# CHAPTER 17

# Ecotoxicology of Selenium

**Harry M. Ohlendorf**

## CONTENTS

## 17.1 INTRODUCTION

Selenium is a naturally occurring semimetallic (also referred to as a metalloid) trace element that is essential for animal nutrition in small quantities but becomes toxic at dietary concentrations that are not much higher than the required levels for good health. Thus, dietary selenium concentrations that are either below or above the optimal range are of concern. This chapter summarizes the ecotoxicology of excessive selenium exposure for animals, especially as reported during the last 15 years. It focuses primarily on freshwater fish and aquatic birds because fish and birds are the groups of animals for which most toxic effects have been reported in the wild. However, information related to bioaccumulation by plants and animals as well as effects in invertebrates, amphibians, reptiles, and mammals also are presented.

Selenium was recognized long ago as a cause of toxicity in domestic poultry and livestock (e.g., References 1–3), although a recent reevaluation of some of the historic samples and related information by O'Toole and Raisbeck[4] indicates that the condition known as "blind staggers" in livestock is probably not caused by selenium. For fish and wildlife, selenium became much more of a concern in the 1970s and 1980s with the discovery of selenium bioaccumulation and severe impacts in fish and aquatic birds. These problems and other general information (such as environmental occurrence, general and biochemical properties, and other characteristics) are summarized in a number of recent reviews and books about selenium (see References 5–19) and are reviewed briefly in this chapter. Readers may wish to consult those references for additional information.

As a result of new information related to food-chain bioaccumulation and effects of selenium in freshwater fish, the ambient water quality criteria were revised in 1987,[20] and they are being reviewed again.[21,22] Water quality criteria for freshwater systems are based on total recoverable selenium (as discussed later in this chapter), whereas the saltwater criteria are based on dissolved selenium. Because of differences in the toxicity of different chemical forms, the current acute criterion (criteria maximum concentration) for freshwater considers selenite and selenate as relative fractions of the total recoverable selenium concentration.

The environmental significance of selenium is a topic of ongoing discussion and debate. A recent issue of the journal *Aquatic Toxicology* (Vol. 57, No. 1) was devoted to papers on selenium. In December 1999, the journal *Human and Ecological Risk Assessment* devoted an issue (Vol. 5, No. 6) to debate and commentary on the topic, "Selenium — A potential time bomb or just another contaminant?" and published several papers on the topic. Needless to say, the debate continues, as not all the issues were resolved. In particular, there are differences of opinion concerning waterborne concentrations of selenium that are protective for fish and wildlife,[23–28] the relative importance of sediment vs. waterborne selenium in affecting aquatic biota,[29–31] and the specific thresholds of selenium in the diet or eggs of birds at which reproduction is adversely affected.[32–34] These topics are discussed in later sections of this chapter.

Selenium concentrations typically are reported as micrograms per liter (μg/L) in fluids and milligrams per kilogram (mg/kg) or micrograms per gram (μg/g) in soil, sediment, plant or animal tissues, and diets. Concentrations in soil, sediment, tissues, and diets can be expressed either on a wet-weight or fresh-weight basis (which are considered to be synonymous), or on dry-weight basis. Conversion from one basis to the other is a function of the moisture content in the sample (which should be reported regardless of which basis is used), as follows:

$$\text{Dry-weight conc.} = \text{Wet-weight conc.} \times \frac{100}{(100 - \text{Moisture percentage})}$$

For example, 10 mg/kg on a wet-weight basis in a sample having 80% moisture is equal to 50 mg/kg on a dry-weight basis. Selenium concentrations in soil, sediment, and tissue throughout this chapter are given on a dry-weight basis, unless otherwise noted.

## 17.2 ENVIRONMENTAL CHEMISTRY

From an ecotoxicological perspective, it is important to have an understanding of selenium's chemical characteristics, environmental sources and occurrence, and cycling in the environment. Those aspects of selenium chemistry can help define geographic areas and environmental settings in which selenium may be a chemical of ecological concern, and they are briefly summarized in this section.

### 17.2.1 Chemical Characteristics

Selenium chemistry is complex, and chemical forms vary in their environmental occurrence, biogeochemistry, and toxicity.[7,15,17,19,35–38] In nature, selenium is generally recognized to occur in four oxidation states: selenide ($Se^{2-}$), elemental selenium ($Se^{0}$), selenite ($Se^{4+}$), and selenate ($Se^{6+}$), although a fifth form (selenium dioxide [$Se^{2+}$]) may occur as a product of combustion of elemental selenium present in fossil fuels or rubbish.[19] Soluble selenates are the predominant form under oxidizing conditions in alkaline soils that are commonly found in arid areas. Selenates are readily available to plants, or they can be slowly reduced to selenites, which also can be taken up by plants and converted into organic forms. The common organic forms in plants include selenomethionine, selenocysteine, dimethylselenide, and dimethyldiselenide. Selenite is the more common soluble form of selenium under reducing conditions and in acidic soils, which occur more typically in higher-rainfall areas. Selenium occurs in various organic and inorganic forms in coal and other fossil fuels. In general, elemental selenium and metallic selenides are not readily bioavailable. However, under oxidizing conditions they can be transformed to bioavailable forms.

Selenium commonly occurs as a mixture of several different chemical forms in surface waters, but selenates and selenites are most common.[10,17,25,38] Metal and organic selenides are common in bottom sediments. Waterborne selenium can be evaluated most reliably as total recoverable concentrations,[17] and total recoverable concentrations are the basis of the ambient water quality criteria for freshwater ecosystems.[20,22] This is because partitioning of selenium between the water column and other compartments of an aquatic ecosystem greatly affects the measurement of waterborne selenium. Low waterborne selenium concentrations can reflect low mass loading of the element, but they also may reflect high biotic uptake[17] (see also Kesterson Reservoir example below). Total recoverable selenium includes suspended detrital particulate matter (a function of biotic uptake) and thus more accurately reflects the total mass load of selenium in the system. Differences between dissolved and total recoverable measurements increase with higher levels of eutrophication of the water body.

During the early and mid-1980s, subsurface agricultural drainage waters from the San Joaquin Valley, California, were disposed of by discharging them to Kesterson Reservoir, which was a series of 12 shallow ponds totaling about 500 ha.[7] Selenium concentrations in water entering the reservoir during 1983 to 1985 averaged about 300 μg Se/L, and aquatic plants and invertebrates contained greatly elevated concentrations of selenium (typically 10 to 100 times those found at a nearby reference site).[7, 39–42] Almost all of the waterborne selenium was in the selenate form. Unlike boron and a number of other constituents, selenium concentrations *decreased* as the water flowed through the series of ponds and evaporated. Similarly, plants and animals as well as sediments in ponds nearer the inflow contained higher concentrations of selenium than those downstream. Thus, bioaccumulation by plants and animals removed substantial amounts of selenium from the water and deposited it into the sediments. Similar findings were reported for another site in Montana.[43]

Selenium is essential for animals and for some plants.[19] Biochemically, it is similar to sulfur; when selenium is present at elevated dietary levels, it replaces sulfur in some metabolic pathways and thereby causes problems[23,36,37,44] (see also, later section on effects in fish). Selenium is an essential component of the enzyme glutathione peroxidase, which, along with vitamin E, serves as an antioxidant to prevent metabolic damage to tissues.

Selenium occurs in various chemical forms in plant and animal tissues, but bioavailability is greater from plant selenium than from animal foods.[5,45] Selenomethionine has been found to be a good surrogate for natural food-chain selenium in experimental studies.[17] In general, the diet is the most important exposure pathway for vertebrate animals and should be included when conducting or evaluating the effects observed in experimental studies.

### 17.2.2 Environmental Sources and Occurrence

Typical concentrations in the Earth's crust are < 0.5 mg Se/kg, but some geologic formations are greatly enriched in selenium.[19,36,39,46–50] Especially when these formations are modified as a result of human activities (e.g., mining, agricultural irrigation), the selenium may be mobilized and become more available to plants and animals, which bioaccumulate selenium and incorporate it into the food chain for other organisms. Selenium tends to be present in large amounts in areas where soils were derived from Cretaceous and Tertiary marine sedimentary rocks, and selenium from these sources is highly mobile and biologically available in arid regions having alkaline soils.[17]

Industrial sources, such as coal and oil combustion, nonferrous metal production (primarily copper and nickel, but also lead, zinc, and cadmium), steel and iron manufacturing, municipal and sewage refuse incineration, and production of phosphate fertilizers, introduce much more selenium into the environment than do natural sources such as volcanic activity or weathering of seleniferous rocks.[51,52] However, there are no ores from which selenium can be mined as a primary product.[53] It is found mainly in sulfide minerals of copper, iron, and lead and is most commonly produced by electrolytic separation from copper during the refining process. Average annual usage is estimated at 1850 metric tons. Selenium is used in a wide variety of industrial applications, with glass (35% of demand) and electrical (30%) applications representing most of the use. Other uses include pigments (10%), metallurgy (10%), agricultural/biological (5%), and miscellaneous applications (10%).

Fossil fuels (coal and oil), as well as associated formations from which they are extracted, may contain elevated concentrations of selenium.[19,25] Concentrations vary considerably depending on the source. The selenium may be mobilized through mining of coal or disposal of fly ash from power plants where the coal was burned. In terrestrial environments, when fly ash is deposited in landfills, it may be taken up in significant amounts by plants growing on the landfill.[18] One well-known example of effects associated with fly ash disposal into water bodies has been reported in a series of studies conducted at Belews Lake, North Carolina (see, for example, Reference 24). Similarly, selenium from crude oil may be discharged with process wastewater from the refinery, thereby contributing to bioaccumulation by organisms in the receiving waters.[25,54]

Mass-loading of selenium into aquatic environments most typically results from disposal of coal fly ash, irrigation wastewater, or oil refinery wastewater.[17,25,54] Mining or smelting of sulfide ores, uranium, phosphate, bentonite, and coal also contributes to mobilization of selenium because of its presence in the mined materials or the overburden that is moved during the mining process.[17,19,47,55]

In freshwater systems, normal background levels of selenium in sediment are 0.2 to 2.0 mg/kg, and in water 0.1 to 0.4 μg/L.[17]

### 17.2.3 Cycling

Selenium reaches the earth's surface through volcanic activity, and some authors have proposed that oceans were enriched in selenium from volcanic activity during Cretaceous times.[7,35,46,50,56] This early enrichment, along with bioaccumulation of selenium in the oceans and later deposition in the sediments,[48] would explain the high selenium concentrations found in many Cretaceous marine rocks. Tertiary marine sedimentary rocks contain selenium resulting from bioaccumulation and deposition of particulate matter derived by erosion of older (Cretaceous) seleniferous sedimentary deposits and also are generally seleniferous.

Irrigation of seleniferous soils can dissolve and mobilize selenium and then transport it in groundwater to irrigation drains.[48–50] Geology, climate, and hydrology were found to be important factors affecting the mobilization and transport of selenium into habitats where biota are adversely affected in the western United States. Although selenium can be transported naturally as a result of rock weathering and drainage into water bodies, the process has been increased through irrigation of soils derived from seleniferous formations, especially Cretaceous and Tertiary marine sedimentary deposits.

Many biogeochemical processes affect the cycling of selenium through different components of the environment.[7,10,35,52,57,58] In wetlands, selenium oxidation-reduction reactions are the most important processes controlling speciation, precipitation/dissolution, sorption/desorption, methylation, and volatilization of selenium. Selenate is stable in well-oxidized environments, but it is converted slowly to selenite or elemental selenium under less-oxidized conditions. It can be further reduced to metal selenides or volatile methylated forms (primarily dimethylselenide) through microbial processes. The metal selenides tend to be deposited in the wetland sediments, whereas volatile forms escape to the atmosphere. Selenides and elemental selenium tend to be very insoluble, although they can be converted to more soluble forms (selenite and selenate) under oxidizing conditions.

The changes described above can be important considerations in the wetting and drying cycles that occur in seasonal wetlands, as well as periodic draw-down of water levels in permanent wetlands. When submerged, especially where large amounts of organic material are present, selenium tends to be present in reduced (and less toxic) forms, and volatilization is favored.[10] If the water level is lowered, the selenium becomes more oxidized and bioavailable. Thus, the selenium present in the wetland sediments and organic matter would become more likely to bioaccumulate into aquatic organisms soon after the wetland is reflooded than when it was previously flooded.

Although selenium in igneous and sedimentary rocks may be insoluble and unavailable to terrestrial plants, chemical weathering and plant and microbial action transform much of it into soluble (and bioavailable) forms.[46,56] Oxidation of this selenium in alkaline soils produces selenate, making it readily available to plants. Selenium uptake by plants is affected by both the concentration and its bioavailability in soil. Selenium uptake by plants is greater in soils having higher pH and in areas of low rainfall. In areas with acidic soils and higher rainfall, selenium is generally less available to plants for uptake.

Selenium enters the atmosphere through volcanic activity, burning of fossil fuels (especially coal), and volatilization from terrestrial or aquatic environments.[35,52] Although concentrations are low (typically around 1 ng/m$^3$ air, but ranging up to 6000 times that level near a smelter[59]), atmospheric dispersion and subsequent deposition of the selenium redistributes it in the environment. For example, deposition is estimated to have contributed to a 15% increase in soil selenium in the United Kingdom during the last century and to contribute between 33 and 82% of plant leaf selenium uptake. Deposition occurs either as "wet deposition" (with precipitation), which accounts for most of the deposition, or "dry deposition" (exchange of particulate and gaseous material between the atmosphere and the global surface).

Numerous studies of selenium transformations, cycling, and volatilization in aquatic and terrestrial ecosystems have been reported (or reviewed) in recent years.[16,18,43,52,60–67] Not surprisingly, rates for these processes vary greatly, depending on temperature, moisture, organic carbon content of soil/sediment, selenium concentration and chemical form, and microbiological activity. Although details of some of the processes are not yet well known, it is clear that microbes are largely responsible for many of the changes that occur.

Plants vary widely in their ability to accumulate and volatilize selenium from contaminated wastewater, soils, and sediments.[7,66,68] Under some circumstances, phytoremediation of these contaminated media may be a viable approach to reducing environmental exposures. However, waterborne selenium can be accumulated by plants and animals living in the wetlands to levels that are harmful to aquatic organisms as well as birds that feed upon them.[7,19] As seleniferous waters enter a wetland, selenium is removed from the water through a combination of bioaccumulation by plants and animals, deposition in the sediments, and volatilization (as relatively low-toxicity dimethylselenide) to the atmosphere.[40,42,69,70] Because of selenium bioaccumulation in the food chain and deposition to sediments, wetlands should not be used for phytoremediation of selenium-contaminated wastewater without conducting an ecological risk assessment to evaluate potential adverse effects to birds and other potentially exposed animals.

The biological half-life of selenium in animals is relatively short, varying from 10 days in pheasants to 64 days in earthworms.[71,72] More detail concerning uptake and loss of selenium in birds is presented in the section on bioaccumulation in birds.

## 17.3 BIOACCUMULATION

Selenium bioaccumulates in aquatic and terrestrial food chains and in higher trophic-level animals that feed on those plants and animals. Bioaccumulation is defined as the combined net accumulation (taking into account any concurrent loss) of a chemical from abiotic media and ingestion of selenium-containing biota (i.e., the food chain). In aquatic systems, bioaccumulation may occur by direct absorption or partitioning from water or sediment or by ingestion. The greatest level of bioaccumulation is typically from water to aquatic plants or invertebrates (often a 1000-fold or greater increase occurs). In terrestrial systems, bioaccumulation is primarily from ingestion, because uptake from air is less than that from water, and airborne concentrations generally are lower than those in water. When concentrations in animals are considered on whole-body basis (which is the appropriate perspective), selenium does not biomagnify through the various trophic levels. (Biomagnification is defined as increasing concentrations through successive trophic levels.) Nevertheless, significant bioaccumulation does occur (as described in this section) and leads to adverse effects in sensitive organisms (especially fish and birds; see next section).

This section presents a general summary of selenium bioaccumulation by plants and animals. Selenium concentrations typically are higher in marine organisms than in those from freshwater ecosystems, and they are higher in areas having seleniferous soils or sediments and in areas receiving industrial, agricultural, or municipal wastes. Additional information, including more detailed tabulations of selenium concentrations in biota, are provided by Jenkins,[73] Wilber,[71] and Eisler.[19]

### 17.3.1 Plants

#### *17.3.1.1 Aquatic*

Background concentrations in freshwater algae are 0.1 to 1.5 mg Se/kg, and in rooted plants they are 0.1 to 2.0 mg Se/kg.[17,19] By comparison, algae and rooted aquatic plants accumulated 20 to 390 mg Se/kg in Kesterson Reservoir, California, which received agricultural drainage waters containing up to 330 μg Se/L.[40–42] Selenium bioaccumulated in aquatic plants 28 to 5100 (mean =

1105) times the concentration in water.[42] Elevated concentrations of selenium also have been found in plants from ponds receiving coal fly ash.[74, 75]

The bioaccumulation and volatilization of selenium by wetland plants (referred to as phytoremediation) can be used to treat waterborne selenium under certain conditions. There was at least a 50-fold variation in selenium accumulation and volatilization among 20 aquatic species tested with water containing selenite or selenate.[66] Several of the aquatic plant species showed selenium volatilization and accumulation rates that were similar to that of Indian mustard (*Brassica juncea*), the best-known terrestrial plant species for phytoremediation. However, the overall implications of using vegetated wetlands to remove selenium from agricultural or industrial wastewaters are not yet well studied. This technology may be appropriate for waters containing moderate levels of selenium (perhaps up to 25 or 30 μg/L), but only if it is designed and managed to limit access by aquatic birds that would feed on plants and invertebrates in the wetland. However, at higher waterborne concentrations of selenium, or with inadequate attention to design and management concerns, the wetland could present undesirable levels of exposure to birds.

Several laboratory microcosm studies (summarized below) have measured selenium bioconcentration (i.e., direct uptake) by algae from water and subsequent transfer of selenium to invertebrates or fish consuming the algae. These studies show that the chemical form of the waterborne selenium is an important factor in affecting uptake by aquatic biota and that sulfate content of the water can influence uptake. These and other similar laboratory studies of aquatic systems show that selenium bioaccumulation is a significant mechanism of concern for higher trophic-level organisms.

Bioconcentration factors (BCFs) from water to macrophytes (rooted plants and filamentous algae), periphyton (algae and associated material attached to the wall of the microcosm), and zooplankton (daphnids and their offspring) were significantly higher for selenomethionine than for selenite, which in turn had higher bioconcentration factors than selenate.[76] Volatilization of selenium from the microcosms followed the same pattern (i.e., highest for selenomethionine at 24% loss during the 28-day exposure). For all three forms of selenium, bioaccumulation was higher in zooplankton than in periphyton, which had greater accumulation than macrophytes.

Another study by Besser et al.[77] with algae (*Chlamydomonas reinhardtii*), daphnids (*Daphnia magna*), and fish (bluegill, *Lepomis macrochirus*) also showed higher bioaccumulation from waterborne selenomethionine than from either selenite or selenate. Algae and daphnids concentrated selenium more strongly from selenite (BCFs = 220 to 3600) than selenate (BCFs = 65 to 500), whereas bluegills concentrated selenium about equally from both inorganic forms (estimated BCFs = 13 to 106). Bioaccumulation of foodborne selenium by daphnids and bluegills was similar in food chains dosed with different forms of selenium, and they did not accumulate selenium concentrations greater than those in their diet (except at very low dietary selenium concentrations). In exposures based on selenite, bluegills accumulated greater selenium concentrations from food than from water, and aqueous and food-chain uptakes were approximately additive. The results indicate that most of the accumulation of inorganic waterborne selenium by bluegills in selenium-contaminated environments is via food uptake. However, organoselenium compounds, such as selenomethionine, may contribute significantly to selenium bioaccumulation by bluegills through both aqueous and food-chain uptake.

Corixids (*Trichocorixa reticulata*), which are common insects in wetlands receiving agricultural drainage waters and are important food-chain organisms for fish and birds, were exposed to waterborne and foodborne selenium.[78] Algae (*Oscillatoria* sp. filaments and a *Microcystis*-type colonial alga), the food for the corixids, were exposed to water containing sodium selenate at selenium concentrations ranging from 7.3 to 870 μg/L for 48-h exposures. Corixids exposed to waterborne selenate did not accumulate selenium above control concentrations. Corixids fed algae exposed to ≥87 μg Se/L selenate had significantly higher selenium concentrations than controls, suggesting that corixids may be isolated from the water and that selenium accumulation is solely through dietary exposure.

Transfer of selenium also was followed through a laboratory food chain (water-algae-rotifers-larval fish), and its effects on the larval fathead minnows (*Pimephales promelas*) were evaluated.[72]

Selenium uptake was measured in rotifers (*Brachionus calyciflorus*) fed algae (*Chlorella pyrenoidosa*) that had been cultured in a selenate-containing medium. The rotifers were then fed to larval fathead minnows, and uptake and loss of selenium as well as effects in the fish were measured. Selenium concentrations in the fish (51.7 and 61.1 µg/g in two experiments) were similar to those in their food (68 and 55 µg/g). Final weights of larvae fed selenium-contaminated rotifers were significantly lower than those of controls, although mortality did not occur. Biological half-life of food-derived selenium in the fish larvae was 28 days.

Selenium concentrations in field-collected widgeongrass (*Ruppia maritima*) varied among samples from four evaporation pond systems in the San Joaquin Valley, California.[79] These plants were used as a substrate to measure selenium bioaccumulation and effects in a laboratory benthic-detrital food chain with midge (*Chironomus decorus*) larvae. Following a 96-h exposure, selenium concentrations had increased significantly in the three groups of larvae exposed to plants having the highest selenium concentrations, but there was no consistent pattern of adverse effects on larvae weights. After a 14-day exposure, selenium bioaccumulation patterns in midge larvae were similar to that observed after 96 h. However, as midge selenium concentrations increased, mean weight decreased, indicating that selenium apparently had a significant effect on growth. The results of this study complement other findings that indicate sediment and the detrital pathway are important in the cycling of selenium in aquatic systems.[80–82]

Waterborne sulfate reduces the uptake of selenium by plants when waterborne selenium is in the selenate form. Increasing sulfate resulted in significantly reduced selenate uptake by algae (*Selenastrum capricornutum*) and increased algal growth.[83] Sulfate also reduced selenium uptake from waterborne selenate by widgeongrass, but selenite and selenomethionine uptake were not affected by sulfate.[84] As in other studies summarized above, uptake of selenomethionine (BCFs up to 21,800) by widgeongrass was much higher than for selenite or selenate, and the highest rates of uptake (i.e., BCFs) occur at the lowest waterborne concentrations. Results of this and other field and laboratory studies show that waterborne selenium occurs under field conditions in a mixture of chemical forms, and BCFs therefore vary depending on the environmental conditions and sources of selenium.

#### *17.3.1.2 Terrestrial*

Background selenium concentrations in terrestrial plants are 0.01 to 0.6 mg/kg.[17,85] Bioaccumulation of selenium by plants depends on the species of plant, environmental conditions, age and phase of plant growth, and the chemical form of selenium present.[3,36,46,85,86] Some plants, such as white clover (*Trifolium repens*), buffalograss (*Buchloe dactyloides*), and grama (*Bouteloua* spp.), growing on seleniferous soils accumulate surprisingly low levels of selenium. In contrast, high sulfur-containing plants like the *Brassica* species (mustard, cabbage, broccoli, and cauliflower) and other Brassicaceae are relatively good accumulators of selenium.

Three groups of plants are generally recognized on the basis of their tendency to accumulate selenium when grown on high-selenium soils (as originally described by Rosenfeld and Beath[3]). The first two groups of plants are referred to as *selenium accumulator* or *indicator* plants. These plants grow well on soils containing high levels of available selenium, and some have been used to locate seleniferous soils. Plants in Group 1 are called primary indicators or accumulators and normally accumulate selenium to very high concentrations (often several thousand mg/kg). Plants in Group 2 are referred to as secondary accumulators and may contain a few hundred mg Se/kg in their tissues. Those in Group 3 include grains, grasses, and many other plants that do not normally accumulate selenium in excess of 50 mg/kg when grown on seleniferous soils. Deep-rooted shrubs and other plants with long tap roots may act as pumps, bringing selenium from the deeper soil profiles to the surface and near-surface soils, where it is available to shallower-rooted plants such as grasses.[87]

Selenium concentrations in plants usually decline with maturity, so the highest concentrations usually occur during the most active growth phase.[3,85] Because selenium is associated with protein in the plant, leaves usually contain higher selenium concentrations than seeds (e.g., see also Schuler et al.[42]). Although plant selenium concentration can generally be used as an indicator of soil selenium status,[85] total selenium concentrations in soil do not necessarily determine whether the plants growing there will induce toxicity or nutritional deficiency in animals.[88]

Following the discovery of significant effects of selenium in aquatic birds feeding at Kesterson Reservoir, California, inflow of agricultural drainage water was halted, the reservoir was dewatered, and lower-elevation portions of the site were filled with soil to prevent groundwater from rising to the ground surface.[89,90] Loss of this wetland habitat was mitigated through wetland enhancement in nearby areas. These actions converted the site to a mosaic of three terrestrial habitats, including areas that were filled, those higher-elevation areas where existing vegetation (mainly cattail, *Typha* sp.) was disked, and other higher-elevation grassland areas where no action was taken. Further descriptions of the site and its early history are provided by Ohlendorf[7] and Ohlendorf and Hothem.[91]

Biological monitoring of Kesterson Reservoir has been conducted annually since 1988 and has included extensive sampling and analyses of soil (for total and water-extractable selenium), plants, invertebrates, birds (mainly their blood or eggs), and mammals.[89,90] The chemical behavior of selenium in the Kesterson environment has been characterized through studies of selenium speciation and fractionation in soils, long-term monitoring of the spatial and depth distribution of selenium in soil, field-measured volatilization rates from several experimental plots, laboratory measurements of selenium reoxidation rates under controlled conditions, modeling studies of the above-mentioned data to determine reoxidation and leaching rates, and bioaccumulation by plants and animals at the site. This information was integrated to conduct ecological risk assessments in 1993 with the limited data set available at that time and again with a much larger data set in 2000.[89,90]

The key factor in the exposure model for the site is the water-extractable selenium concentration in soil, which influences bioaccumulation by plants that form the base of the food chain. Although there is a highly significant ($P < 0.001$) increase in plant selenium with increasing water-extractable soil selenium, the relation between the two matrices is highly variable.[89,90] Selenium concentrations vary somewhat from year to year and among plant species. For example, in 1995, the selenium concentration ranged from 0.20 to 90 mg/kg in 286 samples, with a geometric mean of 3.4 mg/kg; in 1998, the range in 240 samples was 0.20 to 68 mg/kg, with a geometric mean of 3.8 mg/kg.

Uptake and accumulation of selenium by plants growing on fly-ash landfills has been studied extensively and reviewed recently.[18] Concentrations of selenium in vegetation vary by plant species, with legumes typically accumulating more selenium than grasses. Although some other elements (such as boron and molybdenum) also bioaccumulate in these plants, this does not occur to elevated concentrations as consistently as the bioaccumulation of selenium. Application of sulfur in the form of gypsum reduced the bioaccumulation of selenium in several plant species. Plants growing on fly-ash landfills can transport selenium to the surface and make it available to the surrounding environment. Reported concentrations of selenium in plants growing on or near these landfills are typically less than 10 mg/kg, although concentrations exceeding that level have been found in grass (*Phalaris arundinacea*), trees (*Populus deltoides, Salix* spp.), and herbaceous vegetation growing at the margin of an ash-settling pond and in plants growing in fly ash without soil cover.

### 17.3.2 Invertebrates

#### *17.3.2.1 Aquatic/Marine*

Background concentrations in aquatic invertebrates are 0.4 to 4.5 mg Se/kg, although concentrations are typically less than 2 mg Se/kg.[12,17,19] Elevated concentrations have been reported in freshwater and estuarine invertebrates from areas receiving subsurface agricultural drainage,[40–42,82,92] urban/industrial discharges,[93] and in ponds or lakes that received selenium-containing fly ash.[75,94–96]

Among the aquatic invertebrates sampled at Kesterson Reservoir, California, selenium concentrations were highest in benthic species, such as midge larvae (Chironomidae) and dragonfly and damselfly nymphs (Odonata), and lowest in water boatmen (Corixidae).[40–42] Mean concentrations in some of the invertebrates from Kesterson Reservoir exceeded 100 mg Se/kg. Selenium concentrations in adult damselflies were significantly higher than in nymphs.[41] Selenium bioaccumulation factors for invertebrates ranged from 168 to 3700 (mean = 1090).[42] Most aquatic insects contained lower concentrations than rooted aquatic plants but higher concentrations than water, sediment, algae, and diatoms.

Midge larvae show small-scale spatial variability in selenium accumulation,[97] which probably reflects variability in concentration and bioavailability of selenium in sediment and perhaps the ages of the larvae. Grain size and organic carbon content can influence sediment selenium concentration and bioavailability. Consequently, when sampling/monitoring is being conducted to characterize dietary exposure of fish and aquatic birds that feed on midge larvae, it is important to analyze randomly collected, composited samples taken throughout the potential feeding area.

Laboratory experiments have indicated that 96% of selenium in mussels (*Mytilus edulis*) was obtained from ingested food, with much less accumulation directly from water than for several other elements.[98] Several other laboratory studies of the relative bioaccumulation in freshwater aquatic organisms from waterborne selenite, selenate, and selenomethionine are discussed above in the section on plants. At selenate concentrations greater than 5 μg Se/L, selenium uptake by daphnids (*Daphnia magna*) was inversely related to waterborne sulfate concentration.[99] Similar effects were reported for aquatic plants, as described above.[83,84]

#### 17.3.2.2 Terrestrial

Background concentrations in terrestrial invertebrates are 0.1 to 2.5 mg/kg.[17] Earthworms were found to bioaccumulate elevated concentrations of selenium from selenite-enriched soil (up to 7.5 mg Se/kg, fresh weight[100]), and earthworms from soil amended with sewage and those from a control field contained 15 to 22 mg Se/kg (dry weight).[101]

Extensive sampling of terrestrial invertebrates (various kinds of insects as well as spiders) has been conducted at Kesterson Reservoir, California since its conversion to terrestrial habitats in 1988 (described above and in Ohlendorf and Santolo[89]). From 1989 to 1995, samples were collected annually, and sampling was conducted again in 1998.[90] Generally, selenium concentrations in carnivorous invertebrates (such as ambush bugs, scarab beetles, and spiders) were significantly higher than those in noncarnivorous insects (such as herbivorous beetles, crickets, and grasshoppers). For example, the geometric mean for 35 samples of carnivorous invertebrates in 1995 was 18 mg Se/kg (range 4.5 to 48 mg Se/kg), and the geometric mean for 110 samples of noncarnivorous insects was 7.4 mg Se/kg (range 0.60 to 40 mg Se/kg). In 1998, the geometric mean for 35 samples of carnivorous invertebrates was 17.1 mg Se/kg (range 8.1 to 36 mg Se/kg), and the geometric mean for 104 samples of noncarnivorous insects was 12.8 mg Se/kg (range 1.8 to 61 mg Se/kg).

### 17.3.3 Fish

#### 17.3.3.1 Freshwater

Background concentrations in livers of freshwater fish are 2 to 8 mg Se/kg, and in other tissues they are 1.0 to 4.0 mg Se/kg.[17] In the National Contaminant Biomonitoring Program, sampling of freshwater fish was conducted five times throughout the United States during the period 1976 to 1986.[102] Geometric means for whole-body fish were between 0.4 and 0.5 mg Se/kg wet weight (or about 1.6 to 2.0 mg/kg dry weight) in all years, with the highest mean in samples from 1976 to 77. The 85th percentile values were 0.66 to 0.82 mg Se/kg wet weight (about 2.6 to 3.3 mg/kg dry weight). These values were typically exceeded at several stations, most of which are in arid areas of the western United States in basins containing substantial amounts of irrigated agriculture. One exception to that pattern is the Waikele Stream site at Waipahu, Hawaii, where the 85th percentile values were regularly exceeded.

Elevated levels of selenium have been reported in fish primarily from seleniferous areas of the western United States and from reservoirs contaminated by fly ash from combustion of coal.[14,19,25,40,41,96,103–108] Typical patterns found in these fish (and those sampled elsewhere) are that selenium concentrations in muscle tissue are considerably lower than in livers, gonads, or whole-body fish. Selenium concentrations in gonads (and especially the eggs) as well as whole-body fish provide the best basis for evaluating exposure of the fish to selenium and the potential for adverse effects in the fish or their reproduction (as discussed in a later section). However, it is possible, especially if the fish of concern is an endangered species, to evaluate exposure through nonlethal sampling of muscle plugs or eggs so the fish is not sacrificed for analysis.

Mosquitofish (*Gambusia affinis*) from the selenium-contaminated Kesterson Reservoir contained exceptionally high concentrations of selenium.[7,40,41] Fish were analyzed as composite samples of many individuals taken from different portions of the site, and concentrations were compared to those in mosquitofish from a nearby reference site. Mosquitofish from Kesterson typically contained geometric mean concentrations greater than 100 mg Se/kg, in comparison to 1 to 2 mg Se/kg at the reference site. In 1984, one composite sample from Kesterson contained 430 mg Se/kg, and the means for fish from two of the ponds exceeded 300 mg/kg.

Laboratory bioaccumulation studies discussed above describe bioaccumulation of selenium by fish in relation to different waterborne forms of selenium and illustrate the importance of dietary exposure as a route of selenium uptake by fish. Studies report more frequently the rates of uptake of different forms of selenium than depuration after exposure to uncontaminated media (food and water). In one study,[80] cycling and fate of selenite and selenomethionine were compared in a 318-day experiment in which the selenium was introduced as an acute release into two experimental ponds. Biotic components sampled included periphyton, rooted plants (*Elodea canadensis*), snails (*Helisoma* sp.), and mosquitofish. Uptake rates for selenomethionine in the biota were at least an order of magnitude greater than for selenite, but elimination rates were similar (different by less than a factor of two). This study also demonstrated the importance of sediment and organic detritus in the recycling/remobilization of selenium within the ponds.

#### *17.3.3.2 Marine*

Estuarine and marine fish tend to have higher selenium concentrations than do freshwater species, but the differences for most species are generally less than an order of magnitude and sometimes are very similar.[19] Higher concentrations were found in species such as tunas and marlins. Selenium has been shown to accumulate with age/length in several studies of marine fish, and a positive correlation between selenium and mercury has been demonstrated in several species.[109] A positive correlation of selenium with mercury in fish may be a significant factor influencing bioaccumulation. One study has shown that a lower portion of selenium is present as selenate in marine fish (24%) than in freshwater fish (36%).[110]

Kasegalik Lake (on the Belcher Islands, Hudson Bay, in the southern Canadian Arctic) is known to have both salt- and freshwater populations of Arctic char (*Salvelinus alpinus*).[111] Fish of both types were analyzed for mercury and selenium. There was an apparent difference in the two populations for mercury concentrations, but there was no difference between salt- and freshwater groups for selenium (which varied from 0.7 to 1.9 mg Se/kg, wet weight). Thus, some of the apparent differences in selenium levels between marine and freshwater fish may be attributed to species differences as well as relationships between selenium and mercury in larger, long-lived marine species.

### 17.3.4 Amphibians and Reptiles

Background concentrations in livers of amphibians and reptiles are 2.9 to 3.6 mg/kg and in other tissues 1 to 3 mg/kg.[17] Many fewer studies of selenium bioaccumulation in amphibians and reptiles have been conducted than those of other animals,[7,17,19] but information on bioaccumulation is summarized below.

#### 17.3.4.1 Amphibians

Selenium concentrations in adult southern toads (*Bufo terrestris*) from coal-ash-settling basins (17.4 mg/kg, whole-body analysis) were elevated by comparison to toads from a reference site (2.10 mg/kg).[112] When other toads were transferred from the reference site to the settling basin for a period of 7 weeks, their concentrations increased significantly — from 2.10 mg Se/kg to 5.46 mg/kg. Sediments in the ash basin contained 4.4 mg Se/kg, whereas those at the reference site had 0.10 mg Se/kg. Frog (*Rana clamitans*) tadpoles from a fly-ash-contaminated pond contained 4.7 mg Se/kg (wet weight), whereas those from a reference site had 1.5 mg/kg.[75] Red-spotted newts (*Notophthalmus viridescens*) from those same sites contained 4.2 and 1.8 mg Se/kg (wet weight), respectively. Similarly, bullfrog tadpoles (*Rana catesbeiana*) from an ash-waste site had elevated selenium (25.7 mg/kg) by comparison to a reference site (3.4 mg/kg).[113]

Because tadpoles apparently accumulate metals readily from sediments and vegetation they consume, some authors (e.g., References 114, 115) have suggested that they may be good indicators of contaminated environments. To evaluate the effects of gut contents (especially ingested sediment) on analyses of bullfrog tadpoles, wild-caught tadpoles were maintained in clean water and analyzed after 0, 24, 48, and 72 h and then analyzed for metals to test the effects of "clearing" of the gut contents.[115] Metals concentrations in whole bodies and digestive tracts also were evaluated separately. Selenium concentrations in the bodies and tails of the tadpoles were unaffected by clearing, although selenium concentration was higher in the digestive tract than in the body without digestive tract, in the tail, or in the whole body. The tadpoles had higher selenium concentrations than those found in the sediment from the sampling location.

Bullfrogs collected from reference sites in the San Joaquin Valley, California contained 1.0 to 1.9 mg Se/kg when analyzed as whole-body samples.[116] Livers of three bullfrogs from other reference sites in the San Joaquin Valley contained 3.6 to 9.3 mg Se/kg (mean 6.2 mg/kg).[117] The lower end of that range is probably representative of background, whereas the upper value may reflect some elevation above background. Few frogs were observed in nearby Kesterson Reservoir, probably because of the highly saline water it contained. However, ten bullfrogs were collected from the San Luis Drain (which conveyed agricultural subsurface drainage waters to Kesterson); livers of these bullfrogs contained 25 to 88 mg Se/kg (mean 45 mg/kg). These liver concentrations were similar to the selenium concentrations in aquatic insects and mosquitofish in the Drain.

#### 17.3.4.2 Reptiles

Livers of banded water snakes (*Nerodia fasciata*) from ash-settling basins associated with a coal-burning electric power plant were analyzed for selenium and several other elements that were found at elevated levels in sediment within the basins.[118] Selenium concentrations in the livers of snakes from these basins averaged about 140 mg/kg, compared to 3.62 mg Se/kg in livers of snakes from a reference site. Water snakes were found to be feeding on small bullfrogs and bullfrog tadpoles, green treefrogs (*Hyla cinerea*), southern toads, bluegill sunfish (*Lepomis macrochirus*), mosquitofish, red-fin pickerel (*Esox americanus*), and largemouth bass (*Micropterus salmoides*). Samples of those potential prey items were collected at the polluted ash-basin site and at the reference site. Mean selenium concentrations in the amphibians and fish from the polluted site were about 10 to 27 mg/kg, whereas those in amphibians and fish from the reference site were typically less than 2 mg/kg. Thus, the water snakes from the ash-basin site bioaccumulated selenium in their livers to concentrations about tenfold the prey diet concentrations, and those at the reference site had selenium concentrations that were about equal to the prey item levels. Selenium concentrations in livers showed a significant linear relationship with snake body mass.

Water snakes (*Nerodia* sp.) from Florida contained 0.3 to 0.5 mg Se/kg (whole body, wet weight).[119] Whole-body selenium concentrations in lizards and snakes from reference areas in the San Joaquin Valley averaged 0.7 to 2.0 mg/kg.[116] Livers of gopher snakes (*Pituophis melanoleucus*) from other reference sites in the San Joaquin Valley contained mean selenium concentrations of 2.05 and 2.14 mg/kg.[117] Livers of gopher snakes from the nearby Kesterson Reservoir had a mean selenium concentration of 11.1 mg/kg. Selenium bioaccumulation in these snakes at Kesterson and associated reference sites reflected selenium levels in prey species for the snakes at those sites.

Selenium levels in the plasma of wild and farm-reared American alligators (*Alligator mississippiensis*) reflected selenium concentrations in their diet.[120] Plasma selenium levels in captive alligators fed fish (*Micropogon undulatus*) were significantly higher (monthly means of 0.23 to 0.27 mg/kg) than in captive animals fed nutria (*Myocastor coypus*) (0.16 to 0.23 mg/kg) or in wild females (0.15 to 0.20 mg/kg); plasma selenium levels in the alligators fed nutria and in the wild females were not significantly different from each other. Fish contained 2.8 mg Se/kg (dry weight), compared to only 0.04 mg/kg (wet weight) in the nutria. Diet of the wild alligators was composed primarily (~70%) of nutria.

### 17.3.5 Birds

Background concentrations in whole-body birds are < 2 mg/kg.[17] However, bioaccumulation in birds is more frequently (and appropriately) evaluated on the basis of various tissues, as discussed below.

#### *17.3.5.1 Eggs*

Mean background concentrations in eggs of freshwater and terrestrial species are < 3 mg Se/kg (typically 1.5 to 2.5 mg/kg), and maximums are < 5 mg Se/kg.[17,19] In a wide variety of species, selenium concentrations (on a dry-weight basis) in bird eggs range from roughly equivalent to about three or four times the concentrations in the diet of the female at the time of egg-laying.[7,121–127]

When birds feed on selenium-contaminated diets during the laying season, the exposure is quickly reflected by elevated levels of selenium in eggs.[128] Similarly, when the birds are switched to a clean diet, selenium concentrations in eggs decline quickly. When mallard (*Anas platyrhynchos*) hens were fed a diet containing 15 mg Se/kg (as selenomethionine), levels peaked in eggs (to about 13 to 20 mg Se/kg, wet weight) in about 2 weeks on the treated diet and leveled off at a relatively low level (<5 mg Se/kg, wet weight) after about 10 days back on the untreated diet. The findings of this study have important implications for evaluation of field exposures, such as how quickly and for what duration selenium exposure may adversely affect bird reproduction. Concentrations of selenium in eggs are especially important because they provide the best samples for evaluation of potential adverse reproductive effects (see also later section).[129]

Selenium concentrations in the eggs of marine species are variable but may be higher, even in remote areas.[7] For example, eggs of three species (wedge-tailed shearwater [*Puffinus pacificus*], red-footed booby [*Sula sula*], and sooty tern [*Sterna fuscata*]) were sampled at four locations throughout the Hawaiian archipelago, from Oahu to Midway.[130] Mean concentrations (converted from wet weight) varied only slightly by location — from 4.4 to 5.3 mg Se/kg for shearwaters, 5.0 to 6.1 mg Se/kg for boobies, and 4.1 to 5.1 mg Se/kg for terns — but all were higher than typical of freshwater species. Surprisingly low selenium concentrations were found in eggs of white-winged scoters (*Melanitta fusca*) when compared to the concentrations of selenium found in their livers.[131]

Elevated selenium concentrations have been found in eggs of birds nesting in areas affected by agricultural drainage, coal-fly-ash disposal, and discharge from an oil refinery.[7,25,91,129,132,133] At Kesterson Reservoir, California, mean selenium concentrations in bird eggs during 1983 to 1985 were as high as 69.7 mg/kg (in eared grebes [*Podiceps nigricollis*]), with many of the species means exceeding 20 mg Se/kg.[91] Some success in reducing selenium exposures of aquatic birds at evaporation ponds for agricultural drainage water was achieved by placing freshwater mitigation ponds near the contaminated ponds.[134]

#### 17.3.5.2 Livers

Background concentrations in bird livers are < 10 mg Se/kg.[17] In a manner similar to that for eggs, Heinz et al.[135] found that selenium concentrations in the liver respond quickly when birds are placed on or taken off a selenium-contaminated diet (see further description of this study below under Muscle). Thus, selenium concentrations measured in the livers of birds sampled outside the breeding season are not good predictors of potential reproductive effects. In laboratory studies of reproductive effects, livers of male mallards had higher concentrations of selenium than those of females, probably because females excreted part of the selenium they had accumulated through egg-laying.[121,122] Nevertheless, analysis of livers of either male or female field-collected birds can provide a useful indication of the relative level of exposure experienced by the population.

In laboratory studies with birds fed diets containing selenomethionine, selenium concentrations in livers of mallards, black-crowned night-herons (*Nycticorax nycticorax*), and eastern screech-owls (*Otus asio*) ranged from roughly equal to the dietary concentration to about three times the dietary level.[121–126]

At Kesterson Reservoir and a nearby reference site, several species of aquatic birds were sampled early and late in the nesting season as adults, and juveniles were sampled at about the same time as the late-season adults.[136] Selenium concentrations in livers of these birds were consistently much higher at Kesterson than at the reference site for both early and late samplings and seemed to reflect the period of exposure as well as the foraging range of the birds at the two sites. For example, adults of the more mobile species (such as dabbling ducks, *Anas* spp.) showed only three- to fivefold differences in mean selenium concentrations between sites, whereas more sedentary species (such as American coot [*Fulica americana*] and black-necked stilt [*Himantopus mexicanus*]) typically showed 10- to 20-fold differences. Selenium concentrations in juvenile birds were generally similar to those in late-season adults.

Selenium concentrations in the livers of aquatic birds from a number of other areas receiving subsurface agricultural drainage also were elevated and have reflected periods of residence in the contaminated areas.[25,137–139] When use of subsurface agricultural drainage waters for wetland management was curtailed and replaced with better-quality water, selenium concentrations in aquatic birds declined.

Livers of diving ducks (such as scoters [*Melanitta* spp.] and scaups [*Aythya* spp.]) from estuarine habitats have been found to contain higher concentrations of selenium than other aquatic birds in the same habitats.[140–143] The apparent reason for the higher concentrations of selenium in the diving ducks is that they forage on benthic organisms, which bioaccumulate selenium to a higher degree than foods of some of the other aquatic birds.

#### 17.3.5.3 Kidneys

Selenium concentrations in kidneys of birds from selenium-normal areas were somewhat higher than those in the liver (liver/kidney ratios of less than L), but concentrations in the two tissues were similar in birds from selenium-contaminated Kesterson Reservoir and in the Imperial Valley of California.[136,144,145] Selenium concentrations in American coots from Kesterson Reservoir and the reference site were significantly correlated ($r = 0.98$). Significant ratios and significant positive correlations also have been found in other field studies (e.g., Reference 146).

#### 17.3.5.4 Muscle

Background selenium concentrations in muscle tissues of birds are 1 to 3 mg/kg.[17] Concentrations increase and decrease in response to changes in dietary exposure, but the changes occur more slowly than those in eggs or liver.[135] However, for muscle as well as other tissues, when tissue levels of selenium are made very high by elevated dietary exposures and the birds are then placed on a lower-selenium diet, the loss rate for all tissues is fast at first but then slows down.

Mallards were placed on a diet containing 10 mg Se/kg for 6 weeks; predicted equilibrium time for concentrations in the breast muscle of females (81 days) was more than ten times that for time to equilibrium in the liver (7.8 days).[135] The ducks were then provided an untreated diet for 6 weeks, and loss of selenium from liver and muscle was monitored; half-times were 18.7 days for liver and 30.1 days for muscle. Males reached similar levels of selenium in the liver and breast muscle as females and declined to similar levels when treatment ended. In another experiment with mallards, females were fed increasing levels of selenium until some died; survivors were then switched to an untreated diet and selenium was measured in blood, liver, and breast muscle over 64 days. Half-times were 9.8 days for blood and 23.9 days for muscle. Selenium initially decreased in liver by one-half in 3.3 days, with subsequent half-times of 3.9, 6.0, and 45.1 days.

Selenium concentrations in breast muscle from juvenile ducks (*Anas* sp.) at Kesterson Reservoir and a reference site were measured because of concern about human consumption of ducks harvested in the vicinity of Kesterson.[136] Mean selenium concentrations were higher at Kesterson than the reference site and were only slightly lower than those in livers of these birds. However, the relationships between muscle and liver ($r^2 = 0.69$) of the ducks were considerably more variable than those between kidneys and livers of American coots from the two sites ($r^2 = 0.97$).

### 17.3.5.5 Blood

Background concentrations in whole blood are 0.1 to 0.4 mg Se/kg on a wet-weight basis.[17] In experimental studies, selenium concentrations in blood of mallards and American kestrels (*Falco sparverius*) reflected dietary exposure levels.[127,135,147–149] In kestrels, maximal blood concentrations were about the same as those in the selenomethionine-supplemented diet. Mallards receiving selenium (as selenomethionine) at dietary concentrations of 10, 25, or 60 mg/kg had blood-selenium concentrations of 4.5, 8.9, or 16 mg/L (wet-weight basis).[148] The concentration of selenium in blood increased in a time- and dose-dependent manner and reached a plateau after 40 days. (The study by Heinz et al.[135] is described in the previous section.)

Selenium concentrations were measured in terrestrial birds of several species from Kesterson Reservoir, the area surrounding that site, and several reference areas in California from 1994 to 1998.[150] Except for loggerhead shrikes (*Lanius ludovicianus*), blood-selenium was higher in birds from within Kesterson than in birds from other areas. For shrikes, the mean concentrations for birds from Kesterson (13 mg Se/kg [dry weight]) were not significantly different than those from nearby surrounding areas (8.5 mg Se/kg), although the maximum selenium concentration at Kesterson (38 mg/kg) was more than twice the maximum for the surrounding area (16 mg/kg). Among species at Kesterson Reservoir, blood-selenium was higher in loggerhead shrikes and northern harriers (*Circus cyaneus*) than in the other species (hawks and owls) sampled.

### 17.3.5.6 Feathers

Background concentrations of selenium in feathers are 1 to 4 mg/kg and are typically less than 2 mg/kg.[17] Selenium concentrations may be higher in the feathers of birds from areas with elevated levels of mercury because of the interactions between these two elements.

Analysis of feathers may provide useful information concerning exposures of birds to selenium if they are considered carefully. It is important to recognize that the selenium may have been deposited into the feathers at the time they were formed (which may have been months earlier and thousands of miles away from the sampling time and location), or the selenium may be the result of external contamination.[151–154] Concentrations also may have been reduced through leaching. Different kinds of feathers from the same bird may contain different concentrations, depending partly on when and where the feathers were grown during the moult cycle.

Feathers of five species of seabirds were sampled at Johnston Atoll (about 1300 km southwest of Hawaii) and three species at Manana Island, Oahu, Hawaii, and analyzed for selenium and several other elements.[155] Selenium concentrations were generally higher in brown noddies (*Anous stolidus*) than in other species at both locations, and noddies had significantly higher concentrations at Johnston Atoll (11.1 mg Se/kg) than Hawaii (7.59 mg Se/kg).

### 17.3.6 Mammals

Background whole-body selenium concentrations in terrestrial and freshwater mammals are < 1 to 4 mg/kg.[17] In muscle, concentrations are typically < 1 mg Se/kg; in liver, they range from about 1 to 10 mg Se/kg and are typically ≤5 mg/kg; in hair, normal concentrations are <3 mg/kg in individual samples, with population averages ranging from 0.5 to 1.5 mg Se/kg; selenium in whole blood is typically 0.1 to 0.5 mg/L, with concentrations below 0.1 mg/L indicating a deficiency in domestic livestock. Both hair and blood are considered good media for sampling/monitoring selenium status of live mammals.

#### *17.3.6.1 Aquatic/Marine*

Selenium concentrations in raccoons (*Procyon lotor*) collected at Kesterson Reservoir, California averaged 19.9 mg/kg in liver, 28.3 mg/kg in hair, 21.6 mg/kg in feces, and 2.61 mg/L in blood (wet weight in blood, others in dry weight).[156] These concentrations were 12, 30, 21, and 10 times higher, respectively, than those found in raccoons from a nearby reference site. Selenium concentrations in hair provided the strongest statistical separation between the two study areas. Selenium concentrations in raccoons and river otters (*Lontra canadensis*) at sites in Canada were similar to each other, but they were higher than those found in beavers (*Castor canadensis*).[157]

Selenium concentrations in the tissues of marine mammals are highly variable, with some animals having concentrations in the liver or other tissues that are typical of mammals in freshwater or terrestrial environments and others (even of the same species in the same areas) having greatly elevated levels.[19,109] Part of this variation may be due to interactions between selenium and mercury, with much greater bioaccumulation (especially as reflected by concentrations in the liver) when the animals are exposed to elevated levels of mercury.

Selenium concentrations greater than 100 mg/kg have been reported for livers of pilot whales (*Globicephala macrorhynchus*),[158] beluga whales (*Delphinapterus leucus*),[159] striped dolphins (*Stenella coeruleoalba*),[160] bottle-nosed dolphins (*Tursiops truncatus*),[161] ringed seals (*Phoca hispida*),[162] harbor seals (*P. vitulina*),[163] gray seals (*Halichoerus grypus*),[164] and California sea lions (*Zalophus californianus*).[165]

#### *17.3.6.2 Terrestrial*

Mean selenium concentrations in several species of small mammals at Kesterson Reservoir, California, were greatly elevated by comparison to a nearby reference area.[166] For example, means for California voles (*Microtus californicus*) from Kesterson were as much as 522 times those from the reference area (means up to 119 mg Se/kg vs. 0.228 mg Se/kg in livers). There were species-to-species differences at Kesterson; higher selenium concentrations occurred in carnivorous species and species that fed on foods closely linked to the pond water. There were also pond-to-pond differences at Kesterson, with mammals reflecting a pattern similar to that for aquatic organisms (described above).

Following the conversion of Kesterson Reservoir to a terrestrial habitat in 1988 (as described above), selenium concentrations in small mammals have been monitored periodically.[89,90] Mean whole-body selenium concentrations in mice on a site-wide basis are typically 5 to 7 mg Se/kg and tend to be lower in voles than in mice. However, the selenium levels tend to vary spatially within the site, especially as related to habitat type and history of management activities that have altered the site since it served as a reservoir for agricultural drainage waters.

Elevated levels of selenium were found in woodchucks (*Marmota monax*) from the vicinity of a fly-ash landfill,[167] and in American bison (*Bison bison*), elk (*Cervus elaphus*), and mule deer (*Odocoileus hemionus*) that ate forage with elevated levels of selenium in Wyoming.[168] In contrast, analysis of blood from free-ranging mule deer sampled from a number of herds in California indicated the deer were selenium-deficient.[169–173] Several methods of selenium supplementation to the deer were tested; some were more practical and effective than others and resulted in increased reproductive performance of the herds.

## 17.4 EFFECTS IN ANIMALS

### 17.4.1 Nutritional Requirements vs. Toxicity

Selenium is essential for animal nutrition at sub-mg/kg dietary concentrations, but it is toxic to sensitive species when concentrations are only a few mg/kg. For most animals, dietary requirements appear to fall in the range between about 0.05 and 0.5 mg/kg.[5,19,46,56,174–179] Thresholds for toxicity are usually not much more than an order of magnitude higher than the concentrations that cause deficiency (as described below). Selenium deficiency is a more widespread problem for animal health than is excess selenium, so the issue of selenium ecotoxicology is one of balance between meeting the dietary requirements without exceeding toxicity thresholds. Signs of selenium deficiency have been reported in many domesticated species (including chickens, turkeys, pheasants, quail, ducks, cattle, horses, goats, sheep, and swine) as well as wild species (including species such as fish, bighorn sheep [*Ovis canadensis*], pronghorns [*Antilocapra americana*], elk, and deer). In poultry, selenium deficiency causes severe pancreatic atrophy as well as reduced egg production and hatchability; in mammals it causes poor health and reproduction. One of the more common myopathies resulting from selenium deficiency (referred to as "white muscle disease") has been reported in at least 30 of the 50 U.S. states.

Selenium deficiency in some species can be ameliorated through a variety of means including addition of selenium to diets of domestic species, use of intrarumenal boluses that release selenium slowly, and application of selenium-containing fertilizers to grazing lands.[5,176–179] Selenium fertilization and use of intrarumenal boluses have been tested for improvement of reproduction in free-ranging mule deer, sometimes with significant benefit.[170,172,173] However, selenium deficiencies in other kinds of wildlife may not be corrected as easily as those in species such as deer.

### 17.4.2 Toxic Effects

Selenosis and selenium toxicosis are used interchangeably to refer to the pathological conditions resulting from ingestion of high-selenium diets. Those conditions include impaired health and reproduction, as discussed below in relation to vertebrate animals. When water is the only exposure route, selenium is not very toxic to fish or wildlife.[17,20,57] Eggs and larvae of fish and amphibians may be the most sensitive life stages of vertebrate animals to direct effects of waterborne selenium. Eggs and larvae of fish and birds are very sensitive to the lethal or teratogenic effects of selenium transferred to the eggs by the female parent (as discussed below).

#### *17.4.2.1 Invertebrates*

Invertebrates are not particularly sensitive to the toxic effects of selenium, and there is limited information concerning adverse effects to invertebrates under field conditions.[17] In laboratory studies, selenite was found to be more toxic than selenate, regardless of whether the effects were measured as direct toxicity or impairment of reproduction.[20,180,181] Duration of exposure is also an important factor, with toxicity increasing as exposure time increases, at least up to 14 days. For example, the 96-hour $LC_{50}$ for the scud (*Hyalella azteca*) was 340 μg Se/L or 760 μg Se/L in different studies,[182,183] whereas the 14-day $LC_{50}$ was only 70 μg Se/L.[184] Similarly, for the *Daphnia magna* the 96-hour $LC_{50}$ was 710 μg Se/L and the 14-day $LC_{50}$ was 430 μg Se/L.[184]

In a comparison involving the cladoceran *Daphnia magna* and larvae of the midge *Chironomus riparius*, the daphnids were more acutely sensitive than midges to the toxic effects of inorganic selenium.[185] An organic form of selenium (selenomethionine) was toxic to daphnids but relatively nontoxic to midges. Tissue concentrations of 14.7 and 31.7 mg Se/kg in daphnids were associated with reduced growth and reproduction, respectively. Nevertheless, these invertebrates were not adversely affected at waterborne concentrations that would affect fish or would bioaccumulate in the food chain to levels that are harmful to fish or aquatic birds.

In evaluating the potential for selenium to cause toxicity to aquatic invertebrates, it is important to note that other elements can affect the toxicity of selenium. For example, sulfate can reduce the bioavailability and toxicity of selenate selenium (the predominant form in many water bodies) to invertebrates.[99]

Biologically incorporated selenium (presumably in organic forms) in widgeongrass from agricultural drainwater evaporation ponds in the San Joaquin Valley, California, caused significant reductions in growth of midge (*Chironomus decorus*) larvae that were cultured on detrital material from these plants.[79] Significant reductions in growth were observed in midges exposed to plant substrate having more than about 10 mg Se/kg, and relative bioaccumulation (as BAF) was inversely related to selenium concentrations in the plant material (i.e., a relatively smaller amount was taken up when concentrations were greater than 20 mg Se/kg).

#### 17.4.2.2 Fish

Several recent publications provide reviews of the effects of selenium in fish and describe case studies of such effects.[12,14,17,19,23,25,186,187] Those references provide more detailed descriptions of tissue or exposure media concentrations that are associated with no effects or with adverse effects in fish. Some also emphasize that a combination of biological (e.g., condition of larvae/fry, reproductive success, etc.) and chemical (e.g., analysis of ovaries/eggs, whole-body fish, or food-chain organisms) indicators should be used in evaluating risks to fish populations. The mortality of larvae/fry that is associated with excess selenium can have important effects on populations, resulting in lack of recruitment and subsequent declines or disappearance of affected species in a contaminated water body without evidence of "die-offs." Given enough time, once the levels of selenium input are reduced, populations may recover; however, contaminated sediments will serve as a continuing source for selenium cycling.[188]

Excess selenium in the diet of fish leads to substitution of selenium for sulfur during protein synthesis, among other effects.[23] Substitution of selenium for sulfur disrupts normal chemical bonding, resulting in improperly formed and dysfunctional proteins or enzymes. This causes a variety of toxic effects at subcellular, cellular, organ, and system levels.[103,189] These effects are exhibited through effects on reproduction (especially in the form of teratogenesis) and reduced survival of young fish as well as effects on health, physiology, and survival of older fish.

There is general agreement that early life stages (i.e., eggs, larvae, and fry) are more sensitive to the adverse effects of excess selenium than are adult fish, and that some species of fish are more sensitive than others.[14,17,23,186,190] Selenium concentrations in the ovary/eggs as well as those in whole-body fish and the diet are most useful in evaluating risks of adverse effects (i.e., more useful than selenium concentrations in water or sediment, or in livers or other tissues). However, there is no uniform agreement on the selenium concentrations that represent thresholds for adverse effects.

Selenium concentrations in ovaries/eggs are considered to represent a threat of adverse effects when they are greater than 10 mg/kg,[14,23,190] 7 to 13 mg/kg,[17] or 17 mg/kg.[186] At concentrations exceeding the threshold, there is a large increase in the occurrence of teratogenic deformities and a concomitant reduction in survival of the larvae/fry. Although many of the larvae/fry die as a result of the deformities, some fish may survive and have noticeable deformities as adults. Selenium-related deformities observed in fish larvae and fry are due to their absorbing/using the yolk from the egg, which contains elevated levels of selenium deposited there by the female parent for their nourishment and development. Feeding excess but sublethal levels of selenium to fry, juvenile fish, or adult fish does not cause deformities such as those caused through maternal transfer of selenium to the egg.[191–193]

Examples of the kinds of teratogenic deformities that are caused by selenium are provided by Lemly.[23] The typical deformities include lordosis (concave curvature of the lumbar region of the spine), scoliosis (lateral curvature of the spine), kyphosis (convex curvature of the thoracic region of the spine, resulting in a "humpback" condition), missing or deformed fins, missing or deformed gills and gill covers (opercle), abnormally shaped head, missing or deformed eyes, and deformed mouth.

Selenium concentrations in whole-body fish are considered to represent a threat of adverse effects to health or reproductive success when they are greater than 4 mg/kg[14, 190] or greater than 4 to 6 mg/kg;[17] however, DeForest et al.[186] evaluated results of available studies and considered the threshold to be 6 mg/kg for cold-water and anadromous species and 9 mg/kg for warm-water fish. These differences in values result from different interpretations of the available information from the various studies. Despite the differences, however, it is important to note that whole-body selenium concentrations in the range of 4 to 9 mg/kg are recognized to reflect risk of adverse effects in fish.

Threshold selenium concentrations in the diet of fish that are associated with adverse effects have been estimated as 3 mg/kg,[14, 190] as 3–8 mg/kg,[17] or separately as 11 mg/kg for cold-water and anadromous species and 10 mg/kg for warm-water fish.[186] As described in the previous section (Bioaccumulation), different forms of selenium bioaccumulate at very different rates in food-chain organisms, with selenomethionine > selenite > selenate. These differences in bioaccumulation can greatly affect the degree of risk associated with particular waterborne selenium concentrations. However, studies indicate that once the selenium is biologically incorporated in the food organisms, the risk to fish is similar at equivalent dietary concentrations.

In waterborne exposures, organic selenium (selenomethionine) also is much more toxic than selenite, which is more toxic than selenate. For example, the 96-hour $LC_{50}$ values for bluegill (13 µg Se/L) and striped bass (*Morone saxatilis*; 4 µg Se/L) based on selenomethionine were only a small fraction of the $LC_{50}$s for selenite (7800 to 13,000 and 1000 µg Se/L, respectively) and selenate (98,000 and 39,000 µg Se/L).[181] However, as duration of exposure increases, the toxicity of selenium increases. For example, when selenium was introduced to outdoor experimental streams at 10 or 30 µg/L, all adult bluegill sunfish at the higher exposure level died in less than a year, and there also was significantly reduced survival of those at 10 µg Se/L when compared to controls.[194] Exposure of adults at both 10 and 30 µg Se/L for 40 weeks before spawning resulted in reduced embryo and larval survival and produced larvae with a high incidence of edema, lordosis, and internal hemorrhaging.

#### *17.4.2.3 Amphibians and Reptiles*

There is only limited information about the toxic effects of selenium in amphibians, and there is even less information about its effects in reptiles.[17,19,195–198]

In laboratory exposures, amphibian embryos and tadpoles were about as sensitive as aquatic invertebrates and fish larvae/fry to the effects of waterborne inorganic selenium.[199–201] In short-term toxicity tests (typically 96-h exposures, but up to 3 days) with the South African clawed frog (*Xenopus laevis*), $LC_{50}$ values for both selenite and selenate were 1500 µg Se/L or greater. Not surprisingly, the $LC_{50}$ values decrease as exposure duration increases. The 7-day $LC_{50}$ for this species was 456 µg Se/L, but in a 7-day exposure with narrow-mouthed toad (*Gastrophryne carolinensis*) embryos, the $LC_{50}$ values was only 90 μg Se/L.[202, 203] The low value for the narrow-mouthed toad apparently reflects species differences in sensitivity or perhaps differences in exposure conditions.

Adult and larval amphibians exposed to trace elements in coal combustion waste ponds showed physiological and morphological changes as well as bioaccumulation of some elements.[112,113] For example, tadpoles of the bullfrog collected in a coal-ash basin and a downstream drainage swamp had a reduced number of labial teeth and deformities of labial papillae when compared with tadpoles from reference areas. Tadpoles with deformities were less able to graze periphyton than were normal tadpoles, and they had lower growth rates. These tadpoles had significantly elevated concentrations of selenium (25.7 mg/kg) by comparison to a reference site (3.37 mg/kg). They also had significantly greater concentrations of arsenic, barium, cadmium, and chromium than those from the reference site, but they did not significantly bioaccumulate copper or lead in comparison to the reference site. Thus, it is not possible to determine which elements caused the effects.

Banded water snakes from a site contaminated by coal combustion wastes had high concentrations of trace elements, especially selenium and arsenic but also cadmium, chromium, and copper, by comparison to snakes from a nearby reference site.[118] (Results of analyses of livers and potential prey items were summarized in a previous section.) Snakes from the ash-basin site exhibited mean standard metabolic rates 32% higher than snakes from the reference site. The authors concluded that snakes from the contaminated site probably allocate more of their energy to maintenance and theoretically should have less energy available for growth, reproduction, and storage.

### *17.4.2.4 Birds*

Selenium has been identified as the agent responsible for mortality and reproductive failure in birds at a number of sites in the United States (see USDI[17] and Skorupa[25] for a partial list). Much of the research evaluating the effects of selenium in birds has been related primarily to discharges of subsurface agricultural drainage water,[7,129] but effect levels in bird diets and eggs probably are similar once the selenium has been accumulated by food-chain organisms.

Selenium became recognized as a significant environmental contaminant for wildlife in 1983, with the discovery of developmental abnormalities and excessive embryonic and adult mortality in aquatic birds at Kesterson Reservoir, California.[7,91,132,144,204,205] Observations in the field were corroborated through a series of laboratory studies that documented the association between excessive dietary selenium and effects on reproduction and young birds as well as adult health, physiology, and survival. Those effects are described in several other reviews (e.g., References 4, 7, 8, 13, 17, 19, 25) and are summarized below.

#### *17.4.2.4.1 Reproduction and Young Birds*

Prior to the 1980s, most studies of avian reproductive effects in relation to selenium were conducted with domestic poultry.[4,7,13] Studies of wild birds to determine the effects of selenium on reproduction and other aspects of avian biology have been conducted in the field and in the laboratory since 1983. Reproductive effects of selenium reported in wild aquatic birds and domestic poultry include embryo mortality and teratogenesis as well as the failure of adult birds to nest.

Excess selenium in the diet of female birds during the period just before egg-laying results in transfer of selenium to the eggs at harmful levels, although the sensitivity varies among species.[7,8,13,25,128,129,206] Characteristic effects observed in the field and laboratory studies include reduced hatchability of eggs (due to embryo mortality) and high incidence of developmental abnormalities (due to teratogenesis). Selenium-induced abnormalities are often multiple and include defects of the eyes (microphthalmia and possible anophthalmia [i.e., abnormally small or missing eyes]), feet or legs (amelia and ectrodactylia [absence of legs or toes]), beak (incomplete development of the lower beak, spatulate narrowing of the upper beak), brain (hydrocephaly and exencephaly [fluid accumulation in the brain and exposure of the brain]), and abdomen (gastroschisis [an open fissure of the abdomen]). Most of these abnormalities are illustrated through photographs that have been published elsewhere (e.g., References 4, 7, 8, 132, 144).

Reproductive effects have been produced in laboratory dietary studies using both organic and inorganic forms of selenium (e.g., in mallards and domestic chickens [*Gallus gallus*]).[4,121,122,124,125,207–210] Selenomethionine, the most bioavailable and toxic of the different selenium compounds studied, is associated with reduced reproductive effects in mallards at dietary concentrations as low as 7 mg Se/kg.[125] This form of dietary selenium has been found to be the best surrogate for use in feeding studies to parallel the dose-effect levels found in field studies with birds.

Studies at Kesterson Reservoir showed there was poor hatching success of fertile eggs of a number of aquatic bird species[7,204] and that there was also poor survival of those young that did hatch.[7,211] This was attributed to high levels of selenium in the diet of the breeding females as well as the diet of newly hatched birds. Most aquatic insects at Kesterson had mean selenium concentrations greater than 100 mg/kg, and some composite samples exceeded 300 mg/kg.[40–42] Mallard ducklings were fed diets containing either sodium selenite or selenomethionine at concentrations of 0, 10, 20, 40, or 80 mg Se/kg from hatching to 6 weeks of age.[212] Among ducklings that received selenomethionine, 100% of the group receiving 80 mg Se/kg died in less than 6 weeks, and 12.5% of those that received 40 mg Se/kg died during the exposure period.

Predictive criteria for avian selenosis were developed through a broad-scale program that collected response data for avian teratogenesis at selenium-impacted and reference aquatic sites in the San Joaquin Valley, California.[17,206] The largest cumulative sampling effort occurred within the Tulare Basin area in the southern San Joaquin Valley. In the Tulare Basin, evaporative disposal of subsurface irrigation drainage water was accomplished by letting it evaporate from 25 shallow impoundments. Even though the impoundments were not constructed to attract wildlife and were devoid of emergent vegetation (i.e., cattails, bulrush, etc.), large populations of nesting waterbirds used the sites. Two of the most abundant waterbirds that nested at the evaporation basins were American avocets (*Recurvirostra americana*) and black-necked stilts (*Himantopus mexicanus*). Water discharged to the evaporation basins contained from < 1 to > 1000 μg Se/L; consequently, selenium concentrations in eggs from this region spanned four orders of magnitude (< 1 to > 100 mg/kg egg selenium).

Using data on selenium in eggs from the Tulare Basin, combined with data from several other western sites where elevated selenium was found, a detailed exposure-response relationship was documented.[17,25,206] Statistically distinct teratogenesis response functions were delineated for ducks, stilts, and avocets using the Tulare Basin data. The Tulare curves were used to estimate expected frequencies of teratogenesis for ducks, stilts, and avocets using other sites, and the predicted levels were tested against the observed frequencies from the sites. The predicted and observed frequencies of teratogenesis were not significantly different, and therefore the data were combined to generate final response curves. Using this data set, Skorupa[206] developed species-specific response curves for stilts and avocets and a composite duck curve (using combined data from gadwalls [*Anas strepera*], mallards, pintails [*A. acuta*], and redheads [*Aythya americana*]).

Based on the response coefficients and their standard errors, the teratogenesis function for ducks, stilts, and avocets were significantly different.[206] Within this data set, these responses represent "sensitive" (duck), "average" (stilt), and "tolerant" (avocet) species. The probability of overt teratogenesis in stilts increases markedly when selenium concentrations are greater than 40 mg/kg, with an $EC_{10}$ for teratogenic effects of 37 mg/kg. In contrast, the threshold for teratogenesis (expressed as an $EC_{10}$) in mallards is 23 mg/kg, and in avocets it is 74 mg/kg.

A more sensitive measure of avian selenosis is reduced egg hatchability due to embryo inviability.[17,25,206] Egg selenium concentrations that cause embryo inviability are usually below the levels that cause embryo deformities. The threshold for mean egg selenium associated with impaired egg hatchability at the population level for black-necked stilts (and, therefore, the embryotoxicity threshold) was estimated to be 6 to 7 mg/kg. This threshold is approximately equivalent to the $EC_{10}$ on a clutchwise (or henwise) basis and the $EC_{03}$ on an eggwise basis.[33] Skorupa[206] used more than

400 sample eggs from black-necked stilt nests monitored at Tulare Basin, Kesterson Reservoir, and Salton Sea, California to identify the background rate of inviable stilt eggs (8.9%) and developed the following equation to determine the number of clutches containing at least one inviable embryo:

$$Y = \exp(-2.327 + 0.0503X)/(1 + \exp(-2.327 + 0.0503X));\ r^2 = 0.18 \qquad (17.1)$$

where Y is the probability of ≥ 1 inviable egg in a sampled clutch based on selenium concentration in a random sample egg (X). In contrast to the stilt, with a threshold of 6 to 7 mg Se/kg for effects on egg hatchability, the avocet is much less sensitive, with no effects on hatchability at concentrations below 60 mg Se/kg.

In his review of the available information concerning reproductive effects of selenium in birds, Heinz[13] suggested 10 mg Se/kg (converted from 3 mg Se/kg wet-weight basis he reported) in avian eggs as a threshold for effects on hatchability. Conclusions concerning egg-selenium threshold levels for effects in waterfowl depend on which data are included in the analyses, how the effects are expressed, and statistical approaches used in the analyses. For example, Fairbrother et al.[32,34] used some of the data available from experimental studies with mallards and concluded that the threshold for reduced hatchability ($EC_{10}$) is 16 mg Se/kg in the egg. However, Skorupa[33] disagreed with this conclusion.

The data available from six studies in which mallards were fed selenomethionine[121,122,124,125,207,208] are reanalyzed here using a two-parameter logistic regression model to calculate various effect levels for selenium, as follows: $EC_{10}$ = 12.5 mg/kg; $EC_{20}$ = 16.3 mg/kg; $EC_{50}$ = 25.7 mg/kg in the eggs; in the diet, $EC_{10}$ = 4.87 mg/kg; $EC_{20}$ = 5.86 mg/kg; and $EC_{50}$ = 8.05 mg/kg. Table 17.1 provides a brief summary of the relevant studies, including the dietary treatment level of selenium, reported egg hatchability, and mean selenium concentrations in the eggs. In addition, the table provides the egg hatchability results for the treatment groups as a proportion of the controls (to

**Table 17.1 Effects of Dietary Selenium Exposure (as selenomethionine) on Egg Hatchability in Mallards and Associated Selenium Concentrations (mg/kg, dry weight) in Eggs**

| Dietary Selenium (mg/kg)[a] | N | Egg Hatchability (%)[b] | Hatchability as Proportion of Control | Egg Selenium (mg/kg)[c] | Reference |
|---|---|---|---|---|---|
| Control | 11 | 65.7 | 1.0 | 0.16 | 121 |
| 10 | 5 | 30.9 | 0.470 | 15.2 | |
| Control | 32 | 59.6 | 1.0 | 0.59 | 122 |
| 1 | 15 | 70.7 | 1.19 | 2.74 | |
| 2 | 15 | 60.0 | 1.01 | 5.28 | |
| 4 | 15 | 53.4 | 0.896 | 11.2 | |
| 8 | 15 | 36.9 | 0.619 | 36.3 | |
| 16 | 9 | 2.2* | 0.037 | 59.4 | |
| Control | 10 | 41.3 | 1.0 | 1.35 | 207 |
| 10 DL[d] | 15 | 7.6* | 0.184 | 30.4 | |
| 10 L[e] | 15 | 6.4* | 0.155 | 29.4 | |
| Control | 12 | 44.2 | 1.0 | 1.16 | 208 |
| 10 | 11 | 24.0 | 0.543 | 25.1 | |
| Control | 37 | 88.2 | 1.0 | 1.4 | 124 |
| 10 | 35 | 20.0* | 0.227 | 37 | |
| Control | 33 | 62.0 | 1.0 | 0.89 | 125 |
| 3.5 | 29 | 61.0 | 0.984 | 11.6 | |
| 7.0 | 34 | 41.0* | 0.661 | 23.4 | |

[a] Presented as nominal concentration in diet. In some (but not all) studies, control and treated diets were analyzed. Control diets typically contained 0.4 mg/kg selenium.
[b] Asterisks indicate hatchability significantly different than controls.
[c] When mean selenium concentrations in eggs were reported on wet-weight basis, concentrations were converted to approximate dry-weight basis, assuming 70% moisture.
[d] Seleno-DL-methionine.
[e] Seleno-L-methionine.

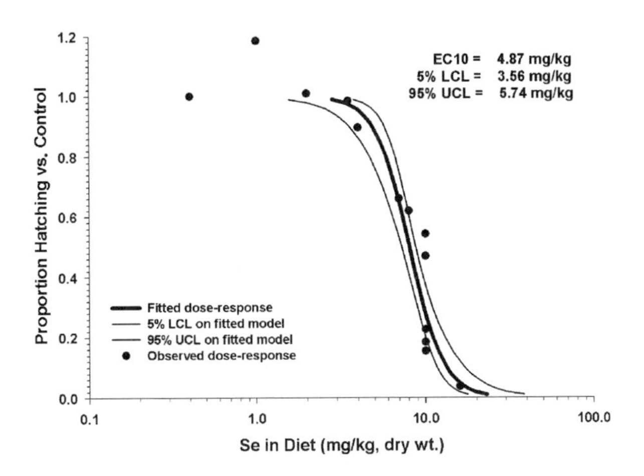


**Figure 17.1** Mallard egg hatchability vs. control as a function of selenium concentration in diet.

take into account the variable hatching success for controls in the various studies). The results of these studies are illustrated graphically in Figures 17.1 and 17.2.

Fertility has been measured in only a small number of selenium studies. In American kestrels fed selenomethionine at 12 mg Se/kg, egg fertility was significantly reduced (by over 14%) compared to kestrels fed 6 mg/kg selenium.[127] Lack of reporting on fertility effects may be due in part to a general practice of simply including infertile eggs as inviable eggs in studies of selenium effects in birds (i.e., "infertility" effects may not be separated from "embryotoxic" effects in the overall measurement of hatchability). Failure to measure infertility as a separate endpoint may be due to the difficulty often associated with distinguishing infertile eggs from those containing embryos that have died very early in development. Nevertheless, decreased fertility is a distinct effect from embryotoxicity, particularly in that it indicates a mechanism acting on adult, rather than embryonic, physiology. Results obtained in kestrels suggest infertility may be a potentially important factor contributing to the overall reproductive impairment in some species. However, in mallards and black-crowned night-herons fed 10 mg Se/kg as selenomethionine, egg fertility was not reduced compared with controls.[121,123,207,208] Similarly, fertility was not affected in mallards fed diets containing selenium at 7 mg/kg[125] or 16 mg/kg[122] as selenomethionine, but hatchability of fertile eggs

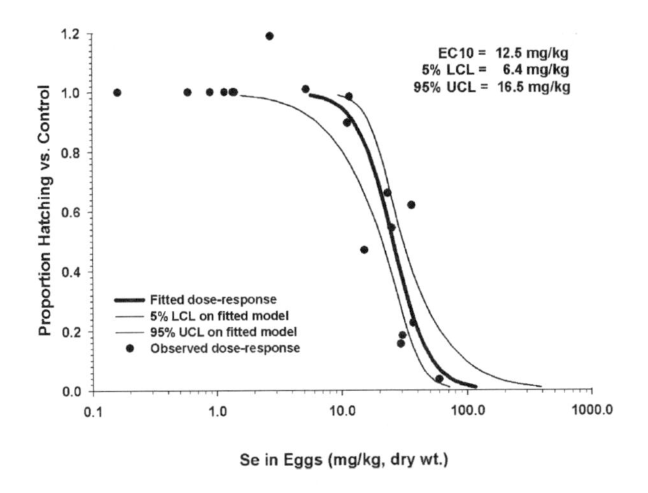


**Figure 17.2** Mallard egg hatchability vs. control as a function of selenium concentration in eggs.

was significantly reduced. Thus, effects on egg fertility in mallards and night-herons are not likely to be as ecologically significant as reduced hatchability.

#### *17.4.2.4.2 Adult Health, Physiology, and Survival*

O'Toole and Raisbeck[4] have recently provided a comprehensive review of the various kinds of lesions associated with acute-subchronic and chronic selenosis in avian species. Other reviews of effects in adult birds also have been published recently.[8,13,19]

Laboratory studies have been conducted with mallards to determine the kinds of lesions and other measurements that can be used for diagnosis of selenium toxicosis in birds.[148,213,214] In general, ducks that received diets containing more than 20 mg Se/kg developed a number of lesions of the liver and integument. Those receiving 40 mg/kg or more selenium in their diets lost weight and had abnormal changes in the integument that involved structures containing hard keratin, such as feathers (alopecia/depterylation [i.e., feather loss]), beaks (necrosis), and nails (onychoptosis [soughed or broken]). When corroborated by elevated selenium concentrations in tissues (especially the liver), these integumentary and hepatic lesions, as well as weight loss, can serve for diagnosis of selenium toxicosis in birds. It should be noted, however, that some birds died without exhibiting any significant morphological lesions, even though they were emaciated.

Selenium toxicosis effects had been described previously in several species of aquatic birds found at Kesterson Reservoir from 1984 to 1986.[7,136,144] Those birds exhibited many of the same signs of selenosis as those later found in mallards (as described above) including hepatic lesions, alopecia, necrosis of the beak, and weight loss. As described in a previous section, most kinds of aquatic invertebrates at Kesterson had mean selenium concentrations greater than 100 mg/kg. Thus, the laboratory studies with mallards provided strong evidence that selenosis was the main factor responsible for morbidities and mortalities among birds at Kesterson Reservoir.[214]

In a 16-week exposure to evaluate susceptibility of mallards to over-winter mortality as a result of selenium exposure, ducks that received dietary concentrations of 20, 40, or 80 mg Se/kg (as selenomethionine) had higher mortality rates than ducks receiving the control or 10 mg Se/kg diets.[147] After 1 week of treatment, body weights were significantly depressed by the 20, 40, or 80 mg Se/kg diet. Ducks that survived for 16 weeks on the 20 mg/kg treatment were returned to a control diet; after 4 weeks, their body weight was similar to that of the control birds. Concentrations of selenium in blood were related to dietary treatment levels, but mortality was not clearly related to a threshold selenium concentration in the blood.

A number of studies (e.g., References 144, 215–218) have described physiological changes that are associated with selenium exposure in field-collected or laboratory-exposed birds. These generally involve changes in measurements associated with liver pathology and glutathione metabolism (e.g., glycogen, protein, total sulfhydryl and protein-bound sulfhydryl concentrations; glutathione peroxidase activity).

A few studies have suggested that effects of selenium on avian immune function are a sensitive endpoint and may affect birds at relatively low exposures.[219–221] However, immunotoxicity of selenium has not been studied sufficiently to provide a threshold-effect level in wild birds.[206]

### *17.4.2.5 Mammals*

Most studies of the effects of excess selenium have focused on livestock and laboratory animals, and there have been no well-documented cases of significant selenosis problems among wild mammals.[17] The most common cause of chronic selenosis is consumption of forage containing more than 5 mg Se/kg. Sublethal effects were observed in dogs fed diets containing about 7 mg Se/kg.[2]

O'Toole and Raisbeck[4] have recently provided a comprehensive review of the various kinds of lesions associated with acute-subchronic and chronic selenosis in mammalian species, and they stress the importance of using multiple lines of evidence (e.g., selenium concentrations in tissues, presence and types of lesions) in diagnosing selenosis. Subchronic and chronic selenosis (sometimes called "alkali disease") commonly involves changes in the integument, especially the hooves, horns, and hair. Loss of hair is a common sign of selenosis in a number of mammalian species and is similar to the bilateral loss of feathers observed in selenium-poisoned birds (described above).

Lesions of the liver occur in some species of mammals (sheep, dogs, cats, and rats), but they are not particularly significant in most livestock.[4] Unlike birds, congenital abnormalities are not commonly associated with selenium toxicosis in mammals. When congenital abnormalities occur, the female parent usually was exposed to near-lethal dietary concentrations of selenium.

At Kesterson Reservoir, which contained seleniferous agricultural drainage waters,[7] small mammals and raccoons were trapped and examined to determine whether selenium adversely affected the health of adult animals or their reproduction.[156,166] Despite very high concentrations of selenium in livers or whole-body small mammals, or in livers, hair, feces, or blood of raccoons, there was no evidence of adverse effects in the mammals. Parameters evaluated for the small mammals included body size, body condition, liver size, and reproduction (including examination for malformations). Those measures as well as histopathology of selected tissues (liver, kidney, and spleen) and blood chemistry were evaluated in the raccoons.

### 17.4.3 Interactions

Many studies have shown interactions between selenium and other inorganic elements (especially arsenic, sulfur, and various metals), vitamins A, C, and E, as well as sulfur-containing amino acids.[7,19,44,46,56,222] The interactions may be synergistic or antagonistic in terms of effects on uptake and metabolic effects, and the degree of interaction is affected by numerous factors. Thus, the topic of interactions is too complex to be addressed in detail in this review, and only a few examples of recent studies are discussed. Readers are encouraged to consult some of the references cited above for more details and a further introduction to this topic. Nevertheless, some of these interactions can be important factors in the design of field or laboratory studies and in the evaluation of results, and they should be taken into consideration.

Interactions between mercury and selenium have probably been studied more intensively in more kinds of animals than most of the others listed above. Mercury and selenium concentrations in the livers of various free-living carnivorous mammals often are highly correlated in a molar ratio of 1:1.[19,109,223,224] However, there is no consistent pattern for such a correlation in the livers of birds. For example, hepatic mercury and selenium were correlated, with an overall mercury:selenium molar ratio of 1:6, in diving ducks (surf scoters [*Melanitta perspicillata*] and greater scaups [*Aythya marila*]) from San Francisco Bay, California, during 1982.[140] In a subsequent study of scoters sampled twice at six locations in the Bay during 1985, mercury and selenium in the livers were not correlated; mercury:selenium ratios were typically between 1:7 and 1:15 for most locations and collection times, but the mean ratio at one site was 1:45.[225]

Elsewhere, mercury and selenium concentrations were positively correlated in some bird livers, but not in others, or they were negatively correlated (see review by Ohlendorf[226]). These relationships may change as birds remain at the sampling location (due to differential accumulation and loss rates for mercury and selenium), they may vary because of differing relative concentrations of the two elements, and other factors (such as the chemical forms present) also may complicate the patterns of bioaccumulation.

In the selenium-mercury studies discussed above, concentrations were reported for total selenium and total mercury. However, recent studies by Henny et al.[227] and Spalding et al.[228] have shown high correlations of selenium with inorganic mercury on a molar basis in fish-eating birds. Those authors suggested that selenium may contribute to the sequestration of inorganic mercury, thereby reducing its toxicity. This conclusion would be consistent with the results of a selenium-mercury interaction study with mallards by Heinz and Hoffman[208] discussed below.

Interactive effects of selenium with arsenic, boron, or mercury have been evaluated in reproductive studies with mallards.[124,125,208] Each of the studies involved varying levels of dietary exposures of breeding mallards to selenium alone, one of the other elements alone, and also selenium in combination with the other chemical. In each study, selenium and the other chemical caused significant adverse effects on reproduction when present alone in the diet at higher treatment levels, but the interactions varied by chemical. Antagonistic interactions between arsenic and selenium occurred whereby arsenic reduced selenium accumulation in duck livers and eggs and alleviated the effects of selenium on hatching success and embryo deformities. There was little evidence of interaction between boron and selenium when ducks were fed the two chemicals in combination. Selenium provided a protective effect that reduced the toxicity of mercury to adult male ducks. However, when the diet contained 10 mg Se/kg plus 10 mg Hg/kg, the effects on reproduction were worse than for either selenium or mercury alone. The number of young produced per female and the frequency of teratogenic effects were significantly affected by the combination of mercury and selenium in the diet, and mercury also enhanced the storage of selenium in duck tissues.

## 17.5 SUMMARY AND CONCLUSIONS

Selenium is a naturally occurring trace element that is essential for animal nutrition, but it becomes toxic to sensitive species of fish or birds at dietary concentrations about two or three times the background levels found in many food-chain organisms. The biogeochemistry of selenium is complex, and it is essential to understand selenium chemistry to enable competent predictions of its ecotoxicology. Selenium concentrations generally are elevated in areas where soils were derived from Cretaceous and Tertiary marine sedimentary rocks in the western United States. Irrigation of such soils has led to contamination of wetlands receiving drainage from the agricultural lands. Industrial sources (such as coal and oil production or combustion and phosphate mining) have contributed selenium to habitats where fish and wildlife can be exposed. Once released to the surface environment, selenium cycling is affected by many biogeochemical processes, especially in aquatic habitats.

Selenium bioaccumulates in aquatic and terrestrial food chains and in higher trophic-level animals that feed on those plants and animals. The greatest level of bioaccumulation is typically from water to aquatic plants or invertebrates (often a 1000-fold or greater increase occurs). Background concentrations have been fairly well established for plants and animals in freshwater and terrestrial ecosystems, and these values can be used in evaluating the selenium status of an area if field-collected samples are analyzed and compared to background values. Such studies have been conducted in a number of areas. Multitrophic-level laboratory studies also have been conducted and help interpret the significance of bioaccumulation of selenium.

Studies of the effects of selenium have focused mainly on domestic livestock and on fish and aquatic birds because those animals have been found to be sensitive to the adverse effects of excess selenium. In fish and birds, the early life stages (fish larvae/fry, bird embryos) are most sensitive, although lethal and sublethal adverse effects also occur in adult animals under field conditions in some areas. Teratogenic effects in fish larvae/fry and in bird embryos may lead to death of young that hatch, but a significant amount of mortality also may occur without visible developmental abnormalities.

Many studies have shown interactions between selenium and other inorganic elements (especially arsenic, sulfur, and various metals), vitamins A, C, and E, as well as sulfur-containing amino acids. These interactions may be synergistic or antagonistic in terms of effects on uptake and metabolic effects. The interactions of selenium and mercury often are of particular interest, and there is no universal rule of thumb as to the net result of these interactions. For example, in mallards, selenium provided a protective effect that reduced the toxicity of mercury in adult ducks, but the effects of mercury plus selenium in the diet caused greater impairment of reproduction than either chemical alone.

Overall, there is enough information to thoroughly evaluate the significance of selenium as an environmental contaminant if it is included among the chemicals of potential concern when ecological risk assessments are planned and conducted.[204] The main difference between selenium and most other contaminants is that it is essential to have a good understanding of selenium's occurrence, complex biogeochemistry, and ecotoxicology to avoid serious errors.

## ACKNOWLEDGMENTS

I appreciate the assistance of Gary M. Santolo in providing some of the material for the section on effects in birds and Bradley E. Sample in performing the statistical analyses of relationships between effects in mallards and the concentrations of selenium in the eggs or diet. Gary H. Heinz reviewed the manuscript and provided useful comments.

## REFERENCES

1. Poley, W. E., Moxon, A. L., and Franke, K. W., Further studies of the effects of selenium poisoning on hatchability, *Poult. Sci.*, 16, 219, 1937.
2. Moxon, A. L. and Rhian, M., Selenium poisoning, *Physiol. Rev.*, 23, 305, 1943.
3. Rosenfeld, I. and Beath, O. A., *Selenium: Geobotany, Biochemistry, Toxicity, and Nutrition*, Academic Press, New York, 1964.
4. O'Toole, D. and Raisbeck, M. F., Magic numbers, elusive lesions: Comparative pathology and toxicology of selenosis in waterfowl and mammalian species, in *Environmental Chemistry of Selenium*, Frankenberger, W. T., Jr. and Engberg, R. A., Eds., Marcel Dekker, New York, 1998, 355 (Chap. 19).
5. Combs, G. F., Jr., and Combs, S. B., *The Role of Selenium in Nutrition*, Academic Press, Orlando, 1986.
6. Jacobs, L. W., *Selenium in Agriculture and the Environment*, American Society of Agronomy and Soil Science Society of America, Madison, WI, 1989.
7. Ohlendorf, H. M., Bioaccumulation and effects of selenium in wildlife, in *Selenium in Agriculture and the Environment*, Jacobs, L. W., Ed., Special Publication No. 23, Soil Science Society of America, Madison, WI, 1989, 133 (Chap. 8).
8. Ohlendorf, H. M., Selenium, in *Noninfectious Diseases of Wildlife, 2nd ed.*, Fairbrother, A., Hoff, G. L., and Locke, L. N., Eds., Iowa State University Press, Ames, 1996, 128 (Chap. 12).
9. Cappon, C. J., Sewage sludge as a source of environmental selenium, *Sci. Total Environ.*, 100, 177, 1991.
10. Masscheleyn, P. H. and Patrick, W. H., Biogeochemical processes affecting selenium cycling in wetlands, *Environ. Toxicol. Chem.*, 12, 2235, 1993.
11. Frankenberger, W. T., Jr. and Benson, S., *Selenium in the Environment*, Marcel Dekker, New York, 1994.
12. Maier, K. J. and Knight, A. W., Ecotoxicology of selenium in freshwater systems, *Rev. Environ. Contam. Toxicol.*, 134, 31, 1994.
13. Heinz, G. H., Selenium in birds, in *Interpreting Environmental Contaminants in Animal Tissues*, Beyer, W. N., Heinz, G. H., and Redmon-Norwood, A. W., Eds., Lewis Publishers, Boca Raton, Florida, 1996, 447 (Chap. 20).
14. Lemly, A. D., Selenium in aquatic organisms, in *Interpreting Environmental Contaminants in Animal Tissues*, Beyer, W. N., Heinz, G. H., and Redmon-Norwood, A. W., Eds., Lewis Publishers, Boca Raton, Florida, 1996, 427 (Chap. 19).
15. U.S. Public Health Service, Toxicological Profile for Selenium (Update), U.S. Department of Health and Human Services, PHS, Agency for Toxic Substances and Disease Registry, Atlanta, 1996.
16. Frankenberger, W. T., Jr. and Engberg, R. A., *Environmental Chemistry of Selenium*, Marcel Dekker, New York, 1998.

17. U.S. Department of the Interior, Guidelines for Interpretation of the Biological Effects of Selected Constituents in Biota, Water, and Sediment, USDI (Bureau of Reclamation, U.S. Fish and Wildlife Service, U.S. Geological Survey, Bureau of Indian Affairs), National Irrigation Water Quality Program Information Report No. 3, Bureau of Reclamation, Denver, 1998, 139 and Errata Sheet.
18. Woodbury, P. B., McCune, D. C., and Weinstein, L. H., A review of selenium uptake, transformation, and accumulation by plants with particular reference to coal fly ash landfills, in *Biogeochemistry of Trace Elements in Coal and Coal Combustion Byproducts*, Sajwan, K. S., Alva, A. K., and Keefer, R. F., Eds., Kluwer Academic/Plenum Publishers, New York, 1999, 309 (Chap. 20).
19. Eisler, R., *Handbook of Chemical Risk Assessment: Health Hazards to Humans, Plants, and Animals*, Vol. 3, Lewis Publishers, Boca Raton, FL, 2000, 1649 (Chap. 31).
20. U.S. Environmental Protection Agency, Ambient Water Quality Criteria for Selenium — 1987, EPA-440/5–87–006, U.S. EPA Office of Water, Washington, D.C., 1987.
21. U.S. Environmental Protection Agency, Report on the Peer Consultation Workshop on Selenium Aquatic Toxicity and Bioaccumulation, EPA 822-R-98–007, U.S. EPA Office of Water, Washington, D.C., 1998.
22. U.S. Environmental Protection Agency, National Recommended Water Quality Criteria — Correction, EPA 822-Z-99–001 U.S. EPA Office of Water, Washington, D.C., 1999.
23. Lemly, A. D., Pathology of selenium poisoning in fish, in *Environmental Chemistry of Selenium*, Frankenberger, W. T., Jr. and Engberg, R. A., Eds., Marcel Dekker, New York, 1998, 281 (Chap. 16).
24. Lemly, A. D., Selenium impacts on fish: An insidious time bomb, *Human Ecol. Risk Assess.*, 5, 1139, 1999.
25. Skorupa, J. P., Selenium poisoning of fish and wildlife in nature: Lessons from twelve real-world examples, in *Environmental Chemistry of Selenium*, Frankenberger, W. T., Jr. and Engberg, R. A., Eds., Marcel Dekker, New York, 1998, 315 (Chap. 18).
26. Chapman, P. M., Invited debate/commentary: Selenium – A potential time bomb or just another contaminant?, *Hum. Ecol. Risk Assess.*, 5, 1123, 1999.
27. Canton, S. P., Acute aquatic life criteria for selenium, *Environ. Toxicol. Chem.*, 18, 1425, 1999.
28. U.S. Department of the Interior and U.S. Department of Commerce, Untitled letter to Ms. Felicia Marcus, Administrator U.S. EPA Region 9, representing the Services' final biological opinion on the effects of the final promulgation of the California Toxics Rule on listed species and critical habitats in California, File Reference 1–1–98-F-21, from U.S. Department of the Interior, Fish and Wildlife Service, Sacramento, and U.S. Department of Commerce, National Marine Fisheries Service, Long Beach, dated March 24, 2000.
29. Canton, S. P. and van Derveer, W. D., Selenium toxicity to aquatic life: An argument for sediment-based water quality criteria, *Environ. Toxicol. Chem.,* 16, 1255, 1997.
30. Van Derveer, W. D., and Canton, S. P., Selenium sediment toxicity thresholds and derivation of water quality criteria for freshwater biota of western streams, *Environ. Toxicol. Chem.*, 16, 1260, 1997.
31. Hamilton, S. J. and Lemly, A. D., Water-sediment controversy in setting environmental standards for selenium, *Ecotoxicol. Environ. Saf.*, 44, 227, 1999.
32. Fairbrother, A. et al., Egg selenium concentrations as predictors of avian toxicity, *Hum. Ecol. Risk Assess.*, 5, 1229, 1999.
33. Skorupa, J. P., Beware of missing data and undernourished statistical models: Comment on Fairbrother et al.'s critical evaluation, *Hum. Ecol. Risk Assess.*, 5, 1255, 1999.
34. Fairbrother, A. et al., Egg selenium thresholds for birds: A response to J. Skorupa's critique of Fairbrother et al., 1999, *Hum. Ecol. Risk Assess.*, 6, 203, 2000.
35. McNeal, J. M. and Balistrieri, L. S., Geochemistry and occurrence of selenium: An overview, in *Selenium in Agriculture and the Environment*, Jacobs, L. W., Ed., Special Publication No. 23, Soil Science Society of America, Madison, WI, 1989, 1 (Chap. 1).
36. Mayland, H. F. et al., Selenium in seleniferous environments, in *Selenium in Agriculture and the Environment*, Jacobs, L. W., Ed., Special Publication No. 23, Soil Science Society of America, Madison, WI, 1989, 15 (Chap. 2).
37. McDowell, L. R., *Minerals in Animal and Human Nutrition*, Academic Press, San Diego, 1992, 294 (Chap. 13).
38. Milne, J. B., The uptake and metabolism of inorganic selenium species, in *Environmental Chemistry of Selenium*, Frankenberger, W. T., Jr. and Engberg, R. A., Eds., Marcel Dekker, New York, 1998, 459 (Chap. 23).

39. Presser, T. S. and Ohlendorf, H. M., Biogeochemical cycling of selenium in the San Joaquin Valley, California, USA, *Environ. Manage.*, 11, 805, 1987.
40. Saiki, M. K. and Lowe, T. P., Selenium in aquatic organisms from subsurface agricultural drainage water, San Joaquin Valley, California, *Arch. Environ. Contam. Toxicol.*, 16, 657, 1987.
41. Hothem, R. L. and Ohlendorf, H. M., Contaminants in foods of aquatic birds at Kesterson Reservoir, California, 1985, *Arch. Environ. Contam. Toxicol.*, 18, 773, 1989.
42. Schuler, C. A., Anthony, R. G., and Ohlendorf, H. M., Selenium in wetlands and waterfowl foods at Kesterson Reservoir, California, 1984, *Arch. Environ. Contam. Toxicol.*, 19, 845, 1990.
43. Zhang, Y. Q. and Moore, J. N., Controls on selenium distribution in wetland sediment, Benton Lake, Montana, *Water Air Soil Pollut.*, 97, 323, 1997.
44. Kishchak, I. T., Supplementation of selenium in the diets of domestic animals, in *Environmental Chemistry of Selenium*, Frankenberger, W. T., Jr. and Engberg, R. A., Eds., Marcel Dekker, New York, 1998, 143 (Chap. 9).
45. Lo, M. T. and Sandi, E., Selenium: Occurrence in foods and its toxicological significance. A review, *J. Environ. Pathol. Toxicol.*, 4, 193, 1980.
46. National Academy of Sciences — National Research Council, *Selenium in Nutrition*, Subcommittee on Selenium, Committee on Animal Nutrition, Board on Agriculture, NRC, National Academy Press, Washington, D.C., 1983.
47. Piper, D. Z. and Medrano, M. D., Geochemistry of the phosphoria formation at Montpelier Canyon, Idaho: Environment and deposition, in The Phosphoria Formation: Its Geochemical and Biological Environment of Deposition, U.S. Geological Survey Bulletin 2023, U.S. Government Printing Office: 1994–583–048/10005, 1994.
48. Presser, T. S., "The Kesterson Effect," *Environ. Manage.*, 18, 437, 1994.
49. Presser, T. S., Sylvester, M. A., and Low, W. H., Bioaccumulation of selenium from natural geologic sources in western states and its potential consequences, *Environ. Manage.*, 18, 423, 1994.
50. Seiler, R. L., Skorupa, J. P., and Peltz, L. A., Areas Susceptible to Irrigation-induced Selenium Contamination of Water and Biota in the Western United States, USGS Circular 1180, U.S. Geological Survey, Carson City, Nevada, 1999.
51. Nriagu, J. O., *Heavy Metals in the Environment, Vol. 1*, CEP Consultants, Edinburgh, 1991.
52. Haygarth, P. M., Global importance and global cycling of selenium, in *Selenium in the Environment*, Frankenberger, W. T., Jr. and Benson, S., Eds., Marcel Dekker, New York, 1994, 1 (Chap. 1).
53. Selenium Tellurium Development Association, Internet site: www.stda.be, Grimbergen, Belgium, 2000.
54. Pease, W. et al., Technical Report: Derivation of Site-Specific Water Quality Criteria for Selenium in San Francisco Bay, San Francisco Bay Regional Water Quality Control Board, Oakland, 1992.
55. Boon, D. Y., Potential selenium problems in Great Plains soils, in *Selenium in Agriculture and the Environment*, Jacobs, L. W., Ed., Special Publication No. 23, Soil Science Society of America, Madison, WI, 1989, 107 (Chap. 6).
56. National Academy of Sciences — National Research Council, *Selenium*, Committee on Medical and Biologic Effects of Environmental Pollutants, NRC, National Academy Press, Washington, D.C., 1976.
57. Maier, K. J. et al. The dynamics of selenium in aquatic ecosystems, in *Trace Substances in Environmental Health-XXI*, Hemphill, D. D., Ed., University of Missouri, Columbia, 1987, 361.
58. Oremland, R. S., Biogeochemical transformations of selenium in anoxic environments, in *Selenium in the Environment*, Frankenberger, W. T., Jr. and Benson, S., Eds., Marcel Dekker, New York, 1994, 389 (Chap. 16).
59. Nriagu, J. O. and Wong, H. K., Selenium pollution of lakes near the smelters at Sudbury, Ontario, *Nature*, 301, 55, 1983.
60. Haygarth, P. M., Jones, K. C., and Harrison, A. F., Selenium cycling through agricultural grasslands in the U.K.: Budgeting the role of the atmosphere, *Sci. Total Environ.*, 103, 89, 1991.
61. Terry, N. and Zayed, A. M., Selenium volatilization by plants, in *Selenium in the Environment*, Frankenberger, W. T., Jr. and Benson, S., Eds., Marcel Dekker, New York 1994, 343 (Chap. 14).
62. Tokunaga, T. K., Pickering, I. J., and Brown, G. E., Jr., Selenium transformations in ponded sediments, *Soil Sci. Soc. Am. J.*, 60, 781, 1996.
63. Zawislanski, P. T. and Zavarin, M., Nature and rates of selenium transformations: A laboratory study of Kesterson Reservoir soils, *Soil Sci. Soc. Am. J.*, 60, 791, 1996.

64. Zhang, Y. Q. and Moore, J. N., Selenium fractionation and speciation in a wetland system, *Environ. Sci. Technol.*, 30, 2613, 1996.
65. Dowdle, P. R. and Oremland, P. R., Microbial oxidation of elemental selenium in soil slurries and bacterial cultures, *Environ. Sci Technol.*, 32, 3749, 1998.
66. Pilon-Smits, E. A. H. et al., Selenium volatilization and accumulation by twenty aquatic plant species, *J. Environ. Qual.,* 28, 1011, 1999.
67. Zhang, Y. Q. and Frankenberger, W. T., Jr., Formation of dimethylselenonium compounds in soil, *Environ. Sci Technol.*, 34, 776, 2000.
68. Terry, N. and Zayed, A. M., Phytoremediation of selenium, in *Environmental Chemistry of Selenium*, Frankenberger, W. T., Jr. and Engberg, R. A., Eds., Marcel Dekker, New York, 1998, 633 (Chap. 31).
69. Cooke, T. C. and Bruland, K. W., Aquatic chemistry of selenium: Evidence of biomethylation, *Environ. Sci. Technol.*, 21, 1214, 1987.
70. Hansen, D. et al., Selenium removal by constructed wetlands: Role of biological volatilization, *Environ. Sci. Technol.,* 18, 773, 1998.
71. Wilber, C. G., *Selenium: A Potential Environmental Poison and A Necessary Food Constituent*, Charles C Thomas, Springfield, Illinois, 1983.
72. Bennett, W. N., Brooks, A. S., and Boraas, M. E., Selenium uptake and transfer in an aquatic food chain and its effect on fathead minnow larvae, *Arch. Environ. Contam. Toxicol.*, 15, 513, 1986.
73. Jenkins, D. W., Biological Monitoring of Toxic Trace Metals, Vol. 2, Toxic Trace Metals in Plants and Animals of the World, Part III, Report 600/3–80–0921, U.S. Environmental Protection Agency, 1980, 1090.
74. Rodgers, J. H., Jr., Cherry, D. S., and Guthrie, R. K., Cycling of elements in duckweed (*Lemna perpusilla*) in an ash settling basin and swamp drainage system. *Water Res.*, 12, 765, 1978.
75. Furr, A. K. et al., Elemental content of aquatic organisms inhabiting a pond contaminated with coal fly ash, *N.Y. Fish Game J.*, 26, 154, 1979.
76. Besser, J. M. et al., Distribution and bioaccumulation of selenium in aquatic microcosms, *Environ. Pollut.*, 62, 1, 1989.
77. Besser, J. M., Canfield, T. J., and La Point, T. W., Bioaccumulation of organic and inorganic selenium in a laboratory food chain, *Environ. Toxicol. Chem.*, 12, 57, 1993.
78. Thomas, B. V., Knight, A. W., and Maier, K. J., Selenium bioaccumulation by the water boatman *Trichocorixa reticulata* (Guerin-Meneville), *Arch. Environ. Contam. Toxicol.*, 36, 295, 1999.
79. Alaimo, J., Ogle, R. S., and Knight, A. W., Selenium uptake by larval *Chironomus decorus* from a *Ruppia maritima*-based benthic/detrital substrate, *Arch. Environ. Contam. Toxicol.*, 27, 441, 1994.
80. Graham, R. V. et al., Comparison of selenomethionine and selenite cycling in freshwater experimental ponds, *Water Air Soil Pollut.*, 62, 25, 1992.
81. Luoma, S. N. et al., Determination of selenium bioavailability to a benthic bivalve from particulate and solute pathways, *Environ. Sci. Technol.*, 26, 485, 1992.
82. Saiki, M. K., Jennings, M. R., and Brumbaugh, W. G., Boron, molybdenum, and selenium in aquatic food chains from the lower San Joaquin River and its tributaries, California, *Arch. Environ. Contam. Toxicol.*, 24, 307, 1993.
83. Williams, M. J. et al., Effects of sulfate on selenate uptake and toxicity in the green alga *Selenastrum capricornutum*, *Arch. Environ. Contam. Toxicol.*, 27, 449, 1994.
84. Bailey, F. C. et al., Effect of sulfate level on selenium uptake by *Ruppia maritima*, *Chemosphere*, 30, 579, 1995.
85. Girling, C. A., Selenium in agriculture and the environment, *Agric. Ecosyst. Environ.*, 11, 37, 1984.
86. Mikkelsen, R. L., Page, A. L., and Bingham, F. T., Factors affecting selenium accumulation by crop plants, in *Selenium in Agriculture and the Environment*, Jacobs, L. W., Ed., Special Publication No. 23, Soil Science Society of America, Madison, WI, 1989, 65 (Chap. 4).
87. James, L. F. et al., Selenium poisoning in livestock: A review and progress, in *Selenium in Agriculture and the Environment*, Jacobs, L. W., Ed., Special Publication No. 23, Soil Science Society of America, Madison, WI, 1989, 123 (Chap. 7).
88. Lakin, H. W., Selenium accumulations in soils and its absorption by plants and animals, *Geol. Soc. Am. Bull.*, 83, 181, 1972.
89. Ohlendorf, H. M. and Santolo, G. M., Kesterson Reservoir — Past, present, and future: An ecological risk assessment, in *Selenium in the Environment*, Frankenberger, W. T., Jr. and Benson, S., Eds., Marcel Dekker, New York, 1994, 69 (Chap. 4).

90. CH2M HILL and Lawrence Berkeley National Laboratory, Ecological Risk Assessment for Kesterson Reservoir, Prepared for U.S. Bureau of Reclamation Mid-Pacific Region, December 2000.
91. Ohlendorf, H. M. and Hothem, R. L., Agricultural drainwater effects on wildlife in central California, in *Handbook of Ecotoxicology*, Hoffman, D. J., Rattner, B. A., Burton, G. A., Jr., and Cairns, J., Jr., Eds., Lewis Publishers, Boca Raton, FL, 1995, 577 (Chap. 26).
92. Leland, H. V. and Scudder, B. C., Trace elements in *Corbicula fluminea* from the San Joaquin River, California, *Sci. Total Environ.*, 97/98, 641, 1990.
93. Johns, C., Luoma, S. N, and Elrod, V., Selenium accumulation in benthic bivalves and fine sediments of San Francisco Bay, the Sacramento-San Joaquin Delta, and selected tributaries, *Estuar. Coastal Shelf Sci.*, 27, 381, 1988.
94. Cherry, D. S. and Guthrie, R. K., Mode of elemental dissipation from ash basin effluent, *Water Air Soil Pollut.*, 9, 403, 1978.
95. Cumbie, P. M. and van Horn, S. L., Selenium accumulation associated with fish mortality and reproductive failure, *Proc. Annu. Conf. Southeast. Assoc. Fish Wildl. Agencies*, 32, 612, 1978.
96. Lemly, A. D., Toxicology of selenium in a freshwater reservoir: Implications for environmental hazard evaluation and safety, *Ecotoxicol. Environ. Saf.*, 10, 314, 1985.
97. Malloy, J. C., Meade, M. L., and Olsen, E. W., Small-scale spatial variation of selenium concentrations in chironomid larvae, *Bull. Environ. Contam. Toxicol.*, 62, 122, 1999.
98. Wang, W. X., Fisher, N. S., and Luoma, S. N., Kinetic determinations of trace element bioaccumulation in the mussel *Mytilus edulis*, *Mar. Ecol. Prog. Ser.*, 140, 91, 1996.
99. Ogle, R. S. and Knight, A. W., Selenium bioaccumulation in aquatic ecosystems: 1. Effects of sulfate on the uptake and toxicity of selenate in *Daphnia magna*, *Arch. Environ. Contam. Toxicol.*, 30, 274, 1996.
100. Gissel-Nielsen, G. and Gissel-Nielsen, M., Ecological effects of selenium application to field crops, *Ambio*, 2, 114, 1973.
101. Helmke, P. A. et al., Effects of soil-applied sewage sludge on concentrations of elements in earthworms, *J. Environ. Qual.*, 8, 322, 1979.
102. Schmitt, C. J. et al., Organochlorine residues and elemental contaminants in U.S. freshwater fish, 1976–1986: National Contaminant Biomonitoring Program, *Rev. Environ. Contam. Toxicol.*, 162, 43, 1999.
103. Sorensen, E. M. B., The effects of selenium on freshwater teleosts, in *Reviews in Environmental Toxicology, Vol. 2*, Hodgson, E., Ed., Elsevier Science Publishers, New York, 1986, 59.
104. Saiki M. K., Jennings, M. R., and May, T. W., Selenium and other elements in freshwater fishes from the irrigated San Joaquin Valley, California, *Sci. Total Environ.*, 126, 109, 1992.
105. Hamilton, S. J. and Waddell, B., Selenium in eggs and milt of razorback sucker (*Xyrauchen texanus*) in the Middle Green River, Utah, *Arch. Environ. Contam. Toxicol.*, 27, 195, 1994.
106. Waddell, B. and May, T., Selenium concentrations in the razorback sucker (*Xyrauchen texanus*): Substitution of non-lethal muscle plugs for muscle tissue in contaminant assessment, *Arch. Environ. Contam. Toxicol.*, 28, 321, 1995.
107. Besser, J. M. et al., Selenium bioaccumulation and hazards in a fish community affected by coal fly ash effluent, *Ecotoxicol. Environ. Saf.*, 35, 7, 1996.
108. Hamilton, S. J., Selenium effects on endangered fish in the Colorado River Basin, in *Environmental Chemistry of Selenium*, Frankenberger, W. T., Jr. and Engberg, R. A., Eds., Marcel Dekker, New York, 1998, 297 (Chap. 17).
109. Furness, R. W. and Rainbow, P. S., *Heavy Metals in the Marine Environment*, CRC Press, Boca Raton, FL, 1990.
110. Cappon, C. J. and Smith, J. C., Chemical form and distribution of selenium in edible seafood, *J. Anal. Toxicol.*, 6, 10, 1982.
111. Hermanson, M. H. and Brozowski, J. R., Assessment of mercury and selenium in Arctic char (*Salvelinus alpinus*) from Kasegalik Lake, Belcher Islands, Northwest Territories, Canada, in Program and Abstracts, *OME 36th Conf. Int. Assoc. for Great Lakes Res.*, June 4–10, 1993, 49.
112. Hopkins, W. A. et al., Elevated trace element concentrations in southern toads, *Bufo terrestris*, exposed to coal combustion waste, *Arch. Environ. Contam. Toxicol.*, 35, 325, 1998.
113. Rowe, C. L. et al., Oral deformities in tadpoles (*Rana catesbeiana*) associated with coal ash deposition: Effects on grazing ability and growth, *Freshwater Biol.*, 36, 723, 1996.

114. Hall, R. J. and Mulhern, B. M., Are anuran amphibians heavy metal accumulators?, in *Vertebrate Ecology and Systematics — A Tribute to Henry S. Fitch*, Seigel, R. A. et al., Eds., Museum of Natural History, Univ. Kansas, Lawrence, 1984, 123.
115. Burger, J. and Snodgrass, J., Heavy metals in bullfrog (*Rana catesbeiana*) tadpoles: Effects of depuration before analysis, *Environ. Contam. Toxicol.*, 17, 2203, 1998.
116. California Department of Fish and Game (CDFG), Selenium in Wildlife from Agroforestry Plantations in the San Joaquin Valley, Final Report to the California Department of Water Resources, CDFG, Bay-Delta and Special Water Projects Div., Stockton, CA, 1993.
117. Ohlendorf, H. M., Hothem, R. L., and Aldrich, T. W., Bioaccumulation of selenium by snakes and frogs in the San Joaquin Valley, California, *Copeia*, 1988, 704, 1988.
118. Hopkins, W. A., Rowe, C. L., and Congdon, J. D., Elevated trace element concentrations and standard metabolic rate in banded water snakes (*Nerodia fasciata*) exposed to coal combustion wastes, *Environ. Toxicol. Chem.*, 18, 1258, 1999.
119. Winger, P. V. et al., Residues of organochlorine insecticides, polychlorinated biphenyls, and heavy metals in biota from Apalachicola River, Florida, 1978, *J. Assoc. Off. Anal. Chem.*, 67, 325, 1984.
120. Lance, V., Joanen, T., and McNeese, L., Selenium, vitamin E, and trace elements in the plasma of wild and farm-reared alligators during the reproductive cycle, *Can. J. Zool.*, 61, 1744, 1983.
121. Heinz, G. H. et al., Reproduction in mallards fed selenium, *Environ. Toxicol. Chem.*, 6, 423, 1987.
122. Heinz, G. H., Hoffman, D. J., and Gold, L. G., Impaired reproduction of mallards fed an organic form of selenium, *J. Wildl. Manage.*, 53, 418, 1989.
123. Smith, G. J. et al., Reproduction in black-crowned night-herons fed selenium, *Lake Reservoir Manage.*, 4, 175, 1988.
124. Stanley, T. R., Jr. et al., Main and interactive effects of arsenic and selenium on mallard reproduction and duckling growth and survival, *Arch. Environ. Contam. Toxicol.*, 26, 444, 1994.
125. Stanley, T. R., Jr. et al., Effects of boron and selenium on mallard reproduction and duckling growth and survival, *Environ. Toxicol. Chem.*, 15, 1124, 1996.
126. Wiemeyer, S. N. and Hoffman, D. J., Reproduction in eastern screech-owls fed selenium, *J. Wildl. Manage.*, 60, 332, 1996.
127. Santolo, G. M. et al., Selenium accumulation and effects on reproduction in captive American kestrels fed selenomethionine, *J. Wildl. Manage.*, 63, 502, 1999.
128. Heinz, G. H., Selenium accumulation and loss in mallard eggs, *Environ. Toxicol. Chem.*, 12, 775, 1993.
129. Skorupa, J. P. and Ohlendorf, H. M., Contaminants in drainage water and avian risk thresholds, in *The Economics and Management of Water and Drainage in Agriculture*, Dinar, A. and Zilberman, D., Eds., Kluwer Academic Publishers, Boston, 1991, 345 (Chap. 18).
130. Ohlendorf, H. M. and Harrison, C. S., Mercury, selenium, cadmium and organochlorines in eggs of three Hawaiian seabird species, *Environ. Pollut. (Ser. B)*, 11, 169, 1986.
131. Henny C. J. et al., Contaminants and sea ducks in Alaska and the Circumpolar Region, *Environ. Health Perspect.(Suppl. 4)*, 103, 41, 1995.
132. Ohlendorf, H. M. et al., Embryonic mortality and abnormalities of aquatic birds: Apparent impacts of selenium from irrigation drainwater, *Sci. Total Environ.*, 52, 49, 1986.
133. King, K. A., Custer, T. W., and Weaver, D. A., Reproductive success of barn swallows nesting near a selenium-contaminated lake in east Texas, USA, *Environ. Pollut.*, 84, 53, 1994.
134. Gordus, A. G., Selenium concentrations in eggs of American avocets and black-necked stilts at an evaporation basin and freshwater wetland in California, *J. Wildl. Manage.*, 63, 497, 1999.
135. Heinz, G. H. et al., Selenium accumulation and elimination in mallards, *Arch. Environ. Contam. Toxicol.*, 19, 374, 1990.
136. Ohlendorf, H. M. et al., Bioaccumulation of selenium in birds at Kesterson Reservoir, California, *Arch. Environ. Contam. Toxicol.*, 19, 495, 1990.
137. Ohlendorf, H. M. et al., Selenium contamination of the Grasslands, a major California waterfowl area, *Sci. Total Environ.*, 66, 169, 1987.
138. Paveglio, F. L., Bunck, C. M., and Heinz, G. H., Selenium and boron in aquatic birds from central California, *J. Wildl. Manage.*, 56, 31, 1992.
139. Paveglio, F. L., Kilbride, K. M., and Bunck, C. M., Selenium in aquatic birds from central California, *J. Wildl. Manage.*, 61, 832, 1997.

140. Ohlendorf, H. M. et al., Selenium and heavy metals in San Francisco Bay diving ducks, *J. Wildl. Manage.*, 50, 64, 1986.
141. Ohlendorf, H. M. et al., Environmental contaminants and diving ducks in San Francisco Bay, in *Selenium and Agricultural Drainage: Implications for San Francisco Bay and the California Environment, Proc. 4th Selenium Symp., Berkeley,* CA, March 21, 1987, Howard, A. Q., Ed., The Bay Institute of San Francisco, Sausalito, California, 1989, 60.
142. Ohlendorf, H. M. and Fleming, W. J., Birds and environmental contaminants in San Francisco and Chesapeake bays, *Mar. Pollut. Bull.*, 19, 487, 1988.
143. Henny C. J. et al., Accumulation of trace elements and organochlorines by surf scoters wintering in the Pacific Northwest, *Northwest Nat.*, 72, 43, 1991.
144. Ohlendorf, H. M. et al., Selenium toxicosis in wild aquatic birds, *J. Toxicol. Environ. Health*, 24, 67, 1988.
145. Koranda, J. J., Biogeochemical Studies of Wintering Waterfowl in the Imperial and Sacramento Valleys, Report UCID-18288, Lawrence Livermore Laboratory, University of California, Livermore, CA, 1979.
146. Goede, A. A., Mercury, selenium, arsenic zinc in waders from the Dutch Wadden Sea, *Environ. Pollut. (Ser. A)*, 37, 287, 1985.
147. Heinz, G. H. and Fitzgerald, M. A., Overwinter survival of mallards fed selenium, *Arch. Environ Contam. Toxicol.*, 25, 90, 1993.
148. O'Toole, D. and Raisbeck, M. F., Experimentally induced selenosis of adult mallard ducks: Clinical signs, lesions, and toxicology, *Vet. Pathol.*, 34, 330, 1997.
149. Yamamoto, J. L., Santolo, G. M., and Wilson, B. W., Selenium accumulation in captive American kestrels (*Falco sparverius*) fed selenomethionine and naturally incorporated selenium, *Environ. Toxicol. Chem.*, 17, 2494, 1998.
150. Santolo, G. M. and Yamamoto, J. T., Selenium in blood of predatory birds from Kesterson Reservoir and other areas in California, *J. Wildl. Manage.*, 63, 1273, 1999.
151. Goede, A. A. and de Bruin, M., The use of bird feather parts as a monitor for metal pollution, *Environ. Pollut. (Ser. B)*, 8, 281, 1984.
152. Goede, A. A. and de Bruin, M., Selenium in a shore bird, the dunlin, from the Dutch Waddenzee, *Mar. Pollut. Bull.*, 16, 115, 1985.
153. Goede, A.A. and de Bruin, M., The use of bird feathers for indicating heavy metal pollution, *Environ. Monit. Assess.*, 7, 249, 1986.
154. Goede, A. A. et al., Selenium, mercury, arsenic and cadmium in the lifecycle of the dunlin, *Calidris alpina*, a migrant wader, *Sci. Total Environ.*, 78, 205, 1989.
155. Burger, J. et al., Lead, cadmium, selenium and mercury in seabird feathers from the tropical Pacific, *Environ. Toxicol. Chem.*, 11, 815, 1992.
156. Clark, D. R., Jr. et al., Selenium accumulation by raccoons exposed to irrigation drainwater at Kesterson National Wildlife Refuge, California, 1986, *Arch. Environ. Contam. Toxicol.*, 18, 787, 1989.
157. Wren, C. D., Distribution of metals in tissues of beaver, raccoon and otter from Ontario, Canada, *Sci. Total Environ.*, 34, 177, 1984.
158. Stoneburner, D. L., Heavy metals in tissues of stranded short-finned pilot whales, *Sci. Total Environ.*, 9, 293, 1978.
159. Mackey, E. A. et al., Bioaccumulation of vanadium and other trace metals in livers of Alaskan cetaceans and pinnipeds, *Arch. Environ. Contam. Toxicol.*, 30, 503, 1996.
160. Itano, K. et al., Mercury and selenium levels in striped dolphins caught off the Pacific coast of Japan, *Agric. Biol. Chem.*, 48, 1109, 1984.
161. Leonzio, C., Focardi, S., and Fossi, C., Heavy metals and selenium in stranded dolphins of the northern Tyrrhenian (NW Mediterranean), *Sci. Total Environ.*, 119, 77, 1992.
162. Kari, T. and Kauranen, P., Mercury and selenium contents of seals from fresh and brackish water in Finland, *Bull. Environ. Contam. Toxicol.*, 19, 273, 1978.
163. Reijnders, P. J. H., Organochlorine and heavy metal residues in harbour seals from the Wadden Sea and their possible effects on reproduction, *Neth. J. Sea Res.*, 14, 30, 1980.
164. Van de Ven, W. S. M., Koeman, J. H., and Svenson, A., Mercury and selenium in wild and experimental seals, *Chemosphere*, 8, 539, 1979.

165. Martin, J. H. et al., Mercury-selenium-bromine imbalance in premature parturient California sea lions, *Mar. Biol.*, 35, 91, 1976.
166. Clark, D. R., Jr., Selenium accumulation in mammals exposed to contaminated California irrigation drainwater, *Sci. Total Environ.*, 66, 147, 1987.
167. Fleming, W. J., Gutenmann, W. H., and Lisk, D. J., Selenium in tissues of woodchucks inhabiting fly ash landfills, *Bull. Environ. Contam. Toxicol.*, 21, 1, 1979.
168. Medeiros, L. C., Belden, R. P., and Williams, E. S., Selenium content of bison, elk, and mule deer, *J. Food Sci.,* 58, 731, 1993.
169. Dierenfeld, E. S. and Jessup, D. A., Variation in serum of alpha-tocopherol, retinol, and selenium of free-ranging mule deer (*Odocoileus hemionus*), *J. Zoo Wildl. Med.*, 21, 425, 1990.
170. Oliver, M. N., Jessup, D. A., and Norman, B. B., Selenium supplementation of mule deer in California, *Trans. Western Sect. Wildl. Soc.*, 26, 87, 1990.
171. Oliver, M. N. et al., Selenium concentrations in blood of free-ranging mule deer in California, *Trans. Western Sect. Wildl. Soc.*, 26, 80, 1990.
172. Flueck, W. T., Whole blood selenium levels and glutathione peroxidase activity in erythrocytes of black-tailed deer, *J. Wildl. Manage.*, 55, 26, 1991.
173. Flueck, W. T., Effect of trace elements on population dynamics: Selenium deficiency in free-ranging black-tailed deer, *Ecology*, 75, 807, 1994.
174. Hodson, P. V. and Hilton, J. W., The nutritional requirements and toxicity to fish of dietary and waterborne selenium, *Ecol. Bull.*, 35, 335, 1983.
175. Gatlin, D. M., III and Wilson, R. P., Dietary selenium requirement of fingerling channel catfish, *J. Nutr.*, 114, 627, 1984.
176. Oldfield, J. E., Selenium: Its Uses in Agriculture, Nutrition & Health, and Environment. Special Publication of Selenium-Tellurium Development Association, Darien, Connecticut, 1990.
177. Oldfield, J. E., Environmental implications of uses of selenium with animals, in *Environmental Chemistry of Selenium*, Frankenberger, W. T., Jr. and Engberg, R. A., Eds., Marcel Dekker, New York, 1998, 129 (Chap. 8).
178. Oldfield, J. E. et al., Risks and Benefits of Selenium in Agriculture. Issue Paper No. 3 and Supplement, Council for Agricultural Science and Technology, Ames, Iowa, 1994.
179. Maas, J., Selenium metabolism in grazing ruminants: Deficiency, supplementation, and environmental implications, in *Environmental Chemistry of Selenium*, Frankenberger, W. T., Jr. and Engberg, R. A., Eds., Marcel Dekker, New York, 1998, 113 (Chap. 7).
180. Brasher, A. M. and Ogle, R. S., Comparative toxicity of selenite and selenate to the amphipod *Hyalella azteca*, *Arch. Environ. Contam. Toxicol.*, 24, 182, 1993.
181. Lemly, A. D., Finger, S. E., and Nelson, M. K., Sources and impacts of irrigation drainwater contaminants in arid wetlands, *Environ. Toxicol. Chem.*, 12, 2265, 1993.
182. Murphy, C. P., Bioaccumulation and toxicity of heavy metals and related trace elements, *J. Water Pollut. Control Fed.*, 53, 993, 1981.
183. Adams, W. J., The toxicity and residue dynamics of selenium in fish and aquatic invertebrates, Ph.D. thesis, Michigan State Univ., East Lansing, 1976.
184. Halter, M. T., Adams, W. J., and Johnson, H. E., Selenium toxicity to *Daphnia magna*, *Hyalella azteca*, and the fathead minnow in hard water. *Bull. Environ. Contam. Toxicol.*, 24, 102, 1980.
185. Ingersoll, C. G., Dwyer, F. J., and May, T. W., Toxicity of inorganic and organic selenium to *Daphnia magna* (Cladocera) and *Chironomus riparius* (Diptera), *Environ. Toxicol. Chem.*, 9, 1171, 1990.
186. DeForest, D. K., Brix, K. V., and Adams, W. J., Critical review of proposed residue-based selenium toxicity thresholds for freshwater fish, *Hum. Ecol. Risk Assess.*, 5, 1187, 1999.
187. Jarvinen, A. W. and Ankley, G. T., *Linkage of Effects to Tissue Residues: Development of a Comprehensive Database for Aquatic Organisms Exposed to Inorganic and Organic Chemicals*, SETAC Press, Pensacola, FL, 1999.
188. Lemly, A. D., Ecosystem recovery following selenium contamination in a freshwater reservoir, *Ecotoxicol. Environ. Saf.*, 36, 275, 1997.
189. Sorensen, E. M. B. et al., Histopathological, hematological, condition-factor, and organ weight changes associated with selenium accumulation in fish from Belews Lake, North Carolina, *Arch. Environ. Contam. Toxicol.*, 13, 153, 1984.

190. Lemly, A. D., Guidelines for evaluating selenium data from aquatic monitoring and assessment studies, *Environ. Monit. Assess.*, 28, 83, 1993.
191. Hamilton, S. J. et al., Toxicity of organic selenium in the diet to chinook salmon, *Environ. Toxicol. Chem.*, 9, 347, 1990.
192. Cleveland, L. et al., Toxicity and bioaccumulation of waterborne and dietary selenium in juvenile bluegill (*Lepomis macrochirus*), *Aquat. Toxicol.*, 27, 265, 1993.
193. Coyle, J. J. et al., Effect of dietary selenium on the reproductive success of bluegills (*Lepomis macrochirus*), *Environ. Toxicol. Chem.*, 12, 551, 1993.
194. Hermanutz, R. O. et al., Effects of elevated selenium concentrations on bluegills (*Lepomis macrochirus*) in outdoor experimental streams, *Environ. Toxicol. Chem.*, 11, 217, 1992.
195. Schuytema, G. S. and Nebeker, A. V., Amphibian Toxicity Data for Water Quality Criteria Chemicals, EPA/600/R-96/124, U.S. Environmental Protection Agency, Corvallis, OR, 1996.
196. Hopkins, W. A., Reptile toxicology: Challenges and opportunities on the last frontier in vertebrate ecotoxicology, *Environ. Toxicol. Chem.*, 19, 2391, 2000.
197. Linder, G. and Grillitsch, B., Ecotoxicology of metals, in *Ecotoxicology of Amphibians and Reptiles*, Sparling, D. W., Linder, G., and Bishop, C. A., Eds., SETAC Press, Pensacola, FL, 2000, 325 (Chap. 7).
198. Meyers-Schöne, L., Ecological risk assessment of reptiles, in *Ecotoxicology of Amphibians and Reptiles*, Sparling, D. W., Linder, G., and Bishop, C. A., Eds., SETAC Press, Pensacola, FL, 2000, 793 (Chap. 14b).
199. Browne, C. L. and Dumont, J. N., Toxicity of selenium to developing *Xenopus laevis* embryos, *J. Toxicol. Environ. Health*, 5, 699, 1979.
200. De Young, D., Bantle, J. A., and Fort, D. J., Assessment of the developmental toxicity of ascorbic acid, sodium selenate, coumarin, serotonin and 13-cis-retenoic acid using FETAX, *Drug Chem. Toxicol.*, 14, 127, 1991.
201. Linder, G. et al., Evaluating amphibian responses in wetlands impacted by mining activities in the western United States, in Fifth Biennial Symposium on Issues and Technology in the Management of Impacted Wildlife, 8–10 April, Boulder, Colorado, Thorpe Ecological Institute, Boulder, 1991, 17.
202. Birge, W. J., Aquatic toxicology of trace elements of coal and fly ash, in *Energy and Environmental Stress in Aquatic Systems*, Thorpe, J. H. and Gibbons, J. W., Eds., U.S. Department of Energy, Technical Information Center, Washington, D.C., 1978, 219.
203. Birge, W. J., Black, J. A., and Westerman, A. G., Evaluation of aquatic pollutants using fish and amphibian eggs as bioassay organisms, in *Animals as Monitors of Environmental Pollutants*, Nielson, S. W., Migaki, G., and Scarpelli, D. G., Eds., National Academy of Sciences, Washington, D.C., 1979, 108.
204. Ohlendorf, H. M., Selenium *was* a time bomb, *Hum. Ecol. Risk Assess.*, 5, 1181, 1999.
205. Ohlendorf, H. M., Hothem, R. L., and Welsh, D., Nest success, cause-specific nest failure, and hatchability of aquatic birds at selenium-contaminated Kesterson Reservoir and a reference site, *Condor*, 91, 787, 1989.
206. Skorupa, J. P., Risk Assessment for the Biota Database of the National Irrigation Water Quality Program, Prepared for the National Irrigation Water Quality Program, USDI, Washington, D.C., April. 1998.
207. Heinz, G. H. and Hoffman, D. J., Comparison of the effects of seleno-L-methionine, seleno-DL-methionine, and selenized yeast on reproduction of mallards, *Environ. Pollut.*, 91, 169, 1996.
208. Heinz, G. H. and Hoffman, D. J., Methylmercury chloride and selenomethionine interactions on health and reproduction in mallards, *Environ. Toxicol. Chem.*, 17, 139, 1998.
209. Gruenwald, P., Malformations caused by necrosis in the embryo: Illustrated by the effect of selenium compounds on chick embryos, *Am. J. Pathol.*, 34, 77, 1958.
210. Ort, J. F. and Latshaw, J. D., The toxic level of sodium selenite in the diet of laying chickens, *J. Nutr.*, 108, 1114, 1978.
211. Williams, M. L., Hothem, R. L., and Ohlendorf, H. M., Recruitment failure in American avocets and black-necked stilts nesting at Kesterson Reservoir, California, 1984–1985, *Condor*, 91, 797, 1989.
212. Heinz, G. H., Hoffman, D. J., and Gold, L. G., Toxicology of organic and inorganic selenium to mallard ducklings, *Arch. Environ. Contam. Toxicol.*, 17, 561, 1988.
213. Albers, P. H., Green, D. E., and Sanderson, C. J., Diagnostic criteria for selenium toxicosis in aquatic birds: Dietary exposure, tissue concentrations, and macroscopic effects, *J. Wildl. Dis.*, 32, 468, 1996.

214. Green, D. E. and Albers, P. H., Diagnostic criteria for selenium toxicosis in aquatic birds: Histologic lesions, *J. Wildl. Dis.*, 33, 385, 1997.
215. Hoffman, D. J. and Heinz, G. H., Effects of mercury and selenium on glutathione metabolism and oxidative stress in mallard ducks, *Environ. Toxicol. Chem.*, 17, 161, 1998.
216. Hoffman, D. J., Heinz, G. H., and Krynitsky, A. J., Hepatic glutathione metabolism and lipid peroxidation in response to excess dietary selenomethionine and selenite in mallard ducklings, *J. Toxicol. Environ. Health*, 27, 263, 1989.
217. Hoffman, D. J. et al., Subchronic hepatotoxicity of selenomethionine ingestion in mallard ducks, *J. Toxicol. Environ. Health*, 32, 449, 1991.
218. Hoffman, D. J. et al., Association of mercury and selenium with altered glutathione metabolism and oxidative stress in diving ducks from the San Francisco Bay region, USA, *Environ. Toxicol. Chem.*, 17, 167, 1998.
219. Fairbrother, A. and Fowles, J., Subchronic effects of sodium selenite and selenomethionine on several immune-functions in mallards, *Arch. Environ. Contam. Toxicol.,* 19, 836, 1990.
220. Whiteley, P. L. and Yuill, T. M., Interactions of environmental contaminants and infectious disease in avian species, in *Acta XX Congressus Internationalis Ornithologici, Christchurch, New Zealand,* 2–9 December 1990, Bell, B. D., Ed., Ornithological Trust Board, Wellington, New Zealand, Vol. 4, 1991, 2338.
221. Schamber, R. A., Belden, E. L., and Raisbeck, M. F., Immunotoxicity of chronic selenium exposure, in *Decades Later: A Time for Reassessment*, Schuman, G. E. and Vance, G. F., Eds., *Proc. 12th Annu. Natl. Meeting*, American Society of Surface Mining and Reclamation, Princeton, WV, 1995, 384.
222. Hill, C. H., Interrelationships of selenium and other trace elements, *Fed. Proc.*, 34, 2096, 1975.
223. Scheuhammer, A. M., The chronic toxicity of aluminium, cadmium, mercury, and lead in birds: A review, *Environ. Pollut.*, 46, 263, 1987.
224. Cuvin-Aralar, M. L. A. and Furness, R. W., Mercury and selenium interaction: A review, *Ecotoxicol. Environ. Saf.*, 21, 348, 1991.
225. Ohlendorf, H. M. et al., Trace elements and organochlorines in surf scoters from San Francisco Bay, 1985, *Environ. Monit. Assess.*, 18, 105, 1991.
226. Ohlendorf, H. M., Marine birds and trace elements in the temperate North Pacific, in *The Status, Ecology, and Conservation of Marine Birds of the North Pacific*, Vermeer, K., Briggs, K. T., Morgan, K. H., and Siegel-Causey, D., Eds., Can. Wildl. Serv. Spec. Publ., Ottawa, 1993, 232.
227. Henny, C. J. et al., Nineteenth century mercury: Hazard to wading birds and cormorants of the Carson River, Nevada, *Ecotoxicology*, in press.
228. Spalding, M. G. et al. Methylmercury accumulation in tissues and effects on growth and appetite in captive great egrets, *J. Wildl. Diseases*, 36, 411, 2000.

CHAPTER 18

# Sources, Pathways, and Effects of PCBs, Dioxins, and Dibenzofurans

**Clifford P. Rice, Patrick W. O'Keefe, and Timothy J. Kubiak**

## CONTENTS

## 18.1 INTRODUCTION

Polychlorinated biphenyls (PCBs), dioxins (PCDDs), and dibenzofurans (PCDFs) all belong to a general class of pollutants labeled polyhalogenated organics. They all share the common distinction of having a benzene backbone that has chlorine substituted on it, and they all have several similarities in their modes of action, environmental pathways, and occurrence as direct products of commerce as opposed to agriculture. Bioconcentration and biomagnification are these compounds' prominent routes of exposure. The atmosphere plays an important role in the transport and fate of these compounds. Each grouping (PCBs and PCDDs/PCDFs) will be dealt with separately under each of the following major headings: Sources and Pathways in the Environment (Sections 18.2 and 18.3), Ambient Levels (Sections 18.4 and 18.5), and Effects (Section 18.6).

### 18.1.1 PCBs

Effective July 1979, the final PCB ban rule was implemented by the U.S. Environmental Protection Agency (EPA), which prohibited the manufacture, processing, distribution in commerce, and use of PCBs except in a totally enclosed system or unless specifically exempted by the EPA.[1] Essentially, the use of PCBs was unabated from their entry in the marketplace in the late 1920s until the early 1970s, when evidence became available that chronic exposure could result in hazard to humans and the environment. The Yusho incident[2] in Japan, in which over 1000 individuals were severely exposed to PCB-contaminated rice oil in 1968, provided a strong impetus to finalize this ban. The total worldwide production through 1976 is estimated to be about $6.1 \times 10^{11}$ g[3], of which about $5.7 \times 10^{11}$ g (93%, 1.25 billion lbs) were produced by Monsanto in the United States. As concern continued to mount through the 1970s, the timing was right for PCBs to become a focus of the new U.S. Toxic Substances Control Act (TSCA), which was formulated in 1976 and pursuant to which all further manufacture of PCBs by U.S. companies was banned as of 1977.

The common names for PCB varied from country to country. In the United States they were called Aroclors, in France, Phenechlor, in Japan, Kaneclor, and in Russia, Sovol.[4] PCB is the acronym for polychlorinated biphenyl. This name accurately depicts its chemical structure. Biphenyl is a compound consisting of two benzene (phenyl) rings connected by a single carbon-to-carbon bond. The polychlorinated modifier signifies that there is at least one or several chlorine atoms substituted at one or more of the ten carbons of the biphenyl backbone. The substitution of chlorine

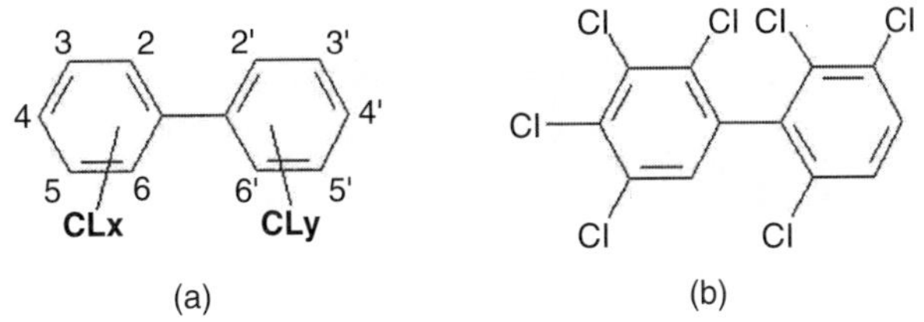

**Figure 18.1** Chemical structures for PCB. (a) PCB congener numbering protocol; (b) congener, designated as BZ# 174, which is a congener that exists as two chiral forms (atropisomers, see text).

**Table 18.1 Distribution of PCB Congeners ($C_{12}H_{10-n}Cl_n$)**

| Homolog Grouping (Based on Chlorine No.) | Possible Congeners |
|---|---|
| Monochloro- | 3 |
| Dichloro- | 12 |
| Trichloro- | 24 |
| Tetrachloro- | 42 |
| Pentachloro- | 46 |
| Hexachloro- | 42 |
| Heptachloro- | 24 |
| Octachloro- | 12 |
| Nonochloro- | 3 |
| Decachloro- | 1 |
| TOTAL | 209 |

on the biphenyl backbone leads to the nomenclature that is typical for the more than 100 congeners that are widely described in the literature (Figure 18.1).

There are actually 209 congeners that are theoretically possible from the base structure (Figure 18.1[a], Table 18.1). The congeners are designated individually with separate BZ numbers that were assigned by Ballschmitter and Zell in 1978.[5] These congener assignments are broadly accepted today with only slight modification by the International Union of Pure and Applied Chemists.[6]

Of the possible 209 compounds, only about 100 to 150 are represented in formulations that have been used and are now widely dispersed in the environment. Besides the great numbers of different chemistries resulting from the varying chlorine substitutions (amounts and positions) that are shown in Table 18.1, there are also configurational variations, or atropisomers, within selected single congeners (nine unique pairs are possible). These configurational differences occur because of restricted rotations of the two phenyl groups when three or more *ortho*-substituted (two or six positions) chlorines are present. Each of the congeners having this property therefore exists in an enatiomeric pair. A prominent example of one of these atropisomers is BZ# 174 (Figure 18.1[b]). As they were synthesized, commercial PCBs had equal quantities of each member of the pair of these atropisomers; however, biological processes will usually degrade one form in preference to the other. Environmental chemists have begun to use measurements of PCB atropisomers in attempts to distinguish chemical from biological processing and to track environmental weathering.[7,8,9]

### 18.1.2 PCDDs and PCDFs

Polychlorinated dibenzo-*p*-dioxins (PCDDs) and polychlorinated dibenzofurans (PCDFs) are two related classes of aromatic heterocyclic compounds (Figure 18.2). In contrast to PCBs, PCDFs and PCDDs have no commercial use and are released into the environment as contaminants of chemical and combustion processes. The number of chlorine atoms can vary between one and eight, which allows for 75 PCDD and 135 PCDF positional isomers, respectively. The toxicity of PCDD/PCDF compounds is associated with substitution at the lateral (2,3,7, and 8) positions, and

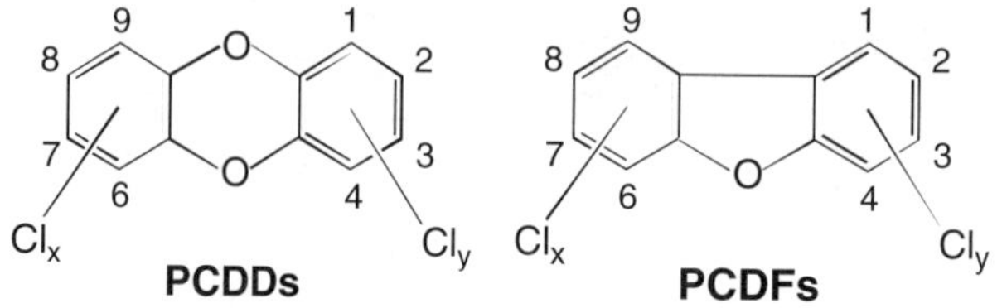

**Figure 18.2** Generalized structures for PCDDs and PCDFs.

the two compounds with all of these positions occupied — 2,3,7,8-TCDD and 2,3,7,8-TCDF — are extremely toxic to certain species of laboratory animals, particularly guinea pigs ($LD_{50}$ approx. 1 µg/kg).[10] However, other species are considerably less sensitive to the two compounds, and consequently there is no clear consensus in the scientific community on the human toxicity of PCDD/PCDF compounds.[11]

PCDD\PCDF compounds are thermally very stable, and extensive decomposition of 2,3,7,8-TCDD, the most toxic PCDD/PCDF compound, does not occur until temperatures exceed 750°C.[12] However, PCDDs and PCDFs do undergo electrophilic substitution reactions,[13] and photochemical degradation occurs in both organic solvents[14] and water.[15] In the gas phase, reactions can occur between photochemically generated hydroxyl (OH) radicals and PCDDs/PCDFs.[16] Since atmospheric PCDDs/PCDFs with six or fewer chlorines exist primarily in the vapor phase, the OH radical reactions could reduce the atmospheric concentrations of all PCDDs/PCDFs except for the hepta- and octaisomers. The 2,3,7,8-TCDD isomer has very low solubility in water, with a value of 20 ng/g reported for dissolution from a thin film,[17] and 483 ng/g reported from generator column studies.[18] As a consequence of its low water solubility, 2,3,7,8-TCDD has a very high carbon-based sediment partition coefficient, with a log $K_{oc}$ value near 7.[19]

## 18.2 SOURCES AND PATHWAYS OF PCBS IN THE ENVIRONMENT

Important concerns for PCBs are their extensive release to the environment and their aquatic and terrestrial fates after release. PCBs are distributed throughout the world, a fact largely resulting from their considerable aquatic and atmospheric mobility. There is a great deal of literature on the environmental fate and properties of PCBs, including numerous books and special reports.[4,20–26] Much of the concern about this group of chemicals arises from their extensive environmental occurrence and from numerous data indicating that members of this group of compounds may be carcinogenic[27] or may cause adverse endocrine and general reproductive effects.[28]

Manufacturing sites that made extended use of the Aroclors as hydraulic fluids or that used them to manufacture electrical devices, especially transformers or fluorescent light ballasts, became the most common sites for environmental buildup. For example, die-casting equipment used great quantities of hydraulic fluids in the pressing out of metal parts for automobiles and related industries. Also, the electrical industries, especially those assembling capacitors and transformers, employed large amounts of the Aroclors as an insulating fluid. Therefore, areas where these facilities occurred, e.g., the Northeast and heavy manufacturing regions in Michigan, Wisconsin, and Illinois, became regions for major PCB buildup. Examples of documented hotspots are the General Electric Plant in Glens Falls, NY;[29] a die-casting operation on the North Branch of the Shiawasee River;[30] electronic manufacturing facilities in New Bedford Harbor;[31] and an automobile foundry and an aluminum manufacturing plant on the St. Lawrence River.[32] There also were specialized uses for PCBs that led to localized buildups. One of these uses was for carbonless paper, and buildups occurred at locations where these papers were deinked and recycled for reuse, e.g., Kalamazoo River, Kalamazoo, MI had several plants in the city and downstream,[33] and there were also plants along the lower Fox River in Wisconsin.[34] As further evidence for this manufacturing localization for PCB contamination, it is noteworthy that even 20 years after cessation of their use, these same areas are now recognized as regions for PCB fishing advisories across the United States[35] — the Northeast, the upper Midwest, and coastal regions near population centers.

With the 1979 U.S. ban, it can now be safely said that most direct sources of PCB releases to surface waters and air have ceased, except for the occasional spills or the increasingly less frequent acts of illegal disposal or dumping.

### 18.2.1 Release of PCBs into the Environment

The primary emission of PCBs into the environment occurs through release to water, with maximum wildlife and human exposure occurring directly or indirectly through aqueous systems. Atmospheric processes are secondary in terms of immediate exposure, even though geographic dispersion through the atmosphere is most likely greater.

#### *18.2.1.1 Emission of PCBs into Water*

Past discharge into rivers and streams has led to significant dispersion of, and exposure to, PCBs. Fortunately, however, most of these direct discharges have ceased, and exposures in aquatic systems generally center around past deposits that reside in sediments. Many of the sites identified in the Great Lakes as "areas of concern" by the U.S./Canadian Great Lakes Water Quality Agreement[36] have this designation because of PCB buildup in their sediments (more than 70% of the 42 sites). Considerable debate continues on the relative merits/costs of dredging these and other areas to reduce exposure from these deposits. Two notable examples of high sediment buildup of PCBs are the Hudson River[37] and New Bedford Harbor.[38] The high level of PCBs in the sediment of the Upper Hudson River site remains from earlier industrial practices that, at the time, were considered acceptable.[39] There are several case studies that could be presented to describe the environmental fate and distribution of PCBs in aqueous systems. One of the most extensively documented is indeed the Hudson River PCB problem.[39,40] Among areas, the PCB spill in Duwamish Harbor in Seattle is of historical interest, as it has long since been corrected.[41] Another more recent focus is an effort underway[42] to define the extent of PCB contamination in Baltimore Harbor.

Some areas have undergone dredging as a mitigative strategy, with sampling built in as a check for improvement in the systems.[30,43] New Bedford Harbor is an estuary severely contaminated with PCBs. It has undergone a multistage Superfund remediation involving a long-term monitoring program. Based on this monitoring, it was found that the major redistributions of contaminated sediments were confined to the immediate vicinity of remedial activities; however, there was evidence that low-molecular-weight PCBs were transported farther.[34,44] In another study by Bremle and co-workers,[45] in which suction dredging was performed on a small Swedish lake, it was determined that 97% of the PCB-contaminated sediment was removed, and little excess PCB was released as a result of the dredging. This may have been aided by the use of a geotextile silt curtain that reduced leakage of PCB to the river downstream. In the early assessment process to determine options to remediate the PCB buildup in the Hudson River, separate sediment transport and resuspension studies were carried out to determine the degree of remobilization of PCBs that might take place during dredging (Tofflemire 1984, as cited in Final Environmental Impact Statement on the Hudson River PCB Reclamation Demonstration Project[46]). Losses were found to be negligible.

#### *18.2.1.2 Emissions of PCBs into the Atmosphere*

While the quantities mobilized through the air are likely similar to losses to aqueous systems, dispersion and dilution into the atmosphere are much more extensive and rapid. The impacts from air dispersion, however, are less obvious than in the aqueous situation. Atmospheric processes are clearly responsible for significant dispersion of PCBs to large isolated waterbodies like the Great

Lakes,[47–50] Lake Baikal,[51,52] and most of the marine portions of the Arctic.[53,54] Furthermore, it has been repeatedly documented that there is a general trend of atmospheric movement of the more volatile components of the Aroclor mixes to colder areas of the world.[55]

Rather than atmospheric emissions emitting directly from source areas, as was the case in the 1960s and 1970s, today's atmospheric levels originate predominantly from recycling of past atmospheric deposits. This recycling tends to maintain higher concentrations in these areas, preventing levels from declining as quickly as was observed right after sources were cut off in the late 1960s.[56] Studies of current atmospheric emission have concluded that reprocessing of soil-sorbed PCBs is a major contributor to loading in urban/industrial areas. Local hot spots will also contribute to these loads; however, estimates using known hot spots vs. general recycling have been carried out in London, and these hot-spot contributions were determined to be rather insignificant.[57]

Indoor levels of PCBs are generally much higher (0.3 $\mu g/m^3$) than levels measured in outdoor air (0.004 $\mu g/m^3$).[58] PCB concentrations were measured in the indoor air of several public buildings in Bloomington, IN.[59] The concentrations were 5 to 300 times higher than outdoor concentrations, and the indoor-air PCB levels were highest in buildings having the earliest construction dates. Indoor air ventilation systems were found to be a short-range source, governed by factors such as the building ventilation rate, and it was hypothesized that this building release even constituted a principal source of tri- and tetrachlorinated PCBs at a Birmingham sampling site.[60] Much of the explanation for higher levels in indoor air appears to be linked to the extensive use of fluorescent light ballasts in the past, especially in schools, office buildings, and institutional building, such as hospitals. Each of these ballasts contains approximately 1.6 kg of PCBs. Many of these ballasts can still be found in use even today.[61]

### 18.2.2 Aquatic and Terrestrial Fate of PCBs

#### *18.2.2.1 Biaccumulation Pathways of PCBs*

PCBs are noted for their tendency to bioaccumulate in aquatic and terrestrial organisms. This characteristic can be depicted by characterizing processes involved in the pharmokinetics of exposed experimental animals.[62] Matthews and Dedrick[62] reviewed this subject and reported their findings for one study in which specific congeners were applied to rats, and the exact chemistry of the compounds and their distribution were carefully followed (Table 18.2). Other rodent studies have shown similar trends for intrabody distribution of PCBs. Notice in Table 18.2 that the more

**Table 18.2 Tissue/Blood Distribution Ratios of PCB Congeners in Rats**

| Compartment | Parent | | | | Metabolite | | | |
|---|---|---|---|---|---|---|---|---|
| | 1-CB | 2-CB | 5-CB | 6-CB | 1-CB | 2-CB | 5-CB | 6-CB |
| Blood | 1 | 1 | 1 | 1 | 1 | 1 | 1 | 1 |
| Gut lumen | 1 | 1 | 1 | 1 | 1 | 1 | 1 | 1 |
| Muscle | 1 | 2 | 1 | 4 | 0.14 | 0.4 | 0.1 | 0.3 |
| Liver | 1 | 3 | 6 | 12 | 2 | 5 | 2 | 4 |
| Skin | 10 | 10 | 7 | 30 | 0.25 | 0.3 | 0.1 | 2 |
| Adipose | 30 | 70 | 70 | 400 | 0.4 | 0.6 | 0.4 | 2 |

| Rate constant | Kinetic Parameters | | | |
|---|---|---|---|---|
| | 1-CB | 2-CB | 5-CB | 6-CB |
| Metabolic clearance ($K_m$,mL/min) | 10.0 | 2.0 | 0.39 | 0.045 |
| Kidney clearance ($K_k$,mL/min) | 0.2 | 0.133 | 0.033 | 0.03 |
| Biliary clearance ($K_B$,mL/min) | 0.2 | 0.35 | 0.3 | 0.3 |
| Gut reabsorption ($K_G$,mL/min$^{-1}$) | 0.00016 | 0.00016 | 0.00016 | 0.00016 |
| Fecal transport ($K_F$min$^{-1}$) | 0.0008 | 0.0008 | 0.008 | 0.0008 |

*Source:* Matthews, H.B. and Dedrick, R.L., *Annu. Rev. Pharmacol. Toxicol.*, 24, 85, 1984. With permission.

lipophilic 5/6-chlorobiphenyls tend to preferentially bioconcentrate in fatty tissues (adipose tissue and skin), whereas the more polar members, especially the polar metabolites, tend to show up in the hydrophilic cell tissues/compartments. Note also that the size of the clearance rate constant (Km) is structure-dependent; e.g., the Km for CB-1 is 10, while the values decrease as the chlorine content increases.

In addition to the gross structural character of the molecule that is established by chlorine content, position of the chlorines on the ring also affects distribution and persistence within organisms. For example among the four hexachlorobiphenyls 2,2',3,3',5,5'-; 2,2',3,3',6,6'-; 2,2',4,4',5,5'-; and 2,2',4,4',6,6', the 2,2',3,3',6,6' was eliminated and metabolized more rapidly than were the other three. This was due to the fact that the 4,5 unsubstituted carbons of this congener were not present in the others. In another set of tests, a group of 6-CBs with no *ortho* and only one *ortho* (2 position) was administered. Levels tested after 29 days showed that there were marked differences in retention of these compounds depending on the animal species (fat levels were higher in the mammalian species than in trout and quail): 8.27, 6.84, and 4.74 in the rat, rabbit, and guinea pig, respectively, vs. 3.02 and 2.15, respectively, in trout and Japanese quail. Furthermore, structure also contributed to variations in retention. The quail retained only the non*ortho* congeners; this was true even when low levels were tested. Rabbits retained the highest levels of the di-*ortho* and mono-*ortho* compounds. Fish retained fairly even amounts of all the components and at lower levels of all of the congeners than for the other organisms.

#### *18.2.2.1.1 Aquatic Bioaccumulative Processes of PCBs*

The aquatic bioaccumulative fate of PCBs has been studied in several ecosystem types and over numerous food-chain pathways. Several examples to support bioaccumulation can be cited. All of these, however, suffer from lack of control of all of the input parameters and also control as the tiers increase. A dramatic example was described by Safe[63] for the Lake Ontario ecosystem. The chain was depicted to start with water at 0.05 ng/g PCB, then progressing through sediment (150 ng/g) to plankton (1880 ng/g) to catfish at 11,580 ng/g to finally the herring gull at 3,530,000 ng/g. Laboratory studies can overcome the lack of input accountability in environmental examples; however, transferring these findings to the field situation is difficult. Eisler[64] provided a relevant review of information on how the sublethal effects of PCBs on aquatic organisms are linked to their high bioaccumulation potential. Briefly, he demonstrated that the high potential for bioaccumulation of PCBs by aquatic organisms is due to their intimate exposure to these compounds and to the highly lipophilic nature of PCBs, causing them to accumulate in the fatty tissues of these organisms.

Bioconcentration factors are used to express this bioaccumulation tendency. Gobas[65] provides an excellent treatise on distinguishing bioaccumulation factors (BAFs) from bioconcentration factors (BCFs) for PCBs based on 1984 data that he and his co-workers generated for individual PCB congeners.[66] The typical values increase by a factor of 10- to 100-fold when ascending major consumption levels in a food chain, i.e., from algae to fish to birds. Depuration of accumulated PCBs is slow. In fish, egg maturation and spawning can, however, result in significant reduction in the body burden of persistent PCBs such as 2,5,2',5'-tetrachloro-biphenyl.[67]

It is largely the bioaccumulative property of PCBs that has caused them to be identified as ubiquitous contaminants. The chemicals tend to concentrate in fatty organisms that often reside at the peak of food chains. Such food chains are especially common in arctic climates, where fats are the most efficient and common means of energy storage. PCBs are found in nearly all marine plant and animal species, fish, mammals, birds (especially fish-eating birds), and, of course humans. Wassermann and co-workers[68] published an extensive review of PCBs in animals that, with the exception of the highest values, is still generally valid today. Specifically, they reported that for marine food webs, zooplankton range from $< 0.003$ μg/g to 1 μg/g, whereas top consumers, such as seals and fish, had ranges of PCB from 0.03 to 212 μg/g. Moessner and Ballschmiter[69] monitored seven indicator congeners of the polychlorinated biphenyls (PCB # 28, 52, 101, 118, 138, 153, and 180) in marine mammals that differed in their geographic distributions. They found that animals from the western North Atlantic were contaminated at levels that were about 15 times higher than for animals from the eastern North Pacific and the Bering Sea/Arctic Ocean.

#### *18.2.2.1.2 Terrestrial Biaccumulative Processes of PCBs*

The terrestrial biaccumulative fate of PCBs is less studied, largely because levels tend to be lower and concern for exposure is less than for aquatic organisms. The lower accumulation in terrestrial organisms is believed to be a function of food chains that are shorter than those in aquatic environments.[54] Levels in terrestrial biota can reach high levels in organisms near PCB landfill sites[70] or in terrestrial communities neighboring regions of high aqueous buildup. For example, for eagles were studied in the Great Lakes region; those nearest the lakes had notably higher levels than those farther inland.[71] Also, tree swallows living near the shores of the Hudson River had higher PCB levels than those from river sites more distant from known PCB pollution.[72]

The most extensive and detailed studies of terrestrial transfer of PCBs exist for the Arctic[73] (several studies were summarized in the AMAP, Assessment Report, Arctic Pollution Issues[54]). Extensive study of caribou and reindeer revealed that, even though concentrations of the PCBs were substantially lower in tissues of these terrestrial herbivores than in marine mammals collected from nearby areas, the importance of these herbivores to the native diets in the Arctic makes this route for human exposure one for concern.[54] Levels of PCBs were higher (~twofold) in Russian reindeer (20 ng/g wet weight) than in Canadian caribou. Elkin and co-workers[74] provided an example of a food-chain transfer by way of caribou in the Northwest Territory of Canada; the transfer was from lichens to caribou to wolves. The pattern of congeners changed as the mixture of PCBs was transferred. The food-chain buildup is obvious, with levels reaching a maximum of about 50 ng/g lipid weight in the wolf, after starting around 0.4 ng/g dry weight for the lichens. The congener shift observed with caribou was similar to that observed by Muir and co-workers[75] in marine organisms — fish to seal to polar bear.

Accumulation of PCBs by dairy cows has been studied by several investigators.[76–79] Thomas and co-workers[78] described the distribution fate of PCBs in cows with a concern for terrestrial exposure through forage and consumption. They developed a model that incorporates degradation, especially for the readily metabolized congeners (e.g., BZ#33). Calamari and co-workers[80] studied plant uptake of PCBs and used this as a measure of the geographic distribution of PCBs. Hermanson and Hites[81] looked at uptake by bark as a means of describing geographic distribution of PCBs and other hydrophobic pollutants.

#### ***18.2.2.2 Abiotic Dispersal of PCBs***

The fate and dispersion of PCBs is greatly influenced by abiotic dispersion processes, volatility, solubility, particle sorption, etc. and these are all important and interactive processes ongoing in the atmospheric and aquatic systems that are the major reservoirs for the world's inventory of PCBs. The concept of inventories is important for an understanding of where likely exposure will occur. One such inventory was conducted by the National Academy of Science in the early 1980s.[21] The accessible PCBs were defined as residing in the mobile environmental reservoir (MER). A major objective of these early assessments was to attempt to balance what was produced with where it had come to rest and to determine how much was still available to contaminate the environment. Much of the existing PCBs at that time (1977) were still in commerce, in storage awaiting destruction, or in reservoirs that were considered inaccessible (landfills, deep sediment, or degraded). Tanabe,[82] using updated information (1987 data), performed a similar exercise and calculated the global distribution of available PCB. Despite the passage of 10 years between the two estimates, there was remarkable similarity between them (Figure 18.3). Tanabe further indi-

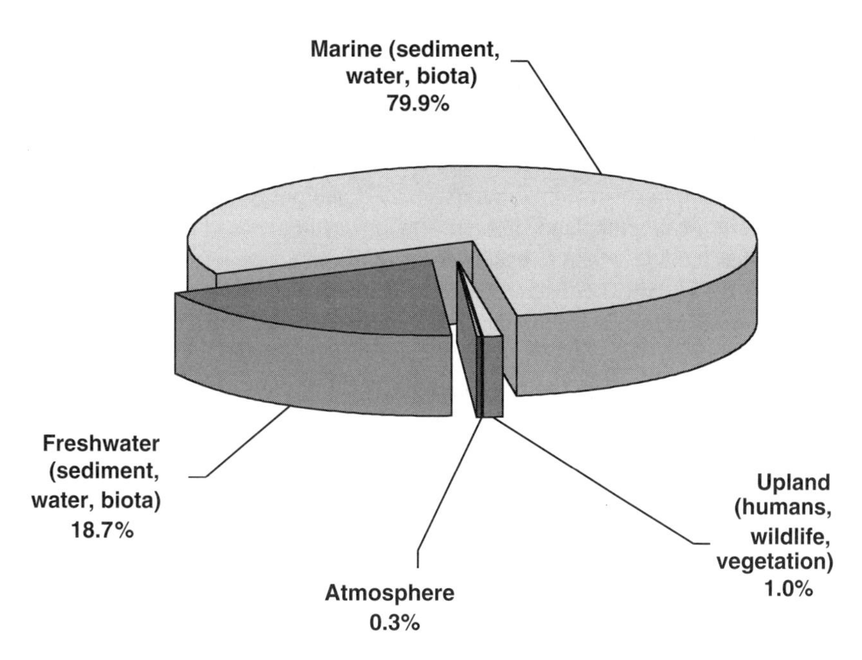


**Figure 18.3** Global distribution of environmentally available PCB, as estimated by Tanabe.[82]

cated the need for increased concern for the marine mammals that would be receiving the greatest exposure according to these predictions. Most recent inventories for PCBs have been developed for specific regions of the country. Harrad and co-workers[57] describe the present United Kingdom environmental levels of PCBs. Only 1% of the amount of PCBs sold in the United Kingdom since 1954 was found to still be present in the U.K. environment. Across the range of congeners, persistence increased with increasing chlorination. The major loss mechanism for PCBs was advection, atmospheric or pelagic, transport from the United Kingdom. There was a dramatic fall in levels in archived soils and vegetation between the mid-1960s and the present. Ninety-three percent of the contemporary U.K. burden is associated with soils, with 3.5% in seawater and 2.1% in marine sediment. Freshwater sediments, vegetation, humans, and sewage sludge combined to account for only 1.4% of the present burden, and PCB loadings in air and freshwater were insignificant. The major loss pathway from the United Kingdom is atmospheric, with sources feeding this atmospheric advection as follows: volatilization from soils (88.1%), leaks from large capacitors (8.5%), production of refuse-derived fuel (RDF) (2.2%), leaks from transformers (0.6%), recovery of contaminated scrap metal (0.5%), and volatilization from sewage sludge-amended land (0.2%).[57]

#### *18.2.2.2.1 Atmospheric Dispersal of PCBs*

The air concentrations of PCBs and other chemicals play an important role in the deposition of these chemicals in terrestrial ecosystems (on leaves/needles, grass, soil).[57,83] Long-range overland transport of PCBs is critically linked to atmospheric routes of exposure; this mechanism was clearly demonstrated for transport of PCBs in Canadian lakes by Muir and co-workers.[84] The conclusion from this study was that atmospheric movement northward, and subsequent fractionation by volatility, led to selective changes in the PCB profile in the liver of burbot that inhabited these lakes. The PCB profiles progressed from a preponderance of heavier (less volatile) to lighter (more volatile) PCB homologs in the burbot livers in lakes dispersed with increasing latitudes north.

The importance of atmospheric processes for dispersal of PCBs has been established on a global scale.[56,85,80] Higher levels occur downwind of known sources, e.g., the Chicago plume[86] and landfill sources.[87] Considerable data, especially on the Great Lakes, have been recorded to indicate that PCBs will exchange across the air/water interface[88,89] and that this process is controlled by temperature, mass balance levels in the air and water, and wind speed.[90] Actually, realization of this process has helped greatly towards reconciling the amount of PCB that could be accounted for by measurements in sediment and water column of the Great Lakes with the predicted amounts based on loadings. Realization of a reverse flux, i.e., gaseous losses out of the water column, has allowed researchers to account for the imbalance in their previous estimates and provided a means for mass balance estimates that include dynamic exchange of PCBs as gases across the air/water interface.

The best predictive chemical constant for describing these fluxes is the compounds' Henry's Law constant (air–water partition coefficient); H-values have been calculated for a considerable number of PCB congeners. Typical values for the estimated Henry's Law constants for Aroclors indicate that water-to-air degassing can be a significant environmental transport process for PCBs when they are in disequilibrium in water vs. overlying air, especially when water temperatures are high and air concentrations are low, e.g., during autumn over the Great Lakes.[90] Reported H-values for the Aroclor mixtures 1242 and 1260, respectively, were 58.5 and 731 Pa $m^3$/mol.[91] Burkhard and co-workers[92] developed a method by which to estimate H-values for the congeners; their estimated values compared favorably with the limited measured values that were available previously. A recent review of H-values for all 209 of the PCB congeners[93] indicates that there is wide range in values (varying from 160 Pa m3/mol for BZ#9, a dichlorobiphenyl, to 1.00 Pa m3/mol for BZ#199, which is an octochlorobiphenyl). Bamford and co-workers[94] recently generated measured H-values for 26 congeners, including direct measurements for their changing values as functions of temperature (–3 to 31°C). This is an important property to consider, especially for environmental modeling.

Wania and Mackay[55] described the relative mobility of PCB homologs on a global scale by using vapor pressure and log octanol–air partition coefficients of PCBs. They grouped PCBs into four categories based on the relative mobilities of the PCBs to move away from sources and toward the poles. These groupings were: 0 to 1 Cl (highly mobile worldwide/no deposition), 1 to 4 Cl (relatively high mobility/deposition in polar latitudes), 4 to 8 Cl (relatively low mobility/deposition in mid-latitudes), and 8 to 9 Cl (low mobility/deposition close to source).

Model predictions of the concentration of PCBs in air can be made by knowing the slope of the log vapor pressure vs. inverse temperature curve (Antoine equation) as well as the expected air concentrations of particulate matter.[95] It has been shown by other researchers that atmospheric PCB concentrations, which are only weakly dependent on transport paths, are strongly dependent on temperature because of the vapor pressures of the compound.[96]

The major source of PCBs to vegetation is transfer of vapor-phase PCBs from air to the aerial aboveground portions of the plants.[97] Harner and co-workers[98] measured atmospheric PCBs near hazardous waste sites that were greater than background. Their conclusion was that PCBs were being emitted from the soils near these sites where previous deposition had occurred. Also, losses from moist soil are greater than from dry soil, due to stronger soil binding in the absence of water.[99]

#### *18.2.2.2.2 Aquatic Dispersal of PCBs*

Abiotic-mediated movement and fate of PCBs in aquatic systems has been monitored extensively and in every conceivable situation: open ocean,[100,101] rivers (see numerous references cited throughout text), large lakes (Lake Baikal,[56,102,51] Great Lakes[48]), small arctic lakes,[103] and embayments and estuaries.[41,104,44]

In water, adsorption to sediment or other organic matter is a major PCB removal process. Experimental and monitoring data have shown that PCB concentrations are higher in sediment and suspended matter than in the associated water column. The low water solubility and, therefore, resulting high octanol–water partition coefficients (expressed by the log $K_{ow}$) range from 4.5 to 8.1 for individual PCB congeners[105] and result in a strong adsorption to soils and sediments, suggesting that leaching should not occur in soil under most conditions. If leaching does occur, it will be

**Table 18.3 Ranges of Physicochemical Properties of Polychlorobiphenyls ($C_{12}H_{10-n}Cl_n$)**

| Homolog Group | Molecular Weight (μg/L) | Solubility (Pa) 20° | Vapor Pressure | Log $K_{ow}$ |
|---|---|---|---|---|
| Monochloro- | 188.7 | $1.3 \times 10^3$–$7 \times 10^3$ | $2.2 \times 10^3$–$9.2 \times 10^2$ | 4.6–4.7 |
| Dichloro- | 223.1 | $0.6 \times 10^2$–$7.9 \times 10^2$ | $3.7 \times 10^2$–$7.5 \times 10$ | 5.2–5.3 |
| Trichloro- | 257.6 | $0.1 \times 10^2$–$6.4 \times 10^2$ | $1.1 \times 10^2$–$1.3 \times 10$ | 5.7–6.1 |
| Tetrachloro- | 292.0 | $0.2 \times 10^2$–$1.7 \times 10^2$ | 1.8–4 | 5.9–6.7 |
| Pentachloro- | 326.4 | 4.5–12 | 5.3–0.88 | 6.4–7.5 |
| Hexachloro- | 360.9 | 0.4–0.9 | 1.9–0.2 | 6.4–7.6 |
| Heptachloro- | 395.3 | 0.5 | 0.53–$4.8 \times 10^{-2}$ | 7.0–7.7 |
| Octachloro- | 429.8 | 0.2–0.3 | $7.8 \times 10^{-2}$–$9 \times 10^{-3}$ | 7.0–7.6 |
| Nonachloro- | 464.2 | 0.1 | $3.2 \times 10^{-2}$–$1.1 \times 10^{-2}$ | 7.7–7.9 |
| Decachloro- | 498.7 | 0.02 | $5.6 \times 10^{-3}$ | 8.4 |

*Source:* Ballschmiter, K. et al., in *Halogenated Biphenyls, Terphenyls, Naphthalenes, Dibenzodioxins and Related Products*, Kimbrough, R.D. and Jensen, A.A., Eds., Elsevier, Amsterdam, 1989. With permission.

greatest for the least-chlorinated congeners. These trends in physical properties are apparent in Table 18.3.

Although adsorption and subsequent sedimentation may immobilize PCBs for relatively long periods of time in aquatic systems, redissolution into the water column has been shown to occur. The substantial quantities of PCBs contained in aquatic sediments can therefore act as a reservoir from which PCBs may be released over long periods of time. Since sorption to soil is proportional to the soil's organic carbon content,[107,108] leaching or loss is expected to be greatest from soils with low organic carbon.

An interesting discovery in remote pristine lakes in the arctic was the differentiation of atmospheric loading from biotransported PCBs. Grayling were monitored to measure background levels in the water. One grayling population was in a salmon-spawning lake, and the other in a nearby lake without salmon and receiving pollutants only via atmospheric deposition. The grayling in the salmon-spawning lake had concentrations of PCBs more than twofold higher than those in the grayling in the salmon-free lake, and the pollutant composition resembled that found in salmon.[103]

### 18.2.3 PCB Removal Processes

Significant removal of PCBs from the MER can occur through natural processes. One process in the aquatic system is simple physical removal due to sediment burial. However, permanent removal by degradation is the preferred method for ultimate protection of the environment. Chemical degradation in natural systems is minimal,[109] except for photolysis. Anderson and Hites[110] measured hydroxyl radical degradation as it occurs during photolytic exposure. They estimated half-lives ranging from 2 to 34 days for several PCB congeners. Atkinson[111] reviewed the literature on the atmospheric chemistry of PCBs and verified that losses due to photolysis can be high. Tropospheric half-lives due to hydroxyl radical removal were listed as follows: mono-, 3.5 to 7.6 days; di-, 5.5 to 11.8 days; tri-, 9.7 to 20.8 days; tetra-, 17.3 to 41.6 days; and penta-, 41.6 to 83.2 days.

Biological degradation is slow, according to most studies performed over a wide range of organisms including higher animals, plants, and several microbial systems. PCBs are very resistant to biological transformation — one of the very features of the group that made them useful in commerce. In spite of the abundant evidence for PCBs' ability to resist biological degradation, much study has gone into attempting to isolate a "bug" or consortium of bugs that will magically clean up the numerous hot spots of PCB known to exist in the environment.[112, 113]

#### *18.2.3.1 Biological Degradation of PCBs*

Metabolism of PCBs by biological systems has been studied extensively. Several reviews of this subject are available.[62,114–116] Invariably, the phenolic products are the major PCB metabolites, although sulfur-containing metabolites (sulfones), *trans* dihydrols, polyhydroxylated PCBs, and their methyl–ether derivatives and ring-degraded microbial oxidation products have been identified. The effects of chlorine substitution patterns on oxidative breakdown seem to fall into the following patterns:[114]

1. Hydroxylation is favored at the *para* position in the least-chlorinated phenyl ring, unless this site is sterically hindered.
2. In the less-chlorinated biphenyls, the *para* position of both biphenyl rings and the carbon that is *para* to the chlorine are all readily hydroxylated.
3. The availability of two vicinal unsubstituted carbon atoms (particularly C-5 and C-4 in the biphenyl nucleus) also facilitates oxidative metabolism of the PCB substrate, but it is not a necessary requirement for metabolism.
4. As the degree of chlorination increases on both phenyl rings, the rate of oxidative metabolism decreases.
5. The metabolism of specific PCB isomers by different species can result in considerable variations in metabolite distribution. Most PCB congeners, particularly those lacking adjacent unsubstituted positions on the biphenyl rings (2,4,5-, 2,3,5-, or 2,3,6- substituted on both rings), are extremely persistent in the environment.

Harner and co-workers[98] have suggested that microbial degradation half-lives on the order of 10 years are possible; however, the actual values are likely congener-specific and are also a function of concentration and microbial condition in the soil. Biodegradation of PCBs in sediment can occur under both aerobic[117] and anaerobic[118] conditions. Biodegradation rates are highly variable because they depend on a number of factors, including the amount of chlorination, concentration, type of microbial population, available nutrients, and temperature. Results show, however, that under aerobic conditions, mono-, di-, and trichlorinated biphenyls generally biodegrade relatively fast; tetrachlorinated biphenyls generally biodegrade slowly; and the more highly chlorinated biphenyls generally resist biodegradation.

Studies of subsurface anaerobic biodegradation of aquatic sediment show that the pattern of congener reactivity is determined by two main factors — reductive potential of the system and molecular shape of the PCBs. PCBs containing chlorines in the *para* positions are preferentially biodegraded, compared to PCBs containing chlorines in other ring positions,[119] and meta chlorines appear to be preferentially lost. Studies from spill sites also show that the highly chlorinated congeners are biotransformed by a reductive dechlorination to less-chlorinated PCBs, which are biodegradable by aerobic processes.[120,121] For highly chlorinated PCBs that have been shown to dechlorinate naturally in anaerobic sediments, there appears to be a concentration threshold (10 μg/g dry weight) below which this anaerobic process will not start.[118,119]

Biodegradation is probably the ultimate loss process for PCBs in sediments (at least in surficial zones); however, diagenesis experiments in deeper sediments need to be carried out in order to assess the lifetime of PCBs that are deeply buried. Biodegradation in soils with high organic carbon content is slower than in less organic-carbon-enriched soils.[122]

## 18.3 SOURCES AND PATHWAYS OF PCDDs AND PCDFs IN THE ENVIRONMENT

### 18.3.1 Release of PCDDs and PCDFs into the Environment

#### *18.3.1.1 Emission of PCDDs and PCDFs into Water*

PCDDs and PCDFs are formed as by-products in several chemical manufacturing processes. In the case of PCBs, a PCDF molecule can be formed by ring closure of a PCB molecule in the presence of oxygen. While industrialized countries ceased producing PCBs in the 1970s, PCDF concentrations in commercial PCB formulations used prior to that time were reported to range from less than 1 mg/kg to more than 10 mg/kg.[123–125] PCBs were used primarily as dielectric fluids in transformers and capacitors (see section on PCBs above), and additional formation of PCDFs has been found to occur in catastrophic events such as arcs and electrical fires, with concentrations of total PCDFs as high as 2000 mg/kg on soot particles from PCB fires.[126] Certain dielectric fluids also contain chlorobenzenes (CBs) in addition to PCBs. While the CBs can also be converted to PCDDs by catastrophic events, the reaction is bimolecular, and therefore the PCDD concentrations are two orders of magnitude lower than the PCDF concentrations.

Emissions into the aquatic environment from PCB fires can be controlled, and any contamination is usually confined to a small area. In contrast, operations associated with the manufacturing and servicing of electrical equipment or with the improper disposal of capacitors and transformers have resulted in more widespread contamination of the aquatic environment by both PCDFs and PCBs.

Chlorophenols (CPs) are contaminated with PCDDs/PCDFs, and the manufacture and utilization of these compounds have resulted in the release of PCDDs/PCDFs into the aquatic environment. Pentachlorophenol (PCP) has been widely used as a fungicide and insecticide in the wood-products industry, and 2,4,5-trichlorophenol (TCP) was used as the precursor of the herbicide 2,4,5-T. At one time, 2,4,5-T was used extensively in the United States and other industrialized countries as a herbicide on grasslands where cattle and sheep were grazed, on rice-growing fields, in the control of deciduous species in conifer forests, and along roadsides. The toxic 2,3,7,8-TCDD is the major PCDD/PCDF contaminant in 2,4,5-T, and concentrations exceeding 10 mg/kg were found in certain 2,4,5-T formulations produced between 1966 and 1970.[127] However, these were unusually high concentrations, since most 2,4,5-T formulations produced in the 1960s did not contain 2,3,7,8-TCDD concentrations greater than 0.2 mg/kg.[128,129]

The production of PCP in the United States was 20 million kg in 1976 and 14.5 million kg in Japan in 1966.[130] After 1971, production ceased in Japan, but the compound is still produced at one plant in the United States. There are no production figures available for Europe. Results from a number of studies have shown that mg/kg concentrations of hexa-, hepta- and octa-CDDs and CDFs are present in commercial formulations of PCP, with reported concentrations varying over several orders of magnitude.[131] One manufacturer produced a purified PCP formulation that contained less than 25 mg/kg of total PCDD/PCDF compounds.[132] The toxic properties of the PCDD/PCDF compounds in PCP can be attributed to the presence of 1,2,3,6,7,8-hexa CDD. This isomer accounts for nearly 50% of the total hexa CDD concentration in some PCP formulations.[132]

The PCDD/PCDF compounds associated with CPs can be released into water either during manufacturing operations or during the use of the products. Sediments in the vicinity of many chlorophenol-manufacturing plants in the United States and Europe have become contaminated with PCDD/PCDF compounds. A plant situated on the Passaic River 3 km upstream from Newark Bay produced approximately 15 million tons of 2,4,5-T between 1948 and 1969. Sediment samples collected from different locations were dated by radiochemical methods and were analyzed for tetra-, hepta- and octa-CDDs and CDFs.[14] Immediately outside the plant a sediment core section from the mid-1960s was contaminated with 7600 pg/g of 2,3,7,8-TCDD, whereas a 1985 section from the same core had a 2,3,7,8-TCDD concentration of 730 pg/g (all sediment concentrations of PCDDs/PCDFs reported in this subsection are expressed on a dry-weight basis). Samples taken further downstream in Newark Bay and in New York Harbor had considerably lower concentrations of 2,3,7,8-TCDD but were contaminated with PCDDs/PCDFs whose patterns were statistically associated with other sources of contamination.[134]

While the utilization of 2,4,5-T has undoubtedly contributed to 2,3,7,8-TCDD contamination in the aquatic environment, it is generally not possible to determine the point sources for this contamination. However, it is possible to determine aquatic contamination by PCDD/PCDF compounds from facilities where wood is treated with PCP. In a wood treatment plant in Pensacola, Florida, that was closed in 1981, it was found in 1984 that sludge and sediments from wastewater impoundments were contaminated with total hexa- to octa-CDD concentrations of 49 and 105 mg/kg, respectively.[135] It was found by taking soil samples at 0.3 and 0.6 m depths that subsurface migration had occurred, with total PCDD concentrations decreasing from 1.1 mg/kg near the impoundment site to 9.9 ng/g at a distance of 275 m from the site. Sediments collected in the harbor of Thunder Bay, Ontario, near a wood-treatment plant also showed evidence of PCDD/PCDF contamination, with a maximum level of 1.3 ng/g.[136]

In 1985, during the course of the U.S. EPA National Dioxin Study, a comprehensive study on PCDD/PCDF contamination in the United States, it was discovered that fish collected downstream from pulp and paper mills that used the kraft process were contaminated with 2,3,7,8-TCDD.[137]

Subsequent investigations showed that the PCDD/PCDF compounds were primarily formed during the chlorination stage of the pulp-bleaching process.[138,139] Only a small number of isomers were formed, with 2,3,7,8-TCDF as the dominant isomer, followed by either 2,3,7,8-TCDD or octa CDD.[140,141] These results are consistent with a mechanism of direct chlorination of dibenzo-*p*-dioxin (DD) and dibenzofuran (DF). It was shown that both DD and DF were present at ng/g levels in petroleum-based defoaming agents formerly used in pulp processing.[142] However, the PCDD/PCDF compounds must also have been formed by other reaction mechanisms, since lignin from which all low-molecular-weight compounds had been removed by methanol extraction formed the PCDD/PCDF compounds in the presence of chlorine.[139] Based on a study of 104 pulp and paper mills in the United States, it was found that the average export of 2,3,7,8-TCDD from a mill site was uniformly distributed in three media: pulp 38%, effluent 29%, and sludge 33%.[143] For the kraft mills involved in the study, mean concentrations of 2,3,7,8-TCDF in softwood pulp, effluent, and sludge were 137, 0.5, and 806 pg/g, respectively, and mean concentrations of 2,3,7,8-TCDD in the same media were 12, 0.06, and 95 pg/g, respectively. Comparable concentrations of the two analytes were found in the effluents of Canadian pulp and paper mills.[141] In the 1990s, defoaming agents containing PCDD/PCDF precursors were eliminated, and steps were taken to replace chlorine as a bleaching agent by other products, primarily chlorine dioxide. Studies conducted in Canada have shown that these process changes have resulted in > 90% reduction in the levels of 2,3,7,8-TCDF and 2,3,7,8-TCDD discharged to the environment from pulp and paper mills.[144–146]

The sludges from pulp and paper mills are frequently used for the fertilization of agricultural lands and therefore can serve as a source of PCDD/PCDF contamination for both terrestrial and aquatic environments. Sludge from sewage treatment plants is frequently used for the same purpose, and results from several investigations have shown that the sewage sludges from industrialized countries are contaminated with PCDD/PCDF compounds, primarily hepta-CDDs and octa-CDD.[147,148] There has been a trend toward decreasing concentrations over time, with current mean PCDD/PCDF concentration in sewage sludges from Catalonia, Spain of 55 pg I-TEQ/g (International Toxic Equivalents[149]), compared to 620 pg I-TEQ/g in the period from 1979 to 1987.[147] For a full discussion of TEQ (dioxin-like toxicity, toxicity equivalency factors and toxicity equivalents), see Section 18.6, Effects of PCBs, PCDDs, and PCDFs.

A predominant source of the PCDD/PCDF compounds in sewage sludge has not been identified. However, there is evidence from the United Kingdom to suggest that combustion and PCP may be major sources of PCDDs/PCDFs in sewage sludges.[148] Furthermore, a study conducted in Germany showed that household wastewater from the washing of PCP-treated textiles contributed between 27 and 94% of the PCDDs/PCDFs in sewage sludge from a nonindustrial area.[147] Additional sources of PCDDs/PCDFs in sewage sludge are textile dyes and formation from precursors during sludge digestion.

Precursor compounds with structures closely related to PCDD/PCDF compounds are present in the chemical processes described above. However, virtually every chemical process in which there is a source of carbon and chlorine has the potential to produce PCDD/PCDF compounds, with the concentrations decreasing as the structures of the precursor compounds deviate from PCDD/PCDF structures. It was found in 1974 that commercial preparations of hexachlorobenzene were contaminated with octa-CDF (0.4–58 mg/kg) and octa-CDD (0.05–212 mg/kg).[150] Only traces (pg/g) of PCDD/PCDF compounds were found in commercial preparations of short-chain chlorinated hydrocarbons.[151] However, these compounds have low boiling points, and high PCDD/PCDF concentrations were found in sediments near waste-discharge points from a vinyl chloride monomer plant in Finland.[152]

#### *18.3.1.2 Emissions of PCDDs and PCDFs into the Atmosphere*

It was originally considered that PCDD/PCDF compounds were only formed as by-products in the manufacture of chlorinated aromatic chemicals. However, in 1977, PCDD/PCDF compounds were found in the flue gases from several municipal incinerators in the Netherlands.[153] A study carried out in 1978 showed that PCDD/PCDF compounds were present in cigarette smoke and in particulate matter from chemical and municipal waste incinerators, automobile and diesel truck mufflers, charcoal-broiled steaks, and fireplaces.[154] Based on these results, it was concluded that PCDD/PCDF compounds were ubiquitous by-products of combustion processes.

Municipal waste incinerators (MWIs) are generally considered to be a major combustion source of PCDD/PCDF compounds, and there are numerous reports in the literature on the determination of PCDDs/PCDFs in both ash by-products (fly-ash and bottom ash) and stack gas emissions. There is considerable variability in the patterns of PCDD/PCDF compounds found in MWI samples, but in general the PCDDs are dominated by octa-CDD, with concentrations of the other congener groups decreasing as the number of chlorine atoms is reduced. In contrast, the PCDFs, which are generally present at higher concentrations than the PCDDs, are dominated by less-highly-chlorinated congeners — usually either penta- or hexa-CDF congeners.[155] The absolute amounts of PCDD/PCDF compounds in stack gas emissions from 32 MWIs in industrialized countries varied from $< 1$ to 33,047 ng/dry standard cubic meter (ng/DSCM).[156] Data compiled by the WHO Regional Office for Europe also showed that there were large differences in the concentrations of 2,3,7,8-substituted PCDD/PCDF compounds emitted by different municipal MWIs.[157,158] The higher concentrations were often associated with abnormal operating conditions during the sampling program. Older incinerators and incinerators with inadequate emission-control equipment also had higher PCDD/PCDF emissions. The stack gas emissions of most MWIs in industrial countries are now close to the desired limit (0.1 ng I-TEQ/$m^3$).[159]

In comparison to MWIs, hazardous waste incinerators (HWIs) consume a much smaller quantity of waste. However, the concentrations and the patterns of PCDD/PCDF compounds emitted by state-of-the-art HWIs are similar to those found for municipal incinerators, although the waste stream in the former often contains large quantities of chlorinated chemicals.[158] While hospital incinerators also consume a relatively small quantity of waste, emissions of PCDD/PCDF compounds from these incinerators can be high.[160,161] In Denmark, it was estimated that 1,750,000 and 16,800 tons of waste were burned annually in MWIs and in hospital incinerators, respectively.[162] However, the total amounts of PCDD/PCDF compounds produced were 2200 g and 900 g, respectively. The high concentrations of PCDD/PCDF compounds in hospital incinerator emissions have been attributed to the presence of a high percentage of plastics in hospital wastes.[163] Also, these incinerators were often poorly designed, and it is now agreed that incinerators of advanced design should be used for the destruction of hospital waste.

Metal-processing plants are also a source of atmospheric PCDD/PCDF compounds. In Sweden in 1992, it was reported that steel mills were the single largest source of PCDD/PCDF compounds.[164] In the case of scrap-metal-reclamation plants, the presence of chlorine-containing plastics, such as polyvinyl chloride (PVC), in the feedstock would appear to be one of the factors responsible for the increased PCDD/PCDF production.[165] However, other factors, such as furnace design, also are important in controlling PCDD/PCDF emissions.[166]

Studies conducted in Sweden[167] and New Zealand[168] showed that automobiles represented an additional source of PCDD/PCDF compounds in the atmosphere when leaded gasoline was used as a fuel, with TCDFs as the major congener group in the emissions. However, no PCDDs/PCDFs were identified in emissions from automobiles using unleaded gasoline, suggesting that the lead scavenger dichloroethane was the source of the PCDDs/PCDFs. While diesel-powered vehicles are still a source of PCDD/PCDF emissions, data from stationary engine tests[169] in Europe and from on-road sampling[170] in the U.S. suggest that these emissions contribute less than 10% of the total atmospheric emissions of PCDDs/PCDFs.

As PCDD/PCDF emissions from incinerators are being reduced by the introduction of new technology, diffuse burning sources, including domestic heating, are making a larger contribution to the total load of atmospheric PCDDs/PCDFs. Based on tests conducted at the U.S. EPA's Open Burning Test facility, it was estimated that between 2 and 40 households burning their trash daily in barrels could produce PCDD/PCDF emissions comparable to a 182,000 kg/day MWI.[171] Bonfires are customarily lit in the United Kingdom during a November festival, and it was determined that PCDD/PCDF emissions during the festival amounted to ~5 to 14% of the United Kingdom's total annual emissions from primary sources.[172]

## 18.3.2 Aquatic and Terrestrial Fate of PCDDs and PCDFs

### *18.3.2.1 Bioaccumulation Pathways of PCDDs and PCDFs*

#### *18.3.2.1.1 Aquatic Bioaccumulation Pathways of PCDDs and PCDFs*

There are three potential reservoirs of PCDDs/PCDFs available to aquatic organisms: water, sediment (suspended and bottom), and food. The importance of each reservoir as a source of PCDDs/PCDFs for fish was studied in a laboratory investigation of 2,3,7,8-TCDD bioaccumulation in Lake Ontario. It was found, using various combinations of contaminated food, sediment, and water, that food was the major source of 2,3,7,8-TCDD for lake trout (*Salvelinus namaycush*).[173] Only 22 to 30% of the 2,3,7,8-TCDD residue in the lake trout was derived from the sediment, and insignificant quantities of 2,3,7,8-TCDD were bioaccumulated from water.

However, the lake trout is a piscivorous species, and results from laboratory experiments conducted with a planktivorous species, guppy (*Poecilia reticulata*), and fly-ash extracts as a PCDD/PCDF source showed that the contribution of food residues to the body burdens of five PCDDs/PCDFs was less than 5% of the contribution from water.[174] A direct comparison with the Lake Ontario study is not possible, since 2,3,7,8-TCDD was not detected in the guppies. An additional bioaccumulation study was carried out with guppies in which one group of fish was exposed to the PCDD/PCDF fly-ash extract dissolved in water and another group was exposed to the same water solution that had been equilibrated with sediment prior to addition to the exposure tanks.[175] While the PCDD/PCDF concentrations in the water were the same in both experimental groups, fish exposed to the sediment-equilibrated water had lower PCDD/PCDF concentrations. It was suggested that PCDD/PCDF compounds in the sediment-equilibrated water were associated with dissolved or colloidal matter and that this association hindered the uptake of the compound across the cell membrane.

For the fish exposed to the sediment-equilibrated water, the bioaccumulation potential for different PCDD/PCDF compounds was evaluated by calculating biota–sediment accumulation factors (BSAFs). The BSAF is defined as the ratio of the concentration of a chemical in tissue lipid to the concentration in sediment organic carbon. This is a very useful parameter since sediment is the major reservoir of PCDDs/PCDFs in the aquatic environment. Also, these hydrophobic compounds associate with organic carbon in sediment and with lipid in biota; therefore, it is possible to compare data from studies in which there are differences in tissue lipid content and organic carbon content. If the affinities of the PCDDs/PCDFs were equal in lipid and in sediment, the BSAFs should approach one and should be independent of a compound's hydrophobicity. However, the BSAFs were all found to be less than one, and they decreased with increasing chlorine content, e.g., from 0.15 for 2,3,7,8-TCDD to 0.003 for OCDD. While the exposures were carried out for the comparatively short period of 30 days, the results could not be attributed to a failure to reach equilibrium, since comparable BSAFs were found when carp (*Cyprinus carpio*) were exposed for 205 days to contaminated Wisconsin River sediments.[176]

In practice, PCDDs/PCDFs may have different affinities for organic carbon and lipids, and these affinities could be influenced by hydrophobicity. Reductions in water solubility or diffusion across biological membranes are other factors that could explain the decrease in BSAF values as molecular weights increase. Sediments in the marine environment can also be a source of PCDDs/PCDFs for aquatic organisms. Sandworms (*Nereis virens*), clams (*Macoma nasuta*), and grass shrimp (*Palaemontes pugio*) were found to bioaccumulate 2,3,7,8-TCDD, 2,3,7,8-TCDF, and PCBs when exposed to sediments from the Passaic River in New Jersey.[177] While clams rapidly accumulated and depurated the compounds, the highest concentrations were found in the worms, which had slower rates of accumulation and depuration.

In their larval stages, aquatic insects are in intimate contact with sediments; it has been found that the emerging insects can serve as a means of transporting contaminants into aquatic and terrestrial food chains. When small areas of Canadian lakes were isolated as mesocosms by polyethylene barriers, it was determined that up to 2% of a dose of tritiated 2,3,7,8-TCDF in the sediment could be exported annually by emerging insects.[178]

Fly ash from incinerators represents a major atmospheric source of PCDDs/PCDFs (see below), and fly-ash particles are undoubtedly present in many sediment-deposition layers of the Industrial Age. Results from a study on the bioavailability of 2,3,7,8-TCDD from municipal incinerator fly ash to carp suggest that PCDD/PCDF bioavailability from fly ash in the aquatic environment can vary over a wide range, depending on the source of the fly ash and on its organic carbon content.[179] It was found that carp exposed to a fly ash from the East Coast with a high organic carbon content (4%) had lower 2,3,7,8-TCDD residues than did carp exposed to fly ash from the Midwest with a lower organic carbon content (1%), even though the East Coast fly ash had 12 times as much 2,3,7,8-TCDD. In a follow-up study with the Midwest fly ash, BSAF values were determined for selected PCDDs/PCDFs.[180] The BSAFs for tetra- to hepta-CDDs decreased from $7.4 \times 10^{-4}$ to $7.8 \times 10^{-5}$, and for tetra- to hepta-CDFs from $5.7 \times 10^{-4}$ to $9.2 \times 10^{-5}$. These BSAFs are considerably lower than the BSAFs reported for sediments. However, certain fly ashes are highly contaminated with PCDDs/PCDFs and therefore could contribute more strongly to PCDD/PCDF body burdens found in fish.

There are also several field investigations of the food-chain bioaccumulation of PCDDs/PCDFs in aquatic birds. In the first study, PCDD/PCDF concentrations in the livers of Dutch fish-eating birds, primarily of cormorants (*Phalacrocorax carbo*), were compared to PCDD/PCDF concentrations in eels (*Anquilla anquilla*), an important food source for cormorants.[181] The major contaminant in the cormorant liver tissue was 2,3,4,7,8-penta-CDF, followed by 2,3,7,8-TCDD and 1,2,3,7,8-hexa-CDD. These compounds were also found at lower concentrations in the eel tissue. Significant correlations (r values from 0.8 to 0.95, $p < 0.05$) were found between the congener concentrations in the cormorants. It was proposed that this pattern was a result of exposure to a relatively constant mixture of PCDDs/PCDFs via a food source such as the eels. Direct uptake from sediments was considered an unlikely bioaccumulation route since sediment PCDD/PCDF concentrations in the Netherlands are highly variable.

In the second study, concentrations of PCDDs/PCDFs, PCBs, and organochlorine pesticides in the eggs, livers, and whole-body tissues of Lake Ontario herring gulls were compared to the PCDD/PCDF concentrations in alewife (*Alosa pseudoharengus*).[182] The alewife and the rainbow smelt (*Osmerus mordax*) constitute the major fish species in the gulls' diet, and both species contain nearly identical concentrations of organochlorine contaminants. Biomagnification factors (BMFs) are ratios of the PCDD/PCDF concentrations in biota at one trophic level and the PCDD/PCDF concentrations in their food from a lower trophic level.

By using BMFs based on wet-tissue weight rather than lipid weight, we can compare results from the Dutch study with the herring gull study. The BMFs for 2,3,7,8-TCDD, 1,2,3,6,7,8-hexa

CDD, and 2,3,4,7,8-penta CDF in the livers of the cormorants were 12, 29, and 340, respectively, compared to 12, 30, and 9, respectively, in the livers of the Lake Ontario herring gulls. The BMFs for the two PCDDs are in close agreement, whereas there is a large discrepancy in the values for 2,3,4,7,8-penta-CDF. The discrepancy could be due to species differences in liver retention, elimination rates, or unidentified food sources. In the herring gull study, liver concentrations were compared to whole-body concentrations. On a lipid-adjusted basis, the liver/whole-body ratios for the PCDDs/PCDFs exceeded those found for other organochlorine compounds and varied from 1.4 for 2,3,7,8-TCDD to 12 for octa-CDD. It has been found in rats that 2,3,4,7,8-penta CDF binds to nonmetabolizing sites on the protein of the microsomal oxidase cytochrome P-448.[183] This was considered a plausible explanation for the increase in liver retention of the PCDDs/PCDFs in the herring gulls as the degree of chlorination increased.

It is now understood that if the tissue levels of certain PCB congeners that share structural similarities with PCDDs/PCDFs ("dioxin-like" PCBs) are high enough, the TEQ contributions of these congeners may be much more significant than those of the PCDDs/PCDFs. The importance of PCBs in contributing to toxicity can be illustrated by results from a study conducted in the western Mediterranean. In this study, eggs were collected from nests of the Audouin's Gull (*Larus audouinii*), a protected piscivorous species, and the yellow-legged gull (*Larus cachinnans*), a scavenging species that feeds mainly at refuse dumps[184]. The total I-TEQs from PCBs and PCDDs/PCDFs were 2955 and 126 pg/g dry weight in the Audouin's gull eggs and in yellow-legged-gull eggs, respectively. However, the PCDD/PCDF contributions to these total TEQ values were less than 3%. The presence of higher total TEQs in Audouin's gulls can be explained by the higher trophic level occupied by this species relative to yellow-legged gulls.

Aquatic biota are exposed to both 2,3,7,8-substituted- and non-2,3,7,8-substituted PCDDs/PCDFs. However, only 2,3,7,8-substituted PCDD/PCDF isomers are bioaccumulated/biomagnified in vertebrates, although in certain cases non-2,3,7,8-substituted PCDDs/PCDFs can be found in invertebrates. The preferential bioaccumulation of 2,3,7,8-substituted PCDDs/PCDFs in vertebrates has been confirmed by a large number of both field and laboratory studies. Metabolism or selective absorption and retention are possible explanations for this effect.

#### *18.3.2.1.2 Terrestrial Bioaccumulation Pathways of PCDDs and PCDFs*

There is a much wider dispersion of PCDD/PCDF compounds in the terrestrial environment than in the aquatic environment, and the potential for terrestrial bioaccumulation is generally considered low. However, industrial accidents have occurred that have resulted in localized contamination by PCDDs/PCDFs. After an explosion at a trichlorophenol manufacturing facility at Seveso near Milan in 1976, severe mortality was found in domestic rabbits raised in areas near the chemical plant. In the group of 341 samples that had positive signals for 2,3,7,8-TCDD, liver concentrations for the analyte ranged from 0.25 ng/g to 1025 ng/g.[185] A fire at a pentachlorophenol wood-treatment facility at Oroville, California, in 1987 contaminated the soil of surrounding farmland with PCDD/PCDF compounds, which were subsequently bioaccumulated by farm animals. There was some suggestion that livestock had been exposed to PCDD/PCDF compounds prior to the fire since a cow slaughtered prior to the fire had the same concentration of PCDD/PCDF compounds as a cow slaughtered after the fire.[186] A controlled laboratory study was carried out in which chickens were fed a formulated diet containing 10% contaminated soil from the area. Approximately 20–50% of the 2,3,7,8-substituted tetra- through hexa-compounds and 50–90% of the hepta-CDD/CDF compounds present in the feed was eliminated in the feces, and 7–54% of all 2,3,7,8-substituted PCDDs/PCDFs was excreted in the eggs.[187] Adipose tissue was the major deposition site for the PCDDs/PCDFs, with less than 0.5% in the liver. BMFs between the soil and the adipose tissue ranged from ~8 for penta-CDDs/CDFs to 0.3 for octa-CDD/CDF.

Several studies have been conducted in which rodent or other small mammals were fed 2,3,7,8-TCDD-contaminated soil or extracts from soil. When results from these studies were compared, large differences in 2,3,7,8-TCDD bioavailability were found. Soil obtained from two areas in Missouri — Times Beach and Minker Stout — was contaminated with approximately 800 ng/g 2,3,7,8-TCDD as a result of the improper use of waste oils for dust control. When the soil was administered to guinea pigs by gavage as an aqueous suspension, severe mortality occurred, with $LD_{50}$ values close to the value for positive controls (1.75 μg/kg body weight).[188] Based on a comparison of liver concentrations in the positive controls and in the soil-treated animals, it can be calculated that ~85% of the 2,3,7,8-TCDD in the soils was bioavailable. In contrast, when guinea pigs were administered soil from a former 2,4,5-T manufacturing site in New Jersey, typical signs of PCDD toxicity were not found, and the calculated bioavailability was only 0.5%.[189] The soil for this study was contaminated with 2200 ng/g 2,3,7,8-TCDD and several other PCDD/PCDF compounds. Two explanations were offered for the observed differences in bioavailability. First, the nature of the soil could alter the bioavailability. The fill at the New Jersey site contained asphalt, and the carbonaceous nature of this matrix could enhance binding of PCDDs/PCDFs. Second, the soil at New Jersey was contaminated over a long period of time, with a generally aqueous medium, whereas the Missouri soils were contaminated with an oil mixture. The presence of oil would tend to reduce the binding of PCDDs/PCDFs to the soil and make them more bioavailable.

The potential for PCDD/PCDF bioaccumulation in ruminants in the wild may be assessed from results of a study on 2,3,7,8-TCDD bioaccumulation in beef cattle.[190] In this study, seven animals were fed a standard cattle ration containing 24 pg/g 2,3,7,8-TCDD for 28 days. Three animals were then sacrificed for tissue analysis, while 2,3,7,8-TCDD elimination in the remaining animals was monitored over a 36-week period by taking fat biopsies at 4-week intervals. The data were examined using a one-compartment kinetic model, and it was extrapolated that a steady-state level would be reached in 500 days, at which time the fat-tissue concentration would be 594 pg/g (all tissue concentrations of PCDDs/PCDFs reported in this subsection are expressed on a wet-weight basis unless noted otherwise). Therefore, a BMF of 25 could be expected for ruminant species in the wild such as deer. This value would be considerably reduced for does with fawns since the carry-over from feed to milk for cows was found to vary from 40% for 2,3,4,7,8-PeCDF to 0.68% for OCDD.[191] The excreted PCDDs/PCDFs would, however, increase the PCDD/PCDF body burden of the fawns.

#### *18.3.2.2 Abiotic Dispersal of PCDDs and PCDFs*

##### *18.3.2.2.1 Atmospheric Dispersal of PCDDs and PCDFs*

Studies carried out to determine the sources of PCDDs/PCDFs in the environment have revealed that the atmosphere is a very important medium for the transport of PCDD/PCDF compounds. With the exception of sediments collected from Lake Ontario, remarkable similarities were found between PCDD/PCDF patterns in urban-air particulates and in sediments from different locations in the Great Lakes.[192] The generalized pattern was dominated by octa-CDD, and it was suggested that atmospheric transport of combustion-derived PCDDs/PCDFs was the common source for each location. By invocation of photolysis as a plausible degradation pathway with the less-chlorinated PCDD/PCDF compounds, the PCDD/PCDF patterns found for municipal incinerators could be transformed into the patterns found in the sediments and the air particulates. In the case of the Lake Ontario sample, which had elevated concentrations of both octa-CDD and octa-CDF, a pattern-recognition technique showed that the sample had a PCDD/PCDF pattern very similar to the PCDD/PCDF pattern in pentachlorophenol. It is, in fact, known that chemical wastes have been transported from the Niagara River into Lake Ontario and that some of these wastes could have included pentachlorophenol and associated PCDDs/PCDFs. In another study of PCDD/PCDF deposition in sediment cores from the Great Lakes, it was determined that the most isolated Great Lake, Lake Superior, receives only 20% of its current input of PCDDs/PCDFs from local sources by atmospheric and aquatic routes. In contrast, Lake Ontario receives >90% of its input from local sources.[193]

A small lake from which sediments were collected, Siskiwit Lake, is situated on a remote island in Lake Superior; it therefore could receive contaminants only via atmospheric transport. By use of radiochemical techniques for dating sediment core sections, annual PCDD/PCDF fluxes to Siskiwit Lake could be plotted against time from 1920 to 1998.[194] It was found that the annual fluxes to the lake were low (< 1 pg/cm$^2$/year) before 1935 and then increased to a maximum of 9.5 pg/cm$^2$/year in the period 1975–1980. Since this increase parallels the increase in production of chlorinated aromatic compounds in the 1940s, it was concluded that combustion of wastes containing these compounds was the most significant source of atmospheric PCDDs/PCDFs. The decline in PCDD/PCDF concentrations that occurred after the 1970s was attributed to a reduction of particulate emissions following the passage of environmental legislation. However, the total decline for the period 1987–1995 was estimated to be 20%, whereas the U.S. EPA has calculated that emissions from combustion sources declined by 75% during the same time period. The lower-than-expected rate of decline of PCDDs/PCDFs in the sediments was attributed to the photochemical conversion of PCP to OCDD and HpCDDs in condensed atmospheric water (see below). Additional studies conducted on sediment cores collected from remote lakes in North America[195] and Europe[196–198] support the pattern of current and historical loading of PCDD/PCDF compounds described for Siskiwit Lake.

It is generally accepted that combustion processes are a major source of PCDDs/PCDFs in the environment. In contrast, a study conducted in 1995 found that, on a global scale, deposition estimates exceeded emission estimates by a factor of approximately four.[199] In this study, depositional fluxes were determined from the PCDD/PCDF concentrations in 107 soil samples. Total global deposition was estimated to be 13,100 ± 2,000 kg/yr. Data from air sampling conducted in the North Atlantic suggested that less than 10% of this total deposition was occurring over the oceans.[200] The deposition estimate was then improved by collection of 63 additional soil samples from areas that had been inadequately sampled in the first study.[201] Since $NO_x$ emissions were highly correlated with PCDD/PCDF fluxes, available $NO_x$ data were used to estimate PCDD/PCDF emissions in areas where no soil samples had been collected. The database for emissions estimates was further expanded after PCDD/PCDF emissions were found to be well correlated with both $CO_2$ emissions and Gross Domestic Product (GDP) of the country. Only 23 countries did not report data for either $CO_2$ or GDP, and these countries were at any rate considered to have very low dioxin emissions.[202] Through the use of all of these additional data, it was determined that global PCDD/PCDF emissions were in the range of 1800 kg/year, compared to a deposition estimate of 3,000–10,000 kg/year.[202] Therefore, although the more recent data show that the mass balance discrepancy may be closer to a factor of two, the findings still point to a deposition rate that exceeds the emissions rate.

Experiments conducted in the laboratory have shown that 1,2,3,4-TCDD can react in the gas phase with OH radicals generated from either the photolysis of ozone ($O_3$) in the presence of water or from the photolysis of hydrogen peroxide.[16] The 1,2,3,4-TCDD-OH reaction rate constant was calculated from experimental data, and the average OH reaction rate constants for PCDD/PCDF congener groups with four to eight chlorines were then calculated from structure–activity considerations. When these average OH rate constants were applied to source data via a simple model incorporating gas/particle partitioning, a reduction in concentrations of lower chlorinated congeners was found, but OCDD and HpCDD were not found to be as predominant as they are in sinks.[202]

In order to reconcile the mass balance data and the congener profile data for sources and sinks, Baker and Hites[202] have suggested that PCP is converted into OCDD and HpCDD by a photochemical reaction in condensed atmospheric water. In support of this hypothesis, it was found that aqueous solutions of PCP could be converted in a photochemical reactor into OCDD, HpCDD, and HxCDD at respective approximate yields of 0.1, 0.01, and 0.003%. Based on current measurements of PCP concentrations in rain, these yields would be sufficient to close the mass balance gap. While the experiments were conducted at environmentally relevant pH (5.5) and irradiation wavelengths

(> 290 nm), the authors have pointed out that the laboratory conditions do not mimic all of the conditions existing in condensed atmospheric water.

#### *18.3.2.2.2 Aquatic Dispersal of PCDDs and PCDFs*

The subject of aquatic dispersal of PCDDs/PCDFs has received considerably less attention than that of atmospheric dispersion. Tyler and co-workers[203] found that PCDD/PCDF concentrations in sediments from the Clyde Estuary decreased with distance downstream from Glasgow. The concentration at a 1-km distance was 6000 pg/g, decreasing to 322 pg/g at a distance of 22 km. Since organic pollutants favor fine sediments with high surface area and high organic carbon content, it was suggested that these fine sediments were being filtered out in their movement downstream, thereby resulting in a decrease in the PCDD/PCDF sediment concentration. However, there was also a decrease in the relative proportion of octa-CDD, suggesting that other physicochemical processes were occurring.

### 18.3.3 PCDD and PCDF Removal Processes in Aquatic Systems

Laboratory studies have shown that PCDDs/PCDFs dissolved in water can be photodegraded on exposure to either sunlight or UV light at wavelengths that are environmentally relevant (> 290 nm).[204,205] The process follows first-order kinetics, and the rates in sunlight are quite rapid for congeners dissolved in pure water (half-lives of 6.4 h and 8.3 h for 1,2,7,8-TCDD and 1,2,7,8-TCDF, respectively). When the PCDDs/PCDFs were dissolved in natural water from streams and lakes, they were photosensitized by humic acids and other organic compounds, increasing the photolysis rates by a factor of two.[15,204] While dechlorination occurs during these reactions to form lower-chlorinated PCDDs/PCDFs, the dechlorination process makes only a minor contribution to the total loss of the parent PCDD/PCDF.

Anaerobic microbial metabolism is another degradation process that has been shown to dechlorinate PCDDs/PCDFs in sediments.[206] In contrast to photodegradation reactions, in which the lateral positions are preferentially dechlorinated, peri positions (the position closest to the ether bridge) are frequently selected for dechlorination by anaerobic microorganisms. This could potentially increase toxicity by leading to the formation of 2,3,7,8-substituted congeners. However, over time, the 2,3,7,8-substituted congeners are subject to degradation by consortia with the capability of removing chlorine atoms from both lateral and peri positions. Studies carried out using sediment inocula from polluted rivers in the United States and Germany suggest that 20 to 30% of PCDDs/PCDFs are dechlorinated over periods varying from 7 months to 2 years during incubations at temperatures varying from 20°C to 30°C.[206–208] There is also evidence that PCDDs/PCDFs can be dechlorinated by abiotic processes involving electron transfer from inorganic (zerovalent iron and zinc) and organic (hydroquinone/quinone and vitamin B12) molecules.[209] Additional data show that peridechlorination may be enhanced by the synergistic actions of the abiotic and microbial dechlorination processes.[210]

While these laboratory studies indicate that PCDDs/PCDFs can be degraded by sunlight, anaerobic microorganisms, or abiotic dechlorination, it is not clear whether any of these degradation processes has a significant impact on the levels of PCDDs/PCDFs in the aquatic environment. Since PCDDs/PCDFs are hydrophobic compounds, they are primarily associated with sediment particles; this association would limit the dissolved concentrations of the compounds available for photodegradation. In the case of anaerobic metabolism, dechlorination rates for OCDD in sediment systems were considerably lower than those found using spiked sediment-free elutriates.[211] Therefore, bioavailability to microorganisms from sediments could be a limiting factor in the anaerobic metabolism of PCDDs/PCDFs. Finally, the anaerobic incubations were carried out in the presence of cosubstrates and nutrients and at temperatures exceeding those generally found in the aquatic environment.

## 18.4 AMBIENT LEVELS OF PCBs

### 18.4.1 Ambient Levels of PCBs in Aquatic Biota

One of the first reports of the environmental persistence of PCB congeners came from the observation of a Swedish chemist, Sören Jensen, who noted their presence in fish. He was looking for evidence of DDT contamination in biological tissues and discovered, in addition to DDT and its metabolites, several other chlorinated hydrocarbons, which were later verified to be chlorinated biphenyls.[212] As accumulators of PCBs, many organisms are among the best monitors of PCB, especially for conducting trend analyses. They integrate exposure over extended time periods, and, since they magnify levels, their concentrations will build up to levels much higher than in their external media, allowing for ease of measurement. Several nations have monitoring programs underway involving analyses of PCBs in sentinel species, usually fish,[29] but also aquatic birds (herring gulls),[36] marine mammals,[213] and shellfish.[214] In the Great Lakes, lake trout have been routinely monitored since 1979; in them, a continuing decline in PCBs has been reported.[36] NOAA Status and Trends data exist for PCBs in coastal areas also showing downward trends of PCBs in mussels.[214] European data, especially from Sweden, have been discussed in several publications, and both terrestrial and aquatic species have been studied.[215] Especially high levels have been reported in porpoises and seals, and several long-term studies have been done on polar bears.[215]

### 18.4.2 Ambient Levels of PCBs in Sediments

Much of the PCBs released into the environment over time are believed to reside in sediment. As clean sediment is delivered (now that external loadings have decreased) these deposited sediments are being buried, and eventually, in the absence of scouring or resuspension, they will be removed from the MER of PCBs. Deviations from this gradual burial process have been documented. In a Lake Erie study of PCBs in biota, investigators found that unusually high wind speeds from 1996 to 1997 led to increased PCB concentrations in biota that reside above the sediment due to resuspension of previously buried sediment deposit.[217] Therefore, deep scouring during flooding in rivers or heavy-storm-induced resuspension in lakes can lead to renewed exposure of PCBs from sediment deposits.

Other additional natural losses such as microbial degradation are also possible, but the rates of such processes in deeper sediments are very slow. Historical assessments of PCB deposition into sediments rely on dating the deposition layers within sediment cores that are removed from the deposit areas and relating the layers with their respective PCB concentrations. The chronology of past depositions of the PCB into undisturbed sediments has been documented for the Hudson River estuary, Great Lakes, and the Baltic.[217–220] Several sediment profiles have provided evidence for anaerobic dechlorination losses, for example, in the Hudson River[121] and in the tributary basin of the Thames River in England.[221] These researchers' evidence for dechlorination was expressed as a change in the ratios of less highly chlorinated (tri- and tetra-) congeners relative to congeners with five or more substituted chlorine atoms. In one of their cores samples, they[221] reported an average ratio of 0.23 from the surface of this core to ratios increasing to 0.32 at 2.3 m depth and finally to 0.62 by 5.8 m below the surface. As long as contaminated sediments reside in place in deposit zones they are subject to remobilization. Careful stewardship of these deposit zones is the preferred option. This stewardship is best provided by careful removal and placement in secure upland disposal facilities. While choosing this option has to be done on a case-by-case basis, it is the most frequently chosen option when costs are not a concern.[39]

### 18.4.3 Ambient Levels of PCBs in Soils

There are few data on the levels of PCBs in soil. The general rule is that ambient air levels of PCBs are greatly affected by soil/air partitioning and that soils and other exposed surfaces in urban communities generally receive influx of atmospheric PCBs. However, now that ambient background levels of PCBs in air are declining, soils in rural areas are releasing more PCBs than they are receiving. Thus, the congener ratios of PCBs in rural soils and at remote sites tend to reflect depletion of the more volatile congeners and enrichment of more highly chlorinated congeners.[96] In tests of several soil types by Ayris and Harrad,[107] the most important influences on the rate constant for soil partitioning were from the log $K_{octanol/air}$ (adjusted for soil temperature) and soil organic carbon.

One of the more careful and complete studies of soil interactions with PCBs has been conducted by Kevin Jones' research group in the United Kingdom.[56,98,222,223] Most of the soils that were the centerpiece of their work came from agricultural plots. The levels of total PCBs ranged from 10 to 670 μg/kg (median 30 μg/kg).[222] Twenty-six congeners were routinely monitored and quantitated in each of the samples. The soils included contemporary samples and historical samples dating back to 1944. All of the contemporary samples had levels similar to those in early 1940s samples; however, there tended to be a higher proportion of more highly chlorinated congeners in the contemporary samples. The soil levels of PCBs increase when atmospheric levels are high by gas-to-soil exchange. Then, as source releases are depleted, the exchange from surface soils is reversed, buffering the levels in the atmosphere.[98] There appears to be a delay (10 years' time lag was observed by Harner and co-workers[98]), and there is a slower release of those accumulated PCBs having half-lives of 10–20 years, an effect that tends to prolong background levels for longer periods of time. Numerous investigators have noted that hotspot buildup often occurred in soils near transformer spills; these remain for long periods of time.

### 18.4.4 Ambient Levels of PCBs in Air

Air concentrations of total PCBs in metropolitan centers generally run about five to ten times higher than background levels and correlate closely with the higher particle concentrations in cities. The total gaseous PCB concentrations at an overlake site 15 km from Chicago ranged from 132 to 1120 pg/m,[3] with higher concentrations occurring in warm periods and when wind directions were from the city. Therefore, Chicago was determined to be a major source of PCBs for Lake Michigan.[86] Because of concern for elevated concentrations of PCBs in the Great Lakes region and because of the sensitivity of this area to PCBs, there is currently an air-monitoring network analyzing PCBs on a regular basis (IADN Network-Integrated Atmospheric Deposition Network).[224,225] Data now available from these collections show that mean yearly values for PCBs range from 89 to 370 pg/m.[3] Concentrations at an urban site near Lake Erie were about two times higher than concentrations at the two more remote sites near Lake Superior.[226]

Five years of data, from 1992 to 1997, were summarized for PCBs measured in air at a site in northwest England. PCB levels declined, with specific half-lives ranging from 2 to 6 years for all of the congeners that were modeled; the decline was fastest for congener 52.[56] Schreitmueller and Ballschmiter[85] measured PCBs in air over the North Atlantic. Values for total PCBs ranged between 48 pg/m$^3$ in 1992 in the eastern North Atlantic and 22 pg/m$^3$ in the central South Atlantic for this year. Several papers confirm that PCB air concentrations vary with season (high concentrations when air temperatures are high and low PCB concentrations when the temperatures are lower). PCB concentrations in air vary with latitude, i.e., they are higher at mid latitudes (most industrial sources exist here) and lower at higher latitudes that are closer to polar regions.[55]

## 18.5 AMBIENT LEVELS OF PCDDs AND PCDFs

### 18.5.1 Ambient Levels of PCDDs and PCDFs in Aquatic Biota

When the first efforts to monitor PCDDs/PCDFs were undertaken in the 1980s, major emphasis was placed on the determination of 2,3,7,8-TCDD in fish from aquatic environments that receive wastes from industrial processes. In 1983, O'Keefe and co-workers[227] found that 2,3,7,8-TCDD was present in fish from all of the Great Lakes. However, the most heavily contaminated fish were found in Lake Ontario and in the Saginaw Bay area of Lake Huron, where concentrations varied from 2 to 162 pg/g and from 2.5 to 29 pg/g, respectively. Fish from the other Great Lakes (Lake Erie, Lake Michigan, and Lake Superior) had < 3pg/g. Later studies[228–230] confirmed these results. Norstrom and co-workers[231] also found that herring gull eggs collected from colonies in Lake Ontario and the Saginaw Bay area of Lake Huron had higher 2,3,7,8-TCDD concentrations (43 to 86 pg/g) than did eggs from colonies in Lakes Erie, Michigan, and Superior (9 to 11 pg/g). PCDDs/PCDFs containing from four to eight chlorine atoms were determined in Great Lakes fish by Stalling and co-workers[232] in 1982 and Zachareweski and co-workers[233] in 1989. Both studies showed that PCDFs were more widely distributed in the Great Lakes than were PCDDs, which suggests that the PCDFs are entering the lakes from other sources, including commercial Aroclors. The major PCDD/PCDF congener in most samples was 2,3,7,8-TCDF, with the exception of samples from Lake Ontario, where 2,3,7,8-TCDD was the major congener, and samples from Saginaw Bay, Lake Huron, where there were significant quantities (> 30 pg/g) of tetra-, penta-, and hexa-CDDs and CDFs.[232] While fish collected from Lake Michigan generally have low concentrations of PCDDs/PCDFs, higher concentrations have been found in biota collected around Green Bay, WI, an area contaminated by effluents from pulp and paper mills and other industrial wastes. Eggs from colonies of Forster's terns located in this area were found by Kubiak and co-workers[234] to contain median 2,3,7,8-TCDD and total PCDD concentrations of 37 and 102 pg/g, respectively.

A national survey conducted by the U.S. EPA demonstrated that contamination of aquatic biota in the U.S. by PCDD/PCDF compounds was not confined to the Great Lakes. In this study, bottom-feeding and predator fish were collected from almost 400 sites including the Great Lakes. It was found that 2,3,7,8-TCDD could be detected in fish from 112 of the 395 sites (28%), with concentrations below 5 pg/g from 74 sites (67%), between 5 and 25 pg/g from 34 sites (32%), and between 25 and 85 pg/g from 4 sites (1%).[137]

Aquatic biota, other than fish and birds, bioaccumulate PCDDs/PCDFs. Bullfrogs (*Rana catesbeiana*) collected near a chemical plant at Jacksonville, Arkansas, that formerly manufactured the herbicide 2,4,5-T had concentrations of 2,3,7,8-TCDD in their fat tissues as high as 68,000 pg/g.[235] For females, concentrations were also elevated in ovaries and oviducts. Snapping turtles (*Chelydra serpentina*) are long-lived organisms consuming both plants and animals, including fish; they therefore tend to bioaccumulate high concentrations of organochlorine compounds. Ryan and co-workers[236] analyzed PCDD/PCDF compounds in fat and liver samples from three snapping turtles from the upper St. Lawrence River. Concentrations of 2,3,7,8-TCDD in the fat tissue varied from 232 to 474 pg/g. It is conceivable that this PCDD congener was transported into the upper St. Lawrence from its source, Lake Ontario. One of the three samples was collected near industrial sites at Massena, New York; this sample contained 3 ng/g 2,3,4,7,8-penta CDF in addition to 2,3,7,8-TCDD. It is now known that this area is heavily contaminated with PCBs and PCDFs.

The aquatic species described above accumulate predominantly 2,3,7,8-substituted isomers. However, crustaceans discriminate to a lesser extent between 2,3,7,8-substituted and non-2,3,7,8-substituted isomers, as was shown in a study conducted by Rappe and co-workers[237] in the New York Bight and the Newark Bay area of New York Harbor. When tissues of blue crabs (*Callinectes sapidus*) and American lobsters (*Homarus americanus*) were analyzed, concentrations were considerably higher in the hepatopancreas than in the meat, with a concentration of over 6000 pg/g 2,3,7,8-TCDD in the hepatopancreas from one crab. While 2,3,7,8-TCDD was the major TCDD isomer, the 2,3,7,8-substituted isomers constituted less than 30% of the total congener group concentrations for the penta- and hexa-CDDs and the tetra- to hexa-CDFs. In general, similar isomer patterns were found in the hepatopancreas and the meat tissue.

In Europe, studies to monitor PCDDs/PCDFs in aquatic biota have focused on major rivers, such as the Rhine and the Elbe, and on the Baltic Sea. In the Elbe, bream (*Abramis brama*) collected in the Hamburg Harbor area near a pesticide manufacturing plant had total 2,3,7,8-substituted PCDD/PCDF concentrations and 2,3,7,8 TCDD concentrations as high as 308 and 102 pg/g, respectively.[238] Lower PCDD/PCDF concentrations (< 50 pg/g) were generally found in bream collected at other locations in Hamburg Harbor and at sampling points from the river mouth to a point 200 km upstream.[239] In all of the samples in this study, the PCDF concentrations exceeded those of PCDD, with 2,3,7,8-TCDF/2.3,7,8-TCDD ratios exceeding 10. Also, comparable levels of 1,2,3,7,8- and 2,3,4,7,8-penta CDF were detected, whereas in both North American and Scandinavian studies the 2,3,4,7,8 isomer has been found to predominate. However, it was also apparent that the capillary gas chromatography column was not capable of separating 1,2,3,7,8-and 2,3,4,7,8-penta CDF standards. Concentrations and isomer patterns of PCDDs/PCDFs in fish from the upper reaches of the Rhine and its tributary, the Neckar, were comparable to those found in the River Elbe fish. The more usual pattern of higher 2,3,4,7,8-penta CDF than 1,2,3,7,8-penta CDF concentrations was found in the Rhine/Neckar samples.[240] Eel samples, however, had quite different patterns that were dominated by octa-CDD, suggesting a more direct influence from sediments.

In 1989, Rappe and co-workers[241] summarized their investigations of PCDD/PCDF concentrations in fish collected from the Baltic Sea and from Lake Vattern in Sweden. Composites of herring from various locations in the Baltic and a control site on the Atlantic coast of Sweden had very similar PCDD/PCDF patterns that were dominated by 2,3,4,7,8-penta CDF (3–19 pg/g), with lower amounts of 2,3,7,8-TCDF (1.7–6.2 pg/g) and no detectable levels of 2,3,7,8-TCDD. Total congener group concentrations varied from 7 pg/g for the Atlantic Coast sample to over 30 pg/g for a sample collected near Stockholm, with intermediate levels for samples from the Gulf of Bothnia, an area in which several pulp and paper mills are located. As salmon are predators, it was not unexpected to find higher PCDD/PCDF concentrations in these fish than in either herrings or hatchery-raised salmon receiving synthetic food. Wild salmon from the Gulf of Bothnia had PCDD/PCDF concentrations approximately five to ten times higher than either the herrings or the hatchery salmon. In samples collected close to pulp and paper mills, the dominant PCDD/PCDF congener was either 2,3,7,8-TCDD, as in the case of perch from Norrsundet in the Gulf of Bothnia (13–19 pg/g), or 2,3,7,8-TCDF for Artic Char from Lake Vattern (20–75 pg/g). For the Artic Char, there was a direct relationship between fish weight and contaminant levels.

Monitoring data are now available showing the time trend of PCDD/PCDF concentrations in aquatic biota. For herring gull eggs from the Great Lakes, concentrations declined between 1981 and 1984.[242] However, there were no obvious temporal trends between 1984 and 1991. In a review of a database on TCDD concentrations in fish and shellfish collected from U.S. waterways between 1979 and 1994, Firestone and co-workers found that there had been a steady decline during the study period.[243] This decline was most apparent at contaminated sites on the Great Lakes. From 1981 to 1982, five of ten fish collected in Lake Ontario had TCDD concentrations exceeding 11 pg/g, whereas in 1994, TCDD was not detected in any of nine fish (≤1–2 pg/g). Significant declines in PCDD/PCDF levels in wildlife have also occurred in British Columbia, Canada as a result of the replacement of chlorine by chlorine dioxide as a bleaching agent for wood pulp and the introduction of regulations to control the use of chlorophenols as antisapstaining agents for undried lumber. In a study of PCDDs/PCDFs in osprey eggs collected in the drainage systems of the Columbia and Fraser Rivers, reductions between four- and tenfold were found in the PCDD/PCDF concentrations in the eggs over the period from 1991 to 1997. These reductions were mainly attributed to regulatory controls on PCDD/PCDF releases from pulp and paper mills introduced between 1991 and 1993.[146,244] However, in many recently industrialized countries and in Central and Eastern European countries, reports published between 1995 and 1997 show that there are locations where wildlife are still contaminated with high concentrations of PCDDs/PCDFs. Fish taken from the Er Jen River in Taiwan, near metal-reclamation plants, had PCDD/PCDF TEQs as high as 2084 pg/g.[245] Blubber tissue from one juvenile and two adult female Baikal seals (*Phoca sibirca*) collected from Lake Baikal in Russia had PCDD/PCDF TEQs of 28, 73, and 93 pg/g, respectively — levels comparable to those found for ringed seals (*Phoca hispida*) in contaminated areas of the Baltic Sea.[246]

### 18.5.2 Ambient Levels of PCDDs and PCDFs in Sediments

Current background PCDD/PCDF concentrations in sediments from various countries in the Northern Hemisphere have been determined from analysis of the top sections of sediment cores from remote lakes. Between 1993 and 1998, it was reported that Siskiwit Lake in the United States,[194] Loch Coire nan Arr in northwest Scotland,[196] and three lakes in Subarctic Finland[197] had recent total PCDD/PCDF sediment concentrations of 382, 488, and 202 ± 92 pg/g, respectively. These data are in remarkable agreement and demonstrate that long-range atmospheric transport can distribute PCDDs/PCDFs over wide areas. Lakes in the Black Forest in Germany, where the prevailing winds can transport PCDDs/PCDFs directly from emission sources in industrial areas, have higher sediment PCDD/PCDF concentrations (8.9–18 ng/g).[198] The highest sediment PCDD/PCDF concentrations occur where there is direct loading to water from point sources. Norwood and co-workers[247] analyzed surface sediments from a number of contaminated coastal areas in the United States and found the highest PCDD/PCDF concentration (44 ng/g) in sediment from Eagle Harbor, Washington, a location that had been contaminated with PCP wastes.

### 18.5.3 Ambient Levels of PCDDs and PCDFs in Soils

Comprehensive studies of PCDD/PCDF contamination in soil samples have been carried out in England. The first study involved the collection of 77 rural soil samples at the intersection points of a 50-km grid.[248] When 12 samples contaminated by local point sources were eliminated from the data set, it was found that the mean concentrations of PCDDs and PCDFs were 311 and 132 pg/g, respectively. Mean concentrations of the PCDDs/PCDFs were considerably higher in soil samples collected from 19 urban areas (11 ng/g for PCDDs and 0.9 ng/g for PCDFs).[249] Principal-components analysis suggested that combustion processes, such as coal burning and municipal incinerators, were the principal sources of PCDD/PCDF compounds in the urban soils. In an investigation of PCDDs/PCDFs in soil samples collected in rural Germany, samples collected from cultivated land and grassland had total PCDD/PCDF concentrations slightly lower than those found in the English study (mean values of 123 pg/g and 170 pg/g for cultivated land and grassland, respectively).[250] However, elevated PCDD/PCDF concentrations were found in soil samples collected from forested areas (mean values of 2.3 ng/g and 6.5 ng/g for deciduous and coniferous forest soils, respectively). It is known that leaves and needles can take up PCDDs/PCDFs, primarily by dry gaseous deposition, and subsequently the falling needles/leaves increase PCDD/PCDF concentrations in the surrounding soil.[251]

### 18.5.4 Ambient Levels of PCDDs and PCDFs in Air

Measurements of PCDD/PCDF compounds in ambient air have been carried out at several locations in Europe, North America, New Zealand, and Japan. In rural areas,[168,252] total PCDD/PCDF concentrations in ambient air generally do not exceed 1 $pg/m^3$ and are often lower (< 0.5 $pg/m^{12}$).[253] In suburban locations, where there is no impact from point sources, PCDD/PCDF concentrations range from 1 to 5 $pg/m^3$.[254] Urban-air concentrations of PCDDs/PCDFs range from 5 to 20 pg/m,[3] but concentrations can exceed 100 $pg/m^3$ where there is a direct impact from industrial emissions.[252] PCDD/PCDF patterns in ambient-air samples from different locations are usually quite similar, with equal amounts of PCDDs and PCDFs. The congener profiles are dominated by 2,3,7,8-TCDF and octa-CDD. The decline in deposition fluxes following the introduction of pollution-control regulations has been discussed previously in this chapter. This decline has also been reflected in ambient-air PCDD/PCDF concentrations, particularly in urban areas. In the German cities of Köln, Duisburg, Dortmund, and Essen, ambient PCDD/PCDF concentrations were in the range of 6.5–17.4 pg/m$^3$ in 1987/1988; these levels had declined to 3.2–9.9 pg/m$^3$ by 1993/1994.[254]

## 18.6 EFFECTS OF PCBs, PCDDs, AND PCDFs

Most of the data describing the ecological effects of PCBs, PCDDs, and PCDFs in the environment deal with direct aquatic exposures (sediments/particulates, food, and water) and, thus, their related aquatic-based foodchains. Over time, especially in the last 20 years, additional toxicological and exposure information has been generated on aquatic species, especially predators, including other aquatic species, terrestrial and marine mammals, and avian species. Before the advent of improved analytical capabilities and availability of standards of the specific congeners of the PCBs, data on toxicity were only available for the commercial mixtures.

Evaluation of the health effects of PCBs is complicated by several factors. Worldwide, each of the commercial PCBs is a mixture of fewer than the possible 209 congeners, and toxicity depends on the precise congener makeup of each of these mixtures that accumulate in the tissues of organisms and the metabolites that arise from these residues. Also, the degree of contamination of commercial PCB mixtures with PCDFs complicates the situation, since PCDFs are considerably more toxic, especially to fish. Furthermore, direct transfer of the unaltered manufactured PCB mixtures to a target organism seldom occurs, except in a laboratory situation. Even then, congener and mixture homolog group patterns change following assimilation due to differential metabolism in various test species. This holds true in the ambient environment as well, where multiple processes change composition and relative proportions of congeners. Of the PCDDs and PCDFs, those that are bioaccumulated and biomagnified are primarily 2,3,7,8-substituted. For PCBs, both dioxin-like and nondioxin-like congeners bioaccumulate and biomagnify in organisms and produce acute and chronic effects.

The earliest tests were done with Aroclor mixtures. Table 18.4, compiled from ATSDR's Toxicological Profile for Polychlorinated Biphenyls,[27] provides an example of these mixture tests, predominantly done for human-health purposes. The lethal dose decreases with length of exposure, and the most serious damage to these organisms occurs through reproductive effects. Reproductive effects occur at doses much lower than mortality levels.

Regulations for PCBs are based on total PCB concentrations (e.g., a sum of the mixed congeners), while PCDD/PCDF regulatory guidelines require individual congener identification and quantification. Mechanistically, this has resulted in an interpretation gap between the PCBs and PCDDs/PCDFs. While specific isomer information has always been available for assessing exposure and toxic dose for PCDD/PCDFs, concentration data for PCBs has always been more complicated to deal with, e.g., even with the congener-specific data for PCBs there is currently no unified and accepted way to integrate all of the multiple congener exposures.

### 18.6.1 Toxicological and Structural Similarities

The general consensus about the mode of action of TCDD is that it involves the activation of the TCDD-receptor complex, and the subsequent translocation of this complex into the cell nucleus is a necessary, but not sufficient, prerequisite for any TCDD-related effect.[255] It is also generally agreed that: (1) animals that possess an aryl hydrocarbon (Ah) receptor respond to TCDD similarly; (2) multiple effects, including enzyme induction, immunotoxicity, reproductive toxicity, developmental toxicity and carcinogenicity, occur in all susceptible species; and (3) chemicals that are

**Table 18.4 Most Sensitive Laboratory Animal Species Tested with Mixed Aroclors**

| Response | Species | Route[b] | Exposure Duration/ Frequency | NOAEL[c] (mg/kg/day) | LOAEL (effect)[d] (mg/kg/day) | Form |
|---|---|---|---|---|---|---|
| | | | **Acute Exposure** | | | |
| Death | Rat | GO | One time | — | 1010 to 4250 ($LD_{50}$) | Mixed |
| Death | Mink | G | One time | — | 750; 4000 ($LD_{50}$) | Mixed |
| Death | Mouse | PCDF | 2 weeks | — | 130 ($LD_{50}$) | A-1254 |
| Reproductive | Rat | GO | 9d | 8 | 32 | A-1254 |
| | | | **Intermediate** | | | |
| Death | Rat | PCDF | 8 months | | 72.4 (80% mort.) | A-1260 |
| Death | Rat | GO | 2.5 weeks | — | 840 (13% lower surv.) | A-1260 |
| Death | Mouse | PCDF | 6 months | — | 48.8 (lower surv.) | A-1221 |
| Death | Monkey | PCDF | 2 months | — | 4 (near 100% mort.) | A-1248 |
| Death | Mink | PCDF | 28–4 months | 8 | 1.9–7.1 ($LD_{50}$) | A-1254 |
| Reproductive | Rat | GO | 1 months | — | 30 | A-1254 |
| Reproductive | Mouse | PCDF | 108 days | 1.25 | 12.5 | A-1254 |
| Reproductive | Monkey | PCDF | 7 months | — | 0.2 | A-1248 |
| Reproductive | Monkey | PCDF | 2 months | — | 4.3 | A-1248 |
| Reproductive | Mink | PCDF | 4–8 months | 0.1 to 0.2 | 0.4 to 0.9 | A-1254 |
| | | | **Chronic** | | | |
| Death | Rat | PCDF | 205 weeks | — | 2.5 (lower surv.) | A-1254 |
| Reproductive | Monkey | PCDF | 16 months | 0.008 | 0.03 | A-1016 |
| Cancer | Rat | PCDF | 14–24 months | — | 1.25 to 5 | A-1254;1260 |

[a] Compiled from ATSDR 1997[27] Toxicological Profile for Polychlorinated Biphenyls (Update) U.S. Department of Health and Human Services, Public Health Service, Washington, D.C., 1997, 429.
[b] Route of exposure, (GO) = gavage, oil; (G) = gavage; (PCDF) = food.
[c] NOEL = no observed effect level.
[d] LOEL = lowest observed effect level.

isostereomers of TCDD (i.e., polychlorinated and polybrominated dibenzo-*p*-dioxins, dibenzofurans, and coplanar biphenyls) act through the same mode of toxic action. This final point is used to justify discussing the mammalian and nonmammalian vertebrate effects of 2,3,7,8-substituted PCDDs and PCDFs together with the effect of coplanar PCBs.

Species-specific factors, such as uptake, disposition, and metabolism of TCDD, as well as interspecies differences in concentration, tissue distribution, and ligand affinity of the Ah receptor, all likely play roles in determining the relative sensitivity of an organism to TCDD. However, the presence of the Ah receptor clearly appears to be a necessary prerequisite for TCDD (and related compounds) to exhibit toxicity. The presence of the Ah receptor in fishes and the lack of the receptor in aquatic invertebrates are consistent with the relative sensitivities of the two groups of species to TCDD and structurally similar compounds. For example, TCDD has been shown to be lethal to a number of fish species when administered through the diet or the water, or by injection (egg or whole organism). Conversely, long-term exposures of a number of different species to TCDD (snails [*Physa* sp.], worms [*Paranais* sp.], and mosquito larvae [*Aedes aegypti*]) failed to result in discernible toxicity.[256]

Congener-specific chemistry advances and toxicological interpretation considerations have consistently driven the need for a normalization protocol based on common mechanisms of action. Today, this protocol is termed Toxicity Equivalency and is now commonly used in risk assessment. However, it is not yet a uniform regulatory requirement, at least in the United States.

For those congeners belonging to this group (all PCDDs, PCDFs and selected planar PCB congeners, Table 18.5), relative toxicity of a congener can be normalized to the potency 2,3,7,8-

TCDD, the prototype, and the most toxic compound of this group of stereoisomers. This normalized value is referred to as the TCDD toxicity equivalent concentration. Considered within a generally accepted model of additivity, these congeners' individual toxicity equivalent concentrations (actual concentration × Toxicity Equivalency Factor, or TEF) are summed to derive a total value of exposure in an organism of interest or in the foods that it consumes.

**Table 18.5 World Health Organization Toxic Equivalency Factors (TEFs) for Humans Mammals, Fish, and Birds[260]**

| | TEF | | |
|---|---|---|---|
| **Congener** | **Humans/Mammals** | **Fish[a]** | **Birds[a]** |
| 2,3,7,8-TCDD | 1 | 1 | 1 |
| 1,2,3,7,8-PentaCDD | 1 | 1 | 1[b] |
| 1,2,3,4,7,8-HexaCDD | 0.1[a] | 0.5 | 0.05[b] |
| 1,2,3,6,7,8-HexaCDD | 0.1[a] | 0.01 | 0.01[b] |
| 1,2,3,7,8,9-HexaCDD | 0.1[a] | 0.01[c] | 0.1[b] |
| 1,2,3,4,6,7,8-Hepta CDD | 0.01 | 0.001 | <0.001[b] |
| OctaCDD | 0.0001[a] | <0.0001 | 0.0001 |
| 2,3,7,8-TetraCDF | 0.1 | 0.05 | 1[b] |
| 1,2,3,7,8-PentaCDF | 0.05 | 0.05 | 0.1[b] |
| 2,3,4,7,8-PentaCDF | 0.5 | 0.5 | 1[b] |
| 1,2,3,4,7,8-HexaCDF | 0.1 | 0.1 | 0.1[b,d] |
| 1,2,3,6,7,8-HexaCDF | 0.1 | 0.1[d] | 0.1[b,d] |
| 1,2,3,7,8,9-HexaCDF | 0.1[a] | 0.1[c,d] | 0.1[d] |
| 2,3,4,6,7,8-HexaCDF | 0.1[a] | 0.1[d,e] | 0.1[d] |
| 1,2,3,4,6,7,8-HeptaCDF | 0.01[a] | 0.01[e] | 0.01[e] |
| 1,2,3,4,7,8,9-HeptaCDF | 0.01[a] | 0.01[c,e] | 0.01[e] |
| OctaCDF | 0.0001[a] | <0.0001[c,e] | 0.0001[e] |
| 3,4,4',5-TetraCB (81) | 0.0001[a,c,d,e] | 0.0005 | 0.1[c] |
| 3,3',4,4'-TetraCB (77) | 0.0001 | 0.0001 | 0.05 |
| 3,3',4,4',5-PentaCB (126) | 0.1 | 0.005 | 0.1 |
| 3,3',4,4',5,5'-HexaCB (169) | 0.01 | 0.00005 | 0.001 |
| 2,3,3',4,4'-PentaCB (105) | 0.0001 | <0.000005 | 0.0001 |
| 2,3,4,4',5-PentaCB (114) | 0.0005[a,d,e] | <0.000005[e] | 0.0001[g] |
| 2,3',4,4',5-PentaCB (118) | 0.0001 | <0.000005 | 0.00001 |
| 2',3,4,4',5-PentaCB (123) | 0.0001[a,d] | <0.000005[e] | 0.00001[g] |
| 2,3,3',4,4',5-HexaCB (156) | 0.0005[d,e] | <0.000005 | 0.0001 |
| 2,3,3',4,4',5'-HexaCB (157) | 0.0005[d,e] | <0.000005[d,e] | 0.0001 |
| 2,3',4,4',5,5'-HexaCB (167) | 0.00001[a] | <0.000005[e] | 0.00001[g] |
| 2,3,3',4,4',5,5'-HeptaCB (189) | 0.0001[a,d] | <0.000005 | 0.00001[g] |

*Abbreviations:* CDD, chlorinated dibenzodioxins; CDF chlorinated dibenzofurans; CB chlorinated biphenyls; QSAR, quantitative structure-activity relationship.

[a] Limited data set
[b] *In vivo* CYP1A induction after *in ovo* exposure
[c] *In vitro* CYP1A induction
[d] QSAR modelling prediction from CYP1A induction (monkey, pig, chicken, or fish)
[e] Structural similarity
[f] No new data from 1993 review (1)
[g] QSAR modelling prediction from class specific TEFs

*Source:* Van den Berg, M. et al., *Environ. Health Perspect.*, 106, 775, 1998. With permission.

The mechanism of action of all planar halogenated hydrocarbons (another designation for the group to which PCDDs and PCDFs and selected PCB congeners belong) is of great importance because the cumulative risks are much greater than their individual toxicities in many instances. Support for this approach was most recently documented by van den Berg and co-workers[257] under the auspices of the World Health Organization. This group reviewed previous work regarding TEFs derived from mammalian species[258, 259] and recommended the establishment of animal-class-specific TEFs for humans/mammals, fish, and birds. The WHO database used to develop these TEFs is available from a U.S. Fish and Wildlife website[260] by clicking on dioxin_information.xls on the cited webpage. Table 18.5 provides a summary of these TEFs.

Most of the effort to isolate these more toxic components has focused on the dioxin-like response from PCBs. This similarity in effect to dioxin, i.e., mixed-function oxidase (MFO) induction, subcutaneous edema, reproductive impairment, weight loss, immune suppression, and hormonal alterations, has now further been narrowed to those compounds in the PCB mix that have a three-dimensional similarity in their structure. The potency and specificity for binding to the Ah receptor, followed by MFO induction (and correlatively, for potential toxicity) of individual PCB congeners can be directly related to how closely they approach the molecular spatial configuration and distribution of forces of 2,3,7,8-TCDD. The most toxicologically active PCB congeners are those having chlorine substitution at the *para* (4 and 4') positions and at least two *meta* (3,3',5,5') substitutions on the biphenyl nucleus, but no *ortho* (2,2',6, and 6') substitutions. The PCB congeners that have no *ortho* substitutions can assume a coplanar configuration, because the *ortho* chlorines are absent that would otherwise prevent rotation of the two opposing phenyl rings from assuming a planar configuration. Because the potential toxicity is enhanced by coplanarity, this rule limits the number of such PCB congeners to four: 77, 81, 126, and 169 (3,3',4,4'-, 3,4,4',5-, 3,3',4,4',5-, 3,3',4,4',5,5'). With the exception of number 81, the non*ortho*-coplanar congeners are potent inducers of aryl hydrocarbon hydroxylase AHH and 7-ethoxyresorufin O-deethylase EROD in *in vitro* rat hepatoma cell preparations. The 81 congener is a very potent avian *in ovo* EROD inducer (see Table 18.5). Virtually no toxicity data are available for this congener. The *in vitro* inductions are correlated with *in vivo* demonstrations of mammalian, avian, and fish toxicity such as thymic atrophy and inhibition of body weight gain.[261] The 81 congener has what is called a "mixed" type response in that it induces both methyl cholanthrene (3-MC)-type reactions and also phenobarbitol (PB)-type enzyme systems. From the data of Tillitt and co-workers[262] regarding mink accumulation and *in vitro* data, PCB 81 appears to be rapidly hydroxylated and is not biomagnified in mammals, but it is present in fish and birds as well as in their egg yolks, which contain lipids. It has not been reported or assessed in terms of human exposures.[263]

The second group of congeners having enzyme-inducing potencies and potential toxicities of high concern are analogs of the four non*ortho*-coplanar congeners that are still relatively coplanar but have a singe *ortho*-chloro substitution. These are congeners 105, 114, 118, 123, 156, 157, 167, and 189. This group of congeners has demonstrated mixed phenobarbitol (PB-)- and 3-methyl cholanthrene (3-MC)-type-inducing properties. Most of the more highly chlorinated dioxins and furans have received relatively little toxicity testing relative to 2,3,7,8-TCDD. Even the TEFs for the pentachloro dioxin and furan analogs are not based on whole-organism toxicity testing but on *in vivo* or *in vitro* MFO induction or quantitative structure–activity relationships (see footnotes in Table 18.5). In fish and birds, most PCDDs and PCDFs have not been reported in the primary literature with *in ovo* data for toxicity, and again the listed TEFs rely more on MFO induction data rather than data sets involving more thorough toxicity screening. However, older data on lethality and developmental anomalies for other PCDDs and PCDFs from the U.S. Food and Drug Administration (FDA) (reported by Verret,[264] and a Verret personal communication cited in Goldstein[265]) are generally consistent with the WHO TEFs for birds. This pioneering work by the FDA is now viewed as having been crucial to stimulating and refining scientific understanding of toxicity effects in birds and other oviparous vertebrates. For instance, the derivation of the WHO TEFs for birds and fish is *in ovo* tissue/residue-based, when such information is available. This stands in contrast to the situation for mammalian TEFs, which are derived from studies of adult dosing with concurrent partial metabolic elimination — despite the acknowledgment that body-burden-based toxicity information is preferable.

### 18.6.2 Structural and Toxicological Dissimilarities

Environmental exposures to PCB congeners and their metabolites that are not mechanistically identical to the PCDDs and PCDFs should still be considered. At a minimum, use of total PCB measurements in the assessment of the compounds' nondioxin-like effects is still prudent until similar equivalency approaches for other PCB congener groups having other modes of action are developed. This approach also provides a framework for interpreting long-term trends in environmental matrices. Giesy and Kannan[266] recently provided a perspective on the relative risks of dioxin-like vs. nondioxin-like PCBs using mink as a model species. Besides this example, little work has been done to assess the effects of the complex mix of dioxin-like and nondioxin-like congeners with their metabolites and across multiple animal classes.

Certain PCB metabolites have been found that may be operating via non-Ah receptor mechanisms or that may provide additional physiologically significant interactions with parent congeners. This fact is a caveat against pursuing only one approach,[268–271] despite the current evidence supporting the dominant toxicological importance of those congeners that act through the Ah-receptor. Conversely, it would be scientifically unsound to disregard the power of the TEF/TEQ approach and not to use it for ecological risk assessment. Based on what is known today, the uncertainties of continuing the present practice of assessing TCDD alone, or using TEQs for the PCDDs and PCDFs when considering total PCBs separately, would be greater than the uncertainty of adopting a TEF/TEQ approach that integrates PCDDs and PCDFs with the dioxin-like PCBs.[272] Given the additional known effects and accompanying uncertainties associated with the remaining PCB congeners, dual-track total PCB-based and TEF/TEQ-based assessments are recommended. More information is being generated and summarized for PCBs that are nondioxin-like.[273,274] Alternative methods for assessing these congeners in functional groups, as is being done for those that act through the Ah receptor, may be forthcoming and may be proposed for consensus adoption.

The primary consideration regarding the nondioxin-like congeners is the need to establish ecological relevance. For this, environmental exposure–effect relationships must be documented. The effects must be demonstrated in a toxicological significance range in which the exposure–effect is equal to or greater than for the dioxin-like congeners in the same species/life stage being assessed. More work with various animal classes will determine whether nondioxin-like congeners are as toxicologically significant as those that are dioxin-like. Invertebrates appear particularly attractive to study because they lack the Ah receptor. Aquatic organisms (fish, amphibians, and reptiles), in particular, require comparative study.

### 18.6.3 Effects Categories

Usually, the earliest sign of general toxicity from exposure to PCBs or PCDDs/PCDFs in controlled laboratory studies is weight loss. Female animals are more sensitive than males, and young animals are more sensitive than adults. Poultry, guinea pigs, mink, and nonhuman primates are the most sensitive.[275–277] Weight loss in chickens — "chicken edema disease" — is accompanied by increased bloating, which is actually an extracellular accumulation of body fluids and not real weight gain.[278,279] Terminally, rodents, mink, and especially birds show hemorrhage in the gastrointestinal tract.

#### *18.6.3.1 Acute Signs of Poisoning*

Acute signs of poisoning that are common include binding to the nuclear Ah receptor and resulting MFO induction of AHH and EROD. Following initial binding, induction or inhibition of other MFOs as well as hormonal and vitamin alterations can occur. Pathologically, subcutaneous, pericardial, and peritoneal edema, eyelid edema, hemorrhage, weight loss or reduced weight gain (termed "wasting" when accompanied by mortality), immune suppression, fatty infiltration of the liver, hepatomegaly, and thymic (lymphoid) and gonadal atrophy have been demonstrated in numerous test species. All of the planar congeners show a degree of toxicity that relates to both the number and the locations of the halogen atoms. Lesser amounts of all of these planar compounds are needed to cause toxicity as the length of exposure increases. As an example, the single oral $LD_{50}$ for 2,3,7,8-TCDD in monkeys was found to be 50 µg/kg, but if 500 ng/kg was administered over 6 months, then the cumulative $LD_{50}$ was 2 µg/kg. Of the polychlorinated dioxins, 2,3,7,8-TCDD is the most studied because of its extreme toxicity to many organisms. Species information reveals the following rank order for adult $LD_{50}$ TCDD sensitivity: guinea pig 1 µg/kg;[275] mink 4.2 µg/kg;[280] rabbit 115 µg/kg; chicken 25–50 µg/kg;[278] male and female rat 22 µg/kg and 45 µg/kg, respectively;[275] mouse 114 µg/kg;[281] bullfrog > 500;[282] and hamster 5000 µg/kg.[283] Comparably, Aulerich and associates reported interperitoneally injected newborn mink mortalities of 62 to 100% with a 1 ng/g TCDD daily dose for 12 days, and a dietary 135-day $LC_{50}$ for mink of 0.05 µg/g for PCB 169.[284,285]

#### *18.6.3.2 Chronic Signs of Poisoning*

Chronic signs of poisoning in laboratory species include ligand binding, MFO induction, and other biochemical alterations, immunotoxicity, developmental defects, chloracne, cancer, and death.[286,287] Reviews by various authors indicate that the dioxin-like congeners correlatively manifest similar effects in a given species when dosages are adjusted to compensate for their lesser potencies relative to 2,3,7,8-TCDD. Interspecies sensitivity is variable, and not all acute and chronic symptoms of poisoning occur in every species. It is also very clear from the literature that life-stage sensitivity is exceedingly important in the full assessment of the effects of these compounds. For instance, the embryonic $LD_{50}$ of 140 pg/g for TCDD in the chicken egg is 140–280 times higher than the $LD_{50}$ in the adult.[278,288] For virtually all species tested, reproductive, developmental, and endocrine-related effects are the most sensitive endpoint indicators.

#### *18.6.3.3 Endocrine and Mixed-Function Oxidase Effects*

Available information suggests that dioxin-like exposures have the potential to compromise basic reproductive and developmental processes that can impair a population through modifying embryonic set points that are normally programmed for future survival and recruitment. The most responsive adverse effects are on the developing immune, nervous, and reproductive systems. In mammals, these effects have been observed at maternal body burdens of 2,3,7,8-TCDD in the range 30–80 pg/g for both nonhuman primates and rodents. While there is a 5000-fold acute adult lethality difference among guinea pigs, rats, mice, and hamsters, the doses associated with fetotoxicity are similar.[287] Among the most striking results of recent ecotoxicological relevant are those involving rats. Prenatal TCDD exposure toward the end of organogenesis produces permanent adverse effects in both male and female rat pups. Males exhibit delayed puberty, altered mating behavior, and decreased spermatogenesis, while females have a unique genital malformation that is characterized by a vaginal thread and cleft phallus, as well as premature reproductive senescence.[289–295] These effects occur at an estimated body burden of 28 pg/g over the background body burden of 4 pg/g of TCDD typically measured in rats and mice.[271]

There is a general relationship between the amount and timing of the dose for these compounds, not unlike "timing windows" for hormones. Lower dosage levels applied over longer periods of time can result in an identical toxic response of a single high dose. Developing embryos and early-life-stage compromise of basic maturation processes are known targets of these compounds. In this context, therefore, these endocrine effects can threaten maintenance of healthy ecological communities by compromising the ability of individuals to survive, breed, and recruit in a population. Because many of these congeners are persistent in the environment and magnify into higher trophic-level biota, full life-cycle exposures in laboratory testing are important for analyses of population effects.

Other noncancer endpoints have been researched and appear associated with endocrine disruption; these have been recently summarized by Birnbaum and Tuomisto.[287] Those with potential ecotoxicological significance that have been found in laboratory species exposed to TCDD and related compounds include: alterations of Leydig cells and normal dentition; induction of cleft palate and hydronephrosis; permanent lowering of core body temperature; onset of hearing deficits associated with circulating T4; learning deficits and impairments possibly associated with brain damage; locomotor and rearing behavior changes; promotion of endometriosis; thymic atrophy and decreases in thymic cellularity and depletion of all four subsets of T-cells; splenic atrophy; thyroid follicular cell hyperplasia; and changes in levels of androgens, estradiol, leutinizing hormone (LH), follicle stimulating hormone (FSH), vitamin A, thyroxine (T4), melatonin, insulin, glucose, serum gastrin, and glucocorticoids. Tissue-specific alterations of several glucocorticoid and estrogen receptor levels have been reported for doses as low as 0.1 of $LD_{50}$s. Alteration of nuclear retinoid receptor expression has been suggested as well as expression of mitogenic growth factors, such as epidermal growth factor (EGF). Many of these effects appear as direct or indirect divergences from hormonal homeostasis.

The most common sublethal biological effect of PCBs, PCDDs, and PCDFs across all vertebrate classes is increased activity in the hepatic microsomal MFO system. Many enzymes are induced or inhibited by exposures either *in vitro* or in the whole organisms. Typical environmental concentrations of TCDD and TCDD-equivalent tissue concentrations are less than 1.0 ng/g. Therefore, most typical effects on organisms will be through sublethal and chronic effects, as adult $LD_{50}$ and $LC_{50}$ values for the species known to be most sensitive are seldom approached in the environment, except at waste sites and in accidental poisoning episodes.

Physiological functions that are controlled by steroid hormones may be altered by exposure of organisms to PCBs PCDDs, and PCDFs. Growth, molting, and reproduction are primary functions that can be affected by exposure of aquatic organisms to PCBs in numerous laboratory investigations. The ability of organisms to eliminate foreign organic compounds or endogenous waste products also may be affected. Steroid biosynthesis and the degradation and biotransformation of foreign compounds are metabolic activities in both fish and higher vertebrates that are strongly influenced by terminal oxidase activities of the microsomal cytochrome P450 system (or MFO system). Some, though not all, PCB congeners are MFO inducers in fish, mammals, and birds and to a lesser extent in aquatic invertebrates. There are excellent historical reviews on this topic, covering mammals, birds, and, more recently, fish.[265,276,277,296,297,299–301]

PCB metabolites can be more toxic than the parent compound.[261] Similarly, some steroids associated with reproduction may be oxidized or altered as a result of the metabolic process used by organisms in the attempt to depurate PCBs and PCDD/PCDFs.[302–304] Effects on reproduction in fish may occur prior to spawning, at spawning and egg deposition, or during egg and larval development. Steroid alteration can effect oogenesis, and the production of toxic metabolites can influence oocyte maturation.[304] Again, the difficulties in isolating these effects in the environment are severely compounded by the fact that they are very likely indirect, complicated by the interaction with nonchemically related environmental variables that also contribute to these responses. Some examples include deficiency syndromes associated with thiamine, vitamin D, or iodine that may be independent of, or interactive with, toxic effects from these compounds.[305–307] The increased sensitivity of organisms to PCBs, PCDDs, and PCDFs when challenged with exposure-independent and exposure-dependent deficiencies needs further elucidation.

### 18.6.4 Wildlife

Generally, the research to date on the effects of the mixed PCBs on mammalian wildlife agree with the data found for traditional laboratory animals. This is partly because few in-depth studies have been carried out on nontraditional laboratory animals, and thus our understanding is far from complete, especially with regard to full-life-cycle testing. For other nonmammalian species, several different types of responses have been observed.

#### 18.6.4.1 Fish

##### 18.6.4.1.1 Acute Effects of Poisoning

Laboratory studies indicate that fish vary in their acute response to PCBs. Eisler,[64] summarizing the $LC_{50}$ responses of fishes, showed quite a considerable range depending on the Aroclor type, the species, and whether the species was marine or freshwater. The general level that produced lethality was in the 10- to 300-μg/g range (the exceptions were two marine fish that tested at 0.1 to 0.9 for long-term lab studies of 12 to 38 days). The longer the exposure period, the lower the $LC_{50}$ value needed to kill the fish species. Early research found PCB toxicities in the range of 8 to 234 μg/L in various fish for 96-hour $LC_{50}$ tests.[308,309] Trout species were found to require 1 to 62 mg/L to show $LC_{50}$ values in 96-hour tests.[310]

In fish, the general characteristic of TCDD-induced toxicity is similar to that observed for mammalian species, i.e., there is a slow wasting-away syndrome leading to mortality. Furthermore, other symptoms are similar, i.e., reproductive toxicity, histopathologic alterations, and possibly immunosuppression and induction of cytochrome-P450-dependent monooxygenases.[256] Generally, concentrations causing observable effects were as follows: for rainbow trout, concentrations as low as 0.038 ng/L in flow-through systems caused mortality,[256] and adult carp (*Cyprinus carpio*) showed mortality after 71 days on 0.06 ng/L in the same study. Exposure of eggs caused the most dramatic effects: some residues as low as 40 pg/g in eggs caused mortality in lake trout (*Salvelinus namaycush*).[312] Other species were not as sensitive, e.g., rainbow trout exhibited mortality at 230 to 488 pg/g.[313] Later stages of growth were also less sensitive. Intraperitoneal injections of juveniles caused 80-day $LD_{50}$s at 5 ng/g in bullheads[314] and 3 ng/g in carp, whereas 16 ng/g of an intraperitoneal dose was required to produce mortality in bluegills (*Lepomis macrochirus*), and 11 ng/g were required in largemouth bass (*Micropterus salmoides*).

##### 18.6.4.1.2 Chronic Effects of Poisoning

For PCBs, long-term toxicity tests on fish indicated little or no effect, even when high water concentrations or dietary exposures were tested. Salmon fingerlings fed 480 mg/kg Aroclor 1254 died after 260 days, while no effect was observed when catfish where fed 2.4–24 mg/kg of four Aroclors for 193 days.[310] Similarly, adult catfish fed 20 mg/kg Aroclor 1242 for 196 days showed no effect.[315] Day-old fathead minnows exposed to 0.9 to 5 μg/L Aroclor 1242 and 0.2 to 5 μg/L of Aroclor 1254 for 260 days showed no effects.[309] However, 10% lower growth of trout was observed when 17-day-old trout were exposed to 0.2 to 2.9 μg/L Aroclors 1254:1260 (1:2) for 90 days.[311]

Oviparous fish appear to show a wide range of reproductive sensitivities. Cold-water species such as lake trout, brook trout, and rainbow trout (*Onchorrhynhus mykiss*) are more sensitive than certain cool-/warm-water species, specifically the fathead minnow (*Pimephales promelas*), channel catfish (*Ictalurus punctatus*), lake herring (*Coregonus artedii*), Japanese medaka (*Oryzias latipes*), white sucker (*Catastomus commersoni*), Northern pike (*Esox lucius*), and zebrafish (*Danio rerio*).[316] The lake trout is the most sensitive ($LC_{50}$ 65 pg/$g_{egg}$), while the least sensitive is the zebrafish ($LC_{50}$ 2610 pg/$g_{egg}$). Thus, tested cool-/warm-water species are 8- to 38-fold less sensitive than lake trout. At the present time, hatching and early-life-stage studies have not been extended to partial and full life-cycle testing, which prolongs exposure, determines long-term survival, and looks for additional functional problems demonstrated in other animal classes. Much like the *in utero* effects in mammals, *in ovo* TCDD exposures in fish result in facial/jaw deformities, blue sac disease (yolk sac and pericardial edema), cranial edema, subcutaneous hemorrhaging, spinal deformities, and MFO induction — developmental symptoms similar to those seen in birds. The species differences in sensitivity may be due to variations in the length of time needed for fry to reach swim-up stage and fill their air bladders, in Ah-receptor toxicodynamics, and in toxicokinetics. Readers are urged to consult the underlying references in the WHO TEF scheme;[257] these references constitute the basis for our understanding of the similarity of chronic effects of related congeners. They also document the diminished potency of the PCBs needed to produce these effects in fish relative to mammals and birds.

Toxicologically, it is important to consider the absorption/absorptive exposure in fish, rather than relying on water exposures. In a series of studies with the brook trout with natural accumulation of TCDD via contaminated food pellets, an adult whole-body concentration of 300 pg/g was sufficient to achieve egg concentrations associated with significant adverse effects on embryo survival and delay in time to initial spawning, suggesting that ovulation was affected. In a high-dose (1200 pg/g) treatment, the hepatosomatic index increases were dose-dependent in both sexes.[318] Companion study results[319] produced an $LC_{50egg}$ of 138 pg/g wet weight at swim-up stage. Eggs constituted 39% of the adult whole-body concentration. Exophthalmia, cranial deformities, and generalized edema and hemorrhaging were observed, consistent with previous egg-injection studies in salmonids. Results from natural deposition of TCDD into the eggs resulted in a lower $LC_{50egg}$ of 138 pg/g, compared to injection concentration of 200 pg/g in a previous study.[319] Threshold $LC_{10egg}$ and $LC_{90egg}$ values were 88 and 184 pg/g.

A physiologically based toxicokinetic model was also produced in the third paper of this series.[320] While affected individuals have not been raised to adulthood, it would obviously be beneficial to have observed these effects in older individuals. This is especially crucial for comparative purposes when wild fish observations are documented. For instance, striped bass (*Morone saxatilis*) have been characterized as having morphological abnormalities. Upon retrospective review of this study, these abnormalities qualitatively resemble certain dioxin-like induced effects (pugheadedness and skeletal) seen in the lower Hudson River and along coastal Long Island in New York during the mid-1970s.[321] Similar symptoms of bone-development abnormalities were seen in brook trout tested with Aroclor 1254; in these, backbone phosphorus and hydroxyproline as well as fry survival were reduced, and calcium backbone levels were increased, at fry whole-body concentrations ranging from 17 to 284 μg/g.[322]

Integration of abiotic and biotic exposure compartments is needed to address risk and to remediate excessive environmental pollution with these compounds. In a modeling paper discussing retrospective analysis of dioxin-like compounds using biota sediment accumulation factors (BSAFs) and dated sediment cores, lake trout from Lake Ontario were assessed. Effects on reproductive impairment were based on egg TEFs and toxicity equivalence concentration (TECs), defined as the sum of the products of the TEFs and the concentration of each congener in a mixture. In this case study based on egg TECs, lake trout suffered some early-life-stage mortality from approximately 1940 to 1990; for three of these five decades, they suffered total mortality. Eggs were contaminated by multiple dioxin-like congeners, with TCDD and PCB 126 exerting the predominant effects. Ecoepidemiological information confirmed this relationship through monitoring of feral, contaminated eggs under laboratory conditions and the appearance of naturally produced 1- to 2-year-old fish in 1994 when the level of the contaminants had declined.[323]

Atlantic salmon have been tested with a complex mixture of Aroclors 1016, 1221, 1254, and 1260. Tissue concentrations that were 5.6 μg/g from the 48-hour water exposure to this mixture produced reduced wet weight and length after 6 months. A higher tissue concentration (14.2 μg/g) was accompanied by altered behavior and reduced predation behavior by the fish.[324]

#### *18.6.4.1.3 Endocrine and Mixed-Function Oxidase Effects*

Endocrine and MFO effects of exposures have been investigated for several decades in fish, but with less firm results than for other animal classes. In early investigations, only thyroid effects were observed at very high-dose (0.4 to 650 mg/kg) Aroclor diets.[310] MFO activity in fish, however, is a major site of response to PCB exposure. Although increased activity of this system alone does not represent a toxicological threat,[325–327] it could have some negative aspects through the formation and function of secondary products and downstream cascade effects from Ah-receptor binding. Thus, reexamination of old data may be warranted.

Reproductive toxicities have been documented in certain species of fish. Baltic flounder (*Platichthys flesus*) exhibited toxicity when ovary concentrations exceeded 0.12 mg/kg fresh weight.[309] Eisler cites several examples in his review chapter on PCBs:[64] among these, rainbow trout at 0.4 mg/kg Aroclor 1242 produced eggs with low survival and numerous fry deformities, rainbow trout at 0.33 mg/kg 1254 had 10 to 28% mortality, and Atlantic salmon at 0.6 to 1.9 mg/kg had 46 to 100% mortality. However, other studies on trout indicated that eggs with residue levels of 1.6 mg/kg did not have diminished survival.[317] Eggs exposed to water with 0.4 to 13 μg/L Aroclor 1254 had 21 to 100% fry mortality.[322] Thyroid hyperplasia has been observed in Great Lakes salmonids, and some attention has been given to this response in laboratory testing with PCBs. However, other than higher thyroid activity, no harmful effects have yet to be documented.[303,309,310]

2,3,4,7,8-pentachlorodibenzofuran, TCDD, and 2,3,7,8-tetrachlorodibenzofuran as well as ß-naphthoflavone inhibited vitellogenin synthesis in 17ß-estradiol-treated rainbow trout liver cells.[331] Potency of inhibition was directly related to strength as an inducer of CYP1A1 protein. PCB congeners 77, 126, and 156 did not inhibit vitellogenin synthesis and induced no levels or only moderate levels of CYP1A1 protein or EROD activity at molar concentrations that did produce vitellogenin effects from the tested furans and dioxin.

The pineal hormone melatonin, an important regulator of endocrine function and circadian rhythms in vertebrates, was investigated in rainbow trout primary hepatocytes. Treatment with 2,3,7,8-TCDD increased the major oxidative metabolite, 6-hydroxymelatonin, about 2.5-fold after 24 h and 1.2-fold after 48 h exposure, relative to the control cultures.[332]

Schmitt and co-workers[333] provided a description of the complexity of relationships in the environment for fish exposures to multiple contaminants, including PCBs, PCDDs, and PCDFs from various regions of the United States, specifically the Mississippi, Columbia, and Rio Grande River watersheds. The yolk protein, vitellogenin, occurrence in plasma and ovarian cells in the testes of male fish from several sites, along with abnormal ratios of sex steroids, suggesting that fish from some sites are exposed to endocrine-modulating substances. Because of the antiestrogenic activity of dioxin-like PCBs, PCDDs, and PCDFs, it is difficult for these compounds to be clearly implicated in male fish vitellogenin induction (a hallmark of an estrogenic substance). However, the abnormal sex steroid ratios might be implicated. Hydroxylated PCBs, which are not dioxin-like, may be important in vitellogenin induction and feminization of male fish; they require further male-specific investigation.

Aroclor 1254 disrupted reproductive/endocrine function in Atlantic croaker (*Micropogonias undulatus*) when they were fed a dietary concentration of 0.5 mg/100g body weight for 17 days. Ovarian growth was impaired, and vitellogenesis decline was associated with a decline in plasma estradiol. *In vitro* spontaneous secretion of gonadotropin from pituitary glands of PCB-treated fish was also demonstrated.[334] More recent work in this species, again with exposure to Aroclor 1254 (dietary dose 1 μg/g for 30 days), demonstrated reduced serotonin, dopamine, testosterone and 11-ketotestosterone, testicular growth inhibition, and reduced testicular gonadal somatic index.[335,336]

Endocrine stress response and abnormal development were demonstrated in carp that were water-exposed to PCB 126.[337] At $10^{-10}$ and $10^{-11}$ mol $L^{-1}$, PCB 126 altered the levels of stress hormones, andrenocorticotropic hormone (ACTH), α-melanocyte-stimulating hormone (MSH), and cortisol in whole-body measurements. Swelling of the yolk sac and pericardium occurred, indicating that edema was induced, as well as PCB-stimulated body pigmentation thought to be mediated by elevated α-MSH at all PCB 126 test concentrations.

TCDD, PCB 126, β-napthoflavone, benzo(a)pyrene, and diindolylmethane caused a concentration-dependent suppression of the secretion of vitellogenin, which was related to the response observed with E2-treated cultured female carp hepatocytes. The order of potency for suppression of vitellogenin secretion was comparable to the order of potency for CYP1A induction. These investigators suggested that the antiestrogenicity of these compounds, while AhR-mediated does not involve CYP1A. This could be relevant for feral fish populations, as they are frequently exposed to Ah-receptor agonists at levels at which AhR-mediated effects are observed.[338]

Study of multiple exposures exemplifies the difficulty in establishing cause-and-effect relationships without complementary laboratory and combined field/laboratory study designs. The joint toxicity of tributyl tin (TBT) and PCB in the diets of Japanese medaka demonstrated additive effects on spawning success but antagonistic effects on embryological success of eggs and swim-up success of larvae.[339]

In a study of zebrafish,[340] 17ß-estradiol; 2,3,3',4,4',5,6-HpCB (PCB-190); 2,3,4,4'-TeCB (PCB-60); and 2,2',4,6,6'-PeCB (PCB-104), which are PCBs considered to be nondioxin-like, increased embryo and larval mortality. The most severe effects on viability were observed following treatment with 17ß-estradiol or the weakly estrogenic PCB-104. Delayed embryo development and hatching were seen. PCB-190 showed only moderate effects on early-life-stage mortality. The fish were reared until sexual maturation, after which they were subjected to gross morphological and histological analyses. Changes in morphology were observed following PCB-104 and PCB-190 treatment. Both substances produced craniofacial malformations, while PCB-104 induced lordosis in females and scoliosis in fish of both sexes. From histological analysis it was found that PCB-104 and 17ß-estradiol resulted in karyorrhexis and karyolysis in the kidney. Ovaries showed atresia and limited failure of testicular spermatogenesis in fish treated with 17ß-estradiol, PCB-104, or PCB-60. This study demonstrated that xenoestrogenic substances are embryotoxic and manifest certain characteristic symptoms in common with the dioxin-like effects, although obviously through a different primary mechanism.

PCB 81 was tested in juvenile rainbow trout and was found to be 0.004-fold as potent as TCDD in the induction of AHH activity.[341] When investigated in medaka embryos in this study, PCB 81 was more toxic than PCB 77, and TEFs of 0.0014 and 0.006 were determined for mortality and for swim bladder inflation, respectively.

Detailed reviews that address PCBs and dioxins have recently covered the effects on the immune, MFO, and endocrine systems in fish.[301,342–346]

### *18.6.4.2 Birds*

#### *18.6.4.2.1 Acute Effects of Poisoning*

As noted by Eisler,[64] birds are more resistant to acutely toxic effects of PCBs than are mammals. $LD_{50}$s ranged from the 60s to > 6000 mg/kg diet for several birds tested in laboratory studies.[62] Little field mortality can be ascribed to PCBs. In southern Ontario, high PCB exposure, as indicated by levels in the brain in excess of 310 μg/g, were believed to have caused mortality in ring-billed gulls in late summer and early autumn 1973.[347] This mortality may also have been caused by high PCDF and PCDD contamination, which is well documented in the Lake Ontario ecosystem.

Birds that forage within aquatic-based foodchains, such as gulls, terns, herons, and mergansers, can accumulate relatively high concentrations of PCBs and PCDD/PCDFs. They differ from true terrestrial species, such as passerine and gallinaceous birds, which generally have lower concentrations of PCBs and PCDD/PCDFs. However, for eagles, falcons, hawks, and owls, which are at the top of their respective food chains, there is a potential for high exposure due to their high biomagnification potentials (i.e., 10,000 and higher) and their unique feeding habits. Those individual species that possess feeding strategies in which they hunt other birds (or their eggs) and mammals — which already may be exposed and also have biomagnified these contaminants — will have the highest exposures and will possibly move past an effects threshold. Additionally, for birds that are scavenger feeders, e.g., foragers at municipal dumps or dead carcasses of seals, the potential for exposure is high, though dependent on the frequency of such behavior.

*Laboratory Species* — There are considerable data for PCB toxicity and more limited data for PCDD toxicity to chickens. This species appears to be unusually sensitive to PCBs and dioxin-like compounds in general. The term "chick edema disease" was first used during investigative studies of unintentional poisonings in young chickens (reviewed in Firestone[348]). Chick edema disease was caused by 1,2,3,7,8,9-HpCDD, which was discovered in chicken feed that was contaminated with a "toxic fat" food supplement. The toxic fat was linked to the degreasing of cattle hides with pentachlorophenol, the ultimate source. More interpretive PCB and PCDD studies have been conducted with chicken (*Gallus gallus*) than with any other bird species.

*Pen and Laboratory Studies* — Adult and juvenile birds of precocial species exhibit varying sensitivities to ingestion of Aroclor and PCB mixtures. In 5-day feeding trials, dietary median lethal concentrations (8-day µg/g-$LC_{50}$s) were 604 with Aroclor 1254 for northern bobwhite (*Colinus virginianus*), 1091 for ring-necked pheasants (*Phasianus colchicus*), 2697 for mallards (*Anas platyrhynchos*), and 2895 for Japanese quail (*Coturnix japonica*).[349] In a related publication on this same Aroclor toxicity study, Heath and co-workers[279] documented the complete absence of fat and the presence of pericardial edema in lethally poisoned bobwhite chicks — characteristics of chick edema disease. This early study also tested the joint lethality of Aroclor 1254 and p,p'-DDE to Japanese quail chicks, using three different but wide-ranging ratios; it found that all combinations "proved to be essentially additive" in the joint action of the two compounds. This has implications for causal determinations of reproductive impairment in the wild where multiple contaminant exposures are common and where statistical significance alone, on a chemical-by-chemical basis, may not account for combined causal associations that are biologically significant.

Dahlgren and co-workers[350] administered capsules daily containing 10, 20, or 210 mg of Aroclor 1254 to 11-week-old hen pheasants until death or sacrifice. PCB ingestion, at all levels in a dose-dependent fashion, killed birds, with those receiving the least PCB taking longer to die. The authors concluded that a brain residue level of 300 to 400 µg/g was indicative of death due to PCB toxicosis. Further studies by these authors[351] revealed that periodic food deprivation increased brain residues, leading to more rapid death.

Young cockerels (day-old at start) that died on dosage of 600 µg/g Aroclor 1260 contained 270 to 420 µg/g of PCBs in the brain.[352] Dosage with phenochlor and chlophen (60% chlorine mixtures) resulted in more variable residues, and these two formulations were more toxic than Aroclor 1260 and produced pathological signs. However, both of these formulations proved to be contaminated with chlorinated dibenzofurans.[353] Pathological examination of these chickens revealed severe edema and lesions.

*Wildlife Species* — Studies conducted with altricial species of birds have included observations with cormorants, herons, finches, and blackbirds. Great cormorants (*Phalacrocorax carbo*) dosed experimentally with clophen A 60 died with brain residues of 76–180 µg/g (mean, 130) whereas herons that died on clophen dosage contained 420–445 µg/g, suggesting that cormorants were more sensitive than herons.[354] These authors concluded that survival time in cormorants was related to the capacity of the birds to store the PCBs in adipose and other tissues, apart from brain, indicating that total-body content is not a good criterion by which to assess mortality.

As observed above, lethality in birds appears to be correlated with brain residue levels. Stickel and co-workers[355] conducted one of the most comprehensive studies designed to establish lethal brain residues of PCBs, using Aroclor 1254, in passerine species. These species included immature male common grackles (*Quiscalus quiscula*), immature female red-winged blackbirds (*Agelaius phoeniceus*), adult male brown-headed cowbirds (*Molothrus ater*), and immature female starlings (*Sturnus vulgaris*). PCB residues in brains of birds that died were distinctly higher than those in sacrificed survivors, providing suitable diagnostic criteria. PCB residues varied from 349 to 763 µg/g wet weight in brains of dead birds and from 54 to 301 µg/g in sacrificed birds. The authors considered a threshold of 310 µg/g to be diagnostic of PCB-induced mortality. PCBs in brains of dead birds for three species did not differ significantly from one another, but residues in starlings (mean 439 μg/g) averaged significantly lower than for red-winged blackbirds and grackles.

#### *18.6.4.2.2 Chronic Effects of Poisoning*

*Egg Injection Studies* — Most chronic-effects studies have focused on the reproductive outcomes of embryonic exposure to PCBs and PCDD/PCDFs. Egg-injection studies have been successfully utilized to mimic natural deposition of these lipophilic contaminants from a laying hen, either by introduction into the air sac or direct yolk-sac injection. Additionally, several methods have been used to observe the reproductive effects of chronic adult exposure, including dietary, water, and injected exposures of the hen to the toxicant, as well as allowing for natural deposition into the egg, in which yolk is the toxicant reservoir. As in other experiments, it appears that (1) the longer the exposure, the greater the effect, and that (2) the earlier in development the compounds are administered (day-zero exposure in an egg), the greater the toxic effect. This makes it difficult to separate out general chronic toxicity from that mediated indirectly through endocrine system disruption. All of the above effects appear to be related for these (PCBs and PCDD/PCDF) compounds.

*Laboratory Species* — In a white leghorn study, hens received diets containing 0 to 80 μg/g Aroclor 1242 for 6 weeks.[356] Egg production, egg weight, shell thickness, and shell weight were not affected, but hatchability was affected within 2 weeks for hens fed as little as 20 μg/g dry weight. Yolks contained 2.4 μg/g (expected whole-egg concentration of 0.87 μg/g). This conversion is based on the chicken yolk comprising 0.364 of the whole-egg content mass.[357] Even 10 μg/g dry weight in the diet caused a small reduction in hatching at the end of 6 weeks (yolk concentration of 3.7 μg/g, or expected whole-egg concentration of 1.3 μg/g). Scott[358] reported that hatchability was decreased after 4 weeks of Aroclor 1248 at 10 μg/g dry weight in the diet, with egg residues of 22.7 μg/g.

Numerous egg-injection studies have been conducted with PCBs in the chicken egg. Blazak and Marcum[359] injected Aroclor 1242 into the air cell of fertile white leghorn eggs. Both 10 and 20 μg/g caused 64 to 67% mortality but did not result in chromosomal breakage, as had been reported in ringed-turtle doves by Peakall and co-workers.[360] Srebocan and co-workers[361] injected Aroclor 1254 into chicken egg air cells and noted decreased activity in key gluconeogenic enzymes. Brunstrom and Orberg[362] injected white leghorn eggs, after 4 days of incubation, into the yolk sacs to attain egg concentrations of 0, 1, 5, and 25 μg/g of Aroclor 1248. Hatchability was 96, 92, 17, and 0%, respectively, with mortality occurring the earliest in the 25 μg/g group (days 6–12). A great number of egg-injection studies have been conducted with specific PCB congeners studying the effects in chickens and other species. Rifkind and co-workers[363] injected chicken eggs at 10 days of incubation with 5 to 1000 nmol/egg for each of three PCB congeners (77, 169, and 2,2',3,3',6,',6'-HCB (PCB 136). PCB 77 caused dose-related decreases in survival from day 10 through 19 days of exposure at 100 to 1000 nmol/egg, and PCB 169 at 500 to 1000 nmol/egg. These decreases in survival were accompanied by decreased thymus weight and increased pericardial and subcutaneous edema in surviving embryos. Another study by these authors revealed that hepatocyte swelling, the major histopathologic change, was apparent within 24 h, after doses as low as 5 nmol/egg for PCB 77.

*Wildlife Species* — Effects of PCB congeners 126 and 77 in chickens, common terns (*Sterna hirundo*), and American kestrels (*Falco sparverius*) were studied through hatching, following air-cell injections on day 4 of incubation.[364] In the chicken, kestrel, and common tern, the $LD_{50}$s for PCB 126 were, respectively, ~0.4 $ng/g_{egg}$, 65 $ng/g_{egg}$, and 104 $ng/g_{egg}$. This study also demonstrated that PCB 77 was about fivefold less toxic than PCB 126 in the chicken model ($LD_{50}$ 2.6 $ng/g_{egg}$). In the kestrel, the $LD_{50}$ for PCB 77 was 316 $ng/g_{egg}$. In common terns, deformities occurred at egg concentrations at which hatching success was affected, whereas malformations occurred below those concentrations affecting hatching success in the chicken and kestrel. Hatching success, malformations, and edema in this study were consistent with the reproductive problems identified in other recent laboratory studies of chickens[288,365] as well as Great Lakes species (Forster's tern, common tern, Caspian tern, double-crested cormorant, bald eagle, and herring gull).[366–369]

Probably the most important finding of the study was on kestrels, considered as a surrogate species for interpreting effects on the bald eagle in the Great Lakes and elsewhere. Eagle eggs from New Jersey[370] and the Great Lakes,[371,372] areas where reproductive impairment including malformations[373] have been documented, contained elevated PCB 126 concentrations at 2–4 ng/g egg to 42 ng/g egg fresh wet weight. PCB 126 is always the largest contributor to TCDD TEQs in birds having significant PCB exposure. In kestrels, malformations begin occurring at 0.23 $ng/g_{egg}$, combined malformations and edema are statistically different from controls at 2.3 $ng/g_{egg}$, and malformations alone are statistically different at 23 $ng/g_{egg}$. These findings present a plausible explanation for the occurrence of bill defects in Great Lakes eaglet nestlings. Growth retardation indicators of toxicity measured in exposed kestrels (hatch weight, liver weight, and femur length) were statistically significant at 0.23 $ng/g_{egg}$. Although these effects have not been studied directly in bald eagles because of their protected status, it seems reasonable to expect growth retardation in more highly exposed individual hatchlings, leading to reduced productivity and reduced survival from other chronic stressors. Kestrel chick developmental toxicity was shown to be lower than for embryos and higher than for adults.[374]

In an exposure study using carp from Saginaw Bay, Lake Huron as a weathered, environmentally relevant source, white leghorn hens were administered between 0.3 (control (0%) carp) and 6.6 (high-dose [35%] carp) mg/kg/day total PCBs, yielding time- and dose-dependent reproductive toxicity.[375] Embryos displayed typical embryo mortality and reduced hatchability, as well as classic deformities including head, neck, and abdominal edema, hemorrhaging, and foot, leg, skull, brain, and yolk-sac deformities. Organ weights were significantly affected including livers, spleens, and bursas. Hatched chicks showed significant effects on body, brain, liver, heart, and bursa weights. These effects are all consistent with the establishment of Koch's fifth postulate, which requires the hypothesized putative factor to be introduced into the test organism and to reproduce symptomatic adverse effects. In this case, the sensitive model species was utilized as a surrogate for fish-eating birds that have demonstrated historical symptoms of PCB and dioxin-like reproductive toxicity.[366–368]

Comparative avian egg-injection studies by Brunstrom and co-workers have shown that chickens are more sensitive than turkeys (*Meleagris gallopavo*), pheasants, ducks (mallards and goldeneyes (*Bucephala clangula*), domestic geese (*Anser anser*), herring gulls, and black-headed gulls (*Larus ridibundus*).[376]

Reproductive impairment has been reported in at least five species of birds that were experimentally dosed with PCBs including chickens, ringed-turtle doves (*Streptopelia risoria*), Japanese quail, mourning doves (*Zenaida macroura*), and ring-necked pheasants. Egg production was reduced with 5 μg/g in the diet, but hatchability was unaffected. However, after 14 weeks, fertility was lower. Hatchability of fertile eggs was only affected when the PCB concentration exceeded 15 μg/g in eggs; at such levels, embryonic mortality was high and occurred at earlier stages. In contrast, Cecil and co-workers[377] reported that dietary levels of 20 μg/g for Aroclor 1254 did not affect hatching success of chickens at the end of 5 weeks with egg residues of 13.2 μg/g. Tumasonis and co-workers[378] exposed white leghorn hens for 6 weeks to 50 mg/L Aroclor 1254 in drinking water. Within 2 weeks, hatching success dropped to 34%. The authors concluded that yolk concentrations > 10–15 μg/g (whole egg concentrations above 4 μg/g) were required to cause this effect. Peakall and Peakall[379] found that 10 μg/g of Aroclor 1254 caused increases in embryonic mortality of ringed-turtle doves; this effect was greatly increased when the eggs were incubated by the parents, but it was decreased by artificial incubation. Monitoring of egg temperatures suggested that mortality increased with decreased parental attentiveness. Further studies with Aroclor 1254 in this species revealed depletions of brain dopamine and norepinephrine that were inversely correlated with brain residues.[380] Other species appear to be less reproductively sensitive to PCB exposure.

Neither Aroclor 1254 nor Aroclor 1242 affected reproductive success of mallards. Other tests showed screech owls (*Otus asio*) and Atlantic puffins (*Fratercula arctica*) to be unresponsive.

In a more recent iteration of the determination of relative species sensitivity to PCB 126,[365] a comparative investigation of yolk-sac-injected eggs demonstrated marked differences in relative embryo mortality sensitivity in the following rank order (most sensitive to least): white leghorn, turkey, common tern, and Japanese quail. For the common tern, the results are consistent with findings by Hoffman and co-workers,[364] although the two studies used different dosage methods. The bobwhite is much more sensitive to PCB 126, based on the reported $LD_{50}$ of 24 $ng/g_{egg}$,[364] than is the Japanese quail, with a no-effect concentration on mortality of 240 $ng/g_{egg}$.[365] For the former, the divergence is consistent with early feeding studies of these species in which bobwhites were three to four times more sensitive in dietary acute exposures with various Aroclors.[279] In contrast, the piscivorous double-crested cormorant is intermediate in sensitivity, with an $LD_{50egg}$ for PCB 126 of 158 ng/g.[381]

Among the various bird species that have been tested with 2,3,7,8-TCDD, the ring-necked pheasant was the most sensitive besides the chicken, with egg injections of 2.1 ng/g producing $LD_{50's}$ and 25 ng/g i.p. for 80% mortality in the adult hen after 77 days. Other studies[382] with ring-necked pheasant determined that the delayed onset of mortality occurred at a single injection of 25 μg/kg. With repeated injections, the cumulative level of 10 μg/kg over 10 weeks caused similar symptoms. Egg-injection studies of ring-necked pheasant eggs demonstrated $LD_{50}$s of 1.35 and 2.2 pg/g, respectively, for albumen and yolk injections..[383] Bobwhite were sensitive to an $LD_{50}$ dose of 15 ng/g in their feed,[384] but mallards and ringed-turtle doves showed no response at concentrations greater than 108 and 810 ng/g of feed (single dose), respectively.

Among the most compelling case histories available to document the reproductive problems caused by PCBs is for Green Bay Forster's terns (*Sterna fosteri*) in the 1980s. Egg-intrinsic and adult behavioral nest-attentiveness problems lead to a variety of measurement endpoints, including wasting syndrome, as documented in a followup[385] to the original study, and provide convincing ecoepidemiological evidence for total PCBs as well as TCDD equivalents as the putative causative factors, as measured in study eggs.[386–390] The above findings did not determine DDT/DDE to be important to Forster's tern reproductive impairment. Additional work on this species, subsequently conducted in Texas, showed that total PCB egg residues were low and that DDT/DDE residues were greater than those at Green Bay; however, reproductive success was normal.[391] This study represents an independent confirmation of the importance of the PCBs in Green Bay, conducted by other investigators in a different time and ecosystem.

Wasting syndrome has been documented to various degrees in fish-eating colonial waterbirds, in at least three Great Lakes species investigations, for Forster's terns,[385] Caspian terns,[392] and herring gulls.[393] A comprehensive review by Hoffman and co-workers[368] should be consulted for historical toxicity information from both the laboratory and field study of birds with these compounds.

#### *18.6.4.2.3 Endocrine and Mixed-Function Oxidase Effects*

*Histological and Biochemical* — Comparison of the toxicities of PCB congeners injected into the yolk sac of chicken eggs at an early stage of development (4 days of incubation) revealed that PCB 126 was the most toxic and also the most potent inducer of MFO (EROD) in chick embryo liver.[394] The EROD-inducing potencies correlated well with the embryolethalities. When injections were administered at the earlier stage of incubation (day 4) via the yolk sac and eggs were incubated until day 18, the congeners were even more toxic. Non*ortho* chlorinated congeners were more potent inhibitors than were *mono-ortho*-chlorinated congeners of lymphoid development in the embryonic bursa of *Fabricius*.[395] The most immunotoxic of the *mono-ortho*-chlorinated analogs of PCB 77 and PCB 126 were about 1000 times less potent than PCB 126.

### *18.6.4.2.4 Other Effects*

Specific congeners have been tested as inducers of porphyria in Japanese quail and American kestrels.[396,397] Liver residue levels for PCB 105 of 2.6 ± 1.7 mg/kg, and for PCB 126 of only 0.091 ±0.08 mg/kg were found to cause porphyria after 2 weeks in Japanese quail.[398] PCB 153 at 52.6 mg/kg in the liver had minimal effects. PCB 126 caused a decrease in thymus weight. All three congeners induced MFO activity as detected by the EROD assay. In American kestrels, residue levels in pooled adipose tissue of 182 mg/kg for PCB 105, of 119 mg/kg for PCB 153, and of 3.3 μg/kg for PCB 126 did not cause porphyria but did induce hepatic MFO activities.[397]

Studies have examined the effects of specific PCB congeners on growth. The effects of feeding 400 μg/g of five hexachlorobiphenyl congeners, including 2,2',4,4',6,6'-HCB (PCB 155), 2,2',3,3',6,6'-HCB (PCB 136), 2,2',4,4',5,5'-HCB (PCB 153), 2,2',3,3',4,4'-HCB (PCB 128), and 3,3',4,4',5,5'-HCB (PCB 169), were examined in growing chicks for 21 days.[398] PCB 169 was the most toxic, causing complete mortality, and it produced the highest accumulation (mean of 203 μg/g) of all congeners in the liver. Growth was reduced by all congeners except PCB 155. Liver-to-body weight ratio was increased by all congeners, with the largest increase produced by PCB 155 and the smallest by PCB 136. Another study examined the effects of PCB congeners 126, 77, and 105 on growth and development of American kestrel nestlings.[374] Dosing with 50 ng/g resulted in pronounced liver enlargement and some mild coagulative necrosis of the liver. Increasing the dose to 250 ng/g resulted in intensification of the above effects as well as lesions of the thyroid, decreased spleen weight, and lymphoid depletion of the spleen and bursa.

In testing of PCB 126 in primary immune organ development,[399] thymus mass decreased at doses between 0.13 and 0.32 $ng/g_{egg}$ and thymic lymphoid cell numbers dropped sharply at doses between 0.051 and 0.13 $ng/g_{egg}$ when injected into chicken egg air cells. Similarly, bursa mass began to decrease at the lowest dose ($0.051 ng/g_{egg}$). The number of viable cells decreased across the dose range and reached a minimum at 0.13 $ng/g_{egg}$. The study concluded that (1) lymphoid cell numbers were more sensitive than organ mass, (2) bursa was more sensitive than the thymus, and (3) the doses to reduce the quantity of viable cells were at least an order of magnitude lower with full-term embryo exposure (day 0–7 injection before incubation, and day-20 egg opening, in a normal 21-day incubation) than with late-stage exposure only. In a related study,[400] thymocyte phenotype $TCR\cdot\beta^+$ decreased with dose as a function of the declines in $TCR\cdot\beta^+$ percentages and total thymocyte numbers. The number of viable lymphoid cells in the bursa decreased to 45% lower than in controls at 0.13 $ng/g_{egg}$ and fell to 76% lower than the controls at 0.8 $ng/g_{egg}$. The authors also calculated that these immunological effects were comparable to those found in Great Lakes herring gull eggs, after correction for interspecies differences in sensitivity to PCB 126. Other related studies on Saginaw Bay Caspian terns have suggested a biologically significant effect on immune function by PHA (phytohemagglutinin), which measures T lymphocyte response.[401] PCB concentrations correlated with reduced immune function to a greater extent than did those of DDE, although both types of compounds were significantly correlated in plasma and eggs. These authors also observed that these and other immune-function tests[393,402] suggest an association with the reduced recruitment of young terns into the contaminated Saginaw Bay ecosystem that has been characterized through studies of breeding reproduction and population dynamics.[403]

*Environmental Extract Egg-Injection Studies and Vitamin D* — In another approach to understanding relative sensitivity, extracts obtained from Great Lakes double-crested cormorants were injected into chicken eggs. One egg-equivalent of cormorant extract injected into chicken eggs produced 77% embryo mortality.[404] Yet at these levels, no malformations occurred, which is unusual, whereas body weight and relative bursa weights were reduced, while relative brain weights were increased.

Egg-injection studies with extracts from wild bird eggs have failed to produce statistically significant increases in deformities in cormorants,[381] chickens,[404] or herring gulls,[405] suggesting that another factor or factors (such as vitamin D deficiency[306]) may be altered during egg development and that this cannot be duplicated in "clean" test eggs. One possibility that has received little attention in ecotoxicological or laboratory studies is that PCBs may alter deposition, in either form or amount, of vitamin D in egg yolks. Vitamin D is lipid-soluble and is critical to proper bone development. Vitamin D is twice hydroxylated, once by hepatic 25-hydroxylase and a second time by renal hydroxylase, to produce the hormone-like activity of 1α,25-dihydroxycholecalciferol that is used for calcium regulation and bone deposition. Additionally in the Great Lakes, where bill defects in cormorant nestlings[406] and dead embryos[407] are relatively common, altered vitamin D deposition into the yolk could be influenced by exposure of cormorant hens to dioxin-like congeners prior to egg laying. This could be a cofactor in the incidence and prevalence of bill defects, along with dioxin-like congener exposure *in ovo*. Under the combined influence of altered enzyme activity and *in ovo* dioxin-like congener contamination, there could be a deficiency, excess, or altered availability of 1α,25-dihydroxycholecalciferol that could be related to bill and skeletal defects or even embryonic mortality.[408]

Embryonic exposure, together with effects in the environment, could result in an enhanced sensitivity of embryos that are simultaneously deficient in available vitamin D and that have altered egg proteins that are associated with reduced vitellogenin synthesis caused by antiestrogenic, dioxin-like congeners. Ascertaining whether such a relationship exists would help to resolve outstanding inconsistencies between extract-based egg injection studies and adult feeding protocols.[375] Investigations along these lines may also help explain the puzzling observation that cormorants exposed to low PCB levels developed bill malformations only after hatching and in conjunction with vitamin D-deficient diets (frozen fish) and sunlight deprivation (resulting from indoor zoological captivity).[306]

Several lines of evidence suggest that during embryonic development, vitamin D, calcium requirements, and vitellogenin formation can be linked to embryotoxic and teratogenic effects caused by certain dioxin-like compounds. Basic bird research on vitamin D has demonstrated that 1α,25-dihydroxycholecalciferol-deficient Japanese quail embryos die of calcium deficiency at day 15 of development when hens produce vitamin D-deficient eggs. Single doses of 125 ng cholecalciferol, 600 ng 24,25-dihydrocholecalciferol, or 100 ng 1α,25-dihydroxycholecalciferol increased hatchability when injected into eggs prior to incubation, but they were toxic when injected into eggs on days 11 and 12. *In vitro* formation of vitellogenin by liver slices of vitamin D-deficient hens is reduced to less than 30% of that occurring in vitamin D-sufficient controls, and vitamin D-deficient hens have lowered Ca2+ in blood plasma.[409] It is not known whether any yolk-related proteins or binding proteins are altered in vitamin D-deficient adult birds. In a review of the primary literature on comparative avian nutrition,[410] Klassing's discussion of vitamins A and D, among others, provides additional evidence for a potential relationship between embryotoxic and teratogenic effects that are consistent with certain dioxin-like effects. For instance, vitamin D-deficient embryos can develop deformed mandibles and beaks and problems pipping the shell due to overall weakness, and they can have poor bone calcification.

*Summary Information* — Besides the studies addressed above, recommended resources on the following selected topics are: MFO effects in birds,[300,411,412] and in-depth development and documentation on TEFs (www.who.nl).

#### 18.6.4.3 Mammals

##### 18.6.4.3.1 Acute Signs of Poisoning

The hallmark responses of MFO induction — wasting and edema — are common to most tested species. In addition, there are species-specific signs of acute toxicity that have been documented. In the monkey, alopecia and swollen eyelids have been demonstrated. In mink, lethargy, anorexia, hemorrhaging, and hemorrhagic gastric ulcers are symptomatic, along with a dose-dependent decrease in feed consumption with corresponding body weight loss. Gross necropsy revealed mottling and discoloration of the liver, spleen, and kidneys. In mink that are exposed to high doses of TCDD, brain, kidneys, heart, and thyroid and adrenal glands all become enlarged relative to the organism's body weight.[280] Fatty infiltration of the liver may or may not occur, depending on the length of the acute exposure. In the European ferret, which is less sensitive than the mink, claws become markedly curled, with an appearance similar to deformed bills in birds.[413] Similarly, hooves of horses that curled upward was a documented effect observed in horses exposed to horse-arena soil that was contaminated by TCDD from a 2,4,5-T manufacturing facility. The episodic horse exposure is not common in the general environment, whereas acute exposures of mink to TCDD and dioxin-like chemicals in contaminated ecosystems are still possible.

#### *18.6.4.3.2 Chronic Signs of Poisoning*

*General Indicators* — Skin lesions are generally observed in various species. The specific class of compounds first ascribed to causing this response was the chlorinated naphthalenes that caused the hyperkeratotic disease of cattle in the United States labeled "X-disease." Later an "X-disease-like symptom" was associated with chlorinated dioxins and chlorinated dibenzofurans;[10] PBBs caused skin lesions like the X-disease in cattle in Michigan,[414] and Old World monkeys developed chloracne and a thickened condition of the eyelids.[415] McConnell[10] listed the following chronic features associated with poisoning by these agents. Chloracne-like lesions are not usually observed in rodents. Internal lesions are quite common with these classes of compounds; such lesions vary with duration of exposure and depend especially on the age of the organism being tested. In all experiments involving immature animals, the size and weight of the thymus are markedly less than in the controls. Chronic effects in blood are quite variable; however, the one common component observed in most long-term studies is a dose-time-related anemia that is postulated to be due to the toxic effects on erythropoiesis from all of these types of compounds. The carcinogenic potential of PCBs and dioxins has been studied, but information on dibenzofurans is not available. Dioxins, specifically TCDD and HxCDD, have been shown to be carcinogenic in rats and mice by several investigators. The reproductive effects from intoxication with PCBs and dioxins have been studied in depth.

The first reported reproductive dysfunction that was suggested to be a result of intoxication with the general class of compounds to which PCBs, dioxins, and dibenzofurans belong was observed in a study of women exposed during 1968 to PCBs in Yusho, Japan. These women exhibited irregularities in their menstrual cycles. As stated earlier, the likely components causing these problems were dibenzofurans and dioxin impurities in the PCBs. TCDD has been reported to affect the testes and accessory male reproductive organs. TCDD also adversely affects spermatogenesis and testosterone levels.[416] Many reports of TCDD developmental toxicity are available. TCDD causes cleft palate in mice and other manifestations of developmental toxicity in other species. Furans also induce cleft palate and hydronephrosis in responsive mice. Additional studies of the effects of TCDD on palate and kidney development have been reported. There are considerable differences in toxicity among members of the chemical classes that make up the dioxins and the furan series of compounds.

Metabolism and excretion are assumed to be important modes of detoxification of these chemicals. Exposure of target tissues is to a great extent closely related to chemical solubility and lipophilicity of the particular compounds. Solubility tends to decrease with increasing degree of halogenation, as do the extent and rate of metabolism (and therefore elimination). Thus, highly halogenated members of the groups tend to be more persistent and to have greater potential for bioaccumulation. Those halogenated dioxins, and furans with adjacent unsubstituted carbon atoms, are more readily metabolized than are those without this feature. The symmetrical tetrachlorinated dibenzo-*p*-dioxins and dibenzofurans have no adjacent unsubstituted carbon atoms and are the most toxic members of their respective classes, including developmental toxicity.

Regarding relative species and strain sensitivities, clear differences among species in susceptibility to developmental toxicity have been reported that are relatable to inherent genetic susceptibility, differences in tissue disposition of the chemical, and perhaps other factors that are currently not understood or appreciated. In general, adverse reproductive effects have been caused in multiple species by 2,3,7,8-substituted dioxins. Greater species differences are noted for developmental toxicity than for adverse reproductive effects. Mice are generally more likely to demonstrate malformations, with or without effects on size of pups or litters. In contrast, rats are more likely to have resorption or decreased fetal weight than malformations. Nonhuman primates are more likely to show an effect on fertility and *in utero* survival than abnormalities of structural development. Observations on humans exposed accidentally to dioxins suggest that humans are not more sensitive to adverse reproductive developmental effects than laboratory animals.[416] Additionally, the recent WHO reevaluation of the "tolerable daily intake" for humans estimated the body burdens of laboratory animals at their respective lowest observed adverse effect levels (LOAELs). Those body burdens were transformed into estimated daily human intakes that on a chronic basis would be expected to lead to similar body burdens in humans. The reevaluation noted that "the estimated human daily intakes are related to the body burdens in animals where adverse effects have been reported."[271]

*Mink* — Hochstein and co-workers[417] determined the $LC_{50}$ dietary concentration in adult female mink fed TCDD for 28 or 125 days. Based on feed consumption of control mink, $LC_{50}$ concentrations approximated 0.264 and 0.047 μg TCDD/kg body weight/day, respectively, for the two lengths of dosing.

In adult female mink exposed to 5 ng/g TCDD for 6 months, the maxilla and mandible were grossly unremarkable, but histologically they had nests of squamous epithelium within the periodontal ligament. There was osteolysis of the adjacent alveolar bone.[418] In a study of 12-week-old male mink fed 24 ng/g PCB 126 for 31–69 days, similar symptoms developed, including thickened gingiva and loose teeth. Squamous cells proliferated into bone and caused jaw osteoporosis.[419] Rhesus macaques (*Macaca mulatta*) also showed these symptoms.[420] This study was one of the first to address equitoxic doses of congeners and Aroclor mixtures that produced similar effects at different doses, including Aroclors 1242, 1248, PCB 77, PCB126, TCDD, 2,3,7,8-TCDF, and 2,3,4,6,7,8-hexachlorodibenzo-*p*-dioxin. The importance of 1,25-dihydroxy-vitamin D3 in the control of tooth development was linked to vitamin D deficiency in rats; it was concluded that the tooth is a target organ for 1,25-dihydroxyvitamin D3.[421] Calcitonin, estrogen, and parathyroid hormone all interact with vitamin D in the control and deposition of bone and for maintaining tooth health. Aroclor 1254 resulted in smaller body weight and increased urinary alkaline phosphatase and lactate dehydrogenase, serum calcium, nephrotoxicity, and abnormal bone morphometry in young, growing male Fisher 344 rats, but serum PTH, phosphorus, alkaline phosphatase, LDH, and 1,25-dihydroxy vitamin D3 were unchanged at any dose.[422] The influence of the dioxin-like congeners on MFO induction/inhibition, and availability of the above hormones, as well as their effect on carrier and other proteins that direct calcium metabolism and toxic effect in general need further elucidation.

A study of multigenerational effects was conducted on mink in which they consumed PCB-contaminated carp from Saginaw Bay, Lake Huron.[423] The adult mink were fed diets formulated to provide 0 (control), 0.25, 0.5, or 1.0 μg/g PCBs through substitution of Saginaw Bay carp for ocean fish in the diet. Continuous F1 exposures to = 0.25 μg/g of PCBs delayed the onset of estrus (as determined by vulvar swelling and time of mating) and lessened the whelping rate. Litters whelped by females continuously exposed to = 0.5 μg/g of PCBs had greater mortality and lower body weights than did controls. Continuous exposure to 1.0 μg/g PCBs had a variable effect on serum thyroxin (T4) and triiodothyronine (T3) concentrations. There were significant differences in kidney, liver, brain, spleen, heart, and thyroid gland weights of the mink continuously exposed to 1.0 μg/g PCBs.

Plasma and liver PCB concentrations of the adult and kit mink were, in general, directly related to the dietary concentration of PCBs and the duration and time of exposure. Short-term parental exposure to PCBs had detrimental effects on survival of subsequent generations of mink, conceived months after the parents were placed on "clean" feed. The LOAEL for use of an environmentally weathered PCB diet in this study was 0.25 μg/g. Using a similar Saginaw Bay carp feeding protocol in a one-generation test, Tillitt and co-workers[262] estimated a total TCDD toxicity equivalent concentration threshold dietary dose of 1.9 pg/g and 60 $pg/g_{liver}$ for adult mink that had been exposed and chemically analyzed. According to earlier WHO mammalian TEFs,[258,259] PCBs contributed approximately 76 to 82% of dietary total TCDD equivalents, while in livers of adult mink, PCBs accounted for 61 to 65% of total TCDD equivalents. Since absorbed concentrations are more relevant in assessment of target organ and reproductive effects, liver concentrations of the latter are more ecotoxicologically important. Rank-ordered contributors to total TCDD equivalent concentrations, from highest to lowest, were PCB126, 2,3,7,8-TCDD, 2,3,7,8-TCDF, 2,3,4,7,8-PeCDF, and PCB 118.

Across dosage groups, PCB 126 was at least fivefold more potent as a contributor to total TEQ concentrations in mink liver than either 2,3,7,8-TCDD or 2,3,7,8-TCDF. These relative importance rankings will vary because ecosystems, on a site-specific scale, have been polluted with different congener ratios from different sources. The new WHO TEFs[257] would increase the overall importance of the PCDDs in this study slightly due to the change of TEF for 1,2,3,7,8-PCDD from 0.5 to 1.0. Chronic effects in adult mink dosed in this study are addressed in a companion paper[424] and are consistent with previous work using known quantities of commercial PCB mixtures. In a study of the risk to Great Lakes fish contaminated with PCBs, virtually every fish species potentially consumed by mink offered some measure of risk.[425]

*Other Mammals* — Common seal populations in the Wadden Sea in The Netherlands appear to have declined due to PCBs.[426,427] Effects on the uteri of Baltic ringed seals have been noted when levels of DDT and PCB are higher than in normal females. The same effect was observed in Baltic grey seals (*Halichaerus grypus*) when concentrations of PCB were > 70 μg/g in blubber fat, and they were not fertile at these concentrations. High PCDD and PCDF levels were also correlated with the high PCBs in these species, suggesting that these compounds may actually be responsible for these reproductive problems.

It has been suggested that PCBs may affect bats.[428,429] Reinhold and associates[430] studied the pond bat (*Myotis dasycneme*) and its major food source, chironomids. They found that PCBs measured in chironomids did not exceed "safe" dietary levels for mammals. Nonetheless, some bats had PCB concentrations of 9, 33, or 76 mg kg-1 lipid weight, which exceed the concentration levels that cause reproductive effects in mink. Further work on bats appear mandatory since several bat species are federally listed species in the United States.

For additional information on the exposure to and the effects of PCBs, PCDDs, and PCDFs on aquatic and marine mammals, some excellent recent reviews can be consulted.[431–433] Another recent review provided analysis of tissue and dietary thresholds for various aquatic mammals.[434] Threshold PCB or TCDD-equivalent concentrations in livers of aquatic mammals that elicited physiological effects ranged from 6.6 to 11 μg PCBs/g (geometric mean: 8.7 μg/g) and 160 to 1400 pg TCDD equivalents (geometric mean: 520 pg/g) for lipid weight, respectively. Biomagnification factors for PCBs and TCDD equivalents varied by species of marine mammal. The dietary threshold concentration ranges for these marine mammals were 10 to 150 ng PCBs/g and 1.4 to 1.9 pg TEQs/g wet weight.

#### 18.6.4.4 Endocrine and Mixed Function Oxidase Effects

There is a dearth of information on wild mammals concerning MFO induction, and little more information is available concerning their immune systems. Controlled studies are extremely difficult to conduct, especially for large marine mammals. Concern over the use of larger mammals in testing is also evident from this paucity of information. As with the preceding discussion on chronic effects, key reviews should be consulted to assess what is available.[431–433]

*Seals* — Harbor seals fed fish from the Wadden Sea (high-level PCB contamination) had significantly lower concentrations of plasma retinol, total thyroxin (TT4) and free thyroxin (FT4), and total triiodothyronine (TT3). The PCB-induced reduction in plasma retinol levels disappeared when seals on a diet of Wadden Sea fish were subsequently fed low-PCB-containing Atlantic Ocean fish. The authors suggested that an increased susceptibility to microbial infections, reproductive disorders, and other pathological alterations were related to reproductive disorders and to the lethal viral infections in seals and other marine mammal populations in the Baltic Sea, North Sea, and Wadden Sea. The reduced plasma retinol and thyroid hormone levels were implicated.[435] Subsequently, a comprehensive review of the seal problem and concerns for their protection was produced in 1992, which included a suggestion to perform feeding experiments with mink as a surrogate species for studying the problem.[436]

In an additional report on these harbor seals, impairment of T-cell-mediated immune responses became apparent during the second year on the Wadden Sea/Atlantic Ocean fish diets used earlier and correlated significantly with TCDD equivalent levels in blubber biopsies taken from the seals after 2 years. Humoral immune responses remained largely unaffected. The authors concluded that cellular immune system suppression response in seals fed a third diet of Baltic herring was induced by the chronic exposure to immunotoxic environmental contaminants accumulated through the food chain and could be of crucial importance in the clearance of morbillivirus (canine distemper virus) infections. Environmental-pollution-related immunosuppression may have contributed to the severity and extent of recent morbillivirus-related mass mortalities among marine mammals.[437] More recently, total PCBs in the blubber and liver of harbor seals were correlated with P450, P420, and MFO activity levels.[438]

Harbor seals assessed from southern San Francisco Bay had a mean whole-blood PCB concentration of 50 ng/mL, about three times the average level reported for blood of captive seals fed exclusively on fish from the Baltic's PCB-contaminated Wadden Sea. These findings support concerns about the ecological effects of PCB contamination in San Francisco Bay.[439] Four-week-old harbor seal pups from coastal California had mean PCB concentrations in blubber of 3.3 μg/g on a lipid basis.[440] This study demonstrated that contaminated pups exhibited relatively strong lymphocyte proliferative response to conconavalin A and moderate responses to poke weed mitogen and phytohemagglutinin. Serum thyroid hormone levels in the pups were typical for levels reported in neonatal seals. Regression analysis revealed associations between reduced immune responses and increasing levels of TCDD equivalents (non-*ortho* PCBs were 46% of the total), suggesting an effect of planar PCBs on T-cell function. Reduced thyroid hormone levels were associated with increasing levels of TCDD equivalents, suggesting the involvement of planar and nonplanar congeners. The estimated threshold for immune- and endocrine-disrupting effects of bioactive PCB congeners was 3 μg/g on a lipid basis for neonatal harbor seals.

The situation for the Baltic grey seal was recently reviewed. It has seen general improvement in reproductive and other parameters between 1977 and 1996, but there is an increase in colonic ulcers in young grey seals, suggesting the possibility of another new contaminant in this ecosystem.[441] The disease complex comprised lesions of claws, skull bone, intestine (colonic ulcers), kidneys (glomerulopathy, tubular cell proliferations), arteries (sclerosis) and adrenal glands (cortical hyperplasia, cortical adenomas). Besides occlusions and stenoses, tumors (leiomyomas) were common in the uterus. A time trend of improving gynecological health with decreasing PCBs and DDT was confirmed.

*Mink* — In a study of mink from the Frazer and Columbia River systems in the Pacific Northwest of North America, juvenile mink demonstrated a significant negative correlation between total PCB concentrations in the liver and baculum length. The association of juvenile baculum length with eventual reproductive success is unknown at the present time.[442] There is supportive evidence for these types of responses occurring in other mammals as follows: dosing of adult rats with PCB 169 produced male offspring that had reduced ventral prostate, seminal vesicles, testes weights, epididymides, and sperm counts and were in general considered to be slow in all maturation events including time to puberty. Delayed development in this sense was not considered an antiandrogenic response because hallmark effects of reduced anogenital distance, male hypospadias, or induction of areolas or nipples did not occur.[443] However, as 600-day-old adults, these PCB 169-exposed offspring had ventral prostates, caudal sperm counts, and ejaculated sperm counts that were all significantly reduced, and there was a significant increase in dorsolateral lobe prostatitis.

*Polar Bear* — In a recent comprehensive review from Canada, the species with the most significant risk of exposure to PCBs and organochlorine (OC) pesticides appeared to be the polar bear (*Ursus maritimus*). In this study, the measured EROD function in the polar bear appeared to have elevated CYP1A-mediated activity, relative to beluga whale (*Delphinapterus leucas*) and ringed seal. The MFO enzyme data for polar bear, beluga, and seal suggest that even the relatively low levels of contaminants present in Arctic animals may not be without biological effects, especially during years of poor feeding.[444] The PCDD/PCDF congener pattern in polar bear milk from Svalbard, Norway was different from that found in polar bear fat from the Canadian Arctic, although non-*ortho*-substituted PCBs in polar bear milk were similar to those found in polar bear fat from the Canadian North.[445] In 1996, two yearling polar bears at Svalbard, Norway were captured that were found to have a normal vaginal opening and a 20-mm penis containing a baculum. The penis was located caudal to the location in a normal male and was concealed within the vaginal opening by a single pair of labia. Neither of the yearlings showed signs of a Y chromosome, so both bears were regarded as female pseudohermaphrodites. Two additional bears with aberrant genital morphology and a high degree of chloral hypertrophy at Svalbard were also classified as female pseudohermaphrodites. The observed rate of female pseudohermaphroditism in this area was 1.5% (4/269). Pseudohermaphroditism in this polar bear population could result from excessive androgen excretion by the mother, be caused by a tumor, or be the result of endocrine disruption from environmental contaminants.[446] A recent compilation of contamination and effects has been produced covering the Svalbard area.[447]

*Whales and Dolphins* — The Gulf of St. Lawrence beluga whales have received considerable attention as a population of which approximately 500 individuals remain. It is not expanding, suggesting reduced reproduction and survival of juveniles. Of all of the tumors reported in cetaceans, 37% were observed in these whales. An adult hermaphrodite beluga has been documented, with two ovaries, two testes, and complete genital tracts of both sexes, with the exception of cervix, vagina, and vulva. Belugas are contaminated with a wide mix of endocrine-disruptive and carcinogenic contaminants. So far no attempts have been made to identify any likely causative agents from the large family of pollutants to which the belugas are exposed.[448] Bottlenose dolphins (*Tursiops truncatus*) have a reduced *in vitro* mitogen response to Con A and PHA that is associated with increased levels of PCBs and DDT in peripheral blood.[449]

### 18.6.4.5 Invertebrates and Other Aquatic Organisms

#### 18.6.4.5.1 Invertebrates

Barber and co-workers[450] found no indication of 2,3,7,8-TCDD toxicity to survival or growth in the amphipod (*Ampelisca abdita*), using spiked sediments from coastal New Jersey waters with a range of 2,3,7,8-TCDD concentrations (0–25 μg/kg dry weight). The maximum 2,3,7,8-TCDD concentration in ambient sediment was 0.62 μg/kg. Ashley and co-workers[451] reported a 2,3,7,8-

TCDD $LD_{50}$ of 30–100 μg/kg for freshwater crayfish (*Pacifastacus leniusculus*), with a delayed mortality of typically 15–40 days after dosing, and anergia. These authors reported that treatment of crayfish with 3 μg/kg of TCDD significantly induced cytochrome P450, as measured spectrally, and that induction and delayed onset of mortality suggested the presence of a receptor-mediated mechanism of TCDD toxicity in crayfish. Hahn and co-workers[452] and Powell and co-workers[453] provide useful information on, respectively, the possible evolutionary precursor of the Ah receptor in invertebrates and the discovery of an ortholog Ah receptor in the invertebrate nematode *Caenorhabditis elegans*.

Acute PCB doses vary by organism, ranging from 10,000 μg/g for hydra[454] to 1.3 μg/L for 121-day exposure of *Daphnia magna*.[455] Crayfish, damselfly, glass shrimp, and stonefly were also tested.[64] Marine invertebrates have been tested; shrimp (Grass, Brown, and Pink) had $LD_{50}$s of 6.1 to 12.5 μg/g.[328,456]

Several reports of decreased growth of aquatic organisms during exposure to PCBs can be found in the literature. For example, diatoms were affected at 0.1 μg/L Aroclor 1254, and oysters were affected at 10 μg/L Aroclor 1016. These concentrations are much higher than are usually encountered in the environment today. When sediment is chemically characterized, various comparative "effects thresholds" have been assessed for invertebrate toxicity.[456] A consensus-based probable effects threshold was calculated to be 676 μg/kg sediment, expressed on a dry-weight basis. This value is relevant for invertebrate species that have been tested and do not apply to other animal classes in which PCBs and related compounds bioaccumulate and biomagnify to higher trophic levels.

#### *18.6.4.5.2 Other Aquatic Organisms*

There is little definitive information on reptiles and amphibians. Snapping turtles have received considerable attention. Studies on the Northern leopard frog have not found the species to be sensitive to dioxin-like exposures; effects studied include PCB 126 MFO induction,[457] EROD activity and feral carcass PCB concentration,[458] PCB 126 weight loss, and growth rate.[459] These results are consistent with the resistance of the bullfrog to TCDD.[235,282] Contaminant reviews that have addressed reptiles and amphibians should be consulted.[460,461]

## 18.7 SUMMARY

PCBs, PCDDs, and PCDFs are all similar in their chemistry, i.e., aromatic rings with varying degrees of chlorination. The dioxins and dibenzofurans differ in the fact that they all have a bridge oxygen substituted across adjoining phenyl rings. The chemistry, fate, and effects of these compounds are very complex. This complexity arises primarily from the fact that these compounds exist as complex mixtures. PCBs are composed of a large number of compounds (usually 100–150 separate congeners) that require specialized methods of analysis. In the dioxin and dibenzofuran class of compounds, there are fewer isomers. However, there is the added problem of the extreme toxicity of these compounds, requiring methods for analysis and detection down to levels of picogram-per-gram concentration.

PCBs are universally distributed throughout the world, a fact largely resulting from their considerable aquatic and atmospheric mobility. Much of the release of these chemicals occurred over the span of their industrial use, from 1930 until the late 1970s, primarily as electrical insulating compounds and as hydraulic fluids. Numerous hot spots exist even today, over 20 years after the compounds were banned by the United States in 1977. Because of the extensive use of PCBs in electrical devices, indoor air levels of PCBs still exceed their levels in most outside air. Initial releases most often occurred into aqueous systems; however, atmospheric losses, especially from aquatic systems, are important for the dispersal of PCBs to remote systems such as the Arctic.

Levels in the air have begun to stabilize, to the point where ambient concentrations in Great Lakes water now degas into the atmosphere rather than continually load from the air, as was the situation in earlier times.

PCBs are noted for their tendency to bioaccumulate in aquatic and terrestrial organisms. This accumulation is a function of the compounds' partitioning behaviors, with increasing buildup accompanying greater chlorine substitution. Empirical data for terrestrial accumulation show the importance of diet and location, especially for caribou in the Arctic. Models for accumulation by dairy cows show the importance of atmospheric accumulation into forage crops and how lipid association and metabolism can be used to explain accumulation levels and possible human exposure. Abiotic dispersal of PCBs is sufficient to explain most of what we know about the existing placement of the world's supply of PCBs. The mobile environmental reservoir for PCBs is largely in the oceans (87%), followed by freshwater sediments, water, and biota (9.9%). How the compounds have distributed themselves in these compartments depends mostly on their physicochemical properties.

The air–water exchange behavior of PCBs has helped modelers to better explain existing levels in large bodies of water, particularly the Great Lakes. The soil reservoir and soil-to-air transfer have been widely recognized as important in maintaining existing levels in rural and metropolitan areas, especially in the United Kingdom. Long-term monitoring of PCBs in air is being used in the Great Lakes and other areas to identify sources. In water, adsorption to sediment or other organic matter is an important fate process for PCBs. Several sites with high- PCB sediment concentrations are being considered for mitigation; one of the most extensively contaminated sites is the upper Hudson River in New York State. Decreasing concentrations of PCBs in biota are beginning to be observed at locations throughout the world. Both aerobic and anaerobic degradation of PCBs have been described. Generally, the overall changes in levels due to degradation are slight, except for processes involving atmospheric photolysis.

During the late 1960s and early 1970s, it was recognized that PCDDs and PCDFs could be formed as by-products during the manufacture of various classes of chlorinated organic chemicals e.g., PCDFs from PCBs and PCDDs/PCDFs from chlorophenols. Since in most industrialized countries the manufacture and use of PCBs have been terminated and chlorophenol use has been severely restricted, emissions from the combustion of municipal wastes are currently the most important sources of PCDDs/PCDFs in the environment. However, diffuse sources such as domestic heating and outdoor trash burning are assuming greater relative importance, as controls are being implemented on municipal incinerators.

On a global scale, estimates of the quantity of deposited PCDDs/PCDFs exceed emissions estimates from combustion by a factor of two to three. Recently it has been proposed that the discrepancy could be accounted for by the photochemical formation of the two major atmospheric PCDDs/PCDFs, heptaCDD and OCDD, from pentachlorophenol in cloud water. While the photochemical process is an alternate pathway for the formation of PCDDs/PCDFs, controls that have been placed on chemical manufacturing and combustion emissions have resulted in significant declines in PCDDs/PCDFs in aquatic, atmospheric, and terrestrial environments. In the 1980s, it was found that biota in large bodies of water receiving industrial discharges, such as the Great Lakes and the Baltic Sea, were contaminated with PCDDs and PCDFs. Concentrations of 2,3,7,8-TCDD in fish from Lake Ontario varied from 2 pg/g for brown bullhead, a planktivorous fish species, to 162 pg/g for brown trout, a piscivorous species. In 1994, no 2,3,7,8-TCDD was detected at limits of detection of 1–2 pg/g in nine fish collected from Lake Ontario, including salmon, walleye, and perch. Similar improvements have been found for other aquatic species. During the same time period, ambient atmospheric concentrations of PCDDs/PCDFs declined by a factor of two in major European cities; current levels generally do not exceed 10 $pg/m.^3$

The ability to address the effects of these three structurally and toxicologically related groupings of halogented organics has greatly improved, especially over the last 10 years. Exposures of test organisms have demonstrated that, for classic measurement endpoints, a given species will respond identically to different dioxin-like congeners when exposure is adjusted to compensate for their different toxic potencies. Use of a combination of relative toxic potency to produce a multitude of measurement endpoints and an additive model of toxicity for the dioxin-like components of the PCBs, PCDDs, and PCDFs have dramatically improved our ability to interpret environmental residues along with representative biological and toxicological responses to a variety of organisms.

Across animal taxa and on a number of levels, there are differences in relative sensitivity to these exposures. Mammal, bird, and fish species may be either highly sensitive or highly resistant to dioxin-like adverse effects, especially chronic reproductive and developmental/endocrine effects. Aquatic food chain species (seals, dolphins, polar bears, piscivorous birds, and cold-water fish species) with high exposure potential through biomagnification are particularly vulnerable to such effects. Invertebrates are relatively insensitive because they lack an Ah receptor. Ecological effects of PCDDs, PCDFs, and PCBs have manifested and continue to manifest in areas where such compounds have been documented historically at high concentrations. Environmental effects can remain for long periods because of ecosystem dynamics and the persistence of these chemicals. Continued study of nondioxin-like congeners, especially with PCBs, is needed across animal taxa. The reporting of total PCBs from the summation of congener-specific analyses, as well as the reporting of individual congeners, is also important because of the large abundance of nondioxin-like PCB congeners in the original commercial mixtures and in environmental samples.

## REFERENCES

1. Bremer, K. E., Dealing with multi-agency and multi-state regulation of polychlorinated biphenyls, in PCBs, in *Human and Environmental Hazards,* D'Itri, M. and Kamrin, M. A., Eds., Butterworth Publishing, Woburn, MA, 1983, 367
2. Kuratsune, M., Yoshimura, T., Matsuzaha, J., and Yamaguchi, A., Epidemiological study of Yusho, a poisoning caused by ingestion of rice oil contaminated with a commercial brand of polychlorinated biphenyls, *Environ. Health Perspect.* 1, 119, 1972.
3. Durfee, R. L., Contos, G., Whitmore, C., Barden, J. D., Hackman, E. E., III, and Westin, R. A., PCBs in the United States — Industrial Use and Environmental Distribution, Versar, Springfield, VA, 1976.
4. Huntzinger, O. S. and Safe, Z. V., *The Chemistry of PCBs,* CRC Press, Boca Raton, FL, 1974.
5. Ballschmiter, K. and Zell, M., Analysis of polychlorinated biphenyls (PCB) by glass capillary gas chromatography, *Fresenius Zeitschrift für Anal. Chem.* 302, 20, 1980.
6. U.S. EPA, *Region v. PCB ID, BZ v. IUPAC*, Comparison of PCB Nomenclature — Past and Present, 2000.
7. Huehnerfuss, H., Pfaffenberger, B., Gehrcke, B., Karbe, L., Koenig, W. A., and Landgraff, O., Stereochemical effects of PCBs in the marine environment: Seasonal variation of coplanar and atropisomeric PCBs in blue mussels (*Mytilus edulis* L.) of the German Bight, *Mar. Pollut. Bull.*, 30, 332, 1995.
8. Jimenez, O., Jimenez, B., Marsili, L., and Gonzalez, M. J., Enantiomeric ratios of chiral polychlorinated biphenyls in stranded cetaceans form the Mediterranean Sea, *Organohalogen Compd.*, 40, 409, 1999.
9. Wong, C. S. and Garrison, A. W., Enantiomer separation of polychlorinated biphenyl atropisomers and polychlorinated biphenyl retention behavior on modified cyclodextrin capillary gas chromatography columns, *J. Chromatogr. A,* 866, 213, 2000.
10. McConnell, E. E., Acute and chronic toxicity and carcinogenesis, in *Halogenated Biphenyls, Terphenyls, Naphthalenes, Dibenzodioxins and Related Compounds*, Kimbrough, R. D. and Jensen, A. A., Eds., Elsevier, Amsterdam, 1989, Chap. 6.
11. Neubert, D., Peculiarities of the toxicity of polyhalogenated dibenzo-*p*-dioxins and dibenzofurans in animals and man, *Chemosphere*, 23, 1869, 1991.
12. Stehl, R. H., Papenfuss, R. R., Bredeweg, R. A., and Roberts, R. W., The stability of pentachlorophenol and chlorinated dioxins to sunlight, heat and combustion, in *Chlorodioxins — Origin and Fate,* Blair, E. H., Ed., Advances in Chemistry Series 120, American Chemical Society, Washington, D.C., 1973, Chap. 13.

13. Pohland, A. E. and Yang, G. C., Preparation and characterization of chlorinated dibenzo-*p*-dioxins, *J. Agric. Food Chem.*, 20, 1093, 1972.
14. Crosby, D. G., Wong, A. S., Plimmer, J. R., and Woolson, E. A., Photodecomposition of chlorinated dibenzo-*p*-dioxins, *Science*, 173, 748, 1971.
15. Friesen, K. J., Muin, C. G., and Webster, G. R. B., Evidence of sensitized photolysis of polychlorinated dibenzo-p-dioxins in natural waters under sunlight conditions, *Environ. Sci. Technol.*, 24, 1739, 1990.
16. Brubaker, W. W., Jr. and Hites, R. A., Gas-phase oxidation products of biphenyl and polychlorinated biphenyls, *Environ. Sci. Technol., 32, 3913, 1998.*
17. Marple, L., Brunck, R., and Throop, L., Water solubility of 2,3,7,8-tetrachloro-dibenzo-*p*-dioxin, *Environ. Sci. Technol.*, 20, 180, 1986.
18. Lodge, K. B., Solubility studies using a generator column for 2,3,7,8-tetrachloro-dibenzo-*p*-dioxin, *Chemosphere*, 18, 933, 1989.
19. Lodge, K. B. and Cook, P. M., Partitioning studies of dioxin between sediment and water: The measurement of $K_{oc}$ between sediment and water, *Chemosphere*, 19, 439, 1989.
20. Erickson, M. D., *Analytical Chemistry of PCBs,* CRC Press/Lewis Publishers, Boca Raton, FL, 1997.
21. NAS/NRC, *Polychlorinated Biphenyls,* National Academy of Sciences/National Research Council, Washington, D.C., 1979.
22. Albaiges, J., *Environmental Analytical Chemistry of PCBs,* Gordon and Breach Science Publishers, Yverdon, Switzerland, 1993.
23. Dobson, S. and van Esch, G. J., *Polychlorinated Biphenyls and Terphenyls,* World Health Organization, Geneva, Switzerland, 1993.
24. Strachan, W. M. J., Polychlorinated Biphenyls (PCBs): Fate and Effects in the Canadian Environment, Environment Canada, Environmental Protection Service, Ottawa, Canada, 1988.
25. American Council on Science and Health, *PCBs, Is the Cure Worth the Cost?*, The Council, Summit, NJ, 1986.
26. Hansen, L. G., *The Ortho Side of PCBs: Occurrence and Disposition*, Kluwer Academic, Boston, 1999.
27. ATSDR, Toxicological Profile for Polychlorinated Biphenyls (Update), U.S. Dept. of Health and Human Services, Public Health Service, Washington, D.C., 1997
28. EDSTAC, Endocrine Disruptor Screening and Testing Advisory Committee (EDSTAC) Final Report, U.S. EPA, Washington, D.C., 1998.
29. Hettling, L. J., Horn, E. G., and Tofflemier, T. J., Summary of Hudson River PCB study results, Dept. of Environmental Conservation, Albany, NY, 1978.
30. Rice, C. P., and White, D. S., PCB availability assessment of river dredging using caged clams and fish, *Environ. Toxicol. Chem.*, 6, 4, 259, 1987.
31. U.S. EPA, Superfund Record of Decision (EPA Region 1): (First Remedial Action), Report, EPA/ROD/R01–90/045 (PB90–254293), New Bedford, MA, 251, March 1990.
32. Hwang, S. A., Gensburg, L. J., Fitzgerald, E., Herzfeld, P. M., and Bush, B., Fingerprinting sources of contamination: Statistical techniques for identifying point sources of PCBs, *J. Occup. Med. Toxicol.,* 2, 4, 365–382, 1993.
33. Stratus Consulting, *Stage I Assessment Plan, Kalamazoo River Environment Site*, Stratus Consulting, Boulder, CO, 2000.
34. Sullivan, J. R. and Delphino, J. J., *A Select Inventory of Chemicals Used in Wisconsin's Lower Fox River Basin*, WIS-SG-82–238, University of Wisconsin Sea Grant Institute, Madison, WI, 1982.
35. U.S. EPA, Office of Water, USEPA Fisheries Advisory Report 1998, EPA-823-PCDF-99–005, 99, 2001.
36. Fuller, K., Shear, H., and Wittig, J., The Great Lakes: An Environmental Atlas and Resource Book, United States Environmental Protection Agency and Government Canada, Great Lakes National Program Office/Environment Canada, Chicago Illinois/Downsview, Ontario, Canada, 1995
37. Bopp, R., Simpson, H. J., Deck, B. L., and Kostyk, N., The persistence of PCB components in sediments of the lower Hudson, *Northeast. Environ. Sci.,* 3, 180, 1984.
38. Ikalainen, A. J. and Allen., D. C., New Bedford Harbor Superfund Project, in *Contaminated Marine Sediments: Assessment and Remediation,* Edited by National Research Council, Committee on Contaminated Sediment, National Academy Press, Washington, D.C., 1989, 365.
39. Sanders, J. E., PCB pollution in the Upper Hudson River, in *Contaminated Marine Sediments: Assessment and Remediation*, Edited by National Research Council, Committee on Contaminated Sediment, National Academy Press, Washington, D.C., 1989, 496.

40. Sanders, J. E., PCB pollution problem in the Upper Hudson River: From environmental disaster to environmental gridlock, *Northeast. Environ. Sci.*, 8, 1, 1989.
41. Pavlou, S. P., Environmental dynamics of trace organic contaminants in estuarine and coastal zone ecosystems, in *Analytical Techniques in Environmental Chemistry*, Albaiges, J., Ed., Pergamon Press, New York, 1980, 387.
42. Ashley, J.T. and Baker, J. E., Hydrophobic organic contaminants in surficial sediments of Baltimore Harbor: Inventories and sources, *Environ. Toxicol. Chem.*, 18, 5, 838, 1999.
43. Anon., Superfund Record of Decision (EPA Region 1): New Bedford Site, Upper and Lower Harbor Operable Units, New Bedford, MA, September 25, 1998. Govt. Reports Announcements & Index (GRA&I) Issue 14 (PB98–963709), 344, 1999.
44. Bergen, B. J., Rahn, K. A., and Nelson, W. G., Remediation at a marine superfund site: Surficial sediment PCB congener concentration, composition, and redistribution, *Environ. Sci. Tech.*, 32, 22, 3496, 1998.
45. Bremle, G., Larsson, P., Hammar, T., Helgee, A., and Troedsson, B., PCB in a river system during sediment remediation, *Water Air Soil Pollut.*, 107, 1–4, 237, 1998.
46. U.S. EPA, Region II, Draft Joint Supplement to the Final Environmental Impact Statement on the Hudson River PCB Reclamation Demonstration Project, USEPA, USEPA, Region II, New York,/New York State Department of Environmental Conservation, Albany, 1987.
47. Eisenreich, S. J., Looney, B. B., and Thornton, B. B., Airborne organic contaminants in the Great Lakes ecosystem, *Environ. Sci. Technol.,* 15, 30, 1981.
48. Swackhamer, D. L., Studies of polychlorinated biphenyls in the Great Lakes, in *Chlorinated Organic Micropollutants,* Hester, R. E. and Harrison, R. M., Eds., London: Royal Society of Chemistry, 1996, 137–153.
49. Strachan, W. M. J. and Eisenreich, S. J., Mass balance accounting of chemicals in the Great Lakes, in *Long Range Transport of Pesticides,* Kurz, J., Ed., Lewis Publishers, Chelsea, MI, 1990, 291–301.
50. Jeremiason, J. D., Hornbuckle, K. C., and Eisenreich, S. J., PCBs in Lake Superior, 1978–1992: Decreases in water concentrations reflect loss by volatilization, *Environ. Sci. Technol.,* 28, 903, 1994.
51. Mamontov, A. A., Mamontova, E. A., Tarasova, E. N., and McLachlan, M. S., Tracing the sources of PCDD/PCDFs and PCBs to Lake Baikal, *Environ. Sci. Technol.*, 34, 741, 2000.
52. Iwata, H., Tanabe, S., Ueda, K., and Tatsukawa, R., Persistent organochlorine residues in air, water, sediments, and soils from the Lake Baikal region, Russia, *Environ. Sci. Technol.,* 29, 3, 792, 1995.
53. Tanaba, S., Distribution, behavior and fate of PCBs in the marine environment by the member awarded the Okada prize of the Oceanographical Society of Japan for 1985, *J. Oceanogr. Soc. Jpn.,* 41, 358, 1985.
54. de March, B. G. E., deWitt, C. A., and Muir, D. C, G., Persistent organic pollutants, in *AMAP Assessment Report: Arctic Pollution Issues*, Wilson, S. J., Murray, J. L., and Huntington, H. P., Eds., Arctic Monitoring and Assessment Programme (AMAP), Oslo, Norway, 1998, 183–371.
55. Wania, F. and Mackay, D., Tracking the persistence of persistent organic compounds, *Environ. Sci. Technol.* 30, 390A, 1996.
56. Sweetman, A. J. and Jones, K. C., Declining PCB Concentrations in the U.K. atmosphere: Evidence and possible causes, *Environ. Sci. Technol.*, 34, 5, 863, 2000.
57. Harrad, S. J., Sewart, A. P., Alcock, R., Boumphrey, R., Burnett, V., Duarte, D. R., Davidson, R., Halsall, C., Sanders, G., Waterhouse, K., Wild, S. R., and Jones, K. C., Polychlorinated biphenyls (PCBs) in the British environment: Sinks, sources and temporal trends, *Environ. Pollut.,* 85, 131, 1994.
58. MacLeod, K. E., Polychlorinated biphenyls in indoor air, *Environ. Sci. Technol.,* 15, 926, 1981.
59. Wallace, J. C., Basu, I., and Hites, R. A., Sampling and analysis artifacts caused by elevated indoor air polychlorinated biphenyl concentrations, *Environ. Sci. Technol.,* 30, 9, 2730, 1996.
60. Currado, G. M. and Harrad, S. A., Factors influencing atmospheric concentrations of polychlorinated biphenyls in Birmingham, U.K., *Environ. Sci. Technol.*, 34, 1, 78, 2000.
61. Gabrio, T., Piechotowski, I., Wallenhorst, T., Klett, M., Cott, L., Friebel, P., Link, B., and Schwenk, M., PCB-blood levels in teachers, working in PCB-contaminated schools, *Chemosphere*, 40, 1055, 2000.
62. Matthews, H. B. and Dedrick, R. L., Metabolism of PCBs, *Annu. Rev. Pharmacol. Toxicol.*, 24, 85, 1984
63. Safe, S., Metabolism, uptake, storage and bioaccumulation, in *Halogenated Biphenyls, Terphenyls, Naphthalenes, Dibenzodioxins and Related Products*, Kimbrough, R. D. and Jensen, A. A., Eds., Elsevier/North Holland Biomedical Press, New York, 1989, Chap. 5.

64. Eisler, R., Polychlorinated biphenyl hazards to fish, wildlife and invertebrates: A synoptic review, Fish and Wildlife Service, U.S. Dept. of the Interior, Washington, D.C., 1986.
65. Gobas, A.P.C., Assessing bioaccumulation factors of persistent organic pollutants in aquatic food-chains, in *Persistent Pollutants: Environmental Behavior and Pathways of Human Exposure,* Harrad, S., Ed., Kluwer Academic Publishers, Norwell, MA, 2001, Chap. 6.
66. Gobas, A. P. C., Graggen, M. N., and Zhang, X., Time response of the Lake Ontario ecosystem to virtual elimination of PCBs, *Environ. Sci. Technol.*, 29, 8, 2038, 1995.
67. Vodicnik, M. J. and Peterson, R. E., The enhancing effect of spawning on elimination of a persistent polychlorinated biphenyl from a female yellow perch, *Fundam. Appl. Toxicol.*, 5, 770, 1985.
68. Wasserman, M., Wassermann, D., Cucos, S., and Miller, J. H., World PCBs map: Storage and effects, in *Man and His Biologic Environment in the 1970s, Ann. N.Y. Acad. Sci.*, 1979, 320.
69. Moessner, S. and Ballschmiter, K., Marine mammals as global pollution indicators for organochlorines, *Chemosphere,* 34, 1285, 1997.
70. Custer, T. W., Sparks, D. W., Sobiech, S. A., Hines, R. K., and Melancon, M. J., Organochlorine accumulation by sentinel mallards at the Winston-Thomas sewage treatment plant, Bloomington, Indiana, *Arch. Environ. Contam. Toxicol.*, 30, 163, 1996.
71. Bowerman, W. W. et al., PCB concentrations in plasma of nestling bald eagles from the Great Lakes Basin, North America, *Organohalogen Compd.,* 4, (Misc. Contrib.), 203, 1990.
72. McCarty, J. P. and Secord, A. L., Reproductive ecology of tree swallows, Tachycineta bicolor) with high levels of polychlorinated biphenyl contamination, *Environ. Toxicol. Chem.*, 18, 1433, 1999.
73. Braune, B., Muir, D., DeMarch, B., Gamberg, M., Poole, K., Currie, R., Dodd, M., Duschenko, W., Eamer, J., Elkin, B., Evans, M., Grundy, S., Hebert, C., Johnstone, R., Kidd, K., Koenig, B., Lockhart, L., Marshall, H., Reimer, K., Sanderson, J., and Shutt, L., Spatial and temporal trends of contaminants in Canadian Arctic freshwater and terrestrial ecosystems: A review, *Sci. Total Environ.,* 230,145, 1999.
74. Elkin, B. T., Organochlorine, heavy metal, and radionuclide contamination transfer through the lichen-caribou wolf food chain, in *Synopsis of research conducted under the Northern Contaminants Program.* Murray, J. L. and Shearer, R. G. Eds., Ottawa: Indian and Northern Affairs Canada, Environmental Studies, 1994, 356–361.
75. Muir, D. C. G., Norstrom, R. J, and Simon, M., Organochlorine contaminants in arctic marine food chains: Accumulation of specific polychlorinated biphenyls and chlordane-related compounds, *Environ. Sci. Technol.*, 22, 1071, 1988.
76. McLachlan, M. S., Model of the Fate of Hydrophobic Contaminants in Cows, *Environ. Sci. Technol.*, 28, 2407, 1994.
77. Sewart, A. and Jones, K. A., A survey of PCB congeners in U.K. cows' milk, *Chemosphere,* 32, 2481, 1996.
78. Thomas, G. O., Sweetman, A. J., and Jones, K. C., Input-output balance of polychlorinated biphenyls in a long-term study of lactating dairy cows, *Environ. Sci. Technol.*, 33, 104, 1999.
79. Sweetman, A. J., Thomas, G. O., and Jones, K. C., Modelling the fate and behaviour of lipophilic organic contaminants in lactating dairy cows, *Environ. Pollut.*, 104, 261, 1999.
80. Calamari, D., Bacci, E., Focardi, S., Gaggi, C., Morosini, M., and Vighi, M., Role of plant biomass in the global environmental partitioning of chlorinated hydrocarbons, *Environ. Sci. Technol,.* 25, 1489, 1991.
81. Hermanson, M. H. and Hites, R. A., Polychlorinated biphenyls in tree bark, *Environ. Sci. Technol.,* 24, 666, 1990.
82. Tanabe, S., PCB problems in the future: Foresight from current knowledge, *Environ. Pollut.* 50, 5, 1988.
83. Boehme, Welsch, P. K., and McLachlan, M. S., Uptake of airborne semivolatile organic compounds in agricultural plants: Field measurements of interspecies variability, *Environ. Sci. Technol.*, 33, 1805, 1999.
84. Muir, D.C., Ford, C. A., Grift, N. P., Metner, D. A., and Lockhart, W. L., Geographic variation of chlorinated hydrocarbons in burbot (*Lota lota*) from remote lakes and rivers in Canada, *Arch. Environ. Contam. Toxicol.*, 19, 530, 1990.
85. Schreitmueller, J. and Ballschmiter, K., Levels of polychlorinated biphenyls in the lower troposphere of the North and South Atlantic Ocean, studies of global baseline pollution XVII, *Fresenius J. Anal. Chem.*, 348, 226, 1994.
86. Zhang, H., Eisenreich, S. J., Franz, T. R., Baker, J. E., and Offenberg, J. H., Evidence for increased gaseous PCB fluxes to Lake Michigan from Chicago, *Environ. Sci. Technol.*, 33, 2129, 1999.

87. Panshin, S. Y. and Hites, R. A., Atmospheric concentrations of polychlorinated biphenyls in Bloomington, Indiana, *Environ. Sci. Technol.*, 28, 2008, 1994.
88. Achman, D. R., Hornbuckle, K. C., and Eisenreich, S. J., Volatilization of polychlorinated biphenyls from Green Gay, Lake Michigan, *Environ. Sci. Technol.*, 27, 75, 1993.
89. Hillery, B. R. and Hites, R. A. Temporal trends in a long-term study of vapor phase PCB concentrations over the Great Lakes, *Organohalogen Compd.*, 28, 536, 1996.
90. Hornbuckle, K. C., Jeremiason, J. D., Sweet, C. W., and Eisenreich, S. J., Seasonal variations in air–water exchange of polychlorinated biphenyls in Lake Superior, *Environ. Sci. Technol.*, 28, 1491, 1994.
91. Thomas, R. G., Volatilization from water, in *Handbook of Chemical Property Estimation Methods,* Lymann, W. J., Reehl, W., and Rosenblatt, D. H., Eds., American Chemical Society, Washington, D.C., 1990, 15-1 to 15-34.
92. Burkhard, L. P., Armstrong, D. E., and Andren, A. W., Henry's law constants for the polychlorinated biphenyls, *Environ. Sci. Technol.*, 19, 590, 1985.
93. Bamford, H. A., Baker, J. E., and Poster, D. L., Review of Methods and Measurements of Selected Hydrophobic Organic Contaminant Aqueous Solubilities, Vapor Pressures, and Air–Water Partition Coefficients, NIST Special Publication 928, National Institute of Standards and Technology, Washington, D.C., 1998.
94. Bamford, H. A., Poster, D. L., and Baker, J. E., Henry's law constants of polychlorinated biphenyl congeners and their variation with temperature, *J. Chem. Eng. Data*, 45, 1069, 2000.
95. Hoff, R. M., Muir, D. C. G., and Grift, N. P., Annual cycle of polychlorinated biphenyls and organohalogen pesticides in air in southern Ontario, 2. Atmospheric transport and sources, *Environ. Sci. Technol.*, 26, 276, 1992.
96. Wania, F., Haugen, J. E., Lei, Y. D., and Mackay, D., Temperature dependence of atmospheric concentrations of semivolatile organic compounds, *Environ. Sci. Technol.*, 32, 1013, 1998.
97. O'Connor, G. A. et al., Plant uptake of sludge-borne PCBs, *J. Environ. Qual.*, 19, 113, 1990.
98. Harner, T., Mackay, D., and Jones, K. C., Model of the long-term exchange of PCBs between soil and the atmosphere in the Southern U.K., *Environ. Sci. Technol.*, 29, 1200, 1995.
99. Taylor, A. W. Volatilization and vapor transport processes, in *Pesticides in the Soil Environment: Processes, Impacts and Modeling,* Cheng, H. H., Ed., Soil Science Society of America Book Series, Madison, WI, 1989, 214–269.
100. Harvey, G. R. and Steinhauer, W. G., Transport pathways of polychlorinated biphenyls in Atlantic water, *J. Mar. Res.*, 34, 561, 1976.
101. Schulz-Bull, D. E., Petrick, G., Bruhn, R., and Duinker, J. C., Chlorobiphenyls (PCBs) and PAHs in water masses of the northern North Atlantic, *Mar. Chem.*, 61, 101, 1998.
102. McConnell, L. L., Kucklick, J. R., Bidleman, T., Ivanov, G. P., and Chernyak, S. M., Air–water gas exchange of organochlorine compounds in Lake Baikal, Russia, *Environ. Sci. Technol.*, 30, 2975, 1996.
103. Ewald, G., Larsson, P., Linge, H., Okla, L., and Szarzi, N., Biotransport of organic pollutants to an inland Alaska Lake by migrating sockeye salmon (*Oncorhynchus nerka*), *Arctic,* 51, 40, 1998.
104. Brownawell, B. J. and Farrington, J. W., Biogeochemistry of PCBs in interstitial waters of a coastal marine sediment, *Geochim. Cosmochim. Acta,* 50, 157, 1986.
105. Hawker, D. W. and Connell, D. W., Octanol–water partition coefficients of polychlorinated biphenyl congeners, *Environ. Sci. Technol.*, 22, 382, 1988.
106. Ballschmiter, K., Rappe, C., and Buser, H. R., Chemical properties, analytical methods and environmental levels of PCBs, PCTs, PCNs and PBBs, in *Halogenated Biphenyls, Terphenyls, Naphthalenes, Dibenzodioxins and Related Products,* Kimbrough, R. D. and Jensen, A. A., Eds., Elsevier, New York, Oxford, Amsterdam, 1989, Chap. 2.
107. Ayris, S. and Harrad, S., The fate and persistence of polychlorinated biphenyls in soil, *J. Environ. Monit.*, 1, 395, 1999.
108. Sklarew, D. S., Attenuation of polychlorinated biphenyls in soils, *Rev. Environ. Contam. Toxicol.*, 98, 1, 1997.
109. Callahan, M. A., Slimak, M. W., and Gabel, N. W. Water-related environmental fate of 129 Priority Pollutants, U.S. Environmental Protection Agency, Washington, D.C., 1979, Chap. 36.
110. Anderson, P. N. and Hites, R. A., OH radical reactions: The major removal pathway for polychlorinated biphenyls from the atmosphere, *Environ. Sci. Technol.*, 30, 1756, 1996.

111. Atkinson, R., Atmospheric chemistry of PCBs, PCDDs and PCDFs, *Issues Environ. Sci. Technol.*, 6, 53, 1996.
112. Maltseva, O. V., Tsoi, T. V., Quensen, J., III, Fukuda, M., and Tiedje, J. M., Degradation of anaerobic reductive dechlorination products of Aroclor 1242 by four aerobic bacteria, *Biodegradation,* 10, 363, 1999.
113. Kas, J., Burkhard, J., Demnerova, K., Kostal, J., Macek, T., Mackova, M., and Pazlarova, J., Perspectives in biodegradation of alkanes and PCBs, *Pure Appl. Chem.*, 69, 2357, 1997.
114. Safe, S., Polyhalogenated aromatics: Uptake, disposition and metabolism, in *Halogenated Biphenyls, Terphenyls, Naphthalenes, Dibenzodioxins and Related Products*. Kimbrough, R. D. and Jensen, A. A., Eds., Elsevier, New York, Oxford, Amsterdam, 1989, Chap. 5.
115. Schnellmann, R. G., Vickers E. M., and Sipes, I. G., Metabolism and disposition of polychlorinated biphenyls, in *Reviews in Biochemical Toxicology,* Vol. 7, Hodgson, E., Bend, J. R., and Philpot, R. M., Eds., Amsterdam, Elsevier, 1985, 247–282.
116. Sundstrom, G., Hutzinger, O., and Safe, S., The metabolism of chlorobiphenyls, *Chemosphere,* 5, 267, 1976.
117. Bedard, D. L., Unterman, R., Bopp, L. H., Brennan, M. J., Haberl, M. L., and Johnson, C., Rapid assay for screening and characterizing microorganisms for the ability to degrade polychlorinated biphenyls, *Appl. Environ. Microbiol.,* 51, 761, 1986.
118. Brown, J., Wagner R. E., Feng, H., Bedard, D. L., Brennan, M. J., Carnahan, J. C., and May, R. J., Environmental dechlorination of PCBs, *Environ. Toxicol. Chem.,* 6, 579, 1987.
119. Rhee, G. Y., Sokol, R. C., Bethoney, C. M., and Bush, B., Dechlorination of polychlorinated biphenyls by Hudson River sediment organisms: Specificity to the chlorination pattern of congeners, *Environ. Sci. Technol.,* 27, 1190, 1993.
120. Rhee, G. Y., Bush, B., Brown, M. P., Kane, M., and Shane, L., Anaerobic biodegradation of polychlorinated biphenyls in Hudson River sediments and dredged sediments in clay encapsulation, *Water Res.,* 23, 957, 1989.
121. Quensen, J., III, Tiedje, J. M., and Boyd, S. A., Reductive dechlorination of polychlorinated biphenyls by anaerobic microorganisms from sediments, *Science,* 242, 752, 1988.
122. Iwata, Y., Westlake, W. E., and Gunther, A., Varying persistence of polychlorinated biphenyls in six California soils under laboratory conditions, *Bull. Environ. Contam. Toxicol.* 9, 204, 1973.
123. Wakimoto, T., Kannan, N., Ono, M., Tatsukawa, R., and Masuda, Y., Isomer-specific determination of polychlorinated dibenzofurans in Japanese and American polychlorinated biphenyls, *Chemosphere,* 17, 743, 1988.
124. Bowes, G. W., Mulvihill, M. J., Simoneit, B. R. T., Burlingame, A. L., and Risebrough, R. W., Identification of chlorinated dibenzofurans in American polychlorinated biphenyls, *Nature,* 256, 305, 1975.
125. Morita, M., Nakagawa, J., Akiyama, K., Mimura, S., and Isono, N., Detailed examination of polychlorinated dibenzofurans in PCB preparations and Kanemi Yusho oil, *Bull. Environ. Contam. Toxicol.,* 18, 67, 1973.
126. O'Keefe, P. W. and Smith, R. M., PCB capacitor/transformer accidents, in *Halogenated Biphenyls, Naphthalenes, Dibenzodioxins and Related Compounds,* Kimbrough, R. D. and Jensen, A. A., Eds., Elsevier, Amsterdam, 1989, Chap. 15.
127. Woolson, E. A., Thomas, R., and Ensor, P. D. J., Survey of polychlorodibenzo-*p*-dioxin content of selected pesticides, *J. Agric. Food Chem.,* 20, 351, 1972.
128. Buser, H.-R. and Bosshardt, H.-P., Determination of 2,3,7,8-tetrachlorodibenzo-*p*-dioxin at parts per billion levels in technical grade 2,4,5-trichlorophenoxyacetic acid, in 2,4,5-T alkyl ester and 2,4,5-T amine salt herbicide formulations by quadrupole mass fragmentography, *J. Chromatogr.*, 90, 71, 1974.
129. Rappe, C., Buser, H.-R., and Bosshardt, H.-P., Identification and quantification of polychlorinated dibenzo-*p*-dioxins (PCDDs) and dibenzofurans (PCDFs) in 2,4,5-T-ester formulations and herbicide orange, *Chemosphere,* 10, 431, 1978.
130. International Program on Chemical Safety, *Environmental Health Criteria 71: Pentachlorophenols,* World Health Organization, 1987.
131. Buser, H.-R. and Bosshardt, H.-P., Determination of polychlorinated dibenzo-*p*-dioxins and dibenzofurans in commercial pentachlorophenols by combined gas chromatography-mass spectrometry, *J. Assoc. Offic. Anal. Chem.,* 59, 562, 1976.
132. Firestone, D., Chemistry and analysis of pentachlorophenol, *FDA By-Lines*, 8, 57, 1977.

133. Tong, H. Y., Monson, S. J., Gross, M. L., Bopp, R., Simpson, H. J., Deck, B. L. and Moser, C., Analysis of dated sediment samples from the Newark Bay area for selected PCDDs/PCDFs, *Chemosphere,* 20, 1497, 1990.
134. Ehrlich, R., Wenning, R. J., Johnson, G. W., Su, S. H., and Paustenbach, D. J., A mixing model for polychlorinated dibenzo-*p*-dioxins and dibenzofurans in surface sediments from Newark Bay, New Jersey, using polytopic vector analysis, *Arch. Environ. Contam. Toxicol.*, 27, 486, 1994.
135. Pereira, W. E., Rostad, C. E., and Sisak, M. E., Geochemical investigations of polychlorinated dibenzo-*p*-dioxins in the subsurface environment at an abandoned wood-treatment facility, *Environ. Chem. Toxicol.,* 4, 629, 1985.
136. McKee, P., Burt, A., and McCurvin, D., Levels of dioxins, furans and other organics in harbor sediments near a wood processing plant using pentachlorophenol and creosote, *Chemosphere,* 20, 1679, 1990.
137. Kuehl, D. W., Butterworth, B. C., McBride, A., Kroner, S., and Bahnick, D., Contamination of fish by 2,3,7,8-tetrachlorodibenzo-*p*-dioxin: A survey of fish from major watersheds in the United States, *Chemosphere,* 18, 1997, 1989.
138. Axegard, P. and Renberg, L., The influence of bleaching chemicals and lignin content on the formation of polychlorinated dioxins and dibenzofurans, *Chemosphere,* 19, 661, 1989.
139. LaFleur, L., Brunck, B., McDonough, T., Ramage, K., Gillespie, W., and Malcolm, E., Studies on the mechanism of PCDD/PCDF formation during the bleaching of pulp, *Chemosphere,* 20, 1731, 1990.
140. Kuehl, D. W., Butterworth, B. C., DeVita, W. M., and Sauer, C. P., Environmental contamination of polychlorinated dibenzo-*p*-dioxins and dibenzofurans associated with pulp and paper mill discharge, *Biomed. Environ. Mass Spectrom.*, 14, 443, 1987.
141. Chement, R. E., Suter, S. A., Reiner, E., McCurvin, D., and Hollinger, D., Concentrations of chlorinated dibenzo-*p*-dioxins and dibenzofurans in effluents and centrifuged particulates from Ontario pulp and paper mills, *Chemosphere*, 19, 649, 1989.
142. Vanness, G., Tiernan, T. O., Hanes, M. S., Reynolds, C. J., Garrett, J. H., Wagel, D. J., and Solch, J. G., Determination of dibenzo-*p*-dioxin (DBD) and dibenzofuran (DBF) in paper pulp mill defoamer additives and in paper pulp samples, *Chemosphere,* 20, 1611, 1990.
143. Whittemore, R. C., LaFleur, L. E., Gillespie, W. J., Amendola, G. A., and Helms, J., USEPA/paper industry cooperative dioxin study: The 104-mill study, *Chemosphere*, 20, 1625, 1990.
144. MacDonald, R. W., Ikonomou, M. G., and Paton, D. W., Historical inputs of PCDDs, PCDFs, and PCBs to a British Columbia interior lake — The effect of environmental controls on pulp mill emissions, *Environ. Sci. Technol.,* 32, 331, 1998.
145. Hagen, M. E., Colodey, A. G., Knapp, W. D., and Samis, S. C., Environmental response to decreased dioxin and furan loadings from British Columbia coastal pulp mills, *Chemosphere,* 34, 1221, 1997.
146. Krahn, P. K., Dioxins and Furans in British Columbia Pulp Mill Effluent, (1987–1995), Short Papers, Vol. 24, *15th Int. Symp. Dioxins Related Compd.*, Edmonton, Canada, 1995, 217.
147. Eljarrat, E., Caixach, J., and Rivera, J., Decline of PCDD and PCDF in sewage sludges from Catalonia, Spain, *Environ. Sci. Technol.*, 33, 2493, 1999.
148. Jones, K. C. and Sewart, A. P., Dioxins and furans in sewage sludges — A review of their occurrence and sources in sludge and of their environmental fate, behavior, and significance in sludge-amended agricultural systems, *Crit. Rev. Environ. Sci. Technol.*, 27, 1, 1997.
149. NATO/CCMS, International Toxicity Equivalent Factor (I-TEF), Method of risk assessment for complex mixtures of dioxins and related compounds, Pilot study on international information exchange on dioxins and related compounds, Report no. 176, 1988.
150. Villaneuva, E. C., Jennings, R. W., Burse, V. W., and Kimbrough, R. D., Evidence of chlorodibenzo-*p*-dioxin and chlorodibenzofuran in hexachlorobenzene, *J. Agric. Food Chem.*, 22, 916, 1974.
151. Heindl, A. and Hutzinger, O., Search for industrial sources of PCDD/PCDF: III. Short-chain chlorinated hydrocarbons, *Chemosphere,* 16, 1949, 1987.
152. Isosaari, P., Kohonen, T., Kiviranta, H., Tuomisto, J., and Vartiainen, T., Assessment of levels, distribution, and risks of polychorinated dibenzo-*p*-dioxins and dibenzofurans in the vicinity of a vinyl chloride production plant, *Environ. Sci. Technol.*, 34, 2684, 2000.
153. Olie, K., Vermuelen, P. L., and Hutzinger, O., Chlorodibenzo-*p*-dioxins and chlorodibenzofurans are trace components of fly ash and flue gas of some municipal incinerators in the Netherlands, *Chemosphere,* 8, 455, 1977.

154. Dow Chemical Co., The Trace Chemistries of Fire — A source of and Routes for the Entry of Chlorinated Dioxins into the Environment, Dow Chemical Company, Midland, MI, 1978.
155. Smith, R. M., O'Keefe, P. W., Hilker, D. R., and Aldous, K. M., Ambient air and incinerator testing for chlorinated dibenzofurans and dioxins by low resolution mass spectrometry, *Chemosphere*, 18, 585, 1989.
156. Siebert, P. C., Alston-Guiden, D., and Jones, K. H., An analysis of worldwide resource recovery emissions and the implications for risk assessment, in *Health Effects of Municipal Waste Incineration*, Hattemer-Frey, H. A. and Travis, C., Eds., CRC Press, Boca Raton, FL, 1991, Chap. 1.
157. W.H.O., PCDD and PCDF Emissions from Incineration, Environmental Health 17, WHO/EURO Copenhagen, Denmark, 1987.
158. Rappe, C. and Buser, H.-R., Chemical and physical properties, analytical methods, sources and environmental levels of halogenated dibenzodioxins and dibenzofurans, in *Halogenated Biphenyls, Triphenyls, Naphthalenes, Dibenzodioxins and Related Compounds*, Kimbrough, R. D. and Jensen, A. A., Eds., Elsevier, Amsterdam, 1989, Chap. 3.
159. Shin, D. H., Choi, S. M., Oh, J. E., and Chang, Y. S., Evaluation of polychlorinated dibenzo-*p*-dioxin/dibenzofuran (PCDD/PCDF) emissions in municipal solid waste incinerators, *Environ. Sci. Technol.*, 33, 2657, 1999.
160. Brown, R. S., Pettit, K., Mundy, K. J., Jones, P. W., and Noble, W., Incineration: The British experience, *Chemosphere,* 20, 1785, 1990.
161. Lindner, G., Jenkins, A. C., McCormack, J., and Adrian, R. C., Dioxins and furans from medical waste incinerators, *Chemosphere,* 20, 1793, 1990.
162. Manscher, O. H., Heidam, N. Z., Vikelsoe, J., Nielsen, P., Blinksbjerg, P., Madsen, H., Pallesen, L., and Tiernan, T. O., The Danish incinerator dioxin study. Part 1, *Chemosphere,* 20, 1779, 1990.
163. Hagenmaier, M., Kraft, M., Brunner, H., and Haag, R., Catalytic effects of fly ash from waste incineration facilities on the formation and decomposition of polychlorinated dibenzo-*p*-dioxins and polychlorinated dibenzofurans, *Environ. Sci. Technol.*, 21, 1080, 1987.
164. Rappe, C., Sources of Exposure, Environmental Levels and Exposure Assessment of PCDDs and PCDFs, Extended Abstracts, Vol. 9, p 5, *12th Int. Symp. Dioxins Related Compd.,* Tampere, Finland, 1992.
165. Tysklind, M., Söderstrom, G., and Rappe, C., PCDD and PCDF emissions from scrap metal melting processes at a steel mill, *Chemosphere*, 19, 705, 1989.
166. Harnly, M., Stephens, R., McLaughlin, C., Marcotte, J., Petreas, M., and Goldman, L., Polychlorinated dibenzo-*p*-dioxin and dibenzofuran contamination at metal recovery facilities, open fire sites, and a railroad car incineration facility, *Environ. Sci. Technol.*, 29, 677, 1995.
167. Marklund, S., Rappe, C., Tysklind, M., and Egeback, K-E., Identification of polychlorinated dibenzofurans and dioxins in exhausts from cars run on leaded gasoline, *Chemosphere*, 16, 29, 1987.
168. Bingham, A. G., Edmunds, C. J., Graham, B. W. L., and Jones, M. T., Determination of PCDDs and PCDFs in car exhaust, *Chemosphere*, 19, 669, 1989.
169. Geueke, K. J., Gessner, A., Quass, U., Broker, G., and Hiester, E., PCDD/PCDF emissions from heavy duty vehicle diesel engines, *Chemosphere*, 38, 2791, 1999.
170. Ryan, J. V. and Gullett, B. K., On-road emission sampling of heavy-duty diesel trucks for polychlorinated dibenzo-*p*-dioxins and dibenzofurans, *Environ. Sci. Technol.*, 34, 4483, 2000.
171. Lamieux, P. M., Lutes, C. C., Abbott, J. A., and Aldous, K. M., Emissions of polychlorinated dibenzo-*p*-dioxins polychlorinated dibenzofurans from the open burning of household waste in barrels, *Environ. Sci. Technol.*, 34, 377, 2000.
172. Lee, R. G. M., Green, N. J. L., Lohmann, R., and Jones, K. C., Seasonal, anthropogenic, air mass and meterolological influences on the atmospheric concentrations of polychlorinated dibenzo-*p*-dioxins and dibenzofurans (PCDDs/PCDFs): Evidence for the importance of diffuse combustion sources, *Environ. Sci. Technol.,* 33, 2864, 1999.
173. Cook, P. M., Batterman, A. R., Butterworth, B. C., and Lodge, K. B., Laboratory Study of TCDD Bioaccumulation by Lake Trout from Lake Ontario Sediments, Food Chain and Water, in *Lake Ontario TCDD Bioaccumulation Study Final Report*, U.S. Environmental Protection Agency, New York, 1990, Chap. 6.
174. Loonen, H., Tonkes, M., Parsons, J. R., and Govers, H. A. J., Relative contributions of water and food to the bioaccumulation of polychlorinated dibenzo-*p*-dioxins and polychlorinated dibenzofurans in guppies, *Sci. Total Environ.,* Suppl. 491, 1993.

175. Loonen, H., Parsons, D., Govers, H. A. J., Effect of sediment on the bioaccumulation of a complex mixture of polychlorinated dibenzo-*p*-dioxins (PCDDs) and polychlorinated dibenzofurans (PCDFs) by fish, *Chemosphere*, 28, 1433, 1994.
176. Kuehl, D. W., Cook, P. M., Batterman, A. R., Lothenbach, D., and Butterworth, B. C., Bioavailability of polychlorinated dibenzo-*p*-dioxins and dibenzofurans from contaminated Wisconsin River sediment to carp, *Chemosphere*, 16, 667, 1987.
177. Rubinstein, N. I., Pruell, R. J., Taplin, B. K., Li Volsi, J. A., and Norwood, C. B., Bioavailability of 2,3,7,8-TCDD, 2,3,7,8-TCDF and PCBs to marine benthos from Passaic River sediments, *Chemosphere*, 20, 1097, 1990.
178. Fairchild, W. L., Muir, D. C. G., Currie, R. S., and Yarechewski, A. L., Emerging insects as a biotic pathway for movement of 2,3,7,8-tetrachlorodibenzofuran from lake sediments, *Environ. Toxicol. Chem.*, 11, 867, 1992.
179. Kuehl, D. W., Cook, P. M., Batterman, A. R., Lothenbach, D. B., Butterworth, B. C., and Johnson, D. L., Bioavailability of 2,3,7,8-tetrachlorodibenzo-*p*-dioxin from municipal incinerator flyash to freshwater fish, *Chemosphere*, 14, 427, 1985.
180. Kuehl, D. W., Cook, P. M., Batterman, A. R., and Butterworth, B. C., Isomer dependent bioavailability of polychlorinated dibenzo-*p*-dioxins and dibenzofurans from municipal incinerator flyash to carp, *Chemosphere*, 16, 657, 1987.
181. Van den Berg, M., Blank, F., Heeremans, C., Wagenaar, H., and Olie, K., Presence of polychlorinated dibenzo-*p*-dioxins and dibenzofurans in fish-eating birds and fish from the Netherlands, *Arch. Environ. Contam. Toxicol.*, 16, 149, 1987.
182. Braune, B. M. and Norstrom, R. J., Dynamics of organochlorine compounds in herring gulls: III. Tissue distribution and bioaccumulation in Lake Ontario gulls, *Environ. Toxicol. Chem.*, 8, 957, 1989.
183. Kuroki, J., Koga, N., and Yoshimura, H., High affinity of 2,3,4,7,8-pentachloro-dibenzofuran to cytochrome P-450 in hepatic microsomes of rats, *Chemosphere*, 15, 731, 1986.
184. Pastor, D., Ruiz, X., Barcelo, D., and Albaiges, J., Dioxins, furans and AHH-active PCB congeners in eggs of two gull species from the Western Mediterranean, *Chemosphere*, 31, 3397, 1995.
185. Fanelli, R., Bertoni, M. P., Bonfanti, M., Castelli, M. G., Chiabrando, C., Martelli, G. P., N'e, M.A., Noseda, A., and Sbarra, C., Routine analysis of 2,3,7,8-tetrachlorodi-benzo-*p*-dioxin in biological samples from the contaminated area of Seveso, Italy, *Bull. Environ. Contam. Toxicol.*, 24, 818, 1980.
186. Chang, R., Hayward, D., Goldman, L., Hanly, M., Flattery, J., and Stephens, R., Foraging farm animals as biomonitors for dioxin contamination, *Chemosphere*, 19, 481, 1989.
187. Stephens, R. D., Petreas, M. X., and Haywood, D. G., Biotransfer and bioaccumulation of dioxins and dibenzofurans from soil, extended abstracts, Vol. 8, p 377, 12$^{th}$ *Int. Symp. Dioxins Related Compds.*, Tampere, Finland, 1992.
188. McConnell, E. E., Lucier, G. W., Rumbaugh, R. C., Albro, P. W., Harvan, D. J., Hass, J. R., and Harris, M. W., Dioxin in soil: Bioavailability after ingestion by guinea pigs, *Science*, 223, 1077, 1984.
189. Umbeit, T. H., Hesse, E. J., and Gallo, M. A., Bioavailability of dioxin in soil from a 2,4,5-T manufacturing site, *Science*, 232, 417, 1986.
190. Jensen, D. J., Hummel, R. A., Mahle, N. H., Kocher, C. W., and Higgins, H. S., A residue study on beef cattle consuming 2,3,7,8-tetrachlorodibenzo-*p*-dioxin, *J. Agric. Food Chem.*, 29, 265, 1981.
191. McLachlan, M. and Richter, W., Uptake and transfer of PCDDs/PCDFs by cattle fed naturally contaminated feedstuffs and feed contaminated as a result of sewage sludge application. 1. Lactating cows, *J. Agric. Food Chem.*, 46, 1166, 1998.
192. Czuczwa, J. and Hites, R. A., Airborne dioxins and dibenzofurans: Sources and fates, *Environ. Sci. Technol.*, 20, 195, 1986.
193. Pearson, R., Swackhamer, D. L., Eisenreich, S. J., and Long, D. T., Concentrations, accumulations and inventories of polychlorinated dibenzo-*p*-dioxins and dibenzofurans in sediments of the Great Lakes, *Environ. Sci. Technol.*, 31, 2903, 1997.
194. Baker, J. I. and Hites, R. A., Siskiwit revisited: Time trends of polychlorinated dibenzo-*p*-dioxin and dibenzofuran deposition at Isle Royale, Michigan, *Environ. Sci. Technol.*, 34, 2887, 2000.
195. Smith, R. M., O'Keefe, P. W., Aldous, K., Briggs, R., Hilker, D., and Connor, S., Measurement of PCDFs and PCDDs in air samples and lake sediments in upstate New York, *Proc. 11$^{th}$ Int. Symp. Dioxins Related Compd.*, Research Triangle Park, Raleigh, NC, 1991.

196. Rose, C. L., Rose, N. L., Harlock, S., and Fernandes, A., A historical record of polychlorinated dibenzo-*p*-dioxin (PCDD) and polychlorinated dibenzofuran (PCDF) deposition to a remote lake site in north-west Scotland, U.K., *Sci. Total Environ.,* 198, 161, 1997.
197. Vartiainen, T., Mannio, J., Korhonen, M., Kinnunen, K., and Strandman, T., Levels of PCDD, PCDF and PCB dated sediments in subartic Finland, *Chemosphere*, 34, 1341, 1997.
198. Juttner, I., Henkelmann, B., Schramm, K. W., Steinberg, C. E. W., Winkler, R., and Kettrrup, A., Occurrence of PCDD/PCDF in dated lake sediments of the Black Forest, Southwestern Germany, *Environ. Sci. Technol.*, 31, 806, 1997.
199. Bruzy, L. P. and Hites, R. A., Global mass balance for polychlorinated dibenzo-*p*-dioxins and dibenzofurans, *Environ. Sci. Technol.*, 30, 1797, 1996.
200. Baker, J. I. and Hites, R. A., Polychlorinated dibenzo-*p*-dioxins and dibenzofurans in the remote North Atlantic marine atmosphere, *Environ. Sci. Technol.*, 33, 14, 1999.
201. Wagrowski, D. M. and Hites, R. A., Insights into the global distribution of polychlorinated dibenzo-*p*-dioxins and dibenzofurans, *Environ. Sci. Technol.*, 34, 2952, 2000.
202. Baker, J. I. And Hites, R. A., Is combustion the major source of polychlorinated dibenzo-*p*-dioxins and dibenzofurans in the environment? A mass balance investigation, *Environ. Sci. Technol.*, 34, 2879, 2000.
203. Tyler, A. O., Millward, G. E., Jones, P. H., and Turner, A., Dioxins in U.K. Estuaries, Extended Abstracts, *12th Int. Symp. Dioxins Related Compd.,* Tampere, Finland, 9, 295, 1992.
204. Dung, M. H. and O'Keefe, P. W., Comparative rates of photolysis of polychlorinated dibenzofurans in organic solvents and in aqueous solutions, *Environ. Sci. Technol.*, 28, 549, 1994.
205. Kim, M. and O'Keefe, P. W., Photodegradation of polychlorinated dibenzo-*p*-dioxins and dibenzofurans in aqueous solutions and in organic solvents, *Chemosphere*, 41, 793, 2000.
206. Adriaens, P., Fu, Q. Z., and Grbicgalic, D., Bioavailability and transformation of highly chlorinated dibenzo-*p*-dioxins in anaerobic soils and sediments, *Environ. Sci. Technol.*, 29, 2252, 1995.
207. Ballerstedt, H., Krauss, A., and Lechner, U., Reductive dechlorination of 1,2,3,4-tetrachlorodibenzo-*p*-dioxin and its products by anaerobic mixed cultures from Saale River sediment, *Environ. Sci. Technol.*, 31, 1749, 1997.
208. Barkovskii, A. L. and Adriaens, P., Microbial dechlorination of historically present and freshly spiked chlorinated dioxins and diversity of dioxin-dechlorinating populations, *Appl. Environ. Microbiol.*, 62, 4556, 1996.
209. Adriaens, P., Chang, P. R., and Barkovskii, A. L., Dechlorination of PCDD/PCDF by organic and inorganic electron transfer molecules in reduced environments, *Chemosphere*, 32, 433, 1996.
210. Barkovskii, A. L. and Adriaens, P., Impact of humic constituents on microbial dechlorination of polychorinated dioxins, *Environ. Toxicol. Chem.* 17, 1013, 1998.
211. Albrecht, I. D., Barkovskii, A. L., and Adriaens, P., Production and dechlorination of 2,3,7,8-tetrachlorodibenzo-*p*-dioxin in historically-contaminated estuarine sediments, *Environ. Sci. Technol.*, 33, 737, 1999.
212. Jensen, S., The PCB Story, *Ambio* 1, 123, 1972.
213. http://www.nwfsc.noaa.gov/issues/Pdfs/EC6504.pdf, National Marine Mammal Contaminant Monitoring, Research Issue Paper, EC 6504 (HQ ID 284/285/286).
214. O'Connor, T. P., *Recent Trends in Coastal Environmental Quality: Results from the First Five Years of the NOAA Mussel Watch Project*, U.S. Printing Office, U.S. Department of Commerce, National Oceanic and Atmospheric Administration, 1992.
215. Bignert, A., Olsson, M., Persson, W., Jensen, S., Zakrisson, S., Litzen, K., Eriksson, U., Haggberg, L., and Alsberg, T., Temporal trends of organochlorines in northern Europe, 1967–1995, Relation to global fractionation, leakage from sediments and international measures, *Environ. Pollut.*, 99, 177, 1998.
216. Muir, D. C. and Norstrom, R. J., Geographical differences and time trends of persistent organic pollutants in the Arctic, *Toxicol. Lett.,* 112–113, 93, 2000.
217. Morrison, H. A., Whittle, M. D., and Haffner, D. G., The relative importance of species invasions and sediment disturbance in regulating chemical dynamics in western Lake Erie, *Ecol. Modeling*, 125, 279, 2000.
218. Bopp, R., Simpson, H. J., Olsen, C. R., Trier, R. M., and Kostyk, N., Chlorinated hydrocarbons and radionuclide chromologies in sediments of the Hudson River estuary, New York, *Environ. Sci. Technol.,* 16, 666, 1982.

219. Golden, K. A., Wong, C. S., Jeremiason, J. D., Eisenreich, S. J., Sanders, G., Hallgren, J., Swackhamer, D. L., Engstrom, D. R., and Long, D. T., Accumulation and preliminary inventory of organochlorines in Great Lakes sediments, *Water Sci. Technol.*, 28, 19, 1993.
220. Broman, D., Naf, C., Axelman, J., and Petersen, H., Time trend analysis of PAHs and PCBs in the northern Baltic proper, *Chemosphere,* 29, 1325, 1994.
221. Scrimshae, M. D. and Lester, J. N., Estimates of the inputs of polychlorinated biphenyls and organochlorine insecticides to the River Thames derived from the sediment record, *Philos. Trans. R. Soc. London, Ser. A,* 355, 189, 1997.
222. Alcock, R. E., Johnston, A. E., McGrath, S. P., Berrow, M. L., and Jone, K. C., Long-term changes in the polychlorinated biphenyl content of United Kingdom soils, *Environ. Sci. Technol.,* 27, 1918, 1993.
223. Jones, K. C., Duarte, D. R., and Cawse, P. A., Changes in the PCB Concentration of United Kingdom air between 1972 and 1992, *Environ. Sci. Technol.*, 29, 272, 1995.
224. Hillery, B. R., Simcik, M., Basu, I., Hoff, R. M., Strachan, W. M. J., Burniston, D., Chan, C. H., Brice, K. A., Sweet, C. W., and Hites, R. A., Atmospheric deposition of toxic pollutants to the Great Lakes as measured by the integrated atmospheric deposition network, *Environ. Sci. Technol.*, 32, 2216, 1998.
225. Simcik, M., Basu, I., Sweet, C. W., and Hites, R. A., Temperature dependence and temporal trends of polychlorinated biphenyl congeners in the Great Lakes atmosphere, *Environ. Sci. Technol.,* 33, 1991, 1999.
226. Hillery, B. R., Basu, I., Sweet, C. W., and Hites, R. A., Temporal and spatial trends in a long-term study of gas-phase PCB concentrations near the Great Lakes, *Environ. Sci. Technol.*, 31, 1811, 1997.
227. O'Keefe, P., Meyer, C., Hilker, D., Aldous, K., Jelus-Tyror, B., Dillon, K., and Donnelly, R., Analysis of 2,3,7,8-tetrachlorodibenzo-*p*-dioxin in Great Lakes fish, *Chemosphere*, 12, 325, 1983.
228. Ryan, J. J., Lau, P-Y., Pillon, J. C., Lewis, D., McLeod, H. A., and Gervais, A., Incidence and levels of 2,3,7,8-tetrachlorodibenzo-*p*-dioxin in Lake Ontario commercial fish, *Environ. Sci. Technol.*, 18, 719, 1984.
229. Fehringer, N. V., Walters, S. M., Kozara, R. J., and Schneider, L., Survey of 2,3,7,8-tetrachlorodibenzo-*p*-dioxin in fish from the Great Lakes and selected Michigan rivers, *J. Agric. Food Chem.*, 33, 626, 1985.
230. Harless, R. L., Oswald, E. O., Lewis, R. G., Dupuy, A. E., Jr., McDaniel, D. D., and Tai, H., Determination of 2,3,7,8-tetrachlorodibenzo-*p*-dioxin in fresh water fish, *Chemosphere*, 11, 193, 1982.
231. Norstrom, R. J., Hallett, D. J., Simon, M., and Mulvihill, M. J., Analysis of Great Lakes herring gull eggs for tetrachlorodibenzo-*p*-dioxins, in *Chlorinated Dioxins and Related Compounds: Impact on the Environment,* Hutzinger, O., Frei, R. W., Merian, E., and Pochiari, F., Eds., Pergamon Press, Oxford, 1982, 173.
232. Stalling, D. L., Smith, L. M., Petty, J. D., Hogan, J. W., Johnson, J. L., Rappe, C., and Buser, H. R., Residues of polychlorinated dibenzo-*p*-dioxins and dibenzofurans in Laurentian Great Lakes fish, in *Human and Environmental Risks of Chlorinated Dioxins and Related Compounds,* Tucker, R. E., Young, A. L., and Gray, A. P., Eds., Plenum Publishers, New York, 1983, 221.
233. Zacharewski, T., Safe, L., Safe, S., Chittim, B., DeVault, D., Wiberg, K., Bergqvist, P-E., and Rappe, C., Comparative analysis of polychlorinated dibenzo-*p*-dioxin and dibenzofuran in Great Lakes fish extracts by gas chromatography-mass spectrometry and *in vitro* enzyme induction activities, *Environ. Sci. Technol.*, 23, 730, 1981.
234. Kubiak, T. J., Harris, H. J., Smith, L. M., Schwartz, T. R., Stalling, D. I., Trick, J. A., Sileo, L., Docherty, D. E., and Erdman, T. C., Microcontaminants and reproductive impairment of the Forster's tern on Green Bay, Lake Michigan — 1983, *Arch. Environ. Contam. Toxicol.*, 18, 706, 1989.
235. Korfmacher, W. A., Hansen, E. B., Jr., and Rowland, K. L., Tissue distribution of 2,3,7,8-TCDD in bull frogs obtained from a 2,3,7,8-TCDD contaminated area, *Chemosphere*, 15, 121, 1986.
236. Ryan, J. J., Lau, B. P-Y., Hardy, J. A., Stone, W. B., O'Keefe, P., and Gierthy, J., 2,3,7,8-Tetrachlorodibenzo-*p*-dioxin and related dioxins and furans from snapping turtle tissues (Chelydra serpentina) from the Upper St. Lawrence River, *Chemosphere,* 15, 537, 1986.
237. Rappe, C., Bergqvist, P-E., Kjeller, L-O., Swanson, S., Belton, T., Ruppel, B., Lockwood, K., and Kahn, P. C., Levels and patterns of PCDD and PCDF contamination in fish, crabs and lobsters from Newark Bay and New York Bight, *Chemosphere*, 22, 239, 1991.

238. Gotz, R., Schumacher, E., Kjeller, L-O., Bergqvist, P-A., and Rappe, C., Polychlorierte Dibenzo-*p*-Dioxine (PCDDs) und Polychlorierte Dibenzofurane (PCDFs) in Sedimenten und Fischen aus dem Hamburger Hafen, *Chemosphere*, 20, 51, 1990.
239. Luckas, B. and Oehme, M., Characteristic contamination levels for polychlorinated hydrocarbons, dibenzofurans and dibenzo-*p*-dioxins in bream (*Abramis brama*) from the River Elbe, *Chemosphere,* 21, 79, 1990.
240. Frommberger, R., Polychlorinated dibenzo-*p*-dioxins and polychlorinated dibenzofurans in fish from South-West Germany: River Rhine and Neckar, *Chemosphere,* 22, 29, 1991.
241. Rappe, C., Bergqvist, P.-A., and Kjeller, L-O., Levels, trends and patterns of PCDDs and PCDFs in Scandinavian environmental samples, *Chemosphere*, 18, 651, 1989.
242. Hebert, C. E., Norstrom, R. J., Simon, M., and Braune, B. M., Temporal trends and sources of PCDDs and PCDFs in the Great Lakes: Herring gull monitoring, 1981–1991, *Environ. Sci. Technol.*, 28, 1268, 1994.
243. Firestone, D., Fehringer, N. V., Walters, S. M., Kozan, R. J., Ayres, R. J., Ogger, J., Schneider, L., Glidden, R. M., Ahlrep, J. R., Brown, P. J., Ford, S. E., Davy, R. A., Gulick, D. J., McCullough, B. H., Sittig, R. A., Smith, P. V., Syvertson, C. N., and Barber, M. R., TCDD residues in fish and shellfish from US waterways, *J. AOAC Int.*, 79, 1174, 1996.
244. Elliott, J. E. and Norstrom, R. J., Chlorinated hydrocarbon contaminants and productivity of Bald Eagle populations on the Pacific coast of Canada, *Environ. Toxicol. Chem.*, 17(6), 1142–1153, 1998.
245. Ling, Y., Soong, D. K., and Lee, M. K., PCDD/PCDFs and coplanar PCBs in sediment and fish samples from the Er-Jen River in Taiwan, *Chemosphere*, 31, 2863, 1995.
246. Tarasova, E. N., Mamontov, A. A., Mamontova, E. A., Klasmeier, J., and McLachlan, M. S., Polychorinated dibenzo-*p*-dioxins (PCDDs) and dibenzofurans (PCDFs) in Baikal seal, *Chemosphere,* 34, 2419, 1997.
247. Norwood, C. B., Hackett, M., Pruell, R. J., Butterworth, B. C., Williamson, K. J., and Nauman, S. M., Polychlorinated dibenzo-*p*-dioxins and dibenzofurans in selected estuarine sediments, *Chemosphere,* 18, 553, 1989.
248. Creaser, C. S., Fernandes, A. R., Al-Haddad, A., Harrad, S. J., Homer, R. B., Skett, P. W., and Cox, E. A., Survey of background levels of PCDDs and PCDFs in U.K. soils, *Chemosphere*, 18, 767, 1987.
249. Creaser, C. S., Fernandes, A. R., Harrad, S. J., and Cox, E. A., Levels and sources of PCDDs and PCDFs in urban British soils, *Chemosphere*, 21, 931, 1990.
250. Rotard, W., Christmann, W., and Knoth, W., Background levels of PCDD/PCDF in soils in Germany, *Chemosphere*, 29, 2193, 1994.
251. Horstman, M., Bopp, U., and McLachlan, M. S., Comparison of the bulk deposition of PCDD/PCDF in a spruces forest and an adjacent clearing, *Chemosphere*, 34, 1245, 1997.
252. Eitzer, B. D. and Hites, R. A., Atmospheric transport and deposition of polychlorinated dibenzo-*p*-dioxins and dibenzofurans, *Environ. Sci. Technol.*, 23, 1396, 1984.
253. Lohmann, R., Green, N. J. L., and Jones, K. C., Atmospheric transport of polychlorinated dibenzo-*p*-dioxins and dibenzofurans (PCDDs/PCDFs) in air masses across the United Kingdom and Ireland: Evidence of emissions and depletion, *Environ. Sci. Technol.*, 33, 2872, 1999.
254. Hiester, E., Bruckman, P., Bohm, R., Eynck, P., Gerlach, A., Mulder, W., Ristow, H., Pronounced decrease of PCDD/PCDF in ambient air, *Chemosphere*, 34, 1231, 1997.
255. Scheuplein, R. J., Gallo, M. A., van der Heijden, K. A., in *Branbury Report 35: Biological Basis for Risk Assessment of Dioxins and Related Compounds (Epilogue)*, Cold Spring Harbor Laboratory Press, Plainview, NY, 1991, 489.
256. Cook, P. M., Erickson, R. J., Spehar, R. L., Bradbury, S. P., and Ankley, G. T., Interim report on data and methods for assessment of 2,3,7,8-tetrachlorodibenzo-*p*-dioxin risks to aquatic life and associated wildlife, EPA/600/R-93/055, Environmental Research Laboratory, Office of Research and Development, U.S. Environmental Protection Agency, 1993.
257. Van den Berg, M., Birnbaum, L., Bosveld, A. T. C., Brunstrom, B., Cook, P., Feeley, M., Giesy, J. P., Hanberg, A., Hasegawa, R., Kennedy, S. W., Kubiak, T., Larsen, J. C, van Leeuwen, X., Liem, A. K., Nolt, C., Peterson, R. E., Poellinger, L., Safe, S., Schrenk, D., Tillitt, D., Tysklind, M., Younes, M., Waern, F., and Zacharewski, T., Toxic equivalency factors (TEFs) for PCBs, PCDDs, PCDFs for humans and wildlife, *Environ. Health Perspect.*, 106, 775, 1998.

258. Ahlborg, U. G., Brouwer, A., Fingerhut, M. A., Jacobson, J. L., Jacobson, S. W., Kennedy, S. W., Kettrup, A. A., Koeman, J. H., Poiger, H., Rappe, C., Safe, S. H., Seegal, R., Tuomisto, J., and van den Berg, M., Impact of polychlorinated dibenzo-*p*-dioxins, dibenzofurans, and biphenyls on human and environmental health, with special emphasis on application of the toxic equivalency factor concept, *Eur. J. Pharmacol. Environ. Toxicol. Pharmacol. Sect.,* 228, 179, 1992.
259. Ahlborg, U. G., Becking, G. C., Birnbaum, L. S., Brouwer, A., Derks, H. J. G. M., Feeley, M., Golor, G., Hanberg, A., Larsen, J. C., Liem, A. K. D., Safe, S. H., Schlatter, C., Younes, M., and Yrjanheikki, E., Toxic equivalency factors for dioxin-like PCBs. Chemosphere, 28, 1049, 1994.
260. U.S. Fish and Wildlife website, http://njfieldoffice.fws.gov/Contaminants/Contaminants.htm
261. McFarland, V. A. and Clarke, J. U., Environmental occurrence, abundance and potential toxicity of polychlorinated biphenyl congeners: Considerations for a congener-specific analysis, *Environ. Health Perspect.*, 81, 225, 1989.
262. Tillitt, D. E., Gale, R. W., Meadows, J. C., Zajicek, J. L., Peterman, P. H., Heaton, S. N., Jones, P. D., Bursian, S. J., Kubiak, T. J., Giesy, J. P., and Aulerich, R. J., Dietary exposure of mink to carp from Saginaw Bay. 3. Characterization of dietary exposure to planar halogenated hydrocarbons, dioxin equivalents, and biomagnification, *Environ. Sci. Technol.*, 30, 283, 1996
263. Liem, A. K. D., Furst, P., and Rappe, C., Exposure of populations to dioxins and related compounds, *Food Additives Contam.*, 17, 241, 2000.
264. Verrett, M. J., Investigation of the Toxic and Teratogenic Effects of Halogenated Dibenzo-*p*-dioxins and Dibenzofurans in the Developing Chicken Embryo, Memorandum Report, U.S. Food and Drug Administration, Washington, D.C., 1976.
265. Goldstein, J. A., Structure-activity relationships for the biochemical effects and the relationship to toxicity, in *Halogenated Biphenyls, Terphenyls, Naphthalenes, Dibenzodioxins and Related Compounds,* Kimbrough, R. D. and Jensen, A. A., Eds., Elsevier, Amsterdam, 1980, Chap. 6.
266. Giesy, J. P. and Kannan, K., Dioxin-like and non-dioxin-like toxic effects of polychlorinated biphenyls (PCBs): Implications for risk assessment, *Crit. Rev. Toxicol.,* 28, 511, 1998.
267. Bergman, A., Athanasiadou, M., Bergek, S., Haraguchi, Jensen, S., and Klasson-Wehler, E., PCB and PCB methyl sulfones in mink treated with PCBs and various PCB fractions, *Ambio,* 21, 570, 1992.
268. Bergman, A., Klasson-Wehler, E., and Kuroki, H., Selective retention of hydroxylated PCB metabolites in blood, *Environ. Health Perspect.* 102, 465, 1994.
269. Porterfield, S. P., Vulnerability of the developing brain to thyroid abnormalities: Environmental insults to the thyroid system, *Environ. Health Perspect.*, 102, Suppl. 2, 125, 1994.
270. Birnbaum, L. S., Endocrine effects of prenatal exposure to PCBs, dioxins and other xenobiotics: Implications for policy and future research, *Environ. Health Perspect.*, 102, 767, 1994.
271. World Health Organization (WHO), Consultation on assessment of the health risk of dioxins: Re-evaluation of the tolerable daily intake (TDI): Executive Summary, *Food Additives Contam.*,17, 223, 2000.
272. Tillitt, D. E., The toxic equivalents approach for fish and wildlife, *Hum. Ecol. Risk Assess.,* 5, 25, 1999.
273. Hansen, L. G., *The Ortho side of PCBs: Occurrence and Disposition*, Kluwer, Boston, 1999.
274. Robertson, L. W. and Hansen, L. G., *PCBs — Recent Advances in the Environmental Toxicology and Health Effects,* University of Kentucky Press, Lexington, 2001.
275. Schwetz, B. A., Norris, J. M., Sparschu, G. L., Rowe, V. K., Gehring, P. J., Emerson, J. L., and Gerbig, P. J., Toxicology of chlorinated dibenzo-*p*-dioxins, *Environ. Health Perspect.,* 5, 87, 1973.
276. Poland, A. and Knutson, J. C., 2,3,7,8-tetrachorodibenzo-*p*-dioxin and related halogenated aromatic hydrocarbons: Examination of the mechanism of toxicity, *Ann. Rev. Pharmacol. Toxicol.*, 22, 517, 1982.
277. Safe, S., Polychlorinated biphenyls (PCBs), dibenzo-*p*-dioxins (PCDDs), dibenzofurans (PCDFs), and related compounds: Environmental and mechanistic considerations which support the development of toxic equivalency factors (TEFs), *CRC Crit. Rev. Toxicol.*, 21, 51, 1990.
278. Greig, J. B., Jones, G., Butler, W. H., and Barnes, J. M., Toxic effects of 2,3,7,8-tetrachlorodibenzo-*p*-dioxin, *Food Cosmet. Toxicol.*, 11, 585, 1973.
279. Heath, R. G., Spann, J. W., Kreitzer, J., and Vance, C., Effects of polychlorinated biphenyls on birds. in *Proc. XV^TH Int. Ornithol. Congr.,* Voous, K. H., Ed., The Hague, The Netherlands, 1972, 475.
280. Hochstein, J. R., Aulerlich, R. J., and Bursian, S. J., Acute toxicity of 2,3,7,8-tetrachlorodibenzo-*p*-dioxin to mink, *Arch. Environ. Contam. Toxicol.,* 17, 33, 1988.

281. Vos, J. G., Moore, J. A., and Zinkl, J. G., Toxicity of 2,3,7,8-tetrachlorodibenzo-*p*-dioxin (TCDD) in C57B1/6 mice, *Toxicol. Appl. Pharmacol.*, 29, 1974, 229.
282. Beatty, P. W., Holscher, M. A., and Neal, R. A., The comparative toxicity of 2,3,7,8-tetrachlorodibenzo-*p*-dioxin in larval and adult forms of *Rana catesbeina, Bull. Environ. Contam. Toxicol.*, 5, 578, 1976.
283. Olson, J. R., Holscher, M. A., and Neal, R. A., Toxicity of 2,3,7,8-tetrachlorodibenzo-*p*-dioxin in the Golden Syrian hamster, *Toxicol. Appl. Phamarcol.*, 59, 67, 1980.
284. Aulerich, R. J., Bursian, S. J., and Napolitano, A. C., Biological effects of epidermal growth factor and 2,3,7,8-tetrachlorodibenzo-*p*-dioxin on developmental parameters of neonatal mink, *Arch. Environ. Contam. Toxicol.,* 17, 1988, 27.
285. Aulerich, R. J., Bursian, S. J., Evans, M. G., Hochstein, J. R., Koudele, K. A., Olson, B. A., and Napolitano, A. C., Toxicity of 3,4,5,3',4',5'-hexachlorobiphenyl to mink, *Arch. Environ. Contam. Toxicol.,* 16, 53, 1987.
286. Birnbaum, L., Re-evaluation of dioxin, Presentation at Great Lakes Water Quality Board 102nd Meeting — Chicago, International Joint Commission, Windsor, Canada, 1993.
287. Birnbaum, L. S. and Tuomisto, J., Non-carcinogenic effects of TCDD in animals, *Food Additives Contam.,* 17, 275, 2000.
288. Powell, D. C., Aulerich, R. J., Meadows, J. C., Tillitt, D. E., Giesy, J. P., Stromborg, K. L., and Bursian, S. J., Effects of 3,3',4,4',5-pentachlorobiphenyl (PCB 126) and 2,3,7,8-tetrachlorodibenzo-*p*-dioxin (TCDD) injected into the yolks of chicken (*Gallus domesticus*) eggs prior to incubation, *Arch. Environ. Contam. Toxicol.*, 31, 404, 1996.
289. Mably, T. A., Moore, R. W., and Peterson, R. E., *In utero* and lactational exposure of male rats to 2,3,7,8-tetrachlorodibenzo-*p*-dioxin. 1. Effects on androgenic status, *Toxicol. Appl. Pharmacol.,* 114, 97, 1992a.
290. Mably, T. A., Moore, R. W., Goy, R. W., and Peterson, R. E., *In utero* and lactational exposure of male rats to 2,3,7,8-tetrachlorodibenzo-*p*-dioxin. 2. Effects on sexual behavior and the regulation of luteinizing hormone secretion in adulthood, *Toxicol. Appl. Pharmacol.*, 114, 108, 1992b.
291. Mably, T. A., Bjerke, D. L., Moore, R. W., Gendron-Fitzpatrick, A., and Petterson, R. E., *In utero* and lactational exposure of male rats to 2,3,7,8-tetrachlorodibenzo-*p*-dioxin: 3. Effects on spermatogenesis and reproductive capability, *Toxicol. Appl. Pharmacol.*, 114, 118, 1992c.
292. Gray, L. E., Jr. and Ostby, J. S., *In utero* 2,3,7,8-tetrachlorodibenzo-*p*-dioxin (TCDD) alters reproductive morphology and function in female rat offspring, *Toxicol. Appl. Pharmacol.*, 133, 285, 1995.
293. Gray, L. E., Jr., Kelce, W. R., Monosson, E., Ostby, J. S., and Birnbaum, L. S., Exposure to TCDD during development permanently alters reproductive function in male Long Evans rats and hamsters: Reduced ejaculated and epididymal sperm numbers and sex accessory gland weights in offspring with normal androgenic status, *Toxicol. Appl. Pharmacol.,* 131, 108, 1995.
294. Gray, L. E., Jr., Ostby, J. S., and Kelce, W. R., A dose-response analysis of the reproductive effects of a single gestational dose of 2,3,7,8-tetrachlorodibenzo-*p*-dioxin in male Long Evans hooded rat offspring, *Toxicol. Appl. Pharmacol.,* 146, 11, 1997a.
295. Gray, L. E., Jr., Wolf, C., Mann, P., and Ostby, J. S., *In utero* exposure to low doses of 2,3,7,8-tetrachlorodibenzo-*p*-dioxin alters reproductive development of female Long Evans hooded rat offspring, *Toxicol. Appl. Pharmacol.,* 146, 237, 1997b.
296. Safe, S., Metabolism, uptake, storage and bioaccumulation, in *Halogenated Biphenyls, Terphenyls, Naphthalenes, Dibenzodioxins and Related Products,* Kimbrough, R. D. Ed., Elsevier/North-Holland Biomedical Press, New York, 1980, 81.
297. Safe, S., Polyhalogenated aromatics: Uptake, disposition and metabolism, in *Halogenated Biphenyls, Terphenyls, Naphthalenes, Dibenzodioxins and Related Products*, Kimbrough, R. D. and Jensen, A. A., Eds., 2nd ed., Elsevier, Amsterdam, New York, Oxford, 1989, 131.
298. Sawyer, T. and Safe, S., PCB isomers and congeners: Induction of arylhydrocarbon hydroxylase and ethoxyresorufin o-deethylase enzyme activities in rat hepatoma cells, *Toxicol. Lett.*, 13, 57, 1982.
299. Goldstein, J. A. and Safe, S., Mechanism of action and structure-activity relationship for the chlorinated dibenzo-*p*-dioxins and related compounds, in *Halogenated Biphenyls, Terphenyls, Naphthalenes, Dibenzodioxins and Related Compounds,* Kimbrough, R. D. and Jensen, A. A., Eds., Elsevier, Amsterdam, 2nd ed., 1989, Chap. 6.
300. Bosveld, A. T. C. (Bart), and van den Berg, M., Effects of polychorinated biphenyls, dibenzo-*p*-dioxins and dibenzofurans on fish-eating birds, *Environ. Rev.*, 2, 147, 1994.

301. Whyte, J. J., Jung, R. E., Schmitt, C. J., and Tillitt, D. E., Ethoxyresorufin-*O*-deethylase (EROD) activity in fish as a biomarker of chemical exposure, *Crit. Rev. Toxicol.*, 30, 347, 2000.
302. Freeman, H. C., Sangalang, G. B., and Fleming, B., The sublethal effects of polychlorinated biphenyl (Aroclor 1254) diet on the Atlantic cod (*Gadus morhua*), *Sci. Total Environ.*, 24, 1, 1982.
303. Klotz, A. V., Stegmann, J. J., and Walsh, C., An aryl hydrocarbon hydroxylating hepatic cytochrome P-450 from the marine fish *Stenotomus chrysops*, *Arch. Biochem. Biophys.*, 226, 578, 1983.
304. Spies, R. B. and Rice, D. W., Effects of organic chemicals on the starry flounder *Platichthys stellatus* in San Francisco Bay. II. Reproductive success of fish captured in San Francisco Bay and spawned in the laboratory, *Mar. Biol.*, 98, 191, 1988.
305. Marcquenski, S. V. and Brown, S. B., Early mortality syndrome in the salmonid fishes from the Great Lakes, in *Chemically Induced Alterations in Functional Development and Reproduction in Fishes*, Rolland, R. M., Gilbertson, M., and Peeterson, R. E., Eds., Society of Environmental Toxicology and Chemistry, Pensacola, FL, 1997, Chap. 10.
306. Kuiken, T., Fox, G. A., and Danesik K. L., Bill malformations in double-crested cormorants with low exposure to organochlorines, *Environ. Toxicol. Chem.*, 18, 2908, 1999.
307. Moccia, R. D., Fox, G. A., and Britton, A., A quantitative assessment of thyroid histopathology of herring gulls (*Larus argentatus*) from the Great Lakes and a hypothesis on the causal role of environmental contaminants, *J. Wildl. Dis.*, 22, 60, 1986
308. Nebeker, A. V. and Puglisi, A., Effect of polychlorinated biphenyls (PCB's) on survival and reproduction of *Daphnia, Gammarus, and Tanytarus*, *Trans. Am. Fish. Soc.*, 103, 722, 1974a.
309. Nebeker, A. V, Puglisi, A., and DeFoe, D. L., Effect of polychlorinated biphenyl compounds on survival and reproduction of the fathead minnows and flagfish, *Trans. Am. Fish. Soc.*, 103, 562, 1974b.
310. Mayer, J., Mehrle, P. M., and Sanders, H. O., Residue dynamics and biological effects of polychlorinated biphenyls in aquatic organisms, *Arch. Environ. Contam. Toxicol.*, 5, 501, 1977.
311. Mayer, K. S., Mayer,.L., and Witt, A., Jr., Waste transformer oil and PCB toxicity to rainbow trout, *Trans. Am. Fish. Soc.*, 114, 869, 1985.
312. Walker, M. K., Spitsbergen, J. M., Olson, J. R., and Peterson, R. E., 2,3,7,8-tetrachlorodibenzo-*p*-dioxin (TCDD) toxicity during early life stage development of lake trout (*Salvelinus namaycush*), *Can. J. Fish. Aquat. Sci.*, 48, 875, 1991a.
313. Walker, M. K. and Peterson, R. E., Potencies of polychlorinated dibenzo-*p*-dioxin, dibenzofuran, and biphenyl congeners, relative to 2,3,7,8-tetrachlorodibenzo-*p*-dioxin, for producing early life stage mortality in rainbow trout (*Oncorhynchus mykiss*), *Aquat. Toxicol.*, 21, 219, 1991b.
314. Kleeman, J. M, Olson, J. R., and Peterson, R. E., Species differences in 2,3,7,8-tetrachlorodibenzo-*p*-dioxin toxicity and biotransformation in fish, *Fundam. Appl. Toxicol.*, 10, 206, 1988.
315. Hansen, L. G., Wiekhotst, W. B., and Simon, J., Effects of dietary Aroclor 1242 on channel catfish (*Ictalurus punctatus*) and the selective accumulation of PCB components, *J. Fish. Res. Bd. Can.*, 33, 1343, 1976.
316. Elonen, G. E., Spehar, R. L., Holcombe, G. W., Johnson, R. D., Fernandez, J. D., Erickson, R. J., Tietge, J. E., and Cook, P. M., Comparative toxicity of 2,3,7,8-tetrachlorodibenzo-*p*-dioxin to seven freshwater fish species during early life-stage development, *Environ. Toxicol. Chem.*, 17, 472, 1998.
317. Tietge, J. E., Johnson, R. D., Jensen, K. M., Cook, P. M., Elonen, G. E., Fernandez, J. D., Holcombe, G. W., Lothenbach, D. B., and Nichols, J. W., Reproductive toxicity and disposition of 2,3,7,8-tetrachlorodibenzo-*p*-dioxin in adult brook trout (*Salvelinus fontinalis*) following a dietary exposure, *Environ. Toxicol. Chem.*, 17, 2395, 1998.
318. Johnson, R. D., Tietge, J. E., Jensen, K. M. Fernandez, J. D., Linnum, A. L., Lothenbach, D. B., Holcombe, G. W., Cook, P. M., Christ, S. A., Lattier, D. L., and Gordon, D. A., Toxicity of 2,3,7,8-tetrachlorodibenzo-*p*-dioxin to early life stage brook trout (*Salvelinus fontinalis*) following parental dietary exposure, *Environ. Toxicol. Chem.*, 17, 2408, 1998.
319. Walker, M. K. and Peterson, R. E., Toxicity of 2,3,7,8-tetrachlorodibenzo-*p*-dioxin to brook trout (*Salvelinus fontinalis*) during early development, *Environ. Toxicol. Chem.*, 13, 817, 1994.
320. Nichols, J. W., Jensen, K. M., Tietge, J. E., and Johnson, R. D., Physiologically based toxicokinetic model for maternal transfer of 2,3,7,8-tetrachlorodibenzo-*p*-dioxin in brook trout (*Salvelinus fontinalis*), *Environ. Toxicol. Chem.*, 17, 2442, 1998.
321. Hickey, C. R., Jr. and Young, B. H., Incidence of morphological abnormalities in striped bass from the Hudson River and coastal Long Island, New York, *N.Y. Fish and Game J.*, 31, 104, 1984.

322. Mauck, W. L., Mehrle, P. M., and Mayer, L., Effects of the polychlorinated biphenyl Aroclor 1254 on growth, survival, and bone development in brook trout (*Salvelinus fontinalis*), *J. Fish. Res. Bd. Can.*, 35, 1084, 1978.
323. Cook, P. M. and Burkhard, L. P., Development of bioaccumulation factors for protection of fish and wildlife in the Great Lakes, in *Proc. Natl. Sediment Bioaccumulation Conf.,* U.S. Environmental Protection Agency, EPA 823-R-98–002, 3–19, 1998.
324. Fisher, J. P., Spitsbergen, B. B., Bush, B., and Janhan, P. G., Effect of embryonic PCB exposure on hatching success, survival, growth, and developmental behavior in landlocked Atlantic salmon, *Salmo salar*, in *Environmental Toxicology and Risk Assessment: 2nd Volume,* Gorsuch, J. W., Dwyer, J., Ingersoll, C. G., and LaPoint, T. W., Eds., ASTM Publications, Philadelphia, 298, 1994.
325. Cantrell, S. M., Lutz, L. H., Tillitt, D. E., and Hannink, M., Embryotoxicity of 2,3,7,8-tetrachloro-dibenzo-*p*-dioxin (TCDD): The embryonic vasculature is a physiological target for TCDD-induced DNA damage and apoptotic cell death in medaka (*Orizia latipes*), *Toxicol. Appl. Pharm.*, 141, 23, 1996.
326. Cantrell, S. M., Joy-Schlezinger, J., Stegeman, J. J., Tillitt, D. E., and Hannink, M., Correlation of 2,3,7,8-tetrachlorodibenzo-*p*-dioxin-induced apoptotic cell death in the embryonic vasculature with embryotoxicity, *Toxicol. Appl. Pharm.*, 148, 24, 1998.
327. Toomey, B.H., Bello, S., Hahn, M.E., Cantrell, S., Wright, P., Tillitt, D.E., and Di Giulio, R.T., 2,3,7,8-tetrachlorodibenzo-*p*-dioxin induces apoptotic cell death and cytochrome P4501A expression in developing *Fundulus heteroclitus* embryos, *Aquat.Toxicol.*, 51, 127, 2001.
328. Ernst, W., Pesticides and technical organic chemicals, in *Marine Ecology,* Vol. V, Part 4, Kinne, O., Ed., John Wiley and Sons, New York, 1984, 1617.
329. Hendricks, J. D., Putman, T. P., and Sinnhuber, R. O., Null effect of dietary Aroclor 1254 on heptacellular carcinoma incidence in rainbow trout (*Salmo gairdneri*) exposed to aflatoxin $B_1$ as embryos, *J. Environ. Toxicol. Pathol.*, 4, 9, 1980
330. Leatherland, J. and Sonstegard, R. A., Lowering of serum thyroxine and triiodothyronine levels in yearling coho salmon, (*Oncorhynchus kisutch*) by dietary mirex and PCBs, *J. Fish. Res. Bd. Can.,* 35, 1285, 1978.
331. Anderson, M. J., Miller, M. R., and Hinton, D. E., *In vitro* modulation of 17-beta-estradiol-induced vitellogenin synthesis: Effects of cytochrome P4501A1 inducing compounds on rainbow trout (*Oncorhynchus mykiss*) liver cells, *Aquat. Toxicol.,* 34, 327, 1996.
332. Pesonen, M., Korkalainen, M., Laitinen, J. T., Andersson, T. B., and Vakkuri, O., 2,3,7,8-Tetrachloro-dibenzo-*p*-dioxin alters melatonin metabolism in fish hepatocytes, *Chem. Biol. Interact.,* 126, 227, 2000.
333. Schmitt, C. J., Bartish, T. M., Blazer, V. S., Gross, T. S., Tillitt, D. E., Bryant, W. L., and Deweese, L. R., Biomonitoring of environmental status and trends (BEST) program: Contaminants and related effects in fish from the Mississippi, Columbia, and Rio Grande basins, Water-Resour. Invest. Rep. (U.S. Geological Survey), 99–4018B, U.S. Geological Survey Toxic Substances Hydrology Program, 2, 437, 1999.
334. Thomas, P., Effects of Aroclor 1254 and cadmium on reproductive endocrine function and ovarian growth in Atlantic croaker, *Mar. Environ. Res.*, 28, 499, 1989.
335. Khan, I. A. and Thomas, P., Disruption of neuroendocrine function in Atlantic croaker exposed to Aroclor 1254, *Mar. Environ. Res.,* 42, 145, 1996.
336. Khan, I. A. and Thomas, P., Aroclor 1254-induced alteration in hypothalamic momamine metabolism in the Atlantic Croaker (*Micopogonia undulatus*): Correlation with pituitary gonaotropin release, *Neurotoxicology,* 18, 553, 1997.
337. Stouthart, W. J. H. X., Huijbregts, M. A. J., Balm, P. H. M., Lock, R. A. C., and Wendleaar Bonga, S. E., Endocrine stress response and abnormal development in carp (*Cyrinus carpio*) larvae after exposure of the embryos to PCB 126, *Fish Physiol. Biochem.*, 18, 321, 1998.
338. Smeets, J. M. W., van Holsteijn, I., Giesy, J. P., and van den Berg, M., The anti-estrogenicity of Ah receptor agonists in carp (*Cyprinus carpio*) hepatocytes, *Toxicol. Sci.*, 52, 178, 1999.
339. Nirmala, K., Oshima, Y., Lee, R., Imada, N., Honjo, T., and Kobayashi, K., Transgenerational toxicity of tributyltin and its combined effects with polychlorinated biphenyls on reproductive processes in Japanese medaka (*Oryzias latipes*), *Environ. Toxicol. Chem.,* 18, 717, 1999.
340. Olsson, P. E., Westerlund, L., Teh, S. J., Billsson, K., Berg, A. H., Tysklind, M., Nilsson, J., Eriksson, L. O., and Hinton, D. E., Effects of maternal exposure to estrogen and PCB on different life stages of zebrafish (*Danio rerio*), *Ambio,* 28, 100, 1999.

341. Harris, G. E., Kiparissis, Y., and Metcalfe, C. D., Assessment of the toxic potential of PCB congener 81 (3,4,4',5-tetrachlorobiphenyl) to fish in relation to other non-ortho-substituted PCB congeners, *Environ. Toxicol. Chem.*, 13, 1405, 1994.
342. Noguchi, G. E., Immunological disorders associated with polychlorinated biphenyls and related halogenated aromatic compounds, in *Fish Diseases and Disorders Volume 2: Non-Infectious Disorders,* Leatherland, J., and Woo, P. T. K., Eds., CABI, Wallingford, U.K., 1998, Chap. 6.
343. Sijm, T. H. M. and Opperhuizen, A., Dioxins: An environmental risk for fish? in *Environmental Contaminants in Wildlife — Interpreting Tissue Concentrations*, Byer, W. N., Heinz, G. H., and Redmon-Norwood, A. W., Eds., CRC Press, Boca Raton, FL, 1995, Chap. 8.
344. Giesy, J. P. and Snyder, E. M., Xenobiotioc modulation of endocrine responses in fishes, in *Principles and Processes for Evaluating Endocrine Disruption in Wildlife,* Kendall, R., Dickerson, R, Giesy, J., and Suk, W., Eds., Lewis Publishers, Chelsea, MI, 1998, Chap. 8.
345. Fairbrother, A., Ankley, G. T., Birnbaum, L. S., Bradbury, S. P., Francis, B., Gray, L. E., Hinton, D., Johnson, L. L., Peterson, R. E., and van der Kraak, G., Reproductive and developmental toxicology of contaminants in oviparous animals, in *Reproductive and Developmental Effects of Contaminants in Oviparous Vertebrates,* Di Giulio, R. T. and Tillitt, D. E., Eds., SETAC, Pensacola, FL, 1997, Chap. 5.
346. Stegeman, J. J., Schlezinger, J. J., Craddock, J. E., and Tillitt, D. E., Cytochrome P450 1A expression in midwater fishes: Potential of chemical contaminants in remote oceanic zones, *Environ. Sci. Technol.,* 35, 54, 2001.
347. Sileo, L., Karstad, L., Frank, R., Holdrinet, M. V. H., Addison, E., and Braun, H. E., Organochlorine poisoning of ring-billed gulls in southern Ontario, *J. Wildl. Dis.*, 13, 313, 1977.
348. Firestone, D., Etiology of chick edema disease, *Environ. Health Perspect.*, 5, 59, 1973.
349. Heath, R. G., Spann, J. W., Hill, E., and Kreitzer, J., Comparative dietary toxicities of pesticides to birds, U.S. Fish Wildlife Service, Spec. Sci. Rep.-Wildlife, 152, 57, 1972.
350. Dahlgren, R. B., Bury, R. J., Linder, R. L., and Reidinger, R., Jr., Residue levels and histopathology in pheasants given polychlorinated biphenyls, *J. Wildl. Manage.*, 36, 524, 1972.
351. Dahlgren, R. B., Linder, R. L., and Tucker, W. L., Effects of stress on pheasants previously given polychlorinated biphenyls, *J. Wildl. Manage.*, 36, 974, 1972.
352. Vos, J. and Koeman, J., Comparative toxicologic study with polychlorinated biphenyls in chickens with special reference to porphyria, edema formation, liver necrosis, and tissue residues, *Toxicol. Appl. Pharmacol.*, 17, 656, 1970.
353. Vos, J. G., Strik, J. J. T. W. A., van Holsteyn, C. W. M., and Pennings, J. H., Polychlorinated biphenyls as inducers of hepatic porphyria in Japanese quail, with special reference to d-aminolevulinic acid synthetase activity, fluorescence, and residues in the liver, *Toxicol. Appl. Pharmacol.*, 20, 232, 1971.
354. Koeman, J. H., van Velzen-Blad, H. C. W., de Vries, R., and Vos, J. G., Effects of PCB and DDE in cormorants and evaluation of PCB residues from an experimental study, *J. Reprod. Fertil. Suppl.*, 19, 353, 1973.
355. Stickel, W. H., Stickel, L., Dyrland, R. A., and Hughes D. L., Aroclor 1254 residues in birds: Lethal levels and loss rates, *Arch. Environ. Contam. Toxicol.*, 13, 7, 1984.
356. Britton, W. M. and Huston, J. M., Influence of polychlorinated biphenyls in the laying hen, *Poult. Sci.*, 52, 1620, 1973.
357. Sotherland, P. R. and Rahn, H., On the composition of bird eggs, *The Condor,* 89, 48, 1987.
358. Scott, M. L., Effects of PCBs, DDT, and mercury compounds in chickens and Japanese quail, *Fed. Proc.*, 36, 1888, 1977.
359. Blazak, W. and Marcum, J. B., Attempts to induce chromosomal breakage in chicken embryos with Aroclor 1242, *Poult. Sci.*, 54, 310, 1975.
360. Peakall, D. B., Lincer, J. L., and Bloom, S. E., Embryonic mortality and chromosomal alterations caused by Aroclor 1254 in ring doves, *Environ. Health Perspect.*, 1, 103, 1972.
361. Srebocan, V., Pompe-Gotal, J., Brmalj, V., and Plazonic, M., Effect of polychlorinated biphenyls (Aroclor 1254) on liver gluconeogenic enzyme activities in embryonic and growing chickens, *Poult. Sci.*, 56, 732, 1977.
362. Brunstrom, B. and Orberg, J., A method for studying embryotoxicity of lipophilic substances experimentally introduced into hens' eggs, *Ambio*, 11, 209, 1982

363. Rifkind, A. B., Sassa, S., Reyes, J., and Muschick, H., Polychlorinated aromatic hydrocarbon lethality, mixed-function oxidase induction, and uroporphyrinogen decarboxylase inhibition in the chick embryo: Dissociation of dose-response relationships, *Toxicol. Appl. Pharmacol.*, 78, 268, 1985.
364. Hoffman, D. J., Melancon, M. J., Klein, P. N., Eisemann, J. D., and Spann, J. W., Comparative developmental toxicity of planar polychlorinated biphenyl congeners in chickens, american kestrel, and common terns, *Environ. Toxicol. Chem.,* 17, 747, 1998.
365. Brunstrom, B. and Halldin, K., EROD induction by environmental contaminants in avian embryo livers, *Comp. Biochem. Physiol. Part C*, 121, 213, 1998.
366. Gilbertson, M., Effects on fish and wildlife populations, in *Halogenated Biphenyls, Terphenyls, Naphthalenes, Dibenzodioxins and Related Compounds*, 2nd ed., Kimbrough, R. D. and Jensen, A. A., Eds., Elsevier, Amsterdam, 1989, Chap. 4.
367. Gilbertson, M., Kubiak T., Ludwig, J., and Fox, G., Great Lakes embryo mortality, edema, and deformities syndrome (GLEMEDS) in colonial fish-eating birds: Similarity to chick-edema disease, *J. Toxicol. Environ. Health*, 33, 455, 1991.
368. Hoffman, D. J., Rice, C. P., and Kubiak, T. J., PCBs and Dioxins in Birds, in *Environmental Contaminants in Wildlife — Interpreting Tissue Concentrations,* Beyer, N. W., Heinz, G. H., and Redmon, A. W., Eds., CRC Press/Lewis Publishers, Boca Raton, FL, 1996a, Chap. 7.
369. Grasman, K. E., Scanlon, P., and Fox, G. A., Reproductive and physiological effects of environmental contaminants in fish-eating birds of the Great Lakes: A review of historic trends, *Environ. Monit. Assess.*, 53, 117, 1998.
370. Clark, K. E., Niles, L. J., and Stansley, W., Environmental contaminants associated with reproductive failure in bald eagle (*Haliaeetus leucocephalus*) eggs in New Jersey, *Bull. Environ. Contam. Toxicol.*, 61, 247, 1998.
371. Schwartz, T. R., Tillitt, D. E., Feltz, K. P., and Peterman, P. H., Determination of mono- and non-o,o'-chlorine substituted polychlorinated biphenyls in aroclors and environmental samples, *Chemosphere,* 26, 1443, 1993.
372. Kubiak, T. J., Personal communication, unpublished data of the Fish and Wildlife Service and U.S. Geological Survey.
373. Bowerman, W. W., Kubiak, T. J., Holt, J.B., Jr., Evans, R. G., Eckstein, R. G., Sindelar, C. R., Best, D. A., and Kozie, K. D., Observed abnormalities in mandibles of nestling bald eagles *Haliaeetus leucocephalus, Bull. Environ. Contam. Toxicol.*, 53, 450, 1994.
374. Hoffman, D. J., Melancon, M. J., Klein, P. N., Rice, C. P., Eisemann, J. D., Hines, R. K., Spann, J. W., and Pendleton, G. W., Developmental toxicity of PCB 126 (3,3',4,4',5-pentachlorobiphenyl) in nestling American kestrels (*Falco sparverius*), *Fundam. Appl. Toxicol.,* 34, 188, 1996b.
375. Summer, C., Giesy, J., Bursian, S., Render, J., Kubiak, T., Jones, P., Verbrugge, D., and Aulerich, R., Effects induced by feeding organochlorine-contaminated carp from Saginaw Bay, Lake Huron, to laying White Leghorn hens. II. Embryotoxic and teratogenic effects, *J. Toxicol. Environ. Health,* 49, 409, 1996.
376. Brunstrom, B., Sensitivity of embryos from duck, goose, herring gull, and various chicken breeds to 3,3',4,4'-tetrachlorobiphenyl, *Poult. Sci.*, 67, 52, 1988.
377. Cecil, H., Bitman, J., Fries, G., Smith, L., and Lillie, R., PCB's in laying hens, American Chemical Society Fall Meeting, 86, 1977.
378. Tumasonis, C., Bush, B., and Baker, D., PCB levels in egg yolks associated with embryonic mortality and deformity of hatched chicks, *Arch. Environ. Contam. Toxicol.*, 1, 312, 1973.
379. Peakall, D. B. and Peakall, M. L., Effects of a polychlorinated biphenyl on the reproduction of artificially and naturally incubated dove eggs, *J. Appl. Ecol.*, 10, 863, 1973.
380. Heinz, G. H., Hill, E., and Contrera, J., Dopamine and norepinephrine depletion in ring doves fed DDE, Dieldrin, and Aroclor 1254, *Toxicol. Appl. Pharmacol.*, 53, 75, 1980.
381. Powell, D. C., Aulerich, R. J., Meadows, J. C., Tillitt, D. E., Powell, J., Restum, J. C., Stromborg, K. L., Giesy, J. P., and Bursian, S. J., Effects of 3,3',4,4',5-pentachlorobiphenyl (PCB 126), 2,3,7,8-tetrachorodibenzo-*p*-dioxin (TCDD), or an extract derived from field-collected cormorant eggs injected into double-crested cormorant (*Phalacrocorax auritus*) eggs, *Environ. Toxicol. Chem.*, 16, 1450, 1997a.
382. Nosek, J. A., Sullivan, J. R., Hurley, S. S., Olson, J. R., Craven, S. R., and Peterson, R. E., Metabolism and disposition of 2,3,7,8-tetrachlorodibenzo-*p*-dioxin in ring-necked pheasant hens, chicks, and eggs, *Toxicol. Environ. Health,* 35, 1992, 153.

383. Nosek, J. A., Sulivan, J. R., Amundson, T. E., Craven, S. R., Miller, L. M., Fitzpatrick, G., Cook, M. E., and Peterson, R. E., Embryotoxicity of 2,3,7,8-tetrachlorodibenzo-*p*-dioxin in ring-necked pheasants, *Arch. Environ. Contam. Toxicol.,* 23, 1993, 250.
384. Hudson, R., Tucker, R., and Haegele, M., *Handbook of Toxicity of Pesticides to Wildlife*, 2nd ed., U.S. Fish and Wildlife Service, Resources Publication No. 153, Washington, D.C.
385. Harris, H. J., Erdman, T. C., Ankley, G. T., and Lodge, K. B., Measures of reproductive success and PCB residues in eggs and chicks of Forster's tern on Green Bay, Lake Michigan — 1988, *Arch. Environ. Contam. Toxicol.*, 25, 304, 1993.
386. Kubiak, T. J., Harris, H. J., Smith, L. M., Schwartz, T. R., Stalling, D. L., Trick, J. A., Sileo, L., Docherty, D. E., and Erdman, T. C., Microcontaminants and reproductive impairment of the Forster's tern on Green Bay, Lake Michigan — 1983, *Arch. Environ. Contam. Toxicol.*, 18, 706, 1989.
387. Hoffman, D. J., Rattner, B. A., Sileo, L., Docherty, D., and Kubiak, T. J., Embryotoxicity, teratogenicity and aryl hydrocarbon hydroxylase activity in Forster's terns on Green Bay, Lake Michigan, *Environ. Res.*, 42, 176, 1987.
388. Smith, L. M., Schwartz, T. R., Feltz, K., and Kubiak, T. J., Determination and occurrence of AHH-active polychlorinated biphenyls, 2,3,7,8-tetrachlorodibenzo-*p*-dioxin and 2,3,7,8-tetrachlorodibenzofuran in Lake Michigan sediment, and biota. The question of their relative significance, *Chemosphere*, 21, 1063, 1990.
389. Schwartz, T. R. and Stalling, D. L., Chemometric comparison of polychlorinated biphenyl residues and toxicologically active polychlorinated biphenyl congeners in the eggs of forster's terns (*Sterna fosteri*), *Arch. Environ. Contam. Toxicol.*, 20, 183, 1991.
390. Schwartz, T. R., Tillitt, D. E., Feltz, K. P., and Peterman, P. H., Determination of mono- and non-o,o'-chlorine substituted polychlorinated biphenyls in aroclors and environmental samples, *Chemosphere*, 26, 1443, 1993.
391. King, K. A., Custer, T. W., and Quinn, J. S., Effects of mercury, selenium, and organochlorine contaminants on reproduction of Forster's terns and black skimmers nesting in a contaminated Texas bay, *Arch. Environ. Contam. Toxicol.*, 20, 32, 1991.
392. Ludwig, J. P., Auman, H. J., Kurita, H., Ludwig, M. E., Campbell, L. M., Giesy, J. P., Tillitt, D. E., Jones, P., Yamashita, N., Tanabe, S., and Tatsukawa, R., Caspian tern reproduction in the Saginaw Bay ecosystem following a 100-year flood event, *J. Great Lakes Res.,* 19, 96, 1993.
393. Grasman, K. A., Fox, G. A., Scanlon, P., and Ludwig, J. P., Organochlorine-associated immunosuppression in prefledgling Caspian terns and herring gulls from the Great Lakes: An ecoepidemiological study, *Environ. Health Perspect.,* 104, 829, 1996.
394. Brunstrom, B., Toxicity of coplanar polychlorinated biphenyls in avian embryos, *Chemosphere*, 19, 765, 1989.
395. Brunstrom, R., Andersson, L., Nikolaidis, E., and Dencker L., Non-ortho- and mono-ortho-chlorine-substituted polychlorinated biphenyls-embryotoxicity and inhibition of lymphocyte development, *Chemosphere*, 20, 1125, 1990.
396. Elliott, J. E., Kennedy, S. W., Peakall, D. B., and Won, H., Polychlorinated biphenyl (PCB) effects on hepatic mixed function oxidases and porphyria in birds. I. Japanese quail, *Comp. Biochem. Physiol.*, 96C, 205, 1990.
397. Elliott, S. E., Kennedy, S. W., Jeffrey, D., and Shutt, L., Polychlorinated biphenyl (PCB) effects on hepatic mixed function oxidases and porphyria in birds. II. American Kestrel, *Comp. Biochem. Physiol.*, 99C, 141, 1991.
398. McKinney, J. D., Chae, K., Gupta, B. N., Moore, J. A., and Goldstein, J. A., Toxicological assessment of hexachlorobiphenyl isomers and 2,3,7,8-tetrachlorodibenzofuran in chicks, *Toxicol. Appl. Pharmacol.*, 36, 65, 1976.
399. Fox, L. L. and Grasman, K. A., Effects of PCB 126 on primary immune organ development in chicken embryos, *J. Toxicol. Environ. Health, Part A,* 58, 233, 1999.
400. Grasman, K. A. and Whitacre, L. L., Effects of PCB 126 on thymocyte surface marker Expression and immune organ development in chicken embryos, *J. Toxicol. Environ. Health, Part A,* 62, 191, 2001.
401. Grasman, K. A. and Fox, G. A., Associations between altered immune function and organochlorine contamination in young Caspian terns (*Sterna caspia*) from Lake Huron, 1997–1999, *Ecotoxicology*, 10, 101, 2001.

402. Grasman, K. A., Armstrong, M., Hammersley, D. L., Scanlon, P., and Fox, G. A., Geographic variation in blood plasma protein concentrations of young herring gull (*Larus argentatus*) and Caspian terns (*Sterna caspia*) from the Great Lakes and Lake Winnipeg, *Compar. Biochem. Physiol., Part C,* 125, 365, 2000.
403. Mora, M. A., Auman, H. J., Ludwig, J. P., Giesy, J. P., Verbrugge, D. A., and Ludwig, M. E., Polychlorinated biphenyls and chlorinated insecticides in plasma of caspian terns: Relationships with age, productivity, and colony site tenacity in the Great Lakes, *Arch. Environ. Contam. Toxicol.,* 24, 320, 1993.
404. Powell, D. C., Aulerich, R. J., Meadows, J. C., Tillitt, D. E., Stromborg, K. L., Kubiak, T. J., Giesy, J. P., and Bursian, S. J., Organochlorine Contaminants in double-crested cormorants from Green Bay, Wisconsin. II. Effects of an extract derived from cormorant eggs on the chicken embryo, *Arch. Environ. Contam. Toxicol.,* 32, 316, 1997b
405. Gilman, A. P., Hallett, D. J., Fox, G. A., Allan, L. J., and Peakall, D. B., Effects of injected organochlorines on naturally incubated herring gull eggs, *J. Wildl. Manage.*, 42, 484, 1978
406. Fox, G. A., Collins, B., Hayaskawa, E., Weseloh, D. V., Ludwig, J. P., Kubiak, T. J., and Erdman, T. C., Reproductive outcomes in colonial fish-eating birds: A biomarker for developmental toxicants in Great Lakes food chains. II. Spatial variation in the occurrence and prevalence of bill defects in young double-crested cormorants in the Great Lakes, *J. Great Lakes Res.* 17, 158, 1991.
407. Yamashita, N., Tanabe, S., Ludwig, J. P., Kurita, H., Ludwig, M. E., and Tatsukawa, R., Embryonic abnormalities and organochlorine contamination in double-crested cormorants (*Phalacrocorax auritus*) and Caspian terns (*Hydroprogne caspia*) from the upper Great Lakes, *Environ. Pollut.*, 79, 163, 1993.
408. Klasing, K. C., Vitamins, in *Comparative Avian Nutrition,* CAB International, New York, 1998, Chap. 11.
409. Elaroussi, M. A., DeLuca, H., Forte, L. R., and Biellier, H. V., Survival of vitamin D-deficient embryos: Time and choice of chlolecalciferol or its metabolites for treatment in ovo, *Poult. Sci.*, 72, 1118, 1993.
410. Joyner, C. J., Peddie, M. J., and Taylor, T. G., The effect of vitamin D deficiency on yolk formation in the hen, *Biochem. Soc. Trans.,* 14, 341, 1986.
411. Tillitt, D. E., Ankley, G. T., Giesy, J. P., Ludwig, J. P., Kurita-Matsuba, H., Weseloh, D. V., Ross, P. S., Bishop, C. A., Sileo, L., Stromborg, K. L., Larson, J., and Kubiak, T. J., Polychlorinated biphenyl residues and egg mortality in double-crested cormorants from the Great Lakes, *Environ. Toxicol. Chem.*, 11, 1281, 1992.
412. Sanderson, J. T., Kennedy, S. W., and Giesy, J. P., *In vitro* induction of ethoxyresorufin-o-deethylase and porphyrins by halogenated aromatic hydrocarbons in avian primary hepatocytes, *Environ. Toxicol. Chem.,* 17, 2006, 1998.
413. Bleavins, M. R., Aulerich, R. J., Ringer, R. K., and Bell, T. G., Excessive nail growth in the European ferret induced by Aroclor 1242, *Arch. Environ. Contam. Toxicol.,* 11, 305, 1982.
414. Jackson, T. and Halbert, L., A toxic syndrome associated with the feeding of polybrominated biphenyl-contaminated protein concentrate to dairy cattle, *J. Am. Vet. Med. Assoc.*, 165, 437, 1974.
415. Allen, J. R., Barsotti, D. A., van Miller, J. P., Abrahamson, L. J., and Lalich, J. J., Morphological changes in monkeys consuming a diet containing low levels of 2,3,7,8-tetrachlorodibenzo-*p*-dioxin, *Food Cosmet. Toxicol.*, 15, 401, 1977.
416. Morrisey, R. E. and Schwetz, B. A., Reproductive and developmental toxicity in animals, in *Halogenated Biphenyls, Terphenyls, Naphthalenes, Dibenzodioxins and Related Products,* 2nd ed., Kimbrough, R. D. and Jensen, A. A., Eds., Elsevier, Amsterdam, 1989, 195.
417. Hochstein, J. R., Bursian, S. J., and Aulerich, R. J., Effects of dietary exposure to 2,3,7,8-tetrachlorodibenzo-*p*-dioxin in adult female mink (*Mustela vison*), *Arch. Environ. Contam. Toxicol.*, 35, 348, 1998.
418. Render, J. A., Hochstein, J. R., Aulerich, R. J., and Bursian, S. J., Proliferation of periodontal squamous epithelium in mink fed 2,3,7,8-tetrachlorodibenzo-*p*-dioxin (TCDD*), Vet. Hum. Toxicol.,* 42, 85, 2000.
419. Render, J. A., Aulerich, R. J., Bursian, S. J., and Nachreiner, R., Proliferation of maxillary and mandibular periodontal squamous cells in mink fed 3,3',4,4',5-pentachlorobiphenyl (PCB 126), *J. Vet. Diagn. Invest.*, 12, 477, 2000.
420. McNulty, W. P., Toxicity and fetotoxicity of TCDD, TCDF and PCB isomers in rhesus macaques (*Macaca mulatta*), *Environ. Health Perspect.*, 60, 77, 1985.

421. Hotton, D., Davideau, J. L., Dupret, J. M., Pike, J. W., Mathieu, H., and Berdal, A., Effects of 1,25-dihydroxyvitamin D3 in the tooth germ. Modulations of the receptor throughout development, *C.R. Seances Soc. Biol. Ses Fil.,* 185, 482, 1991.
422. Andrews, J. E., Polychlorinated biphenyl (Aroclor 1254) induced changes in femur morphometry, calcium metabolism and nephrotoxicity, *Toxicology,* 57, 83, 1989.
423. Restum, J., Bursian, S., Giesy, J., Render, J., Helferich, W., Shipp, E., Verbrugge, D., and Aulerich, R., Multigenerational study of the effects of consumption of PCB-contaminated carp from Saginaw Bay, Lake Huron, on mink. 1. Effects on mink reproduction, kit growth and survival, and selected biological parameters, *J. Toxicol. Environ. Health,* 54, 343, 1998.
424. Heaton, S. N., Bursian, S. J., Giesy, J. P., Tillitt, D. E., Render, J. A., Jones, P. D., Verbrugge, D. A., Kubiak, T. J., and Aulerich, R. J., Dietary exposure on mink to carp from Saginaw Bay, Michigan. 2. Hematology and liver pathology, *Arch. Environ. Contam. Toxicol.*, 29, 411, 1995.
425. Giesy, J. P., Verbrugge, D. A., Othout, R. A., Bowerman, W. W., Mora, M. A., Jones, P. D., Newstead, J. L., Vandervoort, C., Heaton, S. N., Aulerich, R. J., Busian, S. J., Ludwig, J. P., Dawson, G. A., Kubiak, T. J., Best, D. A., and Tillitt, D. E., Contaminants in fishes from Great Lakes-influenced sections and above dams of three Michigan rivers. II. Implications for health of mink, *Arch. Environ. Contam. Toxicol.*, 27, 213, 1994.
426. Paasivirta, J., *Chemical Ecotoxicology,* Lewis Publishers, Chelsea, MI, 1991, 127.
427. Reijnders, P. J. H., Reproductive failure in common seals feeding on fish from polluted coastal waters, *Nature*, 324, 456, 1986.
428. Clark, D. R., Jr. and Lamont, T. G., Organochlorine residues and reproduction in the big brown bat, *J. Wildl. Manag.*, 40, 249, 1974.
429. Clark, D. R. and Krynitsky, A. K., Organochlorine residues and reproduction in the little brown bat, Laurel, Maryland-June 1976, *Pest. Monit. J.*, 12, 113, 1978.
430. Reinhold, J. O., Hendriks, A. J., Slager, L. K., and Ohm, M., Transfer of microcontaminants from sediment to chironomids, and the risk for the Pond bat *Myotis dasycneme* (Chiroptera) preying on them, *Aquat. Ecol.*, 33, 363, 1999.
431. Mason C. and Wren, C. D., Carnivora, in *Ecotoxicology of Wild Mammals,* Shore R. and Rattner, B. A., Eds., Wiley, London, 2001, Chap. 7.
432. Ross, P. S. and Troisi, G. M., Pinnipedia, in *Ecotoxicology of Wild Mammals,* Shore, R. and Rattner, B. A., Eds., Wiley, London, 2001, Chap. 8.
433. O'Shea, T. J. and Aguilar, A., Cetacea and Sirenia. in *Ecotoxicology of Wild Mammals,* Shore, R. and Rattner, B.A., Eds., Wiley, London, 2001, Chap. 9.
434. Kannan, K., Blankenship, A. L., Jones, P. D., and Giesy, J. P., Toxicity reference values for the toxic effects of polychlorinated biphenyls to aquatic mammals, *Hum. Ecol. Risk Assess.,* 6, 181, 2000.
435. Brouwer, A., Reijnders, P. J. H., and Koeman, J. H., Polychlorinated biphenyl (PCB)-contaminated fish induces vitamin A and thyroid hormone deficiency in the common seal (*Phoca vitulina*), *Aquat. Toxicol. (Amsterdam),* 15, 99, 1989.
436. *Ambio* Special Issue: Seals and Seal Protection, 21, 1992.
437. Swart, R. L. De., Ross, P. S., Timmerman, H. H., Vos, H. W., Reijnders, P. J. H., Vos, J. G., and Osterhaus, A. D. M. E., Impaired cellular immune response in harbor seals (*Phoca vitulina*) feeding on environmentally contaminated herring, *Clin. Exp. Immunol.*, 101, 480, 1995.
438. Troisi, G. M. and Mason, C., Cytochromes P450, P420 and mixed-function oxidases as biomarkers of polychlorinated biphenyl (PCB) exposure in harbor seals (*Phoca vitulina*), *Chemosphere*, 35, 1933, 1997.
439. Young, D., Becerra, M., Kopec, D., and Echols, S., GC/MS analysis of PCB congeners in blood of the harbor seal Phoca vitulina from San Francisco Bay, *Chemosphere,* 37, 4, 1998.
440. Shaw, S. D., Brenner, D., Hong, C. S., Bush, B., and Shopp, G. M., Low-level exposure to PCBs is associated with immune and endocrine disruption in neonatal harbor seals (Phoca vitulina) from the California coast, *Organohalogen Compd.,* 42, 11, 1999.
441. Bergman, A. A., Health condition of the Baltic grey seal (*Halichoerus grypus*) during two decades, Gynaecological health improvement but increased prevalence of chronic ulcers, *APMIS*, 107, 270, 1999.
442. Harding, L. E., Harris, M. L., Stephen, C. R., and Elliott, J. E., Reproductive and morphological condition of wild mink (*Mustela vison*) and river otters (*Lutra canadensis*) in relation to chlorinated hydrocarbon contamination, *Environ. Health Perspect.*, 107, 141, 1999.

443. Gray, L. E., Jr., Wolf, C., Lambright, C., Mann, P., Price, M., Cooper, R. L., and Ostby, J., Administration of potentially antiandrogenic pesticides (procymidone, linuron, iprodione, chlozolinate, *p,p'*-DDE, and ketoconazole) and toxic substances (dibutyl- and diethylhexyl phthalate, PCB 169, and ethane dimethane sulphonate) during sexual differentiation produces diverse profiles of reproductive malformations in the male rat, *Toxicol. Ind. Health,* 15, 94, 1999.
444. Muir, D. A., Braune, B., DeMarch, B., Norstrom, R., Wagemann, R., Lockhart, L., Hargrave, L., Bright, D., Addison, R., Payne, J., and Reimer, K., Spatial and temporal trends and effects of contaminants in the Canadian Arctic marine ecosystem: A review, *Sci. Total Environ.*, 230, 83, 1999.
445. Oehme, M., Biseth, A., Schlabach, M., and Wiig, O., Concentrations of polychlorinated dibenzo-*p*-dioxins, dibenzofurans and non-*ortho* substituted biphenyls in polar bear milk from Svalbard Norway, *Environ. Pollut.*, 90, 401, 1995.
446. Wiig, O., Derocher, A. E., Cronin, M. M., and Skaare, J. U., Female pseudohermaphrodite polar bears at Svalbard, *J. Wildl. Dis.*, 34, 792, 1998.
447. Skaare, J. U., Bernhoft, A., Derocher, A., Gabrielsen, G. W., Goksoyr, A., Henriksen, E., Larsen, H. J., Lie, E., and Wiig, O., Organochlorines in top predators at Svalbard: Occurrence, levels and effects, *Toxicol. Lett.*, 112–113, 103, 2000.
448. De Guise, S., Martineau, D., Beland, P., and Fournier, M., Possible mechanisms of action of environmental contaminants on St. Lawrence beluga whales (*Delphinapterus leucas*), *Environ. Health Perspect.,* 103, 73, 1995.
449. Lahvis, G. P., Wells, R. S., Kuehl, D. W., Stewart, J. L., Rhinehart, H. L., and Via, C. S., Decreased lymphocyte responses in free-ranging bottlenose dolphins (*Tursiops truncatus*) are associated with increased concentrations of PCBs and DDT in peripheral blood, *Environ. Health Perspect.*, 103, 67, 1995.
450. Barber, T. R., Chappie, D. J., Duda, D. J., Fuchsman P. J., and Finley, B. L., Using a spiked sediment bioassay to establish a no-effect concentration for dioxin exposure to the amphipod Ampelisca abdita, *Environ. Toxicol. Chem.,* 17, 420, 1998.
451. Ashley, C. M., Simpson, M. G., Holdich, D. M., and Bell, D. R., 2,3,7,8-Tetrachloro-dibenzo-*p*-dioxin is a potent toxin and induces cytochrome P450 in the crayfish, *Pacifastacus leniusculus*, *Aquat. Toxicol. (Amsterdam),* 35, 157, 1996.
452. Hahn, M., Karchner, S., Shapiro, M., and Perera, S., Molecular evolution of two vertebrate aryl hydrocarbon (dioxin) receptors (AHR1 and AHR2) and the PAS family, *Proc. Natl. Acad. Sci. U.S.A.*, 94, 13743, 1997.
453. Powell, C. J., Bradfield, C., and Wood, W., *Caenorhabditis elegans* orthologs of the aryl hydrocarbon receptor and its heterodimerization partner the aryl hydrocarbon receptor nuclear translocator, *Proc. Natl. Acad. Sci.U.S.A.*, 95, 2844, 1998.
454. Adams, J. A. and Haileselassie, H. M., The effects of polychlorinated biphenyls (Aroclors 1016 and 1254) on mortality, reproduction, and regeneration in *Hydra oligactis*, *Arch. Environ. Contam. Toxicol.*, 13, 4913, 1984.
455. EPA, Ambient water quality criteria for polychlorinated biphenyls, U.S. Environmental Protection Agency Rep., 440/5–80–068, 1980, 221.
456. MacDonald, D. D., Ingersoll, C. G., and Berger, T. A., Development and evaluation of consensus-based sediment quality guidelines, *Arch. Environ. Contam. Toxicol.*, 39, 20, 2000.
457. Huang, Y-W., Melancon, M. J., Jung, R. E., and Karasov, W. H., Induction of cytochrome P450-associated monooxygenases in northern leopard frogs, *Rana pipiens,* by 3,3',4,4',5-pentachlorobiphenyl, *Environ. Toxicol. Chem.*, 17, 1564, 1998.
458. Huang, Y-W., Karasov, W. H., Patnode, K. A., and Jefcoate, C. R., Exposure of northern leopard frogs in the Green Bay ecosystem to polychlorinated biphenyls, polychlorinated dibenzo-*p*-dioxins, and polychlorinated dibenzofurans is measured by direct chemistry but not by hepatic ethoxyresorufin-*o*-deethylase activity, *Environ. Toxicol. Chem.,* 18, 2123, 1999.
459. Huang, Y-W., and Karasov, W. H., Oral bioavailability and toxicokinetics of 3,3',4,4',5-pentachlorobiphenyl in northern leopard frogs, *Rana pipiens, Environ. Toxicol. Chem.,* 19, 1788, 2000.
460. Sparling, D. W., Ecotoxicology of organic contaminants to amphibians, in *Ecotoxicology of Amphibians and Reptiles,* Sparling, D. W., Linder, G., and Bishop, C. A., Eds., SETAC Press, Pensacola, FL, 2000, Chap. 8a.

461. Portelli, M. J. and Bishop, C. A., Ecotoxicology of organic contaminants in reptiles: A review of the concentrations and effects of organic contaminants in reptiles, in *Ecotoxicology of Amphibians and Reptiles,* Sparling, D. W., Linder, G., and Bishop, C. A., Eds., SETAC Press, Pensacola, FL, 2000, Chap. 8b.

CHAPTER 19

# Receiving Water Impacts Associated with Urban Wet Weather Flows

**Robert Pitt**

## CONTENTS



## 19.1 INTRODUCTION

The main purpose of treating stormwater is to reduce its adverse impacts on the beneficial uses of receiving waters. Therefore, it is important in any urban stormwater study to assess the detrimental effects that runoff is actually having on a receiving water. Urban receiving waters may have many beneficial-use goals, including:

- stormwater conveyance (flood prevention)
- biological uses (warm water fishery, biological integrity, etc.)
- noncontact recreation (linear parks, aesthetics, boating, etc.)
- contact recreation (swimming)
- water supply

Two joint research projects recently funded by the U.S. Environmental Protection Agency (EPA[145,146]) examined the historical development of stormwater management programs and modifications that should be incorporated into future design procedures. The projects found that, with full development in an urban watershed and with no stormwater controls, it is unlikely that any of the above listed uses can be fully obtained. With less development and with the application of stormwater controls, some uses may be possible. It is important that unreasonable expectations not be placed on urban waters, as the cost to obtain these uses may be prohibitive. With full-scale development and lack of adequate stormwater controls, severely degraded streams will be common. However, stormwater conveyance and aesthetics should be the basic beneficial-use goals for all urban waters. Biological integrity should also be a goal, but with the realization that the natural stream ecosystem will be severely modified with urbanization. Certain basic controls, installed at the time of development, plus protection of stream habitat, may enable partial use of some of these basic goals in urbanized watersheds. Careful planning and optimal utilization of stormwater controls are necessary to achieve these basic goals in most watersheds. Water contact recreation, consumptive fisheries, and water supplies are not appropriate goals for most urbanized watersheds. However, these higher uses may be possible in urban areas where the receiving waters are large and drain mostly undeveloped areas.

*Water Environment & Technology*[1] reported that a recent National Water Quality Inventory released by the EPA only showed a slight improvement in the attainment of beneficial uses in the nation's waters. Urban runoff was cited as the leading source of problems in estuaries, with nutrients and bacteria as the leading problems. Problems in rivers and lakes were caused mostly by agricultural runoff, with urban runoff the third ranked source for lakes and the fourth ranked source for rivers. Bacteria, siltation, and nutrients were the leading problems in the nation's rivers and lakes. Borchardt and Statzner[2] stressed that many conditions may affect receiving waters from stormwater, specifically physical factors (such as shear stress) and chemical factors (such as oxygen depletion and/or nonionized ammonia).

In general, monitoring of urban stormwater runoff has indicated that the biological beneficial uses of urban receiving waters are most likely affected by habitat destruction and long-term exposures to contaminants (especially to macroinvertebrates via contaminated sediment), while documented effects associated with acute exposures of toxicants in the water column are rare.[3–5] Receiving water contaminant concentrations resulting from runoff events and typical laboratory bioassay test results have not indicated many significant short-term receiving-water problems. As an example, Lee and Jones-Lee[6] state that the existence of a beneficial-use impairment is not necessarily implied if numeric criteria are exceeded by short-term discharges. Many toxicologists and water-quality experts have concluded that the relatively short periods of exposures to the toxicant concentrations in stormwater are not sufficient to produce the receiving-water effects that are evident in urban receiving waters, especially considering the relatively large portion of the toxicants that are associated with particulates.[7] Lee and Jones-Lee[7] conclude that the biological problems evident in urban receiving waters are mostly associated with illegal discharges and that the sediment-bound toxicants are of little risk. Mancini and Plummer[8] have long been advocates of numeric water-quality standards for stormwater that reflect the partitioning of the toxicants and the short periods of exposure during rains. Unfortunately, this approach attempts to isolate individual runoff events and does not consider the cumulative adverse effects caused by the frequent exposures of receiving-water organisms to stormwater.[9–11] Recent investigations have identified acute toxicity problems associated with moderate-term (about 10 to 20 days) exposures to adverse toxicant concentrations in urban receiving streams.[12] However, the most severe receiving-water problems are likely associated with chronic exposures to contaminated sediment and to habitat destruction.

Pathogens in stormwater are also a significant concern potentially affecting human health. The use of indicator bacteria is controversial for stormwater, as is the assumed time of typical exposure of swimmers to contaminated receiving waters. However, recent epidemiological studies have shown significant health effects associated with stormwater-contaminated marine swimming areas. Protozoa pathogens, especially associated with likely sewage-contaminated stormwater, is also a public-health concern.

Evaluating a receiving water and understanding the potential role that urban wet-weather flows may have on its beneficial uses is a complex and time-consuming activity. Burton and Pitt[13] have produced a comprehensive book describing the development of effective monitoring strategies including selection of parameters, development of the experimental design, and detailed guidance on sampling, analyses, and data interpretation.

Urban runoff has been found to cause significant receiving water impacts on aquatic life.[3,4,13] The effects are obviously most severe for receiving waters draining heavily urbanized watersheds. However, some studies have shown important aquatic life impacts for streams in watersheds that are less than 10% urbanized.[19,22]

To best identify and understand these impacts, it is necessary to include biological monitoring, using a variety of techniques, and sediment-quality analyses, in a monitoring program. Water-column testing alone has been shown to be very misleading. Most aquatic life impacts associated with urbanization are probably related to long-term problems caused by polluted sediments and food-web disruption. Transient water-column-quality conditions associated with urban runoff probably rarely cause significant aquatic-life impacts.

The underlying theme of these researchers is that an adequate analysis of receiving-water biological impacts must include investigations of a number of biological organism groups (fish, benthic macroinvertebrates, algae, rooted macrophytes, etc.) in addition to studies of water and sediment quality.[13] Simple studies of water quality alone, even with possible comparisons with water-quality criteria for the protection of aquatic life, are usually inadequate to predict biological impacts associated with urban runoff.

Duda et al.[14] presented a discussion on why traditional approaches to assessing water quality — and selecting control options — in urban areas have failed. The main difficulties of traditional approaches when used with urban runoff are: the complexity of contaminant sources, wet-weather monitoring problems, and limitations when using water-quality standards to evaluate the severity of wet-weather receiving-water problems. They also discuss the difficulty of meeting water-quality goals in urban areas that were promulgated in the Water Pollution Control Act.

Relationships between observed receiving-water biological effects and possible causes have been especially difficult to identify, let alone quantify. The studies reported in this paper have identified a wide variety of possible causative agents, including sediment contamination, poor water quality (low dissolved oxygen, high toxicants, etc.), and factors affecting the physical habitat of the stream (high flows, unstable streambeds, absence of refuge areas, etc.). It is expected that all of these factors are problems, but their relative importance varies greatly depending on the watershed and receiving-water conditions. For example, Horner[15] notes that many watershed-, site-, and organism-specific factors must be determined before the best combination of runoff-control practices to protect aquatic life can be determined.

The time scale of biological impacts in receiving waters affected by stormwater must also be considered. Snodgrass et al.[16] reported that ecological responses to watershed changes may take between 5 and 10 years to equilibrate. Therefore, receiving-water investigations conducted soon after disturbances or mitigation may not accurately reflect the long-term conditions that will eventually occur. They found that the first changes due to urbanization will be to stream and groundwater hydrology, followed by fluvial morphology, then water quality, and finally the aquatic ecosystem. They also reported that it is not possible to predict biological responses from in-stream habitat changes or conditions, although they, along with many other researchers,[26,62–67,77,80–82,85–88,99,102,103,105–108] have found that habitat changes are among the most serious causes of the aquatic biological problems associated with urbanization of a watershed.

This chapter contains numerous citations and summaries of many research activities conducted throughout the world concerning receiving-water effects associated with urban wet-weather flows, specifically stormwater. The purpose of these discussions is to demonstrate the widespread nature of the various problems that may occur. During the last few years, the number of technical publications addressing stormwater-quality issues has risen dramatically. The annual literature review issues of *Water Environment Research*[148–152] contain a growing section on urban wet-weather flows. In 1996, about 250 international references were cited in this review, growing to more than 500 in the 2000 review. Many of these references addressed receiving-water impacts and are summarized in this chapter. Obviously, many other interesting papers may exist but have been unintentionally left out of this review.

## 19.2 GROSS INDICATORS OF ACUTE AQUATIC ORGANISM STRESS IN URBAN RECEIVING WATERS

### 19.2.1 Dissolved-Oxygen-Depletion Investigations

Dissolved-oxygen (DO) stream levels have historically been used to indicate receiving-water problems associated with point-source contaminant discharges and with combined sewer overflows. Therefore, early investigations of the effects of stormwater discharges mostly focused on in-stream dissolved oxygen conditions downstream from outfalls. Of course, DO levels are also being evaluated in most current receiving water investigations also, but the emphasis has shifted more toward elevated nutrient and toxicant concentrations, plus numerous other indicators of aquatic organism stress, as described later.

An early study of DO in urban streams affected only by stormwater was conducted by Ketchum[17] in Indiana. Sampling was conducted at nine cities, and the project was designed to detect significant DO deficits in streams during periods of rainfall and runoff. The results of this study indicated that wet-weather DO levels generally appeared to be similar or higher than those observed during dry-weather conditions in the same streams. They found that significant wet-weather DO depletions were not observed, and due to the screening nature of the sampling program, more subtle impacts could not be measured. Heaney et al.,[18] during their review of studies that examined continuous DO monitoring stations downstream from urbanized areas, indicated that the worst DO levels occurred after rainstorms in about one third of the cases studied. This lowered DO could be due to urban runoff moving downstream, combined sewer overflows, or resuspension of benthic deposits. Resuspended benthic deposits could have been previously settled urban runoff settleable solids. They also found that worst-case conditions do not always occur during the low flow periods following storms. As noted below, adverse DO conditions associated with urban runoff are likely to occur a substantial time after the runoff event and downstream from the discharge locations.

Figure 19.1 illustrates a problem that may be common to DO predictions in urban receiving waters. Pitt[19] conducted three long-term biochemical oxygen demand (BOD) experiments with stormwater collected from a residential area in San Jose, California. These were conventional BOD tests, using approved procedures published in the then current version of *Standard Methods*. Basically, many BOD bottles were prepared for each sample, representing replicates for each day for the observations, and for several different dilutions. The bottles were seeded with activated sludge to provide a starting microbial population. As the figure shows, the observed BOD curves do not have a conventional shape. The $BOD_5$ values are about 25 mg/L, typical of what is commonly reported for most stormwater. However, the BOD curves are seen to rapidly increase throughout the 20-day test period, instead of leveling off at about 7 to 10 days, as expected for municipal

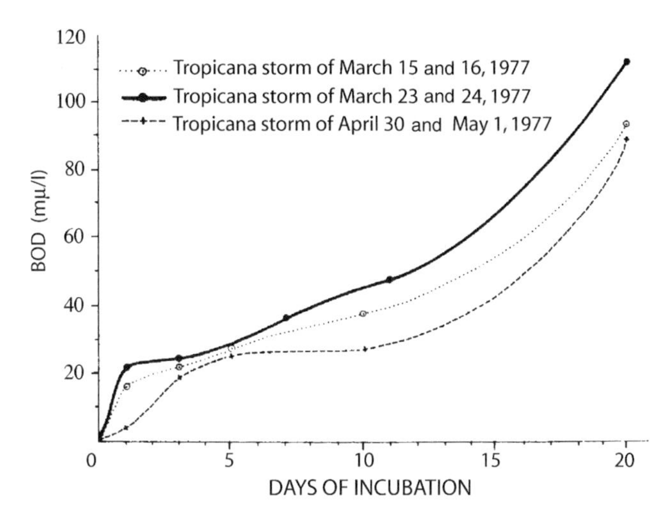


**Figure 19.1** Long-term BOD tests for stormwater.[19]

wastewaters. These curves illustrate the common problem of acclimation of a wastewater to the microorganisms that are present in the test solution. Stormwater has relatively low levels of nutrients and easily assimilated organic material, but moderate levels of toxicants. It is possible that the activated sludge seed requires extra time for the microbial population to shift to a population dominated by organisms capable of effectively degrading the organics in stormwater. Alternatively (or in addition), the more refractory organics in stormwater may simply require a longer period of time for degradation. In any case, the ultimate BOD/$BOD_5$ ratio for stormwater is much greater than for conventional municipal wastewaters, making simple use of observed $BOD_5$ values in receiving water models problematic. Urban-stream sediments are commonly anaerobic, likely caused by the deposition of the slowly decaying stormwater organic compounds. Stormwater effects on short-term stream DO levels may be minimal, but sediment interaction (including scour) with the water can have adverse effects long after the stormwater event that discharged the decaying material. Therefore, the misuse of the classical $BOD_5$ test for stormwater can lead to poor conclusions concerning urban DO conditions, one of the more commonly used indicators of ecological health.

### 19.2.2 Urban Runoff Effects on Receiving Water Contaminant Concentrations

Numerous data are available characterizing stormwater chemical characteristics. This discussion summarizes a few example cases where in-stream measurements found significant changes in quality as a function of land use. These studies usually sampled streams as they passed through urban areas, from upstream relatively uncontaminated areas through and past urban areas. Both wet- and dry-weather sampling was also usually conducted.

In the southeast, many urban lakes in developing areas are typically characterized by high turbidity levels caused by high erosion rates of fine-grained clays. There has been conflicting evidence on the role of these elevated turbidity levels on eutrophication processes and resulting highly fluctuating DO levels. Because of the high sediment loads, these urban lakes are quite different compared to most studied impoundments. Burkholder et al.[20] described a series of enclosure experiments they conducted in Durant Reservoir, near Raleigh, North Carolina. The experimental design allowed investigating the effects of different levels of sediment and nutrients on algal productivity. They found that the effects (reduction of light reduction and coflocculation of clay and phosphate) of low (about 5 mg/L) and moderately high clay (about 15 mg/L) loadings added every 7 to 14 days did not significantly reduce the algal productivity simulation caused by high phosphate loadings. However, they noted that other investigators using higher clay loadings (about

25 mg/L added every 2 days) did see depressed effects of phosphorus enrichment on the test lake. They concluded that dynamically turbid systems, such as represented in southeastern urban lakes, have complex interacting mechanisms between discharged clay and nutrients that make simple predictions of the effects of eutrophication much more difficult than in the more commonly studied clear lakes. In general, they concluded that increased turbidity will either have no effect, or a mitigating effect, on the cultural eutrophication process.

Field and Cibik[21] summarized some potential urban-runoff effects reported in other studies. Two studies of a reservoir near Knoxville, Tennessee, showed that the quality of the contributing streams were degraded to a small extent by urban runoff and that the reservoir itself experienced a significant change in DO, pH, $BOD_5$, conductivity, temperature, total solids, and total coliform bacteria during short storm events. In another study at the Christina River in Newark, Delaware, cadmium and lead concentrations several miles below the urban area remained at elevated values up to 48 h after storm periods. The quality of runoff from similar nonurbanized watersheds was compared with this urbanized area's runoff. They found that concentrations of nitrates, phosphorus, heavy metals, and pesticides were considerably higher in the urbanized areas than in the forested regions. Field and Cibik[21] also reported on a study conducted in Virginia, where water, sediment, detritus, caddisflies, snails, and crayfish were analyzed for iron, manganese, nickel, lead, cadmium, zinc, chromium, and copper. The sampling areas were exposed to wastewater effluent and urban runoff. They found that concentrations increased immediately below stormwater-discharge locations. They also reported on a study from Hawaii that indicated that receiving-water conditions were designated as hazardous because of very high concentrations of suspended solids, heavy metals, and bacterial pathogens.

During the Coyote Creek, San Jose, California study, dry-weather concentrations of many constituents exceeded expected wet-weather concentrations by factors of two to five times.[22] During dry weather, many of the major constituents (e.g., major ions, total solids, etc.) were significantly greater in both the urban and nonurban reaches. These constituents were all found in substantially lower concentrations in the urban runoff and in the rain. The rain and the resultant runoff apparently diluted the concentrations of these constituents in the creek during wet weather. Within the urban area, many constituents were found in greater concentrations during wet weather than during dry weather (chemical oxygen demand, organic nitrogen, and especially heavy metals — lead, zinc, copper, cadmium, mercury, iron, and nickel). Lead concentrations were found to be more than seven times greater in the urban reach than in the nonurban reach during dry weather, with a confidence level of 75%. Other significant increases in urban-area concentrations occurred for nitrogen, chloride, orthophosphate, chemical oxygen demand (COD), specific conductance, sulfate, and zinc. The DO measurements were about 20% less in the urban reach than in the nonurban reach of the creek.

Bolstad and Swank[23] examined the in-stream water quality at five sampling stations in Cowetta Creek in western North Carolina over a 3-year period. The watershed is 4350 ha and is relatively undeveloped (forested) in the area above the most upstream sampling station and becomes more urbanized at the downstream sampling station. Baseflow water quality was good, while most constituents increased during wet weather. Bacteria values increased substantially during wet weather, with total and fecal coliforms and fecal streptococci increasing by two to three times during storms. Water quality was compared to building density for the different monitoring stations, with increasing stormwater contaminant concentrations (especially for turbidity, bacteria, and some inorganic solutes) with increasing building densities. Baseflow concentrations also typically increased with increasing urban density, but at a much lower rate. In addition, the highest concentrations observed during individual events corresponded to the highest flow rates.

### 19.2.3 Reported Fish Kill Information

Urban-runoff impacts are sometimes difficult for many people to appreciate in urban areas. Fish kills are the most obvious indication of water-quality problems for many people. However, because urban receiving-water quality is usually so poor, the aquatic life in typical urban receiving waters is usually limited in abundance and diversity and quite resistant to poor water quality. Sensitive native organisms have typically been displaced, or killed, long ago. It is also quite difficult to identify the specific cause of a fish kill in an urban stream. Ray and White,[24] for example, stated that one of the complicating factors in determining fish kills related to heavy metals is that the fish mortality may lag behind the first toxic exposure by several days and is usually detected many miles downstream from the discharge location. The actual concentrations of the water-quality constituents that may have caused the kill could then be diluted beyond detection limits, making probable sources of the toxic materials impossible to determine in many cases.

Heaney et al.[18] reviewed fish-kill information reported to government agencies from 1970 to 1979. They found that less than 3% of the reported 10,000 fish kills were identified as having been caused by urban runoff. This is less than 30 fish kills per year nationwide. A substantial number of these 10,000 fish kills were not identified as having any direct cause. They concluded that many of these fish kills were likely caused by urban runoff or by a combination of problems that could have been worsened by urban runoff.

During the Bellevue, Washington, receiving-water studies, some fish kills were noted in the unusually clean urban streams.[25] The fish kills were usually associated with inappropriate discharges to the storm drainage system (such as cleaning materials and industrial chemical spills) and not from "typical" urban runoff. However, as noted later, the composition of the fish in the urban stream was quite different, as compared to the control stream.[26]

Fish-kill data have therefore not been found to be a good indication of receiving water problems caused by urban runoff. However, the composition of the fisheries and other aquatic-life taxonomic indicators are sensitive indicators of receiving-water problems in urban streams.

### 19.2.4 Toxicological Effects of Stormwater

Even though acute toxicity of stormwater on most aquatic organisms has been relatively rare, short-term toxicity tests are still commonly conducted as part of some whole-effluent-toxicity (WET) tests required by some state regulatory agencies and by some stormwater researchers.[147]

The need for endpoints for toxicological assessments using multiple stressors was discussed by Marcy and Gerritsen.[27] They used five watershed-level ecological risk assessments to develop appropriate endpoints based on specific project objectives. Dyer and White[28a] also examined the problem of multiple stressors affecting toxicity assessments. They felt that field surveys can rarely be used to verify simple single-parameter laboratory experiments. They developed a watershed approach integrating numerous databases in conjunction with *in situ* biological observations to help examine the effects of many possible causative factors. Toxic-effect endpoints are additive for compounds having the same "mode of toxic action," enabling predictions of complex chemical mixtures in water, as reported by *Environmental Science & Technology*.[28b] They reported that EPA researchers at the Environmental Research Laboratory in Duluth, Minnesota, identified about five or six major action groups that contain almost all of the compounds of interest in the aquatic environment. Much work still needs to be done, but these new analytical methods may enable the in-stream toxic effects of stormwater to be better predicted.

Ireland et al.[29] found that exposure to UV radiation (natural sunlight) increased the toxicity of polycyclic aromatic hydrocarbon (PAH)-contaminated urban sediments to *C. dubia*. The toxicity was removed when the UV wavelengths did not penetrate the water column to the exposed organisms. Toxicity was also reduced significantly in the presence of UV when the organic fraction of the stormwater was removed. Photo-induced toxicity occurred frequently during low-flow conditions and wet weather but was reduced during turbid conditions.

Johnson et al.[30] and Herricks et al.[10,11] describe a structured tier testing protocol to assess both short-term and long-term wet-weather discharge toxicity that they developed and tested. The protocol recognizes that the test systems must be appropriate to the time scale of exposure during the discharge. Therefore, three time-scale protocols were developed — for intraevent, event, and long-term exposures. The use of standard WET tests was found to overestimate the potential toxicity of stormwater discharges.

The effects of stormwater on Lincoln Creek, near Milwaukee, Wisconsin, were described by Crunkilton et al..[12] Lincoln Creek drains a heavily urbanized watershed of 19 mi$^2$ that is about 9 mi long. On-site toxicity testing was conducted with side-stream flow-through aquaria using fathead minnows, plus in-stream biological assessments, along with water and sediment chemical measurements. In the basic tests, Lincoln Creek water was continuously pumped through the test tanks, reflecting the natural changes in water quality during both dry- and wet-weather conditions. The continuous flow-through mortality tests indicated no toxicity until after about 14 days of exposure, with more than 80% mortality after about 25 days, indicating that the shorter-term toxicity tests likely underestimate stormwater toxicity. The biological and physical habitat assessments also supported a definitive relationship between degraded-stream ecology and urban runoff.

Rainbow[31] presented a detailed overview of heavy metals in aquatic invertebrates. He concluded that the presence of a metal in an organism could not indicate directly whether that metal is poisoning the organism. However, if compared to concentrations in a suite of well-researched biomonitors, it is possible to determine if the accumulated concentrations are atypically high, with the possible presence of toxic effects. Allen[32] also presented an overview of metal-contaminated aquatic sediments. Allen's book presents many topics that would enable the user to better interpret measured heavy-metal concentrations in urban-stream sediments.

One of the key objectives of the Chesapeake Bay restoration effort is to reduce the impacts of toxicants, of which stormwater is a recognized major source for the area. Hall et al.[33] describe the *Toxics Reduction Strategy*, based on water-column and sediment-chemical analyses, benthic-community health, and fish-body burdens. More than 40% of the sites have displayed some degree of water-column toxicity, and about 70% of the sites have displayed sediment toxicity. Garries et al.[34] further describe how the list of *Toxics of Concern* is developed for Chesapeake Bay.

Sediment contaminated by stormwater discharges has a detrimental effect on the receiving-water biological community. Schueler[35] summarized *in situ* assessment methods of stormwater-impacted sediments. The use of *in situ* test chambers, using *C. dubia*, eliminates many of the sample-disruption problems associated with conducting sediment toxicity tests in the laboratory. Love and Woolley[36] found that stormwater was alarmingly more toxic than treated sewage and that treatment before reuse of residential-area stormwater may be needed.

Pitt[37] reported a series of laboratory toxicity tests using 20 stormwater and CSO samples. He found that the most promising results are associated with using several complementary tests, instead of any one test method. However, simple screening toxicity tests (such as the Azur Microtox® test) are useful during preliminary assessments or for treatability tests.

Huber and Quigley[38] studied highway construction and repair materials (e.g., deck sealers, wood preservatives, waste-amended pavement, etc.) for their chemical and toxicological properties and leaching characteristics. *Daphnia magna* (a water flea) and the algae *Selenastrum capricornutum* were used for the toxicity tests. Leaching was evaluated as a function of time using batch tests, flat-plate tests, and column tests, as appropriate for the end-use of the highway material. These comprehensive tests identified a number of maintenance and construction materials that should be avoided for use near aquatic environments due to their elevated toxicity.

Kosmala et al.[40] used *C. dubia* in laboratory toxicity tests in combination with field analysis of the *Hydropsychid* life cycle to assess the impact of both the wastewater-treatment-plant effluent and the stormwater overflow on the receiving water. They found that the results seen in the laboratory toxicity tests and in the *in situ* biological measurements were due to nutrient and micropollutant loadings. Marsalek et al.[41] used several different toxicity tests to assess the various types of toxicity in typical urban runoff and in runoff from a multilane highway. The tests included traditional toxicity analysis using *Daphnia magna*, the Microtox toxicity test, submitochondrial particles, and the SOS Chromotest for genotoxicity. Marsalek and Rochfort[42] also investigated the toxicity of urban stormwater and CSO. Acute toxicity, chronic toxicity, and genotoxicity of stormwater and CSO were studied at 19 urban sampling sites in Ontario, Canada, using a battery of seven bioassays. The most frequent responses of severe toxicity were found in stormwater samples (in 14% of all samples), particularly those collected on freeways during the winter months. Compared to stormwater, CSO displayed lower acute toxicity (7% of the samples were moderately toxic, and none of the samples was severely toxic).

Skinner et al.[43] showed that stormwater runoff produced significant toxicity in the early life stages of medaka (*Oryzias latipes*) and inland silverside (*Menidia beryllina*). Developmental problems and toxicity were strongly correlated with the total metal content of the runoff and corresponded with exceedences of water-quality criteria of Cd, Cu, W, and Zn.

Tucker and Burton[44] compared *in situ* vs. laboratory conditions for toxicity testing of nonpoint-source runoff. They found that NPS runoff from urban areas was more toxic to the organisms in the laboratory, while agricultural runoff was more toxic to the organisms exposed *in situ*. The differences seen between the two types of toxicity tests demonstrated the importance of *in situ* assays in examining the effects of NPS runoff. Hatch and Burton,[45] using field and laboratory bioassays, demonstrated the impact of the urban stormwater runoff on *Hyalella azteca*, *Daphnia magna*, and *Pimephales promelas* survival after 48 h of exposure. The significant toxicity seen at the outfall site was attributed to the contaminant accumulation in the sediments and the mobilization of the top layers of sediment during storm events.

Bickford et al.[46] described the methodology developed and implemented by Sydney Water in Australia to assess the risk to humans and aquatic organisms in creeks, rivers, estuaries, and ocean waters from wet-weather flows (WWFs). The model used in this study was designed to predict concentrations of various chemicals in WWFs and compare the values to toxicity reference values. Brent and Herricks[47] proposed a methodology for predicting and quantifying the toxic response of aquatic systems to brief exposures to pollutants such as the contaminants contained in stormwater runoff. The method contains an event-focused toxicity method, a test metric (event toxicity unit, or ETU) to represent the toxicity of intermittent events, and an event-based index that would describe the acute toxicity of this brief exposure. The toxicity metric proposed (PE-$LET_{50}$ [post-exposure lethal exposure time]) was the exposure duration required to kill 50% of the population during a prespecified, postexposure monitoring period. Colford et al.[48] proposed three methods of analytically evaluating the impact of storm-sewer and combined-sewer outflows on public health.

## 19.3 SUBTLE (CHRONIC) EFFECTS OF STORMWATER DISCHARGES ON AQUATIC LIFE

Many studies have shown the severe detrimental effects of urban runoff on receiving-water organisms. These studies have generally either examined receiving-water conditions above and below a city or compared two parallel streams, one urbanized and the other nonurbanized. The researchers usually carefully selected the urbanized streams to minimize contaminant sources other than urban runoff. However, few studies have examined direct cause-and-effect relationships of urban runoff for receiving-water aquatic organisms.[49] The following paragraphs briefly describe a variety of urban receiving-water investigations.

Klein[50] studied 27 small watersheds having similar physical characteristics but varying land uses in the Piedmont region of Maryland. During an initial phase of the study, they found definite relationships between water quality and land use. Subsequent study phases examined aquatic-life relationships in the watersheds. The principal finding was that stream aquatic-life problems were first identified with watersheds having imperviousness areas comprising at least 12% of the watershed. Severe problems were noted after the imperviousness quantities reached 30%.

Receiving-water impact studies were also conducted in North Carolina.[51–53] The benthic fauna occurred mainly on rocks. As sedimentation increased, the amount of exposed rocks decreased, with a decreasing density of benthic macroinvertebrates. Data from 1978 and 1979 in five cities showed that urban streams were grossly polluted by a combination of toxicants and sediment. Chemical analyses, without biological analyses, would have underestimated the severity of the problems because the water-column quality varied rapidly, while the major problems were associated with sediment quality and effects on macroinvertebrates. Macroinvertebrate diversities were severely reduced in the urban streams compared to the control streams. The biotic indices indicated very poor conditions for all urban streams. Occasionally, high populations of pollutant-tolerant organisms were found in the urban streams but abruptly disappeared before subsequent sampling efforts. This was probably caused by intermittent discharges of spills or illegal dumpings of toxicants. Although the cities studied were located in different geographic areas of North Carolina, the results were remarkably uniform.

During the early Coyote Creek, San Jose, California, receiving-water study, 41 stations were sampled in both urban and nonurban perennial-flow stretches of the creek over 3 years. Short- and long-term sampling techniques were used to evaluate the effects of urban runoff on water quality, sediment properties, fish, macroinvertebrates, attached algae, and rooted aquatic vegetation.[22] These investigations found distinct differences in the taxonomic composition and relative abundance of the aquatic biota present. The nonurban sections of the creek supported a comparatively diverse assemblage of aquatic organisms including an abundance of native fishes and numerous benthic macroinvertebrate taxa. In contrast, however, the urban portions of the creek (less than 5% urbanized), affected only by urban runoff discharges and not industrial or municipal discharges, had an aquatic community generally lacking in diversity and was dominated by pollution-tolerant organisms such as mosquitofish and tubificid worms.

A major nonpoint-runoff receiving-water-impact research program was conducted in Georgia (U.S.).[54] Several groups of researchers examined streams in major areas of the state. Benke et al.[55] studied 21 stream ecosystems near Atlanta having watersheds of 1 to 3 square mi each and land uses ranging from 0 to 98% urbanization. They measured stream-water quality but found little relationship between water quality and degree of urbanization. The water-quality parameters also did not identify a major degree of pollution. In contrast, there were major correlations between urbanization and the number of species found. They had problems applying diversity indices to their study because the individual organisms varied greatly in size (biomass). CTA[56] also examined receiving-water aquatic biota impacts associated with urban runoff sources in Georgia. They studied habitat composition, water quality, macroinvertebrates, periphyton, fish, and toxicant concentrations in the water, sediment, and fish. They found that the impacts of land use were the greatest in the urban basins. Beneficial uses were impaired or denied in all three urban basins studied. Fish were absent in two of the basins and severely restricted in the third. The native macroinvertebrates were replaced with pollution-tolerant organisms. The periphyton in the urban streams were very different from those found in the control streams and were dominated by species known to create taste and odor problems.

Pratt et al.[57] used basket artificial substrates to compare benthic population trends along urban and nonurban areas of the Green River in Massachusetts. The benthic community became increasingly disrupted as urbanization increased. The problems were not associated only with times of heavy rain but seemed to be present at all times. The stress was greatest during summer low-flow periods and was probably localized near the streambed. They concluded that the high degree of correspondence between the known sources of urban runoff and the observed effects on the benthic community was a forceful argument that urban runoff was the causal agent of the disruption observed.

Cedar swamps in the New Jersey Pine Barrens were studied by Ehrenfeld and Schneider.[58] They examined 19 wetlands subjected to varying amounts of urbanization. Typical plant species were lost and replaced by weeds and exotic plants in urban-runoff-affected wetlands. Increased uptakes of phosphorus and lead in the plants were found. It was concluded that the presence of stormwater runoff to the cedar swamps caused marked changes in community structure, vegetation dynamics, and plant-tissue-element concentrations.

Medeiros and Coler[59] and Medeiros et al.[60] used a combination of laboratory and field studies to investigate the effects of urban runoff on fathead minnows. Hatchability, survival, and growth were assessed in the laboratory in flow-through and static bioassay tests. Growth was reduced to one half of the control growth rates at 60% dilutions of urban runoff. The observed effects were believed to be associated with a combination of toxicants.

The University of Washington[25,61–67] conducted a series of studies to contrast the biological and chemical conditions in urban Kelsey Creek with rural Bear Creek in Bellevue, Washington. The urban creek was significantly degraded when compared to the rural creek but still supported a productive, albeit limited and unhealthy, salmonid fishery. Many of the fish in the urban creek had respiratory anomalies. The urban creek was not grossly polluted, but flooding from urban developments had increased dramatically in recent years. These increased flows dramatically changed the urban stream's channel by causing unstable conditions with increased streambed movement and by altering the availability of food for the aquatic organisms. The aquatic organisms were very dependent on the few relatively undisturbed reaches. DO concentrations in the sediments depressed embryo salmon survival in the urban creek. Various organic and metallic priority pollutants were discharged to the urban creek, but most of them were apparently carried through the creek system by the high storm flows to Lake Washington. The urbanized Kelsey Creek also had higher water temperatures (probably due to reduced shading) than Bear Creek. This probably caused the faster fish growth in Kelsey Creek.

The fish population in the urbanized Kelsey Creek had adapted to its degrading environment by shifting the species composition from Coho salmon to less sensitive cutthroat trout and by making extensive use of less disturbed refuge areas. Studies of damaged gills found that up to three fourths of the fish in Kelsey Creek were affected with respiratory anomalies, while no cutthroat trout and only two of the Coho salmon sampled in the forested Bear Creek had damaged gills. Massive fish kills in Kelsey Creek and its tributaries were also observed on several occasions during the project due to the dumping of toxic materials down the storm drains.

There were also significant differences in the numbers and types of benthic organisms found in urban and forested creeks during the Bellevue research. Mayflies, stoneflies, caddisflies, and beetles were rarely observed in the urban Kelsey Creek but were quite abundant in the forested Bear Creek. These organisms are commonly regarded as sensitive indicators of environmental degradation. One example of degraded conditions in Kelsey Creek was shown by a species of clams (*Unionidae*) that was not found in Kelsey Creek but was commonly found in Bear Creek. These clams are very sensitive to heavy siltation and unstable sediments. Empty clamshells, however, were found buried in the Kelsey Creek sediments, indicating their previous presence in the creek and their inability to adjust to the changing conditions. The benthic-organism composition in Kelsey Creek varied radically with time and place, while the organisms were much more stable in Bear Creek.

Urban-runoff-impact studies were conducted in the Hillsborough River near Tampa Bay, Florida, as part of the U.S. EPA's Nationwide Urban Runoff Program (NURP).[68] Plants, animals, sediment, and water quality were all studied in the field and supplemented by laboratory bioassay tests. Effects of saltwater intrusion and urban runoff were both measured because of the estuarine environment. During wet weather, freshwater species were found closer to the Bay than during dry weather. In coastal areas, these additional natural factors made it even more difficult to identify the cause-and-effect relationships for aquatic-life problems. During another NURP project, Striegl[69] found that the effects of accumulated contaminants in Lake Ellyn (Glen Ellyn, Illinois) inhibited desirable benthic invertebrates and fish and increased undesirable phytoplankton blooms.

The number of benthic-organism taxa in Shabakunk Creek in Mercer County, New Jersey declined from 13 in relatively undeveloped areas to four below heavily urbanized areas.[70,71] Periphyton samples were also analyzed for heavy metals with significantly higher metal concentrations found below the heavily urbanized area than above.

Stewart et al.[72] collected diatoms (*Bacillariophyta*) and water-quality samples from three streams that drain the Great Marsh in the Indiana Dunes National Lakeshore. They found that diatom species diversity could be used as indicators of water quality, which could then be linked to land use in a watershed. Diatom species diversity was most variable in areas with poorer water quality and was directly correlated to the total alkalinity, total hardness, and specific conductance of the water in the stream. .

A number of papers presented at the 7th International Conference on Urban Storm Drainage, held in Hannover, Germany, described receiving-water studies that investigated organic and heavy metal toxicants. Handová et al.[73] examined the bioavailability of metals from CSOs near Prague. They compared these results with biomonitoring. The metals were ranked according to their mobility as: Cd (95%), Zn (87%), Ni (64%), Cr (59%), Pb (48%), and Cu (45%). The mobile fraction was defined as the metal content that was exchangeable, bound to carbonates, bound to iron and manganese oxides, and bound to organic matter. Boudries et al.[74] and Estèbe et al.[75] investigated heavy metals and organics bound to particulates in the River Seine near Paris. The Paris CSOs caused a significant increase in the aliphatic and aromatic hydrocarbons bound to river sediments. The high flows during the winter were associated with lower heavy metal associations with the sediment, compared to the lower summer flow conditions. These differences were found to be due to dilution of the CSOs in the river and to the changing contributions of rural vs. urban suspended solids during the different seasons.

The Northeastern Illinois Planning Commission[76] compared comprehensive fish survey information from over 40 northeastern Illinois small- to moderate-sized streams and rivers to demographic data for the contributing watershed areas. The streams had watershed areas ranging from about 12 to 222 square mi and had population densities ranging from about 30 to more than 4500 people per square mile. The fish data was used in the index of biotic integrity (IBI) to identify the quality of the fish populations. Table 19.1 lists the fish data that are used in the IBI, and Table 19.2 shows the different scores for the quality categories. Factors necessary for good- and excellent-quality fish communities include the presence of diverse and reproducing fish and other aquatic organisms, including a significant percentage of intolerant species (such as darters and smallmouth bass).

The more commonly used imperviousness-based indicator of development was not used due to a lack of available data and the difficulty of acquiring good-quality current imperviousness data, let alone estimating historical imperviousness data. In contrast, population data was readily available and thought to be an adequate indicator of the extent and density of urbanization in the watersheds. They found that nearly all streams in urban and suburban watersheds having population densities greater than about 300 people per square mile showed signs of considerable impairment to their fish communities (being in fair to very poor condition). In contrast, nearly all rural streams supported fish communities that were rated good or excellent. They identified both point and nonpoint sources

**Table 19.1 Index of Biotic Integrity (IBI) Metrics[76]**

| Category | Metric |
| --- | --- |
| Species richness and composition | Total number of fish species |
| | Number and identity of darter species |
| | Number and identity of sunfish species |
| | Number and identity of sucker species |
| | Number and identity of intolerant species |
| | Proportion of individuals as green sunfish |
| Trophic composition | Proportion of individuals as omnivores |
| | Proportion of individuals as hybrids |
| | Proportion of individuals as piscivores |
| Fish abundance and condition | Number of individuals in sample |
| | Proportion of individuals as hybrids |
| | Proportion of individuals with disease, tumors, fin damage, and skeletal anomalies |

**Table 19.2 Illinois Environmental Protection Agency (IEPA) Biological Stream Characterization (BSC) and Index of Biotic Integrity (IBI) Classifications and Criteria[76]**

| IBI Score | Stream Class | BSC Category | Biotic Resource Quality |
|---|---|---|---|
| 51–60 | A | Unique | Excellent |
| 41–50 | B | Highly valued | Good |
| 31–40 | C | Moderate | Fair |
| 21–30 | D | Limited | Poor |
| ≤ 20 | E | Restricted | Very poor |

as major contributors to these impairments. However, the point-source discharges and CSO discharges have substantially decreased over the past 20 years, while the nonpoint-source discharges have increased significantly with increased development, and the fisheries are still declining in many areas. In stable areas that were mostly affected by point sources and CSOs, documented dramatic improvements in some water-quality indicators (especially DO and ammonia) and the fish populations have occurred. In areas that are similar but that have continued urban development, the fisheries have continued to decline.

The researchers concluded that although rural watersheds have known water-quality problems (especially agricultural chemicals and erosion, plus manure runoff), these issues did not prevent the attainment of mostly high-quality fisheries in these areas. Similar conclusions were noted in the comparison study by the USGS in North Carolina[77] of forested, agricultural, and urban streams. Although the forested streams were of the best quality, the streams in the agricultural areas were of intermediate quality and had significantly better biological conditions than the urban stream (which had poor macroinvertebrate and fish conditions, poor sediment and temperature conditions, and fair substrate and nutrient conditions).

## 19.4 HABITAT EFFECTS CAUSED BY STORMWATER DISCHARGES

Some of the most serious effects of urban runoff are on the aquatic habitat of the receiving waters. These habitat effects are in addition to the pollutant-concentration effects. Numerous papers already referenced found significant sedimentation problems in urban receiving waters. The major effects of urban sediment on the aquatic habitat include silting of spawning and food production areas and unstable bed conditions.[78] Other major habitat-destruction problems include rapidly changing flows and the absence of refuge areas to protect the biota during these flow changes. Removal of riparian vegetation can increase water temperatures and a major source of large organic debris that are important refuge areas. The major references on stream geomorphology that many of the following researchers based their work on were by Leopold et al.,[79] Brookes,[80] and Rosgen.[81] These fundamental references should be consulted for excellent descriptions of the many natural processes affecting streams in transition. Brookes also specifically examines urbanization effects on stream morphology. Knowledge of these basic processes will better enable an understanding of local stream changes occurring with watershed urbanization. This understanding will, in turn, enable more efficient rehabilitation efforts of degraded streams and the use of watershed controls to minimize these effects.

Brookes[80] has documented many cases in the United States and Great Britain of stream-morphology changes associated with urbanization. These changes are mostly responsible for habitat destruction, which is usually the most significant detriment to aquatic life. In many cases, water-quality improvement would result in very little aquatic-life benefits if the physical habitat is grossly modified. The most obvious habitat problems are associated with stream "improvement" projects, ranging from debris removal and stream straightening to channelization projects. Brookes[80,82] presents a number of ways to minimize habitat problems associated with stream-channel projects, including stream restoration.

In Maryland streams affected by urban construction activities, Wolman and Schick[83] observed deposition of channel bars, erosion of channel banks, obstruction of flows, increased flooding, and shifting of channel bottoms, along with concomitant changes in the aquatic life.

Pess and Bilby[84] identified Coho salmon (*Oncorhynchus kisutch*) distribution and abundance in Puget Sound rivers and explained the distribution by using both stream-reach and watershed-scale habitat characteristics, including the influence of urban areas on the habitat. In the Puget Sound region of the U.S. Pacific Northwest, Greenberg et al.[85] developed and evaluated the Urban Stream Baseline Evaluation Method to characterize baseline habitat conditions for salmonids. The methodology, based on assessment of geomorphic suitability, fish distribution, and habitat alteration, was recommended for use to prioritize recovery actions.

Bragg and Kershner[86] investigated the impact on the habitats of aquatic life and they found that coarse woody debris in riparian zones could be used successfully to maintain the integrity of these ecosystems. Larson[87] evaluated the effectiveness in urban areas of these habitat-restoration activities using large woody debris and found that in urban areas, the success of restoration may be hindered by the high sediment loads and increased flow associated with urbanization. Markowitz et al.[88] documented the CSO Long Term Control Plan implemented by the City of Akron, Ohio, which focused on habitat preservation and aquatic-life use of the receiving waters. The plan included these nontraditional alternatives: riparian setbacks in undeveloped areas, stream restoration, linear parks or greenways, and artificial riffles for stream aeration; these alternative approaches were found to cost less than 5% of the typical cost of controlling CSO flows. A methodology to investigate the chronic and cumulative degradation of the river Orne due to CSO and urban runoff was presented by Zobrist et al.,[89] with the results being used to evaluate management activities. Xu et al.[91] reported on the improvement plan being used for a river passing through the downtown area of a city in Western Japan and the problems inherent in developing a compromise strategy between flood control and mitigation and the desire to have an attractive waterway running through the city. The final improvement plan recommended construction of a new flood drain tunnel and a new underground flood-control reservoir.

Cianfrani et al.[92] used a GIS system to document the results of a comprehensive inventory of the natural resources of the Fairmount Park (Philadelphia, Pennsylvania) stream system including vegetation communities, fish, aquatic and terrestrial insects, birds, mollusks, amphibians, reptiles, and streams. The stream assessment also included the characterization of stream reaches by in-stream habitat, geomorphology, and riparian zone. This GIS inventory then was used in planning the restoration of sites in the Fairmount Park system. Derry et al.[93] reported on the habitat management strategies implemented by the City of Olympia, Washington, to control the degradation of aquatic habitats by urban stormwater runoff. These management strategies provided a basis for resolving the conflict between growth and the protection of aquatic resources. Ishikawa et al.[94] reported on the efforts to restore the hydrological cycle in the Izumi River Basin in Yokohama, Japan, while Saeki et al.[95] have documented the efforts of the Tokyo Metropolitan Government and its Basin Committee to restore the natural water cycle in the Kanda River. Kennen[96] investigated the relationship between selected basin and water-quality characteristics in New Jersey streams and the impact on the macroinvertebrate community and stream habitat. He found that urban areas had the greatest probability of having impacted stream areas, with the amount of urban land and the total flow of treated sewage effluent being the strongest explanatory variables for the impact. He also found that levels of impairment were significantly different between the Atlantic Coastal Rivers drainage area and the Lower Delaware River drainage area.

### 19.4.1 Increased Flows from Urbanization

Increased flows are the probably the best known example of impacts associated with urbanization. Most of the recognition has, of course, focused on increased flooding and associated damages. This has led to numerous attempts to control peak flows from new urban areas through the use of regulations that limit postdevelopment peak flows to predevelopment levels for relatively large design storms. The typical response has been to use dry detention ponds. This approach is limited and may actually increase downstream flows. In addition to the serious issue of flooding, high flows also cause detrimental ecological problems in receiving waters. The following discussion presents several case studies where increased flows were found to have serious effects on stream habitat conditions.

The aquatic organism differences in urbanized and control streams found during the Bellevue Urban Runoff Program were probably most associated with the increased peak flows. The increased flows in the urbanized Kelsey Creek resulted in increases in sediment carrying capacity and channel instability of the creek.[61–65] Kelsey Creek had much lower flows than the control Bear Creek during periods between storms. About 30% less water was available in Kelsey Creek during the summers. These low flows may also have significantly affected the aquatic habitat and the ability of the urban creek to flush toxic spills or other dry-weather contaminants from the creek system.[66,67] Kelsey Creek had extreme hydrologic responses to storm. Flooding substantially increased in Kelsey Creek during the period of urban development; the peak annual discharges almost doubled in 30 years, and the flooding frequency also increased due to urbanization.[66,67] These increased flows in urbanized Kelsey Creek resulted in greatly increased sediment transport and channel instability.

Bhaduri et al.[97] also quantified the changes in streamflow and associated decreases in groundwater recharge associated with urbanization. They point out that the most widely addressed hydrologic effect of urbanization is the peak discharge increases that cause local flooding. However, the increase in surface runoff volume also represents a net loss in groundwater recharge. They point out that urbanization is linked to increased variability in volume of water available for wetlands and small streams, causing "flashy" or "flood-and-drought" conditions. In northern Ohio, urbanization at a study area was found to cause a 195% increase in the annual volume of runoff, while the expected increase in the peak flow for the local 100-year event was only 26% for the same site. Although any increase in severe flooding is problematic and cause for concern, the much larger increase in annual runoff volume, and associated decrease in groundwater recharge, likely has a much greater effect on in-stream biological conditions.

Snodgrass et al.[16] reported that in the Toronto, Ontario area, flows causing bankfull conditions occur with a return frequency of about 1.5 years. Storms with this frequency are in general equilibrium with resisting forces that tend to stabilize the channel (such as vegetation and tree root mats), with increased flows overcoming these resisting forces, causing channel enlargement. Infrequent flows can therefore be highly erosive. With urbanization, the flows that were bankfull flows during historical times now occur much more frequently (about every 0.4 years in Toronto). The channel cross-sectional area, therefore, greatly increases to accommodate the increased stream discharges and power associated with the "new" 1.5-year flows that are trying to reestablish equilibrium.

Booth and Jackson[98] examined numerous data from lowland streams in western Washington and concluded that development having about 10% imperviousness caused a readily apparent degradation of aquatic life in the receiving waters. They linked the association between increased imperviousness and biological degradation to increases in flows and sediment discharges. They concluded that conventional methods to size stormwater mitigation measures (especially detention ponds) were seriously inadequate. They felt that without a better understanding of the critical processes that lead to degradation, some downstream damage to the aquatic ecosystem is likely inevitable, without unpopular restrictions to the extent of development in the watershed corresponding to < 10% imperviousness. The stream channels were generally stable if the effective impervious areas remained below 10% of the complete watershed. This level of development corresponds to a 2-year developed condition flow being less than the historical 10-year predeveloped flow condition. They found that the classical goal of detention ponds to maintain predevelopment flows was seriously inadequate because there is no control on the duration of the peak flows. They showed that a duration standard to maintain post development flow durations for all sediment-transporting discharges to predevelopment durations will avoid many receiving water habitat problems associated with stream instability. Without infiltration, the amount of runoff will obviously still increase with urbanization, but the increased water could be discharged from detention facilities at flow rates below the critical threshold, causing sediment transport. The identification of the threshold discharge below which sediment transport does not occur, unfortunately, is difficult and very site-specific. A presumed threshold discharge of about one half of the predevelopment 2-year flow was recommended for gravel-bedded streams. Sand-bedded channels have sediment transport thresholds that are very small, with inevitable bed load transport likely to occur for most levels of urbanization.

### 19.4.2 Channel Modifications Due to Urban Wet Weather Flow Discharges

Changes in physical stream channel characteristics can have a significant effect on the biological health of the stream. These changes in urban streams have been mostly related to changes in the flow regime of the stream, specifically increases in peak flow rates, increased frequencies and durations of erosive flows, and channel modifications made in an attempt to accommodate increased stormwater discharges.

Schueler[99] stated that channel geometric stability can be a good indicator of the effectiveness of stormwater control practices. He also found that once a watershed area has more than about 10 to 15% effective impervious cover, noticeable changes in channel morphology occur, along with quantifiable impacts on water quality and biological conditions. Stephenson[100] studied changes in streamflow volumes in South Africa during urbanization. He found increased stormwater runoff, decreases in the groundwater table, and dramatically decreased times of concentration. The peak flow rates increased by about twofold, about half caused by increased pavement (in an area having about only 5% effective impervious cover), with the remainder caused by decreased times of concentration (related to the increased drainage efficiency of artificial conveyances).

Richey[64] made some observations about bank stabilities in Kelsey and Bear Creeks as part of the Bellevue, Washington, NURP project.[25] She notes that the Kelsey Creek channel width had been constrained during urban development. In addition, 35% of the urbanized Kelsey Creek channel mapped during these projects was modified by the addition of some type of stabilization structure. Only 8% of nonurbanized Bear Creek's length was stabilized. Most of the stabilization structures in Bear Creek were low walls in disrepair, while more than half of the structures observed along Kelsey Creek were large riprap or concrete retention walls. The necessity of the stabilization structures was evident from the extent and severity of erosion cuts and the number of deposition bars observed along the Kelsey Creek stream banks. Bridges and culverts were also frequently found along Kelsey Creek; these structures further act to constrict the channel. As discharges increased and the channel width is constrained, the velocity increases, causing increases in erosion and sediment transport.

The use of heavy riprapping along the creek seemed to worsen the flood problems. Storm flows are unable to spread out onto the flood plain, and the increased velocities are evident downstream along with increased sediment loads. This rapidly moving water has enough energy to erode unprotected banks downstream of riprap. Many erosion cuts along Kelsey Creek downstream of these riprap structures were found. Similar erosion of the banks did not occur in Bear Creek. Much of the Bear Creek channel had a wide flood plain with many side sloughs and back eddies. High flows in Bear Creek could spread onto the flood plains and drop much of their sediment load as the water velocities decreased.

The University of Washington studies also examined sediment transport in urbanized Kelsey and nonurbanized Bear Creeks. Richey[64] found that the relative lack of debris dams and off-channel storage areas and sloughs in Kelsey Creek contributed to the rapid downstream transit of water and materials. Both the small size of the riparian vegetation and the increased stream power probably contributed to the lack of debris in the channel. It is also possible that the channel debris may have been cleared from the stream to facilitate rapid drainage. The high flows from high velocities caused the sediments to be relatively coarse. The finer materials were more easily

**Table 19.3 Hours of Exceedence of Developed Conditions with Zero Runoff Increase Controls Compared to Predevelopment Conditions[102]**

| Recurrence Interval (years) | Existing Flowrate ($m^3/s$) | Exceedence for Predevelopment Conditions (hours per 5 yrs) | Exceedence for Existing Development Conditions, with ZRI Controls (hours per 5 yrs) | Exceedence for Ultimate Development Conditions, with ZRI Controls (hours per 5 yrs) |
|---|---|---|---|---|
| 1.01 (critical mid-bankfull conditions) | 1.24 | 90 | 380 | 900 |
| 1.5 (bankfull conditions) | 2.1 | 30 | 34 | 120 |

transported downstream. Larger boulders were also found in the sediment but were probably from failed riprap or gabion structures.

Maxted[101] examined stream problems in Delaware associated with urbanization. He found an apparent strong correlation between habitat score and biology score from 40 stream study locations. He found that it is not possible to have acceptable biological conditions if the habitat is degraded. The leading contributor to habitat degradation was found to be urban runoff, especially the associated high flows and sediment accumulations.

A number of presentations concerning aquatic habitat effects from urbanization were made at the *Effects of Watershed Development and Management on Aquatic Ecosystems* conference held in Snowbird, Utah, in August 1996, sponsored by the Engineering Foundation and the ASCE. MacRae[102] presented a review of the development of the common zero-runoff-increase (ZRI) discharge criterion, referring to peak discharges before and after development. This criterion is commonly met using detention ponds for the 2-year storm. MacRae shows how this criterion has not effectively protected the receiving-water habitat. He found that streambed and bank erosion is controlled by the frequency and duration of the mid-depth flows (generally occurring more often than once a year), not the bankfull condition (approximated by the 2-year event). During monitoring near Toronto, he found that the duration of the geomorphically significant predevelopment mid-bankfull flows increased by a factor of 4.2 times after 34% of the basin had been urbanized compared to flow conditions before development. The channel had responded by increasing in cross-sectional area by as much as three times in some areas, and was still expanding. Table 19.3 shows the modeled durations of critical discharges for predevelopment conditions, compared to current and ultimate levels of development with "zero-runoff-increase" controls in place. At full development and even with full ZRI compliance in this watershed, the hours exceeding the critical mid-bankfull conditions will increase by a factor of ten, with resulting significant effects on channel stability and the physical habitat.

MacRae[102] also reported other studies that found that channel cross-sectional areas began to enlarge after about 20 to 25% of the watershed was developed, corresponding to about a 5% impervious cover in the watershed. When the watersheds are completely developed, the channel enlargements were about five to seven times the original cross-sectional areas. Changes from stable streambed conditions to unstable conditions appear to occur with basin imperviousness of about 10%, similar to the value reported previously for serious biological degradation. He also summarized a study conducted in British Columbia that examined 30 stream reaches in natural areas, in urbanized areas having peak-flow attenuation ponds, and in urbanized areas not having any stormwater controls. The channel widths in the uncontrolled urban streams were about 1.7 times the widths of the natural streams. The streams having the ponds also showed widening, but at a reduced amount compared to the uncontrolled urban streams. He concluded that an effective criterion to protect stream stability (a major component of habitat protection) must address mid-bankfull events, especially by requiring similar durations and frequencies of stream power (the product of shear stress and flow velocity, not just flow velocity alone) at these depths, compared to satisfactory reference conditions.

Much research on habitat changes and rehabilitation attempts in urban streams has occurred in the Seattle area of western Washington over the past 20 years. Sovern and Washington[103] described the in-stream processes associated with urbanization in this area, as part of a paper describing a recommended approach for the rehabilitation of urban streams. They were concerned that many attempts to "restore" urban streams were destined to failure because of a lack of understanding of the actual changes occurring in streams as the watersheds changed from forested to urban land uses. They presented a concept of the "new urban stream" that attempts to correct several of the most important changes to better accommodate the native Pacific Northwest fish, instead of the unrealistic goal of trying to totally restore the steams to predevelopment conditions. The important factors that affect the direction and magnitude of the changes in a stream's physical characteristics due to urbanization include:

- The depths and widths of the dominant discharge channel will increase directly proportional to the water discharge. The width is also directly proportional to the sediment discharge. The channel width divided by the depth (the channel shape) is also directly related to sediment discharge.
- The channel gradient is inversely proportional to the water discharge rate and directly proportional to the sediment discharge rate and the sediment grain size.
- The sinuosity of the stream is directly proportional to the stream's valley gradient and inversely proportional to the sediment discharge.
- Bed load transport is directly related to the stream power and the concentration of fine material and inversely proportional to the fall diameter of the bed material.

In their natural state, small streams in forested watersheds in western Washington have small low-flow channels (the aquatic habitat channel) with little meandering.[103] The stream banks are nearly vertical because of clayey bank soils and heavy root structures, and the streams have numerous debris jams from fallen timber. The stream widths are also narrow, generally from 3 to 6 feet wide. Stable forested watersheds also support about 250 aquatic plant and animal species along the stream corridor. In contrast, fewer than 50 aquatic plant and animal species are usually found along urban streams. Pool/riffle habitat is dominant along streams having gradients less than about 2% slope, while pool/drop habitat is dominant along streams having gradients from 4 to 10%. The pools form behind large organic debris (LOD) or rocks. The salmon and trout in western Washington have evolved to take advantage of these stream characteristics. Sovern and Washington[103] point out that less athletic fish species (such as chum and pink salmon) cannot utilize the steeper gradient and upper reaches of the streams. Coho, steelhead, and cutthroat can use these upper areas, however.

Urbanization radically affects many of these natural stream characteristics. Pitt and Bissonnette[25] reported that the Coho salmon and cutthroat trout were affected by the increased nutrients and elevated temperatures of the urbanized streams in Bellevue, as studied by the University of Washington under the U.S. EPA's NURP project.[104] These conditions were probably responsible for accelerated growth of the fry that were observed to migrate to Puget Sound and the Pacific Ocean sooner than their counterparts in the control forested watershed that was also studied. However, the degradation of sediments, mainly the decreased particle sizes, adversely affected their spawning areas in streams that had become urbanized.

Sovern and Washington[103] reported that in western Washington frequent high flow rates can be 10 to 100 times the predevelopment flows in urbanized areas, but that the low flows in the urban streams are commonly lower than the predevelopment low flows. They have concluded that the effects of urbanization on western Washington streams are dramatic, in most cases permanently changing the stream hydrologic balance by increasing the annual water volume in the stream, increasing the volume and rate of storm flows, decreasing the low flows during dry periods, and increasing the sediment and contaminant discharges from the watershed. With urbanization, the streams increase in cross-sectional area to accommodate these increased flows, and headwater downcutting occurs to decrease the channel gradient. The gradients of stable urban streams are often only about 1 to 2%, compared to 2 to 10% gradients in natural areas. These changes in width and the downcutting result in very different and changing stream conditions. The common pool/drop habitats are generally replaced by pool/riffle habitats, and the streambed material is comprised of much finer material, for example. The researchers have concluded that once urbanization begins, the effects on stream shape are not completely reversible. Developing and maintaining quality aquatic-life habitat, however, is possible under urban conditions, but it requires human intervention and will not be the same as for forested watersheds.

Other Seattle area researchers have specifically examined the role that large woody debris (LWD) has in stabilizing the habitat in urban streams. Booth et al.[105] found that LWD performs key functions in undisturbed streams that drain lowland forested watersheds in western Washington. These important functions include energy dissipation of the flow energy, channel bank and bed stabilization, sediment trapping, and pool formation. Urbanization typically results in the almost complete removal of this material. They point out that logs and other debris have long been removed from channels in urban areas for many reasons, especially because such debris has the potential to block culverts or to form jams at bridges and can increase bank scour, and many residents favor "neat" stream-bank areas (a lack of woody debris in and near the water and even with mowed grass to the waters edge). Booth et al.[105] present and modify the stream classification system originally developed by Montgomery and Buffington[106] that recognizes LWD as an important component of Pacific Northwest streams that are being severely affected by urbanization.

The role of LWD varies in each channel type, and the effects of its removal also vary. The channel types are described as follows. The upper colluvial channels are wholly surrounded by colluvium (sediment transported by creep or landsliding, and not by stream transport) and generally lie at the top of the channel network. The cascade channels are the steepest of the alluvial channels and are characterized by their tumbling flows around individual boulders that dissipate most of the energy of the flowing water. Only very small pools are in cascade channels. The step-pool channels have accumulations of debris that form a series of steps that are one to four channel widths apart. The steps separate small pools that accumulate fine sediment. The fine sediment can be periodically flushed downstream during rare events.

"Free" step-pool channels are characterized by steps that are made of alluvium that can be periodically transported downstream during high flows, while "forced" step-pool channels are characterized by steps that are made of immovable obstructions (large logs or bedrock). The removal of LWD from a forced step-pool stream in the Cascade Range could be naturally compensated by the common occurrence of large boulders that also form forced steps. However, in the lowlands near Puget Sound, the available sand and gravel stream deposits are too small to form stable steps, and the removal of LWD would have a much more severe effect on the channel stability. Plane-bed channels have long and channel-wide reaches of uniform riffles and do not have pronounced meanders and associated pools. Pool-riffle channels are the most common lowland stream channels in western Washington. These streams have pronounced meanders with pools at the outside of the bends and corresponding bars on the inside of the bends. Riffles form in the relatively straight stretch between the pools. There are also "free" and "forced" pool-riffle channels. Forced riffle-pool channels are typically formed with obstructions, such as LWD, and their removal would generally lead to a plane-bed channel characteristic. Forced riffle-pool channels form due to natural meanders and the inertial forces of the water. Dune-ripple channels have beds mostly made of sand where the character of the bed material changes in response to the flows.

The role of LWD is also highly dependent on the width of the stream. In narrow channels (high gradient colluvial and cascade channels), much of the LWD can be suspended above the flows, rarely being submerged and not available as a fish refuge or sediment trap or to dissipate the water's energy. In wide channels (dune-ripple channels), the LWD may be significantly shorter than the channel width, with minimal stable opportunities to provide steps in the channel. Therefore, LWD plays a much more important role in channels having medium widths (lowland streams having plane-bed and pool-riffle channels), where the timber can become tightly lodged in the common flow channel. The removal of the LWD in these streams, especially in streams having few boulder steps, would have significant effects. Fish populations decline rapidly and precipitously following the removal of LWD in these critical streams.[105]

Horner et al.[107] described an extensive study in the Pacific Northwest where 31 stream reaches were examined beginning in 1994 for a variety of in-stream and watershed characteristics. They concluded that the most severe in-stream biological changes were most likely associated with changes in habitat, especially increased frequencies and magnitudes of high flows. These flow changes were therefore thought to be most related to watershed factors affecting runoff, especially the amount of impervious areas in the watershed. They felt that the most rapid changes in ecological conditions were most likely to occur for urbanizing streams at relatively low levels of development, conditions representing most of the selected study sites.

Horner et al.[107] found a rapid decline in biological conditions as total imperviousness area increases to about 8% in the watershed. The rate of decline is less for higher levels of urbanization. Eight study areas had better biological conditions than expected and were associated with higher amounts of intact wetlands along the riparian corridors than other sites, indicating a possible significant moderating effect associated with preserving stream corridors in their natural condition. The less tolerant Coho salmon is much more abundant than the more tolerant cutthroat trout only for very low levels of urbanization. Stormwater concentrations of zinc were also seen to increase steadily with increasing impervious areas. However, the concentrations are well below the critical water-quality criteria until the impervious cover reaches about 40%, a level much greater than when significant biological effects are noted.

Similar conclusions were made with other metal concentrations and contaminant concentrations in the sediment. They interpreted these findings to imply that contaminant conditions were much less important than habitat destruction when it comes to affecting in-stream biological conditions. They concluded that the preponderance of physical and biological evidence indicated rapid in-stream degradation of biological conditions at early stages of urbanization. However, chemical contaminants did not appear to significantly affect biological conditions in the early stages of urbanization but may have at very high levels of urbanization. Based on their results, they developed a preliminary summary of the conditions that would allow high levels of biological functions in the Puget Sound area:

- Total impervious areas less than 5% of the watershed area, unless mitigated by extensive riparian protection, management efforts, or both;
- 2-year peak flow/winter baseflow ratio of <20;
- Greater than 60% of the upstream buffer should be greater than 30 m wide; and
- Less than 15% of the sediment in the stream bed should be less than 0.85 mm.

Habitat evaluations are commonly and justifiably recognized as critical components of stream and watershed studies. However, Poole et al.[108] caution users concerning their use to quantify aquatic habitat or channel morphology in an attempt to measure the response of individual streams to human activities. Their concern is the subjectivity of habitat surveys and the lack of repeatability, precision, and transferability of the measurement techniques. The measurement parameters are also assigned relatively arbitrary nominal values that are not easily statistically evaluated. According to them, the typical use of habitat unit classifications encourages the focus on direct manipulation or replacement of habitat structures (such as in-stream "restoration" activities) while neglecting the long-term maintenance of habitat-forming biophysical processes (such as controlling the energy distribution of stream discharges and the discharges of sediment into the streams).

Therefore, the use of habitat unit classifications as an indicator of watershed health may be most appropriately used for only very large differences or changes, when conducted over a large portion of a watershed being studied, and only if a sufficiently large number of observations and replicates are made to compensate for the high inherent measurement variations. Many current habitat surveys are being conducted on small scales within a short period of time, with few observations, and without adequate statistical evaluations of the data. The results of these surveys are, therefore, of questionable value. As for all indicators, it is important that methods be developed and tested to improve the accuracy of the tool and that additional supplemental measurement methods also be used to confirm observations and conclusions, especially when evaluating cause-and-effect relationships in watersheds.

## 19.5 STORMWATER CONTAMINATION OF SEDIMENTS AND INCREASED SEDIMENT DISCHARGES IN URBAN STREAMS

Many of the observed biological effects associated with urban runoff may be caused by polluted sediments and associated benthic-organism impacts. There has been a tremendous amount of research focusing on sediment toxicity. The following paragraphs only briefly describe several cases where sediment toxicity was examined in some detail in urban environments in which the major contaminant sources originated from stormwater alone. Burton and Pitt,[13] Allen,[32] Hatch and Burton,[45] among others, provide reviews of other urban-stream-sediment toxicity studies.

The EPA[109] prepared a four-volume report to Congress on the incidence and severity of sediment contamination in the surface waters of the United States. This report was required by the Water Resources Development Act of 1992. This Act defines contaminated sediment as "sediment containing chemical substances in excess of appropriate geochemical, toxicological, or sediment quality criteria or measures; or otherwise considered to pose a threat to human health or the environment." In the national quality survey, the EPA examined data from 65% of the 2111 watersheds in the United States and identified 96 watersheds that contain areas of probable concern. In portions of these waters, benthic organisms and fish may contain chemicals at levels unsafe for regular consumption. Areas of probable concern are located in regions affected by urban and agricultural runoff, municipal and industrial waste discharges, and other contaminant sources. When the fourth volume is completed, much more detailed information will become available concerning the relative role that urban stormwater contributes to national contaminated sediment problems. The development of sediment-quality criteria is an emerging area, with slowly emerging general guidance available to compare locally observed conditions to "standards." In most cases, local reference conditions have been most effectively used to indicate if the observed conditions constitute a problem.[13]

Examples of elevated heavy-metal and nutrient accumulations in urban sediments are numerous. DePinto et al.[110] found that the cadmium content of river sediments can be more than 1000 times greater than the overlying water concentrations, and the accumulation factors in sediments are closely correlated with sediment organic content. They reported that sediments were also able to adsorb phosphorus in proportion to the phosphorus concentrations in the overlaying waters during aerobic periods, but that the sediments released phosphorus during anaerobic periods. Heaney[111] found that long-term impacts of urban runoff related to the resuspension of previously deposited polluted benthos material may be more important than short-term discharges of contaminants from potential "first-flushes."

Another comprehensive study on polluted sediment was conducted by Wilber and Hunter[112] along the Saddle River in New Jersey, where they found significant increases in sediment contamination with increasing urbanization. They found large variations in metal concentrations for different sediment particle sizes in the urban river. The sediment particle-size distribution was the predominant influencing factor for total metal concentrations in the sediments. Areas having fine sediments had a substantially greater concentration of heavy metals than those areas having coarse sediments.

In another study, Pitt and Bozeman[22] observed concentrations for many contaminants in the urban-area sediments of Coyote Creek (San Jose, California) that were much greater than those from the nonurban area. Orthophosphates, TOC, $BOD_5$, sulfates, sulfur, and lead were all found in higher concentrations in the sediments from the urban-area stations, as compared with those from the upstream, nonurban-area stations. The median sediment particle sizes were also found to be significantly smaller at the urban-area stations, reflecting a higher silt content.

Several of the University of Washington projects as well as the Seattle METRO project investigated physical and chemical characteristics of the Kelsey and Bear Creeks sediments as part of the Bellevue, Washington, NURP projects.[25] Perkins[63] found that the size and composition of the sediments near the water interface tended to be more variable and of a larger median size in Kelsey Creek than in Bear Creek. These particle sizes varied in both streams on an annual cycle in response to runoff events. Larger particle sizes were more common during the winter months, when the larger flows were probably more efficient in flushing through the finer materials. Pedersen[61] also states that Kelsey Creek demonstrated a much greater accumulation of sandy sediments in the early spring. This decreases the suitability of the stream substrates for benthic colonization. Scott et al.[26] state that the level of fines in the sediment samples appears to be a more sensitive measure of substrate quality than the geometric mean of the particle-size distribution. Fines were defined as all material less than about 840 μm in diameter. METRO[113] also analyzed organic priority pollutants in 17 creek sediments including several in Kelsey and Bear Creeks. Very few organic compounds were detected in either stream, with the most notable trend being the much more common occurrence of various PAHs in Kelsey Creek, while none were detected in Bear Creek.

The University of Washington project and the Seattle METRO project analyzed interstitial water for various constituents. These samples were obtained by inserting perforated aluminum standpipes into the creek sediment. This water is most affected by the sediment quality and in turn affects the benthic organisms much more than the creek water column. Scott et al.[65] found that the interstitial water pH ranged from 6.5 to 7.6 and did not significantly differ between the two streams but did tend to decrease during the spring months. The lower fall temperatures and pH levels contributed to reductions in ammonium concentrations. The total ammonia and nonionized ammonia concentrations were significantly greater in Kelsey Creek than in Bear Creek. They also found that the interstitial DO concentrations in Kelsey Creek were much below those concentrations considered normal for undisturbed watersheds. These decreased interstitial oxygen concentrations were much less than the water-column concentrations and indicated the possible impact of urban development. The DO concentrations in the interstitial waters and in Bear Creek were also lower than expected, potentially suggesting deteriorating fish spawning conditions. During the winter and spring months, the interstitial oxygen concentrations appeared to be intermediate between those characteristic of disturbed and undisturbed watersheds.

The University of Washington[64] also analyzed heavy metals in the interstitial waters, focusing mostly on the more readily detected lead and zinc measurements compared to the low, or undetectable, copper and chromium concentrations. The urban Kelsey Creek interstitial water had concentrations of heavy metals approximately twice those found in the rural Bear Creek interstitial water. They expect that most of the metals were loosely bound to fine sediment particles. Most of the lead found was associated with the particulates, with very little soluble lead found in the interstitial waters. The interstitial samples taken from the standpipe samplers were full of sediment particles, which could be expected to release lead into solution following the mild acid digestion for exchangeable lead analyses. They also found that the metal concentrations in Kelsey Creek interstitial water decreased in a downstream direction. They thought that this might be caused by stream scouring of the benthic material in that part of the creek. The downstream Kelsey Creek sites were more prone to erosion and channel scouring, while the uppermost stream station was relatively stable.

Variable interstitial water quality may cause variations in sediment toxicity with time and location. Seattle METRO[113] monitored heavy metals in the interstitial waters in Kelsey and Bear Creeks. They found large variations in heavy-metal concentrations, depending on whether the sample was obtained during the wet or the dry season. During storm periods, the interstitial water and creek water heavy-metal concentrations approached the stormwater values (200 μg/L for lead). During nonstorm periods, the interstitial lead concentrations were typically about only 1 μg/L. They also analyzed priority pollutant organics in interstitial waters. Only benzene was found and only in the urban stream. The observed benzene concentrations in two Kelsey Creek samples were

22 and 24 µg/L, while the reported concentrations were less than 1 µg/L in all other interstitial water samples analyzed for benzene.

A number of recent investigations have examined sediment quality in conjunction with biological conditions in urban receiving waters in attempts to identify causative agents affecting the biological community. Arhelger et al.[114] examined conditions in the upper Houston Ship Channel, which receives drainage from the metropolitan Houston area. The channel has been dredged to allow large vessels access to the upper reaches of what used to be a relatively small channel. The dredging has increased the cross-sectional area by about 20 times, with attendant significant decreases in flushing flows. This has caused increased sedimentation of suspended material discharged from the 500-square-mile urban watershed. The sediments have undergone extensive chemical, physical, and toxicity testing, with frequent indications of toxicity. The tests have indicated that the toxicity is most likely caused by the high sediment oxygen demand and associated low DO conditions. Toxicity testing of *Ampelisca* under varied DO conditions showed significant decreases in survival when the bottom DO is less than 3 mg/L, for example. Even though the point-source BOD loads have been reduced by more than 90% since the 1970s, receiving-water and sediment oxygen levels are very low. Arhelger et al.[114] concluded that this remaining DO problem was caused by uncontrolled stormwater discharges.

Previous studies near Auckland, New Zealand have shown that sediment concentrations of many constituents near stormwater outfalls, especially in industrial areas, often exceed guidelines intended to protect bottom-dwelling animals. Guidelines used were as presented by Long et al.[115] and were as follows (along with sediment concentrations from two locations near Auckland):

| mg/kg | Copper | Lead | Zinc |
|---|---|---|---|
| Effects range — low | 34 | 47 | 150 |
| Effects range — median | 270 | 218 | 410 |
| Hellyers/Kaipatiki | 17–36 | 13–95 | 58–192 |
| Pakuranga | 14–65 | 22–112 | 108–345 |

Lead, zinc, and organochlorine were the most widespread potential problems. Field surveys and laboratory toxicity tests had shown circumstantial evidence of chronic toxicity associated with stormwater. Detailed field surveys by Morrisey et al.[116] were therefore conducted to better understand actual toxicity problems in the local marine estuaries that are influenced by complex natural factors. These complicating factors include strong gradients in salinity, sediment texture, currents, and wave action, all radically affecting the natural distribution of benthic fauna. In slowly growing areas or in relatively low-density urban areas, the relatively small rate of accumulation of contaminated sediments from nonpoint-sources could take many years to accumulate to levels that might produce detectable impacts in the receiving waters. In addition, changing urban conditions and changing weather from year to year make the rate of accumulation highly variable. These factors all make it difficult to conduct many types of field experiments that rely on before and after observations or other short-term observations that assume steady conditions. They therefore relied on a "weight-of-evidence" approach that considers many different and reinforcing/confirming procedures (such as the sediment-quality-triad and the effects-range tests, both of which rely on distribution of contaminants and organisms in the field and from laboratory toxicity tests). They also applied their results to the Abundance Biomass Comparison index proposed by Warwick.[117] This index is a relative measure of biomass vs. abundance and has been shown to work well for individual sites when control sites are difficult to identify and study, especially if available "control" sites are already impacted. Pore-water chemistry, sediment quality, and benthic-community composition were included in the field analyses. Statistical analyses identified thc strongest correlations between pH and iron content of the pore water and between the sediment texture and benthic composition. The pH and iron pore-water conditions may affect the bioavailability of the sediment heavy metals. Current and future work includes similar studies in nonurbanized estuaries, the development of chronic toxicity tests using local indigenous organisms, and studies of recolonization of heavily impacted sites.

Watzin et al.[118] examined sediment contamination in Lake Champlain near Burlington, Vermont, to compare several toxicity endpoints with sediment characteristics. They measured sediment pore-water toxicity using *Ceriodaphania dubia*, *Chironomus tentans*, and *Pimephales promelas*, benthic-community composition, and many physical and chemical characteristics at 19 locations. Four major storm drains and the secondary sewage treatment plant all discharged to the harbor. Boat traffic and historical petroleum-handling facilities also affected some of the sampling locations. They found variable levels of toxicity at the different sites, but no correlation was observed between acid-volatile sulfides (AVS), heavy-metal concentrations, and sediment toxicity. However, they did find strong associations between increasing levels of organics and lowering toxicity levels, indicating that possible metal–organic-matter complexation was reducing the metal availability. The sediment toxicity tests did indicate a moderate level of concern, but the macroinvertebrate community was apparently not significantly affected during these tests. They proposed the use of a weight-of-evidence approach that uses multiple indicators of problems and possible sources of the problems, plus repeated observations over seasonal cycles, before management recommendations are developed.

Rhoads and Cahill[120] studied the elevated concentrations of chromium, copper, lead, nickel, and zinc that were found in sediments near storm-sewer outfalls. They noted that copper and zinc concentrations were greater in the bedload compared to the bed material and therefore were more likely to be mobilized during runoff events.

Crabill et al.[121] presented their analysis of the water and sediment in Oak Creek in Arizona, which showed that the sediment fecal coliform counts were on average 2200 times greater than those in the water column. Water-quality standards for fecal coliforms were regularly violated during the summer due to the high recreational activity and animal activity in the watershed as well as the storm surges due to the summer storm season.

Vollertsen et al.[122] characterized the biodegradability of combined-sewer organic matter based on settling velocity. Fast-settling organic matter, which represents the largest fraction of the organic material, was found to be rather slowly biodegradable compared to the slow-settling organic fraction. They stressed that the biodegradability of sewer sediments should be taken into account when evaluating CSO impacts. Vollertsen and Hvitved-Jacobsen[123] studied the stoichiometric and kinetic model parameters for predicting microbial transformations of suspended solids in combined sewer systems.

The effects of large discharges of relatively uncontaminated sediment on the receiving-water aquatic environment were summarized by Schueler.[124] These large discharges are mostly associated with poorly controlled construction sites, where 30 to 300 tons of sediment per acre per year of exposure may be lost. These high rates can be 20 to 2000 times the unit area rates associated with other land uses. Unfortunately, much of this sediment reaches urban receiving waters, where massive impacts on the aquatic environment can result. Unfortunately, high rates of sediment loss can also be associated with later phases of urbanization, where receiving-water channel banks widen to accommodate the increased runoff volume and frequency of high erosive flow rates. Sediment is typically listed as one of the most important pollutants causing receiving water problems in the nation's waters. Schueler[124] listed the impacts that can be associated with suspended sediment:

- Abrades and damages fish gills, increasing risk of infection and disease
- Scours periphyton from streams (plants attached to rocks)
- Causes loss of sensitive or threatened fish species when turbidity exceeds 25 NTU
- Leads to shifts in fish communities toward more sediment tolerant species
- Causes decline in sunfish, bass, chub, and catfish when monthly turbidity exceeds 100 NTU
- Reduces sight distance for trout, with reduction in feeding efficiency
- Reduces light penetration, which causes reduction in plankton and aquatic plant growth

- Reduces filtration efficiency of zooplankton in lakes and estuaries
- Adversely impacts aquatic insects, which are the base of the food chain
- Slightly increases stream temperature in summer
- Carries significant amounts of nutrients and metals
- Increases probability of boating, swimming, and diving accidents resulting from high turbidity
- Increases water treatment to meet drinking water standards
- Increases wear and tear on hydroelectric and water-intake equipment
- Reduces anglers' chances of catching fish
- Diminishes direct and indirect recreational experience of receiving waters

He also listed the impacts that can be associated with deposited sediment:

- Physical smothering of benthic aquatic insect community
- Reduced survival rates for fish eggs
- Destruction of fish spawning areas and redds
- Reduces fish and macroinvertebrate habitat value from 'Imbedding' of stream bottom
- Loss of trout habitat when fine sediments are deposited in spawning or riffle-runs
- Possible elimination of sensitive or threatened darters and dace from fish community
- Depleted DO in lakes or streams due to increases in sediment oxygen demand
- Alarming decline of freshwater mussels
- Reduced channel capacity, exacerbating downstream bank erosion and flooding
- Reduced flood transport capacity under bridges and through culverts
- Loss of storage and lower design life for reservoirs, impoundments, and ponds
- Dredging costs for maintaining navigable channels and reservoir capacity
- Spoiling of sand beaches
- Diminished scenic and recreational value of waterways

### 19.5.1 Sediment Contamination Effects

There is much concern and discussion about contaminated sediments in urban receiving waters. Many historical discussions downplayed the significance of contaminated sediments, based on their assumed "low-availability" to aquatic organisms. However, many of the previously described receiving-water studies found greatly affected benthic-organism populations at sites with contaminated urban sediments, compared to uncontaminated control sites. More specifically, *in situ* sediment toxicity tests in urban receiving waters (such as those conducted by Burton and Stemmer[125]; Burton[126, 128–129]; Burton et al.[127]; Burton and Scott[130]; and Crunkilton et al.[12]) have illustrated the direct toxic effects associated with exposure to contaminated urban sediments, to problems associated with their scour, and to decreases in toxicity associated with their removal from stormwater.

The fate of contaminated sediments, and especially mechanisms that expose contaminants to sensitive organisms, can determine the overall and varied effects that the sediments may have. Scour of fine-grained sediments during periods of high flows in streams and rivers, or due to turbulence from watercraft in shallow waterbodies, has frequently been encountered. In addition, contaminant remobilization may also occur through bioturbation from sediment-dwelling organisms or from nest-building fish. These mechanisms may resuspend contaminants, making them more available to organisms. Burrowing organisms can also transport deeply buried contaminants to surface layers, thereby increasing surface contamination levels, while the surface-scouring mechanisms tend to decrease the concentrations in the surface sediment. Bioturbation has been reported to strongly influence the fate of contaminants and sediment-bound contaminants can be remobilized by this biological activity.[131]

Lee and Jones-Lee[6] reviewed the significance of chemically contaminated sediments and associated impacts. They are especially concerned about the development of sediment-contamination criteria based on simple chemical tests. They concluded that there has been sufficiently demonstrated evidence that the toxic-available form of chemical constituents present in the sediment is the dissolved form present in the interstitial waters. Historically, the EPA assumed that the dissolved form of certain organic toxicants could be estimated based on an equilibrium partitioning model based on the particulate organic carbon present. Likewise, the dissolved forms of heavy metals were assumed to be controlled by metal sulfide precipitates. Lee and Jones-Lee hold that the EPA's overly simplistic two-component-box model used to predict dissolved forms of toxicants should never be used alone without concurrent well-established toxicity measurements. They are also concerned about the use of databases to relate measured sediment chemical conditions with observed biological conditions that are also sometimes used to establish sediment criteria. These databases have not considered some of the most important possible causes of toxicity at the test sites, namely, low DO and high ammonia and hydrogen sulfide concentrations. They outlined the components of sediment toxicity tests they believe are necessary:

- Nonchemically based "toxicity" can be caused by factors such as sediment grain size.
- Natural vs. anthropogenic sediment toxicity also needs to be separated. They mention several instances where sediments are naturally toxic according to laboratory toxicity tests but still support healthy and high-quality sport fisheries in overlying waters. The most obvious natural cause of sediment toxicity is low oxygen levels in the interstitial water. High levels of ammonia and hydrogen sulfide may also then occur. They state that "the presence of highly toxic conditions in sediments from natural causes, which decimates the benthic-organism populations for a considerable part of the year, does not preclude the presence of an outstanding sports fishery."
- The sensitivity of the test organisms to ammonia toxicity should be considered. Several commonly used toxicity test organisms are much less sensitive to ammonia than many naturally occurring aquatic life forms of interest. Some researchers also strip ammonia from sediments before testing, treating ammonia as a test interference. They believe that nutrient-derived toxicity (algal decomposition effects on sediment oxygen demand and the resulting reducing conditions, low DO levels, and high ammonia and hydrogen sulfide levels) may be the most important cause of toxicity in aquatic sediments. An appropriate toxicity investigation evaluation (TIE) should be conducted to identify the cause of any identified toxicity problems. The use of acid-volatile sulfide and heavy-metal concentrations and TOC normalized sediment organic concentrations can be used as part of a TIE to rule out metals or certain organics as the potential cause of toxicity, but the reverse is not reliable (these methods cannot predict toxicity).
- Selecting reference sites is critical. A suite of test toxicity organisms (at least two or three) must be used, along with a suite of reference sites. Multiple references sites is needed to help understand the role of natural causes of toxicity. In addition, investigations should be conducted at least twice in a year during important times for the aquatic organisms.

Many researchers contend that the best approach in developing sediment-quality evaluations would use a best-professional-judgment (BPJ), weight-of-evidence approach. This approach involves an integrated assessment of the aquatic-life-toxicity test results, assessment of the bioaccumulations of hazardous chemicals in edible portions of aquatic life, knowledge of chemical characteristics of the sediments and associated waters, and investigations of the aquatic-life assemblages in the sediments of concern compared to appropriate reference sites.

## 19.6 BIOASSESSMENTS AND OTHER WATERSHED INDICATORS AS COMPONENTS OF RECEIVING WATER EVALUATIONS

Kuehne[132] studied the usefulness of using various aquatic organisms during stream taxonomic surveys as indicators of pollution. He found that invertebrates can reveal pollution for some time after a water pollution event, but they cannot give accurate indications of the nature of the contaminants. He stated that in-stream fish studies had not been employed as biological indicators much before 1975 but that they are comparable in many ways to invertebrates as quality indicators and can be more easily identified. However, because of better information pertaining to invertebrates and due to their limited mobility, certain species may be useful as sensitive indicators of minor changes in water quality. Fish can be highly mobile and cover large sections of a stream as long as their passage is not totally blocked by adverse conditions. Fish-disease surveys were also used during the Bellevue, Washington, urban-runoff studies as an indicator of water-quality problems.[25,65] McHardy et al.[133] also examined heavy-metal uptake in green algae (*Cladophora glomerata*) from urban runoff for use as a biological monitor of specific metals.

Burton et al.,[127] during tests conducted at polluted stream and landfill sites, found that a battery of laboratory and *in situ* bioassay tests were most useful when determining aquatic biota problems. The test series included microbial activity tests, along with exposures of microfaunal organisms, zooplankton, amphipods, and fathead minnows to the test water. The newly developed microbial tests correlated well with *in situ* biological test results. Bascombe et al.[134] also reported on the use of *in situ* biological tests, using an amphipod exposed for 5 to 6 weeks in urban streams, to examine urban-runoff receiving-water effects. Ellis et al.[135] examined bioassay procedures for evaluating urban-runoff effects on receiving-water biota. They concluded that an acceptable criterion for protecting receiving-water organisms should not only provide information on concentration and exposure relationships for *in situ* bioassays but also consider body burdens, recovery rates, and sediment-related effects.

A number of stormwater researchers have recently presented bioassessment and other "watershed indicators" that they have found to be useful tools for quantifying local receiving-water problems. Many of these schemes were presented at the conference *Assessing the Cumulative Impacts of Watershed Development in Aquatic Ecosystems and Water Quality* held in Chicago in March 1996, sponsored by the Northeastern Illinois Planning Commission, and at the conference *Effects of Watershed Development and Management on Aquatic Ecosystems* held in Snowbird, Utah, in August 1996, sponsored by the Engineering Foundation and the ASCE. Several papers from those conferences are summarized by location below.

### 19.6.1 U.S. National Perspective of Bioassessments

Barbour[136] reviewed many of the state programs throughout the United States that are using biological assessments as part of their water-resources programs. Most of the active state bioassessment programs started after 1990, following the publication of the EPA's *Rapid Bioassessment Protocols*[137] and the *Program Guidance for Biocriteria*[138] manuals. By 1996, numeric biocriteria were in place in Ohio and Florida (and promulgated in Maine) and under development in 13 other states. Although the majority of the states had not used biocriteria, nearly three fourths had used bioassessment data to measure the attainment of their aquatic uses. Almost all states were using benthic macroinvertebrates (47 states) and fish (36 states). Seven states were also using algae in their bioassessment programs.

An important aspect of the biocriteria approach is that local and regional expectations be considered in setting specific objectives. In addition, local reference sites representing specific ecoregions are also used to calibrate observations. The basic components of a bioassessment include:

- Study objectives (typically the determination of biological conditions for different watershed characteristics)
- Site classification (identification of homogeneous areas within a watershed, typically using various biological metrics)
- Reference condition (relatively undisturbed areas for comparison and calibration of the metrics)
- Standardized protocols (training and the use of consistent methods)
- Data analysis (selection of several complementary metrics based on local relevancy)
- Habitat assessment (physical habitat structure evaluations, generally a visual technique)
- Quality assurance (assign responsibility, establish protocols, etc. to ensure repeatability)

### 19.6.2 Watershed Indicators of Receiving Water Problems

The EPA[139] published a list of 18 indicators to track the health of the nation's aquatic ecosystems. These indicators are intended to supplement conventional water-quality analyses in compliance-monitoring activities. The use of broader indicators of environmental health is increasing. As an example, 12 states are currently using biological indicators, and 27 states are developing local biological indicators, according to Pelley.[140] Because of the broad nature of the nation's potential receiving-water problems, this list is more general than typically used for specific stormwater issues. These 18 indicators are:[139]

1. Population served by drinking water systems violating health-based requirements.
2. Population served by unfiltered surface water systems at risk from microbiological contamination.
3. Population served by communities by community drinking water systems exceeding lead action levels.
4. Drinking water systems with source-water-protection programs.
5. Fish consumption advisories.
6. Shellfish-growing waters approved for harvest for human consumption.
7. Biological integrity of rivers and estuaries.
8. Species at risk of extinction.
9. Rate of wetland acreage loss.
10. Designated uses: drinking water supply, fish and shellfish consumption, recreation, aquatic like.
11. Groundwater pollutants (nitrate).
12. Surface-water pollutants.
13. Selected coastal surface water pollutants in shellfish.
14. Estuarine eutrophication conditions.
15. Contaminated sediments.
16. Selected point-source loadings to surface water and groundwater.
17. Nonpoint-source sediment loadings from cropland.
18. Marine debris.

These environmental indicators cover a wide range of problems and many are for specific local uses. Most, however, are applicable to stormwater problems in urban areas.

Claytor[141,142] summarized the approach developed by the Center for Watershed Protection as part of their EPA-sponsored research on identifying watershed indicators that can be used to assess the effectiveness of stormwater-management programs.[143] The indicators selected are direct or indirect measurements of conditions or elements that indicate trends or responses of watershed conditions to stormwater-management activities. Categories of these environmental indicators are shown in Table 19.4, ranging from conventional water-quality measurements to citizen surveys. Biological and habitat categories are also represented. Table 19.5 lists the 26 indicators by category. It is recommended that appropriate indicators be selected from each category for a specific area under study. This will enable a better understanding of the linkage of what is done on the land, how the sources are regulated or managed, and the associated receiving-water problems. The indicators were selected to: (1) measure stress or the activities that lead to impacts on receiving waters, (2) assess the resources themselves, and (3) measure the regulatory compliance or program initiatives. Claytor[142] presented a framework for using stormwater indicators, as shown below.

Level 1 (problem identification):

1. Establish management sphere (who is responsible, other regulatory agencies involved, etc.).
2. Gather and review historical data.
3. Identify local uses that may be impacted by stormwater (flooding/drainage, biological integrity, noncontact recreation, drinking-water supply, contact recreation, and aquaculture).
4. Inventory resources and identify constraints (time frame, expertise, funding, and labor limitations)
5. Assess baseline conditions (use rapid assessment methods).

**Table 19.4 Stormwater Indicator Categories[142]**

| Category | Description | Principal Element Being Assessed |
|---|---|---|
| Water quality | Specific water-quality characteristics | Receiving-water quality |
| Physical/hydrological | Measure changes to, or impacts on, the physical environment | Receiving-water quality |
| Biological | Use of biological communities to measure changes to, or impacts on, biological parameters | Receiving-water quality |
| Social | Responses to surveys or questionnaires to assess social concerns | Human activity on the land surface |
| Programmatic | Quantify various nonaquatic parameters for measuring program activities | Regulatory compliance or program initiatives |
| Site | Indicators adapted for assessing specific conditions at the site level | Human activity on the land surface |

**Table 19.5 Environmental Indicators[142]**

| Indicator Category | Indicator Name |
|---|---|
| Water quality indicators | Water-quality pollutant-constituent monitoring |
| | Toxicity testing |
| | Nonpoint source loadings |
| | Exceedence frequencies of water-quality standards |
| | Sediment contamination |
| | Human health criteria |
| Physical and hydrological indicators | Stream widening/downcutting |
| | Physical habitat monitoring |
| | Impacted dry-weather flows |
| | Increased flooding frequency |
| | Stream temperature monitoring |
| Biological indicators | Fish assemblage |
| | Macroinvertebrate assemblage |
| | Single species indicator |
| | Composite indicators |
| | Other biological indicators |
| Social indicators | Public attitude surveys |
| | Industrial/commercial pollution prevention |
| | Public involvement and monitoring |
| | User perception |
| Programmatic indicators | Illicit connections identified/corrected |
| | BMPs installed, inspected, and maintained |
| | Permitting and compliance |
| | Growth and development |
| Site indicators | BMP performance monitoring |
| | Industrial-site-compliance monitoring |

The selection of the indicators to assess the baseline conditions should be based on the local uses of concern, as shown in Table 19.6. Most of the anticipated important uses are shown to require indicators selected for each of the categories.

The Level 2 assessment strategy is for examining the local management program and is outlined below:

1. State goals for program (based on baseline conditions, resources, and constraints).
2. Inventory prior and ongoing efforts (including evaluating the success of ongoing efforts).
3. Develop and implement management program.
4. Develop and implement monitoring program (more quantitative indicators than typically used for the Level 1 evaluations above).

**Table 19.6 Selection of Indicators for Evaluating Baseline Conditions by Receiving-Water Use[142]**

| | Water Quality | Physical/ Hydrological | Biological Indicators | Social Indicators | Programmatic Indicators | Site Indicators |
|---|---|---|---|---|---|---|
| Flooding/drainage | | x | | x | x | x |
| Biological integrity | x | | x | x | x | x |
| Noncontact recreation | x | x | x | x | x | x |
| Water supply | x | | x | x | x | x |
| Contact recreation | x | x | x | x | x | x |
| Aquaculture | x | x | x | x | x | x |

5. Assess indicator results. (Does the stormwater indicator monitoring program measure the overall watershed health?)
6. Reevaluate management program (update and revise management program based on measured successes and failures).

Cave[144] described how environmental indicators are being used to summarize the massive amounts of data being generated by the Rouge River National Wet Weather Demonstration Project in Wayne County (Detroit area), Michigan. This massive project is examining existing receiving-water problems, the performance of stormwater and CSO management practices, and receiving-water responses in a 438-square-mile watershed having more than 1.5 million people in 48 separate communities. The baseline monitoring program now has more than 4 years of continuous monitoring of flow, pH, temperature, conductivity, and DO, supplemented by automatic sampling for other water-quality constituents, at 18 river stations. More than 60 projects are examining the effectiveness of stormwater-management practices, and 20 projects are examining the effectiveness of CSO controls, each also generating large amounts of data. Toxicants are also being monitored in sediment, water, and fish tissue and with semipermeable membranes to help evaluate human-health and aquatic-life effects. Habitat surveys were conducted at 83 locations along more than 200 mi of waterway. Algal diversity and benthic macroinvertebrate assessments were also conducted at these survey locations. Electrofishing surveys were conducted at 36 locations along the main river and in tributaries. Several computer models were also used to predict sources, loadings, and wet-weather flow-management options for the receiving waters and for the drainage systems. A geographic information system was used to manage and provide spatial analyses of the massive amounts of data collected. However, there was still a great need to simply present the data and findings, especially for public presentations. Cave described how they developed a short list of 35 indicators, based on the list of 18 from the EPA and with discussions with state and national regulatory personnel. They then developed seven indices that could be color-coded and placed on maps to indicate areas of existing problems and projected conditions based on alternative management scenarios. These indices are described as follows:

Condition quality indicators:

1. Dissolved oxygen. Concentration and percent saturation values (ecologically important).
2. Fish consumption index. Based on advisories from the Michigan Dept. of Public Health.
3. River flow. Significant for aquatic habitat and fish communities.
4. Bacteria count. *E. coli* counts based on Michigan Water Quality Standards, distinguished for wet and dry conditions.

Multifactor indices:

1. Aquatic biology index. Composite index based on fish and macroinvertebrate community assessments (populations and individuals).
2. Aquatic habitat index. Habitat suitability index, based on substrate, cover, channel morphology, riparian/bank condition, and water quality.
3. Aesthetic index. Based on water clarity, color, odor, and visible debris.

These seven indicators represent 30 physical, chemical, and biological conditions that directly impact the local receiving-water uses (water contact recreation, warmwater fishery, and general aesthetics). Cave presented specific descriptions for each of the indices and gave examples of how they are color-coded for map presentation.

The use of reference sites is common to many bioassessment approaches. As indicated above, reference sites typically are selected as representing as close to natural conditions as possible. However, it is not possible to identify such pristine locations representing varied habitat conditions in most areas of the country. Ohio, for example, has numerous reference sites throughout the state representing a broad range of conditions, but few are completely unimpacted by modifications or human activity in the watersheds. Schueler[77] reviewed a USGS report prepared by Crawford and Leant that examined the differences between streams located in forested, agricultural, and urban watersheds in North Carolina. He points out that in many cases, a completely natural forested area is not a suitable benchmark for current conditions before urbanization. In many areas of the country, agricultural land is being converted to urban land, and the in-stream changes expected may be better compared to agricultural conditions. The USGS study found that the stream impacted by agricultural operations was intermediate in quality, with higher nutrient and worse substrate conditions than the urban stream, but better macroinvertebrate and fish conditions. The forested watershed had the best conditions (good-quality conditions for all categories), except for somewhat higher sediment heavy-metal concentrations than expected. Even though the agricultural watershed had little impervious area, it had high sediment and nutrient discharges, plus some impacted stream corridors. The urban stream had poor macroinvertebrate and fish conditions, poor sediment and temperature conditions, and fair substrate and nutrient conditions.

### 19.6.3 Summary of Assessment Tools

Almost all states using bioassessment tools have relied on the EPA reference documents as the basis for their programs. Common components of these bioassessment programs (in general order of popularity) include:

- Macroinvertebrate surveys (almost all programs, but with varying identification and sampling efforts)
- Habitat surveys (almost all programs)
- Some simple water-quality analyses
- Some watershed characterizations
- Few fish surveys
- Limited sediment-quality analyses
- Limited stream-flow analyses
- Hardly any toxicity testing
- Hardly any comprehensive water quality analyses

Normally, numerous metrics are used, typically only based on macroinvertebrate survey results, which are then assembled into a composite index. Many researchers have identified correlations between these composite index values and habitat conditions. Water-quality analyses in many of these assessments are seldom comprehensive, a possible overreaction to conventional very costly programs that have typically resulted in minimally worthwhile information. Burton and Pitt[13] have recommended a more balanced assessment approach, using toxicity testing and carefully selected water and sediment analyses to supplement the needed biological monitoring activities. A multicomponent assessment enables a more complete evaluation of causative factors and potential mitigation approaches.

## 19.7 SUMMARY OF URBAN-RUNOFF EFFECTS ON RECEIVING WATERS

The effects of urban runoff on receiving-water aquatic organisms or other beneficial uses is very site-specific. Different land-development practices create substantially different runoff-flow characteristics. Different rain patterns cause different particulate washoff, transport, and dilution conditions. Local attitudes also define specific beneficial uses and, therefore, current problems. There is also a wide variety of water types receiving urban runoff, and these waters all have watersheds that are urbanized to various degrees. Therefore, it is not surprising that urban-runoff effects, though generally dramatic, are also quite variable and site-specific.

Previous attempts to identify urban-runoff problems using existing water-quality data have not been conclusive because of differences in sampling procedures and the common practice of pooling data from various sites or conditions.[4] It is therefore necessary to carefully design comprehensive, long-term studies to investigate urban-runoff problems on a site-specific basis.[13] Sediment transport, deposition, and chemistry play key roles in urban receiving waters and need additional research. Receiving-water aquatic biological conditions, especially compared to unaffected receiving waters, should be studied as a supplement to laboratory bioassays. In-stream taxonomic surveys are sensitive to natural variations of pollutant concentrations, flows, and other habitat affects. However, laboratory studies are necessary to help understand potential cause-and-effect relationships because of their ability to better control exposure variables.

These specific studies need to examine beneficial uses directly and not rely on published water-quality criteria and water-column measurements alone. Published criteria are usually not applicable to urban runoff because of the highly variable nature of urban runoff and the unique chemical speciation of its components. Typical natural water-pollutant characteristics (especially chemical mixtures and exposure pulses) are difficult to interpret, compared to simpler artificial systems having continuous discharges of more uniform characteristics.

The long-term aquatic-life effects of urban runoff are probably more important than short-term effects associated with specific events and are related to site-specific conditions associated with dilution, watershed size, and stream size. The long-term effects are probably related to habitat degradation, deposition and accumulation of toxic sediments, or the inability of the aquatic organisms to adjust to repeated exposures to high concentrations of toxic materials or high flow rates.

## REFERENCES


1. *Water Environment & Technology,* News Watch: U.S. water quality shows little improvement over 1992 inventory, 8, 2, 15–16, Feb. 1996a.
2. Borchardt, D. and B. Statzner, Ecological impact of urban stormwater studied in experimental flumes: Population loss by drift and availability of refugial space, *Aquat. Sci.,* 52, 4, 299–314, 1990.
3. Field, R. and R. Pitt, Urban storm-induced discharge impacts: U.S. Environmental Protection Agency research program review, *Water Sci. Technol.,* 22, 10/11, 1990.
4. Pitt, R. E., Biological effects of urban runoff discharges, in *Effects of Urban Runoff on Receiving Systems: An Interdisciplinary Analysis of Impact, Monitoring, and Management,* Engineering Foundation Conference, Mt. Crested Butte, CO, ASCE, New York, 1991.
5. Pitt, R., Effects of urban runoff on aquatic biota., in *Handbook of Ecotoxicology*, Hoffman, D. J., B. A. Rattner, G. A. Burton, Jr., and J. Cairns, Eds., Lewis Publishers/CRC Press, Boca Raton, FL, 1995, 609–630.
6. Lee, G. F. and A. Jones-Lee, Evaluation of the water quality significance of the chemical constituents in aquatic sediments: Coupling sediment quality evaluation results to significant water quality impacts, *WEFTEC '96, Surface Water Quality and Ecology, Parts 1 and 2*, 1996 Water Environment Federation Technical Exposition and Conference, Dallas, WEF, Alexandria, VA, 1996.
7. Lee, G. F. and A. Jones-Lee, Deficiencies in stormwater quality monitoring, *Proc. Eng. Found./ASCE Conf., Stormwater NPDES Related Monitoring Needs*, Torno, H. C., Ed., Mt. Crested Butte, CO, Aug. 1994, ASCE, New York, 1995.
8. Mancini, J. and A. Plummer, Urban runoff and water quality criteria, in *Urban Runoff Quality – Impact and Quality Enhancement Technology,* Urbonas, B. and L. A. Roesner, Eds., Engineering Foundation Conference, Henniker, NH, ASCE, New York, 133–149, Jun. 1986.

9. Davies, P. H., Factors in controlling nonpoint source impacts, *Proc. Eng. Found./ASCE Conf., Stormwater Runoff and Receiving Systems: Impact, Monitoring and Assessment* Mt. Crested Butte, CO, Herricks, E. E., Ed., Lewis Publishers/CRC Press, Boca Raton, FL, 1995, 53–64.
10. Herricks, E. E, I. Milne, and I. Johnson, A protocol for wet weather discharge toxicity assessment, *WEFTEC '96: Proc. of the 69th Annu. Conf. and Exposition*, Dallas, Vol. 4, 1996, 13–24.
11. Herricks, E. E., Ed., *Proc. of Eng. Found./ASCE Conf., Stormwater Runoff and Receiving Systems: Impact, Monitoring and Assessment,* Mt. Crested Butte, CO, 1991, Lewis Publishers/CRC Press, Boca Raton, FL, 1995.
12. Crunkilton, R., J. Kleist, J. Ramcheck, W. DeVita, and D. Villeneueve, Assessment of the response of aquatic organisms to long-term *in situ* exposures to urban runoff, *Proc. of Effects of Watershed Developments and Management on Aquatic Ecosystems Conf.* Snowbird, UT, Aug. 4–9, 1996, Roesner, L. A., Ed., ASCE, New York, 1997.
13. Burton, G. A., Jr. and R. Pitt, *Stormwater Effects Handbook: A Tool Box for Watershed Managers, Scientists, and Engineers*, CRC Press, Boca Raton, FL, 2001.
14. Duda, A. M., D. R. Lenat, and D. Penrose, Water quality in urban streams — What we can expect, *J. Water Pollut. Control Fed.*, 54, 7, 1139–1147, 1982.
15. Horner, R. R., Toward ecologically based urban runoff management, *Proc. of Eng. Found. Conf., Effects of Urban Runoff on Receiving Systems: An Interdisciplinary Analysis of Impact, Monitoring, and Management*, Mt. Crested Butte, CO, ASCE, NY, 1991.
16. Snodgrass, W. J., B. W. Kilgour, L. Leon, N. Eyles, J. Parish, and D. R. Barton, Applying ecological criteria for stream biota and an impact flow model for evaluation of sustainable urban water resources in southern Ontario, in Sustaining Urban Water Resources in the 21st Century, in *Proc. Eng. Found. Conf.,* Rowney, A. C., P. Stahre, and L. A. Roesner, Eds., Malmo, Sweden, Sept. 7–12, 1997, ACSE, New York, 1998.
17. Ketchum, L. H., Jr., Dissolved Oxygen Measurements in Indiana Streams during Urban Runoff, EPA-600/2–78–135, U.S. Environmental Protection Agency, Cincinnati, OH, Aug. 1978.
18. Heaney, J. P., W. C. Huber, and M. E. Lehman, Nationwide Assessment of Receiving Water Impacts from Urban Storm Water Pollution, U.S. Environmental Protection Agency, Cincinnati, OH, Apr. 1980.
19. Pitt, R. E., Demonstration of Nonpoint Pollution Abatement through Improved Street Cleaning Practices, EPA-600/2–79–161, U.S. Environmental Protection Agency, Cincinnati, OH, Aug. 1979.
20. Burkholder, J. M., L. M. Larsen, H. B. Glasgow, Jr., K. M. Mason, P. Gama, and J. E. Parsons, Influence of sediment and phosphorus loading on phytoplankton communities in an urban piedmont reservoir, *J. Lake Reservoir Manage.*, 14, 1, 110–121, 1998.
21. Field, R., and C. Cibik, Urban runoff and combined sewer overflows, *J. Water Pollut. Control Fed.*, 52, 6, 1290–1307, 1980.
22. Pitt, R. and M. Bozeman, Sources of Urban Runoff Pollution and Its Effects on an Urban Creek, EPA-600/52–82–090, U.S. Environmental Protection Agency, Cincinnati, OH, 1982.
23. Bolstad, P. V. and W. T. Swank, Cumulative impacts of land use on water quality in a southern Appalachian watershed, *J. Am. Water Resour. Assoc.*, 33, 3, 519–533, 1997.
24. Ray, S., and W. White, Selected aquatic plants as indicator species for heavy metal pollution, *J. Environ. Sci. Health*, A11, 717, 1976.
25. Pitt, R. E. and P. Bissonnette, Bellevue Urban Runoff Program, Summary Report, PB84 237213, Water Planning Division, U.S. Environmental Protection Agency, Washington, D.C., 1983.
26. Scott, J. B., C. R. Steward, and Q. J. Stober, Effects of urban development on fish population dynamics in Kelsey Creek, Washington, *Trans. Am. Fisher. Soc.*, 115, 4, 555–567, Jul. 1986.
27. Marcy, S. and J. Gerritsen, Developing diverse assessment endpoints to address multiple stressors in watershed ecological risk assessment, Abstract Book: *SETAC 17th Annual Meeting*, Washington, D.C., Nov. 17–21, 1996, 96.
28a. Dyer, S. D. and C. E. White, A watershed approach to assess mixture toxicity via integration of public and private databases, Abstract Book: *SETAC 17th Annual Meeting*. Washington, D.C., Nov. 17–21, 1996, 96.
28b. ES&T *(Environmental Science & Technology),* Toxicity of aquatic mixtures yielding to new theoretical approach, 30, 4, 155a–156a, Apr. 1996a.
29. Ireland, D. S., G. A. Burton, Jr., and G. G. Hess, *In situ* toxicity evaluations of turbidity and photoinduction of polycyclic aromatic hydrocarbons, *Environ. Toxicol. Chem.*, 15, 4, 574–581, 1996.

30. Johnson, I., E. E. Herricks, and I. Milne, Application of a test battery for wet weather discharge toxicity analyses, Vol. 4, 219–229, *WEFTEC '96: Proc. 69th Annu. Conf. and Exposition*, Dallas, 1996.
31. Rainbow, P. S., Heavy metals in aquatic invertebrates, *in Environmental Contaminants in Wildlife: Interpreting Tissue Concentrations*, Beyer, W. N., G. H. Heinz, and A. W. Redmon-Norwood, Eds., CRC Press/Lewis Publishers, Boca Raton, FL, 1996, 405–425.
32. Allen, H. E., Ed., *Metal Contaminated Aquatic Sediments*, Ann Arbor Press, Chelsea, MI, 1996.
33. Hall, L. W., Jr., M. C. Scott, W. D. Killen, Jr., and R. D. Anderson, The effects of land-use characteristics and acid sensitivity on the ecological status of Maryland coastal plain streams, *Environ. Toxicol. Chem.*, 15, 3, 384–394, 1996.
34. Garries, M. J., T. Barron, R. Batiuk, K. Eisenman, J. Gregory, L. Hall, A. Hart, P. Jiapizian, W. Rue, J. Savitz, M. E. Setting, and C. Stoll, Derivation of Chesapeake Bay toxics of concern, Abstract Book: *SETAC 17th Annual Meeting*, Washington, D.C., Nov. 17–21, 1996, 105.
35. Schueler, T., Ed., *In situ*, nonbenthic assessment of stormwater-impacted sediments."*Watershed Prot. Tech.*, 2, 2, 351–353, 1996b.
36. Love, E. and B. Woolley, Environmentally sustainable stormwater quality, *Proc. 8th Int. Conf. on Urban Storm Drainage*, Joliffe, I. B. and J. E. Ball, Eds., The Institution of Engineers of Australia, The International Association for Hydraulic Research, and The International Association on Water Quality, Sydney, Australia, 1575, Aug. 30–Sept. 3, 1999, 50.
37. Pitt, R., Urban stormwater toxic pollutant assessment, sources, and treatability — Closure. *Water Environ. Res.*, 68, 5, 953–955, 1996.
38. Huber, W. C. and M. M. Quigley, Simplified Fate and Transport Model of Runoff from Highway Construction and Repair Materials, *Proc. 8th Int. Conf. on Urban Storm Drainage*, Joliffe, I. B. and J. E. Ball, Eds., The Institution of Engineers of Australia, The International Association for Hydraulic Research, and The International Association on Water Quality, Sydney, Australia, Aug. 30–Sept. 3, 1999, 1209.
39. Doherty, F. G., A. A. Qureshi, and J. B. Razza, Comparison of the *Ceriodaphnia dubia* and Microtox® inhibition tests for toxicity assessment of industrial and municipal wastewaters, *Environ. Toxicol.*, 14, 4, 375, 1999.
40. Kosmala, A., S. Charvet, M. C. Roger, and B. Faessel, Impact assessment of a wastewater treatment plant effluent using instream invertebrates and the *Ceriodaphnia dubia* chronic toxicity test, *Water Res.*, 33, 1, 266, 1999.
41. Marsalek, J., Q. Rochfort, B. Brownlee, T. Mayer, and M. Servos, Exploratory study of urban runoff toxicity, *Water Sci. Technol.*, 39, 12, 33, 1999.
42. Marsalek J. and Q. Rochfort, Toxicity of urban wet-weather pollution sources: Stormwater and CSOs, *Proc. 8th Int. Conf. on Urban Storm Drainage*, Joliffe, I. B. and J. E. Ball, Eds., The Institution of Engineers of Australia, The International Association for Hydraulic Research, and The International Association on Water Quality, Sydney, Australia, Aug. 30–Sept. 3, 1999.
43. Skinner, L., A. de Peyster, and K. Schiff, developmental effects of urban storm water in Medaka (*Oryzias latipes*) and inland silverside (*Menidia beryllina*), *Arch. Environ. Contam. Toxicol.*, 37, 2, 227, 1999.
44. Tucker, K. A., and G. A. Burton, Assessment of nonpoint-source runoff in a stream using *in situ* and laboratory approaches, *Environ. Toxicol. Chem.*, 18, 12, 2797. 1999.
45. Hatch, A. C. and G. A. Burton, Sediment toxicity and stormwater runoff in a contaminated receiving system: Consideration of different bioassays in the laboratory and field, *Chemosphere*, 39, 6, 1001, 1999.
46. Bickford, G., J. Toll, J. Hansen, E. Baker, and R. Keessen, aquatic ecological and human health risk assessment of chemicals in wet weather discharges in the Sydney region, New South Wales, Australia, *Mar. Pollut. Bull.*, 39, 1–2, 335, 1999.
47. Brent, R. N. and E. E. Herricks, A method for the toxicity assessment of wet weather events, *Water Res.*, 33, 10, 2255, 1999.
48. Colford, J., M. I. Tager, L. F. Byers, P. Ricci, A. Hubbard, and R. Horner, methods for assessing the public health impact of outflows from combined sewer systems, *J. Air Waste Manage. Assoc.*, 49, 4, 454, 1999.
49. Heaney, J. P. and W. C. Huber, Nationwide assessment of urban runoff impact on receiving water quality, *Water Resour. Bull.*, 20, 1, 35–42, 1984.
50. Klein, R. D., Urbanization and stream quality impairment, *Water Resour. Bull.*, 15, 4, 1979.

51. Lenet, D. R., D. L. Penrose, and K. Eagleson, Biological Evaluation of Non-Point Sources of Pollutants in North Carolina Streams and Rivers, North Carolina Division of Environmental Management, Biological Series #102, North Carolina Dept. of Natural Resources and Community Development, Raleigh, 1979.
52. Lenet, D. and K. Eagleson, Ecological Effects of Urban Runoff on North Carolina Streams, North Carolina Division of Environmental Management, Biological Series #104. North Carolina Dept. of Natural Resources and Community Development, Raleigh, 1981.
53. Lenet, D. R., D. L. Penrose, and K. W. Eagleson, Variable effects of sediment addition on stream benthos, *Hydrobiologia*, 79, 187–194, 1981.
54. Cook, W. L., F. Parrish, J. D. Satterfield, W. G. Nolan, and P. E. Gaffney, *Biological and Chemical Assessment of Nonpoint Source Pollution in Georgia: Ridge-Valley and Sea Island Streams*, Department of Biology, Georgia State Univ., Atlanta, 1983.
55. Benke, A. C., G. E. Willeke, F. K. Parrish, and D. L. Stites, *Effects of Urbanization on Stream Ecosystems,* School of Biology, Environmental Resources Center, Report No. ERC 07–81, Georgia Institute of Technology, Atlanta, 1981.
56. CTA, Inc., Georgia Nonpoint Source Impact Assessment Study: Blue Ridge/Upland Georgia Cluster, Piedmont Cluster, and Gulf Coastal Plain Cluster, Georgia Environmental Protection Division, Dept. of Natural Resources, Atlanta, 1983.
57. Pratt, J. M., R. A. Coler, and P. J. Godfrey, Ecological effects of urban stormwater runoff on benthic macroinvertebrates inhabiting the Green River, Massachusetts, *Hydrobiologia*, 83, 29–42, 1981.
58. Ehrenfeld, J. G. and J. P. Schneider, The Sensitivity of Cedar Swamps to the Effects of Non-Point Pollution Associated with Suburbanization in the New Jersey Pine Barrens, U.S. Environmental Protection Agency, Office of Water Policy, PB8–4–136779, Washington, D.C., Sept. 1983.
59. Medeiros, C. and R. A. Coler, *A Laboratory/Field Investigation into the Biological Effects of Urban Runoff*, Water Resources Research Center, Univ. of Massachusetts, Amherst, July 1982.
60. Medeiros, C., R. A. Coler, and E. J. Calabrese, A laboratory assessment of the toxicity of urban runoff on the fathead minnow (*Pimephales promelas*), *J. Environ. Sci. Health*, A19. 7, 847–861, 1984.
61. Pedersen, E. R., *The Use of Benthic Invertebrate Data for Evaluating Impacts of Urban Stormwater Runoff*, Masters thesis submitted to the College of Engineering, Univ. of Washington, Seattle, 1981.
62. Richey, J. S., M. A. Perkins, and K. W. Malueg, The effects of urbanization and stormwater runoff on the food quality in two salmonid streams, *Verh. Int. Werein. Limnol.* Vol. 21, 812–818, Stuttgart, Oct. 1981.
63. Perkins, M. A., An Evaluation of Instream Ecological Effects Associated with Urban Runoff to a Lowland Stream in Western Washington, U.S. Environmental Protection Agency, Corvallis Environmental Research Laboratory, Corvallis, OR, Jul. 1982.
64. Richey, J. S., *Effects of Urbanization on a Lowland Stream in Western Washington*, Ph.D. dissertation, Univ. of Washington, Seattle, 1982.
65. Scott, J. B., C. R. Steward, and Q. J. Stober, Impacts of Urban Runoff on Fish Populations in Kelsey Creek, Washington, Contract No. R806387020, U.S. Environmental Protection Agency, Corvallis Environmental Research Laboratory, Corvallis, OR, May 1982.
66. Ebbert, J. C., J. E. Poole, and K. L. Payne, Data Collected by the U.S. Geological Survey During a Study of Urban Runoff in Bellevue, Washington, 1979–82, Preliminary U.S. Geological Survey Open-File Report, Tacoma, WA, 1983.
67. Prych, Edmund A. and J. C. Ebbert, Quantity and Quality of Storm Runoff from Three Urban Catchments in Bellevue, Washington, Preliminary U.S. Geological Survey Water Resources Investigations Report, Tacoma, WA, undated.
68. Mote Marine Laboratory, Biological and Chemical Studies on the Impact of Stormwater Runoff upon the Biological Community of the Hillsborough River, Tampa, FL, Stormwater Management Division, Dept. of Public Works, Tampa, March 1984.
69. Striegl, R. G., Effects of Stormwater Runoff on an Urban Lake, Lake Ellyn at Glen Ellyn, IL, USGS open file report 84–603, Lakewood, CO, 1985.
70. Garie, H. L. and A. McIntosh, Distribution of benthic macroinvertebrates in a stream exposed to urban runoff, *Water Resour. Bull.*, 22, 3, 447–455, 1986.
71. Garie, H. L. and A. McIntosh, Distribution of benthic macroinvertebrates in a stream exposed to urban runoff, *Water Sci. Technol.,* 22, 10/11. 1990.

72. Stewart, P. M., J. T. Butcher, and P. J. Gerovac, Diatom (*Bacillariophyta*) community response to water quality and land use, *Nat. Areas J.,* 19, 2, 155, 1999.
73. Handová, Z., Z. Korícek, M. Liska, J. Marsálek, J. Matena, and J. Sed'a, CSO impacts on receiving waters: Heavy metals in sediments and macrozoobenthos, *Proc. 7th Int. Conf. on Urban Storm Drainage*, Sieker, F. and H.-R. Verworn, Eds., IAHR/IAWQ. SuG-Verlagsgesellschaft, Hannover, Germany, Sept. 9–13, 1996, 485–492.
74. Boudries, H., C. Broguet, J.-M. Mouchel, and D. R. Thévenot, Urban runoff impact on composition and concentration of hydrocarbons in River Seine suspended solids, *Proc. 7th Int. Conf. Urban Storm Drainage*, Sieker, F. and H-R. Verworn, Eds., IAHR/IAWQ. SuG-Verlagsgesellschaft, Hannover, Germany, Sept. 9–13, 1996, 569–574.
75. Estèbe, A., G. Belhomme, S. Lecomte, V. Videau, J-M. Mouchel, and D. R. Thévenot. Urban runoff impacts on particulate metal concentrations in River Seine: Suspended solid and sediment transport, *Proc. 7th Int. Conf. on Urban Storm Drainage*, Sieker, F. and H.-R. Verworn, Eds., IAHR/IAWQ. SuG-Verlagsgesellschaft, Hannover, Germany, Sept. 9–13, 1996, 575–580.
76. Dreher, D. W., Watershed urbanization impacts on stream quality indicators in northeastern Illinois, *Proc. of Assessing the Cumulative Impacts of Watershed Development on Aquatic Ecosystems and Water Quality Conf.,* March 20–21, 1996, Northeastern Illinois Planning Commission, Chicago, IL, 1997, 129–135.
77. Schueler, T., Ed., Comparison of forest, urban and agricultural streams in North Carolina. *Watershed Prot. Tech.,* 2, 4, 503–506, 1997b.
78. Cordone, A. J. and D. W. Kelley, Influences of inorganic sediments on aquatic life of streams, *Californica Fish Game*, 47, 189–228, 1961.
79. Leopold, L. B., M. G. Wolman, and J. P. Miller, *Fluvial Processes in Geomorphology*, W. H. Freeman, San Francisco, 1964.
80. Brookes, A., *Channelized Rivers: Perspectives for Environmental Management*, John Wiley and Sons, 1988.
81. Rosgen, D. L., *A Classification of Natural Rivers*. Catena, Elsevier Science, Amsterdam. 1994.
82. Brookes, A., Design issues, in *Effects of Urban Runoff on Receiving Systems: An Interdisciplinary Analysis of Impact, Monitoring, and Management, Proc. Eng. Found. Conf.*, Mt. Crested Butte, CO, ASCE, New York, 1991.
83. Wolman, M. G. and A. P. Schick, Effects of construction on fluvial sediment, urban and suburban areas of Maryland, *Water Resour. Res.,* 3, 2, 451–464, 1967.
84. Pess, G. R. and R. E. Bilby, Stream-reach and watershed-scale variables and salmonid spawning distribution and abundance in the Puget Sound region, *Proc. AWRA 1999 Annu. Water Resour. Conf. — Watershed Manage. to Protect Declining Species*, Dec. 1999, Seattle, American Water Resources Association, 397, 1999.
85. Greenberg, E. S., M. R. Gagner, and D. R. Reiser, An approach for evaluating baseline habitat conditions within urban/urbanizing streams with application to prioritization of recovery efforts. *Proc. AWRA 1999 Annu. Water Resour. Conf. — Watershed Manage. to Protect Declining Species*, Dec. 1999, Seattle, American Water Resources Association, 409, 1999.
86. Bragg, D. C. and J. L. Kershner, Coarse woody debris in riparian zones — Opportunity for interdisciplinary interaction, *J. For.*, 97, 4, 30, 1999.
87. Larson, M. G., Effectiveness of LWD in stream rehabilitation projects in urban basins. *Proc. AWRA 1999 Annu. Water Resour. Conf. — Watershed Manage. to Protect Declining Species*, Dec. 1999, Seattle, American Water Resources Association, 377, 1999.
88. Markowitz, D. V., P. Gsellman, and J. Bronowsk, Thinking outside the pipe: Non-traditional solutions for water quality improvement. *Proc. Water Environ. Fed. 72nd Annu. Conf. Exposition*, [CD-ROM], New Orleans, 1999.
89. Zobrist, C., N. Cencie, and G. Demortier, Methodological study of CSO's impact: Application to the River Orne, *Proc. 8th Int. Conf. on Urban Storm Drainage*, Joliffe, I. B. and J. E. Ball, Eds., The Institution of Engineers Australia, The International Association for Hydraulic Research, and The International Association on Water Quality, Sydney, Australia, Aug. 30–Sept. 3, 1999, 403.
90. O'Meara, J., J. Murray, and J. Ridgway, Restoration of an urban lake: The Newburgh Lake Project, *Proc. Water Environ. Fed. 72nd Annu. Conf. Exposition*, [CD-ROM], New Orleans, 1999.

91. Xu, S., R. Higashijima, T. Shingu, M. Amakata, and T. Sakai, Improvement plan of Y River going through downtown of F city, *Proc. 8th Int. Conf. on Urban Storm Drainage*, Joliffe, I. B. and J. E. Ball, Eds., The Institution of Engineers Australia, The International Association for Hydraulic Research, and The International Association on Water Quality, Sydney, Australia, Aug. 30–Sept. 3, 1999, 218.
92. Cianfrani, C., W. C. Hession, M. McBride, and J. E. Pizzuto, Urban streams in Fairmount Park, Philadelphia, PA: Assessment and restoration, *Proc. AWRA 1999 Annu. Water Resour. Conf. — Watershed Manage. to Protect Declining Species*, American Water Resources Association, Seattle, Dec. 1999, 393.
93. Derry, W. E., A. Haub, E. Dobey, and J. Hoey, City of Olympia aquatic habitat evaluation and management study, *Proc. AWRA 1999 Annu. Water Resour. Conf. — Watershed Manage. to Protect Declining Species*, American Water Resources Association, Seattle, Dec. 1999, 323.
94. Ishikawa, T., M. Irie, M., T. Terashima, and K. Ozawa, Restoration of the hydrological cycle in the Izumi River Basin in Yokohama, *Proc. 8th Int. Conf. on Urban Storm Drainage*, Joliffe, I. B. and J. E. Ball, Eds., The Institution of Engineers of Australia, The International Association for Hydraulic Research, and The International Association on Water Quality, Sydney, Australia, Aug. 30–Sept. 3, 1999, 530.
95. Saeki, K., Y. Houjoh, I. Inoyae, Y. Shinoda, S. Watanabe, and Y. Nakanishi, Rehabilitation of the water cycle at the basin of the Kanda River, *Proc. 8th Int. Conf. on Urban Storm Drainage*, Joliffe, I. B. and J. E. Ball, Eds., The Institution of Engineers of Australia, The International Association for Hydraulic Research, and The International Association on Water Quality, Sydney, Australia, Aug. 30–Sept. 3, 1999, 522.
96. Kennen, J. G., Relation of macroinvertebrate community impairment to catchment characteristics in New Jersey streams, *J. Am. Water Resour. Assoc.*, 35, 4, 939, 1999.
97. Bhaduri, B., M. Grove, C. Lowry, and J. Harbor, Assessing long-term hydrologic effects of land use change, *J. AWWA*, 89, 11, 94–106, 1997.
98. Booth, D. B., C. R. Jackson, Urbanization of aquatic systems: Degradation thresholds, stormwater detection, and the limits of mitigation, *J. Am. Water Resour. Assoc.,* 33, 5, 1077–1090, 1997.
99. Schueler, T., Ed., Stream channel geometry used to assess land use impacts in the Pacific Northwest, *Watershed Prot. Tech.,* 2, 2, 345–348, 1996a.
100. Stephenson, D., Evaluation of effects of urbanization on storm runoff, *Proc. 7th Int. Conf. on Urban Storm Drainage*, Sieker, F. and H-R. Verworn, Eds., IAHR/IAWQ. SuG-Verlagsgesellschaft, Hannover, Germany, Sept. 9–13, 1996, 31–36.
101. Maxted, J. R., The use of percent impervious cover to predict the ecological condition of wadable nontidal streams in Delaware, *Proc. Assessing the Cumulative Impacts of Watershed Development on Aquatic Ecosystems and Water Quality Conf.*, Northeastern Illinois Planning Commission, Chicago, IL, Mar. 20–21, 1996, 123–127, 1997.
102. MacRae, C. R., Experience from morphological research on Canadian streams: Is control of the two-year frequency runoff event the best basis for stream channel protection? *Proc. of Effects of Watershed Developments and Manage. on Aquat. Ecosys. Conf.*, Roesner, L. A., Ed., Snowbird, UT, Aug. 4–9, 1996, ASCE, New York, 1997.
103. Sovern, D. T. and P. M. Washington, Effects of urban growth on stream habitat, *Proc. of Effects of Watershed Developments and Manage. on Aquat. Ecosys. Conf.*, Roesner, L. A., Ed., Snowbird, UT, Aug. 4–9, 1996, ASCE, New York, 1997, pp. 163–177.
104. EPA (U.S. Environmental Protection Agency), Results of the Nationwide Urban Runoff Program, Water Planning Division, PB 84–185552, Washington, D.C., Dec. 1983.
105. Booth, D.B., D. R. Montgomery, and J. Bothel, Large woody debris in urban streams of the Pacific Northwest, *Proc. of Effects of Watershed Developments and Manage. on Aquatic Ecosystems Conf.*, Roesner, L. A., Ed., Snowbird, UT, Aug. 4–9, 1996, ASCE, New York, 1997.
106. Montgomery, D. R. and J. M. Buffington, Channel Classification, Prediction of Channel Response, and Assessment of Channel Conditions, Washington State Department of Natural Resources, Report TFW-SH10–93–002, 1993.
107. Horner, R. R, D. B. Booth, A. Azous, and C. W. May, Watershed determinants of ecosystem functioning, *Proc. of Effects of Watershed Dev. and Manage. on Aquat. Ecosys. Conf.*, Roesner, L. A., Ed., Snowbird, UT, Aug. 4–9, 1996, ASCE, New York, 1997.

108. Poole, G. C., C. A. Frissell, and S. C. Ralph, In-stream habitat unit classification: Inadequacies for monitoring and some consequences for management, *J. Am. Water Resour. Assoc.*, 33, 4, 879–896, 1997.
109. EPA (U.S. Environmental Protection Agency), Report to Congress: The Incidence and Severity of Sediment Contamination in Surface Waters of the United States, Vol. 1: National Sediment Quality Survey (EPA 823-R-97–006), Vol. 2: Data Summary for Areas of Probable Concern (APC) (EPA 823-R-98–007), Vol. 3: National Sediment Contaminant Point Source Inventory (EPA 823-R-98–008), Vol. 4 (under development): National Sediment Contaminant Nonpoint Source Inventory. National Center for Environmental Publications and Information, Cincinnati, OH, 1998.
110. DePinto, J. V., T. C. Young, and S. C. Martin, Aquatic sediments, *J. Water Pollut. Control Fed.*, 52, 6, 1656–1670, 1980.
111. Heaney, J. P., Nationwide Assessment of Receiving Water Impacts from Urban Storm Water Pollution: First Quarterly Progress Report, Environmental Engineering Sciences, University of Florida, Gainesville, 1978.
112. Wilber, W. G., and J. V. Hunter, *The Influence of Urbanization on the Transport of Heavy Metals in New Jersey Streams*, Water Resources Research Institute, Rutgers University, New Brunswick, NJ, 1980.
113. Galvin, D. D. and R. K. Moore, Toxicants in Urban Runoff, METRO Toxicant Program Report #2, U.S. Environmental Protection Agency Grant #P-000161–01, Lacey, Washington, D.C., Dec. 1982.
114. Arhelger, M., P. Jensen, and Y-C. Su, Houston ship channel sediments and their relation to water quality, *WEFTEC '96: Proc. 69th Annu. Conf. and Exposition*, Dallas, Vol. 4, 1996, 329–337.
115. Long, E. R., D. D. MacDonald, S. L. Smith, and F. D. Calder, Incidence of adverse biological effects within ranges of chemical concentrations in marine and estuarine sediments, *Environ. Manage.*, 19, 81–97, 1996.
116. Morresey, D. J., D. S. Roper, and R. B. Williamson, Biological effects of the build-up of contaminants in sediments in urban estuaries, *Proc. of Effects of Watershed Dev. and Manage. on Aquat. Ecosys. Conf.*, Roesner, L. A., Ed., Snowbird, UT, Aug. 4–9, 1996, ASCE, New York, 1997, 1–20.
117. Warick, R. M., A new method for detecting pollution effects on marine macrobenthic communities, *Mar. Biol.*, 92, 557–562, 1986.
118. Watzin, M. C., A. W. McIntosh, E. A. Brown, R. Lacey, D. C. Lester, K. L. Newbrough, and A. R. Williams, Assessing sediment quality in heterogeneous environments: A case study of a small urban harbor in Lake Champlain, VT, USA, *Environ. Toxicol. Chem.*, 16, 10, 2125–2135, 1997.
119. Driscoll, S. K. and P. F. Landrum, A comparison of equilibrium partitioning and critical body residue approaches for predicting toxicity of sediment-associated fluoranthene to freshwater amphipods, *Environ. Toxicol. Chem.*, 16, 10, 2179–2186, 1997.
120. Rhoads, B. L. and R. A. Cahill, Geomorphological assessment of sediment contamination in an urban stream system, *Appl. Geochem.*, 14, 4, 459, 1999.
121. Crabill, C., R. Donald, J. Snelling, R. Foust, and G. Southam, The impact of sediment fecal coliform reservoirs on seasonal water quality in Oak Creek, Arizona, *Water Res.*, 33, 9, 2163, 1999.
122. Vollertsen, J., T. Hvitved-Jacobsen, I. McGregor, and R. Ashley, Aerobic microbial transformations of pipe and silt trap sediments from combined sewers, *Water Sci. Technol.*, 39, 2, 233, 1999.
123. Vollertsen, J., and T. Hvitved-Jacobsen, Stoichiometric and kinetic model parameters for microbial transformations of suspended solids in combined sewer systems, *Water Res.*, 33, 14, 3127, 1999.
124. Schueler, T., Ed., Impact of suspended and deposited sediment, *Watershed Prot. Tech.*, 2, 3, 443, 1997a.
125. Burton, G. A., Jr., B. L. Stemmer, and K. L. Winks, A multitrophic level evaluation of sediment toxicity in Waukegan and Indiana Harbors, *Environ. Toxicol. Chem.*, 8, 1057–1066, 1989.
126. Burton, G. A., Jr., Evaluation of seven sediment toxicity tests and their relationships to stream parameters, *Toxicity Assess.*, 4, 149–159, 1989.
127. Burton, G. A., Jr., B. L. Stemmer, and K. L. Winks, A multitrophic level evaluation of sediment toxicity in Waukegan and Indiana Harbors, *Environ. Toxicol. Chem.*, 8, 1057–1066, 1989.
128. Burton, G. A., Jr., Assessing freshwater sediment toxicity, *Environ. Toxicol. Chem.*, 10, 12, 1991.
129. Burton, G. A., Jr., Ed., *Sediment Toxicity Assessment,* Lewis Publishers, Boca Raton, FL, 1992.
130. Burton, G. A., Jr. and J. Scott, Sediment toxicity evaluations, *Environ. Sci. Technol.*, 25. 1992.
131. ES&T *(Environ. Sci. Technol.)*, News–Sediments: Pollutant remobilization, 31, 12, 536a, 1997.
132. Kuehne, R. A., *Evaluation of Recovery in a Polluted Creek after Installment of New Sewage Treatment Procedures*, University of Kentucky Water Resources Research Institute, Lexington, May 1975.

133. McHardy, B. M., J. J. George, and J. Salanki, Eds., The uptake of selected heavy metals by the green algae *Cladophora glomerata*, *Proc. of Symp. on Heavy Metals in Water Organisms*· Tihany, Hungary, Akademiai Kiado, Budapest, *Symp. Biol. Hung.*, Vol. 29· 1985, 3–20.
134. Bascombe, A. D., J. B. Ellis, D. M. Revitt, and R. B. E. Shutes, Development of ecotoxicological criteria in urban catchments, *Water Sci. Technol.,* 22, 10/1, 173–179, 1990.
135. Ellis, J. B., R. B. Shutes, and D. M. Revitt, Ecotoxicological approaches and criteria for the assessment of urban runoff impacts on receiving waters, *Proc. Eng. Found. Conf. Effects of Urban Runoff on Receiving Systems: An Interdisciplinary Analysis of Impact, Monitoring, and Management*, Mt. Crested Butte, CO, ASCE, New York, 1991.
136. Barbour, M. T., Measuring the health of aquatic ecosystems using biological assessment techniques: A national perspective, *Proc. of Effects of Watershed Dev. and Manage. on Aquat. Ecosys. Conf.*, Roesner, L. A., Ed., Snowbird, UT, Aug. 4–9, 1996, ASCE, New York, 1997, 18–33.
137. Plafkin, J. L., M. T. Barbour, K. D. Porter, S. K. Gross, and R. M. Hughes, Rapid Bioassessment Protocols for Use in Streams and Rivers: Benthic Macroinvertebrates and Fish. EPA-440–4–89–001, U.S. Environmental Protection Agency, Office of Water Regulations and Standards, Washington, D.C., 1989.
138. EPA (U.S. Environmental Protection Agency), Biological Criteria: National Program Guidance for Surface Waters, EPA-440–5–90–004, U.S. Environmental Protection Agency, Office of Water Regulations and Standards, Washington, D.C., 1990.
139. EPA (U.S. Environmental Protection Agency), Environmental Indicators of Water Quality in the United States, Office of Water, U.S. Environmental Protection Agency, EPA 841-F-96–002, Washington, D.C., June 1996.
140. Pelley, J., National "environmental indicators" issued by EPA to track health of U.S. waters, *Environ. Sci. Technol.*, 31, 9, 381a, 1996.
141. Claytor, R. A., An introduction to stormwater indicators: An urban runoff assessment tool, *Watershed Prot. Tech.,* 2, 2, 321–328, 1996a.
142. Claytor, R. A., An introduction to stormwater indicators: Urban runoff assessment tools, *Proc. Assessing the Cumulative Impacts of Watershed Dev. on Aquat. Ecosys. and Water Qual. Conf.*, Northeastern Illinois Planning Commission, Chicago, IL, March 20–21, 1996, 217–224. Apr. 1997.
143. Claytor, R. A. and W. Brown, Environmental Indicators to Assess the Effectiveness of Municipal and Industrial Stormwater Control Programs, Prepared for the U.S. EPA, Office of Wastewater Management, Center for Watershed Protection, Silver Spring, MD, 1996.
144. Cave, K. A., Receiving water quality indicators for judging stream improvement, *Proc. Eng. Found. Conf. Sustaining Urban Water Resources in the 21st Century,* Rowney, A. C., P. Stahre, and L. A. Roesner, Eds., Malmo, Sweden, Sept. 7–12, 1997, ASCE, New York, 1998.
145. Pitt, R., M. Lilburn, S. Nix, S. R. Durrans, S. Burian, J. Voorhees, and J. Martinson, Guidance Manual for Integrated Wet Weather Flow (WWF) Collection and Treatment Systems for Newly Urbanized Areas (New WWF Systems), U.S. Environmental Protection Agency, 2001, 612.
146. Heaney, J. P., R. Pitt, and R. Field, *Innovative Urban Wet-Weather Flow Management Systems*, Technomics, Lancaster, PA, 2000.
147. Burton, G. A., Jr., R. Pitt, and S. Clark, The role of whole effluent toxicity test methods in assessing stormwater and sediment contamination, *CRC Critical Rev. Environ. Sci. Technol.,* 30, 413–447, 2000.
148. Field, R., R. Pitt, Hsu, K., M. Borst, R. DeGuida, C-Y. Fan, J. Heaney, J. Perdek, and M. Stinson, Urban wet weather flow — 1996 literature review, *Water Environ. Res.,* 69, 4, 426–444, 1997.
149. Field, R., T. O'Connor, C-Y. Fan, R. Pitt, S. Clark, J. Ludwig, and T. Hendrix, Urban wet weather flows — 1997 literature review, *Water Environ. Res.,* 70, 4, 1998.
150. O'Connor, T., R. Field, D. Fischer, R. Rovansek, R. Pitt, S. Clark, and M. Lama. Urban wet weather flows — 1998 literature review, *Water Environ. Res.,* 71, 4, 1999.
151. Fan, C.-Y., R. Field, J. Heaney, R. Pitt, S. Clark, L. Wright, R. Rovansek, and S. Olivera, Urban wet weather flows — 1999 literature review, *Water Environ. Res.,* 72, 5, 2000 (CD-ROM).
152. Clark, S., R. Rovansek, L. Wright, J. Heaney, R. Field, and R. Pitt, Urban wet weather flows — 2000 literature review, *Water Environ. Res.,* 73, 5, 2001 (CD-ROM).

CHAPTER 20

# Nuclear and Thermal

**Linda Meyers-Schöne and Sylvia S. Talmage**

## CONTENTS

## 20.1 INTRODUCTION

Assessing the radiological and physical impact of nuclear-energy-related activities on natural environments is important for the protection of natural systems and their species components. Impacts of nuclear power facilities on the environment result from routine and accidental releases of radionuclides to terrestrial and aquatic ecosystems and the interaction of condenser cooling systems with aquatic ecosystems. Sources of release of radionuclides to the environment include normal operation of nuclear power plants and plutonium-production reactors, nuclear plant accidents, past nuclear weapons testing, and leakage from radioactive waste storage sites. Several excellent reviews of radioecology and radiation effects on the environment have been published.[1–6]

Few conclusions can be drawn about the effects of chronic, low-level doses of radiation on natural populations, communities, and ecosystems. Routine, chronic releases from nuclear reactors are controlled to keep human exposures below specified limits. As a result, natural populations of biota are likely protected.[7–9] However, exposures may be greater as the result of accidents, weapons testing, and habitation at terrestrial and aquatic waste-disposal sites. Numerous studies have documented uptake of radionuclides in field situations without examining potential adverse effects. Discussion in this chapter is focused on studies conducted in the field, with supporting information from relevant laboratory studies. Although radiation doses experienced in the field are difficult to estimate, the selected examples attempt to relate estimated doses or tissue levels of radionuclides to effects.

Power plant condenser cooling systems at both nuclear and fossil-fueled sites may withdraw and discharge large quantities of surface waters. The removal of aquatic organisms by impingement or entrainment and the added load of warm water and chemicals may impact the receiving aquatic ecosystem. These potential impacts have been studied for years[10–17] with no apparent long-term effects observed.[18] Potential problems have been mitigated by redesigning intake structures, regulating intake flow and temperature changes, and building cooling towers or closed-system cooling lakes. The impact of several power plants on the biota of the Hudson River is presented as a case study.

## 20.2 BASIC RADIOLOGICAL CONCEPTS

In order to facilitate a discussion of the effects of radioactivity on the environment, important radiological concepts are briefly presented. In-depth discussions of the chemistry, physics, and acute biological effects of radiation can be found in appropriate textbooks.

Radioactivity is the property of isotopes of elements, termed radionuclides, that spontaneously emit radiation by disintegration (decay) of their unstable nuclei. The radiation can be in the form of particles or electromagnetic waves; the most important of these are alpha ($\alpha$) and beta ($\beta$) particles, neutrons, and gamma ($\gamma$) and related x-rays. Because these types of radiation have sufficient energy to penetrate matter and eject electrons (protons in the case of neutrons) from atoms, thereby producing positively and negatively charged ion pairs, they are referred to as ionizing radiation. Another decay process is electron capture in which the parent nucleus captures an orbital electron and subsequently emits radiation.

### 20.2.1 Types of Ionizing Radiation

The five types of ionizing radiation differ in energy and ability to penetrate living tissues. Penetration ability increases with increasing energy of the radiation. Alpha particles are composed of two protons and two neutrons with a charge of plus two. They are emitted from heavy elements with unstable nuclei such as uranium, plutonium, and radium. Because of their double charge and large mass, they interact strongly with matter, producing many ions; however, they possess a low penetrating ability. For example, alpha particles can be stopped by a sheet of paper; in tissues their penetration is measured in micrometers.

Beta particles are negatively charged electrons ejected from the nuclei of radioactive atoms during the conversion of a neutron to a proton. Energy levels of beta particles are continuous, but maximum energy levels are specific for specific isotopes. Positively charged beta particles, termed positrons ($\beta^+$), are less frequently emitted. Shielding from beta particles can be effected by 3 mm of metal or 6 mm of wood.

Neutrons consist of an electron and a proton. They are released from elements that decay by spontaneous fission. In nuclear reactors, bombardment of elements with high fission probabilities, such as uranium-235 ($^{235}U$), with slow neutrons results in fission products and additional neutrons, which in turn cause a fission "chain reaction." In tissues, neutrons produce ionization indirectly by ejection of protons from the nuclei of hydrogen atoms and by activation of elements via neutron capture, the latter resulting in the release of gamma rays. Due to their large mass and chargeless state, neutrons have great kinetic energy and penetration capacity.

Gamma and x-rays have no mass or electrical charge. Gamma rays are emitted by nuclei, usually in combination with alpha or beta emission, whereas x-rays arise from the electron shells. Both have very short wavelengths and great penetrating ability. Gamma rays can be stopped by 5 to 10 cm of lead or 30 to 60 cm of concrete.

### 20.2.2 Half-Life

By definition, a parent radionuclide is unstable, undergoing spontaneous disintegration or decay by the emission of one or more alpha or beta particles and forming a decay or daughter product. The daughter radionuclide may be stable or radioactive. Very heavy elements such as $^{238}U$ and plutonium-238 and 239 ($^{239-240}Pu$) tend to decay by emission of alpha particles; for lighter elements, beta emission is the most frequent decay process. The time required for half of a given number of atoms to decay is defined as the physical half-life. For example, with the loss of beta particles, half of a given mass of strontium-90 ($^{90}Sr$) decays to its radioactive daughter, yttrium-90 ($^{90}Y$), in approximately 28 years.

### 20.2.3 Units of Measurement

Originally, units of radioactivity and radioactivity measurement were expressed in curies (Ci), Roentgens (R), rads, and rems. Presently, the International System (SI) of Units (Becquerel [Bq], Gray [Gy], and Sievert [Sv]) has been adopted for standard usage. In order to change systems, the following conversion factors are routinely used: 1 Bq = 27 pCi (1 pCi = 0.037 Bq), 1 mGy = 100 mrad, and 10 mSv = 1 rem.

The Gray and Sievert (rad and rem, respectively) apply to all types of ionizing radiations; however, the Sievert takes into consideration the linear energy transfer of the radiation. The Sievert is equal to the absorbed dose in rads multiplied by a quality factor or the relative biological effectiveness of the radiation and other modifying factors. The quality factor for gamma and x-rays and electrons and positrons is approximately 1. A quality factor of ten is used for alpha particles and fast neutrons.

A confounding factor in applying dose measurements is that the external radiation exposure from gamma or x-rays in air cannot be directly translated to absorbed dose. However, for soft tissue, a crude approximation is that a dose of approximately 0.01 Gy will result from exposure to

**Table 20.1 Average Annual Terrestrial Radiation Dose from Natural Sources[a]**

| Source | Dose Rate (mGy/year) |
|---|---|
| Cosmic radiation | 0.4 |
| Terrestrial gamma rays | 0.4 |
| Internal radioactivity | 0.2 |

[a] From Biological Effects of Ionizing Radiation (BEIR) Advisory Committee Report: *The Effects on Populations of Exposure to Low Levels of Ionizing Radiation*, National Academy of Sciences, National Research Council, Washington, D.C., 1972.

1 R ($2.58 \times 10^{-4}$ Coulomb (C)/kg of air from x-rays or gamma rays[12]). The energy deposition in tissue for a dose of 1 R has also been estimated as about 0.0096 J/kg.

## 20.3 SOURCES OF IONIZING RADIATION

Exposure to ionizing radiation may result from background sources as well as radionuclides released during fuel fabrication, the normal operation of nuclear power reactors, nuclear accidents, waste storage sites, and past weapons testing.

### 20.3.1 Background Radiation

To put radiation from anthropogenic sources into perspective, it is important to consider the amount of naturally occurring background radioactivity that all organisms are unavoidably exposed to. Natural background radiation is due to cosmic radiation, terrestrial gamma rays from naturally occurring radionuclides, and internal radioactivity, primarily potassium-40 ($^{40}K$), which occurs naturally in foods. The total dose to man and large terrestrial vertebrates from natural background radiation varies geographically but has been estimated at 1 mGy/year (Table 20.1). The total dose has also been expressed as absorbed dose equivalent: 1 mSv/year. For comparison, global fallout from past nuclear weapons testing adds an additional 0.04 mSv; the contribution of nuclear power is minor, 0.003 μSv/yr.[19] Estimates of background doses received by marine and freshwater biota are presented in Table 20.2. As can be seen from the data, annual doses to aquatic organisms from natural sources are generally less than 5 mGy/year.[2,20]

### 20.3.2 Fuel Fabrication and Production Facilities

The primary fuels for nuclear reactors are uranium dioxide ($UO_2$) enriched in $^{235}U$ to 2 to 3% and, to a lesser extent, thorium-232 ($^{232}Th$). The primary enrichment process has been gaseous diffusion, using $UF_6$ gas. Enrichment plants are presently located at Paducah, Kentucky, and Piketon, Ohio;[21] a third plant in Oak Ridge, Tennessee, was put on standby in 1985 and shut down in 1987. Enrichment by gaseous diffusion is expensive and requires considerable amounts of energy. More efficient enrichment processes, including gas centrifugation and enrichment by lasers, are being developed. Most nuclear reactors utilize $UO_2$ fuel sintered into rods, or the fuel may be in the form of small pellets, clad with a thin metallic coating. Gaseous and liquid releases from such fuel fabrication facilities contain a small amount of U radioisotopes and $^{234}Th$.[22,23] In addition to the gaseous diffusion plants, there are eight additional major uranium fuel fabrication and uranium hexafluoride production facilities licensed to operate in seven states.[21] There are 17 other facilities in the United States that possess or process significant quantities of special nuclear material.

**Table 20.2 Estimates of Annual Doses Received by Marine and Freshwater Organisms from Natural Sources of Radiation (mGy/year)[a]**

| | Phytoplankton | Zooplankton | Mollusca | Crustacea | Fish |
|---|---|---|---|---|---|
| | | | **Freshwater Organisms** | | |
| Cosmic | 0.24 | 0.24 | 0.19 | 0.19 | 0.19–0.24 |
| Water | $6.4. \times 10^{-4}$–0.54 | $9.0 \times 10^{-5}$–$7.4 \times 10^{-2}$ | $4.0 \times 10^{-5}$–$3.1 \times 10^{-2}$ | $4.0 \times 10^{-5}$–$3.1 \times 10^{-2}$ | $4.0 \times 10^{-5}$–$6.1 \times 10^{-2}$ |
| Sediments ($\beta + \gamma$) | 0 | 0 | 0.27–3.2 | 0.27–3.2 | 0–3.2 |
| Internal | — | — | — | — | 0.32–0.42 |
| Total | 0.24–0.78 | 0.24–0.31 | 0.46–3.5 | 0.46–3.5 | 0.51–4.0 |
| | | | **Marine Organisms** | | |
| Cosmic | $4.4. \times 10^{-2}$ | $4.4. \times 10^{-2}$ | $4.4. \times 10^{-2}$ | $4.4. \times 10^{-2}$ | $4.4. \times 10^{-2}$ |
| Water | $3.5. \times 10^{-2}$ | $1.8. \times 10^{-2}$ | $9.0 \times 10^{-3}$ | $9.0 \times 10^{-3}$ | $9.0 \times 10^{-3}$ |
| Sediments ($\beta + \gamma$) | 0 | 0 | 0.27–3.2 | 0.27–3.2 | 0–3.2 |
| Internal | 0.17–0.64 | 0.23–1.4 | 0.65–1.3 | 0.69–1.9 | 0.24–0.37 |
| Total | 0.25–0.72 | 0.29–1.7 | 0.97–4.6 | 1.0–5.2 | 0.29–3.7 |

[a] IAEA, *Effects of Ionizing Radiation on Aquatic Organisms and Ecosystems*, Tech. Rep. Ser. No. 172, International Atomic Energy Agency, Vienna, 1976.

### 20.3.3 Nuclear Power Reactors

As of 1999, there were 104 commercial nuclear power reactors licensed to operate in 31 states.[21] Reactors employed for electric power generation in the United States are light-water reactors, using water as moderator and coolant. Light-water reactors are of two types: pressurized-water reactors and boiling-water reactors. During normal operation of the reactor, new radionuclides are formed by fission of the nuclear fuel and neutron activation of structural materials. Despite containment within the reactor system, small amounts of radionuclides escape through the fuel cladding into the coolant system; from there they are released into liquid and gaseous effluents. Under normal operations, the principal radionuclides present in coolant water of light-water-cooled reactors are tritium ($^{3}H$); sodium-24 ($^{24}Na$); silicon-31 ($^{31}Si$); phosphorus-32 ($^{32}P$); scandium-46 ($^{46}Sc$); chromium-51 ($^{51}Cr$); manganese-56 ($^{56}Mn$); cobalt-58 and cobalt-60 ($^{58}Co$ and $^{60}Co$); copper-64 ($^{64}Cu$); arsenic-76 ($^{76}As$); krypton-85 ($^{85}Kr$); strontium-85, strontium-90, and strontium-91 ($^{85}Sr$, $^{90}Sr$, and $^{91}Sr$); yttrium-90, yttrium-92, and yttrium-93 ($^{90}Y$, $^{92}Y$, and $^{93}Y$); zirconium-97 ($^{97}Zr$); molybdenum-99 ($^{99}Mo$); technecium-99m ($^{99m}Tc$); rhodium-105 ($^{105}Rh$); iodine-130, iodine-131, iodine-133, and iodine-135 ($^{130}I$, $^{131}I$, $^{133}I$, and $^{135}I$); xenon-131 and xenon-133 ($^{131}Xe$ and $^{133}Xe$); cesium-134 and cesium-137 ($^{134}Cs$ and $^{137}Cs$); cerium-143 ($^{143}Ce$); barium-140 ($^{140}Ba$); and neptunium-239 ($^{239}Np$).[24–26] Many of the more than 200 fission products produced are short lived and of no environmental consequence. Each reactor releases an average of $2.22 \times 10^{11}$ Bq/year of mixed fission and neutron-activation products in liquid effluents; greater amounts of $^{3}H$ and noble gases ($^{85}Kr$, $^{131}Xe$, and $^{133}Xe$) are released in gaseous effluents. Biologically important fission and neutron-activation products, based on concentration, half-life, chemical properties, and solubility, are listed in Table 20.3. Also listed are the types of radiation emitted and half-lives of the radionuclides. Of primary concern in environmental studies are $^{90}Sr$ and $^{137}Cs$.

Routine emissions from power plant operation do not contribute significantly to detectable levels in the environment. For example, an off-site survey at the Rancho Seco nuclear plant found primarily $^{137}Cs$ and $^{134}Cs$, with lesser amounts of $^{60}Co$ and $^{58}Co$ along streams receiving liquid effluent and in fields irrigated with water from these streams.[27] Gamma radiation exposures along the creek and in irrigated fields ranged from 8 microroentgen (μR)/hour (background) to 14 μR/hour.

**Table 20.3 Radionuclides of Potential Environmental Concern Produced by Nuclear Reactors[a]**

| Radionuclide | Half-life | Radiation | Source | Element Analog |
|---|---|---|---|---|
| $^{3}H$ | 12.3 years | β | Fission product, neutron activation | H |
| $^{14}C$ | 5568 years | β | Neutron activation | C |
| $^{24}Na$ | 15 h | β, γ | Neutron activation | Na |
| $^{32}P$ | 14 days | β | Neutron activation | P |
| $^{35}S$ | 87 days | β | Neutron activation | S |
| $^{41}Ar$ | 110 min | β, γ | Neutron activation | — |
| $^{45}Ca$ | 164 days | β | Neutron activation | Ca |
| $^{54}Mn$ | 291 days | γ | Neutron activation | Mn |
| $^{55}Fe$ | 2.6 years | X | Neutron activation | Fe |
| $^{59}Fe$ | 45 days | β, γ | Neutron activation | Fe |
| $^{57}Co$ | 270 days | γ | Neutron activation | Co |
| $^{58}Co$ | 71 days | $\beta^{+}$, γ | Neutron activation | Co |
| $^{60}Co$ | 5.2 years | β, γ | Neutron activation | Co |
| $^{65}Zn$ | 245 days | $\beta^{+}$, γ | Neutron activation | Zn |
| $^{85}Kr$ | 10 years | β, γ | Fission product | — |
| $^{89}Sr$ | 51 days | β | Fission product | Ca |
| $^{90}Sr$ | 28 years | β | Fission product | Ca |
| $^{91}Y$ | 58 days | β, γ | Fission product | — |
| $^{95}Zr$ | 65 days | β, γ | Fission product | — |
| $^{103}Ru$ | 40 days | β, γ | Fission product | — |
| $^{106}Ru$ | 1.0 years | β, γ | Fission product | — |
| $^{129}I$ | 1.7 x $10^{7}$ years | β, γ | Fission product | — |
| $^{131}I$ | 8.1 days | β, γ | Fission product | — |
| $^{134}Cs$ | 2 years | β, γ | Fission product | K |
| $^{137}Cs$ | 27 years | β, γ | Fission product | K |
| $^{140}Ba$ | 12.8 days | β, γ | Fission product | Ca |
| $^{143}Ce$ | 33 h | β, γ | Fission product | — |
| $^{144}Ce$ | 285 days | β, γ | Fission product | — |
| $^{147}Nd$ | 11 days | β, γ | Fission product | — |
| $^{239}Pu$ | 24,360 years | α, γ | Neutron activation | — |
| $^{239}Np$ | 2.3 days | β, γ | Neutron activation | — |
| $^{241}Am$ | 4770 years | α, γ | Neutron activation | — |
| $^{242}Cm$ | 163 days | α, γ | Neutron activation | — |

[a] From Whicker, F. W. and Schultz, V., *Radioecology: Nuclear Energy and the Environment*, Vol. II, CRC Press, Boca Raton, Fl, 1982.

In addition to commercial power reactors, there are 37 nonpower reactors licensed to operate in 24 states.[21] These reactors are designed and utilized for research, testing, and educational purposes.

### 20.3.4 Weapons Production

Three U.S. Department of Energy sites have been engaged in the production of weapons-grade plutonium and tritium: the Savannah River Plant, Aiken, South Carolina; the Hanford Reservation, Hanford, Washington; and the Idaho National Engineering Laboratory, Idaho Falls, Idaho. Cooling water for the five reactors at the Savannah River Plant was released to two major aquatic habitats: cooling ponds and natural streams that drain into swamp forest and, ultimately, the Savannah River.[10] At least two of the reactors are no longer in operation. From 1944 to 1971 the Columbia River served as a source of single-pass cooling water for the plutonium production reactors at Hanford. Complete inventories of these releases are not available.

### 20.3.5 Reprocessing Plants

After partial burn-up of the contained fissile material in the reactor, the spent fuel assemblies are withdrawn, stored for 90 days or longer to allow decay of most of the short-lived radioactivity and, in the past, shipped to reprocessing plants.[22] Reprocessing plants in the United States were licensed or scheduled for licensing at West Valley, New York; Morris, Illinois; and Barnwell, South Carolina. Because these sites are not in operation at present, spent fuel is being stored in the United States Reprocessing in connection with production of plutonium for weapons has taken place at the three U.S. Department of Energy weapons production sites. As an example of the amount of radioactivity that may be released to the environment from such activities, reprocessing in connection with the Windscale uranium-fueled reactor in England (now called Sellafield) has released $3.7 \times 10^{14}$ to $5.6 \times 10^{15}$ Bq of $^{137}Cs$ to the Irish Sea annually.[23] Since 1976, discharges have been decreasing as a result of changes in operating procedures. Figures were not available for U.S. plants.

The Mayak Production Association is a nuclear fuel reprocessing facility located in the southern Urals of Russia, approximately 2000 km from Moscow. This facility processed weapons-grade plutonium during 1949 to 1952. Water from the adjacent Techa River was used as a coolant and subsequently discharged to the river. Although large quantities of wastes associated with plutonium production were buried near the facility, the greatest impact to the environment was from discharges to the river. During this period, approximately $10^{17}$ Bq were released into the river. Predominant radionuclides included $^{89}Sr$, $^{90}Sr$, $^{137}Cs$, $^{95}Zr$-$^{95}Nb$, $^{103}Ru$, and isotopes of rare-earth elements. Beta-activity in the river has decreased over time, but $^{90}Sr$ and $^{137}Cs$ remain the major contributors to the radioactivity present.[28]

### 20.3.6 Reactor Accidents

The most serious accident at a U.S.-licensed commercial power plant occurred at Three Mile Island Unit 2 on March 28, 1979. Noble gases and $^{131}I$ were released to the atmosphere, but other fission products were not released in measurable quantities. Detectable levels of radioiodine were found in milk samples of domestic animals and in thyroids of voles (*Microtus pennsylvanicus*) in the immediate area.[18,29]

Several other major accidents — one at Windscale, England, one at Kyshym in the Ural Mountains, Russia, and one at Chernobyl in Ukraine — have occurred. Windscale is a plutonium production reactor located on the northwest coast of England. In October 1957, one of the two air-cooled, graphite-moderated natural-uranium reactors was damaged by fire, resulting in the release of fission products to the surrounding countryside and the Irish Sea. Radioactivity from the principal isotopes released during the fire ($^{131}I$, $^{137}Cs$, $^{89}Sr$, and $^{90}Sr$) was estimated at 6.5 to $7.7 \times 10^{14}$ Bq.[23]

In 1957, a thermal explosion of a tank containing liquid radioactive waste at the Mayak nuclear materials production complex in Russia released $7.4 \times 10^{17}$ Bq, which contaminated the surrounding area.[30,31] Many of the radionuclides were short-lived, but release of $^{90}Sr$ and $^{90}Y$ has resulted in long-term contamination of a 23,000-km$^2$ area. Contributing to the radiation contamination in this region are the direct discharge of radioactive contaminants into the Techa River from 1949 to 1959 and the current discharge and release of radionuclides into the environment by the Mayak Production Association.[31]

In the early morning of April 26, 1986, the 1000-MW water-cooled, pressure-tube, graphite-moderated reactor at the Chernobyl power station in the U.S.S.R. was destroyed by an accident, resulting in heavy local contamination and fallout on an extensive worldwide level. By May 6, 1986, an estimated $8 \times 10^{19}$ Bq of radioactivity from $^{133}Xe$, $^{85}Kr$, $^{131}I$, $^{132}Te$, $^{134}Cs$, $^{137}Cs$, $^{99}Mo$, $^{95}Zr$, $^{103}Ru$, $^{106}Ru$, $^{140}Be$, $^{141}Ce$, $^{144}Ce$, $^{89}Sr$, $^{90}Sr$, $^{238}Pu$, $^{239}Pu$, $^{240}Pu$, $^{241}Pu$, $^{242}Pu$, $^{242}Cm$, and $^{234}Np$ had been released.[23] The fire carried radioactive gases and particulates to relatively high altitudes, resulting in worldwide dispersion. Cesium-137 was the most important contaminant because of its high concentration in the fallout. Fallout over Europe was nonuniform and dependent on rainfall.

For example, from May 5 to 7, ground beta and gamma activity in West Germany, a distance of 1200 km from Chernobyl, ranged from less than 15,000 to more than 240,000 Bq/m$^2$.[32] In Greece, levels ranged from 10 to 137,000 Bq/m$^2$.[33]

### 20.3.7 Low-Level Waste Burial Sites

Low-level waste materials consist of solid trash, laundry wastes, and concentrates from liquid wastes. In the United States, most low-level radioactive wastes have been treated and disposed of by shallow land burial with or without concrete vaults.[21] In addition, liquid wastes at some sites have been discharged to surface ponds. The major federal low-level waste burial sites operated by the U.S. Department of Energy are located at Oak Ridge National Laboratory, Oak Ridge, Tennessee, the Savannah River Plant, Aiken, South Carolina, the Hanford Reservation, Hanford, Washington, Los Alamos Scientific Laboratory, Los Alamos, New Mexico, and the Idaho National Engineering Laboratory, Idaho Falls, Idaho. Environmental contamination from radionuclides has occurred at four of these sites as a result of infiltration into groundwater and surface-water runoff.[1] At some sites, rodents have come into contact with surface contamination or burrowed into the trenches and spread small amounts of radioactivity.[34–37]

In 2000, there were three commercial sites licensed by states under agreements with the Nuclear Regulatory Commission or by the Nuclear Regulatory Commission itself to store radioactive wastes.[21] These sites are located at Barnwell, South Carolina, Hanford, Washington, and Clive, Utah. The sites at West Valley, New York, Maxey Flats, Kentucky, Sheffield, Illinois, and Beatty, Nevada are closed.

High-level wastes, primarily from reprocessing plants, are stored in steel tanks awaiting treatment or development of storage facilities. At West Valley, New York, a high-level waste-vitrification demonstration project has been underway since 1980.[21] The West Valley Demonstration Project began converting liquid high-level waste into glass logs in July 1996. Under the U.S. Department of Energy, a waste-solidification facility project is underway in Hanford, Washington.

### 20.3.8 Weapons Testing

Nuclear testing in the atmosphere was carried out from 1945 until the test moratorium in 1963. The first nuclear detonation took place at the Trinity Site in New Mexico on July 16, 1945. Testing began in the Pacific Ocean in 1946. Fission products and products of neutron activation from materials used for bomb construction are released as radioactive fallout. Some of these are $^{89}$Sr, $^{90}$Sr, $^{95}$Zr, $^{106}$Ru, $^{131}$I, $^{137}$Cs, $^{141}$Ce, $^{144}$Ce, and $^{239}$Pu.[26] Radioactive material introduced into the troposphere was distributed locally, while that introduced into the stratosphere was deposited worldwide. Small amounts continue to be deposited from the stratosphere;[1] the yearly dose to humans is 0.04 mSv.[19] Both of the fission products $^{90}$Sr and $^{137}$Cs are of ecological concern; $^{137}$Cs has a high yield from nuclear fission and is one of the major dose-contributing radionuclides in the environment.

## 20.4 BEHAVIOR OF RADIONUCLIDES IN THE ENVIRONMENT

Radionuclides introduced into the environment become diluted and distributed according to highly complex interactions of physical, chemical, and biological factors.[1,2] The chemical behavior of radionuclides is the same as the stable form of the element and is similar to other elements in the same groupings (vertical columns) of the periodic table of elements. This results in their distribution in various components of the food chain in a manner similar to their stable chemical analogs (Table 20.3). For example, radionuclides such as $^{89}$Sr, $^{90}$Sr, $^{140}$Ba, $^{226}$Ra, and $^{45}$Ca will behave like calcium, and $^{40}$K, $^{86}$Rb, and $^{137}$Cs will have properties similar to potassium. In biota,

$^{137}$Cs will be distributed throughout the soft tissue of the body, whereas $^{90}$Sr will be deposited primarily in bone, shells, or exoskeletons. Tritium ($^{3}$H) will behave like hydrogen, and $^{131}$I will behave like stable iodine. Some radionuclides, such as $^{239}$Pu, lack nutrient analogs, but plutonium accumulates in bone. Inert gases such as $^{85}$Kr tend to stay in the atmosphere until they decay; iodine and radon emitted to the atmosphere also tend to stay in the gaseous phase.[1]

Uptake or concentration ratios of radionuclides, such as strontium and cesium, are dependent on the biota, specific tissues, mode of uptake, environmental source, and the properties of the radionuclide itself. Terrestrial plants tend to discriminate against strontium in favor of calcium; plant/soil concentration ratios average 0.017.[22] Tissue concentrations of strontium do not increase with trophic level, as evidenced by concentration ratios through successive trophic levels of less than 1.0. Turnover is slow, as strontium concentrates in hard tissue such as bone, shell, and exoskeleton. In addition to uptake by ingestion, aquatic organisms adsorb and concentrate radionuclides directly from the water. Because of larger dilution factors and the abundance of competing nutrient elements in seawater, marine organisms do not concentrate radionuclides to the same degree as do freshwater organisms. In freshwater aquatic systems, concentration ratios (concentration in organism/concentration in water) of strontium in algae, mollusks, crustaceans, and fish average 500, 300, 300, and 30, respectively. In marine systems, the respective ratios are 20, 1, 1, and 1.[22,38,39]

Cesium-137 binds very firmly to clay particles in both soil and sediments. Although there is little uptake by plants, with plant/soil concentration ratios of 0.01, cesium does accumulate through food chains in terrestrial systems.[2] In an aquatic system, $^{137}$Cs will move from the water compartment to the sediments, where it is available to detritivores and bottom feeders. From there, cesium moves through the food chain to sessile vegetation and phytoplankton, to herbivorous consumers, to primary consumers such as small fish and carnivorous invertebrates, and finally to secondary consumers (large fish).[40] Successive trophic-level concentration factors may approach 3.0;[2] the biological half-life is weeks to months. Cesium concentration factors for freshwater algae, mollusks, crustaceans, and fish typically average 500, 100, 100, and 2000, respectively; respective values for marine species are 10, 10, 50, and 30.[22,38,39]

Food-chain accumulation of $^{137}$Cs has been followed in the environment because of uptake by large herbivores such as mule deer (*Odocoileus hemionus*) and reindeer (*Rangifer tarandus*), which serve as food for humans. Cesium-137 from fallout is deposited on and accumulated over soil levels by fungi and lichens, which serve as food for these species. Following the Chernobyl accident, lichens (*Cladonia stellaris*) contained 30,000 to 50,000 Bq/kg dry weight; various species of fungi contained 500 to 445,000 Bq/kg dry weight.[41] In mountain pastures of Norway, fresh-weight $^{137}$Cs concentrations in reindeer, which feed primarily on lichens, ranged up to 90,000 Bq/kg of meat.[41,42] Deposition of $^{134,137}$Cs in soil ranged from less than 5 to 200 Bq/m$^2$.

## 20.5 BIOLOGICAL EFFECTS OF RADIATION

Many factors, both intrinsic and extrinsic, may modify the response of living organisms to radiation. In plants, sensitivity to radiation is dependent on chromosome number, volume, and duplication; length and stage of the mitotic cycle; percentage of cells dividing; type of cell or tissue; stage of cell differentiation; stature and structure of organism; age of organism; stage of growth cycle; nutritional condition; concentrations of sensitizing or protective substances; and species.[2]

Depending on total dose, dose rate, type of radiation, and exposure period, exposure to radiation can lead to no observable effects; genetic changes; physiological changes such as effects on the hemopoietic and reproductive systems; effects on growth and development; or life shortening, including cancer or death. In the laboratory, where most studies have focused on response to acute doses, total dose and dose rates can be closely estimated. In natural situations, however, where dose rates are not uniform and where total dose is due to both external and internal irradiation from absorption, ingestion, and inhalation, dose is at best a crude estimate.

Relative sensitivities of plants, animals, and bacteria to acute lethal doses have been summarized by Whicker and Schultz.[2] Although sensitivities vary widely within taxonomic groups, mammals appear more sensitive to radiation than birds, reptiles, amphibians, fish, insects, plants, and microorganisms with acute lethal exposures in mammals ranging from 2 to 13 Gy. Direct mortality is not significant for any taxa until a dose of approximately 2 Gy is reached; disruption of the ecological community would occur at 10 Gy or more. These doses are unlikely to occur except in the event of a catastrophic nuclear accident or nuclear warfare.[2]

Studies have also been conducted using longer-term exposures at much lower doses with observations made on effects other than mortality. Effects on the blood were often studied because acute doses revealed the hemopoietic system as one of the most radiosensitive systems in the body. From the standpoint of survival of the population, reproduction is the most sensitive indicator of radiation impairment.[43] Minimum acute doses required to depress reproduction and growth rates may be less than 10 to 100% of those required to produce direct mortality.[2] According to the National Council on Radiation Protection and Measurements (NCRPM),[8] chronic exposures of $\leq$10 mGy are very unlikely to produce measurable deleterious changes in populations or communities of aquatic plants or animals. Studies reviewed by French et al.[44] indicated that a dose rate of 10 mGy/day was the approximate threshold at which effects became apparent in irradiated natural mammalian populations. The IAEA[9] concluded that limiting the chronic radiation dose rate to 1 mGy/day would be protective of terrestrial plants and animals.

Another sensitive indicator of radiation exposure in vertebrates is the frequency of chromosomal aberrations in bone marrow cells or peripheral lymphocytes. This type of genetic damage appears when the animals are exposed to a dose of 0.25 Gy or more.[45] Observation of chromosome aberrations must be interpreted with care, however, because other agents are capable of causing genetic damage. In addition, DNA repair mechanisms and cellular recovery processes serve to reduce radiation damage. However, even when effects are not observable, there is a possibility of increased risk of cancer or life shortening.[45] The use of stable chromosome translocations as a biodosimetric tool is currently being explored as a method to assess both exposure to radiation and potential risk to biota in the field.[46]

When considering the effects of radiation on humans, concern is focused on individuals, and a mutation or cancer is perceived as a deleterious event. When considering the effects of radiation on nonhuman animals, however, concern is usually not focused on individual animals but on whether the population remains viable and successful. Thus, an increase in the mutation rate or an increase in cancer will have little influence on the population because most organisms serve as food for others, and sick or damaged animals would be rapidly removed.[20] Furthermore, subcellular and cellular lesions that do not result in mortality are not manifested in measurable attributes of populations and communities.[2]

## 20.6 EFFECTS OF RADIATION ON TERRESTRIAL POPULATIONS AND COMMUNITIES

Reviews of responses of populations and communities of terrestrial plants[47] and terrestrial animals[48] to chronic irradiation have been published. In most contaminated areas, levels of radioactivity are too low to detect population- and community-level effects.[2] Experimental field studies using $^{137}Cs$ or $^{60}Co$ as large sources of short-term and chronic gamma radiation have provided data on effects on natural communities of plants. For example, Woodwell's studies[49] with chronic gamma irradiation of an oak-pine forest demonstrated a relationship between plant physiognomy and radiosensitivity. Pine trees were the most sensitive, followed by shrubs and then herbaceous forms. Low and crustose forms such as lichens were the most resistant. A reduction in species diversity was observed at 100 R/day (1 Gy/day).

The use of large gamma sources, such as those used to show changes in plant communities, is not a good method for demonstrating changes in animal populations and communities because many animals, such as invertebrates, are dependent on the presence of vegetation, which may be destroyed by the radiation. In addition, other animals, because of their natural mobility and shielding by vegetation or burrows, do not receive uniform doses.[48] Moreover, radiation doses are difficult to estimate, as dose rate decreases as a power function with distance from the source. In the field, animals confined to enclosures have been irradiated chronically.[2,48] In the laboratory, the acute response of animals is usually measured in terms of acute lethal doses, usually $LD_{50(30)}$ values, the dose lethal to 50% of organisms within 30 days; few laboratory studies involve chronic irradiation.

Many studies conducted in the area of nuclear facilities or testing have documented the uptake of radionuclides by natural biota. No appreciable radiation exposures to natural populations of terrestrial animals has occurred from normal operation of the reactors at nuclear power plant sites in the United States.[18] Higher levels of radiation, either acute or chronic, are necessary to show effects on populations of plants and animals. However, information is available for several other contaminated terrestrial sites where estimated doses were calculated and effects on natural populations studied. These sites include areas contaminated by reactor accidents, several national laboratories, and waste sites. Where applicable, supporting evidence from relevant laboratory studies is included. The most highly radiation-contaminated terrestrial site is the area surrounding the Chernobyl nuclear reactor accident; these studies are noted here and reviewed in Chapter 24 by Eisler.

Effects of the Chernobyl accident on the flora and fauna in the surrounding contaminated area have been reviewed.[50–53] Following the accident, the human population was evacuated from a highly contaminated area within a 30-km radius around the damaged reactor. Wildlife populations have become established in the absence of humans. This area, divided into a 10-km inner zone and an outer zone, now serves as a unique natural ecosystem of enormous radiological interest.[53,54]

Within three months of the Chernobyl accident, Russian scientists began studying the effects of radiation on plants and animals within the highly contaminated zones surrounding the plant and comparing the results with biota from reference areas. Because of the uneven distribution of radiation around the plant, doses to biota were often difficult to estimate. When studies were continued over a 3-year period, 1986 to 1989, recovery of the exposed populations took place, either by immigration of animals into the area or by a decrease in mortality and lethal genetic effects with time. Many of the studies address accumulation of genetic changes in the resident populations, the consequence of which are presently unknown.

### 20.6.1 Plant Populations and Communities

Of the flora in the area surrounding the damaged Chernobyl reactor, the pine and spruce (*Pinus silvestris* and *Picae excelsa*) in the nearby forest were the most radio-sensitive species, and by late summer of 1986, trees in a 4400-ha area were dead or dying. Doses to the trees were estimated at 80 to 100 Gy.[50–53] Doses of 8 to 10 Gy killed younger trees and young shoots. Trees that received sublethal doses, 3 to 4 Gy, lost needles and developed morphological variations such as shortened and curved shoots. At the edge of the 30-km zone, where doses were estimated at 1 to 1.5 Gy, growth was temporarily suppressed. During 1987 to 1989, regenerative processes took over, and damaged trees formed gigantic needles. Mass flowering was observed. The deciduous forest, represented by birch (*Betula tremula*), aspen (*Populus tremula*), alder (*Alnus glutinosa*), and oak (*Quercus robur*), was more radioresistant, although foliage turned yellow and growth was temporarily suppressed in the area that was sublethal to the pine forest. The deciduous species are considered ten times more radioresistant than pines.

Although no visible damage to herbaceous plants was reported in some studies, Savchenko reported increased phenotypic diversity and genetic modifications in several species, including *Plantago lancelolata* (Plantaginaceae) and *Hieracium umbellatum* (Compositae), found throughout the contaminated area.[53] Doses were not estimated. The following changes in plants were additionally reported in the contaminated area: an increase in chlorophyll mutation frequency, a decrease in seed viability, mass gall formation, and the appearance of leaf deformation/asymmetry and abnormal stem branching.

Two species were chosen for genetic toxicity studies. Chromosome aberrations were observed in root meristem cells of *Crepis tectorum* (Compositae), where gamma exposure rates ranged from 0.02 to 20 mR/hour.[55] Beta exposure rates were estimated at ten times higher. For *Arabidopsis thaliana* (Cruciferae), there was a correlation between the radioactive contamination level, which ranged from 0.01 to 240 mR/hour, and frequency of plants with mutations,[56] although there was no effect of these exposures on the germination rate of the plants.[50]

### 20.6.2 Invertebrate Populations and Communities

Species presence and population numbers of pine-forest-litter and soil fauna, such as mites, springtails, beetles, earthworms, and spiders within a 3-km radius of the Chernobyl reactor, were impacted by the fallout.[57] Because of shielding by the soil, soil animals were affected to a lesser degree than pine-litter animals. Also, adult animals were less affected by 30-Gy doses (estimated by thermoluminescent dosimeters distributed on the soil surface) than were eggs and juveniles. Among soil microfauna, first instar nymphs and larvae were absent in the soil following the accident; populations of young earthworms were decreased.[50] The authors estimated that an absorbed dose of about 29 Gy devastated the soil microfauna, while a dose of about 8 Gy led to minor changes. In previously plowed soil, a surface dose of 4000 rad did not affect soil dwellers due to the shielding effect of the soil. Several specific studies were reviewed by Sokolov et al.[50] The soil worm *Aporectodea caliginose,* a diploid species, suffered genetic damage in its male germ cells (chromosome fragments in 20% of the cells), and the population size was smaller in the contaminated zone than in a reference area. In contrast, the hexaploid parthenogenetic species *Dendrobaena octaedra,* which lives in the pine litter, had increased its population size over that of a reference area by 1988. The relative radioresistance of the latter species was attributed to the polyploid genome and the lack of predators.

Following the Chernobyl accident, changes involving greater phenotypic diversity and an increased frequency of rare phenotypes were observed in several species of dragonflies.[53] Fruit flies (*Drosophila melanogaster*), which are known to be radioresistant, showed little difference in mutation frequency following the accident.[58] However, in another study, an increase in dominant lethal mutations in fruit flies collected from an area with a radiation dose of 80.6 mR/hour compared with a reference area was reported.[50]

### 20.6.3 Small-Mammal Populations

Mammals are considered more radiosensitive than other taxonomic groups, with acute lethal doses ranging from 2 to 11 Gy.[2] The mammal population in the Chernobyl area could not be compared pre- and postaccident, as the area was heavily inhabited by humans prior to the accident.[50] But a 90% mortality of small mammals as a consequence of the accident was predicted, based on comparisons with reference plots and given the external radiation doses of up to 6000 R (approximately 60 Gy). By spring of 1987, the populations appeared to increase due to migration from adjacent noncontaminated areas[50] and trapping of small mammals within the contaminated zone 8 years after the accident yielded greater success rates than in uncontaminated areas.[59] Bank voles (*Clethrienemys glareolus*) captured in 1986 showed an increased number of corpora lutea but also an increase in embryonic mortality.[50]

An extensive study of genetic damage to small mammals at the Chernobyl site was carried out with the house mouse, *Mus musculus*.[50,51,60] House mice were captured at three plots with average external radiation doses of 0.1 to 1.5 mR/hour, 1 to 2 mR/hour, and 60 to 100 mR/hour (the latter approximately 0.6 to 1.0 mGy/hour). These animals did not show any signs of radiation sickness. When males were mated with female laboratory mice, only two males (from the maximally contaminated area) of 122 were irreversibly sterile. For the rest of the males caught in this area, only temporary sterility was observed. In both males collected from this area and male progeny of pregnant laboratory female mice caged in the radioactive area, the frequency of chromosome aberrations as indicated by reciprocal translocations in spermatocytes was increased. The frequency of reciprocal translocations in spermatocytes increased linearly with increasing absorbed dose to the testes, which was estimated at 0.1 to 25 Gy for mice caught at the site.[60]

In 1994 and 1995, a study was undertaken to describe the diversity, distribution, and karyotypes of small mammals that live in the most radioactive sites within the exclusion zone near Chernobyl.[59] An examination of karyotypes of 11 species of small mammals from the 30-km exclusion zone with species obtained from outside of the exclusion zone did not document gross chromosomal rearrangements. The diversity and abundance of small-mammal fauna was not reduced at the most radioactive sites, and the trapped specimens did not demonstrate aberrant gross morphological features other than enlargement of the spleen. The most commonly collected species were voles and shrews: *Microtus arvalis*, *M. oeconomus*, *M. rossiaemeridionalis*, and *Sorex araneus*. An examination of the mitochondrial cytochrome b gene of voles from Chernobyl did not reveal an increased mutation rate over that of voles from a reference area.[61,62] In addition to enlargement of the spleen observed in some individuals collected at the site, physiological responses to radiation-induced stress may involve enlarged livers and thymus glands.[63]

Following the Chernobyl accident, several studies documented the genetic impact of radiation on small mammals. Bank voles (*Clethrionomys glareolus*) were extensively studied because they are common inhabitants of the area and have the highest levels of cesium and strontium. Through 1991, Goncharova and Ryabokon found elevated incidences of chromosome aberrations and polyploid cells in voles in the Chernobyl area.[64] In Sweden, bank voles in a Chernobyl-contaminated area were found to have increased numbers of micronucleated polychromatic erythrocytes.[65] However, bank voles collected in 1997 from the most contaminated area at Chernobyl did not have increased frequencies of micronuclei[66] and did not have reduced genetic variation.[67] A possible explanation for these observations is the radioresistance or radioadaptation of bank voles.[68,69] Bank voles trapped in the Chernobyl area and further irradiated with $^{60}Co$ had increased life spans several generations after irradiation. The $LD_{50/30}$ for this species is 966 R.[70] Continuing monitoring studies of voles in the Chernobyl area will help to clarify the long-term genetic effects of radioactive contamination on natural populations.

Compared with Chernobyl, radioactive waste storage sites are minimally contaminated. Following the partial draining of White Oak Lake, a settling basin for low-level radioactive wastes from the Oak Ridge National Laboratory, Dunaway and colleagues[71–74] studied the rodent populations utilizing the site. Doses to cotton rats (*Sigmodon hispidus*) were estimated by placing dosimeters 1 m above ground level and subcutaneously within the animals. It was estimated that the rats were exposed to 0.004 Gy/week internally and 0.025 Gy/week externally based on free-air exposures of 15 mR/hr. The total dose was estimated at 0.03 Gy/week, and lifetime doses were probably less than 2 to 3 Gy. No effects on size, sex ratio, fertility, litter size, blood cell counts, or tissue lesions attributable to radiation were found. In the laboratory, an acute dose of 2 Gy resulted in no blood changes except a transient leucopenia in both cotton rats and rice rats (*Oryzomys palustris*).[75] A dose of 6 Gy was lethal to rice rats but not to cotton rats. Chronic irradiation studies were not conducted with either species.

Since 1952, approximately $3 \times 10^{17}$ Bq of solid radioactive waste has been buried at a 36-ha subsurface disposal site at the Idaho National Engineering Laboratory. Small mammals have burrowed into the site and dispersed some of the radioactivity.[36,76] Radiation dosimeters implanted in deer mice (*Peromyscus maniculatus*) inhabiting the disposal area indicated radiation dose rates of 0.004 to 418 mGy/day (median, 0.02 mGy/day); the dose at a reference site was 0.004 mGy/day.[76] Comparisons with dosimeters placed 15 cm above the ground showed that the doses were primarily external from the buried waste and not from internal emitters ($^{137}Cs$, $^{60}Co$, and $^{95}Nb$). The doses depended to a large degree on radioactivity in the area of capture and season, which in turn reflected animal movement outside the burrow. Implications of the dose were not assessed, but it was noted that the population density within the burial area was not significantly different from the density in an adjacent area. It is likely that the few animals that received the maximum dose rate of 418 mGy/day did not survive. Radiation dose rates received by Ord's kangaroo rats (*Dipodomys ordii*) were much less than those received by deer mice. Doses to either species were less than those received by cotton and rice rats at the Oak Ridge site, where chronic exposure to 0.03 Gy/day did not result in significant effects on fertility.

A concurrent study concluded that radiation doses received by the deer mice inhabiting the area had no effect on body or organ weights and caused no increase in chromosomal aberrations.[77] However, these observations were made two weeks after the mice were transported to an uncontaminated site. Deer mice that were trapped near a radioactive leaching pond and that received dose rates of up to 0.01 Gy/day failed to exhibit any blood changes or pathologic lesions such as tumors.[78] Using a shielded $^{137}Cs$ radiation source to provide a uniform dose rate, French et al.[44] exposed a population of pocket mice (*Perognathus formosus*) enclosed in a desert area to chronic yearly dose rates of 2.11 to 3.60 Gy/year (5.8 to 10 mGy/day). These dose rates were sufficient to increase the death rate of this species.

## 20.6.4 Avian Populations

The early literature on ionizing radiation and wild birds was reviewed by Mellinger and Schultz.[79] Monitoring studies at radioactive sites did not reveal any apparent gross effects on birds. Lethal doses ($LD_{50/30}$) for various species are presented. Birds are only slightly less radiosensitive than mammals, with a range of acute lethal doses of 4.5 to 15 Gy.[2]

Ellegren et al. found an increased frequency of partial albinism among adult and nestling barn swallows (*Hirundo rustica*) captured close to Chernobyl compared to swallows captured in an uncontaminated reference area.[80] As high as 15% of birds captured near Chernobyl in 1996 displayed partial albinism compared with $< 2\%$ in reference areas. Aberrant coloration, such as partial albinism, may be associated with increased risk of predation and reduced mating success; this genotoxic effect appears to have resulted in a loss of fitness in the breeding population that may be associated with a significant decline in breeding population size between 1986 and 1996. The authors proposed that the mutation was associated with germ cells and thus was heritable. The level of radiation in the area of capture at Chernobyl was 300 to 500 μR.

Bird populations were studied at the Idaho National Engineering Laboratory. Populations of barn swallows (*H. rustica*) that nested near the Test Reactor Area radioactive leaching ponds utilized leaching-pond arthropods as a food source and contaminated mud for nest construction.[81] Adult birds contained primarily $^{51}Cr$ (16.1 Bq/g), but 72% of the internal dose rate was due to $^{24}Na$ (0.22 mGy/day). Thermoluminescent dosimeters placed in the nests indicated average external dose rates of 0.84 mGy/day for eggs and 2.20 mGy/day for nestlings, for a total of 54 mGy during the nesting period. Mortality rates for this population did not differ from that of a reference population. The mean growth rate of first-clutch swallows and mean asymptotic weights of immature birds were significantly different ($p < 0.05$) compared with the reference group, but they were within the normal range reported in the literature.

In a related laboratory-field study at the above site, freshly-hatched tree swallows (*Tachycineta bicolor*) acutely exposed to 0.9 Gy showed a normal growth rate, whereas those exposed to 2.7 or 4.5 Gy showed pronounced growth depression by the end of the nestling stage.[82] Earlier, the authors had located a single tree-swallow nest with five eggs chronically exposed to 1.0 Gy/day throughout the nesting period. Fledglings from this nest showed severe growth depression.

## 20.7 EFFECTS OF RADIATION ON AQUATIC POPULATIONS AND COMMUNITIES

Radionuclides have entered aquatic environments from natural sources, nuclear weapons testing, nuclear waste disposal, and accidental releases from nuclear power plants and via runoff from contaminated terrestrial environments. Associated with the presence of radioactive contaminants in aquatic systems is the potential for effects on aquatic biota. Exposure of such organisms may occur externally due to (1) radiation present in water and sediment and (2) the absorption of radionuclides onto the surface of biota, and internally as a consequence of absorption or ingestion. As indicated in Table 20.2, annual doses received by marine and freshwater biota from natural sources of radiation are generally less than 5 mGy/year.[20]

Radiation-induced somatic and genetic effects have been observed in individual organisms following acute exposures in the laboratory.[8,20,83–85] Several general conclusions have been drawn from acute and chronic laboratory investigations on aquatic organisms.[8] Adult fish have radiation sensitivities similar to terrestrial mammals, provided sufficient time is allowed for temperature adjustments in the poikilothermic animals. Aquatic invertebrates tend to be more resistant to effects of radiation than vertebrates. Sensitivities of fish to radiation are dependent on the age of the fish and stage of embryological development. As pointed out by Blaylock and Witherspoon[86] and the NCRP,[8] such studies are often conducted at doses many times greater than that expected in aquatic environments associated with nuclear activities. Surveys of the literature indicate a lack of data on chronic exposures in the environment, especially at the population and community level of organization. [2,8,85]

Included among the freshwater environments that have been studied for radiation effects are three U.S. Department of Energy sites: White Oak Lake at Oak Ridge National Laboratory, seepage basins and a reactor cooling reservoir at the Savannah River Plant, and ponds and streams at the Hanford Site; also studied are aquatic environments adjacent to the Chernobyl accident site and the Mayak plutonium reprocessing facility in Russia as well as marine atolls that have undergone nuclear testing activities. Each of these environments contains radionuclides at above-background concentrations. In some cases, effects have been noted in individual organisms; however, adverse effects have not been observed at the population or community level.[8,20] For the most part, these studies have shown the resilience of populations of freshwater biota to doses of less than 10 mGy/day.

Field studies on the effects of radiation on the marine environment are primarily limited to those that have been conducted near Bikini and Eniwetok atolls and in the North Irish Sea. In both instances, contamination of the marine environment was associated with nuclear-weapons-related activities. Long-term studies conducted in these areas have shown no impact on populations of aquatic biota. Comparisons of freshwater and marine ecosystems indicate that radiation effects on marine systems may be less than those expected in freshwater systems based on the larger size of marine systems and therefore greater potential for contaminants to disperse, and on the greater number of species usually present in marine environments as compared to freshwater ecosystems.[20] The studies discussed below support the resilience of marine ecosystems to low-level ionizing radiation.

### 20.7.1 Plant Populations

The first large-scale introduction of radionuclides into the marine environment as a result of human activities occurred in 1946 on Bikini atoll. Both Bikini atoll and Eniwetok were used as the "Pacific Proving Ground" for test detonations of nuclear and thermonuclear devices. The biological studies that have been conducted in these areas have been summarized by Templeton et al.[83] Intensive studies were conducted from 1950 through the 1960s to determine what effects radiation had on the marine environment. The studies concluded that, although individual organisms perished as a result of nuclear testing, the marine environment became repopulated with the impacted species over time. A few examples of effects on individual organisms from the area have been found. Physiological functions of sessile algae were not noticeably altered one year after an explosion when algae were exposed to 5 to 20 to 30 mR/day.[87] Abnormalities in plant growth have also been noted on Bikini and Eniwetok atolls. Chronic dose rates of gamma radiation of 0.13 to 0.37 Gy/day for 2 to 4 months (total dose 7.8 to 55 Gy) can result in abnormalities similar to those found on these atolls. Abnormalities in plant growth observed in the field at these sites include tumors and changes from vine growth to stalk growth in one plant.[88]

### 20.7.2 Invertebrate Populations

Located on the Oak Ridge National Laboratory's Reservation is a 7-ha impounded embayment known as White Oak Lake, which has been used since 1943 as a settling basin for low-level radioactive waste and nonradioactive contaminants. contribute Most of the radioactivity within the lake is contributed by $^{137}$Cesium, $^{60}$Co, $^{90}$Sr, and $^{3}$H. Transuranic and $^{106}$Ruthenium tradionuclides are also present in the lake.[89,90] Radioactivity within the lake water and sediment has varied over the years. Several species of fish, turtles, aquatic invertebrates, and plants have been reported in the eutrophic lake. Among them, chironomid larvae, snails, mosquitofish, and pond slider and snapping turtles have been studied for possible effects of chronic exposure to low-level radiation on the individuals and populations that reside in the lake. Although genetic effects and reproductive differences were observed between animals from the lake and animals from the reference sites as observed in the following discussions, adverse impacts at the population level were not expected to occur. Populations of aquatic organisms have been present since impoundment of the lake.

Larvae of the midge, *Chironomus tentans*, were collected from White Oak Lake on two separate occasions and examined for chromosome aberrations.[91,92] Chironomids collected from the lake in 1960 received a daily dose of 6.24 mGy, approximately 1000 times normal background level.[93] Comparisons of chromosomes within the salivary gland of these benthic aquatic invertebrates from White Oak Lake and reference populations indicated an increased number of chromosomal aberrations in animals from the contaminated site. The frequency of endemic inversions was, however, similar between the areas. The author attributed the lack of difference to the possible elimination of inversion through natural selection or to genetic drift within the White Oak Lake populations. In a follow-up study conducted 10 years later, when the dose rate received by chironomids was 0.3 mGy/day, no chromosomal aberrations were detected.[94] A laboratory study with a related species, *Chironomus riparius*, revealed that 360 mGy/day was the lowest dose rate at which an increased frequency of chromosome aberrations could be detected in the larval and pupal stages of chironomids reared in tritiated water.[95] A comparative study of chromosomal aberrations induced by chronic gamma radiation from an external $^{60}$Co source and beta radiation from incorporated tritium revealed similarities in the frequency of aberrations.[96]

Chromosomal abnormalities have also been measured in chironomids from the water reservoirs of the Chernobyl atomic power plant.[53] In 1998, two years following the accident, all individuals of *Chironomus balatonicus* had chromosomal abnormalities, with 85.7% having heterozygous inversions, 85.7% with homozygous inversions, and 3.6% of the individuals with "B-chromosomes." The radiation dose to these organisms was not estimated.

A field experiment conducted in the early 1970s examined the effects of chronic irradiation on the population of an aquatic snail, *Physa heterostropha*.[97] White Oak Lake snails, receiving a dose of 6.5 mGy/day, were found to have a significantly lower number of egg capsules per snail than did snails from the control population; however, snails from the contaminated site contained a significantly greater number of eggs per egg capsule. As a consequence, egg production was found to be similar in both populations. Laboratory populations of *Physa heterostropha* receiving a chronic dose of 10 mGy/hour (240 mGy/day) showed no significant differences from the control population in the number of egg capsules per snail, number of eggs per capsule, percent of eggs hatched, mortality, or snail size.[98] A dose rate of 100 mGy/hour (2400 mGy/day), however, significantly decreased each of the population parameters.

Snails have been shown to be good biomonitors of radio-strontium in aquatic environments and to be relatively tolerant of elevated doses of radiation. Mollusks were collected from the Dnieper drainage area and throughout the Kiev administrative region following the Chernobyl nuclear accident.[99] Radioactivity in shells and soft tissue were found to exceed pre-Chernobyl concentrations by factors as great as three orders of magnitude. The highest recorded concentrations were 4 to 5 MBq/kg in shells of *Lymnaea* and *Plaorbarius*. Abnormalities and impacts on snail populations were not recorded in the investigation. Gene frequencies at four polymorphic loci were examined in seven populations of the snail *Dreissena polymorphs* adjacent to the Chernobyl atomic power station.[100] Recorded differences were attributed to conditions of the breeding site, such as substrate characteristics and water flow rate, and not to radiation or thermal differences.

### 20.7.3 Fish Populations

Studies have been conducted that examined the effects of chronic low-level radiation on mosquitofish (*Gambusia affinis*) populations in White Oak Lake.[101,102] Fish collected from the lake in 1965 were found to have significantly larger brood sizes (when adjustments were made to correct for differences in maternal size) than did fish from the reference population. In addition, an increased number of dead and deformed embryos were noted in the White Oak Lake population. The fitness of White Oak Lake mosquitofish was further investigated in 1973 by the rearing of field-collected fish from the lake and from four reference sites.[102] Significant differences were not found in the brood sizes of the $F_1$ generations of the laboratory-reared fish from the five locations. It was therefore concluded that the larger brood size originally observed in White Oak Lake mosquitofish populations was likely attributable to the eutrophic conditions of the lake and not to genetic adaptation as a response to radiation. The increased incidence of dead and deformed embryos did, however, persist in the White Oak Lake stock reared in the laboratory. Measurements of other characteristics revealed differences in the size and variability in the size of female fish and differences in the variability and mean critical thermal maximum of male mosquitofish from the White Oak Lake stock. Although these findings were used to support the explanation of the presence of radiation-induced recessive mutations in the gene pool of White Oak Lake, no significant negative impact on the overall fitness of mosquitofish from the lake was expected.[102]

As an added note, the dose rate to mosquitofish in White Oak Lake was 0.59 mGy/day in 1975 and 3.6 mGy/day in 1965. Both rates were considerably lower than the dose rate found to decrease brood size and increase sterility in the guppy, *Poecilia reticulata*, following chronic exposure in the laboratory.[103] DNA strand breakage was also evaluated in mosquitofish (*Gambusia affinis*) from two sites associated with this facility.[104] Fecundity was found to be negatively correlated with the amount of double-stranded breaks at one radioactively contaminated site. In addition, a higher proportion of strand breaks was found in females with broods containing deformed embryos than in females with normal broods. These data indicate a link between DNA strand breaks and potential population-level effects.

Fish from regions near Chernobyl have been examined for both genetic and morphological abnormalities. Chromosomal aberrations in the corneal epithelium were examined in carp from the cooling reservoir of the Chernobyl Atomic Power Station during 1987. The rate of mutagenesis between carp from the cooling pond and a site 60 km from Chernobyl were not interpreted as significantly different.[50] DNA strand breaks and micronuclei were examined in channel catfish (*Ictalurus punctatus*) from the Chernobyl cooling pond and a fish hatchery 30 km from the Chernobyl site.[105] Catfish from the cooling pond were found with a higher incidence of DNA damage; however, no difference in the number of micronuclei was detected. Radiocesium concentrations in the cooling pond were at least 50% greater than at the U.S. Department of Energy Savannah River Site, where DNA damage was also detected in largemouth bass (*Micropterus salmoides*).[106] The ability of catfish populations within the cooling ponds to withstand sediment exposures of up to approximately 3 Gy/day was attributed to possible acclimation and adaptation over time.[105] This may be supported by the lack of anomalous abnormalities in fish breeding near Chernobyl one year following the nuclear reactor accident.[50]

Activities of the Windscale reprocessing plant were responsible for the release of radionuclides into the marine environment as a consequence of the discharge of low-level radioactive effluent and a nuclear reactor accident in 1957. Fish in the North Irish Sea have been exposed to low-level radiation for over 20 years. Plaice (*Pleuronectes platessa*), a commercial fish species collected from the area in 1968 to 1970, showed no differences in average body length when compared to fish from other areas.[20] In addition, laboratory studies conducted on plaice embryos revealed that a radiation dose rate of 0.05 mGy/day (measured in the area from 1967 to 1969)[85] would not be sufficient to induce a discernible effect on individual plaice or on populations of the fish.[107]

Nuclear testing in the Pacific has also resulted in observable effects in marine fish. Thyroid tissue damage was observed in fish collected from Eniwetok for a period of 8 months following a nuclear detonation. The damage was attributed to the radiation from iodine, despite the absence of radioiodine from the fish thyroids.[108]

### 20.7.4 Turtle Populations

In the late 1980s, two species of turtles were collected from White Oak Lake and a reference site and examined for single-stranded DNA breaks, a nonspecific indicator of possible exposure to genotoxic agents in the environment.[109] Both the pond slider (*Trachemys scripta*) and common snapping turtle (*Chelydra serpentina*) from the lake were found to contain a significantly greater amount of DNA damage than turtles from the reference site. Differences between species were not found. Turtles from White Oak Lake also contained significantly higher tissue concentrations of $^{137}$Cs, $^{90}$Sr, $^{60}$Co, and mercury. Because turtles in the lake were exposed to agents other than radiation that may have induced the DNA damage, radiation could not be singled out as the cause of the damage. The dose received by turtles in the lake was not determined, and comparative laboratory studies have not been conducted to ascertain whether radiation could induce the single-stranded DNA breaks observed in the field.

As with White Oak Lake, the Savannah River Plant is located on a U.S. Department of Energy facility. On the approximately 80,000 ha in South Carolina are five nuclear reactor facilities, two chemical separation plants, a heavy-water separating plant, small test-reactor facilities, and the Savannah River laboratory.[110] Seepage basins on site have received radioactive and nonradioactive wastes from the plant in the past. A reactor cooling pond, designated Pond B, received fission products from a nuclear reactor between 1961 and 1964.[111] A flow cytometric assay was utilized to assess the impact of low-level radiation on the DNA of turtles and ducks from contaminated aquatic environments on the Savannah River Plant. Bickham et al.[112] collected slider turtles from two seepage basin complexes containing primarily $^{134}$Cs, $^{137}$Cs, $^{89}$Sr, $^{90}$Sr, and $^{3}$H. Chromium, mercury, and nitrate were also contaminants within the basins. Turtles from the seepage basins were found to possess significantly greater DNA content in red blood cells than turtles from the reference site. In addition, four turtles from the contaminated basins contained mosaic DNA. The enhanced amount of variation in DNA content was attributed to chromosomal rearrangements in the DNA, which can result in deletions and duplications of genetic material.

The presence of mosaic DNA is related to the proliferation of such cell lines. A correlation was found between the plastron length of the turtle, which is a rough estimate of age, and the amount of variation in cellular DNA. The dose rate to turtles was not determined, and the specific cause of the genetic damage in the Savannah River Plant turtles could not be ascertained. To determine whether low-level radiation could induce genetic damage as observed in the turtles from the seepage basins, slider turtles were collected from Pond B and cellular DNA content in red blood cells measured.[113] As with the turtles from the seepage basins, a significantly greater variation in DNA content was found in turtles from the radioactive site than in turtles from the reference site. DNA from two of the turtles from Pond B showed evidence of aneuploid mosaicism. Although the radiation dose received by the turtles was not determined, Pond B sliders were noted as containing average total body burdens of $^{137}Cs$ and $^{90}Sr$ of 842 Bq and 1879 Bq, respectively. Body burdens of the White Oak Lake pond sliders were, incidentally, within the range of that measured in the Savannah River turtles.[114] Impacts of the variation in DNA content and presence of mosaicism on the turtle populations are unknown.

### 20.7.5 Waterfowl Populations

The same flow cytometric assay used on the Savannah River turtles was used to detect DNA abnormalities in ducks exposed to radionuclides in Pond B.[111] Cesium-137 constituted 99% of the activity present in the reservoir. Game-farm-reared male mallards were released onto Pond B and allowed to roam free during the daylight hours for 12 months. Flow cytometric measurements were conducted on a regular basis by the extraction of blood from exposed and control mallards. Differences were found between the Pond B mallards and the control groups' only after the $^{137}Cs$ concentration in the exposed ducks had reached steady-state equilibrium (mean of 2.52 Bq/g at 8 months). After 9 months, the coefficient of variation in DNA content was significantly greater in the Pond B mallards, with aneuploid mosaicism detected in 2 of the 14 pond ducks. Again, the impacts of these findings on the population level are unknown.

### 20.7.6 Aquatic Communities

The Hanford site near Richland, Washington, contains several aquatic systems with origins that date back to the Manhattan Project in 1943. Hanford ponds and streams were maintained to receive wastewater from nuclear reactor, reprocessing, and laboratory operations. The aquatic systems on site contain a variety of contaminants. Most contain radioactive waste in the form of mixed-fission and activation products with actinides including transuranics in some locations. Chemical waste was also released into some of the systems. In the mid-1970s, several of the aquatic ecosystems on site were studied to determine whether the presence of radioactive wastes could impact the community structure in terms of diversity and productivity.[115] The study characterized limnological and radiological conditions in several ponds, ditches, and a trench. Although some differences in population parameters for algae, macrophytes, and invertebrates were found among the Hanford sites and between the Hanford sites and off-site systems, they could not be correlated with radiological differences. Comparisons of doses from ionizing radiation to aquatic biota in the Hanford ponds and streams to doses reported in the literature to cause minor to intermediate damage in aquatic organisms revealed that dose rates of 50 to 100 R/day in 100-N Trench may have been sufficient to induce limited damage to algae, invertebrates, and amphibians.

Aquatic environments were impacted by the Chernobyl atomic power plant (CAPP) accident in 1986. Many species formerly associated with the CAPP cooling ponds have returned to the ponds.[51] A colony of beaver (*Castor fiber*) storks has returned to the ponds. Bivalve mollusk populations of *Anodonata cygnea* appear to be recovering and are actively growing following the radiation insult; however, populations of *Dreissena* continue to be depressed. Model estimates indicate an external radiation dose to aquatic biota of 2 to 3 mGy/day shortly after the accident. Fish within these ponds also appeared to breed and grow successfully.

## 20.8 RADIOLOGICAL DOSE MODELS

Simplistic models are often used in screening level risk assessments to estimate exposure and risk to ecological receptors. A variety of radiation dose models have been developed to estimate dose to terrestrial[116–119] and aquatic biota.[120–122] Recently, dose models have been developed that allow for a comparison of radionuclide concentrations with screening benchmark values intended to be protective of the ecosystems of concerns. Such models provide guidance on how ecological risks can be assessed by using environmental monitoring data to estimate dose to specific ecological receptors, followed by a comparison of the calculated dose against acceptable dose limits for aquatic[123–126] and terrestrial biota.[123,126,127]

The most comprehensive and user-friendly of these ecological risk-related radiation dose models is that developed by the U.S. Department of Energy.[123] A graded approach, which includes methods for both screening and detailed analyses, was developed as a technical standard to evaluate compliance with specific radiation dose limits to populations of aquatic and terrestrial biota exposed to anthropogenic sources of radiation at U.S. Department of Energy sites. The dose limits used in the analysis of risk are those recommended by the IAEA[9] of 10 mGy/day for aquatic biota, 10 mGy/day for terrestrial plants, and 1 mGy/day for terrestrial animals. A companion tool to the standard is the RAD-BCG Calculator. This tool contains a series of electronic spreadsheets that facilitate the calculation of total radiation dose to each receptor.

Strengths and weaknesses of predictive models in assessing exposure, dose, and effects have been reviewed by radioecologists.[128–131] Basic conclusions are:

- Existing exposure/dose models are often simplistic and utilize generalized default parameters that can result in either over- or underestimation of potential risk to the receptor. Selection of default values is also highly dependent upon professional judgment, which can vary among modelers.
- Existing models should be tested and calibrated with actual data from the field or laboratory.
- Models in the past have been primarily deterministic. Greater technical value would be achieved through stochastic models that incorporate error propagation and probability distributions of specific parameter values.

Existing radiation dose models are, however, useful tools in predicting potential doses and effects to ecological receptors in the absence of site-specific information, as in a screening-level ecological risk assessment.

## 20.9 EFFECTS OF POWER PLANT COOLING SYSTEMS

Over half of the nuclear power plants in the United States use once-through cooling systems. Receiving-water bodies include rivers, lakes and reservoirs, cooling ponds, the Great Lakes, and estuaries and coastal areas.[18] Other units have closed-cycle cooling, primarily mechanical or natural-draft-cooling towers, which results in little water withdrawal. With cooling towers, most of the heat is dissipated to the atmosphere. The number of fossil-fueled plants is considerably greater than the number of nuclear plants.

Operation of once-through cooling systems at power plants, whether nuclear or fossil-fueled, primarily affects the aquatic environment. The impacts on terrestrial resources — construction changes, transmission-line maintenance, and drift from cooling towers — are few and are limited to the immediate area, particularly when considering population-, community-, and ecosystem-level effects.[18] The following discussion concerns both nuclear and fossil-fueled power plants; nuclear plants require larger amounts of cooling water than do fossil-fueled plants. Although the effects of cooling systems on the aquatic environment are site specific, some generalizations can be made.

Once-through condenser cooling systems require large amounts of water that are withdrawn directly from surface waters and returned to the receiving water body with an increased heat and chemical load. During operation of the cooling system, intake screens bar debris and large organisms from entering the plant. Impingement occurs when aquatic organisms are caught on the screens. Impingement results in mortality if the organisms are held against the screens for long periods of time. Entrainment — the pumped passage of small organisms, including eggs and larvae, through the condenser cooling system where they are subjected to heat, mechanical, and pressure stresses

— may also result in mortality. Entrained organisms may suffer heat shock as a result of abrupt exposure to the elevated temperature of the cooling water. On the other hand, organisms acclimated to the temperature of the receiving water may suffer cold shock during winter months when plant shutdowns occur. Periodic condenser chlorination to control fouling also results in mortality of entrained organisms. These impacts of power plant cooling systems on the aquatic environment have been monitored for years,[10–17] with no long-range problems.[18,132,133]

Initially, the concern regarding the impacts of cooling systems on the aquatic environment centered on heated discharges ("thermal effects"), which were regulated under the Clean Water Act (CWA) 316(a) requirements. Depending on the flow, use of biocides, and type or placement of the outlet in the water column, scouring and the absence of benthic invertebrates in the immediate area have been observed. At some plants, seasonal changes in river phytoplankton were amplified in the heated area, with disappearance of some taxa in summer and replacement of green algae with the more heat-tolerant blue-green algae.[134] The response of fish to thermal effluents is seasonal and depends on the thermal preference and temperature tolerance of the individual species.[135] Available data show that thermal plumes in rivers do not usually affect anadromous or resident fish migration; migrating fish have been shown to avoid areas of maximum temperature by passing around or under the plume.[132,136] In some cases, temperatures of 32 to 35°C have not presented a barrier to the presence[137] or movement[138] of fish. A review of the response of fish to heated discharges indicates that some species avoid heated areas, whereas some warm-water species have been observed feeding in thermal plumes at temperatures as high as 41°C.[139] Thus, there have been changes in the distribution of organisms, with replacement of heat-sensitive species of fish, invertebrates, and algae by more resistant species in the immediate vicinity of the discharge; however, none of the changes regarding thermal effluents have impacted the entire ecosystem.

In addition to the direct effects on organisms, there is concern about the impact on fish populations resulting from mortality of fish eggs and larvae carried by the cooling waters through the plant and from impingement of juvenile fish colliding with the intake screens.[14,16] Neither entrainment of early life stages of fish or impingement of fish and shellfish has been found to be a problem at most power plants.[18] Several studies conducted at power plants on the Hudson River indicated that entrainment mortalities are less than 50% for some fish species, and the combined effects of entrainment and impingement would result in 10 to 20% loss of a year class of several fish species.[140] The problem may become significant, however, when many plants are sited on the same river or estuarine system, cooling water intake is considerable compared to natural flow, and early life stages of commercially important anadromous fish are involved.

Five electricity-generating plants with once-through cooling systems are located on the Hudson River estuary between river kilometers 60 and 107. The only nuclear plant among these is the Indian Point power plant located at kilometer 69. In the 1960s, construction of two additional units, both with once-through cooling, was proposed for the Indian Point site. In addition, a proposed pumped-storage hydroelectric plant was proposed for the Cornwall area (kilometer 93). Both of these facilities would draw large volumes of water from an area of the Hudson River estuary that served as a spawning and nursery habitat for several species of commercially and recreationally important fish. It was feared that entrainment and impingement would result in the decline of the striped bass (*Morone saxatilis*) population.

Between 1963 and 1980, an intensive environmental research and assessment program was undertaken, documenting the characteristics of the estuary including geography, hydrology, and physicochemistry of the river; biological productivity; abundance, distribution, and life histories of the fish fauna; and potential impact of additional power plants.[140,141] Mathematical population models were developed to calculate potential impacts. These studies supported a series of legal proceedings (regarding licenses and discharge permits) among scientists, conservationists, and utility companies that ended in an out-of-court settlement in December 1980. As an alternative to building cooling towers, three key measures were agreed upon for mitigating entrainment and impingement loss: flow reductions and power outages at several of the plants during periods of peak entrainment, a barrier net to mitigate fish impingement at one plant, and a program of stocking of hatchery-reared striped bass by the utility companies.

In the late 1990s, concern for the potential adverse environmental effects of power plants focused less on thermal discharges and more on cooling-water intake structures, as the U.S. Environmental Protection Agency revisited the CWA 316(b) requirements for new and existing facilities. The last 30 years of research on impingement and entrainment, i.e., 316(b) issues, were recently summarized.[133] Early assessments of the impacts of cooling-water intakes used models to predict losses to key fish populations. Using worst possible scenarios, these models predicted high annual exploitation rates. However, monitoring of fish populations in the Hudson River (mentioned previously) and the Delaware estuaries has failed to show adverse population-level impacts.[142] A variety of fish-protection technologies have been employed to mitigate impingement and entrainment: fish collection systems, fish diversion systems, physical barriers, and behavioral barriers.[143] Nevertheless, compensation, or the ability of fish populations to offset losses, remains an area of active research. Phytoplankton and zooplankton are less impacted than fish populations.

## 20.10 SUMMARY

This chapter addresses man-made sources of ionizing radiation, the movement of radionuclides in the environment, and potential effects on natural populations and communities of plants and animals. In addition to natural background radiation, irradiation occurs from the normal operation of nuclear power plants and plutonium-production reactors, nuclear plant accidents, nuclear weapons testing, and contact with or leakage from radioactive waste storage sites. No studies of plants or animals exposed chronically to the low doses found in the environment demonstrated a detrimental effect on populations, although some biological effects have been observed. Aquatic organisms associated with environments contaminated by activities at the Mayak Production Association facility in the southern Urals of Russia have been exposed to unusually high levels of radiation for extended periods. Impacts on snail populations have been noted following doses of 2 to 3 Gy/day; however, recovery of some populations over time has also been noted. More common are acute doses at high exposure levels, which are not expected to occur except in the event of a catastrophic nuclear accident — nuclear warfare.

Radiation effects in this chapter focus on field studies, with supporting information from relevant laboratory investigations. Selected examples attempted to relate estimated doses or tissue levels to potential effects; however, dose estimates in the field are often imprecise, and observations are further confounded by the presence of other contaminants or stressors.

Very few studies address the effects of exposure of natural populations of terrestrial plants and animals to chronic, low-level irradiation. Chronic exposure of a population of cotton rats to 0.03 Gy/day at one radioactive site did not result in significant effects on fertility or blood parameters. Results of laboratory studies indicate that adverse effects are not expected at doses below 1 to 2 Gy. Chronic dose rates below 1 mGy/day should be protective of natural terrestrial populations.

Adverse effects of radiation on natural aquatic populations have been demonstrated only in severely contaminated areas. While genetic and reproductive changes have been measured in aquatic biota from areas contaminated with radiation, the overall fecundity of the populations did not appear to be adversely impacted. Changes in DNA have been observed in field-collected turtles and waterfowl, but for the purpose of evaluating the effects of radiation at the population level, these studies are lacking because dose rates were not determined for comparison with laboratory acute radiation data. In addition, impacts on survival, fertility, or fecundity were not measured. In some instances, as with the White Oak Lake, Savannah River Plant, and Hanford studies, observed effects in the field could not be exclusively attributed to radiation because of the presence of chemical contaminants or potential impacts of other environmental stressors.

Responses of aquatic organisms to radiation can be modified by salinity, temperature, and water chemistry in addition to the biological factors such as life stage and metabolic rate. In addition, competition between chemically similar stable elements and radionuclides, such as calcium and $^{90}Sr$ or potassium and $^{137}Cs$, can have an impact on the exposure of aquatic biota to radiation. In cases where dose rates received by organisms in natural aquatic environments have been measured, it appears that chronic dose rates of ≤10 mGy/day do not result in deleterious effects on natural populations of aquatic organisms. Snail populations in adjacent aquatic environments associated with the Mayak nuclear fuel reprocessing plant in Russia, may, however, be an exception. Further field investigations of these chronically exposed aquatic populations are warranted. The general resilience of aquatic species to exposure to low-level radiation are perhaps best illustrated with the marine environments of Bikini and Eniwetok atolls, where populations and communities have survived decades of nuclear weapons testing.

The impacts of power plant cooling systems — impingement, entrainment, elevated water temperatures, heat shock, and cold shock — on aquatic populations and communities have been intensively studied for more than 20 years. Local effects have been observed, but there have been no long-range problems. Procedures such as siting considerations, modifications of intake structures, cooling water flow reductions, and the greater use of cooling towers have helped to mitigate potential environmental impacts. Stocking programs can be used to replenish populations of important fish species.

## REFERENCES

1. Klement, A. W., Jr., Ed., *CRC Handbook of Environmental Radiation*, CRC Press, Boca Raton, FL, 1982.
2. Whicker, F. W. and Schultz, V., *Radioecology: Nuclear Energy and the Environment,* Vol. II, CRC Press, Boca Raton, FL, 1982.
3. Barnthouse, L. W., *Effects of Ionizing Radiation on Terrestrial Plants and Animals: A Workshop Report*, ORNL/TM-13141, Oak Ridge National Laboratory, Oak Ridge, TN, 1995.
4. Eisenbud, M. and Gesell, T. F., *Environmental Radioactivity from Natural, Industrial and Military Sources*, Academic Press, San Diego, 1997, 614.
5. Eisler, R., Radiation Hazards to Fish, Wildlife, and Invertebrates: A Synoptic Review, Contaminant Hazard Reviews Report 29, U.S. Department of the Interior National Biological Service, Washington, D.C.,1994.
6. Kahn, B., Ecological Risks Associated with Radioactive Materials, in *Predicting Ecosystem Risk*, Cairns, J., Jr., Niederlehner, B. R., and Orvos, D. R., Eds, Princeton Scientific Publishing Co., Princeton, NJ, 1992, 347.
7. ICRP, *1990 Recommendations of the International Commission on Radiological Protection*, Publication 60, International Commission on Radiological Protection, Pergamon Press, New York, 1991.
8. NCRPM, *Effects of Ionizing Radiation on Aquatic Organisms*, NCRP Rep. No. 109, National Council on Radiation Protection and Measurements, Bethesda, MD, 1991.
9. IAEA, Effects of Ionizing Radiation on Plants and Animals at Levels Implied by Current Radiation Protection Standards, International Atomic Energy Agency, Vienna, 1992.
10. Gibbons, J. W. and Sharitz, R. R., Eds., *Thermal Ecology,* National Technical Information Service, Springfield, VA, 1974.
11. Esch, G. W. and McFarlane, R. W., Eds., *Thermal Ecology II,* National Technical Information Service, Springfield, VA, 1976.
12. IAEA, Environmental Effects of Cooling Systems at Nuclear Power Plants, International Atomic Energy Agency, Vienna, 1975.
13. IAEA, Combined Effects of Radioactive, Chemical and Thermal Releases to the Environment, International Atomic Energy Agency, Vienna, 1975.
14. Van Winkle, W., Ed., *Assessing the Effects of Power-Plant-Induced Mortality on Fish Populations,* Pergamon Press, New York, 1977.

15. Thorp, J. H. and Gibbons, J. W., Eds., *Energy and Environmental Stress in Aquatic Systems,* CONF-771114, DOE Symposium Series 48, National Technical Information Service, Springfield, NY, 1978.
16. Jensen, L. D., Ed., *Fourth National Workshop on Entrainment and Impingement,* EA Communications, Melville, NY, 1978.
17. Schubel, J. R. and Marcy, B. C., Jr., *Power Plant Entrainment — A Biological Assessment,* Academic Press, New York, 1978.
18. U.S. NRC, Generic Environmental Impact Statement for License Renewal of Nuclear Plants, NUREG-1437, Vol. 1, Washington, D.C., 1991.
19. Biological Effects of Ionizing Radiation (BEIR) Advisory Committee, *Report: The Effects on Populations of Exposure to Low Levels of Ionizing Radiation,* National Academy of Sciences, National Research Council, Washington, D.C., 1972.
20. IAEA, *Effects of Ionizing Radiation on Aquatic Organisms and Ecosystems,* Tech. Rep. Ser. No. 172, International Atomic Energy Agency, Vienna, 1976.
21. U.S. NRC, *Information Digest 2000 Edition*, NUREG 1350, Vol. 12, Washington, D.C., 2000.
22. Eichholz, G. G., *Environmental Aspects of Nuclear Power,* Ann Arbor Science, Ann Arbor, 1976.
23. Eisenbud, M., *Environmental Radioactivity from Natural, Industrial, and Military Sources*, 3rd ed., Academic Press, New York, 1987.
24. Anon, Evaluation of Radiological Conditions in the Vicinity of Hanford for 1962, HW-76526, Hanford Laboratories, 1963.
25. Kahn, B., Blanchard, R. L., Krieger, H. L., Kolde, H. E., Smith, D. B., Martin, A., Gold, S., Averett, W. J., Brinck, W. L., and Karches, G. J., Radiological surveillance studies at a boiling water nuclear power plant, in *Environmental Aspects of Nuclear Power Stations,* IAEA-STI-PUB261, International Atomic Energy Agency, Vienna, 1971.
26. Eisenbud, M., *Environmental Radioactivity from Natural, Industrial, and Military Sources*, 2nd ed., Academic Press, New York, 1973.
27. Miller, C. W., Cottrell, W. D., Loar, J. M., and Witherspoon, J. P., Examination of the impact of radioactive liquid effluent releases from the Rancho Seco nuclear power plant, *Health Phys.,* 58, 263, 1990.
28. Akleyev, A. V., Kostyuchenko, V. A., Peremyslova, L. M., Baturin, V. A., and Popova, I. Ya., Radioecological impacts of the Techa River contamination, *Health Phys.*, 79, 36, 2000.
29. Field, R. W., Field, E. H., Zegers, D. A., and Stenton, G. L., Iodine-131 in thyroids of the meadow vole (*Microtus pennsylvanicus*) in the vicinity of the Three Mile Island nuclear generating plant, *Health Phys.,* 41, 297, 1981.
30. Trabalka, J. R., Eyman, L. D., Auerbach, S. I., Analysis of the 1957–1958 Soviet nuclear accident, *Science*, 209, 345, 1980.
31. Kryshev, I. I., Romanov, G. N., Sazykina, T. G., Isaeva, L. N., Trabalka, J. R., and Blaylock, B. G., Environmental contamination and assessment of doses from radiation releases in the southern Urals, *Health Phys.*, 74, 687, 1998.
32. Hohenemser, C., Deicher, M., Ernst, A., Hofsass, H., Linder, G., and Recknagel, E., Chernobyl: An early report, *Environment,* 28, 6, 1986.
33. Simopoulos, S. E., Soil sampling and $^{137}Cs$ analysis of the Chernobyl fallout in Greece, *Appl. Radiat. Isot.,* 40, 607, 1989.
34. O'Farrell, T. P. and Gilbert, R. O., Transport of radioactive material by jackrabbits on the Hanford Reservation, *Health Phys.,* 29, 9, 1975.
35. Garten, C. T., Jr., Radiocesium uptake by a population of cotton rats (*Sigmodon hispidus*) inhabiting the banks of a radioactive liquid waste pond, *Health Phys.,* 36, 39, 1979.
36. Arthur, W. J., and Markham, O. D., Small mammal burrowing as a radionuclide transport vector at a radioactive waste disposal area in southeastern Idaho, *J. Environ. Qual., 12, 117, 1983.*
37. Talmage, S. S. and Walton, B. T., Small mammals as monitors of environmental contaminants, *Rev. Environ. Contam. Toxicol.,* 119, 47, 1991.
38. IAEA, Generic Models and Parameters for Assessing the Environmental Transfer of Radionuclides from Routine Releases: Exposure of Critical Groups, International Atomic Energy Agency, Vienna, 1982.
39. Till, J. E. and Meyer, H. R., Radiological Assessment: A Textbook on Environmental Dose Analysis, NUREG/CR-333, U.S. Nuclear Regulatory Commission, Washington, D.C., 1983.

40. Pendleton, R. C. and Hanson, W. C., Absorption of cesium-137 by components of an aquatic community, *Proc. 2nd Int. Conf. on the Peaceful Uses of Atomic Energy,* Vol. 18, United Nations, Geneva, 1958, 419.
41. Hove, K., Pedersen, O., Garmo, T. H., and Staaland, H., Fungi: A major source of radiocesium contamination of grazing ruminants in Norway, *Health Phys.,* 59, 189, 1990.
42. Howard, B. J., Beresford, N. A., and Hove, K., Transfer of radiocesium to ruminants in natural and semi-natural ecosystems and appropriate countermeasures, *Health Phys.,* 61, 715, 1991.
43. Turner, F. B., Licht, P., Thrasher, J. D., Medica, P. A., and Lannom, J. R., Jr., Radiation-induced sterility in natural populations of lizards (*Crotaphytus wislizenii* and *Cnemidophorus tigris*), in *Radionuclides in Ecosystems,* Nelson, D. J., Ed., U.S. Atomic Energy Commission, Oak Ridge, TN, 1973, 1131.
44. French, N. R., Maza, B. G., Hill, H. O., Aschwanden, A. P., and Kaaz, H. W., A population study of irradiated desert rodents, *Ecol. Monogr.,* 44, 45, 1974.
45. Hobbs, C. H., and McClellan, R. O., Toxic effects of radiation and radioactive materials in *Casarett and Doull's Toxicology: The Basic Science of Poisons,* 3rd ed., Klaassen, C. D., Amdur, M. O., and Doull, J., Eds., Macmillan, New York, 1986, 669.
46. Ulsh, B. A., Whicker, F. W., Hinton, T. G., Congdon, J. D., and Bedford, J. S., Chromosome translocations in *T. scripta*: The dose-effects and *in vivo* lymphocyte radiation response, *Radiat. Res.*, 155, 63, 2001.
47. Whicker, F. W. and Fraley, L., Jr., Effects of ionizing radiation on terrestrial plant communities, *Adv. Radiat. Biol.,* 4, 317, 1974.
48. Turner, F. B., Effects of continuous irradiation on animal populations, *Adv. Radiat. Biol.,* 5, 83, 1975.
49. Woodwell, G. M., Radiation and the patterns of nature, *Science,* 156, 461, 1967.
50. Sokolov, V. E., Krivolutzky, D. A., Ryabov, I. N., Taskaev, A. I., and Shevchenko, V. A., Bioindication of biological after-effects of the Chernobyl atomic poser station accident in 1986–1987, *Biol. Int.*, 8, 6, 1989.
51. Sokolov, V. E., Rjabov, I. N., Ryabtsev, I. A., Tikhomirov, F. A., Shevchenko, V. A., and Taskaev, A. I., Ecological and genetic consequences of the Chernobyl atomic power plant accident, *Vegetatio,* 109, 91, 1993.
52. Medvedev, Z. A., Chernobyl: Eight years after, *Trends Ecol. Evol.*, 9, 369, 1994.
53. Savchenko, V. K., *The Ecology of the Chernobyl Catastrophe: Scientific Outlines of an International Programme of Collaborative Research*, Man and the Biosphere Series, Vol. 16, United Nations Educational, Scientific and Cultural Organization, Paris, France, and The Parthenon Publishing Group, U.K., 1995.
54. Baker, R. J. and Chessler, R. K., The Chernobyl nuclear disaster and subsequent creation of a wildlife preserve, *Environ. Toxicol. Chem.*, 19, 1231, 2000.
55. Grinikh, L. I. and Shevchenko, V. V., Cytogenetic effects of ionizing radiation in *Crepis tectorum* growing within 30 km of the Chernobyl atomic power station, *Sci. Total Environ.,* 112, 9, 1992.
56. Abramov, V. I., Fedorenko, O. M., and Shevchenko, V. A., Genetic consequences of radioactive contamination of populations of *Arabidopsis*, *Sci. Total Environ.,* 112, 9, 1992.
57. Krivolutzki, D. A. and Pokarzhevski, A. D., Effects of radioactive fallout on soil animal populations in the 30 km zone of the Chernobyl atomic power station, *Sci. Total Environ.,* 112, 69, 1992.
58. Zainullin, V. G., Shevchenko, V. A., Mjasnjankina, E. N., Generalova, M. V., and Rakin, A. O., The mutation frequency of *Drosophila melanogaster* populations living under conditions of increased background radiation due to the Chernobyl accident, *Sci. Total Environ.,* 112, 37, 1992.
59. Baker, R. J., Hamilton, M. J., van den Bussche, R. A., Wiggins, L. E., Sugg, D. W., Smith, M. H., Lomakin, M. D., Gaschak, S. P., Bundova, E. G., Rudenskaya, G. A., and Chessler, R. K., Small mammals from the most radioactive sties near the Chernobyl nuclear power plant, *J. Mammal.*, 77, 155, 1996.
60. Shevchenko, V. A., Pomerantseva, M. D., Ramaiya, L. K., Chekhovich, A. V., Testov, B. V., Genetic disorders in mice exposed to radiation in the vicinity of the Chernobyl nuclear power station, *Sci. Total Environ.*, 112, 45, 1992.
61. Baker, R. J., van den Bussche, R. A., Wright, A. J., Wiggins, L. E., Hamilton, M. J., Reat, E. P., Smith, M. H., Lomakin, M. D., and Chessler, R. K., High levels of genetic change in rodents of Chernobyl, *Nature*, 380, 707, 1996.
62. Baker, R. J., van den Bussche, R. A., Wright, A. J., Wiggins, L. E., Hamilton, M. J., Reat, E. P., Smith, M. H., Lomakin, M. D., and Chessler, R. K., Retraction: High levels of genetic change in rodents of Chernobyl, *Nature*, 390, 100, 1997.

63. Tsiperson, V. P., and Soloviev, M. Y., The impact of chronic radioactive stress on the immunophysiological condition of small mammals, *Sci. Total Environ.*, 203, 105, 1997.
64. Goncharova, R. I. and Ryabokon, N. I., Dynamics of cytogenetic injuries in natural populations of bank vole in the Republic of Belarus, *Radiat. Prot. Dos.*, 62, 37, 1995.
65. Cristaldi, M., Ieradi, L. A., Mascanzoni, D., and Mattei, T., Environmental impact of the Chernobyl accident: Mutagenesis in bank voles from Sweden, *Int. J. Radiat. Biol.*, 59, 31, 1991.
66. Rodgers, B. E. and Baker, R. J., Frequencies of micronuclei in bank voles from zones of high radiation at Chornobyl, Ukraine, *Environ. Toxicol. Chem.*, 19, 1644, 2000.
67. Matson, C. W., Rodgers, B. E., Chessler, R. K., and Baker, R. J., Genetic diversity of *Clethrionomys glareolus* populations from highly contaminated sites in the Chornobyl region, Ukraine, *Environ. Contam. Chem.*, 19, 2130, 2000.
68. Krapivko, T. P. and Il'enko, A. I., First features of radioadaptation in a population of red-backed voles (*Clethrionomys glareolus*) in a radioactive biogeocenosis, *Dokl. Akad. Nauk S.S.R.*, 302, 1272, 1988.
69. Il'enko, A. I., and Krapivko, T. P., Radioresistance of populations of bank voles, *Clethrionomys glareolus* in radionuclide-contaminated areas, *Dokl. Acad. Nauk S.S.R.*, 336, 714, 1994.
70. Il'enko, A. I., Mazheikite, R. B., Nizhnik, G. V., and Ryabtsev, I. A., Radiosensitivity of common redbacked voles inhabiting different geographic regions of European USSR, *Radiobiology*, 17, 95, 1977.
71. Dunaway, P. B. and Kaye, S. V., Studies of small-mammal populations on the radioactive White Oak Lake bed, in *Trans. 26th N. Am. Wildl. Conf.*, 1961, 167.
72. Dunaway, P. B. and Kaye, S. V., Effects of ionizing radiation on mammal populations on the White Oak Lake bed, in *Radioecology, Proc. of the First National Symp. on Radioecology,* Schultz, V. and Klement, A. W., Jr., Eds., Reinhold, New York, 1963, 333.
73. Kaye, S. V. and Dunaway, P. B., Estimation of dose rate and equilibrium state from bioaccumulation of radionuclides by mammals, in *Radioecology, Proc. of the First National Symp. on Radioecology,* Schultz, V. and Klement, A. W., Jr., Eds., Reinhold, New York, 1963, 107.
74. Childs, H. E. and Cosgrove, G. E., A study of pathological conditions in wild rodents in radioactive areas, *Am. Midl. Nat.,* 76, 309, 1976.
75. Kitchings, J. T., Dunaway, P. B., and Story, J. D., Blood changes in irradiated cotton rats and rice rats, *Radiat. Res.,* 42, 331, 1970.
76. Arthur, W. S., Markham, O. D., Groves, C. R., Keller, B. L., and Halford, D. K., Radiation dose to small mammals inhabiting a solid radioactive waste disposal area, *Appl. Ecol.,* 23, 12, 1986.
77. Evenson, L. M., Systemic effects of chronic radiation exposure on rodents inhabiting liquid and solid waste disposal areas, MS thesis, University of Idaho, Moscow, 1981.
78. Evenson, L. M., Olson, D. P., Halford, D. K., and Markham, O. D., Systemic effects of radiation exposure on rodents inhabiting liquid and solid radioactive waste disposal areas, in Ecological Studies on the Idaho Engineering Laboratory Site, Idaho Falls, Markham, O.D., Ed., U.S. DOE Report IDO 12087, U.S. Department of Energy, Washington, D.C., 1978, 99.
79. Mellinger, P. J., and Schultz, V., Ionizing radiation and wild birds: A review, *Critical Reviews in Environ. Control,* CRC Press, 1975, 397.
80. Ellegren, H., Lindgren, G., Primmer, C. R., and Moller, A. P., Fitness loss and germline mutations in barn swallows breeding in Chernobyl, *Nature*, 389, 593, 1997.
81. Millard, J. B., Whicker, F. W., and Markham, O. D., Radionuclide uptake and growth of barn swallows nesting by radioactive leaching ponds, *Health Phys.,* 58, 429, 1990.
82. Zach, R. and Mayoh, K. R., Gamma radiation effects on nestling tree swallows, *Ecology,* 65, 1641, 1984.
83. Templeton, W. L., Nakatani, R. E., and Held, E. E., Radiation effects, in *Radioactivity in the Marine Environment,* National Academy of Sciences, Washington, D.C., 1971, 223.
84. Blaylock, B. G. and Trabalka, J. R., Evaluating the effects of ionizing radiation on aquatic organisms, *Adv. Radiat. Biol.,* 7, 103, 1978.
85. Woodhead, D. S., Contamination due to radioactive materials, in *Marine Ecology: A Comprehensive, Integrated Treatise on Life in Oceans and Coastal Waters,* Kinne, O., Ed., John Wiley and Sons, Chichester, 1984, 1111.
86. Blaylock, B. G. and Witherspoon, J. P., Radiation doses and effects estimated for aquatic biota exposed to radioactive releases from LWR fuel-cycle facilities, *Nucl. Saf.*, 17, 351, 1976.
87. Blinks, L. R., Effects of radiation on marine algae, *J. Cell. Comp. Physiol.,* 39, 11, 1952.

88. Donaldson, L. R., Seymour, A. H., and Nevissi, A. E., University of Washington's radioecological studies in the Marshall Islands, 1946–1977, *Health Phys.*, 73, 214, 1997.
89. Oakes, T. W., Kelly, B. A., Ohnesorge, W. F., Eldridge, J. S., Bird, J. C., Shank, K. E., and Tsakeres, F. S., *Technical Background Information for the Environmental Safety Report,* Vol. 4: White Oak Lake and Dam, ORNL-5681, Oak Ridge National Laboratory, Oak Ridge, TN, 1982.
90. Rogers, J. G., Daniels, K. L., Goodpasture, S. T., and Kimborough, C. W., *Environmental Surveillance of the U.S. Department of Energy Oak Ridge Reservation and Surrounding Environs During 1987,* ES/ESH-4/V2, Oak Ridge National Laboratory, Oak Ridge, TN, 1988.
91. Blaylock, B. G., Chromosomal aberrations in a natural population of *Chironomus tentans* exposed to chronic low-level radiation, *Evolution,* 13, 421, 1965.
92. Blaylock, B. G., Chromosomal polymorphism in irradiated natural populations of *Chironomus, Genetics*, 53, 131, 1966.
93. Nelson, D. J. and Blaylock, B. G., The preliminary investigation of salivary gland chromosomes of *Chironomus tentans* Fabr. from the Clinch River, in *Radioecology,* Schultz, V. and Klement, A. W., Eds., Reinhold, New York, 1963, 367.
94. Auerbach, S. I., Nelson, D. J., and Struxness, E. G., *Environmental Sciences Division Annual Progress Report for Period Ending September 30, 1973,* ORNL-4935, Oak Ridge National Laboratory, Oak Ridge, TN, 1974, 34.
95. Blaylock, B. G., Chromosome aberrations in *Chironomus riparius* developing in different concentrations of tritiated water, in *Radionuclides in Ecosystems,* Vol. 2, Nelson, D. J., Ed., U.S. Atomic Energy Commission, Oak Ridge, TN, 1973, 1169.
96. Blaylock, B. G., The production of chromosome aberrations in *Chironomus riparius* (Diptera:Chironomidae) by tritiated water, *Can. Ent.,* 103, 448, 1971.
97. Cooley, J. L., Effects of chronic environmental radiation on a natural population of the aquatic snail *Physa heterostropha, Radiat. Res.,* 54, 130, 1973.
98. Cooley, J. L. and Miller, F. L., Jr., Effects of chronic irradiation on laboratory populations of the aquatic snail (*Physa heterostropha*), *Radiat. Res.,* 47, 716, 1971.
99. Frantsevich, L., Korniushin, A., Pankov, I., Ermakov, A., and Zakharchuk, T., Application of molluscs for radioecological monitoring of the Chernobyl outburst, *Environ. Pollut.*, 94, 91, 1996.
100. Fetisov, A. N., Rubanovic, A. V., Slichenko, T. S., and Shevchenko, V. A., The structure of *Dreissena polymorpha* populations from basins adjacent to the Chernobyl atomic power station, *Sci. Total Environ.*, 112, 115, 1992.
101. Blaylock, B. G., The fecundity of a *Gambusia affinis affinis* population exposed to chronic environmental radiation, *Radiat. Res.*, 37, 108, 1969.
102. Trabalka, J. R. and Allen, C. P., Aspects of fitness of a mosquitofish *Gambusia affinis* exposed to chronic low-level environmental radiation, *Radiat. Res.*, 70, 198, 1977.
103. Woodhead, D. S., The effects of chronic irradiation on the breeding performance of the guppy (*Poecilia reticulata*) (Osteichthyes:Teleostei), *Int. J. Radiat. Biol.*, 32, 1, 1977.
104. Theodorakis, C. W., Blaylock, B. G., and Shugart, L. R., Genetic ecotoxicology I: DNA integrity and reproduction in mosquitofish exposed *in situ* to radionuclides, *Ecotoxicology*, 6, 205, 1997.
105. Sugg, D. W., Bickham, J. W., Brooks, J. A., Lomakin, M. D., Jagoe, C. H., Dallas, C. E., Smith, M. H., Baker, R. J., and Chesser, R. K., DNA damage and radiocesium in channel catfish from Chernobyl, *Environ. Toxicol. Chem.*, 15, 1057, 1996.
106. Sugg, D. W., Chesser, R. K., Brooks, J. A., and Grasman, B. T., The association of DNA damage to concentrations of mercury and radiocesium in largemouth bass, *Environ. Toxicol. Chem.*, 14, 661, 1995.
107. Woodhead, D. S., The assessment of radiation dose to developing fish embryos due to the accumulation of radioactivity by the egg, *Radiat. Res.,* 43, 582, 1970.
108. Gorbman, A. M. and James, M. S., An exploratory study of the radiation damage in the thyroids of coral reef fishes from Eniwetok Atoll, in *Radioecology,* Schultz, V. and Klement, A. W., Eds., Reinhold and AIBS, Washington, D.C., 1963, 385.
109. Meyers-Schöne, L., Shugart, L. R., Beauchamp, J. J., and Walton, B. T., Comparison of two freshwater turtle species as monitors of radionuclide and chemical contamination: DNA damage and residue analysis, *Environ. Toxicol. Chem.,* 135, 93, 1993.
110. Koli, A. K., Whitmore, R., and Roache, W., Behavior of cesium in fish tissues from Savannah River, *Environ. Int.,* 11, 23, 1985.

111. George, L. S., Dallas, C. E., Brisbin, I. L., Jr., and Evans, D. L., Flow cytometric DNA analysis of ducks accumulating $^{137}$Cs on a reactor reservoir, *Ecotoxicol. Environ. Saf.,* 21, 337, 1991.
112. Bickham, J. W., Hanks, B. G., Smolen, M. J., Lamb, T., and Gibbons, J. W., Flow cytometric analysis of the effects of low-level radiation exposure on natural populations of slider turtles (*Pseudemys scripta*), *Arch. Environ. Contam. Toxicol.,* 17, 837, 1988.
113. Lamb, T., Bickham, J. W., Gibbons, J. W., Smolen, M. J., and McDowell, S., Genetic damage in a population of slider turtles (*Trachemys scripta*) inhabiting a radioactive reservoir, *Arch. Environ. Contam. Toxicol.,* 20, 138, 1991.
114. Meyers-Schöne, L., unpublished data, 1989.
115. Emery, R. M. and McShane, M. C., Nuclear waste ponds and streams on the Hanford site: An ecological search for radiation effects, *Health Phys.,* 38, 787, 1980.
116. Schell, W. R., Linkov, I., Myttenaere, C., and Morel, B., A dynamic model for evaluating radionuclide distribution in forests from nuclear accidents, *Health Phys*., 70, 318, 1996.
117. Thomas, P. A, Radionuclides in the terrestrial ecosystem near a Canadian uranium mill, Part I: Distribution and doses, *Health Phys*., 78, 614, 2000.
118. Antonopoulos-Domis, M., Clouvas, A., Xanthos, S., and Alifrangis, D. A., Radiocesium contamination in a submediterranean semi-natural ecosystem following the Chernobyl accident: Measurements and models, *Health Phys*., 72, 243, 1997.
119. Müller, H. and Pröhl, G., ECOSYS-87: A dynamic model for assessing radiological consequences of nuclear accidents, *Health Phys*., 64, 232, 1993.
120. St-Pierre, S., Chambers, D. B., Lowe, L. M., and Bontoux, J. G., Screening level dose assessment of aquatic biota downstream of the Marcoule Nuclear Complex in southern France, *Health Phys*., 77, 315, 1999.
121. Baker, D. A and Soldat, J. K., Methods for Estimating Doses to Organisms from Radioactive Materials Released into Aquatic Environments, PNL-8150, Pacific Northwest Laboratory, Richland, WA, 1992.
122. Blaylock, B. G., Frank, M. L., and O'Neal, B. R., *Methodology for Estimating Radiation Dose Rates to Freshwater Biota Exposed to Radionuclides in the Environment*, ES/ER/TM-78, Oak Ridge National Laboratory, Oak Ridge, TN, 1993.
123. U.S. Department of Energy, A Graded Approach for Evaluating Radiation Doses to Aquatic and Terrestrial Biota, DOE June 2000 Interim Technical Standard, U.S. Department of Energy, Washington, D.C., 2000.
124. Jones, D. S., Radiological benchmarks for effects on aquatic biota at the Oak Ridge Reservation, *Human Ecol. Risk Assess*., 6, 789, 2000.
125. Bechtel Jacobs Company LLC, Radiological Benchmarks for Screening Contaminants of Potential Concern for Effects on Aquatic Biota at Oak Ridge National Laboratory, Oak Ridge, Tennessee, BJC/OR-80, Oak Ridge National Laboratory, Oak Ridge, TN, 1998.
126. Higley, K. A., and Kuperman R.,. *Radiological Benchmarks for Wildlife a Rocky Flats Environmental Technology Site*, Work sponsored by Kaiser-Hill/Rocky Flats, CO, 1996.
127. IT Corporation, *Predictive Ecological Risk Assessment Methodology,* prepared for Sandia National Laboratories/New Mexico Environmental Restoration Program, Albuquerque, NM, 1998.
128. Whicker, F. W., Shaw, G., Voigt, G., and Holm, E., Radioactive contamination: State of the science and its application to predictive models, *Environ. Pollut*., 100, 133, 1999.
129. Thiessen, K. M., Hoffman, F. O., Rantavaara, A., and Hossain, S., Environmental models undergo international test, *Environ. Sci. Technol*., 31, 358A, 1997.
130. Thiessen, K. M., Thorne, M. C., Maul, P. R., Pröhl, G., and Wheater, H. S., Modelling radionuclide distribution and transport in the environment, *Environ. Pollut*., 100, 151, 1999.
131. Hinton, T. G., Sensitivity analysis of ECOSYS-87: An emphasis on the ingestion pathway as a function of radionuclide and type of deposition, *Health Phys*., 66, 513, 1994.
132. Jensen, L. D., The effects of thermal discharges into surface waters, in *Human and Ecologic Effects of Nuclear Power Plants*, Sagan, L. A. Ed., Charles C Thomas, Springfield, IL, 1974.
133. Wisniewski, J., Power plants and aquatic resources: Issues and assessment, *Environ. Sci. Policy,* 3 (Suppl. 1), 2000.
134. Patrick, R., Effects of abnormal temperatures on algal communities, in *Thermal Ecology,* Gibbons, J. W. and Sharitz, R. R., Eds., National Technical Information Service, Springfield, VA, 1974, 335.

135. Coutant, C. C., Temperature selection by fish — A factor in power-plant impact assessments, in *Environmental Effects of Cooling Systems at Nuclear Power Plants,* International Atomic Energy Agency, Vienna, 1975, 575.
136. Moss, J. L., Boonyaratpalin, S., and Shelton, W. L., Movement of three species of fishes past a thermally influenced area in the Coosa River, Alabama, in *Energy and Environmental Stress in Aquatic Systems,* Thorp, J. H. and Gibbons, J. W., Eds., CONF-771114, DOE Symposium Series 48, National Technical Information Service, Springfield, VA, 1978, 534.
137. Reash, R. J., Seegert, G. L., and Goodfellow, W. L., Experimentally derived upper thermal tolerances for redhorse suckers: Revised 316(a) variance conditions, *Environ. Sci. Policy,* S191, 2000.
138. Wrenn, W. B., Temperature preference and movement of fish in relation to a long, heated discharge channel, in *Thermal Ecology II,* Esch, G. W. and McFarlane, R. W., Eds., National Technical Information Service, Springfield, VA, 1976, 191.
139. Talmage, S. S. and Opresko, D. M., *Literature Review: Responses of Fish to Thermal Discharges,* EPRI-1840, Electric Power Research institute, Palo Alto, CA, 1981.
140. Barnthouse, L. W., Klauda, R. J., Vaughn, D. S., and Kendall, R. L., Eds., *Science, Law, and Hudson River Power Plants: A Case Study in Environmental Impact Assessment,* American Fisheries Society, Monograph 4, Bethesda, MD, 1988.
141. Barnthouse, L., Impacts of power-plant cooling systems on estuarine fish populations: The Hudson River after 25 years, *Environ. Sci. Policy,* 3, S341, 2000.
142. Dey, W. P., Jinks, S. M., and Lauer, G. J., The 316(b) assessment process: Evolution towards a risk-based approach, *Environ. Sci. Policy,* 3, S15, 2000.
143. Taft, E. P., Fish protection technologies: A status report, *Environ. Sci. Policy,* 3, S149, 2000.

# CHAPTER 21

# Global Effects of Deforestation

**Richard A. Houghton**

## CONTENTS



## 21.1 INTRODUCTION

For most of the last 10,000 years, since the development of settled agriculture, the environmental effects of deforestation have been scattered in time and space. Large areas of forest were cleared long ago in Mesoamerica, Europe, and parts of Asia and Africa, some of these clearings lasting to the present and others returning to forests as empires declined.[1] In the last few centuries, however, and particularly in the last several decades, the effects of deforestation have become global, not only in the sense that they occur almost everywhere on the earth, but in the sense that they contribute to global changes in climate through emissions of greenhouse gases. Changes in climate affect all peoples of the planet, whether or not they contributed to the changes. In this sense, the effects are similar to other forms of ecotoxicology: the unintended consequences are often borne by communities that did not benefit from the intended consequences. The consequences of deforestation extend far beyond the borders of forests and thus affect the global commons or, in other words, the public interest.[2] As with other elements of globalization, the balance of private and public rights needs attention.

Deforestation has always had consequences at local and regional scales. Locally, deforestation often increases the frequency and severity of floods and mudslides, increases the day-night range of temperature, reduces the capacity of soils to hold water, and may lead to soil erosion and the silting-in of downstream reservoirs, thus shortening the lifetime of hydroelectric dams. In addition, deforestation results in a loss of fuel, shelter, and other resources for local inhabitants. These effects are particularly troublesome for the people of developing nations, a great number of whom depend directly on the land for their survival. The major cause of deforestation, ironically, is the clearing of new agricultural land for greater production of food. One of the tragedies of deforestation in the tropics is that much of the agricultural land carved from forests is abandoned from production after a few years, having lost its fertility. In many locations, abandoned land may return to forest, but the worldwide increase in degraded lands shows that such reforestation is generally not taking place. Instead, these abandoned agricultural lands often remain degraded, eroded, or waterlogged, with an impoverished, low-statured vegetation of woody shrubs. The net effect of deforestation, with subsequent overuse of the land, is a reduction in the capacity of the earth to support human populations.

Regionally, the effects of deforestation include changes in temperature and moisture. For example, deforestation in the tropics leads to reduced rainfall and elevated temperatures within the region,[3] while deforestation in mid-latitudes leads to cooler temperatures.[4] The changes in surface energy and water budgets may extend beyond the regions deforested and thus have global effects,[5] similar, qualitatively at least, to El Niño and its equivalent in northern mid-latitudes, the North Atlantic Oscillation.

Other effects of deforestation are more clearly global. They include not only the conversion of potentially productive land to land with diminished capacity to support either crops, forests, or people, but also the irreplaceable loss of species, and emissions to the atmosphere of chemically active and heat-trapping trace gases, such as carbon dioxide, methane, nitrous oxide, and carbon monoxide. The last of these global effects is considered here — the effect of deforestation on the emissions of trace gases and, hence, on the earth's atmosphere and climate.

When forests are cleared and replaced with agricultural lands, the carbon originally held in the trees is released as carbon dioxide to the atmosphere, either immediately, if the forests are burned, or more slowly, as the unburned organic matter decays. Cultivation of forest soil also oxidizes organic carbon held in soil and releases it to the atmosphere as carbon dioxide. Reforestation reverses these fluxes of carbon. While forests are regrowing, they withdraw carbon from the atmosphere and accumulate it in trees and soil. Although deforestation itself may not release significant quantities of methane or nitrous oxide to the atmosphere, subsequent use of the land for cattle or other ruminant livestock, for paddy rice, or for agricultural production enhanced by nitrogen fertilizers does release these other heat-trapping gases.

This paper will review the contribution of deforestation and subsequent land use to the increasing concentrations of greenhouse gases in the atmosphere. It will also consider the reverse process: How might forests be managed so as to withdraw carbon from the atmosphere? And how much carbon might be accumulated if forests were reestablished on suitable lands?

## 21.2 GLOBAL WARMING

The scientific community is in agreement about most aspects of global warming. Two scientific assessments of climatic change, involving more than 400 scientists from 26 countries, have been

**Table 21.1 Characteristics of Greenhouse Gases and Their Relative Contributions to Global Warming over a 100-Year Time Horizon**

| Gas | 1990 Emissions (Tg) | Warming Effect of an Emission of 1 kg Relative to that of $CO_2$ | Atmospheric Lifetime (Years) | Relative Contribution over 100 Years |
|---|---|---|---|---|
| Carbon dioxide | 26,000[a] | 1 | 50–200[b] | 61% |
| Methane | 300 | 21[c] | 10 | 15% |
| Nitrous oxide | 6 | 290 | 130 | 4% |
| CFCs and HCFCs[d] | 1 | 1000's | 10's to 100's | 11% |
| Others[e] | | | | 9% |

[a] 26,000 Tg $CO_2$ = 7 Pg C.
[b] The broad range in lifetime for $CO_2$ results from uncertainties in the global carbon cycle and from the fact that the removal of $CO_2$ from the atmosphere depends not only on the amount but on the rate at which $CO_2$ is emitted to the atmosphere.
[c] Includes the indirect effects to concentrations of other greenhouse gases through chemical interactions in the atmosphere.
[d] The radiative properties and atmospheric lifetimes of specific CFCs and HCFCs are known precisely. Only order-of-magnitude averages for the more abundant gases are shown here.
[e] Principally tropospheric ozone, indirectly generated in the atmosphere as a result of other emissions.

*Source:* From Houghton, J. T. et al., *Climate Change 1995: The Science of Climate Change*, Cambridge University Press, Cambridge, U.K., 1996.

carried out by the Intergovernmental Panel on Climatic Change (IPCC),[6,7] and a third assessment was published in 2001.[7a] The IPCC was established by the World Meteorological Organization and the United Nations Environmental Programme. According to these IPCC assessments of climate change, scientists are certain, first, that there is a natural greenhouse effect, and, second, that emissions of carbon dioxide ($CO_2$), methane ($CH_4$), nitrous oxide ($N_2O$), and chlorofluorocarbons (CFCs) from human activities are increasing the concentrations of these greenhouse gases in the atmosphere. "There is new and stronger evidence that most of the warming observed over the last 50 years is attributable to human activities."[7a]

Evidence of the natural greenhouse effect comes from understanding the characteristics of greenhouse gases (Table 21.1), from satellite measurements of radiation at different levels within the earth's atmosphere, and from the high correlation between concentrations of $CO_2$ (and $CH_4$) and the surface temperature of the earth over the last 420,000 years. Without the natural greenhouse effect, the average temperature of the surface of the earth would be 33°C cooler, and life would be very different if it didn't exist at all. Evidence for the increased concentrations of greenhouse gases in the atmosphere comes from direct measurements in the atmosphere over recent decades and, over the last centuries, from measurements of air trapped in the ice of polar glaciers.

Evidence that the 30% increase in atmospheric $CO_2$ since preindustrial times is from human activities comes from several sources. First, the concentration of $CO_2$ fluctuated less than 10 parts per million by volume (ppmv) over the 1000 years preceding 1750. In less than 250 years the concentration has increased from about 285 ppmv to 363 ppmv (in 1997). Nowhere in the last 420,000 years has the concentration been as high as it is now. Second, the releases of $CO_2$ from combustion of fossil fuel and from changes in land use parallel the rise in atmospheric concentrations. Third, the north-south gradient in $CO_2$ concentrations is consistent with the distribution of emissions, predominantly in the northern mid-latitudes. And fourth, the observed changes in isotopes of atmospheric carbon are consistent with the known sources of $CO_2$ and with the global circulation of carbon as described by global carbon models.

The global warming potential of a gas can be calculated from three factors: the amount of gas emitted annually to the atmosphere, the radiative properties of the gas, and its average atmospheric lifetime. Table 21.1 gives these characteristics for the most important greenhouse gases and shows their expected contributions to global warming in the next century if current trends in emissions continue (the IPCC business-as-usual scenario).[6] Over the next 100 years, the relative contributions to a global warming are calculated to be 61% for $CO_2$, 15% for $CH_4$, 4% for $N_2O$, and 11% for CFCs. Other gases, principally tropospheric ozone, are expected to contribute about 9%.

Those greenhouse gases with long atmospheric lifetimes will continue to have a radiative effect for centuries after they are emitted to the atmosphere. The longer the emission rates continue to increase at present rates, the greater will be the reductions in emissions required to obtain a given atmospheric concentration. Reductions of more than 60% in emissions of $CO_2$, $N_2O$, and CFCs will be required immediately to stabilize atmospheric concentrations at today's levels.

This last point bears emphasis. It provides the strongest example of the existing gap between science and public affairs. The Kyoto Protocol calls for developed countries to reduce 1990 levels of emissions by an average of 5% by 2010. For the United States, the reduction is 7%. In fact, emissions in the United States were 11.2% higher in 1998 than they had been in 1990, so the reduction required for compliance with the Kyoto Protocol is no longer 7%, but 18.2%. In contrast to the emissions reductions agreed to in the Kyoto Protocol, however, the reductions required for stabilization of atmospheric concentrations at today's levels are greater than 60%. If $CO_2$ emissions were to stay constant at current levels, they would lead to a nearly constant rate of increase in atmospheric concentrations for at least two centuries, reaching about 500 ppmv by the end of the 21st century.[7]

The mean global surface temperature has increased by 0.3 to 0.6°C since the late nineteenth century, with an acceleration in the warming since 1980. The warmest years to date have been 1998 and 1997. The magnitude of this warming is consistent with the results of global climate models when they include the cooling effect of aerosols, but it is also consistent with natural variability. Nevertheless, the IPCC's latest assessment recognizes a discernible human influence on global climate.[7]

Scientists predict on the basis of General Circulation Models (GCMs) that in the business-as-usual scenario, the mean global temperature will increase by about 2°C (range 1–3.5°C) by the year 2100 (0.2°C per decade). This rate of warming is greater than experienced in the last 10,000 years. However, the rate will vary regionally and seasonally. More of the warming will occur at higher latitudes and in winter. Sea level is predicted to rise about 50 cm by the year 2100 (5 cm per decade with an uncertainty of 1.5 to 9.5 cm per decade). Under alternative scenarios that include some degree of control on emissions, rates of warming are predicted to be one half to one third the rate predicted on the basis of current trends. There are numerous uncertainties in the above predictions. The major ones are the sources and sinks of greenhouse gases, the role of clouds in affecting climate, the role of the oceans in delaying the timing of the warming, and the role of polar ice in affecting sea level and in amplifying the warming.

The prospects of a climatic change associated with business-as-usual may be described qualitatively as follows: the warming will be rapid, 10 to 60 times more rapid than the warming associated with the last retreat of the glaciers, about 10,000 years ago. The warming will be continuous; it will not stop unless deliberate steps are taken to halt it. The combination of a rapid and continuous change will make coping strategies difficult; adjustments will always lag behind the latest climate. The climate will not equilibrate at a doubled $CO_2$ concentration, as modeling experiments suggest, but will keep on changing. The reserves of fossil fuels are sufficiently large to release to the atmosphere on the order of 10 times more carbon than is currently in the atmosphere. Finally, an enhanced greenhouse warming will be irreversible within a human lifetime. The irreversibility results from the long atmospheric residence times of most of the greenhouse gases, on the order of a century or more, and from the thermal lag of the oceans, which commits the earth to a warming of about twice that realized at any point in time. There is no way to remove these gases rapidly from the atmosphere. Although the oceans will eventually absorb the excess $CO_2$ and processes in the atmosphere will eventually consume $N_2O$ and CFCs, reductions in atmospheric concentrations will require centuries.

## 21.3 EFFECTS OF DEFORESTATION ON SOURCES AND SINKS OF GREENHOUSE GASES

### 21.3.1 CARBON

#### *21.3.1.1 Carbon Stored in Vegetation and Soils*

The amount of carbon stored in the living plants of the earth is of the same order as the amount held in the earth's atmosphere. In the eighteenth century, 700–800 Pg (1 Pg = $1 \times 10^{15}$ g) carbon were held in the earth's vegetation, and about 600 PgC were in the atmosphere. Today, trees, grasses, and herbs are believed to hold about 500 PgC, and the atmosphere about 760 Pg. The increase in atmospheric $CO_2$ is accurately known; the decrease in terrestrial biomass is not as well known. The amount of organic carbon stored in the soils of the earth has also decreased (by about 3%) over the last 150 years as a result of cultivation and remains at 1100–1500 PgC. Terrestrial ecosystems (including both the living plants and soils) hold almost three times as much carbon as the atmosphere. Most of this terrestrial carbon is stored in forests. Forests cover about 30% of the land surface and hold almost half of the world's terrestrial carbon. If only vegetation is considered (soils ignored), forests hold about 75% of the living carbon.

In area, tropical forests account for slightly less than half of the world's forests, yet they hold about as much carbon in their vegetation and soils as temperate-zone and boreal forests combined. Considering only vegetation, undisturbed tropical forests hold, on average, about 65% more carbon per unit area than forests outside the tropics. When the carbon in soils is included, temperate zone and tropical forests are more nearly equal. Nevertheless, equivalent rates of deforestation in the two regions will cause considerably more carbon to be released from the tropics than from regions outside the tropics because only a fraction of the soil carbon is lost with cultivation.

The biomass of temperate zone and boreal forests is reasonably well known because foresters have been conducting forest inventories in northern countries for decades. Such inventories are rare in tropical countries, and the distribution of biomass throughout the tropics is poorly known. A recent comparison in the Brazilian Amazon showed that estimates of biomass for the entire region varied by more than a factor of two.[8] Uncertainties result from limited data on belowground biomass and variability in the biomass of trees smaller than those routinely sampled, in vines, in nontree components, in palms, in the shape and density of tree boles, and in the amount of woody debris on the forest floor. There is also considerable uncertainty in the spatial distribution of biomass over large areas. None of the five estimates compared in the Brazilian Amazon showed similar distributions of high- and low-biomass forests.[8]

Per unit area, forests hold 20 to 50 times more carbon in trees than the ecosystems that generally replace them (Table 21.2), and this carbon is released to the atmosphere as forests are transformed to other uses. In the tropics, much of the carbon is released immediately through burning. Afterwards, decay of soil organic matter and woody debris continues to release carbon to the atmosphere, but at lower rates. If croplands are abandoned, regrowth of live vegetation and redevelopment of soil organic matter withdraw carbon from the atmosphere and accumulate it again on land. To calculate the net flux of carbon from deforestation and reforestation, ecologists have documented the changes in carbon associated with different types of land use and different types of ecosystems in different regions of the world. Annual changes in the different reservoirs of carbon (live vegetation, soils, debris, and wood products) determine the annual net flux of carbon between the land and the atmosphere. Because of the variety of ecosystems and land uses, and because the calculations require accounting for cohorts of different ages, bookkeeping models have been developed for the calculations.[9–13]

Table 21.2 compares the relative losses of carbon as a result of converting forests to other uses. The losses in biomass range from 100% for permanently cleared land to 0% for nondestructive

**Table 21.2 Percent of Initial Carbon Stocks Lost to the Atmosphere when Tropical Forests are Converted to Different Kinds of Land Use**

| Land Use | Carbon Lost to the Atmosphere Expressed as % of Initial Carbon Stocks | |
|---|---|---|
| | Vegetation | Soil |
| Cultivated land | 90–100 | 25 |
| Pasture | 90–100 | 12 |
| Degraded croplands and pastures[a] | 60–90 | 12–25 |
| Shifting cultivation | 60 | 10 |
| Degraded forests | 25–50 | <10 |
| Logging[b] | 10–50 | <10 |
| Plantations[c] | 30–50 | <10 |
| Extractive reserves | 0 | 0 |

*Note:* For soils, the stocks are to a depth of 1 m. The loss of carbon may occur within 1 year, with burning, or over 100 years or more, with some wood products. Values are from Houghton et al.[9] unless otherwise indicated.

[a] Croplands and pastures, abandoned because of reduced fertility, may accumulate carbon, but their stocks remain lower than the initial forests.

[b] Based on current estimates of aboveground biomass in undisturbed and logged tropical forests.[61] When logged forests are colonized by settlers, the losses are equivalent to those associated with one of the agricultural uses of land.

[c] Plantations may hold as much or more carbon than natural forests, but a managed plantation will hold, on average, 1/3 to 1/2 as much carbon as an undisturbed forest because it is generally regrowing from harvest.[62]

harvest of fruits, nuts, and latex (extractive reserves). Losses of carbon from soil may also occur, especially if the soils are cultivated.

### *21.3.1.2 The Global Extent of Deforestation*

#### *21.3.1.2.1 Temperate and Boreal Forests*

According to the latest Forest Resources Assessment published by the Food and Agriculture Organization (FAO) of the United Nations,[14] the area of forest in developed countries (largely outside the tropics) increased by an average of $1.76 \times 10^6$ ha/year during the first half of the 1990s. This increase does not include the Russian Federation because the FAO could not determine a reliable estimate. Reported increases in forest area of $2.4 \times 10^6$ ha between 1988 and 1993 are largely a result of reclassifications and definition: forest lands in Russia are defined as lands that trees can grow on and that are not used for other purposes. One estimate finds that the area actually covered by closed canopy forests (more than 30–40% canopy cover) declined by $7.6 \times 10^6$ ha between 1988 and 1993.[15] However, in the 10 years between 1983 and 1993, only about 30% of the forests in Siberia were inventoried, and few of these areas were inventoried twice, so estimates of change are unreliable for the country with the largest share of the world's forests.

The average annual change of $1.76 \times 10^6$ ha included increases of 0.39, 0.76, 0.05, and $0.56 \times 10^6$ ha/year in the developed countries of Europe, North America, Oceania, and countries of the former Soviet Union (not including Russia), respectively, during the period 1990–1995.[14] In China, forest area experienced a net loss of $0.087 \times 10^6$ ha/year. Despite this net loss of forests, China has established the largest area of forest plantations: $34 \times 10^6$ ha by 1995. As mentioned above, it is not clear whether changes in Russian forests would have increased or decreased the average increase of $1.76 \times 10^6$ ha/year in developed (largely temperate-zone and boreal) countries.

**Table 21.3 Average Annual Rates of Deforestation ($10^6$ ha $yr^{-1}$) in Developing (Largely Tropical) Regions**

| | 1980–1990 | 1990–1995 |
|---|---|---|
| Africa | 4.28 | 3.75 |
| Asia-Oceania | 4.41 | 4.17 |
| Latin America and Caribbean | 6.77 | 5.81 |
| Developing world | 15.46 | 13.73 |

*Source:* From FAO, *State of the World's Forests 1997,* FAO, Rome, 1997.

### *21.3.1.2.2 Tropical Forests*

In contrast to the modest increase in developed countries, the average rate of deforestation in developing (largely tropical) countries was $13.7 \times 10^6$ ha/year during the first half of the 1990s (Table 21.3), down somewhat from the $15.5 \times 10^6$ ha/year rate for the 1980s. The recent rate is equivalent to clearing an area about the size of Georgia or Wisconsin each year. The highest rates (in $10^6$ ha/yr) were reported in Brazil (2.554), Indonesia (1.084), Zaire (0.740), Bolivia (0.581), Mexico (0.508), and Venezuela (0.503). For all developing countries, the annual rate of forest loss was about 0.65% of forest area. Relative rates of loss are somewhat smaller in tropical Latin America (0.62%/year) and largest in tropical Asia (0.98%/year). The rates of deforestation in developing countries were offset to some extent by the establishment of forest plantations. Tropical countries reporting the largest areas of plantations in 1995 were India, Indonesia, and Brazil, with 15, 6, and $5 \times 10^6$ ha, respectively. Natural forests and plantations are not readily distinguished in developed countries, and so were combined in the discussion of developed countries above.

The errors in the rates of deforestation reported by the FAO are unknown. The most recent assessment by the FAO[14] revised the 1990 estimate of deforestation from 16.3 to $15.5 \times 10^6$ ha/year (5%) as a result of new estimates of forest cover in 1980 and 1990. At the other extreme, preliminary national communications from Bolivia and Zimbabwe reported rates of deforestation six times less than reported by the FAO.[16] Other countries were more similar to, or used, the FAO estimate. Mexico reported credible rates that varied between 0.370 and $0.858 \times 10^6$ ha/year, a range that is approximately 80% of the mean. It is possible to lower the uncertainty in average rates of deforestation with the use of high-spatial-resolution data from satellites, such as Landsat. However, two estimates of the total area deforested in the Brazilian Amazon, both based on data from Landsat, differed by 25%.[17] The reasons for the difference have not been fully resolved.

One fate of land in the tropics bears special attention from the perspective of deforestation. Most deforestation is thought to be for agriculture. The expansion of human settlements involves relatively little area, and logged areas, if the logging has not been too destructive and if the area is not colonized by farmers, generally return to forests. However, despite the fact that most deforestation is for some form of agriculture, the annual net expansion of agricultural lands is considerably less than the annual net reduction in forest area.[18] For the entire tropics, for example, the expansion of croplands accounted for only 27% of total deforestation. Adding the increase in pasture area accounted for an additional 18% of deforestation. Fully 55% of the deforestation between 1980 and 1985 was explained by an increase in "other land."[19] Although some of this "other land" is urban land, roads, and other settled lands, these uses are unlikely to have accounted for more than a few percent of the area deforested. Most of the increase in "other land" seems likely to be abandoned, degraded croplands and pastures, lands that no longer support crop or livestock production but that do not revert readily to forest, either.

Forests are not converted directly to degraded areas, of course. The transformation of land is from forest to agriculture and, subsequently, to degraded land. The important point is that only about one half of the area of tropical forest lost each year actually expands the area in agriculture. The other half is only temporarily useful. After a few years it is lost, neither agriculturally productive nor forested. If these estimates are correct, making agriculture sustainable, so that agricultural lands can be farmed continuously, may be as effective in halting deforestation as increasing the yields of crops.

The fraction of deforestation used to expand the area in agriculture, as opposed to replacing worn-out land, varies among tropical regions. In Africa, the expansion of croplands accounted for only about 12% of the net area deforested. Eighty-eight percent of the decrease in forest area was matched by the expansion of "other land." In tropical Asia, 40% of the net reduction in forests appeared as an expansion of agricultural lands. In Latin America, about two thirds of the reduction in forests could be accounted for by the expansion of croplands and pastures. If agriculture could be made sustainable throughout the tropics, rates of deforestation could be reduced by about 50% without reducing the expansion of agricultural lands, and large areas of marginal or degraded lands might be reforested

#### *21.3.1.3 A Broader Definition of Deforestation: Land-Use Change*

Because this review is concerned with the effects of forests and forest management on emissions of greenhouse gases, particularly carbon dioxide, it is important to consider changes in the carbon content, or stature, of forests and not just changes in their area. The carbon content of forests, in MgC/ha (1 Mg = $10^6$ g), varies geographically as a function of climate and soil fertility. It also varies temporally as a result of disturbances, including those attributable to human activities, such as logging or shifting cultivation, and as forests respond to management, for example, silviculture or fire suppression (Table 21.2).

##### *21.3.1.3.1 Temperate and Boreal Forests*

In temperate and boreal forests, the small net changes in area do not reflect the changes in carbon stocks. Estimates of recent changes in the carbon content of temperate and boreal zone forests suggest that forests were accumulating 0.7 to 0.8 PgC/yr (1 Pg = $10^{15}$ g) during the 1980s and 1990s.[20–22] The factors responsible for this net accumulation are uncertain. Changes in land use are not thought to have been responsible because releases of carbon from decay of logging debris and wood products approximately balanced accumulations of carbon in growing forests.[9,23] Thus, the accumulation of carbon measured in forest inventories has been attributed largely to changing environmental factors or to recovery from past natural disturbances rather than from human activity (land-use change).

A more recent analysis of the United States alone estimated a net uptake (sink) of 0.15 to 0.35 PgC/year,[24] but less than half of this sink was in forests, and only 25% of the forest sink was attributed to past changes in land use or management. Although forests in the eastern U.S. were accumulating carbon as a result of fire suppression and the abandonment of agricultural lands, increased rates of logging in the northwest were responsible for a net release of carbon in that region. Another recent analysis of forests in the U.S. found that almost all of the carbon accumulating could be attributed to the age structure of the forests, that is, to past disturbances or management rather than to enhanced growth rates.[25] Understanding the reasons for the accumulation of carbon is more than academic. If the present accumulation of carbon is the result of environmental changes (for example, increased concentrations of $CO_2$ in the atmosphere, deposition of nitrogen, changes in climate), then the sink may be expected to continue until other factors limit forest growth. On the other hand, if the current sink is the result of past land-use and management practices, a continuation of this sink into the future is limited, because, once regrown, forests no longer accumulate carbon.

There are other reasons to suspect that the current terrestrial sink may not continue in the future. Indeed, reductions in forest area could be self-amplifying. One of the effects of a global warming, for example, is likely to be an increased rate of respiration (including decomposition of soil organic matter).[26,27] Increased emissions of respiratory $CO_2$ and $CH_4$, in turn, would increase the warming, and the warming might then be beyond human control. Perhaps relatedly, the areas burned annually in Canadian and U.S. forests have increased in recent years. Earlier fire suppression allowed carbon to accumulate in both live and dead plant material, but in the last decades, the frequency of fires in Canada seems to have increased dramatically.[28] The year 2000 was also a year in which an unusually large area of forest burned in the U.S. Whether the change is related to global warming and whether it will continue in the future are unknown at present. The net effect of numerous other global changes currently underway is difficult to predict. The danger, of course, is that the changes will reduce the capacity of the earth to support life and, further, that the changes will initiate processes in the earth's climate system that are irreversible for generations. For the present, the net flux of carbon into or out of northern forests is small relative to the emissions of carbon from deforestation in the tropics. However, these possible feedbacks are not included in any of the models used to predict future climatic change.

#### *21.3.1.3.2 Tropical Forests*

There is considerable evidence that carbon is being lost from tropical forests in addition to the losses resulting from outright deforestation. Selective logging, shifting cultivation, and grazing are widespread within tropical forests, and the increasing practice of these activities for both subsistence and economic interests is reducing the standing stock of carbon in trees.[29–35]

Whether the carbon content is changing in response to factors other than direct human activity is as uncertain in tropical forests as it is in temperate-zone and boreal forests. In fact, changes are more difficult to evaluate in the tropics. In contrast to the national forest inventories of temperate-zone countries, tropical forests have rarely been inventoried. Direct measurement of change is limited to small areas where permanent plots are repeatedly sampled. An analysis of these plots suggests that growth rates may have increased in South America but not in tropical Africa or Asia.[36] The finding may be artificial, however, and related more to methods of measurement than to real change.[37] Evidence for an accumulation of carbon in undisturbed forests also comes from a small number of sites where the flux of $CO_2$ has been measured over a forest. Measurements suggest that undisturbed tropical forests are indeed a sink for carbon.[38,39] However, the method is sensitive to atmospheric conditions, which differ day and night. Because the fluxes also vary day and night, the possibility for a systematic bias in measurement is strong. Indeed, a positive correlation between nighttime respiration and wind speed seems to confirm a methodological, rather than a physiological, basis for variations in respiration and, hence, net uptake. If only those nights with a high wind speed are considered, the calculated net flux of carbon is a source, rather than a large sink (M. Goulden, personal communication).

### ***21.3.1.4 Emissions of Carbon to the Atmosphere Caused by Land-Use Change***

The net flux of carbon to the atmosphere from deforestation and reforestation (in the broad sense, including activities affecting biomass within forests) are calculated from two types of information: historic and current reconstructions of land-use change (for example, annual rates of forest conversion to croplands and annual rates of wood harvest) and a knowledge of the per-hectare stocks of carbon in vegetation and soils and changes in these stocks as a result of land-use change. Between 1850 and 1990, about 120 PgC are calculated to have been released globally into the atmosphere from changes in land use.[40] This value is a net flux; it includes the uptake of carbon in forest growth following harvests as well as the releases of carbon from burning and decay. The total net flux from changes in land use is approximately half of the amount of carbon emitted from combustion of fossil fuels over this period. Before the first part of the twentieth century, the annual net flux from land-use change was greater than annual emissions from fossil fuels. Almost two thirds of the carbon released from terrestrial ecosystems has come from tropical lands; one third has come from temperate and boreal lands. About 90% of the net flux has been from forests, most of the rest arising from cultivation of mid-latitude prairies.

**Table 21.4 Relative Contribution of Deforestation to the Anthropogenic Radiative Effect in the 1980s Relative to Preindustrial Times**

| Gas | Contribution of the Gas to Radiative Forcing[a] | Annual Emissions | Deforestation as Percentage of Total Emissions | Deforestation as Percentage of Radiative Forcing |
|---|---|---|---|---|
| Carbon dioxide | 60% | Pg C | | |
| Industrial | | 5.6 | | |
| Natural | | 0 | | |
| Deforestation | | 2.0 | 25% | 15% |
| Total | | 7.6 | | |
| Methane | 20% | Tg[b] $CH_4$ | | |
| Industrial | | 100 | | |
| Natural | | 150 | | |
| Deforestation | | 250 | 50% | 10% |
| Total | | 500 | | |
| Nitrous oxide | 4% | Tg[b] $N_2O$ | | |
| Industrial | | 1.3 | | |
| Natural | | 7.2 | | |
| Deforestation | | 2.0 | 20% | 1% |
| Total | | 10.5 | | |
| CFCs and HCFCs | 16% | Gg[c] CFC | | |
| Industrial | | 1.0 | | |
| Natural | | 0 | | |
| Deforestation | | 0 | 0% | 0% |
| Total | | 1.0 | | |
| | 100% | | | 26% |

[a] Data from Houghton, J.T. et al., in *Climate Change 2001: The Scientific Basis*, Cambridge University Press, Cambridge, U.K., 2001. With permission.
[b] 1 Tg = $10^{12}$ g.
[c] 1 Gg = $10^9$ g.

Estimates of the current (1980–1995) net flux from land-use change range between 1.6 and 2.4 PgC/year.[17,24,40,41] Outside the tropics, estimates of the current net flux of carbon from land-use change are near zero, or slightly negative (carbon accumulating on land). The areas deforested or reforested are relatively small, and most of the flux results from earlier abandonment of croplands, fire suppression, and logging, with associated regrowth. More than 90% of the global flux of carbon is currently from deforestation in the tropics, especially in Asia and Latin America.[40,41] The countries with the largest sources of carbon are those with the highest rates of deforestation: Brazil, Indonesia, Zaire, Bolivia, Mexico, and Venezuela. Deforestation, in the strict sense, accounts for about 75% of the net flux from the tropics, but reductions in the carbon content of forests (through unsustainable logging and shifting cultivation) are estimated to account for more than 50% of the net carbon loss in some regions of Africa.[35]

If the net release of carbon from deforestation (the broader definition) was 2.0 PgC/yr (midpoint of the range of estimates), deforestation accounted for about 25% of total carbon emissions and about 15% of the enhanced greenhouse effect during the 1980s (Table 21.4). Notice that the relative contributions of different greenhouse gases in Table 21.4 are not the same as those in Table 21.1. Table 21.1 is a prediction for the next 100 years; Table 21.4 is for the 1980s.

Will the management of forests contribute significantly to averting climatic change, or will the next decades see an increase in the rate of deforestation and higher emissions of carbon? It is impossible to know, but one can ask how much carbon might be released into the atmosphere if current trends continue, or how much carbon might be withdrawn from the atmosphere and accumulated on land if incentives favored a reverse of current trends in land use.

On the assumption that current rates of deforestation would continue (or increase), the release of carbon over the next 100 years would range between 120 PgC and 335 PgC.[42] In fact, forests will have been eliminated before 2100 in many countries if today's rates continue. These emissions are equivalent to, and about three times greater than, the amount of carbon estimated to have been released from worldwide deforestation in the last 135 years.

On the other hand, if forests were to be reestablished in areas now degraded, and if woody plants were to be introduced into agricultural lands (agro-forestry), tropical lands might be managed to remove 60 to 160 PgC from the atmosphere over the next 50 to 100 years.[43–45] Peak annual accumulations might reach as much as 1.5 to 3 PgC/year for a few years. Allowing secondary forests to recover to full biomass would also contribute to removing carbon from the atmosphere.

The accumulation of carbon through reforestation, although useful in the short term to withdraw carbon from the atmosphere, does not represent a permanent solution. Once forests have grown to maturity, they are essentially in balance with respect to carbon. They continue to hold the carbon they have accumulated, but the rate of accumulation is low. The most effective use of forests from the perspective of atmospheric carbon dioxide is to substitute wood energy for fossil fuel energy. Gross emissions of carbon will be similar, but as long as next year's wood supply is accumulating in growing forests, the emissions of carbon from combustion of wood will be balanced by these accumulations. Fossil fuels, of course, accumulate much too slowly to be used sustainably in this way.

Future projections can be summarized by the six strategies shown in Table 21.5. The strategies are shown for comparative purposes; all six assume that emissions of carbon from fossil fuels will remain constant at 1990 rates (6.0 PgC/year). In fact, emissions have increased over the 1990s. The first strategy represents the current rate of deforestation and fossil fuel use. A total of 8 PgC are being released each year to the atmosphere, mostly as $CO_2$. The second strategy shows the

**Table 21.5 Potential Annual Fluxes of Carbon (PgC/yr) to or from the Atmosphere from Human Activities**

| Strategy | Fossil Fuel | Deforestation | Reforestation* | Sustainable Harvest of Fuel | Total | Potential Reduction from Present |
|---|---|---|---|---|---|---|
| I | 6.0 | 2.0 | <0.1 | 0 | 8 | 0% |
| II | 6.0 | 2.0 | −1 to −2 | 0 | 6 to 7 | 12 to 25% |
| III | 6.0 | 0.0 | <0.1 | 0 | 6 | 25% |
| IV | 6.0 | 0.0 | −1 to −2 | 0 | 4 to 5 | 35 to 50% |
| V | 0 | 0 | 0 | 0 | 0 | 100% |
| VI | 0 | 0 | −1 to −2 | 0 | −1 to −2 | 112 to 125% |

*Note:* Positive values indicate emissions; negative values indicate a removal of carbon from the atmosphere.

I This strategy represents current (1990) emissions of carbon.

II Reforestation: 100 to 200 PgC might be stored in new plantations, in forests protected from further logging and shifting cultivation, and in agroforestry. The accumulations of carbon are assumed to take place over 100 years.

III No deforestation.

IV Massive reforestation and a halt to deforestation.

V Replacing fossil fuels with wood-based fuels grown sustainably (this assumes the world energy consumption will not increase substantially above 1990 rates). Combustion of wood fuels emits as much or more carbon to the atmosphere as combustion of fossil fuels. The emissions from wood, however, are balanced by accumulations of carbon in the forests growing to provide the next years' fuel.

IV All options are put into effect at once. It is very unlikely in this case, where fuelwood plantations are supplying the world's energy needs, that there will be additional land available to accumulate carbon in new forests.

• This annual accumulation of carbon in growing forests persists only while the forests are growing or while new lands are being reforested. Once forests have matured, in 15 to 100 years, they continue to hold carbon, but they no longer withdraw it from the atmosphere.

effect of massive reforestation: an average of about 1.5 PgC/year might be withdrawn from the atmosphere over the next 100 years and accumulated on land, thereby reducing total annual carbon emissions by 12 to 25%. Halting deforestation (strategy III) has a larger effect; it would reduce total annual emissions by 25%. Halting deforestation and increasing the area of forests through reforestation would reduce current emissions by 35 to 50% (strategies II and III together = strategy IV). Both the efforts together still do not reduce emissions by the 60% necessary for stabilization of atmospheric $CO_2$ concentrations at today's levels. Furthermore, as mentioned above, reforestation is a temporary solution for reducing atmospheric concentrations of $CO_2$. Reductions of emissions by 60% or more require halting deforestation and substituting wood-derived fuels for fossil fuels (strategy V). If massive reforestation were to occur as well (strategy VI), worldwide emissions of carbon would be negative. Only strategies V and VI will halt the rise in atmospheric $CO_2$, and these estimates are wildly optimistic. Clearly, the major reductions in emissions will have to come from a reduced use of fossil fuels.

### 21.3.2 Methane, Nitrous Oxide, and Carbon Monoxide

The major emissions of $CH_4$, $N_2O$, and CO from deforestation do not necessarily occur with the deforestation itself but with subsequent use of the land. For example, paddy rice, cattle, and biomass burning account for 20, 15, and 8%, respectively, of the total annual emissions of $CH_4$, and the burning associated with deforestation releases little of this $CH_4$. Similarly, most $N_2O$ is released from land in the months and years following deforestation, especially if nitrogen fertilizers are applied to enhance or maintain productivity.

Part of the carbon emitted to the atmosphere as a result of deforestation is released through burning, but the majority is released through decay.[32] Biomass burning, on the other hand, is a major source of atmospheric $CH_4$, CO, and other chemically reactive gases that are important because they contribute either directly or indirectly to the heat balance of the earth. The accumulation of $CH_4$ in the atmosphere contributed about 20% of the enhanced radiative forcing in the decade of the 1980s; the contribution from $N_2O$ was about 4% (Table 21.4). While CO is not radiatively important itself, it reacts chemically with hydroxyl radicals (OH) in the atmosphere, thereby affecting the concentration of $CH_4$.

#### *21.3.2.1 Methane*

A small fraction (0.5 to 1.5%) of the carbon released into the atmosphere during biomass burning is $CH_4$. The radiative effect of a molecule of $CH_4$, however, is 25 times that of a $CO_2$ molecule, so if as much as 4% of the carbon were emitted as $CH_4$, the radiative effects of the $CO_2$ and $CH_4$ emissions would be equal in the short term. Because the average residence time of $CH_4$ in the atmosphere is only about 10 years, while that of $CO_2$ is 50 to 200 years, the long-term radiative forcing of $CO_2$ is larger than that for $CH_4$.

If the ratio of $CH_4/CO_2$ emitted in fires associated with deforestation is 1%, and if 40% of the emissions from deforestation results from burning,[46] then only about 10 Tg (1 Tg = $10^{12}$ g) of $CH_4$ were emitted to the atmosphere directly from deforestation. But this flux is based on the net flux of $CO_2$. Gross burning is estimated to release about 40 Tg $CH_4$ annually from burning of pastures, grasslands, and fuelwood.[47] In addition, about 80 Tg $CH_4$ are released from cattle ranching, and about 60 Tg are released from rice cultivation. Because some of these releases are from lands never forested, the contribution from deforestation is somewhat less than the total release of 260 Tg $CH_4$. Overall, about half of the global emissions of $CH_4$ results directly and indirectly from deforestation. More than half of this flux is a result of tropical deforestation and land use, with a lesser fraction released from outside the tropics. The expansion of wetlands through flooding of forests for hydroelectric dams could become a significant new source of $CH_4$ in the future.

### 21.3.2.2 Nitrous Oxide

Nitrous oxide ($N_2O$) is also a biogenic gas emitted to the atmosphere following deforestation. Small amounts of $N_2O$ are released during burning, but most of the release occurs in the months following a fire, especially in new pastures. Fire affects the chemical form of nitrogen in soils and, as a result, favors a different kind of microbial activity (nitrification). One of the by-products of nitrification is the production of NO and $N_2O$.

Estimates of the global emissions of $N_2O$ are uncertain. Industrial sources are thought to contribute about 1.3 Tg $N_2O$-N per year as a result of fossil fuel combustion and the production of adipic acid (nylon) and nitric acid. Biomass burning is estimated to release 0.2 to 1.0 Tg $N_2O$-N per year, and cultivated soils are estimated to release between 0.03 and 3.0 Tg. The role of soils is very uncertain. They may be both a major sink and a major source of atmospheric $N_2O$. Fertilized soils may release ten times more $N_2O$ per unit area than undisturbed soils, and the soils of new pastures may release even higher amounts. Deforestation may be responsible for about 20% of the global increase in $N_2O$ concentrations (Table 21.4).

### 21.3.2.3 Carbon Monoxide

Carbon monoxide (CO) is not a greenhouse gas, but it affects the oxidizing capacity of the atmosphere through interaction with OH and thus indirectly affects the concentrations of other greenhouse gases such as $CH_4$. Increased concentrations of CO in the atmosphere will deplete the concentrations of OH, leave less of the radical available to break down $CH_4$, and thereby increase the concentration and atmospheric lifetime of $CH_4$. Carbon monoxide emissions are generally 5 to 15% of $CO_2$ emissions from burning, depending on the intensity of the burn. More CO is released during smoldering fires than during rapid burning or flaming. The burning associated with deforestation may thus release 40 to 170 Tg C as CO. In addition, the repeated burning of pastures and savannas in the tropics is estimated to release 200 Tg C, as CO.[48] Together, these emissions from the tropics are as large as industrial emissions.

The emissions of $CO_2$, $CH_4$, and $N_2O$ from tropical deforestation and from subsequent use of the land are shown in Table 21.4. Summing the emissions and taking into account their relative contributions to radiative forcing in the decade of the 1980s[6] shows that deforestation directly and indirectly accounts for about 25% of the radiatively active emissions globally. Most of the emissions of $CO_2$ from deforestation are from the tropics, but significant emissions of $CH_4$ and $N_2O$ result from land use outside the tropics.

## 21.4 OPTIONS FOR INCREASING TERRESTRIAL CARBON STORAGE

About 84 nations have signed (although not ratified) the Kyoto Protocol, which stipulates that industrial countries are to reduce their emissions of greenhouse gases by an average of 5% below their 1990 emissions by the first commitment period, 2008–2012. The Protocol specifically recognizes that sources and sinks of carbon attributable to afforestation, reforestation, and deforestation (ARD) are to be counted in achieving these emissions reductions. It also recognizes that sources and sinks of carbon from other activities besides ARD may be counted but does not specify which ones.

The recognition by the Kyoto Protocol that certain terrestrial carbon sinks may be counted in meeting industrialized countries' emissions reductions has met with considerable criticism. Many European nations and some environmentalists argue that counting terrestrial sinks will take the pressure off the more critical need to reduce emissions of carbon from fossil fuels. The United States, Canada, and others, on the other hand, do not want sinks limited to afforestation and reforestation because there are many other activities and management options that may be used to sequester carbon. The recently published IPCC Special Report on Land Use, Land-Use Change, and Forestry[49] provides detailed scientific and technical advice on the feasibility and consequences of defining, counting, and measuring sources and sinks of carbon resulting from ARD and other activities. The issues are complex, both scientifically and politically. The Sixth Conference of the Parties in The Hague in November 2000 failed to reach agreement on how to include terrestrial sinks.

In answer to the question of which sources and sinks should be counted, there are almost as many answers as there are parties to the UN Framework Convention on Climate Change (UNFCCC). The final section of this review addresses this question from the perspective of what is known about the global carbon cycle. The emphasis is on carbon because carbon dioxide has been and is expected to be the most important of the greenhouse gases resulting from human activities. Furthermore, carbon dioxide is the only greenhouse gas that can be accumulated on land and, therefore, withheld, at least temporarily, from the atmosphere. Five broad options are considered:

- Count only those sources and sinks of carbon that result from human activity
- Count only those resulting directly from human activity
- Count only those resulting from afforestation, reforestation, and deforestation
- Count all terrestrial sources and sinks, whatever the cause
- Count no terrestrial sources and sinks of carbon at all

### 21.4.1 Count Only Those Sources and Sinks of Carbon that Result from Human Activity

The first option considered is to include all sources and sinks of carbon that are the result of human activity. This option appears ideal because the UNFCCC and Kyoto Protocol are concerned with reporting and counting anthropogenic sources and sinks. Anthropogenic sources and sinks are defined as those "that are a direct result of human activities or are the result of natural processes that have been affected by human activities."[50] The latter category may also be called "indirect" anthropogenic effects. Sources and sinks not the result of human activity are, by definition, outside of human control and thus not to be penalized or rewarded.

The difficulty with this option is that distinguishing anthropogenic from nonanthropogenic sources and sinks is scientifically difficult. There are at present no direct methods at any scale that enable the net uptake of carbon in forests to be divided between anthropogenic and nonanthropogenic causes. The difficulty in distinguishing anthropogenic and nonanthropogenic factors pertains to almost all measurements of net carbon balance, whether through direct measurement of wood volumes in forest inventories, through direct measurement of $CO_2$ flux over a forest,[38,39] or through model inversions based on atmospheric $CO_2$ concentrations and simulated atmospheric transport.[51–53] If logging or agricultural abandonment were the only activities that led to forest growth, measured growth could be attributed to human activity. But the regrowth of forests may also result from natural disturbances (fire, insects, disease) that are not attributable to human influence, and rates of growth may vary in response to both human-induced and natural variations in the environment (atmospheric $CO_2$, nitrogen deposition, climate).

The global carbon cycle may be summarized by the following equation, where sources (+) and sinks (–) of carbon, in PgC year$^{-1}$, are annual averages for the period 1980–1989:

$$\begin{array}{c}\text{Atmospheric}\\ \text{increase}\end{array} = \begin{array}{c}\text{Fossil}\\ \text{fuel}\\ \text{source}\end{array} + \begin{array}{c}\text{Net source}\\ \text{from changes}\\ \text{in land use}\end{array} - \begin{array}{c}\text{Oceanic}\\ \text{sink}\end{array} - \begin{array}{c}\text{Residual}\\ \text{terrestrial}\\ \text{sink}\end{array}$$

$$3.3(\pm 0.2) = 5.5(\pm 0.5) + 2.0(\pm 0.8) - 2.0(\pm 0.8) - 2.2(\pm 1.3) \qquad (21.1)$$

The net terrestrial sink is only 0.2 PgC year$^{-1}$, resulting from a 2.0 PgC year$^{-1}$ source directly attributable to human activity (deforestation in the broad sense) and a 2.2 PgC year$^{-1}$ (residual) sink attributable to either indirect anthropogenic effects or nonanthropogenic effects. The magnitude of the sink is large — as large as the source attributed to land-use change. Some countries would like to be able to claim credit for this sink, but others insist that credit should only result from anthropogenic-induced changes, in other words, from deliberate management.

The distinction between anthropogenically induced and nonanthropogenically induced carbon storage might be demonstrated with controlled experiments for some processes, such as the accumulation of carbon in soils under alternative management practices. For forest growth, however, controlled experiments would be difficult to carry out and interpret. A recent analysis used successive measurements of wood in U.S. forests to show that the accumulation of carbon could be attributed to the age structure of the forests rather than to changing environmental conditions that may have enhanced growth.[25] But the age structure results from both natural disturbances and management — for example, agricultural abandonment or fire suppression — so a direct anthropogenic cause is not clear from the analysis. Distinctions between the factors responsible for accumulations of carbon might be determined in the future with process-based ecosystem models, but the current variability in the predictions of such models suggests that their use is not yet reliable enough for attribution.

The distinction between anthropogenic and nonanthropogenic sinks can be made politically, of course, and because of the incentives, countries will want to claim all sinks as anthropogenic. The accumulation of carbon in a forest, for whatever reason, can be defined as anthropogenic simply by declaring the forest is managed. Management is a direct anthropogenic activity, and the decision to do nothing (for example, the decision not to suppress fires) is a management decision. Credit might even be claimed for terrestrial sinks resulting from $CO_2$ fertilization, i.e., from fossil fuel $CO_2$ emissions. While there is a certain logic to this claim (higher concentrations of $CO_2$ enhance forest growth and thus store more carbon), again, it is difficult, at present, to determine the fraction of observed growth attributable to elevated $CO_2$. To avoid a bias in counting sinks, and not sources, policies might require that all lands counted as sinks be counted in subsequent commitment periods, whether or not they are still functioning as sinks.

Overall, although this option may be logically the most appealing in terms of counting the appropriate sources and sinks, it is also the most difficult to implement scientifically (that is, it contains the largest number of uncertainties). All or none of the 2.2 PgC/year residual sink might be anthropogenic and, by this option, countable (Table 21.6).

It is worth recognizing that sources and sinks of carbon need not be debited and credited to the countries in which they occur. If reliable methods are eventually developed for distinguishing anthropogenic and nonanthropogenic changes in carbon storage, nonanthropogenic changes might be added to a global pool to be divided proportionally among all participating countries. During periods when atmospheric concentrations increase more than expected, anthropogenic emissions might have to be reduced more than anticipated. Conversely, during periods when the growth rate of atmospheric $CO_2$ is lower than expected, emissions might be allowed to increase. Such an approach would shift the "costs" of nonanthropogenic emissions to the entire community and not penalize (or reward) an individual country. Although such an approach seems worth exploring, balancing incentives for carbon storage within countries with issues of equity among countries will be complex.

### 21.4.2 Count Only Those Resulting Directly from Human Activity

A modification of the first option, that is, including only those terrestrial sources and sinks that result directly from anthropogenic activities, fares little better in terms of attribution. The global flux of carbon attributed to changes in land use, including forestry, averaged about 2.0 PgC $year^{-1}$, or 35% of fossil fuel emissions during the 1980s.[40] But this global perspective hides variation among countries. Almost all of this source was emitted from tropical countries as a result of deforestation. In temperate-zone and boreal countries, carbon sources from decay of wood products

**Table 21.6 Global Terrestrial Sources (+) and Sinks (–) of Carbon ($PgC\ yr^{-1}$) Counted and Not Counted for the Five Options Discussed in the Text**

| Terrestrial Sources and Sinks | Counted | Not Counted[a] |
|---|---|---|
| 1. Changes in carbon directly and indirectly attributable to human activity | –0.2 to 2.0[b] | 0 to –2.2 |
| 2. Changes in carbon directly attributable to human activity | 2.0 | –2.2 |
| 3. Changes in carbon resulting from afforestation, reforestation, and deforestation | 1.7 | –1.9 |
| 4. All changes in carbon | –0.2 | 0 |
| 5. No changes counted | 0 | –0.2 |

[a] The total net flux of carbon always sums to –0.2 PgC $year^{-1}$ (Equation 21.2). This global net sink is composed of a source of 2.0 PgC $year^{-1}$ from changes in land use (directly human induced) and a sink of –2.2 PgC $year^{-1}$ that may have been caused either indirectly from human activity, from non-anthropogenic factors, or from some combination of the two.

[b] The range results from not knowing whether the residual terrestrial sink is human-induced or not. If all terrestrial sources and sinks of carbon result either directly or indirectly from human activity (assuming that methods exist for distinguishing human-induced from non-human-induced changes), then human activity accounts for a net sink of 0.2 PgC $year^{-1}$. At the other extreme, if none of the residual sink results from human activity, then the effect of human activity is equivalent to the source of carbon from land-use change (2.0 PgC $year^{-1}$).

were approximately offset by sinks in regrowing forests.[40] In contrast, the geographic distribution of the residual terrestrial sink is uncertain, as are the mechanisms responsible for it. Under this option, no countries would receive credit for the residual sink of 2.2 PgC $year^{-1}$ (Table 21.6).

To avoid endless discussions of whether sources and sinks of carbon are anthropogenic or not, or direct rather than indirect, there are at least three other options. First, credit might be limited to the sinks resulting from specific activities (option 3, below). Option 4 is to count all changes in terrestrial carbon, ignoring the distinction between anthropogenic and nonanthropogenic causes. And the last option (5) is to ignore terrestrial sources and sinks altogether and consider only changes in fossil fuel emissions.

### 21.4.3 Count Only Those Resulting from Afforestation, Reforestation, and Deforestation

In the current formulation of the Kyoto Protocol, sources and sinks of carbon from afforestation, reforestation, and deforestation (ARD) are to be counted (Article 3.3). These activities are a subset of the changes in land use considered in option 2 above; changes in carbon within existing forests (for example, as a result of logging) do not count. Counting ARD activities creates an incentive to reduce emissions. The elimination of deforestation would reduce global emissions by an estimated 1.7 PgC $year^{-1}$ (Table 21.6). Increased afforestation and reforestation have the potential to reduce emissions (increase the storage of carbon on land) by as much as 1.5 to 2.2 PgC $year^{-1}$.[43,54,55]

Under most definitions of deforestation, this option does not include forestry, that is, the harvest and subsequent regrowth of forests. At a global level, not much is lost in ignoring harvests of wood. Deforestation (conversion of forests to croplands and pastures) accounted for about 85% of the net changes in carbon attributable to land-use change in the 1980s[40] (Table 21.6). Wood harvest, including the fate of harvested products as well as subsequent regrowth of harvested forests, contributed only 15% of the terrestrial source from land-use change. Again, one disadvantage of ignoring harvests is that they may account for much more than 15% of the terrestrial source of carbon in individual countries, while in other countries, such as those in the northern mid-latitudes, past harvests (as well as management) may be a major cause of the current sink.[20,21,24] Limiting the counting of terrestrial sources and sinks to ARD alone reduces the incentive for forestry practices that sequester carbon. It also reduces the incentive to substitute wood-based fuels and products for fossil-based fuels and products, a practice that could reduce emissions of carbon significantly.[56]

On the other hand, there are advantages to limiting the counting of sources and sinks to ARD activities. The advantages result from the magnitude of the changes, their unambiguous nature, and, hence, the ease of measurement. As pointed out above, ARD activities account for 85% of the net emissions from changes in land use. Fossil fuels and ARD together account for more than 95% of the global carbon emissions resulting directly from human activity. Second, because ARD activities involve the replacement of open lands with forests and vice versa, the per-hectare changes in aboveground carbon are large and readily measured. Third, the areas affected by ARD activities can be reasonably well defined through national surveys and remote sensing techniques. And finally, most of the changes in carbon resulting from ARD can be attributed to direct anthropogenic activity. On any particular hectare, some fraction of the observed increase in carbon storage may result from elevated $CO_2$ or N deposition, but the fraction is probably small relative to the increased storage resulting from afforestation or reforestation of that hectare in the first place. In contrast, the environmental enhancement of growth in existing (grown) forests may be large relative to an average accumulation of nearly zero, while at the same time the mechanisms responsible for such an enhancement are difficult to demonstrate (see next option).

### 21.4.4 Count All Terrestrial Sources and Sinks, Whatever the Cause

The fourth option is to ignore the distinction between anthropogenic and nonanthropogenic fluxes and, instead, to measure and count all changes in terrestrial carbon (full carbon accounting).[57] From one perspective, counting all terrestrial sources and sinks of carbon is consistent with the intent of the UNFCCC: stabilization of atmospheric concentrations of greenhouse gases at a level that would prevent dangerous anthropogenic interference with the climate system. Atmospheric concentrations are determined by the sum of all sources and sinks, not just anthropogenic ones. Why not count all? The major obstacle to full carbon accounting is the technical challenge (and cost) of measuring changes in carbon over the entire earth. At present, global analyses based on atmospheric data and models infer a net change in terrestrial carbon storage that, if correct, implies errors as large as 500% in the forest inventories of northern mid-latitudes: 0.6 PgC $year^{-1}$ (see Reference 21) vs. 3.5 PgC $year^{-1}$ (see Reference 51). Recent analyses demonstrate the same relative discrepancy in North America.[24,58] Full carbon accounting offers the potential to resolve the discrepancy, but the methods will have to have an accuracy much higher than the techniques used to date if the resolution is to be successful.

### 21.4.5 Count No Terrestrial Sources and Sinks of Carbon at All

The last option considered here is to ignore terrestrial sources and sinks of carbon altogether. The arguments for ignoring these fluxes are that they are uncertain, small relative to the emissions from fossil fuels, and not worth the effort of measurement, verification, and international negotiation. From the global perspective, the net annual flux of carbon between terrestrial ecosystems and the atmosphere is small and, arguably, not worth measuring or counting for the Kyoto Protocol.

$$\text{Atmospheric increase} = \text{Fossil fuel source} - \text{Oceanic sink} - \text{Terrestrial sink}$$

$$3.3(\pm 0.2) = 5.5(\pm 0.5) - 2.0(\pm 0.8) - 0.2(\pm 1.3) \quad (21.2)$$

The equation shows that terrestrial ecosystems were a small net sink during the 1980s, accumulating less than 5% of the fossil fuel source. The fraction may have increased during the 1990s. Recent estimates include net terrestrial sinks of 1.8 PgC for a single year (1992),[51] 0.7 PgC year$^{-1}$ for the interval 1980–1995,[52] and 1.3 PgC year$^{-1}$ for the period 1985–1995.[53] These three estimates include only atmospheric exchanges, and if an additional 0.4 to 0.7 PgC is transferred each year from land to ocean through rivers,[59] estimates of the net increase in terrestrial carbon storage range between about 0 and 1.4 PgC year$^{-1}$. The latitudinal distribution of this net terrestrial sink is uncertain.

Given the uncertainty, the year-to-year variability, and the small magnitude of this net flux, one might conclude that omitting it from the Kyoto Protocol would be justified. Consideration of fossil fuels alone addresses 75 to 100% of the net anthropogenic emissions of carbon to the atmosphere. Attention to terrestrial sources and sinks dilutes the attention that should be focused on the reduction of fossil fuel use. The counter argument is that there ought to be incentives to halt deforestation and increase the area of forests, thereby reducing net emissions of carbon to the atmosphere. Furthermore, even if the net change in terrestrial carbon storage is small, globally, it is not small for individual countries. In many tropical countries, for example, the emissions of carbon by terrestrial ecosystems are large relative to emissions from fossil fuels.[60] And in countries of the temperate zone, terrestrial sinks may help reduce net emissions significantly. Regardless of the arguments, ignoring terrestrial sinks is not currently an option under the Kyoto Protocol.

## 21.5 SUMMARY AND CONCLUSION

Deforestation releases carbon, principally as $CO_2$, to the atmosphere as the organic carbon stored in trees and soil is oxidized through burning and decay. Emissions of other greenhouse gases, such as $CH_4$ and $N_2O$, are also increased as a result of the conversion of forests to agricultural lands. Current emissions of greenhouse gases from deforestation account for about 25% of the global warming calculated to result from all anthropogenic emissions of greenhouse gases. The greatest rates of deforestation are currently in the tropics, but the greatest emissions of greenhouse gases from all sources, including fossil fuels, are from industrialized countries in the northern mid-latitudes. If current trends continue, tropical forests will be eliminated over the next 100 years or so, and in the process about as much carbon will be released into the atmosphere as was emitted from worldwide combustion of fossil fuels since the start of the industrial revolution. Continued emissions of greenhouse gases from both deforestation and industrial sources will raise global mean temperature by an estimated 1 to 3.5°C by the end of this century.

To avert such a warming, the concentrations of greenhouse gases in the atmosphere must be stabilized. The steps needed for stabilization include (1) a large (= 60%) reduction in the use of fossil fuels, through increased efficiency of energy use and a much expanded use of renewable energy sources; (2) the elimination of deforestation; and (3) reforestation of large areas of land, either to store carbon or to provide renewable fuels to replace fossil fuels. These steps required to curb global warming are good for a host of reasons independent of climatic change. The initial steps may even have negative costs. Eventually, the realization of these steps will be difficult and expensive, but the costs of coping with rapid, continuous, irreversible climatic change will also be expensive.

One of the issues that the nations of the world need to resolve for reducing emissions of greenhouse gases is the accounting of terrestrial sources and sinks of carbon. The ideal option for counting terrestrial sources and sinks of carbon (including only those sources and sinks resulting from human activity) is also the most difficult to implement. Methods cannot yet distinguish between anthropogenic and nonanthropogenic changes in carbon storage. The least expensive option for counting terrestrial sources and sinks is to ignore them altogether. Emissions of carbon from fossil fuels accounted for about 75% of global emissions in the 1980s, and the percentage has increased since then. However, an important incentive is lost in not considering terrestrial sinks. Furthermore, in a world with reduced fossil fuel emissions, the emissions from land-use change may become more important unless they too are reduced. Although counting all sources and sinks would contribute dramatically to scientific understanding of the carbon cycle and thereby increase the confidence of predictions of future atmospheric concentrations from alternative emissions scenarios, the more straightforward (and less expensive) option seems at present to be the one that the Kyoto Protocol initially recognized (ARD), perhaps expanded to include a few other well documented activities that lead to unambiguous changes in carbon storage. Despite numerous arguments to the contrary, the Kyoto Protocol seems to have selected the most feasible, cost-effective, and appropriate option.

Less than full carbon accounting is consistent with the purpose of the UNFCCC, which is to reduce emissions of greenhouse gases so as to avoid dangerous changes in climate. The counting need not be complete, but for precautionary reasons, it must not overestimate sinks (or underestimate sources). In this respect, limiting the counting to a few unambiguous activities is safer than full carbon accounting. The precautionary bias is especially important because of the difference between the common (global) good and the individual (national) good (the tragedy of the commons). Mechanisms for protecting the common good include symmetry and permanence. Symmetry suggests that sinks from reforestation cannot be counted unless sources from prior deforestation are also counted, even if deforestation is accidental and reforestation deliberate. Permanence suggests that parcels of land counted (and thus identified) as sinks must be counted in all subsequent commitment periods. Some forests may switch from sinks to sources in the future (for example, as a result of fire or warming-enhanced respiration), and those sources need to be counted. It is neither consistent with the intention of the UNFCCC nor logical to count sinks (deposits) and not sources (debits). It is important to note, however, that even if some forests change from sinks to sources of carbon, they still hold more carbon than lands without forests.

## ACKNOWLEDGMENTS

The work was supported by the Terrestrial Ecology Program at the National Aeronautics and Space Administration.

## REFERENCES

1. Perlin, J., *A Forest Journey: The Role of Wood in the Development of Civilization*, Harvard University Press, Cambridge, MA, 1989.
2. Woodwell, G. M., Forests: What in the world are they for?, in *World Forests for the Future. Their Use and Conservation*, K. Ramakrishna and G. M. Woodwell, Eds., Yale University Press, New Haven, 1, 1993.
3. Shukla, J., C. A. Nobre, and P. Sellers, Amazonian deforestation and climate change, *Science*, 247, 1322, 1990.
4. Bonan, G. B., Frost followed the plow: Impacts of deforestation on the climate of the United States, *Ecol. Appl.*, 9, 1305, 1999.
5. Brovkin, V., A. Ganopolski, M. Claussen, C. Kubatzki, and V. Petoukhov, Modelling climate response to historical land cover change, *Global Ecol. Biogeogr.*, 8, 509, 1999.
6. Houghton, J. T., G. J. Jenkins, and J. J. Ephraums, Eds., *Climatic Change: The IPCC Scientific Assessment,* Cambridge University Press, Cambridge, U.K., 1990.
7. Houghton, J. T., L. G. Meira Filho, B. A. Callander, N. Harris, A. Kattenberg, and K. Maskell, Eds., *Climate Change 1995: The Science of Climate Change,* Cambridge University Press, Cambridge, U.K., 1996.

7a. Houghton, J. T., Y. Ding, D. J. Griggs, M. Noguer, P. J. van der Linden, X. Dai, K. Maskell, and C. A. Johnson, Eds., *Climate Change 2001: The Scientific Basis*, Cambridge University Press, Cambridge, U.K., 2001.

8. Houghton, R. A., K. T. Lawrence, J. L. Hackler, and S. Brown, The spatial distribution of forest biomass in the Brazilian Amazon: A comparison of estimates, *Global Change Biol.*, 7, 731, 2001.
9. Houghton, R. A., R. D. Boone, J. R. Fruci, J. E. Hobbie, J. M. Melillo, C. A. Palm, B. J. Peterson, G. R. Shaver, G. M. Woodwell, B. Moore, D. L. Skole, and N. Myers, The flux of carbon from terrestrial ecosystems to the atmosphere in 1980 due to changes in land use: Geographic distribution of the global flux, *Tellus*, 39B, 122, 1987.
10. Houghton, R. A., R. D. Boone, J. M. Melillo, C. A. Palm, G. M. Woodwell, N. Myers, B. Moore, and D. L. Skole, Net flux of $CO_2$ from tropical forests in 1980, *Nature*, 316, 617, 1985.
11. Detwiler, R. P. and C. A. S. Hall, Tropical forests and the global carbon cycle, *Science*, 239, 42, 1988.
12. Hall, C. A. S. and J. Uhlig, Refining estimates of carbon released from tropical land-use change, *Can. J. For. Res.*, 21, 118, 1991.
13. Houghton, R. A., J. E. Hobbie, J. M. Melillo, B. Moore, B. J. Peterson, G. R. Shaver, and G. M. Woodwell, Changes in the carbon content of terrestrial biota and soils between 1860 and 1980: A net release of $CO_2$ to the atmosphere, *Ecol. Monogr.*, 53, 235, 1983.
14. FAO, State of the World's Forests 1997, FAO, Rome, 1997.
15. Krankina, O. N., M. E. Harmon, and J. K. Winjum, Carbon storage and sequestration in the Russian forest sector, *Ambio*, 25, 284, 1996.
16. Houghton, R. A. and K. Ramakrishna, A review of national emissions inventories from select non-Annex I countries: Implications for counting sources and sinks of carbon, *Annu. Rev. Energy Environ.*, 24, 571, 1999.
17. Houghton, R. A., D. L. Skole, C. A. Nobre, J. L. Hackler, K. T. Lawrence, and W. H. Chomentowski, Annual fluxes of carbon from deforestation and regrowth in the Brazilian Amazon, *Nature*, 403, 301, 2000.
18. Houghton, R. A., The worldwide extent of land-use change, *BioScience*, 44, 305, 1994.
19. FAO, 1989 Production Yearbook. FAO, Rome, 1990.
20. Houghton, R. A., Terrestrial sources and sinks of carbon inferred from terrestrial data, *Tellus*, 48B, 420, 1996.
21. Houghton, R. A., Historic role of forests in the global carbon cycle, in *Carbon Dioxide Mitigation in Forestry and Wood Industry,* G. H. Kohlmaier, M. Weber, and R. A. Houghton, Eds., Springer-Verlag, Berlin, 1, 1998.
22. Goodale, C. L., M. J. Apps, R. A. Birdsey, C. B. Field, L. S. Heath, R. A. Houghton, J. C. Jenkins, G. H. Kohlmaier, W. Kurz, S. Liu, G.-J. Nabuurs, S. Nilsson, and A. Z. Shvidenko, Forest carbon sinks in the northern hemisphere, *Ecol. Appl.*, 12, 891, 2002.
23. Melillo, J. M., J. R. Fruci, R. A. Houghton, B. Moore, and D. L. Skole, Land-use change in the Soviet Union between 1850 and 1980: Causes of a net release of $CO_2$ to the atmosphere, *Tellus*, 40B, 116, 1988.
24. Houghton, R. A., J. L. Hackler, and K. T. Lawrence, The U.S. carbon budget: Contributions from land-use change, *Science*, 285, 574, 1999.
25. Caspersen, J. P., S. W. Pacala, J. C. Jenkins, G. C. Hurtt, P. R. Moorcroft, and R. A. Birdsey, Contributions of land-use history to carbon accumulation in U.S. forests, *Science*, 290, 1148, 2000.
26. Woodwell, G. M., Biotic effects on the concentration of atmospheric carbon dioxide: A review and projection, in *Changing Climate,* National Academy Press, Washington, D.C., 216, 1983.
27. Woodwell, G. M., F. T. Mackenzie, R. A. Houghton, M. J. Apps, E. Gorham, and E. A. Davidson, Biotic feedbacks in the warming of the earth, *Climatic Change*, 40, 495, 1998.
28. Stocks, B. J., The extent and impact of forest fires in northern circumpolar countries, in *Global Biomass Burning*, J. S. Levine, Ed., MIT Press, Cambridge, MA, 197, 1991.
29. Uhl, C. and I. C. G. Vieira, Ecological impacts of selective logging in the Brazilian Amazon: A case study from the Paragominas region of the State of Para, *Biotropica*, 21, 98, 1989.
30. Woods, P., Effects of logging, drought, and fire on structure and composition of tropical forests in Sabah, Malaysia, *Biotropica*, 21, 290, 1989.
31. Gajaseni, J. and C. F. Jordan, Decline of teak yield in northern Thailand: Effects of selective logging on forest structure, *Biotropica*, 22, 114, 1990.
32. Brown, S., A. J. R. Gillespie, and A. E. Lugo, Biomass of tropical forests of south and southeast Asia, *Can. J. For. Res.*, 21, 111, 1991.

33. Brown, S., L. R. Iverson, and A. E. Lugo, Land use and biomass changes of forests in Peninsular Malaysia from 1972 to 1982: A GIS Approach, in *Effects of Land Use Change on Atmospheric $CO_2$ Concentrations: South and Southeast Asia as a Case Study,* V. H. Dale, Ed., Springer-Verlag, New York, 117, 1994.
34. Flint, E. P. and J. F. Richards, Trends in carbon content of vegetation in south and southeast Asia associated with changes in land use, in *Effects of Land Use Change on Atmospheric $CO_2$ Concentrations: South and Southeast Asia as a Case Study,* V. H. Dale, Ed., Springer-Verlag, New York, 201, 1994.
35. Gaston, G., S. Brown, M. Lorenzini, and K. D. Singh, State and change in carbon pools in the forests of tropical Africa, *Global Change Biol.*, 4, 97, 1998.
36. Phillips, O. L., Y. Malhi, N. Higuchi, W. F. Laurance, P. V. Nunez, R. M. Vasquez, S. G. Laurance, S. V. Ferreira, M. Stern, S. Brown, and J. Grace, Changes in the carbon balance of tropical forests: Evidence from long-term plots, *Science*, 282, 439, 1998.
37. Clark, D. A., Are tropical forests an important global carbon sink? Revisiting the evidence from long-term inventory plots, *Ecol. Appl.*, 12, 3, 2002.
38. Grace, J., J. Lloyd, J. McIntyre, A. C. Miranda, P. Meir, H. S. Miranda, C. Nobre, J. Moncrieff, J. Massheder, Y. Malhi, I. Wright, and J. Gash, Carbon dioxide uptake by an undisturbed tropical rain forest in southwest Amazonia, 1992 to 1993, *Science*, 270, 778, 1995.
39. Malhi, Y., A. D. Nobre, J. Grace, B. Kruijt, M. G. P. Pereira, A. Culf, and S. Scott, Carbon dioxide transfer over a central Amazonian rain forest, *J. Geophys. Res.*, 103, 31,593, 1998.
40. Houghton, R. A., The annual net flux of carbon to the atmosphere from changes in land use 1850–1990, *Tellus*, 51B, 298, 1999.
41. Fearnside, P. M., Global warming and tropical land-use change: Greenhouse gas emissions from biomass burning, decomposition and soils in forest conversion, shifting cultivation and secondary vegetation, *Climatic Change*, 46, 115, 2000.
42. Houghton, R. A., The role of the world's forests in global warming, in *World Forests for the Future: Their Use and Conservation,* K. Ramakrishna and G. M. Woodwell, Ed., Yale University Press, New Haven, 21, 1993.
43. Brown, S., Management of forests for mitigation of greenhouse gas emissions, in *Climatic Change 1995. Impacts, Adaptations and Mitigation of Climate Change: Scientific-Technical Analyses*, J. T. Houghton, G. J. Jenkins, and J. J. Ephraums, Eds., Cambridge University Press, Cambridge, U.K., 773, 1996.
44. Houghton, R. A., The future role of tropical forests in affecting the carbon dioxide concentration of the atmosphere, *Ambio*, 19, 204, 1990.
45. Houghton, R. A., J. D. Unruh, and P. A. Lefebvre, Current land use in the tropics and its potential for sequestering carbon, *Global Biogeochem. Cycles*, 7, 305, 1993.
46. Houghton, R. A., Biomass burning from the perspective of the global carbon cycle, in *Global Biomass Burning*, J. S. Levine, Ed., MIT Press, Cambridge, MA, 321, 1991.
47. Watson, R. T., H. Rodhe, H. Oeschger, and U. Siegenthaler, Greenhouse gases and aerosols, in *Climatic Change. The IPCC Scientific Assessment,* J. T. Houghton, G. J. Jenkins, and J. J. Ephraums, Eds., Cambridge University Press, Cambridge, U.K., 1, 1990.
48. Hao, W. M., M. H. Liu, and P. J. Crutzen, Estimates of annual and regional releases of $CO_2$ and other trace gases to the atmosphere from fires in the tropics, based on the FAO statistics for the period 1975–1980, in *Fire in the Tropical Biota,* J. G. Goldammer, Ed., Springer-Verlag, Berlin, FRG, 440, 1990.
49. Watson, R. T., I. R. Noble, B. Bolin, N. H. Ravindranath, D. J. Verardo, and D. J. Dokken, *Land Use, Land-Use Change, and Forestry,* Cambridge University Press, Cambridge, U.K., 2000.
50. IPCC, IPCC Guidelines for National Greenhouse Gas Inventories, Bracknell, U.K., 1997.
51. Ciais, P., P. P. Tans, J. W. C. White, M. Trolier, R. J. Francey, J. A. Berry, D. R. Randall, P. J. Sellers, J. G. Collatz, and D. S. Schimel, Partitioning of ocean and land uptake of $CO_2$ as inferred by d13C measurements from the NOAA Climate Monitoring and Diagnostics Laboratory global air sampling network, *J. Geophys. Res.*, 100, 5051, 1995.
52. Rayner, P. J., I. G. Enting, R. J. Francey, and R. Langenfelds, Reconstructing the recent carbon cycle from atmospheric $CO_2$, d13 and O2/N2 observations, *Tellus*, 51B, 213, 1999.

53. Bousquet, P., P. Ciais, P. Peylin, M. Ramonet, and P. Monfray, Inverse modeling of annual atmospheric $CO_2$ sources and sinks, 1. Method and control inversion, *J. Geophys. Res.*, 104, 26161, 1999.
54. Houghton, R. A., Converting terrestrial ecosystems from sources to sinks of carbon, *Ambio*, 25, 267, 1996.
55. Schopfhauser, W., World forests: The area for afforestation and their potential for fossil carbon sequestration and substitution, in *Carbon Dioxide Mitigation in Forestry and Wood Industry*, G. H. Kohlmaier, M. Weber, and R. A. Houghton, Eds., Springer-Verlag, Berlin, 185, 1998.
56. Kohlmaier, G. H., M. Weber, and R. A. Houghton, Eds., *Carbon Dioxide Mitigation in Forestry and Wood Industry*, Springer-Verlag, Berlin, 1998.
57. IGBP Terrestrial Carbon Working Group, The terrestrial carbon cycle: Implications for the Kyoto Protocol, *Science*, 280, 1393, 1998.
58. Fan, S., M. Gloor, J. Mahlman, S. Pacala, J. Sarmiento, T. Takahashi, and P. Tans, A large terrestrial carbon sink in North America implied by atmospheric and oceanic $CO_2$ data and model, *Science*, 282, 442, 1998.
59. Sarmiento, J. L. and E. T. Sundquist, Revised budget for the oceanic uptake of anthropogenic carbon dioxide, *Nature*, 356, 589, 1992.
60. Houghton, R. A. and J. L. Hackler, Emissions of carbon from forestry and land-use change in tropical Asia, *Global Change Biol.*, 5, 481, 1999.
61. Brown, S., A. J. R. Gillespie, and A. E. Lugo, Biomass estimation methods for tropical forests with applications to forest inventory data, *For. Sci.*, 35, 881, 1989.
62. Cooper, C. F., Carbon storage in managed forests, *Can. J. For. Res.*, 13, 155, 1982.

# CHAPTER 22

# Pathogens and Disease

**Frederick A. Leighton**

## CONTENTS



## 22.1 INTRODUCTION

Pathogenic organisms are normal components of all ecosystems. Simply defined, pathogenic organisms are life forms that cause disease in other life forms. Most are parasites that live at the expense of their host. Some are free-living but cause disease indirectly, such as by producing toxins to which other life forms become exposed. Ecotoxicology is generally considered to be the study of chemical pollutants in ecosystems. However, pathogenic organisms and the diseases they cause are relevant in this context in several different ways. Pathogens can be regarded as pollutants when they are released by humans into ecosystems for the first time or when they are concentrated in certain areas by human activity. Human activities also can alter ecosystems in ways that change the relationships among pathogens and their hosts and, thus, the occurrence of diseases. The transportation of plants, animals, and human beings around the world has created many new associations of potential pathogens, hosts, vectors, and environments out of which a substantial number of new or newly important diseases have recently emerged. The technology of genetic engineering has made possible the creation of whole new strains of pathogens that are indeed man-made.

This chapter begins with a brief discussion of the nature of pathogenic organisms and the factors that influence the occurrence of the diseases they cause. Four situations in which human activities can alter the occurrence of diseases in the environment are then considered: (1) translocation of pathogens, host species, and vectors to new environments; (2) concentration of pathogens or host species in particular areas; (3) changes in the environment that can alter host-pathogen relationships;

and (4) creation of new pathogens by intentional genetic modification of organisms. While most of the examples used are of diseases in vertebrate animals, the principles exemplified apply to all life forms.

## 22.2 PATHOGENS AND DISEASES

The relationships among pathogens, their hosts, their biological vectors where such exist, and the occurrence of disease are complex.[1] Infection and disease are not synonyms. Infection is entry and growth of a parasitic organism in the body of a host organism, while disease is some departure from a normal state of health. Whether or not infection of a host organism with a pathogen will result in disease usually depends on interactions among specific aspects of the biology of both the host and the pathogen and on environmental factors that interact with both. For example, in vertebrates, many pathogenic bacteria and viruses readily cause disease when they infect individuals for the first time. However, subsequent infections with the same organisms often cause no disease at all. This is the usual situation with the common diseases of childhood, for example, and results from a change in host physiology after the first encounter with the pathogens such that immune defenses against those specific pathogens are greatly enhanced. The eminent ecologist Paul Errington studied the occurrence of epidemic mortality in muskrats due to infection with the bacterium *Clostridium piliforme* and found that, while the causal bacterium appeared to be a constant presence in the muskrats' environment, it caused epidemic disease only under conditions of inadequate food supply, drought, or very high muskrat population densities.[2–4] These examples illustrate that the occurrence of infectious disease can be influenced by many factors; infection with a pathogen is not the only important variable. Rudolph Virchow, a father of modern medicine, recognized this with startling insight in 1848 when he pronounced that the cause of epidemic typhus, then raging in rural Prussia, was poverty, and that it could be prevented by education, liberty and prosperity.[5]

For any particular pathogen, there can be considerable variation in the ability of individual pathogenic organisms to cause disease. Most readers will be aware that, in humans, some strains of influenza virus cause more severe disease than do others. Similarly, there are many strains of the Newcastle Disease virus, an important pathogen of domestic chickens. Some cause rapidly-lethal disease, while others cause no disease at all in chickens but may cause serious disease in other species of birds.[6–7] Examples of eukaryotic pathogens that show great variation in ability to cause disease, include the protozoans *Trichomonas gallinae* in doves and *Trypanosoma brucei* in humans and other African mammals.[8–9]

Many organisms we consider to be pathogens are, in fact, rather benign parasites of their normal host species. This is a common, though not universal, theme among infectious diseases: severe disease may be uncommon in the infected normal host but common when unusual hosts are infected. For example, *Yersinia pestis*, the bacterium that causes Bubonic Plague, is a normal parasite of certain wild rodents and their fleas. In this natural setting, infection is widespread, but lethal disease is rare and even mild disease is unusual. Large outbreaks of severe disease occur when the bacterium is introduced into abnormal but susceptible hosts living at high density such as prairie dogs, urban rats, and urban humans.[10–11] Dutch elm disease in American Elms, and Lassa fever and Lyme disease in humans are all examples of diseases in which parasites that are relatively benign in their normal hosts cause severe disease when they infect new hosts.[12–14] These relationships between pathogens and hosts are not fixed, however. Sometimes there can be coevolution of parasites and hosts that often leads to a degree of mutual adaptation. The pathogen may lose some of its pathogenicity for the new host, while the host may develop some resistance to the pathogen.[15]

There are some similarities between pathogenic organisms and chemical poisons as causes of disease. The effects that both may have on their hosts can vary over a continuum from no detectable effect to rapid death and includes chronic debility and impaired reproduction that can persist indefinitely. Also, for many pathogens, there is a positive relationship between dose and response such that the occurrence and severity of disease are proportional to the number of pathogenic organisms to which a host organism is exposed. Like polluting chemicals, pathogens exert their effects on populations, communities, and ecosystems through their effects on individual host organisms.

However, there also are important differences between chemicals and pathogenic organisms as causes of disease. Pathogenic organisms are alive. They reproduce, pass from host to host, often by intricately complex methods, and they evolve and change over time as do other living things. A pathogenic organism, released in a new environment in which it can survive, cannot be controlled simply by ceasing the practice that led to its release. The ecosystem itself has been infected, and the pathogen is there for good. Furthermore, pathogenic organisms are normal constituents of ecosystems. They play important roles in the population dynamics of their host species and should be regarded, in general, as components of ecosystems that contribute to stability and resilience.[3,16–17] Thus, pathogenic organisms can be viewed in several different ways in the context of ecology. The emphasis here is on situations in which human activity can increase the negative impact of pathogenic organisms on host species.

## 22.3 TRANSLOCATION OF PATHOGENS, HOSTS, AND VECTORS

Movement of people, plants, animals, and products around the world now occurs rapidly and on a vast scale.[18] This movement, or translocation, has the potential to introduce pathogens into ecosystems, to place species of animals and plants into ecosystems that contain pathogens to which the introduced organisms are susceptible, or to introduce into ecosystems species (vectors) that can transmit pathogens from one host to another in new ways. In each scenario, there is potential for a negative impact on populations, communities, and ecosystems, on conservation programs, on human health, and on human economies. Several examples will serve to illustrate these points.

Two parasitic fungi introduced into North America in the 1900s have had major impacts on their host species and associated ecosystems. *Ceratocystis ulmi*, the fungus that causes Dutch elm disease, probably evolved in Asia in elm species with which it had a stable relationship. The fungus arrived in North America with logs imported for veneer manufacture, and the disease was first recognized in American elms (*Ulmus americana*) in Ohio in 1931. Within a few decades, most American elms were dead. The death of the American chestnut (*Castanea dentata*) due to the fungus *Endothia parasitica*, the pathogen of Chestnut blight, also is attributed to importation of a new pathogen from Asia, where it had a stable host-parasite relationship with Asian chestnuts. The American chestnut was the dominant forest tree over much of its former range in eastern North America, and the forest ecosystem it dominated was altered abruptly and completely by the blight; nothing like the primeval forest of eastern North America now exists.[19]

Human populations have been directly and severely affected from time to time by introductions of new pathogens. Aboriginal Americans were overwhelmed and severely depopulated by European diseases, particularly smallpox, well in advance of any military or economic conquests.[20–21] Bubonic plague, caused by the bacterium *Yersinia pestis* and introduced to new human populations through trade routes with Asia, may have killed as many as 100 million people during the first recognized pandemic (wide-spread epidemic) that began in 542 A.D. and killed millions more in two subsequent pandemics and numerous regional epidemics in subsequent years.[11] The current epidemic of Acquired Immunodeficiency Syndrome (AIDS) is an example of pathogens (Human Immunodeficiency Viruses 1 and 2) spreading from their sites of origin in Africa to new populations on that continent and elsewhere. This process appears to be driven by recent and profound alterations in human ecology including urbanization, improved medical care, and travel.[22]

Translocation of human pathogens may have major economic impacts, even if the pathogens' direct effects on human health are small. For example, the epidemic of rabies in raccoons (*Procyon lotor*) that has spread up the east coast of the United States and into Canada over the past 20 years was initiated by translocation of rabid raccoons from a distant endemic area to the state of West Virginia; it has required vast expenditure of public-health funds on postexposure treatments of people bitten by raccoons and on vaccination of wildlife.[23–24] Similarly, the introduction of West Nile virus to the New York City area in the summer of 1999 and its subsequent centrifugal spread evoked massive public expenditure on disease surveillance and mosquito control, severely taxing federal, provincial, state, and municipal resources.[25–26]

Outbreaks of new diseases are most often documented and studied in terms of their effects on a particular affected species rather than in terms of ecosystem-level effects. However, some data are available regarding effects of newly introduced pathogens beyond those on the host species itself. Myxomatosis in England is a case in point. The poxvirus that causes myxomatosis is a New World parasite of rabbits of the genus *Sylvilagus*. It causes severe, lethal disease in 90 to 99% of infected European rabbits (*Oryctolagus cuniculus*) and has been introduced around the world to control rabbit populations. It reached England in 1954 and exterminated rabbits over large areas.[27] An intensive study of the vegetation of several areas affected by myxomatosis was undertaken from 1954–1957 and documented major alterations as a result of the absence of grazing by rabbits. In general, grassland species declined in density and diversity, while woody plants, such as gorse and brambles, grew and multiplied.[28] These effects of rabbit removal are not surprising, but they emphasize the potential ecological effects that new pathogens can have as a result of their effect on their own host species.

Animals or plants translocated to new environments may themselves be at risk from pathogens already present in these environments. This situation can have tragic impacts on conservation efforts such as recovery programs for endangered species. For example, efforts to reintroduce caribou (*Rangifer tarandus*) to Cape Breton Highlands National Park in Nova Scotia (Canada) and to parts of the State of Maine (U.S.), where caribou once were native, have failed because of the presence of a nematode parasite, *Parelaphostrongylus tenuis*, which causes fatal infection in caribou. This parasite had become endemic in both reintroduction sites through expansion of the range of the white-tailed deer (*Odocoileus virginianus*), its normal host.[29–31] Similarly, the recovery program for the endangered whooping crane (*Grus americana*) included establishment of a captive breeding population in the eastern United States, outside of the normal geographic range of the species. In this new environment, the cranes encountered an endemic pathogen, eastern equine encephalitis virus, which is not present in their native habitat and which caused fatal disease.[32]

Some pathogens require a living organism to transmit them efficiently among susceptible hosts. Organisms that carry pathogens from host to host are called vectors; blood-feeding arthropods, such as ticks and mosquitoes, often serve as such vectors. The presence or absence of vectors and the efficiency of the various vectors present as transmitters of a particular pathogen can influence greatly the occurrence of infection and disease among susceptible host species in an ecosystem. Thus, the introduction or elimination of the vector of a pathogen in an ecosystem has the potential to alter the effects of that pathogen on host populations. For example, infection with *Borrelia burgdorferi*, the agent of Lyme disease, occurs in animals only in ecosystems that support the few species of ticks that transmit the disease.[33–34] Similarly, the fungus that causes Dutch elm disease is spread from tree to tree by a bark-dwelling beetle and does not spread in the absence of the beetle. Thus, introduction and establishment of new vector species in ecosystems has the potential to radically alter host-pathogen relationships or to create conditions under which a pathogen introduced from elsewhere may successfully establish in a new ecosystem because a vector necessary for its survival is now present.

The effect of introductions of new pathogens and of vectors on native birds of the Hawaiian Islands is an instructive example.[35] Over half of the native land bird species of these islands became extinct between about 1800 and 1940.[36] There were many causal factors associated with these extinctions, beginning with the arrival of human beings, and, subsequently, of rats, pigs, cats, and mongooses.[36–38] Two introduced diseases, avian malaria and avian pox, appear to have played very significant roles as well, but their introduction was itself a multistep process. The causal pathogens of both diseases — a protozoan and a virus, respectively — are transmitted from bird to bird by mosquitoes that act as vectors. Until 1826, there were no mosquitoes on the Hawaiian Islands. In that year, the crew of the ship Wellington dumped the mosquito-laden remnants of their water kegs into a Hawaiian stream preparatory to filling them with fresh water. Thus, *Culex quinquefasciatus* from Mexico was introduced and a suitable vector for both pox and malaria was established. The pathogens themselves were introduced later, almost certainly with one or more of the many nonnative species of birds brought to the islands for various purposes by settlers. The poxvirus arrived in the late 1800s, while *Plasmodium relictum*, the agent of avian malaria in Hawaii, probably did not affect native birds on a large scale until around 1920. Van Riper et al. (1986) estimated that 16 species of birds became extinct in the late 1800s and early 1900s, while a further nine became extinct after introduction of malaria.[37] Malaria continues to have a negative impact on the native birds of Hawaii, restricting their abundance and distribution while giving a competitive advantage to introduced species that are resistant to the disease.[35]

## 22.4 CONCENTRATION OF PATHOGENS

As noted above, for many pathogens, the dose to which a potential host organism is exposed can determine whether or not infection will occur and, sometimes, whether infection will be unimportant or lead to disease. Thus, human acts that result in the concentration of pathogens within a geographic area can alter the normal ecology of host and parasite in favor of disease. Several well-described diseases caused by bacteria illustrate this point.

*Salmonella* sp. is a genus of bacteria with many different species; it has the ability to infect most vertebrates. Various members of these species cause important diseases in humans and in domestic and wild animals. Two widely different human activities have resulted in increases in prevalence of disease in wild birds due to *Salmonella* sp.: discharge of urban sewage and creation of bird-feeding stations. *Salmonella* sp. thrives in sewage. Muller (1965) found a striking inverse correlation between the prevalence of infection with pathogenic *Salmonella* sp. in gulls and distance from the municipal sewage discharge area of Hamburg, Germany.[39] Lethal infection with *Salmonella* occurs quite commonly among passerine birds clustered around bird feeders in winter.[40–41] The bacterium appears to be carried by certain species, particularly House Sparrows.[42–43] Many birds can become infected when the feeders are contaminated by the carrier birds. The catalyst for such outbreaks appears to be the feeders themselves, which serve as focal points for contamination with bacteria and infection of many individuals.

Concentration of animals around feed sources has resulted in serious disease problems in other species as well. Winter feeding of wild elk (wapiti: *Cervus elaphus*) in western Wyoming (U.S.) is considered responsible for the maintenance of brucellosis in these populations; in the absence of artificial concentrations of animals in winter and spring around feed sources, the disease does not appear to persist in wild elk populations.[44–47] Similarly, infection of wild white-tailed deer with bovine tuberculosis, recently found to be widespread in the State of Michigan (U.S.), has occurred in association with maintenance of very high population densities through provision of food in winter; elsewhere this disease is not maintained in wild white-tail populations.[48] Both these diseases are of major economic importance to agriculture and can be severe diseases in people.

Small wells and ponds created as sources of drinking water for wild or domestic animals also can act as foci of concentration for pathogens and sources of infection for large numbers of animals. Particularly under drought conditions, animals become concentrated around water sources, and the water and surrounding environment can become heavily contaminated with pathogens. In this manner, outbreaks of anthrax (*Bacillus anthracis*) in African wildlife and of necrobacillosis (*Fusibacterium necrophorum*) in elk, deer, and pronghorn (*Antilocapra americana*) in North America have occurred in association with both natural and man-made water sources as well as with feeding stations.[49–52]

## 22.5 DISEASE ASSOCIATED WITH CHANGED ENVIRONMENTS

Since, in general, disease occurs through interactions among a pathogen, its host, and their joint environment, it follows that the prevalence or severity of a disease can be influenced as much by environmental changes as by changes in parameters of pathogen or host. This statement says no more than that alterations made to environments by humans can have indirect and unexpected ramifications throughout the ecosystem. The point made here is that diseases are components of ecosystems that can be affected by such environmental changes.

One instructive example of a disease responding to a changed environment is Sacramento River Chinook Disease.[53] This disease caused extensive losses of juvenile Chinook salmon in the Sacramento and Mokelumne rivers of California beginning in the 1960s. The emergence of this disease as a serious problem in wild and hatchery salmon was associated with the construction of dams on major Sacramento tributaries. Outbreaks of the disease killed up to 80% of young salmon in hatcheries, and a reduction in the annual spawning run of chinook salmon by 60% was attributed, in part, to this disease. The virus that causes the disease is very sensitive to temperature; it replicates at relatively cold temperatures (7–13°C) and does not produce disease in fish held at 15°C or higher. Dams changed the annual temperature regime of the river system from one in which waters passed through the range of 7–13°C in 2 to 3 weeks each spring to one in which this critical temperature range was maintained for 3 to 4 months each year. Thus, dams led to epizootic virus disease by way of altered environmental temperature.

Recent outbreaks of debilitating and lethal disease in balsam fir (*Abies balsamea*) and white spruce (*Picea glauca*) in eastern Canada and the adjacent United States caused by the spruce budworm (*Choristoneura fumiferana*) are another case in point.[54] The increase in frequency, extent, and severity of outbreaks of disease caused by this pathogenic insect have been attributed to major changes in forest ecology as a result of human activities. Over the past 200 years, specific logging practices, suppression of forest fires, and spraying of forests with insecticides all have resulted in huge increases in the land area covered with mature fir and spruce. The spruce budworm is a parasite of mature Balsam Fir and White Spruce, and the size and severity of outbreaks correlate well with the size of stands of mature trees. These are the climax species of normal succession in much of the affected region, but the extent of climax forest is normally greatly restricted by fire and, to a lesser extent, by the budworm. Outbreaks of budworm disease that formerly were small, isolated, and self-limiting now involve vast contiguous areas (for example, 55,000,000 ha in the outbreak that began in 1970). Intervals between outbreaks have become much smaller, and the duration of outbreaks has increased; an outbreak in New Brunswick (Canada), which began in 1949, was still underway in 1983 despite heavy application of insecticides for 30 years.

There is mounting evidence that infectious disease can occur as a result of environmental contaminants causing suppression of normal resistance to infection. This is inherently difficult to document because of the complexity of disease causation under natural conditions. Nonetheless, there is experimental evidence that immune function can be reduced as a toxicological effect of exposure to chemical poisons,[55] and there is also an increasing body of evidence for ecologically significant changes in patterns of disease occurrence as a result of exposure to contaminants. An epidemic of a new virus disease in harbor seals, now called phocine distemper, in northern Europe in 1988 was most severe in seals from the highly-polluted Baltic Sea.[56–57] This observation led to studies in immune function in seals fed contaminant-rich fish from the Baltic and contaminant-poor fish from the Atlantic. Seals fed contaminant-rich fish had distinctly reduced immune function, as assessed by laboratory measures.[58] Additional examples of occurrences of infectious diseases attributed to reduced resistance caused by environmental contaminants are adult mortality and destruction of eggs by protozoan and fungal pathogens observed in crayfish associated with acidification of the aquatic environment and the increased susceptibility to bark beetle damage in ponderosa pines associated with exposure to photochemical air pollutants.[59–60] It should be noted that effects on resistance to infection caused by exposure to environmental pollutants or other environmental disturbances can be very difficult to identify or attribute to the primary cause. A primary reduction in resistance to disease will usually appear to be a problem of a particular infectious disease itself; the pollutant or disturbance that reduced the normal resistance to infection may not be suspected of contributing to the occurrence of disease.

Avian botulism can serve as a final example of a disease that can change in importance when environments are altered.[61] Botulism is a form of food poisoning that occurs when the bacterium *Clostridium botulinum* grows and produces toxins in food material that is subsequently eaten by animals. Botulism is an especially important disease in waterbirds. For these species, the usual source of botulinum toxin is maggots that develop in rotting carcasses of animals that have died for any reason. In most marshes or other wetland areas, resident animals will have spores of the bacterium in their digestive tracts. When such animals die, the bacterium multiplies, spreads within the carcass, and produces toxin. Maggots growing in the same carcasses absorb and concentrate the toxin but are not affected by it. Waterbirds readily feed on maggots and are subsequently killed by the toxin the maggots contain. These birds, in turn, become additional dead carcasses in which more toxin and maggots can develop and, thus, are sources of poisoning for yet more birds.

Many environmental factors can influence whether or not avian botulism will kill large numbers of birds in a wetland. Any cause of animal mortality in a marsh can increase the probability of botulism. For example, construction of electrical power transmission lines across a marsh has resulted in new annual botulism outbreaks because birds killed by collision with lines or towers dropped into the marsh and served as sources of toxin and maggots.[62] An increase in the density of waterbirds on a marsh can increase the likelihood of botulism simply because it will increase the likelihood that maggots from carcasses will be found and eaten. An increase in density of waterbirds on wetlands is a continent-wide phenomenon as the quantity of wetland habitat continues to shrink due to drainage and diversion of water, and birds are crowded onto the residual wetland areas. Large populations of scavengers can reduce the probability of botulism by rapid removal of carcasses, while reduction of scavenger populations can have the opposite effect.[63] Changes in these environmental factors can alter radically the magnitude of mortality from botulism on any particular wetland, resulting in annual fluctuations over several orders of magnitude.

Human actions that fundamentally change ecological relationships also can have unexpected consequences with respect to disease. Bovine spongiform encephalopathy (BSE), also called "mad cow disease," is a case in point. The epidemic of BSE that devastated the British cattle industry in the 1990s and has had major economic and public health ramifications throughout Europe and much of the rest of the world arose from human imposition of carnivorous and cannibalistic feeding on a normally herbivorous animal. It was through feeding cattle with meat and bone meal derived from rendered cattle and sheep that the disease was created and spread, an industrial practice that struck down natural ecological barriers that previously had precluded such an epidemic.[64–66]

## 22.6 GENETICALLY MODIFIED ORGANISMS

It is now possible to alter the genetic makeup of organisms by direct alteration of the genome. The resultant organisms often are called genetically modified organisms (GMOs) or genetically engineered organisms (GEOs). The ecological effects of release of GMO into ecosystems have not yet been evaluated, and more will be learned over the next several decades.[67] Here, it can only be noted that pathogens are among the biota that have been, and certainly will be, subjected to genetic engineering. Already, a poxvirus genetically modified to contain a gene from the very different virus that causes rabies has been developed and widely distributed in nature in baits used for the field vaccination of foxes and raccoons against rabies.[68] This GMO was evaluated extensively prior to use, and no negative impact on the environment has been associated with its use other than the intended reduction or elimination of rabies virus from fox populations and the attendant ecological changes.[69] However, genetic modification of many potential pathogens has been conceived for a wide range of purposes: to use as agents for immuno-contraception, to carry drugs to cancerous cells, to implant normal genes in genetically deficient individuals, and to vaccinate against any of a number of diseases important to people or animals. It seems likely that, in the near future, GMOs derived from pathogens or that are themselves potentially new pathogens will enter the biosphere with increasing frequency. Regulatory processes will identify and prevent dissemination of the most hazardous of these modified pathogens, but some small proportion of GMO released into the environment in the future may create ecological changes by causing disease in various plants or animals in a manner analogous to the translocation of an existing pathogen into a new ecosystem.[70]

## 22.7 PREVENTION AND REMEDIATION

It is immensely more practical, feasible, and affordable to prevent the introduction of new pathogens into ecosystems or to take measures to avoid activities that alter disease occurrence in ecosystems in undesirable ways than it is to pursue remediation after the event. The principal methods of prevention are forethought and risk analysis. It is entirely feasible, for example, to carry out analysis of the risks of disease occurrence prior to taking decisions on translocations of animals and plants or on actions that alter environments in substantial ways. Regulatory procedures that incorporate health risk analysis already are in place in many jurisdictions. Tools and guidelines are available that make such risk analysis readily achievable.[71]

Remediation, on the other hand, is unlikely to succeed. Living pathogens, once established in an ecosystem, are very difficult to eliminate or to control in any substantive way. Small pox was eliminated through unprecedented international effort and massive cost.[72] Rabies has been eliminated from fox populations over large geographic areas by oral vaccine distributed in baits, again at massive cost.[73] But most attempts to eradicate or control living pathogens fail. For most pathogens, neither the technology nor the necessary epidemiological information required for control or eradication programs will exist when they are needed, and each will take years to acquire.

Wobeser (1994) has provided a comprehensive discussion of the options and approaches that may be taken to control or eradicate pathogens in animal populations.[74] While elimination of pathogens from ecosystems often is impossible, specific and modest objectives to reduce specified undesirable effects of pathogens sometimes can be achieved. Measures to achieve these objectives may be directed at the pathogen itself, vector organisms required in the pathogen's life cycle, the host population, the environment, or at human activities and behavior. Success in remediation usually is, at best, modest and limited in scope.

## 22.8 CONCLUSION

From the foregoing, it is clear that human activities can affect the occurrence and significance of many diseases in a variety of ways. This is a lesson only lately learned, but it is a lesson with important implications. The introduction of a pathogen or an environmental change that disturbs the balanced relationships between hosts and parasites can have widespread and unexpected effects. Thus, disease must be included among the many factors to be considered when the environmental impact of human activities is assessed and when regulations to channel or limit these activities are developed.

## REFERENCES

1. Hanson, R. P., Koch is dead, *J. Wildl. Dis.*, 24, 193, 1988.
2. Errington, P. L., Disease cycles in nature — Epizootiology of a disease of muskrats, in *The Problems of Laboratory Animal Disease*, Harris, R. J. C., Ed., Academic Press, New York, 1962, 7.

3. Errington, P. L., The special responsiveness of minks to epizootics in muskrat populations, *Ecol. Monogr.,* 24, 377, 1954.
4. Wobeser, G., Barnes, H. J., and Pierce, K., Tyzzer's disease in muskrats: Re-examination of specimens of hemorrhagic disease collected by Paul Errington, *J. Wildl. Dis.,* 15, 525, 1979.
5. Virchow, R., Report on the typhus epidemic in Upper Silesia, in *Collected Essays on Public Health and Epidemiology*, Rather, L. J., Ed., Science History Publications, Canton, MA, 1985, 205.
6. Alexander, D. J., Historical aspects, in *Newcastle Disease*, Alexander, D. J., Ed., Kluwer Academic Publishers, Boston, 1988, Chap. 1.
7. Hanson, R. P., Heterogeneity in strains of Newcastle disease virus: Key to survival, in *Newcastle Disease*, Alexander, D. J., Ed., Kluwer Academic Publishers, Boston, 1988, Chap.7.
8. Kocan, R. M. and Herman, C. M., Trichomoniasis, in *Infectious and Parasitic Diseases of Wild Birds*, Davis, J. W., Anderson, R. C., Karstad, L., and Trainer, D. O., Eds., Iowa State University Press, Ames, 1971, 282.
9. Acha, P. N. and Szyfres, B., *Zoonoses and Communicable Diseases Common to Man and Animals*, 2nd ed., Pan American Health Organization, Washington, D.C., 1987.
10. Barnes, A. M., Surveillance and control of bubonic plague in the United States, *Symp. Zool. Soc. London,* 50, 237, 1982.
11. Pollitzer, R., *Plague,* W.H.O. Monograph Series No. 22, World Health Organization, Geneva, 1954.
12. Day, P. R., The genetic basis of epidemics, in *Plant Disease: An Advanced Treatise*, Horsfall, J. G. and Cowling, E. B., Eds., Academic Press, New York, 1978, Chap. 13.
13. Fuller, J. G., *Fever*, Ballantine Books, New York, 1974.
14. Anderson, J. F., Epizootiology of *Borrelia* in *Ixodes* tick vectors and reservoir hosts, *Rev. Infect. Dis.,* 11 Suppl. 6, S1451, 1989.
15. Allison, A. C., Co-evolution between hosts and infectious disease agents and its effects on virulence, in *Population Biology of Infectious Diseases*, Anderson, R. M. and May, R. M., Eds., Springer-Verlag, Berlin, 1982, 245.
16. Washburn, J. O., Mercer, D. R., and Anderson, J. R., Regulatory role of parasites: Impact on host population shifts with resource availability, *Science,* 253, 185, 1991.
17. Hornfeldt, B., Synchronous population fluctuations in voles, small game, owls, and tularemia in Northern Sweden, *Oecologia,* 32, 141, 1978.
18. Griffith, B., Scott, J. M., Carpenter, J. W., and Reed, C., Animal translocations and potential disease transmission, *J. Zoo Wildl. Med.,* 24, 231, 1993.
19. Horsfall, J. G. and Cowling, E. B., Some epidemics man has known, in *Plant Disease: An Advanced Treatise*, Horsfall, J. G. and Cowling, E. B., Eds., Academic Press, New York, 1978, Chap. 2.
20. Ramenofsky, A. F., *Vectors of Death*, University of New Mexico Press, Albuquerque, 1987.
21. Diamond, J., *Guns, Germs and Steel: The Fates of Human Societies*, W. W. Norton and Company, New York, 1997.
22. Grmek, M. D., *History of AIDS: Emergence and Origin of a Modern Pandemic*, Princeton University Press, Princeton, 1990.
23. Rupprecht, C. E. and Smith, J. S., Raccoon rabies: The re-emergence of an epizootic in a densely populated area, *Seminars Virol.,* 5, 155, 1994.
24. Meltzer, M. I., Assessing the costs and benefits of an oral vaccine for raccoon rabies: A possible model, *Emerging Infect. Dis.,* 2, 343, 1996.
25. Centers for Disease Control and Prevention, Update: West Nile virus activity — Eastern United States, 2000, *Morbidity and Mortality Weekly Report,* 49, 1044, 2000.
26. Komar, N., West Nile viral encephalitis, *Rev. Sci. Tech. O.I.E.*, 19, 166, 2000.
27. Yuill, T. M., Myxomatosis and fibromatosis, in *Infectious Diseases of Wild Mammals*, Davis, J. W., Karstad, L., and Trainer, D. O., Eds., Iowa State University Press, Ames, 1981, Chap. 13.
28. Thomas, A. S., Changes in vegetation since the advent of myxomatosis, *J. Ecol.,* 48, 287, 1960.
29. Smith, H. J. and Archibald, R. M., Moose sickness, a neurological disease of moose infected with the common cervine parasite, *Elaphostrongylus tenuis*, *Can. Vet. J.,* 8, 173, 1967.
30. Lankester, M. W. and Fong, D., Distribution of elaphostrongyline nematodes (metastrongyloidea: Protostrongylidae) in cervidae and possible effects of moving *Rangifer* spp. into and within North America, *Alces,* 25, 133, 1989.
31. Dauphine, T. C., The disappearance of caribou reintroduced to Cape Breton Highlands National Park, *Can. Field-Nat.,* 89, 299, 1975.

32. Dein, F. J., Carpenter, J. W., Clark, G. G., Montali, R. J., Crabbs, C. L., Tsai, T. F., and Docherty, D. E., Mortality of captive whooping cranes caused by eastern equine encephalitis virus, *J. Am. Vet. Med. Assoc.,* 189, 1006, 1986.
33. Gage, K. L., Ostfeld, R. S., and Olson, J. G., Nonviral vector-borne zoonoses associated with mammals in the United States, *J. Mammal.,* 76, 695, 1995.
34. Gern, L. and Falco, R. C., Lyme disease, *Rev. Sci. Tech. O.I.E.*, 19, 121, 2000.
35. Van Riper, C., III, The impact of introduced vectors and avian malaria on insular passeriform bird populations in Hawaii, *Bull. Soc. Vector Ecol.,* 16, 59, 1991.
36. Warner, R. E., The role of introduced diseases in the extinction of the endemic Hawaiian avifauna, *Condor,* 70, 101, 1968.
37. Van Riper, C., III, van Riper, S. G., Goff, M. L., and Laird, M., The epizootiology and ecological significance of malaria in Hawaiian land birds, *Ecol. Monogr.,* 56, 327, 1986.
38. Olson, S. L. and James, H. F., Fossil birds from the Hawaiian Islands: Evidence for wholesale extinction by man before western contact, *Science,* 217, 633, 1982.
39. Muller, G., Salmonella in bird feces, *Nature,* 207, 1315, 1965.
40. Daoust, P. Y., Busby, D. G., Ferns, L., Goltz, J., McBurney, S., Poppe, C., and Whitney, H., Salmonellosis in songbirds in the Canadian Atlantic provinces during winter-summer 1997–98, *Can. Vet. J.,* 41, 54, 2000.
41. Kirkwood, J. K., Holmes, J. P., and MacGregor, S., Garden bird mortalities, *Vet. Rec.,* 136, 372, 1995.
42. Wobeser, G. A. and Finlayson, M. C., *Salmonella typhimurium* infection in house sparrows, *Arch. Environ. Health,* 19, 882, 1969.
43. Tizard, I. R., Fish, N. A., and Harmeson, J., Free flying sparrows as carriers of salmonellosis, *Can. Vet. J.,* 20, 143, 1979.
44. Thorne, E. T., Morton, J. K., and Ray, W. C., Brucellosis, its effect and impact on elk in western Wyoming, in *North American Elk: Ecology, Behavior and Management*, Boyce, M. S. and Hayden-Wing, L. O., Eds., Univ. of Wyoming, Laramie, 1979, 212.
45. Tessaro, S. V., Bovine tuberculosis and brucellosis in animals, including man, *Alberta,* 3, 207,1992.
46. Meagher, M. and Meyer, M. E., On the origin of brucellosis in bison of Yellowstone National Park: A review, *Conserv. Biol.,* 8, 645, 1994.
47. Cheville, N. F., McCullough, D. R., and Paulson, L. R., *Brucellosis in the Greater Yellowstone Area*, National Academy Press, Washington, D.C., 1998.
48. Schmitt, S. M., Fitzgerald, S. D., Cooley, T. M., Bruning, F. S., Sullivan, L., Berry, D., Carlson, T., Minnis, R. B., Payeur, J. B., and Sikarskie, J., Bovine tuberculosis in free-ranging white-tailed deer from Michigan, *J. Wildl. Dis.,* 33, 749, 1997.
49. Pienaar, U. de V., Epidemiology of anthrax in wild animals and control of anthrax in the Kruger National Park, South Africa, *Fed. Proc.,* 26, 1496, 1967.
50. Ebedes, H., Anthrax epizootics in Etosha National Park, *Madoqua,* 10, 99, 1977.
51. Rosen, M. N., Necrobacillosis, in *Infectious Diseases of Wild Animals*, Davis, J. W., Karstad, L., and Trainer, D. O. Eds., Iowa State University Press, Ames, 1981, Chap. 31.
52. Wobeser, G., Runge, W., and Noble, D., Necrobacillosis in deer and pronghorn antelope in Saskatchewan, *Can. Vet. J.,* 16, 3, 1975.
53. Wingfield, W. H. and Chan, L. D., Studies of the Sacramento River chinook disease and its causative agents, in *A Symposium on Diseases of Fishes and Shellfishes*, Snieszko, S. F., Ed., American Fisheries Society, Washington, D.C., 1970, 307.
54. Blais, J. R., Trends in the frequency, extent, and severity of spruce budworm outbreaks in eastern Canada, *Can. J. For. Res.,* 53, 539, 1983.
55. Shuurman, H.-K., Krajnc-Franken, M. A. M., Kuper, C. F., van Loveren, H., and Vos, J. G., Immune System, in *Handbook of Toxicologic Pathology*, Haschek, W. M. and Rousseaux, C. G., Eds., Academic Press, San Diego, 1991, 421.
56. Dietz, R., Heide-Jorgensen, M.-P., and Harkonen, T., Mass deaths of harbor seals *(Phoca vitulina)* in Europe, *AMBIO,* 18, 258, 1989.
57. Kennedy, S., A review of the 1988 European seal morbillivirus epizootic, *Vet. Rec.,* 127, 563, 1990.
58. Swart, R. D., Ross, P. S., Vedder, L. J., Timmerman, H. H., Heisterkamp, S., Loveren, H. van Vos, J. G., Reijnders, P. J. H., Osterhaus, A. D. M. E., De Swart, R. L., and van Loveren H., Impairment of immune function in harbor seals (*Phoca vitulina*) feeding on fish from polluted waters, *AMBIO,* 23, 155, 1994.

59. Schindler, D. W., Mills, K. H., Malley, D. F., Findlay, D. L., Shearer, J. A., Davies, I. J., Turner, M. A., Linsey, G. A., and Cruikshank, D. R., Long-term ecosystem stress: The effects of years of experimental acidification on a small lake, *Science,* 228, 1395, 1985.
60. Stark, R. W., Miller, P. R., Cobb, F. W., Jr., Wood, D. L., and Parmeter, J. R., Jr., I. Incidence of bark beetle infestation in injured trees, *Hilgardia,* 39, 121, 1968.
61. Wobeser, G. A., Botulism, in *Diseases of Wild Waterfowl*, $2^{nd}$ ed. Plenum Press, New York, 1997, 149.
62. Malcolm, J. M., Bird collisions with a power transmission line and their relation to botulism at a Montana wetland, *Wildl. Soc. Bull.,* 10, 297, 1982.
63. Hamilton, D. B. and Wobeser, G. A., Accumulating evidence of avian botulism risk using certainty factors, *AI Applications,* 3, 1, 1989.
64. Donnelly, C. A., MaWhinney, S., and Anderson, R. M., A review of the BSE epidemic in British cattle, *Ecosys. Health*, 5, 164, 1999.
65. Caskie, P., Davis, J., and Moss, J. E., The economic impact of BSE: A regional perspective, *Appl. Econ.,* 31, 1623, 1999.
66. Zeidler, M. and Ironside, J. W., The new variant of Creutzfeldt-Jakob disease, *Rev. Sci. Tech. O.I.E.*, 19, 98, 2000.
67. Wolfenbarger, L. L. and Phifer, P. R., The ecological risks and benefits of genetically engineered plants, *Science,* 290, 2088, 2000.
68. Rupprecht, C. E., Hanlon, C. A., Koprowski, H., and Hamir, A. N., Oral wildlife rabies vaccination: Development of a recombinant virus vaccine, in *Trans. of the $57^{th}$ North American Wildlife and Natural Resources Conference*, McCabe, R. E., Ed., Wildlife Management Institute, Washington, D.C., 1992, 439.
69. CNEVA, *Red Fox Demography and Management*, Centre National d'Etudes Veterinaires et Alimentaires, Malzeville, France, 1995.
70. Hails, R. S., Cory, J. S., Thomas, M. B., and Kedwards, T., Evaluating risks of genetically modified baculoviruses in the environment, *Aspects Appl. Biol.,* 53, 197, 1999.
71. Two examples of guidelines for disease risk analysis available on the Internet are: <http://wildlife.usask.ca/bookhtml/RiskAnalysis/RSKGUIDINDEX.htm≥ and <http://www.oi.e.,int/eng/normes/MCode/A_00009.htm≥.
72. Fenner, F., Biological control, as exemplified by smallpox eradication and myxomatosis, *Proc. Royal Soc. Lond., B Biol. Sci.,* 218, 259, 1983.
73. Aubert, M. F. A., Masson, E., Artois, M., and Barrat, J., Oral wildlife rabies vaccination field trials in Europe, with recent emphasis on France, *Curr. Topics Microbiol. Immunol.,* 187, 219, 1994.
74. Wobeser, G. A., *Investigation and Management of Disease in Wild Animals,* Plenum Press, New York, 1994.

## CHAPTER 23

# Environmental Factors Affecting Contaminant Toxicity in Aquatic and Terrestrial Vertebrates

**Barnett A. Rattner and Alan G. Heath**

## CONTENTS



## 23.1 INTRODUCTION

Physical and natural factors have long been known to influence the toxicity of environmental contaminants to vertebrates. The majority of data that address this topic have been derived from studies on fish, highly inbred laboratory rodents, and man.[1–3] The degree to which these factors modify toxicity has principally been elucidated by controlled laboratory experiments. Until recently, the significance of such effects to free-ranging vertebrates (Figure 23.1) was frequently overlooked in ecological risk assessments.[4] Drawing upon controlled experiments and observational science, we overview environmental factors that influence pollutant toxicity in fish and wildlife, and present some perspective on their ecotoxicological significance.

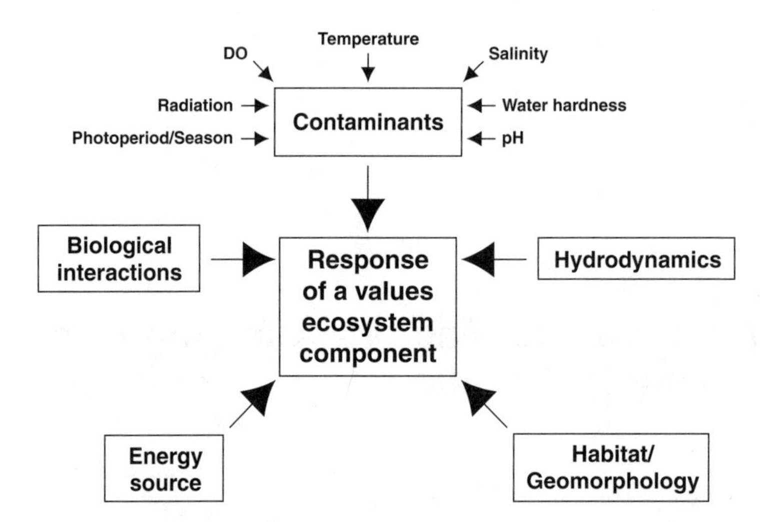


**Figure 23.1** Physical and natural factor interactions with contaminants in an aquatic ecosystem. (Modified from Foran, J.A. and Ferenc, S.A., Eds., *Multiple Stressors in Ecological Risk and Impact Assessment*, SETAC Press, Pensacola, FL, 1997.)

## 23.2 ENVIRONMENTAL FACTORS AFFECTING CONTAMINANT TOXICITY

### 23.2.1 Temperature

Of all environmental variables that affect contaminant toxicity to vertebrates, temperature has received more attention than any other. For example, a 1972 bibliography on the effects of temperature on fish compiled by Raney et al.[5] contains more than 4000 references, and the present number is at least an order of magnitude greater. Guidelines for deriving temperature criteria for freshwater fish were prepared by Brungs and Jones[6] and Houston.[7] The latter reference also addresses the interactions of temperature and chemical toxicity, a subject reviewed in detail by Cairns et al.[8] and Sprague.[9] Information available for terrestrial vertebrates focuses more on effects of drugs than environmental contaminants.[2] Our current understanding indicates that generalizations cannot be made across classes of contaminants. In part, this is because temperature can be both a lethal factor and a controlling factor, acting on a variety of biological processes such as metabolism.

#### *23.2.1.1 Aquatic and Amphibious Forms*

All fish, except large tuna and some sharks, are poikilothermic (ectothermic). Thus, body temperature tracks the environmental temperature rather precisely with little lag, even when temperature changes rapidly.[10,11] The temperature tolerance zone varies greatly among species and to a lesser degree with the age, physiological condition, and temperature to which the fish has been acclimated.[7,12] Exposure to sublethal concentrations of toxicants may reduce the upper and raise the lower lethal temperature thresholds of fish, thereby constricting the tolerance zone.[13–17] Since fish show reduced growth and impaired swimming ability when subjected to the extremes of their temperature tolerance zone, sublethal concentrations of toxicants could have greater effects on these functions at these extremes.[18]

Because body temperature of a fish tracks that of the environment, many metabolic processes exhibit a direct relationship to environmental temperature. As a general rule, energy metabolism increases about twofold with every 10°C rise in temperature. This requires gill ventilation to rise proportionately to meet increased metabolic demands.[19] A rise in water flow over the gills results in more rapid uptake of toxic chemicals through the gills, the major uptake route of waterborne chemicals in freshwater fish and to a lesser extent in marine species.[20,21] It thus could be assumed that higher temperatures would result in greater toxicity at a constant chemical concentration in the water. However, this assumption is unwarranted. Although there are numerous examples of acute toxicity being directly proportional to temperature, there are also cases (especially with chronic exposure) where the opposite occurs, or where temperature has no effect on contaminant toxicity.[22] This is because detoxication and excretion processes also increase with temperature, sometimes at rates equal to or exceeding contaminant uptake. It is important to note that rapid temperature changes coupled with toxicant exposure have more profound effects than responses observed when fish are first acclimated to the extreme temperature and then challenged with the toxicant.[7,20] However, due to limited data on rapid temperature changes, in the discussion that follows, nearly all studies reviewed utilize thermally acclimated fish.

##### *23.2.1.1.1 Acid*

Although literature on acid toxicity to fish is extensive, information on the modulating effect of temperature is relatively scarce.[23] In general, the higher the temperature, the shorter the survival time in a lethal pH for common carp (*Cyprinus carpio*) and brook trout (*Salvelinus fontinalis*).[24,25] Elevated temperature had the same effect on survival when the median lethal pH was estimated for rainbow trout (*Oncorhynchus mykiss*) fingerlings.[26] However, embryos of trout in this study exhibited the opposite response to temperature extremes (i.e., acidity was more harmful at low temperatures). This was attributed to prolonged incubation time at lower temperature, and thus there was more opportunity for the acid to exert effects on the developing embryo.

##### *23.2.1.1.2 Ammonia*

The toxicity of ammonia to fish is influenced greatly by pH and temperature because both factors affect ammonia ionization. Ionized ammonia ($NH_4^+$) has relatively little toxicity (however, see Section 23.2.2.1. on pH effects), whereas the unionized form ($NH_3$) is highly toxic. A 10°C rise in temperature at any given pH results in a threefold increase in formation of unionized ammonia.[27]

In most ammonia toxicity studies, however, the unionized ammonia concentration is the only value reported. Under these circumstances, there is little effect on toxicity at temperatures above 10°C in freshwater fish, but at lower temperatures ammonia is actually more toxic.[28] Hazel et al.[29] found no temperature effect of ammonia on three-spined sticklebacks (*Gasterosteous aculeatus*) in freshwater, but when this euryhaline species was tested in brackish water or seawater, unionized ammonia was much more toxic at the higher temperature. However, in this same report, temperature had no effect on ammonia toxicity at any salinity to another euryhaline species, the striped bass (*Morone saxatilis*), so test species differences appear critical.[29]

##### *23.2.1.1.3 Chlorine*

Chlorine is used extensively as a biocide to prevent fouling of industrial and electric-power-generating cooling systems and as a disinfectant for municipal sewage discharges. In general, it is more toxic at higher temperatures during continuous exposure.[30] However, many electric-power-generating facilities chlorinate intermittently (two or three times each day), thereby exposing the nontarget organisms to pulses of chlorine. Heath,[31] using a laboratory simulation of this regime, found that temperature had little effect on chlorine toxicity for a wide variety of cold- and warmwater fish species. Instead, the fish species and form of chlorine (i.e., free chlorine vs. monochlora-mine) were much more critical in determining relative toxicity.

##### *23.2.1.1.4 Cyanide*

This highly toxic chemical appears to kill fish more rapidly at high temperatures, but only when acutely lethal concentrations are tested.[8] Death occurs more rapidly at high temperatures because uptake is rapid and metabolic demand is accelerated, while aerobic metabolism is blocked by cyanide. At acutely toxic cyanide concentrations, detoxication mechanisms are probably overwhelmed. At lower concentrations, temperature may have no effect on time to death. Fish are able to tolerate moderate concentrations of cyanide at high temperatures, probably because detoxication mechanisms (e.g., rhodanase enzyme activity) are also highly temperature-dependent. Rainbow trout are an important exception; cyanide toxicity (96-hour $LC_{50}$ estimate) decreases at elevated temperatures.[32]

### *23.2.1.1.5 Metals*

Despite the abundance of information on the toxicity of metals to fishes, relatively little work has been conducted on interactions with temperature. Arsenic can exist in waterways in several forms, but the most stable form is arsenate. Rainbow trout are more sensitive to acute doses of this metalloid at 5°C than at 15°C, but the reverse occurs with chronic arsenate exposure.[33] Evidently, a critical body burden of arsenate must be achieved to cause death, which occurs more rapidly at warmer temperatures because of enhanced uptake rate. At lower exposure concentrations (when detoxication mechanisms are not overwhelmed), elevated temperature facilitates detoxication and excretion processes, thereby delaying or preventing accumulation of the lethal body burden.[33]

Zinc toxicity increases or is unaffected at high temperatures.[8] The definitive study by Hodson and Sprague[34] provides the probable explanation for the variety of responses reported in earlier papers. Using Atlantic salmon (*Salmo salar*) as a model, they concluded that fish were more sensitive to zinc at high temperatures, when exposures were at high concentrations and of short duration. As exposure duration was extended, temperature effects were progressively reduced until, at 2-week exposures, the $LC_{50}$ was higher (less toxic) in warm water, a phenomenon similar to that seen with cyanide and arsenate. The mechanisms for the temperature effect may not be the same, however. Zinc causes death by destroying gill tissue. It was found that cold-acclimated fish experienced more gill damage from low doses of zinc than did warm-acclimated fish, even though the latter had greater concentrations of zinc in the gills.[34]

Whereas fish are more sensitive to arsenate and zinc at cold temperatures, cold-acclimated (6°C) trout exhibited a threefold greater 10-day lethal threshold for cadmium compared with fish at 18°C.[35] The elevated toxicity of cadmium at the higher temperature was correlated with enhanced plasma calcium suppression by cadmium. Hypocalcemia is a key mechanism of cadmium toxicity to fish.[21] In addition, Eisler reported a threefold increase in toxicity of cadmium at 20°C compared to 5°C for mummichog (*Fundulus heteroclitus*).[36]

Because cadmium appears to be more toxic at higher temperatures, one might expect that fish would seek cooler water when exposed to cadmium. Contrary to expectations, brown bullhead (*Ictalurus nebulosus*) actually moved to a markedly higher temperature following cadmium exposure.[37] The investigators hypothesized this response to be a mechanism compensating for cadmium-induced suppression of ventilation. Whereas most metals, including cadmium, cause an elevation in gill ventilation, such a response in the bullhead may be an exception.[21]

The effect of temperature on cadmium toxicity in tadpoles (*Bufo* spp.) would appear to be the reverse of that in fish. Ferrari et al.[38] found that a mere 5°C rise in temperature (20–25°C) caused an approximate doubling of the cadmium $LC_{50}$. This was attributed to enhanced detoxication mechanisms in the higher temperature, although no data were presented to test this hypothesis.

Despite the large amount of data available on copper toxicity, there is little information on interactions with temperature. Based on 96-hour $LC_{50}$ estimates, copper-temperature interactions are insignificant for several species of teleosts.[39] However, Dixon and Hilton[40] observed that a high carbohydrate diet in trout increased the chronic toxicity of waterborne copper and that reduced temperature exacerbated this effect.[40] In another study, Felts and Heath[41] found that sublethal copper exposure of bluegills (*Lepomis macrochirus*) delayed metabolic acclimation to temperature change.[41] Further study of copper-temperature interactions in fishes are warranted.

Mercury and lead exhibit elevated toxicity at higher temperatures, apparently due to a direct relation between metal accumulation and temperature.[42,43] Mercury accumulates twice as fast with a 10°C rise in temperature.[44] Metabolism (and therefore gill ventilation) show a similar rate of increase with temperature, which probably facilitates the uptake; however, excretory processes for mercury and lead may fail to rise at a comparable rate. Part of the detoxication process involves binding of the metal to metallothionein.[21] Most measures of tissue accumulation do not differentiate between metal bound to metallothionein and that free in the cells, and thus there may not be a direct correlation between body burden and lethality. The greater accumulation of lead and mercury at high temperatures has relevance for predatory animals, including humans, that consume contaminated fish.

#### *23.2.1.1.6 Organics*

Polychlorinated biphenyl (PCB) accumulation rates are enhanced by increased temperature. Spigarelli et al.[45] explored this relation in adult brown trout (*Salmo trutta*) in both constant temperatures and a diel temperature cycle. About 90% of the PCB was taken in by ingestion of food and the remainder via gills. The accumulation rate followed the growth rate, and both responded to temperature as if the fish had been acclimated to the peak of the diel cycle. Such a finding would be interesting to test with other contaminants, including metals, because diel cycles are common in nature.

Among organic pesticides, organochlorine compounds are generally more toxic to fish than organophosphorus compounds.[46] DDT has been reported to be more toxic at low temperatures in the laboratory. Decreased water temperature in streams two or three months after application can result in delayed mortality of trout.[47,48] DDT also causes trout to choose warmer temperatures, although at high concentrations they may actually prefer cooler temperatures.[49] Eisler[50] found that the 96-hour mortality of *Fundulus heteroclitus* at a given DDT dose was least at 20°C and rose at temperatures above and below 20°C, suggesting the existence of a temperature optimum for minimal toxicity in that species.

Methoxychlor is also more toxic to fish in cold water.[51] Such an interaction with temperature may be limited to salmonids. In tests with lindane, Johnson and Finley report a 2.3-fold decrease in toxicity to rainbow trout with increasing temperature from 2 to 18°C, but the opposite occurred in bluegills (a warmwater centrarchid) over a temperature range of 7 to 29°C.[52] Macek et al.[51] tested effects of four other organochlorine insecticides on bluegills and trout, and found that all the compounds to be more toxic (96-hour $LC_{50}$ estimates) at high temperatures. In European eels (*Anguilla anguilla*), the highest $LC_{50}$ estimate for lindane was at 22°C, while greater lindane toxicity was observed at 15 and 29°C.[53] Thus, temperature effects on organochlorine pesticide toxicity are highly dependent on the species of fish and compound tested.

Although some organochlorine compounds are more toxic at cold temperatures, the opposite occurs for organophosphorus compounds.[47,52,54,55] This is also true for carbaryl (a carbamate), although there are also marked species differences in temperature modulation of its toxicity.[52] The time-course of temperature acclimation may be an important factor because the action of organophosphorus and carbamate insecticides is to inhibit acetylcholinesterase in nervous tissue, and the activity of this enzyme changes during the temperature-acclimation process in fish.[56]

Pyrethroid exposure studies in fish have demonstrated increased toxicity with reduced water temperature.[57,58] One of the most pronounced thermal effects was seen with permethrin (a synthetic pyrethroid) exposure of rainbow trout wherein the 96-hour $LC_{50}$ rose tenfold over a temperature range of 5 to 20°C.[58] Thus, higher temperatures are clearly protective for this compound in trout. Similar types of studies in amphibians have yielded somewhat inconsistent findings. Lethality of several pyrethroids in adult *Rana pipens* was noted to be dramatically enhanced by cold temperature (4 vs. 20°C).[59] Using behavioral endpoints, recovery of newly hatched *Rana clamitan* and *Rana pipiens* tadpoles following sublethal exposures to permethrin and fenvalerate was slower at 15 vs. 20°C.[60] In contrast, a moderately cool temperature (18 vs. 22°C) actually lowered lethality of esfenvalerate in tadpoles (*Rana* spp.).[61]

The cytochrome P4501A enzymes are an important detoxication mechanism for organic contaminants in fish, and these enzymes are inducible by the presence of many toxic substances.[62] Strong induction was detected in both winter- and summer-acclimated arctic char (*Salvelinus alpinus*); however, the response was delayed and longer-lasting in the cold.[63] It should be noted that species differences may be very important here, for Machala et al.[64] found that acclimating carp (a warmwater species) to winter temperatures caused a "highly significant inhibition" of the induction process.

### 23.2.1.2 Terrestrial Forms

Ambient temperature influences the toxicity of contaminants in homeotherms by affecting metabolic rate, energetics, neural control of thermoregulatory function, and contaminant uptake, transformation, and detoxication. For many pollutants, toxicity is lowest within the thermoneutral zone and tends to increase outside this range, thereby exhibiting a "U-shaped" response curve.

#### 23.2.1.2.1 Petroleum Crude Oil

Spillage of oil during transportation and discharge of petroleum hydrocarbons as industrial effluent frequently result in the exposure of birds externally, by the oiling of plumage, and internally, through the ingestion of oil while feeding or preening. Direct external exposure is probably the most important toxic effect because oiling of feathers causes insulating air to escape, resulting in loss of body heat and increased metabolic rate.[65–68] Under such physiological conditions, birds exposed to low ambient temperatures (e.g., cold water or air) rapidly deplete their energy reserves and become hypothermic, which can lead to death. Ingestion of oil may alter metabolic rate in seabirds[67] but does not seem to affect metabolic rate in domestic ducks (*Anas* spp.).[69] Nonetheless, mild cold exposure seems to exacerbate the toxicity of chronically ingested crude oil in mallards (*Anas platyrhynchos*).[70] Studies in England have revealed that avian mortality associated with oil pollution rises from a minimum in the summer to a peak in late winter.[71]

External oiling of wild mammals (e.g., sea otter, *Enhydra lutris*; polar bear, *Urus maritimus*) increases thermal conductance across the insulating fur layer, resulting in substantial heat loss; this is compensated by an elevated metabolic rate.[72] Thermal effects have been predicted to be substantially greater upon exposure to cold air or water, although polar bears have exhibited a marked increase in core temperature following external oil exposure that might reflect a febrile response.

#### 23.2.1.2.2 Organochlorine Contaminants

In Columbiformes and gulls (*Larus fuscus*), ingestion of DDT and PCB were associated with reduced metabolic rates, whereas elevated metabolic rates were reported when these compounds were fed to northern bobwhite (*Colinus virginianus*).[73–76] Cold exposure following ingestion of 100 parts per million (ppm) DDT significantly decreased survival of quail.[76] This response was attributed to the mobilization of stored DDT and its metabolites. Based upon egg loss in a ruffed grouse (*Bonasa umbellus*) population following a DDT spray, cool weather has been suggested to act synergistically on DDT toxicity.[77] A retrospective study of the relationship between organochlorine contaminant concentration in herring gull eggs (*Larus argentatus*) and cold weather in the Great Lakes between 1962 and 1995 indicated that PCB levels were positively correlated with winter severity.[78] This relation appears to reflect altered migratory and feeding behavior, as gulls move to more highly contaminated southerly Great Lakes locations during severe winters and apparently ingest larger quantities of fish.

#### 23.2.1.2.3 Organophosphorus and Carbamate Pesticides

Studies with laboratory rodents show that the toxicity of cholinesterase-inhibiting pesticides are enhanced several fold at reduced and elevated ambient temperature.[79,80] Considerable evidence indicates that this effect is due to disruption of central cholinergic and monoaminergic thermoregulatory function, endocrine dysfunction, and altered pesticide absorption.[81] Furthermore, ambient temperature has long been recognized to affect the environmental half-life of these pesticides when sprayed on crops.[82]

Juvenile northern bobwhite (*C. virginianus*) and mallard ducklings (*A. platyrhynchos*) acutely or subchronically exposed to anticholinesterase compounds (e.g., carbofuran, chlorpyrifos, temephos) exhibit decreased survival rates when maintained at low ambient temperatures compared with thermoneutral controls.[83–85] A twofold increase in parathion lethality has been observed in northern bobwhite (*C. virginianus*) and Japanese quail (*Coturnix japonica*) maintained within environmental chambers at ambient temperatures below or above the thermoneutral zone.[81,86] At sublethal parathion dosages, heat-exposed quail exhibited greater brain acetylcholinesterase inhibition than similarly dosed quail exposed to cold or maintained at thermoneutral temperatures. In contrast, recovery of brain acetylcholinesterase activity in brown-headed cowbirds (*Molothrus ater*) receiving dimethoate was actually slower in cold-exposed birds compared to thermoneutral controls.[83] These alterations in toxicity are attributed to disruption of thermoregulatory function, as evidenced by hypothermia, and other metabolic and endocrine responses to heat and cold but not to effects on the activities of hepatic enzymes involved in anticholinesterase activation or detoxication (e.g., parathion oxidase, paraoxonase, and paraoxon deethylase).[81,85,87,88] The circannual toxicity of parathion was investigated in European starlings (*Sturnus vulgaris*) housed in outdoor pens encountering a wide-range of temperatures during winter, spring, summer and fall.[89] Based upon the acute $LD_{50}$ of parathion, toxicity was greater during hot summer weather compared with cool winter weather ($LD_{50}$ of 118 vs. 160 mg/kg).

In a field setting, low temperatures contributed to the poisoning of about 1500 geese (*Anser anser* and *A. brachyrhynchos*) that fed upon carbophenothion-treated winter wheat, and elevated ambient temperature was suggested to have been an exacerbating component in a sage grouse (*Centrocercus urophasianus*) die-off following dimethoate treatment of alfalfa in Idaho.[90–92]

Temperature-contaminant interaction studies in wild mammals are quite limited. Acute oral exposure to parathion has been demonstrated to evoke hypothermia in white-footed mice (*Peromyscus leucopus*); however, parathion toxicity was not affected in mice maintained at 10°C, possibly because of their ability to undergo facultative torpor.[93]

#### *23.2.1.2.4 Lead Shot*

Lead poisoning from ingested shot is a major cause of waterfowl mortality throughout the world. In North America, most waterfowl that die from ingested lead shot succumb during the cold weather that accompanies the hunting season. Harsh winter weather is generally thought to exacerbate lead toxicity in birds. This has been confirmed in part by studies of Kendall and Scanlon[94] with ringed turtle doves (*Streptopelia risoria*); marked mortality and greater blood and liver lead concentrations were observed in doves exposed to 6°C compared with birds maintained at 21°C. Comparison of lead-shot toxicity (mortality and body weight change) to mallards (*A. platyrhynchos*) and black ducks (*Anas rubripes*) housed outdoors in winter- and summer-dosing trials suggest that shot toxicity was greater in winter.[95] Ambient temperature does not seem to affect the toxicity of alternative shot (e.g., bismuth, tungsten, tin) used for hunting of waterfowl in North America.

#### *23.2.1.2.5 Other Contaminants*

Temperature influences the toxicity of sodium monofluoroacetate (compound 1080), a commonly used rodenticide. When compared to thermoneutral controls, acute toxicity was greater at low ambient temperatures for brushtail possums (*Trichosurus vulpecula*) and also greater at high ambient temperatures for raccoons (*Procyon lotor*).[96,97] It was suggested that target-animal control strategies should include scaling of dosage at various ambient temperatures to minimize impacts on nontarget animals while maintaining overall efficacy.[97]

### 23.2.2 Salinity, Water Hardness, and pH

#### *23.2.2.1 Aquatic and Amphibious Forms*

Although the ionic environment is markedly different between the ocean and freshwater, salinity has relatively little influence on acute toxicity of most chemicals.[9,98] Limited data suggest that chronic responses to toxicants are more readily modulated by salinity than acute responses to toxicants. An important caveat is that tests must be conducted at a salinity to which the species is adapted; lack of ionic acclimation can enhance sensitivity to the chemical.[99] Furthermore, when the same species is acclimated to either seawater or freshwater, sublethal effects on variables such as blood electrolyte concentrations may be far greater in fish acclimated to freshwater.[100] A potential complicating factor related to the toxicity of metals is the enhanced bioavailability of free metal ions (toxic form) at low salinities. This explains why nearly all metals are more toxic at low salinities.[101]

Euryhaline species, such as mummichog (*F. heteroclitus*) and rainbow trout (*O. mykiss*), tend to be most sensitive to toxicants in low salinities; many contaminants are least harmful at intermediate salinity near the isosmotic range of 11 to 15 parts per thousand.[101–104] Because acute doses of a wide variety of chemicals cause ionoregulatory dysfunction, this is not unexpected.[21] For example, Wilson and Taylor[105] found that exposure of freshwater-acclimated rainbow trout to copper at a concentration of 4.9 mmol caused 100% mortality within 24 h due to severe ionoregulatory and respiratory failure. But when trout were acclimated to full strength seawater and exposed to 6.3 mmol copper for 24 h, there were no alterations in ionoregulatory or respiratory function. Note that full strength seawater is actually hypertonic to fish blood; however, the high calcium concentration in the seawater may provide more competition for external binding sites with copper.[106] Even in a freshwater environment, a low level (4 meq/L) of sodium chloride has been found to reduce the ionoregulatory stress (and thus toxicity) of aluminum.[107]

Responses to silver appear to differ in that seawater-acclimated rainbow trout are more sensitive to this metal at higher salinities than at lower ones.[108] The investigators attributed this to the incomplete hypoosmoregulatory ability of the rainbow trout. It is noteworthy that alterations in salinity apparently caused no change in silver speciation.

In their extensive review, Hall and Anderson[101] found no consistent trend of salinity on the toxicity of organic chemicals. Some possible exceptions are the organophosphorus insecticides, which appear to be more toxic as salinity increases.

Water hardness, usually expressed as mg/L $CaCO_3$, substantially affects metal toxicity but seemingly has little effect on the toxicity of organic chemicals.[9,22,109] Where hardness affects toxicity of organics, it can often be attributed to the effect of pH differences (discussed elsewhere in this chapter). The toxicity of common water pollutants, such as ammonia, surfactants, and phenols, are not greatly affected by water salinity and hardness.[9]

The relation of water hardness to metal toxicity has produced extensive literature including the classic 1968 paper by Brown.[110] The almost universal observation is that metals are less toxic in hard water, provided the pH is constant. The effect of hardness varies with the metal, being most dramatic with cadmium, due in part to its precipitation in hard water. Hardness has an almost negligible effect on dissolved lead toxicity.[111]

Copper, zinc, and nickel toxicity are only moderately affected by water hardness.[110,112,113] In the case of zinc, the effect of hardness seems to be due to changes in the fish gills rather than with metal speciation. Increasing hardness reduces the uptake rate of zinc by gill tissue, the prime target organ for zinc lethality.[114] Even at lower doses of zinc, where gill damage is minimal, increased hardness impedes metal uptake into the fish and may facilitate its excretion as well.[115]

The situation is more complex in the case of copper because water pH greatly affects copper speciation.[116] Playle et al.[117] found that calcium (but not $CaCO_3$) reduced gill accumulation of copper at pH 4.8 but not at pH 6.3. Apparently, both calcium and hydrogen ions compete with copper for binding sites on gills.

There has been considerable debate as to whether total hardness or alkalinity are more important in copper toxicity in fish.[118] In part the uncertainty exists because alkalinity and hardness tend to track each other, and copper complexes with both carbonate and hydroxide ions. Furthermore, calcium and magnesium reduce copper complex formation.

Consideration should briefly be given to acid pollution, a topic discussed in the first edition of this text.[119] The calcium content of water (i.e., hardness) has long been known to be protective of the effects of low pH.[120] The lakes and streams most sensitive to acid pollution are those that are poorly buffered and, therefore, almost always characterized by low water hardness. Extensive studies on lakes in northern Norway have shown that those with calcium-to-hydrogen ratios of less than 3:1 are usually devoid of fish, whereas those with calcium-to-hydrogen ratios greater than 4:1 support good fisheries.[120]

An interesting complex-formation phenomenon occurs with silicon. Much of the toxicity from acid pollution is due to the presence of aluminum, which is leached from aluminosilicates of rocks and soils by low pH precipitation.[23,121] The resulting silicic acid in the waters has been shown to greatly reduce or even eliminate the toxicity of aluminum in acid water if excess silicon is present.[122]

It has been claimed that pH has a greater modifying effect on contaminant toxicity to aquatic animals than any other physical factor.[22] This is seen dramatically with ammonia, where a rise of one pH unit in the midrange of that found in surface waters increases the level of $NH_3$, the toxic form, about sixfold over $NH_4^+$, the ionized and less toxic form. Cyanide also becomes more ionized, and thus less toxic, as pH rises, but the effect is less dramatic than that seen with ammonia.[9]

The effect of pH on metal toxicity is complicated because the form of the metal changes greatly with the pH,[116] and the respective toxicities of these forms are quite different. This results in unusual toxicity curves. For example, rainbow trout are most sensitive to copper at pH 6, and this sensitivity decreases as one raises the pH to 7 or lowers it below 6. However, the toxicity increases again as the pH is lowered below 5 or raised above 8.[123] The toxicological interactions among copper and dissolved organic carbon, calcium, and pH have been extensively explored by Welsh and coworkers.[124]

In general, zinc is less toxic at a lower pH presumably due to less accumulation in the gills.[113] This is also true of cadmium and nickel,[125] but at pHs below 5 to 6 (depending on the fish species), acid toxicity alone may be an important factor.

Organic contaminants in the unionized form are more toxic because they more readily penetrate membranes than in the ionic form,[9] and water pH is the major factor influencing ionization.[22] Toxicity of the organophosphorus pesticide chlorpyrifos increased when pH was increased from 7.5 to 9,[52] and the same was seen with trichlorfon over this pH range.[55] However, the pesticide terbufos was least toxic at pH 7.5, with toxicity increasing above and below this point.[55]

#### 23.2.2.2 Terrestrial Forms

Ingestion of hyperosmotic water from alkaline lakes, prairie potholes, and estuarine and marine environments constitutes one stressor frequently encountered by endemic and migratory birds. Regulation of fluid and electrolyte balance under these circumstances is facilitated by the salt gland. Contaminant-induced disruption of osmoregulation by the inhibition of $Na^+$-$K^+$-ATPase activity has been observed in birds.

Friend and Abel[126] reported increased mortality rates in mallard ducklings maintained on 0.5% salt water and challenged with an osmotic stress following dietary exposure to the organophosphorus insecticides bidrin and parathion. Increasing the concentration of NaCl in drinking water also enhances the toxicity of malathion and EPN (alone and in combination). However, other studies failed to confirm this observation in adult birds and even suggested that saltwater may have blunted adverse effects by reducing ingestion rates of test diets containing the organophosphorus insecticide fenthion.[127]

Several studies suggest that salt gland $Na^+$-$K^+$-ATPase activity and salt gland excretory function of birds may be depressed by organochlorine and cholinesterase-inhibiting pesticides.[128–130] Large die-offs and severe population declines of some seabird species were at one time attributed to osmoregulatory failure due to mobilization of organochlorine pesticides and their metabolites during starvation. However, other studies demonstrated that this was unlikely.[131]

The effects of acidity on amphibious forms have recently been reviewed in detail.[132] There is considerable inter- and intraspecific variation in pH tolerance among amphibians. Low pH in amphibian embryos can lead to embryonic mortality by impairing development and by inhibiting enzymes necessary for the expansion of the perivitelline membrane for hatching, causing embryos to curl.[133–137] Acidity can cause the concentration of sodium to drop in larval amphibians, resulting in mortality.[138–139] The toxicity of low pH is ameliorated for embryos of some species (*Ambystoma jeffersonianum, Rana pipiens, Bufo americanus*) by the presence of Ca cation and in larvae by high concentrations of cations such as Ca, Mg, and Na.[137] Little research has been conducted on the effects of pH on adult amphibians. Chronically lethal soil pH for red-backed salamanders (*Plethedon cinereus*) was between 3 and 4, and free-ranging populations were reduced at soil pH 3.7.[140] Sublethal effects of low pH that could reduce long-term survival of amphibians include alteration of time to hatch, abnormal or deformed embryos and hatchlings, altered behavior after hatching, decreased growth rate of larvae, and delayed metamorphosis.[132]

Aluminum (Al) and low pH can act independently or synergistically, and in some species Al can ameliorate the effects of low pH (similar to the effects of Ca cations).[137,141–145] Increasing pH can increase toxicity of Cd and Al to *Rana temporaria* larvae,[146] as Cu can for *R. sylvatia* and *A. jeffersonianum.*[147] Addition of Ca and Mg ameliorated the effects for only *R. sylvatica*. Survival of *A. jeffersonianum* larvae at pH 4.5 and 5.5 was dramatically reduced upon chronic exposure to high concentrations of Al, Cu, Fe, Pb, and Zn in combination.[148]

In contrast to aquatic vertebrates that reside directly in media whose pH can vary dramatically, effects of pH on contaminant toxicity in terrestrial vertebrates are of less concern. Nonetheless, it is well documented that acid precipitation and poor buffering capacity in some habitats can evoke pH-related effects on the bioavailability of several heavy metals and aluminum to birds and mammals. These indirect effects of pH on contaminant toxicity in higher vertebrates have been reviewed in some detail.[119,149]

### 23.2.3 Oxygen Tension

#### *23.2.3.1 Aquatic Forms*

The dissolved oxygen (DO) content of water is quite variable, even under so-called "natural" conditions (e.g., the hypolimnion of thermally stratified lakes is usually low in oxygen). The term *hypoxia* refers to any condition in which oxygen content is below saturation; *anoxia* means no oxygen. Anthropogenic inputs of organic wastes result in algal and microbial blooms, which may cause marked oxygen depletion, especially at night in the absence of photosynthesis. Waters that contain contaminants are often low in oxygen, which can influence their toxicity.

Although the influences of DO on toxicity were of substantial interest in the 1960s and early 1970s, surprisingly little research has been conducted on it until recently. Lloyd[150] found that as DO diminished, acute toxicity of zinc, copper, lead, and phenols increased almost linearly by up to a factor of two. Hydrogen sulfide toxicity to goldfish was increased by 1.4 times by lowering the DO to 10% of saturation, a level of oxygen most species of fish could probably not tolerate.[151,152] Whole pulp mill effluent was about 1.5 times more toxic at 38% oxygen saturation than at 79% saturation.[153] Ammonia was 1.5- to 2-fold more toxic when DO was lowered to 30% saturation.[150,154] Bluegills exposed to 22% oxygen saturation were 1.7 times more sensitive to the surfactant linear alkylate sulfonate (LAS) compared with controls in oxygen saturated water, provided the fish were first acclimated to the low DO.[155]

Acclimation to the low DO could be an important factor deserving more attention. Hokanson and Smith[155] found that the toxicity of LAS increased by an order of magnitude if fish were not acclimated to low DO before exposure. Fish increase their gill ventilation when exposed to hypoxia, which enhances toxicant contact and uptake by gills, a hypothesis first proposed by Lloyd.[21,150] Acclimation to hypoxia reduces ventilation volume and causes other physiological adjustments that facilitate functioning in low DO.[21] Whether acclimation would be relevant to field situations depends on local circumstances. A diurnal DO depletion in a highly eutrophic pond would permit little time for acclimation, whereas acclimation may occur during seasonal cycles.

These data suggest that hypoxia has a relatively small effect on acute toxicity of contaminants in fish. However, Gupta et al.[156] reported that various mixtures of three phenolic compounds exhibited markedly different toxicities when tested with the bronze featherback (*Notopterus notopterus*) maintained at differing DO concentrations. Over a rather limited DO range of 6 to 7.8 ppm, the 96-hour $LC_{50}$s estimates for the mixtures differed by as much as 15-fold.

An unusual interaction between DO with acute anthracene toxicity has been observed in bluegills.[157] The toxicity of this polycyclic aromatic hydrocarbon is induced by ultraviolet light, and this process is oxygen dependent. In this instance, the relation of DO and toxicity is not linear. Instead, a bimodal curve results in which anthracene is least toxic at low and high DO levels. The heightened toxicity at the intermediate DO may, as McCloskey and Oris hypothesize, be due to a complex interaction between ventilation volume and oxygen-dependent toxicity.[157] Ventilation volume of fish increases as DO is lowered, and this brings more toxicant in contact with the gills. At the highest DO, ventilation volume is at its lowest so the gills are exposed to less anthracene.[157]

Studies involving prolonged chemical exposure are even less numerous than those devoted to acute toxicity. In 32-day tests in which fathead minnow (*Pimephales promlas*) embryos and larvae were exposed to 1,2,3-trichlorobenzene, Carlson[158] observed that reducing the DO to 4.5 ppm caused greater effects on survival and weight gain. However, these responses were observed only at a concentration above the toxicity threshold. There did not appear to be any influence of DO on the toxicity threshold of 1,2,3-trichlorobenzene. However, Chapman and Shumway observed that hypoxia caused both a decrease in survival time in lethal concentrations and a lowering of the threshold concentration in steelhead trout embryos and larvae exposed to pentachlorophenol.[159]

The inhibition of acetylcholinesterase in the nervous system of fish can be used to detect organophosphorus and carbamate pesticide poisoning, although other contaminants (e.g., mercury) may also inhibit this enzyme.[160] Hoy et al.[161] found that fish treated with dichlorvos (an organophosphorus compound used to treat for ectoparasite infestation) experienced a severe inhibition of acetylcholinesterase. Enzyme inhibition was greater when the water had low DO. The mechanism may involve rapid uptake of the pesticide due to hyperventilation by the fish in hypoxic water.

The herbicide paraquat causes increased superoxide dismutase (SOD) in cells as a mechanism to remove free radicals during its metabolism. Vig and Nemcsok exposed carp to severe hypoxia (< 1 ppm DO) and found SOD stimulation in the gills, liver, and brain.[162] Paraquat alone stimulated SOD activity only in gill tissue.[162] Combining hypoxia and paraquat produced an additive effect in the gills but no additional enhancement of SOD activity in the other tissues.

Hypoxia can itself be a lethal factor. In combination with sublethal quantities of a toxicant, there is some evidence of exacerbation of the harmful effects of hypoxia. Nitrite causes methemoglobin formation, which effectively reduces the ability of hemoglobin to transport oxygen. It is not surprising that channel catfish (*Ictalurus punctatus*) exposed to nitrite are less able to tolerate low levels of dissolved oxygen.[163] Another mechanism by which a toxic chemical may reduce tolerance to hypoxia is illustrated by the work of Hlohowskyj and Chagnon,[164] who found that phenol, which causes histological damage to the gill tissue, reduces hypoxic tolerance in the stoneroller minnow (*Capostoma anomalum*). Since a wide variety of chemicals damage gill tissue, this phenomenon of reduced hypoxic tolerance resulting from chemical exposure deserves further study.[21]

#### *23.2.3.2 Terrestrial Forms*

Although two reviews have addressed air-quality effects on wildlife, no information is available on the effects of oxygen tension on contaminant toxicity in terrestrial animals.[165,166] This is a fertile research area because some species of birds have been observed at altitudes up to 11,300 m,[167] and many mammals may be exposed to moderate hypoxia in their burrows or during torpor.

### 23.2.4 Nonionizing Radiation

In the past two decades, there has been increased interest in the effects of nonionizing radiation (frequency: 0–300 GHz; wavelength: 10,000 m to 1 mm) on birds with its expanding use by man (e.g., radionavigation; radio, television, and satellite communication; and weather radar). Its pervasiveness has been suggested to result in direct irradiation of birds and their eggs.[168] Direct effects on embryonic development, growth, reproduction, and behavior have been documented in passerines and mallards.[168] Although the potential for interaction with environmental contaminants is great, it has yet to be investigated in free-ranging birds. Because much of the spectrum of nonionizing radiation is rapidly attenuated in the water column, its interaction with contaminants is of limited concern to aquatic vertebrates.

### 23.2.5 Photoperiod and Season

Photoperiod and season are interrelated because the natural cause of photoperiod difference is the change of season. Temperature, which has a profound effect on physiological systems, changes with season. Thus, there is some application of studies on temperature effects to season. Although information is quite limited, photoperiod *per sec* appears to have little effect on toxicant threshold concentrations.[169] However, exposure to ultraviolet light, which is affected by photoperiod and season, has pronounced effects on toxicity of polynuclear aromatic hydrocarbons, carbamates, and some other contaminants through photoactivation mechanisms.[170,171]

There is considerable interest in the use of biochemical, histological, morphological, and physiological biomarkers for evaluating environmental health.[172–174] It must be appreciated that nearly all such parameters change to a greater or lesser degree with season.[175–177] This is not a serious problem provided that the populations under investigation are compared with nearby reference populations sampled at the same time. Unless seasonal fluctuations of parameters are understood in a given species, "baseline" values are difficult to interpret.

#### *23.2.5.1 Aquatic Forms*

Many species of fish are seasonal spawners and thus exhibit marked changes in steroid hormone concentrations and energy stores associated with both temperature and photoperiod. These changes are reflected in alterations in the enzyme systems that metabolize xenobiotic chemicals in fish collected from the field.[178–180] Current data does not permit generalizations as to whether spawning stimulates or inhibits a particular enzyme activity; conflicting results may be due to interspecific differences. Liver metallothionein concentration also appears to be greatly modulated by spawning season, independent of metal exposure.[181] It seems probable that the ability of fish to sequester, transform, and excrete toxic chemicals is affected by season, and these effects are due in part to spawning activity.

#### *23.2.5.2 Terrestrial Forms*

In dosing studies with the avicide 3-chloro-*p*-toluidine hydrochloride, starlings were found to be more sensitive during daylight hours.[182] Reduction of the light cycle in chronic Kepone feeding studies evoked a dramatic increase in toxicity to immature Japanese quail but not adults.[183] Serum B esterases, a key target for cholinesterase-inhibiting pesticides, exhibit diurnal variation, although such changes do not seem to have biologically significant effects on tolerance to anticholinesterase pesticides.[184,185]

Wild birds and mammals exhibit pronounced fluctuations in body mass and composition associated with seasonal changes in diet and physiological condition as well as reproductive and migratory components of their annual cycle. Dramatic alterations in organic and inorganic contaminant tissue burdens have been documented to accompany these changes.[186–194] Furthermore, seasonal variation in enzyme activities (e.g., acetylcholinesterase and monooxygenase activity) have been detected in several species of amphibians, birds, and mammals.[185,195,196] The toxicological significance of these circannual changes remain poorly understood.

## 23.3 CONCLUSIONS

In contrast to toxicity studies conducted in a controlled laboratory setting, free-ranging animals simultaneously encounter a combination of environmental variables that may influence, and even act synergistically to alter, contaminant toxicity. It is not possible to rank these factors, particularly since they are often interrelated (e.g., temperature and seasonal rhythms). Moreover, species differences in toxic chemical sensitivity are often greater than the modulatory effects of environmental variables. However, it is clear that environmental factors (particularly temperature) may alter contaminant exposure and toxicity (accumulation, sublethal effects, and lethality) by more than an order of magnitude in some species. Accordingly, effects of abiotic environmental variables should be considered and factored into ecological risk assessments of anthropogenic pollutants.

## REFERENCES


1. Parrish, P. R., Acute toxicity tests, in *Fundamentals of Aquatic Toxicology*, Rand, G. M. and Petrocelli, S. R., Eds., Hemisphere Publishers, Washington, D.C., 1985, Chap. 2.
2. Hayes, W. J., Jr., Dosage and other factors influencing toxicity, in *Handbook of Pesticide Toxicology*, Hayes, W. J., Jr. and Laws, E. R., Jr., Eds., Academic Press, New York, 1991, Chap. 2.
3. Mayer, F., Marking, I., Pedigo, I., and Brecken, J., Physicochemical Factors Affecting Toxicity: pH, Salinity, and Temperature, Part 1 — Literature Review, EPA 600/X-89–033, U.S. Environmental Protection Agency, Gulf Breeze, FL, 1989.
4. Foran, J. A. and Ferenc, S. A., Eds., *Multiple Stressors in Ecological Risk and Impact Assessment*, SETAC Press, Pensacola, FL, 1997.
5. Raney, E. C., Menzel, B. W., and Weller, E. C., Heated Effluents and Effects on Aquatic Life with Emphasis on Fishes: A Bibliography, U.S. Atomic Energy Commission, TID-3918, 1972.
6. Brungs, W. A. and Jones, B. R., Temperature Criteria for Freshwater Fish: Protocol and Procedures, U.S. Environmental Protection Agency, Ecol. Res. Ser., EPA-600/3–77–061, 1977.
7. Houston, A. H., *Thermal Effects upon Fishes,* National Research Council Canada, Ottawa, 1982.
8. Cairns, J., Heath, A. G., and Parker, B. C., The effects of temperature upon the toxicity of chemicals to aquatic organisms, *Hydrobiologia,* 47, 135, 1975.
9. Sprague, J. B., Factors that modify toxicity, in *Fundamentals of Aquatic Toxicity,* Rand, G. M. and Petrocelli, S. R., Eds., Hemisphere Publishers, Washington, D.C., 1985, Chap. 6.
10. Hughes, G. M. and Roberts, J. L., A study of the effect of temperature changes on the temporary respiratory pumps of the rainbow trout, *J. Exp. Biol.*, 52, 177, 1970.
11. Stauffer, J. R., Dickson, K. L., Heath, A. G., Lane, G. W., and Cairns, J., Body temperature change of bluegill sunfish subjected to thermal shock, *Prog. Fish Cult.*, 37, 90, 1975.
12. Fry, F. E. J., Responses of vertebrate poikilotherms to temperature, in *Thermobiology*, Rose, A. H., Ed., Academic Press, New York, 1967, 375.

13. Paladino, F. V., Spotila, J. R., Schubauer, J. P., and Kowalski, K. T., The critical thermal maximum: A technique used to elucidate physiological stress and adaptation in fishes, *Rev. Can. Biol.*, 39, 115, 1980.
14. Watenpaugh, D. E. and Beitinger, T. L., Se exposure and temperature tolerance of fathead minnows, *Pimephales promelas, J. Therm. Biol.*, 10, 83, 1985.
15. Lydy, M. J. and Wissing, T. E., Effect of sublethal concentrations of copper on the critical thermal maxima (CTMax) of the fantail *(Etheostoma flabellare)* and johny (*E. nigrum*) darters, *Aquat. Toxicol.*, 12, 311, 1988.
16. Heath, S., Bennett, W., Kennedy, J., and Beitinger, T., Heat and cold tolerance of the fathead minnow *Pimephales promelas*, exposed to the synthetic pyrethroid cyfluthrin, *Can. J. Fish. Aquat. Sci.*, 51, 437, 1994.
17. Heath, A. G., Cech, J. J., Brink, L., Moberg, P., and Zinkle, J. G., Physiological responses of fathead minnow larvae to rice pesticides, *Ecotoxicol. Environ. Saf.*, 37, 280, 1997.
18. Smith, L. S., *Introduction to Fish Physiology*, TFH Publications, Neptune City, NJ, 1982.
19. Heath, A. G. and Hughes, G. M., Cardiovascular and respiratory changes during heat stress in rainbow trout, *Salmo gairdneri, J. Exp. Biol.*, 59, 323, 1973.
20. Black, M. C., Millsap, D. S., and McCarthy, J. F., Effects of acute temperature change on respiration and toxicant uptake by rainbow trout, *Salmo gairdneri*, (Richardson) *Physiol. Zool.*, 64, 145, 1991.
21. Heath, A. G., *Water Pollution and Fish Physiology*, 2nd ed., CRC Press, Boca Raton, FL, 1995.
22. Mayer, F. L. and Ellersieck, M. R., Experiences with single-species tests for acute toxic effects on freshwater animals, *Ambio,* 17, 367, 1988.
23. Morris, R., Taylor, E. W., Brown, D. J. A., and Brown, J. A., *Acid Toxicity and Aquatic Animals*, Cambridge University Press, U.K., 1989.
24. Korwin-Kossakowski, M. and Jezierska, B., The effect of temperature on survival of carp fry, *Cyprinus carpio* L., in acidic water, *J. Fish Biol.*, 26, 43, 1985.
25. Robinson, G. D., Dunson, W. A., Wright, J. E., and Mamolito, G. E., Differences in low pH tolerance among strains of brook trout (*Salvelinus fontinalis*), *J. Fish Biol.*, 8, 5, 1976.
26. Kwain, W., Effects of temperature on development and survival of rainbow trout, *Salmo gairdneri,* in acid waters, *J. Fish. Res. Bd. Can.*, 32, 493, 1975.
27. Emerson, K, Russo, R. C., Lund, R. E., and Thurston, R. V., Aqueous ammonia equilibrium calculations: Effect of pH and temperature, *J. Fish. Res. Bd. Can.*, 32, 2379, 1975.
28. EIFAC, Water quality criteria for European freshwater fish, Report on ammonia and inland fisheries, *Water Res.*, 7, 1011, 1973.
29. Hazel, C. R., Thomsen, W., and Meith, S. J., Sensitivity of striped bass and stickleback to ammonia in relation to temperature and salinity, *Calif. Fish Game,* 57, 154, 1971.
30. Burton, D. T., Hall, L. W., Margrey, S. L., and Small, R. D., Interactions of chlorine, temperature change, and exposure time on survival of striped bass (*Morone saxatilis*) eggs and prolarvae, *J. Fish. Res. Bd. Can.*, 36, 1108, 1979.
31. Heath, A. G., Toxicity of intermittent chlorination to freshwater fish: Influence of temperature and chlorine form, *Hydrobiologia*, 56, 39, 1977.
32. Kovacs, T. and Leduc, G., Acute toxicity of cyanide to rainbow trout (*Salmo gairdneri*) acclimated at different temperatures, *Can. J. Fish. Aquat. Sci.,* 39, 1426, 1982.
33. McGeachy, S. M. and Dixon, D. G., Effect of temperature on the chronic toxicity of arsenate to rainbow trout (*Oncorhynchus mykiss*), *Can. J. Fish. Aquat. Sci.*, 47, 2228, 1990.
34. Hodson, P. V. and Sprague, J. B., Temperature-induced changes in acute toxicity of zinc to Atlantic salmon (*Salmo salar*), *J. Fish. Res. Bd. Can.,* 32, 1, 1975.
35. Roch, M. and Maly, E. J., Relationship of cadmium-induced hypocalcemia with mortality in rainbow trout (*Salmo gairdneri*) and the influence of temperature on toxicity, *J. Fish. Res. Bd. Can.*, 36, 1297, 1979.
36. Eisler, R., Cadmium poisoning in *Fundulus heteroclitus* (Pices: Cyprinodontidae) and other marine organisms, *J. Fish. Res. Bd. Can.*, 28, 1225, 1971.
37. Moffitt, B. P., Rae, N., and Rizzi, M., Thermoregulatory and respiratory responses of brown bullhead catfish to cadmium stress, *The Physiologist,* 33, A-105, 1990.
38. Ferrari, L., Salibán, A., and Muino, C., Selective protection of temperature against cadmium acute toxicity to *Bufo arenarum* tadpoles, *Bull. Environ. Contam. Toxicol.*, 50, 212, 1993.

39. Rehwoldt, R., Menapace, L. W., Nerrie, B., and Allessandrello, D., The effect of increased temperature upon the acute toxicity of some heavy metal ions, *Bull. Environ. Contam. Toxicol.*, 8, 91, 1972.
40. Dixon, D. G. and Hilton, J. W., Effects of available dietary carbohydrate and water temperature on the chronic toxicity of waterborne copper to rainbow trout (*Salmo gairdneri*), *Can. J. Fish. Aquat. Sci.*, 42, 1007, 1985.
41. Felts, P. A. and Heath, A. G., Interactions of temperature and sublethal environmental copper exposure on the energy metabolism of bluegill, *Lepomis macrochirus* Rafinesque, *J. Fish Biol.*, 25, 445, 1984.
42. Verma, S. R., Chand, R., and Tonk, I. P., Effects of environmental and biological variables on the toxicity of mercuric chloride, *Water Air Soil Pollut.*, 25, 243, 1985.
43. Somero, G. N., Chow, T. J., Yancey, P. H., and Snyder, C. B., Lead accumulation rates in tissues of the estuarine teleost fish, *Gillichthys mirabilis:* Salinity and temperature effects, *Arch. Environ. Contam. Toxicol.*, 6, 337, 1977.
44. Reinert, R. E., Stone, L. J., and Willford, W. A., Effect of temperature on accumulation of methylmercuric chloride and *p,p'*DDT by rainbow trout *(Salmo gairdneri*), *J. Fish. Res. Bd. Can.*, 31, 1649, 1974.
45. Spigarelli, S., Thommes, M., and Prepejchal, W., Thermal and metabolic factors affecting PCB uptake by adult brown trout, *Environ. Sci. Technol.*, 17, 88, 1983.
46. Murty, A. S., *Toxicity of Pesticides to Fish,* Vol. II, CRC Press, Boca Raton, FL, 1986.
47. Cope, O. B., Sport Fishery Investigations, Laboratory Studies and Toxicology, U.S. Bur. Sport Fish. Wildl. Circ., 226, 51, 1965.
48. Elson, P. F., Effects on wild young salmon of spraying DDT over New Brunswick forests, *J. Fish. Res. Bd. Can.*, 24, 731, 1967.
49. Miller, D. L. and Ogilvie, D. M., Temperature selection in brook trout (*Salvelinus fontinalis)* following exposure to DDT, PCB, or phenol, *Bull. Environ. Contam. Toxicol.*, 14, 545, 1975.
50. Eisler, R., Factors Affecting Pesticide-Induced Toxicity in an Estuarine Fish, U.S. Bur. Sport Fish. Wildl. Tech. Pap., No. 45, 1970.
51. Macek, K. J., Hutchinson, C., and Cope, O. B., The effects of temperature on the susceptibility of bluegills and rainbow trout to selected pesticides, *Bull. Environ. Contam. Toxicol.*, 4, 174, 1969.
52. Johnson, W. W. and Finley, M. T., Handbook of Acute Toxicity of Chemicals to Fish and Aquatic Invertebrates, U.S. Fish Wild. Serv. Resour. Publ., No. 137, Washington, D.C., 1980.
53. Ferrando, M. D., Almar, M. M., and Andreu, E., Lethal toxicity of lindane on a teleost fish, *Anguilla anguilla* from Albufera Lake (Spain): Hardness and temperature effects, *J. Environ. Sci. Health*, B23, 45, 1988.
54. Duangwawasdi, M. and Klaverkamp, J. F., Acephate and fenitrothion toxicity in rainbow trout: Effects of temperature stress and investigations on the sites of action, in *Aquatic Toxicology*, ASTM STP 667, Marking, L. L. and Kimerle, R. A., Eds., American Society for Testing and Materials, Philadelphia, 1979, 35.
55. Howe, G. E., Marking, L. L., Bils, T. D., Rach, J. J., and Mayer, F. L., Effects of water temperature and pH on toxicity of terbufos, trichlorfon, 4-nitrophenol and 2,4-dinitrophenol to the amphipod *Gammarus pseudolimnaeus* and rainbow trout (*Oncorhynchus mykiss), Environ. Toxicol. Chem.*, 13, 51, 1994.
56. Baldwin, J., Adaptation of enzymes to temperature: Acetylcholinesterases in the central nervous system of fishes, *Comp. Biochem. Physiol.*, 40, 181, 1971.
57. Mauck, W. L., Olson, L. E., and Marking, L. L., Toxicity of natural pyrethrins and five pyrethroids to fish, *Arch. Environ. Contam. Toxicol.*, 4, 18, 1976.
58. Kumaraguru, A. K. and Beamish, F. W. H., Lethal toxicity of permethrin (NRDC-143) to rainbow trout, *Salmo gairdneri*, in relation to body weight and water temperature, *Water Res.*, 15, 503, 1981.
59. Cole, L. M. and Casida, J. E., Pyrethroid toxicology in the frog, *Pest. Biochem. Physiol.*, 20, 217, 1983.
60. Berrill, M., Bertram, S., Wilson, A., Louis, S., Brigham, D., and Stromberg, C., Lethal and sublethal impacts of pyrethroid insecticides on amphibian embryos and tadpoles, *Environ. Toxicol. Chem.*, 12, 525, 1993.
61. Materna, E. J., Rabeni, C. F., and LaPoint, T. W., Effects of the synthetic pyrethroid insecticide, esfenvalerate, on larval leopard frogs (*Rana* spp.), *Environ. Toxicol. Chem.*, 14, 613, 1995.
62. Stegeman, J. J. and Hahn, M. E., Biochemistry and molecular biology of monooxygenases: Current perspectives on forms, functions, and regulation of cytochrome P450 in aquatic species, in *Aquatic Toxicology: Molecular, Biochemical, and Cellular Perspectives*, Malins, D. C. and Ostrander, G. K., Eds., Lewis Publishers, Boca Raton, FL, 1993, 87.

63. Jorgensen, E. H. and Wolkers, J., Effect of temperature on the P4501A response in winter-and summer-acclimated Arctic char (*Salvelinus alpinus*) after oral benzo[a]pyrene exposure, *Can. J. Fish. Aquat. Sci.*, 56, 1370, 1999.
64. Machala, M., Nezveda, K., Petrivalsky, M., Jarosova, A., Piacka, V., and Svobodova, Z., Monooxygenase activities in carp as biochemical markers of pollution by polycyclic and polyhalogenated aromatic hydrocarbons: Choice of substrates and effects of temperature, gender and capture stress, *Aquat. Toxicol.*, 37, 113, 1997.
65. Hartung, R., Energy metabolism in oil-covered ducks, *J. Wildl Manage.*, 31, 798, 1967.
66. McEwen, E. H. and Koelink, A. F. C., The heat production of oiled mallards and scaup, *Can. J. Zool.*, 51, 27, 1972.
67. Butler, R. G., Peakall, D. B., Leighton, F. A., Borthwick, J., and Harmon, R. S., Effects of crude oil exposure on standard metabolic rate of Leach's storm-petrel, *Condor*, 88, 248, 1986.
68. Jenssen, B. M. and Ekker, M., Effects of plumage oiling on thermoregulation in common eiders residing in air and in water, in *Trans. 19th Int. Union Game Biol. Cong.*, Trondheim, Norway, 1990, 281.
69. Jenssen, B. M., Effects of ingested crude and dispersed crude oil on thermoregulation in ducks (*Anas platyrhynchos*), *Environ. Res.*, 48, 49, 1989.
70. Holmes, W. N., Gorsline, J., and Cronshaw, J., Effects of mild cold stress on the survival of seawater-adapted mallard ducks (*Anas platyrhynchos*) maintained on food contaminated with petroleum, *Environ. Res.*, 20, 425, 1979.
71. Bourne, W. R. P. and Bibby, D. J., Temperature and the seasonal and geographical occurrence of oiled birds on west European beaches, *Mar. Pollut. Bull.*, 6, 77, 1975.
72. Engelhardt, F. R., Petroleum effects on marine mammals, *Aquat. Toxicol.*, 4, 199, 1983.
73. Jefferies, D. J. and French, M. C., Hyper- and hypothyroidism in pigeons fed DDT: An explanation for the "thin eggshell phenomenon," *Environ. Pollut.*, 1, 235, 1971.
74. Jefferies, D. J. and Parslow, J. L. F., The effect of one polychlorinated biphenyl on size and activity of the gull thyroid, *Bull. Environ. Contam. Toxicol.*, 8, 306, 1972.
75. Torri, G. M. and Mayer, L. P., Effects of polychlorinated biphenyls on the metabolic rates of mourning doves exposed to low ambient temperatures, *Bull. Environ. Contam. Toxicol.*, 27, 678, 1981.
76. Lustick, S., Vos, T., and Peterle, T. J., Effects of DDT on steroid metabolism and energetics in bobwhite quail (*Colinus virginianus*), in *Proc. First Nat. Bobwhite Quail Symp.*, Morrison, J. A. and Lewis, J. C., Eds., Oklahoma State University, Stillwater, 1972, 213.
77. Neave, D. J. and Wright, B. S., The effects of weather and DDT spraying on a ruffed grouse population, *J. Wildl. Manage.*, 33, 1015, 1969.
78. Herbert, C. E., Winter severity affects migration and contaminant accumulation in northern Great Lakes herring gulls, *Ecol. Appl.*, 8, 669, 1998.
79. Keplinger, M. L., Lanier, G. E., and Deichman, W. B., Effects of environmental temperature on the acute toxicity of a number of compounds in rats, *Toxicol. Appl. Pharmacol.*, 1, 156, 1959.
80. Fuhrman, G. J. and Fuhrman, F. A., Effects of temperature on the action of drugs, *Annu. Rev. Pharmacol.,* 1, 65, 1961.
81. Rattner, B. A., Becker, J. M., and Nakatsugawa, T., Enhancement of parathion toxicity to quail by heat and cold exposure, *Pest. Biochem. Physiol.*, 27, 330, 1987.
82. Willis, G. H., McDowell, L. L., Smith, S., and Southwick, L. M., Effects of weather variables on methyl parathion disappearance from cotton foliage, *Bull. Environ. Contam. Toxicol.*, 48, 394, 1992.
83. Maguire, C. C. and Williams, B. A., Response of thermal stressed bobwhite to organophosphorus exposure, *Environ. Pollut.*, 47, 25, 1987.
84. Fleming, W. J., Heinz, G. H., Franson, J. C., and Rattner, B. A., Toxicity of Abate 4E (Temephos) in mallard ducklings and the influence of cold, *Environ. Toxicol. Chem.*, 4, 193, 1985.
85. Martin, P. A. and Solomon, K. R., Acute carbofuran exposure and cold stress: Interactive effects in mallard ducklings, *Pest. Biochem. Physiol.*, 40, 117, 1991.
86. Rattner, B. A., Sileo, L., and Scanes, C. G., Hormonal responses and tolerance to cold of female quail following parathion ingestion, *Pest. Biochem. Physiol.*, 18, 132, 1982.
87. Brunet, R. and McDuff, J., Recovery of brain cholinesterases of brown-headed cowbirds from organophosphorus intoxication: Effect of environmental temperature, *Bull. Environ. Contamin. Toxicol.*, 59, 285, 1997.

88. Rattner, B. A. and Franson, J. C., Methyl parathion and fenvalerate toxicity in American kestrels: Acute physiological responses and effects of cold, *Can. J. Physiol. Pharmacol.*, 62, 787, 1984.
89. Rattner, B. A. and Grue, C. E., Toxicity of parathion to captive European starlings (*Sturnus vulgaris*) — Absence of seasonal effects, *Environ. Toxicol. Chem.*, 9, 1029, 1990.
90. Stanley, P. I. and Bunyan, P. J., Hazards to wintering geese and other wildlife from the use of dieldrin, chlorfenvinphos and carbophenothion as wheat seed treatments, *Proc. R. Soc. London B.*, 205, 31, 1979.
91. Blus, L. J., Staley, C. S., Henny, C. J., Pendleton, G. W., Craig, T. H., Craig, E. H., and Halford, D. K., Effects of organophosphorus insecticides on sage grouse in southeastern Idaho, *J. Wildl. Manage.*, 53, 1139, 1989.
92. Blus, L., unpublished data, 1988.
93. Montz, W. E., Jr. and Kirkpatrick, R. L., Effects of cold ambient temperature on acute mortality of *Peromyscus leucopus* dosed with parathion, *Bull. Environ. Contam. Toxicol.*, 35, 375, 1985.
94. Kendall, R. J. and Scanlon, P. F., The toxicology of lead shot ingestion in ringed turtle doves under conditions of cold exposure, *J. Environ. Pathol. Toxicol. Oncol.,* 5, 183, 1984.
95. Rattner, B. A., Fleming, W. J., and Bunck, C. M., Comparative toxicity of lead shot in black ducks (*Anas rubripes*) and mallards (*Anas platyrhynchos*), *J. Wildl. Dis.,* 25, 175, 1989.
96. Oliver, A. J. and King, D. R., The influence of ambient temperatures on the susceptibility of mice, guinea-pigs and possums to compound 1080, *Aust. Wildl. Res.*, 10, 297, 1983.
97. Eastland, W. G. and Beasom, S. L., Effects of ambient temperature on the 1080-$LD_{50}$ of raccoons, *Wildl. Soc. Bull.*, 14, 234, 1986.
98. Klapow, L. A. and Lewis, R. H., Analysis of toxicity data for California marine water quality standards, *J. Water Poll. Control Fed.*, 51, 2054, 1979.
99. Stickle, W. B., Sabourin, T. D., and Rice, S. D., Sensitivity and osmoregulation of coho salmon *Oncorhynchus kisutch*, exposed to toluene and naphthalene at different salinities, in *Physiological Mechanisms of Marine Pollutant Toxicity*, Vernberg, W. B., Calabrese, A., Thurberg, F. P., and Vernberg, F. J., Eds., Academic Press, New York, 1982, 331.
100. Stagg, R. M. and Shuttleworth, T. J., The accumulation of copper in *Platichthys flesus* L., and its effects on plasma electrolyte concentrations, *J. Fish Biol.*, 20, 491, 1982.
101. Hall, L. and Anderson, R., The influence of salinity on the toxicity of various classes of chemicals to aquatic biota, *Crit. Rev. Toxicol.*, 25, 281, 1995.
102. Levitan, W. M. and Taylor, M. H., Physiology of salinity-dependent naphthalene toxicity in *Fundulus heteroclitus*, *J. Fish. Res. Bd. Can.*, 36, 615, 1979.
103. Herbert, D. W. and Shurben, D. S., The susceptibility of salmonid fish to poisons under estuarine conditions, II. Ammonium chloride, *Int. J. Air Water Pollut.,* 9, 89, 1965.
104. Fulton, M. H. and Scott, G. I., The effect of certain intrinsic and extrinsic variables on the acute toxicity of selected organophosphorus insecticides to the mummichog, *Fundulus heteroclitus*, *J. Environ. Sci. Health*, B26, 459, 1991.
105. Wilson, R. W. and Taylor, E. W., Differential responses to copper in rainbow trout (*Oncorhynchus mykiss*) acclimated to seawater and brackish water, *J. Comp. Physiol.,* 163B, 239, 1993.
106. Lauren, D. D. J. and McDonald, D. G., Effects of copper on branchial ionoregulation in the rainbow trout *Salmo gairdneri*: Modulation by water hardness and pH, *J. Comp. Physiol.*, 155B, 635, 1985.
107. Dietraich, D., Schlatter, C., Blau, N., and Fischer, M., Aluminum and acid rain: Mitigating effects of NaCl on aluminum toxicity to brown trout (*Salmo trutta* Farop) in acid water, *Toxicol. Environ. Chem.*, 19, 17, 1989.
108. Ferguson, E. and Hogstrand, C., Acute silver toxicity to seawater-acclimated rainbow trout: Influence of salinity on toxicity and silver speciation, *Environ. Toxicol. Chem.*, 17, 589, 1998.
109. Inglis, A. and Davis, E. L., Effects of water hardness on the toxicity of several organic and inorganic herbicides to fish, U.S. Fish Wildl. Serv. Tech. Pap. No. 67, Washington, D.C., 1972.
110. Brown, V. M., The calculation of the acute toxicity of mixtures of poisons to rainbow trout, *Water Res.*, 2, 723, 1968.
111. Davies, P. H., Goettl, J. P., Sinley, J. R., and Smith, N. F., Acute and chronic toxicity of lead to rainbow trout, *Salmo gairdneri*, in hard and soft water, *Water Res.*, 10, 199, 1976.
112. Miller, T. G. and Mackay, W. C., The effects of hardness, alkalinity and pH of test water on the toxicity of copper to rainbow trout (*Salmo gairdneri*), *Water Res.*, 14, 129, 1980.

113. Bradley, R. W. and Sprague, J. B., The influence of pH, water hardness, and alkalinity on the acute lethality of zinc to rainbow trout (*Salmo gairdneri), Can. J. Fish. Aquat. Sci.*, 42, 731, 1985.
114. Bradley, R. W. and Sprague, J. B., Accumulation of zinc by rainbow trout as influenced by pH, water hardness and fish size, *Environ. Toxicol. Chem.*, 4, 685, 1985.
115. Everhall, N. C., Macfarlane, N. A. A., and Sedgwick, R. W., The effects of water hardness upon the uptake, accumulation and excretion of zinc in the brown trout, *Salmo trutta* L., *J. Fish Biol.*, 35, 881, 1989.
116. Fleming, C. A. and Trevors, J. T., Copper toxicity and chemistry in the environment: A review, *Water Air Soil Pollut.*, 44, 143, 1989.
117. Playle, R. C., Gensemer, R. W., and Dixon, D. G., Copper accumulation on gills of fathead minnows: Influence of water hardness, complexation and pH of the gill micro-environment, *Environ. Toxicol. Chem.*, 11, 381, 1992.
118. Sorensen, E. M. B., *Metal Poisoning in Fish,* CRC Press, Boca Raton, FL, 1991.
119. Sparling, D. W., Acidic deposition: A review of biological effects, in *Handbook of Ecotoxicology*, Hoffman, D. J., Rattner, B. A., Burton, G. A., Jr., and Cairns, J., Jr., Eds., Lewis Publishers, Boca Raton, GL, 1995, Chap. 14.
120. Brown, D. J. A. and Sadler, K., Fish survival in acid waters, in *Acid Toxicity and Aquatic Animals*, Morris, R., Taylor, E. W., Brown, D. J. A., and Brown, J. A., Eds., Cambridge University Press, Cambridge, 1989, 32.
121. Baker, J. P., Effects on fish of metals associated with acidification, in *Acid Rain/Fisheries*, Johnson, R. E., Ed., American Fisheries Society, Bethesda, MD, 1982, 165.
122. Birchall, J., Exley, C., Chappell, J., and Phillips, M., Acute toxicity of aluminum to fish eliminated in silicon-rich acid waters, *Nature*, 338, 146, 1989.
123. Howorth, R. and Sprague, J., Copper lethality to rainbow trout in waters of various hardness and pH, *Water Res.,* 12, 455, 1978.
124. Welsh, P., Parrott, J., Dixon, P., Hodson, P., Spry, D., and Mierle, G., Estimating acute copper toxicity to larval fathead minnow (*Pimephales promelas*) in soft water from measurements of dissolved organic carbon, calcium, and pH, *Can. J. Fish. Aquat. Sci.*, 53, 1263, 1996.
125. Schubauer-Berigan, M., Dierkes, J., Monson, P., and Ankley, G., pH-dependent toxicity of Cd, Cu, Ni, Pb, and Zn to *Ceriodaphnia dubia, Pimephales promelas, Hyalella azteca and Lumbriculus variegatus, Environ. Toxicol. Chem.*, 12, 1261, 1993.
126. Friend, M. and Abel, J. H., Jr., Inhibition of mallard salt gland function by DDE and organophosphates, in *Wildlife Diseases*, Page, L. A., Ed., Plenum Publishing, New York, 1976, 261.
127. Rattner, B. A., Fleming, W. J., and Murray, H. C., Osmoregulatory function in ducks following ingestion of the organophosphorus insecticide fenthion, *Pest. Biochem. Physiol.*, 20, 246, 1983.
128. Friend, M., Haegele, M. A., and Wilson, R., DDE: Interference with extra-renal salt excretion in the mallard, *Bull. Environ. Contam. Toxicol.*, 9, 49, 1973.
129. Herin, R. A., Suggs, J. E., Lores, E. M., Heiderscheit, L. T., Farmer, J. D., and Prather, D., Correlation of salt gland function with levels of chlorpyrifos in the feed of mallard ducklings, *Pest. Biochem. Physiol.*, 9, 157, 1978.
130. Eastin, W. C., Jr., Fleming, W. J., and Murray, H. C., Organophosphate inhibition of avian salt gland Na, K-ATPase activity, *Comp. Biochem. Physiol.*, 73C, 101, 1982.
131. Miller, D. S., Kinter, W. B., Peakall, D. B., and Risebrough, R. W., DDE feeding and plasma osmoregulation in ducks, guillemots, and puffins, *Am. J. Physiol.*, 231, 370, 1976.
132. Rowe, C. L. and Freda, J., Effects of acidification on amphibians at multiple levels of biological organization, in *Ecotoxicology of Reptiles and Amphibians*, Sparling, D. W., Linder, G., and Bishop, C. A., Eds., SETAC Press, Pensacola, FL, 2000, Chap. 9.
133. Gosner, K. L. and Black, I. H., The effects of acidity on the development and hatching of New Jersey frogs, *Ecology*, 38, 256, 1957.
134. Salthe, S. N., Increase in volume of the perivitelline chamber during development of *Rana pipiens* Schreber, *Physiol. Zool.*, 38, 80, 1965.
135. Pough, F. H., Acid precipitation and embryonic mortality of spotted salamanders, *Ambystoma maculatum, Science*, 192, 68, 1976.
136. Dunson, W. A. and Connell, J., Specific inhibition of hatching in amphibian embryos by low pH, *J. Herpetol*, 16, 314, 1982.

137. Freda, J. and Dunson, W. A., The influence of external cation concentration on the hatching of amphibian embryos in water of low pH, *Can. J. Zool.*, 63, 2649, 1985.
138. Freda, J. and Dunson, W. A., Sodium balance of amphibian larvae exposed to low environmental pH, *Physiol. Zool.*, 57, 435, 1984.
139. Freda, J. and Dunson, W. A., The effect of prior exposure on sodium uptake in tadpoles exposed to low pH water, *J. Comp. Physiol. B*, 156, 649, 1986.
140. Wyman, R. L. and Hawksley-Lescault, D. S., Soil acidity affects distribution, behavior, and physiology of the salamander *Plethedon cinereus, Ecology*, 68, 1819, 1987.
141. Clark, K. L. and Hall, R. J., Effects of elevated hydrogen ion and aluminum concentrations on the survival of amphibian embryos and larvae, *Can. J. Zool.*, 63, 116, 1985.
142. Clark, K. L. and LaZerte, B. D., A laboratory study of the effects of aluminum and pH on amphibian eggs and tadpoles, *Can. J. Fish. Aquat. Sci.*, 44, 1622, 1985.
143. Clark, K. L. and LaZerte, B. D., Intraspecific variation in hydrogen ion and aluminum toxicity in *Bufo americanus* and *Ambystoma maculatum*, *Can. J. Fish. Aquat. Sci.*, 44, 1622, 1987.
144. Freda, J. and McDonald, D. G., Effects of aluminum on the leopard frog, *Rana pipiens*: Life stage comparisons and aluminum uptake, *Can. J. Fish. Aquat. Sci.*, 47, 210, 1990.
145. Freda, J., The influence of aluminum and other metals on amphibians: A review, *Environ. Pollut.*, 71, 305, 1991.
146. Leuven, R. S. E. W., den Hartog, C., Christiaans M. M. C., and Heijligers, W. H. C., Effects of water acidification on the distribution pattern and reproductive success of amphibians, *Experientia*, 42, 495, 1986.
147. Horne, M. T. and Dunson, W. A., Effects of low pH, metals, and water hardness on larval amphibians, *Arch. Environ. Contam. Toxicol.*, 29, 500, 1995.
148. Horne, M. T. and Dunson, W. A., The interactive effects of low pH, toxic metals, and DOC on a simulated temporary pond community, *Environ. Pollut.*, 89, 155, 1995.
149. Lloyd, R., Effect of dissolved oxygen concentrations on the toxicity of several poisons to rainbow trout (*Salmo gairdneri* Richardson), *J. Exp. Biol.*, 38, 447, 1961.
150. Scheuhammer, A. M., Effects of acidification on the availability of toxic metals and calcium to wild birds and mammals, *Environ. Pollut.*, 71, 329, 1991.
151. Adelman, I. R. and Smith, L. L., Toxicity of hydrogen sulfide to goldfish (*Carassius auratus*) as influenced by temperature, oxygen, and bioassay techniques, *J. Fish. Res. Bd. Can.*, 29, 1309, 1972.
152. Doudoroff, P. and Shumway, D. L., Dissolved oxygen requirements of freshwater fishes, FAO Fish. Tech. Pap., 86, 291, 1970.
153. Hicks, D. B. and Dewitt, J. W., Effects of dissolved oxygen on kraft pulp mill effluent toxicity, *Water Res.*, 5, 693, 1971.
154. Thurston, R. V., Phillips, G. R., and Russo, R. C., Increased toxicity of ammonia to rainbow trout (*Salmo gairdneri*) resulting from reduced concentrations of dissolved oxygen, *Can. J. Fish. Aquat. Sci.*, 38, 983, 1981.
155. Hokanson, K. E. F. and Smith, L. L., Some factors influencing toxicity of linear alkylate sulfonate (LAS) to the bluegill, *Trans. Am. Fish. Soc.*, 100, 1, 1971.
156. Gupta, S., Dalela, R. C., and Saxena, P. K., Influence of dissolved oxygen levels on acute toxicity of phenolic compounds to freshwater teleost, *Notopterus notopterus (Pallas)*, *Water Air Soil Pollut.*, 19, 223, 1983.
157. McCoskey, J. T. and Oris, J. T., Effect of water temperature and dissolved oxygen concentration on the photo-induced toxicity of anthracene to juvenile bluegill sunfish (*Lepomis macrochirus*), *Aquat. Toxicol.*, 21, 145, 1991.
158. Carlson, A. R., Effects of lowered dissolved oxygen concentration on the toxicity of 1,2,3-trichlorobenzene to fathead minnows, *Bull. Environ. Contam. Toxicol.*, 38, 667, 1987.
159. Chapman, G. A. and Shumway, D., Effects of sodium pentachlorophenate on survival and energy metabolism of embryonic and larval steelhead trout, in *Pentachlorophenol,* Rao, K. R., Ed., Plenum Press, New York, 1978, 285.
160. Mayer, F. L., Versteeg, D. J., McKee, M. J., Folmar, L. C., Graney, R. L., McCume, D. C., and Rattner, B. A., Physiological and nonspecific biomarkers, in *Biomarkers, Biochemical Physiological, and Histological Markers of Anthropogenic Stress*, Huggett, R. J., Kimerle, R. A., Mehrle, P. M., and Bergman, H. L., Eds., Lewis Publishers, Chelsea, MI, 1992.

161. Hoy, T., Horsberg, T. E., and Wichstrom, R., Inhibition of acetylcholinesterase in rainbow trout following dichlorvos treatment at different water oxygen levels, *Aquaculture*, 95, 33, 1991.
162. Vig, E. and Nemscok, J., The effects of hypoxia and paraquat on superoxide dismutase activity in different organs of carp, *Cyprinus carpio* L., *J. Fish Biol.*, 35, 23, 1989.
163. Watenpaugh, D. E. and Beitinger, T. L., Resistance of nitrite-exposed channel catfish, *Ictalurus punctatus,* to hypoxia, *Bull. Environ. Contam. Toxicol.*, 37, 802, 1986.
164. Hlohowskyj, I. and Chagnon, N., Reduction in tolerance to progressive hypoxia in the central stoneroller minnow following sublethal exposure to phenol, *Water Air Soil Pollut.*, 60, 189, 1991.
165. Newman, J. R., Effects of industrial air pollution on wildlife, *Biol. Conserv.*, 15, 181, 1979.
166. Newman, J. R. and Schreiber, R. K., Animals as indicators of ecosystem responses to air emissions, *Environ. Manage.*, 8, 309, 1984.
167. Laybourne, R. C., Collision between a vulture and an aircraft at an altitude of 37,000 feet, *Wilson Bull.*, 86, 461, 1974.
168. Bryan, T. E. and Gildersleeve, R. P., Effects of nonionizing radiation on birds, *Comp. Biochem. Physiol.*, 89A, 511, 1988.
169. McLeay, D. J. and Gordon, M., Effect of seasonal photoperiod on acute toxic responses of juvenile rainbow trout (*Salmo gairdneri*) to pulpmill effluent, *J. Fish. Res. Bd. Can.*, 35, 1388, 1978.
170. Zaga, A., Little, E. E., Rabeni, C. F., and Ellersieck, M. R., Photoenhanced toxicity of a carbamate insecticide to early life stage anuran amphibians, *Environ. Toxicol. Chem.,* 17, 2543, 1998.
171. Eisler, R., Polycyclic aromatic hydrocarbons, in *Handbook of Chemical Risk Assessment: Health Hazards to Humans, Plants, and Animals, Vol. 2 Organics*, Lewis Publishers, Boca Raton, FL, 2000, 1372, Chap. 25.
172. Heath, A. G., Physiology and ecological health, in *Multiple Stresses in Ecosystems,* Cech, J., Wilson, B., and Crosby, D., Eds., CRC Press, Boca Raton, FL, 59, 1998, Chap. 7.
173. Malins, D. and Ostrander, G., Eds., *Aquatic Toxicology, Molecular, Biochemical, and Cellular Perspectives*, CRC Press, Boca Raton, FL, 1993.
174. Hugget, R. J., Kimerle, R. A., Mehrle, P. M., Jr., and Bergman, H. L., *Biomarkers. Biochemical, Physiological, and Histological Markers of Anthropogenic Stress*, Lewis Publishers, Chelsea, MI, 1992.
175. Stuart, S. and Morris, R., The effects of season and exposure to reduced pH (abrupt and gradual) on some physiological parameters in brown trout (*Salmo trutta*), *Can. J. Zool.*, 63, 178, 1985.
176. Larsson, A., Andersson, T., Forlin, L., and Hardig, J., Physiological disturbances in fish exposed to bleach kraft mill effluents, *J. Water Sci. Technol.*, 20, 67, 1988.
177. Bidwell, J. and Heath, A., An *in situ* study of rock bass, *Ambloplites rupestris*, physiology: Effects of season and mercury contamination, *Hydrobiologia*, 264, 137, 1993.
178. Chambers, J. E. and Yarbrough, J. D., A seasonal study of microsomal mixed-function oxidase components in insecticide-resistant and susceptible mosquitofish, *Gambusia affinis*, *Toxicol. Appl. Pharmacol.*, 48, 497, 1979.
179. Lindstrom-Seppa, P., Seasonal variation of the xenobiotic metabolizing enzyme activities in the liver of male and female vendace (*Coregonus albula* L.), *Aquat. Toxicol.*, 6, 323, 1985.
180. Galgani, F., Bocquene, G., Lucon, M., Grzebyk, D., Letrouit, F., and Claisse, D., EROD measurements in fish from the northwest part of France, *Mar. Pollut. Bull.*, 22, 494, 1991.
181. Olsson, P., Haux, C., and Foerlin, L., Variations in hepatic metallothionein, zinc and copper levels during an annual reproductive cycle in rainbow trout, *Salmo gairdneri*, *Fish Physiol. Biochem.*, 3, 39, 1987.
182. Schwab, R., Eglehoeffer, S., and Osborne, J., Daily and Seasonal Susceptibility of Starlings to the Avicide DRC-1339, Joint Progress Report of the Univ. Calif. Agric. Exp. Stat., U.S. Bureau Sport Fish. Wildl. and Calif. Dept. Agric., 1967.
183. Eroschenko, V. P. and Wolson, W. O., Photoperiods and age as factors modifying the effects of kepone in Japanese quail, *Toxicol. Appl. Pharmacol.*, 29, 329, 1974.
184. Thompson, H. M., Walker, C. H., and Hardy, A. R., Avian esterases as indicators of exposure to insecticides — The factor of diurnal variation, *Bull. Environ. Contam. Toxicol.*, 41, 4, 1988.
185. Rattner, B. A. and Fairbrother, A., Biological variability and the influence of stress on cholinesterase activity, in *Cholinesterase-Inhibiting Insecticides: Their Impact on Wildlife and the Environment*, Mineau, P., Ed., Elsevier, Amsterdam, 1991, Chap. 5.

186. Haarakangas, H., Hyvärinen, H., and Ojanen, M., Seasonal variations and the effects of nesting and moulting on liver and mineral content in the house sparrow (*Passer domesticus* L.), *Comp. Biochem. Physiol.*, 47A, 153, 1974.
187. Osborn, D., Seasonal changes in the fat, protein and metal content of the liver of the starling *Sturnus vulgaris*, *Environ. Pollut.*, 19, 145, 1979.
188. Honda, K., Nasu, T., and Tatsukawa, R., Seasonal changes in mercury accumulation in the Black-eared Kite, *Milvus migrans lineatus*, *Environ. Pollut.*, 42, 325, 1986.
189. Mora, M. A., Anderson, D. W., and Mount, M. E., Seasonal variation of body condition and organochlorines in wild ducks from California and Mexico, *J. Wildl. Manage.*, 51, 132, 1987.
190. Warren, R. J., Wallace, B. M., and Bush, P. B., Trace elements in migrating blue-winged teal: Seasonal-, sex- and age-class variations, *Environ. Toxicol. Chem.*, 9, 521, 1990.
191. Stewart, F. M., Thompson, D. R., Furness, R. W., and Harrison, N., Seasonal variation in heavy metal levels in tissues of common guillemots, *Uria aalge* from northwest Scotland, *Arch. Environ. Contam. Toxicol.*, 27, 168, 1994.
192. Esselink, H., van der Geld, F. M., Jager, L. P., Posthuma-Trumpie, G. A., Zoun, P. E. F., and Baars, A. J., Biomonitoring heavy metals using the barn owl (*Tyto alba guttata*): Sources of variation especially relating to body condition, *Arch. Environ. Contam. Toxicol.*, 28, 471, 1995.
193. Rattner, B. A. and Jehl, J. R., Jr., Dramatic fluctuations in liver mass and metal content of eared grebes (*Podiceps nigricollis*) during autumnal migration, *Bull. Environ. Contam. Toxicol.*, 59, 337, 1997.
194. Shore, R. F. and Rattner, B. A., Eds., *Ecotoxicology of Wild Mammals*, Ecological and Environmental Toxicology Series, John Wiley and Sons, New York, 2001.
195. Hill, E. F. and Murray, H. C., Seasonal variation in diagnostic enzymes and biochemical constituents of captive northern bobwhites and passerines, *Comp. Biochem. Physiol.*, 87B, 933, 1987.
196. Rattner, B. A., Hoffman, D. J., and Marn, C. M., Use of mixed-function oxygenases to monitor contaminant exposure in wildlife, *Environ. Toxicol. Chem.*, 8, 1093, 1989.

SECTION III

# Case Histories and Ecosystem Surveys

# CHAPTER 24

# The Chernobyl Nuclear Power Plant Reactor Accident: Ecotoxicological Update

**Ronald Eisler**

## CONTENTS



## 24.1 INTRODUCTION

The partial meltdown of the 1000 MW reactor at Chernobyl, Ukraine on April 26, 1986 released large amounts of radiocesium and other radionuclides into the environment, causing widespread radioactive contamination of Europe and the former Soviet Union.[1–8] Among the reactors operating in the former Soviet Union, 13 are identical to the one in Chernobyl, Ukraine, including units in Chernobyl, Leningrad, Kursk, and Smolensk.[70] Of the three remaining reactors in Chernobyl, one was closed in 1991 after a fire swept through the turbine hall, causing extensive damage, though not to the reactor.[168] A second unit was closed in 1994 by the Russian president after pledges were received from the Group of Seven to help finance an overall plan to decommission the Chernobyl complex. And the last unit was closed in late 2000.[168]

At least 3,000,000 trillion becquerels (TBq) were released from the fuel during the April 1986 accident, dwarfing — by orders of magnitude — radiation releases from other highly publicized reactor accidents at Windscale (U.K.) and Three-Mile Island (U.S.).[3,9,159] (Note: 1 Bq = 1 disintegration/sec; about 0.037 Bq = 1 picoCurie.) The Chernobyl accident happened while a test was being conducted during a normal scheduled shutdown and is attributed mainly to human error.[3] About 25% of the released radioactive materials escaped during the first day of the accident; the rest over a 9-day period.[3] The initial explosions and heat from the fire carried some of the radioactive materials to an altitude of 1500 m, where they were transported by prevailing winds. Long-range atmospheric transport spread the radioactivity through the northern hemisphere, where it was initially detected in Japan on May 2, in China on May 4, in India on May 5, and in Canada and the United States on May 5–6, 1986. Airborne activity was also detected in Turkey, Kuwait, and Israel in early May. No airborne activity from Chernobyl has been reported south of the equator.[3]

Effective dose equivalents from the Chernobyl accident in various regions of the world were highest in southeastern Europe (1.2 milliSieverts [mSv]), northern Europe (0.97 mSv), and central Europe (0.93 mSv).[3,4] (Note: 1 mSv = 0.1 rem.) In the first year after the accident, whole-body effective dose equivalents were highest in Bulgaria, Austria, Greece, and Romania (0.5–0.8 mSv); Finland, Yugoslavia, Czechoslovakia, and Italy (0.3–0.5 mSv); Switzerland, Poland, U.S.S.R, Hungary, Norway, Germany, and Turkey (0.2–0.3 mSv); and elsewhere (< 0.2 mSv).[3,4] This value was 0.81 mSv in the former Soviet Union, < 0.2 mSv in southwest Asia and western Europe, and < 0.1 mSv elsewhere.[3,4] By comparison, the recommended whole-body annual effective dose equivalent for the general public is < 5 mSv.[3] Thyroid dose equivalents were significantly higher than whole-body effective dose equivalents because of significant amounts of $^{131}I$ in the released materials. Thyroid dose equivalents were as high as 25 mSv to infants in Bulgaria, 20 mSv in Greece, and 20 mSv in Romania; the adult thyroid dose equivalents were usually 80% lower than the infant dose equivalents.[3]

This account briefly summarizes ecological and toxicological aspects of the Chernobyl accident with an emphasis on natural resources; it complements and supplements previous reviews.[8,74,75,121,123,147,149,165,166,172]

## 24.2 LOCAL EFFECTS

The initial contamination rate in the 30-km exclusion zone surrounding the site of the nuclear accident was estimated at 37,000,000 Bq/km$^2$ (1,000 Ci/km$^2$); isotopes included iodine-131, tellurium 127+132, barium-140, lanthanum-140, cerium-141+144, zirconium-95, niobium-95, ruthenium-103+106, praseodymium-144, cesium-134+137, molybdenum-99, strontium-89+90, plutonium-238+239+240+241, silver-110, antimony-125, and others.[89] About 28,000 km$^2$ of land and 2225 settlements in Belarus, Russia, and Ukraine were officially declared contaminated with radiocesium, i.e., levels were > 185,000 Bq $^{134+137}Cs/m^2$ (> 5 Ci/km$^2$).[165] Approximately 850,000 people are still living in these contaminated areas. About 105,000 km$^2$ were contaminated with 37,000 Bq/m$^2$ (1 Ci/km$^2$) or more. In the first years of the catastrophe, 144,000 ha of agricultural land and 492,000 ha of forest were withdrawn from use. More than 4 million people in these three countries were affected by the accident. Current estimates of the eventual toll from cancer deaths as a direct result of Chernobyl range from a minimum of 14,000 to a maximum of 475,000.[165]

### 24.2.1 Acute Effects

At Chernobyl, at least 115 humans received acute bone marrow doses > 1 Grey (Gy), as judged by lymphocyte aberrations. (Note: 1 Gy = 100 rad.) The death toll within 3 months from the accident was at least 30 individuals, usually from groups receiving > 4 Gy, especially the reactor's operating staff and the fire-fighting crew.[3] Humans from highly contaminated areas within 30 km of Chernobyl received mean thyroid doses from radioiodines of 1.6 Gy for adults, 2.1 Gy for those age 7–17 years, and 4.7 Gy for infants.[143] Residents were evacuated from a 30-km exclusion zone surrounding the reactor because of increasing radiation levels; more than 115,000 people, including 27,000 children, were evacuated from the Kiev region, Belarus, and Ukraine.[3] Children evacuated from the contaminated areas showed — in July 1986 — significantly elevated levels of dicentric chromosomal aberrations when compared to resident children of noncontaminated areas.[116] Vitamin E and A deficiencies observed in some children were correlated with high Chernobyl radiation loads of their mothers.[142] Tens of thousands of cattle were also removed from the contaminated area, and consumption of locally produced milk and other foods was banned. Agricultural activities were halted and large-scale decontamination was undertaken.[3] Humans residing in areas where topsoil contamination exceeded 555,000 Bq/m$^2$ were cautioned against inclusion of forest products in their diets and to avoid cattle grazing on wet floodplain meadows without remediation.[146] The forced evacuation and health concerns contributed to severe sociopsychological impacts in all age groups of the displaced population.[126]

The most sensitive ecosystems affected at Chernobyl were the soil fauna and pine forest communities; the majority of the terrestrial vertebrate communities were not adversely affected by released ionizing radiation.[10] Pine forests seemed to be the most sensitive ecosystem. One 400-ha stand of *Pinus silvestris* died and probably received a dose of 80–100 Gy. Other stands experienced heavy mortality of 10- to 12-year-old trees and as much as 95% necrotization of young shoots; these pines received an estimated dose of 8–10 Gy. Abnormal top shoots developed in some *Pinus*, and these probably received 3–4 Gy. In contrast, leafed trees in the Chernobyl Atomic Power Station zone, such as birch, oak, and aspen, survived undamaged, probably because they are about ten times more radioresistant than pines.[10] Extremely high radioresistance was documented in genetically adapted strains of the filamentous fungus *Alternaria alternata* isolated from the reactor of the Chernobyl power plant; other strains of this species are supersensitive to radiation.[167] There was no increase in mutation rate of spiderwort (*Arabidopsis thaliana*), a radiosensitive plant, suggesting that the dose rate was < 0.05 Gy/h in the Chernobyl locale.[10]

Populations of soil mites were reduced in the Chernobyl area, but no population showed a catastrophic drop in numbers. By 1987, soil microfauna — even in the most heavily contaminated plots — was comparable to controls.[10] Flies (*Drosophila* spp.) collected at various distances from the accident site and bred in the laboratory had highest incidences of dominant lethal mutations (14.7%, estimated dose of 0.8 mGy/h) at sites nearest the accident and higher incidences than controls (4.2%).[10]

The most contaminated water body in the Chernobyl emergency zone was the Chernobyl cooling pond ecosystem, an area of about 30 km$^2$.[71,99] On May 30, 1986, the total amount of radioactivity in the water of this system was estimated at 806 TBq and in sediments 5657 TBq.[71] In water, $^{131}$I contributed about 31% of the total radioactivity, $^{140}$Ba-$^{140}$La 25%, $^{95}$Zr-$^{95}$Nb 15%, $^{134}$Cs and $^{137}$Cs 11%, $^{141}$Ce and $^{144}$Ce 10%, $^{103}$Ru and $^{106}$Ru 7%, and $^{90}$Sr < 1%. The distribution pattern in sediments was significantly different: about 41% of the total radioactivity was contributed by $^{95}$Zr-$^{95}$Nb, 27% by $^{141}$Ce and $^{144}$Ce, 16% by $^{103}$Ru and $^{106}$Ru, 12% by $^{140}$Ba-$^{140}$La, 3% by $^{134}$Cs and $^{137}$Cs, 1% by $^{90}$Sr, and 0.5% by $^{131}$I.[71]

Fish populations seemed unaffected in July–August 1987, and no grossly deformed individuals were found; however, $^{134+137}$Cs levels were elevated in young fish. The most heavily contaminated teleost in May 1987 was the carp (*Carassius carassius*). But carp showed no evidence of mutagenesis, as judged by incidence of chromosomal aberrations in cells from the corneal epithelium of carp as far as 60 km from Chernobyl.[10]

In 1986, total radioactivity in muscle of birds near Chernobyl after the accident exceeded the temporarily permitted limits for human food consumption (598 Bq/kg FW) by about 100 times.[89] In 1987, radionuclide concentrations in bird muscle had decreased by a factor of about 7, due in part to physical decay of short-lived isotopes.[89]

Several rodent species comprised the most widely distributed and numerous mammals in the Chernobyl vicinity. It was estimated that about 90% of rodents died in an area receiving 60 Gy and 50% in areas receiving 6 to 60 Gy. Rodent populations seemed normal in spring 1987, and this was attributed to migration from adjacent nonpolluted areas. The most sensitive small mammal was the bank vole (*Clethrionomys glareolus*), which experienced embryonic mortality of 34%. The house mouse (*Mus musculus*) was one of the more radioresistant species. *Mus* from plots receiving 0.6–1.0 mGy/h did not show signs of radiation sickness, were fertile with normal sperm, bred actively, and produced normal young. Some chromosomal aberrations, namely, an increased frequency of reciprocal translocations, were evident.[10] New data on the house mouse suggests that fertility was dramatically reduced in the 30-km zone around the Chernobyl nuclear power plant station in 1986–1987 and that survivors had high frequencies of abnormal spermatozoa heads and dominant lethal mutations.[76] Elevated incidences of germ line mutations were observed in four species of *Apodemus* mice from the vicinity of Chernobyl when compared to control areas; however, variability between species was large.[170]

During the early period after the accident, there was no evidence of increasing mortality, decline in fecundity, or migration of vertebrates as a result of the direct action of ionizing radiation. The numbers and distribution of wildlife species were somewhat affected by the death of the pine stand, the evacuation of people, termination of cultivation of soils (the crop of 1986 remained standing), and the forced transfer of domestic livestock. There were no recorded changes in survival or species composition of game animals and birds. In fact, because humans had evacuated and hunting pressure was negligible, many game species — including fox, hare, deer, moose, wolf, and waterfowl — moved into the zone in autumn 1986–winter 1987 from the adjacent areas in a 50- to 60-km radius.[10]

### 24.2.2 Latent Effects: Humans

The frequency of childhood thyroid cancer in areas of Belarus and Ukraine most affected by the Chernobyl accident is significantly higher than any other region of the world.[114,155–158] Thyroid cancers in children were induced by $^{131+132}I$, although the mechanisms of action are imperfectly understood.[127] During the period 1985–1995, childhood thyroid cancer rates in Belarus were higher after a minimum latent period of 3 years, higher in children under age 10 years than those age 10 to 15 years, and higher in females than males.[114] Between 1990 and 1997, the group most affected were those younger than 5 years old in 1986; the largest number of cases occurred in patients living in areas of thyroid radiation doses > 0.5 Gy.[156,157] The post-Chernobyl thyroid carcinomas were usually papillary, more aggressive at presentation, and frequently associated with thyroid autoimmunity.[155–157] Gene mutations involving the receptor tyrosine kinase (RET) protooncogene were the causative agents specific to papillary cancer.[155,157,158]

Cutaneous radiation syndrome was the primary cause of death in most of the 32 adult patients who died shortly after the accident. Excessive cutaneous fibrosis in eight survivors who participated in the Chernobyl cleanup was treated successfully with interferon over a period of 36 months.[125] There is general agreement that cleanup workers and populations residing in heavily-contaminated areas (> 555,000 Bq $^{137}Cs/m^2$) had an increased frequency of thyroid cancers in the period 1986–1993, but current epidemiological evidence does not conclusively support an increased incidence of other types of cancers.[126]

Increased frequencies of chromosome breakage were evident in plants and small mammals from the Chernobyl-contaminated zone in the period 1986–1991.[164] For humans, chromosome and chromatid aberration frequencies in lymphocytes of Chernobyl evacuees (receiving 330–420 mGy) and cleanup personnel (receiving a maximum of 940 mGy) 1 year after the accident were three to four times higher than in controls.[164] By 1994–1995, aberration frequency in evacuees receiving <250 mGy was the same as the controls; however, this value was 1.7 times higher than controls in cleanup personnel, 1.95 times higher in evacuees receiving 250 mGy, and 2.37 times higher in evacuees receiving > 250 mGy in 1986. In some patients, the increased level of chromosome breakage was associated with high cancer susceptibility.[164]

Russian adult males who participated in Chernobyl cleanup activities (n = 126) received an estimated 0.14–0.15 Gy (maximum of 0.56–0.95 Gy, depending on the statistical model); however, dose estimates could be overestimated if age and smoking status were ignored, particularly for older subjects who smoke.[112] This group had a significantly increased frequency of chromosomal translocations when compared to Russian controls (n = 53).[112] Cancer incidence between 1986 and 1996 for cleanup workers (n = 114,504) with no known oncology before arrival in 1986 at the 30-km zone increased significantly for solid tumors and malignant neoplasms of the digestive tract (but not of the respiratory system) when compared to controls.[110] By 1996, Chernobyl cleanup workers experienced a variety of eye pathologies that seemed to increase in frequency with increasing time postaccident and with high initial doses of absorbed radiation.[140] Eye-pathology patterns are still developing because the latency period for irradiation cataract can exceed 10 years.[140] The incidence of urinary bladder cancer in Ukraine increased from 26.2 per 100,000 population in 1986 to 36.1 in 1997.[139] Individuals from the most severely radiocontaminated areas showed the greatest incidence of early malignant transformation of bladder epithelium.[139] In 1995, the endocrine status and spermatogenesis of Chernobyl cleanup workers were normal except for lower cortisol and higher testosterone.[135] A woman resident of Kiev during the period 1986–1992 who subsequently emigrated to the United States was diagnosed in 1996 with radiation-induced cancer of the ovaries, kidneys, and bile duct.[148]

In certain rural portions of Russia receiving Chernobyl contamination on April 28–29, 1986, $^{137}$Cs concentrations in the human body in 1986–1987 were positively correlated with consumption of meat and dairy products.[160] Domestic livestock were fed clean feed as much as possible for milk production and were fed clean feed for 40 to 120 days before slaughter.[165] Beginning in 1993, however, and persisting to at least 1996, the content of $^{137}$Cs in whole humans correlated positively with the levels of consumption of naturally occurring foodstuffs, such as mushrooms, wild berries, fish, and game.[160]

### 24.2.3 Latent Effects: Plants and Animals

Total radiocesium-137 deposited in soils at Chernobyl sites 2 to 15 km from the reactor was estimated at 1,660,000 Bq/m$^2$, mainly as insoluble fuel particles.[106] The half-time persistence of $^{137}$Cs in surface soils 0–2 cm in depth decreased from 9 years in 1987 to 3 years in 1994; but the residence time of this isotope increased with increasing depth over time. This increase in deeper layers is attributed to the progressive fixation of radiocesium by clay minerals of the soil.[106]

In 1992, bacteria were isolated from soils within 30 km of the power plant. Spore-forming bacilli collected nearest the power plant were more resistant to x-radiation, ultraviolet radiation, and 4-nitroquinoline 1-oxide than were isolates of the same species from control sites.[141] In 1993–1995, populations of cellulolytic, nitrifying, and sulfate-reducing bacteria collected within the 10-km zone from contaminated soils (11,000–629,000 Bq/kg DW soil) were 10 to 100 times lower in abundance and diversity than were populations from control soils (< 222 Bq/kg DW soil).[169]

On heavily contaminated agricultural lands — about 2.4 million ha — about 388,000 ha were withdrawn from use, and the remainder planted with crops of low coefficients of radionuclide transfer from soil, such as grains, potatoes, and corn.[165] In addition, the upper polluted layer of soil (5–6 cm) was ploughed to a depth of 40–50 cm to inhibit transfer to root systems, especially those of grasses. Application of potassium and phosphorus fertilizers in combination with liming further reduced uptake of $^{90}$Sr and $^{137}$Cs by plants.[165] From autumn 1986 to 1991, $^{137}$Cs in vegetable and animal agricultural products from various contaminated areas of Russia decreased at an observed half-time persistence of 0.7 to 1.5 years.[115] Beginning in 1991 and extending through 1995, the average values of transfer factors for $^{137}$Cs from all sources from soil to milk and potatoes were similar to those of the pre-Chernobyl period.[115] Dose rates from soil to the house mouse between 1986 and 1993 ranged from 0.0002 to 2 mGy/h, and these were positively correlated with the frequency of reciprocal translocations in mouse spermatocytes. The frequency of mice heterozygous for recessive lethal mutations decreased over time after the accident.[76,150]

In 1991, pine forests (*Pinus silvestris*) within 10 km of Chernobyl exposed initially to doses of 10–60 Gy had low survival and no regeneration since 1987; mortality was exacerbated by pathogenic insect invaders.[92] Samples of wood and bark from these trees had significant histological changes in resin ducts and radial rays, and this was correlated with radionuclide content of bark.[96] Stands exposed initially to 0.1–1.0 Gy had reduced growth; trees receiving initial doses of less than 0.1 Gy seemed outwardly normal.[92] Between 1986 and 1996, radiocesium-137 concentrated in the bark of pine trees grown in the exclusion zone from about 10,000 Bq/kg DW in 1986 to 37,000 Bq/kg DW in 1996; the high $^{137}$Cs concentrations were associated with growth suppression.[111] In the period 1986–1994, $^{137}$Cs dynamics in forests within the 30-km zone around the Chernobyl reactor were influenced mainly by the size of radioactive particles in the fallout, humidity, soil type, and tree age.[103] In June 1996, abnormal development in three species of plants (black locust tree, *Robina pseudoacacia*; rowan, *Sorbus aucuparia*; camomile, *Matricaria perforata*) was observed between Chernobyl and an uncontaminated area 225 km to the southeast.[128] Abnormalities, including leaf or flower asymmetry, were three to four times more frequent near Chernobyl than more distant sites and were directly related to $^{137}$Cs concentrations in soils.[128] In 1993–1994, about 55% of all young (2- to 9-year-old) plants of pine and spruce growing within 10 km of Chernobyl had abnormal needles, due in part to radiation-induced alterations in protein composition.[144] In 1993–1995, timber products within 30 km of Chernobyl were sufficiently contaminated with radiocesium and other isotopes as to preclude human use, as was the case for berries and mushrooms.[151] Economic damage to forest products within the 30-km exclusion zone is estimated at U.S.$278 million annually with a total estimated loss between 1986 and 2015 of U.S.$8.4 billion.[151]

Concentrations of radioactivity in water, sediments, and biota of the Chernobyl cooling pond ecosystem declined between 1986 and 1990, as judged by $^{137}$Cs concentrations (Table 24.1). Carp and other fish species held in Chernobyl cooling pond waters had reproductive-system anomalies that were more pronounced in males than females.[98,99] Silver carp (*Hypophthalmichthys molitrix*) survivors from the cooling pond of the Chernobyl nuclear power station received 7–11 Gy between 1989 and 1992 and displayed reproductive-system disorders that included sterility, changes in gonadal morphology, and degeneration of reproductive cells.[99] Silver carp born in 1989 from parents reared in Chernobyl cooling pond waters had a marked increase of 17 to 26% above controls in reproductive-system anomalies in 1989–1992.[72] Anomalies included degenerative changes in oocytes, spermatogonia, and spermatocytes, and the appearance of bisexual and sterile fish. The gonadal abnormalities are attributed to the high radiation dose of 7–10 Gy received by the parent fish during gonad formation and the continuing exposure of 0.2 Gy annually to this generation.[72] Mature silver carp dwarf individuals age 4+, descendants of fish irradiated in 1986, were observed in 1991.[100] In 1992, second-generation fish hatched from 3-year-old fish of the first generation. Female reproductive systems were not significantly different from controls; however, males had decreased ejaculate volume and concentration and destructive changes in testes caused by irradiation.[101] Silver carp from this ecosystem also had a dose-dependent decrease in hormonal control over the $Na^+$, $K^+$-pump in erythrocytes, with increased passive permeability of the erythrocyte membrane to radioactive analogs of sodium and potassium.[73]

Radionuclide levels in soil invertebrates, soil, litter, and terrestrial insects within the 30-km exclusion zone declined sharply between 1986 and 1987, usually by a factor of ten or more.[165] Radionuclide concentrations in muscle of fishes from the Kiev Reservoir decreased significantly during the period 1986–1997.[105] Female northern pike, *Esox lucius*, of the 1986 year class from this locale had gonadal abnormalities during the period 1987–1997. Abnormality frequency rate was about 34% and included asymmetry, histopathology, and roe resorption. The internal radiation dose from $^{134+137}$Cs decreased from about 0.1–0.2 Gy in 1986 to 0.001–0.002 Gy during 1993–1997.[105] In the 5-year period following the accident, amphibians and reptiles within the 30-km exclusion zone, had gamma radiation levels that remained stable and sometimes increased due to accumulation of radionuclides.[165] One year after the accident, terrestrial birds within the 30-km exclusion zone had gamma radiation levels five to seven times higher than did conspecifics from uncontaminated areas. Waterfowl, however, during the summer of 1987, had radiation levels that were two to five times higher than a year previously, with highest levels in young and females.[165]

**Table 24.1 Radionuclide Concentrations in Organisms and Abiotic Materials near Chernobyl**

| Locale, Radionuclide, Sample, and Other Variables | Concentration[a] and Reference |
|---|---|
| **Chernobyl cooling pond ecosystem; $^{137}$Cs; 1986 (postaccident) vs. 1990** | |
| Water | 210 FW vs. 14 FW[71] |
| Sediments | 170,000 FW vs. 140,000 FW[71] |
| Algae | 90,000 FW vs. 19,000 FW[71] |
| Fishes; 5 spp.; muscle | 30,000–180,000 FW vs. 8000–80,000 FW[71] |
| **Various ponds; 20–40 km from Chernobyl; $^{137}$Cs; September 9–19, 1992** | |
| Sediments | Usually < 1000 DW; Max. 11,000 DW[171] |
| Crucian carp, *Carassius carassius*; muscle | < 1000–8200 DW[171] |
| **Terrestrial Invertebrates** | |
| Within 30-km exclusion zone, 1986 (postaccident) | |
| Soil invertebrates; gamma vs. beta radiation levels | 21,800–34,800 DW vs. 48,100–166,700 DW[165] |
| Insects; gamma vs. beta radiation levels | 4,800–218,300 DW vs. 200–37,000 DW[165] |
| **Pine forest 80 km from Chernobyl; 1988; $^{134+137}$Cs** | |
| Forest workers, human[b] | 205 FW[162] |
| Soil, up to 5 cm deep | 296 DW[162] |
| Pine needles | 2040 DW[162] |
| Forest substrate | 151,700 DW[162] |
| Mosses | 161,000 DW[162] |
| **Vegetation** | |
| July–August 1988; 6–18 km from power plant | |
| Fungi; 9 spp. | |
| $^{90}$Sr | 1800–93,600 DW[133] |
| $^{134+137}$Cs | 25,900–350,300 DW[133] |
| Lichens, 4 spp. | |
| $^{90}$Sr | 8000–277,300 DW[133] |
| $^{134+137}$Cs | 100,100–1,456,400 DW[133] |
| Mosses, 5 spp. | |
| $^{90}$Sr | 27,100–292,800 DW[133] |
| $^{134+137}$Cs | 118,000–2,602,300 DW[133] |
| **Soils** | |
| Total deposition; sites 2–15 km from reactor; $^{137}$Cs | 1,660,000 Bq/m$^2$ [106] |
| Soils and litter; 1986 (postaccident); within 30-km exclusion zone | 500,000–650,000 DW[165] |
| **Fishes** | |
| Kiev Reservoir (30 km from Chernobyl) | |
| Muscle; 4 spp.; $^{137}$Cs | |
| 1985 | 10 FW[105] |
| 1986 (postaccident) | 54–1998 FW[105] |
| 1987 | 662–1773 FW[105] |
| 1990 | 326–1288 FW[105] |
| 1991 | 224–1431 FW[105] |
| 1992 | 155–1131 FW[105] |
| 1993 | 123–411 FW[105] |
| 1994 | 107–450 FW[105] |
| 1995 | 25–310 FW[105] |
| 1996 | 54–370 FW[105] |
| Muscle; 4 spp.; $^{90}$Sr | |
| 1990 | 20–31 FW[105] |
| 1991 | 23–56 FW[105] |
| 1992 | 31–96 FW[105] |
| 1993 | 18–47 FW[105] |

**Table 24.1 Radionuclide Concentrations in Organisms and Abiotic Materials near Chernobyl *(Continued)***

| Locale, Radionuclide, Sample, and Other Variables | Concentration[a] and Reference |
|---|---|
| Northern pike, *Esox lucius*; muscle; $^{137}Cs$; Kiev Reservoir (30 km from Chernobyl) vs. Lake Kojanoskoye (400 km from Chernobyl); 1994 | 450 FW vs. 28,315 FW[105] |
| Perch, *Perca fluviatilis*; muscle; $^{137}Cs$ | |
| 1994; Kiev Reservoir vs. Lake Kojanoskoye | 327 FW vs. 29,856 FW[105] |
| 1997 | |
| Kiev Reservoir | 326 FW[105] |
| River Teterev (45 km from Chernobyl) | 125 FW[105] |
| Lake Svjatoe (300 km distant) | 90,804 FW[105] |
| **Amphibians and Reptiles** | |
| Within 30-km exclusion zone; 1986 (postaccident); gamma radiation levels; whole body | Max. 370,000 DW (vs. 2400 DW for controls)[165] |
| **Birds** | |
| Summer 1986 vs. summer 1987; gamma radiation; within 30-km exclusion zone; whole body | Max. 300,000 DW vs. 500–8300 DW [165] |
| June 1986; within 30 km of Chernobyl nuclear plant; whole bird; total radioactivity | |
| Mallard, *Anas platyrhynchos* | 80,576 FW[89] |
| Northern shoveler, *Anas clypeata* | 44,400 FW[89] |
| Common teal, *Anas crecca* | 73,177 FW[89] |
| Gerganey, *Anas querquedula* | 62,108 FW[89] |
| Common snipe, *Gallinago gallinago* | 61,420 FW[89] |
| Black grouse, *Lyurus tetrix* | 59,735 FW[89] |
| 1986 (postaccident); within 30 km exclusion zone; muscle; beta emitters; young birds vs. adults | 18,870 FW vs. 7400 FW[89] |
| 1994; mallard; within 10 km of Chernobyl; gamma emitters | |
| Diet | 5000 FW[89] |
| Feathers | 8000 FW[89] |
| Gizzard | 20,000 FW[89] |
| Kidneys | 17,000 FW[89] |
| Liver | 15,000 FW[89] |
| Muscle: breast vs. leg | 28,000 FW vs. 30,000 FW, Max. > 100,000 FW[89] |
| **Mammals** | |
| Lactating dairy cattle fed herbage for 12 days collected from a pasture 3.5 km from Chernobyl; summer 1993; herbage vs. milk | |
| $^{90}Sr$ | 41,900 DW vs. 210 FW[130] |
| $^{137}Cs$ | 28,400 DW vs. 1520 FW[130] |
| Large mammals; 4 spp.; summer 1986; within 30-km exclusion zone; whole body; gamma radiation | 50,000–400,000 DW[165] |
| Small mammals; 8 spp.; skeleton; $^{90}Sr$; near Chernobyl power plant; May 1994–August 1996 | 79,000–497,000 (1200–2,275,000) AW[145] |
| Small mammals; 12 spp.; within 30-km exclusion zone; whole body | |
| Gamma radiation; 1986 (postaccident) vs. 1990 | Max. 225,000 DW vs. Max. 335,000 DW[165] |
| $^{90}Sr$; 1986 (postaccident) vs. 1990 | 500–600 DW vs. 700 DW [165] |
| Within evacuation zone immediately north of Chernobyl near Savichi; 1994–95; initial contamination of 555,000–1,480,000 Bq $^{137}Cs/m^2$ | |
| Moose, *Alces alces* | |
| $^{137}Cs$; muscle | (190–11,700) FW[174] |
| $^{90}Sr$; muscle vs. bone | (4–95) DW vs. (11,420–95,100) DW[174] |
| Roe deer, *Capreolus capreolus* | |
| $^{137}Cs$; muscle | (4840–10,730) FW [174] |
| $^{90}Sr$; muscle vs. bone | (4–23) DW vs. (4250–61,300) DW[174] |
| Wild boar, *Sus scrofa* | |

**Table 24.1 Radionuclide Concentrations in Organisms and Abiotic Materials near Chernobyl *(Continued)***

| Locale, Radionuclide, Sample, and Other Variables | Concentration[a] and Reference |
|---|---|
| $^{137}Cs$; muscle | (3700–61,900) FW[174] |
| $^{90}Sr$; muscle vs. bone | (3–10) DW vs. (4210–30,880) DW[174] |
| Ungulate forage plants; $^{137}Cs$ | |
| Aspen, *Populus* sp.; sprouts vs. bark | 1540 FW vs. 7190 FW[174] |
| Sprouts; 5 species | 1060–4300 FW[174] |
| Within 30 km of Chernobyl vs. 30–50 km southeast of exclusion zone; May 1994–August 1996; $^{134+137}Cs$; muscle | |
| Old World field mouse, *Apodemus agrarius* | 372,300 DW vs. 330 DW[145] |
| Yellow-necked field mouse, *Apodemus flavicollis* | 1,259,000 DW vs. no data[145] |
| Wood mouse, *Apodemus sylvaticus* | 555,200 DW vs. 7500 DW[145] |
| Voles, *Microtus* spp., 3 spp. | 469,400–695,000 DW vs. 780–1000 DW[145] |
| Bank vole, *Clethrionomys glareolus* | 5,062,200 (Max. 73,090,000) DW vs. 1800 (max. 3000) DW[145] |
| Common shrew, *Sorex araneus* | 632,300 (45,800–2,910,000) DW vs. 800 (300–1400) DW[145] |

*Note:* All concentrations are in Bq/kg fresh weight (FW), dry weight (DW), or ash weight (AW) unless indicated otherwise.

[a] Concentrations are shown as mean, range in parentheses, maximum (Max.), and nondetectable (ND).
[b] 65 kg person.

In June 1991, male barn swallows (*Hirundo rustica*) collected within 50 km of Chernobyl — when compared to control areas 100 km distant and with museum samples from both areas — had significant differences in length of tail feathers between the right and left side (fluctuating asymmetry) and in morphology of feathers. The degree of fluctuating asymmetry in male tail length and the frequency of deviant morphology in the tails of male barn swallows were associated with a delay in the start of the breeding season.[117] In 1996, barn swallows near Chernobyl, when compared to conspecifics from distant sites, had decreased lymphocyte and immunoglobulin concentrations, reduced spleen size, and reduced intensity of carotenoid-based coloration.[137] Length of outermost tail feathers of males — important secondary sexual characteristics — was positively related to coloration in controls but not in the Chernobyl population, and this may affect breeding success of Chernobyl swallow populations.[137] In 1994, mallard tissues collected within 10 km of Chernobyl had 8000 to 30,000 Bq/kg FW (Table 24.1). Conspecifics collected 45 km from the reactor in 1994 had tissue concentrations that were 40 to 100 times lower.[89] Population numbers of mallards and teals decreased in the 30-km zone in 1994, perhaps due to decreased food sources of farmlands, and invasion of nest areas by plant communities not used in agriculture.[89] In 1994, there were no cases of radiation-induced pathology in any bird examined.[89] Also in 1994, there were increased sightings of rarely seen species of egrets, cranes, and eagles.[89]

In 1991, 5 years after the accident, a female root vole (*Microtus oeconomus*) with an abnormal karyotype (reciprocal translocation) was found within the 30-km radius of the Chernobyl nuclear power plant. These chromosomal aberrations were probably inherited and did not affect the viability of vole populations.[77] Population density of Chernobyl rodents in 1988–1989 was about twice that predicted from previous cycles and was attributed, in part, to increasing radioresistance and abundance of food supplies.[154]

In 1994–1995, the diversity and abundance of the small mammal population (12 species of rodents) at the most radioactive sites at Chernobyl were the same as reference sites.[78] Rodents from the most radioactive areas did not show gross morphological features other than enlargement of the spleen. There were no gross chromosomal arrangements, as judged by examination of the karyotypes. Also observed within the most heavily contaminated site were red fox (*Vulpes vulpes*), gray wolf (*Canis lupus*), moose (*Alces alces*), river otter (*Lutra lutra*), roe deer (*Capreolus capre-*

*olus*), Russian wild boar (*Sus scrofa*), brown hare (*Lepus europaeus*), and feral dogs.[78] The rich biodiversity and abundance of individuals in the 10-km exclusion zone is attributed, in part, to the absence of human habitation.[163]

In 1994–1996, small mammals collected within 8 km of the Chernobyl reactor were experiencing substantial radiation dose rates from $^{134+137}Cs$ in muscle and $^{90}Sr$ in bone.[145] Radiocesium concentrations averaged 3,200,000 Bq/kg muscle DW and for $^{90}Sr$ in bone 297,000 Bq/kg ash weight. Dose rates from radiocesium averaged 2.4 mGy daily (maximum 60 mGy/day) and from radiostrontium 1.0 mGy daily. Conspecifics captured 30 km southeast of the reactor averaged only 2000 Bq/kg muscle DW and were receiving 0.014 mGy daily from internal radiocesium. Estimated dose rates in certain areas of the Chernobyl region now exceed those reported to interfere with mammalian reproduction.[145]

## 24.3 NONLOCAL EFFECTS

Radioactive material from the Chernobyl accident became widely dispersed throughout Europe and the northern hemisphere (Table 24.2). In Europe alone, about 80,000 TBq of $^{137}Cs$ was deposited as follows: Belarus 33.5%, Russia 24%, Ukraine 20%, Sweden 4.4%, Finland 4.3%, Bulgaria 2.8%, Austria 2.7%, Norway 2.3%, Romania 2%, and Germany 1.1%.[173] It is probable that the full impact of the Chernobyl reactor accident on natural resources will not be known for several decades, primarily because of data gaps on long-term genetic and reproductive effects and on radiocesium cycling and toxicokinetics.

### 24.3.1 Soil and Vegetation

The radiocesium fallout in Sweden was among the highest in western Europe — exceeding 60,000 Bq/m$^2$ on Sweden's Baltic coast — and involved mainly upland pastures and forests.[5–7] In Norway, radiocesium deposition from the Chernobyl accident ranged from < 5000 to > 200,000 Bq/m$^2$, and greatly exceeded deposition from prior nuclear weapons tests.[11] In Italy, heavy rainfall coincided with the passage of the Chernobyl radioactive cloud and caused local high deposition of radionuclides in soil, grass, and plants.[12] The Chernobyl plume reached Greece on May 1, 1986. A total of 14 gamma emitters were identified in Greek soil and vegetation in May 1986, and three ($^{134}Cs$, $^{137}Cs$, $^{131}I$) were also detected in milk of free-grazing animals in the area.[13] Radiocesium-134 and $^{137}Cs$ intake by humans in Germany during the period 1986–1987 was mainly from rye, wheat, milk, and beef.[14] In the United Kingdom, elevated concentrations of radionuclides of iodine, cesium, ruthenium, and other nuclides were measured in air and rainwater during May 2–5, 1986.[1] The background activity concentrations were about three times normal levels in early May, and those of $^{131}I$ approached the derived emergency reference level (DERL) of drinking water of 5 mSv $^{131}I$ (equivalent to a thyroid dose of 50 mSv); however, $^{131}I$ levels were not elevated in foodstuffs or cow's milk.[1] Syria — 1800 km from Chernobyl — had measurable atmospheric concentrations of $^{137}Cs$ and $^{131}I$ and near detection-limit concentrations of $^{144}Ce$, $^{134}Cs$, $^{140}La$, and $^{106}Ru$.[15] The maximum $^{131}I$ thyroid dose equivalent received by Syrians was 116 μSv in adults and 210 μSv in children. One year later, these values were 25 μSv in adults and 70 μSv in a 10 year old.[15]

The amount of fallout radioactivity deposited on plant surfaces depends on the exposed surface area, the developmental season of the plants, and the external morphology. Mosses, which have a comparatively large surface area, showed highest concentrations of radiocesium (Table 24.2). In northern Sweden, most of the radiocesium fallout was deposited on plant surfaces in the forest ecosystem and was readily incorporated into living systems because of browsing by herbivores and cesium's chemical similarity to potassium.[7] In August 1992, the distribution of $^{137}Cs$ fallout from Chernobyl in a Swedish forest was 87% in soils, 6% in the bryophyte layer, and 7% in standing biomass of trees; the mean deposition of $^{137}Cs$ was estimated at 54,000 Bq/m$^2$.[138] The mean $^{137}Cs$

**Table 24.2 Radionuclide Concentrations in Biotic and Abiotic Materials from Various Geographic Locales in Relation to the Chernobyl Nuclear Accident on April 26, 1986**

| Locale, Sample, Radionuclide, and Other Variables | Concentration[a] and Reference |
|---|---|
| **Adriatic Sea** | |
| Limpet, *Patella coerulea*; $^{110m}Ag$; soft parts | |
| June 1986 | 29 FW[86] |
| October 1986 | 14 FW[86] |
| June 1987 | 2 FW[86] |
| Sediments; $^{137}Cs$ | |
| August 1983 (pre-Chernobyl) | 8 DW[86] |
| June 1986 | 10 DW[86] |
| June 1987 | 13 DW[86] |
| June 1990 | 9 DW[86] |
| Sediments; $^{103}Ru$ | |
| June 1986 | 5–19 DW[86] |
| November 1986 | ND[86] |
| **Alaska** | |
| Arctic samples; 1987; $^{134+137}Cs$; radiocesium deposition rate April 1986–August 1986 estimated at 0.5 Bq/m$^2$ | |
| Soil | 2–44 DW[119] |
| Lichens (mostly *Cladonia* spp.), mosses (*Sphagnum* spp.), ledum (*Ledum* spp.) | 16–242 DW[119] |
| Bowhead whale, *Balaena mysticetus*; muscle | 0.6 FW[119] |
| Least cisco, *Coregonus sardinella*; muscle | 0.5 FW[119] |
| Lake whitefish, *Coregonus clupeaformis*; muscle | ND[119] |
| Caribou, *Rangifer tarandus*; muscle | 1.1 FW[119] |
| Subarctic samples, July–August 1988; $^{137}Cs$ vs. $^{134}Cs$ | |
| Lichens | 89 DW vs. 5 DW[119] |
| Mosses | 86 DW vs. 6 DW[119] |
| Ledum | 60 DW vs. 4 DW[119] |
| Mushrooms | 43 DW vs. 5 DW[119] |
| **Alaska/Yukon Territories** | |
| Barren-ground caribou, *Rangifer tarandus granti*, Porcupine herd; $^{137}Cs$; March–November 1987 | |
| Feces | Max. 802 DW[40] |
| Muscle | 133 (26–232) FW[40] |
| Rumen contents | Max. 538 DW[40] |
| **Albania** | |
| $^{137}Cs$; 2–19 May 1986 | |
| Air | Max. 1.8 Bq/m$^3$ [62] |
| Cow, *Bos bovis*; milk | Max. 380 FW[62] |
| Wheat flour | Max. 236 FW[62] |
| $^{131}I$; cow's milk; 2–19 May 1986 | Max. 3500 FW[62] |
| $^{90}Sr$ | |
| Seawater; Adriatic Sea near Durres | |
| 1992 vs. 1993 | 0.0054 FW vs. 0.0032 FW[175] |
| 1994 vs. allowable limit | 0.0022 FW vs. 0.41 FW[175] |
| Fish; 3 species; 1992–1994 | Max. 0.056 FW[175] |
| **Canada** | |
| Red-backed vole, *Clethrionomys gapperi*; southeastern Manitoba; 1986; $^{137}Cs$; spruce bog habitat | |
| Carcass vs. GI tract | |
| April | 159 FW vs. 210 FW[109] |
| July | 76 FW vs. 172 FW[109] |
| October | 603 FW vs. 1095 FW[109] |
| Fungi in diet; 11 spp.; September 1986 | Max. 990 FW–Max. 2476 FW[109] |

**Table 24.2 Radionuclide Concentrations in Biotic and Abiotic Materials from Various Geographic Locales in Relation to the Chernobyl Nuclear Accident on April 26, 1986 *(Continued)***

| Locale, Sample, Radionuclide, and Other Variables | Concentration[a] and Reference |
|---|---|
| Lichens and mosses; July 1986; $^{134+137}Cs$ | |
| Northwest Territories (NWT) | 510–980 DW (none attributed to Chernobyl)[122] |
| Ontario | 250–1140 DW (14–16% attributed to Chernobyl)[122] |
| Alberta | 280–620 DW (12–28% from Chernobyl)[122] |
| Caribou, *Rangifer tarandus*; muscle; $^{134+137}Cs$; NWT | |
| April 1986 (pre-Chernobyl) | 150 FW, 650 DW (0–13% from Chernobyl)[122] |
| January 1987 | 130 FW, 600 DW (28% from Chernobyl)[122] |
| March 1987 | 20 FW, 90 DW (none from Chernobyl)[122] |
| Caribou; northern Quebec; muscle; $^{137}Cs$; 1986–1987 | (166–1129) FW[39] |
| **Croatia** | |
| 1986; annual effective dose in mSv | |
| 1-year-old infants | 1.49[159] |
| Children, age 10 years | 0.93[159] |
| Adults | 0.83[159] |
| Yearly intake in foods by adults, in Bq; $^{137}Cs$ vs. $^{90}Sr$ | |
| 1964 (military tests) | 17,857 vs. 3583[159] |
| 1985 | 36 vs. 71[159] |
| 1986 | 7786 vs. 182[159] |
| 1990 | 138 vs. 99[159] |
| 1995 | 88 vs. 78[159] |
| **Czechoslovakia** | |
| Cow; milk; $^{134+137}Cs$ | |
| May 1986 | 42 FW[50] |
| July 1986 | 10 FW[50] |
| December 1986 | 7 DW[50] |
| Domestic pig, *Sus* sp.; muscle; $^{134+137}Cs$; July 1986 vs. July 1987 | (15–24) FW vs. 22 FW[50] |
| **Danube River, Hungary-Yugoslavia** | |
| Fish; 1986 (postaccident) vs. 1987 | |
| $^{134}Cs$ | 8 FW vs. 4 FW[23] |
| $^{137}Cs$ | 13 FW vs. 12 FW[23] |
| $^{103}Ru$ | 1 FW vs. < 1 FW[23] |
| $^{106}Ru$ | 4 FW vs. 3 FW[23] |
| Sediments; 1986 (postaccident) vs. 1988 | |
| $^{134}Cs$ | 500 DW vs. 80 DW[23] |
| $^{137}Cs$ | 750 DW vs. 200 DW[23] |
| Algae; 1986 (postaccident) vs. 1988 | |
| $^{134}Cs$ | 275 FW vs. 25 FW[23] |
| $^{137}Cs$ | 625 FW vs. 100 FW[23] |
| Seawater; $^{90}Sr$; 1992 | 0.01 FW[175] |
| **Egypt** | |
| Surface soils; Nile delta and north coast; 1988 | |
| $^{137}Cs$ | 4.7 (0.1–17.7) DW; 62 (18–2175) Bq/m$^2$ ash weight[129] |
| $^{90}Sr$ | 5.2 (1.7–17.6) DW; 760 (234–3129) Bq/m$^2$ ash weight[129] |
| **Finland** | |
| Finish Lapland; $^{137}Cs$; 1979–1984 vs. 1986 (postaccident) | |
| Arboreal lichens | 120 DW vs. 590 DW[26] |
| Ground lichens | 230 DW vs. 900 DW[26] |
| Birch, *Betula* sp. | 68 DW vs. 51 DW[26] |
| Horsetails, *Equisetum* sp. | 203 DW vs. 280 DW[26] |
| Bilberry, *Vaccinium* sp. | 120 DW vs. 590 DW[26] |

**Table 24.2 Radionuclide Concentrations in Biotic and Abiotic Materials from Various Geographic Locales in Relation to the Chernobyl Nuclear Accident on April 26, 1986** *(Continued)*

| Locale, Sample, Radionuclide, and Other Variables | Concentration[a] and Reference |
|---|---|
| Lichens, $^{137}Cs$ | |
| From reindeer herding areas; 1986 (postaccident) vs. 1987 | 900 DW vs. 800 DW[30] |
| Isolated areas, 1986–1987 | 3000–10,000 DW[30] |
| Lichens, *Cladonia* spp. vs. surface soil and litter; maximum values; summers 1986/88, decay corrected to May 1, 1986 | |
| $^{238}Pu$ | 5.7 DW vs. 5.1 DW[152] |
| $^{239+240}Pu$ | 1.3 DW vs. 0.6 DW[152] |
| $^{241}Am$ | 1.6 DW vs. 2.3 DW[152] |
| $^{134}Cs$ | 37,000 DW vs. 3660 DW[152] |
| $^{137}Cs$ | 67,000 DW vs. 13,700 DW[152] |
| $^{95}Zr$ | 10,800 DW vs. 19 DW[152] |
| $^{106}Ru$ | 10,000 DW vs. 1170 DW[152] |
| $^{144}Ce$ | 6100 DW vs. 300 DW[152] |
| Gulf of Finland; $^{137}Cs$ | |
| Seawater | |
| 1985 vs. 1986 (postaccident) | 0.01. FW vs. 1.1 FW[71] |
| 1987 vs. 1990 | 0.2 FW vs. 0.05 FW[71] |
| Bottom sediments | |
| 1985 vs. 1986 | 1.2 FW vs. 40 FW[71] |
| 1987 vs. 1990 | 19 FW vs. 5 FW[71] |
| Algae, whole | |
| 1985 vs. 1986 | 4 FW vs. 175 FW[71] |
| 1987 vs. 1990 | 30 FW vs. 14 FW[71] |
| Fish, 2 spp., whole | |
| 1985 | 1–4 FW[71] |
| 1986 (postaccident) | 22–54 FW[71] |
| 1987 | 60–120 FW[71] |
| 1990 | 36–116 FW[71] |
| Lake Paijanne (estimated $^{137}Cs$ Chernobyl loading of 20,000 Bq/m$^2$); $^{137}Cs$; whole fish, 3 spp. (northern pike, *Esox lucius*; yellow perch, *Perca flavescens*; roach, *Rutilus rutilus*) | |
| 1986, pre- vs. post-Chernobyl | 580 FW vs. 1250 FW[24] |
| 1987 vs. 1988 | 1000–2000 FW vs. 160–2000 FW[24] |
| Scotch pine, *Pinus silvestris*; southern Finland; needles; $^{137}Cs$ | |
| June 1986 | |
| Needles formed in 1983 | 20,000 DW[102] |
| Needles formed in 1986 postaccident | 3050 DW[102] |
| Newly formed needles, 1987 vs. 1988 | 2700 DW vs. 4800 DW[102] |
| Reindeer, *Rangifer tarandus*; muscle; $^{137}Cs$ | |
| 1964–1965 (following nuclear tests) | Max. 2600 FW[26,30] |
| 1985–1986 (preaccident) | 300 FW[26,30] |
| 1986 (postaccident)–1987 | 720 FW; Max. 16,000 FW[30] |
| 1987–1988 | 640 FW; Max. 9000 FW[30] |
| Southern Finland; Lakes (n = 52); $^{137}Cs$; 1988 vs. 1992 | |
| Sediments (Bq/m$^2$) | 11,003 DW, Max. 55,700 DW vs. 3010 DW, Max. 20,700 DW[120] |
| Fish; whole less viscera and head | |
| Perch | 2065 FW vs. 530 FW[120] |
| Pike | 2453 FW vs. 687 FW[120] |
| Roach | 565 FW vs. 218 FW[120] |
| **Germany** | |
| Soils; 24 June 1986 | |
| $^{134}Cs$ | Max. 602 Bq/m$^2$ DW[14] |

**Table 24.2 Radionuclide Concentrations in Biotic and Abiotic Materials from Various Geographic Locales in Relation to the Chernobyl Nuclear Accident on April 26, 1986** ***(Continued)***

| Locale, Sample, Radionuclide, and Other Variables | Concentration[a] and Reference |
|---|---|
| $^{137}Cs$ | Max. 1545 Bq/m$^2$ DW[14] |
| $^{103}Ru$ | Max. 808 Bq/m$^2$ DW[14] |
| Pasture vegetation; May 1986 | |
| $^{134}Cs$ | 20 FW[14] |
| $^{137}Cs$ | 40 FW[14] |
| $^{131}I$ | 75 FW[14] |
| Cow; milk; May 1986 | |
| $^{134}Cs$ | 140 FW[14] |
| $^{137}Cs$ | 250 FW[14] |
| $^{131}I$ | 250 FW[14] |
| $^{103}Ru$ | 250 FW[14] |
| Vertebrates; 13 spp.; muscle; $^{137}Cs$; May 1986–January 1987 | |
| Wild boar, *Sus* sp. | 440 (5–5400) FW[113] |
| 12 spp.; maximum values | 21–4300 FW[113] |
| Roe deer, *Capreolus capreolus*; muscle | |
| $^{137}Cs$ | |
| 1986, postaccident | 50 FW; Max. 200 FW[84] |
| 1987 | 30 FW; Max. 700 FW[84] |
| 1988 | < 15 FW; Max. 450 FW[84] |
| 1989 | 4 FW; Max. 140 FW[84] |
| 1990 | 2 FW[84] |
| $^{134}Cs$ | |
| 1986, postaccident | 20 FW; Max. 70 FW[84] |
| 1987 | 10 FW; Max. 210 FW[84] |
| 1988 | < 10 FW; Max. 110 FW[84] |
| 1989 | 0.6 FW; Max. 20 FW[84] |
| 1990 | 0.3 FW[84] |
| **Greece** | |
| Alfalfa, *Medicago sativa*; June 1986 | |
| $^{134}Cs$ | 2303 DW[20] |
| $^{137}Cs$ | 4551 DW[20] |
| $^{103}Ru$ | 358 DW[20] |
| $^{106}Ru$ | 1075 DW[20] |
| Rye grass, *Lolium perenne*; June 1986 | |
| $^{134}Cs$ | 3518 DW[20] |
| $^{137}Cs$ | 7090 DW[20] |
| $^{103}Ru$ | 708 DW[20] |
| $^{106}Ru$ | 1747 DW[20] |
| Fishes | |
| Marine teleosts; muscle; maximum concentrations; July 1986 | |
| $^{140}Ba/^{140}La$ | 20 FW[104] |
| $^{144}Ce$ | 120 FW[104] |
| $^{134+137}Cs$ | 140 FW[104] |
| $^{131}I$ | 55 FW[104] |
| $^{103}Ru$ | 230 FW[104] |
| $^{95}Zr/^{95}Nb$ | 34 FW[104] |
| Freshwater fishes; muscle; $^{134+137}Cs$; June 3–14, 1986 | 205 (5–830) FW[104] |
| Plants, various; measured about 4 months postaccident; $^{137}Cs$; values represent about 9% of initial activity | |
| Aromatic plants, 11 spp. | 22–11,344 FW; 26–22,000 DW[64] |
| Cereals, 4 spp. | 11–2257 FW; 11–2775 DW[64] |
| Fruit bearing trees, 7 spp. | 85–1572 FW; 122–2116 DW[64] |
| Fungi, 4 spp. | 103–5553 FW; 214–11,418 DW[64] |

**Table 24.2 Radionuclide Concentrations in Biotic and Abiotic Materials from Various Geographic Locales in Relation to the Chernobyl Nuclear Accident on April 26, 1986 *(Continued)***

| Locale, Sample, Radionuclide, and Other Variables | Concentration[a] and Reference |
|---|---|
| Marine algae, 4 spp. | 85–139 FW; 529–917 DW[64] |
| Mosses and lichens, 6 spp. | 1184–9413 FW; 1110–18,847 DW[64] |
| Vegetables, 18 spp. | 18–244 FW; 18–299 DW[64] |
| Northern Greece; May 1986; $^{131}$I | |
| Grasses | Max. 1500 FW[13] |
| Milk | |
| Cow | Max. 300 FW[13] |
| Domestic sheep, *Ovis aires* | Max. 800 FW[13] |
| Domestic sheep; thyroid; $^{131}$I; 1986 | |
| 27 June | Max. 4000 FW[61] |
| 2 July | Max. 15,600 FW[61] |
| 3 July | Max. 618,000 FW[61] |
| 5 July | Max. 9000 FW[61] |
| 29 July | Max. 8500 FW[61] |
| 20 August | Max. 600 FW[61] |
| **Ireland** | |
| Migratory birds; muscle; maximum values | |
| 1986/87; waders; 3 spp. | |
| $^{134}$Cs | 206 FW[87] |
| $^{137}$Cs | 565 FW[87] |
| 1986/87; ducks; 9 spp. | |
| $^{134}$Cs | < 34 FW[87] |
| $^{137}$Cs | < 48 FW[87] |
| 1987/88; woodcock, *Scolopax rusticola* | |
| $^{134}$Cs | 32 FW[87] |
| $^{137}$Cs | 119 FW[87] |
| **Italy** | |
| Honey bee, *Apis* spp.; honey; 10 May 1986 | |
| $^{134}$Cs | Max. 171 FW[17] |
| $^{137}$Cs | Max. 363 FW[17] |
| $^{131}$I | Max. 1051 FW[17] |
| $^{103}$Ru | Max. 575 FW[17] |
| Rodent, *Mus musculus domesticus*; carcass, less internal organs; $^{137}$Cs | |
| October–November 1981 | 5 DW[42] |
| May 1986 | 43 DW[42] |
| October–November 1986 | 20 DW[42] |
| May 1987 | 18 DW[42] |
| Northwest Saluggia; May 1986 | |
| $^{137}$Cs | |
| Air | –0.001 Bq/m$^3$ FW[59] |
| Pasture grass | 8000 DW[59] |
| Cow's milk | 180 FW[59] |
| $^{131}$I | |
| Air | –0.005 Bq/m$^3$ FW[59] |
| Pasture grass | 12,000 DW[59] |
| Cow's milk | 870 FW[59] |
| Ligurian Sea | |
| Seafood products (fish, molluscs, crustaceans) of commerce, edible portions; 13 spp.; 1987 vs. 1988 | |
| $^{137}$Cs | 5.9 FW vs. 2.5 FW[118] |
| $^{134}$Cs | 2.7 FW vs. 0.3 FW[118] |
| Littoral areas vs. open sea; 1988 | |
| $^{137}$Cs | 8.4 FW vs. 2.9 FW[118] |
| $^{134}$Cs | 1.7 FW vs. 0.6 FW[118] |

**Table 24.2 Radionuclide Concentrations in Biotic and Abiotic Materials from Various Geographic Locales in Relation to the Chernobyl Nuclear Accident on April 26, 1986** ***(Continued)***

| Locale, Sample, Radionuclide, and Other Variables | Concentration[a] and Reference |
|---|---|
| **Japan** | |
| $^{137}Cs$ | |
| Milk; cow; May 1986 | Max. 0.6 FW[65] |
| Soil, estimated deposition from Chernobyl | 180 Bq/m$^2$ DW[65] |
| $^{131}I$; 10–11 May vs. 30 May 1986 | |
| Grass | 65 FW vs. 14 FW[57] |
| Cow's milk | 4.3 FW vs. ND[57] |
| **Monaco** | |
| Air, Bq/m$^3$, April 26, 1986; Monaco vs. Chernobyl (former Soviet Union) | |
| $^{134}Cs$ | 8.2 vs. 53[81] |
| $^{137}Cs$ | 1.6 vs. 120[81] |
| $^{103}Ru$ | 3.5 vs. 280[81] |
| $^{131}I$ | 4.6 vs. 750[81] |
| $^{106}Ru$ | 3.0 vs. 110[81] |
| $^{140}Ba$ | 9.8 vs. 420[81] |
| $^{99}Mo$ | 3.8 vs. 490[81] |
| $^{141}Ce$ | 3.7 vs. 190[81] |
| $^{144}Ce$ | 2.5 vs. 110[81] |
| $^{95}Zr$ | 1.2 vs. 590[81] |
| Marine copepods, 3 spp.; May 6, 1986; whole organism vs. fecal pellets | |
| $^{103}Ru$ | 280 DW vs. 16,000 DW[79] |
| $^{106}Ru$ | 70 DW vs. 5800 DW[79] |
| $^{134}Cs$ | 22 DW vs. 3400 DW[79] |
| $^{137}Cs$ | 34 DW vs. 6300 DW[79] |
| $^{141}Ce$ | 20 DW vs. 900 DW[79] |
| $^{144}Ce$ | 100 DW vs. 2500 DW[79] |
| Mussel, *Mytilus galloprovincialis*; soft parts; May 6 vs. August 14, 1986 | |
| $^{103}Ru$ | 480 FW vs. 9.6 FW[82] |
| $^{106}Ru$ | 121 FW vs. 11.2 FW[82] |
| $^{131}I$ | 84 FW vs. < 2 FW[82] |
| $^{134}Cs$ | 6 FW vs. 0.1 FW[82] |
| $^{137}Cs$ | 5.2 FW vs. 0.3 FW[82] |
| **Netherlands** | |
| Grass silage; 1986 (postaccident) vs. 1987 | |
| $^{134}Cs$ | Max. 50 DW vs. 2 DW[19] |
| $^{137}Cs$ | Max. 172 DW vs. 9 DW[19] |
| $^{40}K$ | 910 DW vs. 1027 DW[19] |
| Contaminated roughage fed to lactating cows | |
| 10.3 Bq $^{137}Cs$/kg FW grass | (1.0–1.6) FW milk[66] |
| 173–180 Bq $^{137}Cs$/kg FW grass silage | (12–28) FW milk[66] |
| 260–271 Bq $^{137}Cs$/kg DW grass | (5.4–6.2) FW milk[66] |
| **Norway** | |
| Alpine lake and vicinity; $^{134+137}Cs$ | |
| Dwarf birch, *Betula nana*; leaves; August 1986 | 4000 FW[6] |
| Lichens; August 1986 | 60,000 FW[6] |
| Willow, *Salix* spp.; leaves; September 1980 vs. August 1986 | < 50 FW vs. 600 FW[6] |
| Lake sediment; upper 10 cm; July–August 1986 | 1050 FW[6] |
| Aquatic organisms; July–August 1986 | |
| Crustaceans, whole | 5300–6700 FW[6] |
| Aquatic insects, whole | 1300–4120 FW[6] |

**Table 24.2 Radionuclide Concentrations in Biotic and Abiotic Materials from Various Geographic Locales in Relation to the Chernobyl Nuclear Accident on April 26, 1986** ***(Continued)***

| Locale, Sample, Radionuclide, and Other Variables | Concentration[a] and Reference |
|---|---|
| Minnow, *Phoxinus phoxinus*, whole | 8800 FW[6] |
| Brown trout, *Salmo trutta* | |
| Muscle | |
| 1985 | < 100 FW[6] |
| June 1986 | 300 FW[6] |
| August 1986 | 7000 FW[6] |
| June 1988 | 4000 FW[6] |
| Eggs vs. milt; July–August 1986 | 1740–3600 FW vs. 1300 FW[6] |
| Dovrefjell, May 1986 vs. August 1990 | |
| Earthworms (*Lumbricus rubellus*, *Allobophora caliginosa*), whole | 121 FW vs. 74 FW[83] |
| Eurasian woodcock, *Scolopax rusticola*, breast muscle | 737 FW vs. 53 FW[83] |
| Litter | 14,400 DW vs. 2900 DW[83] |
| Pied flycatcher, *Ficedula hypoleuca*; whole nestling; $^{134+137}Cs$; 2 locales; July 1 collections | |
| Luru | |
| 1986 vs. 1989 | 844 FW vs. 607 FW[131] |
| 1992 vs. 1996 | 327 FW vs. 118 FW[131] |
| Lauvsjoen | |
| 1986 vs. 1989 | 102 FW vs. 89 FW[131] |
| 1992 vs. 1996 | 193 FW vs. 32 FW[131] |
| Mushroom, *Lactarius* spp.; post-Chernobyl; $^{134+137}Cs$ | Up to 445,000 FW[11] |
| Lake heavily contaminated by Chernobyl fallout; whole fish less digestive tract; $^{137}Cs$; maximum values | |
| Arctic char, *Salvelinus alpinus* | |
| Autumn 1985 | 32 FW[85] |
| July–September 1986 | 8730 FW[85] |
| May–October 1987 | 9910 FW[85] |
| June–September 1988 | 4980 FW[85] |
| May–October 1989 | 3480 FW[85] |
| Brown trout, *Salmo trutta* | |
| Autumn 1985 | 31 FW[85] |
| July–September 1986 | 13,900 FW[85] |
| May–October 1987 | 7070 FW[85] |
| June–September 1988 | 7150 FW[85] |
| May–October 1989 | 2060 FW[85] |
| Reindeer; muscle; $^{134+137}Cs$ | |
| 1986, after Chernobyl | (10,000–50,000) FW[27] |
| January 1987 vs. September 1988 | Max. 56,000 FW vs. Max. 13,900 FW[29] |
| **Poland** | |
| Freshwater fish, 4 spp.; muscle; January 1987; $^{134+137}Cs$ | 4.5–6.1 FW[22] |
| Southern Baltic Sea; 1986; pre- vs. post-accident; $^{134+137}Cs$ | |
| Water | (0.014–0.020) FW vs. (0.059–0.100) FW[22] |
| Atlantic cod, *Gadus morhua*, muscle | (1.4–2.3) FW vs. (5.0–7.4) FW[22] |
| Flounder, *Pleuronectes flesus;* muscle | (1.1–4.5) FW vs. (3.4–6.7) FW[22] |
| Sawinda Wielka Lake (about 400 km from Chernobyl); 1994/1995; $^{137}Cs$ | |
| Water | 0.008 (0.005–0.011) FW[93] |
| Sediments | 78 (5–223) DW[93] |
| Aquatic plants, whole | |

**Table 24.2 Radionuclide Concentrations in Biotic and Abiotic Materials from Various Geographic Locales in Relation to the Chernobyl Nuclear Accident on April 26, 1986 *(Continued)***

| Locale, Sample, Radionuclide, and Other Variables | Concentration[a] and Reference |
|---|---|
| Waterlily, *Nuphar luteum* | Max. 6 DW[93] |
| Pondweeds, *Potamogeton* spp. | 10 DW; Max. 45 DW[93] |
| Sweet flag, *Acorus calamus* | 8 DW; Max. 15 DW[93] |
| Reed. *Phragmites communis* | 116 DW; Max. 381 DW[93] |
| Molluscs, whole, 3 spp. | Max. 4.6 DW[93] |
| Crayfish, *Astacus* sp., whole | Max. 6.1 DW[93] |
| Fish, muscle, 5 spp. | 3–10 DW; Max. 11.2 DW[93] |
| **Spain** | |
| Seawater; 1988 vs. 1991; Bq/m$^3$ | |
| $^{137}$Cs | (4.1–5.0) vs. (4.1–8.4)[134] |
| $^{134}$Cs | (0.26–0.38) vs. (ND–0.51)[134] |
| Marine flowering plant, *Posidonia oceanica*; March 1987 | |
| $^{60}$Co | (0.3–7.7) DW[134] |
| $^{106}$Ru | (1.5–7.7) DW[134] |
| $^{110m}$Ag | (0.8–8.5) DW[134] |
| $^{144}$Ce | (<0.6–2.5) DW[134] |
| $^{137}$Cs | (0.9–12.2) DW[134] |
| $^{134}$Cs | (0.0–1.3) DW[134] |
| Songthrush, *Turdus philomelos* | |
| October 1986; northeastern Spain; whole bird; $^{134}$Cs vs. $^{137}$Cs; maximum values | 133 FW vs. 420 FW[90] |
| $^{134}$Cs; November 1986 vs. November 1987; muscle | Max. 90 DW vs. Max. 7 DW for adults and 5 DW for young[44] |
| $^{137}$Cs; 1986 (postaccident) vs. 1987; muscle | Max. 208 DW vs. Max. 27 DW for adults and 22 DW for young[44] |
| $^{90}$Sr; 1986 (postaccident) vs. 1987; muscle | Max. 23 DW vs. Max. 7 DW[44] |
| Autumn–winter 1994; muscle vs. bone | . |
| $^{134}$Cs | 90 DW vs. Max. 25 DW[94] |
| $^{137}$Cs | 280 (242–2790) DW vs. (47–90) DW[94] |
| $^{90}$Sr | ND vs. 54 DW[94] |
| **Sweden** | |
| Moose, *Alces alces*; central Sweden; muscle; $^{137}$Cs; adults vs. calves | |
| September 1986 | 300 FW vs. 500 FW[7] |
| September 1987 | 201 FW vs. 401 FW[7] |
| September 1988 | 640 FW vs. 1300 FW[7] |
| 1991, all age groups; May–September; various habitats | Mean 478 FW; Max. 1060 FW; highest in swamps and marshes and lowest in farmlands[80] |
| Moose dietary plants; 1986 (postaccident)–1988; $^{137}$Cs | |
| Birch, *Betula* spp.; leaves | 1200 DW[16] |
| Heather, *Calluna vulgaris*; whole | (13,000–32,000) DW[16,67] |
| Sedge, *Carex* spp. | 12,000 DW[67] |
| Hair grass, *Deschampia flexuosa*; whole | 1900 DW[16] |
| Fireweed, *Epilobium angustifolium* | 400 DW[16] |
| Grasses, various | 2500 DW[16] |
| Buckbean, *Menyanthes trifoliata* | 3800 DW[16] |
| Pine, *Pinus sylvestris*; shoots | 2500 DW[16] |
| Aspen, *Populus tremula*; leaves | 700 DW[16] |
| Willow, *Salix* spp.; leaves | 300 DW[16] |
| Mountain ash, *Sorbus aucuparia*; leaves | 1300 DW[16] |
| Bilberry, *Vaccinium myrtillus* | |
| July 1986 | 2000 FW; 4000 DW[7,16] |
| July 1987 | 1138 FW[7] |
| July 1988 | 600 FW[7] |

**Table 24.2 Radionuclide Concentrations in Biotic and Abiotic Materials from Various Geographic Locales in Relation to the Chernobyl Nuclear Accident on April 26, 1986 *(Continued)***

| Locale, Sample, Radionuclide, and Other Variables | Concentration[a] and Reference |
|---|---|
| Bog whortleberry, *Vaccinium ulgmosum* | 5900 DW[16] |
| Cowberry, *Vaccinium vitisidae* | 7500 DW[16] |
| Boreal coniferous forest in Hille, Sweden; average total deposit of 220,000 Bq $^{137}Cs/m^2$ in 1986; $^{137}Cs$; 1991 | |
| Fungi | (120,000–230,000) DW, (3,000–950,000) DW[124] |
| Lichens | 36,000 (29,000–69,000) DW[124] |
| Mosses, ferns, angiosperms | 8000–19,000 DW (600–37,000) DW[124] |
| Cow's milk; $^{137}Cs$ | |
| July 1986 | Mostly < 250 FW; Max. 375 FW[68] |
| 1987 | Usually < 70 FW; Max. 120 FW[5] |
| Lichen, *Bryoria fuscescens*; $^{137}Cs$; 4 June 1986 | (34,000–120,000) DW[67] |
| Roe deer, *Capreolus* sp.; muscle; $^{137}Cs$; 1986, post-accident | (20–12,000) FW[5] |
| Lichen, *Cladina* spp.; $^{137}Cs$; 1986, postaccident | Max. 40,000 DW[67] |
| Bank vole, *Clethrionomys glareolus*, collected from soil containing various concentrations of $^{134+137}Cs$; voles analyzed less skull and digestive organs | |
| 1800 Bq/m$^2$ soil (control) | Voles had 9 Bq $^{134}Cs$/kg FW, and 39 of $^{137}Cs$; mutation frequency 1.3; total daily irradiation of 4.2 μGy[43] |
| 22,000 Bq/m$^2$ soil | In Bq/kg FW, voles had 279 $^{134}Cs$, and 1031 $^{137}Cs$; mutation frequency 1.5; daily dose rate of 8.8 μGy[43] |
| 90,000 Bq/m$^2$ soil | Voles had 1356 Bq $^{134}Cs$/kg FW and 5119 of $^{137}Cs$; mutation frequency 1.9; daily dose of 26.8 μGy[43] |
| 145,000 Bq/m$^2$ soil | Voles had 2151 Bq $^{134}Cs$/kg FW and 7784 $^{137}Cs$; mutation frequency 2.6; daily dose of 39.4 μGy[43] |
| Northern pike, *Esox lucius*; Lake Storvindeln; muscle $^{137}Cs$ vs. $^{134}Cs$ | |
| Preaccident | Max. 27 FW vs. < 1.2 FW[108] |
| May 12–16, 1986 | Max. 19 FW vs. < 2.4 FW[108] |
| May 21–27, 1987 | 460 FW, Max. 567 FW vs. 156 FW, Max. 199 FW[108] |
| Buckbean, *Menyanthes trifoliata*; $^{137}Cs$; 1985 vs. 1987 | 1000 DW vs. 3880 DW[67] |
| Reindeer dietary lichens; $^{137}Cs$; April 1986 | Usually 40,000–60,0000 DW; Max. 120,000 DW[32] |
| Reindeer | |
| Moved in November 1986 from a highly contaminated area (> 20,000 Bq $^{137}Cs/m^2$) to a less- contaminated area (< 3000 Bq/m$^2$) of natural pasture | $^{137}Cs$ content in muscle declined from 12,000 FW in November 1986 to about 3000 FW in April 1987[25] |
| Muscle; $^{137}Cs$; 1986, post-Chernobyl | (100–40,000) FW[5] |
| Muscle; $^{137}Cs$; 1987; | Max. 96,000 FW[136] |
| Muscle; Ottsjolagret; October 16, 1987; $^{137}Cs$ vs. $^{134}Cs$ | 5300 (2970–8955) FW vs. 2024 (1100–3482) FW[108] |
| Muscle; Ammarnis; $^{137}Cs$ vs. $^{134}Cs$ | |
| Pre-Chernobyl | Max. 140 FW vs. < 3 FW[108] |
| September 1986 | 1670 FW, Max. 2460 FW vs. 610 FW, Max. 930 FW[108] |
| September 1987 | 950 FW, Max. 2884 FW vs. 338 FW, Max. 1020 FW[108] |
| August 1989 | 251 FW, Max. 541 FW vs. no data[108] |
| Common shrew, *Sorex araneus*; July–August 1986; $^{137}Cs$; whole less skull, stomach, and viscera | . |
| From soil containing 1,800 Bq/m$^2$ | 48 FW[69] |
| From soil containing 22,000 Bq/m$^2$ | 751 FW[69] |
| From soil containing 90,000 Bq/m$^2$ | 3233 FW[69] |
| From soil containing 145,000 Bq/m$^2$ | 6289 FW[69] |
| **Syria** | |
| $^{137}Cs$; Air; 7–10 May 1986 | 0.1 Bq/m$^3$ [15] |
| $^{131}I$; 7–10 May 1986 | |
| Air | 4 Bq/m$^3$ [15] |
| Goat, *Capra* sp.; milk | 55 FW[15] |

**Table 24.2 Radionuclide Concentrations in Biotic and Abiotic Materials from Various Geographic Locales in Relation to the Chernobyl Nuclear Accident on April 26, 1986** ***(Continued)***

| Locale, Sample, Radionuclide, and Other Variables | Concentration[a] and Reference |
|---|---|
| Coastal mountains; soil vs. vegetation; 1995 | |
| $^{134}Cs$ | (15–229) DW vs. (18–114) DW[132] |
| $^{137}Cs$ | (500–8011) DW vs. (1019–4535) DW[132] |
| $^{90}Sr$ | (55–235) DW vs. (20–195) DW[132] |
| **United Kingdom** | |
| Upland pastures | |
| Moss, *Sphagnum* sp.; September 1986 | |
| $^{110m}Ag$ | 202 DW[60] |
| $^{144}Ce$ | 202 DW[60] |
| $^{134}Cs$ | 8226 DW[60] |
| $^{137}Cs$ | 17,315 DW[60] |
| $^{106}Ru$ | 1893 DW[60] |
| $^{125}Sb$ | 294 DW[60] |
| Vegetation; $^{134+137}Cs$; June 1986 vs. January 1989 | 6000 DW vs. 1000 DW[51] |
| Marine molluscs, 7 spp.; near nuclear plant; 1984 vs. 1986 (post-Chernobyl) | |
| $^{110m}Ag$ | < 77 FW vs. 13–77 FW[63] |
| $^{60}Co$ | < 29 FW vs. 16–32 FW[63] |
| $^{134}Cs$ | < 14 FW vs. 37–388 FW[63] |
| $^{137}Cs$ | < 139 FW vs. 31–836 FW[63] |
| $^{40}K$ | < 59 FW vs. 57–61 FW[63] |
| $^{238}Pu$ | < 27 FW vs. 11–22 FW[63] |
| $^{239+240}Pu$ | < 107 FW vs. 19–89 FW[63] |
| $^{106}Ru$ | < 632 FW vs. 124–1648 FW[63] |
| $^{125}Sb$ | ND vs. 29 FW[63] |
| Red grouse, *Lagopus lagopus*; muscle; November 1986–February 1987 | |
| $^{134}Cs$; cock vs. hen | 325 FW vs. 602 FW[38] |
| $^{137}Cs$; cock vs. hen | 962 FW vs. 1684 FW[38] |
| Black-headed gull, *Larus ridibundus*; near nuclear reactor; 1980 vs. June 1986 | |
| Egg contents | |
| $^{134}Cs$ | ND vs. 22 FW[63] |
| $^{137}Cs$ | 10 FW vs. 43 FW[63] |
| $^{238}Pu$ | 0.02 FW vs. 0.01 FW[63] |
| $^{239+240}Pu$ | 0.05 FW vs. 0.04 FW[63] |
| Egg shells | |
| $^{134}Cs$ | No data vs. 7 FW[63] |
| $^{137}Cs$ | No data vs. 16 FW[63] |
| $^{238}Pu$ | < 0.2 FW vs. 0.4 FW[63] |
| $^{239+240}Pu$ | 0.6 FW vs. 1.6 FW[63] |
| Woodcock, *Scolopax rusticola*; muscle; November 1986–February 1987 | |
| $^{134}Cs$ | 13 FW[38] |
| $^{137}Cs$ | 42 FW[38] |
| Black grouse, *Tetrao tetrix*; $^{137}Cs$; November 1986–February 1987; diet vs. muscle | 167 FW vs. 270 FW[38] |
| Cow, milk; 5–8 May 1986 | |
| $^{137}Cs$ | Max. 150 FW[2] |
| $^{131}I$ | Max. 127 FW[2] |
| Roe deer, *Capreolus capreolus*; $^{137}Cs$; muscle; November 1986–February 1987 | |
| Calves | 711 FW[38] |
| Hinds | 375–586 FW[38] |
| Stags | 1564 FW[38] |

**Table 24.2 Radionuclide Concentrations in Biotic and Abiotic Materials from Various Geographic Locales in Relation to the Chernobyl Nuclear Accident on April 26, 1986** ***(Continued)***

| Locale, Sample, Radionuclide, and Other Variables | Concentration[a] and Reference |
|---|---|
| Red deer, *Cervus elephus;*, muscle; November 1986–February 1987 | |
| $^{134}$Cs; calf vs. hind | 186 FW vs. 112 FW[38] |
| $^{137}$Cs; calf vs. hind | 535 FW vs. 311 FW[38] |
| Brown hare, *Lepus capensis*; $^{137}$Cs; female; November 1986–February 1987; diet vs. muscle | 198 FW vs. 656 FW[38] |
| Blue hare, *Lepus timidus*; $^{137}$Cs; November 1986–February 1987 | |
| Males; diet vs. muscle | 808 FW vs. 1677 FW[38] |
| Females; diet vs. muscle | 577 FW vs. 1440 FW[38] |
| Rabbit, *Oryctolagus* sp.; muscle; male; November 1986–February 1987 | |
| $^{134}$Cs | 6 FW[38] |
| $^{137}$Cs | 15 FW[38] |
| Domestic sheep | |
| Muscle | |
| September 1986; $^{137}$Cs | 1500 FW[60] |
| July 1987 | |
| $^{134}$Cs | 390 FW[60] |
| $^{137}$Cs | 1170 FW[60] |
| Liver; $^{110m}$Ag; September 1986 vs. July 1987 | |
| Ewes | 34 FW v 55 FW[53] |
| Lambs | 17 FW vs. < 8 FW[53] |
| Diet (rye grass and vegetation); $^{110m}$Ag; 1986 vs. 1987 | 32 DW vs. 10–30 DW[53] |
| Red fox, *Vulpes vulpes*; muscle; November 1986–February 1987; vixen | |
| $^{134}$Cs | 176 FW[38] |
| $^{137}$Cs | (461–643) FW[38] |

*Note:* All concentrations are in Bq/kg fresh weight (FW) or dry weight (DW) unless noted otherwise.

[a] Concentrations are shown as mean, range in parentheses, maximum (Max.), and nondetectable (ND).

concentration in muscle samples from moose shot within a 10,000-ha area around the forest in October 1992 was 810 (51–2133) Bq/kg FW; for roe deer muscle from this area in autumn 1992, it was 4200 (500–12,000) Bq/kg FW.[138]

Forest plants seemed to show less decrease than agricultural crops in $^{137}$Cs activity over time.[16] For example, the effective retention half-time of $^{137}$Cs from Chernobyl was 10 to 20 days in herbaceous plants and 180 days in chestnut trees, *Castanea* spp.[17] In Finland during summer 1986, pine-tree needles lost about 50% of their accumulated $^{137}$Cs activity in 72 days; washout by throughfall accounted for 79% of the decrease and shedding of older needles most of the remainder.[102] The radioactive fallout from the Chernobyl accident also resulted in high $^{137}$Cs levels in Swedish pasture grass and other forage, although levels in grain were relatively low.[18] Radiocesium isotopes were still easily measurable in grass silage harvested in June 1986 and used as fodder for dairy cows in 1988.[19] The rejection of the first harvests of radiocesium-contaminated perennial pasture and in particular of rye grass (*Lolium perenne*) does not constitute a safe practice because later harvests — even 1 year after the contamination of the field — may contain very high values, as in Greece.[20]

### 24.3.2 Aquatic Life

After Chernobyl, the consumption of freshwater fishes by Europeans declined, fishing license sales dropped by 25%, and the sale of fish from radiocesium-contaminated lakes was prohibited.[6] Many remedial measures have been attempted to reduce radiocesium loadings in fish, but none has proven effective.[21]

Bioconcentration factors (BCF) of $^{137}$Cs (Bq/kg FW/Bq/L seawater) measured in plankton and soft-bottom benthos during 1989–1990 in the Adriatic Sea ranged from 29 to 152 for plankton and 100 to 229 for benthos.[86] Radiocesium BCFs in muscle of fishes from the southern Baltic Sea increased three to four times after Chernobyl,[22] and $^{134+137}$Cs and $^{106}$Ru in fishes in the Danube River increased by a factor of five. However, these levels posed negligible risk to human consumers.[23] BCFs of $^{137}$Cs in fishes from Lake Paijanne, Finland — a comparatively contaminated area — ranged between 1250 and 3800; the highest BCF values were measured in the predatory northern pike (*Esox lucius*) a full 3 years after Chernobyl; consumption of these fishes was prohibited.[24] Chernobyl radioactivity, in particular $^{141}$Ce and $^{144}$Ce, entering the Mediterranean as a single pulse, was rapidly removed from surface waters and transported to 200 m in a few days, primarily in fecal pellets of grazing zooplankton.[79]

After the Chernobyl accident, radiocesium isotopes were also elevated in trees and lichens bordering an alpine lake in Scandinavia and in lake sediments, invertebrates, and fishes (Table 24.2). Radiocesium levels in muscle of resident brown trout (*Salmo trutta*) remained elevated for at least 2 years.[6] People consuming food near this alpine lake derived about 90% of their effective dose equivalent from the consumption of freshwater fish, reindeer meat, and milk. The average effective dose equivalent of this group during the next 50 years is estimated at 6–9 mSv with a changed diet and 8–12 mSv without any dietary changes.[6]

Between 1988 and 1992, $^{137}$Cs levels declined in sediments from 52 lakes in southern Finland by 27% and in fish tissues by 26 to 39%.[120] Radiocesium-137 concentrations in whole freshwater fishes from Norwegian lakes contaminated by Chernobyl fallout were quite variable. Major sources of $^{137}$Cs variations included the fish weight and growth rate, and these were related to fish age and diet.[85] Change over time in dissolved-phase $^{137}$Cs concentrations of lake water is significantly related to water residence time and mean lake depth; these variables have been incorporated into models to predict estimates of $^{137}$Cs decline in freshwater systems.[97] Radiocesium-137 activity in muscle of brown trout (*Salmo trutta*) and Arctic char (*Salvelinus alpinus*) from six lakes in Cumbria, England between June 1986 and October 1988 were highest between December 1986 and March 1987, with maximum values of about 1200 Bq/kg FW in trout and 350 in char.[107] Maximum $^{137}$Cs values were related to the initial concentration of $^{137}$Cs in both water and sediments and the maximum monthly geometric mean $^{137}$Cs values obtained from the routine water samples; no similar relationships were found with mean calcium or potassium levels in lake water. Ecological half-life for $^{137}$Cs in fish flesh from Cumbrian lakes was 132 days in char and 180 to 240 days in trout.[107]

### 24.3.3 Wildlife

Reindeer (*Rangifer tarandus*) — also known as caribou in North America — are recognized as a key species in the transfer of radioactivity from the environment to humans because 1) the transfer factor of radioactivity from reindeer feed to reindeer muscle is high; 2) lichens — which constitute a substantial portion of the reindeer diet — are efficient accumulators of strontium, cesium, and actinide radioisotopes; and 3) reindeer feed is not significantly supplemented with grain or other feeds low in contamination.[25–27,29,30,95,136] During the period 1986–1987, about 75% of all reindeer meat produced in Sweden was unfit for human consumption because $^{137}$Cs exceeded 300 Bq/kg fresh weight (FW). In May 1987, the maximum permissible level of $^{137}$Cs in Swedish reindeer and other game was raised to 1500 Bq/kg FW; however, about 25% of slaughtered reindeer in 1987 to 1989 still exceeded this limit.[28] Concentrations in excess of 100,000 Bq $^{134+137}$Cs/kg FW in lichens have been recorded in the most contaminated areas and in the 1986–1987 season was reflected in reindeer muscle concentrations of > 50,000 Bq/kg FW from the most contaminated areas of central Norway.[31] Norwegian reindeer containing 60,000–70,000 Bq $^{137}$Cs/kg FW in muscle receive an estimated yearly dose of 500 mSv.[32] The maximum radiation dose to reindeer in Sweden after Chernobyl was about 200 mSv/year, with a daily dose rate of about 1 mSv during the winter period of maximum tissue concentrations.[25] In general, reindeer calves had higher $^{137}$Cs levels in muscle than adult females (4700 vs. 2700 Bq/kg FW) during September 1988, suggesting translocation to the fetus.[29] Two reindeer herds in Norway that were heavily contaminated with radiocesium had a 25% decline in survival of calves; survival was normal in a herd with low exposure.[27]

Several compounds inhibit uptake and reduce retention of $^{137}$Cs in reindeer muscle from contaminated diets, but the mechanisms of the action are largely unknown. These compounds include zeolite — a group of tectosilicate minerals — when fed at 25–50 g daily;[33] ammonium hexacyanoferrate — also known as Prussian Blue or Giese salt — at 0.3–1.5 g daily;[34–36] bentonite — a montmorillonite clay — when fed at 2% of diet;[33] and high intakes of potassium.[33] In Sweden, radiocesium concentrations in domestic reindeer meat were reduced by early slaughter (to avoid grazing on pastures abundant in fungi and lichens), feeding uncontaminated diets for 8 to 12 weeks prior to slaughter, and adding cesium binders to the feed.[136] Much additional work seems needed on chemical and other processes that will hasten excretion and prevent uptake and accumulation of radionuclides in livestock and wildlife.[95]

Reindeer herding is, at present, the most important occupation in Finnish Lapland and portions of Sweden.[26] Swedish Lapland reindeer herders have experienced a variety of sociocultural problems as a result of the Chernobyl accident. The variability of contamination has been compounded by the variability of expert statements about risk, the change in national limits of Bq concentrations set for marketability of meat, and the variability of the compensation policy for slaughtered reindeer. These concerns may result in fewer Lapps becoming herders and in a general decline in reindeer husbandry.[37]

Caribou in northern Quebec contained up to 1129 Bq $^{137}$Cs/kg muscle FW in 1986–1987, but only 10 to 15% of this amount originated from Chernobyl; the remainder is attributed to fallout from earlier atmospheric nuclear tests.[39] The maximum concentration of $^{137}$Cs in meat of caribou (*Rangifer tarandus granti*) from the Alaskan Porcupine herd after Chernobyl did not exceed 232 Bq/kg FW, and this is substantially below the recommended level of 2260 Bq $^{137}$Cs/kg FW.[40] Radiocesium transfer in an Alaskan lichen-reindeer-wolf (*Canis lupus*) food chain has been estimated. If reindeer forage contained 100 Bq/kg dry weight (DW) lichens and 5 Bq/kg DW in vascular plants, the maximum winter concentrations — at an effective half-life of 8.2 years in lichens and 2.0 years in vascular plants — were estimated at 20 Bq/kg FW in reindeer-caribou skeletal muscle and 24 Bq/kg FW in wolf muscle.[41] The radioactive body burden in exposed reindeer and the character of chromosomal aberrations — which was different in exposed and nonexposed reindeer — indicate a genetic effect of radiation from the Chernobyl accident.[31] Chromosomal aberrations in Norwegian female reindeer positively correlated with increasing radiocesium concentrations in flesh.[27] The frequency of chromosomal aberrations in reindeer calves from central Norway were greatest in those born in 1987, when tissue loadings were equivalent to fetal doses of 70–80 mSv and lower in 1988 (50–60 mSv) and 1989 (40–50 mSv), strongly suggesting a dose-dependent induction. [31]

For many households in Sweden, moose (*Alces alces*) are an important source of meat.[7] Radiocesium concentrations in moose foreleg muscle in Sweden during 1987–1988 were highest in autumn, when the daily dietary intake of the animals was about 25,000 Bq $^{137}$Cs and lowest during the rest of the year, when mean daily intake was about 800 Bq.[16] Cesium-137 levels in moose flesh did not decrease significantly for about 2 years after the Chernobyl accident.[5] Nine years after the accident many moose in Sweden from areas receiving high deposition still showed $^{137}$Cs activity concentrations that exceed the limit for human consumption.[88]

The selection of food by moose is paramount to the uptake of environmental contaminants and the changes in tissue levels over time. Increased foraging on highly contaminated plant species, such as bilberry (*Vaccinium myrtillus*), aquatic plants, and mushrooms, might account for the increased $^{137}$Cs radioactivity in moose.[7,88] Habitat is a useful indicator of $^{137}$Cs radioactivity in moose muscle; radioactivity was highest in moose captured in swamp and marsh habitats and lowest in farmlands.[80,88] For reasons that are not yet clear, transfer coefficients of $^{137}$Cs from diet to muscle were about the same in moose (0.03) and beef cattle (0.02), but were significantly higher in sheep (0.24).[16]

Consumption of game or wildlife in Great Britain after the Chernobyl accident probably does not exceed the annual limits of intake (ALI) based on $^{134+137}$Cs concentrations in game and the numbers of animals that can be eaten in 1 year before the ALI is exceeded.[38] For example, a person who eats hares containing 3114 Bq $^{134+137}$Cs/kg FW in muscle would have to consume 99 hares before exceeding the ALI. For the consumption of red grouse (3022 Bq/kg), this number is 441 grouse; and for the consumption of woodcock (55 Bq/kg), it is 45,455 woodcocks.[38] Rabbits (*Oryctolagus* sp.) from northeastern Italy that were fed Chernobyl-contaminated alfalfa meal (1215 Bq $^{134+137}$Cs/kg diet) had a maximum of 156 Bq/kg muscle FW of $^{134+137}$Cs, a value much lower than the current Italian guideline of 370 Bq/kg FW for milk and children's food and 600 Bq/kg FW for other food.[12] More than 85% of the ingested radiocesium was excreted by rabbits in their feces and urine; about 3% was retained.[12]

Radiocesium-137 contamination of muscle of roe deer in 1986–1987 in Germany was lowest in March and highest in October and is attributed to increased food intake and the high concentrations in food plants such as ferns and mushrooms consumed at this time of the year.[113] Cesium radioactivity in tissues and organs of the wolverine (*Gulo gulo*), lynx (*Felis lynx*), and Arctic fox (*Alopex lagopus*) in central Norway after the Chernobyl accident was highly variable. In general, cesium-137 levels were substantially lower in these carnivores than in lower trophic levels,[45] suggesting little or no food-chain biomagnification, and at variance with results of studies of the omnivore and herbivore food chain.

Mutagenicity tests were used successfully with feral rodents to evaluate the biological effects of the radiation exposure from the Chernobyl accident. Increased mutagenicity in mice (*Mus musculus domesticus*) was evident, as judged by the bone marrow micronucleus test, at 6 months and 1 year after the accident. Rodents with increased chromosomal aberrations also had $^{137}$Cs burdens that were 70% higher 6 months after the accident and 55% higher after 1 year, but elevated radiocesium body burdens alone were not sufficient to account for the increase in mutagenicity.[42] In bank voles, however, mutagenicity (micronucleated polychromatic erythrocytes) correlated well with the $^{137}$Cs content in muscle and in the soil of the collection locale.[43] The estimated daily absorbed doses of 4.2–39.4 μGy were far lower than those required to produce the same effect in the laboratory.[43]

Migratory waterfowl could serve as potential vectors of contamination to the human food chain. For example, cooked meat from migratory waterfowl harvested on their wintering grounds as much as 3000 km from the Chernobyl site may exceed the maximum level of radiocesium generally permitted for human consumption (600 Bq/kg FW), depending on deposition rate.[91] Migratory species of game birds in northern Ireland during the period 1986–1988 were contaminated by radiocesium of Chernobyl origin.[87] The songthrush (*Turdus philomelos*) collected in Spain in November 1986 had elevated concentrations of $^{134}$Cs, $^{137}$Cs, and $^{90}$Sr. The contamination probably occurred in central and northern Europe before the birds' migration to Spain.[44] In 1994, concentrations in songthrush tissues remained elevated.[94] However, Spaniards who ate songthrushes contaminated with radiocesium isotopes usually received about 58 μSv yearly, which is well below current international guidelines.[44] In 1987–1988, ptarmigans (*Lagopus* spp.) from alpine ecosystems in central Norway receiving 20,000–60,000 Bq $^{137}$Cs/m$^2$ in 1986 usually had $<$ 350 Bq $^{137}$Cs/kg FW muscle.[153] Concentrations of $^{137}$Cs in ptarmigan muscle were higher in juveniles than in adults, higher in summer than in winter, and correlated with radiocesium concentrations in food plants.[153]

### 24.3.4 Domestic Animals

Radiocesium isotopes from the Chernobyl accident transferred easily to grazing farm animals.[11] Both $^{134}$Cs and $^{137}$Cs were rapidly distributed throughout the soft tissues of animals after dietary ingestion and were most highly concentrated in muscle.[46,47] Radiocesium activity in milk and flesh of Norwegian sheep and goats, for example, increased three- to fivefold 2 years after the accident and coincided with an abundant growth and availability of fungal fruit bodies with $^{134+137}Cs$ levels as much as 100 times greater than green vegetation.[11] Norway — almost 2000 km from Chernobyl — condemned $70.9 million (U.S.) of domestic animal products in the period 1986–1995 as a result of the accident.[161] In 1986 alone, large amounts of mutton, reindeer meat, and goat's cheese exceeded the radiocesium content set by Norwegian authorities and was destroyed at an estimated loss of $24 million. Economic loss in 1995 was about $2.1 million. In 1996, after 10 years of expensive countermeasures (changes of pasturing areas, special feeding, cesium binders, change in slaughtering time), there was some decline in the size of the areas the numbers of animals involved; however, countermeasures are still required.[161]

In cattle, coefficients of radiocesium transfer from diet to muscle were about 2.5% in adults and 16% in calves. The higher value in calves was probably due to a high availability of cesium from the gastrointestinal tract and to daily uptake of potassium in growing animal muscle.[48] There was no correlation between retention of $^{137}Cs$ and pregnancy stage in cattle.[49] Radiocesium concentrations in pork in Czechoslovakia did not decline between 1986 and 1987 because the feed of pigs during this period contained milk by-products contaminated with $^{134+137}Cs$.[50]

Sheep farming is the main form of husbandry in the uplands of west Cumbria and north Wales, a region that received high levels of radiocesium fallout during the Chernobyl accident. Afterward, typical vegetation activity concentrations were –6000 Bq/kg (down to –1000 Bq/kg in January 1989). But sheep muscle concentrations exceeded 1000 Bq $^{137}Cs$/kg FW, which is the United Kingdom's dietary limit for human health protection.[51] Contaminated lambs — which usually had higher concentrations of $^{137}Cs$ than ewes[38] — that were removed to lowland pastures ($<$ 50 Bq/kg vegetation) rapidly excreted radiocesium in feces and urine, and cesium body burdens had an effective half-life of 11 days.[51] This practice should not significantly increase radiocesium levels in soil and vegetation of lowland pastures.[51] The absorption and retention of radiocesium by suckling lambs is highly efficient, about 66%. Fecal excretion was an important pathway after termination of $^{137}Cs$ ingestion. In weaned animals, the absorption of added ionic cesium was about twice that of cesium fallout after the accident at Chernobyl.[52] Silver-110m was also detected in the brains and livers of ewes and lambs in the United Kingdom. The transfer of $^{110m}Ag$ was associated with perennial rye grass harvested soon after deposition in 1986. Silver-110m was taken up to a greater extent than $^{137}Cs$ in liver; but unlike $^{137}Cs$, the $^{110m}Ag$ was not readily translocated to other tissues. Other than cesium isotopes and $^{131}I$, $^{110m}Ag$ was the only detected nuclide in sheep tissues.[53]

Atmospheric deposition of $^{137}Cs$ from Chernobyl to vegetation and eventually to the milk of sheep, cows, and goats on contaminated silage was reported in Italy, The Netherlands, Japan, and the United Kingdom.[40,54–58] The effective half-life of $^{137}Cs$ was 6.7 days in silage and 13.6 days in milk.[59] The average daily transfer coefficient of $^{134+137}Cs$ from Chernobyl from a 70% grass silage diet to milk of Dutch dairy cows was about 0.25%/liter.[19] In goats (*Capra* sp.), about 12% of orally administered $^{137}Cs$ was collected in milk within 7 days after dosing.[46]

Iodine-131 was one of the most hazardous radionuclides released in the Chernobyl accident because it is easily transferred through the pasture-animal-milk pathway and rapidly concentrated in the thyroid gland to an extent unparalleled by any other organ. Because of its high specific activity, $^{131}I$ can transmit a high dose of radiation to the thyroid.[61] Iodine-131 levels of 618,000 Bq/kg FW in sheep thyroids from northwestern Greece on July 3, 1986 are similar to maximal $^{131}I$ concentrations in sheep thyroids in Tennessee in 1957 after global atmospheric fallout from military weapons tests and in London after the Windscale accident.[61] Iodine-131 has an effective whole-body half-life of about 24 h and is rapidly excreted from sheep and cows.[13] The effective half-life of $^{131}I$ in pasture grass was 3.9 days and 5 days in cow's milk.[59] The daily transfer coefficient of $^{131}I$ from vegetation to cow's milk was 0.007%/L milk. This value was 57 times higher (0.4) in sheep,[58] but the mechanism that accounts for this large interspecies difference is not clear.

## 24.4 SUMMARY

The accident at the Chernobyl, Ukraine nuclear reactor on April 26, 1986 released large amounts of radiocesium and other radionuclides into the environment, contaminating much of the northern hemisphere, especially Europe. In the vicinity of Chernobyl, at least 30 people died, more than 115,000 others were evacuated, and consumption of milk and other foods was banned because of radiocontamination. At least 14,000 human cancer deaths are expected in Russia, Belarus, and Ukraine as a direct result of Chernobyl. The most sensitive local ecosystems, as judged by survival, were the soil fauna, pine forest communities, and certain populations of rodents. Elsewhere, fallout from Chernobyl significantly contaminated freshwater and terrestrial ecosystems and flesh and milk of domestic livestock; in many cases, radionuclide concentrations in biological samples exceeded current radiation protection guidelines. Reindeer (*Rangifer tarandus*) in Scandinavia were among the most seriously afflicted by Chernobyl fallout, probably because their main food during winter (lichens) is an efficient absorber of airborne particles containing radiocesium. Some reindeer calves contaminated with $^{137}$Cs from Chernobyl showed $^{137}$Cs-dependent decreases in survival and increases in frequency of chromosomal aberrations. Although radiation levels in the biosphere are declining with time, latent effects of initial exposure — including an increased frequency of thyroid and other cancers — are now measurable. The full effect of the Chernobyl nuclear reactor accident on natural resources will probably not be known for at least several decades because of gaps in data on long-term genetic and reproductive effects and on radiocesium cycling and toxicokinetics.

## REFERENCES

1. Smith, F. B. and Clark, M. J., Radionuclide deposition from the Chernobyl cloud, *Nature*, 322, 690, 1986.
2. Clark, M. J. and Smith, F. B., Wet and dry deposition of Chernobyl releases, *Nature*, 332, 245, 1988.
3. United Nations Scientific Committee on the Effects of Atomic Radiation, Sources, Effects and Risks of Ionizing Radiation, United Nations, New York, 1988.
4. Aarkrog, A., Environmental radiation and radiation releases, *Int. J. Rad. Biol.*, 57, 619, 1990.
5. Johanson, K. J., The consequences in Sweden of the Chernobyl accident, *Rangifer Spec. Iss.*, 3, 9, 1990.
6. Brittain, J. E., Storruste, A., and Larsen, E., Radiocesium in brown trout (*Salmo trutta*) from a subalpine lake ecosystem after the Chernobyl reactor accident, *J. Environ. Radioact.*, 14, 181, 1991.
7. Palo, R. T., Nelin, P., Nylen, T., and Wickman, G., Radiocesium levels in Swedish moose in relation to deposition, diet, and age, *J. Environ. Qual.*, 20, 690, 1991.
8. Eisler, R., Radiation Hazards to Fish, Wildlife, and Invertebrates: A Synoptic Review, U.S. Natl. Biological Service, Biol. Rep. 26, 1994.
9. Severa, J. and Bar, J., *Handbook of Radioactive Contamination and Decontamination*, Studies in Environmental Science 47, Elsevier, New York, 1991.
10. Sokolov, V. E., Krivolutzky, D. A., Ryabov, I. N., Taskaev, A. E., and Shevchenko, V. A., Bioindication of biological after-effects of the Chernobyl atomic power station accident in 1986–1987, *Biol. Int.*, 18, 6, 1990.
11. Hove, K., Pederson, O., Garmo, T. H., Hansen, H. S., and Staaland, H., Fungi: A major source of radiocesium contamination of grazing ruminants in Norway, *Health Phys.*, 59, 189, 1990.
12. Battiston, G. A., Degetto, S., Gerbasi, R., Sbrignadello, G., Parigi-Bini, R., Xiccato, G., and Cinetto, M., Transfer of Chernobyl fallout radionuclides feed to growing rabbits: Cesium-137 balance, *Sci. Total Environ.*, 105, 1, 1991.
13. Assimakopoulos, P. A., Ioannides, K. G., and Pakou, A. A., The propagation of the Chernobyl $^{131}$I impulse through the air-grass-animal-milk pathway in northwestern Greece, *Sci. Total Environ.*, 85, 295, 1989.
14. Clooth, G. and Aumann, D. C., Environmental transfer parameters and radiological impact of the Chernobyl fallout in and around Bonn (FRG), *J. Environ. Radioact.*, 12, 97, 1990.
15. Othman, I., The impact of the Chernobyl accident on Syria, *J. Radiol. Protect.*, 10, 103, 1990.

16. Bothmer, S. V., Johanson, K. J., and Bergstrom, R., Cesium-137 in moose diet; considerations on intake and accumulation, *Sci. Total Environ.*, 91, 87, 1990.
17. Tonelli, D., Gattavecchia, E., Ghini, S., Porrini, C., Celli, G., and Mercuri, A. M., Honey bees and their products as indicators of environmental radioactive pollution, *J. Radioanal. Nucl. Chem.*, 141, 427, 1990.
18. Andersson, I., Teglof, B., and Elwinger, K., Transfer of $^{137}$Cs from grain to eggs and meat of laying hens and meat of broiler chickens, and the effect of feeding bentonite, *Swed. J. Agric. Res.*, 20, 35, 1990.
19. Voors, P. I. and van Weers, A. W., Transfer of Chernobyl radiocaesium ($^{134}$Cs and $^{137}$Cs) from grass silage to milk in dairy cows, *J. Environ. Radioact.*, 13, 125, 1991.
20. Douka, C. E. and Xenoulis, A. C., Radioactive isotope uptake in a grass-legume association, *Environ. Pollut.*, 73, 11, 1991.
21. Hakanson, L. and Andersson, T., Remedial measures against radioactive caesium in Swedish lake fish after Chernobyl, *Aquat. Sci.*, 54, 141, 1992.
22. Grzybowska, D., Concentration of $^{137}$Cs and $^{90}$Sr in marine fish from the southern Baltic Sea, *Acta Hydrobiol.*, 31, 139, 1989.
23. Conkic, L., Ivo, M., Lulic, S., Kosatic, K., Simor, J., Vancsura, P., Slivka, S., and Bikit, I., The impact of the Chernobyl accident on the radioactivity of the River Danube, *Water Sci. Technol.*, 22, 195, 1990.
24. Korhonen, R., Modeling transfer of $^{137}$Cs fallout in a large Finish watercourse, *Health Phys.*, 59, 443, 1990.
25. Jones, B. E. V., Eriksson, O., and Nordkvist, M., Radiocesium uptake in reindeer on natural pasture, *Sci. Total Environ.*, 85, 207, 1989.
26. Rissanen, K. and Rahola, T., Cs-137 concentration in reindeer and its fodder plants, *Sci. Total Environ.*, 85, 199, 1989.
27. Skogland, T. and Espelien, I., The biological effects of radiocesium contamination of wild reindeer in Norway following the Chernobyl accident, in *Population Dynamics, Vol. 1, Trans. XIXth IUGB Congress*, Myrberget, S., Ed., Int. Union Game Biol., Sept. 1989, Trondheim, Norway, 275, 1990.
28. Ahman, G., Ahman, B., and Rydberg, A., Consequences of the Chernobyl accident for reindeer husbandry in Sweden, *Rangifer Spec. Iss.*, 3, 83, 1990.
29. Eikelmann, I. M. H., Bye, K., and Sletten, H. D., Seasonal variation of cesium 134 and cesium 137 in semidomestic reindeer in Norway after the Chernobyl accident, *Rangifer Spec. Iss.*, 3, 35, 1990.
30. Rissanen, K. and Rahola, T., Radiocesium in lichens and reindeer after the Chernobyl accident, *Rangifer Spec. Iss.*, 3, 55, 1990.
31. Roed, K. H., Eikelmann, I. M. H., Jacobsen, M., and Pedersen, O., Chromosome aberrations in Norwegian reindeer calves exposed to fallout from the Chernobyl accident, *Hereditas,* 115, 201, 1991.
32. Jones, B., Radiation effects in reindeer, *Rangifer Spec. Iss.*, 3, 15, 1990.
33. Ahman, B., Forberg, S., and Ahman, G., Zeolite and bentonite as caesium binders in reindeer feed, *Rangifer Spec. Iss.*, 3, 73, 1990.
34. Hove, K., Staaland, H., Pedersen, O., and Sletten, H. D., Effect of Prussian blue (ammonium-iron-hexacyanoferrate) in reducing the accumulation of radiocesium in reindeer, *Rangifer Spec. Iss.*, 3, 43, 1990.
35. Mathiesen, S. D., Nordoy, L. M., and Blix, A. S., Elimination of radiocesium in contaminated adult Norwegian reindeer, *Rangifer Spec. Iss.*, 3, 49, 1990.
36. Staaland, H., Hove, K., and Pedersen, O., Transport and recycling of radiocesium in the alimentary tract of reindeer, *Rangifer Spec. Iss.*, 3, 63, 1990.
37. Beach, H., Coping with the Chernobyl disaster: A comparison of social effects in two reindeer-herding areas, *Rangifer Spec. Iss.*, 3, 25, 1990.
38. Lowe, V. P. W. and Horrill, A. D., Caesium concentration factors in wild herbivores and the fox (*Vulpes vulpes* L), *Environ. Pollut.*, 70, 93, 1991.
39. Crete, M., Lefebvre, M. A., Cooper, M. B., Marshall, H., Benedetti, J. L., Carriere, P. E., and Nault, R., Contaminants in caribou tissues from northern Quebec, *Rangifer Spec. Iss.*, 3, 289, 1990.
40. Allaye-Chan, A. C., White, R. G., Holleman, D. F., and Russell, D. E., Seasonal concentrations of cesium-137 in rumen content, skeletal muscles and feces of caribou from the Porcupine herd: Lichen ingestion rates and implications for human consumption, *Rangifer Spec. Iss.*, 3, 17, 1990.
41. Holleman, D. F., White, R. G., and Allaye-Chan, A. C., Modelling of radiocesium transfer in the lichen-reindeer/caribou-wolf food chain, *Rangifer Spec. Iss.*, 3, 39, 1990.

42. Cristaldi, M., D'Arcangelo, E., Ieradi, L. A., Mascanzoni, D., Mattei, T., and Castelli, I. V. A., $^{137}$Cs determination and mutagenicity tests in wild *Mus musculus domesticus* before and after the Chernobyl accident, *Environ. Pollut.*, 64, 1, 1990.
43. Cristaldi, M., Ieradi, L. A., Mascanzoni, D., and Mattei, T., Environmental impact of the Chernobyl accident: Mutagenesis in bank voles from Sweden, *Int. J. Rad. Biol.*, 59, 31, 1991.
44. Baeza, A., del Rio, M., Miro, C., Moreno, A., Navarro, E., Paniagua, J. M., and Peris, M. A., Radiocesium and radiostrontium levels in song-thrushes (*Turdus philomelos*) captured in two regions of Spain, *J. Environ. Radioact.*, 13, 13, 1991.
45. Ekker, M., Jenssen, B. M., and Zahlsen, K., Radiocesium in Norwegian carnivores following the Chernobyl fallout, in: *Population Dynamics, Vol. 1, Trans. XIXth IUGB Congress*, Myrberget, S., Ed. Int. Union Game Biol., Sept. 1989, Trondheim, Norway, 275, 1990.
46. Book, S. A., Fallout Cesium-137 Accumulation in Two Subpopulations of Black-Tailed Deer (*Odocoileus hemionus columbianus*), M.A. thesis, University California, Berkeley, 1969.
47. Van den Hoek, J., European research on the transfer of radionuclides to animals — A historical perspective, *Sci. Total Environ.*, 85, 17, 1989.
48. Daburon, F., Fayart, G., and Tricaud, Y., Caesium and iodine metabolism in lactating cows under chronic administration, *Sci. Total Environ.*, 85, 253, 1989.
49. Calamosca, M., Pagano, P., Trent, F., Zaghini, L., Gentile, G., Tarroni, G., and Morandi, L., A modelistic approach to evaluate the factors affecting the $^{137}$Cs transfer from mother to fetus in cattle, *Deutsche Tieraerzt. Wochen.*, 97, 452, 1990.
50. Kliment, V., Contamination of pork by caesium radioisotopes, *J. Environ. Radioact.*, 13, 117, 1991.
51. Crout, N. M. J., Beresford, N. A., and Howard, B. J., The radioecological consequences for lowland pastures used to fatten upland sheep contaminated with radiocaesium, *Sci. Total Environ.*, 103, 73, 1991.
52. Moss, B. W., Unsworth, E. F., McMurray, C. H., Pearce, J., and Kilpatrick, D. J., Studies on the uptake, partition and retention of ionic and fallout radiocaesium by suckling and weaned lambs, *Sci. Total Environ.*, 85, 91, 1989.
53. Beresford, N. A., The transfer of Ag-110m to sheep tissues, *Sci. Total Environ.*, 85, 81, 1989.
54. Belli, M., Drigo, A., Menegon, S., Menin, A., Nazzi, P., Sansone, U., and Toppano, M., Transfer of Chernobyl fall-out caesium radioisotopes in the cow food chain, *Sci. Total Environ.*, 85, 169, 1989.
55. Pearce, J., McMurray, C. H., Unsworth, E. F., Moss, B. W., Gordon, F. J., and Kilpatrick, D. J., Studies of the transfer of dietary radiocaesium from silage to milk in dairy cows, *Sci. Total Environ.*, 85, 267, 1989.
56. Voors, P. I. and van Weers, A. W., Transfer of Chernobyl $^{134}$Cs and $^{137}$Cs in cows from silage to milk, *Sci. Total Environ.*, 85, 179, 1989.
57. Aii, T., Kume, S., Takahashi, S., Kurihara, M., and Mitsuhashi, T., The effect of the radionuclides from Chernobyl on iodine-131 and cesium-137 contents in milk and pastures in south-western Japan, *Jpn. Soc. Zootech. Sci.*, 61, 47, 1990.
58. Monte, L., Evaluation of the environmental transfer parameters for $^{137}$Cs using the contamination produced by the Chernobyl accident at a site in central Italy, *J. Environ. Radioact.*, 12, 13, 1990.
59. Spezzano, P. and Giacomelli, R., Transport of $^{131}$I and $^{137}$Cs from air to cow's milk produced in north-western Italian farms following the Chernobyl accident, *J. Environ. Radioact.*, 13, 235, 1991.
60. Coughtrey, P. J., Kirton, J. A., and Mitchell, N. G., Cesium transfer and cycling in upland pastures, *Sci. Total Environ.*, 85, 149, 1989.
61. Ioannides, K. G. and Pakou, A. A., Radioiodine retention in ovine thyroids in northwestern Greece following the reactor accident at Chernobyl, *Health Phys.*, 60, 517, 1991.
62. Kedhi, M., Aerosol, milk and wheat flour radioactivity in Albania caused by the Chernobyl accident, *J. Radioanal. Nucl. Chem.*, 146, 115, 1990.
63. Lowe, V. P. W., Radionuclides and the birds at Ravenglass, *Environ. Pollut.*, 70, 1, 1991.
64. Sawidis, T., Uptake of radionuclides by plants after the Chernobyl accident, *Environ. Pollut.*, 50, 317, 1988.
65. Imanaka, T. and Koide, H., Radiocesium concentration in milk after the Chernobyl accident in Japan, *J. Radioanal. Nucl. Chem.*, 145, 151, 1990.
66. Vreman, K., van der Struijs, T. D. B., van den Hoek, J., Berende, D. L. M., and Goedhart, P. W., Transfer of Cs-137 from grass and wilted grass silage to milk of dairy cows, *Sci. Total Environ.*, 85, 139, 1989.

67. Eriksson, O., $^{137}$Cs in reindeer forage plants 1986–1988, *Rangifer Spec. Iss.*, 3, 11, 1990.
68. Johanson, K. J., Karlen, G., and Bertilsson, J., The transfer of radiocesium from pasture to milk, *Sci. Total Environ.*, 85, 73, 1989.
69. Mascanzoni, D., Bothmer, S. V., Mattei, T., and Cristaldi, M., Small mammals as biological indicators of radioactive contamination of the environment, *Sci. Total Environ.*, 99, 61, 1990.
70. Mufson, S., G-7 eyes aid for ex-Soviet nuclear plants. *Washington Post*, Washington, D.C., 22 May, 1992, F1.
71. Kryshev, I., Radioactive contamination of aquatic ecosystems following the Chernobyl accident, *J. Environ. Radioact.*, 27, 207, 1995.
72. Makeyeva, A. P., Yemel'yanova, N. G., Belova, N. V., and Ryabov, I.N., Radiobiological analysis of silver carp, *Hypophthalmichthys molitrix*, from the cooling pond of the Chernobyl nuclear power plant since the time of the accident. 2. Development of the reproductive system in the first generation of offspring, *J. Ichthyol.*, 35, 40, 1995.
73. Kotelevtsev, S. V., Nagdaliev, F. F., and Skryabin, G. A., The influence of radionuclides on ion transport and its hormonal regulation in membranes of fish erythrocytes, *J. Ichthyol.*, 36, 109, 1996.
74. Eisler, R., Ecological and toxicological aspects of the partial meltdown of the Chernobyl nuclear power plant reactor, in *Handbook of Ecotoxicology*, Hoffman, D. J., Rattner, B. A., Burton, G. A., Jr., and Cairns, J., Jr., Eds., CRC Press, Lewis Publishers, Boca Raton, FL, 1995, Chap. 24.
75. Eisler, R., Radiation, in *Handbook of Chemical Risk Assessment. Health Hazards to Humans, Plants, and Animals*, Volume 3, Eisler, R., CRC Press, Lewis Publishers, Boca Raton, FL, 2000, Chap. 32.
76. Pomerantseva, M. D., Ramaiya, L. K., and Chekhovich, A. V., Genetic consequences of the Chernobyl disaster for house mice (*Mus musculus*), *Russian J. Genet.*, 32, 264, 1996.
77. Nadzhafova, R. S., Bulatova, N. S., Kozlovskii, A. I., and Ryabov, I. N., Identification of a structural chromosomal rearrangement in the karyotype of a root vole from Chernobyl, *Russian J. Genet.*, 30, 318, 1994.
78. Baker, R. J., Hamilton, M. J., van den Bussche, R. A., Wiggins, L. E., Sugg, D. W., Smith, M. H., Lomakin, M. D., Gaschak, S. P., Bundova, E. G., Rudenskaya, G. A., and Chesser, R. K., Small mammals from the most radioactive sites near the Chernobyl nuclear power plant, *J. Mammal.*, 77, 155, 1996.
79. Fowler, S. W., Buat-Menard, P., Yokoyama, Y., Ballestra, S., Holm, E., and van Nguyen, H., Rapid removal of Chernobyl fallout from Mediterranean surface waters by biological activity, *Nature*, 329, 56, 1987.
80. Nelin, P., Radiocaesium uptake in moose in relation to home range and habitat composition, *J. Environ. Radioact.*, 26, 189, 1995.
81. Whitehead, N. E., Ballestra, S., Holm, E., and Walton, A., Air radionuclide patterns observed at Monaco from the Chernobyl accident, *J. Environ. Radioact.*, 7, 249, 1988.
82. Whitehead, N. E., Ballestra, S., Holm, E., and Huynh-Ngoc, L., Chernobyl radionuclides in shellfish, *J. Environ. Radioact.*, 7, 107, 1988.
83. Kalas, J. A., Bretten, S., Byrkjedal, I., and Njastad, O., Radiocesium ($^{137}$Cs) from the Chernobyl reactor in Eurasian woodcock and earthworms in Norway, *J. Wildl. Manage.*, 58, 141, 1994.
84. Brunn, H., Georgii, S., and Eskens, U., $^{137}$Cesium and $^{134}$cesium in roe deer from north and middle Hesse (Germany) subsequent to the reactor accident in Chernobyl, *Bull. Environ. Contam. Toxicol.*, 51, 633, 1993.
85. Ugedal, O., Forseth, T., Jonsson, B., and Njastad, O., Sources of variation in radiocaesium levels between individual fish from a Chernobyl contaminated Norwegian lake, *J. Appl. Ecol.*, 32, 352, 1995.
86. Marzano, F. N. and Triulzi, C., A radioecological survey of northern and middle Adriatic Sea before and after the Chernobyl event (1979–1990), *Mar. Pollut. Bull.*, 28, 244, 1994.
87. Pearce, J., Radiocesium in migratory bird species in northern Ireland following the Chernobyl accident, *Bull. Environ. Contam. Toxicol.*, 54, 805, 1995.
88. Palo, R. T. and Wallin, K., Variability in diet composition and dynamics of radiocaesium in moose, *J. Appl. Ecol.*, 33, 1077, 1996.
89. Vyazovich, Y., Dynamics of the radionuclide contamination and ecology of wild *anatidae* species in Belarus after the Chernobyl nuclear accident, *Gieber Faune Sauvage, Game Wildl.*, 13, 723, 1996.
90. Ruiz, X., Jover, L., Llorente, G. A., Sanchez-Reyes, A. F., and Febrian, M. I., Song thrushes *Turdus philomelos* wintering in Spain as biological indicators of the Chernobyl accident, *Ornis Scandinavica*, 19, 63, 1987.

91. Brisbin, I. L., Jr., Birds as indicators of global contamination processes: The Chernobyl connection, *Acta XX Congressus Internationalis Ornithologici*, 4, 2503, 1990.
92. Arkhipov, N. P., Kuchma, N. D., Askbrant, S., Pasternak, P. S., and Musica, V. V., Acute and long-term effects of irradiation on pine (*Pinus silvestris*) stands post-Chernobyl, *Sci. Total Environ.*, 157, 383, 1994.
93. Zalewski, M., Mnich, Z., Kapala, J., and Tomczak, M., Radiocaesium in the Sawinda Wielka Lake ecosystem 9 years after the Chernobyl accident, *Polish Arch. Hydrobiol.*, 43, 455, 1996.
94. Navarro, E., Roldan, C., Cervera, J., and Ferrero, J. L., Radioactivity measurements on migrating birds (*Turdus philomelos*) captured in the Comunidad Valenciana (Spain), *Sci. Total Environ.*, 209, 143, 1998.
95. Ahman, B., Contaminants in food chains of Arctic ungulates: What have we learned from the Chernobyl accident?, *Rangifer*, 18, 119, 1998.
96. Skuterud, L., Goltsova, N. I., Naeumann, R., Sikkeland, T., and Lindmo, T., Histological changes in *Pinus silvestris* L. in the proximal-zone around the Chernobyl power plant, *Sci. Total Environ.*, 157, 387, 1994.
97. Smith, J. T., Leonard, D. R. P., Hilton, J., and Appleby, P. G., Towards a generalized model for the primary and secondary contamination of lakes by Chernobyl-derived radiocesium, *Health Phys.*, 72, 880, 1997.
98. Makeyeva, A. P., Belova, N. V., Emel'yanova, N. G., Verigin, B. V., and Ryabov, I. N., Materials on the state of reproductive system of bighead *Aristichthys nobilis* from the cooling pond of Chernobyl nuclear power station in the post-disaster period, *J. Ichthyol.*, 36, 181, 1996.
99. Belova, N. V., Verigin, B. V., Yemel'yanova, N. G., Makeyeva, A. P., and Ryabov, I. N., Radiobiological analysis of silver carp (*Hypophthalmichthys molitrix*) from the cooling pond of the Chernobyl nuclear power station in the post-disaster period. 1. Reproductive system of fish exposed to radioactive contamination, *J. Ichthyol.*, 34, 16, 1994.
100. Belova, N. V., Emel'yanova, N. G., Makeeva, A. P., Verigin, B. V., and Ryabov, I. N., Unique occurrence of dwarf individuals of *Hypophthalmichthys molitrix* in the water cooler of the Chernobyl nuclear power station, *J. Ichthyol.*, 38, 809, 1998.
101. Verigin, B. V., Belova, N. V., Emel'yanova, N. G., Makeyeva, A. P., Vybornov, A. A., and Ryabov, I. N., Radiobiologic analysis of silver carp (*Hypophthalmichthys molitrix*) from the cooling pond of Chernobyl nuclear power station in the post-disaster period. 3. Results of artificial reproduction of irradiated fish, *J. Ichthyol.*, 36, 257, 1996.
102. Nygren, P., Hari, P., Raunemaa, T., Kulmala, M., Luokkanen, S., Holmberg, M., and Nikinmaa, E., Behaviour of $^{137}$Cs from Chernobyl fallout in a Scots pine canopy in southern Finland, *Can. J. For. Res.*, 24, 1210, 1994.
103. Mamikhin, S. V., Tikhomirov, F. A., and Shcheglov, A. I., Dynamics of $^{137}$Cs in the forests of the 30-km zone around the Chernobyl nuclear power plant, *Sci. Total Environ.*, 193, 169, 1997.
104. Kritidis, P., Florou, H., and Synetos, S., The contribution of fish consumption to the dose received by the Greek population due to the Chernobyl accident, *Thallassographika*, 13 (Suppl. 1), 43, 1990.
105. Ryabov, I. N., Belova, N. V., and Polyakova, N. I., Evolution of radiocaesium contamination in fishes after the Chernobyl accident, *Ital. J. Zool.*, 65 (Suppl.), 455, 1998.
106. Bunzl, K., Schimmack, W., Krouglov, S. V., and Alexakhin, R. M., Changes with time in the migration of radiocesium in the soil, as observed near Chernobyl and in Germany, 1986–1994, *Sci. Total Environ.*, 175, 49, 1995.
107. Elliott, J. M., Elliott, J. A., and Hilton, J., Sources of variation in post-Chernobyl radiocaesium in brown trout, *Salmo trutta* L., and Arctic charr, *Salvelinus alpinus* (L.), from six Cumbrian lakes (northwest England), *Annales de Limnologie*, 29, 79, 1993.
108. Forberg, S., Odsjo, T., and Olsson, M., Radiocesium in muscle tissue of reindeer and pike from northern Sweden before and after the Chernobyl accident. A retrospective study on tissue samples from the Swedish Environmental Specimen Bank, *Sci. Total Environ.*, 115, 179, 1992.
109. Mihok, S., Schwartz, B., and Wiewel, A. M., Bioconcentration of fallout $^{137}$Cs by fungi and red-backed voles (*Clethrionomys gapperi*), *Health Phys.*, 57, 959, 1989.
110. Ivanov, V. K., Rastopchin, E. M., Gorsky, A. I., and Ryvkin, V. B., Cancer incidence among liquidators of the Chernobyl accident: Solid tumors, 1986–1995, *Health Phys.*, 74, 309, 1998.

111. Nakajima, H., Ryo, H., Nomura, T., Saito, T., Yamaguchi, Y., and Yekiseeva, K. G., Radionuclides carved on the annual rings of a tree near Chernobyl, *Health Phys.*, 74, 265, 1998.
112. Moore D. H., II and Tucker, J. D., Biological dosimetry of Chernobyl cleanup workers: Inclusion of data on age and smoking provides improved radiation dose estimates, *Radiat. Res.*, 152, 655, 1999.
113. Filli, V. F. and Nievergelt, B., Radiocesium in wild animals after the reactor accident in Chernobyl, *Z. Jagdwiss.*, 42, 249, 1996.
114. Heidenreich, W. F., Kenigsberg, J., Jacob, P., Buglova, E., Goulko, G., Paretzke, H. G., Demidchik, E. P., and Golovneva, A., Time trends of thyroid cancer incidence in Belarus after the Chernobyl accident, *Radiat. Res.*, 151, 617, 1999.
115. Bruk, G. Y., Shutov, V. N., Balonov, M. I., Basalayeva, L. N., and Kislov, M. V., Dynamics of $^{137}Cs$ content in agricultural food products produced in regions of Russia contaminated after the Chernobyl accident, *Radiat. Protect. Dosim.*, 76, 169, 1998.
116. Mikhalevich, L. S., Lloyd, D. C., Edwards, A. A., Perepetskaya, G. A., and Kartel, N. A., Dose estimates made by dicentric analysis for some Belarussian children irradiated by the Chernobyl accident, *Radiat. Protect. Dosim.*, 87, 109, 2000.
117. Moller, A. P., Morphology and sexual selection in the barn swallow *Hirundo rustica* in Chernobyl, Ukraine, *Proc. R. Soc. London B*, 252, 51, 1993.
118. Gallelli, G., Panatto, D., Perdelli, F., and Pellegrino, C., Long-term decline of radiocesium concentration in seafood from the Ligurian Sea (northern Italy) after Chernobyl, *Sci. Total Environ.*, 196, 163, 1997.
119. Baskaran, M., Kelley, J. J., Naidu, A. S., and Holleman, D. F., Environmental radiocesium in subarctic and arctic Alaska following Chernobyl, *Arctic*, 44, 346, 1991.
120. Sarkka, J., Keskitalo, A., and Luukko, A., Temporal changes in concentrations of radiocaesium in lake sediment and fish of southern Finland as related to environmental factors, *Sci. Total Environ.*, 191, 125, 1996.
121. Morris, J. A., Review: After effects of the Chernobyl accident, *Brit. Vet. J.*, 144, 179, 1988.
122. Taylor, H. W., Svoboda, J., Henry, G. H. R., and Wein, R. W., Post-Chernobyl $^{134}Cs$ and $^{137}Cs$ levels at some localities in northern Canada, *Arctic*, 41, 293, 1988.
123. Levi, H. W., Radioactive deposition in Europe after the Chernobyl accident and its long-term consequences, *Ecol. Res*, 6, 201, 1991.
124. Guillitte, O., Melin, J., and Wallberg, L., Biological pathways of radionuclides originating from the Chernobyl fallout in a boreal forest ecosystem, *Sci. Total Environ.*, 157, 207, 1994.
125. Peter, R. U., Gottlober, P., Nadeshina, N., Krahn, G., Braun-Falco, O., and Plewig, G., Interferon gamma in survivors of the Chernobyl power plant accident: New therapeutic option for radiation-induced fibrosis, *Int. J. Rad. Oncol. Biol. Phys.*, 45, 147, 1999.
126. WHO, *Proceedings WHO/HICARE Symposium on Radiological Accidents and Environmental Epidemiology: A Decade after the Chernobyl Accident, 24–25 August 1996, Hiroshima, Japan*, WHO/EHG/98.06, World Health Organization, 1998.
127. Robbins, J. and Schneider, A. B., Radioiodine-induced thyroid cancer: Studies in the aftermath of the accident at Chernobyl, *Trends Endocrinol. Metabol.*, 9, 87, 1998.
128. Moller, A., Developmental instability of plants and radiation from Chernobyl, *Oikos*, 81, 444, 1998.
129. Shawky, S. and El-Tahawy, M., Distribution pattern of $^{90}Sr$ and $^{137}Cs$ in the Nile delta and the adjacent regions after Chernobyl accident, *Appl. Radiat. Isotopes*, 50, 435, 1999.
130. Beresford, N. A., Gaschak, S., Lasarev, N., Arkhipov, A., Chyorny, Y., Atasheva, N., Arkhipov, N., Mayes, R. W., Howard, B. J., Baglay, G., Loginova, L., and Burov, N., The transfer of $^{137}Cs$ and $^{90}Sr$ to dairy cattle fed fresh herbage collected 3.5 km from the Chernobyl nuclear power plant, *J. Environ. Radioact.*, 47, 157, 2000.
131. Lonvik, K. and Thingstad, P. G., Radiocaesium accumulation from Chernobyl fallout in nestlings of two pied flycatcher populations (aves) in central Norway; estimating ecological timelag responses and transfer mechanisms, *J. Environ. Radioact.*, 46, 153, 1999.
132. Al-Rayyes, A. H. and Mamish, S., $^{137}Cs$, $^{134}Cs$ and $^{90}Sr$ in the coastal Syrian mountains after the Chernobyl accident, *J. Environ. Radioact.*, 46, 237, 1999.
133. Nifontova, M. G. and Alexashenko, V. N., Content of $^{90}Sr$ and $^{134,137}Cs$ in fungi, lichens, and mosses in the vicinity of the Chernobyl nuclear power plant, *Soviet J. Ecol.*, 23, 152, 1992.

134. Molero, J., Sanchez-Cabeza, J. A., Merino, J., Mitchell, P. I., and Vidal-Quadras, A., Impact of $^{134}$Cs and $^{137}$Cs from the Chernobyl reactor accident on the Spanish Mediterranean marine environment, *J. Environ. Radioact.*, 43, 357, 1999.
135. Goncharov, N. P., Katsiya, G. V., Kolesnikova, G. S., Dobracheva, G. A. D., Todua, T. N., Vax, V. V., Giwercman, A., and Waites, G. M. H., Endocrine and reproductive health status of men who had experienced short-term radiation exposure at Chernobyl, *Int. J. Androl.*, 21, 271, 1999.
136. Ahman, B., Transfer of radiocaesium via reindeer meat to man — Effects of countermeasures applied in Sweden following the Chernobyl accident, *J. Environ. Radioact.*, 46, 113, 1999.
137. Camplani, A., Saino, N., and Moller, A. P., Carotenoids, sexual signals and immune function in barn swallows from Chernobyl, *Proc. R. Soc. London B*, 266, 1111, 1999.
138. Gee, E. J., Synnott, H. J., Johanson, K. J., Fawaris, B. H., Nielsen, S. P., Horrill, A. D., Kennedy, V. H., Barbayiannis, N., Veresoglou, D. S., Dawson, D. E., Colgan, P. A., and McGarry, A. T., Chernobyl fallout in a Swedish spruce forest ecosystem, *J. Environ. Radioact.*, 48, 59, 2000.
139. Romanenko, A., Lee, C. C. R., Yamamoto, S., Hori, T., Wanibuchi, H., Zaparin, W., Vinnichenko, W., Vozianov, A., and Fukushima, S., Urinary bladder lesions after the Chernobyl accident: Immunohistochemical assessment of p53, proliferating cell nuclear antigen, cyclin D1 and p21, *Jpn. J. Cancer Res.*, 90, 144, 1999.
140. Junk, A. K., Kundiev, Y., Vitte, P., and Worgul, B. V., Eds., *Ocular Radiation Risk Assessment in Populations Exposed to Environmental Radiation Contamination*, Kluwer Academic Publishers, Norwell, MA, 1999.
141. Zavilgelsky, G. B., Abilev, S. K., Sukhodolets, V. V., and Ahmad, S. I., Isolation and analysis of UV and radio-resistant bacteria from Chernobyl, *J. Photochem. Photobiol. B.*, 43, 152, 1998.
142. Neyfakh, E. A., Alimbekova, A. I., and Ivanenko, G. F., Vitamin E and A deficiencies in children correlate with Chernobyl radiation loads of their mothers, *Biochemistry (Moscow)*, 63, 1138, 1998.
143. Gavrilin, Y. I., Khrouch, V. T., Shinkarev, S. M., Krysenko, N. A., Skryabin, A. M., Bouville, A., and Anspaugh, L. R., Chernobyl accident: Reconstruction of thyroid dose for inhabitants of the Republic of Belarus, *Health Phys.*, 76, 105, 1999.
144. Sorochinskii, B. V., Special features of protein composition of anomalous needles of spruce (*Picea abies*) and pine (*Pinus sylvestris*) from the 10-km zone of Chernobyl NPP, *Cytol. Genet.*, 32, 28, 1998.
145. Chesser, R. K., Sugg, D. W., Lomakin, M. D., van den Bussche, R. A., DeWoody, J. A., Jagoe, C. H., Dallas, C. E., Whicker, F. W., Smith, M. H., Gaschak, S. P., Chizhevsky, I. V., Lyabik, V. V., Buntova, E.G., Holloman, K., and Baker, R. J., Concentrations and dose rate estimates of $^{134,137}$cesium and $^{90}$strontium in small mammals at Chernobyl, Ukraine, *Environ. Toxicol. Chem.*, 19, 305, 2000.
146. Korobova, E., Ermakov, A., and Linnik, V., $^{137}$Cs and $^{90}$Sr mobility in soils and transfer in soil-plant systems in the Novozybkov district affected by the Chernobyl accident, *Appl. Geochem.*, 13, 803, 1998.
147. Burlakova, E. B., Ed., *Consequences of the Chernobyl Catastrophe on Human Health*, Nova Science Publishers, Commack, NY, 1999.
148. Holcomb, K., Abulafia, O., Montalto, N., Lee, Y. C., and Mathews, R. P., Multiple primary tumors in a woman exposed to the excess radiation of the Chernobyl nuclear disaster, *Eur. J. Gynae. Oncol.*, 20, 174, 1999.
149. Vargo, G. J., Ed., *The Chernobyl Accident: A Comprehensive Risk Assessment*, Battelle Press, Columbus, OH, 2000.
150. Pomerantseva, M. D., Ramaiya, L. K., and Chekhovich, A. V., Genetic disorders in house mouse germ cells after the Chernobyl catastrophe, *Mutat. Res.*, 381, 97, 1997.
151. Ipatyev, V., Bulavik, I., Baginsky, V., Goncharenko, G., and Dvornik, A., Forest and Chernobyl: Forest ecosystems after the Chernobyl nuclear plant accident: 1986–1994, *J. Environ. Radioact.*, 42, 9, 1999.
152. Paatero, J., Jaakkola, T., and Kulmala, S., Lichen (sp. *Cladonia*) as a deposition indicator for transuranium elements investigated with the Chernobyl fallout, *J. Environ. Radioact.*, 38, 223, 1998.
153. Pedersen, H. C., Nybo, S., and Varskog, P., Seasonal variation in radiocaesium concentration in willow ptarmigan and rock ptarmigan in central Norway after the Chernobyl fallout, *J. Environ. Radioact.*, 41, 65, 1998.
154. Mezhzherin, V. A. and Myakushko, S. A., Strategy of small rodent populations from the Kanev National Park under habitat changes due to technogenic pollution and the accident at the Chernobyl nuclear power station, *Biol. Bull.*, 25, 301, 1998 (translated from *Izves. Akad. Nauk, Seriya Biol.*, 3, 374, 1998).

155. Santoro, M., Thomas, G. A., Vecchio, G., Williams, G. H., Fusco, A., Chiappetta, G., Pozcharskaya, V., Bogdnanova, T. I., Demidchik, E. P., Cherstvoy, E. D., Voscoboinik, L., Tronko, N. D., Carss, A., Bunnel, H., Tonnachera, M., Parma, J., Dumont, J. E., Keller, G., Hofler, H., and Williams, E. D., Gene rearrangement and Chernobyl related thyroid cancers, *Br. J. Cancer*, 82, 315, 2000.
156. Tronko, M. D., Bogdanova, T. I., Komissarenko, I. V., Epstein, O. V., Oliynk, V., Kovalenko, A., Likhtarev, I. Y., Kairo, I., Peters, S. B., and LiVolsi, V. A., Thyroid carcinoma in children and adolescents in Ukraine after the Chernobyl nuclear accident, *Cancer*, 86, 149, 1999.
157. Pacini, F., Vorontsova, T., Molinaro, E., Shavrova, E., Agate, L., Kuchinskaya, E., Elisei, R., Demidchik, E. P., and Pinchera, A., Thyroid consequences of the Chernobyl nuclear accident, *Acta Paediatr. Suppl.*, 443, 23, 1999.
158. Klugbauer, S., Demidchik, E. P., Lengfelder, E., and Rabes, H. M., Molecular analysis of new subtypes of *ELE/RET* rearrangements, their reciprocal transcripts and breakpoints in papillary thyroid carcinomas of children after Chernobyl, *Oncogene*, 16, 671, 1998.
159. Lokobauer, N., Franic, Z., Bauman, A., Maracic, M., Cesar, D., and Sencar, J., Radiation contamination after the Chernobyl nuclear accident and the effective dose received by the population of Croatia, *J. Environ. Radioact.*, 41, 137, 1998.
160. Travnikova, I. G., Bruk, G. Y., and Shutov, V. N., Dietary pattern and content of cesium radionuclides in foodstuffs and bodies of Braynsk oblast rural population after the Chernobyl accident, *Radiochemistry*, 41, 298, 1999.
161. Tveten, U., Brynildsen, L. I., Amundsen, I., and Bergan, T. D. S., Economic consequences of the Chernobyl accident in Norway in the decade 1986–1995, *J. Environ. Radioact.*, 41, 233, 1998.
162. Orel, V. E., Tereschenki, V. M., Mazepa, M. G., and Buzanov, V. A., Physical dosimetry and biological indicators of carcinogenic risk in a cohort of persons exposed to unhealthy ecological factors following the Chernobyl nuclear power plant accident, *Arch. Environ. Health*, 53, 398, 1998.
163. Baker, R. J. and Chesser, R. K., The Chornobyl nuclear disaster and subsequent creation of a wildlife preserve, *Environ. Toxicol. Chem.*, 19, 1231, 2000.
164. Maznik, N. A., Genetic impact of low-dose radiation on human and non-human biota in Chernobyl, Ukraine, *Environ. Monit. Assess.*, 51, 497, 1998.
165. Savchenko, V. K., *The Ecology of the Chernobyl Catastrophe: Scientific Outlines of an International Programme of Collaborative Research*, Man and the Biosphere series: Vol. 16. UNESCO, Paris, and Parthenon Publishing Group, Pearl River, NY, 1995.
166. Sokolov, V. E., Rjbov, I. N., Ryabtsev, I. A., Tikhomirov, F. A., Shevchenko, V. A., and Taskaev, A. I., Ecological and genetic consequences of the Chernobyl atomic power plant accident, *Vegetatio*, 109, 91, 1993.
167. Mironenko, N. V., Alekhina, I. A., Zhadanova, N. N., and Bulat, S. A., Intraspecific variation in gamma-radiation resistance and genomic structure in the filamentous fungus *Alternaria alternata*: A case study of stains inhabiting Chernobyl reactor No. 4, *Ecotoxicol. Environ. Saf.*, 45, 177, 2000.
168. Tyler, P. E., Living in the shadow of Chernobyl's reactors, *New York Times*, New York City, June 4, A1, 2000.
169. Romanovskaya, V. A., Sokolov, I. G., Rokitko, P. V., and Chernaya, N. A., Effect of radioactive contamination on soil bacteria in the 10-km zone around the Chernobyl nuclear power plant, *Microbiology*, 67, 226, 1998.
170. Makova, K. D., Microsatellite Evolution in Mice (*Apodemus*): Origin of Alleles, Multiple Paternity, and Mutation Rate at Chernobyl, Ph.D. thesis, Texas Tech University, 1999.
171. Jagoe, C. H., Chesser, R. K., Smith, M. H., Lomakin, M. D., Lingenfelser, S. K., and Dallas, C. E., Levels of cesium, mercury and lead in fish, and cesium in pond sediments in an inhabited region of the Ukraine near Chernobyl, *Environ. Pollut.* 98, 223, 1997.
172. Karaoglou, A., Desmet, G., Kelly, G. N., and Menzel, H. G., Eds., The Radiological Consequences of the Chernobyl Accident, Publ. EUR 16544, Luxembourg Office for Official Publications of the European Communities, 1996.
173. Izrael, Y. A., De Cort, M., Jones, A. R., Nazarov, I. M., Fridman, S. D., Kvasnikova, E. V., Stukin, E. D., Kelly, G. N., Matveenko, I. I., Pokumeiko, Y. M., Tabatchnyi, L. Y., and Tsaturov, Y., The atlas of caesium-137 contamination of Europe after the Chernobyl accident, Karaoglou A., Desmet G., Kelly, G. N., and Menzel, H. G., Eds., The Radiological Consequences of the Chernobyl Accident, Publ. EUR 16544, Luxembourg Office for Official Publications of the European Communities, 1, 1996.

174. Eriksson, O., Gaichenko, V., Goshchak, S., Jones, B., Jungskar, W., Chizevsky, I., Kurman, A., Panov, G., Ryabtsev, I., Shcherbatchenko, A., Davydchuk, V., Petrov, M., Averin, V, Mikhalusyov, V., and Sokolov, V., Evolution of the contamination rate in game, Karaoglou A., Desmet, G., Kelly, G. N., and Menzel, H. G., Eds., The Radiological Consequences of the Chernobyl Accident, Publ. EUR 16544, Luxembourg Office for Official Publications of the European Communities, 147, 1996.
175. Qafmolla, L., Sinoimeri, M., Grillo, B., and Dollani, K., $^{90}$Sr levels in water samples of Adriatic Sea, Karaoglou, A, Desmet, G., Kelly, G. N., and Meznel, H. G., Eds., The Radiological Consequences of the Chernobyl Accident, Publ. 16544, Luxembourg Office for Official Publications of the European Communities, 169, 1996.

CHAPTER 25

# Pesticides and International Migratory Bird Conservation*

**Michael J. Hooper, Pierre Mineau, María Elena Zaccagnini, and Brian Woodbridge**

## CONTENTS



## 25.1 INTRODUCTION

Since the first occurrences of avian mortality from arsenicals and rodenticides in the early 20$^{th}$ century, concerns over pesticide effects on wildlife have led to the development of regulatory guidelines in many countries. In the United States, the Environmental Protection Agency (U.S. EPA), under the Federal Insecticide, Fungicide and Rodenticide Act of 1947 and its amendments (FIFRA), establishes the criteria for the testing and regulation of pesticides. Under FIFRA, pesticides that pose an unreasonable risk to wildlife should either be restricted in their use or banned altogether. In the face of mounting regulatory requirements, producers have sometimes chosen to withdraw pesticides from the market when the cost of meeting testing requirements exceeds the anticipated value of the product or when the ability of the pesticide to meet environmental safety requirements is doubtful.

Because many countries lack the resources to fully develop pesticide regulatory infrastructures of their own, they frequently rely on decisions made by the U.S. EPA, or equivalent regulatory bodies in other countries, to aid in their own internal decision-making. Restricted-use requirements

* Portions of text from Hooper, M.J., Mineau, P., Zaccagnini, M.E., Winegrad, G.W., and Woodbridge, B., *Pesticide Outlook*, 10(3), 97–102, 1999 have been used, unmodified or modified in this chapter with permission from the Royal Society of Chemistry (http://www.rsc.org/po/).

and outright banning of pesticides are actions that appear in the administrative record and are useful in developing regulatory criteria. Withdrawal of a pesticide, however, leaves a gap in the record and provides little useful data upon which decisions can be made in countries where the chemical is still available. This scenario is precisely what happened in the mid-1980s when the organophosphorus (OP) pesticide monocrotophos (dimethyl (E)-1-methyl-2-(methylcarbamoyl)vinyl phosphate; CAS No. 2157–98–4) was withdrawn from the market in the United States. In the subsequent 10 years, monocrotophos went on to become the second-highest-volume OP pesticide used in the world.[1] Ironically, monocrotophos is one of the most acutely toxic pesticides to birds. It took the death of an estimated 20,000 Swainson's hawks (*Buteo swainsoni*) in Argentina to reawaken the world's interest in monocrotophos. What's been learned since, as demonstrated by the risks that face many transborder avian migrants, has clarified the need for greater international cooperation in confronting pesticide impacts.

## 25.2 THE SWAINSON'S HAWK

A spring and summer inhabitant of western North America, the Swainson's hawk is a medium-sized broad-winged hawk of grassland, sparse shrub, and open woodlands, reaching its highest present-day nesting densities in areas of mixed agriculture and native habitat. During its summer breeding season, the Swainson's hawk nests in solitary trees, shrubs, or groves of trees along streams or agricultural areas, feeding rodents, small birds, reptiles, and insects to its young. The generally monogamous pairs will often return annually to within several km of previous breeding sites. Common farming practices, such as disking, cultivating, flooding, burning, and harvesting, provide hunting opportunities, as prey are disrupted by these actions.[2]

Gregarious by nature, Swainson's hawks can be seen in mixed adult and juvenile groups as large as 100 birds searching for food in the late summer. Flocks of 5000 to 25,000 birds were observed historically in the vicinity of San Antonio, Texas. Such North American assemblages of hawks are now found only in Mexico and Central America, where flocks containing hundreds of thousands of birds pass over known flyways on their southern migration.[2] Though overall population levels are relatively stable, if not increasing slightly, it is generally felt that Swainson's hawk numbers fell with the settling of the west. While populations in some areas remain low and the species is considered threatened (California) or in decline (Nevada, Utah, Oregon, and the prairie provinces of Canada), other areas support relatively abundant and stable populations (Idaho, Montana, Washington, and Colorado).[2,3]

Though scattered banding and field reports document the hawk's presence in Argentina during the northern winter (the southern hemisphere or "austral" summer), the details of their migration weren't well established until 1994, when satellite-transmitter-carrying Swainson's hawks were tracked from their summer territories across western North America to the grasslands, or pampas, of Argentina (Figure 25.1). Hawks from across the western United States and Canada converge in eastern Mexico on the Gulf of Mexico coast. From there they follow a narrow, well-defined path through Central America, cross the Andes Mountains in Columbia, and travel south to Argentina along the eastern slopes of the Andes, moving an average of 188 km/day.[4,5]

The pampas of Argentina comprise about a quarter of the country's area. With slowed beef exports in the latter 20$^{th}$ century, agricultural crop production increased on the pampas, leading to a rise in cultivated acreage, particularly for oil seed such as soybean, corn (maize), and sunflower that provide export crops for the country. Accompanying this production came a general intensification of farming practices including a lower tolerance of pest infestations and greater use of pesticides. Rather than the mammalian, avian, or reptilian prey chosen during the breeding season, the hawk's diet on the pampas is primarily insectivorous, feeding on grasshoppers, dragonflies, grubs, and other invertebrates.[4,6–9]

The first transmitter-fitted birds migrated to La Pampa province in 1994, where large flocks of Swainson's hawks numbering from 2000 to 7000 individuals were found inhabiting the region.

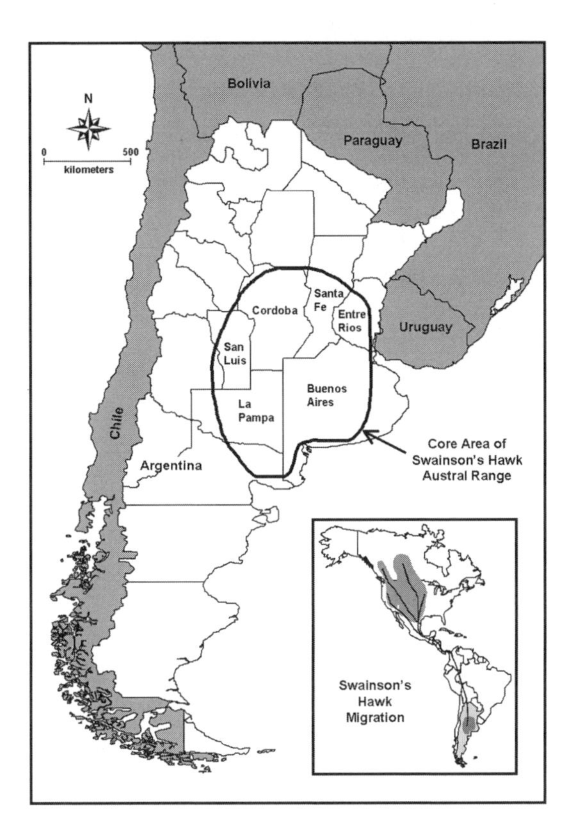


**Figure 25.1** North and South American ranges, migratory pathway and core austral range of Swainson's hawks. (Based on Hooper, M. J. et al., *Pest. Outlook,* 10, 97, 1999. With permission.)

"Aguiluchos langosteros" (translated "locust-eating hawks") foraged on agricultural fields and in the air during the day and roosted in eucalyptus groves (or "montes") that provide windbreaks for local ranches and farms. In early morning, the hawks would glide to nearby fields, waiting for thermals that came with the sun's heat and chasing and consuming "tucuras" (*Dichroplus* sp.), a small grasshopper, in the fields. The first evidence of mortality came with the discovery of over 700 dead hawks under one of the montes.[4] According to the local farmer, hawks died after a neighbor sprayed sunflowers with a pesticide to kill grasshoppers in the young crop. As the carcasses were quite decomposed, no physical data were collected to determine the cause of the deaths. The pesticide used was thought to be monocrotophos. This insecticide would prove to be responsible for the vast majority, if not all, of the Swainson's hawk mortality recorded in Argentina.

## 25.3 MONOCROTOPHOS

Of the pesticides used in U.S. agriculture today, excluding herbicides, the cholinesterase (ChE)-inhibiting OPs and carbamates are the most widely applied.[10] In developing countries, where fewer herbicides are used than in developed ones, cholinesterase-inhibiting insecticides can make up a very high proportion of total pesticide use. Initially developed as military nerve agents in the 1930s, anticholinesterases were put to use as insecticides in the 1940s, competing with the newly discovered DDT and other organochlorines. By the late 1970s, OPs and carbamates had, for the most part, replaced the more persistent organochlorines as the insecticides of choice. Following uptake of a lethal OP dose, the pesticide inhibits the enzyme, acetylcholinesterase (AChE), in the nervous system and its target tissues, eventually causing death by respiratory failure or other direct physi-

**Table 25.1 Acute Oral Toxicity of Monocrotophos to Avian Species**

| Species | Scientific Name | $LD_{50}$[a] | References |
|---|---|---|---|
| Golden eagle | *Aquila chrysaetos* | 0.188 | 45 |
| California quail | *Callipepla californica* | 0. 8 | 45, 46 |
| Northern bobwhite | *Colinus virginianus* | 0.9 | 47, 48 |
| Red-winged blackbird | *Agelaius phoeniceus* | 1.00 | 49 |
| Bengalese finch | *Lonchura domestica* | 1.05 | 50 |
| Red-billed quelea | *Quelea quelea* | 1.3 | 51 |
| House sparrow | *Passer domesticus* | 1.46 | 47, 51 |
| American kestrel | *Falco sparverius* | 1.5 | 48 |
| Eastern screech owl | *Otus asio* | 1.5 | 48 |
| Canada goose | *Branta canadensis* | 1.58 | 47 |
| Ring-necked pheasant | *Phasianus colchicus* | 1.8 | 47, 52, 53 |
| Turkey | *Meleagris gallopavo* | 2.51 | 45 |
| Rock dove | *Columba livia* | 3.45 | 47, 54 |
| Weaver finch | May be *P. domesticus* | 3.5 | 55 |
| Japanese quail | *Coturnix coturnix* | 3.9 | 47, 54, 56, 57 |
| European starling | *Sturnus vulgaris* | 3.9 | 49, 54, 58 |
| Mallard duck | *Anas platyrhynchos* | 4.0 | 47, 59 |
| Common grackle | *Quiscalus quiscula* | 4.21 | 54 |
| Peking duck | *Anas* sp. | 6 | 60 |
| Chukar | *Alectoris chukar* | 6.49 | 47 |
| Chicken | *Gallus gallus* | 6.7 | 55 |
| Grey partridge | *Perdix perdix* | 9.05 | 45 |
| House finch | *Carpodacus mexicanus* | 14.0 | 45 |

*Note:* Values are those studies that used adult birds and technical product only. When multiple values for a species were available, the geometric mean of the $LD_{50}$s was used.

[a] $LD_{50}$ values are in mg active ingredient/kg body weight.

ological consequences of AChE inhibition.[11] Birds are particularly sensitive to OP pesticides, as they have a reduced ability to metabolize and excrete these toxicants from the body.[12]

Monocrotophos was developed in the early 1960s and was introduced in 1965 by Shell and Ciba under the trademarks Azodrin® and Nuvacron,® respectively. It is a broad-spectrum OP with systemic, residual, and contact toxicity to chewing and sucking insects. It has a high water solubility and breaks down relatively quickly in the environment.[13] The avian toxicity of monocrotophos, however, places it among the most toxic pesticides to birds, with $LD_{50}$ values ranging from 0.2 mg active ingredient (a.i.)/kg body weight (b.w.) in the golden eagle (*Aquila chrysaetos*) to 14 mg a.i./kg b.w. in the house finch (*Carpodacus mexicanus*) (Table 25.1).

Because of the U.S. EPA's new avian field testing requirements for pesticides in the mid-1980s, Azodrin,® the commercial monocrotophos product available in the United States, was withdrawn from the market in 1989 by DuPont, who had obtained the U.S. rights to the chemical from Shell 3 years earlier (the international rights being purchased by American Cyanamid and currently retained by BASF). Due to its withdrawal, no official action was taken to further regulate or restrict the use of monocrotophos in the U.S. except to ensure compliance with residue tolerances in imported foodstuffs. In spite of its withdrawal from the United States, monocrotophos remained the second-highest-use OP in the world through the mid-1990s, with production continuing by at least 15 manufacturers worldwide (Figure 25.2).[1]

The U.S. EPA's avian field-testing requirement for monocrotophos was not without cause. Case histories of monocrotophos field trials, posttreatment monitoring efforts, and mortality incidents are numerous.[14] General trends from more than 25 field reports and publications on cotton, sugar cane, sunflower and maize, grain and forage crops, wide-row vegetable crops, and orchards and vineyards suggested that: (1) when monocrotophos was used above 1 kg a.i./ha, avian mortality was readily and predictably seen and (2) mortality occurred frequently below 1 kg a.i./ha, partic-

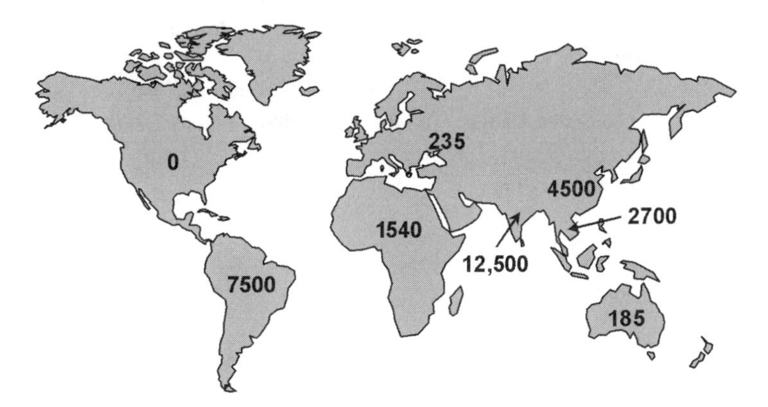


**Figure 25.2** Global monocrotophos use statistics. (From Hooper, M. J. et al., *Pest. Outlook,* 10, 97, 1999. With permission.)

ularly if there were birds present in or near the field or if avian food items (i.e., weed seeds, fruits, or insect prey) or water sources were in or adjacent to the treated field. This meant that the insecticide had the potential to be causing avian mortality — often at rates of application that were no longer efficacious. Few of these reports are available in the open literature. It is only with the recent publication of Australia's National Registration Authority (NRA) monocrotophos risk assessment that the results of many of these studies have become accessible.[14]

It is noteworthy that many of these studies were conducted days to weeks after application and consisted of only quick searches of field perimeters as opposed to formal field studies meeting regulatory requirements for pesticide registration. This both draws into question the validity of negative findings in a few of the studies (notably in orchards) and at the same time demonstrates the obvious magnitude of many of the incidents, given the numbers and ease with which birds were found under the most rudimentary monitoring conditions.

Of particular interest in the list of incidents is the occurrence of a mass mortality of resident, wintering, and migratory raptors that occurred beginning in 1975, when monocrotophos was used to control levant vole (*Microtus guentheri*) populations infesting alfalfa fields in Israel's Hula Valley.[15,16] The voles' disinterest in rodenticide baits in the face of abundant alfalfa crops had led farmers to turn to monocrotophos spray applications. The raptors, already attracted to the vole outbreak, found the dead and dying rodents on the surface of the fields easy prey. By the time authorities finished counting the bodies and collecting and treating the survivors, 150 raptors were dead and 69 were being treated in wildlife care centers. Based on carcass recoveries and census numbers before and after the incident, it was estimated that 300 to 400 raptors were killed in this incident. Species affected in greatest numbers included spotted eagles, common buzzards, black kites, kestrels, and barn owls (Table 25.2). These deaths were particularly significant, as wintering assemblages of several species in the area were decimated to levels from which their numbers did not recover during the winter season.

Israeli agriculture regulators subsequently banned the use of monocrotophos for vole control in alfalfa. The lack of a suitable control agent to replace monocrotophos, however, led to continued unprosecuted use of the compound.[17] By the end of winter 1979, 365 raptors had been documented killed and another 258 treated for OP toxicosis. The barn owl, kestrel, black kite, common buzzard, and spotted eagle made up nearly 80% of the birds affected throughout the four seasons of monocrotophos use (Table 25.2). Numbers decreased during the latter 2 years as decreased alfalfa plantings (due to vole losses) led to decreased area sprayed. Unfortunately, monocrotophos-induced mortality did not end with the raptor deaths of the 1970s. As recently as 1997, applications of monocrotophos to fields in Israel's Bet She'an Valley led to the death or incapacitation of 59 black

**Table 25.2 Raptor Mortality Associated with Monocrotophos Use for Vole Control in the Hula Valley, Israel, from 1975–1979[a]**

| | | Data Summary | | | Percentages | | |
|---|---|---|---|---|---|---|---|
| Common Name | Scientific Name | Killed | Treated | Total | Killed | Treated | Total |
| Spotted eagle | *Aquila clanga* | 30 | 22 | 52 | 8.2 | 8.5 | 8.3 |
| Lesser spotted eagle | *Aquila pomarina* | 10 | 9 | 19 | 2.7 | 3.5 | 3.0 |
| Imperial eagle | *Aquila heliaca* | 5 | 12 | 17 | 1.4 | 4.7 | 2.7 |
| White-tailed sea eagle | *Haliaeetus albicilla* | 1 | — | 1 | 0.3 | 0.0 | 0.2 |
| Long-legged buzzard | *Buteo rufinus* | 9 | 13 | 22 | 2.5 | 5.0 | 3.5 |
| Common buzzard | *Buteo buteo* | 54 | 63 | 117 | 14.8 | 24.4 | 18.8 |
| Sparrowhawk | *Accipiter nisus* | 1 | 0 | 1 | 0.3 | 0.0 | 0.2 |
| Black kite | *Milvus migrans* | 94 | 25 | 119 | 25.8 | 9.7 | 19.1 |
| Marsh harrier | *Circus aeruginosus* | 19 | 7 | 26 | 5.2 | 2.7 | 4.2 |
| Hen or pallid harrier[b] | *Circus cyaneus* or *Circus macrourus* | 14 | 10 | 24 | 3.8 | 3.9 | 3.9 |
| Kestrel | *Falco tinnunculus* | 33 | 32 | 65 | 9.0 | 12.4 | 10.4 |
| Merlin | *Falco columbarius* | 1 | 0 | 1 | 0.3 | 0.0 | 0.2 |
| Short-eared owl | *Asio flammeus* | 12 | 5 | 17 | 3.3 | 1.9 | 2.7 |
| Long-eared owl | *Asio otis* | 5 | 9 | 14 | 1.4 | 3.5 | 2.2 |
| Barn owl | *Tyto alba* | 77 | 50 | 127 | 21.1 | 19.4 | 20.4 |
| Eagle owl | *Bubo bubo* | 0 | 1 | 1 | 0.0 | 0.4 | 0.2 |
| **Total** | | 365 | 258 | 623 | 100 | 100 | 100 |

[a] Data from Mendelssohn, H. and Paz, U., *Biol. Conserv.,* 11, 163, 1977; Mendelssohn, H., Schlueter, P., and Aderret, Y., *Int. Counc. Bird Preservation,* 13, 124, 1979.
[b] Whether hen or pallid harrier, not determined.

kites (*Milvus migrans*), 11 kestrels (*Falco tinnunculus*), and 19 barn owls (*Tyto alba*) as well as many nonraptor migrants such as 168 red-throated pipits (*Anthus cervinus*) and 65 sky larks (*Alauda arvensis*).[18] Ironically, many of the barn owls and kestrels that were killed had been attracted to the area with an extensive nest-box program aimed specifically at controlling local rodent populations.

## 25.4 ARGENTINE MORTALITY INCIDENTS

### 25.4.1 The 1995/1996 Austral Summer

In a follow-up study to the 1994/1995 Swainson's hawk incident, researchers traveled to Argentina, where they worked with the same local farmer and Argentine government wildlife researchers from INTA (Instituto Nacional de Tecnología Agropecuaria) to document the extent of the problem. They discovered four large mortality incidents, totaling over 4100 birds.[19,20] Argentine researchers documented 14 additional mortality incidents from January through March 1996, bringing the estimated total to 5093 dead Swainson's hawks (Table 25.3). In the majority of the incidents, monocrotophos was implicated as the pesticide responsible for the mortalities.[20,21] Preliminary estimates, based on a crude area extrapolation, suggested a total kill of up to 20,000 birds.

Forensic evaluation of the Argentine mortalities used landowner and applicator interviews, analyses of chemical residues from hawk stomach contents, and evaluation of ChE activity in tissues from the dead and incapacitated hawks. Landowner and applicator interviews implicated monocrotophos use (16 of 18 interviews), primarily in alfalfa and sunflowers, for control of the tucura grasshoppers.[21] Though not specifically registered for this use, monocrotophos was registered for use in alfalfa against a number of insect pests. As this type of off-label use was not specifically prohibited in Argentina, farmers were not violating pesticide-use regulations.[22]

Monocrotophos was confirmed in 31 of 45 samples from the sites with recoverable tissues, averaging 0.20 ppm (fresh wt.) in gastrointestinal (GI) tract contents. Monocrotophos is different from most OPs used in agriculture because it is applied in its active form, known as an oxon,

**Table 25.3 Argentine Incidents Involving Primarily Swainson's Hawks and Suspected or Confirmed Pesticide Use**

| Location | Province | Date of Incident | Number of Affected Hawks[a] | Crop | Pesticide Reported Sprayed[b] | Pesticide Confirmed | Certainty Index[c] | References |
|---|---|---|---|---|---|---|---|---|
| Between Hilario Lagos and General Pico | La Pampa | Dec. 94 | 714 | Pasture (alfalfa) | MCP | | 5 | 4, 19, 21 |
| Hilario Lagos and General Pico | La Pampa | Dec. 94 | approx. 1000 | Sunflowers | MCP | | 6 | 19 |
| Near Alta Italia | La Pampa | Jan. 96 | 1 | Unknown | MCP | | 3 | 20 |
| Alta Italia | La Pampa | Jan./Feb. 96 | 387 | Alfalfa | MCP | MCP | 1 | 20, 21, 23 |
| Alta Italia | La Pampa | Jan./Feb. 96 | 103 | Alfalfa | DIM | | 5 | 20, 21, 23 |
| Ingeniero Luiggi | La Pampa | Jan./Feb. 96 | approx. 3024 | Alfalfa | | MCP | 1 | 20, 21, 23 |
| Vertiz | La Pampa | Jan./Feb. 96 | 592–595 | Alfalfa | MCP | MCP | 2 | 20, 21, 23 |
| Parera | La Pampa | Feb. 96 | 2 | | | | 6 | 21 |
| Parera | La Pampa | Jan. 96 | 12–20 | Alfalfa | MCP/CYP | | 5 | 21 |
| Ingeniero Luiggi | La Pampa | Dec. 95 | 14–20 or 80[d] | Alfalfa | MCP | | 6 | 20, 21 |
| Ingeniero Luiggi | La Pampa | Jan./Feb. 96? | approx. 300 | Unknown | MCP | | 6 | 20, 21 |
| Embajador Martini | La Pampa | Dec. 95 | 40–50 | Alfalfa | MCP | | 6 | 20, 21 |
| Hilario Lagos | La Pampa | Mar. 96 | 9 | Oats[e] | CPF/CYP[e] | MCP | 1 | 20, 21 |
| America | Buenos Aires | Feb. 96 | 28 | Alfalfa and Agropiro | MCP | | 6 | 20, 21 |
| Pincén | Córdoba | Jan. 96 | 120 | Alfalfa | MCP | | 6 | 20, 21 |
| Pincén | Córdoba | Feb. 96 | 3 | At roost | MCP | | 6 | 20, 21 |
| Huinca Renancó | Córdoba | Jan. 96 | 4 | At roost | | MCP | 1 | 20, 21 |
| Huinca Renancó | Córdoba | Feb. 1996 | approx. 200 | Pasture (Alfalfa) | MCP/CYP DIM/DEL | | 6 | 20, 21 |
| Huinca Renancó | Córdoba | Dec. 1995 | | Alfalfa | MCP/LAM | | 6 | 20, 21 |
| Huinca Renancó | Córdoba | Nov. 1995 | 30–40 | Alfalfa | MCP | | 6 | 20, 21 |
| Huinca Renancó | Córdoba | Jan. 96 | 20–30 | Alfalfa | MCP | | 6 | 20, 21 |
| Monte de los Gauchos | Córdoba | Jan. 96 | 109 | Sorghum | CPF/CYP[f] | | 6 | 20, 21 |
| Alta Italia | La Pampa | Nov. 96/Jan. 97 | 1[g] | Road | | | 4 | 31 |
| General Pico | La Pampa | Nov. 96/Jan. 97 | 1[g] | Sunflowers | | | 3 | 31 |
| Pozo de Molle | Córdoba | Jan. 97 | 24 | At Roost | CYP/MCP | | 5 | 61 |
| Canals | Córdoba | Nov. 97 | 198 | Soybean | MCP | MCP | 2 | MEZ[h] |

**Table 25.3 Argentine Incidents Involving Primarily Swainson's Hawks and Suspected or Confirmed Pesticide Use** *(Continued)*

| Location | Province | Date of Incident | Number of Affected Hawks [a] | Crop | Pesticide Reported Sprayed[b] | Pesticide Confirmed | Certainty Index[c] | References |
|---|---|---|---|---|---|---|---|---|
| Laspiur | Córdoba | Nov. 97 | 38 | Unknown | FIP | | 5 | MEZ |
| Las Petacas | Santa Fe | Dec. 97 | 154 | Unknown | | | 5 | MEZ |
| Laspiur | Córdoba | Dec. 97 | 97[i] | Unknown | | | 5 | MEZ |
| Bauer y Sigel | Santa Fe | Dec. 97 | 148[j] | sorghum and pasture | MCP | MCP | 2 | MEZ |
| Villa Marcelino | Córdoba | Jan. 98 | 134 | Unknown | MCP[k] | | 5 | MEZ |

*Note:* Known abuses and cases where chemical or biochemical investigation gave negative results are excluded.

[a] Range reflects estimate from one or several observers or different accounts of same incident.
[b] CYP-cypermethrin; DIM-dimethoate; DEL-deltamethrin; CPF-chlorpyrifos; FIP-fipronil; LAM- lambda-cyhalothrin; MCP-monocrotophos.
[c] Determination based on (1) pesticide residues and ChE evidence; (2) residues and circumstances indicative of poisoning; (3) ChE evidence and circumstances indicative of poisoning; (4) ChE evidence; (5) circumstances indicative of poisoning; (6) circumstances indicative of poisoning as related in an interview. Modified from Mineau, P. et al., *J. Raptor Res.*, 33, 1, 1999.
[d] Described as two separate incidents in one reference[20] and as two estimates (from two different individuals) relating to the same incident, by another.[21]
[e] The identity of the pesticide or crop was very tentative. Although the nearest known pesticide treatment to the kill site was a mixture of chlorpyrifos and cypermethrin, this was 3 km from the kill site. The field is described as both wheat[20] and oats.[21]
[f] In this case, the involvement of chlorpyrifos is considered more likely than in the other case reported above. All birds were found on the treated field.
[g] Bird incapacitated only. Held and released. Low cholinesterase values indicative of exposure. Clinical signs in second case. Identity of pesticide unknown but both birds found in zone where MCP was supposed to be excluded.
[h] Zaccagnini, M.E., Unpublished data, 2000.
[i] Another 5 individuals of the following species also found: *Myiopsitta monachus, Colapses melanochloros, Zenaida auriculata, Polyborus plancus.*
[j] Includes a single individual of *Speotyto cunicularia.*
[k] Although MCP is suspected, violent storms could have given rise to this incident. Other incidents in area were storm-related.

**Table 25.4 Cholinesterase Activities of Poisoned and Reference Swainson's Hawks**

| Collection Location | Source | Brain | | Plasma | | | |
|---|---|---|---|---|---|---|---|
| | | N | Total ChE | N | Total ChE | AChE | BChE |
| **California/Colorado, U.S. 1996[b]** | | | | | | | |
| Reference | Field | 1 | 17.2 | 17 | 0.696 ± 0.028 | 0.285 ± 0.017 | 0.410 ± 0.022 |
| | Lit | 2 | 19[a] | — | — | — | — |
| **Argentina 1995–1996[b]** | | | | | | | |
| Site 1 | Field | 3 | 0.20 ± 0.09 (1.2%) | — | — | — | — |
| Site 3 | Field | 8 | 0.79 ± 0.18 (4.6%) | 6 | 0.278 ± 0 087 (39.9%) | 0.023 ± 0.006 (8.4%) | 0.254 ± 0.082 (62.0%) |
| | Euth | 2 | 4.70 (27.3%) | 2 | 0.358 ± 0.074 (51.4%) | 0.032 ± 0.015 (11.2%) | 0.326 ± 0.059 (79.5%) |
| Site 9 | Field | 7 | 1.38 ± 0.22 (8.0%) | — | — | — | — |
| **Argentina 1996–1997[c]** | | | | | | | |
| Various | Field | — | — | 131 | 0.674 ± 0.014 (96.8%) | 0.236 ± 0.006 (82.8%) | 0.438 ± 0.013 (106.8%) |

*Note:* Values are mean ± s.e. in μmoles acetylthiocholine hydrolyzed/min on a per g brain tissue or per mL plasma basis. Values in parentheses are activity as a percent of reference. Field samples were collected from breeding adults in reference areas (U.S.), dead or incapacitated birds (Argentina 1995–1996) or live trapped hawks (1996–1997). Literature values (Lit) are for congeneric red-tailed hawk brain ChE activity. Brain data from Argentine hawks are from mortalities except two birds euthanized (Euth) due to the seriousness of their injuries, providing both plasma and brain samples.

[a] Data from Hill, E.F., *J. Wildlife Dis.*, 24, 54, 1988 and Henny, C.J. et al., *J. Wildl. Manage.*, 49, 648, 1985.
[b] Data from Goldstein, M.I. et al., *Ecotoxicology,* 8, 201, 1999.
[c] Data from Goldstein, M.I. et al., *Ecotoxicology,* 8, 215, 1999.

containing a double bond between the phosphorous and an oxygen atom. Most agricultural OPs are formulated with a similar bond to a sulfur atom, a less reactive (and generally less toxic) form of OPs, requiring metabolic activation to the active oxon and having greater stability in the environment. Monocrotophos levels in GI tract contents were lower than might be expected due to their reactivity with the degrading grasshopper and hawk tissues in the birds. As no other pesticides were detected in any of the samples tested, and the only OP degradates detected were monocrotophos-associated, monocrotophos was implicated in all incidents for which tissues were analyzed.[23]

Brain ChE activities in dead hawks averaged over 90% inhibition (compared to reference values), while the survivors demonstrated over 60% depressions (Table 25.4). Carcasses in Argentina had remained in the field for several days in temperatures reaching daily maxima approaching 40°C. Cholinesterases are, however, particularly resilient to temperature effects, retaining 79% and 60% of their original activities after carcass storage at 36°C for 4 and 8 days postmortem, respectively.[24] Field degradation, therefore, would not have led to such depressed levels. Further, laboratory attempts to reactivate the inhibited enzyme using 2-PAM reactivation techniques were generally unsuccessful and led only to minor increases of activity, indicating that the inhibited ChE had aged to the point where the OP could not be chemically removed. This finding, and the overall lack of spontaneous postmortem ChE reactivation, implicate a dimethylphosphoryl-type OP, such as monocrotophos, that characteristically forms the most stable, quickly aging ChE bond of all the agricultural OPs.[25] Plasma ChE activities were similarly refractory to reactivation attempts. The levels of brain and plasma ChE inhibition documented in Swainson's hawks were consistent with death and incapacitation, respectively, due to anticholinesterase exposure.

Based on the combined findings of the farmer/applicator interviews, chemical analyses, and cholinesterase evaluations, it was concluded that the mortality incidents were due to organophosphate poisoning, and in most cases, the evidence pointed specifically to monocrotophos.

### 25.4.2 Quick Response Leads to Hawk Protection

With the discovery of hawk mortalities in Argentina and the determination that monocrotophos was responsible, the American Bird Conservancy (ABC), representing a number of bird-associated nongovernmental organizations (NGOs) in the United States, requested that the manufacturer, Novartis (formerly Ciba-Geigy, Ltd., then Novartis, currently Syngenta), remove monocrotophos from the global market. The dialog that developed from this interaction between industry and an NGO was joined by Canadian, United States, and Argentine government representatives, academicians, and several other NGOs and monocrotophos producers.

Measures were implemented to protect Swainson's hawks from further impacts. Argentine government agencies established a monocrotophos withdrawal zone that encompassed Swainson's hawk use areas, documented by ground and satellite transmitter studies in the previous year, overlain with the locations of mortality incidents. Pesticide producers and distributors in Argentina withdrew and bought back or exchanged unopened monocrotophos stocks from within this zone. SENASA (Servicio Nacional de Sanidad y Calidad Agroalimentaria), the Argentine pesticide regulatory authority, declared a specific prohibition on any use of monocrotophos on alfalfa or against tucuras on any crop, a necessary action to make these uses of monocrotophos illegal.[22] Labels were modified on monocrotophos containers warning farmers of the new rules.

Intensive farmer information campaigns, focusing on chemical control of grasshoppers and IPM alternatives, were developed by the Argentine government (both INTA and provincial governments), the pesticide industry, and the Argentine NGO Asociación Ornitológica del Plata. Farmers were warned not to use monocrotophos illegally and to refrain from using it altogether when hawks or other birds were present on fields.[26,27] These campaigns were particularly important, as demonstrated by survey findings that up to 33% of pesticide applications were made contrary to label recommendations for crop, pest species, or application rate.[28]

Argentine regulators and scientists worked with scientists from other countries to harmonize the methods used to assess pesticide impacts on wildlife. Four cooperative training workshops led by academic, government, and corporate scientists from the United States, Canada, Switzerland, and Argentina were held in Argentina. Two of the workshops focused on field techniques to monitor and document mortality incidents associated with pesticide use, resulting in the development of a monitoring manual with government-approved methods and procedures.[29] The other two focused on chemical and biochemical analytical techniques to ensure that the methods used in diagnosing exposure and effects were consistent between the primary collaborators. Specialists from each area held extensive training and demonstration sessions. Within 1 year of the occurrence of the large-scale documented kills, a coordinated Argentine program, including a nucleus of trained individuals, was in place to carry out forensic investigations in the core area of the Swainson's hawk wintering range.[30]

The Argentine summer of 1996–1997 was a particularly wet one, and tucura numbers were lower than normal. Intensive monitoring activities were performed, observing Swainson's hawk population movements using satellite and conventional radio-transmitter-fitted hawks.[6] An intensive live-capture and release program that evaluated nonlethal exposures to pesticides demonstrated little evidence of anti-ChE exposures (Table 25.4).[31] The network of field and laboratory personnel was ready to respond quickly to any incident. In the end, there was a single documented mortality incident outside of the intensive control zone, where 24 Swainson's hawks died from an unknown pesticide exposure (Table 25.3). Through a combination of extensive public education and outreach, intensive chemical management, and mild weather, recurrence of the mass Swainson's hawk mortalities of 1996 did not occur in 1997.

Though this good news gave reason for cautious optimism, an intentional misuse of monocrotophos on seed baits in July of 1997 led to the deaths of an estimated 100,000 birds, primarily the eared dove (*Zenaida auriculata*), in Entre Rios Province in eastern Argentina.[27,32] This event intensified the concern for the threat that monocrotophos posed, particularly to flocking bird species.

### 25.4.3 Regulatory Actions since the Incidents

In the time since the activities immediately following the Argentine incidents, a number of producers have initiated phase-outs of monocrotophos, while the number of countries discontinuing registrations has grown. Though they had decided to withdraw Nuvacron® from the entire Argentine market prior to the July 1997 incident, Novartis (now Syngenta) announced in 1998 a phased withdrawal of their monocrotophos products from all world markets as part of a product portfolio review. Monocrotophos would be removed from most countries (except India) by the year 2000, with a complete phase-out by 2003. DowAgrosciences, who recently inherited monocrotophos when it merged with a smaller independent pesticide manufacturer, announced a similar decision in 2000, with a phase-out by 2004, based on concerns for safety to wildlife, particularly migratory birds. The removal of Nuvacron® and the DowAgrosciences product represented approximately 20% of the 30,000 tons of monocrotophos active ingredient marketed annually around the world. No other monocrotophos producers, neither the large multinationals such as BASF (producers of Azodrin®) and Cheminova nor the many national or regional producers in India, China, and Taiwan, have announced similar actions.

A number of national pesticide registration agencies in monocrotophos-using nations have reconsidered their position on the chemical in light of the incidents in Argentina. In 1999, Argentina cancelled all monocrotophos registrations due to the chemical's "... inherent toxicity and the continued occurrence of wildlife mortalities wherever it was used ... even after implementing special restrictions and withdrawals, and establishing an unprecedented public information campaign designed to inform farmers of use restrictions and warn them about treating areas with heavy wildlife use."[33,34]

Australia's National Registration Authority cancelled the Australian registration of monocrotophos after extensive industry and public consultation failed to satisfy the NRA that the OP's continued use under current label guidelines would not harm people or the environment.[14] Of particular note in the NRA review was the fact that, similar to the U.S. EPA reassessment in the mid-1980s, there was no effort made by any producers, distributors, or users to come forward to alleviate the NRA's concerns by performing needed field safety studies. The conclusion of their review was that "The weight of evidence indicates use of monocrotophos poses a high hazard to birds, and it is difficult to defend its continued use."

Concurrent with the implementation of protective measures in Argentina, two additional projects were established to address critical issues associated with the incidents. The first was based on the need for an international ecological risk assessment that covers the variety of uses of monocrotophos. A data review leading to an international ecological risk assessment of monocrotophos is being developed. It examines the available avian data, in the form of laboratory investigations, field studies, and incident reports, that have been produced since the compound's inception over 30 years ago. The goal is to produce an internationally applicable document that provides the background and risks involved with the use of monocrotophos. To eliminate any doubt about the risk posed by monocrotophos, this assessment will be provided to all countries carrying monocrotophos registrations and to producers of the pesticide, so that the repercussions of using the compound are fully understood. Thereafter, decisions to continue selling and using monocrotophos will be made with full understanding of the ecological risk the compound poses. The second initiative, the Avian Incident Monitoring System, is being coordinated by the ABC, with U.S. EPA support, to address the lack of a unified international reporting program for pesticide-related mortality incidents.

Participants are working to create a common format and dissemination plan for incident reporting and information access to improve regulatory decision-making.

## 25.5 CONSERVATION CONSIDERATIONS AND PERSPECTIVE

In their review of the relevance of mortality incidents to raptor populations, Henny and coworkers[35] suggest that the loss of 20,000 birds, estimated to be from 2 to 5% of the total population, cannot be detected within the normally fluctuating numbers of a raptor population. This is particularly true of the Swainson's hawk, as intensive flock observations on the Pampas[6] and the variety of band recoveries in the documented mortality incidents[20] indicate that its movements on its austral range are characterized by substantial mixing of birds from regions throughout its northern range. Had the 20,000 birds been from an isolated regional population, the impacts would have been much greater and easily detectable. The authors conclude that the more pertinent consideration was that the extensive mortality occurred with a compound that was registered and approved for use within the regulatory guidelines of the country.

Though the level of mortality documented on the Argentine pampas does not appear to pose an immediate threat to the overall population of Swainson's hawks, further perspective on the nature of pesticide threats to migratory birds is provided by experiences with wintering dickcissels (*Spiza americana*). Numbers of this neotropical migrant declined steadily during the 1960s and 1970s and have since stabilized at levels 40% below those in 1966.[36]

A North American grassland passerine during the northern summer, the dickcissel migrates every year to the central Venezuelan llanos, where it winters among the rice, sorghum, and sugarcane farms that surround the Orinoco River in the states of Portuguesa, Cojedes, and Guárico (Figure 25.3).[37] Though generally dispersed throughout its northern breeding grounds, the dickcissel is gregarious in the llanos, feeding and roosting in large flocks, much like Swainson's hawks in Argentina. Historically, dickcissels ranged widely across the llanos, feeding on wild grass seed. Conversion of large areas to cereal agricultural has focused their foraging activities on the concentrated food resources of rice and sorghum. Wild grass seed now accounts for less than 20% of their diet. Flocks of greater than 10,000 birds regularly forage on fields of rice and sorghum, affecting approximately 1.5 and 8% of the crops, respectively; damage has been valued at approximately U.S. $1.9 million.[38] The local name for dickcissels in Venezuela is "el pájaro arrocero," the rice bird.

Measures used by farmers to repel the flocks are largely nonlethal (firecrackers, sirens, horns, burning tires), though 14% of farmers indicate that they have used lethal control measures in the past. Large nocturnal roosting flocks, averaging over 500,000 and up to 3 million birds, generally are found in sugarcane fields. OPs such as parathion and monocrotophos were sprayed by ground equipment or from the air to kill the large aggregations.[38] One farmer told of being "knee-deep" in dickcissel carcasses after one roost application.[37] Local water sources and feeding areas are also sprayed in hopes of poisoning dickcissels when they drink and feed.

Based on farmer interviews, documented mass poisonings, and demographic and behavioral characteristics of dickcissel populations on their wintering grounds, the observed population decline from 1966 to 1978 was likely due to intentional poisonings with highly toxic OP pesticides. Dickcissel impacts, when focused on the limited, newly developing farm enterprises of the 1950s and 1960s, were more severe. Established farmers described intensive persecution of dickcissels when their impacts were proportionately the greatest (the 1950s to 1970s). Though dickcissel foraging activities are now dispersed across a greater agricultural landscape, focused crop depredation still impacts farmers at local levels, and based on historical practices in the areas, farmers continue to use lethal control methods. Ongoing use of OP control methods has likely kept dickcissel populations depressed since the late 1970s.[38]

The dickcissel population remains in eminent jeopardy. As the majority of the entire population (6–7 million birds) can be found in 5 or 6 nocturnal roosts, with up to one third of the population

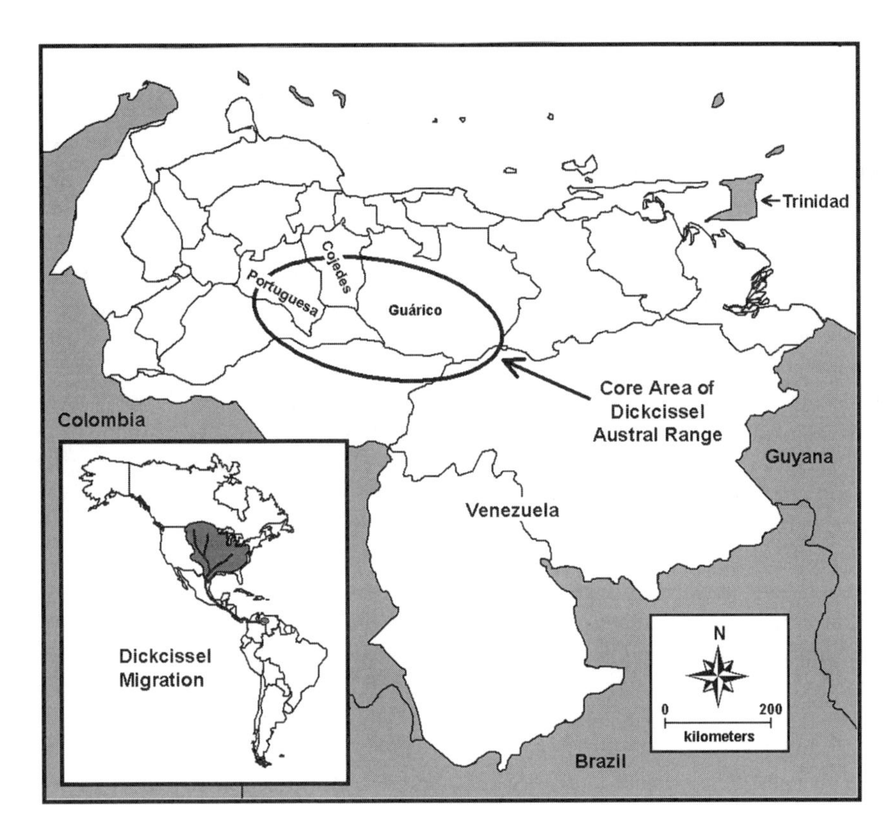


**Figure 25.3** North and South American ranges, migratory pathway and core austral range of dickcissels. (Based on Basili, G. D. and Temple, S. A., *Stud. Avian Biol.*, 19, 289, 1999.)

in one roost, a single poisoning event could have significant impact on the overall population. To avoid such a catastrophe, cooperating government agencies, NGOs, and academicians from the U.S. and Venezuela are attempting, as was done with the Swainson's hawks in Argentina, to find solutions to dickcissel depredation of cereal crops in the llanos. In addition to testing dickcissel response to crop-treatment materials aversive to birds, efforts are focused on modifications of crop timing and mechanisms for compensating farmers for accepting crop losses.[38,39]

## 25.6 WILDLIFE AND PESTICIDES — INTERNATIONAL PROTECTION NEEDS

The experience with Swainson's hawks in Argentina demonstrates a need for improving pesticide risk assessments for migratory wildlife. Pesticide regulation and enforcement, and hence the extent to which wildlife are protected from pesticides, vary greatly from country to country. Unilateral regulation of pesticides within a single country does not provide adequate protection for wildlife that cross international borders. In many countries, regulations are often barely sufficient for the protection of human health, let alone wildlife.[40] As has been seen in the withdrawal and cancellation actions that occurred in the United States and Australia following requests for data on avian safety, pesticide use often follows the path of least regulatory resistance. The resulting lack of regulation can easily lead to chemical applications at rates and under conditions that stray from safe or optimal use. With that said, however, it should be noted that impacts on wildlife are not restricted to countries lacking comprehensive pesticide regulation. Bird kills from labeled pesticide use in the United States and Canada still occur, are often the focus of intense debate, and can result in the reevaluation and regulation of particular pesticide uses.[35,41] Another consideration is that migratory species may be more vulnerable during migration or on their wintering grounds than on the breeding grounds. Because of the Swainson's hawk's high density on the pampas, a handful of Argentine farmers, through no fault of their own, could have significantly more impact on the species than thousands of North American farmers using insecticides of equivalent toxicity.

The development of international assessments of pesticide risk is a new approach to wildlife protection. Current efforts to internationally harmonize pesticide evaluations focus on pesticide toxicity testing methods, primarily as they apply to European and North American regulatory processes.[42] Toxicity and exposure data pertinent to areas outside of these regions are not generally available or incorporated into standard risk assessments. Development of pertinent and regionally applicable toxicity and exposure inputs is needed to begin examining pesticide risks in these countries. The dissemination of these assessments, and not simply a statement on a pesticide's regulatory status, would further provide countries, lacking effective regulatory support, information from which they can make better informed decisions. Monocrotophos was withdrawn from use in the United States in the late 1980s, and because it was never officially banned, its continued use elsewhere did not raise alarms, though its potential toxicity to wildlife was well known to manufacturers and U.S. regulators. A major gap exists in our ability to influence pesticide regulation and protect migratory wildlife when they are outside the bounds of our political borders. Improved communication between similar agencies — across borders — might shorten that gap.

No matter how comprehensive or sophisticated a country's regulatory system may be, it will not benefit wildlife unless chemical-impact assessments are incorporated into conservation efforts. Organized efforts, similar to those used for habitat protection by groups like Partners in Flight, are needed to protect migratory species from pesticides throughout their range. Much of the information necessary for this protection is already being generated; it just needs to be applied more broadly. Habitat use data and detailed natural history accounts of migrants,[43,44] compiled to prioritize habitat protection programs, can be combined with satellite imagery and regional databases that locate significant agricultural areas, intensive cropping practices, and regional pesticide sales. The layering of these data through a geographic information system can identify areas of concern where wildlife assemblages and pesticide use may present particular risk scenarios. The prioritization of resources for further ground-based evaluations can then focus on these areas of concern. An initial effort of this kind is under way for Canadian migrants through the Canadian Wildlife Service.

The Swainson's hawk and dickcissel experiences are a small piece of a complex international puzzle. If it weren't for the curiosity that led to following the species to their austral range, the mortality incidents might never have been documented. In all likelihood, the current efforts to reevaluate the use of monocrotophos and improve international incident reporting would not have proceeded as they have. There are likely similar situations awaiting discovery, not only in the Americas, but throughout the world wherever wildlife habitat and intensive agricultural practices overlap. It is simply a matter of looking. There is a high probability that monocrotophos is continuing to significantly affect bird numbers wherever it is in use. The removal of monocrotophos from the market by two producers will affect less than 20% of the world's supply, and other producers will likely replace the lost supply. Progress forward will depend on continued efforts on the part of biologists to identify pesticide threats when and where they occur, the efforts of international cooperators from all sides of the problem working together to solve them, and the willingness of regulators and pesticide manufacturers to look beyond national boundaries and their own interests to protect migratory wildlife.

## 25.7 SUMMARY

Recent events in Argentina and Venezuela have shown that pesticide issues transcend national borders and that our migratory birds continue to be exposed to potentially damaging pesticides throughout their range. We need to be especially wary of situations, such as that of the Swainson's hawk or the dickcissel, where a large part of a species' population occupies a small geographical area, either in the course of its migration or on wintering grounds. A localized impact, whether or not pesticide-related, could have serious consequences for the population. Mechanisms are needed to put pressure on the pesticide industry to improve its product stewardship and take action on products shown to be environmentally unsound. Currently, the battle against bad pesticides is being fought manufacturer-by-manufacturer and country-by-country, which is wasteful of the meager resources available. On the positive side, the Argentina experience has shown how readily problems can be solved when all stakeholders are willing.

## REFERENCES


1. Voss, G. and Schätzle, P., Special Foreword, *Rev. Environ. Contam. Toxicol.,* 139, xi, 1994.
2. England, A. S., Bechard, M. J., and Houston, C. S., Swainson's hawk (*Buteo swainsoni*), in *The Birds of North America*, No. 265, Pool, A. and Gill, G., Eds., The Academy of Natural Sciences, Philadelphia, and the American Ornithologists' Union, Washington, D.C., 1997.
3. Kirk, D. A. and Hyslop, C., Population status and recent trends in Canadian raptors: A review, *Biol. Conserv.,* 83, 91, 1998.
4. Woodbridge, B., Finley, K. K., and Seager, S. T., An investigation of the Swainson's hawk in Argentina, *J. Raptor Res.,* 29, 202, 1995.
5. Fuller, M. R., Seegar, W. S., and Schueck, L. S., Routes and travel rates of migrating peregrine falcons (*Falco peregrinus)* and Swainson's hawks (*Buteo swainsoni*) in the Western Hemisphere, *J. Avian Biol.,* 29, 433, 1998.
6. Canavelli, S. B., Abundance, Movement and Habitat Use of Swainson's Hawks in their Wintering Grounds, Argentina, M.S. thesis, University of Florida, Gainesville, 2000.
7. Canavelli, S. B., Maceda, J. J., and Bosisio A. C., Dieta del aguilucho langostero (*Buteo swainsoni*) en su área de invernada (La Pampa, Argentina), *Hornero*, 16(2), 89, 2001.
8. Jaramillo, A. P., Wintering Swainson's hawks in Argentina: Food and age segregation, *Condor,* 95, 475, 1993.
9. White, C. M., Boyce, D. A., and Straneck, R., Observations on *Buteo swainsoni* in Argentina, 1984, with comments on food, habitat alteration and agricultural chemicals, in *Raptors in the Modern World,* Meyburg, B. U. and Chancellor, R. D., Eds., World Working Group on Birds of Prey and Owls, Berlin, 1989.
10. Aspelin, A. L., Pesticide Industry Sales and Usage: 1994 and 1995 Market Estimates, Document No. 733-R-97–002, U.S. EPA, 1997.
11. Hill, E. F., Wildlife toxicology of organophosphorus and carbamate pesticides, in *Handbook of Ecotoxicology*, Hoffman, D. J., Rattner, B. A., Burton, G. A., Jr., and Cairns, J., Jr., Eds., Lewis Press, Boca Raton, Fl, 2003, 281, 243.
12. Walker, C. H., Pesticides and birds-mechanisms of selective toxicity, *Agric. Ecosyst. Environ.,* 9, 211, 1983.
13. Guth, J. A., Monocrotophos — Environmental fate and toxicity, *Rev. Environ. Contam. Toxicol.,* 139, 75, 1994.
14. NRA (National Registration Authority for Agricultural and Veterinary Chemicals, Australia), Environmental Assessment (Section 6), in *The NRA Review of Monocrotophos*, NRA Review Series 00.1, Canberra, 2000.
15. Mendelssohn, H. and Paz, U., Mass mortality of birds of prey caused by Azodrin, an organophosphorous insecticide, *Biol. Conserv.,* 11, 163, 1977.
16. Shlosberg, A., Treatment of monocrotophos-poisoned birds of prey with pralidoxime iodide, *J. Am. Vet. Med. Assoc.*, 169, 989, 1976.
17. Mendelssohn, H., Schlueter, P., and Aderret, Y., Report on Azodrin poisoning of birds of prey in the Huleh Valley in Israel, *Int. Counc. Bird Preservation*, 13, 124, 1979.
18. Szabo, J. and Shlosberg, A., Personal communications, 2000.
19. Goldstein, M. I., Woodbridge, B., Zaccagnini, M. E., Canavelli, S. B., and Lanusse, A., An assessment of mortality of Swainson's hawks on wintering grounds in Argentina, *J. Raptor Res.*, 30, 106, 1996.

20. Goldstein, M. I., Lacher, T. E., Jr., Woodbridge B., Bechard M. J., Canavelli S. B., Zaccagnini M. E., Cobb, G. P., Tribolet R., and Hooper M. J., Monocrotophos-induced mass mortality of Swainson's hawks in Argentina, 1995–1996, *Ecotoxicology*, 8, 201, 1999.
21. Canavelli, S. B. and Zaccagnini, M. E., Mortandad de Aguilucho Langostero (*Buteo swainsoni*) en la Region Pampeana: Primera Aproximacion al Problema, Argentine Technical Government Report (INTA) 5/96, 1996.
22. Mineau, P., An assessment of the factors which contributed to the recent pesticide kills of Swainson's Hawks in the Argentine Pampas and recommendations to prevent a re-occurrence, Unpublished report, National Wildlife Research Centre, Canadian Wildlife Service, 1996, http://panna.igc.org/resources/pestis/PESTIS.1997.46.html
23. Scollon, E. J., Goldstein, M. I., Parker, M. E., Hooper, M. J., Lacher, T. E., and Cobb, G. P., Chemical and biochemical evaluation of Swainson's hawk mortalities in Argentina, in *Pesticides and Wildlife,* Johnston, J. J., Ed., ACS Symposium Series No. 771, American Chemical Society, Washington, D.C., 2001, 294.
24. Burn, J. D. and Leighton, F. A., Further studies of brain cholinesterase: Cholinergic receptor ratios in the diagnosis of acute lethal poisoning of birds by anticholinesterase pesticides, *J. Wildl. Dis.,* 32, 216, 1996.
25. Wilson, B. W., Hooper, M. J., Hanson, M. E., and Nieberg, P. S., Reactivation of organophosphate inhibited AChE with oximes, in *Organophosphates: Chemistry, Fate and Effects,* Chambers, J. E. and Levi, P. E., Eds., Academic Press, Orlando, 1992, 107.
26. Zaccagnini, M. E., S.O.S. Aguiluchos, Revista Campo y Tecnología, INTA Año IV, 20, 1996.
27. Krapovickas, S. and Lyons de Perez, J. A., Swainson's hawk in Argentina: International crisis and cooperation, *World Birdwatch*, 19, 12, 1997.
28. Zaccagnini, M. E., Unpublished interview data, 2000.
29. Uhart, M. and Zaccagnini, M. E., Manual de Procedimientos Operativos Estandarizados de Campo para la Documentación de Incidentes de Mortandad de Fauna Silvestre en Agroecosistemas, INTA, 2000.
30. Zaccagnini, M. E., Aguiluchos: Un proyecto que marca un camino hacia la sostenibilidad agrícola, Revista Campo y Tecnología, INTA Año VI, 60, 1998.
31. Goldstein, M. I., Lacher, T. E., Jr., Zaccagnini, M. E., Parker, M., and Hooper, M. J., Monitoring and assessment of Swainson's hawks in Argentina following restrictions on monocrotophos use, 1996–1997, *Ecotoxicology,* 8, 215, 1999.
32. Zaccagnini, M. E. and Canavelli, S. B., Unpublished internal report, INTA, Buenos Aires, Argentina, 1997.
33. Boletin Oficial, 29.173, First Section, Buenos Aires, 17 June 1999.
34. Hooper, M. J., Argentina cancels monocrotophos, *Pest. Outlook,* 10, 174, 1999.
35. Henny, C. J., Mineau, P., Elliott, J. E., and Woodbridge, B., Raptor poisonings and current insecticide use: What do isolated kill reports mean to populations?, in *Proc. 22nd Int. Ornithol. Congr.,* Adams, N. and Slotow, R., Eds., University of Natal, Durban, South Africa, 1998, 1020.
36. Sauer, J. R., Hines, J. E., and Fallon, J., The North American Breeding Bird Survey, Results and Analysis 1966–2000, Version 2001.2, USGS Patuxent Wildlife Research Center, Laurel, 2001.
37. Basili, G. D. and Temple, S. A., Winter ecology, behavior, and conservation needs of dickcissels in Venezuela, *Stud. Avian Biol.,* 19, 289, 1999.
38. Basili, G. D. and Temple, S. A., Dickcissels and crop damage in Venezuela: Defining the problem with ecological models, *Ecol. Appl.,* 9, 732, 1999.
39. Avery, M. L., Tillman, E. A., and Laukert, C. C., Evaluation of chemical repellents for reducing crop damage by dickcissels in Venezuela, *Int. J. Pest. Manage.*, 47, 311, 2001.
40. Wesseling, C., McConnell, R., Partanen, T., and Hogstedt, C., Agricultural pesticide use in developing countries: Health effects and research needs, *Int. J. Health Serv.,* 27, 273, 1997.
41. Mineau, P., Fletcher, M. R., Glazer, L. C., Thomas, N. J., Brassard, C., Wilson, L. K., Elliott, J. E., Lyon, L. A., Henny, C. J., Bollinger, T., and Porter S. L., Poisoning of raptors with organophosphorous pesticides with emphasis on Canada, U.S. and U.K., *J. Raptor Res.,* 33, 1, 1999.
42. Hart A., Balluf, D., Barfknecht, R., Chapman, P. F., Hawkes, T., Joermann, G., Leopold, A., and Luttik, R., *Avian Effects Assessment: A Framework for Contaminants Studies*, SETAC Press, Pensacola, 2001.

43. DeGraaf, R. M. and Rappole, J. H., *Neotropical Migratory Birds: Natural History, Distribution and Population Change*, Comstock Publishing Associates, Ithaca, 1995.
44. Pool, A. and Gill, G., *The Birds of North America,* The Academy of Natural Sciences, Philadelphia, and the American Ornithologists' Union, Washington, D.C., 1997.
45. Hudson, R. H., Tucker, R. K., and Haegle, M. A., Handbook of Toxicity of Pesticides to Wildlife, 2nd ed., Resour. Publ. 153, Fish and Wildlife Service, U.S. Department of Interior, Washington, D.C., 1984.
46. Anon., Azodrin Insecticide/Quail Toxicity Field Study, Shell Report, 7 November 1968, referenced in Guth, J. A., Monocrotophos – Environmental fate and toxicity, *Rev. Environ. Contam. Toxicol.,* 139, 75, 1994.
47. Tucker, R. K. and Crabtree, D. G., Handbook of Toxicity of Pesticides to Wildlife, Resour. Publ. 84, Fish and Wildlife Service, U.S. Department of Interior, U.S. Government Printing Office, Washington, D.C., 1970.
48. Wiemeyer, S. N. and Sparling, D. W., Acute toxicity of four anticholinesterase insecticides to American kestrels, eastern screech owls and northern bobwhites, *Environ. Toxicol. Chem.,* 10, 1139, 1991.
49. Schafer, E. W., Jr., The acute oral toxicity of 369 pesticidal, pharmaceutical and other chemicals to wild birds, *Toxicol. Appl. Pharmacol.*, 21, 315, 1972.
50. Burford, R. G. and Chappel, C., The acute oral toxicity of organophosphate compounds in the finch, Bio-Research Laboratory, Research Report 6708, 1967, referenced in Guth, J.A., Monocrotophos — environmental fate and toxicity, *Rev. Environ. Contam. Toxicol.,* 139, 75, 1994.
51. Schafer, E. W., Jr., Brunton, R. B., Lockyer, N. F., and DeGrazio, J. W, Comparative toxicity of seventeen pesticides to the quelea, house sparrow, and red-winged blackbird, *Toxicol. Appl. Pharmacol.,* 26, 154, 1973.
52. Hanson, R. E. and Kodama, J. K., Toxicological Studies of Azodrin Insecticide in the Pheasant. I, I. Shell-Report M-81–67, 1968, referenced in Guth, J. A., Monocrotophos — Environmental fate and toxicity, *Rev. Environ. Contam. Toxicol.,* 139, 75, 1994.
53. Anon., Azodrin insecticide wildlife investigations in California, Calif. Dept. Fish and Game, Pesticides Investigation Project, December, 1967.
54. Schafer, E. W., Jr. and Brunton, R. B., Indicator bird species for toxicity determinations: Is the technique usable in test method development?, In *Vertebrate Pest Control and Management Materials*, Beck, J. R., Ed., ASTM STP 680, 1979, 157.
55. Brown, V. K. H., Toxicity Studies on the Insecticide Azodrin: Acute Toxicity to Birds, Shell Report TLGR.0032.70, 1970, referenced in Guth, J. A., Monocrotophos — Environmental fate and toxicity, *Rev. Environ. Contam. Toxicol.,* 139, 75, 1994.
56. Lee, J. and Weir, P., Summarization of Toxicity Studies with Azodrin Insecticide Using One Year Old Japanese Quail (*Coturnix coturnix japonicus*), Stanford Research Institute Report No. 48, in Kodama, J. K. and Hanson, R. E., Eds., Toxicological Studies of Azodrin Insecticide in the Pheasant. II, Shell Report M-14–68, 1968, referenced in Guth, J. A., Monocrotophos — environmental fate and toxicity, *Rev. Environ. Contam. Toxicol.,* 139, 75, 1994.
57. Sachsse, K. and Ullman, L., Acute Oral $LD_{50}$ of Technical Monocrotophos (C1414) in the Japanese Quail, Ciba Geigy Report of 21 November, 1974, referenced in Guth, J. A., Monocrotophos — Environmental fate and toxicity, *Rev. Environ. Contam. Toxicol.,* 139, 75, 1994.
58. Schafer, E. W., Jr., Bowles, W. A., Jr., and Hurlbut, J., The acute oral toxicity, repellency, and hazard potential of 998 chemicals to one or more species of wild and domestic birds, *Arch. Environ. Contam. Toxicol.*, 12, 355, 1983.
59. Hudson, R. H., Tucker, R. K., and Haegle, M. A., Effect of age on sensitivity: Acute oral toxicity of 14 pesticides to mallard ducks of several ages, *Toxicol. Appl. Pharm.*, 22, 556, 1972.
60. Sachsse, K. and Ullman, L., Acute $LD_{50}$ of Technical Monocrotophos (C1414) in the Adult Peking Duck, Ciba Geigy Report of 15 April, 1975, referenced in Guth, J. A., Monocrotophos — Environmental fate and toxicity, *Rev. Environ. Contam. Toxicol.,* 139, 75, 1994.
61. Goldstein, M. I., Toxicological Assessment of a Neotropical Migrant on its Non-Breeding Grounds: Case Study of the Swainson's Hawk in Argentina, M.S. thesis, Clemson University, Clemson, SC, 1997.
62. Hill, E. F., Brain cholinesterase activity of apparently normal birds, *J. Wildl. Dis.,* 24, 54, 1988.
63. Henny, C. J., Blus, L. J., Kolbe, E. J., and Fitzner, R. E., Organophosphate insecticide (famphur) topically applied to cattle kills magpies and hawks, *J. Wildl. Manage.*, 49, 648, 1985.

64. Hooper, M. J., Mineau, P., Zaccagnini, M. E., Winegrad, G. W., and Woodbridge, B., Monocrotophos and the Swainson's hawk, *Pest. Outlook*, 10, 97, 1999.

CHAPTER 26

# Effects of Mining Lead on Birds: A Case History at Coeur d'Alene Basin, Idaho

**Charles J. Henny**

## CONTENTS



## 26.1 INTRODUCTION

Mining began in northern Idaho in the 1880s after the discovery of deposits of lead, silver, gold, and other metals.[1] Approximately 75 million tons of metal-enriched sediments associated with mining activities in upstream locations of the Coeur d'Alene (CDA) River Basin have been deposited in the lakebed of Lake Coeur d'Alene.[2] The flood plain, associated wetlands, and lateral lakes of the CDA River contain large volumes of trace-element-rich material. The process of metal enrichment in the flood plain and lake occurs via the CDA River, which transports contaminated material released by mining activity in the upper basin downstream to these areas. Mining activities were sharply curtailed during the 1980s. Average monthly emissions of lead from the main smelter stack were estimated at 8.3 to 11.7 metric tons from 1955 to 1973.[3] Even as recently as 1973, daily discharges of metals into the South Fork included 4400 kg of zinc, 245 kg of lead, and 57 kg of arsenic.[1] From the early 1970s to 1986, levels of lead in sediment were essentially unchanged.[4] Lead concentrations in sediment were generally in the 3000–5000 μg/g dry weight (dw) range in

1-56670-546-0/03/$0.00+$1.50

most lateral lakes during the mid-1990s.[5] Note that all metal concentrations in this report are wet weight (ww) unless stated otherwise.

Waterfowl mortality from lead that originated from mining and smelting activities has occurred in the CDA River Basin since at least the early 1900s.[6] Waterfowl found dead or collected in the area in 1955 had high levels of lead, zinc, and copper in their tissues; soils, plants, and water also were heavily contaminated.[6] Lead toxicosis in North American birds is usually caused by ingestion of spent lead shotgun pellets.[7] However, in the CDA River Basin, environmental lead in sediment, soils, and biota were believed responsible for most lead poisoning of tundra swans (*Cygnus columbianus*).[8] Exposure to lead has resulted in reported die-offs of tundra swans (and other waterfowl) in 1974,[9] 1982,[10] 1984 and 1985,[11] 1987 to 1990,[8] and annual die-offs continue.[12,13] Livers of hunter-killed ducks in the area also contained high lead concentrations.[11,14] From 1987 to 1995, the Idaho Department of Fish and Game collected data on the incidence of ingested lead shot (another source of lead) in hunter-killed mallards (*Anas platyrhynchos*) from the CDA River Basin. The annual incidence of lead shot varied with the following percentages by 3-year periods: 1987–89, 24.1% (n = 224); 1990–92, 18.8% (n = 192); and 1993–95, 19.3% (n = 181).[15] This area was designated a steel shot hunting area in 1986, and 21 of 70 mallards (30.0%) shot in 1987 contained ingested lead shot.[14] The incidence of lead-shot ingestion may be decreasing, but the percentages were not significantly different.[15]

## 26.2 APPROACHES TO EVALUATE EFFECTS OF LEAD ON BIRDS

### 26.2.1 Diagnosis of "Found Dead" Birds

When dead birds were found, a necropsy was performed to determine cause of death. The necropsy protocol at the National Wildlife Health Center in Madison, WI, was designed to confirm the presence or absence of lesions associated with lead poisoning. Other studies[16–18] have shown that emaciation, green-stained perianal plumage, esophageal and proventricular impaction, bile-stained and fractured gizzard lining, enlarged gallbladder, hepatic hemosiderosis, renal degeneration with nuclear inclusion bodies, myocardial necrosis, fibrinoid necrosis of arteries, erythroid hyperplasia, and peripheral neuropathy are associated with lead poisoning of waterfowl. Some of these lesions are more specific (renal inclusion bodies, fibrinoid necrosis) than others (emaciation). A diagnosis of "lead poisoning" in waterfowl found dead or moribund was based on the absence of other apparent causes of morbidity or death and an unequivocally diagnostic combination of one or more of the following criteria: (1) a history compatible with lead poisoning, (2) gross or histopathologic lesions compatible with lead poisoning, and (3) concentrations of lead in samples of liver tissue that ranged from 3.9 to 90 μg/g. A liver lead concentration greater than 2 μg/g indicates a significant environmental exposure; concentrations greater than 6–8 μg/g indicate toxic exposure and are expected in waterfowl that succumb to lead poisoning.[18,19] In the absence of pathology information, 10 μg/g has been suggested as a reliable threshold concentration for diagnosing lead poisoning.[20]

### 26.2.2 Source of Lead (Lead Shot vs. Mining Activity Lead)

From the introduction, it becomes clear that two sources of lead were present in the CDA River Basin: (1) lead from hunters' spent shot (from shotguns) and (2) lead from the mining and smelting industry in the basin. Spent shot ingested by waterfowl, presumably as grit or food particles, has resulted in widespread waterfowl mortality in the United States, Europe, and elsewhere for over a century.[19] Migratory waterfowl in the CDA River Basin with ingested spent shot may have obtained it within the basin or elsewhere, and although the use of lead shot was banned in the basin in 1986, birds containing lead shot were still present during the basin studies summarized here. Although lead shot does not always persist in the gizzard of mallards that die from lead-shot poisoning, nearly 90% retain lead shot at the time of death,[21] compared to only 1 of 8 (12.5%) lead-poisoned mallards found dead in the CDA River Basin.[15] Similarly, for lead-poisoned Canada geese (*Branta canadensis*) found dead, 75–94% contained ingested lead shot when mortality was due to ingested lead shot, while in the CDA River Basin only 6 of 38 (15.8%) contained ingested lead shot.[15] And for tundra swans dying of lead poisoning outside northern Idaho, 95% contained lead shot, while only 5 of 64 (7.8%) contained lead shot from the CDA River Basin.[8,22] The percentage of lead-poisoned birds with lead shot in their gizzards from the CDA River Basin was much lower than found outside the influence of the mining and smelting area, which implies that lead from sources other than lead shot killed most of the ducks, geese, and swans. Thus, a careful evaluation of gizzards for the presence or absence of lead shot (from both collected and found dead birds) was very important for separating lead sources.

### 26.2.3 Evaluation of Living Birds

Live-captured and euthanized birds were initially evaluated based on suggested liver lead criteria (μg/g) of Pain[19] i.e., < 2, background; 2–5.9, subclinical poisoning; 6–15, clinical poisoning; and > 15, severe clinical poisoning. Then, other diagnostic criteria mentioned above for "found dead" birds were evaluated for a final necropsy diagnosis. The necropsy diagnosis "sublethal lead poisoning" for live birds collected was similar to "lead poisoning" for dead or moribund birds, with the difference being that birds were alive when collected. Free-ranging waterfowl that succumb to lead poisoning generally have some combination of the preceding lesions in association with concentrations of lead in the liver ≥6 μg/g.

The live-bird studies, when few birds were collected, basically compared blood lead residues and hematological findings from the CDA River Basin populations (lead from spent shot and sediment) with data from reference areas nearby (lead from spent shot, but no lead or low lead from sediment). All of the bird species studied migrate, but adults (except tundra swans) were in the area for several months nesting before being captured or collected. The potential effects of lead were evaluated by lead concentrations in blood and several blood measurements including δ-aminolevulinic acid dehydratase (ALAD), protoporphyrin, hematocrit, and hemoglobin. Young of different sizes (ages) were captured to evaluate lead accumulation and its effects on blood constituents. ALAD activity in blood is inhibited in the presence of lead, and baseline values were determined from reference areas for comparison with the CDA River Basin. Inhibition of ALAD, an essential enzyme for heme synthesis, has been accepted as a standard bioassay to detect lead exposure in birds.[23–27] In laboratory studies, ALAD inhibition of 80% was associated with other physiological effects, including decreased hemoglobin concentrations and hematocrits in adult birds and precocial young.[25–28] Some live birds were collected to evaluate lead relationships in the liver, kidney, and blood as well as associated histopathology, but data are not summarized here. Lead concentrations in the contents of gizzards and lower gastrointestinal tracts of the euthanized birds were also determined.

### 26.2.4 The Sediment Ingestion Link

With the lead shot source clarified (most birds dying from lead poisoning were not exposed to lead shot), the issue of lead from mining and smelting sources required attention. Inorganic lead in soils and sediment is not readily available to plants. Elias et al.[29] demonstrated that concentrations of lead tend to be lower in small mammals than in their diets, and they suggested that lead concentrations decrease at each successive trophic level in a food chain. Although millions of waterfowl have been killed by lead, the source has generally been lead shot and fishing sinkers[30] rather than contaminated biota or sediments, as reported in the CDA River Basin. Beyer et al.[31] first investigated whether the lead ingested by wood ducks (*Aix sponsa*) from the CDA River Basin was correlated with elements such as aluminum, which are associated with sediments. A good correlation would indicate that lead could be traced directly to sediments ingested incidentally to feeding, rather than plants or invertebrates in the diet. Wood ducks were collected, and digesta from the intestines was analyzed. The acid-insoluble ash content of the digesta was measured, and the values were used to estimate the sediment content of the diet.[32] Wood ducks feeding on aquatic plants in the CDA River Basin and in the reference area ingested relatively little sediment, estimated at less than 2% of the diet (dw).[31] Digesta with little or no sediment also contained low lead concentrations. Beyer et al.[31] concluded that the lead ingested by the wood ducks was derived mainly from sediment. The unexpected result of their study was that even though wood ducks ingested less than 2% sediment in their diets, the sediment was the main source of lead to the wood ducks. Beyer et al.[31] noted that exposure to lead would be expected to be greatest in species like tundra swans that eat tubers and other items associated with sediments. Thus, the parameter insoluble ash content of digesta can be used to estimate lead exposure and related lead concentrations in blood or liver to sediment ingestion.

### 26.2.5 Laboratory Studies

Laboratory studies were also conducted with sediment from the CDA River Basin fed to mallards, Canada geese, and mute swans (*Cygnus olor*) to determine, under laboratory conditions, whether lead from lead-contaminated sediments was bioavailable and toxic to the same or similar species when consumed as part of their diet. The same blood parameters evaluated in the field were evaluated during the laboratory studies.

## 26.3 LEAD EFFECTS ON BIRDS

### 26.3.1 Wild Populations Studied

Perhaps the most useful approach to evaluate lead exposure and effects is to compare residue concentrations and associated blood measurements for a series of living birds sampled in the CDA River Basin between 1986 and 1995 (Table 26.1.) — a period when little or no change occurred in the sediment lead concentrations. The tabular data contrasts information from reference areas with information collected in the CDA River Basin and refers to nesting species, with the exception of tundra swans. The swans stage in the area for 3–5 weeks in the spring before flying north to nest.[6]

The birds of prey (osprey [*Pandion haliaetus*], northern harrier [*Circus cyaneus*], and American kestrel [*Falco sparverius*]) consistently contained the lowest blood lead concentrations in the CDA River Basin, and a limited number of great horned owl (*Bubo virginianus*), western screech-owl (*Otus kennicotti*), and red-tailed hawk (*Buteo jamaicensis*) nestlings and adults also showed low concentrations.[33,34] The studies in the CDA River Basin support the hypothesis that concentrations of heavy metals, such as lead, decline with increasing trophic level, as reported in marine food chains and in detritus-based and grazing food chains in dredge-spoil pond ecosystems.[35,36] Unlike organochlorine pesticides and mercury, lead does not biomagnify. Although lead concentrations in aquatic and terrestrial vertebrates tend to increase with age (Table 26.1), distribution is localized in hard tissues (i.e., bones and teeth).[37]

Honda et al.[38] reported that most lead in poisoned swans was stored in bones (71.4%), followed by muscle (7.86%), liver (6.19%), and kidney (2.38%). By recognizing that most raptors cast pellets (indigestible feathers, fur, and especially bones in owls),[39] and that most lead concentrates in bones, it was anticipated that lead would be less available to raptors. Ospreys feed almost exclusively on fish. For fish, the percentage of lead located in bones was not available, but lower lead concentrations were reported in fish fillets (no bones) than in whole carcasses of the same species.[33] The other raptors have diversified diets, but small mammals like voles and mice were prominently listed for

**Table 26.1 Concentrations of Lead (geo. mean, µg/g ww) in Blood and Associated Mean Hematological Parameters from Birds in the CDA River System and Reference Areas**

| | Blood Lead | | ALAD (Units) | | Hemoglobin (g/dl) | | Protoprophyrin (µg/dl) | | Hematocrit (%) | | |
|---|---|---|---|---|---|---|---|---|---|---|---|
| **Species (N,N)** | **CDA** | **REF** | **CDA** | **REF** | **CDA** | **REF** | **CDA** | **REF** | **CDA** | **REF** | **References[a]** |
| | | | | | **Osprey** | | | | | | |
| Nestling (53,10) | 0.09 | 0.02 | 102.6 | 215.1 | (10.0)[b] | (9.2) | (73.7) | (59.8) | (33.6) | (32.2) | 33 |
| Adult (21,10) | 0.20 | 0.04 | 58.6 | 162.7 | (15.3) | (16.0) | (91.1) | (71.0) | (46.5) | (48.0) | 33 |
| | | | | | **Northern Harrier** | | | | | | |
| Nestling (41,8) | (0.07) | (0.04) | 399.0 | 612.6 | (10.4) | (10.4) | 51.1 | 95.1 | (32.8) | (33.8) | 34 |
| | | | | | **American Kestrel** | | | | | | |
| Nestling (30, 22) | 0.24 | 0.09 | 256.6 | 566.1 | 10.1 | 11.0 | (53.6) | (53.1) | 34.4 | 36.5 | 34 |
| Adult (3, 6) | (0.46) | (0.25) | 58.3 | 314.2 | 13.6 | 14.7 | (70.6) | (46.7) | (47.0) | (44.8) | 34 |
| | | | | | **Canada Goose** | | | | | | |
| HY < 1000g (16, 7) | 0.30 | 0.03 | 86.2 | 237.1 | (12.5) | (13.3) | 202.5 | 13.3 | (36.3) | (39.1) | 15 |
| HY 1000–2000g (14, 18) | 0.26 | 0.01 | 73.8 | 246.0 | (12.6) | (13.4) | 168.3 | 24.5 | 37.8 | 40.4 | 15 |
| HY 2100–3000g (19, 12) | 0.27 | 0.02 | 54.9 | 260.3 | 13.7 | 15.1 | (50.8) | (38.6) | 39.1 | 43.0 | 15 |
| HY ≥ 3100g (9, 7) | 0.34 | 0.01 | 34.8 | 249.7 | 13.7 | 15.5 | (51.0) | (25.7) | (40.2) | (42.6) | 15 |
| Adult (12, 4) | 0.41 | 0.02 | 32.3 | 183.0 | (16.5) | (15.5) | (44.2) | (22.5) | (42.2) | (43.4) | 15 |
| | | | | | **Mallard** | | | | | | |
| HY ≤ 800g (2, 17) | 0.84 | 0.03 | 13.5 | 296.4 | (14.7) | (13.5) | 99.5 | 32.6 | (44.5) | (40.2) | 15 |
| HY > 800g (16, 9) | 1.00 | 0.02 | 18.6 | 199.7 | (15.0) | (15.2) | 188.2 | 26.2 | (42.9) | (42.2) | 15 |
| Adult (44, 13) | 1.77 | 0.03 | 9.2 | 155.8 | (16.0) | (16.5) | 175.0 | 23.5 | (45.2) | (45.1) | 15 |
| | | | | | **Wood Ducks** | | | | | | |
| Adult (20[c], 12) | 1.5 | 0.20 | 11 | 145.0 | 14 | 16 | 219 | 20 | (48) | (48) | 42 |
| Adult Female[d] (25, 0) | 1.1 | — | 10 | — | (18) | — | 76 | — | (51) | — | 42 |
| Adult Male[e] (42, 0) | 0.9 | — | 9 | — | (17) | — | 60 | — | (50) | — | 42 |
| Adult Female[f] (42, 0) | 1.6 | — | 20 | — | 15 | — | 138 | — | (46) | — | 42 |
| | | | | | **Tundra Swan (Adult)[g]** | | | | | | |
| AH[h] (16, 28) | 0.82 | 0.11 | 29 | 138.0 | (18.1) | (16.8) | — | — | (47.8) | (47.4) | 22 |
| S87[i] (4, 0) | 3.3 | — | 3 | — | 10.9 | — | — | — | 36.3 | — | 22 |
| S94–95[i] (15, 0) | 3.3 | — | 9 | — | 11.8 | — | — | — | (41.8) | — | 22 |

*Note:* CDA = Coeur d'Alene, REF = Reference Area, HY = Hatch Year, AH = Apparently Healthy, S = Sick.

[a] Source: see References Cited.
[b] Values in parenthesis ( ), not statistically significant (P > 0.05).
[c] Females 1986 (16 May–20 June).
[d] 1987 (10 April–4 May).
[e] 1987 (7 April–4 May).
[f] 1987 (12 May–17 June).
[g] All at least 6 months old.
[h] CDA in 1987, Reference in 1994–5.
[i] Sick in CDA in 1987 and 1994–5.

northern harriers, American kestrels, and red-tailed hawks.[40] Cast pellets found at a great horned-owl nest along the CDA River included 38 voles.[34] Therefore, mice and voles were collected in the CDA River Basin for lead analysis.

Two pools of *Peromyscus maniculatus* collected at the same site were analyzed.[34] One pool was analyzed intact, and mice in the other pool were dissected and separated into two categories: (1) bones and (2) remainder of carcass, including internal organs. The intact mice contained 45.3 μg/g lead. Bones from the second pool contained 195 μg/g lead, and the remainder of the carcass < 0.10 μg/g. Overall, the second pool (based on weight of bones and remainder of carcass) contained an estimated 59.2 μg/g. Nearly all of the lead was in the bones and, therefore, less available for absorption by species casting bones in their pellets. Jenkins[41] also noted high concentrations of lead in hair, which is also cast in pellets. Osprey, in addition to not eating large bones, also cast pellets like other raptors.

The blood parameters for the Osprey, the largest raptor dataset, showed significantly increased lead concentrations in blood from the CDA River Basin as well as significantly reduced ALAD. However, the blood concentrations were still comparatively low, and mean protoporphyrin, hematocrit and hemoglobin were not significantly changed from the reference areas. The American kestrel and northern harrier also showed significantly reduced ALAD in the CDA River basin. The nestling kestrels, which were sampled in good numbers, also showed significantly reduced hemoglobin (–8.2%) and hematocrits (–5.8%) in the CDA River Basin, which were associated with about twice the blood lead concentrations of the osprey (geo. mean 0.24 vs. 0.09 μg/g). No lead-related raptor deaths were recorded. Ospreys produced young at nearly identical rates in the reference area and the CDA River Basin; these rates were among the highest ever reported in the western United States.[33] Although the number of young per nesting attempt for American kestrels was slightly lower along the CDA River Basin (1.95 young; N = 20 vs. 2.40 young, N = 10), no significant relationship was found between lead and productivity to fledgling.

Of the ducks, geese and swans studied in the CDA River Basin, Canada geese were the least contaminated with lead (Table 26.1). However, all four waterfowl species showed significant and substantial ALAD inhibition: Canada goose, –63.6 to –86.1%, mallard –90.7 to –95.5%, wood duck –86.2 to –93.8%, and the tundra swan –79.0% (but sick birds –93.5% to –97.8%). In spite of the lower lead concentrations in blood, Canada geese goslings showed significantly elevated protoporphyrin in the two smaller age classes, significantly reduced hemoglobin in the two larger age classes, and significantly reduced hematocrits in the two middle age classes.[15] None of these blood parameters in the adult age class differed significantly from reference geese. The mallards, with much higher blood lead concentrations than Canada geese, had elevated protoporphyrin in the CDA River Basin but no significant changes in hemoglobin or hematocrit. Recognizing that interspecific differences exist in response and sensitivity to lead, it appears (at least for hemoglobin and hematocrit) that Canada geese were more sensitive to lead than mallards, i.e., adverse hematologic effects occur at lower blood lead concentrations.

Adult wood duck blood lead concentrations (only age class sampled) were similar to adult mallards, and like the mallards, protoporphyrin was significantly elevated and hematocrit was not significantly different from the reference areas. However, hemoglobin, unlike in mallards, was significantly reduced in female wood ducks sampled later in the season in both 1986 and 1987. It is of interest that these two groups of female wood ducks had higher geometric mean lead concentrations (1.5 and 1.6 μg/g) than those sampled earlier in the season (0.9 and 1.1 μg/g) when no change in hemoglobin was detected. Thus, the amount of time spent in the area and exposed to lead by the migratory wood duck, and probably other species, seems to be a factor.

Lead concentrations in blood of a small series of Canada goose goslings and mallards euthanized during the study were compared with lead concentrations in blood of others bled but not euthanized and showed good agreement.[15] This suggests that inferences could be made for the local populations based on the additional information available from organ residues and histopathology. The series of 22 live mallards collected (10 hatch years [HYs] and 12 adults) between 2–10 August 1994 provided an assessment of the living mallard population nesting in the area. Significantly reduced hematocrits was found in 4 of the 12 (33%) adults and 1 of 10 (10%) HYs, which was similar to 18% and 6%, respectively, for the complete series bled in the CDA River Basin. Likewise, 2 of 12 adults (17%), and 0 of 10 HYs (0%) had reduced hemoglobin concentrations, which compared favorably with the 15% and 0% for the whole series. The series of live mallards captured and necropsied revealed that 5 of 12 adults (42%) and 1 of 10 HYs (10%) were clinically lead poisoned (based upon published classifications[19]). By eliminating the three mallards from the series with lead shot in their gizzards (a different source of lead), none of the HYs, but 36% of the adults, were clinically lead poisoned. Subclinical lead poisoning was found in 50% of the HY mallards and 55% of the adult mallards (again excluding those with ingested lead shot). This prevalence of mallard lead poisoning causes concern for the local breeding population. Again, the presence of lead shot in mallard gizzards in the CDA River Basin is not unexpected but does tend to complicate (but not thwart) the evaluation of lead from the CDA River Basin sediment.

This study provided evidence that breeding mallards in the CDA River Basin accumulated lead, and adults as well as young (only a few months old) accumulated enough lead to adversely affect a series of blood biomarkers responsive to lead. Furthermore, there was no evidence for a decrease in blood lead concentrations in adult mallards between 1987 and 1995.[15] Dead Canada geese and mallards were difficult to locate in the CDA River Basin marshes, but of the lead-poisoned mallards and Canada geese found, few contained ingested lead shot. The incidence rate of ingested lead shot in either euthanized or found dead mallards and Canada geese in the CDA River Basin does not implicate spent lead shot as the principal source of the lead poisoning. Evidence for sediment ingestion in both mallards and Canada geese was provided from fecal data.[20] Therefore, lead-contaminated sediment associated with upstream mining in the CDA River Basin was the lead source responsible for most of the mallard and Canada goose lead poisoning during this study.

The tundra swan only stays in the CDA River Basin for 3–5 weeks in the spring, before continuing the spring migration north to nest in Canada and Alaska. However, swans were routinely found sick or dead during this short time period. This highly visible large white bird is much more easily located than smaller ducks or more neutral colored Canada geese. The “apparently healthy” swans sampled in the CDA River Basin showed significantly higher blood lead concentrations and significantly reduced ALAD when compared to the reference swans, but no significant difference in hemoglobin or hematocrit. Some of the swans sampled may have been in the basin for only a few days. Blus et al.[8] reported that blood lead increased from 0.60 to 0.71 μg/g over 2 days in one swan and from 0.68 to 2.28 μg/g over 5 days in a second swan, and overall blood lead concentrations increased significantly with time. The 19 sick swans bled showed much higher lead concentrations with significant reduction in ALAD, hemoglobin, and hematocrit. These swans were all near death at the time they were bled. The sampling of live swans showed that exposure to and accumulation of lead was rapid and that moribund swans had blood parameters that deteriorated to a critical stage.

Several other species of birds were studied to a lesser extent in 1987 including American robins (*Turdus migratorius*) and tree swallows (*Tachycineta bicolor*). Seven American robin nestlings collected from nests (one young each) in the CDA River Basin contained elevated concentrations of lead in their blood (geo. mean 0.49, range 0.27–0.87 μg/g) and liver (geo. mean 1.19, range 0.19–5.6 μg/g), while young at 11 tree-swallow nests (pooled blood from two young at each nest) contained lower concentrations of lead in their blood (geo. mean 0.19, range no detection to 0.75 μg/g) and liver (geo. mean 0.15, range no detection to 0.40 μg/g).[43] American robins leave the nest at 14–16 days and tree swallows at 16–24 days,[44] and since lead does not transfer appreciably into eggs, it was accumulated to potentially dangerous concentrations in some individuals of both species in about 2–3 weeks. The lead concentrations in the blood of robin nestlings were higher than in any of the raptor species studied. In the spring and early summer, robins eat primarily animal matter including many types associated with sediment (earthworms, beetles, weevils).[45] Adult American robins and song sparrows (*Melospiza melodia*) were also studied in the CDA River Basin in 1995.[46] Adult American robins and song sparrows in the CDA River Basin had significantly reduced ALAD compared to those from reference areas, but no significant differences in hematocrit were found.

### 26.3.2 Field and Laboratory Test Comparisons

The Canada goose and mallard field[15] and laboratory[47–50] studies are summarized in Table 26.2. The CDA River Basin Canada goose goslings (2100–3000 g, i.e., same size as in laboratory study) contained lower lead concentrations than Canada geese in the laboratory exposed to 12% CDA River Basin sediment and generally showed less adverse ALAD and protoporphyrin responses than reported in the laboratory, although protoporphyrin was extremely elevated in younger goslings (≤2000 g) in the field. The reduction of hemoglobin and hematocrit in the field was statistically significant and more severe than in the lowest laboratory treatment. With the Canada geese in the laboratory, the responses of most blood parameters were more adverse as the percentage of CDA River Basin sediment increased in the diet, although ALAD inhibition was already in excess of 95% at the lowest laboratory treatment level. It is noteworthy that two of nine goslings died at the highest dietary treatment rate (48% CDA River Basin sediment). Additional types of information were available from the laboratory study. With the 48% CDA River Basin sediment diet, plasma enzymatic activity for lactate dehydrogenase (LDH-L) increased by about 1.7X, plasma total protein was significantly lower (30–38%), and one gosling exhibited renal tubular nephrosis.[48]

HY mallards in the CDA River Basin field study contained higher concentrations of lead in their blood and liver than HY Canada geese from the same area (Table 26.2) and exhibited a greater increase in protoporphyrin and a greater inhibition of ALAD. However, Canada geese in the field with the lower lead concentrations showed significant reductions in hemoglobin and hematocrit, whereas mallards in the field with higher concentrations of blood lead showed no significant reductions in hemoglobin or hematocrit.

The protoporphyrin increase in HY mallards with a 12% CDA River Basin sediment diet was only 3.7X compared to 5.9X in the field with lower blood lead concentrations in the field, and a similar pattern was shown for adult mallards (Table 26.2). Otherwise, lead values and protoporphyrin responded in a dose-response pattern. Again, ALAD reached in excess of 90% inhibition in the field study, and higher lead concentrations in the laboratory study showed little additional influence. Hemoglobin and hematocrit in mallard ducklings, in contrast with Canada geese, did not show a significant adverse effect, and none of the ducklings died during the laboratory study. Corn in the laboratory diet resulted in higher blood and kidney lead concentrations, as well as a more severe elevation of protoporphyrin, compared to the same percentage sediment in the standard diet. These findings suggest that the more severe protoporphyrin effects in the field with lower lead concentrations may have resulted from a less-than-optimum diet in the field when compared to the standard laboratory diet. The laboratory findings also indicated that duckling brain weight was 7% lower with the 24% CDA River Basin sediment diet, and LDH-L in the 12% CDA River Basin sediment was 24.6% lower than in the clean sediment diet.[49]

The 12% and 24% CDA River Basin sediment fed to the Canada goose goslings and mallards in the laboratory studies seem realistic in view of the blood parameters reported and the fact that Beyer et al.[20] estimated arithmetic means of 11% sediment ingestion for adult Canada geese and 5.9% for adult mallards in the CDA River Basin. The 90$^{th}$ percentile for Canada geese was 23% and for mallards was 16%. These sediment-ingestion estimates were based on acid-soluble ash content in feces, and the latter estimate for mallards assumed a digestibility of 30% for mallard diets.[32] Sediment ingestion is sometimes the principal route of exposure for environmental contaminants like lead that are not readily taken up by plants and invertebrates.[31] The 11% sediment-ingestion rate for Canada geese vs. 5.9% rate for mallards initially seemed to pose a conundrum because HY mallards in the CDA River Basin have much higher blood lead concentrations than HY Canada geese (0.98 vs. 0.28 μg/g). The apparent enigma was solved when Hoffman et al.[48] found that under controlled laboratory conditions Canada goose goslings fed a diet with 12% CDA sediment contained a geometric mean blood lead concentration of 0.68 μg/g, while mallards fed the same percentage sediment in the diet contained 1.41 μg/g (about twice as much), and a similar ratio of lead between species was reported for a diet containing 24% CDA sediment (Canada goose

**Table 26.2 A Comparison of Blood Parameters for CDA River Basin Canada Geese and Mallards and Those Exposed to CDA River Basin Lead in the Laboratory**

| Species | Field/Lab | Age | Geo Mean Lead (μg/g) | | Change from Reference Area or Untreated Control | | | | References[a] |
|---|---|---|---|---|---|---|---|---|---|
| | | | Blood | Liver | Protoporphyrin | ALAD (%) | Hemoglobin (%) | Hematocrit (%) | |
| Canada goose | Field | HY | 0.28 | 0.39 | (1.3X) | −78.0 | −9.3 | −9.1 | 15 |
| | Lab[b] | HY | 0.68 | 2.43 | 4.2X | −96.6 | (−3.7) | (−2.5) | 47, 48 |
| | Lab[c] | HY | 1.61 | 4.49 | 7.2X | −95.8 | (−9.3) | (−7.4) | 47, 48 |
| | Lab[d] | HY | 2.52 | 6.57 | 8.1X | −94.4 | −18.7 | (−10.2) | 47, 48 |
| Mallard | Field | HY | 0.98 | 2.27 | 5.9X | −90.7 to −95.5 | (−1.6) to (+9.5) | (+1.7) to (+10.6) | 15 |
| | Lab[b] | HY | 1.41 | 4.70 | 3.7X | −97.8 | (+0.8) | (+2.6) | 47, 49 |
| | Lab[c] | HY | 2.56 | 7.92 | 6.2X | −97.6 | (−2.4) | (+0.8) | 47, 49 |
| | Lab[e] | HY | 2.45 | 2.45 | 8.8X | −97.4 | (+1.8) | (+0.3) | 47, 49 |
| | Lab[f] | HY | 4.62 | 6.41 | 18.0X | −95.7 | (−3.6) | (−0.3) | 47, 49 |
| Mallard | Field | AD | 1.77 | 4.67 | 7.4X | −94.1 | (−3.4) | (+0.2) | 15 |
| | Lab[g] | AD | 0.99 | 4.4 | 1.9X | −98.3 | (+2.6) | (+4.1) | 50 |
| | Lab[h] | AD | 1.6 | 9.2 | 3.2X | −98.6 | (−1.3) | (0.0) | 50 |
| | Lab[i] | AD | 2.9 | 12 | 6.3X | −98.6 | (−3.2) | (+2.1) | 50 |
| | Lab[j] | AD | 5.8 | 26 | 11.3X | −98.6 | −10.9 | (−0.2) | 50 |

*Note:* CDA = Coeur d'Alene, ( ) not statistically significant from reference area or control ($P > 0.05$). Protoporphyrin X = factor change.

[a] Source: see References Cited.
[b] 12% CDA Sediment in diet (data reflect 6 weeks on diet).
[c] 24% CDA Sediment in diet (data reflect 6 weeks on diet).
[d] 48% CDA Sediment in diet (data reflect 6 weeks on diet).
[e] 12% CDA Sediment in corn diet (data reflect 6 weeks on diet).
[f] 24% CDA Sediment in corn diet (data reflect 6 weeks on diet).
[g] 3% CDA Sediment in pelletized diet (data reflect 10 weeks on diet).
[h] 6% CDA Sediment in pelletized diet (data reflect 10 weeks on diet).
[i] 12% CDA Sediment in pelletized diet (data reflect 10 weeks on diet).
[j] 24% CDA Sediment in pelletized diet (data reflect 10 weeks on diet).

1.61 and mallards 2.56 μg/g). Reasons for accumulation rate differences for lead between the two species remain unknown, but body size and metabolic rate differences may be involved. Thus, sediment-ingestion rates (as determined from feces in the field, or as fed in controlled laboratory studies) may not directly reflect body burdens or blood lead concentrations,[51] although sediment ingestion was believed the most important exposure route.

## 26.4 CONCLUSIONS AND PERSPECTIVES

The CDA River Basin has been the site of considerable research activity on lead during the last 15 years. Diagnostic procedures and techniques developed over 50 years to assess lead poisoning from spent shotgun pellets was used to assess another source of lead (mining activities). The body of waterfowl literature including incidence of lead shot in gizzards of birds "found dead" from lead-shot poisoning was used effectively to separate the two sources of lead (lead shot vs. mining activities) from each other, with the latter being the dominant factor in the CDA River Basin. Canada geese, tundra swans, and mallards continue to die in the CDA River Basin from lead associated with sediment. In addition to the "found dead" birds, live populations were blood-sampled to evaluate lead residues and blood parameters. Waterfowl showed the highest blood residue concentrations and the most adverse effects on blood parameters. It is important to recognize that all species of birds living in the basin were not equally exposed to lead. Beyer et al.[51] concluded that exposure of waterfowl to lead in the CDA River Basin was related to the amount of sediment ingested in the diet and that the relative amount of lead from vegetation and prey was minor. Following the logic of this conclusion (and the fact that raptors do not ingest sediment, and most raptors do not digest bones of prey species [a major storage area for lead]), it becomes clear why the ospreys, hawks, and owls in the CDA River Basin were less contaminated with lead from mining sources. Of the songbirds studied, young American robins, which feed on many invertebrates associated with the soil, were more contaminated than the birds of prey, while the aerial foraging tree swallow was less contaminated than the robin. However, lead concentrations in blood apparently do not represent the complete story because of variability in species sensitivity (e.g., see different hemoglobin and hematocrit responses to lead in blood of Canada geese and mallards [Table 26.2]).

Many of the field and laboratory findings were also used to investigate relationships between sediment lead and dietary lead and then to relate dietary lead to blood lead in waterfowl. This investigation, which was beyond the scope of this review, resulted in a risk assessment related to lead sediment concentrations. However, the lowest effect concentration of sediment lead was estimated as 530 μg/g dw, with some mortality estimated at 1800 μg/g dw.[51] Lead concentrations in sediment are generally in the 3000–5000 μg/g range in most lateral lakes in the CDA River Basin.[5]

## REFERENCES

1. Rabe, F. W. and Flaherty, D. C., The river of green and gold, Idaho Research Foundation, Inc., Natural Resources Series No. 4, Moscow, ID, 1974.
2. Horowitz, A. J., Elrick, K. A., Robbins, J. A., and Cook, R. B., The Effect of Mining and Related Activities on the Sediment-Trace Element Geochemistry of Lake Coeur d'Alene, Idaho, Part II: Subsurface sediments, U.S. Geological Survey, Open-File Report 93-656, 1993.
3. Burrows, G. E., Sharp, J. W., and Root, R. G., A survey of blood lead concentrations in horses in the North Idaho Lead/Silver Belt Area, *Vet. Human Toxicol.*, 23, 328, 1981.
4. Hornig, C. E., Terpening, D. A., and Bogue, M. W., Coeur d'Alene Basin EPA Water Quality Monitoring, 1972–1986, U.S. Environmental Protection Agency, Seattle, WA EPA-910/9-88-216, 1988.
5. Campbell J. K., Audet, D. J., Kern, J. W., Reyes, M., and McDonald, L. L., Metal Contamination of Palustrine and Lacustrine Habitat in the Coeur d'Alene Basin, Idaho, Final Report, U.S. Fish and Wildl. Serv., Spokane, WA, 1999.

6. Chupp, N. R. and Dalke, P. D., Waterfowl mortality in the Coeur d'Alene Valley, Idaho, *J. Wildl. Manage.*, 28, 692, 1964.
7. Anderson, W. L. and Havera, S. P, Blood lead, protoporphyrin, and ingested shot for detecting lead poisoning in waterfowl, *Wildl. Soc. Bull.*, 13, 26, 1985.
8. Blus, L. J., Henny, C. J., Hoffman, D. J., and Grove, R. A., Lead toxicosis in tundra swans near a mining and smelting complex in northern Idaho, *Arch. Environ. Contam. Toxicol.*, 21, 549, 1991.
9. Benson, W. W., Brock, D. W., Gabica, J., and Loomis, M., Swan mortality due to certain heavy metals in the Mission Lake area, Idaho, *Bull. Environ. Contam. Toxicol.*, 15, 171, 1976.
10. Neufeld, J., A Summary of Heavy Metal Contamination in the Lower Coeur d'Alene with Particular Reference to the Coeur d'Alene Wildlife Management Area, Idaho Dept. Fish and Game, Coeur d'Alene, ID, 1987.
11. Krieger, R. I., Toxicity and Bioavailability Studies of Lead and Other Elements in the Lower Coeur d'Alene River, Tech. Bull. 90-3, Bureau Land Management, Boise, ID, 1990.
12. Audet, D. J, Coeur d'Alene Basin Resource Damage Assessment Biological Reconnaissance Investigation, Final Report, U.S. Fish and Wildl. Serv., Spokane, WA, 1997.
13. Audet, D. J., Creekmore, L. H., Sileo, L., Snyder, M. R., Franson, J. C., Smith, M. R., Campbell, J. K., Meteyer, C. U., Locke, L. N., McDonald, L. L., McDonald, T. L., Strickland, D., and Deeds, S., Wildlife Use and Mortality Investigation in the Coeur d'Alene Basin, 1992-97, Final Report, U.S. Fish and Wildl. Serv., Spokane WA, 1999.
14. Casteel, S. W., Nigh, J., Neufeld, J., and Thomas, B. R., Liver lead burden in hunter-killed ducks from the Coeur d'Alene River valley of northern Idaho, *Vet. Human Toxicol.*, 33, 215, 1991.
15. Henny, C. J., Blus, L. J., Hoffman, D. J., Sileo, L., Audet, D. J., and Snyder, M. R., Field evaluation of lead effects on Canada geese and mallards in the Coeur d'Alene River Basin, Idaho, *Arch. Environ. Contam. Toxicol.*, 39, 97, 2000.
16. Wobeser, G. A., Lead and other metals, in *Diseases of Wild Waterfowl,* Wobeser, G. A., Ed., Plenum Press, New York, 1981, pp. 151–159.
17. Friend, M., Lead poisoning, in Field Guide to Wildlife Diseases, Friend, M., Ed., U.S. Fish and Wildl. Serv., Resour. Publ. 167, Washington, D.C., 1987, pp. 175–189.
18. Locke, L. N. and Thomas, N. J., Lead poisoning of waterfowl and raptors, in *Noninfectious Diseases of Wildlife*, Fairbrother, A., Locke, L. N., and Hoff, G. L., Eds., Iowa State University Press, Ames, 1966, pp. 108–117.
19. Pain, D. J., Lead in waterfowl, in *Environmental Contaminants in Wildlife — Interpreting Tissue Concentrations*, Beyer, W. N., Heinz, G. H., and Redmon-Norwood, A. W., Eds., Lewis Publishers, Boca Raton, FL, 1996, pp. 225–264.
20. Beyer, W. N., Audet, D. J., Morton, A., Campbell, J. K., and LeCaptain, L., Lead exposure of waterfowl ingesting Coeur d'Alene River Basin sediments, *J. Environ. Qual.*, 27, 1533, 1998.
21. Bellrose, F. C, Lead poisoning as a mortality factor in waterfowl populations, *Bull. Illinois Nat. Hist. Surv.*, 27, 231, 1959.
22. Blus, L. J., Henny, C. J., Hoffman, D. J., Sileo, L., and Audet, D. J., Persistence of high lead concentrations and associated effects in tundra swans captured near a mining and smelting complex in northern Idaho, *Ecotoxicology*, 8, 125, 1999.
23. Ohi, G., Seki, H., Akiyama, K., and Yagyu, H., The pigeon, a sensor of lead pollution, *Bull. Environ. Contam. Toxicol.*, 12, 92, 1974.
24. Dieter, M. P. and Finley, M. T., δAminolevulinic acid dehydratase enzyme activity in blood, brain, and liver of lead-dosed ducks, *Environ. Res.*, 19, 127, 1979.
25. Hoffman, D. J., Pattee, O. H., Wiemeyer, S. N., and Mulhern, B., Effects of lead shot ingestion on δ-aminolevulinic acid dehydratase activity, hemoglobin concentration, and serum chemistry in bald eagles, *J. Wildl. Dis.*, 17, 423, 1981.
26. Beyer, W. N., Spann, J. W., Sileo, L., and Franson, J. C., Lead poisoning in six captive avian species, *Arch. Environ. Contam. Toxicol.*, 17, 121, 1988.
27. Pain, D. J. and Rattner, B. A., Mortality and hematology associated with the ingestion of one number four shot in black ducks, *Anas rubripes*, *Bull. Environ. Contam. Toxicol.*, 40, 159, 1988.
28. Franson, J. C. and Custer, T. W., Toxicity of dietary lead in young cockerels, *Vet. Human Toxicol.*, 24, 421, 1982.
29. Elias, R., Hirao, Y., and Patterson, C., Impact of present levels of aerosol Pb concentrations on both natural ecosystems and humans, in *Int. Conf. Heavy Metals Environ.*, Vol. 2, Part 1, Hutchinson, T. C., Ed., Institute for Environmental Studies, University of Toronto, 1977, pp. 257–271.

30. U.S. Fish and Wildlife Service, Use of Lead Shot for Hunting Migratory Birds in the United States, Final Environmental Impact Statement, Washington, D.C., FES 86-16, 1986.
31. Beyer, W. N., Blus L. J., Henny, C. J., and Audet, D., The role of sediment ingestion in exposing wood ducks to lead, *Ecotoxicology*, 6, 181, 1997.
32. Beyer, W. N., Connor E. E., and Gerould, S., Estimates of soil ingestion by wildlife, *J. Wildl. Manage.*, 58, 375, 1994.
33. Henny, C. J., Blus, L. J., Hoffman D. J., Grove, R. A., and Hatfield, J. S., Lead accumulation and osprey production near a mining site on the Coeur d'Alene River, Idaho, *Arch. Environ. Contam. Toxicol.*, 21, 415, 1991.
34. Henny, C. J., Blus, L. J., and Grove, R. A., Lead in hawks, falcons and owls downstream from a mining site on the Coeur d'Alene River, Idaho, *Environ. Monit. Assess.*, 29, 267, 1994.
35. International Decade of Oceanographic Exploration (IDOE), Baseline Studies of Pollutants in the Marine Environment and Research Recommendations, Baseline Conference, National Oceanic and Atmospheric Administration, Washington, D.C., 1972.
36. Driftmeyer, J. E. and Odum, W. E., Lead, zinc and manganese, in dredge-spoil pond ecosystems, *Environ. Conserv.*, 2, 39, 1975.
37. Eisler, R., Lead Hazards to Fish, Wildlife and Invertebrates: A Synoptic Review, U.S. Fish Wildl. Serv. Biol. Rep. 85(1.14), 1988.
38. Honda, K., Lee, D. P., and Tatsukawa, R., Lead poisoning in swans in Japan, *Environ. Pollut.*, 65, 209, 1990.
39. Duke, G. E., Jegers, A. A., Loff, G., and Evanson, O. A., Gastric digestion in some raptors, *Comp. Biochem. Physiol.*, 50A, 649, 1975.
40. Sherrod, S. K., Diets of North American falconiformes, *Raptor Res.*, 12, 49, 1978.
41. Jenkins, D. W., Biological Monitoring of Trace Metals, Toxic Trace Metals in Plants and Animals of the World, Vol. 2, Part 2, U.S. Environmental Protection Agency, Rep. 600/3-80-091, 1980, pp. 619-778.
42. Blus, L. J., Henny, C. J., Hoffman, D. J., and Grove, R. A., Accumulation and effects of lead and cadmium on wood ducks near a mining and smelting complex in northern Idaho, *Ecotoxicology*, 2, 139, 1993.
43. Blus, L. J., Henny, C. J., Hoffman, D. J., and Grove, R. A., Accumulation in and effects of lead and cadmium on waterfowl and passerines in northern Idaho, *Environ. Pollut.*, 89, 311, 1995.
44. Harrison, C., *A Field Guide to the Nests, Eggs and Nestlings of North American Birds*, Collins, London, 1978.
45. Martin, A. C., Zim, A. S., and Nelson, A. L., *A Guide to Wildlife Food Habits*, McGraw-Hill, New York, 1951.
46. Johnson, G. D., Audet, D. J., Kern, J. W., LeCaptain, L. J., Strickland, M. D., Hoffman, D. J., and McDonald, L. L., Lead exposure in passerines inhabiting lead-contaminated floodplains in the Coeur d'Alene Basin, Idaho, USA, *Environ. Toxicol. Chem.*, 18, 1190, 1999.
47. Hoffman, D. J., Heinz, G. H., Sileo, L., Audet, D. J., LeCaptain, L. J., and Obrecht, H. H., Toxicity of Lead-Contaminated Sediment to Canada Goose Goslings and Mallard Ducklings, Final Report, Coeur d'Alene Natural Resources Damage Assessment Project, Portland, OR, 1999.
48. Hoffman, D. J., Heinz, G. H., Sileo, L., Audet, D. J., Campbell, J. K., LeCaptain, L. J., and Obrecht, H. H., Developmental toxicity of lead-contaminated sediment in Canada geese (*Branta canadensis*), *J. Toxicol. Environ. Health*, 59A, 235, 2000.
49. Hoffman, D. J., Heinz, G. H., Sileo, L., Audet, D. J., Campbell, J. K., and LeCaptain, L. J., Developmental toxicity of lead-contaminated sediment to mallard ducklings, *Arch. Environ. Contam. Toxicol.*, 39, 221, 2000.
50. Heinz, G. H., Hoffman, D. J., Sileo, L., Audet, D. J., and LeCaptain, L. J., Toxicity of lead-contaminated sediment to mallards, *Arch. Environ. Contam. Toxicol.*, 36, 323, 1999.
51. Beyer, W. N., Audet, D. J., Heinz, G. H., Hoffman, D. J., and Day, D., Relation of waterfowl poisoning to sediment lead concentrations in the Coeur d'Alene River Basin, *Ecotoxicology*, 9, 207, 2000.

CHAPTER 27

# White Phosphorus at Eagle River Flats, Alaska: A Case History of Waterfowl Mortality

**Donald W. Sparling**

## CONTENTS



## 27.1 INTRODUCTION

Eagle River Flats (ERF), Alaska is an 865-ha tidal flat located on the Cook Inlet near Anchorage (Figure 27.1). It is used by waterfowl and shorebirds throughout the spring and summer and is particularly important as a spring and fall staging or stopover area for more than 75 species of migratory ducks, geese, swans, raptors, gulls, shorebirds, and passerines.[1] Massive waterfowl mortalities involving more than 2000 ducks and swans each year were first observed at ERF in 1980,[2] continued unabated through the mid 1990s, and are still occurring, although at substantially lower rates. The most seriously affected species in these die-offs included mallards (*Anas platyrhynchos*), green-winged teal (*A. carolinensis*), pintails (*A. acuta*), greater scaup (*Aythya marila*), and trumpeter (*Cygnus buccinator*) and tundra (*C. columbianus*) swans. Other species, such as widgeon (*Anas americana*), shovelers (*A. clypeata*), snow geese (*Chen caerulescens*), Canada geese

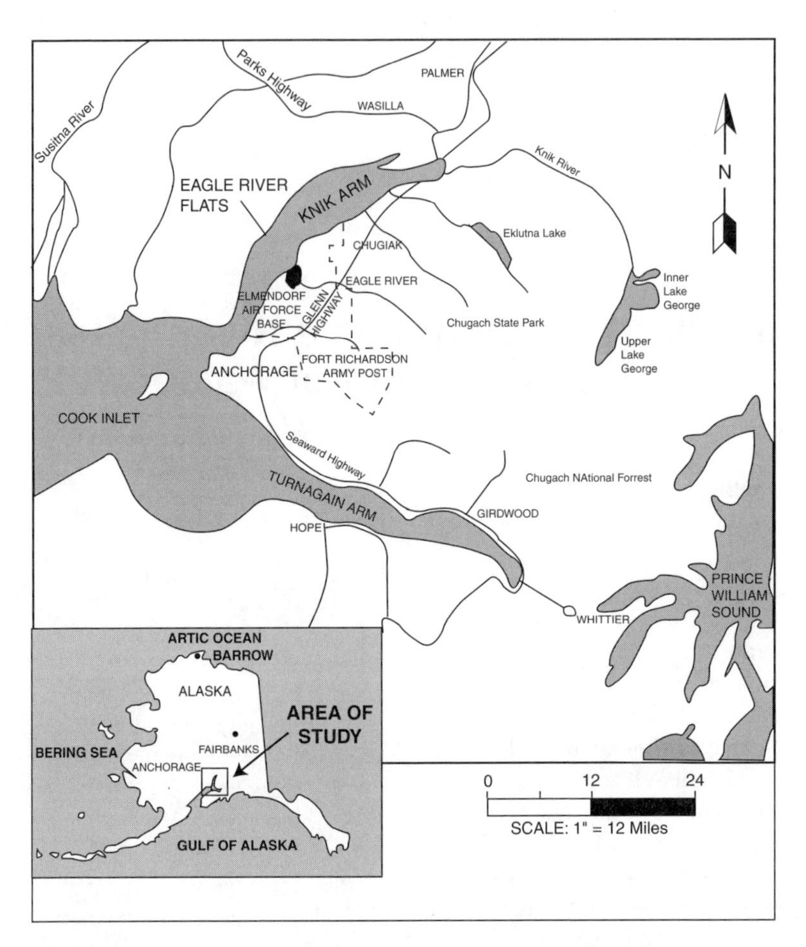


**Figure 27.1** Location of Eagle River Flats on the Knik Arm of the Cook Inlet, Alaska.

(*Branta canadensis*), and several species of shorebirds, utilize the area but experience substantially lower mortality. Avian predators, such as bald eagles (*Haliaeetus leucocephalus*) and gulls (*Larus* sp.), feed on dead and dying waterfowl and sometimes become sick or die (Roebuck et al.[3]). Ten years after the first observation of dead waterfowl, the U.S. Army Corps of Engineers determined that particles of white phosphorus (P4) originating from the firing of munitions into the area were the cause of this mortality.[2,4] Subsequently, teams of scientists and engineers from several agencies have studied the situation using both field and laboratory methods. At present, remediation efforts include dredging and drying areas of high P4 concentration to allow attenuation through vaporization. The site is of particular interest because, unlike most Superfund sites under the Comprehensive Environmental Response, Compensation, and Liability Act (CERCLA), this one has a recognized effect and a known cause. ERF also is a model for the effects and remediation of P4 on at least 71 other sites where the contaminant been found.[5,6] This chapter presents a case history of white-phosphorous-caused waterfowl mortality at ERF. It includes a discussion of the toxicity of white phosphorus, a hazard assessment model, and a description of efforts to remediate the problem.

## 27.2 DESCRIPTION OF EAGLE RIVER FLATS

Eagle River Flats is an estuarine salt marsh located on the southeast side of the Knik Arm of the Cook Inlet (61°'19'N, 149°'44'W). It is separated by a narrow coastal plain from the Chugach

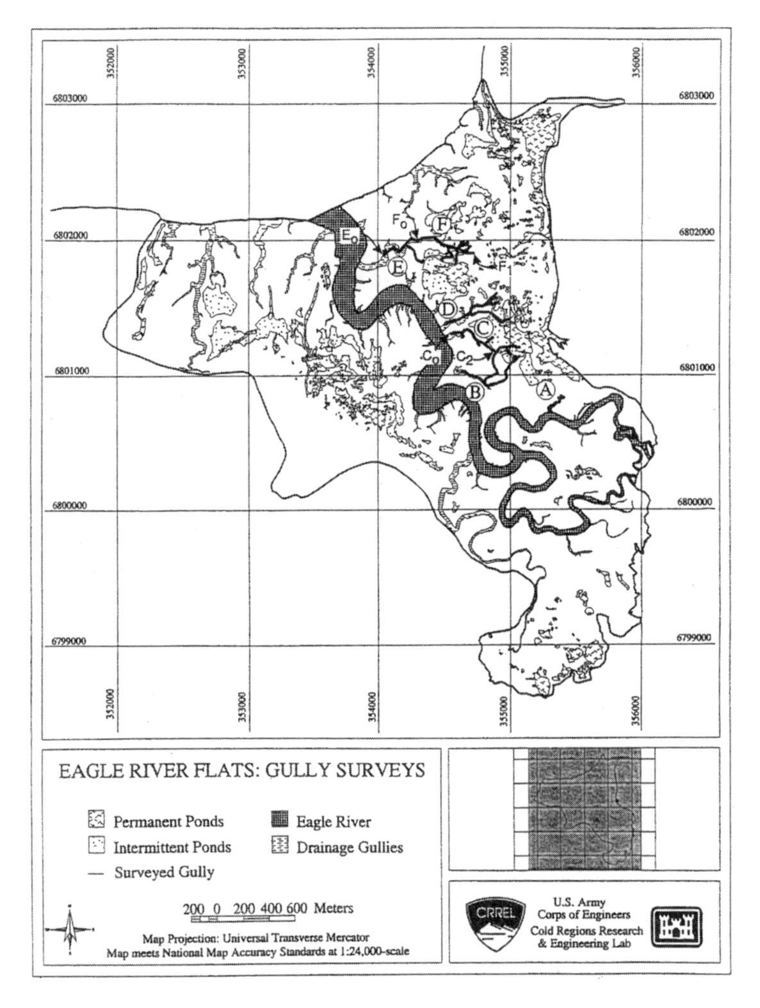


**Figure 27.2** Map of Eagle River Flats, Alaska showing major pools and vegetated areas. (From Racine, C. H. and Brouillette, M., Ecological inventory of Eagle River Flats, U.S. Army, Fort Richardson, AK, 1995.)

Mountains and surrounded on three sides by steep slopes. The Eagle River, a cold, glacial-fed river, flows through the Flats and empties into the Knik Arm (Figure 27.2). The landscape consists of the main river channels, tidal gullies ranging from a few centimeters deep to more than 4 m at low tide, standing pools of a few centimeters to over 1 m deep, open mud flats, and areas of emergent vegetation. The vegetation of ERF is generally characteristic of the Cook Inlet region; 57 species of vascular plants, dominated by sedges and grasses, have been identified.[7] Many of these plants provide habitat or produce seeds that are food for waterfowl and shorebirds, and the area is very attractive to waterbirds, especially in spring because this tidal basin thaws earlier than most of the surrounding habitat. Water in the pools is brackish, ranging in salinity from 4–38 ppt.[7] The central portion of ERF faces the Knik Arm and is most exposed to tidal inundation. Tides are extreme in this area and may exceed 10 m; during even moderately high tides, the entire marsh may be inundated. Soils are generally very fine glacial sediments deposited by back flow from the Eagle River and tidal gullies and by high tides from the sediment-laden Knik Arm.[8] Sedimentation rates range from several millimeters per year on levees, 10–15 mm on mudflats, and up to 20–40 mm per year in ponds. Annually, the daily mean temperature is 1.9°C, and precipitation ranges from 330–508 mm. As will be shown, water conditions and temperatures are very important in any natural attenuation of white phosphorous that might occur on ERF.

### 27.2.1 Biota of Concern

Birds begin to use ERF for feeding and staging as soon as water opens in mid-April to early May. For a few weeks through mid-May to early June, large numbers of migrants come through the area, with individuals staying for a few days to a week or two.[9] During the summer months, usage by ducks declines, but by early August southward-bound migrants begin gathering, and total bird usage is high until freeze-over in October. Over 16 species of waterfowl use ERF during the spring and summer, with mallards, green-winged teal, northern pintail, greater scaup, Canada geese, and swans predominating. Daily totals of waterfowl seen on aerial census routes of ERF vary from 2500 in spring to 100 in midsummer to 1500 in late summer/early autumn.[9] Other species include ravens (*Corvus corax*), shorebirds, gulls, occasional sandhill cranes (*Grus canadensis*), and raptors. Bald eagles are most common during early spring and perch on trees surrounding the marsh, waiting for available prey. Except for voles (*Cleithronymys rutilus*), few mammals occupy the area, although moose (*Alces alces*), wolves (*Canis lupus*), muskrats (*Ondatra zibethicus*), and beaver (*Castor canadensis*) have been seen in the marsh, and brown bears (*Ursa arctos*) frequent the surrounding uplands.

### 27.2.2 Source of Contamination

The root cause of the waterfowl problem stems from the area being used for artillery practice since 1949.[4] The most common artillery has been howitzers and mortars of various sizes. Munitions and compounds used in the area include high explosives (RDX, TNT, HMX), obscurant smoke (white phosphorus, hexachloroethane-zinc), and illumination flares (magnesium). Over a 40-year period, more than 100,000 rounds have been fired into ERF. Accurate records are not available, but of nearly 7000 rounds fired into ERF during most of 1989, about 4% were P4 munitions, and in the 4-year period from 1987–1990, 1217 white-phosphorus rounds with an approximate total P4 mass of 1200 kg impacted the area.[4] After a thorough sediment analysis of the Flats, only P4 was found to be widespread in the shallow marsh ponds used by waterbirds for feeding. P4 also was found in tissues of dead birds.[4] Subsequent toxicity tests showed that P4 was the primary agent of mortality in exposed waterfowl.[2]

## 27.3 CHARACTERISTICS OF WHITE PHOSPHORUS

White phosphorus is elemental phosphorus composed of four phosphorus molecules arranged in a tetrahedron configuration. It is not found in nature and only occurs through manufacturing. P4 is a waxy solid at room temperature but spontaneously combusts between 30 and 40°C. It is highly lipophilic with a high octanol/water-partitioning coefficient (Table 27.1). When in air it reacts with oxygen to form various phosphorous oxides, such as $P_4O_{10}$, that are hygroscopic and form a dense white cloud,[4] a characteristic used by the military for defensive maneuvering. It is relatively inert in water (water solubility is 3 mg/L), where it forms an outer layer or crust that further reduces its solubility. The persistence of P4 in water or sediments is affected by sediment moisture and temperature. When incubated at 20°C in moist but well below saturated sediments, 99.9% of the mass of P4 pellets disappeared after 24 h.[10] However, it took 56 days for 99% of the P4 mass to disappear at 15°C, and at 4°C, 50% of the P4 was still present after 56 days. When soils are saturated, no measurable loss of P4 mass occurred after 60 days, regardless of temperature. Because much of the sediments at ERF either were constantly or periodically saturated and since sediment temperatures tended to remain cool, especially when shaded by vegetation, natural attenuation of P4 is extremely slow.

The distribution of P4 at ERF is very patchy, and sediment concentrations a meter or so apart can vary by two or three orders of magnitude. Highest concentrations are found around craters formed by exploded ordinance. However, P4 can be distributed through suspension of near-colloidal

**Table 27.1 Physical Characteristics of White Phosphorus**

| Property | Data | References |
|---|---|---|
| Molecular weight | 123.89 g/M | 42 |
| Physical state | Waxy solid | 42 |
| Melting point | 44.1°C | 42 |
| Boiling point | 280°C | 42 |
| Density at 20°C | 1.82 g/cm$^3$ | 43 |
| Odor | Garlic-like | 44 |
| Solubility | | |
| Water at 15°C | 3 mg/L | 43 |
| Solvents | Alkali, ether, chloroform, benzene, toluene | 43 |
| Partition coefficients | | |
| Log $K_{ow}$ | 3.08 | 45 |
| Log $K_{oc}$ | 3.05 (estimated) | 45 |
| Vapor pressure at 20°C | 0.026 mm Hg | 44 |
| Henry's Law constant at 20°C | $2.11 \times 10^{-3}$ atm.m$^3$/mol | 45 |
| Autoignition temperature | 30°C | 46 |
| Flashpoint | Spontaneous in air | 47 |
| Oxidation states | +5-$P_4O_{10}$; +3-$P_4O_6$; 0-$P_4$; −2-$P_2H_4$; −3-$PH_3$ (gas) | 5 |

particles in the water column and subsequently transported by tides and currents to other sites. P4 particles have been detected in sediments as deep as 30 cm[4] and have the potential of resurfacing through seismic and hydrological activity. Particles of P4 at ERF range in length from 0.15 mm to 3.5 mm and fall within the size range of naturally occurring seeds. Thus, waterbirds may confuse them as food.

## 27.4 TOXICITY OF WHITE PHOSPHORUS

The toxic signs of white phosphorus in birds can be clearly separated into acute, subacute, and chronic toxicity. Acute toxicity is characterized by stereotypic head bobbing, lethargy, convulsions, and death within a few hours after ingestion.[11,12] At ERF, convulsing waterfowl are attractive to predators, which may ingest P4-contaminated tissues and suffer secondary poisoning.[13,14] Chronic P4 toxicity results in liver and kidney damage, inappetance, multifocal petecia in several tissues, altered blood chemistry, and cessation of reproduction in females.[15,16] White phosphorus can be transported to eggs from exposed hens and may result in teratogenic effects.[16] Egg laying may cease with a single sublethal dose of P4.

### 27.4.1 Acute Toxicity of P4 in Birds

Field observations on free-ranging waterfowl at ERF[2] reported that the first signs of toxicity appeared from 4 to 6 h after ducks (green-winged teal and pintail) arrived on a pond. These signs included rapid head shaking and frequent drinking. They were followed by periods of lethargy and attempts to hide in vegetation. Subsequently, intoxicated birds arched their heads and beaks over their backs, went into convulsions, and died. Many dead waterfowl collected at ERF appeared to have drowned, probably because they lost control of their ability to keep their heads above water.

These signs were consistent with laboratory studies that have been conducted with waterfowl. Coburn et al.[11] dosed wild mallards and black ducks (*Anas rubripes*) with P4 dissolved in almond oil. Time to mortality for birds dosed with ≥ 3 mg/kg body weight ranged from 3.5 to 16 h. They also observed lethargy, increased thirst, leg weakness, rhythmic head bobbing, convulsions, and death. Sparling et al.[12] described three phases of acute toxicity. Within 2 h of dosing, birds demonstrated lethargy, inappetance, increased drinking, unsteady gait, and vomiting. Only 26% of the

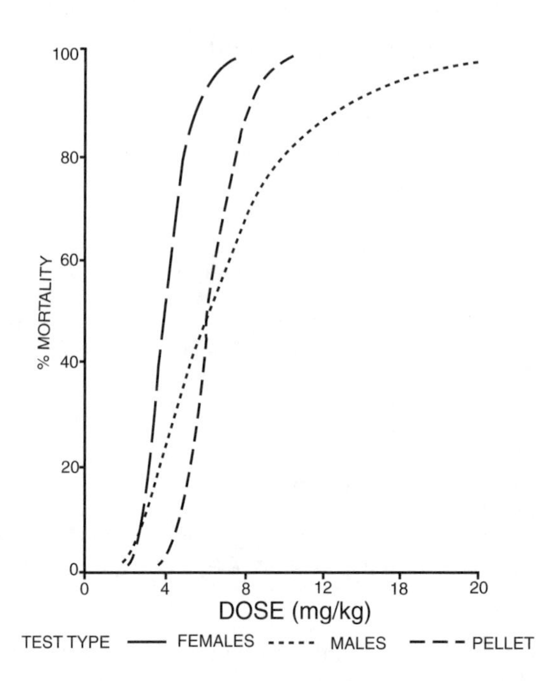


**Figure 27.3** Dose/response curves for male and female mallards given single doses of P4 dissolved in oil and for male mallards given single doses of pelletized P4. (From Sparling, D.W. et al., *J. Wildl. Dis.*, 33, 187, 1997. With permission.)

birds that showed these signs survived. This phase was followed within several hours to 1 day by a short phase of severe tremors and stereotyped head bobbing. All birds seen with these signs died, sometimes without passing into the final phase. The last phase of acute toxicity was characterized by severe convulsions that could last for more than 2 h. The nature of these convulsions was reminiscent of mallards euthanized in an $O_2/CO_2$ chamber but more severe and prolonged, and the authors suggested that the convulsions may have been due to anoxia. The mean time to convulsions was 5 h post dose.

Sparling et al.[12] determined that the $LD_{50}$ for male mallards with P4 dissolved in corn oil was 6.46 mg P4/kg body weight (95% CI = 5.19 to 7.69 mg/kg) (Figure 27.3). Females had an $LD_{50}$ of 6.96 mg/kg (95% CI = 2.66 to 8.96 mg/kg). Although the $LD_{50}$ values were not statistically different between males and females, females had a shallower dose/response curve than males ($p < 0.045$), which might be attributed to the greater fat stores of females. Sparling et al. assessed that the lowest observed adverse effects level (LOAEL) was between 2.6 and 3.7 mg/kg for P4 in oil. They also found that pellets of P4, made by cooling molten P4 in water, had a higher toxicity than dissolved P4 in that the $LD_{50}$ of this more environmentally realistic form of P4 was 3.90 mg/kg (95% CI = 3.24 to 4.69 mg/kg) in male mallards. They did not conduct a separate study with pellets to determine the LOAEL; but, extrapolating from dose/response curves, it lies between 0.3 ($LD_{01}$) and 0.4 ($LD_{10}$) mg/kg.

With P4 pellets, mute swans (*Cygnus olor*) had an $LD_{50}$ of 3.65 mg/kg (95% CI = 1.40 to 4.68 mg/kg), which is very similar to that of mallards.[17] However, the average time to onset of signs was substantially longer than that seen in mallards, often exceeding 1 to 4 days rather than a few hours. The authors attributed the delayed expression of effects to the predominance of small particles of grit in swans that were about the size of P4 pellets and may have delayed passage of P4 past the gizzard; grit size in mallards averages approximately 0.25–0.5 cm in diameter. For comparison, acutely lethal doses in mammals are within the range observed for waterfowl. The $LD_{50}$ with P4 in oil for rats was 3.03 mg/kg in females and 3.76 mg/kg in males.[18] The same study showed an $LD_{50}$ of 4.82–4.85 mg/kg for mice. The lethal dose for humans is around 1–4 g.[5]

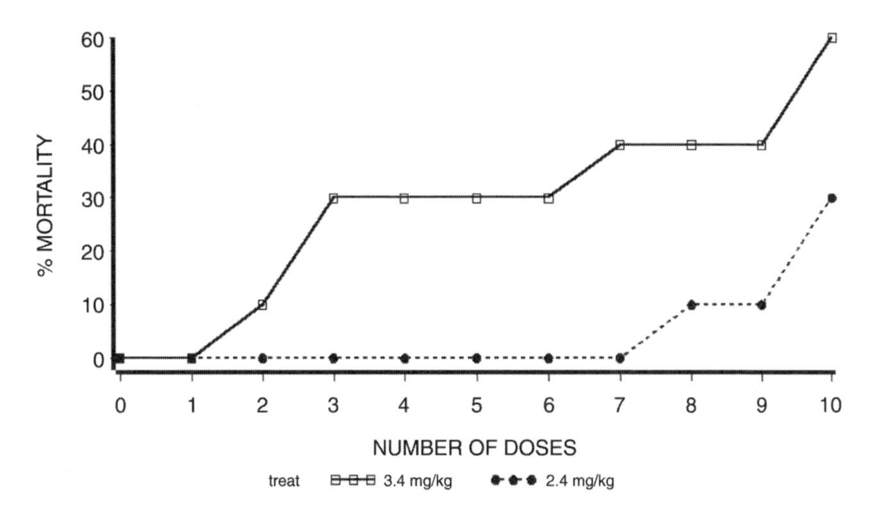


**Figure 27.4** Increased mortality in mallards given repeated doses of pelletized P4 at 2.4 and 3.4 mg P4/kg body weight. (From Sparling, D.W. et al., Toxicological studies of white phosphorus in waterfowl, U. S. Army, Hanover, N.H., 1993.)

Subacute mortality of waterfowl can be exacerbated by repeated exposures over a few days, a situation which is likely to occur given the very patchy distribution of P4 at ERF. Birds can encounter a patch of P4 on one day, ingest a sublethal amount, and then encounter a second patch the next day. Because P4 has some persistence, it can accumulate in tissues through repeated exposure and ultimately result in toxic effects including death. For example, when mallards were experimentally fed pelletized P4 at a rate of 2.4 mg/kg/day for 10 days, no mortality occurred until the seventh day of dosing, and by the tenth day 30% of the birds had died (Figure 27.4). At 3.4 mg/kg, birds began to die by the first day of dosing, and mortality continued until 60% of birds had died by the tenth day.

### 27.4.2 Chronic and Subchronic Effects

A very common chronic effect of white phosphorus is an enlarged, friable, and pale liver (Figure 27.5a,b). This has been frequently reported for birds[11,12,14,19] and mammals.[5,20,21] One of the functions of the liver is to package lipids for excretion by combining them with proteins to make lipoproteins. White phosphorus interferes with this metabolic step, and as a result, lipids accumulate in the organ, enlarging it and altering its texture and density.[22,23] Other liver damage includes necroses or damaged areas, ecchymoses, which appear as blisters on the surface of the liver, and increased vacuolation of hepatocytes, signifying depletion of glycogen reserves.[24] Hepatocellular vacuolations were first observed at acute exposures of P4 in oil at 2 mg/kg. The frequency of hepatic foci did not become appreciable until 6.1 mg/kg, near the $LD_{50}$. Birds that were dosed with dissolved P4 tended to show a greater frequency of these signs than those dosed with P4 pellets.[12]

Kidneys are also frequently damaged by P4. The most consistent renal damage observed in mallards was hyaline droplet nephrosis in the renal convoluted tubules.[12] This became apparent at dosages at and above 4 mg/kg P4 in oil. As with liver, kidney histopathological changes increased in frequency with repeated chronic dosing using pelletized P4. Ancillary necrosis to heart, pancreas, duodenum, and other tissues has been reported, starting at doses around 4 mg/kg of dissolved P4.[11,12] Congested duodena, characterized by swelling and inflamed tissue, was common in birds dosed with pelletized P4. In the lab, 9 to 64% of the dosed birds had congested duodena, depending on dose. In mammals, P4 has reportedly caused damage to cardiovascular,[24] respiratory, renal, muscular, and nervous systems.[5]

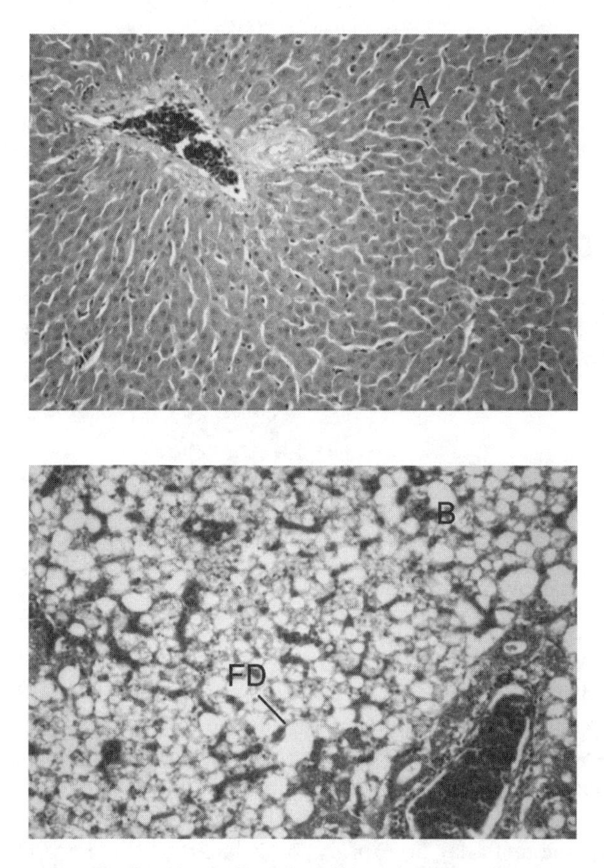


**Figure 27.5** Photomicrographs of a healthy mallard liver (A) and a liver taken from a mallard dosed with P4 dissolved in oil (B). Note areas of fatty degeneration (FD) in photomicrograph B.

In a collection of 30 mallard, pintail, and green-winged teal carcasses taken from ERF in 1993, 40% had congested duodena, 23% had frank liver damage including foci and ecchymoses, 23% had pancreatic petechia, and 43% had coronary petechia.[25] Among ten game-farm mallards released into pens over P4-contaminated pools for 4 h, three died, five had congested duodena, 60% had coronary petechia, and 40% had frank liver damage.

Numerous subacute and chronic changes occur in blood with exposure to P4. Several aspects of blood chemistry respond in a dose-dependent fashion to P4 and may be useful as bioindicators.[14,15] Notably, plasma lactate dehydrogenase (LDH), aspartate aminotransferase (AST), potassium, and uric acid increased with P4, whereas hematocrit, hemoglobin, and total protein decreased.[15] The responses varied between sexes, number of doses, and whether the P4 was in pellet form or dissolved in oil. For example, in nonreproductive females, responses in AST, LDH, potassium, glucose, and hematocrit could be seen after a single dose of 5.4 mg P4/kg body weight in oil, but two doses were needed to see changes in hemoglobin. With pelletized P4 in breeding females (which would be more similar to the conditions seen at ERF), changes were observed in AST, hematocrit, and LDH with repeated doses for 7 days at 0.5 mg/kg/day. At 1.0 mg/kg and 2.0 mg/kg, plasma alanine aminotransferase (ALT), hemoglobin, inorganic P, triglycerides, and uric acid demonstrated changes, and the changes in AST, LDH, and hematocrit became more pronounced. The authors suggested that the ratio of hemoglobin/LDH could provide a good bioindicator of exposure for at least 1 week post exposure (Figure 27.6). The changes in the blood chemistry can be associated with damage in the liver and kidney[26] and with hemolysis. In addition, significant changes in the white blood cell count were noted with a decrease in lymphocytes and an increase in thrombocytes

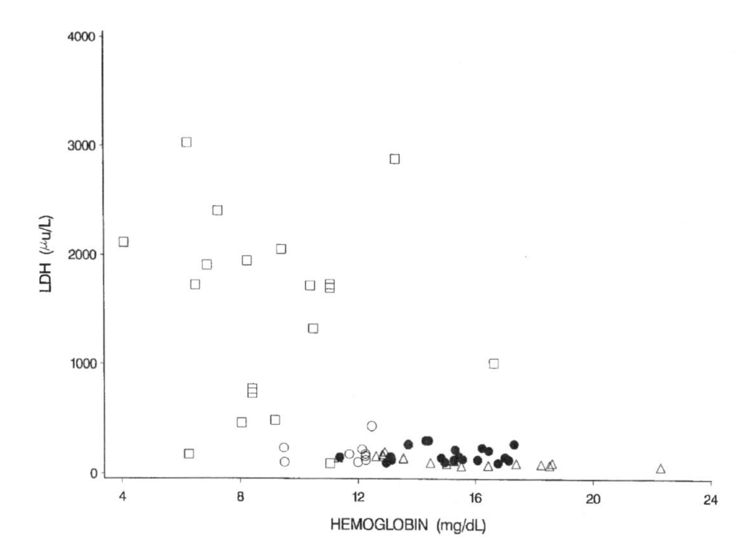


**Figure 27.6** Relationship between plasma LDH and hemoglobin for birds given 0, 0.5 and 1.0 mg/kg pelletized P4 for 7 days. Dot = control birds; circle = 0.5 mg/kg 1 week post dosing; square=1.0 mg/kg 1 week post dosing; triangle=1.0 mg/kg 2 weeks post dosing. (From Sparling, D.W. et al., *Environ. Toxicol. Chem.*, 17, 2521, 1998. With permission.)

with dose.[15] These changes were consistent with differential cell counts in mammals[27] and may have an effect on the immunology of affected animals.

### 27.4.3 White Phosphorus in Tissues

The liver is the primary organ for detoxification of P4 and could be expected to be a center of deposition. However, due to the lipophilic nature of P4, fat and skin are preferred tissues for P4 analysis. Using autoradiography in rabbits, mice, and rats, Cameron and Patrick[28] determined that at 48 h post dose, tissues with intense uptake of $^{32}P$ included liver, renal cortex, epidermis, hair, pancreas, and adrenal cortex. Ovary, renal medulla, spleen, uterus, thymus, lung, and adrenal medulla showed moderate radioactivity. Tissues with low radioactivity included striated muscle, brain, fat, and bone. Goshal et al.[29] demonstrated similar results with autoradiography in rats and determined that within 2 to 3 h the percentage of total dose in the liver reached its maximum of 65–70%. At this time, liver had significantly higher radioactivity than blood, kidney, spleen, pancreas, and brain. Lee et al.[18] also found that liver had the highest radioactivity in rats, whether they were dosed once or multiple times. However, autoradiography does not distinguish between the parent P4 and its metabolites. Because it is a site of metabolism and detoxification, one might expect that the liver would contain a mixture of P4 and associated degradates. However, not much is known about the metabolism of P4 in mammals or other animals.[5]

Fletcher et al.[30] used gas chromatography[31] to examine the uptake and half-life of P4 in Atlantic cod (*Gadus morhua*) and Atlantic salmon (*Salmo salar*). This study is pertinent to the ERF situation because of the discharge and interchange of P4 and sediments into the Cook Inlet by the Eagle River and because of the fishery provided by the Inlet. Fish were exposed to suspended yellow phosphorus (white phosphorus with some impurities) for up to 24 h. Liver, muscle, esophagus, intestine, pyloric caeca, and gill were examined for P4 concentration. For all tissues there was a high degree of relatedness between water and tissue concentrations, but the liver contained the highest concentrations of P4. In cod kept in water ranging from 1.9 to 5780 ppb, liver concentrations ranged from 19 to 519 ppm and muscle concentrations from 0.25 to 30.9 ppm. Between 19 and 55% of the total P4 burden was in liver. Cod, however, are well known for high amounts of oil in their livers, and the P4 may have been suspended in the lipid fraction. In salmon, pyloric caeca had the highest concentrations of P4 (0.08 to 55.5 ppm in water with 0.84 to 1900 ppb concentrations), but this was followed in importance by liver. The biological half-lives for P4 in cod were 5.27 h in liver (with large livers having a half-life of 13.4 h) and 4.71 h in blood. Half-lives were much shorter in salmon and were around 0.5 h for most tissues examined. Little is known about the dynamics of P4 transport into and out of ERF or its dynamics in seawater. Based on the short half-lives found in these fishes and the dilution factors undoubtedly associated with the inflow of Eagle River into the much larger Cook Inlet, it is unlikely that the P4 at ERF poses a threat to fish in the Cook Inlet or to their human consumers.

Nam et al.[32] used gas chromatography to directly quantify P4 concentrations in American kestrels (*Falco sparverius*) fed P4 mixed with food at 5 ppm P4 for an average intake of 6.6 to 131 μg P4/day. Maximum concentrations of P4 were approximately 0.07 ppm in fat and 0.06 ppm in skin. The researchers found that P4 was found primarily in fat and skin, although one bird had trace amounts in thigh muscle. White phosphorus could be detected after the first day of exposure and tended to increase in concentration through time, especially in skin, where the correlation with time was statistically significant. White phosphorus concentrations quickly decreased after P4 was removed from the diet and could not be detected in fat or skin after 3 days. These dosing regimens did not produce signs of toxicity in kestrels.

All of 19 waterfowl carcasses found at ERF in 1990 had P4 in their gizzards.[2] The mass of P4 varied widely, however, from 0.01 μg in a green-winged teal to 3 mg in a mallard to 11 mg in a swan. Mean fat concentrations of P4 were 0.67 ppm (range 0.1 to 2.9) in swans and 0.21 ppm (< 0.01 to 0.43) in ducks. Concentrations in skin were between a tenth and a third of that found in fat.

In a study of waterbirds on ERF,[25] one of five greater yellowlegs (*Tringa tringa*) had measurable levels (0.4 ppm) of P4 in its skin. P4 in fat of 30 ducks (pintails, green-winged teal and mallards) varied from below detection limits of 2 ppb to 19.1 ppm. Skin concentrations ranged from below detectable to 20.5 ppm and gizzard concentrations from below detectable to 628 ppm. Four of the ducks had gizzards that emitted smoke from oxidizing P4 when dissected.

In mallards experimentally dosed with P4 dissolved in corn oil, 18 of 36 given 5.25 to 9.1 mg/kg P4 died. Of those that died, P4 residues ranged from 0.12 to 1.78 ppm in fat and 0.05 to 0.96 ppm in skin.[12] Liver concentrations were much lower, from below detection limits of 1 ppb to 0.02 ppm. Concentrations of P4 in fat differed significantly ($p = 0.003$) across doses but not always in a linear fashion because birds given higher doses died more quickly, probably before all of the P4 could be assimilated. Mallards dosed with P4 pellets at 4 or 6.5 mg/kg had from 0.12 to 3.3 ppm P4 in their fat. Mute swans gavaged with P4 pellets at dosages from 2.98 to 5.28 mg/kg had fat concentrations ranging from below detection limits to 0.38 ppm and skin concentrations from below detection limits to 0.16 ppm.[17] The lower tissue concentrations in swans compared to mallards may have been due to retention of P4 pellets in the gizzards of swans and consequently slower release into the digestive system. In both mallards and swans, survivors that were euthanized after 1 week post dose — even those at the higher doses — had only trace amounts of P4 in fat or skin.

In a study to determine depuration rates of P4 in mallards, game-farm birds were gavaged with a single pellet of P4 at either 1 or 2 mg/kg body weight.[19] Subsamples of birds were then euthanized at 3, 6, 8, 24, 96, and 240 h post dose. Fat, liver, kidney, and breast muscle were analyzed for P4 residues. Concentrations of P4 in fat peaked by 8 h post dose, then declined precipitously by 24 h and more slowly through the next 48 h (Figure 27.7a) The higher concentration at 240 h appears to be due to one bird that may have retained the P4 pellet, because this bird also had elevated P4 concentrations in liver and kidney. During our studies, we came upon a few birds that had smoking gizzards a week or more after dosing. Initial concentrations in fat were 10 to 100 times greater than in other tissues. In the liver and muscle, P4 showed a peak at the first sampling time of 3 h and a rapid loss by 24 h (Figure 27.7b,d). At 2 mg/kg, but not at 1 mg/kg, kidney P4 concentrations were initially high, then decreased rapidly from 3 to 24 h and were below detection at 96 h (Figure

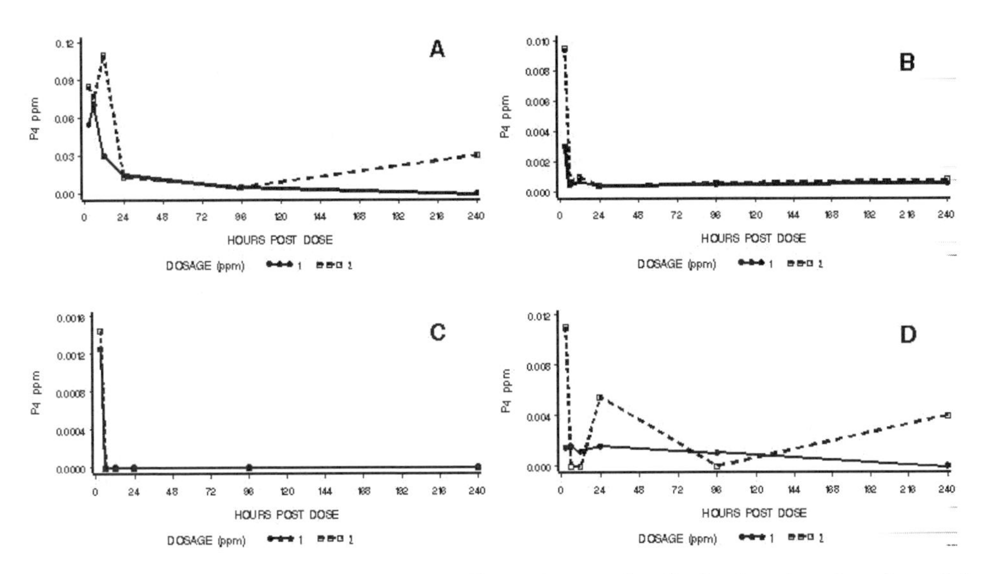


**Figure 27.7** White phosphorus concentrations in different tissues of mallards given 1 or 2 mg/kg pelletized white phosphorus and then sampled at different times to determine uptake and loss; a = fat, b = liver, c = kidney, d = breast muscle. (From Sparling, D.W. et al., Toxicological studies of white phosphorus in waterfowl, U. S. Army, Hanover, NH, 1993.)

27.7c). Nam et al.,[33] using domestic chickens, also reported very rapid loss of P4 from liver, brain, and blood, with a slower decay in fat and skin.

### 27.4.4 Reproductive Effects and Teratogenesis

A proportion of the P4 a female duck ingests may be transferred to her eggs and have a potential for producing teratogenesis in her offspring. White phosphorus has been found in the egg of a gull (*Larus argentatus*), in yolk sacs of some shorebirds at ERF, and in yolk sacs of game-farm mallards treated with P4.[1,3] In one study, domestic chickens were gavaged with P4 dissolved in tricaprylin oil either singly at 1 to 5 mg/kg body weight or daily doses of 1 mg/kg body weight for 5 days.[33] Hens given single doses at 3 or 5 mg/kg and those given repeated doses experienced a reduced rate of egg laying compared to controls or to those given a single 1-mg/kg dose; some hens stopped laying completely. At all dose levels, some P4 was transferred to the yolk of eggs, sometimes as early as the first day of dosing. However, there was considerable variation in the deposition of P4. For the single doses, P4 could be detected at 6, 8, and 9 days post dose, but only 0.003 to 0.007% of the total P4 administered was recovered in eggs. Maximal quantities of P4 in eggs ranged from 0.05 to 0.15 μg and was dose-dependent. For hens given five daily doses, P4 was measurable for 53 days past the end of dosing in stored eggs, reached a maximal mass of 0.12 μg, and accounted for 0.004% of the total administered. None of the hens dosed in that study showed obvious signs of toxicity except for the decreased rate of laying.

Vann et al.[16] gavaged mallard hens with pelletized P4 at 0.5 to 2.0 mg/kg P4 for 7 days to confirm if P4 could be translocated to waterfowl eggs and to determine the effects on mallard reproduction. Although the doses were thought to be sublethal, repeated dosing at 2.0 mg/kg resulted in six of ten hens dying, and this treatment was terminated after four doses. Clinical signs of toxicity, including hepatic necrosis and enlarged spleens, were apparent at all dose levels upon necropsy. White phosphorus measurements were scheduled to be taken on the fourth (presumed to be the first egg to be formed after the onset of dosing) and eleventh (presumed to be after the effects of P4) eggs laid by each bird, but because of the reduced laying rate (see below), measurements frequently could not be taken. For those fourth eggs that could be measured, no P4 was found in control eggs. Four of 17 (23%) of the eggs at 0.5 mg/kg, and 6 of 11 (54%) at 1.0 mg/kg had measurable (> 7 ppb) concentrations of P4. None of the birds at 2.0 mg/kg laid a fourth egg. The highest concentration of P4 detected in eggs was 0.024 mg/kg from a hen given 1.0 mg/kg P4. None of the eleventh eggs had measurable concentrations of P4, but all of these eggs from dosed birds were laid many days after the cessation of dosing.

Vann et al.[16] also examined fertility and hatchability of eggs by using artificial incubation. Fertility of control eggs was 92.3% but dropped significantly ($p < 0.003$) to 86.8% for hens given 0.5 mg/kg P4 and 85.6% for those given 1.0 mg/kg. Of the fertile eggs, 90.1% of the control eggs hatched compared to 78.2% at 0.5 mg/kg and 87.9% at 1.0 mg/kg ($p < 0.01$). Some hens on 0.5 mg/kg continued laying during dosing so their embryos may have had greater exposure to P4 than most of the hens on 1.0 mg/kg who stopped laying shortly after dosing began.

There is some evidence that P4 causes malformations in embryos. Vann et al.[16] dosed mallard hens with five daily doses of pelletized P4 at either 1.3 or 2.6 mg/kg. At the lower dose, 3 of 30 eggs produced deformed embryos that failed to hatch. At 2.6 mg/kg, four unhatched embryos from two hens had malformations. At both dose levels the malformations consisted of spinal defects, microphthalmia, hydrocephaly, and twisted legs and feet. The following year the researchers found that hatching weights of ducklings whose mothers were given 0.5 or 1.0 mg/kg P4 were less than those of control birds ($p < 0.003$). A large proportion, but not all, of the weight difference was due to smaller yolk sacs in treated ducklings. Rates of malformations for controls, 0.5, 1.0, and 2.0 mg/kg groups were, respectively, 1, 7.3, 6.0, and 8.1% ($p < 0.038$). The few control embryos with defects had slight scoliosis and edema, but treated embryos demonstrated more severe scoliosis and edema as well as lordosis, hydrocephaly, and bill curvature.

The greatest effect of white phosphorus on mallard reproduction was a dramatic decrease in frequency of laying. Most of the hens that were given 0.5 mg/kg P4 daily stopped laying by the fifth day of treatment and continued having low to no egg output for at least 20 days (Figure 27.8). Considering that actively laying females have ova at various stages of development, a 5-day lag in response suggests that some of these hens required a brief accumulation of P4 before effects were observed. Hens that were given 1.0 mg/kg and above stopped laying almost immediately, indicating that even partially developed eggs were inhibited from further development. Laying in this group was inhibited for at least 30 days. In the 2.0-mg/kg group, surviving birds stopped

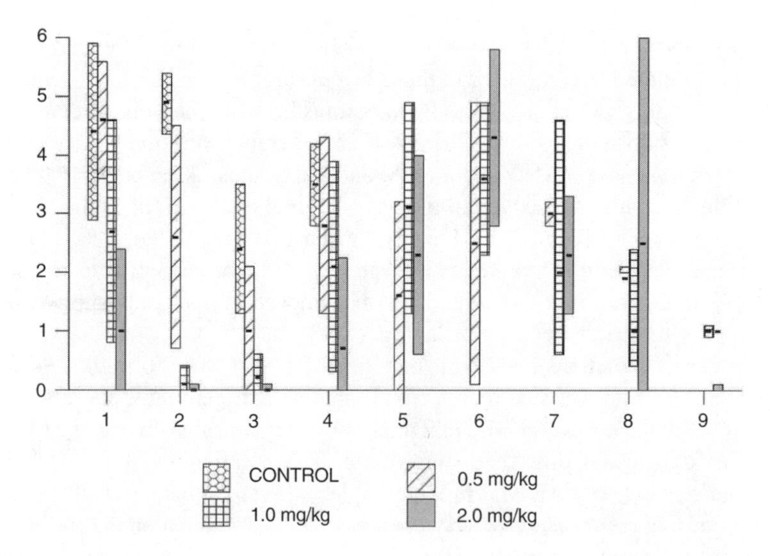


**Figure 27.8** Laying frequency of mallard hens given repeated doses of pelletized P4 daily for 7 days. Each bar marks a 5-day period with mean and standard deviation of number of eggs laid by hens laying eggs during that period. (From Vann, S.L. et al., *Environ. Toxicol. Chem.*, 19, 2525, 2000. With permission.)

laying almost immediately, and only one reinitiated laying weeks later. Control hens took an average of 13 ± 0.4 days to lay a clutch of 12 eggs, those on 0.5 mg/kg that resumed laying took 18 ± 0.7 days, those on 1.0 mg/kg took 28 ± 1.3 days, and the few left from 2.0 mg/kg needed 36.4 ± 0.24 days. Of course, under natural conditions, such abnormal laying patterns would probably lead to abandoned nests. Under the attenuated breeding season of Alaska compared to the contiguous United States, even if hens did not abandon their nests, they would not have had time to successfully produce a brood. The authors suggested that the most likely cause for the curtailment of laying was liver sequestration of lipids necessary for yolk formation. Because of the short breeding season of Alaska, ingestion of even a single dose of P4 at ERF could prevent hens from successfully breeding that year.

Vann et al.[16] also studied the effects of P4 on male mallards. Males that were gavaged repeatedly with 1.0 mg/kg P4 demonstrated reduced testosterone titers in plasma, reduced body weight, and reduced hematocrit, but all these effects were transitory, and no significant differences were detected between controls and treated males at 2 weeks post treatment. There was no difference in the fertility of eggs produced by untreated hens mated with untreated or treated males.

### 27.4.5 Secondary Toxicity

At ERF, the convulsions and thrashings of intoxicated waterfowl attract predators such as bald eagles, ravens, and gulls. These predators feed on various parts of the ducks including gizzards with unassimilated P4 pellets and soft tissues, which may contain P4 residues. Dead eagles collected at ERF had measurable levels of P4 in their fat,[25] and gull eggs in nests at ERF had traces of P4.[3]

The potential for secondary toxicity in predators has been examined in three studies. Roebuck et al.[3] observed 24 predation events on dead or dying waterfowl during a 10-day period at ERF in 1991. Of these events, 15 were by bald eagles, 7 by herring gulls (*Larus argentatus*), and 2 by ravens. During this period, the authors also found 40 fresh feather piles of waterfowl beneath a 0.5-ha tract of trees used as roosts by eagles, demonstrating a high consumption of waterfowl. Fifteen waterfowl carcasses were examined for P4 residues; from 0.02 to > 8000 μg of P4 were found in gizzards of these carcasses, fat concentrations of P4 ranged from 0.04 to 5.92 ppm, and P4 was found in several partially consumed carcasses. Thus, predators could be exposed to chronic or acute toxicity from ingesting contaminated prey.

Roebuck et al.[13] further pointed out that the dynamics of particulate P4 in prey are important to understanding the potential for secondary poisoning. They examined the effects of pellet size on toxicity, and although they did not find any significant difference due to pellet size, they found that substantial amounts of dissolved P4 can be recovered from intestines of dead prey. They raised the question whether predators could become intoxicated only by ingesting gizzards containing intact pellets or if dissolved or assimilated P4 could also be effective. They concluded that P4 dissolved in oil and pelletized P4 were nearly equal in toxic effects and suggested that predators could become ill by ingesting other tissues containing P4.

Sparling et al.[14] examined the problem of pelletized vs. dissolved or incorporated P4 by creating an experimental model of predator and prey. They repeatedly fed poultry chicks sublethal amounts of P4 over a 36-hour period to allow the P4 to pass into the intestines and become incorporated into soft tissues. They then euthanized the chicks, and to one group they added a 1.1-mg pellet of P4 into the gizzard, in another group they dissected out the gizzard to make sure that no pelletized P4 remained, and a third group of chicks were euthanized without being fed P4. The mean mass of P4 in the tissues of treated chicks was 83.9 μg, which, if the soft tissues of the carcass were eaten completely, was equivalent to 0.62 mg/kg for a 136-g kestrel. For those with added pellets, the mean mass was 1184 μg (= 8.7 mg/kg). Captive-bred American kestrels were randomly assigned to the three treatments of pellet, no pellet, and control. Each kestrel was fed a new chick daily, and the amount of food consumed and survival of kestrels were monitored daily. Eight of 15 kestrels fed chicks with P4 pellets died by the end of 7 days, with the first deaths occurring on the second

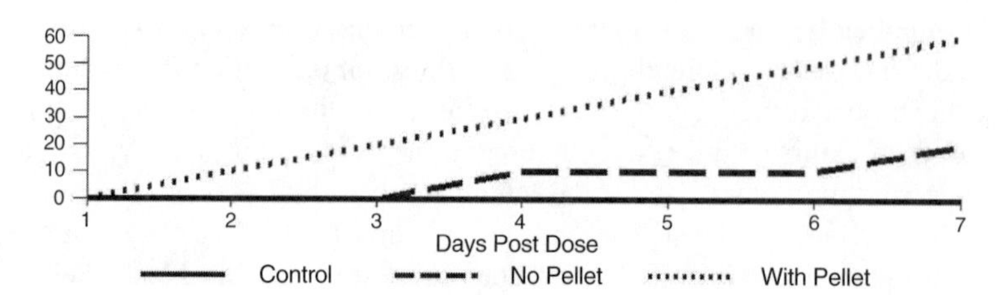


**Figure 27.9** Mortality in American kestrels fed poultry chicks given P4. Control = poultry chicks with no P4; no pellet = chicks with gizzards removed to eliminate any pelletized P4 residue; pellet = chicks with gizzards intact and 1.1 mg P4 pellet added.

day of the study (Figure 27.9). Moreover, 3 of 15 kestrels that were fed treated chicks without an extra pellet died, and the first deaths occurred on the fourth day. None of the kestrels that were fed control chicks died. Food consumption and plasma LDH, ALT, glucose, hematocrit, and hemoglobin differed among treatments. Kestrels fed chicks with the pellet had lower body weight, hemoglobin, and hematocrit and higher LDH, glucose, and ALT than those on the other treatments. This study showed that predators could die from eating contaminated prey, and whereas mortality was greater when pelletized P4 was present, death could also ensue by ingesting dissolved or assimilated P4.

## 27.5 HAZARD ASSESSMENT OF WHITE PHOSPHORUS TOXICITY AT EAGLE RIVER FLATS

Figure 27.10 depicts the daily hazards of P4 to waterfowl at ERF. For each day; the probability of consuming P4 is dependent on the species and location of ERF that the bird uses. Of the five species of ducks that commonly use ERF, pintail mallards and green-winged teal appear to have a higher probability of ingesting P4 than widgeon or shovelers because the first three species consume more benthic invertebrates and seeds than widgeon and lack specialized bill structures of shovelers that facilitate straining and sorting of food.[34]

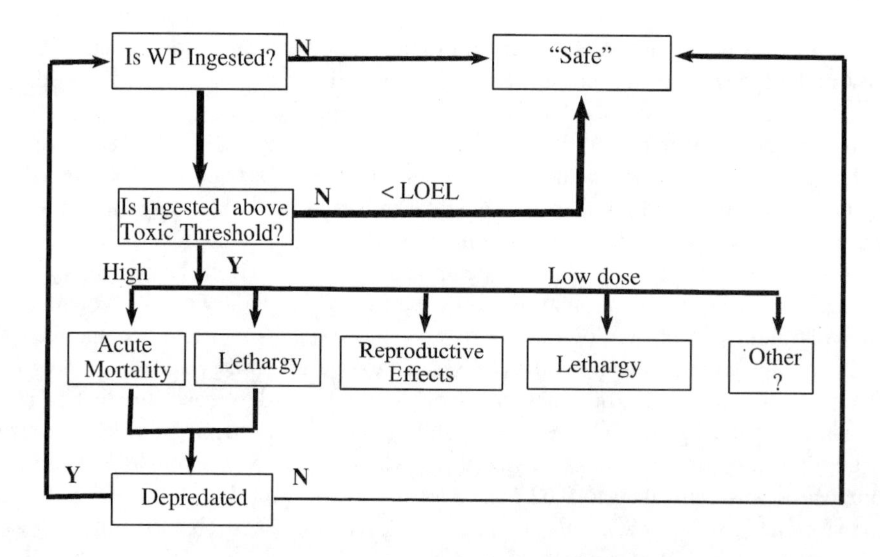


**Figure 27.10** Box diagram representing a daily P4 hazard assessment for waterfowl at Eagle River Flats, Alaska. (From Sparling, D.W. and Federoff, N.E., *Ecotoxicology,* 6, 239, 1997. With permission.)

Although P4 is patchily distributed throughout ERF, some areas have higher concentrations of craters and P4 loci than others; also, where a bird forages makes a substantial difference on its encountering a P4 locus.[4,35] If a duck has encountered a P4 locus and ingests some P4 particles, the subsequent toxicity depends on a number of factors. If the total mass ingested is equivalent or more than 6 mg/kg body weight, the bird will almost assuredly experience acute toxicity and die. If the quantity is around 0.3 mg/kg or less, the bird is likely to survive without ill effects, regardless of whether or not it had a prior dose while at ERF. Males ingesting 0.5 to 1.0 mg/kg may experience no or mild effects, even if P4 had been ingested during the previous day. Females coming into reproductive condition and having consumed these levels of P4, however, may be prevented from breeding successfully that year due to the inhibition of laying. If they continue to lay, they risk decreased fertility and hatchability and increased probability of malformed embryos. The probability of inhibited laying increases with dosage up to 100% at around 2 mg/kg. For either sex, a single exposure of P4 from 1 to 5 ppm will elevate the probability of acute mortality, and the probability will be sharply elevated as the number of exposures increases. However, it is not known to what extent chronic effects of impaired liver and renal function, altered blood chemistry, reduced hematocrit and hemoglobin, and altered white blood cell counts have on the long-term survival of free-ranging waterfowl. To say that a duck is "safe" when it consumes any P4 is only in relative terms.

The total risk encountered by ducks for each day is the sum of the individual daily risks over the period of time that it spends at ERF. Radiotelemetered mallards and green-winged teal spent an average of 6.8 days on ERF during the spring staging period[36] and much longer periods during autumn. Of 32 birds equipped with radiotelemeters during the spring, four — or approximately 12% — died at ERF after visits ranging from 10 to 51 days.[36]

## 27.6 REMEDIATION EFFORTS AT EAGLE RIVER FLATS

Based on the hazard model, the key to reducing the P4 hazard to waterfowl at ERF is to prevent waterfowl from ingesting P4. This can be done in two broadly different ways. The first way is to modify habitat usage and behavior of waterfowl to reduce their use of areas with high concentrations of P4 or restrict their feeding in such areas. At ERF, two methods were used to alter waterfowl behavior. One method employed conventional hazing devices similar to those used at airports.[37] Up to 15 propane cannons were located around open water ponds that were heavily used by waterfowl for foraging. These cannons were accompanied by scarecrows, mylar tape, eagle effigies, strobes, and sirens. In addition, two field technicians fired shell crackers and skyrockets over flocks of birds when they landed near P4 loci. These methods were effective during the brief staging periods in spring and fall and probably reduced avian mortality. However, they required a high commitment of resources and were only effective for the time they were actually in use; as soon as the devices were removed birds resumed their use of the areas. Birds also appeared to habituate to the presence of the noisemakers. Therefore, hazing was never considered a long-term solution to prevent exposure to P4.

Another method used to alter waterfowl behavior and use of an area was broadcasting time-release gelatin beads containing methyl anthranilate (MA) over ponds that had high concentrations of P4.[38,39] MA is a food flavoring that gives grape soda its taste and has been found to be offensive to birds. Various formulations of MA were tested and found to vary in their effectiveness in reducing waterfowl foraging in ponds. However, one formulation (JR930413) appeared to be promising in outdoor laboratory trials and could be effective in reducing waterfowl foraging for up to 10 days post treatment.[39]

The other broad category of methods to reduce P4 hazards is to remove P4 from the environment. The conundrums involved with this category include possible damage or destruction to the ecology of the marsh itself, the risk posed by unexploded ordinance with any excavating technique, the extent of high tides inundating the marsh and preventing soils from drying, and the effects of short seasons and cold temperatures on any natural attenuation of P4.

For relatively small areas, some forms of geotextiles and blocks were considered. Geotextiles are mats composed of various materials that have the intended objective of isolating contaminated sediments from the sediment/water interface and from waterfowl. Blocks are aggregate materials that, when mixed with water, seal the sediment surface. Several geotextiles and blocks were considered and tested[40] including the product AquaBlok™.[6] This block is composed of a blend of calcium, bentonite/organo clays, gravel, and polymers that bind together to form a sealant. The material was mixed on site using cement mixers and applied via helicopter to a 4000-$m^2$ pond where extensive waterfowl mortality had been observed. The barrier successfully reduced the P4 concentration at the sediment-water interface from a mean of $1.59 \pm 1.06$ (SEM) μg/g sediment to $0.01 \pm 0.01$ μg/g. Prior to the application of the barrier, 15 of 24 experimentally confined mallards died within 9 days. After the application, only 3 of 24 mallards died. This compared to a neighboring untreated pond where 21 of 24 mallards died prior to treatment and 24 of 24 died after a simulated but not effected treatment. Foraging behavior declined in the enclosure after treatment with Aqua-Blok™ compared both to the control enclosure before and after a simulated treatment and to the treated one prior to the block. The long-term durability and effectiveness of the AquaBlok™ needs to be determined.[6]

Walsh et al.[41] reviewed the results of three techniques found to be most effective in removing P4 at ERF. Natural attenuation due to sedimentation was considered as being too slow and too uneven compared to the need to quickly reduce waterbird mortality. *In situ* remediation of P4 is the most basic and least intrusive decontamination method examined. This method is appropriate only when sediments have a chance of naturally desaturating; thus, it can only be used in elevated sites or away from the repeated inundation from tides. Its effectiveness can be enhanced by warming the soil through heat-absorbing plastics or other means. A second method augmented natural attenuation by isolating P4 loci in low-lying sites and pumping or draining water to facilitate drying. This method can cause some disturbance in the marsh if ditches are created and can be retarded by groundwater recharge. Dredging was a third method analyzed. This causes more extensive disruption of the marsh and creates deeper pits that may actually be more difficult to dry or drain. After thoroughly evaluating the three methods, the authors concluded that each may be necessary in remediation because of the variability of the environment. Complete remediation of the marsh will take several years, regardless of techniques employed.

## 27.7 SUMMARY

White phosphorus has a limited distribution in the environment because it only occurs where it has been directly used by humans. It is not transported aerially for any distance and, due to its density, has a limited ability to disperse through water. Therefore, it is not a contaminant of broad-scale concern. However, where it does occur, it can cause substantial mortality or critically injure populations of waterfowl. This chronic harm includes impaired liver and kidney functioning, decreased respiratory efficiency, increased susceptibility to predation, loss of body mass, general weakening and malaise, and curtailment of reproductive functioning. Lethal effects occur around 3–4 mg/kg or approximately 3–6 ingested particles; sublethal effects can occur with ingestion of as little as a single particle. The impact of P4 on waterfowl populations nesting around ERF has never been estimated. Even if direct mortality on ERF could be estimated accurately in ducks, the delayed toxicity of P4 in swans (and presumably other species that use small grit size) and the potential for swans to fly away after ingesting a lethal dose of P4 could greatly underestimate the overall mortality. Inhibition of laying, reduced fertility and hatchability, and teratogenesis in hens ingesting even a small amount of P4 could potentially have an effect on populations greater than that exhibited by direct mortality. Predators such as bald eagles and gulls are also at risk due to the toxicity of pelletized, dissolved, and assimilated P4 in prey organisms. Although the U.S. Army stopped using P4 in wetlands in 1993 and remediation efforts have been underway since 1995, waterfowl mortality is expected to continue for several more years. Because Eagle River Flats is only one of several sites where P4 has been found in wetland conditions, further biological investigation is warranted at these other sites.

## REFERENCES

1. Racine, C. H., Walsh, M. E., Collins, C. M., Lawson, D., Henry, K., Reitsma, L., Steele, B., Harris, R., and Bird, S. T., Phase II. Remedial Investigation Report: White Phosphorus Contamination of Salt Marsh Sediments at Eagle River Flats, Alaska, U.S. Army Environmental Center, Aberdeen Proving Ground, MD, 1993.
2. Racine, C. H., Walsh, M. E., Roebuck, B. D., and Collins, C. M., White phosphorus poisoning of waterfowl in an Alaskan salt marsh, *J. Wildl. Dis.,* 28, 669, 1992.
3. Roebuck, B. D., Walsh, M. E., Racine, C. H., and Reitsma, L., Predation of ducks poisoned by white phosphorus: Exposure and risk to predators, *Environ. Toxicol. Chem.,* 13, 1613, 1994.
4. Racine, C. H., Walsh, M. E., Collins, C. M., Calkins, D. J., Roebuck, B. D., and Reitsma, L., Waterfowl Mortality in Eagle River Flats, Alaska: The Role of Munitions Residues, 92–5. U.S. Army Cold Regions Research and Engineering Laboratory, Hanover, NH, 1992.
5. ATSDR, Toxicological Profile for White Phosphorus, Agency for Toxic Substances and Disease Registry, Atlanta, GA, 1994.
6. Pochop, P. A., Cummings, J. L., Yoder, C. A., and Gossweiler, W. A., Physical barrier to reduce WP mortalities of foraging waterfowl, *J. Environ. Eng.,* 126, 182, 2000.
7. Racine, C. H. and Brouillette, M., Ecological inventory of Eagle River Flats, Alaska, in Interagency Expanded Site Investigation: Evaluation of White phosphorus Contamination and Potential Treatability at Eagle River Flats, Alaska, Racine, C. H. and Cate, D., Eds., U.S. Army, Directorate of Public Works, Fort Richardson, AK, 1995.
8. Lawson, D. E., Bigl, S. R., Hunter, L. E., Nadeau, B. M., Weyrick, P. B., and Bodette, J. H., Physical system dynamics, WP fate and transport, remediation and restoration, Eagle River Flats, Ft. Richardson, Alaska, in Interagency Expanded Site Investigation: Evaluation of White Phosphorus Contamination and Potential Treatability at Eagle River Flats, Alaska, Racine, C. H. and Cate, D., Eds., U.S. Army, Directorate of Public Works, Fort Richardson, AK, 1995.
9. Eldridge, W. D., Waterbird utilization of Eagle River Flats: April–October 1994, in Interagency Expanded Site Investigation: Evaluation of White Phosphorus Contamination and Potential Treatability at Eagle River Flats, Alaska, Racine, C. H. and Cate, D., Eds., U.S. Army, Directorate of Public Works, Fort Richardson, AK, 1995.
10. Walsh, M. E., Collins, C. M., and Racine, C. H., Persistence of white phosphorus (P4) particles in salt marsh sediments, *Environ. Toxicol. Chem.,* 15, 846, 1996.
11. Coburn, D. R., DeWitt, J. B., Derby, J. V., Jr., and Ediger, E., Phosphorus poisoning in waterfowl, *J. Am. Pharm. Assoc.,* 3, 151, 1950.
12. Sparling, D. W., Gustafson, M., Klein, P., and Karouna-Renier, N., Toxicity of white phosphorus to waterfowl: Acute exposure in mallards, *J. Wildl. Dis.,* 33, 187, 1997.
13. Roebuck, B. D., Nam, S. I., MacMillan, D. L., Baumgartner, K. J., and Walsh, M. E., Toxicology of white phosphorus (P4) to ducks and risk of their predators: Effects of particle size, *Environ. Toxicol. Chem.,* 17, 511, 1998.
14. Sparling, D. W. and Federoff, N. E., Secondary poisoning of kestrels by white phosphorus, *Ecotoxicology,* 6, 239, 1997.
15. Sparling, D. W., Vann, S., and Grove, R. A., Blood changes in mallards exposed to white phosphorus, *Environ. Toxicol. Chem.,* 17, 2521, 1998.
16. Vann, S. L., Sparling, D. W., and Ottinger, M. A., Effects of white phosphorus on mallard reproduction, *Environ. Toxicol. Chem.,* 19, 2525, 2000.
17. Sparling, D. W., Day, D., and Klein, P., Acute toxicity and sublethal effects of white phosphorus in mute swans, *Cygnus olor, Arch. Environ. Contam. Toxicol.,* 36, 316, 1999.

18. Lee, C. C., Dilley, J. V., and Hodgson, J. R., Mammalian Toxicity of Munition Compounds: Phase I. Acute Oral Toxicity, Primary Skin and Eye Irritation, Dermal Sensitization and Disposition and Metabolism, Environmental Protection Research Division, U.S. Army Medical Research and Development Command, Washington, D.C., 1975.
19. Sparling, D. W., Grove, R., Hill, E., Gustafson, M., and Klein, P., White phosphorus toxicity and bioindicators of exposure in waterfowl and raptors, in Interagency Expanded Site Investigation: Evaluation of White Phosphorus Contamination and Potential Treatability at Eagle River Flats, Alaska, Racine, C. H. and Cate, D., Eds., U.S. Army Cold Regions Research and Engineering Laboratory, Hanover, NH, 1995.
20. Ganote, C. E. and Otis, J. B., Characteristic lesions of yellow phosphorus-induced liver damage, *Lab. Invest.,* 21, 207, 1969.
21. Dianzani, M. U., Liver steatosis induced by white phosphorus, *Morgagni,* 1, 1, 1972.
22. Pani, P., Gravela, E., Mazzarino, C., and Burdin, O. E., On the mechanism of fatty liver in white phosphorus poisoned rats, *Exp. Molec. Pathol.,* 16, 201, 1972.
23. Ghoshal, A. K., Porta, E. A., and Hartcroft, W. S., The role of lipoperoxidation in the pathogenesis of fatty livers induced by phosphorus poisoning in rats, *Am. J. Pathol.,* 54, 275, 1969.
24. Diaz-Rivera, R. S., Ramos-Morales, F., Garcia-Palmieri, M. R., and Ramirez, E. A., The electrocardiographic changes in acute phosphorus poisoning in man, *Am. J. Med. Sci.,* 241, 104, 1961.
25. Sparling, D. W., Grove, R., Hill, E., Gustafson, M., and Comerci, L., Toxicological studies of white phosphorus in waterfowl, in Interagency Site Investigation: Evaluation of White Phosphorus Contamination and Potential Treatability at Eagle River Flats, Alaska, Racine, C. H., Ed., U.S. Army Cold Regions Research and Engineering Laboratory, Hanover, NH, 1993.
26. Franson, J. C., Enzyme activities in plasma, liver, and kidney of black ducks and mallards, *J. Wildl. Manage.,* 18, 481, 1982.
27. Lawrence, J. S. and Huffman, M. M., An increase in the number of monocytes in the blood following subcutaneous administration of yellow phosphorus in oil, *Arch. Pathol.,* 7, 813, 1929.
28. Cameron, J. M. and Patrick, R. S., Acute phosphorus poisoning — The distribution of toxic doses of yellow phosphorus in the tissues of experimental animals, *Med. Sci. Law,* 6, 209, 1966.
29. Ghoshal, A. K., Porta, E. A., and Hartroft, W. S., Isotopic studies on the absorption and tissue distribution of white phosphorus in rats, *Exp. Mol. Pathol.,* 14, 212, 1971.
30. Fletcher, G. L., The dynamics of yellow phosphorus in Atlantic cod and Atlantic salmon: Biological half-times, uptake rates and distribution in tissues, *Environ. Physiol. Biochem.,* 4, 121, 1974.
31. Addison, R. F. and Ackerman, R. G., Direct determination of elemental phosphorus by gas-liquid chromatography, *J. Chromatogr.,* 47, 421, 1970.
32. Nam, S. I., Roebuck, B. D., and Walsh, M. E., Uptake and loss of white phosphorus in American kestrels, *Environ. Toxicol. Chem.,* 13, 637, 1994.
33. Nam, S. I., MacMillan, D. L., and Roebuck, B. D., The translocation of white phosphorus from hen (*Gallus domesticus*) to egg, *Environ. Toxicol. Chem.,* 15, 1564, 1996.
34. Steele, B. B., Reitsma, L. R., Racine, C. H., Burson, I. S. L., Stuart, R., and Theberge, R., Different susceptibilities to white phosphorus poisoning among five species of ducks, *Environ. Toxicol. Chem.,* 16, 2275, 1997.
35. Racine, C. H., Analysis of the Eagle River Flats white phosphorus concentration database, in Interagency Expanded Site Investigation: Evaluation of White Phosphorus Contamination and Potential Treatability at Eagle River Flats, Alaska, Racine, C. H. and Cate, D., Eds., U.S. Army Cold Regions Research and Engineering Laboratory, Hanover, NH, 1995.
36. Cummings, J. L., Clark, L., and Pochop, P. A., Movements, distribution and relative risk of waterfowl, bald eagles and dowitcher using Eagle River Flats, in U.S. Army Eagle River Flats: Protecting Waterfowl from Ingesting White Phosphorus, Cummings, J. L., Ed., USDA, Denver Wildlife Research Center, 1994.
37. Rossi, C., Hazing at Eagle River Flats, in Interagency Expanded Site Investigation: Evaluation of white Phosphorus Contamination and Potential Treatability at Eagle River Flats, Alaska, Racine, C. H. and Cate, D., Eds., U.S. Army, Directorate of Public Works, Fort Richardson, AK, 1995.
38. Clark, L. and Cummings, J. L., U.S. Army Eagle River Flats: Protecting Waterfowl from Ingesting White Phosphorus, Technical Report 93–1, Cummings, J. L., Ed., USDA Denver Wildlife Research Center, Denver, CO, 1993.

39. Cummings, J. L., Clark, L., Pochop, P. A., and Davis, J. E., Jr., Laboratory evaluation of a methyl anthranilate bead formulation on mallard feeding behavior, *J. Wildl. Manage.,* 62, 581, 1998.
40. Henry, K. S., Screening study of barriers to prevent poisoning of waterfowl in Eagle River Flats, Alaska, in Interagency Expanded Site Investigation: Evaluation of White Phosphorus Contamination and Potential Treatability at Eagle River Flats, Alaska, Racine, C. H. and Cate, D., Eds., U.S. Army, Directorate of Public Works, Fort Richardson, AK, 1995.
41. Walsh, M. R., Walsh, M. E., and Collins, C. M., Remediation methods for white phosphorus contamination in a coastal salt marsh, *Environ. Conserv.,* 26, 112, 1999.
42. Budavari, S., O'Neil, M. J., and Smith, A., *The Merck Index*, 11th ed., Merck and Co., Rahway, NJ, 1989.
43. Weast, R. C., Ed., *CRC Handbook of Chemistry and Physics*, 66th ed., CRC Press, Boca Raton, FL, 1985.
44. HSDB, Hazardous Substances Data Base, National Library of Medicine, National Toxicology Information Program, Bethesda, MD, 1993.
45. Spanggord, R. J., Renwick, R., and Chou, T.-W., Environmental Fate of White Phosphorus/Felt and Red Phosphorus/Butyl Rubber Military Screening Smokes, Final Report, DAMD17–82-C-2320, SRI International, Menlo Park, CA, 1985.
46. NSC, Phosphorus (Elemental, White or Yellow), Data sheet I-282, National Safety Council, Chicago, IL, 1990.
47. Sax, N. I., *Dangerous Properties of Industrial Materials*, 6th ed., Van Nostrand Reinhold, New York, 1984.

CHAPTER 28

# A Mining Impacted Stream: Exposure and Effects of Lead and Other Trace Elements on Tree Swallows (*Tachycineta bicolor*) Nesting in the Upper Arkansas River Basin, Colorado

**Christine M. Custer, Thomas W. Custer, Andrew S. Archuleta, Laura C. Coppock, Carol D. Swartz, and John W. Bickham**

## CONTENTS

## 28.1 INTRODUCTION

Hard-rock mining and associated ore processing since the mid-1800s near Leadville, Colorado have contaminated the Arkansas River with lead and other trace elements. Miners worked the watershed extensively for gold, silver, copper, zinc, manganese, and lead causing extensive contamination. Deposits from placer mines and underground mines have disposed large volumes of mining wastes into the Arkansas River. For instance, the 3.4-mile-long Yak tunnel, built between 1895 and 1909, drained numerous underground mines and discharged approximately 210 tons of metals per year into the Arkansas River basin (Environmental Protection Agency [EPA] California Gulch Superfund Web site). Additionally, mercury was used to extract gold from the ore by amalgamation.[1] There are still active, as well as abandoned, mines in the basin. As a result of these many mining activities, a number of contaminated areas designated as EPA Superfund sites exist in the Leadville area.

Water from California Gulch, one such EPA Superfund site, enters the Arkansas River at the south end of Leadville after receiving runoff from numerous tailing piles in the area. This drainage is the primary source of mining wastes in the basin. Lead in sediments of the Arkansas River, sampled in October 1988, increased by 13 times immediately below the confluence with California Gulch;[2] cadmium levels doubled. These sediment levels stayed high for the next 7 km downstream. Lead and cadmium levels then dropped to levels similar to upstream concentrations by approximately 50 km downstream. Concentrations in sediment continued to drop until the final sampling site was reached 190 km downstream.[2] A similar pattern of sediment contamination was present in May 1989 samples as well. Levels of metals at upstream sites are above normal background concentrations, however, because of other mining activities further up the drainage.[3] This highly contaminated segment of the Arkansas River, from the confluence with California Gulch and extending 11 miles downstream, will be referred to in this chapter as the 11-Mile Stretch.

Sediment contamination is transferred into benthic aquatic insects in the Arkansas River. Elevated cadmium levels have been detected in benthic invertebrates immediately downstream of the California Gulch/Arkansas River confluence and for 5 km downstream compared to upstream levels;[4] lead was not measured in that study, however. In the Clark Fork River system in Montana, elevated lead in sediments resulted in elevated lead in benthic invertebrates.[5] Concentrations of trace elements in benthic invertebrates varied by feeding guild; detritivores were the most contaminated, and predators were the least.[4,5]

Animals feeding on benthic aquatic insects may be more or less exposed to sediment-borne contaminants, depending on their food habits. Cadmium concentrations in brown trout (*Salmo trutta*), which feed on benthic aquatic insects from the upper Arkansas River, did not vary between an upstream site and a site 5 km downstream of the confluence with California Gulch,[6] either during the summer or fall of 1991. This may have resulted because cadmium levels in sediment were only slightly elevated below the entry point of California Gulch[2] or because the fish were feeding on less contaminated guilds of benthic invertebrates. No other published data are available on whether elevated trace element concentrations in upper Arkansas River sediments are bioavailable to higher trophic levels or whether they pose a threat to other components of the aquatic food web.

Swallows, especially tree swallows, are now being more-widely used as indicators of local contamination.[7–16] Tree swallows readily use nest boxes, so study sites can be established at specific locations of interest. They feed near their nest box (< 400 m[17]) on emergent aquatic insects,[18] so residues in their tissues reflect sediment contamination for chemicals that transfer into the biota.[19] Data are available on contaminant levels in tree swallows at a number of locations across the U. S. for PCBs,[7,8,14,16] other organochlorines,[9,10,20] metals, including lead,[11,13,21,22] and polycyclic aromatic hydrocarbons.[22]

Contaminant concentrations in tree swallow eggs and nestlings may be a more useful indicator of overall stream contamination than are concentrations in benthic invertebrates or fish. This is especially true over larger geographic areas because fish and invertebrate species composition changes as stream order changes, with the resultant inability to collect the same species at multiple stations even over a fairly small geographic area.[5] Additionally, there may be seasonal variation in availability of benthic aquatic insects and variation in their tolerance to pollutants, which further reduces their utility as bioindicator species.[23] Large variations in metal contamination can occur as a result of differing food habits among aquatic insect taxa,[4,5] which further complicates data interpretation and analysis. Tree swallows integrate contaminant exposure over time and space, are generally found over a large geographic expanse, and can therefore present a much clearer picture of contaminant exposure patterns than invertebrates or fish taxa.

The objectives of this study were to determine whether lead and other trace elements were being transferred from sediments to tree swallows nesting along the Arkansas River, to document contaminant levels of trace metals if they were present, and to determine whether those levels of trace metals were injurious to tree swallows. Our working hypothesis was that there would be a gradient of trace metal contamination, with levels being highest upstream nearer Leadville, CO and lowest farthest downstream.

## 28.2 METHODS

### 28.2.1 Field Collections

In 1997 and 1998, we put up 15 to 35 tree swallow boxes at each of eight sites along the Arkansas River beginning above Leadville on the East Fork of the Arkansas River and extending downstream to Pueblo Reservoir, near Pueblo, CO (Figure 28.1). Locations of study sites (beginning upstream and moving downstream) included Colorado Belle farthest upstream along the East Fork of the Arkansas River; East Fork (near the Bureau of Reclamation's Mine Drainage treatment facility); Docs (at the upper end of the 11-Mile Stretch); Hwy 55 (at the lower end of the 11-Mile Stretch); Buena Vista, Salida, and Pueblo Reservoir. In 1998, tree swallow boxes were not put up at Pueblo Reservoir because of lack of use by tree swallows in 1997. In 1998, we added two new areas: the Shavano Fish Hatchery just downstream from Smeltertown and the Lokey property just downstream from Salida (Figure 28.1); we considered these as one site, called Lokey/Shavano, due to small sample size.

Also in 1998, the Colorado Belle site was expanded to include an area just below the summit of Fremont Pass. We sampled barn swallow (*Hirundo rustica*) nestlings and cliff swallow (*Hirundo*

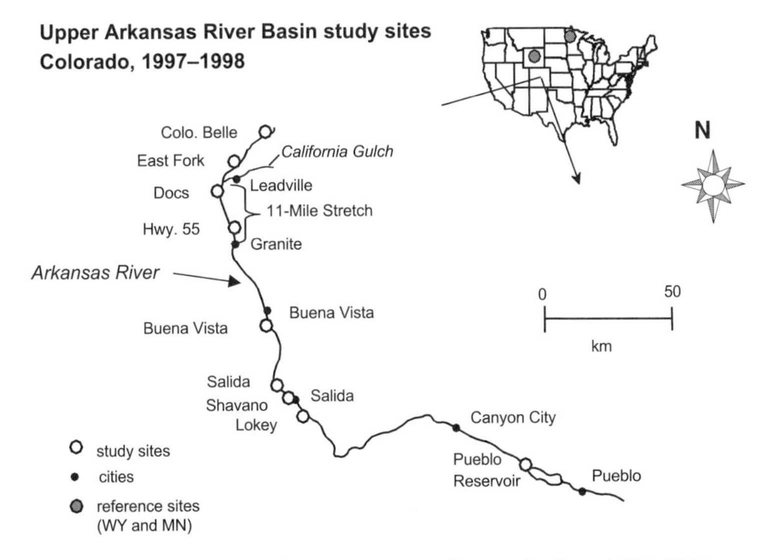


**Figure 28.1** Upper Arkansas River Basin, Colorado tree swallow study sites, 1997–1998.

*pyrrhonota*) eggs at Pueblo Reservoir in lieu of tree swallows in both years because tree swallows did not nest in sufficient numbers at Pueblo Reservoir for an adequate sample. All three species eat aerial insects;[24] however, the similarity or dissimilarity of food habits among the three species at Pueblo Reservoir is unknown. We used two out-of-basin reference sites for comparison of δ-aminolevulinic acid dehydratase (ALAD) activities, Edness Kimball Wilkins State Park, Casper, Wyoming (1997), and Agassiz National Wildlife Refuge, Middle River, Minnesota (1998). These two reference locations were concurrently being studied for trace element exposure and accumulation.[22]

Nest boxes were monitored approximately once per week, and the number of eggs and young present in each box was recorded. We collected two to three eggs from three to five nest boxes at each site approximately mid-way through incubation and collected two 12-day-old nestlings from these same nest boxes for determination of trace metal concentrations in eggs, carcass remainders (entire carcass minus the liver and upper gastrointestinal tract contents), and livers. Unhatched eggs were examined to determine when or why they failed, i.e., if they were infertile, contained dead embryos, were abandoned, etc. These unhatched eggs were not saved for contaminant analysis. Multiple eggs and livers were needed from each nest box to have adequate sample mass for contaminant analyses. Eggs were opened, the contents emptied into a chemically clean jar, and the sample then frozen in a standard freezer. Twelve-day-old nestlings were weighed and then euthanized by decapitation and blood samples collected. The livers were removed, weighed, and placed in a chemically clean jar. Brains were dissected and weighed and then returned to the carcass remainder.

The contents of the stomachs of 12-day-old nestlings were removed and composited by site. Composite diet samples as well as the liver and carcass remainder were submitted to Research Triangle Institute (RTI), Research Triangle Park, North Carolina, for chemical analysis. RTI analyzed for the standard suite of 19 trace elements using standard methodology. Samples were freeze-dried, weighed, and then homogenized in a blender. Subsamples of freeze-dried livers were digested in stages with heat and nitric-perchloric acid and then analyzed for selenium and arsenic by graphite furnace atomic absorption spectrophotometry. Aluminum, barium, beryllium, boron, cadmium, chromium, copper, iron, lead, magnesium, manganese, molybdenum, nickel, strontium, vanadium, and zinc were analyzed by inductively coupled plasma-atomic emission spectrophotometry.

Separate subsamples were digested by nitric acid reflux and analyzed for total mercury by cold vapor atomic absorption spectrophotometry. The nominal lower levels of detection, expressed as μg/g dry weight (dw), were as follows: aluminum (5.0), arsenic (0.5), boron (2.0), barium (0.5), beryllium (0.1), cadmium (0.1), chromium (0.5), copper (0.5), iron (10.0), lead (0.05 in 1997; 0.1 in 1998), mercury (0.1), magnesium (10.0), manganese (0.4), molybdenum (0.5), nickel (0.5), selenium (0.5), strontium (0.2), vanadium (0.5), and zinc (1.0). The number of spikes, duplicates, and blanks was 6 to 8% of the total number of samples analyzed. Recovery rate averaged 106.9% (SE = 0.84, n = 60) for all metals combined; concentrations were not adjusted for recovery. Concentrations of trace elements are reported on a dry-weight basis; approximate wet-weight values can be calculated by using percent moisture values of 81.7% for eggs, 70.3% for livers, and 69.5% for carcass remainders.

Blood samples were taken during decapitation from each 12-day-old nestling for ALAD activity and hematocrit in 1997 and 1998 and for flow cytometry in 1998 only. Blood samples for ALAD were collected in heparinized cryovials and immediately frozen in liquid nitrogen. Samples were analyzed with standard techniques[25,26] by USGS National Wildlife Health Center, Madison, Wisconsin. Each 0.1-mL blood sample was diluted with reagent-grade water and mixed with δ-aminolevulinic acid hydrochloride (ALA) in a sodium phosphorate buffer, pH 6.65. The reaction was stopped after 1 h and the supernatant mixed with Ehrlich's reagent. The absorbance of this mixture was read at 555 nm in a UV-visible spectrophotometer and expressed as nmoles of ALA per minute per mL of red blood cells (RBCs). Hematocrits were determined within 4 h of collection using a microhematocrit capillary tube reader following centrifugation for 5 min at 2000 rpm. Blood for flow cytometry was collected in heparinized capillary tubes, aspirated into freezing media (Hams F10 media with 18% fetal bovine serum and 10% glycerin), inverted several times, and then frozen in liquid nitrogen. Nuclear preparation and staining followed the method of Vindelov.[27] Nuclear suspensions were prepared following lysis with detergent and trypsin digestion. Nuclear suspensions were treated with RNAse and stained with propidium iodide. Nuclear DNA content was analyzed in approximately 10,000 cells per individual on a Coulter Elite flow cytometer by quantification of nuclear fluorescence. The mean half-peak coefficient of variation (HPCV) in the G1 peak was quantified. This value expresses the amount of variation in the DNA content of the 10,000 cells measured. Somatic chromosome damage is suggested when the HPCV in a population is significantly greater than the HPCV in a reference population.

Additionally, food boli were collected from the throats of 8- to 12-day-old nestlings for diet composition in 15 nest boxes in July 1998 using electrical zip ties.[28] Two boxes were sampled twice. The second food sample was collected 4 to 7 days after the first ligature sample was taken. Young from 3 to 4 nest boxes per site were sampled from East Fork, Docs, Buena Vista, and Salida. Lotic, Inc., Unity, Maine, identified insects in one randomly selected bolus per nest box, usually to genus.

### 28.2.2 Statistical Analyses

For most data analyses, trace element concentration data were log-transformed to meet the homogeneity of variance assumption of analysis of variance (ANOVA) and then tested with one- and two-way ANOVA to test for differences between years and among sites within the Arkansas River basin. The antilog and 95% confidence intervals are presented in tables, figures, and text. Only trace elements detected in greater than 50% of samples were statistically analyzed. An exception was lead in eggs, which were statistically analyzed, even though only 41% of eggs had detectable concentrations. Half the detection limit was used in statistical analyses when $< 100\%$ of the samples contained detectable concentrations of a particular chemical. Means were considered significantly different if $P \leq 0.05$ for all statistical tests.

The detection limit for lead differed between years. The percent of livers above the detection limit was calculated based on the detection limit for the year in which the samples were collected. Additionally, the percent above the detection limit was recalculated for 1997 using the detection limit for 1998, the year with the higher detection limit for lead.

HPCV and ALAD were compared among sites and years using ANOVA on the mean values for a nest box, because nest box was our sampling unit. As recommended by Misra and Easton,[29] weighted least squares were used in the ANOVA for HPCV. Analysis of ALAD activity also included comparisons with two out-of-basin reference sites, Edness Kimball Wilkins State Park, Casper, WY (1997) and Agassiz National Wildlife Refuge, Middle River, MN (1998). Simple correlations were run between liver and carcass-remainder concentrations and between liver and diet concentrations for each trace element detected in $> 50\%$ of samples, including only those samples for which an element was detected in both the liver and carcass remainder or in both the liver and diet. Multiple correlations were run between liver and carcass concentrations of trace elements, ALAD, and HPCV for Arkansas River data only. Reproductive data were compiled using Mayfield's[30,31] estimate of daily survival and compared among sites using methods outlined in Hensler and Nichols.[32] Logistic regressions were used to determine the relationship between average lead concentrations in livers and daily survival probabilities.

## 28.3 RESULTS

Occupancy by tree swallows in 1997 ranged from 3% at Pueblo Reservoir to 47% at Buena Vista and East Fork; overall occupancy averaged 24% (Table 28.1). In 1998, the average occupancy averaged 37%, and occupancy ranged from 7 to 66%.

**Table 28.1 Occupancy Rate of Tree Swallow Boxes in the Upper Arkansas River Basin, Colorado, 1997–1998**

| | 1997 | | 1998 | |
|---|---|---|---|---|
| Sites | Number Occupied (# Boxes Available) | Percent Occupied | Number Occupied (# Boxes Available) | Percent Occupied |
| Colorado Belle | 1 (15) | 7 | 2 (30) | 7 |
| East Fork | 7 (15) | 47 | 13 (31) | 42 |
| Docs | 11 (30)[a] | 37 | 20 (35)[b] | 57 |
| Hwy 55 | 3 (29) | 10 | 3 (21) | 14 |
| Buena Vista | 14 (30)[b] | 47 | 17 (32)[b] | 53 |
| Salida | 6 (30) | 20 | 19 (29) | 66 |
| Lokey/Shavano | N/A[c] | | 3 (30) | 10 |
| Pueblo | 1 (30) | 3 | N/A[c] | |
| Overall | 43 (179) | 24 | 77 (208) | 37 |

[a] Includes 3 violet green swallow (*Tachycineta thalassina*) nests.
[b] Includes 1 violet green swallow nest.
[c] N/A = no boxes present.

### 28.3.1 Liver Tissue

Lead was detected in all ten tree swallow livers from Docs (Figure 28.2). The percentages decreased as the distance both downstream and upstream of Docs increased. At Pueblo Reservoir, no lead was detected in the single tree swallow liver (1997) nor was lead detected in the six barn swallow livers in 1997 and 1998. These percentages changed only slightly if the percent above the detection limit for both years were recalculated on the detection limit for 1998 (0.1 μg/g dw) instead. The percent above the detection limit dropped from 83 to 67% at Hwy 55, 50 to 30% at Buena Vista, and 40 to 30% at Salida. This does not change the trend of decreasing percentage containing lead both downstream and upstream of Docs.

Although there was no difference in the concentration of lead in liver between years ($P = 0.970$), both the location ($P < 0.001$) and interaction terms ($P < 0.001$) were significant in the two-way ANOVA. Concentrations of lead were significantly higher at Docs compared to sites farther downstream at Buena Vista, Salida, and Pueblo Reservoir (Table 28.2) and generally declined as the

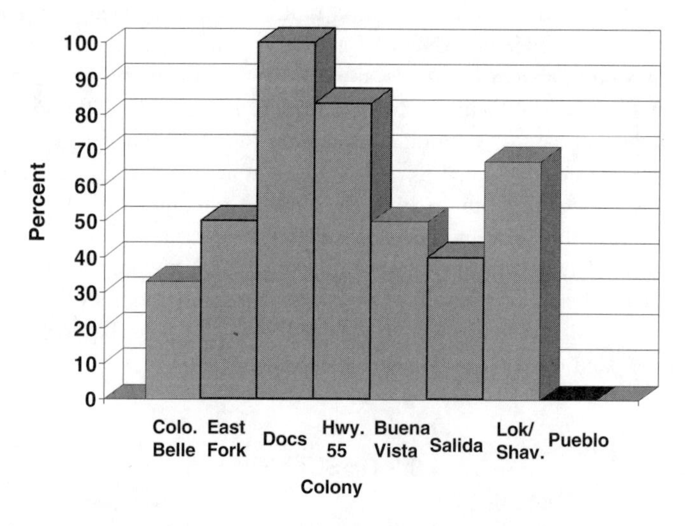


**Figure 28.2** Proportion of liver samples from swallows nesting in the upper Arkansas River basin, Colorado, with detectable concentrations of lead, 1997–1998. Detection limits were 0.05 μg/g dry weight in 1997 and 0.10 in 1998.

**Table 28.2 Trace Element Concentrations (Geometric Mean and 95% Confidence Interval, µg/g dry wt.) in Tree Swallow[b] Livers in the Upper Arkansas River Basin, Colorado, 1997 and 1998[a]**

| Element | Overall Mean | Colo. Belle (*n* = 3)[c] | East Fork (*n* = 10) | Docs (*n* = 10) | Hwy 55 (*n* = 6) | Buena Vista (*n* = 10) | Salida (*n* = 10) | Lokey/Shavano (*n* = 3) | Pueblo[b] (*n* = 7) |
|---|---|---|---|---|---|---|---|---|---|
| Boron-97[a] | 11.01[d]<br>10.26–11.81 | | | | | | | | |
| Boron-98 | 11.98<br>10.3–13.9 | 3.67 b,c[e]<br>1.26–1.34 | 4.44 b<br>2.94–6.71 | 15.31 a,b,c<br>13.2–17.7 | 10.77 a,b,c<br>6.1–19.0 | 21.16 a<br>18.7–24.0 | 18.11 a,c<br>16.9–19.4 | 13.09 a,b,c<br>10.9–15.7 | 25.27 a<br>22.8–28.1 |
| Cadmium | 0.22<br>0.20–0.24 | 0.21 a,b,c<br>0.16–0.27 | 0.14 b,c<br>0.11–0.17 | 0.19 b,c<br>0.16–0.22 | 0.24 a,b<br>0.17–0.33 | 0.52 a<br>0.46–0.58 | 0.25 a,b<br>0.23–0.27 | 0.40 a,b<br>0.31–0.52 | 0.09 c<br>0.07–0.12 |
| Copper | 17.1<br>16.4–17.8 | 24.1 a,b<br>18.0–32.2 | 15.2 b<br>14.2–16.3 | 15.8 b<br>15.0–16.6 | 14.4 b<br>13.5–15.4 | 19.7 a,b<br>17.8–21.9 | 15.9 b<br>14.4–17.5 | 30.4 a<br>25.0–36.9 | 15.9 b<br>15.2–16.7 |
| Lead | 0.12<br>0.10–0.14 | 0.10 a,b,c<br>0.05–0.20 | 0.14 a,b,c<br>0.10–0.20 | 0.54 a<br>0.46–0.65 | 0.14 a,b,c<br>0.11–0.19 | 0.09 b,c<br>0.06–0.13 | 0.06 b,c<br>0.04–0.08 | 0.38 a,b<br>0.14–1.05 | 0.03 c |
| Mercury[f] | 0.067<br>0.064–0.071 | 67% | 50% | 80% | 17% | 20% | 10% | 0% | 0% |
| Selenium | 6.59<br>6.31–6.88 | 7.72 a,b,c<br>7.10–8.40 | 7.86 a,b<br>7.36–8.40 | 5.21 c<br>4.85–5.61 | 5.81 b,c<br>5.01–6.75 | 5.25 c<br>4.95–5.58 | 6.32 b,c<br>5.89–6.77 | 6.77 a,b,c<br>6.36–7.21 | 10.76 a<br>9.55–12.12 |

*Note:* Arsenic, barium, beryllium, nickel, and vanadium were not detected in any liver sample. Aluminum and chromium were detected in one liver each.

[a] Years are presented separately if significantly different.
[b] At Pueblo, livers were from barn and tree swallows.
[c] Number of nest boxes sampled, both years combined.
[d] No significant difference among sites in 1997.
[e] Means sharing same letter are not significantly different.
[f] Percentage with concentrations above the detection limit. Means not compared because fewer than 50% above the detection limit.

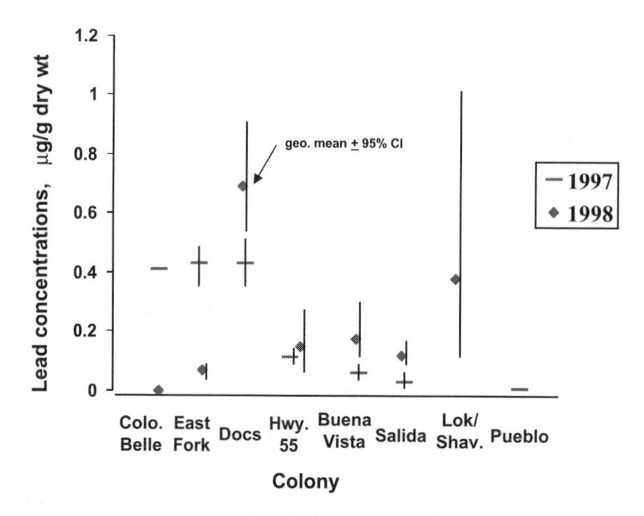


**Figure 28.3** Geometric mean concentrations (horizontal lines and diamonds) and 95% CI (vertical lines) of lead (µg/g dry weight), by year, in swallow livers in the upper Arkansas River basin, Colorado, 1997–1998.

distance from the contamination sourçe increased. An exception was at Lokey/Shavano. The level there did not differ from Docs, although it was about half the concentration (0.38 vs. 0.54 µg/g dw.). The interaction term in the two-way ANOVA was significant mainly because East Fork had marked differences in mean lead concentrations between years (Figure 28.3). Four of four livers in 1997 had detectable concentrations of lead, but only one of six did in 1998.

Arsenic, barium, beryllium, nickel, and vanadium were not detected in any liver samples; aluminum and chromium were detected in only one liver each. For the other 12 trace elements, concentrations in liver tissue (n = 59) did not vary between years except for boron, which was slightly, but significantly, higher in 1998 than in 1997 (Table 28.2). Boron in 1997 (Table 28.2), iron, magnesium, manganese, molybdenum, strontium, and zinc also did not vary among sites within the Arkansas River basin (Table 28.3). Boron in 1998, cadmium, copper, and selenium did vary among locations along the Arkansas River (Table 28.2). The upstream–downstream pattern, i.e., highest in the drainage and decreasing downstream, did not follow for the four metals that varied among sites. For example, cadmium was highest at Buena Vista, copper was highest at Lokey/Shavano, and selenium was highest at Pueblo (Table 28.2). Mercury was detected in too few liver samples to statistically analyze (19 of 59 with detectable concentrations); however, the frequency of occurrence varied among sites ($\chi^2$, P = 0.002). None of the pair-wise site differences were significant (at P < 0.002, n = 28 pair-wise comparisons); however, the frequency of mercury was higher at the three sites near Leadville, (Colorado Belle — 67% with detectable concentrations, East Fork — 50%, and Docs — 80%), than at sites farther downstream (0–20% with detectable concentrations) ($\chi^2$, P < 0.001).

## 28.3.2 Eggs

Concentrations of trace elements in eggs varied among sites in the Arkansas River drainage, and these differences varied slightly between years (i.e., the interaction term in the two-way ANOVA was significant); therefore, data are presented for each year separately for most trace elements (Table 28.4). Lead was detected in only 19 of 46 eggs and did not vary among sites. This was expected because lead generally does not accumulate in eggs and is not considered a tissue suitable for diagnostic purposes.[33] Lead averaged 0.09 µg/g dw (95% CI = 0.08–0.11). Beryllium, cadmium, chromium, nickel, molybdenum, and vanadium were detected in fewer than 50% of the egg samples so were not statistically analyzed. For the three chemicals that varied between years, boron and

**Table 28.3 Trace Element Concentrations (Geometric Mean and 95% Confidence Interval, µg/g dry wgt.) in Tree Swallow Livers That Did Not Vary among Sites or between Years, and in Carcass Remainders and Diet along the Upper Arkansas River Basin, Colorado, 1997 and 1998**

| Element | Liver (*n* = 59) | Carcasses (*n* = 59) | Diet (*n* = 20) |
|---|---|---|---|
| Aluminum | 58ND[a] | 28.8<br>27.2–30.6 | 439<br>273–707 |
| Arsenic | 59ND | 54ND | 0.59<br>0.48–0.71 |
| Barium | 59ND | 3.46<br>3.28–3.67 | 12.9<br>10.9–15.2 |
| Boron | Table 28.2 | 57ND | 21.9<br>19.1–25.1 |
| Cadmium | Table 28.2 | 0.29<br>0.26–0.32 | 2.93<br>2.52–3.41 |
| Copper | Table 28.2 | 7.19<br>7.05–7.35 | 32.1<br>29.9–34.4 |
| Chromium | 58ND | 11.54[b]<br>9.32–14.30 | 16ND |
| Iron | 1332<br>1270–1398 | 258[b]<br>234–284 | 534<br>479–597 |
| Lead | Table 28.2 | 0.41<br>0.36–0.47 | 3.10[c]<br>2.00–4.82 |
| Magnesium | 684<br>676–693 | 812<br>803–822 | 893<br>817–977 |
| Manganese | 4.51<br>4.38–4.63 | 3.93[b]<br>3.60–4.29 | 43.5<br>35.4–53.4 |
| Molybdenum | 2.14<br>2.05–2.23 | 30ND | 1.96<br>1.69–2.26 |
| Nickel | 59ND | 1.14<br>0.94–1.39 | 13ND |
| Selenium | Table 28.2 | 3.11<br>2.96–3.28 | 1.98<br>1.72–2.27 |
| Strontium | 0.21<br>0.20–0.23 | 17.0<br>16.1–17.9 | 35.0<br>25.9–47.5 |
| Zinc | 74<br>73–75 | 144[b]<br>140–148 | 188<br>175–204 |

[a] Number with nondetectable concentrations.
[b] 1998 data only due to inconsistencies with chemical analysis in 1997. Sample size for 1998 data are *n* = 32 for carcasses and 13 for diet.
[c] Contains one sample with 892 µg/g lead.

iron were higher in 1998 than in 1997, strontium was lower in 1998 than in 1997, and no other between-year differences were found. For trace elements, which varied by location, concentrations did not follow a predictable upstream/downstream pattern. Of the trace elements known to be embryotoxic, selenium was significantly higher at Pueblo Reservoir than at the other sites. Mercury was significantly lower at Pueblo Reservoir (1998 only) than at the other sites (Table 28.4). Barium was detected in all eggs and did not vary among sites; the geometric mean concentration and 95% CI were 3.67 µg/g dw, 3.67–4.35.

### 28.3.3 Whole Carcass Remainders

Some trace elements were detected in carcass remainders but not in liver tissue and vice versa. Trace elements that were detected in all 59 carcass remainders, but in one or fewer livers were aluminum, barium, and chromium (Table 283). Detectable concentrations of nickel (35 of 59

**Table 28.4 Trace Element Concentrations (Geometric Mean and 95% Confidence Interval, μg/g dry wgt.) in Tree Swallow[a] Eggs in the Upper Arkansas River Basin, Colorado, 1997 and 1998**

| Element-Year (Year Diff.) | Overall Mean | Colo. Belle (*n* = 2)[b] | East Fork (*n* = 7) | Docs (*n* = 7) | Hwy 55 (*n* = 2) | Buena Vista (*n* = 8) | Salida (*n* = 8) | Lokey/Shavano (*n* = 3) | Pueblo[a] (*n* = 9) |
|---|---|---|---|---|---|---|---|---|---|
| Aluminum-97[c] | 10.21 | | 12.08 a,b[d] | 18.47 a,b | | 5.82 a,b | 2.50 b | | 30.54 a |
| | 7.45–13.99 | | 10.0–14.6 | 17.4–19.6 | | 2.50–13.57 | 2.5–2.5 | | 22.1–42.2 |
| Aluminum-98 | 14.00 | 15.00 a,b | 7.31 b | 9.66 b | 9.54 a,b | 16.67 a,b | 15.49 a,b | 13.53 a,b | 34.19 a |
| | 12.5–15.7 | 12.1–18.6 | 6.9–7.7 | 8.8–10.6 | 7.5–12.2 | 14.2–19.5 | 13.2–18.1 | 13.4–13.7 | 21.5–54.4 |
| Boron-97 (98 > 97)[e] | 11.53 | | 13.41 a,b | 20.46 a | | 14.43 a,b | 13.59 a,b | | 6.00 b |
| | 10.0–13.3 | | 11.6–15.5 | 17.6–23.8 | | 11.3–18.4 | 11.1–16.6 | | 5.16–6.98 |
| Boron-98 | 32.87 | 10.57 b | 31.94 a | 34.35 a | 33.25 a,b | 38.95 a | 33.8 a | 40.58 a | 36.7 a |
| | 30.4–35.6 | 4.3–25.9 | 29.6–34.5 | 32.0–36.9 | 32.7–33.8 | 36.3–41.8 | 31.9–35.9 | 35.0–47.1 | 28.8–46.8 |
| Copper-97 | 3.19 | | 2.81 a,b | 2.09 b | | 2.46 b | 2.52 b | | 6.10 a |
| | 2.78–3.66 | | 2.63–3.00 | 1.82–2.39 | | 2.27–2.67 | 2.28–2.78 | | 4.64–8.03 |
| Copper-98 | 2.80 | 2.70 a | 2.33 a | 3.04 a | 2.63 a | 2.57 a | 3.17 a | 3.25 a | 2.85 a |
| | 2.69–2.91 | 1.99–3.66 | 2.20–2.46 | 2.67–3.46 | 2.56–2.71 | 2.33–2.82 | 2.99–3.36 | 2.88–3.67 | 2.59–3.13 |
| Iron-97 (98 > 97) | 134.4 | | 123.6 a,b | 136.6 a,b | | 101.5 b | 108.4 a,b | | 202.0 a |
| | 122–148 | | 107–142 | 127–147 | | 89–115 | 97–121 | | 170–240 |
| Iron-98 | 153.9 | 157.5 a | 150.6 a | 159.5 a | 146.4 a | 148.3 a | 140.9 a | 151.8 a | 175.1 a |
| | 149–159 | 157–157 | 140–162 | 145–176 | 140–153 | 137–160 | 135–147 | 146–157 | 151–204 |
| Magnesium-97 | 452.3 | | 291.0 b | 317.8 b | | 374.9 b | 382.4 b | | 878.7 a |
| | 398–514 | | 290–292 | 297–340 | | 352–399 | 352–415 | | 750–1030 |
| Magnesium-98 | 399.1 | 476.7 a,b | 329.2 b | 359.6 a,b | 329.7 a,b | 455.9 a,b | 374.4 a,b | 329.8 b | 565.0 a |
| | 381–418 | 460–494 | 617–342 | 342–378 | 316–344 | 406–512 | 345–406 | 323–337 | 480–665 |
| Manganese-97 | 3.07 | | 1.83 b | 2.46 a,b | | 2.43 a,b | 2.41 a,b | | 6.36 a |
| | 2.64–3.58 | | 1.69–1.99 | 2.44–2.48 | | 2.10–2.82 | 2.12–2.75 | | 4.82–8.38 |
| Manganese-98 | 2.85 | 4.00 a | 2.71 a | 2.63 a | 3.10 a | 2.67 a | 2.14 a | 2.63 a | 4.13 a |
| | 2.67–3.05 | 3.82–4.19 | 2.56–2.87 | 2.40–2.87 | 2.50–3.84 | 2.40–2.97 | 1.80–2.53 | 2.18–3.18 | 3.07–5.56 |

| | | | | | | | | | |
|---|---|---|---|---|---|---|---|---|---|
| Mercury-97 | 0.18 | | 0.18 a | 0.24 a | | 0.23 a | 0.27 a | | 0.09 a |
| | 0.15–0.21 | | 0.15–0.22 | 0.20–0.27 | | 0.19–0.29 | 0.24–0.30 | | 0.06–0.14 |
| Mercury-98 | 0.16 | 0.10 a,b | 0.29 a | 0.29 a | 0.11 a,b | 0.22 a | 0.13 a | 0.18 a | 0.05 b |
| | 0.14–0.18 | 0.05–0.19 | 0.23–0.36 | 0.25–0.33 | 0.05–0.23 | 0.21–0.24 | 0.10–0.17 | 0.17–0.19 | 0.05–0.05 |
| Selenium-97 | 3.12 | | 2.22 b | 2.33 b | | 2.23 b | 2.26 b | | 6.97 a |
| | 2.70–3.6 | | 2.10–2.35 | 2.06–2.64 | | 2.18–2.29 | 2.24–2.27 | | 6.24–7.80 |
| Selenium-98 | 2.88 | 2.90 b,c | 2.93 b | 2.74 b,c | 2.77 b,c | 2.12 c,d | 1.98 d | 2.38 b,c,d | 6.72 a |
| | 2.68–3.10 | 2.65–3.18 | 2.81–3.06 | 2.61–2.87 | 2.20–3.49 | 1.99–2.26 | 1.92–2.05 | 2.24–2.52 | 6.57–6.88 |
| Strontium-97 (97 > 98) | 18.8 | | 3.77 b | 17.90 a | | 23.11 a | 17.17 a | | 39.79 a |
| | 15.1–23.49 | | 3.55–4.00 | 17.6–18.2 | | 20.4–26.2 | 11.7–25.2 | | 34.0–46.5 |
| Strontium-98 | 10.95 | 9.94 a,b,c | 4.64 c | 7.83 b,c | 8.74 a,b,c | 16.48 a,b | 12.86 a,b | 8.34 b,c | 27.35 a |
| | 9.74–12.31 | 6.5–15.2 | 4.17–5.17 | 6.92–8.87 | 7.96–9.60 | 13.3–20.4 | 10.0–16.5 | 8.19–8.50 | 22.5–33.2 |
| Zinc-97 | 62.23 | | 51.67 b | 63.29 a,b | | 54.83 b | 47.57 b | | 91.10 a |
| | 57.7–67.2 | | 50.0–53.4 | 62.4–64.2 | | 51.1–58.8 | 46.4–48.7 | | 82.8–100.3 |
| Zinc-98 | 61.95 | 68.55 a | 58.10 a | 66.43 a | 66.07 a | 56.82 a | 51.55 a | 75.13 a | 67.3 a |
| | 59.7–64.3 | 68.1–69.0 | 54.3–62.1 | 63.9–69.1 | 53.7–81.3 | 53.1–60.8 | 48.9–54.4 | 60.8–92.8 | 59.3–76.4 |

*Note:* Beryllium was not detected in any sample. Cadmium was detected in 4 of 46 eggs, chromium in 3 of 46, nickel in 2 of 46, and molybdenum and vanadium were each detected in 1 of 46 eggs.

[a] Pueblo site includes cliff, barn, and tree swallows.
[b] *N* is number of nest boxes from which eggs were collected in 1997 and 1998 combined.
[c] Years were analyzed separately because the interaction term was significant in the two-way ANOVA.
[d] Means sharing same letter are not significantly different. Means only separated with Bonferonni tests if *P* for Bonferonni was set at 0.10 for both copper and iron in 1997. All other means separated at $P < 0.05$.
[e] Indicates when years were significantly different.

carcasses) and vanadium (21 of 59 carcasses) were present in carcass remainders, but neither was detected in liver tissue. Boron, on the other hand, was detected in 58 of 59 livers (Table 28.2) but was detected in only two carcass remainders. For the trace elements detected in > 50% of livers and carcass remainders, there was no correlation between concentrations in the two tissue types for copper ($P = 0.684$), iron ($P = 0.728$), manganese ($P = 0.173$), strontium ($P = 0.705$), or zinc ($P = 0.640$) ($n = 59$, except for strontium with $n = 37$). There were, however, significant positive correlations between liver and carcass-remainder concentrations for cadmium ($P < 0.001$, $R = 0.519$, $n = 55$), lead (0.002, $R = 0.444$, $n = 46$), magnesium ($P < 0.001$, $R = 0.466$, $n = 59$), and selenium ($P < 0.001$, $R = 0.857$, $n = 59$). For the 55 livers and carcasses with detectable concentrations of cadmium, concentrations averaged 25% less in liver (0.29 μg/g dw) compared with carcass remainder (0.38 μg/g dw), lead was 60% less (0.32 and 0.90 μg/g dw in liver and carcasses, n = 46), and magnesium was 16% less (Table 28.3). Selenium concentrations, however, were twice as high in liver compared to the carcass remainder.

### 28.3.4 Diet

Concentrations of trace elements in diet samples generally did not vary among locations within the Arkansas River drainage for most trace elements. This may have been due in part to the small sample sizes for diet, only one or two pooled samples per site per year (Table 28.3 and Appendix 28.1). Although not statistically different, lead concentrations in diet followed the hypothesized upstream/downstream gradient of suspected lead availability. The order of lead concentrations in the diet from highest to lowest was: Docs, Lokey/Shavano, East Fork, Hwy 55, Salida, Buena Vista, Pueblo, and Colorado Belle. Lead concentrations in the diet at East Fork were 19.2 μg/g dw in 1997 and only 2.2 μg/g dw in 1998. One diet sample from Lokey/Shavano contained 892 μg/g dw lead. If this sample is removed from the analysis, then lead concentration in diet averaged 2.30 μg/g dw (95% CI = 1.63–3.24), and the order of descending lead concentrations drops Lokey/Shavano from near the top to the bottom. Concentrations in diet significantly correlated with concentrations in liver tissues for lead ($P = 0.007$, $R^2 = 0.47$), selenium ($P = 0.028$, $R^2 = 0.34$), and boron ($P = 0.031$, $R^2 = 0.333$). There was no relationship between diet and liver concentrations for copper, magnesium, manganese, molybdenum, strontium, or zinc (all Ps > 0.20). The correlation approached significance for cadmium ($P = 0.091$). Trace elements found in fewer than 50% of diet samples ($n = 20$) were beryllium (0 samples), mercury (0 samples), nickel (7 samples), and vanadium (9 samples).

Tree swallows were eating primarily insects of aquatic origin. Ninety-four percent of the boli collected from the throats of nestlings contained aquatic insects (Table 28.5). On average, there were 20 aquatic insects in each food bolus and one terrestrial insect. The two most common insect taxa were mayflies (Ephemeroptera) and true flies (Diptera, Table 28.6). Nearly all of the true flies were aquatic in origin. The most common mayfly genera in swallow boli were *Epeorus* (in seven boli), *Ephemerella* (six boli), *Rithrogena* (four boli) and *Callibaetis* (three boli). The five mayfly genera found in two or fewer boli were *Ameletus, Baetis, Paraleptophlebia, Siphlonurus,* and *Timpanoga*.

**Table 28.5 Food in Boli of Tree Swallow Nestlings along the Upper Arkansas River Basin, Colorado, 1998**

| Origin of Insects | Freq. of Occurrence[a] (*n* = 17 Boli) | Average Number of Insects (per Bolus) | Percent of Diet (by Number of Insects) |
|---|---|---|---|
| Aquatic | 16 (94%) | 20.3 | 86% |
| Terrestrial | 8 (47%) | 1.0 | 14% |

[a] Number and percent of boli that contained each type of insect.

**Table 28.6 Frequency of Insect Taxa in Boli of Tree Swallow Nestlings in the Upper Arkansas River Basin, Colorado in 1998**

| Taxa of Insects (Common Name and Order) | Frequency of Occurrence[a] (*n* = 17 Insect Boli) | Primary Habitat of Insects |
|---|---|---|
| Mayflies (Ephemeroptera) | 88 | aquatic |
| True Flies (Diptera) | 47 | aquatic[b] |
| Caddisflies (Trichoptera) | 18 | aquatic |
| Waterbugs (Hemiptera) | 18 | terrestrial |
| Leafhopper (Homoptera) | 12 | terrestrial |
| Stoneflies (Plecoptera) | 6 | aquatic |
| Beetles (Coleoptera) | 6 | terrestrial |
| Spongillaflies (Neuroptera) | 6 | terrestrial |

[a] Percent of boli that contained each taxon of insect.
[b] 98% of the flies were of aquatic origin.

## 28.3.5 Biomarkers

### *28.3.5.1 ALAD*

ALAD activity at the reference site averaged 74.2 nmol/min/mL of red blood cells. Mean activity of ALAD was significantly reduced along the Arkansas River at East Fork, Docs, Hwy 55, and Buena Vista when compared to the reference sites ($P < 0.001$) (Figure 28.4). Activity of ALAD at Salida, Lokey/Shavano, and Colorado Belle did not differ from the reference values. ALAD values were between 61 and 95% of the reference value. Overall, ALAD at the six Arkansas River basin sites was 74% of the reference value. ALAD values were lower in 1998 than in 1997 ($P < 0.001$), but the interaction term was not significant ($P = 0.171$). ALAD activities at the reference sites were similar between years ($P = 0.661$). The plot of mean ALAD values demonstrates a similar

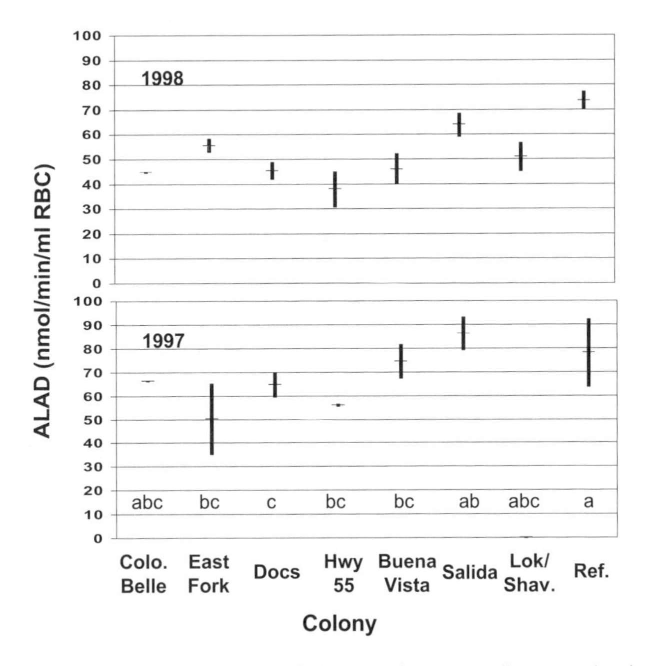


**Figure 28.4** Average ALAD activity (horizontal lines), by year, in tree swallows nesting in the upper Arkansas River basin, CO, 1997–1998 and at the reference sites in Casper, WY (1997) or Agassiz National Wildlife Refuge, Middle River, MN (1998). Vertical lines are ± 1 standard error. Means sharing same letter are not significantly different, and are for both years combined.

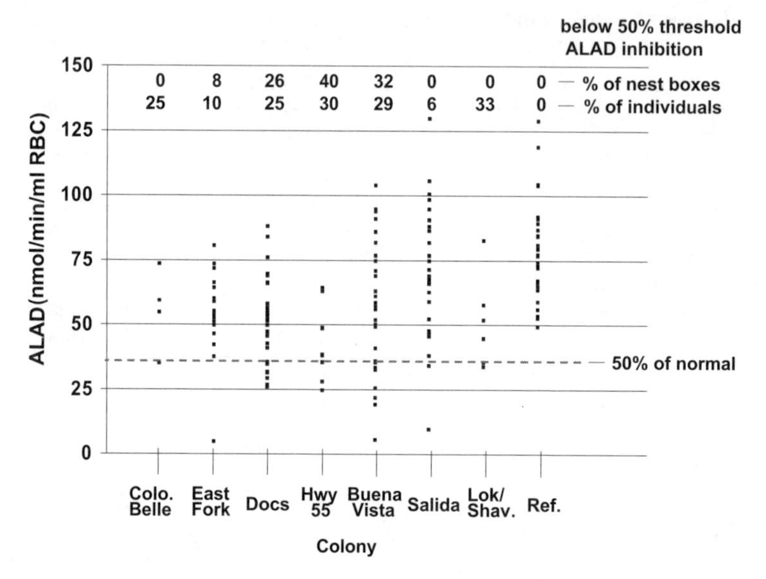


**Figure 28.5** Plot of ALAD values for individual nestlings from the upper Arkansas River Basin, 1997–1998 and from reference locations in WY and MN. The dashed line indicates fifty percent of reference ALAD value.

pattern between years, except for East Fork (Figure 28.4). ALAD values at East Fork were qualitatively, but not significantly, less than Docs in 1997, but not in 1998.

The percent of the samples that had a ≥ 50% reduction in ALAD, when compared to the reference mean, varied between 7 and 40% of nest boxes (Figure 28.5). The plot of individual ALAD values (Figure 28.5) also demonstrated that numerous values were less than the 50% threshold. Overall, 18% of nest boxes (16 of 89) and 20% of individual nestlings (27 of 135) had ALAD reductions ≥ 50% of normal values. There was a significant negative correlation between ALAD and log lead liver concentrations ($P < 0.001$, $n = 48$); however, the amount of variation in ALAD levels explained by this relationship was small ($R^2 = 24\%$).

Hematocrit varied with altitude (Figure 28.6). Hematocrits were significantly higher at higher elevation sites — East Fork, Docs, and Hwy 55 — when compared to lower elevation sites such as Salida, Casper, WY and Agassiz NWR, MN. The slope of altitude did not appear to be linear,

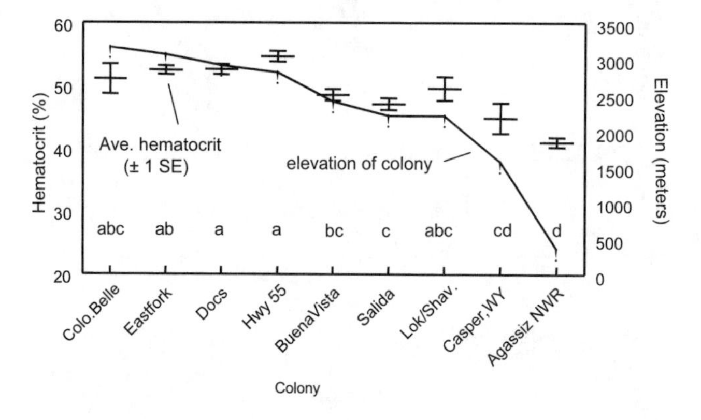


**Figure 28.6** Average hematocrit values (horizontal lines) ± 1 standard error (vertical lines) for tree swallows nesting along the upper Arkansas River, CO, 1997–1998, and at the reference sites at Casper, WY (1997) or Agassiz National Wildlife Refuge, Middle River, MN (1998). Elevations of nesting sites are indicated by the solid connecting line. Hematocrit values sharing the same letter are not significantly different.

**Table 28.7 Correlation Matrix between Half-Peak Coefficient of Variation (HPCV) and Trace Elements That had Significant Correlations for Tree Swallows Nesting along the Upper Arkansas River, Colorado, 1998, *n* = 28**

| | Cadmium (Carcass) | Cadmium (Liver) | Chromium (Carcass) | Magnesium (Liver) |
|---|---|---|---|---|
| HPCV | **0.023** (0.429) | **0.019** (0.440) | **0.049** (–0.376) | **0.017** (–0.449) |
| Cadmium (carcass) | | **0.001** (0.573) | 0.783 (0.054) | **0.005** (–0.511) |
| Cadmium (liver) | | | 0.907 (–0.022) | 0.262 (–0.215) |
| Chromium (carcass) | | | | 0.747 (0.063) |

*Note:* Significant *P*-values are in bold; *R*-values are in parentheses.

i.e., there was a maximum mean hematocrit of approximately 55% and a minimum mean hematocrit of 41%.

#### *28.3.5.2 Flow Cytometry*

Half-peak coefficient of variation (HPCV), which expresses the amount of variation in the quantity of DNA in each red blood cell, between the “a” and “b” nestlings in a box was highly correlated (P = 0.016, R = 0.459). HPCV was positively correlated with carcass and liver cadmium levels (Table 28.7), which were also correlated with one another, as might be expected (P = 0.001). HPCV was negatively correlated with both carcass chromium and liver magnesium (Table 28.7). No other trace elements correlated with HPCV, and no differences were found among sites (P = 0.272, df 6,56) within the Arkansas River basin. Mean ± 1 SE and (n) for each site are as follows: Colorado Belle 3.46 (1); East Fork 3.86 ± 0.08 (10); Docs 3.74 ± 0.10 (16); Hwy55 3.62 ± 0.18 (3); Buena Vista 4.01 ± 0.12 (14); Salida 3.95 ± 0.08 (16); and Lokey/Shavano 4.01 ± 0.21 (3). Overall mean HPCV was 3.87 ± 0.047 (SE), range = 3.07–4.91, n = 63.

### 28.3.6 Reproduction Measures

The modal clutch size was five eggs per clutch and was similar between years. The probability of daily egg survival generally did not differ between the two years of the study, except at Buena Vista (Table 28.8). Buena Vista had a higher probability of egg survival in 1998 than in 1997. In

**Table 28.8 Daily Egg Survival Probabilities (Mayfield Analyses) for Tree Swallows Nesting in the Upper Arkansas River Basin, Colorado, 1997–1998**

| | 1997 | | 1998 | | |
|---|---|---|---|---|---|
| Sites | Probability of Egg Survival | Standard Error | Probability of Egg Survival | Standard Error | Signif. between Years |
| Colo. Belle | 1.0000 a[a] | 0.0000 | 1.0000 a[a] | 0.0000 | 1.000 |
| East Fork | 0.9858 a | 0.0054 | 0.9835 b,c | 0.0050 | 0.755 |
| Docs | 0.9794 a | 0.0057 | 0.9724 c | 0.0047 | 0.343 |
| Hwy. 55 | 0.9861 a | 0.0081 | 0.9903 a,b | 0.0069 | 0.693 |
| Buena Vista | 0.9802 a[b] | 0.0048 | 0.9982 a[b] | 0.0013 | <0.001 |
| Salida | 0.9863 a[b] | 0.0062 | 0.9921 b | 0.0025 | 0.386 |
| Lokey/Shavano | | | 1.0000 a | 0.0000 | |

[a] Means sharing same letter do not vary among locations. Each year analyzed separately.
[b] Mayfield analyses do not include the one nest lost to predation each year at these sites.

1998, the probability of egg survival was significantly less at Docs compared to the other sites, except for East Fork. There was no correlation, however, between average lead concentrations in livers and probability of daily egg survival on a colony basis ($P = 0.214$, $n = 12$). On an individual nest basis, the logistic regression between liver lead and nest success also was not significant ($P = 0.18$, $n = 52$).

The reasons for egg failure varied among sites (Table 28.9). Docs in 1997 had nearly 23% of eggs disappear at hatching. These eggs either failed to hatch or hatched and the nestlings died shortly thereafter. In 1998, lowered hatching success at Docs was due to nest abandonment or death of the adult at the box. Reduced hatching at Buena Vista in 1997 was due to depredation by bears (*Ursus* sp.) and a higher rate of infertility; hatching success at Buena Vista was excellent in 1998.

### 28.3.7 Morphology

Whole-body and liver weights of 12-day-old nestlings did not vary among sites along the Arkansas River, but whole-body weights of Arkansas River tree swallows (21.5 g, $n = 133$) were significantly less than those of the reference birds from WY and MN (22.9 g, $n = 55$) ($P < 0.001$). Nestling whole-body weights in 1997 (22.2 g) were significantly greater than in 1998 (21.8 g) ($P = 0.009$, Arkansas River and reference sites combined). Liver weights followed a similar pattern; livers from the Arkansas River locations (1.13 g, $n = 130$) did not differ from one another. The reference location livers (1.05 g, $n = 55$) weighed significantly less than livers from East Fork (1.21 g) and Salida (1.17 g) in 1997 and 1998 but did not differ from the other Arkansas River sites. In 1997, livers weighed significantly more (1.20 g) than they did in 1998 (1.07 g, $P = 0.003$). The ratio of liver to body weight varied among locations ($P = 0.010$); the reference location had a lower ratio (0.046, $n = 54$) than did Buena Vista (0.056, $n = 32$) or East Fork (0.056, $n = 24$). No other sites differed from one another (0.051, $n = 85$). Brain weights were similar along the Arkansas River (0.399 g, $n = 122$); only brain weights at Colorado Belle (0.341 g) differed from the reference birds (0.428 g, $n = 39$) ($P = 0.004$). When body weight was taken into consideration (brain weight divided by body weight), there were no significant differences among locations or between years ($P = 0.074$).

## 28.4 DISCUSSION

### 28.4.1 Liver Concentrations — Lead

The frequency and concentrations of lead in nestling livers conclusively demonstrated that lead from sediments was bioavailable to tree swallows nesting along the Arkansas River. All livers at Docs, the site closest to the discharges from California Gulch and the area with the highest sediment lead concentrations,[2] contained detectable concentrations of lead, and these concentrations were significantly higher compared to sites farther downstream at Buena Vista, Salida, and Pueblo Reservoir (Table 28.2). Over 80% of livers at Hwy 55, at the lower end of the 11-Mile Stretch, also contained detectable concentrations of lead, but at lower concentrations. Concentrations of lead in tree swallows generally declined as distance from the contamination source increased, just as sediment concentrations did.[2] An exception was at Lokey/Shavano (0.38 μg/g dw in livers), which did not differ from Docs (0.54 μg/g dw). The reason for the relatively high lead concentrations at Lokey/Shavano may be a lead source from an abandoned slag pile just upstream of Lokey/Shavano (ASA, unpublished data). The two sites above Leadville — East Fork and Colorado Belle — had similar concentrations of lead as sites downstream of the 11-Mile Stretch. These upstream sites probably reflect inputs of mining activities farther up the drainage, such as St. Kevins Gulch.[3] The high correlation ($P = 0.007$, $R^2 = 0.47$) between concentrations of lead in the diet and in the liver

**Table 28.9 Fate of Eggs of Tree Swallows Nesting in the Arkansas River Basin, Colorado, 1997–1998**

| | | | Percent Number of Eggs | | | | | | |
|---|---|---|---|---|---|---|---|---|---|
| | | | | Failed to Hatch | | | | | |
| Year & Site | No. of Nest Attempts[a] | Number of Eggs[b] | Hatched | Cause Unknown[c] | Dead Embryo | Infertile | Disappeared at Hatching[d] | Depredated | Abandoned |
| **1997** | | | | | | | | | |
| Colo. Belle | 1 | 6 | 100.0 | | | | | | |
| East Fork | 7 | 31 | 77.4 | 6.5 | 6.5 | | 9.7 | | |
| Docs | 9 | 44 | 70.5 | 2.3 | 2.3 | 2.3 | 22.7 | | |
| Hwy 55 | 3 | 16 | 81.3 | | | 6.3 | 12.5 | | |
| Buena Vista | 15 | 66 | 65.2 | 1.5 | 6.1 | 9.1 | 6.1 | 7.6 | 4.5 |
| Salida | 7 | 24 | 87.5 | | | 8.3 | | 4.2 | |
| **1998** | | | | | | | | | |
| Colo. Belle | 2 | 8 | 100.0 | | | | | | |
| East Fork | 13 | 46 | 76.1 | 2.2 | 6.5 | 13.0 | | | 2.2 |
| Docs | 21 | 101 | 65.3 | 5.0 | 4.0 | 7.0 | 4.0 | | 14.9 |
| Hwy 55 | 3 | 12 | 83.3 | 8.3 | | 8.3 | | | |
| Buena Vista | 16 | 77 | 90.9 | | 1.3 | 1.3 | | 6.5 | |
| Salida | 19 | 87 | 88.5 | 1.1 | 1.1 | 1.1 | 6.9 | | 1.1 |
| Lokey/Shavano | 3 | 8 | 100.0 | | | | | | |

[a] Tree swallow nests only. There may be more than one attempt in a single box.
[b] Does not include eggs collected for contaminant analyses.
[c] Eggs remained in the nest past hatching, but the reason they failed to hatch, either dead embryo or infertile, etc. is unknown.
[d] Eggs remained in the nest until hatching, but whether they died at or before hatching or hatched and then died is unknown.

demonstrated the pathway of lead exposure; tree swallow body burdens were most likely a result of eating insects that had become contaminated with lead from the sediments.

Along the Arkansas River, concentrations of lead in tree swallows were less than in the sympatrically occurring American dipper (*Cinclus mexicanus*), which averaged 0.87 μg/g dw (converted from wet weight [ww]).[34] Concentrations in dippers were about 1.5 times higher than in tree swallows at Docs (0.54 μg/g dw) and 7 times higher than the overall average of tree swallows on the entire Arkansas River (0.12 μg/g dw). This difference is probably due to slightly different food habits of dippers and swallows;[35] dippers eat aquatic insect larvae as well as some fish. Dippers probably do not accumulate much lead via ingestion of sediment because they inhabit high-gradient streams with sand and cobble bottoms.[35]

Trace-element concentrations of lead in livers were generally consistent between years except at East Fork, where concentrations were 0.42 in 1997 but only 0.08 in 1998. This yearly difference in lead exposure is probably real, based on supporting, but independently derived, ALAD activity and diet data. ALAD values at East Fork were less than at Docs in 1997, but not in 1998. Also, lead concentrations in diet from East Fork in 1997 were eight times higher in 1997 than in 1998. All three data sets are consistent and support one another's results.

Concentrations of lead in livers of nestling tree swallows at Docs were similar to concentrations in nestling tree swallow livers from the Coeur d'Alene River basin, Idaho (mean 0.5 μg/g dw, max value 1.3),[21] but lower than adult song sparrow (*Melospiza melodia*) livers (6.7 μg/g dw) from the same drainage.[36] A higher level of lead in adults compared to juveniles was present in barn swallows and starlings,[37,38] but the reverse was true in pied flycatchers (*Ficedula hypoleuca*) in Sweden.[39] The Coeur d'Alene River is one of the most lead-contaminated sites in the United States.[36] The Arkansas River drainage had a higher frequency of livers containing lead than tree swallow study sites in Minnesota (0 of 30, CMC unpubl. data) or Wyoming (0 of 11).[22] Even at the most contaminated Arkansas River site, however, liver concentrations were less than the 10 to 20 μg/g dry-weight value where clinical signs of lead poisoning, such as anemia, tissue lesions, green diarrhea, and muscular incoordination, become evident[40] in falconiformes, columbiformes, and galliformes.

Concentrations of lead in tree swallows were also less than levels in a variety of songbird species contaminated by lead additives in gasoline or other sources of air pollution. Schilderman et al.[41] found between 0.43 and 4.0 μg/g dw (converted from ww using 70% moisture) in urban pigeon livers, and Getz et al.[42] reported 0.6–16.1 μg/g dw in livers of urban house sparrows (*Passer domesticus*), starlings (*Sturnus vulgaris*), grackles (*Quiscalus quiscala*), and American robins (*Turdus migratorius*) in Illinois. Pied flycatcher nestlings next to a sulfide ore smelter in Sweden contained 10 to 24 μg/g dw lead in livers.[39,43]

Studies of the impacts of lead on wildlife populations are limited and have mostly been on acute poisoning associated with lead shot on waterfowl. Few studies have examined reproductive effects on songbirds. Grue et al.[37,38] found no effect of lead from automobile emissions on barn swallow or starling productivity (percent of eggs that produced fledglings) in Maryland. Concentrations of lead in whole carcasses (without feathers, feet, or bills) averaged between 0.7 and 1.5 μg/g dw for barn swallows and 0.78–4.0 μg/g dw for starlings, levels that are both more than those present in tree swallow carcasses in Colorado. Concentrations of lead in barn swallow diet (2.3–3.2 μg/g dw) in Maryland, however, were less than along the Arkansas River. Concentrations in starling diets (7.2–94 μg/g dw) were similar to those in our study. Nyholm[39,43] studied pied flycatchers nesting near and far from a sulfide ore smelter in Sweden. Flycatchers nesting at the two sites closest to the smelter had a reduced clutch size, a higher percentage of unhatched eggs, and lower brain weights than those that nested farthest from the smelter. Lead concentrations in nestling liver tissue was 24.4 and 10 μg/g dw (converted from ww) at the 2 sites nearest the smelter, 20 to 50 times higher than along the Arkansas River in Colorado.

The probability of daily egg survival (0.972) in our study was significantly reduced at the site on the Arkansas River, with the highest lead concentrations compared to other sites in 1998. This trend was also present in 1997 but was not significant. This suggests that lead could be affecting reproduction on the Arkansas River. There was no effect of lead on tree swallow brain weights as was found in Sweden[39,43] or in starlings in Maryland.[38]

### 28.4.2 Biomarkers

#### *28.4.2.1 ALAD*

ALAD analysis is an excellent indicator of lead exposure along the Arkansas River and can aid substantially in the interpretation of tissue lead concentrations. Although mean ALAD activity at Salida did not differ statistically from the reference site, some individual ALAD values at Salida were < 50 nmol/min/mL RBC, the lowest value at the reference sites, or < 37.1 nmol/min/mL RBC, the 50% of threshold indicative of injury according to federal statute (NRDA regulation 43 CFR11.42). This agreed with the actual liver lead concentration data (35% of livers with detectable concentrations of lead) and indicated that biologically significant exposure to lead was occurring, even though mean ALAD values did not differ from reference values at Salida. The slightly reduced ALAD at Lokey/Shavano also supports the finding of lead in two of three livers at that site. There was a significant correlation between ALAD and log lead liver concentrations ($P < 0.001$). The amount of variation in ALAD levels explained by this relationship was small ($R^2 = 24\%$) but similar to the explained variation reported by others.[44]

Inhibition of ALAD was at levels indicative of injury to wildlife (> 50% of normal, NRDA regulation 43 CFR11.42), both for individual values or when averaged by nest box. Between 20 and 40% of nest boxes, in and immediately below the 11-Mile Stretch, had ALAD inhibition indicative of injury to wildlife. This indicates that lead exposure may be having toxic effects on tree swallows.

Levels of ALAD inhibition at our most lead-contaminated site (Docs) was 61% of normal and was similar to the amount of ALAD inhibition in starlings (58% of normal) but more than barn swallows (86% of normal) at the most lead-contaminated freeway site in Maryland.[37,38] On the Coeur d'Alene River, Idaho adult song sparrow and robin ALAD activities were 49% and 25% of normal, a higher inhibition level than we found on the Arkansas River. However, inhibition of ALAD in adult birds is generally greater than in nestlings.[37,38] Urban pigeons in London, England had ALAD values between 4.4% and 21.0% of normal values.[45] Concentrations of lead in livers of these pigeons ranged up to 21.6 μg/g dw, whereas control livers averaged 2 μg/g lead. All were much higher than concentrations in tree swallows along the Arkansas River.

We were not able to test for effects of lead on hematocrit because our data were confounded by the different elevations of the study sites. Field and laboratory studies have shown a decrease in hematocrit with increased exposure to lead.[37,46] Additionally, hematocrit has been shown to increase with altitude,[47] although the variation by altitude is probably not linear, as implied in that paper. Our highest elevation sites were also the sites highest in lead concentrations; there may have been some decrease in hematocrit that we were not able to detect. The maximum hematocrit for tree swallows in this study was reached at approximately 2800 m above sea level; it did not increase above that altitude. The minimum altitude where hematocrit did not decrease further was at approximately 1500 m. It is likely that there is an optimum range within which hematocrits vary according to altitude.

#### *28.4.2.2 Flow Cytometry*

Flow cytometry has identified somatic cell chromosome damage in birds due to petrochemicals, such as the polycyclic aromatic hydrocarbons,[48] and from radiation.[49] Few studies have examined the relationship between trace elements and somatic cell damage in birds. One such study[48] found no correlation between trace elements and HPCV. In that study, petrochemical concentrations, but not trace elements, including cadmium (T. Custer, unpublished data), correlated with HPCV in lesser scaup (*Aythya affinis*). Because the petrochemical contamination in that study was quite high, it may have masked a relationship between trace elements and somatic cell damage. HPCV in tree swallow nestling blood along the Arkansas River was positively correlated with both carcass and liver concentrations of cadmium. This correlation, which was present in both tissue matrices, is intriguing and deserves further study. The negative correlation of HPCV with liver concentrations of magnesium may have resulted from the negative correlation of magnesium with both carcass and liver cadmium levels. The negative correlation of HPCV and carcass chromium levels is problematic. There is no supporting evidence in the literature, although few studies have been done in birds. Further work is needed to understand this result.

### 28.4.3 Diet

Pathway of lead exposure from sediment[2] through insects[4] (Appendix 28.1) to tree swallows (Table 28.2) is clear. Tree swallows fed almost exclusively on aquatic insects, primarily mayflies and aquatic dipterans. Mayflies had higher trace-element contaminant levels than other benthic invertebrate taxa in the upper Arkansas River,[4] which probably contributed to the transfer of contaminants in sediment to the avian community. Lead concentrations in diet and nestling tree swallow livers were highly correlated ($R^2 = 0.47$) and followed the upstream-to-downstream gradient in sediment contamination.[2] There was also a correlation between diet and liver concentrations for selenium and boron; the correlation approached significance for cadmium ($P = 0.09$) as well. Trace elements that did not correlate between diet and liver were generally those that are essential nutrients. Homeostatic processes probably ensured that the optimal amount was present for metabolic processes for these essential elements. Similar results were measured in American dippers along the same stretch of the Arkansas River.[34]

### 28.4.4 Other Trace Elements

Mercury was detected in a larger proportion of tree swallow livers in and around Leadville than at sites farther downstream. Mercury was used in the metallurgy process to extract gold and silver, so the occurrence of mercury higher in the drainage seems reasonable. Average concentrations of mercury along the Arkansas River (0.067 μg/g dw in liver and 0.16–0.18 in eggs) were far lower than concentrations reported in juvenile insectivorous passerines nesting near a mercury production plant (approx. 1.5 μg/g dw in both livers and eggs) in Slovakia and Norway or in nearby reference areas (0.1 μg/g dw or less for both tissue types[50]). Mercury concentrations in eggs and nestling tree swallow livers on the Arkansas River were also less than those in nestling tree swallows near Casper, WY (0.3 and 0.2 μg/g dw, liver and eggs[22]).

Concentrations of cadmium in tree swallow livers (0.22 μg/g) along the Arkansas River were higher than at sites in Wyoming, where all samples were below the detection limit.[22] Nestling dipper livers from locations similar to ours along the Arkansas River averaged 0.497 μg/g dw[34] (converted from ww). This higher concentration may reflect the dipper's slightly higher position in the food web[35] compared to tree swallows. Juvenile finches and tits in Norway, however, had slightly higher cadmium concentrations (0.44–0.66 μg/g dw[51,52]) in liver tissue but similar concentrations of copper and zinc compared with Arkansas River swallows. These areas of Norway are considered to be relatively well protected from long-range transport of heavy metals and have no local pollution sources. Cadmium concentrations in tree swallow livers on the Arkansas River were an order of magnitude below the 40-μg/g dry-weight threshold for cadmium poisoning.[53]

Concentrations of boron, iron, manganese, molybdenum, and zinc were similar to values reported in the literature for tree swallows in Wyoming.[22] Boron levels in red-winged blackbirds (*Agelaius phoeniceus*) and great-tailed grackles (*Quiscalus mexicanus*) in the Mexicali Valley, Mexico (7–13 μg/g dw[54]) were also similar to the levels we found on the Arkansas River as a whole. Average concentrations of boron for the Arkansas River (11–12 μg/g dw) are below the

33–51 μg/g dw found in laboratory mallards (*Anas platyrhynchos*) exhibiting adverse effects.[55] Higher concentrations in boron in swallow livers at Pueblo Reservoir (25.3 μg/g dw) probably reflect inputs from agricultural drainwater.

Selenium concentrations were significantly higher at Pueblo Reservoir than sites farther upstream along the Arkansas River, perhaps due to agricultural drainage inputs farther downstream (ASA, pers. communication). Selenium concentrations in Colorado were less than those present in Wyoming tree swallows, which nested near a major agricultural drainage system[22] and considerably less than the > 10 μg/g dw (eggs) and 33 μg/g dw (liver) thresholds considered harmful to laying females or individuals.[56]

### 28.4.5 Reproduction Measures

Docs had the highest frequency of lead in liver tissue (100%) and also had the highest concentration in the liver (0.54 μg/g dw, average for 1997 and 1998). Although the sites with the highest lead concentrations in liver tissue — Docs in both years and East Fork in 1997 — tended to have a lower probability of egg survival, the correlation between these measurements on a colony basis was not significant ($P = 0.214$). On an individual-nest basis, the logistic regression between liver lead and nest success also was not significant ($P = 0.18$). This is understandable because some nests that were either partially or totally not successful had no nestlings that reached 12 days of age and, hence, were not available to be sampled.

The nationwide average for hatching success in tree swallows is 86.9%.[57] This number was derived only from nests that hatched at least one egg, so it is biased to the high side because it does not include those nests where all eggs failed to hatch. Nevertheless, nests at sites along the Arkansas River with lower lead concentrations either equaled or exceeded this nationwide average; nests with higher lead concentrations tended to be below this nationwide average.

## 28.5 SUMMARY

Lead concentrations in nestling livers conclusively demonstrated that lead from sediments is bioavailable to tree swallows nesting along the Arkansas River. Lead was detected in all tree swallow livers at Docs and in 83% of livers from Hwy 55. Both sites are in the 11-Mile Stretch, the most sediment-contaminated section of the Arkansas River. The proportion of livers with detectable lead was less both downstream and upstream of the 11-Mile Stretch. At Pueblo Reservoir, no lead was detected in the single tree swallow liver or the six barn swallow livers. Actual lead concentrations in tree swallow livers differed significantly among sites and followed the upstream/downstream gradient of sediment contamination. Concentrations of lead were similar between years, except at East Fork where a lower percentage of liver samples contained lead, and at lower concentrations, in 1998 than in 1997. This coincided with lower lead concentrations in the diet in 1998 as well as higher ALAD activity in 1998. The reason for this is unknown, but further sampling at East Fork might be warranted. Food taken from the stomachs of tree swallow nestlings exhibited the same upstream/downstream gradient in lead concentrations. This, along with the correlation between concentrations of lead in the diet and concentrations in liver tissue, demonstrated the pathway from sediment to insects to birds.

ALAD activity indicated that lead exposure was having physiological effects on tree swallow nestlings. Up to 40% of nest boxes at the nesting site nearest the California Gulch input had nestlings with ALAD inhibition > 50% of the mean activity at the reference site. This 50% inhibition threshold for injury determination is that set by federal statute. Sites with > 50% ALAD inhibition were East Fork (8% of boxes), Docs (26%), Hwy 55 (40%), and Buena Vista (32%). Over all, 18% of nest boxes had ALAD inhibition > 50% of normal.

No evidence of chromosomal damage was observed based on flow cytometric analyses. However, HPCV was positively correlated with both liver and carcass cadmium concentrations, indicating that this trace element may cause chromosomal damage in tree swallows.

Concentrations of other trace elements (cadmium, mercury, and selenium) in eggs, livers, and food were generally below known effect levels based on published information of other avian species and were similar to other sites where tree swallows have been studied in the United States.

## ACKNOWLEDGMENTS

We would like to thank Kurt Broderdorp, Craig Miller, Clay Ronish, and Sean Strom of U.S. Fish and Wildlife Service for help in deploying and checking nest boxes; Joe Coogans, Bernard Smith, Department of the Interior, and the Colorado Division of Wildlife for access to their properties; Dwayne Monk and other Leadville Fish Hatchery personnel for logistical support; Milton Smith and Daniel Finley, USGS, National Wildlife Health Center for ALAD analyses; John Tipping for insect identification; and Paul M. Dummer for help with data analysis. We would also like to thank Fred Meyer, Erik I. Nyholm, Cole W. Matson, and two anonymous individuals for review of the manuscript. C.D.S. and J.W.B. were funded in part by NIEHS grants ES 04917.

**Appendix 28.1 Trace Element Concentrations (Geometric Mean and 95% Confidence Interval, μg/g dry wgt.) in Pooled Diet Samples of Tree Swallows[a] along the Upper Arkansas River, Colorado in 1997 and 1998**

| Element | Colo. Belle (*n* = 1)[c] | East Fork (*n* = 3) | Docs (*n* = 3) | Hwy 55 (*n* = 2) | Buena Vista (*n* = 3) | Salida (*n* = 3) | Lokey/Shavano (*n* = 2) | Pueblo[a] (*n* = 3) |
|---|---|---|---|---|---|---|---|---|
| Aluminum | 35.1 | 187.5<br>113–310 | 229.0<br>67–780 | 675.2<br>112–4070 | 511.4<br>290–900 | 1410.0<br>354–5623 | 3963<br>103–152499 | 211.4<br>98–458 |
| Arsenic | 0.25 | 0.51<br>0.36–0.73 | 0.60<br>0.55–0.65 | 0.84<br>0.66–1.08 | 0.38<br>0.25–0.59 | 0.46<br>0.25–0.86 | 1.31<br>0.25–6.85 | 0.77<br>0.43–1.37 |
| Boron | 9.32 | 19.04<br>12.0–30.2 | 16.53<br>12.8–21.4 | 13.62<br>11.9–15.6 | 22.36<br>19.6–25.6 | 22.64<br>20.8–24.65 | 64.75<br>51.5–81.4 | 27.82<br>18.2–42.6 |
| Barium | 4.22 | 13.59<br>11.8–15.6 | 11.90<br>7.8–18.1 | 22.49<br>18.2–27.8 | 6.64<br>5.2–8.5 | 15.49<br>14.0–17.2 | 17.5<br>17.2–17.8 | 17.36<br>7.0–42.9 |
| Cadmium | 4.09 | 3.97<br>3.7–4.3 | 5.13<br>4.3–6.2 | 2.34<br>2.0–2.8 | 3.76<br>2.9–4.9 | 3.25<br>2.9–3.7 | 2.47<br>0.8–7.5 | 1.01<br>0.8–1.2 |
| Copper | 45.9 | 37.59<br>31.5–44.8 | 33.48<br>32.7–34.3 | 23.25<br>19.1–28.3 | 30.6<br>29.0–32.4 | 38.6<br>31.3–47.7 | 30.29<br>25.0–36.7 | 26.13<br>18.9–36.2 |
| Iron | 322.2 | 559.2<br>460–680 | 459.8<br>394–536 | 577.4<br>384–866 | 454.2<br>418–494 | 459.5<br>333–633 | 441.7<br>297–657 | 1038.7<br>683–1579 |
| Lead | 0.74 | 4.52<br>2.16–9.46 | 12.49<br>9.37–16.65 | 4.25<br>4.05–4.47 | 1.83<br>0.75–4.48 | 2.83<br>1.79–4.47 | 6.62[d]<br>0.05–884 | 0.78<br>0.44–1.38 |
| Magnesium | 791 | 1019<br>995–1044 | 1035<br>907–1182 | 633.5<br>610–658 | 751.8<br>612–924 | 998.3<br>710–1403 | 906.6<br>496–1657 | 930.2<br>654–1323 |
| Manganese | 23.6 | 71.84<br>42.8–120.5 | 35.2<br>25.7–48.2 | 25.46<br>21.6–30.0 | 28.06<br>23.9–32.9 | 68.47<br>36.5–128.4 | 29.49<br>12.7–68.5 | 72.39<br>27.0–194.3 |
| Molybdenum[b] | 9.99a | 2.00b<br>1.37–2.94 | 1.30b<br>1.27–1.33 | 1.42b<br>1.12–1.78 | 1.54b<br>1.25–1.89 | 1.04b<br>0.79–1.36 | 3.63a,b<br>3.10–4.25 | 3.29a,b<br>2.98–3.63 |
| Selenium | 3.44 | 2.86<br>2.53–3.24 | 1.88<br>1.79–1.98 | 0.99<br>0.92–1.06 | 1.18<br>1.00–1.40 | 1.56<br>0.99–2.47 | 3.57<br>2.39–5.32 | 2.69<br>1.52–4.77 |
| Strontium | 3.67 | 12.76<br>9.7–16.8 | 22.87<br>14.0–37.3 | 18.63<br>17.7–19.6 | 33.70<br>19.9–57.0 | 88.5<br>86.1–91.0 | 117.0<br>114–120 | 87.57<br>19.0–403.5 |
| Zinc | 162 | 289.5<br>272–308 | 228.9<br>200–262 | 134.5<br>100–181 | 158.0<br>143–175 | 168.2<br>132–214 | 234.8<br>208–265 | 154.8<br>125–192 |

*Note:* Beryllium and mercury were not detected in any diet sample; chromium was detected in 4 of 20 samples; nickel in 6 of 20; and vanadium in 8 of 20 diet samples.

[a] Pueblo site includes barn and cliff swallows only.
[b] Only molybdenum varied among locations ($P < 0.05$).
[c] Sample size is the number of pooled samples for both years combined.
[d] Includes 1 sample from Lokey that contained 892 μg/g lead. If this sample is removed from the analysis then $P = 0.019$ and Docs > Lokey.

## REFERENCES

1. Kirkemo, H., Newman, W. L., and Ashley, R. P., Gold, USGS leaflet, 1996.
2. Kimball, B. A., Callender, E., and Axtmann, E. V., Effects of colloids on metal transport in a river receiving acid mine drainage, upper Arkansas River, Colorado, U.S.A., *Appl. Geochem.*, 10, 285, 1995.
3. Kimball, B. A., Bencala, K. E., and McKnight, D. M., Research on metals in acid mine drainage in the Leadville, Colorado, area, in Proc. Tech. Meet., Phoenix, AZ, Mallard, G. E., and Ragone, S. E., Eds., USGS, Water Resources Investigations Report 88-4220, 1989, 65.
4. Kiffney, P. M. and Clements, W. H., Bioaccumulation of heavy metals by benthic invertebrates at the Arkansas River, Colorado, *Environ. Toxicol. Chem.*, 12, 1507, 1993.
5. Cain, D. J., Luoma, S. N., Carter, J. L., and Fend, S. V., Aquatic insects as bioindicators of trace element contamination in cobble-bottom rivers and streams, *Can. J. Fish. Aquat. Sci.*, 49, 2141, 1992.
6. Clements, W. H. and Rees, D. E., Effects of heavy metals on prey abundance, feeding habits, and metal uptake of brown trout in the Arkansas River, Colorado, *Trans. Am. Fish. Soc.*, 126, 774, 1997.
7. Custer, C. M., Custer, T. W., Allen, P. D., Stromborg, K. L., and Melancon, M. J., Reproduction and organochlorine contamination in tree swallows nesting in the Fox River drainage and Green Bay, Wisconsin, USA, *Environ. Toxicol. Chem.*, 17, 1786, 1998.
8. Custer, C. M., Custer, T. W., and Coffey, M., Organochlorine chemicals in tree swallows nesting in Pool 15 of the Upper Mississippi River, *Bull. Environ. Contam. Toxicol.*, 64, 341, 2000.
9. Shaw, G. C., Organochlorine pesticide and PCB residues in eggs and nestlings of tree swallows, *Tachycineta bicolor*, in central Alberta, *Can. Field-Nat.*, 98, 258, 1983.
10. DeWeese, L. R., Cohen, R. R., and Stafford, C. J., Organochlorine residues and eggshell measurements for tree swallows *Tachycineta bicolor* in Colorado, *Bull. Environ. Contam. Toxicol.*, 35, 767, 1985.
11. Kraus, M. L., Bioaccumulation of heavy metals in pre-fledgling tree swallows, *Tachycineta bicolor*, *Bull. Environ. Contam. Toxicol.*, 43, 407, 1989.
12. Ankley, G. T., Niemi, G. J., Lodge, K. B., Harris, H. J., Beaver, D. L., Tillitt, D. E., Schwartz, T. R., Giesy, J. P., Jones, P. D., and Hagley, C., Uptake of planar polychlorinated biphenyls and 2,3,7,8-substituted polychlorinated dibenzofurans and dibenzo-p-dioxins by birds nesting in the lower Fox River and Green Bay, Wisconsin, USA, *Arch. Environ. Contam. Toxicol.*, 24, 332, 1993.
13. King, K. A., Custer, T. W., and Weaver, D. A., Reproductive success of barn swallows nesting near a selenium-contaminated lake in east Texas, USA, *Environ. Pollut.*, 84, 53, 1994.
14. Bishop, C. A., Koster, M. D., Chek, A. A., Hussell, D. J. T., and Jock, K., Chlorinated hydrocarbons and mercury in sediments, red-winged blackbirds (*Agelaius phoeniceus*) and tree swallows (*Tachycineta bicolor*) from wetlands in the Great Lakes-St. Lawrence River basin, *Environ. Toxicol. Chem.*, 14, 491, 1995.
15. Nichols, J. W., Bioenergetics-based model for accumulation of polychlorinated biphenyls by nestling tree swallows, *Tachycineta bicolor*, *Environ. Sci. Technol.*, 29, 604, 1995.
16. McCarty, J. P. and Secord, A. L., Reproductive ecology of tree swallows (*Tachycineta bicolor*) with high levels of polychlorinated biphenyl contamination, *Environ. Toxicol. Chem.*, 18, 1433, 1999.
17. Quinney, T. E. and Ankney, C. D., Prey size selection by tree swallows, *Auk,* 102, 245, 1985.
18. Blancher, P. J. and McNicol, D. K., Tree swallow diet in relation to wetland acidity, *Can. J. Zool.*, 69, 2629, 1991.
19. Fairchild, W. L., Muir, D. C. G., Currie, R. S., and Yarechewski, A. L., Emerging insects as a biotic pathway for movement of 2,3,7,8-tetrachlorodibenzofuran from lake sediments, *Environ. Toxicol. Chem.*, 11, 867, 1992.
20. Elliott, J. E., Martin, P. A., Arnold, T. W., and Sinclair, P. H., Organochlorines and reproductive success of birds in orchard and non-orchard areas of central British Columbia, Canada, 1990-91, *Arch. Environ. Contam. Toxicol.*, 26, 435, 1994.
21. Blus, L. J., Henny, C. J., Hoffman, D. J., and Grove, R. A., Accumulation in and effects of lead and cadmium on waterfowl and passerines in northern Idaho, *Environ. Pollut.*, 89, 311, 1995.
22. Custer, T. W., Custer, C. M., Dickerson, K., Allen, K., Melancon, M. J., and Schmidt, L. J., Polycyclic aromatic hydrocarbons, aliphatic hydrocarbons, trace elements and monooxygenase activity in birds nesting on the North Platte River, Casper, Wyoming, USA, *Environ. Toxicol. Chem.*, 20, 624, 2001.
23. Clements, W. H., Benthic invertebrate community responses to heavy metals in the Upper Arkansas River Basin, Colorado, *J. N. Am. Benthol. Soc.*, 13, 30, 1994.

24. Ramstack, J. M., Murphy, M. T., and Palmer, M. R., Comparative reproductive biology of three species of swallows in a common environment, *Wilson Bull.*, 110, 233, 1998.
25. Burch, H. B. and Siegel, A. L., Improved method for measurement of delta-aminolevulinic acid dehydratase activity of human erythrocytes, *Clinical Chem.*, 17, 1038, 1971.
26. Scheuhammer, A. M., Erythrocyte δ-aminolevulinic acid dehydratase in birds. II. The effects of lead exposure *in vivo*, *Toxicology*, 45, 165, 1987.
27. Vindelov, L. L., Christensen, I. J., and Nissen, N. I., A detergent-trypsin method for the preparation of nuclei for flow cytometric DNA analysis, *Cytometry*, 3, 323, 1983.
28. Johnson, E. J., Best, L. B., and Heagy, P. A., Food sampling biases associated with the "ligature method," *Condor*, 82, 196, 1980.
29. Misra, R. K. and Easton, M. D. J., Comment on analyzing flow cytometric data for comparison of mean values of the coefficient of variation of the G1 peak, *Cytometry*, 36, 112, 1999.
30. Mayfield, H., Nesting success calculated from exposure, *Wilson. Bull.*, 73, 255, 1961.
31. Mayfield, H., Suggestions for calculating nest success, *Wilson Bull.*, 87, 456, 1975.
32. Hensler, G. L. and Nichols, J. D., The Mayfield methods of estimating nesting success: A model, estimators and simulation results, *Wilson Bull.*, 93, 42, 1981.
33. Pain, D. J., Lead in waterfowl, in *Environmental Contaminants in Wildlife, Interpreting Tissue Concentrations*, Beyer, W. N., Heinz, G. H., and Redmon-Norwood, A. W., CRC Press/Lewis Publishers, Boca Raton, FL, 1996, 251.
34. Strom, S. M., The Utility of Metal Biomarkers in Assessing the Toxicity of Metals in the American Dipper (*Cinclus mexicanus*), M.S. thesis, Colorado State University, Fort Collins, 2000.
35. Kingery, H. E., American dipper (*Cinclus mexicanus*), in *Birds of North America*, No. 229, Poole, A. and Gills, F., Eds., Academy of Natural Sciences, Philadelphia, PA, 1996, 27.
36. Johnson, G. D., Audet, D. J., Kern, J. W., LeCaptain, L. L., Strickland, M. D., Hoffman, D. J., and McDonald, L. L., Lead exposure in passerines inhabiting lead-contaminated floodplains in the Coeur d'Alene River basin, Idaho, USA, *Environ. Toxicol. Chem.*, 18, 1190, 1999.
37. Grue, C. E., O'Shea, T. J., and Hoffman, D. J., Lead concentrations and reproduction in highway-nesting barn swallows, *Condor*, 86, 383, 1984.
38. Grue, C. E., Hoffman, D. J., Beyer, W. N., and Franson, J. C., Lead concentrations and reproductive success in European starlings *Sturnus vulgaris* nesting within highway roadside verges, *Environ. Pollut.*, (Ser. a) 42, 157, 1986.
39. Nyholm, N. E. I., Heavy metal tissue levels, impact on breeding and nestling development in natural populations of pied flycatcher (*Aves*) in the pollution gradient from a smelter, in *Ecotoxicology of Soil Organisms*, Donker, M. H., Eljsackers, H., and Helmback, F., Eds., SETAC Spec. Publ. Ser. Lewis Publishers, Boca Raton, FL, 1994, 373.
40. Franson, J. C., Interpretation of tissue lead residues in birds other than waterfowl, in *Environmental Contaminants in Wildlife, Interpreting Tissue Concentrations*, Beyer, W. N., Heinz, G. H., and Redmon-Norwood, A. W., CRC Press/Lewis Publishers, Boca Raton, FL, 1996, 265.
41. Schilderman, P. A. E. L., Hoogewerff, J. A., van Schooten, F-J., Maas, L. M., Moonen, E. J. C., van Os, B. J. H., van Wijnen, J. H., and Kleinjans, J. C. S., Possible relevance of pigeons as an indicator species for monitoring air pollution, *Environ. Health Perspec.*, 105, 322, 1997.
42. Getz, L. L., Best, L. B., and Prather, M., Lead in urban and rural song birds, *Environ. Pollut.*, 12, 235, 1977.
43. Nyholm, N. E. I, Monitoring of terrestrial environmental metal pollution by means of free-living insectivorous birds, *Ann. di Chimica*, 85, 343, 1995.
44. Henny, C. J., Blus, L. J., Hoffman, D. J., and Grove, R. A., Lead in hawks, falcons and owls downstream from a mining site on the Coeur d'Alene River, Idaho, *Environ. Monit. Assess.*, 29, 267, 1994.
45. Hutton, M., The effects of environmental lead exposure and *in vitro* zinc on tissue δ-aminolevulinic acid dehydratase in urban pigeons, *Comp. Biochem. Physiol.*, 74C, 441, 1983.
46. Hoffman, D. J., Franson, J. C., Pattee, O. H., Bunck, C. M., and Murray, H. C., Biochemical and hematological effects of lead ingestion in nestling American kestrels (*Falco sparverius*), *Comp. Biochem. Physiol.*, 80C, 431, 1985.
47. Keys, G. D., Fleischer, R. C., and Rothstein, S. I., Relationships between elevation, reproduction and the hematocrit level of brown-headed cowbirds, *Comp. Biochem. Physiol.*, 83A, 765, 1986.

48. Custer, T. W., Custer, C. M., Hines, R. K., Sparks, D. W., Melancon, M. J., Hoffman, D. J., Bickham, J. W., and Wickliffe, J. K., Mixed-function oxygenases, oxidative stress, and chromosomal damage measured in lesser scaup wintering on the Indiana Harbor Canal, *Arch. Environ. Contam. Toxicol.*, 28, 522, 2000.
49. George, L. S., Dallas, C. E., Brisbin, I. L., Jr., and Evans, D. L., Flow cytometric DNA analysis of ducks accumulating $^{137}$Cs on a reactor reservoir, *Ecotoxicol. Environ. Saf.*, 21, 337, 1991.
50. Rosten, L. S., Kalas, J. A., Mankovska, B., and Steinnes, E., Mercury exposure to passerine birds in areas close to local emission sources in Slovakia and Norway, *Sci. Total Environ.*, 213, 291, 1998.
51. Hogstad, O., Accumulation of cadmium, copper and zinc in the liver of some passerine species wintering in central Norway, *Sci. Total Environ.*, 183, 187, 1996.
52. Hogstad, O., Winter levels of cadmium, copper and zinc in the liver of passerines from an area with low long range air pollution in central Norway, *Fauna norv. Ser. C. Cinclus*, 19, 83, 1996.
53. Furness, R. W., Cadmium in birds, in *Environmental Contaminants in Wildlife, Interpreting Tissue Concentrations*, Beyer, W. N., Heinz, G. H., and Redmon-Norwood, A. W., CRC Press/Lewis Publishers, Boca Raton, FL, 1996, 389.
54. Mora, M. A. and Anderson, D. W., Selenium, boron, and heavy metals in birds from the Mexicali Valley, Baja California, Mexico, *Bull. Environ. Contam. Toxicol.*, 54, 198, 1995.
55. Howe, P. D., A review of boron effects in the environment, *Biol. Trace Element Res.*, 66, 153, 1998.
56. Heinz, G. H., Selenium in birds, in *Environmental Contaminants in Wildlife, Interpreting Tissue Concentrations*, Beyer, W. N., Heinz, G. H., and Redmon-Norwood, A. W., CRC Press/Lewis Publishers, Boca Raton, FL, 1996, 447.
57. Robertson, R. J., Stutchbury, B. J., and Cohen, R. R., Tree swallow (*Tachycineta bicolor*) in *The Birds of North America*, Poole, A., Stettenheim, P., and Gill, F., Eds., No. 11, Academy of Natural Sciences, Philadelphia, 1992, 26.

CHAPTER 29

# The Hudson River — PCB Case Study

**John P. McCarty**

## CONTENTS



## 29.1 INTRODUCTION

### 29.1.1 The Hudson River — PCB Case History

The Hudson River dominates the history and landscape of eastern New York and surrounding states (Figure 29.1). In environmental terms, the Hudson River has come to symbolize the difficulties associated with finding solutions to the problems of widespread contamination by persistent organic compounds as well as other sources of anthropogenic stress. The case history I describe here does not include many of the complicating factors that make addressing contaminant problems so difficult in the early 21st century such as complex mixtures and diffuse nonpoint sources with no easily recognizable responsible party. Instead, the story of the Hudson River is dominated by a single class of chemicals (PCBs), released from two long recognized point sources, by a still thriving corporation — in short, the very scenario that many environmental laws were written to address. It is informative that solutions to the contamination of the Hudson River have not been reached. In this case study, I review the history of PCB contamination in the context of the Hudson River ecosystem. I describe the patterns of PCB contamination in the biota of the Hudson, the risk contaminants pose to fish and wildlife, and attempts to mitigate those risks.

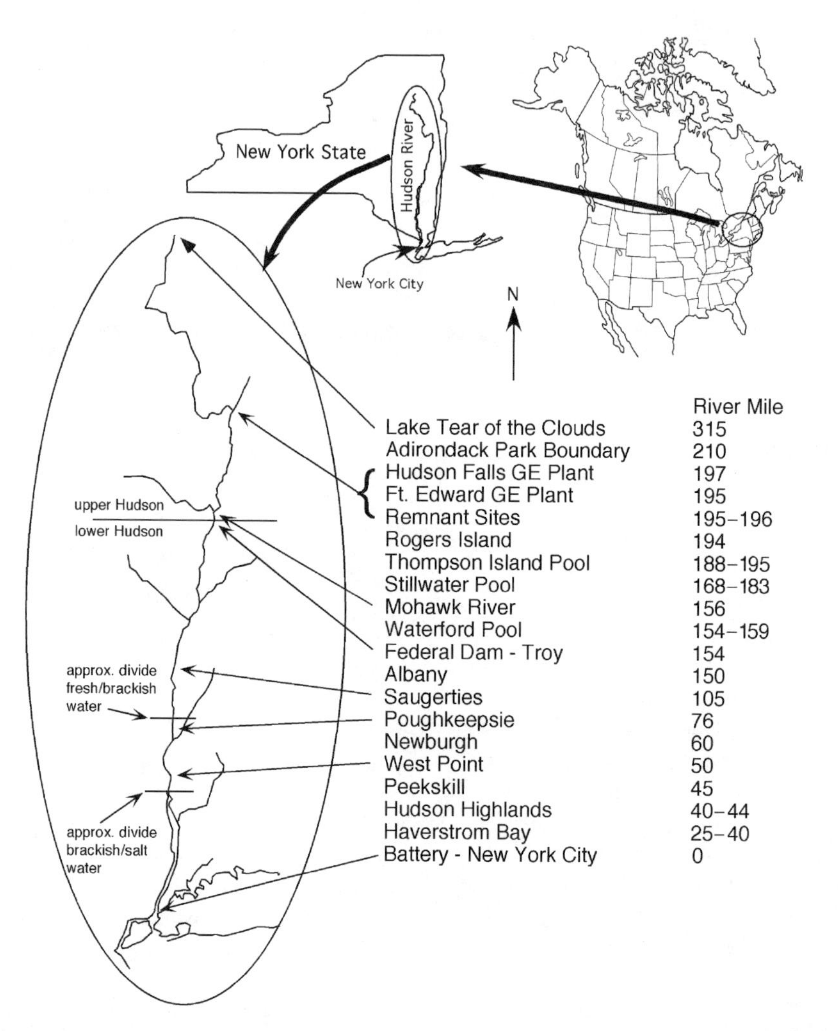


**Figure 29.1** The Hudson River and major landmarks. Locations along the Hudson are traditionally referred to by their distance in miles from the mouth of the river at the Battery in New York City (River Mile [RM]).

Polychlorinated biphenyls, or PCBs, are a group of 209 related molecules (congeners) that were widely used in industry and have become widely distributed in the environment.[1,2] While the broad spread of PCBs throughout the biosphere is significant cause for concern on a global scale, there continue to be significant local and regional problems associated with high concentrations of contaminants remaining in the environment from historical activity.[2] The widely recognized risks to human health and ecosystems posed by PCBs and related persistent organic pollutants (POPs) has led to widespread restrictions on their manufacture and use at the national and international levels.[1,3] Documentation of potential adverse impacts on human health associated with exposure to PCBs dates back many decades and includes skin problems, liver and kidney damage, possible estrogenic activity, and possible increased cancer risk.[4–6] In that past decade, studies of both humans and nonhuman animals provide increasing evidence that children exposed to PCBs *in utero* suffer neurodevelopmental deficits.[7–13] These neurodevelopmental effects of PCBs are expressed as lower IQ scores, deficits in reading ability, delays in development of motor skills, and spatial learning deficits. In addition, adults exposed to PCBs through consumption of contaminated fish score lower on tests of memory and learning.[14]

As the threats to fish, wildlife, and the environment due to bioaccumulation and bioconcentration of organochlorines became widely recognized in the 1960s, high concentrations of PCBs were identified in a variety of organisms. Substantial field and laboratory research has documented a wide array of adverse effects on fish and wildlife associated with PCB exposure.[1,2] Mortality, reproductive failure, porphyria, induction of mixed function oxydase (MFOs), behavioral and developmental abnormalities, and pathologies of the liver, thyroid, spleen, and other organs have been reported in a variety of laboratory and field studies of wildlife.[15–17] As analytical techniques have become more sophisticated, studies of the relative toxicity of individual PCB congeners have added to our understanding of PCB biology. Non-*ortho* planar congeners, such as PCBs 77, 126, and 169, appear to be especially toxic to wildlife.[16,17]

It is far beyond the scope of this chapter to provide a comprehensive review of the thousands of pages of scientific literature on the Hudson River. Instead, it is intended to provide an overview of how the Hudson River case history has progressed, describe some of the scientific generalizations that have emerged, and to point to some of the sources of conflict. The historical background of environmental issues surrounding the Hudson River are first discussed, followed by a brief overview of some of the ecotoxicological studies that are specific to the Hudson River. Finally, I present a brief overview of the progress from identifying the scope of the PCB issue toward deciding on a course of action to mitigate the impacts of the contamination of the river and surrounding ecosystems.

### 29.1.2 Hudson River Ecosystems

The Hudson River watershed encompasses 35,000 km$^2$ of New York and four surrounding states (Figure 29.1). From its origin at Lake Tear of the Clouds near Mount Marcy in the Adirondack Mountains, the Hudson extends for 507 river km to its terminus at the Battery on Manhattan Island, where it empties into New York Bays and ultimately the Atlantic Ocean. The Hudson River is an important historical, cultural, economic, and natural resource for the New York region. The River continues to play a role in transportation, through shipping upriver as far as the Port of Albany and through commercial and recreational traffic on the Champlain Barge Canal, which is contiguous with the Hudson River from Albany to Ft. Edward. The Hudson is also an important source of drinking water for municipalities such as Waterford, Poughkeepsie, and Rhinebeck, NY and is used in times of shortage as a water source for New York City. In addition, the river continues to play a key role in power generation for the region, primarily through use as a source of cooling water for power plants.

The upper third of the Hudson runs through the relatively undisturbed and protected forests of the Adirondack Park, flowing down a relatively steep gradient of 3.1 m/km.[18] The first major settlements on the river — Glens Falls, Hudson Falls, and Fort Edward, New York — lie less than 10 km south of the Adirondack Park boundary, at river mile 200 (320 km — it is traditional to reference locations on the Hudson as river miles (RMs), measured upstream from the terminus of the river in New York City; see Figure 29.1). South of the Adirondacks, the Hudson River valley is a mosaic of farmland and second-growth eastern deciduous forest. The gradient of the river also lessens to 0.62 m/km.[18] From Fort Edward, the Hudson passes through a series of locks and dams of the Champlain Canal system before reaching the Federal Dam at Troy, 80 river km to the south. These locks form a series of pools and form important sediment deposition zones.

The Federal Dam at RM 153 (240 km) marks the transition between the upper Hudson River and the lower Hudson River. The lower Hudson is an estuary, influenced by tidal action and open to shipping. The Hudson estuary is a drowned river valley, with much of the river bottom well below sea level. Over the entire 153 mi of the lower Hudson from Troy to the Battery, the river falls only 1.5 m — an average of 0.006 m/km.[18] Wetlands form an important component of the lower Hudson ecosystem. Both marshes with emergent vegetation and tidal mud flats are important habitat types, especially between RMs 25 and 140.[18]

The tidal estuary of the Hudson has a gradient from fresh water (salinity $< 0.5\%$) to polyhaline (salinity 18–30%). The divide between fresh and saline waters varies both daily with tidal action and seasonally depending on freshwater runoff. Under historically typical levels of runoff, brackish water reaches to 50 to 60 mi above New York City. However, the divide can fluctuate from RM

15 to 100 in times of especially high or low river flows.[18] In recent years, saline water has extended as far as 95 km upstream from the mouth of the river.[19]

Biological productivity of the Hudson River estuary is high. Productivity of the oligohaline and mesohaline reaches of the river between May and October in the early 1990s was estimated at 310–680 g C $m^{-2}$.[20] In the freshwater portions of the estuary, respiration tends to be greater than annual production. Annual phytoplankton productivity averages 300 g C $m^{-2}$, while approximately 40% of the freshwater system's organic carbon is exported directly to the lower river.[21]

### 29.1.3 PCBs and the Hudson Falls/Fort Edward Plants

Recognition of heavy contamination by PCBs in the early 1970s led to the eventual declaration of 320 km of the Hudson River as an EPA Superfund Site (Table 29.1). PCBs were widely used in industry, starting in the 1930s. In the United States, the principal manufacturer of PCBs was the Monsanto Corporation, which sold mixtures of PCB congeners under the trade name Aroclor from 1929 until PCBs were banned in 1977.[22] PCBs are highly stable and were favored as electrical insulators. Between 1946 and 1952, General Electric (GE) began using PCBs in the manufacture of capacitors at plants in Fort Edward and Hudson Falls, New York. It is estimated that these facilities alone purchased 15% of all the PCBs used in the United States.[23] Reported uses of PCBs at these facilities consisted primarily of Aroclors 1016, 1242, and 1254.[22,24] These plants were the largest source of the PCBs in the Hudson River.[22,23] Estimates of total PCB discharges by these plants range from 95,000 to 603,000 kg. However, these estimates were all made prior to the discovery of ongoing inputs of PCBs from the GE plants and surrounding areas, so the true amount of PCBs entering the river is undoubtedly much higher than the commonly cited estimates.[19]

Until 1973, some of the PCBs released by these plants were trapped in sediments held behind a dam at Fort Edward, NY. Removal of the Fort Edward dam because of its deteriorating condition resulted in the release of an estimated 1 million $m^3$ of contaminated sediments.[22] These contaminated sediments tended to accumulate in depositional areas of the Hudson, forming pockets of highly contaminated sediment eventually labeled "hot spots." Although limited data were available, high concentrations of PCBs in fish were in evidence before the removal of the Fort Edward dam (R. Sloan, pers. comm.).[25] Shortly after the dam removal and the redistribution of contaminated sediments via flooding, an increased number of analyses and the subsequent finding of high levels of PCBs in fish in the Hudson helped initiate a long series of studies and assessments of the environmental, ecological, and social impacts of the PCB contamination. These studies continue in the ongoing quest for possible solutions to the PCB problem. By 1976, commercial fishing for striped bass and other species had been banned on the Hudson, and recreational fishing was restricted.[25]

Redistribution of the PCB-contaminated sediments by the removal of the Ft. Edward dam coupled with ongoing inputs of PCBs from the GE site continue to form an important route of entry for PCBs into the food web. Residual deposits of sediments remaining after the removal of the Ft. Edward dam ("remnant deposits") between RMs 195 and 196 were capped in 1991 as part of the 1984 U.S. Environmental Protection Agency (EPA) Record of Decision. The importance of ongoing inputs of PCBs from the GE plants and surrounding areas into the Hudson have been recognized in recent years. One source of these continued inputs was the abandoned Allen Mill site located adjacent to the Hudson Falls plant. Between 1993 and 1995 alone, approximately 45 tons of PCBs were discovered in water conduits within this plant and removed.[19] In addition, PCBs saturate the bedrock under the GE plants. These PCBs continue to seep into the Hudson from the fractured bedrock adjacent to and underlying the river at the Hudson Falls and Ft. Edward plants.[19] In 1994, it was estimated that between 11 and 19 liters (3–5 gallons) per day of PCB product were being collected from seeps at the Hudson Falls GE plant.[25] By 1999, ongoing discharges from the GE plants were estimated at 90 mL per day (3 ounces).[25]

**Table 29.1 Timeline of Major Hudson River–PCB Events**

| Year | Event |
|---|---|
| 1947–1952 | General Electric begins use of large quantities of PCBs in capacitors manufactured at Ft. Edward and Hudson Falls plants adjacent to the Hudson River in New York. For the period 1966–1974, it is estimated that 15% of all PCBs sold in the United States are used by these two plants. [23] An unknown quantity of PCBs eventually reaches the waters of the Hudson River, with early estimates ranging from 95,000 to 603,000 kg. |
| 1969 | Elevated levels of PCBs in Hudson River biota first observed.[23] By 1971 New York State finds PCB levels of 4 to 11 ppm in striped bass. |
| 1971 | Monsanto, the primary manufacturer of PCBs in the United States, voluntarily restricts sales of PCBs to uses in "closed cycle" systems.[23] |
| 1972 | GE applies for an EPA permit to discharge 30 lb/day of chlorinated hydrocarbons. The permit goes into effect 31 January 1975.[23] |
| 1973 | Dam at Ft. Edward removed allowing sediments trapped behind the dam to move downstream. Flow of PCBs downstream is exacerbated by periods of high flow such as observed in 1976. In subsequent years, an estimated 1.3 million cubic yards of PCB-laden sediments are released downstream, redistributing contamination over 322 km (200 mi) of the Hudson River. |
| 1974 | U.S. EPA study finds high levels of PCBs in Hudson River fish. |
| 1976 | Toxic Substance Control Act restricts manufacture and use of PCBs. Monsanto halts production in 1977. U.S. EPA bans all use of PCBs in 1979. |
| 1976 | NYSDEC closes all fishing on the upper Hudson River and restricts commercial fishing along the entire Hudson River. |
| 1976 | GE is ordered to stop discharging PCBs into the Hudson River by July 1977. GE reaches an agreement with the State of New York agreeing to pay $3 million for monitoring of the Hudson River (to be matched by $3 million from NYSDEC), in return for not being held liable by the state. |
| 1977 | Clean Water Act prohibits discharge of PCBs into navigable waters. GE terminates active discharge from plant sites in to the river. NYSDEC implements long-term monitoring of Hudson River fish. |
| 1983 | U.S. EPA includes the Hudson River on the Superfund National Priority List. |
| 1984 | U.S. EPA issues Record of Decision including an interim "No Action" decision regarding contaminated sediments. |
| 1989 | NYSDEC requests that U.S. EPA reconsider their 1984 interim "No Action" decision. U.S. EPA begins the process of reassessment of Hudson River sediments in 1990. Phase One is completed in 1991. Phase Two reports, including the Human Health and Ecological Risk Assessment, are released between 1995 and 2000.[6,19,24] The Phase Three report on the feasibility of remediation is released in December 2000.[28] The U.S. EPA estimates that the cost for the Reassessment will total $21 million. |
| 1991 | New sources of PCB oil flowing from the GE site at Hudson Falls are discovered. |
| 1993 | Cleanup around the GE Hudson Falls plant begins. |
| 1995 | NYSDEC reopens the upper Hudson River to recreational fishing on a catch-and-release basis. |
| 2000 | Human Health and Ecological Risk Assessments are released by the U.S. EPA.[6,19] The U.S. EPA releases the Reassessment Phase 3 Feasibility Study Report and the Agency's (nonbinding) Proposed Plan.[28] The U.S. EPA preferred remedy includes dredging targeted areas in the Upper Hudson River between Fort Edward and Troy, totaling 2.65 million cubic yards. The dredged material will be shipped to existing licensed landfills outside of the Hudson River Valley for disposal. The estimated cost is $460 million. It is planned that the remediation will take five years. The Proposed Plan is open to public comment; by the close of the public comment period in early 2001, 70,000 comments are received. |
| 2001 | Bush administration EPA Administrator Christie Whitman endorses the Clinton-Administration preferred remedy and issues the formal Record of Decision (ROD) from the Reassessment.[55] |
| Ongoing | Implementation of the U.S. EPA proposed plan. Debate will continue on how to best implement and evaluate the limited dredging of the Hudson River "hot spots" called for under the EPA Record of Decision.[26] |
| | Natural Resources Damage Assessment: the U.S. Department of the Interior, National Oceanic and Atmospheric Administration, and New York State Department of Environmental Conservation are responsible for an ongoing evaluation of the degree of damage to the natural resources of the Hudson River.[25,57] This process will evaluate the degree of damage to resources, determine how the resources might be restored, and consider how the public might be compensated for past and future losses of valued resources. |

While other sources of historical PCB inputs exist, it is estimated that the GE-related contamination accounts for close to 100% of the in-place and water-borne contamination of the upper Hudson, declining to 50% in the saline reaches of the lower Hudson.[19]

Extensive sampling of water, sediments, and aquatic biota of the Hudson has been conducted over the last 25 years. Although PCBs move through the system during low-flow periods, high-flow events are a major factor in the transport of PCBs. Contaminated sediments have also entered the surrounding terrestrial ecosystem when flood events deposit contaminated sediments on the flood plain. Earlier navigational dredging projects have also factored into redistribution of sediments within the river and into riparian and upland systems.[19] A general trend toward declining PCB levels had been observed until 1991, when levels in water and the biota again climbed in response to additional, recently documented releases of PCBs from the GE site.[19]

## 29.2 PCBs IN THE HUDSON RIVER VALLEY ECOSYSTEM

### 29.2.1 Water and Sediments

PCBs continue to be transported downstream in the water column both from the continued inputs of PCB oil from the GE plant sites and from the contaminants contained in the sediments of the upstream pools.[19,24] The amount of PCB transported varies greatly with location and among locations depending on seasonal and interannual changes in river flow. For example, at the Thompson Island Dam, total PCB load averaged 1.16 kg/day during periods of low flow and 18.1 kg/day during high-flow periods.[24] The U.S. Geological Survey (USGS) has monitored river flow at several sites along the Hudson since 1977. These data exhibit a general decline in the total PCB load from the initial measurements (2308 kg/year at RM 194.3, 3956 kg/year at RM 168.2 in 1977) through the 1980s, when they fell below 400 kg/year in many years. However, there was significant interannual variation, and a notable increase in total PCB load was evident in 1992, when loads climbed to 951kg/year and 537 kg/year at the two sites.[24] This increase was probably related to high water flows washing fresh PCBs into the Hudson from the Allen Mill at the GE Hudson Falls site.[25]

The sediments of the Hudson River form a reservoir where high levels of PCBs are stored as they cycle through the food web. Sediment surveys conducted by the New York State Department of Environmental Conservation (NYSDEC) starting in 1976–1978 noted the concentration of PCBs in areas of sediment deposition, especially in the pools that form behind the locks of the upper Hudson. Areas of especially high concentrations (mean concentration = 48.4 mg/kg)[24] have become known as "hot spots." This general pattern has persisted since first described in the 1970s and is still recognized in surveys conducted by NYSDEC, GE, and U.S. EPA in the 1980s and 1990s. Sediment concentrations vary on a small spatial scale. The 1984 sediment survey of the Thompson Island Pool sampled 1100 sites and found mean PCB concentrations of 10.9 g/m$^2$ and a median of 3.5 g/m$^2$ (SD = 41.7). However, concentrations ranged from a low of 0.04 g/m$^2$ to a high of 1218 g/m$^2$.[24]

Sediment concentrations also vary on a wider spatial scale. The highest concentrations of PCBs are found in the Thompson Island Pool sediments. Mean concentration in the upper 25 cm of sediment measured in 1991 was 42 mg/kg.[26] Concentrations were lower in pools immediately downstream of Thompson Island Dam, with a mean 26 mg/kg PCB (but a maximum of 4000 mg/kg in one core). Concentrations are lower still in the sediments upstream from the Federal Dam at Troy, where surface sediment concentrations were 9 mg/kg.[26]

### 29.2.2 Fish and Aquatic Invertebrates

The diverse habitats of the Hudson River support several distinct assemblages of fish. Commercially valuable species, such as striped bass (*Morone saxatilis*), American shad (*Alosa sapidissima)*, Atlantic Sturgeon (*Acipenser oxyrhynchus*), and American eel (*Anguilla rostrata*), depend on the estuarine portions of the lower river.[19] Recreational fishers pursue a diversity of these estuarine species as well as largemouth and smallmouth bass (*Micropterus salmoides* and *M. dolomieui*), yellow perch (*Perca flavescens*), pumpkinseed (*Lepomis gibbosus*), and other freshwater species. In all, over 200 fish species are found in the Hudson River.[19]

Until the mid-1970s the lower Hudson supported a commercial fishery, based on striped bass, American shad, American eel, Atlantic sturgeon, and other species, valued at an estimated $2.6 million annually.[18] After high levels of PCBs in commercially exploited species from the Hudson became publicly known, the NYSDEC closed the lower Hudson River to commercial fishing for striped bass in 1976.[25,27] The upper Hudson River was closed to both commercial and recreational fishing in 1976, although a catch-and-release recreational fishery was reopened in 1995.[22,28] Losses to the striped bass fishery alone were estimated at $0.75 to $3.7 million in 1994.[29] In addition, the lower Hudson supports several endangered and threatened species as well as species of special concern such as the federally and state listed shortnose sturgeon (*Acipenser brevirostrum*).

Several groups have collected field data on the exposure of Hudson River fish to PCBs including the National Oceanic and Atmospheric Administration (NOAA), GE, and U.S. EPA. The most comprehensive data have been gathered by NYSDEC. Data collected by NYSDEC in 1998 show high levels of PCBs persist (reported by U.S. EPA[19] and pers. comm. from R. Sloan, NYSDEC). For example, in Thompson Island Pool (Griffin Island at RM 189), average tissue concentrations in 1998 were 32.7 μg/g for carp (*Cyprinus carpio*), 16.2 μg/g for brown bullhead (*Ameiurus nebulosus*), and 18.2 μg/g for largemouth bass (Table 29.2; R. Sloan, NYSDEC, pers. comm.).

Given the social and economic importance of the fishery, much attention has been paid to the suitability of Hudson River fish for human consumption. The closure of the Hudson River fisheries in the 1970's was preceded by health advisories warning residents not to eat fish from the river. Since that time, the U.S. Food and Drug Administration (FDA) has set acceptable limits for human consumption, first at 5 ppm wet mass in 1974, then at the current lower level of 2 ppm in 1984[25] (Table 29.1).

Generalizations about patterns of PCB concentrations in Hudson River fish are difficult to make given the variety of factors that influence contaminant levels. Concentrations of PCBs in fish vary within a species based on the age, fat content, and size of the fish. Among species, those species that are nonmigratory, feed at higher trophic levels, and live further upstream tend to have the highest concentrations. Other factors such as sex, reproductive status and body condition, and habitat use may also contribute significant variation to PCB levels.[30,31]

PCB levels are low in fish from sites upstream of the GE plants (Figure 29.2). Peak levels are observed near the plants and then generally decrease with distance down river. The Mohawk River enters the Hudson at RM 156. Downstream of this point approximately 50% of the river flow (though less than 50% of the sediment load) originated in the Mohawk Valley. This dilution of water from the upper Hudson is in part responsible for the drop in PCB loads in fish from Albany downstream (R. Sloan, NYSDEC, pers. comm.).

Over time, PCB levels in fish have fluctuated. When use of PCBs at the GE facilities stopped in 1977, the levels of PCBs in water and in fish tissues declined rapidly until the early 1980's (Figure 29.3). Since then, levels have changed slowly or stabilized, although they remain subject to significant interyear variation (R. Sloan, NYSDEC, pers. comm.).[19,25,32] As late as 1999, mean concentrations of PCBs in muscle tissue of brown bullheads and largemouth bass remained as high as 13 μg/g and 21 μg/g in the upper river, well above the FDA limit for the edible portion of fish of 2 μg/g.[26,33] Some individual largemouth bass in Thompson Island Pool still had 114 μg/g in 1999 (NYSDEC data).[26]

The social and commercial value placed on striped bass has contributed to this species' becoming an important representative of the Hudson River PCB problem. Striped bass are anadromous, migrating as adults from the Atlantic Ocean to spawning grounds in the Hudson River in the spring, and returning to the Atlantic Ocean in the late spring and summer, although some individuals (particularly males) remain in the Hudson longer (R. Sloan, NYSDEC, pers. comm.). Because

**Table 29.2 Concentration of PCBs in a Representative Range of Hudson River Organisms. Total PCBs (µg/g Based on Wet Mass) Shown for Samples Collected in a Variety of Years and Locations**

| Species | Tissue or Life Stage Analyzed | Total PCBs µg/g (Wet Mass) | Year | Site (River Mile) | Notes |
|---|---|---|---|---|---|
| Mink | liver | 0.6 | 1982–1984 | upper Hudson Valley | a |
| | liver | 0.7 | 1982–1984 | lower Hudson Valley | a |
| | fat | 4.4 | 1982–1984 | upper Hudson Valley | a |
| | fat | 3.5 | 1982–1984 | lower Hudson Valley | a |
| Otter | liver | 7.3 | 1982–1984 | Hudson Valley | a |
| | fat | 19.9 | 1982–1984 | Hudson Valley | a |
| Tree swallow | nestling | 39.8 | 1994–1995 | Special Area 13 (193) | b |
| | nestling | 3.7 | 1994–1995 | Saratoga National Historic Park (173) | b |
| | nestling | 10.0 | 1998 | Special Area 13 (193) | c |
| | nestling | 6.8 | 1998 | Saratoga National Historic Park (173) | c |
| | nestling | 0.4 | 1998 | Lower Hudson | c |
| Wood duck | adult-fat | 0.6 | 1983–1984 | Hudson/Champlain Valley | d |
| | adult-muscle | 0.05 | 1983–1984 | Hudson/Champlain Valley | d |
| | egg | 1.4 | 1995 | Roger's Island (194) | e |
| | adult | 0.3 | 1995 | Roger's Island (194) | e |
| Mallard | adult-fat | 8.9 | 1983–1984 | Hudson/Champlain Valley | d |
| | adult-muscle | 0.12 | 1983–1984 | Hudson/Champlain Valley | d |
| | egg | 1.1 | 1995 | Saratoga National Historic Park (173) | e |
| | adult | 4.0 | 1995 | Roger's Island (194) | e |
| Snapping turtle | fat | 36,080.0 | ca 1983 | Hudson Falls (197) | f |
| Brown bullhead | muscle | 16.2 | 1998 | Griffin Island (189) | g |
| | muscle | 10.1 | 1998 | Stillwater (168) | g |
| | muscle | 1.95 | 1998 | Poughkeepsie (76) | g |
| Carp | muscle | 32.7 | 1998 | Griffin Island (189) | g |
| | muscle | 58.6 | 1998 | Stillwater (168) | g |
| | muscle | 5.61 | 1998 | Poughkeepsie (76) | g |
| Largemouth bass | muscle | 18.2 | 1998 | Griffin Island (189) | g |
| | muscle | 9.6 | 1998 | Stillwater (168) | g |
| | muscle | 2.62 | 1998 | Poughkeepsie (76) | g |
| Striped bass | muscle | 1.90 | 1993 | Geo. Washington Bridge (12) | h |
| | muscle | 1.26 | 1996 | Geo. Washington Bridge (12) | i |
| | muscle | 2.54 | 1994 | Poughkeepsie (74–76) | h |
| | muscle | 1.81 | 1996 | Poughkeepsie (74–76) | i |
| | muscle | 6.41 | 1994 | Albany/Troy (152–153) | h |
| | muscle | 4.89 | 1996 | Albany/Troy (152–153) | i |
| Caddisfly | larvae | 9.6 | 1983 | Roger's Island (194) | j |
| | larvae | 66 | 1983 | Thompson's Island (192) | j |
| | larvae | 10.6–101.1 | 1978–1985 | Upper Hudson | k |
| Midge | larvae | 6.2–7.3 | 1985 | Thompson's Island (192) | l |
| Mixed insects | adults | 17.7 | 1995 | Remnant Site (195) | l |
| | adults | 6.6 | 1995 | Special Area 13 (193) | m |
| Odonata | adults | 1.1 | 1995 | Remnant Site (195) | b |
| | adults | 0.5 | 1995 | Special Area 13 (193) | b |
| | adults | 0.2 | 1995 | Saratoga National Historic Park (173) | b |

[a] Data from Foley et al. 1988.[60]
[b] Data from Secord et al.[40]
[c] USFWS data.[19]
[d] Data from Foley 1992.[61]
[e] USFWS data — A. Secord, pers. comm.
[f] Based on a single specimen.[62]
[g] NYSDEC data, R. Sloan, pers. comm.
[h] NYSDEC data from Sloan et al. 1995.[30]
[i] NYSDEC data, converted to reflect trichlorinated and higher congeners and reported in U.S. EPA 2000.[19]
[j] *Hydropsyche leonardi.*[63]
[k] Unspecified spp. results reported on the basis of dry mass.[64]
[l] *Chironomus tentans* sampled after 96 h in the Hudson, reported on the basis of dry mass.[65]
[m] Include adults of aquatic insect larvae captured by foraging tree swallows and fed to nestlings.[40]

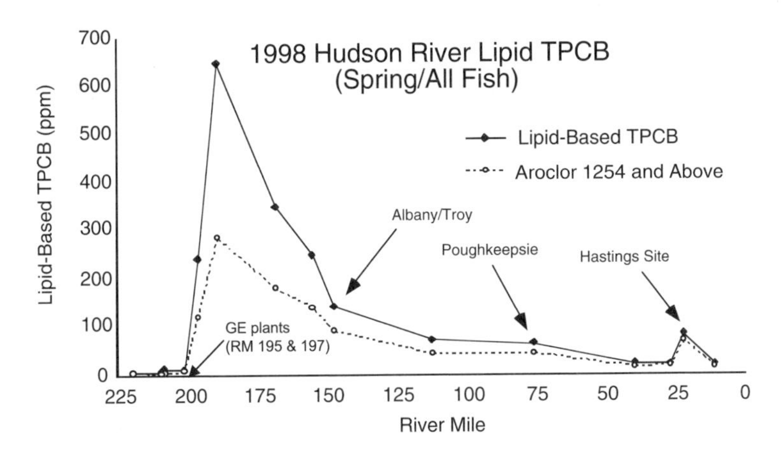


**Figure 29.2** PCB levels in Hudson River fish. River Mile is the distance above the mouth of the river, with the primary source of PCBs being the Hudson Falls and Ft. Edward GE plants at RMs 195–197. Concentrations are given on a lipid-weight basis, with both total PCBs and the higher chlorinated fractions quantified as Aroclor 1254 and higher shown. Figure courtesy of R. Sloan, NYSDEC.

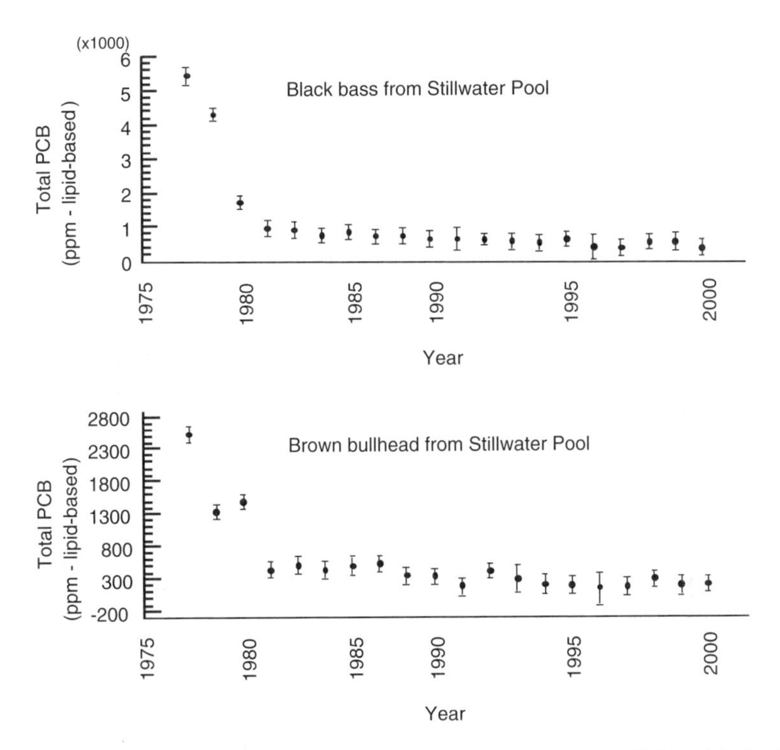


**Figure 29.3** Changes in PCB levels in Hudson River fish over time. Total PCBs (lipid-weight basis) are shown for both black bass (*Micropterus*) and brown bullhead (*Ameiurus nebulosus*) from the Stillwater Pool (RMs 168–183). Figure courtesy of R. Sloan, NYSDEC.

striped bass spend only a portion of their life history in the Hudson River, overall levels of contamination in them tend to be lower than in resident species. Initial observations of PCBs in striped bass date back to at least 1970, when concentrations in excess of 4 μg/g were observed.[22,23] As the PCB contamination came to the public's attention and health warnings were issued, regular sampling of striped bass was initiated.[32] Mean levels of PCBs in the edible portions of striped bass were 18.1 μg/g in 1978.[27] Total PCB levels fell below 10 μg/g by 1980 but showed only a slow decline in subsequent years.[30] In the Hudson River below RM 80, tissue concentrations changed slowly and were around 2 μg/g by the 1992–1994 period.[30] However, in the Albany/Troy area (RM 153) interyear variation was greater than any trend toward decreasing concentrations. Levels were especially high in 1985, 1986, and 1992, when total PCB levels were greater than 15 μg/g (wet mass).[30] Overall, PCB concentrations increase in fish caught further upstream. As in other species, male striped bass have higher levels of contamination than females. For example, males caught in 1997 in the lower estuary averaged 1.25 μg/g PCBs, while females averaged 0.58 μg/g PCB (R. Sloan, NYSDEC, pers. comm.).

While much attention has been paid to the impact of PCBs on the economics of the fishery, the impacts of this contamination on fish health and ecology have been less well studied. Effects on fish attributed to PCBs include declines in physical condition, liver, spleen and kidney cell abnormalities, abnormal bone development, and skewed sex ratios.[34–36]

The U.S. EPA Ecological Risk Assessment evaluated the probable effects of PCB exposure on a variety of representative fish species.[19] The general approach was to compare exposure of Hudson River populations to levels of PCB contamination shown to cause detectable impacts on reproduction, survival, or growth in field and laboratory studies reported in the scientific literature. Based on this approach, the EPA concluded that PCB levels in water and in sediment in the upper Hudson exceeded levels that could be reasonably considered acceptable and that these levels were predicted to remain high through the entire analysis period (e.g., through at least 2018).[19] In the lower Hudson, the U.S. EPA concluded that the risk to fish is primarily limited to piscivorous fish.

### 29.2.3 Wildlife and the Terrestrial Ecosystem

PCBs are known to have been deposited in terrestrial habitats near the Hudson River through dumping of contaminated dredge spoils, deposition along floodplains due to flooding, and exposure of contaminated sediments in the remnant-deposit areas. PCBs enter the terrestrial food chain when primarily terrestrial species eat contaminated aquatic organisms, such as insects and fish, or are exposed to invertebrates associated with contaminated flood-plain soils.

The highly productive waters of the Hudson River attract large numbers of birds and other wildlife. Waterfowl concentrate along the Hudson in large numbers during migration and both peregrine falcons (*Falco peregrinus*) and bald eagles (*Haliaeetus leucocephalus*) have bred along the Hudson since the 1990s. Other species found in the Hudson River watershed and listed as threatened and endangered by New York State or the federal government include several amphibian and reptile species (Blanding's turtle *Emydoidea blandingii*, timber rattlesnake *Crotalus horridus*, bog turtle *Clemmys muhlenbergii*, and northern cricket frog *Acris crepitans*) and two mammals, the Indiana bat *Myotis sodalis* and eastern woodrat *Neotoma magister*.[19]

PCBs are suspected of causing a wide range of negative effects in birds and other wildlife, though few species have been studied under controlled conditions. PCBs are ubiquitous in wild birds, but concentrations above background levels are still found in many areas and are associated with adverse effects.[1,15,17] Among the effects of PCBs observed in birds are developmental abnormalities (especially deformed feet and bills), liver damage, physiological and metabolic changes, reduced reproductive success, and embryonic and adult mortality.[1,2,15,17] Concentrations in brain tissue greater than 300 μg/g have been found in dead or dying adults and are indicative of acute toxicity.[15]

Compared to aquatic habitats, relatively few studies of the implications of PCB contamination for terrestrial species have been conducted along the Hudson. Available data do suggest that wildlife are exposed to high levels of Hudson River PCBs (Table 29.2).[19] However, for most species, sampling has not been systematic or intensive enough to document effects of contamination or the spatial and temporal patterns of PCB concentrations. As a result, most analyses, including the U.S. EPA Ecological Risk Assessment depend on estimates of exposure and use a modeling approach to evaluate risk.[19]

Based on this approach, the U.S. EPA evaluated risk to a variety of wildlife. This risk analysis concluded that PCBs pose significant risk to mammals along the Hudson, including insectivores, such as little brown bats (*Myotis*), and piscivores, such as mink (*Mustela vison*) and river otter (*Lutra canadensis*).[19] Mink and otter along the upper Hudson are especially likely to be exposed to harmful levels of PCBs; the calculated toxicity quotients for these species were up to three orders of magnitude above the acceptable risk quotient of one.[19]

Similar conclusions were reached from the analysis of risk to birds. Based on the U.S. EPA's analysis, exposure of bald eagles along the Hudson to PCBs is likely to result in impaired reproduction.[19] Other piscivores along both the upper and lower Hudson River, such as belted kingfishers (*Ceryle alcyon*) and great blue heron (*Ardea herodias*), are also at risk.[19] Recent empirical data from U.S. Fish and Wildlife Service collecting supports the conclusion that species such as bald eagles and great blue herons are exposed to Hudson River PCBs. Levels of PCBs in eggs and adults of migratory game birds, such as wood ducks (*Aix sponsa*) and mallards (*Anas platyrhychos*), are above the FDA limit for human consumption of poultry and eggs[33] (Table 29.2).

Perhaps the most detailed study of wildlife species along the Hudson River is the tree swallow (*Tachycineta bicolor*) study initiated in 1994 by the U.S. Fish and Wildlife Service (USFWS).[37] The overall goal of the study was to evaluate uptake of PCBs by migratory birds in the Hudson River valley. In addition to evaluating exposure, data on reproductive success and other ecologically relevant endpoints were targeted to begin to establish a basis for evaluating the risk posed by the contamination of the Hudson to migratory birds and other resources.[37] Tree swallows were chosen because they were known to breed along the Hudson River, feed on adults of aquatic insect larvae, and had been the subject of other studies of environmental contaminants.[38]

Breeding colonies of tree swallows were established in nest boxes along the upper Hudson River.[39,40] Exposure was evaluated through analysis of eggs, nestlings, and adults. In 1994–1995, average total PCB levels in tree swallow eggs ranged from 9.3 μg/g at the Saratoga site 45 km downstream of the Hudson Falls GE plant to 29.5 μg/g at Special Area 13 located just 6 km downstream.[40] Nestling contamination levels were higher than those in eggs, ranging from an average of 3.7 μg/g at Saratoga to 62.2 μg/g at Remnant Site 4 (Table 29.2).[40] A single adult from Special Area 13 had 114 μg/g. Levels of PCBs in 1998 tended to be lower, with means of 6.8 μg/g at Saratoga and 10.0 μg/g at Special Area 13 (Table 29.2). Nestlings collected from the lower Hudson in 1998 contained less than 1 μg/g PCB (Table 29.2). The levels of PCBs found in tree swallows from the upper Hudson River are greater than those found from studies at other PCB-contaminated sites and are indicative of the magnitude of the contamination of the Hudson River ecosystem.[38]

The majority of PCBs in Hudson River tree swallows were congeners from the tri-, tetra-, and pentachlorinated homolog groups.[40] In general, PCB congeners 118, 105, and 77 were most prevalent. In most previous studies of PCB congeners in birds, PCBs 138 and 153 have predominated.[16] The presence of high levels of congener 77 is noteworthy because this PCB and related non-*ortho* planar congers, is considered especially toxic to wildlife[16,17] and because congener 77 has not been thought to biomagnify to the same extent as other congeners.[1]

Population dynamics of mobile species such as tree swallows can be dominated by dispersal among different breeding sites. This makes evaluation of the population-level consequences of a stressor such as PCBs difficult or impossible in the absence of quantitative data on dispersal. The simple fact that tree swallows were chosen in part because they are abundant in the Hudson River invalidates any attempt to draw inferences from simple estimates of local population size. While population effects cannot be evaluated directly, data on demographic parameters do indicate some effects of PCB contamination on the reproductive biology of Hudson River tree swallows. Evidence from 1994 and 1995 indicated that Hudson River tree swallows did have an excess number of large clutches (> seven eggs), and in one year the percent of eggs hatching was low.[39] Survival of nestlings that hatched was within normal ranges at all Hudson River sites, but females tended to bury eggs in the nest and abandon nests at a greater rate than has been observed in uncontaminated sites.[39]

Other aspects of tree swallow ecology also appear to be impacted by PCB contamination. Parental behavior, as measured by success at building nests, is impaired along the upper Hudson and is significantly correlated with contaminant levels.[41] Likewise, development of female plumage is abnormal in the upper Hudson River population. Compared to individuals from elsewhere in the species range, the Hudson River population contains a significantly larger proportion of females whose plumage is intermediate in color between subadult and full adult categories.[42] The possible mutagenic effects of high concentrations of PCBs were evaluated by comparing Hudson River tree swallows to individuals from uncontaminated sites using minisatellite DNA fingerprinting.[43] Tree swallows from the Hudson River averaged 0.010 mutations per fragment scored, while swallows from reference sites in Wisconsin, Alberta, and Ontario averaged 0.011 mutations per fragment scored, indicating that PCBs do not result in elevated rates of mutation in this population.[43]

Based in part on these results, the U.S. EPA Ecological Risk Assessment concludes that tree swallows and related insectivorous birds did exceed acceptable exposure limits in the early 1990s along portions of the upper Hudson River. The modeling study indicated that exposure levels were likely to decline in the future (through 2018) and that levels of risk in the future would likely be low.[19] However, the effect levels chosen by the U.S. EPA were not able to take into account negative effects observed on behavioral or developmental endpoints, so the broader implications of the tree swallow field study for the status of migratory birds as a public resource remains to be evaluated.

### 29.2.4 Other Contaminants and Stressors

The Hudson River watershed is heavily populated, and a variety of human-caused stressors are likely to impact fish, wildlife, and ecological communities in the area. Direct impacts of development, shoreline modification, and dredging are all important in portions of the river. The dramatic land use changes seen in the northeast — a period of rapid deforestation and intensive agriculture in the late 18th and 19th centuries, followed by concurrent trends towards reforestation and urbanization in the 20th century — have led to an ecosystem that is heavily influenced by human activities. In particular, inputs of organic carbon, especially from agricultural runoff, are an important influence on the functioning of the Hudson River ecosystem.[21] Presettlement organic carbon inputs into the Hudson are estimated at only 36% of present inputs.[21] The carbon and nutrient cycles of the ecosystem form an important basis for water quality, especially in the lower Hudson. Removal or regulation of point sources of pollution, especially sewage treatment plants, have resulted in significant increases in water quality in recent years.[21,44]

Control of inputs from many point sources has also improved water quality through reductions in heavy metals.[45] Other chemicals of concern, such as mercury, DDT, dioxins, and dibenzofurans, are present in fish and wildlife of the Hudson River, but levels of these contaminants are generally low and judged by NYSDEC to be of secondary importance compared to PCBs.[19] Since 1994, fish-consumption advisories issued by the New York State Department of Health have specifically listed PCBs as the sole chemical of concern in Hudson River fish.[25] Mercury levels in fish from the lower Hudson River were generally less than 1.0 μg/g (R. Sloan, NYSDEC, pers. comm.), though largemouth bass in the upper Hudson had higher levels of mercury than those in the lower river. Concentrations ranged from 0.09 to 0.29 μg/g near Poughkeepsie (RM 74), 0.13 to 0.79 μg/g near Stillwater (RM 168), and 0.15 to 0.92 μg/g near Griffin Island (RM 189; R. Sloan, NYSDEC, pers. comm.).

PCDDs and PCDFs were detected in tree swallow nestlings. Among sites, total concentrations of PCDDs ranged from 0.022 μg/g at Saratoga Battlefield to 0.054 μg/g at Special Area 13.[37] The highest levels (0.077 μg/g) were found at Lock 9 on the Champlain Canal, upstream from the Hudson River. Total concentrations of PCDFs ranged from 0.022 μg/g at the Remnant Site to 0.039 μg/g at Saratoga.[37] Analysis of tree swallow tissues found that PCDDs and PCDFs were both of minor importance when compared to PCB levels using toxic equivalency factors (TEFs), accounting for less than 3% of total TEQs.[37] Other organic compounds (e.g., mirex, dieldrin, DDD, DDE, and DDT) were below levels of detection in tree swallows, as were most metals.[37]

Exotic species abound in the Hudson River ecosystem, and many have exerted significant pressures on the native ecosystems. At least 113 of these species have become established in the freshwaters of the river.[46] Eurasian milfoil (*Myriophyllum spicatum*) and water chestnut (*Trapa natans*) are common in near-shore habitats, as are purple loosestrife (*Lythrum salicaria*) and common reed (*Phragmites australis*) in riparian areas and marshes. Exotic fishes, such as the ubiquitous carp (*Cyprinus carpio*), are also abundant. A diverse assemblage of exotic fishes is supported, including brown trout (*Salmo trutta*), as well as domestic goldfish (*Carassius auratu*).[46] Introduced predators, such as brown trout, can adversely affect native fish communities, while large populations of bottom-dwelling carp can alter both the structure and functioning of aquatic ecosystems.[47–49]

Zebra mussels (*Dreissena polymorpha*) invaded the Hudson River in about 1990 and are abundant where suitable habitat is found. Especially high average densities (17,000/$m^2$) are found on rocks in waters deeper than 5 m.[50,51] Zebra mussels have had a profound effect on the ecosystem. Recent estimates suggest that the zebra mussel population of the lower Hudson River may filter a volume of water equal to the entire volume of the tidal freshwater river every 2 days,[52] altering the planktonic community.[51] In addition, a recent decline in summertime dissolved oxygen levels can be attributed to the arrival and spread of zebra mussels.[53] Much research remains on how or if the zebra mussel influences the dynamics of PCB bioaccumulation and transfer through the Hudson River ecosystem.

## 29.3 MITIGATION ACTIVITIES: PAST AND FUTURE

Any attempt to understand the Hudson River case history that focuses solely on the scientific studies will be incomplete. Much of the focus of the science, such as the emphasis on the biology of striped bass, has been driven by concerns of the public or policymakers. Unfortunately, the link between the products of these scientific efforts and the decisions of the public and policymakers has not always been as direct.

Public awareness and concern about the fate of the Hudson River came early relative to the rise of environmental awareness in many areas of the United States.[54] Regulations and pressure from environmental groups have been successful in bringing controls of many sources of pollution that are most offensive to public senses and have resulted in significant improvements in water quality, especially in the lower Hudson where the majority of the population is concentrated.[44,45]

The PCB issue arose after public support for cleaning up the Hudson had already developed. Indicators of problems emerged by 1970, by which time PCBs had been found in Hudson River fish and Monsanto (manufacturer of PCBs) had warned its customers (including GE) that release of PCBs posed a threat to the environment (Table 29.1). By 1972, GE applied for permission to release 30 lbs of PCBs per day into the Hudson River (13.6 kg/day).

When the public became aware of the magnitude of the PCB problem in the Hudson River, GE and the State of New York were forced into negotiations. The outcome of their negotiations in 1975 resulted in blame being assigned both to GE for release of PCBs and to the state for "regulatory failure" in allowing the releases to continue. The settlement included an agreement by GE to contribute \$3 million for monitoring of the Hudson River, matched by \$3 million from NYSDEC. In addition, GE agreed to devote \$1 million to research on development of an alternative dielectric fluid and to phase out use of PCBs (Table 29.1).[22] The U.S. EPA became involved in 1975. Federal involvement increased when New York State requested federal funds to clean up PCBs under the Clean Water Act (CWA).[22] In 1980, Congress authorized the U.S. EPA to spend \$20 million on the Hudson River cleanup under the CWA. The U.S. EPA was first required to conduct an environmental impact statement, which was completed in 1982.[22] While the environmental impact statement supported remediation and was backed by most stakeholders, U.S. EPA Administrator Ann Gorsuch issued a surprise ruling that the project was not suited for funding under CWA but should be funded under Superfund (Comprehensive Environmental Response, Compensation, and Liability Act [CERCLA]).[22] This began the process of listing the Hudson River site under the National Priorities List of Superfund.

In 1984, the U.S. EPA released its initial Record of Decision (ROD) for the Hudson River Superfund Site. The ROD recognized the problem of PCBs in the Hudson River sediments but selected an interim "No Action" decision because it considered technologies at the time insufficient to safely remediate the site.[26] The ROD did require GE to contain and cap the contaminated sediments and debris in the exposed Remnant Deposits (RMs 195–196).

In 1989, the U.S. EPA announced the reassessment of its 1984 interim "No Action" decision for the upper Hudson River at the request of NYSDEC.[26] In 1991, GE reported an increase in PCBs in the water of the upper Hudson River. This was eventually attributed to releases of contaminated sediment and oil from the Allen Mill located at the GE Hudson Falls site. Between 1993 and 1995, an estimated 45 tons of PCB-saturated sediments were removed from Allen Mill.[26] Additional fresh inputs of PCB oil from the bedrock adjacent to the Hudson Falls site were also discovered in the early 1990s as the Allen Mill site was investigated and stabilized and as the adjacent Bakers Falls was dewatered as part of reconstruction of a hydroelectric plant on the opposite bank of the river.[25]

In the fall of 2000, the U.S. EPA completed its reassessment and released its proposed plan for the upper Hudson River sediments.[26] The reassessment concluded that the PCBs in the sediments of the Hudson River posed significant risk to human health and ecosystem integrity.[6,19] The assessment noted that natural river dynamics prevent the contaminated sediments from being buried by clean sediments in a timely manner, that rates of dechlorination are insufficient to "naturally remediate" the PCBs, and that PCBs from the upper river are transported to the ecosystems of the estuarine Hudson at rate of approximately 214 kg/year.[26] The proposed plan called for the limited dredging of Hudson River sediments, focusing on the "hot spots" of highest concentration.[28] The efficacy of currently available dredging technologies creates a need to weigh risks from allowing PCBs to remain in the river against risks associated with dredging. The U.S. EPA decision is in part based on recent advances in technology and experience with remediation of other PCB sites.[28] While the U.S. Congress ordered the EPA to fund a study by the National Academy of Sciences on alternative remediation strategies (in part because of the controversy over the Hudson River), the final National Research Council Report failed to provide a strong statement for or against dredging on the Hudson River.[1]

Starting in the fall of 2000, the U.S. EPA solicited comments on the plan from the public and other interested parties. After considering the more than 70,000 comments received, the EPA Administrator for the Bush administration, Christie Whitman, supported the preferred remedy of dredging, drafted by the previous Clinton Administration.[55]

Estimates of the cost of implementing the U.S. EPA plan vary from $460 to $500 million. The "targeted dredging" of the hot spots is intended to removed 2.65 million cubic yards ($2.03 \times 10^6$ m$^3$) of sediment.[26] Sediments will then be transported to preexisting disposal facilities outside of the Hudson River valley.[26] This dredging is expected to remove approximately 45,000 kg (100,600 pounds) of PCBs from the river, out of the estimated 603,000 kg of PCBs released by the GE plants prior to 1977.

Currently, there are several routes by which this plan could be implemented. GE could undertake the work described, or the U.S. EPA could seek a court order, forcing GE to undertake the cleanup. If GE were to challenge the court order, the EPA could proceed with the work itself. In that case, Superfund allows the Federal Government to pursue up to triple the cost of the work from GE. Today, it is unclear when this issue will be resolved, and it is likely that the final outcome will involve extensive interactions between GE, the EPA, and the federal courts.

Additional issues remain to be resolved on the Hudson River. There is growing public awareness of the magnitude of PCB contamination in terrestrial habitats along the upper Hudson River. By some estimates, six times more PCBs from the GE plants at Ft. Edward and Hudson Falls were dumped on land than were released directly into the river.[56] The implications of these releases for human and ecological health have received little attention.

Further government action is also likely. CERCLA (and other federal and state laws) grants federal and state agencies the authority to investigate damage to natural resources caused by releases of harmful substances. With respect to the Hudson River, the Department of the Interior (DOI), the National Oceanic and Atmospheric Administration (NOAA), along with the New York State Department of Environmental Conservation, acting on behalf of the public in their roles as trustees of the natural resources, have begun to evaluate damage to natural resources.[57] This process, a Natural Resources Damage Assessment, allows trustees to pursue claims against parties responsible for release of harmful substances. To date, the emphasis of this process has been on the fish resources of the Hudson. The trustees have determined that restrictions and closures of recreational and commercial fisheries and advisories against eating Hudson River fish resulting from PCB contamination have curtailed the public's use of the Hudson's fish resources and constitute an injury to the resource.[25] Future activities will likely focus on other sources of injury, means of restoring the resources, and compensation to the public for past and future losses of resources.[25]

Emphasis on a broader range of scientific research along the Hudson River is called for to support improved risk assessments. There is surprisingly little empirical data on exposure of humans along the Hudson to PCBs or health effects for the residents of the Hudson Valley. While the U.S. EPA's Human Health Risk Assessment did find significant health risks in the absence of remediation, future work on risks under different remediation scenarios is now called for.[6] These assessments would benefit from empirical data on exposure and PCB body burdens of subpopulations living along the Hudson River. Data on human health patterns is also needed, especially noncancer endpoints in children.[8–10]

Exposure data for fish and a few wildlife species do exist. For assessments of ecological risk, the greatest need is for studies of the impact of exposure on individuals and populations of fish and wildlife species. Given the uncertainties about how individual PCB congeners affect different species and how congeners interact within the body, site-specific studies of fish and wildlife are called for. Until recently, terrestrial wildlife have been largely ignored. However, the exposure data from the tree swallow studies clearly indicate that terrestrial species can have very high exposure levels.[40]

There is an especially urgent need for studies of other species of terrestrial animals to assess risk from PCB exposure. Tree swallows appear to be exceptionally resistant to the effects of PCBs.[38] Other, more susceptible species may be at much greater risk but have not been adequately studied.[19,37] A simple set of population surveys is unlikely to be sufficient to answer this question, since even heavily impacted populations can be supplemented by immigration from other populations.[38] PCB levels along the Hudson are high enough to potentially exclude susceptible species,[37,38] so a combination of site-specific lab studies and work with both free-ranging and confined individuals is likely to be needed.

In 1985, one of the first attempts to synthesize the ecological and environmental issues surrounding the Hudson River lamented the fact that over 2 years had passed since the Hudson had become a federal Superfund site, yet efforts at finding a solution to the problem had come to a standstill.[22,58] Earlier reports on the Hudson River noted the difficulties in trying to reach a scientifically sound solution to the Hudson River PCB issue, describing the environmental assessments of the Hudson River as the "illegitimate offspring of an unhappy union between law and science"[59] and noting the delays caused by some scientists "interpreting their analyses with their employers' implicit biases" as their primary consideration.[58] As this chapter is written, 25 years after the Hudson River PCB problem became widely recognized, many of the same issues as described by Limburg are still being debated, and only limited progress has been made toward remediation.[1]

## 29.4 SUMMARY

The amount of PCBs released by GE into the Hudson River will never be known. The total amount is almost certainly greater than the widely cited estimated range of 95,000 to 603,000 kg calculated prior to the discovery of the ongoing inputs of fresh PCB oils from the GE sites at Hudson Falls and Ft. Edward in the early 1990s (estimated at > 11–19 liters, or 3–5 gallons, per day). The pattern that has emerged over the past 25 years of extensive scientific study is that while the levels of PCBs in the ecosystem have declined from the high levels observed in the 1970's, the combination of continued discharges of PCB oil from the GE sites and the continued availability of PCBs in sediments act to ensure that high levels of contamination remain available to the biota.

Much of the attention has focused on the damage to the freshwater and estuarine fisheries, and concentrations of PCBs in most Hudson River species remain above the FDA threshold of 2 ppm. Other wildlife in the Hudson River valley are also exposed to significant levels of PCBs. The USFWS study of tree swallows along the upper Hudson River found PCB concentrations similar to those found in fish living in adjacent areas of the river. These high levels of contamination far exceed those found in other studies using tree swallows and are likely responsible for abnormal reproductive behavior and development observed in the Hudson River population of tree swallows.

While the weight of scientific evidence points to a significant risk of Hudson River PCBs to both human health and the environment, questions remain about the best course of action to remediate that risk and how long the contaminants will remain available if no remediation is attempted.

## ACKNOWLEDGMENTS

I am grateful to the American Association for the Advancement of Science for support through the Science Policy Fellowship program. I extend my gratitude to Dr. Ron Sloan of the NYSDEC for providing access to data on PCBs in fish. Anne Secord and an anonymous reviewer provided helpful comments on an earlier draft.

## REFERENCES

1. National Research Council, *A Risk-Management Strategy for PCB-Contaminated Sediments*, National Academy Press, Washington, D.C., 2001.
2. Rice, C. P., O'Keefe, P., and Kubiak, T., Sources, pathways, and effects of PCBs, dioxins, and dibenzofurans, in *Handbook of Ecotoxicology*, 2nd ed., Hoffman, D. J., Rattner, B. A., Burton, G. A., Jr., and Cairns, J., Jr., Eds., CRC Press, Boca Raton, FL, 2003, 501.
3. Kaiser, J. and Enserink, M., Treaty takes a POP at the dirty dozen, *Science*, 290, 2053, 2000.
4. Brouwer, A., Longnecker, M. P., Birnbaum, L. S., Cogliano, J., Kostyniak, P., Moore, J., Schantz, S., and Winneke, G., Characterization of potential endocrine-related health effects at low-dose levels of exposure to PCBs, *Environ. Health Perspect.*, 107, Suppl. 4, 639, 1999.
5. Safe, S. H., Polychlorinated biphenyls (PCBs): Environmental impact, biochemical and toxic responses, and implications for risk assessment, *Crit. Rev. Toxicol.*, 24, 87, 1994.
6. U.S. Environmental Protection Agency (U.S. EPA), Phase 2 Report, Further Site Characterization and Analysis, Volume 2F — Revised Human Health Risk Assessment: Hudson River PCBs Reassessment RI/FS, Prepared for the U.S. EPA, Region II, New York, by TAMs Consultants, 2000.
7. Faroon, O., Jones, D., and de Rosa, C., Effects of polychlorinated biphenyls on the nervous system, *Toxicol. Ind. Health*, 16, 305, 2001.
8. Jacobson, J. L. and Jacobson, S. W., Intellectual impairment in children exposed to polychlorinated biphenyls *in utero*, *N. Engl. J. Med.*, 335, 783, 1996.
9. Jacobson, J. L. and Jacobson, S. W., Evidence for PCBs as neurodevelopmental toxicants in humans, *Neurotoxicology*, 18, 415, 1997.
10. Ribas-Fito, N., Sala, M., Kogevinas, M., and Sunyer, J., Polychlorinated biphenyls (PCBs) and neurological development in children: A systematic review, *J. Epidemiol. Community Health*, 55, 537, 2001.
11. Rice, D. C., Behavioral impairment produced by low-level postnatal PCB exposure in monkeys, *Environ. Res.*, 80 (Pt. 2), S113, 1999.
12. Schantz, S. L., Moshtaghian, J., and Ness, D. K., Spatial learning deficits in adult rats exposed to *ortho*-substituted PCB congeners during gestation and lactation, *Fundam. Appl. Toxicol.*, 26, 117, 1995.

13. Widholm, J. J., Clarkson, G. B., Strupp, B. J., Crofton, K. M., Seegal, R. F., and Schantz, S. L., Spatial reversal learning in Aroclor 1254 exposed rats: Sex-specific deficits in associative ability and inhibitory control, *Toxicol. Appl. Pharmacol.*, 174, 188, 2001.
14. Schantz, S. L., Gasior, D. M., Polverejan, E., McCaffrey, R. J., Sweeney, A. M., Humphrey, H. E., and Gardiner, J. C., Impairment of memory and learning in older adults exposed to polychlorinated biphenyls via consumption of Great Lakes fish, *Environ. Health Perspect.*, 109, 605, 2001.
15. Eisler, R., Polychlorinated Biphenyl Hazards to Fish, Wildlife, and Invertebrates: A Synoptic Review, U.S. Fish and Wildlife Service Biological Reports, 85(1.7), 1986.
16. Eisler, R. and Belisle, A. A., Planar PCB Hazards to Fish, Wildlife, and Invertebrates: A Synoptic Review, National Biological Service Biological Report 31, 1996.
17. Hoffman, D. J., Rice, C. P., and Kubiak, T. J., PCBs and dioxins in birds, in *Environmental Contaminants in Wildlife: Interpreting Tissue Concentrations*, Beyer, W. N., Heinz, G. H., and Redmon-Norwood, A. W., Eds., Lewis Publishers, Boca Raton, FL, 1996, 165.
18. Moran, M. A. and Limburg, K. E., The Hudson River ecosystem, in *The Hudson River Ecosystem*, Limburg, K. E., Moran, M. A., and McDowell, W. H., Eds. Springer-Verlag, New York, 1986, 6.
19. U.S. Environmental Protection Agency (U.S. EPA), Phase 2 Report, Further Site Characterization and Analysis, Volume 2E — Revised Baseline Ecological Risk Assessment: Hudson River PCBs Reassessment, Prepared for the U.S. EPA, Region II, New York, by TAMs Consultants and Menzie-Cura & Associates, 2000.
20. Swaney, D. P., Howarth, R. W., and Butler, T. J., A novel approach for estimating ecosystem production and respiration in estuaries: Application to the oligohaline and mesohaline Hudson River, *Limnol. Oceanogr.*, 44, 1509, 1999.
21. Howarth, R. W., Schneider, R., and Swaney, D., Metabolism and organic carbon fluxes in the tidal freshwater Hudson River, *Estuaries*, 19, 848, 1996.
22. Limburg, K. E., PCBs in the Hudson, in *The Hudson River Ecosystem*, Limburg, K. E., Moran, M. A., and McDowell, W. H., Eds., Springer-Verlag, New York, 1986, 83.
23. Clesceri, L. S., Case History: PCBs in the Hudson River, in *Introduction to Environmental Toxicology*, Gutherie, F. E. and Perry, J. J., Eds., Elsevier, New York, 1980, 227.
24. U.S. Environmental Protection Agency (U.S. EPA), Phase 2 Report, Further Site Characterization and Analysis, Volume 2C — Data Evaluation and Interpretation Report, Hudson River PCBs Reassessment RI/FS, Prepared for the U.S. EPA, Region II, New York, by TAMs et al., 1997.
25. U.S. Department of the Interior, National Oceanic and Atmospheric Administration, and New York State Department of Environmental Conservation, Injuries to Hudson River Fishery Resources: Fishery Closures and Consumption Restrictions, Hudson River Natural Resources Damage Assessment, Final Report, 2001.
26. U.S. Environmental Protection Agency (U.S. EPA), Hudson River PCBs Superfund Site Proposed Plan, U.S. EPA, Region II, New York, 2000.
27. Bush, B., Streeter, R. W., and Sloan, R. J., Polychlorinated biphenyl (PCB) congeners in striped bass (*Morone saxatilis*) from marine and estuarine waters of New York State determined by capillary gas chromotography, *Arch. Environ. Contam. Toxicol.*, 19, 49, 1989.
28. U.S. Environmental Protection Agency (U.S. EPA), Hudson River PCBs Reassessment RI/FS Phase 3 Report: Feasibility Study, Prepared for the U.S. EPA, Region II, New York and U.S. Army Corps of Engineers, Kansas City District, by TAMs Consultants, 2000.
29. Kahn, J. R. and Buerger, R. B., Valuation and the consequences of multiple sources of environmental deterioration: The case of the New York striped bass fishery, *J. Environ. Manage.*, 40, 257, 1994.
30. Sloan, R., Young, B., and Hattala, K., PCB Paradigms for Striped Bass in New York State, New York State Department of Environmental Conservation, Division of Fish, Wildlife and Marine Resources, Technical Report 95–1, 1995
31. Zlokovitz, E. R. and Secor, D. H., Effect of habitat use on PCB body burden in Hudson River striped bass (*Morone saxatilis*), *Can. J. Fish. Aquat. Sci.*, 56 (Suppl. 1), 86, 1999.
32. Sloan, R. J., Simpson, K. W., Schroeder, R. A., and Barnes, C. R., Temporal trends toward stability of Hudson River PCB contamination, *Bull. Environ. Contam. Toxicol.*, 31, 377, 1983.
33. U.S. Government, Code of Federal Regulations, Title 21 Food and Drugs, Part 109, Unavoidable Contaminants in Food for Human Consumption, Sec. 109.30 Tolerances for Polychlorinated Biphenyls (PCBs), U.S. Government Printing Office, Washington, D.C., 1996.

34. Bowser, P. R., Martineau, D., Sloan, R., Brown, M., and Carusone, C., Prevalence of liver lesions in brown bullheads from a polluted site and a nonpolluted reference site on the Hudson River, New York, *J. Aquat. Animal Health,* 2, 177, 1990.
35. Kim, J. C. S., Chao, E. S., Brown, M. P., and Sloan, R., Pathology of brown bullhead, *Ictalurus nebulosus*, from highly contaminated and relatively clean sections of the Hudson River, *Bull. Environ. Contam, Toxicol.*, 43, 144, 1989.
36. Mehrle, P. M., Haines, T. A., Hamilton, S., Ludke, J. L., Mayer, F. L., and Ribick, M. A., Relationship between body contaminants and bone development in east-coast striped bass, *Trans. Am. Fish. Soc.*, 111, 231, 1982.
37. Secord, A. L. and McCarty, J. P., Polychlorinated Biphenyl Contamination of Tree Swallows in the Upper Hudson River Valley, New York, U.S. Fish and Wildlife Service, New York Field Office, Cortland, NY, 1997.
38. McCarty, J. P., Tree swallows as monitors of environmental stress, *Reviews in Toxicology, Series B: Environmental Toxicology*, in press.
39. McCarty, J. P. and Secord, A. L., Reproductive ecology of Tree Swallows (*Tachycineta bicolor*) with high levels of PCB contamination, *Environ. Toxicol. Chem.,* 18, 1433, 1999.
40. Secord, A. L., McCarty, J. P., Echols, K. R., Meadows, J. C., Gale, R. W., and Tillitt, D. E., Polychlorinated biphenyls and 2,3,7,8-tetrachlorodibenzo-*p*-dioxin equivalents in tree swallows from the upper Hudson River, New York State, USA, *Environ. Toxicol. Chem.,* 18, 2519, 1999.
41. McCarty, J. P. and Secord, A. L., Nest building behavior in PCB contaminated tree swallows (*Tachycineta bicolor*), *Auk,* 116, 55, 1999.
42. McCarty, J. P. and Secord, A. L., Female plumage color, PCB contamination, and reproductive success in Hudson River Tree Swallows (*Tachycineta bicolor*), *Auk,* 117, 987, 2000.
43. Stapleton, M., Dunn, P. O., McCarty, J. P., Secord, A. L., and Whittingham, L. A., Polychlorinated biphenyl contamination and minisatellite DNA mutation rates of tree swallows, *Environ. Toxicol. Chem.,* 20, 2263, 2001.
44. Brosnan, T. M. and O'Shea, M. L., Long-term improvements in water quality due to sewage abatement in the lower Hudson River, *Estuaries*, 19, 890, 1996.
45. Sañudo-Wilhelmy, S. A. and Gill, G. A., Impact of the clean water act on the levels of toxic metals in urban estuaries: The Hudson River estuary revisited, *Environ. Sci. Technol.*, 33, 3477, 1999.
46. Mills, E. L., Scheuerrell, M. D., Strayer, D. L., and Carlton, J. T., Exotic species in the Hudson River basin: A history of invasions and introductions, *Estuaries,* 19, 814, 1996.
47. Moyle, P. B., Fish introductions into North America: Patterns and ecological impact, in *Ecology of Biological Invasions of North America and Hawaii*, Mooney H. A. and Drake, J. A., Eds., Springer-Verlag, New York, 1986, 27.
48. Vitousek, P. M., D'Antonio, C. M., Loope, L. L., and Westbrooks, R., Biological invasions as global environmental change, *Am. Scientist,* 84, 468, 1996.
49. Mooney, H. A. and Hobbs, R. J., Eds., *Invasive Species in a Changing World*, Island Press, Washington, D.C., 2000.
50. Strayer, D. L., Powell, J., Ambrose, P., Smith, L. C., Pace, M. L., and Fischer, D. T., Arrival, spread, and early dynamics of a zebra mussel (*Dreissena polymorpha*) population in the Hudson River estuary, *Can. J. Fish Aquat. Sci.,* 53, 1143, 1996.
51. Strayer, D. L., Caraco, N. F., Cole, J. C., Findlay, S., and Pace, M. L., Transformation of freshwater ecosystems by bivalves: A case study of zebra mussels in the Hudson River, *BioScience,* 49, 19, 1999.
52. Roditi, H. A., Caraco, N. F., Cole, J. J., and Strayer, D. L., Filtration of Hudson River water by the zebra mussel (*Dreissena polymorpha*), *Estuaries,* 19, 824, 1996.
53. Caraco, N. F., Cole, J. J., Findlay, S. E. G., Fischer, D. T., Lampman, G. G., Pace, M. L., and Strayer, D. L., Dissolved oxygen declines in the Hudson River associated with the invasion of the zebra mussel (*Dreissena polymorpha*), *Environ. Sci. Technol.*, 34, 1204, 2000.
54. Boyle, R. H., *The Hudson River: A Natural and Unnatural History,* Norton and Co., New York, 1969.
55. U.S. Environmental Protection Agency (USEPA), Whitman Decides to Dredge Hudson River [Press Release R-125], U.S. EPA, Communications, Education, and Media Relations, Washington, D.C., 2001.
56. Worth, R., PCB Worries are spreading from Hudson to its shores, *New York Times*, 17 April, 2001.

57. State of New York, National Oceanic and Atmospheric Administration, and the U.S. Department of the Interior, Preassessment Screen Determination for the Hudson River, New York, 1 October, 1997.
58. Limburg, K. E. and Moran, M. A., Synthesis and evaluation, in *The Hudson River Ecosystem*, Limburg, K. E., Moran, M. A., and McDowell, W. H., Eds., Springer-Verlag, New York, 1986, 155.
59. McDowell, W. H., Power plant operation on the Hudson River, in *The Hudson River Ecosystem*, Limburg, K. E., Moran, M. A., and McDowell, W. H., Eds., Springer-Verlag, New York, 1986, 40.
60. Foley, R. E., Jackling, S. J., Sloan, R. J., and Brown, M. K., Organochlorine and mercury residues in wild mink and otter: Comparison with fish, *Environ. Toxicol. Chem.*, 7, 363, 1988.
61. Foley, R. E., Organochlorine residues in New York waterfowl harvested by hunters in 1983–84, *Environ. Monit. Assess.*, 21, 37, 1992.
62. Olafsson, P. G., Bryan, A. M., Bush, B., and Stone, W., Snapping turtles — a biological screen for PCBs, *Chemosphere*, 12, 1525, 1983.
63. Bush, B., Simpson, K. W., Shane, L., and Koblintz, R. R., PCB congener analysis of water and caddisfly larvae (Insecta: Trichoptera) in the upper Hudson River by glass capillary chromotography, *Bull. Environ. Contam. Toxicol.*, 34, 96, 1985.
64. Novak, M. A., Reilly, A. A., and Jackling, S. J., Long-term monitoring of polychlorinated biphenyls in the Hudson River (New York) using caddisfly larvae and other macroinvertebrates, *Arch. Environ. Contam. Toxicol.*, 17, 699, 1988.
65. Novak, M. A., Reilly, A. A., Bush, B., and Shane, L., *In situ* determination of PCB congener specific first order absorption/desorption rate constants using *Chironomus tentans* larvae (Insecta: Diptera: Chironomidae), *Water Res.*, 24, 321, 1990.

## CHAPTER 30

# Baseline Ecological Risk Assessment for Aquatic, Wetland, and Terrestrial Habitats along the Clark Fork River, Montana

**Greg Linder, Daniel F. Woodward, and Gary Pascoe**

## CONTENTS

## 30.1 INTRODUCTION

Watersheds in mining districts in the western United States are frequently impacted by metals and other inorganic constituents released during extraction and processing activities. Metal-enriched materials, such as soil eroded from exposed and disturbed landscapes and tailings generated during processing, may be released to the environment and associated with increased metal concentrations in surface water and groundwater. Similarly, dispersed sediments may have been deposited as alluvial materials in riparian areas, yielding soils having metal concentrations much greater than predepositional conditions. The present chapter summarizes ecological risk assessment studies focused on metal-contaminated soil, sediment, and surface water for a series of Superfund sites located in the Clark Fork River (CFR) watershed of western Montana (Figure 30.1).

## 30.2 CASE STUDY BACKGROUND

In the early 1980s, various locations throughout the CFR watershed were recognized as "Superfund" sites, which fell under the regulatory statutes developed as a result of the Comprehensive Environmental Response, Compensation, and Liability Act of 1980 (CERCLA, or Superfund). For

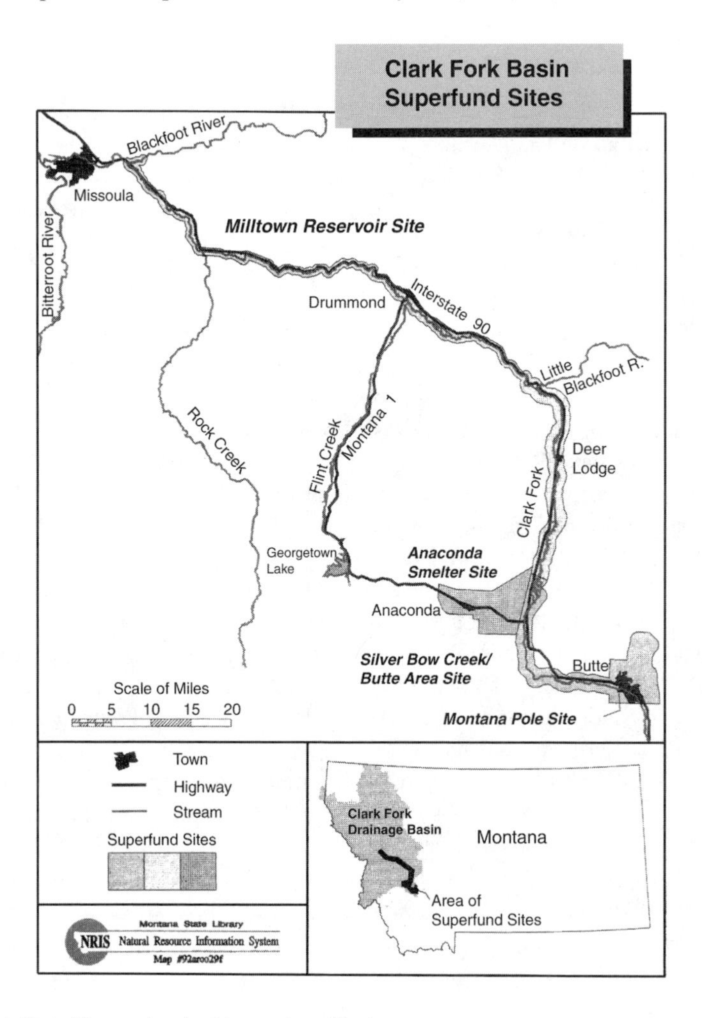


**Figure 30.1** Clark Fork River watershed in western Montana.

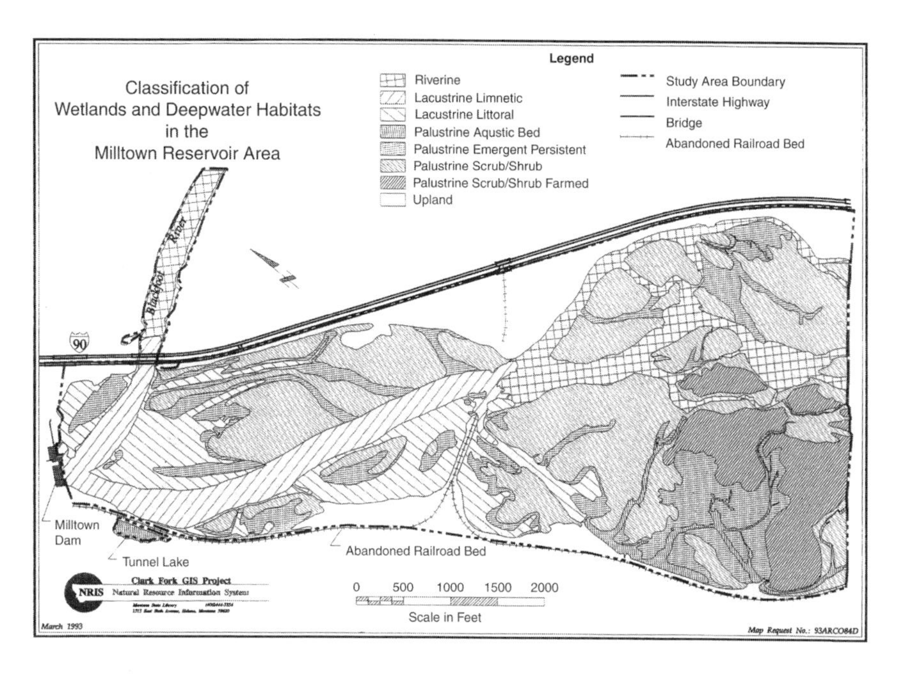


**Figure 30.2** Milltown Reservoir Wetlands at the western terminus of the series of Superfund sites where work was originally completed to support baseline ecological risk assessments for terrestrial, wetland, and aquatic habitats.

example, the reservoir near Milltown, Montana, which is located at the western end of a series of Superfund sites on CFR, was initially listed on the National Priority List, or NPL, in 1982 as a result of a hydrologic investigation that identified reservoir sediments as the likely source of arsenic contamination in drinking water taken from groundwater wells. Subsequently, during the late 1980s and early 1990s, numerous studies contributed to the baseline ecological risk assessment completed for CFR and associated terrestrial and wetland habitats.[1–14] These studies evaluated the aquatic and terrestrial habitats that had been impacted by historical copper mining activities in the Butte and Anaconda areas.[1] Extraction and processing activities in the mining district had released disturbed soils and tailing materials enriched in arsenic, cadmium, copper, lead, and zinc that subsequently contributed to sediment loading of CFR.[15–19] As the primary focus of these baseline studies, the upper CFR consists of approximately 200 km (125 mi) along CFR from Warm Springs, Montana to Milltown Dam, a hydroelectric facility located near Missoula, Montana (see Figures 30.1 and 30.2). The dam lies immediately downstream from the confluence of CFR and Blackfoot River, with Milltown Reservoir occurring behind the dam to a maximum pool elevation of 995 m (3264 ft). The reservoir, associated wetlands, and the reach of CFR from the up-river boundary of Milltown Reservoir (maximum pool elevation) to just down river of Warm Springs Ponds delineates the area of concern discussed in this review.[1]

## 30.2.1 Historic Releases to CFR

Mining had occurred in the Clark Fork Basin since the discovery of gold at Butte in the mid-1860s, and mining-related activities continued until the early 1980s. During that period, an estimated 100 million tons of tailings were lost to CFR and transported down river. With construction of Milltown Dam in 1906–1907, sediments transported down CFR subsequently accumulated behind the dam and contributed to the development of wetlands associated with the reservoir. In 1982, Woessner et al.[20] estimated that 29 ft of sediment were present at the upstream face of the dam. Sediment volume contained within the reservoir was estimated at 3.5 to 4.6 million cubic m (4.5 to 6 million cubic yards, or over 2750 acre-ft).[20–22]

Early mining wastes were primarily coarse-grained materials resulting from placer mining and stamping milling, while later developments in mining technology yielded finer-grained mining wastes, which likely increased mining impacts to the river basin.[23,24] Disposal of tailings, milling and smelting wastes, and mine waste water directly into tributaries of the CFR was common practice throughout the mining districts along CFR until regulatory guidance was established. Waste materials associated with these activities were characterized by concentrations of arsenic, cadmium, copper, lead, and zinc that exceeded expected background concentrations by 10 to 100 times.[17,20,21]

In the present summary of the ecological risk assessment process for the CFR, we briefly summarize the findings and ecological risks to habitats of the upper CFR and Milltown Reservoir Wetlands (MRW). As indicated by the numerous citations in the summary that follows, many studies have been published as a result of these efforts to characterize the baseline ecological risks for the upper CFR-MRW complex, and we encourage the reader to consult these publications for detail beyond our present scope.

## 30.3 ENVIRONMENTAL SETTING

### 30.3.1 The Clark Fork River Basin and Upper Clark Fork River Subbasin

The CFR system drains over 57,000 square km (22,000 square mi) of western Montana as it flows 550 km (340 mi) to Lake Pend Oreille in northern Idaho. The upper Clark Fork subbasin includes the headwaters of the CFR downstream to Milltown Dam, about 200 km to the north and west of the river's origin (125 river miles) (see Figure 30.1). Historically, the CFR was a major corridor and spawning ground for trout migrating out of Lake Pend Oreille in northern Idaho. Migrations ceased, however, with the construction of control structures such as Milltown Dam. The upper river currently supports high-quality game fish in certain river segments, and the basin is valued for its big game, upland bird, and waterfowl hunting. These resources of the basin have created a large recreational and tourism industry in the area.[1]

### 30.3.2 Milltown Reservoir Wetlands

Milltown Reservoir is the first major impoundment on the CFR. It is about 2 km (1.25 mi) in length and is generally shallow except for the two channels of the Clark Fork and Blackfoot Rivers. Sediment accumulation in the impoundment behind Milltown Dam has created a 541-acre wetland, with total jurisdictional wetland acreage at 86% (Figure 30.2).

MRW contains a variety of habitats from riverine to upland. Palustrine wetlands comprise about 56% of the area and have been classified as aquatic bed, emergent persistent, scrub/shrub, and scrub/shrub farmed.[25,26] The major depositional areas of the wetland are lacustrine littoral and palustrine emergent persistent wetlands. In general, the vegetation and wildlife at MRW are typical of riverine wetlands of western Montana, with the wetland habitat characteristically consisting of submerged and emergent herbaceous plants.[2,26,27] The wetland supports a wide variety of terrestrial vertebrate wildlife, including current or former range occupied by threatened or endangered species, e.g., Peregrine falcon (*Falco peregrinus*).[27]

## 30.4 AQUATIC RESOURCES AT RISK

Numerous studies have assessed the potential impact of metals contamination on wildlife and aquatic resources of the CFR, and the results of these investigations suggest that metals contamination in the CFR has degraded the aquatic resource.[14,28]

### 30.4.1 Metals Contamination of Sediments and Effects on Benthic Invertebrates

Sediments of the upper CFR, from Butte and Anaconda to Missoula, are contaminated with arsenic, cadmium, copper, lead, and zinc,[14,28] but sediment from Milltown Reservoir and the CFR was not generally lethal to test organisms in the laboratory.[4] However, Ingersoll et al.[5] found that sediment is a significant source of metals to invertebrates in the CFR. For example, concentrations of metals in amphipods exposed to CFR sediments in relative short-term laboratory exposures resulted in metal bioaccumulation to 50% of that measured in field-collected invertebrates from the CFR. Besser et al.[29] found that sediments from most sites in the Clark Fork and Milltown Reservoir caused (1) significant bioaccumulation of copper and (2) significantly reduced survival and growth in larvae of the midge. Additional studies found that the bioavailability of copper to midges exposed to sediments from the CFR and MRW — as indicated by copper bioaccumulation and growth — were largely controlled by concentrations of acid-volatile sulfides (AVS).[30] These studies also suggested that bioavailability of copper in Clark Fork sediments could vary seasonally, as AVS concentrations change in response to factors such as water temperature, flow regime, and loading of nutrients and organic matter.

Aufwuchs — the biofilm consisting of fine detrital sediment, attached algae, and bacteria — is an important link between particle-bound metals of colloidal size and aquatic invertebrates that feed by grazing and scraping at this trophic level.[31] The accumulation of metals in invertebrates was dependent on functional feeding group, and shredders-scrapers that feed on biofilm accumulated the largest concentration of metals. In a system like the CFR that has received chronic contamination resulting in metals loading in sediment and food-chain organisms, the total load to the biological system is dependent on water, sediment, and food chain. Consequently, any consideration of exposure and toxicity should consider all routes of exposure in a system like the CFR, or the risks associated with exposures in the field would be underestimated.

The adverse effect of metals on aquatic invertebrates may affect benthic community structure and bioaccumulation potential in biota from CFR and MRW.[5,6] Where concentration of metals are highest, community structure will likely be affected, but when biological effects are subtle, community structure may be less affected. Even in the absence of altered community structure, bioaccumulation still occurs, and metals are transferred through the food chain. Poulton et al.[13] identified significant differences in benthic invertebrate community composition (i.e., number of invertebrate taxa and predominance of pollution-intolerant taxa) of metal-contaminated sites in the Clark Fork and recommended that both benthic community structure and metals concentration in invertebrates be used to determine relative impacts in metal-contaminated river systems. This recommendation was based on the close relationships among metal availability, metals accumulation in benthic invertebrates, and transfer to higher trophic levels. The authors stated that, of importance, moderate to high densities of benthic organisms are present in the CFR, even at sites containing high levels of metals in invertebrates, which indicates that, to some degree, the benthic community may be able to tolerate and adjust to metals contamination. Ecologically, this phenomenon may be more important than altered community structure (e.g., changes in taxa richness) and biomass because metal-contaminated invertebrates represent a major source of metal uptake and accumulation in salmonids.

### 30.4.2 Toxicity of Metal-Contaminated Food Organisms to Fish

Benthic macroinvertebrates are an important source of food for the larval trout that inhabit the CFR. Studies conducted between 1992 and 1995 using invertebrates collected from the CFR and fed to fish in the laboratory suggested that such spiked-feeding studies may underestimate dietary exposures in the field.[7,8] In these studies, early-life-stage brown trout and rainbow trout were exposed for 88 to 91 days to simulated CFR water and a diet of benthic invertebrates collected from the river. These exposures resulted in reduced growth and elevated levels of metals in the whole bodies of both species and, in some cases, elevated concentrations in liver. Additionally, these experiments and those of Farag et al.[10] observed that fish fed diets collected from near Warm Springs and from Gold Creek had elevated concentrations of products of lipid peroxidation and histological abnormalities (such as effects on hepatocytes, pancreatic tissue, and the mucosal epithelium of the intestine). These alterations in tissue structure and impaired physiology were also associated with an increase in tissue metals.

In contrast, another feeding study focused on dietary exposures used rainbow trout fed brine shrimp (*Artemia* sp.), which had been exposed to an aqueous metals mixture for 24 h after hatching.[12] Here, metal concentrations in brine shrimp were comparable to metal concentrations in CFR field-collected diets, but rainbow trout fed on the exposed brine shrimp did not show reduced survival or growth. Although additional study would be necessary to focus on mechanisms that contribute to these differences, the duration of exposure in the brine shrimp study was very short relative to exposures in the field, and the metals likely occurred primarily in the free form, adhering to external surfaces of the brine shrimp and not fully incorporated into the biological matrix. Ionic metals are not absorbed as efficiently and may not be as toxic as metals bound to proteins. Exposure differences between laboratory-generated and field-collected diets likely explain differences in experimental outcomes. Similar studies on the Coeur d'Alene River in Idaho, which has metals contamination similar to the CFR, support this explanation for differences noted between studies using field-collected and laboratory-exposed diets. For example, Farag et al.[32] used procedures similar to those of Woodward et al.[7,8] and Farag et al.[10] for their studies on the Coeur d'Alene, and toxicological endpoints similar to those noted from studies on the CFR were observed in feeding studies using field-collected materials from the Coeur d'Alene River (i.e., reduced survival, decreased growth, reduced feeding activity, and histopathological abnormalities). These toxicological endpoints were also associated with accumulations of metals in tissues of dietary-exposed fish.

Observations and measurements on wild fish collected from the CFR are also consistent with dietary exposure pathways and the associated physiological impairment observed in the laboratory.[11] Arsenic, cadmium, copper, and lead accumulated in specific tissues (gill, liver, kidney, pyloric caeca, stomach, large intestine, stomach contents, and whole fish) in a pattern that suggested the dietary source was a major pathway of exposure. An association between tissue metal residues and physiological responses strongly suggested a cause-and-effect relationship between metals in the CFR and physiological impairment of resident fish in the river. Metallothionein, lipid peroxidation, and histological lesions were used as indicators of tissue injury, with damage being present in those tissues associated with digestion (intestine and pyloric caeca) or systemic exposure (liver). Food-chain transfer of metals appears to be a dominant route of exposure for fishes in the upper CFR.

### 30.4.3 Avoidance of Metals by Fish in the Laboratory

Water-quality criteria were formulated to protect survival, growth, and reproduction of sensitive species;[33] however, behavioral avoidance of unfavorable concentrations of metals such as copper and zinc may be an additional cause of reduced fish populations in natural systems. Behavioral avoidance is particularly important because it can occur at concentrations lower than those causing effects on survival and growth. The limiting factor on fish populations may not be acute toxicity or physiological impairment but the displacement of fish from preferred habitats by avoidance of unfavorable conditions.

Copper in the CFR is at or above the concentrations that have previously been avoided by fishes. Woodward et al.[9] and Hansen et al.[34] conducted several experiments to determine if CFR water resulted in an unfavorable behavioral response by brown trout and rainbow trout. These tests were conducted in a cylindrical chamber that received reference water at one end and metal-contaminated water at the other; a distinct boundary formed at the center where the chamber drained. Reference water simulated the CFR without metals and was used for acclimation and as the alternative choice for the fish to the test water. Test water was identical to reference water but was altered by the addition of metals to simulate concentrations existing in the river. Brown trout and rainbow trout avoided conditions simulating the presence of copper and other metals in the CFR, with rainbow trout appearing to be more sensitive to the metals than brown trout. Following 45 days of acclimation to an ambient metals mixture that simulated the CFR, rainbow trout preferred clean water and avoided higher metal concentrations. These controlled laboratory experiments show that both trout species will avoid typical metal concentrations observed in the CFR and may explain, in part, the reduced wild trout population in the CFR.

Snake River cutthroat trout have also been shown to avoid a mixture of metals simulating the Coeur d'Alene Basin, Idaho under similar test conditions as those used for the CFR.[35] Cutthroat trout avoided the mixture of metals, and it was determined that avoidance was due to copper and zinc and not to cadmium and lead. The avoidance response to either copper or zinc alone was similar to the avoidance response to the mixture. Furthermore, fish were raised in water containing metals at water-quality criteria for 90 days. At the end of 90 days, cutthroat detected and preferred a lower zinc concentration and avoided a higher zinc concentration. This study provides further evidence that fish can survive and grow in water containing metals but can detect a gradient in metals concentration, and they prefer the cleaner water.

### 30.4.4 Avoidance of Metals by Wild Fish Populations

Detection and avoidance of undesirable substances in the water has been documented in the laboratory, but wild populations of fish are influenced by numerous other factors. Goldstein et al.[36] used radio telemetry and a migrating population of Chinook salmon to determine the response of this wild population to the presence of metals. In this study, salmon were transferred from their natal stream to a point 2 mi below the confluence of the North Fork and the South Fork of the Coeur d'Alene River. As with CFR, the South Fork of the Coeur d'Alene River has metals contamination from mining (mainly cadmium, lead, and zinc, with some copper), while the North Fork is relatively clean. Fish were tagged with radio transmitters and allowed to move upstream. While other variables were present — flow, temperature, general water quality, and cover — the higher metals concentration in the South Fork tributary was the factor of greatest difference. The majority (70%) of the fish tagged selected the North Fork. The results of this experiment, coupled with the companion laboratory study, indicates that the abundance of free-ranging populations of fish are likely reduced by their unfavorable avoidance reaction to certain metals. Also, the demonstration of an avoidance response in a natural fish population validated results of laboratory avoidance experiments and explained the distribution and movement of fishes in relation to metals contamination.

## 30.5 TERRESTRIAL AND WETLAND RESOURCES AT RISK: BIOLOGICAL ASSESSMENT OF SOILS

In parallel with studies completed on CFR, studies at MRW were also undertaken. Here, designed field and laboratory studies focused on evaluations of soil physicochemical characteristics and biological tests of soils using invertebrates (e.g., earthworm toxicity and bioaccumulation tests and plant tests as currently standardized by ASTM[37,38]). Biological sampling of diet items, such as invertebrates, plants, and wildlife (i.e., small mammals), was also completed, with a primary focus on analyzing dietary exposures to terrestrial and wetland wildlife.[39–42]

### 30.5.1 Soil Physicochemical Characterizations and Metals Analysis

As part of an ecological assessment for MRW, biological effects measured in toxicity tests considered interactions, or potential interactions, between the soil matrix and metals in the soil, particularly with respect to bioavailability of metals.[43] Among other physicochemical characteristics, soil pH, cation-exchange capacity (CEC), and percent organic material (OM) were measured on soil samples collected at MRW. Soil-texture analysis was also completed, and DTPA (diethylenetriaminepentaacetic acid, 0.025M)-extractable metals were analyzed as an indirect estimate of metal bioavailability.[44,45] With these characteristics of the soil matrix available, potential confounding effects associated with soil physicochemical properties could be identified, and interpretations of metal toxicity were subsequently developed with more confidence.[14,46]

Soils throughout MRW were xerofluvents and ranged from loams to silty loams to occasional silty-clay loams that were heterogeneous with respect to soil texture, OM, and CEC.[2,46] For example, OM was variable across MRW but was relatively invariant within defined sampling strata, as were total nitrogen, soil pH, and CEC. Similarly, the distribution of metals — both total and extractable – in soil varied spatially, but variation within sample units identified across the wetland was much less than the variation among all sample units within MRW.[2] The spatial distribution of metals of greatest concern – cadmium, copper, zinc — clearly indicated that sample units could be ranked with respect to concentrations of these metals in soil.[14,46] And when these target analyte rankings were evaluated relative to the biological tests completed on colocated soil samples, an integrated assessment of risk to biological receptors was developed (see Sections 30.5.2 and 30.5.3).[14,46,47]

### 30.5.2 Plant Studies to Evaluate Wetland and Upland Soils

MRW was characterized by a range of terrestrial and wetland habitats and soil types; hence, numerous biological assessment tools were used to evaluate potential adverse effects associated with metal exposures.[2,40]

#### *30.5.2.1 Plant Tests Completed Using Bulk Soils Collected from MRW*

During preliminary studies, seed-germination tests were completed using site soils, with both field and laboratory methods being employed.[48] No statistically significant acute phytotoxicity was observed in these bulk soil tests, which suggested that definitive studies should be focused on the soil-soil interstitial water interface.[2,14,46]

#### *30.5.2.2 Root Elongation and Groundwater Phytotoxicity Assessments*

Groundwater levels at MRW during the definitive investigation generally varied between 24 and 48 in. below ground level, although the upland sampling locations presented groundwater generally characterized by depths to groundwater from 48 to 84 in. below ground level. A limited number of upland locations presented depths to groundwater between 84 and 120 in. below ground level.[2] Phytotoxicity assessments available from root elongation tests with groundwater indicated that samples collected with hand-driven well points were quite heterogeneous with respect to metal concentrations, as would be expected, considering other measurements taken in the rhizosphere (e.g., soil physical structure and metal concentrations in soil). Biological activity associated with groundwater collected from sample units across MRW was relatively low and not statistically significant relative to laboratory controls, but a trend toward reduced root growth was noted across all samples from depositional areas. No samples, however, expressed acute toxicity, as indicated by germination failure in the root-elongation test. From the work with root-elongation testing, reduced growth in plant roots may be linked to elevated metal concentrations in groundwater collected from depositional areas at MRW.[2,14,46]

#### *30.5.2.3 Root Elongation Tests on Soil Eluates*

In conjunction with tests completed with groundwater, standard phytotoxicity toxicity tests using lettuce seeds were also completed using soil eluates.[48] As were groundwater test results, no statistically significant reductions in root length were recorded. Nonetheless, the trend in the data presented biologically significant responses such as consistent reduction in root growth, again in samples collected from depositional locations in MRW. While these biological effects were not statistically significant, these trends were also observed in other tests (e.g., root-elongation tests with groundwater) used for the biological assessment at MRW.[2,46]

### 30.5.2.4 Metals Accumulation in Terrestrial Plants

Vegetation samples collected coincident with soil and groundwater samples indicated that target metals did bioaccumulate in vegetation growing at MRW. The extent of accumulation varied spatially among species and anatomically within plants (i.e., metal residues in roots differed from stems differed from leaves). Emergent wetland vegetation, e.g., cattail and sedge, appeared to accumulate the most metals, while upland vegetation appeared to accumulate metals, especially those having little nutrient value, to relatively low concentrations.[2,41,49,50] Metals with known nutritional value to plants, e.g., zinc, were bioaccumulated many-fold relative to the concentration of metal in soil.

### 30.5.2.5 Summary of Plant Tests

In addition to vegetation tests completed with groundwater, field and laboratory tests with submerged aquatic vascular plants (*Potamogeton pectinatus*[51,52] and *Hydrilla verticillata*[51,53]) were used to evaluate emergent-zone habitats at MRW. In general, emergent-vegetation testing in the laboratory and in the field suggested that biological effects, when evident, were sublethal. Field tests with sago pondweed indicated no adverse effects for growth endpoints (root and shoot length), and physiological markers (peroxidase activity [POD]) indicative of plant stress were unremarkable. Indigenous plant samples (*Elodea* spp.) were also collected concurrent with field tests on sago pondweed and were analyzed for POD activity. For indigenous *Elodea* spp., statistically significant differences among sites were noted across MRW sampling locations, which may be indicative of general plant stress. In parallel laboratory exposures, however, no consistent pattern was noted with respect to growth endpoints (e.g., root and shoot lengths and chlorophyll a) for either test species when tested with bulk sediments collected from MRW sites. While morphologic endpoints related to growth suggested no acute toxicity in laboratory or field exposures with MRW sediments, differences in POD activity across MRW may reflect the spatial variability in metals that are differentially bioavailable.

On the basis of these studies with emergent plants, terrestrial plants exposed to bulk soils, and biological evaluations of groundwater and sediment, rhizosphere exposures to sediment pore water, soil interstitial water, and groundwater coincident were generally not acutely phytotoxic. Yet all groundwater samples collected from deposition zones in the wetlands and upstream braided stretches of CFR appeared to have an inhibitory effect on root elongation as measured using the standard laboratory test species.[14,46] Field surveys completed in conjunction with the sampling and laboratory testing, however, did not suggest that this biological effect was currently being exhibited at these sampling locations during the period of study.[2,26,27]

Soil eluates were also prepared from soils collected at MRW and were tested using root elongation as the toxicity endpoint. While qualitative differences were observed with respect to the inhibition of root growth, no soil-derived eluate showed statistically significant results. As with tests with groundwater, reduced root growth was noted in eluates derived from soils collected in depositional areas; for example, biological effects reflected in reduced root growth appeared to be associated with soils collected from old "oxbow" reaches along the river.[2]

In parallel with these controlled plant tests, field surveys noted no widespread effects on vegetation in the sampled area. For example, from a jurisdictional wetlands delineation[26] and wildlife survey,[27] and from the supporting field surveys completed during the study,[2] the wetlands and adjoining upland habitats at MRW presented a relatively complex plant community, with numerous obligate as well as facultative wetland plant species.[54] In part, the physicochemical characteristics of the soils across MRW may explain, or contribute to, the variation in growth reduction observed in laboratory tests, particularly in those areas occasionally associated with reduced plant vigor in laboratory tests. The physicochemical heterogeneity of soils across MRW was readily apparent, even in sampling units that were similar with respect to texture. The reduced root elongation noted in groundwater evaluations, however, may be an effect directly associated with metal-contaminated groundwater. Again, field surveys indicated that no adverse effects were currently being expressed, though some emergent plants were clearly displaying relatively increased uptake of metal from sediments. CECs varied across soil at MRW and may have influenced the apparent differences in bioavailability of metals in the rhizosphere of these soils, as could the interrelationships among soil texture, organic content, and the geochemical composition of the soils.

These interacting soil-matrix characteristics undoubtedly contribute to the heterogeneity of metal uptake in exposed wetland or upland plants.[2,14,46] Metals did accumulate in plant tissues, with the trend suggesting that roots would accumulate metals to a greater extent than leaves. The pattern of metal accumulation in roots differed, in part, as a function of soil type.[2,46]

### 30.5.3 Animal Studies Completed for Terrestrial and Wetland Evaluations

#### *30.5.3.1 Earthworm Tests in Field and Laboratory*

Earthworms expressed no acute toxicity in either laboratory or field tests;[2,46,55] however, correlation analysis suggested that subtle biological response data (e.g., behavioral, morphologic, and dermatopathologic effects) from field tests were most significantly associated with soils having elevated total and DTPA-extractable copper. The spatial distribution noted in these sublethal-effect biological markers was consistent with other biological assessments completed at MRW (e.g., root-elongation tests).[2,14,46] These observations supported interpretations that subtle biological effects were associated with soils collected from depositional areas having relatively high metal concentrations. For example, exposures to soil associated with sublethal effects in *E. foetida* also presented subtle effects on plant growth (e.g., root elongation reduction) in groundwater and soil eluate tests.

Estimates of the bioaccumulation potential of metal in MRW soils were considered by analyzing earthworm tissues for metals following laboratory exposures.[37] To make these data more relevant for evaluating wildlife risk (that is, earthworms are consumed intact by predators in the field), earthworms were not purged prior to analysis. Bioaccumulation tests clearly suggested that metals in soils from MRW are potentially available to wildlife that prey upon earthworms,[2,46] and the extent to which metals are available for trophic transfer (either as metal incorporated into earthworm tissue through bioaccumulation or transferred via coincidental ingestion of gut contents) varies as a function of metal concentrations in soil and its physicochemical properties, e.g., high vs. low organic material, different levels of extractable metals at relatively similar total metal concentrations. Differential bioavailability of metals may be considered a possible source of these spatially variable expressions of chronic effects associated with metal bioaccumulation.[14,46]

#### *30.5.3.2 Wildlife Surveys and Bioaccumulation Sampling*

Aquatic invertebrates, fish, amphibians, birds, and small mammals inhabiting MRW were surveyed in 1990 following Warren-Hicks et al.[48] and other routine guidance for conducting wildlife surveys.[56–58] Different wetland habitat types were highly interspersed at Milltown Reservoir, resulting in diverse vegetative structure.[26] By design, surveys were not intended to completely assess the presence or absence of species expected to occur in riverine, lake, or wetland habitats; however, 14 taxa of aquatic macroinvertebrates and 20 vertebrate taxa other than birds were observed at the site.

Terrestrial and wetland wildlife were the primary focus of these field surveys. Species of small mammals and other terrestrial vertebrates found on the site were typical of riparian sites in western Montana, and birds observed at MRW from April through October 1990 were reported as either probable or confirmed breeders for this area.[27,59] Water bird and passerine bird species diversity was similar to avian communities occupying similar habitats at nearby Lee Metcalf National Wildlife Refuge, Stevensville, MT, and most of the birds at MRW were characterized as either wetland or edge species.[2,27] For amphibians, preliminary tests following Linder et al.[60] suggested that surface-water exposures would not present acute toxicity. No statistically significant acute toxicity was found in tadpoles exposed to any site samples in the laboratory for 96 h, and, relative to laboratory exposures using defined metal mixtures and single-metal exposures, analysis of these site samples suggested that metal concentrations in surface waters were not sufficient to mediate acute effects at the time the samples were collected and the study was completed. Chronic effects observed following laboratory exposures to some MRW surface-water samples were generally associated with reduced growth, and only a limited number of embryos presented terata. Gross malformations were occasionally noted and included mild abdominal edema and hyperpigmentation. No contaminant-specific malformations were noted in these preliminary tests.[60]

Evaluation of dietary exposure to metals by primary consumers and higher trophic organisms in the wetland habitat was performed by a food-chain evaluation. Exposures, expressed as daily oral intake, were compared to dose-response information (i.e., no observed adverse effects levels, NOAELs, in mg/kg/day) from the scientific literature.[42] The lack of toxicity information on most metals for many of the exposed species necessitated the use of dose-response data from rodents as surrogate toxicity criteria. Exposures were modeled on the results of field measurements in vegetation and lower trophic biota as well as sediments, soils, and water.[41,42] Complete reports of the food-chain analysis are reported in Pascoe et al., [42] but the site-specific data on body burden measurements of arsenic, cadmium, copper, lead, and zinc in the meadow vole (*Microtus pennsylvanicus*) and deer mouse (*Peromyscus maniculatus*) suggested that soil-metal bioavailability was limited to $< 0.1\%$ at the MRW. For example, for small mammals, the highest estimated daily doses for arsenic, cadmium, copper, and zinc were below the toxicity criteria pertinent to the analysis.[14,42] In contrast, for muskrat (*Ondatra zibethicus*), the consumption of aquatic vegetation from the depositional areas of the wetland where metal concentrations were two- to tenfold greater than site-wide concentrations, the calculated lead intake rate (0.23 mg/kg/day) exceeded the toxicity criterion. This was largely influenced by the preference of muskrat for cattails as the primary food source, which had elevated tissue concentrations of lead in these areas.[42]

## 30.6 ECOLOGICAL SIGNIFICANCE OF RISK ASSESSMENT FINDINGS

Evaluating the ecological significance of the results of these analyses was critical to the assessment of risks associated with metal exposures in the CFR-MRW system. Biologically important impacts were defined as those affecting populations, communities, and ecosystems, which acknowledged that measurement of impacts to individuals does not necessarily indicate impacts to the ecosystem.[61,62] The biologically important results would then be the focus in setting remedial goals for the CFR-MWR system.[63]

### 30.6.1 Defining Ecological Significance

For evaluating the studies completed throughout CFR-MRW, the definition of ecological significance focused on a relative ranking of impacts as outlined by Duinker and Beanlands:[62]

- *Major Impact:* affects an entire population or species in sufficient magnitude to cause a decline in abundance or change in distribution beyond which natural recruitment (reproduction, immigration from unaffected areas) would not return that population or species, or any population or species dependent upon it, to its natural level.

- *Moderate Impact:* affects a portion of a population and may bring about a change in abundance and distribution over one or more generations but does not threaten the integrity of that population or any population dependent upon it.
- *Minor Impact:* affects a specific group of localized individuals within a population over a short period of time but does not affect other trophic levels or the population itself.
- *Negligible Impact:* any impacts below the minor category are considered negligible.

For the CRF-MRW ecological risk assessment, the inclusion of quantitative studies on chemical exposures, biological effects assessment, and ecological studies helped characterize the ecological significance of the results.[14]

### 30.6.2 Terrestrial Habitat Findings

Results of the integrated field and laboratory studies for the ecological risk assessment for MRW have been summarized in detail elsewhere.[2,14,46] For the terrestrial wildlife relying on upland and wetland habitats at MRW, the studies found that the bioavailability of metals and arsenic to small mammals was more limited than expected.[41,42] The apparent bioavailability of soil elements to small mammals was less than 0.1%.

The effect of the limited bioavailability of mining waste metals on toxicity tests is illustrated by results of earthworm (*Eisenia foetida*) biological tests.[2,55] Based on the concentrations of total copper in soils collected at MRW, substantial toxicity was predicted for earthworms,[64] yet only subtle effects were observed at the highest concentrations of metals in soil.[2,55] The subtle findings of toxicity tests performed with both terrestrial and aquatic exposure media at MRW generally were consistent with the lack of overt ecological impacts observed in terrestrial and aquatic habitats at the site.[2,14,46] Altered benthic community structure in sediments (e.g., increased number of genera of Chironomidae) was observed in sediments at locations in MRW having the highest extractable metals under oxidizing conditions (i.e., low flow and low acid-volatile-sulfide content).[3,6] However, the ecological significance of the altered benthic community structure was considered of lesser importance to the ecological risk assessment than the potential transfer of metals from benthic invertebrates through the food chain to consumer fish.[14]

The ecological assessment at MRW indicated that no acute toxicity or adverse biological effects were being expressed at the time these studies were completed. Similarly, biological tests and field observations suggested only limited chronic effects were being expressed during the study period.[14] Consistently, and regardless of the field- or laboratory-test methods used, biological assessments at MRW were unable to demonstrate acute toxicity; however, evidence of subtle biological effects was noted in samples collected from depositional areas.[2,14,46]

Sublethal effects observed in biological tests could suggest subtle ecological impacts that could be potentially significant within ecological contexts. For example, the rooting zone tests demonstrated inhibition of root growth in plants exposed to interstitial water collected from the rhizosphere in depositional areas in MRW that presented high metal concentrations. An inhibition in root elongation suggests that lateral root growth might be inhibited in certain plants, which could result in decreased absorption of nutrients including copper, zinc, and iron. Altered root growth, in turn, could affect aboveground biomass, as demonstrated during greenhouse studies with soils from depositional areas, and result in iron-deficiency-induced chlorosis. Thus, the risks to wetland vegetation in depositional areas of MRW may be considered minor to moderate, especially if other stressors became prominent in the multiple-stressor picture characteristic of any riparian wetland field setting (e.g., altered flow regimes, drought).

For the terrestrial habitat, the findings of the exposure and biological effects assessments need to be placed in context with those of the ecological assessment to identify the significance of ecological risk. For example, results of the wildlife survey and wetland delineation study suggested a healthy wetland ecosystem. Thus, integration of the results of the ecological studies with the subtle effects observed in the biological tests suggests that the wetlands did not express significant ecological effects despite possible local impacts related to vegetation stress associated with potential nutrient deficiencies.

In contrast to potential effects expressed by vegetation, very little is available in the scientific literature to form a basis for translating the subtle toxicity test results, such as earthworm morphological changes, to impacts on ecological endpoints related to populations and wetland communities. However, the food-web analysis demonstrated that primary consumers and higher trophic organisms, such as predatory birds, waterfowl, and wild mammals, are not at high risk from exposures to metals in soils at MRW.[41,42] This conclusion was supported by the finding of low bioavailability of metals in soil.[46] The only potential concern was the effect of the modeled intake of metals on muskrat.[14,42] Given the uncertainties in the modeled intake and assumptions related to bioavailability, it is unlikely that the exposures in muskrat would be expressed by adverse effects in their population or ecosystem impacts. In addition, the waterfowl survey at the wetland suggested that the waterfowl populations were healthy and reproducing.[27] No evidence was observed of contaminated sediment invertebrates affecting the health of waterfowl in the wetland. Similarly, the food-chain-transfer study did not suggest that metals in the invertebrates or aquatic plants would impact muskrats, beaver, or mink in the wetland. In summary, the biological investigations completed at MRW assessment studies and food-chain analysis suggested little, if any, acute toxicity from exposures to upland or wetland soils and that the sublethal effects observed would not likely increase ecosystem risks. From this, risks to the terrestrial habitat can best be characterized as minor.[14,62]

In order to develop an integrated snapshot of MRW at the time of these studies and a technical framework for evaluating and managing ecological risks, a relative ranking scheme was used to summarize and integrate exposures and biological effects for the terrestrial habitat[14,46] (Table 30.1). In this procedure, where consistent statistical differences in biological effects were observed in each of the sample units at MRW, a rank score of plus (+) was assigned to units having a consistent "signature" of adverse effects, a mixed signature of adverse effects and no effects was assigned a plus/minus (+/–), and consistently no effects was ranked as a minus (–). Each sample unit was then considered by the sum of ranks for the biological responses and by the concentrations of total copper; for this ranking, a rating of "++" depicts the high-ranked units, "+" the moderate-ranked units, and "–" the low-ranked units. This comparison demonstrates the apparent association between total copper concentrations and a variety of biological effects across the site, with highest rankings for both chemistry and biological responses generally being characteristic of depositional areas at MRW.[2,14,46] The results of this integration of chemical and biological data are depicted geographically on a map of the wetland sampling units in Figure 30.3, where concentrations of total copper and the sum of biological responses are shaded for moderate (+) to high (++) ranks (i.e., "+" is shaded with single lines and "++" with double lines, with the chemistry ranking depicted by vertical lines and the biological ranking by horizontal lines; the triangular representations are described below for sediment ranking). The graphical depiction in Figure 30.3 allows a better illustration of the spatial correlation of biological responses with metal concentrations in the reservoir wetland soils. This type of presentation also more clearly demonstrates the higher concentrations of copper and greater biological effects in sediments that have accumulated directly behind the dam in the old river channel and behind the old railway berm.

Like the soil evaluations in terrestrial and wetland habitats, the sediment evaluations were critical to the evaluation of emergent zone habitats in the wetland. The weight of evidence from the sediment quality triad of chemistry, ecology, and toxicity studies[47] suggests that benthic invertebrate communities in the reservoir are at risk in sediments where metal concentrations are elevated and AVS concentrations are low. Elevated metals in oxidizing environments are found in sediments of the old river channel and upstream of the railroad berm in the MRW (Figure 30.3). Oxidizing environments at risk in the reservoir and upper CFR have been observed in shallow or fast-moving

**Table 30.1 Integration of Biological and Chemical Data for Evaluation of Ecological Risks at MRW***

| | Sample Units[a] | | | | | | | | | |
|---|---|---|---|---|---|---|---|---|---|---|
| **Biological Test[b]** | **7[c]** | **6[d]** | **8[c]** | **2[c]** | **10[e]** | **9[e]** | **11** | **3[e]** | **12** | **4[f]** |
| Earthworm (field) | + | – | + | + | – | – | – | +/– | – | +/– |
| Earthworm (lab) | +/– | + | + | + | +/– | – | – | + | – | ++ |
| Root elongation | + | | + | + | +/– | – | – | + | – | – |
| Seed germination | | | | + | | | | | | – |
| Biomass | + | | | + | | | | | | |
| Sum biological[g] | ++ | + | ++ | ++ | + | + | – | + | – | ++ |
| Cu concentration (μg/g)[h] | 975 | 870 | 761 | 734 | 654 | 595 | 585 | 529 | 470 | 244 |
| Sum chemistry[i] | ++ | ++ | ++ | ++ | + | + | + | + | – | – |

* See Figure 30.3 for locations of sample units (after Linder, Pascoe, and DalSoglio, 1999).

[a] Sample Units rank-ordered by concentration of total Cu; sample Units 1 and 5 were not tested (see Linder et al. 1994).

[b] Consistent statistically significant differences in biological effects observed in various tests completed on soil samples collected in each sample unit were graded as (+), inconsistent but positive responses in some tests as (+/–), and negative responses as (–). Sample Unit 4 was scored (++) for the laboratory earthworm test because of the much greater level of adverse effects observed in these soils relative to laboratory earthworm tests completed with soil samples collected from other sample units.

[c] Biological effects likely associated with metal contamination; sample units included major depositional areas of the wetland.

[d] Inconsistent results in biological tests likely due to low bioavailability of metals; poor association between biological effects and metals concentrations likely occurred due to ameliorating effects of soil physicochemical properties (e.g., cation-exchange capacity, organic material, pH).

[e] Moderate biological responses may be related to low metals contamination.

[f] Biological effects likely due to other factors and not metals.

[g] Relative ranking of biological tests completed as part of the ecological assessment. A value of (++) was assigned to the high-ranked sample units, (+) the moderate-ranked sample units, and (–) the low-ranked sample units.

[h] Mean concentration of total Cu in soil samples.

[i] Relative ranking of concentrations of Cu, Zn, and Cd in soils; soils with Cu concentrations less than 500mg/kg were scored (–), greater than or equal to 500 mg/kg, but less than 700 mg/kg were scored (+), and greater than or equal to 700 mg/kg were scored (++).

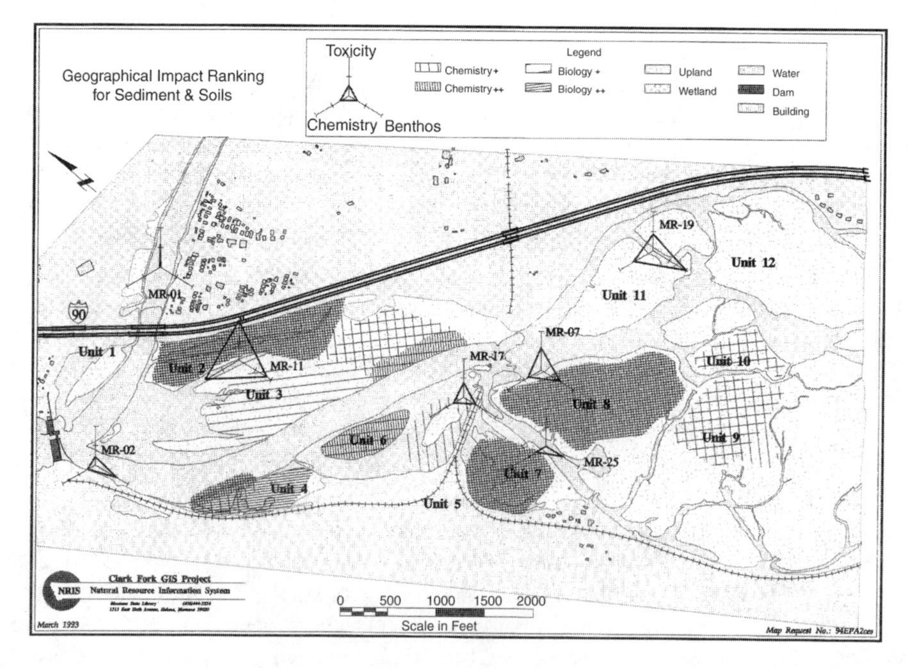


**Figure 30.3** Integration of laboratory and field work completed at MRW yielded a spatial distribution of risks that corresponded to depositional areas on CFR.

waters in depositional areas where metal concentrations are elevated. Sediment toxicity was generally observed where molar SEM Cu/AVS ratios exceeded unity.[5] In contrast, a comparison of sediment-toxicity criteria with no-effect concentrations (NECs) calculated from sediment toxicity tests completed at MRW illustrates the limitations in relying solely on benchmark comparisons for characterizing ecological risks associated with contaminated sediments.[65] For example, at MRW both copper and zinc NECs exceeded commonly applied guidance values,[66] which reflects on the conservativeness inherent in guidance values as pointed out by O'Conner et al.[67]

For evaluation of risks associated with sediments at MRW, the scaled ranking for the chemistry, toxicity, and benthos variables are plotted on maps of the reservoir operable unit as tri-axial graphs in Figure 30.3, as suggested by Chapman et al.[68] The tri-axial graphs help to visualize geographic trends and the magnitude in the differences among station responses. A symmetrical triangle indicates similar chemistry, toxicity, and benthos responses; a larger triangle indicates a more severely degraded station. Here, a large and relatively symmetrical triangle would be representative of a location having elevated chemistry, toxicity, and impacted benthos, suggesting metal-induced degradation of the sediment habitats in this area. In contrast, a smaller symmetrical triangle would represent an area having relatively little, if any, chemical, biological, or ecological data suggestive of adverse effects being characteristic of that location.

### 30.6.3 Aquatic Habitat Findings

Like the sediments collected in MRW, surface waters and sediments collected from CFR presented elevated exposures to sediment metals, which correlated with toxicity to benthic invertebrates and impacted benthic community structure. The finding of increased number of genera of Chironomidae in sediments that exhibited toxicity to benthic invertebrates suggested that benthic-community impacts were likely due to the metal contamination. However, overt effects, such as decreased abundance or loss of taxa in major groups of biota, were not apparent in MRW or CFR sediments. Nonetheless, Canfield et al.[6] note that altered benthic-community structure such as that at CFR and at MRW have been observed in other studies of sediments with high metal concentrations[69,70] and can be explained by a disturbance in population competition dynamics.

A number of lines of evidence indicate severe impacts of contaminant metals on the fishery of the upper CFR. For example, concentrations of various metals in surface water of the upper CFR have exceeded water-quality criteria; metal concentrations are elevated in invertebrates from riffle and depositional areas; toxicity tests have demonstrated that fish health is adversely affected following exposure to water and sediment at metal concentrations typical of the upper CFR; and trout behavior in toxicity tests suggests avoidance of elevated metal concentrations. Fish with reduced growth rate and health could be eliminated from natural populations where metal stresses are present, and their potential avoidance of CFR surface waters may prevent adequate access to suitable habitat and food sources. The weight of evidence from fishery studies indicates that the health of trout populations and the fishery ecology of the upper CFR are at major risk from the metal contamination.[14]

## 30.7 SUMMARY

One of the driving concerns for evaluating the CFR-MRW system was the potential for chronic impacts from metals released during episodic events such as spring snowmelt and summer thunderstorms. Since the ecological surveys and biological assessments (both field and laboratory tests) were performed during summer and fall, the results of the studies could be representative of conditions after the spring snowmelt. Since spring flows of the upper CFR prior to the risk-assessment sampling program were high enough to increase metal concentrations in surface waters, the study results may be reflective of conditions during recovery from acute impacts.[14,46]

Within a risk-assessment context, the adverse effects associated with exposures to sediments can be described as moderate, especially for benthic communities in oxidizing environments. Although sediment biota are impacted at contaminated stations under oxidizing conditions, it is unknown if other ecological parameters such as predator-prey relationships are subsequently affected. The impacted benthic invertebrates in the upper CFR and MRW are likely of greater significance to trout populations through the transfer of metals during feeding where mining wastes have contaminated water, sediment, food-chain organisms, and fish. Because of the potential for metals to affect fish through dietary sources, total load of metals to the system should be evaluated, including concentrations of metals in water, sediment, and biota. Biofilm and benthic invertebrates are important in concentrating and cycling metals up the food chain, and the dependence of early-life-stage fry on invertebrate food sources and their sensitivity to metals in the diet makes these receptors a weak link in the dietary food web. While avoidance of copper by juvenile fishes is the most sensitive endpoint for evaluating degraded water quality in CFR, contributions from all of the studies summarized here clearly indicate that the CFR is a degraded biological resource, particularly in areas impacted by historic deposition of metal-contaminated sediments.

## REFERENCES


1. Pascoe, G. A. and DalSoglio, J. A., Planning and implementation of a comprehensive ecological risk assessment at the Milltown Reservoir–Clark Fork River Superfund Site, Montana, *Environ. Toxicol. Chem.,* 13, 1943, 1994.
2. Linder, G., Hazelwood, R., Palawski, D., Bollman, M., Wilborn, D., Malloy, J., DuBois, K., Ott, S., Pascoe, G., and DalSoglio, J., Ecological assessment for the wetlands at Milltown Reservoir, Missoula, Montana: Characterization of emergent and upland habitats, *Environ. Toxicol. Chem.,* 13, 1957, 1994.
3. Brumbaugh, W. G., Ingersoll, C. G., Kemble, N. E., May, T. W., and Zajicek, J. L., Chemical characterization of sediments and pore water from the upper Clark Fork River and Milltown Reservoir, Montana, *Environ. Toxicol. Chem.,* 13, 1971, 1994.
4. Kemble, N. E., Brumbaugh, W. G., Brunson, E. L., Dwyer, F. J., Ingersoll, C. G., Monda, D. P., and Woodward, D. F., Toxicity of metal-contaminated sediments from the Upper Clark Fork River, Montana, to aquatic invertebrates and fish in laboratory exposures, *Environ. Toxicol. Chem.,* 13, 1985, 1994.
5. Ingersoll, C. G., Brumbaugh, W. G., Dwyer, F. J., and Kemble, N. E., Bioaccumulation of metals by *Hyalella azteca* exposed to contaminated sediments form the Upper Clark Fork River, Montana, *Environ. Toxicol. Chem.,* 13, 2013, 1994.
6. Canfield, T. J., Kemble, N. E., Brumbaugh, W. G., Dwyer, F. J, Ingersoll, C. G., and Fairchild, J. F., Use of benthic invertebrate community structure and the Sediment Quality Triad to evaluate metal-contaminated sediment in the upper Clark Fork River, Montana, *Environ. Toxicol. Chem.,* 13, 1999, 1994.
7. Woodward, D. F., Brumbaugh, W. G., DeLonay, A. J., Little, E. E., and Smith, C. E., Effects on rainbow trout fry of a metals-contaminated diet of benthic invertebrates from the Clark Fork River, Montana, *Trans. Am. Fish. Soc.,* 123, 51, 1994.
8. Woodward, D. F., Farag, A. M., Brumbaugh, W. G., Smith, C. E., and Bergman, H. L., Metals-contaminated benthic invertebrates in the Clark Fork River, Montana: Effects on age-0 brown trout and rainbow trout, *Can. J. Fish. Aquat. Sci.,* 52, 1994, 1995.
9. Woodward, D. F., Hansen, J. A., Bergman, H. L, Little, E. E, and DeLonay, A. J., Brown trout avoidance of metals in water characteristic of the Clark Fork River, Montana, *Can. J. Fish. Aquat. Sci.,* 52, 2031, 1995.
10. Farag, A. M., Boese, C. J., Woodward, D. F., and Bergman, H. L., Physiological changes and tissue metal accumulation in rainbow trout exposed to foodborne and waterborne metals, *Environ. Toxicol. Chem.,* 13, 2021, 1994.
11. Farag, A. M., Stansbury, M. A., Hogstrand, C., MacConnell, E., and Bergman, H. L., The physiological impairment of free-ranging brown trout exposed to metals in the Clark Fork River, Montana, *Can. J. Aquat. Sci.,* 52, 2083–2050, 1995.

12. Mount, D., Barth, A. K, Garrison, T. D., Barten, K. A., and Hockett, J. R., Dietary and waterborne exposure of rainbow trout (*Oncorhynchus mykiss*) to copper, cadmium, lead, and zinc using a live diet, *Environ. Toxicol. Chem.,* 13, 2031, 1994.
13. Poulton, B. C., Monda, D. P., Woodward, D. F., Wildhaber, M. W., and Brumbaugh, W. G., Relations between benthic community structure and metals concentrations in aquatic macroinvertebrates: Clark Fork River, Montana, *J. Freshwater Ecol.,* 10, 277, 1995.
14. Pascoe, G. A., Blanchet, R. J., Linder, G., Brumbaugh, W., Kemble, N., Canfield, T., Ingersoll, C., Farag, A., and DalSoglio, J., Characterization of ecological risks at the Milltown Reservoir-Clark Fork River Sediments Superfund Site, MT, *Environ. Toxicol. Chem.,* 13, 2043, 1994.
15. Andrews, E. D., Longitudinal dispersion of trace metals in the Clark Fork River, Montana, in *Chemical Quality of Water and the Hydrologic Cycle*, Averett, R. C. and McKnight, D. M., Eds., Lewis Publishers, Chelsea, MI, 179, 1987.
16. Axtmann, E. V. and Luoma, S. N., Large-scale distribution of metal contamination in the fine-grained sediments of the Clark Fork River, Montana, *Appl. Geochem.,* 6, 75, 1991.
17. Moore, J. N. and Luoma, S. N., Hazardous waste from large-scale metal extraction: The Clark Fork waste complex, Montana, in *Proc. Clark Fork River Symp.*, Watson, V., Ed., University of Montana, Missoula, MT, 1990.
18. Moore, J. N., Luoma, S. N., and Peters, D., Downstream effects of mine effluent on an intermontane riparian system, *Can. J. Fish. Aquat. Sci.,* 48, 222, 1991.
19. Moore, J. N., Source of Metal Contamination in Milltown Reservoir, Montana: An Interpretation Based on Clark Fork River Bank Sediment, Prepared for U.S. Environmental Protection Agency, Helena, MT, Department of Geology, University of Montana, Missoula, 1985.
20. Woessner, W. W. and Popoff, M. A., *Hydrogeologic Survey of Milltown, Montana and Vicinity*, University of Montana, Dept. of Geology, Missoula, 1982.
21. Woessner, W. W., Moore, J. M., Johns, C., Popoff, M., Sartor, L., and Sullivan, M., *Arsenic Source and Water Supply Remedial Action Study, Milltown, Montana*, Department of Geology, University of Montana, Missoula, 1984.
22. Luoma, S. N., Axtmann, E. V., and Cain, D. J., Fate of mine wastes in the Clark Fork River, Montana, USA, in *Metals and Metalloids in the Hydrosphere: Impact through Mining and Industry, and Prevention Technology in Tropical Environments*, Proceedings of an IHP workshop, International Hydrological Programme, United Nations Educational Scientific and Cultural Organization (UNESCO), Phuket, Thailand, 63, 1989.
23. Axtmann, E. V., Cain, D. J., and Luoma, S. N., Distribution of trace metals in fine-grained bed sediments and benthic insects in the Clark Fork River, Montana, in *Proc. of the Clark Fork River Symp.*, Watson, V., Ed., University of Montana, Missoula, 1990.
24. Moore, J. N., Brook, E. J., and Johns, C., Grain size partitioning of metals in contaminated coarse-grained river floodplain sediment: Clark Fork River, Montana, U.S.A., *Environ. Geol. Water Sci.,* 14, 107, 1989.
25. Cowardin, L. M., Carter, V., Golet, F. C., and Blake, E., Classification of Wetlands and Deepwater Habitats of the United States, FWS/OBS-79/31, Biological Services Program, Fish and Wildlife Service, U.S. Department of the Interior, Washington, D.C., 1979.
26. U.S. Fish and Wildlife Service, Milltown Reservoir Sediments Site Endangerment Assessment. Wetland Delineation and Classification, Prepared for U.S. Environmental Protection Agency, Region 8, Helena, MT, U.S. Fish and Wildlife Service, Helena, MT, 1991.
27. U.S. Fish and Wildlife Service, Milltown Reservoir Sediments Site Endangerment Assessment, Wildlife Survey, U.S. Fish and Wildlife Service, Fish and Wildlife Enhancement, Montana State Office, Helena, MT. Prepared for U.S. Environmental Protection Agency, Region 8, Montana Office, Helena, MT, 1992.
28. Lipton, J., Aquatic Resources Injury Assessment Report, Upper Clark Fork River Basin, Prepared for State of Montana Natural Resource Damage Program, Helena, MT, RCG/Hagler, Bailly, Denver, CO, 1993.
29. Besser, J. M., Kubitz, J. A., Ingersoll, C. G., Braselton, W. E., and Giesy, J. P., Influences on copper bioaccumulation and growth of the midge, *Chironomus tentans*, in metal contaminated sediments, *J. Aquat. Ecosyst. Health,* 4, 157, 1995.
30. Besser, J. M., Ingersoll, C. G., and Giesy, J. P., Effects of spatial and temporal variation of acid-volatile sulfide on the bioavailability of copper and zinc in freshwater sediments, *Environ. Toxicol. Chem.,* 15, 286, 1996.

31. Farag, A. M., Woodward, D. F., Goldstein, J. N., Brumbaugh, W. G., and Meyer, J. S., Concentrations of metals associated with mining waste in sediments, biofilm, benthic macroinvertebrates, and fish from the Coeur d'Alene River Basin, Idaho, *Arch. Environ. Contam. Toxicol.,* 34, 119, 1998.
32. Farag, A. M., Woodward, D. F., Brumbaugh, W., Goldstein, J. N., MacConnell, E., Hogstrand, C., and Barrows, F. T., Dietary effects of metals-contaminated invertebrates from the Coeur d'Alene River, Idaho, on cutthroat trout, *Can. J. Fish. Aquat. Sci.,* 128, 578, 1999.
33. U.S. EPA, Quality Criteria for Water with Subsequent Updates, EPA 440/5–86–001, U.S. Environmental Protection Agency, Office of Water Regulations and Standards, Washington, D.C., 1986.
34. Hansen, J. A., Woodward, D. F., Little, E. E., DeLonay, A. J., and Bergman, H. L., Behavioral avoidance: Possible mechanisms for explaining abundance and distribution of trout species in a metal-impacted river, *Environ. Toxicol. Chem.,* 18, 313, 1999.
35. Woodward, D. F., Goldstein, J. N,, Farag, A. M, and Brumbaugh, W. G., Cutthroat trout avoidance of metals and conditions characteristic of a mining waste site: Coeur d'Alene River, Idaho, *Trans. Am. Fish. Soc.,* 126, 699, 1997.
36. Goldstein, J. N., Woodward, D. F., and Farag, A. M., Movements of adult chinook salmon during spawning migration in a metals contaminated system, Coeur d'Alene River, Idaho, *Trans. Am. Fish. Soc.,* 128, 121, 1999.
37. American Society for Testing and Materials (ASTM), Guide for Conducting Laboratory Soil Toxicity or Bioaccumulation Tests with the Lumbricid Earthworm *Eisenia foetida* (E1676–97), in *ASTM Standards on Biological Effects and Environmental Fate,* 2nd ed., ASTM, West Conshohocken, PA, 1999.
38. ASTM, Practice for Conducting Early Seedling Growth Tests (E1598–94), in *ASTM Standards on Biological Effects and Environmental Fate*, 2nd ed., ASTM, West Conshohocken, PA, 1999.
39. Linder, G., Callahan, C., and Pascoe, G., A strategy for ecological risk assessments for Superfund: Biological methods for evaluating soil contamination, in *Superfund Risk Assessment in Soil Contamination Studies, ASTM STP 1158*, Hoddinott, K. B. and Knowles, G. D., Eds., American Society for Testing and Materials, Philadelphia, 1992.
40. Linder, G., Bollman, M., Gillett, C., King, R., Nwosu, J., Ott, S., Wilborn, D., Henderson, G., Pfleeger, T., Darrow, T., and Lightfoot, D., A framework for field and laboratory studies for ecological risk assessment in wetland and terrestrial habitats: Two case studies, in *Environmental Toxicology and Risk Assessment,* 3rd Vol., ASTM STP 1218, Hughes, J. S., Biddinger, G. R., and Mones, E., Eds., American Society for Testing and Materials, Philadelphia, 1995.
41. Pascoe, G. A., Blanchet, R. J., and Linder, G., Bioavailability of metals and arsenic to small mammals at a mining waste-contaminated wetland, *Arch. Environ. Contam. Toxicol.,* 27, 44, 1994.
42. Pascoe, G. A., Blanchet, R. J., and Linder, G., Food-chain analysis of exposures and risks to wildlife at a metals-contaminated wetland, *Arch. Environ. Contam. Toxicol.,* 30, 306, 1996.
43. Linder, G., Ingham, E., Henderson, G., and Brandt, C. J., Evaluation of Terrestrial Indicators for Use in Ecological Assessments at Hazardous Waste Sites, EPA 600/R-92–183, U.S. Environmental Protection Agency, Environmental Research Laboratory, Corvallis, OR, 1993.
44. Klute, A., Ed., *Methods of Soil Analysis, Part 1. Physical and Mineralogical Methods,* 2nd ed., American Society of Agronomy, Inc. and Soil Science Society of America, Inc., Madison, WI, 1986.
45. Page, A. L., Miller, R. H., and Keeney, D. R., Eds., *Methods of Soil Analysis, Part 2, Chemical and Microbiological Properties*, American Society of Agronomy, Inc. and Soil Science Society of America, Inc., Madison, WI, 1982.
46. Linder, G., Pascoe, G., and DalSoglio, J., An ecological risk assessment for the wetlands at Milltown Reservoir, Missoula, Montana: An overview, in *Ecotoxicology and Risk Assessment for Wetlands,* Lewis, M. A., Powell, R. L., Nelson, M. K., Henry, M. G., Klaine, S. J., Dixon, K. B., and Mayer, F. L., Eds., SETAC Press, Pensacola, FL, 1999, Chap. 5.
47. Pascoe, G. A., Wetland risk assessment, *Environ. Toxicol. Chem.,* 12, 2293, 1993.
48. Warren-Hicks, W., Parkhurst, B., and Baker, S., Jr., Eds. Ecological Assessment of Hazardous Waste Sites, EPA/600/3–89/013, U.S. Environmental Protection Agency, Environmental Research Laboratory, Corvallis, OR, 1989.
49. Johns, C. E., Accumulation and partitioning of arsenic in emergent macrophytes in a reservoir contaminated with mining wastes, in *Heavy Metals in the Environment, Volume 1*, Lindberg, S. E. and Hutchinson, T. C., Eds., 6th Int. Conf., New Orleans, 1987.

50. Johns, C. and Moore, J. N., Copper, zinc, and arsenic in bottom sediments of Clark Fork River reservoirs — preliminary findings, in *Proc. of the Clark Fork River Symp.*, Carlson, C. E. and Bahls, L. L., Eds., Montana Academy of Sciences, Montana College of Mineral Science and Technology, Butte, MT, 1985.
51. Byl, T. D. and Klaine, S. J., Peroxidase activity as an indicator of sublethal stress in the aquatic plant Hydrilla verticillata (Royle), in *Plants for Toxicity Assessment,* 2nd Vol., ASTM STP 1115, Gorsuch, J. W., Lower, W. R., Wang, W., and Lewis, M. A., Eds., American Society for Testing and Materials, Philadelphia, 101, 1992.
52. Fleming, W. J., Ailstock, M. S., Momot, J. J., and Norman, C. M., Response of sago pondweed, a submerged aquatic macrophyte, to herbicides in three laboratory culture systems, in *Plants for Toxicity Assessment,* 2nd Vol., ASTM STP 1115, Gorsuch, J. W., Lower, W. R., Wang, W., and Lewis, M. A., Eds., American Society for Testing and Materials, Philadelphia, 267, 1992.
53. Klaine, S., Milltown Reservoir Sediment Site Wetlands Ecological Risk Assessment: Sediment Phytotoxicity: Final, Prepared for U.S. Environmental Protection Agency, Region 8, Helena, MT, 1992.
54. Federal Interagency Committee for Wetland Delineation, Federal Manual for Identifying and Delineating Jurisdictional Wetlands, U.S. Army Corps of Engineers, U.S. Environmental Protection Agency, U.S. Fish and Wildlife Service, and U.S.D.A. Soil Conservation Service, Washington, D.C. Cooperative technical publication, 1989.
55. Wilborn, D. W., Bollman, M. A., Gillett, C. S., Ott, S. L., and Linder, G. L., A field screening method using earthworms (*Eisenia foetida andrei*) to evaluate contaminated soils, in *Environmental Toxicology and Risk Assessment: Modeling and Risk Assessment,* 6th Vol., ASTM STP 1317, Dwyer, F. J., Doane, T. R., and Hinman, M. L., Eds., ASTM, Philadelphia, 1994.
56. Davis, D. E., Ed., *Handbook of Census Methods for Terrestrial Vertebrates*, CRC Press, Boca Raton, FL, 1982.
57. Gysel, L. W. and Lyon, L. J., Habitat analysis and evaluation, in *Wildlife Management Techniques Manual,* 4th ed., Schemnitz, S., Ed., The Wildlife Society, Washington, D.C., 1980.
58. Ohmart, R. D. and Anderson, B. W., Riparian habitats, in Inventory and Monitoring of Wildlife Habitat, Cooperrider, A. Y., Boyd, R. J., and Stuart, H. R., Eds., U.S. Department of the Interior, Bureau of Land Management Service Center, Denver, CO, 1986.
59. Skaar, D., Flath, D., and Thompson, L. S., *P.D. Skaar's Montana Bird Distribution — Mapping by Latilong*, Montana Acad. Sci. Monogr. No. 3, Missoula, 1985.
60. Linder, G., Wyant, J., Meganck, R., and Williams, B., Evaluating amphibian responses in wetlands impacted by mining activities in the western United States, in *5th Biennial Symposium on Issues and Technology in the Management of Impacted Wildlife*, Thorne Ecological Institute, Boulder, CO, 17, 1991.
61. Barnthouse, L. W., Suter, G. W., and Rosen, A. E., Inferring population-level significance from individual-level effects: An extrapolation from fisheries science to ecotoxicology, in *Aquatic Toxicology and Environmental Fate,* 11th Vol., ASTM STP 1007, Suter, G. W. and Lewis, M. A., Eds., American Society for Testing and Materials, Philadelphia, 1989.
62. Duinker, P. N. and Beanlands, G. E., The significance of environmental impacts: An exploration of the concept, *Environ. Manage.,* 10, 1, 1986.
63. Pascoe, G. A., Role of ecological risk assessment in reducing uncertainties in remedial decisions at mining waste sites, in *Proc. 1st Int. Conf. on Tailings and Mine Waste*, Fort Collins, CO, 147, 1994.
64. Hartenstein, R., Neuhauser, E. F., and Narahara, A., Effects of heavy metal and other elemental additives to activated sludge on growth of *Eisenia foetida*, *J. Environ. Qual.,* 10, 372, 1981.
65. Ingersoll, C. G., Dillon, T., and Biddinger, G. R., *Ecological Risk Assessment of Contaminated Sediments*, SETAC Press, Pensacola, FL, 1997.
66. Long, E. R. and Morgan, L. G., The Potential for Biological Effects of Sediment-Sorbed Contaminants Tested in the National Status and Trends Program, NOAA Technical Memorandum NOS ORCA 52, Seattle, WA, 1991.
67. O'Conner, T. P., Daskalakis, K. D., Hyland, J. L., Paul, J. F., and Summers, J. K., Comparisons of sediment toxicity with predictions based on chemical guidelines, *Environ. Toxicol. Chem.,* 17, 468, 1998.

68. Chapman, P. M., Power, E. A., and Burton, G. A., Jr., Integrative assessments in aquatic ecosystems, in *Sediment Toxicity Assessment*, Burton, G. A., Jr., Ed., Lewis Publishers, Chelsea, MI, Chap. 14, 1992.
69. LaPoint, T. W., Melancon, S. M., and Morris, M. K., Relationships among observed metal concentrations, criteria, and benthic community structural responses in 15 streams, *J. Water Pollut. Control Assoc.,* 56, 1030, 1984.
70. Clements, W. H. and Kiffney, P. M., Integrated laboratory and field approach for assessing impacts of heavy metals at the Arkansas River, Colorado, *Environ. Toxicol. Chem.,* 13, 397, 1994.

SECTION IV

# Methods for Making Estimates, Predictability, and Risk Assessment in Ecotoxicology

CHAPTER 31

# Global Disposition of Contaminants

**Roy M. Harrison, Stuart Harrad, and Jamie Lead**

## CONTENTS



## 31.1 INTRODUCTION

In most instances, pollutant sources are relatively easy to identify. Point sources especially present few problems of quantification, while diffuse sources (e.g., runoff from agricultural land) are more difficult to determine with certainty. The source is, however, only the first part of the picture, and the period that exists between emission/discharge of a pollutant and contact with the receptor may contain many varied and interesting processes. It is the aim of this chapter to describe some of the more important processes involved in pollutant transport and removal from the environment and to demonstrate how such processes influence the distribution of pollutants within the environment. Of particular interest are processes leading to the transfer of chemical substances between environmental compartments, i.e., water to air, air to soil, etc.

Environmental cycles of chemical elements and compounds are generally termed "biogeochemical cycles." Ideally, such cycles include quantitative estimates (however uncertain) of the fluxes between compartments and the total inventory of substance within a given compartment. Such quantification is difficult even for chemical elements, especially in relation to the flux component. For chemical compounds subject in some cases to rather rapid chemical change, estimation of fluxes is even more problematic.

In this chapter, the transport mechanisms responsible for pollutant transfer within and between environmental compartments are first considered. Mathematical treatments allowing calculation of transfer fluxes and lifetimes are then described. Some examples of pollutant distributions are given, indicating where possible the processes responsible. Finally, some examples are provided of the use of present-day environmental measurements to infer historical concentrations of pollutants.

## 31.2 ENVIRONMENTAL TRANSPORT MECHANISMS

### 31.2.1 Atmospheric Transport

Atmospheric motions occur on a number of spatial scales, most notably:

1. Global
2. Synoptic, or large-scale (thousands of kilometers)
3. Mesoscale, or intermediate (tens and hundreds of kilometers)
4. Microscale (ten kilometers and less)

In addition to horizontal transfer processes, movements in the vertical are important, especially for substances with long atmospheric lifetimes. The atmosphere, viewed in the vertical (Figure 31.1), divides readily into discrete regions. The lowermost part, known as the troposphere, is characterized by decreasing temperature with increasing altitude. This region is the most accessible to us and consequently is the part most thoroughly observed in scientific terms. Above the troposphere lies the stratosphere, a region within which temperature increases with altitude. This is the region in which ozone mixing ratios peak, as discussed later. The atmospheric regions above the stratosphere are of little concern in relation to pollution phenomena.

The troposphere is typically also thermally stratified (see Figure 31.1). The main regions are the boundary layer, typically about 1 km in depth during the daytime but often reducing to only around 100 m at night, and the free troposphere, which lies between the boundary layer and the tropopause. These regions are separated by a temperature inversion, which severely limits exchange between them.

The extent to which a pollutant is subject to either vertical or horizontal movement in the atmosphere is a function of its atmospheric lifetime, defined in section 31.4. For *substantial* movement of a substance between compartments, the approximate minimum lifetimes $\tau$ are indicated below:

| | |
|---|---|
| Boundary layer to free troposphere | $\tau > 5$ days |
| Entire tropospheric hemisphere | $\tau > 1$ month |
| Global troposphere | $\tau > 2$ years |
| Troposphere to stratosphere | $\tau > 10$ years |

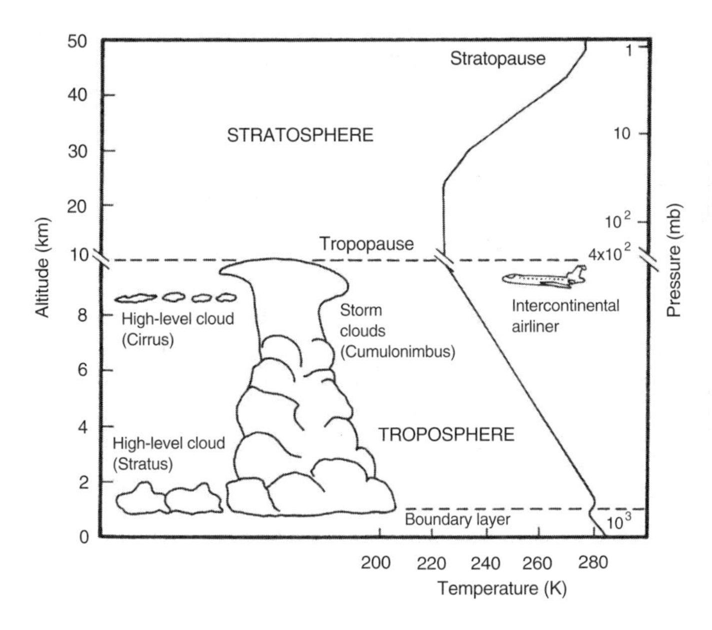


**Figure 31.1** Vertical structure of the atmosphere indicating approximate temperatures and the main regions.

To cite some examples, nitrogen dioxide has a chemical lifetime of about a day, and little of that emitted in the tropospheric boundary layer transfers to the free troposphere. Aerosol emitted from Chernobyl, which had a lifetime of around 1 month, led to contamination of much of the northern hemisphere. Methane, with an atmospheric lifetime of about 9 years, is rather well mixed between northern and southern hemispheres and penetrates in modest amounts into the lower stratosphere. Chlorofluorocarbons and nitrous oxide, with lifetimes in excess of 100 years, have no significant tropospheric sinks and mix appreciably into the stratosphere.

Horizontal atmospheric motions are driven by gradients in temperature and pressure that lead to general circulation, as described in Figure 31.2. Wind speeds in the boundary layer vary greatly

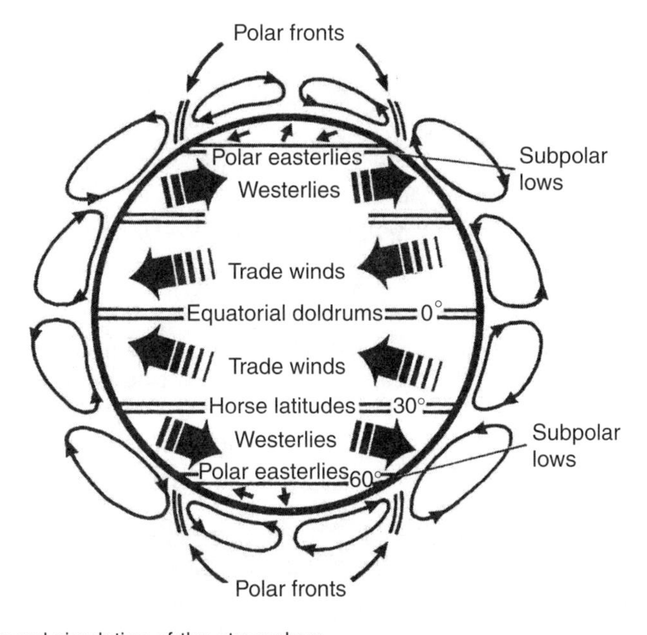


**Figure 31.2** The general circulation of the atmosphere.

from place to place but are typically of the order of 5 m/sec, implying transport distances of around 400 km/day. It is thus possible for a pollutant to travel around the globe at a given latitude in a matter of days, but more substantial north–south mixing takes months. Atmospheric circulation at low latitude is dominated by the Hadley circulation, which involves updrafts of air in equatorial regions, with subsequent movement both north and south at high tropospheric altitudes and subsidence to form the subtropical high–pressure regions. The existence of two such circulatory cells on either side of the equator ensures inefficient mixing between northern and southern hemispheres, and only substances with lifetimes measured in years mix appreciably between the hemispheres. This is particularly important in limiting movement of contaminants from the heavily populated northern hemisphere into the far cleaner southern hemisphere.

Mesoscale circulations involve such processes as land–sea breeze circulations that can profoundly influence pollutant movements in coastal regions. Mountain upslope and downslope winds driven by thermal convective motions may also act as an important transfer route. Processes occurring on the micro- and mesoscale are important in influencing concentrations of locally generated pollutants but play only a minor role in transferring pollutants on a larger scale.

### 31.2.2 Freshwaters

Pollutants entering freshwater may be in the dissolved (< 1nm), colloidal (1 nm – 1 μm), or particulate (> 1 μm) form. The exact nature of the pollutant species has important implications for both bioavailability and environmental transport. For instance, biological impact of trace metals is often related directly to the dissolved (free) metal concentration.[1] In addition, dissolved and colloidal forms of the pollutant will tend to remain in the water column due to mixing processes, while particulate forms will tend to settle out of the water column and be incorporated in the sediments.[2] Once in a river or lake, they may remain in their original form or repartition between these different forms in response to a changed matrix. Important parameters affecting the distribution of pollutants include microbiological activity, concentration, and nature of organic matter, pH, ionic strength, redox potential, and so on. These changes can be illustrated by the following examples:

- *Rain water (pH ~ 5) falling on lake water (pH ~ 7–8).* As the pH increases, metals in the dissolved form will tend to bind to solid phases. Over this pH range many metals change from 0 to 100% bound to solid phases.
- *Freshwater mixing with seawater in estuaries.* Increased ionic strength will cause colloids, and subsequently sediment, to aggregate out of the water column. The fate of associated pollutants will also be affected.
- *Oxic–anoxic boundary in lakes.* The oxidized form of iron (Fe [III]) exists as the solid phase in oxygenated surface waters. The particles sediment from the water column (with any associated pollutants) and come into contact with the deoxygenated bottom waters. At this point the iron is reduced, forming Fe (II), which exists in the dissolved phase. Particulate-phase pollutants may therefore pass into the dissolved phase.

In reality, the processes occurring are much more complex than indicated, with many competing reactions occurring on different spatial and temporal scales. For instance, in estuaries, both pH and chloride concentration as well as ionic strength increase dramatically, with many consequences for pollutant behavior. In the case of lake waters, as the iron is reductively solubilized across the oxic–anoxic interface, released trace pollutants may bind to other solid phases and not be released into the dissolved phase.

Processes such as sorption, precipitation, microbiological activity, and others may affect both organic and inorganic pollutants. Therefore, the assumption that a substance entering a river in an effluent will maintain its original physicochemical form will often prove to be incorrect.

Metals are important contaminants that may undergo a wide range of physical changes such as those outlined above and, in some instances, may also undergo changes in oxidation state leading

| Metals Species | Free Metal Ions | Inorganic Ion Pairs<br>Inorganic Complexes<br>Low Molecular-Weight Organic Complexes | High Molecular-Weight Organic Complexes | Metal Species Adsorbed onto Inorganic Colloids<br>Metals Associated with Detritus | Metals Absorbed into Live Cells<br>Metals Adsorbed onto or Incorporated into Mineral Solids and Precipitates |
|---|---|---|---|---|---|
| Examples | $Mn^{2+}$<br>$Cd^{2+}$ | $NiCl^{+}$<br>$HgCl_4^{2-}$<br>Zn-fulvates | Pb-humates | Co-$MnO_2$<br>Pb-$Fe(OH)_3$ | Cu-clays<br>$PbCO_{3(s)}$ |

**Soluble** **Colloidal** **Particulate**

**Size ← 1 nm —— 10 nm —— 100 nm —— 1000 nm →**

**Figure 31.3** Typical physicochemical forms of trace metals in aquatic systems, as related to size association. (From de Mora, S. J. and Harrison, R. M., *Water Res.*, 17, 723, 1983. With permission.)

to a complete change in physical form. The relevance of such changes to environmental transport is that they affect the size association and hence the mode of movement in the freshwater system. Figure 31.3 exemplifies the possible forms of trace metals and their size association.[3]

Dissolved contaminants will move freely with flowing water and, although subject to diffusive movements caused by turbulence, are predominantly influenced by advective transport (i.e., they are mainly carried along by the water, rather than diffused within it). The behavior of particulate material is more complex, as it may deposit, entering the bottom sediment. This process is controlled by two processes. First, the tendency to deposit is a function of the size and settling velocity of the particles with which the contaminant is associated. However, this tendency is counteracted by turbulence forces that keep particles in suspension and make them available for movement with the water. Thus, in relatively static waters (lakes and ponds), there is a strong tendency for particles to be incorporated into bottom sediments. In fast-moving rivers and estuaries, the turbulent energy of the water keeps the particles in motion and may even lead to resuspension of bottom sediment. In such circumstances, contaminant transport may occur by three main mechanisms:

- As truly suspended particles
- By saltation, or "bouncing" of particles along the sediment surface
- As bed load, or bulk motion, of surface sediment

Changing flow conditions within a river may lead to rapidly altering transport properties. An example is given in Figure 31.4, which shows measurements of total (suspended plus dissolved) concentrations of chromium in the River Thames (U.K.) as a function of discharge (flow). As flow increases from very low values, concentrations of contaminant fall due to dilution of dissolved material in a greater volume of water. They reach a minimum and then begin to increase for flows of greater than about 70 $m^3 s^{-1}$. This flow is known (from other measurements) to correspond to that at which riverbed sediments begin to become resuspended. Concentrations of chromium continue to increase with increasing flow as more bed sediment enters the water.

In slow-moving waters, where much of the suspended sediment enters the bottom sediment, a historical record of inputs to the lake may be preserved in the bottom sediment. For instance, the concentration of lead, zinc, and copper in recently deposited sediments in Lake Erie have increased by several-fold in comparison to sediments deposited ca. 1900.[4] Such records often reflect changes in atmospheric deposition to the lake surface, and for substances such as heavy metals[5] and polynuclear aromatic hydrocarbons, they have been used to reconstruct a record of air quality in the past.

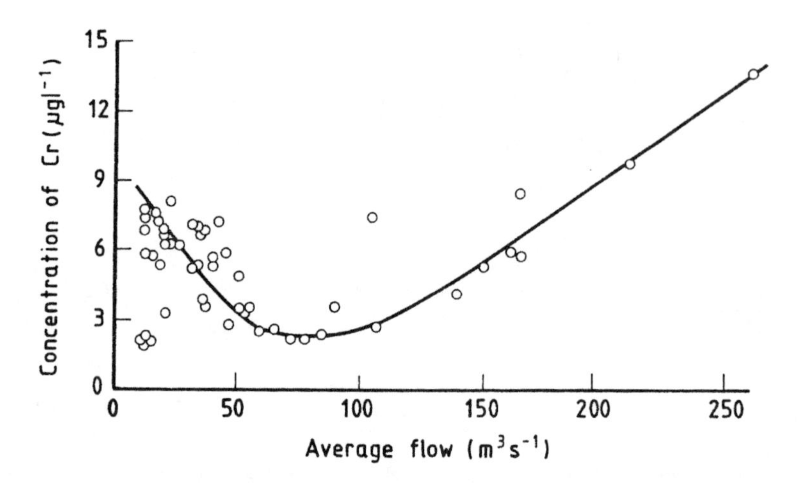


**Figure 31.4** Concentration of chromium (suspended plus dissolved) in the waters of the River Thames as a function of flow rate.

The behavior shown in Figure 31.4 for chromium in the River Thames is not common to all contaminants. In the same study, measurements were made of nitrate and soluble reactive phosphate (SRP). For nitrate, concentrations were almost independent of flow, suggesting the dominance of inputs in waters making up the major part of the river flow (surface runoff and groundwater). In the case of SRP, concentrations of this dissolved species showed a monotonic decrease with flow rate reflective of dilution of effluent inputs (from sewage treatment works) and no input from resuspended sedimentary materials.

### 31.2.3 Marine Transport

While rivers generally take only hours or days to flow to the sea, water has a very long residence time in the ocean. Consequently, transport and transformation processes control the distribution of contaminants. The oceanic surface layers to a depth of around 100 m are driven largely by the wind, the water motion being modified by the Coriolis force that arises from the rotation of the earth. Oceanic circulations in large part consist of gyres constrained by continental boundaries. Faster currents occur along western margins, leading to pronounced circulatory features such as the Gulf Stream, Brazil Current, and Kuroshio Current.

At greater depths within the ocean, circulations are determined by the chemistry and temperature of the water. Dense waters (either cold or more saline) form in cold polar regions, where the sea ice formation leads to an increase in salinity. Antarctic bottom water flows north from the Weddell Sea into the south Atlantic, while North Atlantic deep water moves on a timescale of around 1000 years from the Norwegian Sea through the Indian Ocean into the Pacific.[6]

Surface ocean waters are relatively well mixed and show little variation of temperature with depth. In the layer beneath, termed the "thermocline," temperatures decline rapidly with depth, with the most dramatic temperature changes occurring down to about 1 km in equatorial and temperate latitudes. The deep layers beneath the thermocline show little change in temperature with depth. The thermocline presents a rather sharp boundary and a considerable barrier against mixing of the surface and deep layers.

Owing to the relatively slow time scales of oceanic water movement, even in the surface layers, highest concentrations of contaminants occur in the coastal regions, where inputs are greatest — from rivers, direct coastal discharges, and atmospheric deposition. Much of this contaminant load may deposit to the sediments before mixing processes carry it far from coastal waters.

### 31.2.4 Soils

Soils are by their nature rather immobile. If undisturbed, they may retain a record of contaminant inputs over a very long period. Many pollutants bind strongly to soils and, if there is input from the atmosphere, show a very strong surface enrichment. Mechanical mixing by plowing or other agricultural practices, or bioturbation by burrowing organisms, can cause some vertical and lateral mixing of pollutants in soils.

Following deposition from the atmosphere, some organic pollutants, particularly polychlorinated biphenyls (PCBs), readily undergo volatilization from soil to the extent that for many such compounds, volatilization represents the principal loss mechanism from soil.[7] This has important implications for understanding the origins of the continuing atmospheric presence of PCBs (the manufacture of which was ceased in most western countries in the late 1970s). Specifically, although accidental releases from PCBs remaining in use continue, such volatilization of previously deposited material is widely considered to represent the main contemporary source of PCBs in the atmosphere.

## 31.3 TRANSFER MECHANISMS AND FLUXES BETWEEN ENVIRONMENTAL COMPARTMENTS

Every time it rains, small amounts of highly persistent compounds, such as chlorinated pesticides, are deposited to land and sea. Dry deposition of particles and gases also contributes to inputs. This occurs even for some compounds no longer in large-scale production or use. Residues in soils and waters in parts of the world where heavy usage has occurred are still evaporating into the atmosphere and are carried to locations remote from their sources, where deposition takes place. Exchange between environmental compartments is a major route of transfer for some substances and can lead to unexpected instances of pollution. One such phenomenon occurred in West Cumbria, U.K., where abnormally high levels of plutonium were found in coastal soils. Detailed investigation showed that plutonium discharged from the Sellafield reprocessing plant to the Irish Sea was incorporated into sea spray with an efficiency greater than that expected from its abundance in seawater and carried back to land in marine aerosol.[5,8]

Some of the major processes involved in intercompartmental transfer will now be considered.

### 31.3.1 Atmosphere–Land Surface Exchange

Transfer from air to land can occur by two major routes:

1. Rainfall (termed "wet deposition")
2. As dry particles or gas, without the intervention of rain (termed "dry deposition")

A third pathway, involving "occult" deposition of fog and/or cloudwater droplets, may also be of localized importance.

Concepts of dry deposition were originally developed to describe the transfer of radioactive gases and particles from the atmosphere into terrestrial systems. The process of deposition is first order; that is, the flux to the surface depends linearly upon the atmospheric concentration. The constant of proportionality is termed the "deposition velocity," $v_g$, defined as:

$$v_g = \frac{\text{flux to surface}}{\text{atmospheric concentration at 1 m}}$$

For gases, deposition velocities to soils and vegetation vary greatly according to the affinity of the gas for the individual surface. Thus, for chlorofluorocarbons, deposition velocities are essentially zero, while for highly reactive and water-soluble nitric acid vapor, $v_g$ has a typical value of 2 to 3 cm $s^{-1}$. The deposition velocity is not constant for a given gas or a given surface. While one can make approximate statements about the magnitude of $v_g$ (such as that above for nitric acid), in reality $v_g$ varies with the surface characteristics and atmospheric properties at the time of deposition. This may be expressed as follows for the total resistance to deposition, R.

$$R = \frac{1}{v_g} = r_a + r_b + r_c \tag{31.1}$$

where $r_a$ is the aerodynamic resistance, or the resistance to transfer downward through the atmosphere to within 1 to 2 mm of the surface; $r_b$ is the boundary layer resistance, which is the resistance to transfer through a laminar layer of air of about 1 mm thick over the individual roughness elements of a surface; and $r_c$ is the canopy, or surface resistance, which describes the resistance of the surface itself to take-up of the gas. For a sticky, reactive molecule like nitric acid, $r_c$ is essentially zero, while for chlorofluorocarbons, it is almost infinite and accounts for the differing depositional behavior of these gases. For ozone, $r_c$ over vegetation typically varies with time of day according to the opening of stomatal apertures of the vegetation, necessary for rapid ozone deposition.

Deposition velocities are not as strong a function of chemical properties in airborne particles as in gases. The main determinant of deposition velocity is the particle size (see Figure 31.5), with high values of $v_g$ applicable to large particles due to their inertial properties and very small particles due to their high diffusivities, which cause them to behave more like gases.[9] Between the two lies a region of low dry depositional efficiency. For particles in this size range, around 0.1 to 1 μm in diameter, wet scavenging is also inefficient, and atmospheric lifetimes are long — of the order of 7 to 40 days. Chemical composition may affect dry deposition, as hygroscopic particles are liable to grow adjacent to a humid surface, leading to enhanced deposition.

Some gases are released by soils or vegetation. If their concentration immediately adjacent to the surface exceeds that in the atmosphere above, the concentration gradient will cause them to diffuse upward, and the net flux will be from surface to atmosphere. In this instance, the concept of deposition velocity is of little value. A gas that normally exhibits upward fluxes from soil is

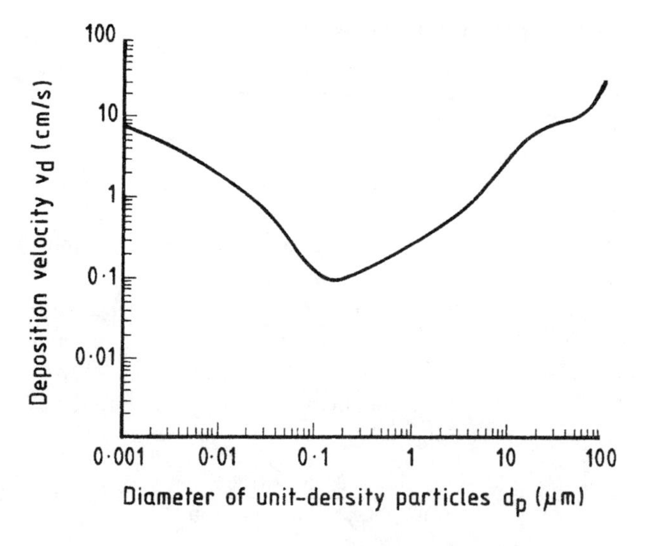


**Figure 31.5** Typical variation of aerosol particle deposition velocity with particle size over land.

nitrous oxide, $N_2O$. Ammonia, $NH_3$, shows ambivalent behavior; net fluxes from fertilized soils are commonly upward into the atmosphere, while over unfertilized soils, within which concentrations of ammonium (and by inference ammonia) are very low, deposition normally occurs.

### 31.3.2 Air–Plant Exchange

Air-to-plant transfer of persistent organic pollutants (POPs), such as dioxins, PCBs, and polycyclic aromatic hydrocarbons (PAH), occurs readily, primarily as a result of vapor-phase deposition onto the leaf surface, either as an equilibrium partitioning process or as a kinetically-controlled process, although particle-bound deposition can be influential for some compounds.[10] Given the facile nature of air–plant transfer of POPs, it is important to evaluate the quantitative role that vegetation plays within the overall biogeochemical cycling of such pollutants. Several authors have speculated that because of the vast surface area covered by vegetation, this role could be significant. Indeed, there is ample evidence that vegetation, particularly forests, remove a substantial quantity of airborne POPs, with consequent reductions in their atmospheric residence times. A steady-state mass-balance model determined that vegetation removed 4% of PAHs from the atmosphere in the northeastern region of the United States. It is evident that vegetation plays a significant role in the biogeochemical cycling of PAHs and other POPs possessing similar propensity for atmosphere-to-foliage transfer. The implications of this phenomenon in terms of human and ecotoxicological impacts are at present unclear, and this research topic is likely to receive much attention in the future.

### 31.3.3 Air–Sea Exchange

For many large bodies of water, the atmosphere is a major route of contaminant input. For example, for the North Sea, which is heavily influenced by adjacent industrialized countries, atmospheric deposition is the major source of some trace metal inputs.

Particles deposit to water surfaces in much the same manner as to land. There are clearly differences arising from the resistance terms in Equation 31.1 that are not identical for the two surfaces or the air above them. In general, similar patterns of behavior are observed, although there is stronger wind-speed dependence in the case of the sea, since surface characteristics change appreciably as high winds break up the sea surface into waves and spray.

Aerosol produced from the sea surface itself can be enriched in trace elements that are transported over land. The extent of enrichment is controversial, with some research indicating a major impact on aerosol and rainwater composition and other papers suggesting little effect in the case of trace metals. One clear example of enrichment is plutonium, cited above. Plutonium is enriched in aerosol and deposition some 30- to 500-fold relative to its abundance in bulk seawater. Fortunately, marine aerosol is of relatively large particle size and deposits within a rather narrow coastal band of land; the majority of deposition occurs within 5 km of the coast.

Incorporation of contaminants in rainwater leads to wet deposition to both land and sea surfaces. Scavenging of contaminants may occur either within the cloud (termed "rainout") or by falling raindrops (termed "washout"). The overall efficiency of the process is often described by the scavenging ratio, W, also known historically as the Washout Factor:

$$W = \frac{\text{concentration in rain (mg kg}^{-1}\text{)}}{\text{concentration in air (mg kg}^{-1}\text{)}}$$

Values of W for particulate species are typically around 200 to 1000, implying crudely that substances present in air at microgram-per-cubic-meter concentrations will occur in rain at milligram-per-liter-levels, as a cubic meter of air weighs 1.2 kg at 25°C and 1 atm pressure. An alternative descriptor of wet removal used by numerical modelers is the washout coefficient, $\lambda$. This is a first-order rate constant in the equation

$$\frac{dc}{dt} = -\lambda t$$

where dc/dt is the rate of change of airborne concentration with time, t. For aerosol species, $\lambda$ typically takes values of around $10^{-4}$ $s^{-1}$.

The sea may act as a source of atmospheric trace gases. A notable example is that of dimethyl sulfide released by marine phytoplankton. It may also act as a sink for reactive gases such as nitric acid and other soluble species.

The direction of flux between air and sea[11] is determined by Henry's Law, which describes the equilibrium between dissolved and gaseous forms of a substance. At equilibrium:

$$C_w(eq) = C_a H^{-1}$$

where $C_w$ and $C_a$ are the concentrations in water and air, respectively, $C_w(eq)$ is the concentration in water at equilibrium with the atmosphere, and H is Henry's Law constant.

If, however, $C_w \neq C_w(eq)$, then a net flux of material to air, or from water to air, will occur. If $C_a H^{-1} > C_w$, net transfer into the water occurs; and if $C_a H^{-1} < C_w$, then the net flux is to the atmosphere.

Suppose that:

$$\Delta C = C_a H^{-1} - C_w$$

The rate of gas transfer is:

$$F = K_{(T)w} \Delta C$$

where $K_{(T)w}$ is termed the "transfer velocity." By analogy with Equation 31.1, one can write:

$$\frac{1}{K_{(T)w}} = \frac{1}{\alpha k_w} + \frac{1}{H k_a} = r_w + r_a \tag{31.2}$$

where $k_a$ and $k_w$ are individual transfer velocities for chemically nonreactive gases in air and water phases, respectively, and $\alpha$ (= $k_{reactive}/k_{inert}$) is a factor that quantifies any enhancement of gas transfer in the water due to chemical reaction. The resistance $r_a$ and $r_w$ are directly analogous to those in Equation 31.1; and for chemically reactive gases, usually $r_a >> r_w$, and atmospheric transfer limits the overall flux. For less reactive gases, $r_w$ is more difficult to predict. It is highly wind-speed dependent, as exemplified by Figure 31.6, which shows the overall transfer velocity for carbon dioxide (for which $k_w$ is dominant) as a function of wind speed at 10 m.[13]

### 31.3.4 Sediment–Water Exchange

The distribution of trace elements in aquatic systems is affected by a number of processes, which are exemplified in Figure 31.7. One important process in the water column is the distribution between the solid-solution interface (see next section). Incorporation of contaminants into suspended solids that subsequently aggregate and deposit to form bottom sediments is a major removal mechanism of the pollutants. Sedimentation rates, as expressed by the rate of increase at the bottom sediment surface, vary greatly from millimeters per 100 years in the deep ocean to centimeters per

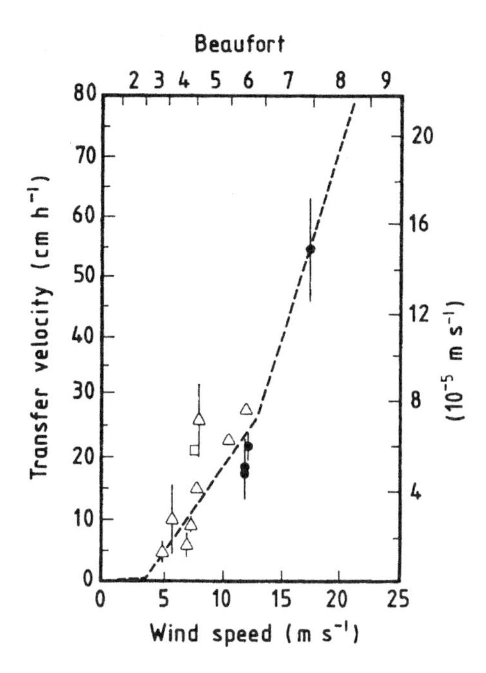


**Figure 31.6** Air–sea transfer velocities for carbon dioxide at 20°C as a function of wind speed at 10 m ($ms^{-1}$ or Beaufort Scale). The graph combines experimental data (points) with a theoretical line. (From Watson, A.J., Upstill-Goddard, R.C., and Liss, P.S., *Nature [London]*, 349, 145, 1991. With permission.)

year in some lakes and coastal environments. Other important controls on the transport of contaminants are the resuspension of particles, the entrapment of water in pore spaces, the upward flow of pore water due to hydrostatic pressure gradients, and diffusional fluxes of dissolved species.[2] Recent work with microelectrodes and other microscale techniques[14] has shown that sediments are extremely heterogeneous over ranges of hundreds of microns and that dissolved concentrations of major and minor species will fluctuate not only vertically and temporally but also laterally, i.e., in three dimensions. These heterogeneities often relate to the effects of biogenic processes.[15]

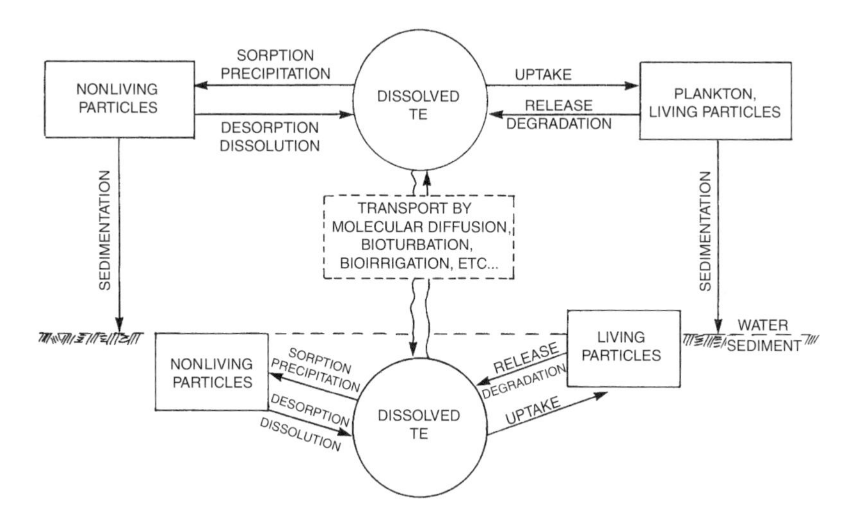


**Figure 31.7** Processes affecting the distribution of trace elements in aquatic systems. (From Tessier, A. et al., in *Chemical and Biological Regulation of Aquatic Systems,* CRC Press, Boca Raton, FL, 1994.)

The net flux of a substance across the water-sediment interface, $F_{z=0}$, is expressed by

$$F_{z=0} = \phi\left(-D\frac{dC}{dz} + UC + U_sC_s\right) \tag{31.3}$$

where

$\phi$ = porosity (volume fraction occupied by water)
D = diffusion coefficient
C = concentration in solution
$C_s$ = concentration in solids
U = rate pore water advection
$U_s$ = sedimentation rate
z = vertical distance

Thus, of the three terms in the right-hand side, the first term represents diffusive transfer, the second term advection in pore water, and the third term transfer in solids.

In a uniform, noncompacting sediment, $U = U_s$, and from above, $C_s = K_dC$. Substitution yields:

$$F_{z=0} = \phi\left[-D\frac{dC}{dz} + U(K_d + 1)C\right]_{z=0} \tag{31.4}$$

The relative importance of the two terms, representing diffusion and sedimentation, depends upon the sedimentation rate and the magnitude of $K_d$. If either of these is large, it is likely that sedimentation processes will dominate the flux. If both are small, diffusive transfer may be dominant, as often occurs in the deep ocean.

A fuller consideration of solute transport would also consider biological processes (bioturbation and bioirrigation by macroinvertebrates) and physical processes (waves, currents, and tides).[12] Although bioturbation appears to be negligible, the other processes can significantly enhance movement. The role of the biofilms may also be important in pollutant dispersion.

Interesting processes can arise in redox–active sediment components. Examples are with iron and manganese. Both elements are present in oxidizing waters in highly insoluble oxidized forms such as $Fe_2O_3$ and $MnO_2$, which become incorporated in surface sediments. Progressively, the sediment becomes buried with freshly sedimenting solids and becomes anoxic as microorganisms consume oxygen more rapidly than it can be supplied from the overlying water. In the anoxic region, chemical reduction takes place, and Fe(III) and Mn(IV) are converted to, respectively, Fe(II) and Mn(II). These are both liable to be more soluble than the more oxidized forms (especially Mn), and Mn(II) diffuses upward through the interstitial water, from where it may either enter the overlying water or reprecipitate in the oxic layer.[16] Figure 31.8 shows typical schematic profiles of solid-associated and dissolved manganese. The subsurface maximum in dissolved manganese occurs where the outward flux and reprecipitation balance the upward supply from greater depth. Solid manganese below this depth is mostly present as $MnCO_3$. Analysis of a core of sediment for manganese is liable to reveal a near-surface maximum for the above reasons. This should not be confused with near-surface enrichment of other sediment components not subject to the same redox chemistry, which in more recent years has often been seen as a result of enhanced rates of contaminant deposition (see Section 31.6). Post-depositional changes in sediments are common and varied and are grouped under the term *diagenesis*.

Diagenetic processes in sediments can lead to the release of contaminants to the overlying water. Thus, for example, microbial decay of organic matter and reduction of iron and manganese in sedimentary solids can lead to rerelease of adsorbed or complexed material that, if not reabsorbed,

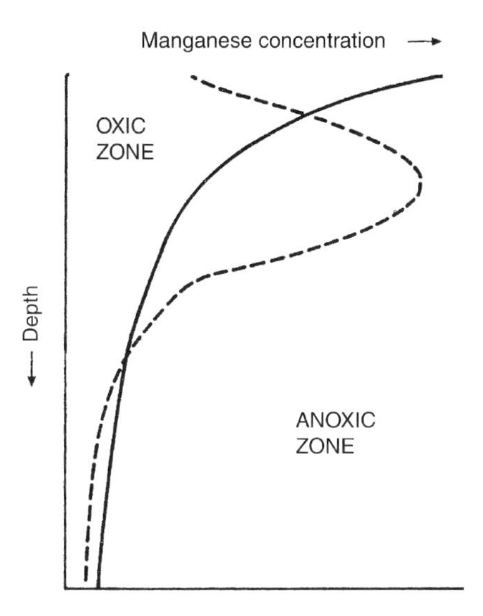


**Figure 31.8** Schematic profiles of manganese in bottom sediment dissolved in the interstitial waters (broken lines) and associated with the solid sedimentary components (solid line).

may be released into the overlying water. This is most likely to occur to an appreciable degree if the bottom water of a lake becomes anoxic for an appreciable period, causing even the surface sediment to be anaerobic.

### 31.3.5 Solid–Solution Exchange

Although these exchanges occur within — rather than between — environmental compartments, e.g., in surface waters, groundwaters, or sediment pore waters, the processes occurring at the solid–solution interface within the water body are essential to understanding the transport of pollutants. Controlling processes, such as biological activity and sedimentation, are highly dependent on reactions occurring at the solid–solution interface. The chemical weathering of minerals that, in large part, determines the composition of natural waters and plays a part in controlling atmospheric carbon dioxide levels through the action of carbonic acid (see below, dissolution of silicates and calcium carbonate) is also dependent on these solid-solution reactions.

$$CaSiO_3 + 2CO_2 + 3H_2O = Ca^{2+} + 2HCO_3^- + H_4SiO_4$$
$$CaCO_3 + 2H^+ = Ca^{2+} + H_2O + CO_2$$

The solid–solution interface is often dominant in regulating the concentrations of dissolved trace elements and can be modelled through the use of a distribution coefficient given by the equation,

$$K_d = \frac{\text{concentration in solid phase material (mg kg}^{-1}\text{)}}{\text{concentration in dissolved phase (mg kg}^{-1}\text{)}}$$

Although the distribution coefficient is useful in modelling processes such as the sorption of organic pollutants on organically coated particles, it is also limited in that it provides a conditional constant, which will change with changing solution conditions. More recently, understanding the partitioning of metals to oxide or natural organic surfaces has been performed in terms of surface complexation theory.[17,18] However, full understanding of these processes will come only in the distant future. The fate of organic pollutants such as PCBs is modified as the "hydrophobic effect." This results in a tendency of organic chemicals to avoid contact with water and results in sorption, particularly on natural organic matter (primarily humic substances). The humic substances themselves tend to sorb to the surface of mineral and microbiological particles, with significant consequences for the fate of the pollutants.

## 31.4 CHEMICAL AND MICROBIOLOGICAL BREAKDOWN

The atmospheric lifetime of many chemical species is dictated by chemical reactions rather than depositional processes. This is also true for some more reactive species in aquatic systems. A simplistic description of relevant processes is given here.

### 31.4.1 Rate Expressions

If contaminant X is removed by reaction with reagent Y in a one-step chemical process,

$$X + Y \rightarrow \text{products}$$

the rate of loss of X is given by:

$$\frac{-d[X]}{dt} = k[X][Y]$$

where k is the rate constant for the reaction. If Y is a species whose concentration is not greatly variable with time, either because it is present in great excess or it is continually replaced after consumption, such that $[Y] \cong \text{constant}$, then

$$\frac{-d[X]}{dt} = k'[X]$$

where $k' = k[Y]$.

This rate expression has a first-order form and is termed "pseudo first-order." The implication of this form of equation, which prevails frequently in the environment, is that a contaminant that ceases to be emitted diminishes in concentration according to a negative exponential form, as with the decay of a radioactive source.

The second-order equation, if applicable, may be used to estimate the rate of removal of a substance by a specific chemical reaction. For example, the main sink for nitrogen dioxide is conversion to nitric acid by reaction with the hydroxyl radical, ·OH. For this reaction,

$$k = 1.1 \times 10^{-11} \text{ cm}^3 \text{ molecule}^{-1} \text{ s}^{-1}$$

and

$$-\frac{d}{dt}[NO_2] = (1.1 \times 10^{-11})(1 \times 10^{6})[NO_2]$$

and

$$\frac{-\frac{d}{dt}[NO_2]}{[NO_2]} = 1.1 \times 10^{-5} \text{ s}^{-1} = 0.04 \text{ h}^{-1}$$

Thus, loss of $NO_2$ occurs at a rate of approximately 4% per hour at the above concentration of OH radical. At twice that OH radical concentration, the loss is 8% per hour.

In aquatic chemistry, many of the reaction processes are not simple one-step reactions, and complex rate expressions result in noninteger reaction orders.

In soils and sediments, many decay and transformations affecting chemicals are microbiologically mediated. For example, in oxidizing sediments, organic materials are oxidized, with consumption of molecular oxygen, $O_2$. When the $O_2$ is depleted, the system becomes anoxic, and anaerobic organisms take over. Initially, nitrate-reducing bacteria are favored. Nitrate is used as an oxygen source and ammonium is produced as product. Subsequently, sulfate is used as an oxygen source and converted to sulfide in the process. Finally, carbohydrate material itself is used as a source of oxygen by methanogenic bacteria, and methane and carbon dioxide are the products.

### 31.4.2 Environmental Lifetimes

In discussing environmental lifetimes, it is useful to think of the analogy of a bath, with a source of input water and removal via an overflow. At equilibrium, the level of water is constant, and the inflow is exactly equaled by the outflow through the overflow. Then, the faster the input, the more rapid the outflow and the shorter the residence time (or lifetime) of water in the bath. This concept is easily related to a mixed reservoir, such as the atmosphere or ocean, but becomes more nebulous when considering solid-earth processes.

To a crude approximation, many environmental contaminants are in some form of equilibrium, and sources (S) equal removal rates (termed "sinks"®):

$$S = R$$

If R and S refer to a given environmental compartment, and A is the total quantity of the contaminant in that compartment, the lifetime (or residence time), τ, is defined as:

$$\tau = \frac{A(\text{kg})}{S(\text{kg s}^{-1})} = \frac{A}{R}$$

In practice, the lifetime defined in this way is the time taken for the concentration of the species to diminish to $^1/e$ of its initial concentration if all source processes ceased.

Alternatively, dividing top and bottom by the volume of the environmental compartment gives:

$$\tau = \frac{[A](\text{kg m}^{-3})}{S'(\text{kg m}^{-3}\text{ s}^{-1})}$$

or ratio of concentration to flux, S′.

If removal is via a first-order or pseudo first-order process, as described above, then:

$$S' = \frac{d[A]}{dt} = k[A]$$

and

$$\tau = \frac{[A]}{k[A]} = \frac{1}{k}$$

where k is the first-order constant for the loss of component A. This is a very useful result and can be used in the context of removal by chemical or depositional pathways. Taking the example of nitrogen dioxide above,

$$\begin{aligned} k' &= k[\cdot OH] \\ &= 0.04\ h^{-1} \end{aligned}$$

and

$$\tau = 25\ h$$

For many atmospheric trace gases, removal is by reaction with the hydroxyl radical, and

$$\tau = \frac{1}{k[\cdot OH]}$$

For methane,

$$k = 5.6 \times 10^{-15}\ cm^3\ molecule^{-1}\ s^{-1}$$

and

$$\tau = 5.7\ \text{years for}\ [\cdot OH] = 1 \times 10^6\ molecule\ cm^{-3}$$

Concentrations of hydroxyl radical do in fact vary diurnally and with latitude and season. The use of long-term average values (as above) for lifetime estimation is therefore justified for the longer-lived substances (e.g., methane); but for short-lived species like nitrogen dioxide, more precise values for the location and time are appropriate.

As mentioned in Section 31.3, washout coefficients are first-order rate constants for wet deposition. Thus, for $\lambda = 10^{-4}$ $sec^{-1}$, the lifetime with respect to removal by this pathway is $10^4$ sec, or 2.8 h. Such a value could obviously apply only during periods of rain and would not be relevant to dry spells.

Lifetimes can also be calculated in relation to dry deposition processes. In this case, the rate constant is given by

$$k = v_g/H$$

where $v_g$ is the deposition velocity and H the mixing depth, or depth of the lowest mixed layer of the atmosphere, normally the boundary layer. Thus, for sulphur dioxide, for which $v_g \cong 1$ cm $s^{-1}$ and a typical mixing depth of 1000 m,

$$\begin{aligned} \tau &= H/v_g \\ &= 28\ h \end{aligned}$$

Within the ocean, dissolved materials may be scavenged by adsorption onto particles, thereby accelerating their deposition. Using the same form of mathematical analysis, deep-water scavenging residence times vary from 0.4 years for particles, to 10 to 50 years for metals such as tin, iron, cobalt, and lead, to over $10^4$ years for cadmium.

## 31.5 SPATIAL DISTRIBUTION OF CONTAMINANTS

As discussed above, spatial distributions of contaminants result from a number of factors, including source locations, dispersal processes, and sink processes. The spatial extent of the distribution from a single source will depend critically on the lifetime of the contaminant, as those of shorter lifetimes will not survive over any great distance.

### 31.5.1 Microscale

Good examples of microscale distributions of contaminants arise from dispersion from point sources. Effective numerical models exist for describing the three-dimensional concentration distribution of pollutants emitted from elevated point sources into the atmosphere. The instantaneous distribution of concentrations is described adequately for most purposes by the Gaussian plume model.[19,20]

Emissions to atmosphere of pollutants such as heavy metals from point sources can lead to distributions of pollutants in local surface soils directly analogous to those for atmospheric concentrations. Many surveys have been carried out of metal concentrations in soils and vegetation around smelters, exhibiting just this kind of distribution, except where low-level, fugitive sources of coarse dusts, such as stockpiles, lead to highly enhanced deposition close to the works. Frequently, this has led to severe contamination, which persists unless subject to cleanup.

A well-known phenomenon of past decades has been the contamination of roadside soils with lead from vehicle exhausts. A typical distribution involves highly elevated concentrations at curbside, falling exponentially to background levels within about 100 m. Such soils typically show a strong surface enrichment of lead, with concentrations declining rapidly in lower layers, reaching local background concentrations by about 15 cm deep. Highly elevated concentrations of lead in soils may also be found in areas of lead mineralization and historical lead mining, where lead-rich soils have contaminated local soils.

Both comparable methods based on Fickian diffusion and systems procedures are available to estimate by numerical means the dispersion of contaminants in rivers.[21]

### 31.5.2 National and Regional Scales

Pollutants with major localized sources may give substantial and significant variations in concentration on the micro- or mesoscale but contribute in only a minor way to national or regional patterns of concentration. Secondary atmospheric pollutants often yield more meaningful patterns on the latter scale, as they are produced more slowly and are less influenced by individual sources; an example is hydrogen ion in rainwater. Strong acidity is generated only rather slowly from oxidation of sulphur dioxide (ca. 1 to 2% per hour) and nitrogen dioxide (ca. 5% per hour). Figure 31.9 shows the distribution of rainwater pH over North America at the height of the acidification problem, which is now less severe. Distinct spatial gradients relate to sources, meteorology, and local geology, since airborne carbonate particles can neutralize acids. Another example of a secondary pollutant showing meaningful spatial patterns in concentration is ozone. Analysis of measurement data in Europe indicates clear north–south gradients in episodic peak hourly ozone, both nationally and across western Europe as a whole.

### 31.5.3 Hemispheric and Global Scales

As mentioned above, the atmospheric circulation provides a strong barrier to tropospheric mixing between northern and southern hemispheres; crossing of the equatorial regions occurs appreciably only for pollutants with lifetimes of a year or more. Ship-borne measurements of air

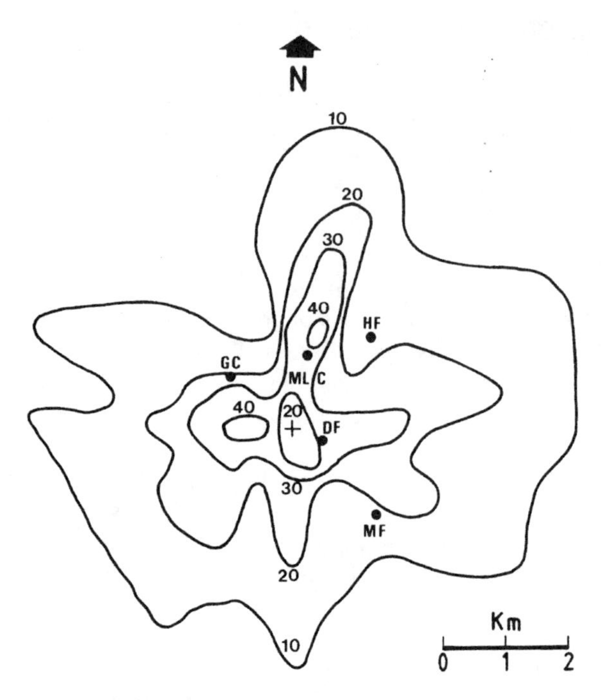


**Figure 31.9** Calculated ground-level concentration distributions (μg $m^{-3}$) of $NO_x$ around an elevated point source within a fertilizer works. + denotes center of works; • denotes sampling sites used to measure ground-level concentrations. (From Boutron, C.F. et al., *Nature,* 353, 153, 1991. With permission.)

composition have revealed north–south gradients of many atmospheric constituents. In some instances, these show sharp discontinuities at the equator, which is consistent with a predominantly northern hemisphere source and little transfer into the southern hemisphere.

As previously mentioned, persistent organic pollutants (POPs) like PCBs are capable of volatilizing from soils and water. Once airborne, they readily undergo long-range atmospheric transport either in the vapor phase or bound to aerosol. It is via such atmospheric transport, which occurs in a repeated cycle sometimes referred to as "the grasshopper effect," that these chemicals distribute in a ubiquitous fashion. This universal distribution is illustrated by the presence of POPs in polar regions, which is a matter of concern, for although inputs to such locations are small, the low temperatures minimize volatilization losses, leading to a steady accumulation of the overall pollutant burden. This accumulation is compounded by the reduced biomass-to-surface-area ratio of polar regions, with the result that the pollutant burden is distributed among a much smaller biomass than in industrial areas. These factors account for the observations of PCB concentrations in the breast milk of Canadian Inuit mothers, which exceed those found in women from urban Canada.

## 31.6 TEMPORAL TRENDS IN CONTAMINANT CONCENTRATIONS

High-quality environmental measurements of trace contaminants have been made only in relatively recent years. There are, therefore, relatively few examples of clear temporal trends discernible from contemporaneous environmental measurements. One very clear example is that of tropospheric carbon dioxide measured at Mauna Loa, Hawaii by Keeling and coworkers[22] since the late 1950s. High-quality data show a very clear upward trend modulated by a seasonal cycle due to varying exchange fluxes with terrestrial biota. Since $CO_2$ has a relatively long tropospheric lifetime, similar general trends are seen at other sites within both northern and southern hemispheres.

It is more difficult to determine trends that predate contemporaneous measurements. However, two methods have been particularly successful. The first involves analysis of ice cores. Gases

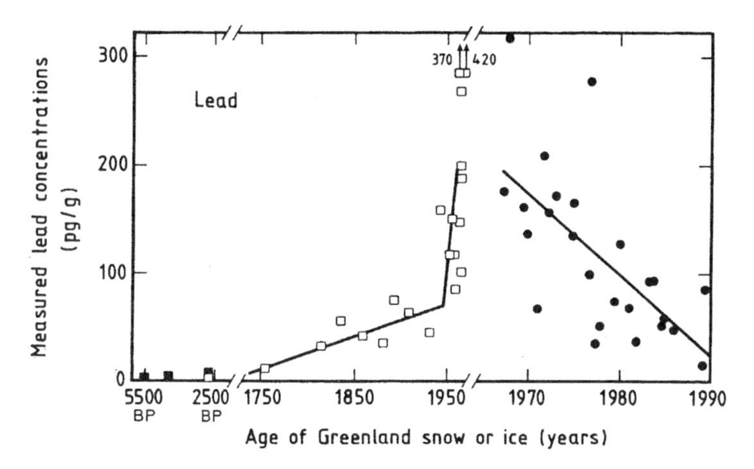


**Figure 31.10** Changes in lead concentrations in Greenland ice and snow from 5500 BP to present. (From Boutron, C.F. et al., *Nature,* 353, 153, 1991. With permission.)

become trapped in ice during snowfall, and careful retrospective analysis of dated ice core samples has revealed valuable information. Examples are carbon dioxide and methane, determined in Antarctic ice[23] up to $1.6 \times 10^5$ years old. Metallic pollutants are also incorporated into snow and ice, and Figure 31.10 shows data for lead in Greenland ice, determined from several studies, showing a massive increase in atmospheric lead deposition, commencing with the Industrial Revolution and accelerating rapidly after 1950 as use of leaded gasoline increased. More recently, deposited ice shows a strong decline from around 1970, when leaded gasoline use was curtailed, first in North America and subsequently in many other developed countries.

Lake sediment cores can also reveal interesting patterns of historical inputs, as the sediments are built up of layers laid down sequentially, the most recent at the water interface. Concentrations of trace contaminants in such sediment cores reflect input processes and subsequent diagenetic changes. In some instances, however, profiles of pollutants have been shown to reflect clearly historical inputs of pollutants such as trace metals from the atmosphere to the water body and hence provide a relative measure of trends in airborne concentrations.

The environmental persistence of POPs means that correctly stored archived samples as well as sediment and snow-pack cores provide an invaluable resource for elucidating past changes in contaminant concentrations.[25] Knowledge of such past trends can be correlated with contemporaneous trends in source activities and used to build further understanding of the likely impact of emission-control strategies on future concentrations. Figure 31.11 shows the temporal variation

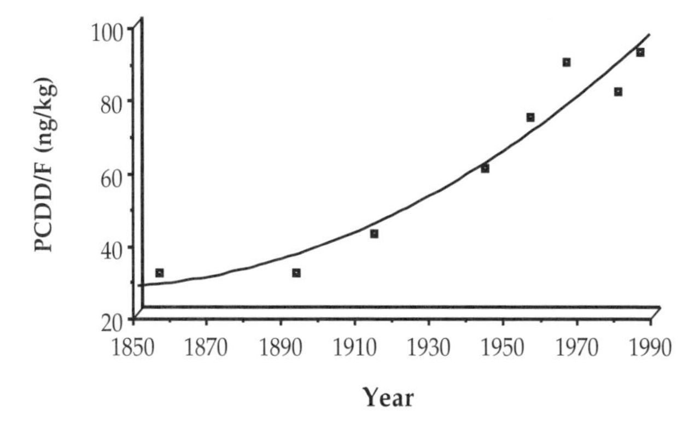


**Figure 31.11** Temporal trends in concentrations of dioxins (PCDD/Fs) in soil measured in the southern United Kingdom.

of dioxin concentrations in archived soil taken in south England between 1850 and 1990.[26] This study shows that a low, essentially constant "background" of dioxins existed in the soil throughout the latter part of the 19th century; since then, dioxin concentrations have risen dramatically. The inference from this is that, while some natural sources do exist, notably forest fires (which may arguably be of significance in some countries), the environmental burden of these compounds has its source in human activity, in particular, the manufacture and use of organochlorine chemicals (such as chlorophenols) and combustion processes (such as waste incineration and coal combustion).

## 31.7 SUMMARY

Distributions of many pollutants are measured on spatial scales varying from the local microscale to the global scale. These may generally be understood in terms of:

1. Distribution of sources
2. Transport mechanisms and intercompartmental transfer
3. Environmental lifetimes
4. Distribution of sinks

While short-lived contaminants are most readily identified close to their source, the more persistent substances, such as heavy metals and PCBs, may achieve a truly global distribution due to atmospheric transport and deposition to soils and surface waters.

Good mathematical treatments are available from many transfer processes and sink mechanisms, although considerable uncertainties still exist in relation to the use of appropriate values of parameters such as diffusion coefficients and transfer velocities. The analysis of pollutant deposits accumulated historically in media such as sediments and polar ice may be used to construct a picture of past pollutant concentrations.

## REFERENCES


1. Campbell, P. G. C., Interactions between trace metals and aquatic organisms: A critique of the free-ion activity model, in *Metal Speciation and Bioavailability*, Tessier, A. and Turner, D. R., Eds., John Wiley and Sons, Chichester, 1995.
2. Stumm, W. and Morgan, J. J., *Aquatic Chemistry*, 3rd ed., John Wiley and Sons, New York, 1996.
3. de Mora, S. J. and Harrison, R. M., The use of physical separation techniques in trace metal speciation studies, *Water Res.,* 27, 723, 1983.
4. Sigg, L., The regulation of trace elements in lakes: The role of sedimentation, in *Chemical and Biological Regulation of Aquatic Systems*, Buffle, J. and de Vitre, R. R., Eds., Lewis Publishers, Boca Raton, FL, 1994.
5. Harrison, R. M., Integrative aspects of pollutant cycling, in *Understanding Our Environment: An Introduction to Environmental Chemistry and Pollution*, Harrison, R. M., Ed., Royal Society of Chemistry, London, 1992.
6. Stowe, K. S., *Ocean Science,* John Wiley and Sons, New York, 1979.
7. Ayris, S. and Harrad, S., The fate and persistence of polychlorinated biphenyls in soil, *J. Environ. Monit.*, 1, 395, 1999.
8. Cawse, P. A., Studies of Environmental Radioactivity in Cumbria, Part 4. Cesium-137 and Plutonium in Soils of Cumbria and the Isle of Man, United Kingdom Atomic Energy Authority, UKAEA Report No. AERE-9851, 1980.
9. Davidson, C. I. and Wu, Y. L., Dry deposition of trace elements, in *Control and Fate of Atmospheric Trace Metals*, Pacyna, J. M. and Ottar, B., Eds., Kluwer Academic, Dordrecht, The Netherlands, 1989.

10. Currado, G. M. and Harrad, S., Transfer of POPs in vegetation: Implications and mechanisms, in *Persistent Organic Pollutants: Environmental Behaviour and Pathways of Human Exposure*, Harrad, S., Ed., Kluwer Academic, Dordrecht, The Netherlands, 2001.
11. Liss, P. S. and Merlivat, L., Air-sea exchange rates: Introduction and synthesis, in *The Role of Air-Sea Exchange in Geochemical Cycling*, in Buat-Menard, P., Ed., Reidel, Dordrecht, The Netherlands, 1986.
12. Tessier, A., Carignan, R., and Belzille, N., Processes occurring near the sediment-water interface: Emphasis on trace elements, in *Chemical and Biological Regulation of Aquatic Systems*, Buffle, J. and de Vitre, R. R., Eds., Lewis Publishers, Boca Raton, FL, 1994.
13. Watson, A. J., Upstill-Goddard, R. C., and Liss, P. S., Air-sea gas exchange in rough and stormy seas measured by a dual-tracer technique, *Nature(London),* 349, 145, 1991.
14. Brendel, P. J. and Luther, G. W., Development of a gold amalgam voltammetric microelectrode for the determination of dissolved Fe, Mn, $O_2$, and S (-II) in porewaters of marine and freshwater sediments, *Environ. Sci. Technol.*, 29, 751–761, 1995.
15. Shuttleworth, S. M., Davison, W., and Hamilton-Taylor, J., Two-dimensional and fine structure in the concentration of iron and manganese in sediment pore-waters, *Environ. Sci. Technol.*, 33, 4169–4175, 1999.
16. Harrison, R. M., de Mora, S. J., Rapsomanikis, S., and Johnston, W. R., *Introductory Chemistry for the Environmental Sciences,* Cambridge University Press, New York, 1991.
17. Dzomabak, D. A. and Morel, F. M. M., *Surface Complexation Modelling — Hydrous Ferric Oxide,* John Wiley and Sons, New York, 1990.
18. Turner, D. R., Problems in trace metal speciation modeling, in *Metal Speciation and Bioavailability,* Tessier, A. and Turner, D. R., Eds., John Wiley and Sons, Chichester, U.K., 1995.
19. Harrison, R. M. and McCartney, H. A., A comparison of the predictions of a simple Gaussian plume dispersion model with measurements of pollutant concentration at ground-level and aloft, *Atmos. Environ.,* 14, 489, 1980.
20. Williams, M. L., Atmospheric dispersal of pollutants and the modelling of air pollution, in *Pollution: Causes, Effects and Control,* 4th ed., Harrison, R. M., Ed., Royal Society of Chemistry, London, 2001.
21. Young, P. C., Quantitative systems methods in evaluation of environmental pollution problems, in *Pollution: Causes, Effects and Control,* 2nd ed., Harrison, R. M., Ed., Royal Society of Chemistry, London, 1990.
22. Keeling, C. D. et al., A three-dimensional model of atmospheric $CO_2$ transport based on observed wind, 4. Mean annual gradients and interannual variations, *Geophys. Monogr.,* 55, 305, 1989.
23. Inter-Governmental Panel on Climate Change, *Climate Change*, Cambridge University Press, New York, 1991.
24. Boutron, C. F., Gorlack, U., Candelone, J.-P., Bolshov, M. A., and Delmas, R. J., Decrease in anthropogenic lead, cadmium and zinc in Greenland snows since the late 1960s, *Nature,* 353, 153, 1991.
25. Sanders, G., Temporal trends in environmental contamination, in *Persistent Organic Pollutants: Environmental Behaviour and Pathways of Human Exposure*, Harrad, S., Ed., Kluwer Academic Publishers, Dordrecht, The Netherlands, 2001.
26. Kjeller, L., Jones, K. C., Johnston, A. E., and Rappe, C., Increases in the polychlorinated dibenzo-*p*-dioxin and -furan content of soils and vegetation since the 1840s, *Environ. Sci. Technol.*, 25, 1619, 1991.

CHAPTER 32

# Bioaccumulation and Bioconcentration in Aquatic Organisms

**Mace G. Barron**

## CONTENTS

## 32.1 INTRODUCTION

Concern for the bioaccumulation of contaminants arose in the 1960s because of incidents such as toxicity from methylmercury residues in shellfish and avian reproductive failure due to chlorinated pesticide residues in aquatic species. Bioaccumulation models were developed in the 1970s to account for processes such as the partitioning of hydrophobic chemicals from water to aquatic organisms.[1] The study of contaminant accumulation and predictive model development expanded in the 1980s to include bioaccumulation from sediments, food chain biomagnification, and carcinogenesis in feral species. Today, determining contaminant bioaccumulation in aquatic organisms is an essential component of assessing risks to piscivorous wildlife. Additionally, current research is evaluating the association between contaminant tissue residues and adverse effects in aquatic organisms.[2]

This chapter presents an overview of the principles and determinants governing bioaccumulation from sediments and water and biomagnification in aquatic-based food webs. Organic and metal contaminants are discussed, with an emphasis on hydrophobic organics. The objective of this chapter is to elucidate concepts relating to bioaccumulation rather than present an exhaustive review of the literature.

## 32.2 BIOACCUMULATION FROM SEDIMENTS

### 32.2.1 Determinants of Bioaccumulation

Aquatic sediments are formed from the deposition of particles and colloids and can act as both a sink and a source of pollutants. Long-term contaminant input leads to sediment concentrations that can exceed the water concentration by several orders of magnitude because of partitioning of chemicals onto sediment-binding sites. Determinants of the bioavailability of sediment-associated contaminants to aquatic organisms include physical/chemical interactions between the chemical and sediment constituents and biological processes determining exposure and uptake. Binding of hydrophobic organics and metals to sediment appears to be controlled by different mechanisms. The majority of information relating to bioaccumulation from sediments relates to hydrophobic organics, with more limited information on metal or organic anion and cation bioaccumulation.

Bioaccumulation of sediment-associated contaminants is highly species-dependent because of the diversity of feeding ecology and living habits of benthic organisms.[3–5] Exposure pathways from sediment-associated contaminants include exposure to sediment pore water (interstitial water), ingestion of sediment particles and dissolved organic matter, direct contact of sediment with body surfaces, and exposure to the boundary layer of water overlying the sediment.[6] Utilization of sediments as a food source by benthic organisms is common.[5] Particle ingestion via filtration of the water column or direct consumption may be a significant route of exposure because benthic organisms may preferentially ingest this fraction of the sediments.[7]

Mode of feeding can significantly influence contaminant bioaccumulation from sediment.[6] For example, deposit-feeding invertebrates may accumulate contaminants to a greater extent than filter feeders. Digestion of sediment by an organism may be required for contaminant absorption because of slow desorption of chemicals bound to sediment.[8] Sediment reprocessing by benthic invertebrates may introduce nonequilibrium conditions in the sediment by removal of rapidly desorbable contaminants.[9] Benthic organisms may also preferentially select higher organic carbon sediments, influencing their exposure sediment-associated contaminants.[9] Types of organic carbon food sources included in sediment can include encrusted mineral grains, organic–mineral aggregates, fecal pellets, plant fragments, microalgae, and fungi.[5] A fraction of organic carbon is digestible, but some organic carbon is present as refractory humic polymers that are not readily digestible.[5] Lee[9] described additional factors that may influence contaminant bioaccumulation. (1) Burrowing organisms (e.g., amphipods) may ventilate pore waters, whereas surface deposit feeders may not, and (2) the formation of tubes and burrows may affect the spatial distribution of contaminants. Ultimately, contaminant bioaccumulation in benthic organisms will be influenced by the heterogenous nature of sediment, which can vary in composition in scales of millimeters.[5]

### 32.2.2 Bioaccumulation Factors

A term used to quantify bioaccumulation from sediments at equilibrium is the bioaccumulation factor (BAF), defined as the chemical concentration in the organism (ng chemical/g tissue) divided by the concentration in sediment (ng/g sediment). Either the wet or dry weight may be used; use of dry weight may eliminate variability due to tissue or sediment hydration. The BAF is a unitless number between 0 and infinity. A biota to sediment accumulation factor (BSAF) can also be calculated from the lipid-normalized chemical concentration (ng chemical/g animal lipid) divided by the organic-carbon-normalized sediment concentration (ng/g sediment organic carbon). The BSAF provides a unitless number between 0 and infinity, with typical values reported for benthic invertebrates ranging from 1 to 10.[4] BAFs and BSAFs are frequently used in ecological risk assessments to estimate contaminant concentrations in benthic organisms.[10] BSAFs for polychlorinated biphenyl (PCB) congeners determined in four species of benthic invertebrates collected from Lake Erie, ranged from 0.1 to 10 and generally increased with increasing log $K_{ow}$ of the congener.[3] BSAFs for polycyclic aromatic hydrocarbons (PAHs) ranged from 0.04 to 8 and generally decreased with log $K_{ow}$, suggesting biotransformation of the larger PAHs.[3] Mean BAFs for eight metals in benthic invertebrates ranged from 0.1 to 3.5 and from 9 to 37 for PCBs.[10]

### 32.2.3 Bioaccumulation of Metals

Concentrations of heavy metals in sediments can exceed those of the overlying water by three to five orders of magnitude.[11] The bioavailability of sediment-associated heavy metals is related to the presence of metal binding sites on the sediment. Increasing the concentration of iron oxides or organic materials in the sediment appeared to reduce metal bioavailability by increasing the number of metal binding sites.[11] Metals can also form insoluble sulfides, and the sulfide fraction of sediments may represent a pool of adsorption sites for metal binding.[12] The toxicity of cadmium has been shown to be related to the acid-volatile sulfide (AVS) content of sediments,[12,13] whereas the binding of cadmium added to oxic sediments was explained by the humic acid content of the sediment.[14] Fu et al.[14] concluded that in oxic sediments, humic substances were the major absorbent for cadmium and possibly other metals. Other factors that may influence metal bioaccumulation from sediments include metal speciation, transformation (e.g., methylation to form hydrophobic alkyl metals), inhibitory interactions of different metals, sediment chemistry (salinity, redox or pH), and binding to dissolved organic matter (DOM).[11] DOM is an important energy source for microbially based aquatic food webs, and the presence of DOM can affect both the distribution of a metal between water and sediments and the bioavailability of pore-water contaminants.[9,15] Benthic organisms may directly affect DOM concentrations through their feeding behaviors and excretory processes.[9]

### 32.2.4 Bioaccumulation of Organic Contaminants

#### *32.2.4.1 Bioavailability*

Binding of organic contaminants to sediments has been attributed to the organic carbon content, clay type and content, cation-exchange capacity, pH, and particle surface area of the sediment.[6] The organic carbon content of sediment can be predictive of contaminant bioavailability because hydrophobic organics can reversibly bind to the pool of organic carbon (typically comprises 0.5 to

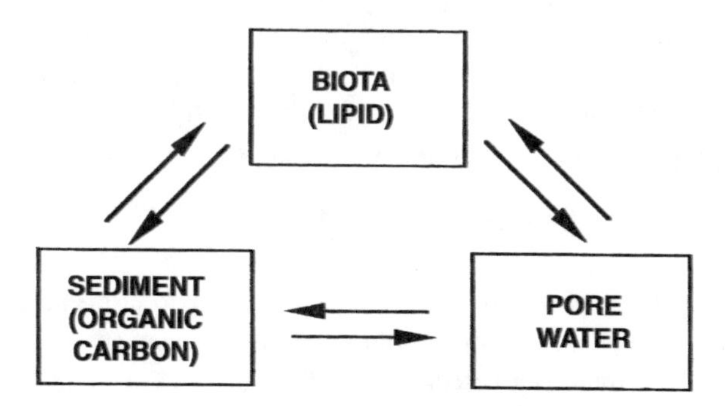


**Figure 32.1** Schematic representation of contaminant bioaccumulation from sediment based on Equilibrium Partitioning theory.

3% of a sediment mass).[16,17] Variability in the bioaccumulation factors for hydrophobic contaminants can be reduced by normalizing for organism lipid content and sediment organic carbon.[16,17] The source of the organic carbon (e.g., mud, plant material) may also influence the bioavailability and bioaccumulation, although De Witt et al.,[18] in a study with fluoranthene, found that only plant material had a significant influence. Contaminant bioavailability has been primarily assessed by the observed increase in toxicity with decreasing organic carbon content of sediments rather than a direct measurement of bioaccumulation.[16,17] Landrum,[19] Lee,[9] and others have assessed bioavailability from the magnitude of the uptake rate or clearance constant. Bioavailability appeared to decrease with increasing contact time between the sediment and contaminant, a phenomenon known as sediment aging.[19]

#### *32.2.4.2 Equilibrium Partitioning Theory*

The Equilibrium Partitioning (EqP) theory of bioaccumulation from sediments (Figure 32.1) is based on thermodynamically driven partitioning between the hydrophobic contaminant and sediment, pore water, and an organism. The EqP theory assumes that a chemical bound to sediment organic carbon is in equilibrium with the chemical dissolved in the aqueous phase of the sediment (pore water) and in the lipid components of the exposed organism. Additional assumptions of EqP theory include absence of stearic hindrance, concentration-independent uptake, and absence of metabolism or degradation of the contaminant.[4] An implicit assumption of EqP theory is that the bioavailable concentration in water is the freely dissolved portion, and the presence of DOM does not affect equilibrium partitioning. The equilibrium level accumulated by benthic organisms is assumed to be independent of the number and types of exposure routes (e.g., sediment ingestion and exposure via pore water).[17] The EqP theory predicts that the concentration in the organism can be accurately estimated from the contaminant concentration freely dissolved in pore water or the organic-carbon-normalized sediment concentration. An extensive review of the data supporting EqP theory (based primarily on halogenated organics resistant to degradation) was presented by Di Toro et al.,[16] who concluded that sediment-to-sediment variation in bioavailability (assessed by toxicity) can be reduced to a factor of two to three by application of the EqP theory, and the influence of particle size appears minimal.

Di Toro and McGrath[20] showed that EqP theory may be applicable to chemicals causing short-term toxicity through a narcosis mode of action, but EqP theory may prove to be less useful at accounting for changes in bioavailability across broad chemical classes.[7] Bierman[17] concluded that EqP theory could predict bioaccumulation within an order of magnitude for specific classes of chemicals, but bioaccumulation varied considerably with the species of organism. Limitations of EqP theory include the assumptions that equilibrium conditions exist and that lipid content is the principal biotic determinant of bioaccumulation. True equilibrium conditions are unlikely to exist because of bioturbation, biodegradation, slow contaminant uptake by organisms, spatial and temporal heterogeneity in pore-water concentrations, variation in organism lipid content and composition over time, and heterogeneity in organic carbon, particle number, and contaminant concentration. Desorption kinetics from the sediment may also become important when pore-water concentrations are depleted.[17] Lee[9] has stated that benthic organisms are not "static players driven solely by thermodynamic partitioning with the external environment." Alternative methods to the EqP approach include estimating contaminant bioaccumulation and bioavailability from (1) direct measurements of chemical concentrations in field-collected sediment and organisms, (2) equilibrium laboratory exposures, and (3) kinetic models that determine the ratio of uptake and elimination-rate coefficients in shorter-term laboratory exposures.

## 32.3 BIOCONCENTRATION FROM WATER

### 32.3.1 Bioconcentration Processes

#### *32.3.1.1 Uptake*

Bioconcentration is the accumulation of waterborne chemicals by aquatic animals through nondietary routes.[21] Bioconcentration can be viewed simply as the result of the competing rates of chemical uptake and elimination. A more complex view of bioconcentration was presented as a conceptual model by Barron (Figure 32.2).[22] Steps in the uptake process include delivery of the waterborne contaminant to the absorbing epithelium (e.g., the gill), contaminant movement across diffusion barriers (e.g., mucus, membranes) into the blood, and internal distribution of the contaminant by the circulatory fluids. In general, hydrophobic organic chemicals will diffuse through cells (permeation), while metals move intercellularly or through intracellular channels (filtration).[23] The permeability of the absorbing membrane (e.g., gill, skin) and other tissues constitutes a series of barriers to absorption and transfer of chemicals.[1,24] The absorption of a chemical from water by an aquatic animal can be viewed as a series of steps, each of which can govern the rate of uptake as well as the extent of accumulation.[22]

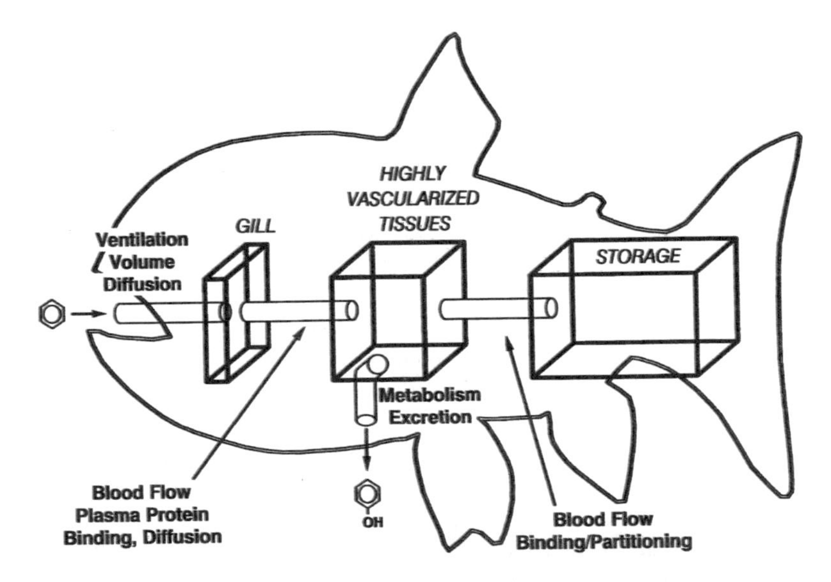


**Figure 32.2** Conceptual model of bioconcentration from water. (From Barron, M.G., *Environ. Sci. Technol.*, 24, 1612, 1990. With permission.)

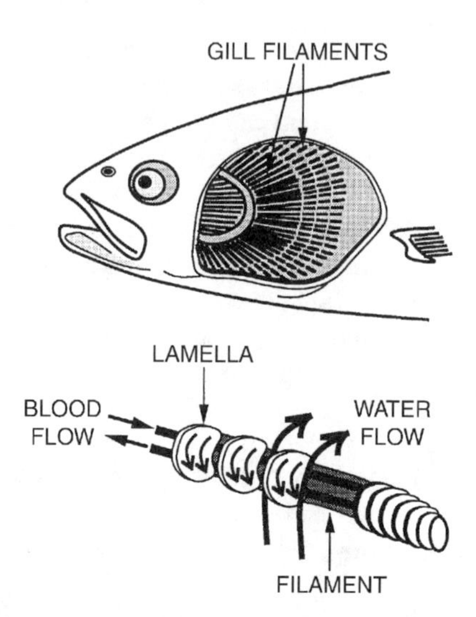


**Figure 32.3** The fish gill. (Adapted from Hickman, C. P., Hickman, C. P., and Hickman, F. M., *Integrated Principles of Zoology,* 5th ed., C.V. Mosby Company, St. Louis, 1974, 473.)

The skin, gill, and digestive tract are potential sites of absorption of waterborne chemicals. In fish, the gill is considered the dominant site of contaminant uptake because of anatomical/physiological properties that maximize absorption efficiency. For example, the gill has high countercurrent flows of blood and water, large absorbing surface, and a small diffusion distance between blood and water (Figure 32.3).[24] Skin may be an important absorption route in very small fish because of the large surface-to-volume ratio.[25] Hayton and Barron[24] concluded that the determining factors of uptake in fish were ventilation volume, blood flow, and membrane permeability. Erickson and McKim[26] incorporated these factors into a simplified model and were able to accurately predict organic-chemical uptake by fish.

#### 32.3.1.2 Distribution and Lipid Partitioning

The tissue distribution of a chemical is dependent on three processes: (1) binding/partitioning with circulatory fluid and tissue components, (2) crossing permeability barriers, and (3) distributing via the species- and tissue-specific blood flows.[27] In aquatic toxicology, it is generally accepted that lipid partitioning of organic chemicals is an important determinant of bioconcentration.[22] Lipids are an important storage site because of large mass, distribution throughout the body, and high affinity and capacity to bind hydrophobic organics. Various authors have shown that normalizing bioconcentration for the lipid content of an organism can reduce interspecific variation in bioconcentration factors. Lipid content and composition of the gill may be more important in controlling uptake than lipid content of the whole animal because hydrophobic chemicals must initially pass through the gill.[28] Bickel[27] concluded that while chemical lipophilicity will determine binding to lipids, structural features of the chemical are important in determining the binding of contaminants to other macromolecules.

The presence of tissue-specific blood perfusion rates causes tissue groups to act as kinetically distinct compartments and prevents many chemicals from rapidly reaching equilibrium.[29] Following absorption, hydrophobic contaminants are redistributed from blood to high-perfusion tissues (e.g., liver), to low-perfusion tissues (skin, muscle), and finally to lipoidal tissues.[27] The prevalence of distributional effects on contaminant accumulation can be seen in the biphasic elimination of a variety of chemicals in both fish and invertebrates.[30] Thus, blood flow as well as binding/partitioning determines the rate of transfer during uptake and elimination of a chemical.[31]

#### *32.3.1.3 Elimination*

Elimination consists of biotransformation and excretion processes. Biotransformation is considered an elimination process because the parent chemical is transformed and thus removed from the organism. Aquatic animals are capable of rapidly and extensively metabolizing a variety of chemical substrates. Biotransformation of organic chemicals acts to decrease bioconcentration by increasing the rate of elimination.[22] Although biotransformation generally acts to increase the excretion rate of chemical residues, some metabolites may be more persistent in tissues than the parent compound.[32]

Excretion processes include passive diffusion across biological membranes and active or facilitated transport in organs of elimination. The hepatobilliary system, gill, and kidney are key excretory systems in fish. Generally, large nonpolar molecules and metabolites formed from hepatic biotransformation will be excreted via the liver. Factors determining renal excretion include: filtration (filtration rate is species specific and only small molecules are filtered) and active secretion (vertebrate kidneys have active organic anion/cation transport systems).[33] Factors determining gill excretion include proper lipid solubility (e.g., rapid excretion of chemicals with log $K_{ow}$ between 1 and 3) and maximum gill clearance is equal to cardiac output.[24,26,34] Excretion of metals may occur by active transport across membranes or by processes such as diapedesis in invertebrates.[35] Slow elimination of hydrophobic organics may ultimately be due to slow distributional processes, such as slow transfer from storage tissues, rather than from limited biotransformation.[24] Some metals may exhibit extremely slow elimination from deep storage compartments such as bone, with half-lives greater than a year.[36]

### 32.3.2 Estimation of Bioconcentration

The term used to quantify the magnitude of bioconcentration is the bioconcentration factor (BCF). The BCF is defined as the proportionality constant relating the concentration of a chemical in the aquatic animal to the concentration in water, under equilibrium conditions (e.g., [fish]/[water]).[21] The BCF is an estimate of a chemical's propensity to accumulate in an aquatic animal at equilibrium, with values between 0 and infinity. BCFs have units of water volume per unit tissue weight (e.g., mL/g) and can be viewed conceptually as the water volume containing the amount of chemical concentrated in 1 g of animal tissue.[30] A BCF can be converted to a unitless number by assuming 1 g of tissue is equal to 1 mL of exposure water.

Procedures used to estimate BCFs are the equilibrium method, kinetic modeling methods, physiologically based methods, and quantitative structure activity relationship (QSAR) methods. The equilibrium method exposes an organism to a constant concentration of waterborne chemical until equilibrium is reached.[37] Kinetic methods apply a compartmental representation of the organism and estimate BCF from uptake- and elimination-rate constants (e.g., $BCF = k_{in}/k_{out}$) or fugacity-based constants.[30,38,39] Physiologically based methods estimate chemical uptake from physiological and morphological parameters such as body weight, gill surface area, and oxygen uptake.[40,41] QSAR methods estimate BCF from physical properties of the chemical. The majority of QSAR models have been linear regression models of the log transformations of BCF and the chemical's hydrophobicity (indexed by the octanol:water partition coefficient).[22] QSAR models for organic chemicals have generally been developed from data for halogenated organics resistant to biotransformation.[22,42]

QSAR models have not been available for estimating the bioconcentration of metals. Implicit in QSAR models based on chemical hydrophobicity are assumptions that (1) the octanol–water system is an appropriate surrogate for a fish lipid system, (2) biotransformation is negligible, and (3) bioconcentration is simply the thermodynamically driven partitioning between water and the lipid phase of the animal.[43] Several research groups have noted that octanol may not be a suitable surrogate for fish lipids,[42] and bioconcentration is dependent upon organism physiology and biochemistry (e.g., biotransformation) as well as lipid content and chemical hydrophobicity.[22,24,39,40]

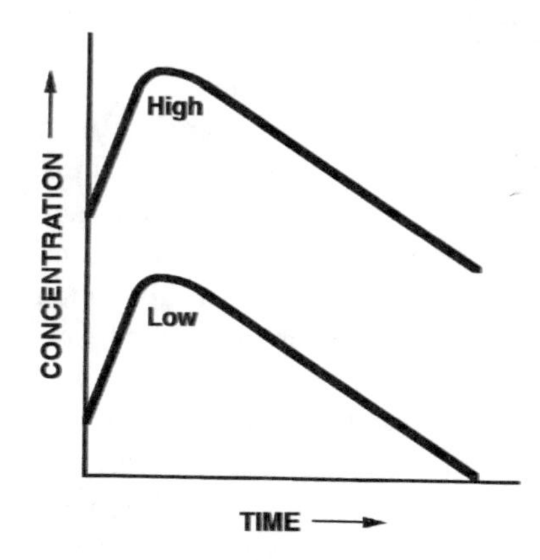


**Figure 32.4** Representation of concentrations over time for chemicals with low and high bioavailability.

### 32.3.3 Bioavailability of Waterborne Contaminants

Bioavailability of waterborne contaminants refers to the rate and extent of contaminant absorption (Figure 32.4). In aquatic toxicology, contaminant bioavailability of waterborne contaminants has usually been assessed using the rate of uptake rather than the extent of contaminant absorption. A decrease in bioavailability indicates that less chemical reaches the systemic circulation of an organism. Factors influencing the bioavailability of waterborne chemicals include the exposure concentration, the presence of particulates or DOM, and stearic hindrance. Generally, chemicals must be in solution (each molecule with a hydration shell) to be transferred across the absorbing epithelium. Thus, some chemicals may not be accumulated because of their insolubility in water. Binding of chemicals to particulates and DOM (e.g., humic acids) results in a decrease in the freely dissolved fraction, which causes a reduction in uptake and bioconcentration.[6,22] Additionally, DOM may interact directly at the cell surface to affect chemical uptake.[44]

A reduction in bioavailability due to decreased absorption of chemicals with large molecular size or shape is termed stearic hindrance. Opperhuizen et al.[45] suggested that the structure of the phospholipid membrane of the gill epithelium can restrict uptake of hydrophobic molecules of long chain length or large cross-sectional area by imposing a physical barrier to diffusion. Slow solvation (cavity formation) in membrane lipids or in blood may also slow or limit accumulation.[45] The bioavailability of metals may be reduced because of high affinity binding to ionized sites on the gill surface, which reduces internal diffusion.[46] Finally, presystemic biotransformation of a chemical during uptake may also reduce bioavailability by lowering the amount of parent chemical reaching the systemic circulation.[47] Ultimately, the temporal and spatial heterogeneity of the environment as well as physiological/biochemical variability in organisms will make an accurate prediction of contaminant bioavailability difficult, particularly for metals.

### 32.3.4 Bioconcentration of Metals

Relative to hydrophobic organics, knowledge of the determinants of metal bioconcentration is limited. Water-quality parameters (e.g., pH, water hardness) may influence metal accumulation by affecting metal speciation and form insoluble organic and inorganic complexes that precipitate and become unavailable for bioconcentration.[48,49] Metals can exist in solution as free ions and inorganic complexes, which will affect bioavailability.[48] Water hardness may influence metal uptake independent of effects on metal speciation. For example, hardness ions may reduce zinc uptake through a chemical mechanism (e.g., inhibition of metal absorption) or biological mechanism (e.g., reduction in membrane permeability).[50] Acclimation of an organism through prior exposure to a metal can also lower metal uptake in subsequent exposures.[51] Formation of organic metals (e.g., alkylation) increases hydrophobicity and causes the metal to bioconcentrate as a hydrophobic organic.[52,53] For example, Besser et al.[52] found that an organic form of selenium had BCFs of 16,000 in algae, 200,000 in daphnids, and 5000 in fish.

The bioconcentration of metals varies greatly with the species of organism and the specific metal.[49,54] Some species can accumulate metals to high levels (e.g., zooplankton), while other species such as fish closely regulate internal concentrations or sequester the metal with cellular binding proteins (e.g., metallothioneins).[49] Uptake of metals has been frequently reported as a zero-order process in both invertebrates and fish.[36,55] Spry and Wood[55] found uptake of zinc to be a saturable process exhibiting Michaelis–Menten kinetics. Thus, saturation of transport processes involved in metal uptake appears responsible for the zero-order uptake of some metals. Size of the organism can also influence metal bioconcentration; there appears to be a negative correlation between body size and metal bioaccumulation.[56] Metals initially accumulate in the gills of both invertebrates and fish.[46,50,51,57] The gill surface is negatively charged and thus provides sites for ionic binding with the metal.[46] Reid and McDonald[46] speculated that lower-binding-affinity metals such as copper have a greater likelihood of entering the fish rather than binding at the gill surface.

### 32.3.5 Bioconcentration of Organic Contaminants

Bioconcentration factors of organic contaminants can vary 100-fold with species and environmental conditions. Some of the species variation in bioconcentration may result from differences in size rather than the species or lipid content of an organism.[58] Lipid content generally varies less than tenfold between species of organisms. Body-size-dependent uptake may be explained by the size dependence of the components of the uptake process.[58] For example, gill surface area and ventilation volume per unit body weight generally decrease with an increase in body size.[22,24]

The effects of environmental conditions on bioconcentration have generally been unpredictable. Temperature can have dramatic effects on bioconcentration because of the effect on biochemistry and physiology of aquatic animals.[22] For example, Karara and Hayton[59] found that the bioconcentration factor for a phthalate ester increased from 45 at 10°C to 6500 at 35°C. The effect of the ionic composition of the water on bioconcentration has received limited investigation.[22] The pH of exposure water will significantly influence the uptake of weak electrolytes by influencing the concentration of the more rapidly absorbed nonionized fraction.[60] Other environmental parameters that can affect bioconcentration include salinity and dissolved oxygen.[61]

The importance of animal physiology and biochemistry in determining organic chemical accumulation was recognized by earlier workers based on drug kinetics in mammalian species.[62] Barron[22] concluded that absorption of hydrophobic organics across the gill was analogous to inhalation of volatile organics in mammals, where ventilation volume may limit the uptake of highly lipid soluble chemicals. Although membrane permeability as a diffusion barrier had been recognized as a significant determinant of bioconcentration, the importance of blood flow in bioconcentration was only recently recognized.[24] Prediction of bioconcentration under various environmental conditions will require an understanding of the influence of physiological and biochemical processes.[22,61]

## 32.4 BIOMAGNIFICATION AND TROPHIC TRANSFER

### 32.4.1 Overview

Biomagnification is the increase in contaminant concentration in an organism in excess of bioconcentration and is caused by the transfer of contaminant residues from lower to higher trophic levels (trophic transfer). Bioconcentration of waterborne chemicals can usually account for contaminant accumulation in aquatic organisms, but many persistent hydrophobic contaminants exhibit increasing concentrations with trophic level in aquatic-based food webs and in piscivorous birds.[63] Further, observations of contaminant concentrations greater than those predicted from BCFs argue for a significant role of biomagnification for chlorinated pesticides, PCBs, dioxins, and many other contaminants.[64]

As in sediment bioaccumulation studies, a BAF can be used to assess the magnitude of accumulation from multiple exposure pathways (e.g., trophic transfer, bioconcentration). The BAF relates concentration in an organism resulting from dietary and aqueous exposures to the concentration in water. To reduce variability, a lipid-normalized BAF can be calculated as: μg contaminant/Kg lipid divided by μg contaminant/liter water. The BAF is generally used to assess the magnitude of site- or study-specific bioaccumulation. A biomagnification factor (BMF; e.g., [fish]/[diet]) can also be computed to evaluate an elevation in fish tissue concentrations above contaminant concentrations in prey species.

### 32.4.2 Dietary Absorption

The fundamental process in trophic transfer of a contaminant is dietary absorption. In fish and mammals, absorption of food occurs primarily in the intestine.[65] Passive diffusion is the most important process in contaminant adsorption for unionized chemicals or ion pairs, whereas some metals may be absorbed through carrier-mediated or facilitated diffusion.[66] In general, chemicals that are water soluble and have moderate lipid solubility are normally absorbed best.[66] Intestinal absorption of hydrophobic contaminants is dependent upon the lipid content of the food source and may exhibit substantial variability in adsorption between individuals and between species.[66,67] Dietary triglycerides and micelles generally increase bioavailability, and contaminant absorption appears to occur concomitantly with lipid absorption.[68] Dissolution from the food source may be the rate limiting determinant of contaminant absorption. For example, desorption half-lives from ingested sediments may be greater than a week, which likely exceeds the intestinal transit time of most organisms.[8] The lipid-associated contaminants are processed into fat vacuoles and chylomicrons (mammals) or very-low-density lipoproteins (fish), which are subsequently released into lymph and the systemic circulation.[68] The process of hydrophobic chemical absorption into blood is presented in Figure 32.5, adapted from Sire et al.[69] The food matrix and physiological parameters will substantially affect the dietary absorption of contaminants including the (1) digestibility and composition of food, (2) drinking water rates, (3) environmental temperature, (4) gastric evacuation rates and pH, and (5) activity of gastrointestinal digestive and metabolic enzymes.[66]

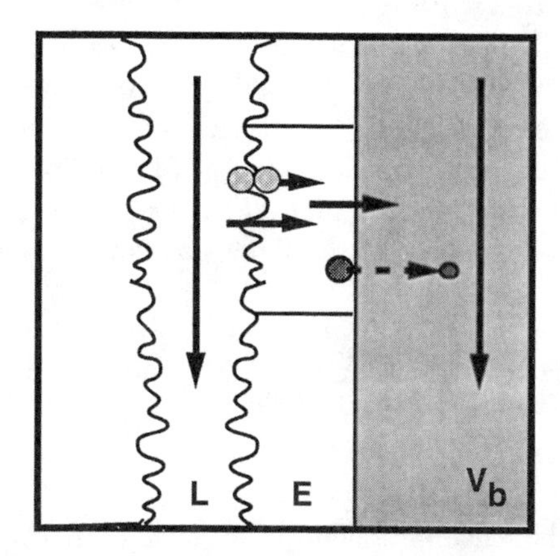


**Figure 32.5** Schematic representation of the intestinal absorption process (adapted from Reference 68). Arrows indicate the movement of food through the intestinal lumen (L), absorption into the epithelium (E), and transport in the circulatory system (Vb) by passive diffusion and cosecretion with lipids.

### 32.4.3 Dietary Bioavailability

Dietary or oral bioavailability, defined as the relative amount of chemical that reaches the systemic circulation, is a unitless number between zero and one. Oral bioavailability may be expressed relative to 100% (complete bioavailability). Pharmacological studies calculate drug bioavailability from the area under the blood-concentration-time curve (AUC), referenced to an intravascular dose.[30] The AUC approach has been applied to aquatic organisms including lobsters, rainbow trout, and channel catfish.[69,70] Oral bioavailability is species- and chemical-specific and in aquatic animals ranges from near zero (e.g., tetracycline) to near 100% (e.g., quinoline).[69,71,72]

In aquatic toxicology, oral bioavailability is frequently termed "absorption efficiency" and is estimated from chemical concentrations in the whole organism.[73] In a study of 17 halogenated organic chemicals fed to rainbow trout, Niimi and Oliver[73] found that most absorption efficiencies were between 0.5 and 0.8 and decreased rapidly at a molecular volume of greater than 0.3 nm.[3] Gobas et al.[74] found a similar relationship for halogenated organics compiled from data for several fish species: absorption efficiencies were approximately 0.5, then declined toward 0 at log $K_{ow}$ greater than 7. In contrast, Fisk et al.[75] reported that adsorption efficiencies of hydrophobic organochlorine compounds exhibited a curvilinear relationship with $K_{ow}$, with maximum values at log $K_{ow}$ 6.9. Causes of reduced bioavailability include incomplete absorption (e.g., poor membrane permeability due to stearic hindrance or incomplete dissolution), and biotransformation during the absorption process (e.g., intestinal lumen, enterocytes, liver).

### 32.4.4 Aquatic-Based Food Webs

#### *32.4.4.1 Biomagnification*

Evidence for biomagnification in aquatic-based food webs has generally been restricted to contaminants that are persistent, halogenated, and very hydrophobic. For example, Thomann and Connolly[76] concluded that 99% of the PCB residues in a Lake Michigan aquatic food web were due to trophic transfer. Based on theoretical modeling, Thomann and coworkers[64] concluded that food-chain biomagnification will not be significant for chemicals with a log $K_{ow}$ less than 5. Of 23 hydrophobic organochlorine compounds fed to rainbow trout, 20 had BMFs ([fish]/[diet]) greater than 1, indicating a potential to biomagnify.[75] Maximum biomagnification occurred at a log $K_{ow}$ of 7.[75] Rasmussen et al.[77] demonstrated that PCBs in lake trout increased with the length of the benthic-based food web and with the lipid content of the tissue. PCB bioaccumulation in algae was consistent with partitioning from water to algal lipids, but uptake was slow relative to growth in biomass.[77] Axelman et al.[78] concluded that equilibrium concentrations of PCBs in plankton do not occur in the environment and should not be used in food-web modeling. Multiple studies have reported biomagnification of chlorinated pesticides in marine mammals and birds.[64,79] Varanasi and Stein[33] concluded that trophic transfer of low-molecular-weight PAHs to higher vertebrates was probable because of low metabolism in fish. Trophic transfer of higher-molecular-weight carcinogenic PAHs — for example, benzo(a)pyrene — was unlikely because of rapid and extensive metabolism by fish.[33] For metals, significant biomagnification in higher vertebrates appears to occur only for hydrophobic alkyl metals because metals are internally regulated by higher organisms.[11] Biomagnification of organic contaminants appears most significant for benthic-based food webs and for very hydrophobic chemicals resistant to biotransformation and biodegradation.[64]

#### *32.4.4.2 Food-Web Models*

Aquatic food-web models describe the transfer of contaminants from surface water and sediment to higher trophic levels and can be used to predict contaminant bioaccumulation in organisms exposed by multiple pathways.[64] The models are composed of a series of equations that calculate contaminant concentrations in predator organisms (e.g., fish) from intake of contaminated food and uptake from water and processes that reduce contaminant concentrations such as growth and excretion.[80] Contaminant concentrations in sediment and water are typically assumed to be in steady-state equilibrium with the aquatic organisms composing the food web.[81]

#### 32.4.4.3 Maternal Transfer

The transfer of contaminants from the maternal parent to eggs and embryos of aquatic organisms represents an additional bioaccumulation pathway. Maternal transfer occurs as contaminants stored in lipids are mobilized and transferred to the gonads, resulting in egg concentrations that can vary substantially with the contaminant. For example, eggs of five species of fish contained 5 to 30% of the total contaminant burden.[82] In contrast, spawning of brook trout fed TCDD did not represent an important route of elimination, and methylmercury concentrations in walleye eggs represented only 0.2 to 2.1% of the total body burden.[83,84]

## 32.5 SUMMARY

Bioaccumulation and bioconcentration are terms describing the transfer of contaminants from the external environment to an organism. In aquatic organisms, bioaccumulation can occur from exposure to sediment (including pore water) or via the food chain (termed trophic transfer). Bioconcentration is the accumulation of waterborne contaminants by aquatic animals through nondietary exposure routes. Biomagnification is defined as the increase in contaminant concentration in an organism in excess of bioconcentration. Biomagnification appears most significant for benthic-based food webs and for very hydrophobic contaminants resistant to biotransformation and biodegradation.

A multitude of factors influence contaminant bioaccumulation and bioconcentration including biological processes (e.g., feeding ecology, living habits, physiology/biochemistry) and the chemistry and composition of the aquatic environment. For organic chemicals, equilibrium partitioning between the organism and the aquatic environment is an important process driving contaminant accumulation. Following absorption of the contaminant, internal distribution and elimination processes as well as partitioning control the rate and extent of contaminant accumulation. Bioaccumulation and bioconcentration of metals is additionally complex because of saturable uptake (even at low exposure concentrations) and regulation of internal concentrations. Estimation and measurement of bioaccumulation has become increasingly important in ecological risk assessment but will remain difficult because of the spatial and temporal heterogeneity of the environment and the biological diversity of aquatic organisms.

## REFERENCES

1. Hamelink, J. L., Waybrant, R. C., and Ball, R. C., A proposal: Exchange equilibria control the degree chlorinated hydrocarbons are biologically magnified in lentic environments, *Trans. Am. Fish. Soc.*, 100, 207, 1971.
2. Barron, M. G., Hansen, J., and Lipton, J., Association between contaminant tissue residues and adverse effects in aquatic organisms, *Rev. Environ. Contam. Toxicol*, 173, 1–37, 2001.
3. Gewurtz, S. B., Lazar, R., and Haffner, G. D., Comparison of polycyclic aromatic hydrocarbon and polychlorinated biphenyl dynamics in benthic invertebrates of Lake Erie, USA, *Environ. Toxicol. Chem.*, 19, 2943, 2000.
4. Lake, J. L., Equilibrium partitioning and bioaccumulation of sediment-associated contaminants by infaunal organisms, *Environ. Toxicol. Chem.*, 9, 1095, 1990.

5. Watling, L., The sedimentary milieu and its consequences for resident organisms, *Am. Zool.*, 31, 789, 1991.
6. Knezovich, J. P., Harrison, F. L., and Wilhelm, R. G., The bioavailability of sediment-sorbed organic chemicals: A review, *Water Air Soil Pollut.*, 32, 233, 1987.
7. Landrum, P. F. and Faust, W. R., Effect of variation in sediment composition on the uptake rate coefficient for selected PCB and PAH congeners by the amphipod, *Diporeia*, sp., *Aquatic Toxicology and Risk Assessment: Fourteenth Volume*, ASTM STP 1124, Mayes, M. A. and Barron, M. G., Eds., American Society for Testing and Materials, Philadelphia, 1991, 263.
8. Schrap, S. M., Bioavailability of organic chemicals in the aquatic environment, *Comp. Biochem. Physiol.*, 100C, 13, 1991.
9. Lee, H., II, A clam's eye view of the bioavailability of sediment-associated contaminants, in *Organic Substances and Sediments in Water, Volume III: Biological*, Baker, R., Ed., Lewis Publishers, Chelsea, MI, 1991, 73.
10. Bechtel Jacobs Company, Biota Sediment Accumulation Factors for Invertebrates: Review and Recommendations for the Oak Ridge Reservation. BJC/OR-112, U.S. Department of Energy, 1998.
11. Bryan, G. W. and Langston, W. J., Bioavailability, accumulation and effects of heavy metals in sediments with special reference to United Kingdom estuaries: A review, *Environ. Pollut.*, 76, 89, 1992.
12. Di Toro, D. M., Mahony, J. U. D., Hansen, D. J., Scott, K. J., Hicks, M. B., Mayr, S. M., and Redmond, M. S., Toxicity of cadmium in sediments: The role of acid volatile sulfide, *Environ. Toxicol. Chem.*, 9, 1487, 1990.
13. Carlson, A. R., Phipps, G. L., Mattson, V. R., Kosian, P. A., and Cotter, A. M., The role of acid-volatile sulfide in determining cadmium bioavailability and toxicity in freshwater sediments, *Environ. Toxicol. Chem.*, 10, 1309, 1991.
14. Fu, G., Allen, H. E., and Yang, C., The importance of humic acids to proton and cadmium binding in sediments, *Environ. Toxicol. Chem.*, 11, 1363, 1992.
15. Farrington, J. W., Biogeochemical processes governing exposure and uptake of organic pollutant compounds in aquatic organisms, *Environ. Health Perspect.*, 90, 75, 1991.
16. Di Toro, D. M., Zarba, C. S., Hansen, D. J., Berry, W. J., Swartz, R. C., Cowan, C. E., Pavlou, S. P., Allen, H. E., Thomas, N. A., and Paquin, P. R., Technical basis for establishing sediment quality criteria for nonionic organic chemicals using equilibrium partitioning, *Environ. Toxicol. Chem.*, 10, 1541, 1991.
17. Bierman, V. J., Equilibrium partitioning and biomagnification of organic chemicals in benthic animals, *Environ. Sci. Technol.*, 24, 1407, 1990.
18. DeWitt, T. H., Ozretich, R. J., Swartz, R. C., Lamberson, J. D., Schults, D. W., Ditsworth, G. R., Jones, J. R. P., Hoselton, L., and Smith, L. M., The influence of organic matter quality on the toxicity and partitioning of sediment associated fluoranthene, *Environ. Toxicol. Chem.*, 11, 197, 1992.
19. Landrum, P. F., Eadie, B. J., and Faust, W. R., Variation in the bioavailability of polycyclic aromatic hydrocarbons to the amphipod *Diporeia* (spp.) with sediment aging, *Environ. Toxicol. Chem.*, 11, 1197, 1992.
20. Di Toro, D. M. and McGrath, J. A., Technical basis for narcotic chemicals and PAH criteria: II. Mixtures and sediments, *Environ. Toxicol. Chem.*, 19, 1971, 2000.
21. Vieth, G. D., DeFoe, D. L., and Bergstedt, B. V., Measuring and estimating the bioconcentration factor of chemicals in fish, *J. Fish. Res. Bd. Can.*, 36, 1040, 1979.
22. Barron, M. G., Bioconcentration, *Environ. Sci. Technol.*, 24, 1612, 1990.
23. Mayer, S. E., Melmon, K. L., and Gilman, A. G., Introduction; the dynamics of drug absorption, distribution, and elimination, *The Pharmacological Basis of Therapeutics*, 6th ed., Goodman, L. S. and Gilman, A., Eds., MacMillan, New York, 1980, Chap. 1.
24. Hayton, W. L. and Barron, M. G., Rate limiting barriers to xenobiotic uptake by the gill, *Environ. Toxicol. Chem.*, 9, 151, 1990.
25. Saarikoski, J., Lindstrom, R., Tyynela, M., and Viluksela, M., Factors affecting the absorption of phenolics and carboxylic acids in the guppy (*Poecilia reticulata*), *Ecotoxicol. Environ. Saf.*, 11, 158, 1986.
26. Erickson, R. J. and McKim, J. M., A model for exchange of organic chemicals at fish gills: Flow and diffusion limitations, *Aquat. Toxicol.*, 18, 175, 1990.
27. Bickel, M. H., The role of adipose tissue in the distribution and storage of drugs, *Prog. Drug Res.*, 28, 273, 1984.

28. Barber, M. C., Suarez, L. A., and Lassiter, R. R., Modeling bioconcentration of nonpolar organic pollutants in fish, *Environ. Toxicol. Chem.*, 7, 545, 1988.
29. Barron, M. G., Tarr, B. D., and Hayton, W. L., Temperature dependence of cardiac output and regional blood flow in rainbow trout (*Salmo gairdneri Richardson*), *J. Fish Biol.*, 31, 735, 1987.
30. Barron, M. G., Stehly, G. R., and Hayton, W. L., Pharmacokinetic modeling in aquatic animals. I. Models and concepts, *Aquat. Toxicol.*, 18, 61, 1990.
31. Anderson, M., A physiologically based toxicokinetic description of the metabolism of inhaled gases and vapors: Analysis at steady state, *Toxicol. Appl. Pharmacol.*, 60, 509, 1981.
32. Varanasi, U. and Stein, J. E., Disposition of xenobiotic chemicals and metabolites in marine organisms, *Environ. Health Perspect.*, 90, 93, 1991.
33. Pritchard, J. B. and Bend, J. R., Relative roles of metabolism and renal excretory mechanisms in xenobiotic elimination by fish, *Environ. Health Perspect.*, 90, 85, 1991.
34. Maren, T. H., Embry, R., and Broder, L. E., The excretion of drugs across the gill of the dogfish, *Squalus acanthias, Comp. Biochem. Physiol.*, 26, 853, 1968.
35. Lobel, P. B., Short-term and long-term uptake of zinc by the mussel, *Mytilus edulis*: A study in individual variability, *Arch. Environ. Contam. Toxicol.*, 16, 723, 1987.
36. Barron, M. G., Schultz, I. R., and Newman, M. E., Pharmacokinetics of zinc$^{65}$ following intravascular administration in channel catfish, *Ecotoxicol. Environ. Saf.*, 45, 304, 2000.
37. Barron, M. G. and Woodburn, K. B., Pesticide bioaccumulation and metabolism: Study requirements, experimental design, and data analysis, in *Xenobiotic Metabolism in Fish*, Smith, D. J., Gingerich, W. H., and Beconi-Barker, M. G., Plenum Press, New York, 1999, Chap. 4.
38. Branson, D. R., Blau, G. E., Alexander, H. C., and Neely, W. B., Bioconcentration of 2,′2′,4,′4′-tetrachlorobiphenyl in rainbow trout as measured by an accelerated test, *Trans. Am. Fish. Soc.*, 104, 785, 1975.
39. Landrum, P. F., Lee, H., II, and Lydy, M., Toxicokinetics in aquatic systems: Model comparisons and use in hazard assessment, *Environ. Toxicol. Chem.*, 11, 1709, 1992.
40. Nichols, J. W., McKim, J. M., Lien, G. L., Hoffman, A. D., Bertelsen, S. L., and Gallinat, C. A., Physiologically based toxicokinetic modelling of three waterborne chloroethanes in channel catfish, *Ictalurus punctatus, Aquat. Toxicol.,* 27, 83, 1993.
41. Yang, R., Thurston, V., Neuman, J., and Randall, D. J., A physiological model to predict xenobiotic concentration in fish, *Aquat. Toxicol.,* 48, 109, 2000.
42. Mackay, D., Correlation of bioconcentration factors, *Environ. Sci. Technol.*, 16, 274, 1982.
43. Petersen, G. I. and Kristensen, P., Bioaccumulation of lipophilic substances in fish early life stages, *Environ. Toxicol. Chem.*, 1998, 1385, 1998.
44. Campbell, P. G. C., Twiss, M. R., and Wilkinson, K. J., Accumulation of natural organic matter on the surfaces of living cells: Implications for the interaction of toxic solutes with aquatic biota, *Can. J. Fish. Aquat. Sci.,* 54, 2543, 1997.
45. Opperhuizen, A., Von der Velde, E. W. , Gobas, F. A. P. C., Liem, D. A. K., Van der Stein, J. M. D., and Hutzinger, O., Relationship between bioconcentration in fish and steric factors of hydrophobic chemicals, *Chemosphere*, 14, 1871, 1985.
46. Reid, S. D. and McDonald, D. G., Metal binding activity of the gills of rainbow trout (*Oncorhynchus mykiss*), *Can. J. Fish. Aquat. Sci.*, 48, 1061, 1991.
47. Barron, M. G., Schultz, I. R., and Hayton, W. L., Presystemic branchial metabolism limits di-2-ethylhexyl phthalate accumulation in fish, *Toxicol. Appl. Pharmacol.*, 98, 49, 1989.
48. Brown, P. L. and Markich, S. J., Evaluation of the free ion activity model of metal-organisms interaction: Extension of the conceptual model, *Aquat. Toxicol.,* 51, 177, 2000.
49. Hodson, P. V., The effect of metal metabolism on uptake, disposition and toxicity in fish, *Aquat. Toxicol.*, 11, 3, 1988.
50. Barron, M. G. and Albeke, S., Calcium control of zinc uptake in rainbow trout, *Aquat. Toxicol.,* 50, 257, 2000.
51. Hogstrand, C., Webb, N., and Wood, C. M., Covariation in regulation of affinity for branchial zinc and calcium uptake in freshwater rainbow trout during adaptation to waterborne zinc, *J. Exp. Biol.*, 201, 1809, 1998.
52. Besser, J. M., Canfield, T. J., and LaPoint, T. W., Bioaccumulation of organic and inorganic selenium in a laboratory food chain, *Environ. Toxicol. Chem.*, 12, 57, 1993.

53. Duvall, S. E. and Barron, M. G., A screening-level probabilistic ecological risk assessment of mercury in Florida Everglades food webs, *Ecotoxicol. Environ. Saf.*, 47, 298, 2000.
54. Ratte, H. T., Bioaccumulation and toxicity of silver compounds: A review, *Environ. Toxicol. Chem.*, 18, 89, 1999.
55. Spry, D. J. and Wood, C. M., Zinc influx across the isolated, perfused head preparation of the rainbow trout (*Salmo gairdneri*) in hard and soft water, *Can. J. Fish. Aquat. Sci.*, 45, 2206, 1988.
56. Newman, M. C. and Heagler, M. G., Allometry of metal bioaccumulation and toxicity, *Metal Ecotoxicology, Concepts and Applications*, Newman, M. C. and McIntosh, A. W., Eds., Lewis Publishers, Chelsea, MI, 1991, Chap. 4.
57. Andersen, J. T. and Baatrup, E., Ultrastructural localization of mercury accumulation in the gills, hepatopancreas, midgut, and antennal glands of the brown shrimp, *Crangon crangon, Aquat. Toxicol.*, 13, 309, 1988.
58. Schultz, I. R. and Hayton, W. L., Body size and the toxicokinetics of trifluralin in rainbow trout, *Toxicol. Appl. Pharmacol.*, 129, 138, 1994.
59. Karara, A. H. and Hayton, W. L., A pharmacokinetic analysis of the effect of temperature on the accumulation of di-2-ethylhexyl phthalate (DEHP) in sheepshead minnow, *Aquat. Toxicol.*, 15, 27, 1989.
60. Lo, I-H. and Hayton, W. L., Effects of pH on the accumulation of sulfonamides by fish, *J. Pharmacol. Biopharm.*, 9, 443, 1981.
61. McKim, J. M. and Erickson, R. J., Environmental impacts on the physiological mechanisms controlling xenobiotic transfer across fish gills, *Physiol. Zool.*, 64, 39, 1991.
62. Hunn, J. B. and Allen, J. L., Movement of drugs across the gills of fishes, *Annu. Rev. Pharmacol.*, 14, 47, 1974.
63. Baumann, P. C. and Whittle, D. M., The status of selected organics in the Laurentian Great Lakes: An overview of DDT, PCBs, dioxins, furans, and aromatic hydrocarbons, *Aquat. Toxicol.*, 11, 241, 1988.
64. Thomann, R. V., Connolly, J. P., and Parkerton, T. F., An equilibrium model of organic chemical accumulation in aquatic food webs with sediment interaction, *Environ. Toxicol. Chem,*. 11, 615, 1992.
65. Honkanen, R. E., Rigler, M. W., and Patton, J. S., Dietary fat assimilation and bile salt absorption in the killifish intestine, *Am. J. Physiol.*, 249, G399, 1985.
66. James, M. O. and Kleinow, K. M., Trophic transfer of chemicals in the aquatic environment, in *Aquatic Toxicology*, Malins, D. C. and Ostrander, G. K., Eds., Lewis Publishers, Boca Raton, FL, 1994, Chap. 1.
67. Van Veld, P. A., Absorption and metabolism of dietary xenobiotics by the intestine of fish, *Rev. Aquat. Sci.*, 2, 185–203, 1990.
68. Sire, M.-F., Lutton, C., and Vernier, J. M., New views on intestinal absorption of lipids in teleostean fishes: An ultrastructural and biochemical study in the rainbow trout, *J. Lipid Res.*, 22, 81, 1981.
69. Stoffregen, D. A., Bowser, P. R., and Babish, J. G., Antibacterial chemotherapeutants for finfish aquaculture: A synopsis of laboratory and field efficacy and safety studies, *J. Aquat. Animal Health*, 8, 181, 1996.
70. Barron, M. G. and James, M. O., Oral bioavailability of sulfadimethoxine in the lobster, *Homarus americanus*, following single and multiple dosing regimes, *Xenobiotica,* 24, 921, 1994.
71. Plakas, S. M., McPhearson, R. M., and Guarino, A. M., Disposition and bioavailability of $H^3$-tetracycline in the channel catfish (*Ictalurus punctatus*), *Xenobiotica*, 18, 83, 1988.
72. Dauble, D. D. and Curtis, L. R., Rapid branchial excretion of dietary quinoline by rainbow trout (*Salmo gairdneri*), *Can. J. Fish. Aquat. Sci.*, 46, 705, 1989.
73. Niimi, A. J. and Oliver, B. G., Influence of molecular weight and molecular volume on dietary absorption efficiency of chemicals by fishes, *Can. J. Fish. Aquat. Sci.*, 45, 222, 1988.
74. Gobas, F. A. P. C., Muir, D. C. G., and Mackay, D., Dynamics of dietary bioaccumulation and faecal elimination of hydrophobic organic chemicals in fish, *Chemosphere*, 5, 943, 1988.
75. Fisk, A. T., Norstrom, R. J., Cymbalisty, C. D., and Muir, D. C. G., Dietary accumulation and depuration of hydrophobic organochlorines: Bioaccumulation parameters and their relationship with the octanol/water partition coefficient, *Environ. Toxicol. Chem.*, 17, 91, 1998.
76. Thomann, R. V. and Connolly, J. P., Model of PCB in the Lake Michigan lake trout food chain, *Environ. Sci. Technol.*, 18, 65, 1984.
77. Rasmussen, J. B., Rowan, D. J., Lean, D. R. S., and Casey, J. H., Food chain structure in Ontario lakes determines PCB levels in lake trout (*Salvelinus namaycush*) and other pelagic fish, *Can. J. Fish. Aquat. Sci.*, 47, 2030, 1990.

78. Axelman, J., Broman, D., and Naf, C., Field measurements of PCB partitioning between water and planktonic organisms: Influence of growth, particle size, and solute-solute interactions, *Environ. Sci. Technol.*, 31, 665, 1997.
79. Kawano, M., Inoue, T., Wada, T., Hidaka, H., and Tatsukawa, R., Bioconcentration and residue patterns of chlordane compounds in marine animals: Invertebrates, fish, mammals and seabirds, *Environ. Sci. Technol.*, 22, 792, 1988.
80. Stehly, G. R., Barron, M. G., and Hayton, W. L., Bioaccumulation and bioavailability in aquatic animals, in *Xenobiotic Metabolism in Fish*, Smith, D. J., Gingerich, W. H., and Beconi-Barker, M. G., Eds., Plenum Press, New York, 1999, Chap. 6.
81. Burkhard, L., Comparison of two models for predicting bioaccumulation of hydrophobic organics in a Great Lakes food web, *Environ. Toxicol. Chem.*, 17, 383, 1998.
82. Niimi, A. J., Biological and toxicological effects of environmental contaminants in fish and their eggs, *Can. J. Fish. Aquat. Sci.*, 40, 306, 1983.
83. Nichols, J. W., Jensen, K. M., Tietge, J. E., and Johnson, R. D., Physiologically based toxicokinetics model for maternal transfer of 2,3,7,8-tetrachlorodibenzo-*p*-dioxin in brook trout (*Salvelinus fontinalis*), *Environ. Toxicol. Chem.*, 17, 2422, 1998.
84. Johnston, T. A., Bodaly, R. A., Latif, M. A., Fudge, R. J. P., and Strange, N. E., Intra-and interpopulation variability in maternal transfer of mercury to eggs of walleye (*Stizostedion vitreum*), *Aquat. Toxicol.*, 52, 73, 2001.

CHAPTER 33

# Structure Activity Relationships for Predicting Ecological Effects of Chemicals

John D. Walker and T. Wayne Schultz

## CONTENTS



## 33.1 INTRODUCTION

Structure activity relationships (SARs) are comparisons or relationships between a chemical structure, chemical substructure or some physical or chemical property associated with that structure or substructure, and a biological (e.g., acute toxicity) or chemical (e.g., hydrolysis) activity.[1]

When the result is expressed quantitatively, the relationship is a quantitative structure activity relationship (QSAR).

Most SARs have been developed for predicting ecological effects of organic chemicals. However, there are some publications that discuss SARs for predicting ecological effects of polymers,[2] mixtures,[3–6] ionic organic chemicals,[7] metals,[8] and organometals.[9–12] The majority of SARs for predicting ecological effects have been developed for fish (fathead minnow, (*Pimephales promelas*) and guppy (*Poecilia reticulata*), aquatic invertebrates (e.g., *Tetrahymena pyriformis* and *Daphnia* spp.), and bacteria (e.g., *Vibrio fischeri*, *Escherichia coli* and *Pseudomonas* spp.). However, a few publications describe SARs for predicting ecological effects of chemicals to terrestrial invertebrates and insects.[13–15]

For SARs that predict ecological effects, this chapter provides some examples, developmental approaches, universal principles, applications, and recommendations for new QSARs. For additional information on QSARs for predicting ecological effects and their applications, readers are referred to other chapters, reviews,[16–20] and several books.[21–33] For readers interested in information on QSARs related to human health or human–health–environment interactions, several books are also available.[33–48]

## 33.2 EXAMPLES OF STRUCTURE ACTIVITY RELATIONSHIPS

A few examples of SARs are provided here to illustrate how they can be used to predict ecological effects. The first example illustrates how SARs can be used to distinguish clusters of benzaldehydes based on type, number, and position of substituents on a benzaldehyde substructure (Figure 33.1).[48,49] The second example illustrates how QSARs account for the effects of hydrophobicity on growth inhibition manifested by alcohols, amines, and aminoalkanols.[50–52] The third example demonstrates how molecular structure within a congeneric series of carboxylic acids affects teratogenicity.[53–54]

### 33.2.1 Distinguishing Clusters of Benzaldehydes

Fish $LC_{50}$ values for benzaldehydes were used to separate clusters of benzaldehydes into two-dimensional space (Figure 33.1). All of the ortho-hydroxy benzaldehydes were more toxic than benzaldehyde (in rectangle), para-hydroxy benzaldehydes, or alkoxy benzaldehydes, because of hydrogen bonding. The ortho-hydroxy benzaldehydes were equally toxic to hexoxybenzaldehyde, which was highly toxic to fish because of its high log octanol/water partition coefficient (log P or log $K_{ow}$).[48,49]

### 33.2.2 Effects of Hydrophobicity on Growth Inhibition Manifested by Aliphatic Saturated Alcohols, Primary Amines, and Aminoalkanols

The classic approach to developing SARs has been to use a congeneric series of chemicals. A congeneric series consists of chemicals with a common functional group or toxicophore but different sizes or shapes of alkyl groups. Thus, the reactivity imparted by the toxicophore is constant, and potency varies with hydrophobicity imparted by the alkyl group. The toxicophores in the examples illustrated here are alcohol, amine, and aminoalkanol functional groups. Examples of QSARs for the congeneric series that consists of alcohols, amines, and aminoalkanols developed with toxicity data for the freshwater ciliate *Tetrahymena pyriformis*[50] can be seen in Equations 33.1–33.3.[51]

Using population growth inhibition of *T. pyriformis* (log $IGC50^{-1}$) and hydrophobicity (log $K_{ow}$) data, the following QSAR was developed for aliphatic alcohols:

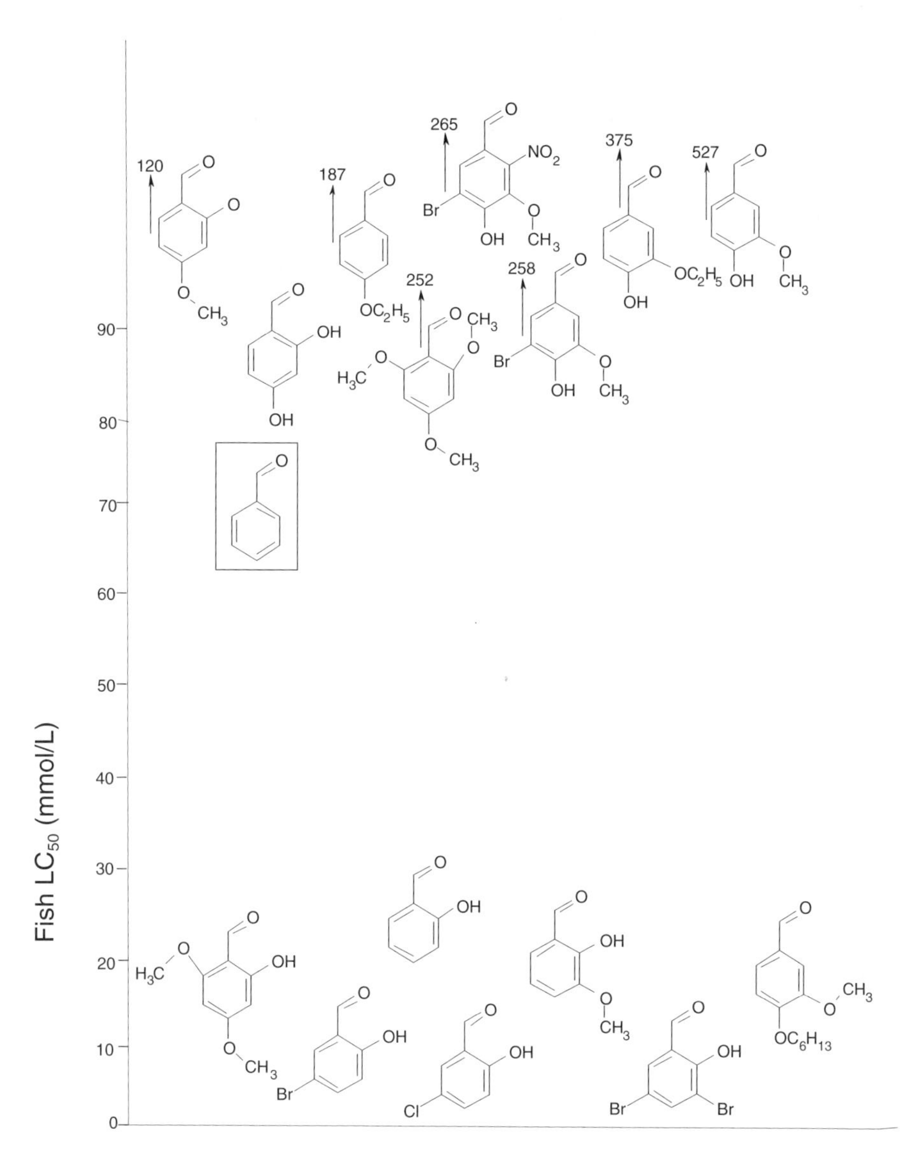


**Figure 33.1** Fish $LC_{50}$ values of hydroxy, alkoxy, and hydroxy-halogen benzaldehydes, compared to benzaldehyde (in rectangle); arrows are used to indicate fish $LC_{50}$ values of alkoxy benzaldehydes that exceed 90 mmol/L.

$$\log (IGC50^{-1}) = 0.80 (\log K_{ow}) - 2.04 \tag{33.1}$$

where $IGC_{50}^{-1}$ is the inverse of concentration (m*M*) resulting in 50% population growth inhibition as compared to controls, and $K_{ow}$ is the 1-octanol/water partition coefficient.[51]

Since the neural narcotic toxicity of saturated alcohols depends only on the uptake of the chemical, the modeling of this mechanism yields a simple hydrophobicity-dependent QSAR. Theoretically, the slope of these models is unity. Deviations from one reflect alterations in bioavailability. In the case of the *T. pyriformis* growth assay, this is the result of two phenomena: volatilization of low-molecular-weight compounds and the binding of high-molecular-weight compounds to hydrophobic components in the medium. To illustrate further, Sinks et al.[52] developed a QSAR for aliphatic primary amines:

$$\log (IGC_{50}^{-1}) = 0.72 (\log K_{ow}) - 1.32 \quad (33.2)$$

A comparison of Equations 33.1 and 33.2 showed that the difference in slopes is minimal, and the difference in intercept is greater than 0.7 of a log unit, with each primary amine being more toxic than its corresponding alcohol. Because of their higher basicity (pKa > 10), the aliphatic amines are almost completely ionized at physiological pH. Thus, it is implied that the protonated form of the chemical is the active toxicant. The above implication may account for the difference in intercept. Regression analysis of hydrophobicity vs. toxicity for the aminoalkanols yielded the QSAR

$$\log (IGC50^{-1}) = 0.17 (\log K_{ow}) - 0.70 \quad (33.3)$$

A comparison of Equation 33.3 with both Equations 33.1 and 33.2 showed both a difference in slopes and a difference in intercepts because each aminoalcohol is more toxic than either corresponding alcohol or primary amine.[52]

### 33.2.3 Effects of Hydrophobicity and Molecular Structure on the Developmental Toxicities Manifested by Carboxylic Acids

Developmental toxicity effects of chemicals, especially teratogenic ones, are among the most difficult to determine experimentally. However, test systems such as the Frog Embryo Teratogenesis Assay–*Xenopus* (FETAX)[55,56] have provided a means of collecting quantitative data for SAR analyses. Since valproic acid (2-propylpentanoic acid) was implicated as a human teratogen,[57] the developmental toxicity and teratogenic effects of carboxylic acids have received noteworthy consideration. Dawson et al.[54] developed a QSAR for acute embryo toxicity for a set of carboxylic acids. The resulting QSAR was:

$$\log (LC_{50}^{-1}) = 0.70 (\log K_{ow}) - 0.39 (pKa) - 0.51 \quad (33.4)$$

where mortality was related to orthogonal descriptors of both hydrophobicity and ionization.

The principal malformation induced in developing frog embryos exposed to valproic acid was microcephaly, with other craniofacial defects also being observed.[54] Dawson and co-workers also applied QSAR techniques to analyze a data set of 45 carboxylic acids with the dependent variable being the 50% teratogenic effect concentration ($EC_{50}^{-1}$). The result was the QSAR:

$$\log (EC50^{-1}) = 0.66 (\log K_{ow}) - 1.47 \quad (33.5)$$

The addition of the ionization constant as a second variable did not improve the fit of the QSAR for malformation.

Hazard index data for selected acyclic carboxylic acids evaluated in FETAX are presented in Table 33.1. Examination of these data reveal a number of molecular structural features associated with a high hazard index. Specifically, these include the presence of (1) a free carboxyl group, (2) an H-saturated C-chain with the optimal length of five C-atoms, and (3) branching of the C-chain, especially in the 2-position.

## 33.3 DEVELOPING QSARS

The development of a QSAR starts with a description of the domain. This comprises the identification of a group of chemicals that behave in the same way with respect to the endpoint to be estimated (e.g., a common mode of action [MOA] for an effect QSAR). This group of chemicals

**Table 33.1 The Hazard Index of Selected Carboxylic Acids Evaluated with *Xenopus* Embryos**

| Compound | Hydrocarbon Moiety | Hazard Index |
|---|---|---|
| Pentane | CH3–(CH2)3–CH3 | not teratogenic |
| Acetic acid | CH3–C | 1.4 |
| Butyric acid | CH3(CH2)2–C | 8.4 |
| Pentanoic acid | CH3(CH2)3–C | 13.2 |
| Hexanoic acid | CH3(CH2)4–C | 12.4 |
| Octanoic acid | CH3(CH2)6–C | 4.5 |
| Decanoic acid | CH3(CH2)8–C | 3.2 |
| Lauric acid | CH3(CH2)10–C | 1.5 |
| 2-Pentenoic acid | CH3–CH2–CH=CH–C | 6.8 |
| 4-Pentenoic acid | CH2–CH–CH2–CH2–C | 5.8 |
| 4-Pentynoic acid | CH2–C–CH2–CH2–C | 7.2 |
| 2-Methyl pentanoic acid | CH3–CH2–CH2–CH(CH3)–C | 13.8 |
| 3-Methyl pentanoic acid | CH3–CH2–CH(CH3)–CH2–C | 8.9 |
| 2-Propyl pentanoic acid | CH3–CH2–CH2–CH(CH2–CH2–CH3)–C | 27.9 |
| 2-Ethyl hexanoic acid | CH3–CH2–CH2–CH2–CH(CH2–CH3)–C | 13.6 |

*Note:* Hazard Index = $LC_{50}/EC_{50}$.

is called the *training set*. The training set is described by identifying a common sub-structure, the class of chemicals to which the training set belongs, structural rules, or MOA information.[58]

The selection of a *training set* is dependent on the skills of the QSAR expert and requires a full range of multidisciplinary scientific knowledge. The domain limits the QSAR's use to the endpoint to be described and the group of chemicals for which it is valid. Furthermore, it limits the valid range of the descriptor within the boundaries of the lowest and highest values of those in the training set. It is also essential that for all the chemicals selected, data on the descriptor (e.g., log P or log $K_{ow}$, water solubility, etc.) are available.

When a QSAR is developed, it must be tested or verified. For this, a further set of substances, with known measured endpoints, that fulfill the requirements of the QSAR domain should have been identified. This group of substances is called the *test set*.

There are three approaches that have been used in the past to develop QSARs. The approach that is used depends on the nature of the scientific data on which the QSAR will be based.

### 33.3.1 Statistical Approach

In many cases, the available test results are correlated with a known descriptor or descriptors by statistical evaluations. This statistical approach is used when the understanding of how the descriptor describes the endpoint is either unknown or poorly understood. In aquatic toxicology, this approach appears to have some application, especially when the independent variables are limited to universal descriptors of hydrophobicity and stereoelectronic interactions.[59,60]

### 33.3.2 Chemical-Class-Based Approach

The chemical-class-based approach is based predominantly on the assumption that compounds from the same two-dimensional chemical class should behave in a toxicologically similar manner. Consequently, homologous series of chemicals have been used to develop structure–toxicity relationships. For each homologous series of chemicals, it was assumed that toxic effects were imparted by common structural components used in two-dimensional chemical-class assignments. This approach has been most commonly applied to neutral nonionic discrete organic chemicals that have no reactive functional groups. The approach has been used by the Toxics Substances Control Act (TSCA) Interagency Testing Committee (ITC) to organize chemicals into structural classes that may be associated with the potential to cause ecological effects to algae, plants, invertebrates, and vertebrates or the potential to cause specific health effects such as acute, carcinogenic, mutagenic,

**Table 33.2 Chemical Classes for Which There are Published QSARs to Predict Ecological Effects**

| Chemical Class | References |
|---|---|
| Acrylates | 68 |
| Alcohols | 69 |
| Aldehydes | 49, 70, 71 |
| Amines | 72–75 |
| Anilines | 76, 77 |
| Benzenes | 60, 78 |
| Chlorophenols | 79 |
| Epoxides | 80, 81 |
| Linear benzene sulfonates | 82 |
| Naphthalenes | 83, 84 |
| Nitrobenzenes | 85, 86 |
| Nitrogenous heterocyclic compounds | 87–89 |
| Organophosphates | 14, 90 |
| Phenols | 77, 91–93 |
| Polyhalogenated aromatics | 94 |

neurotoxic, or reproductive effects.[61–65] The ITC uses this approach to facilitate review of structurally related chemicals and to examine SARs, but not to develop QSARs. A similar approach has been used by the National Toxicology Program to identify "structural alerts" or structural fragments that are associated with a chemical's potential to cause cancer in rodents.[66,67] Chemical classes for which there are published QSARs to predict ecological effects are listed in Table 33.2.

The problem with the chemical-class-based approach is defining a chemical class. This problem is illustrated in Section 33.2.2, where the aquatic toxicity of aminoalkanol (see Equation 33.3) is not predicted well by either the QSAR for amines (see Equation 33.2) or alcohols (see Equation 33.1). Thus, the class-based approach becomes very conservative with models being constructed for a series of toxicants that have the same single functional group or toxicophore and alkyl moieties of varied size. In such series, reactivity is the constant attributable to the functional group, while the methylene groups do not alter reactivity but increase hydrophobicity (i.e., log $K_{ow}$ by approximately 1/2 log unit per group). The net result is a variety of QSARs that predict the toxicity of basic toxicants of varying alkyl chain length but have little application to more complex molecules.

### 33.3.3 Mode-of-Action-Based Approach

The MOA-based approach is the most reliable approach to developing QSARs. It is used when there is clear evidence for understanding the MOA for a group of chemicals and molecular descriptors that impact a specific biological endpoint. Recently, the ITC has combined the chemical-class-based approach with the MOA-based approach.[95]

Research conducted over the past several years addressing the joint toxic action of chemicals and toxicological responses observed in fish have challenged the notion that QSARs can be reliable based on typical chemical classification schemes.[3,96–101] Furthermore, an evaluation of the fathead minnow database by Russom et al.[102] illustrates that toxicological classifications based on typically-used chemical classes can be problematic. Many chemical classes are associated with a baseline narcosis QSAR[103–105] (e.g., ethers, alcohols, ketones, esters, and benzenes) including chemicals acting by a baseline narcosis MOA as well as chemicals acting through an electrophilic-based mode of action (Figure 33.2). Conversely, chemical classes not usually identified as acting by a baseline narcosis mode of action (e.g., the phenols) include chemicals determined to act either through baseline narcosis, polar narcosis, oxidative phosphorylation uncoupling, or electrophilic-based modes of action. Subsequently, QSAR developments and applications have evolved from a chemical-class perspective to one that is more consistent with assumptions regarding MOA.[106,107] The use of MOA-based QSARs, therefore, requires an appreciation of toxic mechanisms, critical structural characteristics, and chemical properties.[19]

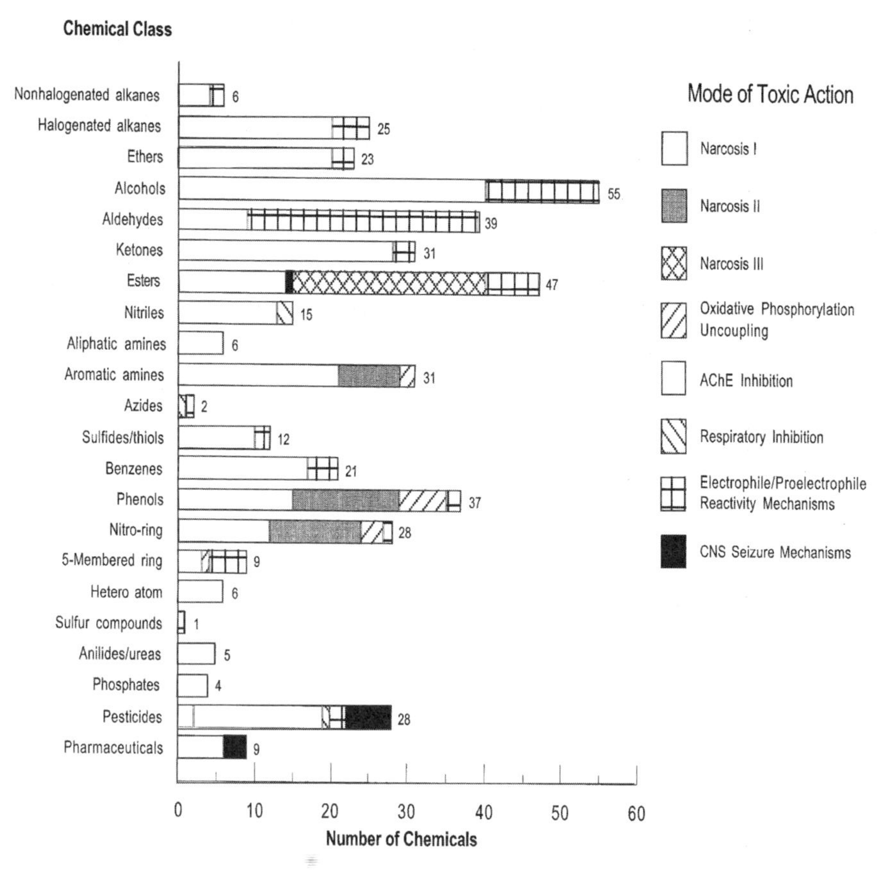


**Figure 33.2** Observed modes of toxic action associated with fathead minnow 96-hour $LC_{50}$ values as a function of chemical classes. (From Russon, C.L. et al., *Environ. Toxicol. Chem.,* 16, 948, 1997. With permission.)

There are at least three advantages of MOA-based QSARs. One advantage is that it is easier to assess a new substance and identify whether it behaves in the same way as those chemicals in the original domain of the QSAR. A second advantage is that the application of MOA-based QSARs is not restricted to congeneric series of chemicals with similar two-dimensional structures; MOA-based QSARs can be applied to chemicals with different two-dimensional structures but identical MOAs. A third advantage is that with MOA-based QSARs, it may be possible to extrapolate QSARs from one species to another.[108]

The problem with the MOA-based approach is determining the specific MOA for a given toxicant. As noted by Bradbury,[109] the assignment of a mode or mechanism of toxic action to a specific toxicant can be a daunting task. This problem is illustrated in Section 33.2.1 and shows how the aquatic toxicity of benzaldehydes varies markedly based on molecular structure. This variability in potency would not have been predicted *a priori*. The reason for such difficulties, as noted by Nendza and Russom,[110] may rest on the fact that the MOA of a given toxicant in a given organism depends upon a plethora of molecular interactions, the vast majority of which are not well understood.

## 33.4 UNIVERSAL PRINCIPLES FOR QSARS TO PREDICT AQUATIC TOXICITY

The establishment of universal principles for predicting aquatic toxicity is far more complicated than for other endpoints. This is in large part due to the availability of four large data sets that were created specifically for predicting adverse effects to aquatic organisms. These databases contain toxicity results for aquatic effects in the fathead minnow,[102] guppy,[106] *pyriformis*,[50,60,111,112] and *V. fischeri* (formerly known as *Photobacterium phosphoreum*).[113,114] In each case, except for *V. fischeri*, the data originated with a single research group using standardized protocols. Moreover, the data derived for a single endpoint (e.g., $EC_{50}$ or $LC_{50}$) includes a wide range of molecular structures representing a variety of MOAs.

While modeling approaches used in SARs can be correlative or pattern recognition, the most commonly used approach in aquatic toxicity is the correlative approach, which employs regression analysis. Regression analyses are quantitative in nature and fit the continuous character of the $EC_{50}$ or $LC_{50}$ data of the four data sets most often used in SAR development in aquatic toxicology. Regression-type analyses relating the similarity or dissimilarity of one or a few molecular properties with a corresponding similarity or dissimilarity in toxicity has its roots in the work of Hansch and Fujita.[115] This methodology is typified by the generic regression model:

$$\log (C^{-1}) = a\,(\text{hydrophobicity}) + b\,(\text{electronic}) + c\,(\text{steric}) \qquad (33.6)$$

where C, the toxicant concentration at which an endpoint (e.g., $LC_{50}$ or $EC_{50}$) is manifested, is related to a parameter quantifying hydrophobic, electronic, and steric properties.

Nowhere in the SAR arena is the Hansch approach more often the approach of choice than in aquatic toxicity. Moreover, an examination of the literature of the past quarter century reveals that the most useful structure–toxicity relationships for aquatic toxicity are those derived using regression analysis, continuous data for both toxic potency and molecular properties, and three-dimensional, whole-molecule-based descriptors.

Absorption, transport, and distribution of the toxicant are thought of as the toxicokinetic phase. These processes are governed largely by a toxicant's ability to partition between lipid and aqueous phases and dissolution in body fluids. Partitioning and dissolution vary with hydrophobicity and water solubility, respectively. Both of these processes are also influenced by the pH of the system and its components. Passive diffusion is the primary means of transport of most xenobiotics within aquatic systems. While the rate of transport depends on a number of properties, such as hydrophobicity, polarity, degree of ionization, and molecular shape and size, it is often assumed in aquatic toxicity modeling that partitioning of a chemical between aquatic organisms and water can be related to hydrophobicity, with $K_{ow}$ being the most commonly used measure of this hydrophobicity.

The typical endpoints used in predictive modeling in aquatic toxicity are not strictly dependent on highly structured interactions; thus, accurate QSARs based on one or a few descriptors are easily constructed. This lack of structured interaction allows for a relaxation of the constraints on molecular structure imposed by modeling a strictly congeneric series without the trepidation of stumbling upon unexpected disruptions in toxic potency. This is indeed the case with narcosis where, in the case of many chemicals, the most important consideration is the absorption characteristics of the toxicant.[73,75]

In aquatic toxicology, the physiological and biochemical characteristics of an organism on structure-toxicity relationships are often limited, especially for nonvertebrate species. However, biotic factors control the rate at which toxicants undergo metabolism and thus control the rate at which a given toxicant reaches the site of action. Metabolism itself is controlled by the energetics of covalent bond cleavage as well as the electronic and steric factors governing interaction with the corresponding enzymes. Steric factors (e.g., size and shape) can shield a bond from metabolic attack. Steric factors also relate to ability of a toxicant to pass into or through biomembranes. These factors are important in the global modeling of aquatic toxicity. An example of this is the proelectrophile mechanism of toxic action.[116]

In aquatic toxicity, interaction with the site of action (i.e., the toxicodynamic phase) is considered to be under electronic or steric control and is typically an electrophilic covalent event.[117] More specifically, the majority of the electrophilic interactions associated with aquatic toxicity are controlled by molecular orbital events.[118] Therefore, quantum chemical descriptors, especially frontier orbital energies, are important global parameters for modeling the toxicodynamic phase of aquatic toxicity.

Utilizing quantitative descriptors for passive transport (i.e., uptake by the biophase) and reactivity (i.e., interaction) with the molecular site of action, especially when coupled with multiple linear regression, is a straightforward approach. This tandem of descriptor limitations and methodology typically leads to a model for aquatic toxicity that is easy to understand and often provides a mechanistic basis. In other words, the overall hydrophobic properties of a toxicant (responsible for its toxicokinetics) as well as its electronic, hydrophobic, and steric properties (determine its toxicodynamics) can be captured, for a wide variety of aquatic toxicants, in a very small number of global descriptors.

The seminal work of Veith and Mekenyan[119] demonstrated the applicability of the toxicokinetic–toxicodynamic approach to predictive modeling of aquatic toxicity. They modeled fish lethality as a plane in a three-dimensional surface by using the log $K_{ow}$ to model the biouptake and the quantum chemical property; also, they used the energy of the lowest unoccupied molecular orbital ($E_{LUMO}$) to model electro(nucleo)philic interaction. The resulting "toxic space" was a continuum, with toxicants showing no evidence of reactive toxicity being distributed along the toxicity-log-$K_{ow}$ facet of the surface. At the same time, toxicants of moderate electrophilicity were distributed in the central area of the plane. Toxicants acting as strong electrophiles or those being respiratory uncouplers were distributed nearer the potency-$E_{LUMO}$ facet of the space.

More recently, the investigations of Karabunarliev et al.[59] and Schultz[60] modeled aquatic toxicity data for *P. promelas* and *T. pyriformis*, respectively. Data, while confined to substituted benzenes, represented the neutral narcosis, polar narcosis, respiratory uncoupling, and weak and strong electrophilic MOAs. The fish toxicity model:

$$\log (LC_{50}^{-1}) = 0.62 (\log K_{ow}) + 9.17 (S_{max}) - 3.21 \qquad (33.7)$$

where hydrophobicity was quantified by log $K_{ow}$ and electrophilic reactivity was quantified by the molecular orbital parameter. Maximum superdelocalizability ($S_{max}$) was similar in terms of slopes, intercept, and statistical fit to the ciliate toxicity model:

$$\log (IGC_{50}^{-1}) = 0.50 (\log K_{ow}) + 9.85 (S_{max}) - 3.47 \qquad (33.8)$$

While on the surface Equation 33.8 may appear to be global in nature, in fact, it too is limited in applicability. Equation 33.8 covers a substantial chemical domain of the aquatic toxicity arena.[120] However, it is not applicable to all quinones,[121] photoactivated toxicants,[122] carboxylic acids[123,124] or many aliphatic chemicals.[112] Thus, in modeling aquatic toxicity, it is necessary to cover the chemical universe with smaller chemical domains. This requires a separate QSAR for each chemical domain and a means of defining the limits of each domain.

One approach to defining the chemical domains of aquatic toxicity is the development of rules-based expert systems. Rules-based systems build associations and generalizations from relatively small groups of chemicals. Typically, the rules-based expert systems are more limited in their application than regression-based approaches. However, they offer greater interpretability. Thus, combining both approaches may provide a means of improving predictability. Such rules-based systems dealing with aquatic toxicology have recently been described.[125]

## 33.5 USES OF QSARS FOR PREDICTING AQUATIC TOXICITY

As noted above, most QSARs for predicting ecological effects have been developed for aquatic organisms. Hence, it is more feasible at this time to describe uses of QSARs for predicting aquatic toxicity than for predicting ecological effects. QSARs for predicting aquatic toxicity have been used by industry and government organizations.

QSARs for predicting aquatic toxicity have been used by industry to promote pollution prevention.[30] QSARs for pollution prevention are driven by simple economics. It is simple and inexpensive to use QSARs at any stage of a chemical's presence in the environment. However, it is more cost-effective to use QSARs in the early stages of a chemical's life cycle to predict chemical toxicity, bioconcentration, or persistence potential than after the chemical is incorporated into a marketplace product. QSARs for pollution prevention are used by chemical manufacturers and the government organizations that regulate those manufacturers.

The use of SARs and QSARs by other U.S. government organizations and by global government organizations has been described recently.[126,127] QSARs for predicting aquatic toxicity have been used by government organizations to screen chemicals for potential toxicity.[30] Government organizations that have used QSARs for predicting aquatic toxicity include the ITC, the U.S. Environmental Protection Agency (U.S. EPA), the European Union (EU), and Environment Canada. The use of QSARs for predicting aquatic toxicity by these government organizations is briefly described below.

### 33.5.1 Toxic Substances Control Act Interagency Testing Committee

The ITC was created to (1) screen and review "existing" chemicals and identify chemicals or classes of chemicals with suspicions of toxicity, but few if any data, (2) add those chemicals or classes of chemicals to the TSCA section 4(e) *Priority Testing List,* and (3) recommend them for testing to the U.S. EPA Administrator. To ensure that the ITC discharged its mandate effectively, the U.S. Congress itemized eight factors in section 4(e) of TSCA that the ITC must consider when recommending a chemical for testing to the Administrator of the U.S. EPA.[63] In fulfilling the ITC's mandate, the members of the 15 U.S. Government organizations represented on the ITC have been careful to meet their obligations under TSCA and consider all relevant factors when making a recommendation. With the fifth factor in this mandatory list — "the extent to which the substance or mixture is closely related to a chemical substance or mixture which is known to present an unreasonable risk of injury to health or the environment" — the U.S. Congress clearly demonstrated its desire that SARs and QSARs be considered during Committee deliberations. With the seventh factor in this mandatory list — "the extent to which testing of the substance or mixture may result in the development of data upon which the effects of the substance or mixture on health or the environment can reasonably be determined or predicted" — the U.S. Congress clearly demonstrated its desire that data be developed to predict effects on human health or the environment (i.e., environment–human-health interactions through development of SARs and QSARs).[95]

The ITC used the substructure-based computerized chemical selection expert system (SuCCSES) and SARs to identify numerous chemicals and structural classes of chemicals that are likely to cause aquatic toxicity.[61–65] A few examples of individual chemicals that the ITC has identified as likely to cause aquatic toxicity include acrylamide, biphenyl, bisphenol A, hexachloro-cyclopentadiene, and tetrabromobisphenol A. A few examples of these structural classes that the ITC has identified as likely to cause aquatic toxicity include aldehydes, alkylphenols, alkylphenol ethoxylates, alkyl phthalate esters, alkyl tins, anilines, antimony compounds, aryl phosphates, benzidene dyes, bromo diphenyl ethers, chlorinated benzenes, chlorinated paraffins, cresols, fluoroalkenes, phenols, siloxanes, toluidine dyes, and xylenes. Data developed as a result of the U.S. EPA's implementation of the ITC's testing recommendations have verified the ITC's predictions.[128–130]

### 33.5.2 U.S. Environmental Protection Agency

The U.S. EPA Office of Pollution Prevention and Toxics uses SARs and QSARs in combination with expert opinions for predicting the aquatic toxicity of new chemicals (chemicals submitted to the U.S. EPA under section 5 of TSCA as Pre-Manufacturing Notices or PMNs). The U.S. EPA has

90 days to predict the aquatic toxicity of PMNs, the annual number of which has ranged from a low of 8 in 1979 to a high of 2991 in 1988. The U.S. EPA developed the ecological effects structure activity relationships (ECOSAR) tool for predicting the aquatic toxicity. (http://www.epa.gov/oppt/exposure/docs).

### 33.5.3 European Union

The EU, through the European Economic Commission (EEC), enacted legislative changes that provided an opportunity to use QSARs for screening and priority setting.[139] These changes in EEC legislation accelerated the need for practical applications of QSARs including the use of QSARs for predicting aquatic toxicity.[140,141] EEC Directive 93/67/EEC provided an opportunity to evaluate existing data and to develop estimates or predictions to fill gaps of missing data during the evaluation of existing chemicals.[58] The EEC Technical Guidance Document provided supporting documentation for Commission Directive 93/67/EEC and guidance on using QSARs to estimate fish 96-hour $LC_{50}$ and daphnid 48-hour $EC_{50}$ values to predict the aquatic toxicity of nonpolar discrete organic chemicals.[142,143]

### 33.5.4 Environment Canada

The recently revised Canadian Environmental Protection Act (CEPA) requires Environment Canada to "categorize" and then "screen" substances listed on Canada's Domestic Substances List (DSL) to determine whether they are "toxic" as defined under CEPA. Since experimental aquatic toxicity data needed to "categorize" and then "screen" substances are only available for a small number of the substances on the DSL, QSARs and other predictive methods are being considered to provide aquatic toxicity estimates needed to "categorize" and then "screen" DSL substances.[144]

## 33.6 RECOMMENDATIONS FOR NEW QSARS TO PREDICT ECOLOGICAL EFFECTS

The major limitation in the further development and validation of QSARs to predict ecological effects is the lack of data of sufficient depth and breadth on biological endpoints of significance. Germane to this problem is limiting the number of compounds known to exhibit the neutral narcosis mechanism of toxic action. This mechanism of toxic action can be modeled well for a wide variety of endpoints and test systems.

### 33.6.1 Organic Chemicals

Since most QSARs for predicting ecological effects have been developed for aquatic organisms, it is recommended that additional QSARs be developed for predicting effects of chemicals on terrestrial and sediment-dwelling organisms. QSARs are needed for such organisms because they account for the bioavailability and effects of chemicals in the soil and sediment environments. As such, predicting the hazard or risk of chemicals to terrestrial and sediment-dwelling organisms and their environments may continue to be resource-intensive until such QSARs are available.

Most QSARs for predicting ecological effects to aquatic organisms have been developed for freshwater aquatic organisms (except for those developed for *V. fischeri*).[114] While some interspecies modeling of toxicity has occurred for aquatic organisms,[145] clearly there is a need for QSARs to predict the effects of chemicals on estuarine and marine organisms to account for the speciation of ionizable chemicals in brackish and salt waters.

There are no QSARs for predicting effects of chemicals on terrestrial wildlife, i.e., birds and wild mammals. There is a critical need for these QSARs, especially given the high exposure potential of terrestrial wildlife to pesticide active and other formulation ingredients that are intentionally dispersed to the environment and the potential exposure to industrial chemicals with potential for long-range transport, persistence, bioconcentration, and toxicity.

### 33.6.2 Organo-Metallic Chemicals

Very few QSAR models have been designed to predict the toxicity of organo-metallic chemicals.[9–12] All of these models were developed to predict the toxicity of organo-tin chemicals and demonstrated that tri-alkyl tin was more toxic than mono-, di- or tetra-alkyl tin. More QSAR models are needed to predict the toxicity of organo-metallic chemicals.

## REFERENCES


1. Walker, J. D., Jaworska, J., Schultz, T. W., Comber, M. H. I., and Dearden, J. C., Guidelines for developing and using quantitative structure activity relationships (QSARs), *Environ. Toxicol. Chem.*, in press, 2003.
2. Boethling, R. S. and Nabholz, J. V., Environmental assessment of polymers under TSCA, in *Ecological Assessment of Polymers*, Hamilton, J. D. and Sutcliffe, R., Eds., Van Nostrand Reinhold, New York, 1997, 187.
3. Broderius, S. J., Kahl, M. D., and Hoglund, M. D., Use of joint toxic response to define the primary mode of toxic action for diverse industrial organic chemicals, *Environ. Toxicol. Chem.*, 14, 1591, 1995.
4. Broderius, S. J. and Kahl, M., Acute toxicity of organic chemical mixtures to the fathead minnow, *Aquat. Toxicol.*, 6, 307, 1985.
5. Hermens, J., Könemann, J. H., Leeuwangh, P., and Musch, A., Quantitative structure-activity relationships in aquatic toxicity studies of chemicals and complex mixtures of chemicals, *Environ. Toxicol. Chem.*, 4, 273, 1985.
6. Schüürman, G., Altenburger, R., and Nendza., M., QSARs for predicting toxicity of organic chemical mixtures, *Environ. Toxicol. Chem.*, in press, 2003.
7. Roberts, D. W., QSAR issues in toxicity of surfactants, *Total Environ.*, 109(110), 557, 1991.
8. Walker, J. D., Enache, M., and Dearden, J. C., QSARs for predicting toxicity of metal ions, *Environ. Toxicol. Chem.*, in press, 2003.
9. Laughlin, R. B., Quantitative structure-activity studies of di- and tri-organotin compounds, in *QSAR in Environmental Toxicology — II*, Kaiser, K. L. E., Ed., D. Reidel Publishing, Dordrecht, The Netherlands, 1986, 189.
10. Laughlin, R. B., French, W., Johannesen, R. B., Guard, H. E., and Brinkman, F. E., Predicting toxicity using computed molecular topologies: The example of triorganotin compounds, *Chemosphere*, 13, 575, 1984.
11. Laughlin, R. B., Johannesen, R. B., French, W., Guard, H. E., and Brinkman, F. E., Structure-activity relationships for organotin compounds, *Environ. Toxicol. Chem.*, 4, 343, 1985.
12. Schüürmann, G. and Röderer, G., Acute toxicity of organotin compounds and its predictability by quantitative structure-activity relationships (QSARs), in *Heavy Metals in the Hydrological Cycle*, Astruc, M. and Lester, J. N., Eds., Selper Ltd., London, 1988, 433.
13. Van Gestel, C. A. M., Ma, W. C., and Smit, E. E., Development of QSARs in terrestrial ecotoxicology: Earthworm toxicity and soil sorption of chlorophenols, chlorobenzenes and dichloroaniline, in *QSAR in Environmental Toxicology — IV*, Hermens, J. L. M. and Opperhuizen, A., Eds., Elsevier, The Netherlands, 1991, 589.
14. Vighi, M., Garlanda, M. M., and Calamari, D., QSARs for toxicity of organophosphorous pesticides to *Daphnia* and honeybees, in *QSAR in Environmental Toxicology — IV.*, Hermens, J. L. M. and Opperhuizen, A., Eds., Elsevier, The Netherlands, 1991, 605.
15. Hirashima, A., Shinkai, K., Kuwano, E., Taniguchi, E., and Eto, M., Quantitative structure-activity studies of octopaminergic agonists and antagonists against *Locusta migratoria* using similarity indexes, *SAR QSAR Environ. Res.*, 11, 45, 2000.
16. Veith, G. D., DeFoe, D., and Knuth, M., Structure-activity relationships for screening organic chemicals for potential ecotoxicity effects, *Drug Metab. Rev.*, 15, 1295, 1984.

17. Lipnick, R. L., Structure-activity relationships, in *Fundamentals of Aquatic Toxicology*, 2nd ed., Rand, G. R., Ed., Taylor & Francis, London, 1995, 609.
18. Schüürmann, G., Ecotoxic modes of action of chemical substances in *Ecotoxicology*, Schüürmann G. and Markert, B., Eds., John Wiley and Spektrum Akademischer Verlag, New York, 1998, 665.
19. Bradbury, S., Russom, C., Ankley, G., Schultz, T. W., and Walker, J. D., QSARs for predicting ecological effects of organic chemicals, *Environ. Toxicol. Chem.*, in press, 2003.
20. Comber, M. H. I., Watts, C., Hermens, J., and Walker, J. D., QSARs for predicting potential ecological risks of organic chemicals, *Environ. Toxicol. Chem.*, in press, 2003.
21. Kaiser, K. L. E., Ed., *QSAR in Environmental Toxicology*, D. Reidel Publishing Co., Dordrecht, The Netherlands, 1984, 406.
22. Kaiser, K. L. E., Ed., *QSAR in Environmental Toxicology — II*, D. Reidel Publishing Co., Dordrecht, The Netherlands, 1986, 465.
23. Turner, J. E., England, M. W., Schultz, T. W., and Kwaak, N. J., Eds., *3rd Int. Workshop on Quantitative Structure-Activity Relationships (QSAR) in Environmental Toxicology*, CONF-880520–9DE88013180, NTIS, Springfield, VA, 1988, 228.
24. Karcher, W. and Devillers, J., Eds., *Practical Applications of Quantitative Structure-Activity Relationships (QSAR) in Environmental Chemistry and Toxicology*, Kluwer Academic Publishers, Dordrecht, The Netherlands, 1990.
25. Hermens, J. L. M. and Opperhuizen, A., Eds., *QSAR in Environmental Toxicology — IV*, Elsevier, The Netherlands, 1991.
26. Hansch, C. and Leo, A., *Exploring QSAR: Fundamentals and Applications in Chemistry and Biology*, American Chemical Society, Washington, D.C., 1995.
27. Hansch, C., Leo, A., and Hoekman., D., *Exploring QSAR: Hydropobic, Electronic and Steric Constants*, American Chemical Society, Washington, D.C., 1995.
28. Chen, F. and Schüürmann, G., Eds., *Quantitative Structure-Activity Relationships (QSAR) in Environmental Sciences — VII*, SETAC Press, Pensacola, FL, 1997.
29. Nendza, M., *Structure-Activity Relationships in Environmental Sciences*, Chapman & Hall, London, 1998.
30. Walker, J. D., Ed., *QSARs for Pollution Prevention, Toxicity Screening, and Risk Assessment*, SETAC Press, Pensacola, FL, 2002.
31. Walker, J. D., Ed., *QSARs for Predicting Biological Activities Related to Endocrine Disruption and Environment–Human Health Interactions*, SETAC Press, Pensacola, FL., 2003.
32. Walker, J. D., Ed., *QSARs for Predicting Ecological Effects of Chemicals*, SETAC Press, Pensacola, FL, 2003.
33. Walker, J. D., Ed., *QSARs for Predicting Physical Properties, Bioconcentration Potential and Environmental Fate of Chemicals*, SETAC Press, Pensacola, FL, 2003.
34. Tichy, M., Ed., *Chemical Structure-Biological Activity Relationships: Quantitative Approaches*, Akademiai Kiado, Budapest, 1976.
35. Franke, R. and Oehms, P., Eds., *Quantitative Structure-Activity Analysis*, Akademie-Verlag, Berlin, 1978.
36. Darvas, F., Ed., *Chemical Structure-Biological Activity Relationships Quantitative Approaches*, Akademiai Kiado, Budapest, Hungary, 1980.
37. Dearden, J. C., Ed., *Quantitative Approaches to Drug Design*, Elsevier Science Publishers, Amsterdam, The Netherlands, 1983.
38. Seydel, J. K., Ed., *QSAR and Strategies in the Design of Bioactive Compounds*, VCH, Bad, 1984.
39. Tichý, M., Ed., *QSAR in Toxicology and Xenobiochemistry*, Elsevier Science Publishers, Amsterdam, The Netherlands, 1985.
40. Hadzi, D. and Jerman-Blazic, B., Eds., *QSAR in Drug Design and Toxicology*, Elsevier Science Publishers, Amsterdam, The Netherlands, 1987.
41. Fauchere, J. L., Ed., *QSAR: Quantitative Structure-Activity Relationships in Drug Design*, Alan R. Liss, Inc., New York, 1988, 291.
42. Silipo, C. and Vittoria, A., Eds., *QSAR: Rational Approaches to the Design of Bioactive Compounds*, Elsevier Science Publishers, Amsterdam, The Netherlands, 1991.
43. Wermuth, C. G., Ed., *Trends in QSAR and Molecular Modelling 92*, ESCOM, Leiden, 1993.

44. Sanz, F., Giraldo, J., and Manaut, F., *QSAR and Molecular Modelling: Concepts, Computational Tools and Biological Applications*, Prous Science Publishers, Barcelona, 1994.
45. Van de Waterbeemd, H., Testa, B., and Folkers, G., Eds., *Computer-Assisted Lead Finding and Optimization: Current Tools for Medicinal Chemistry*, Verlag Helvetica Chimica Acta, Basel, Italy, 1996.
46. Gundertofte, K. and Jørgensen, F. S., Eds., *Molecular Modeling and Prediction of Bioactivity*, Kluwer Academic/Plenum Publishers, New York, 2000.
47. Walker, J. D., Ed., *QSARs for Predicting Biological Activities Related to Endocrine Disruption and Environmental-Human Health Interactions*, SETAC Press, Pensacola, FL, 2003.
48. Walker, J. D. and Printup, H., Using the substructure-based computerized chemical selection expert system (SuCCSES) to analyze aldehydes, I: Developing structure activity relationships (SARs), in *QSARs for Predicting Ecological Effects of Chemicals*, Walker, J. D., Ed., SETAC Press, Pensacola, FL, 2003.
49. Walker, J. D., Printup, H., Dimitrov, S. D., Karabunarliev, S. H., and Mekenyan, O. G., Using the substructure-based computerized chemcial selection expert system (SuCCSES) to analyze aldehydes II: Developing quantitative structure activity relationships (QSARs) for predicting toxicity, in *QSARs for Predicting Ecological Effects of Chemicals*, Walker, J. D., Ed., SETAC Press, Pensacola, FL, 2003.
50. Schultz, T. W., TETRATOX: *Tetrahymena pyriformis* population growth impairment endpoint-A surrogate for fish lethality, *Toxicol. Methods*, 7, 289, 1997.
51. Schultz, T. W. and Tichy, M., Structure-toxicity relationships for unsaturated alcohols to *Tetrahymena pyriformis*: C5 and C6 analogs and primary propargylic alcohols, *Bull. Environ. Contam. Toxicol.*, 51, 681, 1993.
52. Sinks, G. D., Carver, T. A., and Schultz, T. W., Structure-toxicity relationships for aminoalcohols: A comparison with alkanols and alkanamines, *SAR QSAR Environ. Res.*, 9, 217, 1998.
53. Nau, H. and Hendricks, A. G., Valproic acid teratogenesis, ISI Atlas of Science: Institute for Scientific Information, Philadelphia, *Pharmacology*, 52, 1987.
54. Dawson, D. A., Schultz, T. W., and Hunter, R. S., Developmental toxicity of carboxylic acids to *Xenopus* embryos: A quantitative structure-activity relationship and computer-automated structure evaluation, *Carcinog. Mutag. Teratog.*, 16, 109, 1996.
55. Dumont, J. N., Schultz, T. W., Buchanan, M. V., and Kao, G. L., Frog embryo teratogenesis assay: *Xenopus* (FETAX) — a short-term assay applicable to complex environmental mixtures, in *Short-term Bioassays in the Analysis of Complex Environmental Mixtures III,* Waters, M. D., Sandhu, S.S., Lewtas, J., Claxton, L., Chernoff, N., and Nesnow, S., Eds., Penum Publishing Corp., New York., 1983, 393.
56. ASTM, *Standard Guide for Conducting the Frog Embryo Teratogenesis Assay: Xenopus (FETAX)*, American Society for Testing and Materials E1439–91, Philadelphia: ASTM Special Publications, 1991.
57. Robert, E. and Rossa, F., Valproate and birth defects, *Lancet*, 2, 1142, 1983.
58. EEC, Commission Regulation (EC) No 1488/94 of 28 June 1994, laying down the principles for the assessment of risks to man and the environment of existing substances in accordance with Council Regulation (EEC) No 793/93, *Off. J. Eur. Comm.*, L 161, 1994.
59. Karabunarliev, S. H., Mekenyan, O. G., Karcher, W., Russom, C. L., and Bradbury, S. P., Quantum-chemical descriptors for estimating the acute toxicity of electrophiles to the fathead minnow (Pimephales promelas): An analysis based on molecular mechanisms, *Quant. Struct.-Act. Relat.,* 15, 302, 1996.
60. Schultz, T. W., Structure-toxicity relationships for benzenes evaluated with *Tetrahymena pyriformis, Chem. Res. Toxicol.*, 12, 1262, 1999.
61. Walker, J. D. and Brink, R. H., New cost-effective, computerized approaches to selecting chemicals for priority testing consideration, in *Aquatic Toxicology and Environmental Fate: Eleventh Volume, ASTM STP 1007,* Suter, G. W., II and Lewis, M. A., Eds., ASTM, Philadelphia, 1989, 507.
62. Walker, J. D., Chemical selection by the TSCA Interagency Testing Committee: Use of computerized substructure searching to identify chemical groups for health effects, chemical fate and ecological effects testing, *Sci. Total Environ.*, 109/110, 691, 1991.
63. Walker, J. D., The TSCA Interagency Testing Committee, 1977 to 1992: Creation, structure, functions and contributions, in *Environmental Toxicology and Risk Assessment: Second Volume, ASTM STP 1216*, Gorsuch, J. W., Dwyer, F. J., Ingersoll, C. G., and La Pointe, T. W., Eds., ASTM, Philadelphia, 1993, 451.

64. Walker, J. D., The TSCA Interagency Testing Committee's approaches to screening and scoring chemicals and chemical groups: 1977–1983, in *Access and Use of Information Resources in Assessing Health Risks from Chemical Exposure,* Lu, P. Y., Ed., Oak Ridge National Laboratories, Oak Ridge, TN, 1993, 77.
65. Walker, J. D., Estimation methods used by the TSCA Interagency Testing Committee to prioritize chemicals for testing: Exposure and biological effects scoring and structure activity relationships, *Toxicol. Modeling*, 1, 123, 1995.
66. Tennant, R. J., Spalding, J., Stasiewicz, S., and Ashby, J., Prediction of the outcome of rodent carcinogenicity bioassays currently being conducted on 44 chemicals by the National Toxicology Program, *Mutagenesis*, 5, 3, 1990.
67. Ashby, J. and Tennant, R. W., Prediction of rodent carcinogenicity for 44 chemicals: Results, *Mutagenesis*, 9, 7, 1994.
68. Freidig, A. P., Verhaar, H. J. M., and Hermens, J. L. M., Quantitative structure-property relationships for the chemical reactivity of acrylates and methacrylates, *Environ. Toxicol. Chem.*, 18, 1133, 1999.
69. Mekenyan, O. G., Veith, G. D., Bradbury, S. P., and Russom, C. L., Structure-toxicity relationships for α,β-unsaturated alcohols in fish, *Quant. Struct.-Act. Rel.,* 12, 132, 1993.
70. Deneer, J. W., Seinen, W., and Hermens, J. M. L., The acute toxicity of aldehydes to the guppy, *Aquat. Toxicol.,* 12, 185, 1988.
71. Schultz, T. W., Bryant, S. E., and Lin, D. T., Structure-toxicity relationships for *Tetrahymena*: Aliphatic aldehydes, *Bull. Environ. Contam. Toxicol.,* 52, 279, 1994.
72. Dumont, J. N., Schultz, T. W., and Jones, R., Toxicity and teratogenicity of aromatic amines to *Xenopus laevis, Bull. Environ. Contam. Toxicol.,* 22, 159, 1979.
73. Davis, K. R., Schultz, T. W., and Dumont, J. N., Toxic and teratogenic effects of selected aromatic amines on embryos of the amphibian *Xenopus laevis, Arch. Environ. Contam. Toxicol.,* 10, 371, 1981.
74. Newsome, L. D., Johnson, D. E., Lipnick, R. L., Broderius, S. J., and Russom, C. L., A QSAR study of the toxicity of amines to the fathead minnow, in *QSAR in Environmental Toxicology — IV,* Hermens, J. and Opperhuizen, A., Eds., Elsevier, Amsterdam, NL, 1991, 537.
75. De Wolf, W., Yedema, E., Seinen, W., and Hermens, J. L. M., Toxicokinetics of aromatic amines in guppy, *Poecilia reticulata, Sci. Total Environ.*, 109/110, 383, 1991.
76. Hermens, J. L. M., Leeuwangh, P., and Musch, A., Quantitative structure-activity relationships and mixture toxicity studies of chloro- and alkylanilines at an acute lethal toxicity level to the guppy (*Poecilia reticulata*), *Ecotoxicol. Environ. Saf.*, 8, 388, 1984.
77. Damborksy, J. and Schultz, T. W., Comparison of the QSAR models for toxicity and biodegradability of anilines and phenols, *Chemosphere*, 34, 429, 1997.
78. Karabunarliev, S. H., Mekenyan, O. G., Karcher, W., Russom, C. L., and Bradbury, S. P., Quantum-chemical descriptors for estimating the acute toxicity of substituted benzenes to the guppy (*Poecilia reticulata*) and fathead minnow (*Pimephales promelas*), *Quant. Struct.-Act. Relat.,* 15, 311, 1996.
79. Van Gestel, C. A. M. and Ma, W.-C., Toxicity and bioaccumulation of chlorophenols in earthworm in relation to bioavailability in soil, *Ecotoxicol. Environ. Saf.*, 15, 289, 1988.
80. Deneer, J. W., Seinen, W., and Hermens, J. L. M., A quantitative structure-activity relationship for the acute toxicity of some epoxy compounds to the guppy, *Aquat. Toxicol.,* 13, 195, 1988.
81. Purdy, R., The utility of computed superdelocalizability for predicting the $LC_{50}$ values of epoxides to guppies, *Sci. Tot. Environ.*, 109/110, 553, 1991.
82. Roberts, D. W., Aquatic toxicity of linear benzene sulfonates (LAS) — A QSAR analysis, *Comun. J. Com. Esp. Deterg.*, 20, 35, 1989.
83. Schultz, T. W., Dumont, J. N., Sankey, F. D., and Schmoyer, R. L., Structure activity relationships of selected naphthalene derivatives, *Ecotoxicol. Environ. Saf.*, 7, 191, 1983.
84. Schultz, T. W. and Moulton, M. P., Structure-toxicity relationships of selected naphthalene derivatives. II. Principal components analysis, *Bull. Environ. Contam. Toxicol.*, 34, 1, 1985.
85. Deneer, J. W., Sinnige, T. L., Seinen, W., and Hermens, J. L. M., Quantitative structure-activity relationships for the toxicity and bioconcentration factor of nitrobenzene derivatives towards the guppy (*Poecilia reticulata*), *Aquat. Toxicol.*, 10, 115, 1987.
86. Deneer, J. W., van Leeuwen, C. J., Seinen, W., Maas-Diepeveen, J. L., and Hermens, J. L. M., QSAR study of the toxicity of nitrobenzene derivatives towards *Daphnia magna, Chlorella pyrenoidosa* and *Photobacterium phosphoreum, Aquat. Toxicol.*, 15, 83, 1989.

87. Schultz, T. W., Cajina-Quezada, M., and Dumont, J. N., Structure toxicity relationships of selected nitrogenous heterocyclic compounds, *Arch. Environ. Contam. Toxicol.*, 9, 591, 1980.
88. Schultz, T. W. and Cajina-Quezada, M., Structure toxicity relationships of selected nitrogenous heterocyclic compounds II. Dinitrogen molecules, *Arch. Environ. Contam. Toxicol.*, 11, 353, 1982.
89. Schultz, T. W., Kier, L. B., and Hall, L. H., Structure toxicity relationships of selected nitrogenous heterocyclic compounds III. Relations using molecular connectivity, *Bull. Environ. Contam. Toxicol.*, 28, 373, 1982.
90. De Bruijn, J. and Hermens, J., Qualitative and quantitative modelling to toxic effects of organophosphorous compounds to fish, in *QSAR in Environmental Toxicology-IV*, Hermens, J. L. M. and Opperhuizen, A., Eds., Elsevier, Amsterdam, NL, 1991, 441.
91. Saarikoski, J. and Viluksela, M., Relationship between physico-chemical properties of phenols and their toxicity and accumulation in fish, *Ecotoxicol. Environ. Saf.*, 6, 501, 1982.
92. Lipnick, R. L., Charles Ernest Overton: Narcosis studies and a contribution to general pharmacology, *Trends Pharmacol. Sci.*, 7, 161, 1986.
93. Xu, L., Ball, J. W., Dixon, S. L., and Jurs, P. C., Quantitative structure-activity relationships for toxicity of phenols using regression analysis and computational neural networks, *Environ. Toxicol. Chem.*, 13, 841, 1994.
94. Nevalainen, T. and Kolehmainen, E., New QSAR models for polyhalogenated aromatics, *Environ. Toxicol. Chem.*, 13, 1699, 1994.
95. Walker, J. D. and Gray, D. A., The Substructure-based Computerized Chemical Selection Expert System (SuCCSES), *QSARs for Predicting Biological Activities Related to Endocrine Disruption and Environmental–Human Health Interactions*, Walker, J. D., Ed., SETAC Press, Pensacola, FL, 2003.
96. Bradbury, S. P., Carlson, R. W., and Henry, T. R., Polar narcosis in aquatic organisms, in *Aquatic Toxicology and Hazard Assessment,* ASTM STP 1027, Cowgill, U. M. and Williams, L. R., Eds., ASTM, Philadelphia, 1989, 59.
97. McKim, J. M., Schmieder, P. K., Carlson, R. W., Hunt, E. P., and Niemi, G. J., Use of respiratory-cardiovascular responses of rainbow trout (*Salmo gairdneri*) in identifying acute toxicity syndromes in fish: Part 1. Pentachlorophenol, 2,4-dinitrophenol, tricaine methanesulfonate, and 1-octanol, *Environ. Toxicol. Chem.*, 6, 295, 1987.
98. McKim, J. M., Schmieder, P. K., Niemi, G. J., Carlson, R. W., and Henry, T. R., Use of respiratory-cardiovascular responses of rainbow trout (Salmo gairdneri) in identifying acute toxicity syndromes in fish: Part 2. Malathion, carbaryl, acrolein and benzaldehyde, *Environ. Toxicol. Chem.*, 6, 313, 1987.
99. Bradbury, S. P., Henry, T. R., Niemi, G. J., Carlson, R. W., and Snarski, V. M., Use of respiratory-cardiovascular responses of rainbow trout (*Oncorhynchus mykiss*) in identifying acute toxicity syndromes in fish: Part 3, Polar narcotics, *Environ. Toxicol. Chem.*, 8, 247, 1989.
100. Bradbury, S. P., Henry, T. R., Niemi, G. J., Carlson, R. W., and Snarski, V. M., Use of respiratory-cardiovascular responses of rainbow trout (*Oncorhynchus mykiss*) in identifying acute toxicity syndromes in fish: Part 4. Central nervous system seizure agents, *Environ. Toxicol. Chem.*, 10, 115, 1991.
101. Veith, G. D. and Broderius, S. J., Structure-toxicity relationships for industrial chemicals causing Type II narcosis syndrome, in *QSAR in Environmental Toxicology — II*, Kaiser, K. L. E., Ed., D. Reidel Publishing, Dordrecht, The Netherlands, 1987, 385.
102. Russom, C. L., Bradbury, S. P., Broderius, S. J., Hammermeister, D. E., and Drummond, R. A., Predicting modes of toxic action from chemical structure: Acute toxicity in the fathead minnow (*Pimephales promelas*), *Environ. Toxicol. Chem.*, 16, 948, 1997.
103. Könemann, H., Quantitative structure-activity relationships in fish toxicity studies Part 1: Relationship for 50 industrial pollutants, *Toxicology,* 19, 209, 1981.
104. Bobra, A., Shiu, W. Y., and MacKay, D., Quantitative structure-activity for the acute toxicity of chlorobenzenes to *Daphnia magna, Environ. Toxicol. Chem.,* 4, 297, 1985.
105. Veith, G. D., Call, D. J., and Brooke, L. T., Structure-toxicity relationships for the fathead minnow, *Pimephales promelas*: Narcotic industrial chemicals, *Can. J. Fish. Aquat. Sci.,* 40, 743, 1983.
106. Verhaar, H. J. M., van Leeuwen, C. J., and Hermens, J. L., Classifying environmental pollutants. 1: Structure-activity relationships for prediction of aquatic toxicity, *Chemosphere,* 25, 471, 1992.
107. Russom, C. L., Drummond, R. A., and Hoffman, A. D., Acute toxicity and behavioral effects of acrylates and methacrylates to juvenile fathead minnows, *Bull. Environ. Contam. Toxicol.,* 41, 589, 1988.

108. Lipnick, R. L., Baseline toxicity QSAR modeling: A means to assess mechanism of toxicity for aquatic organisms and mammals, in *Environmental Toxicology and Risk Assessment, Second Volume*, ASTM STP 1173, Gorsuch, J. W., Dwyer, F. J., Ingersoll, C. G., and La Point, T. W., Eds., Am. Soc. for Testing and Materials, Philadelphia, 1993, 610.
109. Bradbury, S. P., Predicting modes of toxic action form chemical structure: An overview, *SAR QSAR Environ. Res.*, 2, 89, 1994.
110. Nendza, M. and Russom, C. L., QSAR modeling of the ERL-D fathead minnow acute toxicity database, *Xenobiotica*, 21, 147, 1991.
111. Schultz, T. W., Sinks, G. D., and Bearden, A. P., QSARs in aquatic toxicology: A mechanism of action approach comparing toxic potency to *Pimephales promelas, Tetrahymena pyriformis,* and *Vibrio fischeri*. in Devillers, J., Ed., *Comp. QSAR*, Taylor and Francis, London, 52, 1998.
112. Cronin, M. T. D., Sinks, G. D., and Schultz, T. W., Modelling of toxicity to the ciliate *Tetrahymena pyriformis*: The aliphatic carbonyl domain, in *Forecasting the Environmental Fate and Effects of Chemicals,* Rainbow, P. S., Hopkin, S. P., and Crane, M., Eds., John Wiley and Sons, Ltd., Chichester, U.K., 113, 2001.
113. Kaiser, K. L. E. and Palabrica, V. S., *Photobacterium phosphoreum* toxicity data index. *Water Pollut. Res. J. Can.,* 26, 3, 361, 1991.
114. Kaiser, K. L. E. and Devillers, J., Ecotoxicity of chemicals to Photobacterium phosphoreum, Gordon and Breach Science Publishers, Philadelphia, 1994, 879.
115. Hansch, C. and Fujita, T., A method for the correlation of biological activity and chemical structure, *J. Am. Chem. Soc.*, 86,1616, 1964.
116. Veith, G. D., Lipnick, R. L., and Russom, C. L., The toxicity of acetylenic alcohols to the fathead minnow, Pimephales promelas: Narcosis and proelectrophile activation, *Xenobiotia*, 19, 555, 1989.
117. Hermens, J. L. M., Electrophiles and acute toxicity to fish, *Environ, Health Perspect.*, 87, 219, 1990.
118. Mekenyan, O. G. and Veith, G. D., The electronic factor in QSAR: MO-parameters, competing interactions, reactivity, and toxicity, *SAR QSAR Environ. Res.*, 2, 129, 1994.
119. Veith, G. D. and Mekenyan, O. G., A QSAR approach for estimating the aquatic toxicity of soft electrophiles (QSAR for soft electrophiles), *Quant. Struct.-Act. Relat.*, 12, 349, 1993.
120. Seward, J. R., Cronin, M. T. D., and Schultz, T. W., Structure-toxicity analyses of *Tetrahymena pyriformis* exposed to pyridine – an examination into extension of surface-response domains, *SAR QSAR Environ. Res.*, 11, 489, 2001.
121. Schultz, T. W., Sinks, G. D., and Cronin, M. T. D., Quinone-induced aquatic toxicity to *Tetrahymena*: Structure-activity relationships, *Aquat. Toxicol.*, 39, 267, 1997.
122. Sink, G. D., Schultz, T. W., and Hunter, R. S., UVb-induced toxicity of PAHs: Effects of substituents and heteroatom substitution, *Bull. Environ. Contam. Toxicol.*, 59, 1, 1997.
123. Muccini, M., Layton, A. C., Sayler, G. S., and Schultz, T. W., Aquatic toxicities of halogenated benzoic acids to *Tetrahymena pyriformis*, *Bull. Environ. Contam. Toxicol.*, 62, 616, 1999.
124. Seward, J. R. and Schultz, T. W., QSAR analyses of the toxicity of aliphatic carboxylic acids and salts to *Tetrahymena pyriformis*, *SAR QSAR Environ. Res.*, 10, 557, 1999.
125. Karabunarliev, S. H, Dimitrov, S., Nikolova, N., and Mekenyan, O., Prediction of acute aquatic toxicity of noncongeneric chemicals: Rule-based and quantitative structure-activity relationships, in *Handbook on Quantitative Structure Activity Relationships (QSARs) for Predicting Ecological Effects of Chemicals*, Walker, J. D., Ed., SETAC Press, Pensacola, FL, 2002.
126. Walker, J. D., Applications of QSARs in toxicology: A U.S. government perspective, *J. Molec. Struct. Theochem.*, in press, 2002.
127. Walker, J. D., Carlsen, L., Hulzebos, E., and Simon-Mettich, B., Global government applications of analogs, SARs and QSARs to predict aquatic toxicity, chemical or physical properties, environmental fate parameters and health effects of organic chemicals, *SAR QSAR Environ. Res.*, 13, 607, 2002.
128. Walker, J. D., Relative sensitivity of algae, bacteria, invertebrates, and fish to phenol: Analysis of 234 tests conducted for more than 149 species, *Toxicity Assess. Int. J.*, 3, 415, 1988.
129. Walker, J. D., Review of chemical fate testing conducted under section 4 of the Toxic Substances Control Act: Chemicals, tests and methods, in *Aquatic Toxicology and Risk Assessment: Thirteenth Volume*, ASTM STP 1096, Landis, W. G. and Vanderschaile, W. H., Eds., ASTM, Philadelphia, 1990, 77.
130. Walker, J. D., Bioconcentration, chemical fate and environmental effects testing under section 4 of the Toxic Substances Control Act, *Toxicity Assess. Int. J.*, 5, 61, 1990.

131. Walker, J. D., Bioconcentration, chemical fate and aquatic toxicity testing under the Toxic Substances Control Act: Proposed testing and decision criteria, *Toxicity Assess. Int. J.*, 5,103, 1990.
132. Walker, J. D., Acrylamide aquatic effects: Potential impact of extended exposure, *Environ. Toxicol. Water Qual.*, 6, 363, 1991.
133. Walker, J. D., Review of ecological effects and bioconcentration testing recommended by the TSCA Interagency Testing Committee and implemented by EPA under the Toxic Substances Control Act: Chemicals, tests and methods, in *Environmental Toxicology and Risk Assessment*, Landis, W. G., Hughes, J. S., and Lewis, M. A., Eds., ASTM STP 1179, ASTM, Philadelphia, 1993, 92.
134. Walker, J. D., The TSCA Interagency Testing Committee's role in facilitating development of test methods: Toxicity and bioconcentration testing of chemicals added to sediment, in *Environmental Toxicology and Risk Assessment: Second Volume*, Gorsuch, J. W., Dwyer, F. J., Ingersoll, C. G., and La Pointe, T. W., Eds., ASTM STP 1216, ASTM, Philadelphia, 1993, 688.
135. Walker, J. D., Testing Decisions of the TSCA Interagency Testing Committee for brominated flame retardants: A review of decisions and health and safety data, in *The Future of Fire Retarded Materials: Applications and Regulation*, Fire Retardant Chemicals Association, Lancaster, PA, 1994, 185.
136. Walker, J. D., Recommendations of the TSCA Interagency Testing Committee: Aquatic toxicity, bioconcentration and chemical fate data developed under section 4 of the Toxic Substances Control Act, in *Fundamentals of Aquatic Toxicology II*, Rand, G. M., Ed., Taylor and Francis Publishers, Washington, D.C., 1995, 669.
137. Walker, J. D. and Smock, W. H., Chemicals recommended for testing by the TSCA Interagency Testing Committee: A case study with octamethylcyclotetrasiloxane, *Environ. Toxicol. Chem.*, 14, 1631, 1995.
138. Walker, J. D., Testing decisions of the TSCA Interagency Testing Committee for chemicals on Canada's Domestic Substances List and Priority Substances List: Di-*tert*-butylphenol, ethyl benzene, brominated flame retardants, phthalate esters, chloroparaffins, chlorinated benzenes and anilines in *Environmental Toxicology and Risk Assessment: Fourth Volume*, LaPoint, T. W., Price, F. T., and. Little, E. E, Eds., ASTM STP 1241, ASTM, Philadelphia, 1996, 18.
139. EEC, Commission Directive 93/67/EEC of 20 July 1993, laying down the principles for the assessment of risks to man and the environment of substances notified in accordance with Council Directive 67/548/EEC, *Off. J. Eur. Comm.*, L 227/9, 1993.
140. Feijtel, T. C. J., Evaluation of the use of QSARs for priority setting and risk assessment, *SAR QSAR Environ. Res.*, 3, 237, 1995.
141. Comber, M. H. I. and Feijtel, T. C. J., Assessment and use of quantitative structure activity relationships: A report from two ECETOC task forces, in *Quantitative Structure-Activity Relationships in Environmental Sciences – VII*, Chen, F. and Schüürmann, G., Eds., SETAC, Pensacola, FL, 1997, 25, 365.
142. EEC, Technical Guidance Documents in support of the Commission Directive 93/67/EEC on risk assessment for new notified substances and the Commission Regulation (EC) 1488/94 on risk assessment for existing substances, Ispra, Italy, 1996.
143. ECETOC, QSARs in the Assessment of the Environmental Fate and Effects of Chemicals, Technical Report No. 74, Brussels, 1998.
144. MacDonald, D., Breton, R., Sutcliffe, R., and Walker. J. D., Uses and limitations of quantitative structure-activity relationships (QSARs) to categorize substances on the Canadian Domestic Substance List as persistent and/or bioaccumulative, and inherently toxic to non-human organisms, *SAR QSAR Environ. Res.*, 13, 43, 2002.
145. Dimitov, S. D., Mekenyan, O. G., and Schultz, T. W., Interspecies modeling of narcotic toxicity in aquatic animals, *Bull. Environ. Contam. Toxicol.*, 65, 399, 2000.

# CHAPTER 34

# Predictive Ecotoxicology

**John Cairns, Jr. and B. R. Niederlehner**

## CONTENTS



## 34.1 GENESIS OF PREDICTIVE ECOTOXICOLOGY

When population centers were small and scattered, human activities often did not impair the surrounding environment. However, as human populations grew, increasing sewage loads and harvesting pressure led to local collapses in the ecosystem services upon which human society depended. Waste processing, provision of potable water, and food production are examples of ecosystem services that were pushed beyond sustainability. The inconvenience and hardship resulting from failures in ecosystem services motivated early attempts at proactive management of the environment, including prediction and mitigation of damage before the fact. The ability to predict environmental outcome has improved. However, the development of robust predictive techniques for subtle damage and larger spatial scales remains challenging.

## 34.2 APPRAISAL VS. PREDICTION

Ecotoxicology requires an appreciation for different time frames. Successful management requires not only techniques to appraise existing damage in natural systems but also techniques to predict the environmental outcome of proposed actions. Appraisal defines the problem. Prediction allows for prevention of damage before it occurs. Without prediction, environmental management is doomed to react to each crisis after the fact.

Predictive ecotoxicology is distinct from environmental appraisal because often there is no "real world" to survey. "Reality checks" (i.e., characterizations of the accuracy of predictions) are more difficult and indirect for predictive ecotoxicology than in the case of environmental appraisal. Instead, predictive ecotoxicology must rely on models.

The process of predicting risk involves many steps.[1–4] First, the problem must be defined. What is it that is worthy of protection? While setting goals is the responsibility of society at large, scientists may be called upon both to interpret the significance of changes to the natural world and to operationalize societal goals. Second, a scientific risk assessment compares a prediction of the level of stress that will occur in the environment after some human activity to a prediction of likely biological effects at that level of stress. Third, after the scientific risk assessment has estimated the probability of various adverse effects and the uncertainties of these estimates, these risks are considered along with options, costs, benefits, and political will to arrive at a risk-management plan. Fourth, after a course of action is implemented, the resulting effects in the real world must be compared with the predictions of risk made before the fact. These comparisons serve a quality-control function for environmental management and provide insight into the adequacy of the predictive models and management techniques.

In many regulatory agencies, a commonly accepted framework is used for determining the nature and likelihood of the effects of human actions on the environment. This framework is ecological risk assessment (ERA),[3,4] which is used to organize ecological information and assess its quality so that management decisions can proceed based on the best information available. Solomon and colleagues[5] and Giesy and colleagues[6] provide two excellent examples of the process. Purposes can include predictive applications such as screening new chemicals prior to commercial production, evaluating risks associated with land use changes or exotic species introduction, and ranking the relative environmental consequences of different proposed actions. But ERA is not always predictive.[7] Other applications of ERA are essentially for appraisal: are present uses of chemicals of concern? The question is not what will happen, but what has already happened. The best instances of ERA for this purpose provide the opportunity to assess the accuracy of predictions by comparing predictions of harm to observations in natural systems. Even though predictive ecotoxicology and ERA are not synonymous, the structure and definitions used in ERA provide a useful framework in which to discuss predictive ecotoxicology.

## 34.3 OPERATIONALIZING SOCIETAL GOALS

### 34.3.1 Endpoint Selection

Although society at large sets goals for environmental protection, scientists are called upon to develop efficient ways of measuring change and to interpret the importance of changes in the environment. The relationship between societal goals for environmental protection and scientific methods for assessing the environment have been described by Suter.[8] Concerns about maintaining quality of life have led to the development of amorphous societal goals. For example, current goals in the United States include: "to promote efforts that will prevent or eliminate damage to the environment and biosphere," "to maintain and enhance long-term productivity," to avoid "any irreversible or irretrievable commitments of resources," and "to restore and maintain the chemical,

**Table 34.1 Characteristics Desirable in Endpoints for Appraising, Monitoring, and Predicting Environmental Impact**

1. *Biologically relevant,* i.e., important in maintaining biotic integrity
2. *Socially relevant,* i.e., of obvious value to and observable by shareholders or predictive of an endpoint that is
3. *Sensitive* to hazards with a high signal-to-noise ratio
4. Broadly *applicable* to many hazards and sites
5. *Diagnostic* of the particular hazard causing the problem
6. *Measurable,* i.e., capable of being operationally defined and measured using a standard procedure with documented performance and low measurement error
7. *Interpretable,* i.e., capable of distinguishing acceptable from unacceptable conditions in a scientifically and legally defensible way
8. *Cost-effective,* i.e., inexpensive to measure, providing the maximum amount of information per unit effort
9. *Integrative,* i.e., summarizing information from many unmeasured endpoints
10. One for which *historical data are available* to define normal operating range, trends, and possibly acceptable and unacceptable conditions
11. *Anticipatory,* i.e., capable of providing an indication of degradation before serious harm has occurred, early warning
12. *Nondestructive* of the ecosystem
13. One with potential for *continuity* in measurement over time
14. Of an *appropriate scale* to the management goal
15. *Not redundant* with other measured endpoints, i.e., providing unique information
16. *Timely,* i.e., providing information quickly enough to initiate effective management action before unacceptable damage has occurred

physical, and biological integrity of the nation's water."[8] Scientists are charged with taking these amorphous goals and, based on them, defining assessment endpoints, i.e., formal expressions of actual environmental values worthy of protection. Assessment endpoints may include properties such as sustainable production of various species, retention of soil and nutrients in the ecosystem, maintenance of water supplies, or maintenance of aesthetics or biodiversity. A measurement endpoint — an expression of an observed or measured response to a hazard — is the next step in operationalizing societal goals. While no instrument can measure ecosystem health on a scale from 1 to 100, scientists can determine catch of fish per unit effort or yield per acre, standing crop of preferred species, number of different taxa in an ecosystem, rates of nutrient removal, or rate of flooding. Many examples of measurement endpoints are provided in Section I of this book. These measurable qualities underlie all predictive methods. Obviously, the wider the net is cast in looking for environmental consequences of human actions, the more likely that change can be detected. However, there are inescapable practical limits on how many endpoints can be practically monitored. As such, endpoints must be selected carefully, and criteria for selection of the most efficient and useful endpoints for environmental analyses have been described (Table 34.1).[8–11]

### 34.3.2 Relevance

Among the many desirable characteristics cited is relevance. Relevant endpoints are those closely related to management goals or ecological integrity. While some endpoints are widely valued by society at large (i.e., catchable fish), other endpoints are equally relevant ecologically but less obvious in their value (i.e., status of organisms at the base of the food web). Still other endpoints are clearly irrelevant to many systems because they monitor the response of organisms that do not live in the system or the response to conditions that do not occur in the system. These endpoints are useful only if general properties of living material allow extrapolation from one situation to another.

These extrapolations are the basic tool in predictive ecotoxicology and must be critically evaluated. When endpoints that are far removed from attributes of interest in a particular ecosystem are used because they can be efficiently and crisply measured, a model must then translate the easily measured endpoint into terms of effects on an important attribute. For example, it may not be important to protect the ability of exotic bioluminescent bacteria to produce light in a particular system. However, luminescence may be a useful endpoint because impairment may be an easily measured indication of potential for adverse effects on more important endpoints. In order to be useful as a management tool, the endpoint in the simple test must be related to environmental outcome in a complex natural system. For example, we must predict impairments in biodiversity from impairments in light production.

The nature of this relationship between endpoints and the degree of uncertainty involved must be characterized. Is it a strong relationship? Is it consistent over stresses? What kinds of predictive errors are likely?

Models have been developed to predict the response of one species to a toxicant from the measured response of another, to predict chronic response from measurements of acute response, and to predict the response to a novel toxicant from measured response to structurally similar chemicals.[12] Each extrapolation from one endpoint or hazard to another more closely related to the problem at hand increases the complexity of the predictive model and adds another layer of uncertainty to the predictions. For example, Barnthouse and colleagues[12] estimated that the uncertainty associated with the prediction of a concentration of concern for a population was within an order of magnitude when life-cycle-toxicity test data were available for the species of interest. However, when prediction involved extrapolations across species, length of exposure, or chemicals, uncertainty approached three orders of magnitude. In addition, these estimates of uncertainty are based on variability of data used for model construction. Additional uncertainty will stem from relevant environmental factors (e.g., pH, hardness, or flow changes) or interactions (e.g., competition, harvesting pressure, or disease prevalence) not incorporated into the model. Predictions of adverse impact on clearly relevant endpoints will more likely be accepted as a sufficient basis for decision-making than those based on other endpoints because there is less extrapolation and less uncertainty.

### 34.3.3 Signal-to-Noise Ratios

Another scientific problem with operationalizing management goals is that a point exists at which the discriminatory power demanded by increasingly ambitious management goals outstrips the power of the measurement endpoints. Both laymen and scientists can identify and define characteristics of a grossly damaged environment. However, it is more difficult to identify and measure subtle damage in natural systems (Table 34.2) or the characteristics of healthy ecosystems worthy of protection (Table 34.3).

In order to be useful in appraising or predicting environmental impact, the response to an impact must be large enough to be distinct from normal variations. Either the signal (i.e., the change in response to hazard) must be large or the noise (i.e., normal variation) must be small (Figure 34.1). Appraising, monitoring, and predicting extreme damage are easier than doing so for subtle damage because the signal-to-noise ratio is higher. In the case of extreme damage, the magnitude of the

**Table 34.2 Characteristics of Damaged Ecosystems**

1. Loss of sensitive species, decreased species diversity
2. Increased instability in component populations, greater amplitude of fluctuations
3. Changes in the size spectrum to smaller organisms
4. Retrogression to opportunistic species
5. Decreased life spans, increased turnover of organisms
6. Increased disease and parasite prevalence, decreased condition
7. Shift in trophic dynamics, shortening of food chains, and decline in functional redundancy and diversity
8. Changes in primary production
9. Decreased efficiency of nutrient recycling and increased loss of nutrients
10. Decreased efficiency of energy use, increased respiration
11. Increased circulation of contaminants

*Source:* From Odum, E.P., *Bioscience,* 35, 419, 1985; Rapport, D.J. et al., *Am. Nat.,* 125, 617, 1985; and Schindler, D.W., *Can. J. Fish. Aquat. Sci.,* 44, 6, 1987.

**Table 34.3 Characteristics of Healthy Ecosystems**

1. The maintenance of the community structure and function characteristic of a particular locale
2. The ability to support and maintain a balanced, integrated, adaptive community of organisms having a species composition, diversity, and functional organization comparable to that of natural habitat of the region
3. Habitat for desired diversity and reproduction of organisms
4. Phenotypic and genotypic diversity among organisms
5. Robust food chain to support desired biota
6. Adequate nutrient pool to support desired organisms
7. Adequate nutrient cycling to perpetuate the ecosystem
8. Adequate energy flux for maintaining the trophic structure
9. Feedback mechanisms for damping undesirable oscillations
10. The capacity to temper toxic effects
11. The absence of anthropogenic inputs

*Source:* From Cairns, J. Jr., Integrity of Water, U.S. Government Printing Office, Washington, D.C., 1977, 171; Karr, J.R., *Ecol. Appl.,* 1, 66, 1991; Schaeffer, D.J. et al., *Environ. Manage.,* 12, 445, 1988; and Rapport, D.J., *Persp. Biol. Med.,* 33, 120, 1989.

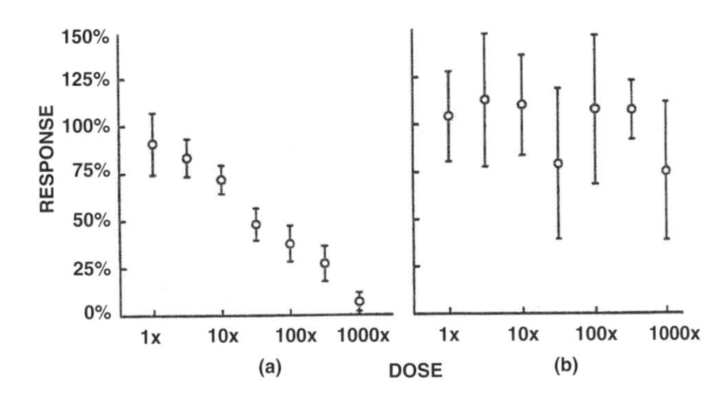


**Figure 34.1** Differences in signal-to-noise ratios of endpoints in biomonitoring. Part (a) illustrates a high signal-to-noise ratio. The change in response after stress is large, and the response does not vary widely. In this case, it is easy to see the impact of the stress. Part (b) illustrates a low signal-to-noise ratio. The signal (i.e., the change in response after stress) is small, but the response varies widely, so there is a large amount of noise.

impairment overwhelms any natural changes in the ecosystems over time and space. In order to interpret the significance of smaller changes, better information must be available on the normal operating range of minimally impacted systems. The significance of a predicted change can be judged only in reference to the normal operating range in similar, minimally impacted systems. Because natural systems are diverse, patchy, and constantly changing diurnally, seasonally, and successionally, some change cannot be attributed to hazard.

Better measurement methods will improve the ability to detect small or early impairments in ecosystem health. Endpoints are being screened in terms of their signal-to-noise ratios. Sampling methods are incorporating blocking, compositing, and other techniques to accommodate natural variability.[13] Multiple relevant variables are being incorporated into dose-response models.[14] In addition, the normal operating ranges for natural systems are being better characterized so that meaningful evaluations of change can be made.[15]

## 34.3.4 Integrating Multiple Effects

Unfortunately, no single endpoint will adequately operationalize all goals for environmental protection. Instead, the multiple effects of a single hazard must be integrated. However, they can be integrated at different stages of a prediction. Some assessments choose one endpoint that depends on the integrity of many responses at lower trophic levels or quicker time scales (e.g., catch per unit effort of predacious, long-lived fish).[16] Other assessments integrate multiple endpoints before the prediction (e.g., the index of biotic integrity [IBI] or other multimetric indices).[15] Alternatively, independent predictive models can be developed for a variety of endpoints and the information integrated at the point of decision making.

## 34.4 DEVELOPMENT OF PREDICTIVE MODELS

### 34.4.1 Dose-Response Models

A scientific risk assessment compares a prediction of the level of stress that will occur in the environment after some human activity to a prediction of likely biological effects at that level of stress. At least two predictive models are involved in this process. In this discussion, predictive models of biological effects are the focus rather than models of environmental fate, but many of the same considerations apply in both cases. Models of environmental fate are well described by Burns and Baughman[17] and Wolfe and Crossland[18] as well as in this book (Chapter 29) and in examples of ERA.[5,6] The most commonly used models for prediction of biological effects are dose-response curves. The pattern of response over increasing levels of hazard is used to predict the probability and magnitude of response to the same hazard under similar circumstances.

Several types of curves are repeatedly found in studies of responses to stress (Figure 34.2). Figure 34.2A shows a distinct change in the rate of response that constitutes a threshold, that is, a point at which an effect begins to be produced. An example of this pattern of response was observed in bacterial numbers in microbial communities exposed to increasing levels of ammonia.[19] Figure 34.2B shows how a stimulation of response at low doses may serve as a practical threshold when impairment is defined as change in one direction, that is, subsidy vs. stress.[20] An example of stimulation of response at low doses has been observed for periphyton production in response to atrazine.[21] Figure 34.2C shows an asymptotic rate of change in response[22] with no absolute threshold; however, at some point, decreases in impairment with decreasing dose may be so small as to be biologically unimportant. In this case, there may be a practical threshold in absence of an absolute threshold. Finally, Figure 34.2D shows a response that changes linearly from background to severe stress levels. There is no discontinuity in the dose-response curve; there is no threshold. This is the pattern described by Woodwell[23] for radiation effects on mutants in *Tradescantia.*

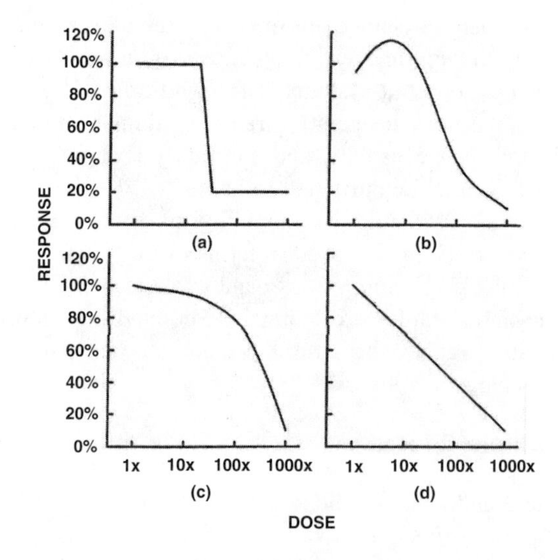


**Figure 34.2** Types of dose-response curves. Curve (a) shows an abrupt change in response with dose, i.e., a threshold. Curve (b) shows a subsidy at low doses that may serve as a practical threshold. Curve (c) is asymptotic with a practical threshold. Curve (d) shows no threshold.

### 34.4.2 Thresholds and Concentrations of Concern

As the power of measurement techniques has improved and the ability to detect small changes has increased, fewer clear thresholds for adverse biological effects are found. Woodwell[23] asked, "Is it reasonable to assume that thresholds for effects of disturbance exist in natural ecosystems or are all disturbances effective, cumulative, and detrimental to the normal functioning of natural ecosystems?" Reports of dose-response curves that are linear down to background levels, like that in Figure 34.2D, undermine sole reliance on estimates of threshold concentrations for environmental prediction and management. If thresholds are an artifact of the statistical power of the measurement methods, then scientists must instead rely on active judgments about the biological importance of the changes involved. This is the basis of the ongoing debate about statistical vs. biological significance of changes and the increasing dependence on regression analyses rather than hypothesis testing to determine concentrations of concern for management.[3,24] While no-observed effect concentrations (NOECs) appear decisive and do provide an indication of signal-to-noise ratio, they are not necessarily objective and accurate estimates of real thresholds. Reliance on professional judgment of the biological importance of changes is required in both methods for deriving concentrations of concern.

### 34.4.3 Data Used to Develop Predictive Models

Most predictive models of biological effects of toxic substances are based on toxicity test data or synoptic surveys of natural systems. A toxicity test is an attempt to simulate, under controlled conditions, some of the chemical or physical conditions to which organisms in natural systems might, at least theoretically, be exposed. Synoptic surveys provide a broad overview of conditions in natural systems. The data used to construct predictive models can be characterized along covarying gradients of complexity, spatial scale, time scale, and similarity to appraisals of natural systems (Table 34.4). In natural systems, complexity, size, and time scale increase together. However, in designed systems, complexity and size can be individually manipulated so that small, complex systems and large, simple systems are possible.

There is a countervailing gradient of complexity in the methods by which toxicity test data are applied to the prediction of environmental outcome. The uncertainty involved in prediction increases with the discrepancy in scale between the measurement endpoint and the assessment endpoint or management goal.

An example of a small extrapolation with relatively little uncertainty is the prediction that repeating a simple toxicity test will yield similar results. Larger extrapolations with greater uncertainty include those from species to species and from chemical to chemical. Still larger extrapolations with larger uncertainty are those from one level of biological organization to another, from one ecosystem type to another, or from one extreme of the temporal or spatial scale to another.

While the simplest toxicity tests require more accessory information and the most complex methods for translation into prediction of environmental outcome, the most complex test methods measure endpoints that are most closely related to management goals and are more simply translated into site-specific effects. At the same time, however, toxicity tests with few components provide simpler and less expensive tools for the comparisons used to rank hazards, to discern trends over time, and to link a causative agent with an observed biological impairment. More data from simple toxicity tests are available to decision makers.

Different objectives require different levels of design complexity; thus, data requirements change from one application to the next. For example, screening in order to prioritize management of various toxins may require little complexity and great generality, but prediction of environmental outcome in site-specific situations with low tolerance for error may dictate greater complexity and little generality.

**Table 34.4 Covarying Gradients in the Design and Application of Toxicity Tests**

| **Design Complexity** | **Low <-> High** | |
|---|---|---|
| Few components | <-> | Many components |
| Few endpoints | <-> | Many endpoints |
| Standardizable | <-> | Site specific |
| Replicable | <-> | Realistic background variation |
| Capable of high precision | <-> | Capable of high accuracy |
| Low cost per experiment unit | <-> | High cost per experimental unit |
| **Design Scale** | **Low <-> High** | |
| Small experimental units | <-> | Large experimental units |
| Small receptors (e.g., bacteria) | <-> | Large receptors (e.g., fish) |
| Receptor low in biological hierarchy (e.g., cells, populations) | <-> | Receptor high, integrating effects at lower levels (e.g., ecosystem dynamics) |
| Receptor low in phylogenetic hierarchy (e.g., procaryotes) | <-> | Receptor high (e.g., vertebrates) |
| Receptor low in food web (e.g., primary producers) | <-> | Receptor high, integrating effects at lower levels (e.g., piscivores) |
| Endpoint early in progression of impairment (e.g., biomarker) | <-> | Endpoint of obvious social and biological importance (e.g., fisheries harvest) |
| Short observation time in human terms | <-> | Long observation time |
| Short observation time in relation to receptor's life history | <-> | Long observation time |
| **Difference in Complexity and Scale between Test and Application** | **High <-> Low** | |
| Endpoints and hazards are qualitatively different from those in assessments of quality in natural systems | <-> | Qualitatively similar to those in assessments of natural systems |
| Bottom-up appraisal of risk | <-> | Top-down appraisal of risk |
| Cause and mechanics known | <-> | Environmental effects known |
| An array of individual tests are combined to provide information to model environmental effects (many experimental units) | <-> | A single test is used to predict environmental effects (few experimental units) |
| Model necessary to translate test results to prediction of environmental effects (with error) | <-> | Information directly applied to site-specific prediction (with error) |

### 34.4.4 Examples of Predictive Models in Ecotoxicology

Although the results of a single toxicity test on a "sensitive species" were previously used to predict general environmental outcome, this approach is no longer commonly used. No one species is reliably most sensitive over toxicant and water-quality conditions.[25] The uncertainty involved in predictions of environmental outcome based on the response of one species has been judged too high. Models predicting environmental outcome from laboratory toxicity tests on individual species now rely on arrays of data. Rather than trying to find one species that is representative of sensitive organisms in the natural systems, several species are tested, and the tolerance of organisms in general is described from this sample. A tolerance distribution for aquatic organisms in general is modeled from an array of data on the $LC_{50}$ values or NOECs of the individual species tested. From this model of the tolerance distribution, the number of taxa affected by some concentration of toxicant can be predicted. An example of this approach can be found in the derivation of numerical water-quality criteria for protection of aquatic life.[26] The same approach, but using different models for distributions of tolerance, has been used by van Straalen and Denneman[27] and van Leeuven.[28]

Using arrays of data from single-species toxicity tests to model the tolerance distributions of species in general is a good way to address the variation in sensitivity among the many species likely to be affected by any discharge; but problems remain. First, the species that can conveniently be tested may not be a random sample.[29] Second, there may be considerable difference between the endpoints measured and qualities that are worthy of management protection, resulting in large uncertainties.[12] Third, this approach usually does not incorporate the interaction between species in determining environmental outcome. The potential importance of interactive factors can be seen in validation studies in which predictions from an array of single-species toxicity tests failed to identify significant features of environmental outcome. LaPoint and Perry[30] list three interactive factors that significantly affect environmental outcome: (1) differences in actual exposure conditions in the laboratory and field (e.g., mobilization of metals under acidification or microbial methylation of mercury), (2) ability of organisms to avoid the hazard, and (3) secondary effects such as loss of a trophic guild or keystone species.

Interactive factors have been incorporated into predictions in two ways. Mathematical models can incorporate interactions into predictions of outcome, or interactions can be included in toxicity tests. An example of a simulation model of toxicant fate and effects that includes interactions is that described by Bartell and colleagues.[31] Simulated patterns of production at different naphthalene exposures support the importance of secondary trophic effects in predicting environmental outcome. This approach to prediction is attractive because it is applicable to many different ecosystems and relies on the type of data that is most readily available. However, each of the many extrapolations in such a model adds to uncertainty. Without validation studies comparing predictions to observed effects, the degree of confidence warranted is unknown and suspect.

More complex toxicity tests empirically include interactions between species and maximize the correspondence between endpoints measured in toxicity tests and characteristics worthy of protection. For example, Tolle and colleagues[32] compared the predictive accuracy of test systems with varying physical complexity. Flower pot and intact soil microcosms were used to predict the yield and uptake of elements by plants in field plots amended with fly ash. They concluded that microcosms were superior to pots for predicting the dose-response curve for alfalfa, but neither type of test accurately predicted dose-response curves for oats. Niederlehner and colleagues[33] used multispecies toxicity tests to simultaneously evaluate the response of many interacting taxa to cadmium, copper, industrial effluents, and changes in pH. They compared predictions from the multispecies test to arrays of conventional toxicity tests, whole-ecosystem studies, and synoptic surveys and concluded that the multispecies test yielded reasonable predictions of chronic tolerance distributions for aquatic organisms in general, with considerable cost savings.

Even more complex test systems have been used in tests for predicting the environmental effects of pesticide usage.[5,6,34,35] When simple tests suggest that adverse effects are possible at estimated environmental concentrations, artificial pond mesocosm tests were used to comprehensively evaluate this presumption of risk. By maximizing physical realism and the correspondence between management goals and measurement endpoints, pond mesocosms can convincingly predict environmental outcome.

Synoptic surveys of natural systems can also be used to develop predictive models. Relationships between chemical, physical, and biological data across many systems are examined. As with all correlative approaches, the ability to identify cause and effect is limited. However, consistent patterns across diverse ecosystems are convincing, especially when supported by toxicity test results. Examples of predictive models based on synoptic surveys include the model relating nutrient loading to eutrophication.[36] Many predictive models have been developed from synoptic surveys of lake acidity.[37] Some of these models predict population-level responses such as the probability of the presence of a valued fish species in waters of various pH. Other models predict community-level effects such as the loss of biotic richness in waters of various pH. In one such study, a model of biotic damage due to lake acidification from a synoptic survey was used to predict total impact on biodiversity of a regional resource resulting from different levels of sulfur emissions.[38] That study provides an excellent example of the progression through problem definition, identification of goal, measurement, modeling, and predictive application.

## 34.5 HOW ACCURATE ARE OUR PREDICTIONS?

A risk assessment depends on at least two predictive models — one for chemical fate and one for biological effects. Obviously, the uncertainty involved in a concatenation of predictions is much greater than in an appraisal of existing damage that relies on direct observation. The degree of uncertainty involved is an important component in decision making based on predictions of environmental outcome.[39] The decision maker must evaluate the probability and magnitude of an impact as well as the margins of error implicit in that estimate and the consequences of being wrong.

Validation is the comparison of predictions to observed effects in the real world or perhaps in a surrogate of the real world such as a mesocosm. Without validation studies, any one predictive model is as useful or useless as any other. There is no criterion for selection. Validation studies characterize the error involved in prediction in order to control and improve the quality of the predictive techniques. If environmental outcome cannot be predicted with sufficient accuracy to prevent excessive costs from underprotection or overprotection of the environment, resources must be devoted to improving predictive models. There are different approaches to the comparisons that constitute validation.[40,41] There are also different judgments about what constitutes sufficient agreement between prediction and appraisal.

A common approach to validation of predictions is a comparison of point estimates or classifications.[42,43] Typically, if an impairment is predicted from toxicity test results (i.e., the concentration exceeds an NOEC value or other threshold estimate) and a statistically significant impairment is observed at a field site similar in exposure, this is taken as evidence of validity. However, point-estimate comparisons rarely examine the magnitude and, thus, significance of impairments. This approach relies on the existence of thresholds that may or may not be an artifact of statistical power.

Another approach is to correlate the magnitude of the toxicity test response with the response in the natural system.[44] Because this approach abandons reliance on point estimates of a threshold and encompasses a gradient of effects, it is a more rigorous test of validity. However, a significant correlation means only that the pattern in response is similar in the toxicity test and the natural system. This is surely to be expected, but it is insufficient for validation. It does not indicate that toxicity tests can predict field effects with accuracy sufficient for management.[45]

Regression analysis makes explicit the intention to predict field effects from toxicity test results. The dose-response curves in the predictive method and natural system can be compared directly. The margins of error can be characterized, and differences in accuracy at different magnitudes of response can be described.

Also important to validation is an analysis of errors in prediction. The frequency and seriousness of the consequences of errors in predictions of biological impairment must be identified. There are three steps in the demonstration of predictive validity.[40] First, predictions of response in the natural system must be made from toxicity tests. Second, these predictions must be compared to actual response. And third, the adequacy of predictions must be evaluated according to a predetermined criterion based on management needs. An example of predictive validation of multispecies toxicity test data using this approach is discussed by Niederlehner and colleagues.[46]

If the system of interest is very large or the biological response occurs over a very long time span, it may not be possible to measure model error in the system of interest. In the absence of empirical determinations of the size of predictive errors, some indication of the relative credibility of alternate models can be obtained from comparisons of the predictive errors of their testable components[47] or from peer review.[8] However, it is important to note that even models with large predictive errors can be useful if they are the best tools available. For example, even though weather forecasts are often wrong, they remain useful.

## 34.6 FUTURE TRENDS

Future developments in predictive ecotoxicology may be focused in several areas. The first area is the development of more robust predictive models for larger spatial scales. For example, the oceans constitute 97% of the planet's water supply. The liquid surface of the planet is substantially larger than the land surface. The spatial scales are enormous, and controls transcend political boundaries to an extent unparalleled for freshwater and terrestrial ecotoxicology problems. Because of the aggregate size of the oceans, the temporal scales are also likely to be equally daunting. Point sources of pollution for the oceans are the world's great rivers, and nonpoint sources include land runoff that goes directly into the oceans and direct inputs from atmospheric deposition. A very readable overview of the challenges involved is provided by Thorne-Miller.[48]

Another major area affecting the development of predictive models in ecotoxicology is the linkage of the precautionary principle with environmental policy. Increasingly, individuals in the United States and elsewhere in the world regard prevention and precaution as "best practices." The precautionary principle calls for preventative and anticipatory measures to be taken when an activity raises threats of harm to the environment or human health, even if significant uncertainty remains in the scientific evidence. The Rio Declaration on environment and development thrust the precautionary principle to the global level in 1992, when the United Nations conference (the Earth Summit) provided a declaration that listed precaution as the fifteenth of 27 principles to guide environmental and development policies. Other countries, such as Germany, introduced what is called "Vorsorgeprinzip,"[49] which roughly translates as the "foresight principle."

In matters of personal health, a precautionary approach might indicate adoption of additional exercise, diets higher in fiber, meditation, and avoiding excessive use or even any use of tobacco and, to a lesser extent, alcohol (since some wine, for example, may have health benefits). In matters of environmental health, a precautionary approach leads to an increased attention to pollution prevention.

The exact role that ecotoxicology will play in the implementation of the precautionary principle in international law is just beginning to emerge.[50] Lyons and colleagues[51] suggest one means of operationalizing the precautionary principle in the management of chemicals. They believe that it is possible to balance the dual societal expectations of a clean and healthy environment and access to technological developments. Such important issues are addressed as the misuse of the precautionary principle to create unjustified barriers to trade and also the unacceptability of a situation in which all countries had to be bound by the lowest standards of human health and environmental protection.

## 34.7 CONCLUSIONS

Prediction of environmental outcome is different from appraisal of existing damage. In prediction, the dependence on a model is absolute. There is often no way to check the accuracy of prediction through field survey. However, accuracy checks are essential in order to make sure that predictive techniques are adequate to meet management needs. Validation studies compare predictions to appraisals of damage in natural systems. Through these comparisons, the magnitude and significance of predictive errors can be evaluated. If the potential costs of acting on imperfect predictions are too high, then predictive methods must be improved.

There are several areas in which methods for predictive models can be improved. In the absence of validated models for extrapolating between endpoints, the correspondence between measurement and assessment endpoints should be maximized. A suite of the most relevant endpoints with good signal-to-noise ratios must be identified. Sole reliance on estimates of thresholds must be replaced with evaluations of biological significance of changes relative to normal operating ranges. Where many societal goals must be addressed, multiple endpoints must be monitored. Techniques for integrating multiple endpoints must be developed. Validation studies must characterize predictive accuracy of various techniques and evaluate the potential costs of errors. The efficiency of test methods must be judged not just on testing costs but also on the costs of acting on that information. If the test method is inexpensive, but predictive inaccuracies lead to environmental damage or unnecessary treatment, the total cost is too high. Areas of extrapolation and errors involved should be clearly communicated to decision makers.

While uncertainty can never be eliminated, society and its representatives must make decisions now on various risks with whatever methodologies are available. Predictions must be made even as the methodologies improve. This is similar to what George Bernard Shaw is reported to have responded when asked what he thought of life: "It is preferable to the alternative." The application of existing methods for predicting environmental outcome has improved environmental quality. Improved methods will reduce costs and answer the needs of increasingly ambitious management goals.

## ACKNOWLEDGMENT

The authors thank Darla Donald for her editorial assistance.

## REFERENCES

1. Cairns, J., Jr., Dickson, K. L., and Maki, A. W., *Estimating the Hazard of Chemical Substances to Aquatic Life*, American Society for Testing and Materials, Philadelphia, 1978.
2. National Research Council, *Risk Assessment in the Federal Government: Managing the Process*, National Academy Press, Washington, D.C., 1983.
3. U.S. EPA (United States Environmental Protection Agency), A Framework for Ecological Risk Assessments, EPA 630/R-92–001, Risk Assessment Forum, Washington, D.C., 1992.
4. U.S. EPA (United States Environmental Protection Agency), Guidelines for ecological risk assessment, *Fed. Regis.*, 63, 93, 26846, 1998.
5. Solomon, K. R. et al., Ecological risk assessment of atrazine in North American surface waters, *Environ. Toxicol. Chem.*, 15, 31, 1996.
6. Giesy, J. P. et al., Chlorpyrifos: Ecological risk assessment in North American aquatic environments, *Rev. Environ. Contam. Toxicol.*, 160, 1, 1999.
7. De Vlaming, V., More on probabilistic risk assessments — A magic bullet for all situations, they are not, *SETAC Globe*, 1, 3, 27, 2000.
8. Suter, G. W., II, Endpoints for regional risk assessments, *Environ. Manage.*, 14, 9, 1990.
9. Macek, K. et al., Discussion session synopsis, in *Estimating the Hazard of Chemical Substances to Aquatic Life,* Cairns, J., Jr., Dickson, K. L., and Maki, A. W., Eds., American Society for Testing and Materials, Philadelphia, 1978, 27.
10. Kelly, J. R. and Harwell, M. A., Indicators of ecosystem response and recovery, in *Ecotoxicology: Problems and Approaches,* Levin, S. A., Harwell, M. A., Kelly, J. R., and Kimball, K. D., Eds., Springer-Verlag, New York, 1989, 9.
11. Hunsaker, C. T. and Carpenter, D. E., Environmental Monitoring and Assessment Program: Ecological Indicators, EPA/600/3–90–060, National Technical Information Service, Springfield, VA, 1990, Chap. 2.
12. Bamthouse, L. W., Suter, G. W., II., and Rosen, A. E., Risks of toxic contaminants to exploited fish populations: Influence of life history, data uncertainty, and exploitation intensity, *Environ. Toxicol. Chem.*, 9, 297, 1990.
13. Cairns, J., Jr., Niederlehner, B. R., and Smith, E. P., The emergence of functional attributes as endpoints in ecotoxicology, in *Sediment Toxicity Assessment,* Burton, G. A., Ed., Lewis Publishers, Chelsea, MI, 1992, Chap. 6.

14. Allus, M. A., Brereton, R. G., and Nickless, G., Chemometric studies of the effect of toxic metals on plants: The use of response surface methodology to investigate the influence of Tl, Cd, and Ag on the growth of cabbage seedlings, *Environ. Pollut.*, 52, 169, 1988.
15. Karr, J. R., Biological integrity: A long-neglected aspect of water resources management, *Ecol. Appl.*, 1, 66, 1991.
16. Edwards, C. J. and Ryder, R. A., Biological Surrogates of Mesotrophic Ecosystem Health in the Laurentian Great Lakes, Report to the Great Lakes Science Advisory Board of the International Joint Commission, Windsor, Ontario, Canada, 1990.
17. Burns, L. A. and Baughman, G. L., Fate modeling, in *Fundamentals of Aquatic Toxicology,* Rand, G. M. and Petrocelli, S. R., Eds., Hemisphere, New York, 1985, Chap. 19.
18. Wolfe, C. J. M. and Crossland, N. O., The environmental fate of organic chemicals, in *Controlling Chemical Hazards,* Cote, R. P. and Wells, P. G., Eds., Urwin Hyman, Boston, MA, 1991, Chap. 2.
19. Niederlehner, B. R. and Cairns, J., Jr., Effects of ammonia on periphytic communities, *Environ. Pollut.*, 66, 207, 1990.
20. Odum, E. P., Finn, J. T., and Franz, E. H., Perturbation theory and the subsidy-stress gradient, *BioScience*, 29, 349, 1979.
21. Pratt, J. R. et al., Effects of atrazine on freshwater microbial communities, *Arch. Environ. Contam. Toxicol.,* 17, 449, 1988.
22. Marcovich, H. and Devoret, R., The effect of small doses of ionizing radiations on *Escherichia coli,* in *Immediate and Low Level Effects of Ionizing Radiations,* Buzzati-Traverso, A. A., Ed., Taylor and Francis, London, 1960, 293.
23. Woodwell, G. M., The threshold problem in ecosystems, in *Ecosystem Analysis and Prediction,* Levin, S. A., Ed., SIAM Institute for Mathematics and Society, Alta, UT, 1974, 9.
24. Stephan, C. E. and Rogers, J. W., Advantages of using regression analyses to calculate results of chronic toxicity tests, in *Aquatic Toxicology and Hazard Assessment: Eighth Symposium,* STP892, Bahner, R. C. and Hansen, D. J., Eds., American Society for Testing and Materials, Philadelphia, 1985, 328.
25. Mayer F. L. and Ellersieck, M. R., Manual of Acute Toxicity: Interpretation and Data Base for 410 Chemicals and 66 Species of Freshwater Animals, Resource Publication 160, U.S. Department of Interior, Washington, D.C., 1986.
26. Stephan, C. E. et al., Guidelines for Deriving Numerical National Water Quality Criteria for the Protection of Aquatic Organisms and Their Uses, PB885–227049, National Technical Information Service, Springfield, VA, 1985.
27. Van Straalen, N. M. and Denneman, G. A. J., Ecological evaluation of soil quality criteria, *Ecotoxicol. Environ. Saf.*, 18, 241, 1989.
28. Van Leeuwen, K., Ecotoxicological effects assessment in the Netherlands: Recent developments, *Environ. Manage.,* 14, 779, 1990.
29. Seegert, G., Fava, J. A., and Cumbie, P. M., How representative are the data sets used to derive national water quality criteria?, in *Aquatic Toxicology and Hazard Assessment: Seventh Symposium,* STP854, Cardwell, R. D., Purdy, R., and Bahner, R. C., Eds., American Society for Testing and Materials, Philadelphia, 1985, 527.
30. La Point, T. W. and Perry, J. A., Use of experimental ecosystems in regulatory decision making, *Environ. Manage.*, 13, 539, 1989.
31. Bartell, S. M., Gardner, R. H., and O'Neill, R. V., An integrated fates and effects model for estimation of risk in aquatic systems, in *Aquatic Toxicology and Hazard Assessment: 10th Volume,* STP971, Adams, W. J., Chapman, G. A., and Landis, W. G., Eds., American Society for Testing and Materials, Philadelphia, PA, 1988, 261.
32. Tolle, D. A., Arthur, M. F., Chesson, J., and van Vorris, P., Comparison of pots versus microcosms for predicting acroecosystem effects due to waste amendment, *Environ. Toxicol. Chem.*, 4, 501, 1985.
33. Niederlehner, B. R. and Cairns, J., Jr., Naturally derived microbial communities as receptors in toxicity tests, in *Ecological Toxicity Testing; Scale, Complexity and Relevance*, Cairns, J., Jr. and Niederlehner, B. R., Eds., Lewis Publishers, Boca Raton, FL, 1995, 123.
34. Touart, L. W., Aquatic Mesocosm Tests to Support Pesticide Registration, EPA 540/09–88–035, National Technical Information Service, Springfield, VA, 1988.

35. Crossland, N. 0. and La Point, T. W., Convenors, Symposium on aquatic mesocosms in ecotoxicology, *Environ. Toxicol. Chem.,* 11, 1, 1992.
36. Vollenweider, R. A., Input-output models with special reference to the phosphorus loading concept in limnology, *Schweiz. Zeitschrift Hydrologie*, 37, 53, 1975.
37. Baker, J. P. and Christensen, S. W., Effects of acidification on biological communities in aquatic ecosystems, in *Acidic Deposition and Aquatic Ecosystems: Regional Case Studies,* Charles, D. F., Ed., Springer-Verlag, New York, 1991, Chap. 4.
38. Minns, C. K. et al., Assessing the potential extent of damage to inland lakes in eastern Canada due to acidic deposition. III. Predicted impacts on species richness in seven groups of aquatic biota, *Can. J. Fish. Aquat. Sci.*, 47, 821, 1990.
39. Crane, M., Grosso, A., and Janssen, C., Statistical techniques for the ecological risk assessment of chemicals in freshwaters, in *Statistics in Ecotoxicology,* Sparks, T., Ed., John Wiley and Sons, New York, 2000, 247.
40. Cairns, J., Jr., Smith, E. P., and Orvos, D., The problem of validating simulation of hazardous exposure in natural systems, in *Proc. 1988 Summer Computer Simulation Conf.,* Barnett, C. C. and Holmes, W. M., Eds., The Society for Computer Simulation International, San Diego, 1988, 448.
41. Sanders, W. M., III, Field validation, in *Fundamentals of Aquatic Toxicology,* Rand, G. M. and Petrocelli, S., Eds., Hemisphere, New York, 1985, 601.
42. Carlson, A. R., Nelson, R. H., and Hammermeister, D., Development and validation of site-specific water quality criteria for copper, *Environ. Toxicol. Chem.*, 5, 997, 1986.
43. Mount, D. I., Steen, A. E., and Norberg-King, T. J., Eds., Validity of Effluent and Ambient Toxicity Testing for Predicting Biological Impact on Five Mile Creek, Birmingham, Alabama, EPA 600/8–85–015, National Technical Information Service, Springfield, VA, 1985.
44. Mount, D. I. et al., Effluent and Ambient Toxicity Testing and Instream Community Response on the Ottawa River, Lima Ohio, EPA 600/3–84–080, National Technical Information Service, Springfield, VA, 1984.
45. Cairns, J., Jr. and Smith, E. P., Developing a statistical support system for environmental hazard assessment, *Hydrobiologia*, 184, 143, 1989.
46. Niederlehner, B. R. et al., Field evaluation of predictions of environmental effects from a multispecies-microcosm toxicity test, *Arch. Environ. Contam. Toxicol.,* 19, 62, 1990.
47. Aber, J. D. et al., A strategy for the regional analysis of the effects of physical and chemical climate change on biogeochemical cycles in northeastern (U.S.) forests, *Ecol. Model.*, 67, 37, 1993.
48. Thorne-Miller, B., *The Living Ocean: Understanding and Protecting Marine Biodiversity,* Island Press, Washington, D.C., 1999.
49. Gullet, W., Environmental protection and the precautionary principle: A response to scientific uncertainty in environmental management, *Environ. Plan. Law J.*, 14, 52, 1997.
50. Saladin, C., Precautionary principle in international law, *Inter. J. Occup. Environ. Health*, 6, 289, 2000.
51. Lyons, G., Ahrens, B. A., and Salter-Green, E., An environmentalist's vision of operationalizing the precautionary principle in the management of chemicals, *Int. J. Occup. Environ. Health*, 6, 70, 2000.
51. Odum, E. P., Trends expected in stressed ecosystems, *Bioscience*, 35, 419, 1985.
52. Rapport, D. J., Regier, H. A., and Hutchinson, T. C., Ecosystem behavior under stress, *Am. Nat.*, 125, 617, 1985.
53. Schindler, D. W., Detecting ecosystem responses to anthropogenic stress, *Can. J. Fish. Aquat. Sci.*, 44, 6, 1987.
54. Cairns, J., Jr., Quantification of biological integrity, in Integrity of Water, 055–001–01068–1, Ballentine, R. K. and Guarraia, L. J., Eds., U.S. Government Printing Office, Washington, D.C., 1977, 171.
55. Schaeffer, D. J., Herricks, E. E., and Kerster, H. W., Ecosystem health. I. Measuring ecosystem health, *Environ. Manage.,* 12, 445, 1988.
56. Rapport, D. J., What constitutes ecosystem health?, *Persp. Biol. Med.*, 33, 120, 1989.

CHAPTER 35

# Population Modeling

**John R. Sauer and Grey W. Pendleton**

## CONTENTS

## 35.1 WHAT IS A POPULATION MODEL?

A model is a simplification of a real system that is used to provide some insights into the system. Population models are often (although not always) formulated as a set of rules or assumptions, expressed as mathematical equations, that describe change in population size over time as a consequence of survival and reproduction, and they include external factors that affect these characteristics. A model simplifies a system, retaining essential components while eliminating parts that are not of interest. Ecology has a rich history of using models to gain insights into populations, often borrowing both model structures and analysis methods from demographers and engineers. Much of the development of the models has been a consequence of mathematicians and physicists seeing simple analogies between their models and patterns in natural systems. Consequently, one major application of ecological modeling has been to emphasize the analysis of dynamics of often complex models to provide insights into theoretical aspects of ecology.[1]

Population models are also an intrinsic part of experimental studies and population management. Prediction of future population dynamics in the context of a model is an essential component of understanding the consequences of changes in demographic parameters such as survival or productivity. The model provides the context for assessing population consequences of environmental influences on the demographic parameters. Obviously, any population manager needs the integrative and predictive ability of models to assess the consequences of management, and the extensive use of models in population viability analysis[2] (PVA) and in management of exploited populations[3] reflects this need. Adaptive management[4] is explicitly model-based in that models provide a basis for evaluating alternative management actions and are revised and updated based on the monitored response to management.

These practical uses of models have created a great deal of interest in both the process of modeling and the estimation of parameters and model structure. Modeling efforts can now draw upon a variety of books and other published works on modeling[5,6,7] and computer software packages for modeling (e.g., Ramas Software[8]). Practical aspects of estimation of survival and reproduction parameters for animals in the field have also been the subject of extensive development,[9] and general estimation programs for survival rates, such as MARK, are now available on the Internet.[10] Finally, papers such as Nichols et al.[11] provide a good example of modeling efforts that integrate existing data into a population model, show how changes in elements of the model affect the theoretical population, and discuss the practical consequences of the model structure in the management of a species. The relevance of population modeling to toxicological studies is demonstrated by increasing numbers of modeling papers appearing in journals specializing in toxicology.[12,13]

In this chapter, we discuss some of the uses of population modeling for toxicological studies and review the basic ideas of population modeling. After reviewing some of the terminology and types of population models, we discuss (1) the use of modeling procedures that assume that populations can be grouped into discernible age classes, and that survival and fecundity rates can be measured over intervals for these groups; (2) methods for analyzing the stable population attributes of these models; (3) methods for assessing the effects of changes in the parameters of the models; and (4) applications of the models to evaluation of the effects of changes in the demographic parameters. These discrete-time, discrete-age class models are commonly used for vertebrates. We also briefly review the analysis of other kinds of models, such as individual-based simulation models and time series models, and discuss the uses of intervention analysis to assess the effects of a given toxin on a population. And, because models are useless without reasonable estimates of parameters, we review methods of obtaining reliable estimates of parameters and relationships used in the model. Our discussion and analysis are somewhat idiosyncratic, and further information and more general expositions on the topic can be found in several textbooks.[5,7]

## 35.2 HOW ARE MODELS OF USE IN ECOTOXICOLOGICAL STUDIES?

For ecotoxicological studies, models have several roles:

1. Field and laboratory studies can provide information on how pesticides or other contaminants affect survival and reproduction.[14] It is often unclear how a specific change in a demographic parameter affects the overall dynamics of the population, as some changes may have little effect on the overall population growth rate while others may have relatively large effects. After survival and reproduction parameters are estimated, models provide a context for examination of population effects of changes in the population parameters. Once the model is developed, it can be tested to see if it does provide a reasonable analogue to the dynamics observed in nature, and the stable population attributes of population growth rates, age distributions, and reproductive values can be estimated.
2. If the model is a realistic mimic of nature, it can be used to predict the possible effects on the population of changes in survival and reproduction rates. Changes in survival and reproductive rates, such as those caused by contaminants, can be hypothesized, and the effects of the changes on growth of the population can be assessed, both analytically from mathematical analysis of the model and through simulation of the model with computers. Direct analysis of models using mathematical procedures can be extremely complex. Analysis of the sensitivity of rates of population growth to changes in the components of the model has traditionally been an important, and sometimes controversial, part of population modeling. PVA seeks to predict extinction probabilities for declining species from models.[2]
3. Modeling provides a logical way of summarizing relevant information regarding the population of interest[15] and is therefore a reasonable way of sorting and summarizing existing information, identifying gaps in knowledge, and directing future research. It often becomes apparent early in the process where additional research is needed to supply information to the model.[16] Developing a quantitative model also requires explicit statement of model assumptions, which can then be tested through appropriate data gathering and experimentation.[15,17]
4. In toxicological studies, it is generally difficult to take information from laboratory and limited field studies and extrapolate to population effects in natural situations. Modeling provides one way of integrating the information and predicting population response. However, uncertainty still exists in translating simulation and analytical results from models to field situations. The adaptive management paradigm is an important philosophical advance that integrates management, modeling, and population monitoring information to increase our understanding of causal factors that influence population change. In adaptive management, it is assumed that uncertainty exists in population response to management and that models can be developed to predict the outcome of management activities (e.g., a demographic model that predicts the population response to a pesticide application).

   Often, the model structure is not well defined, and alternative models can be formulated that provide different predictions of the population response. We use our best information on which model is appropriate to predict the population response and then conduct the management. After management, we monitor the population to determine the response and establish which model best predicted the response. The system is adaptive in that future predictions are then based on the best model, and the management helps us determine which model is most likely. This model-based approach to extracting causal relationships is appealing in that it avoids both the weak inference characteristic of association analyses based simply on monitoring data and the reductionist approaches associated with laboratory or limited field studies. Adaptive management has been implemented in fisheries[4] and in waterfowl management[3] and is advocated as an alternative to PVA, especially when uncertainty occurs in model structure and parameter estimates.[2] It is likely that adequate understanding of the role of toxins in population dynamics of widespread species will require use of a model-based adaptive-management approach in which toxins are modeled in conjunction with other exogenous factors that influence the populations.

5. Many alternative forms of population modeling can be used in toxicological studies, and new modeling methods, such as neural networks,[12] are being applied to toxicology studies. Many monitoring programs are of potential use for correlation studies of toxicology, and in these programs population change is frequently modeled from a series of population estimates. The effects of introducing a pesticide to the population can be directly assessed using time-series analysis methods. This modeling of population changes directly from population size estimates uses estimation procedures from the statistical field of time-series analysis[18] and has only infrequently been used in ecological research.[19]

## 35.3 CONSIDERATIONS IN MODEL BUILDING

### 35.3.1 The Iterative Nature of Modeling

We assume that readers of this chapter will be interested in the general question of how the presence of toxins in the environment modify the population dynamics of a species. Why do we need a model to address this question, when we can evaluate the presence of toxins on individual survival and productivity rates and possibly even detect direct effects of toxins on population sizes of animals? It is easy to detect consequences of toxins for survival and reproduction, but evaluation of these effects at the population level is difficult. A model provides a framework for specifying the effects of the toxins on demographic parameters in the context of other factors that influence population dynamics, allowing investigators to evaluate the possible consequences of toxins for the populations. This integration and extrapolation of information is a critical component of modeling and opens the door to both simulation and analytical studies of the population effects of toxins; in addition, it provides the basis for prediction and information updating needed for adaptive management. However, models are abstractions, and care is needed in developing and updating models to ensure that they are reasonable descriptions of the systems of interest.

To address any question using a model, we must (1) develop a preliminary model of the population dynamics of the species, (2) estimate the parameters of interest, and (3) validate the model using additional information from the population. Obviously, no model is complete, and often the steps will be repeated many times as new information about the population becomes available. Remaining uncertainties in the model can be retained by considering a model set in which alternative models that embody the different assumptions are developed. Adaptive management methods explicitly allow for use of future population information to identify the most likely model from the set.[3] After the model has been shown to be appropriate for the population, we can use the model to evaluate the dynamics of the population. The effects of toxins on the population can be simulated by changing the values of parameters in the model that are likely to be affected by the toxins and observing the consequences to the dynamics of the model or by modeling the effect of the toxin explicitly within the model.

### 35.3.2 What Type of Model Is Appropriate?

A large number of model structures exist. Models can be very simple and designed to give general insights into population processes, or they can be exceedingly complex, if they are needed to evaluate a specific situation. They can be individual-based, in which individual animals are characterized by a vector of state variables and their probability of surviving and reproducing depend on the values of these variables, or aggregated, in which groups of animals are assumed to have similar characteristics and common survival and fecundity rates. Some models have analytical solutions in which an equation can be written for equilibrium population sizes, while other models can be analyzed only through computer simulation. Finally, models can be indexed by discrete increments of time, in which the population is only defined at specific points in time, or continuous, when the population is defined at all times. The appropriate model structure is a consequence of the characteristics of the modeled population, the type of information available for the population, and the desired product.

Selection of the model structure should be guided by both the goals of the modeling exercise and the practical constraints of the existing data. Individual-based models allow each individual to have a series of attributes that could influence its survival, and therefore many opportunities exist for evaluating the influence of the environment (as shown by the values of the attributes) on survival. One difficulty in developing these models, however, is that we often do not know how the attributes influence individuals, so the model becomes a series of guesses with little predictive value. Instead, we usually estimate survival and fecundity for groups of animals, and models that are based upon age classes are more relevant to those data.

Several works provide extensive background into appropriate formulations of model structure.[5,7] The model-building process involves clearly defining (1) the system of interest in space and time; (2) state variables, such as population size, that the model parameters act to change over time; (3) endogenous variables that influence the population and are intrinsic to the system and can be managed; (4) exogenous variables that are outside the system and cannot be managed but also influence the population; (5) a transition equation that is used to predict the change in state variables over time; and (6) a time scale of the modeling effort.

### 35.3.3 Model Output

Once a model is constructed, outputs from the model are used first to verify and validate the model, to judge the success of the modeling effort based on existing information about the population. Once the model is documented to be technically correct and an accurate reflection of the actions of the population under known conditions, it can then be used as a predictive tool to evaluate possible population-level effects of toxins. Several characteristics of the model population that might be affected by toxin-induced changes in survival or reproduction rates might be of interest, depending on the nature of the model and the objectives in building the model. Some of these are applicable to virtually all population models, others to only a few.

In discrete-time demographic models, a characteristic of interest is the asymptotic finite rate of population increase, $\lambda$. Estimates of $\lambda$ indicate whether the population is decreasing, stable, or increasing. Other characteristics that might be of interest, particularly in stochastic models, include probability of extinction or pseudoextinction, average population size over time and the associated variation in population size, and average recovery time following perturbation.[20] Pseudoextinction differs from extinction by being the probability that population size will decline to a specified positive value rather than to 0.[20] Many types of contaminants affect wildlife populations as a short-term source of mortality (e.g., Blus et al.[21]). The time required for a population to return to pre-effect size following a short-term phenomenon could be of interest in toxicological investigations.

### 35.3.4 Range of Applicability

In obtaining reliable estimates of model inputs and their variances, it is important to clearly define the system defined by the model, which includes species, habitat, region, etc. Following this, the sample unit for the various model variables must also be identified. For example, the desired range of applicability for a small mammal population model might be *Microtus* sp. populations in all areas of a specific habitat type (e.g., grasslands) within a physiographic region (e.g., coastal plain). The sampling unit within these grassland areas might be trapping grids where a capture-recapture procedure is used to estimate survival rates.

### 35.3.5 Sources of Variability

Several sources of variation might be considered when estimating precision of model outputs. Total variability associated with estimates of parameters used in models results from process variation (i.e., inherent geographic or temporal variability) within the range of applicability (even if all possible samples within the region were measured), sampling variability (because not all samples are measured), and measurement variability (within a single sample point the parameter might not be known exactly) (Skalski and Robson,[22] Burnham et al.,[23] Wisdom et al.[24]). It is important to identify and, if possible, to estimate the magnitude of all sources of variability affecting model output variables. In the small mammal example, survival rates estimated at each trapping grid have an associated, model-based variance estimate.[25] This is an estimate of the measurement error associated with a single data point. However, this estimate does not contain all of the variability associated with survival-rate estimates for the region. Process and sampling (intergrid) variation in survival rates is likely to be a large component of the variability. For this reason, a single sample grid, although it provides an estimate of one source of variability, would not be sufficient for many purposes. Replicated grids in space and time are necessary for estimating the sample variance component[26] (Burnham et al.,[23] Skalski and Robson[22]). Not all types of variability will be applicable or relevant to all models. For example, if a model were to apply only to a small population covered by a single trapping grid, the model-based variance would be the estimate of sampling variation as well as measurement variation.

An additional source of variability is that associated with estimating relationships between variables. Uncertainty associated with these estimates also affects the variability of model outputs.

### 35.3.6 Sensitivity Analysis

Sensitivity analysis is used to determine which of the model parameters have the greatest effect on the outputs of interest. Sensitivities are important to know, for two reasons. First, if the model is a good approximation of the real population of interest, components in the model that are highly influential in shaping model outputs might also be the factors that influence the actual population, and knowledge of the importance of components will suggest effective management strategies. Second, sensitivity analyses can give insights into how to construct and improve the model. Elements of survival and reproduction that most influence population dynamics should be estimated with the most precision, and relative sensitivities can guide efforts to generate more precise estimates of variables or relationships, to generate parameter estimates from other locations to expand the region of applicability, or to develop improved measurement procedures.

Two general categories of sensitivity analysis are useful in investigating the characteristics of population models, which we call *structural sensitivity* and *statistical sensitivity.*

Structural sensitivity estimates reflects inherent model relationships without regard to variability associated with estimates of model parameters. Often, mathematicians will conduct local sensitivity analyses on models. In these analyses, the effects of adding a very small value to parameters on model output are assessed analytically.[27,28]

Statistical sensitivity determines which model variables have the greatest effect on the model output because of the precision with which they are estimated.[24,29] Even if a component of a model (such as survival rate) does not have great structural sensitivity, variability resulting from sampling and measurement processes could greatly affect the model. Model variables estimated with high variability could produce large changes in model output by chance alone. In some sense, statistical sensitivity measures reliability of model outputs as a function of the data quality of the input variables.

As discussed in the preceding section, variability in estimates of model variables can result from several sources, including measurement variation, sampling variation, and process variation. Examining sensitivity of model output with respect to each source of variation that applies to that particular model could be informative.[24] For example, variables highly sensitive to measurement variation would be good candidates for increased effort to achieve more precise estimates or to develop better measurement procedures, and variables sensitive to sampling variation might suggest larger samples. In addition, sensitivity to process variability could indicate where caution is needed in applying the model to different areas or populations.

All of the sensitivity analyses discussed above are contingent on a fixed model structure and emphasize variation in the magnitude of the variables. However, one could also think of model sensitivity in terms of changes in model structure. Simple models incorporating only survival and fecundity rates are often not accurate, but little is known about the structure of density dependence and other associations among model components. Sensitivity analysis can also investigate variability induced in the model as a consequence of changes in model structure. For example, the consequences for population growth rates of two different models for density dependence could be explored by sensitivity analysis of the two alternative models under different conditions. This model uncertainty is an explicit component of adaptive management. Often, model sets are developed that embody differing assumptions about the model structure. The predictive abilities of each model in the set are "tested" by management, and our understanding of which model best fits the system is updated over time by evaluating the outcome of the management.[3,4,7]

### 35.3.7 Estimating Parameters and Testing Assumptions

When a preliminary model has been developed, reliable estimates of model parameters and relationships must be obtained whenever possible. Using models with only poorly estimated parameters or guesses at relationships is largely an academic exercise. Detailed and complex analyses of such models reveal only information regarding the dynamics of the model, and little or nothing about the dynamics of the population of interest. As was previously discussed, models are a set of assumptions expressed as equations; using data, experimentation, and adaptive management to evaluate the validity of these assumptions is another important component of the modeling process. Testing of assumptions allows for informed refinement of the model structure.

## 35.4 DISCRETE-TIME POPULATION MODELS

We begin our discussion of specific models with the simplest of models, a difference equation model for a population without age structure. In this model, we specify the state of the population at some time (generally denoted as $t$) and a set of rules that govern the transition of the population from its state at time $t$ to that at a future time $t + 1$. This leads to the equation

$$N_{t+1} = \lambda N_t \tag{35.1}$$

where $N_t$ is the number of animals at time $t$, and $\lambda$ is the finite rate of increase. In this simple model, the only information one has on the dynamics of the population is the size of $\lambda$: if $\lambda = 1$, the population is stationary; if $\lambda > 1$, the population is increasing exponentially; and if $\lambda < 1$, the population is declining.

As written in Equation 35.1, this model provides little insight into factors that influence change in population size, and the only way to estimate $\lambda$ is from direct observation of population counts at different times. Although modeling can be done from population counts (we discuss the potential for this in later sections), demographic analyses usually attempt to divide $\lambda$ into parts that represent the four basic causes of population change: births, deaths, immigration, and emigration. We can rewrite $\lambda$ in terms of births and deaths to provide a set of rules for the projection of population size from $t$ to $t + 1$.

For example, we could specify that an individual in the population at $t$ has to survive with probability $P$ to still be in the population at time $t + 1$ and that individuals reproduce right before time $t + 1$ to produce $F$ young/adult in the next period. If this occurs,

$$N_{t+1} = PN_t + PFN_t = P(1 + F)N_t = \lambda N_t \quad (35.2)$$

Alternatively, we could specify that an individual could survive with probability $P$ but reproduce right after count to produce $F$ young, which then survive with probability $P'$ until the next period. In this case,

$$N_{t+1} = PN_t + FP'N_t = (P + FP')N_t = \lambda N_t \quad (35.3)$$

Consideration of these models suggests that we must make several assumptions for these structures to accurately model the population:

1. The interval of time $t$ to $t + 1$ must be defined in terms of the life cycle of the organism of study. Often, this interval refers to the time between annual reproductive events (i.e., generations). This implies that births within the interval must occur at the same time, which is referred to as a birth pulse. For white-tailed deer (*Odocoilus virginianus*) in Maryland, for example, young are born during a relatively restricted time in spring, and the natural interval is 1 year.
2. The model requires that we set a time of census, which specifies the time at which we reference the population relative to when the birth pulse occurs. The two formulations of $\lambda$ presented above differ in time of census relative to when the animals breed. The first case is called a *post-breeding census*, and the second case is the *prebreeding census*. To actually apply these models to a real population, choice of census time is critical in determining the structure of the model and what has to be estimated.
3. To make general statements about population parameters, the model requires that survival and fecundity rates are constant over time. Assumptions of a stable population are critical for the validity of this type of model because analyses of the dynamics of the model are valid only if the model is appropriate for the data. The assumption of a stable population can be tested via estimation of parameters. If $\lambda$ changes over time, it must be indexed by time ($\lambda_t$), which introduces a great deal of complexity for analysis of the model. This can be accommodated by either assuming a model for changes in $\lambda$ over time, treating the survival and fecundity rates as random variables that have variances, or modeling changes in survival and fecundity with covariates.
4. We must also assume that all individuals survive and reproduce at the same rates. In practice, $N$ is often not homogeneous, and the simple model (Equation 35.1) must be extended to allow for age groups or other stages in the population that differ with regard to survival or fecundity. One major source of heterogeneity is sex, and population models are generally constructed for females only.

The ramifications of these assumptions, and the consequences of their violation, have been well described (e.g., Caughley,[30] Caswell[5]). Some of them are relaxed somewhat by developing more complex models that incorporate age-structure or density dependence, and some can be tested by estimation of model parameters. However, adding complexity (and realism) to the model complicates both estimation of the parameters in the model and analysis of the dynamics of the model. The goal of most analyses is to fit the simplest model that accurately describes the population. Ideally, the result is a model complex enough to mimic the population of interest, while being simple enough so that its workings are understandable.

## 35.4.1 Adding Realism

### *35.4.1.1 Density Dependence*

Observation of real populations indicates that population size changes over time, but populations never experience exponential growth or decay for very long. Therefore, it would be reasonable to make $\lambda$ dependent either on time ($\lambda_t$) or on population size $\lambda_{N_t}$. One reasonable elaboration of Equation 35.1 would be to incorporate a density-dependent effect, so that the population would be limited at some level $k$, or

$$N_{t+1} = N_t e^{r(1 - N_t/k)} \tag{35.4}$$

This simple nonlinear difference equation can have extremely complex dynamics for certain values of $r$ and $k$, and the possibility that natural systems with these complex (chaotic) dynamics could be modeled by these simple equations has been the topic of much discussion in recent years.[31] Proponents of chaos theory suggest that other modeling efforts could miss this crucial dynamical feature (e.g., Schaffer[32]). Generally, it is claimed that the presence of these chaotic systems can be detected from long time-series of population counts, and much controversy exists regarding the ability of these methods to detect chaos in the presence of sampling error in the counts. Consequently, this philosophical approach to modeling may be of limited value to applied studies of the effects of toxicants on wildlife populations.

There are ways of incorporating density dependence into models that do not produce chaotic results, and even models that produce chaos under certain parameter values are stable under other conditions. In the models we discuss later in the paper, there are values of parameters for which the model is stable. We also (briefly) discuss density dependence in more complex models and review some of the nuances involved in the estimation of density dependence in nature.

#### *35.4.1.2 Age Specificity*

It is unlikely that all individuals in a population have the same $\lambda$. Different individuals have different probabilities of surviving and reproducing, hence $N_t$ is often divided into groups, such as age classes, members of which have similar survival and fecundity rates. To model this case we use several simultaneous difference equations, which are usually presented as matrices, or

$$N_{t+1} = AN_t \tag{35.5}$$

$N_t$ is a vector of numbers of animals in each age class, or $N'_t = [n_{1,t}, n_{2,t}, \ldots, n_{m,t}]$, (the prime indicates transpose), and the matrix **A** is composed of fecundity and survival elements that project the number of animals in each of the $M$ age class from time $t$ to time $t + 1$. This model is often called the Leslie matrix and was proposed by Lewis[33] and Leslie.[34] Note that **A** is not subscripted, suggesting that it is valid for any time. This is a critical assumption that will be discussed later.

Some care is necessary in developing the elements of the matrix **A**, and a good deal of confusion exists in the literature regarding the definition of age classes and transitions between age classes. The elements determine the contributions of each age class to all other age classes from a specific time (the time of census) to some time in the future. As with $\lambda$ in the simple difference equation, the structure of **A**, and consequently what features must be estimated to construct **A**, depend on when the time of census occurs relative to the birth pulse.

Noon and Sauer[28] discuss the structure of matrix **A** and the numbering of the age classes at time of census for pre- and post-birth-pulse censuses. They suggest that age classes be labeled as the number of intercensus intervals that animals in the model have lived. For prebreeding censuses, animals in the first visible age class have lived one interval, while for post-breeding census the first age class is newborns of age 0. Survival is indexed by the age class at the start of the interval, and fecundity is indexed by the age class at the time births occur. We outline the structure of the matrix **A** for pre- and post-breeding censuses (Figure 35.1).

a. Pre-Breeding Census

$$N_t = \begin{pmatrix} n_1 \\ n_2 \\ n_3 \end{pmatrix}, \quad A \quad \begin{pmatrix} s_0p_1 & s_0p_2 & s_0p_3 \\ s_1 & 0 & 0 \\ 0 & s_2 & 0 \end{pmatrix}$$

b. Post-Breeding Census

$$N_t = \begin{pmatrix} n_0 \\ n_1 \\ n_2 \\ n_3 \end{pmatrix}, \quad A = \begin{pmatrix} s_0p_1 & s_1p_2 & s_2p_3 & 0 \\ s_0 & 0 & 0 & 0 \\ 0 & s_1 & 0 & 0 \\ 0 & 0 & s_2 & 0 \end{pmatrix}$$

c. Post-Breeding Census, Stage-Based

$$N_t = \begin{pmatrix} n_0 \\ n_1 \\ n_2 \\ n_3 \end{pmatrix}, \quad A = \begin{pmatrix} s_0p_1 & s_1p_2 & s_2p_3 & sp_3 \\ s_0 & 0 & 0 & 0 \\ 0 & s_1 & 0 & 0 \\ 0 & 0 & s_2 & s \end{pmatrix}$$

**Figure 35.1** Leslie matrices for (a) pre-breeding and (b) post-breeding censuses. The matrices project the age classes from time $t$ to time $t + 1$, and survival rates ($s_i$) and probabilities of individuals reproducing ($p_i$) apply to the age class $i$ for the interval, with survival always indexed to age class at start of interval, but reproductive rates are indexed by the age class at the time of adjacent census. A simple stage-based variation, in which survival and reproductive rates become constant after age 3, is presented in (c). We also indicate the vectors of age classes, showing the number of visible classes at the time of census.

### 35.4.2 Stable Population Analysis

The population change defined by matrix **A** is dependent on the number of animals in different age classes, and $\lambda$ will differ as the age structure of the population changes. However, under certain circumstances, matrix **A** can be "solved" to provide information on the long-term dynamics of the population. If the matrix **A** is correct in that the structure is appropriately defined and the component survival and fecundity rates correctly estimated, over time the original age structure is lost, and the matrix defines both the long-term growth rate of the population and the asymptotic structure of the population. This property is called *ergodicity*, and if it occurs, we can use matrix theory to analyze the structure of **A**, which provides information on asymptotic rate of increase, stable age distributions, age-specific reproductive values, and the effects of changing elements of **A** on population growth rate.[5]

In a stable population, the rate of change from year to year will become constant. This growth rate is analogous to the $\lambda$ of Equation 35.1 and is a structural component of matrix **A** called the dominant eigenvalue. Corresponding to the dominant eigenvalue is an eigenvector (the right eigenvector), which contains the proportion of the population in each age class. In a stable population growing according to **A**, $\lambda$, the eigenvector (**x**), and **A** are related by:

$$Ax = \lambda x \tag{35.6}$$

Now, we can also analyze the structure of the transpose matrix **A′**, which leads to the same eigenvalue but a different (left) eigenvector **y**

$$A'y = \lambda y \tag{35.7}$$

The rows of the vector **y** represent the reproductive values of each age class in the stable population. These numbers indicate the relative contribution of individuals in each age class to future generations expressed relative to the contribution of a new individual. When these vectors are presented, they are often normalized to set the first value of **y** to 1 ($y_1 = 1$).

The stable-population model can be reparameterized to be the actuarial life table, which is used in human demographic studies.[28] It has provided the basis for a good deal of evolutionary theory, dating back to Fisher,[35] who recognized that it provided a framework for analysis of the effects of natural selection on life histories and suggested that natural selection should act to maximize population growth rate.[35]

#### 35.4.2.1 When Are Population Models Unstable?

There exist circumstances under which the difference equations will not be stable.[36] If this occurs, matrix **A** cannot be solved to provide stable population attributes, and the actual dynamics of the population will be much different than that predicted by the stable-population analysis. Sometimes, the model simply does not meet the structural conditions for stability. A theorem of matrix algebra called the Perron-Frobenius theorem defines structural conditions under which a matrix will have a dominant eigenvalue. Basically, if the matrix is primitive (meaning all elements of the matrix become positive in the matrix $\mathbf{A}^k$ for some $k > 0$) it can achieve stability under appropriate conditions. In some biologically relevant cases, this is not true, for example, if only a single age class is reproductive such as occurs in semelparous organisms.[5]

Often, matrix models are not appropriate because the population is undergoing perturbations in survival and fecundity; hence, the conditions expressed in the model do not apply. For example, attributes of a stable population are generally valid only if the matrix **A** is constant and the population has been growing according to **A** for some time. If the population has only been growing according to **A** for a short time, it may not be stable, and the transient dynamics or rate of approach to the stable age distribution is of interest. And, if elements of the matrix **A** are not fixed but change over time, the population will be stable only under very restrictive conditions. See Caswell[5] for an excellent discussion of stability in population matrices.

### 35.4.3 Sensitivity Analysis

#### 35.4.3.1 Analytical Approaches to Sensitivity Analysis

Caswell[5] observed that the eigenvectors of stable age distributions and reproductive values can be used to define (structural) sensitivities of the dominant eigenvalue of the matrix **A** to small changes in the elements of **A**. A matrix of sensitivities can be defined using

$$\frac{d\lambda}{da_{ij}} = \frac{x_i y_j}{\langle x, y \rangle} \tag{35.8}$$

where lowercase letters represent specific elements of **A** (i,*j* represent row i and column j), **x**, and **y**, and the expression on the left of the equal sign represents the derivative of λ with respect to $a_{ij}$. Because these sensitivities may be difficult to understand due to differences in the magnitudes of fecundity and survival elements, de Kroon et al.[37] suggested estimating a proportional sensitivity (or elasticity) using

$$\frac{d\ln\lambda}{d\ln a_{i,j}} = \frac{a_{i,j} x_i y_j}{\lambda \langle x, y \rangle} \tag{35.9}$$

The sum of the elasticities is 1, allowing direct comparisons of the contributions of matrix elements. See Caswell[5] for examples of this approach.

Goodman[38] and others have discussed analogous sensitivity analyses in the classic life table framework. Meyer and Boyce[39] have also presented methods for sensitivity analysis in the context of toxicology studies.

#### 35.4.3.2 Simulation Approaches to Sensitivity Analysis

Once the matrix is estimated, we can change any of the elements and recalculate the stable-population attributes. For example, if we have evidence that a contaminant decreases survival in an age class, we can change the element in question and find the predicted λ.

### 35.4.4 Deficiencies with Stable Analyses of Simple Matrix Models

Sykes[40] noted that most applications of the matrix model for human and wildlife populations do not provide a reasonable simulation of population dynamics, which he attributed to changes in survival and reproductive rates over time. In general, the assumption of constant survival rates over time is false for most animals. For example, Anderson and Burnham[41] documented that survival rates varied over time in Mallards (*Anas platyrhynchos*), invalidating the use of composite dynamic life-table methods of survival estimation.

Caswell[5] makes a distinction between *projections*, which are defined as population dynamic consequences of demographic parameter values that characterize the system at present, and *forecasts*, which are predictions of the future state of the population. Eigenvalue and simulation analyses are projections in which the present state of the population is defined by **A**, and we evaluate growth and effects of changes conditional on **A**. Clearly, forecasting requires many more assumptions about the structure of the population and knowledge of how the matrix elements interrelate over time.

Generally, matrix models work best when the sources of parameter change are relatively small and chronic, rather than large and abrupt. Asymptotic theory of population growth and stability does not apply to a population with erratic changes in survival and fecundity. Therefore, modeling of abrupt population changes may be best done with simulation models.

#### 35.4.4.1 Incorporating Density Dependence in the Age-Class Model

The assumption that **A** is constant over time is unlikely to be valid in practice. Interactions between population size and survival and fecundity is classically thought to be important in regulating populations, although few quantitative demonstrations of it exist in nature.[28] Density dependence can be incorporated into **A** in several ways, some of which lead to stable populations.

Leslie[34] formulated a time-specific version of **A**, $\mathbf{A}_t$, in which a time-invariant matrix **A** is multiplied by an additional matrix $\mathbf{Q}^{-1}$, in which matrix **Q** is a diagonal matrix with elements

$$q_{i,t} = 1 + aN_{t-i-1} + bN_t \quad (35.10)$$

where the $N_t$ represent the total population size at time $t$ and $a$ and $b$ are constants. This density-dependent formulation will result in a stationary population, i.e., the population size will become constant. Other density-dependent formulations also exist, some of which are considerably less stable.[5]

#### 35.4.4.2 Stochastic Matrices

Another way that **A** can change over time stems from the possibility that the elements of survival and fecundity are not constant but differ as random variables over time. Because the elements are never exactly the same from time to time, the population is never exactly stable. Under certain conditions, the stochastic matrix **A** can provide a stable population (weak ergodicity[5]). See Caswell[5] for a review of the properties of long-term probabilities of extinction based on stochastic matrices.

Two kinds of variability are often discussed when modeling population dynamics. Demographic stochasticity is a consequence of random variation in survival and fecundity, which May[42] discusses as a consequence of animals existing only in integer units. Environmental stochasticity is caused by the action of the environment on population size and in practice can be thought of as treating population size, survival rates, and birth rates as random variables. In simulation models, both sources of stochasticity should be used.

#### 35.4.4.3 Stage-Based Analyses

The age-class model discussed above tends to be restrictive — we often consider survival to be age-dependent only in that young animals' survival is less than that of adults. Therefore, a simplification of the age-specific model that allows only a single adult class is appropriate. More generally, the classes we divide $\mathbf{N}_t$ into can be any grouping (or stage) for which survival and fecundity rates tend to differ in the population. For example, we could place animals into body-mass classes[43] and estimate probabilities of surviving and changing mass classes.[44] These stage-based transition matrices are often called Lefkovitch matrices and are widely used in botanical applications.[45,46] The stable population and sensitivity analysis can also generally be applied to these models.

### 35.4.5 Life Table Response Experiments

Caswell[5] suggested using life table response experiments to assess the effects of contaminants. In these studies, populations are used as replicates and exposed to experimental applications of factors that modify vital rates. Models are constructed for the replicates, and population growth rates are used as summary statistics of the effects of the factors. To evaluate how the factors influence growth, the treatment effects are decomposed to see how the perturbation affects elements of the matrices and hence survival and productivity. These experiments can include analysis of variance, random, and regression designs, and the analysis procedure can also be applied to nonexperimental situations to assess observational results. Caswell provides an informative summary of a two-factor (DDT exposure and food amount) experiment conducted by Rao and Sarma[47] on a rotifer population, extracting effects of food levels and DDT exposure on growth rate and age-specific fertility and survival.

### 35.4.6 Examples of Population Matrix Models

Because the Leslie matrix model has been in the literature since 1942, there have been many examples presented of its use. Unfortunately, most of the early applications do not have realistic estimates of survival and reproductive rates, and a great deal of confusion exists regarding the structure of the matrix. There are many examples of stage-based matrix models. Crouse et al.[48] provide an excellent example of population modeling and sensitivity analyses of a Loggerhead Sea Turtle (*Caretta caretta*) population. Martin et al.[49] analyzed Mallard populations in North America using a stage-based model incorporating age and sex classes.

Van Straalen and Kammenga[50] review 29 studies that apply demographic approaches to ecotoxicology, and Kuhn et al.[13] apply age-structured population models to mysid shrimp to project concentrations of p-nonylphenol at which population-level effects would occur.

## 35.5 INDIVIDUAL-BASED POPULATION MODELS

Another class of models is based on individual animals rather than on homogeneous groups of animals.[51] These models gain additional flexibility, and perhaps realism, through variation in parameter values at the individual level at a cost of greatly increased complexity in the model. Many more variables and relationships often must be estimated to implement individual-based models compared with aggregated models.[17] Individual-based models are usually based on computer simulation and track each individual in the model population among several states.[52] The cycle time at which individuals can change states, die, reproduce, etc. is often short (e.g., 1 day[53]) in relation to between-census (generation) times, unlike many discrete-time models that have cycle and between-census times equal. Survival rate, reproductive rates, and other factors vary among states. Output characteristics of individual-based models are usually determined through computer simulation, while those from aggregated models can often be derived analytically from the model equations.[52] Individual-based models have not been frequently used to investigate the effects of toxic substances, but Samuel[53,54] used an individual-based model of midcontinent North America Mallard populations to examine the effects of lead poisoning and disease.

In aggregated models, external influences, such as contaminant exposure, are included for each age/sex class as the mean level of exposure acting on the mean level of a parameter (survival, fecundity). In individual-based models, a range of values of the external factors can act on the individuals' survival and reproduction probabilities in the model. Individual-based models facilitate incorporation of heterogeneity in parameter values and responses to external factors,[52] nonlinear responses to external factors (e.g., individuals exposed to contaminant levels above the mean have disproportionately higher responses[55]), and rare events, such as contaminant spills.[16] Individual-based models also can be useful for species that have a protracted breeding season and multiple litters or clutches, which violates the assumptions of some aggregated models.

Aggregated models can be thought of as special cases of individual-based models.[52] Because realism is often easier to incorporate into individual-based models, they can serve as a starting point to encourage formation of testable hypotheses.[17] Following testing of hypotheses about model assumptions and structure, extraneous complexity might be rejected, leading to simpler aggregated models and their associated powerful analytical evaluation methods.[52]

### 35.5.1 Sensitivity Analysis

In individual-based population models, virtually all sensitivity analyses are performed through simulation. Input variables are varied singly by a small amount, and the model is then run for a specified number of iterations. Results are summarized as the proportional change in model outputs caused by the change in the model input. These simulations can be based on either sensitivity (change in output for a constant level [i.e., additive] change in input) or elasticity (change in output for a proportional [i.e., multiplicative] change in input), although elasticity is often more easily interpreted when variables have very different measurement scales (e.g., survival rates and clutch sizes in birds). Structural sensitivity and statistical sensitivity (possibly based on several sources of variation) vary only in how the input variables are changed. For structural sensitivity, inputs are varied by a constant amount across variables. For statistical sensitivity, estimates are varied by a constant fraction of the variance estimate for the variable (e.g., 0.5 standard errors). Investigation of sensitivity to different sources of variation would require repeating the analysis with each variance source (such as measurement variance or sampling variance).[29] Most sensitivity analyses are conducted by examining variables one at a time. However, if input variables are correlated, estimates of sensitivity or elasticity could be misleading if the correlation structure among variables is not accounted for. Incorporating covariation among inputs or nonlinear patterns of sensitivity into sensitivity analyses would add greatly to the complexity of the analyses; data are frequently lacking on broad-scale relationships among variables of interest. But if the types of data were available, it would be valuable to incorporate them when estimating sensitivity.

## 35.6 ESTIMATION OF MODEL PARAMETERS AND RELATIONSHIPS

Population change is a consequence of survival, reproduction, immigration, and emigration. Important relationships in the model structure include how these factors influence each other and how they interact with population size (e.g., density dependence, compensatory mortality). Survival and reproductive rates can be estimated, but identifying and quantifying features such as density dependence, compensatory mortality, and the influences of exogenous variables can be extremely difficult.[56,57]

To make models more useful for investigating population dynamics, the effects of external factors on survival and reproductive rates are often included. The effects of the external factors might also interact. For example, the effect of contaminant exposure on survival rate might be affected by exposure to other contaminants or by weather. Obtaining reliable estimates of model parameters and relationships and accompanying variance estimates is an important component of the modeling process. Acquiring these estimates requires experimental work that applies appropriate designs and statistical methods. A variety of books and internet products exist that discuss these methods,[7,10] and their application is prerequisite for successful model development.

One consideration in obtaining reliable parameter estimates is heterogeneity in vital rates, which can cause biased estimates if not recognized.[58–60] A common source of heterogeneity occurs among identifiable groups such as age, sex, or size classes. This possibility should be investigated, and, if necessary, estimates should be produced separately for each group. Unfortunately, other sources of heterogeneity can be more difficult to detect,[58] which is one of many reasons for caution in interpreting model results.

### 35.6.1 Survival

Several approaches are available to estimate survival rates from wildlife populations, depending on the sources of data available. Data from field studies are usually of a different form than data from captive animal studies, and different estimation methods are sometimes used (Table 35.1). Survival rates from all of these methods can be expressed as a period-specific survival rate (i.e., the probability of surviving an interval of a specific length), differing in how the pattern of mortality is modeled within the period. If the period length for the survival estimate differs from that of the model, conversion of the survival to the model interval will be necessary, which is relatively easy with constant survival methods but more difficult for methods that have nonlinear changes in survival rates over time.

**Table 35.1 Methods for Estimating Survival Rates for Several Types of Data, with Selected References and Computer Programs for Implementing the Methods**

| Data Type | Analysis Method | References | Software |
|---|---|---|---|
| Continuous monitoring (includes captive animals and most radio telemetry data) | Direct estimation (i.e., dead/exposed) | | |
| | Parametric models | Kalbfleish and Prentice,[61] Cox and Oakes[62] | |
| | | SAS Inst.[91] | Proc Lifereg |
| | Kaplan-Meier product limit | Pollock et al.[64] | |
| | | White and Garrott[63] | |
| | | Cox and Oakes[62] | |
| | | SAS Inst.[91] | Proc Lifetest |
| | Mayfield | Bart and Robson[66] | |
| | | Heisey and Fuller[67] | MICROMORT |
| | | White[92] | SURVIV |
| Release-recovery | Band recovery models | Brownie et al.[68] | BROWNIE ESTIMATE |
| | | White[92] | SURVIV |
| | | Conroy and Williams[80] | MULT |
| | | White and Burnham[10] | MARK |
| Capture-recapture | Jolly-Seber open population models and extensions | Pollock et al.[25] | JOLLY |
| | | Lebreton et al.[9] | JOLLYAGE SURGE |
| | | Burnham et al.[23] | RELEASE |
| | | White[92] | SURVIV |
| | | White and Burnham[10] | MARK |

If animals in the population can be monitored continually and all enter the study at the same time (e.g., captive animals and some radio telemetry studies), the survival rate for a specific time period can be estimated through direct estimation of the proportion of the study animals that survive the interval. This is a common measure in captive animal toxicology studies. With direct estimation, no assumptions about intraperiod mortality patterns are required unless conversion to a different interval is necessary. Estimation methods for intraperiod survival with this type of data include those that specify a parametric form for the survival rates;[61,62] the parametric methods can have either constant or nonconstant survival rates. Nonparametric methods, such as the Kaplan-Meier product limit,[61,62,63] are also applicable. The Kaplan-Meier estimator has been extended for use when study animals enter the study at different times, which is common with field radio telemetry projects.[63,64] If constant survival is assumed, the Mayfield method is of use,[65–67] particularly when animals are monitored periodically rather than continuously.

If animals are captured, marked, and released and recovered dead at some later time, band-recovery models are useful for estimating survival rates.[68] To use these models there must be a series of equally spaced capture, mark, and release events (e.g., annually) with sufficient animals for a series of recoveries from each of these release times. Period survival rates are produced for the interval between release events. Survival rates can vary among periods, and no assumptions are made about the pattern of intraperiod survival. High recovery rates (or extremely large samples) of the previously marked animals are required to get precise survival estimates. Also, estimates of annual survival rates from band-recovery models are not possible if only young-of-the-year animals are marked.[68,69]

Another set of methods for estimating survival rates is open population capture-mark-recapture models.[9,25] These differ from the band-recovery models in that animals are captured and released on more than one occasion, rather than a single capture and later recovery. "Open population" refers to the fact that deaths, births, and some types of movements can be accommodated by model estimation procedures. Period survival estimates are produced (along with estimates of population size and recruitment) for the intervals between capture occasions, and these rates need not be equal among periods.

An important assumption of these methods is that, on any capture occasion, all animals have the same probability of capture. This assumption can be relaxed by altering sample designs and incorporating closed-population capture-recapture methods in the estimation scheme.[70,71] A further assumption of capture-recapture models is that all emigration is permanent (animals cannot leave the study population and subsequently return). However, this presents a limitation of the survival-rate estimates from these models, as permanent emigration cannot be separated from mortality without additional information[25,72] (but see Nichols et al.[73]). If emigration rates are substantial, survival rates estimated with these methods will be too low. To our knowledge, increased emigration has not been attributed to contaminant exposure, although estimation of emigration rates and causes have not been extensively investigated.[72]

A very general computer package for estimation of survival and other attributes from marked animals is a program called MARK, developed by White and Burnham.[10] This package provides a general structure for estimation from capture data, encompassing band-recovery models, Cormack-Jolly-Seber models, joint live-and-dead-encounter models, modeling of known fates, robust design analyses (incorporating both open- and closed-population models), and multistrata designs that permit migration among strata. Comprehensive documentation of the package is available through the Internet at (http://www.cnr.colostate.edu/class_info/fw663/Mark.html).

### 35.6.2 Reproduction

Approaches for estimating reproductive rates are somewhat more limited than for survival rates (Table 35.2). As with estimates of survival rate, it is essential that a specific definition of reproductive rate be decided upon when building the model and obtaining estimates.[72] Net natality (the number of female young produced per adult female) is a commonly used measure, but other measures might be more appropriate in some situations. What constitutes reproduction is relatively clear-cut for mammals but less so for birds. For birds, reproduction can be defined on the basis of eggs, hatchlings, or fledglings, with the difference that survival rates at these various stages are either combined into the reproductive rate or into the juvenile survival rate.[72] Individual-based models might have a separate survival rate for each of these life stages, and contaminant exposure could be modeled to affect each separately.

It is often assumed with wildlife populations that when animals reach a certain age, all will attempt to reproduce. However, population dynamics can be affected if the proportion of breeders in a population is less than 1,[74] which might result from contaminant exposure. Changes in the age of first reproduction can be thought of as a special case of changes in the yearly proportion of breeders. Estimates of the proportion of the population breeding could be important for some models.[75]

For captive animals or those that can be monitored at will (e.g., some radio telemetry studies with animals marked before the reproductive season), reproductive rate can be estimated directly (e.g., mean brood size, etc.). The proportion of the population breeding can also be estimated directly. With captive animals, assumptions about the applicability of survival-, reproduction-, and breeding-rate estimates to the wild are necessary. It is also necessary to assume that the animals observed to provide data for the estimate are representative of the population of interest, which is best achieved

**Table 35.2 Methods for Estimating Reproductive Rates for Several Types of Data Along with References and Computer Programs**

| Data Type | Analysis Method | References | Software |
|---|---|---|---|
| Continuous monitoring (includes captive animals and most radio telemetry data) | Direction observation (i.e., average number fledging, etc.) | | |
| Capture-recapture | Jolly-Seber open population models and extensions | Pollock et al.[25] Lebreton et al.[9] Clobert et al.[72] | JOLLY JOLLYAGE SURGE |

if they are randomly selected. To ensure unbiased estimates in modeling, random selection of samples should be made in all phases of data gathering whenever possible (see Hurlbert[26]).

For field data, care must also be taken to account for complete reproductive failure, such as complete nest or brood loss. If not detected, complete nest or brood loss will result in biased estimates (too high) of reproductive rates.[66,76]

Another source of recruitment estimates is open-population capture-recapture models. Unlike using direct observation, capture-recapture estimates of recruitment estimate the number of new individuals added to the population between sampling occasions without providing a direct estimate of average production per female or per adult. An estimate of average reproductive rate could be derived using the capture-recapture population estimate along with the recruitment estimate. For species where distinguishing sexes is difficult, a reproductive measure other than net natality will be required. However, one complication associated with capture-recapture estimates of recruitment is that *in situ* reproduction is often indistinguishable from immigration without additional information. Important assumptions for open-population capture-recapture were discussed in the preceding section on estimating survival rates.

Using capture-recapture models to estimate the proportion of the population breeding is a relatively new application. See Lebreton et al.[75] and Clobert et al.[74] for a discussion of these methods.

### 35.6.3 Relationships to External Factors

Period survival rates or reproductive rates estimated from any of the described methods can be related to external variables through the use of regression methods. A critical factor in estimating these relationships is having a sufficient number of survival- or reproductive-rate estimates and explanatory variables to produce an adequate estimate. Multiple explanatory variables and nonlinear relationships require more data (i.e., estimates of survival rates) than linear relationships. Interactions among explanatory variables can also complicate the estimation process, particularly for data from field samples where confounding of variables is a common problem.

For example, levels of contaminants often vary together (e.g., DDE and PCBs), making it difficult to determine which contaminant causes which effect. Or, levels of one contaminant will be high at one location, and levels of another contaminant will be high at another location. Extreme caution should be used before attributing differences in population variables between the two sites to the contaminants; the differences could be due to site-specific habitat differences, an unmeasured contaminant, or many other factors. It should also be remembered that relationships estimated from observational field data rather than from manipulative experimentation (i.e., deliberately applied levels of the explanatory variable) indicate only correlation between the explanatory and response variables rather than cause and effect. The necessity of using such estimated relationships in population models also necessitates caution in use of model outputs. In contrast, manipulative field experiments[77] can be used to estimate cause and effect relationships providing greater confidence in models using these estimates.[22] The adaptive-management approach allows for stronger inference from observed population changes, as the observational data are used to test the predictions from alternative models.[3]

Relating nonconstant survival rates, such as from Kaplan-Meier or some parametric models, to external variables for use in population modeling is difficult and requires carefully designed experiments.[78] Incorporating these types of survival-rate functions would probably be possible only for individual-based models.

Survival probabilities of individuals (or other binomial variables[64]) can be related to explanatory variables using logistic regression[79] or proportional-hazards models.[52,62] More complex capture-recapture and band-recovery models allow direct incorporation of external effects in estimating survival rates.[9,57,78,80]

Modeling of density dependence should be viewed as an analysis of an external factor (population size) on survival or fecundity. If independent estimates of survival and population size exist, analysis of density dependence can proceed as outlined above. However, most of the experimental evidence of density dependence has been derived from nonindependent data sources such as time-series or through key-factor analysis. Kuno[81] and others have shown that these results should be viewed with caution due to the possibility of spurious relationships derived from the sampling procedures. Barker and Sauer[56] review some of the problems associated with analyzing density dependence from a series of annual counts. See Lebreton[82] for a useful discussion of estimating density dependence from population data.

## 35.7 MODELS BASED ON ESTIMATES OF POPULATION SIZE

Occasionally, we only have a time series of population size estimates and want to model the effects of covariates, such as pesticide levels, on population size. Often, these data are collected from a population in the field, and we want to model the effects of the addition of a toxin on subsequent population sizes. However, the methods are also relevant to simulations in which we change the level of survival at a known time and want to see how it affects the output of the model. Although these analyses could also reasonably be discussed in a chapter on statistical models, we have chosen to briefly discuss them here because they are relevant to the analysis of the models described above. The input data for these models are a time series of population sizes generally collected (or calculated from a model simulation) at evenly spaced intervals. Values of the level of toxins or other environmental covariable are also available at each time period, although these data could simply be a dummy variable with value 0 at times when the toxin was not present and 1 at times when it was. The goal of the analysis is to determine if the levels of the covariable were associated with the levels of the population.

A naive analysis of these data would simply use linear regression or correlation to determine if an association existed between the variables. Sauer et al.[83] reviewed how correlation analyses can be used in this context. Unfortunately, this analysis is usually not appropriate because time-series data tend to be autocorrelated, with population sizes dependent upon population sizes in earlier years. This autocorrelation leads to an inflated alpha value in correlations, rejecting the null hypothesis more frequently than is correct.[83,84] To correctly analyze the data, the autocorrelation in the data must be appropriately modeled and incorporated into the analysis.

### 35.7.1 Time-Series Analysis

Time-series analysis provides a means of accurately analyzing these data. Box and Jenkins[18] have provided a framework for modeling autocorrelation in time-series data. These models require that a time series be stationary (i.e., without trend or other features that change the autocorrelation over time), and often the count data are transformed by taking logarithms or differencing the data to remove trend. Then, autocorrelation is modeled by correlation of the data at different lags. The structure of the resulting autocorrelation function (i.e., the correlations between population sizes at a year $t$ and sizes at other years $t + i$, $i = 1,...$) is used to fit an ARMA (autoregressive moving average model). ARMA models have the form:

$$\sum_{j=0}^{p} \alpha_j X_{t-j} = \sum_{k=0}^{q} \beta_k \varepsilon_{t-k} \tag{35.11}$$

where $X$ is the time series of counts, $\alpha$ are the coefficients of the autoregressive part of the model (there are $p$ of them, so it is said to be of order $p$), $\beta$ are the coefficients of the moving average part of the model (there are $q$ of them), and $\varepsilon$ is the error term, which is assumed to be normally distributed with mean 0 and variance $\sigma^2$. The $\alpha$ and $\beta$ terms model the autocorrelation (or correlations between counts at different time differences, or lags). Because this model will only be valid for a stationary time series, often the $X$ will be transformed by subtracting $X_t$ from $X_{t+k}$ (where $k$ is an integer $\geq 1$), which is called $k$th differencing. If the data are differenced, the ARMA model becomes an ARIMA model (for autoregressive *integrated* moving average model).

To actually fit these models to data, one uses a statistical package such as SAS.[85] Inspection of the temporal patterns of autocorrelation vs. the number of lags in the data is performed to evaluate the need for differencing and $p$ and $q$ (the number of $\alpha$ and $\beta$ terms). Based on this inspection, a model is fit and the estimated parameters are tested for statistical significance using $t$ tests, as in a linear regression. Residuals from the model are subjected to autocorrelation analysis to see if additional patterns exist, and new models are fitted. Often, several possible models are fit to the data, and the model that best fits the data (as determined by a minimal value of Akaike's Information Criterion, a measure of goodness of fit discounted by the number of estimated parameters[7]) and provides residuals that do not contain any autocorrelation is selected as the appropriate model.

#### 35.7.1.1 Trend in Time Series

This modeling structure can accommodate many complex sorts of long-term autocorrelation such as seasonality and other patterns. Note, however, that while population growth rate, or trend, is treated as a nuisance parameter in these models, it is the parameter of interest for the stage-structured population models discussed previously. If we are interested in modeling trend, the other time-series components can influence both our view of the trend in the time series and the variance of the trend estimate.[56] For example, if the population has a long-term cycle, estimates of trend from a time period shorter than the cycle will result in a "trend" that is really just a portion of the cycle and is not representative of the long-term dynamics of the system.

Sauer et al.[83] provide a discussion of modeling population trend from time series using linear models. They discuss methods for determining if population trend changed at a specific time or if populations changed over an interval of interest. The crucial result from their work is that it is difficult to model changes in trends from short time series because the time-series components can produce apparent changes in trend. However, if independent replicates of the population time series exist, they can be used to determine if the changes are consistent among the time series.

Although trend is only a nuisance variable in most time-series models, there are cases in which changes in the level of the time series can be modeled in the context of autocorrelation. A covariate can be modeled with the time series to explain some of the variation in population size. We will briefly consider intervention analyses and other transfer-function models as examples of covariate modeling.

#### 35.7.1.2 Intervention Analysis

Single time-series models can be modified to allow assessment of the effects of covariates on the system. Often, we are interested in determining whether the level of a population time series changes at a specified time $t_0$. One could consider the change at time $t_0$ as a permanent change, a step in the level of the series corresponding to the time of initiation of a new pesticide or some other contaminant, or a pulse corresponding to a single episode of exposure. Box and Tiao[86] discuss modeling these changes with indicator variables, where the step variable has value 0 at time $< t_0$ but value 1 at times $\geq t_0$. The indicator for the pulse has value 1 at time $t_0$ but value 0 at all other times.

The effects of these indicator values on the population are determined by a transfer function. The value of this function models the population response to the indicator variable and describes both the change in population associated with the indicator and the return of the population to equilibrium. For example, the transfer function may treat the population response to the change as an instantaneous shift to a different level similar to an indicator variable in regression. If so, the transfer function has a single parameter representing the shift in level. Alternatively, the change to the new level could be gradual, resulting in a function with two parameters. See Box and Tiao[86] for descriptions of some of the possible shapes of the transfer function.

To fit the intervention model, Box and Tiao[86] suggest that an investigator (1) develop a transfer model that best summarizes the effects of the intervention, (2) conduct the data analysis to fit the model, and (3) see if the model adequately fits the data. If not, go back to Step (1) and try alternative models. This procedure has been used to model wildlife populations. Bautista et al.[87] used intervention models to model the effects of hunting and land use changes on Common Crane (*Grus grus*) population changes.

#### 35.7.1.3 Transfer-Function Analyses

If the covariable is itself a time series, a transfer-function model can be developed using time-series methods. Both time series are assumed to be stationary, and if not stationary they must be transformed to stationarity through differencing. Then, time-series components are fit to the covariable using the procedures outlined above. This model transforms the series to a white-noise series. The model is then applied to the population time series. Then, cross-covariances (correlation between the two time series at different lags) are estimated between the two transformed time series and are used to estimate the transfer function.

#### 35.7.1.4 Additional Methods

One of the limitations of standard time-series methods is the requirement of observations equally spaced in time, which might not be available. However, recently developed methods allow appropriate analysis of time-series data with unequal spacing. If relatively few data points are missing (including below detection limit estimates for chemical concentrations), these can be imputed using the correlation structure from the available data and, possibly, auxilliary variables. With the missing data imputed, standard time-series methods are then applied. However, this overestimates the precision of the resulting estimates because some of the points used are the imputed points. To overcome this, the missing data are imputed several times (e.g., 5–10) with the analysis performed for each of the "complete" data sets. These multiple results are then combined to get estimates and variances that include the variation due to the multiple imputations.[88,89] An advantage of this approach is that standard methods can be used while accounting for the uncertainty caused by the missing data. The second recent innovation is the ability to perform regression analyses with a variety of autocorrelated error structures.[90] These include spatial error structures that can be used to model autocorrelation with unequally spaced observations. These regression models also allow incorporation of random (i.e., nested) effects and, unlike standard time-series models, nonnormal error distributions.[90] The flexibility of autocorrelated regression models allows for appropriate analyses, producing valid estimates and variances from many types of commonly encountered field data.

## 35.8 SUMMARY

Population models provide a means for evaluating the effects of a toxin in the context of the life cycle of an organism. By developing a model and estimating demographic parameters, effects of toxins on demographic parameters, population growth rates, and model stability can be assessed. Also, modeling allows us to identify what portions of the life cycle are most sensitive to toxins and can guide future data collection and field experiments. Several types of population models are potentially of use in modeling the effects of toxins on wildlife populations. Choice of model should be based on the available demographic data and characteristics of the population of interest. We describe three classes of models: discrete-time difference equation (Leslie matrix) models, individual-based models, and time-series models. For each class, we discuss some of the nuances of developing, refining, and analyzing the models.

Estimation of model parameters is a crucial component of modeling, and we discuss methods for estimating survival and reproduction of animal populations. Most models have been based on poor estimates of the parameters, and only recently have appropriate methods been developed for estimation. Although interactions among demographic parameters may play a crucial role in the dynamics of the models, remarkably little is known about the actual form of the interactions. We emphasize the importance of the interaction between experiments to estimate demographic parameters and realistic modeling based on these estimates, and of the direct use of models in life table response experiments. The emerging science of adaptive management provides an important new role for modeling as a tool for increasing our understanding of toxicant effects on populations.

## ACKNOWLEDGMENTS

We thank F. Johnson, W. Kendall, and J. D. Nichols for helpful comments on an earlier version of the manuscript.

## REFERENCES

1. Maynard-Smith, J., *Models in Ecology*, Cambridge University Press, New York, 1974.
2. Boyce, M. S., Population viability analysis, *Annu. Rev. Ecol. Syst.*, 23 481, 1992.
3. Johnson, F. A., Moore, C. T., Kendall, W. L., Dubovsky, J. A., Caithamer, D. F., Kelley, J. R., Jr., and Williams, B. K., Uncertainty and the management of mallard harvests, *J. Wildl. Manage.*, 61, 202, 1997.
4. Hilborn, R. and Walters, C. J., *Quantitative Fisheries Stock Assessment: Choice, Dynamics, and Uncertainty*, Chapman and Hall, New York, 1992.
5. Caswell, H., *Matrix Population Models*, Sinauer Associates, Sunderland, MA, 2001.
6. Tuljapurkar, S. and Caswell, H., *Structured Population Models in Marine, Terrestrial, and Freshwater Ecosystems*, Chapman and Hall, New York, 1997.
7. Williams, B. K., Nichols, J. D., and Conroy, M. J., *Analysis and Management of Animal Populations*, Academic Press, San Diego, 2002.
8. Ferson, S., *RAMAS/Stage User Manual: Generalized Stage Based Modeling for Population Dynamics*, Applied Biomathematics, New York, 1990.
9. Lebreton, J-D., Burnham, K. P., Clobert, J., and Anderson, D. R., Modeling survival and testing biological hypotheses using marked animals: A unified approach with case studies, *Ecol. Monogr.*, 62, 67, 1992.
10. White, G. C. and Burnham, K. P., Program MARK: Survival estimation from populations of marked animals, *Bird Study,* 46(Suppl.), 120, 1999.
11. Nichols, J. D., Hensler, G. L., and Sikes, P. W., Jr., Demography of the Everglade kite: Implications for population management, *Ecol. Modeling,* 9, 215, 1980.
12. Kaiser, K. L. E. and Niculescu, S. P., Modeling acute toxicity of chemicals to *Daphnia magna*: A probabilistic neural network approach, *Environ. Toxicol. Chem.*, 20, 420, 2001.
13. Kuhn, A., Munns, W. R., Champlin, D., McKinney, R., Tagliabue, M., Serbst, J., and Gleason, T., Evaluation of the efficacy of extrapolation population modeling to predict the dynamics of *Americamysis bahia* populations in the laboratory, *Environ. Toxicol. Chem.*, 20, 213, 2001.
14. Hoffman, D. J., Rattner, B. A., and Hall, R. J., Wildlife toxicology: Third in a four-part series, *Environ. Sci. Technol.*, 24, 276, 1990.
15. Stormer, F. A. and Johnson, D. H., Introduction: Biometric approaches to modeling, in *Wildlife 2000: Modeling Habitat Relationships of Terrestrial Vertebrates*, Verner, J., Morrison, M. L., and Ralph, C. J., Eds., University of Wisconsin Press, Madison, WI, 1986, 159.

16. Gross, L. J., Rose, K. A., Rykiel, E. J., Jr., Van Winkle, W., and Werner, E. E., Individual-based modeling: Summary of a workshop, in *Individual-Based Models and Approaches in Ecology*, DeAngelis, D. L. and Gross, L. J., Eds., Chapman and Hall, New York, 1992, 511.
17. Murdoch, W. W., McCauley, E., Nisbet, R. M., Gurney, S. C., and de Roos, A. M., Individual-based models: Combining testability and generality, in *Individual-Based Models and Approaches in Ecology*, DeAngelis, D. L. and Gross, L. J., Eds., Chapman and Hall, New York, 1992, 18.
18. Box, G. E. P. and Jenkins, G. M., *Time Series Analysis, Forecasting, and Control*, rev. ed., Holden-Day, Oakland, 1976.
19. Jassby, A. D. and Powell, T. M., Detecting changes in ecological time series, *Ecology*, 71, 2044, 1990.
20. Emlen, J. M., Terrestrial population models for ecological risk assessment: A state-of-the-art review, *Environ. Toxicol. Chem.*, 8, 831, 1989.
21. Blus, L. J., Staley, C. S., Henny, C. J., Pendleton, G. W., Craig, T. H., Craig, E. H., and Halford, D. K., Effects of organophosphorus insecticides on sage grouse in southeastern Idaho, *J. Wildl. Manage.*, 53, 1139, 1989.
22. Skalski, J. R. and Robson, D. S., *Techniques for Wildlife Investigation: Design and Analysis of Capture Data*, Academic Press, New York, 1992.
23. Burnham, K. P., Anderson, D. R., White, G. C., Brownie, C., and Pollock, K. H., Design and analysis methods for fish survival experiments based on release-recapture, *Am. Fish. Soc. Monogr.*, 5, 1987.
24. Wisdom, M. J., Mills, L. S., and Doak, D. F., Life-stage simulation analysis: Estimating vital rate effects on population growth for species conservation, *Ecology*, 81, 628, 2000.
25. Pollock, K. H., Nichols, J. D., Brownie, C., and Hines, J. E., Statistical inference for capture-recapture experiments, *Wildl. Monogr.*, 107, 1990.
26. Hurlbert, S. H., Pseudoreplication and the design of ecological field experiments, *Ecol. Monogr.*, 54, 187, 1984.
27. Emlen, J. M. and Pikitch, E. K., Animal population dynamics: Identification of critical components, *Ecol. Modeling*, 44, 253, 1989.
28. Noon, B. R. and Sauer, J. R., Population dynamics of passerine birds, in *Wildlife 2001: Populations*, McCullough, D. R. and Barrett, R. H., Eds., Elsevier, London, 1992, 441.
29. Wisdom, M. J. and Mills, L. S., Sensitivity analysis to guide population recovery: Prairie chickens as an example, *J. Wildl. Manage.*, 61, 302, 1997.
30. Caughley, G., *Analysis of Vertebrate Populations*, John Wiley and Sons, New York, 1977.
31. May, R. M., Simple mathematical models with very complex dynamics, *Nature*, 261, 459, 1976.
32. Schaffer, W. M., Perceiving order in the chaos of nature, in *Evolution of Life Histories of Mammals: Theory and Pattern*, Boyce, M. S., Ed., Yale University Press, New Haven, 313, 1989.
33. Lewis, E. G., On the generation and growth of a population, *Sankhya*, 6, 93, 1942.
34. Leslie, P. H., On the use of matrices in certain population dynamics, *Biometrika*, 33, 183, 1945.
35. Fisher, R. A., *The Genetical Theory of Natural Selection*, 2nd ed., Dover, New York, 1958.
36. Beddington, J. R., Age distribution and the stability of simple discrete time population models, *J. Theor. Biol.*, 47, 65, 1974.
37. De Kroon, H., Plaisier, A., van Groenendael, J., and Caswell, H., Elasticity: The relative contribution of demographics parameters to population growth rate, *Ecology*, 67, 1427, 1986.
38. Goodman, D., Demographic intervention for closely managed populations, in *Conservation Biology: An Evolutionary-Ecological Perspective*, Soule, M. and Wilcox, B. A., Eds., Sinauer, Sunderland, MA, 1980, 171.
39. Meyer, J. S. and Boyce, M. S., Life historical consequences of pesticides and other insults to vital rates, in *Wildlife Toxicology and Population Modeling: Integrated Studies of Agroecosystems*, Kendall, R. L. and Lacher, T. E., Eds., Lewis, Boca Raton, FL, 1994.
40. Sykes, Z. M., Some stochastic versions of the matrix model for population dynamics, *J. Am. Stat. Assoc.*, 64, 111, 1969.
41. Anderson, D. R. and Burnham, K. P., Population Ecology of the Mallard, VI. The Effect of Eexploitation on Survival, U.S. Fish and Wildl. Serv., Resour. Publ., 128, 1976.
42. May, R. M., *Stability and Complexity in Model Ecosystems*, Monographs in Population Biology, 6, Princeton University Press, Princeton, 1974.
43. Sauer, J. R. and Slade, N. A., Size-based demography of vertebrates, *Annu. Rev. Ecol. Syst.*, 18, 71, 1987.

44. Nichols, J. D., Sauer, J. R., Pollock, K. H., and Hestbeck, J. B., Estimating transition probabilities for stage-based population projection matrices using capture-recapture data, *Ecology*, 73, 306, 1992.
45. Lefkovitch, L. P., The study of population growth in organisms grouped by stages, *Biometrics*, 21, 1, 1965.
46. Werner, P. A. and Caswell, H., Population growth rates and age versus stage-distribution models for teasel (*Dipsacus sylvestris* Huds.), *Ecology*, 58, 1103, 1977.
47. Rao, T. R. and Sarma, S. S., Demographic parameters of *Brachionus patulus* Muller (Rotifera) exposed to sublethal DDT concentrations at high and low food levels, *Hydrobiologia*, 139, 193, 1986.
48. Crouse, D. T., Crowder, L. B., and Caswell, H., A stage-based population model for loggerhead sea turtles and implications for conservation, *Ecology*, 68, 1412, 1987.
49. Martin, F. W., Phospahala, R. S., and Nichols, J. D., Assessment and population management of North American migratory birds, in *Environmental Biomonitoring, Assessment, Prediction, and Management-Certain Case Studies and Related Quantitative Issues*, Cairns, J., Jr., Patil, G. P., and Waters, W. E., Eds., International Cooperative Publishing House, Fairland, MD, 1978, 187.
50. Van Straalen, N. M. and Kammenga, J. E., Assessment of ecotoxicology at the population level using demographic parameters, in *Ecotoxicology*, Schuurmann, G. and Markert, B., Eds, John Wiley and Sons, New York, 1998.
51. DeAngelis, D. L. and Gross, L. J., Eds., *Individual-Based Models and Approaches in Ecology*, Chapman and Hall, New York, 1992.
52. Caswell, H. and John, A. M., From the individual to the population in demographic models, in *Individual-Based Models and Approaches in Ecology*, DeAngelis, D. L. and Gross, L. J., Eds, Chapman and Hall, New York, 1992, 36.
53. Koford, R. R., Sauer, J. R., Johnson, D. H., Nichols, J. D., and Samuel, M. D., A stochastic population model of mid-continental mallards, in *Wildlife 2001: Populations*, McCullough, D. R. and Barrett, R. H., Eds., Elsevier, London, 1992, 170.
54. Samuel, M. D., Influence of disease on a population model of mid-continent mallards, *Trans. N. Am. Wildl. Nat. Resour. Conf.*, 57, 486, 1992.
55. MacKay, N. A., Evaluating the size effects of lampreys and their hosts: Application of an individual-based model, in *Individual-Based Models and Approaches in Ecology*, DeAngelis, D. L. and Gross, L. J., Eds, Chapman and Hall, New York, 1992, 278.
56. Barker, R. J. and Sauer, J. R., Modeling population change from time series data, in *Wildlife 2001: Populations*, McCullough, D. R. and Barrett, R. H., Eds., Elsevier, London, 1992, 182.
57. Barker, R. J., Hines, J. E., and Nichols, J. D., Effect of hunting on annual survival of grey ducks in New Zealand, *J. Wildl. Manage.*, 55, 260, 1991
58. Johnson, D. H., Burnham, K. P., and Nichols, J. D., The role of heterogeneity in animal population dynamics, in *Proc. 13th Int. Biometrics Conf.*, Biometric Society, University of Washington, Seattle, 1986.
59. Nichols, J. D., Stokes, S. L., Hines, J. E., and Conroy, M. J., Additional comments on the assumption of homogeneous survival rates in modern bird banding estimation models, *J. Wildl. Manage.*, 46, 953, 1982.
60. Pollock, K. H. and Raveling, D. G., Assumptions of modern band-recovery models, with emphasis on heterogeneous survival rates, *J. Wildl. Manage.*, 46, 88, 1982.
61. Kalbfleisch, J. D. and Prentice, L., *The Statistical Analysis of Failure Data*, John Wiley and Sons, New York, 1980.
62. Cox, D. R. and Oakes, D., *Analysis of Survival Data*, Chapman and Hall, New York, 1984.
63. White, G. C. and Garrott, R. A., *Analysis of Wildlife Radio-Tracking Data*, Academic Press, New York, 1990.
64. Pollock, K. H., Winterstein, S. R., Bunck, C. M., and Curtis, P. D., Survival analysis in telemetry studies: The staggered entry design, *J. Wildl. Manage.*, 53, 7, 1989.
65. Link, W. A., Efficiency and optimal allocation in the staggered entry design, *Commun. Statist. Theory Methods*, 22, 485, 1993.
66. Bart, J. and Robson, D. S., Estimating survivorship when the subjects are visited periodically, *Ecology*, 63, 1078, 1982.
67. Heisey, D. M. and Fuller, T. K., Evaluation of survival and cause specific mortality rates using telemetry data, *J. Wildl. Manage.*, 49, 668, 1985.

68. Brownie, C., Anderson, D. R., Burnham, K. P., and Robson, D. S., Statistical Inference from Band Recovery Data: A Handbook, U.S. Fish and Wildl. Serv. Resour. Publ. 131. Washington, D.C., 1985.
69. Anderson, D. R., Burnham, K. P., and White, G. C., Problems in estimating age-specific survival rates from recovery data of birds ringed as young, *J. Anim. Ecol.*, 54, 89, 1985.
70. Pollock, K. H., A capture-recapture design robust to unequal probability of capture, *J. Wildl. Manage.*, 46, 757, 1982.
71. Kendall, W. L. and Pollock, K. H., The robust design in capture-recapture studies: A review and evaluation by monte carlo simulation, in *Wildlife 2001: Populations*, McCullough, D. R. and Barrett, R. H., Eds., Elsevier, London, 1992, 31.
72. Clobert, J. and Lebreton, J-D., Estimation of demographic parameters in bird populations, in *Bird Population Studies: Relevance to Conservation and Management*, Perrins, C. M., Lebreton, J. D., and Hirons, G. J. M., Eds., Oxford University Press, Oxford, 1991, 75.
73. Nichols, J. D., Brownie, C., Hines, J. E., Pollock, K. H., and Hestbeck, J. B., The estimation of exchanges among populations or subpopulations, in *Marked Individuals in the Study of Bird Populations*, Lebreton, J-D. and North, P. M., Eds., Advances in Life Sciences, Birkhauser Verlag, Berlin, 1993, 265.
74. Clobert, J., Lebreton, J-D., and Marzolin, G., The estimation of local immature survival rate and age-specific proportions of breeders in bird populations, in *Population Biology of Passerine Birds: An Integrated Approach*, Blondel, J., Gosler, A., Lebreton, J-D., and McCleery, R., Eds, Springer-Verlag, Berlin, 1990, 199.
75. Lebreton, J-D., Hemery, G., Clobert, J., and Coquillart, H., The estimation of age-specific breeding probabilities from recaptures or resightings in vertebrate populations, I. Transversal models, *Biometrics*, 46, 609, 1990.
76. Mayfield, H., Suggestions for calculating nesting success, *Wilson Bull.*, 87, 456, 1975.
77. Stromborg, K. L., Grue, C. E., Nichols, J. D., Hepp, G. R., Hines, J. E., and Bourne, H. C., Postfledging survival of European starlings exposed as nestlings to organophosphorus insecticide, *Ecology*, 69, 590, 1988.
78. Skalski, J. R., Hoffmann, A., and Smith, S. G., Testing the significance of individual- and cohort-level covariates in animal survival studies, in *Marked Individuals in the Study of Bird Populations*, Lebreton, J-D. and North, P. M., Eds., Advances in Life Sciences, Birkhauser Verlag, Berlin, 1993, 9.
79. Agresti, A., *Categorical Data Analysis*, John Wiley and Sons, New York, 1990.
80. Conroy, M. J. and Williams, B. K., A general methodology for maximum likelihood inference from band-recovery data, *Biometrics*, 40, 739, 1984.
81. Kuno, E., Sampling error as a misleading artifact in "key factor analysis," *Res. Pop. Ecol.*, 13, 28, 1971.
82. Lebreton, J-D., Modeling density-dependence, environmental variability, and demographic stochasticity from population counts: An example using Wytham Wood great tits, in *Population Biology of Passerine Birds: An Integrated Approach*, Blondel, J., Gosler, A., Lebreton, J-D., and McCleery, R., Eds, Springer-Verlag, Berlin, 1990, 89.
83. Sauer, J. R., Barker, R. J., and Geissler, P. H., Statistical aspects of modeling population changes from population size data, in *Wildlife Toxicology and Population Modeling: Integrated Studies of Agroecosystems*, Kendall, R. L. and Lacher, T. E., Eds., Lewis Publishers, Boca Raton, FL, 1994.
84. Keller-McNulty, S. and McNulty, M., The independent pairs assumption in hypothesis tests based on rank correlation coefficients, *Am. Stat.*, 41, 40, 1979.
85. SAS Institute, *SAS System for Forecasting Time Series*, SAS Institute, Cary, NC, 1986.
86. Box, G. E. P. and Tiao, G. C., Intervention analysis with applications to economic and environmental problems, *J. Am. Stat. Assoc.*, 70, 70, 1975.
87. Bautista, L. M., Alonso, J. C., and Alonso, J. A., A 20-year study of wintering common crane fluctuations using time series analysis, *J. Wildl. Manage.*, 56, 563, 1992.
88. Rubin, D. B., Multiple imputation after 18+ years, *J. Am. Stat. Assoc.*, 91, 473, 1996.
89. Hopke, P. K., Liu, C., and Rubin, D. B., Multiple imputation for multivariate data with missing below-threshold measurements: Time-series concentrations of pollutants in the arctic, *Biometrics*, 57, 2001.
90. Littell, R. C., Milliken, G. A., Stroup, W. W., and Wolfinger, R. D., *SAS System for Mixed Models*, SAS Institute, Cary, NC, 1996.
91. SAS Institute, *SAS/STAT User's Guide*, Vol. 2., Version 6, 4th ed., Cary, NC, 1989.
92. White, G. C., Numerical estimation of survival rates from band-recovery and biotelemetry data, *J. Wildl. Manage.*, 47, 716, 1983.

CHAPTER 36

# Ecological Risk Assessment: U.S. EPA's Current Guidelines and Future Directions

**Susan B. Norton, William H. van der Schalie, Anne Sergeant, Lynn Blake-Hedges, Randall Wentsel, Victor B. Serveiss, Suzanne M. Marcy, Patricia A. Cirone, Donald J. Rodier, Richard L. Orr, and Steven Wharton***

## CONTENTS



* The views expressed in this chapter are those of the authors and do not necessarily reflect the views or policies of the U.S. Environmental Protection Agency.

## 36.1 INTRODUCTION

Environmental problems facing scientists and decision makers are numerous and varied; they include potential global climate change, spread of pathogens and invasive species, habitat loss, acid deposition, and the release and detection of multiple human-made chemicals in the environment. The types of questions that scientists are asked to inform managers about range from defining the nature and extent of anticipated or observed effects of environmental changes, to prioritizing research and collecting, to comparing management alternatives. Doing this requires a process that flexibly accommodates the diversity of issues yet provides the consistency and rigor needed to have confidence in conclusions.

Ecological risk assessment (ERA) is a process that evaluates the likelihood that adverse ecological effects may occur as a result of exposure to one or more stressors.[1] It is a process for evaluating information, assumptions, and uncertainties to understand the relationships between stressors and ecological effects; its ultimate goal is to inform environmental decisions. It has been used to evaluate a wide variety of environmental issues of interest such as water-regime management in South Florida,[2,3] the chemical and biological stressors used to control gypsy moths,[4] and chemicals that are used in industrial processes for pest control or are present at hazardous waste sites (e.g., References 5–7).

This chapter discusses the ERA process as developed by the U.S. Environmental Protection Agency (EPA); government entities in other countries have also developed guidance and methods (e.g., References 8–13). The field also has benefitted from National Research Council reports,[14] textbooks like this one, and the startup in 1995 of a new journal entitled *Human and Ecological Risk Assessment*. While the term has become more common recently, ERA draws from the experiences of adaptive environmental management,[15] environmental impact statements,[16] hazard evaluation,[17] and biological monitoring of environmental condition.[18] It also builds on procedures used to analyze risks to human health.[19]

ERA has several features that contribute to effective environmental decision-making:

- It can iteratively incorporate new information to improve environmental decision-making. This feature is consistent with adaptive-management principles[15] used in managing natural resources.
- It can be used to express changes in ecological effects as a function of changes in exposure to stressors. This capability may be particularly useful to the decision maker who must evaluate tradeoffs, examine different alternatives, or determine how much stressors must be reduced to achieve a given outcome.
- It explicitly evaluates uncertainty. Uncertainty analysis describes the degree of confidence in the assessment and can help the risk manager focus additional monitoring, modeling, or experiments on those areas that will lead to the greatest reductions in uncertainty.
- It provides a consistent basis for comparing, ranking, and prioritizing risks.
- It considers management goals and objectives as well as scientific issues during planning and problem formulation to ensure that results will be useful to risk managers and other stakeholders.

This chapter describes some recent and ongoing activities in ERA. Section 36.2 summarizes the EPA's Guidelines for Ecological Risk Assessment.[1] Section 36.3 discusses several new technical developments and applications. Section 36.4 discusses ERA's role in broader risk-management activities, including objectives development and economic and human-welfare assessments. These activities can offer additional interpretation of the effects of alternative management options.

## 36.2 THE EPA'S GUIDELINES FOR ECOLOGICAL RISK ASSESSMENT

In 1998, the U.S. EPA published the Guidelines for Ecological Risk Assessment.[1] The Guidelines are intended to promote consistency across EPA risk assessments, improve the quality of the science underlying risk assessments, and inform the public about the EPA's approach to risk assessment. They are the culmination of a long-term effort at the EPA, including the Framework for Ecological Assessment,[20] issue papers,[21] and case studies.[22,23]

The Guidelines describe three phases of ERA: problem formulation, analysis, and risk characterization (Figure 36.1). In problem formulation, risk assessors develop a conceptual model of how ecological effects may result from human activities. A critical component of problem formulation is linking endpoints that can be measured (e.g., survival of fish in a laboratory experiment) with management goals (e.g., preventing fish kills in streams). This link is aided by defining assessment endpoints that are intermediate between broad management goals and specific measurements. During the subsequent analysis phase, the conceptual model guides data evaluation to reach

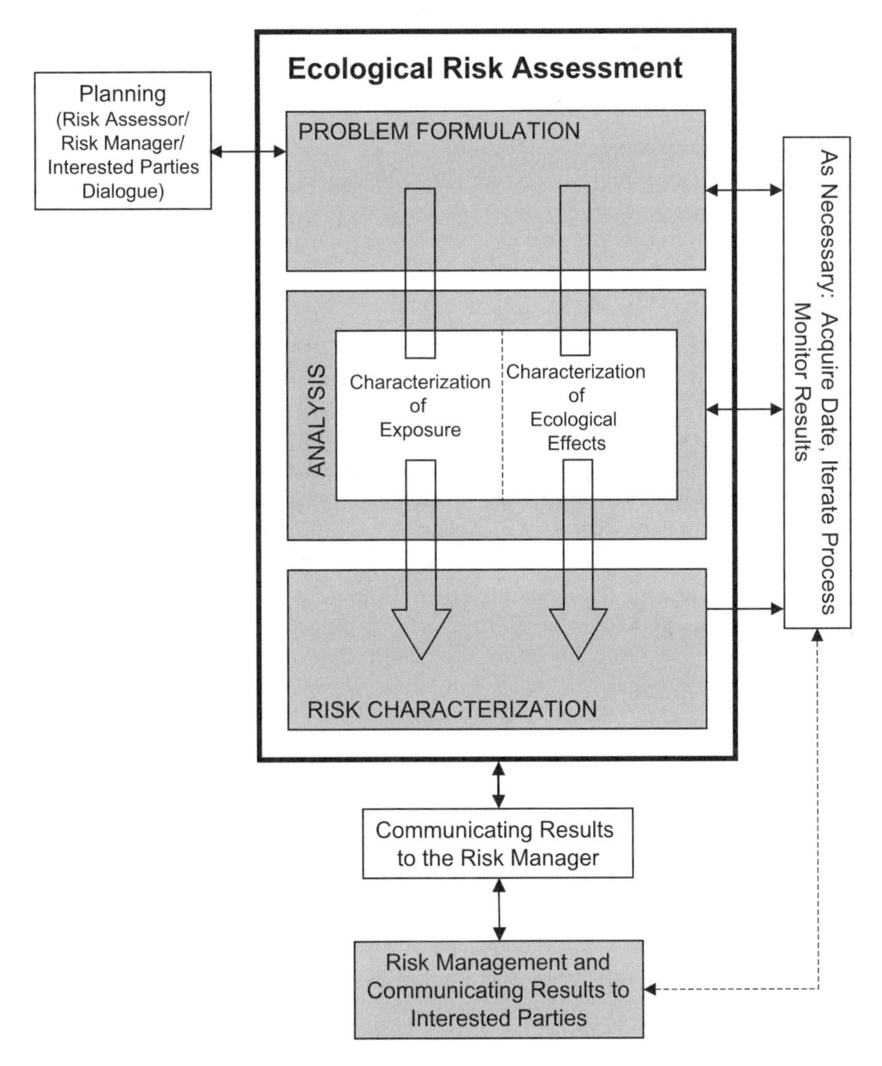


**Figure 36.1** The ecological risk assessment process. (From U.S. Environmental Protection Agency, Guidelines for Ecological Risk Assessment, Washington, D.C., 1988.)

conclusions about exposure and the relationships between stressors and effects. In risk characterization, the information on exposure and stressor-response relationships is integrated to estimate risk, and all of the evidence is brought together to reach final conclusions about the likelihood and magnitude of effects. Effective risk characterizations acknowledge uncertainties and assumptions; in addition, they separate scientific conclusions from policy judgements.[24,25]

The processes for eliciting the views of managers and other stakeholders are critical for ensuring that the results of the assessment can be used to support a management decision.[26] These processes are particularly important in providing input to problem formulation (the "Planning" box shown in the upper left-hand side of Figure 36.1). They include articulating management goals, management alternatives, and the resources available to conduct the assessment. The interactions between risk assessment and risk management during the early phases of the project are discussed further in Section 36.2.1.

The box on the right-hand side of Figure 36.1 represents data collection and monitoring. Risk analysis is a data-driven process. However, the data may have been collected specifically for the risk assessment or developed for another purpose (e.g., routine environmental monitoring). In the latter case, assessors need to critically evaluate the data to ensure that the information can support the assessment (Section 36.2.3). Monitoring data can also be used to evaluate the efficacy of a risk-management decision. For example, if a risk were to be mitigated by reducing exposure, monitoring could help determine whether the desired reductions in exposure and effects were achieved. Monitoring can also document ecological recovery and help determine whether the management approach needs adjustment.[15] Finally, focused monitoring can be used to evaluate the risk assessment approach and improve the process.

### 36.2.1 Planning the Risk Assessment

ERAs are conducted to translate environmental data into information risk managers can use to make informed decisions. To make sure ERAs are useful, risk managers, risk assessors, and, where appropriate, stakeholders plan the process before beginning. Planning generates three outputs: (1) clear management goals, (2) decisions to be made, and (3) the ERA's scope and complexity.

The first step is to decide whether an ERA is really needed. Participants explore what is known about the risk, what can be done to mitigate or prevent it, and whether ERA is the best way to evaluate environmental concerns. An ERA may add little value to the decision process if there are alternatives that circumvent the need for it. Or a discussion with stakeholders may reveal that there is so much opposition to a particular action that there is no point in considering it, much less going to the trouble of preparing an ERA.

#### *36.2.1.1 Involve the Players*

Once planning begins, the next step is to ensure that all the right people are involved. Not all participants have the same roles and responsibilities. Clarifying who the assessment is for, who will judge its worth, who will use it, and who will be affected by any ensuing management actions are critical components of planning an assessment.

*Managers* bring the most in-depth knowledge of the purpose for the assessment, the context for decisions, and the regulatory frameworks that help frame the question. They are the ultimate users of the assessment's results. A manager may be a single individual (e.g., a government manager) or a group of dispersed decision makers (e.g., landowners in a watershed).

*Local stakeholders* contribute in-depth knowledge of area concerns, knowledge about the system, and a historical perspective to the planning process. Although stakeholders may include decision makers, scientists, and analysts, it is important to also include individuals who may be concerned with or impacted by decisions, if possible. In addition to helping to define and articulate the decision context, stakeholders can provide unique knowledge and information for the assessment. Without their involvement, it can be much more difficult to solidify public and political support for assessment activities, particularly over the long term. In addition, environmental-protection strategies built on partnership models are increasingly used to augment the command-and-control methods historically used for managing point sources. If ecological assessment is to inform these new strategies, then the assessment must engage the stakeholders.

For assessments conducted under specific regulations or laws, substantial stakeholder involvement likely occurred during the development of the law or regulation. In these cases, the assessment team may have a starting point that articulates stakeholder concerns.[27]

*Scientists* usually conduct the assessment and are responsible for integrating the decision context and stakeholder concerns with scientific issues. They also gather information on the ecosystem (structure, function, status, and history of system), known or suspected stressors, and susceptible species or communities. They bring knowledge of environmental processes and causal linkages as well as strengths and limitations of different variables and measurement techniques. They are also most likely to understand what and how many resources will be needed to complete the analysis.

#### 36.2.1.2 Identify Management Goals

Risk assessments are conducted to provide information relevant to achieving a management goal. They reflect the decision-maker's overall goals and may address an organism, an ecosystem, or some part or feature of an ecosystem. Early discussions and clear management goals help the risk assessor identify and gather critical information. No matter how they are established, goals that explicitly define what will be protected provide the best foundation for risk-assessment objectives and actions to reduce risk.

It is well worth any time spent to explore goals and make sure they are explicit. A well-defined goal specifies an entity, some attribute, and a desired state. For example, if the management goal is a general statement such as "maintain a sustainable aquatic community," the entity is the aquatic community and one (implicit) attribute is the population size of each member species. The desired state is unclear, probably something to the effect of "enough so they'll still be with us next year," and would be much more helpful if it was expressed as something like "that can sustain a sport fishery for organisms $x$ and $y$ and a commercial fishery for organism $z$." Similarly, "restore a wetland" could be stated more clearly as "restore the prairie pothole wetland at Puddle Prairie State Park (entity) to its areal extent (attribute) before tile drains are installed (desired state)," and "prevent toxicity" could be expressed as "reduce toxic discharges to Lake Serenity (entity; the implicit attribute is the aquatic community structure) to concentrations below Ambient Water-Quality Criteria (desired state)." Explicit goals not only guide the risk assessment and subsequent management decisions; they provide a way to tell whether the action taken was successful.

When goals are very broad, it may be useful to break them down into multiple management objectives. Keeney[28] provides a structured procedure for developing objectives, and Hammond et al.[29] provide a somewhat simplified approach. A series of management objectives can clarify the inherent assumptions within the goal and help determine which ecological entities and attributes best represent each objective.

#### 36.2.1.3 Describe Management Options

ERAs are conducted for a number of purposes including prioritizing stressors, resources, or places for protection or further investigation; identifying causes of observed impacts; predicting the effects of a new or continuing release; and selecting among alternatives for restoration. The investigation's purpose and management goals will influence what options are considered. In turn, a clear understanding of the range of management options under consideration helps to focus the ERA and clarify the degree of confidence that will be needed.

Risk assessors can also provide a scientific perspective on candidate management options. For example, if the management goal is to reestablish spawning habitat for free-living anadromous salmon in a river impounded by multiple dams, a risk assessor might advise the decision maker that several actions will be needed (dam removal, nutrient input reduction, erosion control, pest management) and that none of them will make any difference unless the dams are removed. If this is not among the management options, the planning group might work together to reconsider the goal, abandon it, or present it as a work in progress with interim goals such as water clarity and quality.

ERAs for a region or watershed that address multiple stressors and ecological values are often based on a general goal with multiple potential decisions and many possible options. Such assessments tend to be complex because they involve the multiple stressors, ecological values, and political and economic factors that influence decisions at larger spatial scales (see Section 36.4.2. below). Although time-consuming, ERAs at large spatial scales can be valuable in bringing together diverse stakeholders and focusing management efforts on key stressors.

#### *36.2.1.4 Define Scope and Complexity*

An ERA's purpose determines its geographic scope and complexity (e.g., national, regional, site specific). The scope and complexity are also a function of how sure a decision-maker needs to be about risks to decide what to do about them. The level of confidence needed is influenced by the cost of potential management actions, the level of interest, the degree of difficulty, and the legal context for the decision. ERAs that are completed under legal mandates and likely to be contested or appealed must pay strict attention to uncertainties to ensure that they will stand up in court. The risk manager and risk assessor should be frank with each other about sources of uncertainty, the comfort level needed for the decision, and ways uncertainty can be reduced. For example, successive iterations or tiers of increasing cost and complexity may help reduce uncertainty.

Advice on how to address the interplay of management decisions, study boundaries, data needs, and uncertainty, and specify limits on decision errors may be found in the EPA's guidance on data-quality objectives.[30] The planning group considers the type of decision (e.g., national policy, local impact), available resources, opportunities for increasing resources (e.g., partnering, new data collection, alternative analytical tools), expertise at hand, and what information is needed to make the decision. Detailed analysis is not always needed, and sometimes a simple screening-level assessment is enough.

Planning is complete when the group has decided on (1) management goals, (2) the range of management options the ERA is to address, (3) the ERA's focus and scope, and (4) resource availability. If needed, formal agreements may be used to document the technical approach, spatial scale, temporal scale, and, of course, product deadlines.

### 36.2.2 Problem Formulation

Problem formulation is a process for developing and evaluating preliminary questions about why ecological effects might occur. This step is crucial — it provides the foundation for the ERA. First, participants refine the focus and scope of the ERA, based on the results of the planning step described above. Then, they develop a plan for analyzing data and characterizing risk. Any shortcomings in problem formulation will haunt future work, so it is worth spending time here to refine objectives and plan carefully.

Problem formulation generates three products: assessment endpoints, a conceptual model or models, and an analysis plan. In all but the simplest analyses, the problem-formulation process is interactive and iterative and may require several cycles.

Problem formulation begins with a review of information about the situation that is readily available. Useful information includes stressor sources and characteristics, exposure opportunities, characteristics of the ecosystem potentially at risk, and any potential or observed ecological effects. When information is plentiful, problem formulation goes quickly. If not, the ERA begins with the information at hand, and the problem-formulation process helps identify missing data and provides a framework for further data collection.

### *36.2.2.1 Select Assessment Endpoints*

Assessment endpoints are explicit expressions of the actual environmental value to be protected and provide the focal point for ERA. They are operationally defined by an ecological entity and some attribute. The entity identifies what is being protected. This can be a species (e.g., eelgrass, piping plover), a functional group (e.g., piscivores), a community (e.g., benthic invertebrates), an ecosystem (e.g., lake), a specific habitat (e.g., wet meadows), a unique place (e.g., a remnant of native prairie), or other entity of concern. The second is the attribute or characteristic of that entity that is important to protect and potentially at risk. For piping plovers, it may be reproductive rates; for a lake, frequency and severity of algal blooms; for a wet meadow, the number of native plants. An assessment endpoint needs both an entity and an attribute to serve as a clear link to management goals and the basis for measurement.

Assessment endpoints are similar to management goals but are distinguished by their neutrality and specificity. Assessment endpoints do not describe a desired achievement (i.e., goal), so they do not contain words like "protect," "maintain," or "restore," or indicate a direction for change such as "loss" or "increase."

It can be a challenge to choose the characteristics that will be most useful for the assessment. Ecosystems may be examined at several organizational levels (individual, population, community, ecosystem, landscape) and through multiple ecosystem processes. It is rarely clear which of these attributes are most critical to ecosystem structure or function, and professionals and the public don't always agree on which are most valuable. Three useful criteria to consider are: (1) ecological relevance (the entity should be a current or historical part of the ecosystem being studied), (2) susceptibility (it should be both exposed and sensitive to the effects of the stressor), and (3) relevance to management goals (it should be something the manager can do something about). Ecological relevance and susceptibility are essential for scientifically defensible assessment endpoints. However, endpoints that reflect societal values and management goals are more likely to be used in management decisions. Assessment endpoints that meet all three criteria provide the best foundation for an effective ERA.

The challenge is to find endpoints that meet the selection criteria and are also important to those who make and have to live with the decision — risk managers and the public. Suppose, for example, an assessment is designed to evaluate the risk of applying pesticide around a lake to control insects. It turns out that at this lake midges are susceptible to the pesticide and form the base of a complex food web that supports native fish popular with local anglers. While both midges and fish are key components of the aquatic community in this scenario, evaluating midges alone would not address both ecological and community concerns (and it would probably be difficult to convince local residents that annoying insects should be protected). Selecting the fishery would allow assessors to characterize the risk to the fishery if the midge population is adversely affected and address both sets of issues. (Notice that this may require a trip back to the planning stage to revisit management goals, which on further examination may evolve from something like "eliminate biting insects" to "minimize biting insects while maintaining sport fishing opportunities.") This strategy addresses ecological issues while responding to management concerns. If there is no choice but to use an unpopular assessment endpoint, the risk assessor will need to link it convincingly to values that people do care about.

Practical issues, such as what is required by statute (e.g., the Endangered Species Act) or management options, may influence assessment endpoint selection. Another concern is whether important variables can be measured directly. Assessment endpoint attributes that can be measured directly are best because they avoid the uncertainty introduced by having to extrapolate or estimate values. Nevertheless, sometimes the only data that are available are those measured indirectly or generated by a model. Note also that data availability and measurement convenience are *not* selection criteria: while established measurement protocols may be familiar and easy to use, convenience is no guarantee of appropriateness. For practical reasons, it may be helpful to use assessment endpoints that have well-developed test methods, field measurement techniques, and predictive models.[31] However, it is not necessary to use standardized protocols, and assessment endpoints should not be selected simply because protocols are readily available.

Finally, assessment endpoint selection is an important risk manager–risk assessor checkpoint: all parties should agree that selected assessment endpoints effectively represent the management goals, and the rationale for their selection should be explicit.

#### 36.2.2.2 Conceptual Models

Conceptual models describe key causal relationships that will be evaluated in the risk assessment. They have two components: (1) a set of written hypotheses that propose relationships between ecological effects and stressors and (2) a diagram of the relationships. Conceptual models may include elements such as ecosystem processes that influence receptor responses or exposure scenarios that qualitatively link land-use activities to stressors. They may describe primary, secondary, and tertiary exposure pathways or co-occurrence between exposure pathways, ecological effects, and ecological receptors. The process also identifies what additional information is needed. A carefully prepared conceptual model will provide a helpful reference point throughout the ERA; it can be used to refine models or assumptions, keep the assessment on course, and serve as an outline when preparing outreach materials.

Risk hypotheses (sometimes called assessment questions) articulate how exposure to stressors could cause adverse effects. They are used to guide the analyses conducted in the risk assessment and are not usually designed to be tested statistically. They may predict effects of a stressor prior to its release or explain why observed ecological effects occurred and, ultimately, what caused the effect. Risk hypotheses may range from very simple (e.g., receptor x will be exposed to stressor y at a particular place and time) to very complex (thermal discharge will increase species a's recruitment, decrease species b's recruitment, increase species c's biomass, increase species d's food supply, and increase nutrient levels over time...with possible effects to a predator that happens to be endangered locally, but not regionally).

Diagrams are extremely useful in communicating and exploring conceptual models. They can show important pathways clearly and concisely and be used to refine or generate new questions about relationships. Sometimes they are flow diagrams with boxes and arrows to illustrate relationships (see, for example, Figure 36.2); cartoons are also very helpful. Early conceptual models are usually quite general in that they identify as many potential relationships as possible. As more information is incorporated, risk assessors sort through stressor–effect relationships and the ecosystem processes that influence them to identify the questions most appropriate for the analysis phase. Several documents offer examples and advice on conceptual-model development.[1,32–34]

Conceptual-model development may be one of the most important sources of uncertainty in ERA. If important relationships are missed or incorrectly described, the risk characterization may misrepresent actual risks. Uncertainty may arise from ignorance about how the ecosystem functions, failing to identify and interrelate temporal and spatial parameters, omitting stressors, or overlooking cascading effects. Although it's impossible to avoid simplification and ignorance, conceptual models can help document what is known.[33,35]

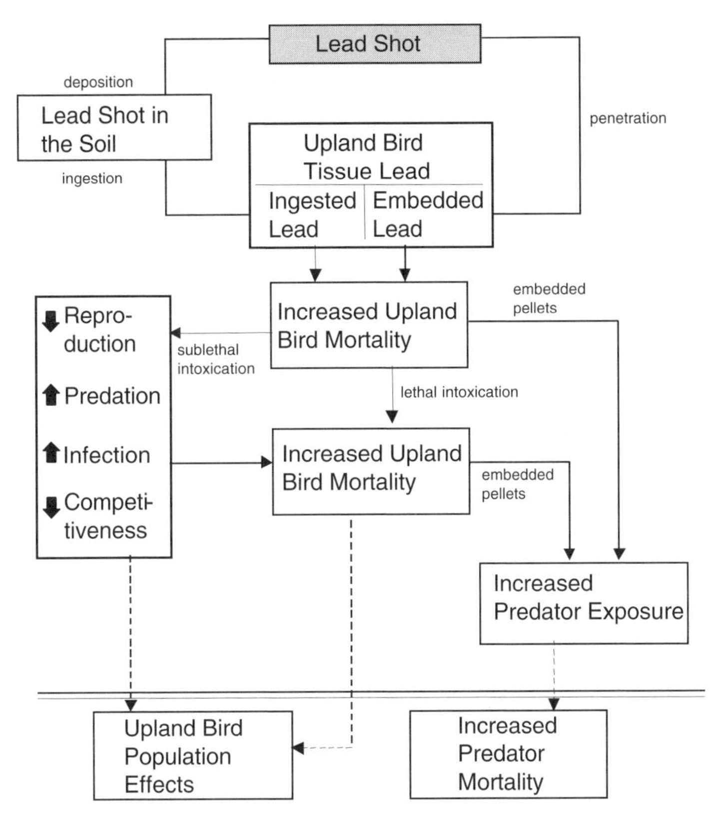


**Figure 36.2** Conceptual model for tracking stress associated with lead shot through upland ecosystems. (From Kendall, R.J. et al., *Environ. Toxicol. Chem.*, 15, 4, 1996. With permission.)

### 36.2.2.3 Analysis Plan

The analysis plan completes problem formulation. Here, participants evaluate the conceptual model to determine how it can be assessed. The plan delineates the assessment design, data needs, analytical methods, and anticipated uncertainties. For assessment approaches that are conducted repeatedly with different stressors (e.g., EPA's new chemical assessments), an analysis plan in a standard format may be available. The more unique and complex the assessment, the more important it is to have a good analysis plan.

The analysis plan identifies the pathways and relationships that will be pursued and emphasizes those most likely to contribute to risk. It also may compare the confidence desired for the management decision with that expected from alternative analyses to help decide what data are still needed and which analytical approach is best. If data are scanty, the analysis plan may recommend data collection or propose a phased or tiered approach. If new data collection is impossible, this should be acknowledged as a source of uncertainty.

When assessment endpoint responses cannot be measured directly, surrogates must be used. The selection of what, where, and how to measure surrogate responses determines whether the ERA is still relevant to the risk-management decisions. For example, an assessment may evaluate the risk of a pesticide used on seeds to an endangered seed-eating bird to help decide whether to register it for use. The assessment endpoint entity is the endangered species. Example attributes include feeding behavior, survival, growth, and reproduction. While it may be possible to directly collect measures of exposure and life-history characteristics, it would not be appropriate to dose the bird with the pesticide to measure sensitivity. So to evaluate susceptibility one would measure (or review the literature for data on) response to the pesticide in an appropriate surrogate. Depending on the chemical, this could be another species with similar life-history characteristics, physiology, or phylogeny.

The analysis plan identifies the pathways and relationships that will be pursued in the assessment and emphasizes those most likely to contribute to risk. It can serve as a checkpoint with managers to compare the results expected from the assessment with the information needs and confidence desired for the management decision (e.g., do they meet Data Quality Objectives[30])? If suitable data are available, the assessment proceeds. If data are scanty, the analysis plan may recommend additional data collection or propose a phased or tiered approach that begins with available information.

### 36.2.3 The Analysis Phase

Analysis is a process that examines the two primary components of risk — exposure and effects and their relationships with the ecosystem. The objective is to provide the ingredients necessary to determine or predict ecological responses to stressors under specific exposure conditions.

The analysis phase is flexible, with substantial interaction between the effects and exposure characterizations as illustrated by the dotted line in Figure 36.1. In particular, when secondary stressors and effects are of concern, analyses of exposure and effects can become intertwined and difficult to differentiate. An assessment of bottomland hardwoods, for example, examined potential changes in the plant and animal communities under different flooding scenarios.[36] Analysts combined the stressor-response and exposure analyses within the FORFLO model for primary effects on the plant community and within the Habitat Suitability Index for secondary effects on the animal community. The model results were used directly in risk characterization and thus also blurred the distinction between analysis and risk estimation.

The nature of the stressor influences the types of analyses conducted. For chemical stressors, exposure estimates emphasize contact and uptake into the organism, and effect estimations often entail extrapolating toxicological data from one test organism to another. For physical stressors, the initial disturbance may cause primary effects on the assessment endpoint (e.g., loss of wetland acreage). But in many cases, secondary effects (e.g., decline of wildlife populations that depend on wetlands) are the main concern; the point of view depends on the assessment endpoints. Because adverse effects can occur even if receptors do not physically contact disturbed habitat, exposure analyses may emphasize co-occurrence with physical stressors rather than contact. For biological stressors, exposure analysis evaluates entry, dispersal, survival, and reproduction[37] (discussed further in Section 36.3.2).

The assessor's first task of the analysis phase is to determine whether available studies are right for the ERA objectives. Data are rarely as complete as one would like, but eventually one must decide whether to make do with what is available, fill gaps with models, or collect new data. Reviewing the data and information that will be used in the risk assessment also provides the opportunity to evaluate how uncertainties in the assessment will be described and, where possible, quantified. Uncertainty analysis can improve credibility because it explicitly describes the size and direction of errors, and it can show where additional data or analytical refinements would do the most good. The development of quantitative methods to evaluate uncertainty is a currently active area, discussed further in Section 36.3.1.

#### *36.2.3.1 Characterize Exposure*

Exposure characterization describes potential or actual contact or co-occurrence of stressors with receptors. The evaluation encompasses stressor sources, receptor behavior, stressor and receptor distribution and movement in the environment, and the extent and pattern of contact or co-occurrence. The objective is to describe how receptors and stressors interact with the environment and each other and the likelihood of any contact or co-occurrence. Exposure analysis describes exposure in terms of intensity, space, and time in units that can be combined with the effects assessment.

#### *36.2.3.1.1 Describe the Source(s)*

A source can be defined in two general ways: as the place from which the stressor either originates or is released (e.g., a smokestack, historically contaminated sediments) or the management practice or action (e.g., dredging) that produces stressors. In some assessments, the original source no longer exists and so is defined as the current location of the stressors. For example, contaminated sediments might be considered a source because the industrial plant that produced the chemicals that ended up there no longer operates. A source influences where and when stressors eventually will be found.

Many stressors have natural counterparts or multiple sources, so it may be necessary to characterize these as well. Many chemicals occur naturally (e.g., most metals), are widespread due to other sources (e.g., polycyclic aromatic hydrocarbons in urban ecosystems), or have significant sources outside the boundaries of the current assessment (e.g., atmospheric nitrogen deposited in Chesapeake Bay). Many physical stressors also have natural counterparts. For instance, construction activities may release sediments into a stream in addition to those already coming from a naturally undercut bank. Human activities may also change the magnitude or frequency of natural disturbance cycles. For example, fire suppression may decrease the frequency but increase the severity of fires because fuel accumulates without small fires to consume it.

#### *36.2.3.1.2 Describe the Distribution of the Stressors or Disturbed Environment*

The second objective of exposure analysis is to describe where and when stressors occur in the environment. For physical stressors that directly alter or eliminate portions of the environment, the assessor describes the disturbed environment. Stressor distribution is examined by evaluating transport pathways from the source as well as the formation and subsequent distribution of secondary stressors.

Secondary stressors may be even more of a concern than the primary stressor. For chemicals, the evaluation usually focuses on metabolites, biodegradation products, or chemicals formed through abiotic processes. As an example, microbial action increases mercury bioaccumulation by transforming inorganic forms to organic species. Ecosystem processes may also form secondary stressors. For example, nutrient inputs into an estuary can increase primary production and subsequent decomposition and deplete dissolved oxygen. Physical disturbances also can generate secondary stressors, and identifying the one that most affects the assessment endpoint can be a difficult task. The removal of riparian vegetation, for example, may increase nutrient levels, stream temperature, sedimentation, and stream flow extremes. However, temperature change may have the greatest effect on adult salmon survival in a particular stream.

If stressors have already been released, direct measurement of environmental media or a combination of mechanistic and empirical modeling and measurement is ideal. If not, models make it possible to either make predictions or interpolate between available measurements. Models also are useful for quantifying relationships between sources and stressors. Models of fate and transport of chemicals are particularly well developed (e.g., Lahlou[38]), and the concept can be extended to other types of sources and stressors. For example, Johnson et al.[39] used wetland extent to predict downstream flood peaks.

#### *36.2.3.1.3 Describe Contact or Co-Occurrence*

The third objective is to describe the intensity and spatial and temporal extent of co-occurrence or contact between stressors and receptors. This is critical — if there is no exposure, there can be no risk. The analysis may consider exposure that may occur in the future, exposure that has already occurred but is not currently evident (e.g., in some retrospective assessments), and exposure of food or habitat resources that might cause a cascade of effects.

People usually think of exposure as actual contact with a stressor, but sometimes a stressor's presence can affect an organism even if there is no contact. Whooping cranes provide a case in point: they use river sandbars as resting areas, and they prefer sandbars with unobstructed views. Without ever actually contacting the birds, dams can modify the flood regime that maintains the sandbars by scouring and redepositing sand, and obstructions such as bridges can interfere with resting behavior. Most stressors must contact receptors to produce an effect. For example, aluminum floc in low pH lakes alters fish gills that it contacts. Finally, some stressors must not only be contacted but also reach the target organ. A toxicant that causes liver tumors in fish, for example, must be absorbed and reach the liver to cause the effect.

Intensity is the most familiar dimension for describing exposure to chemical and biological stressors. Exposure intensity may be expressed as the amount of chemical or number of pathogenic organisms contacted. The temporal dimension comprises duration (the time over which exposure occurs) and frequency (how often it occurs). Finally, exposure timing, including the order or sequence of events, can be an important consideration. Adirondack Mountain lakes receive high concentrations of hydrogen ions and aluminum during snow melt; this period corresponds to the sensitive life stages of some aquatic organisms.

Spatial extent is another exposure dimension. It usually is expressed as area (e.g., hectares of paved habitat, square meters that exceed a particular chemical threshold). At larger spatial scales, the shape or arrangement of exposure may be an important issue, and area alone may not be enough to describe spatial extent for risk assessment. Geographic Information Systems (GISs) provide many options for analyzing and presenting the spatial dimension of exposure (e.g., Pastorok et al.[40]).

#### 36.2.3.2 Characterization of Ecological Effects

Effect characterization describes the stressor's effects, links them to the assessment endpoints, and examines the relationship between the amount of exposure and the magnitude of response. The objective is to describe the relationship between stressor levels and ecological effects and, further, to relate the measurable ecological effects to assessment endpoints when the latter cannot be directly measured.

##### 36.2.3.2.1 Stressor-Response Analysis

To evaluate ecological risks one must examine the relationships between stressors and responses postulated in the analysis plan. For example, an assessor may need point estimates of an effect (such as an $LC_{50}$) to compare the effects of several stressors. The shape of the stressor–response curve may be needed to determine the presence or absence of an effects threshold or for evaluating incremental risks, or stressor–response curves may be used as input for effects models. Assessors usually examine one response (e.g., mortality, mutations), and most quantitative techniques have been developed for univariate analysis. If the response of interest reflects many individual variables (e.g., species abundances in an aquatic community), multivariate techniques may be needed.[41] If it is not possible to quantify stressor–response relationships, they may be described qualitatively.

Stressor–response relationships are described using intensity, time, or space for consistency with exposure estimates. Intensity is most often used for chemicals (e.g., dose, concentration) and pathogens (e.g., spores per milliliter; propagules per unit of substrate). Point estimates and stressor–response curves can be generated for chemicals and some biological stressors. Exposure duration also is commonly used for chemical stressor–response relationships; for example, median acute effect levels are always associated with a time parameter (e.g., 24 h). Space or area is usually used for physical stressors: Thomas et al.[42] related the chance of sighting a spotted owl to the extent and pattern of suitable habitat, and Phipps[43] related tree growth to water-table depth.

If enough data are available, multiple-point estimates can be displayed as cumulative distribution functions. Figure 36.3 shows how this was done for species sensitivity derived from multiple-point

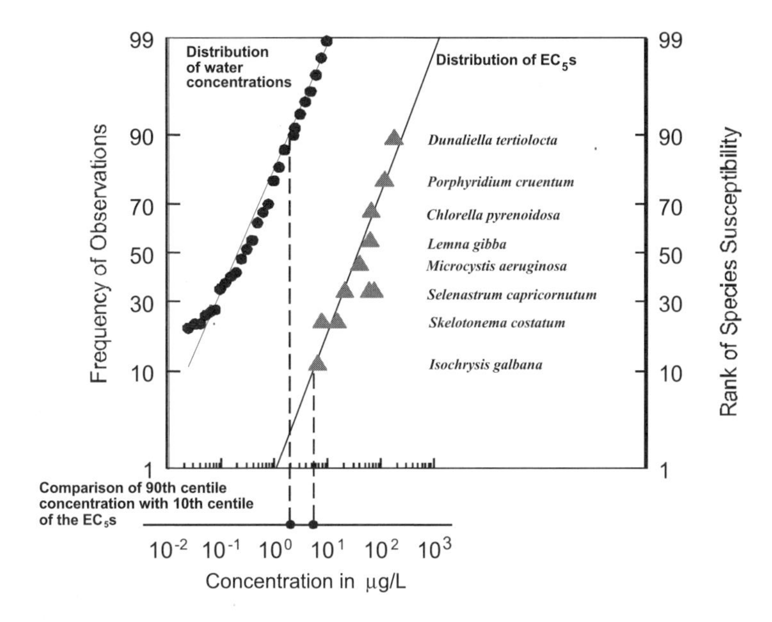


**Figure 36.3** Risk estimation techniques: comparison of exposure distribution of an herbicide in surface waters with freshwater single-species toxicity data. (Redrawn with permission from Baker et al., *Pesticide Risk Assessment and Mitigation*, SETAC, Pensacola, FL, 1994.) (Centile ranks for species $EC_5$ data were obtained using the formula ($100 - n/[N + 1]$), where $n$ is the rank number of the $EC_5$, and $N$ is the total number of data points in the set. (Adapted from Parkhurst, B.R. et al., Methodology for Aquatic Ecological Risk Assessment, Water Research Foundation, Alexandria, VA, 1995.)

estimates ($EC_5$s) for algae (and one vascular plant) exposed to an herbicide. These distributions can help identify stressor levels that affect a chosen proportion of species.

### *36.2.3.2.2 Establishing Cause-and-Effect Relationships (Causality)*

Causality is the relationship between cause (a stressor) and effect (response). Without a clear connection between cause and effect, one cannot place much confidence in the ERA's conclusions. Causal relationships are especially important in risk assessments driven by observed effects such as bird or fish kills or a shift in species composition. When multiple stressors occur, a causal evaluation helps identify the stressors most responsible for effects and eliminate those that are unimportant (Section 36.3.3). Evidence that can be used to support causal inferences may come from observed (e.g., bird kills associated with field application of a pesticide, with death caused by an effect the pesticide is known to elicit) or experimental data (e.g., laboratory tests with the pesticides in question show mortality at tissue concentrations similar to those found in the field). Confidence increases when the effects of different causes can be isolated (e.g., as in controlled experiments) and when multiple lines of evidence support a particular cause. Since we cannot always arrange an experiment, scientists have looked for other ways to support an argument for cause and effect. Human health and ecoepidemiologists have developed a series of considerations that are useful for evaluating causality.[44–47] These include whether the cause preceded the effect in time, whether the cause is consistently associated with an effect, whether effects increase with increasing levels of stress, and whether there is a plausible mechanism.

For chemicals, a modification of Koch's postulates[48–50] has been proposed, and was applied in an evaluation of air pollution impacts on trees:[51]

- The injury, dysfunction, or other putative effect of the toxicant must be regularly associated with exposure to the toxicant and any contributory causal factors.
- Indicators of exposure to the toxicant must be found in the affected organisms.

- The toxic effects must be seen when organisms or communities are exposed to the toxicant under controlled conditions, and any contributory factors should be manifested in the same way during controlled exposures.
- The same indicators of exposure and effects must be identified in the controlled exposures as in the field.

#### *36.2.3.2.3 Linking Effects to Assessment Endpoints*

Stressor–response and causal relationships are most easily evaluated using very specific measures of effects that can be readily measured. These measures may be closely linked with the assessment endpoint. When they are not, extrapolations are used to make this link explicit and preferably quantified. Common extrapolations include those between taxa (e.g., bluegill sunfish mortality to rainbow trout mortality), between responses (from mortality to growth or reproduction) from laboratory to field (from bluegill mortality in laboratory tests to that under field conditions), between geographic areas (an empirical model in one region to another) between spatial scales (local effects to regional effects), and between temporal scales (short-term exposure to longer-term exposure).

Uncertainty factors are commonly used to account for uncertainty associated with different extrapolations. The objective is usually to estimate a stressor level that should not cause adverse effects to the assessment endpoint. Uncertainty factors have been most extensively developed for chemicals because extensive ecotoxicological data are available, especially for aquatic organisms. They are useful when decisions must be made about stressors in a short time and with little information.[52] Despite their usefulness, uncertainty factors can be misused, especially when applied too conservatively such as when several factors are multiplied together without sufficient justification.

Process models for extrapolation are abstractions of a system or process[53] that incorporate causal relationships and can generate predictions even when data are not available.[54] They make it possible to translate data on individual effects (e.g., mortality, growth, and reproduction) to potential alterations in specific populations, communities, or ecosystems. They can be used to evaluate risk questions about a stressor's effect on an assessment endpoint that cannot easily be tested in the laboratory. Population models describe how demographic attributes of a group of individuals of the same species change through time; they have been used in both wildlife and fisheries management and to assess the impacts of power plants and toxicants on specific fish populations.[55–57] Community and ecosystem models (e.g., Jorgensen[32] and Bartell et al.[58]) are useful when the assessment endpoint involves structural (e.g., community composition) or functional (primary production) elements. They can also be useful for examining how effects may cascade through an ecosystem and for estimating changes in ecosystem components such as populations, functional types, feeding guilds, or environmental processes.

The final product of ecological response analysis is a summary of what effects a stressor may elicit in the various receptors of concern, under what conditions these effects occur, and how such effects relate to the assessment endpoints.

### 36.2.4 Risk Characterization

Risk characterization is the final phase of ERA, in which the risk assessor reaches conclusions about the existing or anticipated effects. The assessor uses the results of analysis to estimate risk to the entities included in the assessment endpoints identified in problem formulation (Section 36.2.5.1). Next, the assessor provides context — the significance of any adverse effects and lines of evidence supporting conclusions (Section 36.2.5.2). Finally, the assessor identifies and summarizes the uncertainties, assumptions, and qualifiers in the risk assessment and reports the conclusions (Section 36.2.5.3).

#### *36.2.4.1 Risk Estimation*

Risk estimation is the process of integrating exposure and effects data and evaluating any associated uncertainties. The process uses exposure and stressor–response estimates developed in the analysis phase. The output may be qualitative or quantitative and is used to place the risks in the risk-management context.

A variety of methods are used to combine exposure and effects information into an estimate of risk. Even with limited data, risks may be categorized or ranked using professional judgment. If there are enough data to quantify exposure and effects estimates, the simplest comparative approach is a ratio or quotient; this usually is expressed as an exposure concentration divided by an effects concentration. Quotients are most commonly used for chemical stressors, for which reference or benchmark toxicity values are widely available. The quotient method can be enhanced by using the entire stressor–response curve or by incorporating variability in exposure or effects estimates. Process models can be used to combine empirical estimates with mechanistic data. Even field observations can be used, when a particular exposure scenario and effects co-occur. Because each method has its strengths and limitations, a risk assessment may use more than one approach.

#### *36.2.4.2 Risk Description*

After preparing risk estimates, risk assessors explain the supporting information by evaluating the lines of evidence and the significance of the adverse effects. Ideally, the relationship between the assessment endpoints, measures of effect, and associated lines of evidence will have been established during the analysis phase. If not, the risk assessor can qualitatively link lines of evidence to the assessment endpoints. Regardless of the technique, the technical narrative supporting a risk estimate is as important as the estimate itself.

Confidence in an ERA's conclusions may be increased by using several lines of evidence to interpret and compare risk estimates. These may be derived from different sources or by different techniques, such as quotient estimates, modeling results, or field observational studies. Three factors are used to evaluate lines of evidence: (1) data quality and adequacy, (2) degree and type of uncertainty associated with the evidence, and (3) relationship of the evidence to the risk assessment questions.

This evaluation is more than just a list of the evidence that supports or refutes the risk estimate. The risk assessor should examine each line of evidence and evaluate its contribution in the risk assessment's context. Data or study results often are not reported or carried forward in the risk assessment because they are of insufficient quality. Note that such data may still be valuable if they provide information about topics such as how methodologies could be improved and provide recommendations for further studies.

At this point, the assessment has described potential changes in the assessment endpoints and evaluated lines of evidence. The next step is to explain whether these changes are in fact adverse. For our purposes, adverse effects are changes that are undesirable because they alter valued structural or functional attributes of the entities of interest, that is, the assessment endpoints identified back in problem formulation. The risk assessor evaluates the degree of adversity, a challenging task frequently based on professional judgment. The assessor should also recognize that ecological effects will be considered along with economic, legal, or social factors when a risk-management decision is made and be prepared to discuss adverse effects in that context, if necessary. The risk manager will use all this information to decide whether a particular adverse effect can be tolerated, and may also find it useful when communicating the risk to stakeholders.

The criteria for evaluating the degree of adversity are (1) nature and intensity of effects, (2) spatial and temporal scale, and (3) recovery potential. The nature and intensity of effects help distinguish adverse changes from normal ecosystem variability or those resulting in little or no significant change. For example, for an assessment endpoint involving survival, growth, and reproduction of a species, do predicted effects involve survival and reproduction or only growth? If offspring survival will be affected, by what percentage will it diminish?

Although statistical significance does not necessarily mean ecological significance, risk assessors should consider both ecological and statistical effects when evaluating intensity. For example, a 10% decline in reproduction may be worse for a population of slowly reproducing trees than for rapidly reproducing planktonic algae, even if both can be statistically differentiated from a reference or control condition. Natural ecosystem variation can make it hard to statistically detect stressor-related perturbations. For example, normal intra- and interannual variability in the size of marine fish populations is several orders of magnitude. Furthermore, cyclic events (e.g., migration, tides) are very important in natural systems and may mask or delay stressor-related effects. A field study's inability to detect statistically significant effects does not automatically mean that there are none. Risk assessors should consider the test's power to detect changes, the degree or magnitude of change, and other lines of evidence in reaching their conclusions.

The spatial dimension of risk encompasses both the extent and pattern of effect as well as the context of the effect within the landscape. Factors to consider include the absolute area affected, the extent of critical habitats affected compared with a larger area of interest, and the affected area's role within the landscape. A larger affected area may be (1) subject to a greater number of other stressors and complications from stressor interactions, (2) more likely to contain sensitive species or habitats, or (3) more susceptible to landscape-level changes because many ecosystems may be altered by the stressors. Nevertheless, a smaller area of effect is not always associated with lower risk. An area's function within the landscape may be more important than its absolute area. Destruction of small but unique areas such as critical wetlands may have important effects on local and regional wildlife populations.

The temporal scale for ecosystem change can range from seconds (photosynthesis, prokaryotic reproduction) to centuries (global climate change). Changes within a forest ecosystem can occur gradually over decades or centuries and may be affected by slowly changing external factors such as climate. When interpreting adversity, risk assessors should recognize that any stressor-induced changes occur within the context of multiple natural time scales. In addition, ecosystem changes may involve intrinsic time lags, so observable responses to a stressor may be delayed. Analysts should make an effort to distinguish a stressor's long-term impacts from its immediately visible effects. For example, fish population changes resulting from eutrophication may not become evident for many years after initial increases in nutrient levels and observations of algal blooms.

Considering the temporal scale of adverse effects leads logically to a consideration of recovery. Recovery is the rate and extent of return of a population or community to some aspect of its condition prior to a stressor's introduction.[59] Because ecosystems are dynamic and, even under natural conditions, constantly changing in response to changes in the physical environment (e.g., weather, natural disturbances) or other factors, it is unrealistic to expect that a system will remain static at some level or return to exactly the same state that it was before it was disturbed.[60] Therefore, the assessment should be clear about what attributes define "recovered" in a particular system. Examples might include productivity declines in a eutrophic system, species reestablishment at a particular density, species recolonization of a damaged habitat, or diseased organisms' recovery.

Recovery can be evaluated in spite of the difficulty in predicting events in ecosystems.[61] We can distinguish changes that are usually reversible (e.g., stream recovery from sewage effluent discharge), frequently irreversible (e.g., establishment of introduced species), and always irreversible (e.g., extinction). For example, physical alterations, such as deforestation in the coastal hills of Venezuela in recent history and in Britain during the Neolithic period, changed soil structure and seed sources such that forests cannot easily grow again.[59]

#### *36.2.4.3 Reporting Risks*

When risk characterization is complete, risk assessors should be able to estimate the likelihood and magnitude of ecological effects, indicate the overall degree of confidence in the risk estimates, cite lines of evidence supporting the risk estimates, and interpret the adversity of ecological effects. Usually this information is included in a risk assessment report (sometimes called a risk characterization report).

To ensure that readers can understand the ERA and eventually use its conclusions, its results should be clear (written in ordinary language accessible to others outside the risk-assessment field) and transparent (a reader can tell how analyses and conclusions were developed). It also helps if they are reasonable (do not contain any unusual or unprecedented conclusions) and consistent with other similar assessments.[62]

It is in the assessor's interest to present the results in terms that will be meaningful to the decision maker and other readers. It is a good idea to find out what format the decision maker prefers. If this person only wants the big picture, prepare a summary that tells where supporting information can be found if needed. Or if your reader typically searches for details, by all means provide them — do what it takes to get the assessment read. Although the risk characterization should include enough details somewhere so readers can see clearly how its conclusions were derived, eventually the results need to be distilled into a smaller package such as an executive summary or briefing packet.

Risk managers need to know the major risks and get an idea of whether the conclusions are strongly or weakly supported. Insufficient resources, lack of consensus, or other factors may make it impossible to prepare a detailed and well-documented risk characterization. If this is the case, the risk assessor should carefully explain any issues, obstacles, and correctable shortcomings for the risk manager's consideration.

Risk managers typically decide whether additional follow-up activities are needed. Depending on the importance and visibility of the assessment, confidence in its results, and available resources, it may be advisable to conduct another iteration of the risk assessment to inform a final management decision. Another option is to proceed with the decision, implement management action, and develop a monitoring plan to evaluate the results (see Section 36.1). If the decision is to mitigate risks through exposure reduction, for example, monitoring could help determine whether the action taken really did reduce exposure (and effects).

Finally, risk characterization provides the basis for communicating ecological risks to stakeholders and the general public. This task is usually the risk manager's responsibility, but it may be shared with risk assessors. An even better approach is to use communication specialists familiar with common concerns about risk and effective strategies for addressing them. Although the final version of a risk assessment prepared for the government is usually made available to the public, the communication process is best served by tailoring information to a particular audience. This is the time to refer both to the questions posed by managers and other stakeholders that were involved in planning the assessment and to the conceptual model.

## 36.3 RECENT DEVELOPMENTS AND APPLICATIONS OF ECOLOGICAL RISK ASSESSMENT

As ERA is applied to specific problems, methods and concepts are being refined to meet particular needs. This section explores three evolving applications: probabilistic methods (Section 36.3.1) for pesticides and other chemicals, exposure concepts for biological stressors (Section 36.3.2), and place-based analyses for multiple stressors (Section 36.3.3).

### 36.3.1 Assessing Ecological Risks from Chemicals Using Probabilistic Methods

Deterministic methods are widely accepted by regulatory groups and decision-makers and can produce scientifically sound ERAs.[63] However, distilling toxicity and exposure information into a single value can lead to inconsistency and confusion about the assessment's conclusions. Probabi-

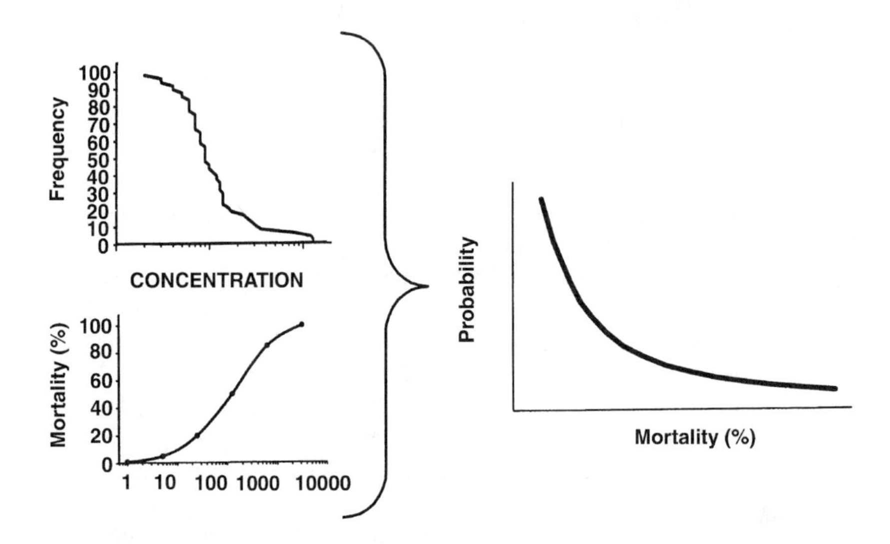


**Figure 36.4** Probabalistic risk assessments combine distributions of effects and exposure to produce estimates of the probability that effects will occur. (From ECOFRAM, Terrestrial Draft Report, U.S. EPA, Washington, D.C., 1999.)

listic ERA methods have been proposed as a way to quantitatively describe natural variability and lack of knowledge and their effect on conclusions.

Probabilistic risk assessments (PRAs) use exposure equations and effects estimates similar to those in deterministic ERAs, but PRAs use distributions rather than single point values for parameters. Distributions for each value used to calculate toxicity or exposure are specified based on available data and the professional judgment of the assessor. Then the distributions can be combined mathematically to yield a range or distribution of risks faced by the endpoint of concern (Figure 36.4). More usually, the distributions are combined using simulations: values for each variable are selected randomly from the distribution and combined, and then the process is repeated thousands of times with programs such as Crystal Ball™. Risk is shown as a curve that combines the exposure and toxicity distributions such as the probability of different levels of effect (e.g., $EC_{10}$).[64] Detailed descriptions of probabilistic risk estimation can be found in several publications.[58,63–67]

Rumbold[68] conducted a PRA on the risk of methylmercury exposure to wading birds in the Everglades. He used an extensive database of fish collected from the Everglades to generate a distribution of the dose of methylmercury from fish consumption to several species of wading birds. From this distribution, he generated curves that showed the fraction of the population that would exceed different toxicological doses.

Probabilistic methods make better use of the information available in dose-response curves, monitoring data, and exposure data. By using ranges or distributions of parameters, probabilistic methods can address both known variability and the lack of knowledge of a parameter's true value. Sensitivity analysis can be conducted to identify the uncertainties that have the greatest impact on the results of the assessment. This can help to focus the ERA on the critical parameters or identify needs for further data collection or research.

Challenges with PRA include gaps in our understanding of how the variables are distributed.[64] Careless treatment of these gaps can create serious flaws in the output. Mathematical assumptions, such as independence of variables, can lead to inaccurate assessments if ignored by assessors. It can also be difficult to describe probabilistic ecological risk assessment results to managers or the public. Deterministic risk assessments typically produce a "bright line" number (concentration or dose) that is considered protective of a given endpoint. In contrast, PRAs typically generate curves that reflect different degrees of protection, for example, 10% of the species has a 1% risk of

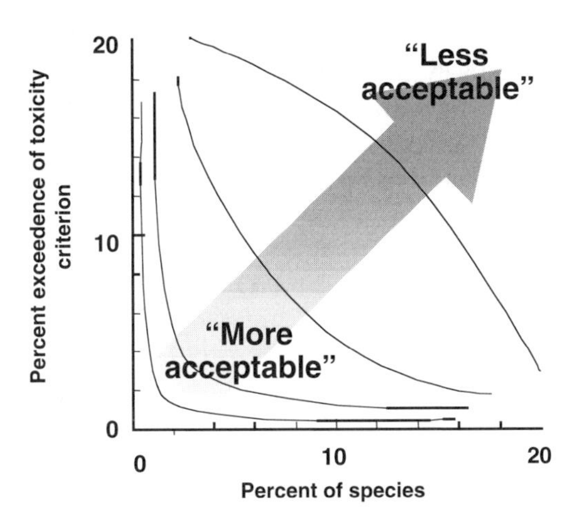


**Figure 36.5** Example curves illustrating possible interpretations of probabilistic risk-assessment results. (From ECOFRAM Terrestrial Draft Report, U.S. EPA, Washington, D.C., 1999.)

exceeding a criterion; 1% has a 10% risk of exceeding it (Figure 36.5). Expressing risks in curves may require additional effort from the risk assessor to effectively communicate results and additional effort from the risk manager to consider the degree of protectiveness required.

Despite the difficulties in calculating and communicating PRA results, these methods provide many benefits. PRA addresses uncertainty and variability by analyzing information available in dose-response curves, exposure distributions, and environmental variability. Probabalistic approaches can increase the realism and credibility of assessments. They can provide insights into low-probability, high-consequence scenarios as well as scenarios that will frequently occur. By providing more information to risk managers, decisions can be better informed, flexible, and appropriately protective.

### 36.3.2 Ecological Risk Assessment of Biological Stressors: Expanding Concepts of Exposure

The most familiar applications of the EPA's Guidelines[1] are for chemical stressors such as pesticides and industrial chemicals, but they also address physical and biological stressors. This section discusses how the general ecological risk paradigm can be adapted to biological stressors, specifically, nonindigenous species. Further discussion and example case studies may be found in Orr.[69]

A significant difference between chemical and biological stressors is that living organisms reproduce. Equally important is that they can adjust to or modify the environment to fit their needs. In addition, a newly established population can, over successive generations, adapt themselves to the new environment. These basic characteristics of life pose a challenge for conducting risk assessments for living organisms. In spite of these difficulties, the principles described in the EPA's Guidelines[1] can be used to evaluate the risks posed by invasive species. This requires a reevaluation of exposure as traditionally applied to chemical stressors.

Exposure is contact or co-occurrence of a stressor with a receptor — a familiar concept for chemical stressors. For biological stressors, exposure analysis evaluates the probability of several factors:[37] (1) organism presence in a pathway leading to their introduction into an area (e.g., insect pests on imported logs); (2) successful entry via the pathway (e.g., the insects survive risk-mitigation measures such as log debarking or fumigation during transport and processing); (3) organism colonization at the entry point (e.g., escape of insect into trees near a pulp mill); and (4) organism potential for spread beyond the initial colonization point.

Evaluating entry pathways is analogous to conducting a source characterization for chemical stressors. It is particularly important for invasive species, since many risk-reduction strategies focus on preventing initial entry. Once the source is identified, the likelihood of entry may be characterized qualitatively or quantitatively or by using a hybrid of both. In their risk analysis of Chilean log importation, for example, the assessment team concluded that the beetle *Hylurgus ligniperda* had a high potential for entry into the United States. Their conclusion was based on the beetle's attraction to freshly cut logs and tendency to burrow under the bark, which would provide protection during transport.[70]

Once the analyst has determined whether a species can enter an area, the next step is to examine whether it can colonize the new habitat and spread beyond the point of introduction. Environmental conditions such as climate, habitat, and suitable hosts determine whether a species becomes established. For example, climate would prevent establishment of the Mediterranean fruit fly in the northeastern United States. Thus, a thorough evaluation of environmental conditions in the area vs. the natural habitat of the stressor is important. Even so, many species can adapt to varying environmental conditions, and the absence of natural predators or diseases may play an even more important role than abiotic factors.

Because of the uncertainty in predicting the effects of potentially invasive species, professional-judgment approaches are commonly used. For example, there may be measures of effect data on a pathogen that attacks a certain tree species not found in the United States, but the assessment endpoint concerns the survival of a commercially important tree found only in the United States. In this case, the analyst would compare the life history and environmental requirements of both the pathogen and the two tree species. Expert panels are typically used for this kind of evaluation.[71] For example, the U.S. Department of Agriculture Forest Service used an expert panel to evaluate the risks of introduced pests to U.S. forests from imported logs.[70] The purpose was to determine whether actions to restrict or regulate the importation of Chilean logs were needed to protect U.S. forests. Stressors included insects, forest pathogens (e.g., fungi), and other pests. The assessment endpoint was the survival and growth of tree species (particularly conifers) in the western United States. Damage that would affect the commercial value of the trees as lumber was also clearly of interest.

The analysis phase was carried out by eliciting professional opinions from a team of six experts. Measures of exposure included distribution information for the imported logs and insect and pathogen attributes such as dispersal mechanisms and life-history characteristics. Measures of ecosystem and receptor characteristics included U.S. climate data, location of geographic barriers, host suitability, and potential host species' ranges. Measures of effect included infectivity of these pests in other countries and the infectivity of similar pests in U.S. hosts (Figure 36.6).

This information was used by the risk-assessment team to evaluate the potential for exposure. They began by evaluating the likelihood of infested logs entering the United States. The distribution of the organism's entry was evaluated by considering the potential for colonization and spread beyond the point of entry as well as its survival and reproduction. Exposure potential was summarized by assigning each of the above elements a judgment-based value of high, medium, or low. Ecological effects were also evaluated by collective professional judgment. Specifically, environmental damage potential was defined as the likelihood of ecosystem destabilization, reduction in biodiversity, loss of keystone species, or reduction or elimination of endangered or threatened species. The team also considered negative impacts beyond environmental damage (e.g., negative economic and social impacts). Again, each factor was assigned a value of high, medium, or low to summarize the potential for ecological effects. The values assigned to the exposure and effect elements were combined into a total risk characterization value using a series of risk tables.[70]

Risk assessment provides a framework for placing information about nonindigenous species into a format that can be used and understood by policy makers for making risk-management decisions. The major difficulty is the high uncertainty associated with predicting the likelihood that a species will establish itself in a new environment and estimating the degree of subsequent impact. This difficulty likely will continue given the lack of information on specific organisms and our current understanding of how ecosystems function. Nevertheless, the degree of uncertainty sur-

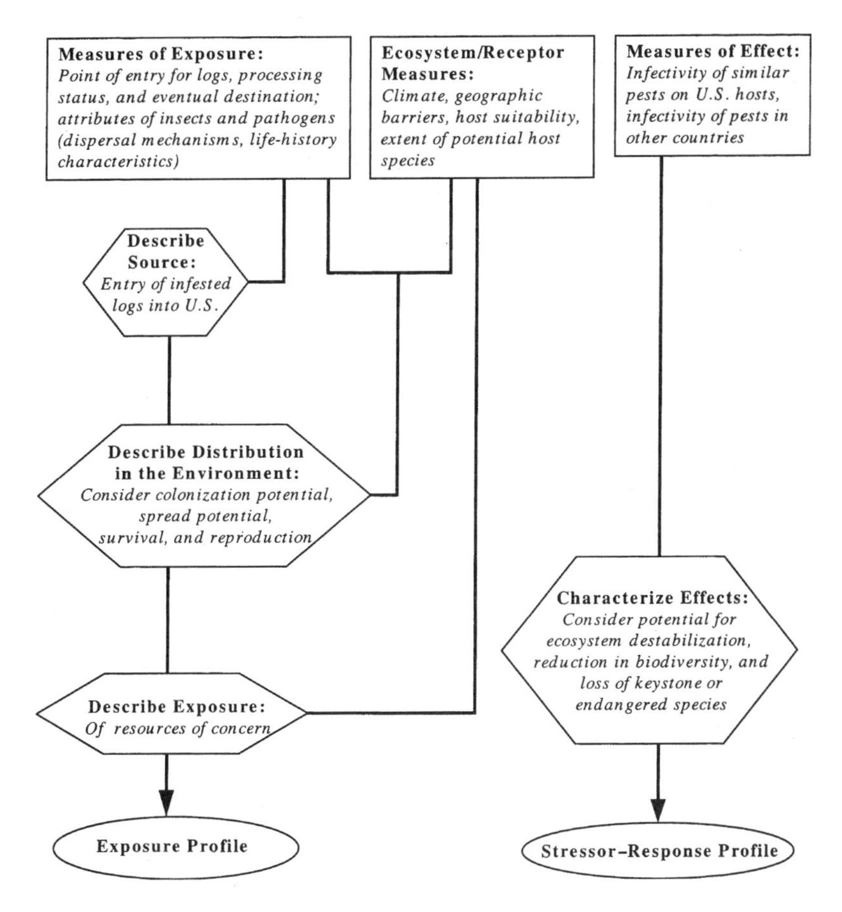


**Figure 36.6** The analysis phase of an ecological risk assessment for invasive species: pest risk assessment of the importation of logs from Chile.[70] Rectangles indicate inputs, hexagons indicate actions, and ovals indicate outputs.

rounding the introduction of nonindigenous organisms only increases the need for careful, unbiased risk assessments before deciding whether to allow an introduction or take other management actions.

### 36.3.3 Evaluating Risks from Multiple Stressors

Risk assessment has a long history of evaluating risks from single (usually chemical) stressors. However, in reality, organisms are exposed to many stressors. Risk assessments that focus on a valued resource (e.g., salmon) or place (e.g., a watershed) must integrate risks across stressors in order to present a complete picture of threats and to identify which stressors may be most urgent or important for management action.

Integrating risks across multiple stressors poses some significant challenges. Stressors usually occur at different temporal and spatial scales. They may operate via different modes of action on different ecosystem components and interact through different physical, chemical, or biological processes. For example, a prototype watershed ERA was performed on a part of the Snake River in Idaho. Demands on water resources have transformed this once free-flowing segment to one with multiple impoundments, flow diversions, and increased chemical and microbiological pollutant loadings. The diversity, reproduction, growth, and survival of representative species from three major trophic levels (fish, invertebrates, and plants) were chosen as assessment endpoints to capture the broad range of potential risks.

Careful selection of assessment endpoints can identify characteristics of the ecosystem that respond to the various stressors under investigation and thus provide a mechanism to integrate across them. Conceptual models provide a useful way to explore and communicate the pathways by which multiple stressors interact. Even these qualitative hypotheses may be helpful for decision-making. As with single-stressor ERAs, they also help identify and select the most important pathways, relationships, and assessment endpoints and help identify uncertainty. For example, Figure 36.7 shows how interactions between siltation and temperature can affect salmon. Cormier et al.[34] used conceptual models to evaluate multiple stressors in the Big Darby Creek of Ohio. This assessment focused on the common endpoint of aquatic community structure to evaluate multiple stressors.

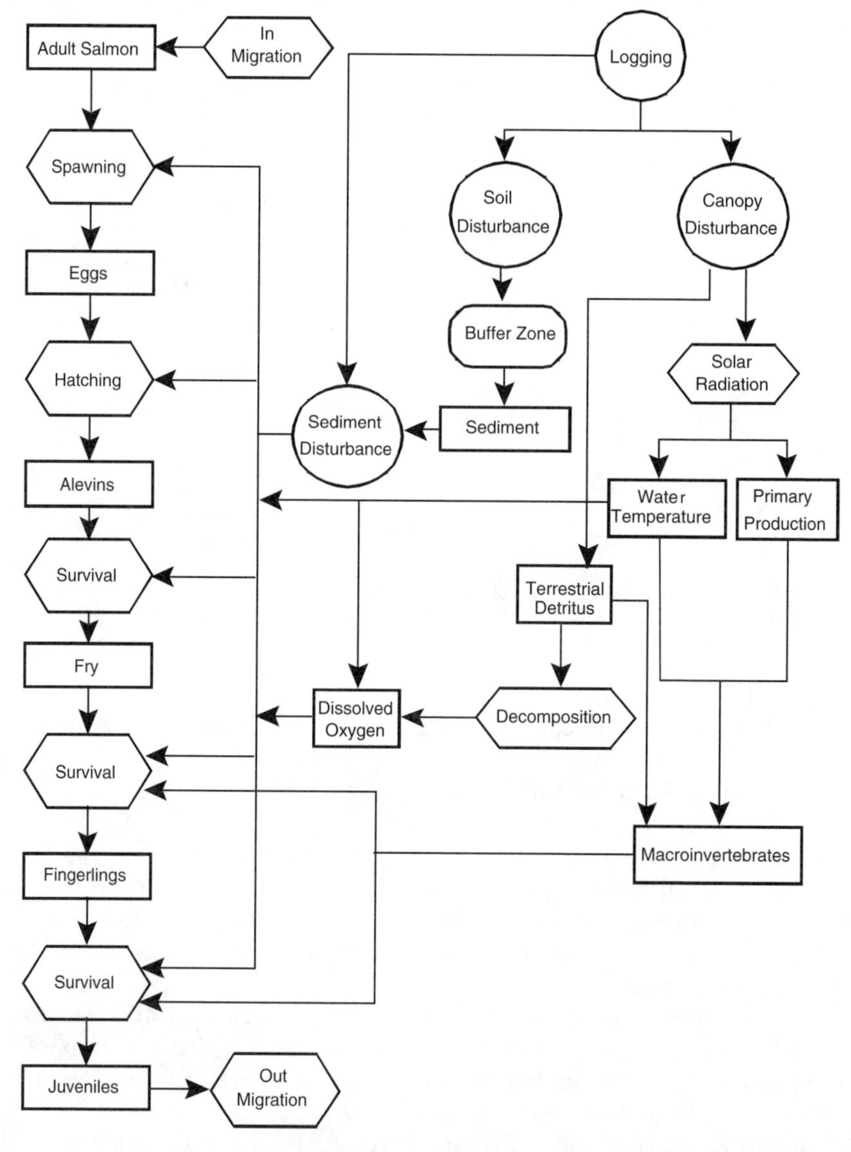


**Figure 36.7** A conceptual model for ecological risk assessment illustrating the effect of multiple stressors on salmon production in a forest stream. (The assessment includes a series of exposures and responses. In the diagram, the circles are stressors, the rectangles are states of receptors, and the hexagons are processes of receptors. The rectangle with rounded corners is an intervention — establishment of buffer zones — that is being considered. (From Suter, G.W. et al., Issue Paper on Characterization of Exposure, U.S. EPA, Washington, D.C., 1994.)

The decision about which pathways and stressors to analyze is best made by an interdisciplinary team's best professional judgment. Foran and Ferenc[72] summarize a variety of techniques, the majority of which involve having team members score the impact of stressors on assessment endpoints (see Reference 73 for an example).

While developing exposure and effects estimates is desirable, it is often difficult for multiple stressors. Different stressors may co-occur in different places, and some may occur only episodically, making them difficult to measure and interpret. Due to the relatively large spatial scale and the multiple stressors that could be present, detailed quantitative exposure information is often unavailable. When sufficient data are available, multivariate statistical models can be used to quantify the associations between sources (e.g., an agricultural field), stressors (e.g., excess nutrients, sediments, and toxic substances), and effects.[41] Process models such as CASM and Aquatox can be useful if they can be parameterized for the system of interest.[74] When models are not available, the system may be simplified by using surrogates such as land-use categories.[75] Or the analyses may need to be limited to the most disruptive stressor.

The Clinch and Powell watershed risk assessment focused on the impacts of multiple stressors on one of the richest assemblages of native fish and freshwater mussels in the world. Nearly half of the species once present in the watershed are now extinct, threatened, or endangered. The ecological assessment used GISs and ERA to structure a multivariate analysis to quantify relationships between instream stressors and land-use patterns and their effects on the fish and mussel communities.[76] One of the most useful analyses in this effort showed how mussel and fish communities declined with proximity to mining, agriculture, and urban land use.

Assessments initiated by an observed effect may need to tease out which of many stressors is the predominant cause. Recently, the EPA has explored how to evaluate causality in situations where an aquatic effect has been observed.[77–80] Methods include eliminating candidate causes, using diagnostic symptoms or protocols, and comparing the strength of evidence for different causes. A formal method may be helpful when the situation is complex or contentious.

## 36.4 ECOLOGICAL RISK ASSESSMENT AND ENVIRONMENTAL MANAGEMENT

The planning phase of the EPA's Guidelines[1] describes interaction between risk assessors and risk managers, albeit in rather general terms and as something that should take place outside the risk assessment process *per se.* This section discusses several ways that the interactions between ERA and management are being explored further. Section 36.4.1 discusses the development of management objectives for ERA, Section 36.4.2 discusses the interface between ERA and management in the context of watershed assessments, and Section 36.4.3 discusses the interactions of ERA with economic benefits analysis and other societal concerns.

### 36.4.1 Developing Management Objectives for Ecological Risk Assessment

ERAs may examine many different species and multiple levels of biological organization, from individual to population, community, and ecosystem. Not everyone values organisms or ecosystems equally, and there is no general agreement on the level of protection they should be afforded. Moreover, a stressor or ecological change may harm some species but benefit others, making it more difficult to decide whether the effects should be avoided. ERAs may also consider species interactions, indirect effects, and nonchemical stressors. Because of these complexities, the process for planning an ERA and deciding on priorities for protection is particularly important.

The bridge between general regulatory goals and specific subjects for protection has been explored from two perspectives. First, the EPA recently collected and organized information about what the EPA had protected in the past.[27] They identified three categories of what they called *Priorities for Ecological Protection*: (1) plants, animals, and their habitats; (2) whole ecosystems, their functions, and services; and (3) special places, and they suggest that these priorities could provide a specific focal point for assessment activities. Second, the processes used to focus broad goals has been further explored, both for assessments prompted by a regulation and those conducted within more flexible partnership processes.

Building on the *Priorities* document, the EPA has described three steps in the planning process: identifying the decision context, developing management objectives, and identifying what is needed to make the decision.[81] These three steps interact substantially with the problem-formulation tasks of integrating available information, describing conceptual models, and identifying endpoints for the assessment. The National Academy of Sciences describes this interaction as an analytic–deliberative process, where the political processes of risk management and the scientific processes of risk assessment are closely intertwined.[26]

The planning process is illustrated in an environmental impact assessment prepared for the Salton Sea restoration effort.[82] The Salton Sea was created by flooding after irrigation water was brought into the southern California desert to support agriculture. Over the years, it became an important recreation area and wetland habitat, although agriculture remains the primary land use in the area. The water level is maintained mainly by agricultural drainwater, which contributes nutrients, pesticides, and other contaminants (e.g., salts) that are concentrated by evaporation, since there is no outlet. Today, the water level is rising and has flooded buildings, roads, and agricultural land. Disease outbreaks among birds have increased in frequency and severity, and recreational use has declined.

Identifying the decision context includes delineating the decision to be made, the context in which it will be made (e.g., public values, legal and policy issues, risk-management options, and geographic and temporal scale), and who needs to be involved (risk managers, risk assessors, and local stakeholders). For the Salton Sea example, two laws were particularly relevant: Public Law 102–575 of 1992 and the Salton Sea Reclamation Act (Public Law 105–372) of 1998. The laws direct the project to, among other things, "provide endangered species habitat, enhance fisheries, and protect human recreational values." The project also needed to comply with the National Environmental Policy Act (NEPA) and the California Environmental Quality Act (CEQA). Two agencies, the Bureau of Reclamation (federal government) and the Salton Sea Authority (state government) led the Environmental Impact Statement (EIS) process and the restoration effort.

The step of developing objectives comprises deciding what is meant by "protect," what is really important and how to achieve it, and choosing objectives. The five major objectives identified in the Salton Sea EIS were to (1) maintain the Sea as a repository for agricultural drainage from the Imperial and Coachella Valleys; (2) provide a safe, productive environment for resident and migratory birds and endangered species; (3) restore recreational uses; (4) maintain a viable sport fishery; and (5) provide opportunities for economic development along the shoreline. Notice that some of these goals contradict each other. While there are no easy remedies to resolving contradictory goals, a facilitated dialog process may be useful. Briefly, the dialog process features (1) equality and the absence of coercive influences (everyone has an equal standing and feels free to speak), (2) listening with empathy (everyone is heard and taken seriously), and (3) bringing assumptions into the open (people explore and explain what's behind their positions).[83] Goals developed by consensus are usually general, so the planning group should be prepared to spend some time making them specific and measurable enough to use in the ERA.

Identifying information needs includes determining what is needed to make the decision (be it from the risk assessment or other analyses), considering what resources are available, deciding what questions the risk assessment should address, and recognizing what the risk manager can expect from the risk assessment. The first and major job of the scientific analysis component of the Salton Sea assessment was to evaluate the potential environmental impacts of five possible restoration alternatives listed in the EIS. The objectives listed above were used to guide this analysis.

A formal planning process to identify management goals and objectives can improve risk assessment by helping decision-makers set specific, measurable objectives that can be used to track progress and document success. It also provides opportunities to incorporate public values into decision-making and to identify potentially conflicting objectives early enough in the process so that the assessment can address them. In addition, it can help focus assessment efforts on resources that are truly valued, ensuring that the assessment provides information relevant to the risk manager's decisions. Finally, a formal process promotes consistency, increases transparency, and aids in communication and improvement of the overall ERA process.

### 36.4.2 Ecological Risk Assessment for Watershed Management

Watershed and other place-based environmental-management approaches use defined geographic areas, partnerships, and science in decision-making.[84] For example, the *Watershed Approach Framework*[85] suggests coordinating environmental management by focusing public- and private-sector efforts on addressing the highest-priority problems within hydrologically defined areas, taking into consideration both surface and ground water. Partnerships with stakeholders are used to achieve the greatest environmental improvements with the resources available. The Watershed Approach has become a key tool for managing over 2000 watersheds in the country.[86]

Assessments conducted in a partnership context are often initiated out of concern for a valued resource or place. A place-based focus facilitates the formation and activities of partnerships that can work together to resolve problems and consider making changes that will improve environmental quality. However, it may be necessary to reconcile multiple values to develop management objectives. In the Big Darby Creek watershed assessment, the partners established three environmental management objectives: (1) establish water-quality criteria for designated uses throughout the watershed, (2) maintain exceptional warm-water criteria for stream segments having that designation between 1990 and 1995, and (3) ensure the continued existence of native species.[34] These were then used to identify information needs that might be fulfilled by the assessment; in the Big Darby, many revolved around better understanding the relationships between human activities throughout the watershed and the diverse aquatic stream community.

A sound plan can form the basis of effective communication. While this is true for all types of risk assessments, place-based assessments generally involve many impacted and interested organizations and individuals, and so the need for more effective communication increases. Many authors have emphasized that stakeholder and manager communication needs to begin during planning, and that recurring discussions are necessary to make the findings most useful.[26,72,87,88] The iterative aspect of the ERA process includes a regularly occurring dialog between the scientists and managers. This feedback loop is intended to incorporate new scientific information and changing risk-management needs into the developing risk assessment. This is especially important in watershed ERA because there are many potential risk managers, with diverse interests, and there is often no legislative mandate for them to take action based on the assessment findings. Instead, their involvement is based on their individual desires to address their respective environmental management objectives. Thus, scientists and managers need to interact regularly to share information, refine the scope of the analysis, and improve the clarity and value of the scientific information to the end users. In the Clinch and Powell Valley assessment, interim results and discussions refined the analyses by focusing work on assessing the impacts of mining, identifying the optimal spatial scale for evaluating riparian land use, and incorporating additional data on toxic-material spills.[76] In the Big Darby Creek assessment, similar discussions stimulated managers and stakeholders to take management actions including removing low-head dams, retargeting erosion-control efforts. The watershed assessment activities also contributed to a proposal to designate a Darby Prairie National Wildlife Refuge, which would protect areas with some of the best aquatic conditions.[89]

ERAs can inform and support watershed management decisions by (1) helping scientists focus on decision makers' needs so managers better understand the ecological implications of their actions and (2) providing watershed-management groups a systematic method for incorporating environmental monitoring and assessment data into decision-making.[90,91]

### 36.4.3 Integrating Ecological Risk Assessment with Economic, Human Health, and Cultural Assessments

Humans and their activities have always been part of the environment, and ecologists and land and resource managers are increasingly asked to provide information to help integrate economic, social, and human-health consequences with ecological issues. Difficult and contentious issues include managing western rivers and natural areas to balance human demands for energy with endangered species' needs for food and habitat. To make good choices, decision-makers must consider ecosystem functions and attributes, human health, and economic and cultural issues. Where economic development compromises ecosystems, this in turn can affect human health and cultural integrity as well as economic opportunities.

ERA information can be used to estimate the value of the services that ecosystems provide (e.g., flood control, nutrient recycling). An economic analysis of ecological benefits is based on the premise that actions affecting the state of an ecological resource will change the goods and services provided by that resource (i.e., the economic benefit endpoints).[92] Typically, these results are presented in terms of dollars, or monetized. Because economic benefits are fundamentally based on ecological resources, economists need to work with ecologists and other scientists to understand what resources are likely to be affected and estimate the magnitude and economic consequences of those changes. In addition, by working with economists to identify and describe resources, ecologists can help ensure that the economic analysis does not overlook significant but less obvious or less direct effects.

Risk assessors and economists begin collaborating during planning, where risk managers, risk assessors, economic analysts, and stakeholders discuss the nature of the problem at hand and alternatives for addressing it. During problem formulation, risk assessors and economists work together to identify and define the linkages between ecological resources and economic benefits (i.e., the goods and services provided or supported by the ecological resource that have economic value to society such as recreation and flood control) (Figure 36.8). During conceptual model development, the economist can explain to the risk assessor how the ERA information will be used in subsequent economic analyses. The ecological risk assessor can identify direct and indirect effects of a risk-management action and explain how the ecological cascade of effects might result in changes that are not immediately apparent. Both viewpoints are needed to extend the conceptual model to include the economic endpoints.

Linking economic benefit endpoints to ecological endpoints can result in several types of findings including good matches between some ecological changes and some economic benefit endpoints as well as ecological changes that economists had not considered and potential economic valuation endpoints for which there are no clear connections to ecological changes identified in the conceptual model. Ecologists and economists can then consider these findings to see if linkages can be further refined or extended. Identification of linkages is an iterative process, and both disciplines should also consider whether there are feedback loops between ecological and economic endpoints. At this point in the process, the goal of collaboration is to be all inclusive and to extend the conceptual diagram to include as many linkages between ecological changes and economic benefit endpoints as is reasonable.

Interactions between the risk assessors and economists may also be useful for prioritizing which endpoints to evaluate and the extent to which relationships will be quantified. ERAs and economic benefit assessments are typically carried out independently. If necessary, analysis plans for the risk assessment and economic assessment can be coordinated to ensure analytical compatibility and to identify any feedback loops whereby economic information may affect the results of the ERA.

The characterization and presentation of ecological risks and the results of the economic benefit analysis are usually also performed independently. Because it may not be possible to monetize or quantify many of the economic benefits of an ecological change, both analysts should make sure the conceptual model allows benefits to be described qualitatively.

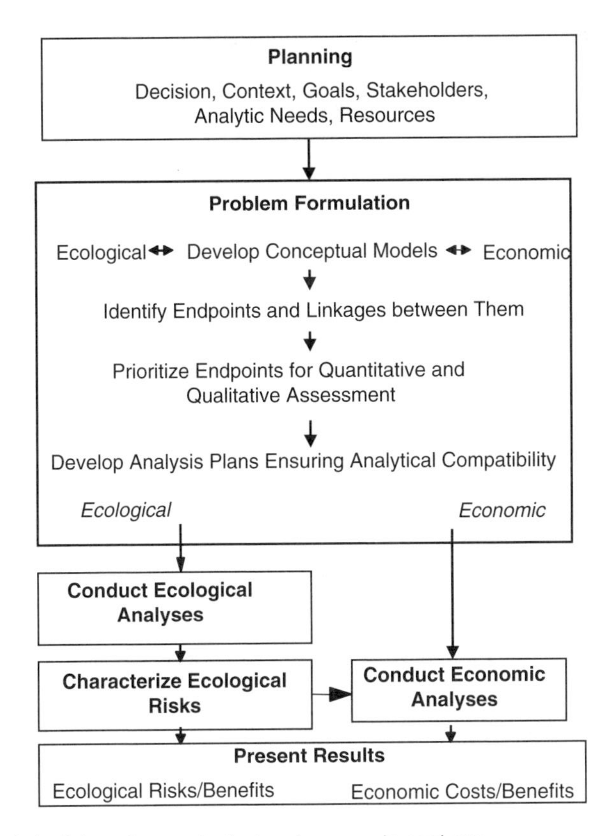


**Figure 36.8** Framework for integrating ecological and economic analyses.

This same general process can be applied to integrate ERA with other analyses: important attributes for human health and cultural values can be defined during planning just as they are for ecosystems. This approach is now being used to evaluate the inter-relationships of human health, economic viability, and ecological and cultural integrity in Arctic ecosystems.[93] Subsistence life-styles serve as a focal point because subsistence is directly linked to all four topics: (1) subsistence is a principal economic base for most communities in Alaska, (2) success in obtaining subsistence foods is a function of ecological condition, (3) community culture centers around the subsistence harvest, and (4) native people's' health directly benefits from eating subsistence foods and is adversely affected by more urban diets. The planning process will be expanded to identify values for all four areas of interest, and the risk assessment process will be used within this broader context. The effort is being approached both on a large regional scale — the Bering Sea — and in small communities.[93]

For example, in the small island village of St. Paul, a village of 750 Aleuts is struggling for economic survival. To broaden its tax base, the community built a deep-sea-fishing port to capitalize on the Bering Sea fishery. Because the current winter fishery is being depleted, the community is considering enlarging the harbor and developing a summer finfish-processing industry. Although this may increase income, finfish processing is not without its own impacts. Preliminary planning activities identified the following scenarios:

The enlarged harbor will allow more ships to use the island, which will increase noise during the breeding season and the risk of fuel spills. Vulnerable wildlife include northern fur seal, kittiwakes and other seabirds, and the endangered Steller sea lion; these organisms, the resources they provide, and an unspoiled environment are integral components of Aleut culture.

Finfish processing uses freshwater. The community's water source is an aquifer, accessed through a well. Although they are currently outside the well's zone of influence, pollutants such as nitrates and petroleum have been introduced into the aquifer. Increased water use may draw the pollutants into the water supply and may also cause salt-water intrusion.

An integrated risk assessment that incorporates ecological, health, economic, and cultural risk can provide managers with a series of potential outcomes for a variety of possible decisions helping them forestall unintended consequences and make reasoned and balanced decisions. As we learn more about how societal goals such as human health, economic benefits, and cultural concerns are influenced by ecological factors, new opportunities for responsible management may become apparent. ERA can provide the cornerstone for these evaluations.

## 36.5 CONCLUSIONS

ERA is a science-based process that allows for a consistent and formal approach for applying ecological information to a broad array of environmental decisions. By providing a common language and process, it increases the reproducibility and clarity of assessments, the credibility of practitioners, and the reasonableness of decisions that it informs. ERA can also provide the framework for examining the interactions between human health, welfare, economics, and ecological condition.

The hallmark of good science is its ability to predict what we will or will not see.[94] ERA has a critical role in predicting when human activities may degrade or improve environmental quality. By reviewing and validating those predictions and using information to improve ERA, we can advance the practical application of ecological and toxicological knowledge. In addition, we can provide managers with means for considering the full range of ecological consequences of their decisions and a sound basis for protecting and improving environmental quality. Finally, as managers increasingly consider the balance between ecological and human systems, ERA can provide a useful foundation for examining the interactions among human health, welfare, the economy, and ecological condition.

## REFERENCES

1. U.S. Environmental Protection Agency, Guidelines for Ecological Risk Assessment, EPA/630/R-95/002F, Risk Assessment Forum, Washington, D.C., 1998.
2. Harwell, M. A., Science and environmental decision making in South Florida, *Ecol. Appl.*, 8, 580, 1998.
3. Harwell, M. A., Long, J. F., Bartuska, A. M., Gentile, J., Harwell, C. C., Myers, V., and Ogden, J. C., Ecosystem management to achieve ecological sustainability: The case of south Florida, *Environ. Manage.*, 20, 497, 1996.
4. Animal and Plant Health Inspection Service, Ecological Risk Assessment: Appendix G of Gypsy Moth Management in the United States: Final Environmental Impact Statement. Volume IV of V., U.S. Department of Agriculture, Forest Service, Northeastern Area State and Private Forestry, Radnor, PA, 1995.
5. Zeeman, M. G., Ecotoxicity Testing and Estimation Methods Developed under Section 5 of the Toxic Substances Control Act (TSCA), in *Fundamentals of Aquatic Toxicology: Effects, Environmental Fate, and Risk Assessment,* Rand, G. M., Ed., Taylor and Francis, Washington, D.C., 1995.
6. Risk Assessment and Management Committee, Generic Non-Indigenous Aquatic Organisms Risk Analysis Review Process, Report to the Aquatic Nuisance Species Task Force, Washington, D.C., 1996.
7. Committee on Environment and Natural Resources, Ecological Risk in the Federal Government, CNER/5–99/001, National Science and Technology Council, Washington, D.C., 1999.
8. Environment Canada, A Framework for Ecological Risk Assessment at Contaminated Sites in Canada: Review and Recommendations, Scientific Series No. 199, Environment Canada, Ottawa, Ontario, 1994.

9. Environment Canada, Ecological Risk Assessments of Priority Substances under the Canadian Environmental Protection Act, Guidance Manual, Chemical Evaluation Division, Commercial Chemicals Evaluation Branch, Ottawa, Canada, 1997.
10. ANZECC, Australian and New Zealand Guidelines for Fresh and Marine Water Quality, Vol. 1, The Guidelines, Paper No. 4, Australia and New Zealand Environment and Conservation Council and Agriculture and Resource Management Council of Australia and New Zealand, Canberra, Australia, 2000.
11. ANZECC, Australian and New Zealand Guidelines for Fresh and Marine Water Quality, Vol. 2, Aquatic Ecosystems — Rationale and Background Information., Paper No. 4, Australia and New Zealand Environment and Conservation Council and Agriculture and Resource Management Council of Australia and New Zealand, Canberra, Australia, 2000.
12. Claassen, M., Strydom, W., Murray, K., and Jooste, S., Ecological Risk Assessment Guidelines, Water Research Commission, Republic of South Africa, Pretoria, 2001.
13. Murray, K. and Claassen, M., An interpretation and evaluation of the U.S. Environmental Protection Agency ecological risk assessment guidelines, *Water SA*, 25, 513, 1999.
14. National Research Council, A paradigm for ecological risk assessment, *Issues in Risk Assessment,* National Academy Press, 1993,
15. Holling, C. S., *Adaptive Environmental Assessment and Management*, John Wiley and Sons, Chichester, U.K., 1978.
16. Beanlands, G. E. and Duinker, P. N., *Ecological Framework for Environmental Impact Assessment in Canada*, Institute for Resources and Environmental Studies, Dalhousie University, Halifax, Nova Scotia, 1983.
17. Cairns, J., Jr., Dickson, K. L., and Maki, A. W., *Estimating the Hazard of Chemical Substances to Aquatic Life, STP 657*, American Society Testing for Testing and Materials, Philadelphia, 1978.
18. Davis, W. S., Biological assessment and criteria: Building on the past, in *Biological Assessment and Criteria,* Davis, W. S. and Simon, T. P., Eds., Lewis Publishers, New York, 1995.
19. Kaplan, S. and Garrick, B. L., On the quantitative definition of risk, *Risk Analysis*, 1, 11, 1984.
20. U.S. Environmental Protection Agency, Framework for Ecological Risk Assessment, EPA/630/R-92/001, Risk Assessment Forum, Washington, D.C., 1992.
21. U.S. Environmental Protection Agency, Summary Report on Issues in Ecological Risk Assessment, EPA/625/3–91/018, Risk Assessment Forum, Office of Research and Development, U.S. Environmental Protection Agency, Washington, D.C., 1994.
22. U.S. Environmental Protection Agency, A Review of Ecological Risk Assessment Case Studies from a Risk Assessment Perspective, EPA/630/R-92/005, Risk Assessment Forum, Office of Research and Development, U.S. Environmental Protection Agency, Washington, D.C., 1993.
23. U.S. Environmental Protection Agency, A Review of Ecological Assessment Case Studies from a Risk Assessment Perspective, EPA/630/R-94/003, Risk Assessment Forum, Office of Research and Development, U.S. Environmental Protection Agency, Washington, D.C., 1994.
24. Browner, C., Policy for Risk Characterization, Policy Memorandum from the U.S. Environmental Protection Agency Administrator Carol Browner, U.S. Environmental Protection Agency, Washington, D.C., 1995.
25. U.S. Environmental Protection Agency, Ecological Risk: A Primer for Risk Managers, EPA/734/R-95/001, U.S. Environmental Protection Agency, Washington, D.C., 1995.
26. National Research Council, *Understanding Risk: Informing Decision in a Democratic Society,* National Academy Press, Washington, D.C., 1996.
27. U.S. Environmental Protection Agency, Priorities for Ecological Protection: An Initial List and Discussion Document for EPA, EPA/600/S-97/002, Office of Research and Development, U.S. Environmental Protection Agency, Washington, D.C., 1997.
28. Keeney, R. L., *Value Focused Thinking: A Path to Creative Decision Making*, Harvard University Press, Cambridge, MA, 1992.
29. Hammond, J. S., Keeney, R. J., and Raiffa, H., *Smart Choices: A Practical Guide to Making Better Decisions*, Harvard Business School Press, Boston, 1999.
30. U.S. Environmental Protection Agency, Guidance for the Data Quality Objectives Process, EPA QA/G-4, Center for Environmental Research Information, U.S. Environmental Protection Agency, Washington, D.C., 1994.

31. Suter, G. W., II, Endpoints for regional ecological risk assessments, *Environ. Manage.*, 14, 19, 1990.
32. Jorgensen, S. E., *Fundamentals of Ecological Modeling*, Elsevier, Amsterdam, 1994.
33. Suter, G. W., II, Developing conceptual models for complex ecological risk assessments, *Human Ecol. Risk Assess.*, 5, 375, 1999.
34. Cormier, S. M., Smith, M., Norton, S. B., and Neiheisel, T., Assessing ecological risk in watersheds: A case study of problem formulation in the Big Darby Creek watershed, Ohio, USA, *Environ. Toxicol. Chem.*, 19, 1082, 2000.
35. Smith, E. P. and Shugart, H. H., Issue Paper on Uncertainty in Ecological Risk Assessment, Ecological Risk Assessment Issue Papers, Risk Assessment Forum, U.S. Environmental Protection Agency, 1994, Chap. 6.
36. Brody, M., Troyer, M. E., and Vallette, Y., Ecological Risk Assessment Case Study: Modeling Future Losses of Bottomland Forest Wetlands and Changes in Wildlife Habitat within a Louisiana Basin, A Review of Ecological Assessment Case Studies from a Risk Assessment Perspective, EPA/630/R-92/005, Risk Assessment Forum, Office of Research and Development, U.S. Environmental Protection Agency, 1993, Chap. 12.
37. Orr, R. L., Cohen, S. D., and Griffen, R. L., Generic Non-Indigenous Pest Risk Assessment Process, Animal and Plant Health Inspection Service, U.S. Department of Agriculture, Riverdale, MD, 1993.
38. Lahlou, M., Shoemaker, L., Choudhury, S., Elmer, R., Hu, A., Manguerra, H., and Parker, A., Basins 2.0 User's Manual, EPA-923-B-98–006, Office of Water, U.S. Environmental Protection Agency, Washington, D.C., 1998.
39. Johnson, C. A., Detenbeck, N. E., and Niemi, G. J., The cumulative effect of wetlands on stream water quality and quantity: A landscape approach, *Biogeochemistry*, 10, 105, 1990.
40. Pastorok, R. A., Butcher, M. K., and Nielson, R. D., Modeling wildlife exposure to toxic chemicals: Trends and recent advances, *Human Ecol. Risk Assess.*, 2, 444, 1996.
41. Fairbrother, A. and Bennett, R. S., Multivariate statistical applications for addressing multiple stresses in ecological risk assessments, in *Multiple Stressors in Ecological Risk and Impact Assessment: Approaches in Risk Estimation,* Ferenc, S. A. and Foran, J. A., Eds., SETAC Press, Pensacola, FL, 2000.
42. Thomas, J. W., Forsman, E. D., Lint, J. B., Meslow, E. C., Noon, B. R., and Verner, J. A., Conservation Strategy for the Spotted Owl, 1990–791/20026, Interagency Scientific Committee to Address the Conservation of the Northern Spotted Owl, U.S. Government Printing Office, Washington, D.C., 1990.
43. Phipps, R. L., Simulation of wetlands forest vegetation dynamics, *Ecol. Modeling*, 7, 257, 1979.
44. Fox, G. A., Practical causal inference for ecoepidemiologists, *J. Toxicol. Environ. Health*, 33, 359, 1991.
45. Hill, A. B., The environment and disease: Association or causation?, *Proc. R. Soc. Med.*, 58, 295, 1965.
46. Susser, M., Rules of inference in epidemiology, *Regulatory Toxicol. Pharmacol.*, 6, 116, 1986.
47. Susser, M., The logic of Sir Carl Popper and the practice of epidemiology, *Am. J. Epidemiol.*, 124, 711, 1986.
48. Yerushalmy, J. and Palmer, C. E., On the methodology of investigations of etiologic factors in chronic disease, *J. Chronic Dis.*, 10, 27, 1959.
49. Hackney, J. D. and Kinn, W. S., Koch's postulates updated: A potentially useful application to laboratory research and policy analysis in environmental toxicology, *Am. Rev. Respiratory Dis.*, 1119, 849, 1979.
50. Woodman, J. N. and Cowling, E. B., Airborne chemicals and forest health, *Environ. Sci. Technol.*, 21, 120, 1987.
51. Suter, G.W., II, *Ecological Risk Assessment*, Lewis Publishers, Boca Raton, FL, 1993.
52. Chapman, P., Fairbrother, A., and Brown, L. C., A critical evaluation of safety (uncertainty) factors for ecological risk assessment, *Environ. Toxicol. Chem.*, 17, 99, 1998.
53. Starfield, A. M. and Bleloch, A. L., *Building Models for Conservation and Wildlife Management*, Burgess International Group, Edina, MN, 1991.
54. Wiegert, R. G. and Bartell, S., Issue Paper on Risk Integration Methods, Ecological Risk Assessment Issue Papers, EPA/630/R-94/009, Risk Assessment Forum, Office of Research and Development, U.S. Environmental Protection Agency, 1994, Chap. 9.
55. Barnthouse, L. W., Suter, G. W., II, Rosen, A. E., and Beauchamp, J. J., Estimating responses of fish populations to toxic contaminants, *Environ. Toxicol. Chem.*, 6, 811, 1986.
56. Barnthouse, L. W., Suter, G. W., II, and Rosen, A. E., Risks of toxic contaminants to exploited fish populations: Influence of life history, data uncertainty, and exploitation intensity, *Environ. Toxicol. Chem.*, 9, 297, 1990.

57. Emlen, J. M., Terrestrial population models for ecological risk assessment: A state-of-the-art review, *Environ. Toxicol. Chem.*, 8, 831, 1989.
58. Bartell, S., Gardner, R., and O'Neill, R., *Ecological Risk Estimation*, Lewis Publishers, Ann Arbor, MI, 1992.
59. Fisher, S. G. and Woodmansee, R., Issue Paper on Ecological Recovery, Ecological Risk Assessment Issue Papers, EPA/630/R-94/009, Risk Assessment Forum, Office of Research and Development, U.S. Environmental Protection Agency, 1994, Chap. 7.
60. Landis, W. G., Matthews, R. A., Markiewicz, A. J., and Matthews, G. B., Multivariate analysis of the impacts of the turbine fuel JP-4 in a microcosm toxicity test with implications for the evaluation of ecosystem dynamics and risk assessment, *Ecotoxicology*, 2, 271, 1993.
61. Niemi, G. J., DeVore, P., Detenbeck, N. E., Taylor, D., Lima, A., Pastor, J., Yount, J. D., and Naiman, R. J., Overview of case studies on recovery of aquatic systems from disturbance, *Environ. Manage.*, 14, 571, 1990.
62. U.S. Environmental Protection Agency, Risk Characterization Handbook, EPA 100-B-00–002, Office of Science Policy, Office of Research and Development, U.S. Environmental Protection Agency, Washington, D.C., 2000.
63. U.S. Environmental Protection Agency, Guiding Principles for Monte Carlo Analysis, EPA/630/R-97/001, Risk Assessment Forum, United States Environmental Protection Agency, Washington, D.C., 1997.
64. ECOFRAM, Terrestrial Draft Report, http://www.epa.gov/oppefed1/ecorisk/, Office of Pesticide Programs, United States Environmental Protection Agency, Washington, D.C., 1999.
65. Warren-Hicks, W. J., Moore, D. R. J., Appling, J. W., Barry, T., Barton, A., Chapman, P., Cirone, P., Clark, J., Cothern, K., Cowan, C., Cox, D., Crocket, T., Dickson, G., Corward-King, E., Farrar, D., Ferson, S., Gilbert, R., Hacker, C., Landis, W., Lavine, M., Levin, L., Macfarlane, M., Mattice, J., Miller, J., Norton, S. B., Parkhurst, B. R., Pastorok, R. A., Power, M., Rodier, D., Schaeffer, D., Smith, E., Suter, G. W., II, Valoppi, L., van Leeuwen, C., Viteri, A., and Williams, B., *Uncertainty Analysis in Ecological Risk Assessment*, Society of Environmental Toxicology and Chemistry, Pensacola, FL, 1995.
66. Moore, D. R. J., Sample, B. E., Suter, G. W., Parkhurst, B. R., and Teed, R. S., A probabalistic risk assessment of the effects of methylmercury and PCBs on mink and kingfishers along East Fork Poplar Creek, Oak Ridge, Tennessee, USA, *Environ. Toxicol. Chem.*, 18, 2941, 1999.
67. Vose, D., *Quantitative Risk Analysis: A Guide to Monte Carlo Simulation Modeling*, John Wiley and Sons, New York, 1996.
68. Rumbold, D., Methylmercury Risk to Everglades Wading Birds: A Probabilistic Ecological Risk Assessment, Everglades Consolidated Report 2000, South Florida Management District, 2000, Appendix 7–3b.
69. Orr, R., Non-Indigenous Species, Ecological Risk Assessment in the Federal Government, White House National Science and Technology Council, Committee on Environment and Natural Resources, CENR/5–99/001, 1999, Chap. 4.
70. U.S. Department of Agriculture, Pest Risk Assessment of the Importation of *Pinus radiata*, *Nothofagus dombeyi* and *Laurelia philippiana* Logs from Chile, Miscellaneous Publication 1517, U.S. Forest Service, 1993.
71. U.S. Environmental Protection Agency, Report on the Shrimp Virus Peer Review and Risk Assessment, EPA/600/R-99/027, National Center for Environmental Assessment, Office of Research and Development, Washington, D.C., 1999.
72. Foran, J. A. and Ferenc, S. A., Multiple stressors in ecological risk and impact assessment, in *Multiple Stressors in Ecological Risk and Impact Assessment,* Foran, J. A. and Ferenc, S. A., Eds., Society of Environmental Toxicology and Chemistry, Pensacola, FL, 1999.
73. Harris, H. J., Wegner, R. B., Harris, V. A., and Devault, D. S., A method for assessing environmental risk: A case study of Green Bay, Lake Michigan, USA., *Environ. Manage.*, 18, 295, 1994.
74. Moore, D. R. J. and Bartell, S., Estimating ecological risks of multiple stressors: Advanced methods and difficult issues, in *Multiple Stressors in Ecological Risk And Impact Assessment: Approaches to Risk Estimation,* Ferenc, S. A. and Foran, J. A., Eds., Society of Environmental Toxicology and Chemistry, Pensacola, FL, 2000, Chap. 4.
75. Gordon, S. I. and Majumder, S., Empirical stressor-response relationships for prospective risk analysis, *Environ. Toxicol. Chem.*, 19, 1055, 2000.

76. Serveiss, V. B., Diamond, J. L., Gowan, D. W., and Hylton, R. E., Watershed ecological risk assessment: The Clinch and Powell valley experience., VWRRC P7 2001, Bosch, D., Eds.,Virginia Polytechnic Institute and State University, Blacksburg, 2001.
77. U.S. Environmental Protection Agency, Stressor Identification Guidance Document, EPA/822/B-00/025, Washington, D.C., Office of Water, 2000.
78. Suter, G. W., II, Norton, S. B., and Cormier, S. M., A method for inferring the cause of observed impairments in aquatic ecosystems, *Environ. Toxicol. Chem.*, 21, 1101, 2002.
79. Norton, S. B., Cormier, S. M., Subramanian, B., Lin, E. L. C., Altfater, D., and Counts, B., Determining probable causes of ecological impairment in the Little Scioto River, Ohio: Part I. Listing candidate causes and analyzing evidence, *Environ. Toxicol. Chem.*, 21, 1112, 2002.
80. Cormier, S., Norton, S. B., Suter, G. W., Lin, E. L. C., Altfater, D., and Counts, B., Determining the causes of impairments in the Little Scioto River, Ohio, Part II. Characterization of causes, *Environ. Toxicol. Chem.*, 21, 1125, 2002.
81. U.S. Environmental Protection Agency, Guidelines for Planning Ecological Risk Assessments, Risk Assessment Forum, Office of Research and Development, U.S. Environmental Protection Agency, Washington, D.C., 2001.
82. Tetra Tech, Salton Sea Restoration Project Draft Environmental Impact Statement/Environmental Impact Report, January 2000, Prepared for Salton Sea Authority and U.S. Department of Interior, Bureau of Reclamation, Lower Colorado Regional Office, Boulder, CO, 2000.
83. Yankelovich, D., *The Magic of Dialogue: Transforming Conflict into Cooperation*, Simon and Schuster, New York, 1999.
84. U.S. Environmental Protection Agency, The Watershed Protection Approach, EPA/503/R-92/002, Office of Water, Washington, D.C., 1991.
85. U.S. Environmental Protection Agency, Watershed Approach Framework, EPA-840-S-96–001, Office of Water, U.S. Environmental Protection Agency, Washington, D.C., 1996.
86. U.S. Environmental Protection Agency, Testimony of J. Charles Fox, Assistant Administrator for Water, U.S. EPA, before the U.S. House of Representatives, 2000.
87. Dover, M. J. and Goulding, D., Communicating with the Public on Ecological Issues: Workshop Report, Cooperative Agreement # CX832519–01–0, Prepared for Office of Sustainable Ecosystems and Communities, Office of Policy, Planning and Evaluation, U.S. EPA, Washington, D.C., 1995.
88. Timmerman, J. G., Ottens, J. J., and Ward, R. C., The information cycle as a framework for defining information goals for water-quality monitoring, *Environ. Manage.*, 25, 229, 2000.
89. U.S. Environmental Protection Agency, Report on the watershed ecological risk characterization workshop, EPA/600/R-99/111, National Center for Environmental Assessment, Office of Research and Development, U.S. Environmental Protection Agency, Washington, D.C., 2000.
90. Serveiss, V. B., Applying ecological risk principles to watershed assessment and management, *Environ. Manage.*, 29, 145, 2002.
91. Serveiss, V. B., Norton, D. J., and Norton, S. B., Watershed Ecological Risk Assessment, http://www.epa.gov/owow/watershed/wacademy/acad2000/ecorisk, The Watershed Academy, Office of Water, U.S. Environmental Protection Agency, 2000.
92. Freeman, M. A., *The Measurement of Environmental and Resource Values Theory and Methods*, Resources for the Future, Washington, D.C., 1993.
93. Interagency Arctic Research Policy Committee, Integrating Science and Management for Sustainable Bering Sea, Arctic Research of the United States: 7th Biennial Arctic Research Plan 2002–2006, National Science Foundation, 2001, Sect. 2.
94. Peters, R. H., *A Critique for Ecology*, University Press, Cambridge, U.K., 1991.
95. Kendall, R. J., Lacher, T. E., Bunck, C., Daniel, B., Grue, C. E., Leighton, F., Stansley, W., Watanabe, P. G., and Whitworth, M., An ecological risk assessment of lead shot exposure in non-waterfowl avian species: Upland game birds and raptors, *Environ. Toxicol. Chem.*, 15, 4, 1996.
97. Baker, J. L., Barefoot, A. C., Beasley, L. E., Burns, L. A., Caulkins, P. P., Clark, J. E., Feulner, R. L., Giesy, J. P., Graney, R. L., Griggs, R. H., Jacoby, H. M., Laskowski, D. A., Maciorowski, A. F., Mihaich, E. M., Nelson, H. P., Jr., Parrish, P. R., Siefert, R. E., Solomon, K. R., and van der Schalie, W. H., *Aquatic Dialogue Group: Pesticide Risk Assessment and Mitigation*, Society of Environmental Toxicology and Chemistry (SETAC) Press, Pensacola, FL, 1994.

98. Parkhurst, B. R., Warren-Hicks, W. J., Etchison, T., Butcher, J. B., Cardwell, R. D., and Voloson, J., Methodology for Aquatic Ecological Risk Assessment, RP91-AER-1, Water Environment Research Foundation, Alexandria, VA, 1995.
99. Suter, G. W., II, Gillett, J. W., and Norton, S. B., Issue Paper on Characterization of Exposure, Ecological Risk Assessment Issue Papers, EPA/630/R-94/009, Risk Assessment Forum, Office of Research and Development, U.S. Environmental Protection Agency, 1994, Chap. 4.

CHAPTER 37

# Ecological Risk Assessment Example: Waterfowl and Shorebirds Feeding in Ephemeral Pools at Kesterson Reservoir, California

**Earl R. Byron, Harry M. Ohlendorf, Gary M. Santolo, Sally M. Benson, Peter T. Zawislanski, Tetsu K. Tokunaga, and Michael Delamore**

## CONTENTS

## 37.1 INTRODUCTION

### 37.1.1 Site Background

Kesterson Reservoir is located in the Grasslands area of northern Merced County, within the San Joaquin Valley of California. The reservoir was constructed between 1968 and 1975 as a series of 12 shallow ponds totaling about 500 ha (1200 acres) that served as a disposal site for subsurface agricultural drainage. Shallow saline groundwater was collected by drainage collection systems from fields in Fresno County and conveyed to the reservoir by the concrete-lined San Luis Drain. Following the discovery of selenium contamination in Kesterson Reservoir, the U.S. Bureau of Reclamation (USBR) halted discharge of agricultural drainage to Kesterson in 1986. The reservoir was subsequently dewatered, and lower elevation portions were filled with soil in 1988 to prevent groundwater from rising to the ground surface. Further description of the site and its early history are provided by USBR et al.,[1] USDI,[2] and Ohlendorf.[3] A map showing the location and configuration of the subdivided, dry "ponds" of the reservoir as well as the three main habitat types is shown in Figure 37.1.

### 37.1.2 Previous Investigations

Beginning in 1983, a series of investigations documented the nature and occurrence of selenium contamination in soils, surface water, and groundwater at Kesterson Reservoir. Early investigations focused on surface water and the physicochemical processes associated with selenium bioaccumulation in the aquatic food chain.[1–5] These studies demonstrated that while selenate was the predominant form of selenium in surface waters, the pond-bottom sediments accumulated very high concentrations (tens to hundreds of mg/kg) of selenium in the elemental and organically associated forms. This accumulation resulted from a combination of microbial transformation of selenate to elemental selenium in the pond-bottom sediments and decomposition of aquatic vegetation such as the macroalgae chara and cattails.[6–8]

In 1988, with the drying out of Kesterson and filling of low-lying areas with soil, studies shifted to focus on the upland environment and the ephemeral pools that form during the rainy winter months. Specific investigations related to the physical and chemical behavior of selenium in the Kesterson environment include the following:

1. **Ephemeral-pool occurrence and waterborne-selenium concentrations**
   - Field measurements of selenium concentrations in ephemeral pools from 1992 through 1999 (see summary by Zawislanski,[9] CH2M HILL[10])
   - Field observations of the inception and duration of ephemeral pools from 1990 through 1999[9]
   - Development of a model relating inception and duration of ponding to the difference between precipitation and cumulative evapotranspiration[9]

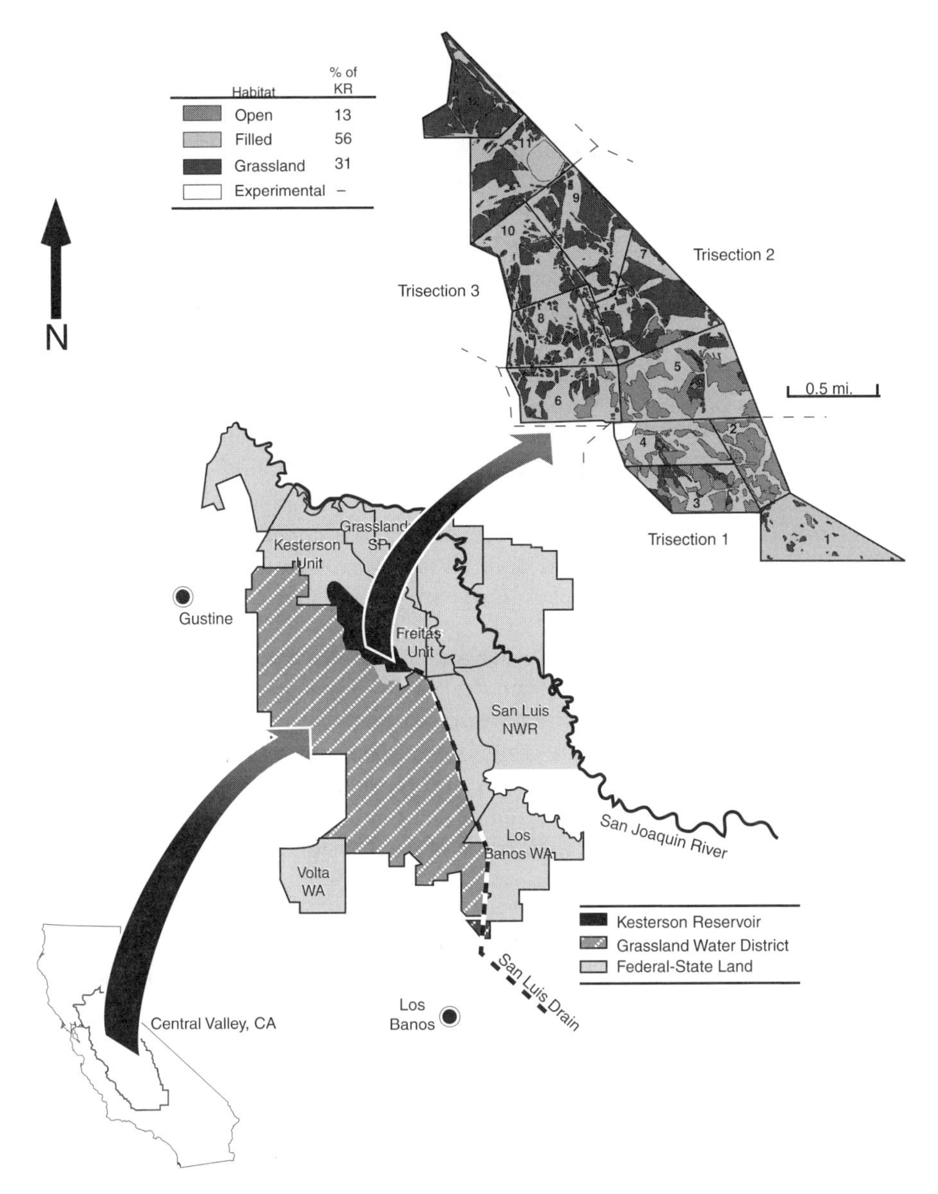


**Figure 37.1** Kesterson Reservoir location and habitats.

- Aerial photographs taken early to mid-spring showing the location and extent of ephemeral pools
- Laboratory studies elucidating the physical and chemical processes contributing to intra-annual trends in selenium concentrations in ephemeral pools[11,12]

2. **Spatial distribution and time trends in soil-selenium concentrations**
   - Selenium speciation and fractionation investigations showing the depth distribution of elemental selenium, selenate, selenite and organically associated selenium at numerous sites throughout Kesterson[6,13]
   - Long-term monitoring of the depth distribution of extractable and total selenium concentrations at several sites from 1988 through 1998[13,14]
   - Synoptic measurements of total and water-extractable-selenium concentrations at 54 locations across Kesterson from 1990 to 1998 in conjunction with the biological monitoring program[15] (for summary see Zawislanski et al.[9])
   - Field-measured volatilization rates from several experimental plots[16,17]

- Laboratory measurements of selenium reoxidation rates under controlled conditions[18–20]
- Modeling studies of the above-mentioned data to determine re-oxidation and leaching rates of selenium[21,22]

Together these studies have been sufficient to develop an understanding of the current status of selenium in soils and ephemeral pools at Kesterson. Most of the selenium is confined to the top 15 cm of soil. Only a small fraction of the selenium in surface soils, typically less than 5%, is currently in extractable or mobile forms (e.g., dissolved selenate and selenite).[9]

Ephemeral pools at Kesterson are formed primarily by the accumulation of rainwater on the soil surface. Selenium concentrations in these pools range from 1 to as much as 1000 μg/L. The highest selenium concentrations occur immediately after the pools have formed and then decrease to relatively stable concentrations in the range of 1 to 200 μg/L after a few weeks (Figure 37.2). The selenium in these pools results from dissolution of salt crusts and diffusion of dissolved selenium from pore waters into the overlying surface waters. Once the pool is formed, anoxic conditions may develop in the underlying soils. Over time, diffusion of selenium from the pools back into the underlying soils depletes the concentration of selenium in surface water. Simultaneously, selenium diffused back into the soils is reduced to immobile forms. As demonstrated by nearly a decade of monitoring, the net effect of these processes is for selenium concentrations in ephemeral pools to remain in the current range for several decades.

**3. Biological monitoring and ecological risk assessment (ERA)**

A series of field and laboratory studies, beginning in 1983, documented the bioaccumulation of selenium in food-chain organisms and its adverse effects on aquatic birds feeding at Kesterson (summarized by Ohlendorf[3]). These birds exhibited high rates of embryo deformities and mortality as well as mortality of adult birds. In contrast, studies of mammals and terrestrial birds using Kesterson showed significant levels of selenium bioaccumulation, but no clear evidence of adverse effects. These findings, among others, contributed to the decision to dewater the reservoir and fill low-lying areas as a means of site remediation.

Selenium monitoring and several focused studies have been conducted at Kesterson since 1988 to determine temporal and spatial trends in selenium levels and to determine whether existing conditions cause adverse effects in wildlife (see, for example, CH2M HILL[10]). Although selenium concentrations are elevated in all environmental media (soils, plants, and animals), the monitoring has not found evidence of embryo deformities, reduced egg viability, or wildlife mortality.

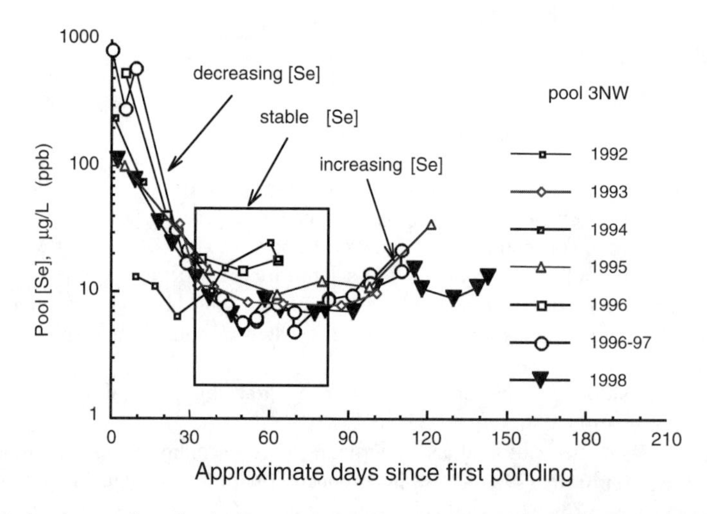


**Figure 37.2** Typical ephemeral-pool-selenium concentration pattern following initial ponding, site 3NW.

An ERA was conducted in 1993 to (1) review pertinent information available at that time, (2) estimate future levels of selenium in various biota (plants and animals) at Kesterson, (3) estimate the risks to animals caused by the site's selenium inventory, (4) assess the significance of the site's selenium toxicosis risks, (5) identify contingency plans and evaluate their effectiveness in eliminating or reducing any potentially significant risks of selenium toxicosis, and (6) recommend research and monitoring that would provide information needed to improve management efficiency for Kesterson or other projects.[23,24]

That risk assessment analyzed data available from the Kesterson Reservoir Biological Monitoring Program (KR BMP) through November 1992 as well as other published and unpublished reports. Total and water-soluble selenium concentrations in the top 15 cm of soil, along with selenium concentrations in a wide range of plants and animals, were used to model potential exposures and predict effects to terrestrial wildlife over a period of 20 years into the future. In addition, waterborne-selenium concentrations and selenium in aquatic invertebrates from ephemeral pools were used to evaluate potential risks to aquatic birds.

The overall conclusions of the risk assessment were that (1) selenium concentrations in the plants and animals at Kesterson were not expected to change markedly during the next 20 years, (2) the selenium concentrations in small mammals were expected to remain below those measured during a study in 1984 when the reservoir was partially flooded with drainwater (and there were no clear effects on small mammal reproduction), (3) selenium concentrations in insectivorous birds were not expected to increase to biologically significant levels, (4) the greatest risk of selenium exposure to terrestrial wildlife appeared to be the possible consumption of mushrooms that contain the highest levels of selenium found at Kesterson (but there was little evidence that birds or mammals ate the mushrooms), (5) much of the reservoir could be covered with standing water during very wet years as a result of pooling of rainwater, and (6) although selenium concentrations in these pools (water and biota) are lower than pre-1989 levels, aquatic birds may feed in pools that persist into spring and be exposed to selenium at levels that could cause adverse reproductive effects. Monitoring of these pools has been included each year during which they occur during late winter and early spring.[10] Waterborne-selenium concentrations in the pools constitute some of the highest potential levels of selenium exposure available to wildlife at Kesterson. The importance of this route of exposure to overall risk could not be understood without conducting a risk assessment with recent data to define exposure and risk from ephemeral pools. This chapter provides a summary of the risk assessment for aquatic exposures completed in 2000 for the USBR by CH2M HILL and Lawrence Berkeley National Laboratory (LBNL).[25] Although that ERA also evaluated terrestrial exposures, they are not included in this chapter.

### 37.1.3 Objectives

The objectives of the current ERA were to:

1. Assess the risks to wildlife from the site's selenium inventory, based on observed and estimated levels of biologically available selenium, particularly as resulting from the site's ephemeral pools.
2. Develop contingency plans for controlling the routes of exposure and risk identified in the assessment.
3. Recommend appropriate monitoring for Kesterson and additional research (if any) that would provide information needed to improve management efficiency for Kesterson and/or other projects.

The basic approach of this study was to assess risk to wildlife using a standard ERA framework, similar to that outlined in U.S. Environmental Protection Agency (EPA) guidance.[26] The primary method of risk determination was to estimate the variability in risk as provided by the natural variation in annual climatic regime. In all cases, Kesterson physical, chemical, and biological monitoring results formed the basis for modeling and risk estimates.

## 37.2 PROBLEM FORMULATION

Problem formulation presents information that was used to develop and focus the Analysis phase of the ERA by describing the ecological characteristics of the ephemeral-pool environment at Kesterson Reservoir, summarizing previously existing data and specifying the goals of the assessment.

### 37.2.1 Physical Setting

Ephemeral pools formed by rainfall ponding reflect conditions where seepage rates into soils lag behind "net" rainfall (rainfall minus evapotranspiration). Thus, primary factors controlling ephemeral-pool formation at Kesterson Reservoir are rainfall, evapotranspiration, soil permeability, and the soil profile hydraulic potential gradient.

The Kesterson area has a semi-arid, Mediterranean climate typical of its location in the San Joaquin Valley. The wet-season months — November through May — receive over 95% of the mean annual rainfall (226 mm) and have relatively low pan evaporation rates. The hot, dry months of June through September have average temperatures ranging from 19 to 26°C.

Field saturated hydraulic conductivity measurements of surface soils at Kesterson Reservoir have shown that permeabilities are typically moderately low to low. The wet season at Kesterson Reservoir coincides with months during which the shallow groundwater rises to nearly reach the soil surface. Under post-closure, prefill conditions, much of the area was topographically low enough to permit surfacing of the groundwater during winter months. This type of groundwater-supported ephemeral-pool formation can be practically independent of rainfall. The recognition that Kesterson ephemeral pools formed by groundwater rise were extremely seleniferous[27] led to the filling of topographically low areas with imported soil.[2] As a result, only a very small portion (< 1%) of Kesterson Reservoir exhibits this mode of pooling. The vast majority of current ephemeral pools form by ponding of rainwater in low-permeability depressions.

### 37.2.2 Ecological Setting: Ephemeral-Pool Habitat

Although the goal of remediation at Kesterson Reservoir in the late 1980s was to dry the reservoir and remove aquatic habitats, pools form at widespread locations in the former reservoir in response to the winter rainy season. In many years the pools are very ephemeral and never achieve significant aquatic invertebrate populations. However, in some years, with higher rainfall or a more extended rainy period, pools persist for several months. Those longest-lasting pools with developed invertebrate populations offer the possibility of exposure from redissolved and bioaccumulated selenium to the Kesterson wildlife community. Waterfowl and shorebirds are of particular concern for exposure from the pools as they begin nesting at the time of pool occurrence in the late winter and early spring.

The ephemeral pools typically form in December or January and may last until March or April in wet years. They vary from a few cm to 1 m in depth and several square meters to several hectares in area. Size, depth, and longevity of the pools are determined by rainfall distribution and regional hydrology. Pools tend to form in the same locations each year when they occur.

### 37.2.3 Chemicals of Potential Ecological Concern

Studies at Kesterson Reservoir have included analyses of water and biota for a number of organic and inorganic constituents that could potentially be found in agricultural drainage.[3,28–31] Although some other chemicals (such as boron) occur at elevated levels at Kesterson, studies at Kesterson, other field sites, and in the laboratory indicate that selenium is responsible for the adverse effects observed in the past in aquatic birds at the site. Therefore, selenium has been the focus of the ongoing monitoring program (as described above) and is the only chemical of potential ecological concern (COPEC) to be considered in this ERA.

Selenium bioavailability and mobility are controlled by selenium speciation. When first introduced to Kesterson Reservoir, selenium was in its most oxidized state of selenate — Se(VI) — which is most mobile due to its high solubility and low sorption. Subsequently, selenium was largely reduced under ponded conditions and very little (< 5%) selenium is currently found as Se(VI). More chemically reduced and less soluble species (elemental selenium, organically-associated selenium, and selenite, Se [IV]), have since dominated. These species can all be oxidized to Se(VI), and both elemental selenium and organoselenium can be oxidized to Se(IV). Long-term monitoring has shown that although selenium may be seasonally reoxidized, it also tends to be reduced during wet months.[32] Because selenium reduction is far more rapid than oxidation,[18,20] the average selenium oxidation state does not change significantly from year to year. The net result is a stable set of species, with roughly equal fractions of elemental selenium, organoselenium, and Se(IV), with minor Se(VI).

The uptake of selenium by biota is limited by soluble selenium concentrations. Despite its high solubility, Se(IV) sorbs strongly onto iron oxides,[33] clay minerals,[34] and soil organic matter.[35] Generally, around 2–10% of total selenium remains in the soil solution of near-surface soils.[18,36] At Kesterson, that translates to an average soluble selenium concentration around 0.1 to 1 mg/kg.[9]

### 37.2.4 Assessment and Measurement Endpoints

The identification of assessment and measurement endpoints is critical to problem formulation because these endpoints structure the assessment to address management concerns and are central to conceptual model development.[26] Assessment endpoints provide a transition between the ecological management goals for the site and measures used to evaluate the endpoints. Ecological management goals for Kesterson include soil, surface-water quality, and food-source conditions capable of supporting small mammal populations that would be typical of similar habitats in the region; and soil, surface-water quality, and food-source conditions capable of supporting migratory birds without significant adverse effects on reproduction.

Although the recent ERA for the site[25] also evaluated risks to mammals (the first goal), the ERA presented in this chapter focuses only on the second goal.

#### *37.2.4.1 Assessment Endpoints*

Assessment endpoints are explicit expressions of the environmental value that is to be protected and are operationally defined by an ecological entity and its attributes.[26] Their relevance is determined by how well they focus on susceptible ecological entities and whether they adequately represent management goals. The following three criteria were used in selection of assessment endpoints:

1. Ecological relevance — migratory birds are integral components of the terrestrial ecosystem of the site, and aquatic species are known to use the ephemeral rainwater pools when pools are present.
2. Susceptibility to the stressor — research has shown that aquatic birds may be adversely affected by exposure to selenium.
3. Relevance to management goals — migratory birds are valued ecological resources for the site, and their protection represents management goals identified above.

The assessment endpoint selected for this ERA is stated as follows: Survival and reproduction of aquatic birds without ecologically significant adverse effects.

#### *37.2.4.2 Measurement Endpoints*

Measurement endpoints (also known as "measures") are the characteristics that are evaluated to provide an indication of whether adverse effects on assessment endpoints have occurred or are

**Table 37.1 Measures of Exposure and Effects and Associated Risk Estimation Used in the Kesterson ERA Model**

| Model | Exposure | Effects | Risk Estimation |
|---|---|---|---|
| Aquatic model for risk from ephemeral pool environments. | Selenium concentrations in water, aquatic invertebrates, and avian eggs. | Literature toxicity values for concentrations of concern for selenium in the avian diet and eggs. Monitoring for teratogenic effects to semiaquatic birds. | Model estimates of the frequency of occurrence of dietary concentrations and potentially exposed eggs of receptor bird species. |

likely to occur. Two categories of measures that are predictive of the assessment endpoints were used for this assessment: (1) measures of exposure were used to evaluate how selenium exposures could be occurring and (2) measures of ecological effects were used to evaluate the response of the aquatic birds when exposed to selenium.

Measures of exposure include measured or modeled (estimated) concentrations of selenium in soil, surface water, and food-chain organisms as well as receptor species. Measures of effects include food-chain bioaccumulation and the responses of aquatic invertebrates and birds to environmental and dietary exposure of selenium. Measures of exposure and effects as well as risk estimation modeling used in the Kesterson Reservoir ERA are shown in Table 37.1.

### 37.2.5 Ecological Conceptual Site Model

The conceptual site model combines information on selenium, potential ecological receptors, potential exposure pathways, assessment endpoints, and measures of exposure and ecological effects. The model provides an overall picture of site-related exposures that can be used to focus the evaluation of selenium in the ERA.

#### *37.2.5.1 Physical/Chemical Parameters*

The physical/chemical parameters used in the ERA models were:

- Soil-selenium concentrations
- Soil pore-water-selenium concentrations
- Pool-water-selenium concentrations
- Estimates of the extent of surface-water ponding for various months

Data sources and synthesis of these parameters are discussed in Section 37.3.1.1. They are critical components of the model, so they are explained in some detail.

#### *37.2.5.2 Identification of Representative Species*

Representative ecological receptors were identified as the aquatic and semiaquatic wildlife that are most likely to be affected by selenium. They include primary and secondary consumers that are aquatic (e.g., macroinvertebrates) and semiaquatic (e.g., shorebirds and waterfowl that feed on aquatic biota). Representative species selected are from communities that are commonly found at Kesterson Reservoir ephemeral pools. The communities potentially exposed directly or indirectly (i.e., through the food chain) to selenium at Kesterson Reservoir are composed of free-swimming or benthic aquatic invertebrates and semiaquatic birds. Representative ecological receptors were selected from these communities to fulfill as many of the following criteria as possible:

- Species that are known to occur or are likely to occur at the site
- Species that relate to the assessment endpoints selected

- Species that are likely to be maximally exposed to selenium
- Sedentary species or species with a small home range
- Species with high reproductive rates
- Species that are known to play an integral role in the ecological community structure at the site
- Species that are known or likely to be especially sensitive to selenium, and thus are an indication of ecological change
- Species that are susceptible to bioaccumulation of selenium from a limited number of food items
- Species that are representative of the foraging guild or serve as food items for higher trophic levels

The aquatic and semiaquatic representative species selected for the Kesterson Reservoir ERA are: (1) free-swimming and benthic insects and microcrustaceans; and (2) semiaquatic birds — killdeer, black-necked stilt, and mallard.

Benthic and free-swimming macroinvertebrates and microcrustaceans are primary consumers that fulfill many of the selection criteria. They would be in direct contact with potentially contaminated sediments and surface water and therefore would be exposed to selenium. They have a small range, have high reproductive rates, and serve an integral role in the aquatic and semiaquatic ecosystem. They serve as important prey items for the semiaquatic bird species chosen for the ERA (i.e., stilt, killdeer, mallard). Various species of aquatic insects (e.g., midge larvae, water boatmen, beetles) and microcrustaceans (ostracods, cladocerans) were collected from ephemeral pools at Kesterson Reservoir during most years from 1990 through 1999.

Semiaquatic birds feed primarily on species associated with aquatic habitats. Black-necked stilts, killdeer, and mallards were chosen as the receptors to assess the effects of selenium exposure through ephemeral aquatic habitats at Kesterson. Their habit of feeding in shallow waters for crustaceans and insects gives them a high potential for exposure to selenium through the ephemeral-pool food-chain pathway. In addition, killdeer consume some terrestrial insects and mallards some terrestrial plants. Representative bioaccumulation levels are available for those terrestrial diet items through ongoing Kesterson Biological Monitoring Program results. Eggs of all three species have been sampled at Kesterson Reservoir.

#### *37.2.5.3 Exposure Pathway Inclusion/Exclusion*

The exposure pathway inclusion/exclusion evaluation is based on information gathered from the problem formulation and the selection of representative species, the probable completeness of each exposure pathway, and the potential for that pathway to be a major or minor route of exposure and risk.

A complete exposure pathway must exist for an exposure to occur, and it must have the following elements, in addition to the presence of suitable habitat for ecological receptors:

- Contaminant source (e.g., selenium in soils and water)
- Mechanism for contaminant release and transport (e.g., solubility in poolwater)
- Exposure point (e.g., ephemeral pool)
- Feasible route of exposure (e.g., ingestion)
- Receptor (e.g., bird)

Contaminant sources and release mechanisms at Kesterson Reservoir consist primarily of on-site source areas (soil-selenium inventory) and availability for movement of solubilized selenium through the soil profile into ephemeral pools. Ecological receptors at the pools can be exposed to chemicals in sediment or surface water via direct or secondary exposure pathways. Direct exposure pathways include ingestion and dermal contact. Secondary exposure pathways are limited to food-chain transfer of bioaccumulated selenium. Potential exposure pathways for representative species are summarized below along with the rationale for inclusion/exclusion in the quantitative and qualitative evaluations to be conducted in the ERA.

Aquatic invertebrates can absorb selenium through their epidermis and can accidentally or purposefully ingest sediment during feeding or burrowing. Benthic organisms are especially prone to exposure to selenium present in sediments as some consume the organic materials from within the sediment (e.g., midge larvae). Aquatic invertebrates serve as a major route of food-chain transfer because they are prey for semiaquatic birds (e.g., waterfowl, shorebirds).

Semiaquatic birds can be exposed to selenium in sediment and surface water from several different behaviors. Birds can inadvertently or purposefully ingest sediment while grooming or while consuming contaminated prey species. Surface water can be ingested as a drinking-water source or during bathing or grooming activities. Dermal contact with sediment or surface water is considered to be a secondary route of exposure for birds and mammals, and it is not considered an important exposure route for selenium (as compared to some organic contaminants).

Food-chain exposure is the primary exposure route of concern for these birds. Food-chain exposure may occur because selenium can be accumulated directly by aquatic invertebrates or in aquatic plants or algae that are consumed by herbivorous invertebrates (and secondarily by predatory insects) that are then eaten by semiaquatic birds. A diagram of the conceptual site model is presented in Figure 37.3.

### 37.2.6 Data Sources and Synthesis

#### *37.2.6.1 Pool Formation Parameters*

This section describes the database and procedures used to estimate probabilities of ephemeral-pool formation and areal pool coverage at Kesterson Reservoir. The estimates to be obtained through analysis of weather data are (1) the probabilities for ephemeral-pool occurrence during individual months and (2) estimates of the fraction of the reservoir that will be ponded. These estimates are based on the following four basic assumptions:

1. Monthly rainfall data from the nearby city of Los Banos can supplement the smaller rainfall record at Kesterson Reservoir.
2. A simple algorithm for predicting pool initiation and duration within a given wet season yields adequate results. This algorithm was found to be successful at calculating the initial date of major ponding to within 10 days of actual dates based on field observations in 10 out of the 11 available calibration years. It also predicted duration of typical ephemeral pools to within 32 days and, in most cases, to within 22 days.
3. The algorithm for predicting ponding can also be extended to estimate areal coverage by ephemeral pools.
4. The 127-year historical rainfall record (Los Banos and Kesterson data) will be representative of future rainfall distributions.

Based on these assumptions, estimated probabilities of pooling and of ponded area are given below.

##### *37.2.6.1.1 Weather Database*

Rainfall records at Kesterson Reservoir begin in 1982, with detailed records available only since 1985. Although this time interval includes drought years as well as the extreme 1997–1998 El Niño year, a larger rainfall record was needed for purposes of estimating rainfall probabilities. Monthly rainfall data are available for the Los Banos weather station from the California Department of Water Resources with records since 1874. Los Banos is located only 20 km (12 mi) away from Kesterson Reservoir, with very little topographic variation between the two locations. Nevertheless, we examined the correlation between their rainfall records before relying on the Los Banos data. Because of the short Kesterson record, this comparison can only be done for the years 1985 through

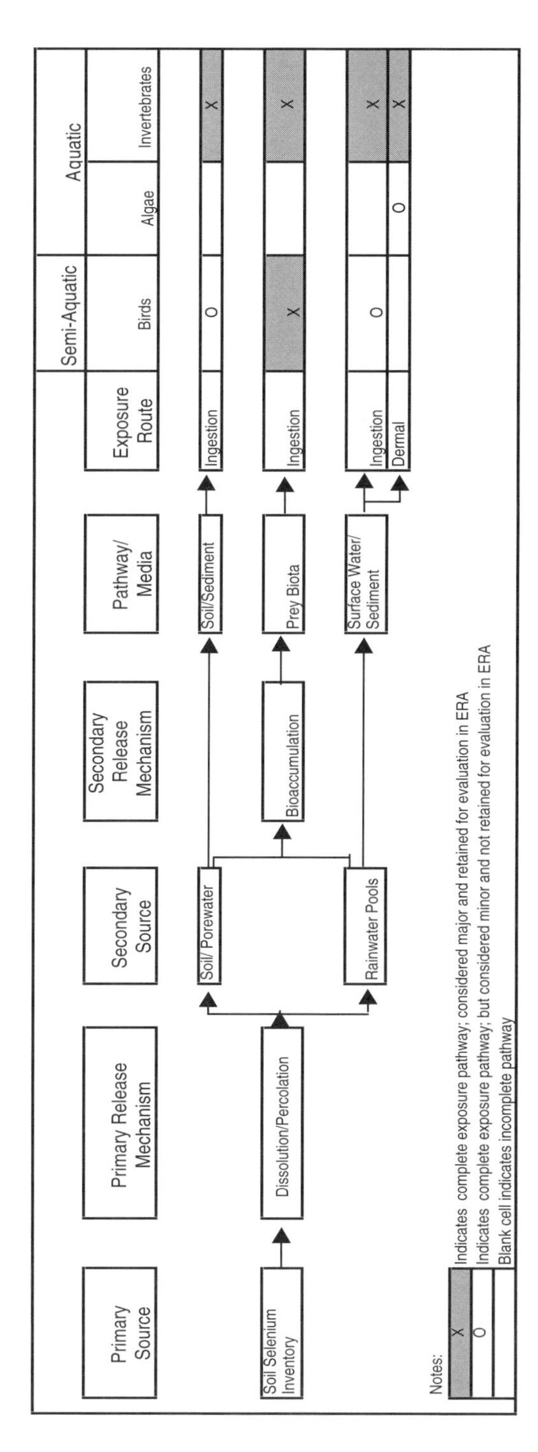


**Figure 37.3** Conceptual site model for Kesterson Reservoir.

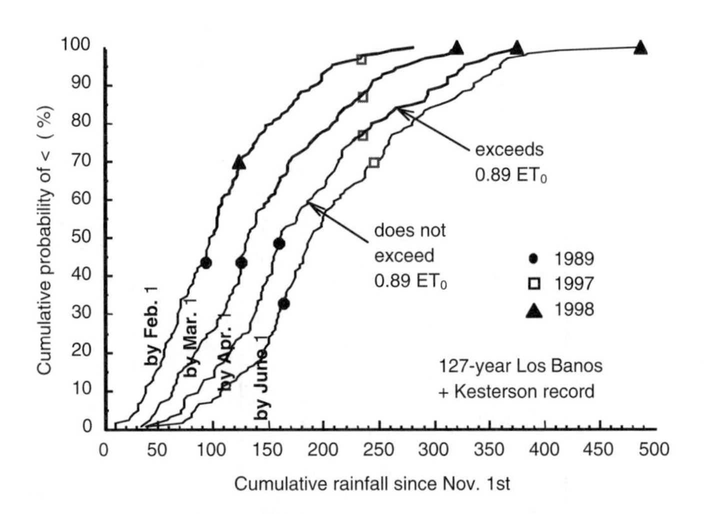


**Figure 37.4** Cumulative probability of exceeding seasonal cumulative rainfall = x mm by Feb. 1, March 1, Apr. 1, and June 1. Cumulative rainfall from the 1989 (drought), 1997 (above average), and 1998 (El Niño) years are indicated on monthly curves. Also shown are segments along the cumulative rainfall probability curves that do not (lighter trace), and that do (darker trace) exceed 0.89 times the cumulative average $ET_0$.

2000. Los Banos monthly rainfall data do provide a fair estimate of monthly rainfall at Kesterson Reservoir. The near-zero (2.4 mm) intercept, near-unit (0.998) slope, and high correlation ($r = 0.946$) indicate that the Los Banos monthly rainfall data can be used directly for predicting Kesterson rainfall. Thus, for purposes of this analysis, the Los Banos monthly rainfall data from 1874 through 1984 were combined with the Kesterson Reservoir rainfall data from 1985 through 2000. The compiled rainfall record shows a mean annual precipitation of 22.6 cm, with a range in annual totals of 5 to 53 cm. The fact that the available data set includes the 1997–1998 El Niño year substantially improves our chances of fairly estimating ponding probabilities because that year yielded the highest rainfall in the available 127-year record. Furthermore, the bulk of the El Niño rainfall occurred within late winter and spring months when ephemeral-pool environments have greatest potential for wildlife impacts (Figure 37.4).

#### *37.2.6.1.2 Predicting the Timing and Duration of Ephemeral Pools at Kesterson Reservoir*

One of the two main inputs into our previously derived ponding model[17] is the CIMIS $ET_0$ parameter. $ET_0$ is the estimated evaporation rate from a reference "crop" (irrigated grass surface) and depends on the local temperature, wind speed, and humidity. Inspection of 9 years of Kesterson Reservoir CIMIS data has shown that while daily $ET_0$ records are quite variable from year to year, the cumulative $ET_0$ record is not. Average $ET_0$ values for Kesterson Reservoir can be used instead of actual daily $ET_0$ values, without much loss of accuracy throughout most of the wet season. For Kesterson Reservoir, cumulative $ET_0$ amounts to 71% of cumulative pan evaporation.

We now consider the simplification that soil water storage has a minor influence on rainfall ponding at Kesterson Reservoir. This simplification is based on two factors previously mentioned. The first is the recognition that the water table is close to the soil surface during the wet season, even during drought years. This factor, combined with the partially saturated soil profile, leaves limited air-filled porosity for storage. In addition, the moderately low permeability of soils at sites most susceptible to rainfall ponding lessens the extent to which infiltration and drainage influences ponding. Thus, for purposes of estimating the inception and duration of wet-season ponding, it appears reasonable as an approximation to overlook hydraulic interactions with the soil profile.

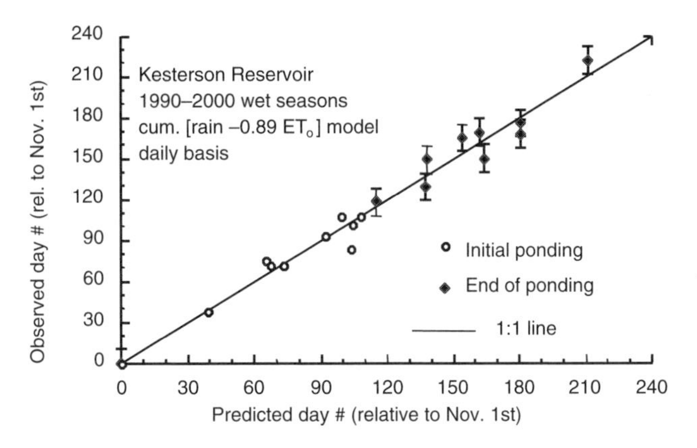


**Figure 37.5** Predicted vs. observed days for initial ponding and cessation of ponding for the years when ponding is predicted.

With these two simplifications (use of the average $ET_o$, and neglecting soil-water exchanges), only information on the temporal variation of rainfall within a given season is needed to estimate timing of ephemeral-pool formation and duration. In the approximation that only rainfall and evaporation control ponding, pools will form at any time when the cumulative rainfall exceeds cumulative evaporation. As an initial approximation, the raw, average cumulative $ET_o$ values were used. However, model calculations were found to underpredict ponding periods. Scaling down the average cumulative $ET_0$ by a constant factor of 0.89 yielded optimized agreement between field-observed and predicted ponding. The optimization criterion selected was that of maximizing the number of correctly predicted minus incorrectly predicted days. This optimization approach yielded a ratio of modeled vs. observed ponding days of 0.965, with 86% of the observed ponding days correctly predicted.

Good agreement is usually obtained between field observations of ponding and model predictions. Predictions vs. observations of ponding for the subset of water years in which ponding is expected by the simple model (9 years out of the 11) are compared graphically in Figure 37.5. The fact that a wide range of rainfall distributions and rainfall totals is included in these results gives support to the model's predictive ability, especially for nondrought years, in which such information is most relevant. The success that this simplified model has had in predicting ephemeral-pool formation and duration indicates that it could be useful in estimating future ponding at Kesterson Reservoir under a range of possible rainfall patterns.

The calculations up to this point have been based on daily rainfall and daily average $ET_0$ data. However, the larger historical weather data only provide monthly rainfall data, so we tested the reliability of calculations based on monthly values. Using the previously described algorithm, but with only the monthly discretized Kesterson Reservoir rainfall data, the ponding model was applied to the 1990–2000 wet seasons. As before, the start of each wet season was defined to begin on November 1. For example, the 1990 wet season begins on November 1, 1989. The cumulative monthly rainfall (CMR) was calculated within a given season relative to November 1. The CMR was compared on a month-by-month basis to 0.89 times the cumulative monthly average $ET_0$ (i.e., $0.89 \times CMET_0$), to determine which months have CMR in excess of $0.89 \times CMET_0$. The whole data set was then evaluated to determine the number of years in which CMR begins to exceed $0.89 \times CMET_0$ in a given month. The number of months within each year (beginning November 1) in which CMR exceeds $0.89 \times CMET_0$ was also calculated. These results provided the basis for estimating probabilities of pond initiation by a given month and probabilities of pond duration, respectively.

### *37.2.6.1.3 Estimating the Areal Extent of Ponding*

Information limited to whether ephemeral pools occur at Kesterson Reservoir within a given month is not very useful without information on the areal extent of the pools. The potential for

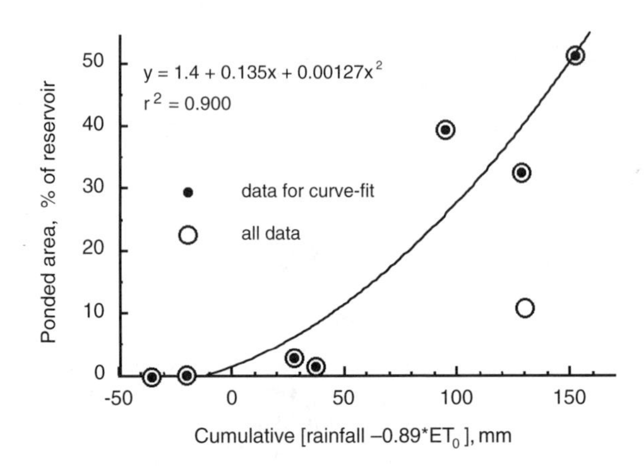


**Figure 37.6** Aerial photograph-based estimates of percent areal ponding at Kesterson Reservoir, correlated to CMR-0.89 × $CMET_0$. This fit is optimized to minimum root mean square deviation, excluding the single point (shown) that would yield low predicted areal ponding.

wildlife exposure and effects of 1% vs. 50% areal ponding within Kesterson Reservoir is clearly very different. Estimating the areal extent of ephemeral pools at Kesterson Reservoir with a simple model is difficult. There is a very limited database on ephemeral-pool areal coverage. Such information, obtained from USBR aerial photographs, was available only for 8 days from 1995 through 1999 (for example, see USDI[37]). However, estimates of ephemeral-pool areal extent are needed as part of the risk-assessment model to provide a link to bird use and exposure.

Because [cumulative rainfall –0.89 × cumulative $ET_0$] (hereinafter referred to as "net rainfall") appears to provide a good estimate of whether pools occur, we tested this parameter for predicting the areal extent of ponding at Kesterson Reservoir. Qualitatively, negative to slightly positive values of net rainfall should be associated with insignificant ponding, while larger positive values should correlate with greater areal ponding. The available ephemeral-pool area data plotted against net rainfall show this rough correlation (Figure 37.6). A second-order polynomial fit, $y = 1.4 + 0.135x + 0.00127x^2$, is shown with these data ($r^2 = 0.90$). The fitting was done with exclusion of the one data point indicative of low ponded area at high net rainfall in order to obtain more conservative predictions. By "conservative" we mean predictions weighted to greater ponded areas. The fit shown in Figure 37.6 was obtained by minimizing the root mean square deviation with respect to all data except the single value mentioned above.

### *37.2.6.2 Estimating Ephemeral-Pool Selenium Concentrations*

Reservoir-wide data collected between 1989 and 1998 were used to define the statistical distribution of selenium concentrations by habitat and trisection. This data set had been collected as part of the ongoing biological and soil monitoring program.[9,10] For a determination of the water-soluble-selenium distribution in surface soils, a subsample of the homogenized soil from each sampling station (20 to 25 g) was used to prepare a 1:5 soil:water extract. Water-soluble-selenium was then analyzed using hydride generation atomic absorption spectroscopy (HGAAS). Selenium concentrations were summarized as standard statistical characterizations (arithmetic, geometric, and harmonic mean, median, standard deviation and error, skewness, and kurtosis) for each combination of habitat and trisection, for both total and soluble selenium.

A transfer factor between soil selenium and ephemeral-pool water was calculated via regression between existing ephemeral-pool-selenium concentration data and 1999 soil-selenium data. Ephemeral-pool water has been collected annually during the wet season. Although some ephemeral-pool sampling began in 1987, regular sampling at selected monitoring sites started during the 1992 wet season. During years when ephemeral-pool formation is significant, pool waters are sampled several

(2 to 4) times per month by the USBR and LBNL for analyses of selenium and other constituents. Within a given site, pool-selenium concentrations typically follow a general seasonal trend. During the initial days of ponding, selenium concentrations are relatively high (Figure 37.2). Over the first 20 to 40 days of ponding, pool-selenium concentrations decrease because of dilution with additional rainwater and because of diffusive transport and reduction in surface sediments. During this interval, pool-selenium concentrations often exhibit more than tenfold decreases. At still longer times, gradual increases in pool-selenium concentrations occur, primarily due to evaporative concentration. This cycle is observed during years when ephemeral pools are persistent, without any evident systematic year-to-year trends. Given the very dynamic nature of pool-selenium concentrations, it is desirable to identify concentrations that are characteristic of each site. For this purpose, a selenium concentration typical of pool waters at 30 to 80 days within each site was selected for estimating soil–pool-water-selenium transfer factors. In addition, as presented below, this age of pool offers the best choice for characterizing ecological exposure.

#### *37.2.6.3 Aquatic Habitats*

Conducting an ERA for Kesterson that incorporates aquatic habitats required the summary of long-term monitoring data on ephemeral-pool and aquatic invertebrate tissue-selenium concentrations. Data from 1990 through 1999 were summarized by individual pool and averaged by trisection. All data were used to develop a regression relationship between waterborne-selenium and aquatic invertebrate tissue-selenium. Terrestrial components of bird diets were taken from the distribution of tissue-selenium results (geometric mean, standard deviation) for those necessary terrestrial components of receptor diets (i.e., plants and terrestrial invertebrates). The relationship between ephemeral-pool waterborne-selenium and aquatic dietary-item-selenium concentrations used as ERA model inputs is shown in Figure 37.7.

In addition to the data on pool water and bird dietary item chemistry, waterfowl and shorebird daily use data from the ongoing Kesterson Reservoir Monitoring Program were summarized by trisection to use in the ERA model. Bird-use data during the late winter and early spring were required to develop a regression relationship to predict bird use based on the extent of ponding by trisection. Bird-use records were chosen for their proximity in timing to the aerial photography estimates of the extent of ephemeral pools. A summary of the regression relationships of bird-use data vs. extent of ponding by trisection is presented below.

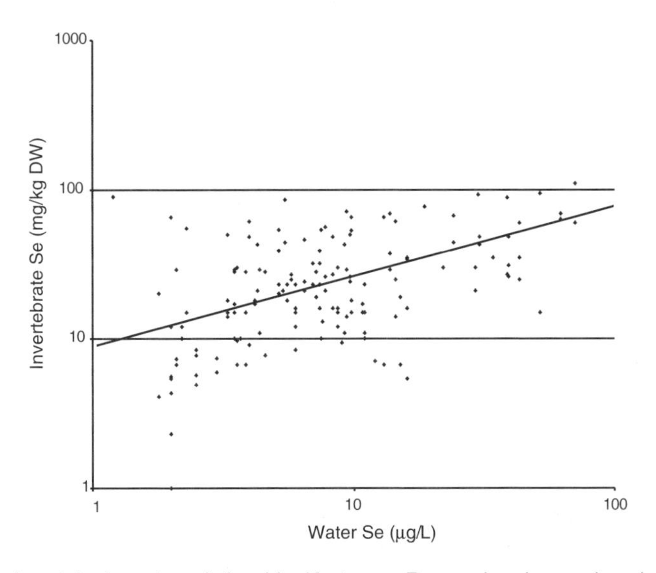


**Figure 37.7** Aquatic invertebrate:water relationship, Kesterson Reservoir ephemeral pools (1990–1998).

### 37.2.7 Modeling

#### *37.2.7.1 Physical/Chemical Model*

The physical/chemical databases and models used to predict water-extractable selenium-transfer factors to pool-selenium concentrations and the extent of ponding by month were discussed previously and are developed into model predictions below.

#### *37.2.7.2 Aquatic Habitat Model*

The aquatic habitats Kesterson ERA model was used to predict exposure of semiaquatic birds to selenium through the ephemeral-pool environment. Terrestrial plants and invertebrates were added to the model for mallards and killdeer (respectively) because these birds do not feed exclusively in the aquatic environment. In contrast, black-necked stilt were assumed to feed exclusively in the ephemeral pools.

The high year-to-year, climate-driven variability of ponding determines all aspects of the exposure of wildlife at Kesterson to selenium available through the ephemeral pools. The Monte Carlo simulation model was structured to represent natural climatic variability as measured at Kesterson for a hypothetical 1000 years of data.

The aquatic habitat model depends on estimates of the areal extent of ephemeral pools. Total month-end ponding was predicted based on the ponding model (Figure 37.5). Ponding by trisection was related to total ponding, as shown in the digital aerial photography record. Regression relationships were then developed that predicted ponding by trisection based on cumulative precipitation. The first step in the aquatic habitat model was to estimate the extent of ponding by trisection at the ends of January, February, and March for any given year (the months where ponding overlaps with bird exposure at Kesterson). A summary of the steps in the aquatic habitat model is as follows:

1. Establish a random distribution of 1000 data points for total ponding at Kesterson Reservoir based on the statistical distribution of ponding as determined by precipitation and ET. All bird use and extent of ponding predictions were calculated based on those hypothetical 1000 years of climatic record.
2. Predict ponding by trisection for the period at the end of January, February, and March (using total to trisection-specific ponding relationships).
3. Predict bird use by trisection based on trisection-specific ponding (using bird use to ponding relationships).
4. Predict ephemeral-pool-selenium concentrations by habitat type based on the transfer factors from soil-water extractable to pool-water-selenium as described above. Create a database of 1000 random selenium-transfer factors based on the measured statistical distribution of transfer factors. Then use the area of habitat types by trisection (as shown in Figure 37.1) to create weighted-average pool-selenium concentrations by trisection (1000 each). Measured habitat type distributions are:

   Trisection 1 = 61.1% Filled, 8.8% Grassland, and 30.1% Open.
   Trisection 2 = 49.9% Filled, 35.1% Grassland, and 15% Open.
   Trisection 3 = 55% Filled, 44.8% Grassland, and 0.2% Open.

5. Predict aquatic invertebrate concentrations by trisection (1000 each) based on the modeled pool-water-selenium concentrations (from Step 4) and the Kesterson-specific regression relationship between those variables. Black-necked stilt were assumed to feed exclusively on these aquatic invertebrates.
6. Estimate the trisection-specific selenium concentrations of the killdeer and mallard diets by adding selected terrestrial dietary items to the aquatic invertebrate diet. Create a random distribution of 1000 data points for terrestrial plants (mallard diet) and insects (killdeer diet) based on the statistical distribution for these items by trisection from the KRBMP. Create average combined diets based

on the assumptions that killdeer diets were 25% pool and 75% terrestrial invertebrates and mallard diets were 75% pool invertebrates and 25% terrestrial plants.

7. Predict average Kesterson-wide dietary concentrations of selenium based on diet weighted by bird use by trisection (1000 average dietary concentrations per species).
8. Predict potential nests and eggs per species by trisection based on predicted bird use and estimates of the average number of eggs per nest (four each for killdeer and black-necked stilt, eight for mallard).
9. Predict the potential egg-selenium concentrations based on Kesterson-specific data on transfer factors for dietary selenium to egg selenium for the three species.
10. For killdeer and stilt only, predict the potential number of hens having at least one egg per nest affected as a result of dietary-selenium exposure (relationship based on egg-selenium concentrations; limited to an extensive database for shorebirds only).

The basic approach of the aquatic habitat model was to predict daily average bird use and dietary exposure from estimates of the extent of ponding. The diet and egg concentrations of selenium can be directly related to literature values on the potential effects of selenium on birds. As described in the literature on black-necked stilts and extended in our model to include killdeer (see Step 10 above), it was possible to predict the number of shorebird hens potentially affected by egg loss over the modeled 1000 years of climate record.

## 37.3 ANALYSIS

### 37.3.1 Exposure Characterization

#### *37.3.1.1 Physicochemical*

##### *37.3.1.1.1 Statistical Distribution of Soil Selenium Concentration*

Total and soluble soil-selenium concentrations have been measured from reservoir-wide synoptic sampling. There are very few year-to-year changes in the Filled and Grassland habitats. In the Open habitat, which comprises only 13% of the reservoir area, there are many year-to-year differences, but an overall trend is absent. Moreover, due to the large degree of spatial variability, discerning statistically meaningful interannual and long-term trends is difficult. As discussed earlier, the most likely change in selenium distribution is one of selenium oxidation to species of higher solubility. It is selenium solubility that largely determines its bioavailability and ecological exposure.

Year-to-year and long-term changes in soluble-selenium concentrations may be neglected with respect to the ERA model. The entire 9-year data set of selenium concentrations, split by habitat and trisection, was used. Differences in selenium concentrations are most pronounced among habitats, with highest levels in the Open habitat and lowest in the Filled habitat. Soil in Trisection 1 generally contains the highest selenium concentrations, except in the Filled habitat, where total selenium does not significantly differ amongst trisections (Table 37.2).

Statistical parameters can be used to determine the type of distribution present. Most data sets are skewed and dominated by low values. In those cases, the distribution is log-normal. In a few cases, skewness is close to 1 and the median and mean values are similar, suggesting an approximately normal distribution (Table 37.2).

##### *37.3.1.1.2 Rainfall Distribution and Pool Formation*

It is necessary to apply the pool-prediction model to the 1874–2000 monthly rainfall data set to obtain estimated probabilities of ponding. Given the fair success of the CMR-($0.89 \times CMET_0$) model for matching the available field observations of ephemeral-pool formation and duration, we

**Table 37.2 Statistical Properties of Soluble-Selenium Distributions in Soil by Habitat and Trisection**

| | Descriptive Statistics: Water-Soluble Se (mg/kg) | | | | | | | | |
|---|---|---|---|---|---|---|---|---|---|
| | Open Habitat (by trisections) | | | Filled Habitat (by trisections) | | | Grassland (by trisections) | | |
| **Statistic** | **T1** | **T2** | **T3** | **T1** | **T2** | **T3** | **T1** | **T2** | **T3** |
| Mean | 0.86 | 0.44 | 0.53 | 0.33 | 0.19 | 0.09 | 0.61 | 0.12 | 0.13 |
| Std. Dev. | 0.69 | 0.40 | 0.75 | 0.75 | 0.20 | 0.13 | 0.60 | 0.14 | 0.12 |
| Std. Error | 0.10 | 0.06 | 0.10 | 0.10 | 0.03 | 0.02 | 0.08 | 0.02 | 0.02 |
| Count | 50 | 54 | 54 | 52 | 54 | 53 | 53 | 54 | 52 |
| Min. | 0.064 | 0.048 | 0.052 | 0.01 | 0.006 | 0.002 | 0.103 | 0.01 | 0.021 |
| Max. | 2.70 | 1.53 | 4.92 | 4.286 | 0.746 | 0.49 | 2.975 | 0.754 | 0.685 |
| Variance | 0.48 | 0.16 | 0.56 | 0.56 | 0.04 | 0.02 | 0.36 | 0.02 | 0.02 |
| Coeff. Var. | 0.81 | 0.91 | 1.40 | 2.28 | 1.05 | 1.50 | 0.99 | 1.17 | 0.97 |
| Range | 2.6 | 1.5 | 4.9 | 4.3 | 0.7 | 0.5 | 2.9 | 0.7 | 0.7 |
| Geom. Mean | 0.59 | 0.31 | 0.34 | 0.11 | 0.09 | 0.04 | 0.42 | 0.07 | 0.09 |
| Harm. Mean | 0.38 | 0.22 | 0.25 | 0.06 | 0.04 | 0.02 | 0.30 | 0.05 | 0.07 |
| Skew | 1.02 | 1.48 | 4.17 | 4.02 | 1.20 | 2.02 | 2.12 | 2.88 | 2.59 |
| Kurtosis | 0.23 | 1.05 | 20.60 | 16.23 | 0.31 | 2.80 | 4.82 | 9.46 | 8.00 |
| Median | 0.59 | 0.31 | 0.32 | 0.10 | 0.12 | 0.04 | 0.40 | 0.06 | 0.10 |

*Note:* Skewness at or below 1 indicates close-to-normal distribution, provided mean and median are similar. All other distributions are log-normal. "T" designates trisection.

estimated probabilities of pond initiation within a given month and probabilities of pool duration. This was done by analysis of the larger (1874–2000) data set using the aforementioned procedure and assuming that the 127-year record is representative of future rainfall distributions. The estimated probability of ponding and estimated probability of pool duration are shown in Figure 37.8. Based on the historical monthly data, no significant ponding is expected for 43% of the wet seasons. Ponding for durations of 1, 2, 3, 4, 5, and 6 months is expected in about 9%, 11%, 14%, 15%, 6%, and 1% of the wet seasons, respectively. Of the years in which ephemeral pools form, they are expected to initially form by the end of November, December, January, February, and March with probabilities of about 27%, 32%, 33%, 7% and 1%, respectively.

Because wildlife use and nesting associated with ephemeral pools occurs during the later months of the wet season, we estimated probabilities of ephemeral-pool occurrence relative to January 1. For this purpose, the CMR-($0.89 \times CMET_o$) model was applied to the 1874–2000 monthly database and examined for expected ponding in the months of February through May. These results show that if February ponding occurs (based on calculations for January 31), there also was ponding in the immediately preceding months. There is a low probability (about 7% of the years in which ponding occurs and about 4% of all years) that ephemeral pools will form in March without ponding in previous months. These results also show that the expected frequency of ponding through the

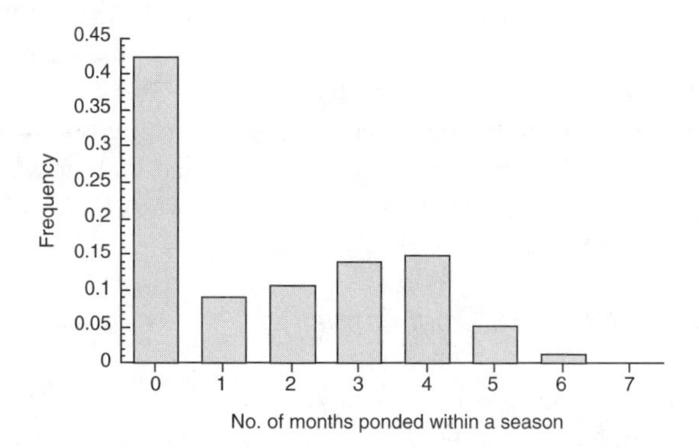


**Figure 37.8** Estimated probability of ponding and estimated probability of pool duration.

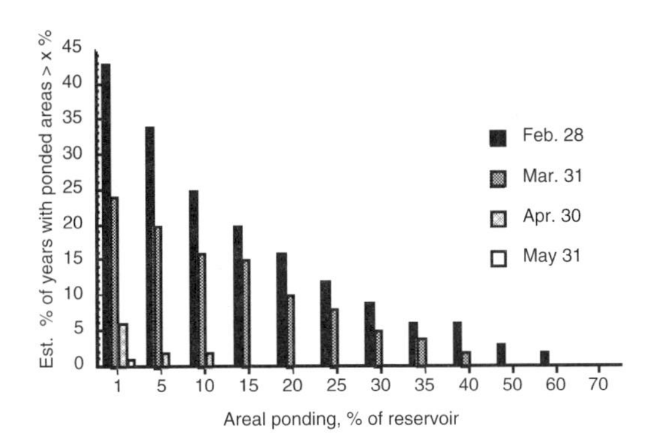


**Figure 37.9** Estimated probability for ephemeral-pool areal coverage exceeding X% at the end of February, March, April, and May.

months of April and May is very low. Based on the historical rainfall record and the CMR-(0.89 $\times$ $CMET_0$) model, only 7 out of the past 127 years would have resulted in ephemeral pools persisting through the end of April. One of these 7 years was 1998, which is consistent with field observations. The model also correctly predicts that no ponding persists through the end of June.

Estimates of ephemeral-pool areal coverage were obtained for periods during the wet season critical to wildlife impacts. Predictions were made for the end of the months of February, March, and April. The 127-year record of monthly CMR-(0.89 $\times$ $CMET_0$) was combined with the second-order polynomial correlation described previously to obtain estimated percent ponding by month. This result indicates that both areally and temporally extensive ephemeral-pool coverage of Kesterson Reservoir is rare. For example, the estimated probabilities of > 10% areal coverage at the end of February, March, and April are about 25%, 16%, and 2%, respectively, based on the historical rainfall data (Figure 37.9).

Pool area by trisection was estimated based on aerial photographic records of pool coverage by trisection as compared to total pool area using the following relationships:

$$\text{T1Pools (m}^2\text{)} = (\text{Total Kesterson pool area (m}^2\text{)} \times 0.304) + 18026 \quad (37.1)$$

$$\text{T2Pools (m}^2\text{)} = (\text{Total Kesterson pool area (m}^2\text{)} \times 0.463) - 1580 \quad (37.2)$$

$$\text{T3Pools (m}^2\text{)} = (\text{Total Kesterson pool area (m}^2\text{)} \times 0.233) - 16446 \quad (37.3)$$

### *37.3.1.1.3 Pool-selenium Concentration and Transfer Factors*

As discussed earlier, ephemeral-pool concentrations typically start at relatively high values, decrease during the main ponding period, and increase to varying degrees towards the later stages of ponding (Figure 37.2). Selenium in ephemeral pools occurs primarily as Se(VI), with up to 30% as Se(IV). The decreases in pool-selenium concentrations during the initial stages of ponding result from transport into shallow sediments, in some cases also by dilution by additional rainfall, and a relatively smaller loss due to volatilization. Year-to-year comparisons in selenium concentrations within individual pools reveal no obvious trends, and over the years the concentrations in individual pools have remained in a consistent range.[17] This also applies to the "stable" concentration range shown in Figure 37.2. This concentration is ecologically more important than the initial high selenium spike, because (a) it is much longer in duration and (b) it is closer in timing to the springtime period coincident with bird nesting. In addition, there is typically a lag of 30 or more days between pool formation and development of invertebrate communities that would provide food for semiaquatic birds. The average "stable" concentrations for each pool were used in calculating a soil-to-pool-water transfer factor.

A ratio of pool selenium to soluble soil selenium was calculated for each of ten sites and subsequently averaged at 0.024. This means that if the concentration of soluble selenium in the

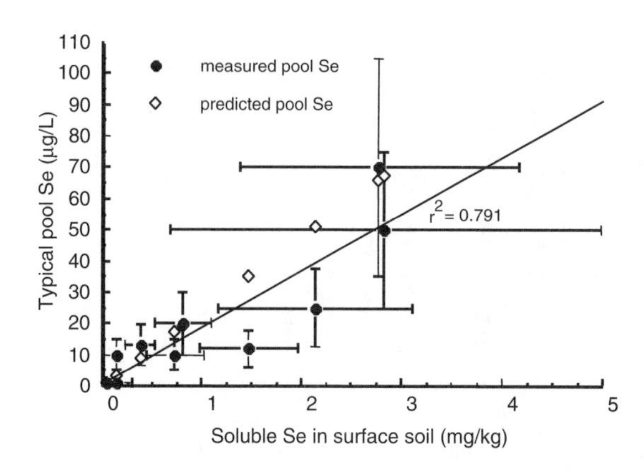


**Figure 37.10** Measured and predicted ephemeral-pool Se based on soluble Se in surface soil.

surface soils is 100 µg/kg, the corresponding concentration in the overlying ephemeral pool will be approximately 2.4 µg/L. In some cases, this approximation slightly underestimates pool concentrations, and in a few cases it substantially overestimates those values (Figure 37.10). This is due to the fact that the partitioning between soil and pool water depends on several variables, some of which are site-specific and include selenium fractionation in surface soils, ground cover, and the amount of decomposing litter, soil permeability, and depth to the water table. Therefore, it is impossible to derive a single transfer factor capable of precisely defining pool selenium levels. A polynomial relationship may better reproduce the actual pool-selenium concentrations, but the linear approximation is accurate within a factor of 2 to 3 and is adequate given the spatial and temporal variability in soil-selenium concentrations.

### *37.3.1.2 Ecological*

#### *37.3.1.2.1 Bird Use*

Bird use is measured at Kesterson Reservoir as part of the long-term monitoring program. Daily use was estimated by trisection for days close to the time of aerial photographs. Those use estimates by trisection and by species were compared to measured flooded area to create regression relationships to be used in the ERA aquatic exposure model. The selected receptor species showed positive, linear relationships between daily use and the extent of ponding. Regression equations are presented in Table 37.3. The relationships were all derived from ponding and bird use in the January-through-March period and were used to predict bird use for those months in the model.

**Table 37.3 Bird Use as a Function of Ponded Area by Trisection***

| Species | Trisection 1 Daily Use | Trisection 2 Daily Use | Trisection 3 Daily Use |
|---|---|---|---|
| Black-necked stilt | [0.048 + 0.00000229 × pool area ($m^2$)] $r^2 = 0.441$ | [–0.007 + 0.00000313 × pool area ($m^2$)] $r^2 = 0.527$ | [–0.041 + 0.00000598 × pool area ($m^2$)] $r^2 = 0.413$ |
| Killdeer | [0.126 + 0.000000917 × pool area ($m^2$)] $r^2 = 0.587$ | [0.125 + 0.00000065 × pool area ($m^2$)] $r^2 = 0.520$ | [0.074 + 0.000000801 × pool area ($m^2$)] $r^2 = 0.637$ |
| Mallard | [–0.085 + 0.00000196 × pool area ($m^2$)] $r^2 = 0.503$ | [0.05 + 0.000000661 × pool area ($m^2$)] $r^2 = 0.723$ | [0.053 + 0.00000112 × pool area ($m^2$)] $r^2 = 0.490$ |

* Regressions based on monitoring data.

**Table 37.4 Geometric Mean Water and Dietary Selenium Concentrations, Kesterson Reservoir (Number of Samples in Parentheses)**

| Parameter | Trisection 1 | Trisection 2 | Trisection 3 |
|---|---|---|---|
| Water, 1992–1999 (total selenium, μg/L) | 8.4 (62) | 6.6 (49) | 7.4 (39) |
| Aquatic Invertebrates, 1992–1999 (whole body concentrations, mg/kg as dry weight) | 27.4 (74) | 17.4 (56) | 14.9 (53) |
| Terrestrial Invertebrates, 1989–1998 (whole body concentrations, mg/kg as dry weight) | 11.6 (400) | 9.9 (409) | 7.4 (365) |
| Terrestrial Plants, 1989–1998 (whole-plant concentrations, mg/kg as dry weight) | 4.4 (690) | 3.4 (691) | 3.3 (749) |

#### *37.3.1.2.2 Pool Water and Aquatic Invertebrate Selenium Concentrations*

Some exposure of wildlife to selenium from ephemeral pools occurs through direct exposure to waterborne selenium, but the primary exposure is through ingestion of aquatic invertebrates along with ingestion of terrestrial invertebrates and plants (for some species). Average concentrations of selenium in water and bird receptor dietary items from the 1990–1999 Kesterson Reservoir Monitoring Program are given in Table 37.4.

Waterborne- and aquatic-invertebrate-selenium concentrations are positively correlated for the Kesterson Reservoir ephemeral pools. Therefore, the following relationship (as shown in Figure 37.7) was used to predict invertebrate tissue concentrations based on predicted poolwater concentrations, where:

$$\text{Pool aquatic invertebrate selenium (mg/kg as DW)} = 10\text{\^{}}(\log(\text{Pool Water Se}) \times 0.393 + 0.986) \qquad (37.4)$$

In addition to their aquatic invertebrate diet when pools are present, killdeer were assumed to consume terrestrial invertebrates 75% of the time, while mallards were assumed to feed 25% of the time on terrestrial plants. Black-necked stilts were assumed to feed solely on aquatic invertebrates in the ephemeral-pool environment and to leave Kesterson when the pools were not present. Terrestrial components of mallard and killdeer diets were estimated based on the mean selenium concentrations for plants and invertebrates, as taken from the terrestrial model. Mean terrestrial component concentrations by trisection are shown in Table 37.4.

Waterborne-selenium concentrations were predicted in the ERA model based on changing climatic regime and estimates of the month-end extent of ponding. The model uses the statistical distribution of transfer factors from water-extractable soil selenium to poolwater selenium to predict an array of possible poolwater-selenium concentrations by trisection. The predictive relationships, expanded in the model to an array of 1000 random average conditions, are shown in Table 37.5. The basic construction of the model was described above.

### 37.3.2 Ecological Effects Characterization

Impacts to waterfowl and shorebirds using Kesterson Reservoir can be estimated by comparing the selenium concentrations in dietary items or eggs to known toxic-effect levels. Recently established guidelines for trace element exposure indicate dietary concentrations of concern from 3 to 7 mg/kg selenium (dry-weight basis) and potential toxic effects when dietary concentrations exceed 7 mg/kg.[38,39] Table 37.6 provides a summary of dietary and egg effect levels that can be used to evaluate potential toxic effects of environmental-selenium concentrations to birds. There are differences in potential dietary exposures and sensitivity among aquatic bird species that may use Kesterson Reservoir in wet years. Mallards are among the most sensitive species, black-necked stilts and killdeer are moderately sensitive, and American avocets are less sensitive.

**Table 37.5 Means and Standard Deviations of Data Expanded in the Monte Carlo Simulations of the Aquatic ERA Model**

| Parameter | Mean | Standard Deviation |
|---|---|---|
| Water-extractable-selenium concentration (μg/kg) to poolwater-selenium concentration (μg/L) transfer factor | 0.0239 | 0.0188 |
| Total percent ponded area, Kesterson Reservoir | End of Jan. = 6.3<br>End of Feb. = 6.6<br>End of Mar. = 3.4 | End of Jan. = 13<br>End of Feb. = 14<br>End of Mar. = 8 |
| Terrestrial plants, selenium concentration | T1 = 5.0<br>T2 = 4.0<br>T3 = 3.5 | T1 = 10.024<br>T2 = 4.845<br>T3 = 8.087 |
| Terrestrial invertebrates, selenium concentration | T1 = 12.5<br>T2 = 9.4<br>T3 = 8.1 | T1 = 8.916<br>T2 = 7.738<br>T3 = 6.768 |

Measured and modeled egg concentrations may be compared to probably toxic-effect levels, as well. Toxicity guidelines suggest that egg-selenium concentrations exceeding 6 mg/kg are of concern because of potential reproductive impairment (Table 37.6).

## 37.4 RISK CHARACTERIZATION

The risk characterization evaluates the evidence linking exposures to COPECs with their potential ecological effects among the representative species identified for the site. Based on the potential risks of adverse effects to those ecological receptors (and similarly exposed species), it leads to conclusions (presented below) and recommendations (which were presented in the original report[25] but are not presented here). The risk characterization consists of three portions: (1) risk estimation, (2) risk description, and (3) uncertainty analysis.

### 37.4.1 Risk Estimation

#### *37.4.1.1 Quantitative Evaluation*

The model output produced predictions of the frequency of given bird use and average dietary concentrations. Figures 37.11 through 37.13 show the potential bird use for each species for the months of interest (as predicted, based on the extent of ponding). The model also produced estimates of the expected frequency for various dietary-selenium concentrations for the three species, which are shown in Figures 37.14 through 37.16. Figures 37.15 and 37.16 also reflect inclusion of terrestrial components in the average diet concentrations for killdeer and mallards, respectively. April was not included in the model due to the low probability of ponding (6%) of a small percentage of the reservoir (< 10%). Bird use and dietary exposure were predicted from ponding and indicated nonexistent risk for April conditions.

**Table 37.6 Toxicity Guidelines for Selenium Exposure in Water, Diet, and Eggs**

| | Level of Effect[a] | | |
|---|---|---|---|
| Criteria | None/Background | Level of Concern | Toxicity Threshold |
| Water (μg/L, total recoverable Se) | <2 | 2 to 5 | >5 |
| Diet (mg/kg dw) | <3 | 3 to 7 | >7 |
| Waterbird eggs (mg/kg dw) | <6 | 6 to 10 | >10 |

[a] Except for avian eggs, these guidelines are intended to be population-based. Thus, trends in means over time should be evaluated. Guidelines for avian eggs are based on individual level response thresholds.[38–41]

*Source:* From Beckon, W.N. and Dunne, M., *Grasslands Bypass Project Annual Report, 1998–99,* San Francisco Estuary Institute, Richmond, CA, 2000; Skorupa, J.P., U.S. Fish and Wildlife Service, pers. comm.

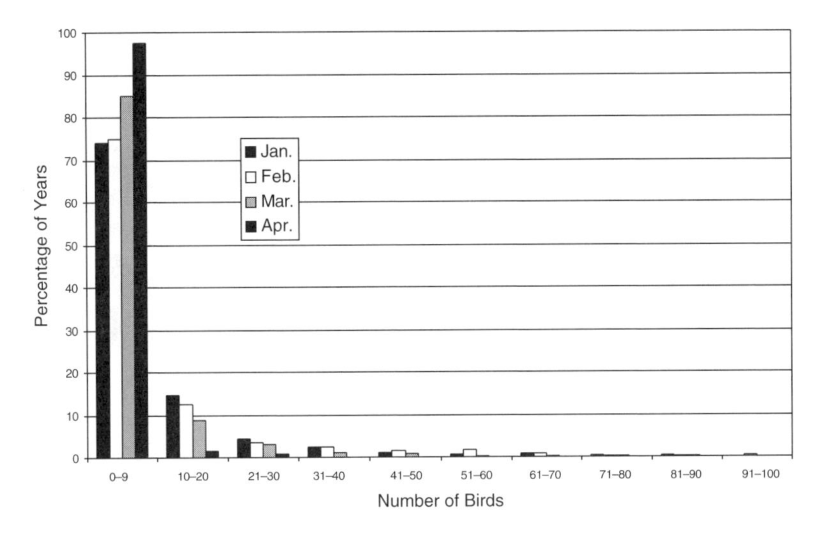


**Figure 37.11** Black-necked stilt daily use estimates.

Estimation of risk based on selenium exposure required the prediction of selenium concentrations in eggs. Monitoring results for selenium concentrations in invertebrate and plant dietary items for birds were compared to egg-selenium concentrations from the 1998 KRBMP data set to yield Kesterson-specific environmental transfer factors. (These site-specific transfer factors are relatively low because they do not represent true diet-to-egg transfers, which can be documented only if 100% of the birds' diets come from within the site and diet composition is well known. Thus, the transfer factors are referred to as "environmental transfer factors.") Extensive ponding and higher-than-normal bird use characterized the year 1998. In addition, we were able to collect a relatively large number of samples of aquatic invertebrates, terrestrial invertebrates, and terrestrial plants that could be used to estimate bird dietary-selenium concentrations for the bird species in the model. The 1998 sampling year was of most value, however, because an unusually high number of semiaquatic birds nested at Kesterson. The average egg-selenium concentrations of black-necked stilts, killdeer, and mallards from 1998 were compared to estimated dietary-selenium concentrations (from measured concentrations of selenium in aquatic invertebrates, terrestrial invertebrates, and terrestrial plants) to yield average environmental transfer factors for selenium. Based on 1998 data, the transfer factors for stilts, killdeer, and mallards were 0.57, 0.51, and 0.48, respectively.

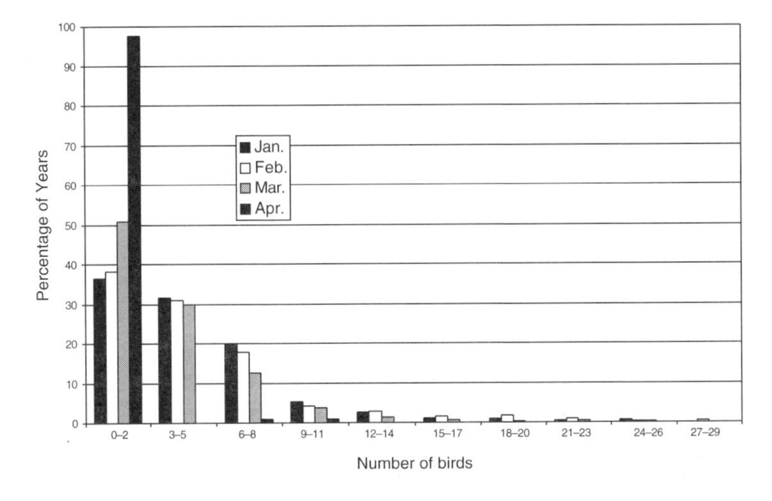


**Figure 37.12** Killdeer daily use estimates.

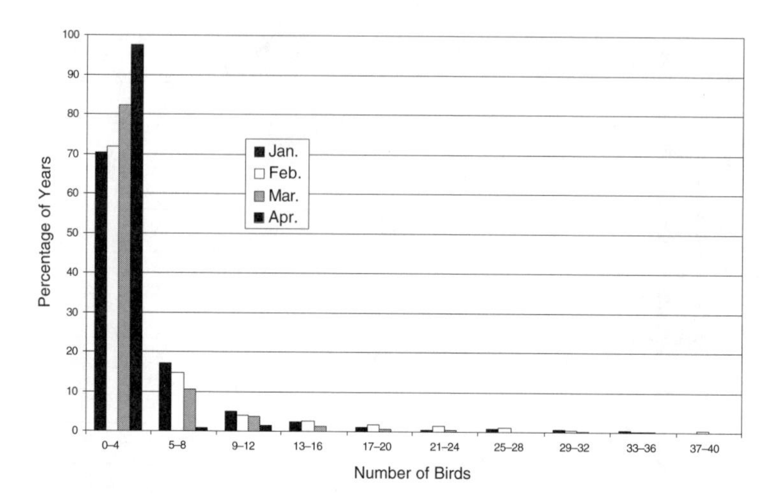


**Figure 37.13** Mallard daily use estimates.

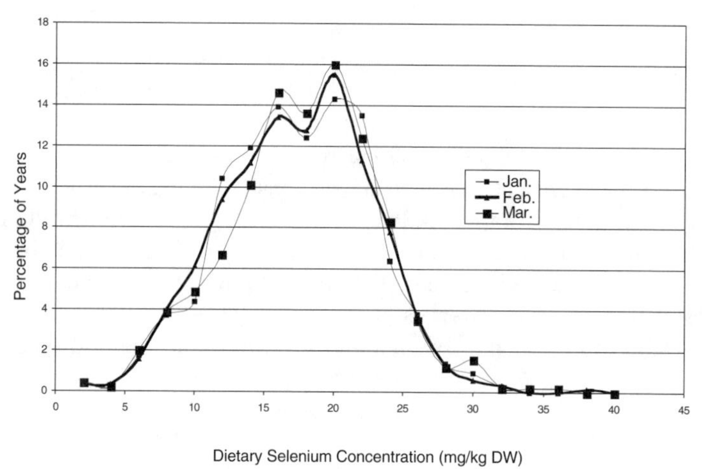


**Figure 37.14** Black-necked stilt dietary selenium.

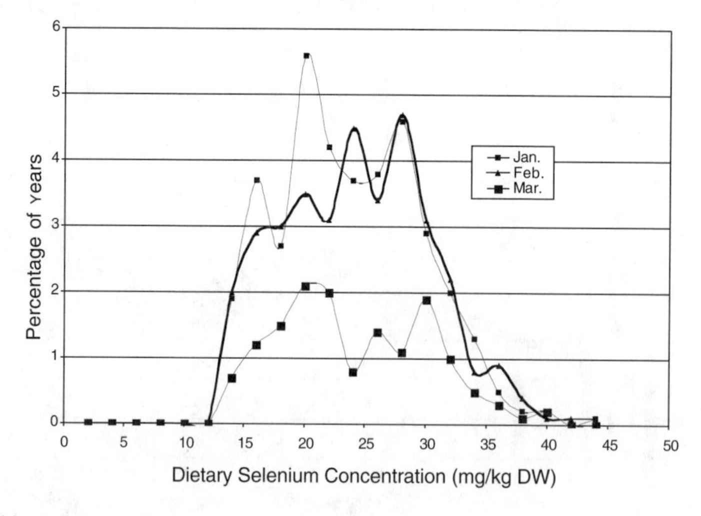


**Figure 37.15** Killdeer dietary selenium.

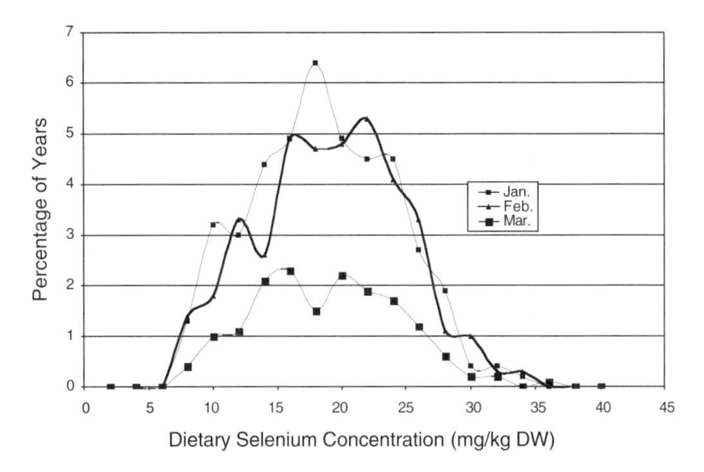


**Figure 37.16** Mallard dietary selenium.

Skorupa[39] developed a relationship between black-necked stilt egg failures and egg-selenium concentrations from a large database of egg chemistry and nesting success. We were able to use Skorupa's equation with our egg-selenium concentration estimates (based on the above environmental transfer factors) to calculate the number of stilt hens with at least one probable egg failure, as follows:

$$\text{Raw percentage of hens affected} = \exp(-2.327+0.0503 \times \text{Egg-selenium conc.})/ (1 + \exp(-2.327 + 0.0503 \times \text{Egg-selenium conc.}) \quad (37.5)$$

$$\text{Actual percentage of hens affected} = [(1\text{-background\%}) - (1\text{-raw\%})]/(1\text{-background\%}) \quad (37.6)$$

$$\text{Number of affected hens} = (\text{No. of birds counted}/2) \times \text{Percentage of hens affected where background\%} = 8.9\% \quad (37.7)$$

The calculation was run as a function of the set of 1000 estimated egg-selenium concentrations for both black-necked stilts and killdeer because of the similarities between the species. No such estimating technique was available for mallards. Although shorebirds may nest in April, April model results were not shown because they demonstrate no exposure. Kesterson is dry 94% of the time in April and shows minimal ponding in years of April flooding.

The probable number of hens affected by egg failure was estimated to exceed the background failure rate of about 8.9% (the calculated background failure rate for black-necked stilts) in a very limited number of cases. The model results indicated less than one killdeer hen would be affected by a loss of eggs due to selenium toxicity (0.1–0.3 hens affected). The results for black-necked stilts showed possible affected hens for the month of January only, with less than three hens affected. However, stilts typically begin nesting later in the season and therefore would rarely be affected. Similarly, the weighted mean selenium concentrations in killdeer eggs predicted by the terrestrial food chain model indicate that the percentage of hens affected (based on purely terrestrial exposure routes) would be less than the background failure rate of 8.9%.

#### *37.4.1.2 Qualitative Evaluation*

Kesterson Reservoir has been managed since 1988 as a terrestrial habitat. Terrestrial exposure to the selenium inventory of the reservoir has not resulted in demonstrable effects on local wildlife and is not expected to cause such effects in the future.

The overall summary of selenium exposure from the ephemeral-pool habitats at Kesterson Reservoir suggests insignificant risk to waterfowl and shorebirds. In most cases, ponding is too limited to represent significant exposure. In the rare cases when ponding persists, a very small number of birds (one or two) may be adversely affected during a small percentage of the years (less than 5%).

### 37.4.2 Risk Description

The evaluation of ecological risk from selenium exposure in the terrestrial and aquatic environments at Kesterson Reservoir is based on several lines of evidence. We have evaluated risks based on a long-term record of water-extractable soil selenium and extensive records of waterborne- and bird dietary-selenium concentrations, combined with an overall low frequency of ponding and the infrequent persistence of ephemeral pools for 1 month or more during the bird nesting season. Risks associated with the ephemeral pools were strongly determined by the balance between exposure (the degree of ponding at critical seasons during the year) and the spatial variation of both waterborne-selenium concentrations and bird use. In particular, bird use was the single most important direct determinant of risk. Although the water and invertebrate fauna of ephemeral pools contain relatively elevated selenium concentrations, risk is most often minimized through the limited ponding and lack of pool use by birds.

There is uncertainty in the modeled predictions of bird use and pool concentrations, but the modeled concentrations and predictions of impairment frequency closely match observed data. As predicted by the model, only a small number of birds are expected to be adversely affected from selenium concentrations in their diet. Monitoring data over 12 years did not show evidence of reproductive impairment to birds nesting at Kesterson attributable to selenium toxicity.

### 37.4.3 Uncertainties/Data Limitations

There are uncertainties and limitations at all levels of this analysis, as is typical of all ERAs. In this section, we describe the sources of uncertainty and specific limitations of the data and analysis.

#### *37.4.3.1 Available Data*

The Kesterson Reservoir database of selenium concentrations in various media, climatic variation, extent of ponding, and bird use is unusually comprehensive for an ERA. Rather than the typical 1 or 2 years of data, most of the summary statistics for selenium are based on 10 to 12 years of data gathered over a full range of seasons and interannual climatic variation. The climate record is based on 127 years of precipitation and estimated evapotranspiration data, incorporating periods of extended drought as well as extreme flooding (such as the 1997–1998 El Niño year). However, in some cases, such as data associated with the aerial photographic record, data sets are more limited. Ranges and measures of variability are presented with individual data summaries and regression relationships in the appropriate sections of this chapter.

It is unclear why the detected rates of egg inviability in killdeer (the most frequently sampled species; 4.3% of nests containing at least one fail-to-hatch [FTH] egg) are somewhat lower than background (8.9% of black-necked stilt nests containing at least one FTH egg), as reported by Skorupa.[39] Since Kesterson Reservoir was filled in 1988, the KRBMP has found seven inviable eggs in the 209 nests that were located during the monitoring. In general, nests were monitored on a weekly basis after they were found and marked. However, eggs in many of those nests were lost to predation, desertion, or destruction of nests (e.g., road grading) before they hatched, and from 1996 to 1999, when 64% of the eggs were collected, over 50% of the nests were lost. Therefore, it was not possible to determine accurately the rate of egg inviability in killdeer. In 1998, when 21 killdeer and 16 black-necked stilt eggs were collected at Kesterson and incubated in the laboratory at UC Davis, there were no killdeer eggs in which embryos died before hatching, and there were two stilt eggs in which embryos did not develop (12.5%). In addition, 41 eggs were collected for examination and selenium analysis; no dead embryos were found in any of those eggs. Thus, it has not been possible to establish relationships between egg-selenium concentrations and reproductive impairment in killdeer, or to document the incidence of egg inviability (due to the high rates of eggs being lost to predation or perhaps other confounding factors).

#### *37.4.3.2 Analytical Techniques and Modeling Assumptions*

Analytical techniques used in the terrestrial and aquatic habitat models are based on a series of linear and log-linear relationships among environmental variables. In each case, assumptions were made about the underlying distribution of the data and the appropriateness of the regression relationship in explaining covariance of the variables. Individual regression equations and bivariate plots are presented in the appropriate sections in this report. As is to be expected from environmental-monitoring data, there is much unexplained variation in predictions, even in cases of statistically significant relationships. However, the basic assumption of the ERA models is that the predictive relationships are all descriptive of underlying causal relationships.

The ERA models are based on Monte Carlo simulations of the distribution of selected input variables. In this case, the underlying assumption is that the true distribution of the environmental variable can be accurately duplicated using random-number sets generated from the mean, standard deviation, and assumed type of distribution of the monitoring data. Most simulated variables (annual extent of ponding, dietary concentrations, bird use) were based on log-normal relationships.

## 37.5 CONCLUSIONS

As a result of this study, several summary conclusions can be made in relation to soil chemistry, the transfer of soluble selenium to surface-water pools, and the extent of ponding on the reservoir:

1. Anticipated future soluble (i.e., bioavailable) soil-selenium levels will fall within the range of concentrations observed over the last decade of soil-selenium monitoring and research.
2. The overall values for soil-to-ephemeral-pool transfer of selenium are not expected to change in the future and can be predicted within a factor of 2 based on historical data.
3. A ponding model based on the historic (127-year) Kesterson-area rainfall record predicts that in 43% of wet seasons, ephemeral pools will not form, or if they do form, they will have a duration of less than 1 month. Ponding for 2 or more months (needed to cause the development of invertebrate pond fauna) is expected to occur in approximately 11% of wet seasons. Ponding for 3 or more months is estimated to occur in about 37% of wet seasons.
4. During the period of spring waterfowl and shorebird nesting, the ponding model predicts that there is a high probability (57% to 94%, depending on the month) that less than 1% of Kesterson Reservoir will be ponded.
5. Ponding during the winter and spring of 1997–1998 represented the maximum ponded condition, based on the historical record and the near worst-case condition, in terms of modeling results for the extent of ephemeral pools.

The Ecological Risk Assessment model yielded the following conclusions:

1. Predicted selenium concentrations in bird dietary items (composed of aquatic invertebrates, terrestrial invertebrates, or terrestrial plants) exceeded dietary levels of concern for reproductive impairment in black-necked stilt and killdeer and exceeded levels of concern 16 to 43% of the time for mallards. However, no evidence of toxicological impact has been observed for these species during the 12 years of the Kesterson Reservoir Biological Monitoring Program, including results for 1997–1998, the year of highest observed ponding at Kesterson.

2. Daily bird use (i.e., potentially exposed individuals) during the months of concern, based on the extent of ponding, was predicted to be less than 20 birds more than 90% of the time for black-necked stilt, killdeer, and mallard.
3. The average number of black-necked stilt and killdeer hens expected to have one or more eggs fail to hatch due to elevated selenium exposure through the Kesterson Reservoir ephemeral pools was expected to be fewer than two hens for 99% of modeled years.

## REFERENCES


1. U.S. Bureau of Reclamation (USBR), Mid-Pacific Region, in cooperation with U.S. Fish and Wildlife Service and U.S. Army Corps of Engineers, Final Environmental Impact Statement, 1986.
2. U.S. Department of the Interior (USDI), Submission to California State Water Resources Control Board in Response to Order No. WQ-88–7: Effectiveness of Filling Ephemeral Pools at Kesterson Reservoir, Kesterson Program Upland Habitat Assessment, and Kesterson Reservoir Final Cleanup Plan, 1989.
3. Ohlendorf, H. M., Bioaccumulation and effects of selenium in wildlife, in *Selenium in Agriculture and the Environment*, Jacobs, L. W., Ed., Soil Science Society of America and American Society of Agronomy, SSSA Special Publ. 23, 1989, pp. 133–177.
4. Ohlendorf, H. M., The birds of Kesterson Reservoir: A historical perspective. *Aquat. Toxicol.*, 57, 1–10, 2002.
5. Presser, T. S. and H. M. Ohlendorf, Biogeochemical cycling of selenium in the San Joaquin Valley, California, USA, *Environ. Manage.*, 11, 805–821. 1987.
6. Weres, O., G. A. Cutter, A. Yee, R. Neal, H. Moehser, and L. Tsao, Section 3500-Se. pp. 3–128 to 3–141, in *Standard Methods for the Examination of Water and Wastewater*, 17th ed., Clesceri, L. S. et al., Eds., Am. Public Health Assoc., Washington, D.C., 1989.
7. Oremland, R. S., J. T. Hollibaugh, A. S. Maest, T. S. Presser, L. G. Miller, and C. W. Culbertson, Selenate reduction to elemental selenium by anaerobic bacteria in sediments and culture: Biogeochemical significance of a novel, sulfate-independent respiration, *Appl. Environ. Microbiol.*, 55, 2333–2343, 1989.
8. Horne, A. J. and J. C. Roth, Selenium Detoxification Studies at Kesterson Reservoir Wetlands: Depuration and Biological Population Dynamics Measured Using an Experimental Mesocosm and Pond 5 under Permanently Flooded Conditions, University of California, Berkeley, Environmental Engineering and Health Sci. Lab. 1989.
9. Zawislanski, P. T., T. K. Tokunaga, S. M. Benson, H. S. Mountford, H. Wong, T. Alusi, R. TerBerg, and K. Olsen, Hydrological and Geochemical Investigations of Selenium Behavior at Kesterson Reservoir, LBNL Report # 43535, Lawrence Berkeley National Laboratory, Berkeley, CA, 1999.
10. CH2M HILL, Kesterson Reservoir 1999 Biological Monitoring, Prepared for U.S. Bureau of Reclamation, Mid-Pacific Region by CH2M HILL, Sacramento, CA, April 2000.
11. Tokunaga, T. K., G. E. Brown, Jr., I. J. Pickering, S. R. Sutton, and S. Bajt, Selenium redox reactions and transport between ponded waters and sediments. *Environ. Sci. Technol.*, 31, 1419–1425, 1997.
12. Tokunaga, T. K., S. R. Sutton, S. Bajt, P. Nuessle, and G. Shea-McCarthy, Selenium diffusion and reduction at the water-sediment boundary: Micro-XANES spectroscopy of reactive transport, *Environ. Sci. Technol.,* 32, 1092–1098, 1998.
13. Tokunaga, T. K., P. T. Zawislanski, P. W. Johannis, S. Benson, and D. S. Lipton, Field investigations of selenium speciation, transformation, and transport in soils from Kesterson Reservoir and Lahontan Valley, in *Selenium in the Environment,* Frankenberger, W. T. and S. Benson, Eds. Marcel Dekker, New York, 1994, pp. 119–138.
14. Zawislanski, P. T., T. K. Tokunaga, S. M. Benson, J. M. Oldfather, and T. N. Narasimhan, Bare soil evaporation and solute movement of selenium in contaminated soils at Kesterson Reservoir, *J. Environ. Qual.,* 12, 447–457, 1992.
15. Wahl, C., S. Benson, and G. Santolo, Temporal and spatial monitoring of soil selenium at Kesterson Reservoir, CA, *Water Air Soil Pollut.,* 74, 345–361, 1994.
16. Frankenberger, W. T. and U. Karlson, Volatilization of selenium from a dewatered seleniferous sediment: A field study, *J. Indust. Microbiol.,* 12, 226–232, 1995.

17. Zawislanski, P. T., T. K. Tokunaga, S. M. Benson, H. S. Mountford, T. C. Sears, H. Wong, D. King, and J. Oldfather, Hydrological and Geochemical Investigations of Selenium Behavior at Kesterson Reservoir, Progress Report, October 1, 1994–September 30, 1996, LBL-41027, Berkeley, CA, 1997.
18. Zawislanski, P. T. and M. Zavarin, nature and rates of selenium transformations in Kesterson Reservoir soils: A laboratory study, *Soil Sci. Soc. Am. J.,* 60, 791–800, 1996.
19. Dowdle, P. R. and R. S. Oremland, Microbial oxidation of elemental selenium in soil slurries and bacterial cultures, *Environ. Sci. Technol.,* 32, 3749–3755, 1998.
20. Losi, M. E. and W. T. Frankenberger, Microbial oxidation and solubilization of precipitated elemental selenium in soil, *J. Environ. Qual.*, 27, 836–843, 1998.
21. Benson, S. M., T. K. Tokunaga, and P. T. Zawislanski, Anticipated Soil Selenium Concentrations at Kesterson Reservoir, Lawrence Berkeley Laboratory Report, LBL-33080, Berkeley, CA, 1992.
22. Wahl, C. and S. Benson, Update to Modeling Soil Selenium Concentrations in the Shallow Soil Profile at Kesterson Reservoir, Merced County, California, 1988–1993, LBNL Report 39215, Lawrence Berkeley Laboratory, Berkeley, CA, 1996.
23. CH2M HILL, Ecological Risk Assessment for Kesterson Reservoir. Prepared for U.S. Bureau of Reclamation, Mid-Pacific Region by CH2M HILL, Sacramento, CA, March 1993.
24. Ohlendorf, H. M. and G. M. Santolo, Kesterson Reservoir — Past, present, and future: An ecological risk assessment, in *Selenium in the Environment,* Frankenberger, W. T., Jr., and S. Benson, Eds., Marcel Dekker, New York, 1994, pp. 69–117.
25. CH2M HILL and Lawrence Berkeley Laboratory, Ecological Risk Assessment for Kesterson Reservoir, Prepared for U.S. Bureau of Reclamation, Mid-Pacific Region, December 2000.
26. U.S. Environmental Protection Agency (USEPA), Guidelines for Ecological Risk Assessment, EPA/630/R-95/002F, USEPA Risk Assessment Forum, Washington, D.C., 1998.
27. Tokunaga, T. K. and S. M. Benson, Selenium in Kesterson Reservoir ephemeral pools formed by groundwater rise. I. A field study, *J. Environ. Qual.,* 21, 246–251, 1992.
28. Saiki, M. K., Concentrations of selenium in aquatic food-chain organisms and fish exposed to agricultural tile drainage water, in *Selenium and Agricultural Drainage: Implications for San Francisco Bay and the California Environment, Proc. of the Second Selenium Symp.*, Howard, A. Q., Ed., 23 March 1985, Berkeley, CA, 1986, pp. 25–33.
29. Ohlendorf, H. M., D. J. Hoffman, M. K. Saiki, and T. W. Aldrich, Embryonic mortality and abnormalities of aquatic birds: Apparent impacts of selenium from irrigation drainwater, *Sci. Total Environ.*, 52, 49–63, 1986.
30. Ohlendorf, H. M., J. P. Skorupa, M. K. Saiki, and D. A. Barnum, Food-chain transfer of trace elements to wildlife, in *Management of Irrigation and Drainage Systems: Integrated Perspectives*, Allen, R. G. and C. M. U. Neale, Eds., American Society of Civil Engineers, New York, 1993, pp. 596–603.
31. Hothem, R. L. and H. M. Ohlendorf, Contaminants in foods of aquatic birds at Kesterson Reservoir, California, 1985, *Arch. Environ. Contam. Toxicol.*, 18, 773–786, 1989.
32. Zawislanski, P. T., G. R. Jayaweera, J. W. Biggar, W. T. Frankenberger, and L. Wu, The Pond 2 Selenium Volatilization Study: A Synthesis of Five Years of Experimental Results, 1990–1995, LBNL Report # 39516, Lawrence Berkeley National Laboratory, Berkeley, CA, 1996.
33. Hamdy, A. A. and G. Gissel-Nielsen, Fixation of selenium by clay minerals and iron oxides, *Z. Pflanzenernaehr. Bodenkd.,* 140, 63–70, 1977.
34. Bar-Yosef, B. and D. Meek, Selenium sorption by Kaolinite and Montmorillonite, *Soil Sci.*, 144, 12–19, 1987.
35. Yläranta, T., Sorption of selenite and selenate in the soil, *Annales Agric. Fenniae.,* 22, 29–39, 1983.
36. Tokunaga, T. K., D. S. Lipton, S. M. Benson, A. Y. Yee, J. M. Oldfather, E. C. Duckart, P. W. Johannis, and K. H. Halvorsen, Soil selenium fractionation, depth profiles and time trends in a vegetated site at Kesterson Reservoir, *Water Air Soil Pollut.*, 57–58, 31–41, 1991.
37. U.S. Bureau of Reclamation (USBR), Mid-Pacific Region, Revised monitoring and reporting program, Number 87–149, 1987.
38. U.S. Department of the Interior (USDI), Guidelines for Interpretation of the Biological Effects of Selected Constituents in Biota, Water, and Sediment, National Irrigation Water Quality Program Information Report No. 2, Denver, CO, 1998.

39. Skorupa, J. P., Risk Assessment for the Biota Database of the National Irrigation Water Quality Program, Prepared for the National Irrigation Water Quality Program, U.S. Department of the Interior, Washington, D.C., April 1998.
40. Beckon, W. N. and M. Dunne, Biological Effects, in *Grasslands Bypass Project Annual Report, 1998–99*, San Francisco Estuary Institute, Richmond, CA, 2000, pp. 63–98.
41. Heinz, G. H., Selenium in birds, in *Environmental Contaminants in Wildlife: Interpreting Tissue Concentrations,* Beyer, W. N., G. H. Heinz, and A. W. Redmon-Norwood, Eds., Lewis Publishers, Boca Raton, FL, 1996, pp. 447–458.

## CHAPTER 38

# Restoration Ecology and Ecotoxicology

**John Cairns, Jr.**

## CONTENTS



Sustainable development is the development that meets the needs of the present without compromising the ability of future generations to meet their own needs.

**World Commission on the Environment and Development as quoted in the U.S. Man and the Biosphere Bulletin, December 1991, 15(4):2**

## 38.1 INTRODUCTION

Since the first edition of the *Handbook of Ecotoxicology* was published, the primary impetus for the developing relationship of restoration ecology and ecotoxicology has been the rapidly developing interest in natural capitalism and industrial ecology (these issues are addressed in detail in Chapter 42 of this volume, "The Role of Ecotoxicology in Industrial Ecology and Natural

Capitalism"). An important related development is the recognition of the environmentally perverse effects of governmental subsidies.[1] While ecological restoration is poorly subsidized, environmental damage and destruction are often heavily subsidized by governmental agencies in all parts of the planet. In addition, the means for more accurate determinations of individual and societal impacts upon natural systems have advanced markedly.[2]

Arguably, the most important rapidly developing trend is the increasing acceptance of the need for implementing the precautionary principle.[3] The precautionary principle essentially states that, when an activity raises threats of harm to human health or the environment, precautionary measures should be taken, even if some cause-and-effect relationships are not fully established scientifically. The importance of this principle to ecotoxicologists is the need to develop methods for predicting toxicological harm before it occurs and to estimate which toxicological thresholds, when crossed, will pose substantial ecological restoration problems. Another development is the establishment by the American Association for the Advancement of Science of the Program in Scientific Freedom, Responsibility and Law: Court Appointed Scientific Experts. This group of advisors may well have profound effects upon the field of ecotoxicology (for details, see http://www.aaas.org/spp/case/case.htm). If successful, this program will ensure a greater role for ecotoxicologists in courts of law in the United States and, concomitantly, a greater responsibility.

The Program in Scientific Freedom, Responsibility and Law: Court Appointed Scientific Experts has launched a demonstration project that will assist federal district judges in obtaining independent scientific and technical experts. Judges wishing to appoint experts under any source of authority will be able to call for assistance in identifying highly qualified scientists and engineers who will serve as experts to the courts, rather than to parties in litigation. An important aspect of the project is a rigorous and independent evaluation that will be conducted by the Federal Judicial Center, the research and education arm of the judiciary. Ecotoxicologists and restoration ecologists will be well advised to keep track of this initiative because it affects not only each profession independently but also the relationship between the two professions. An important volume in this regard is the National Research Council's report *Ecological Indicators for the Nation*.[4] This volume has some useful discussions of thresholds and breakpoints for the types of situations in which both restoration ecologists and ecotoxicologists are likely to be involved. Another useful National Research Council book is *Global Change: Ecosystems Research*.[5] It provides similar but different guidance on assessing ecological change at large temporal and spatial scales.

Humans have three options regarding their relationship to the earth's ecosystems: (1) continue ecological destruction at the present rate, (2) stop ecological damage and achieve a no-net-ecological-loss situation, and (3) maintain ecosystem services per capita, even with a growing population. Most citizens of developed countries would agree that option (1) is unacceptable, although these same people may not yet be prepared to accept the policies that will be essential to embrace either option (2) or option (3). Adopting a no-net-ecological-loss policy will require ecological restoration to restore systems damaged by ecoterrorism (e.g., the Arabian Gulf), by accidental spills (e.g., the Exxon Valdez in Prince William Sound, Alaska), or by cumulative impact of anthropogenic stresses over a long period of time. Ecotoxicology will be essential to ensure that the management practices associated with potentially toxic materials are well understood. Nevertheless, even if the option is fully successful and the population continues to grow, there will be fewer ecosystem services per capita simply because the population is growing in a finite space.

Option (3) will require more vigorous restoration efforts, as well as reduced risk in terms of ecotoxicological effects. Option (3) will be essential to maintain just the present ecosystem services per capita with a growing global population. These ecosystem services are well described by Ehrlich and Ehrlich,[6] but a few illustrative examples include maintaining the atmospheric gas balance, transforming societal wastes into less objectionable compounds, maintaining a genetic inventory that will be helpful during climatic and other large-scale changes, regulating the rainfall pattern and other microclimatic events, and, of course, maintaining water quality.

## 38.2 THE RELATIONSHIP BETWEEN ECOTOXICOLOGY AND RESTORATION ECOLOGY

Environmental damage occurs from ecoaccidents, ecoterrorism, or because an ecological prediction of a safe concentration or course of action was incorrect. In each case, the damaged system should be repaired; when the spill involves hazardous and toxic materials, the fields of ecotoxicology and restoration ecology must be integrated. Additionally, restoration makes use of ecological resiliency — that is, the ability of natural systems to snap back to some semblance of their former condition, either with or without human assistance. Ecological resilience should be included in the safety or application factor used in conjunction with ecotoxicological tests.[7] That is, protection should be greater for ecosystems less likely to "snap back" after disturbance than those that do. Other relationships should be explored by both groups of professionals.

### 38.2.1 Multiple Use and Cumulative Impacts

The news media have numerous examples each day of human impact on the environment. Both the frequency and severity, as well as the complexity, of impacts have been increasing. Multiple use is becoming increasingly favored with a growing population and a finite planet, but multiple use increases the probability of multiple or cumulative impacts. At the same time, the planet's ecological capital is shrinking, thus requiring increasingly effective management of natural resources. The nature of adverse environmental impacts has changed radically since the beginning of the industrial revolution from small, localized effects to the now increasing importance of regional effects (e.g., acid precipitation) and global effects (e.g., climate change). Other factors that have markedly changed include nonpoint sources of stress relative to point-source discharges, the increasing number of potential stressors, and the importance of aggregate or cumulative impacts. In most developed nations, the focus of environmental protection has broadened from the development of stress-specific environmental quality standards to the achievement of broad objectives for restoring self-maintaining ecosystems inextricably coupled with the maintenance of the quality of human life.[8] Coping effectively with the transition in the way society presently affects ecosystem health to restoring and protecting the environment will require substantive improvements in the effectiveness of environmental management strategies.

### 38.2.2 Can Ecosystem Health be Measured?

Ecotoxicologists measure injury to representatives of individual species in laboratory jars or the effects at higher levels of biological organization in laboratory microcosms, but ultimately the desire is to predict concentrations with minimal effects in natural systems (generally referred to as ecosystems). Because humans are health conscious and wish to preserve their health, many wish to do the same thing for ecosystems. However, Glenn W. Suter, II (personal communication) disagrees with treating the metaphor of ecosystem health as a reality because it is not an observable property. Suter feels that the metaphor misrepresents both ecology and health science. Ecosystems are not organisms, he feels, so they do not behave like organisms and do not have properties of organisms, such as health. Additionally, Suter feels that health is not an operational concept for physicians or health-risk assessors because they must predict, diagnose, and treat specific states called diseases or injuries; they do not calculate indices of health. Finally, Suter believes that attempts to operationally define ecosystem health result in the creation of indices of heterogeneous variables. These indices, he says, have no meaning; they cannot be predicted, so they are not applicable to most regulatory problems; they have no diagnostic power; effects on one component are eclipsed by responses of other components; and the reason for high or low index value is unknown. Suter feels

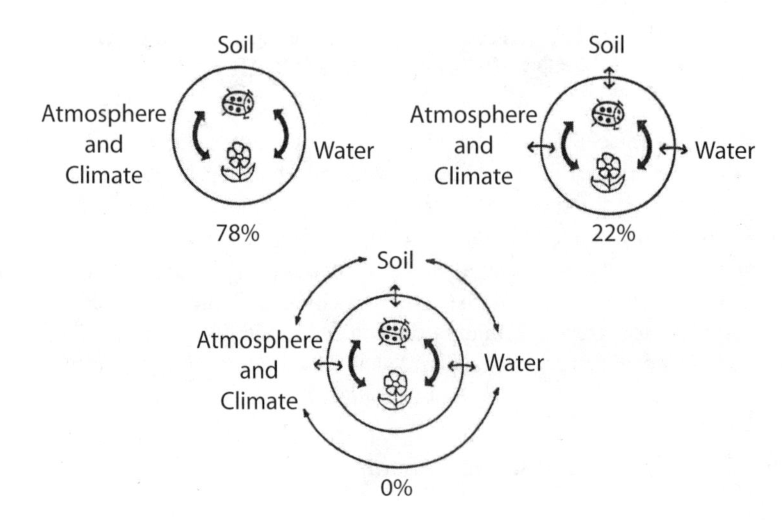


**Figure 38.1** Results of a survey of journal literature in ecology, indicating the dominance in ecology of research that ignores the influence of climate, soil, water, and air on species interactions.

that a better alternative is to assess the real array of ecosystem responses so that causes can be diagnosed, future states can be predicted, and benefits of treatments can be compared.

Unfortunately, perhaps, ecosystem health is a powerful metaphor that focuses desires and aspirations of large numbers of people. It has been frequently advocated as a goal for environmental regulation and management.[9–12] Ecosystem health was the subject of a recent workshop[13] and is used in the title of a new journal (*Journal of Aquatic Ecosystem Health*). Putting aside for the moment the undeniably poetic power of this metaphor, one wonders why it has been so persistent and durable in the field of environmental science. Harte and his colleagues[14] examined 285 articles in four consecutive issues during 1987–1990 of the journals *Ecology, American Naturalist, Oecologia, and Conservation Biology.* They found that:

> If we turn to the current journal literature in ecology and conservation biology, we will not get as much help as we would like. Figure 7 (Figure 38.1 in this discussion), based on our survey of all the articles appearing in some recent volumes of leading journals, illustrates the "Pluto Syndrome" that afflicts ecology today. It shows that most research in ecology might just as well have been carried out on the planet Pluto — climate and the chemical and physical properties of the soil, the water, and the atmosphere of Earth were irrelevant to the research. Even in the 22% of the articles that mentioned at least one of these factors, the reference was generally made as part of a site description and played no role in the interpretation of results. In none of the papers surveyed were the interactions among these factors even mentioned.

In short, most ecologists study populations and not ecosystems; thus, a dearth of terminology and methodology is used to describe and inventory ecosystem condition. Although Whittaker[15] had no doubt that "ecosystems are not organisms," more recently Lovelock,[16] in espousing the Gaia hypothesis, has resurrected the view of the planet as a superorganism.

This discussion explores the relationship between restoration ecology and ecotoxicology. Scientists wish to prevent damage to natural systems, detect damage when it occurs, and restore the damaged system to a close resemblance of its predisturbance condition. Presumably, the same attributes or endpoints could be used for all three purposes. Regrettably, the field of ecotoxicology has yet to achieve a recognizable identity,[17,18] and the field of ecology is presently more concerned with populations than with interactions among the chemical, physical, and biological components of the entities called ecosystems.[14] Both ecotoxicology and restoration ecology purportedly are concerned with the condition (structural and functional) of complex multivariate natural systems. While it is true that ecotoxicology focuses on causes of deterioration from the nominative state and restoration ecology focuses on how to return a disturbed system to its predisturbance nominative state, the number of system attributes shared by these two fields in their diagnoses should be remarkably similar. In fact, a charitable but dispassionate observer would, in examining both fields as they presently exist, have difficulty making any robust connections between the two. However, both groups almost certainly should be using nearly identical ultimate attributes to judge success or failure.

### 38.2.3 Top-Down vs. Bottom-Up Strategies

The top-down and bottom-up approaches appear quite different in the evaluation of environmental degradation at the community and ecosystem levels.[19,20] The top-down method involves directly assessing changes in natural communities and ecosystems and then subsequently diagnosing problems and causative agents. This might also be regarded as a reactive measure, which would include the saprobian approach.[21] The contrasting bottom-up methods use laboratory data that demonstrate effects on simple systems, generally low in environmental realism, and then use the data to model or predict changes in more complex natural ecosystems. The hazard-assessment protocols illustrate this approach.[22] Routine bottom-up procedures for estimating hazard (e.g., laboratory testing with human or ecosystem surrogates, models of fate transport and partitioning) are limited in their ability to predict impacts on natural ecosystems for a number of reasons.[23–27] A list of the most important of these limitations follows.

1. Difficulties involved in the use of effects observed in the laboratory to predict responses in the natural environment.
2. Difficulties involved in the use of different degrees of environmental realism (simple laboratory test systems vs. complex natural systems) and different levels of biological organization (single species laboratory tests vs. complex natural communities and ecosystems).
3. Difficulties in measuring aggregate effects in the laboratory. Most hazard protocols consider the effect of each type of stress individually, even though impacts are inevitably cumulative or aggregative in natural systems. In short, reductionist science is pushed beyond the limits of its capabilities, and integrative science is not given the attention it deserves. Not to be misunderstood, all scientists must be reductionists at times because the scientific method requires this.
4. Difficulties in determining all the chemicals to which organisms are exposed in natural systems. The American Chemical Society Computer Registry of Chemicals has had over 8 million chemicals on the register. Fortunately, only an estimated 76,000 of these are in daily use, including shelf-life extenders for foods, etc. Nevertheless, 76,000 chemicals, when released into the environment following use, can have a bewildering array of interactions. Current testing capabilities cannot test all possible combinations of these chemicals under all possible environmental conditions with all species that are probably exposed. As a consequence, all estimates of hazard or environmental risk will necessarily have a substantial uncertainty involved. Nevertheless, the uncertainty need not be as large as it presently is since methods currently available could be utilized more extensively to reduce the uncertainty substantively.
5. Difficulties of scale and complexity of detail. It seems highly improbable that successful management of sizable ecological landscapes or major ecosystems to achieve broad environmental and socioeconomic objectives is possible without a substantial broadening of the environmental assessment framework to encompass top-down ecosystem objectives. Figure 38.2 provides a simplified flow pattern for an ecosystem quality control system.

## 38.3 RESTORATION OF NATURAL ECOSYSTEMS ALREADY IMPACTED BY HAZARDOUS CHEMICALS AND OTHER ANTHROPOGENIC STRESSES

It is difficult to find any ecosystem in the world, especially aquatic ecosystems, totally unaffected by human activities. Even relatively remote areas have now accumulated trash and other debris

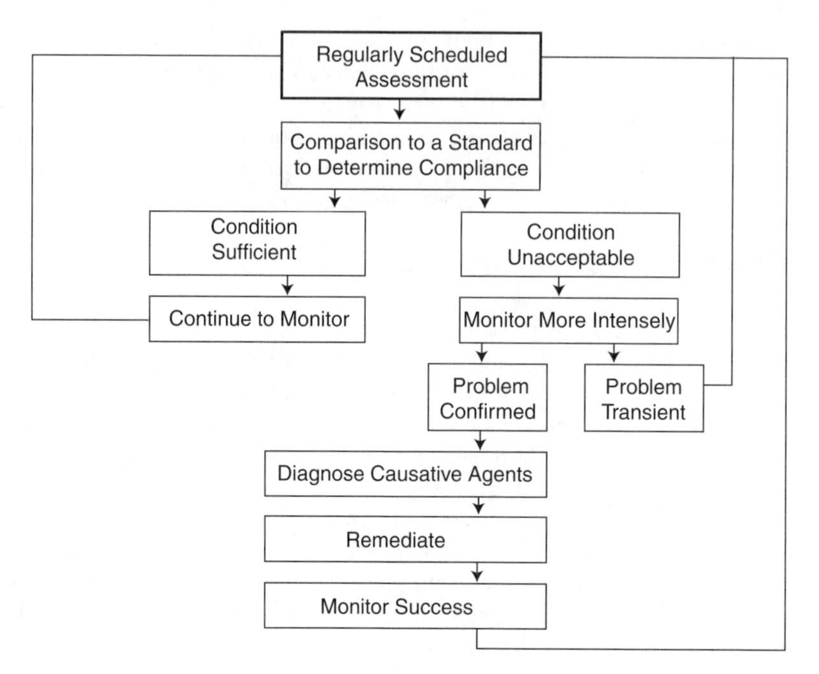


**Figure 38.2** Monitoring the health of an ecosystem over time. (Reproduced with the permission of the International Joint Commission (United States and Canada) from *A Proposed Framework for Developing Indicators of Ecosystem Health for the Great Lakes Region*, a report prepared by Cairns, J.R., Jr., McCormick, P.V., and Niederlehner, B.R., July, 1991.)

from expeditions and may be affected by global climate change or changes in the ozone layer and the like. Furthermore, the human population globally is still increasing dramatically,[28] and considerable restoration is necessary just to retain the per-capita ecosystem services now available.[29] Ecotoxicologists need to determine for restoration ecologists the degree of restoration possible at different levels or concentrations of hazardous materials for ecosystems affected by anthropogenic toxics and hazardous materials.

One of the daunting problems for any professional is attempting to communicate outside the discipline. Can one make the assumption that the specialized jargon, presumably understood by a limited number of initiates, will be equally well understood or even intelligible to those with other backgrounds, however intelligent they might be? Should professionals have a semantic debate about words used in a limited context within the profession (or even avoided entirely) or should they attempt to convey the essence of their beliefs using the common language of the general public? Persuasive evidence indicates that legislators and the general public are disenchanted with scientists. This may well be because scientists avoid simple, direct terms such as *health* in favor of those more scientifically accurate but more easily misunderstood by the general public. I am most sympathetic to Suter's misgivings, but these are outweighed (just barely) by my apprehensions that unfamiliar terminology may cause laypersons to either misunderstand or ignore the message.

An important aspect of restoring ecosystems displaced by anthropogenic or other stresses is relating the development of indicators of ecosystem health to management goals for that particular ecoregion. Figure 38.3 is a crude schematic of the alternative trajectories for an ecosystem for which some management option is contemplated.

All restoration efforts should be accompanied by an effective monitoring program, which is the only way of determining whether the expected events are occurring. Since everything cannot be measured, selected attributes related to objectives for the ecosystem should be developed. "Indicators" should be selected that are useful in determining the extent to which specific objectives have been achieved (e.g., that selected quality control parameters are within the predicted range or

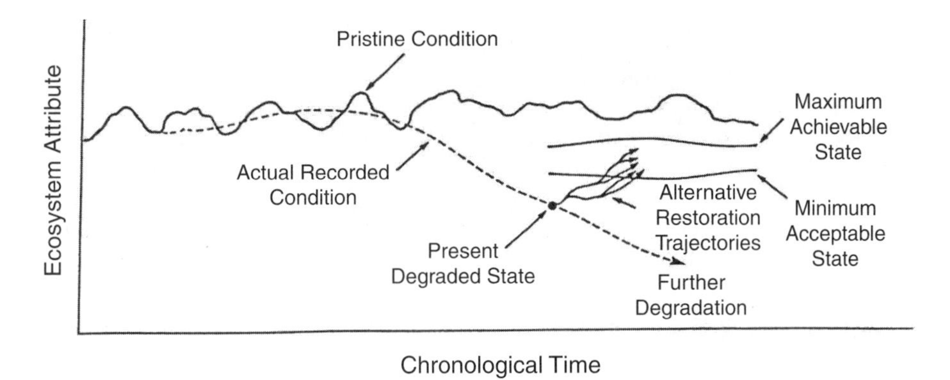


**Figure 38.3** Crude conceptual model of alternative trajectories of an ecosystem. (Reproduced with the permission of the International Joint Commission (United States and Canada) from *A Proposed Framework for Developing Indicators of Ecosystem Health for the Great Lakes Region*, a report prepared by Cairns, J., Jr., McCormick, P.V., and Niederlehner, B.R., July, 1991.)

are approaching the predicted range). However, indicators or attributes cannot be identified until the goals and objectives are explicitly stated. Regrettably, many regulatory agencies frequently select indicators or attributes without explicitly stating ecosystem objectives and goals. Of course, just stating "a healthy ecosystem," while laudable, is not of itself a desirable management objective unless it is specified, both structurally and functionally, what "a healthy ecosystem" means.

Structural attributes are relatively straightforward, although uncertainty about any measurement is always present. Some of the structural attributes involve "critter counting" (such as relationships between different trophic levels), comparison of species array or community structure with others characteristic of that particular habitat, and topographical (slopes, etc.) and hydrologic characteristics. These are the attributes most commonly measured and most easily quantified.

Functional attributes are no less important, but they generally receive less attention. Functional attributes most commonly involve rate processes of one sort or another. Some illustrative examples for a wetlands restoration include: (1) rate of carbon fixation, (2) rate of nutrient spiraling, (3) rate of detritus processing and storage, (4) rate of suspended-solid trapping, and (5) nutrient trapping and storage.

Each wetland, of course, may require a somewhat different mixture of attributes — although a list of attributes may be commonly shared, some will be more important in one system than in another. Success means being on trajectory toward development of the predicted state, and failure means getting a result strikingly different from that predicted.

Whenever a restoration project is underway, it becomes considerably more difficult to admit failure. If an industry is charged with restoration and is bonded to achieve this result (e.g., in association with surface mining), failure to achieve the desired goals or attributes may result in forfeiture of the bond, extension of the period during which bonding is required, or an unexpected expenditure of funds to realign the recovery process with the predicted pathway. In the same sense, a regulatory agency may have to admit that some goals it has set for a restoration project are unachievable and that the organization responsible for the restoration is not accountable for these goals. In this era of reduced public funding, such an admission could be a real threat to the continuance of the regulatory agency at its present level of funding. Finally, for the scientists involved, admission of failure means that their hypothesis was invalid and that all or some of their assumptions were false. While this is painful to admit, even among fellow scientists, it is far, far worse to make this admission to the general public or legislators who have little understanding that failure is a constant part of the scientific process. However, some failures could easily have been prevented by a systematic and orderly approach in project planning. Some worst-case reasons for failure, not involving uncertainty in predictive models, are:

1. No specific goals were set. Instead, vague generalities, such as "fishable, swimmable waters," are utilized instead of more explicit statements of testable objectives. Sometimes, if the objectives are listed, they may describe only the proposed structure and not the proposed functioning of the restored ecosystem. In this case, success in terms of functional attributes may be missed, despite the fact that some of the biological structural attributes do not match the predicted model.
2. In some cases of extreme fragmentation, a fragment may appear to be restored, but in the process of doing so, serious damage may occur in another part of the ecological landscape. For example, increased discharge from a reservoir may enhance wetlands recovery downstream but damage the ecological integrity of wetlands behind the dam.
3. There is no monitoring or follow-up to determine if the project was actually carried to completion, whether it has in fact achieved the stated restoration goals, or whether the mitigation proposed has actually been achieved.
4. Unexpected problems are encountered during the restoration process that make it impossible to follow the original plans. Course corrections may be made on the spur of the moment on site and without consultation with the mixture of organizations that had reached a consensus on the original project. Construction crews may make errors that are difficult to correct because ecologists or hydrologists were not present during critical stages of the restoration process. For example, in Colorado, one mitigation project required that an access road be built through relatively undisturbed wetlands to reach the mitigation site where both restored and constructed wetlands were being developed as a part of the mitigation process.
5. No substantive documentation of the restoration efforts is available that can be given rigorous peer review by appropriate specialists. As a consequence, instead of being required to correct minor flaws in design early in the restoration process, a disaster discloses the inadequacies of the project.
6. Access is often denied to those charged with evaluating a restoration project on private land.
7. Regulatory agencies have insufficient personnel to track the restoration effort or have sufficient personnel but with inadequate professional experience.
8. Cosmetic restoration (greening or revegetation) produces enchanting photographs that can persuade the unknowledgeable that true restoration has occurred. However, when a careful examination is made by trained personnel, the deficiencies are discovered. Nevertheless, course corrections should be made early in the project, and glossing over inadequacies with colorful photographs is all too often substituted for scientifically justifiable evidence.

## 38.4 ESTABLISHING A PROTOCOL FOR INTERACTIONS BETWEEN RESTORATION ECOLOGISTS AND TOXICOLOGISTS

Ecologists and toxicologists have exceedingly limited interactions at present, and establishing a common ground between what appear to be quite distinct professions is difficult. Human society appears to be developing four goals with regard to its relationship with natural systems:

1. Develop predictive models that will accurately estimate effects upon natural systems of various anthropogenic stresses before these potential stressors are utilized, or at least before they are utilized on a wide scale. Lemons[30] identifies a number of important organizational characteristics of ecosystems: ecosystem processes; productivity; decomposition and nutrient cycling; interactions among the biosphere, atmosphere, hydrosphere, and lithosphere; communities; community structure; dynamic networks of interacting individuals and species; symbiotic and mutualistic species; populations; individual organisms; and overall homeostasis. These are illustrative only because any attribute of a natural system might well be used in the assessment process. Naturally, key attributes will vary from one ecosystem to another, and professional judgment will be required to make an appropriate selection.
2. Detect damage in natural systems as soon as possible after its occurrence while the extent of the damage is still minimal. Damage could result from inaccurate predictive models improperly validated or from spurious validation. Additionally, damage frequently occurs from accidental spills and other ecoaccidents, including ecoterrorism. Finally, since natural systems respond to the aggre-

gate of stresses to which they are exposed under conditions not always utilized in the development of the predictive model, aggregate or cumulative impacts may not have been properly evaluated.

3. Restore damaged systems to as close an approximation of predisturbance conditions as possible. Attributes useful in all three of these activities may be quite similar. In short, ecotoxicologists will need system-level attributes to predict and validate anthropogenic effects at higher levels of biological organization such as communities and ecosystems. Second, these same system attributes should be equally useful in providing early warning of damage. The attributes used in resolving the uncertainties posed in both (1) and (2) are now similar at present; this seems to be an accident of the development of the two fields rather than a sound, scientifically justifiable strategy.
4. Have objective quantitative means of determining success or failure in restoration efforts. The system-level attributes used in implementing (1) and (2) should be equally useful in restoration. In fact, if they cannot be used, then an overall management plan integrating these three activities could not be developed. Almost certainly, all three activities would benefit from a dialog discussing suitable attributes among the practitioners of ecotoxicology, biological and ecological monitoring, and ecological restoration.

The crude, illustrative protocol presented in Figure 38.4 is intended to stimulate discussion on the relationship between restoration ecologists and ecotoxicologists. Successful rehabilitation requires information flow between these two specialties. In Phase I, description of the community that is the desired endpoint of restoration efforts serves as the basis for determinations of the necessity for reductions in hazard. In Phase II, tolerance information will guide the choice of species, maximizing success. In Phase III, after species are introduced, observations of their success can be compared to expectations. It would be astonishing if predictions were completely accurate; therefore, Phase III focuses on detecting errors and identifying needs for more robust information where substantial error occurred. This process provides feedback loops that improve the quality of the information used to make decisions. In Phases IV–VI, the new information is incorporated into the rehabilitation plan. Ultimately, an entire book could be allotted to this subject, as was the case for the relationship between those who determined environmental concentrations of chemicals and those who determined the biological effects of these chemicals.[22]

It now seems unthinkable that the two groups whose relationship was explored in detail in the book just cited were not working together from the outset, but the need to develop an effective relationship was not apparent in the 1940s, 1950s, and 1960s. In addition, the relationship between restoration ecologists and ecotoxicologists will not be viewed at the end of this century as it is today. One event that makes this change seem probable was the April 13, 1992 issue of *U.S. News and World Report*, which had extensive documentation of the poisoning of Russia. Problems had already surfaced in various Eastern European bloc countries. Reducing all this damage and its threats to human health and the environment, together with restoration to a closer approximation of predisturbance condition, will definitely require collaboration between ecotoxicologists and restoration ecologists. Since the attributes at the system level for both ecological restoration and ecotoxicology could easily be more robust, energy should be spent on this aspect rather than an extended conflict over which methods of which group should take precedence.

## 38.5 FINANCING ECOSYSTEM RESTORATION

Ecological restoration represents a major shift from increasingly more environmental services that society expects to responsible stewardship. Hazardous-waste sites by definition pose a threat to either human health or the environment or both. The latter is the most likely case since, while living material is not identical in its response to hazardous materials, they are of sufficient similarity to make the assumption that anything adversely affecting environmental health will have some negative impact on human health. In his budget message for 1991, President Bush stated "Today, a consensus is emerging in our society: investments in maintaining and restoring the health of the

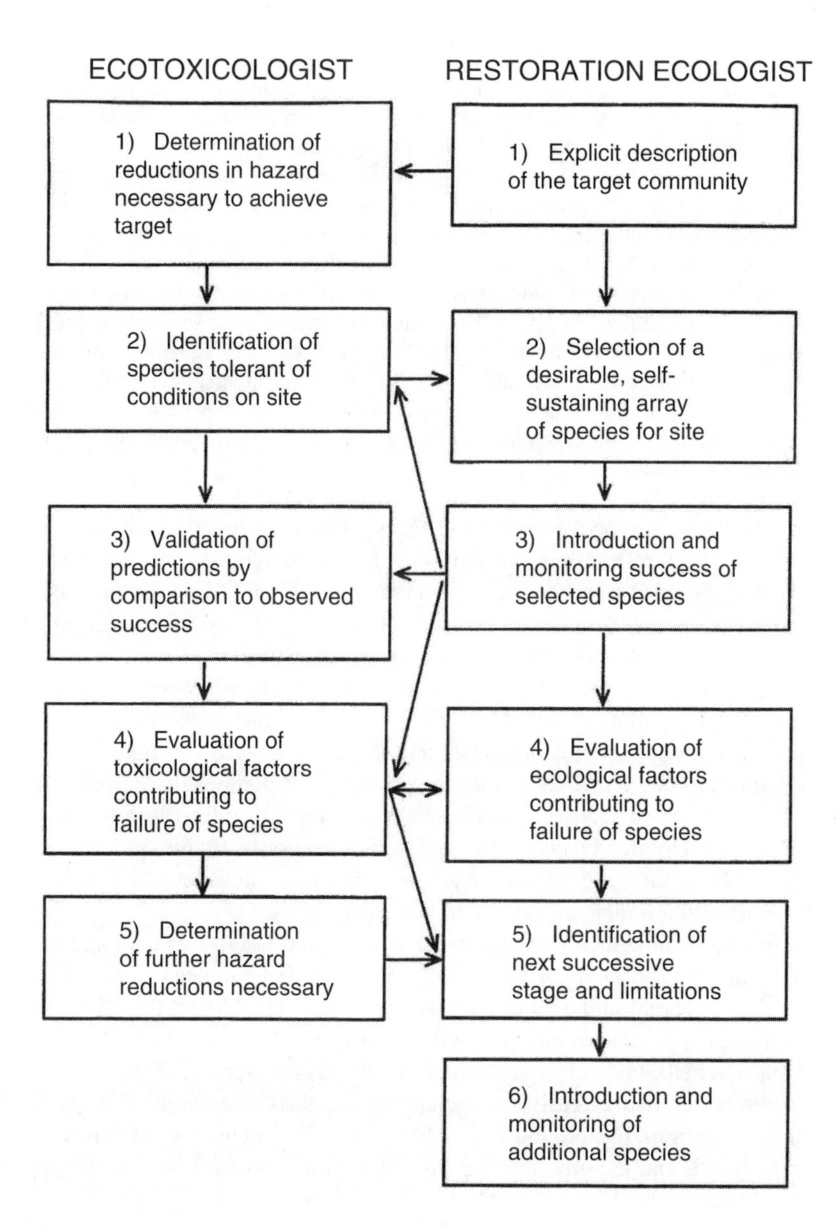


**Figure 38.4** Interactions between restoration ecologists and ecotoxicologists charged with rehabilitating a hazardous waste site.

environment can now be seen as responsible investments for the future." Unfortunately, the present governmental structure of programs and policies related to the environment is fragmented and will probably not adequately approach restoration as a challenge requiring management of rather large systems (i.e., drainage basins, etc.). President Bush's aspiration requires a national, but not solely federal, ecosystem-restoration strategy to meet the changing societal demands on natural resources.

An article in *U.S. News and World Report* (page 40, April 13, 1992 issue) notes that, in the former Soviet Union, economic growth was worth any price. The price has been enormous, even though the order of magnitude is not known. In terms of financing, every country of the world, developed or developing, has sacrificed ecological capital to spur economic development. American citizens decry the loss of the Amazon rain forests but are not equally vocal about the loss of their own forests in the Pacific Northwest and elsewhere. As is the case with medicine, preventing disease is far less costly than curing disease. Ironically, widespread development and implementation of the field of ecotoxicology may only occur when the cost of restoring ecosystems damaged by inattention to this field are fully and robustly documented. The problem, of course, is that, for such things as old-growth forests and a number of other ecosystems, restoration is unlikely to occur in the lifetime of humans now living and may not be possible in the lifetimes of any future generations. Nevertheless, by starting now to restore ecosystems damaged by hazardous materials, especially in those instances such as Russia where extreme contamination has occurred, the economics of restoration will be better understood than it is now. The basic question is where to acquire funds to begin the restoration effort.

After funding is available, a number of elements deserve attention (Chapter 8 of the National Research Council report[8] discusses these for aquatic ecosystems in broad general terms).

1. A national restoration strategy should be directed to broad-based and measurable goals.
2. A national aquatic ecosystem assessment process should monitor the achievements of the nation's goals for wetlands, rivers, and lakes.
3. Policies and programs for aquatic ecosystem restoration should emphasize a landscape perspective.
4. Restoration policies and individual restoration projects should be designed and executed according to the principles of adaptive planning and management.
5. Evaluation and ranking of restoration alternatives should be based on an assessment of opportunity costs rather than a traditional cost-benefit analysis.
6. A single definition of restoration should be incorporated into all appropriate national legislation.
7. A national restoration strategy should allocate leadership to the central government for landscape restoration of national significance and should rely on nongovernmental and governmental units to coordinate restoration programs in local areas.
8. The central government should initiate an interagency and intergovernmental process to develop a unified national strategy for aquatic ecosystem restoration.
9. The development of a unified national program for aquatic ecosystem restoration should be facilitated, and then maintained, under the leadership of a single responsible organizational unit.
10. Current and proposed federal programs should exploit available opportunities for ecosystem restoration.
11. Government should establish a national aquatic ecosystem restoration fund.
12. Reliance should be increased on local and environmental restoration boards for program planning, synthesis, and leadership.
13. Central governments should allow states and local governments to trade governmental funds designated for development, construction, maintenance, and major repair of water projects and to obtain funds instead for aquatic restoration programs.
14. The U.S. government should authorize expansion of the Agricultural Wetlands Reserve Program with funds from Farm Program Cost Savings.
15. The U.S. government should encourage water-pollution credit-trading programs to finance aquatic-ecosystem-restoration programs.
16. The central government and state government should encourage the trading of water rights to promote aquatic ecosystem restoration.
17. Governmental and state agencies should cooperatively design landowner-finance regional restoration projects.

Although these elements relate to aquatic ecosystems, relatively slight modification could include terrestrial ecosystems. Financing, in short, will come from a variety of sources, not the least important of which would be the savings resulting from integrated environmental management as the replacement for the fragmented and relatively uncoordinated present environmental management. Unfortunately, Russia and some other countries are examples of what happens when the life-support system is damaged for short-term gains in the economic life-support system. The vast and expensive military organization maintained by the former Soviet Union contributed to, rather than protected it from, many dangers.

## 38.6 DEVELOPMENT OF AN INTEGRATED ENVIRONMENTAL MANAGEMENT SYSTEM

Ultimately, society will determine that preventing damage is less expensive than restoring damaged ecosystems. Thus, preventative measures will likely become much more important than they now are. Additionally, if the economic evaluation uses opportunity analysis instead of the traditional cost-benefit analysis, many more situations will occur in which ecotoxicologists and restoration ecologists should interact to provide the evidence on which the opportunity-analysis decisions are made.

Integrated environmental management may be defined as proactive or preventative measures that maintain the environment in good condition for a variety of long-range, sustainable uses.[31] Alternatively, integrated environmental management may be regarded as coordinated control, direction, or influence of all human activities in a defined environmental system to achieve and balance the broadest possible range of short- and long-term objectives. Sometimes, a course of action becomes clearer by stating what it is not. Environmental management is NOT fragmented decision making so that only one use is considered at a time. Short-range goals that benefit a single group are not preferred.

The more obvious benefits of integrated environmental management are:

1. Long-term protection of the resource
2. Enhanced potential for nondeleterious multiple use
3. Reduced expenditure of energy and money on conflicts over competing uses and the possibility of redirecting these energies and funds to environmental management
4. More rapid and effective rehabilitation of damaged ecosystems to a more usable condition (more ecosystem services provided)
5. Cost effectiveness

Resource managers have long recognized that the institutional problems associated with managing natural resources are invariably more aggravating and intractable than the scientific and technical problems. Typically, it is not lack of methodology that impedes more effective use of natural systems (although methodology could certainly be improved), but rather the fact that many institutions (each charged with a fragment of resource use or management) fail to integrate system-management responsibilities. This situation was an inevitable consequence of a policy that made specialization a dominant theme in science and engineering. The field of human medicine has recently been redirected toward a more holistic approach, even though specialists still play a critical role. Environmental management might well adopt a similar strategy.

The barriers to integrated environmental management are indeed formidable! The 24 illustrative barriers used by Cairns[30] are listed here in an abbreviated form.

1. Institutions of higher learning are primarily reductionist, not integrative. Students graduating from these educational institutions have had their attention directed primarily toward specialization and reductionist science, with very little, if any, time spent on integrative science.
2. Integrated environmental management takes time, and budgeting this time is often neglected.
3. Turf battles run rampant in many organizations, including universities and colleges.
4. Integrated environmental management is often viewed as a threat to job security.
5. Many individuals and organizations are unwilling to compromise.
6. Short-term profits are too enticing.
7. There is a "what has posterity done for me" attitude.
8. Issues are not simple "good guys/bad guys" for decision makers.
9. The uncertainty of the outcome is often unacceptably high, for example, predicting the precise environmental benefits of reducing stack emissions from fossil fuel power plants.
10. At the global level, developing countries aspire to material benefits per capita now enjoyed by developed countries.

11. Changes in lifestyle (recycling, lower per capita use of energy, etc.) are strongly resisted by some individuals.
12. Specialists feel more comfortable working with "their own kind" rather than alien (from other disciplines) specialists.
13. Environmental despoilers fear the general public will not have the same value system.
14. The present use is considered a "right" or entitlement not open to discussion or compromise (e.g., prior appropriation water rights of western states in the United States).
15. Society is oriented toward growth rather than maintenance.
16. Change is only acceptable in a crisis, not always when the problem is more manageable.
17. There is a general fear that management authority will be abused.
18. Specialists fear criticism resulting from oversimplification sometimes needed for effective communication in integrated environmental management.
19. The belief that all systems are too complex to permit any prescriptive (read: standard methods and procedures) legislation or professional endorsement halts some scientists in their attempts to resolve complex issues.
20. People "turn off" when they face complex issues requiring a substantial effort to understand them properly.
21. Technical information is viewed as inadequate at the present time.
22. Nonspecialists have difficulty determining which evidence is credible, especially when there are specialists taking strong positions that are diametrically opposed.
23. The number of professionals skilled in integrated environmental management or some component of integrated environmental management is inadequate.
24. The political process is oriented toward polarized issues rather than integrated environmental management.

## 38.7 CONCLUSIONS

Ecological restoration in many situations could be facilitated by the interaction of restoration ecologists and ecotoxicologists. It is a *sine qua non* that human society must be motivated to restore at least some of the damaged ecosystems. The National Research Council report[8] provides some case histories for aquatic systems that clearly demonstrate such support. For terrestrial systems, Janzen[32] provides a splendid example for a developing country, and the two-volume set of which the Janzen chapter is a part provides other examples. Illustrative examples of where the relationship between restoration ecology and ecotoxicology might be most effective are rivers and other parts of the hydrologic system where there has been a long-term cumulative impact of hazardous materials or an unexpected spill of hazardous materials. For terrestrial systems, the Superfund sites in the United States, where accumulations of hazardous materials pose a threat to human health and the environment, provide another example.

Although this section of the *Handbook* emphasizes the relationship between restoration ecology and ecotoxicology, it would be most unfortunate if there were a perception that the interaction of these two specialties alone would be adequate for resolving the problems just described. A large number of other disciplines should be involved for a truly integrated problem-solving team capable of providing a long-term solution to a complex multivariate problem. Illustrative areas of competence to be included would be urban and regional planners, economists, sociologists, chemists, groundwater and other types of hydrologists, and a variety of other disciplines. Decision analysis[33] should show how each type of information would affect a given decision, and, of course, if not too difficult or impossible, what type of information is clearly inappropriate for that decision. Regrettably, substantial amounts of data are sometimes gathered for long-term, large-scale projects because certain types of data are thought to be inherently valuable, although no effort is made to determine how the information, once gathered, will be used in the decision-making process. Determining how the information will used in the decision-making process also provides unmistakable indications of what type of working relationship should be developed between and among specialists.

There is persuasive evidence[6] that the earth's population will continue to increase well into the next century. Additionally, individual expectations of a better life are also rising. With a finite planet that is not expandable, this means the single-purpose use for land and water areas acceptable during frontier days is no longer possible. Each individual and nation will have to remember that it is part of a larger system and will often be required to modify its behavior so that ecological capital is not destroyed and, better yet, so that ecological capital is generated. Long-term sustainable use of the planet requires nothing less. Placed in this context, developing a relationship between restoration ecologists and ecotoxicologists is clearly only a necessary first step in the development of a larger integrated environmental management strategy. Nevertheless, if these first steps in developing a relationship between what have been until now more or less independent professions fail, it is highly unlikely that the ultimate complex interdisciplinary teams will be effective. This process should begin in our educational institutions, which, regrettably, have hardly begun to acquaint students with integrative science. Some have made tentative beginnings, often in the face of considerable resistance from the faculty. If the educational system is to change as rapidly as the situation demands, the consumers (i.e., students) and society as a whole must exert some pressure on the educational system.

## REFERENCES


1. Myers, N. and Kent, J., *Perverse Subsidies: Tax $s Undercutting Our Economies and Environments Alike*, International Institute for Sustainable Development, Winnipeg, Canada, 1998.
2. Wackernagel, M. and Rees, W., *Our Ecological Footprint*, New Society Publishers, Gabriola Island, British Columbia, Canada, 1996.
3. Raffensperger, C. and Ticknor, J., *Protecting Public Health and the Environment: Implementing the Precautionary Principle*, Island Press, Washington, D.C., 1999.
4. National Research Council, *Ecological Indicators for the Nation*, National Academy Press, Washington, D.C., 2000.
5. National Research Council, *Global Change: Ecosystems Research*, National Academy Press, Washington, D.C., 2000.
6. Ehrlich, P. R. and Ehrlich, A. H., *Healing the Planet*, Addison-Wesley Publishing, Reading, MA, 1991.
7. Cairns, J., Jr., Editorial: Application factors and ecosystem elasticity: The missing connection, *Environ. Toxicol. Chem.*, 10, 235, 1991.
8. National Research Council, *Restoration of Aquatic Ecosystems: Science, Technology, and Public Policy*, National Academy Press, Washington, D.C., 1991.
9. Karr, J. R., Fausch, K. D., Angermeier, P. L., Yant, P. R., and Schlosser, I. J., *Assessing Biological Integrity in Running Waters; A Method and Its Rationale*, Illinois Natural History Survey Special Publ. 5, Champaign, IL, 1986.
10. Schaeffer, D. J., Herricks, E. E., and Kerster, H. W., Ecosystem health: Measuring ecosystem health, *Environ. Manage.*, 12, 455, 1988.
11. Rapport, D. J., What constitutes ecosystem health? *Persp. Biol. Med.*, 33, 120, 1989.
12. Costanza, R., Toward an operational definition of ecosystem health, in *Ecosystem Health: New Goals for Environmental Management*, Costanza, R., Norton, B., and Haskell, B., Eds., Island Press, NY, 1992.
13. Constanza, R., Norton, B., and Haskell, B., Eds., *Ecosystem Health: New Goals for Environmental Management*, Island Press, NY, 1992.
14. Harte, J., Torn, M., and Jensen, D., The nature and consequences of indirect linkages between climate change and biological diversity, in *Global Warming and Biodiversity*, Peters, R. and Lovejoy, T. E., Eds., Yale University Press, New Haven, CT, 1992.
15. Whittaker, R. H., Recent evolution of ecological concepts in relation to the eastern forests of North America, *Am. J. Bot.*, 44, 197, 1957.
16. Lovelock, J. E., The earth as a living organism, in *Biodiversity*, Wilson, E. O., Ed., National Academy Press, Washington, D.C., 1988, 486.

17. Cairns, J., Jr., Editorial: Will the real ecotoxicologist please stand up? *Environ. Toxicol. Chem.*, 8, 843, 1989.
18. Cairns, J., Jr., Paradigms flossed: The coming of age in ecotoxicology, *Environ. Toxicol. Chem.*, 11, 3, 285, 1992.
19. Norton, S., McVey, M., Colt, J., Durda, J., and Hegner, R., Review of Ecological Risk Assessment Methods, EPA/230–10–88–041, National Technical Information Service, Springfield, VA, 1988.
20. Hunsaker, C. T. and Carpenter, D. E., Eds., Environmental Monitoring and Assessment Program: Ecological Indicators, U.S. Environmental Protection Agency, Office of Research and Development, Research Triangle Park, NC, 1990.
21. Kolkwitz, R. and Marsson, M., Okologie der Pflanzlichen Saprobian, *Ber. dt. Bot. Ges.*, 26, 505, 1908.
22. Cairns, J., Jr., Dickson, K. L., and Maki, A. W., Eds., *Estimating the Hazard of Chemical Substances to Aquatic Life*, STP657, American Society for Testing and Materials, Philadelphia, 1978.
23. National Research Council, *Testing for Effects of Chemicals on Ecosystems*, National Academy Press, Washington, D.C., 1981.
24. Cairns, J., Jr., Are single species toxicity tests alone adequate for estimating environmental hazard?, *Hydrobiologia*, 100, 47, 1983.
25. Cairns, J., Jr., The case for simultaneous toxicity testing at different levels of biological organization, in *Aquatic Toxicology and Hazard Assessment: Sixth Symposium,* STP802, Bishop, W. E., Cardwell, R. D., and Heidolph, B. B., Eds., American Society for Testing and Materials, Philadelphia, 1983.
26. Ryder, R. A. and Edwards, C. J., A Conceptual Approach for the Application of Biological Indicators of Ecosystem Quality in the Great Lakes Basin, Report to the Great Lakes Science Advisory Board of the International Joint Commission (U.S. and Canada). Windsor, Ontario, Canada, 1985.
27. Kimball, K. D. and Levin, S. A., Limitations to laboratory bioassays: The need for ecosystem-level testing, *BioScience*, 35, 165, 1985.
28. Ehrlich, P. R. and Ehrlich, A. H., *The Population Explosion*, Simon and Schuster, New York, 1990.
29. Cairns, J., Jr., Is restoration practical?, *Restor. Ecol.*, March 3–7, 1993.
30. Lemons, J., What is an environmental professional?, *Nat. Assoc. Environ. Prof. Newsl.,* 18, 7, 1993.
31. Cairns, J., Jr., The need for integrated environmental systems management, in *Integrated Environmental Management*, Cairns, J., Jr. and Crawford, T. V., Eds., Lewis Publishers, Chelsea, MI, 1991, 5.
32. Janzen, D. H., Guanacaste National Park: Tropical ecological and biocultural restoration, in *Rehabilitating Damaged Ecosystems, Vol. II*, Cairns, J., Jr., Ed., CRC Press, Boca Raton, FL, 1988, 143.
33. Maguire, L. A., Decision analysis: An integrated approach to ecosystem exploitation and rehabilitation decisions, in *Rehabilitating Damaged Ecosystems, Vol. II,* Cairns, J., Jr., Ed., CRC Press, Boca Raton, FL, 1988, 105.

## SECTION V

# Special Issues in Ecotoxicology

CHAPTER 39

# Endocrine Disrupting Chemicals and Endocrine Active Agents

**Timothy S. Gross, Beverly S. Arnold, María S. Sepúlveda, and Kelly McDonald**

## CONTENTS

## 39.1 INTRODUCTION AND HISTORICAL BACKGROUND

It has been established that a wide variety of anthropogenic (man-made) chemicals in the environment are capable of modulating and adversely affecting or disrupting endocrine function in vertebrate organisms.[1–13] The physiological effects of exposure to these chemicals have been termed "endocrine disruption" and the active compounds labeled as "endocrine-disrupting chemicals" (EDCs) or "endocrine-active-agents." Endocrine disruption has been defined by the U.S. Environmental Protection Agency (EPA)[12] as the action of "an exogenous agent that interferes with the production, release, transport, metabolism, binding, action, or elimination of natural hormones in the body responsible for the maintenance of homeostasis and the regulation of developmental processes." This definition was further expanded by the U.S. EPA Endocrine Disruption Screening and Testing Advisory Committee (EDSTAC)[14] to indicate that these effects are "adverse" and may involve a wide assortment of endocrine-mediated functions and potential receptor-mediated events. Indeed, effects may involve the steroid receptor superfamily, including the sex steroids, thyroid hormones, and adrenal hormones, as well as hypothalamic-pituitary and other protein hormones.

The physiological processes regulated by the endocrine system are diverse and numerous. Likewise, the mechanisms of action and effects of potential EDCs are equally diverse (see Figure 39.1). Receptor-mediated events involve EDCs acting as hormone mimics (agonists or antagonists) and adversely impacting hormone synthesis, catabolism, secretion, transport, and signal transduction. Examples of nonreceptor-mediated modes of EDC action include altered enzyme function and selective toxicities for endocrine-active or target tissues, whereas altered gene expression and induction of oxidative stress are types of receptor mediated events. EDCs may also act by altering developmental processes, often producing multigenerational effects.

Endocrine-active anthropogenic chemicals are also numerous and diverse (see References 1–13). Evidence for endocrine-disrupting effects due to these chemicals comes from a diverse array of

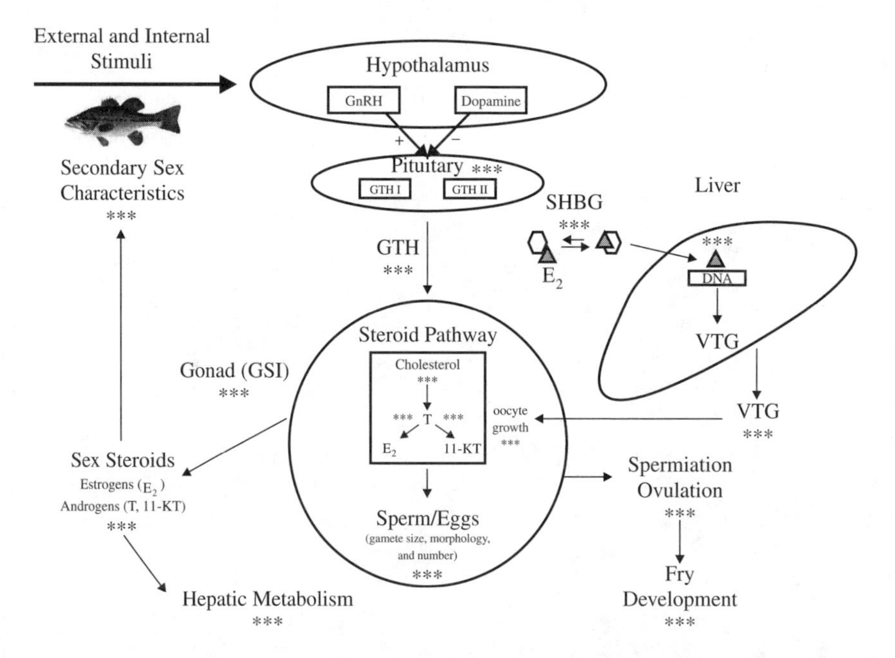


**Figure 39.1** Schematic representation of the hypothalamic-pituitary-gonad-liver axis of teleost fishes. Asterisks denote areas at which EDCs can exert their effects. In general, this model is also applicable for other oviparous vertebrates. Abbreviations: GnRH (gonadotropin releasing hormone); GTH (gonadotropin hormone); GSI (gonadosomatic index); SHBG (serum binding hormone globulin); VTG (vitellogenin); T (testosterone); $E_2$ (17β estradiol); 11KT (11-ketotestosterone).

reports involving multiple vertebrate taxonomic groups, limited invertebrate taxa, and results from both *in vitro* and *in vivo* studies. Reported effects of EDCs have included effects at multiple levels of biological organization including molecular, biochemical, cellular, tissue, and organismal. However, few reports have documented effects at the population level and higher. In addition, most studies have focused upon reproductive effects; however, effects on growth, metabolism, and thyroid and immune function have also been noted. This chapter summarizes the current evidence for the endocrine-disrupting effects of specific chemicals and chemical classes in vertebrate wildlife with a discussion on potential mechanisms/modes of action.

### 39.1.1 General and Comparative Endocrinology

To fully understand the mechanisms by which anthropogenic or natural EDCs may modulate endocrine function, normal functioning of the endocrine system must be understood. Indeed, an assessment of the risk of potential EDC exposures and effects requires critical information from a variety of disciplines, including endocrinology, and an understanding of the variation among and within vertebrate classes. The following section is a brief overview of vertebrate endocrinology and the hormones that may be involved in endocrine modulation or disruption.

The endocrine system is a collection of hormone-secreting cells, tissues, and ductless glands (e.g., pituitary, thyroid, adrenal, and gonads) that play an important role in growth, development, reproduction, and homeostasis. Tissues of the endocrine system synthesize and secrete hormones that influence virtually every stage of the life cycle of an organism, from gametogenesis and fertilization, through development into a sexually mature organism and senescence. Endocrinology is the study of tissues that secrete hormones into the blood and the subsequent effects hormones have on target tissues. Hormones are released into the extracellular environments and affect neighboring cells (paracrine control), the emitting cell (autocrine control), or other target tissues (endocrine control). Some nerve cells also release hormones into the blood (neuroendocrine control) or into extracellular fluid for communication with other nerve cells or nonnerve cells (neurotransmission). Pheromones are hormones secreted into the external environment for communication with other individuals or species. In addition, there are several hormones that act through more than one of these chemical-signaling modes.

Figure 39.1 summarizes the hypothalamic-pituitary-gonadal axis for fish as an example of the endocrine system, its diverse control over reproductive and developmental processes, and sites at which EDCs may exert endocrine-disrupting effects. In general, this model is also applicable to other oviparous vertebrate species including birds, amphibians, and reptiles.

The vertebrate hypothalamus and the pituitary gland (or hypophysis) have an essential role in regulating endocrine and nonendocrine target tissues.[15–17] The hypothalamus and pituitary are functionally and anatomically linked, forming the hypothalamic-pituitary axis. In mammals, the pituitary is composed of four anatomically and functionally distinct regions: the adenohypophysial *pars distalis* and *pars intermedia*, and the neurohypophysial median eminence and *pars nervosa*. In fish, the *pars distalis* is additionally separated into two regions that contain different cell types and produce different hormones.[18] The pituitary gland of amphibians, birds, and reptiles is similar to the mammalian pituitary gland.[16] Indeed, the basic arrangement of the hypothalamic-pituitary axis is essentially the same in all vertebrate groups, with the exception of teleost fishes, which lack a median eminence.[16]

The hypothalamus directly controls pituitary hormone secretion via the production and release of a number of peptide and nonpeptide hormones. These pituitary-tropic hormones are generally categorized as releasing hormones (RH) or release-inhibiting hormones (RIH), depending on their function. Hypothalamic hormones include corticotropin-releasing hormone (CRH), thyrotropin-releasing hormone (TRH), gonadotropin-releasing hormone (GnRH), growth-hormone-releasing hormone (GHRH), growth-hormone release-inhibiting hormone (GHRIH, somatostatin), and prolactin release-inhibiting hormone (PRIH). Other hypothalamic hormones also include critical neurotransmitters such as catecholamine and dopamine.[19]

The principal neurohypophysial (neuropituitary) hormones in mammals are arginine vasopressin and oxytocin. Birds, reptiles, and amphibians have structurally-related peptides: mesotocin and arginine vasotocin,[20] while fish in general have arginine vasotocin and isotocin or mesotocin, depending on the species.[16] These hormones are critical for milk secretion, oviductal and uterine contraction, renal water absorption, and vaso-constriction and dilation. In all vertebrates, these neurohypophysial hormones are produced in the hypothalamus and are transported to the pituitary, where they are stored until release into the bloodstream.

Hormones produced by the mammalian adenohypophysis are the pituitary-derived tropic hormones including growth hormone (GH), adrenocorticotropin (ACTH), melanotropin (MSH), thyroid-stimulating hormone (TSH), prolactin (PRL), and the gonadotropins — follicle-stimulating hormone (FSH), and luteinizing hormone (LH). Secretions of ACTH, TSH, and the gonadotropins (FSH and LH) are each regulated by negative feedback. Although structurally related counterparts for the adenohypophysial hormones have been identified in fish, amphibians, birds, and reptiles,[16] there are important differences in hormone actions across vertebrate groups. For instance, PRL is associated with reproduction and lactation in mammals but is an important osmoregulatory hormone in fish.[21] Although FSH and LH function similarly in mammalian and avian reproduction, reptiles do not synthesize an LH-like gonadotropin and instead utilize FSH to regulate gonadotropin-related functions.[15] In fish and amphibians, two different gonadotropins, GTH-I and GTH-II, have been identified that act similarly to mammalian FSH and LH, respectively.[17] GH generally regulates body and tissue growth; however, in nonmammalian vertebrates, it is also involved in osmoregulation. In mammals and birds, ACTH is responsible for stimulating the production of corticosteroids by the adrenal gland, which in turn plays a role in metabolism, ion regulation, and stress responses. The role of ACTH in fish and amphibians is less clear, however, and MSH may have similar properties in these taxonomic groups. Indeed, similarities in hormone structure may not necessarily represent similar hormone function in nonmammalian vertebrates.

GH is important for bone growth and as an anabolic hormone during development.[22] It has direct effects on a wide variety of tissues as well as indirect effects that are modulated by growth factors such as insulin-like growth factor-I (IGF-I).[22] In conjunction with thyroid hormones, GH is necessary for the development of a wide number of tissues ranging from cardiac[23] and skeletal muscle,[24] to bone[25] and brain development.[26] In nonmammalian species, GH probably functions in a similar manner; however, less is known about growth hormone in fish, amphibians, and reptiles.

The adrenal glands, thyroid gland, and gonads are all directly regulated by the pituitary gland.[16] Thyroid hormones, which are produced by thyroid glands, and steroids produced by the adrenal cortex and gonads can indirectly inhibit their own secretion by inhibiting the release of pituitary and hypothalamic hormones (negative feedback). In response to TSH, the thyroid gland produces two hormones, triiodothyronine ($T_3$) and tetraiodothyronine ($T_4$). In mammals, $T_3$ and $T_4$ have important effects on metabolism and development.[16] Thyroid hormones also play an essential role in fish and amphibian metamorphosis. Indeed, thyroid hormones determine the timing of developmental processes, and metamorphosis is almost entirely controlled directly by thyroid hormones.[16,27–29] Some metamorphic processes that are under the control of thyroid hormones include the migration of the eye and dorsal fin growth in fish,[30,31] amphibian tail and gut resorption,[27,32,33] restructuring of the amphibian head,[34,35] amphibian limb development,[36] and amphibian gill resorption.[37] Thyroid hormones also have important roles during fish smoltification.[38–40]

The mammalian adrenal gland produces two important steroid hormones — aldosterone and corticosterone. Aldosterone plays an important role in the maintenance of sodium concentrations, and corticosterone is primarily involved in regulating blood glucose.[16] Adrenal steroids function similarly in birds but very differently in other nonmammalian vertebrates. In amphibians, aldosterone and corticosterone are equally effective as regulators of blood glucose, whereas in fish and reptiles, corticosterone serves to regulate blood glucose and sodium. While adrenal hormones have critical roles in all vertebrates, characterizations of their functions in nonmammalian vertebrates are limited, and interspecies differences have not been thoroughly evaluated.

In all vertebrate classes, gonadal function is dependent upon the hypothalamic-pituitary axis through the production of GnRH and gonadotropins.[16] In mammals, the gonadotropins include FSH and LH, which control different gonadal events. In females, FSH promotes ovarian follicular growth, and LH induces ovulation. Both gonadotropins are also required for normal estrogen synthesis: LH stimulates the synthesis of androgens, and FSH stimulates aromatization of androgens to estrogen. In males, FSH promotes spermatogenesis, and LH promotes steroidogenesis and spermiation. The mammalian gonad also produces the peptide hormone inhibin, which feeds back to inhibit FSH production. In both males and females, the pulsatile release of GnRH is regulated by the feedback of high circulating levels of androgens and estrogens. In birds, gonadotropins function in a similar manner; however, reptiles do not synthesize an LH-like gonadotropin and utilize FSH to regulate all gonadotropin-related functions.[15] In fish and amphibians, two different gonadotropins — GTH-I and GTH-II — have been identified, and they act similarly to mammalian FSH and LH, respectively. GTH-I is involved in gonadal development, gamete production, and vitellogenesis, a process that involves the hepatic synthesis of yolk protein precursors, vitellogenin (VTG), under the stimulus of estrogens.[16,17] GTH-II stimulates the final stages of oocyte maturation as well as ovulation in females and spermiation in males.

In general, gonadotropins exert effects on vertebrate gonads by binding to specific receptors. The primary gonadal response to gonadotropins is the synthesis and secretion of assorted sex steroids. In all vertebrates, the primary reproductive sex steroids include androgens [e.g., testosterone (T), 11-ketotestosterone (11KT), androstenedione (A), dihydrotestosterone (DHT)], estrogens [estradiol ($E_2$), estrone ($E_1$), estriol ($E_3$)], and progestins [progesterone ($P_4$), dihydroxyprogesterone (DHP)]. Gonadal steroid hormones are involved in every aspect of reproduction, from sex determination to the control of courtship behaviors and the development of secondary sex characteristics.

Sex steroids also play an important role in brain development. For example, in mammals, $E_2$ and DHT are involved in normal sexual differentiation of the brain.[41–43] Although reproductive function is regulated and modulated by sex steroids in all vertebrates,[16,28,44–47] there are distinct functional differences that must be noted. Indeed, functional differences in sex steroids are most evident for fish, amphibians, and reptiles, with significant differences also existing within each of these taxonomic classes.[17] For instance, the primary androgen for spermatogenesis in mammals, birds, and reptiles is T, but in many fish and some amphibians the critical androgen for spermatogenesis is 11KT. Preliminary results from our laboratory would suggest that 11KT might not be the predominant androgen in live-bearing fish (such as mosquito fish *Gambusia holbrooki*). $E_2$ is the sex steroid responsible for oocyte growth and maturation in all vertebrates; however, it also regulates and induces the synthesis of VTG in oviporous vertebrate species.[16,17,48,49]

Progestins are critical to pregnancy in mammals but function in reptiles and birds in post-ovulatory events such as the regulation of eggshell deposition. In fish, progestins are responsible for final egg maturation prior to oviposition. Gonadal sex steroids can also have dramatic effects on sex differentiation in fish, amphibians, and reptiles, effects that are not observed in birds or mammals.[17,28,50] When applied early during development, sex steroids can cause sex reversal in fish, amphibians, and reptiles. Therefore, the genetic sex of the individual can be different from the phenotypic sex. Finally, the effects of sex steroids on gonadal differentiation and sex reversal vary dramatically between species and across developmental stages, and therefore these differences need to be noted and considered in any study of potential EDC effects in vertebrate wildlife.

### 39.1.2 Mechanisms of Endocrine Modulation

There is significant evidence to suggest that a wide variety of anthropogenic chemical contaminants in the environment can disrupt or modulate endocrine function in a wide variety of vertebrate and some invertebrate organisms. However, information regarding the mechanisms that lead to these endocrine modifications is limited. It is, nonetheless, critical that mechanisms and modes of action for EDCs and endocrine-active agents be understood. Mechanisms of action are generally difficult to elucidate and are complicated by multiple factors including chemical properties, routes, timing, and lengths of exposure, as well as endocrine-system and species- and tissue-specific physiological differences. Furthermore, the integration of the nervous, endocrine, reproductive, hepatic, and other target systems, as well as multiple feedback regulatory pathways, adds to the complexity of understanding EDC mechanisms (see, for example, Figure 39.1).

Potential mechanisms of action for EDCs are diverse. EDCs may interrupt multiple pathways along the hypothalmic-pituitary–target-tissue axis, potentially disturbing the normal synthesis, transport, release, binding, action, biotransformation, or elimination of natural hormones in the body. EDCs may alter the hypothalamic-pituitary axis, which can have widespread effects through the disruption of endocrine functions downstream of the hypothalamus. There is increasing evidence that EDCs may disrupt endocrine function by influencing the regulation/release of the pituitary-tropic hormones. Indeed, polychlorinated biphenyls (PCBs) have been shown to interfere with the neurotransmitters that control GnRH secretion, resulting in decreased GnRH production as well as subsequent reductions in gonad size and plasma concentrations of sex steroids.[51] In mammals, neonatal exposure to diethylstilbestrol (DES) or dichlorodiphenyltrichloroethane (DDT) results in both reduced GnRH and LH production.[51] These results demonstrate that interference at one site along the hypothalmic-pituitary axis can affect multiple downstream events. Furthermore, the hypothalamus and pituitary are regulated by the feedback of hormones from several other endocrine-active tissues; therefore, alterations in different hormone concentrations can also affect hypothalmic and pituitary function.

EDCs can exert effects and disrupt the function of other endocrine tissues and hormones downstream of the hypothalamus and pituitary. Hormones are synthesized by specific endocrine tissues, secreted into the bloodstream, and transported by binding proteins to target tissues to interact with receptors, elicit responses, and be metabolized or degraded. EDCs can block or enhance the function of hormones by interfering with any one or several of these critical steps. For instance, EDCs may interfere with hormone synthesis, thereby altering endocrine activity by directly affecting the availability of specific hormones or critical precursors.[28,52] Failure to synthesize appropriate hormones can result from either an alteration in the biosynthetic enzymes and in the availability of precursor molecules. The initial, as well as rate-limiting, step in the biosynthesis of hormones may often be affected. EDCs can inhibit the uptake of critical precursors and the subsequent conversion to hormone products.[53–55]

EDCs can alter the rate at which hormones are metabolized. The cytochrome P450 (CYP450) monooxygenases constitute a super family of enzymes that play essential roles in both the synthesis (steroidogenesis) and metabolism of steroid hormones. Many of these enzymes appear to be sensitive to EDCs.[52,56–59] EDCs can affect the number or activity of specific monooxygenases, thereby affecting the rate of hormone metabolism and clearance. Since specific CYP450 enzymes — like CYP1A — are also responsible for metabolizing foreign compounds — like EDCs — EDC stimulation of CYP1A and other monooxygenases that hydroxylate them prior to their elimination may in turn contribute to increased clearance of sex hormones by inducing other monooxygenase activities.[60] EDCs have also been reported to increase the activity of several other microsomal enzymes including aminopyrine demethylase, glucuronyl transferase, and p-nitroreductase.[61,62] Some EDCs may also induce hormone-like effects due to alternating rates of degradation. For example, many synthetic hormones, such as ethynyl estradiol ($EE_2$), a synthetic estrogen used in birth control pills, are not degraded readily by the enzymes that normally metabolize the endogenous hormones.[63] EDCs can also interfere with the binding of hormones to transport proteins, preventing their delivery to target tissues.[64,65] The absence of available binding proteins may result in both faster uptake or increased degradation of free-circulating hormones.[66–68] For example, the sex-hormone-binding globulin (SHBG) has high affinity for both T and $E_2$, which is necessary to prevent degradation and clearance of these hormones as well as enable their transport to target tissues.[69] EDCs, which mimic estrogens or androgens, may bind to these globulin proteins and displace the endogenous sex steroids, thereby increasing the elimination rates for endogenous hormones. Although several studies suggest that globulins may also facilitate the transport of EDCs to target tissues,[69] the greater binding affinity of globulins for endogenous hormones probably limits this process.[70]

EDCs may bind to hormone receptors and either activate (agonize)[71–73] or inhibit (antagonize)[74] receptor function. Indeed, many studies have focused on EDCs as hormone-mimics and the potential for these compounds to interact with hormone-specific receptors. Potential EDCs have been evaluated extensively for their ability to bind to the estrogen receptor (ER). Estrogens normally bind to the ER located in the nucleus of target cells. The $E_2$-bound ER has a high affinity for DNA sequences called estrogen response elements (ERE). After binding the ERE, the ER-DNA complex interacts with various transcription factors, chromosomal proteins, and regulatory factors in order to induce or inhibit the transcription of specific genes and enable endocrine-specific response. EDCs can block or enhance the function of a hormone or endocrine target tissue by interfering with any one or several of these critical steps. Although the potential estrogenic activities of EDCs have overshadowed studies of other receptor-mediated EDC activities, EDCs that act as androgens or antiandrogens via interaction with the androgen receptor (AR) have also been noted.[74–76]

Unlike the ER, which has an $E_2$ specific response element, the response element for the AR is shared with other steroid receptors including the glucocorticoid (GR), progesterone ($P_4$), and mineralocorticoid (MR) receptors. Therefore, EDCs that have androgenic activities may exert broader effects than those attributed to a simple androgen mimic. EDCs may also interact with a wider variety of receptors important for endocrine function. For example, some EDCs (e.g., 2,3,7,8-tetrachlorodibenzo-dioxin [TCDD] and other planar hydrocarbons) are reported to have antiestrogenic activities by interacting with the aryl hydrocarbon receptor (AhR) rather than by competitively binding to the ER. The AhR is an intracellular receptor that is expressed by many different cell types and that functions as a transcription factor.[77,78] EDC interactions with the AhR may interfere with estrogen responses in a number of ways: by reducing $E_2$ binding to the ER,[79] by blocking the binding of the ER to the ERE,[80] by impairing nuclear translocation,[81] or by suppressing gene transcription.[82] These examples demonstrate the varied receptor-mediated activities of EDCs.

Endocrine-disrupting effects may also occur due to direct or indirect toxicities for specific endocrine-active or target tissues. For example, many lipophilic EDCs will accumulate primarily in fatty tissues, such as the liver and gonads, potentially interfering with the synthesis and mobilization of lipids and thereby inhibiting specific endocrine-related functions such as vitellogenesis. It is important to point out, however, that specific mechanisms or modes of action for most EDCs are not well elucidated or understood. This stems from the fact that mechanisms are often difficult to identify and are complicated by multiple factors including differences in EDC-specific properties, routes of exposure, and vertebrate class and species differences. Nonetheless, it is critical that mechanisms of action for EDCs and endocrine-active agents be understood in order that effects in wildlife be prevented and that appropriate screening and testing methods be developed.

## 39.2 SCREENING AND MONITORING FOR ENDOCRINE DISRUPTING CHEMICALS

Analytical methods have long been used to determine concentrations of chemical residues that persist in the environment (e.g., water, sediment) and accumulate in biota (e.g., tissue and body burdens). Although these approaches are useful for characterizing the presence and distribution of specific EDCs in the environment, they fail to indicate whether chemical exposures have biological consequences. The development of EDC-specific screening and monitoring procedures aid in the establishment of potential relationships between environmental EDC concentrations and biological responses. In the past decade, several *in vitro* and *in vivo* assays have been proposed that can be used to screen or monitor individual EDCs, specific EDC mixtures, or complex environmental mixtures for potential endocrine disrupting or modulating activity.

### 39.2.1 *In Vitro* Assays

Several *in vitro* assays have been described for evaluating potential endocrine-disrupting or modulating activities of EDCs.[75] These assays are based on several specific mechanisms of action for EDCs, including receptor binding, gene expression, cell proliferation, and cell differentiation.[83] Advantages of *in vitro* systems include low cost, high reproducibility, and the rapid analysis of large numbers of samples. These assays are also valuable for studying mechanisms of action of compounds, screening effects of mixtures, and detecting potential interaction effects. Results from these screening procedures can aid in the subsequent development and validation of assays. *In vitro* assays, however, generally lack ecorelevance because pharmacokinetics, biotransformation, and binding to carrier proteins may not be accurately represented. For example, some EDCs are activated or deactivated *in vivo* by enzymatic conversion during metabolism, conjugation, and excretion. These limitations must be considered when interpreting or applying results from *in vitro* screening tests.

Receptor-binding assays can be utilized to screen for and identify potential EDCs (which function via receptor-mediated pathways) since they can evaluate whether specific EDCs can bind to specific receptors. Depending on the receptor of interest, receptor-binding assays utilize either crude cell fractions such as plasma membranes, cytosol, or the nucleus. Cell fractions may be obtained from specific vertebrate organisms or from established cell lines, transformed cells,[84,85] or transfected cells.[86] Although *in vitro* receptor-binding assays are relatively simple and inexpensive to conduct, they do not necessarily reflect binding under *in vivo* conditions and are of very little use in screening for EDCs that operate by nonreceptor-mediated pathways. Finally, these assays do not differentiate between agonist and antagonist properties.

Additional *in vitro* assays have utilized the ability of EDCs to induce target-cell-specific proliferation and differentiation. For instance, MCF-7 cells, derived from human breast cancer cells, have been widely utilized for the development of the E-screen assay, which evaluates the ability of specific EDCs or EDC mixtures to both bind and express the ER[87] and the resultant cell proliferation as a response.[88–94] EDCs are identified as potential $E_2$ agonists if there is a significant increase in cell proliferation, which in turn is quantified by counting cell nuclei[92] or measuring other responses such as metabolic reductions. Although the E-screen assay has been extensively used as a screen for estrogenicity,[76,92,95] a positive response cannot be necessarily interpreted as an indicator for the presence of $E_2$ agonists. In addition, ER antagonists and antiandrogens are not detected using this assay, and thus a significant number of false negatives are common. Before a compound is identified as an EDC, positive responses with the E-screen assay should be confirmed by *in vivo* studies.

A number of additional *in vitro* cell-based expression assays have also been developed to measure receptor-dependent biological responses. Expression assays evaluate the induction or suppression of proteins by specific genes in response to potential receptor-mediated EDCs and mixtures. Measured protein endpoints for these receptor-specific expression assays include: VTG,[71–73,94,96,97] sex-hormone-binding globulins,[98] luciferase,[99] galactosidase,[100] and chloramphenicol acetyltransferase (CAT).[86] However, these assays are general and are not limited to the action of EDCs. Additional cell types/lines that have also been utilized for *in vitro* expression assays include fish hepatocytes,[71,73,94,98] MCF-7,[95,101] HeLA,[86,98] and yeast.[101] The types of cells used in expression assays are critical to any interpretations. Indeed, significant differences in responses between yeast-cell-based assays and mammalian-cell assays have been reported,[98] and sensitivities vary greatly.[100] Nonetheless, expression assays have several advantages as compared to other *in vitro* screening assays. Unlike receptor-binding or cell-proliferation assays, expression assays can be used to detect both agonists and antagonists.[86,99,102] Expression assays can also evaluate potential EDCs that influence many aspects of gene expression in addition to those that operate through receptor-mediated functions. Nonetheless, *in vitro* expression assays generally have high variability and lack ecorelevance.

### 39.2.2 *In Vivo* Assays

The effects of EDCs occur at many biological levels of organization including molecular, biochemical, organelle, cell, tissue, organism, population, community, and ecosystem. The use of a battery of biomarkers that reflect multiple biological levels of organization would enable a more thorough evaluation of both exposure and the potential mechanism of action. Although responses at the population level and higher are the most biologically ecorelevant, they are rarely utilized as biomarkers since these responses are complex, less specific, and require greater effort and time. Indeed, most of the current biomarkers are limited to the measurement of responses at the molecular, biochemical, cellular, and organism levels. *In vivo* assays for the identification of EDCs are not mechanism-dependent and provide results that are more environmentally relevant than *in vitro* assays. Indeed, *in vivo* assays rely upon either natural exposures or controlled exposures based on expected or predicted environmental exposures. *In vivo* assays for EDCs can detect effects on endocrine function, regardless of the mechanism of action, as well as identify a potential EDC that would not necessarily exhibit activity in an *in vitro* screening assay. Most importantly, *in vivo* screening assays both identify potential EDCs and enable the description and evaluation of potential effects.

*In vivo* assays for evaluating EDCs may involve the utilization of specific endocrine biomarkers as a way to evaluate potential effects. Widely used endocrine-endpoint-based *in vivo* assays have included the uterotropic assay, the Hershberger assay, and the thyroid-function assay. Although these assays were not originally designed for the evaluation or identification of EDCs, they have demonstrated the utility of *in vivo* assays for the identification of potential EDCs. The uterotropic assay utilizes prepubertal or adult ovariectomized female rats to assess uterine weight and histological responses to potential EDCs. The Hershberger assay evaluates androgenicity using androgen-dependent tissue (e.g., prostate and seminal vesicles) responses to potential EDCs. The thyroid-gland-function assay evaluates potential EDC exposures and the subsequent evaluation of plasma concentrations of $T_3$, $T_4$, and TSH.

Biomarkers that detect alterations at the biochemical and molecular levels are frequently utilized for *in vivo* EDC-screening assays.[103] Biochemical and molecular responses are generally the first detectable responses to an environmental change or stressor and can serve as early indicators of both exposure and effect. Aside from being highly sensitive changes at the molecular and biochemical level, they can sometimes be predictive of responses at higher levels of organization (tissue and organism levels). Examples of molecular-based *in vivo* EDC-screening assays include receptor analyses, transcriptional-based analyses, and differential display.[14] These assays are, in general, based on an analysis of specific molecular parameters for tissues collected following either natural or experimental exposures to potential EDCs. Although molecular-based *in vivo* assays are highly sensitive, they are difficult to validate and often lack ecorelevance. Examples of current biochemical-based *in vivo* EDC-screening assays include: measurement of VTG production[104] and systemic hormone concentrations (e.g., plasma sex steroids, $T_3$, and $T_4$). In fact, systemic concentrations of various hormones have been frequently utilized as biomarkers for EDCs in fish,[105–110] amphibians,[111,112] reptiles,[113,114] birds,[5,115–120] and mammals.[121] These procedures have broad application to all vertebrate classes since hormones, especially the steroid and thyroid hormones, are evolutionarily conserved across all vertebrate classes. However, it must be noted that the same hormones may differ in function significantly between and within vertebrate classes. For example, the primary androgen for spermatogenesis in mammals, birds, and reptiles is T, but in many fish and some amphibians, the critical androgen for spermatogenesis is 11KT.

VTG has been utilized as a bioindicator of potential exposure and effects of estrogenic EDCs in fish and other oviparous vertebrates.[96,122–124] This phospholipoprotein is produced by the liver under the control of $E_2$ in oviparous female fish, amphibians, reptiles, and birds.[111] Oviparous species have vitellogenic cycles that correspond to egg production. Potential EDCs, which mimic or alter endogenous $E_2$, may induce the expression of VTG. This assay has, in general, focused on responses in males, which do not exhibit clear vitellogenic cycles. However, it must be noted that low background levels of VTG are likely to be normal in males. Thus, an identification of a potential EDC by this method cannot be based solely on the presence of detectable VTG; it must additionally be based on a species-specific VTG response that is significantly increased above background levels.

Additional *in vivo* EDC-screening assays involve endpoints based on responses at the tissue and organism levels. Although these assays may have higher biological and ecological relevance, they are more variable and often specific to vertebrate classes or species. Examples of screening assays that rely on tissue-level responses include tissue somatic indices (e.g., gonadosomatic index-GSI), tissue histopathology, altered secondary sex characteristics,[125–128] and egg- and sperm-quality assessments.[94] *In vivo* assays that rely on organism responses may include assessments of egg numbers/ovarian development,[128–132] sexual maturity,[128] neonatal/embryonic mortality,[129–131,133–135] reproductive impairment,[108,136] and evaluation of egg hatchabilities[129,134,135,137] and nest numbers.[133,134] Population and ecosystem endpoints of reproductive success may include evaluation of pod size, age-class analyses, and population numbers.[135,138,139]

Valid *in vivo* screening procedures should provide information about EDC exposure and be indicative of expected or predicted physiological effects. In addition, *in vivo* biomarkers reflect the complex pharmacokinetic and metabolic factors that can affect EDC uptake and metabolism. It is important to keep in mind, however, that *in vivo* assays and endpoints are influenced by both physiological and environmental variables, which make it difficult to establish clear cause-and-effect relationships between responses and specific EDCs. Nonetheless, these assays are often the most useful for evaluating potential EDC effects and for the identification of environmentally relevant EDCs.

## 39.3 EDC EFFECTS: EVIDENCE FOR SPECIFIC CHEMICALS AND CHEMICAL CLASSES

The previous section reviewed many of the possible mechanisms by which environmental contaminants may alter endocrine function in fish and wildlife. The following section introduces several classes of environmentally relevant contaminants with reported or potential endocrine-disrupting activity in invertebrates and vertebrates. This review presents evidence for EDC effects for several specific chemical classes: polycyclic aromatic hydrocarbons, polychlorinated and polybrominated biphenyls, dioxins, organochlorine and other pesticides, complex environmental mixtures, and metals. For most chemicals, the specific mechanism of action is not well understood, and chemical structure does not necessarily indicate or suggest endocrine functionality, mimicry, or EDC activity (see Figure 39.2 as an example of chemical structures for several environmental estrogens). In fact, direct evidence of endocrine activity is often difficult to demonstrate and thus is generally absent. This review includes reports from a variety of laboratory and field studies that have explored the effects of EDCs in fish and wildlife and discusses the potential or suspected modes of action (MOAs) (see Table 39.1).

### 39.3.1 Polycyclic Aromatic Hydrocarbons (PAHs)

Polycyclic aromatic hydrocarbons (PAHs), whether of natural or anthroprogenic origin, are products of incomplete combustion of organic compounds and enter aquatic environments via oil spills, waste discharge, runoff, and dry or wet deposition. Although they are biodegraded in soils and water within weeks to months, the metabolites are often longer lasting and more toxic.

Birds can be exposed to PAHs through ingestion of contaminated food and water, by preening feathers, or through the skin in cases of oil spills. Petroleum hydrocarbons can also be absorbed through the eggshell.[140] In a review by Hoffman,[141] PAHs applied to the shells of eggs caused mortality and reduced hatchability. In studies reviewed by Fry,[142] exposure to petroleum oil

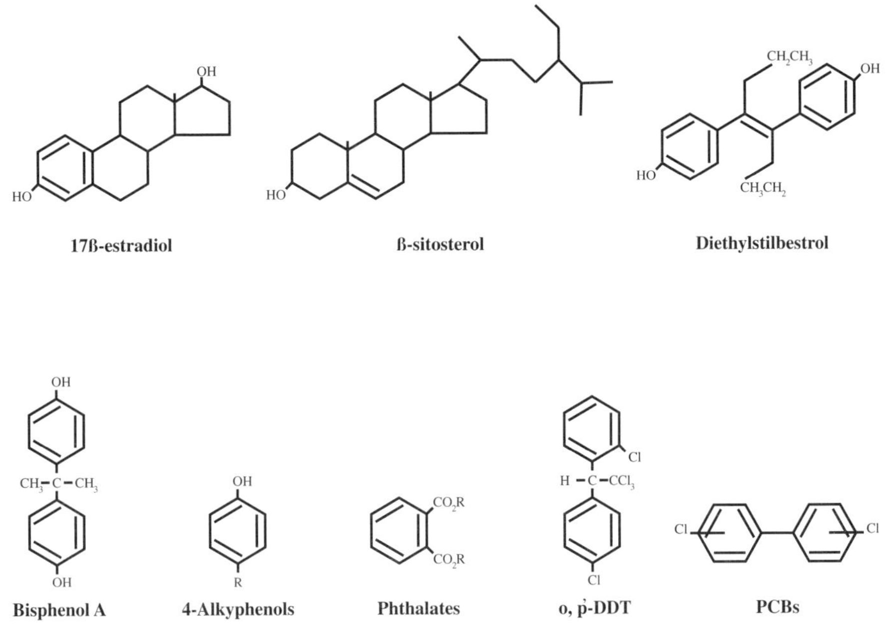

**17ß-estradiol** **ß-sitosterol** **Diethylstilbestrol**

**Bisphenol A** **4-Alkyphenols** **Phthalates** **o, p-DDT** **PCBs**

**Figure 39.2** Structures of some selected natural and environmental estrogens.

increased circulating corticosterone levels and disrupted reproduction through negative feedback to the hypothalamic-pituitary-gonadal system. However other studies[143,144] have shown decreased levels of plasma corticosterone, suggesting that ingested petroleum may interfere with adrenocortical function. Yolk formation may also be depressed after exposure to oil, resulting in a reduction in egg numbers.[145] Exposure to as little as 0.1-mL weathered crude oil (equivalent to 2.5 mL/kg body wt.) interferes with egg production, laying, incubation, and pair bonding. Field exposure of adult storm petrels *(Oceanodroma* sp.) with dependent chicks reduced foraging and feeding of chicks, resulting in reduced growth or death.[146]

Population studies with pigeon guillemots *(Cepphus columba)* after the *Exxon Valdez* spill indicated a decline in numbers for three consecutive years but no effects on reproduction.[147] Reproductive effects in the black oystercatcher (*Haematopus bachmani*) were noted.[148] There was a decrease in nonbreeding pairs, a decrease in egg size, and higher chick mortality, all of which directly related to the amount of oil present in the foraging territory. Birds exposed to oil may exhibit changes in adrenal hormone synthesis and elevated hepatic mixed oxidase activity, which may increase metabolic clearance of corticosterone.[140,149,150] In a laboratory study, female mallards (*Anas platyrhynchos*) that ingested crude oil hatched fewer live ducklings per pair.[116,140] In this study, there was evidence of suppression of follicular development, eggshell thinning, decreased hatchability, and reduced levels of plasma $E_2$, $E_1$, $P_4$, and LH in females. These results suggest that the oil acts on ovarian steroidogenesis, reducing positive feedback to the pituitary and causing a decline in LH, a delay in ovarian maturation, and reduced fertility.

Several field studies have documented altered reproductive activity in fish residing in PAH-contaminated waters. For instance, gonadal development was impaired and $E_2$ concentrations were depressed in English sole (*Parophyrs vetulus*) from highly contaminated areas of Puget Sound, Washington. Reproductive impairment was statistically correlated with elevated PAH concentrations, as measured by the presence of fluorescent aromatic compounds (FACs) in the bile of fish.[151,152] Other examples in which PAH exposure may have been related to endocrine alterations or reproductive dysfunction include altered ovarian development in plaice (*Pleuronectees platessa*) exposed to crude oil,[153] reduced GSI, increased liver size and ethoxyresorufin O-deethylase activity (EROD) in white sucker (*Catostomus commersori*) residing downstream of pulp and paper mills,[154]

**Table 39.1 Summary of Effects and Possible Modes of Action (MOAs) of Endocrine Disrupting Chemicals (EDCs) by Chemical Class and Taxa**

| Chemical | Taxa | Effects | Possible MOA | Sample Reference |
|---|---|---|---|---|
| **PAHs** | Birds | ↓Hatchability | DNA damage<br>Oxidative stress<br>ER agonist | 141 |
| | Fish | ↓GSI<br>Impaired gonadal development | DNA damage<br>Oxidative stress<br>ER agonist | 155<br>152 |
| **PCBs** | Mammals | Abortions & stillbirths | Antiestrogens<br>Act through Ah receptor | 175 |
| | Birds | ↓Eggshell thickness<br>↓Hatching success<br>↑Embryo mortality | Antiestrogens<br>Act through Ah receptor | 189, 195 |
| | Amphibians and Reptiles | ↓Sex hormones<br>↑Mortality & malformation rates | Unknown | 211, 212 |
| | Fish | ↓Spawning<br>↓Hatchability | Antiestrogens<br>Act through Ah receptor | 214, 215 |
| **PBBs** | Mammals | Fetotoxic and teratogenic<br>↑Menstrual cycles<br>↓Sex hormones | Unknown | 234, 606, 607 |
| | Birds | ↓Offspring viability<br>↓Hatchability | Unknown | 235 |
| **Organochlorine pesticides** | | | | |
| Cyclodienes | Birds | ↓Productivity | $E_2$ agonist | 288, 292–294 |
| | Reptiles | Sex reversal<br>↓Sex hormones | $E_2$ agonist | 113, 211 |
| | Fish | ↓Fertilization<br>↓Maturing oocytes<br>Altered spermatogenesis | $E_2$ agonist | 299, 304<br>302 |
| Chlordecone and Mirex | Mammals | Persistent estrus and vaginal changes | Weakly estrogenic | 310 |
| | Birds | ↓Clutch size, egg size, shell thickness, hatchability<br>↑Embryo malformations | | 309 |
| | Fish | Gonadal abnormalities | | 323 |
| DDT and derivatives | Mammals | Persistent vaginal estrus | Androgen antagonist | 325 |
| | Birds | Eggshell thinning<br>Reproductive problems<br>Population reduction | | 338, 343 |

| | | | | |
|---|---|---|---|---|
| | Reptiles and Amphibians | Sex reversal<br>↓Clutch viability sex<br>Altered plasma hormone levels<br>Abnormal gonadal morphology | Hormone mimicry<br>Estrogenicity | 113, 129, 135 |
| | Fish | ↑Oocyte atresia<br>↓Fecundity and fertility<br>↓Sex hormones | Hormone mimicry<br>Estrogenicity<br>Steroid receptors | 136, 346 |
| Hexachlorocyclohexane lindane | Fish | ↓Sex hormones | | 347–349 |
| Vinclozolin | Mammals | Feminization of males | Androgen antagonist | 76 |
| **PCDDs & PCDFs** **TCDD** | Mammals | Impairs sexual differentiation in male rats, delay in testicular descendent and puberty | Antiestrogenic through AhR receptor | 244, 245 |
| | Birds | Developmental alterations<br>Congenital deformities<br>Feminization | Antiestrogenic | 247, 248 |
| | Amphibians and Reptiles | Early metamorphosis<br>↑Frequency of deformities<br>alterations in sex ratios | Antiestrogenic | 264<br>266<br>267 |
| | Fish | Early-life-stage mortality<br>Impaired oocyte development | Antiestrogenic | 273 |
| **Non-organochlorine pesticides** | | | | |
| Organophosphate pesticides (OPs) | Mammals | Depressed reproduction<br>Gonadotrophins | Acts at sites on hypothalamus-pituitary-gonadal-liver axis;<br>Acetylcholinesterase inhibitor | 357 |
| | Birds | Gonadotrophins<br>Developmental defects | | 141 |
| | Amphibians and Reptiles | Altered metamorphosis<br>↑Deformities and delayed development | | 371, 372 |
| | Fish | Retarded ovarian growth<br>↓GSI<br>Arrested spermatogenesis | | 375 |
| Carbamate pesticides | Amphibians | ↑Developmental deformities | Acetylcholinesterase inhibitors<br>Acts on pituitary to alter GnRH and GTH concentrations | 396 |

**Table 39.1 Summary of Effects and Possible Modes of Action (MOAs) of Endocrine Disrupting Chemicals (EDCs) by Chemical Class and Taxa** *(Continued)*

| Chemical | Taxa | Effects | Possible MOA | Sample Reference |
|---|---|---|---|---|
| | Fish | ↑Histopathological alterations in gonads<br>↓GSI (oocyte atresia and spermatogonial necrosis) | | 379, 380 |
| Organometal pesticides (TBT) | Fish | ↓Sperm counts<br>Delayed hatching | May inhibit aromatose | 418<br>412 |
| | Invertebrates | Masculinization of female gastropods<br>Imposex | Competitive inhibitor of aromatase<br>Cytotoxic and genotoxic effects | 402 |
| **Complex Mixtures** | | | | |
| Pulp- and paper-mill Effluents | Fish | ↓Sex hormones<br>↓Gonadal development<br>Delayed sexual maturation<br>Altered secondary sex characteristic expression | Estrogenic<br>ER, AR, AhR agonists | 433<br>453 |
| Sewage-effluents | Fish | ↓Testicular growth and development<br>Altered spermatogenesis<br>↑VTG production in males<br>↓Hatchlings<br>↑Oocyte atresia | Estrogenic<br>ER agonists/antagonist<br>Estrogenic<br>ER binding | 27, 466<br><br>418, 466, 481<br>476, 503 |
| **Metals** | | | | |
| Methylmercury | Mammals | ↓Embryo survival<br>↓Sperm counts | Unknown | 524, 528 |
| | Birds | Impaired reproductive behavior<br>↓Hatchability and nesting success | Unknown | 535, 539 |
| | Amphibians and Reptiles | ↓GSI<br>↓Sperm bundles | | 543 |
| | Fish | ↓GSI<br>↑Gonadal abnormalities<br>Altered gonadal steroidogenesis | Unknown | 546 |

and decreased GSI in bream (*Abramis brama*) inhabiting contaminated areas of the Rhine River.[155] Although PAH concentrations were abnormally high in the field studies described above, they were only one of a group of pollutants that may have caused the observed effects. Furthermore, several histological studies report no differences in the gonads of male and female fish from control and PAH-contaminated sites.[156,157] A combination of field and laboratory experiments is still necessary before the reproductive alterations observed in the wild can be clearly attributed to PAH exposure.

Laboratory and field studies present clear evidence for the adverse affects of PAHs in fish. Thomas et al.[158] have elucidated the impact of benzo(a)pyrene (BaP) on endocrine and reproductive activities in female Atlantic croaker (*Micropogonias undulatus*). Atlantic croaker fed 0.4 mg BaP/70g/day for 30 days during the period of ovarian recrudescence experienced impaired ovarian growth with a concomitant reduction in plasma $E_2$ and T. GSI in control females increased fivefold over the course of the study, whereas GSI of exposed females reached only 66% of controls.

*In vitro* production of sex steroids was not impaired by BaP, and there appeared to be a relationship between the amount of ovarian tissue (i.e., size of ovaries) and steroidogenic capacity. Similar results were reported in a separate study of female Atlantic croaker exposed to BaP via injection for 30 days.[159] In this study, in addition to reduced GSI and plasma sex steroids, a reduction in the number of hepatic ERs and plasma VTG was observed. BaP did not interfere with the binding of $E_2$ to the ER under *in vitro* competition studies, and again there was no clear evidence for a direct effect of BaP on steroidogenesis. *In vitro* competition studies using hepatic ER from spotted seatrout supported earlier results on Atlantic croaker.[160] This is consistent with mammalian studies, and suggests that BaP must undergo metabolic activation in order to interact with the ER. The effects of the PAH 3-methylcholanthrene on endocrine and reproductive function in ricefield eels (*Monopterus albus*) were similar to those observed for BaP-treated Atlantic croaker. Exposure to 4 ppm 3-methylcholanthrene for 7 days resulted in reduced $E_2$, T, VTG, GSI, and altered ovarian histology.[161]

PAHs are known CYP450 inducers. For example, the PAH naphthoflavone induced the expression of CYP4501A1 (the primary xenobiotic-metabolizing enzyme) and inhibited VTG synthesis in $E_2$-stimulated liver cells from rainbow trout (*Oncorhynchus mykiss*).[72] However, naphthoflavone had no effect on vitellogenesis when incubated without $E_2$. The degree of CYP4501A1 induction was directly related to the extent of VTG inhibition, which suggests that naphthoflavone may be acting as an antiestrogen via the AhR, the intracellular receptor involved in CYP4501A1 expression. The effect of naphthoflavone on vitellogenesis *in vivo* appears to be more complicated. When juvenile rainbow trout were treated with 0.5 ppm $E_2$ and 25 or 50 ppm of naphthoflavone, an inhibitory effect on VTG synthesis was observed; however, lower concentrations of naphthoflavone (5 or 12.5 ppm) appeared to potentiate $E_2$-stimulated VTG production.[72] Furthermore, reduced VTG synthesis by higher concentrations of naphthoflavone was correlated with a decrease in radiolabeled $E_2$ binding to the ER. These results suggest that naphthoflavone influences VTG synthesis by regulating ER function, although it is likely that the antiestrogenic activity of PAHs involves multiple mechanisms. Several investigators have proposed that CYP4501A1-inducing compounds affect sex-steroid concentrations by increasing their catabolism.[162] Evidence from a recent study,[163] however, suggests that PAHs may also interfere with steroid biosynthesis. Incubating vitellogenic ovarian tissue from female European flounder (*Platichthys flesus*) with 3 PAHs (phenanthrene, BaP, and chrysene) decreased A and $E_2$ secretion. In addition, phenanthrene inhibited steroid conjugation, and it was concluded that these PAHs inhibited key steroidogenic enzymes, including CYP450 17, 20 lyase, which is responsible for converting C21 to C19 steroids.

### 39.3.2 Polychlorinated and Polybrominated Biphenyls (PCBs and PBBs)

PCBs are a group of synthetic organic chemicals, formed by the chlorination of biphenyls, which include 209 individual compounds (congeners). These substances were manufactured for a wide range of industrial applications including use as hydraulic fluids, lubricants, plasticizers, and coolant/insulation fluids in transformers. Several chemical properties make these compounds both highly useful and potentially hazardous. For instance, their chemical stability makes them ideal for industrial activities involving high temperatures; however, this stability also renders them persistent in the environment. The majority of PCBs that enter the water adsorb to organic particles and sediments, although they are essentially nonbiodegradable in soils and sediments.[164] Furthermore, they are hydrophobic, which makes them excellent lubricants, yet allows them to bioaccumulate in tissues and biomagnify as they are passed along the food chain. Concentrations of PCBs in fish at contaminated sites may range from ppb to ppm. The production of PCBs is currently banned or highly restricted and the use of certain mixtures permitted only under tightly regulated conditions. Nonetheless, PCBs originating from industrial wastes, accidental leaks or spills, and careless disposal continue to be a source of pollution and environmental concern.

Many studies examining the health hazards of PCBs describe the effects of occupational exposure in humans and the physiological responses of mammals and birds that have consumed large quantities of contaminated fish. These studies provide strong evidence that PCB exposure can lead to the development of cancer; disturbances of the immune, hepatic, pulmonary, and nervous systems; and impaired reproduction and development. Many of these abnormalities are enhanced in the offspring, even if exposure occurs prior to conception. Responses are believed to be dependent on species, sex, age, and chemical structure.[165]

Laboratory studies with mink (*Mustela vison*) have established an association between PCB residues and reproductive effects in wildlife,[166] but there are no field studies linking PCBs with reproductive effects.[167] In one study in which mink were fed meat from cows contaminated with Aroclor 1254, concentrations as low as 0.87–1.33 ppm resulted in reproductive failure.[168] Other feeding studies have shown impaired reproduction in mink with fat concentrations of 13.3 ppm and reproductive failure at concentrations of 24.8 ppm.[169]

Field studies with big brown (*Eptisecus fuscus*) and little brown bats (*Myotis lucifugus*) suggested a correlation between PCB residue levels and reproductive toxicity;[170,171] however, captive studies have not supported this link.[172] Studies with ringed seals (*Pusa hispida*) have found a relationship between fat PCB concentrations and uterine-horn occlusions.[173] Later studies with ringed and gray seals (*Halichoerus grypus*), however, failed to detect any relationship between PCB levels and pregnancy or impairment of the uterine horns.[174] Other studies have linked PCBs with abortions and premature pupping in California sea lions (*Zalophus californianus*),[175] tumors and decreased fecundity in Beluga whales (*Delphinapterus leucas*),[176] skeletal lesions in harbor (*Phoca vitulina*) and grey seals,[177,178] and immunosuppression in harbor seals.[179] In a field experiment with harbor seals, animals fed PCB-contaminated fish had a significant reduction in reproductive success; however, in this study it was difficult to separate out the influence of other possible factors and contaminants.[180,181]

PCBs have been associated with embryonic mortality, deformities, and low reproductive success in many species of birds. Laboratory studies with chickens (*Gallus gallus*), ringed turtledoves (*Streptopelia risoria*), and mallards have shown reproductive impairment following ingestion of PCB-laden feed. In three studies, eggs of chickens that received 10–80 ppm Aroclor 1248 in the diet exhibited reduced hatching success;[182–184] however, in another study, a diet of 20 ppm Aroclor 1254 did not affect this parameter.[185] Aroclor 1242 in the drinking water at 50 ppm produced chick embryo mortality and teratogenesis.[186] Aroclor 1254 in the diet of ringed turtledoves has increased embryonic mortality, decreased parental attentiveness,[187] and depleted brain dopamine and norepinephrine.[188] Eggshell thickness was affected in mallard hens,[189] but another study produced no eggshell changes.[190] Studies with screech owls (*Otus asio*)[191] and Atlantic puffins (*Fratercula arctica*)[192] also produced no reproductive effects.

Field studies have indicated PCBs as the cause of mortality of ring-billed gulls (*Larus delawarensis*) in southern Ontario[193] as well as the cause for increased embryo/chick mortality and reduced hatching success.[194,195] PCBs have also been blamed for the low reproductive success and eggshell damage in Lake Michigan herring gulls (*Larus argentatus*).[196,197] Reproductive success of Forsters terns (*Sterna forsteri*) from a Green Bay colony was 52% of that from inland colonies.[198,199] In this study, hatchlings also weighed less, had shorter femurs, exhibited edema, and were malformed. The toxicity was attributed to the PCB congeners 105 and 126, and results indicated PCB congener 77 as accounting for some of the toxicity in the tern eggs.[199–201] PCBs have also been implicated as embryotoxic in eagles,[202] as producing decreased embryonic weight in black-crowned night herons (*Nycticorax nycticorax*),[203] as reducing hatching success of American kestrels (*Falco sparverius*),[204] and as the cause of congenital anomalies and embryonic death in double-crested cormorants (*Phalacrocorax auritus*).[205,206] In cormorants, however, there is discussion as to whether DDT or PCBs are more strongly associated with nest failure.[207,208]

American alligator eggs (*Alligator mississippiensis*) from Lake Apopka, Florida have residues of PCBs as well as a combination of organochlorine pesticides.[209] Alligators from this site have also been documented to have abnormally developed reproductive organs, altered serum hormone concentrations, and decreased egg viability.[135,139,210] However, alligators from Lake Apopka are known to be exposed to a complex mixture of potential EDCs, and therefore it is difficult to pinpoint which compounds are responsible for the observed effects. In red-eared slider turtles (*Trachemys scripta elegans*), males exposed to Aroclor 1242 had significantly lower T concentrations than controls.[211]

The African clawed frog (*Xenopus laevis*) and the European common frog (*Rana temporaria*) were exposed to the PCB mixture Clophen A50 or to PCB 126 for either 10 days or until metamorphosis. Exposed frogs had increased mortality, higher malformation rates, and lower thyroid hormone concentrations.[212] In a similar study, the same frog species were exposed to the mixtures Clophen A50 and Aroclor 1254 or to PCB 126; effects of exposure depended on route and time and length of exposure. This study also indicated a relationship between lowered concentrations of retinoid and PCB exposure.[212] In another study, green frogs (*Rana clamitans*) and leopard frog (*Rana pipiens*) were exposed throughout metamorphosis to PCB 126 at concentrations ranging from 0.005 to 50 ppb. Survival of larvae decreased at the higher concentrations in both species.[213]

In fish, reproductive impairment has been demonstrated under both *in vivo* laboratory studies and in field studies of fish residing in PCB-contaminated waters. Several field studies have attempted to correlate PCB tissue levels with observed reproductive alterations. For instance, PCB levels in the liver and ovarian tissue of female English sole from Puget Sound were associated with the spawning of fewer eggs.[214] Similarly, a negative correlation was found between egg hatchability and total PCB concentrations in the eggs of lake trout (*Salvelinus namaycush*) from the Great Lakes.[215] In a study of the reproductive success of lake trout residing in Lake Michigan, Mac and Edsall[216] suggested that maternally derived PCBs were the cause of reduced egg hatchability and increased fry mortality. Johnson et al.[217] reported decreased egg weight and increased oocyte atresia in female winter flounder (*Pseudopleuronectes americanus*) with high tissue concentrations of PCBs, although there was no evidence that PCBs altered GSI, plasma $E_2$, or fecundity in this species. Interestingly, English sole residing in the same location had reduced plasma $E_2$ concentrations and impaired gonadal development. It was proposed that the different migratory practices of both fish species might have resulted in different susceptibilities to the chemicals, since these behaviors resulted in differences in the timing and duration of exposure to the most highly contaminated waters.

In the laboratory, female Japanese medaka (*Oryzias latipes*) had reduced GSI and were unable to spawn following an injection of 150 ppm of PCB.[218] It was suggested that PCB exposure disrupted $E_2$ metabolism since only control fish, excreted $E_2$ into the water 8 days after the injection. In male goldfish (*Carassius auratus*), PCB exposure resulted in decreased plasma T and 11KT concentrations, while hepatic EROD activity was increased 15-fold.[219] Reduced plasma T in males and $E_2$ and $P_4$ in females, accompanied by an increase in several sex-steroid-metabolizing enzymes, was observed in carp (*Cyprinus carpio*) injected with 250 ppm of the commercial PCB Aroclor 1248.[220] In another study, $E_2$-treated juvenile rainbow trout fed a diet contaminated with PCBs (3, 30, or 300 ppm) for 6 months showed decreased synthesis of VTG.[221]

The decline in estrogens and androgens combined with the elevation of EROD (or other metabolizing enzymes) would suggest that the reduction in sex steroids is related to an increase in metabolism rather than a decreased synthesis. However, this is probably not the only mechanism for PCB-induced damage since several studies also report abnormalities at the organ e.g., GSI, testicular abnormalities) and organism levels (offspring survival, hatchability) in fish showing normal sex-steroid and VTG concentrations.[222–224] In some of these cases, it is believed that the reproductive abnormalities (e.g., delayed spawning, reduced hatchability) may be caused by the accumulation of toxic levels of PCBs in the ovaries and maturing oocytes.[225] Evidence that PCBs bind VTG suggest that lipoproteins are involved in the transport of the contaminants from extra-gonadal tissue into the ovaries.[226] One explanation for the inconsistencies observed between studies might be related to timing of exposure. For instance, the lack of effects of 3,′3′,4,′4′-tetrachloro-biphenyl (TCB) on plasma concentrations of sex steroids and VTG in female striped bass (*Morone saxatilis*) and white perch (*Morone americana*) may have been related to the fact that the fish used in these studies were already vitellogenic and not in a highly active stage of gonadal maturation.[222,223]

The stage of gonadal maturation may also be important in males exposed to PCBs. Atlantic cod (*Gadus morhua*) fed Aroclor 1254 (1–50 ppm) for 5.5 months accumulated significant levels of the PCB in the testes and liver and exhibited considerable testicular damage including fibrosis of lobule walls, necrosis and disintegration of lobule elements, and decreased spermatogenesis.[227] These authors suggest that the stage of gonadal maturation may be related to the degree of chemical sensitivity since only males experiencing rapid spermatogenic proliferation or fully mature males suffered testicular damage (i.e., sexually immature and regressed males were unaffected). In several cases, substantial concentrations of PCBs have been detected in tissues of fish that showed no signs of adverse reproductive effects.[228]

It is not surprising that a wide range of responses has been observed in studies that have differed with regards to species and experimental design. However, the chemical complexity of this class of compounds is an additional factor that complicates interpretation. Slight structural differences in the 209 possible PCB congeners, as well as different compositions of the mixtures, may result in vastly different physiological responses.

There is considerable evidence that PCBs act at multiple sites along the hypothalamic-pituitary-gonadal (HPG) axis,[158,229] and *in vitro* experiments are providing insight into the mechanisms underlying these reproductive alterations. However, extrapolating the actual risks that PCBs impose on the environment and biota are difficult due to the complexity and diversity of the commercial mixtures (of congeners). In addition to understanding the interaction of the PCB mixtures with other environmental pollutants and stressors, consideration must be given to the interaction (additivity, synergism, antagonism) of the individual components that make up these mixtures.

PBBs, formed by the bromination of biphenyls, are similar in structure to PCBs. These chemicals are stable and lipophilic and, therefore, present many of the same environmental hazards as PCBs (e.g., persistence in the environment and long biological half-lives). There are 209 possible PBB congeners, although only 45 have been actually synthesized.[230] FireMaster BP-6, used primarily as a flame-retardant additive in the early 1970s, was the most widely used PBB, although its production was discontinued in 1978.[231,232] The production and distribution of PBBs was insufficient to result in widespread contamination of the environment; however, the accidental contamination of cattle feed by the Michigan Chemical Company in 1973 resulted in the pollution of many Michigan farmlands. Significant concentrations of PBBs were subsequently detected in water and sediment samples and in tissues of fish and ducks residing downstream of the Michigan Chemical Company.[233]

Although information concerning the reproductive effects of PBBs in fish is lacking, there is substantial evidence that these chemicals adversely affect reproductive processes in other species.[234] For instance, feeding adult female chickens a diet contaminated with 45 ppm of the commercial PBB FireMaster FF-1 for 5 weeks resulted in impaired production and hatchability of eggs and in reduced viability of offspring.[235,236] A variety of reproductive effects following PBB exposure have also been reported from other avian (quail) and mammalian (rodents, monkey, cow, and mink) species.[230,234] Although PBBs have been detected in aquatic environments and are known to accumulate in fish tissues,[237] there is very little information regarding the effects of these chemicals on exposed fish. Like PCBs, PBBs are believed to be potent inducers of several monooxygenase enzymes (including EROD), although it is not known whether this induction affects the metabolism of circulating reproductive hormones.

### 39.3.3 Polychlorinated Dibenzo-*p*-Dioxins (PCDDs) and Polychlorinated Dibenzo-*p*-Furans (PCDFs)

PCDDs and PCDFs are structurally related compounds produced during a variety of thermal and chemical reactions including the combustion of PCBs, production of steel and other compounds, and disposal of industrial wastes (via the interaction of chlorophenols). These compounds have also been identified as components of bleached-pulp-mill effluents. PCDDs and PCDFs are halogenated aromatic hydrocarbons with high chemical stability, low water solubility, and limited solubility in many organic solvents. There are 75 possible PCDD congeners and 135 PCDF congeners, although 2,3,7,8-tetrachlorodibenzo-*p*-dioxin, known as TCDD or dioxin, has received the most attention. Concerns regarding TCDD stem from its wide distribution in the environment and extreme toxicity to both humans and wildlife. Although a variety of PCDDs and PCDFs have been detected in fish and wildlife, the 2,3,7,8-substituted congeners are believed to be the most persistent and prevalent in tissue samples analyzed to date, with half-lives of over a year in some fish species.[238] TCDD and related compounds have been implicated in a number of health-related problems including neurotoxicity, hepatoxicity, cardiotoxicity, chloracne, birth defects, immunosuppression, wasting syndrome, and endocrine and reproductive alterations.[240,241,269] In nonhuman primates and rodents, although developmental effects of the immune, reproductive, and nervous systems occur at body burdens in the range of 30–80 pptr, biochemical changes on cytokine expression and metabolizing enzymes are seen at doses ten times lower.[240] Many of the toxic effects associated with exposure to dioxins appear to be dependent on target tissue, species, sex, and age.

Studies on the developmental effects of TCDD in rodents have demonstrated that only transient exposure to relatively low concentrations of TCDD during critical windows of development are capable of eliciting irreversible disruption of organ functioning in offspring. For example, gestational exposure of rats to low concentrations of TCDD (0.064–1.0 ppb) during a critical period of development (day 15 of gestation) causes impaired sexual differentiation in male fetuses including persistence of female traits; decrease in the concentration of T, in the weight of testis and epididymis, and in the production of sperm; and altered sexual behavior during adulthood.[242–244] Similarly, Gray et al.[245] reported a delay in testicular descendent and puberty, with a subsequent reduction in sperm counts and fertility in adult male rodents after a single maternal dose of 1 μg TCDD/kg on day 15 of gestation. Other signs of developmental toxicity in mammals include decreased growth, structural malformations (e.g., cleft palate and hydronephrosis), prenatal mortality, and neurobehavioral changes (e.g., impaired learning in rhesus monkeys).[239]

Laboratory studies on the effects of TCDD in birds have shown significant variation in sensitivity across species, with over 40-fold differences on embryo mortality (reviewed in References 239 and 246). For example, doses of only 20–50 ppt of TCDD in chicken eggs cause mortality and malformations, as opposed to 1000–10,000 ppt in eggs of ring-necked pheasants (*Phasianus colchinus*) and eastern bluebirds (*Sialia sialis*). Chicken embryos are also much more sensitive to the teratogenic effects of TCDD, particularly cardiovascular malformations.

In birds, there is considerable evidence indicating that embryonic exposure to dioxin and dioxin-like compounds can induce developmental alterations. Indeed, several field studies with colonial fish-eating birds from the Great Lakes have implicated dioxin equivalents (the aggregate of AhR-active substances) as the causative factors for the increased incidence of developmental deformities and embryo lethality observed in certain contaminated areas (see reviews in References 247–250).

Together, these epidemiological studies have provided one of the strongest links between contaminant exposure and reproductive/developmental effects in wildlife.

The Great Lakes Embryo Mortality, Edema and Deformities Syndrome (GLEMEDS) was first described in double-crested cormorants but has also been reported in other species including the great blue heron (*Ardea herodias*) and the Caspian tern (*Sterna caspia*). The syndrome is characterized by increased embryo mortality; growth retardation; subcutaneous, pericardial, and peritoneal edema; congenital deformities of the bill and limbs; feminization of embryos; and abnormal parental behavior.[247] This syndrome closely resembles the "chick edema disease" observed in chickens after *in ovo* exposure of hens to PCDDs and PCDFs. Embryotoxicity in piscivorous birds from the Great Lakes has been associated with TCDD concentrations above 100 pg/g (reviewed in Reference 241). Although a reduction in the release of pollutants to the Great Lakes has resulted in significant population improvements for several avian species, particularly double-crested-cormorants and ring-billed gulls (*Larus delawarensis*), reproductive and physiological alterations due to contaminants are still associated with population-level effects in birds that feed on highly contaminated fish (such as Caspian terns and bald eagles, *Haliaeetus leucocephalus*).[249,251]

Similar reproductive and developmental effects due to PCDDs have also been reported from free-ranging populations of great blue herons,[252,253] double-crested cormorants,[254] and wood ducks (*Aix sponsa*)[255] sampled elsewhere. In addition, *in ovo* exposure to dioxins has been associated with the development of asymmetric brains in wild (great blue herons, double-crested cormorants, and bald eagles) and domestic (chickens) species.[256] The behavioral and physiological repercussions of this gross brain deformity, however, are unknown at this time. Other sublethal effects observed in birds exposed to dioxins include decrease bursa and spleen weights in developing embryos[257] and altered thyroid-gland structure, circulating thyroid hormones, and vitamin A (retinoid) status (reviewed in Reference 258). In Belgium and The Netherlands, PCDDs/PCDFs in common tern (*Sterna hirundo*) were correlated with lower yolk sac retinoids and plasma-thyroid concentrations in hatchlings and with unfavorable breeding parameters (delayed laying and smaller eggs and chicks).[259] Similarly, cormorant (*Phalacrocorax carbo*) hatchlings from a PCDD/PCDF-contaminated site in The Netherlands had decreased plasma-thyroid concentrations and an increased *in ovo* respiration rate.[260] Results from laboratory studies, however, have failed to replicate what has been reported from wild avian species, and *in ovo* exposure to TCDD has caused either increases or no changes in thyroid hormones.[120,261,262]

Information on the developmental toxicity effects of TCDD in amphibians and reptiles is scarce. Neal et al.[263] reported no effects in tadpoles and adult bullfrogs (*Rana catesbeiana*) after a single injection of TCDD (500 ppb). Jung and Walker[264] exposed anuran eggs and tadpoles to TCDD for 24 h and observed that American toads (*Bufo americanus*) treated with at least 0.03 ppb appeared to metamorphose earlier than controls and that metamorphosis tended to occur at larger body masses after exposure to higher doses of dioxin. The authors concluded that anuran eggs and tadpoles eliminate TCDD more rapidly and are 100- to 1000-fold less sensitive to its deleterious developmental effects when compared to fish. Differential sensitivity to TCDD and related compounds could be related to differences in metabolism or to different patterns in AhR binding and signal transduction across taxa. In this respect, there is recent information showing a high degree of amino-acid-sequence conservation for the AhR among bird species (97% amino acid identity) but a much lower percent identity across taxa (79 and 74% identity between the amphibian *Necturus maculosus* and bird and mouse sequences, respectively).[265]

In an epidemiological study, developmental abnormalities and hatch rates from eggs of the common snapping turtle (*Chelydra serpentina serpentina*) were assessed in relation to over 70 PAHs, including 8 PCDDs and 14 PCDFs.[266] This study found an increase in the frequency of deformities with increasing contaminant exposure in eggs, particularly PCDDs and PCDFs concentrations. In the laboratory, American alligator eggs (embryo stages 19–22) were treated with TCDD (at doses ranging from 0.1 to 10 ppm) and incubated at male-producing temperatures

(33°C).[267] High doses of TCDD in eggs resulted in dose-dependent alterations in sex ratios, with a higher incidence of female hatchlings.

Results from several field and laboratory studies have established that fish, in particular early life stages, are extremely sensitive to the effects caused by TCDD when compared to other taxa. Effects on fry survival are significant at egg doses ranging from 50 to 5000 pg/TCDD/g, which corresponds to concentrations of 75 to 750 pg TCDD/g in parent fish.[241] Signs of TCDD-induced developmental toxicity resemble blue sac disease, which is an edematous syndrome characterized by yolk sac and pericardial edema, subcutaneous hemorrhages, craniofacial malformations, retarded growth, and death.[268] This syndrome has been well characterized in salmonids after exposure of eggs via water, injection, or maternally derived TCDD. Studies with salmonids have established differential sensitivity to induce sac fry mortality, with $LD_{50}$ values varying from less than 100 pg TCDD/g egg in lake trout (the most sensitive fish species to TCDD developmental toxicity) to 200 and over 300 pg TCDD/g egg in brook trout (*Salvelinus fontinalis*) and in some strains of rainbow trout, respectively.[269–271] TCDD-developmental toxicity has also been reported in several nonsalmonid species including the northern pike (*Esox lucius*), the mumichog (*Fundulus heteroclitus*), the Japanese medaka,[272] and the zebrafish (*Brachydanio rerio*).[273] Regardless of species or egg exposure route, early-life-stage mortality occurs during the sac-fry stage, probably as a consequence of the generalized edema.

Exposure of eggs to TCDD and related compounds may have been responsible for the decline of some fish populations in the Great Lakes since concentrations of dioxin and dioxin equivalent in eggs and fry of salmonids have fallen within the range of those known to induce blue sac disease in the laboratory.[269] The reader is advised to refer to References 269 and 274 for comprehensive reviews on the effects of TCDD and related compounds on fish of the Great Lakes.

Fish reproduction can also be affected after exposure to TCDD. In the laboratory, adult female zebrafish fed 5–20 ng TCDD showed impaired oocyte development with fewer eggs produced and lethal developmental abnormalities in offspring (e.g., malformations of notocord).[273] In a separate study, although TCCD treatment of newly fertilized zebrafish eggs did not affect hatchability, doses of 1.5 ng TCDD/g or more resulted in a variety of structural and physiological abnormalities in larvae.[275]

The mechanisms by which TCDD and structurally related compounds cause endocrine/developmental effects are complex and not completely understood. The relative toxicity of TCDD and other halogenated aromatics is likely dependent on their ability to bind and activate the AhR. Although this AhR mechanism is known for its involvement in the antiestrogenic action of TCDD as well as for its ability to induce structural malformations, its MOA in causing other reproductive and developmental toxicity is less clear.[239] There is substantial evidence that these contaminants induce the expression of certain genes (e.g., translation products comprising Phase I and Phase II enzymes) while altering the transcription of others (ER).

Indeed, the antiestrogenic effects of TCDD in mice have been attributed to its ability to suppress ER gene expression, probably through an inhibition of ER transcription after binding of the TCDD-AhR complex to promoter regions of the ER gene (see Reference 276 for a review of mechanisms of action of dioxins). Antiestrogenic effects of these contaminants have also been documented *in vitro* using fish cell lines. Using carp hepatocytes, Smeets et al.[277] demonstrated that although low concentrations of TCDD caused a suppression of VTG secretion and an induction of CYPIA, the two phenomena were not correlated to each other. From these results, the authors concluded that the antiestrogenic effects of TCDD were probably not caused by increased metabolism of $E_2$ due to induction of CYP1A. Mechanistic studies using mammalian cell lines also support this theory,[80] although increased metabolism of sex steroids may provide an additional or secondary mechanism of antiestrogenicity. In this respect, great blue heron hatchlings and adults exposed to TCDD had increased testosterone hydroxylase activity, a result that was coupled with increased CYP1A1 activity.[278] Changes in hydroxylase activity, however, have not been associated with alterations in circulating sex-steroid concentrations in TCDD-exposed herons.[119,120,261] An additional MOA of

TCDD could be through the pituitary, disrupting normal feedback mechanisms between hormones and LH secretion.[279]

Since vitamin A and thyroid hormones are essential for normal differentiation and development of tissues, alterations in their homeostasis might result in malformations and altered growth. In this respect, there is evidence showing that dioxin may interfere with the metabolism and storage of vitamin A (retinoids)[280] and of thyroid hormones[281] through the co-induction of Phase I (P450) and Phase II (uridine diphosphate-glucoroyltransferase ($T_4$-UDPGT) enzymes. In addition, hydroxy metabolites of PCDDs and PCDFs can compete for thyroxine on the transthyretin (the thyroxine binding prealbumin) binding site.[282] Finally, recent evidence suggests that the hemodynamic and teratogenic effects observed in fish fry affected by blue sac disease could be due to the ability of TCDD to induce oxidative stress and oxidative DNA damage. In the Japanese medaka, oxidative stress to the vascular endothelium of developing embryos induces programmed cell death, or apoptosis.[283] Apoptosis of vascular cells causes alterations in hemodynamics, leading to a generalized loss of function and subsequent mortality.

### 39.3.4 Organochlorine Pesticides and Fungicides

Organochlorine pesticides (OCPs) comprise a large group of structurally diverse compounds used to control agricultural pests and vectors of human disease. Many of these compounds, as well as their metabolites, are environmentally persistent due to their chemical stability, low water solubility, and high lipophilicity. The exact mode of neurotoxicity is not well understood, although OCPs are believed to disrupt the balance of sodium and potassium in nerve cells. The ability of these toxic compounds to bioaccumulate in and often harm unintended species has led to the restricted use of most OCPs. Despite a general reduction in use, several field studies have suggested that OCPs adversely affect endocrine function in fish, indicating that aquatic wildlife is still being exposed to levels capable of altering endocrine and reproductive parameters. For instance, a negative correlation was found between total OCPs and $E_2$ in male carp (*Cyprinus carpio*) in a large-scale field effort to assess the reproductive health of fish in U.S. streams.[106] Similar results were reported in largemouth bass (*Micropterus salmoides*) collected from a contaminated (with OCPs) site in Florida.[284] In the following section, OCPs are discussed according to accepted structural classifications, although effects within the same chemical class may differ drastically. Furthermore, Pickering et al.[285] have suggested that pesticide toxicity is species specific, and a single species may be differentially susceptible to different pesticides. Indeed, the reported effects and mechanisms of action may vary significantly between the various OCPs.

#### *39.3.4.1 Cyclodienes*

The chlorinated cyclodiene pesticides are lipophilic, stable solids with low solubility in water. Although differing from the dichlorodiphenylethanes (i.e., DDT) in their mode of action, they served a similar function in controlling a variety of insect pests. Examples of pesticides in this class include endrin, dieldrin, chlordane, toxaphene, telodrin, isodrin, endosulfan, and heptachlor. Consistent with the nature of organochlorine compounds, cyclodiene pesticides are persistent in soils and sediments with a half-life of 1–14 years in soils following application.

The cyclodienes are believed to produce a wide range of toxic responses in wildlife and adverse effects in laboratory animals. For example, rats exposed to endosulfan had alterations to the nervous, immune, hepatic, renal, and reproductive systems.[286]

Dieldrin levels of 9.4 ppm in purple gallinule (*Porghyrula martinica*) and of 17.5 ppm in the common gallinule (*Gallinula chloropus*) showed no significant effects on percentage of eggs hatched or in the survival of young.[287] Lockie et al.[288] reported that the proportion of successful eyries of the golden eagle (*Aquila chrysaetos*) increased from 31 to 69% as the levels of dieldrin fell from a mean of 0.86 ppm to 0.34 ppm. It has been postulated that dieldrin poisoning of adult birds is the likely mechanism for population decline of bird-of-prey populations, such as the peregrine falcon (*Falco peregrinus*) and Eurasian sparrow hawk (*Accipiter nisus*) in Great Britain and the peregrine falcon in the United States, rather than DDE effects on shell quality.[289–291] Screech owls (*Otus asio*) with egg aldrin concentrations ranging from 0.12 to 0.46 ppm were 57% as productive as controls, with lower clutch sizes, hatch rates, and survival.[292] Heptachlor epoxide reduced nest success in Canada geese (*Branta canadensis*) when eggs contained > 10 ppm[293] and reduced productivity in American kestrels when eggs contained > 1.5 ppm.[294] No relationship was found between heptachlor epoxide residues and shell thickness in eggs of Swainson's hawk (*Buteo swainsoni*), and reproduction was not affected in wild prairie falcons (*Falco mexicanus*) and merlins (*Falco columbarius*).[294,295] Chlordane fed to northern bobwhites (*Colinus virginianus*) in concentrations of 3 and 15 ppm and to mallards at 8 ppm had no effect on reproduction. Similarly, toxaphene fed at 100 ppm to chickens had no significant effect on reproduction.[296] A 2-year study with American black ducks fed a diet containing 1, 10, or 50 ppm toxaphene produced no reproductive effects, although duckling growth, skeletal development, and collagen was decreased in offspring of parents fed 50 ppm.[297]

Chlordane, dieldrin, and toxaphene were tested for their ability to override male-producing incubation temperature in the red-eared slider turtle. Chlordane produced significant sex reversal alone and when administered with $E_2$.[113] In another study, treated male turtles exposed to chlordane had significantly lower testosterone concentrations, and females had significantly lower $P_4$, T, and 5-α-DHT concentrations than controls.[211]

Studies involving the effects of cyclodienes on the reproductive success of fish have produced a range of results, most likely resulting from species- and chemical-specific sensitivities as well as differences in experimental design. For instance, toxaphene at concentrations ranging from 0.02–2.2 ppt did not affect the reproductive success of female zebrafish, as measured by total number of eggs spawned, percentage of fertilized eggs, embryo mortality, and egg hatchability.[298] However, in this species, oviposition appeared to be affected by toxaphene exposure in a dose-dependent manner. Conversely, decreased fertilization has been observed in winter flounder after exposure to 0.001–0.002 ppm dieldrin,[299] reproduction of first-generation flagfish (*Jordanella floridae*) was affected after exposure to 0.3 ppb endrin,[300] and sublethal concentrations of dieldrin and aldrin were reported to induce abortion in mosquitofish.

Although less information is available regarding the effects of cyclodienes on reproductive function in male fish, a laboratory study with tilapia (*Oreochromis mossambicus)* showed disrupted nest-building and decreased reproductive activity.[301] In a study with male striped snakehead (*Channa striatus*), testicular damage and disrupted spermatogenesis were observed after exposure to 0.75–1 ppm of endosulfan for 2–30 days.[302]

Several reports indicate that oocyte development may be a target for cyclodiene-mediated reproductive toxicity. An increase in oocyte atresia was observed in rosy barb (*Barbus conchonius)* exposed to a low dose (46.6 pptr) of aldrin for 2–4 months.[303] Impaired oocyte development and reduced GSI have also been observed in striped snakehead[304] and carp minnow (*Rasbora daniconius*)[305] exposed to endosulfan. Other toxic effects related to endosulfan exposure include reduction in the percentage of maturing and mature oocytes, rupturing of ooctye walls, damage to yolk vesicles, and multiple other histopathological changes in ovarian morphology.[304] Consistent with the observations of oocyte damage and decreased GSI, endosulfan was shown to have an inhibitory effect on vitellogenesis in clarias catfish (*Clarias batrachus*).[306] It is possible that endosulfan directly interferes with VTG synthesis in the liver, a theory that is supported by evidence that endosulfan alters protein synthesis in the liver of clarias catfish. Alternatively, other studies suggest that endosulfan impairs steroidogenesis by interfering with enzymes along the steroid biosynthetic pathway.[307] Likewise, the authors of the later study concluded that endosulfan affected VTG synthesis by interfering with the production or activity of hormones responsible for regulating VTG production. Multiple effects along the hypothalamus-hypophysial-ovarian axis of the Mozambique tilapia (*Sarotherodon mossambicus)* were also observed following an exposure to 0.001 ppm endosulfan for 20 days.[308] In addition to reduced GSI and various histopathological abnormalities associated with ovarian growth and oocyte maturation, degeneration of basophils and acidophils (gonadotrops) in pituitary tissue of endosulfan-treated fish was apparent.

#### 39.3.4.2 Chlordecones (Kepone and Mirex)

Chlordecone, also known as Kepone, and mirex are two structurally similar OCPs that were manufactured and used primarily in the 1960s and 1970s. No longer permitted in the U.S., mirex was used as a pesticide to control fire ants as well as a flame-retardant additive, and chlordecone was used to control insects on a variety of crops and for household purposes. The toxicological effects of chlordecone exposure in humans are well documented as a result of an incident known as the "Kepone Episode," in which many employees and residents in the vicinity of several Kepone manufacturing companies were exposed to intoxicating concentrations of the chemical.[309] The central nervous system, liver, and reproductive organs appeared to be most sensitive to the toxic effects of chlordecone. Comparative studies using laboratory animals have since concluded that the target organs as well as the excretion pathways for chlordecone are similar in humans and rodents, although metabolic pathways differ significantly.

Reproductive impairment in a variety of mammalian and nonmammalian species has been attributed to the estrogenic properties of chlordecone.[309,310] Chlordecone induced constant estrus in mice,[311,312] and neonatal injections in female rodents accelerated vaginal opening and the onset of prolonged vaginal cornification with reductions in ovarian weight.[313] In Japanese quail (*Coturnix coturnix japonica*), Kepone caused oviduct hypertrophy in females[314] and suppressed spermatogenesis in males.[315,316] Mirex fed to mallards at concentrations of 100 ppm decreased duckling survival, and hatch rates were reduced in chickens fed 600 ppm mirex.[317] Hatchability and chick survival were reduced when adults were fed 150 ppm and 75 ppm chlordecone, respectively.[318]

In fish, there is evidence that chlordecone competes with radiolabeled $E_2$ for binding to the hepatic ER in spotted seatrout (*Cynoscion nebulosus*),[160,319] rainbow trout,[320] Atlantic croaker,[321] and channel catfish (*Ictalurus punctatus*). Other alterations attributed to chlordecone exposure include inhibition of oviposition in Japanese medaka,[322] reduced egg production and hatchability in sheephead minnow (*Cyprinodon variegatus*), and histopathological abnormalities in freshwater catfish (*Heteropneustes fossilis*). For instance, exposure of female catfish to chlordecone (0.024 ppm) for 1–2 months resulted in a decrease in the diameter of stage 1–3 oocytes, the formation of interfollicular spaces in the ovaries, and an increase in oocyte atresia.[323] In male catfish, subacute doses (0.024 ppm) over the same time period resulted in significant damage to the seminiferous tubules and cystolysis of spermatids and sperm.

#### 39.3.4.3 Dichlorodiphenylethanes

The dichlorodiphenylethane pesticide reported most often as having endocrine activity is 1,1,1-trichloro-2,2-bis *p*-chlorophenylethane (DDT). Used extensively during World War II to control insect-borne diseases, DDT was released into the environment in substantial quantities and, consequently, accumulated in soil, water, and tissues of many animals including fish. The *p,p′*- and *o,p′*-substituted isoforms of DDT; the dechlorinated analogs, *p,p′*- and *o,p′*-DDD; and the metabolites *o,p′*- and *p,p′*-DDE are various forms that frequently exist in the environment. In highly polluted areas (e.g., Palos Verdes Shelf in southern California), concentrations of DDT (total measured DDT, DDE, and DDD) have exceeded 100 ppm wet weight in the livers of several species. *p,p′*-DDE is one of the most commonly detected and highly persistent OCPs in tissues of aquatic animals, and in a recent study by the U.S. EPA, this metabolite was detected in 98% of fish surveyed at 388 locations in the United States. Over the last two decades, concentrations of DDT and its derivatives have decreased in fish of the United States, Canada, western Europe, and Japan as a result of strict regulation on its use. In addition to DDT, this category of OCPs includes DMC (Dimite), dicofol (Kelthane), methlochlor, methoxychlor, and chlorbenzylate.

Laboratory tests have shown increased uterine weight and persistent vaginal estrus in rats exposed to *o,p′*-DDT.[324,325] Mammals concentrate DDT in adipose tissue, where it can become toxic when fat is lost due to migration or hibernation and the pesticide is unbound.[326] Although concentrations in wildlife have decreased since DDT was banned, little information exists linking DDT exposure in the environment to estrogenic or adverse reproductive effects in wild mammals.[310]

The use of OCPs, especially DDT, was responsible for the declines in populations of many species of predatory birds during the 1950s and 1960s. Most bird declines were due to eggshell thinning, with DDE being responsible for most of this problem; however, some species are more sensitive than others. For example, 3.0 ppm DDE in egg produces eggshell thinning and reduced productivity in the brown pelican (*Pelecanus occidentalis*), with residues > 3.7 ppm leading to total reproductive failure.[327–329] Black-crowned night herons have a gradual decline in productivity as residues increase.[330] Eggshell thinning due to DDE has had a major impact on populations of bald eagles[331,332] and is considered a current risk to double-crested cormorants in Green Bay, Michigan.[207] The mechanism for eggshell thinning, although previously thought to be caused by the estrogenic effects of DDE, is now proposed to involve the inhibition of prostaglandin synthesis in the eggshell gland muscosa.[332,333] Additional reports of DDT effects in birds include decreased egg hatchability in eastern bluebirds (*Sialia sialis*)[334] and the persistence of high concentrations of DDT and metabolites in foodchains in orchard and former orchard areas of the northwest.[335,336] Population declines in raptors, such as the Eurasian sparrowhawk and peregrine falcon, have been linked to dieldrin, which increased adult mortality, in association with the adverse reproductive effects of DDE.[289]

American alligator eggs from Lake Apopka in Florida are known to contain residues of several OCPs including toxaphene, dieldrin, chlordane, DDT and metabolites, as well as PCBs.[209] Several studies have described adverse reproductive effects to the alligator population on this lake including increased embryo mortality[337,338] and morphological and endocrine abnormalities in juvenile alligators (including altered secondary sex characteristics).[135,139] DDE also has been shown to cause sex reversal in alligators and red-eared slider turtles following the treatment of eggs at early embryonic stages and at incubation temperatures necessary for the production of male offspring.[113] DDT has also been reported to induce VTG in the red-eared slider as well as in frogs.[339] Clark et al.[340] reported that technical-grade DDT acted as an antiestrogen and *p,p′*-DDE as an estrogen in tiger salamander (*Ambystoma tigrinum*). In a study with the reed frog (*Hyperolius argus*), *o,p′*-DDT, *o,p′*-DDE and *o,p′*-DDD prematurely induced adult female color patterns in juveniles, but not with *p-p*-substituted isoforms.[341]

A number of studies have demonstrated the effects of DDT and related compounds on the reproductive success of exposed fish. Increased fry mortality has been observed in brown trout (*Salmo trutta*) and brook trout exposed to DDT concentrations ranging from 0.5 to 3.4 ppm/week for 98–308 days.[342] Fry mortality was also reported in a field study of lake trout inhabiting a lake polluted with DDT.[343] In the laboratory, white croaker (*Genyonemus lineatus*) collected from a contaminated site were unable to spawn when total ovarian DDT concentrations exceeded 4 ppm. An increase in oocyte atresia and reduced fecundity and fertility were also observed in white croaker from a DDT-contaminated site.[136] Organism- and population-level effects due to DDT and derivatives include decreased fertilization and embryo deformity in winter flounder,[299] decreased fertility and early oocyte loss in white croaker,[344] and egg mortality in artic char (*Salvelinus alpinus*).[345] Functional male-to-female sex reversal in Japanese medaka has also been reported following the injection of approximately 227 ng *o,p′*-DDT/egg during the course of fertilization. A reduction in viable hatch has also been observed in atlantic herring (*Clupea harengus*) chronically exposed to 0.018 mg DDE/kg ovary.

Substantial evidence from *in vitro* studies (particularly in mammals) suggests that DDT and related OCPs are estrogenic and, thus, mediate endocrine disruption through interaction with the ER. A study by Spies et al.[346] found that female kelp bass (*Paralabrax clathratus*) from a polluted site had lower GTH, T, and $E_2$ concentrations compared to fish from a reference site. Furthermore, the rate of GTH release from the pituitary was enhanced and correlated with hepatic concentrations of DDT. In laboratory studies, an increased rate of GTH release from the pituitary of female kelp bass exposed to DDT was consistent with field observations, yet T production in laboratory exposed fish was enhanced. Receptor binding studies found that $E_2$ binding to the ER was reduced in DDT-exposed fish and, conversely, that DDT was capable of displacing $E_2$ from the ER.[346] While the estrogenic potential of DDT may vary among different species of fish and other vertebrates, the above studies demonstrate the complexities of the reproductive system and the difficulty when interpreting alterations induced by this group of OCPs.

#### 39.3.4.4 Hexachlorocyclohexane

Hexachlorocyclohexane (HCH) is an organochlorine chemical for which there are eight isoforms, and several are used to prepare the technical-grade product. HCH was used primarily as an agricultural pesticide and is, to some extent, still applied as an insecticidal seed dressing. The precise mode of toxicity is unclear, although the α and γ isomers are known convulsants, and the β-isomer is a central nervous system depressant. The technical-grade product, known as HCH or benzene hexachloride (BHC), contains a mixture of alpha (α), beta (β), delta (δ), gamma (γ), and epsilon (ε) isomers; however, most of the insecticidal properties come from the γ-isomer, lindane. This product has received extensive use as an insecticide for fruit, vegetable, and forest crops and as a component of the ointments used to treat head and body lice. Lindane shares many chemical properties with other OCPs, but it has greater polarity and water solubility than most. Although in the U.S. its production has been discontinued since 1977, lindane is still imported and used by U.S. EPA-certified applicators. Lindane can accumulate in the fatty tissues of fish; however, it is easily degraded to less toxic metabolites by algae, fungi, and bacteria inhabiting soil, sediment, and water. Less water-soluble than its gamma counterpart, beta-hexachlorocyclohexane (β-HCH; a by-product generated during the synthesis of γ-HCH) is the most persistent of the HCH isomers and has been known to bioconcentrate in the tissues of invertebrates, fish, birds, and mammals.

Wester et al.[347–349] have examined the effects of β-HCH on the development of the reproductive organs of several fish species. Four-week-old guppies (*Poecilia reticulata*) and post-fertilization Japanese medaka eggs were exposed to a range of concentrations (0.0032–1.0 ppm) for 1–3 m.[349] Female guppies exposed to 0.32–1.0 ppm had a high incidence of premature and abnormal yolk formation despite having no fully mature oocytes. In the absence of developed oocytes, VTG accumulated in the body fluids, including the glomerular filtrate of the kidneys, and may have contributed to several toxic lesions observed in various nonreproductive tissues. VTG was also detected in male guppies exposed to 0.32 and 1.0 ppm β-HCH, and although GTH production by the pituitary was activated, testicular development was delayed and evidence of intersexuality or hermaphroditism was reported. Japanese medaka exposed to β-HCH also exhibited intersexuality or hermaphroditism following exposure to concentrations higher than 0.1 ppm. Overall, these authors concluded that the alterations observed in both species were the direct result of the estrogenic activity of β-HCH or its metabolites.[348,349]

Singh et al.[350–352] conducted a series of *in vivo* studies that examined the effects of γ-HCH (lindane) on steroidogenesis at different phases of the reproductive cycle of goldfish and two species of catfish (*Clarias batrachus* and *Heteropneustes fossilis*). Regardless of reproductive stage, 4 weeks of exposure to 8 ppm γ-HCH resulted in reduced plasma T and $E_2$ concentrations in female clarias catfish, although effects at this dose were more dramatic in females in the later stages of vitellogenesis.[352] Similar results were obtained following exposure of female freshwater catfish to 4 and 8 ppm γ-HCH for 4 weeks.[350] A decrease in T and $E_2$ was observed in all stages of females examined (preparatory, prespawning, spawning, post-spawning, and resting), and sensitivities appeared to increase from the preparatory to the spawning phase. Although the exact mechanism underlying the effects of γ-HCH remains unknown, it has been suggested that this insecticide may inhibit gonadal recrudescence by reducing GTH secretion or the number of GTH receptors, which would likely interfere with steroidogenesis.[350,352]

#### 39.3.4.5 *Vinclozolin*

Vinclozolin (3-(3,5-dichlorophenyl)-5-ethenyl-5-methyl-2,4-oxazolidinedione) is a dicarboximide fungicide used on vegetables and fruits. Two metabolites, M1 and M2, have been reported to be antiandrogens. Pregnant rats that received vinclozolin during gestation produced male offspring with reduced anogenital distance, cleft phallus, retained nipples, and hypospadias.[353,354] Laboratory studies have reported Leydig cell hyperplasia, testicular tubular cell atrophy, penile hypoplasia, as well as hypospadia and infertility in male offspring.[355] Vinclozolin exerts its effects by binding to the androgen receptor.[310,356]

### 39.3.5 Non-Organochlorine Pesticides

#### 39.3.5.1 *Organophosphate Pesticides (OPs)*

Organophosphate chemicals have been used as nerve gas and insecticides. Their effectiveness as chemical-warfare agents and insecticides stems from their ability to inhibit acetylcholinesterase, an enzyme required for basic neuronal function.[327] The majority of OPs are lipophilic liquids, although they tend to have greater polarity and water-solubility than OCPs. Due to their innate instability, most OPs are not believed to bioconcentrate in aquatic species, although select fish species are highly sensitive to the toxic effects of these chemicals. Currently, many OPs are still used in a number of countries to protect crops from insects and farm and domestic animals from endo- and ectoparasites. They may also be used to control disease vectors (e.g., mosquitoes).

There is some indication that OPs might induce endocrine-disrupting effects in wild mammals. Reproductive depression, including decreased percentage of females with embryos and a decreased percentage of births, has been observed in rodents (*Sigmodon hispidus, Microtus ochrogaster*, and *Reithrodontomys fulvescens*) exposed to diazinon.[357] Reproductive activity was also depressed in hispid cotton rats (*S. hispidus*) exposed to carbaryl, a carbamate.[358] A recent study on the population effects of terbufos, however, found no changes on reproductive activity, number of births, or litter size in deer mice (*Peromyscus maniculatus*) and white-footed mice (*P. leucopus*).[359]

In birds, there is some evidence showing altered gonadotrophin release after exposure to OPs. Japanese quail exposed to 10-ppm parathion responded with a decrease in LH levels.[360] Since this effect was observed at near-lethal concentrations (brain cholinesterase activity was inhibited by over 50%), it is unclear whether this represents a case of endocrine disruption or acute toxicity. Bobwhite quail (*Colinus virginianus*) fed 100-ppm parathion exhibited cessation of egg production, inhibition of follicular development, and reduced plasma LH concentrations.[361] Other laboratory studies have reported changes on incubation behavior and egg laying[362–365] and decreased gametogenic function in adult birds exposed to OPs.[366] OPs are also known to induce developmental toxicity effects in birds[141] such as short limbs and parrot beak (known as Type I defects) and skeletal deformities (Type II defects). Results from field studies, however, have generally failed to find effects on several reproductive success parameters (clutch size, hatchability, and number of young fledged/nest) in populations of birds exposed to OPs after agricultural spraying.[367–370]

Exposure of amphibians to OPs may result in altered metamorphosis. Development of bullfrog (*Rana catesbeiana*) tadpoles was significantly delayed after exposures to at least 1000-ppb malathion, possibly because of decreased thyroid function.[371] These concentrations, however, are above those commonly found in wetlands or streams after pesticide application. In a separate study, exposure of premetamorphic northern leopard and green frogs to environmentally relevant concentrations of OPs (< 0.01 ppm of basudin 500EC and technical-grade diazinon) caused deformities and delayed development.[372] Although exposure to the OP methyl parathion (1 ppm) has been associated with bone deformities in amphibians, similar to the Type II defects observed in birds,[373] its effects on metamorphosis are less clear.[374]

Chronic exposure to low environmental concentrations of OPs may lead to a variety of reproductive and developmental effects in fish. For example, Ram and Sathyanesan[375] exposed murrel (*Channa punctatus*) to 20-ppt cythion (50% malathion, 50% organic solvents) for 6 months and observed an increase in oocyte degeneration, which resulted in retarded ovarian growth and lower GSI. These responses were correlated with fewer and less active gonadotropin-producing cells in the pituitary. In males, spermatogenesis was arrested, and some necrotic spermatocytes were apparent. The authors speculated that reduced GTH levels might have contributed to the observed reproductive abnormalities. In freshwater perch (*Anabas testudineus*), although no short-term effects were observed after an exposure to 0.106 ppb Metacid-50 (50% methyl parathion), a reduction in GSI and plasma and ovarian $E_2$ concentrations was evident after 20 days.[376] Exposure to OP pesticides has been documented to elicit a series of histological alterations in ovaries of several fish species. Exposure to 0.1 ppm methyl parathion for 75 days resulted in substantial oocyte damage in carp minnow (*Rasbora daniconius*).[305] Similar effects have been observed in guppies (*Puntius conchonius*) exposed to 53 ppb monocrotophos for 2–4 months.[303] Guppies exposed to fenitrothion have also responded with decreased egg production and abortion.[377] Gonad weights and vitellogenesis were reduced in female striped catfish (*Mystus vittatus*) after 12 weeks of exposure to four different OPs (malathion, birlane, gardona, and phosdrin).[378] Histological abnormalities of the testis has been a common response in male fish exposed to OPs, having been reported in at least ten fish species.[56,161,379–387] Recent evidence indicates that exposure of newly hatched larvae to malathion might also result in developmental alterations (deformed notochord).[388]

Exposure of fish to OPs has also been associated with declines in the concentrations of hormones and VTG. For instance, exposure to malathion and to cythion results in reduced plasma concentrations of $E_2$, T, and VTG.[161] In another study, female catfish (*Heteropneustes fossilis*) had decreased plasma $E_2$ concentrations after an exposure to 1.2-ppm malathion for 72 h.[389] Similar results were reported in studies examining the effect of malathion on sex-steroid concentrations during different phases of the reproductive cycle of the clarias catfish.[352] It was concluded from this study that sensitivity to malathion appeared to increase from the pre- to the post-vitellogenic phases, the latter of which involves ovulation and spawning. Carbaryl-induced thyroid dysfunction has also been reported in this species of freshwater fish.[390]

Evidence suggests that OPs may affect steroidogenesis by acting at multiple sites along the hypothalamic-pituitary-gonadal-liver axis. In fish, exposure to certain OPs reduces GnRH-like factor levels in the hypothalamus and impairs pituitary activity and release of GTH.[391–393] Additional studies have shown reduced 3-hydroxycorticosteroids and 17 hydroxycorticosteroids (3-HSD and 17-HSD) activities in ovaries and testis of exposed fish.[56,57] Singh[384] proposed that malathion reduces $E_2$ in the Asian rice eel by interfering with the enzyme aromatase. Malathion, however, appears not to affect cholesterol biosynthesis, although it has been shown to alter the synthesis and mobilization of other lipids as well as the hydrolysis of esterified cholesterol to free cholesterol.[394,395]

#### 39.3.5.2 Carbamate Pesticides

The carbamate pesticides, many of which are in current use, are derivatives of carbamic acid. Like the OPs, they act as acetylcholinesterase inhibitors, vary with regard to water solubility, and are relatively nonpersistent in the environment. Carbaryl is a carbamate pesticide that controls over 100 species of insects on a variety of crops, agricultural animals, and pets. Although carbaryl has been reported to accumulate in certain species of fish and invertebrates, the risk of biomagnification is low due to its rapid metabolism and degradation. Carbofuran is another carbamate pesticide with wide application, although granular forms were banned in the U.S. in 1994 following a number of bird kills. Carbofuran is used to protect field, fruit, vegetable, and forest crops from insects, mites, and nematodes.

Laboratory studies with amphibians have reported growth inhibition and increased incidence of developmental deformities in tadpoles exposed to carbamate pesticides.[372,373,396] In fish, carbamate pesticides have been shown to induce histopathological alterations in the ovaries and testes. Female fish exposed to carbofuran (range of 1 to 5 ppt) have responded with reductions in GSI, inhibitions in oocyte growth, and increases in oocyte atresia.[379,397] Some of the lesions seen in ovarian tissue of carbofuran-exposed fish have included decreased oocyte diameter, a predominance of immature oocytes, and damage to yolk vesicles and oocyte structure.[305,398] In males, carbofuran causes declines in testes weight and delays in spermatogenesis and induces necrosis of spermatogonia and spermatocytes.[379,380] Although generally less toxic to fish than carbofuran, carbaryl exposure results in many of the same reproductive alterations. Carbaryl-induced alterations, including reduced GSI and plasma $E_2$ concentrations, inhibited oocyte growth, increased oocyte atresia, and damage to yolk vesicles and oocyte structure has been observed in several species of fish after exposures ranging from 2 to 20 ppt.[304,376,387] Increased larval mortality and decreased production and hatchability of eggs are additional responses to carbaryl exposure.[399]

Little information is known regarding the mechanisms by which carbamate pesticides induce reproductive anomalies in fish, although results from both field and laboratory experiments suggest that carbaryl may act at the level of the pituitary by altering GnRH and GTH serum concentrations.[391] Results from the *in vitro* E-screen indicate that neither carbaryl nor carbofuran are estrogenic in nature.[400] Carbofuran also appears to raise cholesterol and phospholipid concentrations in the ovaries and testes of fish, while lowering overall protein, RNA, total lipids, and ascorbic acid.[401]

#### *39.3.5.3 Organometal Pesticides*

Certain metals are highly insoluble in their inorganic form and, therefore, possess little to no toxicity. However, since metal toxicity may be greatly enhanced if binding to an organic ligand occurs, some metals have been modified intentionally to increase their toxicity for use as pesticides. Until 1993, organomercury was used as an antifungal seed dressing in the United Kingdom, organolead has been applied to fruit crops to control caterpillars, and organotin compounds have served a number of functions due to their extreme toxicity. For example, tributyltin (TBT) has served as an algicide, miticide, fungicide, and insecticide, and since the 1960s it has also functioned as a marine antifouling agent. Both agricultural and maritime applications have led to the contamination of aquatic environments. Although the half-life of TBT in water is brief (days to weeks), organotin compounds have the potential to bioaccumulate in aquatic organisms.

One of the best-documented cases of endocrine disruption comes from the work done with marine gastropods exposed to organotin compounds (mainly TBT) contained in antifouling paints. Laboratory and field studies have demonstrated that female gastropods exposed to environmentally relevant doses of TBT develop an irreversible sexual abnormality known as "imposex." This masculinization process involves an increase in T concentrations, which is followed by the imposition of male sex organs (penis and vas deferens) over the oviductal tissues, causing abnormal breeding activity and, in many cases, sterility and population declines.[402–407] Depending on the species and dose attained, oogenesis might be completely supplanted by spermatogenesis. TBT can also induce alterations in the behavior and development of bivalve larvae.[407] It is estimated that about 72 species and 49 genera of prosobranchs have been affected worldwide.[8] In the case of the highly sensitive common dogwhelk (*Nucella lapillus*), imposex is induced at exposures as low as 1–2 pptr, with complete suppression of oogenesis at TBT concentrations above 3–5 pptr.[403] Birds that feed on mollusks have been shown to accumulate butyltins to a greater degree than birds preying on fish, birds, or mammals.[408]

Although the precise mechanisms by which TBT causes endocrine disruption in invertebrates are not entirely known, recent evidence suggests that this compound may act as a competitive inhibitor of CYP450-mediated aromatase.[402,404] TBT may also interfere with sex-steroid metabolism, inhibiting the formation of sulphur conjugates of T and its active metabolites. In addition, TBT is capable of inducing cytotoxic and genotoxic damage to embryonic and larval stages in invertebrates.[409,410]

In fish, organotin compounds are readily bioaccumulated and stored in different tissues, including the gonads.[411] In several fish species, exposure to organotins has been associated with delayed hatching, high embryo and larval mortality, and retarded yolk-sac resorption.[412–417] In guppies, exposure to TBT (11.2–22.3 pptr) and BPA (274–549 ppb) results in significant declines (by 40–75%) in total sperm counts after 21 days.[418] Organotin compounds also induce several alterations in ovaries and testes of fish. Three-spined sticklebacks (*Gasterosteus aculeatus*) exposed to bis(tributyltin)oxide (TBTO) for up to 7.5 months experienced no seasonal increase in GSI, as was apparent in control animals. In addition, ovaries from exposed animals contained 25% resorbing oocytes, as opposed to 0% in controls.[419] In this study, however, several other reproductive endpoints (including spawning behavior, fecundity, hatchability, frequency of deformed fry, and secondary sex characteristics) were not affected by treatment.

#### 39.3.5.4 Triazine Pesticides

The triazine pesticides include some of the most extensively used herbicides in North America. Indeed, atrazine is used to control weeds on more than two thirds of the U.S. acreage containing corn and sorghum as well as 90% of sugarcane acreage. Simazine, another member of this class of herbicides, is currently applied to 30 high-value crops including a variety of fruits, vegetables, nuts, turfgrass, and conifers. Despite the widespread use of these chemicals, relatively little is known regarding their potential health effects to humans and wildlife. Atrazine is only slightly toxic to fish, and the risk for bioaccumulation is extremely low due to its propensity for rapid degradation to less or nontoxic metabolites.

The effects of atrazine on amphibian metamorphosis have been examined in some detail. Tiger salamanders (*Ambystoma tigrinum*) exposed to low concentrations of atrazine (75 ppb) developed at slower rates but were similar in size when compared to controls.[420] In contrast, exposure to high atrazine concentrations (250 ppb) resulted in similar developmental rates but decreased sizes. In addition, $T_4$ was elevated in both groups, whereas corticosterone was depressed in the low-dose group only. These authors hypothesized that the suppression of corticosterone could have resulted in a decreased conversion of $T_4$ to the active form $T_3$, thereby slowing metamorphosis and allowing increased growth. In contrast, Allran and Karasov[421] found no effects on developmental rate and metamorphosis in northern leopard frogs exposed to atrazine (20 and 200 ppb). Atrazine has also been shown to induce teratogenic changes in frog embryos, but at concentrations approaching maximum solubility in water.[422]

Few studies have examined endocrine or reproductive function in fish exposed to atrazine or other triazine pesticides. Channel catfish and gizzard shad (*Dorosoma cepedianum*) maintained for 4.5 months in ponds containing 20-ppb atrazine failed to reproduce, and reproductive success of bluegills (*Lepomis macrochirus*) was reduced by more than 95%.[423] Since the dietary habits of bluegill were largely affected by the herbicide treatment, the authors suggested that impaired reproduction might have been due to impoverishment rather than to direct effect of atrazine exposure. Results from our laboratories have shown that atrazine affects sex steroids in male and female largemouth bass (Gross et al., unpublished data). After 20 days of exposure, plasma 11KT concentrations were elevated in males exposed to 100-ppb atrazine, and $E_2$ concentrations were increased in females exposed to 50- and 100-ppb atrazine. Studies with largemouth bass have also shown that when ovarian follicles are incubated with 10 ppb atrazine, it results in an increased $E_2$ and a decreased T production. Furthermore, *in vitro* T synthesis is greatly reduced when gonads are incubated with a combination of atrazine and floridone or atrazine and chlordane.

Studies with American alligators have shown that atrazine might induce differential responses in developing embryos depending on timing of exposure. Crain et al.[424] reported that atrazine (14 ppb) induced gonadal aromatase activity in male hatchling alligators exposed *in ovo*. In a later study, however, incubation of alligator eggs with atrazine prior to the critical period of gonadal differentiation did not influence sex determination and had no apparent effect on gonadal structure (measured as sex-cord diameter in males, Müllerian duct epithelial cell height, and medullary regression of the ovaries in females) or hepatic aromatase activity.[425] Since most endocrine changes associated with atrazine have been reported in normally organized reproductive systems, the authors hypothesized that the lack of noticeable effects in the latter study was the result of exposing embryos during very early developmental stages, i.e., prior to or during the development of the reproductive system.

Current evidence suggests that atrazine induces endocrine-disruptive effects by acting as a steroid hormone antagonist (antiandrogen or antiestrogen), probably through nonreceptor-mediated mechanisms. Indeed, a number of *in vivo* and *in vitro* studies have failed to detect estrogenic activity for triazines. In two independent studies, oral exposure to atrazine and simazine did not increase uterine weight in immature or ovariectomized female Sprague-Dawley rats,[426,427] and cell-proliferation and binding studies found no evidence for either agonistic or antagonist activity for this herbicide.[426] Furthermore, atrazine and related compounds failed to demonstrate estrogenic activity in human and yeast cells expressing the ER and an estrogen-sensitive reporter gene,[428,429] although the triazines have displaced radiolabeled estradiol from the ER in competition studies.[363,429] Also, a study by Danzo[74] showed that atrazine did not reduce radiolabeled $E_2$ binding to rabbit uterine ER, although it inhibited the binding of DHT to androgen receptor sites in rat testes and reduced the binding of DHT to the androgen-binding protein by 40%. Triazines may also disrupt reproductive function by altering LH and PRL concentrations.[430]

### 39.3.6 Complex Environmental Mixtures

#### *39.3.6.1 Pulp- and Paper-Mill Effluents*

Over the past 15 years, a number of investigators have studied the effects of pulp- and paper-mill effluents on feral and laboratory fish populations. In general, fish exposed to these effluents experience alterations in steroid biosynthesis, gonadal development, sexual maturation, and expression of secondary sex characteristics. Identifying the causative agents in the effluent and establishing cause-and-effect relationships, however, have been challenging tasks, since pulp- and paper-mill effluents are complex mixtures, and the components are not entirely known. Furthermore, variations in wood finish, in the pulping and bleaching process, and in the treatment of effluents between mills lead to different effluent compositions. Nevertheless, all pulping protocols involve the separation and discharge of natural wood components, such as sugars, lipids, resins, and fatty acids, which generally undergo bacteriological treatment in settling and aeration ponds. Depending on the bleaching techniques used, pulp- and paper-mill effluents may also contain different kinds and concentrations of chlorinated organic compounds such as PCDDs and PCDFs.

The most thorough field studies on the reproductive effects of paper-mill effluents have been conducted at Jackfish Bay, Lake Superior. Jackfish Bay has received bleached kraft-mill effluent (BKME) from a nearby pulp mill since 1949 and, therefore, has provided a convenient site for studying the impact of BKME on several fish species. BKME-exposed white suckers show decreased concentrations of several sex-steroid hormones (T, 11KT, and $E_2$).[128,431–435] Declines in steroid concentrations have also been documented in longnose sucker (*Catostomus catostomus*) and lake whitefish (*Coregonus clupeaformis*) from Jackfish Bay,[432,434] in white sucker at other mills,[436–439] and in other effluent-exposed fish species sampled elsewhere.[432,440,441] The consequences of these similar endocrine alterations to whole-animal reproductive fitness and population dynamics, however, have varied greatly among species. For example, longnose sucker exposed to BKME show no organism responses other than an altered age distribution, whereas white sucker and lake whitefish show decreased gonadal sizes, secondary sexual characteristics and egg sizes, and increased age to maturity.[432] In a review of whole-organism responses of fish exposed to different kinds of mill effluents (including unbleached pulps), 80% showed increased age to sexual maturation, and reduced gonadal size was reported in 58% of the studies.[442] These observations provide evidence for species differences in susceptibility to BKME but also show the inherent difficulty when trying to compare biological responses in fish populations inhabiting highly different environments and exposed to complex mixtures likely to vary in chemical composition.

There are relatively few studies on the effects of BKME on egg and fry parameters, and the results from these studies are conflicting. Fertility (as indicated by the percentage of spawned eggs that hatched) was decreased in zebrafish after exposure to chlorinated phenolics from a bleach plant effluent[443] and in brown trout after exposure to BKME.[444] Hatchability was also reduced in pike after exposure of eggs to BKME concentrations as low as 0.5%.[445] Similarly, many field and laboratory studies have reported decline in fecundity in several fish species exposed to paper-mill effluents.[435,436,443,446,447] McMaster et al.,[448] on the other hand, found equal or greater fertilization rates and no effects on hatchabilities of white sucker eggs, despite declines in sex-steroid concentrations, gonad and egg sizes, and sperm motility in BKME-exposed fish. In addition, fecundity and hatchability were not altered after exposures to BKME in several other field[440,449,450] and laboratory studies.[441,451] There is very little information on the developmental effects of BKME. In the laboratory, survival from larvae to adult and growth of fathead minnows (*Pimephales promelas*) were not affected after exposures to up to 20% effluent concentrations.[446,451,452] Studies conducted in our laboratories with largemouth bass, however, have found similar fecundities and hatchabilities but decreased fry growth and survival after exposures to 10% bleached/unbleached kraft-mill effluent (B/UKME).[441] Similarly, Karås et al.[450] reported comparable fecundity and egg mortality in perch (*Perca fluviatilis*) from a BKME-exposed area, but fry hatched from this site were smaller and had an increased frequency of abnormalities, which was translated into lower abundances of fry and young-of-the-year fish. These authors concluded that exposure of perch to BKME had resulted in high mortality rates close to the time of hatching due to either chronic failure of parental reproductive systems or acute toxicity to embryos or early larvae.

Exposure to pulp- and paper-mill effluents has also been associated with alterations in secondary sex characteristics (see Reference 453 for a review of the evidence for masculinization in poeciliids from Florida). Female mosquitofish in Florida inhabiting a stream receiving paper-mill effluents were reported to be strongly masculinized, showing both physical secondary sex characteristics (fully developed gonopodium) and reproductive behavior of males.[454] More recently, masculinization of female fish has been identified from an additional two species (least killifish, *Heterandria formosa* and sailfin molly, *Poecilia latipinna*) collected from an effluent-dominated stream.[455] Masculinization of female fish has been attributed to the action of androgenic hormones that result from the biotransformation of plant sterols (and also cholesterol and stigmasterol) by bacteria such as *Mycobacterium*.[456]

Results from studies on white sucker indicate that several sites within the pituitary-gonadal axis are affected after exposure to BKME. Fish from exposed sites had significantly lower plasma levels of gonadotropin (GTH-II) and showed depressed responsiveness of sex steroids and 17,20ß-dihydroxy-4-pregnen-3-one (a maturation-inducing steroid) after GnRH injections.[457] BKME-exposed fish also had lower circulating levels of testosterone glucoronide, which would be suggestive of altered peripheral steroid metabolism. Similarly to what was observed under *in vivo* conditions, *in vitro* incubations of ovarian follicles collected from BKME-exposed females have also shown reduced steroid production.[433,457] The similarities between both types of studies would suggest that reductions in plasma steroid levels in BKME-exposed fish are mainly due to alterations in ovarian steroid production. Recent studies on white sucker have shown increased apoptotic DNA fragmentation and increased expression of a 70-kDa heat-shock protein in oocytes from prespawning females, which coupled with lower sex steroids may explain the observed decreased gonad weights and delayed sexual maturity.[458]

Although there is extensive literature on the reproductive effects of BKME on fish, very little is known about the chemical compounds that could be held responsible for such changes. Compounds such as dioxins and furans were the first to blame, because of their persistence, bioaccumulative properties, and their known deleterious reproductive and antiestrogenic effects.[434] Recent evidence, however, suggests that the chemicals in pulp-mill effluents responsible for reproductive alterations are relatively short-lived and readily metabolized by fish. For example, mixed-function oxygenase induction and endocrine alterations have also been reported downstream from mills that do not use chlorine bleaching,[432] and these parameters have rapidly returned to normal after cessation of exposure.[434] Indeed, several of the natural wood components in the final effluent, such as sterols, lignans, stilbenes, and resin acids, are believed to be weak estrogens.[459,460] For example, the plant sterol β-sitosterol has demonstrated estrogenic activity by its ability to induce VTG in juvenile rainbow trout[460] and male goldfish[461] and bind to the ER in rainbow trout hepatocytes.[462] Conversely, various phytosterols that survive the treatment process have displayed masculinizing effects under controlled experimental conditions.[463]

### *39.3.6.2 Sewage-Treatment Effluents*

Researchers have shown that effluents coming from sewage-treatment plants might cause estrogenic effects in fish due to their ability to induce the production of VTG (a female specific egg-yolk precursor) in males.[9,464] Recent information has also shown an increase in the incidence of intersex, or hermaphroditism, in populations of wild fish inhabiting rivers contaminated with sewage effluent.[465] Presently, the population-level effects of increased VTG in male fish remain poorly understood, although they are known to be associated with decreased testicular growth.[466,467] Chemical analysis of effluents from sewage-treatment plants has identified several compounds with estrogenic properties, including natural estrogens ($E_2$ and $E_1$), synthetic estrogens widely used in birth-control pills, alkylphenolic chemicals (resulting from the breakdown of nonionic surfactants), plasticizers (bisphenol-A), and phthalates.[466,468,469] The following section reviews the major findings on the endocrine-disrupting effects of the above groups of chemicals in fish.

#### *39.3.6.2.1 Natural and Synthetic Estrogens*

Recent studies using chemical fractionation and biologic screening techniques suggest that natural and synthetic steroidal estrogens may be causing the greatest estrogenic effects in fish inhabiting streams contaminated with sewage effluents.[470] This stems from the fact that both types of estrogens, but especially the synthetic ones, are highly potent hormones, and thus concentrations in the pptr or less are capable of inducing biological effects. For example, $EE_2$ induces VTG synthesis in male rainbow trout at concentrations as low as 0.1 pptr.[471] $EE_2$ concentration in English rivers has ranged from 0.2 to 7 pptr.[472] Although the reproductive consequences of $EE_2$ exposure in fish are mainly unknown at this time, they have been associated with decreased testicular growth and development in immature fish.[466] Altered spermatogenesis has also been reported in fish exposed to natural estrogens.[473] Women are the primary sources of natural and synthetic estrogens in sewage effluents, either because of the excretion of natural and synthetic estrogens as inactive conjugates during menstrual cycling or because of the use of contraceptive pills. During the sewage-treatment process, these conjugates are biotransformed into their parent and biological active compounds.

#### *39.3.6.2.2 Alkyl Phenol Ethoxylates (APEs) and Alkyl Phenols (APs)*

APEs are effective nonionic surfactants serving as components of industrial and domestic detergents, pesticide formulations, cosmetics, and paints. Of all APEs produced, nonylphenol-polyethoxylates and octylphenol-polyethoxylates constitute approximately 80% and 20%, respectively. These chemicals are biodegraded during sewage treatment to form APs such as nonylphenol and octylphenol. Industrial effluents might contain over 100 ppb of nonylphenol, although most streams surveyed in the United Kingdom and in the United States contained equal to or less than

10 and 0.1 ppb, respectively.[474,475] Nonylphenol and octylphenol are hydrophobic and lipophilic and thus can accumulate in sediment and fish adipose tissue. Both APEs and APs are known to have estrogenic properties, as discussed below.

Alkylphenolic chemicals might also be playing an important role as xenoestrogens in sewage effluents. Male rainbow trout exposed to four AP chemicals responded with significant increases in plasma VTG concentrations, particularly after treatment with at least 3 ppb octylphenol.[466] Nonylphenol and two carboxylic acid APE degradation products also induced VTG production in males in this study, but at higher concentrations. Testicular growth was inhibited in response to all four chemicals, with octylphenol having the greatest inhibitory effect. Christianson et al.[476] reported similar effects in male eelpout (*Zoarces viviparous*) exposed to nonylphenol. Twenty-five days after a 10- to 100-ppm nonylphenol injection, a significant increase in plasma VTG with a concomitant decrease in GSI was observed. Histological examination revealed degenerated seminiferous lobules in exposed males as well as decreased guanosine triphosphate (GTP) activity (a marker for Sertoli cell function). Plasma VTG induction has also been reported following nonylphenol exposure in male and immature female rainbow trout,[477] male flounder,[478] male and female Atlantic salmon (*Salmo salar*),[479,480] Japanese medaka,[479,481] and immature channel catfish.[482] Nonylphenol (25 ppm) has also caused a dramatic increase in plasma zona radiata proteins in juvenile female Atlantic salmon.[480] Gray et al.[483] recently reported a reduction in courtship activity in adult Japanese medaka males exposed to octylphenol from 1 day post hatch to 6 months post hatch. In this study, transgenerational effects were also observed (i.e., an increase in fry developmental abnormalities). Similarly, AP-induced developmental toxicity effects have been reported in embryos and larvae of killifish after exposures to octylphenol and 4-*tert*-octylphenol.[484] Disruption of sexual differentiation is yet another effect observed in fish exposed to APs, having been reported in common carp[485] and mosquitofish.[486]

The endocrine-disrupting properties of APs and APEs are mainly related to their ability to bind to the ER. Indeed, APEs have been shown to be estrogenic using several *in vitro* bioassays.[94,104,487] Similarly, APs substituted at position 4 (e.g., 4-nonylphenol) have demonstrated estrogenic activity in various *in vitro* and *in vivo* bioassays.[73,89,91,94,488] P-substituted phenols, such as 4-t-pentylphenol (TPP), are believed to be among the most potent estrogens.[488,489] Ren et al.[490] suggested that nonylphenol may also be involved in the post-transcriptional regulation of VTG mRNA processing. Finally, recent research shows that APs can induce reproductive alterations through an increase in the rate of apoptosis of Sertoli cells, phenomena that can negatively affect the development and release of sperm.[491]

#### *39.3.6.2.3 Bisphenol A*

Bisphenol is the generic name given to a group of diphenylalkanes commonly used in the production of plastics. Bisphenols consist of two phenolic rings joined by a carbon bridge. The bridging carbon has no substituent in bisphenol F and two methyl groups in bisphenol A (BPA). Incomplete polymerization or depolymerization of plastics from heating may result in the release of BPA into the environment and subsequent human and animal exposure. The first reports to document the estrogenic potential of the bisphenols appeared in the 1930s,[492] and a number of investigators employing a variety of techniques have since confirmed those results.[493,494] Evidence suggests that estrogenic potency of these compounds increases with the length of the alkyl substituent at the bridging carbon as well as due to the chemical nature of the substituents.[495] Bisphenols with hydroxyl groups in the para position and an angular conformation are suitable for binding the ER at the acceptor site.

The estrogenicity of BPA has been demonstrated by several *in vitro* assays.[91,95,100,400,489,496,497] More recently, BPA has been shown to induce the synthesis of the VTG protein in rainbow trout liver slices;[498] VTG mRNA in rainbow trout primary hepatocyte cultures;[499] and VTG and zona radiata proteins in Atlantic salmon primary hepatocytes.[500] In the latter study, BPA inhibited the

$E_2$-stimulated induction of VTG and zona radiata proteins, suggesting that the effects of the plasticizer are truly estrogenic in nature. Arukwe et al.[480] observed a dose-dependent increase in plasma VTG and zona radiata proteins following a single intraperitoneal injection of BPA.

Interestingly, recent evidence suggests that the *in vivo* estrogenicity of BPA may be greater than predicted by *in vitro* assays.[501] Male Japanese medaka were exposed to BPA for two weeks and then introduced to a tank with untreated females for spawning studies.[481] In the experimental group, the number of hatchings was reduced, and the concentrations that affected reproduction in this study were lower than concentrations that produced effects in some *in vitro* studies. Furthermore, VTG synthesis was observed at concentrations below those affecting reproduction. BPA is also known to cause significant declines in sperm production.[418,502]

#### *39.3.6.2.4 Other Phenolics*

Several phenolic compounds other than alkyl phenols and BPA have been evaluated for their impact on fish reproduction. For example, polychlorinated phenols are often formed during the chemical reaction of chlorine and phenolic compounds in wood pulp. The polychlorinated phenols are acidic and are chemically reactive compounds of low persistence because of their water solubility. Pentachlorophenol (PCP), a commonly used fungicide in wood preservation, often enters the environment as a component of domestic and industrial effluents, primarily from the forest-products industry.

There is evidence that reproductive effects can be elicited when fish are exposed to PCP. Female rainbow trout exposed for 18 days to sublethal concentrations of PCP (22 and 49 ppm) during the primary ovarian growth phase displayed a significant increase in oocyte atresia and a trend toward decreasing oocyte diameter.[503] The use of purified PCP in this study rebutted the claim that toxicity of technical PCP is due to contamination by PCDDs, PCDFs, or other chlorinated phenols.[504] It has been suggested that in rainbow trout, PCP affects oogenesis by interfering with the production of yolk in the liver.[73] These authors found that PCP may act as an estrogen antagonist since it has shown a slight inhibitory effect on $E_2$-stimulated induction of the ER mRNA and a substantial inhibitory effect on $E_2$-stimulated induction of VTG mRNA. In addition, a study with *Daphnia magna* found that PCP is capable of altering steroid hormone biotransformation and elimination pathways.[505] The potential estrogenicity of several other phenolic compounds was tested by Jobling et al.[468] using a trout ER competition study and several mammalian cell assays. 2,4-dichlorophenol, a component of fungicides and germicides, reduced the binding of radiolabeled $E_2$ to the trout ER. Conversely, 3,4-dimethylphenol and 2-methylphenol, which also serve as fungicides and disinfectants but have no chlorine group, failed to compete for ER binding. In another study, Mukherjee et al.[53] examined the effect of phenol on the steroidogenesis and reproductive activity in sexually maturing carp. After 48 days of exposure to 8-ppt phenol, GSI was reduced, ovarian and liver cholesterol concentrations were increased, and cholesterol conversion to sterol products was inhibited. Previous studies by Kumar and Mukherjee[506] also demonstrated phenol-induced alterations in plasma, ovarian, and hepatic cholesterol concentrations in several species of fish.

### 39.3.7 Metals

Although metals are natural substances, human activity is largely responsible for their abnormal release and accumulation in the environment. Metal toxicity usually results from exposure to high levels of nonessential metals such as mercury (Hg) or cadmium (Cd). Since these and all other metals are nonbiodegradable, the body cannot metabolize them into less toxic forms. Instead, detoxification involves binding to specific proteins (e.g., metallothionein) that function to shield toxic properties or to produce insoluble forms (e.g., intracellular granules) for long-term storage or excretion. If not excreted, some metals can bioaccumulate in tissues, especially if the individual occupies a position at the top of the food chain. Metals probably do not act as classic EDCs, i.e., modulating receptor-mediated effects. Instead, their mechanism of action may involve toxicity of endocrine tissues, altered enzyme binding, and CNS interactions.

### *39.3.7.1 Mercury (Hg)*

Mercury is a nonessential heavy metal found naturally in the environment and used in many industries including battery, paper, paint, chemical, and agriculture, as well as dentistry and medicine. The burning of coal, natural gas, and refining of petroleum products adds 5000 tons of Hg per year to the atmosphere, increasing Hg contamination of aquatic ecosystems worldwide.[507,508] Hg enters aquatic systems either indirectly by atmospheric deposition or from direct discharge of mercuriferous wastes into watersheds.[509,510] Conditions of low pH and high dissolved organic carbon increase the methylation of inorganic Hg to the more toxic methylmercury (MeHg).[511] This methylated form is rapidly bioaccumulated by aquatic species, with body burdens in piscivores increasing with trophic level.[512] While there have been many studies measuring Hg concentrations in wildlife,[513–519] little information is available on its potential effects.

In laboratory tests, Hg has produced stillbirths in dogs and pigs and abortions,[520] abnormal sperm,[521] and low conception rates in the macaque (*Macaca fascicularis*).[522] Hg-laden rats have reduced litter size[523] and decreased survival.[524] In mice, Hg produces decreased fetal survival,[525] fetal malformations,[526] embryo resorption,[527] low sperm counts,[528] and tubular atrophy of testes.[529] Abortions have been reported in guinea pigs (*Cavia porcellus*),[530,531] and Hg in the Florida panther (*Puma concolor coryi*) is thought to reduce kitten survival.[532] These effects in offspring are expected because Hg (both inorganic and organic) is able to cross the placenta, producing behavioral deficits, impaired fertility, and fetal death.[533,534]

Hg impairs reproductive success in birds. Some studies have found a negative correlation between hatching success and Hg concentrations in eggs[518,535,536] or feathers.[537,538] Common terns with liver Hg concentrations between 9 and 21 ppm wet weight showed decreased hatchability and reduced nesting success.[539] Concentrations between 3 and 14 ppm in common loons decreased hatchability and at 52 ppm reduced nesting success.[535] In this same study, brain concentrations >2 ppm reduced egg laying and decreased nest and territory fidelity.[535] Egg concentrations from 0.5 to 1.5 ppm wet weight decreased hatchability in pheasants.[540,541] Mallard eggs with externally applied methylmercury chloride (Me-HgCl) showed decreased embryo weights, developmental abnormalities, and embryonic death.[542] Juvenile survival was also decreased in these studies because of neurological damage. Ducks fed Hg over three generations had decreased reproduction, and ducklings exhibited altered behavior.[542]

There is little information on Hg reproductive toxicity in reptiles and amphibians. Frogs (*Rana cyanophlyctis*) kept in water for less than 3 months had decreased GSI as well as reduced numbers of sperm bundles and increased secondary spermatogonia, indicating a blockage in mitosis and thus in the conversion of spermatogonia into primary spermatocytes.[543] Hg contamination is also reported to have caused a loss in germ cells and sterility in *Rana nigromaculata*[544] and reduced survival in African clawed frog.[545]

In fish, Hg exposure has caused decreased GSI and a variety of gonadal abnormalities.[546] Other responses to exposure involve altered lipid and cholesterol ovarian content,[547] reduced spermatogenesis,[347] and impaired fertilization.[548,549] Studies with tilapia (*Oreochromis niloticus*) have shown reduced plasma $E_2$ in females and plasma 11KT in males with muscle concentrations ranging from 1 to 7 ppm wet weight.[550] Similar effects have been seen in largemouth bass with muscle Hg as low as 0.25 ppm wet weight.[107] These studies suggest a CNS involvement with potential effects on the hypothalamic-pituitary-gonadal axis in fish. Hg may depress hormone production by acting on the gonads and interfering with their development. Testicular atrophy was observed in tilapia with Hg concentrations between 0.4 and 2.7 ppm dry weight and in guppies (*Poecilia reticulata*).[347,550]

### 39.3.7.2 Other Metals

Lead (Pb) is a heavy metal released into the atmosphere from industrial emissions and motor exhaust. It is a nonessential, toxic metal that affects all body systems. Reproductive effects in mammals include alterations in implantation, embryonic development, and reproductive organs.[551,552] Hildebrand et al.[553] found that blood levels of Pb above 390 ppm in male rats induced prostatic hyperplasia, impaired sperm motility, reduced testicular weight, and caused seminiferous tubular damage and spermatogenic cell arrest. Pb is also capable of crossing the blood-brain barrier, interfering with the central nervous system.[554,555] Young can also be exposed through the maternal milk.[556] The fetus and developing young are most sensitive to chronic levels of Pb exposure. Pb blood levels as low as 70–80 ppm can result in neurobehavioral symptoms.[557,558] During embryo development, Pb can cross the placenta and be linked to reduced gestational age and lowered birth weight.[559]

Birds become contaminated with Pb through the consumption of Pb shot or bullets or fishing sinkers. After ingestion, these items gradually dissolve and the birds become progressively weaker and emaciated. Other species, especially raptors, which may prey on these species, are poisoned by ingesting contaminated prey.[560]

There is little research on the reproductive effects of Pb in reptiles. Red-eared slider turtles injected with Pb had reduced righting response,[561] and this heavy metal reduced the rate of development of the Jefferson salamander (*Apystoma jeffersonianum*).[562] Although known to accumulate in fish tissues,[563] few studies have examined the effects of Pb exposure on fish reproduction or endocrine functions. Female Atlantic croaker fed as little as 0.05–0.2 ppm/day for 1 month had reduced GSI, $E_2$, and T.[158,159] Female climbing perch (*Anabas testudineus*) exposed via water to 1.25 ppm Pb for 1 month had lower GSI,[564] and retarded ovarian growth was observed in clarias catfish following long-term exposure (275 days) to 5 ppm.[565] In addition, decreased spermatogenesis and ovarian atresia have been observed in rosy barb exposed to a low dose of Pb nitrate (0.12 ppm) for 60–120 days,[566] and decreased spermatogenesis and testicular hemorrhage were reported in the striped gourami (*Colisa fasciatus*) following 4 days of exposure to 15 ppm Pb nitrate.[567]

Copper (Cu) is an essential metal that is necessary for the activity of various enzymes and for iron utilization. This metal has received widespread use in the preservation and coloring of foods, in brass and copper water pipes and domestic utensils, and in fungicides and insecticides, the latter providing the primary route of exposure to aquatic animals. Egg and larval mortality of the Jefferson salamander were decreased by exposure to Cu.[562] Several studies involving different fish species report a spectrum of reproductive abnormalities following exposure to Cu. Decreased spermatogenesis and ovarian atresia were observed in female rosy barb,[566] whereas testicular abnormalities and arrested spermatogenesis were observed in the male guppy (*Lebistes reticularis*).[568] Although VTG is known to serve as a carrier for many metals, including Cu, the effects of this metal on vitellogenesis are not clear. Cu suppressed vitellogenesis in female mussels (*Mytilis edulis*),[569] whereas cupric acetate appeared to have no effect on vitellogenesis in female clarias catfish.[565] Additional studies involving Cu exposure report decreased egg size and a propensity for deformities on larvae of white sucker,[570] reduced egg viability and hatchability in brook trout,[571] and impaired fertilization and increased larval abnormalities in topsmelt (*Atherinops affinis*).[572]

Cadmium (Cd) is a by-product of copper, lead, and zinc mining and is also found in industrial sludges and phosphate fertilizers. Cd has been reported to accumulate primarily in the liver and kidneys;[573,574] however, several investigators have also detected Cd in the gonads.[575,576] In rats, Cd has been shown to cause Leydig cell tumors[577] and induce prostatic and interstitial cell tumors.[578,579] Cd suppresses egg production in mallards[580] and chickens.[581] Slight gonadal alterations were found in mallards fed Cd and accumulating kidney concentrations up to 50 ppm, while those with kidney concentrations of 100 ppm exhibited testicular atrophy and no sperm production.[582] Again, actively reproducing seabirds have been found with similar kidney concentrations.[583] In amphibians, studies with the African clawed frog showed that females exposed to Cd for 4 weeks produced malformed embryos.[584] Malformations were also seen in *Xenopus* embryos exposed to concentrations ranging from 0.1 to 10 mg $Ca^{2+}$/L. This study showed that embryos were more susceptible from stages 2 to 40, although malformations occurred at all stages.[585]

Diverse effects of Cd exposure have been reported in a number of fish species, although results from different studies are occasionally conflicting. For instance, Cd exposure has led to reduced plasma sex steroids in Asian swamp eel (*Monopterus albus*) and brook trout; reduced VTG in *Monopterus albus*, rainbow trout, bleeker (*Lepidocephalicthys thermalis*), winter flounder, and European flounder; and decreased GSI in Asian swamp eel and winter flounder.[161,586–590] Conversely, Cd has also stimulated steroidogenesis in several fish species.[110,159,591,592] The latter could be explained by evidence showing increased *in vitro* production of GTH after administration of Cd, which is consistent with the enhanced ovarian activity observed in female Atlantic croaker.[591] A wide range of Cd concentrations (0.001–1000 ppm) and durations of exposure (several hours to 90 days) were used in the experiments described above, which could explain the different responses observed.

Cd treatment has also been associated with degenerative changes in the gonads of several fish species.[588,593–595] Adult female guppies exposed to dietary Cd for 30–120 days produced less fry compared with controls, demonstrating the effect of Cd at the organism and possibly at the population level.[596] The mechanisms underlying Cd-induced alterations are poorly understood, although several theories have emerged. There is speculation that vitellogenesis may be impaired because the synthesis of metallothioneins by the liver in response to metal exposure takes priority over the synthesis of VTG.[590] On the other hand, Cd may directly interfere with VTG synthesis at the transcriptional or translational levels. Others suggest that Cd may interfere with the incorporation of VTG into the developing oocyte. In a study by Victor et al.,[588] Cd appeared to impede the transport of VTG across the oolemma into the oocyte. However, Cd-VTG complexes injected into Atlantic croaker were shown to incorporate into the ovaries.[575]

Zinc (Zn) is a component of over 70 metalloenzymes and serves as an important essential metal. Although Zn toxicity is rare, it has been reported in several species. For instance, Zn has been shown to influence hatching success and developmental rates in the Jefferson salamander.[562] Multiple studies also document reproductive alterations in fish following Zn exposure. For example, in a study using clarias catfish, Zn was reported to decrease circulating levels of VTG.[565] Other observed effects in fish include delayed spawning and decreased egg viability in zebrafish,[597] impaired spermatogenesis and increased oocyte atresia in rosy barb,[566] and reduced egg size and increased larval deformities in white sucker.[598]

Selenium (Se) is a natural element/metal required for healthy nutrition in small amounts, but it is toxic at higher concentrations. The processing of fossil fuels releases Se to the environment, where it then accumulates in coal fly ash. Se is also found in high concentrations in certain soils, remaining in wetlands as a by-product of irrigation.[203,599] Reproductive success of birds and fish is more sensitive to Se toxicity than are growth and survival of young or adults.[600–602] In fish, Se exposure can result in subtle but dramatic reproductive failure.[603] In birds, Se egg concentrations of 3 ppm wet weight are considered the threshold for reproductive impairment.[601,604,605]

## 39.4 SUMMARY AND CONCLUSIONS

This chapter has reviewed and selectively summarized the current evidence for potential endocrine-disrupting effects of specific chemicals and chemical classes in vertebrate wildlife and their potential modes of action. Although evidence of endocrine disruption in wild species has accumulated during recent years, most studies are based on indirect evidence rather than defined mechanisms and exposures to specific ECDs. Indeed, most studies of potential EDC effects in wildlife are based on observed adverse reproductive and developmental effects rather than direct evidence of endocrine-modified function or defined endocrine pathways. Nonetheless, a consideration of whether the effects of specific chemicals can be attributed to hormonal properties, mechanisms, or pathways is critical to the identification of a chemical as an EDC or EAA.

This review also evaluated the evidence for endocrine disruption for wildlife and fish in field/natural and control/experimental situations. A wide variety of chemicals have been reported as potential EDCs in wildlife. The major chemical classes summarized here include PAHs, PCBs and PBBs, PCDDs and PCDFs, OCPs, nonorganochlorine pesticides, complex environmental mixtures, and selected metals. In addition, the evidence of potential EDC effects are summarized and reviewed for multiple vertebrate species, with an emphasis on reproductive and developmental effects, which are often modulated by endocrine mechanisms and pathways. Collectively, there is strong evidence of altered reproductive and developmental processes in wildlife exposed to EDCs. Although from most of these studies the mechanisms of action and direct link to endocrine-mediated pathways are often unclear, there is general evidence of an association between effects and chemical/contaminant exposures as well as evidence of effects in multiple vertebrate classes. Much of the evidence for EDC effects in wildlife is derived from observations and studies involving fish. These studies present the clearest link between environmental chemical contaminants and endocrine-disrupting effects. The potential mechanisms of action are diverse (see Figure 39.1), and several endocrine/hormonal mediated pathways are likely.

In recent years, great progress has been made in the development of *in vitro* screening and testing procedures for the identification of potential EDCs. However, these assays have been based primarily on receptor-mediated responses and hormone mimicry. It is important to mention that a wide variety of other potential mechanisms also exist for EDCs (see Figure 39.1 and Table 39.1), and thus there is a strong need for the development of additional screening and testing procedures. On the other hand, *in vivo* studies are more ecorelevant and thus better suited for the assessment of risk in wildlife due to EDCs. However, the interpretation of effects at the organism level and above is difficult and potentially affected by multiple stressors (other than EDCs). Paired studies, involving both field- and laboratory-based exposures as well as *in vitro* assessments of mechanisms are likely needed to adequately identify and evaluate potential EDCs. Nonetheless, studies in wildlife and fish have provided the strongest evidence for accepting the endocrine-disrupting hypothesis and have been critical in the identification and evaluation of potential environmental EDCs.

## REFERENCES

1. Kavlock, R. T., Daston, G. P., DeRosa, C., Fenner-Crisp, P., Gray, L. E., Kaattari, S., Lucier, G., Luster, M., Mac, M. J., Maczka, C., Miller, R., Moore, J., Rolland, R., Scott, G., Sheehan, D. M., Sinks, T., and Tilson, H. A., Research needs for the risk assessment of health and environmental effects of endocrine disruptors: A report of the U.S. EPA sponsored workshop, *Environ. Health Perspect.,* 104, 715, 1996.
2. Kavlock, R. J. and Ankley, G. T., A perspective on the risk assessment process for endocrine-disruptive effects on wildlife and human health, *Risk Anal.,* 16, 731, 1996.
3. Colborn, T., von Saal, F. S., and Soto, A. M., Developmental effects of endocrine-disrupting chemicals in wildlife and humans, *Environ. Health Perspect.,* 101, 378, 1993.
4. Tyler, C. R., Jobling, S., and Sumpter, J. P., Endocrine disruption in wildlife: A critical review of the evidence, *Crit. Rev. Toxicol.,* 28, 319, 1998.
5. Dawson, A., Mechanisms of endocrine disruption with particular reference to occurrence in avian wildlife: A review, *Ecotoxicology,* 9, 59, 2000.
6. Iguchi, T. and Sato, T., Endocrine disruption and developmental abnormalities of female reproduction, *Am. Zool.,* 40, 402, 2000.
7. Harrison, P. T. C., Holmes, P., and Humfrey, C. D. N., Reproductive health in humans and wildlife: Are adverse trends associated with environmental chemical exposure?, *Sci. Total Environ.,* 205, 97, 1997.
8. Depledge, M. H. and Billinghurst, Z., Ecological significance of endocrine disruption in marine invertebrates, *Mar. Pollut. Bull.,* 39, 32, 1999.

9. Sumpter, J. P., Xenoendocrine disrupters — environmental impacts, *Toxicol. Lett.,* 103, 337, 1998.
10. Datson, G. P., Gooch, J. W., Breslin, W. J., Shuey, D. L., Nikiforov, A. I., Fico, T. A., and Gorsuch, J. W., Environmental estrogens and reproductive health: A discussion of the human and environmental data, *Reprod. Toxicol.,* 11, 465, 1997.
11. Kendall, R. J., Dickerson, R. L., Geisey, J. P., and Suk, W. P., *Principles and Processes for Evaluating Endocrine Disruption in Wildlife*, SETAC Press, Pensacola, FL, 1998.
12. U.S. EPA, EPA Special Report on Endocrine Disruption, Fact Sheet, U.S. EPA, Washington, D.C., 1997.
13. National Research Council, *Hormonally Active Agents in the Environment*, Washington, D.C. National Academy Press, 1999.
14. U.S. EPA, Endocrine Disruptor Screening and Testing Committee (EDSTAC) Final Report, U.S. EPA, Washington, D.C., 1998.
15. Licht, P., Reproductive endocrinology of reptiles and amphibians: Gonadotropins, *Annu. Rev. Physiol.,* 41, 337, 1979.
16. Norrism, D. O., *Vertebrate Endocrinology,* Academic Press, San Diego, 1997.
17. Norris, D. O. and Jones, R. E., *Hormones and Reproduction in Fishes, Amphibians and Reptiles,* Plenum Press, New York, 1987.
18. Peter, R. E., Yu, K.-L., Marchant, T. A., and Rosenblum, P. M., Direct neural regulation of the teleost adenohypophysis, *J. Exp. Zool.,* 4, 84, 1990.
19. Ben-Jonathan, N., Arbogast, L. A., and Hyde, J. F., Neuroendocrine regulation of prolactin release, *Prog. Neurobiol.,* 33, 399, 1989.
20. Peter, R. E., Vertebrate neurohormonal systems, in *Vertebrate Endocrinology: Fundamentals and Biomedical Implications,* Pang, P. K. T. and Schreibman, M. P., Eds., Academic Press, New York, 57, 1986.
21. Hirano, T., The spectrum of prolactin actions in teleosts, in *Comparative Endocrinology, Developments and Directives,* Ralph, C. L. and Lilss, A. R., Eds., New York, 53, 1986.
22. Corpas, E., Harman, M., and Blackman, M. R., Human growth hormone and aging, *Endocrinol. Rev.,* 14, 20, 1993.
23. Shimoni, Y., Fiset, C., Clark, R. B., Dixon, J. E., McKinno, D., and Giles, W. R., Thyroid hormone regulates postnatal expression of transient K+ channel isoforms in rat ventricle, *J. Physiol.,* 500, 65, 1997.
24. Muscat, G. E. O., Downes, M., and Dowhan, D. H., Regulation of vertebrate muscle differentiation by thyroid hormone: The role of the *myoD* gene family, *BioEssays,* 17, 211, 1995.
25. Pirinen, S., Endocrine regulation of craniofacial growth, *Acta Odontol. Scand.,* 53, 179, 1995.
26. Leonard, J. L. and Farwell, A. P., Thyroid hormone-regulated actin polymerization in brain, *Thyroid,* 7, 147, 1997.
27. Dodd, M. H. and Dodd, J. M., The biology of metamorphosis, in *Physiology of the Amphibia,* Lofts, B., Ed, Academic Press, New York, 1976.
28. Hayes, T. B., Steroids as potential modifiers of thyroid hormone action in anuran metamorphosis, *Am. Zool.,* 37, 185, 1997.
29. Tata, J. R., Amphibian metamorphosis: An exquisite model for hormonal regulation of postembryonic development in vertebrates, *Dev. Growth Differ.,* 38, 223, 1996.
30. Inui, Y., Yamano, K., and Miwa, S., The role of thyroid hormone in tissue development in metamorphosing flounder, *Aquaculture,* 135, 87098, 1995.
31. Tagawa, M., De Jesus, E. G., and Hirano, T., The thyroid hormone monodeiodinase system during flounder metamorphosis, *Aquaculture,* 135, 128, 1994.
32. Brown, D. D., Wang, Z., Furlow, J. D., Kanamori, A., Schwartzman, R. A., Remo, B. F., and Pinder, A., The thyroid hormone-induced tail resorption program during *Xenopus laevis* metamorphosis, *Proc. Natl. Acad. Sci.,* 93, 1924, 1996.
33. Shi, Y. B., Biphasic intestinal development in amphibians: Embryogenesis and remodeling during metamorphosis, *Curr. Top. Dev. Biol.,* 32, 205, 1996.
34. Hanken, J. and Hall, B. K., Skull development during anuran metamorphosis II. Role of thyroid hormone in osteogenesis, *Anat. Embryol.,* 178, 219, 1988.
35. Hanken, J. and Summers, C. H., Skull development during anuran metamorphosis. III. Role of thyroid hormone in chondrogenesis, *J. Exp. Zool.,* 246, 156, 1988.

36. Buckbinder, L. and Brown, D. D., Thyroid hormone-induced gene expression changes in the developing frog limb, *J. Biol. Chem.*, 267, 25786, 1992.
37. Norris, D. O. and Platt, J. T., $T_3$ and $T_4$-induced rates of metamorphosis in immature and sexually mature larvae of *Ambystoma tigrinum*, *J. Exp. Zool.*, 189, 303, 1974.
38. Dickhoff, W. W., Brown, C. L., Sullivan, C. V., and Bern, H. A., Fish and amphibian models for developmental endocrinology, *J. Exp. Zool.*, 4, 90, 1990.
39. Young, G., Cortisol secretion *in vitro* by the interrenal of coho salmon (*Onchorhynchus kisutch*) during smoltification: Relationship with plasma thyroxine and plasma cortisol., *Gen. Comp. Endocrinol.*, 63, 191, 1986.
40. Young, G. and Lin, R. J., Response of the interrenal to adrenocorticotropic hormone after short-term thyroxine treatment of coho salmon (*Onchorhynchus kisutch*), *J. Exp. Zool.*, 245, 53, 1988.
41. Beyer, C.A. and Hutchison, J.B., Androgens stimulate the morphological maturation of embryonic hypothalamic aromatase-immunoreactive neurons in the mouse, *Dev. Brain Res.*, 98, 74, 1997.
42. Lephart, E. D., A review of brain aromatase cytochrome P450, *Brain Res. Rev.*, 1, 1996.
43. Naftolin, F., Horvath, T. L., Jakab, R. L., Leranth, C., Harada, N., and Balthazart, J., Aromatase immunoreactivity in axon terminals of the vertebrate brain, *Neuroendocrinology*, 63, 149, 1996.
44. Hayes, T. B. and Licht, P., Gonadal involvement in size sex dimorphism in the African bullfrog (*Pyxicephalus adspersus*), *J. Exp. Zool.*, 264, 130, 1992.
45. Hews, D. K. and Moore, M. C., Influence of androgens on differentiation of secondary sex characters in tree lizards, *Urosaurus ornatus*, *Gen. Comp. Endocrinol.*, 97, 86, 1995.
46. Kanamadi, R. D. and Saidapur, S. K., Effect of testosterone on spermatogenesis Leydig cells and thumb pads of the frog *Rana cyanophlyctis*, *J. Karnatak Univ. Sci.*, 31, 157, 1993.
47. Pawar, V. G. and Pancharatna, K., Estradiol-17α-induced oviductal growth in the skipper frog *Rana cyanophlyctis*, *J. Adv. Zool.*, 16, 107, 1995.
48. Rabelo, E. M. and Tata, J. R., Thyroid hormone potentiates estrogen activation of vitellogenin genes and autoinduction of estrogen receptor in adult *Xenopus* hepatocytes, *Mol. Cell. Endocrinol.*, 96, 37, 1993.
49. Rabelo, E. M. L., Baker, B. S., and Tata, J. R., Interplay between thyroid hormone and estrogen in modulating expression of their receptor and vitellogenin genes during *Xenopus* metamorphosis, *Mech. Develop.*, 45, 49, 1994.
50. Van den Hurk, R., Richter, C. J. J., and Janssen-Dommerholt, J., Effects of 17α-methyltestosterone and 11α-hydroxyandrostenedione on gonad differentiation in the African catfish, *Clarias gariepinus*, *Aquaculture*, 83, 179, 1989.
51. Jansen, H. T., Cooke, P. S., Porcelli, J., Liu, T. C., and Hansen, L. G., Estrogenic and antiestrogenic actions of PCBs in the female rat: *In vitro* and *in vivo* studies, *Reprod. Toxicol.*, 7, 237, 1993.
52. Moore, R. W., Jefcoate, C. R., and Peterson, R. E., 2,3,7,8-Tetrachlorodibenzo-*p*-dioxin inhibits steroidogenesis in the rat testis by inhibiting the mobilization of cholesterol to cytochrome P450, *Toxicol. Appl. Pharmacol.*, 109, 85, 1991.
53. Mukherjee, D., Guha D., Kumar V., and Chakrabarty, S., Impairment of steroidogenesis and reproduction in sexually mature *Cyprinus carpio* by phenol and sulfide under laboratory conditions, *Aquat. Toxicol.*, 21, 29, 1991.
54. Singh, P. B., Impact of malathion and γ-BHC on lipid metabolism in the freshwater female catfish, *Heteropneustes fossilis*, *Ecotoxicol. Environ. Saf.*, 23, 22, 1992.
55. McMaster, M. E., Van Der Kraak, G. J., and Munkittrick K. R., Exposure to bleached kraft pulp mill effluent reduces the steroid biosynthetic capacity of white sucker ovarian follicles, *Comp. Biochem. Physiol.*, 112C, 169, 1995.
56. Kapur, K., Kamaldeep, K., and Toor, H., The effect of fenitrothion on reproduction of a teleost fish, *Cyprinus carpio communis* (Linn.): A biochemical study, *Bull. Environ. Contam. Toxicol.*, 20, 438, 1978.
57. Bagchi, P., Chatterjee, S., Ray, A., and Deb, C., Effect of quinalphos, organophosphorous insecticide, on testicular steroidogenesis in fish, *Clarias batrachus*, *Bull. Environ. Contam. Toxicol.*, 44, 871, 1990.
58. Foster, P., Thomas, L., Cook, M., and Walters, D., Effect of di-n-pentyl phthalate treatment on testicular steroidogenic enzymes and cytochrome P-450 in the rat, *Toxicol. Lett.*, 15, 265, 1983.
59. Kirubagaran, R. and Joy, K. P., Toxic effects of mercuric-chloride, methylmercuric chloride, and emisan-6 (an organic mercurial fungicide) on ovarian recrudescence in the catfish *Clarias batrachus* (L.), *Bull. Environ. Contam. Toxicol.*, 41, 902, 1988.

60. Goldstein, J. A. and Safe, S., Mechanism of action and structure-activity relationships for the chlorinated dibenzo-*p*-dioxins and related compounds, in *Halogenated Biphenyls, Terphenyls, Naphthalenes, Dibenzodioxins and Related Products,* Kimbrough, R. D. and Jensen, A., Eds., Elsevier Science Publishers, Amsterdam, 239, 1989.
61. Sivarajah, K., Franklin, C. S., and Williams, W. P., The effects of polychlorinated biphenyls on plasma steroid levels and hepatic microsomal enzymes in fish, *J. Fish Biol.,* 13, 401, 1978.
62. Ankley, G. T., Blazer, V. S., Reinert, R. E., and Agosin, M., Effects of Aroclor-1254 on cytochrome-P-450-dependent monooxygenase, glutathione-S-transferase, and UDP-glucuronosyl transferase activities in channel catfish liver, *Aquat. Toxicol.,* 9, 91, 1986.
63. Goldzieher, J. W., Are low-dose oral-contraceptives safer and better, *Am. J. Obstet. Gynecol.,* 171, 587, 1994.
64. Darnerud, P. O., Morse, D., Klassonwehler, E., and Brouwer, A., Binding of a 3′,3′,4′,4′-tetrachlorobiphenyl (Cb-77) metabolite to fetal transthyretin and effects on fetal thyroid hormone levels in mice, *Toxicology,* 106, 105, 1996.
65. Darnerud, P. O., Sinjari, T., and Jonsson, C. J., Fetal uptake of coplanar polychlorinated biphenyl (PCB) congeners in mice, *Pharmacol. Toxicol.,* 78, 187, 1996.
66. Barter, R. A. and Klaassen, C. D., Reduction of thyroid hormone levels and alteration of thyroid function by four representative UDP-glucuronosyltransferase inducers in rats, *Toxicol. Appl. Pharmacol.,* 128, 9, 1994.
67. Seo, B.-W., Li, M.-H., Hansen, L. G., More, R. W., Peterson, R. E., and Schantz, S. L., Effects of gestational and lactational exposure to coplanar polychlorinated biphenyl (PCB) congeners or 2,3,7,8-tetrachlorodibenzo-*p*-dioxin (TCDD) on thyroid hormone concentrations in weanling rats, *Toxicol. Lett.,* 78, 253, 1995.
68. Van Birgelen, A. P. J. M., Smit, E. A., Kampen, I. M., Groeneveld, C. N., Fase, K. M., van der Kolk, J., Poiger, H., van den Berg, M., Koeman, J. H., and Brouwer, A., Subchronic effects of 2,3,7,8-TCDD or PCBs on thyroid hormone metabolism: Use in risk assessment, *Eur. J. Pharmacol. Environ. Toxicol. Pharmacol. Sect.,* 5, 77, 1995.
69. Rosner, W., The functions of corticosteroid-binding globulin and sex hormone-binding globulin: Recent advances, *Endocrinol. Rev.,* 11, 80, 1990.
70. Arnold, S. F., Klotz, D. M., Collins, B. M., Vonier, P. M., Guillette, L. J., Jr., and McLachlan, J. A., Synergistic activation of estrogen receptor with combinations of environmental chemicals, *Science,* 272, 1489, 1996. Withdrawn from publication, McLachlan, J.A., *Science*, 277, 462, 1997.
71. Anderson, M. J., Olsen, H., Matsumura, F., and Hinton, D. E., *In vivo* modulation of 17ß-estradiol-induced vitellogenin synthesis and estrogen receptor in rainbow trout (*Oncorhynchus mykiss*) liver cells by beta-naphthoflavone, *Toxicol. Appl. Pharmacol.,* 137, 210, 1996.
72. Anderson, M. J., Miller, M. R., and Hinton, D. E., *In vitro* modulation of 17ß-estradiol-induced vitellogenin synthesis: Effects of cytochrome P4501A1 inducing compounds on rainbow trout (*Oncorhynchus mykiss*) liver cells, *Aquat. Toxicol.,* 34, 327, 1996.
73. Flouriot, G., Pakdel, F., Ducouret, B., and Valotaire, Y., Influence of xenobiotics on rainbow trout liver estrogen receptor and vitellogenin gene expression, *J. Mol. Endocrinol.,* 15, 143, 1995.
74. Danzo, B., Environmental xenobiotics may disrupt normal endocrine function by interfering with the binding of physiological ligands to steroid receptors and binding proteins, *Environ. Health Perspect.,* 105, 294, 1997.
75. Gray, L. E. Jr., Kelce, W. R., Wiese, T., Tyl, R., Gaido, K., Cook, J., Klinefelder, G., Desaulniers, D., Wilson, E., Zacharewski, T., Waller, C., Foster, P., Lasky, J., Reel, J., Giesy, J., Laws, S., McLachlan, J., Breslin, W., Cooper, R., DiGiulio, R., Johnson, R., Purdy, R., Mihaich, E., Safe, S., Sonnenschein, C., Weshons, W., Miller, R., McMaster, S., and Colborn, T., Endocrine screening methods. Workshop report: Detection of estrogenic and androgenic hormonal and antihormonal activity for chemicals that act via receptor or steroidogenic enzyme mechanisms, *Reprod. Toxicol.,* 11, 719, 1997.
76. Kelce, W. R., Monosson, E., Gamcsik, M. P., Laws, S. C., and Gray, L. E., Environmental hormone disruptors: Evidence that vinclozolin developmental toxicity is mediated by antiandrogenic metabolites, *Toxicol. Appl. Pharmacol.,* 126, 276, 1994.
77. Hoffman, E. C., Reyes, H., Chu, F. F., Sander, F., Conley, L. H., Brooks, B. A., and Hankinson, O., Cloning of a factor required for activity of the Ah (Dioxin) receptor, *Science,* 252, 954, 1991.

78. Reyes, H., Reisz-Porszasz, S., and Hankinson, O., Identification of the Ah receptor nuclear translocator protein (Arnt) as a component of the DNA binding form of the Ah receptor, *Science,* 256, 1193, 1992.
79. DeVito, M. J., Thomas, T., Martin, E., Umbreit, T. H., and Gallo, M. A., Antiestrogenic action of 2,3,7,8-tetrachlorodibenzo-*p*-dioxin: Tissue-specific regulation of estrogen receptor in CDI mice, *Toxicol. Appl. Pharmacol.,* 113, 284, 1992.
80. Kharat, I. and Saatcioglu, F., Antiestrogenic effects of 2,3,7,8-tetrachlorodibenzo-*p*-dioxin are mediated by direct transcriptional interference with the liganded estrogen receptor — Cross-talk between aryl hydrocarbon- and estrogen-mediated signaling, *J. Biol. Chem.,* 271, 10533, 1996.
81. Zacharewski, T., Harris, M., and Safe, S., Evidence for the mechanism of action of the 2,3,7,8-tetrachlorodibenzo-*p*-dioxin-mediated decrease of nuclear estrogen receptor levels in wild-type and mutant mouse hepa 1c1c7 cells, *Biochem. Pharmacol.,* 41, 1931, 1991.
82. Tian, Y., Ke, S., Thomas, T., Meeker, R. J., and Gallo, M. A., Transcriptional suppression of estrogen receptor gene expression by 2,3,7,8-tetrachlorodibenzo-*p*-dioxin (TCDD), *J. Steroid Biochem. Mol. Biol.,* 67, 17, 1998.
83. McLachlan, J. A., Functional toxicology: A new approach to detect biologically active xenobiotics, *Environ. Health Perspect.,* 101, 386, 1993.
84. Hwang, K. J., Carlson, K. E., Anstead, G. M., and Katzenellenbogen, J. A., Donor-acceptor tetrahydrochrysenes, inherently fluorescent, high affinity, ligands for the estrogen receptor: Binding and fluorescence characteristics and fluorometric assay of receptor, *Biochemistry,* 31, 11536, 1992.
85. Zysk, J. R., Johnson, B., Ozenberger, B.A., Bingham, B., and Gorski, J., Selective uptake of estrogenic compounds by *Saccharomyces cerevesiae*: A mechanism for antiestrogen resistance in yeast expressing the mammalian estrogen receptor, *Endocrinology,* 136, 1323, 1995.
86. Miksicek, R. J., Commonly occurring plant flavonoids have estrogenic activity, *Mol. Pharmacol.,* 44, 37, 1993.
87. Brooks, S. C., Locke, E. R., and Soule, H. D., Estrogen receptor in a human cell line (MCF-7) from breast carcinoma, *J. Biol. Chem.,* 248, 6251, 1973.
88. Soto, A. M., Chung, K. L., and Sonnenschein, C., The pesticides endosulfan, toxaphene and dieldrin have estrogenic effects in human estrogen-sensitive cells, *Environ. Health Perspect.,* 102, 380, 1994.
89. Soto, A. M., Justicia, H., Wray, J. W., and Sonnenschein, C., p-Nonyl phenol: An estrogenic xenobiotic released from "modified" polystyrene, *Environ. Health Perspect.,* 92, 167, 1991.
90. Soto, A. M., Lin, T. M., Justicia, H., Silvia, R. M., and Sonnennschein, C., An "in culture" bioassay to assess the estrogenicity of xenobiotics (E-screen), in *Chemically Induced Alterations in Sexual and Functional Development: The Wildlife/Human Connection,* Colburn, T. and Clement, C., Eds., Princeton Scientific Publishing, Princeton, 295, 1992.
91. Soto, A. M., Sonnenschein, C., Chung, K. L., Fernandez, M. F., Olea, N., and Olea Serrano, F., The E-SCREEN assay as a tool to identify estrogens: An update on estrogenic environmental pollutants, *Environ. Health Perspect.,* 103, 113, 1995.
92. Soto, A. M. and Sonnenshein, C., The role of estrogens on the proliferation of human breast tumor cells (MCF-7), *J. Steroid Biochem.,* 23, 87, 1985.
93. Welshons, W. V., Rottinghaus, G. E., Nonneman, D. J., Dolan-Timpe, M., and Ross, P. F., A sensitive bioassay for detection of dietary estrogens in animal feeds, *J. Vet. Diagn. Invest.,* 2, 268, 1990.
94. White, R., Jobling, S., Hoare, S. A., Sumpter, J. P., and Parker, M. G., Environmentally persistent alkylphenolic compounds are estrogenic, *Endocrinology,* 135, 175, 1994.
95. Olea, N., Pulgar, R., Perez, P., Olea-Serrano, F., Rivas, A., Novillo-Fertrell, A., Pedraza, V., Soto, A. M., and Sonnenschein, C., Estrogenicity of resin-based composites and sealants used in dentistry, *Environ. Health Perspect.,* 104, 298, 1996.
96. Folmar, L. C., Denslow, N. D., Wallace, R. A., LaFleur, G., Gross, T. S., Bonomelli, S., and Sullivan, C. V., A highly conserved N-terminal sequence for teleost vitellogenin with potential value to the biochemistry, molecular biology and pathology of vitellogenesis, *J. Fish Biol.,* 46, 255, 1995.
97. Denslow, N. D., Chow, M., Kroll, K., Wieser, C., Wiebe, J., Johnson, B., Shoeb, T., and Gross, T. S., Determination of baseline seasonal information on vitellogenin production in female and male largemouth bass collected from lakes in Central Florida, *Abstr. 18$^{th}$ Annu. Meeting SETAC,* 1997.
98. Zacharewski, T., *In vitro* bioassays for assessing estrogenic substances, *Environ. Sci. Technol.,* 31, 613, 1997.

99. Pons, M., Gagne, D., Nicolas, J. C., and Mehtali, M., A new cellular model of response to estrogens: A bioluminescent test to characterize (anti) estrogen molecules, *BioTechniques,* 456, 1990.
100. Gaido, K. W., Leonard, L. S., Lovell, S., Gould, J. C., Babai, D., Portier, C. J., and McDonnell, D. P., Evaluation of chemicals with endocrine modulation activity in a yeast-based steroid hormone receptor gene transcription assay, *Toxicol. Appl. Pharmacol.,* 143, 205, 1997.
101. Ramamoorthy, K., Wang, F., Chen, I. C., Norris, J. D., McDonnell, D. P., Leonard, L. S., Gaido, K. W., Bocchinfuso, W. P., Korach, K. S., and Safe, S., Estrogenic activity of a dieldrin/toxaphene mixture in the mouse uterus, MCF-7 human breast cancer cells, and yeast-based estrogen receptor assays: No apparent synergism, *Endocrinology,* 138, 1520, 1997.
102. Demirpence, E., Pons, M., Balaguer, P., and Gagne, D., Study of an antiestrogenic effect of retinoic acid in MCF-7 cells, *Biochem. Biophys. Res. Commun.,* 183, 100, 1992.
103. Stegeman, J. J., Brouwer, M., Di Giulio, R. T., Förlin, L., Fowler, B. A., Sanders, B. M., and van Veld, P. A., Molecular responses to environmental contamination: Enzyme and protein systems as indicators of chemical exposure and effect, in *Biomarkers: Biochemical, Physiological, and Histological Markers of Anthropogenic Stress, Proceedings of the 8th Pellston Workshop,* Huggett, R. J., Kimerle, R. A., Mehrle, P. M., and Bergman, H. L., Eds., Lewis Publishers, Chelsea, MI, 1992.
104. Sumpter, J. P. and Jobling S., Vitellogenesis as a biomarker for estrogenic contamination of the aquatic environment, *Environ. Health Perspect.,* 103, 173, 1995.
105. Down, N. E., Peter, R. E., and Leatherland, J. F., Seasonal changes in serum gonadotropin, testosterone, 11-ketotestosterone, and estradiol-17ß levels and their relation to tumor burden in gonadal tumor-bearing carp x goldfish hybrids in the Great Lakes, *Gen. Comp. Endocrinol.,* 77, 192, 1990.
106. Goodbred, S. L., Gilliom, R. J., Gross, T. S., Denslow, N. P., Bryant, W. L., and Schoeb, T. R., Reconnaissance of 17ß-Estradiol, 11-Ketotestosterone, Vitellogenin, and Gonad Histopathology in Common Carp of United States Streams: Potential for Contaminant-Induced Endocrine Disruption, USGS Open-File Report #96–627, U.S. Geological Survey, Sacramento, 1997.
107. Gross, T. S., Wiebe, J. J., Wieser, C. M., Ruessler, D. S., and Aikens, K., An evaluation of methyl mercury as an endocrine disruptor in largemouth bass, in *Metal Ions in Biology and Medicine,* John Libbey Eurotext, Paris, 2000.
108. Hontela, A., Dumont, P., Duclos, D., and Fortin R., Endocrine and metabolic dysfunction in yellow perch (*Perca flavescens*) exposed to organic contaminants and heavy metals in the St. Lawrence River, *Environ. Toxicol. Chem.,* 14, 725, 1995.
109. Leatherland, J., Endocrine and reproductive function in Great Lakes salmon, in *Chemically Induced Alterations in Sexual and Functional Development: The Wildlife/Human Connection,* Colborn, T. and Clement, C., Eds., Princeton Scientific Publishing, Princeton, 129, 1992.
110. Sangalang, G. B. and Freeman, H. C., Effect of sublethal cadmium on maturation and testosterone, and 11-ketotestosterone production in brook trout (*Salvelinus fontinalis*), *Biol. Reprod.,* 11, 429, 1974.
111. Carnevali, O. and Belvedere P., Comparative studies of fish, amphibian, and reptilian vitellogenins, *J. Exp. Zool.,* 259, 18, 1991.
112. Cheek, A. O., Ide, C. F., Bollinger, J. E., Rider, C. V., and McLachlan, J. A., Alteration of leopard frog (*Rana pipiens*) metamorphosis by the herbicide acetochlor, *Arch. Environ. Con. Tox.,* 37, 70, 1999.
113. Willingham, E. and Crews, D., Sex reversal effects of environmentally relevant xenobiotic concentrations on the red-eared slider turtle, a species with temperature-dependent sex determination, *Gen. Comp. Endocrinol.,* 113, 429, 1999.
114. Arnold, S. F., Vonier, P. M., Collins, B. M., Klotz, D. M., Guillette, L. J., and McLachlan, J. A., *In vitro* synergistic interaction of alligator and human estrogen receptors with combinations of environmental chemicals, *Environ. Health Perspect.,* 105, 615, 1997.
115. Cavanaugh, K. P. and Holmes, W. N., Effects of ingested petroleum on the development of ovarian endocrine function in photostimulated mallard ducks (*Anas platyrhynchos*), *Arch. Environ. Contam. Toxicol.,* 16, 247, 1987.
116. Cavanaugh, K. P. and Holmes, W. N., Effects of ingested petroleum on plasma-levels of ovarian-steroid hormones in photostimulated mallard ducks, *Arch. Environ. Contam. Toxicol.,* 11, 503, 1982.
117. Janz, D. M. and Bellward, G. D., *In ovo* 2,3,7,8-tetrachlorodibenzo-*p*-dioxin exposure in three avian species, 2. Effects on estrogen receptor and plasma sex steroid hormones during the perinatal period, *Toxicol. Appl. Pharmacol.,* 139, 292, 1996.

118. Capdevielle, M. C. and Scanes C. G., Effect of dietary acid or aluminum on growth and growth-related hormones in young chickens, *Toxicol. Appl. Pharmacol.,* 133, 164, 1995.
119. Ludholm, C. E., Inhibition of prostaglandin synthesis in eggshell gland mucosa as a mechanism for *p,p*'-DDE-induced eggshell thinning in birds: A comparison of ducks and domestic fowls, *Comp. Biochem. Physiol.*, 106C, 389, 1993.
120. Janz, D. M. and Bellward, G. D., *In ovo* 2,3,7,8-tetrachlorodibenzo-*p*-dioxin exposure in three avian species. 1. Effects on thyroid hormones and growth during perinatal period and the perinatal period, *Toxicol. Appl. Pharmacol.,* 139, 281, 1996.
121. Golden, R. J., Noller, K. L., Titus-Ernstoff, L., Kaufman, R. H., Mittendorf, R., Stillman, R., and Reese, E. A., Environmental endocrine modulators and human health: An assessment of the biological evidence, *Crit. Rev. Toxicol.,* 28, 109, 1998.
122. Copeland, P. A. and Thomas P., The measurement of plasma vitellogenin levels in a marine teleost, the spotted seatrout (*Cynoscion nebulosus*) by homologous radioimmunoassay, *Comp. Biochem. Physiol.,* 91B, 17, 1988.
123. Denison, M. S., Chambers, J. E., and Yarbrough, J. D., Persistent vitellogenin-like protein and binding of DDT in the serum of insecticide-resistant mosquitofish (*Gambusia affinis*), *Comp. Biochem. Physiol.,* 69C, 109, 1981.
124. Denslow, N. D., Chow, M., Chow, M. M., Bonomelli, S., Folmar, L. C., Heppell, S. A., and Sullivan, C. V., Development of biomarkers for environmental contaminants affecting fish, in *Chemically Induced Alterations in Functional Development and Reproduction of Fishes,* Rolland, R., Gilbertson, M., and Peterson, R. E., Eds., SETAC Press, Pensacola, FL, 73, 1997.
125. Crews, D., Bergeron, J. M., and McLachlan, J. A., The role of estrogen in turtle sex determination and the effect of PCBs, *Environ. Health Perspect.,* 103, 73, 1995.
126. Davis, W. P. and Bortone, S. A., Effects of kraft mill effluent on the sexuality of fishes: An environmental early warning?, in *Chemically Induced Alterations in Sexual and Functional Development: The Wildlife/Human Connection,* Colborn, T. and Clement, C., Eds., Princeton Scientific Publishing, Princeton, 113, 1992.
127. Gimeno, S., Gerritsen, A., Bowmer, T., and Komen, H., Feminization of male carp, *Nature,* 384, 221, 1996.
128. Munkittrick, K. R., Van Der Kraak, G. J., McMaster, M. E., and Portt, C. B., Response of hepatic MFO activity and plasma sex steroids to secondary treatment of bleached kraft pulp mill effluent and mill shutdown, *Environ. Toxicol. Chem.,* 11, 1427, 1992.
129. Gross, T. S., Guillette L. J., Percival H. F., Masson G. R., Matter J. M., and Woodward, A. R., Contaminant-induced reproductive anomalies in Florida alligators, *Comp. Pathol. Bull.,* 26, 1, 1994.
130. Sepúlveda, M. S., Johnson, W. E., Higman, J. C., Denslow, N. D., Schoeb, T. R., and Gross, T. S., Evaluation of biomarkers of reproductive function and potential comtaminant effects in Florida largemouth bass (*Micropterus salmoides floridanus*) sampled from the St. Johns River, *Sci. Tot. Environ.*, 289, 133, 2002.
131. Sepúlveda, M. S., Wiebe, J. J., Harvey, A., Basto, J., Ruessler, D. S., Roldan, E., and Gross, T. S., An evaluation of environmental contaminants and developmental toxicity for the American alligator in central Florida, in *Abstracts: 2001 Society of Toxicology Annual Meeting*, 2001.
132. Gross, T. S., Shrestha, S., Wieser, C., Wiebe, J., Denslow, N., Chow, C., Johnson, W. E., and Stout, R., Evaluation of potential endocrine disrupting effects of water-soluble herbicides in largemouth bass, *Abstr. 18$^{th}$ Annu. Meeting SETAC*, 1997.
133. Giroux, D., Correlative Evaluation of Chlorinated Hydrocarbon Residues and Reproductive Anomalies in Alligators, Masters thesis, University of Florida, 1997.
134. Masson, G. R., Environmental Influences on Reproductive Potential, Clutch Viability and Embryonic Mortality of the American Alligator in Florida, Ph.D. dissertation, University of Florida, 1995.
135. Guillette, L. J., Jr., Gross, T. S., Masson, G. R., Matter, J. M., Percival, H. F., and Woodward, A. R., Developmental abnormalities of the gonad and abnormal sex hormone concentrations in juvenile alligators from contaminated and control lakes in Florida, *Environ. Health Perspect.*, 102, 680, 1994.
136. Hose, J. E., Cross, J. N., Smith, S. G. S., and Diehl, D., Reproductive impairment in a fish inhabiting a contaminated coastal environment off of southern California, *Environ. Pollut.*, 57, 139, 1989.
137. Gross, D. A., Thymus, Spleen and Bone Marrow Hypoplasia and Decreased Antibody Responses in Hatchling Lake Apopka Alligators. Masters thesis, University of Florida, 1997.

138. Percival, H. F., Rice, K., Woodward, A., Jennings, M., Masson, G., and Abercombie, C., Depressed alligator clutch viability on Lake Apopka, Florida, 5th Annual Lake Management Society Symposium, World Meeting Number 942, 5020, 1994.
139. Gross, T. S., Guillette, L. J., Percival, H. F., Masson, G. R., Matter, J. M., and Woodward, A. R., Contaminant induced reproductive anomalies in Florida alligators, *Comp. Path. Bull.,* 4, 2, 1995.
140. Fairbrother, A., Ankley, G. T., Birnbaum, L. S., Bradbury, S. P., Francis, B., Gray, L. E., Hinton, D., Johnson, L. L., Peterson, R. E., and Van Der Kraak, G., Reproductive and developmental toxicology of contaminants in oviparous animals, in *Reproductive and Developmental Effects of Contaminants in Oviparous Vertebrates, SETAC Pellston Workshop on Reproductive and Developmental Effects of Contaminants in Oviparous Vertebrates,* Di Giulio, R. T. and Tillitt, D., Eds., SETAC Press, Pensacola, FL, 1999.
141. Hoffman, D. J., Embryotoxicity and teratogenicity of environmental contaminants to bird eggs, *Rev. Environ. Contam. Toxicol.,* 115, 39, 1990.
142. Fry, D. M., Reproductive effects in birds exposed to pesticides and industrial chemicals, *Environ. Health Perspect.,* 103, 165, 1995.
143. Harvey, S., Klandorf, H., and Phillips, J. G., Reproductive performance and endocrine responses to ingested petroleum in domestic ducks (*Anas platyrhynchos*), *Gen. Comp. Endocrinol.,* 45, 372, 1981.
144. Rattner, B. A. and Eastin, W. C., Plasma corticosterone and thyroxine concentrations during chronic ingestion of crude oil in mallard ducks (*Anas platyrhynchos*), *Comp. Biochem. Physiol.,* 68C, 103, 1981.
145. Fry, D. M., Swenson, J., Addiego, L. A., Grau, C. R., and Kang, A., Reduced reproduction of wedge-tailed shearwaters exposed to weathered Santa-Barbara crude-oil, *Arch. Environ. Contam. Toxicol.,* 15, 453, 1986.
146. Trivelpiece, W. Z., Butler, R. G., Miller, D. S., and Peakall, D. B., Reduced survival of chicks of oil-dosed adult Leach's storm petrels, *Condor,* 86, 81, 1984.
147. Oakley, K. L. and Kuletz, K., Population, reproduction, and foraging of pigeon guillemots at Naked Island, Alaska, before and after the *Exxon Valdez* oil spill, *Am. Fish. Soc. Symp.,* 18, 759, 1996.
148. Sharpe, B. E., Cody, M., and Turner R., Effects of the *Exxon Valdez* oil spill on the black oystercatcher, *Am. Fish. Soc. Symp.,* 18, 748, 1996.
149. Holmes, W. N., Petroleum pollutants in the marine environment and their possible effects on seabirds, in *Reviews of Environmental Toxicology,* I, Hodgson, E., Ed., Elsevier Science Publications, Amsterdam, 251, 1984.
150. Rattner, B. A., Eroschenko, V. P., Fox, G. A., Fry, D. M., and Gorsline, J., Avian endocrine responses to environmental pollutants, *J. Exp. Zool.,* 232, 683, 1984.
151. Johnson, L. L., Casillas, E., Collier, T. K., McCain, B. B., and Varanasi, U., Contaminant effects on ovarian development in English sole (*Parophyrs vetulus*) from Puget Sound, Washington, *Can. J. Fish. Aquat. Sci.,* 45, 2133, 1988.
152. Johnson, L., Casillas, E., Sol, S., Collier, T., Stein, J., and Varanasi, U., Contaminant effects on reproductive success in selected benthic fish, *Mar. Environ. Res.,* 35, 165, 1993.
153. Stott, G. G., Haensly, W. E., Neff, J. M., and Sharp, J. R., Histopathologic survey of ovaries of plaice, *Pleuronectes platessa* L, from Aber Wrach and Aber Benoit, Brittany, France — Long-term effects of the *Amoco Cadiz* crude-oil spill, *J. Fish Dis.,* 6, 429, 1983.
154. McMaster, M. E., Munkittrick, K. R., Luxon, P. L., and Van Der Kraak, G. J., Impact of low-level sampling stress on interpretation of physiological responses of white sucker exposed to effluent from a bleached kraft pulp mill, *Ecotoxicol. Environ. Saf.,* 27, 251, 1994.
155. Slooff, W., Dezwart, D., and Vandekerkhoff, J. F. J., Monitoring the rivers Rhine and Meuse in the Netherlands for toxicity, *Aquat. Toxicol.,* 4, 189, 1983.
156. Stott, G. G., McArthur, N. H., Tarpley, R., Sis, R. F., and Jacobs, V., Histopathological survey of male gonads of fish from petroleum production and control sites in the Gulf of Mexico, *J. Fish Biol.,* 17, 593, 1980.
157. Stott, G. G., Mcarthur, N. H., Tarpley, R., Jacobs, V., and Sis, R. F., Histopathologic survey of ovaries of fish from petroleum production and control sites in the Gulf of Mexico, *J. Fish Biol.,* 18, 261, 1981.
158. Thomas, P., Reproductive endocrine function in female Atlantic croaker exposed to pollutants, *Mar. Environ. Res.,* 24, 179, 1988.

159. Thomas, P., Teleost model for studying the effects of chemicals on female reproductive endocrine function, *J. Exp. Zool.,* 4, 126, 1990.
160. Thomas, P. and Smith, J., Binding of xenobiotics to the estrogen receptor of spotted seatrout: A screening assay for potential estrogenic effects, *Mar. Environ. Res.,* 35, 147, 1993.
161. Singh, H., Interaction of xenobiotics with reproductive endocrine functions in a protogynous teleost, *Monopterus albus, Mar. Environ. Res.,* 28, 285, 1989.
162. Safe, S., Astroff, B., Harris, M., Zacharewski, T., Dickerson, R., Romkes, M., and Biegel, L., 2,3,7,8-Tetrachlorodibenzo-*p*-dioxin (TCDD) and related compounds as antiestrogens: Characterization and mechanism of action, *Pharmacol. Toxicol.,* 69, 400, 1991.
163. Rocha Moneiro, P. R., Polycyclic aromatic hydrocarbons inhibit *in vitro* ovarian steroidogenesis in the flounder (*Platichthys flesus* L.), *Aquat. Toxicol.,* 48, 549, 2000.
164. Mackay, D., Sang, S., Vlahos, P., Gobas, F., Diamond, M., and Dolan, D., A rate constant model of chemical dynamics in a lake ecosystem; PCBs in Lake Ontario, *J. Great Lakes Res.,* 20, 625, 1994.
165. Safe, S. H., Polychlorinated-biphenyls (PCBs) — environmental-impact, biochemical and toxic responses, and implications for risk assessment, *Crit. Rev. Toxicol.,* 24, 87, 1994.
166. Wren, C. D., Cause-effect linkages between chemicals and populations of mink (*Mustela vison*) and otter (*Lutra canadensis*) in the Great Lakes basin, *J. Toxicol. Environ. Health,* 33, 549, 1991.
167. Foley, R. E., Jackling, S. J., Sloan, R. J., and Brown, M. K., Organochlorine and mercury residues in wild mink and otter — comparison with fish, *Environ. Toxicol. Chem.,* 7, 363, 1988.
168. Platonow, N. S. and Karstad, L. H., Dietary effects of polychlorinated biphenyls on mink, *Can. J. Comp. Med.,* 37, 391, 1973.
169. Hornshaw, T. C., Aulerich, R. J., and Johnson, H. E., Feeding Great Lakes fish to mink — Effects on mink and accumulation and elimination of PCBs by mink, *J. Toxicol. Environ. Health,* 11, 933, 1983.
170. Clark, D. R., Jr. and Lamont, T. G., Organochlorine residues and reproduction in big brown bat, *J. Wildlife Manage.,* 40, 249, 1976.
171. Clark, D. R., Jr. and Krynitsky, A., Organochlorine residues and reproduction in the little brown bat, *Pest. Monitor. J.,* 12, 113, 1978.
172. Clark, D. R. and Stafford, C. J., Effects of DDE and PCB (Aroclor 1260) on experimentally poisoned female little brown bats (*Myotis lucifugus*) — Lethal brain concentrations, *J. Toxicol. Environ. Health,* 7, 925, 1981.
173. Helle, E., Olsson M., and Jensen S., PCB levels correlated with pathological changes in seal uteri, *Ambio,* 5, 261, 1976.
174. Perttila, M., Stenman, O., Pyysalo, H., and Wickstrom, K., Heavy metals and organochlorine compounds in seals in the Gulf of Finland, *Mar. Environ. Res.,* 18, 43, 1986.
175. DeLong, R. L., Gilmartin W. G., and Simpson J. G., Premature births in California sea lions: Association with high organochlorine pollutant residue levels, *Science,* 181, 1168, 1973.
176. Martineau, D., Beland, P., Desjardins, C., and Lagace, A., Levels of organochlorine chemicals in tissues of beluga whales, *Arch. Environ. Contam. Toxicol.,* 16, 137, 1987.
177. Bergman, A., Olsson, M., and Reiland, S., Skull-bone lesions in the Baltic grey seal (*Halichoerus grypus*), *Ambio,* 21, 517, 1992.
178. Mortensen, P., Bignert, A., and Olsson, M., Prevalence of skull lesions in harbor seals (*Phoca vitulina*) in Swedish and Danish museum collections: 1835–1988, *Ambio,* 21, 520, 1992.
179. Casteel, S. W., Cowart, R. P., Weis, C. P., Henningsen, G. M., Hoffman, E., Brattin, W. J., Guzman, R. E., Starost, M. F., Payne, J. T., Stockham, S. L., Becker, S. V., Drexler, J. W., and Turk, J. R., Bioavailability of lead to juvenile swine dosed with soil from the Smuggler Mountain NPL site of Aspen, Colorado, *Fundam. Appl. Toxicol.,* 36, 177, 1997.
180. Reijnders, P. J. H., Reproductive failure in common seals feeding on fish from polluted coastal waters, *Nature,* 324, 456, 1986.
181. Kamrin, M. A. and Ringer, R. K., Toxicological implications of PCB residues in mammals, in *Environmental Contaminants in Wildlife: Interpreting Tissue Concentrations,* Beyer, W. N., Heinz, G. H., and Redmon-Norwood, A., Eds., Lewis Publishers, Boca Raton, FL, 153, 1996.
182. Platonow, N. S. and Reinhart, B. S., Effects of polychlorinated biphenyls (Aroclor 1254) on chicken egg-production, fertility and hatchability, *Can. J. Comp. Med.,* 37, 341, 1973.
183. Britton, W. M. and Huston, T. M., Influence of polychlorinated biphenyls in laying hen, *Poult. Sci.,* 52, 1620, 1973.

184. Scott, M. L., Effects of PCBs, DDT, and mercury-compounds in chickens and Japanese quail, *Fed. Proc.,* 36, 1888, 1977.
185. Cecil, H. C., Bitman, Fries, G. F., Denton, C. A., Lillie, R. J., and Harris, S. J., Dietary *p,p′*-DDT, *o,p′*-DDT or *p,p′*-DDE and changes in egg-shell characteristics and pesticide accumulation in egg contents and body fat of caged white leghorns, *Poult. Sci.,* 51, 130-&, 1972.
186. Tumasonis, C. F., Bush, B., and Baker, F. D., PCB levels in egg yolks associated with embryonic mortality and deformity of hatched chicks, *Arch. Environ. Contam. Toxicol.,* 1, 312, 1973.
187. Peakall, D. B. and Peakall, M. L., Effect of a polychlorinated biphenyl on reproduction of artificially and naturally incubated dove eggs, *J. Appl. Ecol.,* 10, 863, 1973.
188. Heinz, G. H., Hill, E. F., and Contrera, J. F., Dopamine and norepinephrine depletion in ring doves fed DDE, dieldrin, and Aroclor-1254, *Toxicol. Appl. Pharmacol.,* 53, 75, 1980.
189. Haseltine, S. D. and Prouty, R. M., Aroclor-1242 and reproductive success of adult mallards (*Anas platyrhynchos*), *Environ. Res.,* 23, 29, 1980.
190. Custer, T. W. and Heinz, G. H., Reproductive success and nest attentiveness of mallard ducks fed Aroclor-1254, *Environ. Pollut. Series A-Ecol. Biol.,* 21, 313, 1980.
191. Mclane, M. A. R. and Hughes, D. L., Reproductive success of screech owls fed aroclor-1248, *Arch. Environ. Contam. Toxicol.,* 9, 661, 1980.
192. Harris, M. P. and Osborn, D., Effect of a polychlorinated biphenyl on the survival and breeding of puffins, *J. Appl. Ecol.,* 18, 471, 1981.
193. Sileo, L., Karstad, L., Frank, R., Holdrinet, M. V. H., Addison, E., and Braun, H. E., Organochlorine poisoning of ring-billed gulls in southern Ontario, *J. Wildl. Dis.,* 13, 313, 1977.
194. Gilbertson, M., Pollutants in breeding herring gulls in the lower Great Lakes, *Can. Field Nat.,* 88, 273, 1974.
195. Gilbertson, M. and Hale, R., Early embryonic mortality in a herring gull colony in Lake Ontario, *Can. Field Nat.,* 88, 354, 1974.
196. Keith, J. A., Reproduction in a population of herring gulls (*Larus argentatus*) contaminated by DDT, *J. Appl. Toxicol.,* 3, 57, 1966.
197. Ludwig, J. and Tomoff, C., Reproductive success and insecticide residues in Lake Michigan herring gulls, *Jack-Pine Warbler,* 44, 77, 1966.
198. Hoffman, D. J., Rattner, B. A., Sileo, L., Docherty, D., and Kubiak, T. J., Embryotoxicity, teratogenicity, and aryl-hydrocarbon hydroxylase-activity in Forsters terns on Green-Bay, Lake, Michigan, *Environ. Res.,* 42, 176, 1987.
199. Kubiak, T. J., Harris, H. J., Smith, L. M., Schwartz, T. R., Stalling, D. L., Trick, J. A., Sileo, L., Docherty, D. E., and Erdman, T. C., Microcontaminants and reproductive impairment of the Forsters tern on Green Bay, Lake, Michigan — 1983, *Arch. Environ. Contam. Toxicol.,* 18, 706, 1989.
200. Brunstrom, B., Broman, D., and Naf, C., Embryotoxicity of polycyclic aromatic-hydrocarbons (PAHs) in 3 domestic avian species, and of PAHs and coplanar polychlorinated-biphenyls (PCBs) in the common eider, *Environ. Pollut.,* 67, 133, 1990.
201. Hoffman, D. J., Melancon, M. J., Eisemann, J. D., and Klein, P. N., Comparative toxicity of planar PCB congeners by egg injection, *Abstr. 16th Annu. Meeting SETAC,* SETAC Press, Pensacola, FL, 207, 199, 1995.
202. Colborn, T., Epidemiology of Great Lakes bald eagles, *J. Toxicol. Environ. Health,* 33, 395, 1991.
203. Ohlendorf, H. M., Hoffman, D. J., Saiki, M. K., and Aldrich, T. W., Embryonic mortality and abnormalities of aquatic birds: Apparent impacts of selenium from irrigation drainwater, *Sci. Total Environ.,* 52, 49, 1986.
204. Fernie, K. J., Smits, J. E., Bortolotti, G. R., and Bird, D. M., Reproduction success of American kestrels exposed to dietary polychlorinated biphenyls, *Environ. Toxicol. Chem.,* 20, 776, 2001.
205. Fox, G. A., Collins, B., Hayakawa, E., Weseloh, D. V., Ludwig, J. P., Kubiak, T. J., and Erdman, T. C., Reproductive outcomes in colonial fish-eating birds — A biomarker for developmental toxicants in Great Lakes food-chains, 2. Spatial variation in the occurrence and prevalence of bill defects in young double-crested cormorants in the Great Lakes, 1979–1987, *J. Great Lakes Res.,* 17, 158, 1991.
206. Yamashita, N., Tanabe, S., Ludwig, J. P., Kurita, H., Ludwig, M. E., and Tatsukawa, R., Embryonic abnormalities and organochlorine contamination in double-creasted cormorants (*Phalacrocorax auritus*) and Caspian terns (*Hydroprogne caspia*) from the upper Great Lakes, *Environ. Pollut.,* 79, 163, 1993.

207. Custer, T. W., Custer, C. M., Hines, R. K., Gutreuter, S., Stromborg, K. L., Allen, P. D., and Melancon, M. J., Organochlorine contaminants and reproductive success of double-crested cormorants from Green Bay, Wisconsin, USA, *Environ. Toxicol. Chem.,* 18, 1209, 1999.
208. Voogt, P. D., Dirksen, S., Boudewijn, T. J., Bosveld, A. T. C., and Murk, A. J., Do polychlorinated biphenyls contribute to reproduction effects in fish-eating birds?, *Environ. Toxicol. Chem.,* 20, 1149–1150, 2001.
209. Heinz, G. H., Percival, H. F., and Jennings, M. L., Contaminants in American alligator eggs from Lake Apopka, Lake Griffin, and Lake Okeechobee, Florida, *Environ. Monitor. Assess.,* 16, 277, 1991.
210. Guillette, L. J., Jr., Pickford, D. B., Crain, D. A., Rooney, A. A., and Percival, H. F., Reduction in penis size and plasma testosterone concentrations in juvenile alligators living in a contaminated environment, *Gen. Com. Endocrinol.,* 101, 32, 1996.
211. Willingham, E., Rhen, T., Sakata, J. T., and Crews, D., Embryonic treatment with xenobiotics disrupts steroid hormone profiles in hatchling red-eared slider turtles (*Trachemys scripta elegans*), *Environ. Health Perspect.,* 108, 329, 2000.
212. Gutleb, A. C., Appelman, J., Bronkhorst, M., van den Berg, J. H. J., and Murk, A. J., Effects of oral exposure to polychlorinated biphenyls (PCBs) on the development and metamorphosis of two amphibian species (*Xenopus laevis* and *Rana temporaria*), *Sci. Total Environ.,* 262, 147, 2000.
213. Rosenshield, M. L., Jofre, M. B., and Karasov, W. H., Effects of polychlorinated biphenyl 126 on green frog (*Rana clamitans*) and leopard frog (*Rana pipiens*) hatching success, development, and metamorphosis, *Environ. Toxicol. Chem.,* 18, 2478, 1999.
214. Johnson, L. L., Sol, S. Y., Lomax, D. P., Nelson, G. M., Sloan, C. A., and Casillas, E., Fecundity and egg weight in English sole, *Pleuronectes vetulus*, from Puget Sound, Washington: Influence of nutritional status and chemical contaminants, *Fish. Bull.,* 95, 231, 1997.
215. Mac, M. J. and Schwartz, T. R., Investigations into the effects of PCB congeners on reproduction in lake trout from the Great Lakes, *Chemosphere,* 25, 189, 1992.
216. Mac, M. J. and Edsall, C. C., Environmental contaminants and the reproductive success of lake trout in the Great Lakes — An epidemiologic approach, *J. Toxicol. Environ. Health,* 33, 375, 1991.
217. Johnson, L. L., Stein, J. E., Collier, T. K., Casillas, E., and Varanasi, U., Indicators of reproductive development in prespawning female winter flounder (*Pleuronectes americanus*) from urban and non-urban estuaries in the northeast United States, *Sci. Total Environ.,* 141, 241, 1994.
218. Ando, H. and Yano, T., Effects of PCB on sexual maturation in female medaka (*Oryzias latips*), *Sci. Bull. Fac. Ag., Kyushu University,* 36, 79, 1982.
219. Kidd, K., Van Der Kraak, G., and Munkittrick, K., The effects of PCB 126 on hepatic mixed function oxygenase activity and steroidogenic capacity of the goldfish (*Carassius auratus*), *Can. Tech. Report Fish. Aquat. Sci.,* 1942, 470, 1993.
220. Yano, T. and Matsuyama, H., Stimulatory effect of PCB on the metabolism of sex hormones in carp hepatopancreas, *Bull. Jap. Soc. Sci. Fish.,* 52, 1847, 1986.
221. Chen, T. T. and Sonstegard, R. A., Development of a rapid, sensitive and quantitative test for the assessment of the effects of xenobiotics on reproduction in fish, *Mar. Environ. Res.,* 14, 429, 1984.
222. Monosson, E., Fleming, W. J., and Sullivan, C. V., Effects of the planar PCB 3′,3′,4′,4′-tetrachlorobiphenyl (TCB) on ovarian development, plasma levels of sex steroid hormones and vitellogenin, and progeny survival in the white perch (*Morone americana*), *Aquat. Toxicol.,* 29, 1, 1994.
223. Monosson, E., Hogson, R. G., Fleming, W. J., and Sullivan, C. V., Blood plasma levels of sex steroid hormones and vitellogenin in striped bass (*Morone saxatilis*) exposed to 3′, 3′,4′,4′- tetrachlorobiphenyl (TCB), *Bull. Environ. Contam. Toxicol.,* 56, 782, 1996.
224. Freeman, H. C. and Idler, D. R., Effect of polychlorinated biphenyl on steroidogenesis and reproduction in brook trout (*Salvelinus fontinalis*), *Can. J. Biochem.,* 53, 666, 1975.
225. Ankley, G., Tillitt, D., Giesy, J., Jones, P., and Verbrugge, D., Bioassay-derived 2,3,7,8-tetrachlorodibenzo-*p*-dioxin equivalents in PCB-containing extracts from the flesh and eggs of Lake Michigan chinook salmon (*Oncorhynchus tshawytscha*) and possible implications for reproduction, *Can. J. Fish. Aquat. Sci.,* 48, 1685, 1991.
226. Ungerer, J. and Thomas, P., Transport and accumulation of organochlorines in the ovaries of Atlantic croaker (*Micropogonias undulatus*), *Mar. Environ. Res.,* 42, 167, 1996.

227. Sangalang, G. B., Freeman, H. C., and Crowell, R., Testicular abnormalities in cod (*Gadus-morhua*) fed Aroclor-1254, *Arch. Environ. Contam. Toxicol.,* 10, 617, 1981.
228. Suedel, B. C., Dillon, T. M., and Benson, W. H., Subchronic effects of five di-ortho PCB congeners on survival, growth and reproduction in the fathead minnow *Pimephales promelas*, *Environ. Toxicol. Chem.,* 16, 1526, 1997.
229. Thomas, P., Changes in the plasma-levels of maturation-inducing steroids in several perciform fishes during induced ovulation, *Am. Zool.,* 28, A53, 1988.
230. Dicarlo, F. J., Seifter, J., and Decarlo, V. J., Assessment of hazards of polybrominated biphenyls, *Environ. Health Perspect.,* 23, 351, 1978.
231. Roboz, J., Greaves, J., and Bekesi, J. G., Polybrominated biphenyls in model and environmentally contaminated human-blood — Protein-binding and immunotoxicological studies, *Environ. Health Perspect.,* 60, 107, 1985.
232. Roboz, J., Greaves, J., McCamish, M., Holland, J. F., and Bekesi, G., An *in vitro* model for the binding of polybrominated biphenyls in environmentally contaminated blood, *Arch. Environ. Contam. Toxicol.,* 14, 137, 1985.
233. Hesse, J. L. and Powers, R. A., Polybrominated biphenyl (PBB) contamination of Pine River, Gratiot, and Midland Counties, Michigan, *Environ. Health Perspect.,* 23, 19, 1978.
234. Damstra, T., Jurgelski, W., Posner, H. S., Vouk, V. B., Bernheim, N. J., Guthrie, J., Luster, M., and Falk, H. L., Toxicity of polybrominated biphenyls (PBBs) in domestic and laboratory-animals, *Environ. Health Perspect.,* 44, 175, 1982.
235. Polin, D. and Ringer, R. K., Polybrominated biphenyls in chicken eggs vs. hatchability, *Proc. Soc. Exp. Biol. Med.,* 159, 131, 1978.
236. Polin, D. and Ringer, R. K., PBB fed to adult female chickens — its effect on egg-production, reproduction, viability of offspring, and residues in tissues and eggs, *Environ. Health Perspect.,* 23, 283, 1978.
237. Zitko, V., Accumulation of polybrominated biphenyls by fish, *Bull. Environ. Contam. Toxicol.,* 17, 285, 1977.
238. Kuehl, D., Butterworth, B. C., McBride, A., Kroner, S., and Bahnick, D., Contamination of fish by 2,3,7,8-tetrachlorodibenzo-*p*-dioxin — A survey of fish from major watersheds in the United States, *Chemosphere,* 18, 1997, 1989.
239. Peterson, R. E., Theoblad, H. M., and Kimmel, G. L., Developmental and reproductive toxicity of dioxins and related compounds: Cross-species comparisons, *Crit. Rev. Toxicol.,* 23, 283, 1993.
240. Birnbaum, L. S. and Tuomisto, J., Non-carcinogenic effects of TCDD in animals, *Food Add. Contam.,* 17, 275, 2000.
241. Boening, D., Toxicity of 2,3,7,8-tetrachlorodibenzo-*p*-dioxin to several ecological receptor groups: A short review, *Ecotoxicol. Environ. Saf.,* 39, 155, 1998.
242. Mably, T. A., Bjerke, D. L., Moore, R. W., Gendronfitzpatrick, A., and Peterson, R. E., *In utero* and lactational exposure of male rats to 2,3,7,8-tetrachlorodibenzo-*p*-dioxin, 3. Effects on spermatogenesis and reproductive capacity, *Toxicol. Appl. Pharmacol.,* 114, 118, 1992.
243. Mably, T. A., Moore, R. W., Goy, R. W., and Peterson, R. E., *In utero* and lactational exposure of male rats to 2,3,7,8-tetrachlorodibenzo-*p*-dioxin, 2. Effects on sexual behavior and the regulation of luteinizing hormone secretion in adulthood, *Toxicol. Appl. Pharmacol.,* 114, 108, 1992.
244. Mably, T. A., Moore, R. W., and Peterson, R. E., *In utero* and lactational exposure of male rats to 2,3,7,8-tetrachlorodibenzo-*p*-dioxin. 1. Effects on androgenic status, *Toxicol. Appl. Pharmacol.,* 114, 97, 1992.
245. Gray, L., Kelce, W., Monosson, E., Ostby, J., and Birnbaum, L., Exposure to TCDD during development permanently alters reproductive function in male Long Evans rats and hamsters-reduced ejaculated and epididymal sperm numbers and sex accessory-gland weights in offspring with normal androgenic status, *Toxicol. Appl. Pharmacol.,* 131, 108, 1995.
246. Hoffman, D., Rice, C., and Kubiak, T., PCBs and dioxins in birds, in *Environmental Contaminants in Wildlife. Interpreting Tissue Concentrations,* Beyer, W., Heinz, G., and Redmon-Norwood, A., Eds., CRC Lewis Publishers, Boca Raton, FL, 165, 1996.
247. Gilbertson, M., Kubiak, T., and Ludwig, J., Great Lakes embryo mortality, edema and deformities syndrome (GLEMEDS) in colonial fish-eating birds: Similarity to chick-edema disease, *J. Toxicol. Environ. Health,* 455, 1991.

248. Giesey, J., Ludwig, J., and Tillitt, D., Deformities in birds of the Great Lakes region, *Environ. Sci. Technol.*, 28, 128, 1994.
249. Grasman, K., Scanlon, P., and Fox, G., Reproductive and physiological effects of environmental contaminants in fish-eating birds of the Great Lakes: A review of historical trends, *Environ. Monitor. Assess.*, 53, 117, 1998.
250. Ludwig, J., Kurita-Matsuba, H., Auman, H., Ludwig, M., Summer, C., Glesy, J., Tillitt, D., and Jones, P., Deformities, PCBs, and TCDD-Equivalents in double-crested cormorants (*Phalacrocorax auritus*) and Caspian terns (*Hydroprogne caspia*) of the upper Great Lakes 1986–91: Testing a cause-effect hypothesis, *J. Great Lakes Res.*, 22, 172, 1996.
251. Heinz, G., Contaminant effects on Great Lakes fish-eating birds: A population perspective, in *Principles and Processes for Evaluating Endocrine Disruption in Wildlife,* Kendall, R. J., Dickerson, R. L., Giesy, J. P., and Suk, W. P., Eds., SETAC Press, Pensacola, FL, 141, 1998.
252. Hart, L., Cheng, K., Whitehead, P., Shah, R., Lewis, R., Ruschlowski, S., Blair, R., Bennett, D., Bandiera, S., Norstrom, R., and Bellward, G., Dioxin contamination and growth and development in great blue heron embryos, *J. Toxicol. Environ. Health,* 32, 331, 1991.
253. Elliott, J., Butler, R., Norstrom, R., and Whitehead, P., Environmental contaminants and reproductive success of great blue herons (*Ardea herodias*) in British Columbia, 1986–1989, *Environ. Pollut.*, 59, 91, 1989.
254. Sanderson, J., Nosrtrom, R., Elliott, J., Hart, L., Cheng, K., and Bellward, G., Biological effects of polychlorinated dibenzo-*p*-dioxins, dibenzofurans, and biphenyls in double-crested cormorant chicks (*Phalacrocorax auritus*), *J. Toxicol. Environ. Health,* 41, 247, 1994.
255. White, D. and Seginak, J., Dioxins and furans linked to reproductive impairment in wood ducks at Bayou Meto, Arkansas, *J. Wildlife Manage.,* 58, 100, 1994b.
256. Henshel, D., Developmental neurotoxic effects of dioxin and dioxin-like compounds on domestic and wild avian species, *Environ. Toxicol. Chem.,* 17, 88, 1998.
257. Powell, D., Aulerich, R., Meadows, J., Tillitt, D., Kelly, M., Stromborg, K., Melancon, M., Fitzgerald, S., and Bursian, S., Effects of 3′,3′,4′,4′,5-pentachlorobiphenyl and 2,3,7,8-tetrachlorodibenzo-*p*-dioxin injected into the yolks of double-crested cormorant (*Phalacrocorax auritus*) eggs prior to incubation, *Environ. Toxicol. Chem.,* 17, 2035, 1998.
258. Rolland, R., A review of chemically-induced alterations in thyroid and vitamin A status from field studies of wildlife and fish, *J. Wildl. Dis.,* 36, 615, 2000.
259. Murk, A., Boudewijn, T., Meininger, P., Bosveld, A., Rossaert, G., and Ysebaert, P., Effects of polyhalogenated aromatic hydrocarbons and related contaminants on common tern reproduction: Integration of biological, biochemical, and chemical data, *Arch. Environ. Contam. Toxicol.,* 31, 128, 1996.
260. Van der Berg, M., Craane, B., Sinnige, T., Van Mourik, S., Dirksen, S., Boudewijn, T., van der Gaag, M., Lutke-schipolt, I., Spenkelink, B., and Brouwer, A., Biochemical and toxic effects of polychlorinated biphenyls (PCB's), dibenzo-*p*-dioxins (PCDDs) and dibenzofurans (PCDFs) in the cormorant (*Phalacrocorax carbo*) after *in ovo* exposure, *Environ. Toxicol. Chem.,* 13, 803, 1994.
261. Janz, D. and Bellward, G., Effects of 2,3,7,8-tetrachloro-*p*-dioxin exposure on plasma thyroid and sex steroid hormone concentrations and estrogen receptor levels in adult great blue herons, *Environ. Toxicol. Chem.,* 16, 985, 1997.
262. Gould, J., Cooper, K., and Scanes, C., Effects of polychlorinated biphenyl mixtures and three specific congeners on growth and circulating growth-related hormones, *Gen. Comp. Endocrinol.,* 106, 221, 1997.
263. Neal, R., Beatty, P., and Gasiewicz, T., Studies of the mechanisms of toxicity of 2,3,7,8-tetrachlorodibenzo-*p*-dioxin (TCDD), *Annals N.Y. Acad. Sci.,* 320, 204, 1979.
264. Jung, R. and Walker, M., Effects of 2,3,7,8-tetrachlorodibenzo-*p*-dioxin (TCDD) on development of anuran amphibians, *Environ. Toxicol. Chem.,* 16, 230, 1997.
265. Karchner, S., Kennedy, S., Trudeau, S., and Hahn, M., Towards molecular understanding of species differences in dioxin sensitivity: Initial characterization of Ah receptor cDNAs in birds and an amphibian, *Mar. Environ. Res.,* 50, 51, 2000.
266. Bishop, C. A., Ng, P., Petit, K. E., Kennedy, S. W., Stegeman, J. J., Norstrom, R. J., and Brooks, R. J., Environmental contamination and developmental abnormalities in eggs and hatchlings of the common snapping turtle (*Chelydra serpentina serpentina*) from the Great Lakes St. Lawrence River basin (1989–91), *Environ. Pollut.,* 101, 143, 1998.

267. Matter, J., McMurry, C., Anthony, A., and Dickerson, R., Development and implementation of endocrine biomarkers of exposure and effects in American alligators (*Alligator mississippiensis*), *Chemosphere,* 37, 1905, 1998.
268. Cooper, K., The effects of polychlorinated dibenzo-*p*-dioxins and polychlorinated dibenzofurans on aquatic organisms, *CRC Crit. Rev. Aquat. Sci.,* 1, 227, 1989.
269. Walker, M. K. and Peterson, R. E., Aquatic toxicity of dioxins and related chemicals, in *Dioxins and Health,* Schecter, A., Ed., Eds., Plenum Press, New York, 347, 1994.
270. Walker, M. and Peterson, R., Potencies of polychlorinated dibenzo-*p*-dioxin, dibenzofuran, and biphenyl congeners, relative to 2,3,7,8-tetrachlorodibenzo-*p*-dioxin, for producing early life stage mortality in rainbow trout (*Oncorhynchus mykiss*), *Aquat. Toxicol.,* 21, 219, 1991.
271. Walker, M. K., Spitsbergen, J. M., Olson, J. R., and Peterson, R. E., 2,3,7,8-tetrachlorodibenzo-*p*-dioxin (TCDD) toxicity during early life stage development of lake trout (*Salvelinus namaycush*), *Can. J. Fish. Aquat. Sci.,* 48, 875, 1991.
272. Wisk, J. and Cooper, K., The stage specific toxicity of 2,3,7,8-tetrachlorodibenzo-*p*-dioxin in embryos of the Japanese medaka (*Oryzias latipes*), *Environ. Toxicol. Chem.,* 9, 1159, 1990.
273. Wannemacher, R., Rebstock, A., Kulzer, E., Schrenk, D., and Bock, K. W., Effects of 2,3,7,8-tetrachlorodibenzo-dioxin on reproduction and oogenesis in zebrafish (*Brachydanio rerio*), *Chemosphere,* 24, 1361, 1992.
274. Giesy, J. P. and Snyder, E. M., Xenobiotic modulation of endocrine function in fishes, in *Principles and Processes for Evaluating Endocrine Disruption in Wildlife*, SETAC Press, Pensacola, FL, 155, 1998.
275. Henry, T. R., Spitsbergen, J. M., Hornung, M. W., Abnet, C. C., and Peterson, R. E., Early life stage toxicity of 2,3,7,8-tetrachlorodibenzo-*p*-dioxin in zebrafish (*Danio rerio*), *Toxicol. Appl. Pharmacol.,* 142, 56, 1997.
276. Safe, S. H., 2,3,7,8-Tetrachlorodibenzo-*p*-dioxin (TCDD) and related environmental antiestrogens: Characterization and mechanism of action, in *Endocrine Disruptors. Effects on Male and Female Reproductive Systems*, CRC Press, Boca Raton, FL, 187, 1999.
277. Smeets, J., van Holsteijn, I., Giesy, J., and van den Berg, M., The anti-estrogenicity of Ah receptor agonists in carp (*Cyprinus carpio*) hepatocytes, *Toxicol. Sci.,* 52, 178, 1999.
278. Sanderson, J., Janz, D., Bellward, G., and Giesy, J., Effects of embryonic and adult exposure to 2,3,7,8-tetrachloro-*p*-dioxin on hepatic microsomal testosterone hydroxylase activities in great blue herons (*Ardea herodias*), *Environ. Toxicol. Chem.,* 16, 1304, 1997.
279. Bookstaff, R., Moore, R., and Peterson, R., 2,3,7,8-tetrachlorodibenzo-*p*-dioxin in male rats: Different effects of in utero versus lactational exposure, *Toxicol. Appl. Pharmacol.,* 104, 212, 1990.
280. Spear, P., Bilodeau, A., and Branchaud, A., Retiboids: From metabolism to environmental monitoring, *Chemosphere,* 25, 1733, 1992.
281. Brouwer, A., Morse, D., Lans, M., Schuur, A., Murk, A., Klasson-Wehler, E., Bergman, A., and Visser, T., Interactions of persistent environmental organohalogens and the thyroid hormone system: Mechanisms and possible consequences for animal and human health, in *Endocrine-Disrupting Chemicals: Neural, Endocrine, and Behavioral Effects,* Colborn, T., vom Saal, F., and Short, P., Eds., Stockton Press, Basingstoke, U.K., 1998, 59.
282. Lans, M., Klassonwehler, E., Willemsen, M., Meussen, E., Safe, S., and Brouwer, A., Structure-dependent, competitive interaction of hydroxy-polychlorinated biphenyls, hydroxy-dibenzo-*p*-dioxins and hydroxy-dibenzofurans with human transthyretin, *Chem.-Biol. Interact.,* 88, 7, 1993.
283. Cantrell, S. M., Lutz, L. H., Tillitt, D. E., and Hannink, M., Embryotoxicity of 2,3,7,8-tetrachlorodibenzo-*p*-dioxin (TCDD): The embryonic vasculature is a physiological target for TCDD-induced DNA damage and apoptotic cell death in medaka (*Orizias latipes*), *Toxicol. Appl. Pharmacol.,* 141, 23, 1996.
284. Gross, D. A., Gross, T. S., Johnson, B., and Folmar, L., Characterization of endocrine-disruption and clinical manifestations in largemouth bass from Florida lakes, in *Abstr. 16th Annu. Meeting SETAC,* SETAC Press, Pensacola, FL, 185, 1995.
285. Pickering, O. H., Henderson, C., and Lemke, A. E., The toxicity of organic phosphorus insecticides to different species of warmwater fishes, *Trans. Am. Fish. Soc.,* 91, 175, 1962.
286. Naqvi, S. M. and Vaishnavi, C., Bioaccumulative potential and toxicity of endosulfan insecticide to non-target animals, *Comp. Biochem. Physiol.,* 105C, 347, 1993.

287. Fowler, J. F., Newsom, L. D., Graves, J. B., Bonner, F. L., and Schillin, P. E., Effect of dieldrin on egg hatchability, chick survival and eggshell thickness in purple and common gallinules, *Bull. Environ. Contam. Toxicol.*, 6, 495, 1971.
288. Lockie, J. D., Ratcliff, D. A., and Balharry, R., Breeding success and organo-chlorine residues in golden eagles in west Scotland, *J. Appl. Ecol.*, 6, 381, 1969.
289. Blus, L. J., Organochlorine pesticides, in *Handbook of Ecotoxicology,* Hoffman, D. J., Rattner, B. A., Burton, G. A., Jr., and Cairns, J., Jr., Eds., Lewis Publishers, Boca Raton, FL, 275, 1995.
290. Newton, I. and Bogan, J., Organochlorine residues, eggshell thinning and hatching success in British sparrowhawks, *Nature,* 249, 582, 1974.
291. Newton, I. and Bogan, J., The role of different organochlorine compounds in the breeding of British sparrowhawks, *J. Appl. Ecol.*, 15, 105, 1978.
292. Fleming, W. J., McLane, M., and Cromartie, E., Endrin decreases screech owl productivity, *J. Wildl. Manage.*, 46, 462, 1982.
293. Blus, L. J., Henny, C. J., Lenhart, D. J., and Kaiser, T. E., Effects of heptachlor- and lindane-treated seed on Canada geese, *J. Wildl. Manage.*, 48, 1097, 1984.
294. Henny, C. J., Blus, L. J., and Stafford, C. J., Effects of heptachlor on American kestrels in the Columbia Basin, Oregon, *J. Wildl. Manage.*, 47, 1080, 1983.
295. Fyfe, R. W., Risebrough, R. W., and Walker, W. I., Pollutant effects on the reproduction of the prairie falcons and merlins of the Canadian prairies, *Can. Field Nat.*, 90, 346, 1976.
296. Bush, P. B., Kiker, J. T., Page, R. K., Booth, N. H., and Fletcher, O. J., Effects of graded levels of toxaphene on poultry residue accumulation, egg-production, shell quality, and hatchability in white leghorns, *J. Agric. Food Chem.*, 25, 928, 1977.
297. Heinz, G. H. and Finley, M. T., Toxaphene does not affect avoidance-behavior of young black ducks, *J. Wildl. Manage.*, 42, 408, 1978.
298. Faahraeus-Van Ree, G. E. and Payne, J. F., Effect of toxaphene on reproduction of fish, *Chemosphere,* 34, 855, 1997.
299. Smith, R. M. and Cole, C. F., Effects of egg concentrations of DDT and dieldrin on development in winter flounder (*Pseudopleuronectes americanus*), *J. Fish. Res. Bd. Can.*, 30, 1894, 1973.
300. Hermanutz, R. O., Endrin and malathion toxicity to flagfish (*Jordanella floridae*), *Arch. Environ. Contam. Toxicol.*, 7, 159, 1978.
301. Matthiessen, P. and Logan, J. W. M., Low concentration effects of endosulfan insecticide on reproductive-behavior in the tropical cichlid fish *Sarotherodon mossambicus*, *Bull. Environ. Contam. Toxicol.*, 33, 575, 1984.
302. Arora, N. and Kulshrestha, S.K., Comparison of the toxic effects of 2 pesticides on the testes of a fresh-water teleost *Channa striatus*, *Acta Hydrochem. Hydrobiol.*, 12, 435, 1984.
303. Kumar, S. and Pant, S. C., Comparative sublethal ovarian pathology of some pesticides in the teleost, *Puntius conchonius* Hamilton, *Bull. Environ. Contam. Toxicol.*, 41, 227, 1988.
304. Kulshrestha, S. K. and Arora, N., Impairments induced by sublethal doses of 2 pesticides in the ovaries of a fresh-water teleost *Channa striatus* Bloch, *Toxicol. Lett.*, 20, 93, 1984.
305. Rastogi, A. and Kulshrestha, S. K., Effect of sublethal doses of 3 pesticides on the ovary of a carp minnow *Rasbora daniconius*, *Bull. Environ. Contam. Toxicol.*, 45, 742, 1990.
306. Chakravorty, S., Lal, B., and Singh, T. P., Effect of endosulfan (thiodan) on vitellogenesis and its modulation by different hormones in the vitellogenic catfish *Clarias batrachus*, *Toxicology,* 75, 191, 1992.
307. Inbaraj, R. M. and Haider, S., Effect of malathion and endosulfan on brain acetylcholinesterase and ovarian steroidogenesis of *Channa punctatus* (Bloch), *Ecotoxicol. Environ. Saf.*, 16, 123, 1988.
308. Shukla, L. and Pandey, A. K., Effects of endosulfan on the hypothalamo hypophysial complex and fish reproductive physiology, *Bull. Environ. Contam. Toxicol.*, 36, 122, 1986.
309. Guzelian, P. S., Comparative toxicology of chlordecone (kepone) in humans and experimental animals, *Annu. Rev. Pharmacol. Toxicol.*, 22, 89, 1982.
310. Yamamoto, J. T., Donohoe, R. M., Fry, D. M., Golub, M. S., and Donald, J. M., Environmental estrogens: Implications for reproduction in wildlife, in *Noninfectious Diseases of Wildlife,* Fairbrother, A., Locke, L. N., and Hoff, G. L., Eds., Iowa State University Press, Ames, 31, 1996.
311. Huber, J., Some physiologic effects of the insecticide Kepone in the laboratory mouse, *Toxicol. Appl. Pharmacol.*, 7, 516, 1965.

312. Hammond, B., Katzenellenbogen, B. S., Krauthammer, N., and McConnel, J., Estrogenic activity of the insecticide chlordecone (Kepone) and interaction with uterine estrogen receptors, *Proc. Natl. Acad. Sci.,* 76, 6641, 1979.
313. Gellert, R. J., Uterotrophic activity of polychlorinated biphenyls (PCB) and induction of precocious reproductive aging in neonatally treated female rats, *Environ. Res.,* 16, 123, 1978.
314. McFarland, L. Z. and Lacy, P. B., Physiologic and endocrinologic effects of the insecticide Kepone in the Japanese quail, *Toxicol. Appl. Pharmacol.,* 15, 441, 1969.
315. Eroschenko, V. P., Alterations in testes of Japanese quail during and after ingestion of insecticide Kepone, *Toxicol. Appl. Pharmacol.,* 43, 535, 1978.
316. Eroschenko, V. P., Morphological alterations in testes of Japanese quail during and after ingestion of estrogenic insecticide Kepone, *Anat. Rec.,* 190, 391, 1978.
317. Romanoff, A. L. and Romanoff, A. J., *The Avian Egg*, John Wiley and Sons, New York, 1949.
318. Naber, E. C. and Ware, G. W., Effect of Kepone and Mirex on reproductive performance in laying hen, *Poult. Sci.,* 44, 875, 1965.
319. Thomas, P. and Khan, I. A., Mechanisms of chemical interference with reproductive endocrine function in sciaenid fishes, in *Chemically Induced Alterations in Functional Development and Reproduction of Fishes*, Rolland, R. M., Gilbertson, M., and Peterson, R. E., Eds., SETAC Press, Pensacola, FL, 29, 1997.
320. Donohoe, R. M. and Curtis, L. R., Estrogenic activity of chlordecone, *o,p′*-DDT and *o,p′*-DDE in juvenile rainbow trout: Induction of vitellogenesis and interaction with hepatic estrogen binding sites, *Aquat. Toxicol.,* 36, 31, 1996.
321. Thomas, P., Gosh, S., and Smith, J. S., Interactions of organochlorines with steroid hormone receptors in Atlantic croaker, *Mar. Environ. Res.,* 42, 176, 1996.
322. Curtis, L. R. and Beyers, R. J., Inhibition of oviposition in teleost *Oryzias latipes*, induced by subacute kepone exposure, *Comp. Biochem. Physiol. C.,* 61, 15, 1978.
323. Srivastava, A. K. and Srivastava, A. K., Effects of chlordecone on the gonads of freshwater catfish, *Heteropneustes fossilis*, *Bull. Environ. Contam. Toxicol.,* 53, 186, 1994.
324. Bitman, J., Cecil, C., Jarros, S. J., and Fries, G. F., Estrogenic activity of *o,p′*-DDT in the mammalian uterus and avian oviduct, *Science,* 162, 371, 1969.
325. Clement, J. G. and Okey, A. B., Estrogenic and antiestrogenic effects of DDT administered in the diet to immature female rats, *Can. J. Physiol. Pharmacol.,* 50, 971, 1972.
326. Aguilar, A. and Borrell, A., Reproductive transfer and variation of body load of organochlorine pollutants with age in fin whales (*Balaenoptera physalus*), *Arch. Environ. Contam. Toxicol.,* 27, 548, 1994.
327. Clement, J. G., Hormonal consequences of organophosphate poisoning, *Fundam. Appl. Toxicol.,* 5, S61, 1985.
328. Blus, L. J., Neeley, B. S., Belisle, A. A., and Prouty, R. M., Organochlorine residues in brown pelican eggs: Relation to reproductive success, *Environ. Pollut.,* 7, 81, 1974.
329. Blus, L. J., Further interpretation of the relation of organochlorine residues in brown pelican eggs to reproductive success, *Environ. Pollut.,* A28, 15, 1982.
330. Blus, L. J., DDT, DDD, and DDE in birds, in *Environmental Contaminants in Wildlife: Interpreting Tissue Concentrations,* Beyer, W. N., Heinz, G. H., and Redmon-Norwood, A. W., Eds., Lewis Publishers, Boca Raton, FL, 49, 1996.
331. Wiemeyer, S. N., Bunck, C. M., and Stafford, C. J., Environmental contaminants in bald eagle eggs 1980–84 and further interpretations of relationships to productivity and shell thickness, *Arch. Environ. Contam. Toxicol.,* 24, 213, 1993.
332. Bowerman, W. W., Best, D. A., Grubb, T. G., Sikarskie, J. G., and Giesy, J. P., Assessment of environmental endocrine disruptors in bald eagles of the Great Lakes, *Chemosphere,* 41, 1569, 2000.
333. Lundholm, C. E., DDE-induced eggshell thinning in birds: Effects of *p,p′*,-DDE on the calcium and prostaglandin metabolism of the eggshell gland, *Comp. Biochem. Physiol. C.,* 118, 113, 1997.
334. Bishop, C. A., Collins, B., Mineau, P., Burgess, N. M., Read, W. F., and Risley, C., Reproduction of cavity-nesting birds in pesticide-sprayed apple orchards in southern Ontario, Canada, 1988–1994, *Environ. Toxicol. Chem.,* 19, 588, 2000.
335. Elliott, J. E., Martin, P. A., Arnold, T. W., and Sinclair, P. H., Organochlorines and reproductive success of birds in orchard and non-orchard areas of central British-Columbia, Canada, 1990–91, *Arch. Environ. Contam. Toxicol.,* 26, 435, 1994.

336. Hebert, C. E., Weseloh, D. V., Kot, L., and Glooschenko, V., Organochlorine contaminants in a terrestrial foodweb on the Niagara Peninsula, Ontario, Canada, 1987–89, *Arch. Environ. Contam. Toxicol.,* 26, 356, 1994.
337. Woodward, A. R., Jennings, M. L., and Percival, H. F., Egg collecting and hatch rates of American alligator eggs in Florida, *Wildl. Soc. Bull.,* 17, 124, 1989.
338. Woodward, A. R., Jennings, M. L., Percival, H. F., and Moore, J. F., Low clutch viability of American alligators on Lake Apopka, *Florida Sci.,* 56, 52, 1993.
339. Palmer, B. D. and Palmer, S. K., Vitellogenin induction by xenobiotic estrogens in the red-eared turtle and the African clawed frog, *Environ. Health Perspect.,* 103, 19, 1995.
340. Clark, E. J., Norris, D. O., and Jones, R. E., Interactions of gonadal steroids and pesticides (DDT, DDE) on gonaduct growth in larval tiger salamanders, *Ambystoma tigrinum, Gen. Comp. Endocrinol.,* 109, 94, 1998.
341. Noriega, N. C. and Hayes, T. B., DDT congener effects on secondary sex coloration in the reed frog *Hyperolius argus*: A partial evaluation of the *Hyperolius argus* endocrine screen, *Comp. Biochem. Physiol. B.,* 126, 231, 2000.
342. Macek, K. J., Reproduction in brook trout (*Salvelinus fortinalis*) fed sublethal concentrations of DDT, *J. Fish. Res. Bd. Can.,* 25, 1787, 1968.
343. Burdick, G. E., Harris, E. J., Dean, H. J., Walker, T. M., Skea, J., and Colby, D., The accumulation of DDT in lake trout and the effect on reproduction, *Trans. Am. Fish. Soc.,* 93, 127, 1964.
344. Cross, J. N. and Hose, J. E., Evidence for impaired reproduction in white croaker (*Genyonemus lineatus*) from contaminated areas off southern California, *Mar. Environ. Res.,* 24, 185, 1988.
345. Monod, G., Egg mortality of Lake Geneva charr (*Salvelinus alpinus* L.) contaminated by PCB and DDT derivatives, *Bull. Environ. Contam. Toxicol.,* 35, 531, 1985.
346. Spies, R. B., Thomas, P., and Matsui, M., Effects of DDT and PCB on reproductive endocrinology of *Paralabrax clathratus* in southern California, *Mar. Environ. Res.,* 42, 75, 1996.
347. Wester, P. W., Histopathological effects of environmental pollutants β-HCH and methyl mercury on reproductive organs in freshwater fish, *Comp. Biochem. Physiol.,* 100C, 237, 1991.
348. Wester, P. W., Canton, J. H., and Bisschoop, A., Histopathological study of *Poecilia reticulata* (guppy) after long-term β-hexachlorocyclohexane exposure, *Aquat. Toxicol.,* 6, 271, 1985.
349. Wester, P. W. and Canton, J. H., Histopathological study of *Oryzias latipes* (Medaka) after long-term ß-hexachlorocyclohexane exposure, *Aquat. Toxicol.,* 9, 21, 1986.
350. Singh, P. B. and Singh, T. P., Impact of γ-BHC on lipid class levels and their modulation by reproductive hormones in the freshwater catfish, *Heteropneustes fossilis, Bull. Environ. Contam. Toxicol.,* 48, 23, 1992.
351. Singh, P., Kime, D., Epler, P., and Chyb, J., Impact of hexachlorocyclohexane exposure on plasma gonadotropin levels and *in vitro* stimulation of gonadal steroid production by carp hypophyseal homogenate in *Carassius auratus, J. Fish Biol.,* 44, 195, 1994.
352. Singh, S. and Singh, T. P., Impact of malathion and hexachlorocyclohexane on plasma profiles of three sex hormones during different phases of the reproductive cycle in *Clarias batrachus, Pest. Biochem. Physiol.,* 27, 301, 1987.
353. Gray, L. E., Ostby, J., and Kelce, W. R., Developmental effects of an environmental antiandrogen: The fungicide vinclozolin alters sex differentiation of the male rat, *Toxicol. Appl. Pharmacol.,* 129, 46, 1994.
354. Gray, L. E., Tiered screening and testing strategy for xenoestrogens and antiandrogens, *Toxicol. Lett.,* 103, 677, 1998.
355. Hellwig, J., Reproduction Study with Reg No. 83–258 (Vinclozolin) in Rats Continuous Dietary Administration over 2 Generations (2 Litters in the First and 2 Litters in the Second Generation), Report #92/11251, Department of Toxicology, BASF AG, Ludwigshafen, 1992.
356. Knuppen, R., Final Report: Study of the Binding of 3-(3,5-Dichlorophenyl)-5-Methylvinyl-2,4-Dion to the Androgen Receptor in MCF-7 Cells, Institute of Biochemical Endocrinology, Medical University of Lübeck, 1990.
357. Sheffield, S. R., Effects of Field Exposure to Diazinon on Small Mammals Inhabiting a Semi-Enclosed Prairie Grassland Ecosystem, I. Ecological and Reproductive Effects, II. Sublethal Effects, Ph.D. dissertation, Oklahoma State University, 1996.
358. Pomeroy, S. and Barrett, G., Dynamics of enclosed small mammal populations in relation to an experimental pesticide application, *Am. Mid. Nat.,* 93, 91, 1975.

359. Block, E. K., Lacher, T. E., Brewer, L. W., Cobb, G. P., and Kendall, R. J., Population responses of *Peromyscus* resident in Iowa cornfields treated with the organophosphorus pesticide COUNTER®, *Ecotoxicology,* 8, 189, 1999.
360. Rattner, B. A., Clarke, R. N., and Ottinger, M. A., Depression of plasma luteinizing hormone concentration in quail by the anticholinesterase insecticide parathion, *Comp. Biochem. Physiol.,* 83C, 451, 1986.
361. Rattner, B. A., Sileo, L., and Scanes, C. G., Oviposition and the plasma concentrations of LH, progesterone and corticosterone in bobwhite quail (*Colinus virginianus*) fed parathion, *J. Reprod. Fertil.,* 66, 147, 1982.
362. Bennett, R. S., Williams, B. A., Schmedding, D. W., and Bennett, J. K., Effects of dietary exposure to methyl parathion on egg laying and incubation in mallards, *Environ. Toxicol. Chem.,* 10, 501, 1991.
363. Vonier, P. M., Crain, D. A., McLachlan, J. A., Guillette, L. J., and Arnold, S. F., Interaction of environmental chemicals with the estrogen and progesterone receptors from the oviduct of the American alligator, *Environ. Health Perspect.,* 104, 1318, 1996.
364. White, D., Mitchell, C., and Hill, E., Parathion alters incubation behavior of laughing gulls, *Bull. Environ. Contam. Toxicol.,* 31, 93, 1983.
365. Stromborg, K., Seed treatment of pesticide effects on pheasant reproduction at sublethal doses, *J. Wildl. Manage.,* 41, 632, 1977.
366. Maitra, S. K. and Sarkar, R., Influence of methyl parathion on gametogenic and acetylcholinesterase activity in the testis of whitethroated munia (*Lonchura malabarica*), *Arch. Environ. Contam. Toxicol.,* 30, 384, 1996.
367. Knapton, R. and Mineau, P., Effects of granular formulations of terbufos and fonofos applied to cornfields on mortality and reproductive success of songbirds, *Ecotoxicology,* 4, 138, 1995.
368. Pascual, J., Peris, S., and Robredo, F., Efectos de tratamientos forestales con cipermetrina y malation sobre el exito de cria del herrerillo comun (*Parus caeruleus*), *Ecologia,* 359, 1991.
369. Robinson, S. C., Kendall, R. J., Robinson, R., Driver, C. J., and Lacher, T. E., Effects of agricultural spraying of methyl parathion on cholinesterase activity and reproductive success in wild starlings (*Sturnus vulgaris*), *Environ. Toxicol. Chem.,* 7, 343, 1988.
370. Powell, G. V. N., Reproduction by an altricial songbird, the red-winged blackbird, in fields treated with the organophosphate insecticide fenthion, *J. Appl. Ecol.,* 21, 83, 1984.
371. Fordham, C., Tessari, J., Ramsdell, H., and Keefe, T., Effects of malathion on survival, growth, development, and equilibrium posture of bullfrog tadpoles (*Rana catesbeiana*), *Environ. Toxicol. Chem.,* 20, 179, 2001.
372. Harris, M., Bishop, C., Struger, J., Ripley, B., and Bogart, J., The functional integrity of northern leopard frog (*Rana pipiens*) and green frog (*Rana clamitans*) populations in orchard wetlands. II. Effects of pesticides and eutrophic conditions on early life stage development, *Environ. Toxicol. Dev.,* 17, 1351, 1998.
373. Alvarez, R., Honrubia, M., and Herraez, M., Skeletal malformations induced by the insecticides ZZ-Aphox and Folidol during larval development of *Rana perezi, Arch. Environ. Contam. Toxicol.,* 28, 349, 1995.
374. Pauli, B., Coulson, D., and Berrill, M., Sensitivity of amphibian embryos and tadpoles to Mimic® 240 LV insecticide following single or double exposures, *Environ. Toxicol. Chem.,* 18, 2538, 1999.
375. Ram, R. N. and Sathyanesan, A. G., Ammonium sulfate induced nuclear changes in the oocyte of the fish, *Channa punctatus* (BI)., *Bull. Environ. Contam. Toxicol.,* 36, 871, 1986.
376. Choudhury, C., Ray, A., Bhattacharya, S., and Bhattacharya, S., Non-lethal concentrations of pesticide impair ovarian function in the freshwater perch, *Anabas testudineus, Environ. Biol. Fishes,* 36, 319, 1993.
377. Yasuno, M., Hatakeyama, S., and Miyashita, M., Reproduction in the guppy (*Poecilia reticulata*) under chronic exposure to mephos and fenitrothion, *Bull. Environ. Contam. Toxicol.,* 25, 29, 1980.
378. Haider, S. and Upadhyaya, N., Effect of commercial formulation of four organophosphorus insecticides on the LH-induced germinal vesicle breakdown in the oocytes of a freshwater teleost, *Mystus vittatus* — a preliminary *in vitro* study, *Ecotoxicol. Environ. Saf.,* 161, 1986.
379. Saxena, P. and Mani, K., Quantitative study of testicular recrudescence in the freshwater teleost, *Channa punctatus* exposed to pesticides, *Bull. Environ. Contam. Toxicol.,* 34, 597, 1985.
380. Saxena, P. and Mani, K., Effect of safe concentrations of some pesticides on testicular recrudescence in the freshwater murrel, *Channa punctatus*: A morphological study, *Ecotoxicol. Environ. Saf.,* 14, 56, 1987.

381. Pandey, A. and Shukla, L., Effect of an organophosphorous insecticide malathion on the testicular histophysiology in *Sorathoradon mossambicus*, *Natl. Acad. Sci. Lett.*, 5, 141, 1982.
382. Kling, D., Total atresia of the ovaries of *Tilapia leucosticta* (Cichlidae) after intoxication with the insecticide Lebaycid, *Experientia,* 37, 73, 1981.
383. Shukla, L., Shrivastava, A., Merwani, D., and Pandey, A., Effect of sublethal malathion on ovarian histophysiology in *Sarotherodon mossambicus*, *Comp. Physiol. Ecol.*, 9, 13, 1984.
384. Singh, H., Effects of malathion on steroidogenesis and sex reversal in *Monopterus albus*, *Mar. Environ. Res.*, 35, 159, 1993.
385. Pawar, K. and Katdare, M., Effects of sublethal and lethal concentrations of fenitrothion, BHC and carbofuran on behaviour and oxygen consumption of the freshwater prawn *Macrobrachium kistnensis* (Tiwari), *Archiv für Hydrobiologie,* 99, 398, 1984.
386. Mani, K. and Saxena, P., Effect of safe concentrations of some pesticides on ovarian recrudescence in the frehwater murrel, *Channa punctatus* (Bi): A quantitative study, *Ecotoxicol. Environ. Saf.*, 9, 241, 1985.
387. Saxena, P. K. and Garg, M., Effect of insecticidal pollution on ovarian recrudescence in the fresh water teleost *Channa punctatus*, *Indian J. Exp. Biol.*, 16, 689, 1978.
388. Lien, N., Adriaens, D., and Janssen, C., Morphological abnormalities in African catfish (*Clarias gariepinus*) larvae exposed to malathion, *Chemosphere,* 35, 1475, 1997.
389. Dutta, H., Nath, A., Adhikari, S., Roy, P., Singh, N., and Datta Munshi, J., Sublethal malathion induced changes in the ovary of an air-breathing fish, *Heteropneustes fossilis*: A histological study, *Hydrobiologia,* 215, 1994.
390. Sinha, N., Lal, B., and Singh, T., Carbaryl-induced thyroid dysfunction in the freshwater catfish *Clarias batrachus*, *Ecotoxicol. Environ. Saf.*, 213, 240, 1991.
391. Ghosh, P., Bhattacharya, S., and Bhattacharya, S., Impact of nonlethal levels of metacid-50 and carbaryl on thyroid function and cholinergic system of *Channa punctatus*, *Biomed. Environ. Sci.*, 2, 92, 1989.
392. Moore, A. and Waring, C., Sublethal effects of the pesticide diazinon on olfactory function in mature male Atlantic salmon parr, *J. Fish Biol.*, 48, 758, 1996.
393. Singh, H. and Singh, T., Thyroid activity and TSH potency of the pituitary gland and blood serum in response to cythion and hexadrin treatment in the freshwater catfish, *Heteropneustes fossilis* (Bloch), *Environ. Res.*, 22, 184, 1980.
394. Singh, P., Impact of malathion and γ-BHC on lipid metabolism in the freshwater female catfish, *Heteropneustes fossilis*, *Ecotoxicol. Environ. Saf.*, 23, 22, 1992.
395. Lal, B. and Singh, T., Impact of pesticides on lipid metabolism in the freshwater catfish, *Clarias batrachus*, during the vitellogenic phase of its annual reproductive cycle, *Ecotoxicol. Environ. Saf.*, 13, 13, 1987.
396. Bridges, D. W., Cech J. J., Jr., and Pedro, D. N., Seasonal hematological changes in winter flounder, *Pseudopleuronectes americanus*, *Trans. Am. Fish. Soc.*, 5, 596, 1976.
397. Chatterjee, S., Dutta, A., and Ghosh, R., Impact of carbofuran in the oocyte maturation of catfish, *Heteropenustes fossilis* (Bloch), *Arch. Environ. Contam. Toxicol.*, 32, 426, 1997.
398. Sukumar, A. and Karpagaganapathy, P., Pesticide-induced atresia in ovary of a fresh water fish, *Colisa lalia*, *Bull. Environ. Contam. Toxicol.*, 48, 357, 1992.
399. Carlson, A., Effects of long-term exposure to carbaryl (sevin) on survival, growth, and reproduction of the fathead minnow (*Pimephales promelas*), *J. Fish. Res. Bd. Can.*, 29, 583, 1972.
400. Sonnenschein, C. and Soto, A., An updated review of environmental estrogen and androgen mimics and antagonists, *J. Steroid Biochem.*, 65, 143, 1998.
401. Saxena, P., Mani, K., and Kondal, J., Effect of a safe application rate (SAR) concentrations of some biocides on the gonads of the fresh water murrel, *Channa punctatus* — A biochemical study, *Ecotoxicol. Environ. Saf.*, 12, 1, 1986.
402. Matthiessen, P. and Gibbs, P. E., Critical appraisal of the evidence for tributyltin-mediated endocrine disruption in mollusks, *Environ. Toxicol. Chem.*, 17, 37, 1998.
403. Gibbs, P., Pascoe, P., and Burt, G., Sex change in the female dog-whelk, *Nucella lapillus*, induced by tributyltin from antifouling paints, *J. Mar. Biol. Assoc. U.K.*, 68, 715, 1988.
404. Spooner, N., Gibbs, P., Bryan, G., and Goad, L., The effect of tributiltin upon steroid titres in the female dogwhelk, *Nucella lapillus*, and the development of imposex, *Mar. Environ. Res.*, 32, 37, 1991.

405. Bryan, G., Gibbs, P., Burt, G., and Hummerstone, L., The effects of tributyltin (TBT) accumulation on adult dog-whelks, *Nucella lapillus*: Long-term field and laboratory experiments, *J. Mar. Biol. Assoc. U.K.,* 67, 525, 1987.
406. Gibbs, P., Spencer, B., and Pascoe, P., The American oyster drill, *Urosalpinx cinerea* (Gastropoda): Evidence of decline in an imposex-affected population (R. Blackwater, Essex), *J. Mar. Assoc. U.K.,* 71, 827, 1991.
407. Ruiz, J., Bryan, G., and Gibbs, P., Effects of tributyltin (TBT) exposure on the veliger larvae development of the bivalve *Scrobicularia plana* (da Costa), *J. Exp. Mar. Biol. Ecol.,* 186, 53, 1995.
408. Kannan, K., Senthilkumar, K., Elliott, J. E., Feyk, L. A., and Giesey, J. P., Occurrence of butyltin compounds in tissues of water birds and seaducks from the United States and Canada, *Arch. Environ. Contam. Toxicol.,* 35, 64, 1998.
409. Jha, A. N., Hagger, J. A., Hill, S. J., and Depledge, M. H., Genotoxic, cytotoxic and developmental effects of tributyltin oxide (TBTO): An integrated approach to the evaluation of the relative sensitivities of two marine species, *Mar. Environ. Res.,* 50, 565, 2000.
410. Jha, A. N., Hagger, J. A., and Hill, S. J., Tributyltin induces cytogenetic damage in the early life stages of the marine mussel, *Mytilus edulis*, *Environ. Mol. Mutagen.,* 35, 343, 2000.
411. Martin, R., Dixon, D., Maguire, R., Hodson, P., and Tkacz, R., Acute toxicity, uptake, depuration and tissue distribution of tri-n-butyltin in rainbow trout, *Salmo gairdneri*, *Aquat. Toxicol.,* 15, 37, 1989.
412. Strmac, M. and Braunbeck, T., Effects of triphenyltin acetate on survival, hatching success, and liver ultrastructure of early life stages of zebrafish (*Danio rerio*), *Ecotoxicol. Environ. Saf.,* 44, 25, 1999.
413. Nirmala, K., Oshima, Y., Lee, R., Imada, N., Honjo, T., and Kobayashi, K., Transgenerational toxicity of tributyltin and its combined effects with polychlorinated biphenyls on reproductive processes in Japanese medaka (*Oryzias Latipes*), *Environ. Toxicol. Chem.,* 18, 717, 1999.
414. Manning, C., Lytle, T., Walker, W., and Lytle, J., Life-cycle toxicity of bis(tributyltin) oxide to the sheepshead minnow (*Cyprinodon variegatus*), *Arch. Environ. Contam. Toxicol.,* 37, 258, 1999.
415. Seinen, W., Helder, T., Vernij, H., Penninks, A., and Leeuwangh, P., Short term toxicity of tri-n-butyltin chloride in rainbow trout (*Salmo gairdneri* Richardson) yolk sac fry, *Sci. Total Environ.,* 19, 155, 1981.
416. Pinkney, A., Matteson, L., and Wright, D., Effects of tributyltin on survival, growth, morphometry, and RNA-DNA ratio of larval striped bass, *Morene saxatilis*, *Arch. Environ. Contam. Toxicol.,* 1990.
417. Fent, K. and Meier, W., Tributyltin-induced effects on early life stages of minnows *Phoxinus phoxinus*, *Arch. Environ. Contam. Chem.,* 428, 1992.
418. Haubruge, E., Petit, F., and Gage, M. J. G., Reduced sperm counts in guppies (*Poecilia reticulata*) following exposure to low levels of tributyltin and bisphenol A, *Proc. R. Soc. London B,* 267, 2333, 2000.
419. Holm, G., Norrgren, L., and Linden, O., Reproductive and histopathological effects of long-term experimental exposure to bis(tributyltin)oxide (TBTO) on the three-spined stickleback, *Gasterosteus aculeatus* Linnaeus., *J. Fish Biol.,* 38, 373, 1991.
420. Larson, D., McDonald, S., Fivizzani, A., Newton, W., and Hamilton, S., Effects of the herbicide atrazine on *Ambystoma tigrinum* metamorphosis: Duration, larval growth, and hormonal response, *Physiol. Zool.,* 71, 671, 1998.
421. Allran, J. and Karasov, W., Effects of atrazine and nitrate on northern leopard frog (*Rana pipiens*) larvae exposed in the laboratory from posthatch through metamorphosis, *Environ. Toxicol. Chem.,* 19, 2850, 2000.
422. Morgan, M., Scheuerman, P., Bishop, C., and Pyles, R., Teratogenic potential of atrazine and 2,4-D using FETAX, *J. Toxicol. Environ. Health,* 48, 151, 1996.
423. Kettle, W., de Noyelles, F., Jr., Heacock, B., and Kadoum, A., Diet and reproductive success of bluegill recovered from experimental ponds treated with atrazine, *Bull. Environ. Contam. Toxicol.,* 38, 47, 1987.
424. Crain, D. A., Guillette, L. J., Rooney, A. A., and Pickford, D. B., Alterations in steroidogenesis in alligators (*Alligator mississippiensis*) exposed naturally and experimentally to environmental contaminants, *Environ. Health Perspect.,* 105, 528, 1997.
425. Crain, D. A., Spiteri, I. D., and Guillette, L. J., The functional and structural observations of the neonatal reproductive system of alligators exposed *in ovo* to atrazine, 2,4-D, or estradiol, *Toxicol. Ind. Health,* 15, 180, 1999.

426. Connor, K., Howell, J., Chen, I., Liu, H., Berhane, K., Sciaretta, C., Safe, S., and Zacharewski, T., Failure of chloro-s-triazine-derived compounds to induce estrogen receptor-mediated responses *in vivo* and *in vitro*, *Fundam. Appl. Toxicol.*, 30, 93, 1996.
427. Tennant, M., Hill, D., Eldridge, J., Wetzel, L., Breckenridge, C., and Stevens, J., Possible antiestrogenic properties of chloro-s-triazines in rat uterus, *J. Toxicol. Environ. Health,* 43, 183, 1994.
428. Balaguer, P., Francois, F., Comunale, F., Fenet, H., Boussioux, A., Pons, M., Nicolas, J., and Casellas, C., Reporter cell lines to study the estrogenic effects of xenoestrogens, *Sci. Total Environ.*, 233, 47, 1999.
429. Tran, D., Kow, K., McLachlan, J., and Arnold, S., The inhibition of estrogen receptor-mediated responses by chloro-s-triazine-derived compounds is dependent on estradiol concentration in yeast, *Biochem. Biphys. Res. Commun.*, 227, 140, 1996.
430. Cooper, R., Stoker, T., Tyrey, L., Goldman, J., and McElroy, W., Atrazine disrupts the hypothalamic control of pituitary-ovarian function, *Toxicol. Sci.*, 53, 297, 2000.
431. McMaster, M. E., Van Der Kraak, G. J., Portt, C. B., Munkittrick, K. R., Sibley, P. K., Smith, I. R., and Dixon, D. G., Changes in hepatic mixed-function oxygenase (MFO) activity, plasma steroid levels, and age at maturity of a white sucker (*Catostomus commersoni*) population exposed to bleached kraft pulp mill effluent, *Aquat. Toxicol.*, 21, 199, 1991.
432. McMaster, M. E., Van Der Kraak, G. J., and Munkittrick, K. R., An epidemiological evaluation of the biochemical basis for steroid hormonal depressions in fish exposed to industrial wastes, *J. Great Lakes Res.*, 22, 153, 1996.
433. McMaster, M. E., Van Der Kraak, G. J., and Munkittrick, K. R., Biochemical basis for hormonal dysfunctions of fish exposed to organic wastes, in *Proc. 38th Conf. Int. Assoc. of Great Lakes Res.*, International Association for Great Lakes Research, East Lansing, MI, 30, 1995.
434. Munkittrick, K. R., McMaster, M. E., Portt, C. B., Van Der Kraak, G. J., Smith, I. R., and Dixon, D. G., Changes in maturity, plasma sex steroid levels, hepatic mixed-function oxygenase activity, and the presence of external lesions in lake whitefish (*Coregonus clupeaformis*) exposed to bleached kraft mill effluent, *Can. J. Fish. Aquat. Sci.*, 49, 1560, 1992.
435. Munkittrick, K. R., Portt, C. B., Van Der Kraak, G. J., Smith, I. R., and Rokosh, D. A., Impact of bleached kraft mill effluent on population characteristics, liver MFO activity, and serum steroid levels of a Lake Superior white sucker (*Catostomus commersoni*) population, *Can. J. Fish. Aquat. Sci.*, 48, 1371, 1991.
436. Gagnon, M. M., Dodson J. J., Hodson P. V., Van Der Kraak, G., and Carey, J. H., Seasonal effects of bleached kraft mill effluent on reproductive parameters of white sucker (*Catostomus commersoni*) populations of the St. Maurice River, Quebec, Canada, *Can. J. Fish. Aquat. Sci.*, 51, 337, 1994a.
437. Munkittrick, K. R., Van Der Kraak, G. J., McMaster, M. E., Portt, C. B., van den Heuvel, M. R., and Servos, M. R., Survey of receiving-water environmental impacts associated with discharges from pulp mills 2. Gonad size, liver size, hepatic EROD activity and plasma sex steroid levels in white sucker (*Catostomus commersoni*), *Environ. Toxicol. Chem.*, 13, 1089, 1994.
438. Hodson, P. V., McWhirter, M., Ralph, K., Gray, B., Thivierge, D., Carey, J. H., Van Der Kraak, G., Whittle, D. M., and Levesque, M. C., Effects of bleached kraft mill effluent on fish in the St. Maurice River, Quebec, *Environ. Toxicol. Chem.*, 11, 1635, 1992.
439. Gagnon, M. M., Dodson, J. J., and Hodson, P. V., Ability of BKME (bleached kraft mill effluent) exposed white suckers (*Catostomus commersoni*) to synthesize steroid hormones, *Comp. Biochem. Physiol.*, 107C, 265, 1994.
440. Adams, M. S., Crumby W. D., Greeley M. S., Shugart L. R., and Saylor C. F., Responses of fish populations and communities to pulp mill effluents: A holistic assessment, *Ecotoxicol. Environ. Saf.*, 24, 347, 1992.
441. Sepúlveda, M. S., Quinn, B. P., Denslow, N. D., Holm, S. E., and Gross, T. S., Effects of paper mill effluents on the reproductive success of Florida largemouth bass, *Environ. Toxicol. Chem.*, in press.
442. Sandstrom, O., *In situ* assessments of the impact of pulp mill effluent on life-history variables in fish, in *Environmental Fate and Effects of Pulp and Paper Mill Effluents,* Servos, M. R., Munkittrick, K.R., Carey, J., and Van Der Kraak, G., Eds., St. Lucie Press, Boca Raton, FL, 449, 1996.
443. Landner, L., Neilsson, A. H., Sorensen, L., Tarnholm, A., and Viktor, T., Short-term test for predicting the potential of xenobiotics to impair reproductive success in fish, *Ecotoxicol. Environ. Saf.*, 9, 282, 1985.

444. Vuorinen, P. J. and Vuorinen, M., Effects of bleached kraft mill effluent on reproduction of brown trout (*Salmo trutta* L.) on a restricted diet, *Finn. Fish. Res.,* 6, 92, 1985.
445. Tana, J. J., Sublethal effects of chlorinated phenols and resin acids on rainbow trout (*Salmo gairdneri*), *Water Sci. Technol.,* 20, 77, 1988.
446. Kovacs, T. G., Gibbons, J. S., Tremblay, L. A., O'Conner, B. I., Martel, P. H., and Voss, R. I., Effect of a secondary-treated bleached kraft pulp mill effluent on aquatic organisms as assessed by short-term and long-term laboratory tests, *Ecotoxicol. Environ. Saf.,* 31, 7, 1995.
447. Gagnon, M. M., Bussieres, D., Dodson J. J., and Hodson, P. V., White sucker (*Catostomus commersoni*) growth and sexual maturation in pulp mill-contaminated and reference rivers, *Environ. Toxicol. Chem.,* 14, 317, 1995.
448. McMaster, M. E., Portt, C. B., Munkittrick, K. R., and Dixon, D. G., Milt characteristics, reproductive performance, and larval survival and development of white sucker exposed to bleached kraft mill effluent, *Ecotoxicol. Environ. Saf.,* 23, 103, 1992.
449. Swanson, S., Shelast, R., Schryder, R., Kloepper-Sams, P., Marchant, T., Kroeber, K., Bernstein, J., and Owens, J. W., Fish populations and biomarker responses at a Canadian bleached kraft mill site, *Tappi J.,* 75, 139, 1992.
450. Karås, P., Neuman, E., and Sandström, O., Effects of a pulp mill effluent on the population dynamics of perch, *Perca fluviatilis*, *Can. J. Fish. Aquat. Sci.,* 48, 28, 1991.
451. Kovacs, T. G., Gibbons, J. S., Martel, P. H., and Voss, R. H., Improved effluent quality at a bleached kraft mill as determined by laboratory biotests, *J. Toxicol. Environ. Health,* 49, 533, 1996.
452. Kovacs, T. G. and Megraw, S. R., Laboratory responses of whole organisms exposed to pulp and paper mill effluents: 1991–1994, in *Environmental Fate and Effects of Pulp and Paper Mill Effluents,* Servos, M. R., Munkittrick, K. R., Carey, J., and Van Der Kraak, G., Eds., St. Lucie Press, Delray Beach, FL, 459, 1996.
453. Bortone, S. A. and Davis, W. P., Fish intersexuality as indicator of environmental stress: Monitoring fish reproductive systems can serve to alert humans to potential harm, *BioScience,* 44, 165, 1994.
454. Howell, W. M., Black, D. A., and Bortone, S. A., Abnormal expression of secondary sex characters in a population of mosquitofish, *Gambusia affinis holbrooki*: Evidence for environmentally induced masculinization, *Copeia,* 4, 676, 1980.
455. Bortone, S. A. and Cody, R. P., Morphological masculinization in Poecillid females from a paper mill effluent receiving tributary of the St. John's River, Florida, USA, *Bull. Environ. Toxicol. Chem.,* 63, 150, 1999.
456. Howell, W. M. and Denton T. E., Gonopodial morphogenesis in female mosquitofish, *Gambusia affinis affinis*, masculinized by exposure to degradation products from plant sterols, *Environ. Biol. Fishes,* 24, 43, 1989.
457. Van Der Kraak, G. J., Munkittrick, K. R., McMaster, M. E., Portt, C. B., and Chang, J. P., Exposure to bleached kraft pulp mill effluent disrupts the pituitary-gonadal axis of white sucker at multiple sites, *Toxicol. Appl. Pharmacol.,* 115, 224, 1992.
458. Janz, D. M., McMaster, M. E., Munkittrick, K. R., and Van Der Kraak, G., Elevated ovarian follicular apoptosis and heat shock protein-70 expression in white sucker exposed to bleached kraft pulp mill effluent, *Toxicol. Appl. Pharmacol.,* 147, 391, 1997.
459. Van Der Kraak, G., Munkittrick, K. R., McMaster, M. E., and MacLatchy, D. L., A comparison of bleached kraft mill effluent, 17β-estradiol and β-sitosterol effects on reproductive function in fish, in *Principles and Processes for Evaluating Endocrine Disruption in Wildlife,* Kendall, R., Dickerson, R., Giesy, J., and Suk, W., Eds., SETAC Press, Pensacola, FL, 249, 1998.
460. Mellanen, P., Petanen, T., Lehtimaki, J., Makela, S., Bylund, G., Holmbom, B., Mannila, E., Oikari, A., and Santti, R., Wood-derived estrogens: Studies *in vitro* with breast cancer cell lines and *in vivo* in trout, *Toxicol. Appl. Pharmacol.,* 136, 381, 1996.
461. MacLatchy, D. L. and Van Der Kraak, G. J., The phytoestrogen ß-sitosterol alters the reproductive endocrine status of goldfish, *Toxicol. Appl. Pharmacol.,* 134, 305, 1995.
462. Tremblay, L. and Van Der Kraak, G. V., Use of a series of homologous *in vitro* and *in vivo* assays to evaluate the endocrine modulating actions of ß-sitosterol in rainbow trout, *Aquat. Toxicol.,* 43, 149, 1998.
463. Denton, T. E., Howell, W. M., Allison, J. J., McCollum, J., and Marks, B., Masculinization of female mosquitofish by exposure to plant sterols and *Mycobacterium smegmatis*, *Bull. Environ. Contam. Toxicol.,* 35, 627, 1985.

464. Tyler, C. and Routledge, E., Natural and anthropogenic environmental oestrogens: The scientific basis for risk assessment. Oestrogenic effects in fish in English rivers with evidence of their causation, *Pure Appl. Chem.,* 70, 1795, 1998.
465. Jobling, S., Nolan, M., Tyler, C., Brighty, G., and Sumpter, J., Widespread sexual disruption in wild fish, *Environ. Sci. Technol.,* 32, 2498, 1998.
466. Jobling, S., Sheahan, D., Osborne, J., Matthiessen, P., and Sumpter, J., Inhibition of testicular growth in rainbow trout (*Oncorynchus mykiss*) exposed to environmental estrogens, *Environ. Toxicol. Chem.,* 15, 194, 1996.
467. Harries, J., Sheahan, D., Jobling, S., Matthiessen, P., Neall, M., Sumpter, J., Taylor, T., and Zaman, N., Estrogenic activity in five United Kingdom rivers detected by measurement of vitellogenesis in caged male trout, *Environ. Toxicol. Chem.,* 16, 534, 1997.
468. Jobling, S., Reynolds, T., White, R., Parker, M. G., and Sumpter, J. P., A variety of environmentally persistent chemicals, including some phthalate plasticizers, are weakly estrogenic, *Environ. Health Perspect.,* 103, 582, 1995.
469. Nimrod, A. and Benson, W., Environmental estrogenic effects of alkylphenol ethoxylates, *Crit. Rev. Toxicol.,* 26, 335, 1996.
470. Desbrow, C., Routledge, E., Brighty, G., Sumpter, J., and Waldock, M., Identification of estrogenic chemicals in STW effluent. I. Chemical fractionation and *in vitro* biological screening, *Environ. Sci. Technol.,* 32, 1549, 1998.
471. Purdom, C., Hardiman, P., Bye, V., Eno, N., Tyler, C., and Sumpter, J., Estrogenic effects of effluents from sewage treatment works, *Chem. Ecol.,* 8, 275, 1994.
472. Routledge, E., Sheahan, D., Desbrow, C., Brighty, G., Waldock, M., and Sumpter, J., Identification of estrogenic chemicals in STW effluent. II. *In vitro* responses in trout and roach, *Environ. Sci. Technol.,* 32, 1559, 1998.
473. Billard, R., Breton, B., and Richard, M., On the inhibitory effect of some steroids on spermatogenesis in adult rainbow trout (*Salmo gairdneri*), *Can. J. Zool.,* 59, 1479, 1981.
474. Naylor, C., Environmental fate and safety of nonylphenol ethoxylates, *Test. Chem. Col.,* 57, 1577, 1985.
475. Blackburn, M. and Waldock, M., Concentrations of alkylphenols in rivers and estuaries in England and Wales, *Water Res.,* 29, 1623, 1995.
476. Christiansen, T., Korsgaard, B., and Jespersen, Å., Effects of nonylphenol and 17ß-oestradiol on vitellogenin synthesis, testicular structure and cytology in male eelpout *Zoarces viviparus*, *J. Exp. Biol.,* 201, 179, 1998.
477. Lech, J. J., Lewis, S. K., and Ren, L., *In vivo* estrogenic activity of nonylphenol in rainbow trout, *Fundam. Appl. Toxicol.,* 30, 229, 1996.
478. Christensen, L., Korsgaard, B., and Bjerregaard, P., The effect of 4-nonylphenol on the synthesis of vitellogenin in the flounder *Platichthys flesus*, *Aquat. Toxicol.,* 46, 211, 1999.
479. Madsen, S., Mathiesen, A., and Korsgaard, B., Effects of 17β-estradiol and 4-nonylphenol on smoltification and vitellogenesis in Atlantic salmon (*Salmo salar*), *Fish Physiol. Biochem.,* 17, 303, 1997.
480. Arukwe, A., Celius, T., Walther, B. T., and Goksoyr, A., Effects of xenoestrogen treatment on zona radiata protein and vitellogenin expression in Atlantic salmon (*Salmo salar*), *Aquat. Toxicol.,* 49, 159, 2000.
481. Shioda, T. and Wakabayashi, M., Effect of certain chemicals on the reproduction of medaka (*Oryzias latipes*), *Chemosphere,* 40, 239, 2000.
482. Nimrod, A. C. and Benson, W. H., Estrogenic responses to xenobiotics in channel catfish (*Ictalurus punctatus*), *Mar. Environ. Res.,* 42, 155, 1996.
483. Gray, M., Teather, K., and Metcalfe, C., Reproductive success and behavior of Japanese medaka (*Oryzias latipes*) exposed to 4-*tert*-octylphenol, *Environ. Toxicol. Chem.,* 18, 2587, 1999.
484. Kelly, S. A. and Di Giulio, R. T., Developmental toxicity of estrogenic alkylphenols in killifish (*Fundulus heteroclitus*), *Environ. Toxicol. Chem.,* 19, 2564, 2000.
485. Gimeno, S., Komen, H., Venderbosch, P. W. M., and Bowmer, T., Disruption of sexual differentiation in genetic male common carp (*Cyprinus carpio*) exposed to an alkylphenol during different life stages, *Environ. Sci. Tech.,* 31, 2884, 1997.
486. Dreze, V., Monod, G., Cravedi, J. P., Biagianti-Risbourg, S., and Le Gac, F., Effects of 4-nonylphenol on sex differentiation and puberty in mosquitofish (*Gambusia holbrooki*), *Ecotoxicology,* 9, 93, 2000.
487. Routledge, E. J. and Sumpter, J. P., Estrogenic activity of surfactants and some of their degradation products assessed using a recombinant yeast screen, *Environ. Toxicol. Chem.,* 15, 241, 1996.

488. Jobling, S. and Sumpter, J. P., Detergent components in sewage effluent are weakly estrogenic to fish: An *in vitro* study using rainbow trout (*Oncorhynchus mykiss*) hepatocytes, *Aquat. Toxicol.,* 27, 361, 1993.
489. Smeets, J., Rankouhi, T., Nichols, K., Komen, H., Kaminski, N., Giesy, J., and van den Berg, M., *In vitro* vitellogenin production by carp (*Cyprinus carpio*) hepatocytes as a screening method for determining (anti)estrogenic activity of xenobiotics, *Toxicol. Appl. Pharmacol.,* 157, 68, 1999.
490. Ren, L., Lewis, S., and Lech, J. J., Effects of estrogen and nonylphenol on the post-transcriptional regulation of vitellogenin gene expression, *Chem.-Biol. Interact.,* 100, 67, 1996.
491. Raychoudhury, S. S., Blake, C. A., and Millette, C. F., Toxic effects of octylphenol on cultured spermatogenic cells and Sertoli cells, *Toxicol. Appl. Pharmacol.,* 157, 192, 1999.
492. Dodds, E. and Lawson, W., Molecular structure in relation to oestrogenic activity compounds without a phenanthrene nucleus, *Proc. R. Soc. London B,* 125, 222, 1938.
493. Morrissey, R., George, J., Price, C., Tyl, R., Marr, M., and Kimmel, C., The developmental toxicity of bisphenol A in rats and mice, *Fundam. Appl. Toxicol.,* 8, 571, 1987.
494. Lutz, I. and Kloas, W., Amphibians as a model to study endocrine disruptors: I. Environmental pollution and estrogen receptor binding, *Sci. Total Environ.,* 225, 49, 1999.
495. Perez, P., Pulgar, R., Olea-Serrano, F., Villalobos, M., Rivas, A., Metzler, M., Pedraza, V., and Olea, N., The estrogenicity of bisphenol A-related diphenylalkanes with various substituents at the central carbon and the hydroxy groups, *Environ. Health Perspect.,* 106, 167, 1998.
496. Krishnan, A., Stathis, P., Permuth, S., Tokes, L., and Feldman, D., Bisphenol-A: An estrogenic substance is released from polycarbonate flasks during autoclaving, *Endocrinology,* 132, 2279, 1993.
497. Milligan, S., Khan, O., and Nash, M., Competitive binding of xenobiotic oestrogens to rat alpha-fetoprotein and to sex steroid binding proteins in human and rainbow trout (*Oncorhynchus mykiss*) plasma, *Gen. Comp. Endocrinol.,* 112, 89, 1998.
498. Shilling, A. and Williams, D., Determining relative estrogenicity by quantifying vitellogenin induction in rainbow trout liver slices, *Toxicol. Appl. Pharmacol.,* 164, 330, 2000.
499. Islinger, M., Pawlowski, S., Hollert, H., Volkl, A., and Braunbeck, T., Measurement of vitellogenin-mRNA expression in primary cultures of rainbow trout hepatocytes in a non-radioactive dot blot/RNAse protection-assay, *Sci. Total Environ.,* 233, 109, 1999.
500. Celius, T., Haugen, T., Grotmol, T., and Walther, B., A sensitive zonagenetic assay for rapid *in vitro* assessment of estrogenic potency of xenobiotics and mycotoxins, *Environ. Health Perspect.,* 107, 63, 1999.
501. Steinmetz, R., Brown, N., Allen, D., Bigsby, R., and Ben, J., The environmental estrogen bisphenol A stimulates prolactin release *in vitro* and *in vivo*, *Endocrinology,* 138, 1780, 1997.
502. VomSaal, F., Cooke, P., Buchanan, D., Palanza, P., Thayer, K., Nagel, S., Parmigiani, S., and Welshons, W., A physiologically based approach to the study of bisphenol A and other estrogenic chemicals on the size of reproductive organs, daily sperm production, and behavior, *Toxicol. Ind. Health,* 14, 239, 1998.
503. Nagler, J. J., Aysola, P., and Ruby, S. M., Effect of sublethal pentachlorophenol on early oogenesis in maturing female rainbow trout (*Salmo gairdneri*), *Arch. Environ. Contam. Toxicol.,* 15, 549, 1986.
504. Cleveland, L., Buckler, D., Mayer, F., and Branson, D., Toxicity of three preparations of pentachlorophenol to fathead minnows — A comparative study, *Environ. Toxicol. Chem.,* 1, 205, 1982.
505. Parks, L. and Leblanc, G., Reductions in steroid hormone biotransformation/elimination as a biomarker of pentachlorophenol chronic toxicity, *Aquat. Toxicol.,* 34, 291, 1996.
506. Kumar, V. and Mukherjee, D., Phenol and sulfide induced changes in the ovary and liver of sexually maturing common carp, *Cyprinus carpio*, *Aquat. Toxicol.,* 13, 53, 1988.
507. Slemr, F. and Langer, E., Increase in global atmospheric concentrations of mercury inferred from measurements over the Atlantic Ocean, *Nature,* 355, 434, 1992.
508. Swain, E., Engstrom, D. R., Brigham, M. E., Henning, T. A., and Brezonik, P. L., Increasing rates of atmospheric mercury deposition in midcontinental North America, *Science,* 257, 784, 1992.
509. Livett, E. A., Geochemical monitoring of atmospheric heavy-metal pollution — Theory and applications, *Adv. Ecol. Res.,* 18, 65, 1988.
510. Zillioux, E. J., Porcella, D. B., and Benoit, J. M., Mercury cycling and effects in freshwater wetland ecosystems, *Environ. Toxicol. Chem.,* 12, 2245, 1993.
511. Spry, D. J. and Wiener J. G., Metal bioavailability and toxicity to fish in low-alkalinity lakes: A critical review, *Environ. Pollut.,* 71, 243, 1991.

512. Huckabee, J. W., Elwood, J. E., and Hildebrand, S. G., Accumulation of mercury in freshwater biota, in *The Biogeochemistry of Mercury in the Environment,* Nriagu, J. O., Ed., Elsevier, Amsterdam, 277, 1979.
513. Lange, T. R., Royals, H. E., and Connor, L. L., Influence of water chemistry on mercury concentration in largemouth bass from Florida lakes, *Trans. Am. Fish. Soc.,* 122, 74, 1993.
514. Lange, T. R., Royals, H. E., and Connor, L. L., Mercury accumulation in largemouth bass (*Micropterus salmoides*) in a Florida lake, *Arch. Environ. Contam. Toxicol.,* 27, 466, 1994.
515. Bodaly, R. A., Hecky, R. E., and Fudge, R. J. P., Increases in fish mercury levels in lakes flooded by the Churchill River diversion, northern Manitoba, *Can. J. Fish. Aquat. Sci.,* 41, 682, 1984.
516. Cumbie, P. M., Mercury Accumulation in Wildlife in the Southeast, Ph.D. dissertation, University of Georgia, 1975.
517. Cumbie, P. M., Mercury in hair of bobcats and raccoons, *J. Wildl. Manage.,* 39, 419, 1975.
518. Fimreite, N., Mercury contamination of aquatic birds in northwestern Ontario, *J. Wildl. Manage.,* 38, 120, 1974.
519. Sheffy, T. B. and St. Amant, J. R., Mercury burdens in furbearers in Wisconsin, *J. Wildl. Manage.,* 46, 1117, 1982.
520. Khera, K. S., Teratogenic and genetic effects of mercury toxicity, in *The Biogeochemistry of Mercury in the Environment,* Nriagu, N. O., Ed., Elsevier/North Holland Biomedical Press, New York, 501, 1979.
521. Mohamed, M. K., Burbacher, T. M., and Mottet, N. K., Effects of methyl mercury on testicular functions in *Macaca fascicularis* monkeys, *Pharm. Toxicol.,* 60, 29, 1987.
522. Burbacher, T. M., Mohamed, M. K., and Mottet, N. K., Methylmercury effects on reproduction and offspring size at birth, *Reprod. Toxicol.,* 1, 267, 1988.
523. Khera, K. S., Reproductive capability of male rats and mice treated with methylmercury, *Toxicol. Appl. Pharmacol.,* 24, 167, 1973.
524. Lee, J. H. and Han, D. H., Maternal and fetal toxicity of methylmercuric chloride administered to pregnant Fischer 344 rats, *J. Toxicol. Environ. Health,* 45, 415, 1995.
525. Hughes, J. A. and Annau, Z., Postnatal behavioral effects in mice after prenatal exposure to methylmercury, *Pharmacol., Biochem. Behav.,* 4, 385, 1976.
526. Fuyuta, M., Fujimoto, T., and Hirata, S., Embryotoxic effects of methylmercuric chloride administered to mice and rats during organogenesis, *Teratology,* 18, 353, 1978.
527. Fuyuta, M., Jujimoto, T., and Kiyofuji, E., Teratogenic effects of a single oral administration of methylmercuric chloride in mice, *Acta Anatomy (Basel),* 104, 356, 1979.
528. Hirano, M., Mitsumori, K., Maita, K., and Shirasu, Y., Further carcinogenicity study on methylmercury chloride in ICR mice, *Nippon Juigaku Zasshi, Jpn. J. Vet. Sci.,* 48, 127, 1986.
529. Mitsumori, K., Hirano, M., Ueda, H., Maita, K., and Shirasu, Y., Chronic toxicity and carcinogenicity of methylmercury chloride in B6C3F1 mice, *Fundam. Appl. Toxicol.,* 14, 179, 1990.
530. Inouye, M. and Kajiwara, Y., Developmental disturbances of the fetal brain in guinea pigs caused by methylmercury, *Arch. Toxicol.,* 62, 15, 1988.
531. National Research Council, *Toxicological Effects of Methylmercury,* National Academy Press, Washington, D.C., 2000.
532. Roelke, M. E., Schultz, D. P., Facemire, C. F., Sundlof, S. F., and Royals, H. E., Mercury Contamination in Florida Panthers, Florida Panther Interagency Committee, 1991.
533. Wolfe, M. F., Schwarzbach, S., and Sulaiman, R. A., Effects of mercury on wildlife: A comprehensive review, *Environ. Toxicol. Chem.,* 17, 146, 1998.
534. Von Burg, R. and Greenwood, M. R., Mercury, in *Metals and Their Compounds in the Environment,* Merian, E., Ed., VCH, Weinheim, Germany, 1991.
535. Barr, J. F., Population dynamics of the common loon (*Gavia immer*) associated with mercury-contaminated waters in northwestern Ontario, Can. Wildl. Serv. Occasional Pap. No., 56, 1, 1986.
536. Wiemeyer, S. N., Lamont, T. G., Bunck, C. M., Sindelar, C. R., Gramlich, F. J., Fraser, J. D., and Byrd, M. A., Organochlorine pesticide, polychlorobiphenyl, and mercury residues in bald eagle eggs — 1969–79 — and their relationships to shell thinning and reproduction, *Arch. Environ. Contam. Toxicol.,* 13, 529, 1984.
537. Furness, R. W., Muirhead, S. J., and Woodburn, M., Using bird feathers to measure mercury in the environment: Relationships between mercury content and moult, *Mar. Pollut. Bull.,* 17, 27, 1986.

538. Furness, R. W. and Hutton, M., Pollutants and impaired breeding of Great Skuas (*Catharacta-skua*) in Britain, *Ibis,* 122, 88, 1980.
539. Finley, M. T. and Stendell R. C., Survival and reproductive success of black ducks fed methyl mercury, *Environ. Pollut.,* 16, 1, 1978.
540. Newton, I. and Haas, M. B., Pollutants in merlin eggs and their effects on breeding, *Br. Birds,* 81, 258, 1988.
541. Heinz, G. H., Methylmercury: Reproductive and behavioral effects on three generations of mallard ducks, *J. Wildl. Manage.,* 43, 394, 1979.
542. Heinz, G., Effects of low dietary levels of methyl mercury on mallard reproduction, *Bull. Environ. Contam. Toxicol.,* 11, 386, 1974.
543. Kanamadi, R. D. and Saidapur, K. S., Effects of exposure to sublethal mercuric chloride on the testis and fat body of the frog *Rana cyanophlyctis*, *J. Herpetol.,* 26, 499, 1992.
544. Harnfenist, A., Power, T., Clark, K. L., and Peakall, D. B., A Review and Evaluation of the Amphibian Toxicological Literature, Technical Report Series No. 61, Canadian Wildlife Service, Quebec, 1989.
545. Ide, C., Jelaso, A., and Austin, C., Effects of methylmercury chloride on development of the frog *Xenopus laevis*, *Neurotoxicology,* 16, 763, 1995.
546. Friedmann, A. S., Watzin, M. C., Brinck-Johnson, T., and Leiter, J. C., Low levels of dietary methylmercury inhibit growth and gonadal development in juvenile walleye (*Stizostedion vitreum*), *Aquat. Toxicol.,* 35, 265, 1996.
547. Verma, S. R. and Tonk, I. P., Effect of sublethal concentrations of mercury on the composition of liver, muscles and ovary of *Notopterus notopterus*, *Water Air Soil Pollut.,* 20, 287, 1983.
548. Billard, R. and Roubaud, P., The effect of metals and cyanide on fertilization in rainbow trout (*Salmo-gairdneri*), *Water Res.,* 19, 209, 1985.
549. Khan, A. T. and Weis, J. S., Toxic effects of mercuric-chloride on sperm and egg viability of 2 populations of mummichog, *Fundulus heteroclitus*, *Environ. Pollut.,* 48, 263, 1987.
550. Arnold, B. S., Distribution of Mercury within Different Trophic Levels of the Okefenokee Swamp, within Tissues of Top Level Predators, and Reproductive Effects of Methyl Mercury in the Nile Tilapia (*Oreochromis Niloticus*), Ph.D. dissertation, University of Georgia, 2000.
551. Rom, W. N., Effects of lead on female and reproduction, *Mount Sinai J. Med.,* 43, 542, 1976.
552. Lee, I. P., Effects of environmental metals on male reproduction, in *Reproductive and Developmental Toxicity of Metals,* Clarkson, T. W., Nordberg, G. F., and Sager, P. R., Eds., Plenum Press, New York, 253, 1983.
553. Hildebrand, D. C., Der, H., Griffin, W., and Fahim, F. S., Effect of lead acetate on reproduction, *Am. J. Obstet. Gynecol.,* 115, 1058, 1973.
554. Barltop, D., Transfer of lead to the human fetus, in *Mineral Metabolism in Pediatrics,* Barltop, D. and Barland, W. L., Eds., Davis, Philadelphia, 135, 1969.
555. Davis, J. M., Risk assessment of the developmental neurotoxicity of lead, *Neurotoxicology,* 11, 285, 1990.
556. Beach, J. R. and Henning, S. J., The distribution of lead in milk and the fate of milk lead in the gastrointestinal-tract of suckling rats, *Pediatr. Res.,* 23, 58, 1988.
557. Hogstedt, C., Hane, M., Agrell, A., and Bodin, L., Neuropsychological test results and symptoms among workers with well-defined long-term exposure to lead, *Br. J. Ind. Med.,* 40, 99, 1983.
558. Pocock, S. J., Shaper, A. G., Ashby, D., Delves, H. T., and Clayton, B. E., The relationship between blood lead, blood-pressure, stroke, and heart-attacks in middle-aged British men, *Environ. Health Perspect.,* 78, 23, 1988.
559. Davis, J. M. and Svendsgaard, D. J., Lead and child-development, *Nature,* 329, 297, 1987.
560. Davidson, W. R. and Nettles, V. F., Field Manual of Wildlife Diseases in the Southeastern United States, Southeastern Cooperative Wildlife Disease Studies, Athens, GA, 1997.
561. Burger, J., Carruth-Hinchey, C., Ondroff, E., McMahon, M., Gibbons, J. W., and Gochfeld, M., Effects of lead on behavior, growth, and survival of hatchling slider turtles, *J. Toxicol. Environ. Health,* 55, 495, 1998.
562. Horne, M. T. and W. A. Dunson, Toxicity of metals and low pH to embryos and larvae of the jefferson salamander, *Ampystoma jeffersonianum*, *Arch. Environ. Con. Chem.,* 29, 110, 1995.
563. Reichert, W. L., Federighi, D. A., and Malins, D. C., Uptake and metabolism of lead and cadmium in coho salmon (*Oncorhynchus kisutch*), *Comp. Biochem. Physiol.,* 63C, 229, 1979.

564. Tulasi, S. J., Reddy, P. U. M., and Rao, J. V. R., Effects of lead on the spawning potential of the freshwater fish, *Anabas testudineus*, *Bull. Environ. Contam. Toxicol.,* 43, 858, 1989.
565. Panigrahi, A., Dasmahapatra, A. K., and Medda, A. K., Effect of lead, zinc, mercury, and copper with and without estrogen on serum vitellogenin level in Magur fish (*Clarias batrachus* L.)., *Gegenbaurs Morphologisches Jahrbuch,* 136, 775, 1990.
566. Kumar, S. and Pant, S. C., Comparative effects of the sublethal poisoning of zinc, copper and lead on the gonads of the teleost *Puntius conchonius* Ham, *Toxicol. Lett.,* 23, 189, 1984.
567. Srivastava, A. K., Changes induced by lead in fish testis, *J. Environ. Biol.,* 8, 329, 1987.
568. Sehgal, R., Tomar, V., and Pandey, A. K., Comparative effects of 2 heavy metallic salts on the testis of viviparous teleost, *Lebistes reticulatus* (Peters), *J. Environ. Biol.,* 5, 185, 1984.
569. Myint, U. M. and Tyler, P. A., Effects of temperature, nutritive and metal stressors on the reproductive biology of *Mytilis edulis*, *Mar. Biol.,* 67, 209, 1982.
570. Munkittrick, K. R. and Dixon, D. G., A holistic approach to ecosystem health assessment using fish population characteristics, *Hydrobiologia,* 188/189, 123, 1989.
571. McKim, J. M. and Benoit, D. A., Effects of long-term exposures to copper on survival, growth, and reproduction of brook trout (*Salvelinus fontinalis*), *J. Fish. Res. Bd. Can.,* 28, 655, 1971.
572. Anderson, B. S., Middaugh, D. P., Hunt, J. W., and Turpen, S. L., Copper toxicity to sperm, embryos and larvae of topsmelt *Atherinops affinis*, with notes on induced spawning, *Mar. Environ. Res.,* 31, 17, 1991.
573. Busacker, G. P. and Chavin, W., Uptake, distribution and biological half-life of Cd-109 in tissues of *Carassius auratus* L., *Am. Zool.,* 17, 865, 1977.
574. Joensen, J. E. and Korsgaard, B., Uptake, time-course distribution and elimination of cadmium in embryos and tissues of the pregnant *Zoarces viviparus* (L.) after intraovarian loading, *J. Fish Biol.,* 28, 61, 1986.
575. Ghosh, P. and Thomas, P., Binding of metals to red drum vitellogenin and incorporation into oocytes, *Mar. Environ. Res.,* 39, 165, 1995.
576. Shackley, S. E., King, P. E., and Gordon, S. M., Vitellogenesis and trace metals in a marine teleost, *J. Fish Biol.,* 18, 349, 1981.
577. Gunn, S. A., Gould, T. C., and Anderson, W. A. D., Cadmium-induced interstitial cell tumors in rats and mice and their prevention by zinc, *J. Natl. Cancer Inst.,* 31, 745, 1963.
578. Waalkes, M. P. and Rehm, S., Carcinogenicity of oral cadmium in the male wistar (WF/NCR) rat — Effect of chronic dietary zinc deficiency, *Fundam. Appl. Toxicol.,* 19, 512, 1992.
579. Waalkes, M. P. and Misra, R. R., Cadmium carcinogenicity and genotoxicity, in *Toxicology of Metals,* Chang, L. W., Ed., Lewis Publishers, Boca Raton, 231, 1996.
580. White, D. H. and Finley, M. T., Uptake and retention of dietary cadmium in mallard ducks, *Environ. Res.,* 17, 53, 1978.
581. Sell, J. L., Cadmium and the laying hen: Apparent absorption, tissue distribution, and virtual absence of transfer into eggs, *Poult. Sci.,* 54, 1674, 1975.
582. Richardson, M. E. and Fox, M. R. S., Dietary cadmium and enteropathy in the Japanese quail — Histochemical and ultrastructural studies, *Lab. Invest.,* 31, 722, 1975.
583. Furness, R. W., Cadmium in birds, in *Environmental Contaminants in Wildlife: Interpreting Tissue Concentrations,* Beyer, W. N., Heinz, G. H., and Redmon-Norwood, A. W., Eds, Lewis Publishers, Boca Raton, FL, 389, 1996.
584. Kotyzova, D. and Sundeman, F. W., Maternal exposure to Cd(II) causes malformations of *Xenopus laevis* embryos, *Ann. Clin. Lab. Sci.,* 28, 224, 1998.
585. Herkovits, J., Cardellini, P., Pavanati, C., and PerezColl, C. S., Susceptibility of early life stages of Xenopus laevis to cadmium, *Environ. Toxicol. Chem.,* 16, 312, 1997.
586. Haux, C., Bjornsson, B. T., Förlin, L., Larsson, A., and Deftos, L. J., Influence of cadmium exposure on plasma calcium, vitellogenin and calcitonin in vitellogenic rainbow trout, *Mar. Environ. Res.,* 24, 199, 1988.
587. Sangalang, G. B. and O'Halloran, M. J., Cadmium-induced testicular injury and alterations of androgen synthesis in brook trout, *Nature,* 240, 460, 1972.
588. Victor, B., Mahalingam, S., and Sarojini, R., Toxicity of mercury and cadmium on oocyte differentiation and vitellogenesis of the teleost, *Lepidocephalichtyhs thermalis* (Bleeker), *J. Environ. Biol.,* 7, 209, 1986.

589. Pereira, J. J., Ziskowski, J., Mercaldo-Allen, R., Kuropat, C., Luedke, D., and Gould, E., Vitellogenin in winter flounder (*Pleuronectes americanus*) from Long Island Sound and Boston Harbor, *Estuaries Estudo,* 15, 289, 1992.
590. Povlsen, A. F., Korsgaard, B., and Bjerregaard, P., The effect of cadmium on vitellogenin metabolism in estradiol-induced flounder (*Platichtys flesus* L.) males and females, *Aquat. Toxicol.,* 17, 253, 1990.
591. Thomas, P., Effects of Aroclor 1254 and cadmium on reproductive endocrine function and ovarian growth in Atlantic croaker, *Mar. Environ. Res.,* 28, 499, 1989.
592. Kime, D. E., The effect of cadmium on steroidogenesis by testes of the rainbow-trout, *Salmo gairdneri, Toxicol. Lett.,* 22, 83, 1984.
593. Pundir, R. and Saxena, A. B., Seasonal changes in the testes of fish *Puntius ticto*, and their relation to heavy metal toxicity, *Bull. Environ. Contam. Toxicol.,* 45, 288, 1990.
594. Kumari, M. and Dutt, N. H. G., Cadmium-induced histomorphological changes in the testis and pituitary gonadotropic hormone secreting cells of the cyprinid *Puntius sarana*, *Bollettino di Zoologia,* 58, 71, 1991.
595. Sehgal, R., Tomar, V., and Pandey, A. K., Comparative effects of 2 heavy metallic salts on the testis of viviparous teleost, *Lebistes reticulatus* (Peters), *J. Environ. Biol.,* 5, 185, 1984.
596. Hatakeyama, S. and Yasuno, M., Chronic effects of Cd on the reproduction of the guppy (*Poecilia reticulata*) through Cd-accumulated midge larvae (*Chironomus yoshimatsui*), *Ecotoxicol. Environ. Saf.,* 14, 191, 1987.
597. Speranza, A. W., Seeley, R. J., Seeley, V. A., and Perlmutter, A., Effect of sublethal concentrations of zinc on reproduction in zebrafish, *Brachydanio rerio* Hamilton-Buchanan, *Environ. Pollut.,* 12, 217, 1977.
598. Munkittrick, K. R. and Dixon, D. G., Use of white sucker (*Catostomus commersoni*) populations to assess the health of aquatic ecosystems exposed to low-level contaminant stress, *Can. J. Fish. Aquat. Sci.,* 46, 1455, 1989.
599. Ohlendorf, H. M., Kilness, A. W., Simmons, J. L., Stroud, R. K., Hoffman, D. J., and Moore, J. F., Selenium toxicosis in wild aquatic birds, *J. Toxicol. Environ. Health,* 24, 67, 1988.
600. Lemly, A. D., Assessing the toxic threat of selenium to fish and aquatic birds, *Environ. Monitor. Assess.,* 43, 19, 1996.
601. Heinz, G. H., Selenium in birds, in *Environmental Contaminants in Wildlife: Interpreting Tissue Concentrations,* Beyer, W. N., Heinz, G. H., and Redmon-Norwood, A. W., Eds., Lewis Publishers, Boca Raton, FL, 447, 1996.
602. Heinz, G. H., Mercury poisoning in wildlife, in *Noninfectious Diseases of Wildlife*, Fairbrother, A., Locke, L. N., and Hoff, G. L., Eds., Lewis Publisher, Boca Raton, FL, 118, 1996.
603. Lemly, A. D., Selenium in aquatic organisms, in *Environmental Contaminants in Wildlife: Interpreting Tissue Concentrations,* Beyer, W. N., Heinz, G. H., and Redmon-Norwood, A. W., Eds., Lewis Publishers, Boca Raton, FL, 427, 1996.
604. Heinz, G. H., Hoffman, D. J., and Gold, L. G., Impaired reproduction of mallards fed an organic form of selenium, *J. Wildl. Manage.,* 53, 418, 1989.
605. Heinz, G. H., Hoffman, D. J., Krynitsky, A. J., and Weller, D. M. G., Reproduction in mallards fed selenium, *Environ. Toxicol. Chem.,* 6, 423, 1987.

CHAPTER 40

# A Review of the Role of Contaminants in Amphibian Declines

**Donald W. Sparling**

## CONTENTS



## 40.1 INTRODUCTION

For the past decade, there has been growing concern about worldwide declines in amphibian populations,[1,2] and a general phenomenon of declining populations was recognized in the mid-1990's. Subsequent research has validated this concern.[3,4] These population declines have been defined either as decreases in numbers of individuals in an area or, preferably because of greater reliability, a decrease in the number of sites occupied by breeding amphibians. Widespread population declines have occurred in North America,[5–7] Europe,[3,8,9] Australia,[10] and Central and South America.[11,12] Population declines in eastern Europe, Asia, and Africa have been suggested but are not as well documented. Worldwide, more than 500 populations of frogs and salamanders have been listed as declining or of concern.[4,13] In the United States, a third of known amphibian species are thought to be in trouble.[14] While the most severely affected populations are in the mountains of the western United States, serious declines have also been observed among some species in the Midwest and Southeast.[2]

Several reasons for these population declines have been proposed including habitat destruction or degradation, disease, climate change, introduction of exotic predators or competitors, increased solar radiation, and contaminants. These factors can operate in isolation of others, or two or more factors may interact to impact a population. The purpose of this chapter is to examine how contaminants may be involved with amphibian population declines, including their possible interaction with other factors. This is an overview, not an exhaustive summary of the literature; a thorough review of the literature is in the recent book edited by Sparling, Linder, and Bishop.[15]

Habitat destruction or degradation is a primary factor for many population declines. Because many species of amphibians require both land and water to complete their life cycles, draining of wetlands, clear-cutting of forests, and fragmentation and isolation of suitable patches of habitat all contribute to decreased breeding and survival.[7,16] Many other factors, such as grazing of riparian shorelines, discharge of effluents, and channelization, can degrade streams and rivers, making them unsuitable as amphibian habitat.

Little is known about amphibian diseases in general, but they are apparent problems in some areas.[17–19] Amphibians are subject to many bacterial, fungal, and viral diseases,[19] some of which have been implicated in epizootics. According to one hypothesis, disappearance of several Australian frogs has been associated with epizootics of an iridovirus that traveled in a northward wave of contagion, causing widespread extinctions.[10] Lips[11] proposed that a combination of disease and environmental contamination led to the decline of six species of frogs and salamanders in Costa Rica and suggested that these factors were responsible for declines in most tropical upland sites. Many of the tadpoles she collected had damaged and missing labial tooth rows. This type of damage has been subsequently attributed to chytrid fungi (*Phylum Chytridiomycota*). Not much is known about these fungi except that they typically digest chitin, but one or more species may have adapted to digesting keratin, which composes tadpole mouth parts. They have been linked to amphibian population declines in Australia and Central America[20] and seem to be very widespread among tadpoles in the United States (D. E. Green, G. Fellers, personal communication).

In a 7-year study of the endangered Wyoming toad (*Bufo baxteri*), the most common mortality factor in captive and free-ranging adults was mycotic dermatitis with a secondary infection of the bacterium *Basidiobolus ranarum*.[21] Of particular interest was the observation that the bacterium has been very common in the toad's environment for a long time, yet the toad apparently has become highly susceptible to infection, which begs the question of whether other factors have reduced its tolerance to the bacterium.

While diseases are potentially very important, additional epidemiological work is required to satisfy Koch's postulates of infectious diseases for most of the proposed agents and populations, and there is a strong likelihood that other factors like environmental contaminants are interacting with disease agents to increase their virility or decrease amphibian immune responses.

Another proposed cause for amphibian population declines is exotic species — especially introduced predators and competitors. Populations of the Cascades toad (*Bufo cascadae*), red-legged frogs (*Rana aurora*), mountain yellow-legged frogs (*R. muscosa*), and foothill yellow-legged frogs (*R. boylii*) are declining in the western Sierra Nevada Mountains of California.[5,6] Many of the lakes that were inhabited by these amphibians have been stocked with nonnative salmonids that are predacious on eggs, tadpoles, and young frogs. The presence of these exotic predators showed a clear discontinuity with the occurrence of native amphibian species.[22,23] Also, a survey using direct-sight locations of trout and long-toed salamanders (*Ambystoma macrodactylum*) in the Bitterroot Mountains of Montana demonstrated a high degree of correlation between the presence of trout and the absence of salamander larvae.[24] Direct-observation surveying methods are more limited than seining or electrofishing techniques, but the relationship between trout and salamanders was strengthened through historical data — 10 years earlier six lakes that had trout also had no salamander larvae, but the trout had been extirpated in the interim, and salamanders had repopulated the lakes.

Bullfrogs (*R. catesbeiana*), which feed on adult and larvae of other amphibians, have also been introduced into western lakes and have been blamed for amphibian population declines.[25,26] A study using experimental ponds inoculated with *R. catesbeiana* tadpoles, *R. aurora* tadpoles, and mosquitofish (*Gambusia affinis*) showed that survival of *R. aurora* tadpoles was less than 5% in ponds stocked with *R. catesbeiana*. Whereas mosquitofish had no measurable effect on *R. aurora* survival, there was significant depression of *R. aurora* tadpole growth and an increase in the frequency of injuries among frogs in ponds stocked with mosquitofish compared to controls.[27] A similar relationship between the occurrence of frogs and fish has been recorded in Spain.[28]

Over the past several years, scientists have been concerned about increased ultraviolet radiation, especially UV-B due to thinning of the atmospheric ozone layer. Because this thinning is related to air pollutants, including $CO_2$, $NO_x$, and difluorocarbons, increased UV-B radiation is indirectly anthropogenic. Direct exposure to ultraviolet radiation in the laboratory reduces hatchability and increases mortality of tadpoles. Ankley et al.[29,30] demonstrated that northern leopard frog (*R. pipiens)* tadpoles exhibited a significantly higher frequency and severity of hindlimb malformations when exposed to UV-B light than did controls. These malformations, however, were not like those found under field conditions because the experimental ones were usually bilaterally symmetrical, whereas those observed in the field are typically unilateral.[31] An understanding of the role of ultraviolet radiation in amphibian population declines is confounded by several factors: (1) many of the older laboratory experiments that subjected amphibians to ultraviolet radiation did not emulate natural conditions or did not see an effect;[32–34] (2) there are important species-specific and life-stage differences in the sensitivity of amphibians; for example, exposure to UV-B decreased larval and egg survival of *R. aurora* but affected only larval Pacific treefrogs (*Hyla [Pseudacris] regilla*);[35] (3) unless the water is exceptionally clear and colorless, UV-B penetration into the water column is rapidly attenuated, and tadpole shelter-seeking behavior may further reduce exposure;[36] and (4) tadpoles of several species of amphibians produce photolyase, an enzyme that repairs photodamaged tissues under low light conditions.[37,38]

Contaminants have been frequently cited as contributing to amphibian population declines.[7,16,39,40] Arguments used to support the role of contaminants include multiple life stages that may expose amphibians to a variety of contaminants in the water and on land, a highly permeable skin that facilitates uptake of contaminants, a moderate to high sensitivity in acute and chronic toxicity tests compared to other aquatic organisms, reproductive seasons that overlap with agrochemical applications that can affect reproduction, and varying degrees of mobility that may affect the probability of exposure, particularly to pesticides in agricultural landscapes. Prior to 1998, most of the information on contaminant exposure in amphibians was residue information with little information on the effects of these contaminants. Most of the available information on effects comes from laboratory toxicity studies, with little information from field experiments. To date, no studies have been published that show a clear cause-and-effect relationship between contaminants and long-term population declines in amphibians. This absence of evidence is most likely due to the lack of research, not because contaminants are unimportant.

## 40.2 CRITERIA FOR DETERMINING A POPULATION EFFECT

Any attempt to determine if a contaminant, or any stressor, has an effect on a population of amphibians is constrained by several factors. First, local groups of many species exist as metapopulations — adults that inhabit nearby ponds, sometimes separated by a kilometer or more, move from one pond to another and breed with those in other ponds, resulting in genetic mixing and opportunities to recolonize depopulated ponds.[41] Thus, local events that reduce a group of amphibians inhabiting one pond may be compensated by neighboring ponds. Alternatively, an impacted pond may serve as a sink for immigrating individuals. Second, generally it is not known if mortality caused by contaminants is additive or compensatory. Many species of anurans (frogs and toads) have high fecundity, so that eggs, embryos, or tadpoles that die from exposure to contaminants could have died anyway from other factors, and the overall insult is no different. Conversely, long-term population recruitment and mortality may be delicately balanced so that the addition of a contaminant that restricts reproduction or increases mortality may be more than a population can tolerate and decline occurs.

Therefore, to determine whether a particular factor such as a contaminant has a negative effect on a population two conditions must be met: (1) the ecologically relevant levels or concentrations (i.e., the levels that could be expected to occur under typical conditions) of the contaminant must result either in significantly reduced recruitment via immigration and impaired reproduction or increased loss above that caused by other stressors through emigration or mortality rates within a free-ranging population and (2) the effect on the population must be additive to other stressors such as predation, parasitism, or energy requirements. That is, decreases in population recruitment or increases in population losses cannot be totally compensated by other factors.

In reality, these criteria are seldom, if ever, met in ecotoxicological studies; thus, only indirect evidence is available to determine the role contaminants may have in amphibian population declines.

## 40.3 LEVELS OF EVIDENCE

Although there are no studies to date that show an unambiguous relationship between environmental contaminants and long-term amphibian population declines, we can use less direct evidence to investigate if such relationships can occur. Based on immediacy and ecological relevance, this evidence can be categorized into three groups:

1. Data collected under field conditions that document at least temporary declines in amphibian numbers when they are exposed to environmental contaminants.
2. Data obtained from laboratory studies that demonstrate lethal or severe sublethal responses to contaminants at or below concentrations typically expected or observed in the field; ideally populations should increase when these contaminants are remediated.
3. Laboratory or field studies that show that contaminants can interact with other hypothesized causes for amphibian declines and increase the probability of population effects.

The most direct and reliable evidence for contaminant-related population declines would come from field studies that reflect or report natural situations. However, the majority of contaminant effects studies on amphibians have occurred in the laboratory under highly artificial conditions. In nature, amphibians are exposed to a host of potentially interacting forces of abiotic factors and water chemistry, predators, parasites, competitors, and habitat that are absent from laboratory investigations. Often, laboratory studies use single, technical-grade chemicals, whereas natural populations are exposed to formulations (e.g., pesticides) or to mixtures of contaminants (e.g., metals, polychlorinated biphenyls [PCBs], polycyclic aromatic hydrocarbons [PAHs]). Under laboratory situations, stressors such as predators and competition are ameliorated or absent, but the stark habitat provided by glass aquaria and fluorescent lighting adds undetermined stresses to organisms. Many other factors that affect response to contaminants (e.g., dissolved organic carbons, temperature and oxygen fluctuations, refugia) are absent in laboratories. Thus, the effective or lethal concentrations determined under laboratory conditions, although frequently our only clue to response, may be at best rough approximations of toxic exposures in natural wetlands.

Ecological relevance can be increased by the use of outdoor laboratories, which allow natural variation in weather and lighting conditions. Relevancy increases as other organisms such as predators, competitors, or vegetation are incorporated in mesocosm and macrocosm studies, but the ability to control all relevant variables decreases as size and complexity increase. A few attempts have been made to strike a balance between experimental control and ecological relevance in amphibian tests. Experimental manipulation of natural wetlands may be difficult due to size and logistics, legal restrictions on applying contaminants to existing wetlands, lack of true controls or replicates, or uncontrollable factors related to land-use practices.

### 40.3.1 Evidence for Direct Involvement

Contaminants can have a direct effect on amphibian populations by impairing physiological processes or resulting in mortality. In contrast to indirect effects, discussed later, direct effects are often easier to identify and to establish causal links.

#### *40.3.1.1 Field Studies Showing Population Effects*

Most of the studies that have examined the direct effects of contaminants on amphibian populations have dealt with acid deposition. Some of these studies also examined correlates of acidification including sulfates, dissolved aluminum and other metals, alkalinity, and cations. Many have been correlative in nature, i.e., drawing relationships between wetland chemistry and presence of species or relative abundance, but several have been experimental.

One correlative study examined 35 wetlands in Pennsylvania.[42] The number of egg masses deposited by Jefferson salamanders (*Ambystoma jeffersonianum*) was positively correlated with pH and alkalinity and negatively correlated with aluminum concentration. In the same study, egg mass deposition by spotted salamanders (*A. maculatum*) correlated positively with pH and negatively with total cations (Na, K, Mg, Ca) and specific conductance. Another correlative study in Ontario, Canada found only weak relationships among species richness of amphibians and water characteristics among 180 wetlands.[43] Although species richness correlated negatively with chloride, magnesium, and turbidity, multiple regression analysis only accounted for 19% of the variance in richness. However, those wetlands were well buffered and not prone to acidification.

In a field experiment in England, variation in egg fertility could be explained primarily by the concentration of monomeric aluminum with additional explanation provided by the concentration of zinc, silicon, and molybdenum.[44] Monomeric aluminum also accounted for the greatest amount of variance in the frequency of abnormal yolk plugs in eggs. Aluminum is the most common metal in the earth's crust, but acidic conditions cause it to dissolve and form toxic species of which monomeric ($Al^{3+}$) is among the most toxic to amphibians.

Horne and Dunson[45] used outdoor mesocosms to test the effects of low pH, dissolved organic carbons (DOC), and metals on survivorship in three species of amphibians. Exposure concentrations were set to reflect typical conditions in central Pennsylvania (pH 4.5 to 5.5, DOC 15–30 mg/L). The time required for wood frog (*R. sylvatica*) larvae to metamorphose increased with high DOC and low pH. Delayed metamorphosis can be disadvantageous to species such as *R. sylvatica* that inhabit temporary wetlands by increasing the risk of dessication. Survival of *A. jeffersonianum* and *A. maculatum* was reduced by pH and metal concentrations in the same study.

The relative abundances of northern cricket frogs (*Acris crepitans*) and gray treefrogs (*Hyla versicolor*) were significantly lower in experimental wetlands that had been acidified to a pH of 5 to 5.5 than in wetlands with a pH range of 6 to 6.8.[46] In this study, all the experimental wetlands were naturally colonized by four species of anurans from surrounding wetlands. For *H. versicolor*, there was a soil × pH interaction in that the greatest effects of pH were seen in wetlands with a clay base rather than a loam base. The soil effect could be explained by the fact that clay soils had approximately twice the concentration of metals as the loam soils and that acidification of sediment and water facilitates metal dissolution. No pH or soil effects were observed in green frogs (*R. clamitans*) or southern leopard frogs (*R. sphenocephala*).

Some amphibians, such as the salamander genus *Plethodon,* spend their entire life cycle on land. Yet these species may also be limited by acid deposition. The distribution and population density of the red-backed salamander *(P. cinereus)* was restricted by soil pH $\leq$ 3.7.[47] While this was a very acidic soil, it is not uncommon in coniferous and deciduous forests in the Northeast where humic acids combine with atmospheric deposition of acid-causing anions.

The effects of acidification on amphibian health and populations are well known, and acid deposition certainly may be a contributing factor to population declines in some areas. However, as stated by Rowe and Freda,[48] no studies have documented specifically that acid deposition has been involved with any long-lasting decline in amphibian populations. Hazards due to acid deposition would be greatest where subsoils are granitic, poorly buffered, with low acid-neutralizing capacity (ANC) and where anthropogenic sources of acid-causing anions such as sulfates and nitrates are downwind. Within North America, these areas tend to be most abundant in the northeastern United States and eastern Canada. The Rocky Mountains have poorly buffered soils but lack an abundant source of acid-causing anions. Florida has many ponds and wetlands with low pH, but the source of the acidity appears to be organic acids rather than anthropogenic sources,[49] and the amphibians have probably adapted to low pH over long evolutionary periods. Bradford et al.[50] did not find any relation between ANC and populations of anurans in the Sierra Mountains, and they did not find any regional differences in ANC or other water-chemistry parameters related to acid deposition. They concluded that acid deposition was not important in the decline of these anurans.

There are fewer experimental or correlative studies with other contaminants and amphibians under natural conditions. Another pervasive potential source of contaminants is nitrogen fertilizers. Nitrogen is used or found in nature in various forms from ammonium ion to ammonia, nitrite, and nitrate. Of these, ammonia and ammonium are the most toxic forms to amphibians, followed by nitrites and nitrates. The use of nitrogen fertilizers in North America is tremendous, with application rates sometimes exceeding 8 metric tons/km$^2$ on agricultural lands and more than 72 million tons worldwide;[51] greatest use in the United States is in the East and in the Midwest. Aqueous concentrations may exceed 100 mg/L in ponds and between 2 and 40 mg/L in streams.

In one study in Britain, Oldham et al.[52] released adult common frogs (*R. temporaria*) into plastic arenas situated in the middle of a field that had been treated with ammonium nitrate at 10.8 g/m.$^2$ Adults began displaying signs of toxicity — increased rate of breathing — and one died; although these signs were considered predictive of death, frogs were removed before they succumbed. Based on laboratory data, the authors estimated that the $EC_{50}$ for *R. temporaria* was 6.9 g/m,$^2$ well below the application rate. Fortunately, the toxicity of ammonium nitrate disappears rapidly as the solution is absorbed by moist soil. However, laboratory toxicity tests (see below) frequently conclude that lethal concentrations are below anticipated ambient levels.

A 20-year decline in amphibian numbers in Poland was ascribed to high levels of nitrogen in surface waters coming from the use of fertilizers on surrounding agricultural lands.[53] In an agricultural study area in Ontario, Bishop et al.[54] determined that habitat loss and nitrogen from fertilizers were more important than pesticides in restricting amphibian survival and species richness. A recent review of the subject[51] concluded that the problem of nitrogen-based fertilizers is extensive and toxic enough to represent one of the most pervasive contaminant threats to amphibian survival in North America and perhaps elsewhere. Of more than 8500 water samples collected from states and provinces around the Great Lakes, 19.8% had nitrate concentrations that exceeded those found to produce sublethal effects in amphibians. Sublethal and lethal effects can be caused by nitrate concentrations ranging from 2.5 to 100 mg/L, depending on species and study.

Chlorpyrifos is a highly toxic but widely used pesticide. In June 2000, the U.S. Environmental Protection Agency ordered a phase-out of chlorpyrifos use in several applications because of human health risks.* The pesticide is also very highly toxic to aquatic organisms, having an $LC_{50}$ of around 9–15 μg/L. A field study by Moulton[55] found adverse effects, including mortality in adult and larval *Hyla femoralis,* after exposure to environmentally realistic concentrations of chlorpyrifos.

Mazanti[56] conducted a combined field and laboratory study on the effects of two commercial pesticide formulations: Lorsban 4E (44.9% chlorpyrifos) and Bicep II (27.4% atrazine, 35.6% metolachlor). Both formulations are used extensively in agriculture, with Bicep II sprayed as a preemergent herbicide for broadleaved plants and Lorsban 4E applied about 3 weeks later to control insects. In the laboratory, a combination of both pesticides was far more toxic than either pesticide alone, and a high dose (2.0 mg/L atrazine, 2.5 mg/L metolachlor, 1.0 mg/L chlorpyrifos) killed

* http://www.epa.gov/pesticides/op/chlorpyrifos-methyl/summary.htm.

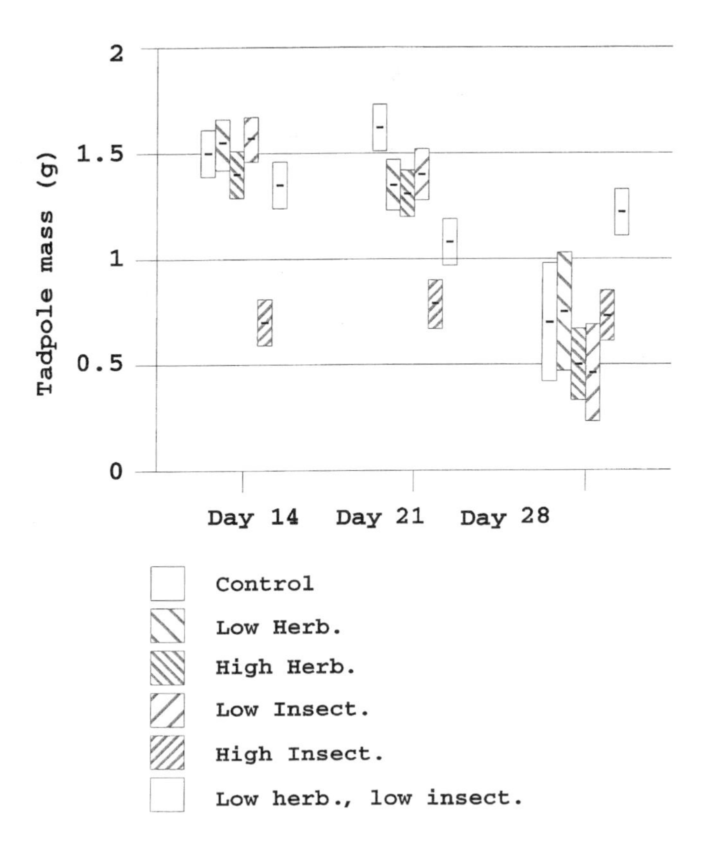


**Figure 40.1** Growth (x ± SD) of gray treefrog (*Hyla versicolor*) tadpoles through time when exposed to low and high concentrations of the insecticide Lorsban (active ingredient is chlorpyrifos), low and high concentrations of the herbicide Bicep II (active ingredients atrazine and metalochlor) and low concentrations of both insecticide and herbicide. See text for concentrations. The primary size difference at 28 days is due to a lack of metamorphosis in high and low insecticide rather than to growth per se. All tadpoles exposed to combined high herbicide and high insecticide died by 11 days.

100% of *H. versicolor* tadpoles within 11 days. A low combined dose and high individual doses of insecticide or herbicide alone caused lethargy and reduced growth but did not increase mortality over controls. In naturally colonized, experimental wetlands, low-herbicide/low-insecticide (0.2 mg/L atrazine, 0.25 mg/L metoachlor, 0.1 mg/L chlorpyrifos) and high-herbicide/low-insecticide treatments resulted in almost complete, short-term extirpation of *H. versicolor* and *R. clamitans* tadpoles compared to controls (Figure 40.1). By the end of the summer, however, these prolonged breeding species had repopulated the treated wetlands. If species with contracted or explosive breeding seasons had been similarly affected, their reproduction would have been severely curtailed for the year.

A recent study implicates chlorpyrifos and other insecticides in the declines of ranid populations in the Sierra Mountains of California. The Central Valley of California is an intensely agricultural region that sits between the Pacific Coast and the Sierra Mountains. The most severely impacted sites for *R. aurora, R. mucosa,* and *R. boylii* are in the mountains east and above the Central Valley in what appear as pristine wetlands. However, 60% of California pesticides used in agriculture, thousands of kilograms of active ingredient pesticides, are aerially sprayed on the croplands of the Central Valley every year.* A large proportion of these pesticides are organophosphorus insecticides such as chlorpyrifos, malathion and diazinon which inhibit cholinesterase.

* http://www.cdpr.ca.gov/docs/pur/purmain.htm.

Other highly toxic pesticides such as endosulfan and trifluralin are also sprayed in large quantities. *Pseudacris regilla* occupies many of the same habitats as the declining ranid species but its numbers are not nearly as depressed as the ranids. Using *P. regilla* as a sentinel species, the authors found that cholinesterase activity was significantly depressed in tadpoles in the Sierra Nevada downwind of the Central Valley compared to coastal populations.[199] The samples with the lowest cholinesterase values came from Yosemite and Sequoia National Parks where mean activity level was often less than 50% of reference means on the coast. In the mountains, more than 83% of the tadpoles and adults collected from Lake Tahoe region had measurable concentrations of endosulfan and 50% of those sampled in Yosemite and Sequoia National Parks contained diazinon or chlorpyrifos residues (Figure 40.2). Both endosulfan and the organophosphorus pesticides have half-lives measured in days so detection of these pesticides in *P. regilla* signified very recent exposure.

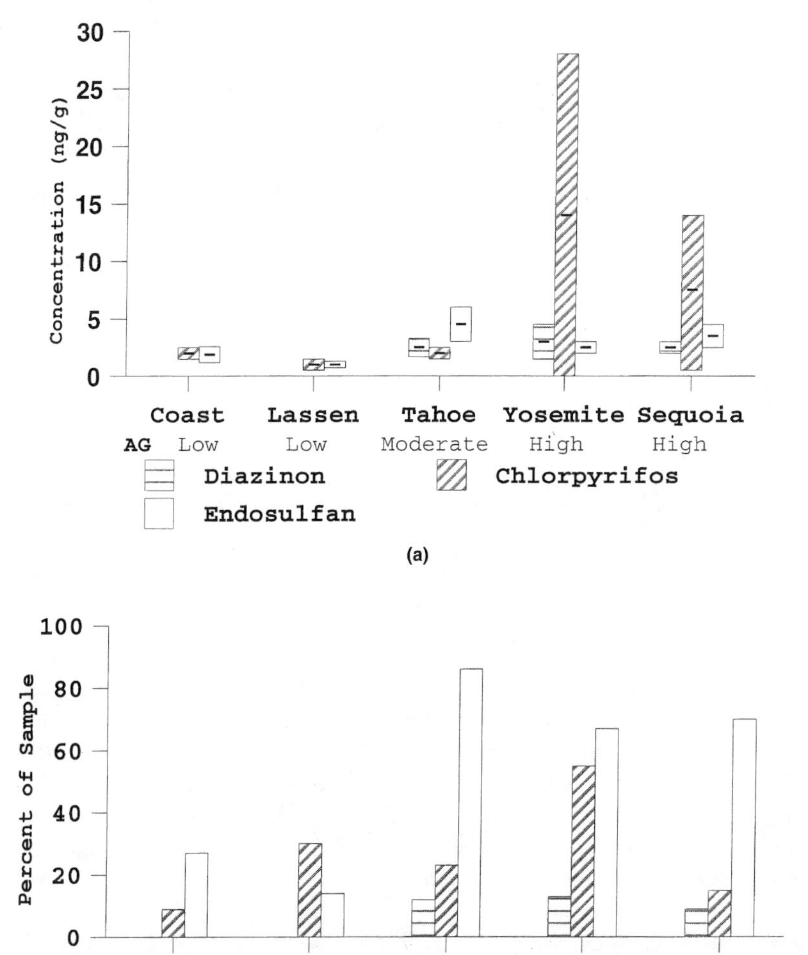


**Figure 40.2** Mean (± SD) concentrations (a) and frequencies of detection (b) for chlorpyrifos, diazinon and endosulfan in Pacific treefrog (*Hyla [Pseudacris] regilla*) adults and tadpoles collected from California, 1999.

Fenitrothion, an organophosphorus insecticide, used to be sprayed widely to control spruce budworm (*Choristoneura fumiferana*). The relative abundance of mink frogs (*R. septentrionalis*) among New Brunswick ponds was inversely related to the intensity of spraying in forest ponds.[57] Ponds that had been sprayed in 3 of the preceding 4 years had significantly lower relative abundances of frogs than ponds that had been sprayed less often. In addition, the percent cover of submerged vegetation related positively to frog abundance, perhaps because it provided shelter from direct exposure to the insecticide.

Although DDT and most other chlorinated hydrocarbon pesticides are no longer used in developed countries, these compounds have been associated with die-offs of amphibians and could still be having some effect in Third World countries, where they are being used. For example, DDT was applied to forests to control tussock moths (*Orgyia pseudotsugata*). A die-off of western spotted frogs (*R. pretiosia*) occurred following a spraying of DDT with fuel oil as solvent at a rate of 0.72 kg DDT/ha. An unknown percentage of the adult frogs were dead, and those that were found dead had 5–10 times the concentration of DDT derivatives on a lipid basis than those that were still alive.[58] There was 100% egg mortality in *R. pipiens* inhabiting a wetland adjacent to cropland that had been sprayed with aldrin.[59] A mixture of endrin, aldrin, dieldrin, and toxaphene at 2.7 kg/ha used in mosquito control killed all the *R. catesbeiana* inhabiting a wetland.[60]

Whereas the number of documented cases of die-offs in amphibians due to contaminants is sparse, certainly the vast majority of cases either are unreported, or causal factors are not identified. Both acid deposition, with the elevated concentrations of metals produced by increased acidity, and pesticides can cause local reductions or extirpations of amphibian communities, and in some cases, they may be related to regional declines.

#### *40.3.1.2 Toxicity in the Laboratory at Ecologically Relevant Concentrations*

As with field studies, more research on amphibian ecotoxicology has been conducted in conjunction with acid deposition than with any other contaminant. Studies have examined direct effects of elevated hydrogen ion concentration plus associated effects due to the dynamics of dissolved aluminum; mitigating effects of water hardness, dissolved organic content, and organic acids; effects of temperature; and other factors. Endpoints in these studies have included survivorship at various life stages (egg, embryo, tadpoles), ion regulation, metal uptake, respiration and energy conversion, and behavioral responses. More complete reviews of the effects on aquatic biota in general have been published.[61,62] Here we focus on the direct effects of acid deposition and pesticides on factors that affect amphibian population dynamics — survivorship and reproduction.

Readers are referred to several reviews of the effects and tolerance limits of amphibians to acidity.[48,61,63–65] Sensitivity to pH varies among species. Among the most sensitive species are the natterjack toad (*Bufo calamatia*), with a lower pH limit (at which substantial or total mortality occurs) of 4.5 and an upper effects level (where sublethal effects or mild mortality can be seen) of pH 4.7. In comparison, the more tolerant pine barrens treefrog (*Hyla andersoni*) can tolerate pH as low as 3.1.[61,65] Most amphibian species are sensitive to pH = 4.5 or 4.6. In general, the toxicity of pH decreases as water hardness increases. Dissolved organic carbon can ameliorate the toxic effect of pH and Al but can itself be toxic to amphibians[66,67] and can complex with Al to alter its availability. The speciation and toxicity of aluminum is complex as pH drops below 5.5, and three response categories to Al have been identified.[68] These range from sensitivity only to pH below 4.5 to sensitivity to pH and aluminum between 4.5 and 5.5, to Al sensitivity only at pH $> 4.5$ and even an amelioration of pH toxicity by Al at pH $< 4.5$. Because of these dynamics, toxicity due to pH, Al, and DOC often cannot be totally distinguished under natural conditions.[61]

As a very brief synopsis of the relatively extensive literature on acid deposition and amphibians, embryos tend to be the most sensitive life stage, especially with elevated concentrations of Al[69] that induce a characteristic curling of the tail due to alterations in the vitelline membrane of the egg. Serious impairment of hatching due to ecologically realistic levels of pH and Al are well documented in several species.[42,70–72] Similarly, tadpole mortality increases significantly at pH below 5.0.[73–76] Acute or episodic exposure to reduced pH and elevated Al, such as what might occur with a rapid snowmelt, tends to be more toxic than chronic exposure.[76] There is some evidence that amphibians may adapt physiologically or genetically to long-term exposure to reduced pH.[77] Hardly anything is known about the effects of pH or Al on adult amphibians.[48]

Among the plethora of other contaminants, many occur in the environment at lethal concentrations as determined by laboratory $LC_{50}$ tests (Table 40.1). In this table, ambient concentrations for most of the chemicals come from Eisler.[78] However, organophosphorus, carbamate, and pyrethroid pesticides are not persistent, and environmental fate studies of these compounds often result in measurements below instrument detection levels. Thus, when measurable concentrations are detected, they may be a fraction of the maximum to which the amphibians were exposed. Because few reliable data exist, I used label application instructions[79] to derive a rough approximation for some of these; amount per area was converted to concentrations in water based on wetlands with average depth of 0.5 m, under the presumption that most amphibians lay their eggs in shallow water. Realized concentrations can vary considerably from these estimates due to actual depth of eggs masses, applicator deviations from label application rates, and additional pesticides entering a wetland through runoff or wind. Lethal toxicity data come from many sources including some recent reviews.[80–82]

The table has several limitations and is intended only to illustrate that there are many compounds that potentially are found in the environment at higher than toxic concentrations. Limitations include the previously stated weaknesses in relying on laboratory toxicity tests to reflect natural exposures and differences in testing methodology among labs that may affect results. Life stages and duration of tests also are important because of differential sensitivity among stages and because length of exposure may greatly alter $LC_{50}$ values. Whereas many laboratory toxicity tests end after 24 or 96 h, free-ranging animals may be exposed for much longer durations or be subject to repeated exposures through time. Third, the table does not report the many instances of nondetects for specific contaminants. The relative frequency of nondetects is biased against totally anthropogenic and short-lived chemicals compared to molecules such as metals that are persistent and also naturally occurring. Fourth, the presence of a substance in sediment may not reflect its bioavailability; it may be bound to sediment particles. Fifth, acute lethality, as demonstrated by the $LC_{50}$ values, is the ultimate and strongest effect of contaminant exposure. Sublethal or chronic effects, which may reduce the survivorship or reproductive potential of an organism, can occur at concentrations much lower than those necessary for acute toxicity. In view of these caveats, perhaps the greatest value of the table is to clearly indicate the great need for ecologically relevant, cause-and-effect data on amphibian populations and contaminant exposure.

Despite these limitations, a few salient points may be developed:

- In contrast to the 75,000 chemicals manufactured in the United States each year, not including pesticides and the 3000 that are considered high-production-volume chemicals (> 1 million pounds produced),[83] toxicity data for amphibians are known for only a very, very small proportion.
- Relationships between ambient concentrations of conventionally used pesticides and effects are poorly known.
- Ambient concentrations are often several times greater than the lethal concentrations, as determined by lab tests, and this difference extends across all categories of contaminants for which there are data.
- Formulations can make a difference. For example, at a given concentration of azinphos-methyl, the formulation Guthion 2S may result in amphibian die-offs, whereas use of the alternative formulation Guthion might not.
- Different life stages of amphibians vary in their sensitivity to the same contaminant. For example, tadpole *Bufo arenarum* are more sensitive than embryos to parathion, and adult *R. pipiens* are less sensitive to TFM than either tadpoles or embryos.

**Table 40.1 Median Lethal Concentrations (LC50s) for Various Contaminants in Amphibians that are Near Below Reported Concentrations in the Environment[a]**

| Compound | Ambient Levels | | Species | Stage | Dosage | Effect | Reference |
|---|---|---|---|---|---|---|---|
| | Matrix | Concentration | | | | | |
| Azinphos-methyl | Water | 0.71 mg/L | *Rana catesbeiana* | Tadpole | 1.8 kg/ha | 100% mortality | 165 |
| As Guthion | | | *Hyla regilla* | Tadpole | 8.7–9.7 mg/L | 100% mortality | 166 |
| Guthion | | | *H. regilla* | Tadpole | 4.14 mg/L | 96 h $LC_{50}$ | 166 |
| As Guthion 2S | | | *H. regilla* | Tadpole | 0.8–1.5 m/L | 100% mortality | 165 |
| Guthion 2S | | | *H. regilla* | Tadpole | 0.46–0.84 mg/L | 96 h $LC_{50}$ | 166 |
| Guthion 2S | | | *Bufo woodhousei fowleri* | Tadpole | 0.13 mg/L | 96 h $LC_{50}$ | 167 |
| Guthion 2S | | | *Ambystoma gracile* | Larvae | 1.67 mg/L | 96 h $LC_{50}$ | 168 |
| Guthion 2S | | | *A. maculatum* | Larvae | 1.9 mg/L | 96 h $LC_{50}$ | 168 |
| Chlorpyrifos | Water | 0.68 mg/L | *Rana tigrina* | Tadpole | 10 µg/L | 6 days $LC_{50}$ | 169 |
| Fenitrothion | Water | 75.5 µg/L | *R. catesbeiana* | Tadpole | 4 mg/L | field mortality | 170 |
| Parathion | Water | 0.34 mg/L | *Bufo arenarum* | Embryos | 20.2 mg/L | 7d $LC_{50}$ | 171 |
| Parathion | | | *B. arenarum* | Tadpoles | 4.5 mg/L | 7d $LC_{50}$ | 171 |
| Temephos | | | *R. clamitans* | Tadpoles | 4.24 µg/L | 96 h $LC_{50}$ | 100 |
| Carbaryl | Water | 0.45 mg/L | *R. tigrina* | Tadpoles | 10 mg/L | MRT[a] 110 min. | 172 |
| Deltamethrin | Water | 2.6 µg/L | *B. arenarum* | Tadpoles | 4.4 µg/L | 96 h $LC_{50}$ | 173 |
| Esfenvalerate | Water | 1 µg/L | *R. blairi* | Tadpoles | 7.29 µg/L | 96 h $LC_{50}$ | 174 |
| Esfenvalerate | | | *R. sphenocephala* | Tadpoles | 7.79 µg/L | 96 h $LC_{50}$ | 174 |
| Atrazine | Water | 0.1–480 µg/L | *R. pipiens* | Tadpoles | 14.5–47.5 mg/L | 96 h $LC_{50}$ | 175 |
| | Sediment | 0–7.3 mg/L | *B. americana* | Tadpoles | 10.7–26.5 mg/L | 96 h $LC_{50}$ | 175 |
| Triclopyr | | 0.1 mg/L | *R. clamitans* | Tadpoles | 2.4 mg/L | 100% mortality | 170 |
| TFM | | | *Hyla versicolor* | Tadpoles | 1.98 mg/L | 96 h $LC_{50}$ | 176 |
| TFM | | | *R. pipiens* | Tadpoles | 2.76 mg/L | 96 h $LC_{50}$ | 176 |
| TFM | | | *R. catesbeiana* | Tadpoles | 3.55 mg/L | 96 h $LC_{50}$ | 176 |
| TFM | | | *R. pipiens* | Tadpoles | 0.95 mg/L | 96 h $LC_{50}$ | 177 |
| TFM | | | *R. pipiens* | Adults | 12.99 mg/L | 96 h $LC_{50}$ | 177 |
| TFM | | | *R. pipiens* | Embryos | 1.0 mg/L | arrested development | 177 |
| Paraquat | Water | 0.17 mg/L | *B. woodhousei fowleri* | Tadpoles | 15 mg/L | 96 h $LC_{50}$ | 178 |
| Paraquat | | | *Pseudacris triserata* | Tadpoles | 28 mg/L | 96 h $LC_{50}$ | 178 |
| Rotenone | Water | 0.05 mg/L | *R. pipiens* | Tadpoles | 5 µg/L | 24 h $LC_{50}$ | 179 |
| Rotenone | | | *R. sphenocephala* | Tadpoles | 30 µg/L | 24 h $LC_{50}$ | 179 |
| Arsenic | Water | 0–243 mg/L | *Adelotus brevis* | Tadpoles | 55 mg/L | 96 h $LC_{50}$ | 180 |
| Arsenic | Sediment | 5–3500 ppm | *Ambystoma opacum* | Embryo | 4.45 mg/L | 8d $LC_{50}$ | 181 |
| Arsenic | | | *Gastrophyrne carolinensis* | Embryo | 0.04 mg/L | 7d $LC_{50}$ | 181 |
| Arsenic | | | *R. hexadactyla* | Tadpole | 0.249 mg/L | 96 h $LC_{50}$ | 182 |

**Table 40.1 Median Lethal Concentrations ($LC_{50}$s) for Various Contaminants in Amphibians that are near below Reported Concentrations in the Environment[a] *(Continued)***

| Compound | Ambient Levels | | Species | Stage | Dosage | Effect | Reference |
|---|---|---|---|---|---|---|---|
| | Matrix | Concentration | | | | | |
| Boron | Water | 0–0.3 mg/L | *B. woodhousei fowleri* | Embryo | 123 mg/L | 8d $LC_{50}$ | 183 |
| Boron | Sediment | 10–500 ppm | *R. pipiens* | Embryo | 130 mg/L | 8d $LC_{50}$ | 183 |
| Cadmium | Water | 0.5 μg/L | *G. carolinensis* | Embryo | 0.04 mg/L | 7d $LC_{50}$ | 181 |
| Cadmium | Sediment | 0.16 ppm | *A. opacum* | Embryo | 0.15 mg/L | 8d $LC_{50}$ | 181 |
| Cadmium | | | *B. arenarum* | Tadpole | 6.77 mg/L | 96 h $LC_{50}$ | 184 |
| Chromium | Water | 0–780 mg/L | *G. carolinensis* | Embryo | 0.03 mg/L | 7d $LC_{50}$ | 181 |
| Chromium | Sediment | 1–25000 ppm | *R. hexadactyla* | Tadpole | 10 mg/L | 96 h $LC_{50}$ | 182 |
| Copper | Water | 0.001–0.1 mg/L | *A. jeffersonianum* | Embryo | 0.315 mg/L | 96 h $LC_{50}$ | 185 |
| Copper | Sediment | 7–6500 ppm | *B. melanostictus* | Tadpole | 0.320 mg/L | 96 h $LC_{50}$ | 186 |
| Copper | | | *R. hexadactyla* | Tadpole | 0.039 mg/L | 96 h $LC_{50}$ | 182 |
| Copper | | | *R. pipiens* | Embryo | 0.05 mg/L | 8d $LC_{50}$ | 187 |
| Copper | | | *R. pipiens* | Adult | 6.37 mg/L | 72 h $LC_{50}$ | 188 |
| Lead | Water | 0–5 μg/L | *B. arenarum* | Embryo | 0.47 mg/L | 48 h $LC_{50}$ | 189 |
| Lead | Sediment | 0–7 ppm | *A. opacum* | Embryo | 1.46 mg/L | 8d $LC_{50}$ | 181 |
| Lead | | | *R. hexadactyla* | Tadpole | 33.28 mg/L | 96 h $LC_{50}$ | 182 |
| Mercury | Water | 0.4–10.7 μg/L | *Acris crepitans* | Embryo | 0.01 mg/L | 7d $LC_{50}$0 | 190 |
| Mercury | Sediment | 0.1–140 ppm | *Ambystoma opacum* | Embryo | 0.107 mg/L | 7d $LC_{50}$ | 190 |
| Mercury | | | *B. fowleri* | Embryo | 0.065 mg/L | 7d $LC_{50}$ | 190 |
| Mercury | | | *B. fowleri* | Tadpole | 0.025 mg/L | 7d $LC_{50}$ | 183 |
| Mercury | | | *G. carolinensis* | Embryo | 0.001 mg/L | 7d $LC_{50}$ | 181 |
| Mercury | | | *H. chyrsocelis* | Embryo | 0.002 mg/L | 7d $LC_{50}$ | 190 |
| Mercury | | | *R. pipiens* | Embryo | 0.007 mg/L | 96 h $LC_{50}$ | 190 |
| Nickel | Water<br>Sediment | 0–183 μg/L<br>0.1–500 ppm | *A. opacum* | Tadpole | 0.42 mg/L | 8d $LC_{50}$ | 191 |
| Selenium | Water<br>Sediment | 0–1.4 μg/L<br>0.07–4.6 ppm | *G. carolinensis* | Embryo | 0.09 mg/L | 7d $LC_{50}$ | 181 |
| Silver | Water<br>Sediment | .08–760 μg/L<br>0–150 ppm | *R. pipiens* | Embryo-larval | 10 μg/L | 96 h $LC_{50}$ | 192 |
| Zinc | Water | .03–3.3 mg/L | *A. opacum* | Embryo | 2.38 mg/L | 8d $LC_{50}$ | 190 |
| Zinc | Sediment | 15–2800 ppm | *B. melanostictus* | Tadpole | 19.86 mg/L | 96 h $LC_{50}$ | 186 |

| | | | | | | | |
|---|---|---|---|---|---|---|---|
| Zinc | | | *G. carolinensis* | Embryo | 0.001 mg/L | 7d $LC_{50}$ | 181 |
| Zinc | | | *R. hexadactyla* | Tadpole | 2.1 mg/L | 96 h $LC_{50}$ | 182 |
| Tributyltin | Water | 0.2–60 μg/L | *R. temporaria* | Embryo | 0.028 mg/L | 5d $LC_{50}$ | 193 |
| | Sediment | 0–1.3 ppm | | | | | |
| Toxaphene | Water | 0.05 μg/L | *B. woodhousei* | Tadpole | 0.14 mg/L | 96 h $LC_{50}$ | 167 |
| Toxaphene | Sediment | .07–4.6 mg/L | *Pseudacris triserata* | Tadpole | 0.50 mg/L | 96 h $LC_{50}$ | 167 |
| Toxaphene | | | *R. sphenocephala* | Tadpole | 0.43 mg/L | 96 h $LC_{50}$ | 194 |
| Toxaphene | | | *R. sphenocephala* | Juvenile | 0.80 mg/L | 96 h $LC_{50}$ | 194 |
| Endosulfan | Water | 0–0.7 μg/L | *R. tigrina* | Tadpoles | 1.8 μg/L | 96 h $LC_{50}$ | 195 |
| Endosulfan | | | *B. melanostictus* | Tadpoles | 123 μg/L | 96 h $LC_{50}$ | 196 |
| Endosulfan | | | *R. sylvatica* | Tadpoles | 0.07–0.36 mg/L | 80–100% mortality 12 days post exposure | 197 |
| Endosulfan | | | *B. americanus* | Tadpoles | 0.04–0.25 mg/L | 60–90% mortality 12 days post exposure | 197 |
| Endosulfan | | | *R. clamitans* | Tadpoles | 0.05–0.34 mg/L | 30–80% mortality 12 days post exposure | 197 |
| Endrin | Water | 0.4 mg/L | *B. woodhousei* | Tadpole | 0.12 mg/L | 96 h $LC_{50}$ | 167 |
| Endrin | | | *P. triserata* | Tadpole | 0.18 mg/L | 96 h $LC_{50}$ | 167 |
| Endrin | | | *R. sphenocephala* | Embryo | 0.03 mg/L | 96 h $LC_{50}$ | 194 |
| Endrin | | | *R. sphenocephala* | Tadpole | 0.01 mg/L | 96 h $LC_{50}$ | 194 |
| Endrin | | | *R. sphenocephala* | Juvenile | 0.008 mg/L | 96 h $LC_{50}$ | 194 |
| No. 2 fuel oil | | | *R. sylvatica* | Tadpole | 26 mg/L | 96 h $LC_{50}$ | 198 |
| No. 2 fuel oil | | | *A. maculatum* | Larvae | 86 mg/L | 96 h $LC_{50}$ | 198 |
| Nitrate | Streams | 2–40 mg/L | *B. americanus* | Tadpole | 13.6 mg/L | 96 h $LC_{50}$ | 51 |
| Nitrate | Ponds | 1–100 mg/L | *Pseudacris triserata* | Tadpole | 17 mg/L | 96 h $LC_{50}$ | 51 |
| Nitrate | | | *R. pipiens* | Tadpole | 22.6 mg/L | 96 h $LC_{50}$ | 51 |
| Nitrate | | | *Bufo bufo* | Tadpole | 9 mg/L | 96 h $LC_{50}$ | 51 |
| Nitrate | | | *Hyla regilla* | Tadpole | 32.4 mg/L | 96 h $LC_{50}$ | 88 |

[a] For empty ambient concentration cells use the preceding concentration for that compound.

- Amphibians are exquisitely sensitive to certain contaminants including chlorpyrifos, temephos, rotenone, silver, zinc, tributyltin, TFM, and endosulfan; laboratory toxicity tests for synthetic pyrethroids had indicated that they also are very toxic, but field trials that allowed for natural attenuation of pyrethroids indicate that the pesticides are less toxic than had been believed.
- For other contaminants, sediment toxicity may be more important than water toxicity due to concentrations in sediment; but very few studies have examined sediment toxicity in amphibians.

#### 40.3.1.3 Do Criteria Developed from Fish Toxicity Data Protect Amphibians?

At present, there are no federal regulatory criteria on toxicants for amphibians. Instead, data from fish studies are often assumed to provide comprehensive coverage and safety. This implies that fish and amphibians are equally sensitive to the toxicants for which criteria have been established. Birge et al.[84] compared the toxicity of a variety of amphibians and organic compounds or metals to that of fish species commonly used in toxicity tests such as the rainbow trout (*Oncorhynchus mykiss*), fathead minnow (*Pimephales promelas*), and largemouth bass (*Micropterus salmoides*). Comparisons were standardized as much as possible including comparable life stages, water chemistry, and durations. Twenty-eight species of native amphibians from the families of Ambystomatidae, Microhylidae, Hylidae, Ranidae, and Bufonidae and the African clawed frog (*Xenopus laevis*) were included in these tests. Median lethal toxicity values for metals in amphibians varied by 100-fold in amphibians (e.g., Hg — 1.0 μg/L in *Gastrophyrne carolinensis* to 103 μg/L in *Ambystoma opacum*). In all, 50 metals and inorganics were tested, as were 13 organic compounds, for a total of 694 amphibian/fish comparisons. Their results showed that amphibians had lower $LC_{50}$ values than fishes in:

- 64% of all the tests
- 74% of the comparisons among the 15 most sensitive amphibian species and fishes
- 80% of the comparisons involving amphibians and warm-water fishes
- 66% of the 13 most metal-sensitive amphibians and rainbow trout, and
- 74% for all amphibian species vs. the fathead minnow

The researchers' overall conclusions were that there was great variation among amphibian species in their sensitivity to metal and organic contaminants, that amphibians generally were more sensitive than fish, and that water-quality criteria established for fish may not be protective of amphibians. From the perspective of population declines, these data also demonstrate that amphibian responses to contaminants may not be predictable from responses of fishes.

Bridges and Semlitsch[85] recently added another conundrum into the attempt to simplify understanding the effects of contaminants on amphibian populations. Through a well-designed study, they demonstrated that significant differences in sensitivity to contaminants such as the pesticide carbaryl can occur at all demographic levels — among closely related species, among populations of the same species (*R. sphenocephala*), and even among families within the same population. Thus, using a standard toxicity value for amphibians as a whole can be very misleading.

### 40.3.2 Sublethal Effects and Interactions between Contaminants and Other Factors

Contaminants have a host of sublethal effects that, while not overtly killing an animal, may weaken it and make it more vulnerable to other factors. Thus, they interact with the other factors that have been related to amphibian population declines. Figure 40.3 graphically illustrates this. High concentrations of a chemical induce acute mortality. Lower concentrations, if sufficiently persistent in the environment or presented as episodic events, may induce mortality but over a longer period of time, often resulting in inconspicuous die-offs. At still lower concentrations, the effects of the contaminants may present themselves in one or more sublethal mechanisms (on the

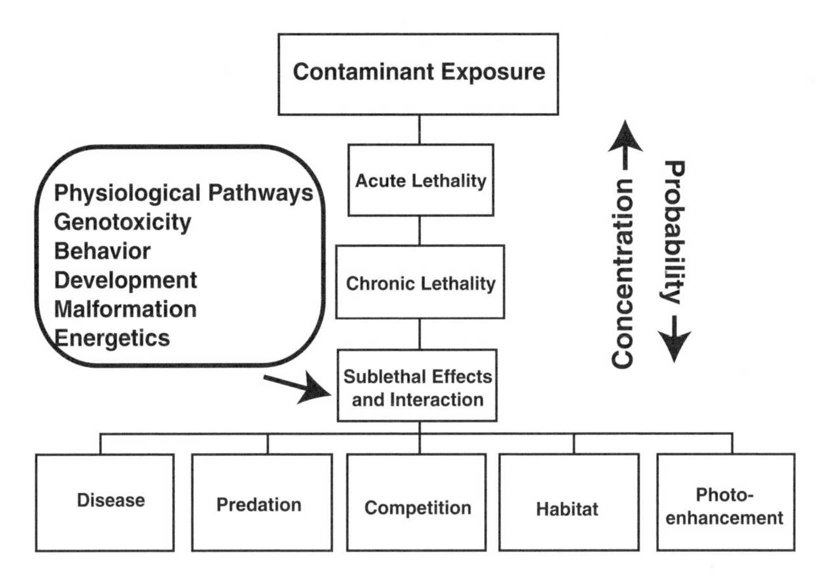


**Figure 40.3** Schematic demonstrating the decreasing direct effects of contaminants on amphibian populations and the increasing potential for interactions with other factors. At high exposure concentrations the potential for interactions is low because animals die quickly. As concentrations decrease death or ill effects take longer to manifest and the possibility of interactions with other limiting factors increases. Mechanisms through which sublethal effects and interactions take place are shown to the left of the chart.

left side of the chart) that can interact with other factors (bottom of chart) to induce population declines, often with ambiguous and confusing causes.

### *40.3.2.1 A Brief Synopsis of Sublethal Mechanisms in Amphibians*

There is not sufficient space in this chapter to discuss all the ways contaminants may affect amphibians. Many of these effects are consistent with other, better studied species or are presented in more exhaustive reviews.[15] This section briefly covers physiological effects, genetic damage, behavioral effects, problems with development including malformations and decreased energy efficiency.

#### *40.3.2.1.1 Effects on Expected Physiological Pathways*

Contaminants that have been studied in other vertebrates affect predictable physiological pathways. For example, organophosphorus and carbamate pesticides block cholinesterase and disrupt normal functioning of the nervous system, which results in loss of neural control. Overt signs of this disruption may include loss of coordination, listlessness, and convulsions. Available data show that amphibians are affected in similar ways as other vertebrates by these pesticides. Heavy metals chelate with enzymes to disrupt physiological pathways and cause histological damage in some organs;[86] low pH affects ion balance[73,87] as does nitrate and ammonia compounds;[88] aluminum binds with phosphorus and prevents its uptake, resulting in rickets.[89] In these cases, the responses of amphibians should be qualitatively similar to those of other vertebrates, especially fish. In contrast to birds and particularly mammals, however, the mixed-function oxidase system of amphibians appears to be less developed and induction of P450 responses less pronounced.[90,91] Mixed-function oxidases are involved in the metabolism of cellular toxins and the hydrolysis of some organic contaminants such as certain pesticides, PAHs, and PCBs. Because the degradates of these compounds may be more toxic than their parent form, diminished ability to metabolize the parent form may not necessarily be bad, and the responses of amphibians to these compounds may differ from that of homeotherms. In general, anurans have relatively low sensitivity to PCBs compared to other vertebrates, especially mammals.[82] In other cases, scientists can expect amphibians to respond to contaminants in ways similar to fish and birds, and existing knowledge of effects and bioindicators is transferable.

#### 40.3.2.1.2 Genotoxicity

Organic contaminants can induce chromosomal and DNA mutations, at least in somatic cells.[92] These mutations can alter cellular functioning and impair survival. Genotoxic effects in amphibians have been reported for herbicides,[93] other pesticides,[94] PCBs,[95] wastewater,[96] and PAHs.[97] The effects of some PAHs can be enhanced by 100-fold or more by irradiation by ultraviolet light [98,99] (see Section 40.3.2.2.5, Photoenhanced Toxicity).

#### 40.3.2.1.3 Effects on Behavior and Development

There are several laboratory studies that have shown a decrease in activity related to contaminants. For example, temephos, an organophosphorus insecticide, reduced the activity rate of *R. sphenocephala* tadpoles and produced distinct diphasic activity cycles in relation to dose.[100] Maintenance of equilibrium in tadpole *R. catesbeiana* was upset by the pesticide malathion at the lowest concentration tested (500 μg/L).[101] Development was also delayed in this species at 1000 μg/L. Chlorpyrifos (0.1 mg/L) reduced swimming speed and startle responses in *Hyla versicolor* when prodded; lethargy and abnormal feeding behavior were even more pronounced when low doses of atrazine (0.2 mg/L) and metolachlor (0.25 mg/L) were added to reflect a common exposure paradigm of herbicides and insecticides.[56] In a classical conditioning experiment, *R. catesbeiana* and *R. clamitans* avoidance–preference responses were impaired by exposure to 750 μg Pb/L for 120 h, indicating that tadpoles had poorer learning and retention when exposed to lead.[102] The authors also noted increased variation in locomotor activity with animals at 625 μg Pb/L or more, demonstrating spontaneous spurts of activity not seen in controls.

Statistically slower developmental times, which could be caused by many physiological and behavioral factors, have been observed in response to atrazine in *H. versicolor*,[103] tiger salamanders (*Ambystoma tigrinum*),[104] and other species. Similarly, developmental effects occur with chromium,[105] triphenyltin,[106] dioxins,[107] acidification and aluminum,[48,68] and a host of other contaminants.[15] Reduced developmental rates may expose tadpoles to drying wetland conditions and may affect survival in other ways.[108–110] Some contaminants such as the herbicide acetochlor may accelerate development, presumably by stimulating the thyroid or interacting with corticosterone.[111]

Amphibians that undergo complete metamorphosis are potentially excellent candidates to study and screen for endocrine-disrupting chemicals, especially those that affect the thyroid.[112] Thyroid stimulating hormone (TSH) and tri-iodothyronine ($T_3$) are the overall regulators of the climax stage of metamorphosis,[113] although their actions may be modified by other hormones.[114] Because thyroid action can be affected by several contaminants,[111,115] metamorphosis can be altered by exposure to these chemicals and can also serve as a useful endpoint in screening potential thyroid disruptors. Sex hormones also may be affected by contaminants in amphibians, as they are in other organisms,[116] and these effects may directly impinge on reproduction. Recently, Hayes et al.[200] showed that atrazine may be a potent endocrine disruptor; concentrations as low as 0.1 ppb significantly increase the incidence of ovitestes in *Xenopus laevis.*

#### 40.3.2.1.4 Contaminants and Amphibian Malformations

Considerable research attention has been given to amphibian malformations over the past few years because they have recently attracted public attention and become more widespread than when they were first seen by a group of students in 1995, but the problem does not appear to be a completely new phenomenon. In fact, malformations in amphibians have been described for more than 300 years.[117] Clearly, one cause of missing and supernumerary limbs and portions of limbs is parasitism from trematodes.[118] However, parasitism does not explain all the incidents of high frequencies of malformations, and considerable evidence also points to waterborne chemicals as another cause.[119–121] Although the malformations of greatest current interest primarily involve the hindlimbs, there are many different kinds of malformations including missing and supernumerary forelimbs, abnormal mouths, spinal problems, and internal abnormalities.[117] Several contaminants have been linked to malformations including chlorinated hydrocarbon pesticides,[122,123] boron,[124] atrazine,[125] ammonia,[126] coal ash effluent,[127–129] methyl-parathion and pirimicarb,[130] methoprene derivatives,[100,131] maneb,[132] several pharmaceuticals and organic contaminants,[133,134] and metals (see Reference 135 for review). Many of these compounds can cause malformations at environmentally realistic concentrations and decrease the survival of affected individuals. In a recent study, Kiesecker[201] found that environmentally realistic concentrations of malathion interacted with trematode infestations to significantly increase the incidence of malformations in native amphibians compared to trematodes alone. However, the malformations seen in the wetlands of Minnesota and elsewhere often differ in important details from the types of malformations produced in laboratories.

##### *40.3.2.1.5 Effects on Energetics*

Rowe et al.[136] found that *R. catesbeiana* tadpoles collected from wetlands contaminated by effluent from coal ash retaining ponds had standard metabolic rates that were 40–175% greater than similarly aged tadpoles collected from clean sites. Assuming that an organism's energy allocations are finite and divided among growth, storage, maintenance, and reproduction, higher metabolic rates could leave less energy for growth or reproduction. Alternatively, tadpoles would be required to spend greater time foraging and less time on other activities such as predator avoidance and resting.

Adult *R. ridibunda* that were exposed to 200 mg/L Cd in solution for 30 days showed signs of metabolic stress in the liver for the first 10 days, after which protective mechanisms, including the synthesis of metallothioneins and glutathione, were initiated; by 30 days of exposure, most liver parameters returned to normal.[137]

An indirect relationship was developed between toxicity of carbaryl and metabolism in *R. clamitans* tadpoles.[138] Because amphibians are poikliothermic, barring behavioral changes, metabolic rate is directly related to ambient temperature. Carbaryl was significantly more toxic at 27°C than at 17 or 22°C. At 27°C, lethal concentrations approached expected ambient concentrations, whereas the concentration needed to significantly increase mortality at 17°C was appreciably higher than anticipated environmental levels. Mudpuppies (*Necturus maculosus*) living in streams contaminated with a variety of organochlorines including PCBs and pesticides had lower liver glycogen concentrations than those inhabiting cleaner streams; liver glycogen can be a major energy reserve for vertebrates and is influenced by corticosterone, which, in turn, was experimentally shown to be altered by PCB exposure.[139]

#### ***40.3.2.2 Evidence for Interactions between Contaminants and Other Factors Associated with Amphibian Population Declines***

Except in very simple ecosystems, it is very rare for only one factor to restrict the growth or survival of a population. Whereas a factor may be limiting for a certain period, through the course of time many constraints may affect a population, and their effects may be interactive. Thus, contaminants may cause added stress to individuals in a population, which makes them more vulnerable to parasites or disease. Similarly, a contaminant may alter behavior to make animals more vulnerable to predation or less competitive. Stress due to sublethal effects of an herbicide may increase metabolic demands at the same time the herbicide is decreasing food supply. Therefore, the actual causes of population declines may be difficult to discern. This section examines the potential interaction of biological factors of disease, predation, and competition and nonbiological aspects of habitat composition and solar radiation with contaminants to negatively affect amphibian populations.

#### *40.3.2.2.1 Interaction with Disease Agents*

It is clear that disease is associated with amphibian population declines. Not clear, however, is why disease organisms, which presumably have been associated with amphibians for many years, should suddenly be significant in global declines of these animals. The possibility that contaminant-caused stress may reduce the immunological defenses of amphibians periodically has been raised.[17,140] Unfortunately, the relationship between contaminants and immunosuppression in poikliothermic vertebrates has been studied very little.[141] Captive, adult male Woodhouse's toads (*Bufo woodhousi*) were experimentally exposed to sublethal concentrations (0.001 and 0.011 mg/g) of commercial-grade, formulated malathion and the bacterium *Aeromonas hydrophila*, the organism responsible for red leg disease.[141] No deaths occurred over the 30-day test period among controls injected with saline or among animals injected with the low dose of malathion alone. When toads were exposed only to *A. hydrophila,* mortality rose to 20%, and at the high rate of malathion alone it was 40%. Toads that were exposed to both *A. hydrophila* and malathion had 80% mortality at the low dose and 100% at the high. Moreover, occurrence of hepatomegaly (enlarged liver) went from 40% for the high dose of malathion alone (0% for control, low dose, and *A. hydrophila* alone) to 80 and 100% when both malathion and *A. hydrophila* were presented.

#### *40.3.2.2.2 Interaction with Predation*

Because of interspecific differences in sensitivity to contaminants, predator/prey relationships within a polluted wetland may be affected due to changes in species composition, to lethargy and impaired escape responses by prey, or to other factors. For example, fish species tend to be more sensitive to acidification than many invertebrates. Thus, a reduction in the abundance and species richness of fish may open a wetland for greater densities of predacious invertebrates. While reduction in the abundance of fish might seem of benefit to amphibians, many invertebrates, such as odonates, notonectids, corixids and dytiscids, are important predators on amphibian eggs and small larvae. Henrikson[142] discussed the effects of a change from fish to large invertebrates as chief predators on amphibian eggs and early larvae and showed that the predator/prey relationships in acidified wetlands are complex. Insects tended to avoid eggs of *R. arvalis* because of a thick jelly. *Bufo bufo* eggs were avoided by insects with chewing mouth parts but not those with sucking mouth parts. Tadpole *R. arvalis* were readily depredated by insects and by the newt *Triturus vulgaris* but *B. bufo* were not because of lower palatability. Fish, on the other hand, might not be as particular in their selection of eggs.

Tadpole Columbia spotted frogs (*R. luteiventris*) that were exposed to water from an aquarium that held tadpole-eating fish sought cover more often and reduced movements compared to those exposed to water with fish that had not eaten tadpoles.[143] Reduced movements and seeking cover were considered as antipredator behaviors. When zinc (7.4 ppm), lead (0.11 ppm), or high levels of metal-laden soil taken from a Superfund site were added to the test chamber, tadpoles did not show antipredatory behavior. Simultaneous presentation of lead and zinc had greater effects on the tadpoles than either metal alone.

Coal ash, which is the by-product of coal combustion in energy production, is a common source of pollution in the United States.[144] Because all of the combustible carbon has been burned off, coal ash has high concentrations of inorganic metals and minerals. Typically, coal ash is pumped into basins where it settles, but the surface waters eventually enter natural waterways. Exposure to water coming from settling basins decreased movement and escape responses in *R. catesbeiana* tadpoles. When exposed tadpoles were experimentally housed with a snapping turtle (*Chelydra serpentina*), their survival was only 6% after 3 days compared to 97–100% among controls.

Tadpole green treefrogs (*Hyla cinerea*) showed a higher mortality rate when aluminum was added to pH 4.5 water compared to controls or aluminum at pH 5.5; growth of the tadpoles was reduced at both pH levels.[202] When dragonfly nymphs were added to the test units, they depredated at a higher rate in water with 150 μg/L aluminum and pH 4.5 than without aluminum. The authors attributed the higher rate of predation to a slower growth rate of tadpoles, which kept them at a vulnerable size longer than controls.

A recent paper[145] clearly demonstrated both the longer-term effects of pesticide exposure in *H. versicolor* tadpoles and a relationship between predation and pesticide effects. The 96-hour $LC_{50}$ of carbaryl in amphibians ranges from 2.5 to 20.6 mg/L. However, after 10 days, 0.09 mg/L carbaryl resulted in 90% mortality and 0.05 mg/L in 60% mortality in laboratory tests. Moreover, at the lower dose, the presence of a caged larval salamander (*Ambystoma maculatum*), which is a predator on tadpoles, resulted in 100% mortality of *H. versicolor* tadpoles after 8 days. Activity and growth rates were also suppressed both by the low and high concentrations of carbaryl and by the presence of a predator under the low concentration. At the low carbaryl concentration, there was an interaction between presence of predator and chemical treatment. No predator effect was observed at the high concentration of carbaryl because the pesticide alone was sufficient to reduce growth and survival.

#### *40.3.2.2.3 Interactions between Contaminants and Competition*

As with predation, differences in sensitivity to contaminants among amphibian species and between amphibians and nonamphibian competitors may alter community composition and affect species richness. Poor, relatively tolerant competitors may outperform more sensitive species under polluted conditions. For instance, *R. clamitans* tadpoles, which did not show a response to acidification or the addition of aluminum in field tests, were more abundant in ponds that had been treated with aluminum and acid than in control ponds, possibly due to release from competition with other sympatric species that showed significant decreases in relative abundance to acidification.[46]

As an example, *H. gratiosa* and *H. femoralis* are sympatric species in the southeast United States. Of the two, *H. femoralis* seems to be more acid-tolerant than *H. gratiosa*.[146] *H. gratiosa* also tends to show pronounced responses to intraspecific competition in that growth rates and age at metamorphosis are density-dependent. At circumneutral (6.0) pH, the presence of both species resulted in reduced growth and increased time to metamorphosis for *H. gratiosa*. At this pH, the survival rate of *H. gratiosa* was also lower when *H. femoralis* was present than when it was not. In general, *H. femoralis* tended to do better at low pH, where *H. gratiosa* may have been at a competitive disadvantage. Pehek[147] saw that competition between *H. andersonii* and either *R. sphenocephala* or *H. versicolor* resulted in decreased survival and body mass of *H. andersonii*, but that pH (3.9 vs. 5.9–8.4) had no effect on any of the species or on competitive effects.

Heavy-metal concentrations may alter the competitive balance between snails and tadpoles, both of which feed on detritus and algae.[148] Snails (*Lymnaea pulustris*) were not measurably affected by the presence of zinc, cadmium, or lead, but tadpole *R. lutriventris* took longer to metamorphose. In the absence of heavy metals, the presence of tadpoles resulted in decreased snail recruitment. Competition between snails and tadpoles led to shifts in feeding locations and higher heavy-metal concentrations in tadpole tissues than when competition did not occur. Interspecies interactions were even more complex when a fish predator (*Lepomis macrochirus*) was added.

#### *40.3.2.2.4 Interactions with Habitat Characteristics*

Habitat characteristics for aquatic stages of amphibians include water chemistry and presence of multiple contaminants, food, and shelter, some of which have been discussed above. For terrestrial amphibians, habitat may involve landscape features such as travel lanes, juxtaposition and interspersion of breeding ponds, forests, cropland, and other factors. The type of habitat a species occupies may provide clues to the types of contaminants it's likely to encounter. For example, amphibians inhabiting agricultural areas may be more subjected to pesticides than those inhabiting forested vernal pools. Amphibians such as *R. clamitans* or *R. catesbeiana* that breed in semipermanent or permanent ponds may be subjected to different forms of pollution than stream-breeding *Ambystoma* species or *R. sylvatica* that breed in vernal wetlands. The study of interactions among contaminant exposure, survival, and landscape features is in its infancy, but several researchers are beginning to work in this area.

Perhaps the most pertinent study of this kind to date was conducted by Bunnell and Zampella.[149] Using detrended-correspondence analysis and principal-component analysis, they found a primary gradient predicting the distribution of anurans in New Jersey that was formed by increasing pH and emergent herbaceous vegetation and decreasing specific conductance. The authors concluded that landscape patterns influence the distribution of adult anurans, whereas pond chemistry limited recruitment. Other studies have shown that movement patterns between water and terrestrial habitats are nonrandom,[41] that there are important species differences in use patterns of aquatic and terrestrial habitats,[150] and that terrestrial habitat characteristics are important factors in predicting amphibian occurrence, even for highly aquatic species.[151,152] Given that terrestrial and habitat landscape features are important to the distribution of amphibians, additional research on how contaminants interact with these habitat characteristics becomes critical.

#### *40.3.2.2.5 Photoenhanced Toxicity*

Recent findings on the interactions between ultraviolet radiation and contaminant toxicity raise very interesting and alarming possibilities. Most acute toxicity tests have occurred under laboratory conditions with artificial lighting. However, studies have shown that the toxicity of some chemicals, including pesticides and PAHs, may increase by 400 times in the presence of ultraviolet radiation, particularly UV-B.[153] Photoenhanced toxicity can occur either through photodegradation of parental forms of contaminants into more toxic degradates or through sensitizing organisms to contaminants.[154] Photodegradation can occur, for example, with several organic contaminants commonly found in petroleum such as PAHs.[155] Photoenhanced toxicity has been identified in fish, aquatic insects, and macrophytes.[156] Due to atmospheric depletion of ozone, some parts of the world are now exposed to greater ultraviolet radiation than they have been historically.[157] Thus, there could be an interaction between ultraviolet radiation and contaminant toxicity. This is particularly relevant to amphibians whose population declines are occurring in montane areas, where the strength of ultraviolet radiation is greater than at sea level.

The toxicity of at least one carbamate insecticide — carbaryl — is enhanced under UV-B exposure in frogs,[158] but most other pesticides have not been tested. Mortality and swimming activity of *H. versicolor* were reduced by exposure to either carbaryl and realistic doses of UV-B compared to controls, but when both were present, the effects increased dramatically. Without UV-B (but UV-A present) 96-hour mortality in *H. versicolor* embryos ranged from 0 to 76.7% after 4 days with carbaryl concentrations ranging from 0 to 15 mg/L. No mortality was observed with just UV-B and no carbaryl. When UV-B and carbaryl were combined, mortality soared to 93% after 2 days. Significant interactions between UV-B and carbaryl were also seen in *H. versicolor* tadpoles; for example, mortality increased from 20% after 4 days at 1.24 mg/L carbaryl alone to 93.3% after 2 days with the same amount of carbaryl and low (4 $\mu W/cm^2$) UV-B. Increased toxicity of carbaryl in *Xenopus laevis* tests appeared to be due more to photoactivation than to photosensitization in that pre-irradiated carbaryl was more toxic than nonirradiated carbaryl, and there was no difference in mortality between embryos that had been exposed to carbaryl and then irradiated and those that were not exposed to UV-B.

Larval *R. sphenocephala* were exposed to water-soluble fractions of weathered petroleum distillates collected from an abandoned oil field and then either to UV-B or lighting without UV-B.[159]

All of those that were exposed to both distillates and UV-B died within 4 days, even at 5% distillate concentration; in contrast, none of the animals died when exposed to the distillates without UV-B or with UV-B but no distillates.

The genotoxicity of PAHs, as determined by the frequency of micronucleated red blood cells, can increase by more than 100 times when UV-B is present. Mature amphibian red blood cells are nucleated and certain genotoxic chemicals result in small nuclei due to chromosomal breakage and loss. Benzo[a]pyrene, a common constituent of oil and incompletely combusted fuels, had no apparent toxicity in the newt *Pleurodeles waltl*, even at 500 ppb when UV-B was not present. When the newts were tested in daylight, however, significant increases in micronuclei frequency resulted at 500 ppb, and even at 25 ppb, with the addition of UV-A radiation. Photosensitization may have been the primary mechanism for this increased toxicity in that pre-irradiation of benzo[a]pyrene reduced the toxicity of the chemical, but animals that had been exposed to nonirradiated benzo[a]pyrene and then post-irradiated showed significant increases in toxicity compared to newts that did not receive post-irradiation. Photoenhancement of toxicity was also observed for benz[a]anthracene, 7,12-dimethyl-benz[a]anthracene and benz[a]anthraquionne.[98,99] Similarly, UV-B-irradiated fluoranthene, a common PAH in sediments, is significantly more toxic to amphibian embryos and tadpoles than fluoranthene without UV-B.[160,161]

Photoenhanced toxicity is potentially more important to amphibian population declines than direct effects of UV-B. Because most toxicity studies have occurred under laboratory conditions without the presence of UV-B and only low amounts of UV-A, published $LC_{50}$ values for chemicals that can be photoactivated or to which photosensitization occurs may be an order of magnitude or greater than lethal concentrations in the field. Pesticide concentrations in pristine areas of the Sierra Nevada,[162,163] for example, were thought to be sublethal. If photoenhanced toxicity is occurring, the measured values may be well within the actual lethal range for the amphibians that are declining there. While tadpoles can alter behavior to avoid direct sunlight,[36,38] avoidance of water contaminated with photoactivated chemicals may be more difficult. Also, some species of tadpoles have mechanisms to repair photodamaged cells[164] but might not have the ability to recover from chemical exposure.

## 40.4 CONCLUSIONS

Although there are no published studies that demonstrate beyond all doubt that contaminants are involved in long-term population declines of amphibians, there is ample evidence and reason to encourage active research and concern about effects. Many contaminants are lethal to amphibians at environmentally realistic concentrations. Acute mortality from these compounds may be difficult to detect because investigators would have to be present shortly after exposures. Chronic mortality may be masked by metapopulation phenomena, so that areas that serve as population sinks may be repeatedly recolonized and difficult to identify. Metapopulation dynamics also make it more difficult to define discrete populations. Contaminants also have many sublethal effects on behavior, energetics, malformations, and diverse effects on physiological pathways which by themselves might not lead to overt death but could alter reproduction or interact with other factors to result in gradual declines in populations. Scientific understanding of these interactions, and of the ecotoxicology of amphibians in general, is far behind what is known about birds, fish, and mammals, and research is desperately needed in this area. Some specific suggestions for critically needed research include:

- Determination of lethal concentrations of common contaminants — pesticides, PAHs, metals — under environmentally realistic conditions of light, temperature, and water chemistry
- Better understanding of the effects of long-term (weeks, months), low-concentration exposure of persistent pesticides and stable contaminants on amphibians

- Development and refinement of bioindicators in amphibians to use in monitoring and screening for potential effects of contaminants in declining amphibian populations
- Further studies on the interaction between contaminants and disease agents including immunosuppression in amphibians
- Additional research on the interaction between ultraviolet radiation — both UV-A and UV-B — and a broader range of contaminants
- Development of models to assess risk in amphibians as related to landscape factors and especially in the aquatic–terrestrial interface
- Population survival requires that organisms survive through their entire life cycle; to date, there have been no studies examining the responses of amphibians to contaminants in all phases (egg, embryo, larvae and adult) of their life cycles.

## REFERENCES


1. Blaustein, A. R., Chicken little or Nero's fiddling? A perspective on declining amphibian populations, *Herpetologica,* 50, 85, 1994.
2. Wake, D. B., Declining amphibian populations, *Science,* 253, 860, 1991.
3. Houlahan, J. E., Findlay, C. S., Schmidt, B. R., Meyers, A. H., and Kuzmin, S. L., Quantitative evidence for global amphibian population declines, *Nature,* 404, 752, 2000.
4. Alford, R. A. and Richards, S. J., Global amphibian declines: A problem in applied ecology, *Annu. Rev. Ecol. Syst.,* 30, 133, 1999.
5. Drost, C. A. and Fellers, G. M., Collapse of a regional frog fauna in the Yosemite area of the California Sierra Nevada, USA, *Conserv. Biol.,* 10, 414, 1996.
6. Fellers, G. M. and Drost, C. A., Disappearance of the Cascades frog *Rana cascadae* at the southern end of its range, California, USA, *Biol. Conserv.,* 65, 177, 1993.
7. Corn, P. S., Amphibian declines: A review of some current hypotheses, in *Ecotoxicology of Amphibians and Reptiles*, Sparling, D. W., Linder, G., and Bishop, C. A., Eds., SETC Press, Pensacola, FL, 2000. 663.
8. Beebee, T. J. C., Changes in dewpond numbers and amphibian diversity over 20 years on chalk downland in Sussex, England, *Biol. Conserv.,* 81, 215, 2001.
9. Beebee, T. J. C., Flower, R. J., Stevenson, A. C., Patrick, S. T., Appleby, P. G., Fletcher, C., Marsh, C., Natkanski, J., Rippey, B., and Batterbee, R. W., Decline of natterjack toad, *Bufo calamita*, in Britain: Palaeoecological, documentary and experimental evidence for breeding site acidification, *Biol. Conserv.,* 53, 20, 1990.
10. Laurance, W. F., McDonald, K. R., and Speare, R., Epidemic disease and the catastrophic decline of Australian rain forest frogs, *Conserv. Biol.,* 10, 406, 2001.
11. Lips, K. R., Decline of a tropical montane amphibian fauna, *Conserv. Biol.,* 12, 106, 1998.
12. Crump, M. L., Hensley, F. R., and Clark, K. L., Apparent decline of the golden toad: Underground or extinct?, *Copeia,* 1992, 413, 1992.
13. Vial, J. L. and Saylor, L., The status of amphibian populations: A compilation and analysis, Working Document No. 1, IUCN/SSC Declining Amphibian Populations Taskforce, 1993.
14. Bury, R. B., Corn, P. S., Dodd, C. K., McDiramid, R. W., and Scott, N. J., Amphibians, in Our Living Resources, LaRoe, E. T., Farris, G. S., Puckett, C. E., Doran, P. D., and Mac, M. J., Eds., United States Department of the Interior National Biological Service, Washington, D.C., 1995, 124.
15. Sparling, D. W., Linder, G., and Bishop, C. A., Eds., *Ecotoxicology of Amphibians and Reptiles*, SETAC, Pensacola, FL, 2000.
16. Pechmann, J. H. K. and Wilbur, H. M., Putting declining amphibian populations in perspective: Natural fluctuations and human impacts, *Herpetologica,* 50, 65, 1994.
17. Carey, C., Cohen, N., and Rollins-Smith, L., Amphibian declines: An immunological perspective, *Develop. Comp. Immunol.,* 23, 459, 1999.
18. Carey, C., Infectious disease and worldwide declines of amphibian populations, with comments on emerging diseases in coral reef organisms and in humans, *Environ. Health Perspect.,* 108, 143, 2000.
19. Crawshaw, G. J., Diseases and pathology of amphibians and reptiles, in *Ecotoxicology of Amphibians and Reptiles*, Sparling, D. W., Linder, G., and Bishop, C. A., Eds., SETAC, Pensacola, FL, 199, 2000.

20. Berger, L., Speare, R., Daszak, P., Green, D. E., Cunningham, A. A., Goggin, C. L., Slocombe, R., Ragan, M. A., Hyatt, A. D., and McDonald, K. R., Chytridiomycosis causes amphibian mortality associated with population declines in the rain forests of Australia and Central America, *Proc. Natl. Acad. Sci.*, 95, 9031, 1998.
21. Taylor, S. K., Williams, E. S., Thorne, E. T., Mills, K. W., Withers, D. I., and Pier, A. C., Causes of mortality of the toad, *J. Wildl. Dis.*, 35, 49, 1999.
22. Bradford, D. F., Allopatric distributions of native frogs and introduced fishes in high Sierra Nevada lakes of California: Implications of the negative impact of fish introductions, *Copeia,* 1989, 775, 1989.
23. Bradford, D. F., Tabatabai, F., and Graber, D. M., Isolation of remaining populations of the native frog, *Rana muscosa*, by introduced fishes in Sequoia and Kings Canyon National Parks, California, *Conserv. Biol.*, 7, 882, 1993.
24. Funk, W. C. and Dunlap, W. W., Colonization of high-elevation lakes by long-toed salamanders (*Ambystoma macrodactylum*) after the extinction of introduced trout populations, *Can. J. Zool.*, 77, 1759, 1999.
25. Hayes, M. P. and Jennings, M. R., Decline of ranid frog species in western North America: Are bullfrogs (*Rana catesbeiana*) responsible?, *J. Herpetol.*, 20, 490, 1986.
26. Adams, M. J., Correlated factors in amphibian decline: Exotic species and habitat change in Western Washington, *J. Wildl. Manage.*, 63, 1162, 1999.
27. Lawler, S. P., Dritz, D., Strange, T., and Holyoak, M., Effects of introduced mosquitofish and bullfrogs on the threatened California red-legged frog, *Conserv. Biol.*, 13, 613, 1999.
28. Brana, F., Frechilla, L., and Oriazola, G., Effect of introduced fish on amphibian assemblages in mountain lakes of northern Spain, *Herpetol. J.*, 6, 145, 1996.
29. Ankley, G. T., Tietge, J. E., DeFoe, D. L., Jensen, K. M., Holcombe, G. W., Durhan, E. J., and Diamond, S. A., Effects of ultraviolet light and methoprene on survival and development of *Rana pipiens*, *Environ. Toxicol. Chem.*, 17, 2530, 1998.
30. Ankley, G. T., Tietge, J. E., Holcombe, G. W., DeFoe, D. L., Diamond, S. A., Jensen, K. M., and Degitz, S. J., Effects of laboratory ultraviolet radiation and natural sunlight on survival and development of *Rana pipiens*, *Can. J. Zool.*, 78, 1092, 2000.
31. Meteyer, C. U., Loeffler, I. K., Fallon, J. F., Converse, K. A., Green, E., Helgen, J. C., Kersten, S., Levey, R., Eaton-Poole, L., and Burkhart, J. G., Hind limb malformations in free-living northern leopard frogs (*Rana pipiens*) from Maine, Minnesota, and Vermont suggest multiple etiologies, *Teratology,* 62, 151, 2000.
32. Grant, K. P. and Licht, L. E., Effects of ultraviolet radiation on life-history stages of anurans from Ontario, Canada, *Can. J. Zool.*, 73, 2292, 1995.
33. Starnes, S. M., Kennedy, C. A., and Petranka, J. W., Sensitivity of embryos of southern Appalachian amphibians to ambient solar UV-B radiation, *Conserv. Biol.*, 14, 277, 2000.
34. Corn, P. S., Effects of ultraviolet radiation on boreal toads in Colorado, *Ecol. Appl.*, 8, 18, 1998.
35. Ovaska, K., Davis, T. M., and Flamarique, I. N., Hatching success and larval survival of the frogs *Hyla regilla* and *Rana aurora* under ambient and artificially enhanced solar ultraviolet radiation, *Can. J. Zool.*, 75, 1081, 1997.
36. Van de Mortel, T. F. and Buttemier, W. A., Avoidance of ultraviolet-B radiation in frogs and tadpoles of the species *Litoria aurea, L. dentata,* and *L. peronii.*, *Proc. Linn. Soc., N.S. W.,* 119, 173, 1998.
37. Van de Mortel, T. F., Buttemer, W., Hoffman, P., Hays, J., and Blaustein, A., A comparison of photolyase activity in three Australian tree frogs, *Oecologia,* 115, 366, 1998.
38. Blaustein, A., Hoffman, P. D., and Kiesecker, J. M., DNA repair activity and resistance to solar UV-B radiation in eggs of the red-legged frog, *Conserv. Biol.,* 10, 1398, 1996.
39. Bury, R. B., A historical perspective and critique of the declining amphibian crisis, *Wildl. Soc. Bull.*, 27, 1064, 1999.
40. Carey, C. and Bryant, C. J., Possible interrelations among environmental toxicants, amphibian development, and decline of amphibian populations, *Environ. Health Perspect.,* 103, 13, 1995.
41. Dodd, C. K., Jr. and Cade, B. S., Movement patterns and the conservation of amphibians breeding in small, temporary wetlands, *Conserv. Biol.,* 12, 331, 1998.
42. Rowe, C. L. and Dunson, W. A., Relationships among abiotic parameters and breeding effort by three amphibians in temporary wetlands of central Pennsylvania, *Wetlands,* 13, 237, 1993.

43. Hecnar, S. J. and M'Closkey, R. T., Amphibian species richness and distribution in relation to pond water chemistry in south-western Ontario, Canada, *Freshwater Biol.,* 36, 7, 1996.
44. Beattie, R. C. and Tyler-Jones, R., The effects of low pH and aluminum on breeding success in the frog *Rana temporaria*, *Herpetology,* 26, 353, 1992.
45. Horne, M. T. and Dunson, W. A., The interactive effects of low pH, toxic metals, and DOC on a simulated temporary pond community, *Environ. Pollut.,* 89, 155, 1995.
46. Sparling, D. W., Lowe, T. P., Day, D., and Dolan, K., Responses of amphibian populations to water and soil factors in experimentally treated aquatic macrocosms, *Arch. Environ. Contam. Toxicol.,* 29, 455, 1995.
47. Wyman, R. L. and Hawksley-Lescault, D. S., Soil acidity affects distribution, behavior, and physiology of the salamander *Plethodon cinereus*, *Ecology,* 68, 1819, 1987.
48. Rowe, C. L. and Freda, J., Effects of acidification on amphibians at multiple levels of biological organization, in *Ecotoxicology of Amphibians and Reptiles*, Sparling, D. W., Linder, G., and Bishop, C. A., Eds., SETAC, Pensacola, FL, 545, 2000.
49. Pollman, C. D. and Canfield, D. E., Florida, in *Acidic Deposition and Aquatic Ecosystems*, Charles, D. F., Ed., Springer-Verlag, New York, 367, 1991.
50. Bradford, D. F., Gordon, M. S., Johnson, D. F., Andrews, R. D., and Jennings, W. B., Acidic deposition as an unlikely cause for amphibian population declines in the Sierra Nevada, California, *Biol. Conserv.,* 69, 155, 1994.
51. Rouse, J. D., Bishop, C. A., and Struger, J., Nitrogen pollution: An assessment of its threat to amphibian survival, *Environ. Health Perspect.,* 107, 799, 1999.
52. Oldham, R. S., Latham, D. M., Hilton-Brown, D., Towns, M., Cooke, A. S., and Burn, A., The effect of ammonium nitrate fertiliser on frog (*Rana temporaria*) survival, *Ag. Ecosyst. Environ.,* 61, 69, 1997.
53. Berger, L., Disappearance of amphibian larvae in the agricultural landscape, *Ecol. Int. Bull.,* 17, 65, 1989.
54. Bishop, C. A., Mahnony, N. A., Struger, J., Ng, P., and Pettit, K. E., Anuran development, density and diversity in relation to agricultural activity in the Holland River watershed, Ontario, Canada (1990–1992), *Environ. Monitor. Assess.,* 59, 21, 1999.
55. Moulton, C. A., Assessing Effects of Pesticides on Frog Population, Ph.D. dissertation, North Carolina State University, Raleigh, 1996.
56. Mazanti, L. E., The effects of atrazine, metolachlor and chlorpyrifos on the growth and survival of larval frogs under laboratory and field conditions, Ph.D. Dissertation, University of Maryland, College Park, 1999.
57. McAlpine, D. F., Burgess, N. M., and Busby, D. G., Densities of mink frogs, *Rana septentrionalis,* in New Brunswick forest ponds sprayed with the insecticide fenitrothion, *Bull. Environ. Contam. Toxicol.,* 60, 30, 1998.
58. Kirk, J. J., Western spotted frog (*Rana pretiosa*) mortality following forest spraying of DDT, *Herpetol. Rev.,* 19, 51, 1988.
59. Hazelwood, E., Frog pond contaminated, *Br. J. Herpetol.,* 4, 177, 1970.
60. Mulla, M. S., Toxicity of new organochlorine insecticides to mosquito fish and some other aquatic organisms, *Mosquito News,* 23, 299, 1966.
61. Sparling, D. W., Acidic deposition: A review of biological effects, in *Handbook of Ecotoxicology,* Hoffman, D. J., Rattner, R. A., Burton, G. A., Jr., and Cairns, J., Jr., Eds., Lewis Publishers, Boca Raton, FL, 301, 1994.
62. Baker, J. P., Bernard, D. P., Christensen, S. W., Sale, M. J., Freda, J., Heltcher, K., Marmorek, D., Rowe, L., Scanlon, P., and Suter, G., Biological effects of changes in surface water acid-base chemistry, in National Acid Precipitation Assessment Program, Acidic Deposition: State of Science and Technology, Vol. II., Report 13, NAPAP, Government Printing Office, Washington, D.C., 1990.
63. Pierce, B. A., The effects of acid precipitation on amphibians, *Ecotoxicology,* 2, 65, 1996.
64. Freda, J., Sadinski, W. J., and Dunson, W. A., Long term monitoring of amphibian populations with respect to the effects of acidic deposition, *Water Air Soil Pollut.,* 55, 445, 1991.
65. Freda, J. and Dunson, W. A., The effect of acidic precipitation on amphibian breeding in temporary ponds in Pennsylvania, U.S. Fish Wildlife Service EELUT, Washington, D.C., 1985.
66. Saber, P. A. and Dunson, W. A., Toxicity of bog water to embryonic and larval anuran amphibians, *J. Exp. Biol.,* 204, 33, 1978.

67. Freda, J., Cavdek, V., and McDonald, D. G., Role of organic complexation in the toxicity of aluminum to *Rana pipiens* embryos and *Bufo americanus* tadpoles, *Can. J. Fish. Aquat. Sci.,* 47, 217, 1989.
68. Freda, J., The effects of aluminum and other metals on amphibians, *Environ. Pollut.,* 71, 305, 2001.
69. Pierce, B. A., Hoskins, J. B., and Epstein, B., Acid tolerance in Connecticut wood frogs (*Rana sylvatica*), *J. Herpetol.,* 18, 159, 1984.
70. Andren, C., Henrikson, L., Olsson, M., and Nilson, G., Effects of pH and aluminium on embryonic and early larval stages of Swedish brown frogs *Rana arvalis, R. temporaria* and *R. dalmatina, Holoarctic Ecol.,* 11, 127, 1988.
71. Whiteman, H. H., Howard, R. D., and Whitten, K. A., Effects of pH on embryo tolerance and adult behavior in the tiger salamander, *Ambystoma tigrinum tigrinum, Can. J. Zool.,* 73, 1529, 1995.
72. Rowe, C. L., Sadinski, W. J., and Dunson, W. A., Predation on larval and embryonic amphibians by acid tolerant caddisfly larvae (*Ptilostomis postica*), *J. Herpetol.,* 28, 356, 1994.
73. Clark, K. L. and LaZerte, B. D., Intraspecific variation in hydrogen ion and aluminum toxicity in *Bufo americanus* and *Ambystoma maculatum, Can. J. Fish. Aquat. Sci.,* 44, 1622, 1987.
74. Clark, K. L., Responses of spotted salamander, *Ambystoma maculatum* populations in Central Ontario to habitat acidity, *Can. Field-Nat.,* 100, 463, 1986.
75. Sadinski, W. J. and Dunson, W. A., A multilevel study of effects of low pH on amphibians of temporary ponds, *J. Herpetol.,* 26, 413, 1992.
76. Rowe, C. L., Sadinski, W. J., and Dunson, W. A., Effects of acute and chronic acidification on three larval amphibians that breed in temporary ponds, *Arch. Environ. Contam. Toxicol.,* 23, 339, 1992.
77. Andren, C., Marden, M., and Nilson, G., Tolerance to low pH in a population of moor frogs, *Rana arvalis,* from an acid and a neutral environment: A possible case of rapid evolutionary response to acidification, *Oikos,* 56, 215, 1988.
78. Eisler, R., *Handbook of Chemical Risk Assessment: Health Hazards to Humans, Plants, and Animals*, Lewis Publishers, Boca Raton, FL, 2000.
79. American Crop Protection Association, *Crop Protection Reference*, 15th ed., C and P Press, New York, 1999.
80. Linder, G. and Grillitsch, B., Ecotoxicology of metals, in *Ecotoxicology of Amphibians and Reptiles*, Sparling, D. W., Linder, G., and Bishop, C. A., Eds., SETAC, Pensacola, FL, 325, 2000.
81. Cowman, D. F. and Mazanti, L. E., Ecotoxicology of "new generation" pesticides to amphibians, in *Ecotoxicology of Amphibians and Reptiles*, Sparling, D. W., Linder, G., and Bishop, C. A., Eds., SETAC, Pensacola, FL, 233, 2000.
82. Sparling, D. W., Ecotoxicology of organic contaminants to amphibians, in *Ecotoxicology of Amphibians and Reptiles*, Sparling, D. W., Linder, G., and Bishop, C. A., Eds., SETAC, Pensacola, FL, 461, 2000.
83. Roe, D., Pease, W., Florini, K., and Silbergeld, E., *Toxic Ignorance*, Environmental Defense Fund, New York, 1997.
84. Birge, W. J., Westerman, A. G., and Spromberg, J. A., Comparative toxicology and risk assessment of amphibians, in *Ecotoxicology of Amphibians and Reptiles*, Sparling, D. W., Linder, G., and Bishop, C. A., Eds., SETAC, Pensacola, FL, 727, 2000.
85. Bridges, C. M. and Semlitsch, R. D., Variation in pesticide tolerance of tadpoles among and within species of Ranidae may explain patterns of amphibian declines, *Arch. Environ Contam Toxicol.*, in Press.
86. Domingo, J. L., Metal-induced developmental toxicity in mammals: A review, *J. Toxicol. Environ. Health,* 42, 123, 1994.
87. Freda, J. and McDonald, D. G., Effects of aluminum on the leopard frog, *Rana pipiens*: Life stage comparisons and aluminum uptake, *Can. J. Fish. Aquat. Sci.,* 47, 210, 1990.
88. Schuytema, G. S. and Nebeker, A. V., Comparative effects of ammonium and nitrate compounds on Pacific treefrog and African clawed frog embryos, *Arch. Environ. Contam. Toxicol.,* 36, 200, 1999.
89. Sparling, D. W. and Lowe, T. P., Environmental hazards of aluminum to plants, invertebrates, fish, and wildlife, *Rev. Environ. Contam. Toxicol.,* 145, 1, 1996.
90. Walker, C. H. and Ronis, M. J. J., The monooxygenases of birds, reptiles and amphibians, *Xenobiotica,* 1111, 1989.
91. Schwen, R. J. and Mannering, G. J., Hepatic cytochrome P-450-dependent monooxygenase systems of the trout, frog and snake. III. Induction, *Comp. Biochem. Physiol.,* 71, 445, 1982.

92. Ferrier, V., Gauthier, L., Zoll-Moreux, C., and l'Haridon, J., Genotoxicity tests in amphibians — A review, in *Microscale Testing in Aquatic Toxicology: Advances, Technologies and Practice*, Wells, P. G., Lee, K., Blaise, C., and Gauthier, J., Eds., CRC Press, Boca Raton, FL, 1999.
93. Clements, C., Ralph, S., and Petras, M., Genotoxicity of select herbicides in *Rana catesbeiana* tadpoles using the alkaline single-cell gel DNA electrophoresis (Comet) assay, *Environ. Mol. Mutagen.,* 29, 277, 1997.
94. Harris, M. L., Bishop, C. A., Struger, J., van den Heuvel, M., van der Kraak, G. J., Dixon, G., Ripley, B., and Bogart, J. P., The functional integrity of northern leopard frog (*Rana pipiens*) and green frog (*Rana clamitans*) populations in orchard wetlands. I. Genetics, physiology and biochemistry of breeding adults and young-of-the-year, *Environ. Toxicol. Chem.,* 7, 1338, 1998.
95. Fernandez, M., Gauthier, L., and Jaylet, A., Use of newt larvae for *in vivo* genotoxicity testing of water: Results on 19 compounds evaluated by the micronucleus test, *Mutagenesis,* 4, 17, 1989.
96. Gauthier, L., van der Gaag, M. A., l'Haridon, J., Ferrier, V., and Fernandez, M., *In vivo* detection of waste water and industrial effluent genotoxicity: Use of the newt micronucleus test (Jaylet test), *Sci. Total Environ.,* 138, 249, 1993.
97. Djomo, J. E., Ferrier, V., Gauthier, L., Zoll-Moreux, C., and Marty, J., Amphibian micronucleus test *in vivo*: Evaluation of the genotoxicity of some major polycyclic aromatic hydrocarbons found in crude oil, *Mutagenesis,* 10, 223, 1995.
98. Fernandez, M. and l'Haridon, J., Effects of light on the cytotoxicity and genotoxicity of benzo(a)pyrene and an oil refinery effluent in the newt, *Environ. Mol. Mutagen.,* 24, 124, 1994.
99. Fernandez, M. and l'Haridon, J., Influence of lighting conditions on toxicity and genotoxicity of various PAH in the newt *in vivo, Mut. Res.,* 298, 31, 1992.
100. Sparling, D. W., Effects of Altosid and Abate-4E on deformities and survival in southern leopard frogs under semi-natural conditions, *J. Iowa Acad. Sci.,* 107, 90, 2000.
101. Fordham, C. L., Tessari, J. D., Ramsdell, H. S., and Keefe, T. J., Effects of malathion on survival, growth, development, and equilibrium posture of bullfrog tadpoles (*Rana catesbeiana*), *Environ. Toxicol. Chem.,* 20, 179, 2001.
102. Steele, C. W., Strickler-Shaw, S., and Taylor, D. H., Effects of sublethal lead exposure on the behaviours of green frog (*Rana clamitans*), bullfrog (*Rana catesbeiana*) and American toad (*Bufo americanus*) tadpoles, *Mar. Fresh Behav. Physiol.,* 32, 1, 1999.
103. Diana, S. G., Resetartits, W. J., Schaeffer, D. J., Beckman, K. B., and Beasley, V. R., Effects of atrazine on amphibian growth and survival in artificial aquatic communities, *Environ. Toxicol. Chem.,* 19, 2961, 2000.
104. Larson, D. L., McDonald, S., Fivizzani, A. J., Newton, W. E., and Hamilton, S. J., Effects of the herbicide atrazine on *Ambystoma tigrinum* metamorphosis: Duration, larval growth and hormonal response, *Physiol. Zool.,* 71, 671, 1998.
105. Anusuya, D. and Christy, L., Effects of chromium toxicity on hatching and development of tadpoles of *Bufo melanostictus, J. Environ. Biol.,* 20, 321, 1999.
106. Fioramonti, E., Semlitsch, R. D., Reyer, H.-U., and Fent, K., Effects of triphenyltin and pH on the growth and development of *Rana lessonae* and *Rana esculenta* tadpoles, *Arch. Environ. Contam. Toxicol.,* 16, 1940, 1997.
107. Jung, R. E. and Walker, M. K., Effects of 2,3,7,8-tetrachlorobenzo-*p*-dioxin (TCDD) on development of anuran amphibians, *Environ. Toxicol. Chem.,* 16, 230, 1998.
108. Semlitsch, R. D., Scott, D. E., and Pechmann, J. H. K., Time and size at metamorphosis related to adult fitness in *Ambystoma talpoideum, Ecology,* 69, 184, 1988.
109. Smith, D. C., Adult recruitment in chorus frogs: Effects of size and date at metamorphosis, *Ecology,* 68, 344, 1987.
110. Beck, C. W. and Congdon, J. D., Effects of age and size at metamorphosis on performance and metabolic rates of southern toad, *Bufo terrestris*, metamorphs, *Func. Ecol.,* 14, 32, 2000.
111. Check, A. O., Ide, C. F., Bollinger, J. E., Rider, C. V., and McLachlan, J. A., Alteration of leopard frog (*Rana pipiens*) metamorphosis by the herbicide acetochlor, *Arch. Environ. Contam. Toxicol.,* 37, 70, 1999.
112. Tata, J. R., Amphibian metamorphosis as a model for studying the developmental actions of thyroid hormone, *Ann. D'Endocrinol.,* 59, 433, 1998.

113. Wolffe, A. P. and Shi, Y-B., A hypothesis for the transcriptional control of amphibian metamorphosis by the thyroid hormone receptor, *Am. Zool.*, 39, 807, 1999.
114. Rosenkilde, P., The role of hormones in the regulation of amphibian metamorphosis, *in Metamorphosis, 8th Symp. Br. Soc. Dev. Biol.*, Balls, M. and Bounes, M., Eds., Clarendon Press, Oxford, 222, 1985.
115. Miranda, L. A., Pisano, A., and Paz, D., Effect of potassium perchlorate on thyroid activity of *Bufo arenarum* larvae, *Comm. Biol.*, 10, 125, 1992.
116. Kloas, W., Lutz, I., and Einspanier, R., Amphibians as a model to study endocrine disruptors: II. Estrogenic activity of environmental chemicals *in vitro* and *in vivo*, *Sci. Total Environ.*, 225, 59, 1999.
117. Ouellet, M., Amphibian deformities: Current state of knowledge, in *Ecotoxicology of Amphibians and Reptiles*, Sparling, D. W., Linder, G., and Bishop, C. A., Eds., SETAC, Pensacola, FL, 617, 2000.
118. Johnson, P. T. J., Lunde, K. B., Ritchie, E. G., and Launer, A. E., The effect of trematode infection on amphibian limb development and survivorship, *Science*, 284, 802, 1999.
119. Fort, D. J., Rogers, R. L., Copely, H. F., Bruning, L. A., Stover, E. L., Helgen, J. C., and Burkhart, J. G., Progress toward identifying causes of maldevelopment induced in *Xenopus* by pond water and sediment extracts from Minnesota, USA, *Environ. Toxicol. Chem.*, 18, 2316, 1999.
120. Ouellet, M., Bonin, J., Rodrigue, J., and DesGranges, J. L., Hindlimb deformities (ectromelia, ectrodactyly) in free-living anurans from agricultural habitats, *J. Wildl. Dis.*, 33, 95, 1997.
121. Bonin, J., Ouellet, M., Rodrigue, J., DesGranges, J.-L., Gagne, F., Sharbel, T. F., and Lowcock, L. A., Measuring the health of frogs in agricultural habitats subjected to pesticides, in *Amphibians in Decline: Canadian Studies of a Global Problem*, Green, D. M., Ed., Society of Amphibians and Reptiles, Saint Louis, 246, 1997.
122. Cooke, A. S., The effects of DDT, dieldrin and 2,4-D on amphibian spawn and tadpoles, *Environ. Pollut.*, 3, 51, 1972.
123. Osborn, D., Cooke, A. S., and Freestone, S., Histology of a teratogenic effect of DDT on *Rana temporaria* tadpoles, *Environ. Pollut.*, 25, 305, 1981.
124. Laposata, M. M. and Dunson, W. A., Effects of boron and nitrate on hatching success of amphibian eggs, *Arch. Environ. Contam. Toxicol.*, 35, 615, 1998.
125. Britson, C. A. and Threlkeld, S. T., Abundance, metamorphosis, developmental, and behavioral abnormalities in *Hyla chyrsoscelis* tadpoles following exposure to three agrichemicals and methyl mercury in outdoor mesocosms, *Bull. Environ. Contam. Toxicol.*, 61, 154, 1998.
126. Jofre, M. B. and Karasov, W. H., Direct effects of ammonia on three species of North American anuran amphibians, *Environ. Toxicol. Chem.*, 18, 1806, 1999.
127. Rowe, C. L., Kinney, O. M., and Congdon, J. D., Oral deformities in tadpoles of the bullfrog (*Rana catesbeiana*) caused by conditions in a polluted habitat, *Copeia*, 1998, 244, 1998.
128. Rowe, C. L., Kinney, O. M., Fiori, A. P., and Congdon, J. D., Oral deformities in tadpoles (*Rana catesbeiana*) associated with coal ash deposition: Effects on grazing ability and growth, *Freshwater Biol.*, 36, 723, 1996.
129. Hopkins, W. A., Congdon, J., and Ray, J. K., Incidence and impact of axial malformations in larval bullfrogs (*Rana catesbeiana*) developing in sites polluted by a coal-burning power plant, *Environ. Toxicol. Chem.*, 19, 862, 2000.
130. Alvarez, R., Honrubia, M. P., and Herráez, M. P., Skeletal malformations induced by the insecticides ZZ-Aphox® and Folidol during larval development of *Rana perezei*, *Arch. Environ. Contam. Toxicol.*, 28, 349, 1995.
131. La Clair, J. J., Bantle, J. A., and Dumont, J., Photoproducts and metabolites of a common insect growth regulator produce developmental deformities in *Xenopus*, *Environ. Sci. Technol.*, 32, 1453, 1998.
132. Zavanella, T., Pacces-Zaffaroni, N., and Arias, E., Abnormal limb regeneration in adult newts exposed to the fungicide Maneb 80: A histological study, *J. Toxicol. Environ. Health*, 13, 753, 1984.
133. Dawson, D. A., Joint action of benzoic hydrazide and β-aminopropionitrile on *Xenopus* embryo development, *Toxicology*, 81, 123, 1993.
134. Schultz, T. W., Dumont, J. N., and Epler, R. G., The embryotoxic and osteolathyrogenic effects of semicarbazide, *Toxicology*, 36, 183, 1985.
135. Harfenist, A., Power, T., Clark, K. L., and Peakall, D. B., A review and evaluation of the amphibian toxicological literature, in Technical Report Series, 61, Canadian Wildlife Service, Ottawa, ON, 1989.

136. Rowe, C. L., Kinney, O. M., Nagle, R. D., and Congdon, J. D., Elevated maintenance costs in an anuran (*Rana catesbeiana*) exposed to a mixture of trace elements during the embryonic and early larval periods, *Physiol. Zool.*, 71, 27, 1998.
137. Vogiatzis, A. K. and Loumbourdis, N. S., A study of glycogen, lactate, total fats, protein, and glucose concentration in the liver of the frog, *Rana ridibunda,* after exposure to cadmium for 30 days, *Environ. Pollut.,* 104, 335, 1999.
138. Boone, M. D. and Bridges, C. M., The effect of temperature on the potency of carbaryl for survival of tadpoles of the green frog (*Rana clamitans*), *Environ. Toxicol. Chem.,* 18, 1482, 1999.
139. Gendron, A. D., Bishop, C. A., Fortin, T. S., and Greenberg, B. M., *In vivo* testing of the functional integrity of the corticosterone-producing axis in mudpuppy (*Amphibia*) exposed to chlorinated hydrocarbons in the wild, *Environ. Toxicol. Chem.*, 16, 1694, 1997.
140. Daszak, P., Berger, L., Cunningham, A. A., Hyatt, A. D., Green, D. E., and Speare, R., Emerging infectious diseases and amphibian population declines, *Emerging Infect. Dis. J.,* Centers for Disease Control and Prevention, Atlanta, 1999.
141. Taylor, S. K., Williams, E. S., and Mills, K. W., Effects of malathion on disease susceptibility in Woodhouse's toads, *J. Wildl. Dis.*, 35, 536, 1999.
142. Henrikson, B. I., Predation on amphibian eggs and tadpoles by common predators in acidified lakes, *Holoarctic Ecol.,* 13, 201, 1990.
143. Lefcort, H., Meguire, R. A., Wilson, L. H., and Ettinger, W. F., Heavy metals alter the survival, growth, metamorphosis and antipredatory behavior of Columbia spotted frog (*Rana luteiventris*) tadpoles, *Arch. Environ. Contam. Toxicol.,* 35, 447, 1998.
144. Raimondo, S. M., Rowe, C. L., and Congdon, J. D., Exposure to coal ash impacts swimming performance and predator avoidance in larval bullfrogs (*Rana catesbeiana*), *J. Herpetol.,* 32, 289, 1998.
145. Relyea, R. A. and Mills, N., Predator-induced stress makes the pesticide carbaryl more deadly to gray treefrog tadpoles, *Proc. Natl. Acad. Sci.,* www.pnas.org/cgi/doi/10.1073/pnas.o31076198, 1, 2001.
146. Warner, S. C., Travis, J., and Dunson, W. A., Effect of pH on variation on interspecific competition between two species of tadpoles, *Ecology,* 74, 183, 1993.
147. Pehek, E. L., Competition, pH, and the ecology of larval Hyla andersonii, *Ecology,* 76, 1786, 1995.
148. Lefcort, H., Thomson, S. M., Cowles, E. E., Harowicz, H. L., Livaudais, B. M., Roberts, W. E., and Ettinger, W. F., Ramifications of predator avoidance: Predator and heavy-metal-mediated competition between tadpoles and snails, *Ecol. Appl.,* 9, 1477, 1999.
149. Bunnell, J. F. and Zampella, R. A., Acid water anuran pond communities along a regional forest to agro-urban ecotone, *Copeia,* 1999, 614, 1999.
150. Kolozsvary, M. B. and Swihart, R. K., Habitat fragmentation and the distribution of amphibians: Patch and landscape correlates in farmland, *Can. J. Zool.,* 77, 1288, 1999.
151. Marnell, F., Discriminant analysis of the terrestrial and aquatic habitat determinants of the smooth newt (*Triturus vulgaris*) and the common frog (*Rana temporaria*) in Ireland, *J. Zool. (London),* 244, 1, 1998.
152. Vos, C. C. and Chardon, J. P., Effects of habitat fragmentation and road density on the distribution pattern of the moor frog (*Rana arvalis*), *J. Appl. Ecol.,* 35, 44, 1998.
153. Bowling, J. W., Leversee, G. J., Landrum, P. F., and Giesy, J. P., Acute mortality of anthracene contaminated fish exposed to sunlight, *Aquat. Toxicol.,* 3, 79, 1983.
154. Newsted, J. L. and Giesy, J. P., Jr., Predictive models for photoinduced acute toxicity of polycyclic aromatic hydrocarbons to *Daphnia magna,* Strauss (Cladocera, Crustacea), *Environ. Toxicol. Chem.,* 6, 445, 1987.
155. Zepp, R. G. and Schlotzhauer, P. F., Photoreactivity of selected aromatic hydrocarbons in water, in *Polynuclear Aromatic Hydrocarbons, 3rd Int. Symp. Chem. Biol. — Carcinogen. Mutagen.*, Jones, P. W. and Leber, P., Eds., Ann Arbor Science Publishers, Ann Arbor, 141, 1979.
156. Arfsten, D. P., Schaeffer, D. J., and Mulveny, D. C., The effects of near ultraviolet radiation on the toxic effects of polycyclic aromatic hydrocarbons in animals and plants: A review, *Ecotoxicol. Environ. Saf.,* 33, 1, 1996.
157. Herman, J. R., Bhartia, P. K., Ziemke, J., Ahmad, Z., and Larko, D., UV increases (1972–1992) from decreases in total ozone, *Geophys. Res. Lett.,* 23, 2117, 1996.
158. Zaga, A., Little, E. E., Rabeni, C. F., and Ellersieck, M. R., Photoenhanced toxicity of a carbamate insecticide to early life stage anuran amphibians, *Environ. Toxicol. Chem.,* 17, 2543, 1998.

159. Little, E. E., Calfee, R., Cleveland, L., Skinner, R., Zaga-Parkhurst, A., and Barron, M. G., Photo-enhanced toxicity in amphibians: Synergistic interactions of solar ultraviolet radiation and aquatic contaminants, *J. Iowa Acad. Sci.,* 107, 67, 2000.
160. Hatch, A. C. and Burton, G. A., Jr., Effects of photoinduced toxicity of fluoranthene on amphibian embryos and larvae, *Environ. Toxicol. Chem.,* 17, 1777, 1998.
161. Walker, S. E., Taylor, D. H., and Oris, J. T., Behavioral and histopathological effects of fluoranthene on bullfrog larvae (*Rana catesbeiana*), *Environ. Toxicol. Chem.,* 17, 734, 1998.
162. McConnell, L. L., LeNoir, J. S., Datta, S., and Seiber, J. N., Wet deposition of current-use pesticides in the Sierra Nevada Mountain Range, California, USA, *Environ. Toxicol. Chem.,* 17, 1908, 1998.
163. LeNoir, J. S., McConnell, L. L., Fellers, G. M., Cahill, T. M., and Seiber, J. N., Summertime transport of current-use pesticides from California's Central Valley to the Sierra Nevada Mountain Range, *Environ. Toxicol. Chem.,* 18, 2715, 1999.
164. Crump, D., Berrill, M., Coulson, D., Lean, D., McGilivray, L., and Smith, A., Sensitivity of amphibian embryos, tadpoles and larvae to enhanced UV-B radiation in natural pond conditions, *Can. J. Zool.,* 1956, 1999.
165. Mulla, M. S., Frog and toad control with insecticides!, *Pest Control,* 30, 20, 1962.
166. Schuytema, G. S., Nebeker, A. V., and Griffis, W. L., Comparative toxicity of Guthion and Guthion 2S to *Xenopus laevis* and *Pseudacris regilla* tadpoles, *Arch. Environ. Contam. Toxicol.,* 27, 250, 1994.
167. Sanders, H. O., Pesticide toxicities to tadpoles of the western chorus frog *Pseudacris triserata* and Fowler's toad *Bufo woodhousii fowleri*, *Copeia,* 1970, 246, 1970.
168. Nebeker, A. V., Schuytema, G. S., Griffis, W. L., and Cataldo, A., Impact of guthion on survival and growth of the frog *Pseudacris regilla* and the salamanders *Ambystoma gracile* and *Ambystoma maculatum*, *Arch. Environ. Contam. Toxicol.,* 35, 48, 1998.
169. Barron, M. G. and Woodburn, K. B., Ecotoxicology of chlorpyrifos, *Rev. Environ. Contam. Toxicol.,* 144, 1, 1995.
170. Berill, M. and Bertram, S., Effects of low concentrations of forest use pesticides on frog embryos and tadpoles, *Environ. Toxicol. Chem.,* 13, 657, 1994.
171. Anguiano, O. L., Montagna, C. M., Chifflet de Llamas, M., Gauna, L., and Perchen de D'Angelo, A. M., Comparative toxicity of parathion in early embryos and larvae of the toad, *Bufo arenarum* Hensel, *Bull. Environ. Contam. Toxicol.,* 52, 649, 1994.
172. Marian, M. P., Arul, V., and Pandian, T. J., Acute and chronic effects of carbaryl on survival, growth, and metamorphosis in the bullfrog (*Rana tigrina*), *Arch. Environ. Contam. Toxicol.,* 12, 271, 1983.
173. Salibian, A., Effects of deltamethrin on the South American toad, *Bufo arenarum*, tadpoles, *Bull. Environ. Contam. Toxicol.,* 48, 616, 2001.
174. Materna, E. J., Rabeni, C. F., and LaPoint, T. W., Effects of the synthetic pyrethroid insecticide, esfenvalerate, on larval leopard frogs (*Rana* spp.), *Environ. Toxicol. Chem.,* 14, 613, 1995.
175. Howe, G. E., Gillis, R., and Mowbray, R. C., Effect of chemical synergy and larval stage on the toxicity of atrazine and alachlor to amphibian larvae, *Environ. Toxicol. Chem.,* 17, 519, 1998.
176. Chandler, J. H. and Marking, L. L., Toxicity of the lampricide 3-trifluoromethyl-4-nitrophenol (TFM) to selected aquatic invertebrates and frog larvae, *Invest. Fish Contam.,* 62, 1, 1975.
177. Kane, A. S., Wesley, W. D., Reimschuessel, R., and Lipsky, M. M., 3-Trifluoromethyl-4-nitrophenol (TFM) toxicity and hepatic microsomal UDP-glucuronyltransferase activity in larval and adult bullfrogs, *Aquat. Toxicol.,* 27, 51, 1993.
178. Mayer, F. L. and Ellersieck, M. R., Manual of Acute Toxicity: Interpretation and Data Base for 410 Chemicals and 66 Species of Freshwater Animals, Resource Publication 160, U.S. Fish and Wildlife Service, Washington, D.C., 1986.
179. Fontenot, L. W., Noblet, G. P., and Platt, S. G., Rotenone hazards to amphibians and reptiles, *Herpetol. Rev.,* 25, 150, 1994.
180. Johnson, W. W. and Finley, M. T., Handbook of Acute Toxicity of Chemicals to Fish and Aquatic Invertebrates, Resource Publication 137, U.S. Fish Wildlife Service, Washington, D.C., 1980.
181. Birge, W. J., Aquatic toxicology of trace elements of coal and fly ash, in Energy and Environmental Stress in Aquatic Systems, Thorp, J. H. and Gibbons, G. W., Eds., U.S. Dept. Energy, Technical Information Center, Washington, D.C., 219, 1978.
182. Khangarot, B. S., Sehgal, A., and Bhasin, M. K., Man and the biosphere, studies on the Sikkim Himalayas, *Acta Hydrochim. Hydrobiol.,* 13, 259, 1985.

183. Birge, W. J. and Black, J. A., A continuous flow system using fish and amphibian eggs for bioassay determinations of embryonic mortality and teratogenesis, in Final Technical Report, EPA 560/5–77–002, U.S. Environmental Protection Agency, Office of Toxic Substances, Washington, D.C., 1977.
184. Ferrari, L., Slaibian, A., and Muino, C. V., Selective protection of temperature against cadmium acute toxicity to Bufo arenarum tadpoles, *Bull. Environ. Contam. Toxicol.,* 50, 212, 1993.
185. Horne, M. T. and Dunson, W. A., Effects of low pH, metals, and water hardness on larval amphibians, *Arch. Environ. Contam. Toxicol.,* 27, 323, 1995.
186. Khangarot, B. S. and Ray, P. K., Sensitivity of toad tadpoles, *Bufo melanostrictus* (Schneider), to heavy metals, *Bull. Environ. Contam. Toxicol.,* 38, 532, 1987.
187. Birge, W. J. and Black, J. A., Effects of copper on embryonic and juvenile stages of aquatic animals, in *Copper in the Environment. Part II,* Nriagu, J. O., Ed., John Wiley and Sons, New York, 374, 1979.
188. Kaplan, H. M. and Yoh, L., Toxicity of copper for frogs, *Herpetologica,* 17, 131, 21961.
189. Perez-Coll, C. S., Herkovits, J., and Salibian, A., Embryotoxicity of lead to *Bufo arenarum, Bull. Environ. Contam. Toxicol.,* 41, 247, 1988.
190. Birge, W. J., Black, J. A., and Westerman, A. G., Evaluation of aquatic pollutants using fish and amphibian eggs as bioassay organisms, in Animals as Monitors of Environmental Pollutants, National Research Council, National Academy of Sciences, Washington, D.C., 108, 1978.
191. U.S. Environmental Protection Agency, Ambient water quality criteria for nickel, 440/5–80–060, U.S. Environmental Protection Agency Report, 1980.
192. Birge, W. J. and Zuiderveen, J. A., The comparative toxicity of silver to aquatic biota, in *Transport, Fate and Effects of Silver in the Environment, 3rd Int. Conf.*, August 6–9, Washington, D.C., Andren, A. W. and Bober, T. W., Eds., University Wisconsin Sea Grant Institute, Madison, WI, 1995.
193. Lauglin, R. and Linden, O., Sublethal responses of the tadpoles of the European frog *Rana temporaria* to two tributyltin compounds, *Bull. Environ. Contam. Toxicol.,* 28, 494, 1982.
194. Hall, R. J. and Swineford, D., Toxic effects of endrin and toxaphene on the southern leopard frog *Rana sphenocephala, Environ. Pollut.,* 23, 53, 1980.
195. Gopal, K., Khanna, R. N., Anand, M., and Gupta, G. S. D., The acute toxicity of endosulfan to freshwater organisms, *Toxicol. Lett.,* 7, 453, 1981.
196. Vardia, H. K., Rao, P. S., and Durve, V. S., Sensitivity of toad larvae to 2,4-D and endosulfan pesticides, *Arch. Hydrobiol.,* 100, 395, 1984.
197. Berrill, M., Coulson, D., McGillivary, L., and Pauli, B., Toxicity of endosulfan to aquatic stages of anuran amphibians, *Environ. Toxicol. Chem.,* 17, 1738, 1998.
198. Hedtke, S. F. and Puglisi, F. A., Short-term toxicity of five oils to four freshwater species, *Arch. Environ. Contam. Toxicol.,* 11, 425, 1982.
199. Sparling, D. W., Fellers, G. M., and McConnell, L. L., Pesticides and amphibian population declines in California, USA, *Environ. Contam. Toxicol.*, 20, 1591, 2001.
200. Hayes, T. B., Collins, A., Lee, M., Mendoza, M., Noreiga, N., Stuart, A. A., and Vonk, A., Hermaphroditic, demasculanized frogs after exposure to the herbicide, atrazine, at low ecologically relevant doses, *Proc. Natl. Acad. Sci.*, 99, 5476, 2002.
201. Kiesecker, J. M., Synergism between trematode infections and pesticide exposure: A link to amphibian limb deformities in nature? *Proc. Natl. Acad. Sci.*, 99, 9900, 2002.
202. Jung, R. E. and Jagoe, C. H., Effects of low pH and aluminum on body size, swimming performance, and susceptibility to predation of green tree frog (*Hyla cinerea*) tadpoles, *Can. J. Zool.*, 73, 2171, 1995.

CHAPTER 41

# Genetic Effects of Contaminant Exposure and Potential Impacts on Animal Populations

Lee R. Shugart, Christopher W. Theodorakis, Amy M. Bickham, and John W. Bickham

## CONTENTS

"... it is important to take genetics into account in understanding if and how chemical contaminants impact populations."

**Peter Calow**[1]

## 41.1 INTRODUCTION

"Understanding changes to the genetic apparatus of an organism exposed to contaminants in the environment is essential to demonstrating an impact on parameters of ecological significance such as population effects. That field of environmental science that attempts to (a) identify changes in the genetic material of natural biota that may be induced by exposure to genotoxicants in their environment and (b) the consequences at various levels of biological organization (molecular, cellular, individual, population, etc.) that may result from this exposure is termed genetic ecotoxicology.[2]

Within genetic ecotoxicology, it is critical to realize that there are two possible classes of effects. First, there are effects that occur in the somatic or reproductive tissues of an organism. These effects are the result of direct exposure to a genotoxicant and have the potential to lead to somatic or heritable (genotoxicological) disease states. Another class of effects results indirectly from contaminant stress on a population and leads to alterations in the genetic makeup of populations, a process termed evolutionary toxicology.[3] These latter types of effects alter the inclusive fitness of populations, such as by the reduction of genetic variability, and can potentially have profound impact on biomarker studies.[4] For example, populations inhabiting contaminated and reference sites might be adapted to different environmental conditions and thus respond differently than expected in such studies.

With respect to genotoxicology ("a" above), it is now possible to identify molecular targets of genotoxicants with extreme sensitivity and to determine how chemical modifications of these targets affect function at a precise molecular level.[5] For several reasons, approaches and studies related to "b" above are not as far advanced. First, a major challenge has been to develop assays with the sensitivity to demonstrate those subtle changes in the genetic material of organisms exposed to genotoxicants that may be genetic markers of population effects.[6,7] However, recent advances in the discipline of molecular biology may provide the experimental tools with which to investigate those key biological mechanisms at the genetic level that regulate and limit responses of ecological relevance.[8,9] Second, most studies performed *in situ* are often burdened by complicating environmental factors. Individual genetic variability within a population, population size, and exposure to complex mixtures are just a few of the many problems that must be addressed in order to interpret data generated by the sophisticated methodologies currently in use.

This chapter is divided into three main sections. The importance of understanding contaminant-induced DNA damage and related effects in relation to population-level studies is covered in Section 41.2, Genetic Effects. Section 41.3, Environmental Population Genetics, focuses mainly on new methodologies applicable to population genetic studies. Finally, Section 41.4, Case Histories, details several different investigations that provide insight on how chemical contaminants may impact populations.

## 41.2 GENETIC EFFECTS

### 41.2.1 Introduction

Within a cell, the structural integrity of the DNA molecule is in a constant state of flux between a functionally stable double-stranded entity without discontinuity and some intermediate, unstable state. This latter state is a transient phenomenon triggered by normal cellular processes. However,

**Table 41.1 Cellular Responses after Exposure to Genotoxicants**

| Biological Response | Expression in Cell | Temporal Occurrence[a] |
|---|---|---|
| Detoxication | Protein induction: P450 enzyme system and metallothionine | Early |
| DNA Structural Modification | | |
| Adduct | Covalent attachment of genotoxicant to DNA | Early |
| Strand Breaks | Breakage of DNA phosphodiester linkages | Early |
| Base Modification | Hypomethylation and chemical modification of bases | Early/Middle |
| Repair | Induction of DNA repair enzymes | Early |
| Abnormal DNA | Apoptosis | Early/Middle |
| | Chromosomal aberrations, micronuclei, aneuploidy, mutations | Middle/Late |
| Pathological Conditions | Neoplasia, tumors, and protein dysfunction | Late |

[a] Temporal occurrence subsequent to exposure will depend on species of and type of genotoxicant. Early: hours to days; Middle: days to weeks/months; Late: weeks/months to years. (*Source:* From Shugart, L.R., *Ecotoxicology,* 9, 329, 2000. With permission.)

these processes can be disrupted when exposure to a genotoxicant occurs, often with the concomitant loss of structural integrity of the DNA molecule. Some of the cellular responses that may be expressed after exposure to genotoxicants are given in Table 41.1. The organism's inability (whether transient or permanent) to cope with loss of structural integrity provides the investigator the opportunity to detect environmental exposure to a genotoxicant. In addition, the occurrence of DNA damage provides a means to investigate the qualitative and quantitative relationships between the formation of DNA damage, subsequent DNA processing, appearance of deleterious lesions, and irreversible effects on reproduction and fitness. The reader is referred to the scientific literature for current reviews on this topic, in particular those of Shugart,[10–12] Shugart et al.,[13] Dixon and Wilson,[14] and Wirigin and Theodorakis.[15]

### 41.2.2 Types of DNA Modifications

A summary of some of the more common DNA structural modifications that occur when a genotoxicant becomes bioavailable and interacts with cellular DNA is recorded in Table 41.2. Two general classes of structural modifications can be inferred from the information contained therein. First, there are those modifications that identify the specific genotoxicant responsible for the structural modification. For example, ultraviolet light in the 290–300 nm range (UV-B) causes specific dimerization of pyrimidine bases within the DNA. Also, many chemicals, such as the polycyclic aromatic hydrocarbon (PAHs) and benzo[a]pyrene (BaP), can form an adduct with the

**Table 41.2 DNA Structural Modifications Caused by Genotoxicants**

| Genotoxicant | Type of Modification | Mechanism |
|---|---|---|
| Physical | Thymine-Thymine Dimer | Dimerization of pyrimidine bases by UV-B light |
| | Strand Breakage | Breakage of phosphodiester linkages due to formation of free radicals by ionizing radiation |
| Chemical | Adduct | Covalent attachment of genotoxicant to DNA molecule |
| | Altered Bases | Chemical modification of existing bases |
| | Abasic Site | Loss of chemically unstable adduct or damaged base |
| | Strand Breaks | Breakage of phosphodiester linkages due to formation of free radicals and abasic sites |
| | Hypomethylated DNA | Improper postreplication |
| | Mutation | Improper DNA repair |

*Source:* From Shugart, L.R., *Ecotoxicology,* 9, 329, 2000. With permission.

DNA. After metabolic activation, the BaP becomes covalently attached to the DNA. In both examples, the structural modification represents a specific fingerprint of the responsible genotoxicant. Second, there are those structural modifications that, although not specific to a particular genotoxicant, nevertheless suggest that exposure has occurred (e.g., breakage of the phosphodiester backbone of the DNA molecule). Strand breakage of the DNA can result when a genotoxicant produces free radical or forms an abasic site or sites. Also, many genotoxicants are known to interfere with normal DNA processing activities such as replication, methylation, and repair, which in turn may result in mutations (e.g., base addition/deletion). The detection of nonspecific structural modification (e.g., strand breaks, abasic sites, hypomethylation and mutations) may imply genotoxicant exposure, especially if the level or degree of these types of modification to the DNA molecule is not what might be anticipated (e.g., when compared to controls).

### 41.2.3 Detection of DNA Modifications

#### *41.2.3.1 DNA Adducts*

Detection of structural damage to DNA such as adducts is not an easy task, for several reasons. First, environmental genotoxicants are usually present at low concentrations and once they become bioavailable are readily detoxified.[16] Therefore, the potential for *in situ* DNA damage is not high, and the amount that is found is often on the order of one adduct per $10^7$ nucleotides or less. Second, until recently, the analytical technologies with the required selectivity and sensitivity to detect extremely low levels of DNA damage were not readily available. However, the application of modern techniques from the scientific disciplines of biochemistry and molecular biology has begun to alleviate this problem.[5,14] In this regard, the possible application of DNA fingerprinting using PCR methodologies for the detection of structural DNA damaged, including adducts, caused by exposure to genotoxic environmental agents has been addressed.[17]

The adverse health effects of most environmental chemicals are the result of their covalent binding to physiologically important receptor molecules. Identification of the interactive products with DNA, especially adducts, can represent the most direct and biologically relevant indicator of exposure to a genotoxicant.[18] Numerous analytical methods to detect and quantify DNA adducts are available,[12,17] with the $^{32}P$-postlabeling technique being the most used. The methodology is described in detail elsewhere.[12,14,18] Because the salient features of this technique are sensitivity and selectivity, it is finding increased application in environmental monitoring studies where genotoxic contaminants exist.[14,18,19] Lists of recent investigations that used the $^{32}P$-postlabeling technique to screen for DNA adducts in organisms taken from contaminated environments are available.[12,20] It should be noted, however, that this technique is subject to problems that may interfere with the interpretation of the data generated.[21] Several laboratories from Europe and North America are currently participating in a project to determine the extent of variability with this technique.[22]

#### *41.2.3.2 DNA Strand Breaks*

Because both physical and chemical genotoxicants have the potential to cause DNA strand breaks,[12] recent environmental studies have included this structural modification as an indicator of genotoxicant exposure. Several of the popular strand-break assays are based on the observation that under *in vitro* denaturation conditions of high pH, the rate of conversion of double-stranded DNA to the single-stranded moiety is proportional to the number of strand breaks in the DNA molecule.[23] Among these are the alkaline elution assay,[24] the alkaline unwinding assay,[25] the gel electrophoresis method,[26] and the comet assay.[27] A list of investigations where these techniques have been applied are found in Shugart.[12]

### 41.2.4 Cytogenetic Effects

DNA damage that is not corrected or is improperly processed may potentiate irreversible cellular events[14,15] that result in the appearance after cell division of abnormally processed DNA (e.g., chromosomal aberrations, micronuclei, somatic mutations etc., Table 41.1.).

Such cytogenetic effects result in alteration of the chromosome structure or chromosome number. The traditional approach microscopically analyzes condensed chromosomes in metaphase cells to determine the karyotype (i.e., number and appearance of chromosomes). Less laborious and time consuming methods than karyological examination include micronucleus analysis and detection of variation of DNA content among cells by flow cytometry. Micronuclei result from acentric fragments of whole chromosomes that lag at anaphase and subsequently do not become incorporated into either daughter nuclei after cell division but form their own small nucleus.[28] Flow cytometry is used to measure the differences in total DNA content among cells that result from the unequal assortment of fragmented or rearranged chromosomal material after cell division.[29] Obviously, cytogenetic, micronucleus, and flow cytometric analyses are measuring related phenomena.

### 41.2.5 Mutations

In addition to cytogenetic effects, faulty repair of genotoxic-induced DNA damage can result in the occurrence of mutations in the DNA molecule (i.e., point mutations, additions/deletions, translocations, etc.). In somatic tissue, mutations in oncogenes and tumor suppressor genes have been associated with the initiation of chemical carcinogenesis. Because these genes are involved in the regulation of cell growth, differentiation, and DNA repair, mutational events in these genes can be correlated with aberrant cellular function, which can then be related to individual- and, it is hoped, population-level effects.[2,9] Wirgin and Theodorakis[15] discuss recent application of this approach in relation to somatic and heritable effects of environmental contaminants on fish.

### 41.2.6 Protein Induction

The genetic apparatus of an organism can interact with a genotoxicant in a variety of ways that may not result in structural modification to its DNA (Table 41.1.). The most common response is that which results in the induction of a protein, or sets of proteins, involved with cellular detoxication processes. For example, the organism may perceive[7] the genotoxicant and modify its physiology, as is found with the induction of the P4501A1 detoxication system.[15,16,30] The induction of the P4501A1 system can be detected by an increase in enzyme activity or enzyme protein, and the magnitude of induction provides a measure of the degree of interaction of the inducing agent with the aryl hydrocarbon hydroxylase receptor (Ah-receptor) in the cytoplasm of the exposed cell.

Metallothionein is a constitutive protein associated with the maintenance of homeostasis of the trace metals zinc and copper. It is known to play a role in the detoxication of the genotoxic metals cadmium and mercury, and upregulation of the metallothionein gene can serve as an early warning signal of metal-induced toxicity.[15,16]

A wide range of genotoxic agents can act as inducers (see discussion below).

### 41.2.7 Genotoxic Agents

A variety of contaminants can induce genotoxic responses. Some chemicals, including PAHs and their nitrogenated or chlorinated derivatives, mycotoxins such as aflotoxins and related compounds, and vinyl chloride typically exert their genotoxicity via formation of bulky adducts.[18,31–35] However, induction of oxidative damage may be a secondary mechanism of genotoxicity.[36] A second class is comprised of those genotoxic chemicals that cause derivatization of nucleotide bases via transfer of methyl or ethyl moieties and includes the potent carcinogens diethynitrosamine and methynitrosurea.[37] Another class of genotoxic agents includes metals such as arsenic, cadmium, chromium, mercury, nickel, and lead.[38,39] There are three possible mechanisms whereby metals may induce genotoxicity. First, some metals, chromium in particular, may adduct nucleotide bases.[40] Second, there is growing evidence that metals may inhibit repair of DNA damage induced by chemicals or endogenous metabolism.[41] Third, metals may increase levels of oxidative stress via redox cycling and Fenton reactions.[42] There are also many organic chemicals that can potentiate genotoxicity via oxidative stress induction, and these include cyclic or aliphatic chlorinated hydrocarbons and several classes of pesticides.[43–46]

Besides chemical agents, there are physical agents that can also lead to DNA damage, most notably several types of radiation. For example, ionizing radiation in the form of high-energy photons (γ- and x-rays), electrons (β-rays), or helium nuclei (α-rays) may be genotoxic. One mechanism by which ionizing radiation can induce DNA damage is via direct interaction of the radioactive particles with the DNA molecule.[47] This can result in base alterations or breaks in the sugar phosphate backbone. Alternatively, the radioactive particles may interact with water or oxygen molecules, producing oxyradicals.[48] These radicals may also produce base alterations or DNA strand breaks. Another type of physical genotoxicant is ultraviolet radiation, specifically UV-B. Irradiation of DNA with UV-B may result in covalent attachment of adjacent pyrimidine bases, resulting in so-called cyclobutane dimers.[49] A secondary genotoxic effect of UV radiation is the production of oxyradicals.[50] There have been suggestions that other types of electromagnetic radiation (e.g., radio and microwaves) and magnetic fields may also produce genotoxic effects, but this research is equivocal.[51,52]

## 41.3 ENVIRONMENTAL POPULATION GENETICS

### 41.3.1 Introduction

Chemical contamination can cause population reduction by the effects of somatic and heritable mutations as well as nongenetic modes of toxicity.[2,3,9,16] Although the original damage caused by chemical contaminants may be at the molecular level, there are emergent effects at the level of populations, such as the loss of genetic diversity, that are not predictable based solely on knowledge of the mechanism of toxicity of the chemical contaminants. In this regard, population genetic diversity has been proposed as a bioindicator of a population's vulnerability to natural and anthropogenic stressors, as a record of genotypic variation in the population history and the effects of genotypic changes on the spatial distribution and abundance of populations in a geographic region.

Even though there is an extensive scientific literature in regards to classical Mendelian genetics, protein polymorphism, and DNA-marker studies in the field of population genetics, it is only recently that studies of the effects of pollution on population genetics have come to the forefront.[53,54]

### 41.3.2 Genetic Markers

The oldest and most classical genetic marker is the *phenotype,* the visible traits or characters of individuals within a biological species. Phenotypic traits, such as mortality, developmental abnormalities, DNA strand breakage, physiology, and metabolism, stand as valid characters for population studies.

Another approach to population genetic analysis is to examine protein polymorphisms. An electrophoretic methodology, known as allozyme analysis, detects charge characteristics of enzymatic proteins produced by amino acid substitution. Allozyme analysis has been used in the past 20 years to assess the relationship between allozyme genotype and exposure to chemical compounds.[55] The importance of the methodology to population genetic studies has been reviewed.[53–55]

**Table 41.3 A Comparison of the DNA Marker Methods**

| Marker | Polymorphic Markers per Reaction | Markers per Genome | Marker Type |
|---|---|---|---|
| RFLP | 1–2 | 1000 | Co-dominant |
| RAPD | 4–6 | 10,000 | Dominant/Co-dominant |
| SSR | 3–10 | 10,000 | Co-dominant |
| AFLP | 10–50 | >100,000 | Co-dominant/Dominant |

*Source:* From D'Surney, S.J., Shugart, L.R., and Theodorakis. C.W., *Ecotoxicology,* 10, 201, 2001. With permission.

**Table 41.4 A Comparison of the Targets, Resolutions and Costs of Various Methods for Surveying Genetic Diversity in Natural Populations[a]**

| Method | Target | Resolution | Cost ($) | Development |
|---|---|---|---|---|
| Allozyme | Nucleus | Low | 0.1 | Low |
| Microsatellite | Nucleus | High | 3.0 | High |
| RAPD | Nucleus | High | 3.0 | Low |
| Single gene sequence | Nucleus | Low-High | 30.0 | High |
| RFLP | mtDNA | Medium | 100.0+ | Low |
| Single gene sequence | mtDNA | Low-High | 30.0 | Low |
| Single gene mutation screen | mtDNA | Low-High | 0.5 | Low |

[a] The costs are those involved in screening a single gene for variation, while development refers to the relative effort which must be expended before the collection of information on a new species.

*Source:* From Bickham, J.W., Sandhu, S., Hebert, P.D.N., Chikhi, L. and Athwal, R., *Mut. Res.,* 463, 33, 2000. With permission.

Recently, the application of DNA sequencing and the polymerase chain reaction (PCR)-based technologies has revolutionized the science of generating high-throughput genetic markers. New genetic-marker systems generated by the PCR methodology with applications to environmental genetics include RFLPs (restriction fragment length polymorphism), RAPD (random amplified polymorphic DNA), SSRs (simple sequence repeats such as mini- and microsatellites), and AFLP (amplified fragment length polymorphism).[17,53,54] These PCR-derived methods provide the potential to encompass large genomic regions, both coding and noncoding. A limited comparison of the capabilities of the various types of genetic markers is given in Table 41.3.

Since the various methodologies discussed above target different segments of the genome, possess differing resolution (Table 41.3), and involve varied operating and developmental costs, there is no single optimal technique (Table 41.4). Instead, methodological selection for environmental population genetic studies is guided by the problem under investigation.[53]

## 41.4 CASE HISTORIES

This section is not intended to review all relevant literature but rather to present several case histories to demonstrate and illustrate (a) the types of techniques and methodologies that were applied to a particular study (b) that exposure to environmental contamination is a possible or likely cause of the population impacts observed, and (c) the kinds of genetic alterations seen or suspected to have occurred.

### 41.4.1 Allozymes

Differences in allozyme allele frequencies between contaminated and reference sites have been found to occur in many species,[56,57] and this may suggest that there is a selective advantage to certain genotypes over others in contaminated populations. For example, Gillespie and Guttman[56] reported that some allozyme alleles were present at a higher frequency in contaminated stoneroller (*Campostoma anomalum*) populations than in reference populations. Fish with these alleles also had longer survival times when exposed to copper in the laboratory.[58] *In vitro* enzymatic assays indicated that the enzymatic activity of these particular alleles was less inhibited by copper than that of the alleles that were more prevalent in noncontaminated populations.[59] This not only linked genotype frequencies with survival (a component of selection) but also demonstrated a biochemical basis for differential susceptibility. On the other hand, selection may not act directly on the allozyme loci themselves, but rather these loci may be closely linked to other genes (e.g., detoxification enzymes, etc.) that impart a selective advantage.

In another series of studies, Newman et al.[60] found that survival time of eastern mosquitofish (*Gambusia holbrooki*) exposed to heavy metals was correlated with allozyme genotype, particularly glucose–phosphate isomerase (GPI) alleles. Mulvey et al. (1995) went on to find that reproductive performance (number of gravid females and developing embryos per female) in these fish exposed to mercury was dependant on GPI genotype. Such differences in reproductive performance were in accordance with differences in survival among GPI genotypes. However, there was no evidence that such differences in survival and reproduction were related to differential susceptibility among GPI genotypes to enzymatic inhibition by mercury, either *in vitro* or *in vivo*.[62,63]

Correlations between laboratory exposures and natural populations are not necessarily straightforward. Diamond et al.[64] found that survival time of metal-exposed *G. holbrooki* was dependent upon allozyme genotype, but this pattern was not consistent among different populations or for fish collected from different years. Indeed, Lee et al.[65] argued that correlations among broods or other subunits of a structured population may influence observed differences between polluted and reference populations. Consequently, the effectiveness of demonstrating contaminant-induced selection using allozyme may depend on the life-history characteristics, behavior, or local population structure.

In order to address other variables besides contaminant selection, Newman and Jagoe[66] simulated mercury-driven selection for *G. holbrooki* GPI genotypes. They used simple and complex models to quantify the relative effects of viability selection, random genetic drift and migration on the GPI-allele frequencies, and sexual and fecundity selection. A simple suggested viability selection was a greater determinant than mortality-driven genetic drift, sexual selection, or fecundity selection. They also found that gene flow could abolish the effects of mercury selection on genetic differentiation among populations. In general, their model simulations indicated that changes in allele frequencies may reflect population-level effects of pollution, provided that the system under study is properly understood.

### 41.4.2 Puget Sound, Washington

It has been known for a long time that exposure to genotoxic agents may lead to neoplastic and preneoplastic lesions, and such patterns have been found in natural populations exposed to high levels of genotoxic contaminants, primarily in fish. Tumor incidence in fish and other aquatic organisms associated with exposure to genotoxic contaminants at a variety of sites throughout the United States including Boston Harbor,[67–71] the Hudson River,[72] Elizabeth River, Virginia,[73] the Black River in Ohio, and the Great Lakes.[74] However, perhaps one of the best-studied systems is in Puget Sound, Washington, which includes Eagle Harbor, a site heavily contaminated with PAH-laden creosote.

This system has been found to be contaminated with high levels of PAHs, and PAH adducts were associated with hepatic carcinomas in populations of English sole *Pleuronectes vetulus*.[75,76] In addition, levels of hepatic DNA adducts have corresponded with known sediment or tissue-contaminant concentrations. In laboratory experiments using sole exposed to sediment collected from Eagle Harbor, PAH adducts demonstrated a linear dose-response function for both PAH concentration and length of exposure.[77] In native fish populations, these adducts were associated with not only neoplastic lesions but also degenerative and preneoplastic lesions, and such lesions have shown significant associations with other biomarkers of PAH exposure and effect such as elevated cytochrome P450 and biliary PAH metabolite levels.[75,76] Additionally, Reichert et al.[76] have used a molecular epizootiological approach to provide definitive evidence that exposure to PAHs was the etiological agent in development of neoplasms. They found that levels of hepatic DNA adducts were a significant risk factor in the development of neoplasia in feral sole populations.

Further studies employ sequencing of the K-ras oncogene in order to study mutational events associated with neoplastic and preneoplastic lesions in this species.[78] Hepatic lesion frequencies as well levels of DNA adducts and other PAH-indicative biomarkers were lower in fish collected after the cessation of discharge than were historical data collected when contaminant-generating activities were ongoing. The site has been capped with uncontaminated sediment, and these biomarker and histological endpoints are being used to assess the efficacy of remediation activates of Eagle Harbor.[79]

Besides PAH adducts, other measures of DNA damage have also been examined in English sole from Puget Sound. For example, Malins and Haimanot[80] found that oxidative DNA damage was highest in livers from sole that were tumorous from contaminated sites, least in sole livers from references areas, and intermediate in tumor-free fish from contaminated sites. This method also exposed positive associations between levels of oxidized bases and severity of preneoplastic and nonneoplastic lesions in livers of fish from the same area.[81]

### 41.4.3 Sunfish

In 1987, the measurement of DNA strand breaks (see Section 41.2, Genetic Effects) in sunfish was implemented as a biological monitoring technique for environmental genotoxicity.[82,83] Sunfish were initially collected and analyzed over a period of several years (1987–1992) from a contaminated stream (primarily mercury) and reference stream[84] as part of a Biological Monitoring and Abatement Program for the U.S. Department of Energy (USDOE) in Oak Ridge, Tennessee. Analyses for DNA strand breaks in sunfish inhabiting these same streams were performed again in 1994–1995 by Nadig et al.[85] and finally in 1997 by Theodarakis et al.[86] Data collected indicated that the DNA structural integrity of sunfish from the reference stream was good (few DNA strand breaks) and remained relatively constant over the entire 10-year sampling period. However, levels of DNA strand breaks of sunfish from the contaminated stream fluctuated and varied with time of sampling. DNA damage was high in 1987 but started to decline in 1988. By 1992, the levels of DNA strand breaks were comparable to that found in the sunfish from the reference stream. The data from the 1994–1995 sampling period[85] indicated a return to the high level of DNA strand breaks observed in 1987. However, by the 1997 sampling period,[86] no significant DNA strand breakage was noted.

These data, in conjunction with other indicators of stress and toxicity,[84] suggested that sunfish in the contaminated stream were being exposed to genotoxicants in a recurring manner. An improving aquatic environment, due to the effects of remedial actions implemented by the USDOE during the early years of sampling, was thought to be responsible for the diminution in DNA strand breakage that returned to normal levels observed in 1992. Subsequent release of contaminants into the stream after 1992 resulted in a return to the high levels of DNA strand breaks, which was documented in the 1994–1995 sampling period.[85] Correction of the problem saw a return to the low levels of DNA strand breaks in the sunfish during the 1997 sampling period.[86]

Theodorakis et al.[86] extended the investigation of genetic effects in the sunfish to include both DNA strand breaks and chromosomal damage (measured by flow cytometry). In general, chromosomal damage in sunfish appeared to be correlated with mutagenicity of the sediment in the stream and was related to community-level responses (e.g., community diversity and percent pollution-tolerant species). Because responses at several levels of biological organization showed similar patterns of downstream effects, the authors suggested a causal relationship between contamination and observable biological effects.

The studies just described focused mainly on individual-level genetic effect (DNA strand breaks and chromosomal damage) from exposure of sunfish to contamination in their environment.

However, the studies of Nadig et al.[85] extended this investigation beyond genetic effects at the individual level and examined potential alteration of population genetics. Using DNA markers produced by the RAPD technique (see Section 41.3.2, Genetic Markers), specific and unique genotypes were identified. Two measures of genetic diversity — the band-sharing index and the nucleon diversity index — showed that the sunfish from the contaminated and reference sites were different. Difference in genetic distance between populations was attributed to selection pressure of contaminants. This conclusion was supported by the finding that frequencies of certain unique genotypes in sunfish from the contaminated site correlated with a downstream gradient of mercury.

Taken together, these several studies[67–86] show that sunfish were experiencing genotoxic stress as a result of exposure to contaminants in their environment. Analysis of DNA structural integrity reflected the level of insult from exposure at the time of sampling, while chromosomal damage data revealed the occurrence of irreversible cellular events as a result of this exposure. The observation that genetic diversity was altered in sunfish populations from contaminated sites compared with those from reference sites suggests that genetic selection occurred in the resident population and was probably due to contaminant effects.

The USDOE Biological Monitoring and Abatement Program in Oak Ridge has collected and archived a wealth of scientific information over the years on such topics as contaminant effects on biological species, waste management, and risk assessment.[87] This program has, by design, the potential to advance our knowledge in the science of environmental population genetics, but to date it has been noticeably underutilized in this respect.

### 41.4.4 Mosquitofish

Beginning in 1992, a series of studies was initiated to determine the effects of ionizing radiation on DNA integrity and population genetics of western mosquitofish (*Gambusia affinis*) living in radionuclide-contaminated ponds on the Oak Ridge National Laboratory in Oak Ridge. In the first phase of these studies, DNA strand breakage was measured in mosquitofish exposed to ionizing radiation *in situ*.[88] This was done by examination of four populations of mosquitofish, two from sites contaminated with radionuclides (Pond 3513 and White Oak Lake) and two from clean sites (Crystal Springs and Wolf Creek). The results of this study[88] demonstrated that the double-stranded MML (median molecular length of DNA fragments detected by gel electrophoresis) of DNA of the fish from White Oak Lake and Pond 3513 was lower than from either of the two reference sites, indicating a higher degree of DNA strand breakage. Also, the single-stranded MML in the DNA of fish from Pond 3513 was lower than in any other population. It was also found that there was a direct correlation of DNA integrity (i.e., MML) with fecundity at least for single-stranded MML. There were no such relationships observed in the reference sites. These observations imply that resistance to DNA damage carries a fitness component, in that individuals that are better able to prevent or repair DNA damage are at a selective advantage in their environment. However, it could also be argued that this relationship is due to environmental factors. Therefore, the population genetics of these fish were examined to determine if this correlation had a genetic, rather than environmental, etiology.

In the next phase of these studies,[89] the RAPD technique was employed in order to determine if the certain genotypes could impart a selective advantage in contaminated environments. A total of 142 RAPD bands were identified, and of these 16 were found to be present at a higher frequency in the contaminated sites relative to the reference sites ("contaminant-indicative bands"). The differences in frequency of the contaminant-indicative bands between contaminated and reference populations suggests that these bands may be genetic markers of loci that provide some sort of selective advantage in radionuclide-contaminated habitats. If this were true, it should be reflected in some component of fitness. To test this hypothesis, fecundity was examined in fish from each of the four populations with and without the contaminant-indicative bands. It was found that for seven of the contaminant-indicative bands in Pond 3513 and White Oak Lake, females that displayed these bands had a higher fecundity than those that did not. This was true for only one band in the Crystal Springs population.[89] Another component of fitness is survival. Thus, if there is differential fitness between genotypes, then survival should be dependent on genotype for those fish exposed to radiation. To test this hypothesis, mosquitofish were collected from a noncontaminated pond and caged in another noncontaminated pond or in Pond 3513. It was found that for nine of the contaminant-indicative bands, the percent survival of fish with the band was greater than that for fish without the band.[90]

These data imply that the contaminant-indicative bands may be genetic markers of loci that confer some sort of selective advantage in contaminated populations, in this case a higher degree of relative radioresistance. If the amount of DNA damage is a reflection of relative radioresistance, then the relative amount of DNA damage should be dependent on RAPD genotype. Therefore, the MMLs were compared for individuals with and without the contaminant-indicative bands. In order to do this, three separate experiments were performed. The first experiment used fish collected from the four populations described previously and used in determination of band frequencies.[89] In the second experiment,[91] 30 fish were collected from a noncontaminated pond and exposed to 20 Gy (approximately 12 min exposure time) of x-rays in the laboratory. The third experiment[90] used the fish from the caging experiment described above. The results from these experiments indicated that for many of the contaminant-indicative bands, the fish that displayed the bands had higher DNA integrity than fish that did not display the bands.

If these bands are indeed genetic markers of loci that confer relative radioresistance, then this should also be reflected in other species exposed to radionuclides. To test this hypothesis, samples of a closely related species, *G. holbrooki*, were collected from two radionuclide-contaminated and two reference sites on the USDOE Savannah River Site (SRS). The population genetic structure of these mosquitofish was examined by the RAPD technique, using the same primers as were used in the Oak Ridge studies.[92] It was revealed that the frequency of three RAPD markers (i.e., PCR-amplified DNA fragments) was greater in the DNA of fish from contaminated than the reference sites, and the frequency of two markers was greater in the reference than in the contaminated sites. These DNA fragments were the same size and amplified by the same PCR primers used in the ORNL study. Southern blot analysis, using labeled *G. affinis* RAPD bands as probes, revealed that the SRS *G. holbrooki* contaminant-indicative markers were homologous to the ORNL *G. affinis* contaminant-indicative markers.

If these RAPD fragments are genetic markers of selective advantage to fish in contaminated habitats, then it is possible that they are being amplified from a physiologically important locus. Thus, their DNA sequences may be conserved across taxa. To test this possibility, probes were made from 3 of the *G. affinis* RAPD primers described above. They were then hybridized to RAPD amplification products obtained from human, herring gull (*Larus argentatus*), and sea urchin (*Strongylocentrotus droebachiensis*) DNA, using the same RAPD primers as were used to produce the *G. affinis* RAPD bands described above. Southern blot analysis revealed that these markers were conserved in DNA sequence and molecular length in all species examined. The *G. affinis* bands were also cloned and sequenced, but the results of DNA sequencing efforts did not provide definitive evidence as to the identity of these loci.[92] Although the identity of these bands is still unknown, the high degree of conservatism suggests that these loci might play an important role in molecular processes such as DNA repair, fitness, and survival.

These studies are significant for two reasons. First, genetic differences between populations may suggest selection for specific genotypes, but to validate this hypothesis, differential fitness and possible biochemical/molecular mechanisms for differential responses to toxicants must be shown. Second, integration of genotoxic or other molecular biomarkers (e.g., DNA strand breakage) into population genetic analyses could provide valuable insight as to the etiology and consequences of population genetic alterations. The concordance of all these results indicates that radiation exposure selects for certain genotypes, and the contaminant-indicative bands are markers of genes or other elements that confer a selective advantage in contaminated environments.[93]

Due to ongoing restoration activities at the Oak Ridge National Laboratory, Pond 3513 is scheduled for remediation. To facilitate future scientific research initiatives with mosquitofish from this contaminated environment, samples were taken and are currently being maintained in laboratory aquaria at the Environmental Sciences division. Also, some carcasses have been archived and preserved in liquid nitrogen. Interested investigators should contact Dr. Mark Greeley.[94]

### 41.4.5 Kangaroo Rats

The Nevada Test Site (NTS) is a nuclear weapons testing facility operated by the USDOE. Between 1951 and 1963, there were 105 aboveground tests of atomic weapons conducted at the NTS or its associated bombing range. In some sites, towers were located upon which bombs were placed for detonation. These ground-zero, or T, sites were used multiple times, and the surrounding areas received considerable radioactive contamination.

Theodorakis et al.[95] conducted studies of the genotoxic effects of radiation from aboveground atomic bomb tests on Merriam's kangaroo rat (*Dipodomys merriami*) at two of the T sites (T1 and T4). Initially, they used flow cytometry and the micronucleus assay to detect the somatic effects of radiation. These studies were inconclusive because, although cytogenetic analysis suggested genotoxic effects (means were higher in the contaminated sites than in the control sites), the differences were not statistically different. This is in spite of the fact that previous studies of heteromyid rodents exposed to chronic low-level radioactivity had revealed ecological effects.[96] Theodorakis et al.[95] subsequently conducted molecular genetic analyses to better characterize the populations and search for population-level genetic effects.

Two molecular genetic analyses — RAPDs and mtDNA control-region sequences — were employed in their study. Although the nuclear RAPDs did not reveal any differences among the four localities (two reference, R1, R2, and two ground-zero sites, T1 and T4), the maternally inherited mtDNA showed significant differences among populations. This was interpreted to mean that males disperse at a greater rate than females; thus, the nuclear markers reflect panmixia, but the maternal markers show population differentiation. This is consistent with behavioral studies on kangaroo rats in which males have been shown to disperse at a greater rate than females.

It was found that some mtDNA haplotypes were shared among sites (potential migrant haplotypes), and others were restricted to only a single site (potential resident haplotypes). Theodorakis et al.[95] surmised that the unique haplotypes represented long-term residents and the shared haplotypes represented potential recent immigrants. To test this hypothesis, the flow-cytometry and micronucleus data were reanalyzed. It was found that when the animals with migrant haplotypes were excluded from the analysis, one of the contaminated sites had significantly increased DNA damage compared to one of the control sites. Furthermore, when animals from the contaminated sites were considered alone, individuals with resident haplotypes had significantly greater chromosome damage compared with animals with migrant haplotypes. This study shows that molecular genetic data can be used to better interpret biomarker data and that it is possible for genotoxic effects to be masked by high immigration from uncontaminated sites into contaminated sites. MtDNA is a potentially valuable genetic marker for differentiating among potential immigrants and residents.

The kangaroo rat data led Theodorakis et al.[95] to hypothesize a specific demographic pattern of movement among populations. Reference area 2 (R2) proved to be significantly different in the biomarker analyses from contaminated T4; R1 and T1 were not different. They hypothesized that R1 was more likely to be exchanging migrants with the contaminated sites than was R2. This could be investigated by long-term field studies using mark–recapture techniques, but such data would take several years to obtain. To test this hypothesis using the genetics data, they conducted a phylogenetic analysis of the haplotypes and plotted the localities at which each haplotype occurred. Using the method of Slatkin and Madison,[97] the hypothetical immigration events needed to explain the topology of the tree (which itself reflects the geneological history or genetic relatedness of the haplotypes) was reconstructed. Using this analysis, Theodorakis et al.[95] found that 27 migration events were needed to explain the tree, for 23 of which the direction of migration could be determined. Of these, 13 migration events involved movement of animals from the reference areas into the contaminated areas, and 6 involved migration from the contaminated areas into the reference areas. This is consistent with their conclusion that the shared haplotypes represented migrant individuals. The greatest number of migration events involved animals migrating from R1 $\rightarrow$ T4 (n = 7) and R1 $\rightarrow$ T1 (n = 5).

Therefore, this analysis supports the hypothesis that R1 serves as a significant source of migrant individuals for the contaminated areas. Furthermore, it changed the perception of the ecology of the ground-zero sites. The data are suggestive that the ground-zero sites are in fact sinks that are populated with a relatively high proportion of migrant individuals. For purposes of ecological risk assessment and ecotoxicological studies using biomarkers, the population genetic data in this case proved critical in obtaining a clear assessment of effects.

### 41.4.6 Bank Voles

The meltdown at Chornobyl caused the worst nuclear power plant disaster, highly contaminating the area surrounding the reactor. Unfortunately, the impacts of this contamination upon wildlife remain largely undetermined. Two studies have been published, however, that shed light on the genetic effects of chronic exposure to radiation in natural populations near Chornobyl.

Matson et al.[98] studied genetic effects of radiation on the bank vole, *Clethrionomys glareolus*, because it exhibits the highest internal levels of $^{134,137}$Cesium and $^{90}$Strontium among rodent species living in this area. Samples were collected over time from two contaminated sites, Glyboke Lake and the Red Forest (which has the highest levels of radiation of any area studied). Samples were also taken from one reference area, Oranoe, located outside the 30-km restriction zone. From these samples, a 291-base-pair region of the highly variable mtDNA control region, the D-loop, was sequenced and used to identify haplotypes. This study showed significantly higher genetic diversity in contaminated sites in comparison to the reference site.

Baker et al.[99] continued the previous study, monitoring spatial and temporal dynamics of haplotype frequencies and genetic diversity. In addition to sampling the same reference and experimental sites used in Matson et al.,[98] two additional reference sites were added, Chista and Nedanchichy (which has the lowest levels of radiation of any area studied). Sequential sampling of populations consistently showed significantly higher genetic diversity in experimental sites as compared to the reference sites. However, based on these data alone, the cause of increased variation in animals from contaminated sites was not determined. Two possible explanations to explain this observation were offered. First, increased genetic diversity could have resulted from mutations induced by exposure to radiation. Alternatively, it could be that the populations of *C. glareolus* were extirpated as a result of the meltdown of the reactor, causing an ecological sink. Consequently, multiple founder effects of animals emigrating from different areas resulted in an increase in genetic diversity in this area. To distinguish between these two hypotheses, monitoring studies are now being conducted at these sites. These studies include establishing pedigrees for resident bank voles using microsatellite analyses and monitoring changes in genetic diversity through time. Such data should reveal if new variants are evolving within the populations or are introduced by immigration.

## 41.5 SUMMARY

Laboratory studies have identified as genotoxic a large number of chemicals that are commonly found in contaminated environments. The adverse health effects of genotoxic chemicals on organisms often result from the consequence of direct DNA damage. While the majority of chemical-induced alterations to DNA are repaired, some are either not repaired or improperly repaired, leading to mutations and changes in the genetic make-up of affected individuals.[13] Genetic alterations in somatic tissue of an individual may not only have a number of immediate effects on the cells involved, but they may also provide an important clue as to the nature of the stress experienced by a population. Nevertheless, the most profound and long-lasting environmental effects occur at higher levels of biological organization.[3]

When mutations occur in germ cells, they can potentially be passed to the offspring. Extrapolation of observations made at the somatic-cell level of biological organization to events occurring in germ cells in the same organism is difficult due to the inherent difference in sensitivity of these types of cells to genotoxicants. Individuals carrying harmful mutations are often eliminated from the population due to a strong selection against less fit and less well-adapted individuals.[9,100] However, the main concern for induced heritable mutations is that they will lower the reproductive output of an affected population since affected individuals have relatively low viability and fertility.

In addition, toxic chemicals, which do not interact directly with DNA, can also cause genetic effects on a population due to the selection or elimination of resistance or sensitive individuals in a population. Thus, adaptation can result in a narrowing of genetic diversity, which in turn is exasperated by associated ecological influences such as genetic drift, bottlenecks, and inbreeding, as well as the risk of producing the fixation of deleterious alleles.[3]

Distinctive groups that differ genetically exist within natural wildlife populations. Variation in responses of organisms within these groups to toxic stress can be attributed in part to their genetic variations. In addition, contamination may influence the genetic composition of individuals within these populations and impose new or additional selection pressures. Stressed organisms are even more vulnerable to additional stressors, which may further jeopardize the survival of the population. Thus, the degree of genetic variation maintained by a population may be evidence of its capacity to survive future environmental alterations by tempering or modulating the stress-related effects of pollution. Genotypes that survive pollutant exposure may represent those individuals that are most tolerant to environmental stressors. For more detailed discussions on this topic, the reader should consult the scientific literature.[2,3,9,53] The work of Belfiore and Anderson[104] on distinguishing between genetic alterations caused by natural processes and contaminants is especially relevant.

Genetic markers offer the most direct approach for measurement of genetic diversity. Two approaches pertaining to selection by anthropogenic stressors are found in the literature. The first is to identify genetic markers linked to either resistance or sensitivity to particular stressors or combination of stressors in select species, and the second is to employ a suite of genetic markers to examine population-level responses. Despite its limited resolution, allozyme analysis remains the simplest and most rapid technique for surveying genetic diversity in single-copy nuclear genes. The appeal of the PCR-based technologies is based on several factors including the simplicity of the procedure, the requirement for small amounts of DNA, and the potential to access many genetic loci. Employing genetic markers to assess genetic diversity of natural populations appears to be a promising and useful approach for determining the effects of environmental pollution on ecosystems.[3]

Since the more significant ecological effects of contamination usually occur at the population or higher levels of biological organization, monitoring changes in population genetic structure will become a valuable component of ecological risk assessments.[101] Research efforts in genetic ecotoxicology that deal with the ecological significance of exposure are rapidly expanding, as evidenced by the publication of a Special Issue on Environmental Population Genetics[102] in the scientific journal *Ecotoxicology*. This special issue is a compilation of several current scientific research endeavors employing different approaches including classical allozyme analysis and genetic markers for studying the diversity (genetic variation) of population. These studies describe approaches and methodologies for the detection of stressor-induced effects on genetic diversity of populations, and several of them detail important case studies that demonstrate the usefulness of a particular approach to a given environmental problem.

## ACKNOWLEDGMENTS

Opinions expressed or implied in this chapter are solely those of the authors and are not those of any institution or federal agency that participated in or sponsored the studies described herein. LRS is president of LR Shugart and Associates, Inc., a consulting firm specializing in ecotoxicological issues. Funding for research on bank voles was provided in part by a contract (DE-FC09–96SR18546) between the United States Department of Energy (DOE) and the University of Georgia. One of the participants of the bank vole project, AMB, is supported by a Howard Hughes Medical Institute grant through the Undergraduate Biological Sciences Education Program to Texas Tech University. We thank R. J. Baker for his comments. JWB is presently funded by grant ES04917 from NIEHS. Studies of kangaroo rats and sunfish were funded in part by DOE under Cooperative Agreement No. DE-FC04–95AL85832. This chapter is contribution no. 103 of the Center for Biosystematics and Biodiversity at Texas A&M University.

## REFERENCES

1. Calow, P., Foreword, in *Genetics and Ecotoxicology,* Forbes, V. E., Ed., Taylor and Frances, Philadelphia, 1998, p. ix.
2. Anderson, S., Sadinski, W., Shugart, L., Bussard, P., Depledge, M., Ford, T., Hose, J., Stegeman, J., Suk, W., Wirgin, I., and Wogan, G., Genetic and molecular ecotoxicology: A research framework, *Environ. Health Perspec.,* 102, 3, 1994.
3. Bickham, J. W. and Smolen, M. J., Somatic and heritable effects of environmental genotoxins and the emergence of evolutionary toxicology, *Environ. Health Perspect.,* 102, 25, 1994.
4. Peakall, D. B. and Shugart, L. R., Biomarkers, in *Encyclopedia of Environmental Analysis and Remediation,* Meyers, R. A., Ed., John Wiley and Sons, New York, 1998, 132.
5. Marnett, L. J., Frontiers in molecular toxicology, *Chem. Res. Toxicol.,* 6, 739, 1993.
6. Dieter, M. P., Identification and quantification of pollutants that have the potential to affect evolutionary processes, *Environ. Health Perspect.*, 101, 278, 1993.
7. Thaler, D. S., The evolution of genetic intelligence, *Science*, 264, 224, 1994.
8. Chasan, R., Molecular biology and ecology: A marriage of more than convenience, *Plant Sci. News,* 1143, 1991.
9. Depledge, M. H., Genetic ecotoxicology: An overview, *J. Exp. Mar. Biol. Ecol.,* 200, 57, 1996.
10. Shugart, L. R., Biological monitoring: Testing for genotoxicity, in *Biological Markers of Environmental Contaminants*, McCarthy, J. F. and Shugart, L. R., Eds., Lewis Publishers, Boca Raton, FL, 1990, 205.
11. Shugart, L. R., Structural damage to DNA in response to toxicant exposure, in *Genetics and Ecotoxicology,* Forbes, V. E., Ed., Taylor and Frances, Philadelphia, 1998, p. 151.
12. Shugart, L. R., DNA damage as a biomarker of exposure, *Ecotoxicology*, 9, 329, 2000.
13. Shugart, L. R., Bickham, J., Jackim, G., McMahon, G., Ridley, W., Stein, J., and Steiner, S., DNA alterations, in *Biomarkers: Biochemical, Physiological, and Histological Markers of Anthropogenic Stress,* Huggett, R., Kimerie, R., Mehrle, P., and Bergman H., Eds., Lewis Publishers, Boca Raton, FL, 1992, 127.
14. Dixon, D. R. and Wilson, J. T., Genetics and marine pollution, *Hydrobiologia,* 420, 29, 2000.
15. Wirgin, I. and Theodorakis, C. W., Molecular biomarkers in aquatic organisms: DNA- and RNA-based endpoints, in *Biological Indicators of Aquatic Ecosystem Health,* Adams, S. M., Ed., American Fisheries Society, Bethesda, MD, 2002, 43.
16. Shugart, L. R., Molecular markers to toxic agents, in *Ecotoxicology a Hierarchical Treatment*, Newman, M. C. and Jagoe, C. H., Eds., Lewis Publishers, Boca Raton, FL, 1996, p. 131.
17. Savva, D., The use of arbitrarily primed PCR(AP-PCR) fingerprinting to detect exposure to genotoxic chemicals, *Ecotoxicology,* 9, 341, 2000.
18. Qu, S.-X., Bai, C.-L., and Stacey, N. H., Determination of bulky DNA adducts in biomonitoring of carcinogenic chemical exposures: Features and comparison of current techniques, *Biomarkers*, 2, 3, 1997.

19. Jones, N. J. and Parry, J. M., The detection of DNA adducts, DNA base changes and chromosome damage for the assessment of exposure to genotoxic pollutants, *Aquat. Toxicol.*, 22, 323. 1992.
20. Pfau, W., DNA adducts in marine and freshwater fish as biomarkers of environmental contamination, *Biomarkers,* 2, 145, 1997.
21. Harvey, J. S. and Parry, J. M., Application of the $^{32}$P-postlabeling assay for the detection of DNA adducts: False positives and artifacts and their implications for environmental biomonitoring, *Aquat. Toxicol.*, 40, 293, 1998.
22. Balk, L., (http://www.cefas.co.uk/bequalm),
23. Rydberg, B., The rate of strand separation in alkali of DNA of irradiated mammalian cells, *Radiat. Res.*, 61, 274, 1975.
24. Kohn, K. W., Erickson, L. C., Ewig, A. G., and Friedman, C. A., Fractionation of DNA from mammalian cells by alkaline elution, *Biochemistry,* 15, 4629, 1976.
25. Shugart, L. R., Quantitation of chemically induced damage to DNA of aquatic organisms by alkaline unwinding assay, *Aquat. Toxicol.*, 13, 43, 1988.
26. Theodorakis, C. W., D'Surney, S. J., and Shugart, L. R., Detection of genotoxic insult as DNA strand breaks in fish blood cells by agarose gel electrophoresis, *Environ. Toxicol. Chem.*, 7, 1023, 1994.
27. Fairbairn, D. W., Olive, P. L., and O'Neill, K. L., The comet assay: A comprehensive review, *Mut. Res.,* 339, 37, 1995.
28. Schmid, W., The micronucleus test for cytogenetic analysis, in *Chemical Mutagens, Principles and Methods for Their Detection, Vol. 6.*, Hollander, A., Ed., Plenum Press, New York, 1976, p. 31.
29. Bickham, J. W., Flow cytometry as a technique to monitor the effects of environmental genotoxins on wildlife populations, in *In Situ Evaluation of Biological Hazard of Environmental Pollutants,* Sandhu, S., Lower, W. R., DeSerres, F. J., Suk, W. A., and Tice, R. R., Eds., Environmental Research Series Vol. 38., Plenum Press, New York, 1990, p. 97.
30. Guengerich, F. P., Cytochrome P450 enzymes, *Am. Sci.*, 81, 440, 1993.
31. Kriek, E., Rojas, M., Alexandrov, K., and Bartsch, H., Polycyclic aromatic hydrocarbon-DNA adducts in humans: Relevance as biomarkers for exposure and cancer risk, *Mut. Res.*, 400, 215, 1998.
32. Fu, P. P. and Herreno-Saenz, D., Nitro-polycyclic aromatic hydrocarbons: A class of genotoxic environmental pollutants, *J. Environ. Sci. Health, Part C: Environ. Carcinogen. Ecotoxicol. Rev.*, C17, 1, 1999.
33. Fu, P. P., von Tungeln, L. S., Chiu, L-H., and Own, Z. Y., Halogenated-polycyclic aromatic hydrocarbons: A class of genotoxic environmental pollutants, *J. Environ. Sci. Health, Part C: Environ. Carcinogen. Ecotoxicol. Rev.*, C17, 71, 1999.
34. Swenberg, J. A., Bogdanffy, M. S., Ham, A., Holt, S., Kim, A., Morinello, E. J., Ranasinghe, A., Scheller, N., and Upton, P. B., Formation and repair of DNA adducts in vinyl chloride- and vinyl fluoride-induced carcinogenesis, *IARC Sci. Publ. (FRANCE), 1999*, 150, 29, 1999.
35. Wang, J. S. and Groopman, J. D., DNA damage by mycotoxins, *Mut. Res.*, 424, 167, 1999.
36. Pickering, R. W., A toxicological review of polycyclic aromatic hydrocarbons, *J. Toxicol. Cutaneous Ocular Toxicol.*, 18, 101, 1999.
37. Van Zeeland, A. A., Molecular dosimetry of chemical mutagens: Relationship between DNA adduct formation and genetic changes analyzed at the molecular level, *Mut. Res.*, 353, 123, 1996.
38. Christie, N. T. and Costa, M., *In vitro* assessment of the toxicity of metal compounds. III. Effects of metals on DNA structure and function in intact cells, *Biol. Trace Element Res.*, 555, 1983.
39. Snow, E. T., Metal carcinogenesis: Mechanistic implications, *Pharmacol. Ther. J.*, 53, 31, 1992.
40. Singh, J. J., McLean, A., Pritchard, D. E., Montaser, A., and Patierno, S. R., Sensitive quantitation of chromium-DNA adducts by inductively coupled plasma mass spectrometry with a direct injection high-efficiency nebulizer, *Toxicol. Sci.*, 46, 260, 1998.
41. Hartwig, A., Role of DNA repair inhibition in lead- and cadmium-induced genotoxicity: A review, *Environ. Health Perspec.*, 102, 45, 1994.
42. Stohs, S. J. and Bagchi, D., Oxidative mechanisms in the toxicity of metal ions, *Free Radical Biol. Med.*, 18, 321, 1995.
43. Muscarella, D. E., Keown, J. F., and Bloom, S. E., Evaluation of the genotoxic and embryotoxic potential of chlorpyrifos and its metabolites *in vivo* and *in vitro*, *Environ. Mutagen.*, 6, 13, 1984.
44. Vijayaraghavan, M. and Nagarajan, B., Mutagenic potential of acute exposure to organophosphorus and organochlorine compounds, *Mut. Res.*, 321, 103, 1994.

45. Huang, Q., Wang, X., Liao, Y., Kong, L., Han, S., and Wang, L., Discriminant analysis of the relationship between genotoxicity and molecular structure of organochlorine compounds, *Bull. Environ. Contam. Toxicol.*, 55, 796, 1995.
46. Campana, M. A., Panzeri, A. M., Moreno, V. J., and Dulout, F. N., Genotoxic evaluation of the pyrethroid lambda-cyhalothrin using the micronucleus test in erythrocytes of the fish *Cheirodon interruptus interruptus*, *Mut. Res. Genet. Toxicol. Environ. Mutagen.*, 438, 159, 1999.
47. Little, J. B., Radiation carcinogenesis, *Carcinogenesis*, 21, 397, 2000.
48. Shulte-Frohlinde, D. and Vonsonntag, C., Radiolysis of DNA and model systems in the presence of oxygen, in *Oxidative Stress*, Seiss, H., Ed., Academic Press, London, 1985, p. 11.
49. Ananthaswamy, H. A. and Pierceall, W. E., Molecular mechanisms of ultraviolet radiation carcinogenesis, *Photochem. Photobiol.*, 52, 1119, 1990.
50. Scharffetter-Kochanek, K., Wlaschek, M., Brenneisen, P., Schauen, M., Blaudschun, R., and Wenk, J., UV-induced reactive oxygen species in photocarcinogenesis and photoaging, *Biol. Chem. Hoppe-Seyler*, 378, 1247, 1997.
51. Malyapa, R. S., Ahern, E. W., Straube, W. L., Moros, E. G., Pickard, W. F., and Roti, J. L., Measurement of DNA damage after exposure to 2450 MHz electromagnetic radiation, *Radiat. Res.*, 148, 608, 1997.
52. McCann, J., Kheifets, L., and Rafferty, C., Cancer risk assessment of extremely low frequency electric and magnetic fields: A critical review of methodology, *Environ. Health Perspect.*, 106, 701, 1998.
53. Bickham, J. W., Sandhu, S., Hebert, P. D. N., Chikhi, L., and Athwal, R., Effects of chemical contaminants on genetic diversity in natural populations: Implications for biomonitoring and ecotoxicology, *Mut. Res.*, 463, 33, 2000.
54. D'Surney, S. J., Shugart, L. R., and Theodorakis, C. W., Genetic markers and genotyping methodologies: An overview, *Ecotoxicology*, 10, 201, 2001.
55. Gillespie, R. B. and Guttman, S. I., Chemical-induced changes in the genetic structure of populations, effects on allozymes, in *Genetic Ecotoxicology*, Forbes, V. E., Ed., Taylor and Frances, Philadelphia, 1998, p. 55.
56. Gillespie, R. B. and Guttman, S. I., Effects of contaminants on the frequencies of allozymes in populations of central stonerollers, *Environ. Toxicol. Chem.*, 8, 309, 1988.
57. Guttman, S. I., Population genetic structure and ecotoxicology, *Environ. Health Perspect.*, 102, 97, 1994.
58. Changon, N. L. and Guttman, S. I., Differential survivorship of allozyme genotypes in mosquitofish populations exposed to copper or cadmium, *Environ. Toxicol. Chem.*, 8, 319, 1989.
59. Changon, N. L. and Guttman, S. I., Biochemical analysis of allozyme copper and cadmium tolerance in fish using starch gel electrophoresis, *Environ. Toxicol. Chem.*, 8, 1141, 1989.
60. Newman, M. C., Diamond, S. A., Mulvey, M., and Dixon, P., Allozyme genotype and time to death of mosquitofish, *Gambusia affinis* (Baird and Girard) during acute toxicant exposure: A comparison of arsenate and inorganic mercury, *Aquat. Toxicol.*, 15, 141, 1989.
61. Mulvey, M., Newman, M. C., Chazal, A., Keklak, M. M., Heagler M. G., and Hales, L. S., Jr., Genetic and demographic responses of mosquitofish (*Gambusia holbrooki* Girard 1859) populations stressed by mercury, *Environ. Toxicol. Chem.*, 14, 1411, 1995.
62. Kramer, V. J. and Newman, M. C., Inhibition of glucose phosphate isomerase allozymes of the mosquitofish, *Gambusia holbrooki*, by mercury, *Environ. Toxicol. Chem.*, 13, 9, 1994.
63. Kramer, V. J., Newman, M. C., Mulvey, M., and Ultsch, G. R., Glycolysis and Krebs cycle metabolites in mosquitofish, *Gambusia holbrooki*, Girard 1859, exposed to mercuric chloride: Allozyme genotype effects, *Environ. Toxicol. Chem.*, 11, 357, 1992.
64. Diamond, S. A., Newman, M. C., Mulvey, M., and Guttman, S. I., Allozyme genotype and time-to-death of mosquitofish, *Gambusia holbrooki*, during acute inorganic mercury exposure: A comparison of populations, *Aquat. Toxicol.*, 21, 119, 1991.
65. Lee, C. J., Newman, M. C., and Mulvey, M., Time to death of mosquitofish (*Gambusia holbrooki*) during acute inorganic mercury exposure: Population structure effects, *Arch. Environ. Contam. Toxicol.*, 22, 284, 1992.
66. Newman, M. C. and Jagoe, R. H., Allozymes reflect the population-level effect of mercury: Simulations of the mosquitofish (*Gambusia holbrooki* Girard) GPI-2, *Ecotoxicology*, 7, 141, 1998.
67. McMahon, G., Huber, L. J., Moore, M. J., Stegeman, J. J., and Wogan, G. N., Mutations in c-Ki-ras oncogenes in diseased livers of winter flounder from Boston Harbor, *Proc. Natl. Acad. Sci. U.S.A.*, 87, 841, 1990.

68. Moore, M. J., Shea, D., Hillman, R. E., and Stegeman, J. J., Trends in hepatic tumours and hydropic vacuolation, fin erosion, organic chemicals and stable isotope ratios in winter flounder from Massachusetts, USA, *Mar. Pollut. Bull.*, 32, 458, 1996.
69. Murchelano, R. A. and Wolke, R. E., Neoplasms and nonneoplastic liver lesions in winter flounder, *Pseudopleuronectes americanus*, from Boston Harbor, Massachusetts, *Environ. Health Perspect.*, 90, 17, 1991.
70. Smolowitz, R. and Leavitt, D., Neoplasia and other pollution associated lesions in *Mya arenaria* from Boston Harbor, *J. Shellfish Res.*, 15, 520, 1996.
71. Varanasi, U., Reichert, W. L., and Stein, J. E., $^{32}$P-postlabeling analysis of DNA adducts in liver of wild English sole (*Parophrys vetulus*) and winter flounder (*Pseudopleuronectes americanus*), *Cancer Res.*, 49, 1171, 1989.
72. Wirgin, I. and Waldman, J. R., Altered gene expression and genetic damage in North American fish populations, *Mut. Res.*, 399, 193, 1998
73. Cooper, P. S., Vogelbein, W. K., and van Veld, P. A., Altered expression of the xenobiotic transporter P-glycoprotein in liver and liver tumors of mummichog (*Fundulus heteroclitus*) from a creosote-contaminated environment, *Biomarkers,* 4, 48, 1999.
74. Baumann, P. C., Epizootics of cancer in fish associated with genotoxins in sediment and water, *Mut. Res. Rev. Mut. Res.,* 411, 227, 1998.
75. Myers, M. S., Johnson, L. L., Hom T., Collier, T. K., Stein, J. E., and Varanasi, U., Toxicopathic hepatic lesions in subadult English sole (*Pleuronectes vetulus*) from Puget Sound, Washington, USA: Relationships with other biomarkers of contaminant exposure, *Mar. Environ. Res.,* 45, 47, 1998.
76. Reichert, W. L., Myers, M. S., Peck-Miller, K., French, B., Anulacion, B. F., Collier, T. K., Stein, J. E., and Varanasi, U., Molecular epizootiology of genotoxic events in marine fish: Linking contaminant exposure, DNA damage, and tissue-level alterations, *Mut. Res. Rev. Mut. Res.*, 411, 215, 1998.
77. French, B. L., Reichert, W. L., Hom, T., Nishimoto, M., Sanborn, H. R., and Stein, J. E., Accumulation and dose-response of hepatic DNA adducts in English sole (*Pleuronectes vetulus*) exposed to a gradient of contaminated sediments, *Aquat. Toxicol.,* 36, 1, 1996.
78. Peck-Miller, K. A., Meyers, M., Collier, T. K., and Stein, J. E., Complete cDNA sequence of the Ki-*ras* proto-oncogene in the liver of wild English sole (*Pleuronectes vetulus*) and mutation analysis of hepatic neoplasms and other toxicopathic liver lesions, *Mol. Carcinogen.*, 23, 207, 1998.
79. Myers, M., Anulacion, B., French, B., Hom, T., Reichert, W., Hufnagle, L., and Collier, T., Biomarker and histopathologic responses in flatfish following site remediation in Eagle Harbor, WA, *Mar. Environ. Res.*, 50, 435, 2000.
80. Malins, D. C. and Haimanot R., The etiology of cancer: Hydroxyl radical-induced DNA lesions in histologically normal livers of fish from a population with liver tumors, *Aquat. Toxicol.*, 20, 123, 1991.
81. Malins, D. C., Polisar, N. L., Garner, M. M., and Gunselman, S. J., Mutagenic DNA base modifications are correlated with lesions in nonneoplastic hepatic tissue of the English sole carcinogenesis model, *Cancer Res.,* 56, 5563, 1996.
82. Shugart, L. R., DNA damage as an indicator of pollutant-induced genotoxicity, in *Aauatic Toxicology and Risk Assessment: Sublethal Indicators of Toxic Stress, Vol. 13, STP 1096*, Landis, W. G. and van der Schalie, W. H., Eds., American Society for Testing and Materials, Philadelphia, 1990, p. 348.
83. Shugart, L. R., Environmental genotoxicology, in *Fundamentals of Aquatic Toxicology, 2nd ed.*, Rand, G., Ed., Taylor and Francis, Washington, D.C., 1995, p. 405.
84. Shugart, L. R. and Theodorakis, C. W., Environmental genotoxicity: Probing the underlying mechanisms, *Environ. Health Perspect.,* 102, 13, 1994.
85. Nadig, S. G., Lee, K. L., and Adams, S. M., Evaluating alterations of genetic diversity in sunfish populations exposed to contaminants using RAPD assay, *Aquat. Toxicol.,* 43, 163, 1998.
86. Theodorakis, C. W., Swartz, C. D., Rogers, W. J., Bickham, J. W., Donnely, K. C., and Adams, S. M., Relationship between genotoxicity, mutagenicity, and fish community structure in a contaminated stream, *J. Aquat. Ecosyst. Stress Recovery*, 7, 131, 2000.
87. Sutter, G. and Loar, J., Weighing the ecological risk of hazardous waste sites: The Oak Ridge case, *Environ. Sci. Technol.,* 26, 432, 1992.
88. Theodorakis, C. W., Blaylock, G. B., and Shugart, L. R., Genetic ecotoxicology, I. DNA integrity and reproduction in mosquitofish exposed *in situ* to radionuclides, *Ecotoxicology*, 5, 1, 1996.

89. Theodorakis, C. W. and Shugart, L. R., Genetic ecotoxicology. II. Population genetic structure in radionuclide-contaminated mosquitofish (*Gambusia affinis*), *Ecotoxicology*, 6, 335, 1997.
90. Theodorakis, C. W., Elbl, T., and Shugart, L. R., Genetic ecotoxicology. IV: Survival and DNA strand breakage is dependant on genotype in radionuclide-exposed mosquitofish, *Aquat. Toxicol.*, 45, 279, 1999.
91. Theodorakis, C. W. and Shugart, L. R., Genetic ecotoxicology. III: The relationship between DNA strand breaks and genotype in mosquitofish exposed to radiation, *Ecotoxicology*, 7, 227, 1998.
92. Theodorakis, C. W., Bickham, J. W., Elbl, T., Shugart, L. R., and Chesser, R. K., Genetics of radionuclide-contaminated mosquitofish populations and homology between *Gambusia affinis* and *G. holbrooki*, *Environ. Toxicol. Chem.*, 10, 1992, 1998.
93. Theodorakis, C. W. and Shugart, L. R., Natural selection in contaminated habitats: A case study using RAPD genotypes, in *Genetics and Ecotoxicology*, Forbes, V. E., Ed., Taylor and Francis, Philadelphia, 1998, p. 123.
94. Greeley, M., (http://www.esd.gov/people/greeley/greeley.html).
95. Theodorakis, C. W., Bickham, J. W., Lamb, T., Medica, P. A., and Lyne, T. B., Integration of genotoxicity and population genetic analyses in kangaroo rats (*Dipodomys merriami*) exposed to radionuclide contamination at the Nevada Test Site, *Environ. Toxicol. Chem.*, 20, 317, 2001.
96. French, N. R., Maza, B. G., Hill, H. O., Aschwanden, A. P., and Kaaz, H. W., A population study of irradiated desert rodents, *Ecol. Monogr.*, 44, 45, 1974.
97. Slatkin, M. and Maddison, W. P., A cladistic measure of gene flow inferred from the phylogenies of alleles, *Genetics*, 123 603, 1989.
98. Matson, C. W., Rodgers, B. E., Chesser, R. K., and Baker, R. J., Genetic diversity of *Clethrionomys glareolus* populations from highly contaminated sites in the Chernobyl Region, Ukraine, *Environ. Toxicol. Chem.*, 19, 2130, 2000.
99. Baker, R. J., Bickham, A. M., Bondarkov, M., Gaschak, A., Matson, C. W., Rodgers, B. E., Wickliffe, J. K., and Chesser, R. K., Consequences of highly polluted environments on population structure: The Bank Vole (*Clethrionomys glareolus*) at Chernobyl, *Ecotoxicology*, 10, 211, 2001.
100. Wurgler, F. E. and Kramers, P. G. N., Environmental effects of genotoxins (ecogenotoxicology), *Mutagenesis*, 7, 321, 1992.
101. Belfiore, N. M. and Anderson, S. L., Genetic patterns as a tool for monitoring and assessment of environmental impacts: The example of genetic ecotoxicology, *Environ. Monitor. Assess.*, 51, 465, 1998.
102. Shugart, L. R., Special issue of *Ecotoxicology* on environmental population genetics, *Ecotoxicology*, 10, 199, 2001.

CHAPTER 42

# The Role of Ecotoxicology in Industrial Ecology and Natural Capitalism

**John Cairns, Jr.**

## CONTENTS

## 42.1 INTRODUCTION

Achieving sustainable use of the planet will require a new view of the relationship between human society and the environment and a concomitant acceptance of responsibility by those now living to provide a quality life for their descendants for an indefinite period of time. Both industrial ecology and natural capitalism provide useful guidelines and case histories on how these two changes might be achieved. Industrial ecology is the study of the flows of materials and energy in the industrial environment and the effects of those flows on natural systems.[1,2] Natural capitalism refers to the increasingly critical relationship between natural capital (i.e., natural resources), living systems and the ecosystem services they provide, and human-made capital.[3]

The idea of sustainability, which includes concepts from both industrial ecology and natural capitalism, regards the environment as an inclusive system that serves as the ecological life-support system upon which humans depend. Both industrial ecology and natural capitalism espouse mimicking natural cyclic phenomena so that wastes can be immediately reincorporated into natural systems in ways that benefit natural systems and preserve their integrity. The ecological life-support system upon which humans depend provides services, such as the maintenance of the atmospheric gas balance, that are frequently not replaceable by present technologies and that the present economic system does not value appropriately because they are "free." However, since the most important ecosystem services cannot, at present, be replaced by technology and are essential to human survival, they are priceless. Both industrial ecology and natural capitalism advocate the preservation of natural capital (old growth forests, soils, biotic diversity from which many medicines are developed, etc.) so that it can continue to provide "interest" — i.e., ecosystem services.

These new approaches will require a global ethos or set of guiding values based on equity and fairness in the use and protection of natural systems, not only among members of the human species but in relationships with other species as well. These approaches are basically ethical constructs, so implementing them will require sound science that assures a relationship between human society and natural systems and that protects the health and integrity of natural systems and facilitates the accumulation, rather than the destruction, of natural capital. Since sound science is essential for implementation, ecotoxicologists have the opportunity to play a dominant role by determining how to reincorporate the wastes of human society into natural systems to the benefit of natural systems and to reduce or eliminate the stress that this process presently produces. Essentially, since natural systems cannot speak for themselves, ecotoxicology will provide the informational feedback loops that guide the new relationship between human society and natural systems.

Because both human society and natural systems are dynamic and multivariate, the future is inherently uncertain. As additional evidence accumulates, the uncertainty will be reduced, and precautionary measures will reduce risks of major catastrophes.

Natural capitalism and two of its subdisciplines — industrial and municipal ecology — are essential components in developing a sustainable relationship with natural systems and protecting both natural capital and the delivery of ecosystem services. The role of ecotoxicologists in this sustainability approach includes: (1) shifting goals and endpoints from an absence of harm to evidence of health, (2) increasing both temporal and spatial scales of ecotoxicological studies, (3) achieving a critical mass of qualified personnel, (4) including demographic change in ecotoxicological analysis and judgment, (5) developing new ecological thresholds, (6) being prepared for environmental surprises, (7) focusing on design for a quality environment, (8) using ecosystem services as endpoints in ecotoxicological studies, and (9) being prepared for climate change and other events that might destabilize the biosphere and require major adjustments to the process of ecotoxicological testing.

## 42.2 THE INHERENTLY UNCERTAIN FUTURE

McNeill[4] in his excellent book comments that the 20th century qualifies as a peculiar century in environmental history because of the acceleration of so many processes that brought ecological change. In an era when sustainable practices are being widely discussed, although rarely implemented, McNeill[4] notes that China had approximately 3000 years of unsustainable development. It then changed, not to sustainability, but to some new and different kind of unsustainability. In fact, economist Simon[5] actually advocates changing from one unsustainable concept to another, although not expressing it the same way, i.e., resources are infinitely substitutable and the ultimate resources are human ingenuity, creativity, and technology. Thus, resources, including natural systems, can be utilized in an unsustainable way because, when they are depleted, substitutes will inevitably be found through human ingenuity. McNeill[4] notes that imperial China was always adopting new food crops, new technologies, and shifting its trade relations with its neighbors. Thus, it was constantly adapting and then surviving crises. If change is necessary, why not move from an unsustainable system to a sustainable one?

The new factor is the appearance of globalization and increased unsustainable practices on a planetary scale. Even during that 3000-year period just mentioned, Chinese society suffered horrendous periods of starvation and mortality. At present, a different perception of human "rights" has surfaced, that is, the "right" to adequate nourishment, adequate housing, and freedom of assembly and expression. Exercising these rights requires not only an enlightened social contract respected by a substantial majority of persons in human society, but also an adequate resource base per capita that permits going beyond mere survival.

Yet compelling evidence indicates that crucial thresholds of ecological damage have already been crossed and are being crossed regularly at local levels in many parts of the planet. The field of restoration ecology[6,7] will provide much useful information, as will books that cover a variety of ecological indicators.[8]

One major difficulty in using the information will be estimating the quality of the vast quantity of information that will be generated if industrial ecology and natural capitalism acquire even a modest-sized group of practitioners. Regrettably, some data generated will be considered proprietary and not made generally available, either because it might be harmful to the company that generated the information or because of general fears that the information might be used in some way against the organization in the future. Some information may come from unknown sources, from organizations or individuals not generally known to the ecotoxicological community, or from groups that do not have the methodology or other information depicted with sufficient detail for the reader to make an informed judgment on its quality, etc. Even information in peer-reviewed journals may not always be reliable if qualified reviewers cannot be found. Alternatively, high-quality information may suffer inordinate delays because one or more reviewers did not fulfill the professional responsibility of completing a review in a timely manner.

Those who feel that the present societal view is invulnerable would do well to read works that focus on the early empires.[9] It is prudent, not subversive, to consider when a shift in approach is advisable and what the new approach should be. Changes in approach can be thrust upon society by catastrophic events such as the terrorist attacks of September 11, 2001 on the World Trade Center in New York City and the Pentagon in Washington, D.C. Alternatively, they might occur as a result of reason guided by intelligence and the desire to take precautionary measures despite uncertainty because the consequences of no action might well be severe. In the first case, events cause the change in approach. In the second, a management decision must be made despite some degree of uncertainty of the need because the scientific evidence is not yet robust. The National Academy of Engineering[10] is one of a number of organizations that examine technological trajectories to facilitate transitions from one stage to another. Ironically, it is often the cumulative impact of decisions that seem small in isolation from other decisions but, in the aggregate, exert a surprising tyranny over human lives.[11,12]

The perfection of resource partitioning in natural systems is literally the result of millions of years of evolution. The extraordinary diversity of some ecosystems, usually accompanied by low numbers of individuals per species and infrequent dominance by any one species, illustrates how finely tuned this process can become. Keeping in mind that many of the species on the planet have yet to be named and that even those named have functional roles that are poorly understood, it is presumptuous, arguably arrogant, to assume that humans can couple their social system to natural systems in ways that enhance the integrity of both systems. Nevertheless, even modest improvements in this relationship between humans and natural systems should produce dramatic benefits, and the increased knowledge about complex ecosystems should lead to further improvements. As the President's Acid Rain Review Panel of the Office of Technology Assessment[13] noted: "Recommendations based on imperfect data run the risk of being in error; recommendations for inaction pending collections of all the desirable data entail even greater risk of damage." But, as Likens[14] points out, even when reasonably precise information is available, its interpretation can be strongly influenced by human value judgments. He gives the following example. A spokesperson for the proponents of a multimillion-dollar tourist and residential canal estate development stated: "... even though the development would mean the destruction of 53% of the wetland (in the Bateman's Bay area), that would be balanced by the construction of a wetlands education center to teach people of the need to protect wetlands!"

Haeckel's[15] original definition of ecology, despite its focus on animals, is quite relevant today:

> By ecology we mean the body of knowledge concerning the economy of nature; the investigation of the total relations of the animal both to its inorganic and to its organic environment; including, above all, its friendly and inimical relations with those animals and plants with which it comes directly or indirectly into contact.

Note Haeckel's use of the phrase "economy of nature," which gets to the heart of human society's present dilemma. Human society's economy is quite different from nature's economy.[16] The materials that human society produces are all too frequently unsuitable for nature's economy and sometimes, arguably most times, a threat to its integrity.[17] Industrial ecology as well as the quest for sustainable use of the planet is an attempt to develop a harmonious relationship between the relatively new human economy and the ancient economy of nature. After noting that human society's present course is unsustainable, Thompson[18] states: "To reverse that course, we need to emulate the economics of nature. It's the straightest path toward stimulating the crucial changes needed to make sustainable development work."

## 42.3 ACKNOWLEDGING THE OBVIOUS AND ABANDONING DENIAL

Bartlett[19] gives an elegant analysis of the "Hubbert curve,"[20] which empirically approximates the full cycle of the growth, peaking, and subsequent decline to zero of the "production" [(quantity/year) vs. year] of a finite, nonrenewable resource. Bartlett notes that each increase of one billion barrels in the size of the world estimated ultimate oil recovery beyond the value of $2.0 \times 10^{12}$ bbl can be expected to result in a delay of approximately 5.5 days in the date of maximum production. Enormous uncertainties also exist in the amount of natural gas (which has been suggested as an alternative fuel for vehicles) remaining in the United States.[21] The World Resources Institute[22] scorecard shows that many of the important ecological life-support systems are declining in capacity, while some, such as forests, are increasing, and some, such as wood fuel production and coastal recreation, are unknown. Myers,[23] despite an acute awareness of the plethora of bad environmental news, nevertheless feels that these unprecedented environmental problems also invoke opportunities of parallel scope! Myers[23] notes: "The vital question is not 'How can we afford to do the necessary?' it is 'How can we afford not to do it?' The biggest cost will not be to our pocketbooks; it will be to our philosophies."

Aside from the obvious evidence of abused resources, there is also a strong element of denial. In the United States in spring of 2000, gasoline prices increased markedly, as did the number of gas-guzzling sport utility vehicles. An alternative hypothesis is that humans are not in denial; they have just decided that if they are "on the Titanic," they might as well fulfill their dreams as long as there is time to do so. However, human society, although aware of the risks, may very well cross one or more crucial ecological thresholds and provoke a sudden nonlinear effect. In the past, such an event seemed unimaginable. For example, McNeill[4] calls attention to the irony in the 1930 statement of American physicist and Nobel Prize winner Robert Millikan who stated that there was no risk that humanity could do real harm to anything as gigantic as Earth. In the same year, American chemical engineer Thomas Midgley invented chlorofluorocarbons, the chemicals now known to be responsible for thinning the stratospheric ozone layer. Midgley certainly had no idea of the profound global consequences of the widespread use of this new compound.

For ecotoxicologists, the prospect is daunting. New chemical compounds are being produced at an unprecedented rate, and the time permitted for testing has not lengthened, due in large part to intense global economic competition as well as the cost of lengthy testing. Furthermore, testing procedures and predictive modeling provide no absolute guarantees that events, such as eggshell thinning caused by some pesticides in some bird eggs, would be a threat, nor is it likely that ozone depletion would initially be predicted to have major effects. Such revelations require perfect hindsight. Even if the precautionary principle (which asserts that when an activity raises threats of harm to human health or the environment, precautionary measures should be taken, even if some cause-and-effect relationships are not fully established scientifically[24]) is applied and is widely accepted in practice, there must be some persuasive, but not necessarily conclusive, evidence regarding the consequences that will result if precautionary measures are not taken. A substantial portion of the burden of the proof of likely consequences will certainly fall on ecotoxicologists, as it should. At present, political and judicial systems worldwide are simply not prepared to cope with scientific uncertainty, as is quite evident from discussions of global warming, stratospheric ozone depletion, and biotic impoverishment caused by the extinction of species. Clear and compelling evidence exists that human society has the ability to change the biospheric life-support system of the planet — for both better or worse.

## 42.4 INDUSTRIAL AND MUNICIPAL ECOLOGY

Allenby[25] discusses eight unifying themes concerning industrial ecology.

1. "... a globalized economy and society is evolving that will not necessarily be homogenous."
2. "The earth is increasingly ... engineered" and "... human choice and technology determine the structure not only of human lives and environments, but "..." other life forms as well.
3. "... reductionist science ..." must "... be augmented ... by more systems-based, comprehensive approaches."
4. "Policy generally functions in the short term ..." temporally and a limited area spatially.
5. Modern institutions are "... changing in both unparalleled and little recognized ..." ways.
6. Both sustainability and sustainable development are ambiguous terms both "... because of a lack of knowledge ..." and "... because they involve social choice."
7. "Evolution toward an economically and environmentally efficient economy will differentially favor certain industrial sectors and technological systems and disfavor others."
8. Although "... environmental issues are occasionally framed in apocalyptic terms ... [w]hat is threatened from a human perspective ... is the stability of global economic and social systems."

Graedel[26] notes that "technological resources should cycle just as nature's resources do." He gives a superb discussion of the basics of municipal ecology or, as Newcombe[27] names it, "the metabolism of a city." Graedel[26] uses the term *ecocity* and maintains that cities be "... regarded as organisms, and analyzed as such, in an attempt to improve their current environmental performance and long-term sustainability." This view provides a clear opportunity for ecotoxicologists since they are accustomed to determining response of complex systems to chemical and other stresses. However, it will require an array of methods, procedures, and endpoints suited to this new undertaking.

### 42.4.1 Mimicking Natural Cycles

Nature's wastes are most often used by some other part of nature (e.g., dung by the African dung beetle),[28] and there are few waste products produced by any species that are not of immediate value to one or more other species. Some plants and animals produce particularly toxic materials, but they are not usually widely distributed and are not usually persistent. Other species have varied means of avoiding exposure, and, in fact, many toxics are produced in order to avoid predation. Industrial wastes, on the other hand, often have no value to other species or organisms and may be fatal to them.

Although some materials do accumulate, such as fossil fuels, most of nature keeps moving. Industrial ecology attempts to mimic natural cycles, which mostly keep materials moving in a way that does not disrupt the biosphere, and to modify industrial wastes so that they are more amenable to cycling in natural systems. The concept of industrial ecology or industrial symbiosis with natural systems is to model industrial systems after the cycles characteristic of natural systems. A more specific definition of industrial ecology is provided by Graedel and Allenby:[2]

> Industrial ecology is the means by which humanity can deliberately and rationally approach and maintain a desirable carrying capacity, given continued economic, cultural and technological evolution. The concept requires that an industrial system be viewed not in isolation from its surrounding systems, but in concert with them. It is a systems view in which one seeks to optimize the total materials cycle from virgin material, to finished material, to component, to product, to obsolete product, and to ultimate disposal.

Similarly, municipal ecology proposes to model municipal systems (towns, cities, and other concentrations of human population)[26] after the cycles characteristic of natural systems. Both industrial and municipal systems should be designed to cycle materials in such a way that they are used serially, rather than in a once-through system in which raw materials are extracted from natural systems and transformed into products, and then ultimately both the product and the waste are discarded into natural systems without regard for the ability of the systems to assimilate and cycle them. Recycling is not an afterthought, as it now is in most human societies; but, rather, the concept is incorporated into all aspects of a cyclic process. It is not enough that the waste products do no harm to natural systems; rather, the essence of industrial and municipal ecology is that the wastes are amenable to incorporation into the natural systems in a way that will promote ecological integrity and health.

If the waste products of anthropogenic activities are not harmful to natural systems but of benefit to them, there is no justification for transporting the wastes considerable distances from the point of generation. If human society mimicked the economies of nature and produced wastes that are not dangerous to natural systems and a threat to their integrity but, rather, are beneficial to natural systems and improve their health and condition, there would be no need to transport such wastes or store them in containers for long periods of time. If wastes were indeed beneficial to natural systems, then producers of the wastes would be losing a valuable resource by exporting them. In fact, the only reason for exporting wastes is the recognition that they are harmful to human health and the environment. The only justification for doing so is that the exporter is so exceptional that the health of its citizens and its environment are more important than the areas to which the wastes are being exported. The fact that other areas are willing, even eager, to accept these wastes does not in any way diminish the moral dilemma of the exporter.

Regrettably, this dumping of harmful wastes on others has been discussed in economic terms and not in ethical and moral terms, which should make the action unthinkable. Even for those favoring economic thinking, unethical and immoral practices are not good for either the exporter or the recipient. Resolving these ethical issues entails determining how to develop wastes that can be reintroduced into natural systems without harm to the natural systems or to human health. There is no moral or ethical dilemma if the wastes are effectively resources to the recipient. However, there is a dilemma if the wastes are harmful to the exposed organisms or effectively block access to essential resources.

All aspects of each industrial processing system must be examined in terms of its compatibility with natural systems wherever and whenever the two interface. Within this conceptual framework, industrial and natural processes are designed to be more compatible and, ultimately, symbiotic, rather than isolated from each other conceptually, as they now are. Human society must operate in such a way that its waste products are readily and beneficially reincorporated into natural systems. If a judgmental error is made in the ecosystem assimilative capacity for anthropogenic wastes, ecosystem stress should be apparent in one or more of a series of departures from the nominative state, such as ecological integrity, resilience,[29] biotic impoverishment,[30] and increased variability, all of which are exceedingly difficult but not impossible to measure.

The example most often cited for industrial ecology is the Danish industries of Kalundborg.[28,31] In this small city, the industries act as if they are an integrated system or a web linking the "metabolism" of one company with that of the others. For example, the "waste" energy in the form of spent steam from a power plant is used to heat the town, to heat fermentation vats for a pharmaceutical company, and to heat water for aquaculture. Thus, the spent steam does not become an environmentally harmful waste discharge, but rather an economic asset cycled through a system that consists of a web of previously isolated components.

On March 10, 2000, the Physicians for Social Responsibility (PSR) held a conference on drinking water entitled "Drinking water and disease: what every healthcare provider should know." The news media have given enormous attention to disease problems associated with poor drinking water, food processing, and the like. What has received less attention is the fact that chlorination for disinfection of drinking water supplies, although it has helped to control cholera and other diseases that once menaced much of human society, has a downside in that the chemical by-products formed as a result of chlorination may be associated with increased risk of bladder cancer, colorectal cancer, and adverse reproductive and developmental effects.[32] The report also notes that anthropogenic pollutants contaminate ground and surface water that supply drinking water. Leaking underground storage tanks and hazardous waste sites, for example, can contaminate groundwater use for individual and community drinking water supplies. Agricultural and livestock production introduce pesticides, fertilizers, animal waste, and antibiotics into the hydrologic cycle. All sorts of contaminants have been identified in drinking water including trace amounts of caffeine.

Tests to identify complex chemicals with precision are costly and are well beyond the analytical capabilities of most water-treatment facilities, except possibly those of large metropolitan areas. In addition, even when the chemicals have been identified, targeting sources is still problematic because wastes are now moved considerable distances from their site of production to sites within the United States and to some other developed countries. Quality control monitoring of safety of underground storage sites has often been inadequate. Industrial ecology can do much to alleviate this problem by minimizing the production of wastes that are unsuitable for release into the environment and by treating wastes as valuable resources and reusing them in a variety of ways. Spreading wastes over wide areas throughout the environment primarily by means of the transportation system and faulty and poorly monitored long-term storage systems should be remedied.

### 42.4.2 Toxicants in Human Behavioral Problems

New research suggests that millions more children than previously thought might have lead-linked mental impairment,[33] while another study supports a strong link between that exposure and juvenile delinquency. Other such findings have also been reported elsewhere.[34]

Suppose there is significant evidence that these contaminants are capable of producing behavioral changes, particularly in the young, and that they are ubiquitous in the environment. This scenario means, first of all, that reference specimens are impossible to obtain or, at best, are exceedingly rare. Even the investigators studying the problem are likely to have altered behaviors. In addition, if humans are as unique as some believe, tests on surrogate species, however scientifically reproducible, may not extrapolate well to the human species.

Even if the possibility of contaminant-altered behavior is remote (and it appears not to be), the precautionary principle requires that precautionary action be taken, especially when the means for such action, namely industrial ecology, shows considerable promise and has been tested to a significant degree.

Even if the principles, concepts, and procedures of industrial ecology are immediately implemented, the environmental concentrations of persistent chemicals and their various transformation products will remain for some time. Since these chemicals are numerous and likely to have at least some interactions, there are two quite obvious tasks for ecotoxicologists — determine routes of exposure and means to avoid them and determine the rates of degradation and the ways in which particularly important chemicals partition in the environment, which will provide both the means to avoid exposure and the possibility of *in situ* treatment of the highest concentrations.

### 42.4.3 Quality Control Monitoring

Environmental monitoring is carried out for the purpose of determining that previously established quality-control conditions have been met.[35] The quality-control conditions will be those developed by environmental scientists.[36] At the very least, monitoring should provide an early warning that quality-control conditions are not being met and provide a degree of validation of the predictive models developed from ecotoxicological testing. In the early development of this area, both false negatives and false positives are likely to be common, possibly even frequent, especially if multiple lines of evidence are not gathered because of cost-cutting measures. A false positive is a signal that quality-control conditions are not being met when in fact they are, and a false negative is an indication that no adverse effects are occurring when in fact they are. These false positives and negatives will be a source of frustration to the general public and its representatives if their ecotoxicological literacy is not raised well above present levels!

In order for quality control monitoring of ecosystem conditions to be effective, the information systems must, above all, be rapid and comprehensive. Decision makers will likely insist that monitoring systems also be economical, but, as is the case for preventative medicine, the cost of ignorance often far exceeds the cost of the preventative measures. The need for speed in effective biological monitoring systems has been recognized for years.[37,38] Major problems will arise in spatial and temporal complexity at the landscape and larger levels; investigators may initially cope with these problems through the use of echelon analysis[39,40] and various other types of system-level assessment.[41,42] Although methods and procedures for the determination of ecosystem health are still in the early development stages, a considerable body of literature already exists.[43–47]

In any widespread activity, the sampling methods, procedures, and protocols become standardized. There are organizations whose mission may be entirely or substantially devoted to such activities (such as the American Society for Testing and Materials, the American Public Health Association, the American Water Works Association, the European Island Fisheries Committee). When any activity becomes increasingly common, it is important to know how a particular number is derived and whether it is from a biological, chemical, or physical sampling. In addition, when any activity becomes particularly common, measurements are made by persons with less formal training than research investigators. Even research investigators should be obliged to collect some data in the same way in order to facilitate comparisons between studies made in widely differing geographic areas or measurements made over considerable spans of time at the same place.

Researchers tend to be attracted to the latest methodologies and technologies; but for studies covering large temporal and spatial spans, consistency in making the measurements, especially in highly variable systems, results in a marked reduction in uncertainty. "Unknowns" in "round robin" testing evaluations are often used to determine the reliability of the laboratory or individual's analytical procedure. It is often a shock to many environmental professionals when they first find how important standard methods are in courts of law. Courts of law are particularly fond of standard methods for three important reasons: (1) they usually represent a strong consensus by practitioners in a particular field as to how a particular measurement should be made; (2) each step of the analytic process is described in painstaking detail so that it is extraordinarily difficult to inadvertently deviate from the established methodology; and (3) because of their widespread use, especially in compliance to regulatory requirements, they are subjected to continual intense scrutiny, reevaluations, and descriptions of situations, which would produce spurious results.

Standard methods already exist for many chemical and physical parameters, and they already have a well-established role in many other areas that affect the daily lives of humans and the environment. Biological standardized methods exist, but they are not nearly as numerous as those in other categories. Before a methodology has achieved the status of a standard method, it is often termed a provisional method, which has less status than a standard method. However, the provisional method is already attracting more attention than would otherwise be the case had it not entered a standardization process. The disadvantage of standard methods is their extreme rigidity. Rigidity has notable advantages in replication, etc. but is often frustrating when adjustments must be made for conditions of a specific site, many of which are unique.

Protocols are produced by a consensus of carefully selected professionals and have a certain stature in courts of law. Even protocols are regularly revised and subcomponents of them explored in great detail. Standard methods and procedures give confidence to the general public, decision makers, and regulatory agencies that appropriate methods are being used. Modification of protocols gives confidence in the process of decision making, integration of information, and predetermining the normal state that, if not met, calls for immediate corrective action.

In historic times, environmental monitoring was practiced in an informal sense. The nobility in many countries, from India to many parts of Europe and, more recently, in parts of North America, had large nature preserves predominantly used for recreation, such as hunting. They employed gamekeepers, river wardens, and the like to keep a watchful eye on the condition of the preserve and also, especially if it was very large, to drive off poachers. The people charged with preserving the system often had little or no formal education and were definitely not acquainted with the complex ecological and statistical analyses at the system level. Nevertheless, these keepers had direct observation of the condition of the system based on a life-long association with it and an encyclopedic knowledge of its components and processes. If they used poor management skills, they suffered severely and, in some primitive cultures, often died. Today, when most inhabitants of the planet have only a fragmentary understanding of natural systems and very little personal involvement with them, society must rely upon rather complex assessments. In an era when many people have a distrust of science, communicating this complex information to the general public will indeed be a challenge. However, society will suffer if its relationship with natural systems is not managed skillfully.

Espousing keepers is in sharp contrast to most of this discussion. At worst, it might be construed as an argument against science and informed management. It does, however, recognize two realities: (1) there are, at present, insufficient professionals to implement these practices on a large scale and (2) even if they become available, many countries will be too poor to employ the professionals and provide them with adequate equipment. One hopes this situation will change; but, in the meantime, it will probably be essential to implement some older practices that may alert management of the need to acquire, temporarily or permanently, more skilled professionals with better equipment.

### 42.4.4 Cessation of Production of Persistent, Non-Degradable Compounds

The most common justification for producing persistent chemicals, many of which are known to be toxic and all of which have the potential to be toxic or environmentally harmful, is that technology will be developed in the future that will transform them into nonpersistent, harmless chemicals. However, the existence of a need is not likely to create the funding necessary to develop this new technology unless there is evidence that the storage systems are not working. The monitoring costs to assure that storage systems are working are too high. There must be unmistakable, or at least persuasive, evidence that faulty storage sites present an immediate threat to human health or the environment before such practices are halted. The many proponents of a future technological solution never mention an alternative that is more attractive in terms of reducing risk. The alternative is to use human ingenuity, creativity, and technology to produce nonpersistent chemicals that need no storage and do not depend on the development of some future, unknown technology in order to be rid of them. In addition, society is running out of storage space because most people resist such sites in their backyard, near their groundwater supply, or near other aquatic ecosystems.

When the United States Congress developed legislation to identify and decontaminate hazardous waste sites (Toxic Substances Control Act), the number that already existed was startling to most people. The difficulty of determining who was responsible for the hazardous wastes was beyond the capability of the justice system in practically every instance. Instead of the allocated funds going toward cleanup of the storage sites, which would have reduced their potential for harm, most of the money went into legal fees, which regrettably have neither led to substantive reductions in risk nor identified a satisfactory means of addressing the problem. Scientists are the ones who must make the most critical judgments on risks to human health and the environment from toxic substances. Ecotoxicologists, of course, have a role to play, both in the development of nonpersistent chemicals and in the detoxification of persistent chemicals. Since there are no courts of science and technology that are comparable to courts of law, the work of ecotoxicologists is directed by, and all too often suppressed or ignored by, the legal system.

In one view, science is supposed to be value-free; however, personal values of scientists determine what research is carried out, how teaching is designed, and the choice of words and phrases in professional publications and presentations at professional meetings. Most scientists attempt to maintain objectivity, and some even carry the effort to the extreme limit of not expressing any opinion about even the possible extinction of their favorite species. The quest for objectivity is greatly assisted by the peer-review system, but even this rigorous quality control system permits value judgments to appear in many journals, even if they are not clearly labeled as editorials, commentaries, or speculations.

The quest for sustainable use of the planet is basically a value-laden goal. A central value is that it would be a good goal for human society to be able to use the planet indefinitely. Other species and the natural systems they inhabit are viewed as valuable as the ecological life-support system of human society and the source of natural capital upon which other forms of human capital depend. Those who use the term "sustainable development" add a further value, that is, development as it is now understood in human society is good and should never cease. Scientists have an ethical and moral responsibility to speak up when damage being done by toxicants to natural systems destroys or impairs the integrity of those systems. Using the excuse of scientific objectivity to avoid speaking out when systems are being destroyed is far worse than stating a value judgment that is clearly identified as such and based on evidence and reason. It is instructive to examine the litigation on the harmful effects of cigarette smoking to see how the public views the absence of ethical values.

### 42.4.5 Industrial Ecology Buffer Zones

The buffer-zone approach is applicable to both private and public structures and, when used in an ecological context, is appropriate when there is an interest in protecting the integrity of both natural and hybrid ecological systems from anthropogenic damage. Each buffer zone requires an array of quality-control conditions and a monitoring system to confirm that they are being met and to provide an early warning if they are not being met.

In the context of sustainable use of the planet, ecotoxicologists will be asked to design and monitor buffer zones around each industry or industrial complex. Ideally, aggregates of industries in which the waste products of one can be used beneficially by another[28] might have a single communal buffer zone. The advantage of this design is cost savings through shared ecotoxicological and monitoring resources. The disadvantage is that incompetent work in one industry may lead to problems for the entire group, particularly if the guilty industry is not willing to admit mistakes and errors. On the positive side, the costs of accidental-spill-contaminant teams for the entire complex could be shared, thereby reducing the cost and, perhaps, increasing the innovative approaches to spill containment.

For those industries willing to share a single group of equipment inspectors for such things as wiring, pump performance, and the like, cost could be substantially reduced. In addition, if an effective industrial-ecological buffer zone is designed, workers should be willing to live closer to their place of employment and, perhaps, in a very well designed buffer zone, even use portions of it for recreation. This design would significantly reduce commuting time, require less area for parking lots, and, since the workers would live closer to the industry, provide an additional incentive for environmentally compatible activities and better maintenance of equipment to prevent environmentally damaging accidental spills, etc. Four major advantages, other than the ones just briefly described, exist to establishing such buffer zones.

1. These zones would serve to validate and confirm models regarding the impact of waste products on ecosystems.
2. By maintaining ecological integrity, buffer zones would provide evidence of protection of adjacent ecosystems. Inevitably, instances would occur where failure to meet standards and criteria would be covered up due to fear of reprisals, fear of losing jobs, or fear of the consequences of poor professional performance. The fate of "whistle blowers" is already well known in society. Generally, their former colleagues turn on them, administrators punish them, and society forgets their courageous acts all too quickly. It is a tribute to human society that, given these daunting obstacles, people still continue to "blow the whistle" when it is abundantly clear that something is wrong. One safeguard against coverups would be to have all buffer-zone plans around industry become an integral part of a bioregional plan with professionals who would check the validity of the conclusions drawn, the interpretation of the data, and the adequacy of the data.
3. Buffer zones can provide an attractive workplace, i.e., a naturalistic setting. This concept is often dismissed as impractical, noncompetitive, and an unnecessary expenditure of capital funds. A Netherlands bank in southeastern Amsterdam, on a site chosen by the workers because of its proximity to their homes, is such a facility.[3] Although a buffer zone is absent because of the highly urban location, this "organic" building integrates art, natural and local materials, sunlight, green plants, energy conservation, quiet, and order, all of which please the employees and did not cost one guilder more than the market average. The information on maintenance is even better since the money spent to put the energy-saving system in place paid for itself in the first 3 months. After an initial start-up period, the bank complex used 92% less energy than an adjacent bank constructed at the same time, representing a saving of 2.9 million guilders per year and making it one of the most energy-efficient buildings in Europe. The energy and time saved by employees because of the short distance to the workplace are additional benefits not included in the figures just mentioned. There are numerous other examples of resource savings. Large urban areas have cut traffic by using metro stations (subway and surface high-speed transport) as development foci. In addition, clustering of housing permits the preservation of substantial natural areas for hiking forest trails

or for footpaths. Residents in some areas do without an automobile by using inexpensive, communal transportation for shopping, cultural events, and the like. Cluster housing can work at an optimal level if a high level of civility and civic responsibility exists. However well cluster housing works in providing some respite to natural areas, it is unlikely to be popular in the United States due to individual choices and, therefore, even less likely for industrial choices. Nevertheless, the success in efforts to conserve resources through cluster housing and communal transportation in Europe, parts of South America, and parts of the United States indicates that one should not abandon hope.

4. Buffer zones provide sites for determining the threshold responses in complex systems. Inevitably, accidents will occur at some industrial sites such as equipment failures, spills, errors in the predictive models, and the like. Thus, any aggregate industrial ecology buffer zones will provide some sites for determining the threshold responses in complex ecosystems (or at least some naturalistic assemblages of plants and animals bearing some resemblance to natural ecosystems, thereby having some useful responses, particularly with regard to determining thresholds). These buffer zones are not great substitutes for natural systems, but there are limitations to the complexities that can be mimicked in laboratory tests, including micro- and mesocosms. One does not wish to place natural systems at risk by carrying out controlled experiments in them; and yet it is desirable to have some evidence of responses at this level of biological organization. While accidents are not to be welcomed, one could take full advantage of them when they happen. Therefore, any test should go well beyond normal routine monitoring so as to gather as much useful information as possible resulting from accidents.

## 42.5 NATURAL CAPITALISM

Hawken and colleagues[3] note that

> … natural capitalism recognizes the critical interdependency between the production and use of human-made capital and the maintenance and supply of natural capital. The traditional definition of capital deals with accumulated wealth in the form of investments, factories, and equipment.

They[3] also note that

> … in actual fact an economy needs four types of capital to function properly:
>
> - Human capital, in the form of labor and intelligence, culture and organization
> - Financial capital, consisting of cash, investments, and monetary instruments
> - Manufactured capital, including infrastructure, machines, tools, and factories
> - Natural capital, made up of resources, living systems, and ecosystem services

The present industrial system uses the first three forms of capital to transform natural capital into the materials that are currently taken for granted daily such as cars, highways, food, television sets, etc. Under the current concept of economic growth, which positions development of human artifacts at a much higher priority level than the preservation of natural systems, the ecological price, as is already becoming evident, is severe. The destruction and degradation of natural systems is occurring at a rate unprecedented in human history.

While stock markets boom worldwide and economic growth continues at rates as much as a 10-year doubling time, the natural capital upon which this economic prosperity depends is speedily declining. Natural capital includes resources that are so familiar that they are taken for granted; examples include water, a suitable atmospheric gas balance, minerals, forests, fisheries, soil, and models for pharmaceutical chemicals. All of these resources in the aggregate form the planet's biosphere, a "skin" on Earth's surface so thin that it cannot be detected from outer space. The biosphere has for all of human history been humankind's ecological life-support system. Only in relatively recent times has the technological life-support system been added that originated from the agricultural revolution first and, subsequently, the industrial revolution. These two support systems have made possible the concomitant dramatic rate of increase in human population size and affluence. However, infinite growth on a finite planet is simply not possible, despite economic claims to the contrary. Common sense indicates that continued increases are simply not rational, and even if they were, they are not desirable. Natural capitalism demands that the supply of natural capital and the services it performs be given the same attention that has been given to the first three types of capital.

Hawken and colleagues[3] note that what might be called "industrial capitalism" does not fully conform to its own accounting principles. It liquidates its capital and calls it income. It neglects to assign any value to the largest stocks of capital it employs, natural resources and living systems, as well as the social and cultural systems that are the basis of human capital. Valuing natural capital is a difficult and imprecise exercise at best,[3] but the value of its services has been calculated as at least U.S. $33 trillion, using 1994 dollars as a reference point.[48] This valuation is undoubtedly a conservative figure, given the fact that things human society cannot live without and cannot replace at any price transcend customary value systems. The Hope Diamond is unique and therefore irreplaceable, but still has value. However, there are many diamonds, none of which are an essential component of the ecological life-support system. Humans have benefited from the latter for their entire existence as a species. If it changed to a far less suitable set of conditions, human society would suffer — possibly even collapse. How does one value an essential life-support system? Arguably, it is undervalued at present.

From another perspective, humans are not an essential component of the ecological life-support system and, therefore, add little value to it. Natural systems functioned without humans for billions of years and, doubtless, would do so if *Homo sapiens* became extinct. How does one value such a system? Certainly it is in human society's enlightened self-interest to take precautionary measures to protect its integrity.

Since the advent of the agricultural and industrial revolutions and the increasing concentration of humans into urban and suburban areas, society's life-support system has become both technological and ecological.[49] When technological services are disrupted for even short periods of time by earthquakes, hurricanes, or snow and ice storms, humans are made well aware of their necessity, even if they take them for granted during normal times. Urban and suburban dwellers, and even those in rural areas, are far less sensitized to ecosystem services because these systems rarely fail so dramatically. Furthermore, damage to the technological system can be repaired much more quickly than damage to the ecological systems. Technological systems in Japanese and European cities that suffered enormous damage during WWII enjoyed rapid restoration. On the other hand, destruction of old-growth forests may require recovery times far longer than an individual human life span, and the recovery is not guaranteed to the extent that it is for technological systems.

Natural capitalism is an attempt to redesign technological systems so that they not only do not threaten ecological systems but actually enhance them and, therefore, increase natural capital. Some ecologists oppose the strategy of interfacing industrial product life cycles and ecological or biological assimilative capacity of natural systems for this purpose.[50] Assimilative capacity is defined as the ability of natural systems to absorb and transform anthropogenic waste without damaging the system's integrity. All of these strategies are based on (1) the assumption that human society's life-support system is both technological and ecological and (2) the abundant evidence that the technological system as presently designed threatens the ecological system, while the reverse is true only to a minor extent. Thus, sustainable use of the planet requires achieving a balance between the technological and ecological systems since the ecological systems are being destroyed and degraded at a rate greater than they are being repaired, and technological systems and other human artifacts are constantly encroaching on the physical space utilized by ecological systems.

Natural capital must be accumulated and preserved. The role of preservation is not a new one, but the accumulation of natural capital requires that ecosystems be moved closer to optimal condition than they now are. For badly degraded or moderately stressed systems, of which there are many on the planet, this restoration is comparatively easy; when conditions begin to approach optimum more closely, much professional judgment and additional knowledge will be required.

## 42.6 NEW ROLES FOR ECOTOXICOLOGISTS

In the context of establishing sustainable use of the planet through emphasis on industrial ecology and natural capitalism, the current roles or activities of ecotoxicologists will probably be abandoned. Furthermore, it is virtually certain that most of these areas of activity will continue to be developed and refined. Major changes and additions that will accommodate industrial ecology and natural capitalism will dramatically enlarge the activities of ecotoxicologists.

### 42.6.1 Shift from Absence of Harm to Presence of Health

A shift from a concentration on absence of harm to a focus on presence of health will probably follow the developments in the field of human health, which has moved from symptoms of disease and malfunction to attributes of well-being and health. Even so, symptoms of ill health and malfunction will continue to be extremely important, just as they presently are in human medicine. Literally billions of humans on the planet are malnourished, afflicted by one or more diseases, or need substantive improvement in their health. Similarly, many of the world's ecosystems are in comparably poor condition and desperately need remedial action. So, the first shift for ecotoxicologists will be in endpoints from those whose presence shows harm or degradation to those whose presence affirms good condition and well-being. Some of these shifts have already occurred at the species level in ecotoxicological testing.

The concept of "health" in ecology, whether at the individual, population, or ecosystem level, necessarily involves value judgments.[51] As Nielsen[52] states: "In the final analysis, what is considered healthy must be reasonable from biological, physical, ethical, and aesthetic points of view, as determined by people." Therefore, health is not a scientific determination per se but a social construct, and its defining characteristics will evolve with time and circumstances.

Ecosystems are more than aggregations of individual species. At each level of biological organization, starting from subcellular levels on up to communities and ecosystems, new properties not apparent at lower levels emerge at each level. These emerging attributes will require much developmental effort in the field of ecotoxicology. However, the search will not be random, and it should not be based on the personal preferences of individual investigators or on those attributes most easily measured and amenable to standardization.

The "interest" from natural capital is ecosystem services, which collectively make up the planet's ecological life-support system. Therefore, natural systems cannot merely continue to exist; they must be in condition for optimal and reliable delivery of these services. More important, each natural system is delivering not one but, typically, a long list of services that ultimately must be assessed in the aggregate. So it has been for human health and so it must be for ecosystem health.

### 42.6.2 Increased Temporal and Spatial Scales

Another change for ecotoxicologists will occur in increased testing at higher levels of biological organization.[53] Tests at levels of biological organization higher than single species, for example, in microcosms, mesocosms, and field enclosures, will continue to be extremely valuable because investigators identify good indications of response thresholds and patterns and can subject test systems to a degree of stress that would be unacceptable in natural systems.

Micro- and mesocosms are used in monitoring complex systems; they are not miniature ecosystems but are constructs that replicate significant cause-and-effect pathways of natural systems. The process of developing an ecosystem-based capability for ecological risk assessments has been described to illustrate how such units can furnish useful information as well as other sources of system-level information.[54] Ecological function and resilience have been espoused as important criteria for environmental impact assessment and ecological risk analysis.[55] Microcosms need not be inordinately expensive, especially if naturally derived microbial communities are used as receptors in toxicity testing.[56] The uncertainties associated with extrapolating from toxicological responses in laboratory systems to the responses of natural systems have also been described.[57] A key issue in this context is the correspondence of a microscale toxicity test to responses to toxicants in natural systems. Various endpoints and thresholds in the field of ecotoxicology are useful in this regard.[59] These preliminary developments need to be expanded rapidly and the robustness of the methods and procedures confirmed and validated since this is a key issue for ecotoxicologists in both industrial ecology and natural capitalism.

It seems abundantly clear that, while ecotoxicologists may continue the types of testing already developed, they must be prepared to move outside the laboratory and beyond field enclosures to engage in bioregional assessment and planning. However, for sustainable use of the planet,[60] the physical and biological basis for the services provided by nature should not be systematically diminished. To achieve sustainability, the life-support system's integrity cannot be impaired. At the very least, the ability of natural systems to assimilate societal wastes without themselves being degraded or damaging their ecological integrity must be used but not abused.

Ecological integrity has been defined as the maintenance of the structure and function characteristic of a locale.[61] Meeting this condition requires that assimilative capacity be quantified and that human society adjust its waste disposal into natural systems so that they remain healthy and suitable for sustained use. An ecological landscape usually consists of a mosaic of habitat types that are interactive and dynamic. Dynamic means that they are constantly changing and they most likely have cyclic behaviors dependent on climate cycles and the like. Ecotoxicologists must, therefore, incorporate more ecological principles into their activities than they presently do, and ecologists must be more aware of this need and the frames of reference used by toxicologists in general. Details on recognizing needs in increased temporal and spatial scales must also be considered.[62–64]

### 42.6.3 Achieving a Critical Mass of Qualified Personnel

When the military produces a new tank or plane, it also takes great pains to ensure that appropriate personnel are available to run it, maintain it, repair it, and use it as part of a larger strategic plan. If a public consensus existed for implementing suggestions for ecological restoration and made funds available for doing so, an adequate number of prepared professional personnel would still be lacking.[65] The educational system is still a series of petty fiefdoms and tribal units that are often excellent for quality control and teaching activities of specialized disciplines but not for the integration of disparate types of information of the type needed for both ecotoxicology and sustainable use of the planet. Educational courses are often taught and disciplines operate as if each were the only flower facing the sun, and students are badly underprepared for the integration of a wide variety of social, physical, and biological sciences, not to mention economics, engineering, law, and the like. Even major research universities are not particularly aware of designing for quality environments where their interactions with natural systems are concerned. This disparity definitely needs to be changed.

A survey of environmental management systems at North American universities[66] has been conducted by considering two components: (1) key areas and major challenges and (2) the posture and behavior of environmental management systems toward environmental issues. Interestingly, only 50 responses were received from an initial mailing of 269, although 18 of these were returned as undeliverable. This figure alone shows how unaware of, or indifferent to, the environmental crisis these institutions are. Alternatively, one might also assume that many North American universities are quite dependent on extramural funding from sources that might be offended by almost any strong environmental position taken by the university. Responses on the areas of posture and behavior were divided into four major components:

1. Environmental leaders: these people feel that environmental problems do affect the university, and they know where the problems are; as a consequence, they have developed the necessary programs and performance to mitigate environmental problems.
2. Environmental strugglers: these people feel that environmental problems do not affect the university, and they are not sure where the problems are until they arise; as a consequence, they are struggling to develop an effective environmental management system.
3. Accidental "greens": these people do not see environmental management systems as a necessary tool because they are not aware of the environmental problems that affect the university; as a consequence, they have not yet considered preventative programs.
4. Environmental dinosaurs: these people feel that environmental issues do not affect their institutions, and they are therefore not aware that there are environmental problems; they do not see environmental management systems or any other environmental program as necessary.

Although the survey is more extensive than reported here, these survey responses do not inspire confidence that university graduates will be as prepared as they should be for facing environmental management issues.

An international conference was held in December of 1997 in Thessaloniki, Greece to celebrate the 20th anniversary of the Tbilisi Doctrines and to reorient education for *sustainability* (italics mine) in the 21st century.[67] The culmination of this event was a Thessaloniki Declaration — a charter for the future of education for sustainability. In only two of the 29 statements made in the Declaration was the term *environmental education* mentioned. One of those references suggested that environmental education be referred to as education for environment and sustainability. Scant use of the term *environmental education* indicates that the term is finding decreasing support in the international community.[67] It is noteworthy that *sustainability* is the word of choice in this Declaration, rather than *sustainable development*.

Despite the call for total integration of environmental education into all subject areas, few holistic approaches can be found in the United States.[67,68] Criticisms[69] include assertions that environmental education (1) is based on emotionalism rather than facts, (2) tends to be issue driven rather than information driven, and (3) is politically motivated. The public perception of the environmentalists and environmental education is too frequently associated with emotional activism rather than reasoned argument. People must become better informed about environmental issues and become more literate.

#### 42.6.3.1 Professional Certification

Scientists will almost certainly have an increased role in the societal decision-making processes as long as they use reason and information, even when major catastrophes occur. Professionals must insist that those who label themselves ecotoxicologists must have appropriate credentials. The process of certification is long overdue! Societies must begin the process of certifying professionals in every country at the earliest possible time. As a preliminary step, one might consider three levels of certification: (1) gathering data, (2) grouping data and carrying out quality control practices, and (3) interpreting and analyzing data and making professional judgments, together with design for investigative projects intended to furnish information useful in the decision-making process. Those sufficiently skilled to carry out particular experiments, especially using standard methods, may not have the background for making professional judgments, especially in areas where extrapolations are necessary and much professional knowledge is required.

The task of assembling, integrating, and condensing all the information needed for both industrial ecology and natural capitalism is unprecedented. For routine measurements, production of standard methods with the usual great detail required by such organizations as the American Society for Testing and Materials will be extremely helpful. However, most ecosystems and ecological landscapes are mosaics and will require different mixtures of standard methods and different temporal and spatial units. Finally, each system will have some unique properties for which standard methods are inappropriate. For these situations, certification of the organizations or individuals generating the information will reduce the uncertainty about its quality. If the certification process is as rigorous as it should be, relatively few individuals and organizations will qualify, certainly not enough to generate the masses of information that now seems necessary.

#### *42.6.3.2 Obtaining Hands-On Experience*

The educational system should provide the formal courses necessary in chemistry, statistics, biology, and the like. However, the practical operational experience should be in a laboratory certified in the field of ecotoxicology staffed by experienced professionals in that field. This step is a practice of some other professions and would provide the necessary hands-on experience under supervision. Funding agencies could assist this process by providing fellowships, grants-in-aid, etc. for students who aspire to become ecotoxicologists.

#### *42.6.3.3 Funding*

At present, it appears extremely unlikely that the average person would be willing to pay the costs of ensuring that healthy ecosystems exist to provide the services upon which human society depends. An environmental catastrophe might well be the only means of causing such a massive shift.

The areas of industrial ecology, natural capitalism, and other terms referring to the same concepts have two major components: (1) preserving the planet's ecological life-support system and its services to human society and (2) maintaining and preserving natural capital, which is the source of the planet's economic and societal well-being and which is responsible for the life-support services. Once the need for preserving the planet's ecological life-support system is recognized and ecotoxicology and other necessary fields are given the respect that is justified by the enormity of this undertaking, funding will become available. The exact means of doing so will be a function of the perceived urgency, which will, in turn, be closely related to the size of the catastrophe that results from crossing the threshold. In the meantime, prudence dictates allotting more significant funding to education at all levels so that the basics of science, engineering, and other necessary fields are well established in the generations to come.

### 42.6.4 Demographic Change

Demographers expect the human population to grow to at least 9 billion by 2050, and resource depletion is inevitable unless restoration of natural capital becomes a major societal goal globally. Ecotoxicologists are more accustomed to preventing damage than repairing ecological damage, but now they must have their mission increased to encompass both activities. Ecotoxicology studies the effects of stress and restoration ecology rehabilitation after stress; thus, the two fields of study have more in common than is initially apparent.

Even if human society embraces the quest for sustainable use of the planet to the degree that (1) major behavioral changes occur in the use of energy and the establishment of a robust energy policy, (2) demographic transition to lower fertility is hastened, and (3) less per-capita resource use is embraced, ecotoxicologists will still be challenged to estimate what is likely to happen before it is too late to do anything about it. On a more positive note, present methodology, predictive modeling, validation of predictive models, and determination of ecotoxicological effects at the ecosystem and landscape level can clearly be vastly improved. These improvements will substantially reduce but not eliminate uncertainty, but they will not occur until those politicians who control research budgets understand that these efforts require long-term, consistent funding and interdisciplinary research teams accustomed to working together and able to do so for long periods of time. Finally, it is worth reemphasizing that although the present system of higher education has made remarkable strides toward multidimensional integration of knowledge, students are still woefully unprepared for the new challenges they will encounter after graduation.

### 42.6.5 Ecological Thresholds

The goal of natural capitalism is sustainable use of the planet. Industrial ecology also has a sustainability goal, but it is focused primarily on industrial systems, which will play an important role in achieving sustainability. Toxicologists in general and ecotoxicologists in particular have long recognized that the only way to determine a precise threshold is to cross it. Most laboratory toxicity tests are designed with this determination in mind, so that a range of concentrations producing a 100% effect (e.g., death or mortality) to 0% effect (e.g., 100% survival) are customary. This strategy is also designed to find the critical range for more complicated test systems, such as micro- and mesocosms. However, the biosphere is a complex, multivariate system not amenable to the type of testing with which ecotoxicologists are familiar. Societal decisions on the use of fossil fuels and the like are the equivalent of a major global experiment in which human society is a part of the test system.

In a statement that will be all too familiar to ecotoxicologists, but less so to politicians and citizens, Costanza[70] has written: "The problem is that one knows one has a sustainable system only after the fact." Stated in terms more familiar to ecotoxicologists, researchers can estimate these thresholds and, using the precautionary principle, attempt to avoid approaching them too closely. Although the precautionary principle has been endorsed by the United Nations, Germany, and a number of other political entities, it has not been widely implemented. Arguably, the most probable reason for this lack of implementation is that it is viewed as an obstacle to economic growth; as a consequence, the most probable scenario is that the various biospheric thresholds will be found only by crossing them. Crossing some of these thresholds is likely to be catastrophic; crossing others may cause minor, local, or regional damage.

Illustrative examples can be taken from history of crossing sustainability thresholds.[71] One key lesson from history is that most irrigation-based civilizations fail. Irrigation is a cornerstone of present-day agriculture and has been a major force in human advancement for at least 6000 years. Will these systems escape the fate of their defunct antecedents?[71] In addition to this example, other interesting information has been gathered about the ecological collapse of ancient civilizations.[72–74] Such information should alert human society to the dangers of overconfidence.

### 42.6.6 Environmental Surprises

In the present enlightened state of computer technology, the Internet, and global communications, will there be environmental surprises? Do not count on political leaders to be saviors rather than managers! As historian Edward Gibbon[9] stated at the end of the first volume of *Decline and Fall of the Roman Empire*: "The people became indifferent or inured to the Emperor's debauches so long as he paved the roads and remitted taxes." If everyone were interested in both natural capitalism and industrial ecology, would there be any environmental surprises likely to push the environment across a crucial threshold? Bright[75] discusses three types of environmental surprises:

1. A *discontinuity* is an abrupt shift in a trend or previously stable state; the abruptness is not necessarily apparent on a human scale; what counts is the time frame of the processes involved.
2. A *synergism* is a change in which several phenomena combine to produce an effect that is greater than would have been expected from adding up individual or separate effects.
3. An *unnoticed trend,* even if it produces no discontinuities or synergisms, may still do a surprising amount of damage before it is discovered.

As a consequence, there is a high probability, almost a certainty, that one or more crucial environmental thresholds will be crossed, even if the precautionary principle is implemented to a much greater degree than it is now. Ecotoxicologists must be prepared to detect evidence that a threshold has been crossed — not in a laboratory, but in natural systems. This role is quite different from the one most ecotoxicological professionals now assume. Natural systems are generally characterized by considerable variability, and one must be able to distinguish between normal variability and the onset of a new trend. In addition, when there is persuasive evidence that a threshold has been crossed, one must use all the evidence available to estimate the probable location of the threshold.

Ecotoxicologists must be prepared to detect the crossings of such thresholds at every level, from local to global, and must also be prepared to give some advice about how the system may be brought below the adverse response threshold. Ecotoxicologists might justifiably complain that this responsibility is a huge expansion of their present role, which will bring them into an area in which there is even higher uncertainty than the present range of responsibilities. They can also properly claim that they are presently unprepared for this new role and need much time to do so. No professional group is well prepared for this particular role, even though ecotoxicologists are particularly well prepared for thresholds involving ecosystems. Arguably no other group, even classical ecologists, is so well acquainted with stress thresholds, especially those involving hazardous chemicals and other comparable stressors. Collaboration with those working at the systems (rather than the single-species) level is definitely advisable.

### 42.6.7 Design for Quality Environment

For the past 100 years, both human population growth and economic growth have accelerated, and, although the former shows some signs of moderating (but by no means stopping, with 81 million additional living persons annually), economic growth with increased consumption of resources is accelerating even further with globalization. The dimensions of these changes are difficult even for the environmentally literate to comprehend, and for the affluent, life is good and change unthinkable. The world's millions who are living in poverty, who are ill-housed and ill-fed, have insufficient political clout or resources to initiate the major changes required. Still, there are grounds for cautious optimism as a consequence of industrial ecology, natural capitalism, and the moderate portions of the green environmental movement.

Society would do well to reflect on who or what will drive the engines of change from the present practices to the new sustainability paradigm. Lemann[76] discusses power elites, and in his concluding statement, he notes that rapid economic change and concentration of power have provoked strong, unpredictable reactions. People feel left out and try to use politics and government to slow down the pace of change and to build intermediate structures that will protect them. However, the economic boom, by so profoundly streamlining America's operations, is creating propitious conditions for a powerful counter reaction. Alternatively, severe economic crisis, such as droughts in North America's grain-producing areas, total depletion of surface water and groundwater for agricultural purposes, salinization of soils, or some other environmental changes, may produce consequences so severe that human society's relationship with natural systems is reexamined and restated. In some cases, citizens' groups have filed lawsuits to delay or halt environmental destruction. If the consequences of environmental damage reach catastrophic levels in the minds of enough people, then gross environmental damage, and even moderate environmental damage, may become taboo,[77] as is the case with nuclear weapons. If this aversion to nuclear weapons can grow in strength and become locked into military doctrine, without being fully appreciated or even acknowledged by society, the possibility exists for a comparable situation with regard to environmental damage. Scientists would do well to note both existing trends and highly probable trends because they often indicate that much more time should be spent raising "grass roots" literacy about both environmental matters and ecotoxicology than they are now doing.

### 42.6.8 Ecosystem Services

Ecosystem services can be used as toxicological endpoints,[78] and the effects of toxicants on ecosystem services must be considered.[79] Abundant literature exists on ecosystem services themselves[80] and their astonishing value,[48] but very little has been written about the effect of toxicants on their delivery. Some of the services will be global in importance, such as maintaining atmospheric gas balance, and others will be highly site-specific, such as the protection of specific endangered species. Reduced to its essence, however, the problem becomes: what effect does a toxicant have upon a particular function? The challenge is to first identify breakpoints or thresholds in natural systems or in micro- and mesocosms and then find the point at which toxicants approach or cross the threshold.

### 42.6.9 Climate Change

The probability of significant global climate change is fairly high, although the rate and amount of change are uncertain. Not only does temperature often affect toxicity, particularly in cold-blooded organisms, but it affects chemical reactions, partitioning, and transformation of various chemicals as well. Finally, climate change will of itself affect communities of organisms and habitat conditions — sometimes in the same ways that a toxicant does. For example, climate change may adversely affect some species that are sensitive to either increases or decreases in temperature and, thus, increase the dominance of those that either are or are not favored by the newer temperatures. When natural systems are under multidimensional stress effects, separating the effects of one particular stress from the others will not be easy. As a consequence, the role of microcosms and mesocosms will increase dramatically since they provide the opportunity of controlling the variables in a way that might lead to separating their effect. Industrial/ecological hybrid systems will also play an important role because they are multidimensional, but they will be of necessity less amenable to deliberate exposure to drastic stress.

Possibly one of the most important effects of climate change, if it occurs gradually, will be the expansion or contraction of the ranges of many species, alteration in cyclic events, and interactions between species. For example, a flowering plant may produce flowers primarily in response to temperature changes, while the pollinators, such as hummingbirds, might be more affected by photoperiod. As a result, a temperature change without a corresponding photoperiod change would disrupt the synchronization that has developed over a substantial period of time. Both pollinator and pollinated would suffer, and the effects would have nothing to do with whatever toxicants are present.

## 42.7 SUMMARY

The trends described here seem almost inevitable if sustainable use of the planet is to be achieved. The world ecological crisis is of sufficient dimensions that the rate of change for ecotoxicologists will make the rapid growth of recent years seem as if it had occurred at a snail's pace. Because of the need of human society to interface in an enhancing way with ecosystems, general principles will have to be modified for each locale, because no two ecosystems are alike, although there will be areas of correspondence in basic structure and function. Furthermore, although microcosms and mesocosms — now the primary tools for evaluating effects on complex systems — are dynamic, they are not nearly as dynamic as natural systems. Since they are controlled, to a significant degree, they are not as likely to be vulnerable to such disruptions as invasion by exotics, which may be especially likely to occur when natural systems are weakened by toxicants and their resistance to invasion diminished.

These trends will almost certainly become so powerful, because they involve the planet's life-support system, that if scientists do not accept these new responsibilities and challenges, they will undoubtedly be accepted by other, often less qualified, people. Disciplines exist in isolation from each other, although the degree of isolation is far less than it was a decade ago. Both industrial ecology and natural capitalism require a far greater degree of interaction between disciplines than presently exists. The benefits of any interactions will not fully be realized until the needs described in this chapter are recognized by funding organizations and implemented by developing new funding initiatives. Furthermore, an entirely new funding approach must be developed. Since many, if not all, of the applications will be highly site-specific, local responsibility from business, political units, and the citizenry needs to be developed. Environmental activists and industries must work together and avoid the polarization that now exists. Political leaders and legislatures will need to be held more accountable for their election-campaign rhetoric and their failure to follow through. Sufficient examples of successful interactions exist, which indicate that all of these goals can be achieved. Successful implementation of sustainable use of the planet requires that all groups benefit, including industries. Natural capitalism shows that environmental sensitivity can go hand in hand with substantial profits.

There are enough case histories to illustrate that both industrial ecology and natural capitalism are feasible and workable. There is persuasive evidence that there will be a shift to these new ideas in the first half of the 21st century. There is also persuasive evidence that many of human society's present practices are unsustainable for multigenerational time spans. The past century has shown a remarkable array of changes, and it seems probable that there will be even more in the next century. Paradigm shifts often occur suddenly as a result of episodic events for which human society was either unprepared or initially reluctant to accept. But a major paradigm shift is essential before future needs can be met!

## ACKNOWLEDGMENTS

I am indebted to Eva Call for transcribing the dictation of the first draft of this manuscript and for subsequent alterations. My editorial assistant, Darla Donald, provided her usual exemplary assistance in preparing the manuscript for publication. I am indebted to Alan Heath, B. R. Niederlehner, and David Orvos for comments on an early draft of this manuscript and to B. R. Niederlehner for assistance in collecting some references. Correspondence with Peter Leigh provided some very useful insights for parts of this manuscript. The Cairns Foundation paid for the transcription and processing costs for this manuscript.

## REFERENCES

1. White, R. M., Preface, in *The Greening of Industrial Ecosystems*, Allenby, B. R. and Richards, D. J., Eds., National Academy Press, Washington, D.C., 1994, v.
2. Graedel, T. E. and Allenby, B. R., *Industrial Ecology,* Prentice-Hall, Upper Saddle River, NJ, 1995.
3. Hawken, P., Lovins, A., and Lovins, H., *Natural Capitalism: Creating the Next Industrial Revolution*, Little, Brown, and Company, New York, 1999.
4. McNeill, J. R., *Something New under the Sun: An Environmental History of the Twentieth-Century World*, W.W. Norton and Co., New York, 2000.
5. Simon, J. L., *The Ultimate Resource*, Princeton University Press, Princeton, 1983.
6. National Research Council, *Restoring Aquatic Ecosystems: Science Technology and Public Policy,* National Academy Press, Washington, D.C., 1992.
7. Holl, K. D. and Cairns, J., Jr., Monitoring and appraisal, in *Handbook of Restoration Ecology,* Davy, A. J. and Perrow, M. R., Eds., Blackwell Science, Oxford, 2002.

8. National Research Council, *Ecological Indicators for the Nation*, National Academy Press, Washington, D.C., 2000.
9. Gibbon, E., *The History of the Decline and Fall of the Roman Empire, 1776–1788.*
10. National Academy of Engineering, *Technological Trajectories and the Human Environment*, National Academy Press, Washington, D.C., 1997.
11. Kahn, A. E., The tyranny of small decisions: Market failures, imperfections, and the limits of economics, *Kyklos*, 19, 23, 1966.
12. Odum, W. E., Environmental degradation and the tyranny of small decisions, *BioScience*, 32, 9, 728, 1982.
13. President's Acid Rain Review Panel, Acid Rain and Transported Air Pollutants: Implications for Public Policy, Report OTA-o-204, Office of Technology Assessment, U.S. Congress, Washington, D.C., 1984.
14. Likens, G. E., *The Ecosystem Approach: Its Use and Abuse*, Ecology Institute, Oldendorf, Luhe, Germany, 1992.
15. Haeckel, E., *Generelle Morphologie der Organismen: Allegemeine durch die von Charles Darwin reformiente Descendeny Theorie*, 2 Vols., Reimer Publishers, Berlin, Germany, 1866.
16. Tullock, G., *The Economics of Non-Human Societies*, Pallas Press, Tucson, 1994.
17. Robèrt, K.-H., Daly, H., Hawken, P., and Holmberg, J., A Compass for Sustainable Development, The Natural Step Environmental Institute, Stockholm, undated pamphlet.
18. Thompson, A. F., Sustainable development: "Depoliticking" the environment and making it a matter of natural economics, in *Sustainable Development: Rules of the Game*: *Natural Economics*, Roy F. Weston, West Chester, PA, 1995, 2.
19. Bartlett, A. A., An analysis of U.S. and world oil production patterns using Hubbert-style curves, *Math. Geol.*, 32(1), 1, 2000.
20. Hubbert, M. K., U.S. Energy Resources, A Review as of 1972, A Background Prepared at the Request of Henry M. Jackson, Chairman, Committee on Interior and Insular Affairs, United States Senate, Pursuant to Senate Resolution 45, a National Fuels and Energy Policy Study; Serial No. 93–40 (92–75), Part I, U.S. Government Printing Office, Washington, D.C., 1974.
21. Bartlett, A. A. and Ristinen, R. A., Natural gas and transportation, *Phys. Soc.*, 24(4), 2, 1995.
22. World Resources Institute, *A Guide to World Resources 2000–2001*, World Resources Institute, Washington, D.C., 2000.
23. Myers, N., The new millennium: An ecology and an economy of hope, *Current Sci.*, 78(6), 686, 2000.
24. Raffensperger, C. and Tickner, J., *Protecting Public Health and the Environment: Implementing the Precautionary Principle*, Island Press, Washington, D.C., 1999.
25. Allenby, B. R., *Industrial Ecology: Policy Framework and Implementation*, Prentice Hall, New York, 1999.
26. Graedel, T. E., Industrial ecology and the ecocity, *The Bridge*, 29(4), 10, 1999.
27. Newcombe, K., Energy use in Hong Kong: Socioeconomic distribution patterns of personal energy use, and the energy slave syndrome, *Urban Ecol.*, 4, 179, 1979.
28. Hawken, P., *The Ecology of Commerce: How Business Can Save the Planet*, Weidenfeld and Nicolsen Publishers, London, 1993.
29. Holling, C. S., Resilience and stability of ecological systems, *Annu. Rev. Ecol. Syst.*, 4, 1, 1973.
30. Woodwell, G. M., Ed., *The Earth in Transition: Patterns and Processes of Biotic Impoverishment*, Cambridge University Press, Cambridge, MA, 1990.
31. Tibbs, H. B. C., Industrial ecology: An environmental agenda for industry, *Whole Earth Rev.*, 77, 4, 1992.
32. Anonymous, PSR's drinking water and disease conference makes waves, *Physicians for Social Responsibility Reports*, 21(2), 3, 2000.
33. Associated Press, Lead-exposure problems more pervasive than once thought, *The Roanoke Times*, May 16, A10, 2000.
34. Colborn, T. and Clement, C., *Chemically-Induced Alterations in Sexual and Functional Development: The Wildlife/Human Connection*, Princeton Scientific Publishing Co., Princeton, 1997.
35. Cairns, J., Jr., Critical species, including man, within the biosphere, *Naturwissenschaften*, 62, 5, 193, 1975.
36. Jungwirth, M., Muhar, S., and Schmutz, S., *Assessing the Ecological Integrity of Running Waters*, Kluwer Academic Publishers, Dordrecht, The Netherlands, 2000, Reprinted from *Hydrobiologia*, 422/423, 2000.

37. Cairns, J., Jr. et al., A preliminary report on rapid biological information systems for water pollution control, *J. Water Pollut. Control Fed.*, 42(5), 685, 1970.
38. Cairns, J., Jr. et al., Coherent optical spatial filtering of diatoms in water pollution monitoring, *Archiv. Mikrobiol.*, 83, 141, 1972.
39. Myers, W. L., Patil, G. P., and Joly, K., Echelon approach to areas of concern in synoptic regional monitoring, *Environ. Ecol. Stat.*, 4(2), 131, 1997.
40. Myers, W. L., Patil, G. P., and Taillie, C., Conceptualizing pattern analysis of spectral change relative to ecosystem health, *Ecosyst. Health*, 5(4), 285, 1999.
41. Rapport, D. J., Regier, H. A., and Hutchinson, T. C., Ecosystem behavior under stress, *Am. Nat.*, 125, 617, 1995.
42. Rapport, D. J. and Whitford, W. G., How ecosystems respond to stress: Common properties of arid and aquatic systems, *BioScience,* 49(3), 193, 1999.
43. De Soyza, A. G., Whitford, W. G., and Herrick, J. E., Sensitivity testing of indicators of ecosystem health, *Ecosyst. Health*, 3(1), 44, 1997.
44. Frohn, R. C., *Remote Sensing for Landscape Ecology: New Metric Indicators for Monitoring, Modeling and Assessment of Ecosystems*, Lewis Publishers, Boca Raton, FL, 1998.
45. Hargis, C. D., Bissonette, J. A., and David, J. L., Understanding measures in landscape pattern, in *Wildlife and Landscape Ecology: Effects of Pattern and Scale*, Bissonette, J. A., Ed., Springer-Verlag, New York, 1997, 231.
46. Johnson, G. D., Myers, W. L., and Patil, G. P., Stochastic generating models for simulating hierarchically structured multi-cover landscapes, *Landscape Ecol.*, 14, 413, 1999.
47. Pearson, S. M. and Gardner, R. H., Understanding neutral models: Useful tools for standing landscape models, in *Wildlife and Landscape Ecology: Effects of Pattern and Scale*, Bissonette, J. A., Ed., Springer-Verlag, New York, 1997, 215.
48. Costanza, R. et al., The value of the world's ecosystem services and natural capital, *Nature*, 387, 253, 1997.
49. Cairns, J., Jr., Determining the balance between technological and ecosystem services, in *Engineering within Ecological Constraints*, Schulze, P. C., Ed., National Academy Press, Washington, D.C., 1996, 13.
50. Cairns, J., Jr., Interfacing product life cycles and ecological assimilative capacity, in *Interconnections between Human and Ecosystem Health*, Di Giulio, R. T. and Monosson, E., Eds., Chapman and Hall Publishers, London, 1996, 115.
51. Rapport, D. J. et al., Reply to Calow: Critics of environmental health misrepresented, *Environ. Health*, 6, 1, 5, 2000.
52. Nielson, N. O., The meaning of health, *Ecosyst. Health*, 5, 65, 1999.
53. Cairns, J., Jr., Bidwell, J. R., and Arnegard, M. E., Toxicity testing with communities: Microcosms, mesocosms, and whole-system manipulations, *Rev. Environ. Contam. Toxicol.,* 147, 45, 1996.
54. Cairns, J., Jr. and McCormick, P. V., Developing an ecosystem-based capability for ecological risk assessments, in *Readings from the Environmental Professional: Risk Assessment*, Lemons, J., Ed., Blackwell Science, Cambridge, MA, 1995, 25.
55. Cairns, J., Jr. and Niederlehner, B. R., Ecological function and resilience: Neglected criteria for environmental impact assessment and ecological risk analysis, in *Readings from the Environmental Professional: National Environmental Policy Act*, Lemons, J., Ed., Blackwell Science, Cambridge, MA, 1995, 110.
56. Niederlehner, B. R. and Cairns, J., Jr., Naturally derived microbial communities as receptors in toxicity tests, in *Ecological Toxicity Testing: Scale, Complexity, and Relevance*, Cairns, J., Jr. and Niederlehner, B. R., Eds., Lewis Publishers, Boca Raton, FL, 1995, 123.
57. Cairns, J., Jr. and Smith, E. P., Uncertainties associated with extrapolating from toxicologic responses in laboratory systems to the responses of natural systems, in *Scientific Uncertainty and Environmental Problem Solving*, Lemons, J., Ed., Blackwell Science, Cambridge, MA, 1996, 188.
58. Cairns, J., Jr., Niederlehner, B. R., and Smith, E. P., Correspondence of a microscale toxicity test to responses to toxicants in natural systems, in *Microscale Testing in Aquatic Toxicology: Advances, Techniques, and Practice*, Wells, P. G., Lee, K., Blaise, C., and Gauthier, J., Eds., CRC Press, Boca Raton, FL, 1998, 539.
59. Cairns, J., Jr., Endpoints and thresholds in ecotoxicology, in *Ecotoxicology: Ecological Fundamentals, Chemical Exposure, and Biological Effects*, Schuurmann, G. and Markerdt, B., Eds., John Wiley and Sons, New York, 1998, 751.

60. Cairns, J., Jr., Commentary: Defining goals and conditions for a sustainable world, *Environ. Health Perspect.*, 105, 11, 1997, 1164.
61. Cairns, J., Jr., Quantification of biological integrity, in Integrity of Water, Ballentine, R. K. and Guarria, L. J., Eds., U.S. Environmental Protection Agency, Office of Water and Hazardous Materials, Washington, D.C., 055–001–010680–1, 1977, 171.
62. Lawton, J. H., *Community Ecology in a Changing World*, Ecology Institute, Oldendorf/Luhe, Germany, 2000.
63. Mooney, H. A., *The Globalization of Ecological Thought*, Ecology Institute, Oldendorf/Luhe, Germany, 1998.
64. Ehrlich, P., *A World of Wounds: Ecologists and the Human Dilemma*, Ecology Institute, Oldendorf/Luhe, Germany, 1997.
65. National Research Council, Report of the Committee for Study of Environmental Manpower, Commission on Human Resources, *Manpower for Environmental Pollution Control*, National Academy Press, Washington, D.C., 1977.
66. Herremans, I. and Allwright, D. E., Environmental management systems at North American universities: What drives good performance? *The Declaration*, Association of University Leaders for a Sustainable Future, 3(3), 14, 2000.
67. Knapp, D., The Thessaloniki Declaration: A wake-up call for environmental education? *J. Environ. Ed.*, 31(3), 32, 2000.
68. Simmons, B., More infusion confusion: A look at environmental education curriculum materials, *J. Environ. Ed.*, 20(4), 26 1989.
69. Wilke, R., EE criticism: Challenge and opportunity, *Environ. Ed. Advocate*, Fall, 3, 1996.
70. Costanza, R., *Ecological Economics: The Science and Management of Sustainability*, Columbia University Press, New York, 1991.
71. Postel, S., *Pillar of Sand: Can the Irrigation Miracle Last?* W.W. Norton and Co., New York, 1999.
72. Diamond, J., Ecological collapses of ancient civilizations: The golden age that never was, *Bull. Am. Acad. Arts Sci.*, XLVII, 5, 37, 1994.
73. Diamond, J., Paradises lost, *Discover*, 18(11), 68, 1997.
74. Diamond, J., *Guns, Germs, and Steel: The Fates of Human Societies*, W.W. Norton and Co., New York, 1997.
75. Bright, C., Anticipating environmental "surprise," in *State of the World 2000*, Brown, L. R., Flavian, C. R., and French, H., Eds, W.W. Norton and Co., New York, 2000, 22.
76. Lemann, N., No man's town: The good times are killing off America's local elites, *The New Yorker*, June 3, 43, 2000.
77. Schelling, T. C., The legacy of Hiroshima: A half-century without nuclear war, *The Key Reporter*, 65(3), 3, 2000.
78. Cairns, J., Jr., The case for ecosystem services as toxicological endpoints, *Human Ecol. Risk Assess.*, 1(3), 171, 1995.
79. Cairns, J., Jr. and Niederlehner, B. R., Estimating the effects of toxicants on ecosystem services, *Environ. Health Perspect.*, 102(11), 936, 1994.
80. Daily, G., Ed., *Nature's Services: Societal Dependence on Natural Ecosystems*, Island Press, Washington, D.C., 1997.

CHAPTER 43

# Indirect Effects of Pesticides on Farmland Wildlife

**Nick Sotherton and John Holland**

## CONTENTS



## 43.1 INTRODUCTION AND HISTORICAL BACKGROUND

### 43.1.1 Wildlife Declines

In recent decades, there has been growing concern that many species of wildlife inhabiting farmland are in severe and widespread decline throughout much of Europe and North America. The best-documented evidence comes from bird data collected in Britain[1–4] and throughout Europe.[5] In North America, birds classified as those found nesting in grasslands, including those found in small-grain cereals, have also shown widespread steep declines over wide areas of their range since the 1960's.[6–10] This was considered to be a loss greater than that measured for neotropical migrants.[11] However, declining trends in other wildlife groups are also apparent including butterflies,[12] beneficial insects,[13,14] annual arable wildflowers,[15] and amphibians.[16] The intensification of agriculture over the last 50 years, and especially the more widespread use of pesticides, is often regarded as being responsible for the decline in wildlife. Reports on the direct effects of pesticides on wildlife are numerous, but the indirect effects are also considered to be substantial, although the link to declining wildlife is more difficult to detect.

Among species of farmland birds, the grey partridge (*Perdix perdix*) has been one of the best-studied species, and this has led to its description as the barometer of countryside change in Britain. Its decline has been documented, the population dynamics of this species are well understood, the precise mechanisms responsible for its decline have been identified, and recovery programs to reverse declines and restore former distributions and abundance have been recommended.[17–21]

This review will focus on field-based studies examining the indirect effects of pesticides on farmland wildlife, although some groups of pesticides, because of their persistence and dispersal in the food chain and water, accumulate in environments far beyond agricultural areas.

### 43.1.2 The Intensification of Agriculture

The spatial and temporal declines in farmland wildlife have coincided with changing patterns of agricultural development. There has been intense speculation as to the causes of these declines, but a consensus has developed that associates them with the increased production of food.[5,22] Such intensification has been expressed in many ways, but all have led to more intensively managed farming systems involving increased inputs, simplified rotations and monoculture cropping, and the polarization of specific farming enterprises in distinct areas of production.[23] For example, in Britain, mixed farming has been largely replaced by specialist areas of livestock production or arable farming.[24] In addition, noncropped cover, such as hedgerows and fence lines, has been removed to increase field size,[25] thereby removing habitat used for foraging, nesting, and escape cover.[26]

Agrochemical inputs have also increased. Average levels of nitrogen fertilizer applied to crops of winter wheat and spring barley in Britain have increased by 900 and 500%, respectively, since the early 1950s.[27,28] This followed a massive improvement in plant-breeding technology that produced cereal varieties that would produce yield responses to inorganic fertilizers.[29]

The use of pesticides has also increased. In the United Kingdom, the number of pesticide active ingredients approved for use increased from 11 to 105 between 1957 and 1995 (Figure 43.1). Use of pesticides expressed as tonnage applied has fallen, but when expressed as area treated, the increase was from 1,931,000 ha sprayed in 1974 to 16,590,000 ha in 1998 (Table 43.1). Also, over a similar time period, the spectrum of activity of the herbicides, expressed as the number of weed taxa classified as susceptible to herbicides, increased from 22 in 1970 to 38 in 1995.[30] Similar increases in the use of pesticides occurred in North America. In the United States herbicide usage has increased by 180% in the period 1971–1987,[31] with 96% of pesticide-treated areas receiving a herbicide and 30% insecticides.[32] In Canada from 1971 to 1986, there was a threefold increase in herbicide use, while insecticide use increased fivefold, resulting in 1986 in 34 and 7% of total farmland being treated with herbicides and insecticides, respectively.[33] In addition, the spiralling cost of developing and registering new pesticides has led manufacturers to concentrate their efforts on broad-spectrum products for use in a world market at the expense of more selective ones that offer environmental benefits through the preservation of some nontarget species but that may have more limited markets.

These many changes in agricultural production systems are now considered responsible for the decline in much of farmland wildlife. However, individual factors alone cannot often be identified as the cause because the functioning of agroecosystems is relatively complex, and thus a number of interacting factors will be responsible.[4] The response of the grey partridge provides one of the best researched examples, but the principles apply to a much wider range of wildlife.

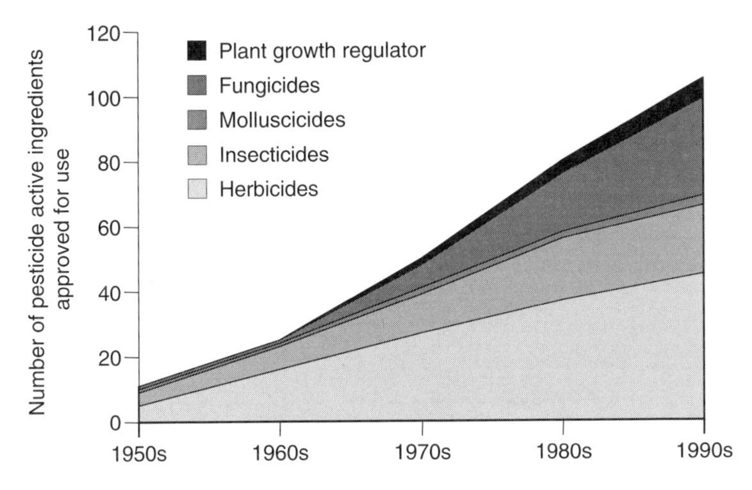


**Figure 43.1** The number of active pesticide ingredients approved for use in U.K. cereal fields by decade.

**Table 43.1 The Use of Pesticides on Winter Wheat in England and Wales 1974–1998**

| | 1974 | 1977 | 1982 | 1988 | 1990 | 1992 | 1994 | 1998 |
|---|---|---|---|---|---|---|---|---|
| Total area grown (x $10^3$ ha) | 1139 | 1025 | 1620 | 1780 | 1895 | 2067 | 1811 | 2045 |
| Area of crop treated (x $10^3$ ha) | | | | | | | | |
| Insecticides | 44 | 491 | 345 | 1004 | 1952 | 1585 | 1462 | 2068 |
| Fungicides | 94 | 367 | 3274 | 4815 | 6044 | 7016 | 6145 | 8053 |
| Herbicides | 1793 | 1647 | 4161 | 4385 | 4297 | 4608 | 4879 | 6469 |

### 43.1.3 The Decline of the Grey Partridge in the United Kingdom

As a hunted species, the grey partridge represents a valuable source of historical data for bird numbers stretching back many decades because of the bag data (hunting records) it generates. In the United Kingdom, bag records reveal that about 2 million birds were killed annually on a sustainable basis between 1870 and 1930.[34] Between 1940 and 1990, the same source of bag data showed a decline of about 85%[35] (Figure 43.2). As a result of this national decline, The Game Conservancy Trust began an intensive study of grey partridges on farmland in southern England in 1968, and it was still collecting data in 2002.[36] Details of this Sussex study may be found

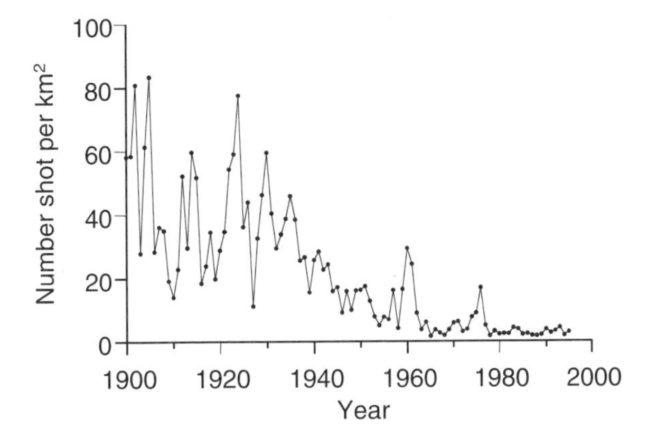


**Figure 43.2** Mean annual number of grey partridges shot per km$^2$ from 12 estates in southern and eastern England for which records extend back to 1900. (From: Aebischer, N.J., in *Conservation and the Use of Wildlife Resources,* Chapman & Hall, London, 1997. With permission.)

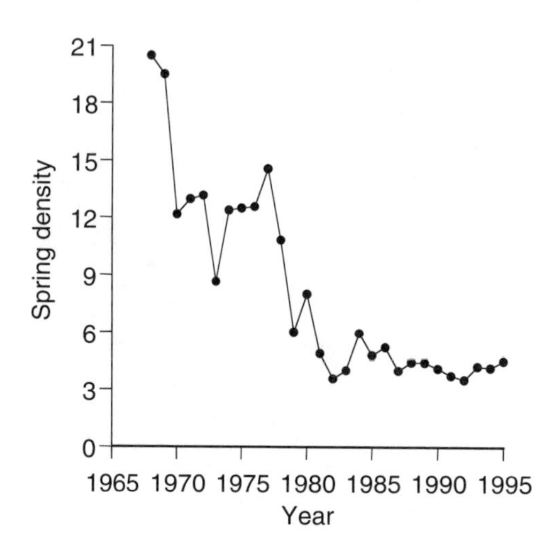


**Figure 43.3** Changes in the annual abundance of grey partridges in the U.K. Spring density (pairs/km$^2$) on The Game Conservancy Trust's main study area in Sussex, 1968–1993. (From: Aebischer, N.J., in *Conservation and the Use of Wildlife Resources,* Chapman & Hall, London, 1997. With permission.)

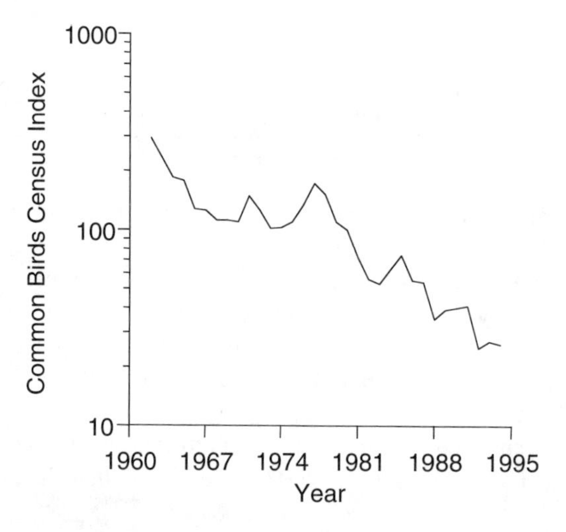


**Figure 43.4** Changes in the annual abundance of grey partridges in the U.K. National index of abundance based on the Common Bird Census of the British Trust for Ornithology, 1962–1994. (From: Aebischer, N.J., in *Conservation and the Use of Wildlife Resources,* Chapman & Hall, London, 1997. With permission.)

elsewhere.[37] Monitoring this study area showed that in the late 1960s, spring pair density on the 62-km$^2$ study area was 20 pairs/km$^2$. This has fallen to less than five pairs in recent years (Figure 43.3). This trend is very similar to the national situation provided by the British Trust for Ornithology's database whose index of abundance for the grey partridge declined between 1962 and 1994 by 90%[35] (Figure 43.4).

### 43.1.4 The Impact of Pesticides

If pesticide use is implicated in this decline of a once common and widespread farmland bird in the United Kingdom, then a mechanism for such a decline must be identified, a hypothesis formulated as to its impact and then tested in the field.

**Table 43.2 Incidence of Mortality in Gamebirds in Britain**

| Decade | DNOC (Herbicide) | Cyclodienes (Mainly Dieldrin) | Other Insecticides | Total |
|---|---|---|---|---|
| 1950s | 31 | >85 | 34 | >150 |
| 1960s | 0 | 228 | 26 | 254 |
| 1970s | 1 | 18 | 13 | 32 |
| Total | 32 | >330 | 73 | >436 |

*Source:* From Potts, G.R., *The Partridge: Pesticides, Predation and Conservation,* Collins, London, 1986. With permission.

#### 43.1.4.1 Direct Effects

In the early 1950s, many birds were found dead on farmland in the U.K. The use of DDT and cyclodiene insecticides like aldrin and dieldrin caused nationwide declines in numbers of sparrow hawks (*Accipiter nisus*) and peregrine falcons (*Falco peregrinus*)[38,39] and killed large numbers of graminivorous birds.[40] For partridges and pheasants, a similar trend emerges whereby deaths attributable to the herbicide DNOC, the cyclodiene insecticides, and other insecticides were recorded in the 1950s, 1960s, and 1970s.[17] (Table 43.2). These peaked in the 1960s but declined after reductions in use following government-imposed regulatory restrictions. Although the amounts of pesticides used have increased since the 1950s, the rate of increase was greatest after this period, when direct toxicity was a problem. Furthermore, those products available during the 1950s (e.g., organochlorines), although generally less toxic than those used in the 1970s and more recently (e.g., carbamates and organophosphates), they were more likely to bioaccumulate.

The withdrawal of these products, which began in the mid 1960s, greatly reduced the exposure, so that today this direct effect is not considered to play a major role in the decline of farmland birds in the United Kingdom.[41] Moreover, raptors, which were especially vulnerable to bioaccumulation, have recovered in those countries where organochlorines are no longer used.[1,5,42] In Spain, however, levels of DDT and lindane are present at a sufficient level to pose a threat to the survival of red-legged partridges (*Alectoris rufa*).[43] Unfortunately, banned[44] or poor-quality products containing contaminants[45] are still made available in developing countries[46] and as a consequence can be detected in species feeding in agricultural areas such as raptors[47] and neotropical migrants.[48] The persistence of organochlorines in the soil has also enabled them to accumulate in higher organisms, leading to secondary poisoning. In Canadian orchards, DDT is still being detected in American robins (*Turdus migratorius*) 20 years after the last application at levels sufficient to cause mortality or reproductive effects.[49] This has occurred through the food chain, as the robins feed predominantly on earthworms, which readily accumulate DDT.

Direct mortality may still occur with modern insecticides. When the toxicological appraisal as given on the product label of 100 commonly used pesticides was examined, a relatively high proportion of the insecticides and other products (principally rodenticides and molluscides) were classified as very toxic to birds (Table 43.3). Pesticide incidents involving birds are still frequently recorded in the U.K. via routes described as following approved or unspecified uses.[50] Species routinely involved include gamebirds, ducks, corvids, geese, pigeons, gulls, and birds of prey. Recent concern has been expressed over mortality rates among owls found dead from residues of rodenticides; their deaths are thought to have been caused by secondary poisoning.[51] Similarly, buzzards (*Buteo buteo*) were dying from feeding on earthworms in sugar beet fields treated with carbofuran.[52]

Granular insecticides and nematicides may also cause direct mortality because they are readily consumed by ground-feeding birds.[53,54] For these birds, the method of application may reduce exposure. In a comparison of liquid vs. granular applications of the organophosphorus insecticide diazinon, survival of bobwhite quail (*Colinus virginianus*) was only reduced by the granular form.[55] In the same experiment, gray-tailed voles (*Microtus canicaudus*) were unaffected, indicating that quail feeding habits increased their susceptibility. However, some liquid formulations can still

**Table 43.3 Relatively Toxicity to Nontarget Organisms for 100 of the Most Commonly Used Pesticides in the United Kingdom (Percentages are Given for Each Product Grouping within Each Organism)**

| Organism | Category | Herbicide | Insecticide | Fungicide | Other[a] | Total Products |
|---|---|---|---|---|---|---|
| Birds | Very toxic | 3 | 17 | 0 | 17 | 7 |
| | Toxic | 11 | 33 | 12 | 28 | 19 |
| | Harmful | 39 | 17 | 20 | 17 | 26 |
| | Nontoxic | 47 | 33 | 68 | 39 | 48 |
| Fish | Very toxic | 17 | 67 | 39 | 39 | 36 |
| | Toxic | 25 | 24 | 43 | 11 | 27 |
| | Harmful | 17 | 6 | 18 | 17 | 15 |
| | Nontoxic | 42 | 6 | 0 | 33 | 22 |
| Invertebrates | Very toxic | 15 | 78 | 44 | 47 | 40 |
| | Toxic | 29 | 17 | 32 | 12 | 24 |
| | Harmful | 18 | 0 | 20 | 12 | 14 |
| | Nontoxic | 38 | 6 | 4 | 29 | 21 |

[a] Includes rodenticides, molluscides, growth regulators, desiccants and wood preservatives.

produce mortality among birds. Numerous examples can, however, be found in the literature concerning bird mortality caused by modern pesticides from either direct consumption or secondary transfer. For example, the consumption of prey debilitated by insecticides can also increase the risk of consuming toxic doses of pesticides, and this has been shown to occur with insectivorous and vermivorous raptors.[56] However, the data show that many pesticide-related mortalities go undetected or are not reported. Moreover, the exposure–effect relationship may be difficult to detect if multiple stressors are involved.[57]

### *43.1.4.2 Sublethal Effects*

Many insecticides can adversely affect birds without killing them, for example by increasing their susceptibility to disease or predation, by lowering their productivity, the hatchability of their eggs, the viability of their chicks or by altering their behavior patterns. There has been no evidence, however, to suggest that sub-lethal effects are important in the partridge decline. Recent research with red-legged partridges has shown that some pesticides may increase a bird's susceptibility to later applications of other pesticides. Two separately harmless pesticides suddenly became toxic when used together or in rapid succession. The same synergistic effect has been detected in pigeons, but not in starlings.[58,59] However, these effects have not yet been demonstrated outside the laboratory.

There has been a failure, so far, to detect large-scale lethal or sublethal effects of pesticides on partridges at a level that would equate to the estimated millions of birds[60] represented by the scale of the loss (85%) over the last four decades. This is the reason why gamebird research in the United Kingdom has concentrated on the indirect effects following the reduction of weeds and insects on gamebird chick food chains.

### *43.1.4.3 Indirect Effects*

Earlier studies[61,62] had found that the key factor causing changes in a grey partridge population in southern England was chick survival (Figure 43.5); in addition, those studies had clearly linked the observed national decline with the availability of sufficient quantities of the preferred insects that are essential in the diet of young chicks[17,63] (Figure 43.6). Moreover, it has been shown that pesticides appeared to be a major factor reducing populations of preferred insects. These insects include a group of Coleoptera (Chrysomelidae, small diurnal Carabidae and Curculionidae), caterpillars of Lepidoptera and Tenthredinidae (sawflies: especially species of the genus *Dolerus*), and many members of the Heteroptera (bugs) especially species of the genus *Calocoris*.[64] The data

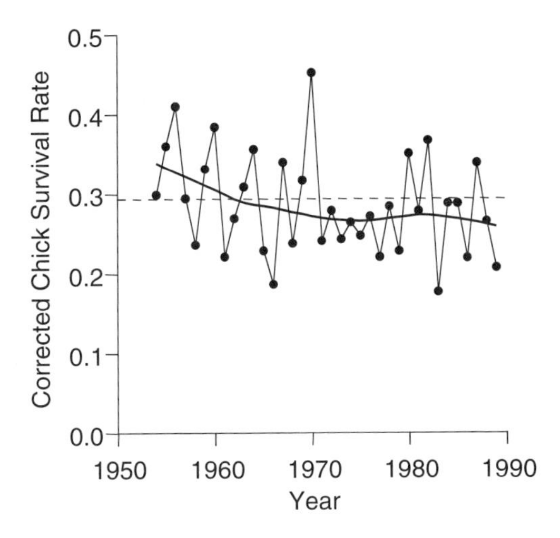


**Figure 43.5** Annual chick survival rate (to 6 weeks) of grey partridge in Sussex, 1953–1990, corrected for weather and releasing of hand-reared birds. The thick line indicates the long-term trend; the dotted horizontal line is the minimum chick survival rate required to maintain the population. (From: Aebischer, N.J., in *Conservation and the Use of Wildlife Resources,* Chapman & Hall, London, 1997. With permission.)

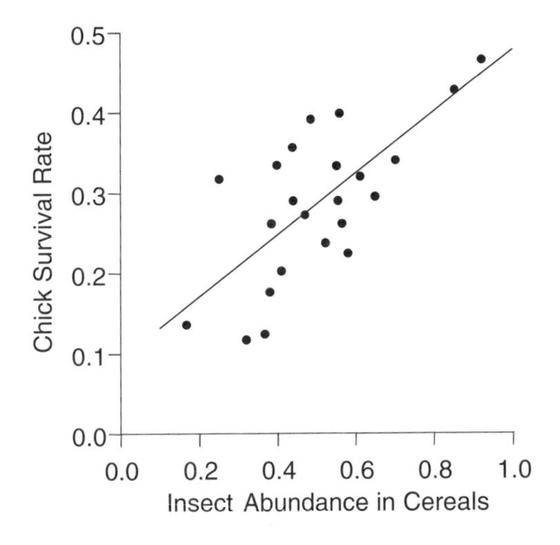


**Figure 43.6** Relationships between grey partridge chick survival (to 6 weeks) and chick-food insects. Annual survival rates in Sussex in relation to abundance indices of chick-food insects sampled in cereals in mid-June (time of peak chick hatch) 1970–1992. (From: Aebischer, N.J., in *Conservation and the Use of Wildlife Resources,* Chapman & Hall, London, 1997. With permission.)

supporting the importance of insects to grey partridge chicks were reviewed in 2000.[65] Many of these preferred insects were found to be more abundant at the edges of cereal fields, where grey partridge broods preferentially foraged.[66]

## 43.2 CONSTRUCTING THE HYPOTHESIS FOR THE INDIRECT EFFECTS OF PESTICIDES

There is now an enormous amount of data documenting that various types of pesticides can reduce the numbers of nontarget chick-food insects that are crucial in the diet of young gamebird chicks in their first two weeks of life.

### 43.2.1 The Impact of Pesticides

#### *43.2.1.1 Insecticides*

The use of insecticides has been shown to significantly reduce densities of chick-food insects in cereal crops at the time chicks hatch and actively begin foraging for food. Such studies have either compared groups of commonly used insecticides,[67,68] or they have focused on particular groups of insecticides like the organophosphates[69–73] or the pyrethroids.[74,75] There was evidence from feeding trials that some insecticides can cause taste aversion in birds, which would ultimately have the same impact as invertebrate mortality because the food resource were made unavailable.[76] This remains to be proven for birds foraging under natural conditions. Alternatively, the direct impact of insecticides to chick-food insects[77] has been reported and specifically for the sawflies (Hymenoptera: Tenthredinidae)[78] and the Heteroptera (Hemiptera).[79]

All these studies document short-term reductions of invertebrates following insecticide application. With respect to chick survival, such reductions are crucial as a continual supply of food is required; lack of food, even for relatively short periods, in combination with inclement weather may be sufficient to cause mortality or adversely influence development. However, long-term trends are also important, as demonstrated in the link between declining chick-food invertebrates and the grey partridge. The declines found for many insect groups in Sussex[13] have also been mirrored elsewhere, although they are not always so clearly demonstrable.[80] In Germany, a 50 to 81% reduction in the activity/density of carabids and formicids was found over a 30-year period, with a 48 to 85% decrease in species numbers.[14] Similarly, an 81% decrease in carabid beetle trapping rates and a 90% decrease in biomass were found from 1971–1974 compared to 1978–1983.[81] Indeed, monitoring conducted within the LINK Integrated Farming System project in the U.K. from 1992–1997 on six farms revealed that the numbers and diversity of invertebrates were very low at half of the sites, suggesting that intensive arable farming was having adverse effects.[82] Species dominance may also change,[83,84] which can also alter food availability for birds. In general, invertebrate species with the poorest dispersal power have declined the most.[85]

Whether an insecticide has an impact on invertebrates may vary, depending on a complex of interacting factors such as the insecticide's characteristics, the invertebrate species and its associated biology, the application technique, and environmental conditions.[86,87] Consequently, the impact on individual invertebrates in the crop will vary considerably. Taking many of these factors into consideration by calculating a hazard index, insects inhabiting the top of the crop were found to be most vulnerable to insecticides.[67] Moreover, the duration of the effect will depend on the chemical's persistence and the capacity for reinvasion from untreated refuges, which again will be species-dependent.[71] The speed of recovery appears to be inversely related to the size of the treated area.[88] Thus, the impact on species dependent on invertebrates as a food resource will vary considerably, the extent of effect varying with their foraging range, the scale of the insecticide application, and weather conditions. In addition, insecticides may also influence the breeding condition of insectivorous birds: those with lower fat reserves as a result of poorer feeding conditions produce fewer broods and are less successful with each one, although there is only evidence for this with respect to granivorous birds and seed availability (see below).

#### *43.2.1.2 Fungicides*

The use of fungicides has also increased in U.K. cereal fields, and the insecticidal properties of these fungicides and their impact on chick-food insects have also been reported.[68,89,90] Modern fungicides are, however, still relatively harmful to invertebrates (Table 43.3) and may also decrease food availability for saprophytic species; this may explain the decline in the Sussex study of *Tachyporus* species (Coleoptera: Staphylinidae), a group of fungal-feeding rove beetles.[13]

### *43.2.1.3 Herbicides*

Herbicides have only rarely been shown to cause direct mortality of farmland arthropods, other soil organisms, birds, or mammals, especially under field conditions,[91–93] although some are classified as very toxic (Table 43.3). However, they influence the density, distribution, and composition of weeds present in cropped fields. These provide foliage and seeds for phytophagous and polyphagous species, influence the microclimate and thereby soil moisture, and govern the degree of physical protection from predators.[93] Thus, the environment created after herbicide applications will favor xerophobic or euryhygric species, which prefer more open habitat, while phytophagous, xerophilic species will be confined to weedy patches and field margins.[94] Arable weed abundance and diversity have declined in the long-term with the intensification of agriculture and in particular the use of more efficient herbicides,[15] which is considered to be important in the decline of invertebrates.[30,93]

Many chick-food insects are also phytophagous or associated with high densities of weed cover.[95–98] The use of herbicides has also been an important factor in the dynamics of grey partridge chick survival because of their ability to remove weeds that are the host plants of many of these phytophagous beneficial insects.[63,96,99–101] Weed seeds are also an important dietary component of many adult farmland birds, especially in autumn and winter when they feed upon cereal stubbles.[26,41] Low seed availability during this period as a result of intensive herbicide use in the preceding crop may lead to a situation in which birds start breeding with lower fat reserves and consequently have a lower reproductive capability. Evidence that spring seed supplies are insufficient was demonstrated with ring-necked pheasants. Supplementary feeding increased their breeding success.[102] Similarly, the number of clutches produced by Turtle doves (*Streptopelia turtur*) has declined since the 1960s from 2.9 to 1.6 because their preferred food, the broad-leaved weed common fumitory (*Fumaria officinalis*), is not sufficiently available.[103]

### 43.2.2 The Hypothesis

As partridge numbers declined, the annual rate of chick survival, governed by the availability of their chick-food insects, also fell. At the same time, herbicide and insecticide use increased. Experimental evidence proved the powerful impact of insect-host-plant removal by herbicides as well as the impact of nontarget-insect removal by insecticides. Routes of exposure that would produce direct toxicities to partridges were discounted, and sublethal impacts provided no plausible explanation for the magnitude of the observed declines. Therefore, in the early 1980s in the United Kingdom, The Game Conservancy Trust developed the hypothesis that the grey partridge decline was caused by the use of insecticides killing insects and herbicides removing host plants, both of which reduced chick-food insects.

## 43.3 TESTING THE HYPOTHESIS

A series of large-scale, replicated experiments was set up in the mid-1980s to test this hypothesis. A large farming estate (about 3000 ha) was divided into three areas or beats and each area divided into two blocks. The six trial plots contained a total of 37 cereal fields. In each field, in one block of land in each area, a 6-meter-wide strip of the cereal crops was not sprayed with pesticides after January 1. Fields in the other block of land per area were fully sprayed right up to the crop edge. The delineation between blocks of land either sprayed or with unsprayed edges was chosen to coincide with natural barriers to partridge movement such as woods, shelter belts, and blocks of break crops such as oilseed rape known to minimize movement of broods between spray treatments. Details of the experiment, the cropping, and the pesticide regime are described elsewhere.[20] After

**Table 43.4 Mean Grey Partridge Brood Sizes (± SE) on Blocks of Cereal Fields with Sprayed and Selectively Sprayed Headlands in Southern and Eastern England (Sample Sizes in Parentheses)**

| | | Grey Partridge, Mean Brood Size | | |
|---|---|---|---|---|
| Study Area | Year | Sprayed Headlands | Selectively Sprayed Headlands | P |
| Principal study | 1983 | 4.7 ±1.1 (39) | 8.4 ± 1.2 (29) | <0.01 |
| Farm | 1984 | 7.5 ± 0.8 (34) | 10.0 ± 0.6 (34) | <0.01 |
| Southern England | 1985 | 3.3 ± 0.7 (9) | 5.7 ± 0.6 (34) | <0.05 |
| | 1986 | 5.9 ± 1.6 (17) | 6.2 ± 1.0 (21) | NS |
| Eastern England | 1984 | 4.7 ± 0.4 (71) | 7.8 ± 0.6 (57) | <0.001 |
| | 1985 | 2.7 ± 0.4 (19) | 4.0 ± 0.7 (19) | <0.05 |
| | 1986 | 4.8 ± 0.6 (32) | 8.7 ± 1.5 (6) | <0.001 |

harvest, field counts of partridges were made from vehicles at dawn and dusk, when family parties or coveys of partridges were actively feeding.

After 1984 in southern England, the cereal crop edge pesticide recommendation changed from no sprays after January 1 to a selective spray regime whereby certain autumn-applied insecticides, noninsecticidal fungicides, and certain graminicides were used. These selectively sprayed edges became known as Conservation Headlands and were selected for practical agronomic reasons. However, such a spray regime still withheld all spring and summer insecticides and all broad-leaved weed herbicides, the group of plants most used as insect host plants. Between 1984 and 1986, a series of experiments was set up on eight farms in eastern England to further test this hypothesis.

Without herbicides, weeds grew in the unsprayed and, later, the Conservation Headlands. Individual weed densities, total numbers of species, total biomass, and percent weed cover were all found to be significantly higher in crop edges with reduced or no herbicides compared to those fully sprayed.[100,101,104,105] The numbers of chick-food insects were also significantly increased by the provision of host plants provided by the absence of herbicides. Two- and threefold increases in density were recorded both between fields[68] and within a field.[105]

In response to increased densities of chick-food insects in areas of crops not receiving full inputs of pesticides, mean grey partridge brood sizes were significantly greater compared to brood sizes in areas of farmland that were fully sprayed[20,21,106,107] (Table 43.4).

Using radiotelemetry with broods foraging in spring barley fields, survival rates to 21 days of age and mean distances between successive roost sites were significantly higher and lower, respectively, in unsprayed headlands compared to those that were fully sprayed. Home range sizes were smaller for broods in fields with unsprayed edges, but the proportion of the 6-meter-wide crop edge in their home range was significantly greater for such broods.[21]

From these studies it has been concluded that grey partridge declines in the United Kingdom have been caused primarily by the indirect effects of pesticides acting on food chains.[41,108] The Conservation Headland hypothesis was similarly tested in the United States. The density and diversity of plants were increased with associated increases in arthropods. The improved food availability and habitat modification subsequently increased densities and diversity of small mammals. Gamebirds were also expected to be encouraged by the mechanism given for grey partridge.[109] Further evidence is now appearing that this scenario is occurring elsewhere. Poor chick survival along with poor hen survival during the breeding period were identified in France as being responsible for the decline of grey partridge.[110] In South Africa, decline of helmeted guinea fowl (*Numida meleagris*) was linked to a reduction in insect and seed food and loss of nesting habitat in intensive agricultural areas.[111] The grey partridge, however, is likely to be one of the bird species most vulnerable to indirect pesticide effects because it resides year round within agricultural fields, feeding predominantly within crops. Similarly, birds feeding predominantly within orchards may similarly suffer because of the high number of insecticide applications and the extended period over which applications occur. Thus, the birds are more exposed to direct toxicity through ingestion of treated prey and indirectly by removal of their food.[112] Indeed, avian abundance and diversity was lower in conventional compared to organic orchards but no mechanisms was identified.[113] Most other species make some use of noncrop areas within agricultural areas or utilize other habitats during the year and are consequently at lower risk of contamination.[112,114] Even so, in Canada, those species that spent more time within crops showed population declines.[112] The other changes associated with intensive agriculture may, however, also be contributing to their decline.

Other changes in agriculture and game management have also contributed to the decline of the grey partridge and other farmland wildlife. Some of these changes act synergistically with pesticide effects, leading to more rapid and extensive reductions. These key factors include removal or damage of edge habitat, which is used for foraging and serves to supplement that available in the field. This also provides essential cover for nesting and escape from predation.[17,115] The deliberate or misapplication of herbicides leading to herbicide drift can damage the functioning of such habitats.[116,117] Finally, the level of mammalian and avian predator control has shown to be important for many farmland bird species,[118] and an interaction exists with noncrop habitat. Poor cover appears to increase the chance that nests, chicks, or adults will be predated and may lead to premature nest abandonment.[115,119] In addition, the time spent foraging is a function of prey availability and consequently the time exposed to predation.[17,115]

## 43.4 INDIRECT EFFECTS OF PESTICIDES ON OTHER FARMLAND WILDLIFE

### 43.4.1 Benefits of Conservation Headlands for Other Farmland Wildlife

Evidence for the indirect effects of pesticides on grey partridges can be readily extended to declines of other farmland species of wildlife. Certainly, in our large-scale experiments in southern and eastern England, other species of gamebirds responded to the presence of increased food resources provided in cereal crop edges where pesticide regimes were modified. Although not in decline, the red-legged partridge also responded to the increased food resource. Mean brood sizes were generally larger with increased chick-food insect densities, but differences become statistically significant between treatment only when data from all study areas were pooled.[21] The benefits of Conservation Headlands have been similarly demonstrated for a wide range of species, not just in the United Kingdom but also in Sweden[120] and The Netherlands.[121] Mean brood sizes of ring-necked pheasant (*Phasianus colchicus*) also increased significantly in the presence of crop edges receiving reduced levels of pesticides.[21,106]

Data on the effects of Conservation Headlands on other species of birds are more variable. In an extensive survey of passerine birds in hedgerows and adjacent crops during the breeding season, the presence of three species of bird — greenfinch (*Carduelis chloris*), robin (*Erithacus rubecula*), and song thrush (*Turdus philomelos*) — was lower in hedgerows adjacent to autumn-sown cereals, which had received pesticides, compared to the presence of those species adjacent to autumn crops which were fully sprayed. Most other species showed similar, but insignificant, differences. However, most species showed a tendency towards higher incidences in hedgerows adjacent to spring-sown cereals with reduced pesticide inputs than in those adjacent to fully sprayed spring-sown crops. There was no manipulative or experimental aspect to this study, and no comment as to mechanisms involved was made.[122] Insecticide drift into hedgerows poses a threat to birds, both directly as found for nesting great tits (*Parus major*) and also indirectly through the removal of invertebrate food.[123] In the Netherlands, Conservation Headlands were used in preference to standard crop edges by the insectivorous bird, *Motacilla flava flava,*[124] probably because phytophagous insect densities were higher. However, skylarks (*Alauda arvensis),* which feed on plants as well as insects at ground level, avoided the Conservation Headlands, probably because of their extensive weed growth.[124]

In another survey of the use made by a farmland bird in an arable farmland habitat, it was found that corn buntings (*Miliaria calandra*), a species in steep decline in the United Kingdom, also responded to the use of pesticides and the density of insect chick-food items thereby produced.[125] Adults preferentially foraged in areas of high chick-food-insect density, and density was negatively correlated with the number of insecticide applications made that season. Low chick-food-insect abundance close to nests resulted in greater distances travelled from the nest by foraging parents and longer trip durations. Nestling weight was positively correlated with insect abundance. Finally, the probability of nest survival was negatively correlated with the abundance of chick-food insects close to the nest, apparently as a result of increased predation.[125]

During the course of the experiments on crop margins, the recommendations for pesticide use changed from a no-spray moratorium after January 1 to a selective regime called Conservation Headlands, whereby broad-leaved weed herbicides and summer insecticides were restricted. At this time, the study on nongame species began. Benefits of Conservation Headlands to butterflies (Lepidoptera) started in the mid-1980s with some simple censuses using standard techniques. Over a 5-year period, significantly more individuals representing 13 species were found in Conservation Headlands compared to headlands that were fully sprayed. When pesticide use was reduced, Lepidoptera numbers increased.[126–128]

Detailed observations of butterfly behavior in the two headland treatments have shown that flight speed and duration were lower in Conservation Headlands.[129] Activity patterns also changed, with some species showing different behaviors between spray management regimes. Small white (*Pieris rapae*) males spent almost all of their time in flight in field margins that had fully sprayed headlands, with little time devoted to either feeding or resting. In field margins with Conservation Headlands, the proportion of time spent in flight was radically reduced, with a much higher proportion of time spent in feeding and resting.[129] *Pieris rapae* males observed in fully sprayed field margins also spent most of their time in the hedgerow area, whereas the hedgerows were virtually ignored in field margins where there were Conservation Headlands.

For many species of strongly flying butterflies (Pieridae and Nymphalidae), the nectar provided by weeds such as charlock (*Sinapsis arvensis*), field pansy (*Viola arvensis*), and creeping thistle (*Cirsium arvense*) attracted them into Conservation Headlands. The larval host plants of some of the Pieridae (*S. arvensis* and rape, *Brassica napus*) were found in Conservation Headlands[126] and were exploited as larval food plants. Increased numbers of such species as green-veined white (*Pieris napi*), *P. rapae* and orange tip (*Anthocharis cardamines*) may also have been caused by their presence. However, the host plants of the Satyridae, Hesperiidae, Nymphalidae, and Lycaenidae were not found in the Conservation Headlands. Indeed, those of the satyrids and hesperiids (*Graminea*) were selectively controlled using herbicides. This was fortunate, as any eggs laid in Conservation Headlands could constitute lost reproductive effort following destruction by harvest and tillage.

For some species, it was not possible to make sufficient observations in fully sprayed headlands because of low numbers. However, such an explanation cannot be applied to all species recorded on farmland in southern England. No such speculation of time spent between the sprayed and selectively sprayed headland treatments could be found for meadow brown (*Maniola jurtina*) males,[129] while other species such as the gatekeeper (*Pyronica tighonus*) spent their time in the hedgerow regardless of the spray regime in the adjacent cereal crop edge. However, the abundance of both these species was clearly greater adjacent to Conservation Headlands,[126] despite our inability to explain these differences in terms of their behavior. It may have been that when hedgerow nectar was scarce, unsuitable, or unavailable, headland nectar was used to "top-up," or act as the major nectar source.

Another possible explanation for the increased abundance of hedgerow species could be the reduced pesticide drift into field boundaries where the spray boom was switched off 6 m away from the field edge. Conservation Headlands may be acting as buffer zones, protecting hedgerows from summer insecticides and broad-leaved residual herbicides throughout the year. The amount of pesticide deposition into field boundaries adjacent to Conservation Headlands and fully sprayed crop edges has been measured in both the autumn and summer. The crop in the Conservation

Headland buffer zone absorbed a significant amount of pesticide before it reached the noncropped area of the field boundary, thereby reducing spray drift.[130] Levels of mortality of *P. brassicae* larvae exposed to grass upon which insecticide deposits had drifted from fully sprayed and selectively sprayed crops were significantly different. Very high mortality (100%) was found in larvae exposed to vegetation collected from fully sprayed sections of crop compared to much lower levels of mortality of larvae (18%) exposed to vegetation collected from vegetation adjacent to an nonsprayed headland.[130] Conservation Headlands were also shown to reduce pesticide drift by 95%, and thus contamination of watercourses and risk to aquatic wildlife decreased.[121]

So, for the Lepidoptera, evidence for both direct and indirect routes of exposure to pesticides was found. Longer-term trends in relative abundance have been established on the study farm over a 5-year period and comparisons made with data derived from the U.K.'s National Butterfly Monitoring Scheme. For two satyrid species (*M. jurtina* and *P. tithonus*), populations were either maintained or increased relative to 1984 levels at the study farm. This compares with the south–southeast region of the National Butterfly Monitoring Scheme (many of the incorporated sites being National Nature Reserves), where these species showed declines over the same period.[126]

Other groups of beneficial insects, including the stenophagous and polyphagous natural enemies of cereal pests, may also benefit from the resources provided in Conservation Headlands or from the degree of exclusion of the adverse effects of pesticides that they provide. Reinvasion of a dimethoate-treated field by epigeal invertebrates was quicker and more extensive in the half surrounded by unsprayed outer 6-m border compared to areas fully sprayed.[71] The authors concluded that this was because the margins from which the reinvading arthropods originated were protected from spray drift. Indeed, arthropod abundance is usually greatest at the field edges, declining into the field.[94,131] Preliminary work has also shown how some economically important species of hover fly (Diptera: Syrphidae) such as *Episyrphus balteatus* made use of the nectar and pollen of flowering weeds in Conservation Headlands such as *S. arvensis*, *Matricaria* spp., and *C. arvense*.[132,133] Similarly, bumblebees made use of the pollen resources provided by arable weeds in Conservation Headlands.[134] Again, analysis of the behavior of adult flies showed that they were retained in the weedy strips because they were foraging on the weeds.[135]

Overall numbers and, in some years, diversity of carabid beetles were higher in unsprayed edges; this was especially true of the spermophagous species, which were responding to the greater abundance of weeds.[136] Microclimate will change if weed cover increases, and this may influence species composition, as carabids differ in their environmental requirements.[137] Polyphagous species, such as the carabid beetles *Pterostichus melanarius* and *Agonum dorsale* were also better fed (i.e., a significantly higher proportion of males, gravid females, and nongravid females had all four portions of their digestive tracts full of solid food remains) in Conservation Headlands compared to individuals from fully sprayed crop edges. It is possible that the greater prey availability (larger numbers and more diverse alternative prey items) may increase predator numbers through improved diet and fecundity.[100] For all these groups of nontarget insects, there is evidence of benefit from the increased provision of food resources (pollen, nectar invertebrates) as a result of the reduced use of pesticides.

#### 43.4.1.1 Farmland Birds

In a recent review of the indirect effects of pesticides on birds, the authors concluded that many species of birds fed on a varied diet of plant and invertebrate species found on farmed land. In most groups of farmland birds, populations are either declining or stable and increasing, but in two — the thrushes and the buntings — all species are in decline.[41] The authors conclude from the available data that their food items are also in decline on U.K. farmland. They found that in the majority of studies, the use of pesticides resulted in short-term reductions in the abundance of bird food items and that such effects could persist for weeks or months after initial applications.[41] Many cite these short-term reductions as being unimportant and of little impact on breeding success of birds. For example, when the impact of two insecticides, Deltamethrin and Furadan, on the chestnut-collared longspurs (*Calcarius ornatus*), which feeds predominantly on grasshoppers, were compared, both insecticides decreased grasshopper populations by more than 90%, but the number of grasshoppers in nestling diets was only decreased in nests within the deltamethrin-sprayed plots.[138] However, the insecticide spraying did not decrease the total biomass of arthropods delivered to the nestlings, and nestling weight and size were unaffected. In this case, birds overcame the effect of the insecticide spraying by foraging further, although previously this was shown not to occur. Instead, their diet changed.[138] In contrast, another species present, the Baird's sparrow (*Ammodramus bairdii*), abandoned many of its nests in plots treated with the other insecticide, Furadan.[138] This difference between species occurred because the design of experiments, the pesticides, and bird species involved can all influence the outcome of field based studies.[139] Thus, the timing of such insect reductions during the stage when nestlings need feeding could be crucial.

Strong temporal associations between the start of the declines and the widespread use of pesticides in cereals were also found.[41] Strong associations (more than 50% of cereals treated at the beginning of the population decline) of 11 bird species to the use patterns of herbicides were cited. For fungicides, the researchers found associations with six species but no associations for insecticides or molluscicides. For these two pesticide groups, they only refer to probable associations where between 10 and 50% of cereals were treated. However, for 11 out of 12 species, they identify the year the decline started, and those were during a period of increasing pesticide use.[41]

For other declining species, there is evidence, but not experimental proof, that the cause is related to pesticide use and the resulting reductions of bird food, both plant and insect material, in the summer to feed chicks and as seeds to feed birds over the winter. However, for these species, other hypotheses linking the declines to changing agriculture practices are equally compelling to explain these declines in the United Kingdom and Canada.[41,93]

Where pesticides use is eliminated, as in organic production, the benefits would be expected to be greatest. A survey of farms in the United Kingdom revealed higher numbers of birds on organic compared to conventional farms.[140] However, the organic farms were characterized by smaller fields with a greater proportion of hedgerows and more diverse crop rotations, and the effect of eliminating pesticide use could consequently not be separated out as the determining factor in the bird increases. Invertebrate food sources for birds have been found to be higher in Danish organic cereal fields,[141] but the differences were lower at the edge compared to the field center, which explains why fewer differences were found between organic compared to conventional cereal fields in the United Kingdom, where only edge samples were taken.[142] In Canada, birds typical of wetland areas were more abundant in the wetland areas on organic farms and in wild areas compared to on conventional farms or those using conservation tillage.[143] Overall differences in avifaunal density and diversity between the farming systems were small in comparison to differences between farms and wild sites, again emphasizing the importance of noncrop habitats.

#### *43.4.1.2 Other Wildlife*

Evidence for the direct and indirect effects of pesticides on other wildlife is less comprehensive than for birds and invertebrates. The pyrethroid insecticides commonly used in temperate farmland areas are not regarded as toxic to mammals,[144] but organophosphorus insecticides are highly toxic and can also reduce activity levels.[145] Insecticides have also been shown to directly reduce the reproductive ability of small mammals, the impact varying between species leading to a change in their competitive ability.[146] As a consequence, the duration of effect lasted longer than the insecticide (diazinon) persisted in the environment.

Amphibians are in worldwide decline, but the causes behind these declines are complex,[147] although pesticides are considered to be the most important factor in agricultural areas.[16] Insecticides originating from intensive agricultural areas were responsible for the decline in riverine amphibian populations,[148] and this may occur through direct toxicity[149] or morphological changes that reduce survival.[16,150] The herbicide atrazine has also been shown to cause morphological changes.[151] However, there is no evidence that indirect effects through reduced food availability or habitat modification are contributing to these declines, although this is a possibility.

## 43.5 SUMMARY

Many species of farmland birds are in decline in the United Kingdom, and there is considerable evidence that the indirect effects of pesticides are the cause. Of the 24 species listed as in decline in the United Kingdom, pesticides were thought to have a possible link to 19 species, 11 with reasonably good evidence. For three species, it was thought there was no evidence, but for only one — the grey partridge — was there conclusive evidence for the indirect effect of pesticides.[41] For the grey partridge, survival of chicks is strongly correlated with pesticide use, including insecticides, insecticidal fungicides, and herbicides, which have all been shown to reduce the insect foods of a partridge chick.[17] The timing of the partridge decline coincides with the increased use of pesticides, and the effects of pesticides have not been changed by the effects of other changes to farming practice. Changes to chick survival drives the population size of the grey partridge,[17] and it was concluded that the timing and magnitude of changes in population size and chick survival are consistent with the increased use of pesticides as the cause.[41] However, the experimental manipulation of pesticides alone (the Conservation Headlands experiments), when all other variables remained constant, showed a clear link between the indirect effects of pesticides and the decline of this once common U.K. farmland bird.

Indirect effects are also likely to impact a wide variety of farmland wildlife that are dependent on the same food chain as the grey partridge, and evidence of this is starting to appear for some passerines. Small mammals and amphibians may similarly suffer from a decline in their food and the creation of simplistic habitats through the creation of monocultures. Wildlife may also decline if pesticides cause sublethal effects, for example, by changing reproductive success or by altering behavior. However, these are seldom investigated in the field because their impact may be confounded by other factors. Insects especially have been shown in the laboratory to exhibit a range of sublethal effects,[87] which may then be exacerbated through an interaction with other species, habitat, or the environment.

The development of more efficient pesticides combined with intensive farming methods has allowed farmers to substantially increase yields, to such an extent that set-aside was introduced to control production in the U.S. and in the European Union (EU). This has provided an opportunity to provide areas of habitat able to support a greater abundance of wildlife than could ever be achieved in a crop, however low the inputs. Moreover, the areas under set-aside are massive in comparison to nature reserves, and, because they are located in farmed areas, they have the potential to create a diversity not previously possible. The debate over whether widespread use of extensive farming systems or intensive production with a higher proportion of noncrop areas has yet to be resolved. There is no doubt that set-aside, if sympathetically managed, can support a diverse wildlife.[152] Unfortunately, in the past, the management options for set-aside were restricted by regulation or expense, and the full potential has not been realized. Evidence is now available that wildlife can recover using conventional pesticide inputs, if set-aside along with predator control is managed to provide food and cover for wildlife.[153] Further evidence on the benefits of set-aside were revealed in an ornithological survey comparing set-aside with arable crops and pasture.[154] Nonpasserines, passerines, insectivores, and granivores all exhibited a preference for rotational set-aside over any farmed crops, probably because of better foraging.

The alternative approach of using more extensive farming systems utilizing lower inputs of pesticides through the greater use of natural control mechanisms has been studied extensively in the EU[155] and in the United States.[30] Their impact on wildlife, however, has often been limited to studies of invertebrates because of the resources available or the limitations of scale. Whether the response was direct or indirect was also not ascertained. The results from these studies have been variable, with some reporting increases in arthropod numbers and diversity where lower inputs of pesticides were used[156–164] but others finding no clear benefit. One of the earliest of these studies, the Boxworth project,[165] also examined the overall impact of conventional and lower-input approaches on small mammals and farmland birds. There was, however, no conclusive evidence of an overall effect of the pesticide regimes on the breeding performance of the most common bird species.[166] Pesticide use has declined considerably in the U.K. since the project finished (Table 43.1), with conventional inputs now being comparable to the lower-input approach used, and consequently the impact on wildlife should be lower.

Pesticides have been identified as contributing to the decline of farmland wildlife, although the impact is often exacerbated by other farm practices associated with intensive agriculture. Their impact may, however, often not be detected because the direct and indirect effects of pesticides are complex and have been studied in detail for only a few species. Costs will inhibit further detailed studies such as those conducted on the grey partridge, but if key indicators can be identified from the grey partridge studies that relate to the population response of other species, then it may be possible to predict the impact of pesticide applications. The abundance of chick-food insects may be one such variable.

## REFERENCES


1. Marchant, J. H., Hudson, R., Carter, S. P., and Whittington, P., *Population Trends in British Breeding Birds,* BTO, Tring, 1990.
2. Fuller, R. J., Gregory, R. D., Gibbons, D. W., Marchant, J. H., Wilson, J. D., Baillie, S. R., and Carter, N., Population declines and range contractions among lowland farmland birds in Britain, *Conserv. Biol.,* 9, 1425, 1995.
3. Gibbons, D. W., Reid, J. B., and Chapman, R. A., *The New Atlas of Breeding Birds in Britain and Ireland: 1988–1991*, Spencer, R., Poyser, T., and Gibbons, D. W., Eds., T and A. D. Poyser, London, 1993.
4. Chamberlain, D. E., Fuller, R. J., Bunce, R. G. H., Duckworth, J. C., and Shrubb, M., Changes in the abundance of farmland birds in relation to the timing of agricultural intensification in England and Wales, *J. Appl. Ecol.*, 37, 771, 2000.
5. Tucker, G. M. and Heath, M. F., *Birds in Europe: Their Conservation Status*, Birdlife International, Cambridge, 1994.
6. Labinsky, R. F., Midwest pheasant abundance declines, *Wild. Soc. Bull.*, 4, 182, 1976.
7. Vance, D. R., Changes in land use and wildlife populations in south-eastern Illinois, *Wild. Soc. Bull.*, 4, 11, 1976.
8. Klimstra, W. D., Bobwhite quail and changing land use, *Proc. 2nd Natl. Bobwhite Quail Symp.*, 1982.
9. Jobin, B., DesGranges, J. L., and Boutin, C., Population trends in selected species of farmland birds in relation to recent developments in agriculture in the St. Lawrence valley, *Agric. Ecosyst. Environ.*, 57, 103, 1996.
10. Downes, C. and Collins, B. T., The Canadian Breeding Bird Survey, 1966–1994, *Can. Wildl. Serv. Progr. Note,* No. 210, 1996.
11. Knopf, F., Avian assemblages on altered grasslands, in *Studies in Avian Biology,* Cooper Ornithological Society, Camarillo, CA, 1994, 247.
12. Heath, J., Pollard, E., and Thomas, J., *Atlas of Butterflies in Britain and Ireland,* Viking, Harmondsworth, 1984.
13. Aebischer, N. J., Twenty years of monitoring invertebrates and weeds in cereal fields in Sussex, in *The Ecology of Temperate Cereal Fields,* Firbank, L. G., Carter, N., Darbyshire, J. R., and Potts, G. R., Eds., Blackwell Scientific Publications, Oxford, 1991, 305.
14. Heydemann, B. and Meyer, H., Auswirkungen der Intensivkultur auf die Fauna in den Agrarbiotopen, *Schr. Reihe. Dtsch. Rat Landespfl. Wirtsch.*, 42, 174, 1983.
15. Schumacher, W., Measures taken to preserve arable weeds and their associated communities in Central Europe, in *Field Margins*, Way, J. M. and Greig-Smith, P. W., Eds., British Crop Protection Conference Monograph No. 35, London, 1987, 11.

16. Berrill, M., Bertram, S., McGillivray, L., Kolohon, M., and Pauli, B., Effects of low concentrations of forest-use pesticides on frog embryos and tadpoles, *Environ. Toxicol. Chem.*, 13, 657, 1994.
17. Potts, G. R., *The Partridge: Pesticides, Predation and Conservation*, Collins, London, 1986.
18. Potts, G. R. and Aebischer, N. J., Modelling the population dynamics of the grey partridge: Conservation and management, in *Bird Population Studies: Their Relevance to Conservation Management,* Perrins, C. M., Lebreton, J. D., and Hirons, G. J. M., Eds., Oxford University Press, Oxford, 1991, 373.
19. Potts, G. R. and Aebischer, N. J., Population dynamics of the grey partridge *Perdix perdix* 1793–1993: Monitoring, modelling and management, *Ibis,* 137, 29, 1995.
20. Rands, M. R. W., Pesticide use on cereals and the survival of partridge chicks: A field experiment, *J. Appl. Ecol.,* 22, 49, 1985.
21. Rands, M. R. W., The survival of gamebird chicks in relation to pesticide use on cereals, *Ibis,* 128, 57, 1986.
22. Murphy, M., Economic implications of supply and environmental control of U.K. – EC agriculture, Brighton Crop Protection Conference – Weeds, 533, 1989.
23. O'Connor, R. J. and Shrubb, M., *Farming and Birds,* Cambridge University Press, Cambridge, 1986.
24. Tapper, S. C. and Barnes, R. F. W., Influence of farming practice on the ecology of the brown hare (*Lepus europaeus*), *J. Appl. Ecol.,* 23, 39, 1986.
25. Chapman, J. and Sheail, J., Field margins – an historical perspective, in *Field Margins: Integrating Agriculture and Conservation,* Boatman, N., Ed., British Crop Protection Conference Monograph No. 58, Farnham, 3, 1994, 3.
26. Landers, J. L. and Mueller, B. S., Bobwhite Quail Management: A Habitat Approach, Misc. Pub. No. *6 Tall Timbers Res. Stn.*, Tallahassee, FL, 39, 1986.
27. Church, B. M., Use of fertiliser in England and Wales, 1980, *Report Rothamstead Experimental Station for 1980,* 2, 115, 1981.
28. Chalmers, A., Kershaw, C., and Leach, P., Fertiliser use on farm crops in Great Britain: Results from the survey of fertilising practice, 1969–88, *Outlook Agric.,* 19, 269, 1990.
29. Fischbeck, G., The evolution of cereal crops, in *The Ecology of Temperate Cereal Fields,* Firbank, L. G., Carter, N., Darbyshire, J. F., and Potts, G. R., Eds., Blackwell Scientific Publications, Oxford, 1991, 31.
30. National Research Council, Alternative Agriculture, National Research Council, Washington, D.C., 1989.
31. Pimentel, D. and Levitan, L., Pesticides: Amounts applied and amounts reaching pests, *BioScience*, 36, 86, 1986.
32. Statistics Canada, 1951–1986 (quinquennial), Census of Canada, *Agriculture*, Statistics Canada, Ottowa, Ontario.
33. Ewald, J. A. and Aebischer, N. J., Trends in pesticide use and efficacy during 26 years of changing agriculture in southern England, *Environ. Monitor. Assess.*, 64, 493, 2000.
34. Tapper, S. C., Ed., *A Question of Balance,* The Game Conservancy Trust, Fordingbridge, 1999.
35. Aebischer, N. J., Gamebirds: Management of the grey partridge in Britain, in *Conservation and the Use of Wildlife Resources,* Bolton, M., Ed., Chapman and Hall, London, 1997, 131.
36. Potts, G. R. P., Grey partridges, in *The Game Conservancy Trust Review of 2000*, Miles, S., Ed., The Game Conservancy Trust, Fordingbridge, U.K., 2001, 22.
37. Aebischer, N. J., Twenty years of monitoring invertebrates and weeds in cereal fields in Sussex, in *The Ecology of Temperate Cereal Fields,* Firbank, L. G., Carter, N., Darbyshire, J. F., and Potts, G. R., Eds., Blackwell Scientific Publications, Oxford, 1991, 305.
38. Newton, I. and Hass, M. B., The return of the Sparrowhawk, *Br. Birds*, 77, 47, 1984.
39. Ratcliffe, D. A., *The Peregrine Falcon*, T and A. D. Poyser, Calton, U.K., 1980.
40. Cramp, S., Conder, P. J., and Ash, J. S., *The Deaths of Birds and Mammals from Toxic Chemicals*, BTO/RSPB/Game Research Association, U.K., 1963.
41. Campbell, L. H., Avery, M. I., Donald, P., Evans, A. D., Green, R. E., and Wilson, J. D., *A Review of the Indirect Effects of Pesticides on Birds,* JNCC Rep. No. 227, Joint Nature Conservation Committee, Peterborough, U.K., 1997.
42. Kirk, D. A. and Hyslop, C., Population status and recent trends in Canadian raptors: A review, *Biol. Conserv.*, 83, 91, 1998.

43. Herrera, A., Arino, A., Conchello, M. P., Lazaro, R., Bayarri, S., Yague, C., Peiro, J. M., Aranda, S., and Simon, M. D., Red-legged partridges (*Alectoris rufa*) as bioindicators for persistent chlorinated chemicals in Spain, *Arch. Environ. Contam. Toxicol.*, 38, 114, 2000.
44. Anonymous, Banned DDT on sale in Tanzania, *Pesticide News*, 35, 10, 1997.
45. Anon., Risky pesticides on sale in Africa, *Pest. News*, 51, 3, 2001.
46. Anon., Exporting risk – U.S. hazardous trade 1995–96, *Pest. News*, 40, 4, 1998.
47. Smith, I. and Bouwman, H., Levels of organochlorine pesticides in raptors from the North-West Province, South Africa, *Ostrich*, 71, 36, 2000.
48. Klemens, J. A., Harper, R. G., Frick, J. A., Capparella, A. P., Richardson, H. B., and Coffey, M. J., Patterns of organochlorine pesticide contamination in Neotropical migrant passerines in relation to diet and winter habitat, *Chemosphere*, 41, 1107, 2000.
49. Harris, M. L., Wilson, L. K., Elliott, J. E., Bishop, C. A., Tomlin, A. D., and Henning, K. V., Transfer of DDT and metabolites from fruit orchard soils to American robins (*Turdus migratorius*) twenty years after agricultural use of DDT in Canada, *Arch. Environ. Contam. Toxicol.*, 39, 205, 2000.
50. Fletcher, M. R., Hunter, K., Barnett, E. A., and Sharp, E. A., Pesticide Poisoning of Animals 1997: Investigations of Suspected Incidents in the United Kingdom, Ministry of Agriculture, Fisheries and Food, London, 1998.
51. Newton, I., Dale, L., Finnie, J. K., Freestone, P., Malcolm, H., Osborn, D., Wright, J., Wyatt, C., and Wylies, I., Wildlife and Pollution: 1996/97, JNCC Annual Report No. 271, Joint Nature Conservation Committee, Peterborough, 1997.
52. Dietrich, D. R., Schmid, P., Zweifel, U., Schlatter, C., Jennieiermann, S., Bachmann, H., Buhler, U., and Zbinden, N., Mortality of birds of prey following field application of granular carbofuran — A case-study, *Arch. Environ. Contam. Toxicol.*, 29, 140, 1995.
53. Stafford, T. R. and Best, L. B., Bird response to grit and pesticide granule characteristics: Implications for risk assessment and risk reduction, *Environ. Toxicol. Chem.*, 18, 722, 1999.
54. Mineau, P., Avian mortality in agro-ecosystems 1. The case against granular insecticides in Canada, in Field Methods for the Study of the Environmental Effects of Pesticides, BCPC Monograph No. 40, Greaves, M. P. and Greig-Smith, P. W., Eds., British Crop Protection Council, Thronton Heath, U.K., 1988, 3.
55. Wang, G. M., Wolff, J. O., and Edge, W. D., Gray-tailed voles do not move to avoid exposure to the insecticide Guthion® 2S, *Environ. Toxicol. Chem.*, 20, 406, 2001.
56. Mineau, P., Fletcher, M. R., Glaser, L. C., Thomas, N. J., Brassard, C., Wilson, L. K., Elliott, J. E., Lyon, L. A., Henny, C. J., Bollinger, T., and Porter, S. L., Poisoning of raptors with organophosphorus and carbamate pesticides with emphasis on Canada, U.S. and U.K., *J. Raptor Res.*, 33, 37, 1999.
57. Vyas, N. B., Factors influencing estimation of pesticide-related wildlife mortality, *Toxicol. Ind. Health*, 15, 186, 1999.
58. Johnston, G., Collett, G., Walker, C., Dawson, A., Boyd, I., and Osborn, D., Enhancement of malathion toxicity to the hybrid red-legged partridge following exposure to prochloraz, *Pest. Biochem. Physiol.*, 35, 107, 1989.
59. Johnston, G., The study of interactive effects of pollutants: A biomarker approach, *Sci. Total Environ.*, 171, 1, 1995.
60. Potts, G. R., Cereal farming, pesticides and grey partridges, in *Farming and Birds in Europe*, Pain, D. J. and Pienkowski, M. W., Eds., Academic Press, London, 1997, 151.
61. Blank, T. H., Southwood, T. R. E., and Cross, D. J., The ecology of the partridge. I. Outline of populations processes with particular reference to chick mortality and nest density, *J. Anim. Ecol.*, 36, 549, 1967.
62. Potts, G. R., Recent changes in the farmland fauna with special reference to the decline of the Grey Partridge (*Perdix perdix*), *Bird Study*, 17, 145, 1970.
63. Southwood, T. R. E. and Cross, D. J., The ecology of the partridge, III. Breeding success and the abundance of insects in natural habitats, *J. Anim. Ecol.*, 38, 497, 1969.
64. Sotherton, N. W. and Moreby, S. J., Beneficial arthropods other than natural enemies in cereals, in *Set-aside, Br. Crop Protect. Conf. Monogr.,* Clarke, J., Ed., No. 50, 1992, 223.
65. Sotherton, N. W., The development of a gamebird research strategy: Unravelling the importance of arthropod populations, in *Quail IV: Proc. 4th National Quail Symp.*, Brennan, L. A., Palmer, W. E., Burger, L. W., Jr., and Pruden, T. L., Eds., Tall Timbers Research Station, Tallahassee, FL, 158, 2000.

66. Green, R. E., The feeding ecology and survival of partridge chicks (*Alectoris rufa* and *Perdix perdix*) on arable farmland in East Anglia, *J. Appl. Ecol.*, 21, 817, 1984.
67. Alford, J., Miller, P. H. C., Goulson, D., and Holland, J. M., Predicting susceptibility of non-target insect species to different insecticide applications in winter wheat, *1998 Brighton Crop Protect. Conf. - Pests Dis.*, 2, 599, 1998.
68. Sotherton, N. W., Farming methods to reduce the exposure of non-target arthropods to pesticides, in *Pesticides and Non-Target Invertebrates,* Jepson, P. C., Ed., Intercept, Wimborne, U.K., 1989, 195.
69. Vickerman, G. P. and Sunderland, K. D., Some effects of dimethoate on arthropods in winter wheat, *J. Appl. Ecol.*, 14, 767, 1977.
70. Powell, W., Dean, G. S., and Bardner, R., Effects of pirimicarb, dimethoate and benomyl on natural enemies of cereal aphids in winter wheat, *Ann. Appl. Biol.,* 106, 235, 1985.
71. Holland, J. M., Winder, L., and Perry, J. N., The impact of dimethoate on the spatial distribution of beneficial arthropods and their reinvasion in winter wheat, *Ann. Appl. Biol.,* 136, 93, 2000.
72. Jepson, P. C., Efe, E., and Wiles, J. A., The toxicity of dimethoate to predatory Coleoptera: Developing an approach to risk analysis for broad-spectrum pesticides, *Arch. Environ. Contam. Toxicol.,* 28, 500, 1995.
73. Duffield, S. J., Jepson, P. C., Wratten, S. D., and Sotherton, N. W., Spatial changes in invertebrate predation role in winter wheat following treatment with dimethoate, *Entomol. Exp. Appl.,* 78, 9, 1996.
74. Wiles, J. A. and Jepson, P. C., The susceptibility of a cereal aphid and its natural enemies to deltamethrin, *Pest. Sci.,* 36, 263–272, 1992.
75. Wiles, J. A. and Jepson, P. C., Substrate-mediated toxicity of deltamethrin residues to beneficial invertebrates: Estimation of toxicity factors to aid risk assessment, *Arch. Environ. Contam. Toxicol.,* 27, 384, 1994.
76. Nicolaus, L. K. and Lee, H. S., Low acute exposure to organophosphate produces long-term changes in bird feeding behaviour, *Ecol. Appl.*, 9, 1039, 1999.
77. Moreby, S., Southway, S., Barker, A., and Holland, J. M., A comparison of the effect of new and established insecticides on non-target invertebrates of winter wheat fields, *Environ. Toxicol. Chem.,* 20, 2243, 2001.
78. Sotherton, N. W., The effects of six insecticides used in U.K. cereal fields on sawfly larvae (Hymenoptera: Tenthredinidae), *Brighton Crop Protect. Conf. — Pests Dis.*, 3, 999, 1990.
79. Moreby, S. J., Sotherton, N. W., and Jepson, P. C., The effects of pesticides on species of non-target heteroptera inhabiting cereal fields in southern England, *Pest. Sci.*, 51, 39, 1997.
80. Lebrun, Ph., Baguette, M., and Dufrêne, M., Species diversity in a carabid community: Comparison of values estimated at 23 years interval, *Acta Phytopathol. Entomol. Hung.,* 22, 165, 1987.
81. Basedow, T., Der Einfluß Gesteigerter Bewirtschaftungsintensität im Getreidebau auf die Laufkäfer (Coleoptera: Carabidae), *Mitt. Biol. Bundesanst. Land-Forstwirtsch. Berl.-Dahl.,* 235, 123, 1987.
82. Holland, J. M., Cook, S. K. Drysdale, A., Hewitt, M. V., Spink J., and Turley, D., The impact on non-target arthropods of integrated compared to conventional farming: Results from the LINK Integrated Farming Systems project, *Brighton Crop Protect. Conf. — Pests Dis.*, 2, 625, 1998.
83. Croy, P., Faunistisch-ökologische Untersuchungen der Carabiden im Emfeld Eines Industriellen Balungsgebeites, *Entomol. Nachr. Ber.*, 31, 1, 1987.
84. Körner, H., Der Einfluß der Pflanzenschutzmittel auf die Faunnenvielfalt der Agrarlandschaft (unter Besonderer Berücksichtigung der Gliederfüßler der Oberfläche der Felder), *Bayer. Landwirtsch. Jahrb.*, 67, 375, 1990.
85. Desender, K. and Turin, H., Loss of habitats and changes in the composition of ground and tiger beetle fauna in four West European countries since 1950 (Coleoptera: Carabidae, Cicinelidae), *Biol. Conserv.*, 48, 277, 1989.
86. Holland, J. M. and Luff, M. L., The effects of agricultural practices on Carabidae in temperate agroecosystems, *Int. Pest. Manag. Rev.*, 5, 109, 2000.
87. Jepson, P. C., The temporal and spatial dynamics of pesticide side-effects on non-target invertebrates, in *Pesticides and Non-Target Invertebrates*, Jepson, P. C., Ed., Intercept, Wimborne, U.K., 1989, 95.
88. Duffield, S. J. and Aebischer, N. J., The effect of spatial scale of treatment with dimethoate on invertebrate population recovery in winter wheat, *J. Appl. Ecol.*, 31, 263, 1994.
89. Sotherton, N. W. and Moreby, S. J., The effects of foliar fungicides on beneficial arthropods in wheat fields, *Entomophaga*, 33, 87, 1988.

90. Sotherton, N. W., Moreby, S. J., and Langley, M. G., The effects of the foliar fungicide pyrazophos on beneficial arthropods in barley fields, *Ann. Appl. Biol.*, 111, 75, 1987.
91. Brust, G. E., Direct and indirect effects of four herbicides on the activity of carabid beetles (Coleoptera: Carabidae), *Pest. Sci.*, 30, 309, 1990.
92. Zhang, J. X., Drummond, F. A., and Leibman, M., Effect of crop habitat and potato management practices on the population abundance of adult *Harpalus rufipes* (Coleoptera: Carabidae) in Maine, *J. Agric. Entomol.*, 15, 63, 1997.
93. Freemark, K. and Boutin C., Impacts of agricultural herbicide use on terrestrial wildlife in temperate landscapes: A review with special reference to North America, *Agric. Ecosyst. Environ.*, 52, 67, 1995.
94. Holland, J. M., Perry, J. N., and Winder, L., The within-field spatial and temporal distribution of arthropods within winter wheat, *Bull. Entomol. Res.*, 89, 499, 1999.
95. Sotherton, N. W., Observations on the biology and ecology of the chrysomelid beetle *Gastrophysa polygoni* in cereals, *Ecol. Entomol.,* 7, 197, 1982.
96. Sotherton, N. W., Effects of herbicides on the chrysomelid beetle *Gastrophysa polygoni* (L.) in laboratory and field, *Z. Angew. Ent.*, 94, 446, 1982.
97. Barker, A. M. and Maczka, C. J. M., The relationships between host selection and subsequent larval performance in three free-living graminivorous sawflies, *Ecol. Entomol.*, 21, 317, 1996.
98. Moreby, S. J. and Southway, S. E., The importance of the crop edge compared to the mid-field, in providing invertebrate food for farmland birds, *Asp. Appl. Biol.*, 54, 217, 1999.
99. Vickerman, G. P., Some effects of grass weed control on the arthropod fauna of cereals, *12th Br. Weed Control Conf. Brighton Crop Protect. Conf.,* 929, 1974.
100. Chiverton, P. A. and Sotherton, N. W., The effects on beneficial arthropods of the exclusion of herbicides from cereal crop edges, *J. Appl. Ecol.*, 28, 1027, 1991.
101. Moreby, S. J. and Southway, S. E., Influence of autumn applied herbicides on summer and autumn food available to birds in winter wheat fields in southern England, *Agric. Ecosyst. Environ.,* 72, 285, 1999.
102. Draycott, R. A. H., Hoodless, A. N., Ludiman, M. N., and Robertson, P. A., Effects of spring feeding on body condition of captive-reared ring necked pheasants in Great Britain, *J. Wildl. Manage.*, 62, 557, 1998.
103. Browne, S. and Aebischer, N. J., Turtle doves in the modern agricultural environment, in *The Game Conservancy Trust Review of 2000*, Miles, S., Ed., The Game Conservancy Trust, Fordingbridge, 2000, 85.
104. Sotherton, N. W., Rands, M. R. W., and Moreby, S. J., Comparison of herbicide treated and untreated headlands on the survival of game and wildlife, *Br. Crop Protect. Conf. — Weeds,* 3, 991, 1985.
105. Sotherton, N. W., Conservation Headlands: A practical combination of intensive cereal farming and conservation, in *The Ecology of Temperate Cereal Fields,* Firbank, L. G., Carter, N., Darbyshire, J. R., and Potts, G. R., Eds., Blackwell Scientific Publications, Oxford, 1991, 373.
106. Sotherton, N. W., Robertson, P. A., and Dowell, S. D., Manipulating pesticide use to increase the production of wild gamebirds in Britain, in *Quail III: 3rd Natl. Quail Symp.,* Church, K. E. and Dailey, T. V., Eds., Missouri Department of Conservation, Jefferson City, MO, 1993, 92.
107. Sotherton, N. W. and Robertson, P. A., Indirect impacts of pesticides on the production of wild gamebirds in Britain, in *Perdix V, Grey Partidge and Ring-Necked Pheasant Workshop,* Church, K. E., Warner, R. E., and Brady, S. J., Eds., Kansas Department of Wildlife and Parks, Emporia, KS, 1990, 84.
108. Burn, A., Pesticides and their effects on lowland farmland birds, in *Ecology and Conservation of Lowland Farmland Birds,* Aebischer, N. J., Evans, A. D., Grice, P. V., and Vickery, J. A., Eds., British Ornithologists' Union, Tring, 2000, 89.
109. Vance, A. S., Habitat Mitigation: The Use of Unsprayed Field Borders to Reduce the Indirect Effect of Herbicides on Selected Farmland Wildlife in Southern Iowa, unpublished M.Sc. thesis, Clemson University, Clemson, SC, 1996.
110. Bro, E., Sarrazin, F., Clobert, J., and Reitz, F., Demography and the decline of the grey partridge *Perdix perdix* in France, *J. Appl. Ecol.*, 37, 432, 2000.
111. Malan, G. and Benn, G. A., Agricultural land-use patterns and the decline of the helmeted guineafowl *Numida meleagris* (Linnaeus 1766) in KwaZulu-Natal, South Africa, *Agric. Ecosyst. Environ.*, 73, 29, 1999.

112. Boutin, C., Freemark, K. E., and Kirk, D. A., Farmland birds in southern Ontario: Field use, activity patterns and vulnerability to pesticide use, *Agric. Ecosyst. Environ.*, 72, 239, 1999.
113. Fluetsch, K. M. and Sparling, D. W., Avian nesting success and diversity in conventionally and organically managed apple orchards, *Environ. Toxicol. Chem.*, 13, 1651,1994.
114. Wilson, J. D., Morris, A. J., Arroyo, B. E., Clark, S. C., and Bradbury, R. B., A review of the abundance and diversity of invertebrate and plant foods of granivorous birds in northern Europe in relation to agricultural change, *Agric. Ecosyst. Environ.*, 75, 13, 1999.
115. Roseberry, D. L. and Klimstra, W. D., *Population Ecology of the Bobwhite*, Southern Illinois University Press, Carbondale, 1984.
116. Boatman, N. D., Blake, K. A., Aebischer, N. J., and Sotherton, N. W., Factors affecting the herbaceous flora of hedgerows on farms and its value as wildlife habitat, in *Hedgerow Management and Nature Conservation*, Watt, T. A. and Buckley, G. P., Eds., Wye College Press, Wye, U.K., 1994, 33.
117. Stoate, C., The influence of field boundary structure on breeding territory establishment of whitethroat *Sylvia communis* and yellowhammer *Emberiza citrinella*, *Asp. Appl. Biol.*, 54, 12, 1999.
118. Stoate, C. and Thomson, D. L., Effects of predation on lowland farmland bird populations, in *Ecology and Conservation of Lowland Farmland Birds*, Aebischer, N. J., Evans, A. D., Price, P. V., and Vickery, J. A., Eds., British Ornithologists' Union, Tring, 2000, 134.
119. Rands, M. R. W., The effect of nest site selection on nest predation in grey partridge *Perdix perdix* and red-legged partridge *Alectoris rufa*, *Ornis Scand.*, 19, 35, 1988.
120. Chiverton, P. A., The benefits of unsprayed cereal crop margins to grey partridges *Perdix perdix* and pheasants *Phasianus colchicus* in Sweden, *Wildl. Biol.*, 5, 83, 1999.
121. De Snoo, G. R., Unsprayed field margins: Effects on environment, biodiversity and agricultural practice, *Landscape Urban Plann.*, 46, 151, 1999.
122. Green, R. E., Osborne, P. E., and Sears, E. J., The distribution of passerine birds in hedgerows during the breeding season in relation to characteristics of the hedgerow and adjacent farmland, *J. Appl. Ecol.*, 31, 677, 1994.
123. Cordi, B., Fossi, C., and Depledge, M., Temporal biomarker responses in wild passerine birds exposed to pesticide spray drift, *Environ. Toxicol. Chem.*, 16, 2118, 1997.
124. De Snoo, G. R., Dobbelstein, R. T. J. M., and Koelewijm, S., Effects of unsprayed crop edges on farmland birds, in *Field Margins — Integrating Agriculture and Conservation*, Boatman, N. D., Ed., Br. Crop Protect. Conf. Monogr., No. 58, Farnham, 1994, 221.
125. Brickle, N. W., Harper, D. G. C., Aebischer, N. J., and Cockayne, S. H., Effects of agricultural intensification on the breeding success of corn bunting *Miliaria calandra*, *J. Appl. Ecol.*, 37, 742, 2000.
126. Dover, J. W., Sotherton, N. W., and Gobbett, K., Reduced pesticide inputs on cereal field margins: The effects on butterfly abundance, *Ecol. Entomol.* 15, 17, 1990.
127. Rands, M. R. W. and Sotherton, N. W., Pesticide use on cereal crops and changes in the abundance of butterflies on arable farmland, *Biol. Conserv.*, 36, 71, 1986.
128. Dover, J. W., The conservation of insects on arable farmland, in *The Conservation of Insects and their Habitats*, Collins, N. W. and Thomas, J., Eds., Academic Press, New York, 12, 1991, 293.
129. Dover, J. W., Conservation headlands: Effects on butterfly distribution and behaviour, *Agric. Ecosys. Environ.*, 63, 31, 1997.
130. Longley, M., Cilgi, T., Jepson, P. C., and Sotherton, N. W., Measurements of pesticide spray drift deposition into field boundaries and hedgerows. 1. Summer applications, *Environ. Toxicol. Contam.*, 16, 165, 1997.
131. Moreby, S. J., Heteroptera distribution and diversity within the cereal ecosystem, in *Integrated Crop Protection: Towards Sustainability*, McKinlay, R. G. and Atkinson, D., Eds., Br. Crop Protect. Conf. Monogr., No. 63, Farnham, 1995, 151.
132. Cowgill, S. E., Wratten, S. D., and Sotherton, N. W., The selective use of floral resources by the hoverfly *Episyrphus balteatus (Diptera: Syrphidae)* on farmland, *Ann. Appl. Biol.*, 122, 223, 1993.
133. De Snoo, G. R. and de Leeuw, J., Non-target insects in unsprayed cereal edges and aphid dispersal to the adjacent crop, *J. Appl. Entomol.*, 120, 501, 1996.
134. Holland, J. M., Protecting field edges and boundaries from pesticides: The benefits for farmland wildlife, *Proc. BBA-Workshop Risk Assess.*, Risk Mitigation Measures for Plant Protection Products (WORMM), September 27–29, 1999 (in press).

135. Cowgill, S. E., Wratten, S. D., and Sotherton, N. W., The effects of weeds on the numbers of hoverfly *(Diptera:Syrphidae)* adults and the distribution and composition of their eggs in winter wheat, *Ann. Appl. Biol.*, 123, 499, 1993.
136. De Snoo, G. R., van der Poll, R. J., and de Leeuw, J., Carabids in sprayed and unsprayed crop edges of winter wheat, sugar beet and potatoes, *Acta Jutl.*, 70, 199, 1995.
137. Thiele, H. U., *Carabid Beetles in Their Environments*, Springer-Verlag, Berlin, 1977.
138. Martin, P. A., Johnson, D. L., Forsyth, D. J., and Hill, B. D., Effects of two grasshopper control insecticides on food resources and reproductive success of two species of grassland songbirds, *Environ. Toxicol. Chem.*, 19, 29, 87, 2000.
139. Blus, L. J. and Henny, C. J., Field studies on pesticides and birds: Unexpected and unique relations, *Ecol. Appl.*, 7, 1125, 1997.
140. Anon., The Effect of Organic Farming Regimes on Breeding and Winter Bird populations, Parts I – IV, BTO Research Report 154, British Trust for Ornithology, Thetford, Norfolk, 1995, 175.
141. Reddersen, J., The arthropod fauna of organic versus conventional cereal fields in Denmark, *Biol. Agric. Hortic.*, 11, 61, 1997.
142. Moreby, S. J., Aebischer, N. J., Southway, S. E., and Sotherton, N. W., A comparison of the flora and arthropod fauna of organically and conventionally grown winter wheat in southern England, *Ann. Appl. Biol.*, 125, 13, 1994.
143. Shutler, D., Mullie, A., and Clark, R. G., Bird communities of prairie uplands and Wetlands in relation to farming practices in Saskatchewan, *Conserv. Biol.*, 14, 1441, 2000.
144. Sonderlund, D. M., Metabolic considerations in pyrethroid design, *Xenobiotica*, 22, 1185, 1992.
145. Block, E. K., Lacher, T. E., Brewer, L. W., Cobb, G. P., and Kendall, R. J., Population responses of *Peromyscus* resident in Iowa cornfields treated with the organophosphorus pesticide COUNTER®, *Ecotoxicology*, 8, 189, 1999.
146. Sheffield, S. R. and Lochmiller, R. L., Effects of field exposure to diazinon on small mammals inhabiting a semi-enclosed prairie grassland ecosystem, I. Ecological and reproductive effects, *Environ. Toxicol. Chem.*, 12, 284, 2001.
147. Kiesecker, J. M., Blaustein, A. R., and Belden, L. K., Complex causes of amphibian population declines, *Nature*, 410, 681, 2001.
148. Bishop, C. A., Mahony, N. A., Struger, J., Ng, P., and Pettit, K. E., Anuran development, density and diversity in relation to agricultural activity in the Holland River watershed, Ontario, Canada (1990–1992), *Environ. Monitor. Assess.*, 27, 51, 1999.
149. Bridges, C. M. and Semlitsch, R. D., Variation in pesticide tolerance of tadpoles among and within species of ranidae and patterns of amphibian decline, *Conserv. Biol.*, 14, 1490, 2000.
150. Bridges, C. M., Long-term effects of pesticide exposure at various life stages of the southern leopard frog (*Rana sphenocephala*), *Arch. Environ. Contam. Toxicol.*, 39, 91, 2000.
151. Diana, S. G., Resetarits, W. J., Schaeffer, D. J., Beckmen, K. B., and Beasley, V. R., Effects of atrazine on amphibian growth and survival in artificial aquatic communities, *Environ. Toxicol. Chem.*, 19, 2961, 2000.
152. Sotherton, N. W., Land use changes and the decline of farmland wildlife: An appraisal of the set-aside approach, *Biol. Conserv.*, 83, 259, 1998.
153. Boatman, N. D. and Stoate, C., Integrating biodiversity conservation into arable agriculture, *Asp. Appl. Biol.*, 62, 21, 2000.
154. Henderson, I. G., Cooper, J., Fuller, R. J., and Vickery, J., The relative abundance of birds on set-aside and neighbouring fields in summer, *J. Appl. Ecol.*, 37, 335, 2000.
155. Holland, J. M., Frampton, G. K., Cilgi, T., and Wratten, S. D., Arable acronyms analysed — A review of integrated farming systems research in Western Europe, *Ann. Appl. Biol.*, 125, 399, 1994.
156. El Titi, A., The Lautenbach project 1978–89: Integrated wheat production on a commercial arable farm, south-west Germany, in *The Ecology of Temperate Cereal Fields*, Firbank, L. G., Carter, N., Darbyshire, J. R., and Potts, G. R., Eds., Blackwell, Oxford, 1991, 399.
157. Cárcamo, H. A., Niemala, J. K., and Spence, J. R., Farming and ground beetles: Effects of agronomic practice on populations and community structure, *Can. Entomol.*, 127, 123, 1995.

158. Büchs, W., Harenberg, A., and Zimmermann, J., The invertebrate ecology of farmland as a mirror of the intensity of the impact of man? — An approach to interpreting results of field experiments carried out in different crop management intensities of a sugar beet and an oil seed rape rotation including set-aside, *Biol. Agric. Hortic.*, 15, 83, 1997.
159. Basedow, T. H., The species composition and frequency of spiders (Araneae) in fields of winter wheat grown under different conditions in Germany, *J. Appl. Entomol.*, 122, 598, 1998.
160. Ellsbury, M. M., Powell, J. E., Forcella, F., Woodson, W. D., Clay, S. A., and Riedell, W. E., Diversity and dominant species of ground beetle assemblages (Coleoptera: Carabidae) in crop rotation and chemical input systems for the Northern Great Plains, *Ann. Entomol. Soc. Am.*, 91, 619, 1998.
161. Huusela-Veistola, E., Effects of pesticide use and cultivation techniques on ground beetles (Col, Carabidae) in cereal fields, *Ann. Zool. Fenn.*, 33, 197, 1996.
162. Booij, C. J. H. and Noorlander, J., Farming systems and insect predators, *Agric. Ecosyst. Environ.*, 40, 125, 1992.
163. Winstone, L., Iles, D. R., and Kendall, D. J., Effects of rotation and cultivation on polyphagous predators in conventional and integrated farming systems, *Asp. Appl. Biol.*, 47, 111, 1996.
164. Gardner, S. M., Luff, M. L., Riding, A., and Holland, J. M., Evaluation of Carabid Beetle Populations as Indicators of Normal Field Ecosystems, MAFF Project Report, MAFF, London, 1999.
165. Greig-Smith, P. W., Frampton, G. K., and Hardy, A. R., Eds., *Pesticides, Cereal Farming and the Environment*, HMSO, London, 1992.
166. Fletcher, M. R., Jones, S. A., Greig-Smith, P. W., Hardy, A. R., and Hart, A. M. D., Population density and breeding success of birds, in *Pesticides, Cereal Farming and the Environment: The Boxworth Project*, Greig-Smith, P. W., Frampton G.K., and Hardy, T., Eds., HMSO, London, 1992, 160.

CHAPTER 44

# Trace Element and Nutrition Interactions in Fish and Wildlife

**Steven J. Hamilton and David J. Hoffman**

## CONTENTS

## 44.1 INTRODUCTION

Nutrition of test animals is one of the most important variables in any biological experiment. Animals with poor nutrition can be physiologically abnormal, which can result in biased results in experiments.[1] In the environment, natural feeds typically supply essential vitamins, minerals, trace elements, and other nutrients needed by fish and wildlife to maintain homeostatic functions. However, when fish and wildlife species are cultured and used in experiments, they must have adequate artificial nutrition to ensure survival and growth and, in some experiments, reproductive success including survival and growth of young. Much effort has been expended to document the essential vitamin, mineral, and other nutritional requirements of fish[2,3] and wildlife.[4,5] Deficiencies of vitamins, minerals, and other nutrients in prepared fish foods can result in skeletal deformities, cataracts, histological lesions, abnormal behavior, and many other abnormalities; in addition, excessive amounts of vitamins and minerals have also resulted in abnormalities.[2] Similar findings have been reported in wildlife species in response to nutrient deficiencies and excesses.[4–6] Even in routine fish and invertebrate culture activities, monocultures of live foods such as brine shrimp nauplii can vary in their nutritional quality due to genetics, amino acids, fatty acid composition, caloric content, carotenoids, and residues of chlorinated hydrocarbons and inorganic elements.[7]

The quality of commercial or experimentally prepared diets used in fish toxicology studies can influence the acute and chronic toxicity of test compounds.[8] Mehrle et al.[8] showed that a high-protein (45%) diet increased a fish's tolerance of a test chemical compared to a low-protein (23%) diet. Furthermore, the type of protein incorporated in a diet can influence a fish's tolerance to toxicants, i.e., protein sources casein and gelatin resulted in a greater tolerance of a test chemical than fishmeal and soybean meal as protein sources. Phillips and Buhler[9] and Hilton[10] also reported that diet quality was an important consideration in interpreting experimental results in fish studies. As with prepared fish diets, live diets can influence the results of toxicity tests with fish. For example, the geographical source of the brine shrimp can alter the toxicity exhibited by fish exposed to test compounds.[11,12]

Several studies have compared the uptake of inorganic elements from water and food either simultaneously or within a series of experiments.[13–19] Although exposures were generally in the μg/L range for water and in the μg/g range for food, body burdens of toxicant were mostly derived from food exposures. Thus, the nutrition and dietary content of trace elements can play an important role in the culture and use of fish and wildlife in research activities.

### 44.1.1 Overview of Trace Elements Essential to Fish and Wildlife Nutrition

Trace elements, considered essential micronutrients in most animals, include arsenic, cobalt, copper, iron, manganese, molybdenum, selenium, vanadium, and zinc.[5,6,20,21] For these elements, there are three levels of biological activity: trace concentrations are required for normal growth and development, moderate concentrations can be stored without disrupting homeostatic functions, and elevated concentrations can result in toxic effects. Numerous other trace elements are not considered essential and at elevated environmental concentrations can result in toxic effects. Human industrial and agricultural activity has hastened the release of inorganic elements from geologic sources and made them available to fish and wildlife in aquatic and terrestrial ecosystems around the globe.

Uptake of essential trace elements can be from water or diet. Uptake of water-soluble trace elements by fish can be either by gills or epidermis. However, dietary exposure of fish to inorganic elements is usually the dominant pathway of uptake because fish are typically at higher trophic levels in the aquatic food web,[22,23] as are birds and mammals that feed in aquatic ecosystems.

### 44.1.2 Overview of Trace Elements Toxic to Fish and Wildlife

Reviews of trace elements in human health have been reported by Goyer.[21] For fish, Leland and Kuwabara[24] reviewed the effects of copper, lead, mercury, and zinc, and Sorensen[25] reviewed the effects of arsenic, cadmium, copper, lead, mercury, selenium, zinc, and mixtures of elements. Since the publication of those two studies, many additional dietary toxicity studies with fish have been conducted. The effects on wildlife of arsenic, boron, cadmium, chromium, copper, iron, mercury, lead, manganese, molybdenum, nickel, selenium, silver, tin, and zinc have been reviewed by Klasing[5] and Eisler.[6] Further reviews of effects of lead, mercury, and selenium are provided within other chapters of this book. Ohlendorf et al.[26] reported on the uptake and potential interaction in wildlife of selenium with arsenic, boron, and molybdenum from food chain organisms in areas in central California impacted by agricultural irrigation activities.

Diet is recognized as a highly relevant route of inorganic element uptake, especially in chronically exposed feral fish.[27–30] Dallinger et al.[23] coined the phrase “the food chain effect” to describe the movement of trace elements along the food chain in polluted aquatic environments to fish and the resulting adverse effects. They noted that two factors affect the food chain effect: (1) high concentrations of contamination of the food and (2) the reduction of species diversity that occurs when susceptible species are eliminated and metal-tolerant food organisms become dominant. The dominance of the dietary pathway for uptake of inorganic elements in fish is particularly true for bioaccumulative elements like selenium. For example, fish species have disappeared from aquatic ecosystems due to food chain contamination by selenium at Belews Lake in North Carolina,[31,32] Martin Lake in Texas,[33,34] and Kesterson Reservoir in California.[35,36]

Few trace elements have been studied in aquatic food chains, but there is some information on arsenic, cadmium, vanadium, and mercury and a substantial amount of literature on selenium. Chen and Folt[37] reported that arsenic was not biomagnified to fish in an arsenic- and lead-contaminated aquatic ecosystem. Biomagnification is defined as the processes of both bioconcentration (uptake from water) and bioaccumulation (uptake of a contaminant from both water and dietary sources) that result in increased tissue concentrations of a contaminant as it passes through two or more trophic levels. Devi et al.[38] reported that cadmium was bioaccumulated in a simple aquatic food chain comprised of duckweed (*Lemna minor*) and red swamp crayfish (*Procambarus clarkii*). The information on vanadium is mostly for marine ecosystems because vanadium is an element of concern in seawater. Reports by Unsal[39] and Miramand and Fowler[40] showed that vanadium bioaccumulates in marine food chains, and it was assumed by the authors that a similar bioaccumulation occurred in freshwater ecosystems. Suedel et al.[41] report that methylmercury and possibly inorganic mercury biomagnify in aquatic ecosystems. Cherry and Guthrie[42] reported that selenium and zinc biomagnified in an aquatic ecosystem contaminated with coal ash. Suedel et al.[41] acknowledged that dietary uptake of inorganic elements was an important route of exposure in higher trophic levels.

Several papers document and review selenium in aquatic food chains from various viewpoints such as those of an aquatic toxicologist,[43] wildlife toxicologists,[44,45] an ecologist,[46] a research chemist,[47] a modeler,[48] a national selenium expert,[49] and other experts.[25,50,51]

Bioaccumulation of trace elements in food chain components such as aquatic invertebrates and fish has been documented in aquatic ecosystems contaminated with mixtures of elements. Cherry and Guthrie[42] examined contaminated food chains in settling basins for coal ash and reported that several elements were elevated in higher tropic levels including arsenic, copper, manganese, selenium, and zinc. These elements were biomagnified by at least one biotic component compared to sediment concentrations of the elements. They reported that aquatic plants had high concentrations of arsenic, manganese, mercury, and titanium; invertebrates had biomagnified concentrations of arsenic, cadmium, chromium, cobalt, copper, and mercury; and fish had biomagnified selenium and zinc.[42]

Furr et al.[52] also examined contaminated food chains in coal-ash-settling basins and reported that aluminum, cobalt, europium, iron, lanthanum, lutetium, samarium, selenium, and titanium were elevated in biota from a fly-ash pond compared to a reference pond. They concluded that, from a toxicological standpoint, selenium was the only element of concern to biota. Patrick and Loutit[53] also showed accumulation of chromium, copper, iron, lead, manganese, and zinc in fish fed tubificid worms, which in turn had ingested inorganic-enriched heterotrophic bacteria. Dallinger and Kautzky[22] concluded that uptake of inorganic elements in fish in a polluted river in Italy was primarily from food organisms because the waterborne concentrations of elements were low, which suggested to the authors that absorption through the gills was of secondary importance.

Of course, not all trace elements accumulate in fish and wildlife primarily from dietary exposure. For example, Suzuki et al.[54] reported that dietary exposure was the primary route for manganese and zinc, water exposure was the primary route for strontium and rhodium-ruthenium, and diet and water had equal roles in the accumulation of cesium, cobalt, and cerium-praseodymium in yellowtail (*Seriola quinqueradiata*).

There is a substantial database of information on the toxicity of dietary selenium in fish and wildlife but a more limited database on the dietary toxicity of other elements. Consequently, the following sections emphasize selenium and interactions with selenium. The paucity of information on the dietary toxicity of other elements should not be misinterpreted to mean that they are not important environmental considerations.

## 44.2 TOXICITY OF SELENIUM TO FISH AND ITS EFFECTS

Several early dietary studies with selenite and rainbow trout (*Oncorhynchus mykiss*) were published in the early 1980s by Canadian researchers.[55–59] These investigations were not apparently connected to the fish population disappearances in Belews Lake, North Carolina, which were believed to be due to selenium toxicity.[31] This same group of researchers also conducted two waterborne studies with selenite and rainbow trout to complement their dietary studies.[14,60] These early selenium studies explored dietary requirements, elimination and uptake rates from water and diet sources, the minimum dietary requirement for rainbow trout (between 0.15 and 0.38 μg/g in dry feed) for maximal storage, half-life period, influence of dietary carbohydrate, and toxic concentrations in water and diet. It was shown that plasma glutathione peroxidase homeostasis was maintained at up to 1.25 μg/g dry-feed activity; toxicity occurred at 13 μg/g dry feed, but the authors speculated that dietary concentrations in excess of 3 μg/g in dry feed over long time periods might be toxic; liver and kidney were the primary tissues of storage; and excess dietary carbohydrate enhanced dietary selenium toxicity in rainbow trout. Channel catfish (*Ictalurus punctutus*) have similar responses to dietary selenium (requirement between 0.1 and 0.5 μg/g and toxicity at 15 μg/g) to those of rainbow trout.[61]

Two other early investigations of dietary selenite toxicity to rainbow trout were conducted by the Colorado Division of Wildlife.[62,63] Their investigations were prompted, in part, by two field investigations in Colorado and Wyoming that concluded selenium toxicity was occurring in fish via the food chain.[64,65] Barnhart[64] was the first publication to suggest that selenium in the food chain was causing fishery problems in Sweitzer Lake in western Colorado. It is interesting to note that these publications and those of the Hilton group[55–59] did not mention selenium problems in Belews Lake, North Carolina, which was occurring during the same time period. Goettl and Davies[63] reported that dietary selenium toxicity occurred between 5 and 10 μg/g dry diet, which is remarkably close to that reported by the Hilton/Hodson group and also to later dietary studies conducted with selenomethionine and other fish species in the late 1980s.

Sandholm et al.[66] were the first to show that selenium accumulation in fish was greater from dietary sources such as phytoplankton or zooplankton than from water. They also showed that there was little difference in fish accumulation of selenite or selenomethionine in the food chain. Besser et al.[67] further reported that selenate and selenite were accumulated in fish primarily via the food chain, whereas selenomethionine was accumulated via both aqueous and food chain uptake. Kleinow and Brooks[68] reported that selenite and selenate were efficiently absorbed from the gastrointestinal tract of fish, thus resulting in a high assimilation efficiency.

In general, dietary studies with selenomethionine have reported that toxic responses in fish were similar to those in fish fed diets containing naturally incorporated selenium compounds such as fishmeal made from western mosquitofish (*Gambusia affinis*)[69] or zooplankton.[70] However, Bell and Cowey[71] reported that the digestibility and availability of selenium to Atlantic salmon (*Salmo salar*) were the least for fishmeal (source not given) and followed the order from greatest to least: selenomethionine > selenite > selenocystine > fishmeal. The comparability of selenium-laden fishmeal diet and a selenomethionine-fortified diet in three studies with chinook salmon (*Oncorhynchus tshawytscha*)[69] compared to differential digestibility reported in Atlantic salmon[71] may reflect species differences or fishmeal differences. In the study by Hamilton et al.,[69] the authors noted in both their freshwater and brackish-water studies with chinook salmon that fish growth was significantly reduced at lower concentrations and in shorter exposure periods in fish fed the diet made with the western mosquitofish fishmeal compared to the selenomethionine diet, which contained a comparable amount of clean fishmeal from western mosquitofish. They suggested that the slightly greater toxic effect in fish fed the fishmeal diet could have been caused by three factors: (1) additional toxic elements accumulated in the western mosquitofish inhabiting the San Luis Drain, (boron, chromium, and strontium); (2) other forms of organoselenium such as selenocystine present in the western mosquitofish; or (3) differential uptake, distribution, or elimination of the protein-bound organoselenium in the fish fed the western mosquitofish fishmeal diet compared to fish fed the free amino acid selenomethionine diet that contained a comparable amount of clean fishmeal.

Due to the similarity between selenomethionine and naturally selenium-laden food organisms, selenomethionine-fortified diets have been used in studies with bluegill (*Lepomis macrochirus*) to determine toxic effects on reproduction.[72–74] The study by Woock et al.[72] incorporated a waterborne exposure to 10 μg/L selenite and reported that larvae survival was reduced in the adult dietary exposure to 13 μg/g diet. The Cleveland et al.[73] study with juvenile fish reported effects at some of the lowest whole-body selenium residues (~4–5 μg/g, wet weight) that have been linked to adverse effects in fish. The Coyle et al.[74] study incorporated a 10 μg/L waterborne selenium exposure and reported adverse effects on fry when adults were exposed to 33 μg/g in the diet. Ogle and Knight[75] conducted a reproduction study with adult fathead minnow (*Pimephales promelas*) and progeny exposed to a dietary mixture of selenate, selenite, and selenomethionine. They reported reduced growth of adults exposed to 20 μg/g selenium after 56, 70, 84, and 96 days of exposure, but no effects on progeny.

As was reported with selenomethionine, Bryson et al.[70] reported that selenocystine incorporated into a fish food diet also produced adverse effects comparable to selenium-laden zooplankton. Bryson et al.[70] and Woock et al.[72] reported that diets incorporating selenite were not as toxic to fish as diets incorporating selenomethionine. This finding may be related to those of Lorentzen et al.[76] who reported that at low dietary selenium concentrations (~1–2 μg/g), dietary selenite was accumulated primarily in liver, whereas dietary selenomethionine accumulated primarily in whole body and muscle. This differentiation of storage compartments for different selenium compounds may influence the toxic effects observed.

Studies conducted in experimental outdoor streams at the U.S. Environmental Protection Agency's Monticello Research Station have demonstrated adverse effects on reproduction of fathead minnow and bluegill exposed for about a year to a selenite concentration of 10 μg/L.[77–79] Although these studies did not measure selenium concentrations in the food chain, selenium was rapidly accumulated during the growing season (i.e., May–September) in sediments and emergent, floating, and submerged aquatic plants in wetlands that comprised a portion of the experimental streams.[80] Consequently, accumulation of selenium in food organisms no doubt occurred and contributed to a dietary exposure of the fish.

Several dietary selenium studies have been conducted with food organisms collected from selenium-contaminated environments. Woock[81] demonstrated in a cage study with golden shiners (*Notemigonus crysoleucas*) that fish in cages with access to bottom sediments accumulated more selenium than fish held in cages suspended about 1.5 m above the sediments. This study showed that effects in fish were linked to selenium exposure via sediment, benthic organisms, detritus, or a combination of sediment compartments. A similar finding was reported by Barnhart[64] who reported that "numerous species of game fish" lived at least 4 months when held in a livebox, which limited access to food organisms and sediment, but fish lived less than 2 months when released in selenium-contaminated Sweitzer Lake, Colorado. The highly toxic nature of benthic organisms from selenium-contaminated Belews Lake, North Carolina was shown by Finley[82] in an experiment where bluegill-fed *Hexagenia* nymphs died in 17 to 44 days. In another study, selenium-contaminated red shiner (*Notropos lutrensis*) collected from Belews Lake, North Carolina were highly toxic to striped bass (*Morone saxatilis*).[83] In a series of experiments with the endangered razorback sucker (*Xyrauchen texanus*), larvae were fed selenium-laden food organisms from sites in the upper Colorado River and the Green River, both in the upper Colorado River basin.[84–86] Larvae consistently showed adverse effects at 4–5 μg/g in the diet (zooplankton collected from backwater areas).

Another important aspect influencing the toxic effects in fish resulting from dietary selenium exposure is season.[87] The toxicity to bluegill of combined low dietary (5.1 μg/g) and low waterborne (4.8 μg/L) selenium at low water temperature (4°C) resulted in significantly increased mortality of fish.[88] The combination of a stress-related elevated energy demand from selenium exposure and reductions in feeding due to cold temperature and short photoperiod led to a severe depletion of stored body lipid and an energetic drain that resulted in the death of about a third of the fish tested. Heinz and Fitzgerald[89] also suggested that stress from winter conditions might have increased the harmful effects of dietary selenium in adult mallards (*Anas platyrhynchos*).

## 44.3 INTERACTIONS OF SELENIUM WITH OTHER TRACE ELEMENTS IN FISH

Selenium interacts with several trace elements in fish, birds, and mammals.[25,90–92] These interactions can be additive, antagonistic, or synergistic, and in some cases the interaction was reversed, i.e., antagonism changed to synergism. In general, selenium toxicity was alleviated by antimony, arsenic, bismuth, cadmium, copper, germanium, mercury, silver, and tungsten,[90–93] whereas chromium, cobalt, fluorine, molybdenum, nickel, tellurium, uranium, vanadium, and zinc apparently have no effect on selenium toxicity.[94,95] Dietary studies conducted with fish have assessed the interaction between selenium and arsenic, copper, and mercury, as discussed below.

### 44.3.1 Arsenic Interaction with Selenium

An arsenic interaction with selenium seems to have occurred in a toxicity test conducted with endangered larval razorback sucker (Table 44.1).[85] The study involved exposing larvae to several food treatments and water treatments at locations near Grand Junction, CO. This discussion will be limited to one water treatment and two food treatments: nauplii of brine shrimp (reference food) or natural zooplankton collected from a low-selenium wetland (termed HTEW: Horsethief east wetland) located near the Colorado River, but up-gradient from irrigation activities.

Selenium residues in razorback sucker larvae after 10 days of exposure were 6.3–6.7 μg/g in the brine shrimp treatment and 6.1–7.0 μg/g in the HTEW treatment. However, survival was very different between the two food treatments: 87% in the brine shrimp treatment, but 15–20% in the HTEW treatment. The difference may have been due in part to the selenium concentrations in food at 10 days of exposure because the brine shrimp treatment had lower selenium concentrations (3.2 μg/g) than the HTEW treatment (5.0 μg/g). Nevertheless, selenium concentrations in both food treatments were above the selenium dietary toxic threshold (3 μg/g),[96,97] and whole-body residues in larvae from both food treatments were above the whole-body adverse effect threshold (4 μg/g).[55,69,73,75,88,98–101]

The same scenario occurred at day 30 in razorback sucker larvae fed the brine shrimp treatment, where whole-body residues were 5.2 μg/g, and in larvae fed the HTEW treatment, where whole-body residues were 8.2 μg/g. Survival of larvae was different between the food treatments: 81–83% in the brine shrimp treatment, but 0–10% in the HTEW treatment. Considering that the whole-body residues were somewhat similar, and selenium concentrations in food were relatively close, but survival was very different between the food treatments, it seems that the selenium residue in larvae fed the brine shrimp treatment was somehow inactivated from having a toxic effect, whereas in the HTEW treatment no inactivation occurred.

Hamilton et al.[86] discussed the potential confounding factors, such as difference between zooplankton and nauplii of brine shrimp (i.e., caloric content, nutrition value, trace element content, suitability as fish food), and concluded that an interaction probably occurred between arsenic and selenium in the brine shrimp treatment. Arsenic concentrations in brine shrimp nauplii were 24 μg/g, and in the HTEW treatment they were 6 μg/g. Arsenic concentrations in brine shrimp nauplii were not elevated sufficiently to cause dietary toxicity[102] but may have ameliorated the toxic stress of dietary selenium. Arsenic compounds have been shown to protect against the toxicity of a variety of forms of selenium including selenite, selenocystine, and selenomethionine.[93] The protective effect of arsenic has been observed in rats, dogs, swine, cattle, and birds.[93] In general, arsenic exposure in water or diet protected against dietary selenium toxicity,[103–110] but combined arsenic and selenium waterborne exposure did not.[111,112] Dubois et al.[104] and Klug et al.[105] reported that the toxicity of selenite, selenomethionine, selenocystine, and seleniferous grain was reduced in rats by exposure to arsenic as either arsenite or arsenate, but not as arsenic sulfides. Klug et al.[106] exposed rats for 12 weeks to arsenic in water and selenium in the diet and reported that arsenic protected against selenium-induced mortality, reduced growth, and reduced feeding, even though selenium residues were increased in liver (28%), kidney (141%), and muscle (52%) compared to exposure to only

**Table 44.1 Interactive Effects of Nutritional Factors and Other Elements on Selenium Toxicity in Fish**

| Species, Age | Form of Se, Dietary (Conc.) | Interactive Factor | Observation Period | Effects | Ref. |
|---|---|---|---|---|---|
| Rainbow trout, juvenile | Sodium selenite (5 or 10 ppm) | Low (0.8%) vs. elevated (24%) carbohydrate | Juvenile through 16 weeks | Excess dietary carbohydrate enhanced selenium toxicity (reduced body weight, enlarged glycogen-filled livers, food avoidance) | 57 |
| Chinook salmon, swim-up larvae | Diets with clean fishmeal fortified with seleno-D,L-methionine vs. fishmeal using selenium-laden western mosquitofish (5.3–9.6 ppm) | Organic forms of selenium or trace elements or nutritional factors | Swim-up through 90 days | Fish growth reduced at slightly lower selenium concentrations and in shorter exposure period fed diets with fishmeal using selenium-laden western mosquitofish compared to clean fishmeal fortified with seleno-D,L-methionine | 69 |
| Razorback sucker, larvae | Natural zooplankton from selenium-contaminated sites vs. brine shrimp from a commercial source (6–7 ppm) | Elevated arsenic in nauplii of brine shrimp (24 ppm) vs. low arsenic in zooplankton (6 ppm) | 5-day-old through 30 days | Elevated arsenic in brine shrimp counteracted selenium to prevent toxic effects, whereas low arsenic in zooplankton did not. Similar selenium residues in larvae but greatly different effects | 86 |
| Atlantic salmon, fry | Organic selenium in fish (unknown) | Dietary copper (500 ppm) | Fry through 12 weeks | Reduced growth, depleted energy stores, and selenium-copper antagonism | 115, 116 |
| Pearl dace (adult)<br>Yellow perch (~2 g)<br>Northern pike (~35 g) | Natural zooplankton exposed to waterborne selenium (6, 100 ppb) | Mercury-contaminated lakes | 3, 6, and 8 weeks | Dietary selenium, but not waterborne selenium, reduced mercury uptake in fish; at elevated concentrations, selenium reduced survival of yellow perch but not pearl dace | 118–120 |
| Yellow perch (adult)<br>Northern pike (adult) | Selenite uptake in the food chain (5 ppb in water) | Mercury-contaminated lakes | Adults for 1–2 years | Reduced mercury residues below level of human health, but selenium reduced fish survival and reproduction | 122–125 |

selenium in the diet (no arsenic exposure). Klug et al.[106] concluded that arsenic counteracted selenium toxicity in some way other than increasing selenium elimination. Levander and Argrett[107] also showed that arsenic protected against selenosis in rats and that selenium residues were increased in carcass over animals in the selenium-only exposure. Others have concluded that arsenic exposure increased the elimination of selenium in bile in short-term (1–10 h) injection experiments.[90–93] In a reciprocal manner, selenium has been reported to reduce arsenic-induced teratogenic deformities in hamsters.[113]

Considering that (1) the whole-body residues of larvae were elevated in the brine shrimp and HTEW dietary treatments, (2) selenium concentrations in both diet treatments were above the selenium dietary toxicity threshold, (3) arsenic concentrations in the brine shrimp treatment were elevated (24 μg/g) and in the HTEW treatment were low (6.0 μg/g), (4) survival was high in the brine shrimp treatments but low in the HTEW treatment, and (5) no confounding factors seemed to be present between the two dietary treatments such as caloric content, nutritional value, trace element content (other than arsenic), and suitability as fish food, it seems likely that arsenic interacted with selenium in the fish larvae and inhibited mortality in larvae fed the brine shrimp treatment but did not inhibit mortality in the HTEW treatment.[86] These results were similar to observations of Klug et al.[106] and Levander and Argrett.[107]

### 44.3.2 Copper Interaction with Selenium

Lorentzen et al.[114] and Berntssen et al.[115,116] both reported that elevated dietary copper reduced concentrations of selenium in liver of Atlantic salmon. Lorentzen et al.[114] suggested that reduced selenium concentrations were due to the formation of insoluble copper–selenium complexes in the intestinal lumen, reducing selenium bioavailability or the excretion of copper–selenium complexes from the liver through the bile. Berntssen et al.[116] reported that dietary copper exposure significantly reduced selenium concentrations in intestine and liver, which in turn reduced glutathione concentrations (selenium is a component of glutathione). However, dietary copper did not affect antioxidant glutathione peroxidase enzyme activity. In contrast, Hilton and Hodson[57] reported that increasing dietary selenium exposure of rainbow trout significantly increased copper concentrations in liver. They speculated that dietary selenium interfered with copper transport or excretion.

### 44.3.3 Mercury Interaction with Selenium

Perhaps one of the most published interactions between inorganic elements is that between mercury and selenium. Pelletier[117] reviewed the literature for aquatic organisms and concluded that many authors reported simultaneous bioaccumulation of mercury and selenium, but there was no evidence of natural joint bioaccumulation of mercury and selenium in fishes, crustaceans, and mollusks. He noted that many results were unrelated and sometimes contradictory.

A series of experiments by Rudd, Turner, and others[118–120] investigated the ability of selenium to ameliorate the toxic effects on fish inhabiting a mercury-contaminated lake in the English-Wabigoon River system in Ontario, Canada. They conducted enclosure experiments in the lake and reported that selenium additions reduced mercury accumulation in fish. Selenium interfered with mercury mobilization through the food web rather than mercury accumulation directly from water. They cautioned that selenium amelioration of mercury should be approached with caution because selenium readily and efficiently accumulated in the food organisms and fish, especially through the food chain, and recommended that selenium additions be limited to 1 μg/L. Klaverkamp et al.[121] reported that exposure of northern pike (*Esox lucius*) to waterborne selenium at 1 μg/L reduced mercury accumulation in carcass, but exposure to 100 μg/L selenium increased mercury accumulation in carcass.

Another series of experiments in mercury-contaminated lakes in Sweden also tested the ameliorating effects of selenium.[122–124] Like the studies by Rudd, Turner, and others, these studies also confirmed that selenium readily reduces mercury accumulation in fish, but selenium bioaccumulated in fish via the food chain if waterborne selenium concentrations were > 3–5 μg/L. Fish kills of yellow perch (*Perca flavescens*) occurred in 4 of the 11 lakes in their study, which prompted Lindqvist et al.,[125] who reviewed mercury concerns in the Swedish environment, to recommend against the use of the selenium amelioration technique in mercury-contaminated lakes.

## 44.4 EFFECTS OF TOXIC DIETARY TRACE ELEMENTS (OTHER THAN SELENIUM) TO FISH

### 44.4.1 Aluminum

Dietary-aluminum studies with fish showed that aluminum accumulated in fish, but apparently did not result in toxic effects.[126–128] Interestingly, Poston[126] reported no effects on growth, survival, or feed conversion in Atlantic salmon fed up to 2000 μg/g dietary aluminum. These studies contrast with waterborne studies that showed that, under acid rain conditions (i.e., low pH), certain aluminum species can be toxic to fish, although they did not seem to accumulate substantially.[128] Poston[126] reported that trace amounts of aluminum had some nutritional benefits to Atlantic salmon, thus suggesting that aluminum may be an essential inorganic element and most likely metabolically regulated.

### 44.4.2 Arsenic

Dietary inorganic arsenic compounds (arsenic trioxide [AT] and disodium arsenate heptahydrate [DSA]) were more toxic to rainbow trout than organic arsenic compounds (dimethylarsinic acid [DMA] and arsanilic acid [AA]).[129] The NOEC (no-observed-effect concentration) for inorganic arsenic was between 1 and 137 μg/g for DSA, between 1 and 180 μg/g for AT, and at least 1497 μg/g for DMA and AA. In a more detailed dietary-arsenic study with rainbow trout, Cockell et al.[102] reported adverse effects in fry at 33 μg/g of pentavalent arsenic as DSA. In another dietary arsenic study with rainbow trout exposed to trivalent arsenic (sodium arsenite), Oladimeji et al.[130] reported adverse effects at 20 and 30 μg/g, thus suggesting that the dietary toxicity of trivalent and pentavalent forms of arsenic were similar. However, in aquatic ecosystems, arsenic does seem to biomagnify in invertebrates but not in fish.[42] Chen and Folt[37] reported that in the arsenic- and lead-contaminated Aberjona Watershed of Massachusetts, arsenic accumulated in lower food-chain components such as algae and zooplankton, but not in upper trophic levels including fish. Woolson[131] also reported that although arsenic was bioconcentrated from water to aquatic plants, it was not biomagnified to invertebrates or fish. Although arsenic bioaccumulated in plants and zooplankton,[37,42] no biological effects were measured in fish.[37,42,131] Furthermore, Chen and Folt[37] noted that although arsenic was bioaccumulated in lower trophic levels from the arsenic-contaminated site, fish from that site had tissue arsenic concentrations similar to reference areas.

### 44.4.3 Cadmium

There is a controversy as to the importance of exposure route in the accumulation of cadmium in fish. Studies have reported that (1) a greater proportion of cadmium was accumulated from food than from water,[19,22,132] (2) about equal amounts of cadmium were accumulated from diet or water exposures,[133] or (3) more cadmium was accumulated from water than dietary exposures.[134–136] Several studies have reported cadmium concentrations in food that were related to adverse effects. Handy[137] reported that dietary cadmium at 10 mg/g caused increased mortality of subadult rainbow trout in a 28-day study. Hatakeyama and Yasuno[138] reported that the reproduction of the guppy (*Poecilia reticulata*) was adversely affected, i.e., the cumulative number of fry produced was reduced by exposure to 210 μg/g in midge used as a food source. Kumada et al.[135] reported that 100 μg/g cadmium as cadmium stearate in the diet caused liver and kidney damage in rainbow trout. Sublethal dietary concentration of 10 μg/fish/day fed to tilapia (*Oreochromis mossambicus*) resulted in disruption of plasma calcium, magnesium, and phosphate concentrations, which was ameliorated somewhat by high waterborne calcium concentrations.[139] Crespo et al.[140] reported that sublethal exposure of rainbow trout to dietary cadmium concentrations of 5 μg/g resulted in morphological disorders in the middle and posterior intestine, which the authors concluded would impair intestinal absorption of nutrients. Lundebye et al.[141] also reported intestine damage in Atlantic salmon exposed to dietary cadmium. In a comparison of the toxicity of cadmium and copper to rainbow trout, Handy[137] reported that dietary cadmium was more toxic than dietary copper, which is an essential element to animals.

In the extensive review of cadmium in food webs by Kay[142] and the review by Handy,[30] the authors cited several studies that reported cadmium concentrations to be generally higher in aquatic invertebrates than in fish, thus suggesting that cadmium was not biomagnified through the food web. Nevertheless, the authors concluded that food, as a source of cadmium, might be relatively more important in nature than implied from laboratory studies. Dallinger and Kautzky[22] and Dallinger et al.[23] also concluded that dietary exposure of fish to cadmium was especially important in aquatic ecosystems where waterborne cadmium concentrations were low, but sediment concentrations could cycle cadmium into the food chain.

Harrison and Curtis[143] compared natural foods to commercial diets fortified with cadmium and reported that five times more cadmium was accumulated by fish from natural foods than from a fortified commercial diet. This observation is the opposite of Merlini et al.[144] and Pentreath,[145] who reported that zinc was more efficiently accumulated from a prepared diet than from a natural food. This difference between cadmium and zinc accumulation from different types of foods may be due to the fact that zinc is an essential element to animals, whereas cadmium is not. Likewise, fish mortality in dietary cadmium exposures may not be due exclusively to direct dietary cadmium toxicity because investigators have reported disruption of copper and zinc metabolism that may have contributed to the observed mortality.[137,146]

A concern raised in dietary-cadmium studies reviewed by Kay[142] was that a major problem in cadmium-treated artificial diets was the use of soluble chloride and nitrate forms of cadmium, which could rapidly leach from the diet, thus resulting in a partial aqueous cadmium exposure. He further noted that rapid-exchange flow-through exposure systems and immediate removal of uneaten food were only partial solutions to the problem of waterborne exposure. These two concerns — soluble trace elements added to diets and rapid flushing or rapid removal — apply to any study where a trace element or other toxic is added to a prepared diet.

### 44.4.4 Cesium

Aoyama et al.[147] reported that predator fish accumulate cesium from prey fish. The cesium concentration in predator fish increased with ration size, but altering the feeding interval did not influence the accumulation of the trace element. Fish growth, however, diluted the body residue when the dietary concentration of cesium was constant.

### 44.4.5 Chromium

One study investigated the effects of dietary chromium in fish using trivalent chromium.[148] Rainbow trout did not accumulate chromium in carcass, fin, vertebrae, skin, gill, muscle, or intestine, even at the highest concentration tested (8.2 μg/g). Nevertheless, chromium apparently caused reduced growth at this concentration. These results were consistent with those of Patrick and Loutit[53] and Dallinger and Kautzky,[22] who investigated aquatic ecosystems contaminated with a mixture of inorganic elements and reported that chromium in the food chain accumulated less in fish than other elements such as cadmium, copper, lead, and zinc. Part of the reason for a lack of adverse effects from chromium may be due to it's metabolic regulation as an essential element in animals.

### 44.4.6 Cobalt

Baudin and Fritsch[149] investigated the relative contribution of cobalt from food and water with common carp (*Cyprinus carpio*) in a 63-day exposure. They concluded that the uptake and accumulation of cobalt, which is an essential element for plants and animals, was greater from water exposure than from food. However, the exposure concentrations were very low (0.2 μg/L in water and 0.13 μg/g in food), which resulted in low accumulation in fish (total cobalt 0.0033 μg/g). Suzuki et al.[54] reported that cobalt accumulated about equally in fish exposed by either waterborne or dietary routes of exposure.

### 44.4.7 Copper

The dietary concentration of copper that was toxic to rainbow trout as reported by Lanno et al.[150] was 664 μg/g. This elevated toxic concentration, relative to the lower toxic dietary concentrations of arsenic, cadmium, and mercury that cause adverse effects in fish, may be due, in part, to copper being an essential inorganic element to fish and other animals. Miller et al.[151] reported that rainbow trout exposed to dietary copper as high as 684 μg/g showed no adverse effects on survival, growth, condition factor, or food conversion efficiency, but they noted that diet seemed to be the dominant source of copper in their water and dietary exposure study. Similarly, Handy et al.[152] reported no effects on survival or growth of rainbow trout exposed to 490 μg/g of dietary copper, but lipids were reduced in liver, and swimming activity was altered. Berntssen et al.[115] reported adverse effects on growth and whole-body energy stores of protein and glycogen in Atlantic salmon fed 500 μg/g or greater copper in the diet. Lundebye et al.[141] also reported that growth was reduced in Atlantic salmon exposed to dietary copper at concentrations of 700 μg/g or greater. Most investigators hypothesized that biochemical regulation of copper prevents toxic effects at low concentrations (i.e., $< 100$ μg/g) because copper is an essential element to fish and other animals. In a comparison of the toxicity of copper and cadmium to rainbow trout, Handy[137] reported that dietary copper was less toxic than dietary cadmium. This differential toxicity probably reflects the regulation of copper and the limited regulation of cadmium via metallothionein.[137,146]

### 44.4.8 Lead

Hodson et al.[153] reported that dietary lead was not accumulated by rainbow trout, although the highest concentration tested in the diet was only 1 μg/g. Chen and Folt[37] reported that elevated lead concentrations in the arsenic- and lead-contaminated Aberjona Watershed of Massachusetts were accumulated in lower food-chain components, such as algae and zooplankton, but not in upper trophic levels including fish. Leland and McNurney[154] also reported that the highest lead concentrations were in periphyton and macrophytes, and lead was not biomagnified through the aquatic food chain. They showed that lead accumulation in fish was lowest in piscivores, intermediate in predators of macroinvertebrates, and highest in grazers or detritus feeders. Patrick and Loutit[53] and Dallinger and Kautzky[22] investigated aquatic ecosystems contaminated with a mixture of inorganic elements and reported that lead was accumulated in fish through the food chain, but to a lesser extent than other important elements such as copper and zinc. Nevertheless, a dietary lead concentration of 10 μg/g fed to rainbow trout caused morphological disorders in the middle and posterior intestine and altered chlorine and sodium fluxes and sodium–potassium ATPase activity.[140] Vighi[15] reported that more lead was accumulated via the food chain than from water in an algae-daphnia-guppy food-chain exposure. However, the accumulation factors showed high accumulation in algae, lower concentrations in daphnia, and still lower in fish. Exposure of rainbow trout to very elevated dietary lead (7 mg/g) caused a 12% incidence of black-tail, a well-established symptom of lead toxicosis in fish.[155]

### 44.4.9 Manganese

Manganese accumulates in fish primarily from dietary sources.[54,145] Dallinger and Kautzky[22] studied a trace-element-contaminated (copper, cadmium, chromium, lead, manganese, nickel, zinc) river ecosystem and reported that rainbow trout accumulated a substantial amount of manganese from dietary sources. No one has reported adverse effects in fish associated with dietary manganese exposure.[22,54,145]

### 44.4.10 Mercury

Mercury has no known essential function in vertebrate organisms.[156] Diet was the primary route of methylmercury uptake by fish and methylmercury the dominant mercury form in fish.[156]

For example, Phillips and Buhler[13] reported that 70% of dietary methylmercury was assimilated fish, whereas only 10% of waterborne methylmercury was assimilated. Handy[30] and Wiener and Spry[156] noted that the trophic transfer of mercury to fish has received a substantial amount of attention because of mercury poisoning in humans from consumption of contaminated fish. However, there is limited information on dietary mercury toxicity to fish. Dietary mercury exposure of fish to about 20–40 μg/g resulted in adverse effects such as gastrointestinal damage and inhibition of enzymes in the gut.[30] Zhou and Wong[157] reported a fourfold difference in mercury accumulation from the diet in six species of fish due to differences in feeding behavior.

### 44.4.11 Vanadium

One study investigated the toxicity of vanadium in the diet to fish.[158] Vanadium concentrations of 10.2 μg/g and greater in the diet of rainbow trout reduced growth and feeding responses. It was concluded that rainbow trout were extremely sensitive to dietary vanadium, and dietary vanadium toxicity was similar to that of dietary selenium toxicity. In contrast, waterborne exposure of fish to vanadium in freshwater ecosystems does not seem to be of major concern.[159] Hamilton et al.[84] suggested that elevated vanadium residues in aquatic invertebrates fed to larval endangered razorback suckers in an on-site toxicity test contributed to the mortality of larvae. Vanadium is one of the least-studied trace elements from a dietary-toxicity standpoint but could potentially be a very important dietary contaminant. Most research with dietary vanadium has been done to determine its essentiality.[160]

### 44.4.12 Uranium

One dietary study has been conducted with uranium in which lake whitefish (*Coregonus clupeaformis*) exposed for 100 days to 100 μg/g of dietary uranium had concentration- and duration-dependent histopathologies in liver and posterior kidney.[161,162] Those effects were chemotoxic rather than radiotoxic because α-radiation rates were negligible.

### 44.4.13 Zinc

In a comparison of waterborne vs. dietary uptake of zinc, Merlini et al.[144] reported a two- to fourfold greater uptake in pumpkinseed (*Lepomis gibbosus*) of zinc from food than from water. Likewise, Willis and Sunda[17] compared waterborne vs. dietary uptake in western mosquitofish and spot (*Leiostomus xanthurus*) and reported that 78 to 82% of the accumulated zinc came from the dietary uptake. Investigations have reported that zinc was more efficiently accumulated from surficially incorporated prepared diets than from zinc incorporated into live foods.[144,145] However, others have reported that zinc proteinate (i.e., a chelated amino acid complex of zinc) accumulated to greater concentrations in fish than did dietary zinc supplemented as unbound zinc as zinc sulfate.[163–165] Wekell et al.[163] reported that exposure of rainbow trout to up to 1.7 mg/g dietary zinc did not cause effects on growth or survival, which was similar to Brafield and Koodie,[165] who found no effects in common carp exposed to up to 5 mg/g dietary zinc. These results were probably due to the fact that zinc is an essential element to animals and is efficiently regulated in fish from a dietary exposure route.[16] However, exposure of rainbow trout to very elevated dietary zinc (8.2 mg/g) has been reported to reduce growth, which may have been due to feed avoidance.[155] In general, waterborne exposure to zinc seems to lead to greater toxic effects than from dietary exposure.[18,166] Elevated concentrations of calcium or phosphorus in the diet can cause a functional zinc deficiency in fish.[30] Similarly, Satoh et al.[167] reported that fishmeal diets deficient in zinc reduced growth of rainbow trout.

### 44.4.14 Mixtures of Inorganic Elements

It is rare in the aquatic environment that one inorganic contaminant stands out alone as the sole source of pollutant stress on fish, with the possible exceptions of mercury and selenium. Rather, a mixture of contaminants is often present at generally low concentrations in water but elevated concentrations in sediment and food organisms.[22,23,53] Woodward, Farag, and coworkers have reported two good examples of mixtures of elements contaminating food organisms and the resulting adverse effects in fish in the Clark Fork River of Montana[168–172] and the Coeur d'Alene River of Idaho.[170,173,174] Both rivers were Superfund sites contaminated by wastes from mining activities and have depressed trout populations. The Clark Fork River is contaminated dominantly by copper and the Coeur d'Alene River by lead.

Food organisms from the Clark Fork River contained elevated concentrations of arsenic, cadmium, copper, lead, and zinc and when fed to brown trout (*Salmo trutta*) or rainbow trout resulted in adverse physiological, metabolic, behavioral, and histopathological effects.[168–172] Woodward et al.[171] reported that dietary exposure to the mixture of trace elements was more important than waterborne exposure in causing reduced growth and survival of fish. One interesting aspect of these studies was that the benthic invertebrates collected from the Clark Fork River and used in the experimental diets were either homogenized and refrozen for later feeding or pasteurized and supplemented with vitamins and minerals. Fish growth was reduced in fish fed either the unsupplemented diet or the supplemented diet. However, survival was reduced in fish fed the unsupplemented diet but not in those fed the supplemented diet, which suggested that either pasteurization (eliminated diseases in the wild caught invertebrates) or addition of vitamins and minerals (which may have produced healthier fish) may have enabled the fish to better withstand stresses from the dietary inorganic elements. A large number of studies were conducted as part of the Clark Fork River investigations including chemistry studies of terrestrial and aquatic abiotic and biotic components; ecological and population studies of terrestrial and aquatic organisms; and *in situ*, laboratory, and field studies of organisms. The weight of evidence in the risk assessment gave great weight to the findings in the fish studies with dietary exposure to inorganic elements.[175]

One interesting study conducted as part of the Clark Fork River investigation was a dietary study with nauplii of brine shrimp. Nauplii were enriched with cadmium, copper, lead, and zinc, individually and as a mixture including arsenic, then fed to rainbow trout fry for 60 days starting at 11-days post swim-up.[176] Fish exposures also included simultaneous exposure to a mixture of waterborne elements at sublethal concentrations. Nauplii were exposed to aqueous element concentrations for about 24 h before hatch and about 24 h post hatch to achieve residues close to those in invertebrates from the Clark Fork River; then they were fed to fish. No adverse effects on survival or growth of fish were observed. In citing this study, Woodward et al.[172] noted that the duration of the brine shrimp exposure was short; thus, the elements were probably attached to the external surfaces of the nauplii and in the free form instead of incorporated protein, as in the studies by Woodward et al.[171,172] and Farag et al.[168,169] They also noted that ionic inorganic elements were not absorbed as efficiently in the gut and may not be as toxic as elements bound to proteins.[143,165,170,177] Farag et al.[170] reported that inorganic elements incorporated naturally into invertebrates collected from the Clark Fork River were processed differently during digestion by cutthroat trout (*Oncorhynchus clarki*) than those from diets made up of brine shrimp nauplii exposed to a mixture of elements in the laboratory for ~24 h, as was done by Mount et al.[176] They cautioned that the results from studies that incorporate inorganic elements into live foods as part of a dietary exposure may result in toxicological effects different than those using natural foods with elevated inorganic elements.

Results similar to those in the Clark Fork River studies were found in the Coeur d'Alene River studies, where the elements of concern included arsenic, cadmium, copper, lead, mercury, and zinc.[170,173,174] In general, inorganic elements were highest in sediments and *aufwuchs*, intermediate in invertebrates, and lowest in fish. Although the elements did not biomagnify, they were bioavailable and did biotransfer to fish. Fish fed diets incorporating benthic invertebrates collected from contaminated sites in the Coeur d'Alene River had reduced survival, growth, and feeding activity and increased histopathological abnormalities.[174] One interesting aspect reported by Farag et al.[173] was that smaller invertebrates accumulated greater concentrations of elements than larger invertebrates, which the authors hypothesized would expose early life stages of fish to larger doses of elements than adults.

## 44.5 INTERACTIVE EFFECTS OF NUTRITION AND OTHER ELEMENTS ON SELENIUM TOXICITY IN BIRDS

Fewer trace element/nutrition interaction studies have been conducted with wildlife species than with fish. Most of these studies have been conducted with birds and virtually none with wild mammals. This most likely is due to the fact that the human health effects literature, as well as human and agricultural nutrition literature, is abundant with interaction studies conducted with laboratory rodents as well as with other species of domestic mammals that are viewed by some as surrogate species for mammalian wildlife. Most of the studies conducted with avian wildlife species appear to fall into three categories: (1) those related to interactions between selenium, nutritional factors, and other trace elements; (2) those related to interactions between lead, nutritional factors, and other trace elements; and (3) a smaller number of studies related to interactions among other trace elements. The findings from studies in these three categories are summarized and discussed below for selenium and in subsequent sections for lead, other trace elements, and field studies.

Agricultural drainwater, sewage sludge, fly ash from coal-fired power plants, and mining of phosphates and metal ores are all sources of selenium contamination of the aquatic environment.[178,179] High concentrations of selenium from agricultural drainwater accumulated in the aquatic food chain at the Kesterson National Wildlife Refuge in the San Joaquin Valley of California. These included mean selenium concentrations of 73 μg/g in submerged rooted aquatic plants and over 100 μg/g in aquatic insects consumed by ducks and other aquatic birds.[44] Similar problems of elevated selenium have been reported elsewhere in the western United States in at least a dozen different locations[45] but most notably in the Tulare Basin in California.[180] Resulting adverse biological and physiological effects of exposure have included avian mortality, impaired reproduction and teratogenesis, and histopathological lesions with alterations in hepatic glutathione metabolism and oxidative stress.[181–184] As a consequence of these findings, a series of laboratory studies was conducted primarily with mallards to help interpret the potential toxicity of different forms of selenium, dietary sources of selenium, and interactions with other dietary components including methionine, protein, and various trace elements that might be encountered in nature.

### 44.5.1 Different Dietary Forms of Selenium and Comparative Effects

The form of selenium in the diet was found to play an important role in the subsequent uptake and resulting toxicity. Selenomethionine is a major form of selenium found in wheat and soybean protein and may be a major form of selenium in other plants.[185,186] This form of selenium is presumably an important exposure source for wild aquatic birds since the toxic thresholds in eggs for decreased hatching success and teratogenicity from lab studies with mallards proved to be nearly identical to those derived from field studies.[179] Selenomethionine becomes particularly embryotoxic and teratogenic when it exceeds 4 ppm Se in the laboratory diet of mallards. At higher concentrations, it becomes toxic to other stages of the mallard life cycle.

#### *44.5.1.1 Reproductive Effects in Mallards*

Findings from reproductive studies have indicated that selenium as selenomethionine is considerably more teratogenic and generally more embryotoxic than sodium selenite or selenocystine, due to much higher accumulation in eggs when provided in this form in the diet. When mallards were fed 10 ppm selenium as selenomethionine, 25 ppm as sodium selenite was required to produce a similar decrease of 40–44% in the total number of eggs that hatched compared to controls.[187,188] Furthermore, 10 ppm Se as selenomethionine was more teratogenic than sodium selenite at 25 ppm, and resulted in an incidence of 13.1% malformations that were often multiple, whereas sodium selenite (10 and 25 ppm Se) resulted in 3.6 and 4.2% malformations. The teratogenicity of selenomethionine was confirmed in a second experiment in which mallards received 1, 2, 4, 8, or 16 ppm Se as selenomethionine, resulting in 0.9, 0.5, 1.4, 6.8, and 67.9% malformations, respectively.[189] Selenocystine at 16 ppm in the diet did not impair reproduction or result in significant malformations. In a subsequent reproductive study with mallards, seleno-DL-methionine and seleno-L-methionine were found to be of similar toxicity, and both forms were more toxic than selenium derived from selenized yeast, which is most likely due to the presence of less toxic forms of selenium in the yeast.[190]

#### *44.5.1.2 Survival and Growth in Mallard Ducklings*

Selenium as selenomethionine resulted in generally greater oxidative stress than did selenium as selenite in the diet of mallard ducklings, as reflected by effects on hepatic glutathione metabolism and lipid peroxidation.[191] These findings reflect the much greater accumulation of selenium in the liver and other organs as dietary selenomethionine than as selenite. However, both forms of selenium reduced duckling growth and survival.[192] When day-old ducklings were fed 10, 20, or 40 ppm Se as seleno-DL-methionine or sodium selenite for six weeks, selenium from selenomethionine accumulated in a dose-dependent manner in the liver. Hepatic and plasma GSH peroxidase activity was initially elevated at 10 ppm Se as selenomethionine, whereas GSSG reductase activity was elevated at higher dietary concentrations of selenium. A decrease in the concentration of hepatic-reduced glutathione (GSH) and total hepatic thiols (SH) at 20 ppm Se in the diet was accompanied by an increase in the ratio of oxidized glutathione (GSSG) to GSH and an increase in thiobarbituric acid reactive substances (TBARS) concentration as evidence of lipid peroxidation.[191] Selenium from selenite accumulated in the liver to an apparent maximum at 10 ppm in the diet, resulting in an increase in hepatic GSH and GSSG, accompanied by a small decrease in hepatic total SH. Sodium selenite resulted in increased hepatic GSSG reductase activity at 10 ppm and in plasma GSSG reductase activity at 40 ppm. A small increase in lipid peroxidation occurred at 40 ppm.

#### *44.5.1.3 Subchronic Effects on Immune Function and Hepatotoxicity in Adult Mallards*

The subchronic effects of selenomethionine and sodium selenite were compared with respect to several immunologic, hematologic, and serologic parameters in adult male mallards, using concentrations in drinking water of 0, 0.5, or 3.5 mg/L Se as sodium selenite or 2.2 mg/L Se as selenomethionine for 12 weeks.[193] A battery of *in vivo* and *in vitro* immunologic assays was performed on each bird throughout the study. The selenomethionine-treated birds displayed an impaired delayed-type hypersensitive (DTH) response to tuberculin (*M. bovis*), as measured by the number of positive reactions present 24 h post purified protein derivative (PPD) challenge. This group also exhibited increased serum alanine aminotransferase (ALT) and plasma GSH peroxidase activities. The selenium concentration in the liver and breast muscle was significantly elevated 4- and 14-fold, respectively, over controls. Sodium-selenite-treated birds did not display any detectable differences in immune function or selenium accumulation in tissues above that of controls. Serum ALT activity was increased in the 3.5 mg/L group, although to a lesser extent than in selenomethionine-treated birds. Concentrations of selenium as sodium selenite did not affect the immune system, whereas low concentrations of selenomethionine (2.2 mg/L Se) appeared to suppress certain aspects of the mallard immune response.

#### 44.5.1.4 Nutrition of Diet, Source of Selenium, and Bioavailability

Since the composition of the diet of wild ducklings may vary considerably in selenium-contaminated environments, a study was conducted to compare seleno-DL-methionine (DL, previously used in many lab studies), seleno-L-methionine (L, a form found in nature), selenized yeast (Y), and selenized wheat (W).[194, 195] Day-old mallard ducklings received an untreated diet (controls) containing 75% wheat (22% protein) or the same diet containing 15 or 30 ppm Se in the above forms for 2 weeks (Table 44.2). All forms of selenium caused significant increases in plasma and hepatic glutathione peroxidase activities. Selenium as L at 30 ppm in the diet was the most toxic form, resulting in high mortality (64%) and impaired growth (> 50%) in survivors and the greatest increase in ratio of hepatic GSSG:GSH. Selenium as both L and DL decreased the concentrations of hepatic GSH and total thiols. Selenium as Y accumulated the least in liver (approximately 50% of other forms) and had less effect on GSH and total thiols. In contrast, when a different and commercially based basal diet, rather than the 75% wheat diet, was provided, survival of ducklings was not affected by 30 ppm Se. Again, selenium as selenized yeast was found to be less toxic than seleno-D,L-methionine, seleno-L-methionine, or selenized wheat, as reflected by the fewer effects on growth, plasma alkaline phosphatase activity, and hepatic oxidative stress. Here, greater oxidative stress for DL and L diets was manifested by significant decreases in hepatic GSH and total SH. This may be related, in part, to the nearly double tissue accumulation of selenium from selenomethionine diets than from selenized yeast.

#### 44.5.1.5 Dietary Methionine and Protein

Since both quantity and composition of dietary protein for wild ducklings may vary in selenium-contaminated environments, several studies were conducted to examine effects of methionine and protein concentration on selenium toxicity. Findings from these studies have shown the potential for antagonistic effects of selenium, methionine, and protein on duckling survival and oxidative stress.

In one study, day-old mallard ducklings received one of the following diets containing 22% protein: unsupplemented controls, 15 ppm Se (as selenomethionine), 60 ppm Se, methionine-supplemented controls, 15 ppm Se with methionine supplement, or 60 ppm Se with methionine supplement for 4 weeks.[196] In a second concurrent experiment, the above sequence was repeated with a protein-restricted (11%) but isocaloric diet. In a third concurrent experiment, protein was increased to 44% with 0, 15, or 60 ppm Se added. With 22% protein and 60 ppm Se in the diet, duckling survival and growth were reduced, and histopathological lesions of the liver occurred. Antagonistic interactive effects occurred between supplementary methionine and selenium including complete to partial alleviation of the following selenium effects: mortality, hepatic lesions, and altered glutathione and thiol status.

**Table 44.2 Interactive Effects of Nutritional Factors and Other Elements on Selenium Toxicity in Birds**

| Species, Age | Form of Se, Dietary (conc., ppm) | Interactive Factor | Observation Period | Effects | Ref. |
|---|---|---|---|---|---|
| Mallard, duckling | Compared these forms: seleno-D,L-methionine (DL), seleno-L-methionine (L), selenium from yeast (30 ppm) | Wheat diet vs. standard diet | Day-old through 2 weeks | With wheat diet: L was most toxic with respect to mortality, decreased growth, and increased oxidative stress | 194, 195 |
| | | | | With standard diet: all forms of Se less toxic than with the wheat diet | |
| Mallard, duckling | Seleno-D,L-methionine (15 ppm) | Dietary methionine (0.42% added) | Day-old through 4 weeks | Methionine decreased Se-related histopathological lesions and oxidative stress | 196 |
| | (60 ppm) | | | Methionine decreased Se-related mortality, histopathological lesions, and oxidative stress | |
| Mallard, duckling | Seleno-D,L-methionine (15 ppm) | Restricted dietary protein (7 or 11% vs. 22% protein controls) | Day-old through 4 weeks | Restricted dietary protein increased Se-related toxicity with respect to growth, plasma chemistries, hepatic Se accumulation, and oxidative stress | 110, 196 197 |
| | (60 ppm) | | | Se caused complete mortality in combination with restricted dietary protein | |
| Mallard, duckling | Seleno-D,L-methionine (15 ppm) | Excess dietary protein (44 vs. 22% protein controls) | Day-old through 4 weeks | Excess dietary protein increased Se-related toxicity with respect to growth and oxidative stress | 196 |
| | (60 ppm) | | | Excess dietary protein increased Se-related toxicity with respect to mortality, growth, hematocrit, hemoglobin, plasma chemistries, and oxidative stress | |
| Mallard, embryo through post-hatching (reproduction study) | Seleno-D,L-methionine (10 ppm) | Sodium arsenate (25, 100, or 400 ppm As) | Embryo through 14 days post-hatching | Sodium arsenate decreased the accumulation of Se in liver and egg, and Se-related hatching failure, and teratogenesis | 198 |
| Mallard, duckling | Seleno-D,L-methionine (15 ppm) | Sodium arsenate (200 ppm As) | Day-old through 4 weeks | Sodium arsenate decreased Se-related oxidative stress | 110 |
| | (60 ppm) | | | Sodium arsenate decreased Se-related mortality, impaired growth, histopathological lesions, hepatic Se concentration, and oxidative stress | |

| | | | | | |
|---|---|---|---|---|---|
| Mallard, embryo through post-hatching (reproduction study) | Seleno-D,L-methionine (7 ppm) | Dietary B (boric acid) (900 ppm B) | Embryo through 14 days post hatching | No interactive effects observed | 199 |
| Mallard, duckling | Seleno-D,L-methionine (15 ppm) | Dietary B (boric acid) (1000 ppm) with and without restricted dietary protein | Day-old through 4 weeks | With normal protein (22%) diet, B and Se in combination decreased liver weight and altered several plasma chemistries<br>With restricted protein (7%) diet, B and Se in combination increased mortality and increased hepatic Se accumulation | 197 |
| | (60 ppm) | | | With normal protein (22%) diet B and Se in combination decreased liver weight and altered several plasma chemistries<br>With restricted protein (7%) diet Se alone caused complete mortality | |
| Japanese quail, chicks | Sodium selenite (5 or 8 ppm) | Methylmercury chloride (20) | 23 days | Se protected from Hg effects including mortality and tremors | 200, 201 |
| Japanese quail, embryos and chicks (reproductive study) | Sodium selenite (6 or 12 ppm) | Methylmercury chloride (15) | Through hatching | Se and Hg were mutually protective from adverse reproductive effects including decreased egg productivity, hatching success, and teratogenesis | 202 |
| Mallard, adult | Seleno-D,L-methionine (10 ppm) | Dietary methylmercury (10 ppm Hg) | 10 weeks | Se protected from Hg effects including mortality, paralysis of the legs, and oxidative stress | 203, 204 |
| Mallard, embryo (reproductive study) | Seleno-D,L-methionine (10 ppm) | Dietary methylmercury (10 ppm Hg) | Embryo through 7 days post hatching | Combined Se and Hg was more toxic than either alone with respect to teratogenesis and duckling production | 203 |

With 11% protein, selenium toxicity was greater than with 22% protein; growth of controls was less than that with 22% protein, selenium (60 ppm) caused 100% mortality, and methionine supplementation, although protective, afforded less protection than it did with 22% protein. Two other studies also reported similar interactive effects between selenium and restricted dietary protein (7%) including mortality, impaired growth, and hepatic oxidative stress.[110,197]

With 44% protein in the diet, ducklings experienced oxidative and renal stress, and selenium was more toxic than with methionine-supplemented diets containing 22% protein.

## 44.5.2 Other Trace Elements

### *44.5.2.1 Arsenic and Selenium*

High concentrations of arsenic (As) may be found in combination with selenium in aquatic food chains associated with irrigation drainwater. Studies with mallard ducklings examined the potential for interactive effects between Se and As on duckling survival, growth, and oxidative stress.[110] Day-old ducklings received diets containing 22% protein as an untreated diet (controls) or diets containing 15 ppm Se (as selenomethionine), 60 ppm Se, 200 ppm As (as sodium arsenate), 15 ppm Se with 200 ppm As, or 60 ppm Se with 200 ppm As for 4 weeks. With 22% protein and 60 ppm Se in the diet, duckling survival and growth was reduced, and livers had histopathological lesions. However, when selenium was provided with arsenic in the diet, antagonistic interactive effects occurred and included complete to partial alleviation of the following selenium effects: mortality, impaired growth, hepatic lesions, and oxidative stress. In another study, 99 pairs of breeding mallards were fed diets supplemented with arsenic (sodium arsenate) at 0, 25, 100, or 400 ppm in combination with selenium (seleno-DL-methionine) at 0 or 10 ppm in a replicated factorial experiment.[198] Ducklings that hatched from parents on the various diets were placed on the same treatment combination as their parents. Arsenic accumulated in adult livers and eggs, reduced adult weight gain and liver weight, delayed the onset of egg laying, decreased whole-egg weight, and caused eggshell thinning. Arsenic did not affect hatching success and was not teratogenic. In ducklings, arsenic accumulated in the liver and reduced body weight, growth, and liver weight. Arsenic decreased overall duckling production. Selenium accumulated in adult liver and eggs, was teratogenic, and decreased hatching success. In ducklings, selenium accumulated in the liver and reduced body weight and growth, and increased liver weight. Selenium increased duckling mortality and decreased overall duckling production. When selenium and arsenic were combined, antagonistic interactions between arsenic and selenium occurred, where arsenic decreased the accumulation of selenium in liver and egg and alleviated the effects of selenium on hatching success and embryo deformities.

### *44.5.2.2 Boron and Selenium*

Boron (B) and selenium sometimes occur together in high concentrations in the environment and can accumulate in plants and invertebrates consumed by waterfowl. A reproductive study with mallards examined toxicity and interaction between boron and selenium in the diet.[199] This study demonstrated that boron and selenium can individually affect mallard reproduction and duckling growth, but with apparent minimal interaction between the two. Breeding mallards were fed diets supplemented with boron (as boric acid) at 0, 450, or 900 ppm, in combination with selenium (as seleno-DL-methionine) at 0, 3.5, or 7 ppm, in a replicated factorial experiment. Ducklings hatched received the same treatment combination as their parents for 2 weeks. Boron and selenium accumulated in adult livers, eggs, and duckling livers. In adults, boron and selenium caused weight loss, and boron decreased hemoglobin concentration, egg weight, and egg fertility. Both boron and selenium reduced hatching success and duckling weight, and boron reduced duckling growth and duckling production and caused several alterations in duckling liver biochemistry, including increased lipid peroxidation (TBARS), and decreased total thiol and protein concentrations, but with few apparent interactive effects.

Another study examined boron and selenium for effect on duckling survival and growth for a longer duration of 4 weeks.[197] Day-old mallard ducklings received diets containing 22% protein as an untreated diet (controls) or diets containing 15 ppm Se (as selenomethionine), 60 ppm Se, 1000 ppm B (as boric acid), 15 ppm Se with 1000 ppm B, or 60 ppm Se with 1000 ppm B for 4 weeks. This same study was repeated, but with only 7% protein in the diet. With 22% protein and 60 ppm Se in the diet, duckling survival and growth was reduced and histopathological lesions of the liver occurred. Boron alone caused some reduction in growth. Several interactive effects occurred between boron and selenium; the combination decreased liver weight and affected several plasma chemistries. However, with a restricted protein (7%) diet, boron and selenium (15 ppm) in combination increased mortality and increased hepatic selenium accumulation. These findings suggest the potential for more severe toxicological effects of selenium and boron independently and interactively on duckling survival and development when dietary protein is diminished.

#### *44.5.2.3 Mercury and Selenium*

Earlier studies have shown that selenium as selenite was protective against the effects of methylmercury to quail; day-old Japanese quail (*Coturnix coturnix*) fed 5 or 8 ppm Se as sodium selenite and 20 ppm Hg as methylmercury chloride were protected from mercury effects including high mortality and tremors.[200,201] In another study, the addition of 3 or 6 ppm Se as sodium selenite reduced the toxicity of 20 ppm Hg as methylmercury hydroxide to young Japanese quail, even though brain levels of mercury were higher when sodium selenite was included in the diet.[202] Selenium levels in the brain were also higher, suggesting that selenium was somehow protecting the brain from methylmercury damage. Mercury and selenium in these forms were mutually protective to reproduction; mercury alone reduced egg production and hatchability, but the addition of selenium lessened these effects, whereas selenium alone reduced hatchability and increased the incidence of deformed embryos, but the addition of mercury reduced these effects.

Recent findings with environmentally relevant forms of mercury and selenium have also shown that mercury and selenium may be antagonistic to each other in adult mallards but synergistic to the reproductive process and embryos. In one study, the effects of selenium (seleno-D,L-methionine) and mercury (methylmercury chloride) were examined separately and in combination.[203] Mallard drakes received one of the following diets: untreated feed (controls), or feed containing 10 ppm Se, 10 ppm Hg, or 10 ppm Se in combination with 10 ppm Hg for 10 weeks. One of 12 ducks fed 10 ppm Hg died, and eight others suffered paralysis of the legs. However, when the diet contained 10 ppm Se in addition to the 10 ppm Hg, none of these effects were apparent. The presence of methylmercury in the diet greatly enhanced the storage of selenium in tissues. The livers of males fed 10 ppm Se contained a mean of 9.6 ppm Se, whereas the livers of males fed 10 ppm Se plus 10 ppm Hg contained a mean of 114 ppm Se. However, selenium did not enhance the storage of mercury. The following clinical and biochemical alterations occurred in response to the mercury exposure: hematocrit and hemoglobin concentrations decreased; activities of the enzymes glutathione (GSH) peroxidase (plasma and liver), glutathione-S-transferase (liver), and glucose-6-phosphate dehydrogenase (G-6-PDH) (liver and brain) decreased; hepatic oxidized glutathione (GSSG) concentration increased relative to reduced glutathione (GSH); and lipid peroxidation in the brain was evident as detected by increased TBARS.[204] Effects of selenium alone included increased hepatic GSSG reductase activity and increased brain TBARS concentration. However, selenium in combination with mercury partially or totally alleviated the effects of mercury on GSH peroxidase, G-6-PDH, and GSSG. It was concluded that since both mercury and excess selenium can affect thiol status, measurement of associated enzymes in conjunction with thiol status may be a useful bioindicator to discriminate between mercury and selenium effects. The ability of selenium to restore the activities of G-6-PDH, GSH peroxidase, and GSH status involved in antioxidative defense mechanisms may be crucial to biological protection from the toxic effects of methylmercury.

In contrast to the protective effect of selenium against mercury poisoning in adult mallards, selenium plus mercury was worse than selenium or mercury alone for some measurements of reproductive success.[203] Both selenium and mercury lowered duckling production through reductions in hatching success and survival of ducklings, but the combination was worse than either mercury or selenium alone. Controls produced an average of 7.6 young per female, females fed 10 ppm Se produced an average of 2.8 young, females fed 10 ppm Hg produced 1.1 young, and females fed both mercury and selenium produced 0.2 young. Teratogenic effects were notably greater for the combined treatment; deformities were recorded in 6.1% of the embryos of controls, 16.4% for those fed methylmercury chloride, 36.2% for those fed selenomethionine, and 73.4% for those fed methylmercury chloride and selenomethionine.

## 44.6 EFFECTS OF NUTRITION AND OTHER ELEMENTS ON LEAD TOXICITY IN BIRDS

### 44.6.1 Diet and Lead Shot Ingestion in Waterfowl

Jordan[205] concluded after controlled experiments with mallards, lesser scaup (*Aytha affinis*), and Canada geese (*Branta canadensis*) that diet was the most important variable modifying the toxic effects of ingested lead-shot pellets. A corn diet resulted in 50% mortality and 34% net weight loss by 25 days, following a dose of four No. 6 lead shot (Table 44.3). However, commercial duck food in the form of pellets was the most effective in lessening these toxic effects, followed by aquatic plants and grains. Corn is deficient in essential minerals, including calcium, several vitamins, and low in protein, hence enhancing the effects of lead. However, the author concluded that these nutritional factors alone were not entirely responsible. The size and hardness of the diet was also a factor. The physical form of the diet appeared to determine the rate of food intake of captive waterfowl, where commercial duck pellets were taken in greatest amounts, followed by small grains and wild seeds, with rice more so than mixed grains, and corn the least. The smaller softer feeds, including duck pellets, which essentially became mash after ingestion, were more easily reduced to a softer form in the gizzard and hence passed along more readily to the intestine, hastening the rate of elimination of lead. Thus, both nutritional benefits and movement of a greater volume of food are likely to facilitate the elimination of lead removed from the surface of ingested lead shot. Others have also reported greater toxicity of lead-shot ingestion in conjunction with a corn diet.[206,207]

### 44.6.2 Calcium and Lead Shot Ingestion

Carlson and Nielsen[208] studied the role of dietary calcium on lead-shot poisoning in adult mallards. In this study, there were three dietary groups consisting of: Group I, fed pelleted calcium-supplemented corn (1.2% Ca); group II, fed a pelleted commercial duck ration with 1.2% Ca; and Group III, fed cracked corn (0.03% Ca). Ten ducks from each group were given four No. 4 lead shot via an esophageal tube, and five ducks from each group were kept as pair-fed controls. Anorexia and weight loss were most severe in the lead-shot-dosed Group III ducks. Group III had a maximum reduction in food consumption of 87%, followed by a slight improvement in appetite but a loss of 35% of the initial body weight. Group I treated ducks had a reduction in food consumption of 64% that eventually returned to the quantities consumed at the start of the experiment; Group I ducks lost 18% of their initial body weight. Group II treated ducks maintained healthy appetites during the experiment and had a weight gain of 2% of their initial body weight. The number of ducks that became moribund and were euthanized differed significantly among the treated groups, with 100% of Group III, 50% of Group I, and 0% of Group II treated ducks becoming moribund.

**Table 44.3 Interactive Effects of Nutritional Factors and Other Elements on Lead Toxicity in Birds**

| Species, Age | Form of Pb, Dose or Dietary Concentration | Interactive Factor | Observation Period | Effects | Ref. |
|---|---|---|---|---|---|
| Mallard, Lesser scaup Canada goose (adults) | Four No. 6 lead shot pellets | Commercial feed vs. corn and other diets | 25 days post dosing | Lead-shot ingestion most toxic with corn diet and least toxic with softer diets with respect to mortality and weight loss | 205–207 |
| Mallard, adult | Four No. 4 lead shot pellets | Dietary calcium supplementation | 9 weeks | Calcium supplementation of a corn diet decreased lead-related mortality by one half and decreased weight loss as well as histopathological lesions | 208 |
| Mallard, adult | Lead contaminated mining sediment (1000 ppm Pb in diet) | Ground corn diet vs. commercial diet | 15 weeks | Ground corn diet containing lead sediment resulted in 80% mortality, greatest weight loss, lowest hemoglobin concentration, and threefold greater hepatic lead accumulation than with commercial diet containing lead sediment | 209 |
| Mallard, ducklings | Lead contaminated mining sediment (830 ppm Pb in diet) | Two thirds ground corn/one third commercial diet mixture vs. commercial diet | 6 weeks | Ground corn diet mixture containing lead sediment resulted in highest blood lead concentration, reduced growth, most oxidative stress, and increased prevalence of acid-fast inclusion bodies in the kidney compared to commercial diet containing lead sediment | 210–212 |
| Japanese quail, young adults | Lead acetate in drinking water at 100 or 400 µg/mL | Poultry feed vs. ground corn | 7 days | Corn diet resulted in lead-related mortality, greatest weight loss and suppressed antibody-mediated immunity | 214 |
| Zebra finch | 10 µg/mL of lead as lead acetate and of cadmium as cadmium chloride in drinking water | 0.3 vs. 3% calcium | 51 days | Low dietary calcium enhanced hepatic and renal accumulation of lead (~400%) and cadmium (~200%), increased metallothionein production, and inhibited ALAD the most compared to controls | 215 |
| Ringed turtle dove | 50 ppm of lead as lead acetate, 20 ppm cadmium as cadmium chloride, and 1500 ppm aluminum as aluminum sulfate | 0.4 vs. 2% calcium | 6 months | Low dietary calcium enhanced the accumulation of lead and cadmium but not aluminum, increased metallothionein production, and inhibited ALAD the most compared to controls | 215 |

All three groups of treated ducks had increased protoporphyrin IX concentrations compared with controls. Groups I and III lead-treated ducks had significant reductions in erythrocyte counts, PCV, hemoglobin concentrations, and mean corpuscular hemoglobin concentrations as compared with controls. Group II had reduced hemoglobin concentration and mean corpuscular hemoglobin concentration compared with controls. However, none of the above changes differed statistically among the three dietary groups of lead-poisoned ducks. Gross pathologic lesions of treated ducks (pectoral muscle atrophy and green discoloration of gizzard mucosa) were most pronounced in Group III, followed by Group I and Group II, in that order. Histopathologic lesions consisting of neuropile vacuolization, intranuclear inclusion bodies, neuronal degeneration, and myocardial and pectoral muscle degeneration, necrosis, and atrophy occurred most frequently in the Group III treated ducks, less in the Group I treated ducks, and least in the Group II treated ducks. These findings suggested that the addition of calcium to the corn diet had an ameliorating effect, but other nutrients found in the Group II commercial duck diet were also important in mollifying lead absorption and metabolism in mallard ducks.

### 44.6.3 Diet and Lead-Contaminated Mining Sediment in Waterfowl

A series of experiments was conducted to measure the toxicity of lead-contaminated mining sediment from the Coeur d'Alene River basin (CDARB) in Idaho to waterfowl, including mallards, Canada geese, and mute swans (*Cygnus olor*), under optimal and less-than-optimal nutritional conditions.

#### *44.6.3.1 Adult Mallards*

In adult mallards, mortality of 80% occurred from lead poisoning within 15 weeks when fed a less than optimal diet of ground corn containing 24% CDARB sediment (with 3954 μg/g lead in this sample of sediment), but none died when 24% CDARB sediment was mixed into a nutritionally balanced commercial duck diet.[209] Birds fed the corn diet containing 24% CDARB sediment lost weight. Mallards fed 24% CDARB sediment in the commercial diet and corn diet had elevated levels of protoporphyrin (424 and 689 μg/dl) compared to controls fed the commercial diet and corn diet (45 and 25 μg/dl). Four of five mallards fed 24% CDARB sediment in a commercial diet and all five fed the contaminated sediment in a corn diet had renal intranuclear inclusion bodies. Lead accumulation was higher in the livers of mallards fed 24% lead-contaminated sediment in the corn diet (38 μg/g) than in the commercial diet (13 μg/g). These laboratory findings with mallards demonstrated that ingestion of realistic amounts of sediment from the Coeur d'Alene River Basin caused lead poisoning in waterfowl, as had been reported in the field, and a less-than-optimal diet such as corn could severely intensify many aspects of these effects.

#### *44.6.3.2 Duckling Growth and Survival*

Other studies were conducted to assess the effects of the same sediment on posthatching development of mallard ducklings for 6 weeks.[210–212] Day-old ducklings received untreated control diet, clean-sediment-supplemented (24%) control diet, or CDARB-sediment-supplemented (3449 μg/g lead) diets at 12% or 24%. The 12% CDARB diet resulted in a blood lead concentration of 1.4 ppm (wet weight) with over 90% depression of red blood cell ALAD activity and over a threefold elevation of free erythrocyte protoporphyrin concentration. The 24% CDARB diet resulted in blood lead of 2.6 ppm with over sixfold elevation of protoporphyrin. In this group, the liver lead concentration was 7.9 ppm (wet weight), and there was a 40% increase in hepatic reduced glutathione concentration. The kidney lead concentration in this group was 8.0 ppm and acid-fast inclusion bodies were present in the kidneys of four of nine ducklings. When ducklings were on a less-than-optimal diet (two thirds corn and one third standard diet), CDARB sediment was more toxic; blood lead levels were 1.7-fold higher, body growth and liver biochemistry (TBARS) were more affected, and the prevalence of acid-fast inclusion bodies increased.

In addition, significant effects on the brain were seen due to lead accumulation and included oxidative stress, altered metabolites, and decreased brain weights.[211,212] Significant increases in brain-reduced glutathione (GSH), calcium, and triglyceride concentrations occurred, with significant decreases in brain protein, ATP, and brain weights. Nutrient level of the diets significantly affected brain GSH, triglyceride, inorganic phosphorus, and protein concentrations, acetylcholinesterase activity, and brain and body weights. Nutrient levels marginally affected lead accumulation in the brain. The incidence and duration of ten behaviors was recorded (resting, standing, moving, drinking, dabbling, feeding, pecking, preening, bathing, and swimming). Contaminated sediment significantly affected the proportion of time spent swimming yet did not significantly affect any of the other recorded behaviors. There were also signs of disruption of balance and mobility. Nutrient level affected the amount of time spent in water-related behaviors.

#### *44.6.3.3 Mute Swans*

In another study with captive mute swans, CDARB sediments were combined with either a commercial avian maintenance diet or a less optimal diet of ground rice and fed to captive mute swans for 6 weeks.[213] Experimental treatments consisted of maintenance or rice diets containing 0, 12 (no rice group), or 24% lead-contaminated (3950 μg/g lead) sediment or 24% uncontaminated (9.7 μg/g lead) sediment. Swans fed the less optimal rice diet containing 24% lead-contaminated sediment were the most severely affected, experiencing a 24% decrease in mean body weight, including three birds that became emaciated. All birds in this treatment group had significant reductions in hematocrit and hemoglobin concentrations, nephrosis, and abnormally dark, viscous bile. This group also had the greatest mean concentrations of lead in blood (3.2 μg/g), brain (2.2 μg/g), and liver (8.5 μg/g). These birds had significant increases in mean plasma ALT, cholesterol, and uric acid concentrations and decreased plasma triglyceride concentrations. This study revealed that mute swans consuming environmentally relevant concentrations of CDARB sediment developed severe sublethal lead poisoning and that toxic effects were more pronounced when the birds were fed lead-contaminated sediment combined with rice, which closely resembles the diet of swans in the wild.

### 44.6.4 Diet and Acute Lead Exposure on Immunity of Quail

Grasman and Scanlon[214] investigated the interacting effects of acute lead exposure and different diets on antibody and T-cell-mediated immunity in 9-week-old male Japanese quail. The treatments were: (1) a positive control group fed 20 μg/g corticosterone, (2) a negative control group given no lead or corticosterone, (3) a low-lead group, and (4) a high-lead group. The low- and high-lead groups received 100 and 400 μg/mL lead as lead acetate in drinking water for 7 days. The two diets were poultry feed and ground corn. Control quail fed corn lost 13–14% of initial body mass, but lead-dosed quail fed corn lost 23–24%. All quail fed poultry feed gained body mass. On the corn diet, three high-lead and one low-lead quail died of lead poisoning. Corn increased the percentage of heterophils in white blood cells and decreased lymphocytes and monocytes. There was marginal evidence that lead increased the heterophil/lymphocyte ratio in corn-fed quail. Corticosterone suppressed this response more than lead. In corn-fed quail, lead suppressed the primary total antibody response to immunization with chukar partridge (*Alectoris graeca*) erythrocytes. However, lead suppressed antibody-mediated immunity only at dosages that also caused clinical lead poisoning.

### 44.6.5 Interactive Effects of Calcium on Lead, Cadmium, and Aluminum in Finches and Doves

Scheuhammer[215] examined the influence of low dietary calcium on the accumulation and effects of dietary lead, cadmium, and aluminum in zebra finches (*Taeniopygia guttata*) and ring doves

(*Streptopelia risoria*). In zebra finches fed a diet containing low Ca (0.3%), the hepatic and renal accumulation of lead was enhanced approximately 400% and of cadmium about 150–200%, compared to birds fed a 3.0% Ca diet. Low dietary calcium also caused bones of female finches to lose an average of about 60% of their normal calcium content. Loss of bone-calcium was also observed in male finches but was less than in females. In reproductively active ring doves, low (0.4%) dietary calcium enhanced the accumulation of lead and cadmium but not of aluminum, compared with accumulation in doves consuming a 2.0% Ca diet. Enhanced accumulation of lead and cadmium was accompanied by an increased synthesis of the metal-binding protein metallothionein and by a greater inhibition of delta-aminolevulinic acid dehydratase activity. These results indicated that under conditions of reduced dietary calcium availability, such as can occur in acid-impacted environments, wild birds risk increased uptake of certain toxic metals more rapidly.

## 44.7 OTHER INTERACTIVE EFFECTS OF NUTRITIONAL FACTORS AND TRACE ELEMENTS IN BIRDS

### 44.7.1 Calcium, Phosphorus, and Aluminum in Waterfowl and Passerines

Acidification may alter the species composition and abundance of aquatic organisms that serve as food, reduce caloric content of aquatic insects, or elevate metal levels in food.[216] Calcium-rich taxa such as snails and freshwater clams are usually scarce in acidified waters, and calcium deficiencies may limit the distribution of certain aquatic birds. In order to further test this hypothesis, Sparling[217,218] fed day-old black ducks (*Anas rubripes*) and mallards diets varying in calcium, phosphorous, and aluminum for 10 weeks. Ducklings on low-calcium, low-phosphorus diets experienced the greatest problems with elevated aluminum. At the highest level of aluminum (5000 μg/g), all ducklings died within 2 weeks and mortality exceeded 60% for those on 1000 μg/g Al. In contrast, no mallards died and mortality did not exceed 30% for black ducks on normal levels of calcium and phosphorus or on low-calcium, high-phosphorus diets, regardless of aluminum diet. Femur and liver aluminum levels correlated highly with dietary aluminum. Many of the differences among treatments diminished or disappeared by 10 weeks of age. It was concluded that rapidly growing young birds may have greater problems coping with elevated aluminum than adult birds, especially when high aluminum is combined with low levels of calcium and phosphorous. Because acidified lakes and streams often have higher aluminum and lower calcium levels than circumneutral surface waters, the possibility of aluminum toxicity in aquatic birds should not be dismissed.

Miles et al.[219] fed adult European starlings (*Sturnus vulgaris*) diets varying in levels of calcium, phosphorous, and aluminum to determine if aluminum interacted with the other elements to affect reproductive capability.

Starlings on normal levels of calcium and phosphorous but with elevated levels of aluminum (1000 or 5000 ppm) laid significantly more eggs than those on a control level (200 ppm) of aluminum. Reduced levels of calcium, however, resulted in smaller clutch sizes than the normal level. Low-calcium and low-phosphorus reduced food consumption and body weights of the adult starlings. Mortality was higher in adults on low-calcium and low-phosphorus and normal levels of these nutrients when aluminum concentrations were increased, but differences were not statistically significant. Eggshells of starlings on elevated aluminum were thicker but not stronger than those on low aluminum. Eggshell and bone characteristics were affected more by calcium and phosphorus levels in the diets than by aluminum levels.

### 44.7.2 Arsenic and Dietary Protein in Mallards

Day-old mallard ducklings received untreated diets of either 22% protein (controls) or restricted dietary protein (7%), with or without 200 ppm As (as sodium arsenate).[110] With restricted dietary protein, arsenic was considerably more toxic with respect to effects on survival, growth, histopathological lesions, and oxidative stress. Plasma LDH and GSSG reductase activities were elevated. Greater accumulation of arsenic in the liver occurred, with greater oxidative stress recorded as increased hepatic TBARS.

### 44.7.3 Boron and Dietary Protein in Mallards

Day-old mallard ducklings received untreated diets of either 22% protein (controls) or restricted dietary protein (7%), with or without 1000 ppm B (as boric acid).[197] With restricted dietary protein, boron had a greater negative effect on duckling growth than with adequate dietary protein.

### 44.7.4 Mercury and Vitamin E in Quail

Mercury as methylmercury chloride at 30 ppm in the diet of day-old Japanese quail caused mortality and signs of mercury toxicity, decreased growth rate, hematocrit, and bone calcification, whereas the addition of varying amounts of supplemental vitamin E significantly decreased these effects.[220] The effects of vitamin E and five synthetic antioxidants, including ethoxyquin, BHT, DPPD, Antabuse and BHA, were examined on methylmercury-induced mortality in day-old Japanese quail.[221] At 0.5% in the diet, both ethoxyquin and DPPD reduced mercury-related mortality. However, vitamin E was equal or superior to all synthetic antioxidants tested in alleviating mercury-related toxicity.

### 44.7.5 Mercury, Lead, and Cadmium in Pekin Ducks

Adult female Pekin ducks (*Anas platyrhynchos*) were fed diets with no metals added (controls), 8 mg of methylmercury chloride/kg feed, 80 mg of lead acetate/kg feed, 80 mg of cadmium chloride/kg of feed, or all combinations of two and three of the above metals for 13 weeks.[222] Cadmium, when administered alone or in combination, caused a 60-fold increase in kidney metallothionein levels. A combination of lead and mercury significantly increased the concentration of lead in kidney above that of lead alone. Cadmium, when administered alone or in combination, caused an increase in the levels of zinc and copper in kidney. Administration of cadmium alone or in combination caused degenerative changes in kidney proximal tubules. However, the severity of the degenerative changes increased when cadmium was administered with the other metals.

## 44.8 FIELD STUDY INTERACTIONS OF TRACE ELEMENTS IN WILDLIFE

In field studies, interactions between trace elements and between other nutritional factors and trace elements are harder to demonstrate due to a lack of controlled conditions and due to the presence of often multiple variables including trace elements themselves. However, in both wild birds and mammals, significant correlations between mercury and selenium levels in tissues have been shown, and high levels of inorganic mercury are found in tissues such as liver and kidney in an apparently bound form with mercury. Mercury and selenium concentrations in the livers of wild mammals, particularly piscivorous and carnivorous ones, are often highly correlated in a molar ratio of 1:1.[178,223–225] The relationship between total mercury and selenium concentrations was found to be more variable for wild aquatic birds, depending on the species.[226–229] However, more recent studies that were able to differentiate between methylmercury and demethylated mercury in livers of aquatic birds have revealed that demethylated mercury exists as inorganic mercury protectively sequestered with selenium and correlates in a molar ratio of 1:1 with selenium.[230,231]

When Moller[232] compared the concentrations of Al, As, B, Ba, Be, Cd, Cr, Cu, Fe, Pb, Li, Mg, Mn, Hg, Mo, Ni, Se, Ag, V, and Zn in aquatic bird liver, fish liver, whole bivalves, insects, and waters in several aquatic ecosystems in northern California, there was evidence of *in vivo* and environmental interactions including the observation of manganese as a possible cofactor or indicator in selenium bioaccumulation. The nearest neighbor selenium correlation in aquatic bird liver tissue that resulted was Cd-Mn-Se-Hg-As. The correlation of liver selenium to manganese *in vivo,* and the finding that the majority of the variance in liver selenium concentration was contained in the manganese term of the regression model relating selenium to cadmium, manganese, and mercury was new knowledge in the study of aquatic birds. A linear relationship between liver selenium and environmental manganese (water and sediment) suggested a water-chemistry compartmentalization or activation of toxicants, or, alternatively, the hepatic concentrations of selenium, manganese, and iron suggested induction of enzymes in response to oxidative stress.

The possibility that aluminum may reach toxic levels in the foods of aquatic birds nesting near acidified ponds was first raised in Sweden.[233] Scheuhammer reviewed and studied the effects of acidification on wildlife inhabiting aquatic or semiaquatic environments, with particular reference to the possibility of increased dietary exposure to mercury, cadmium, lead or aluminum and decreased availability of essential dietary minerals such as calcium.[234,235] It was concluded that: (1) piscivores risk increased exposure to dietary methylmercury in acidified habitats, with concentrations in prey becoming potentially high enough to cause reproductive impairment in birds and mammals; (2) piscivores risk minimal increased exposure to dietary cadmium, lead, or aluminum because either these metals are not increased in fish due to acidification or increases tend to be trivial from a toxicological perspective; (3) insectivores and omnivores may, under certain conditions, experience increased exposure to toxic metals in some acidified environments. However, of greater impact on these species is the decrease in availability of dietary calcium due to the pH-related extinction of high-calcium aquatic invertebrate taxa (molluscs, crustaceans). Egg laying and eggshell integrity in birds and the growth of hatchling birds and neonatal mammals are all highly dependent upon calcium. Acidification-related changes in the dietary availability of other essential elements, such as magnesium, selenium, and phosphorus, have not been established and require further investigation; and (4) herbivores may risk increased exposure to aluminum and lead, and perhaps cadmium, in acidified environments because certain macrophytes can accumulate high concentration of these metals under acidic conditions. The relative importance of pH in determining the metal concentrations of major browse species and the toxicological consequences for herbivorous wildlife are not well established and require further study. A decreased availability of dietary calcium is also likely for herbivores inhabiting acidified environments.

## 44.9 CONCLUDING REMARKS AND FUTURE RESEARCH NEEDS

It is clear from a review of the literature that there is only limited information on the effects of foodborne trace elements and other nutritional factors on fish and wildlife. There is an extensive database on foodborne selenium toxicity to a variety of fish species compared to information available for other trace elements. Yet, the information on dietary-selenium toxicity to fish is incomplete because few of the possible interactions of selenium with nutritional factors or with other trace elements have been examined. In comparison, the single dietary toxicity test with one species of fish and vanadium, which was thought by researchers to have dietary toxicity comparable to that of selenium, is a glaring information gap that would hinder a hazard assessment where vanadium might be a potential contributor to stresses in the aquatic ecosystem. The few dietary tests with other elements, such as arsenic, cesium, chromium, cobalt, lead, mercury, and uranium, and limited tests with cadmium, copper, and zinc are also major information gaps in need of research.

Adequate hazard assessments must incorporate information from not only waterborne exposures, but also dietary exposures. In-depth studies with foodborne mixtures of trace elements, such as those conducted by Woodward, Farag, and coworkers in the Clark Fork River and Coeur d'Alene River, provide valuable information about specific ecosystem hazard assessments. However, dietary studies gave conflicting results when fish were fed nauplii of brine shrimp briefly exposed to a mixture of trace elements in an effort to simulate trace-element concentrations in food organisms in the Clark Fork River. This latter study shows the need for careful experimental design if environmental conditions are to be adequately simulated in laboratory studies.

The emphasis in aquatic toxicology over the past few decades has been on waterborne exposures. Consequently, national water-quality standards have been propagated based on a waterborne approach. Recent studies have shown that single-element criteria do not account for interactions with other waterborne factors such as other trace elements or water characteristics. Limited information from dietary studies with trace elements, especially selenium, show that diet can be an important contributor to toxic effects observed in contaminated ecosystems, yet water-quality standards do not consider potential effects from dietary exposures. Incorporation of dietary criteria into national criteria for trace elements will occur only after a sufficient database of information is generated from dietary toxicity studies.

Fewer trace element/nutrition interaction studies have been conducted with wildlife species than with fish, and most of these studies have been conducted with avian wildlife. Nearly all of the studies conducted with birds fall into three categories: (1) selenium-related interactions between nutritional factors and other trace elements, (2) lead-related studies of nutritional factors including trace elements, and (3) only a few studies with other trace elements such as aluminum. Recent findings with environmentally relevant forms of mercury (methylmercury) and selenium (selenomethionine) in birds have shown that mercury and selenium may be antagonistic to each other in adult birds but synergistic to the reproductive process in embryos. Therefore, further investigations in birds are warranted to compare the interactions of various forms of mercury and selenium, particularly with respect to effects on reproduction and teratogenesis. The ability of selenium to restore the activities of G-6-PDH, GSH peroxidase, and GSH status involved in antioxidative defense mechanisms may be crucial to biological protection from the toxic effects of methylmercury in adults. Further studies are needed to examine the relationship between selenium and other trace elements that may be toxic by compromising cellular antioxidative defense mechanisms.

Virtually no trace element/nutrition interaction studies have been conducted with mammalian wildlife, probably because the human-health-effects literature and the agricultural-nutrition literature are abundant with interaction studies conducted with laboratory rodents as well as other species of domestic mammals, which are viewed by some as surrogate species for mammalian wildlife. However, there are many unique species of wild mammals with varied sensitivities to certain classes of contaminants including trace elements, as reviewed by Shore and Rattner.[236] Therefore, there is a need for comparative interaction studies in these species, particularly with respect to locations where multiple trace-element contaminants exist and other nutritional conditions may be suboptimal.

## ACKNOWLEDGMENTS

The authors thank Aida Farag and Gary Heinz for reviewing the chapter.

## REFERENCES

1. NAS (National Academy of Sciences), Control of diets in laboratory animal experimentation, *Inst. Lab. Anim. Res. News*, 21, A1, 1979.
2. Piper, R. G., McElwain, I. B., Orme, L. E., McCraren, J. P., Fowler, L. G., and Leonard, J. R., Fish Hatchery Management, U.S. Fish and Wildlife Service, Washington, D.C., 1982.
3. Lim, C. and Webster, C. D., Eds., *Nutrition and Fish Health*, Haworth Press, Binghamton, NY, 437, 2001.

4. Kirkpatrick, R. L., Ed., *Comparative Influence of Nutrition on Reproduction and Survival of Wild Birds and Mammals: An Overview*, Caesar Kleberg Wildlife Research Institute, Kingsville, TX, 60, 1988.
5. Klasing, K. C., *Comparative Avian Nutrition,* A CAB International Publication, Oxford University Press Incorporated, New York, 360, 1998.
6. Eisler, R., *Handbook of Chemical Risk Assessment: Health Hazards to Humans, Plants, and Animals*, Vol. 1: *Metals*, Lewis Publishers, CRC Press, Boca Raton, FL, 738, 2000.
7. Leger, P., Bengtson, D. A., Simpson, K. L., and Sorgeloos, P., The use and nutritional value of *Artemia* as a food source, *Oceanogr. Mar. Biol. Ann. Rev.,* 24, 521, 1986.
8. Mehrle, P. M., Mayer, F. L., and Johnson, W. W., Diet quality in fish toxicology: Effects on acute and chronic toxicity, in *Aquatic Toxicology and Hazard Evaluation*, Mayer, F. L. and Hamelink, J. L., Eds., American Society for Testing and Materials, Special Technical Publication 634, Philadelphia, 269, 1977.
9. Phillips, G. R. and Buhler, D. R., Influences of dieldrin on the growth and body composition of fingerling rainbow trout (*Salmo gairdneri*) fed Oregon moist pellets or tubificid worms (*Tubifex* sp.), *J. Fish. Res. Bd. Can.*, 36, 77, 1979.
10. Hilton, J. W., Influence of diet on fish toxicology studies, in Canadian Technical Report of Fisheries and Aquatic Sciences, No. 975, Winnipeg, Manitoba, 150, 1980.
11. Beck, A. D. and Bengtson, D. A., International study on *Artemia.* XXII: Nutrition in aquatic toxicology — diet quality of geographical strains of the brine shrimp, *Artemia*, in *Aquatic Toxicology and Hazard Assessment: Fifth Conference*, Pearson, J. G., Foster, R. B., and Bishop, W. E., Eds., American Society for Testing and Materials, Special Technical Publication 766, Philadelphia, 161, 1982.
12. Bengtson, D. A. S., Beck, A. D., Lussier, S. M., Migneault, D., and Olney, C. E., International study on *Artemia.* XXXI: Nutritional effects in toxicity tests: Use of different *Artemia* geographical strains, in *Ecotoxicological Testing for the Marine Environment, Volume 2*, Persoone, G., Jaspers, E., and Claus, C., Eds., State University Ghent and Institute of Marine Scientific Research, Bredene, Belgium, 399, 1984.
13. Phillips, G. R. and Buhler, D. R., The relative contributions of methylmercury from food or water to rainbow trout (*Salmo gairdneri*) in a controlled laboratory environment, *Trans. Am. Fish. Soc.*, 107, 853, 1978.
14. Hodson, P. V., Spry, D. J., and Blunt, B. R., Effects on rainbow trout (*Salmo gairdneri*) of a chronic exposure to waterborne selenium, *Can. J. Fish. Aquat. Sci.*, 37, 233, 1980.
15. Vighi, M., Lead uptake and release in an experimental trophic chain, *Ecotoxicol. Environ. Saf.*, 5, 177, 1981.
16. Milner, N. J., The accumulation of zinc by 0-group plaice, *Pleuronectes platessa* (L.), from high concentrations in sea water and food, *J. Fish Biol.*, 21, 325, 1982.
17. Willis, J. N. and Sunda, W. G., Relative contributions of food and water in the accumulation of zinc by two species of marine fish, *Mar. Biol.*, 80, 273, 1984.
18. Spry, D. J., Hodson, P. V., and Wood, C. M., Relative contributions of dietary and waterborne zinc in the rainbow trout, *Salmo gairdneri.*, *Can. J. Fish. Aquat. Sci.*, 45, 32, 1988.
19. Harrison, S. E. and Klaverkamp, J. F., Uptake, elimination and tissue distribution of dietary and aqueous cadmium by rainbow trout (*Salmo gairdneri* Richardson) and lake whitefish (*Coregonus clupeaformis* Mitchill), *Environ. Toxicol. Chem.*, 8, 87, 1989.
20. Hopkins, L. L. and Mohr, H. E., Vanadium as an essential nutrient, in *Federation Proceedings*, 33, 1773, 1974.
21. Goyer, R. A., Toxic effects of metals, in *Casarett and Doull's Toxicology: The Basis Science of Poisons*, Amdur, M. O., Doull, J., and Klaassen, C. D., Eds., Pergamon Press, New York, 623, 1991.
22. Dallinger, R. and Kautzky, H., The importance of contaminated food for the uptake of heavy metals by rainbow trout (*Salmo gairdneri*): A field study, *Oecologia*, 67, 82, 1985.
23. Dallinger, R., Prosi, F., Segner, H., and Back, H., Contaminated food and uptake of heavy metals by fish: A review and a proposal for further research, *Oecologia*, 73, 91, 1987.
24. Leland, H. V. and Kuwabara, J. S., Trace metals, in *Fundamentals of Aquatic Toxicology: Methods and Applications*, Rand, G. M. and Petrocelli, S. R., Eds., Hemisphere Publishing, Washington, D.C., 374, 1985.
25. Sorensen, E. M. B., *Metal Poisoning in Fish*, CRC Press, Boca Raton, FL, 1991.

26. Ohlendorf, H. M., Skorupa, J. P., Saiki, M. K., and Barnum, D. A., Food-chain transfer of trace elements to wildlife, in *Management of Irrigation and Drainage Systems: Integrated Perspectives*, Allen, R. G., Ed., American Society of Civil Engineers, New York, 596, 1993.
27. Harrison, S. E., Klaverkamp, K. F., and Hesslein, R. H., Fates of metal radiotracers added to a whole lake: Accumulation in fathead minnow (*Pimephales promelas*) and lake trout (*Salvelinus namaycush*), *Water Air Soil Pollut.*, 52, 277, 1990.
28. Miller, P. A., Munkittrick, K. R., and Dixon, D. G., Relationship between concentrations of copper and zinc in water, sediment, benthic invertebrates, and tissues of white sucker (*Catostomus commersoni*) at metal-contaminated sites, *Can. J. Fish. Aquat. Sci.*, 49, 978, 1992.
29. Campbell, K. R., Concentrations of heavy metals associated with urban runoff in fish living in stormwater treatment ponds, *Arch. Environ. Contam. Toxicol.*, 27, 352, 1994.
30. Handy, R. D., Dietary exposure to toxic metals in fish, in *Toxicology of Aquatic Pollution*, Taylor, E. W., Ed., University Press, Cambridge, MA, 29, 1996.
31. Cumbie, P. M. and van Horn, S. L., Selenium accumulation associated with fish mortality and reproductive failure, in *Proc. Annu. Conf. Southeast. Assoc. Fish Wildl. Agencies,* 32, 612, 1978.
32. Lemly, A. D., Toxicology of selenium in a freshwater reservoir: Implications for environmental hazard evaluation and safety, *Ecotoxicol. Environ. Saf.*, 10, 314, 1985.
33. Garrett, G. P. and Inman, C. R., Selenium-induced changes in fish populations of a heated reservoir, in *Proc. Annu. Conf. Southeast. Assoc. Fish Wildl. Agencies*, 38, 291, 1984.
34. Sorensen, E. M. B., Selenium accumulation, reproductive status, and histopathological changes in environmentally exposed redear sunfish, *Arch. Toxicol.*, 61, 324, 1988.
35. Harris, T., The selenium question, *Defenders*, March/April, 10, 1986.
36. Vencil, B., The Migratory Bird Treaty Act — protecting wildlife on our national refuges — California's Kesterson Reservoir, a case in point, *Nat. Res. J.*, 26, 609, 1986.
37. Chen, C. Y. and Folt, C. L., Bioaccumulation and diminution of arsenic and lead in a freshwater food web, *Environ. Sci. Technol.*, 34, 3878, 2000.
38. Devi, M., Thomas, D. A., Barber, J. T., and Fingerman, M., Accumulation and physiological and biochemical effects of cadmium in a simple aquatic food chain, *Ecotoxicol. Environ. Saf.*, 33, 38, 1996.
39. Unsal, M., The accumulation and transfer of vanadium within the food chain, *Mar. Pollut. Bull.*, 13, 139, 1982.
40. Miramand, P. and Fowler, S. W., Bioaccumulation and transfer of vanadium in marine organisms, in *Vanadium in the Environment, Part 1: Chemistry and Biochemistry*, Nriagu, J. O., Ed., Wiley & Sons, New York, 167, 1998.
41. Suedel, B. C., Boraczek, J. A., Peddicord, R. K., Clifford, P. A., and Dillon, T. M., Trophic transfer and biomagnification potential of contaminants in aquatic ecosystems, *Rev. Environ. Contam. Toxicol.*, 136, 21, 1994.
42. Cherry, D. S. and Guthrie, R. K., Toxic metals in surface waters from coal ash, *Water Resour. Bull.*, 13, 1227, 1977.
43. Saiki, M. K., A field example of selenium contamination in an aquatic food chain, in *Selenium in the Environment, Proc. Calif. Agric. Technol. Inst.*, Publication Number CAT1/860201, California State University, Fresno, 67, 1986.
44. Ohlendorf, H. M., Bioaccumulation and effects of selenium in wildlife, in *Selenium in Agriculture and the Environment*, Jacobs, L. W., Ed., Soil Science Society of America Special Publication Number 23, Madison, WI, 133, 1989.
45. Skorupa, J. P., Selenium poisoning of fish and wildlife in nature: Lessons from twelve real-world examples, in *Environmental Chemistry of Selenium,* Frankenburger, W. T., Jr. and Engberg, R. A., Eds., Marcel Dekker, New York, 315, 1998.
46. Maier, K. J., Foe, C., Ogle, R. S., Williams, M. J., Knight, A. W., Kiffney, P., and Melton, L. A., The dynamics of selenium in aquatic ecosystems, in *Trace Substances in Environmental Health-XXI*, Hemphill, D. D., Ed., University of Missouri, Columbia, 361, 1987.
47. Presser, T. S., Sylvester, M. A., and Low, W. H., Bioaccumulation of selenium from natural geologic sources in western states and its potential consequences, *Environ. Manage.*, 18, 423, 1994.
48. Bowie, G. L., Sanders, J. G., Riedel, G. F., Gilmour, C. C., Breitburg, D. L., Cutter, G. A., and Porcella, D. B., Assessing selenium cycling and accumulation in aquatic ecosystems, *Water Air Soil Pollut.*, 90, 93, 1996.

49. Lemly, A. D., Selenium transport and bioaccumulation in aquatic ecosystems: A proposal for water quality criteria based on hydrological units, *Ecotoxicol. Environ. Saf.*, 42, 150, 1999.
50. Davis, E. A., Maier, K. J., and Knight, A. W., The biological consequences of selenium in aquatic ecosystems, *Calif. Agric.*, 42, 18–20, 1988.
51. Hamilton, S. J., Selenium effects on endangered fish in the Colorado River basin, in *Environmental Chemistry of Selenium*, Frankenberger, W. T., Jr. and Engberg, R. A., Eds., Marcel Dekker, New York, 297, 1998.
52. Furr, A. K., Parkinson, T. F., Youngs, W. D., Berg, C. O., Gutenmann, W. H., Pakkala, I. S., and Lisk, D. J., Elemental content of aquatic organisms inhabiting a pond contaminated with coal fly ash, *N.Y. Fish Game J.*, 26, 154, 1979.
53. Patrick, F. M. and Loutit, M. W., Passage of metals to freshwater fish from their food, *Water Res.*, 12, 395, 1978.
54. Suzuki, Y., Nakahara, M., Nakamura, R., and Euda, T., Roles of food and sea water in the accumulation of radionuclides by marine fish, *Bull. Jpn. Soc. Sci. Fish.*, 45, 1409, 1979.
55. Hilton, J. W., Hodson, P. V., and Slinger, S. J., The requirement and toxicity of selenium in rainbow trout (*Salmo gairdneri*), *J. Nutr.*, 110, 2527, 1980.
56. Hilton, J. W., Hodson, P. V., and Slinger, S. J., Absorption, distribution, half-life and possible routes of elimination of dietary selenium in juvenile rainbow trout (*Salmo gairdneri*), *Comp. Biochem. Physiol.*, 71C, 49, 1982.
57. Hilton, J. W. and Hodson, P. V., Effect of increased dietary carbohydrate on selenium metabolism and toxicity in rainbow trout (*Salmo gairdneri*), *J. Nutr.*, 113, 1241, 1983.
58. Hodson, P. V. and Hilton, J. W., The nutritional requirements and toxicity to fish of dietary and waterborne selenium, *Environ. Biogeochem. Ecol. Bull. (Stockholm)*, 35, 335, 1983.
59. Hicks, B. D., Hilton, J. W., and Ferguson, H. W., Influence of dietary selenium on the occurrence of nephrocalcinosis in the rainbow trout, *Salmo gairdneri* Richardson, *J. Fish Dis.*, 7, 379, 1984.
60. Hodson, P. V., Hilton, J. W., and Slinger, S. J., Accumulation of waterborne selenium by rainbow trout (*Salmo gairdneri*), eggs, fry and juveniles, *Fish Physiol. Biochem.*, 1, 187, 1986.
61. Gatlin, D. M., III and Wilson, R. P., Dietary selenium requirement of fingerling channel catfish, *J. Nutr.*, 114, 627, 1984.
62. Goettl, J. P. and Davies, P. H., Water Pollution Studies, Job Progress Report Federal Aid Project F-33-R-12, Colorado Division of Wildlife, Fort Collins, CO, 1977.
63. Goettl, J. P. and Davies, P. H., Water Pollution Studies, Job Progress Report Federal Aid Project F-33-R-13, Colorado Division of Wildlife, Fort Collins, CO, 1978.
64. Barnhart, R. A., Chemical factors affecting the survival of game fish in a western Colorado reservoir, Master's thesis, Colorado State University, Fort Collins, CO, 1957.
65. Birkner, J. H., Selenium in aquatic organisms from seleniferous habitats, Ph.D. dissertation, Colorado State University, Fort Collins, CO, 121, 1978.
66. Sandholm, M., Oksanen, H. E., and Pesonen, L., Uptake of selenium by aquatic organisms, *Limnol. Oceanogr.*, 18, 496, 1973.
67. Besser, J. M., Canfield, T. J., and LaPoint, T. W., Bioaccumulation of organic and inorganic selenium in a laboratory food chain, *Environ. Toxicol. Chem.*, 12, 57, 1993.
68. Kleinow, K. M. and Brooks, A. S., Selenium compounds in the fathead minnow (*Pimephales promelas*). II: Quantitative approach to gastrointestinal absorption, routes of elimination and influence of dietary pretreatment, *Comp. Biochem. Physiol.*, 83C, 71, 1986.
69. Hamilton, S. J., Buhl, K. J., Faerber, N. L., Wiedmeyer, R. H., and Bullard, F. A., Toxicity of organic selenium in the diet to chinook salmon, *Environ. Toxicol. Chem.*, 9, 347, 1990.
70. Bryson, W. T., MacPherson, K. A., Mallin, M. A., Partin, W. E., and Woock, S. E., Roxboro Steam Electric Plant Hyco Reservoir 1984 Bioassay Report, Carolina Power and Light Company, New Hill, NC, 1985.
71. Bell, J. G. and Cowey, C. B., Digestibility and bioavailability of dietary selenium from fishmeal, selenite, selenomethionine and selenocystine in Atlantic salmon (*Salmo salar*), *Aquaculture*, 81, 61, 1989.
72. Woock, S. E., Garrett, W. R., Partin, W. E., and Bryson, W. T., Decreased survival and teratogenesis during laboratory selenium exposures to bluegill, *Lepomis macrochirus*, *Bull. Environ. Contam. Toxicol.*, 39, 998, 1987.

73. Cleveland, L., Little, E. E., Buckler, D. R., and Wiedmeyer, R. H., Toxicity and bioaccumulation of waterborne and dietary selenium in juvenile bluegill (*Lepomis macrochirus*), *Aquat. Toxicol.*, 27, 265, 1993.
74. Coyle, J. J., Buckler, D. R., Ingersoll, C. G., Fairchild, J. F., and May, T. W., Effect of dietary selenium on the reproductive success of bluegills (*Lepomis macrochirus*), *Environ. Toxicol. Chem.*, 12, 551, 1993.
75. Ogle, R. S. and Knight, A. W., Effects of elevated foodborne selenium on growth and reproduction of the fathead minnow (*Pimephales promelas*), *Arch. Environ. Contam. Toxicol.*, 18, 795, 1989.
76. Lorentzen, M., Maage, A., and Julshamn, K., Effects of dietary selenite or selenomethionine on tissue selenium levels of Atlantic salmon (*Salmo salar*), *Aquaculture*, 121, 359, 1994.
77. Schultz, R. and Hermanutz, R., Transfer of toxic concentrations of selenium from parent to progeny in the fathead minnow (*Pimephales promelas*), *Bull. Environ. Contam. Toxicol.*, 45, 568, 1990.
78. Hermanutz, R. O., Malformation of the fathead minnow (*Pimephales promelas*) in an ecosystem with elevated selenium concentrations, *Bull. Environ. Contam. Toxicol.*, 49, 290, 1992.
79. Hermanutz, R. O., Allen, K. N., Roush, T. H., and Hedtke, S. F., Effects of elevated selenium concentrations on bluegills (*Lepomis macrochirus*) in outdoor experimental streams, *Environ. Toxicol. Chem.*, 11, 217, 1992.
80. Allen, K. N., Seasonal variation of selenium in outdoor experimental stream-wetland systems, *J. Environ. Qual.*, 20, 865, 1991.
81. Woock, S. E., Accumulation of selenium by golden shiners *Notemigonus crysoleucas*: Hyco Reservoir N.C. Cage Study 1981–1982, Final Report, Carolina Power and Light Company, New Hill, NC, 1984.
82. Finley, K. A., Observations of bluegills fed selenium-contaminated *Hexagenia* nymphs collected from Belews Lake, North Carolina, *Bull. Environ. Contam. Toxicol.*, 35, 816, 1985.
83. Coughlan, D. J. and Velte, J. S., Dietary toxicity of selenium-contaminated red shiners to striped bass, *Trans. Am. Fish. Soc.*, 118, 400, 1989.
84. Hamilton, S. J., Buhl, K. J., Bullard, F. A., and McDonald, S. F., Evaluation of Toxicity to Larval Razorback Sucker of Selenium-Laden Food Organisms from Ouray NWR on the Green River, Utah, National Biological Service, Yankton, SD, Final Report to the Recovery Implementation Program for the Endangered Fishes of the Colorado River Basin, Denver, CO, 1996.
85. Hamilton, S. J., Holley, K. M., Buhl, K. J., Bullard, F. A., Weston, L. K., and McDonald, S. F., The Evaluation of Contaminant Impacts on Razorback Sucker Feld in Flooded Bottomland Sites near Grand Junction, Colorado — 1996, U.S. Geological Survey, Columbia Environmental Research Center, Yankton, SD, Final Report, 2001a.
86. Hamilton, S. J., Holley, K. M., Buhl, K. J., Bullard, F. A., Weston, L. K., and McDonald, S. F., The Evaluation of Contaminant Impacts on Razorback Sucker Held in Flooded Bottomland Sites near Grand Junction, Colorado — 1997, U.S. Geological Survey, Columbia Environmental Research Center, Yankton, SD, Final Report, 2001b.
87. Lemly, A. D., Role of season in aquatic hazard assessment, *Environ. Monit. Assess.*, 45, 89, 1997.
88. Lemly, A. D., Metabolic stress during winter increases the toxicity of selenium to fish, *Aquat. Toxicol.*, 27, 133, 1993.
89. Heinz, G. H. and Fitzgerald, M. A., Overwinter survival of mallards fed selenium, *Arch. Environ. Contam. Toxicol.*, 25, 90, 1993.
90. Diplock, A. T., Metabolic aspects of selenium action and toxicity, *CRC Crit. Rev. Toxicol.*, 5, 271, 1976.
91. Whanger, P. D., Selenium and heavy metal toxicity, in *Selenium in Biology and Medicine*, Spallholz, J. E., Martin, J. L., and Ganther, H. E., Eds., AVI Publishing, Westport, CT, 230, 1981.
92. Marier, J. R. and Jaworski, J. F., Interactions of Selenium, National Research Council Canada Report No. 20643, Ottawa, Canada, 1983.
93. Levander, O. A., Metabolic interrelationships between arsenic and selenium, *Environ. Health Perspect.*, 19, 159, 1977.
94. Hill, C. H., Interrelationships of selenium with other trace elements, in *Fed. Proc. Fed. Am. Soc. Exp. Biol.*, 34, 2096, 1975.
95. Ewan, R. C., Toxicology and adverse effects of mineral imbalance with emphasis on selenium and other minerals, in *Toxicity of Heavy Metals in the Environment, Part 1*, Oehme, F. W., Ed., Marcel Dekker, New York, 445, 1978.
96. Maier, K. J. and Knight, A. W., Ecotoxicology of selenium in freshwater systems, *Rev. Environ. Contam. Toxicol.*, 134, 31, 1994.

97. Lemly, A. D., Selenium in aquatic organisms, in *Environmental Contaminants in Wildlife: Interpreting Tissue Concentrations*, Beyer, W. N., Heinz, G. H., and Redmon-Norwood, A. W., Eds., CRC Press, Lewis Publishers, Boca Raton, FL, 427, 1996.
98. Hamilton, S. J., Palmisano, A. N., Wedemeyer, G. A., and Yasutake, W. T., Impacts of selenium on early life stages and smoltification of fall chinook salmon, *Trans. N. Am. Wildl. Nat. Res. Conf.*, 51, 343, 1986.
99. Hunn, J. P., Hamilton, S. J., and Buckler, D. R., Toxicity of sodium selenite to rainbow trout fry, *Water Res.*, 21, 233, 1987.
100. Hamilton, S. J. and Wiedmeyer, R. H., Concentrations of boron, molybdenum, and selenium in chinook salmon, *Trans. Am. Fish. Soc.*, 119, 500, 1990.
101. Hamilton, S. J., Rationale for a tissue-based selenium criterion for aquatic life, *Aquat. Toxicol.*, 57, 85, 2002.
102. Cockell, K. A., Hilton, J. W., and Bettger, W. J., Chronic toxicity of dietary disodium arsenate heptahydrate to juvenile rainbow trout (*Oncorhynchus mykiss*), *Arch. Environ. Contam. Toxicol.*, 21, 518, 1991.
103. Moxon, A. L., The effect of arsenic on the toxicity of seleniferous grains, *Science*, 88, 81, 1938.
104. Dubois, K. P., Moxon, A. L., and Olson, O. E., Further studies on the effectiveness of arsenic in preventing selenium poisoning, *J. Nutr.*, 19, 477, 1940.
105. Klug, H. L., Petersen, D. F., and Moxon, A. L., The toxicity of selenium analogues of cystine and methionine, in *Proc. S.D. Acad. Sci.*, 28, 117, 1949.
106. Klug, H. L., Lampson, G. P., and Moxon, A. L., The distribution of selenium and arsenic in the body tissues of rats fed selenium, arsenic, and selenium plus arsenic, in *Proc. S.D. Acad. Sci.*, 29, 57, 1950.
107. Levander, O. A. and Argrett, L. C., Effects of arsenic, mercury, thallium, and lead on selenium metabolism in rats, *Toxicol. Appl. Pharmacol.*, 14, 308, 1969.
108. Thapar, N. T., Guenthner, E., Carlson, C. W., and Olson, O. E., Dietary selenium and arsenic additions to diets for chickens over a life cycle, *Poult. Sci.*, 48, 1988, 1969.
109. Howell, G. O. and Hill, C. H., Biological interaction of selenium with other trace elements in chicks, *Environ. Health Perspect.*, 25, 147, 1978.
110. Hoffman, D. J., Sanderson, C. J., LeCaptain, L. J., Cromartie, E., and Pendleton, G. W., Interactive effects of arsenate, selenium, and dietary protein on survival, growth, and physiology in mallard ducklings, *Arch. Environ. Contam. Toxicol.*, 22, 55, 1992.
111. Cabe, P. A., Carmichael, N. G., and Tilson, H. A., Effects of selenium, alone and in combination with silver or arsenic, in rats, *Neurobehav. Toxicol.*, 1, 275, 1979.
112. Frost, D. V., Selenium and vitamin E as antidotes to heavy metal toxicities, in *Selenium in Biology and Medicine*, Spallholz, J. E., Martin, J. L., and Ganther, H. E., Eds., AVI Publishing, Westport, CT, 490, 1981.
113. Holmberg, R. E. and Ferm, V. H., Interrelationships of selenium, cadmium, and arsenic in mammalian teratogenesis, *Arch. Environ. Health*, 18, 873, 1969.
114. Lorentzen, M., Maage, A., and Julshamn, K., Supplementing copper to a fish meal based diet fed to Atlantic salmon parr affects liver copper and selenium concentrations, *Aquaculture Nutr.*, 4, 67, 1998.
115. Berntssen, M. H. G., Lundebye, A.-K., and Maage, A., Effects of elevated dietary copper concentrations on growth, feed utilisation and nutritional status of Atlantic salmon (*Salmo salar* L.) fry, *Aquaculture,* 174, 167, 1999.
116. Berntssen, M. H. G., Lundeye, A. K., and Hamre, K., Tissue lipid peroxidative responses in Atlantic salmon (*Salmo salar* L.) parr fed high levels of dietary copper and cadmium, *Fish Phys. Biochem.*, 23, 35, 2000.
117. Pelletier, E., Mercury-selenium interactions in aquatic organisms: A review, *Mar. Environ. Res.*, 18, 111, 1985.
118. Rudd, J. W. M., Turner, M. A., Townsend, B. E., Swick, A., and Furutani, A., Dynamics of selenium in mercury-contaminated experimental freshwater ecosystems, *Can. J. Fish. Aquat. Sci.*, 37, 848, 1980.
119. Turner, M. A. and Rudd, J. W. M., The English-Wabigoon River system, III. Selenium in lake enclosures: Its geochemistry, bioaccumulation, and ability to reduce mercury accumulation, *Can. J. Fish. Aquat. Sci.*, 40, 2228, 1983.
120. Turner, M. A. and Swick, A. L., The English-Wabigoon river system, IV. Interaction between mercury and selenium accumulated from waterborne and dietary sources by northern pike (*Esox lucius*), *Can. J. Fish. Aquat. Sci.*, 40, 2241, 1983.

121. Klaverkamp, J. F., Hodgins, D. A., and Lutz, A., Selenite toxicity and mercury-selenium interactions in juvenile fish, *Arch. Environ. Contam. Toxicol.*, 12, 405, 1983.
122. Paulsson, K. and Lundbergh, K., The selenium method for treatment of lakes for elevated levels of mercury in fish, *Sci. Total Environ.*, 87/88, 495, 1989.
123. Paulsson, K. and Lundbergh, K., Treatment of mercury contaminated fish by selenium addition, *Water Air Soil Pollut.*, 56, 833, 1991.
124. Paulsson, K. and Lundbergh, K., Selenium treatment of mercury-contaminated water systems, in *Proc. Selenium-Tellurium Dev. Assoc. 5th Int. Symp.*, Brussels, 287, 1994.
125. Lindqvist, O., Johansson, K., Aastrup, M., Andersson, A., Bringmark, L., Hovsenius, G., Hakanson, L., Iverfeldt, A., Meili, M., and Timm, B., Mercury in the Swedish environment: Recent research on causes, consequences and corrective methods, *Water Air Soil Pollut.*, 55, xi, 1991.
126. Poston, H. A., Effects of dietary aluminum on growth and composition of young Atlantic salmon, *Prog. Fish-Culturist*, 53, 7, 1991.
127. Handy, R. D., The accumulation of dietary aluminum by rainbow trout, *Oncorhynchus mykiss*, at high exposure concentrations, *J. Fish Biol.*, 42, 603, 1993a.
128. Haines, T. A., Acidic precipitation and its consequences for aquatic ecosystems: A review, *Trans. Am. Fish. Soc.*, 110, 669, 1981.
129. Cockell, K. A. and Hilton, J. W., Preliminary investigations on the comparative chronic toxicity of four dietary arsenicals to juvenile rainbow trout (*Salmo gairdneri* R.), *Aquat. Toxicol.*, 12, 73, 1988.
130. Oladimeji, A. A., Qadri, S. U., and deFreitas, A. S. W., Long-term effects of arsenic accumulation in rainbow trout, *Salmo gairdneri*, *Bull. Environ. Contam. Toxicol.*, 32, 732, 1984.
131. Woolson, E. A., Bioaccumulation of arsenicals, in *Arsenical Pesticides*, Woolson, E. A., Ed., American Chemical Society, Washington, D.C., 97, 1975.
132. Pentreath, R. J., The accumulation of cadmium by the plaice, *Pleuronectes platessa* L. and the thornback ray, *Raja clavata* L., *J. Exp. Mar. Biol. Ecol.*, 30, 223, 1977.
133. Kraal, M. H., Kraak, M. H. S., DeGroot, C. J., and Davids, C., Uptake and tissue distribution of dietary and aqueous cadmium by carp (*Cyprinus carpio*), *Ecotoxicol. Environ. Saf.*, 31, 179, 1995.
134. Williams, D. R. and Giesy, J. P., Relative importance of food and water sources to cadmium uptake by *Gambusia affinis* (Poeciliidae), *Environ. Res.*, 16, 326, 1978.
135. Kumada, H., Kimura, S., and Yokote, M., Accumulation and biological effects of cadmium in rainbow trout, *Bull. Jpn. Soc. Sci. Fish.*, 46, 97, 1980.
136. Hatakeyama, S. and Yasuno, M., Accumulation and effects of cadmium on guppy (*Poecilia reticulata*) fed cadmium-dosed Cladocera (*Moina macrocopa*), *Bull. Environ. Contam. Toxicol.*, 29, 159, 1982.
137. Handy, R. D., The effect of acute exposure to dietary Cd and Cu on organ toxicant concentrations in rainbow trout, *Oncorhynchus mykiss*, *Aquat. Toxicol.*, 27, 1, 1993b.
138. Hatakeyama S. and Yasuno, M., Chronic effects of Cd on the reproduction of the guppy (*Poecilia reticulata*) through Cd-accumulated midge larvae (*Chironomus yoshimatsui*), *Ecotoxicol. Environ. Saf.*, 14, 191, 1987.
139. Pratap, H. B., Fu, H., Lock, R. A. C., and Wendelaar Bonga, S. E., Effect of waterborne and dietary cadmium on plasma ions of the teleost *Oreochromis mossambicus* in relation to water calcium levels, *Arch. Environ. Contam. Toxicol.*, 18, 568, 1989.
140. Crespo, S., Nonnotte, G., Colin, D. A., Leray, C., Nonnotte, L., and Aubree, A., Morphological and functional alterations induced in trout intestine by dietary cadmium and lead, *J. Fish Biol.*, 28, 69, 1986.
141. Lundebye, A.-K., Berntssen, M. H. G., Wendelaar Bonga, S. E., and Maage, A., Biochemical and physiological responses in Atlantic salmon (*Salmo salar*) following dietary exposure to copper and cadmium, *Mar. Pollut. Bull.*, 39, 137, 1999.
142. Kay, S. H., Cadmium in aquatic food webs, *Res. Rev.*, 96, 13, 1985.
143. Harrison, S. E. and Curtis, P. J., Comparative accumulation efficiency of $^{109}$cadmium from natural food (*Hyalella azteca*) and artificial diet by rainbow trout (*Oncorhynchus mykiss*), *Bull. Environ. Contam. Toxicol.*, 49, 757, 1992.
144. Merlini M., Pozzi, G., Brazzelli, A., and Berg, A., The transfer of $^{65}$Zn from natural and synthetic foods to freshwater fish, in *Radioecology and Energy Resources*, Cushing, C. E., Ed., Dowden, Hutchinson, and Ross, Stroudsburg, PA, 226, 1976.
145. Pentreath, R. J., The accumulation and retention of $^{65}$Zn and $^{54}$Mn by the plaice, *Pleuronectes platessa* L., *J. Exp. Mar. Biol. Ecol.*, 12, 1, 1973.

146. Weber, D. N., Eisch, S., Spieler, R. E., and Petering, D. H., Metal redistribution in largemouth bass (*Micropterus salmoides*) in response to restrainment stress and dietary cadmium: Role of metallothionein and other metal-binding proteins, *Comp. Biochem. Physiol.*, 101C, 255, 1992.
147. Aoyama, I., Inoue, Y., and Inoue, Y., Experimental study on the concentration process of trace element through a food chain from the viewpoint of nutrition ecology, *Water Res.*, 12, 831, 1978.
148. Tacon, A. G. J. and Beveridge, M. M., Effects of dietary trivalent chromium on rainbow trout, *Nutr. Rep. Int.*, 25, 49, 1982.
149. Baudin, J. P. and Fritsch, A. F., Relative contributions of food and water in the accumulation of $^{60}Co$ by a freshwater fish, *Water Res.*, 23, 817, 1989.
150. Lanno, R. P., Slinger, S. J., and Hilton, J. W., Maximum tolerable and toxicity levels of dietary copper in rainbow trout (*Salmo gairdneri* Richardson), *Aquaculture*, 49, 257, 1985.
151. Miller, P. A., Lanno, R. P., McMaster, M. E., and Dixon, D. G., Relative contributions of dietary and waterborne copper to tissue copper burdens and waterborne-copper tolerance in rainbow trout (*Oncorhynchus mykiss*), *Can. J. Fish. Aquat. Sci.*, 50, 1683, 1993.
152. Handy, R. D., Sims, D. W., Giles, A., Campbell, H. A., and Musonda, M. M., Metabolic trade-off between locomotion and detoxification for maintenance of blood chemistry and growth parameters by rainbow trout (*Oncorhynchus mykiss*) during chronic dietary exposure to copper, *Aquat. Toxicol.*, 47, 23, 1999.
153. Hodson, P. V., Blunt, B. R., and Spry, D. J., Chronic toxicity of water-borne and dietary lead to rainbow trout (*Salmo gairdneri*) in Lake Ontario water, *Water Res.*, 12, 869, 1978.
154. Leland, H. V. and McNurney, J. M., Lead transport in a river ecosystem, in *Proc. Int. Conf. Transport Persistent Chem. Aquat. Ecosyst.*, National Research Council of Canada, Ottawa, Canada, 1974.
155. Goettl, J. P. and Davies, P. H., Water Pollution Studies, Job Progress Report F-33-R-11, Colorado Division of Wildlife, Fort Collins, CO, 1976.
156. Wiener, J. G. and Spry, D. J., Toxicological significance of mercury in freshwater fish, in *Environmental Contaminants in Wildlife: Interpreting Tissue Concentrations*, Beyer, W. N., Heinz, G. H., and Redmon-Norwood, A. W., Eds., CRC Press, Boca Raton, FL, 297, 1996.
157. Zhou, H. Y. and Wong, M. H., Mercury accumulation in freshwater fish with emphasis on the dietary influence, *Water Res.*, 34, 4234, 2000.
158. Hilton, J. W. and Bettger, W. J., Dietary vanadium toxicity in juvenile rainbow trout: A preliminary study, *Aquat. Toxicol.*, 12, 63, 1988.
159. Holdway, D. A., Sprague, J. B., and Dick, J. G., Bioconcentration of vanadium in American flagfish over one reproductive cycle, *Water Res.*, 17, 937, 1983.
160. Fishbein, L., Trace and ultra trace elements in nutrition: An overview, *Toxicol. Environ. Chem.*, 14, 73, 1987.
161. Cooley, H. M. and Klaverkamp, J. F., Accumulation and distribution of dietary uranium in lake whitefish (*Coregonus clupeaformis*), *Aquat. Toxicol.*, 48, 477, 2000.
162. Cooley, H. M., Evans, R. E., and Klaverkamp, J. F., Toxicology of dietary uranium in lake whitefish (*Coregonus clupeaformis*), *Aquat. Toxicol.*, 48, 495, 2000.
163. Wekell, J. C., Shearer, K. D., and Houle, C. R., High zinc supplementation of rainbow trout diets, *Prog. Fish-Culturist*, 45, 144, 1983.
164. Hardy, R. W. and Shearer, K. D., Effect of dietary calcium phosphate and zinc supplementation on whole body zinc concentration of rainbow trout (*Salmo gairdneri*), *Can. J. Fish. Aquat. Sci.*, 42, 181, 1985.
165. Brafield, A. E. and Koodie, A. V., Effects of dietary zinc on the assimilation efficiency of carp (*Cyprinus carpio* L.), *J. Fish Biol.*, 39, 893, 1991.
166. Kock, G. and Bucher, F., Accumulation of zinc in rainbow trout (*Oncorhynchus mykiss*) after waterborne and dietary exposure, *Bull. Environ. Contam. Toxicol.*, 58, 305, 1997.
167. Satoh, S., Takeuchi, T., Narabe, Y., and Watanabe, T., Effects of deletion of several trace elements from fish meal diets on growth and mineral composition of rainbow trout fingerlings, *Bull. Jpn. Soc. Sci. Fish.*, 49, 1909, 1983.
168. Farag, A. M., Boese, C. J., Woodward, D. F., and Bergman, H. L., Physiological changes and tissue metal accumulation in rainbow trout exposed to foodborne and waterborne metals, *Environ. Toxicol. Chem.*, 13, 2021, 1994.

169. Farag, A. M., Stansbury, M. A., Hogstrand, C., MacConnell, E., and Bergman, H. L., The physiological impairment of free-ranging brown trout exposed to metals in the Clark Fork River, Montana, *Can. J. Fish. Aquat. Sci.*, 52, 2038, 1995.
170. Farag, A. M., Suedkamp, M. J., Meyer, J. S., Barrows, R., and Woodward, D. F., Distribution of metals during digestion by cutthroat trout fed benthic invertebrates contaminated in the Clark Fork River, Montana and the Coeur d'Alene River, Idaho, U.S.A., and fed artificially contaminated *Artemia*, *J. Fish Biol.*, 56, 173, 2000.
171. Woodward, D. F., Brumbaugh, W. G., DeLonay, A. J., Little, E. E., and Smith, C. E., Effects on rainbow trout fry of a metals-contaminated diet of benthic invertebrates from the Clark Fork River, Montana, *Trans. Am. Fish. Soc.* 123, 51, 1994.
172. Woodward, D. F., Farag, A. M., Bergman, H. L., DeLonay, A. J., Little, E. E., Smith, C. E., and Barrows, F. T., Metals-contaminated benthic invertebrates in the Clark Fork River, Montana: Effects on age-0 brown trout and rainbow trout, *Can. J. Fish. Aquat. Sci.*, 52, 1994, 1995.
173. Farag, A. M., Woodward, D. F., Goldstein, J. N., Brumbaugh, W., and Meyer, J. S., Concentrations of metals associated with mining waste in sediments, biofilm, benthic macroinvertebrates, and fish from the Coeur d'Alene River basin, Idaho, *Arch. Environ. Contam. Toxicol.*, 34, 119, 1998.
174. Farag, A. M., Woodward, D. F., Brumbaugh, W., Goldstein, J. N., MacConnell, E., Hogstrand, C., and Barrows, F. T., Dietary effects of metals-contaminated invertebrates from the Coeur d'Alene River, Idaho, on cutthroat trout, *Trans. Am. Fish. Soc.*, 128, 578, 1999.
175. Pascoe, G. A., Blanchet, R. J., Linder, G., Palawski, D., Brumbaugh, W. G., Canfield, T. J., Kemble, N. E., Ingersoll, C. G., Farag, A., and DalSoglio, J. A., Characterization of ecological risks at the Milltown Reservoir — Clark Fork River Sediments Superfund site, Montana, *Environ. Toxicol. Chem.*, 13, 2043, 1994.
176. Mount, D. R., Barth, A. K., Garrison, T. D., Barten, K. A., and Hockett, J. R., Dietary and waterborne exposure of rainbow trout (*Oncorhynchus mykiss*) to copper, cadmium, lead and zinc using a live diet, *Environ. Toxicol. Chem.*, 13, 2031, 1994.
177. Hodson, P. V., The effect of metal metabolism on uptake, disposition and toxicity in fish, *Aquat. Toxicol.*, 11, 3, 1988.
178. Eisler, R., *Handbook of Chemical Risk Assessment: Health Hazards to Humans, Plants, and Animals, Vol. 3: Metalloids, Radiation, Cumulative Index to Chemicals and Species*, Lewis Publishers, Boca Raton, FL, 1829, 2000.
179. Heinz, G. H., Selenium in birds, in *Environmental Contaminants in Wildlife: Interpreting Tissue Concentrations,* Beyer, W. N., Heinz, G. H., and Redmon-Norwood, A. W., Eds., Lewis Publishers, Boca Raton, FL, 447, 1996.
180. Skorupa, J. P. and Ohlendorf, H. M., Contaminants in drainage water and avian risk thresholds, in *The Economics and Management of Water and Drainage in Agriculture*, Kluwer, Boston, 1991.
181. Ohlendorf, H. M., Hoffman, D. J., Saiki, M. K., and Aldrich, T. W., Embryonic mortality and abnormalities of aquatic birds: Apparent impacts of selenium from irrigation drainwater, *Sci. Total Environ.*, 52, 49, 1986.
182. Ohlendorf, H. M., Kilness, A. W., Simmons, J. L., Stroud, R. K., Hoffman, D. J., and Moore, J. F., Selenium toxicosis in wild aquatic birds, *J. Toxicol. Environ. Health,* 24, 67, 1988.
183. Hoffman, D. J., Heinz, G. H., and Krynitsky, A. J., Hepatic glutathione metabolism and lipid peroxidation in response to excess dietary selenomethionine and selenite in mallard ducklings, *J. Toxicol. Environ. Health*, 27, 263, 1989.
184. Ohlendorf, H. M. and Skorupa, J. P., Selenium in relation to wildlife and agricultural drainage water, in *4th Int. Symp. Uses Selenium Tellurium*, Selenium-Tellurium Development Association, Darien, CT, 314, 1989.
185. Olson, O. E., Novacek, E. J., Whitehead, E. I., and Palmer, I. S., Investigation on selenium in wheat, *Phytochemistry,* 9, 1181, 1970.
186. Yasumoto, K., Suzuki, T., and Yoshida, M., Identification of selenomethionine in soybean protein, *J. Agric. Food Chem.,* 36, 463, 1988.
187. Heinz, G. H., Hoffman, D. J., Krynitsky, A. J., and Weller, D. M. G., Reproduction in mallards fed selenium, *Environ. Toxicol. Chem.*, 6, 423, 1987.
188. Hoffman, D. J. and Heinz, G. H., Embryotoxic and teratogenic effects of selenium in the diet of mallards, *J. Toxicol. Environ. Health,* 24, 477, 1988.

189. Heinz, G. H., Hoffman, D. J., and Gold, L. G., Impaired reproduction of mallards fed an organic form of selenium, *J. Wildl. Manage.,* 53, 418, 1989
190. Heinz, G. H. and Hoffman, D. J., Comparison of the effects of seleno-L-methionine, seleno-D,L-methionine, and selenized yeast on reproduction of mallards, *Environ. Pollut.,* 91, 169, 1996.
191. Hoffman, D. J., Heinz, G. H., and Krynitsky, A. J., Hepatic glutathione metabolism and lipid peroxidation in response to excess dietary selenomethionine and selenite in mallard ducklings, *J. Toxicol. Environ. Health,* 27, 263, 1989.
192. Heinz, G. H., Hoffman, D. J., and Gold, L. G., Toxicity of organic and inorganic selenium to mallard ducklings, *Environ. Toxicol. Chem.,* 17, 561, 1988.
193. Fairbrother, A. and Fowles, J., Subchronic effects of sodium selenite and selenomethionine on several immune functions in mallards, *Arch. Environ. Contam. Toxicol.,* 19, 836, 1990.
194. Hoffman, D. J., Heinz, G. H., LeCaptain, L. J., Eisemann, J. D., and Pendleton, G. W., Toxicity and oxidative stress of different forms of organic selenium and dietary protein in mallard ducklings, *Arch. Environ. Contam. Toxicol.,* 31, 120, 1996.
195. Heinz, G. H., Hoffman, D. J., and LeCaptain, L. J., Toxicity of seleno-L-methionine, seleno-D,L-methionine, high selenium wheat, and selenized yeast to mallard ducklings, *Arch. Environ. Contam. Toxicol.,* 30, 93, 1996.
196. Hoffman, D. J., Sanderson, C. J., LeCaptain, L. J., Cromartie, E., and Pendleton, G. S., Interactive effects of selenium, methionine and dietary protein on survival, growth, and physiology in mallard ducklings, *Arch. Environ. Contam. Toxicol.,* 23, 163, 1992.
197. Hoffman, D. J., Sanderson, C. J., LeCaptain, L. J., Cromartie, E., and Pendleton, G. S., Interactive effects of boron, selenium and dietary protein on survival, growth, and physiology in mallard ducklings, *Arch. Environ. Contam. Toxicol.,* 20, 49, 1991.
198. Stanley, T. R., Jr., Spann, J. W., Smith, G. J., and Rosscoe, R., Main and interactive effects of arsenic and selenium on mallard reproduction and duckling growth and survival, *Arch. Environ. Toxicol.,* 26, 444, 1994.
199. Stanley, T. R., Jr., Smith, G. J., Hoffman, D. J., Heinz, G. H., and Rosscoe, R., Effects of boron and selenium on mallard reproduction and duckling growth and survival, *Environ. Toxicol. Chem.,* 15, 1124, 1996.
200. Stoewsand, G. S., Bache, C. A., and Lisk, D. J., Dietary selenium protection of methylmercury intoxication of Japanese quail, *Bull. Environ. Contam. Toxicol.,* 11, 152, 1974.
201. Sell, J. L. and Horani, F. G., Influence of selenium on toxicity and metabolism of methylmercury in chicks and quail, *Nutr. Reports Int.,* 14, 439, 1976.
202. El-Begearmi, M. M., Sunde, M. L., and Ganther, H. E., A mutual protective effect of mercury and selenium in Japanese quail, *Poult. Sci.,* 56, 313, 1977.
203. Heinz, G. H. and Hoffman D. J., Methylmercury chloride and selenomethionine interactions on health and reproduction in mallards, *Environ. Toxicol. Chem.,* 17, 139, 1998.
204. Hoffman, D. J. and Heinz, G. H., Effects of mercury and selenium on glutathione metabolism and oxidative stress in mallard ducks, *Environ. Toxicol. Chem.,* 17, 161, 1998.
205. Jordan, J. S., Influence of diet in lead poisoning in waterfowl, *Trans. N.E. Sec. Wildl. Soc.,* 25, 143, 1968.
206. Longcore, J. R., Andrews, R., Locke, L. N., Bagley, G. E., and Young, L. T., Toxicity of Lead and Proposed Substitute shot to Mallards, U.S. Fish Wildl. Serv. Spec. Sci. Rep.- Wildl., 183, 1, 1974.
207. Finley, M. T. and Dieter, M. P., Toxicity of experimental lead-iron shot versus commercial lead shot in mallards, *J. Wildl. Manage.,* 42, 32, 1978.
208. Carlson, B. L. and Nielsen, S. W., Influence of dietary calcium on lead poisoning in mallard ducks (*Anas platyrhynchos*), *Am. J. Vet. Res.,* 46, 276, 1985.
209. Heinz G. H., Hoffman, D. J., Sileo, L., Audet, D. J., and LeCaptain, L. J., Toxicity of lead-contaminated sediment to mallards, *Arch. Environ. Contam. Toxicol.,* 36, 323, 1999.
210. Hoffman, D. J., Heinz, G. H., Sileo, L., Audet, D. J., and LeCaptain, L. J., Developmental toxicity of lead-contaminated sediment to mallard ducklings, *Arch. Environ. Contam. Toxicol.,* 39, 22, 2000.
211. Douglas-Stroebel, E., Brewer, G. L., and Hoffman, D. J., Effects of lead-contaminated sediment on mallard duckling behavior, in *Proc. SETAC 19th Annu. Meeting*, Charlotte, NC, 19, 1998.
212. Douglas-Stroebel, E., Brewer, G. L., and Hoffman, D. J., Effects of lead-contaminated sediment and nutrient level on mallard duckling brain growth and biochemistry, in *Proc. SETAC 22nd Annu. Meeting*, Baltimore, 2001.

213. Day, D. D., Beyer, W. N., Hoffman, D. J., Morton, A., Sileo, L., Audet, D. J., and Ottinger, M. A., Toxicity of lead contaminated sediments to mute swans, *Arch. Environ. Contam. Toxicol.* (in press), 2002.
214. Grasman, K. A. and Scanlon, P. F., Effects of acute lead ingestion and diet on antibody and T-cell-mediated immunity in Japanese quail, *Arch. Environ. Contam. Toxicol.,* 28, 161, 1995.
215. Scheuhammer, A. M., Influence of reduced dietary calcium on the accumulation and effects of lead, cadmium, and aluminum in birds, *Environ. Pollut.,* 94, 337, 1996.
216. Sparling, D. W., Acidic deposition: A review of biological effects, in *Handbook of Ecotoxicology,* Hoffman, D. J., Rattner, B. A., Burton, G. A., Jr., and Cairns, J., Jr., Eds., Lewis Publishers, Boca Raton, FL, 1995, 301.
217. Sparling, D. W., Acid precipitation and food quality: Inhibition of growth and survival in black ducks and mallards by dietary aluminum, calcium, and phosphorus, *Arch. Environ. Contam. Toxicol.,* 19, 457, 1990.
218. Sparling, D. W., Acid precipitation and food quality: Effects of dietary Al, Ca, and P on bone and liver characteristics in American black ducks and mallards, *Arch. Environ. Contam. Toxicol.,* 21, 281, 1991.
219. Miles A. K., Grue, C. E., Pendleton, G. W., and Soares, J. H., Jr., Effects of dietary aluminum, calcium, and phosphorus on egg and bone of European starlings, *Arch. Environ. Contam. Toxicol.*, 24, 206, 1993.
220. Welsh, S. O. and Soares, J. H., Jr., The protective effect of vitamin E and selenium against methyl mercury toxicity in the Japanese quail, *Nutr. Rep. Int.,* 13, 43, 1976.
221. Kling, L. J. and Soares, J. H., Jr., Vitamin E deficiency in the Japanese quail, *Poult. Sci.*, 59, 2353, 1980.
222. Prasada Rao, P. V. V., Jordan, S. A., and Bhatnagar, M. K., Combined nephrotoxicity of methylmercury, lead, and cadmium in Pekin ducks: Metallothionein, metal interactions, and histopathology, *J. Toxicol. Environ. Health,* 26, 327, 1989.
223. Koeman, J. H., Peeters, W. H. M., Koudstaal-Hol, C. H. M., Tjioe, P. S., and de Goeij, J. M., Mercury-selenium correlations in marine mammals, *Nature,* 245, 385, 1973.
224. Scheuhammer, A. M., The chronic toxicity of aluminum, cadmium, mercury, and lead in birds: A review, *Environ. Pollut.,* 46, 263, 1987.
225. Rainbow, P. S. and Furness, R. W., Heavy metals in the marine environment, in *Heavy Metals in the Marine Environment,* Furness, R. W. and Rainbow, P. S., Eds., CRC Press, Boca Raton, FL, 1990.
226. Ohlendorf, H. M., Lowe, R. W., Kelly, P. R., and Harvey, T. E., Selenium and heavy metals in San Francisco Bay diving ducks, *J. Wildl. Manage.,* 50, 64, 1986.
227. Elliott, J. E., Scheuhammer, A. M., Leighton, F. A., and Pearce, P. A., Heavy metal and metallothionein concentrations in Atlantic Canadian seabirds, *Arch. Environ. Contam. Toxicol.,* 22, 63, 1992.
228. Kim, E. Y., Saeki, K., Tanabe, S., Tanaka, H., and Tatrsukawa, R., Specific accumulation of mercury and selenium in seabirds, *Environ. Pollut.*, 94, 261, 1996.
229. Bischoff, K., Pichner, J., Braselton, W. E., Counard, C., Evers, D. C., and Edwards, W. C., Mercury and selenium concentrations in livers and eggs of common loons (*Gavia immer*) from Minnesota, *Arch. Environ. Contam. Toxicol.,* 42, 71, 2002.
230. Spalding, M. G., Frederick, P. C., McGill, H. C., Bouton, S. N., and McDowell, L. R., Methylmercury accumulation in tissues and effects on growth and appetite in captive great egrets, *J. Wildl. Dis.,* 36, 411, 2000.
231. Henny, C. J., Hill, E. F., Hoffman, D. J., Spalding, M. G., and Grove, R. A., Nineteenth century mercury: Hazard to wading birds and cormorants of the Carson River, Nevada, *Ecotoxicology,* 11, 213, 2002.
232. Moller, G., Biogeochemical interactions affecting hepatic trace element levels in aquatic birds, *Environ. Toxicol. Chem.*, 5, 1025, 1996.
233. Nyholm, N. E. I., Evidence of involvement of aluminum in causation of defective formation of eggshells and of impaired breeding in wild passerine birds, *Environ. Res.,* 26, 363, 1981.
234. Scheuhammer, A. M., Effects of acidification on the availability of toxic metals and calcium to wild birds and mammals, *Environ. Pollut.,* 71, 329, 1991.
235. Scheuhammer, A. M., McNicol, D. K., Mallory, M. L., and Kerekes, J. J., Relationships between lake chemistry and calcium and trace metal concentrations of aquatic invertebrates eaten by breeding insectivorous waterfowl, *Environ. Pollut.,* 96, 235, 1997.
236. Shore, R. F. and Rattner, B. A., Eds., *Ecotoxicology of Wild Mammals,* John Wiley and Sons, Chichester, U.K., 730, 2001.

## CHAPTER 45

# Animal Species Endangerment: The Role of Environmental Pollution

**Oliver H. Pattee, Valerie L. Fellows, and Dixie L. Bounds**

## CONTENTS

## 45.1 INTRODUCTION

During the last 100 years, human activities have contributed to an extinction rate rivaling the meteor strike that ended the dinosaur era. The speed, severity, and taxonomic diversity of declining species is a major concern of the world's ecologists.[1] Extinctions are taking place at a rate of approximately 100 species per day.[2] In 1989, Wilson[3] projected the loss of species at more than 20% of the planet's total biodiversity in 20 years.

The Nature Conservancy compiled a list of 20,481 plants and animals within the United States and estimated, within arbitrary parameters, the persistence of each species.[4–6] There were 4325 animals on the list: 2524 vertebrates and 1801 invertebrates. Seventy-five percent (1893) of the vertebrates and 61% (1098) of the invertebrates were secure or apparently secure and could be expected to persist. The remaining animal species (1334) are likely to go extinct. This chapter discusses the role of contaminants/pollution in the decline of species. A subset of the pertinent information represented by recovery plans and listing packages provides data on species response and contaminant impact.

### 45.1.1 What Is Threatened and Endangered: Definitions

The U.S. Endangered Species Preservation Act of 1966 granted authorization to protect listed species and their habitat and gave approval to acquire land for imperiled species. The Endangered Species Conservation Act of 1969 broadened the law to include international species, prohibited importation of listed species, and called for an international convention on the conservation of endangered species.

Outside the United States, the protection of threatened and endangered species is ambiguous and poorly defined. Legislation that formally lists and protects species moves slowly along many legislative pathways. Ray and Ginsberg[7] evaluated the status of protective legislation in Australia, Canada, and the European Union. In this chapter, we will use their evaluations on the status of species protection and legislative turmoil. The Australian Endangered Species Protection Act of 1992 allows the designation of endangered communities and the remediation of activities that could threaten designated communities. Progress in Canada has been slow, especially at the national level, where the legislation that has passed has been weakened significantly, apparently reflecting a lack of interest by the federal government in assuming a leadership position.[7] The European Union is handicapped by the cultural and linguistic differences of its member states. Consequently, they rely on a disparate set of national endangered species legislation and regional environmental laws that may have provisions for species protection. A unified approach to endangered species protection appears unlikely within the European Union. The most probable legislation will deal with the question on a subregional level.[7]

#### *45.1.1.1 The Convention on International Trade in Endangered Species of Wild Fauna and Flora and the Endangered Species Act (U.S.)*

In 1973, a convention was developed on the worldwide conservation of endangered species. More commonly known as CITES, the Convention on International Trade in Endangered Species of Wild Fauna and Floral restricts international commerce in plant and animal species harmed by trade or believed to be harmed by trade. The United States' 1973 Endangered Species Act (ESA)

**Table 45.1 Listing and Recovery Plans (as of June 2002)**

| | Endangered | | Threatened | | U.S. Species with Recovery Plans |
|---|---|---|---|---|---|
| | U.S. | Foreign | U.S. | Foreign | |
| Mammals | 65 | 251 | 9 | 17 | 53 |
| Birds | 78 | 175 | 14 | 6 | 75 |
| Reptiles | 14 | 64 | 22 | 15 | 32 |
| Amphibians | 11 | 8 | 9 | 1 | 12 |
| Fishes | 71 | 11 | 44 | 0 | 95 |
| Snails | 21 | 1 | 11 | 0 | 27 |
| Clams | 62 | 2 | 8 | 0 | 56 |
| Crustaceans | 18 | 0 | 3 | 0 | 12 |
| Insects | 35 | 4 | 9 | 0 | 29 |
| Arachnids | 12 | 0 | 0 | 0 | 5 |
| **Total** | 387 | 516 | 129 | 39 | 396 |

was established to provide protection for threatened and endangered species (nationally and internationally) and their habitats (nationally). The ESA stands as the most powerful law in the United States, perhaps the world, protecting flora and fauna.

The U.S. Fish and Wildlife Service[8] lists 1071 animal species (516 in the United States, 555 foreign species) under the ESA as endangered or threatened as of June 2002 (Table 45.1). By definition, a species under ESA includes any species or subspecies of fish, wildlife, or plant; any variety of plant; and any distinct population segment of any vertebrate species that interbreeds when mature. Excluded from the ESA are any species of the class Insecta determined to be pests whose protection would present an overwhelming and overriding risk to humans. A species is classified as *endangered* when it is in danger of extinction within the foreseeable future throughout all or a significant portion of its range. A determination of *threatened* is made when the plant or animal so designated is likely to become endangered within the foreseeable future throughout all or a significant portion of its range.

#### *4.5.1.1.2 Listing Packages and Recovery Plans*

*Listing packages* describes the process as well as the documents required under ESA to declare a species threatened or endangered. The determination of status is made based only on the biological and trade information, with no consideration of economic impact or other possible impacts. Criteria for listing are quite specific:

- The present or threatened destruction, modification, or curtailment of habitat or range (pollution events figure prominently here).
- Overutilization for commercial, recreational, scientific, or education purposes.
- Disease or predation.
- The inadequacy of existing regulatory mechanisms. Biocides are regulated, but the full range of their impacts are poorly understood and rarely mentioned or included.
- Other natural or human-made factors affecting a species' continued existence. Human-made problems such as depletion of groundwater, infiltration of contaminants, and careless handling of chemicals create additional hazards that could have been avoided.

The proposed rule is published in the Federal Register, opening a comment period of 6 months. Within 1 year (1) a final listing rule must be published, (2) the proposal must be withdrawn, or (3) a determination may be made to extend the comment period for up to 6 months. Once the proposed rule has been approved, recovery planning starts. Causes of the species' decline must be identified and a recovery plan developed. The recovery plan may be developed by a panel of experts or a knowledgeable consultant but remains the responsibility of the Fish and Wildlife Service or National Marine Fisheries Service. The recovery plan describes and prioritizes the tasks essential to the species recovery. A public review period takes place prior to finalizing the plan and the final approval of the regional director. Approval of the plan does not insure funds are available to carry out the plan, and years may go by before the priority tasks are addressed.

### 45.1.2 Factors Driving Species Decline

Multiple factors contribute to the decline of a species. Habitat destruction is the primary factor that threatens species[1,9–15] and is listed as a significant factor affecting 73% of endangered species.[16–17] The second major factor causing species decline is the introduction of nonnative species.[15,17] This affects 68% of endangered species.[16,17] Pollution and overharvesting were identified as impacting 38% and 15% of endangered species, respectively.[13,15–18] Other factors affecting species decline include hybridization, competition, disease, and other interspecific interactions.[15,17]

Although frequently espoused, environmental pollution as a significant cause of the initial decline is problematic. Effects of DDE/DDT on avian reproduction are well documented; that a major contaminant also bioaccumulated, biomagnified, and targeted an enzyme system universal to birds (eggshell formation) was unexpected and unanticipated. No other contaminant has impacted animal survival to such an extent, and DDT remains one of the few examples of pollution actually extirpating animal species over a significant portion of their range.

However, once a species is reduced to a remnant of its former population size and distribution, its vulnerability to catastrophic pollution events increases, frequently exceeding or replacing the factors responsible for the initial decline. Large-scale environmental events, such global warming, acid rain, and sea-level rise, attract considerable attention as speciation events — both good (species numbers increase) and bad (species numbers decrease).

## 45.2 CHARACTERISTICS OF POLLUTION

The terms *pollution* and *contamination* remain ambiguous, even in the scientific community. For example, in the literature reviewed, each study started the same way — by defining terms. Definitions were applied and derived from a publication with different definitions or categories of pollution for each study. In general, authors need to move toward a universal definition. We classify pollution in two main categories: physical pollution and chemical contamination.

### 45.2.1 Physical Pollution

Physical pollution is most frequently caused by habitat modification such as channelization, dredging, dam and impoundment construction, urbanization, logging, and agricultural operations.[19–21] Aquatic fauna are extremely sensitive to environmental variations, especially pollution and contamination.[10–12,14,22] Habitat destruction was ranked the first and pollution the second most important factor affecting species decline for aquatic animals including amphibians, fish, mussels, crayfish, dragonflies, and damselflies.[15] Lakes, streams, rivers, and freshwater ponds across the nation are adversely affected by various forms of pollution and contamination.

The effects of physical pollution include siltation, sedimentation, altered thermal regimes, altered water-flow regimes, changes in water depth, altered oxygen availability, and changes in pH, salinity, and erodability. Altered hydrologic regimes affected 34% of all species historically and currently affect 28% of all organisms.[22] Channel modification results from flood-control devices, improper agricultural practices, grazing, logging, and mining operations; these activities remove the substrate that organisms live in and lead to increased siltation.[23,24] Siltation is the most important water-quality problem across 45% of U.S. river miles and 28% of U.S. stream miles.[19] Siltation covers spawning sites for fish and destroys benthic food sources.[25] Freshwater clams and other bottom-dwelling filter feeders suffocate when covered by a layer of silt because they are relatively immobile as adults.[24,26]

Water development projects, including dams, impoundments, river channelization, navigational barriers, and agricultural diversion, contribute to the decline of 30% of all threatened and endangered species. Over 91% of endangered freshwater fish and 99% of freshwater mussels are affected by water development.[15] Large, multiple impoundments are the most prominent factor affecting freshwater mussels in any given watershed.[27] These watershed changes are estimated to impact 36% of the species affected by this class of habitat damage.[22] Damming dramatically alters substrate composition, water chemistry, water temperature, and the amount of dissolved oxygen in the water.[24,28,30] Altering free-flowing water contributes to erosion, modifies the transport of organic matter and sediment, changes the temperature regimes, and destabilizes bankments.[12,31] Dams degrade fish-spawning habitat, leading to decreased densities and species richness of fish communities.[29,31] Larval freshwater mussels depend on a host fish to attach to, so they are indirectly threatened by human actions that affect host-fish populations,[24] as are many benthic organisms. The benthic community, especially mussels, is most likely affected indirectly.

### 45.2.2 Chemical Contamination

Chemical contamination includes wastewater discharge, agricultural runoff, nutrient enrichment, pesticide/herbicide use, heavy metals, and organic chemicals. Contaminants originate from both point and nonpoint sources including municipal, industrial, and agricultural runoff and discharges.[19] Over 17,365 waterways in the U.S. have been altered by a multitude of point- and nonpoint-source discharges of pollutants.[32] Approximately 37% of U.S. rivers are affected by nutrient pollution, which is followed by pathogen indicators (27%), pesticides (26%), and organic enrichments combined with low dissolved oxygen (24%). Agricultural practices impact 72% of streams, followed by municipal point-source pollution (15%), urban runoff (11%), resource extraction (11%), and industrial point-source pollution (7%).[19]

## 45.3 SOURCES OF POLLUTION

Venues that allow pollutants to enter the environment are numerous, and articles listing all the potential contaminants are frequent. Basically, all of the factors are derivatives of three sources.

1. Agricultural
2. Municipal
3. Industrial

The path is twisted and contains many false trails, but eventually one or more of these three sources cover the introduced/anthropomorphic-derived pollutants. Compounds derived from municipal and industrial sources are difficult to separate. Frequently, industrial-derived pollutants are discharged into municipal treatment plants, from where they pass on into the environment. Rather than try to follow the trail to its source, industrial and municipal will be treated as a single category.

### 45.3.1 Agricultural

Agricultural practices are ranked as the most influential factor affecting water quality for rivers, streams, and lakes.[19,22,29,33] Nonpoint agricultural pollutants, such as silt, nutrients, farm waste, and fertilizers, are detrimental to aquatic species.[15,22,25,29] Commercial fertilizer is the primary agricultural nonpoint source of nitrates and phosphates,[33] which serve as catalysts for the development of anoxic water, toxic algal blooms, and the eutrophication of plant production. Organic nutrients enter waterways via municipal, industrial, and agricultural pathways including feedlot and pasture runoff.[19] Nitrogen is a common contaminant also released from industrial activities, transportation sources, sewage, and atmospheric acid deposition.[20] Breakdown of accumulated toxic nutrients uses up large quantities of dissolved oxygen. This results in low levels of dissolved oxygen in the water, rendering the habitat unsuitable for many aquatic species.[19]

Agricultural runoff also contains pesticides/insecticides and herbicides. Agricultural pollution is most likely to cause sublethal toxic effects in a species, affecting hormone regulation, reproduction, and embryonic development.[25,29,34–38] Pesticides affect not only aquatic species larval development, but also adult immune systems, rendering organisms more susceptible to disease. With fewer healthy adults in the breeding population, fewer young will be produced, and of those produced, more offspring will not develop normally.[8] Constant pesticide applications that affect immune system development can only suppress an already small population characterized by an endangered or threatened species.

Herbicides produce a wide range of detrimental effects in endangered species. Common herbicides are Atrazine, Paraquat, Diquat, Alachlor, Propham, and Barban.[36,38] Most effects are indirect, such as the loss of plants that act as food sources and provide shelter from predators.[37] Most endangered and threatened species are habitat specialists, and alteration of their habitat may reduce their chances of survival.

### 45.3.2 Municipal and Industrial Pollution

Many of the same pesticides and nutrients used in large-scale agricultural practices are also used by municipal entities including urban and rural households. Runoff from fertilized lawns, household termite control, and organic nutrients from grass clippings may alter the soil makeup and can render a habitat unsuitable for certain plant and animal species.[19] Such municipal sources produce substantial amounts of runoff that is discharged into surface waters, serving as the largest source of pollution for rivers, streams, and estuaries. Municipal discharges affect 16% of rivers and streams, 17% of lakes and reservoirs, and over 35% of estuaries. Runoff from city streets, construction sites, storm sewers, household waste, and leachate from septic tanks and landfills alters the sediment load in aquatic systems and deposits contaminants directly into surface- and groundwater sources.

Industrial sources of pollution include plants, steel mills, paper mills, mining establishments, and chemical manufacturers, whose effluent is usually categorized as point-source pollution. In the U.S., over 160,000 industrial facilities discharge 92,000 metric tons of hazardous wastes into municipal treatment plants annually, and as much as 18% of that remains untreated when it is discharged into surface water.[39] Municipal and industrial effluents carry toxic organic chemicals, toxic nutrients, heavy metals, salts, and pathogens.[19] Runoff and discharges from municipal and industrial effluents eliminate opportunities for threatened species to recolonize local populations, destroy benthic food sources for aquatic species, and isolate populations resulting in lower genetic variability and, ultimately, viability.[39]

Polycyclic aromatic hydrocarbons (PAHs) are one of the more toxic human-generated wastes. PAHs are carcinogens generated by the steel and iron industries, heating and power generation, and fossil fuel production and use.[36,37] PAHs are released into the environment through petroleum spills, sewage effluents, and other accidental spills and are strongly implicated in the formation of liver neoplasms and tumors in fish and amphibians.[36,37,40] PAHs are particularly hazardous to small, isolated, at-risk populations because spills high in PAHs are highly toxic.

A major group of contaminants associated almost exclusively with human activities is the heavy metals. They have been implicated in negative impacts affecting over 48% of U.S. lakes and reservoirs, 15% of rivers and streams, 10% of estuaries.[19] Sources of toxic heavy metals and metalloids include atmospheric deposition, industrial activities, mining leachate, and stormwater runoff.[19,20] Aquatic organisms exposed to high levels of metals in their environment suffer from a decreased metabolic rate and a decreased ability to siphon for food.[41,42] Aquatic organisms like freshwater mussels are sedentary and cannot move from an environment that is becoming increasingly toxic from the deposition of heavy metals. Exposed mussels ingest metals during their siphoning activities and concentrate metals until their metabolic rate is decreased to a level where they cannot siphon for food any longer. Some of the most common pollutants found in municipal and industrial runoff are copper, lead, zinc, mercury, aluminum, and cadmium.[19,37] Mercury and lead are the most studied, but others, such as cadmium and selenium, are known hazards, even if they do require more research before their true impact can be accurately predicted. The dynamics of how each metal affects an organism depends on pH, alkalinity, organic enrichment, water movement, and biomagnification in the food web.[37] For example, mercury enters the environment through its use in mining, paints, fungicides, batteries, and thermometers. Metallic mercury is easily methylated by microorganisms in the environment to form methylmercury or dimethyl mercury, which are hazardous to aquatic animals.[20] Methylmercury attacks an organism's central nervous system but is not excreted from organisms readily because it is fat soluble, so it tends to bioaccumulate in the food chain.[20,37] The process of mercury methlyation is enhanced by the rate of decomposition of organic matter and the availability of organic nutrients, and it remains a consistent problem in newly flooded reservoirs.

## 45.4 CASE HISTORIES

Only 75% (n = 396) of the species listed in the United States have approved recovery plans, ranging from a low of 42% for arachnids to a high of 86% for reptiles (Table 45.1). Using the recovery plans as our data source, we tallied by taxonomic category (Table 45.1) the number of contaminants or pollution identified as a cause of the decline, contributed to the decline, or potentially posed a threat. Although incomplete because it only utilizes a portion of the listed species and none of those proposed for listing, this approach suffices for illustrative purposes.

### 45.4.1 Terrestrial Organisms

There were 194 terrestrial species with recovery plans. This corresponds roughly to mammals, birds, reptiles, insects, and arachnids (Table 45.1). These species had 400 cumulative threats listed in recovery plans (Figure 45.1). Municipal sources were listed most frequently (n = 105), followed by industrial sources (n = 100), agricultural sources (n = 80), waterway/navigation (n = 51), industrial/municipal (n = 42), and agriculture/municipal/industrial combination (n = 22). The main stressors affecting terrestrial species were discharges and effluents (n = 129), pesticides and poisons (n = 112), hydrologic modifications (n = 99), heavy metals (n = 25), and waste runoff (n = 14) (Figure 45.2). The following species are offered as case-history examples.

#### 45.4.1.1 *American Burying Beetle,* Nicrophorus americanus *(USFWS 1991)*[43]

Listed as endangered with a recovery plan approved in 1991, this coleopteran has a disjunct population, with subpopulations in Rhode Island and Oklahoma.[43] Estimated total population is less than 1000 individuals. Initial declines are believed to be due to the extensive use of DDT in agriculture, although there is historical evidence that populations were decimated by disease. However, extensive habitat alteration, habitat loss, and conversion to other uses have resulted in severe habitat fragmentation. In all likelihood, DDT in combination with habitat losses/alteration caused the decline and now inhibits the surviving populations from reoccupying habitat. Additionally, pesticides and herbicides present a severe threat to the beetle because a significant portion of the population could be killed in a single incident.

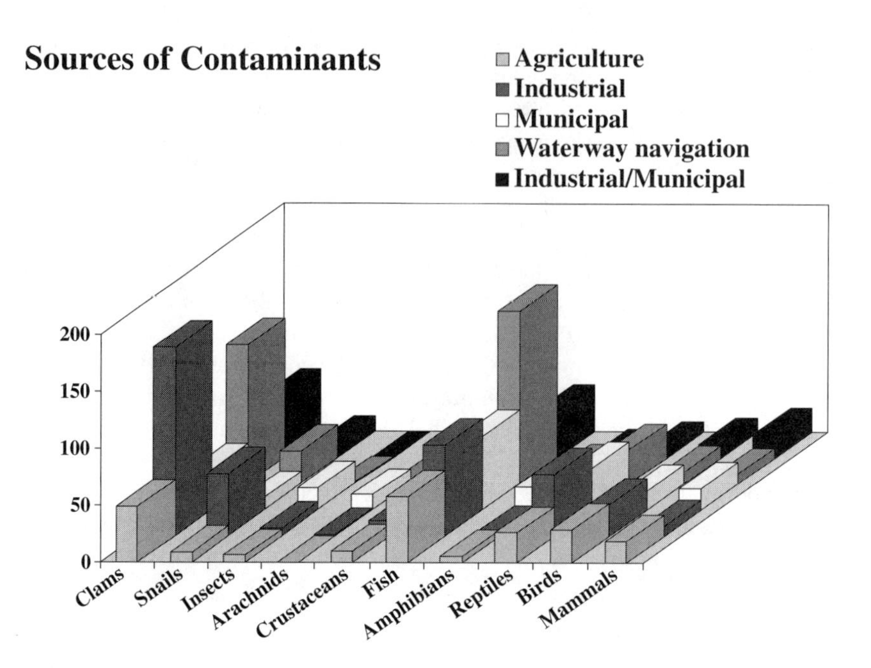


**Figure 45.1** Sources of contaminants affecting U.S. endangered and threatened species.

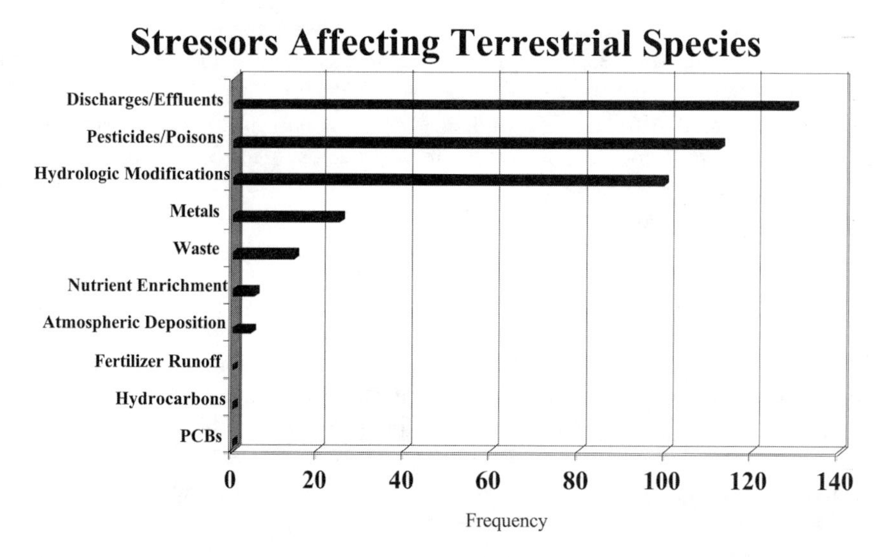


**Figure 45.2** Contaminant and pollution stressors affecting U.S. endangered and threatened terrestrial species.

### *45.4.1.2 Desert Tortoise (Mojave Population),* Gopherus agassizi *(USFWS 1994)*[44]

Populations of this reptile occur in California, Nevada, Arizona, and Utah. Throughout their range, tortoises are being impacted by similar problems, but the impact is most acute in the six Mojave subpopulations, especially those in California. The Mojave segment was listed as threatened in the most recently approved recovery plan (1994).[44]

Overgrazing by exotics (horses and burros) and off-road-vehicle use has reduced the undisturbed and undamaged desert scrub preferred by tortoises. Compounding these population-wide threats is the proliferation of roads and urbanization, putting more tortoises in harm's way from vehicles, domestic pets (especially dogs), and the large commercial and amateur pet trade.

The widespread occurrence of a lung disease appears to affect populations already stressed. A new issue is raven predation. Supported by the proliferation of dumps and landfills, raven populations are increasing; a recently learned behavior involves dropping juvenile tortoises from high altitude onto hard-surface roads to crack the shell, kill and consume the tortoise.

#### *45.4.1.3 Florida Panther,* **Felis concolor coryi** *(USFWS 1995)*[45]

The second revision of the recovery plan for this endangered feline was approved in 1995.[45] Before it was extirpated through most of its range, it occupied most of the southeast from Texas to North Carolina. It now consists of a highly inbred population of 20–30 animals in a single location within Florida.

Isolated from contact with other mountain lion populations, gene flow has been eliminated, and the remnant population suffers the fate of all limited gene pools — disease and parasitism take their toll as do stochastic events such as road kills, fires, and inbreeding depression. Human presence continues to increase, with more panther habitat being converted to uses detrimental to the panther's well-being.

Tissues from panthers and their prey items show mercury levels that may be toxic as well as constitute a reproductive hazard.[46] Sugar cane is a major crop in the area utilizing substantial amounts of pesticides; maintenance of habitat as grasslands for cattle grazing contributes to these hazards. Runoff from these fields exposes the panther to a suite of biotoxins, which may or may not have an effect.

#### *45.4.1.4 Eastern Brown Pelican,* **Pelecanus occidentalis carolinensis** *(USFWS 1979)*[47]

Listed as endangered in 1979, this subpopulation was delisted in 1985. The wild population nested along the Gulf coast from Mexico to Florida, then along the Atlantic coast, from Florida north.[47] The brown pelican's decline on the east coast bore little resemblance to that of the Pacific coast. On the Pacific coast, the decline was much slower and could be explained by the shell thinning and subsequent reproductive failure attributed to DDE/DDT. On the east and Gulf coasts, the decline started in the 1930's and was attributed to a combination of factors including weather, predation, starvation, disturbance, vandalism, and DDE/DDT. Louisiana's breeding population disappeared between 1957 and 1961, leaving a remnant population in Texas, none along the Gulf coast until Florida, and South Carolina, where 20,000–30,000 pelicans (mostly in Florida) continued to breed with no apparent problem. This precipitous decline suggests a toxicant rather than failure to produce viable young. Louisiana lost 40% of a population it was seeking to reestablish from an endrin spill, and oil spills pose a constant and familiar hazard. Brown pelicans nationwide have benefited from the reduction of organochlorine pesticide use.

#### *45.4.1.5 California Condor,* **Gymnogyps californianus** *(USFWS 1996)*[48]

The California condor was listed as an endangered species in 1967, but did not have a recovery plan until 1975; the current plan was approved in 1996 and represents the third revision.[48] Historically (1700–1800), the California condor was believed to have been a common species along the Pacific coast from Baja California, Mexico, to British Columbia, Canada, where they fed on marine mammals washed up on shore. Their distribution during the Pleistocene was apparently continent-wide. With a wingspan approaching 3 m, this wide-ranging soaring scavenger depends on its vision to locate primarily mammal carcasses in grasslands and other open environments. Ridgelines and hilltops are especially liked because they make it easier for the condor to resume flight from a running start. The declines in marine mammals hurt condor populations, as did the conversion of the Central Valley from marshes and grasslands to agriculture and the subsequent loss of Tule elk, mule deer, and pronghorn antelope. As agriculture expanded in the valley floor, both the condors and the cattle industry retreated into the surrounding foothills and mountains of the Coast range, Transverse range, and Sierra Nevada.

The causes of the decline may never be completely understood, but some aspects of the decline are obvious. Loss of foraging range and declines in the primary food sources made foraging more difficult, as the birds searched for carcasses that were increasingly less common. Assembling personal natural-history-specimen collections was quite popular during the Victorian era and carried on into the 1900s; eggshell collecting was one of the more popular themes. Consequently, condor eggs were much sought after and 'hundreds were taken from the wild for formal collections as well as for amateur collections. Also known to have killed condors are power transmission lines, compound 1080, radiator fluid, cyanide, and lead. Elevated organochlorine residues are reported from eggs in the 1980s, although no evidence of significant shell thinning was found.

Despite conservation efforts, the species was in decline throughout the 1900s, culminating in the removal from the wild of the last condor in April 1987. The decision to move the entire population into captivity was the result of the losses incurred by the wild population during the winter of 1984–1985. Four of the five pairs suffered losses, three pairs had one member disappear, and the fourth pair vanished. Faced with a situation in which half of the wild population vanished, bringing the wild population into captivity and planning to restore this population through the release of captive-reared birds appeared to be the only available tactic.

Lead poisoning appears to be a significant problem. Four out of the last five condors recovered after death had died of lead poisoning. Elevated blood-lead levels and mortality have been reported in all of the released condor populations including those from the remote Arizona site. All the evidence indicates that the source of the lead causing raptor mortality is not organic, biologically incorporated, or derived from sediments, but is metallic lead in the form of shot or fragments from expanding bullets.

### 45.4.2 Aquatic Organisms

Aquatic organisms serve as excellent indicators of environmental quality because they are often specialized for such restricted habitats. Freshwater environments are susceptible to degradation caused by changes in biogeochemical processes and changes in the local watershed including the surrounding terrestrial habitat.[25] For example, because mussels are relatively immobile and their reproductive cycle is fairly complex, they are exposed to changes in a watershed at various stages of their life cycle. Good water quality is critical for survival, making mussels extremely sensitive to environmental pollution. Mussels ingest and concentrate heavy metals, contaminants, and pesticides from water, making their absence or presence an indication of water quality and chemical contamination.[24,29]

Freshwater aquatic organisms comprise a substantial number of threatened species listed under the ESA.[4] Aquatic fauna are proportionately more threatened than terrestrial species.[4,13,21] Freshwater mussels, crayfish, amphibians, and fish have the largest numbers of species at risk. More than 60% of freshwater mussels and crayfish are at risk of endangerment or extinction.[13] Various scientists have developed similar values estimating species imperilment. Approximately 37% of freshwater fish and 35% of amphibians are imperiled.[21] Over 20% of the world's fish species are either extinct or endangered,[17,25] and one third of all U.S. fish species are threatened or endangered.[12] Of all freshwater organisms, mussels are the most rapidly declining group,[27] with estimates as high as 10% extinction in the past century.[21]

There were 202 aquatic species that had recovery plans, divided into the categories of amphibians, fish, clams, and crustaceans (Table 45.1). Recovery plans for aquatic species totaled 1089 contaminant and pollution threats (Figure 45.1). Waterway navigation was the main source of

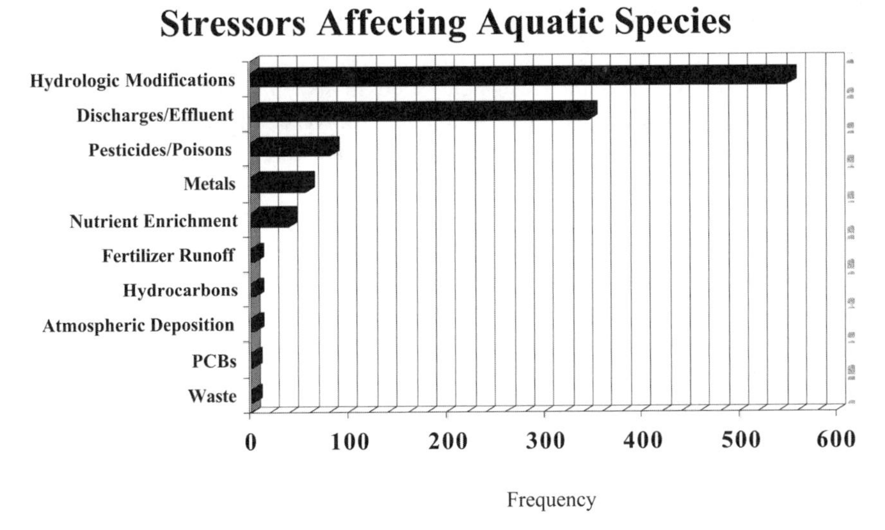


**Figure 45.3** Contaminant pollution stressors affecting U.S. threatened and endangered aquatic species.

contaminant and pollution threats affecting aquatic species (n = 326), followed by industrial sources (n = 318), municipal sources (n = 151), agricultural sources (n = 131), industrial/municipal sources (n = 117), and agricultural/municipal/industrial combinations (n = 46). The stressors listed most frequently were hydrologic modifications (n = 548), discharges and effluents (n = 344), poisons and pesticides (n = 80), heavy metals (n = 56), and nutrient enrichment (n = 39) (Figure 45.3). The following species are offered as case history examples.

#### 45.4.2.1 *San Diego Fairy Shrimp,* Branchinecta sandiegonenis *(USFWS 1998)*[49]

This endangered freshwater crustacean had its recovery plan approved in 1998.[49] Found only in small, shallow ephemeral ponds in southern California, this fairy shrimp numbers less than 1000 individuals. Historically, these vernal pools were subjected to the impacts of overgrazing, trampling, and waste dumping (industrial, municipal, and agricultural). A significant number of pools were lost when they were inundated by reservoirs created for flood control and domestic water supply.

Since the shrimp can tolerate only minimal changes in water quality, maintaining suitable aquatic habitat is the main issue they face. Joining cattle as a significant degrader of habitat quality are off-road vehicles. Runoff from agricultural fields, highways, and urban areas degrades water quality and further threatens this native shrimp.

#### 45.4.2.2 *Utah Valvata Snail,* Valvata utahensis *(USFWS 1995)*[50]

The endangered Utah valvata snail, whose recovery plan was approved in 1995, is restricted to cold-water springs in Utah and Idaho.[50] Although numbering about 12,000 individuals, its specialized and unique habitat requirements make it quite vulnerable to extirpation from acute pollution events. It is known to occur at four sites.

It has been severely impacted by impoundments intended for power generation. The loss of critical habitat and the severe habitat fragmentation have isolated populations with large stretches of still, hostile impounded water. The remaining populations are threatened by water-quality issues including runoff from feedlots and farm fields as well as excessive water withdrawals that expose critical habitat to dessication and increased water temperatures. The restricted distribution makes the remaining populations quite vulnerable to catastrophic pollution events.

#### *45.4.2.3 Wyoming Toad,* Bufo hemiophrys baxteri *(USFWS 1991)*[51]

The Wyoming toad was classified as endangered in its 1991 recovery plan.[51] It is a relic population of a species that was widespread when the climate was wetter but is now reduced to a single population of 100–150 animals inhabiting the wetlands around Laramie, Wyoming.

Climate has been the major cause of the decline, reducing the toad's preferred habitat to wetlands associated with the widely scattered rivers and creeks of the high plains short grass prairies. The result was isolated subpopulations that could not interact and had no suitable habitat for expansion. These are also the areas that have been converted to homesteads, pastures, and agricultural lands. Virtually all of these areas lie in private ownership and are used as pastures and hay meadows, with irrigation and mowing schedules impacting the toad to varying degrees. Populations disappeared without notice until the species was reduced to a single population.

The surviving population is further threatened by disease, predation, development, and ongoing agricultural practices. Of particular concern has been the use of fenthion as a mosquito-control agent. Efforts to reduce mosquito populations are the greatest hazard the toad population faces, with extinction in a single spray event possible, depending on the toxicity, persistence, and exposure levels the population experiences.

#### *45.4.2.4 Blue Pike,* Stizostedion vitreum glaucum *(USFWS 1975)*[52]

The blue pike was a very popular food fish and quite abundant at one time throughout the Lake Huron/Lake Erie/Lake Ontario ecosystem. The recovery plan was signed August 1975, designating the blue pike as an endangered species.[52] The dramatic drop in the commercial and recreational catch year to year warned of the problem, but it was not addressed soon enough to save the blue pike. Further surveys failed to find any fish, and it was declared extinct. The primary factors leading to extinction were over-fishing by commercial and recreational users in concert with a rapidly increasing level of pollution that led to severe oxygen depletion. The introduction of smelt accelerated the downturn because smelt exploited the same food resources as the blue pike and were more tolerant of the deteriorating conditions. As the population levels declined, hybridization became more prominent as individuals failed to find conspecifics and mated with related but nonconspecifics.

#### *45.4.2.5 Devil's Hole Pupfish,* Cyprinodon diabolus *(USFWS 1980)*[53]

Listed as endangered in July 1980, the Devil's hole pupfish is one of many pupfish found in the southwestern U.S. with a restricted range — relics of a wetter, cooler climate.[53] As the climate changed, pupfish populations become isolated and evolved into separate species. The Devil's Hole pupfish are restricted to one water-filled limestone cavern in Nevada. The maximum population estimate was 300–400 fish. Additionally, they are also restricted by the shallow shelf that provides the only source of plant and invertebrate resources for the entire population. Water depletion, particularly falling groundwater, jeopardizes the species through loss of its foraging habitat.

Groundwater depletion has been the most immediate concern. Runoff from adjacent fields into the Devil's Hole watershed and groundwater has raised concerns that toxic substances would move into the cavern. Pesticides pose an immediate threat to the fish, as herbicides do to the plants growing on the foraging rock shelf. The influx of surface water also carries silt and nonpoint-source contaminants into the cavern with unknown effects. Finally, the limited habitat and small population places the species in immediate threat of extinction. Any perturbation of the system, any spill that flows into the cavern or reduces water flow into the system, is capable of driving the Devil's Hole pupfish to extinction.

#### *45.4.2.6 Curtis' Pearly Mussel,* **Epioblasma florentina curtis** *(USFWS 1986)*[54]

The Curtis' pearly mussel was listed as an endangered species at the time the recovery plan was approved in February 1986.[54] This freshwater mussel historically occurred in the basins of the White and St. Frances Rivers in Missouri and Arkansas. Present distribution is 6.1 mi of the Little Blue River and 7 mi of the Castor River, all in Akansas. Like so many of its endangered cousins, the Curtis' pearly mussel requires stable substrates of sand, gravel, cobble, or boulder in riffles and runs of moderate water volume.

Much of its habitat has been covered by impoundments, destroying its suitability to this riverine species. Channelization and gravel dredging has destroyed additional miles of stable river habitat. Impoundments have also impacted downstream water quality, altering water chemistry, temperatures, and flow regimes. Encroaching agricultural fields and building construction degrades water quality by releasing nutrients, pesticides, herbicides, and sediments. The sediments clog occupied habitat, suffocating adult mussels and driving host fish from the area. From the agricultural fields also come low levels of a variety of agricultural chemicals of unknown toxicity or sublethal effects. Finally, the restricted distribution makes the surviving populations vulnerable to spills that might sterilize the river system.

## 45.5 SUMMARY AND CONCLUSIONS

Multiple factors contribute to the decline of species. Habitat destruction is the primary factor that threatens species, affecting 73% of endangered species. The second major factor causing species decline is the introduction of nonnative species, affecting 68% of endangered species. Pollution and overharvesting were identified as impacting, respectively, 38 and 15% of endangered species. Other factors affecting species decline include hybridization, competition, disease, and other interspecific interactions.

Once a species is reduced to a remnant of its former population size and distribution, its vulnerability to catastrophic pollution events increases, frequently exceeding or replacing the factors responsible for the initial decline. Small, isolated populations are particularly vunlerable to catastrophic loss by an acute event, such as a chemical spill or pesticide application. However, when it comes to surviving a single disaster, widespread subpopulations of a species are far more resilient and ensure genetic survival. Hypothesizing theoretical concerns of potential factors that could affect an endangered species could predispose the scientific and political communities to jeopardizing threats.

The user of recovery plans as a data source must be aware of the bias within the data set. These data should be used with the caveat that the source of information in recovery plans is not always based on scientific research and rigorous data collection. Over 58% of the information identifying species threats is based on estimates or personal communication, while only 42% is based on peer-reviewed literature, academic research, or government reports. Many recovery plans were written when a species was initially listed in the 1970s or 1980s. Politics, human disturbance, and habitat demand issues evolve over a 20- to 30-year period, leaving much of the threats facing endangered species outdated and inadequate. These data are most valuable when used to facilitate reviews of Section 7 consultations and environmental impact statements, review permit applications, conduct environmental risk assessments, prioritize research needs, and identify limiting factors affecting species health. These data are also useful in identifying potential threats to species' health. Without properly identifying threats to endangered species based on sound, scientific research, there is little hope to successfully recover an endangered species.

## REFERENCES


1. National Research Council, *Science and The Endangered Species Act,* National Academy Press, Washington, D.C., 1995.
2. Clark, T. W., Reading, R. P., and Clarke, A. L., *Endangered Species Recovery: Finding the Lessons, Improving the Process,* Island Press, Washington, D.C., 1994.
3. Wilson, E. O., Threats to biodiversity, *Sci. Amer.*, 261, 108, 1989.
4. Master, L. L., Assessing threats and setting priorities for conservation, *Conserv. Biol.,* 5, 559, 1991.
5. Morse, L. E., Standard and alternative taxonomic data in the multi-institutional Natural Heritage Data Center, in *Designs for a Global Plant Species Information System*, Bisby, F. A., Russell, G. F., and Pankhurst, R. J., Eds., Oxford University Press, Oxford, 1993, 69.
6. Stein, B. A., Towards common goals: Collections information in conservation databases, *Assoc. Syst. Coll. Newslett.,* 21,1, 1993.
7. Ray, J. C. and Ginsberg, J. R., Endangered species legislation beyond the borders of the United States, *Conserv. Biol.,* 13, 956, 1999.
8. U.S. Fish and Wildlife Service, Box score, Endang. Spec. Tech. Bull. XXVIII, 32, 2002.
9. Ehrlich, P. R., The loss of diversity: Causes and consequences, in *Biodiversity,* Wilson, E. O. and Peter, F. M., Eds., National Academy Press, Washington, D.C., 1988, 21.
10. Fischman, R. L., Biological diversity and environmental protection: Authorities to reduce risk, *Environ. Law,* 22, 435, 1992.
11. Wilson, E. O., *The Diversity of Life,* Belknap Press, Cambridge, MA, 1992.
12. Allan, J. D. and Flecker, A. S., Biodiversity conservation in running waters, *BioScience,* 43, 32, 1993.
13. Stein, B. A. and Chipley, R. M., Eds., Priorities for Conservation: 1996 Annual Report Card for U.S. Plant and Animal Species, The Nature Conservancy, Arlington, VA, 1996.
14. Foin, T. C., Riley, S. P. D., Pawley, A. L., Ayres, D. R., Carlsen, T. M., Hodum, P. J., and Switzer, P. V., Improving recovery planning for threatened and endangered species, *Bioscience,* 48, 177, 1998.
15. Wilcove, D. S., Rothstein, D. R., Dubow, J., Phillips, A., and Losos, E., Quantifying threats to imperiled species in the United States, *Bioscience,* 48, 607, 1998.
16. Miller, R. R., Williams, J. D., and Williams, J. E., Extinctions of North American fishes during the past century, *Fisheries,* 14, 22, 1989.
17. Wilson, E. O., *The Diversity of Life,* Harvard University Press, Cambridge, MA, 1992, 424.
18. Flather, C. H., Joyce, L. A., and Bloomgardem, C. A., Species Endangerment Patterns in the United States, U.S. Department of Agriculture Forest Service, General Technical Report RM-241, Rocky Mountain Forest and Range Experiment Station, 1994.
19. Environmental Protection Agency, The Quality of Our Nation's Waters: 1992, EPA Office of Water, Washington, D.C., 1994.
20. Naiman, R. J., Magnuson, J. J., McKnight, D. M., and Stanford, J. A., Eds., *The Freshwater Imperative: A Research Agenda*, Island Press, Washington, D.C., 1995.
21. Flack, S. and Chipley, R., Eds., *Troubled Waters: Protecting Our Aquatic Heritage,* The Nature Conservancy, Arlington, VA, 1996.
22. Richter, B. D., Braun, D. P., Mendelson, M. A., and Master, L. L., Threats to imperiled freshwater fauna, *Conserv. Biol.,* 11, 1081, 1997.
23. Fuller, S. L. H., Clams and mussels, in *Pollution Ecology of Freshwater Invertebrates,* Hart, C. W., Jr. and Fuller, S. L. H., Eds., Academic Press, New York, 1974, 215.
24. Bogan, A. E., Freshwater bivalve extinctions (Mollusca: Unionoidae): A search for causes, *Am. Zool.,* 33, 599, 1993.
25. Moyle, P. B. and Leidy, R. A., Loss of biodiversity in aquatic ecosystems: Evidence from fish faunas, in *Conservation Biology: The Theory and Practice of Nature Conservation, Preservation, and Management*, Fielder, P. L. and Jain, S. K., Eds., Chapman and Hall, New York, 1992, 127.
26. Anderson, R. M., Layzer, J. B., and Gordon, M. E., Recent catastrophic decline of mussels (Bivalvia, Unionidae) in the Little South Fork Cumberland River, Kentucky, *Brimleyana,* 17, 1, 1991.
27. Vaughn, C. C. and Taylor, C. M., Impoundments and the decline of freshwater mussels: a case study of an extinction gradient, *Conserv. Biol.,* 13,912, 1999.

28. Williams, J. D., Fuller, S. L. H., and Grace, R., Effects of impoundment on freshwater mussels (Mollusca:Bivalvia:Unionidae) in the main channel of the Black Warrior and Tombigbee rivers in western Alabama, *Bull. Alab. Mus. Natl. Hist.,* 13, 1, 1992.
29. Wilcove, D. S. and Bean, M. J., Eds., *The Big Kill,* Environmental Defense Fund, Washington, D.C., 1994.
30. Petts, G. E., Long-term consequences of upstream impoundment, *Environ. Conserv.*, 7, 325, 1980.
31. Bain, M. B., Finn, J. T., and Booke, H. E., Streamflow regulation and fish community structure, *Ecology,* 69, 382, 1988.
32. EPA, Can our coasts survive more growth?, *EPA J.,* 15, 5, 1989.
33. Puckett, L. J., Identifying the major sources of nutrient water pollution, *Environ. Sci. Technol.,* 29, 408, 1995.
34. Freed, V. H., Haque, R., Schmedding, D., and Kohnert, R., Physicochemical properties of some organophosphates in relation to their chronic toxicity, *Environ. Health Perspect.,* 13, 77, 1976.
35. Hall, R. J. and Kolbe, E., Bioconcentration of organophosphorus pesticides to hazardous levels by amphibians, *J. Toxicol. Environ. Health,* 6, 853, 1980.
36. Eisler, R., Contaminant Hazard Reviews, Patuxent Wildlife Research Center, Laurel, MD, 1994.
37. Diana, S. G. and Beasley, V. R., Amphibian toxicology, in *Status and Convservation of Midwestern Amphibians,* Lannoo, M. J., Ed., University of Iowa Press, Iowa City, 1998, 266.
38. Ecobichon, D. J., The toxic effect of pesticides, in *Casarett and Doull's Toxicology: The Science of Poisons,* Klaassen, C. D., Ed., McGraw-Hill, New York, 1996, 643.
39. EPA, Report to Congress on the Discharge of Hazardous Wastes to Publicly Owned Treatment Works, U.S. Environmental Protection Agency, Washington, D.C., 1986.
40. Sarokin, D. and Schulkin, J., The role of pollution in large-scale population disturbances, *Environ. Sci. Technol.*, 26, 1476, 1992.
41. Salanki, J., Behavioural status in mussels under changing environmental conditions, *Symp. Biol.*, 19, 169, 1979.
42. U.S. Fish and Wildlife Service, Fine-Rayed Pigtoe Pearly Mussel Recovery Plan, U.S. Fish and Wildlife Service, Atlanta, 1984.
43. U.S. Fish and Wildlife Service, American Burying Beetle Recovery Plan, U.S. Fish and Wildlife Service, Albuquerque, 1991.
44. U.S. Fish and Wildlife Service, Desert Tortoise (Mojave Population) Recovery Plan, U.S. Fish and Wildlife Service, Portland, OR, 1994.
45. U.S. Fish and Wildlife Service, Second Revision, Florida Panther Recovery Plan, U.S. Fish and Wildlife Service, Atlanta, 1995.
46. Roelke, M. E., Schultz, D. P., Facemire, C. F., and, Sundolf, S. F., Mercury contamination in the free-ranging endangered Florida panther (*Felis concolor coryi*), *Proc. Am. Assoc. Zoo Vet.,* 20, 277, 1991.
47. U.S. Fish and Wildlife Service, Eastern Brown Pelican Recovery Plan, U.S. Fish and Wildlife Service, Atlanta, 1979.
48. U.S. Fish and Wildlife Service, California Condor Recovery Plan, 3rd revision, U.S. Fish and Wildlife Service, Portland, OR, 1996.
49. U.S. Fish and Wildlife Service, San Diego Fairy Shrimp Recovery Plan, U.S. Fish and Wildlife Service, Portland, OR, 1998.
50. U.S. Fish and Wildlife Service, Utah Valvata Snail Recovery Plan, U.S. Fish and Wildlife Service, Denver, 1995.
51. U.S. Fish and Wildlife Service, Wyoming toad recovery plan, U.S. Fish and Wildlife Service, Denver, 1991.
52. U.S. Fish and Wildlife Service, Blue Pike Recovery Plan, U.S. Fish and Wildlife Service, Twin Cities, MN, 1975.
53. U.S. Fish and Wildlife Service, Devil's Hole Pupfish Recovery Plan, U.S. Fish and Wildlife Service, Portland, OR, 1980.
54. U.S. Fish and Wildlife Service, Curtis' Pearly Mussel Recovery Plan, U.S. Fish and Wildlife Service, Twin Cities, MN, 1986.

# Index

## A

## B

## C

## D

## E

## F

## G

## H

## I

## K



## L

## M

## N



## O

## P

## Q



## R

## S

## T

## U

## V



## W

## X



## Z